Theory & Performance of

ELECTRICAL MACHINES

(DC Machines and AC Machines)

15th Thoroughly Revised Edition

J.B. Gupta

Theory & Performance of

ELECTRICAL MACHINES

(DC Machines and AC Machines)

For

B.Sc. (Engineering); B.Tech. B.E. A.M.I.E. (India) Section B; U.P.S.C. I.E.S. (Electrical Engineering); GATE (Electrical Engineering) and Other Equivalent Engineering Examinations

By

J.B. Gupta

B.Sc. (Elec. Engg.); C. Eng. (I);
F.I.E. (India); M.I.S.T.E.

S.K. KATARIA & SONS®

Publisher of Engineering Books

4885/109, Prakash Mahal, Dr. Subhash Bhargav Lane,
Opposite Delhi Medical Association, Daryaganj, New Delhi–110002 (India)
Phone: +91-11-43551243; Mobile: +91-9871775858, 7678425209
e-mail: katariabook@gmail.com, katariabooks@yahoo.com
Website: www.skkatariaandsons.com
(e-books, subscriptions and bulk buying also available on our website)

Published by:

S.K.Kataria&Sons®
4885/109, Prakash Mahal, Dr. Subhash Bhargav Lane,
Opposite Delhi Medical Association, Daryaganj, New Delhi-110002 (India)
Phone: +91-11-43551243; Mobile: +91-9871775858, 7678425209
e-mail: katariabook@gmail.com, katariabooks@yahoo.com

Edited By: **Rajeev Manglik**
B.E. (Elec.) Pb. University
and
Rohit Manglik
B.E. (Elec.) University of Roorkee

Send all correspondence to: M/s S.K. Kataria & Sons, New Delhi
e-mail: katariabook@gmail.com

First Edition: 1967; Fifteenth Edition: January, 2015; Reprint: 2016; Reprint: 2017 (With minor Corrections and amendments); Reprint: 2026-2027

Laser Type Setting by:
SARA ASSIGNMENTS, Delhi

Printed at:
AL-AZIZ BOOKS BINDING HOUSE

PREFACE TO THE FIFTEENTH EDITION

The author feels great pleasure in presenting the 15th edition of the book with marked improvement over the previous one. The book has been thoroughly revised, rewritten and designed so as to meet the needs of Degree (Engineering/Technology) and A.M.I.E. Sec B students and of those preparing for U.P.S.C. I.E.S. (Electrical Engineering) and GATE (Electrical Engineering). The subject matter has been presented in an easily comprehensible manner and to provide a detailed exposition of the fundamentals. A large number (almost all) of old diagrams have been replaced by new ones and numerous new diagrams have been added to make the subject more illustrative. A large number of worked-out examples, theoretical exercises and numerical problems have been deleted and replaced from the latest examination papers of different Indian Universities/Technical Universities; GATE (Electrical Engineering); U.P.S.C. I.E.S. (Electrical Engineering) and other equivalent examining bodies. An earnest effort has been made to provide a comprehensive coverage of the courses as prescribed by the Institution of Engineers (India), Indian and Foreign Universities, Technical Universities and various agencies conducting competitive examinations.

The hardest thing about revision of this book was deciding what to include and what not to include. This was more important than anything else. This is because obsolete topics is a waste of time and effort. So, my first effort and topmost priority in revision of this book was to exclude obsolete topics and include such topics which are of vital importance but could not be included in the previous edition.

The book consists of 22 chapters divided into 6 sections detailed below:

SECTION I : Chapter on "AC Circuits" has been included. Additional topic included is "Matching Characteristics of Electric Machine and Load".

SECTION II : *Additional topics included are:* Comparison Between Simplex Lap and Simplex Wave Windings; and Speed Control of PMDC Motors.

SECTION III : Chapter Titled "Special Transformers" has been added. It includes topics on Phase-shifting Transformer; Isolation Transformers; Audio-frequency Transformers; Radio-frequency (RF) Transformers; Variable-frequency Transformers; Pulse Transformers; Grounding Transformers, Traction Transformers in addition to existing topics.

Other additional topics included are: Transformation of Impedance; 3-Phase-To-6-Phase Conversion; 3-Phase-To-Twelve-Phase Conversion; Frequency Response of Transformers.

SECTION IV : *Additional topics included are:* Excitation Systems; Factors Controlling Size of Alternators; ASA Method of Determination of Voltage Regulation of an Alternator; Motoring and Generating Operations; Capability Curves of Alternator; Synchronising Power and Synchronising Torque Coefficients; Significance of Synchronising Power Coefficient; Electric Braking of Synchronous Motors.

SECTION V : Chapter Titled "Special Electrical Machines" has been added. It covers the topics covered in the chapters "Special Types of DC Machines" and "Special Induction and Polyphase Commutator Machines" and, *in addition, the following topics:*

AC Tachometers; Multistack VR Stepper Motor; Stepper Motor Drive Circuits; Servomotors Vs Stepper Motors; Switched Reluctance Motor, Brushless Vs Brushed Motors; Applications of Brushless DC Motors; Low Inertia Motors; Synchros and Resolvers.

Other topics included are: Relation Between Rotor Torque and Power Factor; Analogy With a DC Motor and Maximum Torque.

SECTION VI : Additional topics included are: Classification of Power Converters; DC Motor Control Through Converters and Chopper Control of DC Motors.

Chapters of "Synchronous Convertors" and "Rectifiers", being obsolete now, have been deleted.

Chapters of Polyphase Circuits (being covered in the Chapter of AC Circuits); Special Types of DC Machines and Special Induction and Polyphase Commutator Machines (being covered in the chapter of Special Electrical Machines) and Rotating Magnetic Fields (being covered in the chapter of Basic Concept of Rotating Electrical Machines) have been deleted.

The solution to illustrative examples covering almost all topics and subtopics of different chapters makes the entire presentation easy to follow. Selected exercises and numerical problems with answers given at the end of almost each chapter help the students in evaluating their understanding of the subject. Short Answer Type Questions With Answers given at the end of each chapter will certainly help the students in preparing for viva voce/theory examinations and interviews.

Most of the agencies conducting examinations nowadays follow the pattern of objective evaluation. Even Institution of Engineers (India) and other examining bodies (Universities and State Boards) are adopting the pattern of objective evaluation (partly) for their regular examinations. The book contains 1,076 objective type questions (Multiple Choice Type: 634; Matching Type: 20; True/False Type: 240 and Fill in The Blanks Type: 182) designed to bring out the hidden aspects and also test knowledge, understanding and application aspect of learning. Most of the objective type questions included in the book are from examination papers of various universities, U.P.S.C. I.E.S. and GATE or are of equivalent standard. This will help the students to score better in their regular as well as competitive examinations.

It is earnestly hoped that with the extensive additions, revision, modifications and improvement, the present edition will facilitate the students not only in preparing for their regular examinations but also in preparing themselves for competitive examinations and prove more useful than the earlier edition.

Every care has been taken to eliminate misprints, omissions and errors but it is too much to expect that no inaccuracy/misprint/error has crept in and the author would be grateful to the readers for bringing to his notice any such errors/misprints/omissions they may come across while going through the book.

The author will like to thank the readers (both teachers and students) who sent their valuable suggestions that became the basis for the revision of the book.

The author shall feel satisfied if the book meets the needs of the students for whom it is meant. Suggestions and criticism for the improvement of the book are always welcomed.

The author will like to thank all the Scientists, Engineers, Authors and Publishers whose works and text have been the source of enlightenment, inspiration and guidance to him in the presentation.

Lastly, but not the least, the author is thankful to his publishers for bringing out the revised edition of the book in such a short period and for pricing it moderately despite heavy cost of paper and printing.

J.B. Gupta

January 9, 2015

CONTENTS

SECTION I

PRELIMINARIES & PRINCIPLES OF ELECTROMECHANICAL ENERGY CONVERSION

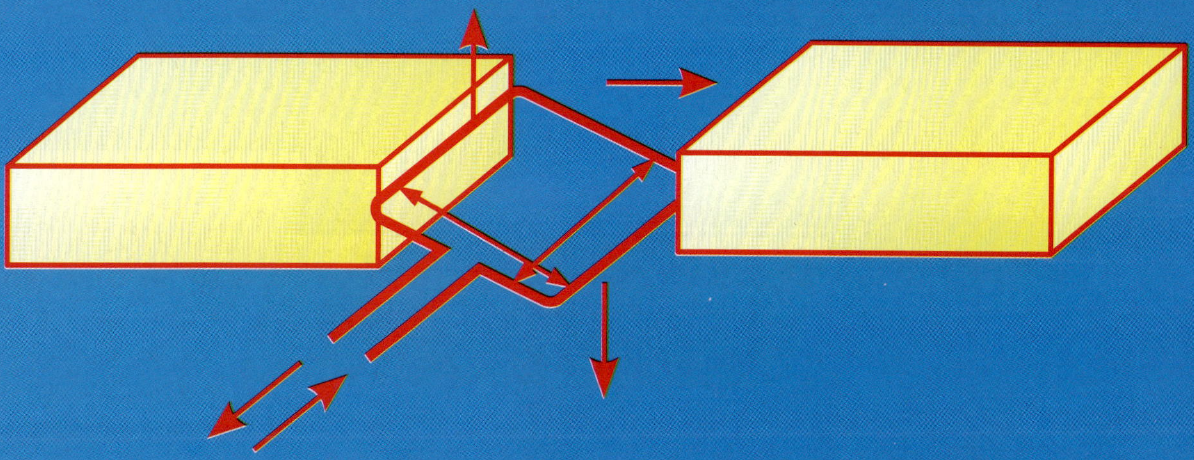

CHAPTERS COVERED IN THIS SECTION

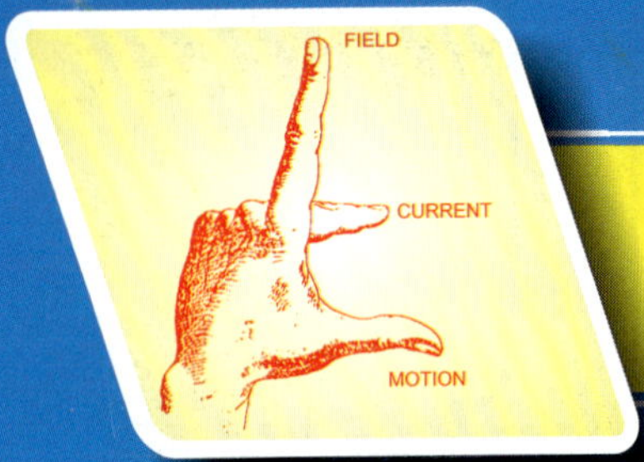

MAGNETIC CIRCUITS AND INDUCTION

AC CIRCUITS

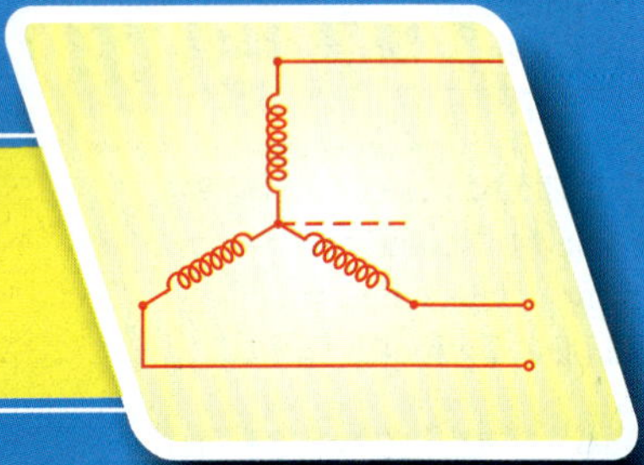

PRINCIPLES OF ELECTROMECHANICAL ENERGY CONVERSION

BASIC CONCEPTS OF ROTATING ELECTRICAL MACHINES

1 CHAPTER

Magnetic Circuits and Induction

INSIDE THIS CHAPTER

The basics of electromagnetism, electromagnetic induction, single-phase and three-phase circuits are very significant to the study of electrical machines. First two chapters will introduce the basics which are the backbone of electrical machines.

1.1 ELECTROMAGNETIC SYSTEM

It is hard fact that the *electromagnetic system* is an essential element of all rotating electrical machines/electromechanical devices and static devices such as transformers. Practically, the operation of all electric generators and motors (may be fractional kilowatt units used in domestic appliances, gigantic several thousand kW motors used in heavy industries or several thousand MW generators installed in power stations) depend upon the electromagnetic field as the coupling medium which permits interchange of energy in either direction between the electrical system and mechanical system and vice versa. Similarly static transformers provide the means for transferring electrical energy from one system to another through the medium of a magnetic field. Though the static transformer is not an electromechanical conversion device, but its analysis is similar to that of rotating machinery. Thus it is seen that all the electrical machinery including transformers make use of the medium of magnetic field for energy conversion/transfer and study of such devices essentially involves electric and magnetic circuit analysis and their interaction.

Hence the review of physical laws governing magnetic fields, production of emf and mechanical force and brief study of simple magnetic circuits becomes essential.

1.2 ELECTRIC AND MAGNETIC FIELDS

The space around an electric charge is always under stress, and experiences a force on another electric charge when placed there. The region or space in which the stress exists is called the *electric* (or *dielectric* or *electrostatic*) *field*.

Similarly, the space around the poles of a magnet is under stress and is known as the *magnetic field*.

The force in the space around a magnet can be pictured by examining the pattern made by iron fillings, sprinkled on a card board placed over the magnet (Fig. 1.1). Each little splinter of iron acts like a compass needle, attracting other fillings, at its ends and repelling those lying parallel to it. These chains of fillings led to the assumption that the region (field) around a magnet contains invisible "lines of force". The total number of lines of force surrounding a magnet, as illustrated in Fig. 1.2, is called the total *magnetic flux.*

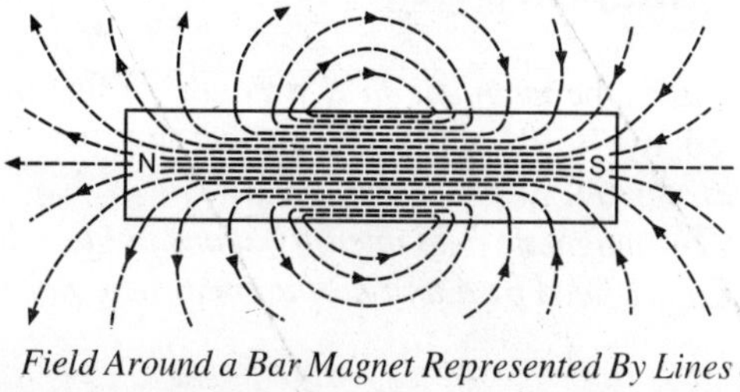

Fig. 1.1 *Pattern of Iron Fillings Around a Bar Magnet*

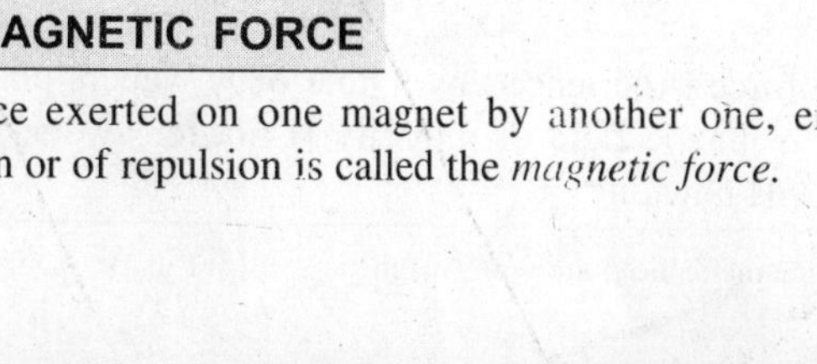

Fig. 1.2 *Field Around a Bar Magnet Represented By Lines of Force*

1.3 MAGNETIC FORCE

The force exerted on one magnet by another one, either of attraction or of repulsion is called the *magnetic force.*

According to Coulomb's first law unlike poles attract each other and like poles repel each other.

According to Coulomb's second law force between two magnetic poles is directly proportional to the product of their pole strengths and inversely proportional to the square of the distance between them.

i.e. $F \propto \frac{m_1 m_2}{d^2}$ where m_1 and m_2 are pole strengths, d is the distance between the two poles and F is the force.

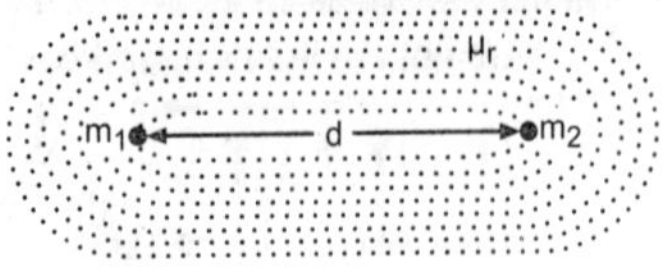

Fig. 1.3

In SI system, magnetic force between two magnetic poles is given as

$$F = \frac{m_1 m_2}{4\pi\mu_0\mu_r d^2} \quad ...(1.1)$$

where m_1 and m_2 are in webers, d in metres and F in newtons.

1.4 RELATIVE PERMEABILITY (μ_r)

The force of attraction or repulsion between the two poles of strengths m_1 and m_2 webers placed d metres apart in any medium of relative permeability μ_r is given by

$$F_{medium} = \frac{m_1 m_2}{4\pi\mu_0\mu_r d^2}$$

If the same poles *i.e.* of strengths m_1 and m_2 are placed in air medium at the same distance *i.e.* d metres apart then force acting between them is given by,

$$F_{air} = \frac{m_1 m_2}{4\pi\mu_0 d^2}$$

$$\text{So} \quad \frac{F_{air}}{F_{medium}} = \frac{\dfrac{m_1 m_2}{4\pi\mu_0 d^2}}{\dfrac{m_1 m_2}{4\pi\mu_0\mu_r d^2}} = \mu_r$$

Hence relative permeability of the medium (μ_r) is defined as the ratio of the force between two magnetic poles placed at a certain distance in air to the force between them placed at the same distance in the medium.

1.5 MAGNETIC FIELD STRENGTH

The force experienced on a unit (one weber) N-pole placed d metres apart from another pole of strength m webers in a medium of relative permeability μ_r is given by $F = \frac{m}{4\pi\mu_0\mu_r d^2}$ newtons, and is known as the *magnetic field strength** at that point. It is represented by H. Its units are either newtons/weber or ampere-turns/metre.

Hence the magnetic field intensity at any point is defined as the force experienced by a unit north pole, when placed at that point.

$$i.e. \quad H = \frac{m}{4\pi\mu_0\mu_r d^2} \text{ newtons/Wb} \quad ...(1.2)$$

The force experienced by a pole of m webers placed in a uniform magnetic field of intensity H newtons/weber, will be equal to mH newtons.

Magnetic field intensity is a vector quantity *i.e* it has both magnitude and direction.

Magnetic field strength (H) at any point is said to be *unit magnetic field strength* if the force experienced by a unit north pole placed at that point is one newton.

1.6 LINES OF MAGNETIC FLUX AND LINES OF INDUCTION

Inspite of the fact that lines of magnetic flux have no physical existence, they do form a very convenient and useful basis for explaining various magnetic effects and for calculating their magnitudes.

Magnetic line of force is defined as the curve along which a unit N-pole would move if free to do so, when placed in a magnetic field. The tangent to these curves at any point gives the direction of magnetic field at that point.

Lines of magnetic flux are assumed to have the following properties:

(*i*) Each line of magnetic flux forms a closed loop, as shown by dotted lines in Fig. 1.4. This means that a line of flux emerging from any point at N-pole end of a magnet passing through the surrounding space back to the S-pole end and is then assumed to continue through the magnet to the point at which it is emerged at the N-pole end.

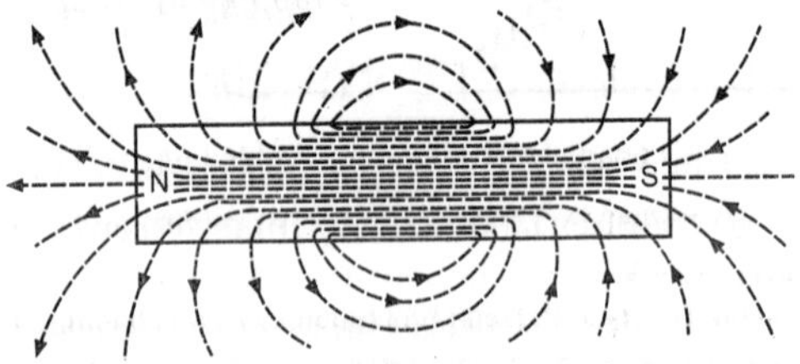

Fig. 1.4

(*ii*) The lines of magnetic flux never intersect.

(*iii*) Lines of magnetic flux tend to contract lengthwise and repel another sideways.

(*iv*) Lines of force emanate from and terminate on a surface normally.

Lines of magnetic flux travel from N-pole to S-pole externally, and continue to travel through the magnet and finally they reach the N-pole again forming closed curves, as illustrated in Fig. 1.4. The portion of the curves within the magnetic material are called the *lines of induction.*

1.7 MAGNETIC FLUX AND MAGNETIC FLUX DENSITY

Magnetic flux is the total number of lines of force comprising the magnetic field. In SI unit system unit north pole i.e. north pole of strength of one weber is supposed to radiate out the flux of one weber. It is represented by symbol Φ and is measured in webers.

$\therefore$ Flux emanated from north pole of strength m webers $= m$ Wb

* Terms magnetic field strength, magnetic field intensity, magnetizing force, magnetic intensity, field intensity are interchangeably used.

Magnetic flux density is defined as the magnetic flux passing per unit area through any material through a plane at right angles to the direction of flux. This is also known as *magnetic induction.* It is represented by English letter B and is measured in Tesla (Weber/m^2)

$$i.e. \quad B = \frac{\Phi}{a} \qquad ...(1.3)$$

Magnetic flux density is a scalar quantity.

1.8 RELATION BETWEEN MAGNETIC FIELD INTENSITY (H) AND INDUCTION DENSITY (B)

At any point in a magnetic field, field strength or magnetic field intensity H is the force that maintains the magnetic flux and produces a particular value of flux density B at that point. Thus the field intensity H is the cause and the flux density B is the effect. Thus the flux density can be assumed to be proportional to the field intensity in a magnetic field, *i.e.* in free space

$$\mathbf{B} = \mu_0 \mathbf{H} \qquad ...(1.4)$$

where B is in webers per square metre, H is in newtons per weber (or ampere-turns per metre) and μ_0 is a magnetic space constant, called the *permeability of free space,* and is equal to $4\pi \times 10^{-7}$ H/m.

It has been found that if the same field intensity H, which was applied in free space as mentioned above is now applied in a solid of relative permeability μ_r, the flux density B will be μ_r times that was in free space

$$i.e. \quad \mathbf{B} = \mu_0 \mu_r \mathbf{H} \qquad ...(1.5)$$

$$\text{or} \quad \mathbf{B} = \mu \mathbf{H} \qquad \text{where } \mu = \mu_0 \mu_r \qquad ...(1.6)$$

Comparing Eqs. (1.4) and (1.5) the *relative permeability* (μ_r) of a magnetic material may be defined as the ratio of flux density produced in the material to the flux density produced in vacuum or in a non-magnetic core, provided the field intensity is the same in both cases.

The value of μ_r varies from material to material and also depends on the degree to which the material is magnetized. It may have a value as high as 2,500.

1.9 MAGNETIC FIELD DUE TO A CURRENT CARRYING CONDUCTOR

In 1819 it was discovered by a Danish Physicist, Hans Christian Oersted that an electric current is always accomplished by certain magnetic effects.

Oersted found that when current is passed through a conductor placed above the magnetic needle, the needle turns in a certain direction, as shown in Fig. 1.5. He also found that when the direction of flow of current is reversed the magnetic needle also deflects in opposite direction.

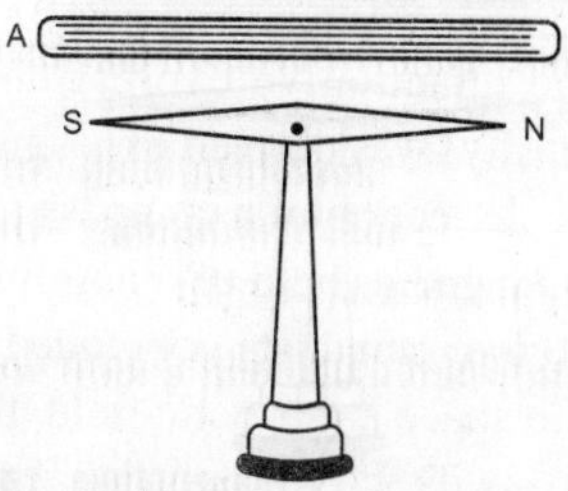

(a) No Current Flow

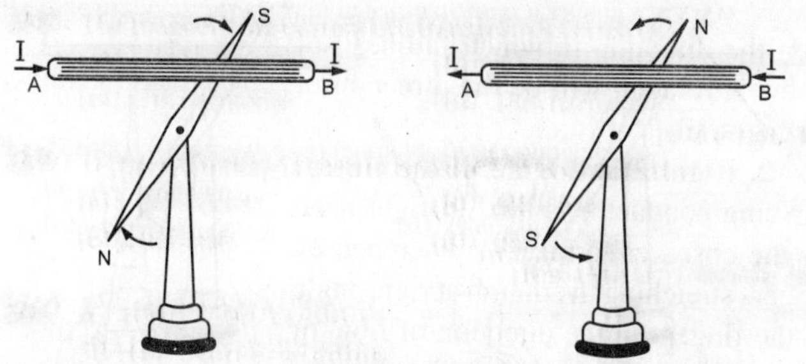

(b) Current Flowing From A to B *(c) Current Flowing From B to A*

Fig. 1.5

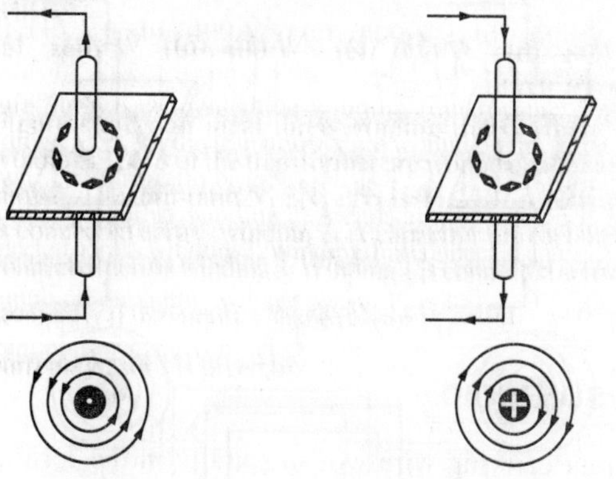

(a) *(b)*

Fig. 1.6

From Figs. 1.6 (*a*) and 1.6 (*b*) it is observed that when the current is passed through conductor in upward direction, the direction of lines of force is counter-clockwise direction (observed from the top of the conductor) and when the current is passed through the conductor in downward direction, the direction of lines of force is clockwise (observed from the top of the conductor).

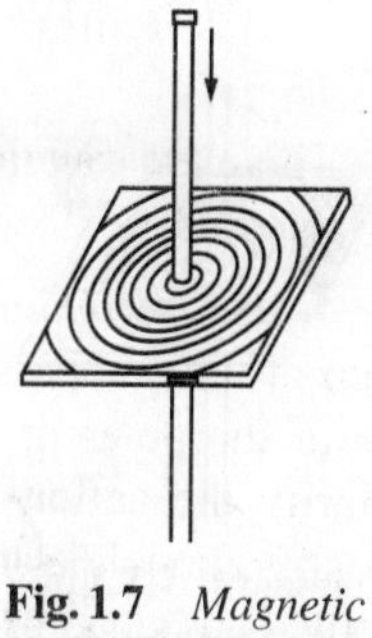

Fig. 1.7 *Magnetic Field Around A Current Carrying Conductor*

The properties of the lines of magnetic induction around a current carrying conductor are summarized as below:

(*i*) Lines of magnetic induction are circles, symmetrical about, and concentric with, the axis of the conductor.

(*ii*) The spacing between the lines of induction decreases as we move closer to the conductor (Fig. 1.7).

(*iii*) The direction of lines of magnetic induction depends on the direction of flow of current through the conductor.

(*iv*) Magnetic induction or flux density depends upon the strength (or magnitude) of the current flowing through the conductor.

The direction of lines of force (magnetic field) around a straight current carrying conductor may be determined by any of the following rules:

1. Corkscrew Rule. If the right handed corkscrew is held with its axis parallel to the conductor pointing the direction of flow of current and the head of the screw is rotated in such a direction that the screw moves in the direction of flow of current then the direction in which the head of screw is rotated, will be the direction of magnetic lines of force.

2. Right Hand Rule. If the current carrying conductor is held in right hand by the observer so that it is encircled by fingers stretching the thumb at right angle to the fingers in the direction of flow of current then finger tips will point the direction of magnetic lines of force, as shown in Fig. 1.8 (*b*).

(*a*) *Corkscrew Rule*

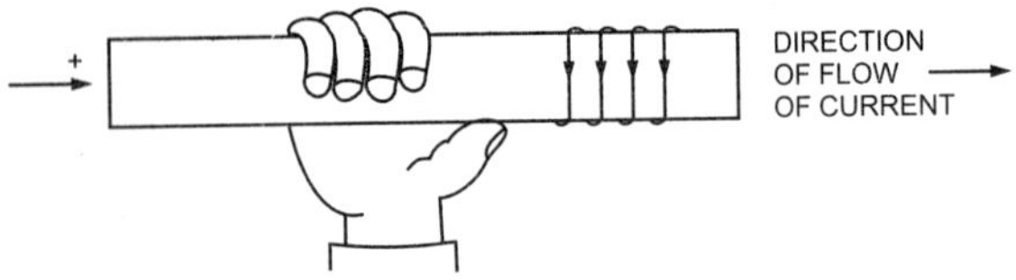

(*b*) *Right Hand Rule*

Fig. 1.8

1.10 SOLENOID

The current carrying wire wound spirally in the form of helix about an axis, as shown in Fig. 1.9, is known as *solenoid or coil.* Magnetic field produced due to current carrying solenoid is fairly uniform over a small region in the middle of the coil. It acts just like a bar magnet having north and south poles.

Fig. 1.9 *Solenoid*

There are several methods used to determine the polarity of the solenoid.

1. By Use of Compass Needle. If one of the poles (say north pole) of the compass needle be brought into close proximity to one of the poles of the current carrying solenoid of unknown polarity the action of the compass needle will immediately classify the pole as north or south depending upon whether the needle is repelled or attracted.

2. Helix Rule. If the helix is held in right hand in such a manner that the finger tips point in the direction of flow of current and thumb is outstretched longitudinally along the coil, it will point towards north pole (Fig. 1.10).

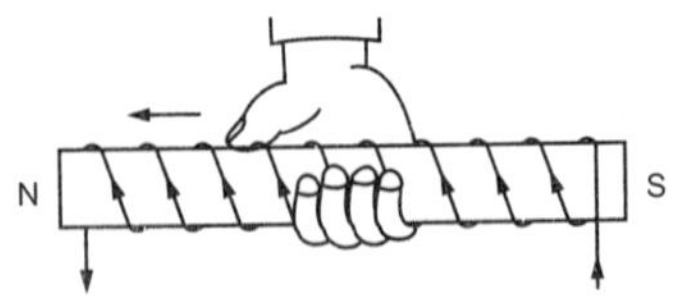

Fig. 1.10 *Helix Rule*

1.11 BASIC LAW OF MAGNETIC FIELD

A magnetic field always accompanies electric currents. Knowing the magnitude and paths of the currents, the corresponding magnetic fields can be computed. In doing so, account must be taken not only of the currents in the conductors of the electric circuit but also of the intermolecular currents that cause magnetization of the medium surrounding the current carrying conductors. In practice, it is necessary only when there are ferro-magnetic substances (such as iron, steel, cast iron, nickel, cobalt and certain alloys) near the conductors.

Let us imagine a magnetic field with currents I_1, I_2, and I_3 an arbitrary closed contour *l* and an arbitrary surface S supported by it (Fig. 1.11). At least a part of these currents intersect the surface: some in one direction and others in the other direction. Let us choose an arbitrary direction of transverse of the contour *l* as positive and consider all currents as positive if they intersect the surface S in the same direction as the end of a corkscrew handle moving along the contour *l* in a positive direction. Thus the current I_1 is negative and current I_2 is positive.

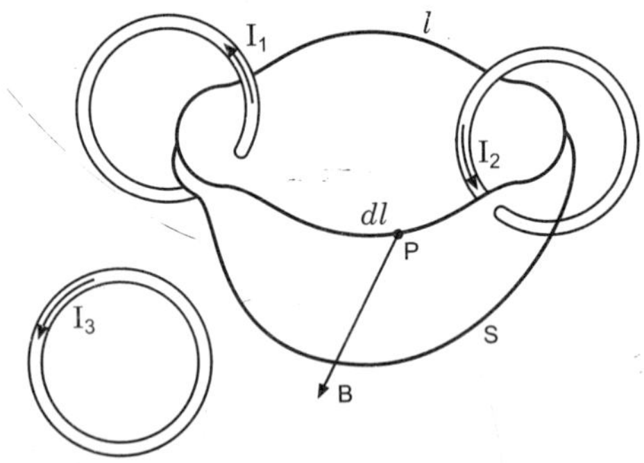

Fig. 1.11 *Magnetic Field Basic Law*

The vector of magnetic field intensity **H** at any point of the contour P depends on the currents both intersecting and not intersecting the surface S, on the shape and dimensions of the conductors carrying these currents, and on the position of point P relative to these conductors. However, the line integral of the vector **H** over the whole contour depends only on the algebraic sum of the currents intersecting the surface S, and is equal to that sum i.e.

$$\oint \mathbf{H}\, dl = \Sigma I = I_2 - I_1 = I \qquad ...(1.7)$$

The above relation is known as *Ampere's circuital law*, sometimes referred as ***Ampere's work law***, which states that the line integral of the magnetic field intensity **H** over a closed path is equal to the current enclosed.

The above law is very comprehensive and provides a basis for calculation of magnetic circuits and makes it possible in some cases (such as in the cases of a long current carrying conductor or a long solenoid) to determine readily the strength of the magnetic field.

1.12 BIOT-SAVART LAW

The quantitative expression for magnetic induction (or flux density) was derived experimentally by Biot and Savart. Taking the deflection of a compass needle as a measure of the intensity of magnetic field of a current, varying in magnitude and shape, the two scientists concluded that any current element projects into space a magnetic field, the magnetic flux density of which, dB, is directly proportional to the length of the element d*l*, the current I, the sine of the angle θ between the direction of the current and the vector joining a given point of the field and the current element, and is inversely proportional to the square of the distance of the given point from the current element, *r* i.e $dB \propto \dfrac{I\,dl \sin\theta}{r^2}$ or $dB = K\dfrac{I\,dl \sin\theta}{r^2}$

where K is a proportionality constant, which depends upon the magnetic properties of the medium and the system of units employed. In SI system of units

$K = \dfrac{\mu_0 \mu_r}{4\pi}$.

$$\text{So } dB = \frac{\mu_0 \mu_r}{4\pi} \cdot \frac{I\,dl \sin\theta}{r^2} \quad ...(1.8)$$

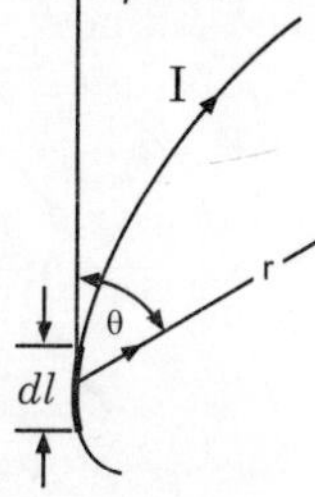

Fig. 1.12

1.13 FORCE ON A CURRENT CARRYING CONDUCTOR LYING IN THE MAGNETIC FIELD

In Fig. 1.13 (*a*) a uniform magnetic field between the two opposite poles is shown.

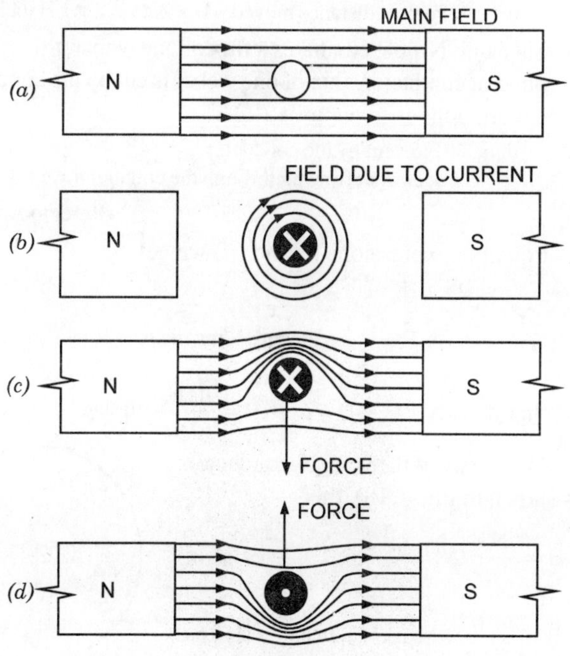

Fig. 1.13

In Fig. 1.13 (*b*) the cross section of a conductor carrying current in inward direction placed between the two magnets, the field being temporarily removed, is shown. By applying the right-hand thumb rule, the direction of the field around the conductor is found to be clockwise.

If the current carrying conductor shown in Fig. 1.13 (*b*) is placed in the magnetic field shown in Fig. 1.13 (*a*), the resultant magnetic field would be similar to that shown in Fig. 1.13 (*c*).

The lines of force above the conductor are strengthened, since they are in same direction but the lines of force below the conductor are weakened because the two fields below the conductor are opposite in directions and hence tend to destroy each other. Magnetic lines like rubber bands have a tendency to strengthen out and, therefore, a force is experienced on the conductor in the downward direction, as shown in Fig. 1.13 (*c*).

If the direction of current is reversed in the conductor, as shown in Fig. 1.13 (*d*), the direction of force experienced is reversed. In this case the lines of force above the conductor are weakened while those below the conductor are strengthened.

Hence it is observed that when a current carrying conductor is placed at right angle to the direction of magnetic field, a mechanical force is experienced on the conductor in a direction perpendicular to both the direction of magnetic field and flow of current.

The direction of this force can be determined by applying *Fleming's Left Hand Rule* which states that if the thumb, forefinger and middle finger of the left hand are stretched in such a way that they are at right angles to each other mutually and forefinger points towards the direction of the magnetic field, middle finger towards the direction of the flow of current then thumb will point the direction of force acting on the conductor [Fig. 1.14 (*a*)].

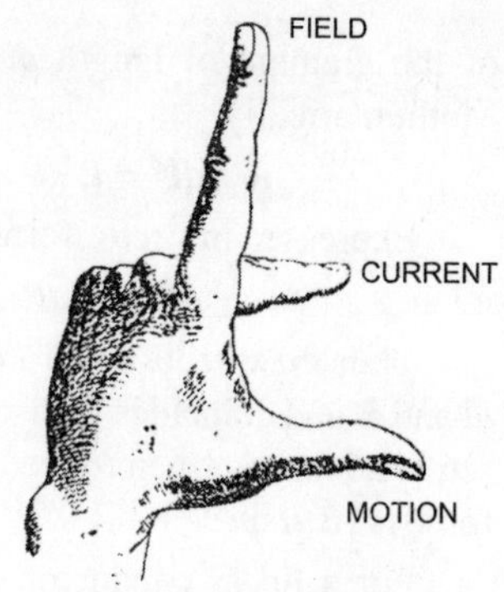

(*a*) *Fleming's Left Hand Rule*

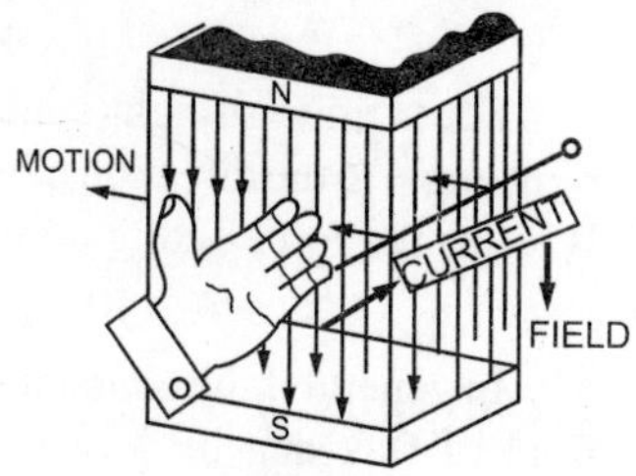

(*b*)

Fig. 1.14

If the current in the conductor is reversed, keeping the direction of magnetic field unchanged the direction of force will reverse.

Similarly, if the direction of the magnetic field is reversed, keeping the direction of flow of current in the conductor unchanged, the direction of force will reverse.

Similarly, if the direction of the magnetic field is reversed, keeping the direction of flow of current in the conductor unchanged, the direction of force will reverse.

It should be noted that no force is exerted on a conductor when it lies parallel to the magnetic field.

Figure 1.14 (*b*) shows another method of finding the direction of force acting on a current carrying conductor placed in a magnetic field. It is known as *Flat left hand rule*. If the palm of the left hand is so placed that the lines of induction leaving the N-pole enter the palm and the four fingers are in the direction of flow of current in the conductor, then the thumb will indicate the direction of force acting on the conductor.

The force *d*F acting upon an element of length *dl* of the conductor is proportional to the magnetic induction B, the current I, the length of element *dl* and sine of angle θ between the direction

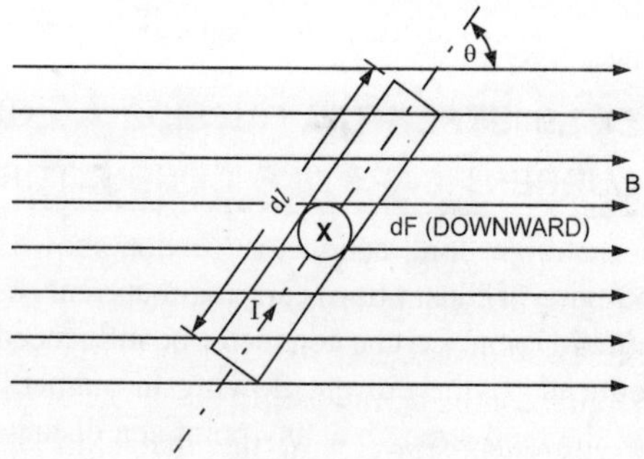

Fig. 1.15

of the element of length dl and direction of magnetic field. Mathematically

$$d\mathbf{F} = \mathrm{B I}\, dl \sin\theta \text{ newtons} \quad ...(1.9)$$

Expressed in vector form

$$d\mathbf{F} = \mathrm{I}\,(dl \times \mathbf{B}) \quad ...(1.10)$$

Thus the direction of force $d\mathbf{F}$ is perpendicular to the vector dl and $\mathbf{B}$ and coincides with the direction in which a right handed screw advances when rotated from dl to B. The magnitude of the force is $\mathrm{IB}dl\sin\theta$.

For a linear conductor of length l in a uniform field B is given as

$$\mathbf{F} = \mathrm{I}\,(l \times \mathbf{B})$$

For a finite straight conductor of length l metres placed at right angle to the direction of a uniform magnetic field of strength B teslas the force acting upon the conductor will be

$$\mathrm{F} = \mathrm{BI}l \text{ newtons} \quad ...(1.11a)$$

In general if the conductor lies at an angle θ with a magnetic field of flux density B webers/metre^2, then mechanical force experienced on a current carrying conductor is given by

$$\mathrm{F} = \mathrm{BI}l\sin\theta \text{ newtons} \quad ...(1.11b)$$

where l is the length of the conductor in metres and I is the current carried by the conductor in amperes.

1.14 WORK DONE IN MOVING A CURRENT CARRYING CONDUCTOR IN THE MAGNETIC FIELD

Consider a conductor of length l metres, carrying a current of I amperes flowing downward, placed at right angle to a uniform magnetic field of flux density B webers/metre^2 (lines of force coming out of the plane of the paper shown by dots in Fig. 1.16).

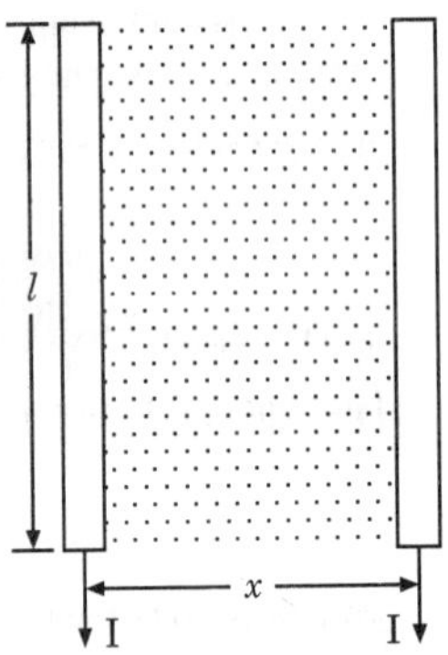

Fig. 1.16

The mechanical force experienced on the conductor,

$$\mathrm{F} = \mathrm{BI}l \text{ newtons}$$

If the above conductor is moved against this force F through a distance of x metres, then

Work done = Force × distance moved = F x Joules = BIlx joules

Since lx is the area swept out by the conductor

$\therefore$ Work done = B I a joules = (B a)I = Φ I joules ...(1.12)

(since total flux, Φ = B × a)

= Flux cut by the conductor × current

1.15 FIELD STRENGTH OUTSIDE A LONG CURRENT CARRYING CONDUCTOR

Consider a straight long conductor (section shown is magnified for the purpose of illustration) carrying a current of I amperes in downward direction. Let the conductor be influenced only by the field produced by the current flowing through it (no external field). Let the field strength at any point at a distance of r metres from the centre of the conductor due to its own field be H newtons/wb.

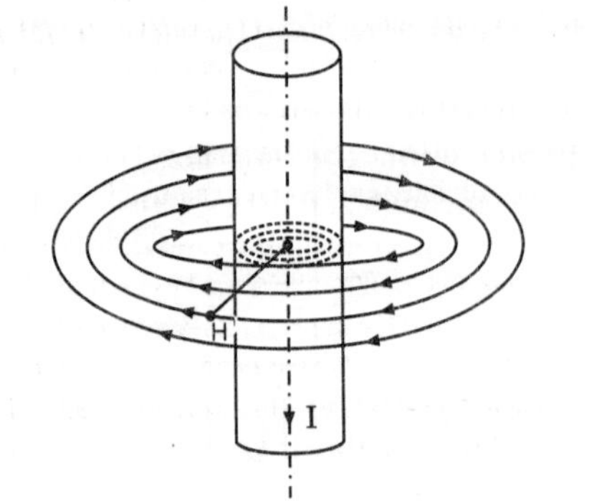

Fig. 1.17

Force acting on a unit N-pole placed at this point
= H newtons, tangential to the lines of force

Work done in moving a unit N-pole along the circumference of a circle of radius r metres against this force

= Force × distance moved = H × 2 π r = 2 π r H joules ...(i)

But a unit N-pole emanates a flux of one weber, therefore, in one round of conductor, flux of one weber is cut by the conductor, hence work done is given as

Work done = Flux cut by the conductor × current flowing through the conductor = 1 × I ...(ii)

(Refer to Art. 1.14)

Equating expressions (i) and (ii) we get

$$2\pi r \mathrm{H} = \mathrm{I}$$

$$\text{or } \mathrm{H} = \frac{\mathrm{I}}{2\pi r} \text{ ampere-turns/metre} \quad ...(1.13)$$

$$\text{and flux density, } \mathrm{B} = \mu \mathrm{H} = \mu_0 \mu_r \mathrm{H} = \frac{\mu_0 \mu_r \mathrm{I}}{2\pi r} \text{ teslas} \quad ...(1.14)$$

Obviously, if there are N-conductors (as shown in Fig. 1.18), then

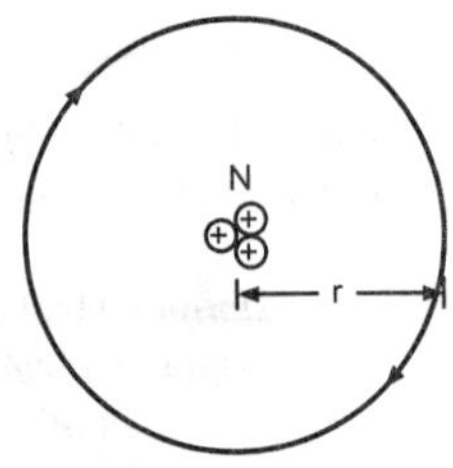

Fig. 1.18

$$\mathrm{H} = \frac{\mathrm{NI}}{2\pi r} \text{ AT/m} \quad ...(1.15)$$

$$\mathrm{B} = \frac{\mu_0 \mathrm{NI}}{2\pi r} \text{ teslas..........in air} \quad ...(1.16a)$$

$$= \frac{\mu_0 \mu_r \mathrm{NI}}{2\pi r}$$

teslas.......in a substance of relative permeability, μ_r ...(1.16b)

1.16 FORCE BETWEEN TWO PARALLEL CURRENT CARRYING CONDUCTORS

When two parallel conductors carrying current in the same direction (say in downward direction), as shown in Fig. 1.19 (a) the lines of force encircle each conductor in the same direction and the resultant field is an envelop of acting like elastic bands tending to pull the conductors together.

When two parallel conductors carry current in the opposite directions, as shown in Fig. 1.19 (b), the lines of force encircle but not concentric either with another or with the conductors. The lines of force are crowded between the conductors tending to repel the conductors further apart.

Thus we see that the force of attraction or repulsion is experienced between two parallel conductors carrying current in the same or in opposite directions respectively.

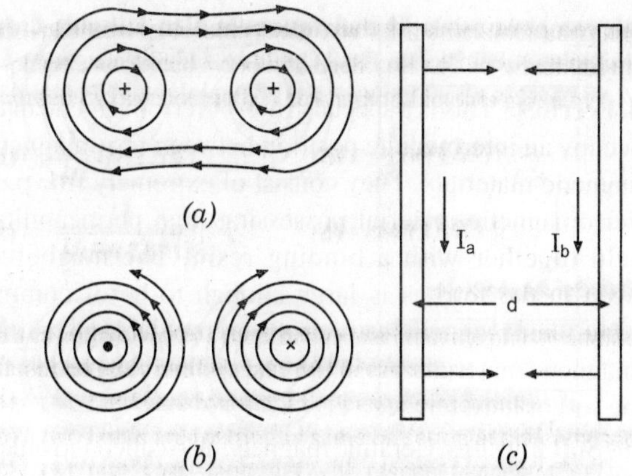

Fig. 1.19

In order to determine the magnitude of force between two parallel current carrying conductors let us consider two parallel conductors A and B each of length *l* metres and placed *d* metres apart in air.

Let the current flowing through the conductors A and B be I_a and I_b respectively.

Field produced due to one of the two conductors experiences a force on another conductor, therefore, to determine the force between two parallel current carrying conductors one of them (say conductor A) be considered placed in the magnetic field due to another one (due to conductor B.)

$$\text{Magnetic field due to conductor B, } H = \frac{I_b}{2\pi d} \text{ N/Wb}$$

$$\text{Force acting on conductor A, } F = B\, I_a\, l$$

where B is the flux density of the field due to conductor B

$$\text{or } F = \mu_0 \mu_r H \times I_a \times l \qquad \text{Since } B = \mu_0 \mu_r H$$

$$= 4\pi \times 10^{-7} \times 1 \times \frac{I_b}{2\pi d} \times I_a \times l$$

$$= \frac{2 I_a I_b\, l}{d} \times 10^{-7} \text{ N or } \frac{2 I_a I_b}{d} \times 10^{-7} \text{ N/m} \quad ...(1.17)$$

Example 1.1. Two long, parallel conductors are laid at a centre to centre distance of 1.5 m in air. Calculate and state the nature of force between them when they are carrying a curent of 1,000 A each in opposite directions. The length of each conductor is 120 m.

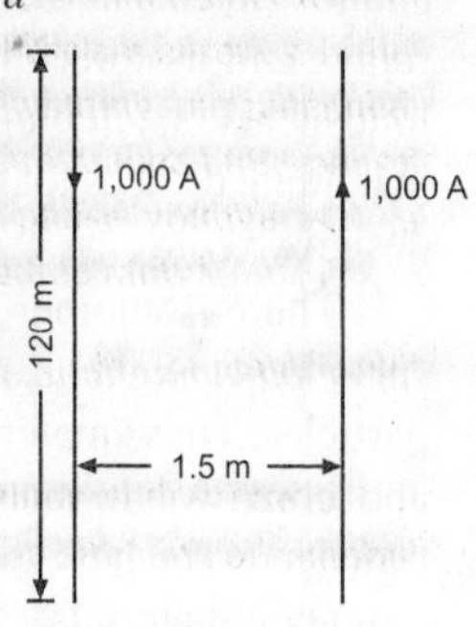

Fig. 1.20

Solution: Length of each conductor,

$$l = 120 \text{ m}$$

Current in each conductor,

$$I_a = I_b = 1{,}000 \text{ A}$$

Centre to centre distance between conductors,

$$d = 1.5 \text{ m}$$

$$\text{Force experienced, } F = \frac{2 I_a I_b}{d} \times l \times 10^{-7} \text{ newtons}$$

$$= \frac{2 \times 1{,}000 \times 1{,}000}{1.5} \times 120 \times 10^{-7}$$

$$= 16 \text{ newtons } \textbf{Ans.}$$

Since the conductors carry current in the opposite directions, conductors will be pulled apart (*i.e.* repulsive force).

Example 1.2. Two long parallel wires A and B, 12 cm apart carry currents of 750A and 500A respectively in opposite directions. Determine the flux density at the mid point of the perpendicular line between the wires.

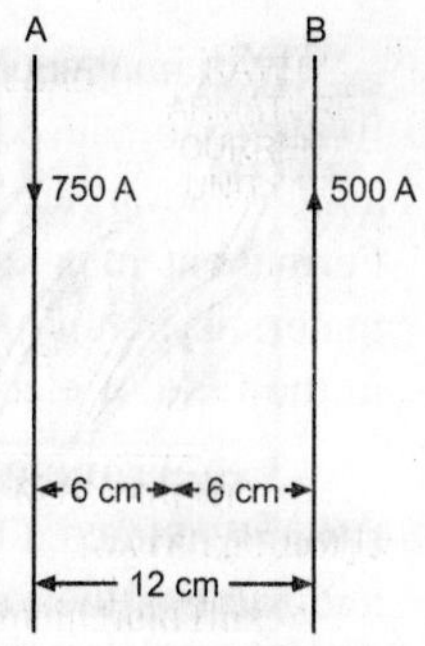

Fig. 1.21

Solution: Current I_A = 750 A

Current I_B = 500 A

Flux density at mid point between two conductors, due to conductor A,

$$B_1 = \frac{\mu_0 \mu_r I_A}{2\pi r}$$

$$= \frac{4\pi \times 10^{-7} \times 1 \times 750}{2\pi \times 0.06}$$

$$= 2.5 \times 10^{-3} \text{ T}$$

Flux density at mid point between two conductors due to conductor B,

$$B_2 = \frac{\mu_0 \mu_r I_B}{2\pi r} = \frac{4\pi \times 10^{-7} \times 1 \times 500}{2\pi \times 0.06} = 1.667 \times 10^{-3} \text{ T}$$

As seen from Fig. 1.19 (*b*), when the two conductors carry current in opposite directions, the fields are additive, so

Net flux density, $B = B_1 + B_2 = 2.5 + 1.667 = 4.167 \times 10^{-3}$ T **Ans.**

1.17 MAGNETIC MATERIALS

Although all materials have magnetic properties of some kind being either *diamagnetic*, *paramagnetic* or *ferromagnetic*, the term "magnetic material" is customarily applied only to substances which exhibit ferromagnetism.

1. Paramagnetic Materials. The materials, which are not strongly attracted by a magnet, such as aluminium, tin, platinum, magnesium, manganese etc. are known as *paramagnetic materials*. Their relative permeability is small but positive. For example, the relative permeabilties for aluminium, air and platinum are 1.00000065, 1.0000031 and 1.00036 respectively. Such materials are slightly magnetized when placed in a strong magnetic field and act in the direction of the magnetic field.

In paramagnetic materials the individual atomic dipoles are oriented in a random fashion, as shown in Fig 1.22. The resultant magnetic field is, therefore, negligible. When an external magnetic field is applied, the permanent magnetic dipoles orient themselves parallel to the applied magnetic field and give rise to a positive magnetization. Since the orientation of the dipoles parallel to the applied magnetic field is not complete, the magnetization is small. These materials have little application in the field of electrical engineering.

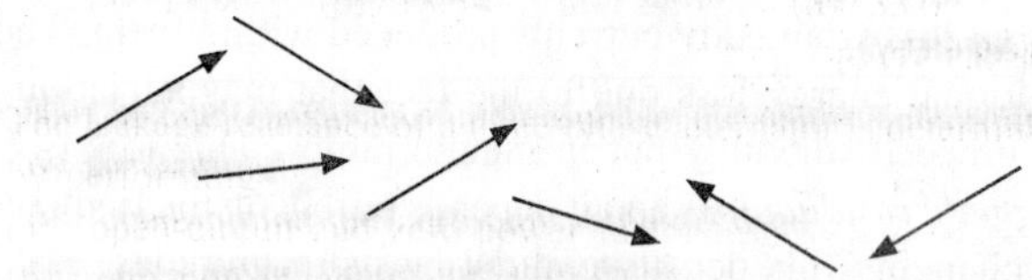

Fig. 1.22

2. Diamagnetic Materials. The materials which are repelled by a magnet such as zinc, mercury, lead, sulphur, copper, silver, bismuth, wood etc., are known as *diamagnetic materials*. Their permeability is slightly less than unity. For example the relative permeabilities of bismuth, copper and wood are 0.99983, 0.999995 and 0.9999995 respectively. They are slightly magnetized when placed in a strong magnetic field and act in the direction opposite to that of applied magnetic field.

In diamagnetic materials, the two relatively weak magnetic fields (one caused due to orbital revolution and other due to axial rotation) are in opposite directions and cancel each other. Permanent magnetic dipoles are absent in them. Diamagnetic materials are unimportant from the point of views of application in the field of electrical engineering.

3. Ferromagnetic Materials. The materials, which are strongly attracted by a magnet such as iron, steel, nickel, cobalt and some of their alloys, are known as *ferromagnetic materials.* Their permeability is very high (varying from several hundreds to many thousands).

The opposing magnetic effects of electron orbital motion and electron spin do not eliminate each other in an atom of such a material. There is a relatively large contribution from each atom which aids in the establishment of an internal magnetic field, so that when the material is placed in a magnetic field, its value is increased many times the value that was present in the free space before the material was placed there.

For the purpose of electrical engineering it will suffice to classify the materials simply as non-ferromagnetic and ferromagnetic materials. The former includes materials of relative permeabilities practically equal to unity while the latter have relative permeabilities many times greater than unity. Paramagnetic and diamagnetic materials fall in the former (*i.e.* non-ferromagnetic) class.

Ferromagnetic materials are of two types: (*a*) those easily magnetized, called the *soft magnetic materials* and (*b*) those retaining their magnetism with great tenacity, designated as *hard magnetic materials.*

(*a*) **Soft Ferromagnetic Materials.** They have high relative permeability, low coercive force, easily magnetized and demagnetized and have extremely small hysteresis. Soft ferromagnetic materials are iron and its alloys with nickel, cobalt, tungsten and aluminium. Ease of magnetization and demagnetization makes them highly suitable for applications involving changing magnetic flux as in electromagnets, electric motors, generators, transformers, inductors, telephone receivers, relays etc. They are also useful for magnetic screening. Their properties may be greatly enhanced through careful manufacture, and by heating and slow annealing so as to achieve a high degree of crystal purity. Large magnetic moment at room temperature makes soft ferromagnetic materials extremely useful for magnetic circuits but ferromagnetics are very good conductors and suffer energy loss from eddy currents produced within them. There is additional energy loss due to the fact that magnetization does not proceed smoothly but in minute jumps. This loss is called *magnetic residual loss* and it depends purely on the frequency of the changing flux density and not on its magnitude.

(*b*) **Hard Ferromagnetic Materials.** They have relatively low permeability, and very high coercive force. These are difficult to magnetize and demagnetize. Typical hard ferromagnetic materials include cobalt steel and various ferromagnetic alloys of nickel, aluminium and cobalt. They retain high percentage of their magnetization and have relatively high hysteresis loss. They are highly suited for use as permanent magnets in loudspeakers etc. Sometimes they are also called the *permanent magnetic materials.* During manufacturing, a high degree of dislocation is introduced. Hard ferromagnetic materials are frequently heated to high temperature and then quenched in suitable liquid to introduce strains.

4. Ferrites. This is a special group of ferromagnetic materials that occupy an intermediate position between ferromagnetic and nonmagnetic materials. They consist of extremely fine particles of a ferromagnetic material possessing high permeability, and are held together with a binding resin. The magnetization produced in the ferrites is large enough to be of commercial value but their magnetic saturations are not as high as those of ferromagnetic materials. As in the case of ferromagnetics, ferrites may be soft or hard ferrites.

(*a*) **Soft Ferrites.** Ceramic magnets, also called ferromagnetic ceramics and ferrites, are made of an iron oxide, Fe_2O_3, with one or more divalent oxides such as NiO, MnO or ZnO. These magnets have a square hysteresis loop and high resistance to demagnetization and are valued for magnets for computing machines where a high resonance is desired. The great advantage of ferrites is their high resistivity. Commercial ferrites can have resistivities as high as 10^9 ohm-cm. Eddy currents resulting from alternating fields are therefore, reduced to a minimum, and the range of application of these magnetic materials is extended to high frequencies, even to microwaves. Ferrites are carefully made by mixing powdered oxides, compacting and sintering at high temperatures. High-frequency transformers in television and frequency modulated receivers are almost always made with ferrite cores. Ferrites, with large magnetostrictive effects, are sometimes used in electromechanical transducers. In high frequency application, magnetostriction in ferrites can lead to undesirable noise.

(*b*) **Hard Ferrites.** These are ceramic permanent magnetic materials. The most important family of hard ferrites has the basic composition $MO.Fe_2O_3$, where M is a barium (Ba) ion or a strontium (Sr) ion. These materials have a hexagonal structure and low in cost and density. Hard ferrites are used in generators, relays and motors. Electronic applications include magnets for loudspeakers, telephone ringers and receivers. They are also used in holding devices for door closers, seals, latches and in several toy designs.

There is another category of magnetic materials, known as *super-paramagnetic,* made from powdered iron or other magnetic particles. These materials are used in transformers for electronics and cores for inductors. Permalloy (molybdenum-nickel-iron powder) is the best known example of this important category of magnetic materials.

Applications of Hard Magnetic Materials. The high-energy hard magnetic materials are employed in a host of different devices in a variety of technological fields. One common application is in motors. Permanent magnets are far superior to electromagnets in that their magnetic fields are continuously maintained and without the necessity of having to expand electrical power; furthermore, no heat is generated during operation. Motors using permanent magnets are much smaller than their electromagnet counterparts and are utilized extensively in fractional kW units.

Permanent magnet (or hard magnetic) materials employed for permanent magnet dc (PMDC) motors are of three types viz Alnicos, ceramics (ferrites) and rare-earth materials.

ALNICO (aluminium-nickel-iron cobalt) are commercially the most important of the hard magnetic materials. Alnico is an exceptionally hard magnetic material, and therefore, Alnico magnets are used in motors up to 200 kW.

The most important family of hard ferrites has the basic composition $MO.Fe_2O_3$, where M is a barium (Ba) iron or a strontium (Sr) iron. These materials have a hexagonal structure and low in cost and density. Ceramic magnets are most economical in fractional kW motors.

The rare-earth magnetic materials are very costly, but are the most economic choice in very small motors. The high permeance and coercivity allow marked reductions in motor frame size for the same output compared to motors using ceramic (ferrite) magnets. For very high temperature applications Alnico or rare-earth cobalt magnets must be used.

Familiar motor applications include the following: in cordless drills and screw drivers; in automobiles (steerings, window winder, wiper, washer, and fan motors); in audio and video recorders; and clocks. Other common devices that employ these magnetic materials are speakers in audio systems, light-weight earphones, hearing aids, and on computer peripherals. The permanent magnets are also used in instruments like galvanometer, ammeters, voltmeters, wattmeters, compasses, electron beam focussing and positioning, telephones, loudspeakers, eddy current brakes etc.

1.18 PROPERTIES OF MAGNETIC MATERIALS

In the context of electromechanical energy-conversion devices, the importance of magnetic materials is twofold. Through their use it is possible to obtain large magnetic flux densities with relatively low levels of magnetizing force. Since magnetic forces and energy density are increased with increasing flux density, this effect plays a large role in the performance of energy-conversion devices.

In addition, magnetic materials can be used to constrain and direct magnetic fields in well-defined paths. In a transformer they are used to maximize the coupling between the windings as well as to lower the excitation current needed for transformer operation. In electric machinery they are used to shape the fields to maximize the desired torque-producing characteristics. Thus magnetic materials can be used to provide specific desirable device characteristics.

Ferromagnetic materials, composed of iron and alloys of iron with cobalt, tungsten, nickel, aluminium and other metals, are by far the most common magnetic materials. Although these materials are characterized by a wide range of properties, the basic phenomena responsible for their properties are common to them all.

Ferromagnetic materials are composed of a large number of domains, *i.e.* regions in which the magnetic moments of all the atoms are parallel, giving rise to a net magnetic moment for that domain. In an unmagnetized sample of material, the domain magnetic moments are randomly oriented and the net magnetic flux in the material is zero.

When an external magnetizing force is applied to this material, the domain magnetic moments tend to align with the applied magnetic field. As a result, the dipole magnetic moments add to the applied field, resulting in a much larger value of flux density than would exist due to magnetizing force alone. Thus the effective permeability μ, equal to the ratio of total magnetic flux density to the applied magnetizing force, is large compared with the permeability of free space μ_0. This behaviour continues until all the magnetic moments are aligned with the applied field, at this point they can no longer contribute to increasing the magnetic flux density, and the material is said to be fully saturated.

In the absence of an externally applied magnetizing force, the domain magnetic moments naturally align along certain directions associated with the crystal structure of the domain, known as *axes of easy magnetization*. Thus if the applied magnetizing force is now reduced, the domain magnetic moments relax to the direction of easy magnetism nearest to that of the applied field. As a result, when applied field is reduced to zero, the magnetic dipole moments will no longer be totally random in their orientation; they will retain a net magnetization component along the applied field direction. It is this effect which is responsible for the phenomenon known as *magnetic hysteresis*.

The relationship between B and H for a ferromagnetic material is both nonlinear and multivalued. In general, the characteristics of the material cannot be described analytically. The most common curve used to describe a magnetic material is the *B-H curve* or *hysteresis loop*. It is usual to plot H as abscissas and B as ordinates.

Another interesting characteristic of magnetic materials is revealed when the field intensity, having been increased to some value, say H_1, is subsequently reduced. It is found that the material opposes demagnetization and, accordingly does not retrace along

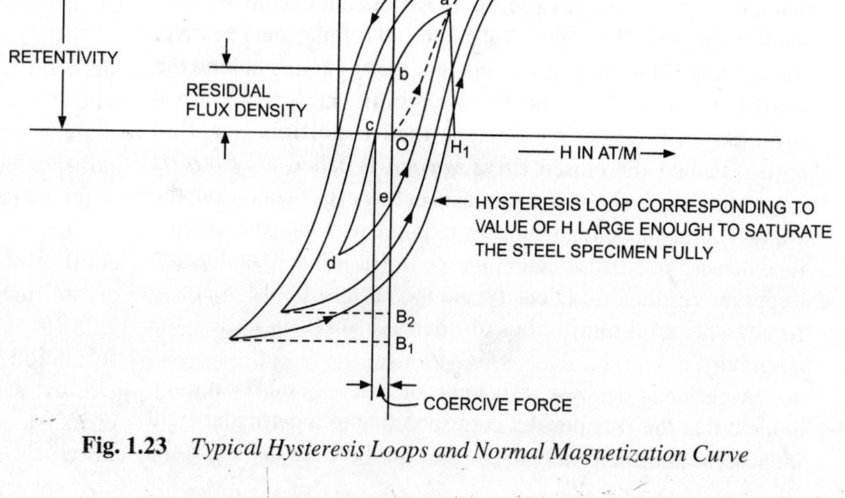

Fig. 1.23 *Typical Hysteresis Loops and Normal Magnetization Curve*

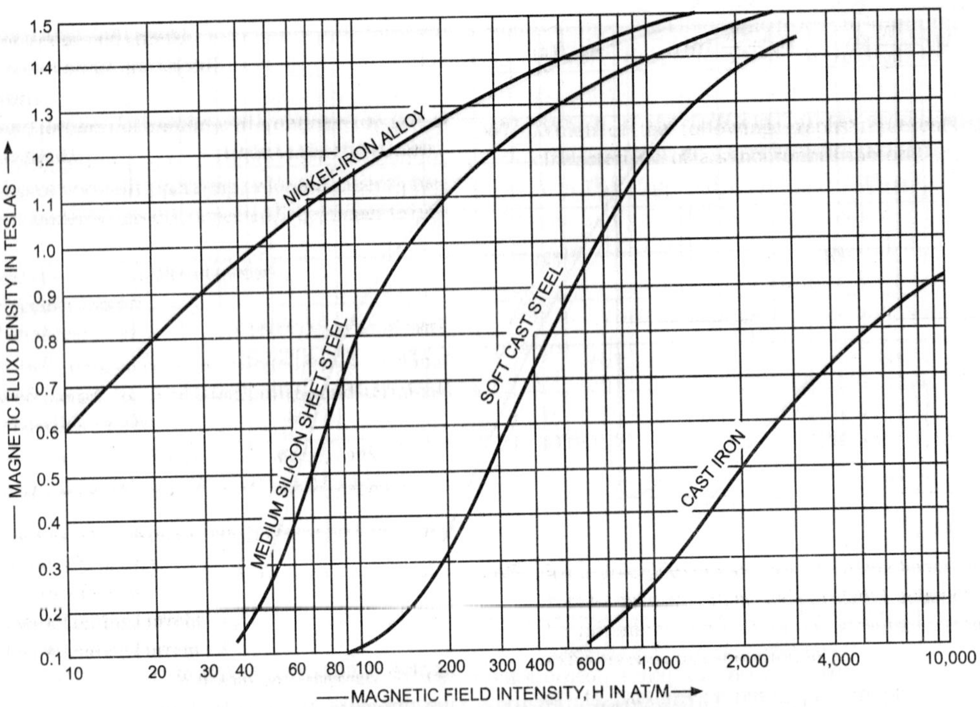

Fig. 1.24 *Magnetization Curves of Typical Ferromagnetic Materials*

the magnetizing curve *Oa* but rather along a curve located above *Oa* (curve *ab* in Fig. 1.23). Furthermore, it is seen that when the field intensity *H* is returned to zero, the flux density *B* is no longer zero as was the case with virgin sample. This happens because some of the domains remain oriented in the direction of the originally applied field, as discussed above. The value of flux density B that remains after the field intensity H is removed is called the ***residual flux density, remanent or residual magnetism***. Moreover, its value varies with the extent to which the material is magnetized. The maximum possible value of the residual flux density is called *retentivity* and results whenever values of H are used that cause complete saturation.

Frequently, in engineering applications of ferromagnetic materials the steel is subjected to cyclically varying values of H having the same positive and negative limits. As magnetizing force H varies through many identical cycles, the graph of flux density B versus magnetizing force H gradually approaches a fixed closed curve as shown in Fig. 1.23. The loop is always traversed in the direction indicated by the arrows. Since time is the implicit variable for these loops, note that B is always lagging behind H. Thus, when H is zero, B is finite and positive, as at point *b*, and when *B* is zero, as at point *c*, *H* is finite and negative, and so forth. This tendency of the flux density to lag behind the field intensity when the ferromagnetic material is in a symmetrically cyclically magnetized condition is called *hysteresis* and the closed curve *abcdea* is called a *hysteresis loop*. Moreover, when the material is in this cyclic condition, the amount of magnetic field intensity required to reduce the residual flux density to zero is called the *coercive force*. Usually, the larger the residual flux density, the larger must be the coercive force. The minimum value of coercive force is called the *coercivity*.

A glance at the hysteresis loops of Fig. 1.23 makes it quite evident that the flux density corresponding to a particular field intensity is not single valued. Its value lies between certain limits depending upon the previous history of the ferromagnetic material. However, since in many situations involving magnetic devices this previous history is unknown, a compromise procedure is used in making magnetic calculations by working with a single-valued curve called the *normal magnetization curve*. This curve is found by drawing a curve through the tips of a group of hysteresis loops generated while in a cyclic condition. Such a curve is *oafg* in Fig. 1.23.

Typical normal magnetization curves of commonly used ferromagnetic materials are shown in Fig. 1.24.

1.19 MAGNETOSTRICTION

When a ferromagnetic material is magnetized, its dimensions change slightly, and the sample being magnetized either expands or contracts in the direction of magnetization. This magnetically induced reversible elastic strain ($\delta l/l$) is called *magnetostriction* and is of the order of 10^{-6}. The energy due to the mechanical stresses created by magnetostriction is called *magnetostrictive energy*. For iron, the magnetostriction is positive at low fields and negative at high fields.

The cause of magnetostriction is attributed to the change in the bond length between the atoms in a ferromagnetic metal when their electron-spin dipole moments are rotated into alignment during magnetization. The fields of the dipoles may attract or repel each other, leading to the contraction or expansion of the metal during magnetization.

Let us now consider the effect of magnetostriction on the equilibrium configuration of the domain structure of cubic crystalline materials, such as shown in Figs. 1.25 (*a*) and 1.25 (*b*). Because of the cubic symmetry of the crystals, the formation of triangular-shaped domains, called *domains of closure*, at the ends of the crystal eliminates the magnetostatic energy associated with an external magnetic field and hence lowers the energy of the material. It might appear that very large

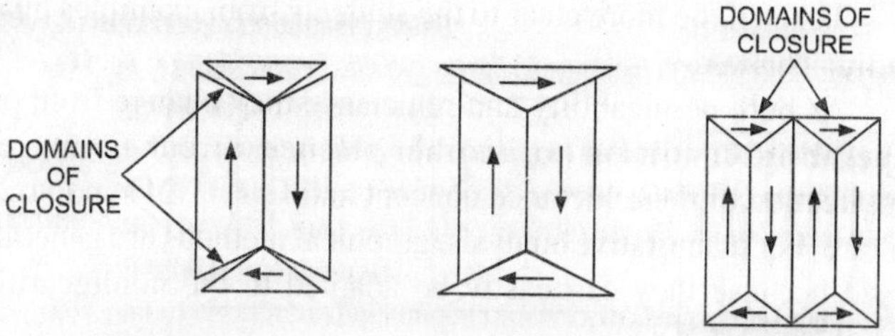

(a) Negative Magnetostriction (b) Positive Magnetostriction (c) Reduced Domain Size

Fig. 1.25 *Magnetostriction in Cubic Magnetic Materials*

domains such as those shown in Figs. 1.25 (*a*) and 1.25 (*b*) would be the lowest energy and most stable configuration since there is minimum wall energy. However, this is not the case since magnetostrictive stresses introduced during magnetization tend to be larger domains. Smaller magnetic domains, such as those in the Fig. 1.25 (*c*) reduce magnetostrictive stresses but increase domain wall area and energy. Thus, the equilibrium domain configuration is reached when the sum of the magnetostrictive and domain wall energies is a minimum.

In summary, the domain structure formed in ferromagnetic materials is determined by the various contributions of exchange, magnetostatic, magnetocrystalline anisotropic, domain wall, and magnetostrictive energies to its total magnetic energy. The equilibrium or most stable configuration is that for which the total magnetic energy is the lowest.

1.20 MAGNETIC CIRCUITS

Practically all electric power machinery (*e.g.* transformers, generators, motors) depend for their operation upon the magnetism produced by the magnetic circuits. The closed path followed by magnetic flux is called a *magnetic circuit* (Fig. 1.26) just as the closed path followed by current is called an electric circuit.

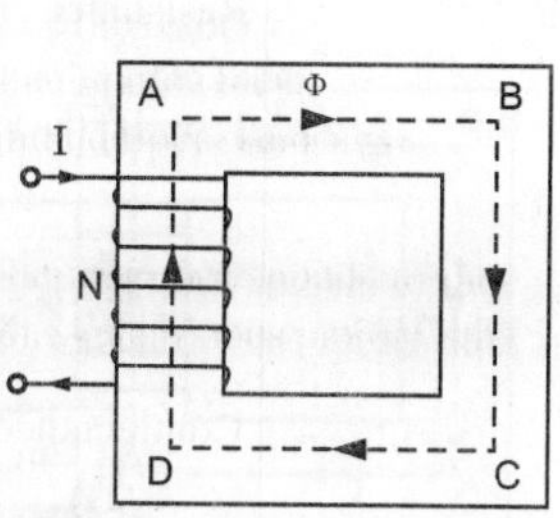

Fig. 1.26 *Simple Magnetic Circuit*

There is a lot of similarity between the magnetic and electric circuits, and will be shown later on during discussion.

A magnetic circuit consists of a structure composed for the most part of high permeability magnetic material. The presence of high permeability material causes the magnetic flux to be confined to the paths defined by the structure, much as currents are confined to the conductors of an electric circuit. A simple example of a magnetic circuit is shown in Fig. 1.26. The core is assumed to be composed of magnetic material whose permeability is much greater than that of the surrounding air. The core is of uniform cross section and is excited by a winding having N turns and carrying a current of I amperes. This winding develops a magnetic field in the core, as illustrated in the figure. The magnetic field can be visualized in terms of flux lines, which form closed loops interlinking with the winding. According to basic law of magnetic field, called the *Ampere's circuital law* (sometimes referred to as *Ampere's work law*) the line integral of **H** around a closed path is equal to the net current enclosed by that path *i.e.*

$$\oint \mathbf{H}\, dl = \Sigma\, \mathrm{I} \qquad \text{...(1.18)}$$

The above law is very comprehensive and provides a basis for calculation of magnetic circuits and makes it possible in some cases (such as in the case of a long current carrying conductor or a long solenoid) to determine readily the strength of the magnetic field.

1.20.1. Magnetic Circuit Concepts. The flux producing ability of the coil in Fig. 1.26 or of a coil on any other magnetic circuit is proportional to the number of turns N and the current I. The product NI is called the *magnetomotive force* (mmf) and determines the amount of flux developed in the magnetic circuit.

MMF of the magnetic circuit is the magnetic potential difference that tends to force flux around the magnetic circuit and is analogous to the electromotive force (emf) in an electric circuit.

$$\text{MMF} = \text{NI ampere-turns} \qquad \text{...(1.19)}$$

The resulting flux Φ in the magnetic circuit, besides being dependent on the mmf, is also a function of the opposition of the iron to carrying flux. This opposition is called the *reluctance* S of the magnetic circuit.

As in the case with resistance in the electric circuit, reluctance S of the magnetic circuit is directly proportional to length l, inversely proportional to cross-sectional area a, and dependent on the nature of material of the magnetic circuit. The reluctance of the magnetic circuit is given as

$$\text{S} = \frac{l}{\mu\, a} \text{ ampere-turns/weber} \qquad \text{...(1.20)}$$

when the flux Φ is constant over the length and uniform over the area. The quantity μ expresses the property of the magnetic material called its *permeability*. Permeability is a measure of the receptiveness of the material having magnetic flux developed in it. For free space, the permeability μ_0 equals $4\pi \times 10^{-7}$ H/m in the SI system. The relative permeability of magnetic material μ_r may range up to thousands.

Total flux developed in the circuit is given as

$$\Phi = \frac{\text{MMF}}{\text{S}} \text{ webers} \qquad \text{...(1.21)}$$

The above Eq. (1.21) is sometimes referred to as *Ohm's law* for the magnetic circuit. It serves to emphasize the mathematical analogy between the magnetic circuit and the electric circuit. Analogous quantities in the two circuits are listed in Art. 1.20.3.

Magnetic circuits differ from electric circuits in one important respect. The reluctance of the magnetic circuit containing iron or ferromagnetic material depends upon the flux carried by it. With the increase in flux, a large change in mmf is required to develop the same change in flux.

1.20.2. Determination of Ampere-Turns. In any magnetic circuit, flux created is given as

$$\Phi = \frac{\text{MMF}}{\text{Reluctance}} = \frac{\text{AT}}{l/\mu a} = \frac{\text{AT}}{l/\mu_0\, \mu_r\, a}$$

$$\text{or AT required} = \frac{\Phi}{a} \times \frac{1}{\mu_0\, \mu_r} \times l = \frac{\text{B}}{\mu_0\, \mu_r} \times l$$

$$= \text{H} \times l \text{ ampere-turns} \qquad \text{...(1.22)}$$

Hence for determination of AT for a magnetic circuit

(*i*) First find field strength H in each part of the magnetic circuit, (*ii*) find the length of various parts of magnetic circuit, (*iii*) find the number of ampere-turns required for the various parts of magnetic circuit from the relation AT = Hl where l is the length of the part in metres and lastly, (*iv*) find total number of ampere-turns for the whole series magnetic circuit by adding ampere-turns determined for various paths in magnetic circuit.

This will be more clear to the students from examples given in this chapter.

As both permeability and reluctance may change from one operating condition to another. Hence direct numerical application of the reluctance concept and Eqs. (1.21) and (1.22) is rare. For quantitative analysis, graphical methods are generally used because they can easily be adapted to the nonlinearities involved (Art. 1.18).

1.20.3. Comparison Between Magnetic and Electric Circuits

	Magnetic Circuit	Electric Circuit
	SIMILARITIES	
Circuit	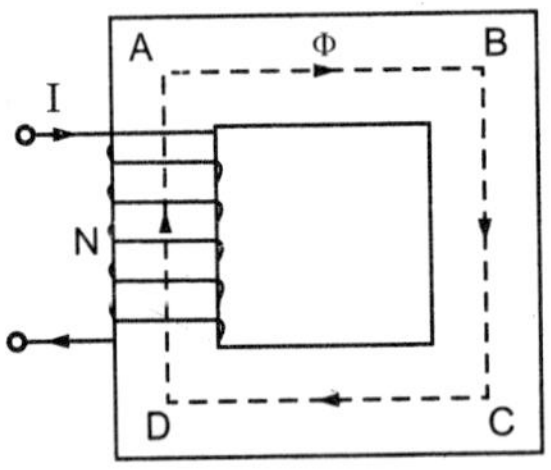	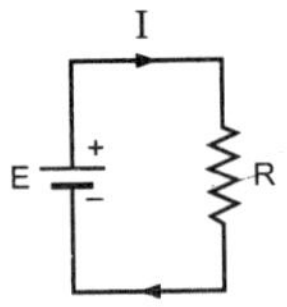
1. Definition	The closed path followed by magnetic flux is called a magnetic circuit	The closed path followed by an electric current is called an *electric circuit*
2. Driving force	**MMF** is required to establish flux Φ in the magnetic circuit and is measured in ampere-turns (AT) or amperes	**EMF** is required to cause flow of current in an electric circuit and is measured in volts
3. Response	Flux, $\Phi = \dfrac{\text{Driving force}}{\text{Magnetic reluctance}} = \dfrac{AT}{S}$ webers	Current, $I = \dfrac{\text{Driving force}}{\text{Electric resistance}} = \dfrac{E}{R}$ amperes
4. Impedance	Reluctance, $S = \dfrac{l}{\mu_0 \mu_r a} = \dfrac{l}{\mu a}$ AT/Wb	Resistance, $R = \rho \dfrac{l}{a}$ ohms
(*a*) For series circuits	$S = S_1 + S_2 + S_3 + ...$	$R = R_1 + R_2 + R_3 + ...$
(*b*) For parallel circuits	$S = \dfrac{1}{\dfrac{1}{S_1}+\dfrac{1}{S_2}+\dfrac{1}{S_3}+...}$	$R = \dfrac{1}{\dfrac{1}{R_1}+\dfrac{1}{R_2}+\dfrac{1}{R_3}+...}$
5. Admittance	Permeance $= \dfrac{1}{\text{Reluctance}}$ in Wb/AT	Conductance $= \dfrac{1}{\text{Resistance}}$ in siemens
6. Proportionality constant	Reluctivity $= \dfrac{1}{\text{Permeability}} = \dfrac{1}{\mu}$ in m/H	Resistivity, $\rho = \dfrac{1}{\text{Conductivity}}$ in ohm-metre
7. Density	Flux density, $B = \mu \times$ magnetic field intensity $= \mu H$ tesla or Wb/m^2	Current density, $J = \dfrac{\text{Electric field intensity}}{\rho}$ in A/m^2
8. Field intensity	Magnetic field intensity, $H = \dfrac{MMF}{l} = \dfrac{NI}{l}$ in AT/m	Electric field intensity $= \dfrac{E}{l}$ in volts/metre
9. Drop	MMF drop $= \Phi \times$ reluctance $= \Phi S$	Voltage drop $= I \times$ Resistance $= IR$

DISSIMILARITIES	
1. Flux does not actually flow in a magnetic circuit.	Current does flow in an electric circuit.
2. Permeability does not largely vary from material to material and there is hardly any material which can act as an insulator to the magnetic flux.	Conductivity varies largely from material to material so much so that some materials are insulators to electric current and some are very good conductors.
3. For a particular temperature the permeability depends up on the flux density (or total flux).	For a particular temperature, conductivity (or resistivity) is constant and independent of current strength.
4. Flux can pass through air.	Current would not flow through the air until an arc is struck.
5. Residual flux persists after removal of mmf.	The current is reduced to zero after removal of source of emf.
6. There is no waste of energy due to reluctance in the magnetic circuit and, therefore, energy is required only to create the magnetic flux but not to maintain it.	In an electric circuit, resistance causes heat to be generated resulting in waste of energy and, therefore, energy is required as long as the current flows.

1.20.4. Magnetic Circuits With Air Gaps. Energy-conversion devices which incorporate a moving element have necessarily air gaps in their magnetic circuits. Air gaps are also provided in the magnetic circuits to avoid saturation. A magnetic circuit with an air gap is shown in Fig. 1.27. An air gap is nothing else but a volume of air between two magnetic surfaces. The length of air gap l_g equals the distance between the two magnetic surfaces. The area of x-section of any one of the surfaces gives the air-gap area, a_g. When the air-gap length l_g is much smaller than the dimensions of the adjacent core faces, the magnetic flux Φ is constrained essentially to reside in the core and the air gap and is continuous throughout the magnetic circuit. Thus the configuration shown in Fig. 1.27 can be analysed as a magnetic circuit with two series components, a magnetic or iron core of permeability μ and mean length l_i and an air gap of permeability μ_0 and length l_g. Since the permeability of air is constant, the air gap is a linear part of the magnetic circuit and the flux density in the air gap is proportional to the mmf across the air gap. The necessary mmf is calculated separately for the air gap and the iron portions and then added to determine the total mmf.

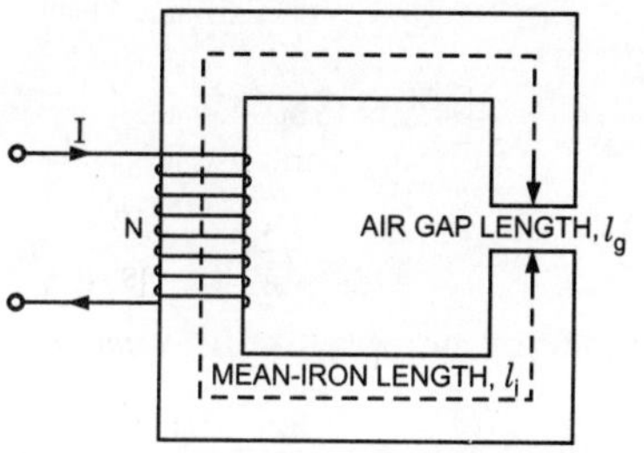

Fig. 1.27 *Magnetic Circuit With Air Gap*

1.20.5. Composite Magnetic Circuits. Consider a circular ring made from different materials of lengths l_1, l_2 and l_3, cross-sectional areas a_1, a_2 and a_3 and relative permeability μ_{r1}, μ_{r2} and μ_{r3} respectively with a cut of length l_g known as air gap. The total reluctance is the arithmetic sum of individual reluctances as they are joined in series.

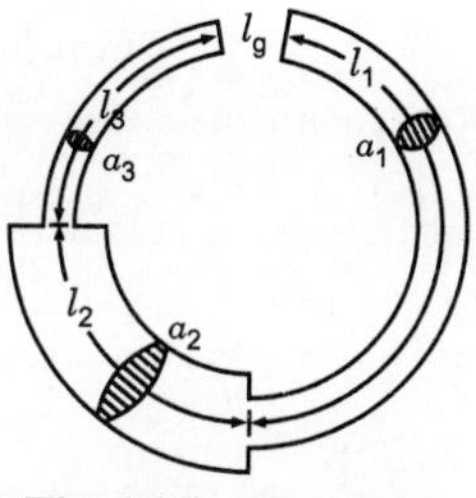

Fig. 1.28 *Composite Magnetic Circuit*

∴ Total reluctance,

$$S = \frac{l_1}{\mu_0 \mu_{r_1} a_1} + \frac{l_2}{\mu_0 \mu_{r_2} a_2} + \frac{l_3}{\mu_0 \mu_{r_3} a_3} + \frac{l_g}{\mu_0 a_g}$$

$$\text{Total mmf} = \Phi \times S$$

$$= \Phi \left[\frac{l_1}{\mu_0 \mu_{r_1} a_1} + \frac{l_2}{\mu_0 \mu_{r_2} a_2} + \frac{l_3}{\mu_0 \mu_{r_3} a_3} + \frac{l_g}{\mu_0 a_g} \right]$$

or Total ampere-turns required

$$= H_1 l_1 + H_2 l_2 + H_3 l_3 + H_g l_g \qquad ...(1.23)$$

= Sum of ampere-turns required for individual parts of the magnetic circuit.

1.20.6. Parallel Magnetic Circuits. In series circuits, all parts of the magnetic circuit carry same flux and total ampere-turns required to create a given flux is the arithmetic sum of the ampere-turns required for individual parts of the circuit.

But if the various paths of the magnetic circuit are in parallel, as shown in Fig. 1.29 the ampere-turns required for the combination is equal to the ampere-turns required to create the given flux in one path.

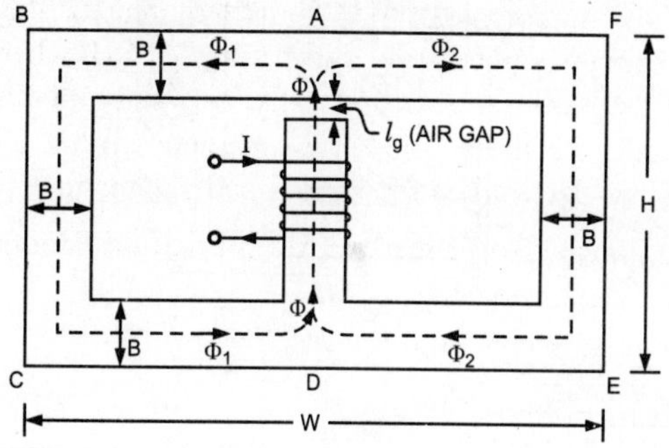

Fig. 1.29

For example for the circuit shown in Fig. 1.29 paths ABCD and AFED are in parallel, so ampere-turns required to create flux Φ_1 in path ABCD is equal to ampere-turns required to create flux Φ_2 in path AFED and also equal to the ampere-turns required for both of the paths.

Hence total ampere-turns required for magnetic circuit shown in Fig. 1.29.

= AT for path DA + AT for path ABCD

= AT for path DA + AT for path AFED

1.20.7. Magnetic Leakage, Fringing and Stacking Factor. Leakage flux is the flux, which follows a leaking path, as shown in Fig. 1.30. Flux in the air gap is known as *useful flux* which is utilized for various useful purposes. For the purpose of calculations the iron is supposed to carry whole of the flux throughout its entire length.

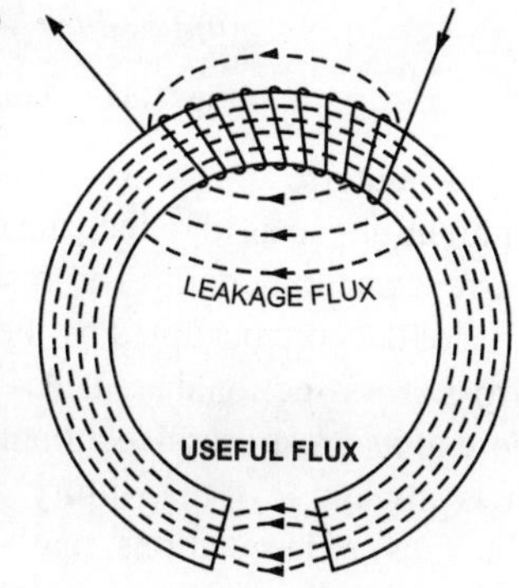

Fig. 1.30

The ratio of total flux (flux in the iron path) to the useful flux (flux in the air) is known as *leakage factor*.

$$i.e. \text{ Leakage factor} = \frac{\text{Total flux}}{\text{Useful flux}} = \frac{\text{Flux in the iron path}}{\text{Flux in the air gap}}$$

It is also seen from Fig. 1.30 that the useful flux passing across the gap tends to bulge outwards, thereby increasing the effective area of gap and reducing the flux density in the gap. This effect is referred to as *fringing*; and the longer the air gap, the greater is the fringing.

In electrical machines (such as in generators and motors), magnetic leakage is undesirable as it causes increase in weight (not decrease in their power efficiency) and cost of manufacture. Though magnetic leakage cannot be avoided completely but can be reduced to the minimum by placing the magnetizing or exciting coils as close as possible to the air gap or to the points in the magnetic circuit where the flux is to be utilized for useful purposes.

The value of leakage factor for modern electric machinery is about 1.2.

Two examples of leakage flux given in Fig. 1.31 are of a relay and of a two-pole electrical machine.

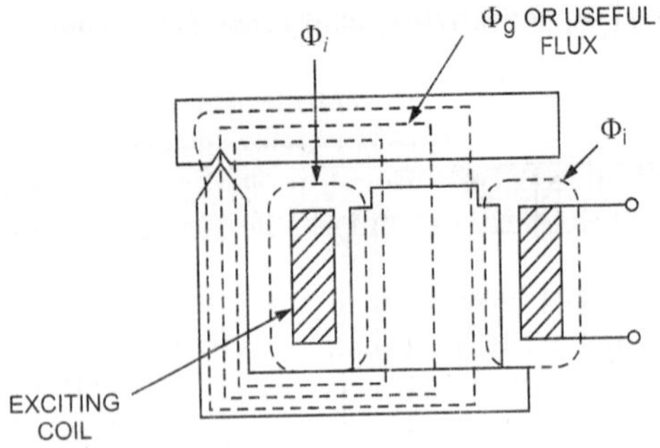

(a) *Electromagnetic Relay*

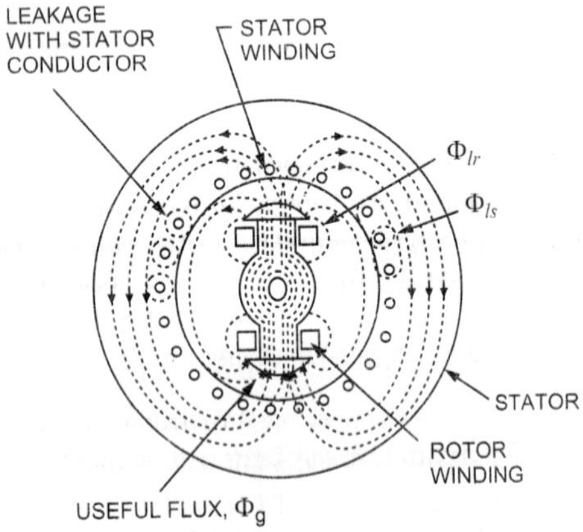

(b) *Two-Pole Electrical Machine*

Fig. 1.31 *Examples of Leakage*

Magnetic cores are made up of thin, lightly insulated laminations so as to reduce the power loss in cores owing to eddy current phenomenon.* The result is that the net cross-sectional area of the core occupied by the magnetic material is less than its gross cross-sectional area. The ratio of net cross-sectional of the core to its gross cross-sectional area is known as *stacking factor*. It is always less than unity and its value depends upon the thickness of laminations, may vary from 0.5 to 0.95 approaching unity with the increase in thickness of laminations.

1.20.8. Flux Linkage. Consider a N turn coil of any shape. On one side of the coil, all turns would carry current in one direction and those on the other side in opposite direction. As the number of turns are more, each turn of the coil will carry the same current, the mmf produced would be more and therefore more flux. If the total flux is Φ webers, the flux linkage with each turn is Φ weber turn. Hence, the flux linkage with the coil having N turns is

$$\psi = \Phi N \text{ weber-turns} \qquad ...(1.24)$$

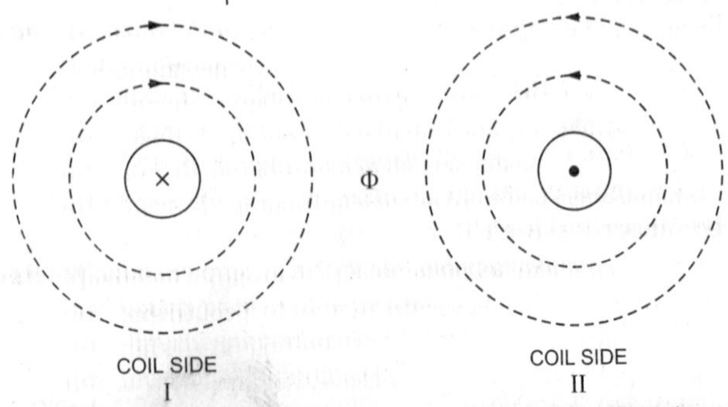

Fig. 1.32 *Flux Linkage With a Closed One-Turn Circuit*

Example 1.3. An electromagnet has an air gap of 4 mm and flux density in the gap is 1.3 Wb/m². Determine the ampere-turns for the gap. [*U.P. Technical Univ. Even Semester 2006-07; Mahamaya Technical Univ. Odd Semester 2011-12*]

Solution: Magnetising force, $H_g = \dfrac{B}{\mu_0 \mu_r} = \dfrac{1.3}{4\pi \times 10^{-7} \times 1}$

$= 1.035 \times 10^6$ AT/m $\quad \because \mu_r$ for air = 1

Ampere-turns required for the gap $= H_g l_g = 1.035 \times 10^6 \times 4 \times 10^{-3}$

$= 4{,}140$ **Ans.**

Example 1.4. A wrought iron bar 30 cm long and 2 cm in diameter is bent into a circular shape as given in Fig. 1.33. It is then wound with 500 turns of wire. Calculate the current required to produce a flux of 0.5 mWb in magnetic circuit with an air gap of 1 mm; μ_r (iron) = 4,000 (assume constant).

[*U.P. Technical Univ. Even Semester 2004-05*]

Solution: Flux to be produced, $\Phi = 0.5$ mWb $= 0.5 \times 10^{-3}$ Wb

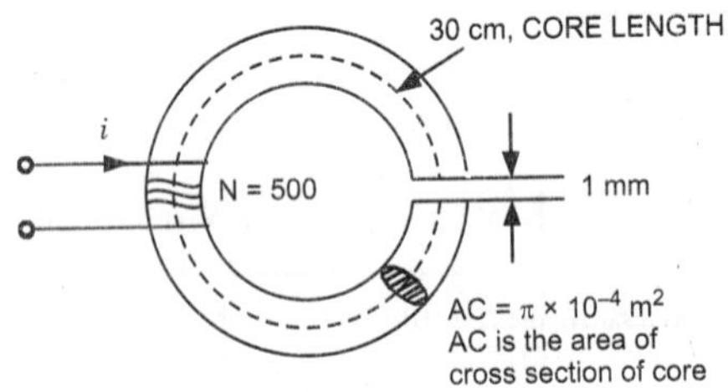

Fig. 1.33

Area of cross section of core, $a = \pi \times 10^{-4}$ m²

Flux density required, $B = \dfrac{\Phi}{a} = \dfrac{0.5 \times 10^{-3}}{\pi \times 10^{-4}} = \dfrac{5}{\pi} = 1.5915$ T

Air-gap length, $l_g = 1$ mm $= 0.001$ m

Core length, $l_i = 30$ cm $= 0.3$ m

Total ampere-turns required,

$$AT = \frac{B}{\mu_0} l_g + \frac{B}{\mu_0 \mu_r} l_i$$

$$= \frac{1.5915}{4\pi \times 10^{-7}} \times 0.001 + \frac{1.5915}{4\pi \times 10^{-7} \times 4{,}000} \times 0.3$$

$$= 1{,}266 + 95 = 1{,}361$$

Current required, $I = \dfrac{AT}{N} = \dfrac{1{,}361}{500} = 2.72$ A **Ans.**

Example 1.5. A flux density of 2.25 Wb/m² is required in 3 mm air gap of an electromagnetic having an iron path 1 m long. Calculate the magnetizing force and current required if the electromagnet has 1,275 turns. Assume relative permeability of iron to be 1,200.

[*M.D. Univ. Electromechanical Energy Conversion December-2013*]

Solution: Flux density to be produced in the air gap,

$B_g = 2.25$ Wb/m²

$\mu_0 = 4\pi \times 10^{-7}$ H/m

Magnetising force for air gap,

$$H_g = \frac{B_g}{\mu_0} = \frac{2.25}{4\pi \times 10^{-7}} = 1{,}790 \times 10^3 \text{ AT/m} \quad \textbf{Ans.}$$

Flux density required in iron path,

$B_i = B_g = 2.25$ Wb/m² neglecting magnetic leakage

* Refer to Art. 1.29.2.

Relative permeability of iron,

$\mu_r = 1{,}200$

Magnetising force for iron path,

$$H_i = \frac{B_i}{\mu_0 \mu_r} = \frac{2.25}{4\pi\times10^{-7}\times1{,}200} = 1{,}492 \text{ AT/m} \quad \textbf{Ans.}$$

Length of air gap, $l_g = 3$ mm $= 0.003$ m

Length of iron path, $l_i = 1$ m

Total ampere-turns required,

$$AT = H_g\, l_g + H_i\, l_i$$
$$= 1{,}790\times10^3\times0.003 + 1{,}492\times1 = 6{,}862$$

$$\text{Current required, } I = \frac{AT}{N} = \frac{6{,}862}{1{,}275} = 5.38 \text{ A} \quad \textbf{Ans.}$$

Example 1.6. A ring of ferromagnetic material has a rectangular cross section. The inner diameter is 7.4 in., the outer diameter is 9 in., and the thickness is 0.8 in. There is a coil of 600 turns wound on the ring. When the coil carries a current of 2.5 A, the flux produced in the ring is 1.2×10^{-3} Wb. Find: (i) Magnetic field intensity (ii) Reluctance (iii) Permeability.

[U.P. Technical Univ. Even Semester 2007-08; Odd Semester 2013-14]

Solution: Number of turns in the coil, $N = 600$

Current carried by coil, $I = 2.5$ A

Flux produced, $\Phi = 1.2\times10^{-3}$ Wb

Mean length of magnetic flux path is that of a circle midway between the inside and outside diameter i.e.

$$l = \frac{\pi(7.4+9)}{2} = 25.76 \text{ in} = 25.76\times0.0254 = 0.6543 \text{ m}$$

Cross-sectional area of the toroidal core,

$$a = \frac{9-7.4}{2}\times0.8 = 0.64 \text{ in}^2 = 4.13\times10^{-4} \text{ m}^2$$

(*i*) Magnetic flux density, $B = \dfrac{\Phi}{a} = \dfrac{1.2\times10^{-3}}{4.13\times10^{-4}} = 2.9 \text{ Wb/m}^2$ **Ans.**

(ii) Reluctance, $S = \dfrac{MMF}{\Phi} = \dfrac{NI}{\Phi} = \dfrac{600\times2.5}{1.2\times10^{-3}}$

$= 1{,}250\times10^3$ AT/Wb **Ans.**

(iii) Permeability, $\mu = \dfrac{l}{Sa} = \dfrac{0.6543}{1{,}250\times10^3\times4.13\times10^{-4}}$

$= 1.267\times10^{-3}$ H/m **Ans.**

Relative permeability, $\mu_r = \dfrac{\mu}{\mu_0} = \dfrac{1.267\times10^{-3}}{4\pi\times10^{-7}} = 1{,}008$ **Ans.**

Example 1.7. A ring of iron has a mean diameter of 15 cm, a cross section of 1.5 cm² and has a radial air gap of 0.5 mm cut in it. It is uniformly wound with 1, 500 turns of insulated wire and a current of 1.2 A produces a flux of 0.1 mWb across the air gap. Calculate the relative permeability of iron on the assumption that there is no magnetic leakage.

[M.D. Univ. Electromechanical Energy Conversion December 2005]

Solution: Flux to be produced in the air gap, $\Phi_g = 0.1$ mWb

$= 0.1\times10^{-3}$ Wb

Flux to be produced in the iron path, $\Phi_i = 0.1$ mWb

∵ there is no magnetic leakage

Area of cross section, $a = 1.5 \text{ cm}^2 = 1.5\times10^{-4} \text{ m}^2$

Length of air gap, $l_g = 0.5 \text{ mm} = 0.5\times10^{-3}$ m

Length of iron path, $l_i = \pi\times0.15 - 0.5\times10^{-3} = 0.4707$ m

Number of turns provided, $N = 1{,}500$

Current flowing through the coil, $I = 1.2$ A

Total ampere-turns provided, $AT = N\times I = 1{,}500\times1.2 = 1{,}800$

Ampere-turns required for air gap,

$$AT_g = \frac{\Phi_g / a}{\mu_0}\, l_g$$

$$= \frac{0.1\times10^{-3}}{4\pi\times10^{-7}\times1.5\times10^{-4}}\times0.5\times10^{-3} = 265$$

Ampere-turns provided for iron path,

$$AT_i = AT - AT_g = 1{,}800 - 265 = 1{,}535$$

Thus ampere-turns provided for iron path,

$$AT_i = \frac{\Phi_i / a_i}{\mu_0\mu_r}\, l_i = AT - AT_g$$

or $$\frac{0.1\times10^{-3}}{4\pi\times10^{-7}\times\mu_r\times1.5\times10^{-4}}\times0.4707 = 1{,}535$$

or Relative permeability of iron, $\mu_r = \dfrac{0.1\times10^{-3}\times0.4707}{4\pi\times10^{-7}\times1.5\times10^{-4}\times1{,}535}$

$= 163$ **Ans.**

Example 1.8. A steel ring 12 cm mean radius and of circular cross section 1 cm in radius has an air gap 2 mm length. It is wound uniformly with 550 turns of wire core carrying 3 A. Neglecting magnetic leakage. The air gap takes 60% of total mmf. Find total reluctance.

[M.D. Univ. Electromechanical Energy Conversion December-2013]

Solution: Area of cross section of steel ring,

$$a = \pi r^2 = \pi\times(0.01)^2 = \pi\times10^{-4} \text{ m}^2$$

Total ampere-turns provided,

$$AT = NI = 550\times3 = 1{,}650$$

Ampere-turns provided for air gap,

$$AT_g = 60\% \text{ of total mmf} = 0.6\times1{,}650 = 990$$

Ampere-turns provided for iron path,

$$AT_i = AT - AT_g = 1{,}650 - 990 = 660$$

Flux density in air gap,

$$B_g = \frac{AT_g\times\mu_0}{l_g} = \frac{990\times4\pi\times10^{-7}}{2\times10^{-3}} = 0.622 \text{ T}$$

Magnetic flux in air gap,

$$\Phi_g = B_g\times a = 0.622\times\pi\times10^{-4} = 0.195 \text{ mWb}$$

Magnetic flux in iron path,

$\Phi_i = \Phi_g = 0.195$ mWb neglecting magnetic leakage

Total reluctance,

S = Air-gap reluctance + reluctance for iron path

$$= \frac{AT_g}{\Phi_g} + \frac{AT_i}{\Phi_i} = \frac{AT}{\Phi} = \frac{1{,}650}{0.195\times10^{-3}} = 0.846\times10^7 \text{ AT/Wb} \quad \textbf{Ans.}$$

Example 1.9. A circular iron ring having a cross-sectional area of 10 cm² and a length of 4π cm in iron has an air gap of 0.4π mm made a saw cut. The relative permeability of iron is 10^3 and permeability of free space is $4\pi\times10^{-7}$ H/m. The ring is wound with a coil of 2,000 turns and carries 2 mA. Determine the air-gap flux neglect leakage and fringing. *[Pb. Technical Univ. December-2013]*

Solution: Let the flux density in iron as well as air gap be B Wb/m²

Total ampere-turns provided,

$$AT = N\times I = 2{,}000\times2\times10^{-3} = 4$$

Total ampere-turns required

= AT required for air gap

+ AT required for iron path

or Total ampere-turns provided

$$= \frac{B}{\mu_0} l_g + \frac{B}{\mu_0 \mu_r} l_i$$

$$\text{or } 4 = \frac{B}{\mu_0}\left(l_g + \frac{l_i}{\mu_r}\right)$$

$$\text{or Flux density, } B = \frac{4 \times \mu_0}{l_g + \dfrac{l_i}{\mu_r}}$$

$$= \frac{4 \times 4\pi \times 10^{-7}}{0.4\pi \times 10^{-3} + \dfrac{4\pi \times 10^{-2} - 0.4\pi \times 10^{-3}}{1{,}000}}$$

$$= \frac{16\pi \times 10^{-7}}{0.4\pi \times 10^{-3} + 0.0396\pi \times 10^{-3}}$$

$$= \frac{16 \times 10^{-7}}{0.4396 \times 10^{-3}} = 3.64 \times 10^{-3} \text{ T} \quad \textbf{Ans.}$$

Example 1.10. An iron ring is made up of three parts: l_1 = 10 cm, A_1 = 8 cm², l_2 = 8 cm, A_2 = 3 cm², l_3 = 6 cm, A_3 = 2.5 cm².

It is wound with a coil of 250 turns. Calculate current required to produce flux of 0.4 mWb.

μ_1 = 2,670, μ_2 = 1,050, μ_3 = 600.

[*U.P. Technical Univ. Even Semester 2007-08*]

Solution: Total reluctance,

$$S = \frac{l_1}{\mu_0 \mu_{r_1} a_1} + \frac{l_2}{\mu_0 \mu_{r_2} a_2} + \frac{l_3}{\mu_0 \mu_{r_3} a_3}$$

$$= \frac{0.1}{4\pi \times 10^{-7} \times 2{,}670 \times 8 \times 10^{-4}} + \frac{0.08}{4\pi \times 10^{-7} \times 1{,}050 \times 3 \times 10^{-4}} + \frac{0.06}{4\pi \times 10^{-7} \times 600 \times 2.5 \times 10^{-4}}$$

$= 0.37255 \times 10^5 + 2.021 \times 10^5 + 3.183 \times 10^5$

$= 5.57655 \times 10^5$ AT/Wb

MMF = Flux Φ × Reluctance S = $0.4 \times 10^{-3} \times 5.57655 \times 10^5$

= 223 ampere-turns

Current required,

$$I = \frac{\text{MMF}}{N} = \frac{223}{250} = 0.892 \text{ A} \quad \textbf{Ans.}$$

Example 1.11. A cast steel ring has a circular cross section of 3 cm in diameter and a mean circumference of 80 cm. A 1 mm air gap is cut in the ring which is wound with a coil of 600 turns. Estimate the current required to establish a flux of 0.75 mWb in the air gap. Neglect leakage.

Magnetization data:

H (AT/M)	200	400	600	800	1,000	1,200	1,400	1,600	1,800
B (T)	0.1	0.32	0.6	0.9	1.08	1.18	1.27	1.32	1.36

[*U.P. Technical Univ. Electromechanical Energy Conversion-I, 2004-2005*]

Solution: Area of x-section, $a = \frac{\pi}{4}(d)^2 = \frac{\pi}{4}(0.03)^2 = 0.0007 \text{ m}^2$

Flux density in the air gap, $B_g = \frac{\Phi_g}{a} = \frac{0.75 \times 10^{-3}}{0.0007} = 1.07$ T

Ampere-turns required for air gap,

$$AT_g = \frac{B_g}{\mu_0} \times l_g = \frac{1.07}{4\pi \times 10^{-7}} \times 1 \times 10^{-3} = 850$$

Flux density in iron path,

$B_i = B_g = 1.07$ T $\quad\because$ leakage in negligible.

From Fig. 1.34, magnetising force, H corresponding to flux density B of 1.07 T is 980 AT/m

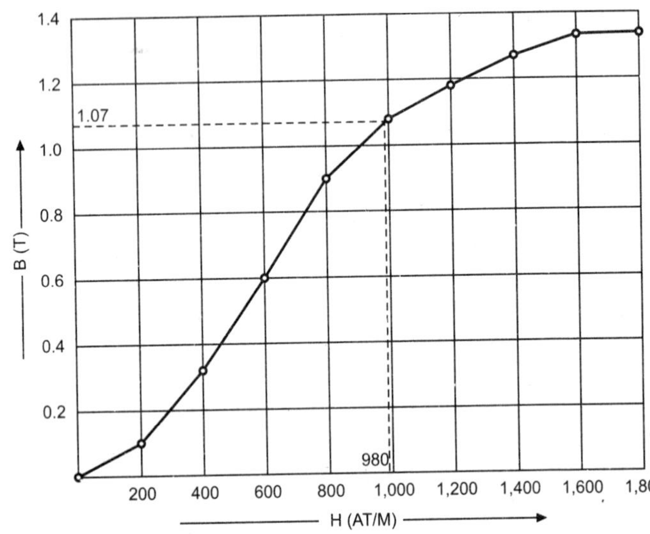

Fig. 1.34 *Magnetization Curve (Example 1.11)*

AT required for iron path, $AT_i = H \times l_i = 980 \times (80 - 0.1) \times 10^{-2}$

= 783

Total ampere-turns required, $AT = AT_g + AT_i = 850 + 783 = 1{,}633$

$$\text{Current required, } I = \frac{AT}{N} = \frac{1{,}633}{600} = 2.72 \text{ A } \textbf{Ans.}$$

Example 1.12. A circular iron ring has a mean circumference of 1.5 m and a cross-sectional area of 100 cm². A saw cut of 0.4 cm wide is made in the ring. Calculate the magnetizing current required to produce a flux of 0.8 mWb in the air gap if the ring is wound with a coil of 350 turns. Assume relative permeability of iron as 400 and leakage factor 1.25.

[*M.D. Univ. Electromechanical Energy Conversion, May-2007*]

Solution: Area of cross section, a = 100 cm² = 0.01 m²

Flux to be produced in the air gap, Φ_g = 0.8 mWb = 0.8×10^{-3} Wb

$$\text{Flux density in the air gap, } B_g = \frac{\Phi_g}{a} = \frac{0.8 \times 10^{-3}}{0.01} = 0.08 \text{ T}$$

Ampere-turns required for air gap,

$$AT_g = \frac{B_g}{\mu_0} \times l_g = \frac{0.08}{4\pi \times 10^{-7}} \times 0.4 \times 10^{-2} = 255$$

Flux through iron path, $\Phi_i = \Phi_g$ × leakage factor

$= 0.8 \times 10^{-3} \times 1.25 = 1 \times 10^{-3}$ Wb

Flux density in the iron path,

$$B_i = \frac{\Phi_i}{a} = \frac{1 \times 10^{-3}}{0.01} = 0.1 \text{ T}$$

Ampere-turns required for iron path,

$$AT_i = \frac{B_i}{\mu_0 \mu_r} l_i = \frac{0.1 \times (1.5 - 0.004)}{4\pi \times 10^{-7} \times 400} = 298$$

Total ampere-turns required,

$$AT = AT_g + AT_i = 255 + 298 = 553$$

$$\text{Current required, } I = \frac{AT}{N} = \frac{553}{350} = 1.58 \text{ A } \textbf{Ans.}$$

Example 1.13. A rectangular magnetic core shown in Fig. 1.35 (*a*) has square cross section of area 16 cm². An air gap of 2 mm is cut across one of its limbs. Find the exciting current needed in the coil having 1,000 turns wound on the core to create an air-gap flux of 4 m Wb. The relative permeability of the core is 2,000.

[*U.P. Technical Univ. February-2002*]

Solution: Flux to be created, $\Phi = 4 \text{ mWb} = 0.004 \text{ Wb}$

Area of x-section, $a = 16 \text{ cm}^2 = 16 \times 10^{-4} \text{ m}^2$

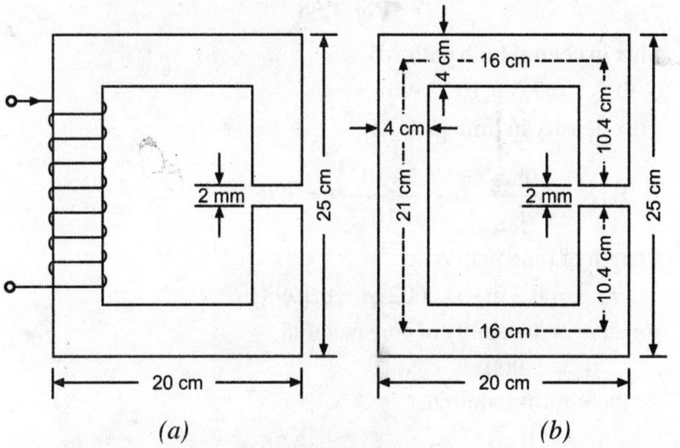

Fig. 1.35

$$\text{Fluse density required, } B = \frac{\Phi}{a} = \frac{4\times10^{-3}}{16\times10^{-4}} = 2.5 \text{ T}$$

Length of air gap $l_g = 2 \text{ mm} = 2 \times 10^{-3} \text{ m}$

Each side of cross section $= \sqrt{16} = 4 \text{ cm}$

$$\text{Length of iron path, } l_i = \left(25 - 2\times\frac{4}{2} + 20 - 2\times\frac{4}{2}\right)\times 2 - 0.2$$

$$= 73.8 \text{ cm} = 0.738 \text{ m}$$

Total ampere-turns required,

$$AT = \frac{B}{\mu_0} l_g + \frac{B}{\mu_0 \mu_r} l_i$$

$$= \frac{2.5\times2\times10^{-3}}{4\pi\times10^{-7}} + \frac{2.5\times0.738}{2{,}000\times4\pi\times10^{-7}}$$

$$= 3{,}979 + 734 = 4{,}713$$

Number of turns on the coil = 1,000

$$\text{Exciting current required, } I = \frac{AT}{N} = \frac{4{,}713}{1{,}000} = 4.713 \text{ A } \textbf{Ans.}$$

Example 1.14. The magnetic circuit frame shown in Fig. 1.36 is built of iron of square cross section of 3 cm. Each air gap is 2 mm wide and each of the coil is wound with 1,000 turns. The relative permeability of part A and B may be taken as 1,000 and 1,200 respectively. Calculate (a) reluctance of part A (b) reluctance of two air gaps (c) total reluctance of complete magnetic path (d) total mmf (e) total flux and (f) flux density.

Solution: Cross-sectional area,

$$a = (0.03)^2 = 0.0009 \text{ m}^2$$

Mean length flux path in part A,

$$l_A = 20 - (1.5 + 1.5) = 17 \text{ cm or } 0.17 \text{ m}$$

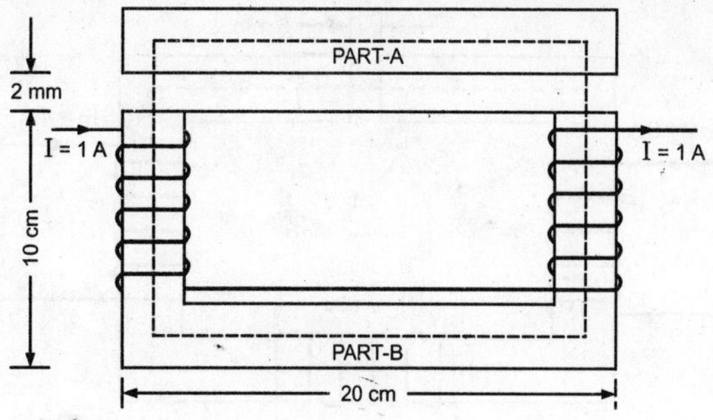

Fig. 1.36

Mean length flux path in part B,

$$l_B = (20 - 1.5 - 1.5) + (10 - 1.5) + (10 - 1.5)$$

$$= 34 \text{ cm or } 0.34 \text{ m}$$

Relative permeability of part A, $\mu_{r\,A} = 1{,}000$

Relative permeability of part B, $\mu_{rB} = 1{,}200$

Length of two air gaps, $l_g = 2 \times 2 = 4 \text{ mm} = 0.004 \text{ m}$

(*a*) (*i*) Reluctance of part A, $S_A = \dfrac{0.17}{4\pi\times10^{-7}\times1{,}000\times0.0009}$

$$= 1{,}50{,}313 \text{ AT/Wb } \textbf{Ans.}$$

(*ii*) Reluctance of part B, $S_B = \dfrac{l_B}{\mu_0\, \mu_{rB} \times a}$

$$= \frac{0.34}{4\pi\times10^{-7}\times1{,}200\times0.0009}$$

$$= 2{,}50{,}521 \text{ AT/Wb}$$

(*b*) Reluctance of two air gaps,

$$S_g = \frac{l_g}{\mu_0 a} = \frac{0.004}{4\pi\times10^{-7}\times0.0009}$$

$$= 35{,}36{,}776 \text{ AT/Wb } \textbf{Ans.}$$

(*c*) Total reluctance, $S = S_A + S_B + S_g$

$$= 1{,}50{,}313 + 2{,}50{,}521 + 35{,}36{,}776$$

$$= 39{,}37{,}610 \text{ AT/Wb } \textbf{Ans.}$$

(*d*) Total mmf = 2 × 1,000 × 1 = 2,000 AT **Ans.**

because mmf is produced by two coils on part B each having 1,000 turns and carrying a current of 1 A

$$(e) \text{ Total flux} = \frac{\text{MMF}}{\text{Reluctance}} = \frac{2{,}000}{39{,}37{,}610}$$

$$= 5.08 \times 10^{-4} \text{ Wb. } \textbf{Ans.}$$

$$(f) \text{ Flux density, } B = \frac{\Phi}{a} = \frac{5.08\times10^{-4}}{0.0009}$$

$$= 0.564 \text{ Wb/m}^2 \text{ (or Tesla) } \textbf{Ans.}$$

Example 1.15. A rectangular iron core shown in the figure (Fig. 1.37) has mean length of a magnetic circuit of 100 cm, cross section of 2 cm × 2 cm and relative permeability of 1,450. A cut of size 5 mm in the core has been made. The three coils A, B and C on the core have number of turns $N_A = 335$, $N_B = 600$ and $N_C = 600$, and the respective currents flowing are 1.5 A, 4 A and 3 A. The direction of currents are as shown. Find the air-gap flux, neglecting fringing of flux.

Solution: It is observed that flux produced by coils B and C are in the same direction but that by coil A is in opposite direction.

Hence net available ampere-turns,

$$= AT_B + AT_C - AT_A$$

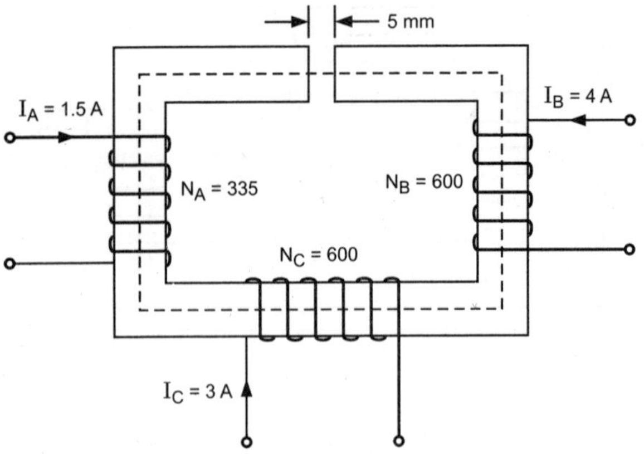

Fig. 1.37

$$= 600 \times 4 + 600 \times 3 - 335 \times 1.5$$
$$= 3{,}697.5$$

Mean length of iron path, l_i = 100 cm = 1 m

Area of cross section, $a = 2 \times 10^{-2} \times 2 \times 10^{-2} = 4 \times 10^{-4}$ m^2

Relative permeability of iron path,

$$\mu_r = 1{,}450$$

$$\text{Reluctance of iron path, } S_i = \frac{l_i}{\mu_0 \mu_r a}$$

$$= \frac{1}{4\pi \times 10^{-7} \times 1{,}450 \times 4 \times 10^{-4}}$$
$$= 1.372025 \times 10^6 \text{ AT/Wb}$$

Length of air gap $l_g = 5 \text{ mm} = 5 \times 10^{-3}$ m

$$\text{Reluctance of air gap, } S_g = \frac{l_g}{\mu_0 \times a} = \frac{5 \times 10^{-3}}{4\pi \times 10^{-7} \times 4 \times 10^{-4}}$$
$$= 9.947184 \times 10^6 \text{ AT/Wb}$$

Total reluctance of the circuit,

$$S = S_i + S_g$$
$$= (1.372025 \times 10^6 + 9.947184 \times 10^6) \text{ AT/Wb}$$
$$= 11.31921 \times 10^6 \text{ AT/Wb}$$

$$\text{Flux in the air-gap, } \Phi = \frac{\text{Total available AT}}{\text{Total reluctance}} = \frac{3{,}697.5}{11.31921 \times 10^6}$$

$$= 0.3267 \times 10^{-3} \text{ Wb or } 0.3267 \text{ mWb}$$ **Ans.**

Example 1.16. The magnetic circuit of the figure given below has cast steel core with dimensions as shown:

Mean length from A to B through either outer limb = 0.5 m.

Mean length from A to B through the central limb = 0.2 m.

In the magnetic circuit shown it is required to establish a flux of 0.75 mWb in the air gap of the central limb. Determine the mmf of the exciting coil if for the core material relative permeability μ_r = 5,000. Neglect fringing. *[RGPV December-2003]*

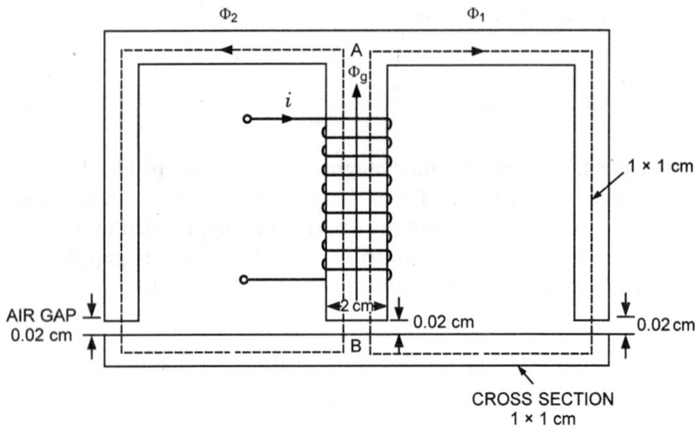

Fig. 1.38

Solution: Flux in the air gap of central limb,

$$\Phi_{g_1} = 0.75 \text{ m Wb} = 7.5 \times 10^{-4} \text{ Wb}$$

Area of cross section,

$$a_{g_1} = 2 \times 1 \times 10^{-4} = 2 \times 10^{-4} \text{ m}^2$$

Flux density,

$$B_{g_1} = \frac{\Phi_{g_1}}{a_{g_1}} = \frac{7.5 \times 10^{-4}}{2 \times 10^{-4}} = 3.75 \text{ T}$$

Ampere-turns required,

$$AT_{g_1} = H_g l_g = \frac{B_{g_1}}{\mu_0} l_{g_1} = \frac{3.75}{4\pi \times 10^{-7}} \times 0.02 \times 10^{-2} = 596$$

Flux in central limb path AB,

$$\Phi_{AB_1} = 7.5 \times 10^{-4} \text{ Wb}$$

Flux density in limb path AB,

$$B_{AB_1} = \frac{\Phi_{AB}}{a_{AB}} = \frac{7.5 \times 10^{-4}}{2 \times 10^{-4}} = 3.75 \text{ T}$$

Length of limb path AB,

$$l_{AB_1} = 0.2 \text{ m} - 0.0002 \text{ m} = 0.1998 \text{ m}$$

Relative permeability of core material,

$$\mu_r = 5{,}000$$

Ampere-turns required,

$$AT_{AB_1} = \frac{B_{AB_1}}{\mu_o \mu_r} \times l_{AB_1} = \frac{3.75}{4\pi \times 10^{-7} \times 5{,}000} \times 0.1998 = 120$$

Flux in the outer limb path,

$$\Phi_{AB_2} = \frac{\Phi_{g_1}}{2} = \frac{7.5 \times 10^{-4}}{2} = 3.75 \times 10^{-4} \text{ Wb}$$

Area of cross section,

$$a_2 = 1 \times 1 \times 10^{-4} = 1 \times 10^{-4} \text{ m}^2$$

Flux density in the outer limb path

$$= \frac{3.75 \times 10^{-4}}{1 \times 10^{-4}} = 3.75 \text{ T}$$

Ampere-turns required for gap in the outer limb,

$$AT_{g_2} = \frac{B_{g_2}}{\mu_o} l_{g_2} = \frac{3.75}{4\pi \times 10^{-7}} \times 0.02 \times 10^{-2} = 596$$

Ampere-turns required for the outer limb,

$$AT_{AB_2} = \frac{B_{AB_2}}{\mu_o \mu_r} \times l_{AB_2} = \frac{3.75 \times (0.5 - 0.0002)}{4\pi \times 10^{-7} \times 5{,}000} = 300$$

Total ampere-turns (mmf) required

$$= AT_{g_1} + AT_{AB_1} + AT_{g_2} + AT_{AB_2}$$
$$= 596 + 120 + 596 + 300 = 1{,}612$$ **Ans.**

Example 1.17. Fig. 1.39(*a*) shows a magnetic circuit formed by an ideal core material. Determine the magnetic flux density in the air gap. *[GATE 1997]*

Solution: The equivalent circuit is given in Fig. 1.39(*b*) where S_s is reluctance of nonmagnetic sleeve and S_g is the reluctance of air gap.

Magnetising force, $F_1 = F_2 = 50 \times 10 = 500$ AT

Length of nonmagnetic sleeve, l_s = 1 mm

Length of air gap, l_g = 4 mm

Since core is ideal, its reluctance is zero.

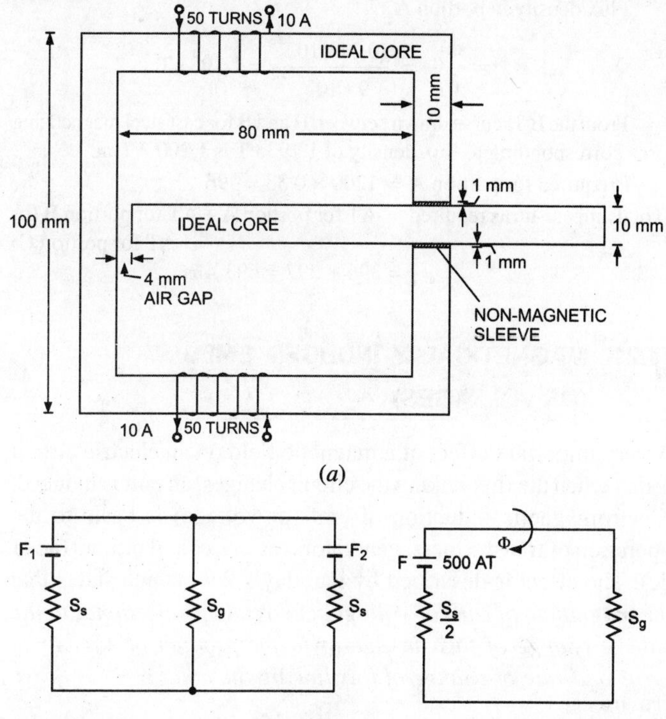

(*a*)

(*b*)

Fig. 1.39

Reluctance of nonmagnetic sleeve, $S_s = \dfrac{l_s}{\mu_0 \mu_r a}$

where $l_s = 1$ mm, $\mu_0 = 4\pi \times 10^{-7}$, and $\mu_r = 1$ being nonmagnetic

So $S_s = \dfrac{1 \times 10^{-3}}{4\pi \times 10^{-7} a}$

and $S_g = \dfrac{l_g}{\mu_0 a_g} = \dfrac{4 \times 10^{-3}}{4\pi \times 10^{-7} \times a}$

Total reluctance of magnetic circuit,

$$S = \frac{S_s}{2} + S_g = \frac{1 \times 10^{-3}}{2 \times 4\pi \times 10^{-7} a} + \frac{4 \times 10^{-3}}{4\pi \times 10^{-7} a} = \frac{4.5}{4\pi \times 10^{-4} a}$$

Magnetic flux density,

$$B = \frac{\Phi}{a} = \frac{\text{MMF}}{S \times a} = \frac{500}{\dfrac{4.5}{4\pi \times 10^{-4} a} \times a} = \frac{500 \times 4\pi \times 10^{-4}}{4.5}$$

$$= 0.14 \text{ Wb/m}^2 \quad \textbf{Ans.}$$

Example 1.18. A cast steel magnetic structure made of a bar of section 2 cm × 2 cm is shown in Fig. 1.40. Determine the current that the 500 turn magnetizing coil on the left limb should carry so that a flux of 2 mWb is produced in the right limb. Take μ_r = 600 and neglect leakage. *[Chhatisgarh Vivekanand Technical Univ. 2006-07]*

Solution: Flux created by magnetizing coil, say Φ, is divided at junction point into two paths depending upon the reluctances of portions B and C respectively.

Reluctance of portion B, $S_B = \dfrac{l_B}{\mu_0 \mu_r a} = \dfrac{0.15}{\mu_0 \mu_r a}$

Reluctance of portion C, $S_C = \dfrac{l_C}{\mu_0 \mu_r a} = \dfrac{0.25}{\mu_0 \mu_r a}$

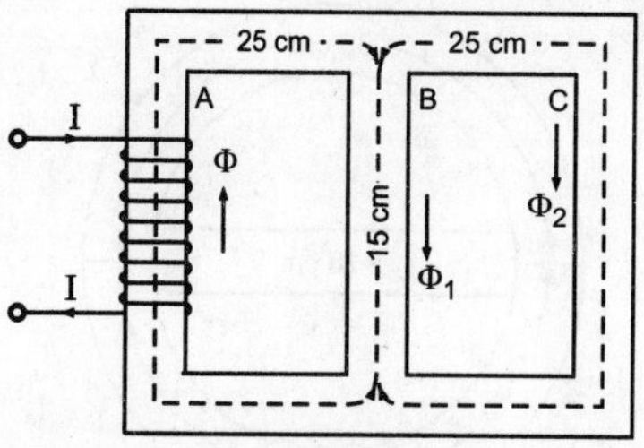

Fig. 1.40

Flux in portion C,

$$\Phi_C = 2 \text{ m Wb} = 0.002 \text{ Wb (given)}$$

Flux in portion B,

$$\Phi_B = \Phi_C \times \frac{S_C}{S_B} = 2 \times 10^{-3} \times \frac{0.25/\mu_0 \mu_r a}{0.15/\mu_0 \mu_r a} = 3.33 \times 10^{-3} \text{ Wb}$$

Total flux in portion A,

$$\Phi = \Phi_B + \Phi_C = 3.33 \times 10^{-3} + 2 \times 10^{-3} = 5.33 \times 10^{-3} \text{ Wb}$$

Flux density in portion A,

$$B_A = \frac{\Phi}{a} = \frac{5.33 \times 10^{-3}}{4 \times 10^{-4}} = 13.33 \text{ Wb/m}^2$$

AT required for portion A,

$$AT_A = \frac{B_A \times l_A}{\mu_0 \mu_r} = \frac{13.33 \times 0.25}{4\pi \times 10^{-7} \times 600} = 4{,}416$$

Flux density in path B,

$$B_B = \frac{3.33 \times 10^{-3}}{4 \times 10^{-4}} = 8.33 \text{ Wb/m}^2$$

AT required for portion B,

$$AT_B = \frac{B_B \, l_B}{\mu_0 \mu_r} = \frac{8.33 \times 0.15}{4\pi \times 10^{-7} \times 600} = 1{,}658$$

Total ampere-turns required,

$$AT = AT_A + AT_B = 4{,}416 + 1{,}658 = 6{,}074 \textbf{ Ans.}$$

Required current,

$$I = \frac{AT}{N} = \frac{6{,}074}{500} = 12.15 \text{ A} \textbf{ Ans.}$$

Example 1.19. A ring of cast steel has an external diameter of 24 cm and a square x-section of 3 cm side. Inside and across the ring an ordinary steel bar 18 × 3 × 0. 4 cm is fitted with negligible gap. Calculate the number of ampere-turns required to be applied to one half of the ring to produce a flux density of 1 Wb/m² in the other half. Neglect leakage.

The B-H characteristics area follows:

	For cast steel			For ordinary steel		
B(Wb/m²)	**1.0**	**1.1**	**1.2**	**1.2**	**1.4**	**1.45**
H(AT/m)	**900**	**1,020**	**1,230**	**590**	**1,200**	**1,650**

Solution: External diameter of the ring = 24 cm = 0.24 m

Internal diameter of the ring = 24 – 6 = 18 cm = 0.18 m

Mean diameter of the ring, $D = \dfrac{0.24 + 0.18}{2} = 0.21$ m

Mean length of the ring = $\pi D = \pi \times 0.21 = 0.66$ m

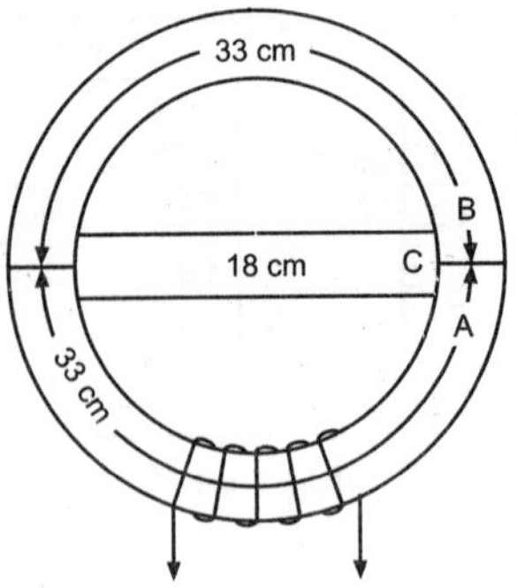

Fig. 1.41

Length of upper portion of ring, l_A = length of lower portion of ring l_A

$$= \frac{0.66}{2} = 0.33 \text{ m}$$

Area of x-section of ring $a_A = a_B = 0.03 \times 0.03 = 9 \times 10^{-4}$ m^2

Length of cross-bar, l_C = 18 cm = 0.18 m

Area of cross section of cross-bar, $a_C = 0.03 \times 0.004 = 12 \times 10^{-5}$ m^2

This is a case of parallel magnetic circuit.

Flux created in A divides itself when it comes to the junction point depending upon the reluctance of portions B and C respectively.

$$i.e.\ \Phi_A = \Phi_B + \Phi_C$$

Flux density in the upper half, B_B = 1 T

For cast steel, against B = 1 T ; H = 900 AT/m

∴ Total ampere-turns required for portion B = 900 × 0.33 = 297

Since portions B and C are in parallel

∴ AT required for portion C

= AT required of portion B = 297

Magnetising force for the portion C,

$$H = \frac{297}{0.18} = 1{,}650 \text{ AT/m}$$

For ordinary steel density,

B_C = 1.45 T when H = 1,650 AT/m

Flux in portion B,

$$\Phi_B = B_B \times a_B = 1 \times 9 \times 10^{-4} = 9 \times 10^{-4} \text{ Wb}$$

Flux in portion C,

$$\Phi_C = B_C \times a_C = 1.45 \times 12 \times 10^{-5} = 1.74 \times 10^{-4} \text{ Wb}$$

Flux in portion A,

$$\Phi_A = \Phi_B + \Phi_C = 9 \times 10^{-4} + 1.74 \times 10^{-4} = 10.74 \times 10^{-4} \text{ Wb}$$

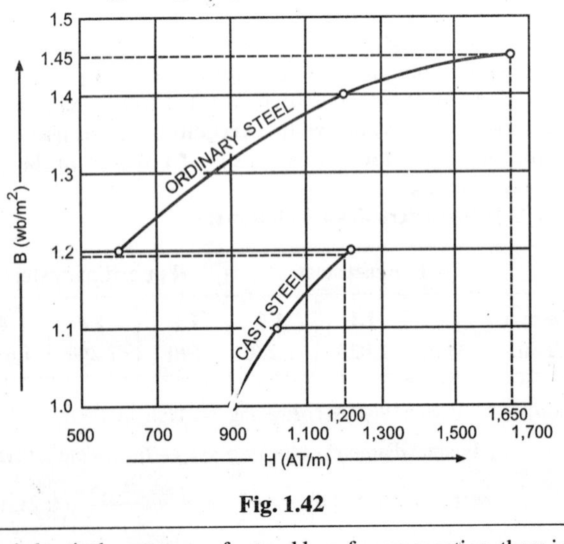

Fig. 1.42

Flux density in portion A,

$$B_A = \frac{\Phi_A}{a_A} = \frac{10.74 \times 10^{-4}}{9 \times 10^{-4}} = 1.1933 \text{ T}$$

From the B-H curve drawn between H and B for cast steel magnetising force corresponding to flux density of 1.1933 T is 1,200 AT/m.

AT required for portion A = 1200 × 0.33 = 396

Total ampere-turns required = AT for portion A + AT for portion B (or AT for portion C)

= 396 + 297 = 693 **Ans.**

1.21 MAGNETICALLY INDUCED EMFS (OR VOLTAGES)

A very important effect of a magnetic field on an electric circuit is that when the flux linking the circuit changes, an emf is induced. Electromagnetic induction of emf (or voltage) is basic to the operation of transformers, generators (ac or dc) and motors (ac or dc). The effect is described by Faraday's law, which states that *the magnitude of emf (or voltage) is directly proportional to the rate of change of flux linkage or to the product of number of turns and rate of change of flux linking the coil.*

$$i.e. \text{ induced emf, } e = N \frac{d\Phi}{dt} \quad ...(1.25)$$

where $N \frac{d\Phi}{dt}$ is the product of number of turns and rate of change of linking flux and is termed as *rate of change of flux linkage.*

The direction of induced emf is governed by Lenz's law* which states that *the direction of induced emf or voltage is such that the current produced by it sets up a magnetic field opposing the cause that produces it.*

A minus (–) sign is required to be placed before the right hand side quantity of Eq. (1.25) just to indicate the phenomenon explained by Lenz's law. Thus the expression for induced emf becomes as

$$e = -N \frac{d\Phi}{dt} \text{ volts} \quad ...(1.26)$$

where Φ is in webers and time *t* is in seconds.

The induced emf or voltage appears in any circuit which may be linked by the changing flux, including the circuit giving rise to flux. Thus, current in a circuit produces a magnetic field linking the circuit and permeating the medium around it. Growth, decay or any other change of current with time causes corresponding changes of the magnetic flux and induces an emf in the circuit.

1.21.1. Dynamically Induced EMF. We have learnt that when the flux linking with the coil or circuit changes, an emf is induced in the coil or circuit.

EMF can be induced by changing the flux linking in two ways :

(*i*) By increasing or decreasing the magnitude of the current producing the linking flux. In this case there is no motion of the conductor or of coil relative to the field and, therefore, emf induced in this way is known as *statically induced emf.*

* Lenz's law is the outcome of natural law: for every action, there is an equal and opposite reaction.

(*ii*) By moving a conductor in a uniform magnetic field and emf produced in this way is known as *dynamically induced emf.*

Consider a conductor of length l metres placed in a uniform magnetic field of density B Wb/m^2, as shown in Fig. 1.43 (*a*).

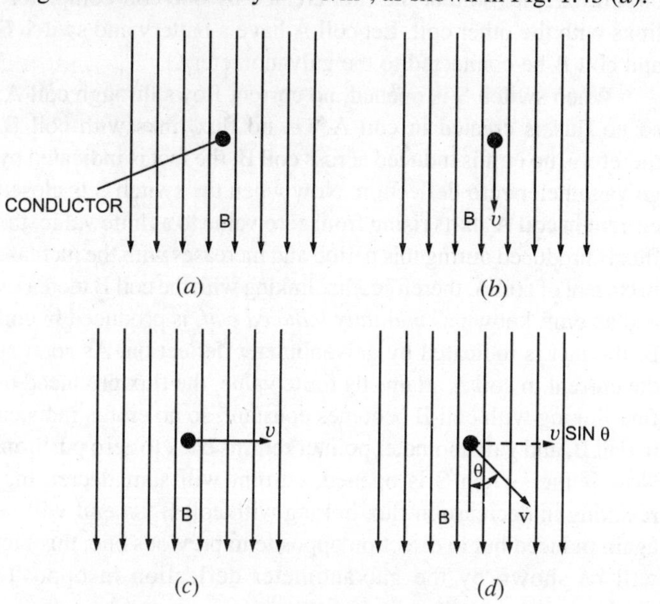

Fig. 1.43

Let this conductor be moved with velocity v m/s in the direction of the field, as shown in Fig. 1.43 (*b*). In this case no flux is cut by the conductor, therefore, no emf is induced in it.

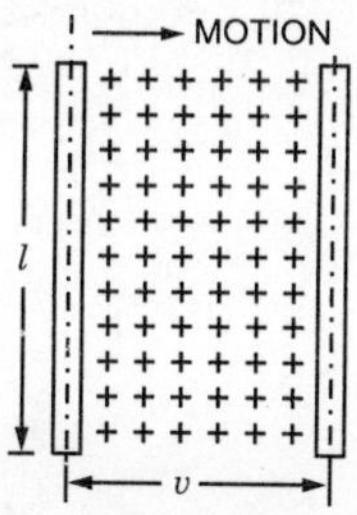

Fig. 1.44

Now if this conductor is moved with velocity v m/s in a direction perpendicular to its own length and perpendicular to the direction of the magnetic field, as shown in Fig. 1.43 (*c*) flux is cut by the conductor, therefore, an emf is induced in the conductor.

Area swept per second by the conductor = $l \times v \ m^2/s$

Flux cut per second = Flux density × area swept per second = Blv

Rate of change of flux, $\frac{d\Phi}{dt}$ = Flux cut per second = B$l\,v$ Wb/s

$$\text{Induced emf, } e = \frac{d\Phi}{dt} = Blv \text{ volts}$$

If the conductor is moved with velocity v metres per second in a direction perpendicular to its own length and at an angle θ to the direction of magnetic field, as shown in Fig. 1.43 (*d*), the magnitude of emf induced is proportional to the component of the velocity in a direction perpendicular to the direction of the magnetic field and induced emf is given by

$$e = Blv \sin\theta \text{ volts} \qquad ...(1.27)$$

The direction of this induced emf is given by Fleming's right hand rule.

If the thumb, forefinger and middle finger of right hand are held mutually perpendicular to each other, forefinger pointing into the direction of the field and thumb in the direction of motion of conductor then the middle finger will point in the direction of the induced emf as shown in Fig. 1.45 (*a*).

Figure 1.45 (*b*) illustrates another way of determination of induced emf, known as *right hand flat palm* rule. This law states

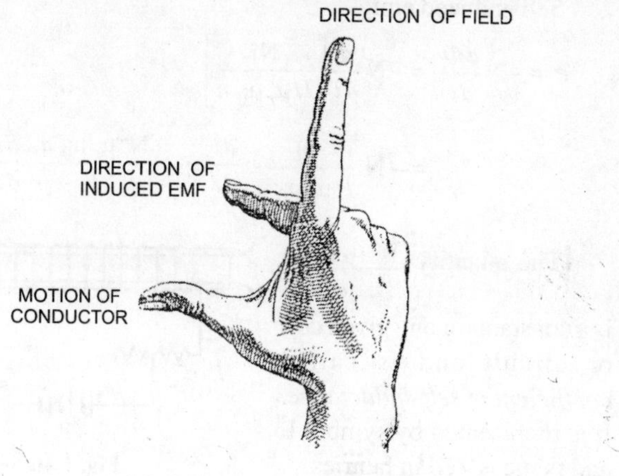

(*a*) *Fleming's Right Hand Rule*

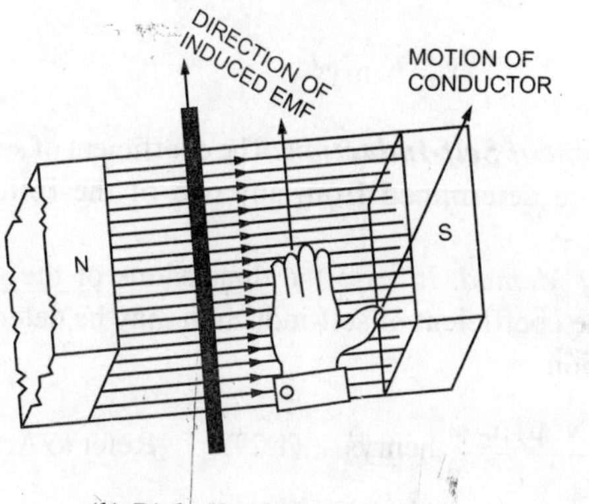

(*b*) *Right Hand Flat Palm Rule*

Fig. 1.45 *Determination of Direction of Induced EMF*

that if right hand is so placed in a magnetic field along the conductor that the magnetic lines of force emerging from the north pole enter the palm and the thumb points in the direction of motion of conductor, the other four fingers will give the direction of induced emf or current.

1.21.2. Statically Induced EMF. *Statically* induced emf may be (*a*) self-induced emf or (*b*) mutually induced emf.

(a) Self-Induced EMF. When the current flowing through the coil is changed, the flux linking with its own winding changes and due to the change in linking flux with the coil, an emf, known as *self-induced emf, is induced.*

Since according to Lenz's law, an induced emf acts to oppose the change that produces it, a self-induced emf is always in such a direction as to oppose the change of current in the coil or circuit in which it is induced. This property of the coil or circuit due to which it opposes any change of the current in the coil or circuit, is known as *self-inductance.*

Consider a solenoid of N turns, length l metres, area of x-section a square metres and of relative permeability μ_r. When the solenoid carries a current of i amperes, a magnetic field of flux $\frac{Ni}{l/\mu_0 \mu_r \, a}$ webers is set up around the solenoid and links with it.

If the current flowing through the solenoid is changed, the flux produced by it will change and, therefore, an emf will be induced.

Self-induced emf,

$$e = -N\frac{d\Phi}{dt} = -N\frac{d}{dt}\left[\frac{Ni}{l/\mu_r\mu_0 a}\right]$$

$$= -N\frac{N}{l/\mu_r\mu_0 a}\frac{di}{dt} = -\frac{N^2\mu_r\mu_0 a}{l}\cdot\frac{di}{dt} \quad ...(1.28)$$

The quantity $\frac{N^2\mu_r\mu_0 a}{l}$ is a constant for any given coil or circuit and is called *coefficient of self-inductance.* It is represented by symbol L and is measured in henries.

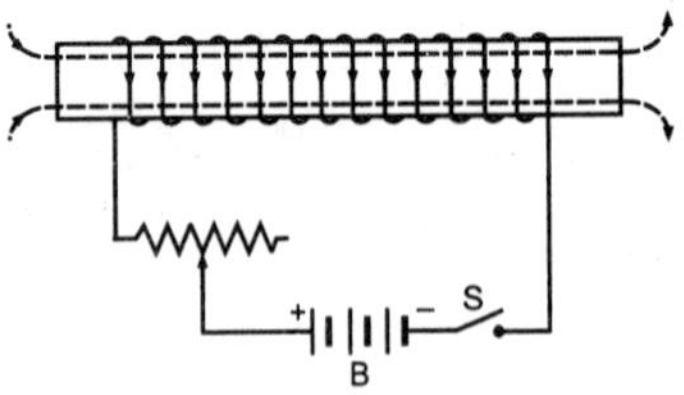

Fig. 1.46

Hence self-induced emf, $e = -L\frac{di}{dt}$

where $L = \frac{N^2\mu_r\mu_0 a}{l}$ henrys

Coefficient of Self-Induction. The coefficient of self-induction (L) can be determined from any one of the following three relations.

First Method. In case the dimensions of the solenoid are given, the coefficient of self-induction may be determined from the relation

$$L = \frac{N^2\mu_r\mu_0 a}{l} \text{ henrys} \quad ...(1.29) \quad \text{[Refer to Art. 1.21.2 (a)]}$$

Second Method. In case the magnitude of induced emf in a coil for a given rate of change of current in the coil is known, self-inductance of the coil may be determined from the following relation.

$$e = L\frac{di}{dt} \quad \text{...[Refer to Art. 1.21.2 (a)]}$$

$$\text{or } L = \frac{e}{di/dt} \quad ...(1.30)$$

Third Method. In case the number of turns of the coil and flux produced per ampere of current in the coil is known, the self-inductance of the coil may be determined from the following relation

$$L = \frac{N\Phi}{i} \quad ...(1.31)$$

The above relation can be derived as follows:

Magnetic flux produced in a coil of N turns, length l metres, area of x-section a metres^2 and relative permeability μ_r when carrying a current of i amperes is given by

$$\Phi = \frac{Ni}{l/\mu_r\mu_0 a} = \frac{N}{l/\mu_r\mu_0 a}i$$

and self-inductance of the coil,

$$L = \frac{N^2\mu_r\mu_0 a}{l} = \frac{N}{i}\frac{N\mu_r\mu_0 a}{l}i = \frac{N\Phi}{i}$$

From the above relation it is obvious that the self-inductance of a coil or circuit is equal to weber-turns per ampere in the coil or circuit.

In the above relation if $N\Phi = 1$ Wb-turn and $i = 1$ A then L = 1 H.

Hence a coil is said to have a self-inductance of one henry if a current of 1 A, when flowing through it, produces flux linkage of 1 Wb-turn in it.

(b) *Mutually Induced EMF.* Consider two coils A and B placed closed together so that the flux created by one coil completely links with the other coil. Let coil A have a battery and switch S and coil B be connected to the galvanometer G.

When switch S is opened, no current flows through coil A, so no flux is created in coil A, *i.e.* no flux links with coil B, therefore, no emf is induced across coil B, the fact is indicated by galvanometer zero deflection. Now when the switch S is closed current in coil A starts rising from zero value to a finite value, the flux is produced during this period and increases with the increase in current of coil A, therefore, flux linking with the coil B increases and an emf, known as *mutually induced emf,* is produced in coil B, the fact is indicated by galvanometer deflection. As soon as the current in coil A attains its finite value, the flux produced or flux linking with coil B becomes constant, so no emf is induced in coil B, and galvanometer pointer returns back to zero position. Now if the switch S is opened, current will start decreasing, resulting in decrease in flux linking with coil B, an emf will be again induced but in direction opposite to previous one, this fact will be shown by the galvanometer deflection in opposite direction.

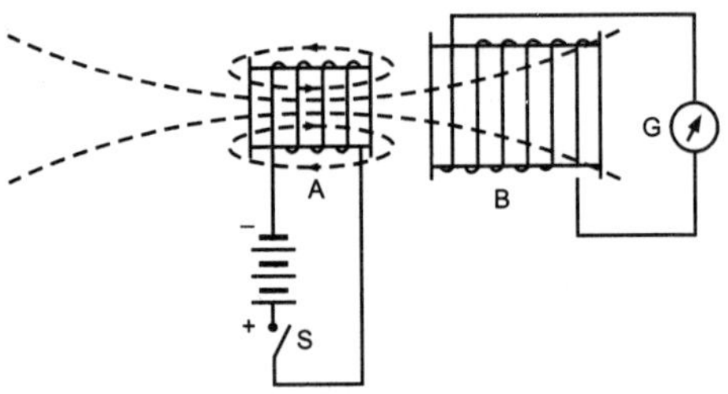

Fig. 1.47

Hence whenever the current in coil A changes, the flux linking with coil B changes and an emf, known as mutually induced emf, is induced in coil B.

Consider coil A of turns N_1 wound on a core of length l metres, area of cross section a square metres and relative permeability μ_r. When the current of i_1 amperes flows through it, a flux of $\frac{N_1 i_1}{l/\mu_r\mu_0 a}$ webers is set up around the coil. Let whole of the flux produced due to flow of current in coil A be linked with the coil B having N_2 turns and placed near by coil A.

Mutually induced emf,

e_m = – Rate of change of flux linkage of coil B

= – N_2 × rate of change of flux in coil A

$$= -N_2\frac{d}{dt}\left[\frac{N_1 i_1}{l/\mu_r\mu_0 a}\right] = -\frac{N_1 N_2\mu_r\mu_0 a}{l}\cdot\frac{di_1}{dt} \quad ...(1.32)$$

The quantity $\frac{N_1 N_2\mu_r\mu_0 a}{l}$ is called the *coefficient of mutual induction* of coil B with respect to coil A. It is represented by symbol M and is measured in henrys.

Hence mutually inducd emf,

$$e_m = -M\frac{di_1}{dt} \quad \text{where } M = \frac{N_1 N_2\mu_r\mu_0 a}{l} \text{ henrys}$$

Coefficient of Mutual Induction. Mutual inductance may be defined as the ability of one coil or circuit to induce an emf in a nearby coil by induction when the current flowing in the first coil is changed. The action is also reciprocal *i.e.* the change in current flowing through second coil will also induce an emf in the first coil. The ability of reciprocal induction is measured in terms of the coefficient of mutual induction M.

The coefficient of mutual induction (M) can be determined from any one of the following three relations.

First Method. In case the dimensions of the coils are given, the coefficient of mutual induction may be determined from the relation

$$M = \frac{N_1 N_2 \mu_r \mu_0 a}{l} \text{ henrys} \quad ...(1.33) \quad \text{[Refer to Art. 1.21.2 (b)]}$$

Second Method. In case the magnitude of induced emf in the second coil for a given rate of change of current in the first coil is known, mutual inductance between the coil may be determined from the following relation

$$e_m = M\frac{di_1}{dt} \quad \text{[Refer to Art. 1.21.2 (b)]}$$

$$\text{or } M = \frac{e_m}{di_1/dt} \quad ...(1.34)$$

Third Method. In case the number of turns of the coil and flux linking with this coil per ampere of current in another coil is known, the mutual inductance of the coil may be determined from the following relation

$$M = N_2 \frac{\Phi_2}{i_1} \text{ Henry} \quad ...(1.35)$$

1.21.3. Leakage Reactance. A coil current varying sinusoidally may be expressed as $i = I \sin \omega t$ and therefore, change of current w.r.t. time is given as

$$\frac{di}{dt} = \omega I \cos \omega t$$

and induced emf is given as

$$e = L\frac{di}{dt} = \omega L I \cos \omega t$$

From the above relationship it is observed that ωL must be measured in ohms ωL is known as *self-inductive reactance* and is symbolized as X. If there is mutual inductance, the reactance symbol used will X_m, which is proportional to ωM.

When leakage flux is associated with a winding carrying "steady state" alternating current, a reactive voltage is induced. The magnetic reluctance offered to leakage flux is predominantly due to the air path under normal conditions and therefore, this flux varies directly as the current, providing a constant inductance. Consequently the reactive voltage may be considered as because of constant leakage reactance and will be symbolized as X. It is determining factor in the machine performance.

Example 1.20. Find the induced emf in a conductor of length 150 cm moving at an angle of 30° to the direction of uniform magnetic field of flux density 1.2 Wb/m² with a velocity of 60 m/s.

Solution: Flux density B = 1.2 T

Length of conductor, l = 150 cm = 1.5 m

Velocity of conductor, v = 60 m/s

Angle of movement of conductor from the direction of magnetic field,

$\theta = 30°$

EMF induced, $e = Blv \sin\theta = 1.2 \times 1.5 \times 60 \times \sin 30° = 108 \times 0.5 = 54$ V **Ans.**

Example 1.21. A conductor of active length 30 cm carries a current of 100 A and lies at right angles to a magnetic field of strength 0.4 Wb/m². Calculate the force in newtons exerted on it.

If the force causes the conductor to move at a velocity of 10 m/s, calculate (*i*) the emf induced in it and (*ii*) the power in watts developed by it.

Solution: Length of conductor, l = 30 cm = 0.3 m

Current flowing through conductor,

I = 100 A

Field strength, B = 0.4 Wb/m²

Force exerted on the conductor, $F = B I l = 0.4 \times 100 \times 0.3 = 12$ N **Ans.**

Conductor velocity, v = 10 m/s

Induced emf, $e = B l v = 0.4 \times 0.3 \times 10 = 1.2$ V **Ans.**

Power developed, $P = F \times v = 12 \times 10 = 120$ watts **Ans.**

Example 1.22. A square coil of 10 cm side and with 100 turns is rotated at a uniform speed of 500 rpm about an axis at right angle to a uniform field of 0.5 T. Calculate the instantaneous value of induced emf when the plane of coil is (*i*) at right angle to the plane of the field (*ii*) at 30° to the plane of the field and (*iii*) in the plane of the field.

Solution: As there are two active sides in the coil, emf induced in the coil is given by $e = 2Blv \sin\theta$ volts

where θ is the angle of the coil movement with the direction of the field.

(*a*) *When plane of the coil is at right angle to the plane of the field.*

The coil moves at an angle of 0° with the direction of the field *i.e.* moves just parallel to the lines of force or θ = 0 or sin θ = 0°

∴ emf induced, e = 0 **Ans.**

(*b*) *When the plane of the coil makes an angle of 30° with the plane of the field*

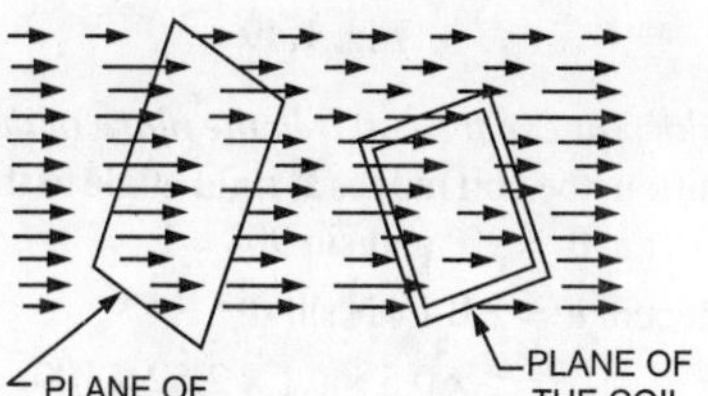

Fig. 1.48

The coil moves at an angle of (90° – 30°) with the direction of the field *i.e.* θ = 60° and $\sin 60° = \frac{\sqrt{3}}{2}$

Flux density, B = 0.5 T

Length of coil, l = 10 cm = 0.1 m

Number of turns N = 100

$$\text{Velocity, } v = \omega r \text{ m/s} = \frac{2\pi rpm}{60} r \text{ m/s}$$

$$= \frac{2\pi \times 500}{60} \times 0.05 = 2.62 \text{ m/s}$$

EMF induced, $e = 2 Blv N \sin\theta$

$$= 2 \times 0.5 \times 0.1 \times 2.62 \times 100 \times \frac{\sqrt{3}}{2}$$

= 22.65 volts **Ans.**

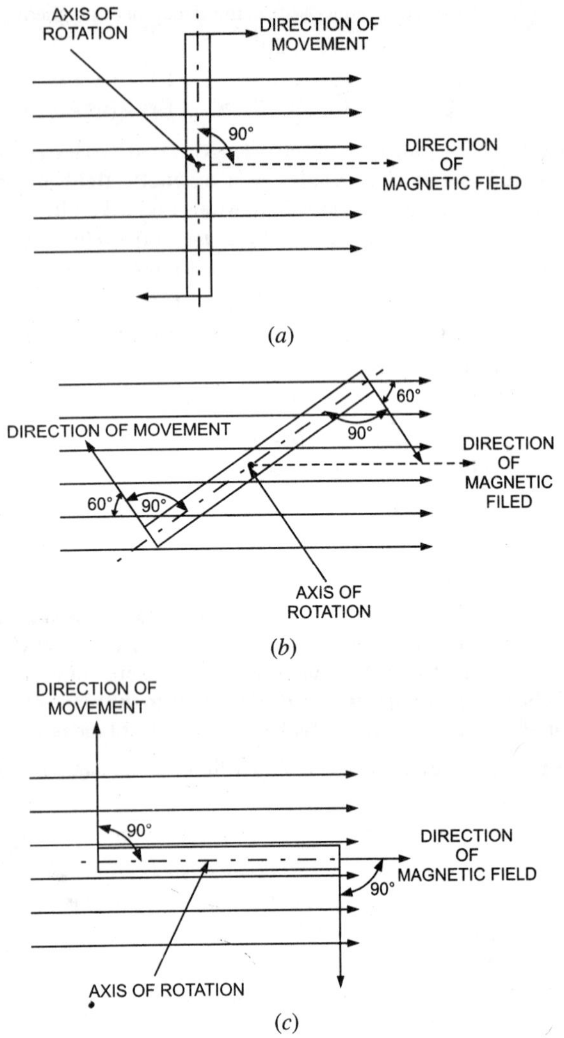

Fig. 1.49

(c) When the plane of the coil is in the plane of the field

In this position, the coil moves at right angle to the field direction

$i.e.\ \theta = 90°$ and $\sin 90° = 1$

$\therefore$ EMF induced, $e = 2\,B\,l\,v\,N \sin\theta$

$= 2 \times 0.5 \times 0.1 \times 2.62 \times 100 \times 1 = 26.2$ V **Ans.**

Example 1.23. A square coil of 10 turns and 10 cm side is moved through a steady magnetic field of 1 Wb/m² at a constant velocity of 2 m/s with its plane perpendicular to the field as shown in Fig. 1.50. Plot the variation of induced emf as the coil moves along the field.

Solution: Number of turns of coil, N = 10

Magnetic field strength B = 1 Wb/m²

Movement velocity, v = 2 m/s

Coil side, a = 10 cm = 0.1 m

When the coil enters, let x length of the loop lie inside the field, therefore,

$$\text{Flux } \Phi = B \times A = B \times (x \times a) = B\,x\,a$$

$$\text{Rate of change of flux, } \frac{d\Phi}{dt} = Ba\frac{dx}{dt} = Bav$$

$$\text{Induced emf, } e = N\frac{d\Phi}{dt} = N\,Bav$$

$$= 10 \times 1 \times 0.1 \times 2 = 2 \text{ V}$$

Plot of variation of induced emf is given in Fig. 1.51.

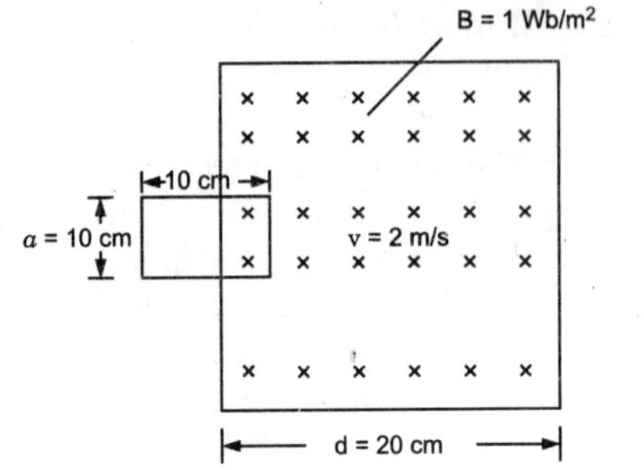

Fig. 1.50

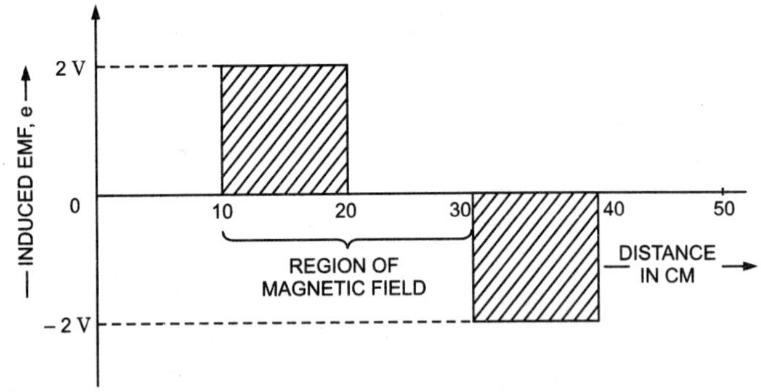

Fig. 1.51

Example 1.24. A rectangular loop of sides *a* and *b* has its plane normal to a magnetic flux density of strength $B_0 \sin \omega t$. What is the voltage induced in the above loop ? *[GATE 1997]*

Solution: Area of rectangular loop,

$$A = a \times b$$

Magnetic flux density, $B = B_0 \sin \omega t$

$$\text{Voltage induced, } v = -\int \frac{\delta B}{\delta t}\, dS$$

$$= -\int_0^a \int_0^b B_0\,\omega \cos \omega t\, dx\, dy$$

$$= B_0\,\omega \cos \omega t \times a \times b = B_0\,\omega\,A \cos \omega t \quad \textbf{Ans.}$$

Example 1.25. A solenoid has 1,200 turns and carries a current of 2 A. The iron core has a length of 0.4 m and cross section of 80 cm², the relative permeability is 1,000. Calculate the self-induced emf in the solenoid, if the current is switched off in 0.01 second.

Solution: Area of x-section of iron core,

$a = 80 \text{ cm}^2 = 0.008 \text{ m}^2$

Self-inductance of coil,

$$L = \frac{N^2 \mu_r \mu_0 a}{l} = \frac{(1{,}200)^2 \times 1{,}000 \times 4\pi \times 10^{-7} \times 0.008}{0.4}$$

$$= 36.191 \text{ H}$$

Rate of change of current,

$$\frac{di}{dt} = \frac{2-0}{0.01} = 200 \text{ A/s}$$

Self-induced emf,

$$e = L\frac{di}{dt} = 36.191 \times 200 = 7{,}238 \text{ V} \quad \textbf{Ans.}$$

Example 1.26. A coil of 300 turns, wound on a core of nonmagnetic material, has an inductance of 10 mH. Calculate *(i)* the flux produced by a current of 5 A and *(ii)* the average value of emf induced when a current of 5 A is reversed in 8 milliseconds.

Solution: (*i*) Flux produced, $\Phi = \frac{iL}{N} = \frac{5 \times 0.01}{300} = 0.1667$ mWb **Ans.**

$$\because \ L = \frac{N\Phi}{i}$$

Rate of change of current, $\frac{di}{dt} = \frac{5-(-5)}{0.008} = 1{,}250$ A/s

Induced emf, $e = L\frac{di}{dt} = 0.01 \times 1{,}250 = 12.5$ V **Ans.**

Example 1.27. A coil of resistance 90 Ω is placed in a magnetic field of 1 mWb. The coil has 100 turns and a galvanometer of 405 Ω resistance is connected in series with it. Find the average emf and current if the coil is moved in 1/10th second from the given field to a field of 0.2 mWb.

[M.D. Univ. Electromechanical Energy Conversion, December-2010]

Solution: Number of turns on the coil,

$$N = 100$$

Change of flux in 1/10 second,

$$d\Phi = 0.001 - 0.0002 = 0.0008 \text{ Wb}$$

Rate of change of flux,

$$\frac{d\Phi}{dt} = \frac{0.0008}{0.1} = 0.008 \text{ Wb/second}$$

Induced emf,

$$e = N\frac{d\phi}{dt} = 100 \times 0.008 = 0.8 \text{ V} \textbf{ Ans.}$$

Current through coil,

$$I = \frac{e}{\text{Coil resistance + galvanometer resistance}}$$

$$= \frac{0.8}{90+405} = \frac{0.8}{495} = 0.00162 \text{ A or } 1.62 \text{ mA} \textbf{ Ans.}$$

Example 1.28. A coil of 1,000 turns is wound on a laminated core of steel having a cross section of 5 cm^2. The core has an air gap of 2 mm cut at right angle. What value of current is required to have an air gap flux density of 0.5 T ? Permeability of steel may be taken as infinity. Determine the coil inductance.

Solution: Total ampere-turns required,

$$AT = \frac{B}{\mu_0 \mu_r} l_i + \frac{B}{\mu_0} l_g = 0 + \frac{0.5}{4\pi \times 10^{-7}} \times 2 \times 10^{-3} = 796 \quad \because \mu_r = \infty$$

Current required,

$$i = \frac{AT}{N} = \frac{796}{1{,}000} = 0.796 \text{ A} \textbf{ Ans.}$$

Inductance of coil,

$$L = \frac{N\Phi}{i} = \frac{1{,}000 \times 0.5 \times 5 \times 10^{-4}}{0.796} = 0.314 \text{ H} \textbf{ Ans.}$$

Example 1.29. The resistance and inductance of a coil are 3 Ω and 0.1 m H respectively. What potential difference exists at the terminals of the solenoid at the instant when the current is 1 A, but increasing at the rate of 10,000 A per second?

Solution: Resistance of the coil, R = 3 Ω

Inductance of the coil, $L = 0.1$ m H $= 1 \times 10^{-4}$ H

Current flowing through the coil, I = 1.0 A

Rate of increase of current, $\frac{di}{dt} = 10{,}000$ A/s

Potential difference across the solenoid,

V = Voltage drop in resistance + induced voltage in the coil

$$= IR + L\frac{di}{dt} = 1 \times 3 + 1 \times 10^{-4} \times 10{,}000 = 4 \text{ V} \textbf{ Ans.}$$

Example 1.30. Find the inductance of a coil in which a current of 0.2 A increasing at the rate of 0.4 A per second represents a power flow of 0.4 watt.

Solution: Power flow, $p = 0.4$ watt

Current, $i = 0.2$ A

Induced emf, $e = \frac{p}{i} = \frac{0.4}{0.2} = 2$ V $\quad \because p = ei$

Rate of increase of current, $\frac{di}{dt} = 0.4$ A/s

Inductance of coil, $L = \frac{e}{di/dt} = \frac{2}{0.4} = 5$ H **Ans.**

Example 1.31. Two coils having 100 and 150 turns respectively are wound side by side on a closed iron circuit of section 125 cm^2 and mean length 200 cm. If the permeability of iron is 2,000, calculate (*a*) self-inductance of each coil (*b*) mutual inductance between them (*c*) the emf induced in the second coil if current in first coil changes from 0 to 5 A in 0.02 *s*.

Solution: Area of x-section of iron circuit, a = 125 cm^2 = 0.0125 m^2

Length of iron circuit, l = 200 cm = 2.0 m

(*a*) Self-inductance of first coil,

$$L_1 = \frac{N_1^2 \mu_0 \mu_r a}{l} = \frac{(100)^2 \times 4\pi \times 10^{-7} \times 2{,}000 \times 0.0125}{2.0}$$

$$= 157.1 \text{ m H} \textbf{ Ans.}$$

Self-inductance of second coil,

$$L_2 = \frac{N_2^2 \mu_0 \mu_r a}{l} = \frac{(150)^2 \times 4\pi \times 10^{-7} \times 2{,}000 \times 0.0125}{2.0}$$

$$= 353.4 \text{ mH} \textbf{ Ans.}$$

(*b*) Mutual inductance between coil,

$$M = \frac{N_1 N_2 \mu_0 \mu_r a}{l}$$

$$= \frac{100 \times 150 \times 4\pi \times 10^{-7} \times 2{,}000 \times 0.0125}{2.0} = 235.6 \text{ mH} \textbf{ Ans.}$$

Rate of change of current in first coil,

$$\frac{di_1}{dt} = \frac{5-0}{0.02} = 250 \text{ A/s}$$

(*c*) EMF induced in second coil,

$$e_2 = M\frac{di_1}{dt} = 235.6 \times 10^{-3} \times 250 = 58.9 \text{ V} \textbf{ Ans.}$$

Example 1.32. Two coils have a mutual inductance of 0.3 H. If the current in one coil is varied from 5 A to 2 A in 0.4 s, calculate (*i*) the average emf induced in the second coil (*ii*) the change of flux linked with the second coil assuming that it is wound with 200 turns.

Solution: Mutual inductance,

$$M = 0.3 \text{ H}$$

Rate of change of current in one coil,

$$\frac{di_1}{dt} = \frac{5-2}{0.4} = 7.5 \text{ A/s}$$

(*i*) EMF induced in second coil,

$$e_2 = M\frac{di_1}{dt} = 0.3 \times 7.5 = 2.25 \text{ V} \textbf{ Ans.}$$

(*ii*) Since mutually induced emf,

$$e_2 = N_2 \frac{d\Phi_2}{dt}$$

∴ Change of flux linked with the second coil,

$$\frac{d\Phi_2}{dt} = \frac{e_2}{N_2} = \frac{2.25}{200} = 0.01125 \text{ Wb/s } \textbf{Ans.}$$

Example 1.33. A long single layer solenoid has an effective diameter of 10 cm and is wound with 2,500 T/m. There is a small concentrated coil having its plane lying in the centre cross-sectional plane of the solenoid. Calculate the mutual inductance between the two coils if the concentrated coil has 120 turns on an effective diameter of (*a*) 8 cm and (*b*) 12 cm.

Solution: (*a*) When the effective diameter of concentrated coil is 8 cm (less than that of solenoid)

Flux density at the centre of the solenoid,

$$B = \mu_0 H = \frac{\mu_0 N I_1}{l} = 2{,}500 \times 4\pi \times 10^{-7} \times I_1$$

$$= 0.00314\, I_1 \text{ Wb/m}^2$$

where I_1 is the current flowing through the solenoid and N/*l* is equal to number of turns per metre length of solenoid *i.e.* 2,500

Area of search coil,

$$a_S = \frac{\pi}{4}(0.08)^2 = 0.005 \text{ m}^2$$

Flux linking with search coil,

$$\Phi_2 = B\, a_S = 0.00314\, I_1 \times 0.005 = 15.79 \times 10^{-6}\, I_1 \text{ Wb}$$

Mutual inductance between the coil,

$$M = \frac{N_2 \Phi_2}{I_1} = \frac{120 \times 15.79 \times 10^{-6}\, I_1}{I_1} = 1.895 \text{ mH } \textbf{Ans.}$$

(*b*) When the effective diameter of the concentrated coil is 12 cm (more than that of the solenoid)

In this case the effective area of the search coil will be equal to that of long solenoid as the field strength outside the solenoid is negligible.

$$\text{So, } \Phi_2 = 0.00314\, I_1 \times \frac{\pi}{4}(0.10)^2 = 24.67\, I_1 \times 10^{-6} \text{ Wb}$$

$$\text{and } M = \frac{N_2 \Phi_2}{I_1} = \frac{120 \times 24.67 \times I_1 \times 10^{-6}}{I_1} = 2.96 \text{ mH } \textbf{Ans.}$$

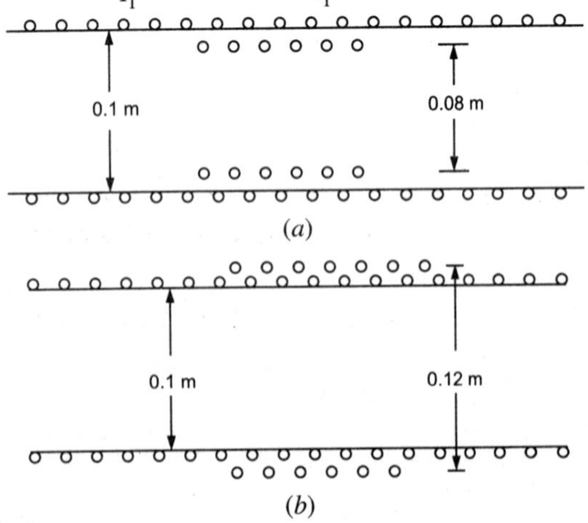

Fig. 1.52

Example 1.34. The self-inductance of a coil of 500 turns is 0.25 H. If 60% of the flux of this coil is linked with a second coil of 10,000 turns, calculate (*i*) The mutual inductance of the two coils, (*ii*) EMF induced in the second coil, when current in the first coil changes at the rate of 100 A/second. [*Rajasthan Technical Univ. 2007*]

Solution: Self-inductance of the first coil,

$$L_1 = 0.25 \text{ H}$$

Number of turns on first coil,

$$N_1 = 500$$

Flux per ampere of current in first coil,

$$\frac{\Phi_1}{i_1} = \frac{L_1}{N_1} = \frac{0.25}{500} = 0.5 \text{ mWb} \qquad \because L_1 = N_1 \frac{\Phi_1}{i_1}$$

Flux linked with second coil per ampere of current in first coil

$$\frac{\Phi_2}{i_1} = \frac{0.6\,\Phi_1}{i_1} = 0.6 \times 0.5 = 0.3 \text{ mWb}$$

(*i*) The mutual inductance of the two coils,

$$M = \frac{N_2 \Phi_2}{i_1} = 10{,}000 \times 0.3 \times 10^{-3} = 3 \text{ H } \textbf{Ans.}$$

(*ii*) EMF induced in the second coil,

$$e_2 = M \frac{di_1}{dt} = 3 \times 100 = 300 \text{ V } \textbf{Ans.}$$

Example 1.35. A coil with axial length 25 cm and diameter of 20 cm has 200 turns. It is placed in a uniform flux of 0.002 Wb/m² (*a*) If the coil is rotated at 25 revolutions per second, find the voltage induced in the coil. (*b*) What will be the force on each conductor and torque acting on the coil if it carries a current of 10 A.

[*Rajasthan Technical Univ. Electrical Machines-I, 2008*]

Solution: Length of coil, l = 25 cm = 0.25 m

Number of turns, N = 200

$$\text{Velocity, } v = \omega r \text{ m/s} = 2\pi \times rps \times r \text{ m/s} = 2\pi \times 25 \times \frac{10}{100}$$

$$= 5\pi \text{ m}$$

$$\because r = \frac{20}{2} = 10 \text{ cm} = \frac{10}{100} \text{ m}$$

Flux density, B = 0.002 Wb/m²

(*a*) Induced e.m.f., $e = NBlv = 200 \times 0.002 \times 0.25 \times 5\pi = 1.57$ V **Ans.**

(*b*) Force on each conductor,

$$F = BIl = 0.002 \times 10 \times 0.25 = 0.005 \text{ N} \quad \textbf{Ans.}$$

Torque acting on the coil,

$$T = 2 \times F \times r \times N = 2 \times 0.005 \times 0.1 \times 200$$

$$= 0.2 \text{ N-m} \quad \textbf{Ans.}$$

1.22 MUTUAL COUPLING IN A SIMPLE MAGNETIC CIRCUIT

The self-inductance of a circuit is intimately associated with the magnetic field linking the circuit. The self-inductance emf may be thought of as the emf induced in the circuit by a magnetic field produced by the circuit current.

Since a magnetic field exists in the region around the current that develops it, there is also a possibility that an emf be induced in the other circuits linked by the field. Two circuits linked by the same magnetic field are said to be *coupled* to each other. The circuit element used to represent *magnetic coupling* is shown in Fig. 1.53 (*a*) and is called *mutual inductance*. It is represented by symbol M and is measured in henrys, already stated in Art. 1.21.2 (*b*). The volt-ampere relationship is one which gives the induced emf in one circuit by a current in another and is given as

$$e_2 = M \frac{di_1}{dt}$$

A similar equation can, of course, be written giving an emf e_1 induced by a current i_2. The two dots, called *polarity markings,* in Fig. 1.53 (*b*) are used for indicating the direction of the magnetic coupling between the two coils : note from the figure and from equation given above that by matching the dots, the directions of currents and voltage drops are made to correspond with those of Fig. 1.53 (*b*) and Eq. (1.33) for mutual inductance.

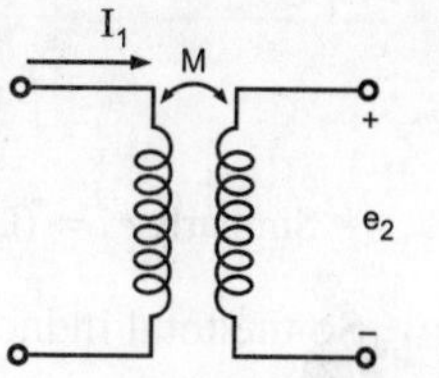

(*a*) *Schematic Representation of Mutual Inductance*

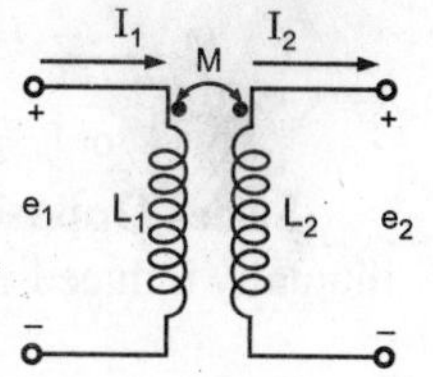

(*b*) *Mutual Inductance With Current in Both Coils*

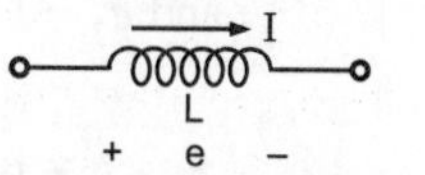

(*c*) *Schematic Representation of Self-Inductance*

Fig. 1.53

If currents are present in both coupled circuits, emfs of self-inductance and mutual inductance are induced in each circuit. The self-induced emf, have the directions shown in Fig. 1.53 (*c*) the mutually induced emfs, follow the pattern of Fig. 1.53 (*a*). Figure 1.53 (*b*) shows two such elements. Each coil is characterized by its own self-inductance; the combination has a mutual inductance indicating the coupling between the coils. The volt-ampere relationships for this case are given below :

$$e_1 = L_1 \frac{di_1}{dt} + M\frac{di_2}{dt} \qquad ...(1.36)$$

$$\text{and } e_2 = L_2 \frac{di_2}{dt} + M\frac{di_1}{dt} \qquad ...(1.37)$$

Coupling between two closed circuits permits the transfer of energy between the circuits through the medium of the mutual magnetic field. This phenomenon is the basis on which transformers operate.

A dot marking convention can be used to denote the relative instantaneous polarities of the two mutually coupled inductive coils such as between transformer windings. The noteworthy points about dot marking conventions are:

1. The instantaneous polarities of the voltages across each inductor with respect to the dotted terminals are the same, as shown in Fig. 1.53 (*b*).

2. The instantaneous directions of current entering the primary inductor at its dotted end and the current leaving the secondary inductor at its dotted end are the same, as shown in Fig. 1.53 (*b*).

1.23 COEFFICIENT OF COUPLING

When two coils are placed near each other, all the flux produced by one coil does not link with the other coil, as assumed in Art. 1.21.2 (*b*), only a certain portion (say, K) of flux produced by one coil links with other coil, K being less than unity. K is called the *coefficient of coupling.*

Flux created in coil A due to current of i_1 amperes,

$$\Phi_1 = \frac{N_1 i_1}{l/\mu_0 \mu_r a} \text{ Wb}$$

Flux linking with coil B,

$$\Phi_2 = K\Phi_1 = \frac{K N_1 i_1}{l/\mu_0 \mu_r a} \text{ Wb}$$

Coefficient of mutual inductance,

$$M = \frac{N_2 \Phi_2}{i_1} = \frac{K N_1 N_2}{l/\mu_0 \mu_r a} \qquad ...(i)$$

Coefficient of self-inductance of coil A,

$$L_1 = \frac{N_1 \Phi_1}{i_1} = \frac{N_1^2}{l/\mu_0 \mu_r a} \qquad ...(ii)$$

Coefficient of self-inductance of coil B,

$$L_2 = \frac{N_2 \Phi_2}{i_2} = \frac{N_2^2}{l/\mu_0 \mu_r a} \qquad ...(iii)$$

Multiplying Eqs. (*ii*) and (*iii*) and taking square root of both sides we get

$$\sqrt{L_1 L_2} = \frac{N_1 N_2}{l/\mu_0 \mu_r a} \qquad ...(iv)$$

Comparing Eqs. (*i*) and (*iv*) we get

$$M = K\sqrt{L_1 L_2}$$

$$\text{or } K = \frac{M}{\sqrt{L_1 L_2}} \qquad ...(1.38)$$

When the coils are tightly coupled *i.e.* when the flux due to one coil links with the other coil completely, the coefficient of coupling, K is unity and the coefficient of mutual inductance M is given as

$$M = \sqrt{L_1 L_2} \qquad ...(1.39)$$

When the flux due to one coil does not link with the other coil at all, the value of coefficient of coupling, K is zero.

Example 1.36. Two coils A of 1,200 turns and B of 800 turns lie near each other so that 60 percent of the flux produced in one links the other. It is found that a current of 5 A in A produces a flux of 0.25 mWb while the same current in B produces a flux of 0.15 mWb. Determine the mutual inductance and coefficient of coupling between the coils.

Solution: Number of turns on coil A, N_1 = 1,200

Number of turns on coil B, N_2 = 800

Flux per ampere of current in coil A,

$$\frac{\Phi_1}{i_1} = \frac{0.25 \times 10^{-3}}{5} = 0.05 \times 10^{-3} \text{ Wb}$$

Self-inductance of coil A,

$$L_1 = N_1 \frac{\Phi_1}{i_1} = 1{,}200 \times 0.05 \times 10^{-3} = 0.06 \text{ H}$$

Flux per ampere of current in coil B,

$$\frac{\Phi_2}{i_2} = \frac{0.15 \times 10^{-3}}{5} = 0.03 \times 10^{-3} \text{ Wb}$$

Self-inductance of coil B,

$$L_2 = N_2 \frac{\Phi_2}{i_2} = 800 \times 0.03 \times 10^{-3} = 0.024 \text{ H}$$

Flux linking coil B per ampere of current in coil A,

$$\frac{\Phi_2}{i_1} = \frac{0.25 \times 10^{-3}}{5} \times 0.6 = 0.03 \times 10^{-3} \text{ Wb}$$

Mutual inductance between the coils,

$$M = N_2 \frac{\Phi_2}{i_1} = 800 \times 0.03 \times 10^{-3} = 0.024 \text{ H } \textbf{Ans.}$$

Coefficient of coupling,

$$K = \frac{M}{\sqrt{L_1 L_2}} = \frac{0.024}{\sqrt{0.06 \times 0.024}} = 0.6325 \textbf{ Ans.}$$

Example 1.37. Define coefficient of coupling and calculate it for two coils having self-inductances of 60 mH and 80 mH. The mutual inductance between them is 40 mH

Solution: Self-inductance, $L_1 = 60 \text{ mH} = 0.06 \text{ H}$

Self-inductance, $L_2 = 80 \text{ m H} = 0.08 \text{ H}$

Mutual inductance, $M = 40 \text{ mH} = 0.04 \text{ H}$

$$\text{Coefficient of coupling, } K = \frac{M}{\sqrt{L_1 L_2}} = \frac{0.04}{\sqrt{0.06 \times 0.08}}$$

$$= 0.577 \textbf{ Ans.}$$

1.24 INDUCTORS IN SERIES

Figure 1.54 (*a*) represents two inductors (inductance coils) in series with the axis of one coil perpendicular to the axis of the other. In this case there is no mutual inductance *i.e.* M = 0. If the coils are connected is series, represented as L_1 and L_2 in Fig. 1.54 (*b*), so that when the current enters the dot end of coil L_1 and leaves, it must enter L_2 at its dotted end, the fluxes of two coils will add. Such a series connection is known as *series aiding.* If the connections to inductor L_2 are reversed so that the current must enter its undotted end, their fluxes will oppose each other. This series connection is known as *series opposing.* It has been assumed that the axes of the inductance coils are on the same straight line.

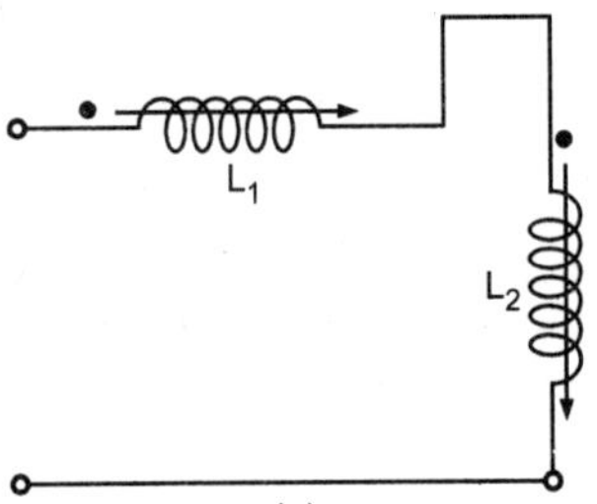

(*a*)

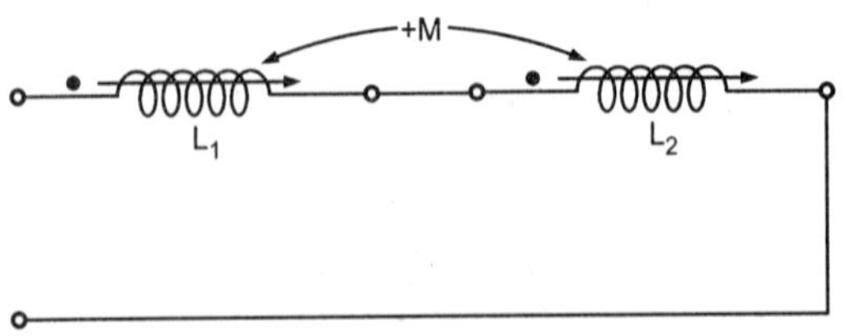

(*b*) *Series Aiding Connections*

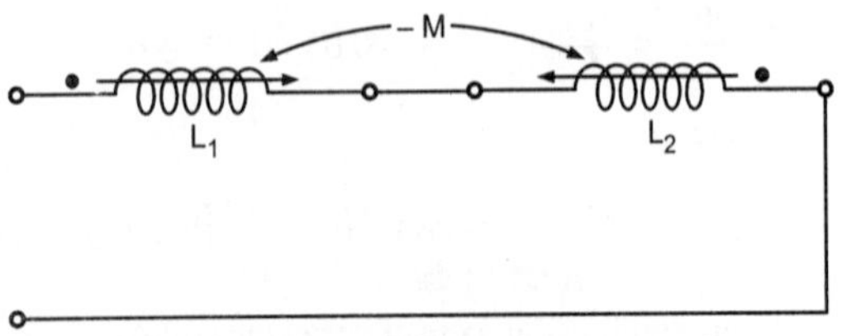

(*c*) *Series Opposing Connections*

Fig. 1.54

Series Aiding. For the series aiding connections, the total emf induced in each of the coils L_1 and L_2 is due to coil's self-inductance and the emf induced by the other coil. So

$$e_1 = L_1 \frac{di}{dt} + M\frac{di}{dt} = (L_1 + M)\frac{di}{dt}$$

$$\text{Similarly } e_2 = (L_2 + M)\frac{di}{dt}$$

So the total induced emf in the circuit is given as

$$e = e_1 + e_2 = (L_1 + L_2 + 2M)\frac{di}{dt}$$

$$\text{or } \frac{e}{di/dt} = L_1 + L_2 + 2M = \text{Total inductance of the circuit, L}$$

$$\text{or } L = L_1 + L_2 + 2M \qquad \text{...(1.40 } a)$$

Series Opposing. For the series opposing connections the mutually induced emf opposes the self induced emf. So

$$e_1 = L_1 \frac{di}{dt} - M\frac{di}{dt} = (L_1 - M)\frac{di}{dt}$$

$$\text{and } e_2 = L_2 \frac{di}{dt} - M\frac{di}{dt} = (L_2 - M)\frac{di}{dt}$$

$$e = e_1 + e_2 = (L_1 + L_2 - 2M)\frac{di}{dt}$$

$$\text{or } \frac{e}{di/dt} = (L_1 + L_2 - 2M)$$

$$\text{or } L_1 = (L_1 + L_2 - 2M) \qquad \text{... (1.40 } b)$$

Example 1.38. A primary coil having an inductance of 100 μH is connected in series with a secondary coil of 240 μH and the total inductance of the combination is measured to be 146 μH. Determine the coefficient of coupling.

Solution: Self-inductance of first coil, $L_1 = 100 \times 10^{-6}$ H

Self-inductance of second coil, $L_2 = 240 \times 10^{-6}$ H

Total inductance of combination, $L = 146 \times 10^{-6}$ H

Since the total inductance of the combination is less than the arithmetic sum of the self-inductances of coils, the coils have been connected in such a way that fluxes are opposing each other. Hence equivalent inductance of combination is given by the relation

$$L = L_1 + L_2 - 2M$$

or Mutual inductance between coils,

$$M = \frac{L_1 + L_2 - L}{2}$$

$$= \frac{100 \times 10^{-6} + 240 \times 10^{-6} - 146 \times 10^{-6}}{2} = 97 \times 10^{-6} \text{ H}$$

Coefficient of coupling K

$$= \frac{M}{\sqrt{L_1 L_2}} = \frac{97 \times 10^{-6}}{\sqrt{100 \times 10^{-6} \times 240 \times 10^{-6}}} = 0.626 \textbf{ Ans.}$$

Example 1.39. Two identical coils with terminals T_1, T_2 and T_3, T_4 respectively are placed side by side. The inductances measured under different sets of connections are as follows ;

When T_2 is connected to T_3 and inductance measured between T_1 and T_4 is 4 H.

When T_2 is connected to T_4 and inductance measured between T_1 and T_3 is 0.8 H.

Determine the self-inductance of each coil and the coefficient of coupling.

Solution: Self-inductance of coil with terminals T_1, T_2

= Self-inductance of coil B with terminals T_3, T_4

= L (say) $\quad\because$ the coils are identical

Let the mutual inductance between the coils be M

Equivalent inductance between T_1 and T_4 (T_2 connected to T_3),

$$L + L + 2M = 4 \text{ H}$$

$$\text{or } L + M = 2\text{H} \quad ...(i)$$

Equivalent inductance between T_1 and T_3 (T_2 connected to T_4)

$$L + L - 2M = 0.8 \text{ H}$$

$$\text{or } L - M = 0.4 \text{ H} \quad ...(ii)$$

Solving Eqs. (*i*) and (*ii*) we have

Self inductance of each coil,

$$L = \frac{2 + 0.4}{2} = 1.2 \text{ H} \textbf{ Ans.}$$

Mutual inductance between coil,

$$M = \frac{2 - 0.4}{0.2} = 0.8 \text{ H}$$

$$\text{Coefficient of coupling, } K = \frac{M}{\sqrt{L_1 L_2}} = \frac{0.8}{\sqrt{1.2 \times 1.2}} = 0.667 \text{ or } 66.7\% \textbf{ Ans.}$$

1.25 INDUCTORS IN PARALLEL

Let us consider two coils of inductances L_1 and L_2 connected in parallel, as shown in Fig. 1.55.

The supply current divides into two components i_1 and i_2 flowing through two coils.

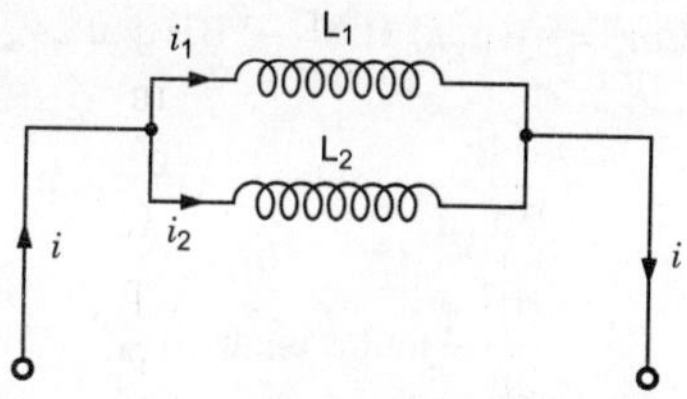

Fig. 1.55

i.e. $i = i_1 + i_2$

$$\text{or} \quad \frac{di}{dt} = \frac{di_1}{dt} + \frac{di_2}{dt}$$

Self-induced emf in coil A,

$$e_{LA} = -L_1 \frac{di_1}{dt}$$

Mutually induced emf in coil A due to change of current in coil B,

$$e_{MA} = -M \frac{di_2}{dt}$$

where M is the mutual coefficient of inductance

Resultant emf induced in coil A

$$= e_{LA} + e_{MA} = -\left(L_1 \frac{di_1}{dt} + M \frac{di_2}{dt}\right)$$

Similarly resultant emf induced in coil B

$$= -\left(L_2 \frac{di_2}{dt} + M \frac{di_1}{dt}\right)$$

As both coils are connected in parallel, therefore, resultant emf induced in both of the coils must be equal

$$\therefore \quad L_1 \frac{di_1}{dt} + M \frac{di_2}{dt} = L_2 \frac{di_2}{dt} + M \frac{di_1}{dt}$$

$$\text{or} \quad (L_1 - M) \frac{di_1}{dt} = (L_2 - M) \frac{di_2}{dt}$$

$$\text{or} \quad \frac{di_1}{dt} = \left(\frac{(L_2 - M)}{(L_1 - M)}\right) \frac{di_2}{dt} \quad ...(i)$$

$$\text{or} \quad \frac{di_1}{dt} + \frac{di_2}{dt} = \left(\frac{(L_2 - M)}{(L_1 - M)} + 1\right) \frac{di_2}{dt}$$

$$\text{or} \quad \frac{di}{dt} = \left(\frac{L_1 + L_2 - 2M}{L_1 - M}\right) \frac{di_2}{dt} \quad ...(ii)$$

If L is the equivalent inductance of the combination then induced emf

$$= -L \frac{di}{dt}$$

Since induced emf in parallel combination

= Induced emf in either of the coils,

$$\therefore L \frac{di}{dt} = L_1 \frac{di_1}{dt} + M \frac{di_2}{dt}$$

$$\text{or } L = \frac{L_1 \frac{di_1}{dt} + M \frac{di_2}{dt}}{\frac{di}{dt}}$$

Substituting $\frac{di_1}{dt} = \left[\frac{L_2 - M}{L_1 - M}\right] \frac{di_2}{dt}$ from Eq. (*i*) and

$\frac{di}{dt} = \left[\frac{L_1 + L_2 - 2M}{L_1 - M}\right] \frac{di_2}{dt}$ from Eq. (*ii*) we get

$$L = \frac{L_1 \left[\frac{L_2 - M}{L_1 - M}\right] \frac{di_2}{dt} + M \frac{di_2}{dt}}{\left[\frac{L_1 + L_2 - 2M}{L_1 - M}\right] \frac{di_2}{dt}}$$

$$\text{or } L = \frac{L_1 L_2 - L_1 M + L_1 M - M^2}{L_1 + L_2 - 2M} = \frac{L_1 L_2 - M^2}{L_1 + L_2 - 2M} \quad ...(1.41)$$

when mutual flux helps the individual flux

$$\text{or } L = \frac{L_1 L_2 - M^2}{L_1 + L_2 + 2M}$$

when mutual flux opposes the individual flux. ...(1.42)

Example 1.40. Two coils of self-inductances 150 mH and 250 mH and of mutual inductance 120 mH are connected in parallel. Determine the equivalent inductance of the combination if (*i*) mutual flux helps the individual flux (*ii*) mutual flux opposes the individual flux.

Solution: Self-inductance of coil A, L_1 = 150 mH = 0.15 H

Self-inductance of coil B, L_2 = 250 mH = 0. 25 H

Mutual inductance, M = 120 mH = 0.12 H

(*i*) Equivalent inductance of the combination when mutual flux helps the individual flux

$$L = \frac{L_1 L_2 - M^2}{L_1 + L_2 - 2M} = \frac{0.15 \times 0.25 - (0.12)^2}{0.15 + 0.25 - 2 \times 0.12} = 0.144 \text{ H} \textbf{ Ans.}$$

(*ii*) Equivalent inductance of the combination when mutual flux opposes the individual flux.

$$L = \frac{L_1 L_2 - M^2}{L_1 + L_2 + 2M} = \frac{0.15 \times 0.25 - (0.12)^2}{0.15 + 0.25 + 2 \times 0.12} = 0.036 \text{ H} \textbf{ Ans.}$$

1.26 ENERGY STORED IN A MAGNETIC FIELD

When a coil is connected to an electric source, the current flowing in the circuit gradually increases from zero to its final value, and a magnetic field is established. Consequently, a portion of the electrical energy supplied by the electric source is stored as magnetic field, while remainder, supplied during the rise of current, is dissipated from the magnetising coil as heat. After the magnetic field has been established, and the current has attained its maximum or steady value, any more energy given to it will be dissipated as heat. In other words, no additional energy is required to maintain the magnetic field, once the steady-state has reached. If, however, the circuit of a steady-state established magnetic field is broken, the energy stored in it will be spent in generating an induced emf or current.

Figure 1.56 depicts an iron-cored coil when the resistance r of the coil is lumped outside so that the exciting coil is devoid of any resistance (pure, lossless). Let the inductance of the coil be L Henrys and a current of i amperes be flowing through it at any instant t. At this instant the current is rising at the rate of $\frac{di}{dt}$ amperes per second. This induces an emf e in the coil. The induced emf in the coil is given by the expression

$$e = L\frac{di}{dt} \quad ...(1.43)$$

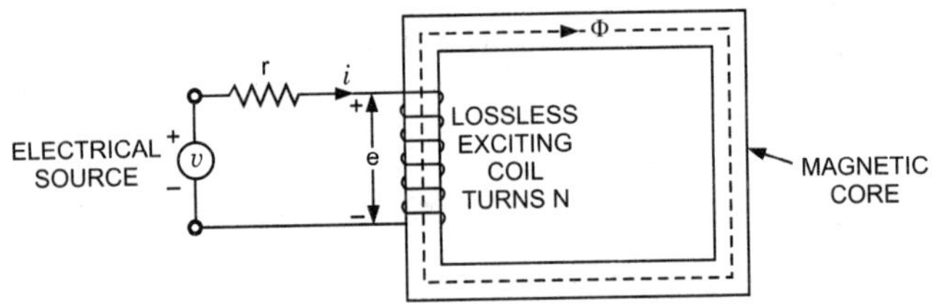

Fig. 1.56

Multiplying both sides of above equation by i, we have the power input to the coil,

$$p = ei = Li\frac{di}{dt} \quad ...(1.44)$$

which is positive when both i and $\frac{di}{dt}$ have the same sign, else it is negative. If the coil current were zero at $t = 0$ and has attained the value of I amperes at $t = T$, the energy input to the coil during this interval of T seconds is

$$W_e = \int_o^T p\,dt = \int_o^I Li\frac{di}{dt}dt = \int_o^I Li\,di = \frac{1}{2}LI^2 \text{ joules} \quad ...(1.45)$$

It should be noted that the total stored energy in the magnetic field depends upon the final or steady-state value of the current and is independent of the manner in which the current has increased or time it has taken to grow.

Equation (1.43) can be written as

$$e = \frac{d\psi}{dt}$$

Correspondingly the stored energy is

$$W_f = W_e = \int ei\,dt = \int \frac{d\psi}{dt} i\,dt = \int_o^{\psi} i\,d\psi = \psi\, i \text{ joules} \quad ...(1.46)$$

Equation (1.43) can also be written as

$$e = N\frac{d\Phi}{dt}$$

and energy stored $W_f = W_e = \int ei\,dt = \int Ni\frac{d\Phi}{dt} \times dt = \int Ni\,d\Phi$ joules

$$= \int \Phi S\,d\Phi = \frac{1}{2}S\Phi^2 \text{ joules} \quad ...(1.47)$$

where S is the reluctance of the magnetic circuit and Φ is the flux established in that magnetic circuit.

In case of an air gap in the core, air gap reluctance being far larger than that of the core, major portion of the field energy would reside in the air gap.

1.26.1. Energy Stored Per Unit Volume. Energy stored in a magnetic field of self-inductance L and carrying a current of I amperes

$$= \frac{1}{2}LI^2$$

$$\text{Energy stored in magnetic field} = \frac{1}{2}\frac{N^2\mu_0\mu_r a}{l} I^2 \text{ joules}$$

$$= \frac{1}{2}\mu_r\mu_0\left(\frac{NI}{l}\right)^2 \times al \text{ joules}$$

$$\text{Since } L = \frac{N^2\mu_0\mu_r a}{l} \text{ joules}$$

Now since $\frac{NI}{l} = H$, the magnetising force and al = volume of the magnetic field in m^3.

$$\therefore \text{ Energy stored/m}^3 = \frac{1}{2}\mu_r\mu_0 H^2 = \frac{1}{2} BH \text{ joules}$$

$$= \frac{B^2}{2\mu_0\mu_r} \text{ joules in a medium} \quad ...(1.48)$$

$$= \frac{B^2}{2\mu_0} \text{ joules in air} \quad ...(1.49)$$

Example 1.41. Find the inductance of a coil in which (*i*) current of 0.1 A yields an energy storage of 0.05 J (*ii*) a current of 0.1 A increasing at the rate of 0.5 A/s represents a power flow of 0.5 watt.

Solution: (*i*) Energy stored = 0.05 J

Current through coil, I = 0.1 A

$$\text{Inductance of coil, } L = \frac{\text{Energy stored}}{(1/2)I^2}$$

$$= \frac{0.05}{\left(\frac{1}{2}\right)(0.1)^2} = 10 \text{ H } \textbf{Ans.}$$

$$(ii) \text{ Power flow, } p = e\,i = \left(L\frac{di}{dt}\right)i = Li\frac{di}{dt}$$

$$\text{or } L = \frac{p}{i\frac{di}{dt}} = \frac{0.5}{0.1 \times 0.5} = 10 \text{ H } \textbf{Ans.}$$

Example 1.42. A coil of 100 turns is wound on a torroidal magnetic core having a reluctance of 10^4 AT per Wb. When the coil current is 5 A and is increasing at the rate of 200 A/s, determine (*i*) energy stored in the magnetic circuit (*ii*) voltage applied across the coil. Assume coil resistance as zero. (*iii*) How are your answers affected if the coil resistance is 2 Ω. [*Hyderabad Univ. Electrical Technology 1991*]

Solution: Number of turns on the coil, N = 100

Reluctance of the magnetic circuit,

$$S = \frac{l}{\mu_0\mu_r a} = 10^4 \text{ AT/Wb}$$

Inductance of coil, $L = \frac{N^2 \mu_0 \mu_r a}{l} = \frac{N^2}{l/\mu_0 \mu_r a} = \frac{(100)^2}{10^4} = 1$ H

(*i*) Energy stored $= \frac{1}{2} LI^2 = \frac{1}{2} \times 1 \times (5)^2 = 12.5$ J **Ans.**

(*ii*) Voltage applied across the coil

= Self-induced emf in the coil

$= L \frac{di}{dt} = 1 \times 200 = 200$ V **Ans.**

(*iii*) Voltage across the coil

$= L \frac{di}{dt} + IR = 1 \times 200 + 5 \times 2 = 210$ V **Ans.**

There will be no change in the stored energy of the coil. However, there will be a loss of energy of I^2R *i.e.* $(5)^2 \times 2 = 50$ watts.

Example 1.43. The field winding of a dc electromagnet is wound with 960 turns and has resistance of 50 Ω. The exciting voltage is 230 V and the magnetic flux linking the coil is 5 mWb. Find (*i*) self-inductance of the coil, (*ii*) the energy stored in the magnetic field.

[M.D. Univ. Electromechanical Energy Conversion, December-2011]

Solution: Current through field winding,

$$I = \frac{\text{Exciting voltage}}{\text{Field winding resistance}} = \frac{230}{50} = 4.6 \text{ A}$$

Magnetic flux produced,

$$\Phi = 5 \text{ mWb} = 0.005 \text{ Wb}$$

(*i*) Inductance of coil, $L = \frac{N\Phi}{I} = \frac{960 \times 0.005}{4.6} = 1.043$ H **Ans.**

(*ii*) Energy stored in the magnetic field

$= \frac{1}{2} LI^2 = \frac{1}{2} \times 1.043 \times (4.6)^2 = 11.04$ J **Ans.**

Example 1.44. A 10 kW, 1440 rpm dc shunt generator has a time constant $\frac{L_f}{r_f}$ of 0.2 sec for its field winding. Under normal operating conditions, the $I_f^2 r_f$ loss in the field winding is 400 watts. Compute the energy stored in the magnetic field produced by the field winding, under normal operating conditions.

[U.P. Technical Univ. Electromechanical Energy Conversion-I, 2005-06]

Solution: Field winding loss $i_f^2 r_f = 400$ watts

Field winding time constant, $\tau = \frac{L_f}{r_f} = 0.2$

or $L_f = 0.2 \, r_f$ ...(*i*)

Energy stored in the magnetic field produced by field winding under normal operating conditions,

$W_{field} = \frac{1}{2} L_f i_f^2 = \frac{1}{2} \times 0.2 \, r_f i_f^2$ $\because$ From Eq. (*i*) $L_f = 0.2 \, r_f$

$= 0.1 \, i_f^2 r_f = 0.1 \times 400 = 40$ watts **Ans.**

Example 1.45. A toroidal iron ring has a uniform cross-sectional area of 50 mm², and a mean magnetic path length of 100 mm. The ring has an air gap of 1 mm. The ring is excited with a dc current of 1 A through a coil of 100 turns wound uniformly along its length. The iron may be assumed to be perfect magnetic material. The effect of fringing at the gap may be assumed to increase the effective area of magnetic flux at the gap by 10%. Evaluate

(*a*) the exciting mmf of the coil,
(*b*) the effective reluctance of the magnetic circuit,
(*c*) the magnetic flux in the air gap,
(*d*) the inductance of the coil,
(*e*) the energy stored in the magnetic field under the above excitation.

[GATE 1991]

Solution: (*a*) Exciting mmf of the coil

= Number of turns on coil × current flowing through coil

$= NI = 100 \times 1 = 100$ AT **Ans.**

(*b*) Reluctance of the magnetic circuit,

S = Reluctance of iron path + Reluctance of air gap

$$= 0 + \frac{l_g}{\mu_0 a} = \frac{1 \times 10^{-3}}{4\pi \times 10^{-7} \times 50 \times 10^{-6} \times 1.1}$$

$= 14.4686 \times 10^6$ AT/Wb **Ans.**

(*c*) Magnetic flux in the gap,

$$\Phi = \frac{MMF}{S} = \frac{100}{14.4686 \times 10^6}$$

$= 6.91 \times 10^{-6}$ Wb **Ans.**

(*d*) Inductance of coil,

$$L = \frac{N^2 a \mu_r \mu_o}{l} = \frac{N^2}{l / a \mu_r \mu_o} = \frac{N^2}{S}$$

$$= \frac{100 \times 100}{14.4686 \times 10^6} = 0.69 \text{ mH } \textbf{Ans.}$$

(*e*) Energy stored $= \frac{1}{2} LI^2 = \frac{1}{2} \times 0.69 \times 10^{-3} \times 1^2$

$= 3.45 \times 10^{-4}$ J **Ans.**

1.27 MAGNETIC PULL BETWEEN TWO PARALLEL FACES

Consider two parallel faces, each of area a square metre and an air gap of length x metre between them. There is flux* of Φ webers crossing the air gap. There is an energy storage in the air gap, which can be computed from Eq. (1.47).

$W_f = \frac{1}{2} S\Phi^2$ joules where S is the reluctance of the magnetic circuit which is assumed to reside in the air gap only. This implies that the permeability of the rest of the magnetic circuit is infinity.

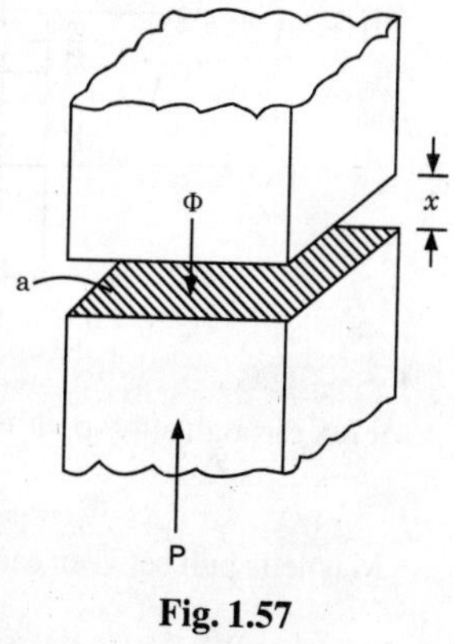

Fig. 1.57

If one of the pole moves through a distance dx, there is a change in the stored energy. With the decrease in air-gap length, the reluctance reduces and therefore, some energy must release from the storage in magnetic field. If P is the force of attraction between two poles in newtons, the mechanical work is done at the expense of energy storage in the field.

* The flux Φ might have been produced by putting a current carrying coil around the magnetic circuit of which these poles are the part or might have been excited by a permanent magnet.

$$\text{Thus } P\,dx = d\,W_f = -d\left[\frac{1}{2}\Phi^2 S\right]$$

$$\text{or } P = -\frac{d}{dx}\left[\frac{1}{2}\Phi^2 S\right] = -\frac{1}{2}\Phi^2\frac{dS}{dx} \qquad ...(1.50)$$

$$\text{or } P = -\frac{1}{2}\Phi^2\frac{d}{dx}\left[\frac{x}{\mu_o a}\right] \qquad \because \; S = \frac{x}{\mu_o a} \text{ and } \Phi = Ba$$

$$= -\frac{1}{2}\frac{\Phi^2}{\mu_o a} = -\frac{1}{2}(B \times a)^2 \times \frac{1}{\mu_o a}$$

$$= -\frac{1}{2}\frac{B^2 a}{\mu_o} \text{ newtons.} \qquad ...(1.51)$$

Negative sign in the above expression indicates that the magnetic pull tends to decrease the air gap.

Example 1.46. A relay coil has 200 turns and is excited by a current of 2.5 A. The pole area is 2 cm^2. Determine the total pull on the armature when

(*i*) the relay is open and air gap is 2 mm

(*ii*) the relay is closed and the air gap is 0.1 mm

Assume permeability of the core material as infinity.

Solution: Exciting mmf of the coil = NI = 200 × 2.5 = 500 AT

Reluctance of the core material is zero, because its permeability is assumed to be infinity.

Hence the total reluctance of the magnetic circuit,

$$S = S_g,$$

$$\text{Reluctance of air gaps} = \frac{2 \times g}{\mu_o a} \qquad \because \text{ There are two air gaps.}$$

$$= \frac{2g}{4\pi \times 10^{-7} \times 2 \times 10^{-4}} = \frac{10^{11}}{4\pi} g \text{ AT/Wb}$$

When the relay is open, g = 2 mm = 2 × 10^{-3} m

$$\text{and } S = \frac{10^{11}}{4\pi} \times 2 \times 10^{-3} = \frac{10^8}{2\pi} \text{ AT/Wb}$$

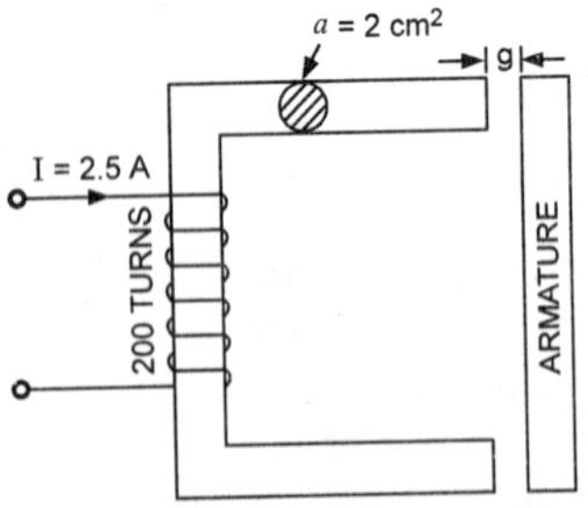

Fig. 1.58

$$\text{Flux through air gap, } \Phi = \frac{\text{MMF}}{S} = \frac{500 \times 2\pi}{10^8}$$

$$= 31.42 \times 10^{-6} \text{ Wb}$$

Magnetic pull between each pole face,

$$P = -\frac{1}{2}\Phi^2\frac{dS_g}{dg} = -\frac{1}{2}\left(\frac{\pi}{10^5}\right)^2 \times \frac{d}{dg}\left(\frac{10^{11}}{4\pi}g\right)$$

$$= -\frac{1}{2}\frac{\pi^2}{10^{10}} \times \frac{10^{11}}{4\pi} = -3.927 \text{ Newtons}$$

Hence the total pull = 2 × 3.927 = 7.854 Newtons **Ans.**

(*ii*) When the relay is closed i.e. g = 0.1 mm = 0.1 × 10^{-3} m

$$\text{Reluctance of the air gap} = \frac{10^{11}}{4\pi} \times 0.1 \times 10^{-3} = \frac{10^7}{4\pi} \text{ AT/Wb}$$

$$\text{Flux through air gap, } \phi = \frac{\text{MMF}}{S} = \frac{500 \times 4\pi}{10^7} = 628.32 \times 10^{-6} \text{ Wb}$$

$$\text{and } \frac{dS_g}{dg} = \frac{10^{11}}{4\pi}$$

So magnetic pull for each air gap,

$$P = -\frac{1}{2}\phi^2\frac{dS_g}{dg} = -\frac{1}{2}(628.32 \times 10^{-6})^2 \times \frac{10^{11}}{4\pi}$$

$$= 1,570.8 \text{ Newtons}$$

Hence the total pull = 2 × 1,570.8 = 3,141.6 Newtons **Ans.**

It is observed that $\frac{dS_g}{dg}$ does not change. It is the flux which has increased and has contributed to the increase in the magnetic pull.

Example 1.47. Determine the pull if the permeability of the core material is 2,500. Mean length of the path through the core may be assumed to be 12.5 cm for the relay shown in Fig. 1.58.

Solution: Reluctance of core material,

$$S_C = \frac{l_c}{\mu_o \mu_r a} = \frac{12.5 \times 10^{-2}}{4\pi \times 10^{-7} \times 2{,}500 \times 2 \times 10^{-4}}$$

$$= 0.198944 \times 10^6 \text{ AT/Wb}$$

(*i*) When relay is open

Total reluctance,

$$S_T = 0.198944 \times 10^6 + 15.9155 \times 10^6$$

$$= 16.11394 \times 10^6 \text{ AT/Wb}$$

$$\text{Flux, } \Phi = \frac{500}{16.11394 \times 10^6} \text{ Wb} = 31.029 \times 10^{-6} \text{ Wb}$$

$$\text{Total pull} = 2 \times P = 2 \times \frac{1}{2}(31.029 \times 10^{-6})^2 \times \frac{10^{11}}{4\pi}$$

$$= 7.6617 \text{ Newtons } \textbf{Ans.}$$

(*ii*) When the relay is closed

Total reluctance,

$$S_T = 0.198944 \times 10^6 + \frac{10^7}{4\pi} = 0.99472 \times 10^6 \text{ AT/Wb}$$

$$\text{Flux, } \Phi = \frac{500}{0.99472 \times 10^6} = 502.655 \times 10^{-6} \text{ Wb}$$

$$\text{Total pull} = 2 \times P = 2 \times \frac{1}{2}(502.655 \times 10^{-6})^2 \times \frac{10^{11}}{4\pi}$$

$$= 2{,}010.6 \text{ Newtons } \textbf{Ans.}$$

1.28 AC OPERATION OF MAGNETIC CIRCUITS

The magnetic circuits of ac machines and many other electromagnetic devices including transformers are excited from ac rather than dc sources. With dc excitation, the steady-state current is determined by the impressed voltage and the resistance of the circuit, the inductance entering only into transient processes (building up and decaying processes of current at the switching on/off instants). The magnetic flux in the magnetic circuit then adjusts itself in accordance with this value of current so that the relationship imposed by B-H or magnetisation curve is satisfied.

With ac excitation, however, inductance enters into the steady-state performance as well; the result for most magnetic circuits, although not for all, is that, to a close approximation, the flux is determined by the impressed voltage and frequency, and the magnetizing current must adjust itself in accordance with this flux so that the relationship imposed by the magnetisation curve is satisfied.

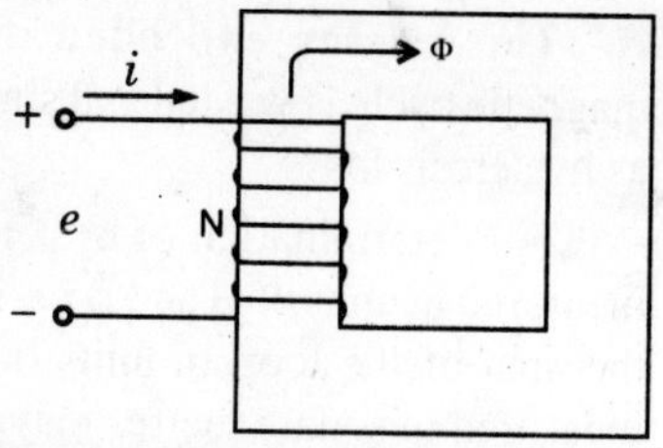

Fig. 1.59 *Magnetic Circuit With AC Excitation*

Except where preservation of linear relationship is of great importance, the normal working flux density in a magnetic circuit is kept beyond the linear portion of the magnetisation curve for the overall circuit (*i.e.* partially saturating the circuit). This is done so as to affect the economic utilisation of magnetic material. Thus accurate analysis cannot be predicted on constant self-inductance. Equivalent circuits containing parameters that do remain substantially constant are used instead. The reactive effect of the time-varying flux on the exciting circuit can readily be shown from Faraday's law,

$$e = N\frac{d\Phi}{dt}$$

Consider an iron core excited by a winding having N turns and carrying a current of i amperes (Fig. 1.59). A magnetic flux Φ is produced by the exciting current i. Let the magnetic flux Φ vary sinusoidally with time t as in

$$\Phi = \Phi_{max} \sin 2\pi f t \qquad ...(1.52)$$

Φ_{max} being the maximum value of flux in the cycle and f is the supply frequency. The induced emf in accordance with Faraday's law is

$$e = N\frac{d\Phi}{dt} = 2\pi f N \Phi_{max} \cos 2\pi f t \qquad ...(1.53)$$

and its effective or rms value is

$$E_{rms} = \frac{2\pi}{\sqrt{2}} f N \Phi_{max} = 4.44 f N \Phi_{max} \text{ volts} \qquad ...(1.54)$$

The polarity of the emf must, in accordance with Lenz's law, oppose the change in flux and, therefore, is as shown in Fig. 1.59 when the flux is increasing. Since the current produces the flux, the two may be considered in phase. From Eq. (1.53), the induced voltage leads the flux, and hence the exciting current by $\pi/2$ radians or 90°. The induced emf and the coil resistance drop oppose the applied voltage. The resistance drop does not exceed a few per cent of the applied voltage in ac machines, most transformers, and many other electromagnetic devices. To a close approximation, resistance drop may be neglected and the induced emf E and applied voltage V may be considered equal in magnitude. The flux Φ_{max} is then determined by the applied voltage V in accordance with Eq. (1.54), even if maintenance of this flux requires a magnetising current far in excess of rated current for the device.

Example 1.48. For the ac excited magnetic circuit of Fig. 1.60, calculate the excitation current and induced emf of the coil to produce a core flux of 0.6 sin 314 mWb. [*U.P. Technical Univ. September-2001*]

Solution: Maximum value of flux to be created,

$$\Phi_{max} = 0.6 \text{ mWb} = 6 \times 10^{-4} \text{ Wb}$$

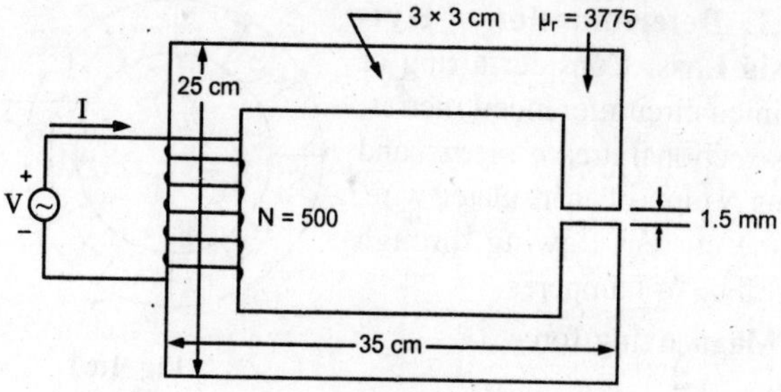

Fig. 1.60

Area of x-section, $a = 3 \times 3 = 9 \text{ cm}^2 = 9 \times 10^{-4} \text{ m}^2$

Flux density required,

$$B_{max} = \frac{\Phi_{max}}{a} = \frac{6\times10^{-4}}{9\times10^{-4}} = 0.667 \text{ T}$$

Length of air gap,

$$l_g = 1.5 \text{ mm} = 1.5 \times 10^{-3} \text{ m}$$

Length of iron path,

$$l_i = \left(25 - 2\times\frac{3}{2} + 35 - 2\times\frac{3}{2}\right)\times 2 - 0.15 = 107.85 \text{ cm}$$

Total ampere-turns required,

$$AT_{max} = \frac{B_{max}}{\mu_0} l_g + \frac{B_{max}}{\mu_0\mu_r} l_i$$

$$= \frac{0.667\times1.5\times10^{-3}}{4\pi\times10^{-7}} + \frac{0.667\times1.0785}{3{,}775\times4\pi\times10^{-7}}$$

$$= 796 + 152 = 948$$

Number of turns on the coil, N = 500

Maximum value of excitation current required,

$$I_{max} = \frac{AT_{max}}{N} = \frac{948}{500} = 1.896 \text{ A}$$

RMS value of excitation current,

$$I_{rms} = \frac{I_{max}}{\sqrt{2}} = \frac{1.896}{1.414} = 1.34 \text{ A} \quad \textbf{Ans.}$$

RMS value of induced emf in the coil,

$$E_{rms} = 4.44 f N \Phi_{max} = 4.44 \times \frac{314}{2\pi} \times 500 \times 0.6 \times 10^{-3}$$

$$= 66.6 \text{ V} \quad \textbf{Ans.}$$

1.29 HYSTERESIS AND EDDY CURRENT LOSSES

When magnetic circuits are subjected to time-varying flux densities, there are two causes of power loss in the form of heat in the iron core. These losses are significant in determining the heating, rating, and efficiency of rotating electrical machines, transformers, and ac operated devices.

The first loss is associated with the phenomenon of hysteresis, discussed in Art. 1.18, and is an expression of the fact that when ferromagnetic material is involved, not all the energy of the magnetic field is returned to the circuit when the mmf is removed. It is known as *hysteresis loss.* When the flux varies from + B, to – B, [Fig. 1.23] at the frequency f, the hysteresis loss per unit volume of material may be shown to be proportional to the area of the hysteresis loop and to the number of loops traversed per second.

1.29.1. Determination of Hysteresis Loss. Consider a ring of specimen circumference l metres, cross-sectional area a metre^2 and having N turns of an insulated wire. Let the current flowing through the coil be of I amperes.

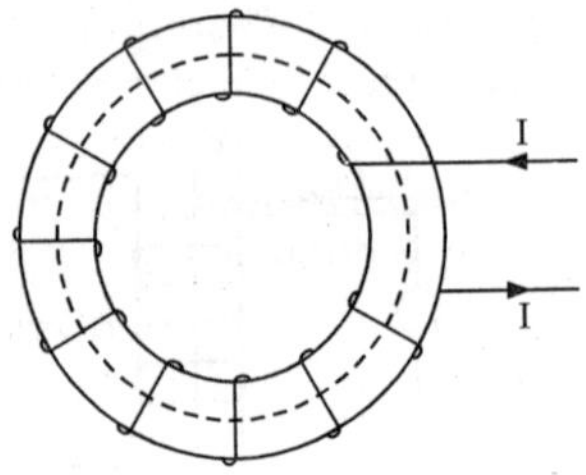

Fig. 1.61

Magnetizing force,

$$H = \frac{NI}{l} \quad \text{or} \quad I = \frac{Hl}{N}$$

Let the flux density at this instant be B.

Total flux through the ring, $\Phi = Ba$ webers.

When the current flowing through the solenoid alters, the flux produced in the iron ring also alters, so the emf (e') is induced, whose value is given by

$$e' = -N\frac{d\Phi}{dt} = -N\frac{d(Ba)}{dt} = -Na\frac{dB}{dt}$$

According to Lenz's law, this induced emf will oppose the flow of current, therefore, in order to maintain the current I in the coil, the source of supply must have an equal and opposite emf.

Hence applied emf,

$$e = -e' = Na\frac{dB}{dt}$$

Energy consumed in short time dt, during which flux density has changed

$$= e\,I\,dt = Na\frac{dB}{dt} \times I \times dt$$

$$= Na\frac{dB}{dt} \times \frac{Hl}{N} \times dt = alHdB \text{ joules} \qquad \because I = \frac{Hl}{N}$$

Thus total work done or energy consumed during one complete cycle of magnetization,

$$W = al\int_o^{B_{max}} HdB$$

Now al is the volume of the ring and HdB is the area of elementary strip of B-H curve shown in Fig. 1.62 and $\int HdB$ is the total area enclosed by hysteresis loop.

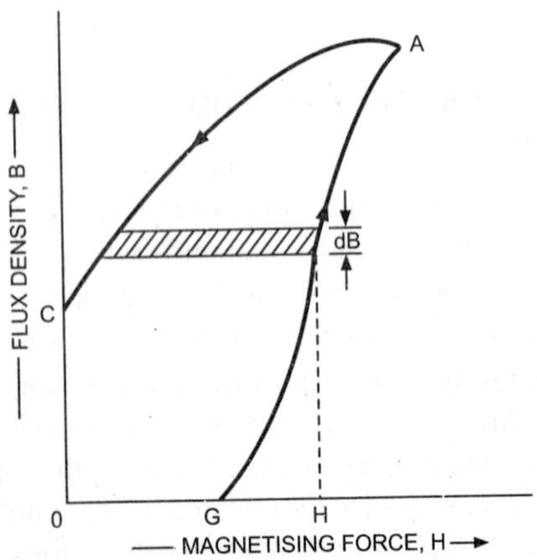

Fig. 1.62

$\therefore$ Energy consumed per cycle

= Volume of ring × area of the hysteresis loop ...(1.55)

This energy expended in taking specimen through a magnetic cycle is wasted and since it appears as heat, it is termed as hysteresis loss.

For determination of hysteresis loss, the area of the loop is measured in units of H and B *i.e.* in AT/m and Wb/m^2. To measure the area of the loop in units of H and B, the area of the loop measured in square metres may be multiplied by the scales of H and B.

For example if one metre represents x AT/m on H-axis and y Wb/m^2 on B-axis, then

$$\text{Hysteresis loss, } W_h = \text{Area of hysteresis loop in } m^2 \times x \times y \text{ joules/m}^3\text{/cycle} \qquad ...(1.56)$$

Hysteresis loss can also be determined by *Steinmetz method*, according to which hysteresis loss per cubic metre per cycle of magnetization of a magnetic material depends upon (*i*) the maximum value of the flux density and (*ii*) the magnetic quality of the material.

$\therefore$ Hysteresis loss $\alpha\ (B_{max})^{1.6}$ J/m^3/cycle

$= \eta\ (B_{max})^{1.6}$ J/m^3/cycle

$= \eta\ (B_{max})^{1.6} f\,v$ J/s or watts ...(1.57)

where η is a constant for a given specimen and a given range of flux density and is known as *Steinmetz hysteresis coefficient*, f is the frequency of reversals of magnetization and v is the volume of magnetic material in m^3.

1.29.2. Eddy Current Losses. The second loss arises from the fact that the core itself is composed of conducting material, so that the voltage induced in it by the varying flux produces circulating currents in the iron. These are called *eddy currents* and are accompanied by i^2r loss in the core called the *eddy current loss*. Since the eddy currents depend upon the rate of change of flux as well as the resistance of the path, it is reasonable to expect this loss to vary as the square of both the maximum flux density and the frequency. Eddy current loss is given by the expression

$$P_e = K_e f^2 (B_{max})^2 t^2 v \text{ watts or joules/second} \qquad ...(1.58)$$

where K_e is the *eddy current coefficient* and depends upon the type of core material, f is the number of complete magnetization cycles per second, B_{max} is the maximum flux density in teslas (Wb/m^2), t is the thickness of laminations in metres and v is the volume of core material in m^3.

To increase the core resistance and thereby minimize eddy currents, magnetic cores subjected to alternating fluxes are assembled from thin sheets with an insulating layer (surface oxide or varnish) between successive laminations. Eddy currents also have a magnetic effect, tending to make the flux density at the centre lower than at the surface. This effect, known as *screening effect*, is negligible in properly laminated cores at power frequencies but may be significant at higher frequencies.

Hysteresis and eddy current losses, taken together, are known as *core or iron loss*. Core loss is present in dc machines as well as ac machines, for the rotor iron of a rotating dc machine contains flux which varies cyclically in both magnitude and direction. Many machines are operated at constant voltage and constant frequency or speed and, as a result, have substantially constant core losses regardless of load being supplied. The operating voltage and frequency of a power transformer, for example, are normally constant, by virtue of Eq. (1.58), the maximum flux and flux density are, therefore, constant, and the core loss remains the same whether the transformer is loaded or not.

Example 1.49. The hysteresis loop of a sample of sheet steel subjected to a maximum flux density of 1.3 Wb/m² has an area of 93 cm², the scales being 1 cm = 0.1 Wb/m² and 1 cm = 50 AT/m. Calculate the hysteresis loss in watts when 1,500 cm³ of the same material is subjected to an alternating flux of 1.3 Wb/m² peak value of a frequency of 65 Hz. [*Allahabad Univ. Electromechanics, 1992*]

Solution: Hysteresis loss/m³/cycle

= Area of hysteresis loss in B-H units

$= 93 \times 0.1 \times 50 = 465$ J/m³/cycle

Number of reversals per second,

$f = 65$

Volume of the material, $v = 1{,}500 \text{ cm}^3 = 0.0015 \text{ m}^3$

Hysteresis loss, p_h = Hysteresis loss /m³/cycle $\times f \times v$

$= 465 \times 65 \times 0.0015 = 45.34$ W **Ans.**

Example 1.50. Calculate the hourly loss of energy in kWh in a specimen of iron, the hysteresis loop of which is equivalent in area to 250 J/m³, frequency of 50 Hz, specific gravity of iron 7.5; weight of specimen 10 kg.

Solution: Volume of iron $= \dfrac{\text{Mass}}{\text{Density}} = \dfrac{10 \times 1{,}000}{7.5}$

$= \dfrac{10^5}{75} \text{ cm}^3 = \dfrac{1}{750} \text{ m}^3$

Number of cycles per hour $= 50 \times 60 \times 60 = 180{,}000$

Energy loss per cubic metre of iron volume per cycle

= Area of hysteresis loop = 250 J

Total energy loss per hour

= Area of hysteresis loop

× volume of iron × number of cycles per hour

$= 250 \times \dfrac{1}{750} \times 180{,}000 = 60{,}000 \text{ J} = \dfrac{60{,}000}{36 \times 10^5}$ kWh

$= 0.0166$ kWh **Ans.**

Example 1.51. Calculate the loss of energy caused by hysteresis in one hour in 50 kg of iron if the peak flux density reached is 1.3 T (Wb/m²) and the frequency is 25 Hz. Assume Steinmetz coefficient as 628 J/m³ and density of iron as 7.8 × 10³ kg/m³. What will be the area of B-H curve of this specimen if 1 cm = 12.5 AT/m and 1 cm = 0.1 T (Wb/m²)?

Solution: Hysteresis loss/second $= \eta \, (B_{max})^{1.6} f v$

$= 628 \times (1.3)^{1.6} \times 25 \times \dfrac{50}{7.8 \times 10^3} = 153$ J/s

Hysteresis loss per hour

$= 153 \times 60 \times 60 = 550{,}800$ J **Ans.**

Hysteresis loss per m³ per cycle as per Steinmetz law

$= \eta \, (B_{max})^{1.6} = 628 \times 1.3^{1.6}$

Also hysteresis loss per m³ per cycle

$= xy \times$ area of B-H loop

$\therefore$ Area of B-H loop $= \dfrac{\text{Hysteresis loss/ m}^3\text{/ cycle}}{xy}$

$= \dfrac{628 \times 1.3^{1.6}}{12.5 \times 0.1} = 764 \text{ cm}^2$ **Ans.**

Example 1.52. The flux in a magnetic core is alternating sinusoidally at a frequency of 600 Hz. The maximum flux density is 2 Tesla and the eddy current loss is 15 watts. Find the eddy current loss in the core if the frequency is raised to 800 Hz and the maximum flux density is reduced to 1.5 Tesla.

[*U.P. Technical Univ; Electromechanical Energy Conversion-I, 2004-05*]

Solution: From Eq. (1.58)

Eddy current loss, $P_e \propto B_{max}^2 \times f^2$

$$\therefore \frac{P_{e_2}}{P_{e_1}} = \left(\frac{B_{max_2}}{B_{max_1}}\right)^2 \times \left(\frac{f_2}{f_1}\right)^2$$

$$\text{or } P_{e_2} = P_{e_1} \times \left(\frac{B_{max_2}}{B_{max_1}}\right)^2 \times \left(\frac{f_2}{f_1}\right)^2$$

$$= 15 \times \left(\frac{1.5}{2}\right)^2 \times \left(\frac{800}{600}\right)^2 = 15 \text{ watts } \textbf{Ans.}$$

Example 1.53. A coil wound on a magnetic core is excited by the following voltage sources:

(*a*) 100 V, 50 Hz (*b*) 110 V, 60 Hz

Compare the hysteresis losses and eddy current losses with these two different sources.

For hysteresis loss, consider Steinmetz constant η = 2.

[*U.P.S.C. I.E.S. Electrical Engineering-II, 2009*]

Solution: From Eqs. (1.57), (1.58) and (1.54) respectively we have

Hysteresis loss, $P_h \propto (B_{max})^\eta f$

$$\text{i.e. } \frac{P_{h_1}}{P_{h_2}} = \left(\frac{B_{max_1}}{B_{max_2}}\right)^2 \times \frac{f_1}{f_2} \quad \text{...(i)} \quad \because \eta = 2$$

Eddy current loss, $P_e \propto (B_{max})^2 f^2$

$$\text{i.e. } \frac{P_{e_1}}{P_{e_2}} = \left(\frac{B_{max_1}}{B_{max_2}} \times \frac{f_1}{f_2}\right)^2 \quad \text{...(ii)}$$

and induced emf in coil, $E \propto \phi_{max} f \propto B_{max} f$

$\because B_{max} = \dfrac{\Phi_{max}}{a}$ where a is constant for a given core

or Excitation voltage, $V \propto B_{max} f$ $\quad \because V \simeq E$

$$\text{i.e. } \frac{V_1}{V_2} = \frac{B_{max_1} f_1}{B_{max_2} f_2} \quad \text{...(iii)}$$

From Eqs. (i) and (iii) we have

$$\frac{P_{h_1}}{P_{h_2}} = \left(\frac{V_1}{V_2}\right)^2 \times \frac{f_2}{f_1} = \left(\frac{100}{110}\right)^2 \times \frac{60}{50} = 0.992 \textbf{ Ans.}$$

From Eqs. (ii) and (iii) we have

$$\frac{P_{e_1}}{P_{e_2}} = \left(\frac{V_1}{V_2}\right)^2 = \left(\frac{100}{110}\right)^2 = 0.826 \textbf{ Ans.}$$

Example 1.54. If in a single-phase transformer core, hysteresis and eddy current losses are 80 W and 50 W at normal voltage and frequency, then calculate losses when voltage and frequency are increased by 20%. [*U.P. Technical Univ. Odd Semester, 2013-14*]

Solution: When voltage and frequency are increased in the same ratio, flux density remains unchanged. Thus

Hysteresis loss, $P_h' = P_h \times \dfrac{f_2}{f_1} = 80 \times 1.2 = 96$ W **Ans.**

Eddy current loss, $P_e' = P_e \times \left(\dfrac{f_2}{f_1}\right)^2 = 50 \times (1.2)^2 = 72$ W **Ans.**

Example 1.55. The area of the hysteresis loop of a given magnetic material is 50 cm^2 with the two axes scaled as 1 cm = 20 AT/m and 1 cm = 50 mWb/m^2, for 50 Hz frequency, find the total hysteresis loss.

[*Rajasthan Technical Univ. Electrical Machines-I, 2011, 2013*]

Solution: Hysteresis loss/m^3/cycle = Area of hysteresis loss in B-H units

$= 50 \times 20 \times 50 \times 10^{-3}$

$= 50$ J/m^3/cycle

Total hysteresis loss/m^3 = Hysteresis loss/m^3/cycle $\times f$

$= 50 \times 50 = 2{,}500$ watts **Ans.**

1.30 FORCE ON A MOVING CHARGE

In an electric field, the *electric field strength* or *electric field intensity,* E is defined as the vector force on a unit charge.

Obviously the force on a charged particle will be as given by the expression

$$F = QE \qquad ...(1.59)$$

This force is in the same direction as the electric field intensity (for a positive charge) and is directly proportional to both **E** and Q. If the charge is in motion, the force at any point in its trajectory is then given by Eq. (1.59).

A charged particle in motion in a magnetic field of flux density **B** is found experimentally to experience a force whose magnitude is proportional to the product of the charge Q, its velocity **v**, and the flux density **B**, and to the sine of the angle between the vectors **v** and **B**. The direction of the force is perpendicular to both **v** and **B** and is given by a unit vector in the direction of **v** × **B**. The force may, therefore, be expressed as

$$F = Q(\mathbf{v} \times \mathbf{B}) \qquad ...(1.60)$$

A fundamental difference in the effect of the electric and magnetic fields on the charged particles is now apparent, for a force which is always applied in a direction at right angles to the direction in which the particle is proceeding can never change the magnitude of the particle velocity. In other words, the *acceleration* vector is always normal to the velocity vector. The kinetic energy of the particle remains unchanged, and it follows that the steady magnetic field is incapable of tansferring energy to the moving charge. The electric field, on the other hand, exerts a force on the particle which is independent of the direction in which the particle is progressing and, therefore, affects an energy transfer between field and particle in general.

The force on a moving particle due to combined electric and magnetic fields is obtained easily by superposition

$$F = Q\,\mathbf{E} + Q(\mathbf{v} \times \mathbf{B}) = Q(\mathbf{E} + \mathbf{v} \times \mathbf{B}) \qquad ...(1.61)$$

This equation is known as the *Lorentz force equation*, and its solution is required in determining electron orbits in the magnetron, proton paths in the cyclotron, plasma characteristics in a *magneto-hydrodynamic* (MHD) generator, or, in general, charged particle motion in combined electric and magnetic fields.

HIGHLIGHTS

1. The space around the poles of a magnet is called the magnetic field, and is represented by magnetic lines of force.
2. The force between two magnetic poles is given by
$$F = \frac{m_1 m_2}{4\pi\mu_0\mu_r d^2} \text{ newtons}$$
3. Magnetic field intensity at any point is given by
$$H = \frac{m}{4\pi\mu_0\mu_r d^2} \text{ newtons per weber}$$
4. Magnetic flux density is given by
$$B = \mu_0 \mu_r H \text{ Wb/m}^2$$
5. Ampere's circuital law or Ampere's work law is given by
$$\oint H\,dl = \Sigma I$$
6. According to Biot-Savart law
$$dB = \frac{\mu_0\mu_r}{4\pi}\,\frac{I\,dl \sin\theta}{r^2}$$
7. Force on a current carrying conductor lying in the magnetic field is given by
$$F = BIl \sin\theta \text{ newtons}$$
8. Work done in moving a current carrying conductor in the magnetic field is given by the product of flux cut by the conductor and current flowing through it *i.e.* WD = ΦI joules.
9. The field strength at a point lying at a perpendicular distance of r metres from a long straight current carrying conductor is given by
$$H = \frac{I}{2\pi r} \text{ AT/m and } B = \frac{\mu_0\mu_0 I}{2\pi r} \text{ Wb/m}^2$$
10. Force between two long current carrying conductors is given by
$$F = \frac{2I_a I_b}{d} \times l \times 10^{-7} \text{ Newtons}$$
11. Although all materials have magnetic properties of some kind being either *diamagnetic*, *paramagnetic* or *ferromagnetic*, the term "magnetic material" is customarily applied only to substances which exhibit ferromagnetism. Ferromagnetic materials are of two types: (*a*) those easily magnetized, called the soft magnetic materials and (*b*) those retaining their magnetism with great tenacity, designated as hard magnetic materials.
12. The curve drawn providing relationship between induction density B and magnetizing force H is known as B-H or magnetization curve.
13. The phenomenon of lagging of magnetization or induction flux density B behind the magnetizing force H is known as magnetic hysteresis.
14. When a ferromagnetic material is magnetized, its dimensions change slightly, and the sample being magnetized either expands or contracts in the direction of magnetization. This magnetically induced reversible elastic strain ($\delta l/l$) is called *magnetostriction* and is of the order of 10^{-6}. The energy due to the mechanical stresses created by magnetostriction is called *magnetostrictive energy.* For iron, the magnetostriction is positive at low fields and negative at high fields.
15. The closed path followed by magnetic flux is called a *magnetic circuit.* There is a lot of similarity between the magnetic and electric circuits, as mentioned in Art. 1.20.3.
16. **MMF** of the magnetic circuit is a magnetic potential that tends to force flux around the magnetic circuit and is analogous to the emf in an electric circuit.
$$\text{MMF} = \text{NI ampere-turns}$$

17. Reluctance is the name given to that property of material which opposes the creation of magnetic flux in it. It is measured in ampere-turns/Wb or amperes/Wb. It may be given as

$$S = \frac{l}{\mu_0 \mu_r a} \text{ AT/Wb}$$

where l is the length of magnetic path, a is x-sectional area, μ_0 is the permeability of the evacuated space or air and is equal to $4\pi \times 10^{-7}$ H/m, and μ_r is the relative permeability of the medium with respect to evacuated space or air.

18. Flux produced Φ, mmf and reluctance are related as follows

$$\Phi = \frac{\text{MMF}}{\text{Reluctance}} = \frac{\text{NI}}{l} \times \mu_0 \mu_r a$$

19. Permeance of a magnetic circuit is the reciprocal of reluctance and is given as

$$\text{Permeance} = \frac{1}{\text{Reluctance}} = \frac{\mu_0 \mu_r a}{l}$$

20. For a series magnetic circuit

Total ampere-turns required $= H_1 l_1 + H_2 l_2 + H_3 l_3 + \ldots$

where H_1 is the magnetizing force required to create flux through part l_1 of the magnetic circuit and so on.

21. For parallel magnetic circuit,

AT required = AT required for either parallel magnetic path.

22. Leakage factor is the ratio of total flux (flux in the iron path) and useful flux (flux in the air gap).

23. The magnitude of dynamically induced emf in a conductor moving at right angles to a uniform magnetic field is given as $e = Blv$ volts

If the conductor moves at an angle θ to the magnetic field, then

$$e = Blv \sin\theta \text{ volts}$$

The direction of dynamically induced emf is given by Fleming's right hand rule. According to Fleming's right hand rule:

If the thumb, forefinger and middle finger of right hand are held mutually perpendicular to each other, forefinger pointing into the direction of the field and thumb in the direction of motion of conductor then the middle finger will point in the direction of the induced emf as shown in Fig 1.45 (*a*).

24. When the current flowing through the coil is changed, the flux linking with its own winding changes and due to the change in linking flux with the coil, an emf, known as *self-induced emf, is induced.* The magnitude of self-induced emf, e in a coil of N turns is given by

$$e = N\frac{d\Phi}{dt} = L\frac{di}{dt}$$

where L is known as *self-inductance* of the coil and is given as

$$L = \frac{e}{di/dt} = \frac{N\Phi}{i} = \frac{N^2 \mu_o \mu_r a}{l} = \frac{N^2}{\text{Reluctance, S}}$$

25. Whenever the current in coil A (placed near another coil B) changes, the flux linking with coil B changes and an emf, known as mutually induced emf, is induced in coil B.

The magnitude of mutually induced emf is given as

$$e_m = M\frac{di_1}{dt} \text{ volts}$$

where $\frac{di_1}{dt}$ is the rate of change of current in coil A and M is mutual inductance of the two coils in henry and is given as

$$M = \frac{e_m}{di_1/dt} = \frac{N_2 \Phi_2}{i_1} = \frac{N_1 N_2 a \mu_o \mu_r}{l} = \frac{N_1 N_2}{\text{Reluctance, S}}$$

26. The coefficient of coupling between two coils is given as

$$K = \frac{M}{\sqrt{L_1 L_2}}$$

where L_1 and L_2 are the individual inductances of the coils and M is the mutual inductance between the coils.

27. Energy stored in an inductive coil is given by

$$W_e = \frac{1}{2} LI^2 \text{ joules}$$

Energy stored in a magnetic circuit is given by

$$W_f = \frac{1}{2} S\Phi^2 \text{ joules}$$

28. When two coils of inductance L_1 and L_2 are connected in series and M is the mutual inductance between them, the total inductance of the circuit is given as

$L = L_1 + L_2 + 2M$ in series aiding

and $L = L_1 + L_2 - 2M$ in series opposing

29. Magnetic pull between two parallel faces is given by

$$P = \frac{B^2 a}{2\mu_0} \text{ newtons}$$

30. The emf induced in a N-turn coil excited by ac is given by equation

$$E_{rms} = 4.44 f N \Phi_{max} \text{ volts}$$

where Φ_{max} is the maximum value of flux, linking with the coil, and varying sinusoidally with time and f is the supply frequency.

31. The energy loss associated with the phenomenon of hysteresis is called the hysteresis loss and is given as

Hysteresis loss, $P_h = \eta (B_{max})^{1.6} f v$ joules/second or watts

where η = Steinmetz hysteresis coefficient, v is the volume of core in m^3, B_{max} is the peak magnitude of flux density and f is the frequency of magnetic cycles per second.

Hysteresis loss is also given as

$$P_h = xy \times (\text{area of loop in mm}^2) \times f v \text{ joules}$$

where x and y = Scale factors of H and B

32. If the magnetic circuit is made up of iron and if the flux in the circuit is variable, currents, known as *eddy currents,* will be induced in the iron circuit itself.

Power loss due to eddy currents is given as

$$P_e = K_e (B_{max})^2 f^2 t^2 \text{ v watts}$$

where K_e is the eddy current coefficient, f is the number of complete magnetization cycles per second, t is the thickness of laminations, v is the volume of material and B_{max} is the maximum value of flux density.

EXERCISES

1. State and explain the following laws:
   (*a*) Coulomb's law. (*b*) Basic law of magnetic field.
   (*c*) Biot-Savart law.
2. Establish the relationship between magnetic field density and magnetic field intensity.
3. Show that the magnetic flux density at a distance r from an infinitely long wire of radius $a < r$ carrying a current I is given by $B = \frac{\mu_0 I}{2\pi r}$.
4. State Biot-Savart law. Apply this law to derive magnetic field due to an infinitely long straight current carrying thin conductor.

5. Derive an expression for the electrodynamic force on a conductor carrying current in a magnetic filed.
6. Explain the behaviour of a current carrying conductor when placed in a magnetic field.
[M.D. Univ. Electromechanical Energy Conversion, December-2007]
7. Obtain an expression for the force between two parallel conductors carrying currents. **[Pb. Technical Univ. May-2009]**
8. What is a magnetic material? Explain the terms :
(*i*) Magnetic intensity (*ii*) Magnetic flux density (*iii*) Permeability and (*iv*) Magnetic flux.
[M.D. Univ. Electromechanical Energy Conversion, December-2007]
9. Classify the materials on the basis of their magnetic properties.
[M.D. Univ. Electromechanical Energy Conversion, May-2008]
10. Describe the classification of materials. Give examples of each class.
[M.D. Univ. Electromechanical Energy Conversion, December-2012]
11. Distinguish between diamagnetic, paramagnetic and ferromagnetic materials. **[U.P.S.C. I.E.S. Electrical Engineering-I, 2007]**
12. What are different types of magnetic materials? Distinguish between hard and soft magnetic materials.
[U.P.S.C. I.E.S. Electrical Engineering-I, 2008]
13. Distinguish between 'hard and soft' magnetic materials. List the properties and applications of 'hard' magnetic materials giving two examples of such materials. **[U.P.S.C. I.E.S. Electrical Engineering-I, 2009]**
14. With respect to magnetic behaviour, give the classification of magnetic materials. Discuss general electric and magnetic characteristics of ferrites and their applications. **[U.P.S.C. I.E.S. Electrical Engineering-I, 2011]**
15. How is B-H curve of a ferromagnetic material different from that of a nonmagnetic material ? Name all the salient regions of B-H curve of a magnetic material.
16. Explain the phenomenon of (*i*) Magnetostriction and (*ii*) hysteresis in ferromagnetic materials. **[U.P.S.C. I.E.S. Electrical Engineering I, 1992]**
17. What do you understand by magnetic hysteresis? Differentiate between hard and soft magnetic materials.
[U.P.S.C. I.E.S. Electrical Engineering-I, 2005]
18. Define the following terms as applied to magnetic circuits. (*i*) MMF (*ii*) Flux density (*iii*) Reluctance (*iv*) Permeability.
[U.P. Technical Univ. Even Semester, 2013-14]
19. Clearly define mmf and emf.
[Anna Univ. Electrical Machines-I, May/June-2012]
20. Explain magnetic leakage and fringing.
[M.D. Univ. Electromechanical Energy Conversion, December-2013]
21. Compare and contrast magnetic circuit with electric circuit.
[M.D. Univ. Electromechanical Energy Conversion, May-2009]
22. What are the similarities and dissimilarities between electric and magnetic circuits? Enlist any five applications of magnetic circuits in engineering fields. **[U.P. Technical Univ. Odd Semester 2013-14]**
23. Explain the following:
(*i*) Faraday's laws of electromagnetic induction.
(*ii*) Lenz's law.
(*iii*) Fleming's right and left hand rules.
Also mention the significances of above laws and rules in electric machines.
[U.P. Technical Univ. Electromechanical Energy Conversion-I, 2013-14]
24. State and explain laws of electromagnetic induction and explain static and dynamic emf.
[M.D. Univ. Electromechanical Energy Conversion, December-2013]
25. Explain clearly the statically and dynamically induced emfs.
[Anna Univ. Electrical Machines-I, May/June-2012]
26. What is meant by dynamically induced emf and statically induced emf? On what factors do these depend? Derive equations for these two emfs. **[Rajasthan Technical Univ. Electrical Machines-I, 2008]**
27. Explain coefficient of self-induction and coefficient of mutual induction. Also derive the equations for these two inductances.
[M.D. Univ. Electromechanical Energy Conversion, December-2011]
28. Give the expression for the inductance of an 'N' turn coil wound on a core having length '*l*', cross-sectional area 'A' and magnetic permeability 'μ'. Explain the significance of different terms involved.
[Rajasthan Technical Univ. Electrical Machines-I, 2011]
29. Explain the concept of mutual inductance. Define coefficient of coupling and derive the expression for relationship between self-inductances of two coils, mutual inductance between them and the coefficient of coupling.
30. Define coefficient of coupling and show that $k = \dfrac{M}{\sqrt{L_1 L_2}}$. Also distinguish between self-inductance and mutual inductance.
[M.D. Univ. Electromechanical Energy Conversion, December-2012]
31. Two coils of self-inductances L_1 and L_2 are placed side by side so that the mutual inductance between them is M. If they are connected in series addition, derive for the expression for the net inductance of the coils.
32. Derive an expression for energy stored in a magnetic circuit.
[V.T.U. Belgaum Karnataka Univ. Winter 2003]
33. What are the uses of electromagnets? Deduce an expression for the lifting power of an electromagnet in terms of the flux density and the area of the x-section.
34. Discuss ac operation of magnetic circuits.
[Anna Univ. Electrical Machines-I, May/June-2012]
35. A coil wound on an iron core is excited from an ac source at voltage $V_{(rms)}$. Derive the expression for maximum flux in the core. Why is it independent of the core reluctance?
[R.G. Technical Univ. M.P., June-2014]
36. Explain hysteresis and eddy current losses. Prove that hysteresis loss in magnetic material is equal to the area of hysteresis loop.
[M.D. Univ. Electromechanical Energy Conversion, December-2013]
37. What do you understand by (*i*) B-H curve and (*ii*) hysteresis loss? Show that hysteresis loss in volume V of material when operated at f Hz is $P_h = W_h Vf$ watts, when W_h is the hysteresis loss in one cycle. **[M.D. Univ. Electromechanical Energy Conversion, December-2004; December-2012]**
38. What are eddy current losses? How these losses can be reduced?
[M.D. Univ. Electromechanical Energy Conversion, May-2006]
39. What are eddy currents? Explain their nature. Suggest a remedy for reducing such currents. Give a brief account of some of the usual applications of the eddy current effects.
[M.D. Univ. Electromechanical Energy Conversion, December-2006]
40. Write short notes on the following :
(*i*) Classification of magnetic materials.
(*ii*) Soft and hard ferrites.
(*iii*) Magnetostriction. **[U.P.S.C. I.E.S. 1992]**
(*iv*) Magnetic leakage and fringing.
(*v*) AC operation of magnetic circuits.
(*vi*) Hysteresis and eddy current losses.

SHORT ANSWER TYPE QUESTIONS WITH ANSWERS

Q. 1. How will you find the direction of magnetic field around a current carrying conductor ?

Ans. By Corkscrew rule or right hand rule.

Q. 2. How will you determine the polarity of a solenoid ?

Ans. By using compass needle or by using helix rule.

Q. 3. What is basic law of magnetic field ?

Ans. The line integral of the vector H over the whole contour depends only on the algebraic sum of the currents intersecting the surfaces, and is equal to that sum i.e.

$$\oint H\,dl = \Sigma I$$

The above equation is known as *Ampere's circuital law* or sometimes referred to *Ampere's work law*.

Q. 4. On what factors does the force experienced by a current carrying conductor placed in a magnetic field depend ?

Ans. The force experienced on a current carrying conductor placed in a magnetic field is directly proportional to (*i*) flux density (field strength), B (*ii*) the magnitude of current flowing through the conductor (*iii*) the length of the conductor and (*iv*) sine of the angle at which the conductor lies with the magnetic field *i.e.* $F = BIl \sin\theta$ newtons.

Q. 5. What is Fleming's left hand rule ?

Ans. Fleming's left hand rule is used to determine the direction of mechanical force experienced by a current carrying conductor placed in a magnetic field.

Fleming's left hand rule states that if the thumb, forefinger and middle finger of the left hand are stretched in such a way that they are at right angles to each other mutually and forefinger points towards the direction of the magnetic field, middle finger towards the direction of flow of current then thumb will point in the direction of force acting on the conductor [Fig. 1.14(*a*)].

Q. 6. On what factors does the force acting between two parallel current carrying conductors depend ?

Ans. Force acting between two parallel current carrying conductors depend on the magnitude of currents carried by them and spacing between the conductors.

The force experienced between two parallel conductors spaced *d* metres apart in air and carrying currents of I_1 and I_2 amperes respectively is given by the expression

$$F = \frac{2I_1I_2}{d} \times 10^{-7} \text{ Newtons per metre}$$

Q. 7. What is magnetic material? **[M.D. Univ. December-2007]**

Ans. Materials in which a state of magnetisation can be induced are known as magnetic materials.

Q. 8. Give the relative permeabilities of paramagnetic materials, diamagnetic materials and ferromagnetic materials.

Ans.

Type of Materials	*Relative Permeability*
Paramagnetic materials	Small but positive. Slightly higher than unity; say 1.00036 for platinum.
Diamagnetic materials	Slightly less than unity; say 0.999995 for copper.
Ferromagnetic materials	Very high (varying from several hundreds to thousands)

Q. 9. What is the value of permeability of free space ?

Ans. The permeability for free space is $4\pi \times 10^{-7}$ H/m.

Q. 10. Give the relationship between magnetic field intensity H and magnetic flux density.

Ans. Field intensity H is the cause and flux density B is the effect. Flux density will be μ times the field intensity where μ is the permeability of the magnetic material.

Q. 11. What are ferromagnetic materials ?

Ans. The materials, which are strongly attracted by a magnet such as iron, steel, nickel, cobalt and some of their alloys, are known as ferromagnetic materials. Their permeability is very high (varying from several hundreds to many thousands.)

Q. 12. What are Ferrites ?

Ans. A special group of ferromagnetic materials, that occupy an intermediate position between ferromagnetic and nonmagnetic materials, are called ferrites. They consist of extremely fine particles of a ferromagnetic material possessing high permeability and are held together with a binding resin. The magnetization produced in the ferrites is large enough to be of commercial value but their magnetic saturations are not as high as those of ferromagnetic materials.

Q. 13. Why are hard steel and alnico alloys well suited for permanent magnets ?

Ans. Hard steel and alnico alloys are well suited for permanent magnets because of their high retentivity power and large coercive force.

Q. 14. Why are wrought iron and cast steel most suitable materials for cores of electromagnets.

Ans. Because they have high permeability and good retentivity.

Q. 15. Why are low carbon steel, silicon alloys, mumetal sheets most suitable for transformer and armature cores ?

Ans. Because these materials have very high permeability and very low hysteresis loss.

Q. 16. What do you understand by B-H curve? **[M.D. Univ. December-2004]**

Ans. The curve drawn between induction density B and magnetizing force H is known as B-H or magnetisation curve.

Q. 17. What is meant by magnetic hysteresis? **[M.D. Univ. May-2004]**

Ans. The phenomenon of lagging of magnetization or induction flux density behind the magnetising force is known as magnetic hysteresis.

Q. 18. What is magnetomotive force and write its unit. **[U.P. Technical Univ. Odd Semester, 2013-14]**

Ans. **MMF** of the magnetic circuit is a magnetic potential that tends to force flux around the magnetic circuit and is analogous to the emf in an electric circuit.

MMF = NI ampere-turns

Q. 19. On what factors does the magnetic reluctance depend ?

Ans. The reluctance of the magnetic circuit is directly proportional to length *l*, inversely proportional to cross-sectional area and dependent on the nature of material of the magnetic circuit.

Q. 20. Is magnetic reluctance constant for all flux densities ?

Ans. No.

Q. 21. What is meant by leakage and fringing ?

Ans. Leakage flux is the flux, which follows a leaking path, as shown in Fig. 1.30.

Useful flux passing across the gap tends to bulge outwards, as shown in Fig. 1.30, thereby increasing the effective area of gap and reducing the flux density in the gap. This effect is referred to as fringing; and the longer the air gap, the greater is the fringing.

Q.22. What is leakage coefficient? Give its effects in magnetic circuit. **[P.T.U. B.E.E.E. May-2009]**

Ans. Leakage flux is the flux, which follows a leaking path, as shown in Fig. 1.30 (page 13). Flux in the air gap is known as *useful flux* which is utilized for various useful purposes. For the purpose of calculations the iron is supposed to carry whole of the flux throughout its entire length.

The ratio of total flux (flux in the iron path) to the useful flux (flux in the air gap) is known as *leakage factor*.

$$i.e.\ \text{Leakage factor} = \frac{\text{Total flux}}{\text{Useful flux}}$$

$$\text{or Leakage factor} = \frac{\text{Flux in the iron path}}{\text{Flux in the air gap}}$$

In electrical machines (such as in generators and motors), magnetic leakage is undesirable as it causes increase in weight (not decrease in their power efficiency) and cost of manufacture. Though magnetic leakage cannot be avoided completely but can be reduced to the minimum by placing the magnetizing or exciting coils as close as possible to the air gap or to the points in the magnetic circuit where the flux is to be utilized for useful purposes.

The value of leakage factor for modern electric machinery is about 1.2.

Q. 23. What information can be derived from the B-H loop? **[M.D. Univ. May-2004]**

Ans. The area enclosed by the hysteresis loop gives the following information regarding hysteresis loss occurring in the sample under testing. Hysteresis loss/m^3/cycle = Area of hysteresis loss in B-H scales.

Q. 24. Draw hysteresis loops for (*i*) permanent magnets (*ii*) transformer and (*iii*) Ferrite.

Ans. Hysteresis loops for permanent magnets, transformer core and ferrite are shown below in Figs. 1(*a*), (*b*), (*c*) respectively.

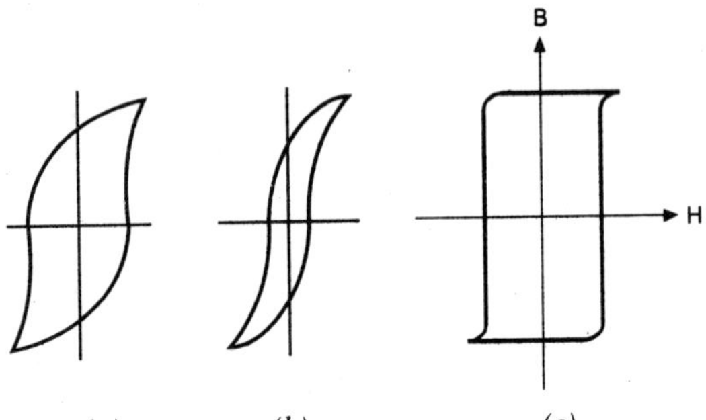

(*a*) (*b*) (*c*)

Fig. 1

Q. 25. What is meant by the terms: Remanence and coercivity?

Ans. While decreasing the magnetising force H, when H is brought to zero the induction density B is represented by *ob* in Fig. 1.23 and is called residual magnetism or remanence. The power of retaining the residual magnetism is called the retentivity of the material.

The demagnetizing force required to neutralize the residual magnestism is known as coercive force.

Q. 26. Explain practical significance of coercivity and retentivity. **[M.D. Univ. May-2005]**

Ans. The demagnetizing force required to neutralize the residual magnetism is known as coercive force or coercivity, while the power of retaining the residual magnetism is called retentivity of the material.

Retentivity, coercivity, magnetic permeability and hysteresis loss are the properties of magnetic materials which decide the utility of the material. Materials of high retentivity and large coercivity, such as hard steel, alnico alloys, are well suited for permanent magnets.

Materials with high magnetic permeability and good retentivity, such as wrought iron and cast steel, are suitable for cores of electromagnet.

Materials with very high permeability and very low hysteresis loss, such as iron, low carbon steel, silicon alloys, permalloy or mumetal sheets, are most suitable for transformer cores and armatures.

Q. 27. State Faraday's laws of electromagnetic induction. **[P.T.U., December 2008; December-2009]**

Ans. A very important effect of a magnetic field on an electric circuit is that when the flux linking the circuit changes, an emf is induced. Electromagnetic induction of emf (or voltage) is basic to the operation of transformers, generators (ac or dc) and motors (ac or dc). The effect is described by Faraday's law, which states that *the magnitude of emf (or voltage) induced is directly proportional to the rate of change of flux linkage or to the product of number of turns and rate of change of flux linking the coil.*

$$i.e.\ \text{induced emf, } e = N\frac{d\Phi}{dt}$$

where $N\frac{d\Phi}{dt}$ is the product of number of turns and rate of change of linking flux and is termed as rate of change of flux linkage.

Q. 28. What is indicated by the negative sign before $N\frac{d\Phi}{dt}$ in expression $e = -N\frac{d\Phi}{dt}$?

Ans. The negative sign before $N\frac{d\Phi}{dt}$ in expression $e = -N\frac{d\Phi}{dt}$ indicates the phenomenon explained by Lenz's law.

Q. 29. What is Lenz's law ?

Ans. The direction of emf induced in a conductor or coil is governed by Lenz's law which states that the direction of induced emf or voltage is such that the current produced by it sets up a magnetic field opposing the very cause which produces it.

Q. 30. Distinguish between statically and dynamically induced emfs. Give examples of each. **[M.D. Univ. May-2004]**

Ans. When flux linking with any coil or circuit changes, an emf is induced in the coil or circuit, called the *electromagnetically induced emf.*

EMF induced in a coil or circuit by increasing or decreasing magnitude of the linking flux (no relative motion between coil or circuit and magnetic field) is known as statically *induced emf.*

The emf induced in a conductor or coil due to relative motion between it and the magnetic field, that produces flux, is known as dynamically induced emf.

EMFs induced in transformer windings are statically induced emfs while emf induced in armature winding of dc and ac generators are dynamically induced emfs.

Q. 31. What is Fleming's right hand rule ?

Ans. According to Fleming's right hand rule if the thumb, forefinger and middle finger of the right hand are held mutually perpendicular to each other, forefinger pointing into the direction of the field and thumb in the direction of motion of conductor then middle finger will point in the direction of the induced emf, as shown in Fig. 1.45 (*a*).

Q. 32. What is the application of Fleming's right hand rule ?

Ans. Fleming's right hand rule is used in determination of direction of emf induced in a conductor moving in a uniform magnetic field.

Q. 33. A conductor of effective length $l(m)$ moves in the direction of a uniform magnetic field B (Tesla) at a constant speed v (ms^{-1}). What is the magnitude of the emf induced in the conductor ?

Ans. No, as no flux is cut by the conductor.

Q. 34. A circular coil suspended by a light inextensible string is lowered at a uniform rate in a constant magnetic field such that the plane of the coil is always perpendicular to the direction of induction. Will the tension in the string increase, decrease or remain the same in the process ?

Ans. Will decrease.

Q. 35. What is mutual inductance ? Give its unit. **[P.T.U., December-2007]**

Ans. Mutual inductance may be defined as the ability of one coil or circuit to induce an emf in a nearby coil by induction when the current flowing in the first coil is changed. The action is also reciprocal *i.e.* the change in current flowing through second coil will also induce an emf in the first coil. The ability of reciprocal induction is measured in terms of the coefficient of mutual induction M. Its unit is Henry.

Q. 36. Define coefficient of coupling.

[U.P. Technical Univ. Odd Semester, 2013-14]

Ans. When the two coils are placed near each other, all the flux produced by one coil does not link with the other coil, only a certain portion (say, K) of flux produced by one coil link with the other coil, K being less than unity. K is called the coefficient of coupling.

Q. 37. What is hysteresis loss? On what factors does it depend ?

Ans. The energy expended in taking a specimen through a magnetic cycle is known as hysteresis loss.

Hysteresis loss per cubic metre per cycle of magnetization of a magnetic material depends upon (*i*) the maximum value of flux density and (*ii*) the magnetic quality of the material.

Q. 38. What are eddy current losses? How these losses can be reduced?

[M.D. Univ. May-2006]

Ans. Voltage induced in the core material (conducting material) by the alternating flux causes circulating currents, known as eddy currents, in the iron these eddy currents are associated with i^2r loss, called the eddy current loss.

Eddy current losses can be reduced by making core of thin laminations.

Q. 39. Mention the parameters on which the eddy current loss depend.

[M.D. Univ. May-2005]

Ans. The eddy current loss depends upon (*i*) the maximum value of the flux density (*ii*) number of complete magnetization cycles per second (*iii*) thickness of the laminations and (*iv*) volume of the core material and is given by the expression

$$P_e = K_e (B_{max})^2 f^2 t^2 v \text{ watts}$$

where K_e is the eddy current coefficient and depends upon the type of magnetic material used.

Q. 40. What are the undesirable effects of eddy current and hysteresis loss ? **[R.G.T.U. Bhopal, June-2002]**

Ans. Eddy current and hysteresis loss appears in the form of heat and so raises the temperature of that portion of magnetic circuit, in which they occur.

Q. 41. How can eddy current and hysteresis losses be minimised ?

Ans. Hysteresis loss can be minimised by choosing a core material with low hysteresis coefficient such as low carbon steel, silicon alloys, alloy steel.

Eddy current loss can be minimised by using thin laminations.

Q. 42. Mention some applications of eddy currents.

[R.G.T.U. Bhopal, June-2002]

Ans. Use of eddy currents is made in (*i*) eddy current heating (*ii*) eddy current braking and (*iii*) to develop damping torque in measuring instruments.

PROBLEMS

1. What will be the current passing through the ring shaped air-cored coil when number of turns is 800 and ampere-turns are 3,200.

[Pb. Technical Univ., May-2006]

[**Ans.** 4 A]

2. A flux density of 1.2 Wb/m^2 is required in the 2 mm air gap of an electromagnet having an iron path 1 m long. Calculate the magnetising force and current required if the electromagnet has 1,273 turns. Assume relative permeability of iron to be 1,500.

[M.D. Univ. Electromechanical Energy Conversion, December-2006-07]

[**Ans.** 955×10^3 AT/m, 637×10^3 AT/m; 2 A]

3. An iron ring of mean diameter 22 cm and cross section 10 cm^2 has an air gap 1 mm wide. The ring is wound uniformly with 200 turns of wire. The permeability of ring material is 1,000. A flux of 0.16 mWb is required in the gap. What current should be passed through the wire ? [**Ans.** 1.075 A]

4. An iron ring has cross section 3 cm^2 and a mean diameter of 25 cm. An air gap of 0.4 mm has been made by a saw cut across the section. The ring is wound with 200 turns through which a current of 2 A is passed. If the total flux is 21×10^{-5} webers, find μ for iron assuming no leakage. [**Ans.** 2,470]

5. An iron ring has a mean circumferential length of 60 cm with an air gap of 1 mm and a uniform winding of 300 turns. When a current of 1 A flows through the coil, find the flux density. The relative permeability of iron is 300. Assume $\mu = 4\pi \times 10^{-7}$ H/m.

[**Ans.** 0.1256 T]

6. An iron ring of mean circumference 80 cm is made from round iron of area 8 cm^2. It has a saw cut of 2 mm wide and is wound with 500 turns. Find the current required to produce a flux of 0.8 m Wb across the air gap. Assume a relative permeability for iron as 625 and leakage factor of 1.25. [**Ans.** 5.73 A]

7. Explain the following terms:

Magnetomotive force, magnetic flux, reluctance and flux density.

Estimate their respective values in the following cases:

A steel ring, 30 cm mean diameter and of circular section 2 cm in diameter, has an air gap 1 mm long. It is wound uniformly with 600 turns of wire carrying a current of 2.5 A. Neglect magnetic leakage. The iron path takes about 40% of the total mmf.

[M.D. Univ. Electromechanical Energy Conversion, May-2006]

[**Ans.** 1500 AT, 0.355 mWb, 422.5×10^4 AT/Wb, 1.13 T]

8. A steel ring 10 cm mean radius and of circular cross section 1 cm in radius has an air gap of 1 mm length. It is wound uniformly with 500 turns of wire carrying 3 A. Neglect magnetic leakage. The air gap takes 60% of total mmf. Find the total reluctance.

[M.D. Univ. Electromechanical Energy Conversion, May-2008]

[**Ans.** 422.5×10^4 AT/Wb]

9. A 0.5 m long wire moves at right angles to its length at 40 m/s in uniform magnetic field of 1Wb/m^2. Calculate the emf induced in the conductor when the direction of motion is (*i*) perpendicular to field, (*ii*) inclined at 30° to the direction of field. [**Ans.** (*i*) 20 V (*ii*) 10 V]

10. If a current through a coil of inductance 1 H is reduced from 10 A to –10 A in 0.05 s, calculate the mean value of emf induced in the coil. [**Ans.** 400 V]

11. A coil of 300 turns, wound on a core of nonmagnetic material has an inductance of 10 mH. Calculate (*i*) the flux produced by a current of 5 A and (*ii*) the average value of emf induced when a current of 5 A is reversed in 8 milli-seconds.[**Ans.** (*i*)0.1667m Wb (*ii*)12.5 V]

12. An iron ring or toroid, 0.2 m in diameter and 10 cm^2 sectional area of core, is uniformly wound with 250 turns of wire. If the flux density in the core is to be 1 Wb/m^2 and relative permeability of iron, $\mu_r = 500$, what is the exciting current required to be passed in the winding ? Determine also the self-inductance. [**Ans.** 4 A, 62.5 mH]

13. Two coils having 30 and 600 turns respectively are wound side by side on a closed iron circuit of area of cross section 100 cm^2 and mean length 200 cm. Estimate the mutual inductance between the coils if the relative permeability of the iron is 2,000. If a current of zero ampere grows to 20 A in a time of 0.02 second in the first coil, find the emf induced in the second coil. [**Ans.** 226 mH; 226 V]

14. The self-inductance of a coil of 500 turns is 0.25 H. If 60% of the flux is linked with a second coil of 6.000 turns, calculate mutual inductance of two coils and emf induced in the second coil when the current in first coil changes at the rate of 100 A/second. [**Ans.** 0.3 H; 30 V]

15. A solenoid 70 cm in length and 45 mm in radius is wound with 2,100 turns. Another coil of 700 turns is wound over this solenoid. Determine the self-inductances of the coils and the mutual inductance between the coils. [**Ans.** 50.36 mH; 5.6 mH; 16.8 mH]

16. Two coils A and B of 600 and 100 turns respectively are wound uniformly around a wooden ring of mean circumference 80 cm. The x-sectional area of the ring is 4 cm^2. Calculate (*a*) self-inductance of each coil (*b*) mutual inductance between the coils (*c*) emf induced in the coil B when a current of 2A in coil A is reversed in 0.01 second.

[**Ans.** (*a*) 226.2 μ H; 6.283 μH; (*b*) 37.7 μH; (*c*) 15.08 mV]

17. A primary coil having an inductance of 100 μH is connected in series with a secondary coil of 240 μH and the total inductance of the combination is measured to be 146 μH. Determine the mutual inductance between the two coils and the coefficient of coupling.

[**Ans.** 97 μH; 0.626]

18. A solenoid 1 m in length and 10 cm in diameter has 5,000 turns. Calculate the energy stored in the magnetic field when a current of 2 A flows in the solenoid. [**Ans.** 0.493 J]

19. An iron ring 10 cm in diameter and 8 cm^2 in cross section is wound with 300 turns. An exciting current of 2 A is necessary to produce a flux density of 1.2 Wb/m^2. Calculate the energy stored in the magnetic field. [**Ans.** 0.288 J]

20. An air cored coil has 1,000 turns of copper wire whose diameter is 0.5 mm. The length of the coil is 1 m and the diameter of the coil is 5 cm. The coil carries 5 A. Determine the following:

(*i*) The resistance of the coil.

(*ii*) The inductance of the coil.

(*iii*) The energy stored in the magnetic field.

[**Ans.** (*i*) 13.76 Ω, (*ii*) 2.47 mH; (*iii*) 0.031J]

21. An electromagnet wound with 2,000 turns produces a flux of 0.015 Wb when excited by a current of 2 A. Determine (*i*) self–inductance of this system, (*ii*) energy stored when current is 2 A, and (*iii*) the induced voltage in the winding when the current is broken in 0.04 second. [**Ans.** (*i*) 15 H (*ii*) 30 J (*iii*) 750 V]

22. Calculate the hourly loss of energy in a specimen of iron, the hysteresis loop of which is equivalent in area to 250 J/m^3. Frequency 25 Hz, density of iron 7.5, weight 10 kg. [**Ans.** 30,000 J]

23. The hysteresis loop of a sample of sheet steel subjected to a maximum flux density of 1.3 T has an area of 93 cm^2, the scales being 1 cm = 0.1 T and 1cm = 50 AT/m. Calculate the hysteresis loss in watts when 1,500 cm^3 of the same material is subjected to an alternating flux density of 1.3 T peak value at a frequency of 65 Hz.

[**Ans.** 45.34 watts]

24. The hysteresis loop for a specimen of iron weighing 12 kg is equivalent in area to 300 J/m^3. Find the loss of energy per hour at 25 Hz. Take density of iron as equal to 7.5 g/cc. [**Ans.** 43,200 J]

25. The hysteresis loop of a certain magnetic material is drawn at a maximum flux density of 1.0 Wb/m^2 with the following scales.

1 cm = 250 AT/cm 1 cm = 0.1 Wb/m^2

The area of the loop is found to be 50 cm^2. If the density of the material is 7.8 gm/cm^3, calculate the loss in watts per kg due to hysteresis, when the material is subjected to a maximum alternating flux density of 1 Wb/m^2 at 50 Hz. [**Ans.** 800 watts]

26. The area of the hysteresis loop obtained with a certain specimen of iron was 9.3 cm^2. The coordinates were such that 1 cm = 1,000 AT/m and 1 cm = 0.2 Wb/m^2. Calculate (*a*) the hysteresis loss per m^3 per cycle and (*b*) the hysteresis loss per m^3 at a frequency of 50 Hz if the maximum flux density were 1.5 Wb/m^2 (*c*) the hysteresis loss per m^3 for a maximum flux density of 1.2 Wb/m^2 and a frequency of 30 Hz, assuming the loss to be proportional to.

[**Ans.** (*a*) 1,860 J/m^3/cycle (*b*) 93,000 W/m^3 (*c*) 37.36 kW]

27. Determine the force in kg necessary to separate two surfaces with 100 cm^2 of contact area when the flux density normal to the surface is 1 Wb/m^2. [**Ans.** 405.6 kg]

28. The two halves of a cast iron ring of mean diameter 40 cm and cross section area of 10 cm^2 are placed together with an effective air gap of 0.2 mm at each end. If the upper half of the ring is wound with 500 turn coil carrying a current of 2 A, calculate the magnetic pull between the two halves. Assume relative permeability of cast iron to be constant at 250. [**Ans.** 45.8 newtons]

❑❑❑

2 CHAPTER

AC Circuits

INSIDE THIS CHAPTER

2.1 TYPES OF SUPPLY

There are two types of supply viz dc and ac.

The voltage or current available from batteries or solar cells is *direct* in the sense that the polarity remains the same. Such sources are called direct current sources. A plot of output of such sources with respect to time is a straight line parallel to time axis as illustrated in Fig. 2.1. This shows that the voltage/current output of a dc source is constant with respect to time, unless the chemicals in the battery are exhausted or the light incident on solar cell varies.

Power obtainable from batteries or solar cells is very small as compared to the total power we need. The dc power in bulk can be generated by using dc generators, and will be discussed in detail later on. Standard dc voltages are 230 and 460 V.

A current (or voltage) is called *alternating* if it periodically changes its direction and magnitude whereas direct currents are steady and in one direction. In more restricted sense, alternating current is a periodically varying current, the average value of which, over a period, is zero.

At present a large percentage of electrical energy (nearly all) being used for domestic, commercial and industrial purposes is generated as alternating current because of technical and economical reasons.

For industrial applications of direct current like electrolytic processes, welding processes and variable speed motor drives, the present trend is to generate ac and then convert it into dc by converters. For conversion of ac into dc the dc generator as a part of motor (ac)- generator (dc) set has to compete the SCR rectifiers and other various power controlled devices which usually are cheaper in cost, compact in size, relatively noise-free in operation and require minimum maintenance, though they suffer from the disadvantages of having poor power factor, harmonic generation, poor braking etc.

In our country the standard voltage for low tension (*lt*) distribution is 240 / 415 V (240 V between phase and neutral wires and 415 V from line-to-line). This is a reasonably safe voltage. Higher the voltage used, lower is the installation cost i.e. economy is achieved but increase of voltage means more hazard to human life and property. Hence, a compromise is made. All over the world the standard voltage used for *lt* distribution is in the range of 200 / 350 V to 250 / 440 V except in American continent where 110 /190 V is the distribution voltage. Generation, transmission and *ht* distribution voltages are governed by economy and operation.

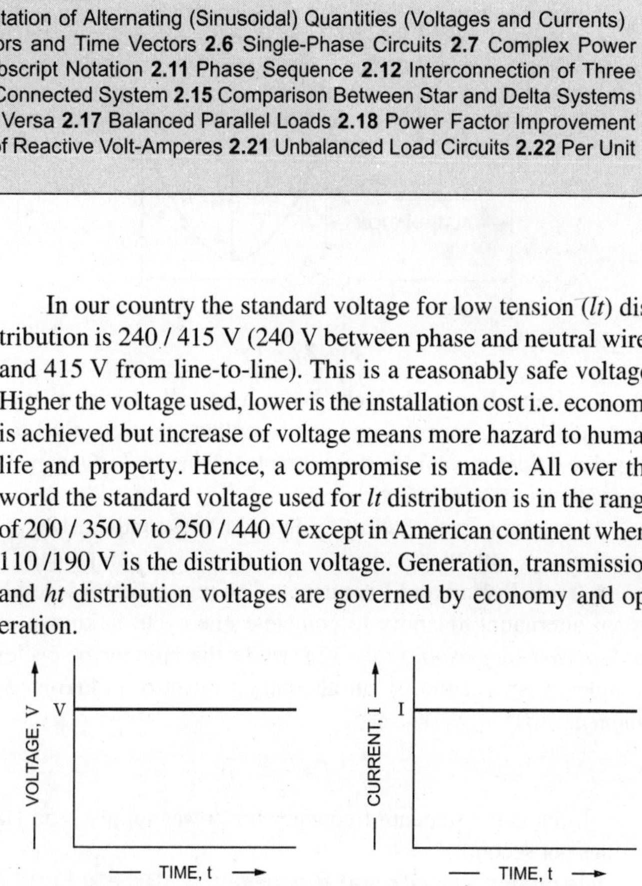

(*a*) *Output Characteristic of Battery* (*b*) *Output Characteristic of Solar Cell*

Fig. 2.1

2.2 TERMINOLOGY

An *alternating quantity* (voltage or current) is one which changes continuously in magnitude and alternates in direction at regular intervals of time.

An alternating quantity that varies sinusoidally (*i.e.* according to the sin of angle θ) is called the *sinusoidal quantity.* Sinusoidal quantities are expressed as

$$e = E_{max} \sin \theta = E_{max} \sin \omega t = E_{max} \sin 2\pi ft$$

$$\text{and } i = I_{max} \sin \theta = I_{max} \sin \omega t = I_{max} \sin 2\pi ft$$

The important ac terms are defined below:

1. Waveform. The shape of the curve of the voltage or current when plotted against time as abscissa (base) is called the *waveform*.

2. Instantaneous Value. The value of alternating quantity (emf, voltage or current) at any particular instant is called the *instantaneous value* and is designated by a small italic letter (*e* for emf, *v* for voltage and *i* for current). The instantaneous values of an alternating quantity can be determined either from the curve or from an equation of the alternating quantity. For example, the instantaneous values of emf represented by the curve shown in Fig. 2.2 at 0, $\frac{\pi}{2}$, π and $\frac{3\pi}{2}$ are zero, $+E_{max}$, zero and $-E_{max}$ respectively.

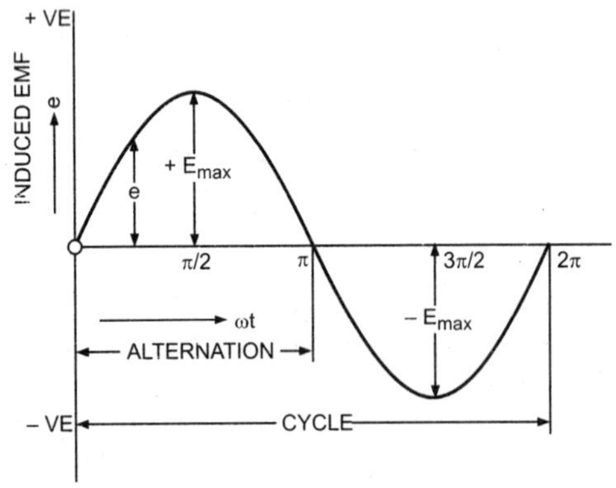

Fig. 2.2

3. Alternation and Cycle. When a periodic wave, such as sinusoidal wave, goes through one complete set of positive and negative values it is said to have completed one *cycle*. One cycle corresponds to 360° or 2 π radians.

Alternation is one half of cycle and corresponds to 180° or π radians.

4. Time Period and Frequency. The time taken in seconds by an alternating quantity to complete one cycle is known as *time period* or *periodic time* (T), while the number of cycles completed per second by an alternating quantity is known as *frequency* (*f*).

$$f = \frac{1}{T}$$

In India the standard frequency for power supply is 50 Hz (cycles per second).

5. Angular Velocity and Frequency. A glance at Fig. 2.2 indicates that each cycle spans 2π radians. Hence, if this quantity is divided by the time period, *angular velocity* of the sine function is obtained. It is denoted by ω and is expressed in radians per second.

∴ Angular velocity,

$$\omega = \frac{\text{Angle turned}}{\text{Time taken}} = \frac{2\pi}{T} = \frac{2\pi}{1/f} = 2\pi f \text{ radians per second} \quad ...(2.1)$$

6. Electrical Time Degrees and Mechanical Degrees. It is seen that the coil must revolve past a pair of poles in order to carry the generated emf through one complete cycle. In circuit work, one complete cycle of voltage or current is designated as 360 electrical degrees or 2 π electrical radians. To correlate with this designation the arc through which a coil of dynamo must rotate in order to generate one cycle of emf is called 360 electrical degrees. In a 2-pole machine one complete revolution of coil produces one cycle of emf. But in a multipolar machine, such as four, six or eight pole machine, the emf completes one cycle or 360 electrical degrees or 2 π electrical radians as soon as the coil passes a pair of poles and a mechanical degree will be equal to as many electrical degrees as there are pairs of poles in the dynamo structure. In a multipolar machine, the number of cycles completed per second by generated emf,

f = Pair of poles × number of revolutions made per second

$$= \frac{P}{2} \times \frac{N}{60} = \frac{PN}{120} \quad ...(2.2)$$

where N is the speed of rotation of the coil in rpm.

7. Amplitude. The maximum value, positive or negative, which an alternating quantity attains during one cycle is called the *amplitude* of the alternating quantity.

8. General Expression For AC Quantities. The general expression for an alternating voltage and current are given as

$$v = V_{max} \sin(\omega t \pm \phi)\,; \quad i = I_{max} \sin(\omega t \pm \phi)$$

where φ is the phase angle in degrees or radians.

The coefficient of the sine of the time angle gives maximum or peak value of the alternating quantity (emf, voltage or current)

Coefficient of time *t* divided by 2 π gives the frequency of the periodic wave.

9. Average or Mean Value of Alternating Current. The *average (or mean) value* of an alternating current is expressed by that steady current (dc) which transfers across any circuit the same charge as transferred by that alternating current during a given time duration.

Average or mean value of alternating current is given as

$$I_{av} = \frac{i_1 + i_2 + i_3 + ...+ i_n}{n} = \frac{\text{Area of one alternation}}{\text{Length of base over one alternation}}$$

For sinusoidal current $I_{av} = \frac{2}{\pi} I_{max} = 0.637\, I_{max}$.

10. Effective or RMS Value of Alternating Current. *Effective* or *virtual value* of alternating current or voltage is equal to the square root of the mean of the squares of successive ordinates and that is why it is known as *root mean square* (rms) value.

$$\textit{i.e.}\ I_{eff} \text{ or } I_{rms} = \sqrt{\frac{i_1^2 + i_2^2 + i_3^2 + ... + i_n^2}{n}}$$

The *rms* or *effective value* of an alternating current or voltage is given by that steady current or voltage which when flows or applied to a given resistance for a given time produces the same amount of heat as when the alternating current or voltage flowing or applied to the same resistance for the same time.

For sinusoidal current, $I_{rms} = \frac{I_{max}}{\sqrt{2}} = 0.707\, I_{max}$

11. Form Factor and Peak Factor. The ratio of rms value to the average value of a waveform is known as *form factor*. For sinusoidal wave its value is 1.11.

The ratio of maximum value to the rms value of a waveform is known as peak factor. For sinusoidal wave its value is $\sqrt{2}$ or 1.414.

The effective and average values of different waves are given, on next page, in tabular form.

Waveform	*Sinusoidal*	*Half Wave Rectified AC*	*Square*	*Triangular*	*Sawtooth*	*Circular*
Effective Value	0.707 I_{max}	0.5 I_{max}	I_{max}	$\frac{1}{\sqrt{3}} I_{max}$	$\frac{1}{\sqrt{3}} I_{max}$	$\sqrt{\frac{2}{3}} I_{max}$
Average Value	0.637 I_{max}	0.318 I_{max}	I_{max}	0.5 I_{max}	0.5 I_{max}	0.785 I_{max}

2.3 GRAPHICAL OR PHASOR REPRESENTATION OF ALTERNATING (SINUSOIDAL) QUANTITIES (VOLTAGES AND CURRENTS)

It has already been pointed out that it is assumed that alternating voltages and currents follow sine law and generators are designed to give emfs having sine waveform. The above said assumption makes the calculations simple. The method of representing alternating quantities by waveform or by the equations giving instantaneous values is quite cumbersome. For solution of ac problems it is advantageous to represent a sinusoidal quantity (voltage or current) by a line of definite length rotating in counter-clockwise* direction with the same angular velocity as that of the sinusoidal quantity. Such a rotating line is called the *phasor.***

Consider a line OA (or phasor as it is called) representing to scale the maximum value of an alternating quantity, say emf *i.e.* OA = E_{max} and rotating in counter-clockwise direction at an angular velocity ω radians/second about the point O, as shown in Fig. 2.3. An arrow head is put at the outer end of the phasor, partly to indicate which end is assumed to move and partly to indicate the precise length of the phasor when two or more phasors happen to coincide.

Figure 2.3 shows OA when it has rotated through an angle θ, being equal to ω*t*, from the position occupied when the emf was passing through its zero value. The projection of OA on Y-axis, OB = OA sin θ = E_{max} sin ω*t* = *e*, the value of the emf at that instant

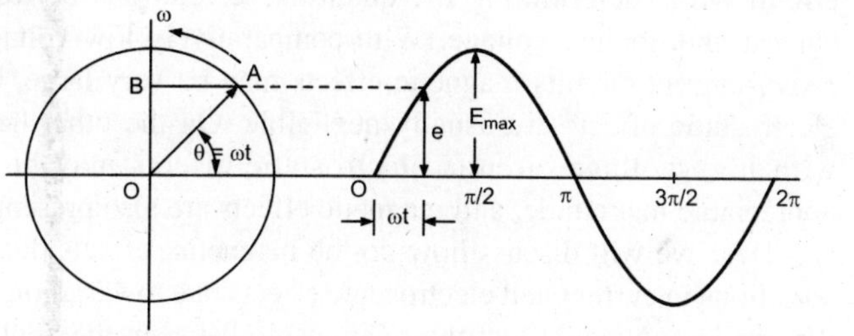

Fig. 2.3 *Phasor Representation of An Alternating Quantity*

Thus, the projection of OA on the vertical axis represents to scale the instantaneous value of emf.

It will be seen that the phasor OA rotating in counter-clockwise direction will represent a sinusoidal quantity (voltage or current) if

(*i*) its length is equal to the peak or maximum value of the sinusoidal voltage or current to a suitable scale.

(*ii*) it is in horizontal position at the instant the alternating quantity (voltage or current) is zero and increasing and

(*iii*) its angular velocity is such that it completes one revolution in the same time as taken by the alternating quantity (voltage or current) to complete one cycle.

A vector or phasor may be represented by a line of length proportional to its magnitude, oriented at the appropriate angle to a chosen reference direction, and drawn from an origin positioned at any convenient point.

2.4 ADDITION AND SUBTRACTION OF ALTERNATING QUANTITIES

A simpler and direct method is by adding the sinusoidal quantities as phasors i.e., two or more alternating quantities may be added in the same way as forces are added, namely by constructing parallelograms or closed polygons and either measuring or calculating the length of the diagonals or closing sides and the magnitudes of the phase angles.

Illustration: The way in which the two given currents i_1 and i_2 can be added by the parallelogram rule of phasor addition is illustrated in Fig. 2.4 where the currents are shown as phasor drawn from the origin O of the system of coordinates. The resultant phasor is the diagonal of the parallelogram formed by the phasors $I_{1\ max}$ and $I_{2\ max}$.

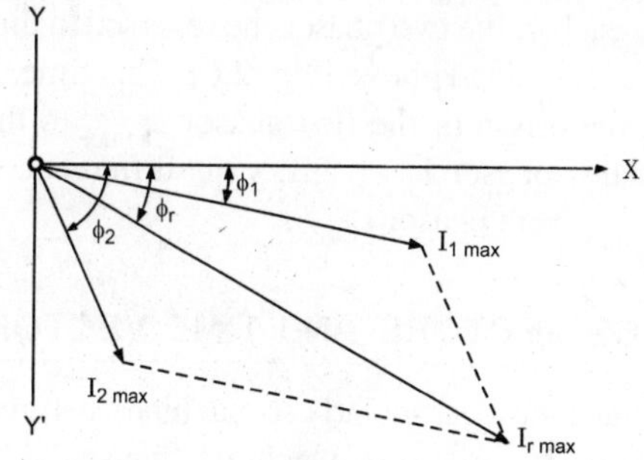

Fig. 2.4

This method is more convenient when more than two phasors are to be added, as shown in Fig. 2.5. From the end point of $I_{1\ max}$ a phasor is constructed parallel to $I_{2\ max}$ of the same magnitude and direction as the latter; then from the end point of $I_{2\ max}$ a

* It is a standard convention that the phasor is rotated in counter-clockwise direction — a convention that is in harmony with the general use of polar coordinates.

** Strictly speaking, sinusoidal quantities are scalar quantities varying periodically with time and, according to the definition of vector quantities, they are not true vectors. Voltage is simply energy or work per coulomb and cannot be classified as a vector. Current is also not a vector quantity because it is merely the flow of electrons through a wire. But keeping in mind that any sinusoidal quantity at a given frequency is completely specified by its amplitude and phase angle, its similarity to a vector quantity is evident, since the amplitude may be considered as the magnitude and the phase angle as the direction of a vector. To account for the difference the term phasor has been adopted, instead of the term vector, for representing graphically the magnitude and phase of a sinusoidal current or voltage.

phasor is constructed parallel to $I_{3\ max}$ and so on. Phasor $I_{r\ max}$ from the origin of the first phasor ($I_{1\ max}$) to the end point of the last phasor ($I_{5\ max}$) represents the sum of all the phasors.

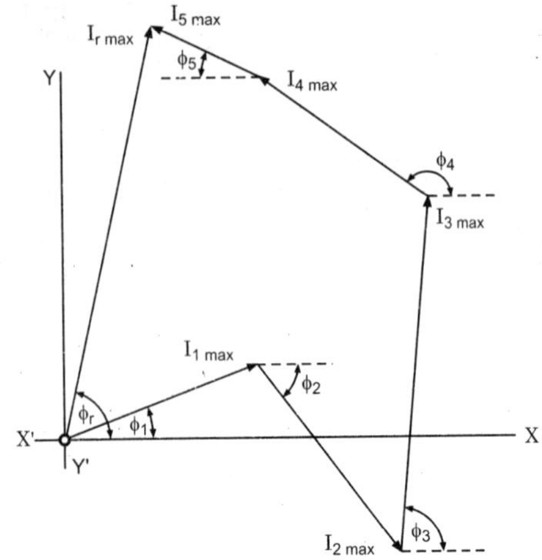

Fig. 2.5

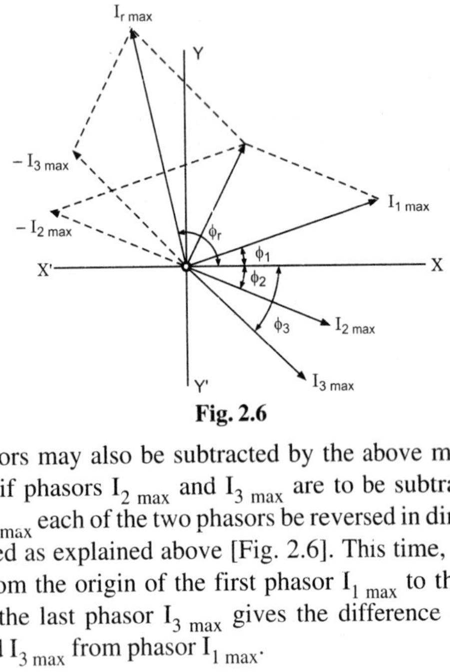

Fig. 2.6

Phasors may also be subtracted by the above method. For example if phasors $I_{2\ max}$ and $I_{3\ max}$ are to be subtracted from phasor $I_{1\ max}$ each of the two phasors be reversed in direction and then added as explained above [Fig. 2.6]. This time, the phasor drawn from the origin of the first phasor $I_{1\ max}$ to the terminal point of the last phasor $I_{3\ max}$ gives the difference of phasors $I_{2\ max}$ and $I_{3\ max}$ from phasor $I_{1\ max}$.

2.5 SPACE VECTORS AND TIME VECTORS

In studying machines, particularly ac machines, certain quantities are found to have variations which are sinusoidal in time, or sinusoidally distributed in space. For addition of two or more such quantities of the same kind, it is convenient to represent them as phasors/vectors and adopt the technique discussed above.

The distinction between time and space variations is quite significant and this can be done by assigning the term *time vector* or *space vector*. However, the term '*phasor*' has been accepted as a suitable designation for a time vector and for sinusoidal space variations, the term *space vector* is commonly used as there is no commonly accepted alternative.

Figure 2.7 depicts an instance where the two types of vectors are closely related. From the rotating pole, a line drawn along the flux axis, (at the point of maximum flux density assumed sinusoidally distributed) is of length proportional to the mutual

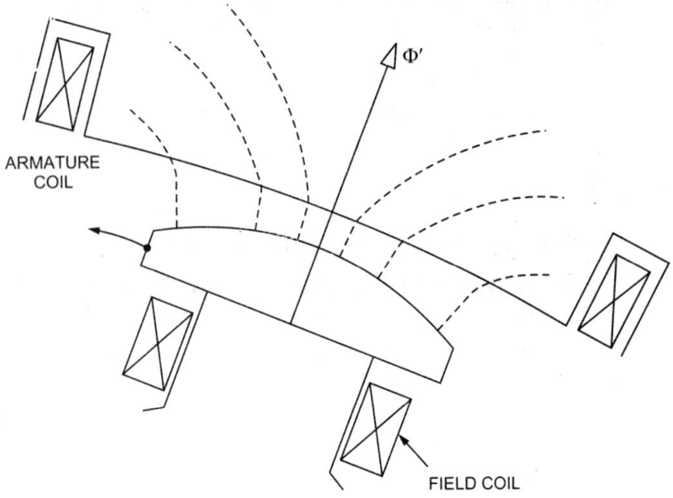

Fig. 2.7 *Time Vector and Space Vector of Flux*

flux **Φ** per pole. This is a *space vector* and has no time variation w.r.t. the field coil provided the excitation current is constant and the bore is smooth. Relative to the coil on stator, however, same flux is experienced as a time variation of flux linkage. When the space vector is intended to be understood, a bold-face phi with prime, **Φ′**, is used. In case it is the time varying, an unprimed **Φ** is used to represent the maximum value of flux per pole crossing the air gap to link the armature coils and induce an emf. A small ϕ will be used for an instantaneous value of flux.

2.6 SINGLE-PHASE CIRCUITS

In a dc circuit the relationship between the applied voltage V and current flowing through the circuit I is a simple one and is given by the expression I = V/R but in an ac circuit this simple relationship does not hold good. Variations in current and applied voltage set up magnetic and electrostatic effects respectively and these must be taken into account with the resistance of the circuit while determining the quantitative relations between current and applied voltage. With comparatively low-voltage, heavy-current circuits magnetic effects may be very large, but electrostatic effects are usually negligible. On the other hand with high-voltage circuits electrostatic effects may be of appreciable magnitude, and magnetic effects are also present.

Here we will discuss how do the magnetic effects due to variations in current and electrostatic effects due to variations in the applied voltage affect the relationship between the applied voltage and current.

2.6.1. Pure Resistive Circuit. A pure resistive or a non-inductive circuit is a circuit which has inductance so small that at normal frequency its reactance is negligible as compared to its resistance. Ordinary filament lamps, water resistances etc. are the examples of non-inductive resistances. If the circuit is non-inductive, no reactance emf (*i.e.* self- induced or back emf) is set up and whole of the applied voltage is utilised in overcoming the ohmic resistance of the circuit.

When an alternating voltage given by expression $v = V_{max} \sin \omega t$ is applied across a pure resistive circuit, the alternating current flowing through the circuit is given as

$$i = I_{max} \sin \omega t \quad \text{where } I_{max} = \frac{V_{max}}{R}$$

Power in pure resistive circuit, $P = V_{rms} \times I_{rms}$ or VI watts

Power factor of pure resistive circuit is unity.

2.6.2. Pure Inductive Circuit. An inductive circuit is a coil with or without an iron core having negligible resistance. Practically pure inductance can never be had as the inductive coil has always small resistance. However, a coil of thick copper wire wound on a laminated iron core has negligible resistance and is known as a *choke coil.**

When an alternating voltage given by expression $v = V_{max} \sin \omega t$ is applied across a pure inductive circuit (a coil with or without an iron core having negligible resistance), the alternating current flowing through the circuit is given as

$$i = I_{max}\left(\omega t - \frac{\pi}{2}\right)$$

where $I_{max} = \dfrac{V_{max}}{X_L}$; Inductive reactance X_L being equal to $2\pi f L$ or ωL ohms and power absorbed by a pure inductor = 0

PF of a pure inductive circuit is zero.

2.6.3. Pure Capacitive Circuit. When a dc voltage is impressed across the plates of a perfect condenser, it will become charged to full voltage almost instantaneously. The charging current will flow only during the period of "build up" and will cease to flow as soon as the capacitor has attained the steady voltage of the source. This implies that for a direct current, a capacitor is a break in the circuit or an infinitely high resistance.

In a pure capacitive circuit, if $v = V_{max} \sin \omega t$

$$i = I_{max} \sin\left(\omega t + \frac{\pi}{2}\right)$$

where $I_{max} = \dfrac{V_{max}}{X_C}$; the capacitive reactance X_C being equal to $\dfrac{1}{\omega C}$ or $\dfrac{1}{2\pi f C}$ ohms

Power absorbed, P = 0

PF is also zero.

An ac circuit can have three elements, namely R, L and C. Power is only consumed in R. The power consumed in elements L and C are zero.

The current through pure inductor lags behind the applied voltage by $\frac{\pi}{2}$ radians while the current through a pure capacitor leads the applied voltage by $\frac{\pi}{2}$ radians.

2.6.4. Series R-L Circuit. This is the case most generally met within practice, nearly all circuits contain both resistance and inductance.

In an R-L series ac circuit, if

$$v = V_{max} \sin \omega t$$

$$i = I_{max} \sin(\omega t - \phi)$$

$$I_{max} = \frac{V_{max}}{Z};\ Z = \sqrt{R^2 + \omega^2 L^2};\ \phi = \text{Tan}^{-1}\frac{\omega L}{R};$$

$$P = VI\cos\phi = I^2 R$$

$$\text{Power factor, } \cos\phi = \frac{R}{Z} \text{ (lagging)}$$

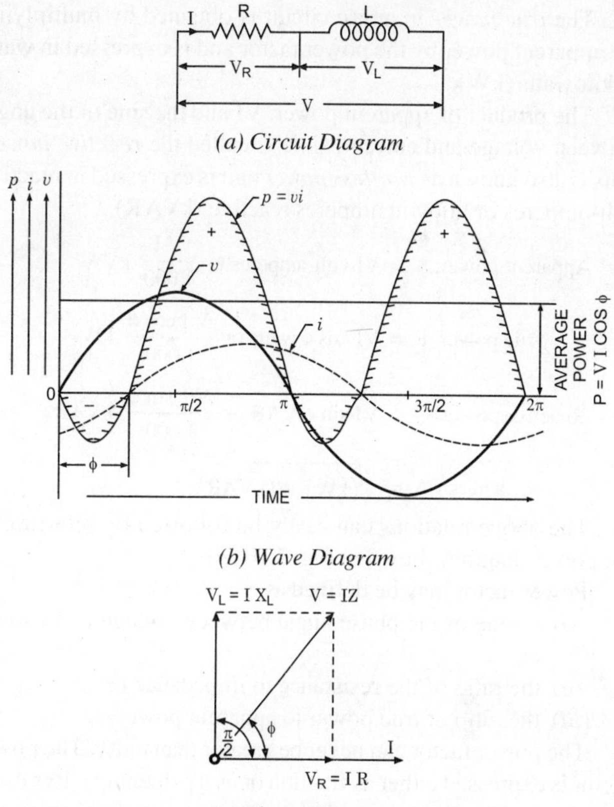

(a) Circuit Diagram

(b) Wave Diagram

(c) Phasor Diagram

Fig. 2.8 *Resistance and Inductance in Series*

2.6.5. Impedance Triangle. If a triangle ABC is drawn so that $AB = \frac{V_R}{I} = R$, $BC = \frac{V_L}{I} = X_L$ and $AC = \frac{V}{I} = Z$, it is a triangle similar to the voltage triangle shown in Fig. 2.9 (*a*). Such a triangle is called an *impedance triangle*, which is most useful in letting one see at a glance how R, X, and Z are related to each other. The angle between Z and R sides of the impedance triangle is known as *phase angle of the circuit* and cos of this angle is known as *power factor of the circuit.*

$$\text{Power factor} = \text{Cos}\,\phi = \frac{R}{Z}$$

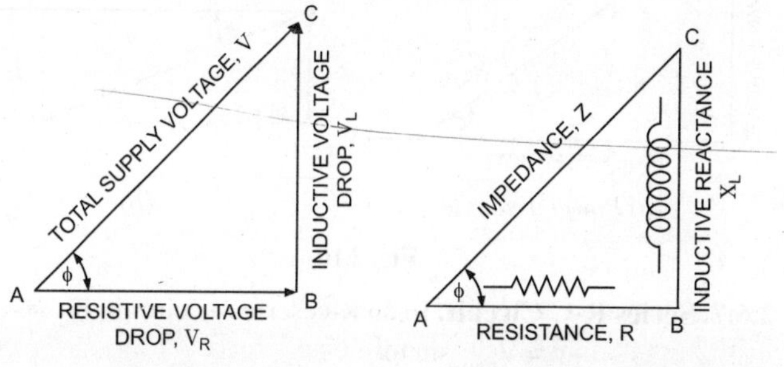

(a) Voltage Triangle *(b) Impedance Triangle*

Fig. 2.9

2.6.6. Apparent power, True Power, Reactive Power AND Power Factor. The product of rms values of current and voltage, VI is called the *apparent power* and is measured in volt-amperes or kilovolt amperes (kVA).

* Inductors or choke coils are made of iron core because large valued flux densities can be produced in iron cores and so inductances of large values can be had. Air-cored inductors become too much bulky to give an inductance of a given value.

The *true power* in an a c circuit is obtained by multiplying the apparent power by the power factor and is expressed in watts or kilowatts (kW).

The product of apparent power, VI and the sine of the angle between voltage and current, sin ϕ is called the *reactive power*. This is also known as *wattless power* and is expressed in reactive volt-amperes or kilovolt amperes reactive (kVAR).

i.e Apparent power, $S = VI$ volt-amperes or $\dfrac{VI}{1{,}000}$ kVA

True power, $P = VI \cos\phi$ watts or $\dfrac{VI\cos\phi}{1{,}000}$ kW

Reactive power, $Q = VI \sin\phi$ VAR or $\dfrac{VI\sin\phi}{1{,}000}$ kVAR

and $\text{kVA} = \sqrt{(\text{kW})^2 + (\text{kVAR})^2}$

The above relations can easily be followed by referring to the power diagram shown in Fig. 2.10 (*a*).

Power factor may be defined as

(*i*) cosine of the phase angle between voltage and current or

(*ii*) the ratio of the resistance to impedance or

(*iii*) the ratio of true power to apparent power.

The power factor can never be greater than unity. The power factor is expressed either as fraction or as a percentage. It is usual practice to attach the word 'lagging' or 'leading' with the numerical value of power factor to signify whether the current lags behind or leads the voltage.

Active Component of Current. The current component which is in phase with circuit voltage (*i.e.* I cos ϕ) and contributes to active or true power of the circuit is called the *active* (*wattful* or *in-phase*) *component* of current.

Reactive Component of Current. The current component which is in quadrature (or 90° out of phase) to circuit voltage (*i.e.* I sin ϕ) and contributes to reactive power of the circuit, is called the *reactive* (or *wattless*) *component* of current.

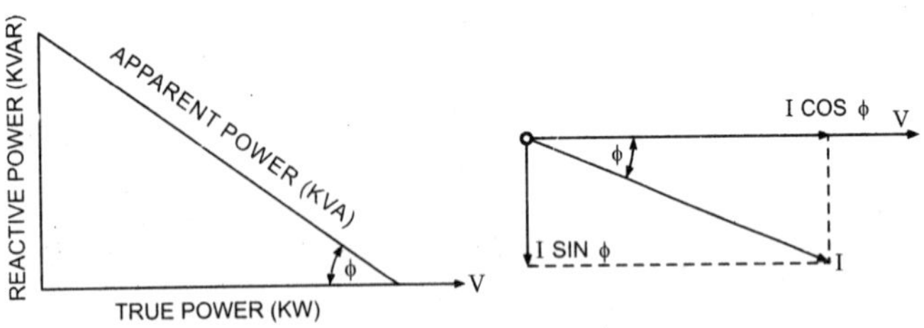

(*a*) Power Triangle (*b*)

Fig. 2.10

2.6.7. Series R-C Circuit. In an R-C series ac circuit, if

$$v = V_{max} \sin \omega t$$
$$i = I_{max} \sin(\omega t + \phi)$$

$$I_{max} = \frac{V_{max}}{Z};\ Z = \sqrt{R^2 + \left(\frac{1}{\omega C}\right)^2} = \sqrt{R^2 + X_C^2}$$

$$\phi = \text{Tan}^{-1}\frac{X_C}{R} = \text{Tan}^{-1}\frac{1}{\omega CR}$$

$$P = VI\cos\phi = I^2 R$$

$$\text{Power factor, } \cos\phi = \frac{R}{Z} \text{ (leading)}$$

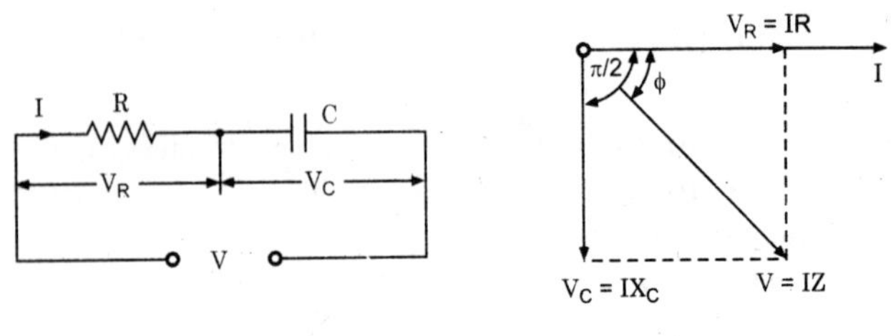

(*a*) Circuit Diagram (*b*) Phasor Diagram

Fig. 2.11

2.6.8. Series R-L-C Circuit. In an R-L-C series ac circuit

$$\text{Impedance, } Z = \sqrt{R^2 + (X_L - X_C)^2}$$

Circuit behaves as an inductive circuit if $X_L > X_C$; as a capacitive if $X_L < X_C$ and as a purely resistive if $X_L = X_C$

$$\text{Circuit current, } I = \frac{V}{Z}$$

$$\text{Phase angle, } \phi = \text{Tan}^{-1}\frac{X_L - X_C}{R}$$

$$\text{Power consumed, } P = VI\cos\phi = I^2 R$$

$$\text{Power factor, } \cos\phi = \frac{R}{Z}$$

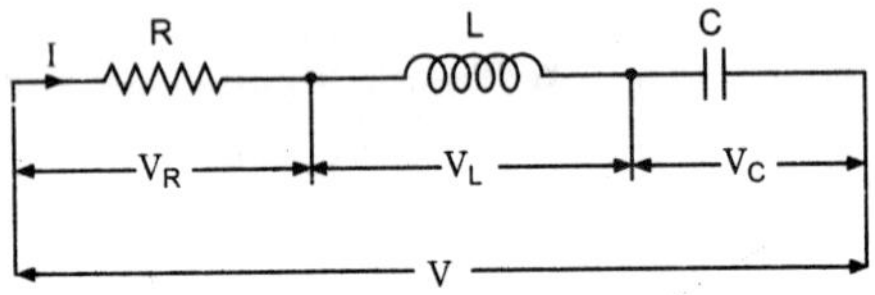

(*a*) Circuit Diagram

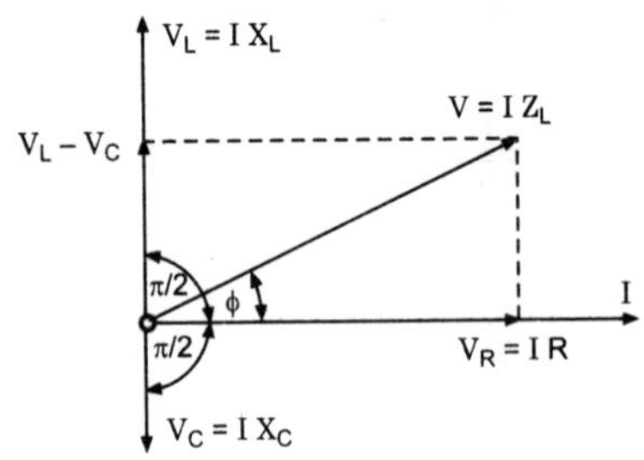

(*b*) Phasor Diagram

Fig. 2.12

2.7 COMPLEX POWER

Apparent power S, true power P and reactive power Q can be determined by *j* method but for it, it is necessary to change the sign of the imaginary part of either V or I.

Complex quantity so obtained is known as *conjugate* of the original complex quantity.

Now $\mathbf{V}_C\mathbf{I} = V(\cos\alpha - j\sin\alpha) \times I(\cos\beta + j\sin\beta)$
$= VI[\cos(\alpha - \beta) - j\sin(\alpha - \beta)]$.

The real part of the above expression VI cos $(\alpha - \beta)$ gives the true power and imaginary part VI sin $(\alpha - \beta)$ gives the reactive power. It should, however, be noted that the real power obtained by the conjugate method is the same regardless of whether V or I is reversed but sign of reactive

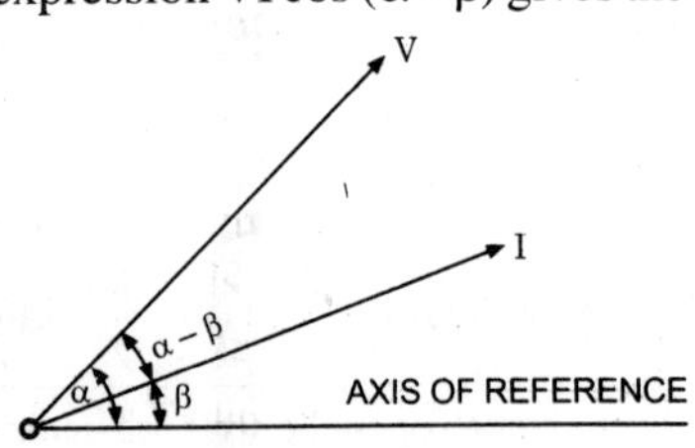

Fig. 2.13

volt-amperes will depend upon it. When the conjugate of voltage is used, capacitive VARs have a positive sign and inductive VARs a negative sign. If the conjugate of the current rather than of the voltage is used the capacitive VARs have a negative sign and inductive VARs a positive sign.

The product of V and I_C is called the *complex power* where I_C is the conjugate of I.

2.8 POLYPHASE CIRCUITS

The type of alternating currents and voltages discussed so far are termed as *single-phase currents and voltages* as they consist of a single alternating current and voltage waves. Single-phase systems involving single-phase currents and voltages are quite satisfactory for domestic applications. Even the motors employed in domestic applications are mostly single-phase, for example, motors for mixers, coolers, fans, air-conditioners, refrigerators. However, the single-phase system has its own limitations and, therefore, has been replaced by polyphase system. For supplying power to electric furnaces 2-phase system is generally employed. Six-phase system is usually employed in connection with converting machinery and apparatus. For general supply three-phase system is universally used. For generation, transmission and distribution of electric power 3-phase system has been universally adopted. The two-phase supply and six-phase supply are obtained from 3-phase supply.

Polyphase system means the system which consists of numerous (poly means numerous or multiple) windings or circuits (phase* means winding or a circuit).

A polyphase system is essentially a combination of several single-phase voltages having same magnitude and frequency but displaced from one another by equal angle (electrical), which depends upon the number of phases and can be determined from the following relation

$$\text{Electrical displacement} = \frac{360 \text{ electrical degrees}}{\text{Number of phases}} \qquad ...(2.3)$$

The above relation does not hold good for two-phase windings, which are displaced by 90 electrical degrees apart.

A supply system is said to be symmetrical when several voltages of the same frequency have equal magnitude and are displaced from one another by equal time angle. 3-phase, 3-wire or 4-wire supply system will be symmetrical when the line to line voltages are equal in magnitude and displaced in phase by 120 electrical degrees with respect to each other. Further in a four-wire system, the voltage with respect to neutral of the three phase wires should be equal to one another in magnitude and displaced in phase by 120° with respect to each other.

A 3-phase supply will be unbalanced when either of the three-phase voltages are unequal in magnitude or the phase angle between these phases is not equal to 120°.

A load circuit is said to be *balanced* when the loads (impedances) connected in various phases are same in magnitude as well as in phase. Any three-phase load in which the impedances in one or more phases differ from the impedances of other phases is called an *unbalanced three-phase load.*

In case one phase out of three phases of a 3-phase supply connected to the 3-phase load is not available, then such a condition is called the *single phasing*.

2.8.1. Merits and Demerits of Polyphase System Over Single-Phase System. The advantages of a polyphase system over single-phase system are enumerated below:

(*i*) In a single-phase circuit the power delivered is pulsating. Even when current and voltage are in phase, the power is zero twice in each cycle, and when the current leads or lags behind the voltage, the power is negative twice and zero four times during each cycle. This is not objectionable for lighting and small motors but with large motors it causes excessive vibrations. In polyphase system the total power delivered is constant if loads are balanced though the power of any one phase or circuit may be negative. So polyphase system is highly desirable particularly for power loads.

(*ii*) The rating of a given machine increases with the increase in number of phases. For example output of a 3-phase motor is 1.5 times the output of a single-phase motor of same size.

(*iii*) Single-phase induction motors have no starting torque and so it is necessary to provide these motors with an auxiliary means of starting , but in case of three-phase motors except synchronous motors, there is no need of providing an auxiliary means for starting.

(*iv*) Power factor of a single-phase motor is lower than that of a polyphase motor of the same rating (output and speed). The efficiency of a polyphase motor is also higher than that of a single-phase motor.

(*v*) Three-phase system requires 3/4th weight of copper of that required by single-phase system to transmit the same amount of power at a given voltage and over a given distance.

(*vi*) Rotating magnetic field can be set up by passing polyphase currents through stationary coils.

(*vii*) Polyphase system is more capable and reliable than single-phase system .

(*viii*) Parallel operation of polyphase alternators is simple as compared to that of single-phase alternators because of pulsating reaction in single-phase alternators.

However, three-phase operation is not as practical for domestic applications where motors are usually smaller than 1 kW and where lighting circuits supply most of the load.

2.8.2. Commonly Used Polyphase System. Although there are several polyphase systems such as two-phase, three-phase but 3-phase system is invariably adopted because of its inherent advantages over all other polyphase systems.

The demand for two-phase system has almost disappeared because that does not have any advantage that is not equalled or surpassed by three-phase system in generation, transmission or utilization. Three-phase system is universally employed for generation, transmission and distribution of electric power. Two-phase supply and six-phase supply, when required, is obtained from 3-phase supply.

* Each emf of a polyphase system is referred to as a *phase*.

Systems with number of phases, more than three, increase complexity and cost of transmission and utilisation hardware and becomes uneconomical.

A knowledge of 3-phase power systems is, therefore, essential for understanding of power technology. Fortunately, the basic circuit technique used in solving single-phase circuits are directly applicable to 3-phase circuits because three phases are identical and one phase represents the behaviour of all the above.

2.9 GENERATION OF 3-PHASE EMF

When three coils fastened rigidly together and 120° (electrical) apart rotate about the same axis in a uniform magnetic field, the induced emf in each of them will have a phase difference of 120° or $\frac{2}{3}\pi$ radians.

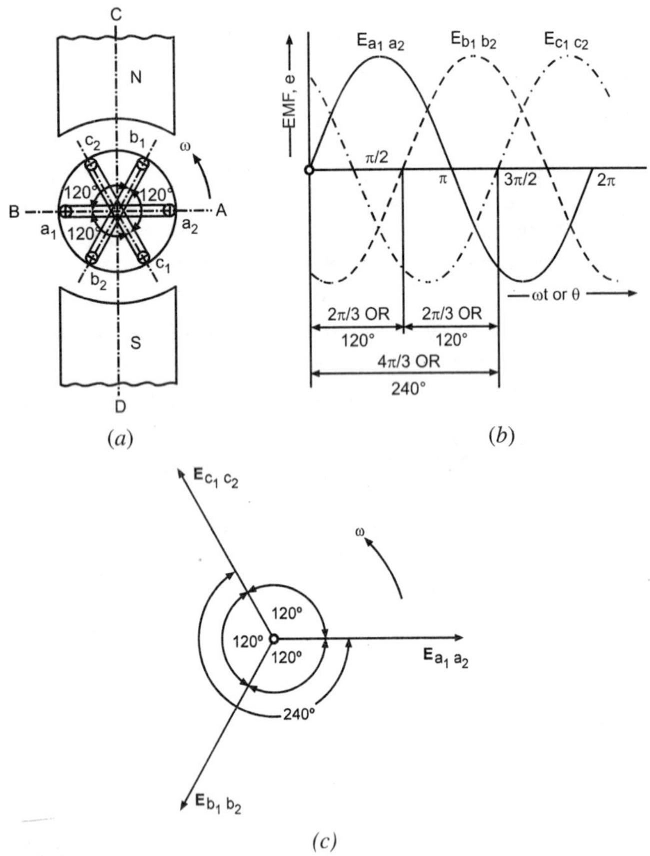

Fig. 2.14 *Generation of Three-Phase EMF*

Consider three identical coils a_1a_2, b_1b_2 and c_1c_2 mounted on the same axis but displaced from each other by 120° rotating in counter-clockwise direction in a bipolar magnetic field, as shown in Fig. 2.14(*a*). Here a_1, b_1 and c_1 are the start terminals and a_2, b_2 and c_2 are the finish terminals of the three coils. When the coil a_1a_2 is in position AB the induced emf in this coil is zero and is increasing in positive direction, the coil b_1b_2 is 120° behind coil a_1a_2 so the emf induced in this coil is approaching its maximum negative value and the coil c_1c_2 is 240° behind coil a_1a_2 so as the emf induced in the coil has passed its positive maximum value and is decreasing.

Since each coil being identical has an equal number of turns and is wound with the wire of the same type and same x-section, the induced emfs in each of the coils are of same magnitude. The induced emf in each coil is also of same frequency and same waveform (sinusoidal in this case)but displaced from each other by $\frac{2\pi}{3}$ radians or 120°, as illustrated in Fig. 2.14 (*b*) by waveforms .

Accordingly, the instantaneous values of the emfs induced in coils a_1a_2, b_1b_2 and c_1c_2 may be given as

$$\left.\begin{aligned} e_{a_1a_2} &= E_{max} \sin \omega t \\ e_{b_1b_2} &= E_{max} \sin\left(\omega t - \frac{2\pi}{3}\right) \quad \text{or } E_{max} \sin(\omega t - 120°) \\ e_{c_1c_2} &= E_{max} \sin\left(\omega t - \frac{4\pi}{3}\right) \quad \text{or } E_{max} \sin(\omega t - 240°) \\ &\qquad \text{or } E_{max} \sin(\omega t + 120°) \end{aligned}\right\} \qquad ...(2.4)$$

if $t = 0$ corresponds to the instant when the voltage or emf of coil a_1a_2 passes through zero and increases in positive direction.

In symbolic notation, the real axis is usually made to coincide with the emf (or voltage) of coil a_1a_2 (say phase A). Then the effective values of the symmetrical system are given in complex numbers as

$$\left.\begin{aligned} E_{a_1a_2} &= E_P \\ E_{b_1b_2} &= E_P \angle -120° \text{ or } E_P e^{-j\,120°} \\ &\quad \text{or } E_P(-0.5 - j\,0.866) \\ E_{c_1c_2} &= E_P \angle +120° \text{ or } E_P e^{+j\,120°} \\ &\quad \text{or } E_P(-0.5 + j\,0.866) \end{aligned}\right\} \qquad ...(2.5)$$

It is easy to prove that, at any instant, the algebraic sum of the three emfs is zero.

This result can be established analytically with the aid of the relations in Eq. (2.4)

$$\begin{aligned} \text{Resultant} &= e_{a_1a_2} + e_{b_1b_2} + e_{c_1c_2} \\ &= E_m\,[\sin \omega t + \sin(\omega t - 120°) + \sin(\omega t - 240°)] \\ &= E_m\,[\sin \omega t - 2\sin(\omega t - 180°)\cos 60°] \\ &= E_m\,[\sin \omega t - 2\sin \omega t \cos 60°] = 0 \end{aligned}$$

It is easier to establish the result by phasor diagram shown in Fig. 2.15.

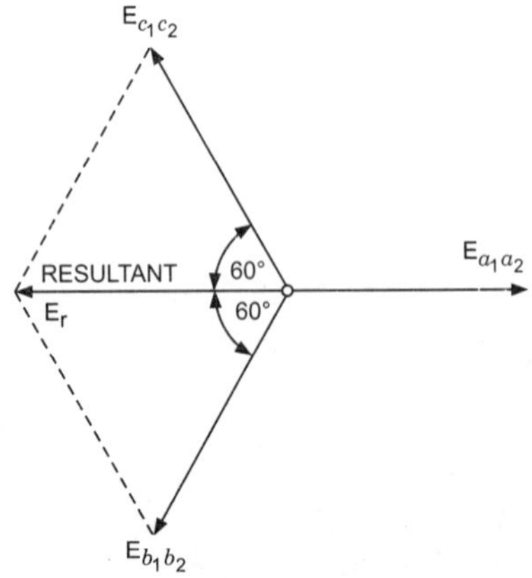

Fig. 2.15

Indeed, combining the phasors $E_{b_1b_2}$ and $E_{c_1c_2}$ we find that, due to symmetry, the phasor $\mathbf{E}_{b_1b_2} + \mathbf{E}_{c_1c_2}$ is directed horizontally to the left, *i.e.* it is opposite to phasor $E_{a_1a_2}$.

$$i.e.\ \mathbf{E}_{a_1a_2} + \mathbf{E}_{b_1b_2} + \mathbf{E}_{c_1c_2} = \mathbf{E}_{a_1a_2} - (\mathbf{E}_{a_1a_2}) = 0$$

The result can be established by using complex notation–relations (2.5)

$$\mathbf{E}_{a_1a_2} + \mathbf{E}_{b_1b_2} + \mathbf{E}_{c_1c_2} = E_P + E_P(-0.5 - j\,0.866) + E_P(-0.5 + j\,0.866) = 0$$

The result is also obvious from wave diagram shown in Fig. 2.14 (*b*).

2.10 DOUBLE SUBSCRIPT NOTATION

The solution of problems involving circuits and systems containing a number of voltages and currents is simplified and less susceptible to error if the voltage and current phasors are designated by some systematic notation. The double subscript notation is a very useful concept from this point of view. In this notation two letters are placed at the foot of the symbol for voltage or current, the order in which the subscripts are written, indicates the direction in which the voltage acts or current flows. For example, if a voltage in a circuit acts in such a direction as to cause a current to flow from A to B, the positive direction of voltage is from A to B, and the voltage may be represented by V_{AB} or E_{AB}, the order of the subscripts denoting that the voltage or emf is acting from A to B. If the voltage is indicated by V_{BA} or E_{BA} it means that point B is positive w.r.t. point A (during its positive half cycle) and the voltage causes the current to flow from B to A *i.e.* V_{BA} or E_{BA} indicates that voltage or emf acts in a direction opposite to that in which V_{AB} or E_{AB} acts.

So $\mathbf{V}_{BA} = -\mathbf{V}_{AB}$ ...(2.6)

Similarly I_{AB} indicates that current is flowing from A to B but I_{BA} indicates that current is flowing in direction from B to A *i.e.* $\mathbf{I}_{BA} = -\mathbf{I}_{AB}$

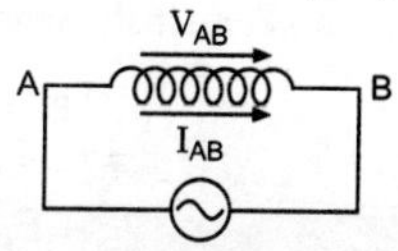

Fig. 2.16

2.11 PHASE SEQUENCE

Phase sequence is the order or sequence in which the currents or voltages in different phases attain their maximum values one after the other.

In Fig. 2.14 (*a*) three coils a_1a_2, b_1b_2 and c_1c_2 rotating in counter-clockwise direction are shown. Since the coil b_1b_2 is 120º behind coil a_1a_2 and coil c_1c_2 is 240º behind coil a_1a_2, therefore, first coil a_1a_2 attains maximum or peak value of induced emf, the coil b_1b_1 attains maximum or peak value of induced emf, when the coils rotate further 120º (electrical) and coil c_1c_2 attains peak value of induced emf when the coils rotate through 240º (electrical). Since the induced emfs in three coils a_1a_2, b_1b_2 and c_1c_2 attain maximum values in order *a*, *b*, *c*, phase sequence is *a b* c.

If the direction of rotation of the coils is reversed *i.e.* clockwise, the three-phase emfs would attain their maximum values in order *a*, *c*, *b* and, therefore, phase sequence would be *acb*. Since the coils can be rotated in clockwise or in counter-clockwise direction, there can be only two possible phase sequences.

The knowledge of phase sequence of a 3-ϕ supply is essential for making connections for alternators and transformers to operate them in parallel. Reversal of phase sequence of a 3-phase alternator which is to be paralleled with another similar alternator may cause extensive damage to both the machines. The direction of rotation of an induction motor depends on the phase sequence. If the phase sequence is reversed by interchanging any two terminals of the 3-ϕ supply, the motor would rotate in the opposite direction.

The phase sequence of the voltage applied to the load, in general, is determined by the order in which three phase lines are connected. In the case of 3-phase unbalanced loads, the effect of reversal of phase sequence, in general, causes a completely different set of values of the currents. So while working on such systems it is essential to specify the phase sequence clearly to avoid unnecessary confusion.

2.12 INTERCONNECTION OF THREE PHASES

Each coil of three phases has two terminals (one 'start' and another 'finish') and if individual phase is connected to a separate load circuit, as illustrated in Fig. 2.17, we get a non-interlinked three-phase, six-wire system. Then, by Ohm's law, the current in individual phase will be

$$I_P = E_P / Z_P \qquad ...(2.7)$$

where I_P is the phase current, E_P is the phase emf or voltage.

In such a system, each circuit will require two conductors, therefore, 6 conductors in all.

Hence such a system will be very complicated and expensive.

In practice non-interlinked 3-phase, 6-wire system is not employed but the phases are interconnected to provide 3 or 4-wire, 3-phase system. The two principal methods of connecting the separate phases of three-phase system are the star (or wye) connections and the delta (or mesh) connections.

The interconnections must be carried out in such a way that if closed circuits are formed the sum of instantaneous emfs in them must be zero in order that there shall be no circulating current in these circuits.

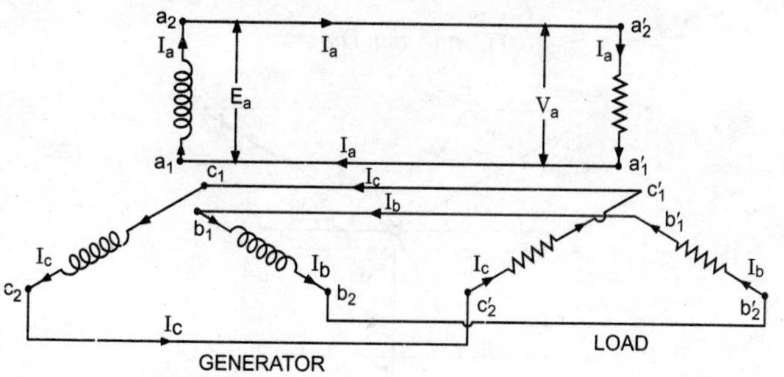

Fig. 2.17 *Non-interlinked Three-Phase System*

2.13 STAR OR WYE (Y)-CONNECTED SYSTEM

This system is obtained by joining together similar ends, either the start or the finish, the other ends are joined to the line wire, as shown in Fig. 2.18 (*a*). The common point N at which similar (start or finish) ends are connected is called the *neutral or star point*. Ordinarily only three wires are carried to the external circuit giving 3-phase 3-wire star-connected system but sometimes a fourth wire, known as *neutral wire* is carried to the neutral point of the external load circuit giving 3-phase, 4-wire star-connected system. The voltage between any line and the neutral point *i.e.* voltage across the phase winding is called the *phase voltage*;

while the voltage between any two outers is called the *line voltage*. The neutral point is usually earth connected.

In Fig. 2.18 (*a*) positive directions of emfs are taken from star point outwards. The arrowheads on emfs and currents indicate the positive direction and not their directions at any instant. The three phases be either numbered as 1, 2, 3, or a, b, c or R, Y, B. In Fig. 2.18 (*a*) the three phases are numbered as usually done; R, Y and B indicate the three natural colours red, yellow and blue respectively.

It is worthwhile to state here that there are only two possible phase sequences, namely RYB and RBY. By convention sequence RYB is taken as positive and RBY as negative.

In Fig. 2.18 (*b*) the emfs induced in 3 phases are shown by phasors. In star connection, there are two windings between each pair of outers and due to joining of similar ends together the emf induced in them are in opposition.

Hence the potential difference between the two outers, known as *line voltage,* is the phasor difference of phase emfs of the two phases concerned.

For example, the potential difference between outers R and Y or line voltage E_{RY} is the phasor difference of phase emfs E_R and E_Y or phasor sum of phase emfs E_R and $(-E_Y)$.

i.e. $\mathbf{E}_{RY} = \mathbf{E}_R - \mathbf{E}_Y$ (phasor difference)

or $\mathbf{E}_{RY} = \mathbf{E}_R + (-\mathbf{E}_Y)$ (phasor sum)

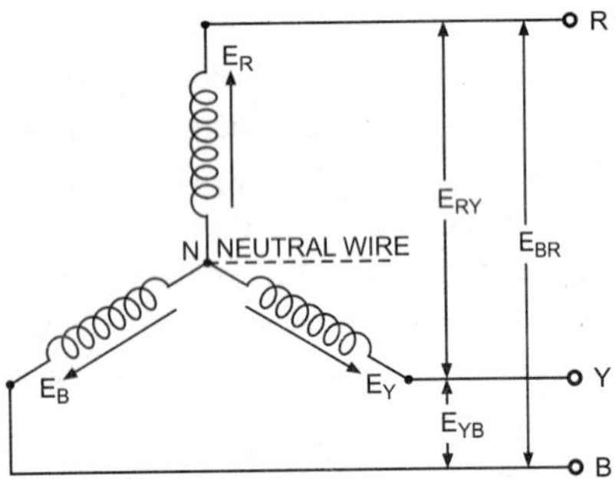

(*a*) *Connection Diagram*

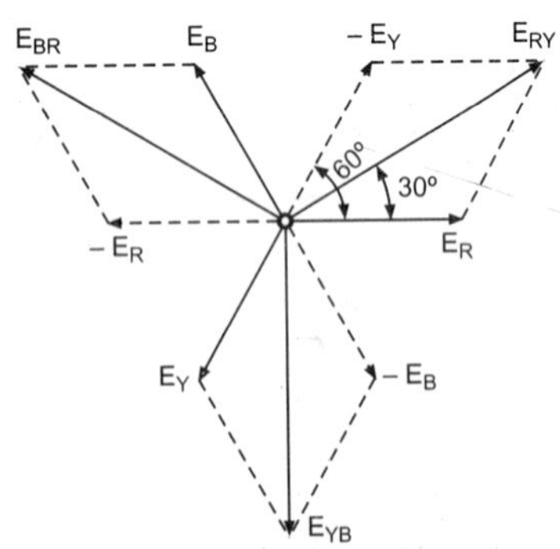

(*b*) *Phasor Diagram of Line and Phase Voltages*

Fig. 2.18 *Three-Phase Star-Connected System*

Since phase angle between phasors E_R and $(-E_Y)$ is 60°, From phasor diagram shown in Fig. 2.18 (*b*)

$$E_{RY} = \sqrt{E_R^2 + E_Y^2 + 2\,E_R\,E_Y \cos 60^\circ}$$

Let $E_R = E_Y = E_B = E_P$ (phase voltage) (Assuming balanced system) then line voltage

$$E_{RY} = \sqrt{E_P^2 + E_P^2 + 2E_P\,E_P \times (1/2)} = \sqrt{3}\,E_P$$

Similarly potential difference between outers Y and B

or line voltage $\mathbf{E}_{YB} = \mathbf{E}_Y - \mathbf{E}_B = \sqrt{3}\ E_P$ in magnitude

and potential difference between outers B and R

or line voltage, $\mathbf{E}_{BR} = \mathbf{E}_B - \mathbf{E}_R = \sqrt{3}\ E_P$ in magnitude

In balanced star (Y) system E_{RY}, E_{YB} and E_{BR} are equal in magnitude and are known as *line voltages*. Let each of them be equal to E_L

$$\therefore\ E_L = \sqrt{3}\ E_P \text{ (in balanced star-connected system)} \quad ...(2.8)$$

Since in star-connected system each line conductor is connected to separate phase, current flowing through the lines and phases are same

i.e. Line current, I_L = Phase current, I_P ...(2.9)

2.13.1. 3-Phase Power. Phase voltages and currents in a balanced three-phase circuit (star or delta) may be written in the instantaneous form as below :

$$e_R = \sqrt{2}\ E_P \sin \omega t \quad ...(2.10\ a)$$

$$e_Y = \sqrt{2}\ E_P \sin (\omega t - 120^\circ) \quad ...(2.10\ b)$$

$$e_B = \sqrt{2}\ E_P \sin (\omega t + 120^\circ) \quad ...(2.10\ c)$$

$$\text{and } i_R = \sqrt{2}\ I_P \sin (\omega t - \phi) \quad ...(2.11\ a)$$

$$i_Y = \sqrt{2}\ I_P \sin (\omega t - \phi - 120^\circ) \quad ...(2.11\ b)$$

$$i_B = \sqrt{2}\ I_P \sin (\omega t - \phi + 120^\circ) \quad ...(2.11\ c)$$

where ϕ is the phase angle between phase voltage and phase current.

Total instantaneous power,

$$\begin{aligned} p &= e_R\, i_R + e_Y\, i_Y + e_B\, i_B \\ &= E_P\, I_P\, [2 \sin \omega t \sin (\omega t - \phi)] \\ &\quad + E_P\, I_P\, [2 \sin (\omega t - 120^\circ) \sin (\omega t - \phi - 120^\circ) \\ &\quad + E_P\, I_P\, [2 \sin (\omega t + 120^\circ) \sin (\omega t - \phi + 120^\circ] \\ &= E_P\, I_P\, [\cos \phi - \cos (2\,\omega t - \phi) + \cos \phi \\ &\quad - \cos (2\,\omega t - \phi - 240^\circ) + \cos \phi - \cos (2\,\omega t - \phi + 240^\circ)] \end{aligned}$$

or Average power,

$$P = 3\, E_P\, I_P \cos \phi \quad ...(2.12)$$

$\because$ the sum of the three second harmonic oscillating terms which have a progressive phase difference of 120° is zero.

Thus the instantaneous power in a 3-phase balanced system is constant and equal to three times the average power per phase. This is in contrast to power in a single-phase system which has a second harmonic oscillating component. Constancy of power in a balanced 3-phase system affords the unique advantage of uniform torque in 3-phase electric machines.

Substituting $E_P = \dfrac{E_L}{\sqrt{3}}$ and $I_P = I_L$ for a 3-phase star-connected balanced system in Eq. (2.12) we have

Total circuit power,

$$P = 3\frac{E_L}{\sqrt{3}} \times I_L \cos \phi = \sqrt{3}\ E_L\, I_L \cos \phi \quad ...(2.13)$$

i.e. Total power,

$$P = \sqrt{3} \times \text{line voltage} \times \text{line current} \times \text{power factor.}$$

Total apparent power,

$$S = 3 \times \text{apparent power per phase}$$

$$= 3\, E_P\, I_P = 3\frac{E_L}{\sqrt{3}}\, I_L = \sqrt{3}\ E_L\, I_L \quad ...(2.14)$$

Total reactive power,

$$Q = 3\, E_P\, I_P \sin\phi = 3 \frac{E_L}{\sqrt{3}} I_L \sin\phi$$

$$= \sqrt{3}\, E_L\, I_L \sin\phi = \frac{P}{\cos\phi} \sin\phi = P \tan\phi \quad ...(2.15)$$

It should be noted that in a balanced star-connected system

(*i*) Line voltages are 120º apart.

(*ii*) Line voltages are 30º ahead of the respective phase voltages.

(*iii*) Line voltages are $\sqrt{3}$ times of phase voltages.

(*iv*) Line currents are equal to phase currents.

(*v*) The angle between line currents and the corresponding line voltages is $(30° \pm \phi)$; + ve for lagging currents and –ve for leading currents.

(*vi*) True power output = $\sqrt{3}\, E_L\, I_L \cos\phi$, where ϕ is the angle between the respective phase current and phase voltage (not between line current and line voltage).

(*vii*) Apparent power = $\sqrt{3}\, E_L I_L$

(*viii*) In balanced system, the potential of neutral or star point is zero because potential at neutral or star point,

$$E_{NR} + E_{NY} + E_{NB} = 0$$

Example 2.1. An rms voltage in a 3-phase star circuit is given by 231 V (ph-N). Write the instantaneous voltage expressions. If the current in each phase lag the corresponding phase voltage by 30º, what are the expressions of instantaneous currents.

[*M.D. Univ. Electrical Technology, May-2010*]

Solution: Let the pd across phase RN be the reference phasor and phase sequence be R Y B. Hence the phase voltage be expressed as

$$V_{RN} = 231\sqrt{2} \sin \omega t = 326.7 \sin \omega t \text{ volts}$$

$$V_{YN} = 326.7 \sin (\omega t - 120°) \text{ volts}$$

and $V_{BN} = 326.7 \sin (\omega t - 240°) = 326.7 \sin (\omega t + 120°)$ volts

As the phase currents lags the corresponding phase voltage by 30º, the phase currents be expressed as

$$I_R = \frac{326.7}{Z} \sin (\omega t - 30°) \text{ amperes}$$

$$I_Y = \frac{326.7}{Z} \sin (\omega t - 30° - 120°) = \frac{326.7}{Z} \sin (\omega t - 150°) \text{ amperes}$$

and $I_B = \frac{326.7}{Z} \sin (\omega t - 30° + 120°) = \frac{326.7}{Z} \sin (\omega t + 90°)$ amperes

where Z is the impedance connected in each phase in ohms.

Example 2.2. A balanced 3-phase star-connected load of 18 kW taking a leading current of 60 amperes when connected across a 3-phase 440 V, 50 Hz supply. Find the values and nature of the load.

[*Electrical Engineering U.P. Technical Univ. Even Semester 2007-08; G.B. Technical Univ. Odd Semester 2012-13*]

Solution: Load supplied, P = 18 kW = 18,000 W

Line voltage, V_L = 440 V

Line current, I_L = 60 A

Phase voltage, $V_P = \frac{V_L}{\sqrt{3}} = \frac{440}{\sqrt{3}} = 254$ V

Phase current, $I_P = I_L = 60$ A

Load impedance, $Z = \frac{V_P}{I_P} = \frac{254}{60} = 4.234\ \Omega$

Load power factor,

$$\cos\phi = \frac{P}{\sqrt{3}\, V_L\, I_L} = \frac{18{,}000}{\sqrt{3} \times 440 \times 60} = 0.39365 \text{ (leading) } \textbf{Ans.}$$

$\because$ Current is leading

Resistance of load,

$R = Z \cos\phi = 4.234 \times 0.39365 = 1.6667\ \Omega$ **Ans.**

Reactance of load,

$$X = Z \sin\phi = 4.234 \sqrt{1 - 0.39365^2}$$

$= 4.234 \times 0.9193 = 3.89242\ \Omega$ (capacitive) **Ans.**

$\because$ Current is leading

Example 2.3. A balanced star-connected load of (8 + *j*6) Ω per phase is connected to a 3-phase, 230 V, 50 Hz supply. Find the line current, pf, power, volt-amperes and reactive power. Draw the phasor diagram for the above circuit.

[*Pb. Technical Univ. Electrical Engineering, June-1999*]

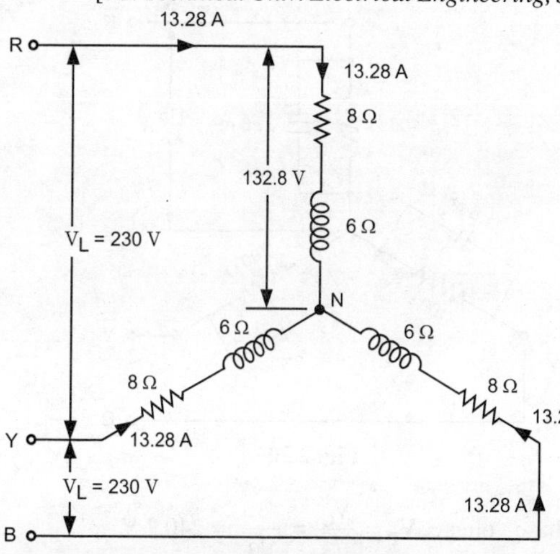

(*a*) *Circuit Diagram*

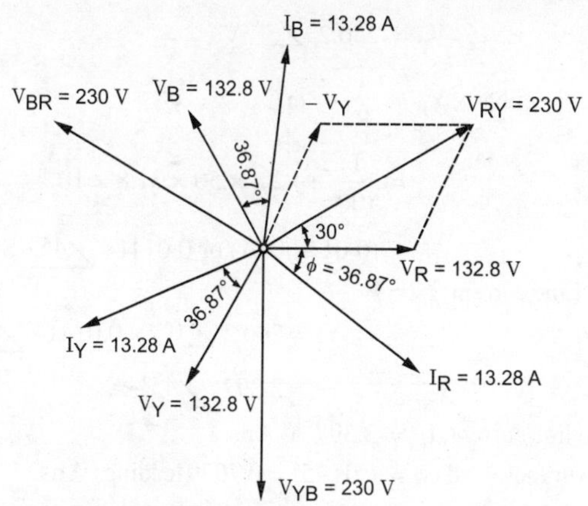

(*b*) *Phasor Diagram For Currents and Voltages*

Fig. 2.19

Solution: Phase voltage, $V_P = \frac{\text{Line voltage, } V_L}{\sqrt{3}} = \frac{230}{\sqrt{3}} = 132.8$ V

Impedance per phase,

$$Z_P = \sqrt{R_P^2 + X_P^2} = \sqrt{8^2 + 6^2} = 10\ \Omega$$

Phase current, $I_P = \frac{V_P}{Z_P} = \frac{132.8}{10} = 13.28$ A

Line current, $I_L = I_P = 13.28$ A **Ans.**

$$\text{Power factor, } \cos\phi = \frac{R_P}{Z_P} = \frac{8}{10} = 0.8 \text{ (lagging)} \quad \textbf{Ans.}$$

$$\text{Power, } P = \sqrt{3}\, V_L I_L \cos\phi = \sqrt{3} \times 230 \times 13.28 \times 0.8$$
$$= 4{,}232 \text{ watts or } 4.232 \text{ kW} \quad \textbf{Ans.}$$

Total volt-amperes,

$$S = \sqrt{3}\, V_L I_L = \sqrt{3} \times 230 \times 13.28 = 5{,}290 \text{ VA} \quad \textbf{Ans.}$$

$$\text{Reactive power, } Q = \sqrt{3}\, V_L I_L \sin\phi$$
$$= \sqrt{3} \times 230 \times 13.28 \sin\cos^{-1}(0.8)$$
$$= \sqrt{3} \times 230 \times 13.28 \times 0.6$$
$$= 3{,}174 \text{ VAR or } 3.174 \text{ kVAR} \quad \textbf{Ans.}$$

Phasor diagram for currents and voltages is shown in Fig. 2.19 (*b*).

Example 2.4. Each phase of a star-connected load consists of a resistance of 100 Ω in parallel with a capacitance 31.8 μF. Calculate the line current, power absorbed, total kVA and power factor of the load when connected to a 416 V, 3ϕ, 4 wire 50 Hz supply.

[*Pb. Technical Univ. Basic Electrical Engineering, December-2003*]

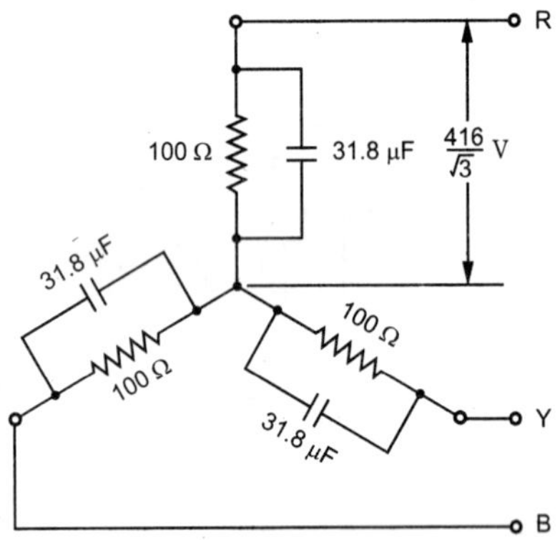

Fig. 2.20

Solution: Phase voltage, $V_P = \frac{V_L}{\sqrt{3}} = \frac{416}{\sqrt{3}} = 240.2$ V

Taking phase voltage as reference phasor,

$$\mathbf{V_P} = 240.2 \angle 0^\circ \text{ V}$$

$$\text{Admittance per phase, } \mathbf{Y_P} = \frac{1}{R} + j\,\omega C$$
$$= \frac{1}{100} + j\,2\,\pi \times 50 \times 31.8 \times 10^{-6}$$
$$= (0.01 + j0.01) \text{ or } 0.01414 \angle 45^\circ \text{ S}$$

$$\text{Line current, } \mathbf{I_L} = \mathbf{I}$$
$$= \mathbf{V_P}\,\mathbf{Y_P} = 240.2 \angle 0^\circ \times 0.01414 \angle 45^\circ$$
$$= 3.397 \angle 45^\circ \text{ A}$$

So line current, I_L = 3.397 A **Ans.**

Power factor, $\cos\phi$ = Cos 45° = 0.707 (leading) **Ans.**

$$\text{Power absorbed, } P = \sqrt{3}\, V_L I_L \cos\phi$$
$$= \sqrt{3} \times 416 \times 3.397 \times 0.707$$
$$= 1{,}730 \text{ watts or } 1.73 \text{ kW} \quad \textbf{Ans.}$$

$$\text{Total kVA, } S = \frac{\sqrt{3} V_L I_L}{1{,}000} = \frac{\sqrt{3} \times 416 \times 3.397}{1{,}000} = 2.45 \quad \textbf{Ans.}$$

Example 2.5. Three-phase star-connected load when supplied from a 400 V, 50 Hz source takes a line current of 10 A at 66.86° w.r.t. to its line voltage. Calculate : (*i*) impedance parameters (*ii*) pf and active power consumed. Draw the phasor diagram.

[*Nagpur University, April-1998*]

Solution: As mentioned in Art 2.13, line voltages are 30° ahead of respective phase voltage *i.e.* phase voltages lag behind respective line voltages by 30°. It is also illustrated in phasor diagram shown in Fig. 2.21. Thus phase angle between phase current and its phase voltage is (66.86° – 30°) = 36.86° (lagging). This is because a current differing in phase with respect to its line voltage by 66.86° can only be lagging.

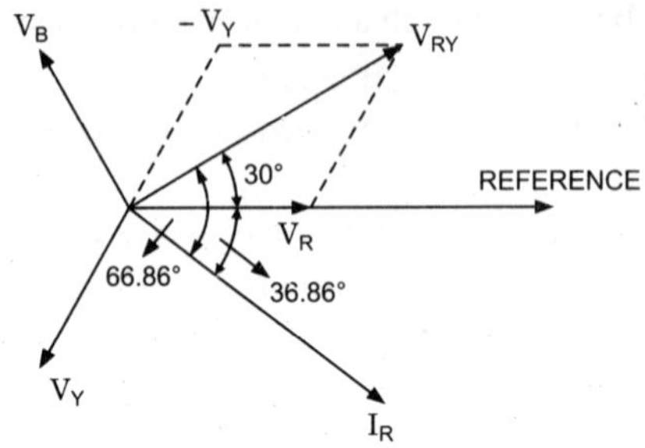

Fig. 2.21

$$(i) \text{ Impedance } Z = \frac{V_P}{I_P} = \frac{V_L/\sqrt{3}}{I_L} = \frac{400}{\sqrt{3} \times 10} = 23.1\ \Omega \quad \textbf{Ans.}$$

Resistance R = Z cos ϕ = 23.1 cos 36.86° = 18.48 Ω **Ans.**

Reactance X_L = Z sin ϕ = 23.1 sin 36.86° = 13.86 Ω **Ans.**

(*ii*) Power factor = Cos ϕ = Cos 36.86° = 0.8 (lagging) **Ans.**

Total active power consumed

$$= \sqrt{3}\, V_L I_L \cos\phi = \sqrt{3} \times 400 \times 10 \times 0.8$$
$$= 5{,}544 \text{ W or } 5.544 \text{ kW} \quad \textbf{Ans.}$$

Example 2.6. A balanced star-connected load is supplied from a 3-phase 50 Hz, 415 V supply. Current in each phase is $20 \angle -30^\circ$ A, the 30° angle being w.r.t. phase voltage. Determine the total power supplied, (ii) the overall power factor, (iii) Draw the phasor diagram.

[*U.P.S.C. I.E.S. Electrical Engineering-I, 2013*]

Solution: Line voltage, V_L = 415 V

Line current, I_L = Phase current, I_P = 20 A

Phase angle, ϕ = –30° or 30° (lagging)

$$\text{(i) Total power supplied, } P = \sqrt{3}\, V_L I_L \cos\phi$$
$$= \sqrt{3} \times 415 \times 20 \times \cos 30^\circ$$
$$= 12{,}450 \text{ W or } 12.45 \text{ kW} \quad \textbf{Ans.}$$

(ii) Overall power factor = Cos ϕ = Cos 30° = 0.866 (lagging)

(iii) Phasor diagram is shown in Fig. 2.22.

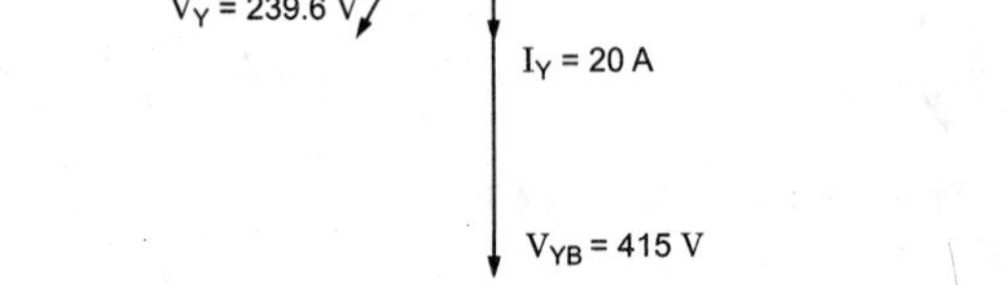

Fig. 2.22 *Phasor Diagram*

2.14 MESH OR DELTA (Δ)-CONNECTED SYSTEM

When the starting end of one coil is connected to the finishing end of another coil, as shown in Fig. 2.23 (*a*), the *delta or mesh connection is obtained*. The direction of emfs in the coils have been taken as positive from start end to finish end in Fig. 2.23 (*a*).

From Fig. 2.23 (*b*) it is obvious that line current is phasor difference of phase currents of two phases concerned. For example the line current in red outer I_R will be equal to the phasor difference of phase currents I_{YR} and I_{RB}. The current phasors are shown in Fig. 2.23 (*b*).

From Figs. 2.23 (*a*) and 2.23 (*b*) we have

Line current, $\mathbf{I_R} = \mathbf{I_{YR}} - \mathbf{I_{RB}} = \mathbf{I_{YR}} + (-\mathbf{I_{RB}})$

Since phase angle between phase current phasors $\mathbf{I_{YR}}$ and $-\mathbf{I_{RB}}$ is 60°,

$$I_R = \sqrt{I^2_{YR} + I^2_{RB} + 2\, I_{YR}\, I_{RB} \cos 60°}$$

Assuming balanced load

The phase current in each winding is equal and let each be equal to I_P

$$\therefore \text{Line current } I_R = \sqrt{I^2_P + I^2_P + 2\, I_P\, I_P \times \frac{1}{2}} = \sqrt{3}\, I_P \quad ...(2.16)$$

Similarly line current $\mathbf{I_Y} = \mathbf{I_{BY}} - \mathbf{I_{YR}} = \sqrt{3}\, I_P$ in magnitude

and line current $\mathbf{I_B} = \mathbf{I_{RB}} - \mathbf{I_{BY}} = \sqrt{3}\, I_P$ in magnitude

Since in delta-connected system, only one phase is included between any pair of line outers, potential difference between the line outers, called the *line voltage*, is equal to phase voltage.

i.e. Line voltage, E_L = Phase voltage, E_P ...(2.17)

Power output per phase = $E_P\, I_P \cos\phi$ where cos φ is the power factor of the load

Total power output = $3\, E_P\, I_P \cos\phi$

$$= 3\, E_L \frac{I_L}{\sqrt{3}} \cos\phi = \sqrt{3}\, E_L\, I_L \cos\phi \quad ...(2.18)$$

i.e. Total power output = $\sqrt{3}$ × line voltage × line current × power factor

Apparent power of 3-phase delta-connected system

= 3 × apparent power per phase

$$= 3\, E_P\, I_P = 3\, E_L \frac{I_L}{\sqrt{3}} = \sqrt{3}\, E_L\, I_L \quad ...(2.19)$$

Total reactive power, Q = $3\, E_P\, I_P \sin\phi$

$$= \sqrt{3}\, E_L\, I_L \sin\phi = P \tan\phi \quad ...(2.20)$$

It should be noted that in balanced delta-connected system

(*i*) Line currents are 120° apart.

(*ii*) Line currents are 30° behind the respective phase currents.

(*iii*) Line currents are $\sqrt{3}$ times phase currents.

(*iv*) The angle between line currents and corresponding line voltage is (30° ± φ) as in star system.

(*v*) Line voltages are equal to phase voltages.

(*vi*) Power = $\sqrt{3}\, E_L\, I_L \cos\phi$, where φ is the phase angle between respective phase current and phase voltage.

(*vii*) Apparent power = $\sqrt{3}\, E_L\, I_L$

(*viii*) In balanced system the resultant emf in the closed circuit will be zero.

i.e. $\mathbf{E_{RY}} + \mathbf{E_{YB}} + \mathbf{E_{BR}} = 0$

Hence there will be no circulating current in the mesh if no load is connected to the lines.

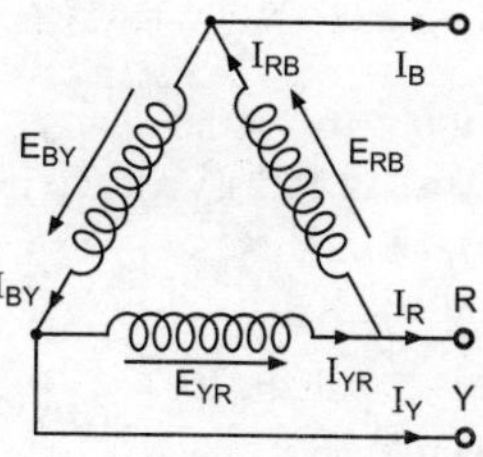

(*a*) Connection Diagram

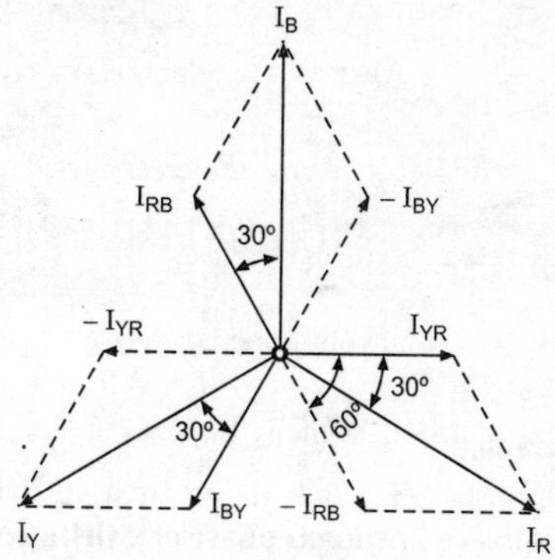

(*b*) Phasor Diagram

Fig. 2.23 *Three-Phase Delta-Connected System*

Example 2.7. When the balanced impedances are connected to a delta across 3-φ, 500 V, 50 Hz supply, the line current drawn is 20 A at 0.3 pf lagging. Calculate resistance and inductance per phase.

[*M.D. Univ. Electrical Technology December-2007*]

Solution: Phase voltage, $V_P = V_L = 500$ V

$$\text{Phase current, } I_P = \frac{I_L}{\sqrt{3}} = \frac{20}{\sqrt{3}} = 11.547 \text{ A}$$

Power factor, cos φ = 0.3 (lagging)

$$\text{Impedance per phase, } Z_P = \frac{V_P}{I_P} = \frac{500}{11.547} = 43.3\ \Omega$$

Resistance per phase, $R_P = Z_P \cos\phi = 43.3 \times 0.3 = 12.99\ \Omega$ **Ans.**

Inductive reactance per phase,

$$X_P = \sqrt{Z_P^2 - R_P^2} = \sqrt{(43.3)^2 - (12.99)^2}$$
$$= 41.3\ \Omega$$

$$\text{Inductance per phase, } L = \frac{X_P}{2\pi f} = \frac{41.3}{2\pi \times 50} = 0.1315 \text{ H} \quad \textbf{Ans.}$$

Example 2.8. A balanced delta connected load of (12 + *j*9) Ω/phase is connected to 3-phase 400 V supply. Find:

(*i*) Line current (*ii*) Power factor (*iii*) Power drawn (*iv*) Reactive Volt-Amperes (*v*) Total Volt-Amperes.

[*Electrical Engineering Mahamaya Technical Univ. Second Semester 2010-11; U.P. Technical Univ. Second Semester 2013-14*]

Solution: Circuit resistance, R = 12 Ω

Circuit inductive reactance, X_L = 9 Ω

$$\text{Circuit impedance, } Z = \sqrt{R^2 + X_L^2} = \sqrt{12^2 + 9^2} = 15\ \Omega$$

Phase angle of the circuit,

$$\phi = \text{Tan}^{-1}\frac{X_L}{R} = \text{Tan}^{-1}\frac{9}{12} = 36.87° \text{ (lagging)}$$

$$\text{Phase current, } I_P = \frac{V_P}{Z} = \frac{V_L}{Z} = \frac{400}{15} = 26.667 \text{ A}$$

∵ in Δ system, $V_P = V_L$

(*i*) Line current,

$$I_L = \sqrt{3}\, I_P = \sqrt{3} \times 26.667 = 46.2 \text{ A} \quad \textbf{Ans.}$$

(*ii*) Power factor,

cos φ = Cos 36.87° = 0.8 (lagging) **Ans.**

(*iii*) Power drawn,

$$P = 3\,I_P^2\,R = 3 \times 26.667^2 \times 12$$
$$= 25{,}600 \text{ W or } 25.6 \text{ kW} \quad \textbf{Ans.}$$

$$\text{Also, } P = \sqrt{3}\,VI\cos\phi = \sqrt{3} \times 400 \times 46.2 \times 0.8$$
$$= 25{,}600 \text{ W} \quad \text{Same as above}$$

(*iv*) Reactive volt amperes,

$$Q = \sqrt{3}\,VI\sin\phi = \sqrt{3} \times 400 \times 46.2 \times 0.6$$
$$= 19{,}200 \text{ VAR or } 19.2 \text{ kVAR} \quad \textbf{Ans.}$$

(*v*) Total volt-amperes,

$$S = \sqrt{3} \times VI = \sqrt{3} \times 400 \times 46.2$$
$$= 32{,}000 \text{ VA} \quad \text{or} \quad 32 \text{ kVA} \quad \textbf{Ans.}$$

Example 2.9. Each phase of a delta-connected load has a resistance of 25 Ω, an inductance of 0.15 H and a capacitance of 120 μF in series. The load is connected across a 400 V, 50 Hz, 3-phase supply. Determine the line current, active power and reactive volt-amperes.

[*U.P. Technical Univ. Electrical Engineering, July-2002*]

Solution: Resistance per phase, $R_P = 25\ \Omega$

Inductive reactance per phase, $X_{LP} = 2\pi f L = 2\pi \times 50 \times 0.15 = 47.124\ \Omega$

Capacitive reactance per phase,

$$X_{CP} = \frac{1}{2\pi f C} = \frac{1}{2\pi \times 50 \times 120 \times 10^{-6}} = 26.526\ \Omega$$

Net reactance per phase,

$$X_p = X_{LP} - X_{CP} = 47.124 - 26.526 = 20.6\ \Omega$$

Impedance per phase

$$= Z_P = \sqrt{R_P^2 + X_P^2} = \sqrt{25^2 + 20.6^2} = 32.4\ \Omega$$

$$\text{Phase angle, } \phi = \text{Tan}^{-1}\frac{X_p}{R_p} = \text{Tan}^{-1}\frac{20.6}{25} = 39.5^\circ \text{ (lag)}$$

$$\text{Phase current, } I_P = \frac{V_P}{Z_P} = \frac{V_L}{Z_P} = \frac{400}{32.4} = 12.35 \text{ A}$$

$$\text{Line current, } I_L = \sqrt{3}\,I_p = \sqrt{3} \times 12.35 = 21.38 \text{ A} \quad \textbf{Ans.}$$

$$\text{Active power, } P = \sqrt{3}\,V_L I_L \cos\phi$$
$$= \sqrt{3} \times 400 \times 21.38 \times \cos 39.5^\circ$$
$$= 11{,}430 \text{ W or } 11.43 \text{ kW} \quad \textbf{Ans.}$$

Reactive volt-amperes,

$$Q = \sqrt{3}\,V_L I_L \sin\phi = \sqrt{3} \times 400 \times 21.38 \times \sin 39.5^\circ$$
$$= 9{,}420 \text{ VAR or } 9.42 \text{ kVAR} \quad \textbf{Ans.}$$

2.15 COMPARISON BETWEEN STAR AND DELTA SYSTEMS

1. In star-connected system similar ends are joined together whereas in delta-connected system dissimilar ends are joined together.
2. In a star-connected system phase voltage is equal to $\frac{1}{\sqrt{3}}$ times the line voltage, but in delta-connected system phase voltage is equal to line voltage, therefore, star-connected alternator needs less number of turns per phase and less insulation than a delta-connected alternator for the same line voltage. These are the reasons that 3-phase alternators are usually star-connected.
3. In star-connected system, phase current is equal to line current *i.e.* $I_P = I_L$ but in delta-connected system, the phase current is $\frac{1}{\sqrt{3}}$ times of line current *i.e.* $I_P = \frac{I_L}{\sqrt{3}}$
4. It is possible to carry neutral wire to the load in star-connected system only *i.e.* only star-connected system can give 3-phase 4-wire arrangement and hence only star-connected system can be used for lighting as well as power loads *i.e.* for an unbalanced load. Lighting loads are connected between neutral and on any one of the outers and power circuits can be connected to the lines.
5. Neutral point of a star-connected system can be connected to earth, so relays and protective devices can be provided in star-connected system for protection of the system against ground faults and it is usually done in practice.
6. Delta connections are mostly used in transformers for operating small low-voltage 3-phase motors and are best suited for rotary convertors. Most of the three-phase induction motors are delta-connected, 3-phase loads are also usually delta-connected as it provides flexibility for addition or removal of a load on a single phase in case of an unbalanced load.

2.16 CONVERSION OF BALANCED LOAD SYSTEM FROM STAR TO DELTA AND VICE VERSA

Any balanced star-connected system can be completely replaced by an equivalent delta-connected system or vice versa because of their relationships between phase and line voltages and currents. For example, a balanced star-connected load having an impedance of magnitude Z with a power factor cos ϕ (or $Z\angle\phi$) in each phase can be replaced by an equivalent delta-connected load having an impedance of magnitude 3Z and power factor cos ϕ (*i.e.* $3Z\angle\phi$) in each phase.

This may be established as follows:

For a balanced star-connected load, let the line voltage be V_L, line current be I_L and impedance per phase be Z

$$\text{Phase voltage, } V_P = \frac{V_L}{\sqrt{3}}$$

$$\text{Phase current, } I_P = I_L$$

$$\text{So impedance per phase, } Z_S = \frac{V_P}{I_P} = \frac{V_L/\sqrt{3}}{I_L} = \frac{1}{\sqrt{3}}\frac{V_L}{I_L} \quad ...(2.21)$$

Now in equivalent delta-connected system for the same line values of voltage and current as in case of star-connected system

$$\text{Phase voltage, } V_P = V_L$$

$$\text{Phase current, } I_P = \frac{I_L}{\sqrt{3}}$$

$$\text{So impedance per phase, } Z_D = \frac{V_P}{I_P} = \frac{V_L}{I_L/\sqrt{3}} = \sqrt{3}\frac{V_L}{I_L} \quad ...(2.22)$$

Comparing Eqs. (2.21) and (2.22) we have

$$Z_S = \frac{1}{3}Z_D \text{ or } Z_D = 3\,Z_S \quad ...(2.23)$$

2.17 BALANCED PARALLEL LOADS

A combination of balanced 3-ϕ loads connected in parallel may be solved by any one of the following methods.

1. All the given loads may be converted into either equivalent Y or Δ-connected loads and then combined together according to the law governing parallel circuits.
2. Alternative method is of working out volt-amperes. The real power and reactive power of various loads may be added arithmetically and algebraically respectively in order to give total volt-amperes according to expression

$$VA = \sqrt{W^2 + VAR^2}$$

$$\text{or } S = \sqrt{P^2 + Q^2}$$

where P is the power in watts (or kW), Q is the reactive power in reactive volt-amperes (or kVAR) and S is the volt-amperes (or kVA)

Example 2.10. A three-phase, three-wire CBA system, with an effective line voltage 200 V, has balanced delta-connected load with impedance $Z = 15\angle 30°\ \Omega$. Obtain the line and phase currents by the single line equivalent method.

[U.P.S.C. I.E.S. Electrical Engineering-I, 2008]

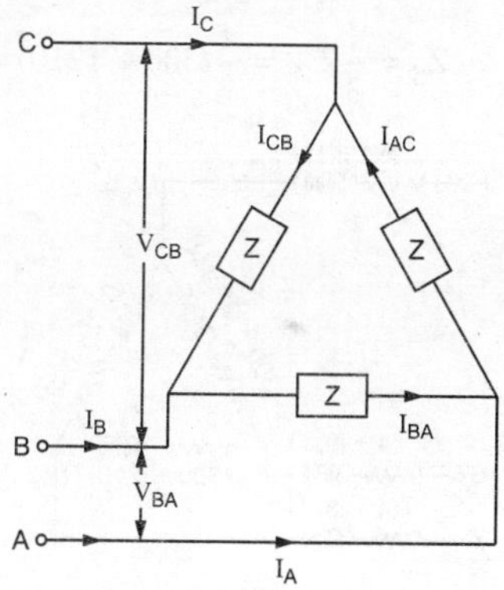

Fig. 2.24

Solution: Taking V_{CB} as the reference phasor we have

$$\mathbf{V}_{CB} = 200\angle 0°\ V$$

$$\mathbf{V}_{BA} = 200\angle -120°\ V$$

$$\mathbf{V}_{AC} = 200\angle 120°\ V$$

Load impedance,

$$Z = 15\angle 30°\ \Omega$$

Different phase currents are:

$$\mathbf{I}_{CB} = \frac{\mathbf{V}_{CB}}{\mathbf{Z}} = \frac{200\angle 0°}{15\angle 30°} = 13.33\angle -30°\ A \quad \textbf{Ans.}$$

$$\mathbf{I}_{BA} = \frac{\mathbf{V}_{BA}}{\mathbf{Z}} = \frac{200\angle -120°}{15\angle 30°} = 13.33\angle -150°\ A \quad \textbf{Ans.}$$

$$\mathbf{I}_{AC} = \frac{\mathbf{V}_{AC}}{\mathbf{Z}} = \frac{200\angle 120°}{15\angle 30°} = 13.33\angle 90°\ A \quad \textbf{Ans.}$$

Different line currents are:

$$\mathbf{I}_A = \mathbf{I}_{AC} - \mathbf{I}_{BA} = (13.33\angle 90° - 13.33\angle -150°)\ A$$
$$= (0 + j\,13.33) - (11.544 - j\,6.665)$$
$$= (11.544 + j\,19.995) = 23.088\angle 60°\ A \quad \textbf{Ans.}$$

$$\mathbf{I}_B = \mathbf{I}_{BA} - \mathbf{I}_{CB} = (13.33\angle -150° - 13.33\angle -30°)\ A$$
$$= 23.088\angle -180°\ A \quad \textbf{Ans.}$$

$$\mathbf{I}_C = \mathbf{I}_{CB} - \mathbf{I}_{AC} = (13.33\angle -30° - 13.33\angle 90°)\ A$$
$$= 23.088\angle -60°\ A \quad \textbf{Ans.}$$

Example 2.11. A 3-phase load consisting of resistance 25 Ω, inductance of 0.15 H and capacitance of 100 μF is connected to 400 V, 50 Hz. Calculate line current, power factor, total power (*i*) when connected in star, (*ii*) when connected in delta.

[U.P. Technical Univ. Odd Semester 2013-14]

Solution: Resistance per phase, $R_p = 25\ \Omega$

Inductive reactance per phase, $X_{Lp} = 2\pi fL = 2\pi \times 50 \times 0.15 = 47.124\ \Omega$

Capacitive reactance per phase,

$$X_{Cp} = \frac{1}{2\pi fC} = \frac{1}{2\pi \times 50 \times 100 \times 10^{-6}} = 31.83\ \Omega$$

Net reactance per phase,

$$X_p = X_{Lp} - X_{Cp} = 47.124 - 31.83 = 15.294\ \Omega$$

Impedance per phase,

$$Z_p = \sqrt{R_p^2 + X_p^2} = \sqrt{25^2 + 15.294^2} = 29.3\ \Omega$$

$$\text{Power factor, } \cos\phi = \frac{R_p}{Z_p} = \frac{25}{29.3} = 0.853 \text{ (lagging)} \quad \textbf{Ans.}$$

(*i*) When connected in star

$$\text{Line current, } I_L = \text{Phase current, } I_P = \frac{V_P}{Z_P} = \frac{400/\sqrt{3}}{29.3}$$
$$= 7.88\ A \quad \textbf{Ans.}$$

$$\text{Total power, } P = \sqrt{3}\, V_L I_L \cos\phi$$
$$= \sqrt{3} \times 400 \times 7.88 \times 0.853$$
$$= 4{,}658\ W \text{ or } 4.658\ kW \quad \textbf{Ans.}$$

(*ii*) When connected in delta

$$\text{Line current, } I_L = \sqrt{3} \times I_P = \sqrt{3} \times \frac{V_P}{Z_P}$$
$$= \sqrt{3} \times \frac{V_L}{Z_P} = \frac{\sqrt{3} \times 400}{29.3} = 23.64\ A \quad \textbf{Ans.}$$

$$\text{Total power, } P = \sqrt{3}\, V_L I_L \cos\phi = \sqrt{3} \times 400 \times 23.64 \times 0.853$$
$$= 13{,}974\ W \text{ or } 13.974\ kW \quad \textbf{Ans.}$$

Note: *For same load in delta-connection line current and power drawn are three times when connected in star.*

Example 2.12. A 415 V, three-phase generator supplies power to both a delta and a star-connected load in the manner shown in Fig. 2.25. All the phase impedances are identical and specifically equal to (5 + *j* 8.66) Ω. Compute the total generator current in each line. *[U.P.S.C. I.E.S. Electrical Engineering-I, 2005]*

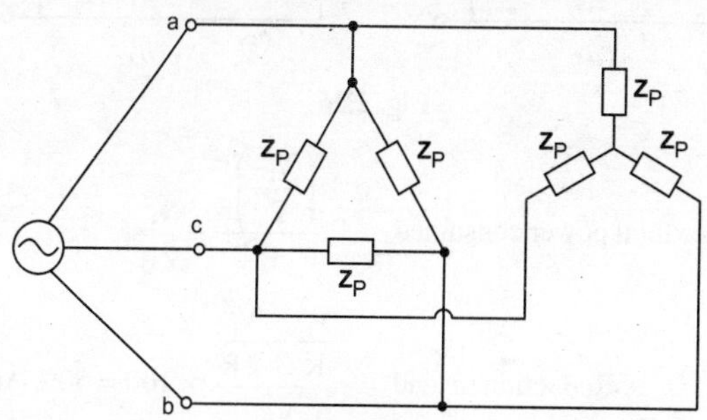

Fig. 2.25

Solution: Line voltage, $V_L = 415$

$$\text{Phase impedance, } \mathbf{Z}_P = (5 + j\,8.66)\ \Omega = 10\angle 60°\ \Omega$$

Line current in delta-connected load,

$$\mathbf{I}_{L\,DELTA} = \sqrt{3}\, \mathbf{I}_{P\,DELTA} = \sqrt{3}\, \frac{\mathbf{V}_P}{\mathbf{Z}_P} = \frac{\sqrt{3} \times 415\angle 0°}{10\angle 60°}$$
$$= 71.88\angle -60°\ A = (35.945 - j\,62.26)\ A$$

Line current in star-connected load,

$$\mathbf{I}_{L\,STAR} = \mathbf{I}_{P\,STAR} = \frac{\mathbf{V}_P}{\mathbf{Z}_P} = \frac{\frac{V_L}{\sqrt{3}}}{Z_P} = \frac{415\angle 0°}{\sqrt{3} \times 10\angle 60°}$$

$= 23.96 \angle -60° \text{ A} = (11.982 - j\,20.75) \text{ A}$

Total line current,

$$\mathbf{I}_L = \mathbf{I}_{L\,DELTA} + \mathbf{I}_{L\,STAR}$$
$$= (35.945 - j\,62.26) + (11.98 - j\,20.75) \text{ A}$$
$$= (47.925 - j\,83.01) = 95.84 \angle -60° \text{ A}$$

Line current in phase *a*,

$$\mathbf{I}_{La} = 95.84 \angle -60° \text{ A} \quad \textbf{Ans.}$$

Line current in phase *b*,

$$\mathbf{I}_{Lb} = 95.84 \angle -60° - 120° = 95.84 \angle -180° \text{ A} \quad \textbf{Ans.}$$

Line current in phase *c*,

$$\mathbf{I}_{Lc} = 95.84 \angle -60° + 120° = 95.84 \angle 60° \text{ A} \quad \textbf{Ans.}$$

Example 2.13. A 3- wire, 3-phase supply feeds a load consisting of three equal resistors. By how much is the load reduced if one of the resistor be removed ? (*i*) When the load is star-connected (*ii*) When the load is delta-connected.

[*R.G.P.V. Bhopal Basic Electrical Engineering, June-2002*]

Solution: Let the line voltage of 3-phase supply be V_L volts and resistance of each phase be of R ohms.

(*i*) **In Star**

$$\text{Power consumed, } P = 3 \times \frac{V_P^2}{R} = 3 \times \left(\frac{V_L}{\sqrt{3}}\right)^2 \times \frac{1}{R} = \frac{V_L^2}{R}$$

After disconnection or removal of one of the resistances, the circuit is no longer 3-phase but converted into single phase circuit, having two resistors, each of R ohms connected in series across supply of V_L volts, as shown in Fig. 2.26(*b*).

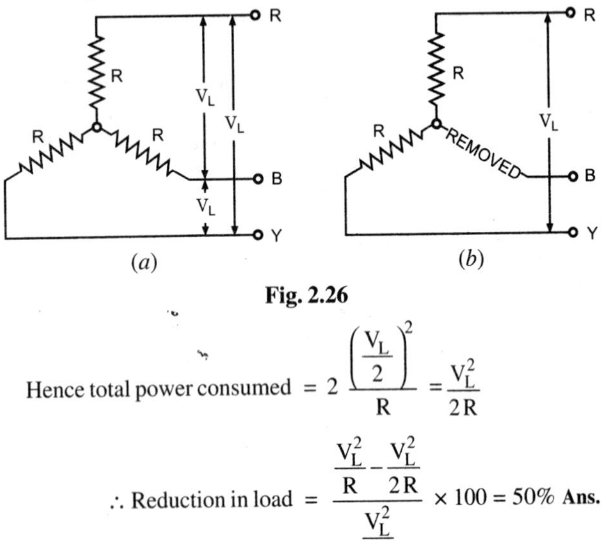

(*a*) (*b*)

Fig. 2.26

$$\text{Hence total power consumed } = 2\,\frac{\left(\dfrac{V_L}{2}\right)^2}{R} = \frac{V_L^2}{2R}$$

$$\therefore \text{ Reduction in load } = \frac{\dfrac{V_L^2}{R} - \dfrac{V_L^2}{2R}}{\dfrac{V_L^2}{R}} \times 100 = 50\% \textbf{ Ans.}$$

(*ii*) **In Delta**

$$\text{Power consumed } = 3 \times \frac{V_P^2}{R} = 3 \times \frac{V_L^2}{R} = 3\,\frac{V_L^2}{R}$$

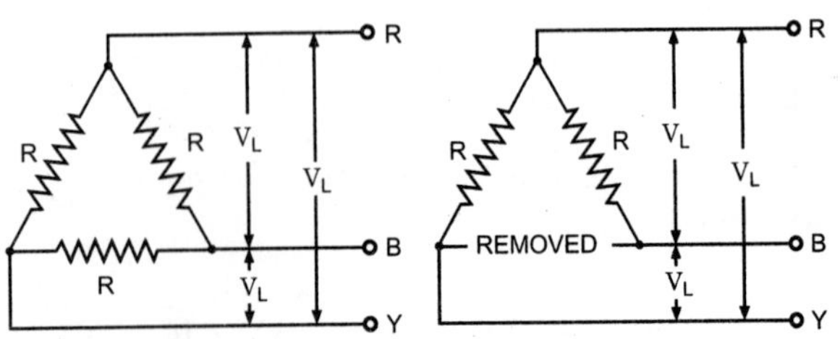

(*a*) (*b*)

Fig. 2.27

After removal of one of the resistance, each of the remaining two resistances is connected across supply of V_L volts as shown in Fig. 2.27(*b*).

$$\therefore \text{ Power consumed } = 2 \times \frac{V_L^2}{R} = \frac{2V_L^2}{R}$$

$$\text{So reduction in load } = \frac{\dfrac{3V_L^2}{R} - \dfrac{2V_L^2}{R}}{\dfrac{3V_L^2}{R}} \times 100 = 33.33\% \textbf{ Ans.}$$

Example 2.14. A balanced delta-connected load having an impedance Z_L = (300 + *j*210) Ω in each phase is supplied from 400 V, 3-φ supply through a 3-φ line having an impedance of Z_S = (4 + *j*8) Ω in each phase. Find the total power supplied to the load as well as the current and voltage in each phase of the load.

[*R.G.T.U. Basic Elec. Engineering, June-2007*]

Solution: Converting delta-connected load into star equivalent, we have

$$Z_S = \frac{1}{3}Z_D = \frac{1}{3}(300 + j210) = (100 + j70)\ \Omega \text{ Refer to Eq. (2.23)}$$

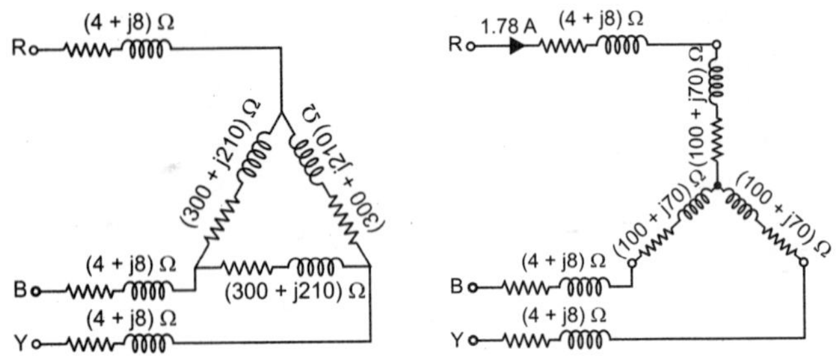

(*a*) (*b*)

Fig. 2.28

Line impedance in each line,

$$Z_{Line} = (4 + j8)$$

Total impedance across each phase,

$$Z_T = Z_{Load} + Z_{Line} = (100 + j\,70) + (4 + j\,8)$$
$$= (104 + j78) = 130\ \Omega$$

Current in each phase of equivalent star-connected load,

$$I_P = I_L = \frac{V_P}{Z_T} = \frac{V_L/\sqrt{3}}{Z_T} = \frac{400/\sqrt{3}}{130} = 1.78 \text{ A}$$

Total power supplied to the load,

$$P = 3I_P^2R = 3 \times (1.78)^2 \times 100 = 947 \text{ W} \quad \textbf{Ans.}$$

Current in each line of delta-connected load

= Line current of equivalent star-connected load = 1.78 A

Current in each each phase of delta-connected load

$$= \frac{1.78}{\sqrt{3}} = 1.03 \text{ A} \quad \textbf{Ans.}$$

Voltage across each phase of delta-connected load

$$= I_P Z_{delta} = 1.03 \times \sqrt{300^2 + 210^2} = 377.2 \text{ V} \quad \textbf{Ans.}$$

2.18 POWER FACTOR IMPROVEMENT

The power factor for lighting and heating loads supplied from 3-phase power supply ranges from 0.95 to unity but the power factor for motor loads ranges from 0.5 to 0.9. Power factor for single-phase motors may be as low as 0.4 (lagging) and for electric welding may be even 0.2 or 0.3 (lagging).

2.18.1. Disadvantages of Low Power Factor. The current for a given load supplied at a constant voltage will be higher at lower power factor and lower at higher power factor. For example if load P is supplied at terminal voltage V and at power factor cos ϕ by a 3-phase balanced system then load current is given by

$$I_L = \frac{P}{\sqrt{3}\, V \cos\phi}$$

If P and V are constant, the load current I_L is inversely proportional to power factor, cos ϕ *i.e.* lower the power factor, higher the current and vice versa. The higher current due to poor power factor affects the system and results in the following disadvantages:

(*i*) Rating of generators and transformers are proportional to their output current hence inversely proportional to power factor, therefore, large generators and transformers are required to deliver same load but at a low power factor.

(*ii*) The x-sectional area of the bus-bar and the contact surface of switchgear is required to be enlarged for the same power to be delivered but at a low power factor.

(*iii*) For the same power to be transmitted but at low power factor, the transmission line or the distributor or cable will have to carry more current. Thus more conductor material is required for transmission lines, distributors and cables to deliver the same load but at a low power factor.

(*iv*) Copper losses are proportional to the square of the current hence inversely proportional to the square of the power factor *i.e.* more copper losses incur at low power factor, which results in poor efficiency.

(*v*) Low lagging power factor results in large voltage drop in generators, transformers, transmission lines and distributors which results in poor regulation. Hence extra regulating equipment is required to keep the voltage drop within permissible limits.

Thus we see that the low power factor leads to a high capital cost for the alternators, switchgears, transformers, transmission lines, distributors and cables etc.

2.18.2. Causes of Low Power Factor

(*i*) All ac motors (except overexcited synchronous motors and certain types of commutator motors) and transformers operate at lagging power factor. The power factor decreases with the decrease in load. For example an induction motor has a reasonable higher power factor of 0.85 at full load, 0.8 at 75% of full load, 0.7 at half-full load, 0.5 at 25% of full load and as low as 0.1 on no load.

(*ii*) Arc lamps operate at low power factor (lagging) due to typical characteristic of the arc.

(*iii*) Industrial heating furnaces, such as arc and induction furnaces, operate on very low lagging power factor.

(*iv*) Due to increase in supply main voltage, which usually occurs during low load periods, such as lunch hours, night hours etc. the magnetizing current of inductive reactances increases and the power factor of the electrical plant as a whole comes down.

2.18.3. Methods of Power Factor Improvement. The following methods are usually adopted for improving the power factor:

(*i*) *By Use of Induction Motors With Phase Advancers.* The power factor of an induction motor falls mainly due to its exciting current drawn from the ac supply mains, because exciting current lags behind the voltage by $\frac{\pi}{2}$. It may be improved by equipping the set with an ac exciter or phase advancer which supplies this exciting current to the motor circuit at slip frequency. Such an exciter may be mounted on the same shaft as the main motor or may be suitably driven from it. Use of phase advancer is not generally economical in conjuction with motors below 150 kW but above this size, phase advancers are frequently employed.

(*ii*) *By Use of Static Capacitors.* Power factor can be improved by connecting the capacitors in parallel with the equipment operating at lagging power factor such as induction motors, fluorescent tubes. Static capacitors have small losses, require almost no maintenance and are very convenient in the small sizes. The current drawn by induction motors and fluorescent tubes can be resolved into two components, the active component which is in phase with the supply voltage and the quadrature or wattless component of constant magnitude. The capacitors draw current leading the supply voltage by $\frac{\pi}{2}$ and neutralize the quadrature or wattless component of current drawn by the equipment across which these are connected. Capacitors can be connected either in star or in delta, as shown in Figs. 2.29 and 2.30 respectively.

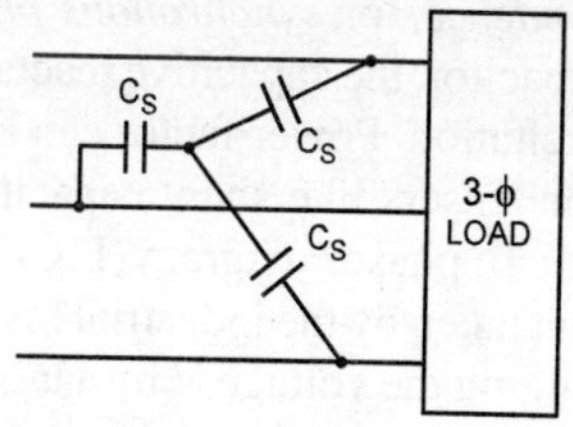

Fig. 2.29 *Star-Connected Capacitors*

These capacitors remain connected permanently in parallel with the equipment and are across the supply mains whenever the equipment is switched on.

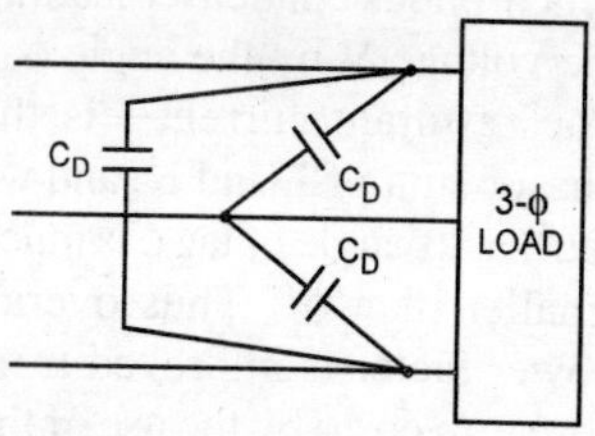

Fig. 2.30 *Delta-Connected Capacitors*

The value of static capacitors for improvement of power factor can be determined as follows:

Let the current drawn by an inductive load be of I amperes lagging behind the applied voltage V by an angle ϕ_1 and pf be improved to cos ϕ_2 (lagging) by connecting a capacitor across the inductive load, as illustrated in Fig. 2.31 (*a*).

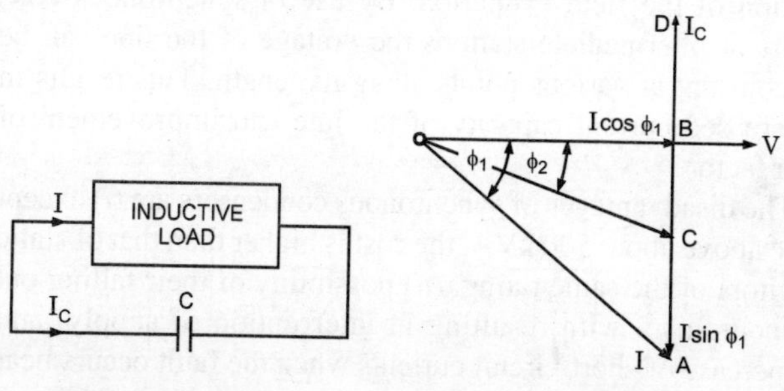

(*a*) *Circuit Diagram* (*b*) *Phasor Diagram*

Fig. 2.31

Active component of current drawn by load = I cos ϕ_1 represented by phasor OB in phasor diagram. [Fig. 2.31(*b*)]

Reactive (lagging) component of current drawn by load $= I \sin \phi_1 = I \cos \phi_1 \times \tan \phi_1$, represented by phasor AB in phasor diagram.

Let the current drawn by capacitor C be of I_C amperes leading the applied voltage V by $\pi/2$, as represented by phasor BD in Fig. 2.31 (*b*).

Lagging component of current drawn from supply mains, as represented by phasor BC in phasor diagram $= OB \tan \phi_2$

$$= I \cos \phi_1 \tan \phi_2.$$

The current drawn by capacitor C, I_C should neutralize the reactive component (lagging), represented by phasor AC in phasor diagram *i.e.* I_C should be equal to current represented by phasor AB less current represented by phasor BC. So

I_C = Current represented by phasor AB – current represented by phasor BC

$$= I \cos \phi_1 \tan \phi_1 - I \cos \phi_1 \tan \phi_2 = I \cos \phi_1 (\tan \phi_1 - \tan \phi_2)$$

The value of capacitance required, $C = \dfrac{I_C}{2\pi f V}$

where V is supply voltage and f is the supply frequency.

(iii) By Use of Synchronous Condensers. An overexcited synchronous motor operating on no load is called the *synchronous condenser* or *synchronous phase advancer* and behaves like a capacitor, the capacitive reactance of which depends on the motor excitation. Power factor can be improved by using synchronous condensers like shunt capacitors connected across the supply.

In phasor diagram (Fig. 2.32), phasor I_L represents the current taken by the industrial load, lagging the voltage V by a large angle ϕ_L and phasor I_M represents the current drawn by the synchronous condenser leading the voltage V by the angle ϕ_M. The resultant current I is the phasor sum of I_L and I_M and we see that its angle of lag ϕ is much smaller than ϕ_L. Thus overall power factor is improved from $\cos \phi_L$ to $\cos \phi$ by the use of the synchronous motor. By this way the power factor can be made unity even.

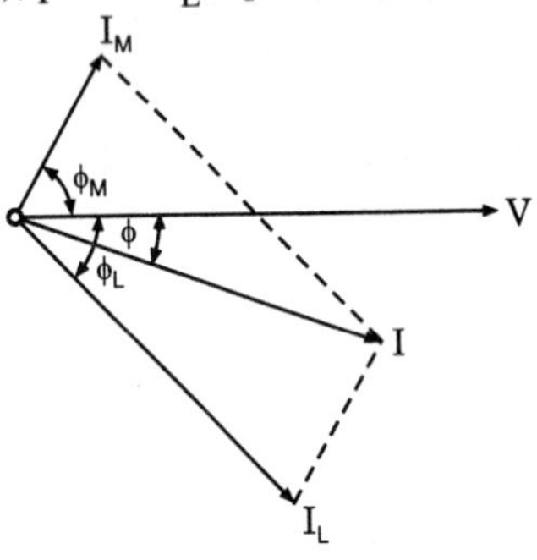

Fig. 2.32

The advantage of synchronous condenser over static capacitor is that power factor can very easily be controlled by variation of the field excitation. By use of synchronous condensers at intermediate stations the voltage of the line can be kept constant at various points along its length. This results in an increased current capacity of the line and improvement of power factor.

The disadvantages of synchronous condensers are (*i*) except in size above about 500 kVA, the cost is higher than that of static capacitors of the same rating, (*ii*) possibility of their falling out of synchronism with resulting in interruption of supply, and (*iii*) increase of short-circuit currents when the fault occurs near the synchronous condenser.

2.18.4. Advantages of Power Factor Improvement. If the power factor of the supply or power station is raised to unity, the current for the same amount of power to be supplied is reduced to minimum. This results in reduction of transmission line copper losses and reduction of voltage drop in transmission line and in alternator windings as copper losses are directly proportional to the square of supply current and voltage drop is directly proportional to the supply current.

The advantages of good power factor are that (*i*) load output of a given plant is better utilized (*ii*) line losses are reduced to minimum, therefore, efficiency of the plant is improved (*iii*) voltage drop or voltage regulation is considerably reduced.

Example 2.15. A 46 mH inductive coil has a resistance of 10 Ω.

(*a*) How much current will it draw if connected across a 100 V, 60 Hz, source?

(*b*) Determine the value of the capacitance that must be connected across the coil to make the power factor of the overall circuit unity.

[*U.P. Technical Univ. Electrical Engineering First Semester 2008-09*]

Solution: (*a*) Resistance of coil,

$$R = 10\ \Omega$$

Inductive reactance of coil,

$$X_L = 2\pi f L = 2\pi \times 60 \times 46 \times 10^{-3} = 17.34\ \Omega$$

Impedance of the circuit,

$$Z = \sqrt{R^2 + X_L^2} = \sqrt{10^2 + 17.34^2} = 20\ \Omega$$

Current drawn, $I = \dfrac{V}{Z} = \dfrac{100}{20} = 5\text{A}$ **Ans.**

(*b*) Phase angle, $\phi = \text{Tan}^{-1} \dfrac{X_L}{R} = \text{Tan}^{-1} \dfrac{17.34}{10} = 60°$ (lagging)

Reactive component of current

$$= I \sin \phi = 5 \sin 60° = 4.33\ \text{A}$$

The power factor of the overall circuit will be unity if the current drawn by the capacitor, connected across the coil, is equal to the reactive (lagging) component of current drawn by coil

i.e. when $2\pi f CV = I \sin \phi$

or $C = \dfrac{4.33}{2\pi \times 60 \times 100} = 114.86\ \mu\text{F}$ **Ans.**

Example 2.16. A circuit consisting of a single resistor R and an inductor L in series is driven by a 25 V_{rms} 50 Hz sinusoidal voltage source. A capacitor is to be placed in parallel with the source to improve the power factor. Given that the average power dissipated in the R is 100 W and the reactive power delivered to the L is 75 VAR, what value of C will yield a 0.9 pf lagging as seen by the source? [*GATE 1998*]

Solution: Total power drawn, P = 100 W

Reactive power drawn, Q_1 = 75 VAR

Improved pf, $\cos \phi_2$ = 0.9 (lagging)

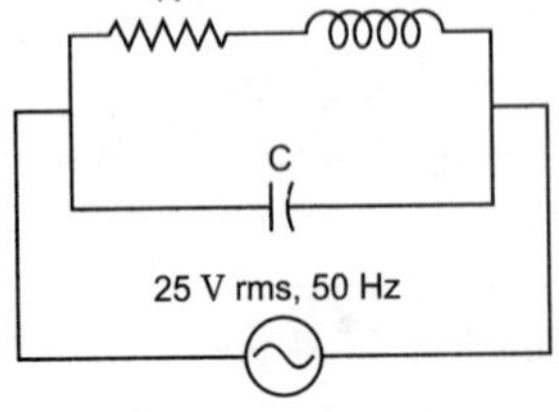

Fig. 2.33

$$\phi_2 = \text{Cos}^{-1}\ 0.9 \text{ (lagging)} = 25.842° \text{ (lagging)}$$

Reactive power drawn with improved pf,

$$Q_2 = P \tan 25.842° = 100 \times 0.4843 = 48.43\ \text{VAR}$$

Leading VAR to be supplied by the capacitor C to be connected across the load for power factor improvement

$$= VI_C = V \times 2\pi f C V = 2\pi f C V^2 = Q_1 - Q_2$$

$$\text{or} \quad C = \frac{Q_1 - Q_2}{2\pi f V^2} = \frac{75 - 48.43}{2\pi \times 50 \times 25^2} = 135.32\ \mu F \ \textbf{Ans.}$$

Example 2.17. An industrial consumer is operating 3-phase, 1.0 kW induction motor at a lagging pf of 0.8 and a source voltage of 400 V_{rms}. He wishes to raise the pf to 0.95 lagging by connecting a circuit element in parallel to the load (motor). Indicate the type of circuit element (inductive or capacitive) and find its value if the operating frequency is 50 Hz. Give your comments on the new line and motor currents. [*U.P.S.C. I.E.S. Electrical Engineering-I, 2005*]

Solution: Motor input, P = 1 kW = 1,000 W

Power factor, $\cos \phi_1 = 0.8$ (lagging),

$\phi_1 = \text{Cos}^{-1}\ 0.8 = 36.87°$ (lagging)

Improved pf, $\cos \phi_2 = 0.95$ (lagging),

$\phi_2 = \text{Cos}^{-1}\ 0.95 = 18.195°$ (lagging)

Reactive power to be supplied by the pf improving device

$$= P(\tan \phi_1 - \tan \phi_2)$$
$$= 1{,}000\ (\tan 36.87° - \tan 18.195°)$$
$$= 1{,}000\ (0.75 - 0.3287) = 421.3 \text{ VAR}$$

Since leading VARs are to be supplied by the power factor improving device, it must be capacitive.

If C is the capacitance of each capacitor then leading VARs supplied will be equal to

$$3 \times 2\pi f V_P C \times V_P = 6\pi f C V_P^2$$
$$= 6\pi f C V_L^2 \text{ (assuming delta-connection)}$$

$$\text{or} \quad C = \frac{421.3}{6\pi f V_L^2} = \frac{421.3}{6\pi \times 50 \times (400)^2} = 2.794\ \mu F \ \textbf{Ans.}$$

Example 2.18. A large consumer of electricity requires 10 kW, 230 V_{rms} of power at a pf angle of 60° lagging through a transmission line of resistance 0.1 Ω. Determine the power produced by the utility to meet the load demand. If the consumer corrects his pf from 0.5 to 0.9 lagging, what power the utility has to produce to meet the load demand? [*U.P.S.C. I.E.S. Electrical Engineering-I, 2004*]

Solution: Load current, $I_L = \dfrac{P \text{ in kW} \times 1{,}000}{V \cos \phi} = \dfrac{10 \times 1{,}000}{230 \cos 60°}$

$= 86.96$ A

Line losses $= I_L^2 R = (86.96)^2 \times 0.1 = 756$ W

Power to be supplied = 10,000 + 756.14

= 10,756 W or 10.756 kW

kVAR to be supplied = Load kVAR = $P \tan \phi$

= 10 tan 60° = 17.32 kVAR

Power produced by the utility,

$S = (10.756 + j\,17.32)$ kVA $= 20.4 \angle 58.16°$ kVA **Ans.**

Improved power factor,

$\cos \phi = 0.9$ lagging,

$\phi = \text{Cos}^{-1}\ 0.9 = 25.84°$ (lagging)

Load current, $I_L' = \dfrac{10 \times 1{,}000}{230 \times 0.9} = 48.31$ A

Lines losses $= (48.31)^2 \times 0.1 = 233.4$ W or 0.2334 kW

Power to be supplied = 10 + 0.2334 = 10.2334 kW

kVAR to be supplied = Load kVAR = 10 tan 25.84° = 4.84 kVAR

Power produced by the utility,

$S' = (10.2334 + j\,4.84) = 11.32 \angle 25.31°$ kVA **Ans.**

Example 2.19. A 3-phase, star-connected motor, connected across a 3-phase, star-connected supply of 400 V, 50 Hz takes a current of 20 A at 0.8 pf lagging. Determine the capacitance of capacitors per phase that are to be connected in delta across the terminals of the motor to raise the pf to unity and the line current with capacitors connected. [*U.P.S.C. I.E.S. Electrical Engineering-I, 2013*]

Solution: Line voltage, $V_L = 400$ V

Line current, $I_L = 20$ A

Power factor, $\cos \phi_1 = 0.8$ (lagging)

Phase angle, $\phi_1 = \text{Cos}^{-1}\ 0.8 = 36.87°$ (lagging)

Phase angle, $\phi_2 = \text{Cos}^{-1}\ 1.0 = 0°$

Power supplied to motor, $P = \sqrt{3}\, V_L I_L \cos \phi$

$= \sqrt{3} \times 400 \times 20 \times \cos 36.87 = 11{,}085$ W

Reactive power to be supplied by capacitors connected in delta across the motor terminals,

$$Q = P(\tan \phi_1 - \tan \phi_2)$$
$$= 11{,}085(\tan 36.87° - \tan 0°)$$
$$= 8{,}313.75 \text{ VAR}$$

If C is the capacitance of each capacitor then leading VARs supplied will be equal to $3 \times 2\pi f V_P C \times V_P = 6\pi f C V_P^2 = 6\pi f C V_L^2$

∵ Capacitors are connected in delta

$$\text{So, } C = \frac{8{,}313.75}{6\pi f V_L^2} = \frac{8{,}313.75}{6 \times \pi \times 50 \times (400)^2}$$
$$= 55.13\ \mu F \ \textbf{Ans.}$$

Line current with capacitors connected across motor terminals,

$I_2 = I_1 \cos \phi_1 = 20 \times 0.8 = 16$ A **Ans.**

2.19 MEASUREMENT OF POWER IN THREE-PHASE CIRCUITS

Wattmeter consists of a low resistance coil, called the *current coil* (*CC*) which is inserted in series with line carrying current to be measured and a high resistance coil, called the *pressure coil* (*PC*), which is connected in series with a high resistance between the two points across which potential difference is to be measured, as shown is Fig. 2.34. The wattmeter gives reading which is proportional to the product of the current flowing through its current coil, potential difference across its pressure coil and cosine of the phase angle between voltage and current.

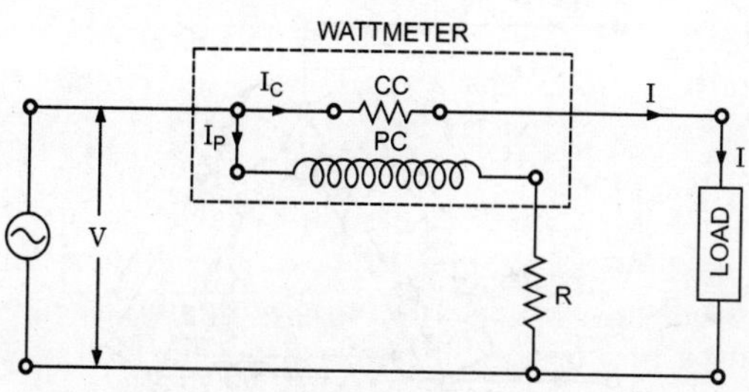

Fig. 2.34 *Wattmeter Connections*

2.19.1. Three-Wattmeter Method. This method is employed for measurement of power in 3-phase, 4-wire circuits, the connections being as shown in Fig. 2.35 (*a*). As the neutral wire is common to the three phases, each wattmeter reads power in its own phase, and the total power of the load circuit is given by the sum of the readings of the three wattmeters

i.e. Total power of load circuit, $P = W_1 + W_2 + W_3$

The above method is only useful for measuring power in 3-phase, 4-wire load circuits. In case of 3-phase, 3-wire star-connected circuits difficulty is experienced in getting neutral. In special cases when it is necessary to employ this method for measurement of power in 3-phase, 3-wire circuits an "artificial star" can be formed by connecting three equal high resistances in star to the three line conductors, as shown in Fig. 2.35 (*b*). (In case of low-voltage circuits three potential coils may be connected to form the common star).

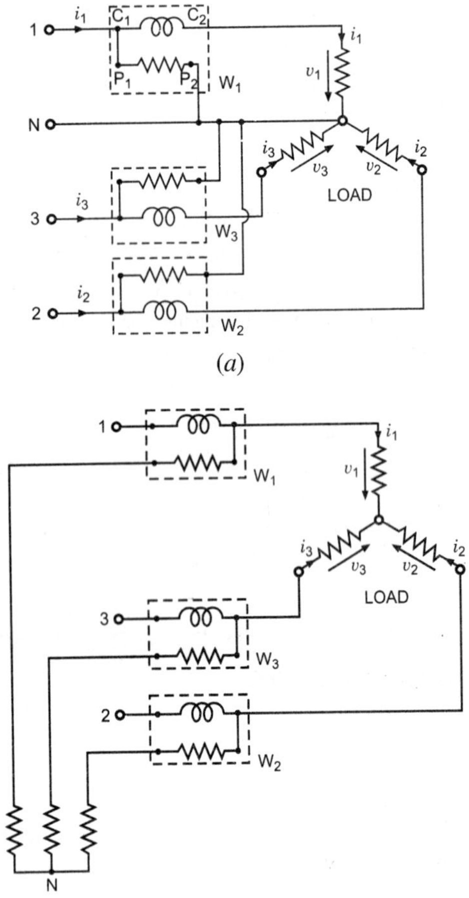

Fig. 2.35

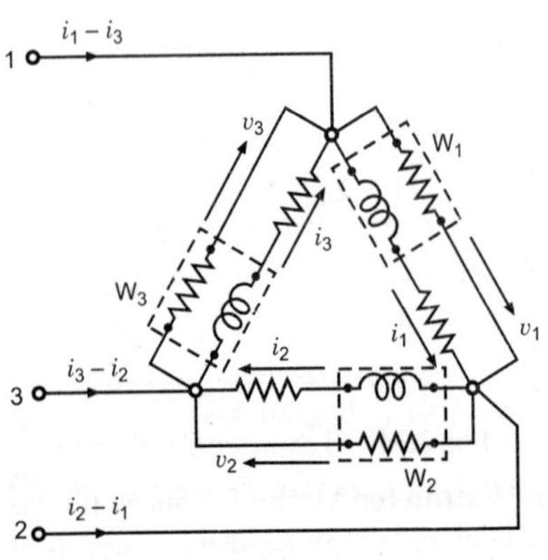

Fig. 2.35 *Three-Wattmeter Method of Measuring 3-Phase Power*

In case of delta-connected circuits the difficulty in adopting the above mentioned method for measurement of power is due to the fact that the phase coils are required to be broken for inserting current coils of wattmeters, as shown in Fig. 2.35 (*c*).

2.19.2. One-Wattmeter Method. In a balanced 3-wire, 3-phase load circuit the power in each phase is equal and, therefore, total power of the circuit can be determined by multiplying the power measured in any one phase by three.

Figure 2.36 (*a*) shows connections for measurement of power in one phase of the 3-phase, 3-wire star-connected balanced load circuit. Total power of the load circuit will be 3 times the reading of wattmeter.

Figure 2.36 (*b*) shows connections for measurement of power in 3-phase, 3-wire delta-connected balanced load circuits by one-wattmeter method. In this method of measurement of power, resistance coils of resistance equal to that of pressure coil is connected in each of remaining two phases, as shown in Fig. 2.36 (*b*). The pressure coil and resistance coils form a balanced star-connection.

It should be noted that the one-wattmeter method has the disadvantage that even a slight degree of unbalance in the loading produce a large error in the measurement.

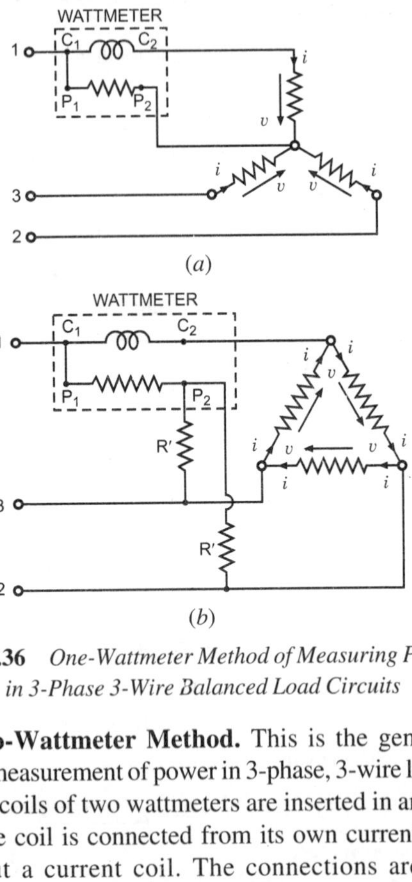

Fig. 2.36 *One-Wattmeter Method of Measuring Power in 3-Phase 3-Wire Balanced Load Circuits*

2.19.3. Two-Wattmeter Method. This is the generally used method for measurement of power in 3-phase, 3-wire load circuits. The current coils of two wattmeters are inserted in any two lines and pressure coil is connected from its own current coil to the line without a current coil. The connections are shown in Fig. 2.37.

Let v_1, v_2, v_3 and i_1, i_2, i_3 be the voltages and currents of the three loads connected across three different phases at any particular instant. These being instantaneous values, the power at the instant under consideration is equal to the sum of their products regardless of power factor

i.e. instantaneous power,

$$p = v_1 i_1 + v_2 i_2 + v_3 i_3 \text{ watts} \quad ...(2.24)$$

(*a*) **Star-Connected System.** Since all the three phases meet at a star point so according to Kirchhoff's first law, the algebraic sum of three instantaneous currents is zero.

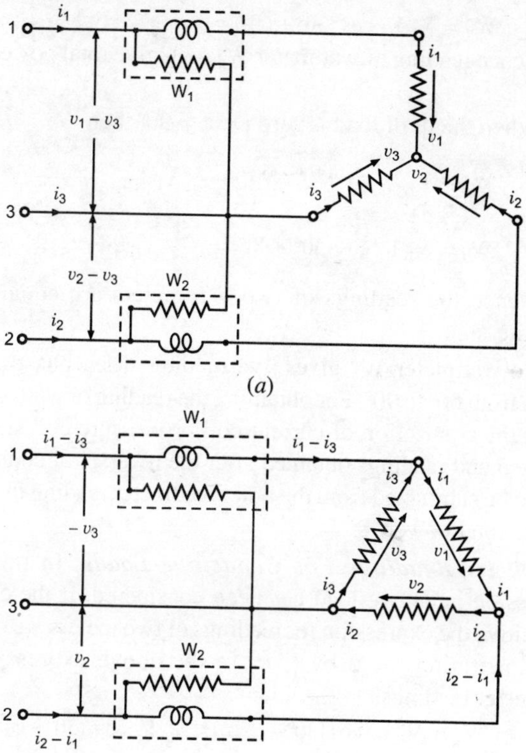

Fig. 2.37 *Two-Wattmeter Method of Measuring 3-Phase 3-Wire Power*

$$i.e.\ i_1 + i_2 + i_3 = 0 \text{ or } i_3 = -(i_1 + i_2)$$

Substituting $i_3 = -(i_1 + i_2)$ in Eq. (2.24) we get

$$\text{Instantaneous power, } p = v_1 i_1 + v_2 i_2 - v_3 (i_1 + i_2)$$
$$= i_1 (v_1 - v_3) + i_2 (v_2 - v_3)$$

Since i_1 is the instantaneous current flowing through the current coil and $(v_1 - v_3)$ is the instantaneous potential difference across pressure coil of wattmeter W_1, $(v_1 - v_3)\, i_1 = w_1$, the instantaneous power measured by wattmeter W_1.

Similarly i_2 is the instantaneous current flowing through the current coil and $(v_2 - v_3)$ is the instantaneous potential difference across pressure coil of wattmeter W_2, therefore, $(v_2 - v_3)\, i_2 = w_2$ the instantaneous power measured by wattmeter W_2.

$$\text{Hence total instantaneous power, } p = w_1 + w_2$$
$$\text{or Total average power, } P = W_1 + W_2 \quad ...(2.25)$$

Hence the algebraic sum of two wattmeter readings gives the total power in the 3-phase, 3-wire star-connected load circuits whether the load is balanced or unbalanced.

(*b*) **Delta-Connected System.** In delta-connected system the three phases form a closed loop and according to Kirchhoff's second law.

$$v_1 + v_2 + v_3 = 0 \text{ or } v_1 = -(v_2 + v_3)$$

$$\text{Instantaneous power, } p = v_1 i_1 + v_2 i_2 + v_3 i_3$$
$$= -(v_2 + v_3)\, i_1 + v_2 i_2 + v_3 i_3$$
$$= -v_3 (i_1 - i_3) + v_2 (i_2 - i_1) \quad ...(2.26)$$

Since $-v_3$ is the instantaneous pd across pressure coil and $(i_1 - i_3)$ is the instantaneous current flowing through current coil of wattmeter W_1, wattmeter W_1 reads average of $-v_3 (i_1 - i_3)$ and similarly wattmeter W_2 reads average of $v_2 (i_2 - i_1)$

$$\text{Hence total power, } P = W_1 + W_2 \quad ...(2.27)$$

Hence the algebraic sum of two wattmeter readings gives the total power of the circuit irrespective of the fact that the circuit is balanced or unbalanced and star-connected or delta-connected.

Determination of Power Factor From Wattmeter Readings. If load is balanced then pf of the load can also be determined from the wattmeter readings.

The phasor diagram for a balanced star-connected inductive load is shown in Fig. 2.38. Let V_1, V_2, and V_3 be the rms values of phase voltages and I_1, I_2 and I_3 be the rms values of phase currents.

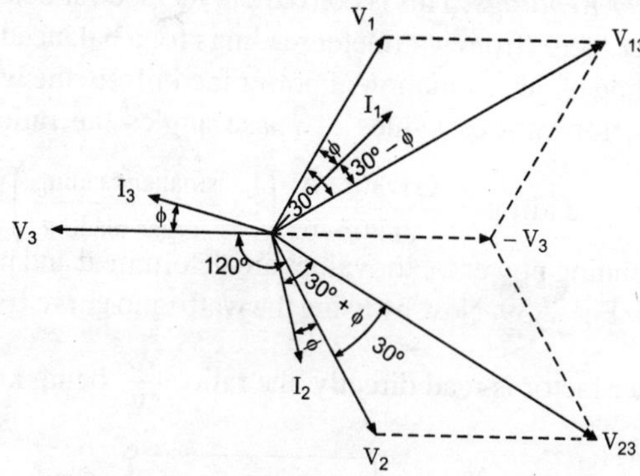

Fig. 2.38 *Phasor Diagram For Balanced Star-connected Inductive Load*

Since load is balanced,

(*i*) Phase voltages V_1, V_2 and V_3 will be equal (say, equal to V_P)

(*ii*) Phase currents I_1, I_2 and I_3 will be equal (say, equal to I_P or I_L)

(*iii*) Phase angles between respective phase voltages and phase currents will be equal, say ϕ

The current in current coil of wattmeter $W_1 = I_1 = I_L$ lagging behind V_1 by ϕ.

The pd across pressure coil of wattmeter $W_1 = V_{13} = \sqrt{3}\ V_P = V_L$ lagging behind V_1 by 30°.

Therefore, phase angle between voltage across potential coil and current through current coil of wattmeter W_1 is $(30° - \phi)$ and reading of wattmeter

$$W_1 = V_L I_L \cos (30° - \phi) \quad ...(2.28)$$

The current in current coil of wattmeter W_2

$= I_2 = I_L$ lagging behind V_2 by ϕ.

The pd across pressure coil of wattmeter W_2

$= V_{23} = \sqrt{3}\ V_P = V_L$ leading V_2 by 30°.

Therefore, phase angle between pd across potential coil and current through current coil of wattmeter W_2 is $(30° + \phi)$.

$$\text{Hence reading of wattmeter } W_2 = V_L I_L \cos (30° + \phi) \quad ...(2.29)$$

The sum of two wattmeter readings

$$W_1 + W_2 = V_L I_L \cos (30° - \phi) + V_L I_L \cos (30° + \phi)$$
$$= V_L I_L \times 2 \cos 30° \cos \phi = \sqrt{3}\, V_L I_L \cos \phi$$
$$= \text{True power of load} \quad ...(2.30)$$

$$\text{and } W_1 - W_2 = V_L I_L \cos (30° - \phi) - V_L I_L \cos (30° + \phi)$$
$$= V_L I_L \times 2 \sin 30° \sin \phi = V_L I_L \sin \phi \quad ...(2.31)$$

Dividing Eq. (2.31) by Eq. (2.30) we get

$$\frac{W_1 - W_2}{W_1 + W_2} = \frac{\text{Tan}\,\phi}{\sqrt{3}}$$

$$\text{or } \phi = \text{Tan}^{-1}\left(\frac{W_1 - W_2}{W_1 + W_2}\right) \times \sqrt{3}$$

and pf of load circuit,

$$\cos\phi = \text{Cos}\tan^{-1}\frac{\sqrt{3}(W_1 - W_2)}{W_1 + W_2} \quad \text{...(2.32)}$$

Hence phase angle ϕ and pf, $\cos\phi$ can be determined from readings of two wattmeters.

Alternative Method of Determination of Power Factor From Wattmeter Readings. This is convenient method for determining the power factor from wattmeter readings for a balanced load. In this method of determination of power factor from the wattmeter readings, for various values of phase angles the ratio of two wattmeter readings, $\frac{\text{Cos } 30° + \phi}{\text{Cos } 30° - \phi}$ $\left(i.e.\ \frac{\text{smaller reading}}{\text{larger reading}}\right)$ and the corresponding power factor values are determined and plotted as shown in Fig. 2.39. Now by using the watt-ratio curve (Fig. 2.39) the power factor is read directly, the ratio, $\frac{W_2}{W_1}$ being known.

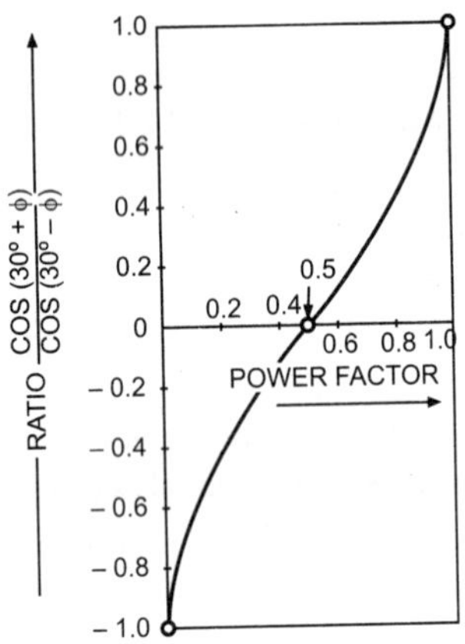

Fig. 2.39 *The Watt-ratio Curve*

It is seen that, when $\frac{W_2}{W_1} = 1.0$, the power factor is unity; when $\frac{W_2}{W_1} = 0$, the power factor is 0.5; when $\frac{W_2}{W_1}$ is negative i.e. when it becomes necessary to reverse W_2, the power factor is less than 0.5.

Variation in Wattmeter Readings. Wattmeter readings for inductive loads

$$W_1 = V_L I_L \cos(30° - \phi)$$

$$W_2 = V_L I_L \cos(30° + \phi)$$

(*i*) When pf of load is unity *i.e.* $\phi = 0$ then

$$W_1 = V_L I_L \cos 30° = \frac{\sqrt{3}}{2} V_L I_L = \text{Half of total power}$$

$$W_2 = V_L I_L \cos 30° = \frac{\sqrt{3}}{2} V_L I_L = \text{Half of total power}$$

Therefore, readings of both wattmeters are same, positive and equal to half the total power of the circuit.

(*ii*) When pf of load is 0.5 *i.e.* $\phi = 60°$ then

$$W_1 = V_L I_L \cos(30° - 60°) = \frac{\sqrt{3}}{2} V_L I_L = \text{Total power}$$

$$W_2 = V_L I_L \cos(30° + 60°) = 0$$

Hence reading of wattmeter W_1 will give total power of the load.

When the pf of load is zero *i.e.* $\phi = 90°$ then

$$W_1 = V_L I_L \cos(30° - 90°) = \frac{1}{2} V_L I_L$$

$$W_2 = V_L I_L \cos(30° + 90°) = -\frac{1}{2} V_L I_L$$

Hence the readings of two wattmeters are equal but of opposite sign.

So wattmeter W_2 gives –ve reading when phase angle ϕ varies from 60° to 90°. For obtaining the reading of wattmeter W_2 either the connection of current coil or pressure coil should be changed and readings obtained after the reversal of connections should be subtracted from the other wattmeter reading in order to get the total power.

Wattmeter Readings For Capacitive Loads. In the above discussion inductive load has been considered. If the circuit is capacitive the expression for readings of two meters are obtained by substituting in ϕ by $(-\phi)$ in the above expressions for wattmeter readings.

$$W_1 = V_L I_L \cos[30° - (-\phi)] = V_L I_L \cos(30° + \phi)$$

$$W_2 = V_L I_L \cos[30° + (-\phi)] = V_L I_L \cos(30° - \phi)$$

i.e. readings of wattmeters are interchanged.

Example 2.20. Explain two wattmeter method to measure three phase power. If in a 2 wattmeter method, readings of two wattmeters are 1,200 W and 300 W, find the power factor of the load.

[*U.P. Technical Univ. Electrical Engineering Odd Semester 2013-14*]

Solution: W_1 = 1,200 watts and W_2 = 300 watts

Power factor of load,

$$\cos\phi = \text{Cos}\tan^{-1}\frac{W_1 - W_2}{W_1 + W_2}\sqrt{3}$$

$$= \text{Cos}\tan^{-1}\frac{1{,}200 - 300}{1{,}200 + 300}\sqrt{3} = 0.6934 \text{ (lag)}$$ **Ans.**

Example 2.21. In a 2-wattmeter method power measured was 30 kW at 0.7 pf lagging. Find the reading of each wattmeter.

[*U.P. Technical Univ. Electrical Engnieering Second Semester 2006-07; Mahamaya Technical Univ. First Semester 2011-12*]

Solution: Total load supplied, P = 30 k W

Power factor, $\cos\phi$ = 0.7 (lagging)

Phase angle, $\phi = \text{Cos}^{-1}\ 0.7 = 45.57°$ (lagging)

$$\text{So } V_L I_L = \frac{\text{P in kW} \times 1{,}000}{\sqrt{3}\cos\phi}$$

$$= \frac{30 \times 1{,}000}{\sqrt{3} \times 0.7} = 24{,}743.6 \text{ V A}$$

Reading of wattmeter $W_1 = V_L I_L \cos(30° - \phi)$

= 24,743.6 cos (30° – 45.57°)

= 23,835 W or 23.835 k W **Ans.**

Reading of wattmeter, $W_2 = V_L I_L \cos(30° + \phi)$

= 24,743.6 cos (30° + 45.57°)

= 6,165 W or 6.165 kW **Ans.**

Example 2.22. A 3-phase balanced load connected across a 3-ϕ, 400 V ac supply draws a line current of 10 A. Two wattmeters are used to measure input power. The ratio of two wattmeter readings is 2 : 1. Find the readings of the two wattmeters.

[U.P. Technical Univ. Electrical Engineering February 2002; Odd Semester 2013-14]

Solution: Let the ratio of wattmeter readings be *r i.e.* $\frac{W_2}{W_1} = r$

and $\tan \phi = \sqrt{3}\left(\frac{W_1 - W_2}{W_1 + W_2}\right) = \sqrt{3}\left(\frac{1 - \frac{W_2}{W_1}}{1 + \frac{W_2}{W_1}}\right) = \sqrt{3}\left(\frac{1-r}{1+r}\right)$

and power factor,

$$\cos \phi = \frac{1}{\sec \phi} = \frac{1}{\sqrt{1 + \tan^2 \phi}} = \frac{1}{\sqrt{1 + 3\left(\frac{1-r}{1+r}\right)^2}}$$

Substituting $r = \frac{W_2}{W_1} = \frac{1}{2} = 0.5$ in above expression we get

Power factor of the load,

$$\cos \phi = \frac{1}{\sqrt{1 + 3\left(\frac{1-0.5}{1+0.5}\right)^2}} = 0.866$$

or phase angle, $\phi = \text{Cos}^{-1}\, 0.866 = 30^\circ$

Wattmeter reading, $W_1 = V_L I_L \cos (30^\circ - 30^\circ)$

$= 400 \times 10 \cos 0^\circ = 4{,}000$ W **Ans.**

Wattmeter reading, $W_2 = V_L I_L \cos (30^\circ + 30^\circ)$

$= 400 \times 10 \times \cos 60^\circ = 2{,}000$ W **Ans.**

Example 2.23. Three identical coils are connected in star to a 200 V, 3-phase supply and each taken 400 W. The pf is 0.8 lagging. What will be the magnitude of line currents and total power if the same coils are connected in delta to the same supply? If the total power in delta-connected coils is measured by two-wattmeter method, what will be the readings of the wattmeters?

[U.P.S.C. I.E.S. Electrical Engineering-I, 2010]

Solution: Phase voltage, $V_P = \frac{V_L}{\sqrt{3}} = \frac{200}{\sqrt{3}} = 115.47$ V

Power consumed per phase,

$P = 400$ W

Phase current, $I_P = \frac{P}{V_P \cos \phi} = \frac{400}{115.47 \times 0.8} = 4.33$ A

$\because$ pf = 0.8

Impedance, $Z_P = \frac{V_P}{I_P} = \frac{115.47}{4.33} = 26.667\ \Omega$

When same coils are connected in delta

Phase voltage, $V_P = V_L = 200$ V

Phase current, $I_P = \frac{V_P}{Z_P} = \frac{200}{26.667} = 7.5$ A

Line current, $I_L = \sqrt{3}\, I_P = \sqrt{3} \times 7.5 = 13$ A **Ans.**

Power factor, $\cos \phi = 0.8$ (lagging) $\because$ Coils are same

Power consumed $= \sqrt{3}\, V_L I_L \cos \phi = \sqrt{3} \times 200 \times 13 \times 0.8$

$= 3{,}600$ W **Ans.**

Phase angle, $\phi = \text{Cos}^{-1}\, 0.8 = 36.87^\circ$ (lagging)

Reading on wattmeter 1, $W_1 = V_L I_L \cos (30^\circ - \phi)$

$= 200 \times 13 \cos (30^\circ - 36.87^\circ)$

$= 200 \times 13 \times 0.99282 = 2{,}580$ W **Ans.**

Reading on wattmeter 2, $W_2 = V_L I_L \cos (30^\circ + \phi)$

$= 200 \times 13 \times \cos (30^\circ + 36.87^\circ)$

$= 200 \times 13 \times 0.39282 = 1{,}020$ W **Ans.**

Example 2.24. When the three identical star-connected coils are supplied with 440 V, 50 Hz, 3-ϕ supply, the 1-ϕ wattmeter whose current coil is connected in line R and pressure coil across the phase R and neutral reads 6 kW and the ammeter connected in R-phase reads 30 A. Assuming RYB phase sequence find (*i*) resistance and reactance of the coil (*ii*) the power factor of the load (*iii*) reactive power of 3-ϕ load. *[Nagpur Univ. Electrical Engineering I, 1993]*

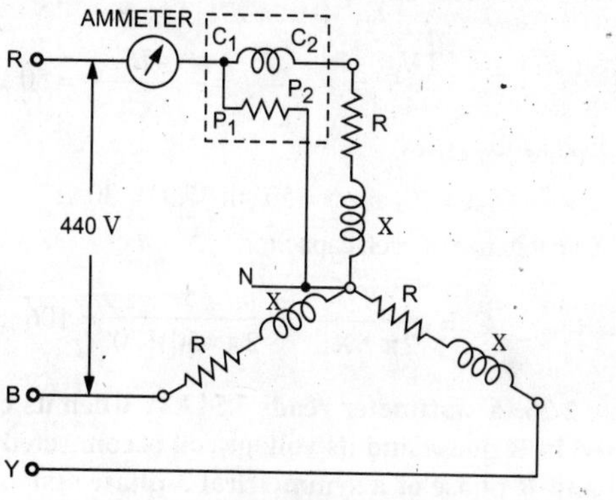

Fig. 2.40

Solution: Line voltage, $V_L = 440$ V

Phase voltage, $V_P = \frac{V_L}{\sqrt{3}} = \frac{440}{\sqrt{3}} = 254$ V

Power drawn per phase, P = Reading of wattmeter

= 6 kW or 6,000 W

Phase current, I_P = Line current

= Ammeter reading = 30 A

Coil impedance, $Z_P = \frac{V_P}{I_P} = \frac{254}{30} = 8.468\ \Omega$

Power factor of load, $\cos \phi = \frac{P}{V_P I_P} = \frac{6{,}000}{254 \times 30}$

= 0.7874 (lag) **Ans.**

Phase angle $\phi = \text{Cos}^{-1}\, 0.7874 = 38.06^\circ$

(*i*) Coil resistance, $R_P = Z_P \cos \phi = 8.468 \times 0.7874$

$= 6.668\ \Omega$ **Ans.**

Coil reactance, $X_P = Z_P \sin \phi = 8.468 \times \sin 38.06^\circ$

$= 5.22\ \Omega$ (inductive) **Ans.**

Reactive power of load, $Q = \sqrt{3}\, V_L I_L \sin \phi$

$= \sqrt{3} \times 440 \times 30 \times \sin 38.06^\circ$

= 14,095 VAR or 14.095 kVAR **Ans.**

Example 2.25. A star-connected balanced load is supplied from a 3-ϕ balanced supply with a line voltage of 416 V at a frequency of 50 Hz. Each phase of the load consists of a resistance and a capacitor joined in series and the readings on two wattmeters connected to measure the load power supplied are 782 W and 1,980 W, both positive. Calculate (*i*) power factor of the circuit (*ii*) line current (*iii*) the capacitance of each capacitor.

[Nagpur Univ. Elec. Engineering I, 1993]

Solution: Total power of load, $P = W_1 + W_2 = 782 + 1{,}980 = 2{,}762$ W

(*i*) Load power factor,

$$\cos\phi = \cos \tan^{-1}\left[-\sqrt{3}\frac{(W_1 - W_2)}{W_1 + W_2}\right]$$

$$= \cos \tan^{-1}\left[-\sqrt{3}\frac{(782 - 1{,}980)}{782 + 1{,}980}\right]$$

$= \cos 36.9° = 0.8$ (lead) **Ans.**

(*ii*) Line current, $I_L = \dfrac{P}{\sqrt{3}\,V_L \cos\phi} = \dfrac{2{,}762}{\sqrt{3}\times 416 \times 0.8} = 4.8$ A **Ans.**

Impedance per phase,

$$Z_P = \frac{V_P}{I_P} = \frac{V_L/\sqrt{3}}{I_L} = \frac{416}{\sqrt{3}\times 4.8} = 50\ \Omega$$

Reactance per phase,

$X_P = Z_P \sin\phi = 50 \sin 36.9° = 30\ \Omega$

(*iii*) Capacitance of each capacitor,

$$C = \frac{1}{2\pi f X_P} = \frac{1}{2\pi \times 50 \times 30} = 106\ \mu\text{F}$$ **Ans.**

Example 2.26. A wattmeter reads 5.54 kW when its current coil is connected in R phase and its voltage coil is connected between the neutral and R phase of a symmetrical 3-phase system supplying a balanced load of 30 A at 400 V. What will be the reading on the instrument if the connections to the current coil remain unchanged and the voltage coil is connected between B and Y phase ? Take phase sequence RYB. Draw corresponding phasor diagram.

Solution: Phase current, I_P = Line current = 30 A

Phase voltage, $V_P = \dfrac{V_L}{\sqrt{3}} = \dfrac{400}{\sqrt{3}} = 231$ V

Power supplied per phase, P = Reading of wattmeter = 5.54 kW or 5,540 W

Load power factor, $\cos\phi = \dfrac{P}{V_P I_P} = \dfrac{5{,}540}{231 \times 30} = 0.8$

or $\phi = \cos^{-1} 0.8 = 36.87°$

In the second case when the pressure coil is connected across phases B and Y, the phase angle between current in phase R and voltage across voltage coil is 90°– Φ as obvious from phasor diagram, shown in Fig. 2.41.

So wattmeter reading $= I_R V_{YB} \cos(90° - \phi)$

$= 30 \times 400 \cos(90° - 36.87°)$

$= 7{,}200$ watts or 7.2 kW **Ans.**

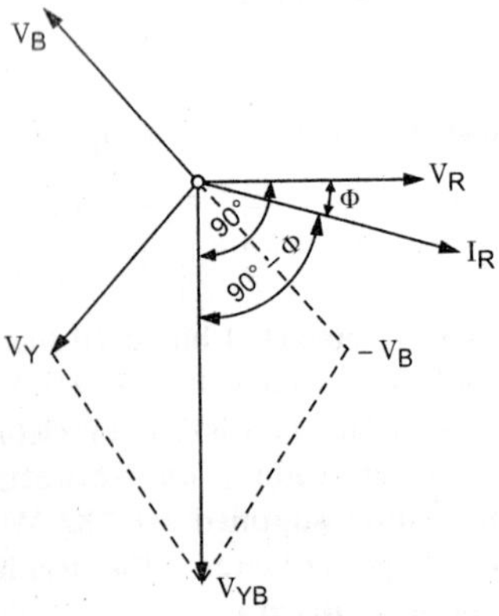

Fig. 2.41

Example 2.27. Three equal star-connected inductors take 8 kW at power factor 0.8 when connected to a 460 V, 3-φ, 3-wire supply. Give the connection diagram of 2 single phase wattmeters to measure the power in the circuit. Allocate the reading of each wattmeter. Find the line currents if one of the inductor is short-circuited. Draw the phasor diagrams of the currents and voltages under the condition.

Solution: The connection diagram of two single phase wattmeters to measure the power in the circuit is shown in Fig. 2.37 (*a*).

Power, P = 8 kW = 8,000 watts

Line voltage, V_L = 460 volts

Power factor, $\cos\phi$ = 0.8 (lag)

Phase angle, $\phi = \cos^{-1} 0.8 = 36.9°$ (lag)

Line current, $I_L = \dfrac{P}{\sqrt{3}\,V_L \cos\phi} = \dfrac{8{,}000}{\sqrt{3}\times 460 \times 0.8} = 12.55$ A

Reading of wattmeter $W_1 = V_L I_L \cos(30° - \phi)$

$= 460 \times 12.55 \cos(30° - 36.9°)$

$= 5{,}732$ watts **Ans.**

Reading of wattmeter $W_2 = V_L I_L \cos(30° + \phi)$

$= 460 \times 12.55 \cos(30° + 36.9°)$

$= 2{,}268$ watts **Ans.**

Phase current, $I_P = I_L = 12.55$ A

Phase impedance, $Z_P = \dfrac{V_P}{I_P} = \dfrac{400/\sqrt{3}}{12.55} = 21.16\ \Omega$

Phase resistance, $R_P = Z_P \cos\phi = 21.16 \times 0.8 = 16.93\ \Omega$

Phase reactance, $X_P = Z_P \sin\phi = 21.16 \times 0.6 = 12.7\ \Omega$

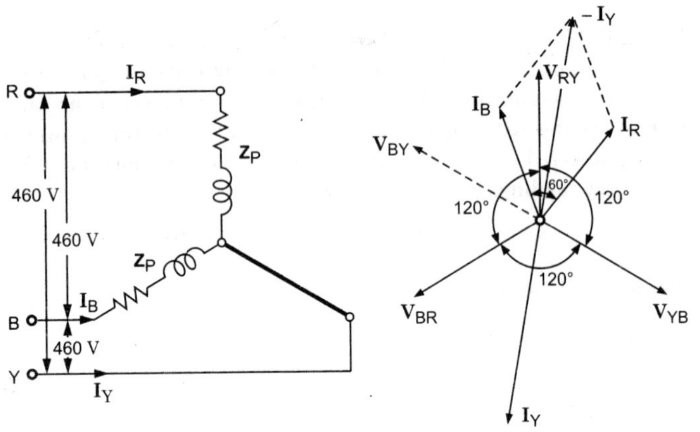

(*a*) *Circuit Diagram* (*b*) *Phasor Diagram*

Fig. 2.42

The circuit diagram when one of the inductor is short circuited is shown is Fig. 2.42 (*a*)

Current in line R, $I_R = \dfrac{V_{RY}}{Z_P} = \dfrac{460}{21.16} = 21.74$ A

Current in line B, $I_B = \dfrac{V_{BY}}{Z_P} = \dfrac{460}{21.16} = 21.74$ A

The phasor diagram is shown in Fig. 2.42 (*b*).

The currents in line R and line B are equal in magnitude but differ in phase by 60°, as shown in phasor diagram.

Current in line Y, $\mathbf{I_Y} = -(\mathbf{I_R} + \mathbf{I_B})$

Magnitude of current flowing in line Y,

$I_Y = 2\, I_R \cos 30° = 2 \times 21.74 \times 0.866 = 37.65$ A **Ans.**

Example 2.28. Two wattmeters A and B give readings as 5,000 W and 1,000 W respectively during the power measurement of 3-ϕ, 3-wire balanced load system. (*a*) Calculate the power and power factor (*i*) both meters read direct and (*ii*) one of them reads in reverse (*b*) if the voltage of the circuit is 400 V, what is the value of the capacitance which must be introduced in each phase to cause the whole of the power to appear on A. The frequency of supply is 50 Hz. [*Nagpur Univ. Elec. Engineering- I, 1992*]

Solution: (*a*) (*i*) When both meters read direct

$$\text{Power factor, } \cos\phi = \text{Cos}\tan^{-1}\frac{\sqrt{3}(W_1 - W_2)}{W_1 + W_2}$$

$$= \text{Cos}\tan^{-1}\frac{(5{,}000 - 1{,}000)}{5{,}000 + 1{,}000}$$

$$= 0.655 \text{ (lag) } \textbf{Ans.}$$

$$\text{Total power, } P = 5{,}000 + 1{,}000 = 6{,}000 \text{ W or } 6 \text{ kW } \textbf{Ans.}$$

(*ii*) When one of the meters reads in reverse *i.e.* when $W_2 = -1{,}000$ W

$$\text{Total power, } P = 5{,}000 + (-1{,}000) = 4{,}000 \text{ W or } 4 \text{ kW } \textbf{Ans.}$$

$$\text{Power factor, } \cos\phi = \text{Cos}\tan^{-1}\sqrt{3}\left[\frac{5{,}000 - (-1{,}000)}{5{,}000 + (-1{,}000)}\right]$$

$$= 0.359 \text{ (lag)}$$

(*b*) Reading of wattmeter B will be zero and whole of the power would appear on wattmeter A when power factor $\cos\phi' = 0.5$ or $\phi' = 60°$. It means pf is to be improved from 0.359 to 0.5.

$$\text{Load current per phase, } I = \frac{P}{\sqrt{3}V_L\cos\phi} = \frac{4{,}000}{\sqrt{3}\times 400\times 0.359}$$

$$= 16.07 \text{ A}$$

Load impedance per phase,

$$Z_P = \frac{V_P}{I_P} = \frac{400/\sqrt{3}}{16.07} = 14.36\ \Omega$$

Load resistance per phase,

$$R_P = Z_P\cos\phi = 14.36 \times 0.359 = 5.156\ \Omega$$

Load reactance per phase,

$$X_P = Z_P\sin\phi = 14.36\sin\cos^{-1}0.359 = 13.4\ \Omega$$

Since there is no change in resistance

∴ Reactance per phase in the circuit,

$$X'_P = R_P\tan\phi' = 5.156\tan 60° = 8.93\ \Omega$$

Capacitive reactance introduced in each phase,

$$X_C = X_P - X'_P = 13.4 - 8.93 = 4.47\ \Omega$$

Capacitance introduced in each phase,

$$C = \frac{1}{2\pi f X_C} = \frac{1}{2\pi\times 50\times 4.47} = 712\ \mu\text{F } \textbf{Ans.}$$

Example 2.29. A 3-ϕ, 434-V, 50 Hz supply is connected to a 3-ϕ, star-connected induction motor and a synchronous motor. Impedance of each phase of induction motor is (1.25 + *j* 2.17) Ω. The synchronous motor is overexcited and it draws a current of 120 A at 0.87 leading pf. Two wattmeters are connected in usual manner to measure the power drawn by the motors. Calculate (*i*) the reading of each wattmeter (*ii*) combined power factor.

[*Hyderabad Univ. Electrical Technology 1992*]

Solution: Supply voltage, $V_L = 434$ V

$$\text{Phase voltage, } V_P = \frac{434}{\sqrt{3}} = 250 \text{ V}$$

Phase impedance of 3-ϕ induction motor,

$$\mathbf{Z}_{P1} = (1.25 + j\,2.17) = 2.5\angle 60°\ \Omega$$

Taking phase voltage as reference phasor it is expressed as $250\angle 0°$ V

$$\text{Phase current, } \mathbf{I}_{P1} = \frac{\mathbf{V}_P}{\mathbf{Z}_{P1}} = \frac{250\angle 0°}{2.5\angle 60°} = 100\angle -60° \text{ A}$$

$$\text{Line current, } \mathbf{I}_{L1} = \mathbf{I}_P = 100\angle -60° \text{ A} = (50 - j\,86.6) \text{ A}$$

Phase angle of synchronous motor, $\phi_2 = \text{Cos}^{-1}\,0.87 = 29.54°$

So line current drawn by synchronous motor,

$$\mathbf{I}_{L2} = 120\angle 29.54° = (104.4 + j\,59.17) \text{ A}$$

$$\text{Total current, } I_L = \mathbf{I}_{L1} + \mathbf{I}_{L2}$$

$$= (50 - j\,86.6) + (104.4 + j\,59.17)$$

$$= (154.4 - j\,27.43) \text{ or } 156.82\angle -10.1° \text{ A}$$

(*i*) Reading of wattmeter

$$W_1 = V_L I_L\cos(30° - \phi)$$

$$= 434 \times 156.82\cos[30° - 10.1°]$$

$$= 63{,}996 \text{ W or } 63.996 \text{ kW } \textbf{Ans.}$$

Reading of wattmeter

$$W_2 = V_L I_L\cos(30° + \phi)$$

$$= 434 \times 156.82\cos[30° + 10.1°]$$

$$= 52{,}060 \text{ W or } 52.06 \text{ kW } \textbf{Ans.}$$

(*ii*) Combined power factor

$$= \text{Cos}(-10.1°) = 0.9845 \text{ (lag) } \textbf{Ans.}$$

2.20 MEASUREMENT OF REACTIVE VOLT-AMPERES

$$\frac{\text{Reactive power}}{\text{True power}} = \frac{VI\sin\phi}{VI\cos\phi} = \text{Tan}\,\phi$$

$$\text{or } \phi = \text{Tan}^{-1}\frac{\text{Reactive power}}{\text{True power}}$$

For measurement of reactive power in a single-phase circuit use of specially compensated wattmeter is made. In compensated wattmeter the voltage applied to the pressure coil is 90° out of phase with the actual voltage, obviously the wattmeter will read V I cos (90° – ϕ) *i.e.* V I sin ϕ, the reactive power. In a single-phase measurement this is affected by making the field of pressure coil lag behind by 90° when it measures actual power.

In the case of balanced three-phase circuit the reactive power can be determined by use of one wattmeter. The connections of wattmeter for measurement of reactive power in 3-ϕ balanced circuit are shown in Fig. 2.43 (*a*). The current coil of the wattmeter is inserted in one line and the pressure coil of it is connected across the other two lines.

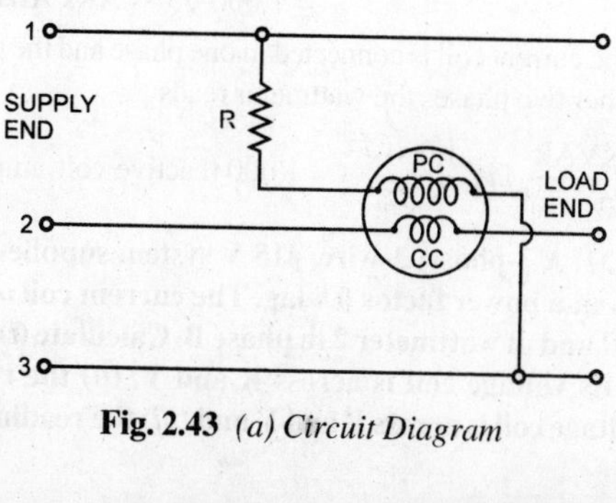

Fig. 2.43 *(a) Circuit Diagram*

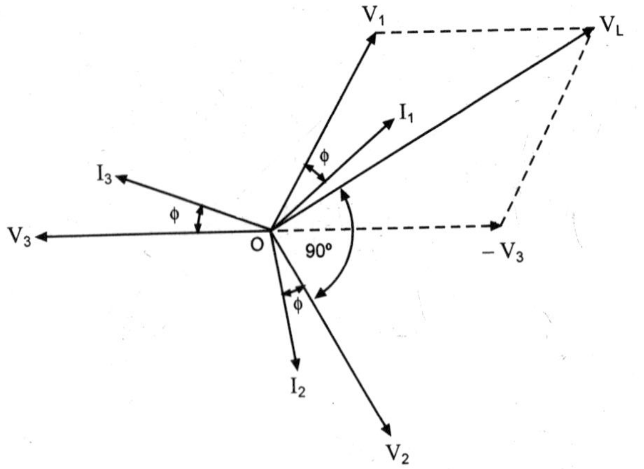

(b) Phasor Diagram

Fig. 2.43

The current flowing through the current coil of wattmeter is $I_2 = I_P$ (say)

The pd applied across the potential coil of wattmeter

$$\mathbf{V} = \mathbf{V}_1 - \mathbf{V}_3$$

$$= \sqrt{3}\, V_P \text{ leading phasor } V_2, \text{ by } 90^\circ$$

$$= \sqrt{3}\, V_P \text{ leading phasor of current } I_2, \text{ by } (90^\circ + \phi)$$

∴ Wattmeter indicates

$$= \sqrt{3}\, V_P I_P \cos(90^\circ + \phi) = -\sqrt{3}\, V_P I_P \sin\phi = Q$$

Total reactive power of the circuit,

$$= 3\, V_P I_P \sin\phi = -\sqrt{3}\, Q$$

Reactive power can also be determined from two wattmeter readings connected for measurement of power

$$W_1 - W_2 = V_L I_L \cos(30^\circ - \phi) - V_L I_L \cos(30^\circ + \phi)$$

$$\text{or } W_1 - W_2 = V_L I_L \times 2 \sin 30^\circ \sin\phi$$

$$\text{or } V_L I_L \sin\phi = W_1 - W_2$$

Reactive power of load circuit, Q

$$= \sqrt{3}\, V_L I_L \sin\phi$$

$$= \sqrt{3}\,(W_1 - W_2) \text{ VARs} \quad ...(2.33)$$

Example 2.30. Two wattmeters are connected to measure power in a 3-phase network. The two readings are 2,000 watts and 1,000 watts respectively. If another wattmeter be connected such that its current coil is in one phase and the potential coil is across the other two phase terminals, what will it read ? Also estimate the reactive power of the network.

Solution: Reading of wattmeter, W_1 = 2,000 Watts

Reading of wattmeter, W_2 = 1,000 watts

Reactive power of the network,

$$Q = \sqrt{3}\,(W_1 - W_2) = \sqrt{3}\,(2{,}000 - 1{,}000)$$

$$= 1{,}000\sqrt{3} \text{ VARs } \textbf{Ans.}$$

When the current coil is connected in one phase and the potential coil across the other two phases the wattmeter reads

$$\frac{\text{Total KVAR}}{\sqrt{3}} \quad i.e. \quad \frac{1{,}000\sqrt{3}}{\sqrt{3}} = 1{,}000 \text{ reactive volt-amperes } \textbf{Ans.}$$

Example 2.31. A 3-phase, 3-wire, 415 V system supplies a balanced load of 20 A at a power factor 0.8 lag. The current coil of wattmeter is in phase R and of wattmeter 2 in phase B. Calculate (*i*) the reading on 1 when its voltage coil is across R and Y, (*ii*) the reading on 2 when its voltage coil is across B and Y and (*iii*) the reading on 1 when its voltage coil is across Y and B. Justify your answer with relevant phasor diagram.

Solution: Supply voltage, V_L = 415 V

Line current, I_L = 20 A

Phase angle, $\phi = \cos^{-1} 0.8 = 36.87^\circ$ (lag)

(*i*) Reading on wattmeter 1 when its voltage coil is across R and Y and current coil in R phase (Fig. 2.37), as usual

$$= V_L I_L \cos(30^\circ - \Phi)$$

$$= 415 \times 20 \times \cos(30^\circ - 36.87^\circ) = 8{,}240 \text{ W } \textbf{Ans.}$$

(*ii*) Reading on wattmeter 2 when its current coil is in phase B and the voltage coil across B and Y (Fig. 2.37), as usual

$$= V_L I_L \cos(30^\circ + \phi)$$

$$= 415 \times 20 \times \cos(30^\circ + 36.87^\circ) = 3{,}260 \text{ W } \textbf{Ans.}$$

(*iii*) When the current coil is connected in one phase (phasor R) and the voltage coil is across other two phases (Y and B), the wattmeter reads $\frac{\text{total kVAR}}{\sqrt{3}}$ and

$$\text{Total kVAR} = \sqrt{3}\,(W_1 - W_2)$$

$$\text{So wattmeter reading} = \frac{\text{Total K V A R}}{\sqrt{3}} = \sqrt{3}\,\frac{W_1 - W_2}{\sqrt{3}}$$

$$= W_1 - W_2 = 8{,}240 - 3{,}260$$

$$= 4{,}980 \text{ VAR } \textbf{Ans.}$$

2.21 UNBALANCED LOAD CIRCUITS

The load circuits dealt with so far were *balanced circuits*. In balanced polyphase circuits the impedances in different phases are same (*i.e.* of same magnitude and same phase). Now we will consider the methods of solving unbalanced delta-connected and star-connected load circuits *i.e.* the circuits consisting of impedances in one or more phases different from those in other phases.

2.21.1. Unbalanced Delta-Connected Load Circuits. In Δ-connected load circuits the voltage across each load phase is fixed, independent of the nature of the load and is equal to the line voltage so different phase currents can be determined in the usual way (as in balanced load circuits) and the line currents are determined by taking phasor difference of phase currents in pairs. It is to be noted that in this case (unlike with balanced loads) the line currents will neither be equal in magnitude nor in phase.

The method for solution of unbalanced delta-connected load circuits will be more clear from the following workedout example.

Example 2.32. A delta load of $Z_{AB} = 52\angle 45^\circ\,\Omega$, $Z_{BC} = 52\angle -30^\circ\,\Omega$ and $Z_{CA} = 10\angle 0^\circ\,\Omega$ are connected to a 230 volts, 3-phase source. Find the magnitude of the line currents, when the phase sequence is ABC. [*Rajasthan Technical Univ., March-2009*]

Solution: Phase voltage $V_P = V_L = 230$ V

Let the pd across phase AB be the reference phasor. Since phase sequence is ABC,

$$\mathbf{V}_{AB} = 230\angle 0^\circ \text{ V}$$

$$\mathbf{V}_{BC} = 230\angle -120^\circ \text{ V}$$

$$\mathbf{V}_{CA} = 230\angle 120^\circ \text{ V}$$

$$\mathbf{Z}_{AB} = 52\angle 45^\circ\ \Omega$$

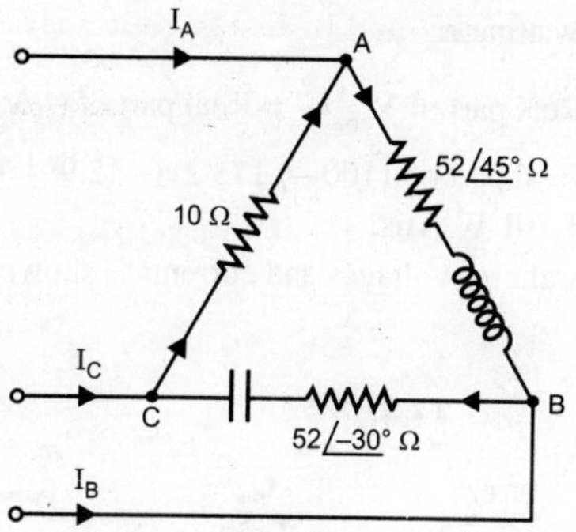

Fig. 2.44

$$Z_{BC} = 52 \angle -30° \ \Omega$$

$$Z_{CA} = 10 \angle 0° \ \Omega$$

Phase current,

$$I_{AB} = \frac{V_{AB}}{Z_{AB}} = \frac{230 \angle 0°}{52 \angle 45°} = 4.423 \angle -45° \text{ A} = (3.13 - j3.13) \text{ A}$$

Phase current,

$$I_{BC} = \frac{V_{BC}}{Z_{BC}} = \frac{230 \angle -120°}{52 \angle -30°} = 4.423 \angle -90° \text{ A} = (0 - j4.23) \text{ A}$$

Phase current,

$$I_{CA} = \frac{V_{CA}}{Z_{CA}} = \frac{230 \angle 120°}{10 \angle 0°} = 23 \angle 120° = (-11.5 + j19.9) \text{ A}$$

Line current,

$$I_A = I_{AB} - I_{CA} = (3.13 - j3.13) - (-11.5 + j19.9)$$
$$= (14.63 + j16.77) = 22.25 \text{ A} \quad \textbf{Ans.}$$

Line current,

$$I_B = I_{BC} - I_{AB} = (0 - j4.23) - (3.13 - j3.13)$$
$$= (-3.13 - j1.10) = 3.32 \text{ A} \quad \textbf{Ans.}$$

Line current,

$$I_C = I_{CA} - I_{BC} = (-11.5 + j19.9) - (0 - j4.23)$$
$$= -11.5 + j24.13 = 26.73 \text{ A} \quad \textbf{Ans.}$$

Example 2.33. Three loads 31 + *j* 59, 30 – *j* 40 and 80 + *j* 60 are connected in delta to a 3-phase, 200 V supply. Find the phase currents, line currents and power.

Solution: Phase voltage, $V_P = V_L = 200$ V

Let the pd across phase RY be the reference phasor and phase sequence be RYB.

Hence the phase voltages may be written as

$$V_{RY} = 200 \angle 0° \text{ V}$$

$$V_{YB} = 200 \angle -120° \text{ V}$$

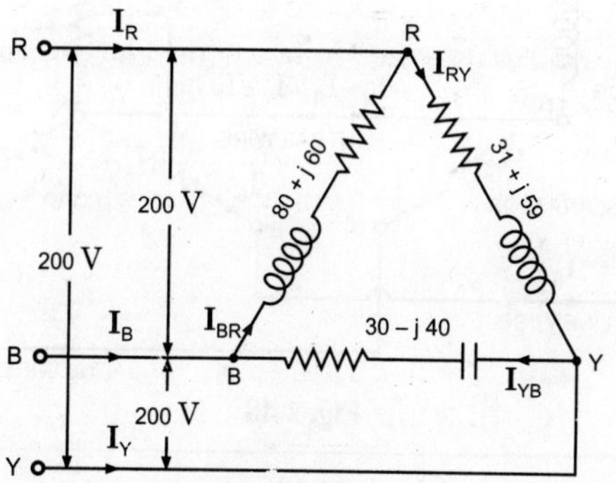

Fig. 2.45

and $$V_{BR} = 200 \angle 120° \text{ V}$$

Impedance $$Z_{RY} = 31 + j\,59 = 66.648 \angle 62.28° \ \Omega$$

Impedance $$Z_{YB} = 30 - j\,40 = 50 \angle -53.13° \ \Omega$$

Impedance $$Z_{BR} = 80 + j\,60 = 100 \angle 36.87° \ \Omega$$

Phase current, $$I_{RY} = \frac{V_{RY}}{Z_{RY}} = \frac{200 \angle 0°}{66.648 \angle 62.28°} \text{ A}$$

$$= 3 \angle -62.28° \text{ A or } (1.395 - j\,2.656) \text{ A} \quad \textbf{Ans.}$$

Phase current, $$I_{YB} = \frac{V_{YB}}{Z_{YB}} = \frac{200 \angle -120°}{50 \angle -53.13°} \text{ A}$$

$$= 4 \angle -66.87° \text{ A or } (1.571 - j\,3.678) \text{ A} \quad \textbf{Ans.}$$

Phase current, $$I_{BR} = \frac{V_{BR}}{Z_{BR}} = \frac{200 \angle 120°}{100 \angle 36.87°} \text{ A}$$

$$= 2 \angle 83.13° \text{ A} = (0.239 + j\,1.986) \text{ A} \quad \textbf{Ans.}$$

Line current in phase R,

$$I_R = I_{RY} - I_{BR} = (1.395 - j\,2.656) - (0.239 + j\,1.986)$$
$$= (1.156 - j\,4.642) \text{ A or } 4.784 \angle -76° \text{ A}$$

Line current is phase Y,

$$I_Y = I_{YB} - I_{RY} = (1.571 - j\,3.678) - (1.395 - j\,2.656)$$
$$= (0.176 - j\,1.022) \text{ A or } 1.037 \angle -80.23° \text{ A} \quad \textbf{Ans.}$$

Line current in phase B,

$$I_B = I_{BR} - I_{YB} = (0.239 + j\,1.986) - (1.571 - j\,3.678)$$
$$= (-1.332 + j\,5.664) \text{ A or } 5.819 \angle 103.23° \text{ A} \quad \textbf{Ans.}$$

Power absorbed,

$$P = I_{RY}^2 \times R_{RY} + I_{YB}^2 R_{YB} + I_{BR}^2 R_{BR}$$
$$= 3^2 \times 31 + 4^2 \times 30 + 2^2 \times 80 = 1{,}079 \text{ W} \quad \textbf{Ans.}$$

Now let the phase sequence be RBY. Now the phase voltages will be written as

$$V_{RY} = 200 \angle 0° \text{ V},$$

$$V_{YB} = 200 \angle 120° \text{ V and } V_{BR} = 200 \angle -120° \text{ V}$$

Phase current, $$I_{RY} = \frac{200 \angle 0°}{66.648 \angle 62.28°} \text{ A}$$

$$= 3 \angle -62.28° \text{ A or } (1.395 - j\,2.656) \text{ A} \quad \textbf{Ans.}$$

Phase current, $$I_{YB} = \frac{200 \angle 120°}{50 \angle -53.13°} \text{ A}$$

$$= 4 \angle 173.13° \text{ A or } (-3.971 + j\,0.478) \text{ A} \quad \textbf{Ans.}$$

Phase current, $$I_{BR} = \frac{200 \angle -120°}{100 \angle 36.87°} \text{ A}$$

$$= 2 \angle -156.87° \text{ A or } (-1.839 - j\,0.786) \text{ A} \quad \textbf{Ans.}$$

Line current in phase R,

$$I_R = I_{RY} - I_{BR} = (1.395 - j\,2.656) - (-1.839 - j\,0.786)$$
$$= (3.234 - j\,1.87) \text{ A or } 3.736 \angle -30.04° \text{ A} \quad \textbf{Ans.}$$

Line current in phase Y,

$$I_Y = I_{YB} - I_{RY} = (-3.971 + j\,0.478) - (1.395 - j\,2.656)$$
$$= (-5.366 + j\,3.134) \text{ A or } 6.214 \angle 149.71° \text{ A} \quad \textbf{Ans.}$$

Line current in phase B,

$$\mathbf{I}_B = \mathbf{I}_{BR} - \mathbf{I}_{YB} = (-1.839 - j\,0.786) - (-3.971 + j\,0.478)$$

$$= (2.132 - j\,1.264) \text{ A or } 2.479 \angle -30.66^\circ \text{ A } \textbf{Ans.}$$

Power absorbed will remain the same as the magnitude of currents flowing different phases are the same.

Thus we see that when phase sequence is reversed

(*i*) Phase currents are of same magnitude but differ in phase.

(*ii*) Line currents change both in magnitude and angle.

Example 2.34. A balanced 3-phase inductive load of 6 kW at 0.8 pf is connected across abc and a 4 kW resistive load is connected across ab as shown in the diagram. If $V_{ab} = V_{bc} = V_{ca} = 200$ V in magnitude, find the wattmeter readings W_1 and W_2 for phase sequence ab - ca - bc.

Solution: Let the voltage V_{ab} be the reference phasor

$$\text{the } \mathbf{V}_{ab} = 200 \angle 0^\circ \text{ V} = (200 + j\,0) \text{ V}$$

$$\mathbf{V}_{bc} = 200 \angle 120^\circ \text{ V} = (-100 + j\,173.2) \text{ V}$$

$$\mathbf{V}_{ca} = 200 \angle -120^\circ \text{ V} = (-100 - j\,173.2) \text{ V}$$

$$\text{Currents in phases ab, ca and ba} = \frac{6{,}000}{3 \times 200 \times 0.8} = 12.5 \text{ A}$$

lagging behind their respective phase voltage by $\cos^{-1} 0.8$ (36.87°)

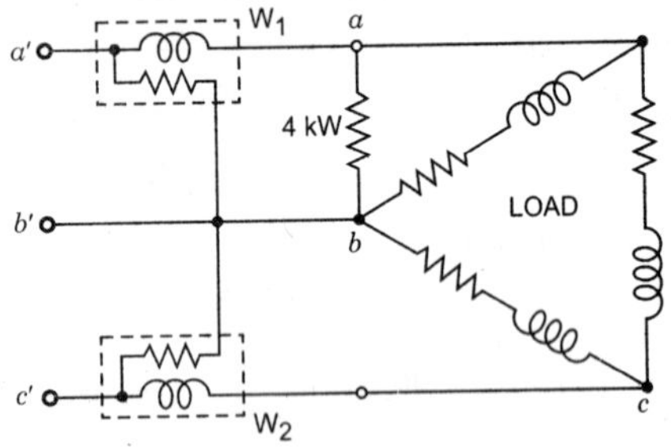

Fig. 2.46

Current in phase ab due to 4 kW resistive load

$$= \frac{4{,}000}{200} = 20 \text{ A in phase with } V_{ab}$$

$$\text{Phase current } \mathbf{I}_{ab} = 12.5 \angle -36.87^\circ + 20 \angle 0^\circ$$

$$= (10.0 - j\,7.5) + 20 = (30 - j\,7.5) \text{ A}$$

$$= 30.92 \angle -14.04^\circ \text{ A}$$

$$\text{Phase current } \mathbf{I}_{bc} = 12.5 \angle 120^\circ - 36.87^\circ \text{ A}$$

$$= 12.5 \angle 83.13^\circ \text{ A} = (1.495 + j\,12.41) \text{ A}$$

$$\text{Phase current } \mathbf{I}_{ca} = 12.5 \angle -120^\circ - 36.87^\circ \text{ A}$$

$$= 12.5 \angle -156.87^\circ \text{ A} = (-11.495 - j\,4.91) \text{ A}$$

$$\text{Line current } \mathbf{I}_a = \mathbf{I}_{ab} - \mathbf{I}_{ca} = (30 - j\,7.5) - (-11.495 - j\,4.91)$$

$$= (41.495 - j\,2.59) \text{ A}$$

$$\text{Line current } \mathbf{I}_c = \mathbf{I}_{ca} - \mathbf{I}_{bc}$$

$$= (-11.495 - j\,4.91) - (1.495 + j\,12.41)$$

$$= (-12.991 - j\,17.32) \text{ A}$$

Reading of wattmeter

$$W_1 = \text{Real part of } \mathbf{V}_{ab}\,\mathbf{I}_a^*$$

$$= \text{Real part of } (200 + j\,0)\,(41.495 + j\,2.59)$$

$$= 8{,}299 \text{ W } \textbf{Ans.}$$

Reading of wattmeter

$$W_2 = \text{Real part of } \mathbf{V}_{cb}\,\mathbf{I}_c^* = \text{Real part of } (-\mathbf{V}_{bc}\,\mathbf{I}_c^*)$$

$$= \text{Real part of } (100 - j\,173.2)\,(-12.991 + j\,17.32)$$

$$= 1{,}701 \text{ W } \textbf{Ans.}$$

Phasor diagram for voltages and currents is shown in Fig. 2.47.

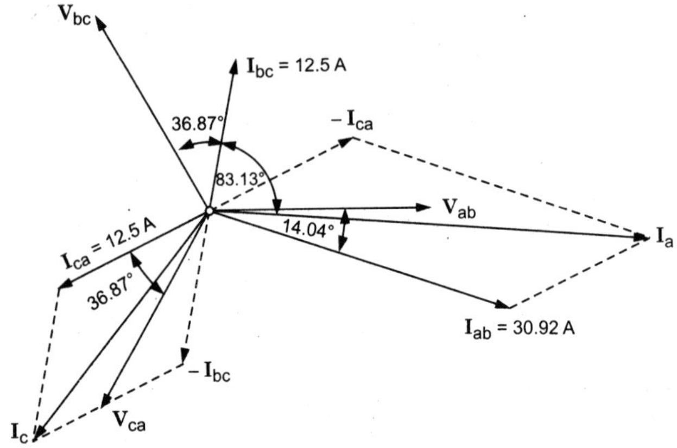

Fig. 2.47

2.21.2. 3-Phase, 4-Wire Star-Connected Unbalanced Load Circuits. In a 3-phase, 4-wire distribution system the lines are supplied with 415 volts (line to line) and voltage from any line to neutral is 240 volts. Lighting loads and 240 V-appliances are connected between line and neutral. This will give unbalanced star-connected load. Again, line currents are unbalanced and due to unbalance, the neutral wire carries some current. In such circuits the star points of load and the generator are tied together through neutral wire (assumed to be having zero impedance), therefore, they are at the same potential and voltage across each impedance is same and equal to phase voltage irrespective of the fact that the circuit is balanced or unbalanced. The three-phase currents or line currents can be determined by dividing the phase voltage by the impedance of the concerned phase,

$$\text{i.e. } \mathbf{I}_R = \frac{\mathbf{V}_R}{\mathbf{Z}_R};\ \mathbf{I}_Y = \frac{\mathbf{V}_Y}{\mathbf{Z}_Y} \text{ and } \mathbf{I}_B = \frac{\mathbf{V}_B}{\mathbf{Z}_B}$$

The current in neutral wire can be determined by applying Kirchhoff's first law (or current law) at star or neutral point N according to which

$$\mathbf{I}_R + \mathbf{I}_Y + \mathbf{I}_B + \mathbf{I}_N = 0$$

$$\text{or current in neutral wire, } \mathbf{I}_N = -(\mathbf{I}_R + \mathbf{I}_Y + \mathbf{I}_B)$$

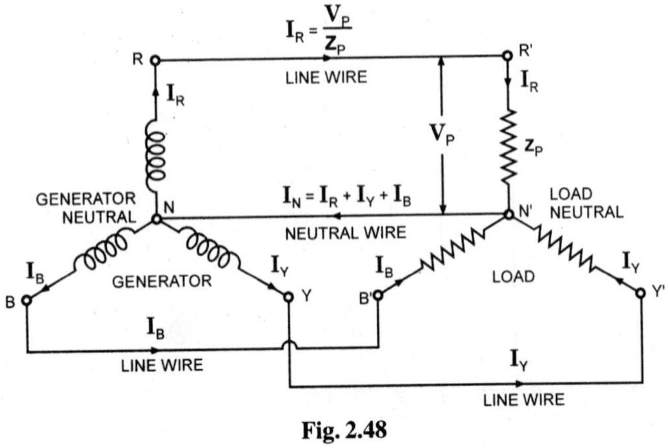

Fig. 2.48

* Means conjugate of

In unbalanced load circuits, the line currents are unequal and do not have a 120° phase difference. But in balanced load circuits the line currents are equal and have the same phase difference with their respective phase voltages, so their phase sum is zero and current in the neutral wire is zero. Since in balanced star-connected load circuits, neutral wire carries no current and, it is usually omitted.

Example 2.35. A 3-phase, 4-wire, supply is obtained from a star-connected alternator. The output voltage V_{RN} is represented by 250 $\angle 0°$ V. Phase-sequence is RYB:

(*i*) What are V_{YN} and V_{BN}?

(*ii*) Express V_{RY}, V_{YB} and V_{BR}.

(*iii*) If the phase sequence is reversed, would your answers change ? Explain. [*R.G.P.V. Bhopal, December-2004*]

Solution: The output voltage V_{RN} is represented by

$$V_{RN} = 250 \angle 0° \text{ V}$$

(*i*) Since phase voltage V_{YN} lags behind the phase voltage V_{RN} by 120° and phase voltage V_{BN} lags behind the phase voltage V_{YN} by 120°, phase sequence being RYB, the output voltages V_{YN} and V_{BN} are represented by

$$V_{YN} = 250 \angle -120° \text{ V} \quad \textbf{Ans.}$$

$$V_{BN} = 250 \angle -240° \text{ V or } 250 \angle 120° \text{ V} \quad \textbf{Ans.}$$

(*ii*) Since line voltages are $\sqrt{3}$ times of phase voltages and lead respective phase voltage by 30°, we have output line voltages as

$$V_{RY} = 250\sqrt{3} \angle 0° + 30° \text{ V} = 250\sqrt{3} \angle 30° \text{ V} \quad \textbf{Ans.}$$

$$V_{YB} = 250\sqrt{3} \angle -120° + 30° \text{ V} = 250\sqrt{3} \angle -90° \text{ V} \quad \textbf{Ans.}$$

and $$V_{BR} = 250\sqrt{3} \angle 120° + 30° \text{ V} = 250\sqrt{3} \angle 150° \text{ V} \quad \textbf{Ans.}$$

(*iii*) When phase sequence is reversed, *i.e.* phase sequence becomes RBY, the phase voltage V_{BN} lags behind the phase voltage V_{RN} by 120° and phase voltage V_{YN} lags behind the phase voltage V_{BN} by 120° and the results become as follows:

$$V_{BN} = 250 \angle -120° \text{ V} \quad \textbf{Ans.}$$

$$V_{YN} = 250 \angle 120° \text{ V} \quad \textbf{Ans.}$$

$$V_{RB} = 250\sqrt{3} \angle 30° \text{ V} \quad \textbf{Ans.}$$

$$V_{BY} = 250\sqrt{3} \angle -90° \text{ V} \quad \textbf{Ans.}$$

and $$V_{YR} = 250\sqrt{3} \angle 150° \text{ V} \quad \textbf{Ans.}$$

Example 2.36. A 3-phase, 4-wire CBA system, with an effective line voltage of 169.7 V, has three impedances of 20 $\angle -30°$ Ω in a Y-connection (Fig. 2.49). Determine the line currents and draw the voltage-current phasor diagram.

[*U.P.S.C. I.E.S. Electrical Engineering-I, 2007*]

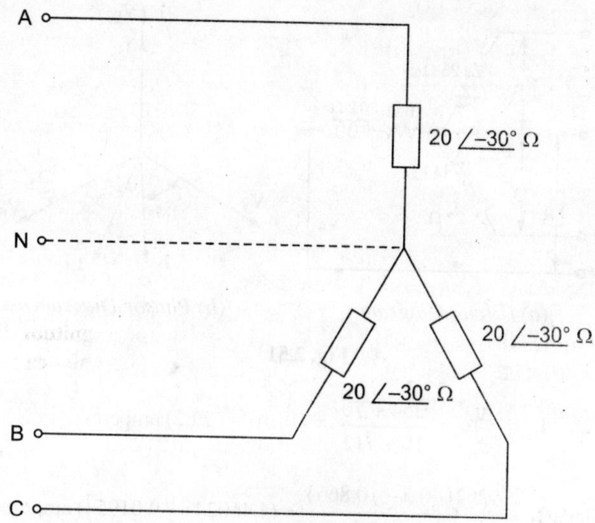

Fig. 2.49

Solution: Line voltage, V_L = 169.7 V

$$\text{Phase voltage, } V_P = \frac{V_L}{\sqrt{3}} = \frac{169.7}{\sqrt{3}} = 98 \text{ V}$$

Taking phase voltage V_C as reference phasor, phase voltages can be expressed as

$$V_C = 98 \angle 0° \text{ V}$$

$$V_B = 98 \angle -120° \text{ V}$$

$$V_A = 98 \angle 120° \text{ V}$$

Line current, being equal to phase currents are:

$$\mathbf{I}_A = \frac{V_A}{\mathbf{Z}} = \frac{98 \angle 120°}{20 \angle -30°} = 4.9 \angle 150° \text{ A } \textbf{Ans.}$$

$$\mathbf{I}_B = \frac{V_B}{\mathbf{Z}} = \frac{98 \angle -120°}{20 \angle -30°} = 4.9 \angle -90° \text{ A } \textbf{Ans.}$$

$$\mathbf{I}_C = \frac{V_C}{\mathbf{Z}} = \frac{9.8 \angle 0°}{20 \angle -30°} = 4.9 \angle 30° \text{ A } \textbf{Ans.}$$

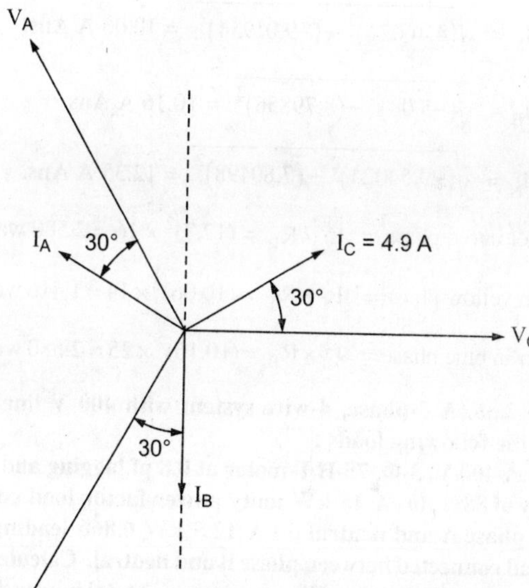

Fig. 2.50 *Phasor Diagram*

Example 2.37. A 440/254 volt, 3 phase, 4-core supplies an unbalanced load represented by the following impedances in ohms connected between R, Y and B phases respectively and the neutral 16 + *j* 12; 14 − *j* 21 and 25. The phase sequence is RYB. Calculate the current in each conductor of the cable and the readings on each of the three wattmeters connected in each line and neutral.

Solution: The potential differences between the three lines and neutral are given by:

$$V_R = 254\,(1 + j\,0)$$

$$V_Y = 254\,(-0.5 - j\,0.866)$$

$$V_B = 254\,(-0.5 + j\,0.866) \quad \text{taking } V_R \text{ as the reference phasor}$$

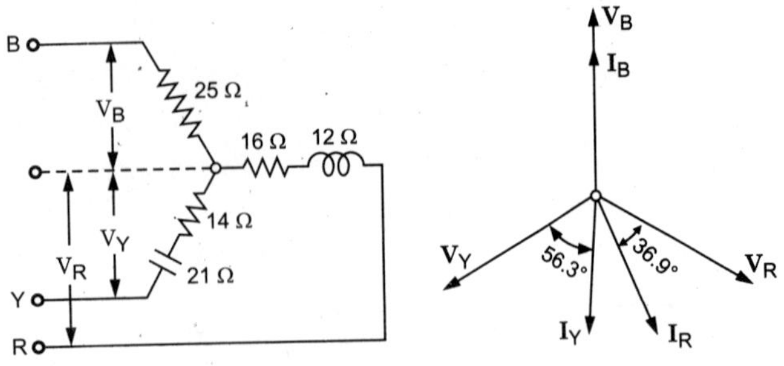

(*a*) *Circuit Diagram* (*b*) *Phasor Diagram*

Fig. 2.51

$$I_R = \frac{V_R}{Z_R} = \frac{254 + j0}{16 + j12} = (10.16 - j\,7.62) \text{ amperes}$$

$$\text{Similarly } \mathbf{I}_Y = \frac{254(-0.5 - j0.866)}{14 - j21} = (4.46024 - j\,9.01954) \text{ amperes}$$

$$\mathbf{I}_B = \frac{254(-0.5 + j0.866)}{25} = (-\,5.08 + j\,8.79856) \text{ amperes}$$

Now current in neutral wire,

$$\mathbf{I}_N = -\,(\mathbf{I}_R + \mathbf{I}_Y + \mathbf{I}_B) = -\,[(10.16 - j\,7.62) + (4.46024 - j\,9.01954) + (-\,5.08 + j\,8.79856)] = -\,9.54024 + j\,7.80498$$

Evaluating these currents we get

$$I_R = \sqrt{(10.16)^2 + (-7.62)^2} = 12.7 \text{ A } \textbf{Ans.}$$

$$I_Y = \sqrt{(4.46024)^2 + (-9.01954)^2} = 10.06 \text{ A } \textbf{Ans.}$$

$$I_B = \sqrt{(-5.08)^2 + (8.79856)^2} = 10.16 \text{ A } \textbf{Ans.}$$

$$I_N = \sqrt{(-9.54024)^2 + (7.80498)^2} = 12.35 \text{ A } \textbf{Ans.}$$

$$\text{Power in red phase} = I_R^2 \times R_R = (12.7)^2 \times 16 = 2{,}580 \text{ watts } \textbf{Ans.}$$

$$\text{Power in yellow phase} = I_Y^2 \times R_Y = (10.06)^2 \times 14 = 1{,}416 \text{ watts } \textbf{Ans.}$$

$$\text{Power in blue phase} = I_B^2 \times R_B = (10.16)^2 \times 25 = 2{,}580 \text{ watts } \textbf{Ans.}$$

Example 2.38. A 3-phase, 4-wire system with 400 V line voltage supplies the following loads :

(*a*) A 400 V, 3-Φ, 75-H P motor at 0.8 pf lagging and with an efficiency of 88%, (*b*) A 15 kW unity power factor load connected between phase A and neutral (*c*) A 12.5 kW 0.866 leading power factor load connected between phase B and neutral. Calculate (*i*) the neutral current (*ii*) the three line currents (*iii*) total power delivered by the supply. Phase sequence being ABC.

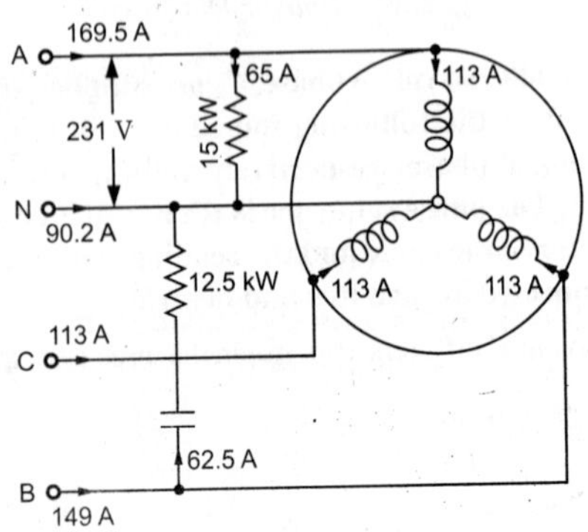

Fig. 2.52 *Circuit Diagram*

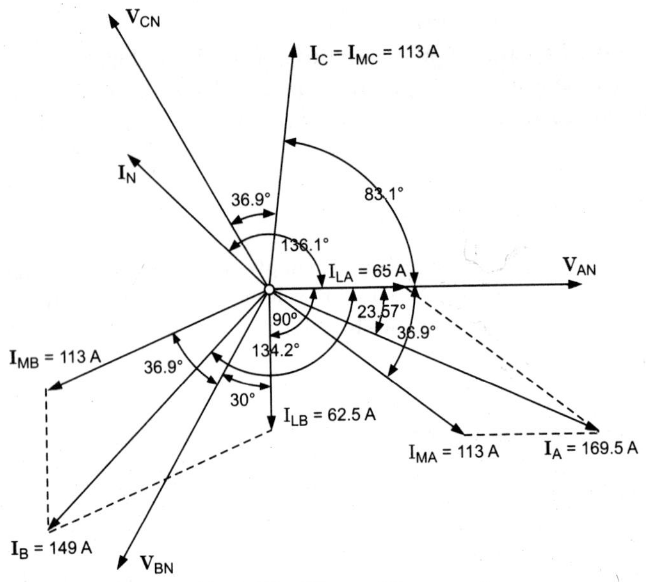

Fig. 2.53 *Phasor Diagram*

Solution: The potential differences between the three lines and neutral are given as:

$$\mathbf{V}_{AN} = \frac{400}{\sqrt{3}}(1 + j\,0) \text{ or } \frac{400}{\sqrt{3}}\angle 0^\circ \text{ volts}$$

$$\mathbf{V}_{BN} = \frac{400}{\sqrt{3}}(-\,0.5 - j\,0.866) \text{ or } \frac{400}{\sqrt{3}}\angle -120^\circ \text{ volts}$$

$$\text{and } \mathbf{V}_{CN} = \frac{400}{\sqrt{3}}(-\,0.5 + j\,0.866) \text{ or } \frac{400}{\sqrt{3}}\angle 120^\circ \text{ volts}$$

Current drawn by 15 kW load at unity power factor,

$$I_{LA} = \frac{\text{Load in watts}}{V_P \times \cos\Phi} = \frac{15{,}000}{\frac{400}{\sqrt{3}} \times 1} = 65 \text{ A in phase with } V_{AN}$$

$$\therefore I_{LA} = 65\angle 0^\circ \text{ or } (65 + j\,0) \text{ A}$$

Current drawn by 12.5 kW load at 0.866 power factor leading,

$$I_{LB} = \frac{\text{Load in watts}}{V_P \times \cos\Phi} = \frac{12{,}500}{\frac{400}{\sqrt{3}} \times 0.866}$$

$$= 62.5 \text{ A leading } V_{BN} \text{ by an angle } \cos^{-1} 0.866 \text{ or } 30^\circ$$

$$\therefore \mathbf{I}_{LB} = 62.5\angle -120 + 30^\circ = 62.5\angle -90^\circ \text{ or } (0 - j\,62.5) \text{ A}$$

Current drawn by the motor per phase

= Line current drawn by motor

$$\therefore I_L = I_P = \frac{\text{Motor output in watts}}{\sqrt{3} \times V_L \times \cos\Phi \times \eta}$$

$$= \frac{75 \times 735.5}{\sqrt{3} \times 400 \times 0.8 \times 0.88} = 113 \text{ A}$$

∴ Motor will draw 113 A lagging the respective phase voltage by an angle $\cos^{-1}$ 0.8 or 36.9°

(*i*) Current in neutral wire,

$$\mathbf{I}_N = -\,(\mathbf{I}_{LA} + \mathbf{I}_{LB}) = -\,[(65 + j\,0) + (0 - j\,62.5)]$$

$$= (-\,65 + j\,62.5) \text{ A} = 90.2\angle 136.1^\circ \text{ A}$$

(*ii*) Current in line A will be phasor sum of I_{LA} and I_{MA}

$$\text{or } \mathbf{I}_A = \mathbf{I}_{LA} + \mathbf{I}_{MA} = (65 + j\,0) + 113\,(0.8 - j\,0.6)$$

$$= 169.5\angle -23.57^\circ \text{ A } \textbf{Ans.}$$

Current in line B will be phasor sum of I_{LB} and I_{MB}

$$\text{or } I_B = 62.5 \angle -90^\circ + 113 \angle -120^\circ - 36.9^\circ$$

$$= 149 \angle -134.2^\circ \text{ A } \textbf{Ans.}$$

Current in line C will be due to motor only and will be equal to 113 A lagging V_{CN} by 36.9° or lead V_{AN} by 83.1°

(*iii*) Total power delivered

$$= 15 + 12.5 + \frac{75 \times 0.7355}{0.88} = 90.185 \text{ kW } \textbf{Ans.}$$

Phasor diagram is shown in Fig. 2.53.

2.21.3. Consequences of Disconnecting Neutral in a 3-phase, 4-Wire System. In case of a balanced three-phase, four-wire system when the neutral is disconnected, no change is produced. But in case of unbalanced 3-phase, 4-wire system when the neutral is disconnected, the loads which are connected between any two line conductors and the neutral are connected in series and potential difference across the combined load becomes equal to line voltage. The potential difference across each load is thus changed as per rating of the load. The changes in voltages and currents are usually very large (in most cases inadmissible). It is because of this reason that no fuses and circuit breakers are ever used in the neutral wire of such a 3-phase system.

The effect of disconnecting the neutral in a 3-phase 4-wire system will be clear from the following solved example.

Example 2.39. In a 3-phase, 4-wire, 400/230 V system a lamp of 100 watts is connected to one phase and neutral and a lamp of 200 watts is connected to the second phase and the neutral. Calculate the voltage across each lamp when the neutral wire is disconnected.

Solution: Rating of lamp L_1 = 100 watts

Rating of lamp L_2 = 200 watts

$$\text{Resistance of lamp } L_1, R_1 = \frac{(V_P)^2}{\text{Wattage of lamp } L_1}$$

$$= \frac{(230)^2}{100} = 529\ \Omega$$

$$\text{Resistance of lamp } L_2, R_2 = \frac{(230)^2}{200} = 264.5\ \Omega$$

When the neutral wire is disconnected, as shown in Fig. 2.54 (*b*) the lamp L_1 (a resistance of 529 Ω) and lamp L_2 (a resistance of 264.5 Ω) are connected in series and the potential difference across them would be equal to line voltage *i.e.* 400 V.

Current flowing through lamps,

$$I = \frac{V_L}{R_1 + R_2} = \frac{400}{529 + 264.5} = 0.504 \text{ A}$$

Voltage across lamp L_1 = $I R_1 = 0.504 \times 529 = 266.67$ V **Ans.**

Voltage across lamp L_2 = $I R_2 = 0.504 \times 264.5 = 133.33$ V **Ans.**

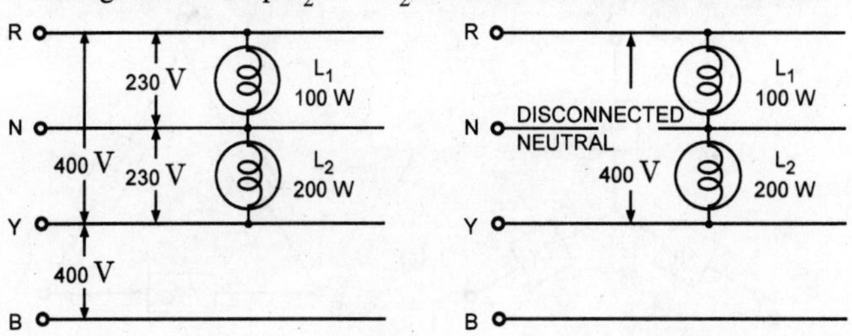

(*a*) (*b*)

Fig. 2.54

Note: *From the above illustration it is clear that if the neutral in 3-phase, 4-wire unbalanced system is disconnected, the potential difference across high resistance is increased and that across low resistance is decreased.*

2.21.4. 3-Phase, 3-Wire Star-Connected Unbalanced Load Circuits. Problems of unbalanced Y-connected load without neutral are purely academic and have little practical importance. Such a load situation would arise only when a star-connected load is unbalanced and its star point is not connected to the neutral of the generator. This is really a faulty or improper operation and needs repairs.

In case of balanced loads, the phasor sum of three line currents at the load star point is zero and operation remains the same with or without the neutral connection. But in case of unbalanced star-connected load without neutral there is no neutral wire to permit the flow of current caused due to unbalancing of loads. There is only one possibility—the individual line currents change so that their phasor sum becomes zero. This does not mean that currents become identical in all lines, for the phasor sum of different currents can be zero even with different line currents (different in magnitude as well as in phase). The change in currents consequently, causes change in voltage across each load. Thus the three-phase voltages are different (or unbalanced) in magnitude and in phase and it is quite possible that one phase voltage may exceed the line voltage. Such a condition is undesirable, since some loads may operate inefficiently due to lowering of the voltage and the other equipment may get damaged due to overvoltage.

As mentioned above, if the neutral point of the load is not connected to the neutral of the generator, the phase voltages do not remain the same across each phase but vary according to the unbalancing of load and under certain conditions of loading a considerable voltage may exist between the two star points. Such an isolated star point is known as a *floating neutral* because its potential is always changing.

The phasor diagram for an unbalanced 3-phase, 3-wire star-connected load circuit is shown in Fig. 2.55 in which line voltages are represented by an equilateral triangle RYB. As the load is unbalanced, instead of neutral point N, N' will be the neutral point and the potential difference across different phases will be V'_R, V'_Y and V'_B instead of V_R, V_Y and V_B and potential existing between two star points will be represented by NN'.

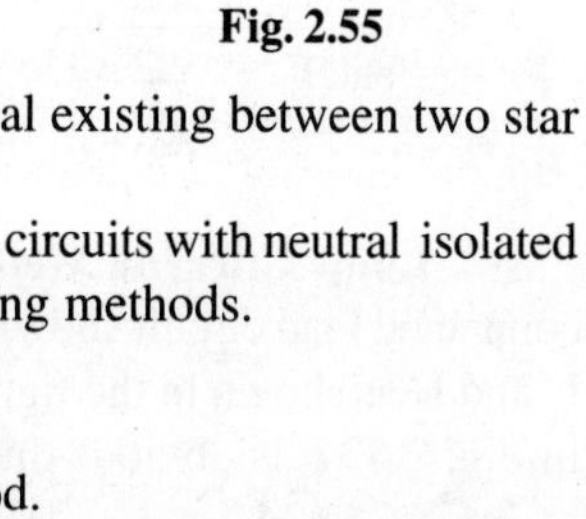

Fig. 2.55

Unbalanced star-connected load circuits with neutral isolated can be solved by any of the following methods.

1. By Kirchhoff's law
2. Loop current method and
3. Star-delta conversion method.
4. By Millman's theorem.

1. By Kirchhoff's Laws: Consider an unbalanced star-connected load circuit shown in Fig. 2.56.

By Kirchhoff's laws we get

$$\mathbf{V}_{RY} = \mathbf{I}_R \mathbf{Z}_R - \mathbf{I}_Y \mathbf{Z}_Y \quad ...(2.34)$$

$$\mathbf{V}_{YB} = \mathbf{I}_Y \mathbf{Z}_Y - \mathbf{I}_B \mathbf{Z}_B \quad ...(2.35)$$

$$\mathbf{V}_{BR} = \mathbf{I}_B \mathbf{Z}_B - \mathbf{I}_R \mathbf{Z}_R \qquad ...(2.36)$$

$$\text{and } \mathbf{I}_R + \mathbf{I}_Y + \mathbf{I}_B = 0 \qquad ...(2.37)$$

Adding Eqs. (2.34) and (2.36) we get

$$\mathbf{V}_{RY} + \mathbf{V}_{BR} = \mathbf{I}_R \mathbf{Z}_R - \mathbf{I}_Y \mathbf{Z}_Y + \mathbf{I}_B \mathbf{Z}_B - \mathbf{I}_R \mathbf{Z}_R = \mathbf{I}_B \mathbf{Z}_B - \mathbf{I}_Y \mathbf{Z}_Y$$

$$\text{or } \mathbf{I}_B = \frac{\mathbf{V}_{RY} + \mathbf{V}_{BR} + \mathbf{I}_Y \mathbf{Z}_Y}{\mathbf{Z}_B} \qquad ...(2.38)$$

Substituting $-\mathbf{I}_R = \mathbf{I}_Y + \mathbf{I}_B$ from Eq. (2.37) in Eq. (2.36) we get

$$\mathbf{V}_{BR} = \mathbf{I}_B \mathbf{Z}_B + (\mathbf{I}_Y + \mathbf{I}_B) \mathbf{Z}_R$$
$$= \mathbf{I}_B (\mathbf{Z}_B + \mathbf{Z}_R) + \mathbf{I}_Y \mathbf{Z}_R \qquad ...(2.39)$$

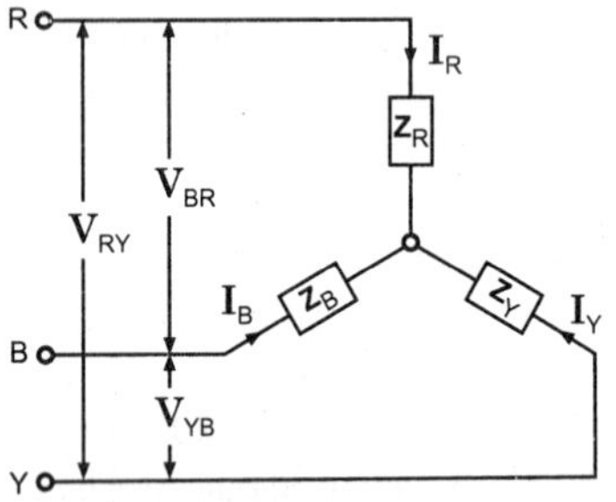

Fig. 2.56

Substituting $\mathbf{I}_B = \frac{\mathbf{V}_{RY} + \mathbf{V}_{BR} + \mathbf{I}_Y \mathbf{Z}_Y}{\mathbf{Z}_B}$ from Eq. (2.38) in Eq. (2.39) we get

$$\mathbf{V}_{BR} = \frac{\mathbf{V}_{RY} + \mathbf{V}_{BR} + \mathbf{I}_Y \mathbf{Z}_Y}{\mathbf{Z}_B} (\mathbf{Z}_B + \mathbf{Z}_R) + \mathbf{I}_Y \mathbf{Z}_R$$

$$\text{or } \mathbf{V}_{BR} \mathbf{Z}_B = \mathbf{V}_{RY} \mathbf{Z}_B + \mathbf{V}_{BR} \mathbf{Z}_B + \mathbf{I}_Y \mathbf{Z}_Y \mathbf{Z}_B + \mathbf{V}_{RY} \mathbf{Z}_R + \mathbf{V}_{BR} \mathbf{Z}_R + \mathbf{I}_Y \mathbf{Z}_Y \mathbf{Z}_R + \mathbf{I}_Y \mathbf{Z}_R \mathbf{Z}_B$$

$$\text{or } \mathbf{I}_Y (\mathbf{Z}_R \mathbf{Z}_Y + \mathbf{Z}_Y \mathbf{Z}_B + \mathbf{Z}_B \mathbf{Z}_R) = -\mathbf{Z}_R (\mathbf{V}_{RY} + \mathbf{V}_{BR}) - \mathbf{V}_{RY} \mathbf{Z}_B$$

$$\text{or } \mathbf{I}_Y (\mathbf{Z}_R \mathbf{Z}_Y + \mathbf{Z}_Y \mathbf{Z}_B + \mathbf{Z}_B \mathbf{Z}_R) = \mathbf{Z}_R \mathbf{V}_{YB} - \mathbf{Z}_B \mathbf{V}_{RY}$$

$$\because \mathbf{V}_{RY} + \mathbf{V}_{YB} + \mathbf{V}_{BR} = 0$$

$$\text{or } \mathbf{I}_Y = \frac{\mathbf{V}_{YB} \mathbf{Z}_R}{\mathbf{Z}_R \mathbf{Z}_Y + \mathbf{Z}_Y \mathbf{Z}_B + \mathbf{Z}_B \mathbf{Z}_R} - \frac{\mathbf{V}_{RY} \mathbf{Z}_B}{\mathbf{Z}_R \mathbf{Z}_Y + \mathbf{Z}_Y \mathbf{Z}_B + \mathbf{Z}_B \mathbf{Z}_R}$$

$$\text{or } \mathbf{I}_Y = \frac{\mathbf{V}_{YB}}{\mathbf{Z}_Y + \mathbf{Z}_B + \frac{\mathbf{Z}_Y \mathbf{Z}_B}{\mathbf{Z}_R}} - \frac{\mathbf{V}_{RY}}{\mathbf{Z}_R + \mathbf{Z}_Y + \frac{\mathbf{Z}_R \mathbf{Z}_Y}{\mathbf{Z}_B}} \qquad ...(2.40)$$

$$\text{Similary, } \mathbf{I}_B = \frac{\mathbf{V}_{BR}}{\mathbf{Z}_B + \mathbf{Z}_R + \frac{\mathbf{Z}_B \mathbf{Z}_R}{\mathbf{Z}_Y}} - \frac{\mathbf{V}_{YB}}{\mathbf{Z}_Y + \mathbf{Z}_B + \frac{\mathbf{Z}_Y \mathbf{Z}_B}{\mathbf{Z}_R}} \qquad ...(2.41)$$

$$\text{and } \mathbf{I}_R = \frac{\mathbf{V}_{RY}}{\mathbf{Z}_R + \mathbf{Z}_Y + \frac{\mathbf{Z}_R \mathbf{Z}_Y}{\mathbf{Z}_B}} - \frac{\mathbf{V}_{BR}}{\mathbf{Z}_B + \mathbf{Z}_R + \frac{\mathbf{Z}_B \mathbf{Z}_R}{\mathbf{Z}_Y}} \qquad ...(2.42)$$

2. Loop Current Method. Consider an unbalanced star-connected load circuit shown in Fig. 2.57. Take the loop currents I_1 and I_2 as shown in the figure. From the circuit diagram shown in Fig. 2.57 it is obvious that

$$\mathbf{I}_R = \mathbf{I}_1 ; \mathbf{I}_Y = \mathbf{I}_2 - \mathbf{I}_1 \text{ and } \mathbf{I}_B = -\mathbf{I}_2$$

Now considering Kirchhoff's voltage law we have

$$\mathbf{V}_{RY} = \mathbf{I}_R \mathbf{Z}_R - \mathbf{I}_Y \mathbf{Z}_Y = \mathbf{I}_1 \mathbf{Z}_R - (\mathbf{I}_2 - \mathbf{I}_1) \mathbf{Z}_Y$$
$$= \mathbf{I}_1 (\mathbf{Z}_R + \mathbf{Z}_Y) - \mathbf{I}_2 \mathbf{Z}_Y \qquad ...(2.43)$$

$$\text{and } \mathbf{V}_{YB} = \mathbf{I}_Y \mathbf{Z}_Y - \mathbf{I}_B \mathbf{Z}_B$$
$$= (\mathbf{I}_2 - \mathbf{I}_1) \mathbf{Z}_Y - (-\mathbf{I}_2 \mathbf{Z}_B) = \mathbf{I}_2 (\mathbf{Z}_Y + \mathbf{Z}_B) - \mathbf{I}_1 \mathbf{Z}_Y \qquad ...(2.44)$$

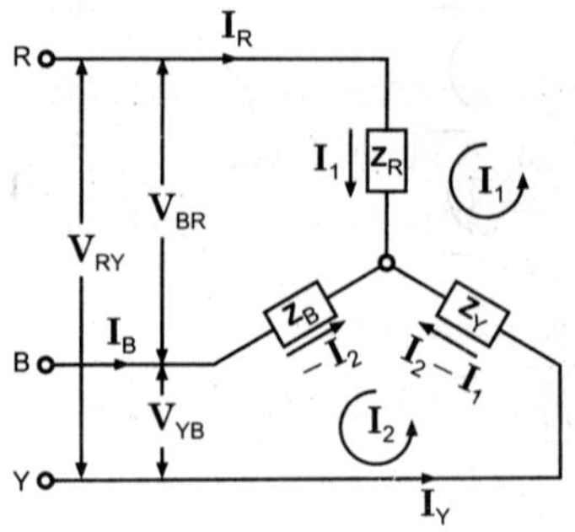

Fig. 2.57

Solving Eqs. (2.43) and (2.44) for $\mathbf{I}_1$ and $\mathbf{I}_2$ we have

$$\mathbf{I}_1 = \frac{\mathbf{V}_{RY}(\mathbf{Z}_Y + \mathbf{Z}_B) + \mathbf{V}_{YB}\mathbf{Z}_Y}{(\mathbf{Z}_R + \mathbf{Z}_Y)(\mathbf{Z}_Y + \mathbf{Z}_B) - \mathbf{Z}_Y^2} \qquad ...(2.45)$$

$$\text{and } \mathbf{I}_2 = \frac{\mathbf{V}_{YB}(\mathbf{Z}_R + \mathbf{Z}_Y) + \mathbf{V}_{RY}\mathbf{Z}_Y}{(\mathbf{Z}_R + \mathbf{Z}_Y)(\mathbf{Z}_Y + \mathbf{Z}_B) - \mathbf{Z}_Y^2} \qquad ...(2.46)$$

3. Star-Delta Conversion Method. In case of unbalanced star-connected load it is sometimes more convenient to convert it into an equivalent delta-connected load and solve it.

Let $\mathbf{Z}_R$, $\mathbf{Z}_Y$ and $\mathbf{Z}_B$ be the impedances of the star-connected load and $\mathbf{Z}_{RY}$, $\mathbf{Z}_{YB}$ and $\mathbf{Z}_{BR}$ be the impedances of the equivalent delta-connected load then

$$\mathbf{Z}_{RY} = \mathbf{Z}_R + \mathbf{Z}_Y + \frac{\mathbf{Z}_R \mathbf{Z}_Y}{\mathbf{Z}_B} \qquad ...(2.47)$$

$$\mathbf{Z}_{YB} = \mathbf{Z}_Y + \mathbf{Z}_B + \frac{\mathbf{Z}_Y \mathbf{Z}_B}{\mathbf{Z}_R} \qquad (2.48)$$

$$\text{and } \mathbf{Z}_{BR} = \mathbf{Z}_B + \mathbf{Z}_R + \frac{\mathbf{Z}_B \mathbf{Z}_R}{\mathbf{Z}_Y}$$

Now the phase currents and line currents can be determined as discussed in Art. 2.21.1 *i.e.* from the following relations.

i.e. $$\mathbf{I}_{RY} = \frac{\mathbf{V}_{RY}}{\mathbf{Z}_{RY}} ; \mathbf{I}_{YB} = \frac{\mathbf{V}_{YB}}{\mathbf{Z}_{YB}} ; \mathbf{I}_{BR} = \frac{\mathbf{V}_{BR}}{\mathbf{Z}_{BR}}$$

4. By Millman's Theorem. As per this theorem, the voltage of the load star point with respect to earth or neutral of the generator is given as (Fig. 2.55)

$$\mathbf{V}_{NN'} = \frac{\mathbf{V}_R \mathbf{Y}_R + \mathbf{V}_Y \mathbf{Y}_Y + \mathbf{V}_B \mathbf{Y}_B}{\mathbf{Y}_R + \mathbf{Y}_Y + \mathbf{Y}_B}$$

where $\mathbf{V}_R$, $\mathbf{V}_Y$ and $\mathbf{V}_B$ are the phase voltages of generator or 3-phase supply.

As seen from Fig. 2.55 the voltage across each of the load is given as

$$\mathbf{V}'_R = \mathbf{V}_R - \mathbf{V}_{NN'} ; \mathbf{V}'_Y = \mathbf{V}_Y - \mathbf{V}_{NN'} \text{ and } \mathbf{V}'_B = \mathbf{V}_B - \mathbf{V}_{NN'}$$

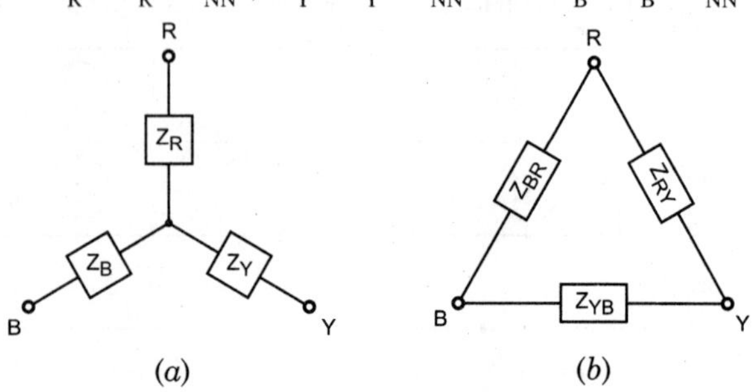

(*a*) (*b*)

Fig. 2.58 *Delta-Star Transformation*

Now the phase current can be determined from the relations

$$I_R = \frac{V'_R}{Z_R}\ ;\ I_Y = \frac{V'_Y}{Z_Y} \text{ and } I_B = \frac{V'_B}{Z_B}$$

Example 2.40. Calculate the voltage drop across each of the three star-connected impedances shown here (Fig. 2.59) when they are connected to a 400 V, 3-φ system (phase sequence RYB). Also find the potential of star point above ground.

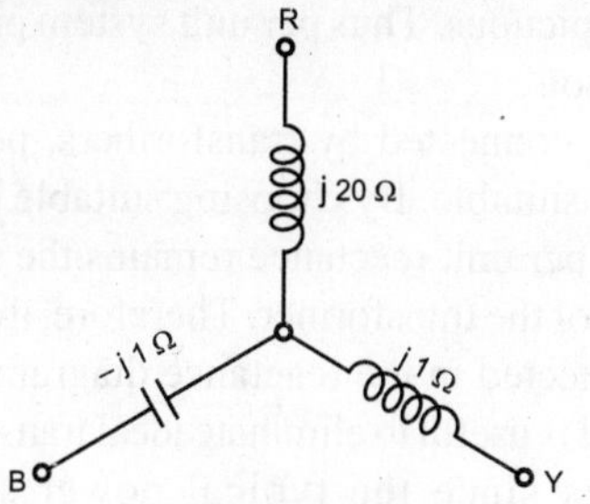

Fig. 2.59

Solution: Taking $\mathbf{V}_{RY}$ as reference phasor

$$\mathbf{V}_{RY} = 400\,\angle 0^\circ \text{ or } 400\,(1 + j\,0) \text{ volts}$$

$$\mathbf{V}_{YB} = 400\,\angle -120^\circ \text{ or } 400\,(-0.5 - j\,0.866) \text{ volts}$$

$$\mathbf{V}_{BR} = 400\,\angle 120^\circ \text{ or } 400\,(-0.5 + j\,0.866) \text{ volts}$$

$$\mathbf{Z}_R = 20\,\angle 90^\circ \text{ or } (0 + j\,20)\ \Omega$$

$$\mathbf{Z}_Y = 1\,\angle 90^\circ \text{ or } (0 + j\,1)\ \Omega$$

$$\mathbf{Z}_B = 1\,\angle -90^\circ \text{ or } (0 - j\,1)\ \Omega$$

$$\text{Now } \mathbf{I}_R = \frac{\mathbf{V}_{RY}}{\mathbf{Z}_R + \mathbf{Z}_Y + \dfrac{\mathbf{Z}_R\mathbf{Z}_Y}{\mathbf{Z}_B}} - \frac{\mathbf{V}_{BR}}{\mathbf{Z}_B + \mathbf{Z}_R + \dfrac{\mathbf{Z}_B\mathbf{Z}_R}{\mathbf{Z}_Y}}$$

$$= \frac{400\,\angle 0^\circ}{(0 + j20) + (0 + j1) + \dfrac{20\,\angle 90^\circ \times 1\,\angle 90^\circ}{1\,\angle -90^\circ}} - \frac{400\,\angle 120^\circ}{(0 - j1) + (0 + j20) + \dfrac{1\,\angle -90^\circ \times 20\,\angle 90^\circ}{1\,\angle 90^\circ}}$$

$$= \frac{400\,\angle 0^\circ}{0 + j21 - j20} - \frac{400\,\angle 120^\circ}{0 + j19 - j20}$$

$$= 400\,\angle -90^\circ - 400\,\angle 210^\circ$$

$$= (346.41 - j\,200) \text{ A or } 400\,\angle -30^\circ \text{ A}$$

$$\mathbf{I}_Y = \frac{\mathbf{V}_{YB}}{\mathbf{Z}_Y + \mathbf{Z}_B + \dfrac{\mathbf{Z}_Y\mathbf{Z}_B}{\mathbf{Z}_R}} - \frac{\mathbf{V}_{RY}}{\mathbf{Z}_R + \mathbf{Z}_Y + \dfrac{\mathbf{Z}_R\mathbf{Z}_Y}{\mathbf{Z}_B}}$$

$$= \frac{400\,\angle -120^\circ}{(0 + j1) + (0 - j1) + \dfrac{1\,\angle 90^\circ \times 1\,\angle -90^\circ}{20\,\angle 90^\circ}} - \frac{400\,\angle 0^\circ}{(0 + j20) + (0 + j1) + \dfrac{20\,\angle 90^\circ \times 1\,\angle 90^\circ}{1\,\angle -90^\circ}}$$

$$= 8{,}000\,\angle -30^\circ - 400\,\angle -90^\circ = (6{,}928 - j\,4{,}000) + j\,400$$

$$= (6{,}928 - j\,3{,}600) \text{ or } 7{,}807.5\,\angle -27.46^\circ \text{ A } \textbf{Ans.}$$

$$\mathbf{I}_B = \frac{\mathbf{V}_{BR}}{\mathbf{Z}_B + \mathbf{Z}_R + \dfrac{\mathbf{Z}_B\mathbf{Z}_R}{\mathbf{Z}_Y}} - \frac{\mathbf{V}_{YB}}{\mathbf{Z}_Y + \mathbf{Z}_B + \dfrac{\mathbf{Z}_Y\mathbf{Z}_B}{\mathbf{Z}_R}}$$

$$= \frac{400\,\angle 120^\circ}{(0 - j1) + (0 + j20) + \dfrac{1\,\angle -90^\circ \times 20\,\angle 90^\circ}{1\,\angle 90^\circ}} - \frac{400\,\angle -120^\circ}{(0 + j1) + (0 - j1) + \dfrac{1\,\angle 90^\circ \times 1\,\angle -90^\circ}{20\,\angle 90^\circ}}$$

$$= 400\,\angle 210^\circ - 8{,}000\,\angle -30^\circ$$

$$= -346.41 - j\,200 - (6{,}928 - j\,4{,}000)$$

$$= (-7{,}274.41 + j\,3{,}800) \text{ or } 8{,}207\,\angle 152.42^\circ \text{ A}$$

Voltage drop

$$\mathbf{V}'_R = \mathbf{I}_R\,\mathbf{Z}_R = 400\,\angle -30^\circ \times 20\,\angle 90^\circ = 8{,}000\,\angle 60^\circ \text{ V } \textbf{Ans.}$$

Voltage drop

$$\mathbf{V}'_Y = \mathbf{I}_Y\,\mathbf{Z}_Y = 7{,}807.5\,\angle -27.46^\circ \times 1\,\angle 90^\circ$$

$$= 7{,}807.5\,\angle 62.54^\circ \text{ V } \textbf{Ans.}$$

Voltage drop

$$\mathbf{V}'_B = \mathbf{I}_B\,\mathbf{Z}_B = 8{,}207\,\angle 152.42^\circ \times 1\,\angle -90^\circ$$

$$= 8{,}207\,\angle 62.42^\circ \text{ V } \textbf{Ans.}$$

Potential of star above ground,

$$\mathbf{V}_{N'N} = -\frac{1}{3}\,(\mathbf{V}'_R + \mathbf{V}'_Y + \mathbf{V}'_B)$$

$$= -\frac{1}{3}\,[8{,}000\,\angle 60^\circ + 7{,}807.5\,\angle 62.54^\circ + 8{,}207\,\angle 62.42^\circ]$$

$$= 8{,}005\,\angle -118.34^\circ \text{ V } \textbf{Ans.}$$

Example 2.41. Three loads (10 + *j* 15), (3 – *j* 4) and (20 + *j* 10) Ω are connected in star to a 3-φ, 200 V supply. Find phasor currents, line currents and power. [*M.D. Univ. Electrical Technology May-2008*]

Solution: Assuming phase sequence R, Y, B and taking V_{RY} as reference phasor we have

$$\mathbf{V}_{RY} = 200\,\angle 0^\circ \text{ V}, \mathbf{V}_{YB} = 200\,\angle -120^\circ$$

$$\text{and } \mathbf{V}_{BR} = 200\,\angle 120^\circ \text{ V}$$

$$\text{Impedances, } \mathbf{Z}_R = (10 + j\,15) = 18.028\,\angle 56.31^\circ\ \Omega$$

$$\mathbf{Z}_Y = (3 - j\,4) = 5\,\angle -53.13^\circ\ \Omega$$

$$\mathbf{Z}_B = (20 + j\,10) = 22.36\,\angle 26.56^\circ\ \Omega$$

By delta-star conversion

$$\mathbf{Z}_{RY} = \mathbf{Z}_R + \mathbf{Z}_Y + \frac{\mathbf{Z}_R\,\mathbf{Z}_Y}{\mathbf{Z}_B}$$

$$= 18.028\,\angle 56.31^\circ + 5\,\angle -53.13^\circ + \frac{18.028\,\angle 56.31 \times 5\,\angle -53.13^\circ}{22.36\,\angle 26.56^\circ}$$

$$= (10 + j\,15) + (3 - j\,4) + 4.031\,\angle -23.38^\circ$$

$$= (10 + j\,15) + (3 - j\,4) + (3.7 - j\,1.6) = (16.7 + j\,9.4) = 19.16\,\angle 29.37^\circ\ \Omega$$

$$\mathbf{Z}_{YB} = \mathbf{Z}_Y + \mathbf{Z}_B + \frac{\mathbf{Z}_Y\,\mathbf{Z}_B}{\mathbf{Z}_R}$$

$$= (3 - j\,4) + (20 + j\,10) + \frac{5\,\angle -53.13^\circ \times 22.36\,\angle 26.56^\circ}{18.028\,\angle 56.31^\circ}$$

$= (3 - j\,4) + (20 + j\,10) + 6.2 \angle -82.88^\circ$

$= (3 - j\,4) + (20 + j\,10) + (0.77 - j\,6.15) = (23.77 - j\,0.15)$

$= 23.77 \angle -0.36^\circ\ \Omega$

$$\mathbf{Z}_{BR} = \mathbf{Z}_B + \mathbf{Z}_R + \frac{\mathbf{Z}_B \mathbf{Z}_R}{\mathbf{Z}_Y}$$

$$= (20 + j\,10) + (10 + j\,15) + \frac{22.36 \angle 26.56^\circ \times 18.028 \angle 56.31^\circ}{5 \angle -53.13^\circ}$$

$= (20 + j\,10) + (10 + j\,15) + 80.62 \angle 136^\circ$

$= (20 + j\,10) + (10 + j\,15) + (-5.8 + j\,56)$

$= (24.2 + j\,81) = 84.54 \angle 73.37^\circ\ \Omega$

$$\mathbf{I}_{RY} = \frac{\mathbf{V}_{RY}}{\mathbf{Z}_{RY}} = \frac{200 \angle 0^\circ}{19.16 \angle 29.37^\circ}$$

$= 10.44 \angle -29.37^\circ = (9.097 - j\,5.12)$ A

$$\mathbf{I}_{YB} = \frac{\mathbf{V}_{YB}}{\mathbf{Z}_{YB}} = \frac{200 \angle -120^\circ}{23.77 \angle -0.36^\circ}$$

$= 8.414 \angle -119.64^\circ = (-4.16 - j\,7.31)$ A

$$\mathbf{I}_{BR} = \frac{\mathbf{V}_{BR}}{\mathbf{Z}_{BR}} = \frac{200 \angle 120^\circ}{84.54 \angle 73.37^\circ} = 2.37 \angle -73.37^\circ$$

$= (0.677 - j\,2.27)$ A

$\mathbf{I}_R = \mathbf{I}_{RY} - \mathbf{I}_{BR} = (9.097 - j\,5.12) - (0.677 - j\,2.27)$

$= (8.42 - j\,2.85)$ A $= 8.89 \angle -18.7^\circ$ A **Ans.**

$\mathbf{I}_Y = \mathbf{I}_{YB} - \mathbf{I}_{RY} = (-4.16 - j\,7.31) - (9.097 - j\,5.12)$

$= (13.257 - j\,2.19)$ A $= 13.437 \angle -9.38^\circ$ A

$\mathbf{I}_B = \mathbf{I}_{BR} - \mathbf{I}_{YB} = (0.677 - j\,2.27) - (-4.16 - j\,7.31)$

$= (4.837 + j\,5.04) = 6.98 \angle 46.18^\circ$ A **Ans.**

Total power consumed $= I_R^2 R_R + I_Y^2 R_Y + I_B^2 R_B$

$= (8.89)^2 \times 10 + (13.437)^2 \times 3 + (6.98)^2 \times 20$

$= 790.3 + 541.65 + 974.41 = 2{,}306.36$ W **Ans.**

2.22 PER UNIT (PU) METHOD OF REPRESENTING QUANTITIES

The quantities involved in power system are kVA, voltage, current and impedance of the equivalent circuits of the various system components. The equivalent circuits are at different voltages and are connected together in the system by means of transformers and interconnections. Each apparatus is rated in kVA and its impedance in actual ohms or in percentage value* referred to its rated kVA and rated voltage. In power system analysis, it is usual to express voltage, current, kVA and impedance in per unit of base or reference values of these quantities. Such a method simplifies the calculations.

The per unit value of any quantity is defined as

$$\frac{\text{the actual value of the quantity in any unit}}{\text{the base or reference value in the same unit}}$$

2.22.1. Advantages and Drawbacks of Per Unit (PU) System

Advantages. 1. Calculations are simplified.

2. The characteristics of machines (generators, transformers, motors etc.) when described in per unit system are specified by almost the same number, regardless of the rating of the machines. In other words, the characteristics (or parameters) tend to fall in relatively narrow range, making erroneous values conspicuous. Thus per unit system provides a method of comparison.
3. For circuits connected by transformers, per unit system is particularly suitable. By choosing suitable base kV's for the circuits the per unit reactance remains the same, referred to either sides of the transformer. Therefore, the various circuits can be connected in the reactance diagram.
4. This method is useful to eliminate ideal transformers as circuit components since the typical power system contains hundreds, if not thousands of transformers, and this is a non-trivial savings.

Drawbacks. 1. Some equations that hold in the unscaled cases are modified when scaled into per unit. Factors such as $\sqrt{3}$ and 3 are removed or added by this method.

2. Equivalent circuits of the components are modified, making them somewhat more abstract. Sometimes phase shifts that are clearly present in the unscaled circuit vanish in the per unit circuit.

2.22.2. Selection of Bases. For a common representation, base kVA and base voltage are to be chosen. Then the base current and base impedance can be expressed as follows :

$$\text{Base current, } I_B = \frac{\text{Base kVA}\,(\text{kVA}_B)}{\text{Base kV}\,(\text{kV}_B)} \text{ amperes} \quad ...(2.49)$$

$$\text{Base impedance, } Z_B = \frac{\text{Base voltage in volts, } V_B}{\text{Base current in amperes, } I_B} \text{ ohms}$$

$$= \frac{V_B \times V_B}{I_B \times V_B}$$

$$= \frac{V_B^2}{V_B\, I_B}$$

$$= \frac{\left(\dfrac{V_B}{1{,}000}\right) \times \left(\dfrac{V_B}{1{,}000}\right) \times 1{,}000}{\dfrac{V_B\, I_B}{1{,}000}}$$

$$= \frac{(\text{kV}_B)^2 \times 1{,}000}{\text{kVA}_B}$$

$$= \frac{(\text{kV}_B)^2}{\text{MVA}_B} \text{ ohms} \quad ...(2.50)$$

For a single-phase, phase-to-neutral voltage, kVA per phase are taken as bases while in a three-phase system, three-phase line-to-line voltage and three-phase kVA are used as bases. This simplifies the calculations.

* Per cent value = Per unit value × 100

Per cent value is not convenient for use as the factor of 100 has to be carried in computations.

The base values in a system are so selected that the per unit voltages and currents in the system are approximately unity. Sometimes the base kVA is chosen equal to the sum of the ratings of the various equipments in the system or equal to the kVA rating of the largest unit connected in the system.

If in the network there is no transformer present, the same base voltage is used throughout, but if the transformers are present, the rule is to change the base voltage in proportion to the transformation ratio of the transformer when transformer is reached. Thus, all impedances in the network before the transformer is reached, including the transformer primary leakage impedance, are expressed in per unit, to the voltage base chosen for the primary side; all impedances beyond the transformer, including the transformer secondary leakage impedance, are expressed in per unit to a new base voltage which is the primary side base voltage multiplied by the transformer transformation ratio. This is very important.

2.22.3. Determination of Base Impedance. Voltage, current, volt-amperes and impedances are so related that selection of base values for any two of them determines the base values of the remaining two. For example for a single-phase circuit we have the following relations

$$S_B = V_B I_B \quad ...(2.51)$$

$$Z_B = \frac{V_B}{I_B} \quad ...(2.52)$$

where S_B, V_B, I_B and Z_B represent the base power, base voltage, base current and base impedance expressed in volt-amperes, volts, amperes and ohms respectively. Thus from above two equations, if any two of the four quantities are specified, the remaining two may be determined without any problem. As already mentioned in Art. 2.22.2, base kVA and base kV are usually specified. Let kVA_B and kV_B be the base kVA and base kV, then

$$\text{Per unit kV} = \frac{\text{Actual kV}}{\text{Base kV}} = \frac{kV_{actual}}{kV_B} \quad ...(2.53)$$

$$\text{Base current, } I_B = \frac{\text{Base kVA}}{\text{Base kV}} = \frac{kVA_B}{kV_B} \text{ amperes} \quad ...(2.54)$$

$$\text{Per unit current, } I_{pu} = \frac{\text{Actual current}}{\text{Base current}}$$

$$= \frac{\text{Actual current}}{kVA_B} \times kV_B \quad ...(2.55)$$

Base impedance,

$$Z_B = \frac{\text{Base kV} \times 1{,}000}{\text{Base current}} = kV_B \times \frac{kV_B}{kVA_B} \times 1{,}000$$

$$= \frac{(kV_B)^2 \times 1{,}000}{kVA_B} \text{ ohms}$$

...already worked out [Eq. (2.50)]

Base power,

$$S_B = \text{Base kVA} \quad ...(2.56)$$

Per unit impedance,

$$Z_{pu} = \frac{\text{Actual impedance}}{\text{Base impedance}}$$

$$= \text{Actual impedance} \times \frac{kVA_B}{(kV_B)^2 \times 1{,}000} \quad ...(2.57)$$

In a **3-phase system** rather than obtaining the per unit values using per phase quantities, the per unit values can be obtained directly by using three-phase base quantities (total kVA, line-to-line voltage and line currents). Let kVA_B be the base kVA (total output of three phases) and kV_B be the base line-to-line voltage in kV.

Assuming star connection (equivalent star can always be found)

$$\text{Base current, } I_B = \frac{kVA_B}{\sqrt{3}\, kV_B} \text{ amperes}$$

Base impedance,

$$Z_B = \frac{kV_B \times 1{,}000}{\sqrt{3} \times I_B} = \frac{kV_B}{\sqrt{3}} \times \frac{\sqrt{3}\, kV_B}{kVA_B} \times 1{,}000$$

$$= \frac{(kV_B)^2 \times 1{,}000}{kVA_B} \text{ ohms} \quad ...(2.58)$$

Per unit impedance,

$$Z_{pu} = \frac{\text{Actual impedance}}{\text{Base impedance}}$$

$$= \frac{\text{Actual impedance} \times kVA_B}{(kV_B)^2 \times 1{,}000} \quad ...(2.59)$$

It should be noted that the same expression is obtained for a single-phase and three-phase systems. In power system problems the data are given in terms of 3-phase kVA, line-to-line kV, the machine kVA and machine phase-to-phase (or line-to-line) kV rating. Further the direct-axis synchronous reactance is also known as positive sequence reactance of the machine.

2.22.4. Change of Base. Normally the per unit impedance of various components corresponding to its own rating voltage and kVA are given and since we choose one common base kVA and base kV for the whole system, therefore, it becomes imperative to determine the per unit impedance of the various equipments corresponding to the common base kV and base kVA. If the individual quantities are $Z_{pu\,old}$, kVA_{old}, kV_{old} and the common base quantities are $Z_{pu\,new}$, kVA_{new} and kV_{new}, then making use of Eq. (2.57), according to which per unit impedance is directly proportional to the base kVA and inversely proportional to the square of the base kV, we have

Per unit impedance to new base,

$$Z_{pu\,new} = Z_{pu\,old} \times \frac{kVA_{new}}{kVA_{old}} \times \frac{(kV_{old})^2}{(kV_{new})^2} \quad ...(2.60)$$

This is very important relation used in power system analysis.

Example 2.42. Base voltage = 1,100 volts. Base kVA = 10^6. What is base impedance ?

Solution: Base kVA $= 10^6$

$$\text{Base kV} = \frac{1{,}100}{1{,}000} = 1.1 \text{ kV}$$

$$\text{Base current, } I_B = \frac{kVA_B}{kV_B} = \frac{10^6}{1.1}$$

$$\text{Base impedance, } Z_B = \frac{kV_B \times 1{,}000}{I_B} = \frac{1.1 \times 1{,}000 \times 1.1}{10^6}$$

$$= 0.00121\ \Omega \quad \textbf{Ans.}$$

Base impedance can also be determined directly by substituting the values of base kVA and base kV in Eq. (2.50) and we get

$$Z_B = \frac{(kV_B)^2 \times 1{,}000}{kVA_B} = \frac{(1.1)^2 \times 1{,}000}{10^6}$$

$$= 0.00121\ \Omega \quad \textbf{Ans.}$$

Same as above.

Example 2.43. If the resistance in ohms is 5 Ω, find the per unit value. Given base kVA = 10 and base kV = 11.

Solution: Resistance, R = 5 Ω

$$\text{Base resistance, } R_B = \frac{(kV_B)^2 \times 1{,}000}{kVA_B} = \frac{(11)^2 \times 1{,}000}{10}$$

$$= 12{,}100\ \Omega$$

$$\text{Per unit resistance, } R_{pu} = \frac{R}{R_B} = \frac{5}{12{,}100} = 0.000413 \text{ pu} \quad \textbf{Ans.}$$

Example 2.44. A single-phase transformer is rated as 2.5 kVA, 11/0.4 kV. If the leakage reactance is 0.96 Ω when referred to low-voltage side, then determine its leakage reactance in per unit.

Solution: Base voltage, $kV_B = 0.4$ kV

Base kVA, $kVA_B = 2.5$ kVA

Low-voltage side base impedance,

$$Z_{BLV} = \frac{(kV_B)^2 \times 1{,}000}{kVA_B} = \frac{(0.4)^2 \times 1{,}000}{2.5} = 64\ \Omega$$

Leakage reactance in per unit,

$$Z_{pu} = \frac{\text{Actual reactance}}{\text{Base impedance}} = \frac{0.96}{64} = 0.015 \text{ pu} \quad \textbf{Ans.}$$

Example 2.45. A 30 MVA, 11 kV generator has a reactance of 0.2 pu referred to its ratings as bases. Determine the per unit reactance when referred to base kVA of 50,000 kVA and base kV of 33 kV.

Solution: Old base kVA = 30,000 kVA

Old base kV = 11 kV

Old per unit impedance,

$Z_{pu\ old}$ = 0.2 pu

New base kVA = 50,000 kVA

New base kV = 33 kV

New per unit impedance,

$$Z_{pu\ new} = Z_{pu\ old} \times \frac{\text{New base kVA}}{\text{Old base kVA}} \times \left(\frac{\text{Old base kV}}{\text{New base kV}}\right)^2$$

$$= 0.2 \times \frac{50{,}000}{30{,}000} \times \left(\frac{11}{33}\right)^2 = 0.037 \text{ pu} \quad \textbf{Ans.}$$

HIGHLIGHTS

1. An *alternating quantity* (voltage or current) is one which changes continuously in magnitude and alternates in direction at regular intervals of time. An alternating quantity that varies sinusoidally (*i.e.* according to the sin of angle θ) is called the *sinusoidal quantity.*

   Sinusoidal quantities are expressed as

   $$e = E_{max} \sin\theta = E_{max} \sin\omega t = E_{max} \sin 2\pi f t$$

   $$\text{and } i = I_{max} \sin\theta = I_{max} \sin\omega t = I_{max} \sin 2\pi f t$$

2. The effective and average values of different waves are given below in tabular form.

Waveform	*Sinusoidal*	*Half Wave Rectified AC*	*Square*	*Triangular*	*Sawtooth*	*Circular*
Effective Value	$0.707\ I_{max}$	$0.5\ I_{max}$	I_{max}	$\frac{1}{\sqrt{3}} I_{max}$	$\frac{1}{\sqrt{3}} I_{max}$	$\sqrt{\frac{2}{3}}\ I_{max}$
Average Value	$0.637\ I_{max}$	$0.318\ I_{max}$	I_{max}	$0.5\ I_{max}$	$0.5\ I_{max}$	$0.785\ I_{max}$

3. Two or more alternating quantities may be added in the same way as the forces are added.
4. An ac circuit can have three elements, namely R, L and C. Power is only consumed in R. The power consumed in elements L and C are zero.

   The current through pure inductor lags behind the applied voltage by $\frac{\pi}{2}$ radians while the current through a pure capacitor leads the applied voltage by $\frac{\pi}{2}$ radians.
5. In an R-L-C series ac circuit

   $$\text{Impedance, } Z = \sqrt{R^2 + (X_L - X_C)^2}$$

   Circuit behaves as an inductive circuit if $X_L > X_C$; as a capacitive if $X_L < X_C$ and as a purely resistive if $X_L = X_C$

   $$\text{Circuit current, } I = \frac{V}{Z}$$

   $$\text{Phase angle, } \phi = \tan^{-1}\frac{X_L - X_C}{R}$$

   $$\text{Power consumed, } P = V I \cos\phi = I^2 R$$

   $$\text{Power factor, } \cos\phi = \frac{R}{Z}$$
6. *Polyphase system* means the system which consists of numerous windings or circuits.
7. A supply system is said to be *symmetrical* when different voltages of the same frequency have equal magnitude and are displaced from one another by equal time angle. 3-phase, 3-wire supply system will be symmetrical when line-to-line voltages are equal in magnitude and displaced by 120 electrical degrees w.r.t. each other.
8. A 3-phase supply will be *unbalanced* when either of three phase voltages are unequal in magnitude or the phase angles between these phases is not equal to 120°.
9. A load circuit is said to be balanced when the loads (impedances) connected in different phases are same in magnitude as well as in phase.

10. For a balanced system, the phasor sum of the three phase voltages is always zero.
11. In case one phase out of three phases of a 3-phase supply connected to the three-phase load is not available, then such a condition is called the single phasing.
12. Commonly used polyphase system is 3-phase system.
13. *Phase sequence* is the order in which the currents or voltages in different phases attain their maximum values.
14. The three-phase system is connected in two ways namely star connection and delta connection.
15. Star-connected system is obtained by joining together similar ends, either the start or the finish, the other ends are joined to the line wires. The common point at which similar ends are connected is called the neutral or star point.
16. In a balanced star-connected system
    (*i*) Line voltages are 120° apart.
    (*ii*) Line voltages are 30° ahead of the respective phase voltages.
    (*iii*) Line voltages are $\sqrt{3}$ times of phase voltages.
    (*iv*) Line currents are equal to phase currents.
    (*v*) The angle between line currents and the corresponding line voltages is (30° ± ϕ); + ve for lagging currents and –ve for leading currents.
    (*vi*) True power output = $\sqrt{3}\ E_L I_L \cos\phi$, where ϕ is the angle between the respective phase current and phase voltage (not between line current and line voltage).
    (*vii*) Apparent power = $\sqrt{3}\ E_L I_L$
    (*viii*) In balanced system, the potential of neutral or star point is zero because potential at neutral or star point,

$$\mathbf{E}_{NR} + \mathbf{E}_{NY} + \mathbf{E}_{NB} = 0$$

17. When the starting end of one coil is joined to the finishing end of another coil, the delta or mesh connection is obtained. The direction of emf in the coils are taken as positive from start to finished ends.
18. In balanced delta connected system
    (*i*) Line currents are 120° apart.
    (*ii*) Line currents are 30° behind the respective phase currents.
    (*iii*) Line currents are $\sqrt{3}$ times phase currents.
    (*iv*) The angle between line currents and corresponding line voltage is (30° ± ϕ) as in star system.
    (*v*) Line voltages are equal to phase voltages.
    (*vi*) Power = $\sqrt{3}\ E_L I_L \cos\phi$, where ϕ is the phase angle between respective phase current and phase voltage.
    (*vii*) Apparent power = $\sqrt{3}\ E_L I_L$
    (*viii*) In balanced system the resultant emf in the closed circuit will be zero.

$$i.e.\ \mathbf{E}_{RY} + \mathbf{E}_{YB} + \mathbf{E}_{BR} = 0$$

    Hence there will be no circulating current in the mesh if no load is connected to the lines.
19. Two-wattmeter method is generally used to measure 3-phase power. In this method, the current coils of the two wattmeters are connected in any two lines and potential coil is connected from its own current coil to the line without a current coil. The sum of the two wattmeter readings gives the total power of the circuit.
    This method of power measurement is applicable to the 3-phase, 3-wire circuit irrespective of the fact that the circuit is balanced or unbalanced, and star-connected or delta-connected.
    If load is balanced, then its power factor can also be calculated from the readings of the two wattmeters, which are given as

$$W_1 = V_L I_L \cos(30° - \phi)$$

$$\text{and } W_2 = V_L I_L \cos(30° + \phi) \text{ in case of inductive loads}$$

    and power factor cos ϕ

$$= \text{Cos}\left[\tan^{-1}\frac{\sqrt{3}\,(W_1 - W_2)}{W_1 + W_2}\right]$$

$$W_1 = V_L I_L \cos(30° + \phi)$$

$$\text{and } W_2 = V_L I_L \cos(30° - \phi) \text{ in case of capacitive circuits.}$$

20. In Δ-connected load circuits the voltage across each load phase is fixed, independent of the nature of the load and is equal to the line voltage so different phase currents can be determined in the usual way (as in balanced load circuits) and the line currents are determined by taking phasor difference of phase currents in pairs.
21. In three-phase 4-wire star-connected unbalanced load circuits the three phase currents or line currents can be determined by dividing the phase voltage by the impedance of concerned phase

$$\text{i.e. } \mathbf{I}_R = \frac{\mathbf{V}_R}{\mathbf{Z}_R};\ \mathbf{I}_Y = \frac{\mathbf{V}_Y}{\mathbf{Z}_Y} \text{ and } \mathbf{I}_B = \frac{\mathbf{V}_B}{\mathbf{Z}_B}$$

    Current in neutral wire is given by $\mathbf{I}_N = -(\mathbf{I}_R + \mathbf{I}_Y + \mathbf{I}_B)$
22. In case of unbalanced 3-phase, 3-wire star-connected load circuits, three phase voltages are different in magnitude and in phase and it is quite possible that one phase voltage may exceed the line voltage.
    Unbalanced star-connected load circuits with neutral isolated can be solved by any of the following methods :
    1. Kirchhoff's laws method. 2. Loop current method
    3. Star-delta conversion method. 4. Milliman's theorem method.
23. The per unit value of any quantity is defined as

$$\frac{\text{the actual value of the quantity in any unit}}{\text{the base or reference value in the same unit}}$$

EXERCISES

1. What do you understand by the following terms.
   (*i*) Three phase balance supply
   (*ii*) Three phase balance load.
   (*iii*) Three phase unbalance load
   (*iv*) Advantages of 3-phase over single-phase.
   [M.D. Univ. Electrical Technology, May-2002]
2. Discuss the advantages of 3-phase ac system over single-phase.
   [M.D. Univ. Electrical Technology, Ist Semester 2006-07]
3. What is a 3-phase system? Give its necessity and advantages. What is meaning of phase sequence and how can it be changed?
   [G.B. Technical Univ. Electrical Engineering Odd Semester 2009-10]
4. Show that power in 3-phase balanced system is constant at every instant and is given by 3 $V_P I_P \cos\phi$.
   [U.P. Technical Univ. Even Semester, 2013-14]
5. Describe star-connection method for interconnection of 3-phase supply. **[R.G. Technical Univ., December-2011]**

6. Show that for star connections of three-phase circuit, the line voltage is equal to 1.732 times the phase voltage whereas line current is equal to phase current. Also draw phasor diagram to support your answer.
7. Show that in a 3-ϕ star-connected system, the line voltage is $\sqrt{3}$ times of phase voltage. **[U.P. Technical Univ. Even Semester, 2013-14]**
8. Obtain the relationship between phase current and line current, phase voltage and line voltage in the 3-phase system in (*i*) star (Y connection), (*ii*) Delta (Δ connection). **[Rajasthan Technical Univ. February-2010]**
9. Find out the relation between line and phase quantities in delta connection. Also draw the phasor diagram and find the relation for power in delta connection. **[Rajasthan Technical Univ. June 2010]**
10. Derive the relation between line voltage and phase voltage, line current and phase current for delta connection in 3-phase system. **[M.D. Univ., May-2013]**
11. Explain the significance of power factor in ac system. **[M.D. Univ. Electrical Technology, December-2011; May-2012]**
12. What are the causes and disadvantages of low power factor? **[U.P. Technical Univ. Odd Semester, 2013-14]**
13. Explain the various methods to improve the power factor. **[Rajasthan Technical Univ., February-2010]**
14. The power in a balanced 3-phase, 3-wire, star-connected system is to be measured by two wattmeters. Draw neat connection and phasor diagrams. Do not derive the expression for the power measurement by each wattmeter. **[U.P.S.C. I.E.S. Electrical Engineering-I, 2013]**
15. Explain two wattmeter method of power measurement in 3-phase ac system at balanced load. **[M.D. Univ., May-2013]**
16. Discuss two-wattmeter method for power measurement in three-phase system and obtain a relation for power factor. **[DCRUSTM, Electrical Technology, December-2011/January-2012]**
17. Derive and explain the two-wattmeter method of measurement of three-phase power for a balanced star-connected load. How is the three-phase power determined ? Discuss the variations in readings for different power factors of loads from unity to zero. **[U.P. Technical Univ. Electrical Engineering Second Semester 2007-08]**
18. Prove that the power in a three-phase balanced circuit can be deduced from the readings of two wattmeters with the help of suitable vector diagrams. Discuss the nature of power factor when, (*i*) the two wattmeters read equal and positive values, (*ii*) the two wattmeters read equal and opposite values, (*iii*) the one wattmeter reads zero value. **[U.P. Technical Univ., February-2001]**
19. Distinguish between 3-phase balanced and unbalanced supply. What is the impact of unbalanced load on the power supply? **[R.G. Technical Univ., June-2014]**
20. A 3-phase load operate with balanced voltages applied to its terminals, and draws balanced currents. The potential coil of a moving coil wattmeter is connected from R to Y terminals of the load. The current coil of the meter is connected in series with phase B. By appropriate derivation, show that the quantity indicated by this wattmeter is proportional to the reactive power drawn by the load. **[GATE 2000]**
21. Star-connected unbalanced load is normally not used on 3-phase, 3-wire system. Explain why ?
22. Write short notes on the following :
    (*i*) Advantages of 3-phase system over single-phase system.
    (*ii*) Double subscript notation.
    (*iii*) Interconnection of 3-phases.
    (*iv*) Comparison of 3-phase star-and delta-connected systems.
23. What do you understand by per unit representation of a power system? Discuss. **[U.P. Technical Univ., 2005-06]**
24. What is meant by percentage reactance? Why is it preferable to express the reactances of various elements in percentage values for short-circuit calculations?
25. Explain the terms base kVA, base kV and percentage reactance. State the relationship between them.

SHORT ANSWER TYPE QUESTIONS WITH ANSWERS

Q. 1. What are unbalanced circuits? **[M.D. Univ. December-2009]**

Ans. A load circuit is said to be *balanced* when the loads (impedances) connected in various phases are same in magnitude as well as in phase. Any three-phase load in which the impedances in one or more phases differ from the impedances of other phases is called an *unbalanced three-phase load.*

Q. 2. Differentiate between balanced and unbalanced circuits. **[M.D. Univ., December-2010]**

Ans. A circuit is said to be balanced when the loads (impedances) connected in different phases are same in magnitude as well as in phase. But in unbalanced circuits impedances in one or more phases differ from those in other phases.

Q. 3. Why is an unbalanced load not normally used on a 3-phase 3-wire system ? Any line/phase voltages equal in such situations ? **[P.T.U. 1999]**

Ans. Normally, unbalanced load is not employed on 3-phase, 3-wire ungrounded star-connected system because unbalanced loading may cause different voltage drops in lines. Consequently the three-phase voltages are different (or unbalanced) in magnitude as well as in phase and it is quite possible that one phase voltage may exceed the line voltage. Such a condition is undesirable, since some loads may operate inefficiently due to lowering of the voltage and the other equipment may get damaged due to overvoltage.

Q. 4. Differentiate between balanced and unbalanced three-phase supply and balanced and unbalanced three-phase load. **[R.G.T.U., June-2002]**

Ans. 3-phase supply will be balanced when line-to-line voltages are equal in magnitude and displaced in phase by 120 electrical degrees with respect to each other. On the other hand a 3-phase supply will be unbalanced when either of the three-phase voltages are unequal in magnitude or the phase angle between these phase voltages is not equal to 120°.

A 3-phase load circuit is said to be balanced when the loads (impedances) connected in the different phases are the same in magnitude as well as in phase. On the other hand any three-phase load in which the impedances in one or more phases differ from the impedances of other phases is called the unbalanced three-phase load.

Q. 5. Show by a phasor diagram that the sum of the three-phase balanced currents is zero. **[R.G.T.U., June-2002]**

Ans. For a balanced 3-phase system, all phases should have equal maximum values and should be displaced by 120 electrical degrees with respect to each other, as illustrated in phasor diagram (Fig. 1)

i.e. $I_R = I_Y = I_B = I_P$ (say)

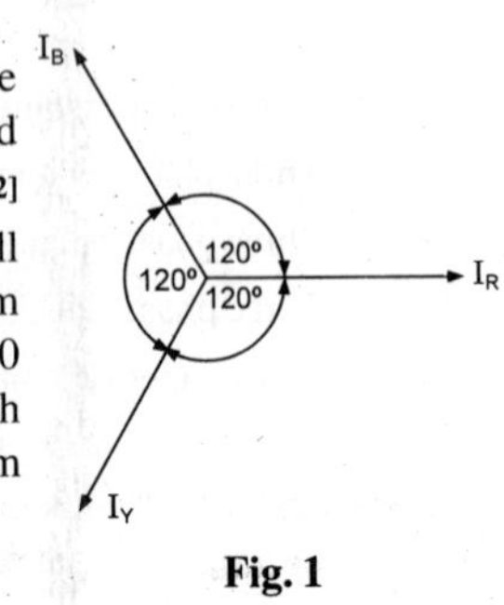

Fig. 1

The resultant of three-phase balanced currents = $\mathbf{I_R} + \mathbf{I_Y} + \mathbf{I_B}$

$= I_R \angle 0^\circ + I_B \angle 120^\circ + I_Y \angle -120^\circ$

$= I_R + I_B \cos 120^\circ + I_Y \cos(-120^\circ) + jI_B \sin 120^\circ + jI_Y \sin(-120^\circ)$

$= I_R - \frac{1}{2} I_B - \frac{1}{2} I_Y + j\left(\frac{\sqrt{3}}{2} I_B - \frac{\sqrt{3}}{2} I_Y\right) = 0 \quad \because I_R = I_Y = I_B$

Q. 6. What do you mean by phase sequence of three-phase supply ? **[R.G.T.U., June-2002]**

Ans. Phase sequence is meant the order in which the currents or voltages in different phases attain their maximum values one after the other.

Q. 7. Why interconnection of 3-phase system is necessary ?

Ans. Non-interlinked three-phase system will require 6 wires while interconnected 3-phase system will require 3 or 4 wires. So non-interlinked system will be very complicated and expensive.

Q. 8. Differentiate between star and delta connections. **[P.T.U., June-2003]**

Ans. In star-connected system similar (either start or finish) terminals of the three-phases are connected together to provide star or neutral point [refer to Fig. 2.18 (*a*)] while in delta-connected system start terminal of one phase is connected to the finish terminal of the second phase and start terminal of second phase is connected to finish terminal of the third phase and so on to provide a closed circuit in 3-phase system [refer to Fig. 2.23 (*a*)].

Q. 9. Differentiate between phase and line voltages in 3-phase circuits. **[M.D. Univ., May-2010]**

Ans. The voltage between any line and the neutral point or voltage across any phase winding is called the phase voltage whereas the voltage between any two outers is called the line voltage.

Q. 10. Give the relation between phase and line values of voltage and current for star connection. **[P.T.U., December-2009]**

Ans. In star-connection,

Line current, I_L = Phase current, I_P

Line voltage, $V_L = \sqrt{3}$ times phase voltage.

Q. 11. Discuss, why the stator windings of alternator are generally star-connected ? **[P.T.U., December-2000]**

Ans. In a star-connected system phase voltage is equal to $1/\sqrt{3}$ times the line voltage, but in a delta-connected system phase voltage is equal to line voltage, therefore, star-connected alternator needs less number of turns per phase and less insulation than a delta-connected alternator for the same line voltage.

Neutral point of a star-connected system can be connected to earth, so relays and protective devices can be provided in star-connected system for protection of the system against ground faults and it is usually done in practice. These are the reasons that 3-phase alternators are usually star-connected.

Q. 12. Why delta-connected system cannot be employed for supplying lighting as well as power loads ?

Ans. Delta-connected system cannot provide 3-phase 4-wire system (*i.e.* neutral wire cannot be carried) which is essential for supplying lighting as well as power loads.

Q. 13. Discuss, why the phase spread of three-phase winding is 60° and not 120° ? **[P.T.U., May-2010]**

Ans. The phase spread is to be taken in two dimensions. Total spread available in two dimensions is 180° and, therefore, the spread of three phase winding is $\frac{180^\circ}{3}$ *i.e.* 60°.

Q. 14. Name two electrical equipments in each case having unity power factor and lagging power factor. **[P.T.U., June-2003]**

Ans. Electrical equipments operating

at unity power factor: Incandescent lamps, electric heater.

at lagging power factor: Arc lamps, induction furnaces.

Q. 15. Draw the circuit diagram for 3-phase power measurement by two wattmeters. Under what operating conditions, one of the wattmeters will show a negative reading ? **[P.T.U., December-2001]**

Ans. The circuit diagrams for 3-phase power measurement by two wattmeters are shown in Figs. 2.37 (*a*) and 2.37 (*b*).

One of the wattmeters will show a negative reading when the power factor of the load circuit is below 0.5.

Q. 16. While measuring power in a balanced three-phase circuit, one of the wattmeters keeps showing zero as the load is varied. How do you account for it ? **[P.T.U., May-2002]**

Ans. One of the two wattmeters, in a two-wattmeter method of measuring power in a balanced 3-phase circuit, will keep showing zero as the load is varied if and only if the power factor of the load remains constant at 0.5. The reading of other wattmeter will vary with the variation in load.

Q. 17. What will happen if the neutral gets disconnected in a 3-phase 4-wire unbalanced circuit ?

Ans. In case of unbalanced 3-phase, 4-wire system when the neutral is disconnected, the loads which are connected between any two line conductors and the neutral are connected in series and potential difference across the combined load becomes equal to line voltage. The potential difference across each load is thus changed as per rating of the load. The changes in voltages and currents are usually very large (in most cases inadmissible).

Q. 18. What is meant by floating neutral in a 3-phase ac system ?

Ans. If the neutral point of the load is not connected to the neutral of the generator, the phase voltages do not remain the same across each phase but vary according to the unbalancing of load and under certain conditions of loading a considerable voltage may exist between the two star points. Such an isolated star-point is known as a *floating neutral* because its potential is always changing.

Q. 19. Why star-connected unbalanced load is not normally used on 3-phase, 3-wire system ?

Ans. In case of unbalanced 3-phase, 3-wire star-connected load circuits, three-phase voltages are different in magnitude and in phase and it is quite possible that one phase voltage may exceed the line voltage. Such a condition is undesirable, since some loads may operate inefficiently due to lowering of the voltage and the other equipment may get damaged due to overvoltage.

Q. 20. What is meant by percentage reactance ?

Ans. Percentage reactance can be defined as the voltage drop across the given reactance expressed as percentage of normal voltage when carrying full-load current pertaining to normal rating.

Q. 21. Which of the electrical quantities are chosen as base values ?

Ans. For common representation, base kVA (or MVA) and base voltage (kV) are chosen as base values.

Q. 22. What is meant by per unit value of any quantity ?

Ans. Per unit value of any quantity is defined as the ratio of the actual value of the quantity in any unit to the base or reference value in the same unit.

Q. 23. What is the relationship between base kVA, base kV and percentage reactance ?

Ans. Percentage reactance

$$= \text{Actual reactance} \times \frac{\text{base kVA}}{(\text{base kV})^2 \times 1{,}000} \times 100$$

Q. 24. Write expression for per unit impedance (Z_{pu}) in a 3-phase system. **[M.D. Univ. 2010-11]**

Ans. $Z_{pu} = \dfrac{Z \text{ in ohms} \times MVA_B}{(kV_B)^2}$

Q. 25. Why the per unit method is considered superior to per cent method for short-circuit calculations ?

Ans. Per cent value is not convenient for use as the factor of 100 has to be carried in computations.

Q. 26. Define base impedance of a power system element and write an expression for it in terms of base MVA and kV.

Ans. Base impedance is defined as the ratio of base voltage in volts and base current in amperes and it is expressed in terms of base MVA and kV as follows :

$$\text{Base impedance in ohms} = \frac{(\text{Base kV})^2}{\text{Base MVA}}$$

Q. 27. What is meant by base quantities ? [Bharathiyar Univ., April-1997]

Ans. For common representation, the quantities (voltage, current, power or impedance) choosen as common or reference quantities are called the base quantities.

Q. 28. Define pu impedance. [Bharathiyar Univ., November-1996]

Ans. Per unit impedance is defined as the ratio of voltage drop across the given impedance to the normal voltage when carrying full-load current pertaining to normal rating.

PROBLEMS

1. If the phase voltage of a 3-phase star-connected system is 200 V, what will be the line voltages (a) when the phases are correctly connected (b) when connections to one of phases are reversed ?

[U.P. Technical Univ. Odd Semester 2000-01; Odd Semester 2003-04]

[**Ans.** 200 V, 346.41 V, 200 V]

2. A balanced star-connected load of (8 + *j*6) ohms is connected across three-phase, 50 Hz, 440 V supply system. Calculate :

(*i*) Line current

(*ii*) Power absorbed

(*iii*) Reactive volt-ampere. [R.G.T.U., Jan./Feb.-2008]

[**Ans.** (*i*) 25.4 A, (*ii*) 15.486 kW, (*iii*) 11.615 kVAR]

3. A balanced star-connected load is supplied from a symmetrical 3-phase, 410 V system. The current in each phase is 30 A and lags 30° behind the phase voltage. Find (*i*) phase voltage (*ii*) total power (*iii*) reactive power drawn by load. [R.G.T.U., 2001]

[**Ans.** (*i*) 236.7 V, (*ii*) 18.45 kW, (*iii*) 10.65 kVAR (lagging)]

4. A Y (star)-connected alternator supplies a delta (Δ)-connected load. The impedance of the load branch is (8 + *j*6) Ω/ph. The line voltage is 230 V. Determine :

(*i*) current in the load branch.

(*ii*) power consumed by the load.

(*iii*) pf (cos φ) of the load.

(*iv*) reactive power of the load.

[Rajasthan Technical Univ., February-2010]

[**Ans.** (*i*) 23 A, (*ii*) 12.696 kW, (*iii*) 0.8 leading (*iv*) 9.5222 kVA (leading)]

5. A balanced Δ-connected load having an impedance (20 – *j*15) Ω is connected to a Δ-connected source having $V = 300\angle 0°$ V. Calculate the phase currents of the load.

[DCRUST Electrical Technology, December-2011/January-2012]

[**Ans.** $12\angle 36.87°$]

6. Three impedances of 20 + *j* 15 ohms each are connected in delta across 400 V, 3-phase supply. Determine: (*i*) Line current (*ii*) power factor and (*iii*) total power consumed.

[M.D. Univ. Electrical Technology, December-2005-06]

[**Ans.** (*i*) 27.7 A, (*ii*) 0.8 (lagging), (*iii*) 15.36 kW]

7. Three identical coils connected in delta across 400 V, 50-Hz, 3-phase ac supply, take a line current of 17.32 A at power factor of 0.8 lagging. Calculate (*i*) the phase current (*ii*) the resistance and inductance of each coil (*iii*) the power drawn by each coil.

[U.P. Technical Univ. Electrical Engineering, February-2002]

[**Ans.** (*i*) 10 A, (*ii*) 32 Ω, 76.4 mH, (*iii*) 3.2 kW]

8. A balanced Δ-connected load of impedance 16 + *j* 12 Ω/phase is connected to a 3-phase 400 V supply. Find the phase current, line current, power factor, power, reactive VA and total VA.

[U.P. Technical Univ. Electrical Engineering Second Semester 2008-09]

[**Ans.** 20 A, 34.64 A, 0.8 (lagging) 19.2 kW, 14.4 kVAR, 24.0 kVA]

9. A balanced delta connected load of (12 + *j*9) Ω/phase is connected to 3-phase 400 V supply. Find:

(*i*) Line current

(*ii*) Power factor

(*iii*) Power drawn

(*iv*) Reactive Volt-Amperes

(*v*) Total Volt-Amperes.

[Electrical Engineering Mahamaya Technical Univ. Even Semester 2010-11; U.P. Technical Univ. Even Semester 2013-14]

[**Ans.** (*i*) 46.2 A, (*ii*) 0.8 (lagging), (*iii*) 25.6 kW, (*iv*) 19.2 kVAR, (*v*) 32 kVA]

10. Three 100 Ω resistors are connected in star across a 440 V, 50 Hz, 3-phase supply. One of the phase is disconnected. What will be the line current and total absorbed power. [Pb. Technical Univ., Dec.-2005]

[**Ans.** 2.2 A, 968 W]

11. A 3-phase balanced load draws 10 kW power from a 400 V, 3-phase, 50 Hz, 4-wire supply at 0.8 lagging power factor.

(*i*) Determine the line current.

(*ii*) What would be the line current if power factor were raised to unity, power remaining the same?

(*iii*) How can the power factor be improved to unity?

[R.G.T.U., December-2004]

[**Ans.** (*i*) 18.04 A, (*ii*) 14.43 A]

12. Calculate the active and reactive current components in each phase of a star-connected 10,000 V, 3-phase alternator supplying 5,000 kW at a power factor of 0.8. If the total current remains the same when the load power factor is raised to 0.9, find the new output.

[R.G.T.U., December-2001]

[**Ans.** 288.7 A, 216.5 A, 5,625 kW]

13. In the two wattmeter method of power measurement in a 3 phase circuit, the readings of the wattmeters are 1,200 W and 300 W. What is the pf of the load: [M.D. Univ. Rohtak Electrical Technology, May-2003]

[**Ans.** 0.6934 (lag)]

14. Two wattmeters connected to measure the total power in a 3-phase balanced circuit. One measures 4,800 W, while the other reads backwards. On reversing the latter it is found to read 400 W. What is the total power and power factor ? [R.G.P.V., 2001]

[**Ans.** 4.4 kW, 0.438 (lag)]

15. A star-connected three-phase load has a resistance of 8 Ω and an inductive reactance of 6 Ω in each phase. It is fed from a 400 V, 3-phase balanced supply. Determine the line current, power factor, active and reactive powers. Draw phasor diagram showing phase and line voltages and currents. If power measurement is made using two-wattmeter method, what will be the readings of both wattmeters ?
[U.P. Technical Univ. Odd Semester 2005-06]

[**Ans.** 23.1 A, 0.8 (lagging), 12.8 kW, 9.6 kVA, 9.171 kW, 3.629 kW]

16. A three-phase balanced system supplies 110 V to a delta-connected load whose impedances are equal to $(3.54 + j3.54)$ ohms. Determine the line currents and draw the phase diagram.
[Rajasthan Technical Univ. 2008]

[**Ans.** $38.1\angle -75°$ A, $38.1\angle 165°$ A, $38.1\angle 45°$ A]

17. Three identical coils, each having a reactance of 20 Ω and resistance of 20 Ω are connected in (*a*) star (*b*) delta across a 440 V, 3-phase line. Calculate for each method of connection the line current and readings on each of two wattmeters connected to measure the power.
[Allahabad University Electro-mechanics 1992]

[**Ans.** (*a*) 8.98 A; 3,817 W; 1,023 W (*b*) 26.94 A, 11,451; 3,069 W]

18. Three inductive coils each having a resistance of 20 Ω and reactance of 15 Ω are connected in star across a 3-phase, 400 V, 50 Hz supply. Calculate the readings of two wattmeters connected in the circuit to measure the power input.
[R.G.P.V., February-2005]

[**Ans.** 3.6685 kW, 1.4515 kW]

19. A star-connected balanced 3-balanced 3-phase circuit containing $Z = 20\angle 30°$ Ω in each branch is connected across 400 V, 3-phase, 50 Hz, 4-wire supply. Calculate

(*i*) current flowing through each line,

(*ii*) total power drawn,

(*iii*) readings of two wattmeters connected for the measurement of total power.
[R.G.P.V., December-2002]

[**Ans.** 11.55 A, 6.93 kW, 4.62 kW, 2.31 kW]

20. Calculate the reading of each wattmeter in the circuit shown in Fig. 1. The load impedance $Z = 40\angle -30°$ Ω.
[U.P. Technical Univ. Even Semester 2004-05]

[**Ans.** 7.48246 kW, 3.74123 kW]

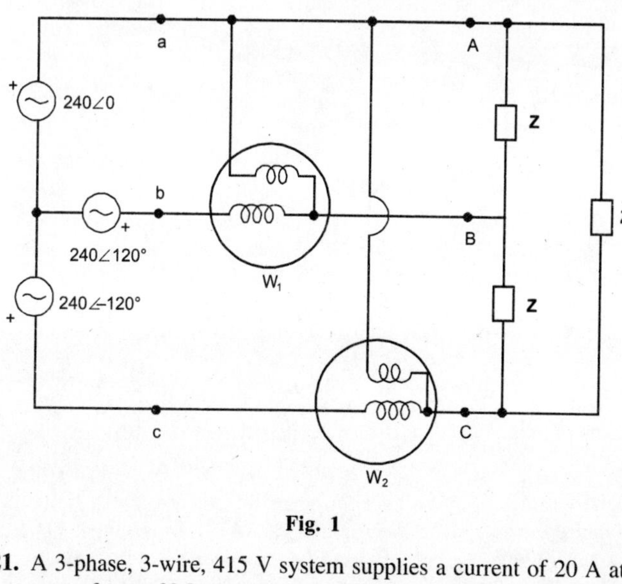

Fig. 1

21. A 3-phase, 3-wire, 415 V system supplies a current of 20 A at a power factor of 0.8 to a star-connected balanced load. Two wattmeters are used to measure total power. The current coil of wattmeter 1 is in the phase R and its coil is across R and Y. The current coil of wattmeter 2 is in the phase B and its voltage coil is across B and Y. Draw the relevant connection and vector diagrams, and (*b*) Calculate the reading on wattmeter W_1 and the reading on wattmeter W_2.

[**Ans.** 8,240 W, 3,260 W]

22. Power in a balanced 3-phase system is measured by the two wattmeter method and it is found that the ratio of the two readings is 2 to 1. What is the power factor of the system?
[Allahabad Univ. Electrical Science-I 1991]

[**Ans.** 0.866 (lag)]

23. The power input to a 2,000 V, 50 Hz, 3-phase motor running on full load at an efficiency of 90% is measured by two wattmeters which indicate 300 kW and 100 kW respectively. Calculate (*i*) input (*ii*) power factor (*iii*) line current and (*iv*) hp output.

[**Ans.** (*i*) 400 kW (*ii*) 0.756 (lag) (*iii*) 152 A (*iv*) 490 (metric)]

24. A 3-phase motor draws a line current of 50 A from 220 V source while starting. The pf is 0.4. Find the readings of the two wattmeters connected to measure power.
[Pb. Univ., June 1993]

[**Ans.** 8.85 kW ; (–) 1.23 kW]

25. A 400 V, 50 Hz, 3-Φ supply of phase sequence RYB is connected to a delta-connected load in which $\mathbf{Z}_{RY} = 100$ Ω, $\mathbf{Z}_{YB} = (20 + j\,60)$ Ω and $\mathbf{Z}_{BR} = -j\,106$ Ω. Take $\mathbf{V}_{RY}$ as the reference and (*a*) calculate the line currents (magnitudes and angles) (*b*) find the wattmeter readings when the two wattmeter method is used with meters in lines R and B

[**Ans.** $7.5\angle 14.54°$ A; $10.27\angle -172.29°$ A; $4.3\angle -47.11°$ A; 2,904 W, – 506 W]

26. A 3-phase, mesh connected load is supplied through a 11/0.4 kV star-delta connected transformer. The load impedances are respectively $\mathbf{Z}_1 = 20 + j\,0$, $\mathbf{Z}_2 = 10 - j\,5$ and $\mathbf{Z}_3 = 15 + j\,10$. Determine the line currents, the line impedance being negligible. Determine also the currents in the windings of the transformer.

Hint: The transformer secondary currents will be the same as those in the corresponding phases of the load.

[**Ans.** $28.87\angle -50°$ A, $42.01\angle -121.8°$ A and $57.93\angle 86.5°$ A; $I_{12} = 20\angle 0°$ A, $I_{23} = 35.78\angle -93.4°$ A and $I_{31} = 22.14\angle 86.3°$ A]

27. Three resistors 10, 20 and 20 ohms are connected in star to the terminals A, B and C of a 3-phase, 3-wire supply through two single phase wattmeters for measurement of total power with current coils in lines A and C and pressure coils between A and B and C and B. Calculate (*i*) the line currents (*ii*) the reading of each wattmeter. The line voltage is 400 V.

[**Ans.** $17.32\angle -29.97°$ A ; $13.24\angle -160.81°$ A ; $13.24\angle -100.9°$ A; 6 kW and 4 kW]

28. Three impedances Z_R, Z_Y and Z_B are connected is star across a 440 V, 3-Φ supply. If the voltage of the star point relative to supply neutral is $200\angle 150°$ V and Y and B line currents are $10\angle -90°$ A and $20\angle 90°$ A respectively, all with respect to the voltage between the supply neutral and the R-line, calculate the values of Z_R, Z_Y and Z_B.
[Nagpur Univ. Electrical Circuits 1991]

[**Ans.** $43.85\angle 76.8°$ Ω, $32.3\angle 8.4°$ Ω, $6.43\angle -21°$ Ω]

29. Three impedances $\mathbf{Z}_1 = 4 + j\,3$ ohms, $\mathbf{Z}_2 = 8 + j\,6$ ohms and $\mathbf{Z}_3 = 6 + j\,4.5$ ohms are star-connected across symmetrical 3-phase, 400 V mains. Determine the current in each line, the total power supplied and the reading on each of two wattmeters connected to read the total power. The current coils are in lines 1 and 2 and the phase sequence is 1 2 3.

[**Ans.** 37.44 A, 26.81 A, 32.56 A, 14,670 W, 3,055 W, 17,725 W]

30. A star-connected load is supplied by a symmetrical 3-phase 415 V ac system. The load consists of the following admittances; RS = (0.05 + *j* 0) mho, YS = (0.06 + *j* 0.03) mho, BS = (0.04 – *j* 0.03) mho. The phase sequence is RYB.

Calculate the potential of star point S with respect to terminal R and the magnitude of the current in the R line. The star point S is connected to the system neutral.

[**Ans.** $155 \angle -169.88°$ V ; $7.75 \angle -169.88°$ A]

31. If the system shown below $\mathbf{Z}_a = \mathbf{Z}_c = 100 \angle 0°\ \Omega$, $\mathbf{Z}_b = 100 \angle 90°\ \Omega$ and $|V_{ab}| = |V_{bc}| = |V_{ca}| = 100$ volts. Find the reading of voltmeter connected across Z_a when the phase sequence is *ab-bc-ca*. What will be the reading if the phase sequence is reversed?

[**Ans.** 86.76 V; 32.85 V]

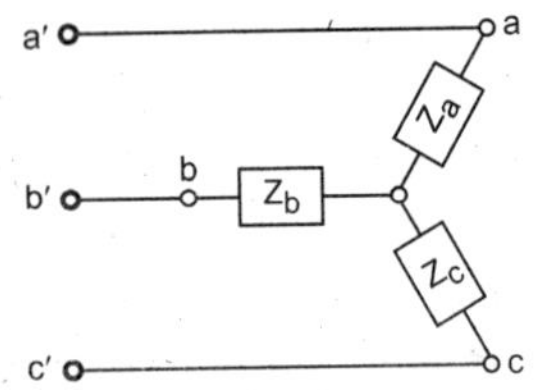

Fig. 2

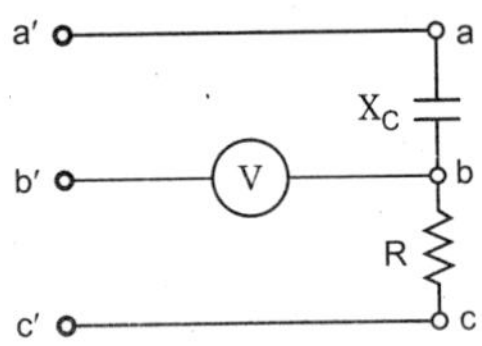

Fig. 3

32. If $X_C = 100\ \Omega$, $R = 100\ \Omega$ and $V_{ab} = V_{bc} = V_{ca} = 141.14$ volts in magnitude then show that the reading of the voltmeter exceeds 141.4 volts if the phase sequence is *ab-bc-ca*. Neglect current drawn by the voltmeter.

[**Hint:** Voltage across ca will be $141.4 \angle 120°$ V, R and X_C are in series across voltage ca *i.e.* $141.4 \angle 120°$ V. Determine the potential of terminal *b* with respect to terminal a by voltage divide rule. Potential of terminal *b'* w.r.t.o terminal *a'* or *a* (terminals *a'* and *a* being at the same potential) is 141.4 V. Thus the pd between terminals *b'* and *b* can be determined]

33. A Y-connected load is supplied from a 400 V, 3-phase, 3-wire symmetrical system. The branch circuit impedances are:

$Z_R = 10\sqrt{3} + j10\ \Omega$; $Z_Y = 20 + j\,20\sqrt{3}\ \Omega$; $Z_B = 0 - j\,10\ \Omega$

Determine the current in each branch. Phase sequence is RYB

[Nagpur Univ. Network Analysis 1993]

[**Ans.** $26.7 \angle -57.75°$ A; $6.47 \angle -133.7°$ A; $28.95 \angle -109.7°$ A]

34. Convert an impedance of (4 + *j* 24) into per unit impedance with base voltage of 11 kV and base kVA of 500.

[**Ans.** (0.0165 + *j* 0.0992) pu]

35. The primary and secondary windings of a single-phase 1,000 kVA, 2,000/1,000 V transformer have leakage reactances each of 2 Ω. Find the pu reactance of the transformer. **[Rajasthan Univ., 2004]**

[**Ans.** 2.5 pu]

3 CHAPTER

Principles of Electromechanical Energy Conversion

INSIDE THIS CHAPTER

3.1 INTRODUCTION

The role of electricity in modern technology is that of an extremely versatile intermediary. Although energy is seldom directly available in electrical form, and ultimately it is seldom required in electrical form, yet conversion of other forms of energies into electrical energy is a common practice. The chief advantages of this conversion are that energy in electrical form can be transmitted, controlled and utilized with relative simplicity, reliability and efficiency. Energy conversion devices are required first to convert energy in non-electrical form to energy in electrical form and then to convert electrical energy into the desired useful form, such as light, heat, sound or mechanical energy. Thus energy conversion devices are needed at both ends of the electrical system. One typical example is generation of electrical energy (conversion of non-electrical form energy into electrical form) at nuclear power station and then transmission and distribution over lines and finally conversion into mechanical energy by an electric motor for final use. Another example is the conversion of sound energy into electrical energy at the talker's end, its transmission in electrical form over lines and then its final conversion to sound waves at the listener's end.

We are concerned here with the **electromechanical energy-conversion process**, which takes place through the medium of the electric or magnetic field of the conversion device. Although the various conversion devices operate on similar principles, their structures depend upon their function. Various devices operating on the principles of electromechanical energy conversion may be classified as follows:

1. Transducers-devices for measurement and control. Such devices usually operate under linear input-output conditions and with relatively small signals and include torque motors, microphones, loudspeakers, gramophone pick-ups etc.

2. Force-producing devices involving restricted mechanical movements and include electromagnets, solenoid-actuators, relays, moving coil instruments, moving iron instruments etc.

3. Continuous energy conversion devices such as generators and motors. These devices are mainly employed for bulk energy conversion and utilization.

A *dynamo* is a machine that converts either mechanical energy into electrical energy or electrical energy into mechanical energy. When a dynamo is driven mechanically by a prime mover such as by a diesel engine, steam engine, steam turbine or water turbine and supplies electrical energy, it is called a *generator* [Fig. 3.1 (*a*)]. It means that in generation operation, the mechanical power is supplied by the prime mover through the shaft of the generator against the mechanical torque produced by the electrical load current. When the dynamo draws electrical energy from the supply mains and drives mechanical devices such as line shafts and machine tools it is called a *motor* [Fig. 3.1 (*b*)]. In other words it can be stated that the generator converts mechanical power ωT into electrical power EI and motor converts electrical power EI into mechanical power ωT.

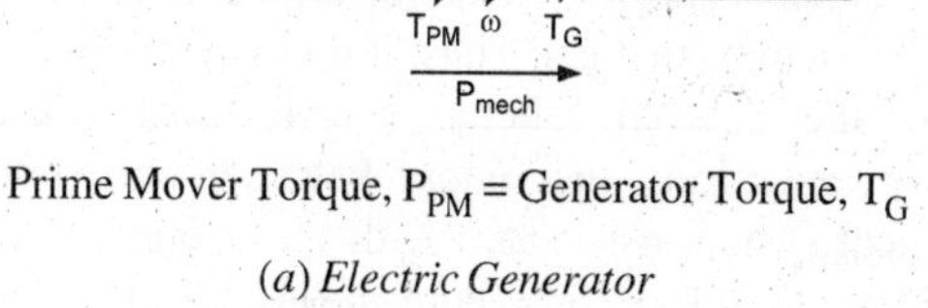

Prime Mover Torque, P_{PM} = Generator Torque, T_G

(*a*) *Electric Generator*

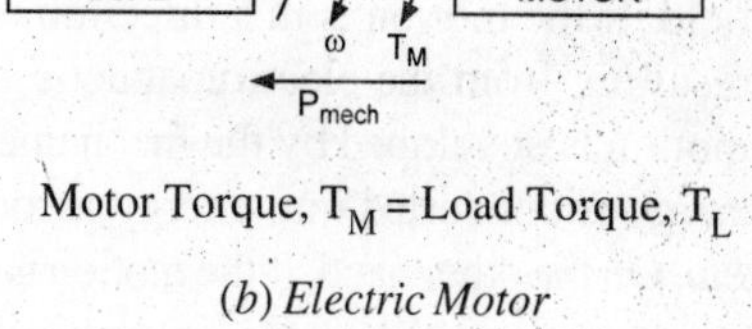

Motor Torque, T_M = Load Torque, T_L

(*b*) *Electric Motor*

Fig. 3.1

The motoring and generating action is basically a reversible process *i.e.* the same machine is capable of operating either as a generator or as a motor depending upon whether the supplied power is mechanical or electrical. This is illustrated in Fig. 3.2. However, in practice, devices may be designed and constructed to suit one particular mode of conversion or the other.

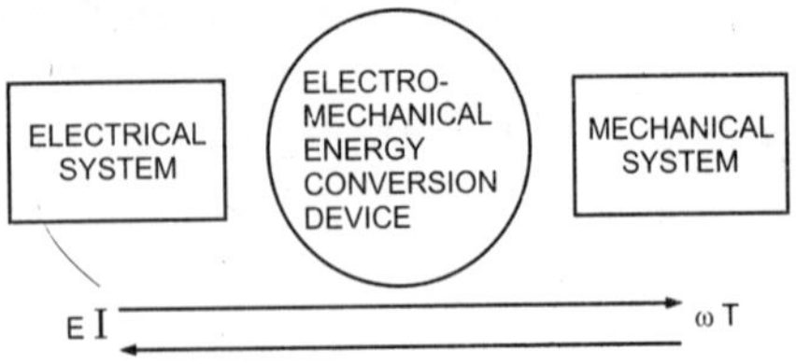

Fig. 3.2 *Block Representation of Electromechanical Energy Conversion*

An *electromechanical energy conversion device* is a link between an electrical and mechanical system and *electromechanical energy conversion* needs the presence of natural phenomena which interrelate electric and magnetic fields on one hand and mechanical force and motion on the other.

The most important phenomena which make this conversion possible are:

1. Whenever a current carrying conductor is placed in a magnetic field in such a way that a component of the length of the conductor is perpendicular to the field, a force will exist between the conductor and the field.
2. Whenever a piece of ferromagnetic material is placed in a magnetic field, a force is experienced upon the material, tending to bring it into the position of the densest portion of the field. The force tends to align the material specimen so that the reluctance of the magnetic path passing through the material will be at minimum.
3. A force of attraction is exerted upon the charged plates of a capacitor.
4. Whenever a piece of dielectric material is located in an electrostatic field, a force is experienced upon the material tending to bring it into the densest portion of the field. In case the specimen of material is of unsymmetrical configuration the force tends to align its longer axis with the electrostatic field.

All of these phenomena involve a force relation that exists between either a magnetic field or an electrostatic field and a material structure (matter). Thus, if relative motion between the field and the material structure is associated with this force, energy conversion must occur between the field and the mechanical system associated with the structure. All of these phenomena, therefore, provide a means for electromechanical energy conversion i.e. interchange of energy between a magnetic field (or an electrostatic field), and a mechanical system. The medium for the energy transfer is either a magnetic field or an electrostatic field. If the motion is in a direction opposite to that of the force resulting from the electromagnetic or electrostatic relations the motion is developed by the mechanical system, and mechanical energy is converted into energy associated with the electrical system. On the other hand, if the motion is in the direction of the force resulting from the electromagnetic or electrostatic relations, the motion is caused by the associated electrical system, and electrical energy is converted into mechanical energy.

For instance in phenomenon 1, when the motion is in the direction of the electromagnetic force, electrical energy from the electric source which causes the current to flow in the conductor is transferred through the associated magnetic field into mechanical energy. While in case the motion is opposite to the direction of the electromagnetic force, mechanical energy is transferred through the associated magnetic field into electrical energy of the system connected to the current carrying conductor.

In this text we are primarily interested in *electromechanical energy conversion by means of continually rotating machines.* The energy conversion in most of these machines with their present day construction is produced through the presence of both phenomenon 1 and phenomenon 2. In both of these phenomena, the magnetic field is the medium for the energy transfer. Also, in all of our present day continually rotating electrical machines, the energy of the magnetic field is directly associated with the energy of an electrical system external to the machine. Thus, the energy transfer involved always is between a mechanical system and an electrical system through the medium of a magnetic field.

Note: *The energy storing capacity of the magnetic field is much larger (about 25,600 times) than that of electric field. In view of this fact, electromechanical energy conversion devices with magnetic field as the coupling medium between electrical and mechanical systems are more common.*

3.2 MOTION (OR SPEED) EMF AND TRANSFORMER EMF

The vast majority of electrical machines are electromagnetic devices and their operation, design and construction are based upon the principles relating mechanical force and motion to current and electromotive force (emf). These principles are associated with the names of Faraday and Ampere.

Two different processes of developing emfs are termed as *motion or speed emf* and *transformer emf.*

According to Faraday's law relating to generation of motion or speed emf, if a conductor of active length* l metres is moved with a velocity of v m/s in a magnetic field of constant flux density B teslas (Wb/m^2), as illustrated in Fig. 3.3 then the emf induced in the conductor will be equal to the rate of flux cutting and given by the expression

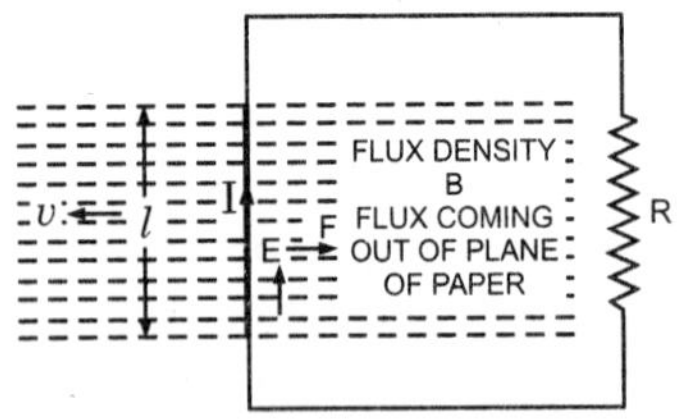

Fig. 3.3 *A Conductor Moving With Velocity v, Perpendicular to Magnetic Flux Coming Out of Paper (Generator Action)*

$$E = Blv \text{ volts} \quad ...(3.1)$$

The direction of emf induced is upward, as determined by Fleming's right-hand rule or right hand flat palm rule.

The other process referred to as transformer emf relates to the emf that is set up in coil when the magnetic flux linking the coil

* That part of the conductor which cuts the magnetic flux.

is varying with time. The expression for transformer emf is given as

$$E = -\text{ Rate of change of flux linkage}$$

$$= \frac{-d\Psi}{dt} = -N\frac{d\Phi}{dt} \text{ volts} \quad ...(3.2)$$

where Ψ is the flux linkage of the coil and is equal to the product of number of turns of the coil N, and flux Φ in webers, linking the coil.

Minus sign placed before the right hand quantity of Eq. (3.2) takes into account the sense of induced emf dictated by Lenz's law*.

The Eqs. (3.1) and (3.2) for induced emf expressed in different ways based on two different processes–motion emf and transformer emf give identical results as proved below.

If in Fig. 3.3, x is the distance moved by the conductor in time t, then change in flux linkage in the coil (single turn),

$$d\Psi = Bl dx$$

and rate of change of flux linkage,

$$\frac{d\Psi}{dt} = Bl\frac{dx}{dt} = Blv \quad ...(3.3)$$

As regards the direction of induced emf in the conductor, it is directed upward in the conductor, as determined by Fleming's right hand rule. The upward direction of induced emf is also indicated by minus sign because the flux produced by the current in this direction tends to reduce the flux in the coil which is being increased by its motion, as according to Maxwell's screw rule upward direction of current produces flux toward the paper in the area of the coil. So any of the two Eqs. (3.1) and (3.2) for induced emf can be used but Eq. (3.1) is convenient for computation of emf in electromagnetic devices where flux cutting by conductors takes place due to relative motion between conductors and field system.

If the circuit of the generator illustrated in Fig. 3.3 is closed, a current, say of I amperes, will flow through this circuit and a mechanical force given by the expression $F = BIl$ newtons will be experienced on the conductor by Ampere's law. The direction of force F, as determined by Fleming's left hand rule** or left hand flat palm rule**, is opposite to that of movement of conductor. As the force experienced on the conductor opposes the movement of conductor, an equal and opposite driving force, therefore, must be applied to the conductor and work is required to be done at the rate of $Fv = BIlv = EI$ watts. Thus in a generator the mechanical power Fv supplied by the prime mover reappears in the form of electrical power EI *i.e.* mechanical power supplied is converted into electrical power.

Now consider a conductor of effective length l metres, carrying a current of I amperes and placed in a magnetic field of flux density B teslas (Wb/m^2), as illustrated in Fig. 3.4. A mechanical force F given by the expression $F = BIl$ newtons is experienced on the conductor. The direction of this force F, as determined by Fleming's left hand rule or left hand flat palm rule, is towards left, as illustrated in the figure. Let the conductor move with a velocity of v metres/second under the influence of the force. Now when the conductor moves with a velocity of v m/s under the influence of force F, it cuts the magnetic lines of force and so an emf $E = Blv$ volts is induced in the conductor. This induced emf is known as *counter emf* or back *emf,* as this is in the direction opposite to that of current or applied voltage V, as determined by Fleming's right hand rule or right hand flat palm rule. To maintain the flow of current through the conductor, a voltage V is required to be applied which should be sufficient to overcome the back emf and voltage drop in the conductor. If R is the resistance of the conductor, then voltage applied V must be equal to E + I R. If R is negligible, the electrical power supplied by the source = $VI = EI = BlvI = Fv$ watts i.e electrical power supplied is converted into mechanical power.

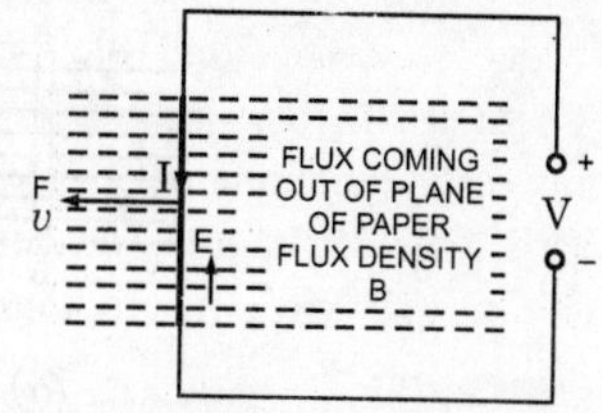

Fig. 3.4 *Motor Operation*

General Case of Induced EMF. In case there is a relative movement between the magnetic field and coil and simultaneously there is change in linking flux with time, then both speed and transformer emfs would be induced in the coil. Thus, the total emf induced in the coil is given as

$$e = \text{Speed or rotational emf } e_r + \text{transformer emf } e_t$$

3.3 EMF INDUCED IN A COIL ROTATING IN A MAGNETIC FIELD

Consider a single turn rectangular coil rotating in counter-clockwise direction with angular velocity ω radians per second in a uniform magnetic field of flux density B teslas (Wb/m^2), as illustrated in Fig. 3.5.

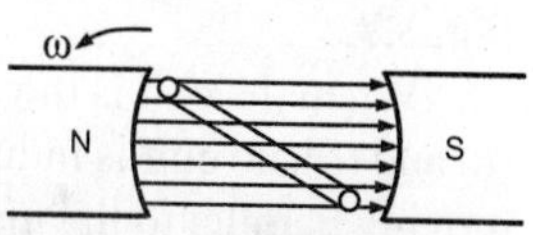

Fig. 3.5

When the coil is rotated in a uniform magnetic field with uniform angular velocity, an emf is induced in its coil sides. The magnitude of emf induced in the coil depends on (i) the flux density of the main magnetic field, B teslas (ii) the effective length of the coil side in the magnetic field, l metres (iii) number of turns on the coil, N (iv) velocity of the coil side perpendicular to the field v_p metres/second and is given by the expression

$$e = 2NBlv_p \text{ volts} \quad ...(3.4)$$

($\because$ two coil sides are connected in series)

The coil sides do not always cut the lines of flux in perpendicular direction while rotating in a magnetic field. The component of velocity of the coil perpendicular to the field is $v_p = v \sin \theta$, where v is the peripheral velocity of the coil and θ is the angle between the conductor and line perpendicular to the magnetic axis (Fig. 3.6).

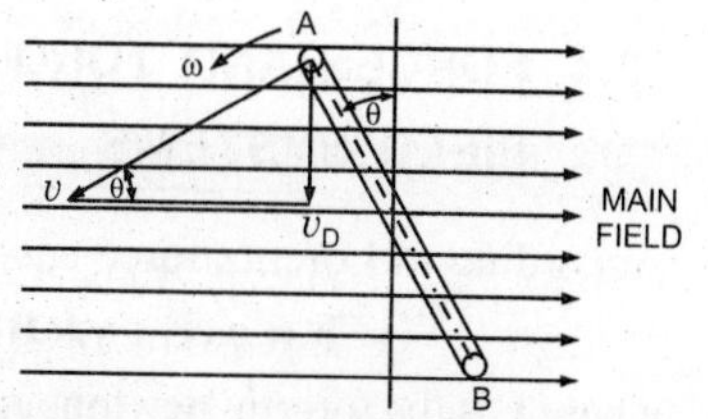

Fig. 3.6

* Already explained in Art. 1.21.

** Already explained in Art. 1.13.

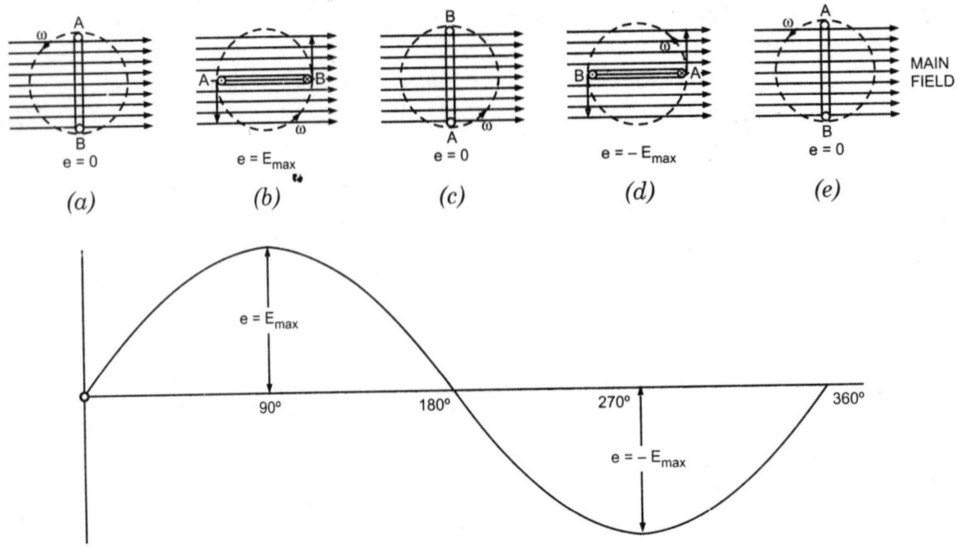

Fig. 3.7

So the emf equation for the generated voltage becomes

$$e = 2\,N\,B\,l\,v \sin\theta \text{ volts} \qquad ...(3.5)$$

The magnitude of induced emf can be determined if the values of N, B, l, v and sin θ are known.

The value of the induced emf varies in magnitude and direction according to the instantaneous position of the coil relative to the magnetic field. The cross-sectional view of the coil and its different position at different instants are shown in Fig. 3.7.

When the coil is in vertical position as shown in Fig. 3.7 (*a*), (*c*) and (*e*), no emf is induced in the coil sides as the coil sides are moving parallel to the magnetic field and the rate of cutting flux is zero. When the coil is horizontal and moving at right angles to the magnetic field, as shown in Fig. 3.7 (*b*) and (*d*), the magnitude of emf induced in the conductors A and B is maximum because the rate of cutting flux is maximum. The direction of emf induced in conductor A at instant shown in Fig. 3.7 (*b*) is outward and at instant shown in Fig. 3.7 (*d*) is inwards, as determined by Fleming's right hand rule. The direction of emf induced in conductor B is opposite to that of conductor A, as obvious from figs. The wave shape of the emf induced in the coil is also shown in Fig. 3.7.

From the above discussion as well as from Eq. (3.5) it is obvious that the waveform of the induced emf is sinusoidal.

3.4 FORCES AND TORQUES IN MAGNETIC FIELD SYSTEMS

According to Lorentz force equation

$$\mathbf{F} = q\,(\mathbf{E} + \mathbf{v} \times \mathbf{B}) \qquad ...(3.6)$$ Refer to Art 1.30

where **F** is the force in newtons on a particle of charge q coulombs in the presence of electric and magnetic fields, **E** being in volts per metre, **B** in Teslas and **v** is the velocity of the particle relative to the magnetic field, in m/s.

Thus in a pure electric field system, the force is determined simply by the charge on the particle and the electric field

$$\mathbf{F} = q\,\mathbf{E} \qquad ...(3.7)$$

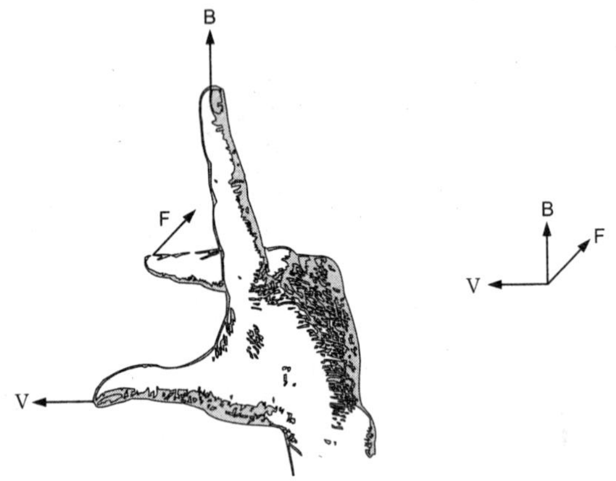

Fig. 3.8 *Right Hand Rule For Determination of Direction of Lorentz Force*

The force **F** acts in the direction of the electric field and is independent of the particle motion.

In magnetic field systems the situation is somewhat more complex. Here the force is given as

$$\mathbf{F} = q\,(\mathbf{v} \times \mathbf{B}) \qquad ...(3.8)$$

and is thus determined by the magnitude of the charge on the particle and the magnitude of induction density **B** as well as the velocity of the particle. In fact the direction of force is always perpendicular to both the direction of particle motion and the direction of magnetic field. Mathematically, this is given by the vector cross product **v** × **B** in Eq. (3.8). The magnitude of this cross product is equal to the product of the magnitudes of **v** and **B** and the sine of the angle between them; its direction can be determined from the right hand rule, which states that when the thumb of the right hand points in the direction of **v** and the index finger points in the direction of **B**, the force points in the direction normal to the palm of the hand, as depicted in Fig. 3.8.

For situations where large number of charged particles are in motion, it is convenient to rewrite Eq. (3.8) in terms of the current density **J**, in which case the force is a force density

$$\mathbf{F} = \mathbf{J} \times \mathbf{B} \text{ N/m}^3 \qquad ...(3.9)$$

For currents flowing in conducting media, Eq. (3.9) can be employed for determination of force density acting on the material itself.

For situations in which the forces act only upon current carrying elements and which are of simple geometry, Eq. (3.9) is usually the simplest and easiest way to determine the forces acting on the system. Very few practical situations fall into this class, however. In fact, most electromechanical energy conversion devices contain magnetic material; in such systems forces act directly on the magnetic material and clearly cannot be calculated from Eq. (3.9).

Techniques for calculating the detailed, localized forces acting on the magnetic materials are extremely complex and need detailed knowledge of the field distribution throughout the structure. Fortunately, most electromechanical energy conversion devices are constructed of rigid, non-deforming structures. In these devices, it is the net force or torque that is of importance, and the details of the localized force distribution are of secondary interest. For example, in a properly designed motor the net accelerating torque acting on the rotor determines the motor's characteristics; accompanying forces, which act to squash or make the solid rotor oval, play no significant role and generally are not even computed.

For understanding the behaviour of rotating machinery, a simple physical picture is quite useful. Associated with the rotor structure is a magnetic field, and similarly with the stator; one can picture them as a set of north and south poles associated with each structure. Just as a compass needle tries to align with the earth's magnetic field, these two sets of fields attempt to align, and torque is associated with their displacement from alignment. Thus in a motor, the stator magnetic field rotates ahead of that of the rotor, pulling on it and performing work. The opposite is true for a generator, here the rotor does the work on the stator.

Various techniques have been evolved for calculation of the net forces of concern in electromechanical energy conversion. The technique usually employed is known as the *energy method* and is based upon the principle of *conservation of energy.* The basis for this method can be understood with reference to Fig. 3.9 (*a*) where a magnetic field-based electromechanical energy conversion device is indicated schematically as a lossless magnetic energy storage system with two terminals. The electric terminal has the terminal variables voltage e and current i and the mechanical terminal has the terminal variables force f_{field} and position x. This sort of representation is valid in situations where the loss mechanism can be separated out from the energy storage mechanism; in these cases the electrical losses such as ohmic losses can be included as external elements connected to the electrical terminals and the mechanical losses such as frictional losses can be included external to the mechanical systems. Fig. 3.9 (*b*) shows a typical example of such a system; a simple force-producing device with a single coil forms the electrical terminal, and a movable plunger serves as the mechanical terminal.

The interaction between the electric and mechanical terminals, i.e., the electromechanical energy conversion, occurs through the medium of the magnetic stored energy. Since the energy storage system is lossless, it is a simple matter to write that the time rate of change of W_{field}, the stored energy in the

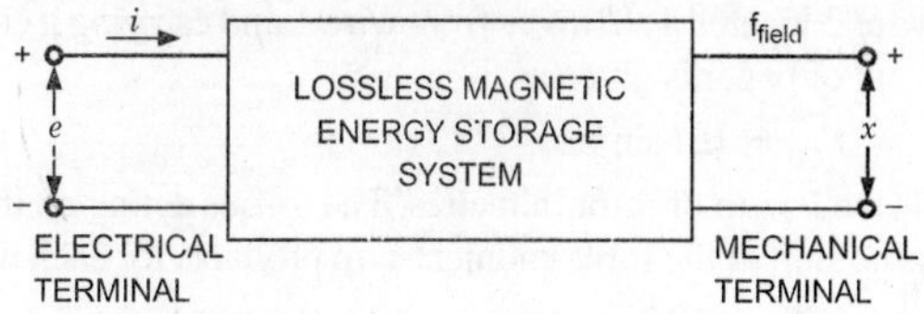

(a) Schematic Magnetic Field Electromechanical Energy Conversion Device

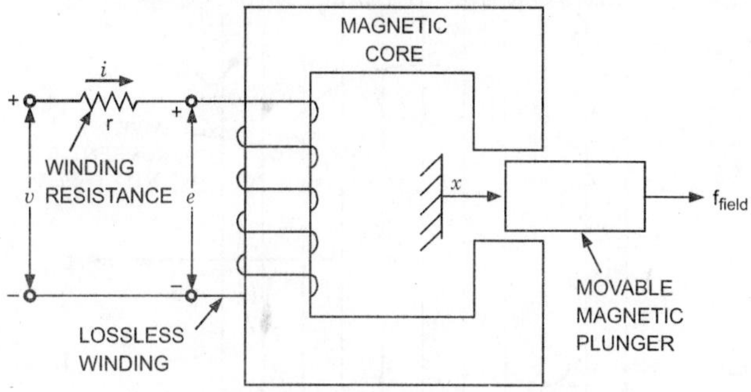

(b) Simple Force-Producing Device

Fig. 3.9

magnetic field, is equal to the electric power input less the mechanical power output of the energy storage system i.e.,

$$\frac{dW_{\text{field}}}{dt} = ei - f_{\text{field}}\frac{dx}{dt} = i\frac{d\psi}{dt} - f_{\text{field}}\frac{dx}{dt} \qquad ...(3.10)$$

$$\because \; e = N\frac{d\Phi}{dt} = \frac{d\psi}{dt}$$

$$\text{or } dW_{\text{field}} = i\,d\psi - f_{\text{field}}dx \qquad ...(3.11)$$

Equation (3.11) permits us to solve the force simply as a function of the flux linkage ψ and the mechanical terminal position x. This result comes about as a consequence of our ability to separate the losses out of the physical problem, resulting in a lossless energy storage element, as in Fig. 3.9 (*a*).

Equations (3.10) and (3.11) form the basis for the energy method. This technique is quite powerful in its ability to compute forces and torques in complex situations of electromechanical energy conversion systems. It is to be noted here that this power comes at the expense of a detailed picture of the force-producing mechanism. The forces themselves are produced by such well known physical phenomena as the Lorentz force on current carrying element, described in Art 1.30, or the interaction of the magnetic fields with the dipoles in the magnetic material.

Example 3.1. A nonmagnetic rotor having a single-turn coil is placed in a uniform magnetic field of magnitude 0.8 T, as shown in Fig 3.10. The coil sides are at a radius of 0.125 m, and the coil carries a current of 12 A, as shown. Determine the θ-directed torque as a function of rotor position α. Rotor may be assumed to be 0.5 m long.

Solution: The force per unit length on a wire carrying a current of I amperes can be determined by multiplying Eq. (3.9) by cross-sectional area of the wire. As the product of the current density and the cross-sectional area of the wire is simply the current **I**, the force per unit length acting on the wire is given as

$$\mathbf{F} = \mathbf{I} \times \mathbf{B}$$

Thus, for wire 1 carrying a current of I amperes into the paper, the θ-directed force is given as

$$F_{1\theta} = IBl \sin \alpha$$

and for wire 2 located 180° away from wire 1 and carrying a current of I amperes out of paper is given as

$$F_{2\theta} = IBl \sin \alpha$$

where l is the length of rotor in metres. The torque acting on the rotor is given by the sum of the force-moment-arm products for each wire i.e.

$$T = 2IBRl \sin \alpha = 2 \times 12 \times 0.8 \times 0.125 \times 0.5 \sin \alpha$$
$$= 1.2 \sin \alpha \text{ N-m} \quad \textbf{Ans.}$$

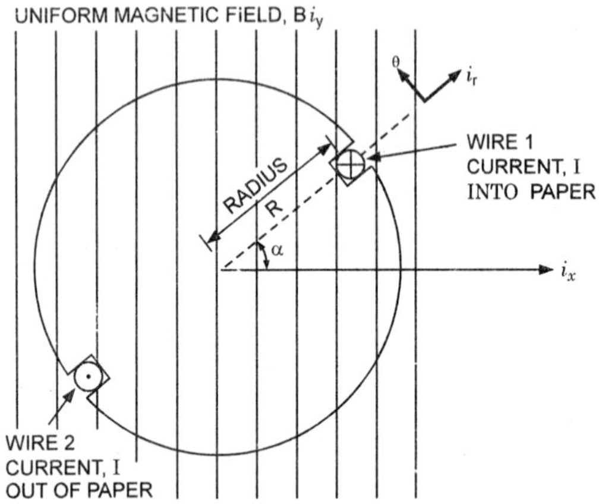

Fig. 3.10 *Single Coil Rotor*

3.5 ENERGY BALANCE

Since, for all practical purposes, the mass of the materials used in the construction of an electrical machine remains constant under the conditions of operation, the principle of conservation of energy can be applied in the analysis of the energy conversion. The input energy must, therefore, be equal to the summation of the useful output energy, the energy converted into heat, and the change in the energy stored in the magnetic field. Thus energy balance equation may be written as

(Energy input from electric source)
= (Mechanical energy output)
+ (Increase in energy stored in the magnetic field)
+ (Energy converted into heat) ...(3.12)

Equation (3.12) is applicable to all electromechanical energy conversion devices; it is written so that the electrical and mechanical energy terms have positive values for motor action. The equation applies equally well to the generator action; these terms then simply have negative values. For generator action

(Total mechanical energy input)
= (Electrical energy output)
+ (Increase in energy stored in the magnetic field)
+ (Energy converted into heat)

Irreversible conversion of energy to heat occurs due to three reasons; part of electrical energy is converted directly to heat in the resistances of the current paths, part of the mechanical energy developed within the device is absorbed in the friction and windage and converted into heat, and part of the energy absorbed by the coupling field is converted into heat in magnetic core loss (for magnetic coupling) or dielectric loss (for electric coupling). If the energy losses in the electrical system, the mechanical system, and the coupling field are grouped with the corresponding terms in Eq. (3.12), the energy balance equation may be rewritten as

$$\begin{pmatrix}\text{Electrical energy}\\ \text{input less resistance}\\ \text{or ohmic losses}\end{pmatrix} = \begin{pmatrix}\text{Mechanical energy output}\\ \text{plus friction and windage}\\ \text{losses}\end{pmatrix} + \begin{pmatrix}\text{Increase in energy}\\ \text{stored in coupling field}\\ \text{plus associated losses}\end{pmatrix} \quad ...(3.13)$$

The left-hand side of Eq. (3.13) may be expressed in terms of the currents and voltages in the electric circuits of the coupling device. Consider, for instance, the energy conversion device shown schematically in Fig. 3.11. The differential energy input from the electrical source in time dt is $vidt$, where v is the instantaneous terminal voltage and i is the instantaneous current. The energy loss in the resistance of the device is $i^2r\,d\,t$, where r is the resistance of the current path. Hence the left-hand side of Eq. (3.13) is

$$dW_{elec} = vi\,dt - i^2\,r\,dt = (v - ir)\,i\,dt \quad ...(3.14)$$

where $d\,W_{elec}$ is the net electrical energy input to the coupling device. For absorption of energy, by the coupling device, from the electric circuit, the development of reaction by the coupling field in the circuit is essential. In Fig. 311 induced emf e in the reaction of the coupling field. Reaction on the input is an essential part of the process of interconversion of energy between an electric circuit and any other medium. The coupling field may be like a reservoir of energy, that gives energy to the output system and being replenished through the reaction of the field on the input system. For the voltages to balances properly, the emf e must be

$$e = v - ir \quad ...(3.15)$$

and, therefore, Eq. (3.14) becomes

$$dW_{elec} = (v - ir)\,i\,dt = ei\,dt \quad ...(3.16)$$

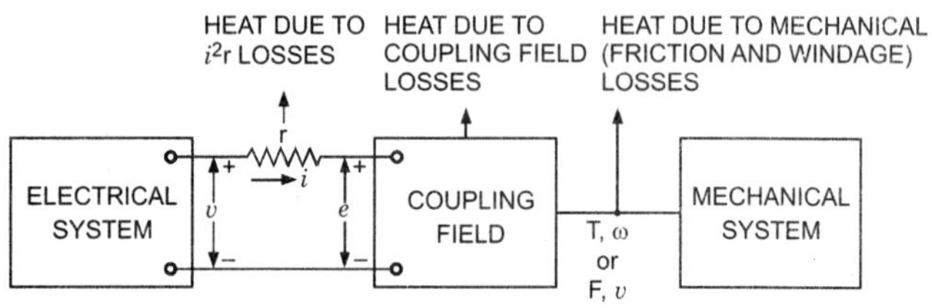

Fig. 3.11 *General Representation of Electromechanical Energy Conversion Device*

If electrical energy is supplied to the coupling field from more than one circuit, the total electrical energy input is the sum of terms of the form of Eq. (3.16).

The first term on the right-hand side of Eq. (3.13) is the total energy converted into mechanical form. It is the useful mechanical energy plus the mechanical (friction and windage) losses caused due to motion of the mechanical parts of the energy conversion device. The second term on the right-hand side of Eq. (3.13) is the total energy absorbed by the coupling field, including both stored energy and losses.

From the above discussion, it is evident that electrical losses (ohmic and coupling field losses) as well as mechanical (friction and windage) losses, though always present, play no basic role in the energy conversion process. They can be accounted for as losses in the electrical and mechanical systems on the two sides of the coupling element, as indicated in Fig. 3.11. The basic energy conversion process is that which involves the coupling field and its action and reaction on the electrical and mechanical systems. For motor action, the sum of energy absorbed by the coupling field and the internal energy converted into mechanical form is always equal to the internal electrical energy associated with the flow of electric current against the emf e caused by the coupling field. In differential form, the Eq. (3.13) may be written as

$$dW_{elec} = ei\,dt = dW_{field} + dW_{mech} \quad ...(3.17)$$

where dW_{mech} is the differential energy converted into mechanical form and dW_{field} is the differential energy absorbed by the coupling field.

Equation (3.17) along with Faraday's law for induced emf $\left(e = -\frac{d\psi}{dt}\right)$ is the fundamental basis for analysis of energy conversion devices.

3.6 COUPLING FIELD REACTION

The interaction between electrical and mechanical terminals is through the medium of energy stored in the coupling field. In other words it can be said that coupling field is the link between electrical and mechanical systems. For rotation or movement of a moving member with respect to a stationary member, presence of air gap between the stationary and the moving members is necessary. The energy stored in the coupling field must develop action and reaction on the electrical and mechanical systems for the conversion of energy from electrical to mechanical form (motoring mode) or from mechanical to electrical form (generating mode).

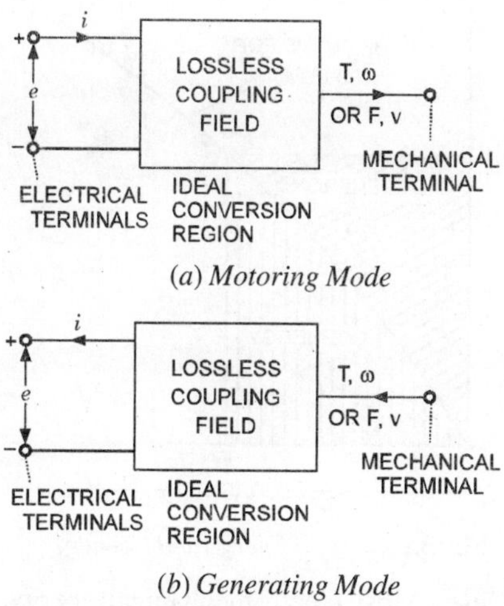

Fig. 3.12 *Representation of Lossless Electromechanical Energy Conversion Device*

In case the output is mechanical, as in a motoring mode, the coupling field must react with the electrical system as to absorb electrical energy from electrical system. In such a case i.e. in a motoring mode this reaction is counter emf e, as shown in Fig. 3.12 (*a*), similar to the emf induced in the primary winding of a transformer. The coupling field extracts energy proportional to ei from the electrical system, converts it into mechanical form and delivers energy proportional to $T\omega$ (or Fv) to the mechanical system.

In case the output is electrical, as in a generating mode, the coupling field must react with mechanical system so as to absorb mechanical energy from mechanical system. In such a case i.e. in a generating mode, this reaction is the counter-torque T (opposite to the applied mechanical torque of the prime mover) developed due to interaction of current i and the coupling field. Thus the coupling field extracts mechanical energy proportional to $T\omega$ from the mechanical system, converts it into electrical form and delivers energy proportional to ei to the electrical system.

Thus, it may be concluded from the above discussion that the coupling field serves as the energy conversion region and the reaction of the coupling magnetic field on the electrical or mechanical system is essential for electromechanical energy conversion process. Induced emf e and torque T, being associated with the coupling field, are usually known as electromechanical coupling terms.

3.7 SINGLY-EXCITED MAGNETIC FIELD SYSTEMS

3.7.1. Energy in Magnetic System. Energy can be stored or retrieved from a magnetic system by means of an exciting coil connected to an electrical source. Consider, for instance, the magnetic field system of an attracted armature relay shown in Fig. 3.13. The resistance r of the exciting coil is shown by a series resistance lumping outside the coil which then is considered as an ideal lossless coil. The magnetic flux is produced in the magnetic circuit due to flow of current i in the exciting coil. It is assumed that all the flux so created is confined to the iron core and, therefore, links all the N turns producing the coil flux linkages ψ which is equal to NΦ.

The instantaneous voltage equation for the electric circuit is written, according to Kirchhoff's voltage law.

$$v = ir + e = ir + \frac{d\psi}{dt} \quad ...(3.18)$$

Multiplying both sides of above Eq. (3.18) by $i\,dt$ we have

$$vi\,dt = i^2 r\,dt + i\,d\psi$$

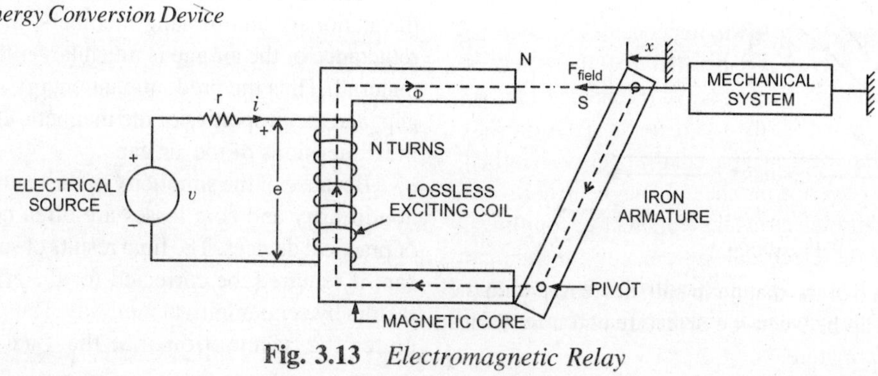

Fig. 3.13 *Electromagnetic Relay*

or $(v - ir)\, i\, dt = i\, d\psi$

or $ei\, dt = i\, d\psi$ ...(3.19)

The electrical energy input to the ideal coil due to flow of current i in time dt is given as

$$dW_{elec} = ei\, dt = i\, d\psi \quad ...(3.20)$$

Assuming for the time being that armature is held fixed at position x, all the input energy is stored in the magnetic field. Thus

$$dW_{elec} = ei\, dt = dW_{field} \quad ...(3.21)$$

where dW_{field} is the change in field energy in time dt. Substituting $ei\, dt = i\, d\psi$ from Eq. (3.20) in above Eq. (3.21) we have

$$dW_{elec} = i\, d\psi = iN\, d\Phi = Md\Phi = dW_{field} \quad ...(3.22)$$

where M = Ni, the *mmf*.

The relationship $i - \psi$ or $M - \psi$ is a functional one corresponding to the magnetic circuit which, in general, is nonlinear and also it exhibits hysteresis. The energy absorbed by the field for finite change in flux linkages or flux is given, from the above Eq. (3.22), as

$$\Delta W_{field} = \int_{\psi_1}^{\psi_2} i(\psi)\, d\psi = \int_{\Phi_1}^{\Phi_2} M(\Phi)\, d\Phi \quad ...(3.23)$$

As the flux in the magnetic circuit undergoes a cycle $\Phi_1 \rightarrow \Phi_2 \rightarrow \Phi_1$, an irrecoverable loss of energy occurs owing to hysteresis and eddy currents in the iron. Assuming here that these losses are separated out and are met directly by the electric source, the ideal coil and the magnetic circuit becomes a *conservative system* with energy interchange between themselves so that the net energy is conserved.

The energy absorbed by the magnetic system in establishing flux linkages ψ or flux Φ from initial zero value,

$$W_{field} = \int_0^{\psi} i(\psi)\, d\psi = \int_0^{\Phi} M(\Phi)\, d\Phi \quad ...(3.24)$$

This is the energy of the magnetic field with given mechanical configuration when its state corresponds to flux Φ or flux linkages ψ.

The i-ψ relationship is actually the magnetization curve and this curve varies with the configuration value x, as illustrated in Fig. 3.14.

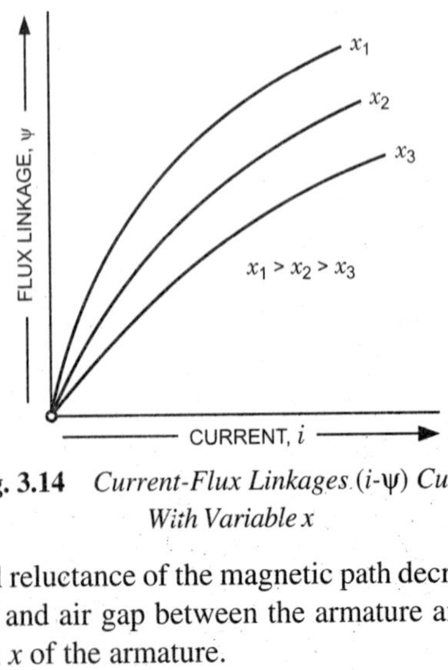

Fig. 3.14 *Current-Flux Linkages (i-ψ) Curves With Variable x*

The total reluctance of the magnetic path decreases with the increase in x and air gap between the armature and core varies with position x of the armature.

Thus i-ψ relationship may be expressed as

$i = i(\psi, x)$ if ψ is the independent variable

or $\psi = \psi(i, x)$ if i is the independent variable.

Thus, the field energy [Eq. (3.24)] is, in general, a function of two variables i.e.

$$\left.\begin{aligned} W_{field} &= W_{field}(\psi, x) \quad ...(i) \\ W_{field} &= W_{field}(i, x) \quad ...(ii) \end{aligned}\right\} \quad ...(3.25)$$

According to above equations field energy is determined by instantaneous values of the system states (ψ, x) or (i, x) and is independent of the path followed by these states in attaining these values.

It means that the field energy at any instant is history independent (does not exhibit hysteresis).

A change in ψ with fixed x causes electromagnetic energy interchange which is governed by Eqs. (3.18) and (3.22). Similarly a change in x with fixed ψ causes interchange of energy between the magnetic circuit and the mechanical system. The general case of such energy interchanges (electrical-magnetic-mechanical) will be discussed in Art 3.7.2.

According to Eq. (3.24), the area between the ψ-axis and i-ψ curve represents the field energy as illustrated in Fig. 3.15. A new term, *coenergy* may be now defined as

$$W'_{field}(i, x) = i\psi - W_{field}(\psi, x) \quad ...(3.26)$$

wherein by expressing flux linkages ψ as $\psi(i, x)$, the independent variables of W'_{field} become i and x. The coenergy on Fig. 3.15 is shown to be complementary area of i-ψ rectangle.

Thus, $$W'_{field} = \int_0^i \psi\, di \quad ...(3.27)$$

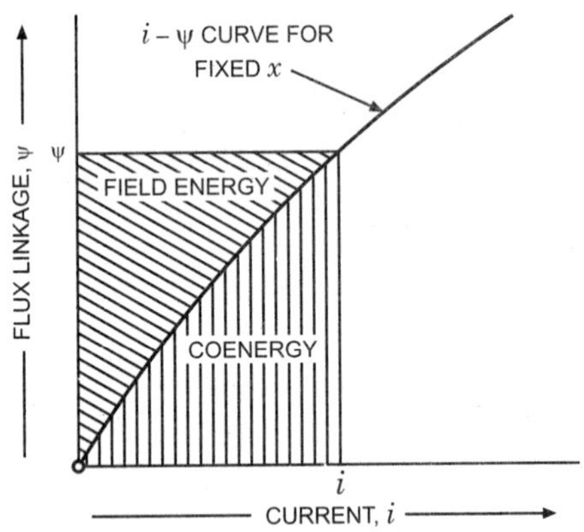

Fig. 3.15 *Field Energy and Coenergy*

Linear Case. Most electromechanical energy conversion devices contain air gaps in their magnetic circuits so as to separate the stationary and moving members. In most of such cases the reluctance of the air gap is much larger than that of the magnetic material. Thus the predominant energy storage occurs in the air gap, and the properties of the magnetic circuit are determined by the dimensions of the air gap.

Because of the simplicity of the resulting relations, magnetic nonlinearity and core losses are often neglected in the analysis of practical devices. The final results of such approximate analysis can, if required, be corrected for the effects of these neglected factors by semiempirical methods. Thus, analysis are carried out under the assumption that the flux and *mmf* are directly proportional, as in air, for the entire magnetic circuit. With the linearity assumption the analysis is greatly simplified.

Assuming linearity, Eq. (3.24) becomes

$$W_{field} = \frac{1}{2}\, i\psi = \frac{1}{2}\, M\Phi = \frac{1}{2}\, S\Phi^2 \qquad ...(3.28)$$

where S is the reluctance of the magnetic circuit and is equal to $\frac{mmf}{\Phi}$ i.e. $\frac{M}{\Phi}$

Since coil inductance is given as

$$L = \frac{N\Phi}{i} = \frac{\psi}{i},$$

or $i = \frac{\psi}{L}$ and the field energy can be expressed as

$$W_{field} = \frac{1}{2}\frac{\psi^2}{L} \qquad ...(3.29) \qquad \because \frac{1}{2}\, i\psi = \frac{1}{2}\frac{\psi}{L} \times \psi$$

In the linear case the inductance L is independent of current i but is a function of configuration x. Thus the field energy is a special function of two independent variables ψ and x i.e.

$$W_{field}(\psi, x) = \frac{1}{2}\frac{\psi^2}{L(x)} \qquad ...(3.30)$$

The field energy is distributed throughout the space occupied by the field. Neglecting losses and assuming permeability to be constant, the energy density of the field is obtained from Eq. (3.24) i.e.

$$w_{field} = \frac{W_{field}}{a_{field}} = \int_0^{\psi} \frac{Ni}{l}\, d\,\frac{\psi}{Na} = \int_0^{B} H\, dB$$

$$= \frac{1}{2}\, HB = \frac{1}{2}\frac{B^2}{\mu}\ J/m^3 \qquad ...(3.31)$$

where H is magnetic field intensity in AT/m and is equal to $\frac{Ni}{l}$ and B is magnetic field density in Wb/m^2 or Tesla.

The above expression for energy density is significant from the point of view of design wherein the capability of the material is to be fully utilized in arriving at the gross dimensions of the device.

For linear case it may be easily followed from Eq. (3.27) or Fig. 3.15 that the coenergy is numerically equal to energy i.e.

$$W'_{field} = W_{field} = \frac{1}{2}\, \psi i = \frac{1}{2}\, M\Phi = \frac{1}{2}\, S\Phi^2 \qquad ...(3.32)$$

Also in terms of coil inductance

$$W'_{field} = \int_0^i \psi\, di$$

$$= \int_0^i Li\, di = \frac{1}{2}\, Li^2 \qquad \because \psi = N\Phi = \frac{N\Phi}{i} \times i = Li$$

or in general

$$W'_{field}(i, x) = \frac{1}{2}\, L(x)\, i^2 \qquad ...(3.33)$$

The expression for coenergy density is given as

$$w'_{field} = \int_0^H B\, dH \qquad ...(3.34)$$

For linear case the above expression becomes

$$w'_{field} = \frac{1}{2}\, BH = \frac{1}{2}\, \mu H^2 = \frac{1}{2}\frac{B^2}{\mu}\ J/m^3 \qquad ...(3.35)$$

The fact that field energy can be expressed in terms of circuit parameter L, clears the way for electric circuit approach to the analysis of electrical machines i.e. the generalized theory of electrical machines. Thus the field energy approach serves as the physical basis for the generalized theory of electrical machines.

3.7.2. General Analysis of Electromagnetic System (Field Energy and Mechanical Force). Consider once again the attracted armature relay excited by an electrical source, as illustrated in Fig. 3.13. The field develops a magnetic force F_{field} acting upon the external mechanical system (which may be composed of passive and active mechanical elements) in the direction indicated in the figure. The mechanical work done by the field in moving the armature by a distance dx in the positive direction is given as

$$dW_{mech} = F_{field}\, dx \qquad ...(3.36)$$

This energy is drawn from the field by virtue of change dx in field configuration. According to the principle of energy conservation

Mechanical energy output

= Electrical energy input – increase in field energy

or $F_{field}\, dx = dW_{elec} - dW_{field}$

or $F_{field}\, dx = i\, d\psi - dW_{field}$ ...(3.37)

$\because$ from Eq. (3.22) $dW_{elec} = i\, d\psi$

Obviously $F_{field}\, dx$ is the gross mechanical output and a part of it will be lost in mechanical friction.

From Eq. (3.26) we have

$$W_{field} = i\psi - W'_{field}(i, x)$$

$$\text{So } d\,W_{field} = d(i\psi) - d\,W'_{field}(i, x)$$

$$= i\, d\psi + \psi\, d\,i - \left[\frac{\partial W'_{field}}{\partial i}\, di + \frac{\partial W'_{field}}{\partial x}\, dx\right] \qquad ...(3.38)$$

Substituting value of dW_{field} from Eq. (3.38) in Eq. (3.37) we have

$$F_{field}\, dx = i\, d\psi - \left[i\, d\psi + \psi\, d\,i - \left\{\frac{\partial W'_{field}}{\partial i}\, di + \frac{\partial W'_{field}}{\partial x}\, dx\right\}\right]$$

$$\text{or } F_{field}\, dx = \left\{\frac{\partial W'_{field}}{\partial i} - \psi\right\} di + \frac{\partial W'_{field}}{\partial x}\, dx \qquad ...(3.39)$$

Because of the incremental variations di and dx being independent and absence of term i in the left-hand side of above Eq. (3.39), its coefficient on the right hand must be equal to zero

$$\text{i.e.} \quad \frac{\partial W'_{field}}{\partial i} - \psi = 0$$

$$\text{or } \psi = \frac{\partial W'_{field}}{\partial i} \qquad ...(3.40)$$

Thus field from Eq. (3.39) becomes

$$F_{field} = \frac{\partial W'_{field}(i, x)}{\partial x} \qquad ...(3.41)$$

The above expression for magnetic or mechanical force produced applies when i is an independent variable i.e. when it is a current excited system.

In case (ψ, x) are independent variables

$$W_{field} = W_{field}(\psi, x)$$

$$\text{and } d\,W_{field} = \frac{\partial W_{field}}{\partial \psi} d\psi + \frac{\partial W_{field}}{\partial x} dx \quad ...(3.42)$$

Substituting for $d\,W_{field}$ from above Eq. (3.42) in Eq. (3.37) we have

$$F_{field}\,dx = i\,d\psi - \left[\frac{\partial W_{field}}{\partial \psi} d\psi + \frac{\partial W_{field}}{\partial x} dx\right]$$

$$\text{or } F_{field}\,dx = -\frac{\partial W_{field}}{\partial x} dx + \left\{i - \frac{\partial W_{field}}{\partial \psi}\right\} d\psi \quad ...(3.43)$$

Since $d\psi$, the independent differential is absent on the left-hand side of the above Eq. (3.43), its coefficient on the right hand must be equal to zero

$$\text{i.e. } i - \frac{\partial W_{field}}{\partial \psi} = 0$$

$$\text{or} \quad i = \frac{\partial W_{field}(\psi, x)}{\partial \psi} \quad ...(3.44)$$

Hence from Eq. (3.43)

$$F_{field} = \frac{-\partial W_{field}(\psi, x)}{\partial x} \quad ...(3.45)$$

In the above form of expression for the magnetic force, ψ is independent variable i.e. it is a voltage controlled system as voltage is derivative of ψ.

Equations (3.41) and (3.45) are equivalent and can both be used to calculate the force for a given system, user preference and convenience often dictate the choice.

In linear systems where inductances are specified it is more convenient to use coenergy for determination of force produced — refer to Eq. (3.41).

In case of voltage controlled system, the current can be determined by writing the necessary circuit equations.

In a rotary case, the Eqs. (3.41) and (3.45) for force will apply for torque by replacing x with angular rotation θ.

Direction of Mechanical Force Developed. From Eq. (3.45) it is obvious that mechanical force developed F_{field} is positive* when $\frac{\partial W_{field}(\psi, x)}{\partial x}$ is negative. It means that stored energy of the field is reduced with the increase of x if flux linkages ψ are kept fixed. In the particular case of electromagnetic relay shown in Fig. 3.13, as the armature moves towards left (i.e. as x increases), the field energy for fixed flux linkages ψ is reduced. This is due to reduced air gap. Thus it can be concluded that the mechanical force developed by the field acts in a direction to reduce field energy i.e., the system seeks a position of minimum field energy. Similary from Eq. (3.42) it follows that the system seeks a position of maximum coenergy.

Also in case of magnetic relay shown in Fig. 3.13, the mechanical force developed acts in a direction to increase x thereby reducing the reluctance of the magnetic circuit and increasing the inductance of coil.

3.7.3. Determination of Magnetic or Mechanical Force. *(a) Nonlinear Case.* Mechanical force developed by the magnetic field is given by partial derivatives of coenergy or energy [Eqs. (3.42) and (3.45)]. In the general nonlinear case, the derivative must be found numerically or graphically by assuming a small increment Δx. Thus

$$F_{field} = \left.\frac{\Delta W'_{field}}{\Delta x}\right|_{i = \text{constant}}$$

$$\text{or } F_{field} = \left.\frac{-\Delta W_{field}}{\Delta x}\right|_{\psi = \text{constant}} \quad ...(3.46)$$

The above two expressions will provide slightly different numerical values of F_{field} as Δx is finite. Obviously, F_{field} is the same in each case as $\Delta x \to 0$.

(b) Linear Case (Neglecting Reluctance of Iron Path)

$$W'_{field}(i, x) = \frac{1}{2} L(x)\,i^2 \qquad \text{Refer to Eq. (3.33)}$$

$$\text{or } \partial W'_{field} = \frac{1}{2} i^2\,\partial L(x)$$

and mechanical force developed,

$$F_{field} = \frac{\partial W'_{field}}{\partial x} = \frac{1}{2} i^2 \frac{\partial L(x)}{\partial x} \quad ...(3.47)$$

From above equation it is obvious that the mechanical or magnetic force developed acts in a direction to increase the exciting coil inductance, as already mentioned.

Also from Eq. (3.30) we have

$$W_{field}(\psi, x) = \frac{1}{2}\frac{\psi^2}{L(x)}$$

$$\text{So } F_{field} = \frac{-\partial W_{field}(\psi, x)}{\partial x} = \frac{1}{2}\left[\frac{\psi}{L(x)}\right]^2 \frac{\partial L(x)}{\partial x} \quad ...(3.48)$$

Since $i = \frac{\psi}{L}$, Eqs. (3.47) and (3.48) are equivalent.

Also from Eq. (3.28), $W_{field} = \frac{1}{2} S\Phi^2$

$$\therefore F_{field} = \frac{-\partial W_{field}}{\partial x} = -\frac{1}{2}\Phi^2 \frac{\partial S(x)}{\partial x} \quad ...(3.49)$$

It is to be noted here that ψ and Φ do not differ as independent variables because these are related by a constant N ($\psi = N\Phi$). From above Eq. (3.49) it is obvious that the force acts in a direction to reduce the magnetic system reluctance, as already mentioned.

Also from Eq. (3.28)

$$W_{field}(\psi, x) = \frac{1}{2}\psi i(x)$$

$$\text{or } F_{field} = -\frac{\partial W_{field}}{\partial x} = -\frac{1}{2}\psi \frac{\partial i(x)}{\partial x} \quad ...(3.50)$$

* The direction of F_{field} is taken as positive when it acts in the positive reference direction of x.

Similarly in a rotary case

$$T_{field} = \frac{-\partial W_{field}(i, \theta)}{\partial \theta} \qquad ...(3.51)$$

The above cases are applicable to motion through a limited distance or rotation through a definite angle and are the equations for singly-excited systems such as in electromagnetic relays, reluctance motors. In such systems the electromechanical energy conversion takes place through only one source (ac or dc) and that is why they are called *singly-excited magnetic field systems.*

3.7.4. Mechanical Work Done. Figure 3.16 (*a*) shows the magnetic circuit of a relay initially with air gap between the core and the armature. On the closure of switch *S*, the exciting current i increases from zero to $i_1 = \frac{V}{r}$ where r is the resistance of the coil. As explained earlier, the energy stored in the magnetic field is represented by the cross-hatched area *oabo* in Fig. 3.16 (*b*).

In Fig. 3.17 (*a*), the armature is assumed in the closed position i.e. the air gap g is zero. On closure of switch *S*, the exciting current again increases from zero to $i_1 = \frac{V}{r}$ whereas the flux linkages increase from zero to ψ_2. It is to be noted here that the flux linkages ψ_2 will be greater than the flux linkages ψ_1, because reluctance of the magnetic path is considerably reduced for the given mmf. The energy stored in the magnetic field is now given by the cross-hatched area *ocdo*.

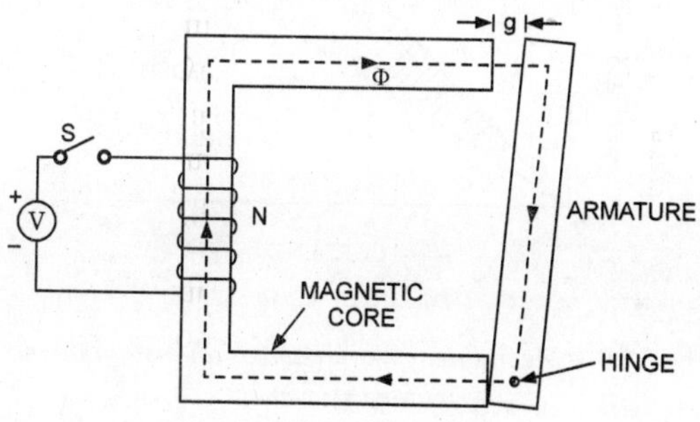

(a) Magnetic Circuit

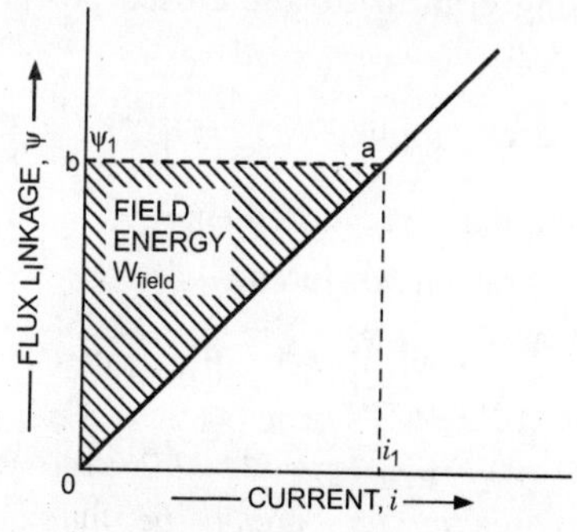

(b) Current-Flux Linkages Curve

Fig. 3.16 *Electromagnetic Relay in Open Position*

When the armature is in open position, as the magnetic field builds up, magnetic pull between the core and the armature develops. Under the action of the magnetic pull (a mechanical force), the armature moves to the closed position. During the armature movement, from its open position, the reluctance of the magnetic path reduces, thereby increasing the flux linkages from its initial value of ψ_1. Because of these increments in flux linkages, an emf, known as counter emf, is induced in the coil. The induced emf opposes the flow of exciting current i, i.e.,

$$\text{Current } i = \frac{V - E}{\text{Coil impedance}}$$

The magnitude of counter emf induced in the exciting coil depends on how fast the armature moves. We will consider the two extreme cases. One, the current is allowed to establish and gap g is reduced to zero very slowly (i.e. at almost zero speed). Second when the armature is held in position till the current i establishes and then the armature is allowed to move and reduce the gap g to zero in zero time (i.e. instantly). Neither of these two would be followed in practice.

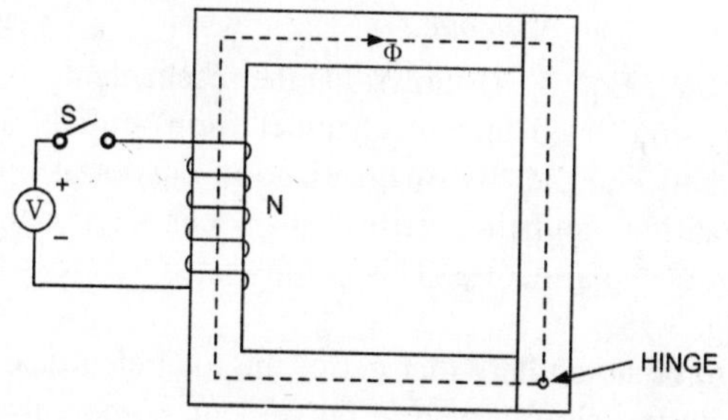

(a) Magnetic Circuit

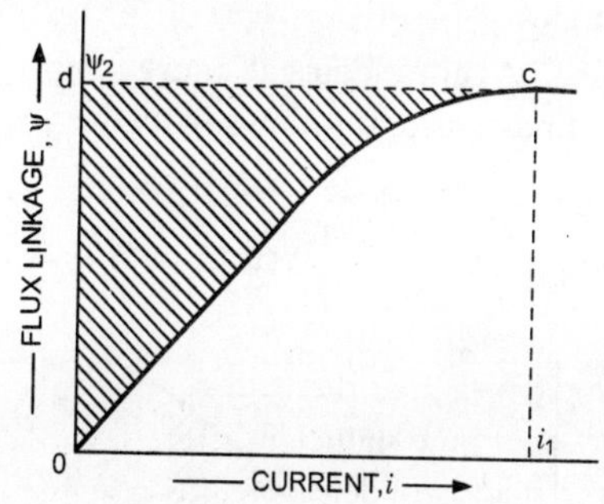

(b) Current-Flux Linkages Curve

Fig. 3.17 *Electromagnetic Relay in Closed Position*

(*a*) **Slow Movement of Armature.** When the armature is in the open position, the exciting current is i_1, the flux linkages are ψ_1 and the operating point is *a*. [Refer to Fig. 3.18] When the armature is allowed to move very slowly, the flux changes at negligibly low rate, the induced emf is, therefore, negligible, the current i_1 stays substantially constant $\left(i_1 = \frac{V}{r}\right)$. Operating point *a* must move in such a way that i_1 is maintained. Thus it moves along the vertical dotted line ac in Fig. 3.18. Consider the significance of the various areas in Fig. 3.18. Area *oaa'bo* to the left of the original magnetization curve (i.e. area *oaa'bo*) represents the energy drawn from the electrical source and absorbed by the magnetic field during the initial excitation of the coil while the armature is in open position. For simplicity, hysteresis and eddy currents are neglected. This energy then is stored in the magnetic field. After the armature has moved into the closed position, the energy stored in the field is represented by the area *ocdo* to the left of the final magnetization curve (i.e. area *oa'cdbo*). Consequently the increase in energy stored in the field,

$$\Delta W_{field} = \text{Magnetic energy stored in the closed position} - \text{magnetic energy stored in the open position}$$

$$= \text{Area } oa'cdbo - \text{area } oaa'bo \qquad ...(3.52)$$

While the armature is moving, the flux linkages are increasing from ψ_1 to ψ_2 and an emf is induced in the coil. The corresponding energy drawn from the electrical source during this time period is, from Eq. (3.20).

$$W_{elec} = \int_{\psi_1}^{\psi_2} i_1 \, d\psi = i_1(\psi_2 - \psi_1) = \text{Area } acdba'a \qquad ...(3.53)$$

Substituting Eqs. (3.52) and (3.53) in Eq. (3.17) we have

$$W_{elec} = \Delta W_{field} + W_{mech}$$

or $\text{Area } acdba'a = (\text{Area } oa'cdbo - \text{area } oaa'bo) + W_{mech}$

or $W_{mech} = \text{Area } acdba'a + \text{area } oaa'bo - \text{area } oa'cdbo$

$$= \text{Area } oaca'o \qquad ...(3.54)$$

The above Eq. (3.54) shows that the mechanical work done (or energy converted into mechanical form) equals the area enclosed between the two magnetization curves at open and closed positions and the vertical ψ-i locus during the slow movement of the armature. This is illustrated by cross-hatched area in Fig. 3.18.

It is to be noted here that out of the total electrical energy supplied during slow armature movement, some energy gets stored in the magnetic field and the remaining is converted into mechanical energy.

Linear Case. Again electrical energy supplied, $W_{elec} = i_1(\psi_2 - \psi_1)$. For linear case, it follows from the geometry of the Fig. 3.19 that

$$\Delta W_{field} = \Delta W'_{field} = W_{mech} = \frac{1}{2} i_1(\psi_2 - \psi_1) = \frac{1}{2} W_{elec} \qquad ...(3.55)$$

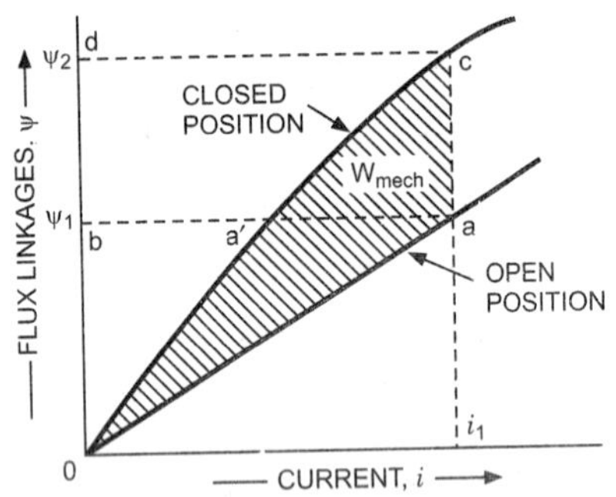

Fig. 3.18 *Mechanical Work Done With Slow Armature Movement (Nonlinear Case)*

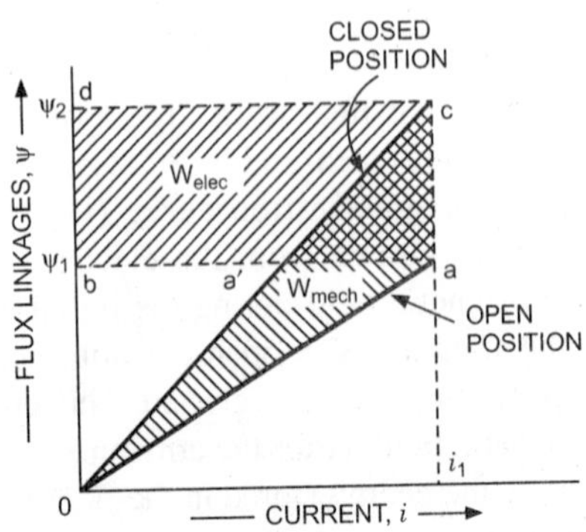

Fig. 3.19 *Mechanical Work Done With Slow Armature Movement (Linear Case)*

It means that half the electrical energy input gets stored in the field and the other half is output as mechanical energy.

(*b*) **Instantaneous Movement of Armature.** When the armature moves to close the air gap instantaneously (i.e. in zero time), change in flux linkages ψ will be negligible and will remain constant at ψ_1. This is because the flux linkages with an inductive circuit cannot change abruptly. The operating point, therefore, moves along the horizontal dotted line from a to a', and after the armature has closed, the operating point moves along the closed position magnetization curve $a'c$ as illustrated in Fig. 3.20 (*a*), because the final operating point has to be c.

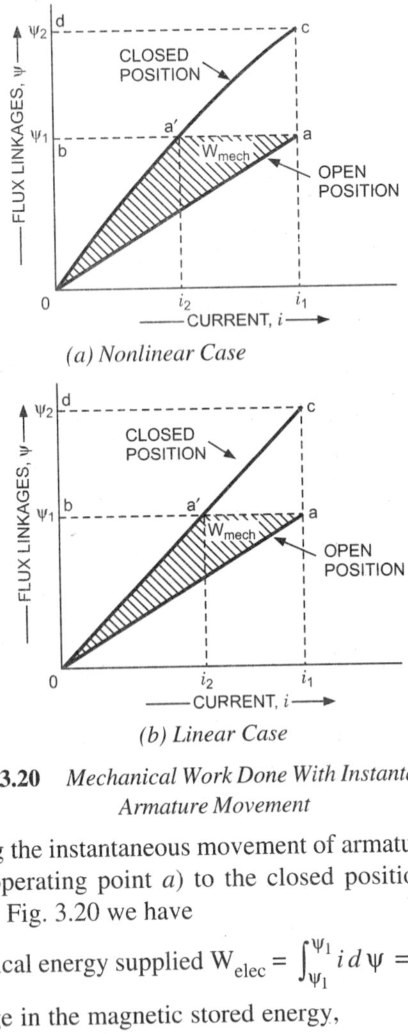

Fig. 3.20 *Mechanical Work Done With Instantaneous Armature Movement*

During the instantaneous movement of armature from open position (operating point a) to the closed position (operating point a') in Fig. 3.20 we have

$$\text{Electrical energy supplied } W_{elec} = \int_{\psi_1}^{\psi_1} i \, d\psi = 0$$

Change in the magnetic stored energy,

$$\Delta W_{field} = \text{Area } oa'bo - \text{area } oaa'bo$$

But $W_{elec} = \Delta W_{field} + W_{mech}$

i.e. $0 = \Delta W_{field} + W_{mech}$ $\qquad \because W_{elec} = 0$

or $W_{mech} = -\Delta W_{field} = -[\text{area } oa'bo - \text{area } oaa'bo]$

$$= \text{area } oaa'bo - \text{area } oa'bo = \text{area } oaa'o \qquad ...(3.56)$$

From the above Eq. (3.56) it is obvious that the mechanical work done is equal to the area enclosed between the two magnetization curves at open and closed positions and the horizontal ψ-i locus during the instantaneous movement of the armature. Since there is no electrical input, the mechanical energy output is at the expense of the field energy stored.

i.e. Mechanical energy output = Reduction in the energy stored in the field.

For linear case [Fig. 3.20 (*b*)]

$$W_{mech} = \frac{1}{2}\psi_1(i_1 - i_2) \qquad ...(3.57)$$

(*c*) **Transient Movement of Armature.** The armature movement will neither be too slow nor too fast, but will be somewhere in between the two extreme limits discussed above. The flux linkage-current (ψ-i) relationship is a general path from *a* to *c* as illustrated in Figs. 3.21 (*a*) and 3.21 (*b*).

In general case

$$\Delta W_{elec} = \text{Area } ba'acd$$

$$\Delta W_{field} = \text{Area } oa'cdo - \text{area } oaa'bo$$

and

$$W_{mech} = \Delta W_{elec} - \Delta W_{field}$$

$$= \text{Area } ba'acd - \text{area } oa'cdo + \text{area } oaa'bo$$

$$= \text{Area } oaca'o \qquad ...(3.58)$$

Equation (3.58) again shows that the mechanical work done is equal to the area enclosed between the two magnetization curves at open and closed positions and the (ψ-i) locus during the transient movement of armature.

Since ac is a general movement, the area *oaca'o*, which represents the mechanical energy output has to be computed graphically or numerically.

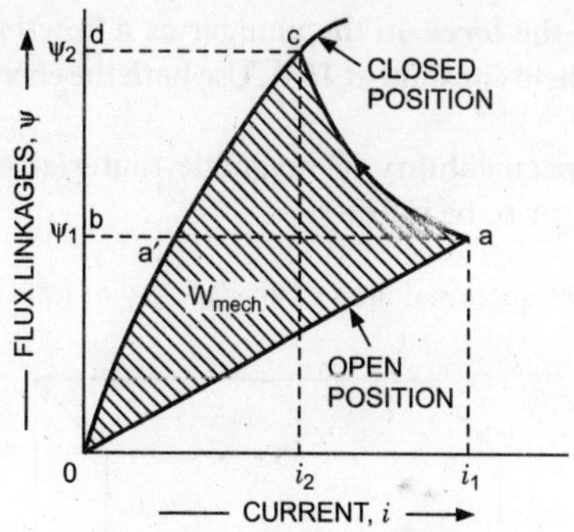

(*a*) *Nonlinear Case*

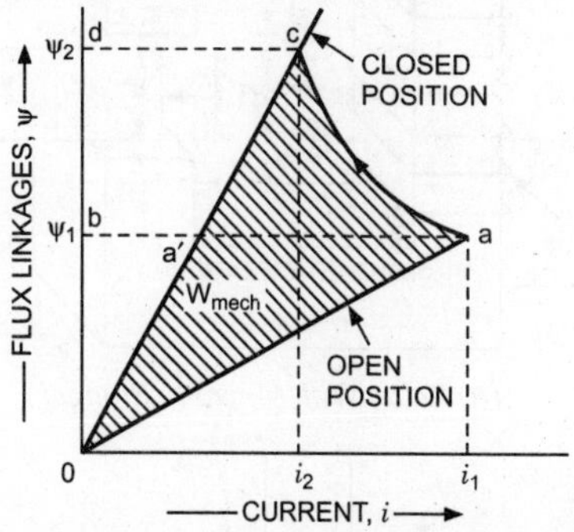

(*b*) *Linear Case*

Fig. 3.21 *Mechanical Work Done During Transient Movement of Armature*

3.8 ENERGY FLOW IN ELECTROMECHANICAL DEVICES

Electromechanical energy conversion process is reversible one and the development of mechanical force is governed by Eqs. (3.42) and (3.45). If the armature of the electromagnetic relay shown in Fig. 3.13 is allowed to move in the positive *x* direction (i.e. towards left) under the influence of mechanical force developed by magnetic field, electrical energy is converted into mechanical energy through the coupling field. When the armature is moved toward right under the influence of external force, mechanical energy is converted into electrical energy through the coupling field.

Electromechanical energy conversion process is not limited to translatory devices as illustrated but is equally applicable to rotary devices. Electrical and mechanical losses cause irreversible flow of energy out of a practical conversion device. The flow of energy in electromechanical conversion in either direction along with irrecoverable energy losses is shown in Figs. 3.22 (*a*) and 3.22 (*b*).

Example 3.2. The relay shown in Fig. 3.23 (*a*) is made from magnetic material with a movable plunger. The height of the plunger is much greater than the air-gap length ($h >> g$). Determine the magnetic stored energy as a function of plunger position ($0 < x < d$) for $N = 2{,}000$, $g = 2$ mm, $d = 0.2$ m, $l = 0.1$ m and $i = 10$ A.

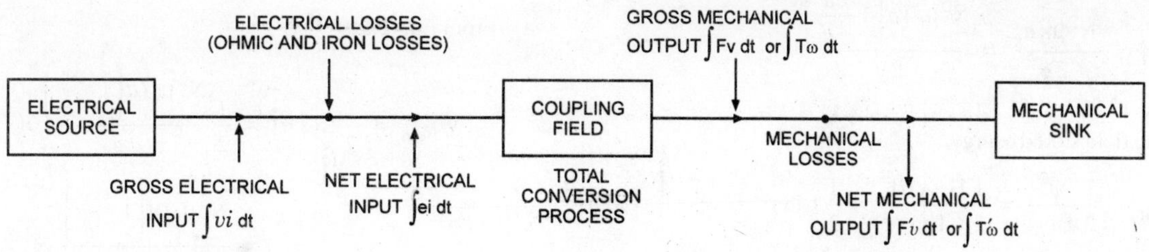

(*a*)

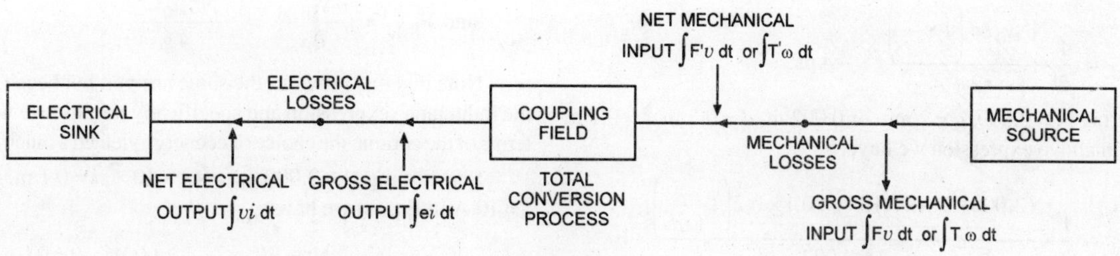

(*b*)

Fig. 3.22

Also find the force on the plunger as a function of x when the coil current is held constant at 15 A. Use both the energy and coenergy expressions.

Assume permeability of magnetic material of the relay and movable plunger to be infinity.

Solution: Cross-sectional area of air gap, $a_g = l\,(d-x) = l d\left(1-\frac{x}{d}\right)$

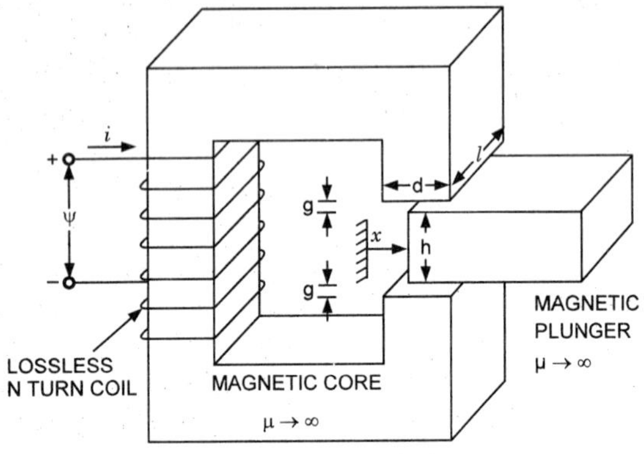

(a) Relay With Movable Plunger

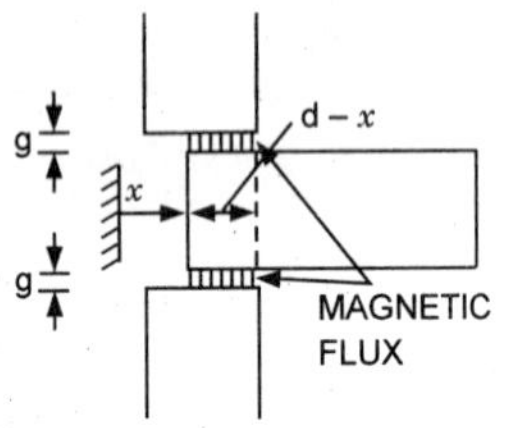

(b) Detail Showing Air gap Configuration With Plunger Partially Moved

Fig. 3.23

Reluctance of magnetic path = 0, as the permeability of the magnetic path is infinite.

$$\text{Reluctance of air gap} = \frac{\text{Length of air gap}}{\mu_o a_g} = \frac{2g}{\mu_o l d\left(1-\frac{x}{d}\right)}$$

So inductance is given as

$$L(x) = \frac{N^2 \mu_0 a_g}{2g} = \frac{N^2 \mu_0 l d\left(1-\frac{x}{d}\right)}{2g} \quad ...(i)$$

From Eq. (3.30)

Magnetic field stored energy,

$$W_{field} = \frac{1}{2}\frac{\psi^2}{L(x)} = \frac{1}{2}\frac{[L(x)i]^2}{L(x)} = \frac{1}{2}L(x)i^2 \quad ...(ii)$$

$$\because \quad \psi = L(x)\, i$$

Substituting the value of $L(x)$ from Eq. (*i*) into above Eq. (*ii*) we have

$$W_{field} = \frac{1}{2}\frac{N^2\mu_0 l d\left(1-\frac{x}{d}\right)}{2g} i^2 \quad \textbf{Ans.}$$

Substituting N = 2,000, g = 2 mm or 0.002 m, d = 0.2 m, l = 0.1 m and i = 10 A in above expression we have

$$W_{field} = \frac{1}{2}\times\frac{(2{,}000)^2\times 4\pi\times 10^{-7}\times 0.1\times 0.2\left(1-\frac{x}{d}\right)}{2\times 0.002}\times 10^2$$

$$= 1{,}256\left(1-\frac{x}{d}\right) \text{ joules} \quad \textbf{Ans.}$$

From Eq. (3.30)

$$\text{Magnetic field stored energy, } W_{field} = \frac{1}{2}\frac{\psi^2}{L(x)}$$

Substituting the value of L (x) from Eq. (*i*) in above expression we have

$$W_{field} = \frac{1}{2}\frac{\psi^2\times 2g}{N^2\mu_o l d\left(1-\frac{x}{d}\right)} = \frac{g\psi^2}{N^2\mu_0 l d\left(1-\frac{x}{d}\right)} \quad ...(iii)$$

and from Eq. (3.45)

$$F_{field} = -\frac{\partial W_{field}}{\partial x} = -\frac{g\psi^2}{N^2\mu_0 l d^2\left(1-\frac{x}{d}\right)^2} \quad ...(iv)$$

Substituting $\psi = L(x)\, i$ in above Eq. we have

$$F_{field} = -\frac{g L^2(x) i^2}{N^2\mu_0 l d^2\left(1-\frac{x}{d}\right)^2}$$

$$= -\frac{g i^2}{N^2\mu_0 l d^2\left(1-\frac{x}{d}\right)^2}\times\left(\frac{N^2\mu_0 l d\left(1-\frac{x}{d}\right)}{2g}\right)^2$$

$$= -\frac{N^2\mu_0 l}{4g} i^2 \quad ...(v)$$

$$\because \quad \text{From Eq. } (i)\ L(x) = \frac{N^2\mu_o l d\left(1-\frac{x}{d}\right)}{2g}$$

Thus it is seen that for constant current, the magnetic force is independent of the plunger position and acts to pull the plunger back into the relay.

Alternatively, the force can be determined from the coenergy. From Eq. (3.41)

$$F_{field} = \frac{\partial W'_{field}}{\partial x}$$

From Eq. (3.33)

$$W'_{field} = \frac{1}{2}L(x)i^2 = \frac{1}{2}\frac{N^2\mu_0 l d\left(1-\frac{x}{d}\right)}{2g} i^2$$

$$= \frac{N^2\mu_0 l d\left(1-\frac{x}{d}\right) i^2}{4g}$$

$$\text{and } F_{field} = \frac{\partial W'_{field}}{\partial x} = -\frac{N^2\mu_0 l i^2}{4g}$$

Note that this provides the same answer, but because of the form of the inductance expression and specifically because we want a solution in terms of the current, the choice of coenergy yielded a much simpler solution.

Substituting N = 2,000, $\mu_0 = 4\pi\times 10^{-7}$, l = 0.1 m, g = 0.002 m and i = 10 A in Eq. (*v*) we have

$$F_{field} = -\frac{N^2\mu_0 l i^2}{4g} = -\frac{(2{,}000)^2\times 4\pi\times 10^{-7}\times 0.1\times 10^2}{4\times 0.002}$$

$$= -6{,}283 \text{ N} \quad \textbf{Ans.}$$

Example 3.3. For the electromagnetic system shown in Fig. 3.24, each limb of the core has cross-sectional area of 20 cm^2. The coil has 1,200 turns. Assume that the iron part of the magnetic circuit has $\mu_r = \infty$.

(*a*) For a coil current of 2.5 A and air-gap length x = 1 cm; calculate (*i*) the coenergy of the magnetic field, and (*ii*) the force on the movable component.

(*b*) Calculate (*i*) mechanical work done (*ii*) electrical energy if the armature moves from x = 1 cm to x = 0.5 cm while the coil current is maintained constant at 2.5 A.

Solution: Since magnetic circuit is assumed to have relative permeability μ_r equal to infinity, the reluctance of the iron path is zero.

Inductance of the magnetic circuit,

$$L(x) = \frac{N^2 \mu_0 a}{2x} \qquad \because \mu_r = 1$$

$$= \frac{(1{,}200)^2 \times 4\pi \times 10^{-7} \times 20 \times 10^{-4}}{2x}$$

$$= \frac{1.152\pi \times 10^{-3}}{2x}\ \text{H}$$

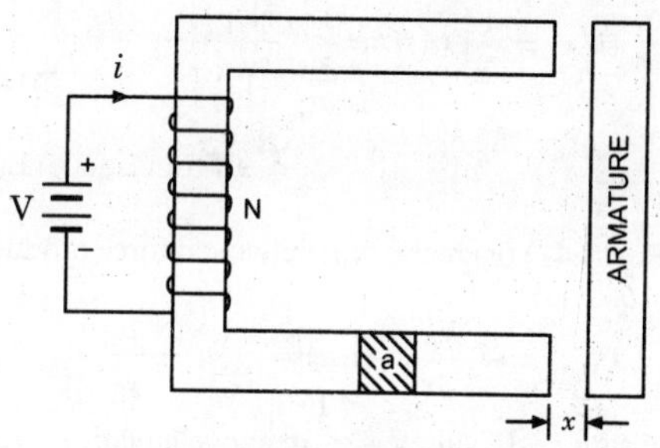

Fig. 3.24

$$\text{For } x = 1 \text{ cm} = 0.01 \text{ m, } L = \frac{1.152\pi \times 10^{-3}}{2 \times 0.01} = 0.181 \text{ H} \quad \textbf{Ans.}$$

(*i*) Coenergy of the magnetic field

$$= \frac{1}{2} L(x) i^2 = \frac{1}{2} \times 0.181 \times (2.5)^2 = 0.565625 \text{ J} \quad \textbf{Ans.}$$

Since $W'_{field} = \frac{1}{2} L(x)\, i^2$, Force on the movable component,

$$F_{field} = \frac{\partial W'_{field}}{\partial x} = \frac{1}{2} i^2 \frac{\partial}{\partial x}\left(\frac{1.152\pi \times 10^{-3}}{2x}\right)$$

$$= \frac{-1.152\pi \times 10^{-3}}{4x^2} i^2$$

$$= \frac{-1.152\pi \times 10^{-3} \times (2.5)^2}{4 \times (0.01)^2}$$

$$= -56.5625 \text{ N } \textbf{Ans.}$$

Negative sign implies that force acts in a direction to reduce x (i.e. (air-gap length))

(*b*) Electrical energy,

$$\delta W_{elec} = i(\psi_2 - \psi_1) \qquad \because \partial W_{elec} = i\, d\psi$$

$$= i(L_2 i - L_1 i) \qquad \because \psi = Li$$

$$= i^2 (L_2 - L_1)$$

$$= (2.5^2)^2 \times \frac{1.152\pi \times 10^{-3}}{2}\left(\frac{1}{0.005} - \frac{1}{0.01}\right) = 1.131 \text{ J } \textbf{Ans.}$$

Mechanical work done,

$$\partial W_{mech} = \frac{1}{2} \partial W_{elec} = \frac{1}{2} \times 1.131 \text{ J} = 0.565 \text{ J} \quad \textbf{Ans.}$$

Also mechanical work done

$$= \int_{x_1}^{x_2} F_{field}\, dx = \int_{0.01}^{0.05} \frac{-1.152\pi \times 10^{-3} \times i^2}{4x^2} dx$$

$$= \frac{1.152\pi \times 10^{-3} \times (2.5^2)^2}{4}\left[\frac{1}{x}\right]_{0.01}^{0.005}$$

$$= 1.8\,\pi \times 10^{-3}\left[\frac{1}{0.005} - \frac{1}{0.01}\right] = 0.565 \text{ J} \quad \textbf{Ans.}$$

Example 3.4. The magnetic flux density on the surface of an iron face is 1.6 T, which is a typical saturation level value for ferromagnetic material. Find the force density on iron face.

[U.P. Technical Univ. Electromechanical Energy Conversion-I, 2004-05]

Solution: Let a be the area in m^2 of the magnetic surfaces which are separated by a distance g. For a displacement x in the direction of force F_{field}, distance between the two surfaces is reduced to $(g - x)$.

Force between parallel faces is given as

$$F_{field} = -\frac{1}{2}\Phi_g^2 \frac{dS_g}{dx} \qquad \text{...(i) Refer to Eq. (3.49)}$$

$$\text{Reluctance of gap length, } S_g = \frac{g - x}{\mu_o a} \qquad \text{...(ii)}$$

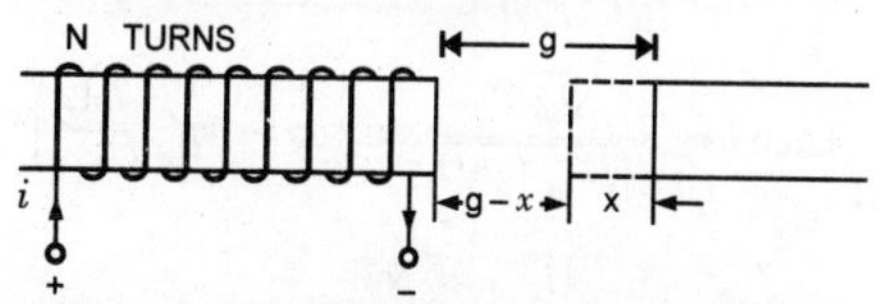

Fig. 3.25

Substituting value of S_g from Eq. (*ii*) in Eq. (*i*) we have

$$F_{field} = -\frac{1}{2}\Phi_g^2 \frac{d}{dx}\frac{(g - x)}{\mu_o a}$$

$$= -\frac{1}{2}\Phi_g^2 \times \frac{1}{\mu_o a} \times (-1)$$

$$= \frac{1}{2}(B_g a)^2 \times \frac{1}{\mu_o a} = \frac{B_g^2 a}{2\mu_o}$$

The force density = Force per unit area

$$= \frac{F_{field}}{a} = \frac{B_g^2 a}{2\mu_o} \times \frac{1}{a} = \frac{1}{2}\frac{B_g^2}{\mu_o}$$

$$= \frac{1}{2} \times \frac{1.6^2}{4\pi \times 10^{-7}}$$

$$= 1.0186 \times 10^6 \text{ N/m}^2 \textbf{ Ans.}$$

Example 3.5. The electromagnetic relay shown in Fig. 3.26 is excited from a voltage source $v = \sqrt{2}$ V sin ωt.

Find the expression for the average force on the armature, when the armature is held fixed at distance x. *[J.U. December-1999]*

Solution : Let the reluctance of iron path be constant c.

$$\text{Reluctance of air path} = \frac{\text{Gap length}}{\mu_0 a} = \frac{2x}{\mu_0 a} = bx$$

Thus the total reluctance of magnetic path, $S = bx + c$

$$W_{field}(\psi, x) = \frac{1}{2} S(x)\,\Phi^2 \qquad \text{Here } \psi = N\Phi$$

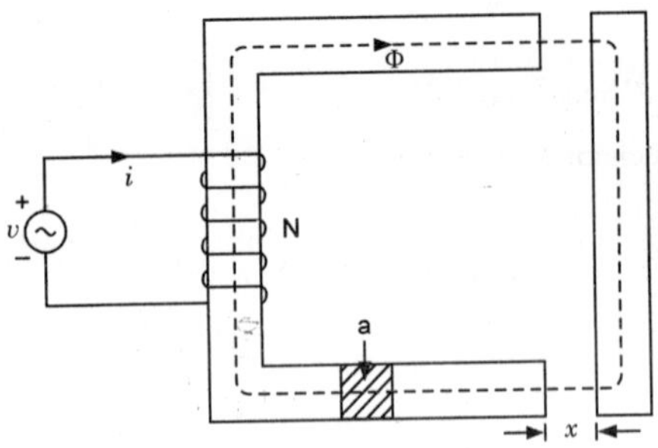

Fig. 3.26

$$F_{field} = \frac{\partial W_f(\Phi, x)}{\partial x} = -\frac{1}{2}\Phi^2 \frac{\partial S}{\partial x}$$

$$= -\frac{1}{2}\Phi^2 \frac{\partial}{\partial x}(bx + c) = -\frac{1}{2}\Phi^2 b \quad ...(i)$$

The derivation of S w.r.t. x is positive so that F_{field} is negative. It means that the force acts in a direction to reduce x.

Now in a circuit,

$$\text{Applied voltage, } v = S + L\frac{di}{dt}$$

The steady-state solution of the above equation is

$$i = \frac{v}{\sqrt{r^2 + \omega^2 L^2}} \angle \tan^{-1}\frac{\omega L}{r} \quad ...(ii)$$

$$\text{Then } i = \frac{\sqrt{2}\,V}{\sqrt{r^2 + \omega^2 L^2}} \sin\left(\omega t - \tan^{-1}\frac{\omega L}{r}\right) \quad ...(iii)$$

$$\text{and } \Phi = \frac{Ni}{l/\mu_r\mu_0 a} = \frac{Ni}{S} \quad ...(iv)$$

Substituting the value of i from Eq. (*iii*) in above Eq. (*iv*) we have

$$\Phi = \frac{\sqrt{2}\,NV}{S\sqrt{r^2 + \omega^2 L^2}} \sin\left(\omega t - \tan^{-1}\frac{\omega L}{r}\right)$$

$$= \frac{\sqrt{2}\,NV}{\sqrt{(rS)^2 + (\omega N^2)^2}} \sin\left(\omega t - \tan^{-1}\frac{\omega N^2}{Sr}\right) \quad ...(v) \because SL = N^2$$

Substituting the value of Φ from Eq. (*v*) in Eq. (*i*) we have

$$F_{field} = -\frac{1}{2}\frac{2N^2V^2 b}{(rS)^2 + (N^2\omega)^2} \sin^2\left(\omega t - \tan^{-1}\frac{\omega N^2}{Sr}\right)$$

$$= -\frac{N^2V^2 b}{(rS)^2 + (N^2\omega)^2} \sin^2\left(\omega t - \tan^{-1}\frac{\omega N^2}{Sr}\right)$$

Time average value of F_{field},

$$F_{field(av)} = \frac{1}{T}\int_0^T F_{field}\, dt = -\frac{1}{2}\frac{bN^2V^2}{(rS)^2 + (N^2\omega)^2} \quad \textbf{Ans.}$$

T being equal to $\frac{2\pi}{\omega}$.

Example 3.6. For the electromagnetic device shown in Fig. 3.27, the cross-sectional area normal to the flux is A and the reluctance is offered by air gap alone. Compute the average force on the movable member in terms of N, x, A etc. when (i) $i = I_{max} \sin(\omega t + \alpha)$ (ii) $v = V_{max} \cos \omega t$.

[*G.B. Technical Univ. Electromechanical Conversion-I, 2010-11*]

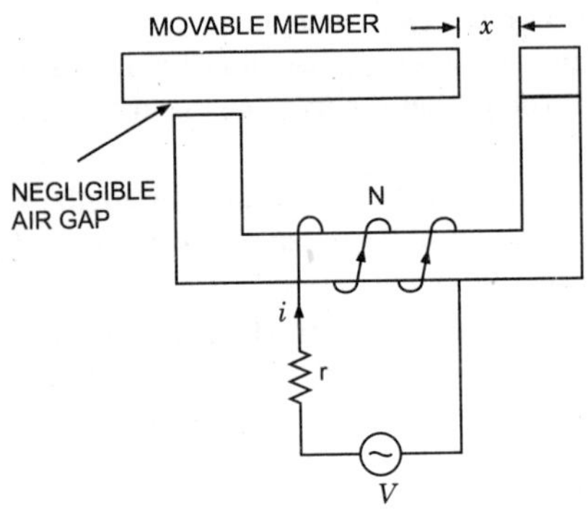

Fig. 3.27

Solution: (*i*) Reluctance offered by air gap,

$$S = \frac{x}{\mu_0 A} \quad ...(i)$$

$$L(x) = \frac{N^2}{S} = \frac{N^2\mu_0 A}{x} \quad ...(ii) \because \text{From Eq. } (i)\ S = \frac{x}{\mu_0 A}$$

From Eq. (3.33)

$$W'_{field}(i, x) = \frac{1}{2}L(x)i^2 = \frac{1}{2}i^2\frac{N^2\mu_0 A}{x} \quad ...(iii)$$

$$\because \text{From Eq. } (ii)\ L(x) = \frac{N^2\mu_0 A}{x}$$

From Eq. (3.41) magnetic or mechanical force produced

$$F_{field} = \frac{\partial W'_{field}}{\partial x}(i, x) = \frac{-1}{2}i^2\frac{N^2\mu_0 A}{x^2}$$

Substituting $i = I_m \sin(\omega t + \alpha)$ in above equation, we have

$$F_{field} = \frac{-1}{2}I_m^2 \sin^2(\omega t + \alpha)\frac{N^2\mu_0 A}{x^2}$$

$$= \frac{-1}{2}\frac{N^2\mu_0 A}{x^2}I_m^2\left[\frac{1 - \cos 2(\omega t + \alpha)}{2}\right]$$

$$= -\frac{1}{4}\frac{N^2\mu_0 A}{x^2}I_m^2 + \frac{1}{4}\frac{N^2\mu_0 A}{x^2}I_m^2 \cos 2(\omega t + \alpha)$$

$$\text{Average magnetic or mechanical force} = \frac{-1}{4}\frac{N^2\mu_0 A}{x^2}I_m^2 \quad \textbf{Ans.}$$

$\because$ the average of second term over a complete cycle is zero

(*ii*) From Eq. (3.49)

$$F_{field} = \frac{-1}{2}\Phi^2\frac{\partial S(x)}{\partial x} = -\frac{1}{2}\Phi^2\frac{1}{\mu_0 A} \quad ...(i)$$

For electric circuit, $v = ir + L(x)\frac{di}{dt}$

Substituting $v = V_m \cos \omega t$ in above equation, we have

$$V_m \cos \omega t = ir + L(x)\frac{di}{dt}$$

Its steady state solution is

$$I = \frac{V_m}{\sqrt{2}\sqrt{r^2 + [\omega L(x)]^2}} \angle -\tan^{-1}\frac{\omega L(x)}{r}$$

$$\text{So } i = \frac{V_m}{\sqrt{r^2 + [\omega L(x)]^2}} \cos\left[\omega t - \tan^{-1}\frac{\omega L(x)}{r}\right]$$

$$\text{Now } \Phi = \frac{Ni}{S} = \frac{N V_m \mu_0 A}{x\sqrt{r^2 + [\omega L(x)]^2}} \cos\left[\omega t - \tan^{-1}\frac{\omega L(x)}{r}\right]$$

Substituting value of Φ from above equation in the expression for field [Eq. (*i*)], we have

$$F_{field} = \frac{-1}{2}\frac{N^2 V_m^2 \mu_0^2 A^2 \cos^2\left[\omega t - \tan^{-1}\frac{\omega L(x)}{r}\right]}{\mu_0 A\left[x^2\{r^2 + (\omega L(x)^2\}\right]}$$

Substituting $L(x) = \frac{N^2 \mu_0 A}{x}$ in above expression, we have

$$F_{field} = \frac{-1}{2}\frac{N^2 V_m^2 \mu_0 A}{r^2 x^2 + (\omega N^2 \mu_0 A)^2}\cos^2\left[\omega t - \tan^{-1}\frac{\omega L(x)}{r}\right]$$

$$\text{or } F_{field\ av} = \frac{-1}{4}\frac{N^2 V_m^2 \mu_0 A}{r^2 x^2 + \omega^2 N^4 \mu_0^2 A^2} \quad \textbf{Ans.}$$

Example 3.7. A movable part of an electromagnet is displaced axially by 1.2 mm, as depicted in Fig. 3.28. Pole width is 5 mm, flux density is 1 Wb/m² and gap length is 4 mm. If the gap length is kept constant, determine the force that acts in a direction to bring them into axial alignment.

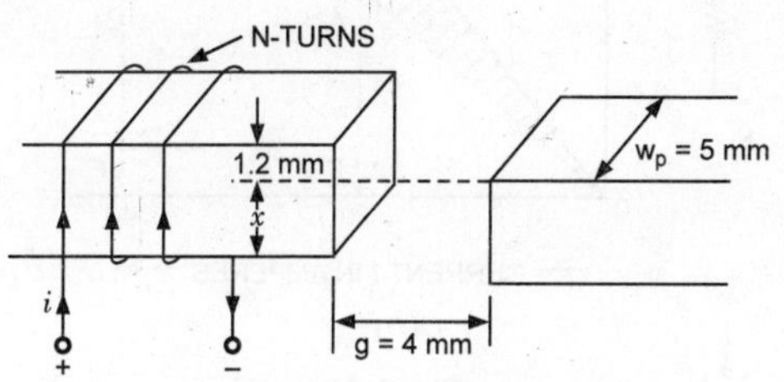

Fig. 3.28 *Electromagnet*

Solution: Let all the field energy be in the overlapping surfaces of the pole faces and overlapping height be *x*, as depicted in Fig. 3.28. Then reluctance of the overlapping region

$$S_g = \frac{g}{\mu_o a} = \frac{g}{\mu_o x\, w_p} \qquad \because\ a = x\, wp$$

$$\text{and} \quad W_{field} = \frac{1}{2}S_g \Phi_g^2 = \frac{1}{2}\frac{g}{\mu_o x w_p}\Phi_g^2 \qquad \text{Refer to equation (3.28)}$$

$$F_{field} = -\frac{\partial W_{field}(\Phi, x)}{\partial x}$$

$$= -\frac{1}{2}\frac{g\Phi_g^2}{\mu_0 w_p}\frac{\partial}{\partial x}\left(\frac{1}{x}\right) = \frac{1}{2}\Phi_g^2 \frac{g}{\mu_0 w_p x^2}$$

Substituting $\Phi_g = B_g w_p x$ in above expression we have

$$F_{field} = \frac{1}{2}(B_g w_p x)^2 \times \frac{g}{\mu_o w_p x^2} = \frac{1}{2}\frac{B_g^2 w_p g}{\mu_o}$$

Substituting B_g = 1 Wb/m² (or Tesla), w_p = 5 mm or 0.005 m, g = 4 mm or 0.004 m and $\mu_0 = 4\pi \times 10^{-7}$ in the above expression we have

$$F_{field} = \frac{1}{2} \times \frac{(1)^2 \times (0.005)(0.004)}{4\pi \times 10^{-7}} = 7.96 \text{ Newtons} \quad \textbf{Ans.}$$

Example 3.8. The cross-sectional view of a cylindrical plunger is shown in Fig. 3.29. The plunger (or armature) is free to move inside a non-ferromagnetic guide around which the coil is surrounded by a cylindrically shaped steel shell. The plunger is separated from the shell by an air gap of length *g*.

(*i*) Derive the expression for the magnetic force exerted on the plunger when it is in the position shown in Fig. 3.29. The length of the plunger is at least equal to that of the shell. Also it has a radius of *a* metres. Reluctance of the steel may be considered negligible.

(*ii*) Determine the magnitude of force when mmf is 1,500 AT and the dimensions of plunger magnet are as below:

a = 25 mm; g = 1.25 mm; x = 25 mm; and h = 50 mm.

Solution : For the direction of coil current assumed a typical flux path is that shown in Fig. 3.29. It is noteworthy that the reluctance seen by the flux Φ as it crosses the bottom gap is less than that seen by the same flux crossing the gap at the upper part of the plunger. This is because the cross-sectional area of the magnetic circuit is larger at the lower part than it is at the upper part of the plunger. As a matter of fact, at the lower part of the plunger the cross-sectional area seen by the flux is fixed for all positions of plunger lying in the range $0 \le x \le h$ and specifically is equal to $2\pi a h$ (assuming *g* very small in comparison to *a*). Accordingly, a force is exerted on the plunger so directed as to reduce the reluctance. This means that the magnetic force acts to move the plunger upward.

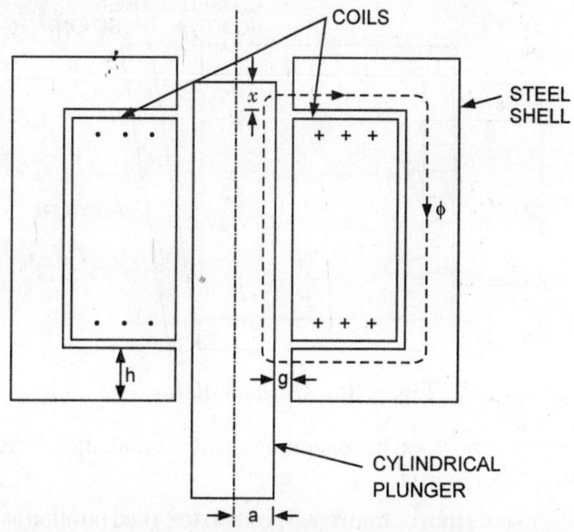

Fig. 3.29

(*i*) Since the reluctance of the steel is negligible, the reluctance associated with the typical flux shown in Fig. 3.29 is merely the sum of the reluctances with each air gap. Thus,

$$S = \frac{g}{\mu_0 \times 2\pi a h} + \frac{g}{\mu_0 \times 2\pi a x}$$

$$= \frac{g}{2\pi a \mu_o}\left(\frac{1}{h} + \frac{1}{x}\right) \qquad \text{...(i)}$$

$$\text{and} \quad \frac{dS}{dx} = \frac{g}{2\pi a \mu_0}\left(\frac{-1}{x^2}\right) \qquad \text{...(ii)}$$

$$\text{Also flux, } \Phi = \frac{MMF}{S} = \frac{MMF}{g/2\pi a \mu_0} \times \frac{xh}{x+h} \qquad \text{...(iii)}$$

Magnitude of magnetic force is given as

$$F = -\frac{1}{2}\Phi^2 \frac{dS}{dx}$$

$$= -\frac{1}{2} \times \left[\frac{2\pi a \mu_0}{g}\right]^2 \times \frac{(MMF)^2 x^2 h^2}{(x+h)^2} \times \frac{g}{2\pi a \mu_0} \times \left(\frac{-1}{x^2}\right)$$

$$= \frac{\pi a \mu_0 \times (MMF)^2 h^2}{g} \times \frac{1}{(x+h)^2}$$

$$= \frac{3.948\times10^{-6}}{g}\times\frac{ah^2}{(x+h)^2}\times(\text{MMF})^2 \quad ...(iv)\ \textbf{Ans.}$$

(*ii*) Substituting MMF = 1,500 AT; a = 25 mm = 0.025 m, g = 1.25 mm = 0.00125 m, x = 25 mm = 0.025 m and h = 50 mm = 0.05 m in above expression (*iv*) we have

$$F = \frac{3.948\times10^{-6}}{0.00125}\times\frac{0.025\times(0.05)^2}{(0.025+0.05)^2}\times(1,500)^2 = 78.96\text{ N } \textbf{Ans.}$$

Example 3.9. For the magnetic relay given in Fig. 3.30, the magnetization curves for open and closed positions of the armature are linear. If the armature of the relay moves from open to closed position at constant current, show that the electrical energy input is shared equally between field energy stored and the mechanical work done.

[U.P. Technical Univ. Electromechanical Energy Conversion-I, 2001-02]

Solution: When the relay is in open position, assume the operating point to be a, so that the current and flux linkage are i_1 and ψ_1 respectively (Fig. 3.19).

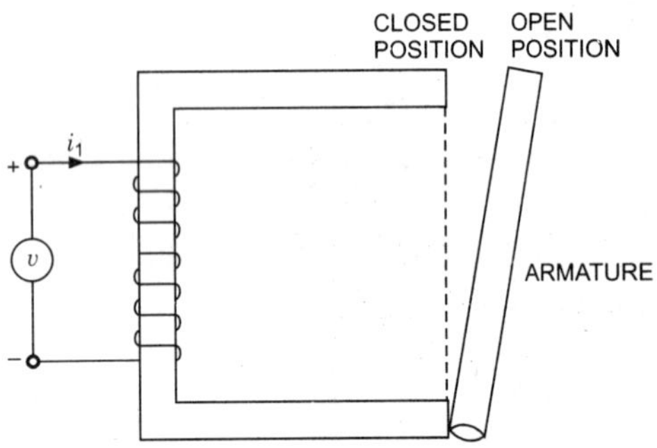

Fig. 3.30 *Magnetic Relay*

In the closed position the operating point is c and, therefore, current is i_1 and flux linkage is ψ_2.

Since armature moves from open position to closed position at constant current, $\psi - i$ locus is along the vertical line ac.

During the movement of the armature from open position to closed position, the electrical energy supplied,

$$W_{elec} = i_1(\psi_2 - \psi_1) \quad ...(i)$$

The mechanical work done, W_{mech} is given by the area *oaca′o* and is equal to

$$\frac{1}{2}i(\psi_2 - \psi_1) = \frac{1}{2}W_{elec} \quad ...(ii)$$

Magnetic stored energy,

W_{field} = Triangular area *oa′cdbo* – triangular area *oaa′bo*

$$= \frac{1}{2}\psi_2 i_1 - \frac{1}{2}\psi_1 i_1 = \frac{1}{2}(\psi_2 - \psi)i_1 = \frac{1}{2}W_{elec} \quad ...(iii)$$

From Eqs. (*ii*) and (*iii*) it is obvious that electrical energy supplied during movement of the armature at constant current from open to closed position is shared equally between field energy stored and the mechanical work done.

Example 3.10. The simple magnetic relay of Fig. 3.31 (*a*) gave the following ψ-*i* characteristics.

(*i*) Open position, $\psi = 0.4\,i$ Wb-turn for all values of current.

(*ii*) Closed position, $\psi = 0.06\,i$ for $0 \le i \le 20$

$\psi = 1.2 + 0.03\,(i - 20)$ $i > 20$

For an armature movement from open to closed position, find the magnitude of average magnetic force. The air gap length is 2 cm and the current during armature movement is constant at 40 A.

[U.P. Technical Univ. Electromechanical Energy Conversion-I, 2007-08]

Solution: The variations of flux linkage with current are depicted in Fig. 3.31 (*b*), both for open and closed positions of the armature, B is the operating point for open position and C is the operating point for closed position. The ψ-*i* locus during the movement of armature is along the vertical line BC, since current *i* remains constant at 40 A.

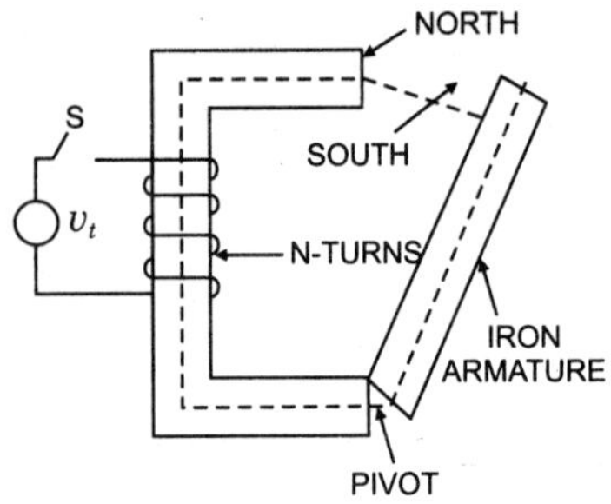

(*a*)

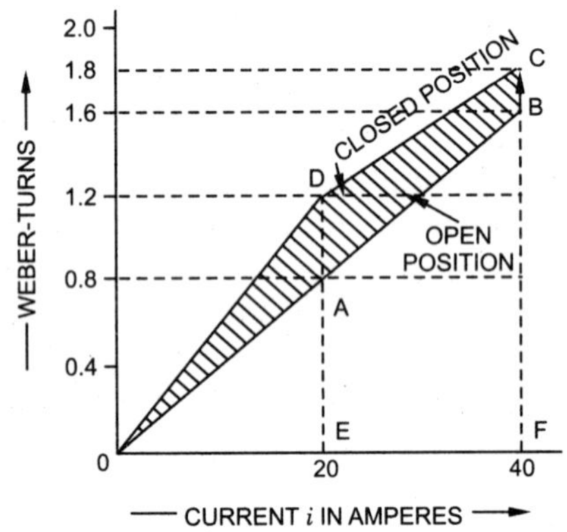

(*b*)

Fig. 3.31

Mechanical workdone during the movement of armature

= Area OABCDO = Area ODCFEO – Area OABFEO

$$= \left[\frac{1}{2}\times20\times1.2+\left(\frac{1.2+1.8}{2}\right)\times20-\frac{1}{2}\times40\times1.6\right]$$

= 42 – 32 = 10 joules

Average electromagnetic force,

$$F_{av} = \frac{\text{Mechanical workdone}}{\text{Air-gap length}} = \frac{10}{2\times10^{-2}} = 500\text{ N} \quad \textbf{Ans.}$$

Example 3.11. For the relay shown in Fig. 3.32 (*a*), the air gap x = 50 mm. The magnetization curve of the relay is shown in Fig. 3.32 (*b*).

(*a*) Calculate the energy stored in the field with the relay in open position.

(*b*) Assuming the relay armature moving very fast (i.e. under the condition of constant flux), calculate the work done in going from open position to the closed position.

(*c*) Where does the energy of part (*b*) comes from ?

(*d*) Calculate the average force exerted on the armature in part (*b*).

(*e*) Assuming the relay armature moving very slowly (i.e. at constant mmf), calculate the work done in going from open position to the closed position.

(*f*) Does the energy part of (*e*) come from the original stored field energy ? Explain.

Solution : Flux in closed position,

$$\Phi_2 = OG = BD = \frac{OD}{OE}\times AD = \frac{1,200}{750}\times0.012 = 0.0192\text{ Wb}$$

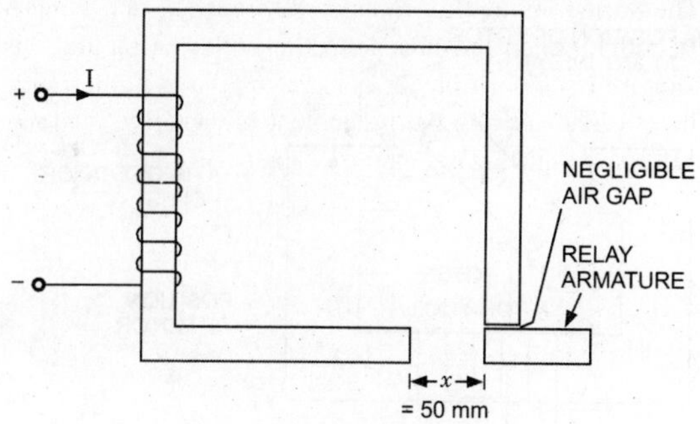

(a) Magnetic Relay

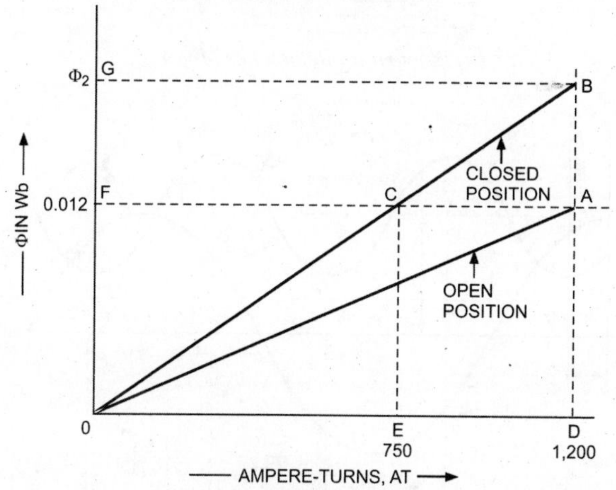

(b) Magnetization Curve

Fig. 3.32

(*a*) Energy stored in the field with the relay in open position,

$$W_{field} = \frac{1}{2} i_1 \psi_1 = \frac{1}{2} i_1 \times N \times \Phi_1$$

$$= \frac{1}{2} AT_1 \times \Phi_1 = 1{,}200 \times 0.012 = 7.2 \text{ J} \quad \textbf{Ans.}$$

(*b*) Mechanical work done in the armature movement from open to closed position under the condition of constant flux

$$W_{mech} = \frac{1}{2} \psi_1 (i_1 - i_2) = \frac{1}{2} N\Phi_1 \left(\frac{AT_1}{N} - \frac{AT_2}{N} \right)$$

$$= \frac{1}{2} \Phi_1 (AT_1 - AT_2)$$

$$= \frac{1}{2} \times 0.012 \times (1{,}200 - 750) = 2.7 \text{ J} \quad \textbf{Ans.}$$

(*c*) Mechanical energy in doing work in part (*b*) is drawn from the field energy which is reduced by an equal amount.

(*d*) Average force exerted on armature on fast movement of armature from open to closed position,

$$F = \frac{\text{Work done}}{\text{Distance moved}} = \frac{2.7}{0.05} = 54 \text{ N} \quad \textbf{Ans.}$$

(*e*) Mechanical work done in the armature movement from open to closed position under the condition of constant mmf,

$$W_{mech} = \frac{1}{2} i_1 (\psi_2 - \psi_1) = \frac{1}{2} N i_1 \left(\frac{\psi_2}{N} - \frac{\psi_1}{N} \right)$$

$$= \frac{1}{2} AT_1 (\Phi_2 - \Phi_2)$$

$$= \frac{1}{2} \times 1{,}200\, (0.0192 - 0.012)$$

$$= 4.32 \text{ J} \quad \textbf{Ans.}$$

(*f*) The mechanical energy comes from the electrical energy supplied during armature movement at constant mmf. Half of the input energy gets stored in the field and the other half is utilised in doing mechanical work.

Example 3.12. The magnetic circuit shown in Fig. 3.33 is made of cast steel. The rotor is free to turn about a vertical axis. The dimensions are shown in the figure.

(*a*) Derive an expression for the torque acting on the rotor in terms of the dimensions and the magnetic field in the two air gaps. Neglect the fringing effect.

(*b*) The maximum flux density in the overlapping portions of the air gap is limited to approximately 2 T, because of saturation in the steel. Determine the maximum torque in Nm for *r* = 25 mm, *l* = 25 mm and *g* = 2.5 mm.

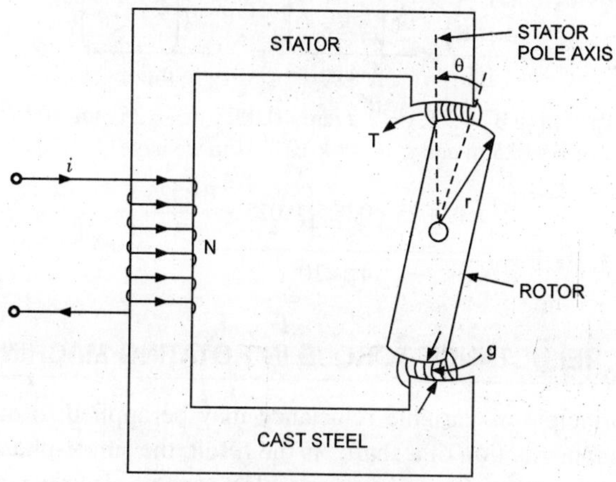

AXIAL LENGTH (NORMAL TO PAPER) = l

Fig. 3.33

Solution: Average gap radius $= \left(r + \frac{1}{2} g \right)$

Air-gap area normal to flux, $A_g = \left(r + \frac{1}{2} g \right) \theta l$

Total air-gap length along the flux path $= 2g$

Reluctance offered by two air gaps,

$$S_g = \frac{2g}{\mu_0 \times \text{area of air gap normal to flux}}$$

$$= \frac{2g}{\mu_0 \left(r + \frac{1}{2} g \right) \theta l}$$

$$\text{Flux established, } \Phi_g = \frac{\text{MMF}}{S_g} = \frac{N \times i \times \mu_0 \left(r + \frac{1}{2} g \right) \theta l}{2g}$$

$$\text{Flux linkages, } \psi_g = \Phi_g \times N = \frac{N^2 i \mu_0 \left(r + \frac{1}{2} g \right) \theta l}{2g}$$

$$\text{Coenergy, } W'_{field}(i, \theta) = \frac{1}{2} \psi_g i = \frac{1}{2} \frac{N^2 i^2 \mu_0 \left(r + \frac{1}{2} g \right) \theta l}{2g}$$

$$\text{Torque developed, } T_e = \frac{\partial W'_{field}}{\partial \theta} = \frac{N^2 i^2 \mu_0 \left(r + \frac{1}{2} g \right) l}{4g} \quad \text{...(}i\text{)}$$

The developed torque acts to cause an increase in coenergy for a given coil current *i*. This happens with θ increasing i.e. when the rotor tends to align itself with the stator. The noteworthy point in this case is that developed electromagnetic torque, T_e is independent of angle θ.

From the design point of view, the torque expression must be in terms of the flux density B_g rather than the exciting coil current, because the limiting maximum value of developed electromagnetic torque depends upon the limiting maximum value of flux density (i.e. when the magnetic material is stressed to its saturation value of 1.6 T).

$$\text{Flux density, } B_g = \frac{\Phi_g}{a_g} = \frac{\mu_0 Ni}{2g}$$

Substituting the value of $Ni = \dfrac{2gB_g}{\mu_0}$ in Eq. (*i*) we have

$$T_e = \frac{4g^2 B_g^2}{\mu_0^2} \times \frac{\mu_0\left(r+\frac{1}{2}g\right)l}{4g} = \frac{B_g^2 gl\left(r+\frac{1}{2}g\right)}{\mu_0} \qquad \text{...(ii)}$$

Substituting $B_g = 2$ T, $g = 2.5$ mm $= 0.0025$ m; $l = 25$ mm $= 0.025$ m; $r = 25$ mm $= 0.025$ m and $\mu_o = 4\pi \times 10^{-7}$ H/m we have

$$T_e = \frac{2^2 \times 0.0025 \times 0.025\left(0.025 + \frac{0.0025}{2}\right)}{4\pi \times 10^{-7}} = 5.22 \text{ Nm } \textbf{Ans.}$$

3.9 RELUCTANCE TORQUE IN ROTATING MACHINES

The principle of variable reluctance may be applied to cause continuous rotation of a shaft. As the result, the single-phase ac reluctance motor, is one of the simplest type of electric motor. Such a motor is the synchronous one i.e. it operates at a speed proportional to the supply frequency.

An elementary form of a single-phase reluctance motor is shown in Fig. 3.34 (*a*). Such a machine has saliency (salient poles) both on the stator and rotor. The reluctance or permeance of the magnetic circuit depends on the relative angular position of rotor and stator. Thus in Fig. 3.34 (*a*) the reluctance S is a periodic function of angle θ_r between the long axis of the rotor and the axis of stator poles, as indicated in Fig. 3.34 (*a*). When the axis of the rotor is directly in line with the axis of the stator poles, called the *direct or d-axis*, ($\theta_r = 0, \pi, 2\pi, 3\pi,...$), the reluctance offered to the stator flux, by two small air gaps in series with high permeability iron, is minimum. This minimum reluctance is called the *direct-axis reluctance* S_d. On the other hand when the axis of rotor is perpendicular to the axis of stator poles called the *quadrature or q-axis*, $\left(\theta_r = \frac{\pi}{2}, \frac{3\pi}{2}, \frac{5\pi}{2}, ...\right)$, the reluctance offered to the stator flux, by two very large air gaps in series with high permeability iron is maximum. This maximum reluctance is called the quadrature-axis reluctance S_q. The stator winding is connected to a source of single-phase alternating voltage. The flux, therefore, is also alternating, as illustrated by the curve in Fig. 3.34 (*b*).

In accordance with the rotational equivalent of Eq. (3.49)

$$T_e = -\frac{1}{2}\Phi^2 \frac{dS}{d\theta_r} \qquad \text{...(3.59)}$$

where T_e is the instantaneous torque acting in the direction to increase the space angle θ_r. The corresponding curves of Φ^2 and $\dfrac{dS}{d\theta_r}$ are shown in Fig. 3.34 (*c*).

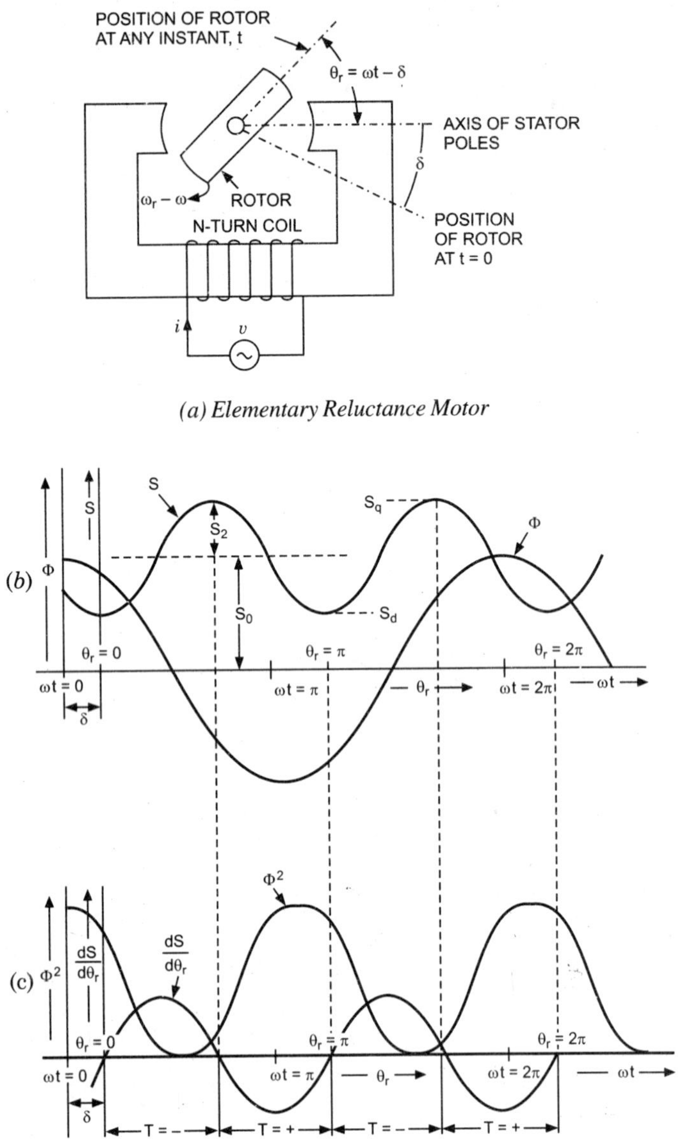

(*a*) *Elementary Reluctance Motor*

(*b*) *and* (*c*) *Flux and Reluctance Variations*

Fig. 3.34 *Single Phase Reluctance Motor*

The direction of torque is determined by the sign of $\dfrac{dS}{d\theta_r}$ from Eq. (3.59). It is positive while the reluctance is decreasing.

From Eq. (3.59) and Fig. 3.34 (*c*) it is observed that, if the rotor speed and the phase relation between the reluctance variation and the flux wave are such that the average square of flux is greater while the reluctance is decreasing than it is while the reluctance is increasing. The average torque acts in the direction to increase θ_r and thereby to keep the rotor revolving. These requirements can be met only when the rotor speed is such that the reluctance goes through one cycle of its variation in each half cycle of the flux wave; i.e., the rotor must revolve one-half a revolution in one-half a cycle of the flux wave. Its average angular velocity in radians per second, therefore, must be equal the time angular velocity ω of the flux wave, whence the rotor must make one revolution for each cycle of the supply voltage. The rotor speed satisfying these requirements is known as *synchronous speed*.

For the simple reluctance motor shown in Fig. 3.34 (*a*),

Synchronous speed, $N = f$ revolutions per second

where f is the supply frequency in Hz. Ignoring the usually negligible effect of harmonics, no average torque is developed at any other speed. The motor is, therefore, not self starting. It requires auxiliary means for its starting. In the simplest types of clock motors, the motor is started manually by spinning the rotor at or above synchronous speed; as the rotor coasts through synchronous speed, it "pulls into step" and continues to operate at this speed. In case of large synchronous motors and self-starting motors are provided with auxiliary windings which develop induction motor action at start.

The basic theory of the single-phase reluctance motor can be developed on the assumption of sinusoidal variations of the flux and reluctance. No doubt, the variation of reluctance depends on the geometry of the magnetic circuit, and the waveform of flux depends on the waveform of the applied voltage. The assumption of sinusoidal waveforms is made primarily because of convenience, it is also a fairly realistic assumption. Let the instantaneous value of flux be given as

$$\phi = \Phi_{max} \cos \omega t \quad \text{...(3.60)}$$

where Φ_{max} is the maximum value of flux.

Assume that the flux produced by coil mmf Ni is confined to the stator core. The stator winding resistance is negligible, because in well designed electromechanical energy conversion devices the winding resistances are usually small. As a consequence of these assumptions, the magnitude of counter or reaction emf e must always be equal to applied voltage v. For sinusoidal variations

$$\text{Counter or reaction emf, } e = N\frac{d\phi}{dt} = N\frac{d}{dt}(\Phi_{max} \cos\omega t)$$

$$= -N\omega\Phi_{max} \sin \omega t \quad \text{...(3.61)}$$

Hence the relation between rms value of counter emf in volts and the maximum value of the flux in webers is

$$E_{rms} = \frac{2\pi}{\sqrt{2}} f N\Phi_{max} = 4.44 f N\Phi_{max} \quad \text{...(3.62)}$$

Thus, when a sinusoidally varying voltage is applied across a winding having negligibly small resistance, a sinusoidally varying flux must be established whose maximum value Φ_{max} satisfies Eq. (3.62). The applied voltage and the induced counter emf are then equal, and Kirchhoff's law is satisfied.

The instantaneous value of reluctance is a function of the instantaneous angle θ_r and assuming sinusoidal variation the reluctance may be expressed as

$$S = \frac{1}{2}(S_q + S_d) - \frac{1}{2}(S_q - S_d)\cos 2\theta_r \quad \text{...(3.63)}$$

Differentiating above Eq. (3.63) we have

$$\frac{dS}{d\theta_r} = (S_q - S_d)\sin 2\theta_r \quad \text{...(3.64)}$$

For the instantaneous position of rotor at any instant t, as shown in Fig. 3.34 (*a*), the space angle θ_r may be expressed as

$$\theta_r = \omega_r t - \delta \quad \text{...(3.65)}$$

where ω_r is the shaft angular velocity in radians per second and δ is the instantaneous position of rotor at zero time when the flux is passing through its maximum value. Actually the torque developed by a singly-excited reluctance motor fluctuates considerably, therefore, the instantaneous speed is not absolutely constant. The inertia of the rotor and its mechanical load, however, maintain speed substantially constant.

For convenience, δ is taken as the lag angle, as shown in Fig. 3.34 (*a*).

Substituting the value of θ_r from Eq. (3.65) in Eq. (3.64) we have

$$\frac{dS}{d\theta_r} = (S_q - S_d)\sin(2\omega_r t - 2\delta) \quad \text{...(3.66)}$$

Substituting the value of ϕ and $\frac{dS}{d\theta_r}$ from Eqs. (3.60) and (3.66) respectively in basic torque Eq. [Eq. (3.59)] we have

$$T_e = -\frac{1}{2}\Phi_{max}^2 \cos^2 \omega_r t\,(S_q - S_d)\sin(2\omega_r t - 2\delta)$$

$$= -\frac{1}{4}\Phi_{max}^2 (1 + \cos 2\omega t)(S_q - S_d)\sin(2\omega_r t - 2\delta)$$

$$\because \quad \cos^2 \omega t = \frac{1 + \cos 2\omega t}{2}$$

$$= -\frac{1}{4}\Phi_{max}^2 (S_q - S_d)[\sin(2\omega_r t - 2\delta) + \sin(2\omega_r t - 2\delta)\cos 2\omega t \quad \text{...(3.67)}$$

But $\sin A \cos B = \frac{1}{2}[\sin(A + B) + \sin(A - B)]$

Therefore, Eq. (3.67) can be written as

$$T_e = -\frac{1}{4}\Phi_{max}^2 (S_q - S_d)\Big[\sin(2\omega_r t - 2\delta) + \frac{1}{2}\sin(2\omega_r t + 2\omega t - 2\delta) + \frac{1}{2}\sin(2\omega_r t - 2\omega t - 2\delta)\Big] \quad \text{...(3.68)}$$

Above Eq. (3.68) provides the instantaneous value of the torque. The average value of the torque can be obtained by determining the average value of Eq. (3.68). When the shaft angular velocity ω_r is not equal to the time angular velocity ω of the flux wave, the three sine terms in Eq. (3.68) are functions of time and the average value of each of them over a complete cycle is zero. Hence no average torque is produced

i.e. $T_{e\,(av)} = 0 \quad$ when $\omega_r \# \omega$

However, if the rotor speed is such that $\omega_r = \omega$, then instantaneous torque is given as

$$T_e = -\frac{1}{4}\Phi_{max}^2 (S_g - S_d) \times \Big[\sin(2\omega t - 2\delta) + \frac{1}{2}\sin(4\omega t - 2\delta) + \frac{1}{2}\sin(-2\delta)\Big] \quad \text{...(3.69)}$$

In above Eq. (3.69), first two terms are sine functions of time, and, therefore, their average values are zero. They represent pulsating components of the torque of twice and four times supply frequency. The last sine term is independent of time. The average value of the torque, therefore, is given as

$$T_{e\,(av)} = -\frac{1}{4}\Phi_{max}^2 (S_q - S_d)\Big[\frac{1}{2}\sin(-2\delta)\Big]$$

$$= \frac{1}{8}\Phi_{max}^2 (S_q - S_d)\sin 2\delta \quad \text{...(3.70)}$$

This shows that the reluctance motor can produce torque only at one speed, called the *synchronous speed.* If the applied voltage and frequency are constant and the resistance drop is negligible, Φ_{max} must be constant, as given by Eq. (3.62). The reluctances S_q and S_d depend primarily on the configuration of the magnetic circuit and are, therefore, constant for a given motor. Thus the rotor phase angle δ is the only variable on RHS of Eq. (3.70). A plot providing the variation of torque $T_{e\,(av)}$ with angle δ is illustrated in Fig. 3.35. When the load torque on motor changes, then the angle δ adjusts itself till reluctance torque becomes equal to the torque required to drive the mechanical load coupled to its shaft plus the torque required in overcoming the motor rotational losses. If the load torque increases, the motor momentarily slows down, thereby increasing the lag angle δ until sufficient electromagnetic torque is produced to bear the increased load. After the brief transient required for adjustment of the torque angle δ, operation is resumed at synchronous speed. The noteworthy point is that the maximum value of the average electromagnetic torque the motor can develop occurs when δ = 45° and is given as

$$T_{e\,(av)\,maxm} = \frac{1}{8}\Phi^2_{max}\,(S_q - S_d) \qquad ...(3.71)$$

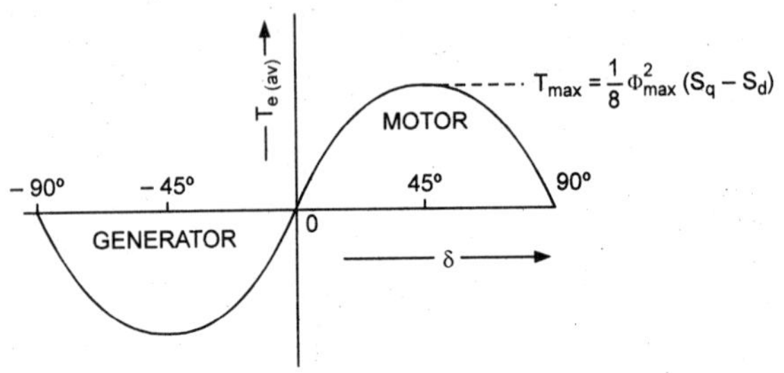

Fig. 3.35 *Torque-Angle Characteristic of a Reluctance Motor*

The motor stalls in case a mechanical load coupled to the motor shaft exerts a torque exceeding this value.

As in case of any electromechanical energy conversion device, the process is essentially reversible. If mechanical power is supplied to the shaft, the rotor advances in phase, i.e. the angle δ becomes negative (or leading). The average electromagnetic torque changes sign and, therefore, represents mechanical power absorbed and converted into electrical power, as illustrated by the portion of the torque-angle characteristic to the left of origin in Fig. 3.35. As with motor action, there is a definite limit to the mechanical power that can be absorbed, occurring when δ = – 45°. Any further increase in driving torque will cause excessive speed and loss of synchronism.

Reluctance motors are widely used for absolute constant speed applications, such as in timing devices, signalling devices, recording instruments, phonograph turntables, control apparatus etc.

In the singly-excited motor [Fig. 3.34(*a*)], the reluctance torque is produced because the reluctance seen by the stator flux varies with the movement of rotor. If the salient pole rotor of Fig. 3.34(*a*) is replaced by a cylindrical rotor, the reluctance offered to the stator flux remains constant for all positions of rotor and therefore, no reluctance torque is produced.

Example 3.13. Determine approximately the maximum torque that the motor can develop when it is connected to 230 V, 50 Hz supply. It is given that the exciting winding has 1,500 turns; the inductance of the winding in direct-axis position is 0.8 H and when the rotor is in quadrature-axis position, the inductance is 0.4 H.

Solution: Inductance of exciting winding

when the rotor is in direct-axis position,

$$L_d = 0.8 \text{ H}$$

when the rotor is in quadrature-axis position,

$$L_q = 0.4 \text{ H}$$

$$\Phi_{max} = \frac{E}{4.44\,fN} = \frac{V}{4.44\,fN}$$

$$= \frac{230}{4.44 \times 50 \times 1{,}500} = 0.00069 \text{ weber}$$

$$S_q = \frac{N^2}{L_q} = \frac{(1{,}500)^2}{0.4} = 5.625 \times 10^6 \text{ AT/Wb}$$

$$\text{Since } L = \frac{N^2}{\text{Reluctance}}$$

$$\text{and } S_d = \frac{N^2}{L_d} = \frac{(1{,}500)^2}{0.8} = 2.8125 \times 10^6 \text{ AT/Wb}$$

The value of the maximum torque that the motor can develop,

$$T_{e\,(av)\,max} = \frac{1}{8}\Phi^2_{max}\,(S_q - S_d)$$

$$= \frac{1}{8} \times (0.00069)^2 \times (5.625 - 2.8125) \times 10^6$$

$= 0.1677$ N-m **Ans.**

3.10 DOUBLY-EXCITED MAGNETIC FIELD SYSTEMS

Singly-excited devices discussed earlier, are generally employed for motion through a limited distance or rotation through a prescribed angle. In singly-excited devices, electromechanical conversion takes place through either dc or ac source and that is why they are called singly- excited devices. Examples of singly-excited devices are electromagnets, relays, moving-iron instruments, reluctance motors etc.

A doubly-excited magnetic system is one which has two independent sources of excitation. Most of the electromagnetic devices belong to doubly-excited or multiply-excited magnetic systems. In measurement systems it is often desirable to obtain torques proportional to two electric signals; power as the product of voltage and current is a common example. Similarly, most continuous energy conversion devices - generators and motors need multiple excitation (multiple windings). One group of windings is mounted on a stationary member and the other group on a movable member. The tendency for the magnetic field energy to change when one group of windings moves with respect to the other gives rise to mechanical forces. The theory of numerous types of energy conversion devices, including most rotating machines, is based on this general principle. DC shunt machines, synchronous machines, loudspeakers, tachometers etc. fall in this category of the devices.

A simple model of a doubly-excited magnetic system is shown in Fig. 3.36. The model consists of stator iron having N_1 turn coil energised from source 1 and rotor iron having N_2 turn coil energised from source 2. The mmfs produced by both the stator and rotor windings are in the same direction and the electromagnetic torque T_e is in the counter-clockwise direction, as illustrated in Fig. 3.36. Magnetic saturation and hysteresis, for convenience may be neglected.

From Eq. (3.20), the electrical energy supplied from two sources is given as

$$d\,W_{elec} = i_1\,d\,\psi_1 + i_2\,d\,\psi_2 \qquad ...(3.72)$$

where ψ_1 and ψ_2 are the instantaneous flux linkages of stator and rotor windings respectively. Since magnetic saturation is assumed to be negligible, the flux linkages ψ_1 and ψ_2 may be expressed in terms of self- and mutual inductances

$$\left.\begin{aligned}\psi_1 &= L_1 i_1 + M_{12} i_2\\ \text{and}\quad \psi_2 &= L_2 i_2 + M_{21} i_1\end{aligned}\right\} \qquad ...(3.73)$$

where L_1 is self-inductance of stator winding, L_2 is self-inductance of rotor winding and M_{12}(or M_{21}) is mutual inductance between stator and rotor windings.

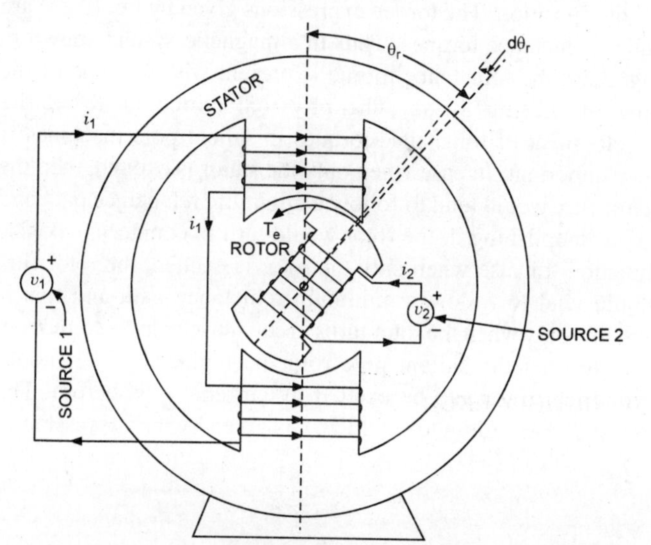

Fig. 3.36 *Doubly-Excited Magnetic System*

Let initially both of the windings be unexcited (i.e. both i_1 and i_2 are zero) and space angle between stator and rotor axes be θ_r. Now let the stator and rotor windings be switched on to their respective supply sources, so that the currents rise from 0 to i_1 and i_2 respectively.

If the rotor is held fixed, then mechanical work done $d\,W_{mech}$ is zero

$$\text{and } d\,W_{elec} = dW_{mech} + dW_{field} = 0 + W_{field}$$

i.e. with rotor held fixed, all the electrical energy input by the two supply sources is stored in the magnetic field and from Eq. (3.72)

$$\begin{aligned}d\,W_{field} = dW_{elec} &= i_1 d\psi_1 + i_2 d\psi_2\\ &= i_1\,d\,(L_1\,i_1 + M_{12}\,i_2) + i_2\,d\,(L_2\,i_2 + M_{21}\,i_1) \quad ...(3.74)\end{aligned}$$

$$\text{Since } L = \frac{N^2 \mu_o \mu_r a}{l} = \frac{N^2}{l/\mu_o \mu_r a} = \frac{N^2}{S}$$

$$\text{and } M = \frac{N_1 N_2 \mu_o \mu_r a}{l} = \frac{N_1 N_2}{l/\mu_o \mu_r a} = \frac{N_1 N_2}{S},$$

$$L_1 = \frac{N_1^2}{S_1};\quad L_2 = \frac{N_2^2}{S_2} \quad \text{and} \quad M_{12} = M_{21} = \frac{N_1 N_2}{S_{12}}$$

where S_1, S_2 and S_{12} are the respective reluctances seen by the stator flux, rotor flux and by the resultant of stator and rotor flux.

Since the rotor is held fixed (not allowed to move), the reluctances and, therefore, inductances are constant. Thus the differential changes in inductances $d\,L_1$, $d\,L_2$ and $d\,M_{12}$ are all zeros and Eq. (3.74) is reduced to

$$\begin{aligned}d\,W_{field} &= i_1\,L_1\,di_1 + i_2\,L_2\,di_2 + i_1\,M_{12}\,di_2 + i_2\,M_{21}\,di_1\\ &= i_1\,L_1\,di_1 + i_2\,L_2\,di_2 + M_{12}\,d\,(i_1\,i_2)\end{aligned}$$

The magnetic field energy stored in establishing the currents from zero to i_1 and i_2 is given as

$$\begin{aligned}W_{field} &= L_1\int_0^{i_1} i_1\,di_1 + L_2\int_0^{i_2} i_2\,di_2 + M_{12}\int_0^{i_1 i_2} d(i_1 i_2)\\ &= \frac{1}{2}L_1 i_1^2 + \frac{1}{2}L_2 i_2^2 + M_{12} i_1 i_2 \qquad ...(3.75)\end{aligned}$$

Let us assume that the rotor moves through a virtual displacement $d\theta_r$ in the direction of magnetic torque T_e, as illustrated in Fig. 3.36. With the movement of rotor, reluctances S_1, S_2 and S_{12} and, therefore, inductances L_1, L_2, M_{12} must change.

Differentiation of Eq. (3.74) with respect to inductances and currents provides for electrical energy input

$$\begin{aligned}d\,W_{elec} = {}& L_1 i_1 d i_1 + M_{12} i_1 d i_2 + i_1^2 d\,L_1 + i_1 i_2 d\,M_{12}\\ &+ L_2 i_2 d i_2 + M_{21} i_2 d i_1 + i_2^2 d\,L_2 + i_1 i_2 d\,M_{12} \qquad ...(3.76)\end{aligned}$$

Similarly from Eq. (3.75), the increment in field energy is

$$\begin{aligned}d\,W_{field} = {}& L_1 i_1 d i_1 + \frac{1}{2} i_1^2 d\,L_1 + L_2 i_2 d i_2 + \frac{1}{2} i_2^2 d\,L_2\\ &+ M_{12} i_1 d i_2 + M_{12} i_2 d i_1 + i_1 i_2 d\,M_{12} \qquad ...(3.77)\end{aligned}$$

The differential mechanical work done during virtual displacement $d\,\theta_r$

$$d\,W_{mech} = T_e\,d\,\theta_r$$

From the law of conservation of energy

$$d\,W_{elec} = T_e\,d\,\theta_r + d\,W_{field} \qquad ...(3.78)$$

When the values of $d\,W_{elec}$ and $d\,W_{field}$ are substituted from Eqs. (3.76) and (3.77) in Eq. (3.78), the four terms $L_1\,i_1\,d\,i_1 + L_2\,i_2\,d\,i_2 + M_{12}\,i_1\,d\,i_2 + M_{12}\,i_2\,d\,i_1$ expressing the electrical energy input caused by current increments di_1 and di_2 are balanced by identical terms expressing the corresponding increments in field energy. These terms, therefore, cancel. In other words, differential changes in currents have no effect on the mechanical forces. Thus we get

$$\begin{aligned}&i_1^2 d\,L_1 + i_2^2 d\,L_2 + 2 i_1 i_2 d\,M_{12}\\ &= \frac{1}{2} i_1^2 d\,L_1 + \frac{1}{2} i_2^2 d\,L_2 + i_1 i_2 d\,M_{12} + T_e\,d\,\theta_r \qquad ...(3.79)\end{aligned}$$

$$\text{or } T_e = \frac{1}{2} i_1^2 \frac{d\,L_1}{d\theta_r} + \frac{1}{2} i_2^2 \frac{d\,L_2}{d\theta_r} + i_1 i_2 \frac{d\,M_{12}}{d\theta_r} \qquad ...(3.80)$$

The translational equivalent of Eq. (3.80) is obtained if torque T_e is replaced by translational force f_e and angular displacement $d\theta_r$ by linear displacement dx in the direction of force. The extension of Eq. (3.80) to a situation involving several circuits whose self- and mutual inductances depend upon the angular position of some member should be obvious from consideration of the line of reasoning followed in the derivation of Eq. (3.80). For example, with three circuits,

$$\begin{aligned}T_e = {}& \frac{1}{2} i_1^2 \frac{d\,L_1}{d\theta_r} + \frac{1}{2} i_2^2 \frac{d\,L_2}{d\theta_r} + \frac{1}{2} i_3^2 \frac{d\,L_3}{d\theta_r}\\ &+ i_1 i_2 \frac{d\,M_{12}}{d\theta_r} + i_2 i_3 \frac{d\,M_{23}}{d\theta_r} + i_3 i_1 \frac{d\,M_{31}}{d\theta_r} \qquad ...(3.81)\end{aligned}$$

From Eq. (3.79) it is obvious that the sum of the three terms on the RHS expressing the increment in magnetic field energy at constant currents is half of the LHS expressing the electrical energy input at constant currents. The other half of the electrical energy input at constant currents is delivered as mechanical energy. The increment in field energy at constant currents, therefore, equals the mechanical energy provided. Thus, the torque or force equals the rate of increase of field energy with respect to motion at constant current.

Thus differentiation of Eq. (3.75) provides

$$T_e = \frac{\partial W_{field}}{\partial \theta_r}(i_1, i_2, \theta_r) \quad ...(3.82)$$

in which the field energy must be explicitly expressed in terms of the currents and angle θ_r and the partial derivatives must be taken with the currents treated as constants. The translational equivalent of Eq. (3.82) is

$$f_e = \frac{\partial W_{field}}{\partial x}(i_1, i_2, x) \quad ...(3.83)$$

where f_e is the electromagnetic force acting in the positive direction of x.

Integration of Eq. (3.74) for $d\,W_{elec}$, with constant currents, gives

$$W_{elec} = i_1^2 L_1 + i_2^2 L_2 + 2 i_1 i_2 M_{12} \quad ...(3.84)$$

For a linear magnetic circuit, $W_{field} = W'_{field}$, therefore, Eq. (3.82) can be written as

$$T_e = \frac{\partial W'_{field}}{\partial \theta_r}(i_1, i_2, \theta_2) \quad ...(3.85)$$

3.10.1. Doubly Excited Linear Magnetic Field System. Fig. 3.37 illustrates a simple model of doubly-excited linear magnetic system.

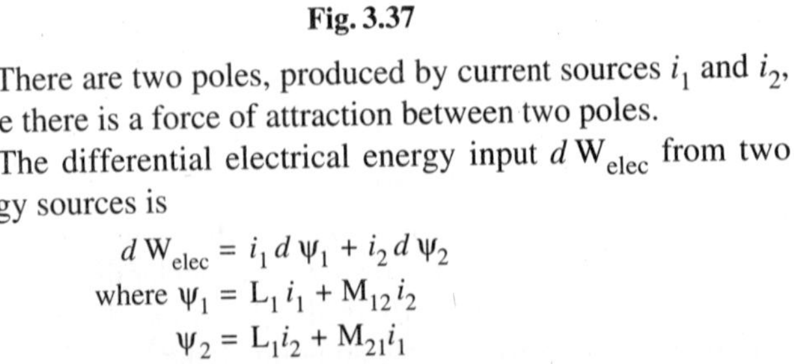

Fig. 3.37

There are two poles, produced by current sources i_1 and i_2, hence there is a force of attraction between two poles.

The differential electrical energy input $d\,W_{elec}$ from two energy sources is

$$d\,W_{elec} = i_1\, d\psi_1 + i_2\, d\psi_2$$

$$\text{where } \psi_1 = L_1 i_1 + M_{12} i_2$$

$$\psi_2 = L_1 i_2 + M_{21} i_1$$

$$\text{and } M_{12} = M_{21}$$

The magnitude of magnetic force f_e can be obtained by adopting the procedure as followed for magnetic torque. The translational equivalent of Eq. (3.80) is

$$f_e = \frac{1}{2} i_1^2 \frac{dL_1}{dx} + \frac{1}{2} i_2^2 \frac{dL_2}{dx} + i_1 i_2 \frac{dM_{12}}{dx} \quad ...(3.86)$$

$$\text{or } f_e = \frac{\partial W_{field}}{\partial x}(i_1, i_2, x) = \frac{\partial W'_{field}}{\partial x}(i_1, i_2, x) \quad ...(3.87)$$

3.10.2. Electromagnetic and Reluctance Torques. The total torque produced by the doubly-excited magnetic system is given by Eq. (3.80)

In Fig. 3.36, if rotor current $I_2 = 0$, then from Eq. (3.80)

$$T_e = \frac{1}{2} i_1^2 \frac{dL_1}{d\theta_r} \quad ...(3.88\ a)$$

and if stator current $I_1 = 0$, then from Eq. (3.80)

$$T_e = \frac{1}{2} i_2^2 \frac{dL_2}{d\theta_r} \quad ...(3.88\ b)$$

Equation (3.88 *a*) reveals that even with rotor unexcited (i.e. $i_2 = 0$), the torque is developed. This is because reluctance S_1 seen by stator flux changes with rotor position. A change in S_1 causes the stator self-inductance L_1 to change with θ_r. Similarly Eq. (3.88 *b*) reveals that with stator unexcited (i.e. $i_1 = 0$) the torque can be produced, because rotor inductance L_2 is a function of rotor position. The torque expressions given by Eq. (3.88) are called reluctance torques. Thus in a magnetic system shown in Fig. 3.36, the reluctance torque is present with any one of the currents acting alone. The physical concepts about the development of reluctance torques in salient pole machines is very important. In case when only the stator is excited, then the stator flux would tend to follow a minimum reluctance path and for accomplishing it, the rotor would turn in counter-clockwise direction. In case when only the rotor is excited, the rotor flux would tend to follow a minimum reluctance path and this is possible only when the rotor turns in counter-clockwise direction.

Now let the salient pole rotor be replaced by a smooth cylindrical rotor and be excited by current i_2, as before. The torque T_e, according to Eq. (3.80), is given by the expression

$$T_e = \frac{1}{2} i_2^2 \frac{dL_2}{d\theta_r} + i_1 i_2 \frac{dM_{12}}{d\theta_r} \quad ...(3.89)$$

The reluctance torque term $\frac{1}{2} L_1^2 \frac{dL_1}{d\theta_r}$ is absent. This is because the reluctance S_1 seen by the stator flux does not vary with the movement of rotor, consequently L_1 is constant and $\frac{dL_1}{d\theta_r}$ is zero.

Thus we conclude that the first two terms of Eq. (3.80), involving the change of self-inductances L_1 and L_2 represent reluctance torques for the mechanical structure shown in Fig. 3.36 while the last term $i_1 i_2 \frac{dM_{21}}{d\theta_r}$ represents that component of total torque, which is dependent on the currents in both the stator and rotor windings and also on the angular rate of change of mutual inductance M_{12}. This component of torque i.e. the torque represented by the last term $i_1 i_2 \frac{dM_{12}}{d\theta_r}$ in Eq. (3.80) is commonly called the *electromagnetic torque* of electromagnetic energy conversion devices. It means that electromagnetic torque can be developed only when both of the stator and rotor windings are mutually coupled and carry currents.

The physical concept about development of electromagnetic torque for the mechanical structure shown in Fig. 3.36 is as follows :

The north and south poles developed on stator by stator current i_1 and south and north poles developed on rotor by rotor current i_2 attract each other tending to align their fields. The torque produced in this way by the interaction of stator and rotor magnetic fields is called *electromagnetic* or *interaction torque.*

The noteworthy point is that the reluctance torque component is independent of direction of flow of currents through stator and rotor windings while the interaction torque component does depend on the direction of currents i_1 and i_2. In Fig. 3.36, with the indicated direction of currents i_1 and i_2, the electromagnetic or interaction, torque is counter-clockwise, as illustrated. If the direction of flow of current through either stator or rotor is reversed, the interaction torque would be opposite to that shown in Fig. 3.36 i.e. in clockwise direction, but the direction of reluctance torque would still be in the same direction as before the reversal of direction of current flow i.e. counter-clockwise.

Example 3.14. Two coils with self-inductances 1 H and 4 H have a mutual inductance 1 H. The rms values of currents flowing in the two coils is 4 A and 1 A respectively. Find (i) the coupling factor (ii) the energy stored in the magnetic system.

[Rajasthan Technical Univ. Electrical Machines-I, 2011]

Solution: Inductances of coils, $L_1 = 1$ H

$L_2 = 4$ H

Mutual inductance, $M_{12} = 1$ H

(*i*) Coupling factor, $k = \dfrac{M}{\sqrt{L_1 L_2}} = \dfrac{1}{\sqrt{1 \times 4}} = 0.5$ **Ans.**

(*ii*) Energy stored in magnetic field

$$= \frac{1}{2} L_1 I_1^2 + \frac{1}{2} L_2 I_2^2 + M_{12} I_1 I_2 \qquad \text{...Refer to Eq. (3.75)}$$

$$= \frac{1}{2} \times 1 \times 4^2 + \frac{1}{2} \times 4 \times 1^2 + 1 \times 1 \times 4 = 8 + 2 + 4 = 14 \text{ watts } \textbf{Ans.}$$

Example 3.15. Two coupled coils have self- and mutual inductance in henrys as follows :

$$L_{11} = 2 + \frac{1}{2x};\ L_{22} = 1 + \frac{1}{2x};\ L_{12} = L_{21} = \frac{1}{2x}$$

over a certain range of linear displacement *x*. The first coil is excited by a constant current of 20 A and the second by a constant current of –10 A. Calculate the energy supplied by each source, if *x* changes from 0.5 m to 1 m.

[U.P. Technical Univ. Electromechanical Energy Conversion-I, 2005-06]

Solution: Self-inductance of coil 1, $L_{11} = 2 + \dfrac{1}{2x}$

Self-inductance of coil 2, $L_{22} = 1 + \dfrac{1}{2x}$

Mutual inductance between the coils, $L_{12} = L_{21} = \dfrac{1}{2x}$

Flux linkages, $\psi_1 = (L_{11} i_1 + L_{12} i_2)$

$$= \left(2 + \frac{1}{2x}\right) \times 20 + \frac{1}{2x} \times (-10) = \frac{5}{x} + 40$$

Flux linkages, $\psi_2 = (L_{22} i_2 + L_{21} i_1)$

$$= \left(1 + \frac{1}{2x}\right) \times (-10) + \frac{1}{2x} \times 20 = \frac{5}{x} - 10$$

Electrical energy supplied by source 1,

$$dW_{elec_1} = \int_{\psi_1(x=0.5)}^{\psi_1(x=1.0)} i_1\, d\psi_1$$

$$= i_1[\psi_1 (x = 1.0) - \psi_1 (x = 0.5)]$$

$$= 20\left[5 + 40 - \frac{5}{0.5} - 40\right] = -100 \text{ J} \quad \textbf{Ans.}$$

i.e. the source 1 draws an energy of 100 J.

Electrical energy supplied by source 2,

$$dW_{elec2} = \int_{\psi_2(x=0.5)}^{\psi_2(x=1.0)} i_2\, d\psi_2$$

$$= i_2 [\psi_2 (x = 1.0) - \psi_2 (x = 0.5)]$$

$$= -10\left[5 - 10 - \frac{5}{0.5} + 10\right] = 50 \text{ J} \quad \textbf{Ans.}$$

3.11 TORQUE IN SINGLE-PHASE CYLINDRICAL ROTOR MACHINE

Consider a basic form of single-phase synchronous machine of smooth cylindrical rotor type, as shown in Fig. 3.38 (*a*). A single-phase supply is connected to stator and a voltmeter across the rotor winding to determine the variation of stator to rotor mutual inductance M_{12}. With $\theta_r = 0$, the value of linkage flux is maximum, which is indicated by the high voltmeter reading. It means that for $\theta_r = 0$, the mutual inductance between stator and

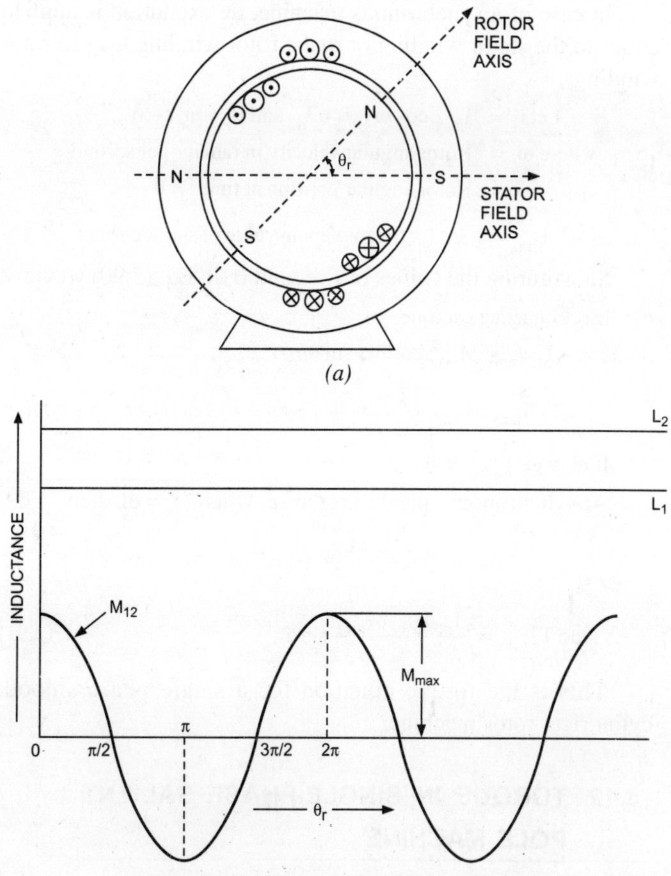

(b) Inductance Variation With Space Angle θ_r

Fig. 3.38 *Single Phase Cylindrical Rotor Synchronous Machine* rotor is maximum i.e. $M_{12} = M_{max}$. When the rotor angle is $\frac{\pi}{2}$, the value of flux linkage is zero, which is indicated by the zero reading on voltmeter i.e. for $\theta_r = \frac{\pi}{2}$, the mutual inductance M_{12} is zero. This is because the stator winding axis is perpendicular to the rotor winding axis. Again with rotor angle π, the rotor flux linkages are maximum but in opposite direction, $M_{12} = -M_{max}$.

The variation of mutual inductance between stator and rotor M_{12} is to be assumed to be a cosine function of θ_r, as illustrated in Fig. 3.38 (*b*).

Inductances L_1 and L_2 are constant as they are independent of movement of rotor. The mutual inductance between stator and rotor at any rotor angle θ_r, can be expressed as

$$M_{12} = M_{max}\cos\theta_r$$

Let i_1 and i_2 be the instantaneous currents in the stator and rotor windings respectively, then

$$W_{field}(i_1, i_2, \theta_r) = \frac{1}{2} i_1^2 L_1 + \frac{1}{2} i_2^2 L_2 + i_1 i_2 M_{12}$$

$$= \frac{1}{2} i_1^2 L_1 + \frac{1}{2} i_2^2 L_2 + i_1 i_2 M_{max}\cos\theta_r$$

$$\text{Torque, } T_e = \frac{\partial W_{field}}{\partial \theta_r} = 0 + 0 - i_1 i_2 M_{max}\sin\theta_r$$

$$= -i_1 i_2 M_{max}\sin\theta_r \qquad ...(3.90)$$

Negative sign indicates that the torque T_e acts to reduce the space angle θ_r.

In case of a synchronous machine, dc excitation is applied either to the stator winding or to the rotor winding (say to rotor winding).

Let $i_1 = I_{max}\cos\omega t$, $i_2 = I_{dc}$ and $\theta_r = \omega_r t - \delta$

where ω_r = Rotor angular velocity in radians per second

δ = Rotor angular position at time $t = 0$

$I_{max} = \sqrt{2}I_1$, the maximum value of stator current.

Substituting the values of i_1, i_2 and θ_r in Eq. (3.90) we have

Electromagnetic torque,

$$T_e = -I_{dc} I_{max} M_{max}\cos\omega t \sin(\omega_r t - \delta)$$

$$= -I_{dc} I_{max} M_{max} \times \frac{1}{2}[\sin(\omega_r t + \omega t - \delta) + \sin(\omega_r t - \delta - \omega t)]$$

If $\omega_r \# \omega$, $T_{e(av)} = 0$

At synchronous speed of rotor i.e. when $\omega_r = \omega$, then

$$T_e = -\frac{I_{dc} I_{max} M_{max}}{2}[\sin(2\omega t - \delta) + \sin(-\delta)]$$

$$\text{and } T_{e(av)} = \frac{I_{dc} I_{max} M_{max}}{2}\sin\delta \qquad ...(3.91)$$

This is the torque equation for a single phase smooth cylindrical rotor machine.

3.12 TORQUE IN SINGLE-PHASE SALIENT POLE MACHINE

Figure 3.39 (*a*) illustrates an elementary form of a single-phase salient pole synchronous machine.

Let L_1, L_2 be the self-inductances of stator and rotor windings and M_{12} be mutual inductance and θ_r be the space angle. The variations of L_1, L_2 and M_{12} with space angle θ_r are shown in Fig. 3.39 (*b*).

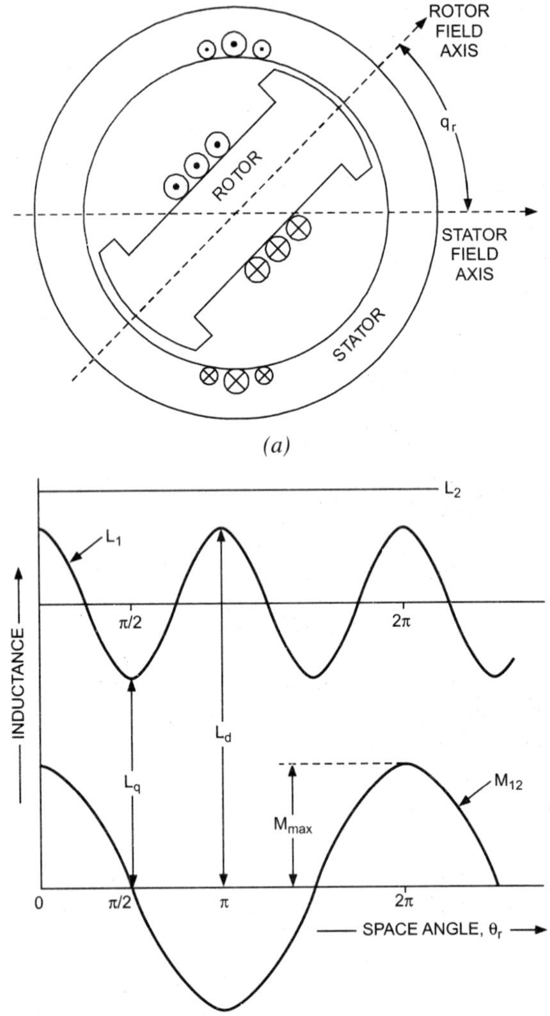

(*a*)

(*b*) *Inductance Variation With Space Angle* θ_r

Fig. 3.39 *Single-Phase Salient Pole Synchronous Machine*

Rotor self-inductance, L_2 = Constant

Stator self-inductance,

$$L_1 = \frac{1}{2}(L_d + L_q) + \frac{1}{2}(L_d - L_q)\cos 2\theta_r$$

and Mutual inductance,

$$M_{12} = M_{max}\cos\theta_r$$

Magnetic field energy stored, $W_{field}(i_1, i_2, \theta_r)$

$$= \frac{1}{2} i_1^2\left[\frac{1}{2}(L_d + L_q) + \frac{1}{2}(L_d - L_q)\cos 2\theta_r\right] + \frac{1}{2} i_2^2 L_2 + i_1 i_2 M_{max}\cos\theta_r$$

The equation for instantaneous magnetic torque is

$$T_e = -\frac{1}{2} i_1^2\left[\frac{1}{2}\times 2(L_d - L_q)\sin 2\theta_r\right] - i_1 i_2 M_{max}\sin\theta_r$$

$$= -\frac{1}{2} i_1^2 (L_d - L_q)\sin 2\theta_r - i_1 i_2 M_{max}\sin\theta_r$$

In salient pole machine, rotor winding carries direct current.

If rotor current, $i_2 = I_f$

Stator current, $i_1 = I_{max} \cos \omega t$

and space angle, $\theta_r = \omega_r t - \delta$, as before

$$\therefore T_e = -\frac{1}{2} I_{max}^2 \cos^2 \omega t (L_d - L_q) \sin(2\omega_r t - 2\delta) - I_f I_{max} \cos\omega t . M_{max} \sin(\omega_r t - \delta)$$

$$= -\frac{1}{4} I_{max}^2 (1 + \cos 2\omega t)(L_d - L_q) \sin(2\omega_r t - 2\delta) - I_f I_{max} \cos\omega t . M_{max} \sin(\omega_r t - \delta)$$

$$= -\frac{1}{4} I_{max}^2 (L_d - L_q) \times \left[\sin(2\omega_r t - 2\delta) + \frac{1}{2}\sin(2\omega_r t + 2\omega t - 2\delta) + \frac{1}{2}\sin(2\omega_r t - 2\omega t - 2\delta)\right]$$

$$-I_f I_{max} M_{max} \times \frac{1}{2}[\sin(\omega_r t + \omega t - \delta) + \sin(\omega_r t - \omega t - \delta)]$$

For $\omega_r \# \omega$, $T_{e\,(av)} = 0$

But when $\omega_r = \omega$,

$$T_e = -\frac{1}{4} I_{max}^2 (L_d - L_q) \times \left[\sin(2\omega t - 2\delta) + \frac{1}{2}\sin(4\omega t - 2\delta) + \frac{1}{2}\sin(-2\delta)\right]$$

$$-\frac{I_f I_{max} M_{max}}{2}[\sin(2\omega t - \delta) + \sin(-\delta)]$$

$$\text{So, } T_{e\,(av)} = -\frac{1}{4} I_{max}^2 (L_d - L_q)\left[\frac{1}{2}\sin(-2\delta)\right] - \frac{I_f I_{max} M_{max}}{2}\sin(-\delta)$$

$$= \frac{1}{8} I_{max}^2 (L_d - L_q)\sin 2\delta + \frac{1}{2} I_f I_{max} M_{max} \sin\delta \quad \text{...(3.92)}$$

If field current is reduced to zero, then

$$T_{e\,(av)} = \frac{1}{8} I_{max}^2 (L_d - L_q)\sin 2\delta \quad \text{...(3.93)}$$

Above Eq. (3.93) is the reluctance torque equation for single-phase salient pole machine. The other component of Eq. (3.92) i.e. $\frac{1}{2} I_f I_{max} M_{max} \sin\delta$ represents the interaction or electromagnetic torque of single phase salient pole synchronous machine.

Example 3.16. In the system shown in Fig. 3.40, the inductances in henrys are given by $L_{11} = (3 + \cos 2\theta) \times 10^{-3}$; $L_{22} = 30 + 10 \cos 2\theta$ and $L_{12} = 0.2 \cos\theta$. Find the torque for currents $i_1 = 1.2$ A and $i_2 = 0.01$ A.

Solution: The magnetic torque is given as

$$T_e = \frac{1}{2} i_1^2 \frac{dL_{11}}{d\theta} + \frac{1}{2} i_2^2 \frac{dL_{22}}{d\theta} + i_1 i_2 \frac{dL_{12}}{d\theta} \quad \text{...refer to Eq. (3.80)}$$

$$= \frac{1}{2} i_1^2 \frac{d}{d\theta}(3 + \cos 2\theta) \times 10^{-3} + \frac{1}{2} i_2^2 \frac{d}{d\theta}(30 + 10\cos 2\theta) + i_1 i_2 \frac{d}{d\theta}(0.2\cos\theta)$$

$$= \frac{1}{2} i_1^2 \times (-2 \times 10^{-3})\sin 2\theta + \frac{1}{2} i_2^2 (-20 \sin 2\theta) + i_1 i_2 (-0.2 \sin\theta)$$

$$= -\left[0.001 i_1^2 \sin 2\theta + 10 i_2^2 \sin 2\theta + 0.2 i_1 i_2 \sin\theta\right]$$

The torque for currents

$i_1 = 1.2$ A and $i_2 = 0.01$ A,

$$T_e = -[0.001 \times (1.2)^2 \sin 2\theta + 10 \times (0.01)^2 \sin 2\theta + 0.2 \times 1.2 \times 0.01 \sin\theta]$$

$$= -2.44 \times 10^{-3} \sin 2\theta - 2.4 \times 10^{-3} \sin\theta \text{ N-m} \quad \textbf{Ans.}$$

Notice that the torque expression consists of terms of two types. One term is proportional to the product of two currents and to the sine of the angular displacement. This torque is due to the mutual interaction between the rotor and stator currents; it acts in a direction to align the rotor and stator so as to maximize the coenergy. Alternately, it can be thought of as being due to the tendency of two magnetic fields to align.

The torque expression also consists of two terms each proportional to the sine of twice the angle and to the square of one of the coil currents. These terms correspond to the torques one sees in singly-excited systems. Here the torque is due to the fact that the coenergy is a function of rotor position and the torque acts in a direction to align the rotor so as to maximize the coenergy. The 2θ variation is due to the corresponding variation in self inductances, which in turn is due to the variation of the air-gap reluctance; notice that rotating the rotor by 180° or π radians from any given position gives the same air-gap reluctance and hence twice the angle variation. This torque component is known as the reluctance torque.

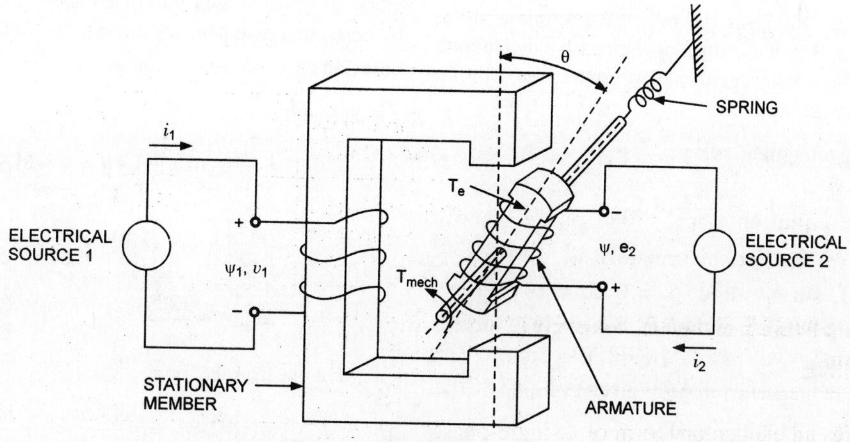

Fig. 3.40

Example 3.17. For doubly excited magnetic field system, various inductances are:

$$L_{11} = (4 + \cos 2\theta) \times 10^{-3}\text{ H}$$
$$L_{12} = 0.15 \cos \theta \text{ H}$$
$$L_{22} = (20 + 5 \cos 2\theta)\text{ H}$$

Determine torque developed if $i_1 = 5$A and $i_2 = 0.10$ A.

[G.B. Technical Univ. Electromechanical Energy Conversion-I, 2011-12]

Solution: Torque developed,

$$T_e = \frac{1}{2}i_1^2\frac{dL_{11}}{d\theta} + \frac{1}{2}i_2^2\frac{dL_{22}}{d\theta} + i_1 i_2\frac{dL_{12}}{d\theta}$$

$$= \frac{1}{2}i_1^2\frac{d}{d\theta}(4+\cos 2\theta)\times 10^{-3} + \frac{1}{2}i_2^2(20+5\cos 2\theta)$$

$$+ i_1 i_2\frac{d}{d\theta}(0.15\cos\theta)$$

$$= \frac{1}{2}i_1^2\times(-2\sin 2\theta)\times 10^{-3} + \frac{1}{2}i_2^2(-10\sin 2\theta) + i_1 i_2(-0.15\sin\theta)$$

$$= -\left[i_1^2(\sin 2\theta)\times 10^{-3} + 5(\sin 2\theta)i_2^2 + 0.15\, i_1 i_2 \sin\theta\right]$$

$$= -[5^2\times 10^{-3}\sin 2\theta + 5\times 0.1^2\sin 2\theta + 0.15\times 5\times 0.1\sin\theta]$$

$$= -0.075\sin 2\theta - 0.075\sin\theta \quad \textbf{Ans.}$$

The first term $-0.075 \sin 2\theta$ is the reluctance torque and the second term $-0.075 \sin \theta$ is the torque developed by the mutual component. The negative sign indicates that the torque is restoring in nature i.e. it opposes the displacement θ.

Example 3.18. For a doubly-excited magnetic field system, various inductances are: $L_{11} = (4 + \cos 2\theta) \times 10^{-3}$ H, $L_{12} = 0.15 \cos \theta$ H and $L_{22} = (20 + 5 \cos 2\theta)$ H. Find the torque developed if $i_1 = 1$A and $i_2 = 0.02$ A. Also explain the significance of each term in torque.

[U.P. Technical Univ. Electromechanical Energy Conversion-I, 2004-05]

Solution: Torque developed,

$$T_e = \frac{1}{2}i_1^2\frac{dL_{11}}{d\theta} + \frac{1}{2}i_2^2\frac{dL_{22}}{d\theta} + i_1 i_2\frac{dL_{12}}{d\theta}$$

$$= \frac{1}{2}i_1^2\frac{d}{d\theta}(4+\cos 2\theta)\times 10^{-3} + \frac{1}{2}i_2^2\frac{d}{d\theta}(20+5\cos 2\theta)$$

$$+ i_1 i_2\frac{d}{d\theta}(0.15\cos\theta)$$

$$= \frac{1}{2}i_1^2\times(-2\sin 2\theta)\times 10^{-3} + \frac{1}{2}i_2^2(-10)\sin 2\theta + i_1 i_2(-0.15)\sin\theta$$

$$= -[1\times 10^{-3}\sin 2\theta + 5(0.02)^2\sin 2\theta + 1\times 0.02\times 0.15\sin\theta]$$

$$= -3\times 10^{-3}\sin 2\theta - 3\times 10^{-3}\sin\theta \quad \textbf{Ans.}$$

The first term $-3 \times 10^{-3} \sin 2\theta$ is the reluctance torque and the second term $-3 \times 10^{-3} \sin \theta$ is the torque developed by the mutual component. The negative sign indicates that the torque is restoring in nature i.e. it opposes the displacement θ.

Example 3.19. In an electromagnetic relay

$$L_{11} = \frac{K_1}{x},\quad L_{22} = \frac{K_3}{x},\quad L_{12} = \frac{K_2}{x}$$

Find the expression for the force on armature, if

$$i_1 = I_1 \sin \omega_1 t \quad \text{and} \quad i_2 = I_2 \sin \omega_2 t$$

For what relationship between ω_1 and ω_2, the average force is (*i*) maximum (*ii*) minimum? *[RGPV, December-2002]*

Solution: The magnitude of magnetic force f_e is given by Eq. (3.86) i.e.

$$f_e = \frac{1}{2}i_1^2\frac{dL_{11}}{dx} + \frac{1}{2}i_2^2\frac{dL_{22}}{dx} + i_1 i_2\frac{dL_{12}}{dx}$$

$$= \frac{1}{2}I_1^2\sin^2\omega_1 t\frac{d}{dx}\left(\frac{K_1}{x}\right) + \frac{1}{2}I_2^2\sin^2\omega_2 t\left(\frac{K_3}{x}\right)$$

$$+ I_1 I_2\sin\omega_1 t\sin\omega_2 t\frac{d}{dx}\left(\frac{K_2}{x}\right)$$

$$= -\frac{1}{2}\frac{K_1}{x^2}I_1^2\sin^2\omega_1 t - \frac{1}{2}\frac{K_3}{x^2}I_2^2\sin^2\omega_2 t - \frac{K_2}{x^2}I_1 I_2\sin\omega_1 t\sin\omega_2 t$$

$$= -\frac{1}{2}\frac{K_1}{x^2}I_1^2\frac{(1-\cos 2\omega_1 t)}{2} - \frac{1}{2}\frac{K_3}{x^2}I_2^2\frac{(1-\cos 2\omega_2 t)}{2}$$

$$-\frac{K_2}{x^2}I_1 I_2\left[\frac{\cos(\omega_1-\omega_2)t - \cos(\omega_1+\omega_2)t}{2}\right]$$

$$= -\frac{1}{4}\frac{K_1}{x^2}I_1^2 + \frac{1}{4}\frac{K_1}{x^2}I_1^2\cos 2\omega_1 t - \frac{1}{4}\frac{K_3}{x^2}I_2^2 + \frac{1}{4}\frac{K_3}{x^2}I_2^2\cos 2\omega_2 t$$

$$-\frac{K_2}{2x^2}I_1 I_2\cos(\omega_1-\omega_2)t + \frac{K_2}{2x^2}I_1 I_2\cos(\omega_1+\omega_2)t$$

The above expression for magnetic force f_e consists of mixed frequency terms and average force is given as

$$f_{e\,(av)} = \lim_{T\to\infty}\frac{1}{T}\int_0^T f_e(t)\,dt$$

For $\omega_1 = \omega_2$

$$f_{e\,(av)} = -\frac{K_1}{4x^2}I_1^2 - \frac{K_3}{4x^2}I_2^2 - \frac{K_2}{2x^2}I_1 I_2, \text{ the maximum force } \textbf{Ans.}$$

For $\omega_1 \# \omega_2$

$$f_{e\,(av)} = -\frac{K_1}{4x^2}I_1^2 - \frac{K_3}{4x^2}I_2^2, \quad \text{the minimum force.} \quad \textbf{Ans.}$$

Example 3.20. The self- and mutual inductances of two coupled coils are

$$L_1 = 4 + \frac{1}{x};\quad L_2 = 2 + \frac{1}{x};\quad \text{and}\quad M_{12} = M_{21} = \frac{1}{x}$$

where x is a certain linear displacement. For constant currents of I_1 = 10 A and I_2 = – 5 A determine :

(*i*) the time-average force at x = 0.6 m

(*ii*) the mechanical work done if x increases from 0.6 m to 1.2 m.

(*iii*) the electrical energy supplied by each electrical source in part (*ii*)

(*iv*) the change in magnetic field energy in part (*ii*).

Hence verify that the sum of the energies associated with mechanical work done and field energy is equal to the energy supplied by both electrical sources during the motion from x = 0.6 to x = 1 m.

Solution: Coenergy is given as

$$W'_{field}(i_1, i_2, x)$$

$$= \frac{1}{2}L_1 i_1^2 + \frac{1}{2}L_2 i_2^2 + i_1 i_2 M_{12}$$

$$= \frac{1}{2}\left(4+\frac{1}{x}\right)\times 10^2 + \frac{1}{2}\left(2+\frac{1}{x}\right)\times(-5)^2 + \frac{1}{x}\times 10\times(-5)$$

$$= 225 + \frac{12.5}{x}$$

(*i*) Time average force,

$$F_{e(av)} = \frac{\partial W'_{field}}{\partial x} = -\frac{12.5}{x^2} = -\frac{12.5}{(0.6)^2} = 34.72\text{ N} \quad \textbf{Ans.}$$

(*ii*) Mechanical work done when x increases from 0.6 m to 1.2 m,

$$W_{mech} = \int_{0.6}^{1.2} f_e\,dx$$

$$= \int_{0.6}^{1.2} -\frac{12.5}{x^2}\,dx = 12.5\left|\frac{1}{x}\right|_{0.6}^{1.2} = -10.4166 \text{ J} \quad \textbf{Ans.}$$

(*iii*) Flux linkages,

$$\psi_1 = (L_1 i_1 + M_{12} i_2) = \left(4+\frac{1}{x}\right)\times 10 + \frac{1}{x}\times(-5) = 40+\frac{5}{x}$$

Electrical energy supplied by source 1,

$$d W_{elec1} = \int_{\psi_1(x=0.6)}^{\psi_1(x=1.2)} i_1\,d\psi_1$$

$$= i_1\,[\psi_1(x=1.2)-\psi_1(x=0.6)]$$

$$= 10\left[40+\frac{5}{1.2}-40-\frac{5}{0.6}\right] = -41.666 \text{ J}$$

Flux linkages,

$$\psi_2 = L_2 i_2 + M_{12} i_1 = \left(2+\frac{1}{x}\right)\times(-5) + \frac{1}{x}\times 10 = -10+\frac{5}{x}$$

Electrical energy supplied by source 2,

$$d W_{elec\,2} = \int_{\psi_2(x=0.6)}^{\psi_2(x=1.2)} i_2\,d\psi_2$$

$$= i_2\,[\psi_2(x=1.2)-\psi_2(x=0.6)]$$

$$= -5\left[-10+\frac{5}{1.2}+10-\frac{5}{0.6}\right] = 20.833 \text{ J}$$

Electrical energy supplied by the two sources,

$$d W_{elec} = d W_{elec\,1} + d W_{elec\,2}$$

$$= (-41.666+20.833) \text{ J} = -20.833 \text{ J} \quad \textbf{Ans.}$$

(*iv*) In linear case with constant current excitation

$$W_{field} = W'_{field} = 225+\frac{12.5}{x}$$

Change in field energy,

$$d W_{field} = \left[\left(225+\frac{12.5}{x}\right)(x=1.2) - \left(225+\frac{12.5}{x}\right)(x=0.6)\right]$$

$$= -10.4166 \text{ J} \quad \textbf{Ans.}$$

Here $d W_{field} + W_{mech} = -10.4166 - 10.4166 = -20.833 \text{ J} = d W_{elec}$

Hence it is verified that the sum of energies associated with mechanical work done and field energy is equal to the energy supplied by both electrical sources during the motion from $x = 0.6$ to $x = 1.2$ m.

Example 3.21. The self- and mutual inductances of a doubly-excited magnetic system are

$$L_1 = L_2 = \frac{2}{1+x}; \quad M_{12} = M_{21} = \frac{1}{1+x}$$

Calculate the time-average force and coil currents at $x = 0.6$ m if

(*a*) both the coils are connected in parallel and excited from a voltage source of 200 cos 314 *t*.

(*b*) both the coils are connected in series and excited from a voltage source of 200 cos 314 *t*.

(*c*) coil 1 is shorted and while coil 2 is excited from a voltage source of 200 cos 314 *t*

(*d*) both coils are connected in series and a current of 2 cos 314 *t* flows through them.

Solution: Coenergy is given as

$$W'_{field}(i_1, i_2, x) = \frac{1}{2}L_1 i_1^2 + \frac{1}{2}L_2 i_2^2 + i_1 i_2 M_{12}$$

$$= \frac{1}{2}\times\frac{2}{x+1}i_1^2 + \frac{1}{2}\times\frac{2}{x+1}i_2^2 + i_1 i_2\times\frac{1}{x+1}$$

$$= \frac{1}{x+1}\left(i_1^2 + i_1 i_2 + i_2^2\right)$$

$$f_e = \frac{\partial W'_{field}}{\partial x} = \frac{d}{dx}\cdot\frac{1}{x+1}\left(i_1^2+i_1 i_2+i_2^2\right)$$

$$= \frac{-1}{(x+1)^2}\left(i_1^2+i_1 i_2+i_2^2\right) \quad ...(i)$$

The force acts in a direction to reduce x.

(*a*) Both the coils are connected in parallel across a voltage source of 200 cos 314 *t*

$$L_1 = L_2 = \frac{2}{1+x} = \frac{2}{1+0.6} = \frac{5}{4}$$

$$M_{12} = M_{21} = \frac{1}{1+x} = \frac{1}{1+0.6} = \frac{5}{8}$$

From Eq. (3.73) we have,

$$e_1 = \frac{d\psi_1}{dt} = L_1\frac{di_1}{dt} + M_{12}\frac{di_2}{dt}$$

$$= \frac{5}{4}\frac{di_1}{dt} + \frac{5}{8}\frac{di_2}{dt} = v = 200\cos 314\,t \quad ...(ii)$$

$$e_2 = \frac{d\psi_2}{dt} = L_2\frac{di_2}{dt} + M_{21}\frac{di_1}{dt}$$

$$= \frac{5}{4}\frac{di_2}{dt} + \frac{5}{8}\frac{di_1}{dt} = v = 200\cos 314\,t \quad ...(iii)$$

Solving Eqs. (*ii*) and (*iii*) we have

$$\frac{di_1}{dt} = \frac{di_2}{dt} = \frac{320}{3}\cos 314\,t$$

Integrating above equation we have

$$i_1 = i_2 = \frac{320}{942}\sin 314\,t \quad \textbf{Ans.}$$

Substituting $x = 0.6$ m and $i_1 = i_2 = \frac{320}{942}\sin 314t$ in Eq. (*i*) we have

$$f_e = \frac{-1}{(0.6+1)^2}\left[\left(\frac{320}{942}\right)^2\sin^2 314t + \frac{320}{942}\sin 314t \times\frac{320}{942}\sin 314t + \left(\frac{320}{942}\right)^2\sin^2 314t\right]$$

$$= \frac{-1}{2.56}\times\frac{3\times 1{,}02{,}400}{(942)^2}\sin^2 314t$$

$$f_{e\,(av)} = \frac{-1}{2.56}\times\frac{3\times 1{,}02{,}400}{(942)^2}\times\frac{1}{2} = -0.0676 \text{ N} \quad \textbf{Ans.}$$

$$\because \frac{1}{T}\int_o^T \sin^2\omega t\,dt = \frac{1}{2}$$

(*b*) Both the coils are connected in series across a voltage source of 200 cos 314 *t*

$$v = \frac{d\psi_1}{dt} + \frac{d\psi_2}{dt}$$

$$= \left(\frac{5}{4}\frac{di_1}{dt} + \frac{5}{8}\frac{di_2}{dt}\right) + \left(\frac{5}{4}\frac{di_2}{dt} + \frac{5}{8}\frac{di_1}{dt}\right)$$

$$= \frac{15}{8}\frac{di_1}{dt} + \frac{15}{8}\frac{di_2}{dt} = 3.75\frac{di}{dt} \because \quad i_1 = i_2 = i \text{ (say)}$$

So $3.75 \dfrac{di}{dt} = 200 \cos 314\, t$

Integrating above expression we have

$$i = \frac{200}{3.75 \times 314} \sin 314t \quad \textbf{Ans.}$$

Substituting the value of x and currents i_1 and i_2 in Eq. (*i*) we have

$$f_e = \frac{-1}{(0.6+1)^2} \times 3 \times \left(\frac{200}{3.75 \times 314}\right)^2 \sin^2 314t$$

$$f_{e\,(av)} = \frac{-1}{(1.6)^2} \times 3 \times \left(\frac{200}{3.75 \times 314}\right)^2 \times \frac{1}{2} = -\,0.0169 \text{ N} \quad \textbf{Ans.}$$

(*c*) Coil 1 is shorted and coil 2 is connected across a voltage source of 200 cos 314 *t*

$$\frac{5}{4}\frac{di_1}{dt} + \frac{5}{8}\frac{di_2}{dt} = 0 \quad \text{and} \quad \frac{5}{4}\frac{di_2}{dt} + \frac{5}{8}\frac{di_1}{dt} = 200 \cos 314\, i$$

Solving above equations we have

$$\frac{di_1}{dt} = \frac{-320}{3} \cos 314t \quad \text{and} \quad \frac{di_2}{dt} = \frac{640}{3} \cos 314t$$

Integrating above expressions we have

$$\left.\begin{aligned} i_1 &= \frac{-320}{942} \sin 314t \\ \text{and } i_2 &= \frac{640}{942} \sin 314t \end{aligned}\right\} \quad \textbf{Ans.}$$

Substituting for i_1, i_2 and x in Eq. (*i*) we have

$$f_e = \frac{-1}{(1.6)^2}\left[\left(\frac{-320}{942}\sin 314t\right)^2 + \left[\frac{640}{942}\sin 314t\right]^2 + \frac{-320}{942}\sin 314 \times \frac{640}{942}\sin 314t\right]$$

$$= \frac{-1}{2.56} \times \left(\frac{320}{942}\right)^2 \times 3 \times \sin^2 314t$$

$$f_{e\,(av)} = \frac{-1}{2.56} \times \left(\frac{320}{942}\right)^2 \times 3 \times \frac{1}{2} = -\,0.0676 \text{ N} \quad \textbf{Ans.}$$

(*d*) Both coils are in series and currents through them

$i_1 = i_2 = 2 \cos 314\, t$ **Ans.**

Substituting for i_1, i_2 and x in Eq. (*i*) we have

$$f_e = \frac{-1}{(1.6)^2} \times 3 \times (2 \cos 314)^2 = -\,4.6875 \cos^2 314\, t$$

$$f_{e\,(av)} = \frac{-1}{2} \times 4.6875 = -\,2.34375 \text{ N} \quad \textbf{Ans.}$$

Example 3.22. A doubly-excited magnetic field system has self- and mutual inductances of

$$\mathbf{L_{11} = L_{22} = 2}$$
$$\mathbf{L_{12} = L_{21} = \cos\theta}$$

where θ is the angle between the axes of the coils.

(*i*) The coils are connected in parallel to a voltage source $v = V_m \sin \omega t$. Derive an expression for the instantaneous torque as a function of the angular position θ. Find therefrom the time-average torque. Evaluate for θ = 30°, $v = 100 \sin 314\, t$.

(*ii*) If coil 2 is shorted while coil 1 carries a current of $i_1 = I_m \sin \omega t$, derive expressions for instantaneous and time average torques. Compute the value of time average torque when θ = 45° and $i_1 = \sqrt{2} \sin 314t$.

[U.P. Technical Univ. Electromechanical Energy Conversion-I, 2003-04]

Solution: Coenergy is given as

$$W'_{\text{field}}(i_1, i_2, \theta) = \frac{1}{2} L_{11} i_1^2 + \frac{1}{2} L_{22} i_2^2 + i_1 i_2 L_{12}$$

$$= \frac{1}{2} \times 2^2 \times i_1^2 + \frac{1}{2} \times 2^2 i_2^2 + i_1 i_2 \cos\theta$$

$$= 2i_1^2 + 2i_2^2 + i_1 i_2 \cos\theta$$

$$\text{Magnetic torque, } T_e = \frac{\partial}{\partial\theta} W'_{\text{field}}(i_1, i_2, \theta)$$

$$= \frac{\partial}{\partial\theta}\left[2i_1^2 + 2i_2^2 + i_1 i_2 \cos\theta\right]$$

$$= -\,i_1 i_2 \sin\theta \quad \ldots(i)$$

(*i*) From circuit equations

$$e_1 = \frac{d\psi_1}{dt} = L_{11}\frac{di_1}{dt} + L_{12}\frac{di_2}{dt}$$

$$= 2\frac{di_1}{dt} + \cos\theta \frac{di_2}{dt} = v = V_m \cos \omega t$$

$$e_2 = \frac{d\psi}{dt} = L_{22}\frac{di_2}{dt} + L_{21}\frac{di_1}{dt}$$

$$= 2\frac{di_2}{dt} + \cos\theta\frac{di_1}{dt} = v = V_m \cos \omega t$$

Solving above equations we have

$$\frac{di_1}{dt} = \frac{di_2}{dt} = \frac{V_m \cos \omega t}{2 + \cos\theta}$$

Integrating above expression we have

$$i_1 = i_2 = \frac{V_m \sin \omega t}{\omega(2 + \cos\theta)}$$

Substituting for i_1 and i_2 in Eq. (*i*) we have

$$T_e = -\left[\frac{V_m \sin \omega t}{\omega(2+\cos\theta)}\right]^2 \sin\theta = -\frac{V_m^2 \sin^2 \omega t}{\omega^2 (2+\cos\theta)^2} \sin\theta$$

$$T_{e\,(av)} = -\frac{V_m^2 \sin\theta}{2\omega^2 (2+\cos\theta)^2}$$

For θ = 30° and $v = 100 \sin 314t$ we have

$$T_{e\,(av)} = -\frac{(100)^2 \times \sin 30^\circ}{2 \times (314)^2 \times (2 + \cos 30^\circ)^2}$$

$$= -\frac{10{,}000 \times \frac{1}{2}}{2 \times (314)^2\,(2.866)^2} = -\,0.003087 \text{ N-m} \quad \textbf{Ans.}$$

(*ii*) From circuit equations

$$L_{22}\frac{di_2}{dt} + L_{21}\frac{di_1}{dt} = 2\frac{di_2}{dt} + \cos\theta\frac{di_1}{dt} = 0$$

$$\text{or} \quad \frac{di_2}{dt} = -\frac{1}{2}\cos\theta\frac{di_1}{dt}$$

Integrating above expression we have

$$i_2 = -\frac{1}{2}\cos\theta\, i_1$$

and $i_1 = I_m \sin\omega t$ ...(given)

$$i_2 = -\frac{1}{2}\cos\theta \times I_m \sin\omega t = -\frac{1}{2} I_m \cos\theta \sin\omega t$$

Substituting for i_1 and i_2 in Eq. (*i*) we have

$$T_e = -I_m \sin\omega t \times \left(-\frac{1}{2}\right)\cos\theta \times I_m \sin\omega t \sin\theta$$

$$= \frac{1}{2} I_m^2 \sin\theta \cdot \cos\theta \sin^2\omega t = \frac{1}{4} I_m^2 \sin 2\theta \sin^2 \omega t$$

$$T_{e\,(av)} = \frac{1}{8} I_m^2 \sin 2\theta$$

For $\theta = 45°$ and $i_1 = \sqrt{2}\sin 314t$ we have

$$T_{e\,(av)} = \frac{1}{8} \times (\sqrt{2})^2 \sin 2 \times 45° = 0.25 \text{ N-m} \quad \textbf{Ans.}$$

Example 3.23. Determine, in part (*ii*) of Example 3.22, at what angle the rotor will come to rest, if it is made free to move.

Solution: The average torque will be zero and change sign at $\theta = 0°, 90°, 180°$ .... The rotor can attain position of rest at any of these values of θ but it will be in stable equilibrium position only at $\theta = 90°, 270°$, ...

Example 3.24. A doubly-excited rotating machine has following self- and mutual inductances :

$$\boldsymbol{r_s = 40\ \Omega,\ r_r = 2\ \Omega,\ M_{sr} = 0.08 \cos\theta_r,}$$
$$\boldsymbol{L_s = 0.16\text{ H},\ L_r = 0.04 + 0.02\cos 2\theta_r}$$

θ_r is the space angle between coil axes. The rotor is revolving at a speed of 100 rad/sec. For i_s = 10 A dc and i_r = 2 A dc, derive expressions for the instantaneous voltages applied to the stator and rotor windings.

[*U.P. Technical Univ. Electromechanical Energy Conversion-I, 2005-06*]

Solution:
Stator resistance, $r_s = 40\ \Omega$
Rotor resistance, $r_r = 2\ \Omega$
Stator self inductance, $L_s = 0.16$ H
Rotor self inductance, $L_r = 0.04 + 0.02 \cos 2\theta_r$
Mutual inductance $M_{sr} = 0.08 \cos\theta_r$
Rotor speed, $\omega_r = 100$ radians per second
Stator current, $i_s = 10$ A dc
Rotor current, $i_r = 2$ A dc

Flux linking with stator winding,

$$\psi_s = L_s i_s + M_{sr} i_r = 0.16 \times 10 + 0.08 \cos\theta_r \times 2$$
$$= 1.6 + 0.16 \cos\theta_r$$

Flux linking with rotor winding,

$$\psi_r = L_r i_r + M_{sr} i_s$$
$$= (0.04 + 0.02\cos 2\theta_r) \times 2 + 0.08 \cos\theta_r \times 10$$
$$= 0.08 + 0.04 \cos 2\theta_r + 0.8 \cos\theta_r$$

Stator voltage equation is

$$e_s = r_s i_s + \frac{d\psi_s}{dt} = 40 \times 10 + \frac{d}{d\theta_r}(1.6 + 0.16\cos\theta_r) \times \frac{d\theta_r}{dt}$$

$$= 400 - 0.16 \sin\theta_r \frac{d\theta_r}{dt} = 400 - 0.16 \sin\theta_r \times 100$$

$$\because \frac{d\theta_r}{dt} = \omega_r = 100 \text{ radians/second}$$

$$= 400 - 16 \sin\theta_r \quad \textbf{Ans.}$$

Rotor voltage equation is

$$e_r = r_r i_r + \frac{d\psi_r}{dt} = 2 \times 2 + \frac{d}{dt}(0.08 + 0.04 \cos 2\theta_r + 0.8 \cos\theta_r)$$

$$= 4 + \frac{d\theta_r}{dt} \times \frac{d}{d\theta_r}(0.08 + 0.04 \cos 2\theta_r + 0.8 \cos\theta_r)$$

$$= 4 + 100 \times [(-0.04 \sin 2\theta_r) \times 2 - 0.8 \sin\theta_r)]$$

$$= 4 - 8 \sin 2\theta_r - 80 \sin\theta_r \quad \textbf{Ans.}$$

3.13 FORCES AND TORQUES IN PERMANENT MAGNET SYSTEMS

Consider a permanent magnet moving armature relay shown in Fig. 3.41. The dc magnetization curve for a permanent magnet is shown in Fig. 3.42. If it is linearly explorated at the lower end it may be expressed as

Magnetic flux density in permanent magnet,

$$B_m = \mu_R (H_m - H'_C) = \mu_R H_m - B_r \quad ...(3.94)$$

where μ_R is the recoil permeability of the permanent magnetic (PM) material, H_m is the magnetic field intensity in the permanent magnet and B_r is the residual flux density.

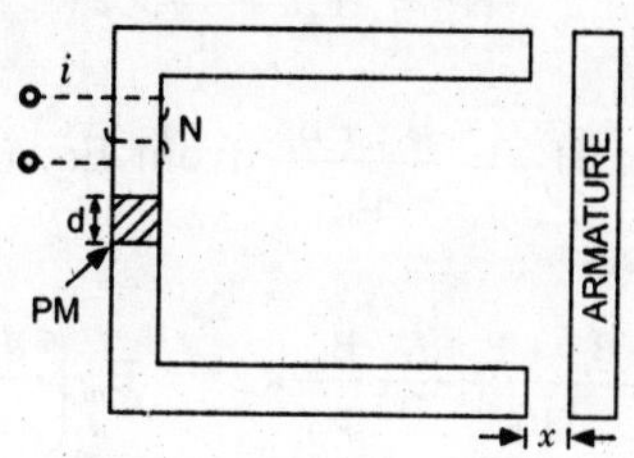

Fig. 3.41 *Permanent Magnet Moving Armature Relay*

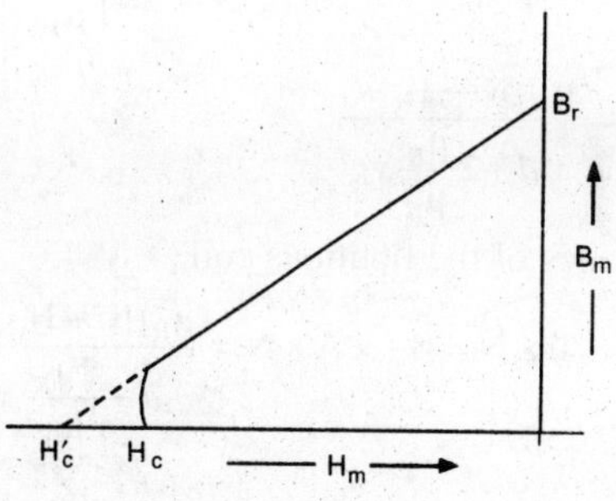

Fig. 3.42 *B-H Curve For Permanent Magnet*

Recoil permeability of the PM material,

$$\mu_R = \frac{B_r}{H'_C} \text{ on linearized basis}$$

Assume iron portion of magnetic circuit including the armature to have infinite permeability i.e. negligible reluctance.

According to Eq. (3.41)

$$F_{\text{field}} = \frac{\partial W'_{\text{field}}(i, x)}{\partial x}$$

Since there is no exciting current (i.e. $i = 0$) in the system, system coenergy is independent of current and coenergy will be a function of the space variable x only i.e.

$$W'_{\text{field}}(i, x) \Rightarrow W'_{\text{field}}(x)$$

Coenergy is given by Eq. (3.27) i.e.

$$W'_{\text{field}} = \int_o^i \psi\, di$$

The above expression needs careful interpretation in case of a PM excited systems because in this case excitation current is zero. Here the limits of integration mean from state of zero flux to a state of certain flux (but with no current). The state of zero flux may be imagined by means of a fictitious N turn coil carrying excitation current i, as depicted by dotted lines in Fig. 3.41. The current is assumed to be adjusted to a value I causing the core flux to reduce to zero and the original state is then attained by imaging the current I to reduce to zero. Thus expression for coenergy in system with PM is modified to

$$W'_{field} = \int_I^o \psi\, di \qquad ...(3.95)$$

For any value of current,

$$Ni = H_m d + H_g \times 2x \qquad ...(3.96)$$

If A is uniform cross-sectional area,

$$B_m A = B_g A \qquad \text{Since flux is same}$$

$$\text{or} \quad B_m = \mu_R H_m = B_g = \mu_0 H_g \qquad ...(3.97)$$

Substituting $H_g = B_m/\mu_0$ from above equation in Eq. (3.96) we have

$$Ni = H_m d + \frac{B_m}{\mu_o} \times 2x$$

Substituting $H_m = \dfrac{B_m + B_r}{\mu_R}$ from Eq. (3.94) in above equation, we have

$$Ni = \frac{(B_m + B_r)d}{\mu_R} + \frac{B_m}{\mu_0} \times 2x = B_m\left(\frac{d}{\mu_R} + \frac{2x}{\mu_0}\right) + \frac{B_r d}{\mu_r}$$

$$= B_m\left(\frac{d}{\mu_R} + \frac{2x}{\mu_0}\right) + H'_C d$$

$$\text{or } B_m = \frac{\mu_R (Ni - H'_C d)}{d + 2\dfrac{\mu_R}{\mu_0} x} \qquad ...(3.98)$$

Flux linkages of the fictitious coil,

$$\psi = \Phi \times N = B_m \times A \times N = \frac{\mu_R (Ni - H'_C d)}{d + \dfrac{2\mu_R}{\mu_0} x} \times A \times N \qquad ...(3.99)$$

For flux and flux linkages to be zero

$$i = \frac{H'_C d}{N} \qquad ...(3.100)$$

Substituting in Eq. (3.95) we have

$$W'_{field}(x) = \int_i^0 \frac{\mu_R NA (Ni - H'_C d)}{d + \dfrac{2\mu_R}{\mu_0} x} di = \frac{\mu_R A (H'_C)^2 d^2}{2\left[d + 2\dfrac{\mu_R}{\mu_0} x\right]}$$

The force on the armature is given as

$$F_{field} = \frac{d W'_{field}}{dx} = \frac{\mu_R^2 (H'_C)^2 d^2 A}{\mu_0 \left[d + 2\dfrac{\mu_R}{\mu_0} x\right]^2}$$

Substituting $\mu_R H'_C = B_r$ in above expression we have

$$F_{field} = \frac{AB_r^2}{\mu_0 \left[1 + 2\dfrac{\mu_R}{\mu_0}\left(\dfrac{x}{d}\right)\right]^2} \qquad ...(3.101)$$

Example 3.25. Determine the magnitude of force in a system with permanent magnet for air gaps of (*i*) $x = 0$ and (*ii*) $x = 6$ mm. The other datas are as follows:

$$B_r = 0.95 \text{ T}; \quad H'_C = 7.2 \times 10^5 \text{ AT/m};$$

$$d = 30 \text{ mm}; \quad A = 500 \text{ mm}^2; \quad \mu_0 = 4\pi \times 10^{-7} \text{ H/m}$$

Solution: Recoil permeability of the permanent magnet material,

$$\mu_R = \frac{B_r}{H'_C} = \frac{0.95}{7.2 \times 10^5} = 13.2 \times 10^{-7}$$

For $x = 0$,

$$\text{Magnetic force, } F_{field} = -\frac{AB_r^2}{\mu_0\left[1 + \dfrac{2\mu_r}{\mu_0}\left(\dfrac{x}{d}\right)\right]^2}$$

$$= -\frac{500 \times 10^{-6} \times (0.95)^2}{4\pi \times 10^{-7}} = -359 \text{ N} \quad \textbf{Ans.}$$

For $x = 6$ mm,

$$F_{field} = -\frac{500 \times 10^{-6} \times (0.95)^2}{4\pi \times 10^{-7}\left[1 + \dfrac{2 \times 13.2 \times 10^{-7}}{4\pi \times 10^{-7}} \times \dfrac{6}{30}\right]^2}$$

$$= -178 \text{ N} \quad \textbf{Ans.}$$

– *ve* sign indicates that the force is restoring in nature i.e. it acts in a direction to reduce the air gap.

3.14 SINGLY-EXCITED ELECTRIC FIELD SYSTEMS

The electric field as a coupling medium between electrical and mechanical systems can be treated in the same way as the magnetic field discussed earlier. Charge and voltage associated with electric field are analogous to flux linkages and current respectively in a magnetic field.

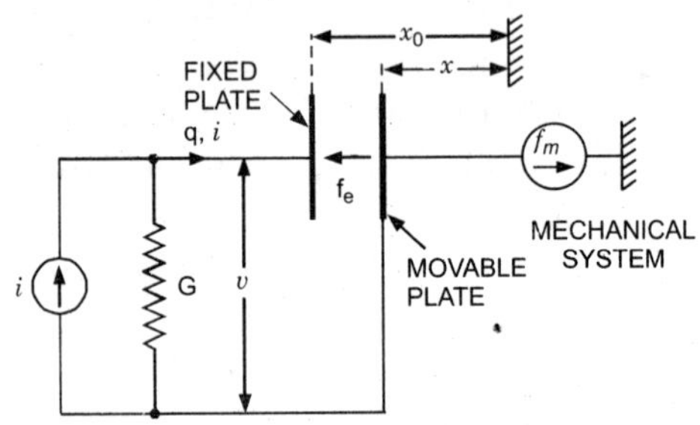

Fig. 3.43 *Singly-Excited Electric Field Energy Conversion Device*

Figure 3.43 shows a parallel plate capacitor with fixed and movable plates representing the electric field as the coupling field. The capacitor is fed with a current source. The leakage current of the capacitor is represented outside by the conductance G so that the capacitor field is conservative. It means that all the energy supplied is recoverable from the device. The energy flow in an electric field system is also governed by Eq. (3.17) as in a magnetic field system.

The value of differential electrical energy input to the capacitor in time dt is

$$d W_{elec} = v\, i\, dt = v\, dq \qquad ...(3.102) \qquad \because \; dq = i\, dt$$

The mechanical work done in time dt by the electric field force f_e,

$$d\,W_{mech} = f_e\,dx \qquad ...(3.103)$$

where dx is the displacement of movable plate in the direction of f_e during time dt.

According to the principle of energy conservation

Mechanical work done by the electric field force

= Electric energy input – increase in field energy

$$\text{or } f_e\,dx = v\,d\,q - d\,W_{field} \qquad ...(3.104)$$

When the movable plate of the capacitor is held fixed in a position,

$f_e\,dx = 0$ and Eq. (3.104) becomes

$$d\,W_{elec} = d\,W_{field} = v\,dq \qquad ...(3.105)$$

In above Eq. (3.105) v remains constant and charge builds up from 0 to q.

The total field energy,

$$W_{field} = \int_0^q v\,dq$$

In a capacitor v and q are linearly related as $C = q/v$, assuming linear electric field

$$W_{field} = \frac{1}{2}\frac{q^2}{C} = \frac{1}{2}qv \qquad ...(3.106)$$

The capacitance of the device, C is a function of configuration (distance between the plates) and be expressed as

$$C = \frac{\varepsilon_0 A}{x_0 - x}$$

where ε_0 is the permittivity of free space and A is the plate area.

Thus the field energy W_{field} is a function of two independent variables q and x, i.e.,

$$W_{field}(q, x) = \frac{1}{2}\frac{q^2}{C(x)} \qquad ...(3.107)$$

The above equation reveals that the electrical field energy can be changed electrically by varying q or mechanically by varying x (i.e. by moving the movable plate).

The electric field energy can also be expressed as

$$W_{field}(v, x) = \frac{1}{2}vq = \frac{1}{2}C(x)v^2 \qquad ...(3.108)$$

The energy density in electric field is expressed as

$$W_{field} = \int_0^D E\,dD = \frac{1}{2}\frac{D^2}{\varepsilon_0} \qquad ...(3.109)$$

where E is electric field intensity or potential gradient in volts per metre and D is electric flux density in Coulombs/m^2.

Coenergy. For a linear case, the variation of q with variation in v is illustrated in Fig. 3.44. As in a magnetic system, coenergy W'_{field} is the area OABO.

$$W'_{field} = \text{Area OABO} = \int_0^{v_1} q\,dv = C\int_0^{v_1} v\,dv = \frac{1}{2}Cv_1^2$$

$$\text{In general coenergy, } W'_{field}(v, x) = \frac{1}{2}C(x)v^2 \qquad ...(3.110)$$

$$\text{and coenergy density, } w'_{field} = \frac{1}{2}\varepsilon_0 E^2 \qquad ...(3.111)$$

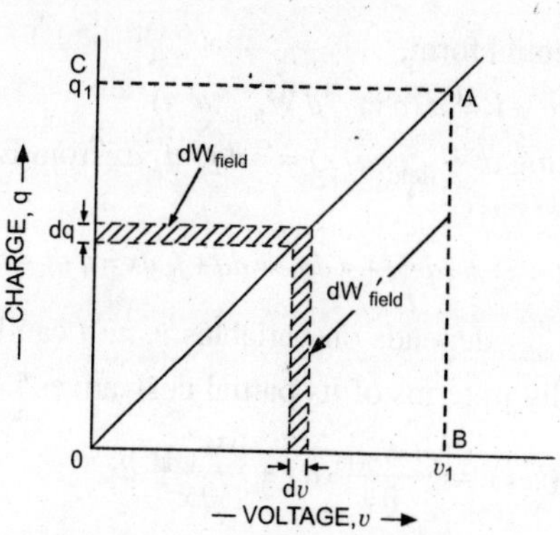

Fig. 3.44 *Electric Field Energy and Co-energy in a Linear Circuit*

Determination of Mechanical Force. The stored electric field energy is dependent on variables q and x and so the expression for it can be written as

$$d\,W_{field}(q, x) = v\,dq - f_e\,dx \qquad ...(3.112) \quad \text{Refer to Eq. (3.104)}$$

The electric field energy $W_{field}(q, x)$ can also be expressed mathematically in terms of its two partial derivatives as

$$d\,W_{field}(q, x) = \frac{\partial W_{field}}{\partial q}dq + \frac{\partial W_{field}}{\partial x}dx \qquad ...(3.113)$$

The field energy in Eqs. (3.112) and (3.113) is the same, therefore, the coefficients associated with dq and dx must be equal in both of the above equations i.e.

$$v = \frac{\partial W_{field}(q, x)}{\partial q} \qquad ...(3.114)$$

$$\text{and } f_e = -\frac{\partial W_{field}(q, x)}{\partial x} \qquad ...(3.115)$$

Alternatively the expression for mechanical force due to electric field can be determined from the principles of virtual work and conservation of energy. Let the plates of the capacitor in Fig. 3.45 be allowed to move closer together a differential distance x under the influence of force of attraction between the oppositely charged plates. Mechanical work done $f_e\,dx$ will then be done by the electric field. The force depends only on the strength of the electric field and is independent of the way in which the field variables may change.

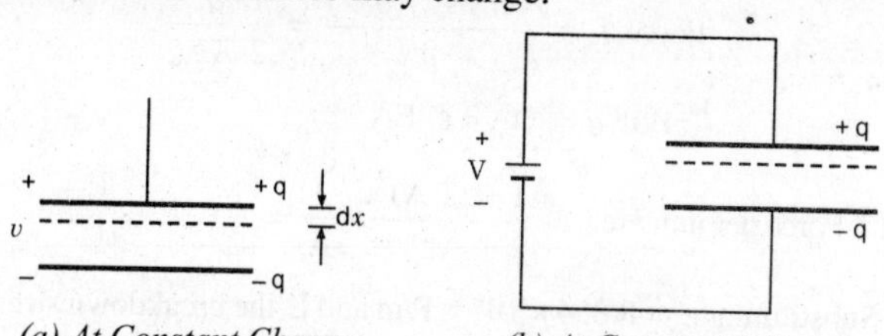

(a) At Constant Charge *(b) At Constant Voltage*

Fig. 3.45 *Charged Capacitor*

First assume that the capacitor has been charged and the source disconnected, as in Fig. 3.45 (*a*). Then assume the virtual displacement dx. The charge q is constant, and the electrical energy input is zero. Equation (3.17) then reduces to

$$d\,W_{mech} = f_e\,dx = -\partial\,W_{field}(q, x)$$

$$\text{or } f_e = -\frac{\partial W_{field}}{\partial x}(q, x)$$

From Fig. 3.44

$$\text{Coenergy, } W'_{field} = q_1v_1 - W_{field}$$

$$\text{In general, } W'_{field}(v, x) = qv - W_{field}(q, x)$$

In differential form,

$$d W'_{\text{field}}(v, x) = d(q v) - d W_{\text{field}}(q, x) \quad ...(3.116)$$

Substituting $d W_{\text{field}}(q, x) = v\, dq - f_e\, dx$ from Eq. (3.112) in Eq. (3.116) we have

$$d W'_{\text{field}}(v, x) = q\, dv + v\, dq - v\, dq + f_e\, dx = q\, dv + f_e\, dx \quad ...(3.117)$$

Since W'_{field} depends on variables v, x; it can be expressed mathematically in terms of its partial derivatives i.e.

$$d W'_{\text{field}}(v, x) = \frac{\partial W'_{\text{field}}}{\partial v} dv + \frac{\partial W'_{\text{field}}}{\partial x} dx \quad ...(3.118)$$

Comparing coefficients of dv and dx in Eqs. (3.117) and (3.118) for $d W'_{\text{field}}(v, x)$ we have

$$q = \frac{\partial W'_{\text{field}}(v, x)}{\partial v} \quad ...(3.119)$$

$$\text{and } f_e = \frac{\partial W'_{\text{field}}(v, x)}{\partial x} \quad ...(3.120)$$

Substituting $W'_{\text{field}}(v, x) = \frac{1}{2} C(x) v^2$ from Eq. (3.110) in above Eq. (3.120) we have

$$f_e = \frac{1}{2} v^2 \frac{dC(x)}{dx} \quad ...(3.121)$$

The above expression for mechanical force indicates that the force acts in such a direction as to increase the capacitance of the configuration.

Example 3.26. Determine an expression for the force per unit area between the plates of a parallel plate air capacitor using both the energy and coenergy methods.

Solution: For a distance x_o between the two plates, movement of a plate by distance x in the direction reduces the distance between the plates from x_o to $(x_o - x)$ (Fig. 3.43). Electric field energy stored in this space,

$$W_{\text{field}\,(q,\,x)} = \frac{1}{2}\frac{q^2}{C} = \frac{1}{2} q^2 \frac{(x_o - x)}{A\varepsilon_o}$$

From Eq. (3.115)

$$\text{Force, } f_e = -\frac{\partial W_{\text{field}}(q, x)}{\partial x} = \frac{1}{2}\frac{q^2}{A\varepsilon_o}$$

$$\text{But } q = DA = \varepsilon_o EA$$

$$\text{and Force per unit area} = \frac{1}{2}\frac{(\varepsilon_o E A)^2}{A\varepsilon_o} \times \frac{1}{A} = \frac{1}{2}\varepsilon_o E^2$$

Substituting $\varepsilon_o = 8.854 \times 10^{-12}$ F/m and E, the breakdown strength of air $= 3 \times 10^6$ V/m in above expression we have

$$\text{Force per unit area} = \frac{1}{2} \times 8.854 \times 10^{-12} \times (3 \times 10^6)^2$$

$$= 39.84 \text{ N/m}^2 \quad \textbf{Ans.}$$

From Eq. (3.110)

$$\text{Coenergy, } W'_{\text{field}}(v, x) = \frac{1}{2} C v^2 = \frac{1}{2} v^2 \frac{A\varepsilon_o}{(x_o - x)}$$

$$\text{and from Eq. (3.120), } f_e = \frac{\partial W'_{\text{field}}(v, x)}{\partial x} = \frac{1}{2} v^2 \frac{A\varepsilon_o}{(x_o - x)^2}$$

$$\text{Force per unit area} = \frac{1}{2} v^2 \frac{\varepsilon_o}{(x_o - x)^2}$$

$$\text{but } v = E(x_o - x)$$

$$\text{So force per unit area} = \frac{1}{2}[E(x_o - x)]^2 \times \frac{\varepsilon_o}{(x_o - x)^2} = \frac{1}{2}\varepsilon_o E^2$$

Both the methods give the same expression for the force, as expected.

In example 3.4, the force density on the bounding surface in a magnetic field near saturation was found to be 1.0186×10^6 N/m² while it has a value of only 39.84 N/m² in electric field with electric field intensity at its breakdown value. The ratio of the two is found to be $\frac{1.0186 \times 10^6}{39.84} =$ 25,600 i.e. the force density in the magnetic field is about 25,600 times the value in electric field at the maximum possible field strength assumed. This is because of low value of ε_o being equal to 8.854×10^{-12} F/m compared to $\mu_o = 4\pi \times 10^{-7}$ A/m.

3.15 DYNAMIC EQUATIONS OF ELECTROMECHANICAL SYSTEMS

We have derived expressions for the forces developed in electromechanical energy conversion devices as functions of electrical variables and the mechanical displacement. These expressions were derived for a conservative energy conversion system; any losses were assigned to the electrical and mechanical sources. Such conversion devices always operate as a coupling means between an electric system and a mechanical system. Thus our interest should be not only in how the conversion device itself operates but also in how the complete electromechanical system operates with the conversion device contained within it.

The model of a simple conversion system illustrated in Fig. 3.46 depicts the basic portions, the details of which may vary from system to system. The system shown consists of three parts: an external electric portion, the energy conversion means, and the external mechanical portion. The electric portion is merely represented by a voltage source v and resistance.

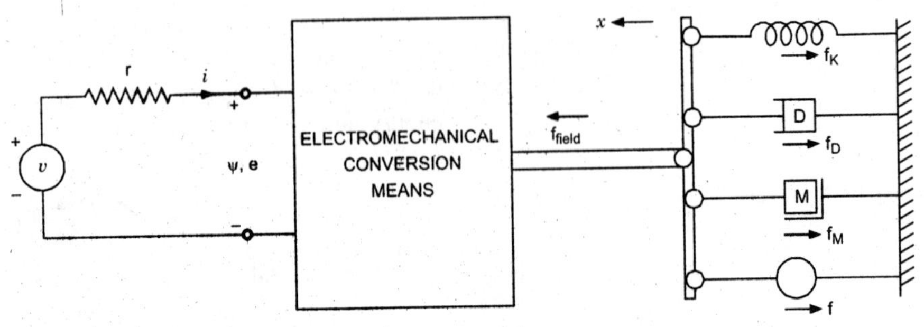

Fig. 3.46 *Model of a Singly-Excited Electromechanical System*

Alternatively it could be represented by a current source i and a parallel conductance G.

Let the electromechanical conversion means has an inductance $L(x)$.

The voltage equation for the electrical system may be written as

$$v = ir + \frac{d\psi}{dt} = ir + \frac{d}{dt}[L(x)i] = ir + L(x)\frac{di}{dt} + i\frac{dL(x)}{dx}\cdot\frac{dx}{dt} \quad ...(3.122)$$

The second term on RHS of Eq. (3.122) is the self-inductance or transformer voltage term while the last or third term is the speed voltage term, which is responsible for the energy transfer between the external electric system and the energy conversion means.

The mechanical portion of the system shown in Fig. 3.46 shows symbols for a spring of spring constant K, a damper of damping constant D, a mass M and external mechanical force f.

The magnetic force f_e is opposed by the inertia force $f_M\left(M\frac{d^2x}{dt^2}\right)$, damping force $f_D\left(D\frac{dx}{dt}\right)$, restraining spring force $f_K(Kx)$ and external force f. Thus force balance equation is

$$f_e = M\frac{d^2x}{dt^2} + D\frac{dx}{dt} + Kx + f \quad ...(3.123)$$

The magnetic force f_e can be determined from the partial derivative of W_{field}, as already explained.

Thus both the Eqs. (3.122) and (3.123) describe the total behaviour of a linear electromechanical system.

Example 3.27. Figure 3.47 shows the cross section of a cylindrical solenoid magnet in which the plunger of mass M moves in brass guide rings of thickness *t* and mean diameter *d*. The motion of plunger is simple harmonic one. The plunger is supported by a spring whose elastance is k newtons per unit displacement. The coil has N turns and is excited from sinusoidal voltage source so that the air-gap flux at any time is given as

$$\phi = \Phi_{max} \cos \omega t$$

(a) Derive expressions for (i) the magnetic force acting on the movable steel plunger in terms of Φ_{max}, ω and t and (ii) the coil voltage as the function of Φ_{max}, N, ω and t.

(b) Obtain the differential equation for the movement of the steel plunger in terms of Φ_{max}, ω and t.

Assume coil resistance, effects of magnetic leakage and fringings and reluctance of steel to be negligible and the movement of steel plunger frictionless. Take relative permeability of brass to be unity.

Solution: (*a*) (*i*) Let the cylindrical steel plunger move a small distance x under the influence of magnetic force f_e. This will reduce the air-gap length from x_0 to $(x_0 - x)$. The reluctance of this reduced air-gap length is given as

$$S_1 = \frac{x_0 - x}{\mu_0 A} \quad ...(i)$$

Reluctance of the brass guide rings between the steel plunger and cylindrical steel shell is given as

$$S_2 = \frac{2t}{\mu_0 A} \quad ...(ii)$$

$$\text{Total reluctance, } S = S_1 + S_2 = \frac{x_0 - x}{\mu_0 A} + \frac{2t}{\mu_0 A} \quad ...(iii)$$

Magnetic force, f_e is given as

$$f_e = -\frac{1}{2}\phi^2\frac{dS}{dx} = -\frac{1}{2}\phi^2\frac{d}{dx}\left[\frac{x_0 - x}{\mu_0 A} + \frac{2t}{\mu_0 A}\right]$$

$$= \frac{1}{2}\phi^2 \times \frac{1}{\mu_0 A} \quad ...(iv)$$

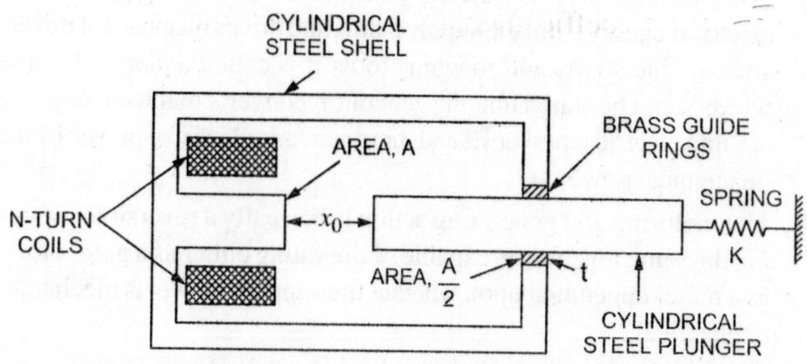

Fig. 3.47 *Solenoid Magnet*

Substituting $\phi = \Phi_{max} \cos \omega t$ in above expression for f_e, we have expression for magnetic force acting on the movable steel plunger and so

$$f_e = \frac{\Phi_{max}^2 \cos^2 \omega t}{2\mu_0 A} = \frac{\Phi_{max}^2(1 + \cos 2\omega t)}{4\mu_0 A} \quad ...(v) \textbf{ Ans.}$$

$$(ii) \text{ Coil voltage, } e = -\frac{d\psi}{dt} = -N\frac{d\phi}{dt} = -\frac{d}{dt}(N\,\Phi_{max} \cos \omega t)$$

$$= \omega N\,\Phi_{max} \sin \omega t \quad ...(vi) \textbf{ Ans.}$$

(*b*) The magnetic force f_e is utilised to overcome (*i*) spring tension (*ii*) friction and (*iii*) mass M.

For any displacement x,

$$f_k = kx \text{ newtons} \quad ...(vii)$$

$$\text{Damping force, } f_D = D\frac{dx}{dt} \text{ newtons} \quad ...(viii)$$

where D is the coefficient of friction of the movable part in newtons per metre per second and $\frac{dx}{dt}$ is the velocity of the movable part in m/s.

$$\text{Inertia force, } f_M = M\frac{d^2x}{dt^2} \text{ newtons} \quad ...(ix)$$

where M is the mass of the movable steel plunger in kg and $\frac{d^2x}{dt^2}$ is the acceleration in m/s^2.

So, the force-balance equation is given as

$$f_e = f_M + f_D + f_k$$

Substituting the values of f_e, f_M, f_D and f_k from Eqs. (*v*), (*ix*), (*viii*) and (vii) respectively in above equation we have

$$\frac{\Phi_{max}^2(1 + \cos 2\omega t)}{4\mu_0 A} = M\frac{d^2x}{dt^2} + 0 + kx$$

$$\text{or } M\frac{d^2x}{dt^2} + kx = \frac{\Phi_{max}^2}{4\mu_0 A}(1 + \cos 2\omega t) \textbf{ Ans.}$$

HIGHLIGHTS

1. An *electromechanical energy conversion device* is a link between an electrical and mechanical system and *electromechanical energy conversion* needs the presence of natural phenomena which interrelate electric and magnetic fields on one hand and mechanical force and motion on the other.
2. A *dynamo* is a machine that converts either mechanical energy into electrical energy or electrical energy into mechanical energy. When a dynamo is driven mechanically by a prime mover such as by a diesel engine, steam engine, steam turbine or water turbine and supplies electrical energy, it is called a *generator*. When the dynamo draws electrical energy from the supply mains and drives mechanical devices such as line shafts and machine tools it is called a *motor*. In other words it can be stated that the generator converts mechanical power ωT into electrical power EI and motor coverts electrical power EI into mechanical power ωT.
3. The motoring and generating action is basically a reversible process *i.e.* the same machine is capable of operating either as a generator or as a motor depending upon whether the supplied power is mechanical or electrical.
4. The energy balance equation is given by:

$$\int dW_{input} = \int dW_{output} + \int dW_{magnetic\ field} + \int dW_{heat}$$

5. For steady-state operation, the energy and power equations are:

$$W_{input} = W_{output} + W_{heat}$$
$$P_{input} = P_{output} + P_{heat}$$

6. The expression for coenergy density is given as

$$W'_{field} = \int_0^H B\,dH$$

7. For linear case the above equation becomes

$$W'_{field} = \frac{1}{2}BH = \frac{1}{2}\mu H^2 = \frac{1}{2}\frac{B^2}{\mu}\ \text{J/m}^3$$

8. In an electromagnetic system, the equation for magnetic or mechanical force is given by

$$F_{field} = \frac{\partial W'_{field}(i, x)}{\partial x} \quad \text{when the system is current excited}$$

and $$F_{field} = \frac{-\partial W_{field}(\psi, x)}{\partial x}$$

when the system is voltage controlled

In a rotary case, the above equations for force will apply for torque by replacing x with angular rotation θ.

9. During slow movement of electromagnetic relay armature, half the electrical energy input gets stored in the field and the other half is output as mechanical energy.

During instantaneous movement of electromagnetic relay armature, there is no electrical input and the mechanical energy output is at the expense of the field energy stored.

10. The principle of variable reluctance may be applied to cause continuous rotation of shaft, and the result, the single-phase ac reluctance motor, is one of the simplest type of electric motor.

The expression for electromagnetic torque in a reluctance motor is given by

$$T_e\,(av) = \frac{1}{8}\Phi^2_{max}(S_q - S_d)\sin 2\delta.$$

11. A doubly excited magnetic system is one which has two independent sources of excitation.

The torque equation in a doubly excited magnetic system is given by

$$T_e = \frac{1}{2}i_1^2\frac{dL_1}{d\theta_r} + \frac{1}{2}i_2^2\frac{dL_2}{d\theta_r} + i_1 i_2\frac{dM_{12}}{d\theta_r}$$

12. The torque equation for a single-phase smooth cylindrical rotor machine is given by

$$T_e\,(av) = \frac{I_{dc}\,I_{max}\,M_{max}}{2}\sin\delta$$

13. The torque equation for a single-phase salient pole machine is given by

$$T_e\,(av) = \frac{1}{8}I^2_{max}(I_d - I_q)\sin 2\delta$$

14. Voltage equation for electrical system may be written as

$$v = ir + L(x)\frac{di}{dt} + i\frac{dL(x)}{dx}\cdot\frac{dx}{dt}$$

and the force balance equation is

$$f_e = M\frac{d^2x}{dt^2} + D\frac{dx}{dt} + Kx + f$$

Both of the above two equations describe the total behaviour of a linear electromechanical system.

EXERCISES

1. In every electromagnetic conversion device, both generator and motor action takes place simultaneously. Explain.
[Rajasthan Technical Univ. Electrical Machines-I, 2008]
2. Discuss the principle of "electromechanical energy conversion". Also explain its applications and limitations.
[U.P. Technical Univ. Electromechanical Energy Conversion-I, 2013-14]
3. Explain briefly an electromechanical energy conversion device with the help of a block diagram.
[J.N. Technological Univ. Electrical Machines-I, March-2014]
4. Describe the principle of energy conversion. From a consideration of the various energies involved, develop the model of an electromechanical energy conversion device.
[U.P. Technical Univ. Electromechanical Energy Conversion-I, 2008-09; J.N. Technological Univ. Electrical Machines-I, March-2014]
5. Define and explain the concept of transformer emf and motional emf in dc machines. **[Pb. Univ. Elec. Machines-I December 1991]**
6. Explain flow of energy in electromechanical device with a suitable model and write energy balance equation.
[G.B. Technical Univ. Electromechanical Energy Conversion-I, 2011-12]
7. Draw a schematic diagram indicating flow of energy in the conversion of mechanical energy to electrical energy.
[Rajasthan Technical Univ. Electrical Machines-I, January-2013]
8. Draw neat labelled schematic diagrams to show flow of energy in electromechanical energy conversion via a coupling magnetic field, both for motors as well as for generators.
[U.P. Technical Univ. Electromechanical Energy Conversion-I, 2009-10]

9. What are the principles of field energy concepts for analyzing energy conversion devices?
[G.G.S.I.P. Univ. Electromechanical Energy Conversion-I, December-2011]

10. Explain principle of conservation of energy in electromechanical system and give its general representation.
[Rajasthan Univ. Electrical Machines-I, 2006]

11. Deduce the energy balance equation for the physical phenomenon involved in energy conversion.
[Electrical Machines-I, Rajasthan Univ. 2001, 2003; Rajasthan Technical Univ. 2010]

12. Draw a schematic diagram indicating flow of energy in the conversion of mechanical energy into electrical energy.
[Rajasthan Technical Univ. Electrical Machines-I, 2011]

13. State and briefly explain the various phenomenon useful for electromechanical energy conversion in rotating machines.
[Rajasthan Technical Univ. Electrical Machines-I, 2007, 2010]

14. Describe the principle of energy conversion. Show that the reaction of coupling magnetic field on the electrical or mechanical system is essential for the electromechanical energy conversion.
[Rajasthan Technical Univ. Electrical Machines-I, 2007; January-2013]

15. Explain "singly-excited systems". Also mention its examples.
[G.B. Technical Univ. Electromechanical Energy Conversion-I, 2012-13]

16. Derive expressions for field energy and coenergy in a singly-excited electromechanical unit.
[J.N. Technological Univ. Electrical Machines-I, March-2014]

17. For a singly-excited magnetic system, establish relationship between magnetic field energy and coenergy. Develop also the relation for the magnetic stored energy and coenergy in terms of reluctance and permeance.
[G.B. Technical Univ. Electromechanical Energy Conversion-I, 2011-12]

18. What do you mean by "energy" and "coenergy"? What is the importance of these terms in electrical machine?
[G.B. Technical Univ. Electromechanical Energy Conversion-I, 2012-13]

19. What is the concept of coenergy? Explain it with the help of an example. [M.D. Univ. Electromechanical Energy Conversion, May-2009]

20. Define energy and coenergy. What is the significance of coenergy ? Show that the field energy in a linear magnetic system is given by

$$W_f = \frac{1}{2}Li^2 = \frac{1}{2}\psi i = \frac{1}{2L}\psi^2$$

[U.P. Technical Univ. Electromechanical Energy Conversion-I, 2006-07]

21. For the electromechanical system shown in Fig. 1, the air gap flux density under steady operating condition is $B(t) = B_m \sin \omega t$.
Find (*i*) The induced coil voltage.
(*ii*) The force of field origin as a function of time, and
(*iii*) The motion of armature as a function of time.
[U.P. Technical Univ. Electromechanical Energy Conversion-I, 2006-07]

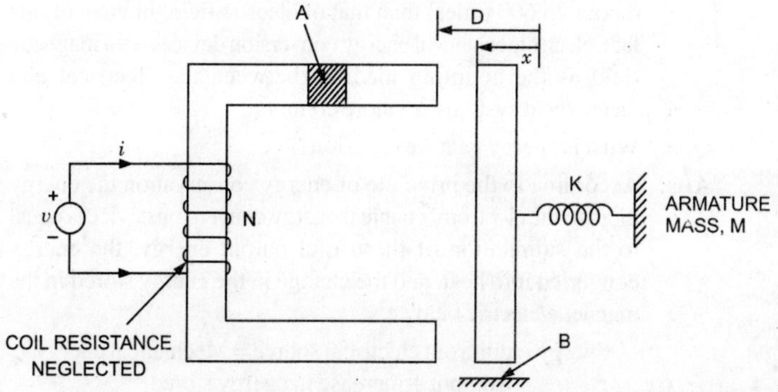

Fig. 1

22. The $\lambda - i$ characteristic of a magnetic circuit are frequently with straight line segments as shown in Fig. 2. The circuit is considered linear up to point A and in saturation from point A to B.

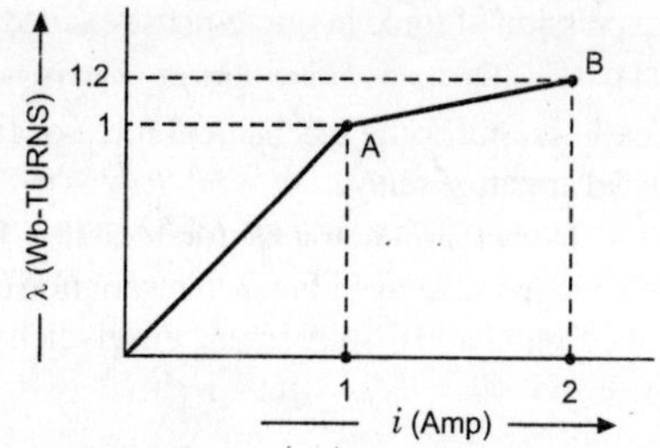

Fig. 2

Find the energy stored in magnetic field for magnetic circuit at point A and at point B.
[U.P. Technical Univ. Electromechanical Energy Conversion-I, 2005-06]

23. Explain the meaning of conservative system and discuss its properties. Also show that the energy stored (W_{fld}) in the singly-excited magnetic system with instantaneous value of the flux linkage (λ) of the circuit at constant inductance L is given by

$$W_{fld} = \frac{1}{2}\frac{\lambda^2}{L}$$

[U.P. Technical Univ. Electromechanical Energy Conversion-I, 2005-06]

24. Explain the term 'co-enegy' in electromechanical energy conversion and show that coenergy is given by:

$W_f = \frac{1}{2} PF^2$ where P = permeance of the magnetic circuit and F = mmf in coil of magnetic circuit.
[U.P. Technical Univ. Electromechanical Energy Conversion-I, 2004-05]

25. For a linear magnetic circuit, derive the following relations for the stored magnetic energy W_{field} and coenergy W_{field}.

$$W_{field} - W'_{field} = \frac{1}{2}F\Phi = \frac{1}{2}\psi i = \frac{1}{2}\Phi^2 R l = \frac{F^2}{R l} = \frac{1}{2}LI^2 \text{ joules}$$

where F = Instantaneous mmf, ψ = Magnetic flux linkage
f = Instantaneous flux = R l = Reluctance
i = Instantaneous current, L = Self-inductance.
[U.P. Technical Univ. Electromechanical Energy Conversion-I, 2008-09]

26. Derive the expression for energy stored in the magnetic field and show that the same is equal to the area between ϕ-F curve and the flux axis.
[U.P. Technical Univ. Electromechanical Energy Conversion-I, 2009-10]

27. For an electromagnetic system, show that the energy stored in a magnetic field is equal to the area enclosed between magnetization curves for open and closed positions of the armature and flux linkage-current during the armature movement.
[Rajasthan Technical Univ. Electrical Machines-I, 2007]

28. Describe the principle of virtual work and hence show that the magnetic force f_e is given by

$$f_e = \frac{-\partial W_{field}}{\partial x}(\Phi, x) = \frac{-\partial W_{field}}{\partial x}(\Phi, x)$$

Above terms have usual meanings.
[Electromechanical Energy Conversion-I, U.P. Technical Univ., 2008-09; G.B. Technical Univ., 2011-12]

29. Explain the mechanical energy and work done in a singly-excited system when actual displacement occurs.
[J.N. Technological Univ. Electrical Machines-I, March-2014]

30. For a linear magnetic circuit derive the relation for stored energy and coenergy. Also show that the magnetic energy density can be given by $\frac{1}{2}B^2/\mu$ joules/m^3.
[G.G.S.I.P. Univ. Electromechanical Energy Conversion-I, December-2011]

31. Derive the expression of force in single-phase excited magnetic field system. [M.D. Univ. Electromechanical Energy Conversion, December-2012]

32. Deduce an expression for the mechanical force of field origin in a typical attracted armature relay.
[Anna Univ. Chennai Electrical Machines-I, May/June-2012]

33. For the simple magnetic relay of Fig. 3, the variation of flux linkages ψ in terms of current i and displacement x is given by the relation.

$$\psi = ix^{1/2}.$$

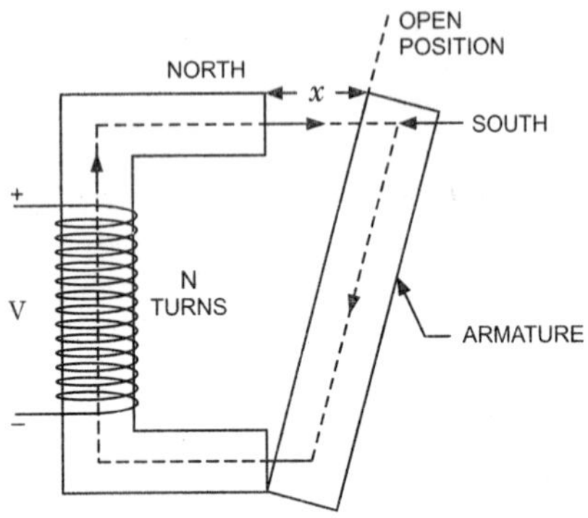

Fig. 3

Develop expression for the magnetic force.
[G.B. Technical Univ. Electromechanical Energy Conversion-I, 2011-12]

34. For a certain relay, the magnetization curves for open and closed positions of the armature are linear. If the armature of the relay moves from open to closed position at constant current (very slowly), show that the mechanical work done is equal to the field energy stored and is half of the electrical energy input.
[U.P. Technical Univ. Electromechanical Energy Conversion-I, 2002-03]

35. Distinguish between singly-excited and multiply-excited magnetic field systems. [J.N. Technological Univ. Electrical Machines-I, March-2014]

36. Derive an expression for the torque in a doubly-excited system having salient pole type of stator as well as rotor. State the assumptions made. [U.P. Technical Univ. Electromechanical Energy Conversion-I, 2006-07]

37. Find an expression for the magnetic force developed in a multiply-excited magnetic systems.
[Anna Univ. Chennai Electrical Machines-I, May/June-2012]

38. Show that the torque developed in a doubly-excited magnetic system is equal to the rate of increase of field energy with respect to displacement at constant current.
[J.N. Technological Univ. Electrical Machines-I, March-2014]

39. Derive torque equation for a single phase smooth cylindrical rotor machine.

40. Derive the expression for torque in a doubly-excited system. Explain the significance of reluctance torques and co-alignment torque in the same. Also derive the expression for torque of a singly-excited system from that of a doubly excited system.
[U.P. Technical Univ. Electromechanical Energy Conversion-I, 2009-10]

41. State and explain the terms (*i*) direct-axis reluctance (*ii*) quadrature-axis reluctance and (*iii*) reluctance torque.

Show that the reluctance motor can develop torque only at one speed.

42. Explain the schematic arrangement of an element reluctance machine. Obtain expression for torque acting on the rotor as a function of current input to the exciting coil and the angle of stator and rotor overlap.
[U.P. Technical Univ; Electromechanical Energy Conversion-I, 2005-06]

43. An electric machine has cylindrical stator and salient pole rotor. Justify the correctness of the following statements:

(*i*) Reluctance torque is produced when exciting winding is on rotor.

(*ii*) When stator and rotor both carry exciting windings, electromagnetic as well as reluctance torques are produced.
[U.P. Technical Univ. Electromechanical Energy Conversion-I, 2005-06]

44. Develop torque equation in an electric machine with cylindrical air gap. [U.P. Technical Univ. Electromechanical Energy Conversion-I, 2011-12]

45. Explain the concept of torques and forces in a system with permanent magnets. [M.D. Univ. Electromechanical Energy Conversion, May-2009]

46. Give schematic representation of singly-excited electromechanical system model. Also write down the dynamical equation in differential form for the overall system.
[U.P. Technical Univ. Electromechanical Energy Conversion-I, 2005-06]

47. Derive the dynamic equation for electromechanical energy conversion.
[M.D. Univ. Electromechanical Energy Conversion, December-2012]

48. Write short notes on the following:

(*i*) Principle of energy conversion

(*ii*) Energy stored in magnetic field. [G.B. Technical Univ. 2012-13]

(*iii*) Singly and multiply excited magnetic field systems.

(*iv*) Reluctance and interaction torques.

SHORT ANSWER TYPE QUESTIONS WITH ANSWERS

Q. 1. What are the advantages of analysing energy conversion devices by field energy concept ?
[U.P. Technical Univ. Electromechanical Energy Conversion-I, 2008-09]

Ans. The analysis of energy conversion devices by field energy concept is more general and broad based, as it is applicable to all devices having linear, rotational or vibratory motions. Its other advantages are:

1. This approach leads physically to the generalized theory of electrical machines.
2. It can deal with both steady-state and transient analysis of all electromechanical energy conversion devices.
3. It provides more physical insight into the operation of all such devices, which is indispensable.
4. This conventional approach can be introduced at any stage, to study the effect of saturation, commutation etc.

Q. 2. Why are magnetic field systems employed in practical electromechanical energy conversion devices rather than electric field systems? [Rajasthan Technical Univ. 2011]

Ans. The energy storing capacity of magnetic field is much larger (about 25,600 times) than that of electric field. In view of this fact, electromechanical energy conversion devices with magnetic field as the coupling medium between the electrical and mechanical systems are more common.

Q. 3. What is energy balance equation?

Ans. According to the principle of energy conservation the energy input to the electromechanical energy conversion device is equal to the summation of the useful output energy, the energy converted into heat, and the change in the energy stored in the magnetic/electric field i.e.

Energy input from electrical source = Mechanical energy output + increase in energy stored in the magnetic/electric field + energy converted into heat

Q. 4. Predominant energy storage occurs in the air gap of an electromechanical energy conversion device. Is this statement correct ? Give reason in support of your answer.

Ans. The statement is correct. This is because the reluctance of air gap is much larger than that of the magnetic material.

Q. 5. For an electromechanical energy conversion process, reaction of coupling magnetic field on electrical or mechanical system is essential. **[Pb. Technical Univ. 2001]**

Ans. In case the output is mechanical, as in a motoring mode, the coupling field must react with the electrical system as to absorb electrical energy from electrical system. In such a case i.e. in a motoring mode this reaction is counter emf *e*, as shown in Fig. 3.12 (*a*), similar to the emf induced in the primary winding of a transformer. The coupling field extracts energy proportional to *ei* from the electrical system, converts it into mechanical form and delivers energy proportional to Tω (or Fv) to the mechanical system.

In case the output is electrical, as in a generating mode, the coupling field must react with mechanical system so as to absorb mechanical energy from mechanical system. In such a case i.e. in a generating mode, this reaction is the counter-torque *T* (opposite to the applied mechanical torque of the prime mover) developed due to interaction of current *i* and the coupling field. Thus the coupling field extracts mechanical energy proportional to *T*ω from the mechanical system, converts it into electrical form and delivers energy proportional to *ei* to the electrical system.

Q. 6. What do you mean by coenergy? **[Anna Univ. May/June-2012]**

Ans. Coenergy has no physical significance. It is given as input electrical energy less the stored energy in the field.

i.e. Coenergy, W'_{field} = Input electrical energy – stored energy in the field, W_{field}.

Q. 7. Give the expression for the energy density of magnetic field.

Ans. Neglecting losses and assuming permeability to be constant, the energy density of the magnetic field is given as

$$w_{field} = \frac{1}{2}\frac{B^2}{\mu} \text{ J/m}^3$$

where *B* is the magnetic field density in Wb/m^2.

Q. 8. When the armature of an electromagnetic relay moves to close the air gap instantaneously, where does the energy come from for doing mechanical work?

Ans. In instantaneous movement of armature to close the air-gap flux linkages ψ remains almost constant and there is no electrical input. The energy required in doing mechanical work comes from the field energy which is reduced by an equal amount.

Q. 9. When the armature of an electromagnet relay moves to close the air gap slowly, where does the energy come from for doing mechanical work?

Ans. In slow movement of armature to close the air gap, current *i* stays substantially constant, energy is drawn from electrical source, half of the input energy is utilised in doing mechanical work and rest is stored in the magnetic field.

Q. 10. Give the examples of singly-excited and doubly-excited electromechanical energy conversion devices.

Ans. **Singly-excited electromechanical energy conversion devices:** Electromagnets, relays, moving iron instruments, reluctance motors etc.

Doubly-excited electromechanical energy conversion devices: DC shunt machines, synchronous machines, loudspeakers, tachometers etc.

Q. 11. How is the reluctance torque produced in a rotating machine? **[Pb. Technical Univ. 2001, 2002]**

Ans. In rotating electrical machines, when the reluctance seen by the working flux varies with the movement of rotor, the reluctance torque is developed.

Q. 12. Give expressions for energy stored in a capacitor.

Ans. Energy stored in a capacitor is given by expressions.

$$W_{field} = \frac{1}{2}q^2/C = \frac{1}{2}CV^2 = \frac{1}{2}qV.$$

Q. 13. Give an expression for the energy density in an electric field.

Ans. Energy density in electric field is given as $w_{field} = \frac{1}{2}\frac{D^2}{\varepsilon_o}$

where D is the electric flux density in C/m^2.

Q. 14. Why energy storing capacity of magnetic field is much larger than that of electric field ?

Ans. Because the value of permittivity of free space is very small compared to permeability of free space.

PROBLEMS

1. For the electromagnetic relay shown in Fig. 3.24, calculate the maximum force on the armature if saturation flux density in the iron part is 1.8 T. Number of coil turns = 1,200 and cross-sectional area of the core = 20 cm^2 **[Ans. – 5,156.6 N]**

2. For the electromagnet shown in Fig. 1, obtain the expression for the magnetic force. Neglect leakage, fringing and reluctance of the magnetic core.

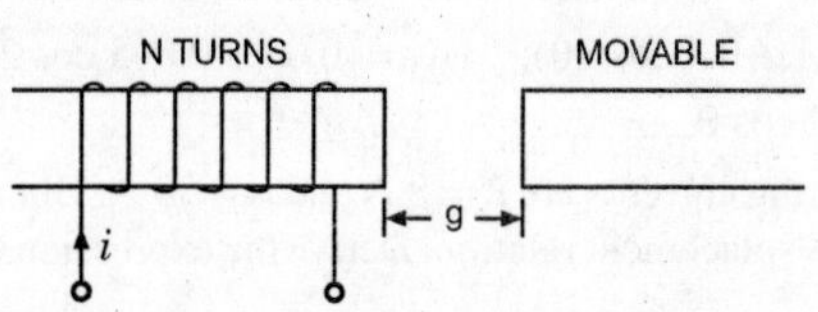

Fig. 1

[**Ans.** $F_{field} = \frac{B_g^2 a}{2\mu_0}$ where *a* is the area of x-section of the electro-magnet]

3. A coil and plunger magnet is shown in Fig. 2. The coil has 1,000 turns and carries a current of 4.5 A.

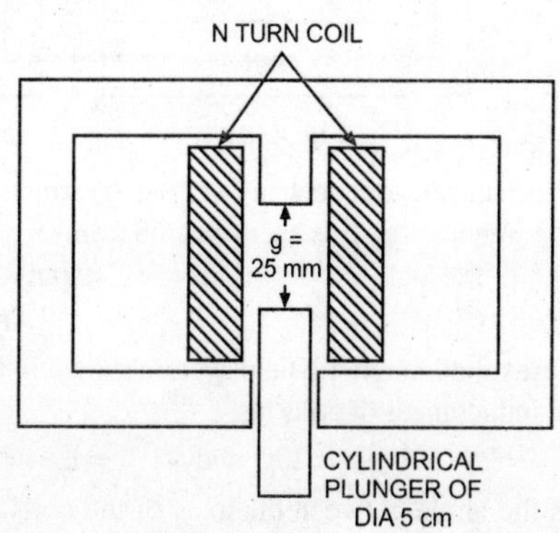

Fig. 2

Determine (*i*) flux density and coil inductance for gap length of 15 mm (*ii*) energy stored in the magnetic field and electromagnetic force for the gap length of 10 mm.

Magnetic leakage, fringing and magnetic core reluctance may be neglected. [**Ans.** (*i*) 0.5655 T; 0.2475 H (*ii*) 1.67 W; 185.625 N]

4. For the coil and plunger magnet shown in Fig. 2 determine the mechanical work done, if the plunger is allowed to move slowly from $g = 25$ mm to $g = 10$ mm.

Find the electrical energy supplied by the source for the above condition.

Assume N = 1,000 and coil current $i = 4.5$ A [**Ans.** 2.5 W; 5 W]

5. For the magnetic circuit shown in Fig. 3, length of iron path is 1.0 m. Air gap length is 4 mm and area of x-section of iron = 50 × 50 mm^2; Relative permeability of iron μ_r = 1,000. Exciting current, I = 2.5 A. Number of turns on the coil, N = 1,200. Calculate and compare the field energy stored and field energy density in iron as well as in air gap. Fringing and leakage flux are negligible.

[**Ans:** 4.0 times; 1,000 times]

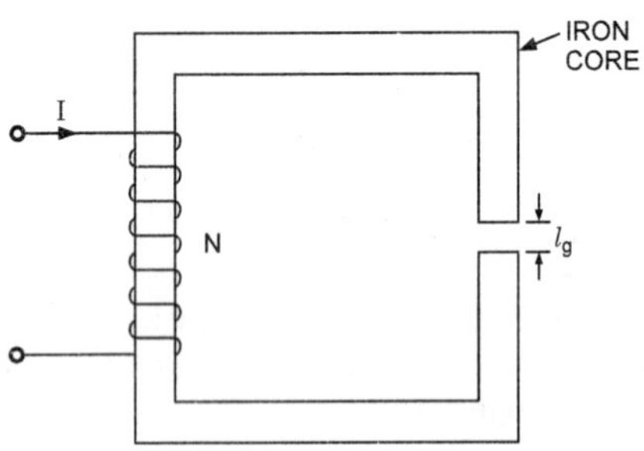

Fig. 3

6. For the relay shown in Fig. 4 (*a*) the air gap x = 0.05 m. The magnetisation curve of the relay for open and closed positions is given in Fig. 4 (*b*). Calculate

(*i*) The mechanical work done in the armature moving from open to closed position, if the relay armature moves very fast. Where does this energy come from?

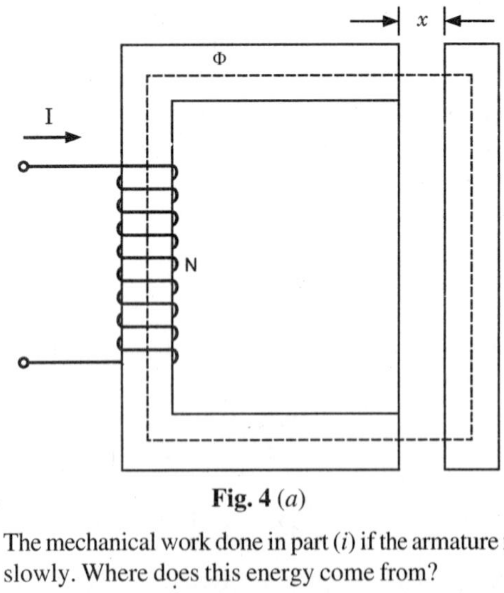

Fig. 4 (*a*)

(*ii*) The mechanical work done in part (*i*) if the armature moves very slowly. Where does this energy come from?

[R.G.P.V. M.P., June-2003]

[**Ans:** 2.7, 4.32J]

7. The doubly-excited magnetic field system shown in Fig. 5 has self- and mutual inductances of coils as

$$L_{11} = L_{22} = 1 + \cos 2\theta \quad \text{and} \quad L_{12} = L_{21} = \cos\theta$$

where θ is the angle between the axes of the coils. The coils are

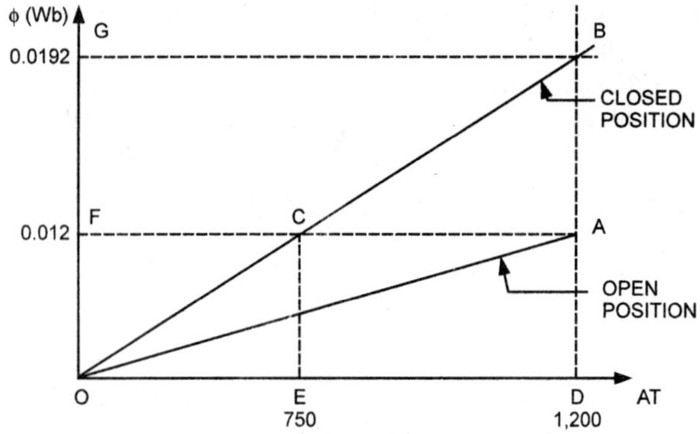

Fig. 4 (*b*)

connected in series and carries a current of I sin ω *t*.

Obtain an expression for the time average torque as a function of angle θ.

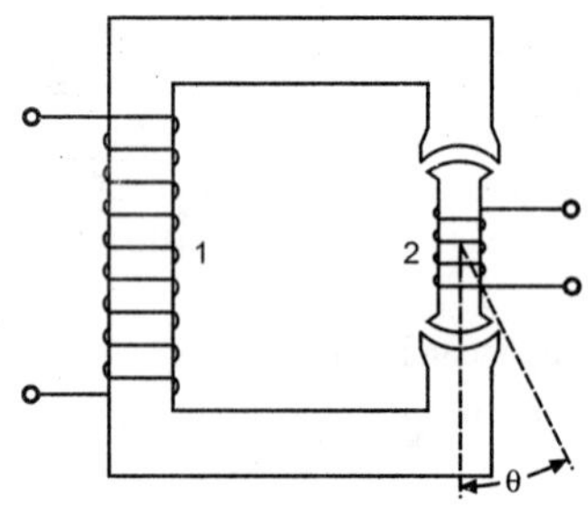

Fig. 5

$$\left[\textbf{Ans. } T_{\text{field } av} = -\frac{I^2}{2}(2\sin 2\theta + \sin\theta)\right]$$

8. An electromechanical device has the following stator and rotor coil inductances

$$L_{11} = L_{22} = 1.6 + \cos 2\theta \quad \text{and} \quad L_{12} = L_{21} = 1 + 0.4\cos\theta$$

Determine the torque developed as function of angle θ when both coils are connected in parallel to a voltage source 160 sin 314 *t* volts. Find the time average torque for θ = 30°. [**Ans.** – 0.00892 Nm]

9. Self- and mutual inductances of two coupled coils are:

$$L_{11} = 3 + \frac{1}{2x}, \quad L_{22} = 2 + \frac{1}{2x}, \quad L_{12} = L_{21} = \frac{1}{2x}$$

where x is a certain linear displacement. For constant currents of 10 and – 5 A determine

(*i*) time average force at x = 0.5 m.

(*ii*) the mechanical work done in increasing x from 0.5 m to 1 m.

(*iii*) the electrical energy supplied by each electrical source in part (*ii*).

(*iv*) the change in magnetic field energy in part (*ii*)

[**Ans.** (*i*) 25 N (*ii*) – 6.25 J (*iii*) – 25 J, 12.5 J, 12.5 J (*iv*) 6.25 J]

10. In a doubly-excited rotary machine, the inductance coefficients are :

$L_{11} = (1.1 + 0.4\cos 2\theta)$; $L_{22} = (0.03 + 0.005\cos 2\theta)$;

$L_{12} = 0.2\cos\theta$

The exciting currents are i_1 = 8 A and i_2 = 50 A. Obtain the torque/angular displacement relation. Derive the expression used, if any.

[I.A.S. 1994]

[**Ans.** – 38.1 sin 2θ – 80 sin θ Nm]

❑❑❑

CHAPTER 4

Basic Concepts of Rotating Electrical Machines

INSIDE THIS CHAPTER

4.1 EMF GENERATION IN ROTATING ELECTRICAL MACHINES

A very important effect of magnetic field on an electric circuit is inducement of emf on variation of flux linking a circuit. As already explained in Art 3.2 whenever flux is cut by the conductor, *motion* or *speed* emf is induced in it and if the circuit is closed, flow of electric current will be caused. For production of motion or speed emf, three things are necessary – a magnetic field, conductor and motion of conductor with respect to field. Cyclic change in flux linkage of a coil results in generation of alternating emf. The field windings are provided to give cyclic north-south space distribution of poles. Armature winding, which is an arrangement* of conductors, is provided to develop desired emfs in a heteropolar** magnetic field. A cylindrical structure is natural choice for such a machine.

Since it is immaterial for generation of an induced emf whether a conductor moves across a magnetic field or vice versa, the rotating machines may be constructed with either the armature or field structure as the revolving member. In all medium and large size synchronous machines, the armature winding is housed in suitably shaped slots cut in the stator while the field winding is provided on the revolving member called the *rotor*. The advantages of such an arrangement will be discussed later on (refer to Art 15.1.2). The field winding is excited from an external dc supply source through a pair of slip rings. However, small rating synchronous generators can be built with armature as rotating member because the voltage generated is comparatively low and the current to be collected by the brushes is small.

4.2 PHYSICAL CONCEPT OF PRODUCTION OF TORQUE

4.2.1. Torque Due to Alignments of Two Magnetic Fields and The Concept of Torque Angle. If a piece of soft iron , free to rotate, is placed between two opposite magnetic poles, as shown in Fig. 4.1, the piece gets magnetised due to magnetic induction. The magnetic lines of force pass through the soft iron piece, the molecules of the soft iron line up parallel with the magnetic lines of force and with their N-pole pointing in the direction that the lines of force are travelling through the iron piece, as illustrated in Fig. 4.1. A force of attraction or repulsion, which has a tendency to align both the fields, would act on the soft iron piece depending on the polarity of the pole formed on the soft iron piece with respect to the main field. This may result in the rotation of the iron piece, depending upon its relative position with respect to the main field. The angle between the axis of the main magnetic field and the magnetic axis of the soft iron piece (or rotor winding) is called the *torque angle*. This angle is measured with respect to the direction of rotation of the soft iron piece (or rotor).

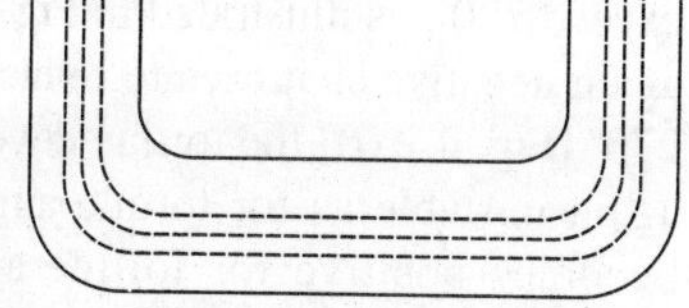

Fig. 4.1

In Fig. 4.1, the two fields are aligning in the same direction and, therefore, no torque is experienced on the

* In practical machines group of conductors are distributed in different ways in slots all over the periphery of the armature. The conductors may be connected in series/parallel and in star/delta combinations depending upon the current and voltage rating of the machine.

** Field system consisting of a number of poles arranged alternately N and S–commonly employed field system. Machines employing such a field system is called the *heteropolar machine*.

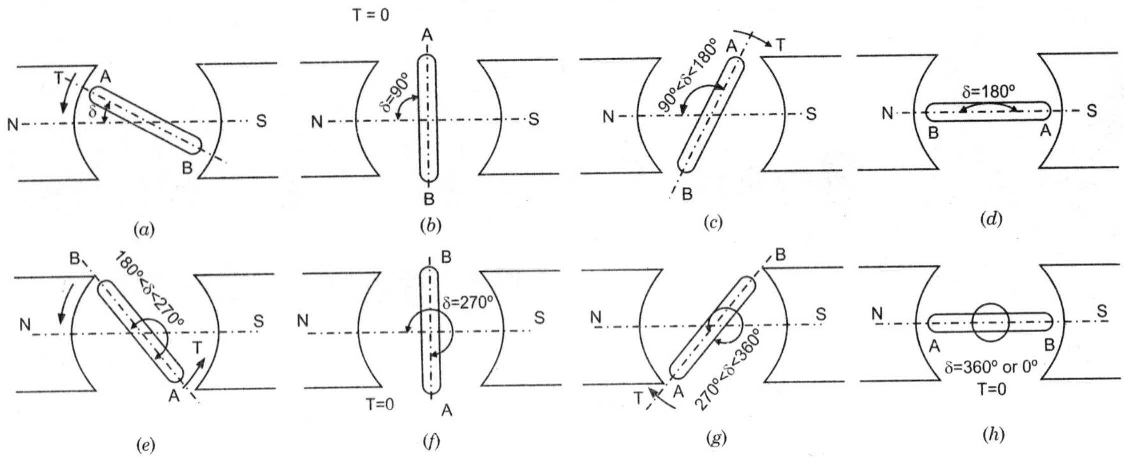

Fig. 4.2

soft iron piece. However, there is only force of attraction at ends A and B of soft iron piece. The forces acting at the two ends A and B being equal in magnitude and opposite in direction cancel each other. In this case torque angle is zero and torque acting on the iron piece is also zero.

The different positions of soft iron piece with respect to main field are shown in Fig. 4.2.

In Fig. 4.2 (*a*), the magnetic axis of the soft iron piece makes an angle δ (0 < δ < 90°) with the axis of the main magnetic field and in this position of the soft iron piece, it tends to align itself in the direction of the field owing to the force of attraction at its both ends. This torque is considered to be negative as it tries to reduce the torque angle. Thus, for any value of δ less than 90°, the torque is negative.

When torque angle becomes 90°, as illustrated in Fig. 4.2 (*b*), the torque developed is zero. This is unstable position of the soft iron piece as slight change in δ in either direction will develop the torque in that direction.

When torque angle δ lies between 90° and 180°, as illustrated in Fig. 4.2 (*c*), the ends of soft iron piece A and B (now north and south poles) are attracted toward south and north poles of the main magnet respectively and thus torque developed tends to increase the torque angle. So the torque developed is taken as positive.

When the torque angle δ becomes 180°, as shown in Fig. 4.2 (*d*), the torque again becomes zero. But this is a stable position as any change in the value of torque angle will create a torque causing restoration of the position of the iron piece.

When torque angle δ becomes more than 180° but does not exceed 270°, as illustrated in Fig. 4.2 (*e*), the torque produced is again negative as it tries to reduce the torque angle. For δ to be 270° [Fig. 4.2 (*f*)] the torque developed is zero. This position is again unstable as for torque angle δ = 90°. The torque again becomes positive for torque angle between 270° and 360° [Fig. 4.2 (*g*)]. When δ becomes 360° *i.e.* 0°, the torque developed is zero and the position is stable.

From the above discussion it may be concluded that the torque developed is the function of torque angle δ. The torque becomes alternately negative and positive, as illustrated in Fig. 4.3, and correspondingly the iron piece cannot respond to torque variations and, therefore, it remains standstill. The torque so produced is called the *reluctance torque*. This principle of developing torque is used in the reluctance motors.

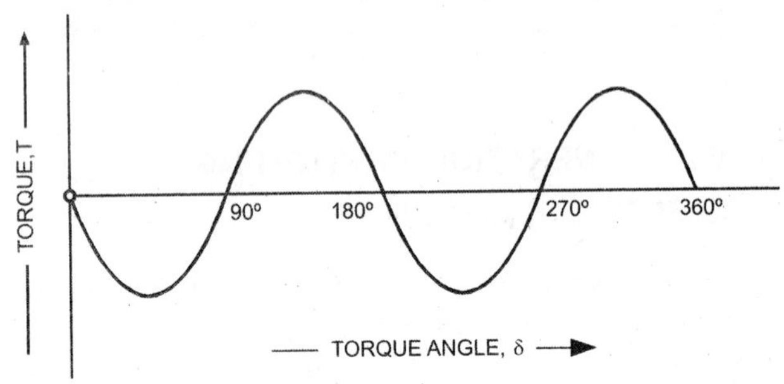

Fig. 4.3

4.2.2. Electromagnetic (or Interaction Torque). Torque considerations in machines may be based on simple formulation of the fact that a mechanical force is exerted on a current carrying conductor placed in a magnetic field. A two-pole cylindrical rotor with one conductor is shown in Fig. 4.4. Only the stator magnetic flux path due to field excitation of field coils has been shown in Fig. 4.4(*a*) and there is no current in rotor conductor. Figure 4.4(*b*) illustrates the distribution of flux due to current only in rotor conductor and there is no excitation in stator field coils. Figure 4.4(*c*) illustrates the resultant flux distribution with both field coils and rotor conductors excited. Magnetic lines like rubber bands have a tendency to strengthen out and, therefore, a force is experienced on the conductor in the downward direction, as illustrated in Fig. 4.4 (*c*).

Now consider one current carrying coil on the rotor. The direction of the current in the coil is in inward direction indicated by cross under stator north pole and in outward direction indicated by dot under south pole, as illustrated in Fig. 4.4 (*d*). The magnetic flux created by the coil interacts with the flux created by the stator coils and provides the resultant magnetic flux distribution, as illustrated in Fig. 4.4 (*d*). The torque so developed is counter-clockwise and called the *interaction* or *electromagnetic torque*.

The development of *electromagnetic* (or interaction) torque may be physically understood by reference to Fig. 4.5 (*a*), in which one permanent magnet, free to rotate, is placed in the field of a stationary magnet. The tendency of the two fields to align themselves in the same direction is called the *interaction torque*.

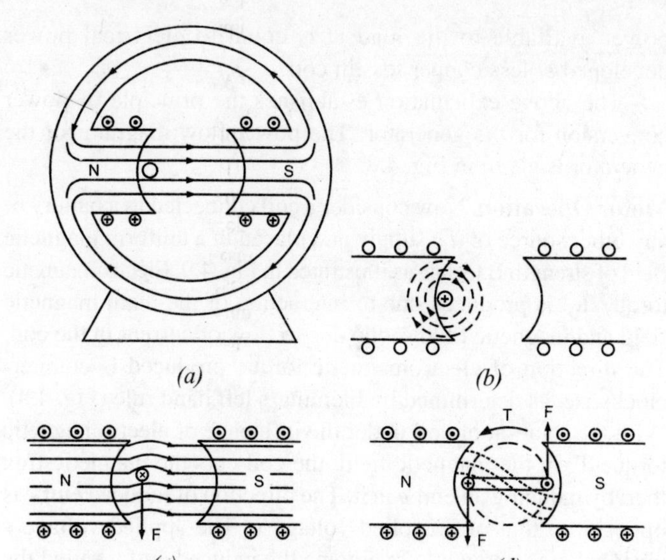

Fig. 4.4

The angle between stator field axis and rotor field axis is called the *torque angle*, δ.

The different position of permanent magnet with respect to main field are shown in Fig. 4.5.

When torque angle δ is less than 90°, the torque developed due to alignment of the two magnetic fields is negative as it tends to reduce the torque angle. When δ is 90° the torque developed is maximum. For torque angle δ exceeding 90° but not exceeding 180° as shown in Fig. 4.5 (*b*), the fields tend to align due to production of repulsive force between similar poles and as such δ tends to reduce and the torque developed is, therefore, considered negative. For δ = 180° [Fig. 4.5 (*c*)], the fields are aligned and so the torque produced is zero. When the

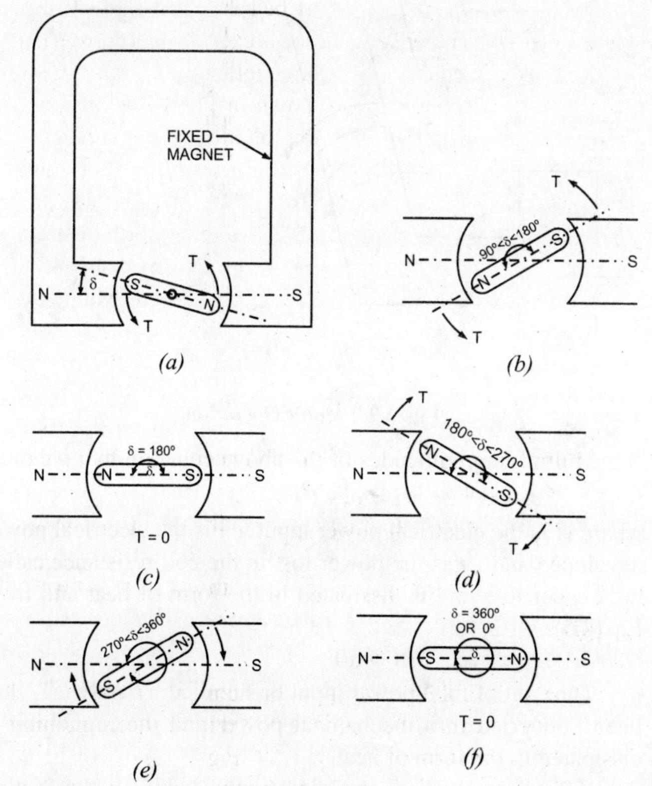

Fig. 4.5

torque angle δ becomes more than 180° but does not exceed 270° [Fig. 4.5 (*d*)] the torque developed is positive as it tends to increase torque angle δ. When torque angle δ varies between 270° and 360° [Fig. 4.5 (*e*)], the torque developed is again positive. However, it remains zero for δ = 360° or 0° [Fig. 4.5 (*f*)].

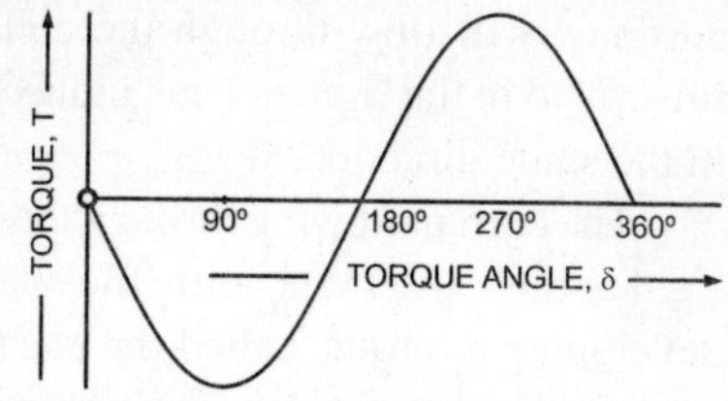

Fig. 4.6

From the above discussion it is revealed that the torque is negative for the first half rotation, positive for the next half rotation and is maximum for torque angles of 90° and 270°, as shown in Fig. 4.6. The magnitude of electromagnetic or interaction torque in all rotating machines is given as

$$T_e \propto \text{Stator field strength} \times \text{rotor field strength} \times \sin \delta \quad ...(4.1)$$

4.3 ELEMENTARY CONCEPT OF AN ELECTRICAL MACHINE

Generator Operation. Consider a coil being rotated in clockwise direction with angular velocity ω radians per second under the influence of driving torque T_d in a uniform magnetic field of flux density B teslas, as illustrated in Fig. 4.7. An emf is induced in the coil due to change in flux linkages (or due to cutting of magnetic lines of force). This induced emf will be of magnitude $e = 2Blv \sin(90° - \delta)$ or $2Blv \cos \delta$ and its direction can be determined by Lenz's law or Fleming's right hand rule, as already discussed. From the Fig. 4.7, it is obvious that the coil sides are cutting the flux causing the flux linkages to reduce and, therefore, emf is induced in the coil in the direction indicated in the figure. Since the coil circuit is closed through load resistance R_L, a

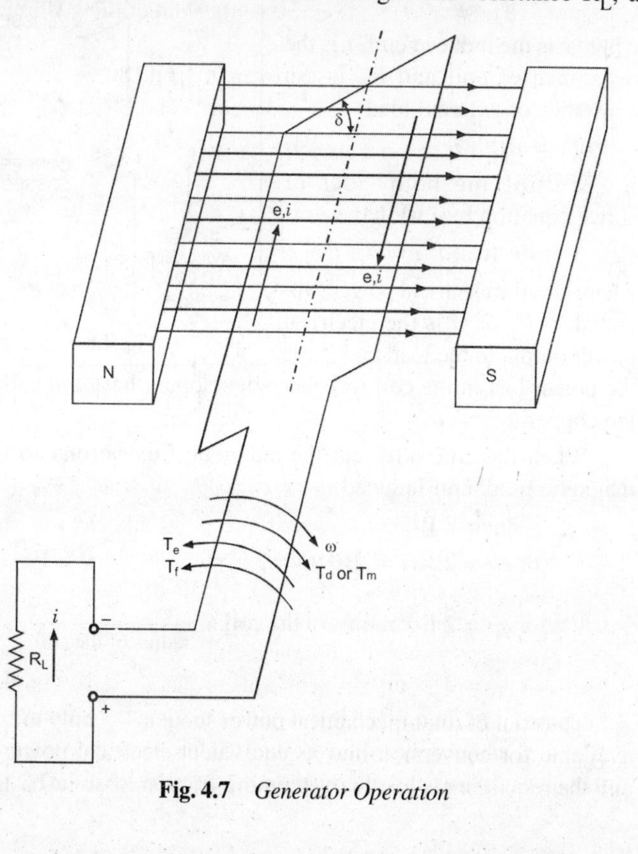

Fig. 4.7 *Generator Operation*

current i will flow through the coil and load resistance R_L, as illustrated in the figure. Thus induced emf e and the current i are in the same direction.

The current flowing through the coil creates its own magnetic field which interacts with the main magnetic field thereby developing a torque, called the *electromagnetic torque*, T_e. The magnitude of this torque will be proportional to F or $BIl \sin\delta$ where δ is the torque angle.

As obvious from the figure, the torque angle is leading and as such the torque developed opposes the rotation of the machine. The direction of the torque developed can also be determined by Fleming's left hand rule.

With coil circuit open (*i.e.* R_L being removed), the current flowing through the coil is zero and, therefore, electromagnetic torque produced $T_e = 0$. The driving torque T_d (or mechanical torque T_m) is required to be sufficient to overcome the friction torque T_f and no-load loss (being ignored for the time being). As the direction of friction is opposite to that of motion of the coil and as such the friction torque opposes the driving or mechanical torque. When the coil circuit is closed through load resistance R_L current i flows through the coil developing an electromagnetic torque T_e and the driving or mechanical torque must, therefore, be enhanced to overcome the developed electromagnetic torque T_e in addition to friction torque T_f.

$$i.e.\ T_d = T_e + T_f \qquad ...(4.2)$$

Neglecting friction torque (assuming friction torque negligibly small in comparison to electromagnetic torque) we have $T_d = T_e$.

The current flowing through the coil.

$$i = \frac{e}{R_L + r} \qquad ...(4.3)$$

where e is the induced emf, r is the resistance of coil and R_L is the resistance of external load

$$\text{or } e = iR_L + ir \qquad ...(4.4)$$

Multiplying both sides of above equation by i we have

$$e\,i = i^2 R_L + i^2 r = vi + i^2 r$$

where ei is the electrical power generated, $i^2 R_L$ or vi is the electrical power output to the load and $i^2 r$ is the power lost in the coil resistance developing heat and called the copper loss.

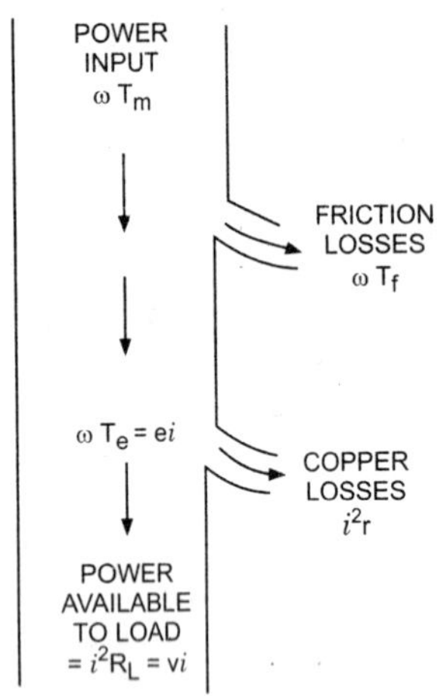

Fig. 4.8

When the coil sides cut the magnetic flux normal to the magnetic field, emf induced

$$e = 2\,Blv$$

$$\text{or } e\,i = 2Blvi = 2Bil \times v = 2\,F \times v \qquad \because F = Bil$$

$$\text{or } e\,i = 2\,F \times \text{radius of the coil} \times \frac{v}{\text{radius of the coil}}$$

$$= T_e \times \omega \qquad ...(4.5)$$

Thus out of total mechanical power input ωT_m, only ωT_e is available for conversion into its equivalent electrical power ei and the remaining is lost in overcoming friction losses ωT_f. The power available to the load vi is equal to electrical power developed ei less copper loss in coil.

The above explanation establishes the principle of power conversion for the generator. The power flow diagram for the generator is given in Fig. 4.8.

Motor Operation. Now consider a coil connected to a battery or any other source of $d\,c$ supply and placed in a uniform magnetic field of strength B teslas, as illustrated in Fig. 4.9. Electromagnetic torque T_e is produced due to interaction of the main magnetic field and magnetic field set up due to flow of current in the coil. The direction of electromagnetic torque produced is counter-clockwise, as determined by Fleming's left hand rule (Fig. 4.9). With the rotation of coil under the influence of electromagnetic torque T_e in the magnetic field, the coil cuts the magnetic flux thereby inducing an emf e in it. The direction of induced emf e is opposite to that of a applied voltage v. The applied voltage v must be large enough to overcome the induced emf e called the *counter or back emf* and voltage drop in the coil resistance $i\,r$ *i.e.*

$$v = e + ir \qquad ...(4.6)$$

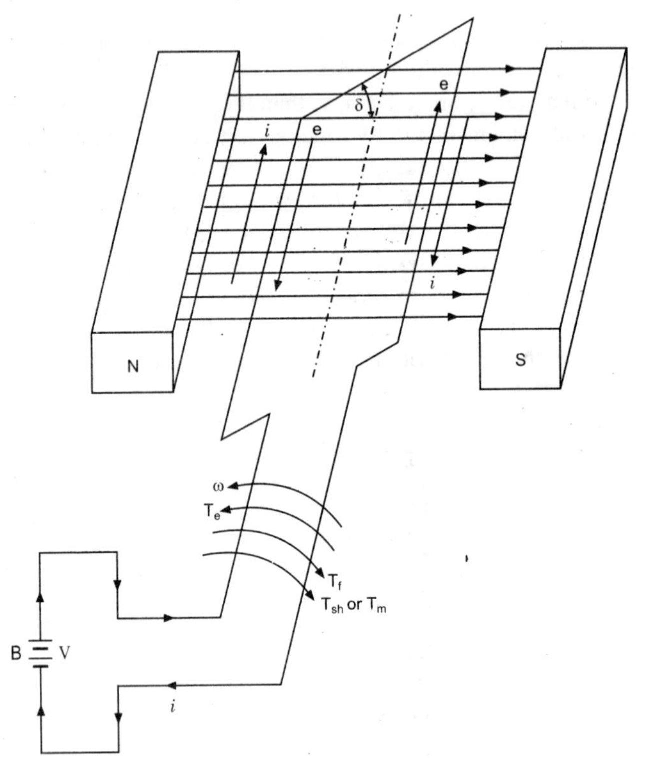

Fig. 4.9 *Motor Operation*

Multiplying both sides of the above equation by i we have

$$v\,i = ei + i^2 r \qquad ...(4.7)$$

where vi is the electrical power input, $e\,i$ is the electrical power developed and $i^2 r$ is the power lost in the coil resistance called the copper loss and is dissipated in the form of heat and from Eq. (4.5)

$$ei = \omega T_e$$

Thus out of total power input or supplied $v\,i$, only ωT_e has been converted into mechanical power and the remaining is dissipated in the form of heat.

If T_m is mechanical opposing torque and T_f is the friction torque then

$$T_e = T_m + T_f \quad ...(4.8)$$

$$\text{or } \omega T_e = \omega T_m + \omega T_f \quad ...(4.9)$$

i.e. electrical power converted into mechanical power (ωT_e) is equal to the sum of the power available at the shaft for driving the load and the power required to overcome the friction loss. Sometimes the mechanical torque T_m is also called the *shaft torque* T_{sh}. The power flow diagram is given in Fig. 4.10.

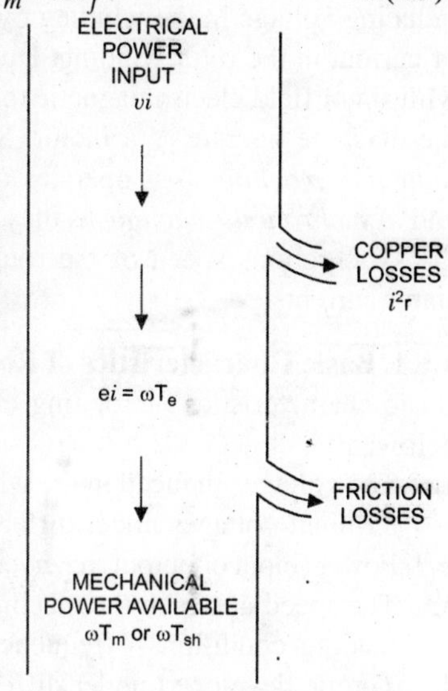

Fig. 4.10

The essential difference between the generator and motor operation is given below with reference to Figs. 4.7 and 4.9.

1. In generator operation induced emf *e* and current *i* are in the same direction whereas in motor operation induced emf *e* and current *i* are in opposition to each other.
2. In case of generator, electromagnetic torque developed opposes the rotation *i.e.* T_e and ω are in opposite directions whereas in a motor T_e and ω are in the same direction.
3. In case of a generator, electromagnetic torque developed T_e and friction torque T_f are in the same direction whereas in a motor T_e and T_f are in opposite directions.

4.4 COMMON FEATURES OF ROTATING ELECTRICAL MACHINES

All the rotating electrical machines employed for generation of electrical energy, electric drives and for control systems have many common essential features from the point of view of construction. The essential features of such machines are illustrated in Fig. 4.11.

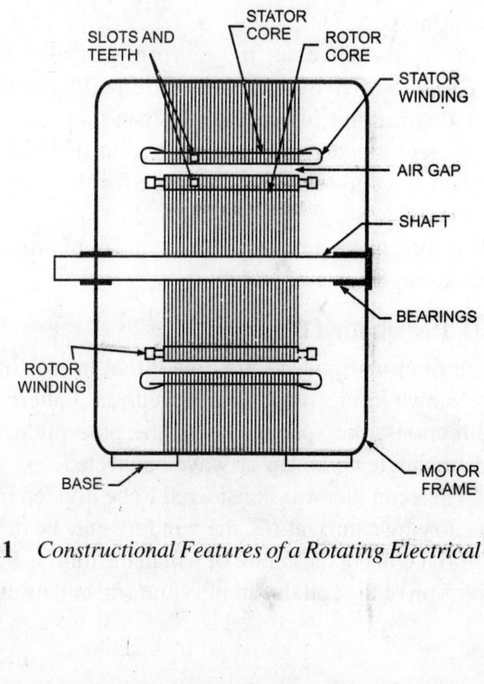

Fig. 4.11 *Constructional Features of a Rotating Electrical Machine*

1. Every rotating electrical machine has an outer stationary member called the *stator* and inner rotating member, called the *rotor*. The latter is mounted in bearings which are fixed on the stationary member.
2. Both the stator and rotor carry concentric cylindrical cores. These are made of ferromagnetic material and are usually fabricated from laminated steel sheets in order to provide optimum electric and magnetic characteristics.
3. There is a clearance between the two cylindrical ferromagnetic cores, called the *air gap*.
4. Every rotating electrical machine usually has exciting or field windings to creat the working flux and armature windings in which the working emf is induced by the working flux. The current in the field or excitation windings does not vary with the load on the machine. The current flowing through the field winding is known as field, magnetising or exciting current. The power rating of field windings is 0.5 to 2% of the rated power of the machine. In both dc and ac machines, the field windings are energized from dc source.
5. A common magnetic flux Φ crosses the air gap from one core to the other in a combined magnetic circuit. This magnetic flux Φ, in crossing the air gap, provides alternate N and S poles on both stator and rotor. Such a magnetic system with alternate N and S poles along with its air gap periphery is called the *heteropolar system*.
6. The important feature of all the rotating electrical machines is that there is a relative motion between the two cylindrical ferromagnetic surfaces for electromechanical energy conversion.
7. The current flowing through armature windings of both the dc and ac machines is ac and varies with the load. Since the armature winding handles all the power, the rating of the armature winding is equal to the power rating of the machine. Since in both dc and ac machines, the armature windings deal with ac only, armature structure of all rotating electrical machines are laminated to reduce eddy current losses. To reduce the hysteresis loss the core material is selected so as to have small hysteresis loop.
8. All types of large rotating electrical machines are provided with the radial and axial ventilating ducts for cooling.

The cylindrical surface on either element (stator or rotor) may be continuous (neglecting slot openings) or may have salient poles. Near the air-gap surface of each or either element, conductors run parallel to the axis of cylinders. These conductors are suitably connected to give windings, called the *rotor/stator windings*.

If power is supplied to / taken out from the rotor, sliding contacts are essential obviously.

Thus the general arrangement is the same for all types of rotating electrical machines. The basic difference between different types of rotating electrical machines lies in the distribution of the conductors forming windings and in whether the two members have continuous cores or salient poles. The operation of any given machine depends on the nature of excitation given to its windings.

4.5 TYPES OF ROTATING ELECTRICAL MACHINES

As already mentioned above, in rotating electrical machines there are two major parts such as stator and rotor. The rotating electrical machines are of two types viz. dc machines and ac machines. In dc machines stator is used as field and the rotor is used as an armature. AC machines, on the basis of principle of operation, are of two types viz. synchronous machines and asynchronous (or non-synchronous) machines. AC machines, on the basis of number of phases employed, are also of two types viz. single-phase and three-phase machines. The rotating electrical machines may be classified as follows:

- Rotating Electrical Machines
  - DC Machines
  - AC Machines
    - Synchronous
      - Single Phase
      - Three Phase
    - Asynchronous
      - Single Phase
      - Three Phase

1. DC Machines. In a dc machine, the field winding is provided on the stationary part and provides a field stationary in space in the air gap as it is energized from dc supply source. The armature (or rotor) has commutator type winding with stationary brushes for connection to external (load/supply) circuit (load circuit in case of generator and supply circuit in case of motor).

2. Synchronous Machines. Synchronous machines are of two types namely synchronous generators and synchronous motors. In synchronous machines, synchronous generators may be constructed with either the armature or field structure as the revolving member. Practically all medium and large size synchronous machines (both generators as well as motors) are always constructed with revolving field. The field winding provided on the rotor is energized from dc supply source and therefore develops a magnetic field moving in space at the speed of the rotation of the rotor. An emf will be induced in the three-phase winding provided on the stator when the rotor rotates. The magnitude of emf induced in the 3-phase winding due to rotation of rotor depends on the strength of dc field and the frequency corresponds to the speed of rotation of rotor in case of generator. In case of a three-phase synchronous motor, three-phase winding is energized from 3-phase supply source. This develops a rotating field in the air gap. The magnetic field developed by the 3-phase winding provided on the stator interacts with the dc field produced and makes the machine run as a motor at synchronous speed.

3. Induction (or Asynchronous) Machines. Like other electrical machines, the asynchronous machine is reversible, i.e., it can operate as both a motor and a generator. The mode of the machine is determined by the speed of the rotating field in relation to the rotor. The distinguishing feature of a such a motor is that it is a singly-excited machine, although such machines are provided with both field (rotor) and armature (stator) windings. In such a machine, the three-phase winding provided on the stator, when connected to three-phase supply, develops a rotating magnetic field which is cut by the closed circuit of the rotor, thereby inducing voltage in the winding by induction. This causes flow of current in the rotor winding. Due to interaction of rotor field with stator field electromagnetic torque is developed and makes the machine operate as a motor. Such a machine is called the *induction machine* as it operates on the principle of induction and *asynchronous machine* as they operate at a speed other than the synchronous speed of the rotating field developed by the stator currents.

4.5.1. Basic Characteristics of Rotating Electrical Machines. Basic characteristics of rotating electrical machines are given below:

1. The voltages induced in the windings, the load currents and terminal voltages under different loading conditions.
2. Power (input or output) received or delivered by the machine.
3. The speed at which the machine is running under different loading conditions—frequency.
4. Torque developed under different loading conditions.

Interrelations between the above quantities for a machine operating as a generator or motor can be worked out for different types of electrical machines. These relations decide the suitability of a particular type of machine for a particular application.

4.6 MMF AND FLUX DISTRIBUTION IN DC AND AC MACHINES

It is important to be able to find the value of the flux created by current in distributed windings and to determine the way in which flux density varies over the core circumference.

The air gap, in the case of induction motors, round rotor synchronous machines and many of the commutator machines, is uniform, except for the slotting, and therefore the reluctance of this gap does not vary from point to point around the periphery. But with other machines, such as dc machines and the salient pole synchronous machines, though one side of the gap is plain, the other side has two regions (*i*) the pole arc where the air-gap length is small, and (*ii*) interpolar arc, where the gap may be considered infinite.

Now, in the former case, the uniform-gap machine, the variation of flux density over the armature periphery depends only on the distribution of mmf of the windings. In case of machines of second category, the distribution of mmf is first found and then the due allowance is made for the variation of reluctance.

It is, therefore, necessary to study first the mmf produced by the different arrangements of windings.

4.6.1. MMF Distribution Diagrams.

1. Single Conductors Spaced One Pole Pitch. In each of single conductors, shown in Fig. 4.12, carries a current I alternately in opposite directions, the spacing being the pole pitch τ. This corresponds to single coils, lap or wave connected.

Now if each conductor is considered to be divided into two parts, each carrying a current I/2, the winding may be described as a series of flat coils in the centre of which the mmf is I/2. Over the centre portion of the coil this mmf is uniform and its direction changes from one coil to next. The mmf distribution diagram is thus a series of rectangles, the mmf remaining constant between conductors and changing by the value of the conductor current from one side of the conductor to the other.

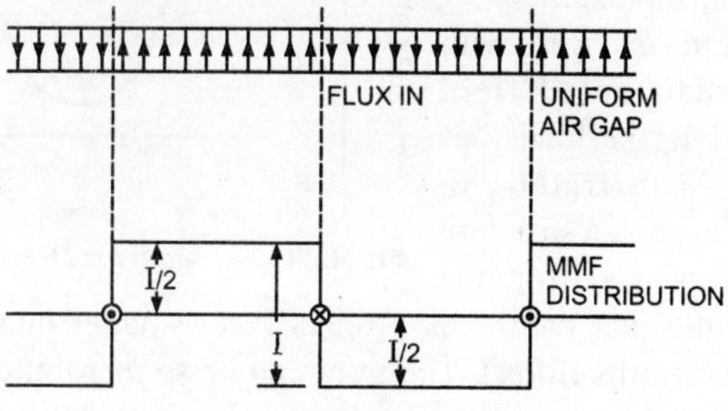

Fig. 4.12

2. Distributed Winding—Isolated Coils. In Fig. 4.13, three sets of conductors similar to those shown in Fig. 4.12 over lapping each other are shown, there being one conductor every 60°. The mmf diagrams similarly overlap, and the total mmf will thus be a stepped wave as illustrated. This may be traced by starting at conductor A which is the centre of the group and moving to the right, noting how the mmf changes at each conductor by the amount of current in it.

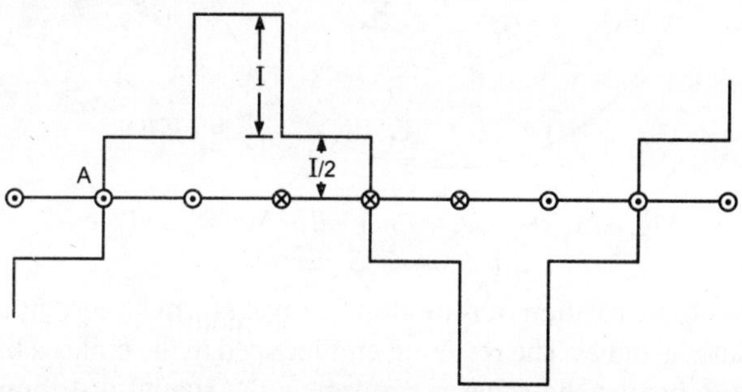

Fig. 4.13

3. Distributed Winding–Uniformly Spread. In case of a large number of coils, instead of the mmf wave being stepped, a smooth transition may be assumed, and the mmf diagram becomes triangular, as illustrated in Fig. 4.14. The maximum value of the mmf—the apex of the triangle will be Σ I/2, Σ I being then sum of currents in the conductors spread over one pole.

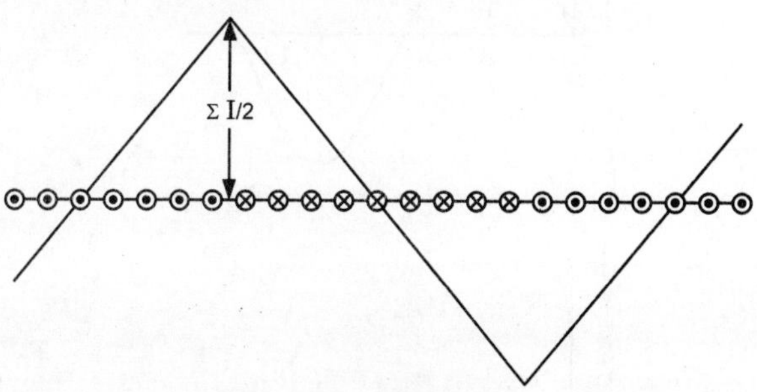

Fig. 4.14

4. Distributed Winding–Limited Spread. If the conductors only occupy a portion of the armature surface, the mmf wave will be trapezoidal, as illustrated in Fig. 4.15. Such mmf wave is found with the rotor winding of a smooth rotor alternator. Another example is due to the instantaneous current in a single phase of an alternator stator winding.

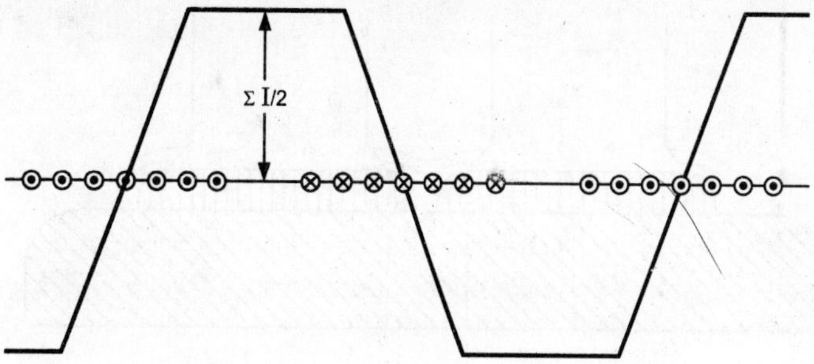

Fig. 4.15

4.6.2. Flux Waveform in DC Machine. The actual flux distribution of a machine depends to a considerable extent upon the shape of the pole shoe, the relative magnitude of the arc subtended by the shoe, and upon the relative value of the air gap and the clearance between the tips of adjacent pole shoes.

In Fig. 4.16 is represented the flux distribution obtained for rectangular poles, without pole shoes, and with a uniform clearance between the faces of the poles and armature. The field mmf is a rectangular distribution, and this, combined with the uniform gap over the pole arc, results in an rectangular flux wave. Fringing occurs at the pole tips making the flux to spread into the interpolar arc, but the flux density there is rapidly reduced so that the flux waveform takes the shape of the dotted line of the diagram. As a result, the flux density in air around the armature is a flat-topped wave.

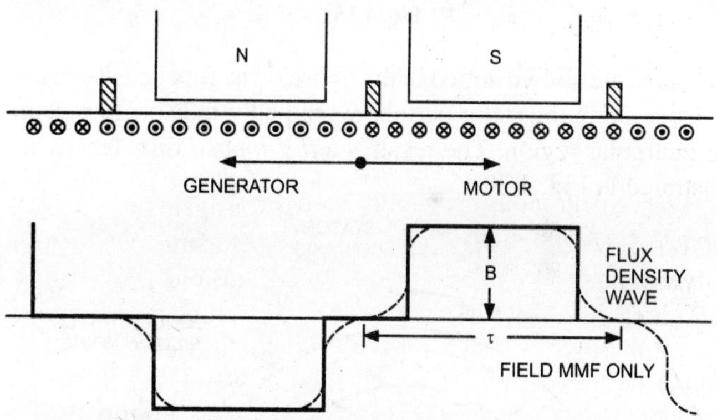

Fig. 4.16 *Flux Distribution Curve For a 2-Pole DC Machine*

4.6.3. Flux Distribution Curve of Synchronous Generator. The flux distribution curve in the air gap of a synchronous generator is not usually so rectangular as in a dc machine (Fig. 4.17), but it is never perfectly sinusoidal unless the machine has a salient pole rotor with specially shaped pole shoes graduating the air gap as illustrated in Fig. 4.18, or a cylindrical rotor with a sinusoidally distributed field winding.

4.7 ELEMENTARY MACHINES

4.7.1. Elementary Synchronous Machines. Elementary salient pole synchronous generator with 2 pole rotor is shown in Fig. 4.19. The magnetic neutral regions are located in the interpolar gaps. Without any significant loss of accuracy, the reluctance of the iron path can be neglected. Assuming the air gap over the pole shoes uniform, the flux density in the air gap over the pole shoes is constant, as mmf acting along any flux path is constant for the concentrated field winding (magnetic

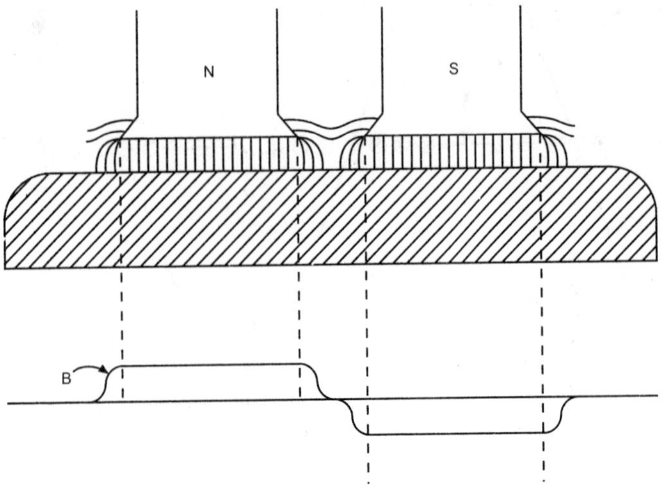

Fig. 4.17

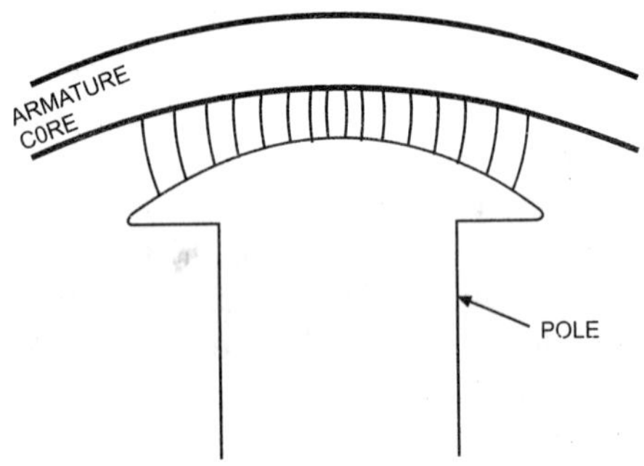

Fig. 4.18

flux paths are shown dotted in the figure). The flux density in the air layer along the stator periphery gradually falls off to zero in the interpolar region. The result is a *flat-topped* flux density as illustrated in Fig. 4.20.

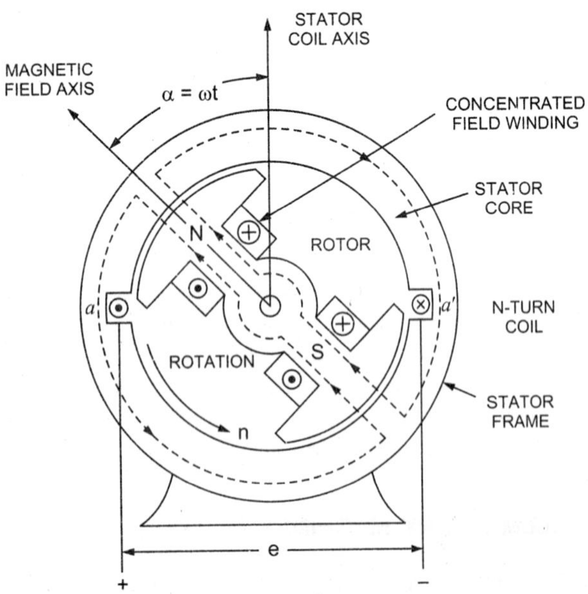

Fig. 4.19 *Elementary Salient Pole Synchronous Generator With 2-Pole Rotor*

In ac machines flux distribution is required to be perfectly sinusoidal so that emf induced in armature conductors is sinusoidal. This is achieved in salient pole machines by providing nonuniform air gap above the pole shoes, as shown in Fig. 4.18 (minimum air gap in the middle of the pole shoe progressively increasing towards outer edges).

Sinusoidal flux distribution in the air gap can also be had by using *nonsalient* or *smooth cylindrical rotor* with suitably distributed field winding along periphery of rotor as illustrated in Fig. 4.21. In such a structure as one moves away from the pole axis, the flux paths link progressively smaller number of field ampere-turns (mmf). The mmf can be so distributed as to provide a nearly sinusoidal flux distribution in the air gap.

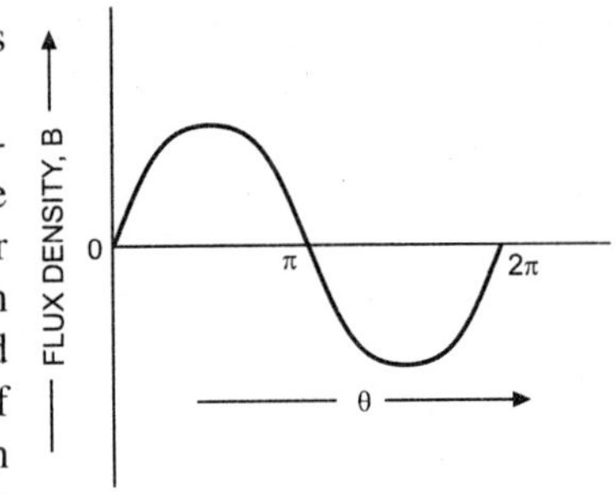

Fig. 4.20 *Air-gap Flux Density Wave*

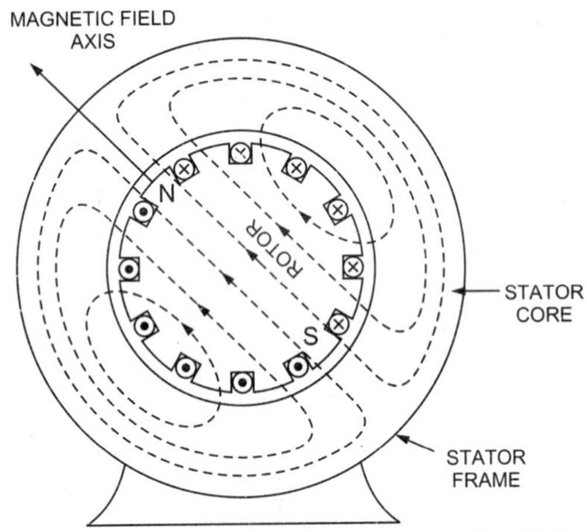

Fig. 4.21 *Smooth Cylindrical or Nonsalient Pole Synchronous Generator*

With the rotation of rotor, the flux waveform sweeps by the coil sides a and a′. The resulting emf induced in the coil is a time function having the same waveform as the spatial distribution flux density B [Figs. 4.22 (*a*) and 4.22 (*b*)].

(*a*) *Space Distribution of Flux Density*

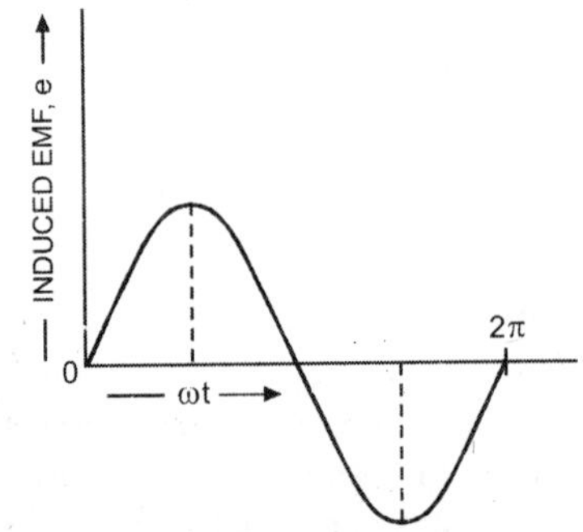

(*b*) *Corresponding Waveform of the Induced EMF*

Fig. 4.22

The coil voltage completes a cycle of values for each revolution of the two-pole machine shown in Fig. 4.19. Its frequency in Hz (cycles per second) is equal to the speed of rotor in revolutions per second (*rps*) i.e. the electric frequency is synchronised with the mechanical speed of rotor and that is why, the machine is designated *synchronous machine.* Thus a 2-pole synchronous machine must revolve at a speed of 50 rps or 3,000 rpm to produce a 50-Hz emf.

A large number of synchronous machines have more than two poles. For instance, the most elementary 4-pole single-phase synchronous generator is shown in Fig. 4.23. The field coils are connected so as to produce poles of alternate north and south polarity. There are two complete wavelengths or cycles in the flux distribution around the periphery, as illustrated in Fig. 4.24(*a*). The armature winding now consists of two coils a_1a_1' and a_2a_2' connected in series by their end connections. The span of each coil is 1/2 wavelength of flux. The induced emf (or generated voltage) now passes through two complete cycles per revolution of the rotor. The frequency in Hz then is twice the speed in *rps*.

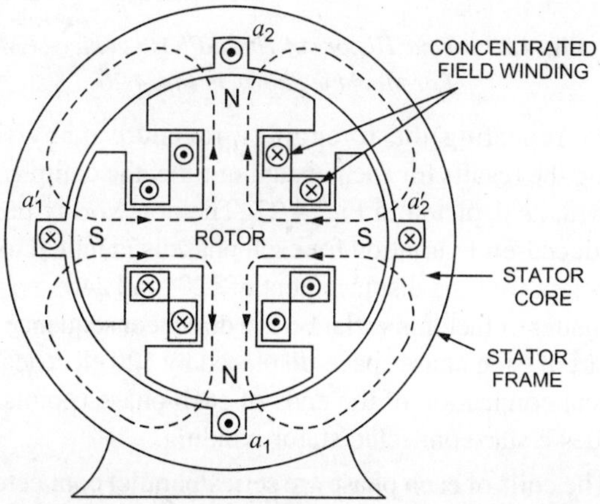

Fig. 4.23 *Elementary 4-pole Synchronous Generator*

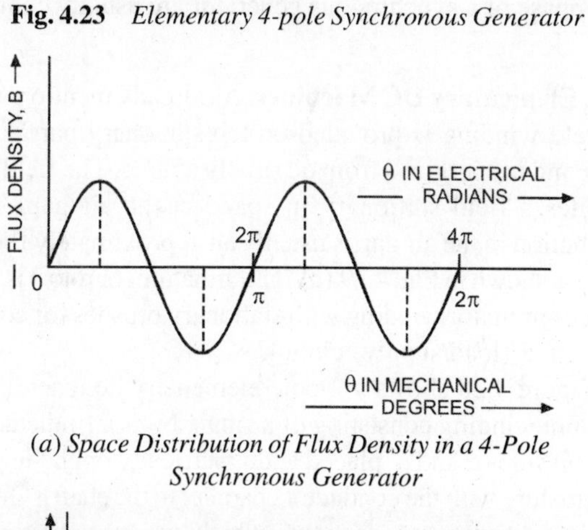

(*a*) *Space Distribution of Flux Density in a 4-Pole Synchronous Generator*

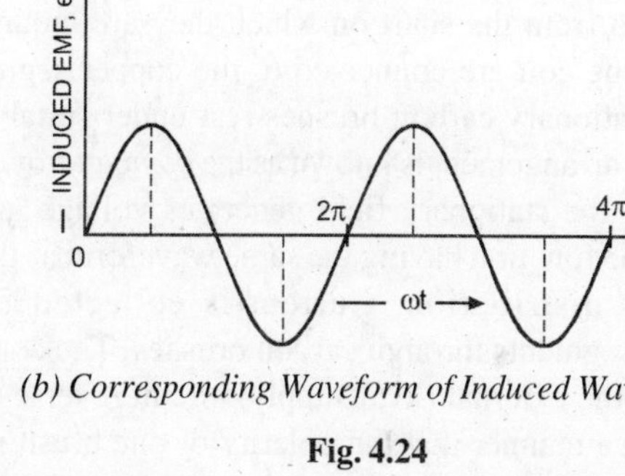

(*b*) *Corresponding Waveform of Induced Waveform*

Fig. 4.24

In a multipolar machine, it is convenient to concentrate on a single pair of poles and to recognize the electric, magnetic, and mechanical conditions associated with every other pole pair, repetitions of those for the pair under consideration. That is why, it is convenient to express angles in *electrical degrees* or *electrical radians* rather than in mechanical units. The total angular span of one pole pair is therefore taken as one cycle of flux distribution equals 360 electrical degrees or 2π electrical radians. As there are $\frac{P}{2}$ complete wavelengths or cycles in one complete revolution,

$$\theta_{elec} = \frac{p}{2}\theta_{mech} \qquad ...(4.10)$$

The induced emf in the coil of a P-pole machine passes through a complete cycle every time a pair of poles sweeps by, or $\frac{P}{2}$ times each revolution. The frequency of the induced emf in the coil is therefore given by the expression

$$f = \frac{P}{2} \times n = \frac{P}{2} \times \frac{N}{60} = \frac{PN}{120} \qquad ...(4.11)$$

where P is the number of poles on the machine and N is the mechanical speed of rotation in *rpm*. The radial frequency ω_{elec} of the wave of induced emf is given as

$$\omega_{elec} = \frac{P}{2}\omega_m \qquad ...(4.12)$$

where ω_m is the mechanical speed in radians per second.

The two coils of the 4-pole synchronous generator shown in Fig. 4.23 are seats of identical emfs and can be connected either in series or parallel as shown in Figs. 4.25 (*a*) and 4.25 (*b*). The series connection provides double the voltage of one coil but can handle the same maximum current as any one coil. On the other hand the parallel connection has twice the maximum current-carrying capacity of one of the coils but has the same voltage as that of each coil. Thus the machine can be designed to provide desired voltage and current rating by adopting series-parallel arrangement of coils groups.

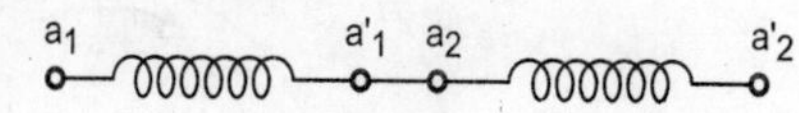

(*a*)

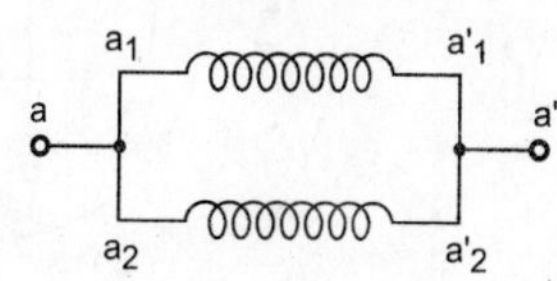

(*b*)

Fig. 4.25

Three-Phase Synchronous Generator. The synchronous machines may be *single, two* or *3-phase* types. For the single-phase type, the armature has all the coils connected for addition of individual voltages; this provides a pair of output terminals and a single ac output. For the two-phase type, the armature has two sets of windings placed so the outputs from three terminals (one common) are 90° out of phase. For the three-phase type, the armature has three sets of windings placed so that three outputs with a mutual phase difference of 120° are available. Three windings may be connected either in star or delta. Large ac machines are invariably 3-phase type. This is due to inherent advantages of 3-phase system over single-phase system.

The synchronous machines may be rotating armature type or rotating field type. Practically all medium and large machines are always constructed with revolving field due to inherent advantages of rotating field construction as discussed in Art. 15.1.2.

The generation of three-phase voltage can be readily understood from a study of Fig. 4.26(*a*). The machine shown in Fig. 4.26(*a*) is a 2-pole rotor machine, the field winding of which is assumed excited from a dc source to produce the pole flux. It may also be assumed that the pole pieces are shaped to provide a sinusoidal flux field. Appearing in the stator is a balanced three-phase winding with the axis of each phase displaced by 120° or $\frac{2}{3}\pi$ radians. The complete winding of each phase is represented by a single coil in Fig. 4.26(*a*). Now the rotor may be considered to be driven by the prime mover in a counter-clockwise direction at synchronous speed. According to Fleming's right hand rule for the field direction shown, it is revealed that the instantaneous emf induced in coil sides a, b', c' is in outward direction while in coil sides a', b, c it is in the inward direction. Furthermore, since coil sides a and a' are located beneath the maximum value of the flux-density wave, the induced emf in phase a is at its maximum value. The corresponding induced emf in phase *b*, as identified by coil *b*, is seen to be of opposite sign and of smaller magnitude. In fact, since coil sides *b* and b' are both displaced from the maximum flux-density position by 60°, it follows that this instantaneous induced emf is at one-half (cos 60°) its maximum value. A similar reasoning gives the same result for phase c.

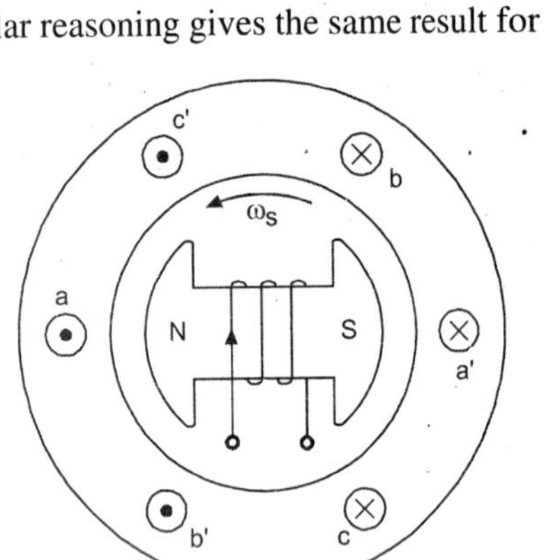

(*a*) *Phase* a *at a Positive Maximum-Time t_1*

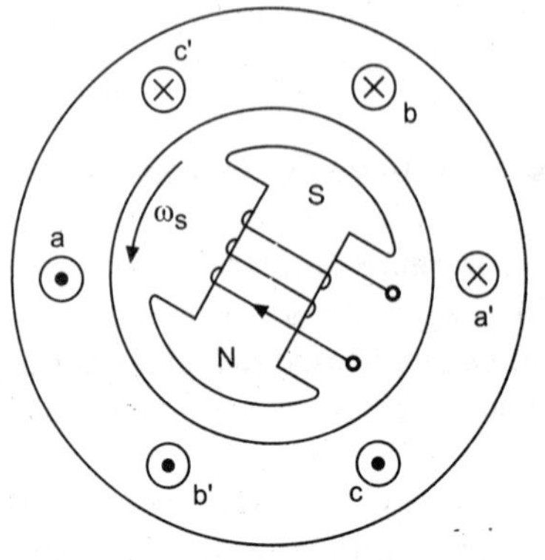

(*b*) *Phase b at a Negative Maximum - Time t_2*

Fig. 4.26 *Induced-Voltage Distribution in the Stator Winding of a 3-Phase Synchronous Generator*

Now if the rotor is considered to have advanced by 60° in the counter-clockwise direction, this puts the amplitude of the north-pole flux directly beneath b', indicating that the induced emf is now a negative maximum in phase *b*. Also, the coil *c-c'* now finds itself under the influence of flux of reverse polarity as shown in Fig. 4.26 (*b*). Hence the direction of induced emf in coil side c is now outward and in coil side c' it is inward. The instantaneous values of the three-phase voltages for the first and second time instants are shown in Fig. 4.27 and are identified at t_1 and t_2, respectively.

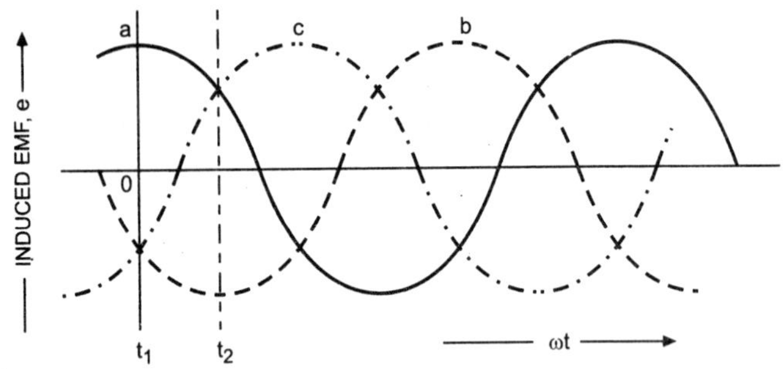

Fig. 4.27 *Time History of Three-Phase Induced EMFs For Machine Shown in Fig. 4.26*

By repeating the foregoing procedure many times and plotting the results for each instant of time, the complete curve is obtained, as depicted in Fig. 4.27. The noteworthy point is that the induced-emf variation for each phase is identical to the other two except for time displacement of 120° and 240°, respectively. As a matter of fact this variation is a direct consequence of having the start of each phase space-displaced by 120 electrical degrees. Physical connection of the ends of each phase (points a', b', c') provides a star-connected stator winding.

The coils of each phase are series/parallel connected and the three phase of a synchronous generator are usually connected in star.

4.7.2. Elementary DC Machines. As already mentioned earlier, the field winding is provided on the stationary part, known as *stator* and is energized from dc supply source. The field winding provides a field stationary in space in the air gap. The flux distribution in the air gap is usually an approximately flat-topped wave, as shown in Fig. 4.29 (*a*). The armature (or rotor) is provided with commutator winding with stationary brushes for connection to external (load/supply) circuit.

Figure 4.28 shows a 2-pole elementary dc machine having armature winding consisting of a single N-turn [indicated by the two coil sides *a* and a' placed at diametrically opposite points on the armature with the conductors parallel to the shaft]. One copper ring is split into two portions, which are insulated from each other and also from the shaft on which they are mounted. The two ends of the coil are connected to the copper segments, on which two stationary carbon brushes rest under suitable spring pressure. The arrangement is known as the *commutator*. Rotation of the coil in the stationary field generates voltage in the coil which is a time function having the same waveform as the spatial flux-density distribution. Current is collected from the commutator segments through carbon brushes. The connections of the coil to the external (load/supply) circuit reverse each half cycle in such a manner that the polarity of one brush is always positive and the other negative. This, in fact, is the rectification action of the commutator-brush arrangement.

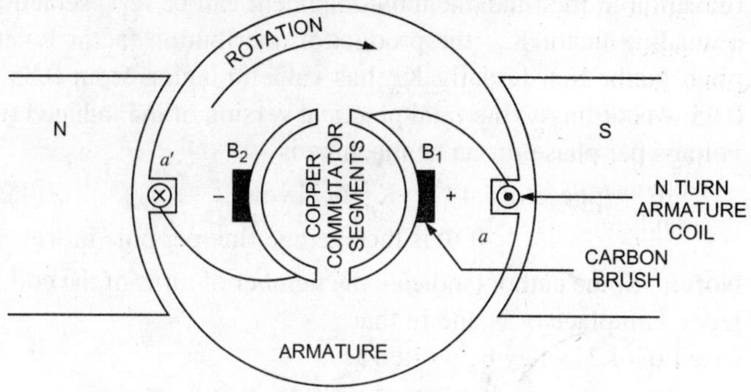

Fig. 4.28 *Elementary DC Machine With Commutator*

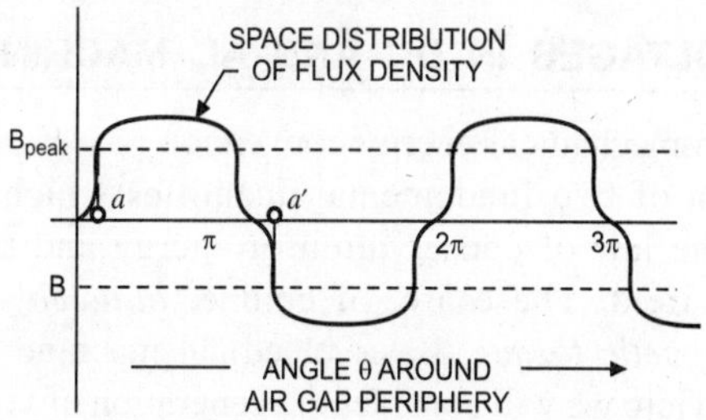

(a) Space Distribution of Air gap Flux Density in an Elementary DC Machine

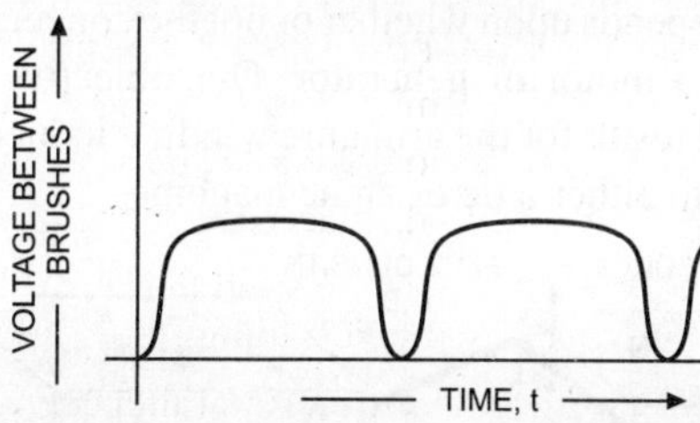

(b) Waveform of Voltage Between Brushes

Fig. 4.29

The commutator provides full-wave rectification, transforming the voltage waveform between brushes to that shown in Fig. 4.29 (*b*) and making available a unidirectional voltage to the external circuit. The dc machine shown in Fig. 4.28 is, of course, simplified but it is unrealistic in the practical sense. The actual machine has a large number of multi-turn coils and commutator segments, with the coils evenly distributed around the surface of the armature. The coils may be connected in the form of lap or wave winding.

From Fig. 4.28, it is revealed that the effect of direct current in the field winding of a dc machine is to produce a magnetic flux distribution which is stationary with respect to the stator. Similarly, the effect of the commutator is such that when direct current flows through the brushes, the armature develops a magnetic flux distribution which is also fixed in space and whose axis, determined by the design of the machine and the position of the brushes, is typically perpendicular to the axis of the field flux. Just as in the ac machines, it is the interaction of these two flux distributions that develops the torque of the dc machine. In case of machine is acting as a generator, the torque opposes the rotation. If it is acting as a motor, the electromagnetic torque acts in the direction of rotation. Similarly an emf is generated in both generator and motor i.e. there is a generation action in both motor and generation operation. In generator operation, the generated emf produces the armature current, whereas, in motor operation the generated emf opposes the conduction.

Since in dc machines, the commutator makes necessary that the armature should be rotating one, or brushes should revolve with the field, it is convenient to have armature as a rotating member in dc machines.

4.8 GENERATED VOLTAGE OF AC WINDING

An elementary ac machine is depicted in cross section in Fig. 4.30. The coils on both the rotor and stator have been shown as single multi-turn full pitch coils. The analysis can readily be extended to distributed windings. The field winding on the rotor is assumed to create a sinusoidal space wave of flux density at the stator surface. The rotor is rotating at a constant angular velocity of ω electrical radians per second. Although a 2-pole machine is shown, the derivations presented here are also applicable to a P-pole machine.

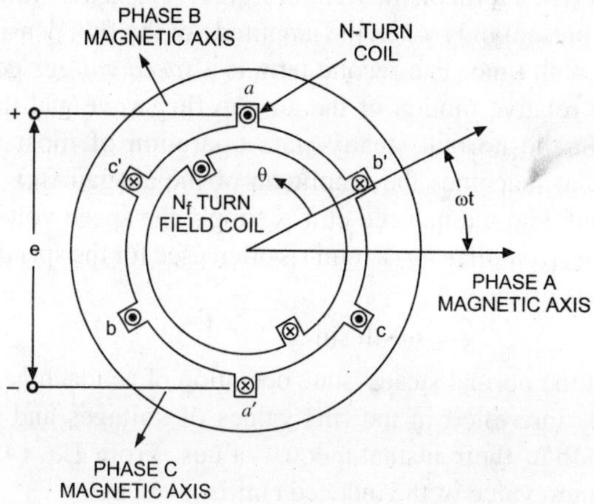

Fig. 4.30 *Cross-Sectional View of an Elementary 3-Phase AC Machine*

When the rotor poles are in line with the magnetic axis of the stator coil the flux linkage with the stator N-turn coil is NΦ, where Φ is the air-gap flux per pole. For the assumed sinusoidal flux density wave

$$B = B_{peak} \cos \theta \qquad ...(4.13)$$

where B_{peak} is the peak value of flux density B at the rotor pole center and θ is measured in electrical radians from the pole axis of rotor. The air-gap flux per pole is the integral of the flux density over the pole area. Thus, for a 2-pole machine

$$\Phi = \int_{-\pi/2}^{\pi/2} B_{peak} \cos \theta \, lr \, d\theta = 2 B_{peak} \, lr \qquad ...(4.14)$$

where l = active coil-side length (axial stator length)

and r = mean radius of the stator at the air gap.

For a P-pole machine

$$\Phi = \frac{2}{P} \times 2 B_{peak} \, lr \qquad ...(4.15)$$

$\because$ the pole area is $\frac{2}{P}$ times that of a 2-pole machine of the same length and diameter

With the rotation of rotor, the flux linkage varies as the cosine of the angle α between the magnetic axes of the stator coil and rotor. With the rotor revolving at a constant angular velocity ω, the flux linkage with the stator coil is given as

$$\psi = N\Phi \cos \omega t \qquad ...(4.16)$$

where time t is arbitrarily chosen as zero when the peak of the flux-density wave coincides with the magnetic axis of the stator coil. According to Faraday's law, the emf induced in the stator coil is given as

$$e = \frac{d\psi}{dt} = N\frac{d\Phi}{dt} \cos \omega t - \omega N\Phi \sin \omega t \qquad ...(4.17)$$

The polarity of this induced emf is such that if the stator coils were short circuited, it would result in flow of current in the direction that would oppose any change in the flux linkage of the stator coil. Although Eq. (4.17) has been derived assuming that only the field winding was creating air-gap flux, this equation applies equally well to the general situation where Φ is the net air-gap flux per pole produced by currents on both the stator and the rotor.

The first term on the RHS of Eq. (4.17) is a transformer emf and is present only when the amplitude of the flux density wave varies with time. The second term is a *speed voltage* generated due to relative motion of the air-gap flux wave and the stator coil. In the normal steady-state operation of most rotating electrical machines the amplitude of the air-gap flux wave is constant, and the induced emf is simply the speed voltage. The term *electromotive force* (emf) is often used for the speed voltage which is given as

$$e = \omega N\Phi \sin \omega t \qquad ...(4.18)$$

In the normal steady-state operation of ac machines we are usually interested in the rms values of voltages and currents rather than their instantaneous values. From Eq. (4.18) the maximum value of the induced emf is given as

$$E_{max} = \omega N\Phi = 2\pi f N\Phi \qquad ...(4.19)$$

$\because \quad \omega = 2\pi f$ where f is the frequency in Hz.

and the rms value is given as

$$E_{rms} = \frac{E_{max}}{\sqrt{2}} = \frac{2\pi}{\sqrt{2}} f N\Phi = 4.44 f N\Phi \qquad ...(4.20)$$

A comparison of above Eq. (4.20) and expression for emf induced in transformer windings $E = 4.44 f \Phi_{max} N$ reveals that the equations have identical form. There is a difference, however, and it lies in the meaning of flux. In a transformer, Φ_{max} is the maximum flux corresponding to peak magnetising current consistent with the magnitude of applied voltage. In an ac electromechanical energy conversion device, Φ is the maximum flux per pole that links with a coil having N turns and spanning the full pitch.

In any practical machine the total turns per phase are not concentrated in a single coil but rather are distributed over one-third of a pole pitch (or 60 electrical degrees for each of the three phases). In addition, the individual coils that make up N turns are designed not to span the full pitch but rather only about 80 to 85% of the pole pitch. Such a coil is called a *fractional pitch coil*. The use of a distributed winding employing fractional-pitch coils has the advantage of virtually eliminating the effects of all harmonics that may be present in the flux density wave while only slightly reducing the fundamental component. The reduction in the fundamental component can be represented by a winding factor K_w (the product of distribution factor K_d and pitch factor K_p). Usually K_w has values ranging from 0.85 to 0.95. Accordingly, the final practical version of the induced rms voltage per phase for an ac machine is

$$E_{rms}/\text{phase} = 4.44\, K_d K_p \Phi f N \text{ volts} \qquad ...(4.21)$$

if Φ is the average flux per pole in webers.

Note: Some authors indicate the number of turns of the coil by letter T in place of N and in that case Eq. (4.21) may be written as

$$E_{rms}/\text{phase} = 4.44\, K_d K_p \Phi f T \text{ volts}$$

4.9 VOLTAGES IN DC AND AC MACHINES

The electromechanical energy conversion process involves the interaction of two fundamental quantities which are related through the law of conservation of energy and the coupling magnetic field. These are, of course, *induced voltage* and *electromagnetic torque*. Basic electrical machine is shown in Fig. 4.31. Here we will consider the generation of voltages in the armature winding brought about by changing the flux linkages. Whether or not this generated voltage is a reaction voltage (*counter emf*) depends upon whether or not the conversion device is behaving as a motor or generator. Our objective now is to obtain a general result for the armature winding induced emf that can be applied to either a dc or an ac machine.

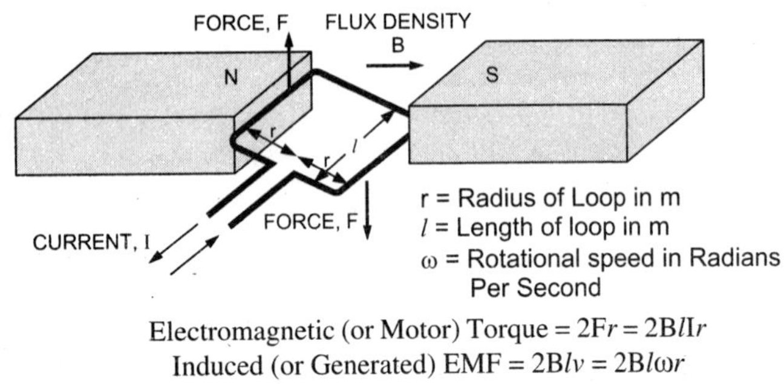

Fig. 4.31 *Basic Electrical Machine*

There are different ways of deriving the emf equation and several ways of expressing it. The following is an attempt to get an expression which is easy to remember, easy to develop and of general application.

An emf induced in a conductor of length l moving with linear velocity v in a non-time varying magnetic field of flux density B is given by the equation.

$$e = Blv \qquad ...(4.22)$$

where B, l, and v are mutually perpendicular. Properly interpreted, this equation also applies to rotating machines and provides a useful basis for visualizing induced voltages.

The relative field-conductor velocity,

$$v = \omega r = \frac{2\pi N}{60} \times \frac{D}{2} = \frac{\pi DN}{60} \qquad ...(4.23)$$

where D is the diameter of armature and N is the relative rotational speed in revolutions per minute (rpm).

The average flux density,

$$B_{av} = \frac{\text{Flux per pole} \times \text{number of poles}}{\text{Cylindrical area of armature surface}}$$

$$= \frac{\Phi P}{\pi D l} \qquad ...(4.24)$$

where l is the axial length of armature.

EMF induced per conductor, e

$$= B_{av} \times l \times v = \frac{\Phi P}{\pi D l} \times l \times \frac{\pi D N}{60}$$

$$= \Phi P \times \frac{N}{60} = \Phi P n \qquad ...(4.25)$$

Since ΦP is the flux cut by one conductor in one revolution and n is the number of revolutions per second, the induced emf can be expressed as

e = Flux cut by one conductor in one revolution × revolutions per second (rps)

= Flux cut per second

If number of conductors in series per circuit or per phase $= Z_s$, then average emf induced per circuit or per phase is given as

E_{av} = Average emf induced per conductor × number of conductors in series per circuit or per phase

$$= \Phi P \frac{N}{60} \times Z_s \qquad ...(4.26)$$

In case of a dc machine, the number of conductors in series between positive brush and negative brush is equal to the total number of conductors divided by the number of parallel paths $= \frac{Z}{A}$

∴ Total emf generated between the positive and negative terminals,

$$E = \Phi P \frac{N}{60} \times \frac{Z}{A} = \Phi \frac{ZN}{60} \times \frac{P}{A} \qquad ...(4.27)$$

If Φ is in webers, generated voltage given by above equation will be in volts.

DC machines are either lap wound or wave wound. In case of lap winding A = P while in wave wound machine A = 2.

For an ac machine, and considering only the fundamental for the moment, the average emf per conductor from Eq. (4.25) is $\Phi P \frac{N}{60}$ where Φ is now the fundamental flux per pole. The rms value, applying the form factor for a sine wave $\left(\frac{\pi}{2\sqrt{2}}\right)$ is, therefore, equal to $\frac{1.11 \times \Phi PN}{60}$ per conductor.

If there are Z_P conductors in series per phase, the arithmetic sum of their emfs would be

$$1.11 \ \Phi P Z_P \frac{N}{60} \qquad ...(4.28)$$

In rotating a coil in a two-pole field at a rate of 1 revolution per second results in a frequency of 1 Hz. For a 4-pole field each revolution per second yields 2 Hz. For a P-pole field, the frequency generated in Hz is

$$f = \frac{P}{2} \times n = \frac{P}{2} \times \frac{N}{60} = \frac{PN}{120}$$

Now substituting $\frac{PN}{120} = f$ in the Eq. (4.28) for rms emf induced per phase we have

$$E_{rms}/\text{phase} = 1.11 \ \Phi Z_P P \times \frac{N}{60}$$

$$= 2.22 \ \Phi Z_P f$$

$$= 2.22 \ \Phi \times 2T \times f = 4.44 \ \Phi f T \qquad ...(4.29)$$

since $Z_P = 2T$ (turns)

As already discussed in Art 4.8, the Eq. (4.29) and expression for transformer emf are identical and the practical version of the induced rms voltage per phase for an ac machine is

$$E_{rms}/\text{phase} = 4.44 \ k_d \ k_p \ \Phi f T \text{ volts} \qquad ...(4.30)$$

if Φ is the average flux per pole in webers.

4.10 TORQUE PRODUCTION AND TORQUE BALANCE

Torque considerations in machines may be based on simple formulation of the fact that a mechanical force is exerted on a current carrying conductor in a magnetic field. Quantitatively, the force on a straight section of conductor length l metres perpendicular to a magnetic field of density B webers per square metre and carrying the current i amperes is given as

$$F = Bli \text{ newtons}$$

If D is the diameter of the rotor, the force F acts at a radial distance of $\frac{D}{2}$ metres and the associated torque per conductor given as

$$T_c = Bli \frac{D}{2}$$

The average flux density, $B_{av} = \frac{\Phi P}{\pi D l}$

So torque developed per conductor, $T_c = \frac{\Phi P}{\pi D l} \times l \times i \times \frac{D}{2}$

$$= \frac{1}{2\pi} \Phi P i \qquad ...(4.31)$$

The torque associated with the entire winding is the summation of the torques for the individual conductors or coil sides. Torque developed by such electromagnetic action is called the *electromagnetic torque* and is given as

$$T_e = \frac{1}{2\pi} \Phi Z P i \qquad ...(4.32)$$

where Z is the number of conductors on the rotor surface.

A glance at Eq. (4.26) and (4.32) shows that both the induced voltage and the developed electromagnetic torque depend upon the coupling magnetic field. The factor that determines whether the induced emf, e and the developed torque T are actions or reactions depends whether generator or motor action is being developed.

The two mechanical quantities T and ω_m, and the two electrical quantities e and i, which are coupled through the medium of magnetic field B, are related further by the law of conservation of energy. This is readily demonstrated by dividing Eq. (4.26) by Eq. (4.32). Thus

$$\frac{e}{T_e} = \frac{\Phi P n Z}{\frac{1}{2\pi}\Phi Z P i} = \frac{2\pi n}{i} = \frac{\omega_m}{i} \quad ...(4.33)$$

where $\omega_m = 2\pi n$ = Mechanical angular velocity of the rotor.

Rearranging above Eq. (4.33) we have

$$ei = T_e \omega_m \quad ...(4.34)$$

The above equation is called the *torque balance equation.*

In a generator, the resulting torque is a *counter torque* opposing the rotation. It is the torque which the prime mover must overcome, and it is the mechanism through which greater electric power output calls for greater mechanical power input. In a motor, the rotational or speed voltage in the armature acts in opposition to the applied voltage and is termed as *counter* or *back emf.* All these statements are specific, tangible expressions of the reversibility which is a characteristic of all electro-mechanical energy conversion processes.

In case of a dc motor, the current per conductor,

$i = \frac{I_a}{A}$ where I_a is the armature current and A is the number of parallel paths.

Substituting $i = \frac{I_a}{A}$ in Eq. (4.32) we get torque developed by the armature of a dc motor. Thus

$$T_a = \frac{1}{2\pi}\Phi Z P \frac{I_a}{A} \text{ N-m} \quad ...(4.35)$$

4.11 ROTATING MAGNETIC FIELDS

A rotating magnetic field is that field which is constant in magnitude but whose axis of direction rotates in space as field system of a dc machine. With a stationary field system the magnetic field is stationary in space, but if the field system is rotated at a certain speed, its magnetic field will rotate with it at the same speed.

The magnetic field produced by single-phase alternating current is an *alternating magnetic field*, the field acting along a fixed axis, varying in magnitude periodically and changing its direction alternately positive and negative (Refer to Fig. 4.32).

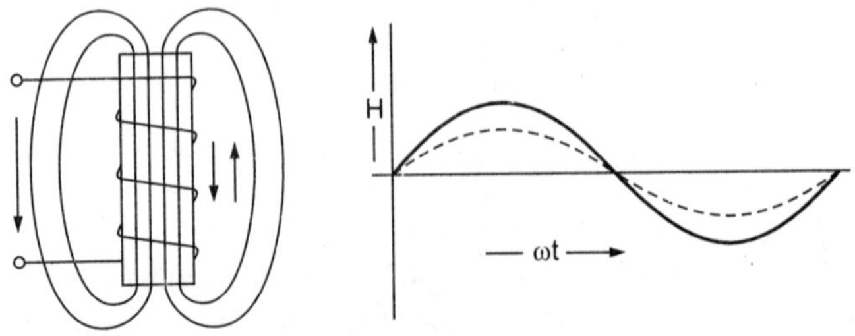

Fig. 4.32 *Alternating Magnetic Field*

4.11.1. Ferraris Principle. The magnetic field produced by single-phase alternating current is an alternating magnetic field which, by Ferraris principle, can be resolved into two rotating fields of half its amplitude and rotating in opposite directions at synchronous speed.

Consider two rotating magnetic fields OA_1 and OA_2, each having magnitude of H units and travelling in opposite directions with angular velocity ω, as shown in Fig. 4.33.

Let both the fields start travelling from axis OX at time $t = 0$.

After time t seconds the angle through which fields OA_1 and OA_2 have rotated, $\theta = \omega t$ radians.

i.e. Field represented by vector OA_1 has moved in counter-clockwise direction through an angle $\theta = \omega t$ from axis OX and field represented by vector OA_2 has moved in clockwise direction through same angle θ, because both fields are travelling with same angular velocity ω.

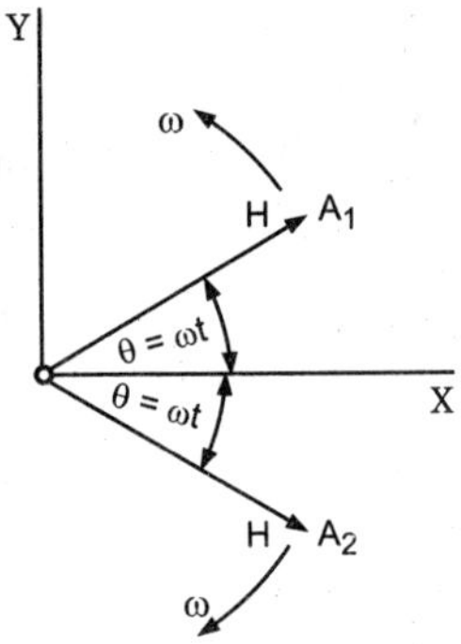

Fig. 4.33

Resolving the magnetic fields represented by vectors OA_1 and OA_2 along axis X and axis Y, we get

$$\text{X-component, } H_x = OA_1 \cos\theta + OA_2 \cos\theta$$
$$= H\cos\theta + H\cos\theta$$
$$= 2H\cos\theta = 2H\cos\omega t$$
$$\text{Y-component, } H_y = OA_1 \sin\theta - OA_2 \sin\theta$$
$$= H\sin\theta - H\sin\theta = 0$$

Hence resultant magnetic field is 2H cos ω*t* along X-axis. It is, therefore, obvious that two rotating magnetic fields, travelling in opposite directions with the same angular velocity, result in an alternating field of twice their amplitude. *Conversely an alternating field can be replaced exactly by two rotating fields of half its amplitude travelling in opposite directions at synchronous speed.*

4.12 PRODUCTION OF TWO-PHASE ROTATING MAGNETIC FIELD

Consider two similar coils placed 90° (electrical) apart and connected to two phase supply, as shown in Fig. 4.34. An alternating field is set up by each coil along its own axis i.e. coil A sets up an alternating field along X-axis and coil B sets up an alternating field along Y-axis. Let the fields set up by coils A and B, each be resolved into two components rotating in opposite directions with angular velocity $\omega = 2\pi f$, f being the supply

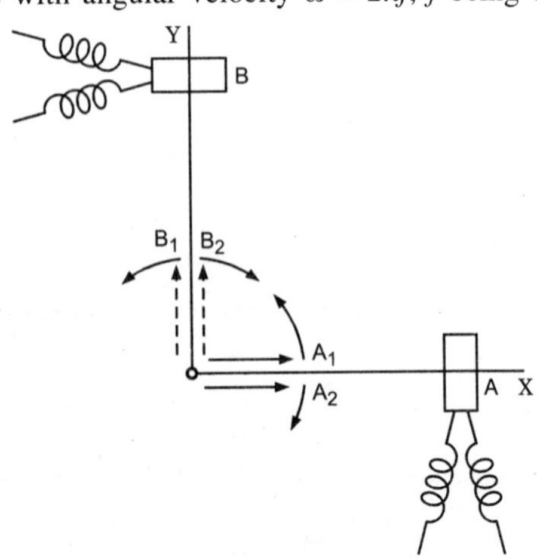

Fig. 4.34

frequency. Let the components of fields due to coil A be OA_1 and OA_2 and the components of field due to coil B be OB_1 and OB_2, as shown in Fig. 4.34.

Consider the instant when current in coil A is maximum. *i.e.* I_A is maximum. At this instant magnetic field due to coil A is maximum, therefore, both the components OA_1 and OA_2 are along the axis of the coil A *i.e.* along OX, as shown in Fig. 4.35.

When the current in coil A is maximum, the current in coil B is 90 electrical degrees behind the instant of maximum value, therefore, its components OB_1 and OB_2 will have to rotate through 90° to come along the axis of the coil B *i.e.*, axis Y. Hence at this instant component OB_1 is along OX and component OB_2 is along OX′ as shown in Fig. 4.35.

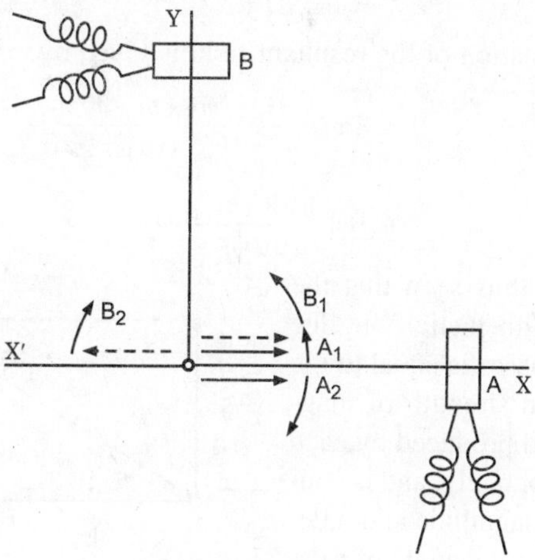

Fig. 4.35

From Fig. 4.35. it is obvious that rotating components OA_2 and OB_2 are rotating in clockwise direction with the same angular velocity but opposing each other, therefore, their resultant is always zero. Components OA_1 and OB_1 are rotating in counter-clockwise direction with the same angular velocity. Since their direction is also same so they result in a pure rotating magnetic field.

It is obvious that the direction of resultant rotating field is same as the direction of phase sequence in the two coils. Thus to reverse the direction of the rotating field, phase sequence must be reversed by reversing the connections of one of the coils.

The manner in which the rotating magnetic field is established by a two-phase stator winding can also be illustrated by considering the direction of current flow through the two phases at several successive instants.

Consider a simplified stator core with two equally spaced coils, as shown in Fig. 4.36. Normally the conductors are evenly spread round the core, but they are here shown in separate groups to enable the individual phases to be distinguished. The dots and crosses show the directions of the currents in the conductors when they are positive.

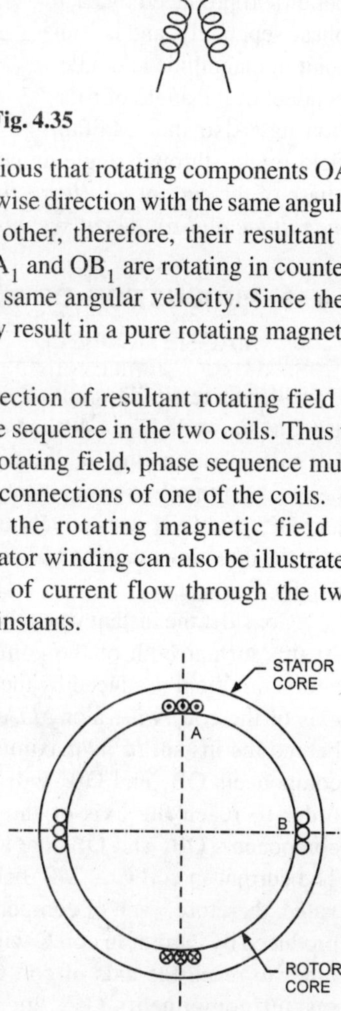

Fig. 4.36

Of course, they are never all positive at the same instant, nor are they ever all equal at any one instant. Phase relationship of the currents flowing through the coils is shown in Fig. 4.37. The different times are marked off at 45° intervals on the current waves in Fig. 4.37.

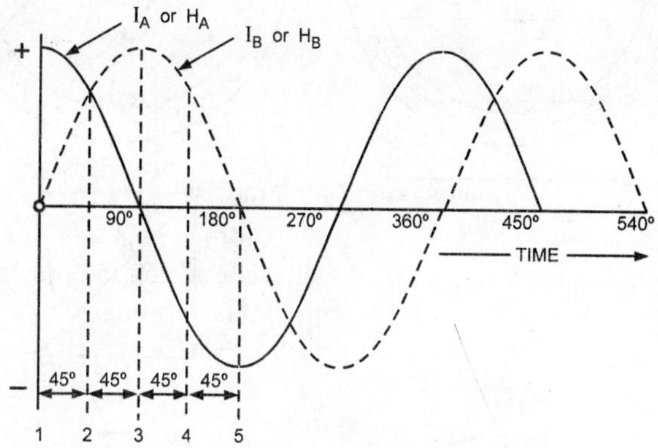

Fig. 4.37

At instant 1, the current in phase A, I_A is positive and at its maximum value and the current in phase B, I_B is zero. The directions of the currents in the conductors are indicated by crosses and dots. The field produced is towards the right, creating a north pole on the stator surface at the left and the south pole on the right, as shown in Fig. 4.38 (1).

At instant 2, the current in each phase is positive and is 0.707 of the maximum value and the resultant field due to these currents is shown in Fig. 4.38 (2). It will be seen that the axis of the resultant field has turned through 45° from that of 4.38 (1). At instant 3, the current in phase A has fallen to zero and the current in phase B is positive and at its maximum value; the resultant field is vertically downward as shown. Thus we see that the resultant field has rotated further through 45°. The remaining Figs in Fig. 4.38 show the resultant field at the instants 4 and 5.

It is evident from Fig. 4.38 that the resultant field due to the currents flowing in the two-phase windings is of the magnitude equal to the peak value of the field produced by individual phase windings and shifts around the stator surface, moving through a definite distance in each of the time interval considered.

From Fig. 4.38, it is also evident that in a 2-pole winding the field makes 1/2 revolution in 1/2 cycle of the current, and, therefore, one revolution in each cycle of current. In a 4-pole winding in which each phase has two separate groups of coils connected in series, it may be shown that the rotating field makes one revolution in two cycles of the current. In general, the field makes one revolution in P/2 cycles where P is the number of poles produced by the stator winding. If N is the speed of the resultant magnetic field then

$$\frac{N}{60} = \frac{f}{P/2}$$

$$\text{or } N = \frac{60 \times 2f}{P} = \frac{120f}{P}$$

The speed at which the rotating field revolves is called the *synchronous speed.*

By redrawing Figs. 4.38 (2), (3), (4) and (5) with the current in, say, phase B reversed, it can be shown that the direction of rotation of the resultant magnetic field is then counter-clockwise

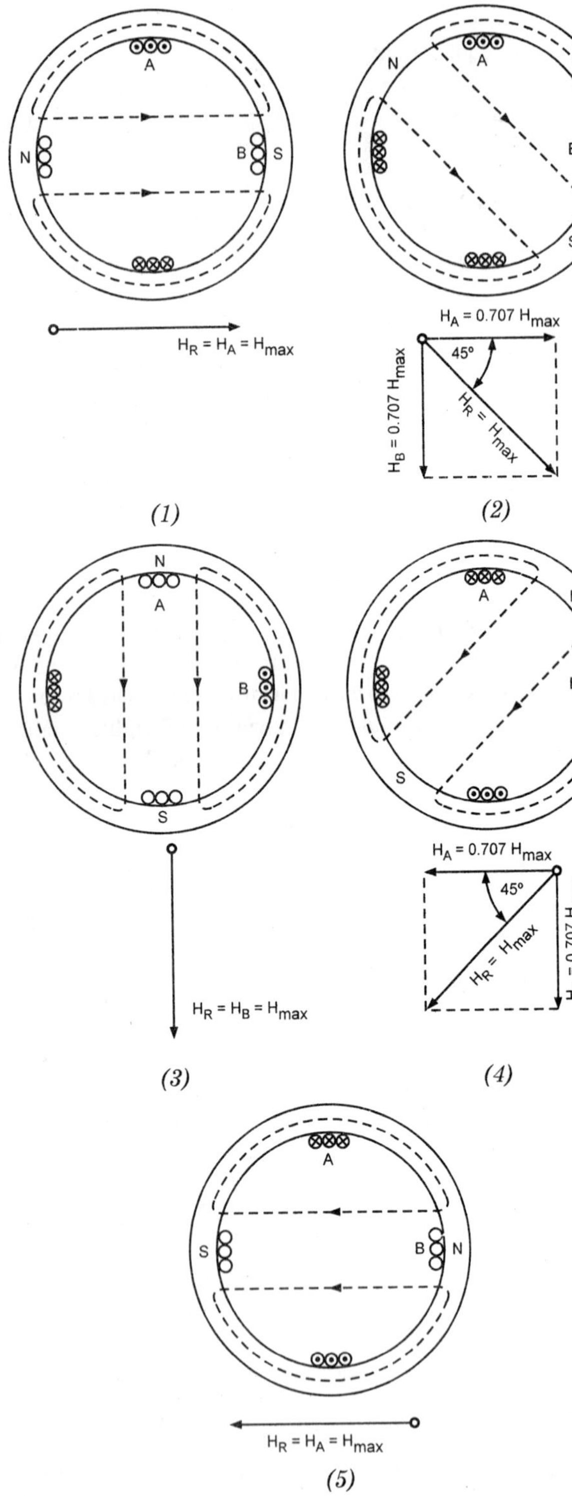

Fig. 4.38

i.e. the direction of the rotation of the resultant magnetic field can be reversed by changing over the connection to any one of the two phases.

Mathematical Proof. The above fact can be proved mathematically as below:

The time diagrams for currents in coils A and B are shown in Fig. 4.37. Coils A and B establish alternating fields along their respective axis *i.e.* along X-axis and Y-axis respectively. Since phase difference between the currents I_A and I_B flowing through coils A and B respectively is 90°, hence phase difference between the alternating fields H_A and H_B produced is 90°.

Let each of the coil produce an alternating field of maximum strength H_{max} along its own axis. Considering an instant when the current in coil A has advanced through an angle ωt from its maximum value, the magnetic field due to this coil, H_A is represented by H_{max} cos ωt and lies along X-axis. At the same instant the magnetic field dut to the other coil H_B, which is lagging behind H_A by 90°, is represented by H_{max} cos (ωt – 90°) and lies along vertical axis.

$$\text{The resultant field, } H = \sqrt{\{H_{max} \cos \omega t\}^2 + \{H_{max} \cos(\omega t - 90)\}^2}$$

$$= H_{max} \sqrt{\cos^2 \omega t + \sin^2 \omega t} = H_{max}$$

Inclination of the resultant field is given by

$$\therefore\ \theta = \text{Tan}^{-1} \frac{H_{max} \cos(\omega t - 90°)}{H_{max} \cos \omega t}$$

$$= \text{Tan}^{-1} \frac{\sin \omega t}{\cos \omega t} = \omega t$$

It is thus seen that the resultant magnetic field due to two phases is equal to the maximum strength of magnetic field produced by each phase separately, and is constant in magnitude and independent of the angle of rotation ωt. Also the resultant field rotates through angle ωt corresponding to an advance in phase of the current ωt. *Hence the resultant field is of constant magnitude and rotates at synchronous speed.*

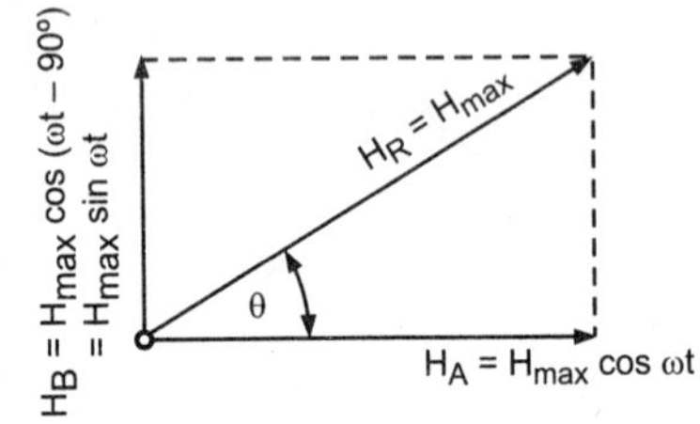

Fig. 4.39

4.13 PRODUCTION OF 3-PHASE ROTATING MAGNETIC FIELD

Consider three similar coils A, B and C displaced in space by 120°, as shown in Fig. 4.40, and connected to 3-phase ac supply. Let each of alternating fields due to currents in coils A, B and C be resolved into two components OA_1 and OA_2, OB_1 and OB_2 and OC_1 and OC_2 respectively travelling with angular velocity ω in the directions shown in Fig. 4.40, where ω = 2π*f*, *f* being the supply frequency.

Consider the instant when the current in coil A is maximum. At this instant both of the components OA_1 and OA_2 of the alternating field produced by the current in coil A are along the axis of the coil A *i.e.* along OA. The current in coil B is 120° behind the instant of its maximum value, therefore, each of its components OB_1 and OB_2 will have to rotate through 120° in order to reach the axis of the coil B *i.e.* along OB. Hence components OB_1 and OB_2 are along OC and OA respectively. The current in coil C is 240° being the instant of its maximum value, therefore, each of components OC_1 and OC_2 of the field produced by current in coil C will have to rotate through 240° in order to reach the axis of coil C *i.e.* along OC. Hence at this instant components OC_1 and OC_2 are along OB and OA respectively.

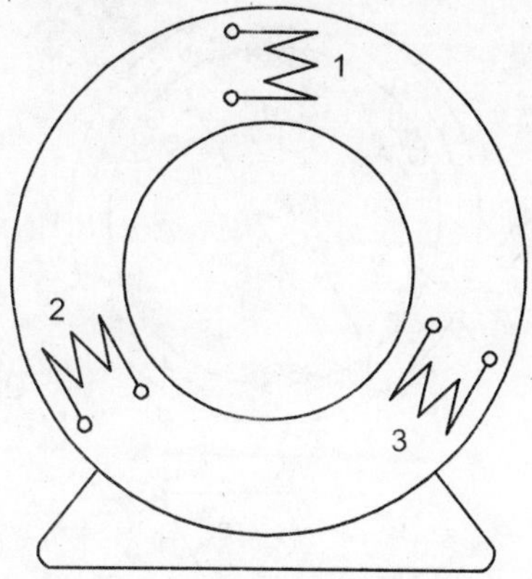

(*a*) *Simple Stator of a Rotating Machine Featuring Three Spatially Positioned Coils at 120° Degrees*

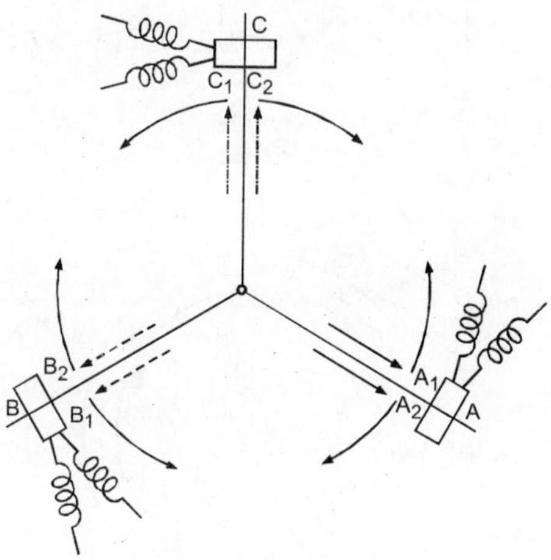

(*b*)

Fig. 4.40

From Fig. 4.41 it is clear that components OA_1, OB_1 and OC_1 are rotating in counter-clockwise direction with same angular velocity ω but are 120° apart, therefore, their resultant is always zero. Components OA_2, OB_2 and OC_2 are rotating in clockwise direction with same angular velocity ω and also points in same direction, therefore, resultant gives a pure rotating field rotating at synchronous speed.

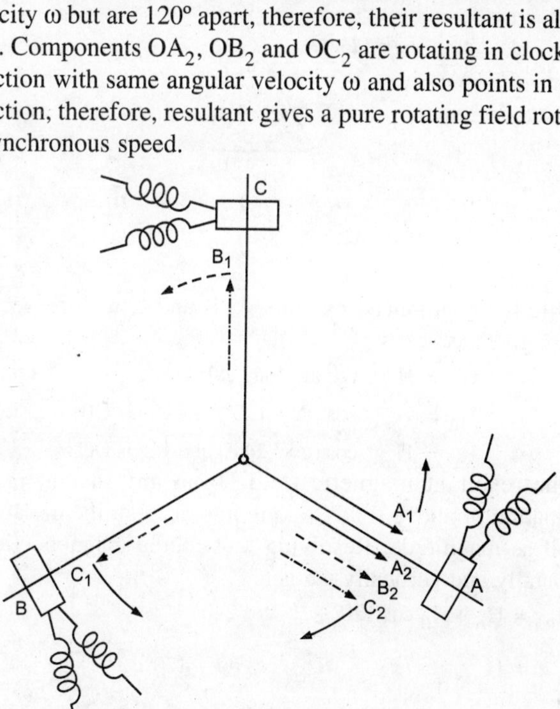

Fig. 4.41

The direction of rotation is the same as the direction of phase sequence *i.e.* first current in coil A attains its maximum value, then current in coil B attains its maximum value and then current in coil C attains its maximum value, as shown in Fig. 4.42. Hence to reverse the direction of rotation of the magnetic field, it is necessary to change the phase sequence which can be done by interchanging any two of the terminals.

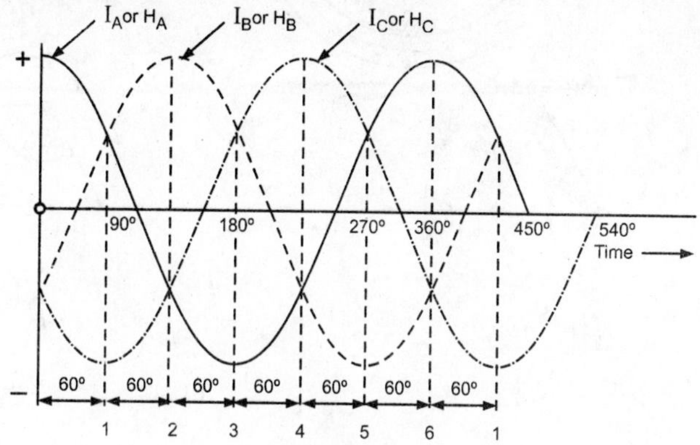

Fig. 4.42

For proving the establishment of rotating magnetic field by three phase currents by considering the direction of flow of currents through three phase windings at several successive instants, let us again consider a simplified stator core with three equally spaced phase windings. The time diagrams for currents in phase windings A, B and C are shown in Fig. 4.42. The different instants are marked off at 60° intervals on the current waves in Fig. 4.42.

Figure 4.43 shows the directions of currents in the stator conductors and the resulting magnetic fields at each of six different instants chosen in Fig. 4.42. The dots and crosses show the directions of currents in the conductors when they are positive. At instant 1, the current is outward in the three coil sides on the upper half of the stator and is inward in three coil sides on the lower half of the stator. The resulting field is established towards the right, creating a north pole on the stator surface at the left and a south pole at the right, as shown. At instant 2, the current is still in the same direction in phase B and C but has reversed in phase A. This establishes a field of the same strength as at instant 1 but in a position 60° in a clockwise direction from instant 1. At instant 3, the current in phase C has reversed, causing the resultant field to be moved another 60° in the clockwise direction. The remaining diagrams in Fig. 4.43 show the resultant field at instant 4, 5 and 6.

It is thus evident from Fig. 4.43 that the resultant field established by the currents flowing in the three phase windings is of the magnitude 1.5 times of the maximum strength produced by individual phase windings and shifts around the stator surface, moving through a definite distance in each of the time intervals considered. The resultant field makes one revolution in one cycle. If the stator is wound for P poles, the field makes one revolution in $\frac{P}{2}$ cycles or $\frac{2f}{P}$ revolutions in one second or $\frac{120f}{P}$ revolutions in one minute.

If any two of the three supply lines to the stator winding are reversed, thereby reversing the phase sequence of the stator

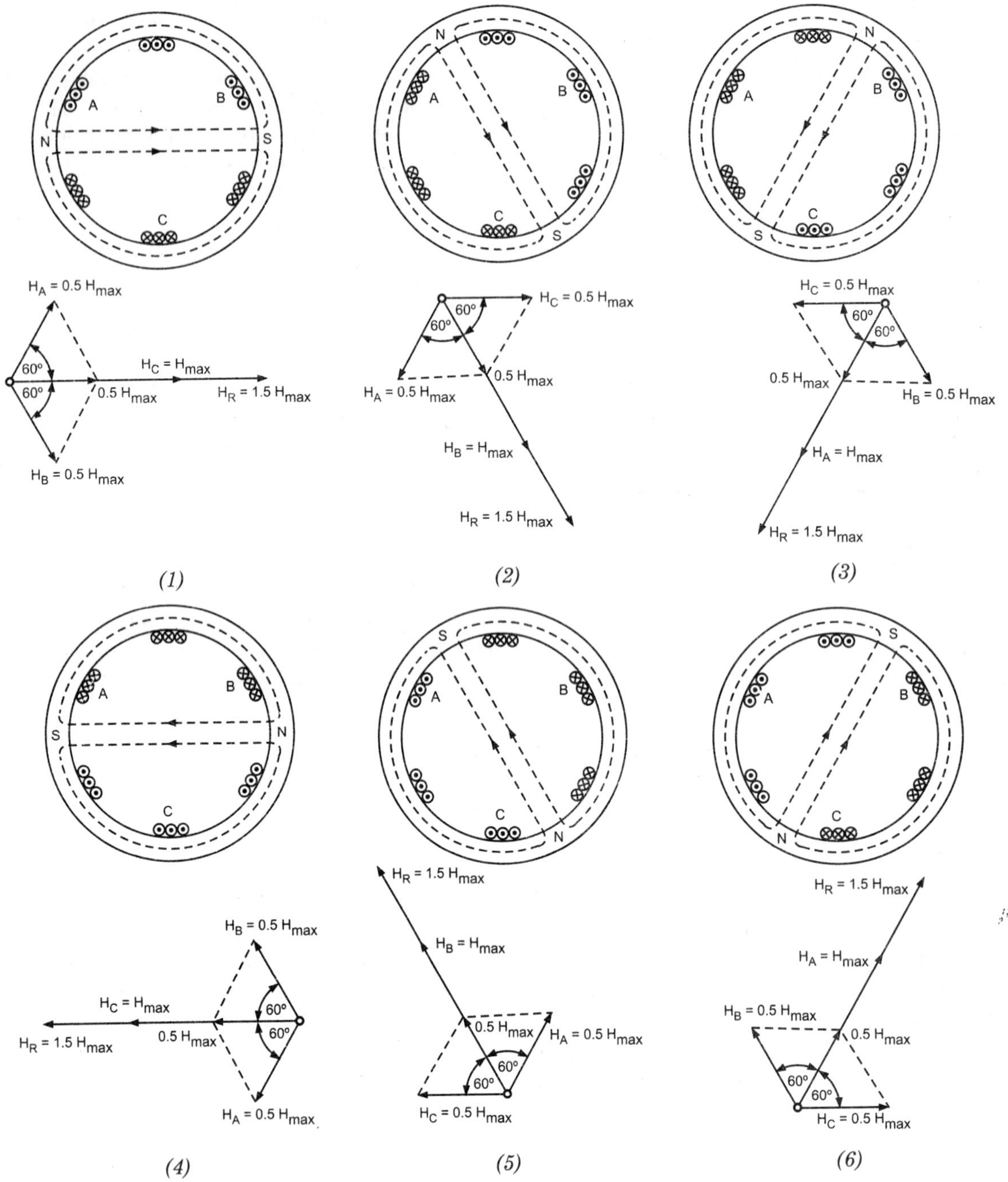

Fig. 4.43

currents, the analysis will show the rotation of the resultant magnetic field to be in the reverse direction, or counter-clockwise.

Mathematical Proof. The above fact can also be proved mathematically as below:

The time diagrams for currents in coils A, B and C are shown in Fig. 4.42. Coils A, B and C establish alternating (pulsating) fields along their respective axis *i.e.* along axis OA, OB and OC respectively. Since phase difference between the currents flowing through coils A, B and C respectively is 120°, therefore, phase difference between the magnetic fields H_A, H_B and H_C produced is 120° (electrical).

Let each of the coil produce an alternating field of maximum strength H_{max} along its own axis then instantaneous values of magnetic fields produced by coils A, B and C with respect to time are given as below:

$$H_A = H_{max} \cos \omega t \text{ along axis OA}$$

$$H_B = H_{max} \cos (\omega t - 120°) \text{ along axis OB}$$

$$H_C = H_{max} \cos (\omega t - 240°) \text{ along axis OC}$$

The resultant magnetic field H_R at any instant is the combination of above fields, taking into account the directions as well as magnitude. Resolving these three magnetic fields horizontally and vertically we get

$$H_{Rh} = H_A + H_B \cos 120° + H_C \cos 240°$$

$$= H_{max} \cos \omega t - \frac{1}{2} H_{max} \cos (\omega t - 120°)$$

$$- \frac{1}{2} H_{max} \cos (\omega t - 240°)$$

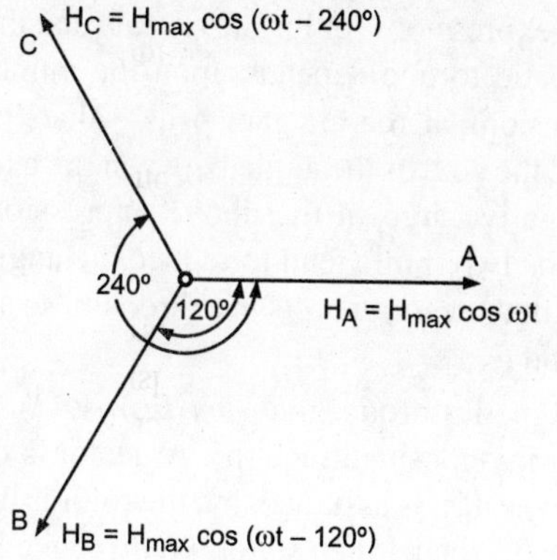

Fig. 4.44

$$= H_{max} \cos \omega t$$

$$- \frac{1}{2} H_{max} [\cos (\omega t - 120°) + \cos (\omega t - 240°)]$$

$$= H_{max} \cos \omega t - \frac{1}{2} H_{max} [2 \cos (\omega t - 180°) \cos 60°]$$

$$= H_{max} \cos \omega t + \frac{1}{2} H_{max} \cos \omega t = \frac{3}{2} H_{max} \cos \omega t$$

$$H_{Rv} = H_A \sin 0° + H_B \sin (-120°) + H_C \sin (-240°)$$

$$= 0 - \frac{\sqrt{3}}{2} H_{max} \cos (\omega t - 120°)$$

$$+ \frac{\sqrt{3}}{2} H_{max} \cos (\omega t - 240°)$$

$$= \frac{\sqrt{3}}{2} H_{max} [\cos (\omega t - 240°) - \cos (\omega t - 120°)]$$

$$= \frac{\sqrt{3}}{2} H_{max} [2 \sin (\omega t - 180°) \sin 60°] = \frac{3}{2} H_{max} \sin \omega t$$

The resultant field, H_R

$$= \sqrt{(H_{Rh})^2 + (H_{Rv})^2}$$

$$= \sqrt{\left(\frac{3}{2} H_{max} \cos \omega t\right)^2 + \left(\frac{3}{2} H_{max} \sin \omega t\right)^2} = \frac{3}{2} H_{max}$$

Inclination of the resultant field is given by

$$\theta = \text{Tan}^{-1} \frac{H_{Rv}}{H_{Rh}} = \text{Tan}^{-1} \frac{\frac{3}{2} H_{max} \sin \omega t}{\frac{3}{2} H_{max} \cos \omega t} = \omega t$$

It shows that the resultant magnetic field due to three phases is of constant magnitude (3/2 times the maximum strength of magnetic field produced by each phase separately) and travels in space with a uniform angular velocity ω radians per second. *Hence resultant field is of constant magnitude and is rotating at synchronous speed.*

The above principle is not only applicable to synchronous machines but to all machines with polyphase stator windings.

4.14 TORQUE IN ROUND ROTOR MACHINE

When the currents flow through both of the stator and rotor windings, they develop their own magnetic fields along their respective axes which are sinusoidally distributed along the air gap. Let F_1 and F_2 be the peak values of the mmfs per pole (or per air gap crossing) respectively of stator and rotor fields. In case the stator or rotor has more than one winding, then F_1 is the resultant stator mmf per pole produced by all stator windings and F_2 is the resultant rotor mmf per pole produced by all rotor windings. These two mmfs F_1 and F_2, in effect, cause the appearance of stator and rotor poles, along their respective magnetic axis—this is illustrated in Fig. 4.45 (*a*). The stator and rotor poles so developed, have a tendency to align their magnetic axis and in this way, electromagnetic torque is produced. The torque tending to align the two fields is developed only when the two fields have the same number of poles and are stationary with respect to each other. Two relatively rotating fields will develop alternating torque as they cross each other so that the average torque is zero. All types of electrical machines (dc, synchronous, induction) are, therefore, devised to develop interacting fields with zero relative velocity.

For deriving a general torque expression, the following assumptions are made:

1. The mmfs due to stator and rotor currents are both sinusoidally distributed in the air gaps. For achieving it distributed windings are used.
2. The air gap (clearance between stator and rotor) is assumed to be uniform throughout i.e. rotor is assumed smooth cylindrical.
3. The tangential (along the air-gap periphery) component of the magnetic field is negligible compared to the radial (along the radius) component. In other words, the flux goes straight across the gap.
4. The radial length of the air gap is small compared to the radii of stator and rotor. This effectively means that there is negligible difference between the flux density at the stator surface, at the rotor surface or at any radial distance in the air gap.
5. The iron parts (stator and rotor iron) are assumed to have infinite permeability. This means that the saturation and hysteresis are negligible.
6. Only the fundamental sinusoidal components of the stator and rotor mmf waves are considered.

Since the mmfs F_1 and F_2 are spatial sinusoidal waves, these can be represented by space phasors $\mathbf{F}_1$ and $\mathbf{F}_2$ directed along their respective magnetic axis in Fig. 4.45 (*a*). The resultant mmf $\mathbf{F}_R$ acting across the air gap is equal to the phasor sum of $\mathbf{F}_1$ and $\mathbf{F}_2$. The magnitude of the resultant mmf F_R is given as

$$F_R = \sqrt{F_1^2 + F_2^2 + 2F_1F_2 \cos \alpha} \qquad ...(4.36)$$

where α is the space angle between peak values of stator mmf F_1 and rotor mmf F_2.

Resultant field intensity H, developed by resultant mmf wave F_R, is sinusoidal along the air-gap periphery. As mentioned above, F_R acts across one air gap, therefore, peak value of magnetic field H_{peak} is given as

$$H_{peak} = \frac{F_R}{l_g} \qquad ...(4.37)$$

The coenergy density at a point in the air gap, where magnetic field intensity is H, is given by $\frac{1}{2} \mu_0 H^2$.

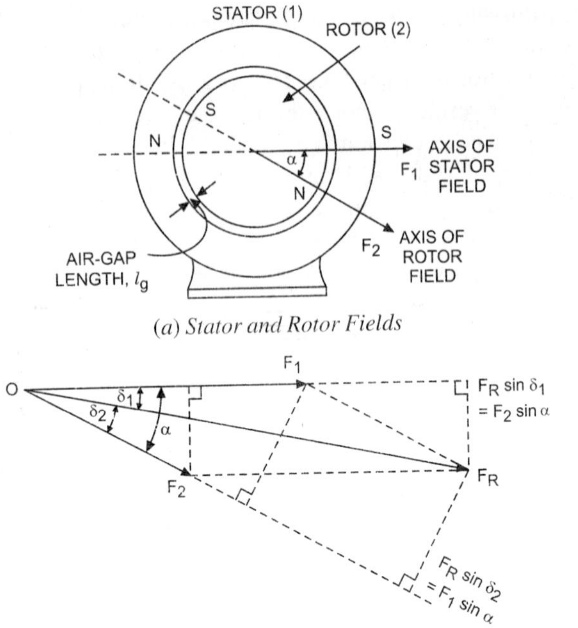

(a) Stator and Rotor Fields

(b) Space Phasor Diagram For MMFs F_1 and F_2.

Fig. 4.45 *Production of Torque in Non-Salient Pole Machine.*

So peak coenergy density in the air gap

$$= \frac{1}{2}\mu_0 H^2_{peak} = \frac{1}{2}\mu_0\left(\frac{F_R}{l_g}\right)^2 = \frac{1}{2}\frac{\mu_0}{l_g^2}F_R^2 \quad ...(4.38)$$

Since mmfs F_1 and F_2 are sinusoidal waves, their resultant mmf F_R must also be a sinusoidal wave. The average value of the coenergy density over the air gap volume, is

$$W'_{field} = \frac{1}{2}\frac{\mu_0}{l_g^2} \text{ (average value of } F_R^2)$$

$$= \frac{1}{2}\frac{\mu_0}{l_g^2}\left(\frac{1}{2}F_R^2\right) = \frac{1}{4}\mu_0\left(\frac{F_R}{l_g}\right)^2 \quad ...(4.39)$$

If D and l are the average diameter of the core (at the air gap) and axial length of the core respectively, then

Volume of air gap = $\pi D l l_g$

The total energy of the field is then

$$W'_{field} = \frac{1}{4}\mu_0\left(\frac{F_R}{l_g}\right)^2 \times \pi D l l_g$$

$$= \frac{\pi}{4}\mu_0\left(\frac{F_R}{l_g}\right)^2 D l l_g$$

$$= \frac{\pi\mu_0 D l}{4 l_g}F_R^2 \quad ...(4.40)$$

$$= \frac{\pi\mu_0 D l}{4 l_g}(F_1^2 + F_2^2 + 2F_1F_2 \cos\alpha) \quad ...(4.41)$$

$\because$ From Eq. (4.36) $F_R^2 = F_1^2 + F_2^2 + 2F_1F_2 \cos\alpha$

Torque produced is given by

$$T = +\frac{\partial W'_{field}}{\partial \alpha} = -\frac{\pi\mu_0 D l}{2 l_g}F_1F_2 \sin\alpha \quad ...(4.42)$$

For a P-pole machine

$$T = -\left(\frac{P}{2}\right)\frac{\mu_0 \pi D l}{2 l_g}F_1F_2 \sin\alpha \quad ...(4.43)$$

The above expressions for torque reveals that the magnitude of electromagnetic torque depends upon the number of poles, machine dimensions at the air gap, peak values of stator and rotor mmfs and the sine of the angle between the axes of the two fields. The negative sign in the above expression for torque indicates that the two mmfs tend to reduce the angle α between them. It means that the torque acts in a direction so as to align the two mmfs F_1 and F_2.

Electromagnetic torque given by Eq. (4.43) acts on both stator and rotor in opposite directions. As stator is fixed, it is the rotor that revolves. If it is assumed that the rotor is held fixed and the stator is free to rotate, then stator would rotate in a direction opposite to that in which rotor had rotated with stator held fixed. It means that stator and rotor experience equal and opposite torques as given by Eq. (4.43). In practice, electromagnetic torque acting on the stator is transmitted to ground through its foundation and the rotor rotates.

With reference to the geometry of Fig. 4.45 (*b*)

$$F_1 \sin\alpha = F_R \sin\delta_2$$

and $F_2 \sin\alpha = F_R \sin\delta_1$

The torque expression given by Eq. (4.43) can, therefore, be expressed in two more alternative forms

$$T = -\left(\frac{P}{2}\right)\frac{\mu_0 \pi D l}{2 l_g}F_1 F_R \sin\delta_1 \quad ...(4.44)$$

$$T = -\left(\frac{P}{2}\right)\frac{\mu_0 \pi D l}{2 l_g}F_2 F_R \sin\delta_2 \quad ...(4.45)$$

An examination of Eqs. (4.44) and (4.45) reveals that the electromagnetic torque is proportional to the product of one mmf (here F_1 or F_2) and a normal component of the resultant mmf F_R.

In Eq. (4.43) torque is proportional to the products of two mmfs (or fields) produced by their respective currents and sine of the angle between them. In Eqs. (4.44) and (4.45), the torque is proportional to the product of resultant mmf (or field), one of the two mmfs and sine of the angle between them.

Alternatively, the expression for torque can also be written in terms of resultant flux Φ_R,

$$\Phi_R = \text{Average flux density over a pole} \times \text{pole area}$$

$$= B_{av} \times \text{pole area} = \frac{2}{\pi}B_R \times \frac{\pi D l}{P} = \frac{2 B_R D l}{P} \quad ...(4.46)$$

where B_R is the peak value of the resultant flux density wave.

$\because$ Average value of sinusoid is $\frac{2}{\pi}$ times its peak value

and pole area $= \frac{\pi D l}{P}$

$$B_R = \mu_o H_g = \frac{F_R}{l_g} \quad ...(4.47)$$

Combining Eqs. (4.45), (4.46) and (4.47) we have

$$T = -\frac{\pi}{8}P^2 F_2 \Phi_R \sin\delta_2 \quad ...(4.48)$$

The above expression provides a very useful expression for the torque in machines with cylindrical air gaps.

4.15 MATCHING CHARACTERISTICS OF ELECTRIC MACHINE AND LOAD

An electromechanical energy conversion system consists of two main components, viz., the machine and the mechanical load. Mechanical load may be either rotating in nature, or it may have a translational motion. A rotating load is either directly coupled to the shaft of the motor or through some gearing mechanism. The characteristics of the machine, generally, play a vital role in the operating behaviour of the complete system.

For selection of reliable and efficient motor it is essential that the conditions of service are well known. It is not sufficient to simply specify the output power in kW and the speed but it is also necessary to know the following additional particulars :

(*i*) Torque at the shaft during running, starting and at different loads.
(*ii*) Accelerating torque and braking torque.
(*iii*) Switching frequency.
(*iv*) Efficiency of motor at different loads.
(*v*) Other working requirements.

In studying the behaviour of a motor selected for a particular driven unit, one of the first problems involved is to determine whether the speed-torque characteristic of the motor suits the requirements imposed by the speed-torque characteristic of the driven unit. Drive behaviour during the transient period of a start-up, braking, or speed change-over also depends upon how the speed-torque characteristics of the motor and the driven unit vary with speed. It is, therefore, imperative to study these characteristics in order to be able to select correctly the motor and obtain an economical drive.

4.15.1. Speed-Torque Characteristics of Machines or Mechanisms. *The speed-torque characteristic of a machine or mechanism* given by the relation $\omega = f(T_L)$ is defined as the relationship between the speed at which it is operated and the resisting or load torque it develops.

Different kinds of mechanisms and machines exhibit different speed-torque characteristics. However, several general conclusions may be drawn if we use the following empirical equation for the speed-torque characteristic of some driven unit of industrial equipment

$$T_L = T_0 + (T_{rn} - T_0)\left(\frac{\omega}{\omega_n}\right)^x \quad ...(4.49)$$

where T_L is gross load (or resisting) torque developed by the unit at speed ω, T_0 is the resisting torque developed by the unit due to friction in its moving parts, T_{rn} is the resisting torque developed by the unit when it is driven at its nominal rated speed ω_n and x is the exponential coefficient characterising the change in resisting torque with the change in speed.

The above Eq. (4.49) permits the speed-torque characteristics of different kinds of machines and mechanisms to be roughly divided into the following categories :

1. ***Loads Requiring Constant Torque At All Speeds.*** Such a load presents to the motor a passive torque which is essentially independent of speed. It is characterized also by the requirement of an extra torque at very near zero speed. For this characteristic $x = 0$ and the load torque T_L does not depend upon speed. The speed-torque characteristic for such loads is illustrated by vertical line in Fig. 4.46. Such loads are dry friction, cranes during hoisting, hoist winches, machine tool feed mechanism, piston pumps operating against a constant pressure head, and conveyors handling a constant weight of material per unit time. In power applications it is usually called the breakaway torque and in control systems, it is referred to as *stiction* (derived from sticking friction).

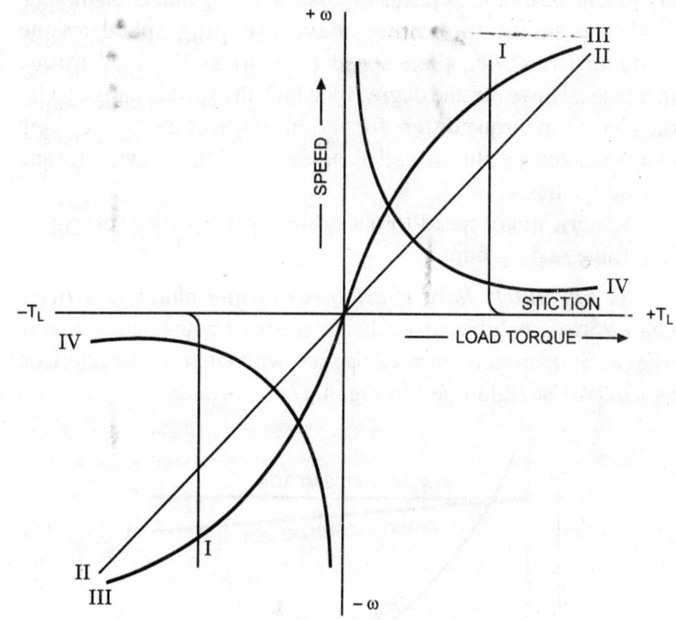

Fig. 4.46 *Speed-Torque Characteristics of Typical Different Types of Machines and Mechanisms*

Since it changes sign with reversal of rotation, the dry friction torque characteristic is discontinuous, as illustrated in Fig. 4.46.

2. ***Loads With Linear-Rising Characteristic.*** Such speed-torque characteristics, illustrated by straight line II in Fig. 4.46, are exhibited by calendering machines, eddy current brakes, separately-excited dc generators supplying fixed ohmic resistance loads and fluid or viscous friction. In this case $x = 1$ and the load torque T_L rises in direct proportion to the speed.

3. ***Loads With Nonlinear-Rising (Parabolic) Characteristic.*** For such a characteristic $x = 2$ and the load torque T_L is proportional to the square of the speed. Such a characteristic is illustrated by curve III in Fig. 4.46. A load with considerable windage, of which a fan is the extreme example, has a torque which varies almost as the square of the speed. Blowers, centrifugal pumps, propellers in ships or aeroplanes, waterwheels, pipe friction, velocity head of pumps etc. also have the same type of speed-torque characteristics.

4. ***Loads With Nonlinear Falling (Hyperbolic) Characteristic.*** For such a characteristic $x = -1$ and the load torque T_L is inversely proportional to the speed, while the power required to drive the given unit remains unchanged. Such a characteristic is illustrated by curve IV in Fig. 4.46. Certain types of lathes, boring machines, milling machines and other kinds of metal-cutting machine tools, steel-mill coilers fall under this category of loads.

The categories of loads listed above do not cover all the cases that may be met with in practice but give a good idea of the characteristics typical of a great many kinds of industrial equipment. In actual practice we may come across loads which are combination of these basic types of loads.

4.15.2. Motor Speed-Torque Characteristics. A motor speed-torque characteristic is defined as the relationship between the speed at which it operates and the torque it developes, *i.e.*, $\omega = f(T)$.

Practically all electric motors–shunt wound, series wound, compound dc motors, squirrel cage and slip-ring induction motors, and ac commutator motors, have drooping speed-torque characteristics, *i.e.*, their speed falls off as the load torque increases. However, the degree by which the speed changes with the change in torque differs for various types of motors, it being characterized by the so-called hardness of their speed-torque characteristics.

Electric motor speed-torque characteristics may be classified into three main groups.

1. ***Absolutely hard (flat) speed-torque characteristic,*** a characteristic exhibiting no change in speed with change in load torque. Synchronous motors operate with such a characteristic (horizontal straight line I in Fig. 4.47).

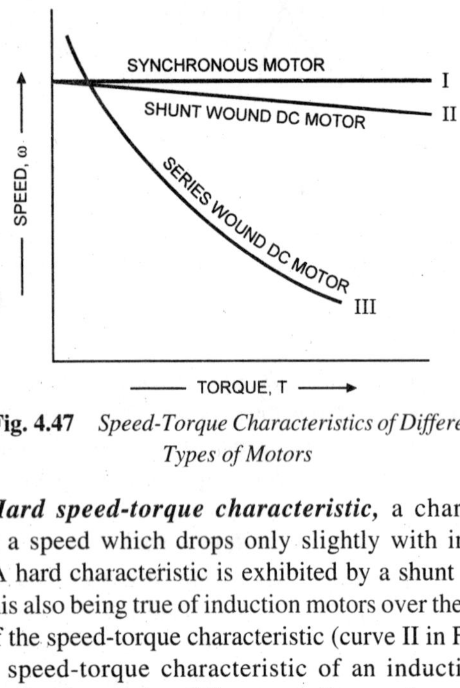

Fig. 4.47 *Speed-Torque Characteristics of Different Types of Motors*

2. ***Hard speed-torque characteristic,*** a characteristic showing a speed which drops only slightly with increase in torque. A hard characteristic is exhibited by a shunt wound dc motor, this also being true of induction motors over the operating region of the speed-torque characteristic (curve II in Fig. 4.47).

The speed-torque characteristic of an induction motor exhibits a "hardness" that differs according to what point along the characteristic is taken into consideration (Fig 4.48). Between the points of the maximum torque in motor operation $T_{max\,M}$ and the maximum torque in generator operation $T_{max\,G}$ an induction machine will exhibit a rather hard characteristic.

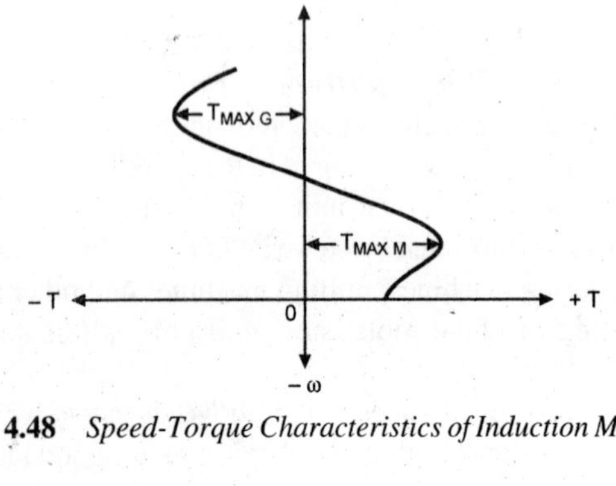

Fig. 4.48 *Speed-Torque Characteristics of Induction Motors*

3. ***Soft speed characteristic,*** a characteristic showing considerable drop in speed with rise in torque. Series wound dc motor possesses such characteristic, especially along the low-torque portion of the characteristic (curve III in Fig. 4.47). For such motors, the degree of hardness of the characteristic varies all along the curve.

Compound wound dc motors, depending on the degree of hardness their speed-torque characteristics display, may be considered as hard-or soft-characteristic motors.

For given speed-torque characteristic of load, motor should have speed-torque characteristic such that

(*i*) The characteristics of both of the load and the motor should intersect as far as possible at 90° as illustrated in Fig. 4.49. This gives very sharp point of intersection i.e. point of operation is well defined. It is further observed that slight change in motor characteristic from thick to dotted line due to variation in supply voltage or frequency gives only a slight change in speed. However if the two characteristics intersect at an angle much lesser than 90° (Fig. 4.50), slight variation in the motor characteristic will shift the operating point considerably.

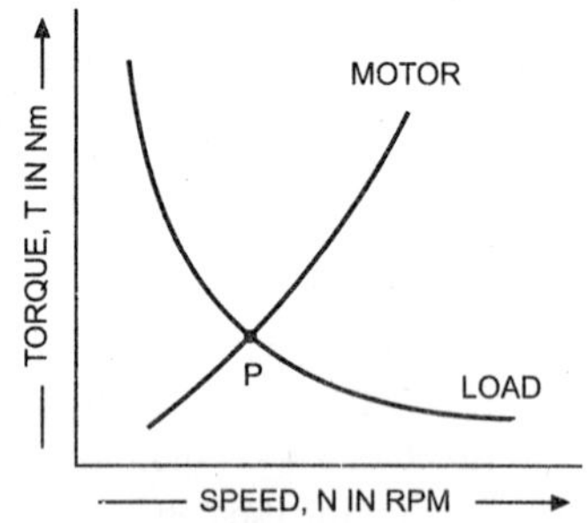

Fig. 4.49

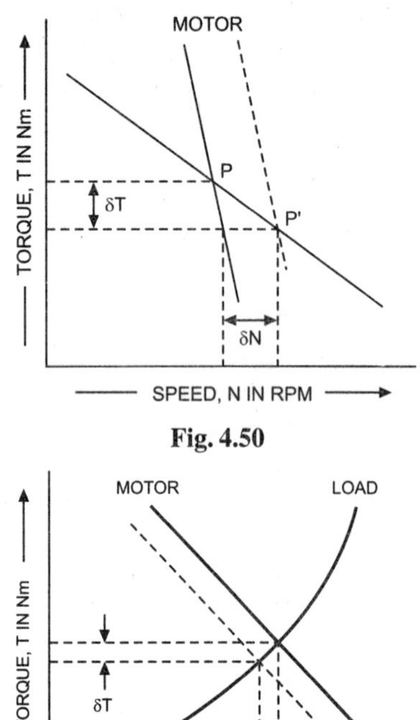

Fig. 4.50

(*ii*) **Stable Operation:** For stable operation of the drive it is necessary that after the point of intersection of the two characteristics, the motor torque should decrease with the increase in speed and vice versa, as depicted in Fig. 4.51. However if the two characteristic curves intersect as shown in Fig. 4.49, operation will not be stable, because slight increase in speed gives rise to acceleration as the motor torque exceeds the load torque. Similarly decrease in speed produces retardation because the motor torque becomes lesser than that of load.

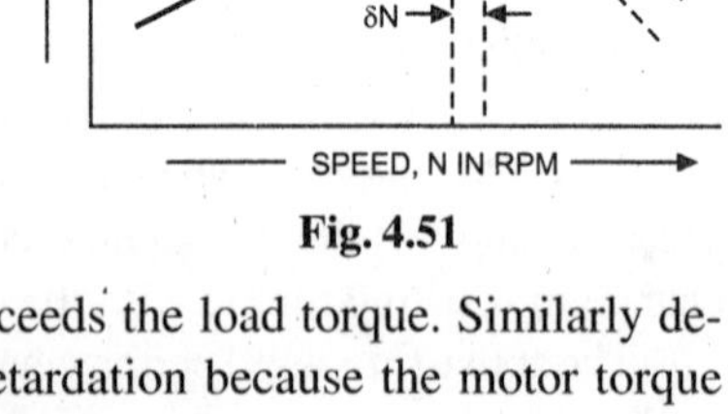

Fig. 4.51

4.16 INSULATING MATERIALS

Insulating materials are employed for providing electrical insulation between parts at different potentials. These materials begin to deteriorate at relatively small temperatures. For reliable operation, it is essential that the temperature rise in electrical machines and equipment do not exceed the permissible maximum temperature of the insulating materials used therein.

The three fundamental electrical properties of insulating materials of utmost importance in the operation of electrical machines or equipment are resistivity (insulation resistance), dielectric strength and dielectric loss angle. In addition to the above, other important properties of insulating materials are mechanical strength, heat resistance, hygroscopicity.

An insulating material should have high resistivity, high dielectric strength, low dielectric loss, good heat conductivity, sufficient mechanical strength to withstand vibrations etc., capability of withstanding a repeated heat cycle without deterioration and should be non-hygroscopic.

The materials used for insulation in electrical machines and apparatus are numerous (such as insulating varnishes, mineral waxes, synthetic waxes, varnish, cloth, silk, cotton, rubber, vulcanized rubber, hard rubber, bakelite, paper, wood, mica, asbestos, glass, porcelain, marble, slate, insulating oils etc.). Some of the most important insulating materials are briefly described below:

1. Cotton. Cotton cloth or tape, made from cotton fibre is flexible, mechanically strong and thin. It gives a high space factor when used for wrapping of coil windings. Cotton has low dielectric strength and is hygroscopic. It must be impregnated with varnish after winding. Varnished cotton clothes have high dielectric strength and great flexibility.

Cotton (cotton tape, empire cloth etc.) is mainly employed for covering bare wire used in winding of small magnet coils, armature windings of small machines, chokes and coils of small transformers.

2. Silk. It is a protein fibre having a long chain structure similar to that of cellulose. The fibre is solid and from 0.01 to 0.015 mm in diameter. It has a higher dielectric strength and is less hygroscopic than cotton. Space factor is better than for cotton, but it is more expensive. Silk is coated with a waxy form of lubricant having good insulating properties. It is mainly used for windings of small machines.

3. Asbestos. It is a fibrous crystalline mineral. Asbestos fibres are from a few tenths of a mm to several cm long. It is made into yarn tape, cloth, paper, board etc. The properties of asbestos as an electrical insulator are not very good.

4. Glass. In electrical machines, glass is mainly used as a glass tape and varnished glass cloth. It has high electrical and mechanical strength and is capable of withstanding higher temperatures. Glass tapes are used for insulating the windings and winding ends of electrical machines operating at high temperatures. Varnished glass cloth is used mainly for insulating the stator windings in high voltage electrical machines.

5. Insulating Varnishes. Insulating varnishes are solutions of resinous materials (such as gums, asphalts etc.) in volatile liquids (such as petrol, benzene, alcohols etc.). There are three types of insulating varnishes namely (*i*) impregnating (*ii*) finishing or coating and (*iii*) sticking or bonding:

Impregnating varnishes serve to treat porous fibrous or solid insulators used in electrical machines and apparatus. *Finishing* or *coating varnishes* produce a hard lustrous coating on materials which are more or less resistant to air moisture. *Sticking* or *bonding* varnishes are used to cement cloth, mica, paper etc.

6. Bakelite. It is synthetic product produced by condensation of phenolic materials with formaldehyde. Its insulating properties are good and is used in manufacturing of all kinds of small electric fittings such as lamp holders, switch covers, terminal boards. Its relative permittivity is 4.5 to 6 and electric strength is 10 to 20 kV/mm. It is also used as an insulating cylinder in transformers.

7. Paper. It is prepared from wood pulp, rags or plant fibre by a suitable chemical process. It is quite cheap and hygroscopic. It has low capacitance, high dielectric strength and high insulation resistivity when dry or impregnated with oil. A small amount of moisture lowers considerably its insulation resistance. Hence before using paper as an insulation it must be impregnated in an insulating oil. It is used for condensers, cables, laminates, bakelite sleeves and mica paper.

8. Mica. It is a mineral product and of finely laminated structure. It is easily split into thin sheets, which are very flexible, tough and high resistance to heat. Its electric strength is very high (80 to 200 kV/mm). It is fireproof and does not absorb moisture. Mica splitting can be pressed into square plates to be used for capacitors, washers for electrical appliances etc. Most commonly mica splittings are bonded together by means of bonding varnish and built up into sheets, plates and various commercial products. It is greatly used in commutators for separating its segments and in condensers as dielectric.

4.16.1. Classification of Insulating Materials. The temperature rise which electric motor can safely withstand is determined by the thermal stability (limiting temperature) of the insulating materials used in them. By using materials of higher thermal stability, motors of the same physical size may be rated for larger power output.

Different insulating materials have different limiting temperatures. According to their limiting temperature all insulating materials employed in electric machines are classified as per IS 1271 as below:

1. Class Y- Cotton yarn and fabrics, paper and fibrous materials of cellulose or silk neither impregnated with nor immersed in liquid dielectric. The limiting temperature is 90 °C.

2. Class A- The same materials as mentioned above, but impregnated with or immersed in liquid dielectric. Limiting temperature is 105 °C.

3. Class E- Certain synthetic organic films and other materials having the same thermal stability. Limiting temperature is 120 °C.

4. Class B- Mica, asbestos or glass-fibre base materials combined with an organic binding agent. The limiting temperature is 130 °C.

5. Class F- Mica, asbestos or glass -fibre base materials combined with suitable synthetic binding and impregnating agents. The limiting temperature is 155 °C.

6. Class H- Mica, asbestos or glass fibre combined with silicon binding and impregnating agents. The limiting temperature is 180 °C.

7. Class C- Mica, ceramic materials, glass, or quartz used without or with inorganic binding agents. For this class of insulating materials, the limiting temperature is above 180 °C and is restricted only by the specific physical, chemical or electrical properties of the material concerned.

The limiting temperatures are never exceeded in practice, otherwise it will result in deterioration and breakdown of the insulation and will shorten the life of the motors.

4.17 POWER LOSSES AND EFFICIENCY

Electromechanical energy conversion is necessarily accompanied by a certain amount of irreversible conversion of energy to heat in the conversion device or machine. These energy losses arise because of circuit resistances (copper or electrical losses), because of the existence of alternating or fluctuating magnetic fields (core or iron losses), and from mechanical factors (friction and windage). Although they play essentially no basic role in the energy conversion process, losses are nevertheless important factors in the practical application of machines.

Treatment of machines losses is important for three reasons :

(*i*) Power losses determine the efficiency of the machine and appreciably influence its operating cost. For example, a machine with low efficiency has more power losses and, therefore, increased operating cost.

(*ii*) Power losses determine the heating of the machine and hence fix the rating or power output that can be obtained without deterioration of the insulation because of overheating.

(*iii*) Voltage drops are associated with ohmic or electrical losses while current components, like core-loss current, pertains to the iron loss in electrical machines. Obviously, the losses associated with voltage drops or current components must be properly accounted for in a machine representation so that the analysis of the electrical machine can be carried out as desired.

The efficiency (η) of an electrical machine, like that of any other apparatus, is defined as the ratio of useful power output to the input power, the two being measured in the same unit (either in watts or kW).

$$\text{i.e. Machine efficiency, } \eta = \frac{\text{Output}}{\text{Input}} \quad ...(4.50)$$

$$= \frac{\text{Output}}{\text{Output + losses}} \quad ...(4.51)$$

$$= \frac{\text{Input-losses}}{\text{Input}} = 1 - \frac{\text{Losses}}{\text{Input}} \quad ...(4.52)$$

Rotating electrical machines in general operate efficiently at light loads. The full-load efficiency of average motors, for example, is about 74 per cent for 0.75 kW size, 89 per cent for 37 kW, 93 per cent for 375 kW, and 97 per cent for 3,750 kW. The efficiency of slow speed motors is usually lower than that of high speed motors, the total spread being 3 or 4 per cent.

The forms of Eqs. (4.51) and (4.52) are often used for electrical machines, since their efficiency is most commonly determined by measurement of losses instead of directly measuring the output and input under load. Loss measurements have the advantage of convenience and economy and of yielding more accurate and precise values of efficiency because a given percentage error in measuring losses causes only one-tenth of that percentage error in the efficiency. Efficiencies determined from measurements of losses can be used in comparing machines if exactly the same methods of measurement and computation are used in each case.

Consideration of machine parts in which power losses can occur will show that the following list includes all the possibilities within the machine itself.

1. Copper or Electrical (I^2R) Losses. These losses occur in the rotor and stator (armature and field) windings. In synchronous and dc shunt machines (except where variable speed is obtained by control of shunt field current), the field copper loss is constant because such machines are normally operated with constant field current. In series and induction machines, the losses in both windings vary as the square of the line current.

2. Core or Iron Losses. These losses consist of eddy current and hysteresis losses. In synchronous and induction machines, these losses are confined essentially to the stator iron, and in dc machines essentially to the rotor (or armature) iron, although in both cases a small core loss will be present in the other member because of small flux variations caused by the slots. The iron in the offending member is laminated to reduce the eddy current loss. In all except series machines, variable-speed shunt motors, and to a lesser degree, compound motors, the air-gap flux and hence the core losses are sensibly constant regardless of load.

3. Friction and Windage Losses. These losses are constant unless the speed varies appreciably. The sum of the friction and windage and core losses is called the *rotational losses.*

4. Stray Load Losses. These are additional hysteresis and eddy current losses arising from any distortion in flux distribution caused by the load currents. These losses are difficult to measure and are usually about 1 per cent of the machine output.

4.18 TEMPERATURE RISE IN ELECTRICAL MACHINES

Heat is developed in all electrical machines due to the losses in the various parts, principally copper loss (I^2R loss) in conductors, eddy current and hysteresis loss in iron and mechanical losses (in rotating machines only) due to friction of the bearing, air friction or windage causing the temperature of that part to rise. This temperature rise continues until all the heat generated is dissipated to the surroundings by one or more of the natural modes of heat transfer, viz. conduction, convection and radiation.

Ultimately then, under steady load, each part achieves a final temperature, the magnitude of which depends on the balance between the rate at which heat is developed in that part (or received by conduction from a hotter part), and the rate at which the heat can be dissipated which is determined by effectiveness of the cooling method.

The temperature rise depends upon (*i*) the amount of heat produced and (*ii*) the amount of heat dissipated per 1°C rise of the surface of a machine.

According to Newton's law of cooling the rate of loss of energy of a hot body is proportional to the difference in temperature between that body and its surroundings. This law is approximately true for moderate temperature difference (up to 100° C) and for bodies dissipating heat by radiation and natural convection. It means that the amount of heat dissipated per 1° C rise of the surface of a machine depends on the surface area of cooling.

The size of motor for any service is governed by the maximum temperature rise when operating under the given load conditions and the maximum torque required. The former is more important because if the motor operates satisfactorily for maximum temperature rise, it will usually provide the required maximum torque, except in special cases where the load consists of heavy peaks followed by relatively long intervals of no load. Electrical machines are, therefore, designed for a limited temperature rise. In fact, the continuous rating of a machine is that rating for which the final temperature rise is equal to or just below the permissible value of temperature rise for the insulating material used in protection of motor windings. When the machine is overloaded for such a long time that its final temperature rise exceeds the permissible limit, it is likely to be damaged. In worst cases, it will result in an immediate thermal breakdown of the insulating material which will cause a short circuit in the motor, thus putting a stop to its functioning. The short circuit may also lead to a fire. In less severe cases immediate thermal breakdown of the insulating material may not occur, but the quality of the insulation will deteriorate such that thermal breakdown with future overloads or even normal loads might soon occur, thus shortening the useful life of the machine.

The temperature rise to which a motor be allowed to rise is limited by the insulation. The maximum temperature rises which should not be exceeded by different types of motor are fully set out in the relevant ISS. Since temperature rise is one of the chief features in fixing the size of a motor, its calculations become a matter of importance and may be carried out as detailed below.

Heating-Time Curve. For determination of an expression for the temperature rise of an electrical machine after time t seconds from the instant of switching it on, let

Power converted into heat = P joules/s or watts

Mass of active parts of machine = m kg

Specific heat of material = C_p joules/kg/°C

Surface area of cooling = S metres^2

Coefficient of cooling = α in watts per metre^2 of surface per °C of difference between surface and ambient cooling temperatures.

Assumptions Made. (*i*) Losses or heat produced remains constant during the temperature rise.

(*ii*) Temperature of the cooling medium remains unchanged.

and (*iii*) Heat dissipated is directly proportional to the difference in temperature of the motor and the cooling medium.

Suppose a machine attains a temperature rise of θ°C above ambient temperature after t seconds of switching on the machine and further rise of temperature by $d\theta$ in very small time dt.

Energy converted into heat = Pdt joules

Heat absorbed = $mC_p d\theta$ joules

Heat dissipated = $S\theta\alpha dt$ joules

Since energy converted into heat

= Heat absorbed + heat dissipated

$$\therefore \; P dt = mC_p d\theta + S\theta\alpha dt \quad ...(4.53)$$

$$\text{or } (P - S\alpha\theta)dt = mC_p\, d\theta$$

$$\text{or } \frac{dt}{mC_p/S\alpha} = \frac{d\theta}{\frac{P}{S\alpha} - \theta} \quad ...(4.54)$$

When final temperature is reached, there is no absorption of heat. Whatever heat is generated, has to be dissipated.

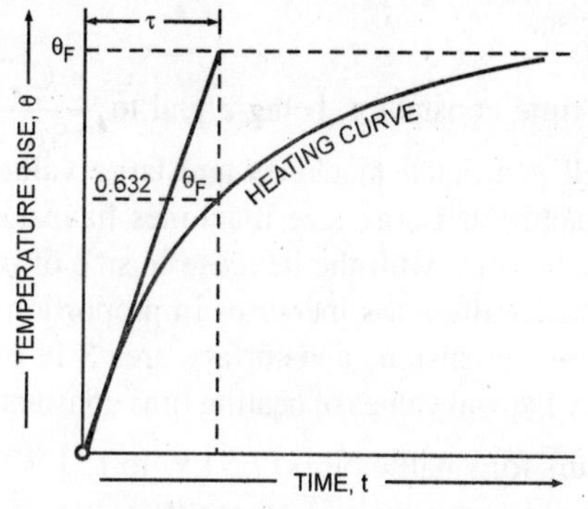

Fig. 4.52(*a*) *Temperature Rise-Time Curve*

$$\therefore \; P dt = S\theta_F \alpha\, dt$$

or Final temperature rise,

$$\theta_F = \frac{P}{S\alpha} \quad ...(4.55)$$

Substituting $\frac{P}{S\alpha} = \theta_F$ in Eq. (4.54), we get

$$\frac{S\alpha\, dt}{mC_p} = \frac{d\theta}{\theta_F - \theta}$$

Integrating both sides, we get

$$\frac{S\alpha}{mC_p} t = -\log_e(\theta_F - \theta) + K_1 \quad ...(4.56)$$

where K_1 is a constant of integration.

Substituting $t = 0, \theta = \theta_1$, initial temperature rise from initial conditions, we have

$$0 = -\log_e(\theta_F - \theta_1) + K_1$$

$$\text{or } K_1 = \log_e(\theta_F - \theta_1)$$

Substituting $K_1 = \log_e(\theta_F - \theta_1)$ in Eq. (4.56), we get

$$\frac{S\alpha}{mC_p} t = -\log_e(\theta_F - \theta) + \log_e(\theta_F - \theta_1) = \log_e\left(\frac{\theta_F - \theta_1}{\theta_F - \theta}\right)$$

$$\text{or } e^{\frac{S\alpha t}{mC_p}} = \frac{\theta_F - \theta_1}{\theta_F - \theta}$$

$$\text{or } \theta_F - \theta = (\theta_F - \theta_1)\, e^{-S\alpha t/m C_p}$$

$$\text{or } \theta = \theta_F - (\theta_F - \theta_1)\, e^{-S\alpha t/m C_p} = \theta_F - (\theta_F - \theta_1)\, e^{-t/\tau} \quad ...(4.57)$$

where $\tau = \frac{mC_p}{S\alpha}$ and is known as *heating time constant.*

If motor is started from ambient temperature, $\theta_1 = 0$, we have

$$\theta = \theta_F - \theta_F e^{-t/\tau} = \theta_F(1 - e^{-t/\tau}) \quad ...(4.58)$$

Heating time constant is defined as the time taken by the motor in attaining the final steady value if the initial rate of rise of temperature were maintained throughout the operation.

Substituting $t = \tau$ in Eq. (4.58), we have

$$\theta = \theta_F (1 - e^{-1}) = 0.632\, \theta_F \qquad ...(4.59)$$

So *heating time constant may also be defined as the time duration during which the machine will attain 63.2% of its final temperature rise above ambient temperature.*

From Eq. (4.55) it is obvious that θ_F is directly proportional to the power losses and inversely proportional to the surface area S and specific heat dissipation α.

For poorly ventilated machines, it will attain a higher final temperature rise.

Heating time constant τ, being equal to $\frac{mC_p}{S\alpha}$, has small value for well ventilated machines and large value for poorly ventilated machines. Large size machines have large heating time constant because with the increase in size of machine, the volume and hence the mass increases in proportion to the third power of linear dimensions and surface area S in proportion to second power. Typical values of heating time constant lie between about $1\frac{1}{2}$ hours for small motors (7.5 kW to 15 kW) up to about 5 hours for motors of several hundred kW.

Cooling-Time Curve. Let the machine be switched off after reaching final steady-temperature rise of θ_F. When the machine is switched off, no heat is produced, therefore,

$$0 = \text{Heat absorbed} + \text{heat dissipated}$$

or $0 = mC_p d\theta + S\theta\alpha' dt$ where α' is the rate of heat dissipation during cooling

$$\text{or } dt = \frac{-mC_p\, d\theta}{S\alpha'\theta}$$

Integrating both sides, we have

$$t = \frac{-mC_p}{S\alpha'} \log_e \theta + K_2 \qquad ...(4.60)$$

where K_2 is a constant of integration

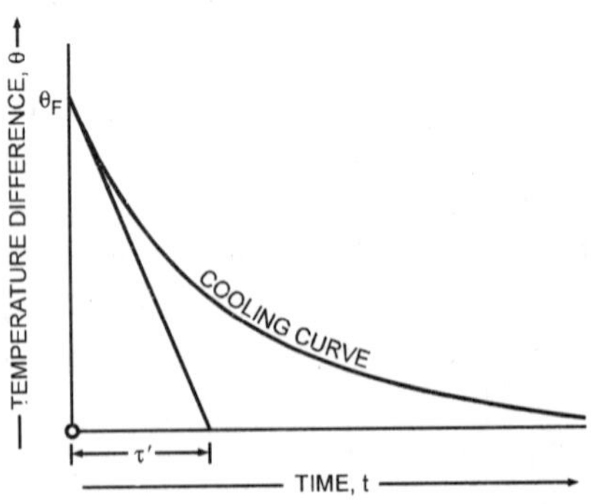

Fig. 4.52(*b*)

Substituting $t = 0$, $\theta = \theta_F$ from initial conditions in above equation, we have

$$K_2 = \frac{mC_p}{S\alpha'} \log_e \theta_F$$

Substituting $K_2 = \frac{mC_p}{S\alpha'} \log_e \theta_F$ in Eq. (4.60), we have

$$t = \frac{-mC_p}{S\alpha'} \log_e \theta + \frac{mC_p}{S\alpha'} \log_e \theta_F = \frac{-mC_p}{S\alpha'} \log_e \frac{\theta}{\theta_F}$$

$$\text{or } \theta = \theta_F\, e^{-S\alpha' t/mC_p} = \theta_F\, e^{-t/\tau'} \qquad ...(4.61)$$

where $\tau' = \frac{mC_p}{S\alpha'}$, called the *cooling time constant.*

Substituting $t = \tau'$ in Eq. (4.61), we have

$$\theta = \theta_F\, e^{-1} = 0.368\, \theta_F \qquad ...(4.62)$$

i.e., cooling time constant may also be defined as the time required to cool the machine down to 0.368 times the initial temperature rise above ambient temperature.

The heating and cooling curves follow an exponential law. Heating time constant and cooling time constant may be different for the same machine, as the ventilation conditions in the two cases may not be the same. The cooling time constant of a rotating machine is usually larger than its heating time constant owing to poor ventilation conditions when the machine cools. In self-cooled rotating machines the cooling time constant is about 2-3 times greater than its heating time constant because cooling conditions are worse at standstill.

4.19 DUTY CYCLES

The nominal duty of a drive motor is the duty corresponding to the service conditions and performance marked on its name plate.

There are three types of duties viz continuous duty, short-time duty and intermittent duty.

The heating and cooling curves for *continuous duty motor* are shown in Fig. 4.53 (*a*). Continuous duty is that duty when the on-period is so long that the motor attains a steady-state temperature rise.

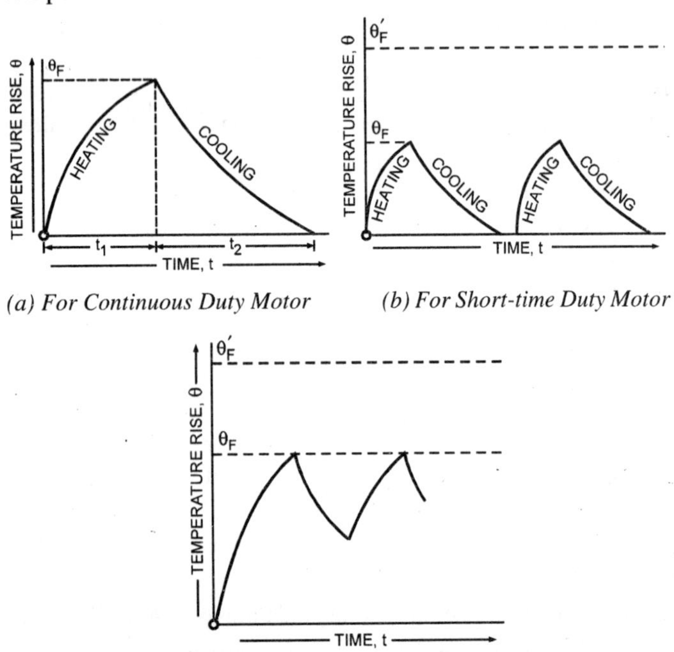

(*a*) *For Continuous Duty Motor* (*b*) *For Short-time Duty Motor*

(*c*) *For Intermittent Periodic Duty Motor*

Fig. 4.53 *Heating and Cooling Curves*

The heating and cooling curves for *short-time duty motor* are given in Fig. 4.53 (*b*). The short-time duty motor operates at a constant load for some specified periods which is then followed by a period of rest. The period of run (or load) is so short that machine cannot attain its steady temperature rise while the period of rest is too long that the motor temperature drops to the ambient temperature.

The heating and cooling curves for *intermittent periodic duty motors* are illustrated in Fig. 4.53 (*c*). On intermittent duty the periods of constant load and rest with machine de-energized alternate. The loading periods are too short to allow the motor to attain its final steady-state value while periods of rest are too small to allow the motor to cool down to the ambient temperature. *Intermittent rating of a machine is defined as the load which is applied during a certain fraction of time of a load cycle and the temperature rise limit is not exceeded.*

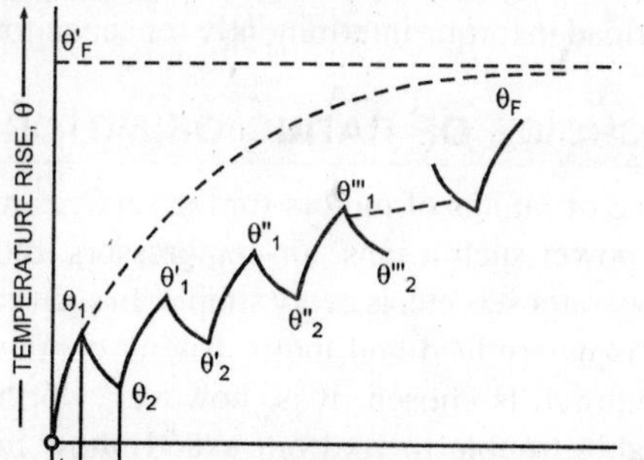

Fig. 4.54

When a machine is intermittently loaded, it will cool down during the time it is off, and temperature will rise when it is on, as illustrated in Fig. 4.54.

Continuous duty motors are employed to drive fans, compressors, generators etc. They may be in operation for many hours and even days in succession. *Short-time duty motors* are used in navigation-lock gates, railway turntables, bascule bridges, and the like. *Intermittent-duty motors* are employed in cranes, hoists, lifts, rolling mills, some metal-working machines, etc.

The duty cycles for various motor duties are shown in Fig. 4.55.

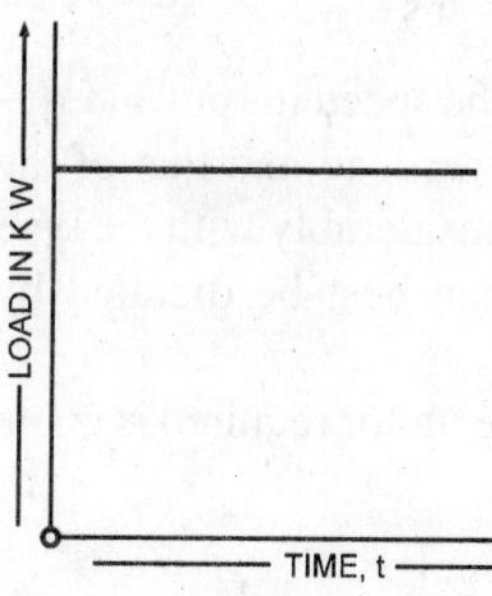

(*a*) *Continuous Duty Cycle*

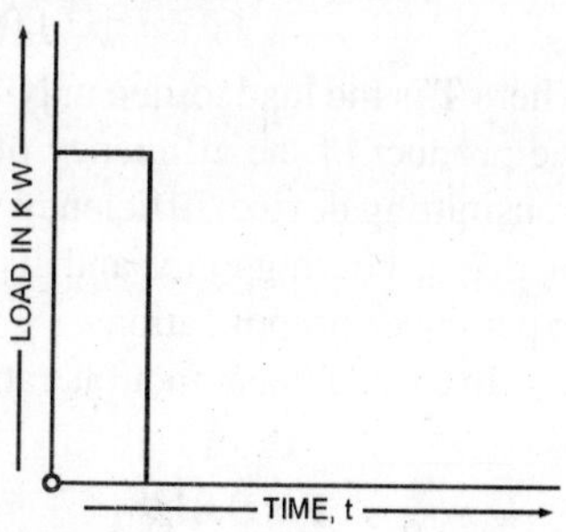

(*b*) *Short-Time Duty Cycle*

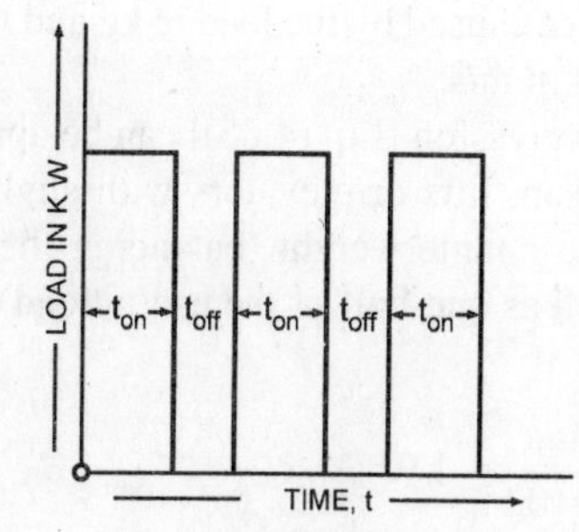

(*c*) *Intermittent Duty Cycle*

Fig. 4.55

4.19.1. Maximum Temperature Attained With Intermittent Loading. When a machine is intermittently loaded it will cool down during the time duration it is off and its temperature will rise when it is on, as illustrated in Fig. 4.54. Let θ_1, θ'_1, θ''_1, be θ'''_1 be the temperature rise after heating and θ_2, θ'_2, θ''_2, θ'''_2 be the temperature rise after cooling, as illustrated in Fig. 4.54.

$$\text{Now } \theta_1 = \theta'_F(1 - e^{-t_1/\tau}) = \theta'_F(1 - e^x) \text{ where } x = \frac{-t_1}{\tau}$$

$$\text{and } \theta_2 = \theta_1 e^{-t_2/\tau'} = \theta_1 e^y = \theta'_F(1 - e^x)e^y \text{ where } y = \frac{-t_2}{\tau'}$$

$$\begin{aligned}
\theta'_1 &= \theta'_F - (\theta'_F - \theta_2)e^x \\
&= \theta'_F(1 - e^x) + \theta_2 e^x \\
&= \theta'_F(1 - e^x) + \theta'_F(1 - e^x)\, e^y e^x \\
&= \theta'_F(1 - e^x)\,(1 + e^x e^y) \\
\theta'_2 &= \theta'_1 e^y = \theta'_F(1 - e^x)\,(1 + e^{x+y}).e^y \\
\theta''_1 &= \theta'_F(1 - e^x) + \theta'_2 e^x \\
&= \theta'_F(1 - e^x) + \theta'_F e^x e^y (1 - e^x)\,(1 + e^x e^y) \\
&= \theta'_F(1 - e^x)\,[1 + e^x e^y + e^{2x} \cdot e^{2y}]
\end{aligned}$$

For *n* times intermittency, we have

$$\begin{aligned}
\theta_1^{1-n} &= \theta'_F(1 - e^x)[1 + e^x e^y + e^{2x} e^{2y} + \ldots e^{(n-1)x} \times e^{(n-1)y}] \\
&= \theta'_F(1 - e^x)\left[\frac{1 - e^{nx} e^{ny}}{1 - e^x e^y}\right]
\end{aligned}$$

When $n = \infty$ both e^{nx} and e^{ny} will be zero as *x* and *y* are negative. If θ_F is the maximum temperature rise with intermittent load then

$$\theta_F = \theta'_F\left(\frac{1 - e^x}{1 - e^x e^y}\right) = \theta'_F \frac{1 - e^x}{1 - e^{x+y}} = \theta'_F \frac{1 - e^{-t_1/\tau}}{1 - e^{-(t_1/\tau + t_2/\tau')}} \quad \ldots(4.63)$$

4.20 RATING OF MACHINES

A name plate fixed to the outside frame of an electrical machine records the data pertaining to its rating. A machine rating specifies the voltage, current, speed, excitation, power factor, efficiency, power output etc.

The rating of a machine should give all information necessary so that, if the machine is operated within the limits of all factors specified in its rating, the machine will operate satisfactorily and safely and will give reasonable length of service. Therefore, the rating of a machine must give the necessary information to safeguard the application of the machine from conditions of operation which (*i*) would result in unsafe mechanical or electrical strains upon any part of its structure or (*ii*) would result in excessive deterioration of the mechanical or electrical characteristics of the materials of which the machine is constructed, To give this information, the rating of an electric machine should include the output, voltage, speed, and any other information that may necessary for the proper operation of the machine.

Generators and motors are rated in terms of kVA or kW output at a given speed and voltage. The size and rating of an electric machine for any service is mainly governed by the factor 'temperature rise'. The maximum temperature, to which an electric machine is allowed to reach, is limited by the type of insulation used.

The maximum temperature rise permissible with insulation A and insulation B are 40°C and 50°C respectively. Overloads are generally permissible for short period of time but when machines are required to carry greater loads than those specified, they must be kept under inspection to see that the temperature does not rise too much and that severe sparking at the commutator does not occur.

The type of service to which a machine is subjected is of great importance. The machines operating continuously at rated (or near rated) load are physically larger than those working at intermittent loads. Also electric machines that are not enclosed and are, in addition, well cooled by fans are likely to have higher ratings than covered up machines or those machines located where air does not circulate freely through and over them. High speed machines and machines employing mica, glass tapes, and the new silicons as insulation can generally be physically smaller, in given ratings, than the low speed machines employing standard insulations. The output of dc machines is also limited by the factor 'commutation sparking'. This factor often limits the output, even though heating may not have proceeded to permissible values.

The machines according to ISI specifications are classified as follows.

Continuous Rating. This is an output, which a machine delivers continuously without exceeding the permissible temperature rise. It can deliver 25 per cent overload for two hours.

Continuous Maximum Rating. Similar to continuous rating but not allowing any overload. It is used for motors of capacity larger than 1.84 kW (2.5 hp) per rpm.

Short-Time Rating. This is an output which an electric machine can deliver for a specified period (say 1 hr., $\frac{1}{2}$ hr., $\frac{1}{4}$ hr. etc.) without exceeding the specified temperature rise.

4.20.1. Effect of Altitude on Rated Output. At high altitudes, the density of cooling air decreases. As a result, cooling capacity is reduced. The effect is negligible for elevations not exceeding 1,000 m. The output is reduced as stated below in tabular form.

Altitude above sea level in metres	*Permissible output in percentage of the rated value for motor speeds*	
	up to 1,000 rpm	*above 1,000 rpm*
0–1,000	100	100
2,000	95	92
3,000	90	85
4,000	87	80
5,000	85	77

4.20.2. Overload Capacity of Induction Motors. Motors are designed to carry overloads, without any adverse effect, as mentioned below.

Sustained overloads	*Time duration*	*Motor kW output per 1,000 rpm*
25% overload in torque	2 hours	7.5 and above
	30 minutes	3 to 6.8
	15 minutes	0.75 to 3
Momentary overloads		
50% overload in torque	1 minute for all sizes	
100% overload in torque	15 seconds for all sizes	

Single-phase induction motors are usually not designed to carry overload. However, 20-25% overload in torque intermittently may be permitted in split-phase ac motors and 40-45% overload in torque intermittently for capacitor-start motors.

4.21 CHOICE OF RATING OF MOTORS

The choice of ratings of motors for the services requiring fairly constant power such a fans, air-compressors, blowers, pumps, motor-generator sets etc. is every simple. In such cases the power required is determined and motor having continuous rating of power required is chosen. It is, however, worthwhile to take considerable trouble to find out exactly how much power is actually required by the driven machinery, since if the motor chosen be too small it will be overloaded and will, therefore, overheat and deteriorate, while if it is too large the efficiency and power factor will probably be less than those of a motor of the correct size. If there is no possibility of overloading then a motor with a continuous maximum rating may be selected, which will be comparatively smaller in size and cheaper in cost. Such a motor may operate at a higher temperature (55°C in place of 40°C).

For the constant torque loads, the size of the motor can be determined as follows :

In case of rotary motion, rating of motor required is given as

$$P = T \times \frac{2\pi N}{60} \times \frac{1}{\eta} \times \frac{1}{1{,}000} \text{ kW} = \frac{TN}{9{,}550\eta} \text{ kW} \quad \text{...(4.64)}$$

where T is the load torque in N-m, N is the speed in rpm and η is the product of the efficiency of the driven load and that of the transmitting device. Efficiency varies considerably with the type of drive, bearings etc. and its value can best be dictated by experience in applications.

In case of linear motion, rating of the motor required is given as

$$P = \frac{F \times 9.81 \times v}{\eta} \times \frac{1}{1{,}000} \text{ kW} = \frac{F \times v}{102\,\eta} \text{ kW} \quad \text{...(4.65)}$$

where F is the force caused by the load in kg and v is the velocity of motion of load in m/s.

The above expression [Eq. (4.65)] can be applied in case of hoisting mechanism, lifts or elevators with slight modification and that is due to counterweight (balancing the weight of the cage or car as well as one-half of the useful load). The modified expression is

$$P = \frac{Fv}{2 \times 102\,\eta} \text{ kW} \quad \text{...(4.66)}$$

v in case of normal passenger lift cabins lies in the range of 0.5–1.5 m/s.

In case of pumps, the rating of the motor required,

$$P = \frac{Q \times \rho \times h}{102\,\eta} \text{ kW} \qquad \text{...(4.67)}$$

where Q is the delivery of pumps in m^3/s, ρ is the density of liquid pumped in kg/m^3 and h is the gross head (static head + friction head) in m. Efficiency η lies in the range of 80 to 90% in case of reciprocating pumps and 40 to 80% in case of centrifugal pumps.

Similarly, the rating of a fan motor is given as

$$P = \frac{Qh}{102\,\eta} \text{ kW} \qquad \text{...(4.68)}$$

where Q is the volume of air or any other gas in m^3/s and h is the pressure in mm of water or in kg/m^2. Efficiency may be taken to be 0.8 for larger power fans and 0.6 in case of smaller power fans.

The selection of size of motor is also simple for the services where the motor is run for a short time so that the temperature rise is less than the maximum permissible value and then it is allowed to cool down to the ambient temperature. For such services a motor rated at the required output power for the given period is chosen.

In cases where the load fluctuates over a given cycle, as in rolling mills or colliery winders, the size of the motor is determined accurately by finding the heating and cooling curves of motors under consideration and plotting heating curve for each motor, when working on the given cycle. The smallest motor which delivers the load according to given cycle without exceeding the specified temperature rise may be choosen.

Other methods for determination of rating of a motor for continuous duty and variable load are given below.

1. *Method of Average Losses.* In this method it is assumed that the temperature rise attained by the motor with fluctuating loading conditions over a certain period of time will be the same as that attained by the motor with a certain load of constant magnitude. This will be true provided that the average losses in the motor for the same time are the same for both the conditions of operation.

This method consists of determining average losses in the motor when it operates according to the given load diagram and then comparing with the losses corresponding to the continuous duty of the machine when operated at its nominal rating.

2. *Equivalent Current Method.* This method is based upon the assumption that the actual variable current may be replaced by an equivalent current I_{eq} which produces the same losses in the motor as the actual current.

$$\therefore\ I_{eq} = \sqrt{\frac{I_1^2 t_1 + I_2^2 t_2 + I_3^2 t_3 + \ldots + I_n^2 t_n}{t_1 + t_2 + t_3 + \ldots + t_n}} \qquad \text{...(4.69)}$$

The heating and cooling conditions in self-ventilated machines depend upon its speed. At low speeds the cooling conditions are poorer than at normal speeds. Therefore, if the work cycle involves slow-speed operation, it must be taken into consideration when using Eq. (4.69).

The equivalent current is compared with the rated current of the motor selected. The equivalent current may be less than or equal to the rated current of the machine.

3. *Equivalent Torque Method.* This method is based upon the assumption that the motor current is proportional to the torque and heating is proportional to the square of current (*i.e.,* heating is proportional to the square of torque).

The equivalent torque is determined in the same manner as the equivalent current *i.e.,*

$$T_{eq} = \sqrt{\frac{T_1^2 t_1 + T_2^2 t_2 + T_3^2 t_3 + \ldots + T_n^2 t_n}{t_1 + t_2 + t_3 + \ldots + t_n}} \qquad \text{...(4.70)}$$

A motor having torque nearly equal to T_{eq} is selected.

4. *Equivalent Power Method.* At constant speed or where the variations in speed are small, the equivalent power is given by the relationship,

$$P_{eq} = \sqrt{\frac{P_1^2 t_1 + P_2^2 t_2 + P_3^2 t_3 + \ldots + P_n^2 t_n}{t_1 + t_2 + t_3 + \ldots + t_n}} \qquad \text{...(4.71)}$$

A motor having power rating P_{eq} is selected.

The equivalent current method is the most accurate out of the four methods discussed above. This method may be employed for determining the capacity for all applications except where it is necessary to take into account the variations in so called 'constant losses'.

The equivalent torque method cannot be employed for applications where equivalent current method cannot be applied or in cases in which flux does not remain constant like dc series motors.

The equivalent power method cannot be used for motors whose speed varies considerably under load, especially when dealing with starting and braking conditions.

Example 4.1. The outside surface of a 12 hp (metric) motor is equivalent to a cylinder of 65 cm and 1 metre length. The motor weighs 400 kg and has a specific heat of 700 J/kg/°C. The outer surface is capable of heat dissipation of 12 W per metre square per degree centigrade. Find the final temperature rise and thermal constant of motor when operating at full load with an efficiency of 90 %. [*J.N. University Hyderabad, April-2003*]

Solution: Power output of machine = 12 × 735.5 = 8,826 W

$$\text{Power input} = \frac{\text{Power output}}{\eta} = \frac{8{,}826}{0.9} = 9{,}807 \text{ W}$$

Power losses, P = Power input – power output = 9,807 – 8,826 = 980.7 W

Mass of active parts of machine, m = 400 kg

Specific heat of material, C_p = 700 J/kg/°C

Coefficient of cooling, α = 12 W/m²/°C

Surface area of cooling, $S = \pi dl = \pi \times 0.65 \times 1 = 2.042 \text{ m}^2$

$$\text{Final temperature rise, } \theta_F = \frac{P}{S\alpha} = \frac{980.7}{2.042 \times 12} = 40\ °C \textbf{ Ans.}$$

$$\text{Heating time constant, } \tau = \frac{mC_p}{S\alpha} = \frac{400 \times 700}{2.042 \times 12} = 11{,}427 \text{ seconds}$$

or 3 hours 10 minutes **Ans.**

Example 4.2. The mean temperature rise of a machine to run on full load is :

Time in hours	**0.2**	**0.4**	**0.6**	**1.0**	**1.5**	**2.0**	**3.0**
Temperature rise in °C	**7.0**	**13.5**	**19.3**	**29.1**	**39.1**	**46.7**	**57.3**

Estimate the final steady-state temperature and heating time constant.

Solution: For determination of steady-state temperature plot temperature-rise against time from the given datas, as shown in Fig. 4.56. Now plot increments of temperature for equal intervals of time against temperature rise to the left of Y-axis, as shown in Fig. 4.56.

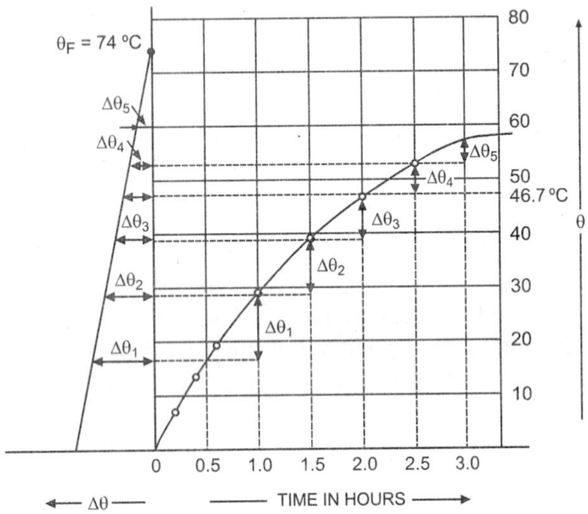

Fig. 4.56

Straight line drawn through the plotted points cuts the Y-axis, where the rate of change of temperature rise is zero. Therefore, the point of intersection of the line drawn, as mentioned above, with the Y-axis gives steady-state temperature. From Fig. 4.56, $\theta_F = 74$ °C. **Ans.**

From definition, heating time constant is the time at which temperature rise is 0.632 times of final steady state temperature, i.e., 0.632 × 74 = 46.7 °C.

From Fig. 4.56, time corresponding to temperature rise of 46.7 °C is 2 hours. Hence heating time constant is 2 hours. **Ans.**

Example 4.3. The full-load temperature rise of totally enclosed motor is 20 °C after one hour and 34 °C after two hours. Find (*i*) final steady-temperature rise on full load, (*ii*) the heating time constant (*iii*) one hour rating of the motor for the same final temperature rise. Take no-load losses = copper losses at rated continuous load.

Solution : From Eq. (4.58)

$$\theta = \theta_F(1 - e^{-t/\tau}) \quad ...(i)$$

Now from given datas when t = 1 hour, θ = 20 °C and when t = 2 hours, θ = 34 °C. Substituting the above values of t and θ in Eq. (*i*) we get

$$20 = \theta_F\left(1 - e^{-1/\tau}\right) \quad ...(ii)$$

$$\text{and } 34 = \theta_F\left(1 - e^{-2/\tau}\right) \quad ...(iii)$$

Dividing equation (*iii*) by Eq. (*ii*) we get

$$\frac{34}{20} = \frac{1 - e^{-2/\tau}}{1 - e^{-1/\tau}}$$

$$\text{or } 1 + e^{-1/\tau} = 1.7$$

$$\text{or } e^{-1/\tau} = 0.7$$

$$\text{or Heating time constant, } \tau = \frac{-1}{\log_e 0.7} = 2.8 \text{ hours } \textbf{Ans.}$$

Substituting τ = 2.8 hours in equation (*ii*) we get

$$\text{Final steady temperature rise, } \theta_F = \frac{20}{1 - e^{-1/2.8}} = 66.67 \text{ °C } \textbf{Ans.}$$

If temperature rise after one hour is 66.67°C, final temperature rise θ_F will be given as,

$$\theta'_F = \frac{\theta_F}{1 - e^{-t/\tau}} = \frac{66.67}{1 - e^{-1/2.8}} = 222 \text{ °C}$$

$$\begin{aligned}\text{Total losses at full load} &= \text{Iron loss + copper loss} \\ &= 2 \times \text{copper loss} = 2P_c\end{aligned}$$

where P_c is full-load copper loss

$\because$ Iron loss = no-load loss = full-load copper loss

Let one hour rating be x times full load then

$$\begin{aligned}\text{Total losses} &= \text{Iron losses} + x^2 P_c \\ &= P_c + x^2 P_c = (1 + x^2) P_c\end{aligned}$$

$$\therefore \frac{\text{Total losses at one-hour rating}}{\text{Total loss at full-load}} = \frac{\theta'_F}{\theta_F}$$

$$\text{or } \frac{\left(1 + x^2\right)P_c}{2P_c} = \frac{222}{66.67}$$

$$\text{or } x = 2.38$$

So one-hour rating of the motor is 2.38 times full-load rating **Ans.**

Example 4.4. Determine the thermal model of motor for heating. A motor is continuously rated at 50 kW. It has a heating time constant of 100 minutes. Determine the one hour rating of the motor. The motor losses can be expressed as 0.6 full-load copper losses + x^2 full-load copper loss where x is a load as a fraction of full-load.

[*U.P. Technical Univ. Electric Drives, 2013-14*]

Solution: Assume that the motor cools down completely between each load period.

Let P kW be the one hour rating, θ_F be the final temperature rise in case of continuous rating and W be the full-load copper loss.

Final temperature rise in case of one hour rating

$$\theta'_F = \frac{\theta_F}{1 - e^{-t/\tau}} = \frac{\theta_F}{1 - e^{-60/100}} = \frac{\theta_F}{1 - e^{-0.6}}$$

$$= 2.21637\, \theta_F$$

$$\text{Losses at 50 kW} = 0.6 \text{ W} + 1^2 \text{ W} = 1.6 \text{ W} \qquad \because x = 1$$

$$\text{Losses at P kW} = 0.6 \text{ W} + \left(\frac{P}{50}\right)^2 \text{W} \qquad \because x = \frac{P}{50}$$

$$\frac{\text{Total losses at P kW}}{\text{Total losses at 50 kW}} = \frac{\theta'_F}{\theta_F}$$

$$\frac{0.6 \text{ W} + \left(\frac{P}{50}\right)^2 \text{W}}{1.6 \text{ W}} = 2.21637$$

$$\text{or } P = \sqrt{(2.21637 \times 1.6 - 0.6)} \times 50 = 85.82 \text{ kW } \textbf{Ans.}$$

Example 4.5. A motor running continuously on full load has a temperature rise of 20 °C. The heating time constant is 60 minutes. How long the motor should be run at twice the continuously rated output without overheating ? The maximum efficiency of motor at full load is 80%.

[*Pb. Technical Univ. Electric Drives and Utilization May 2008*]

Solution: Heating time constant, τ = 60 minutes

Final temperature rise, θ_F = 20°C in case of continuous rating

Let P kW be the continuous rating, and t be the time duration in minutes for which motor can be run at twice the continuously rated output without overheating.

Since maximum efficiency occurs 80% of full load *i.e.*, 0.8 P, therefore at 0.8 P load copper loss is equal to constant loss, let each be equal to W watts.

$$\text{Total losses at P kW} = W + \left(\frac{P}{0.8P}\right)^2 W = 2.5625\ W$$

$$\text{Total losses at 2P kW} = W + \left(\frac{2P}{0.8P}\right)^2 W = 7.25\ W$$

Since copper loss varies as the square of load

$$\frac{\text{Total losses at 2P kW}}{\text{Total losses at P kW}} = \frac{\theta'_F}{\theta_F}$$

$$\text{or } \theta'_F = \theta_F \times \frac{\text{Total losses at 2P kW}}{\text{Total losses at P kW}}$$

$$= 20 \times \frac{7.25}{2.5625} = 56.585\ °C$$

$$\text{Also } \theta'_F = \frac{\theta_F}{1 - e^{-t/\tau}}$$

$$\text{or } t = \tau \log_e \frac{1}{1 - \dfrac{\theta_F}{\theta'_F}} = 60 \log_e \frac{1}{1 - \dfrac{20}{56.585}}$$

$= 26.17$ minutes say, 26 minutes **Ans.**

Example 4.6. If one hour rating of a machine is $\sqrt{3}$ times the continuous rating and if the steady temperature rise for one hour rating is twice for normal load, find the ratio of iron losses to I^2R losses at full load.

Solution: Losses on continuous rating $= P_c + P_i$

Losses one one-hour rating $= x^2 P_c + P_i = 3P_c + P_i$

$\because$ one hour rating is $\sqrt{3}$ times the continuous rating *i.e.* $x = \sqrt{3}$

Ratio of final steady temperature rises with 1 hour and continuous rating is given as

$$\frac{\theta'_F}{\theta_F} = \frac{P_i + x^2 P_c}{P_i + P_c}$$

$$\text{or } 2 = \frac{3P_c + P_i}{P_c + P_i}$$

$$\text{or } P_i = P_c$$

Thus iron losses are equal to full-load copper losses at full load **Ans.**

Example 4.7. The 15 minute rating of a motor is 400 watts. The heating time constant is 60 minutes. Determine the continuous rating of the motor if the maximum efficiency occurs at 80 percent of full load.

Solution: Since maximum efficiency occurs at 80% of full load so at 80% of full-load, iron loss is equal to copper loss, let each equal to P_c watts.

$$\text{Losses on full load} = P_c + \left(\frac{1}{0.8}\right)^2 P_c = 2.5625\ P_c$$

$$\text{Losses at load of 400 watts} = P_c + \left(\frac{400}{0.8 \times \text{full load}}\right)^2 P_c$$

Ratio of final temperature rises with 15 minute rating and continuous rating is given as,

$$\frac{\theta'_F}{\theta_F} = \frac{1}{1 - e^{-t/\tau}}$$

$$= \frac{\text{Total loss on 15-minute rating}}{\text{Total loss on full load}}$$

$$\text{or } \frac{1}{1 - e^{-15/60}} = \frac{P_c + \left(\dfrac{400}{0.8 \times \text{full load}}\right)^2 P_c}{2.5625\ P_c}$$

or Continuous rating $= 153.7$ watts **Ans.**

Example 4.8. A motor has a thermal time constant of 45 minutes. When the motor runs continuously on full load, its final temperature rise is 80 °C. (*i*) What is the temperature rise after 1 hour if the motor runs continuously on full load? (*ii*) If the temperature on one-hour rating is 80 °C, find the maximum steady-state temperature at this rating. (*iii*) How much time does the motor take for its temperature to rise from 50 °C to 80 °C if it is working at its 1 hour rating ?

Solution: (*i*) From given datas:

Final temperature rise, $\theta_F = 80$ °C

Thermal time constant, $\tau = 45$ minutes $= 0.75$ hour

Temperature rise after 1 hour (*i.e* when $t = 1$) can be had by substituting $\theta_F = 80$ °C; $\tau = 0.75$ hour and $t = 1$ hour in Eq. (2.47)

$$i.e.\ \theta = \theta_F (1 - e^{-t/\tau})$$

$$= 80 (1 - e^{-1/0.75}) = 58.9\ °C\ \textbf{Ans.}$$

(*ii*) Let the maximum steady-temperature rise at one hour rating be θ'_F

Now from given datas $\tau = 0.75$ hour, $\theta = 80$ °C when $t = 1$ hour

$$\therefore \theta'_F = \frac{\theta_F}{1 - e^{-t/\tau}} = \frac{80}{1 - e^{-1/0.75}} = 108.64\ °C\ \textbf{Ans.}$$

(*iii*) Time taken to attain temperature of 80 °C (assuming initial temperature as 0 °C) in one hour or 60 minutes,

Final temperature rise, $\theta'_F = 108.64$ °C

Now let the time taken to attain temperature of 50 °C from 0 °C be t hours then

$$50 = 108.64 (1 - e^{-t/0.75})$$

or $t = 0.4625$ hour or 27.75 minutes

Hence time taken to increase temperature from 50 °C to 80 °C $= 60 - 27.75 = 32.25$ minutes **Ans.**

Example 4.9. A motor has a continuous rating of 100 kW. The heating and cooling time constants are 50 and 70 minutes respectively. The motor has a maximum efficiency at 80% of full-load and is employed in an intermittent periodic load cycle consisting of a load period of 10 minutes followed by no-load period of 10 minutes. Calculate the value of load in kW during the load period.

[U.P. Technical Univ. Electric Drives 2009-10]

Solution: Heating time constant, $\tau = 50$ minutes

Cooling time constant, $\tau' = 70$ minutes

On-load period, $t_1 = 10$ minutes

Off-load period, $t_2 = 10$ minutes

From Eq. (4.63)

$$\frac{\theta'_F}{\theta_F} = \frac{1 - e^{-\left(\frac{t_1}{\tau} + \frac{t_2}{\tau'}\right)}}{1 - e^{-t_1/\tau}} = \frac{1 - e^{-\left(\frac{10}{50} + \frac{10}{70}\right)}}{1 - e^{-10/50}} = \frac{1 - 0.70974}{1 - 0.81873} = 1.6$$

Let P be rating of motor when subjected to a duty cycle of 10 minutes on certain load followed by 10 minutes on no load.

Since maximum efficiency occurs at 80 per cent of full load *i.e.* at a load of 80 kW, therefore, at 80 kW load copper loss is equal to constant loss, let each be equal to W watts.

$$\text{Losses at 100 kW load} = W + \left(\frac{100}{80}\right)^2 W = 2.5625\ W$$

$$\text{Losses at P kW load} = W + \left(\frac{P}{80}\right)^2 W = W\left(1 + \frac{P^2}{6{,}400}\right)$$

Thus $$\frac{\text{Total losses at P kW output}}{\text{Total losses at 100 kW output}} = \frac{\theta'_F}{\theta_F}$$

So $$\frac{W\left(1+\frac{P^2}{6{,}400}\right)}{2.5625\ W} = 1.6$$

or $P = \sqrt{(2.5625 \times 1.6 - 1) \times 6{,}400} = 140.85$ kW **Ans.**

Example 4.10. The temperature rise of a motor after operating for 30 minutes on full load is 20°C. After another 30 minutes on the same load the temperature rise becomes 30°C. Assuming that the temperature increases according to an exponential law, determine the final temperature rise and the heating time constant.

[*U.P. Technical Univ. Electric Drives 2007-08*]

Solution: From Eq. (4.58)

$$\theta = \theta_F(1 - e^{-t/\tau})$$

Now from given datas when $t_1 = 30$ minutes, $\theta_1 = 20$°C and when $t_2 = 30 + 30 = 60$ minutes, $\theta_2 = 30$°C. Substituting the values for t and θ in above equation, we have

$$20 = \theta_F(1 - e^{-30/\tau}) \quad ...(i)$$

$$30 = \theta_F(1 - e^{-60/\tau}) \quad ...(ii)$$

Dividing Eq. (*ii*) by Eq. (*i*), we have

$$\frac{1 - e^{-60/\tau}}{1 - e^{-30/\tau}} = \frac{30}{20} = 1.5$$

or $$1 + e^{-30/\tau} = 1.5$$

or Heating time constant, $$\tau = \frac{-30}{\log_e 0.5} = 43.28 \text{ minutes}$$

Substituting $\tau = 43.28$ minutes in Eq. (*i*), we have

$$\theta_F = \frac{20}{1 - e^{-30/43.28}} = 40\text{ °C} \quad \textbf{Ans.}$$

Example 4.11. Calculate the maximum overload that can be carried by a 25 kW motor if the temperature rise is not to exceed 50°C after one hour on overload. The temperature rise on full-load after one hour is 30°C and after 2 hours is 40°C. The losses vary as square of the load.

Solution: The expression for temperature rise is given as

$$\theta = \theta_F(1 - e^{-t/\tau}) \quad ...(i)$$

From given datas when $t = 1$ hour, $\theta = 30$°C and when $t = 2$ hours, $\theta = 40$°C.

Substituting these values of t and θ in above expression (*i*), we have

$$30 = \theta_F(1 - e^{-1/\tau}) \quad ...(ii)$$

and $$40 = \theta_F(1 - e^{-2/\tau}) \quad ...(iii)$$

Solving above equations (*ii*) and (*iii*), we have

$$e^{-1/\tau} = \frac{40}{30} - 1 = \frac{1}{3}$$

Substituting $e^{-1/\tau} = \frac{1}{3}$ in equation (*ii*), we have

$$\theta_F = \frac{30}{1 - \frac{1}{3}} = 45\text{°C}$$

If temperature rise after 1 hour is not to exceed 50°C, final temperature rise, θ'_F will be given as

$$\theta'_F = \frac{50}{1 - e^{-1/\tau}} = \frac{50}{1 - \frac{1}{3}} = 75\text{°C}$$

$$\because t = 1 \text{ and } e^{-1/\tau} = \frac{1}{3}$$

Let the maximum overload that can be carried by a 25 kW motor be P kW.

Since losses vary as the square of load

$$\frac{\theta'_F}{\theta_F} = \frac{\text{Losses at P kW}}{\text{Original losses}}$$

$$\frac{75}{45} = \left(\frac{P}{25}\right)^2$$

or $P = 25\sqrt{75/45} = 32.3$ kW **Ans.**

Example 4.12. A motor has a heating time constant of 70 minutes and cooling time constant of 90 minutes when stationary. When run continuously on full load of 18.5 kW, the total temperature rise is 40 °C. What load could the motor deliver for 10 minutes if this is followed by a shut down period long enough for it to cool without the temperature rise exceeding 40 °C. If it is on intermittent load of 10 minutes on load and followed by 15 minutes of shut down, what is the maximum value of load which it could supply during the on-load period without maximum temperature rise exceeding 40 °C ? Assume losses to be proportional to square of load.

Solution: Final temperature rise, $\theta_F = 40$ °C

Heating time constant, $\tau = 70$ minutes

Cooling time constant, $\tau' = 90$ minutes

Continuous rating of motor = 18.5 kW

(*i*) When load period is followed by shut down period long enough for its cooling, let the load supplied by the motor for 10 minutes without temperature rise exceeding 40 °C be P kW.

$$\text{Final temperature rise, } \theta'_F = \frac{\theta_F}{1 - e^{-t/\tau}} = \frac{40}{1 - e^{-10/70}} = 300$$

Since losses are proportional to the square of load,

$$\frac{\text{Total losses at P kW output}}{\text{Total losses at 18.5 kW output}} = \left(\frac{P}{18.5}\right)^2 = \frac{\theta'_F}{\theta_F} = \frac{300}{40}$$

or $P = 18.5\sqrt{7.5} = 50.66$ kW **Ans.**

When the motor is on intermittent load of 10 minutes on load followed by 15 minutes shut down

From Eq. (4.63)

$$\theta'_F = \theta_F \frac{1 - e^{-\left(\frac{t_1}{\tau} + \frac{t_2}{\tau'}\right)}}{1 - e^{-t_1/\tau}} = 40 \times \frac{1 - e^{-\left(\frac{10}{70} + \frac{15}{90}\right)}}{1 - e^{-10/70}}$$

$$= 40 \times \frac{1 - 0.7388}{1 - 0.8669} = 80\text{ °C}$$

Let maximum value of load that can be supplied of be P' kW

$$\therefore \left(\frac{P'}{18.5}\right)^2 = \frac{\theta'_F}{\theta_F} = \frac{80}{40}$$

or $P' = 18.5\sqrt{2} = 26.163$ kW **Ans.**

Example 4.13. A motor has a heating time constant of 90 minutes and a cooling time constant of 120 minutes and final steady temperature rise on full load of 60°C. The motor has repeated load cycle of full load for 30 minutes followed by a stationary period of 30 minutes. Determine the maximum and minimum temperatures. Also determine the overload on the motor that can be allowed on this cycle such that the maximum temperature rise does not exceed the permissible value of 60°C.

[*G.B. Technical Univ. Electric Drives, 2012-13*]

Solution: Heating time constant, $\tau = 90$ minutes

Cooling time constant, $\tau' = 120$ minutes

On-load period, $t_1 = 30$ minutes

Off-load period, $t_2 = 30$ minutes

Final steady temperature rise on full load, $\theta'_F = 60°C$

Minimum temperature rise

$$\theta_2 = \theta'_F\left(1 - e^{\frac{-t_1}{\tau}}\right)e^{\frac{-t_2}{\tau'}} \quad \text{(refer to Art. 4.19.1 and Fig. 4.54),}$$

$$= 60\left(1 - e^{\frac{-30}{90}}\right)e^{\frac{-30}{120}}$$

$$= 60\left(1 - e^{\frac{-1}{3}}\right)e^{-0.25} = 13.24°C \textbf{ Ans.}$$

Maximum temperature rise,

$$\theta_F = \theta'_F \frac{1 - e^{-t_1/\tau}}{1 - e^{-\left(\frac{t_1}{\tau} + \frac{t_2}{\tau'}\right)}} \quad \text{Refer to Eq. (4.63)}$$

$$= 60 \times \frac{1 - e^{-30/90}}{1 - e^{-\left(\frac{30}{90} + \frac{30}{120}\right)}} = 60 \times \frac{1 - e^{-1/3}}{1 - e^{-0.583333}}$$

$$= \frac{60 \times 0.28347}{0.44196} = 38.48°C \textbf{ Ans.}$$

Let the overload on the motor be x times of the motor rating so that the motor can be allowed on this cycle such that the maximum temperature rise does not exceed the permissible value of 60°C. Let the losses vary as the square of the load

$$\frac{\theta'_F}{\theta_F} = \frac{\text{Losses at overload}}{\text{Losses at original load}}$$

$$\frac{60}{38.8} = \left(\frac{1 + x}{1}\right)^2$$

$$\text{or } x = \sqrt{\frac{60}{38.8}} - 1 = 0.2487 \textbf{ Ans.}$$

Example 4.14. A totally enclosed motor has a temperature rise of 20 °C after half an hour and 35 °C after 1 hour on full load. Determine the temperature rise after 2 hours on load. If a fan is fitted driving cool air over the external surface of the machine which reduces the final steady-temperature rise of the machine to 80 percent of the original value, determine the temperature rise from cold after 1 hour on full load.

Solution : From Eq. (4.58) $\theta = \theta_F (1 - e^{-t/\tau})$ ...(*i*)

Now from given datas when $t = \frac{1}{2}$ hour, $\theta = 20$ °C and when $t = 1$ hour, $\theta = 35$ °C

Substituting the above values for t and θ in Eq. (*i*) we get

$$20 = \theta_F\left(1 - e^{-0.5/\tau}\right) \quad ...(ii)$$

$$35 = \theta_F\left(1 - e^{-1/\tau}\right) \quad ...(iii)$$

Dividing equation (*iii*) by equation (*ii*) we get

$$\frac{1 - e^{-1/\tau}}{1 - e^{-0.5/\tau}} = \frac{35}{20} = 1.75$$

$$\text{or } 1 + e^{-0.5/\tau} = 1.75$$

$$\text{or } \tau = \frac{-0.5}{\log_e 0.75} = 1.738 \text{ hours}$$

Substituting $e^{-0.5/\tau} = 1.75 - 1 = 0.75$ in equation (*ii*) we get

$$\theta_F = \frac{20}{1 - 0.75} = 80 \text{ °C}$$

Temperature rise after 2 hours,

$$\theta' = \theta_F\left(1 - e^{-t/\tau}\right) = 80\left(1 - e^{-2/1.738}\right) = 54.7 \text{ °C } \textbf{Ans.}$$

When a fan is fitted driving cool air over the external surface of the machine, coefficient of cooling is increased resulting in decrease in steady final temperature rise θ_F and heating time constant τ. Let the new coefficient of cooling, steady final temperature rise and heating time constant be α_1, θ_{F_1} and τ_1 respectively.

$$\theta_{F_1} = 0.8\,\theta_F$$

$$\text{Since } \theta_{F_1} = \frac{P}{S\alpha}$$

$$\frac{\alpha_1}{\alpha} = \frac{\theta_F}{\theta_{F_1}}$$

$$\text{or } \alpha_1 = \frac{\theta_F}{\theta_{F_1}}\alpha = \frac{\alpha}{0.8} = 1.25\,\alpha$$

Heating time constant is given as

$$\tau = \frac{mC_p}{S\alpha}$$

$$\text{So } \frac{\tau_1}{\tau} = \frac{\alpha}{\alpha_1}$$

$$\text{or } \tau_1 = \frac{\alpha}{1.25\alpha}\tau = 0.8\,\tau = 0.8 \times 1.738 = 1.39 \text{ hours}$$

$$\theta_{F_1} = 0.8\,\theta_F = 0.8 \times 80 = 64 \text{ °C}$$

Temperature rise from cold after 1 hour at full load

$$\theta_1 = \theta_{F_1}\left(1 - e^{-t/\tau_1}\right)$$

$$= 64\left(1 - e^{-1/1.39}\right) = 32.8 \text{ °C } \textbf{Ans.}$$

Example 4.15. A certain motor has to perform the following duty cycle: 100 kW for 10 minutes, no load for 5 minutes, 50 kW for 8 minutes and no load for 4 minutes. The duty cycle is repeated indefinitely. Draw the curve for the load cycle. Assuming that the heating is proportional to the square of the load, determine the suitable size of a continuously rated motor.

Solution: The load cycle is illustrated in Fig. 4.57.

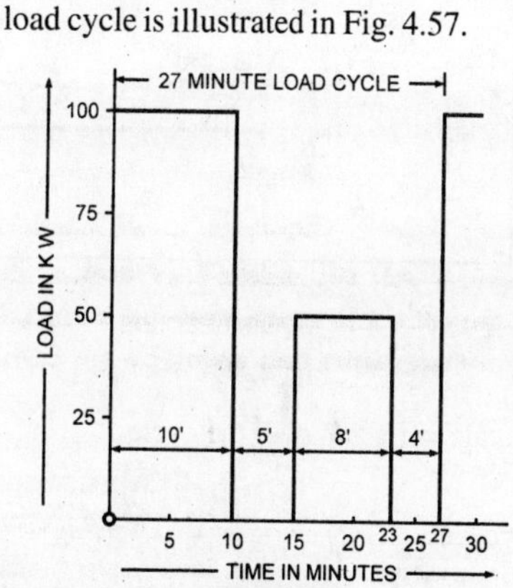

Fig. 4.57

Since heating is proportional to the square of the load,

Rating of continuously rated motor

$$= \text{RMS kW} = \sqrt{\frac{(100)^2 \times 10 + (50)^2 \times 8}{27}} = 66.67 \text{ kW}$$

i.e A motor of 70 kW will be adequate.

Example 4.16. A motor has a cyclic loading as given below:

(i) 250 N-m for 15 minutes

(ii) 350 N-m for 20 minutes

(iii) 100 N-m for 15 minutes

(iv) No load for 10 minutes.

The motor runs at a constant speed of 500 rpm. Determine the rating of suitable motor.

[*U.P. Technical Univ. Electric Drives, 2013-14*]

Solution: Equivalent load torque, from Eq. (4.70)

$$T_{eq} = \sqrt{\frac{T_1^2 t_1 + T_2^2 t_2 + T_3^2 t_3 + T_4^2 t_4}{t_1 + t_2 + t_3 + t_4}}$$

$$= \sqrt{\frac{250^2 \times 15 + (350)^2 \times 20 + (100)^2 \times 15 + (0)^2 \times 10}{15 + 20 + 15 + 10}}$$

= 244 N-m

Speed of motor,

N = 500 rpm

Power rating of suitable motor,

$$P = T_{eq} \times \frac{2\pi N}{60} = 244 \times \frac{2\pi \times 500}{60}$$

= 12,776 W or 12.776 kW **Ans.**

Example 4.17. Based on the rms torque, estimate the kW rating of a 750 rpm motor used for driving an equipment having the following load torque curve :

(*i*) For first 10 seconds, the torque is constant at 40 kg-m;

(*ii*) For next 30 seconds the torque varies linearly with time from 35 kg-m to 15 kg-m;

(*iii*) For the last 50 seconds, the torque is constant and equal to 10 kg-m.

Solution: The load cycle is shown in Fig. 4.58.

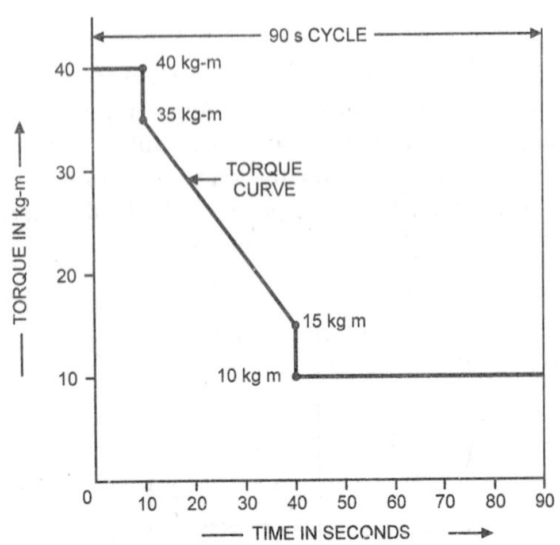

Fig. 4.58

$$(T_{eq})^2 \times \text{Time of cycle} = \int T^2\, dt \text{ over the cycle}$$

$$= \int_0^{10} 40^2\, dt + \int_{10}^{40} \left[35 - \frac{(35-15)}{30}(t-10)\right]^2 dt + \int_{40}^{90} 10^2\, dt$$

$$= \int_0^{10} 1,600\, dt + \int_{10}^{40} \left[1,225 - \frac{140}{3}(t-10) + \frac{4}{9}(t-10)^2\right] dt + \int_{40}^{90} 100\, dt$$

$$= [1,600]_0^{10} + \left[1,225t - \frac{140}{6}(t-10)^2 + \frac{4}{27}(t-10)^3\right]_{10}^{40} + [100]_{40}^{90}$$

$$= 16,000 + 1,225 \times 30 - \frac{70}{3} \times 900 + \frac{4}{27} \times 27,000 + 5,000$$

$$= 16,000 + 36,750 - 21,000 + 4,000 + 5,000 = 40,750$$

$$\text{or } T_{eq} = \sqrt{\frac{40,750}{90}} = 21.278 \text{ kg-m}$$

$$\text{Power rating of motor, } P = T_{eq} \text{ in kg-m} \times 9.81 \times \frac{2\pi N}{60}$$

$$= 21.278 \times 9.81 \times \frac{2\pi \times 750}{60}$$

= 16,400 W or 16.4 kW **Ans.**

Example 4.18. A motor has to deliver a load rising uniformly from zero to a maximum of 1,500 kW in 20 s during the acceleration period, 1,000 kW for 50 s during the full-load period and during the deceleration period of 10 s when regenerative braking takes place the kW returned to the supply falls from an initial value of 500 kW to zero uniformly. The interval for decking before the next load cycle starts is 20 s. Estimate the rating of motor.

[*J.N. University Hyderabad, April-2003*]

Solution: The load cycle is shown in Fig. 4.59.

Assuming losses to be proportional to (power)^2

(kW rating of motor)^2 × time of cycle

$$= \int (\text{kW})^2\, dt \quad \text{over the cycle}$$

$$= \int_0^{20} \left(\frac{1,500}{20} t\right)^2 dt + \int_{20}^{70} 1,000^2\, dt + \int_{70}^{80} [50\,(t-70)]^2\, dt$$

$$= \int_0^{20} (75t)^2\, dt + \int_{20}^{70} 1,000^2\, dt + \int_{70}^{80} [50\,(t-70)]^2\, dt$$

$$= [1,875t^3]_0^{20} + [1 \times 10^6 t]_{20}^{70} + \left[\frac{2,500}{3}(t-70)^3\right]_{70}^{80}$$

$$= 15 \times 10^6 + 50 \times 10^6 + \frac{2.5}{3} \times 10^6 = 65.833 \times 10^6$$

$$\text{kW rating of motor} = \sqrt{\frac{65.833 \times 10^6}{100}} = 811.4 \text{ kW} \quad \textbf{Ans.}$$

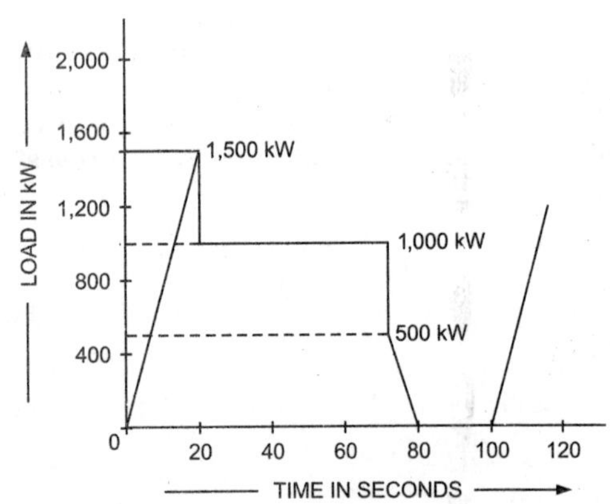

Fig. 4.59

Hence, the rating of continuously rated motor on the basis of rms value should be 850 kW (nearest size available) in order to supply the given load without exceeding the specified temperature.

Note: During regenerative braking, even though the power is returned to the line, the machine will be carrying current. So far as heating is concerned, it is immaterial whether machine is drawing current or returning it to the line.

4.22 TYPES OF ENCLOSURES

In order to protect machines against the ingress of dirt and dust or larger foreign bodies like spanners, vermin etc. into it, it is desirable that some suitable enclosure be provided. But the enclosure provided for this protection will interfere with the free entry of cool air into the machine with the result that an enclosed machine does not give the same output as does an open type machine of the same dimensions for a given temperature rise.

The purpose of providing enclosure is two fold—firstly providing protection to persons against contact with live or moving parts inside the enclosure and protection to machines against the ingress of solid foreign bodies, and secondly providing protection to machines against harmful ingress of water. Various types of enclosures giving various degrees of protection and cooling facilities for use in different working conditions can be built. ISI 4691 provides an ingress protection code which consists of the letter IP followed by two numbers—the first characteristic numeral designates the extent of protection of persons against contact and of the machine against ingress of solid foreign bodies, while the second characteristic numeral indicates the extent of protection to the machine against harmful ingress of water. Further addition of letter S or M indicates that the degree of protection against ingress of water is applicable only while the machine is stationary or rotating respectively. Absence of both the letters means that protection is provided under both the conditions. For example enclosure marked IP 22 will provide protection against ingress bodies of diameter more than 12 mm and against ingress of water falling up to 15° from vertical under both the conditions. Similarly enclosures marked IP 44 and IP 56 will provide protection against ingress of bodies of diameter more than 1 mm and against water splashed under both the conditions and protection against ingress of bodies in powder form and water projected by a nozzle under both the conditions respectively. For further details refer to ISI code 4691.

The various types of enclosures, as specified by relevant ISI are as follows:

1. *Open Type.* The bearings are supported on simple pedestals or end brackets. This type of enclosure does not protect the machines against dust, dirt and larger foreign bodies but facilitates in cooling. The machines of this type are usually large dc generators and dc motors, which are installed in a neat and clean separate room such as in case of power stations, substations, collieries, rolling mills. Because all the moving parts are exposed, hand rail and other suitable rail guards are provided to prevent unauthorized persons coming into contact specially with moving parts.

2. *Protected Type.* Protected type machines are such in which access to live and rotating parts is so limited as to prevent accidental contact with such parts or ingress of foreign bodies while ventilation is not materially obstructed. Protected type motors are used mostly for industrial purposes.

3. *Screen-Protected Type.* In this type of machines, openings provided for ventilation are covered with wire screen, expanded metal or other suitable perforated covers having apertures not exceeding 3.25 cm^2 in area but not less than 0.6 cm^2 in area. Though such an enclosure does not protect the motor against dirt and dust but larger bodies and big insects cannot enter into the machine . Such motors are very widely employed for most of industrial applications where dry clean conditions prevail and gas or chemical fumes are not present. Fig. 4.60 (*a*) illustrates a screen-protected motor in which no attempt has been made to exclude dust or fluff from the windings. It is suitable for most forms of electric drive and permits ample ventilation.

4. *Drip-Proof Type.* In such machines openings provided for ventilation are so protected by overhanging cowls that liquid drops or dust particles falling on the machine vertically cannot enter the machine. Such motors are used in damp situations such as for pumping sets. Fig. 4.60 (*b*) illustrates a drip-proof motor, similar to the screen-protected type, but having protection against dripping moisture.

5. *Splash-Proof Type.* In such machines the ventilating openings are so constructed that liquid drops or dust particles falling on the machine or coming towards it in a straight line at any angle not greater than 100° from the vertical cannot enter the machine. Such motor can be used in rains.

6. *Totally Enclosed Type.* Such machines have solid frames and end shields but no opening for ventilation. They get cooled by surface radiation only. In such machines no dirt or foreign matter can enter the motor and clog the air passage. These machines are used in situations where dust, water, oil, chemicals, gases or other injurious substances are likely to enter the machine

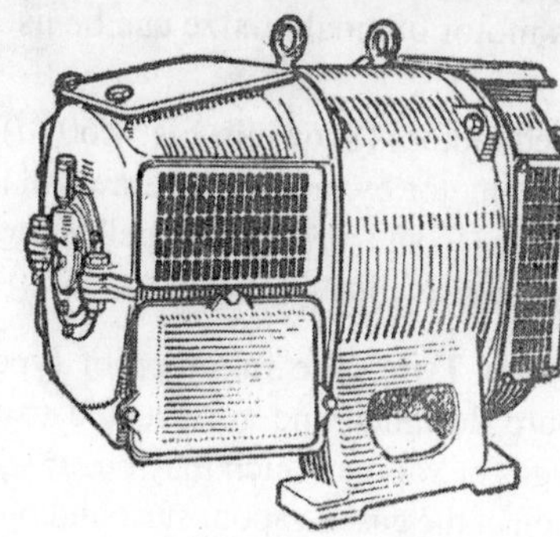

(a) Screen-Protected Motor

(b) Drip-Proof Motor

Fig. 4.60

but natural cooling is sufficient such as for driving coal handling plant, sawmills etc. In these machines due to cooling difficulties the output obtainable is only about 50 to 70% of that obtainable from a protected type of same dimensions and for same temperature rise. Totally enclosed motors are seldom furnished in sizes above 1.5 or 2 kW because higher rating motors require frames of much larger sizes than fan cooled motors of corresponding rating.

7. *Totally Enclosed, Fan Cooled Type.* In order to assist in heat dissipation, most of the manufacturers have developed totally enclosed fan cooled motors in which an outer casing is built around the motor and air is blown by means of a fan mounted on the shaft between the casing and external surface of the motor. To increase cooling rate, motor is provided with external fans. Such motors are used in sawmills, flour mills, cement works etc. These motors are not waterproof and should not be used in places likely to be submerged. Fig. 4.60 (*c*) illustrates a totally enclosed motor in which the main means of cooling is by radiation and internal air circulation.

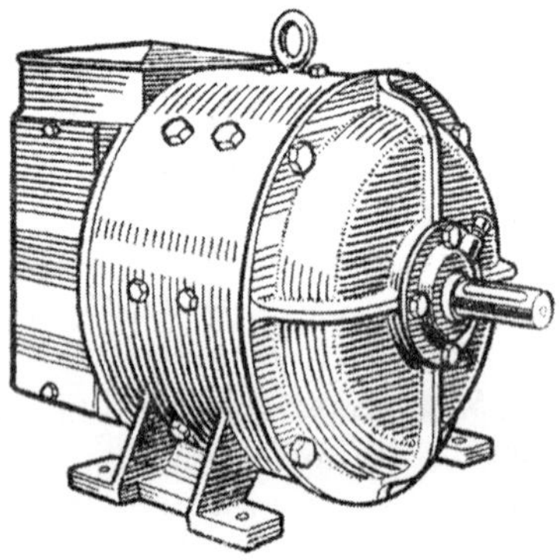

(c) Totally-Enclosed Motor

Fig. 4.60

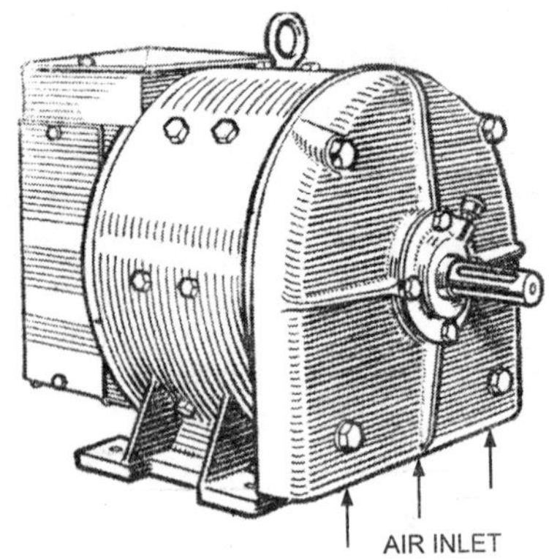

(*i*)

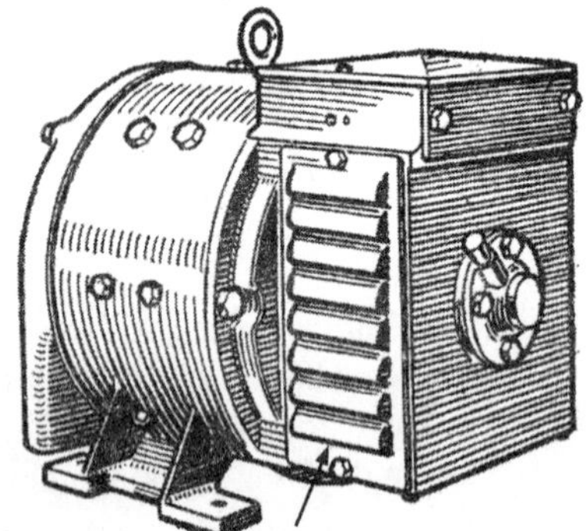

(*ii*)

(d) Pipe-Ventilated Motor

Fig. 4.60

8. *Pipe Ventilated Type.* These are totally enclosed type machines in which to overcome the heat dissipation difficulties, a duct or pipe is provided through which clean air is led to the motor from outside the dust laden building or area. The cost of the piping may be overweighed by the saving in cost of motor, as motor of smaller size can be used with same output.

Pipe ventilated motor is shown in Fig. 4.60 (*d*). Fresh clean air is brought to the motor by piping and drawn in through a big inlet, as illustrated in (*i*), and air is then expelled through louvres or taken away by pipe, as illustrated in (*ii*).

9. *Flame-Proof Type.* The flame-proof type machine is provided enclosure designed and constructed to withstand an explosion of the gas or vapour which may occur within it and to prevent the ignition of the gas or vapour surrounding the machine. The ventilating openings in the machine are made long enough for the flame to be contained within the frame. Flame-proof type motors are used in collieries, petroleum stations or chemical plants.

4.23 METHODS OF VENTILATION AND COOLING OF MACHINES

It is necessary to provide suitable ventilation and cooling for the machines so that the temperature rise at any part of the machine does not exceed the permissible limit governed by the type of insulation employed in the construction. The cooling of electrical machine by means of an air stream is called *ventilation of the machine*.

The circulation of air in the machine can be arranged by the use of a built-in fan or fan with a separate drive. Either a centrifugal type fan or an axial type fan is employed for rotating electrical machines. A centrifugal fan forces air from the centre to flow outwards and operates the same way irrespective of the direction of rotation of the machine. An axial or propeller type fan moves the air in the opposite direction when the direction of rotation of the machine reverses.

According to IS : 4722-1968, the cooling system for rotating electrical machines are classified into three types depending up the origin of cooling.

1. *Natural Cooling.* The machine is cooled by air movements in the machine due to its rotation or due to temperature difference between inside parts and the outside temperature of the air. The machine thus is cooled without the use of a fan, by the movement of air and radiation. This method of cooling is employed only in small fractional kW machines with rating within several hundred watts, since the condition of cooling are comparatively favourable.

2. *Self Cooling.* The machine is cooled by air, blown by fan integrally built with the motor or mounted on the shaft.

3. *Separate Cooling.* The machine is cooled either by a fan not driven by its shaft, or it is cooled by a cooling medium other than air put into motion by means not belonging to machine.

According to the manner of cooling the ventilation systems are classified as

1. *Open-Circuit Ventilation.* The heat is given up directly to the cooling air flowing through the machine which is being continuously replaced. The open-circuit ventilation can be further divided into two types in accordance with how the air is brought into the machine.

(a) Induced Ventilation. Induced ventilation method used in an electrical machine is illustrated in Fig. 4.61. In this arrangement, a fan produces a reduced pressure of air inside the machine and air from outside is sucked into the machine owing to the atmospheric pressure. The air is circulated through the machine and then pushed out by the fan into the atmosphere. The fan is normally arranged inside the machine, as illustrated in the figure. The fan can be provided externally also.

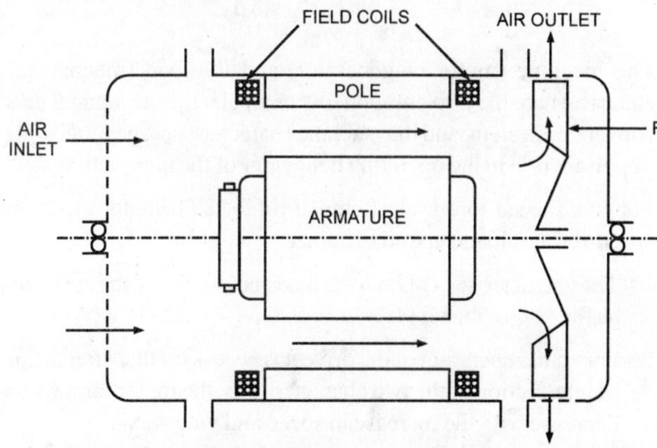

Fig. 4.61 *Induced Ventilation in Electrical Machine*

(b) Forced Ventilation. In this method of ventilation, illustrated in Fig. 4.62, the fan sucks the air from outside and forces it into the machine and finally pushed out into the atmosphere. For forced ventilation, the pressure of air inside the machine is greater than the atmospheric pressure outside. The fan may be arranged internally, as illustrated in the figure or externally,

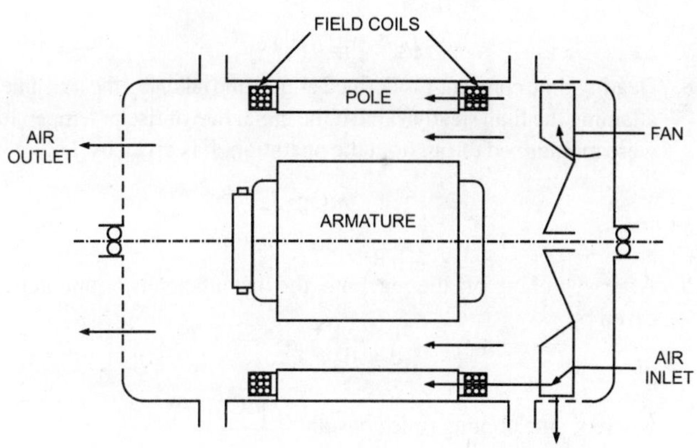

Fig. 4.62 *Forced Ventilation in Electrical Machine*

2. ***Closed-Circuit Ventilation.*** In this system of cooling, heat is transferred to the cooling medium such as air or hydrogen through an intermediate cooling medium circulating in a closed circuit between the interior of the machine and the cooling device, *e.g.*, a water cooler can be used to cool the cooling medium such as air.

For increasing the surface of contact for cooler, stator and rotor cores of electrical machines are provided with ducts. These ducts may be radial or axial type depending on the direction of air flow through them.

3. Surface Ventilation. The heat is transferred from inside of the machine, by cooling medium, to the external surface of a totally enclosed machine. The external surface is being cooled by natural means or mainly by air blown by fan.

4.24 FRAME SIZE

A frame is the mechanical structure required to house a stator along with its bearings, end covers and terminal box. The standardization of frame size leads to economy. All modern motors of small size are built with standard frame sizes as specified in IEC 72 which lists a coherent range of main structural dimensions with centre heights between 56 and 1,000 mm.

A frame size is designated by a number which is its centre height (H) expressed in mm. Thus frame designated 132 has a centre height-(H) of 132 mm. The frame sizes with heights between 56 and 1,000 mm as per recommendations of IEC are 56, 63, 71, 80, 90, 100, 112, 132, 160, 180, 200, 225, 250, 280, 315, 355, 400, 450, 500, 630, 710, 800, 900, 1,000 mm.

The outputs from the standard frames are periodically assessed to take into account the latest technological advancements.

The frames may be diecast or fabricated. Machines of ratings up to 50 kW have their frames diecast in a strong silicon aluminium alloy and in some cases with the stator core cast in. The frames of larger machines are fabricated by welding steel plates. The advantage of fabrication is its adaptability to new designs and modifications.

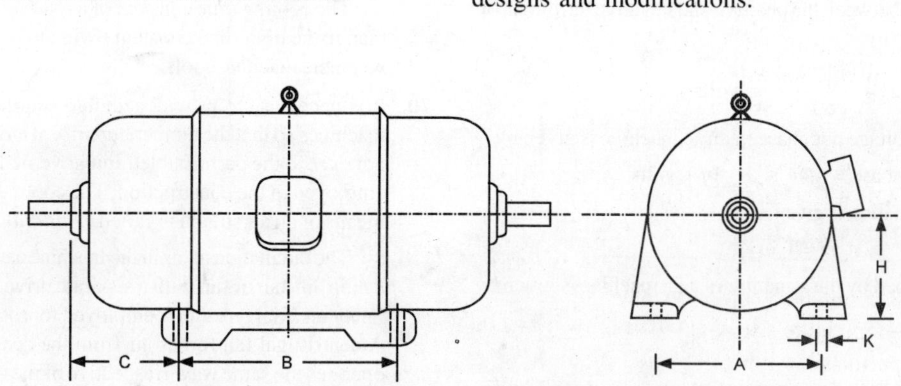

Fig. 4.63 *Structural Dimensions of Standard Frames*

HIGHLIGHTS

1. There are three basic types of electrical rotating machines viz. dc machines, synchronous machines and induction machines. In all of these machines the basic principles are essentially the same. Voltages are generated by relative motion of magnetic field with respect to armature winding, and torques are produced by interaction of stator and rotor magnetic fields. The characteristics of the different types of machines are determined by the methods of excitation and connection of windings, but the basic principles are essentially the similar. The basic analytical tools for studying rotating machines are expressions for the generated voltages and for the electromagnetic torque.
2. There are essentially two types of forces/torques developed in electromechanical devices viz. electromagnetic force/torque and reluctance force/torque. The force/torque developed due to interaction of the fields produced by the currents in the two windings is known as electromagnetic force/torque. The reluctance force/torque, dependent on the current in only one winding, is due to the variations in the reluctance of the air gap in the magnetic circuit carrying the flux linking with the winding. Both of these phenomena usually occur simultaneously.
3. In a multipolar machine one pair of poles or one cycle of flux distribution equals 360 electrical degrees or 2π electrical radians. Since there P/2 complete wavelengths or cycles in one complete revolution, it follows that
$$\theta_{elec} = \frac{P}{2}\theta_{mech}$$
4. The frequency of induced emf in a coil in a multipolar machine is given by the expression
$$f = \frac{PN}{120} \text{ Hz}$$
where P is number of poles and N is the mechanical speed of rotation in rpm.
5. The radial frequency ω_{elec} of the wave of emf induced in P-pole machine is given as
$$\omega_{elec} = \frac{P}{2}\omega_{mech}$$
6. The essential elements for generating emf in alternators are prime mover, stator, rotor and exciter.
7. Practically all medium and large ac machines are always constructed with revolving field due to its inherent advantages.
In dc machines, the commutator makes necessary either the armature should be rotating, or brushes should revolve with the field, so it is convenient to have armature as a rotating member in dc machines.
8. The emf generated between the positive and negative terminals of a dc machine is given by
$$E = \frac{\Phi ZN}{60} \times \frac{P}{A} \text{ volts}$$
The induced rms voltage per phase in an ac machine is given by
$$E_{rms}/\text{phase} = 4.44\, K_d K_p \Phi f T \text{ volts}$$
9. Torque balance equation is given as
$$ei = T_e \omega_m$$
The torque developed by the armature of a dc machine is given by
$$T_a = \frac{1}{2\pi}\Phi ZP \frac{I_a}{A} \text{ N-m}$$
10. A rotating magnetic field is that field which is constant in magnitude but whose axis of direction rotates in space as field system of a dc machine.
11. When two identical coils placed at 90° in space are excited from 2-phase balanced supply, the resultant mmf will be rotating in nature with an amplitude equal to that of each phase.
12. When three identical coils placed at 120° in space are excited from 3-phase balanced currents, the resultant mmf will be rotating in nature with an amplitude 1.5 times that of each phase.
13. The expression for the torque developed in machines with cylindrical air gap is given by
$$T = -\frac{\pi}{8}P^2 F_2 \Phi_R \sin\delta_2$$
14. The machine (motor or generator) and the load (mechanical or electrical) are the two components of an electromechanical energy conversion system, and the machine characteristics, generally, play a important role in the operating behaviour of the complete system.
15. For given speed-torque characteristic of load, the motor should have speed-torque characteristic such that
    1. The characteristics of both the load and the motor should intersect as far as possible at 90°.
    2. For stable operation of the drive it is necessary that after the point of intersection of the two characteristics, the motor torque should decrease with the increase in speed and vice versa.
16. Final temperature rise in an electrical machine is given by
$$\theta_F = \frac{P}{S\alpha}$$
17. Instantaneous temperature after start of an electric machine is given by
$$\theta = \theta_F - (\theta_F - \theta_1)\, e^{-t/\tau}$$
18. Heating time constant τ is defined as the time taken by the machine in attaining the final steady value if the initial rate of rise of temperature were maintained throughout the operation. It is given by
$$\tau = \frac{mC_p}{S\alpha}$$
19. After switching off the machine, the instantaneous temperature is given by
$$\theta = \theta_F\, e^{-t/\tau'}$$
where τ', the cooling time constant = $\dfrac{mC_p}{S\alpha'}$
The cooling time constant of a rotating machine is usually larger than its heating time constant owing to poor ventilation conditions when the machine cools.
20. It is necessary to provide suitable ventilation and cooling for the machines so that the temperature rise at any part of the machine does not exceed the permissible limit governed by the type of insulation employed in the construction. The cooling of electrical machine by means of an air stream is called *ventilation of the machine.*
The circulation of air in the machine can be arranged by the use of a built -in-fan or fan with a separate drive. Either a centrifugal type fan or an axial type fan is employed for rotating electrical machines. A centrifugal fan forces air from the centre to flow outwards and operates the same way irrespective of the direction of rotation of the machine. An axial or propeller type fan moves the air in the opposite direction when the direction of rotation of the machine reverses.

EXERCISES

1. How two magnetic fields interreact to produce torque? Hence define torque angle.
2. Explain how torque is produced in a rotating electrical machine. Differentiate between reluctance and electromagnetic torque.
3. (*a*) Give elementary concepts of generator and motor.
   (*b*) Explain how the torque is produced in electrical machines and give the concept of torque angle.
4. Draw and explain the power flow diagram for a dc machine working as a generator and as a motor.
5. Show the direction of electromagnetic torque, frictional torque and mechanical torque and direction of rotation (*i*) in a motor and (*ii*) in a generator.
6. From the construction point of view, enumerate the common essential features of rotating electrical machines.
7. Deduce expressions for the (*i*) voltage induced (*ii*) torque developed in a dc machine armature. Show that both are always existing irrespective of the mode of operation i.e. generator or motor. **[Pb. Univ. Elec. Machines-II, December-1994]**
8. Derive an expression for the emf generated in a dc machine by using Blv concept.
9. Derive from first principles, the emf equation of a 3-phase synchronous machine. **[R.G.P.V. M.P., 2002]**
10. Write down the voltage equation for an electromechanical device and explain the term speed voltage in the same. **[U.P. Technical Univ. Electromechanical Energy Conversion-I, 2009-10]**
11. Derive the expressions for the emf and the electromagnetic torque of a dc machine. **[W.B. Univ. of Technology Electrical Machine-I, 2013]**
12. Discuss in brief about "torque production and torque balance". **[R.G.P.V. M.P., 2001]**
13. Derive the general torque relation for a rotating electrical machine in terms of two magnetic fields. **[R.G.P.V. M.P., 2001]**
14. Prove that a single-phase motor winding when excited by a single-phase supply produces two equal and opposite revolving fields.
15. Show that when two identical coils displaced in space by 90° are connected across 2-phase ac supply, a uniformly rotating field of constant amplitude rotating at synchronous speed is produced.
16. When a 3-ϕ supply is given to a 3-phase winding, a rotating field of constant magnitude will be produed. Justify the above statement. **[U.P.S.C. I.E.S. Elec. Engineering II, 1993]**
17. Prove that a balanced three-phase supply results in a rotating magnetic field. **[G.B. Technical Univ. Electrical Machines, 2010-11]**
18. For a non-salient machine derive the general torque expression in terms of the interacting stator and rotor mmfs. State the various assumptions made. **[R.G.P.V. Bhopal, June-2005]**
19. Classify electrical insulating materials on the basis of Indian standards.
20. Classify the losses in a rotating electrical machine stating all possible losses. **[R.G.P.V. Bhopal, 2001]**
21. Enumerate the sources of heat generation in rotating electrical machines and name the parts where it is generated. **[Pb. Univ. Electrical Machine Design Dec., 1995]**
22. Derive a temperature-rise versus time relationship for an electrical machine considering it to be a homogeneous body. Define the terms (*i*) heating time constant (*ii*) cooling time constant and (*iii*) maximum temperature rise. List the factors on which they depend. **[Pb. Univ. Electrical Machine Design, June-1992]**
23. Obtain the expression for temperature rise of an electrical machine. State the assumptions made if any. **[J.N. University Hyderabad, December 2002/January 2003]**
24. With the help of heating and cooling curves define and explain the terms:
    (*i*) Heating time constant (*ii*) Cooling time constant. **[Nagpur University Summer 2004]**
25. Classify the ratings of motor based on duty. Also enumerate the various methods in detail for the calculation of rating of motor. **[U.P. Technical Univ. Electric Drives, 2013-14]**
26. What are the limiting temperatures of motors when the class of insulation used is :
    (*i*) Y, (*ii*) A, (*iii*) E, (*iv*) B, (*v*) F, (*vi*) H and (*vii*) C ?
    Give typical examples of loads demanding continuous duty, intermittent duty and short-time duty.
27. Discuss different classes of motor duty cycles. Give industrial application for each class of duty cycle.
    Discuss the equivalent power method for determining the rating of an electric motor which is subjected to intermittent periodic duty.
28. Draw the pattern of temperature rise characteristic under steady-state for (*i*) short-time duty (*ii*) intermittent duty.
    What is the equivalent current method of estimating motor rating ? Explain. Can this method be used if motor speed varies during a load cycle?
29. Discuss 'rms current', 'rms torque' and 'rms power' methods for evaluation and rating of motor for continuous duty variable load conditions. Give the merits and limitations. **[U.P. Technical Univ; Electric Drives 2005-06]**
30. Explain how would you estimate the rating of a motor for intermittent duty cycle.
31. Specify the different types of enclosures of motors, highlighting the special features of each type. Mention the area of application of each.
32. Briefly explain various types of cooling systems for rotating electrical machines.
33. Write briefly on ventilation and cooling of dc machines.
34. Write short notes on the following:
    (*i*) Basic concepts of rotating electrical machines.
    (*ii*) MMF and flux distribution in dc and ac machines.
    (*iii*) Generated emf in machines. **[G.B. Technical Univ. 2012-13]**
    (*iv*) Production of 3-phase rotating magnetic fields.
    (*v*) Matching characteristics of electrical machine and load.
    (*vi*) Thermal model of motor for heating.
    (*vii*) Types of enclosures for electrical machines.

SHORT ANSWER TYPE QUESTIONS WITH ANSWERS

Q. 1. Figure 1 shows cross-sectional view of a coil rotating in a magnetic field. The direction of rotation and the direction of flow of current in the coil sides is also shown in the figure.

(*i*) Find the direction of emf induced in the coil sides.

(*ii*) Determine the direction of electromagnetic torque and torque angle.

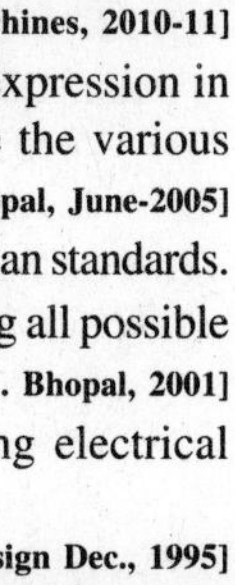

Fig. 1

(*iii*) Mark the directions of mechanical torque and frictional torque.

(*iv*) This device is working as a motor or generator ?

(*v*) Give answers to all the above questions when the coil has rotated through 180° but the direction of the current in the coil is unchanged.

(*vi*) Does the device change its mode of action (becomes motor to generator or generator to the motor) when the coil turns through 180° from the initial position ? If yes, what should be done to the current in the coil to get continuous generating or motoring action.

Ans. (*i*) Opposite to that of direction of flow of current.

(*ii*) Clockwise.

(*iii*) Counter-clockwise, 330°.

(*iv*) working as a motor.

(*v*) in the direction of flow of current; counter-clockwise; 30°; clockwise, counter-clockwise; working as a generator.

(*vi*) Yes, action becomes motor to generator; by providing a commutator in case of a dc machine or by driving the main poles in case of a synchronous machine.

Q. 2. For the fig. shown (Fig. 2) mark the direction of emf induced in the stator coils when the rotor rotates in the clockwise direction. Also mark the coil sides which belong to the same stator coils.

Ans. The direction of the emf induced is illustrated in Fig. 3; coil sides belonging to the same coil are as 1-5, 2-6, 3-7, 4-8, 9-13, 10-14, 11-15, 12-16.

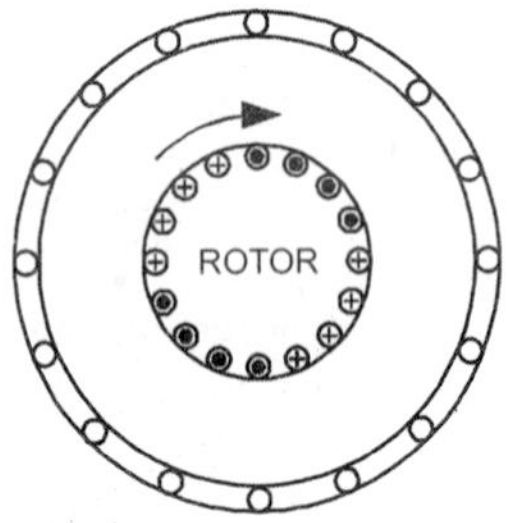

Fig. 2

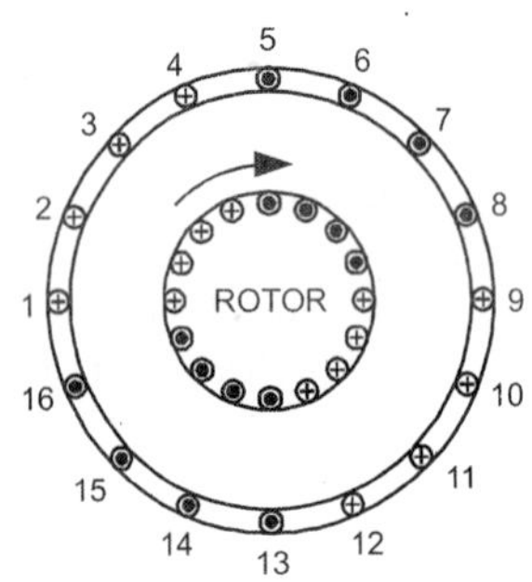

Fig. 3

Q. 3. What is meant by back emf in a dc motor ?

Ans. When the coil rotates under the influence of electromagnetic torque in the magnetic field (motor action) the coil cuts the magnetic flux thereby inducing an emf in it which opposes the applied voltage. This induced emf is called the *back emf.*

Q. 4. An aeroplane is flying horizontally with a velocity v m/s cutting earth's vertical lines of force, B Wb/m^2. A voltmeter is connected across its two wings l metre apart. What voltage will the voltmeter indicate ?

Ans. Blv volts

Q. 5. State at what value of torque angle the torque produced is (*a*) maximum (*b*) minimum?

Ans. (*a*) ± 90° (*b*) 0 and 180°.

Q. 6. In fig. (Fig. 4) mark the currents in the conductors to obtain clockwise torque.

Ans. Refer to Fig. 5

Q. 7. Why is the armature core of a dc generator laminated?

Ans. In order to reduce eddy current losses

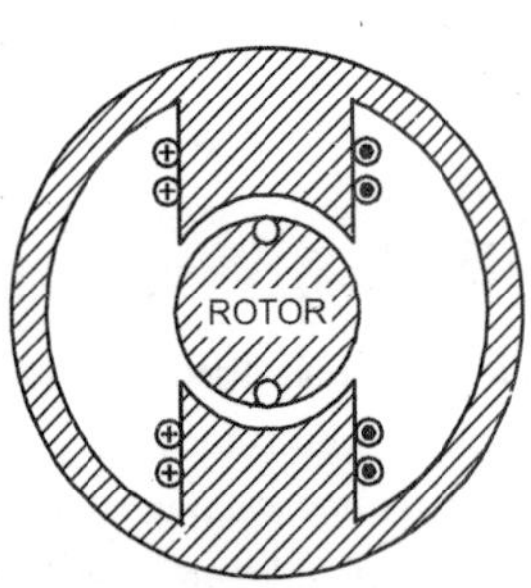

Fig. 4

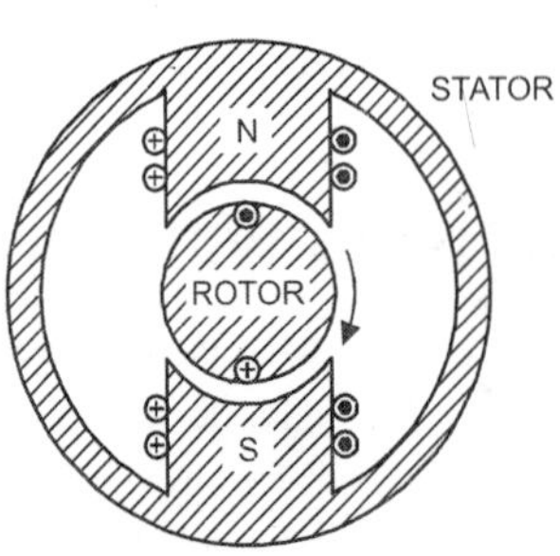

Fig. 5

Q. 8. A conductor of effective length l (m) moves in the direction of a uniform magnetic field B (tesla) at a constant speed v ($m\ s^{-1}$). What is the magnitude of the emf induced in the conductor ?

Ans. Zero.

Q. 9. A circular coil suspended by a light inextensible string is lowered at a uniform rate in a constant magnetic field such that the plane of the coil is always perpendicular to the direction of induction. Will the tension in the string increase, decrease or remain the same in the process?

Ans. Decrease.

Q. 10. Name the different types of rotating electrical machines.

Ans. The rotating electrical machines are of two types namely 1. *dc* machines 2. *ac* machines. AC machines, on the basis of principle of operation, are of two types viz synchronous machines and induction machines (non-synchronous machines). AC machines, on the basis of number of phases employed, are also of two types viz single-phase and three-phase machines.

Q. 11. What is meant by torque angle?

Ans. The angle between the axis of the main (or stator) magnetic field and the magnetic axis of the rotor is called the torque angle.

Q. 12. What is meant by interaction or electromagnetic torque?

Ans. The torque developed due to interaction of stator and rotor magnetic fields is called the interaction or electromagnetic torque.

Q. 13. What is relationship between driving torque, friction torque and electromagnetic torque in case of a generator?

Ans. Driving torque, T_d = Electromagnetic torque, T_e + friction torque, T_f.

Q. 14. What is a rotating magnetic field ?

Ans. A rotating magnetic field is that field which is constant in magnitude but whose axis of direction rotates in space like field system of a dc machine.

Q. 15. How can the direction of the stator revolving field be reversed ? **[Bharathiyar Univ; April-1996]**

Ans. The direction of the stator revolving field can be reversed by reversing any two of the three supply lines to the stator winding.

Q. 16. What is meant by "energy loss" in a machine ?

Ans. Whole of the input energy to the electrical machine is not converted into useful output energy but a part of the energy is converted into heat and is lost for the useful purpose of the machine. The energy converted into heat is called the energy loss.

Q. 17. What is the basic power equation for a dynamo ?

Ans. $P_{input} = P_{output} + P_{losses}$

Q. 18. Why do copper losses occur in a dc machine ?

Ans. Copper losses occur in dc machine due to power consumed in forcing a current against the resistance of windings and connections.

Q. 19. How do the various losses occurring in a dc machine vary with the load ?

Ans. The copper losses occurring in armature winding, series field winding, compensating winding and interpole winding, if provided, almost vary as the square of the load current while that occurring in dc shunt field winding remain practically constant. Rotational losses (iron and mechanical losses) are almost independent of load.

Q. 20. Does core loss occur in armature or in the poles of a dc machine ?

Ans. Core loss occurs in armature core and pole faces.

Q. 21. On what factors do hysteresis and eddy current losses depend ?

Ans. Hysteresis loss depends upon the peak magnitude of the flux density attained in a magnetic cycle, frequency of magnetic cycles, volume of magnetic material and type of magnetic material.

Eddy current loss depends also upon the thickness of magnetic material laminations in addition to factors mentioned above.

Q. 22. How the eddy current losses will change if the thickness of laminations is increased ?

Ans. Eddy current losses vary as the square of the thickness of magnetic material laminations.

Q. 23. Why mechanical losses occur in a dc machine ?

Ans. Mechanical losses occur in dc machines due to motion of rotating parts through the surrounding medium and the friction between brushes and commutator.

Q. 24. Give the condition when the rate of rise of temperature in a machine will be zero.

Ans. Heat generated = Heat dissipated to the surroundings by one or more of the natural modes of heat transfer.

Q. 25. On what factors does the rate of temperature depend ?

Ans. The temperature rise depends upon (*i*) the amount of heat produced and (*ii*) the amount of heat dissipated per °C rise of the surface of a machine.

Q. 26. Define heating time constant of the machine.

Ans. Heating time constant may be defined as the time during which the machine will attain 63.2% of its final steady temperature.

Q. 27. Why the cooling time constant of a rotating machine is usually larger than its heating time constant ?

Ans. The cooling time constant of a rotating machine is usually larger than its heating time constant owing to poor ventilation conditions, when the machine cools.

Q. 28. What is meant by "rating of motor" ? **[P.T.U., December-2009]**

Ans. A name plate fixed to the outside frame of an electrical machine records the data pertaining to its rating. A machine rating specifies the voltage, current, speed, excitation, power factor, efficiency, power output etc.

The rating of a machine should give all information necessary so that, if the machine is operated within the limits of all factors specified in its rating, the machine will operate satisfactorily and safely and will give reasonable length of service. Therefore, the rating of a machine must give the necessary information to safeguard the application of the machine from conditions of operation which (*i*) would result in unsafe mechanical or electrical strains upon any part of its structure or (*ii*) would result in excessive deterioration of the mechanical or electrical characteristics of the materials of which the machine is constructed, To give this information, the rating of an electric machine should include the output, voltage, speed, and any other information that may necessary for the proper operation of the machine.

Q. 29. What informations should the rating of a machine give ?

Ans. The rating of an electrical rotating machine should include the output, voltage, speed and any other information that may be necessary for the proper operation of the machine.

Q. 30. In what terms are dc machines rated ?

Ans. The dc machines are rated in terms of kW output at a given speed and voltage.

Q. 31. What is the main factor which governs the size and rating of an electric machine ?

Ans. "Temperature rise" is the main factor which governs the size and rating of an electric machine.

Q. 32. What are the standard period for short-time rating of a machine ?

Ans. Standard short-time ratings are 10, 30, 60 and 90 minutes.

Q. 33. A motor of smaller rating can be selected for a short time duty. Why? **[U.P. Technical Univ. Electric Drives, 2009-10]**

Ans. The heating and cooling curves for *short-time duty motor* are given in Fig. 4.53 (*b*). The short-time duty motor operates at a constant load for some specified periods which is then followed by a period of rest. The period of run (or load) is so short that machine cannot attain its steady temperature rise while the period of rest is too long that the motor temperature drops to the ambient temperature. Thus a motor of smaller rating can be operated with a larger power output for the maximum temperature rise in a short duration.

Q. 34. Why a totally enclosed type of motor is more expensive than a screen protected type ?

Ans. In totally enclosed type motor, due to cooling difficulties, the output obtainable is only about 50 to 70% that obtained from a protected type of same dimensions and for same temperature rise.

Q. 35. When will you recommend radial ventilating system ?

Ans. Radial ventilating system is used for machines of small and medium ratings.

Q. 36. Why machines having long core length are provided with radial ventilating ducts while machines having deep cores are provided with axial cooling ?

Ans. Radial ventilating ducts provided with long core machines take away heat at a faster rate. Also in case of deep cores axial ventilating ducts does. Otherwise the central portion of long core and inner portion of deep core may attain a temperature higher than permissible value.

Q. 37. The outside of the machine should be cleaned frequently and the air passages blown out regularly. Why ?

Ans. Otherwise dust deposit upon the external surface of a machine will act as a heat insulant and in air ducts will restrict the flow of cooling air and consequently temperature will rise.

PROBLEMS

1. The heating and cooling time constants of a motor are 1 hour and 2 hours respectively. Final temperature rise attained is 100 °C. The motor runs at full load for 30 minutes and then kept idle for 12 minutes and the cycle is repeated indefinitely. Determine the temperature rise for motor after one cycle. **[Nagpur University Winter 1997]**

   [**Ans.** 35.6 °C]

2. Determine the one-hour rating of a 15 hp motor having heating time constant of 2 hours. The motor attains the temperature rise of 40 °C on continuous run at full load. Assume that the losses are proportional to the square of load and the motor is allowed to cool down to the ambient temperature before being loaded again. **[Nagpur University Summer 2001]**

   [**Ans.** 24 hp]

3. Determine the half-hour rating of a 25 kW motor having a time constant of 1.5 hours. Assume that the motor cools down completely between each load period. **[U.P. Technical Univ. Electric Drives 2005-06]**

   [**Ans.** 60 kW]

4. A 100 hp motor has temperature rise of 50 °C when running continuously on full load. It has a time constant of 90 minutes. Determine half-hour rating of the motor for same temperature rise. Assume that losses are proportional to the square of load and motor cools completely between each load period. **[J.N. University Hyderabad, December 2002/January 2003]**

   [**Ans.** 188 hp]

5. The following temperature rises were observed in a temperature rise test on a dc machine at full loads.

After one hour — 15 °C

After two hours — 25 °C

Find out (*i*) Final temperature rise and time constant.

(*ii*) The steady temperature rise after 1 hour at 50 % overload from cold.

Assume that the final temperature rise on 50 % overload is 90°.

[Nagpur University Summer 1998]

[**Ans.** 45 °C; 2.4663 hours; 30 °C]

6. At full load of 10 hp the temperature rise of motor is 25 °C after 1 hour and 40 °C after 2 hours. Find the final temperature rise on full load. Assume that the iron losses are 80 % of full-load copper losses.

[J.N. University Hyderabad, April-2003]

[**Ans.** 62.5 °C]

7. The outside of a 12.5 kW dc motor is equivalent to a cylinder of a 100 cm diameter and 140 cm in length. The machine may be considered to be a homogeneous material weighing 650 kg and of specific heat 700 J/kg/°C. Assuming that the outer cylindrical surface dissipates heat at 0.0012 W/cm^2/°C, estimate the thermal time constant of the machine and its final temperature rise when operating on full load with an efficiency of 85%. [Ans. 2.394 Hours; 41.8 °C)

8. In a temperature rise test on a dc machine at full load, the following temperature rises were observed; 16 °C after one hour and 28 °C after 2 hours. Find (*i*) final steady temperature rise and the time constant (*ii*) the steady temperature rise after one hour at 50% overload from cold. Assume that the final temperature rise on 50% overload is 92 °C.

[**Hint:**Heating time constant of a motor is independent of rate of heat production so the heating time constant at 50% overload is the same as on full load.] [**Ans.** (*i*) 64 °C; 3.481 hours (*ii*) 123 °C]

9. A 50 kW motor with a heating time constant of 100 minutes has a final temperature rise of 50 °C on continuous rating. Find the half-hour rating of the motor for this temperature rise assuming that it cools down completely, between each load period. The motor has maximum efficiency of 80% at its full load.

[**Ans.** 120 kW]

10. The heating and cooling time constants of 100 kW motor are 90 and 120 minutes respectively. Find the rating of motor when subjected to a duty cycle of 18 minutes on certain load and 30 minutes on no load. Assume that the losses are proportional to the square of load.

[**Ans.** 141.4 kW]

11. An induction motor has a temperature rise of 25 °C after one hour and 40 °C after two hours while operating at full-load output of 7.5 kW. The constant losses of the motor are 80% of full-load copper losses. Determine how long the motor can be operated at twice the rated output without overheating.

[**Ans.** 55 minutes]

12. The heating time constant of a 22 kW motor is 90 minutes. If the motor is to be run continuously on full load, it attains a maximum steady temperature rise of 40 °C. Calculate the half-hour rating of the motor for this temperature rise. Assume that losses are proportional to the square of the load and the motor cools down completely between each load cycle. [**Ans.** 41.32 kW]

13. A motor driving some haulage equipment for a mine has to deliver a load which follows the following duty cycle:

100 kW for 10 minutes, no load for 4 minutes, 50 kW for 10 minutes, no load for 6 minutes. The cycle being repeated indefinitely. Estimate the suitable size of continuously rated motor for the purpose.

[**Ans.** 65 kW]

14. A machine on a short test run gave the following mean temperature rise :

Time hours	0	0.25	0.5	0.75	1.0	1.25	1.5
Temp. rise °C	42.5	45.0	47.4	49.5	51.4	53.0	54.4

Estimate graphically the final temperature rise and the time constant. Justify the procedure adopted by you. [**Ans.** 62 °C, 1.6 hours]

15. An electric motor has load variation as given below: Torque 240 N-m for 20 minutes, 140 N-m for 10 minutes, 300 N-m for 10 minutes, 200 N-m for 20 minutes. If the speed of the motor is 720 rpm, find the power rating of motor. [**Ans.** 17 kW]

16. A motor working in a coal mine has to exert power starting from zero and rising uniformly to 100 hp in 5 minutes after which it works at a constant rate of 50 hp for 10 minutes. Then, a no-load period of 3 minutes the cycle is repeated indefinitely, estimate suitable size of motor. [Nagpur University Summer 2000]

[**Ans.** 48.11 hp; say 50 hp]

17. A motor has the following duty cycle :

Load rising from 200 to 400 hp ... 4 minutes

Uniform load 300 hp ... 2 minutes

Regenerative braking hp returned to supply from 50 to zero ... 1 minute

Remaining idle for ... 1 minute

Estimate the hp of motor.

[J.N. University Hyderabad November 2003]

[**Ans.** 178 hp]

❑❑❑

SECTION II

DC MACHINES

CHAPTERS COVERED IN THIS SECTION

CONSTRUCTION OF DC MACHINES

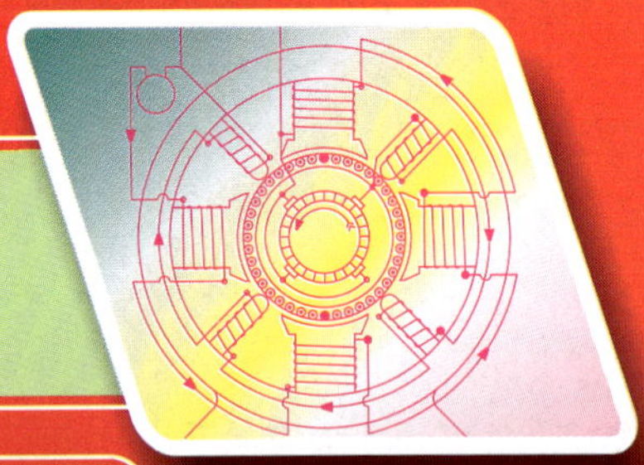

ARMATURE REACTION AND COMMUTATION IN DC MACHINES

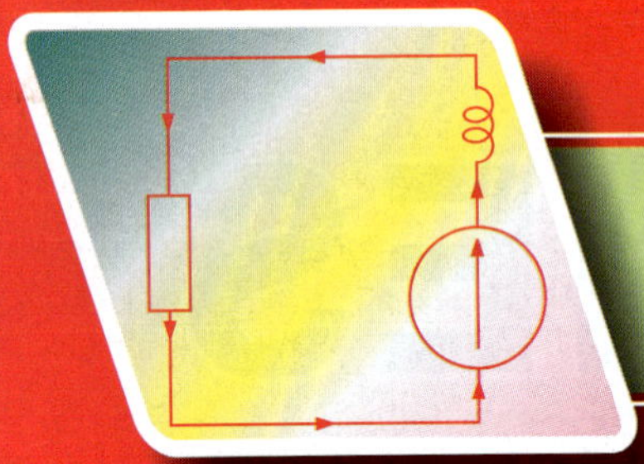

OPERATING CHARACTERISTICS AND APPLICATIONS OF DC GENERATORS

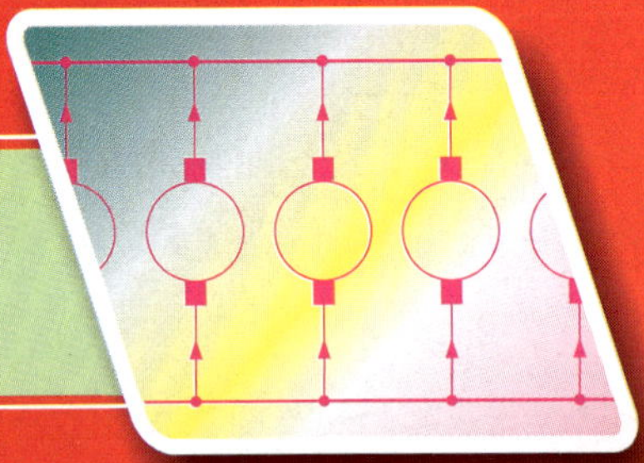

PARALLEL OPERATION OF DC GENERATORS

OPERATING CHARACTERISTICS AND APPLICATIONS OF DC MOTORS

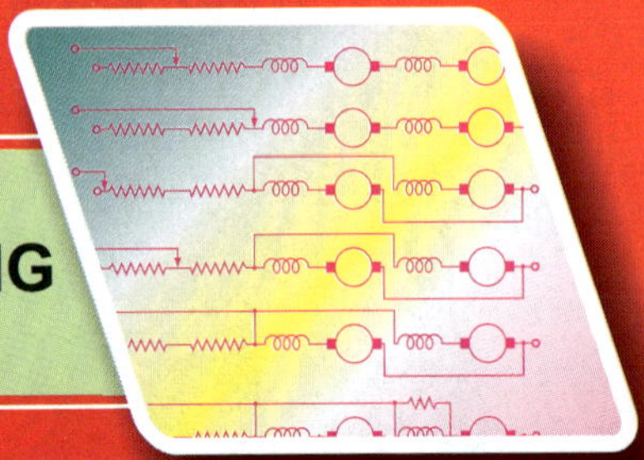

SPEED CONTROL, STARTING AND BRAKING OF DC MOTORS

LOSSES, TESTING AND EFFICIENCY OF DC MACHINES

5 CHAPTER

Construction of DC Machines

5.1 INTRODUCTION

DC machine is actually an *alternating current* machine, but furnished with a special device, called the *commutator*, which under certain conditions converts ac into dc and vice versa.

Inspite of the fact that the commutator has made the operating conditions of a dc machine complicated, it is a highly versatile energy converting device. By means of various combinations of shunt, series, and separately excited field windings they can be designed to give a wide variety of voltage-current or speed-torque characteristics for both dynamic and steady-state operation. Because of the ease with which they can be controlled, dc motors are often used in applications requiring a wide range of motor speeds or precise control of motor output— rolling mills, overhead cranes and traction; drives for process industry, battery driven vehicles, machine tools requiring precise speed control, in miniature range in tape recorders, cameras etc. Small dc motors are widely employed in control applications. Small dc generators are used for power supply in ships, aircrafts, automobiles and other vehicles isolated from inland ac network system.

Both induced emfs and mechanical forces are developed in a machine whether it is a generator or motor. As such a dc generator and motor have identical construction. In this chapter dc machines used as generators, although much of what is said concerning generators is equally applicable to motors, are dealt with.

The term "generator" denotes that it generates electrical energy but actually it does not. It simply converts mechanical energy supplied to it into electrical energy. A dc generator may be compared with a water force pump to make operation of a dc generator more clear.

As the force pump does not produce water but causes a mechanical pressure which forces the existing water into an elevated reservoir against the back pressure due to its weight, in the same way the generator does not produce electricity but creates potential difference, which causes the electric current to flow from low pressure terminal to high pressure terminal in the machine and from high pressure terminal to low pressure terminal in the external circuit against the resistance of the circuit.

The generator operates on the principle of the production of dynamically induced emf *i.e.* whenever flux is cut by the conductor, dynamically induced emf is produced in it according to the laws of electromagnetic induction, which will cause a flow of current in the conductor if the circuit is closed.

For production of dynamically induced emf, three things are necessary—a magnetic field, a conductor and motion of the conductor with respect to the field. In dc generators the field is produced by the field magnets which are stationary. Permanent magnets are used for very small capacity machines and electromagnets are used for large machines to create magnetic flux. The conductors are situated on the periphery of the armature being rotated by the prime mover.

5.1.1. Classification of Electrical Machines By Constructional Features. Electrical machines may be classified by constructional features and their sub-division into power output ranges and speed ranges can be done.

Electrical machines as per their power outputs may be classified as

(*i*) small size electrical machines with an output up to 0.6 kW.

(*ii*) medium size electrical machines with power outputs ranging from 0.6 kW to 250 kW.

(*iii*) large size electrical machines with outputs exceeding 250 kW but not beyond about 5,000 kW.

Electrical machines as per their operating speeds may be classified as

(*i*) low speed machines— speed ranges from 250 to 400 rpm.

(*ii*) medium speed machines—speed ranges from 400 to 1,500 rpm.

(*iii*) high speed machines—speed more than 1,500 rpm.

5.2 SIMPLE LOOP GENERATOR

A single turn rectangular copper conductor loop ABCD rotating in clockwise direction about its own axis XX′ in the uniform magnetic field is shown in Fig. 5.1. As the coil is rotated in the magnetic field by some mechanical means (not shown in the Fig.) the flux linking with the loop changes continually, therefore, an emf is induced in it. The magnitude of emf induced at any instant is proportional to the rate of change of linking flux at that instant and its direction is given by Fleming's right hand rule.

While rotating at the instant the loop of wire ABCD assumes the position shown in Fig. 5.1 (*a*) *i.e.* when the loop ABCD is just parallel to the faces of field magnets N and S, the flux linking with the loop is maximum but the rate of change of linking flux is zero, as at this instant no flux is cut by the coil sides AB and CD which are just moving parallel to them. Hence induced emf is zero when the loop is in this position. This position of the loop is known as *neutral position* and let it be taken starting position and angle of rotation be measured from this position.

As the coil is turned in clockwise direction at a constant speed, the coil sides begin to cut across the field, slowly at first but at gradually increasing rate. Thus the magnitude of emf induced gradually increases as the loop moves and becomes maximum when the loop assumes the position shown in Fig. 5.1(*b*). During this period the loop rotates through $\pi/2$ radians or 90° and induced emf in the loop increases following the sine law, as shown in Fig. 5.2. The direction of emf induced in the coil, as given by Fleming's right hand rule, is from B to A and from D to C. In the position of the loop shown in Fig. 5.1 (*b*) the coil sides are moving at right angles to the field and are, therefore, cutting across the field at maximum rate, consequently, the emf induced at this instant is of maximum value. In the next quarter of the revolution of the loop *i.e.* between $\pi/2$ radians and π radians (or 90° to 180°), the rate at which the conductors cut across the magnetic field gradually decreases, causing the magnitude of induced emf to fall gradually with the angular movement of the loop and becomes zero at the instant the loop becomes again parallel to the faces of the field magnets but with the sides AB and CD's position interchanged with respect to zero position [Fig. 5.1 (*c*)].

In the third quarter of the revolution of loop *i.e.* between π and $3\pi/2$ radians (or 180° to 270°) the rate at which the conductors cut across the magnetic field hence induced emf gradually increases as the loop moves and becomes maximum at the instant, the loop assumes the position shown in Fig. 5.1 (*d*). However, the direction of emf in the loop is now from A to B and from C to D *i.e.* opposite to that in the first two quarters.

In the fourth quarter of the revolution of the loop *i.e.* between $3\pi/2$ and 2π radians (or 270° to 360°), the induced emf decreases as the coil moves and becomes zero when it completes 2π radians or 360° from the starting instant. At this instant the loop comes to its absolute original position and hence the loop is said to have completed one cycle. The cycle then repeats for each revolution of the armature.

The emf thus generated in the loop is of the form shown in Fig. 5.2 from which it is obvious that emf induced in armature conductors is of pulsating nature. Such an emf is known as an *alternating emf.*

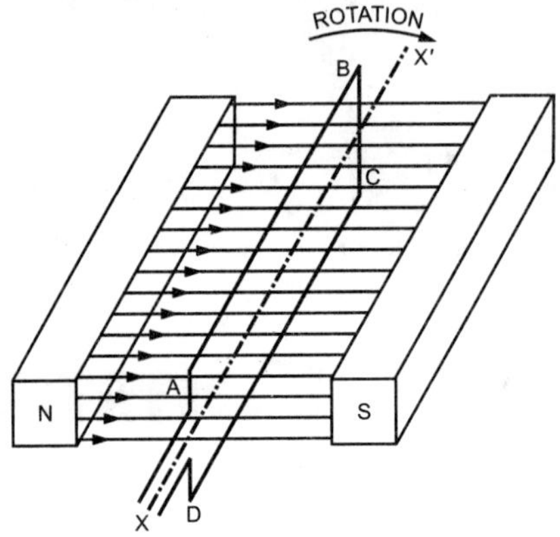

(*a*) *Armature Coil in Neutral Plane*

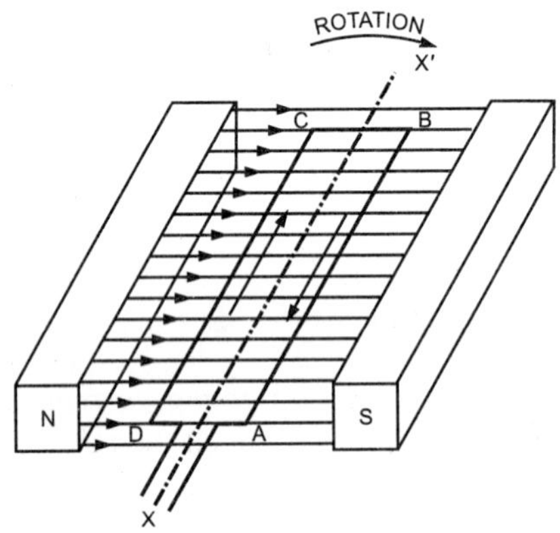

(*b*) *Armature Coil in The 90° Position*

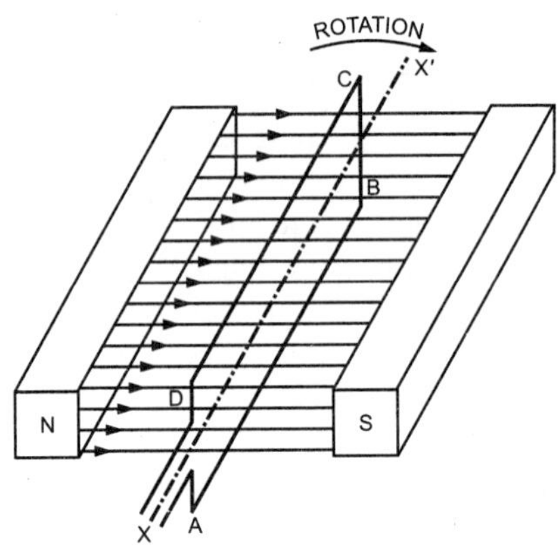

(*c*) *Armature Coil in The 180° Position*

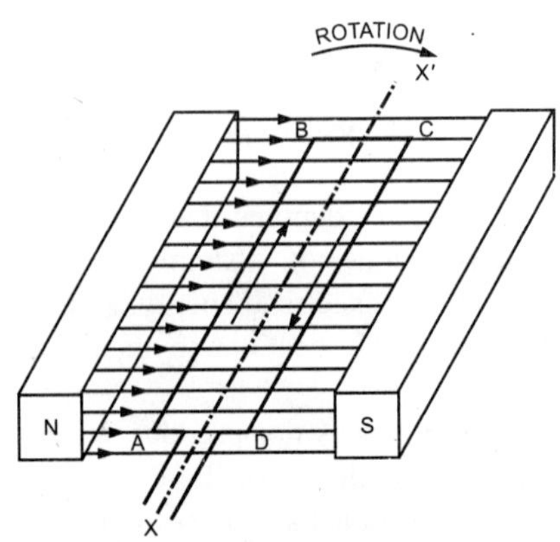

(*d*) *Armature Coil in The 270° Position*

Fig. 5.1

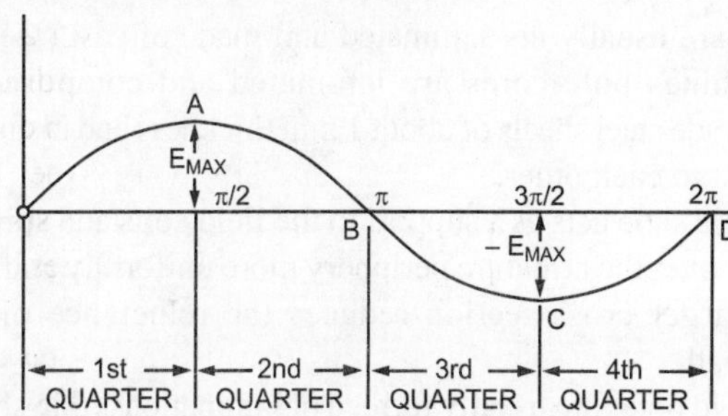

Fig. 5.2 *Curve Showing Variation of Induced EMF in a Coil Rotated in a Uniform Magnetic Field*

The current induced in the coil is collected and conveyed to the external load circuit by connecting the coil terminals to two continuous and insulated rings, known as *slip rings or collector rings*, mounted on the generator shaft and making the two stationary brushes pressing against the slip rings, one brush bearing on each ring, as shown in Fig. 5.3. When the coil is rotated, the generated alternating emf causes a current to flow first in one direction and then in the other through the coil and external circuit. Such a current is called an *alternating current.*

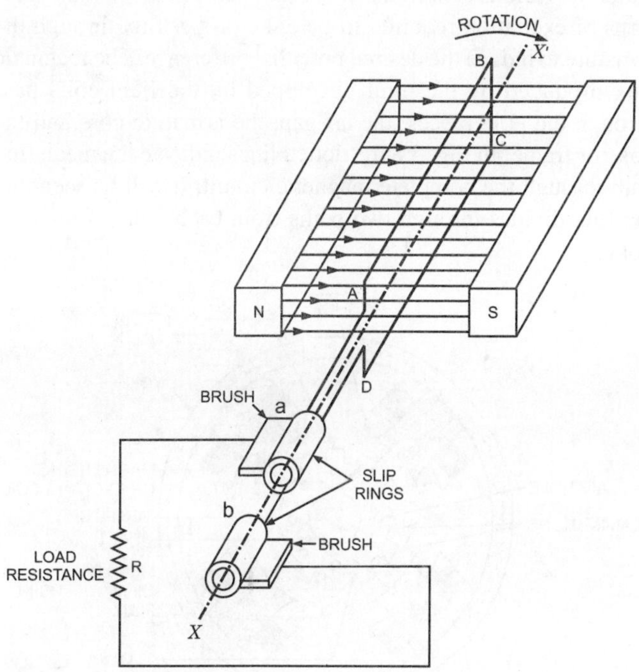

Fig. 5.3 *Elementary Generator With Brushes and Collector Rings*

To obtain the unidirectional or direct current in the external circuit, the collecting arrangement is modified as shown in Fig. 5.4. In this arrangement, the slip rings are replaced by split ring made of a conducting material and splitted into two halves separated from each other by insulation and brushes are placed diametrically opposite instead of being side by side.

It will be observed that in the first half revolution current flows along ABMLCDA *i.e.* the brush M in contact with segment '*a*' acts as the + ve pole of the supply and brush L in contact with segment '*b*' acts as the – ve pole.

In the next half revolution, the direction of induced currents in the coil is reversed, but at the same time the positions of segments '*a*' and '*b*' are also reversed, with the result that brushes M and L again come in contact with + ve segment '*b*' and – ve segment '*a*' respectively. Thus the direction of current in the external load circuit remains the same.

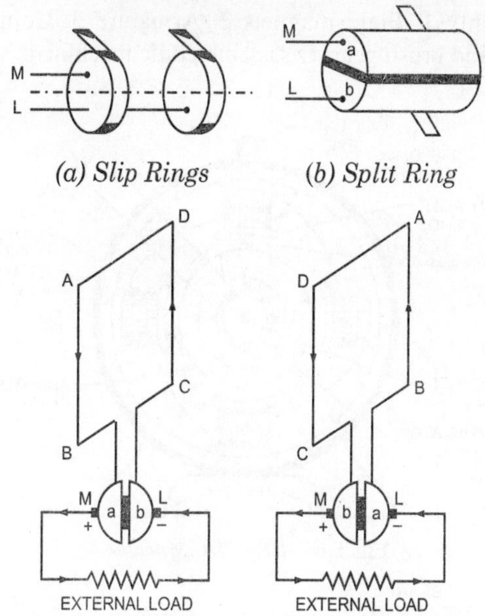

Fig. 5.4

The position of the brushes shall be so arranged that the change of segments from one segment to the other takes place when the plane of the rotating coil is perpendicular to the plane of field, since in this position the induced emf or induced current in the coil will be zero.

Thus the emf or current induced in the armature conductors of a dc generator is alternating which is rectified by the split ring known as *commutator*.

Although after rectification the current through the load circuit is always in the same direction, but it is not a steady current since the emf generated in the armature coil and applied to the brushes, varies from zero to maximum and back to zero twice each revolution. The variation in brush voltage is shown in Fig. 5.5.

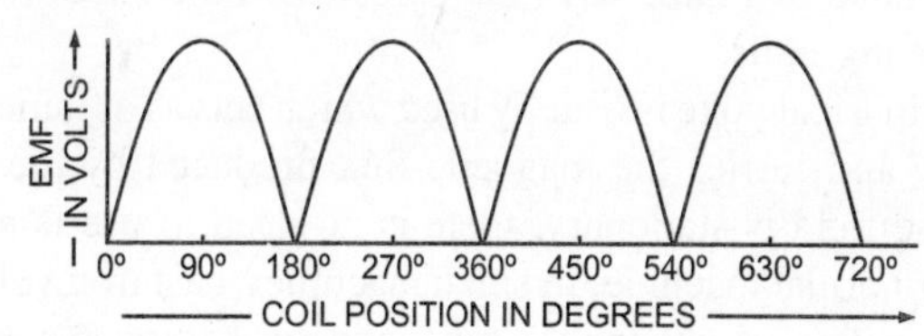

Fig. 5.5 *EMF Curve of a Simple DC Generator*

A pulsating direct current such as given by a single coil generator is not suitable for most commercial purposes. However, by employing a large number of coils and commutator segments, with the coils evenly distributed around the surface of the armature the brush voltage may be made practically constant.

The voltage generated by a single turn armature coil is small. For this reason, the coils employed in commercial generators consist of several turns in series thereby increasing the magnitude of generated emf in direct proportion to the number of the turns in the coil.

5.3 ESSENTIAL PARTS OF A DC MACHINE

DC machine (whether a generator or motor) with four poles is shown in Fig. 5.6. In construction, dc machine consists of four parts mainly 1. Field magnets 2. Armature 3. Commutator 4. Brush and brush gear. Disassembled dc machine is shown in Fig. 5.7.

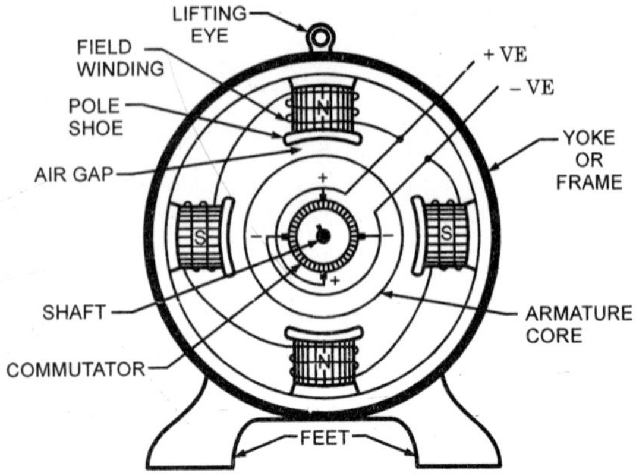

Fig. 5.6 *4-Pole DC Machine*

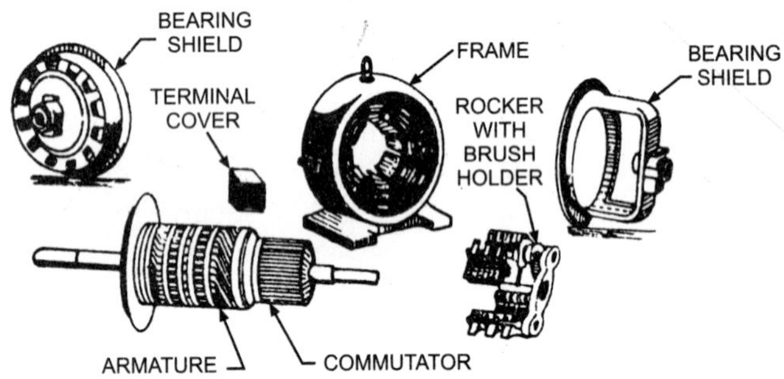

Fig. 5.7 *DC Machine (Disassembled)*

1. Field System - The object of the field system is to create a uniform magnetic field, within which the armature rotates.

Electromagnets are preferred in comparison with permanent magnets on account of their greater magnetic effects and field strength regulation, which can be achieved by controlling the magnetising current.

Field magnet consists of four parts given below :

(*i*) Yoke or Frame (*ii*) Pole cores (*iii*) Pole shoes and (*iv*) Magnetising coils.

Cylindrical yoke is usually used which acts as a frame of the machine and carries the magnetic flux produced by the poles. Since the field is stationary, there is no need to use laminated yoke for normal machine. In small machines, cast iron yokes are used, because of cheapness but yoke of a large machine is invariably made of fabricated steel due to its high permeability. In case of small machines the cheapness is the main consideration, and not the weight but in large machines, weight is the main consideration. Since the permeability of cast steel is about twice of cast iron, the weight of cast steel required will be only half of the cast iron if used for the same reluctance. The manufacturing process consists of rolling a steel slab around a cylindrical mandrel and then welding it at the bottom. The lifting eye, feet and the terminal box etc. are welded to the frame afterwards. Such yokes possess sufficient mechanical strength and have high permeability. With small machines the yoke may be one piece but for larger machines it is usual to employ a split yoke, the two halves being bolted together. This promotes ease in handling and convenience of assembly.

Pole core is usually of circular section and is used to carry the coils of insulated wires carrying the exciting (or field) current. Pole cores are usually not laminated and made of cast steel. In some machines pole cores are laminated and composed of electrical grade steel sheets of about 1 mm thickness and insulated with respect to each other.

The pole shoe acts as a support to the field coils and spreads out the flux over the armature periphery more uniformly and also being of larger cross section reduces the reluctance of the magnetic path.

The field poles are usually formed of laminations (thin sheets of steel) and are bolted to the frame or yoke to which are also fastened the end bells with their *bearings* and the *brush rigging*. In small machines the poles are cast integral with the yoke from cast iron due to its low cost and less machining required by individual parts. In some machines the yoke and pole cores are made in single casting and laminated pole shoes are attached to the pole cores. The pole faces or pole shoes are always laminated to avoid heating and eddy current losses caused by the fluctuations in the flux distribution on the pole face due to movement of armature slots and teeth.

The object of the magnetising or field coils is to provide, under the various conditions of operation, the number of ampere-turns of excitation required to give the proper flux through the armature to induce the desired potential difference. The magnetic flux produced by the mmf developed by the field coils pass through the pole pieces, the air gap, the armature core and the yoke or frame. In Fig. 5.8 the dotted lines indicate the mean flux path through the complete magnetic circuit. It will be seen that the flux divides through two paths from each pole through the yoke.

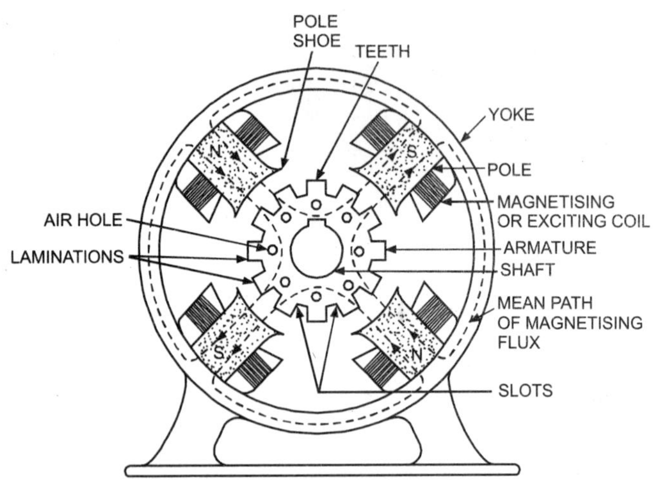

Fig. 5.8 *Cross Section of Field System of a DC Generator*

There are several field constructions adopted according to the type of excitation. In shunt field, many turns of fine wire are used, in series field few turns of large cross-sectional area are used and in compound field both shunt and series windings are used. Shunt coils are usually wound with double cotton covered wires. The field coils, after proper winding, are dipped in an insulating varnish and baked in an oven, which provides stiffness, mechanical strength and good insulating properties to the windings.

In the design of a generator, the number of poles required by the field structure depends on the speed of the armature and the output for which the machine is designed. In a two pole machine there are two voltage maximums per revolution of the armature, while in a four-pole machine there are four voltage maximums for one revolution. If the armature speed is kept constant, the number of poles determines the rate at which the individual coils cut the magnetic flux. Hence the output voltage increases with the increase in number of poles for a constant armature speed. In any generator the field poles are always produced in pairs, since a pair is necessary to produce a set of magnetic poles.

More number of poles reduces the weight of the core and yoke, overall diameter and length of the machine, length of commutator and cost of copper in the field and the armature. With more number of poles distortion of field form is not excessive. But with more number of poles (*i*) frequency of flux reversals is increased thereby increasing iron losses (*ii*) labour charges are increased and (*iii*) tendency of flash-over between brush arms is increased.

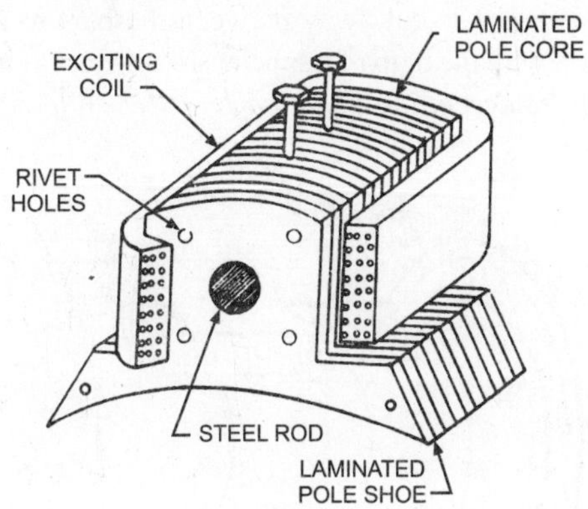

Fig. 5.9 *Laminated Pole Core and Pole Shoe*

The upper limit to the number of poles is imposed by frequency which is of the order of 20-30 Hz for large machines and not exceeding 50 Hz for small ones. The lower limit is due to the current per brush arm, which should not exceed about 400 A otherwise an excessively long commutator will be required.

2. Armature. It is a rotating part of a dc machine and is built up in a cylindrical or drum shape. The purpose of armature is to rotate the conductors in the uniform magnetic field. It consists of coils of insulated wires wound around an iron and so arranged that electric currents are induced in these wires when the armature is rotated in a magnetic field. In addition, its most important function is to provide a path of very low reluctance to the magnetic flux. The armature core is made from high permeability silicon-steel stampings, each stamping, being separated from its neighbouring one by thin paper or thin coating of varnish as insulation.

A small air gap exists between the pole pieces and the armature so that there will be no rubbing in the machine. However, this gap is kept as small as possible, since larger the air gap greater is the mmf required to create the required flux. The air-gap length is about 1.0 mm to 6 mm (say 1 mm for a 1 kW machine, 1.5 to 1.75 mm for medium size machines and 6 mm for 800 kW machine).

The use of high grade steel is made (*a*) to keep hysteresis loss low, which is due to cyclic change of magnetisation caused by rotation of the core in the magnetic field and (*b*) to reduce the eddy currents in the core which are induced by the rotation of the core in the magnetic field. By using stampings or laminations, the path of the eddy currents is cut into several units. The laminations must be in such a direction that they are perpendicular to the paths of eddy currents and parallel to the flux. Each lamination is about 0.3 to 0.6 mm thick.

The slots are either die-cut or punched on the outer periphery of the circular core stamping and the key way is located on the inner diameter, as shown in Fig. 5.10.

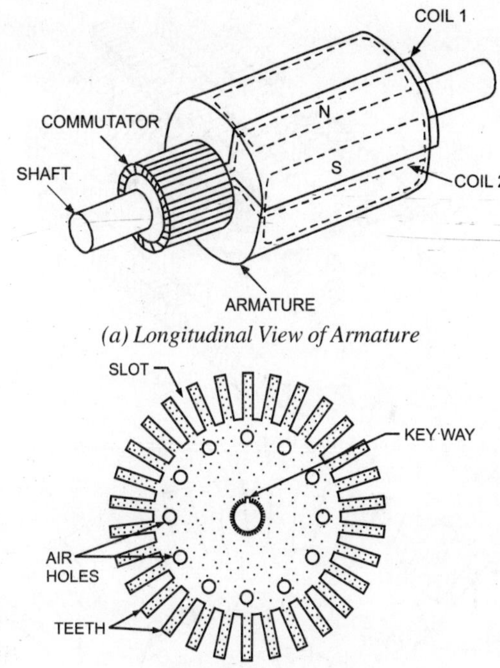

(a) Longitudinal View of Armature

(b) Armature Lamination

Fig. 5.10 *Armature*

The slots are normally open type and usually placed parallel to axis of the armature but are also sometimes skewed at a small angle to the axis to avoid vibration of teeth. The width and depth of the slots are made to accommodate the conductors and the insulation.

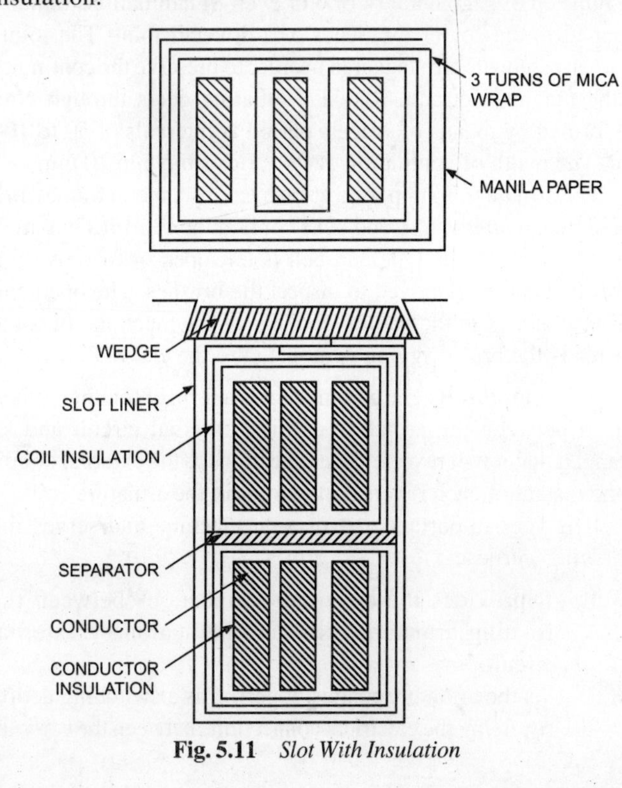

Fig. 5.11 *Slot With Insulation*

Core punchings up to diameter of 0.5 m are generally made in one piece as shown in Fig. 5.10. These core punchings are usually keyed directly to the shaft and punched with holes near the shaft to give longitudinal ventilating ducts. By the fanning action of the armature, air is drawn in through these ducts, thus producing efficient ventilation.

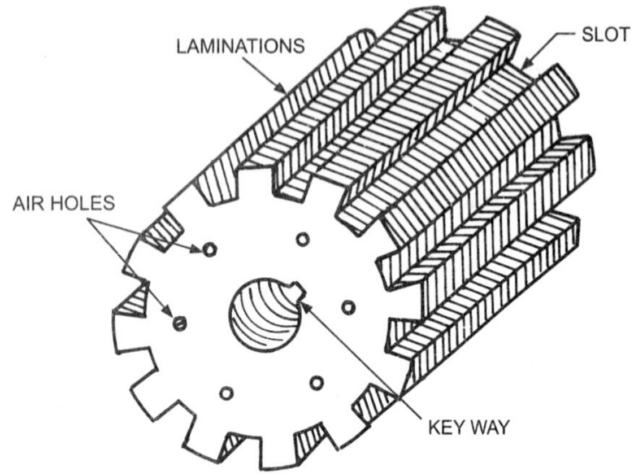

Fig. 5.12 *Assembled View of Armature Laminations*

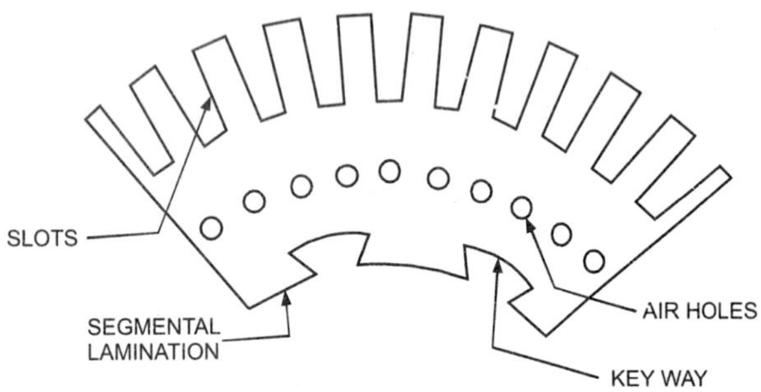

Fig. 5.13 *Segmental Lamination*

For larger machines having large armature diameter it is not economical to punch out a complete ring. Core of larger diameter are built up of segmental (4 or 6 or even 8) laminations that are attached to the spider by means of a dovetail joint. The joints between segments are staggered in order to preserve the continuity of the magnetic circuit. Radial ventilating ducts through core are formed by means of spacers placed at intervals of 50 to 100 mm. The width of ventilating ducts varies from 5 to 10 mm.

The armature is supported at each end by a metal framework called the *end bells*. The end bells contain the bearings in which the armature rotates. One end bell is left open or made with a cover that can be removed to inspect the brushes. The open end bell also assists in the natural cooling of the machine. In some machines the brush rigging is mounted to the end bell.

3. Commutator. The commutator is a form of rotating switch placed between the armature and the external circuit and so arranged that it will reverse the connections to the external circuit at the instant of each reversal of current in the armature coils.

It is very important part of a dc machine and serves the following purposes:

1. It provides the electrical connections between the rotating armature coils and the stationary external circuit.
2. As the armature rotates, it performs a switching action reversing the electrical connections between the external circuit and each armature coil in turn so that the armature coil voltages add together and result in a dc output voltage.
3. It also keeps the rotor or armature mmf stationary in space.

The commutator is essentially of cylindrical structure and is built up of wedge shaped segments of high conductivity hard drawn copper or drop forged copper. These segments are insulated from each other by thin layers of mica (usually of 0.5 to 1mm thickness). Mica is to be preferred but cannot be used for large commutator because of the difficulty of obtaining large sheets, making the cost of large mica segments prohibitive. On account of cost also micanite is often used for small commutators. The segments are held together by means of two V-shaped rings that fit into the V- grooves cut into the segments. Nowadays usually the mica insulation is cut away between the bars to a depth of about 1.5 mm with the help of a special slotting tool. This process is called the *'undercutting the mica'*.

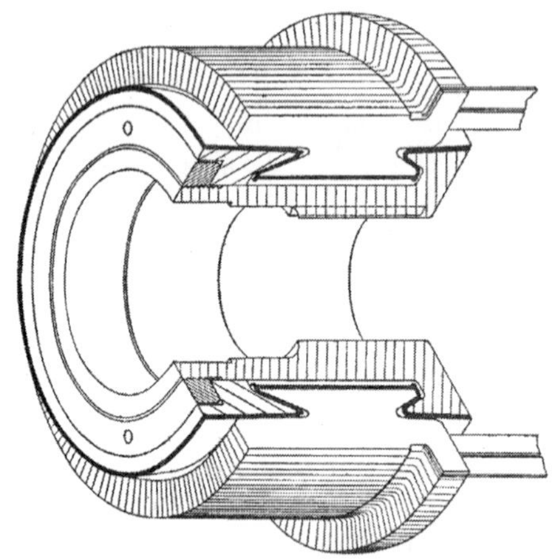

(a) Commutator

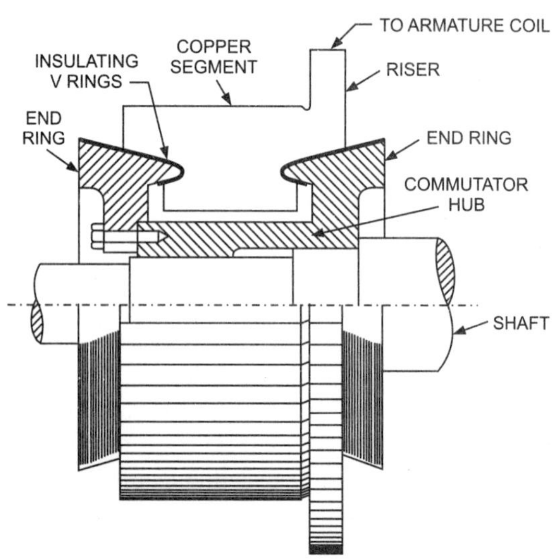

(b) Section View of Commutator Segments

Fig. 5.14

The copper is insulated from the Vee-rings and the hub by micanite carefully moulded to the exact shape required. Frequently the hub is not insulated, but sufficient clearance is left to avoid the need of this insulation.

If the armature and commutator diameters do not differ much, the winding ends are directly soldered to the commutator bars. Otherwise they are soldered with copper lugs or risers. The risers have air spaces between them so that air is drawn across the commutator thereby keeping the commutator cool.

The commutator is pressed on to the armature shaft, and the outer periphery is then machined to provide a smooth surface with which a stationary carbon (or graphite or copper) brush can maintain continuous contact as the armature and commutator rotate. Great care is taken in building the commutator because even slight eccentricity will cause the brushes to bounce, causing undue sparking.

4. Brushes. The function of brushes is to collect current from the commutator and supply it to the external load circuit (the armature of the machine being connected to the external load circuit via the commutator and brushes). The brushes are rectangular in shape and rest on the commutator. Brushes are manufactured in a variety of compositions and degrees of hardness to suit the commutation requirements. They may be classified roughly as carbon, carbon graphite, graphite, metal graphite and copper. The allowable current density at the brush contact varies from 5 A per square cm in case of carbon to 23 A per square cm in the case of copper.

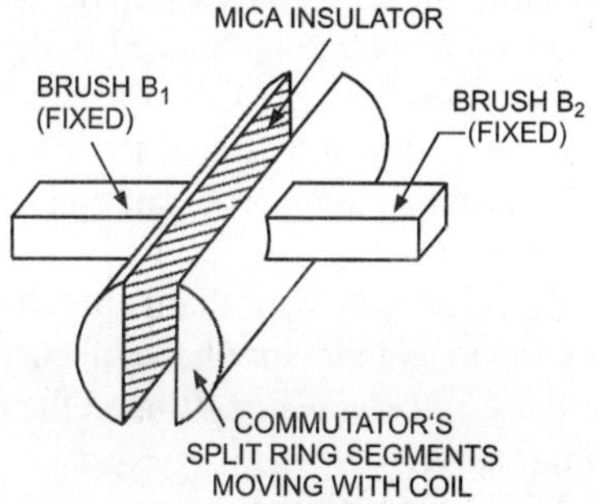

(a) Brushes

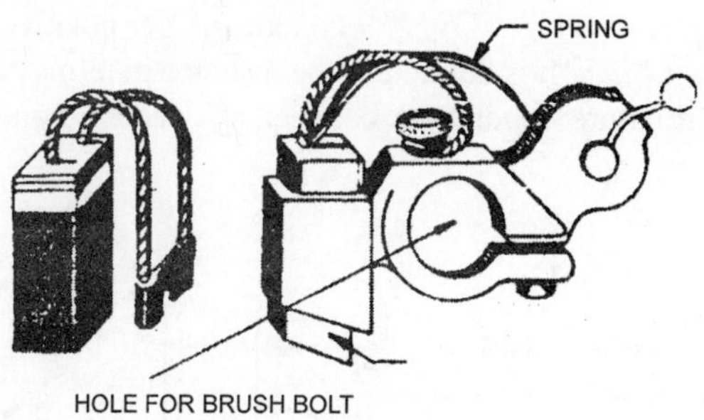

(b) Brush Holder

Fig. 5.15 *Brush and Brush Holder*

Copper brushes are employed only for machines designed for large currents at low voltages. Unless very carefully lubricated they cut the commutator very quickly and, in any case, the wear is rapid. Graphite and carbon graphite brushes are self lubricating and are, therefore, widely used. Even with the softest brushes, however, there is a gradual wearing away of the commutator, and if the mica between the commutator segments does not wear down so rapidly as the segments do, the high mica will cause the brushes to make poor contact with the segments, and sparking will result, with consequent damage to the commutator. To prevent this, the mica is frequently ''undercut'' to a level below the commutator surface by means of a narrow milling cutter.

Sooner or later, commutators generally wear out of true and must be turned down in a lathe, but interpole machines with soft brushes and undercut mica will run for a long time without any maintenance on the commutator. No lubrication is applied to the surface of a commutator with undercut mica, because the lubricant will collect carbon or graphite dust from the brushes and hold it in the grooves above the mica and thus provide leakage paths for the current from segment to segment, which may ultimately develop into a short-circuit or "flash-over" on the commutator. All modern brushes, except the copper ones, contain enough graphite to provide adequate lubrication.

The brushes are housed in brush-holders (usually of the box type) which are mounted on the brush-holder studs or brackets. In turn, the brush-holder studs are mounted on a brush yoke or rocker arm. The brush-rocker can be rotated so as to change the position of the whole brush system in relation to the machine poles. The brush-holder studs are insulated from the brush yoke by means of insulating sleeves and discs. The brush yoke, brush holders and brushes make the brush gear.

Most of the motors have brushes radially placed, that is, their centre line is radial to the commutator. This permits operation in both directions. Even in non-reversing machines, in both directions the brushes are usually radial. However, in some machines the brushes are set in inclined position to avoid vibration which may result in sparking at the brushes.

The brushes are held under pressure over the commutator by a combination of brush holders and springs whose tension may be adjusted. It is important that the desired pressure be applied to the brushes (say 1.5 – 2.5 N/cm^2) to ensure satisfactory commutation–large pressure will cause heating of the commutator and brushes owing to friction whereas the low pressure may cause sparking owing to imperfect contact between the brushes and the commutator.

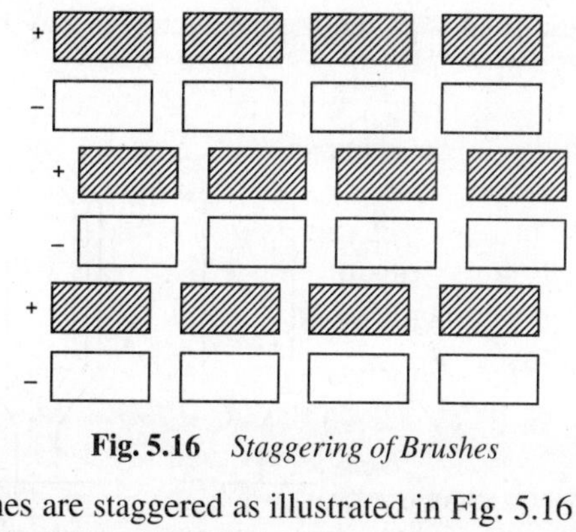

Fig. 5.16 *Staggering of Brushes*

Brushes are staggered as illustrated in Fig. 5.16 in order to prevent ridge formation on the surface of the commutator. Each set of positive brushes being staggered with relation to the previous set of positive brushes, and the negative brushes similarly arranged. This way each track on the commutator is covered by equal number of positive and negative brushes.

5. Bearings. With small machines, ball-bearings may be used at both ends. For larger machines, roller-bearings are used at the driving end, and ball-bearings may be used at the non-driving end, *i.e.* at the commutator end. Thrust-bearings are used where excessive end thrust is anticipated. Sleeve-bearings, with ring lubrication are used for motors when very silent running is required. For large machines pedestal-bearings are generally used.

6. Shaft. The shaft is made of mild steel with a maximum breaking strength. The shaft is used to transfer mechanical power from or to the machine. The rotating parts such as armature core, commutator, cooling fan etc. are keyed to the shaft.

Sectional view of dc rotor consisting of armature shaft, armature core, armature winding and commutator is illustrated in Fig. 5.17.

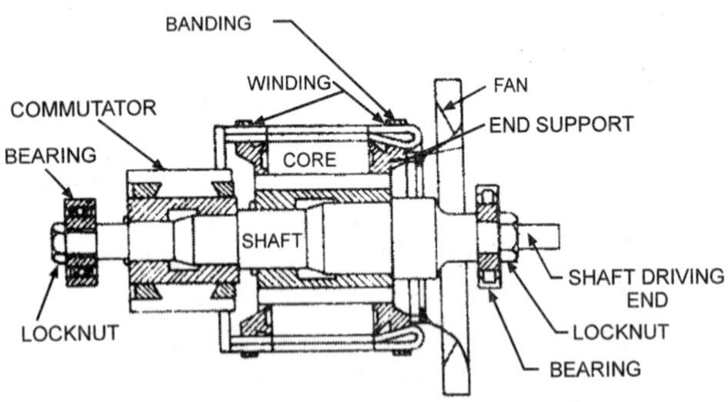

Fig. 5.17 *Sectional View of Rotor Assembly of a DC Machine*

5.4 ARMATURE WINDINGS

Armature winding is an arrangement of conductors to develop desired emfs by relative motion in a heteropolar* magnetic field. In winding, conductor or group of conductors are distributed in different ways in slots all over the periphery of the armature. The conductors may be connected in series and parallel combinations depending upon the current and voltage rating of the machine.

5.4.1. Terminology. It is necessary to understand some of the important terms before discussing windings. The most commonly used terms in describing a winding are given below :

1. *Conductor.* Each individual length of wire lying within the magnetic field is called the conductor. It may be made of one or two or more parallel strands. Symbol Z will be used for the total number of conductors in the armature winding.
2. *Turn.* When the two conductors lying in a magnetic field are connected in series, so that emf induced in them help each other or the resultant induced emf becomes double (assuming full pitch coil) of that due to one conductor, it is known as a *turn*.
3. *Coil.* When one or more turns are connected in series and the two ends of it are connected to adjacent commutator segments (in lap winding) it is termed as a *coil*. Single turn and multi-turn coils are shown in Fig. 5.18 (*a*) and 5.18 (*b*) respectively. Multi-turn coils are used when total number of armature conductors is excessive and it is not feasible to use single turn coils, because it will require large number of commutator segments and if used it will not give sparkless commutation and it will not be economical due to use of more copper in end connections. Symbol C will be used for the total number of coils in the armature winding.
4. *Coil side.* Each coil, single-turn or multi-turn, has two sides called the *coil sides*, embedded in two different slots nearly a pole pitch apart.
5. *Coil Group.* A coil group may have one or more single coils.
6. *Winding.* Number of coils arranged in coil groups is called the *winding*.
7. *Inductor.* One of the wires making up the coil side is called the *inductor* and a voltage is induced in the inductor.
8. *Front End Connector.* A wire that connects the end of a coil to a commutator segment, is called the *front end connector*. This wire is located at that part of the coil that is nearest the commutator.
9. *Back End Connector.* A wire that connects an inductor on one side of coil to an inductor on the other side of the coil is called the *back end connector*. It is on the end opposite to that of commutator.
10. *Pole Pitch.* It is defined as the number of conductors per pole. If there are 54 conductors for 6 poles then pole pitch will be equal to 54/6 *i.e.* 9 conductors per pole.
11. *Front Pitch.* It is defined as the distance in terms of number of armature conductors between the second conductor of

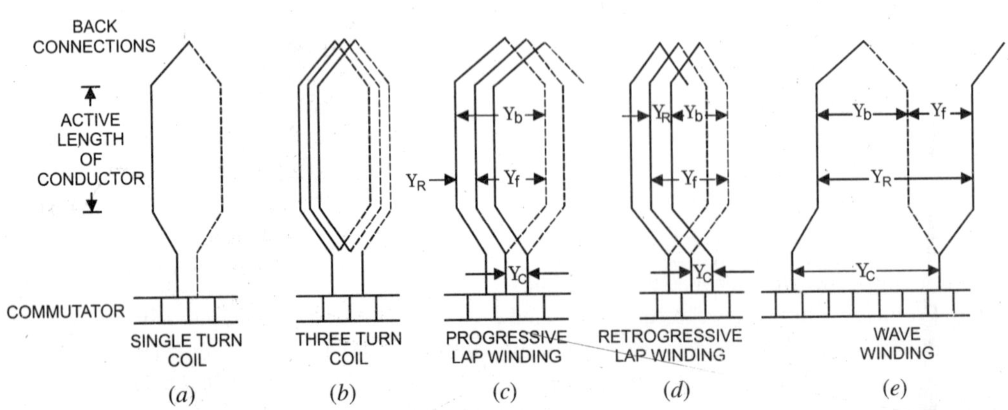

Fig. 5.18

* Field system consisting of a number of poles, arranged alternately N and S—commonly employed field system. Machine empolying such a field system is called the heteropolar machine. The machine empolying a field in which there is only one active pole, so arranged that the active conductors move through a field of constant intensity is called the homopolar machine.

one coil and the first conductor of next coil which are connected to the same commutator segment on the front or commutator end. It is represented by Y_f, as shown in Figs. 5.18 (*c*), (*d*) and (*e*).

12. *Back Pitch.* It is defined as the distance in terms of number of armature conductors between the last and the first conductors of the coil. It is also called the *coil span* or *coil spread.* It is denoted by Y_b as illustrated in Figs. 5.18 (*c*), (*d*)), and (*e*).
13. *Resultant Pitch.* It is defined as the distance in terms of number of conductors between the start of one coil and start of next coil to which it is connected. It is denoted by Y_R, as illustrated in Figs. 5.18 (*c*), (*d*), and (*e*).
14. *Commutator Pitch.* It is defined as the distance measured in terms of commutator segments between the segments to which the two ends of a coil are connected. It is denoted by Y_c as shown in Figs. 5.18 (*c*), (*d*) and (*e*).
15. *Coil Span or Coil Pitch.* When the coil span of the winding is equal to pole pitch, the coils are called the *full-pitched coils*. Full-pitched coils have the advantage of giving maximum emf induced in the coil. Coils 1, 2 and 3 shown in Fig. 5.19 are short-pitched, full-pitched and over-pitched coils respectively.

    Since coil No. 1 has span less than the pole pitch, during the revolution of armature there will be an instant during which both sides of such coils will be under the influence of one and the same pole so emf induced in these two sides of the coil will be in opposition. Similarly there will be an instant during which both sides of an over-pitched coil will be under the influence of two similar poles and emf induced in the two sides of such a coil will be in opposition.

    The advantages of fractional pitch windings are that substantial saving is affected in the copper of end connections and commutation is improved, owing to lesser mutual inductance between the coils.

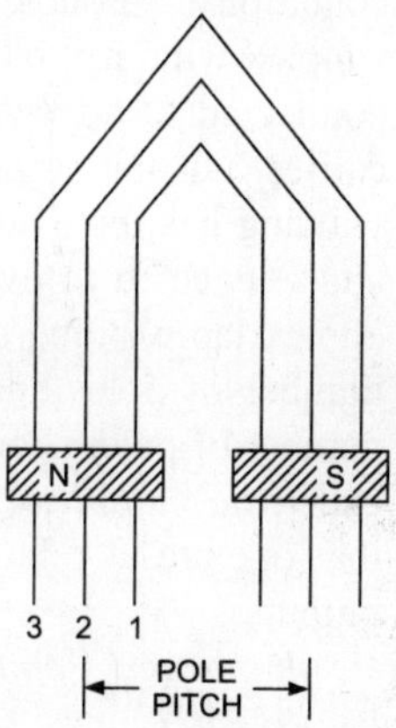

Fig. 5.19

16. *Degree of Re-entrancy of an Armature Winding.* A winding is said to be *single re-entrant* if on tracing through it once, all armature conductors are included on returning to the starting point.

    A winding is said to be *double re-entrant* if only half the conductors are included in tracing the winding once and so on.

5.4.2. Types of Armature Windings. Armature windings are always of the non-salient pole type and usually are symmetrically distributed in slots around the complete circumference of the armature. The armature windings, according to the degree of closure produced by the winding, are of two types namely:

(*a*) Open coil winding (*b*) Closed coil winding.

Open coil winding is that winding which does not close on itself *i.e.* a close circuit will not be formed until some external connection is made to a source or load. The open coil winding is never employed in dc machines, but is usually employed in ac machines.

Closed coil winding is that winding which closes on itself. In such a winding if one starts tracing through it, one will come back to the starting point without passing through any external connection. DC machines employ only closed coil windings in order to provide for the commutation of the coils. Though closed type winding can be used in ac machines also, but usually open coil windings are used.

The continuous or closed armature windings are of two types: (*a*) Gramme-ring Winding (*b*) Drum Winding.

The gramme-ring type of armature winding is an early form of winding which has been replaced by the more efficient winding known as *drum type winding*. Hence our discussion will be restricted to drum winding.

5.5 DRUM WINDING

In this type of winding the conductors are housed in slots over the armature surface and connected to one another by front and back connectors or coil ends. The drum type winding has the following two principal advantages.

1. All armature copper, except end connections, is active *i.e.* it cuts flux and, therefore, is active in generating emf.
2. The coils can be pre-formed and insulated before placing on the armature, hence cost is reduc

The drum winding may be either single layer or double layer winding. Single layer winding, in which one conductor or one coil side is placed in each armature slot, is rarely used. The two layer winding, in which there are two conductors or coil sides per slot arranged in two layers, is usually used for economy reasons. The two layer winding permits the use of coils pre-formed and insulated before being placed on the armature. The coils usually are wound on machine driven forms which give them the proper shape and the turns are then bound together usually with cotton tape. The ends are left bare so that they may be soldered later to the commutator bars. The coils are then dipped into some insulating compound such as asphaltum and are then dried. For very high temperature operation, other materials, such as mica and paper tape, fibre glass tape, and silicon impregnated insulation are used.

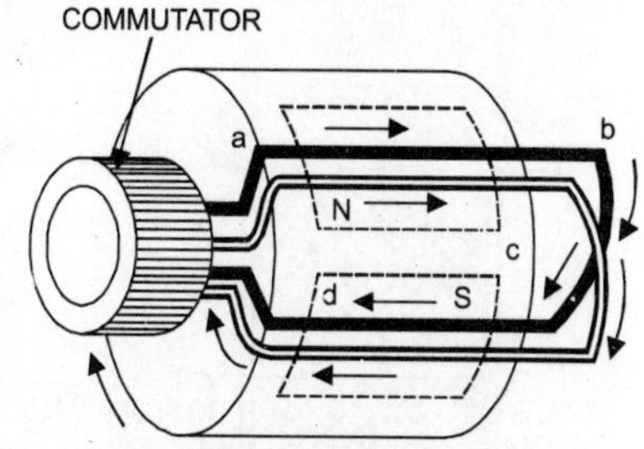

Fig. 5.20 *Drum Winding*

Before placing the coils in the slots, the slots are given U-shaped slot liners of leatheroid or presspahn to ensure mechanical protection of the coils. After the coils have been dropped into the slots, wedges of wood or hard fibre are driven to hold them in place.

In order that the emf induced in the coil be a maximum, other conditions being fixed, the span of the coil should be equal to pole pitch. However, the span may be reduced to as much as eight-tenths the pole pitch without any serious reduction in the induced emf. When this is done, the winding is called a *fractional pitch winding*. The advantages of fractional pitch windings are that substantial saving is affected in the copper of end connections and commutation is improved, owing to lesser mutual inductance between the coils.

Usually, one side of every coil lies in the top of one slot and the other side lies in the bottom of some other slot at a distance of approximately one pole pitch along the armature. Thus at least two coil sides occupy each slot. Such windings in which two coil sides occupy each slot are most commonly used for all medium sized machines. In dynamos of larger ratings it is often necessary to place several coil sides or elements in a single slot usually four, six or eight. More than eight elements in a single slot is rarely used. The coil sides lying at the upper half of the slots are numbered odd 1, 3, 5, 7 etc. while those at the lower half are numbered even. It is to be noted that the coil side 2 lies below coil side 1, 4 below 3 etc. progresses clockwise. Placing of several elements in a single slot gives fewer slots than segments which have got following advantages.

(*i*) As the number of slots is reduced the armature core teeth become mechanically stronger, so that, from the standpoint of handing in manufacture, there is less damage to laminations and coils.

(*ii*) As the number of commutator segments is increased, the voltage between those that are adjacent to each other decreases and the number of turns of wire in the coil or coils connected to adjacent segments also decreases. The result is that there is less sparking at the commutator because of the improved commutation.

(*iii*) Assuming that a comparatively large number of segments has been selected so that good commutation will result, the choice of an armature core with one-half, one-third, one-fourth etc. as many slots means that fewer coils will have to be constructed, this reduces the cost of manufacture.

Two layer winding permits the end connections to be easily made as the coil ends can be bent around one another in a systematic manner, passing from the bottom to the top layer by means of the peculiar twist in the ends of the coils.

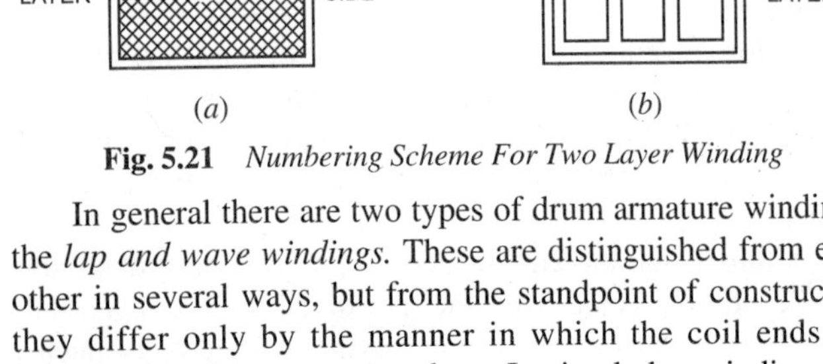

Fig. 5.21 *Numbering Scheme For Two Layer Winding*

In general there are two types of drum armature windings: the *lap and wave windings*. These are distinguished from each other in several ways, but from the standpoint of construction they differ only by the manner in which the coil ends are connected to the commutator bars. In simple lap windings the coil ends are connected to adjacent commutator segments while the ends of the wave winding coils are connected to the commutator segments at points separated by approximately twice the distance between adjacent poles.

There is also a third type of drum winding called the *frog-leg winding*. This winding consists of a lap winding and wave winding placed on the same armature. The wave winding is connected to the commutator segments at equipotential points so that this winding part can be employed as an equaliser for the lap winding.

5.6 LAP WINDING

Single turn lap winding is shown in Figs. 5.18 (*c*) and (*d*). In lap winding finish end of one coil is connected to a commutator segment and to the start end of the adjacent coil situated under the same pole and similarly all coils are connected. This winding is known as *lap winding* because the sides of successive coils overlap each other. Lap winding may be further classified as being a *simplex* (single) or *multiplex* (double or triple) windings.

In the simplex lap winding there are as many parallel paths or circuits through the winding as there are field poles on the machine.

Double and triple windings are used on armature designed for supply of large currents at low voltage. The sole purpose of such a winding is to increase the number of parallel paths enabling the armature to carry a large total current, at the same time reducing the conductor current to improve commutation conditions. A double or duplex winding consists of two similar simplex windings placed in alternate slots on the armature and connected to alternate commutator segments. Each winding carries half the armature current. Likewise, a triple or triplex winding has three similar windings occupying every third slot and connected to every third commutator segment. Hence in duplex lap winding the number of parallel circuits is twice the number of poles and in triplex lap winding, three times the number of poles. For this reason the lap winding is sometimes called the *multiple or parallel winding* and is suited for machines that operate at relatively low voltages but with high current outputs.

Important Point Regarding Lap Winding. 1. The coil or back pitch Y_b must be approximately equal to the pole pitch *i.e.*

$Y_b \approx \frac{Z}{P}$ where Z is the number of conductors on armature and P is the number of poles.

2. The back pitch Y_b should be either lesser or greater than front pitch Y_f by $2m$ where m is the multiplicity of the winding.

$$i.e.\ Y_b = Y_f \pm 2m$$

where, m = 1 for simplex winding

m = 2 for duplex winding

m = 3 for triplex winding and so on.

When Y_b is greater than Y_f the winding progresses from left to right and so known as *progressive winding*. When Y_b is lesser than Y_f the winding progresses from right to left and, therefore, such a winding is known as *retrogressive winding*.

3. The back pitch and front pitch must be odd.

4. The average pitch is given by

$$Y_{av} = \frac{Y_b + Y_f}{2}$$

and should be equal to pole pitch *i.e.* $\frac{Z}{P}$

5. The resultant pitch Y_R is always even, being the difference of two odd numbers and is equal to $2m$ where m is the multiplicity of the winding.

i.e. Resultant pitch, Y_R = 2 for simplex lap winding

Y_R = 4 for duplex lap winding

and Y_R = 6 for triplex lap winding.

6. The commutator pitch, $Y_c = m$ *i.e.*, Y_c is equal to 1, 2, 3, 4 etc., respectively for simplex, duplex, triplex, quadruplex etc. lap windings.

7. Number of parallel paths in lap winding = mP *i.e.* number of parallel paths is equal to P, 2P, 3P, 4P etc. respectively, for simplex, duplex, triplex, quadruplex, etc. lap windings.

Example 5.1. Draw the developed winding diagram of progressive lap winding for 4 poles, 24 slots with one coil side per slot, single layer showing therein position of the poles, direction of motion, direction of induced emfs and position of brushes.

Solution: Developed winding diagram is obtained by imagining the armature surface removed and so laid out flat that the slots and conductors can be viewed without the necessity of turning round the armature in order to trace out the armature winding. Such a developed winding diagram is shown in Fig. 5.22.

Number of poles, P = 4

Number of coil sides, Z = 24

$$\text{Average pitch, } Y_{av} = \frac{Y_b + Y_f}{2} = \frac{Z}{P} = \frac{24}{4} = 6$$

$$\text{or } Y_b + Y_f = 12 \qquad ...(i)$$

and for progressive simplex lap winding

$$Y_b = Y_f + 2 \qquad ...(ii)$$

Solving Eqs. (*i*) and (*ii*) we get $Y_b = 7$ and $Y_f = 5$

Drawing of Winding: In order to draw the winding diagram, first of all draw the coil sides and number them. Now to make connections start from any coil side, say with first coil side.

In order to get the coil side to which the 1st coil side is to be connected at the back, add back pitch to it. So first coil side will be connected to the 1 + 7 = 8th coil side at the back by means of end connections.

In order to get the coil side to which 8th coil side is to be connected on the front or commutator side of the winding, deduct front pitch from it, and, therefore, 8th coil side will be connected to the 8 – 5 = 3rd coil at the front or commutator end.

Now repeat the process, according to which 3rd coil side will be connected to 10th coil side at the back and 10th coil side will be connected to 5th coil side at the next commutator segment.

Thus a table can be made indicating the way in which various coil sides will be connected. The coil sides are connected according to table of the end connections given below.

At Back End		*At Front End*		
Coil Side No.	*Connected to Coil Side No.*	*Coil Side No.*	*Connected to Coil Side No.*	*Through Segment No.*
1	8	8	3	2
3	10	10	5	3
5	12	12	7	4
7	14	14	9	5
9	16	16	11	6
11	18	18	13	7
13	20	20	15	8
15	22	22	17	9
17	24	24	19	10
19	2	2	21	11
21	4	4	23	12
23	6	6	1	1

Position of Brushes. Brush positions can be located by determining the direction of currents flowing in the various coil sides. Assuming the position of poles as shown in the developed winding diagram and the motion of coil sides from right to left, the direction of induced emf in the coil sides under the influence of N-pole will be downwards and in coil sides under influence of S-pole will be upwards. There may be some coil sides in the interpolar space for which there may be doubt about the directions of the induced emf. In such cases the directions of emf should not be marked. Since conductors 1, 2, 3, 4, 5, 6, 13, 14, 15, 16, 17 and 18 are under the influence of N-pole, these conductors will have current flowing in downward direction while conductors 7, 8, 9, 10, 11, 12, 19, 20, 21, 22, 23 and 24 being under the influence of S-pole will have current flowing in upward direction. Now draw the equivalent ring winding diagram, as shown in Fig. 5.23 and put the arrows as per main winding diagram. From the winding diagram it is found that commutator segments 1 and 7 are meeting point of emfs and currents are flowing outwards from the conductors, so + ve brushes be fixed at these segments. Similarly commutator segments 4 and 10 are separating points of emfs and currents are flowing inwards so –ve brushes be fixed at these segments.

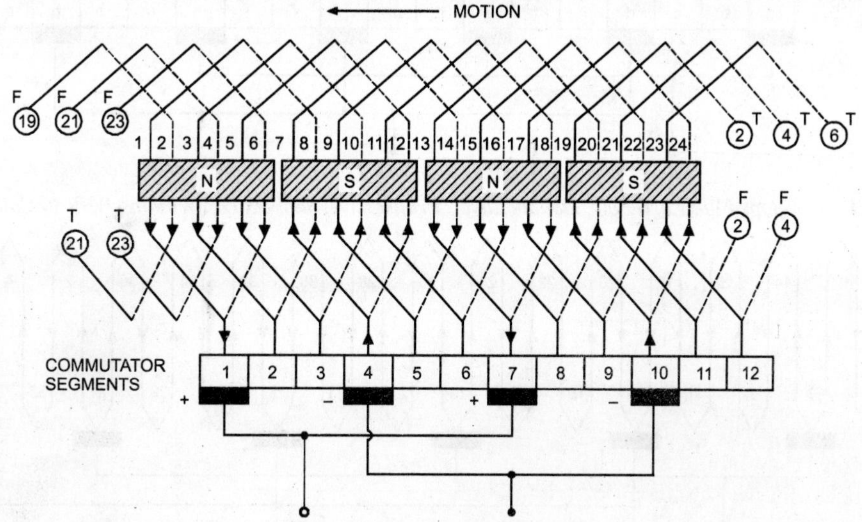

Fig. 5.22 *Developed View of 4-Pole, Single Layer, Progressive Simplex Lap Winding With 24 Coil Sides*

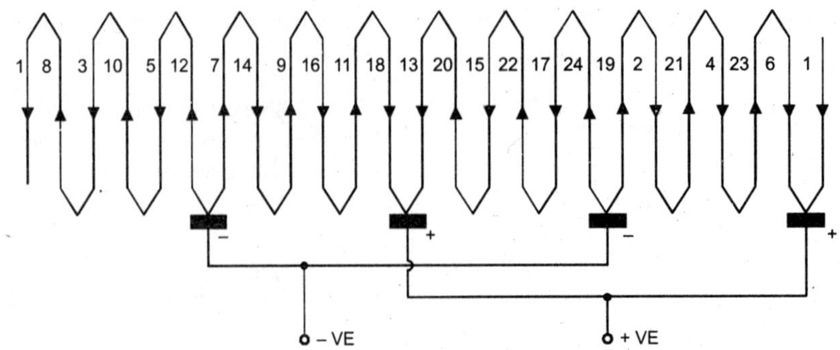

Fig. 5.23 *Equivalent Ring or Spiral Winding*

The brushes of same polarity are connected together and, therefore, the armature winding is divided in four paths in parallel.

Example 5.2. Draw the developed winding diagram of lap winding for 6 poles, 18 slots with two coils sides per slot, double layer showing therein position of the poles, direction of motion, direction of induced emfs and position of brushes.

Solution : Number of poles, P = 6

Number of slots = 18

Numbers of coil sides, Z = Number of slots × number of coil sides per slot

= 18 × 2 = 36

$$\text{Average pitch, } Y_{av} = \frac{Y_b + Y_f}{2} = \frac{36}{6} = 6 \qquad ...(i)$$

For progressive simplex lap winding

$$Y_b = Y_f + 2 \qquad ...(ii)$$

Solving Eqs. (*i*) and (*ii*) we have

$$Y_b = 7 \text{ and } Y_f = 5$$

The developed winding and equivalent ring winding diagrams are shown in Figs. 5.24 and 5.25 respectively.

5.6.1. Characteristics of Simplex Lap Winding. From the winding diagrams shown in Figs. 5.22 and 5.24 the following conclusions regarding simplex lap winding are arrived at:

1. The winding starting from any coil side returns back to that very coil side after connecting all the coil sides once and only once hence it is singly re-entrant closed winding.
2. Back pitch and front pitch of all the coils remain the same and odd number.
3. The total number of brushes is equal to the number of poles, and the brushes are connected to the coil sides,

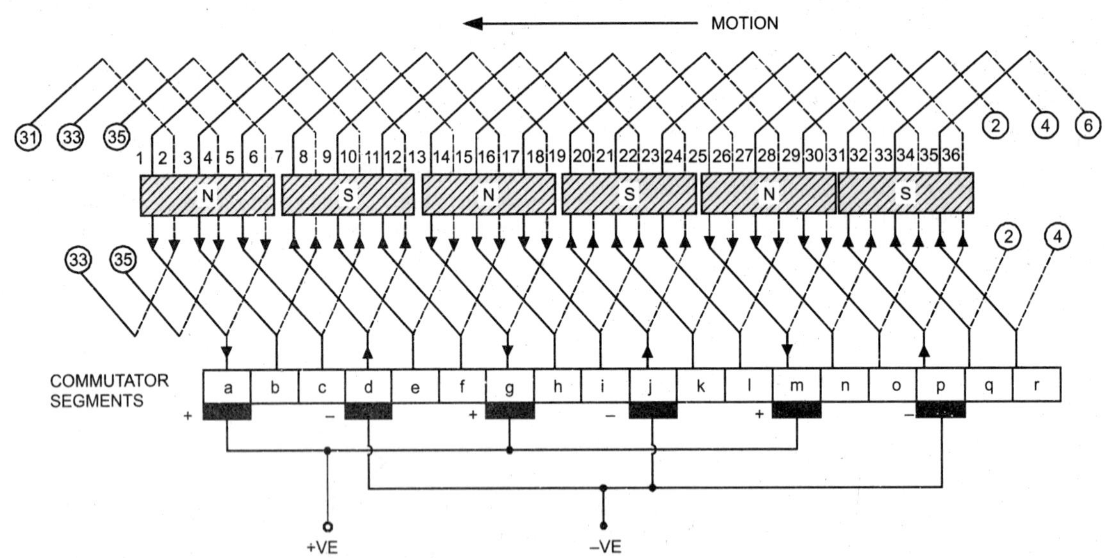

Fig. 5.24 *Developed View of 6 Pole, Double Layer, Progressive, Simplex Lap Winding With 36 Coil Sides*

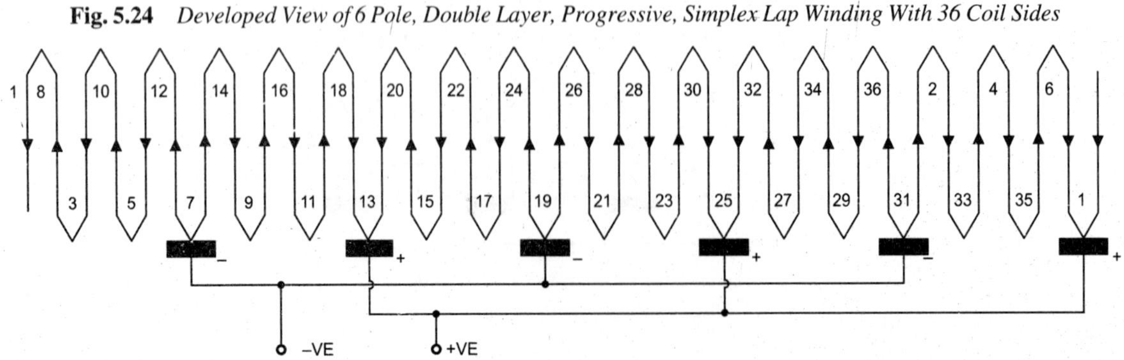

Fig. 5.25 *Ring Winding Diagram*

which instantaneously lie between the poles and have no emf induced in them.

4. There are as many parallel paths in the armature as the number of poles.
5. The emf between + ve and – ve brushes is equal to the emf generated in any of the parallel paths. If Z is the total number of armature coil sides and P is the number of poles, then armature coil sides connected in series in each path = Z/P

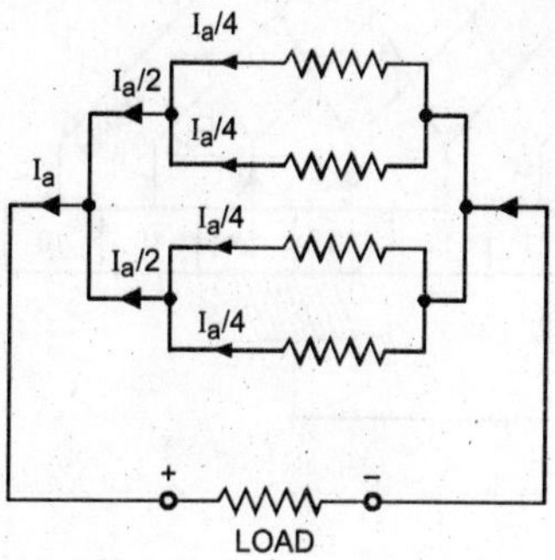

Fig. 5.26 *Current Distribution in Parallel Paths*

Total generated emf,

E = EMF generated per path

$$= \text{Average emf per coil side} \times \frac{Z}{P} = e_{av} \times \frac{Z}{P}$$

6. Round the complete armature winding, the resultant emf is zero, so that there will be no circulating current around the closed winding.
7. If I_a is the total armature current, then current per parallel path (or carried by each coil side) is given by

$$I_c = \frac{I_a}{P}$$

8. If l is the length of each armature conductor, a is its area of x-section and ρ is the resistivity of the material of conductor, then resistance of each conductor (or coil side) $= \rho \frac{l}{a}$

Number of conductors connected in series

$$= \frac{Z}{A} \text{ in each parallel path}$$

Resistance of each parallel path $= \rho \frac{l}{a} \cdot \frac{Z}{A}$

Since the number of paths in parallel is A,

Equivalent armature resistance, $R_a = \dfrac{\rho \frac{l}{a} \cdot \frac{Z}{A}}{A} = \rho \frac{l}{a} \cdot \frac{Z}{A^2}$

Example 5.3. Design and draw a 2 layer progressive duplex lap winding for a 4 pole dc generator with 16 slots, each slot having 2 coil sides. Indicate the position of brushes and their polarities.

Solution: Number of poles, P = 4

Number of coil sides, Z = No of slots × coil sides in each slot = 16 × 2 = 32

$$\text{Average pitch, } Y_a = \frac{Y_b + Y_f}{2} = \frac{Z}{P} = \frac{32}{4} = 8 \quad ...(i)$$

For duplex progressive lap winding

$$Y_b = Y_f + 2 \times 2 \quad ...(ii)$$

$$\text{or } Y_b = Y_f + 4 \quad ...(iii)$$

The back pitch and front pitch must be odd.

The nearest values of Y_b and Y_f satisfying the above three conditions are 9 and 5 respectively.

The scheme of connections for the winding is given below :

At Back End		*At Front End*		
Coil Side No.	*Connected to Coil Side No.*	*Coil Side No.*	*Connected to Coil Side No.*	*Through Segment No.*
Winding No. 1				
1	10	10	5	3
5	14	14	9	5
9	18	18	13	7
13	22	22	17	9
17	26	26	21	11
21	30	30	25	13
25	2	2	29	15
29	6	6	1	1
Winding No. 2				
3	12	12	7	4
7	16	16	11	6
11	20	20	15	8
15	24	24	19	10
19	28	28	23	12
23	32	32	27	14
27	4	4	31	16
31	8	8	3	2

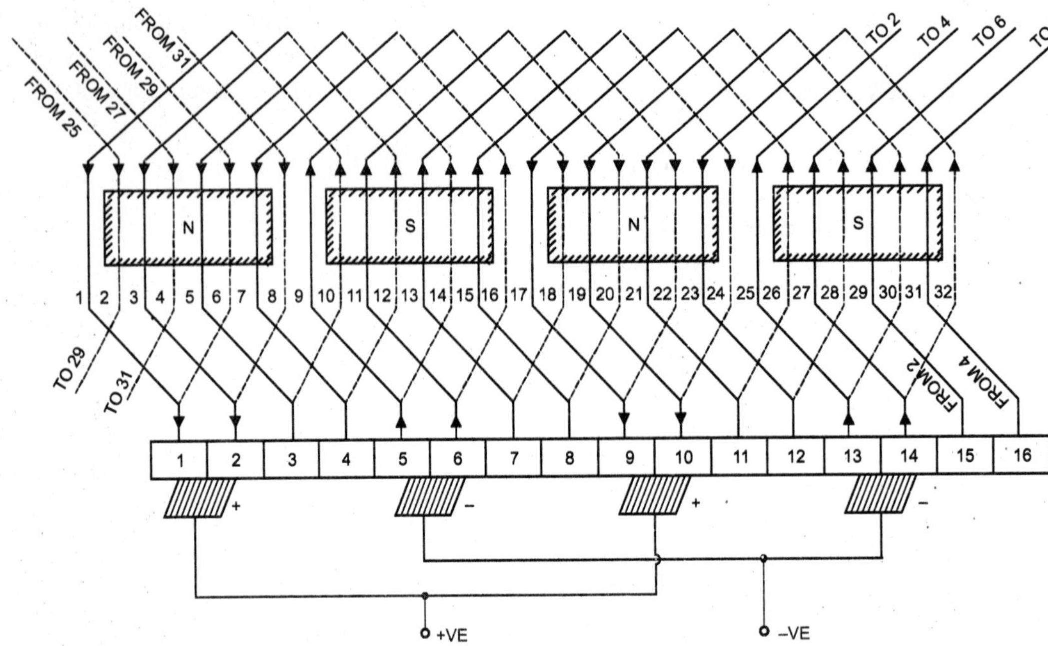

Fig. 5.27 *Developed Diagram of 4 Pole, Duplex Lap Winding With 32 Coil Sides*

The developed winding and equivalent ring winding diagrams for the above winding are shown in Figs. 5.27 and 5.28 (*a*) and (*b*) respectively with the position and polarity of the brushes marked as discussed in Example 5.1

From the winding diagram drawn for duplex lap winding (shown in Fig. 5.27) the following conclusions are arrived at:

1. The total number of brushes is equal to the number of poles and each brush spans two commutator segments. Similarly in triplex winding each brush will span three commutator segments. Hence in a multiplex winding the brush spans as many segments as the degree of multiplicity.

2. The number of parallel paths in each circuit (or winding) is 4 and there are two such circuits (or windings), therefore, number of parallel paths = 2 × 4 = 8.

In general the number of parallel paths for multiplex lap winding is given by A = *m*P where *m* is the multiplicity of the winding and P is the number of poles.

3. There are two independent windings, each closing on itself. As each of the winding closes on itself, the winding is *doubly re-entrant duplex winding*.

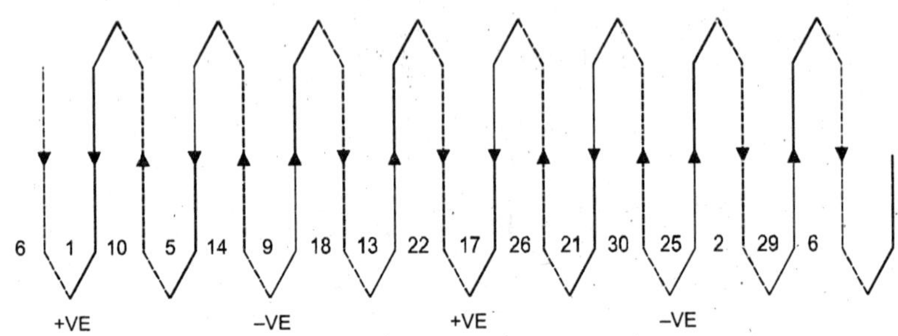

Fig. 5.28 (*a*) *Equivalent Ring or Spiral Winding For First Circuit*

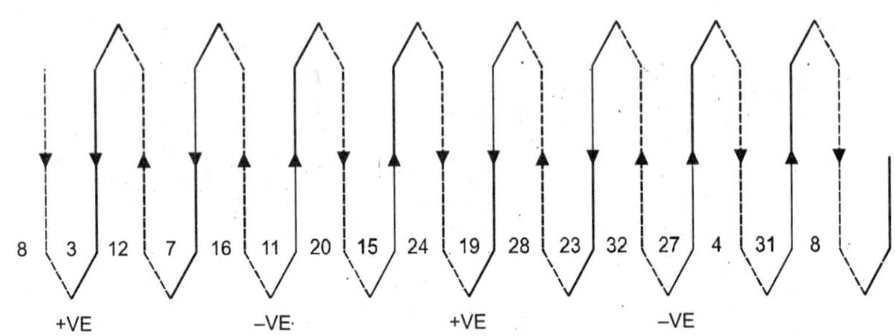

Fig. 5.28 (*b*) *Equivalent Ring or Spiral Winding For Second Circuit*

It will be observed that if in duplex winding the coils are not multiple of 2 (as is in the above case) there will not be two independent windings and in such a case winding will close only once and so will be known as *singly re-entrant duplex winding*.

Similarly if the number of coils in triplex winding are not multiple of 3, the triplex winding will close once and will be called as *singly re-entrant triplex winding*.

In general the *degree of re-entrance of multiplex lap winding* is equal to the great common factor between the number of commutator segments and the multiplicity of the winding.

5.6.2. Equalizer Connections in Lap Windings. In lap winding all the conductors in any parallel path lie under one pair of poles. If flux from all the poles is exactly the same, the induced emf in each path hence current carried by each path will be equal. But inspite of best efforts this condition is not practicable either due to variation in gap length or due to different magnetic properties of steel. The currents flowing in various paths may also differ due to difference in their resistances. Hence there is always slight difference in the generated voltage in the various parallel paths, and as a result, large circulating currents flow in the armature winding. These circulating currents not only tend to heat the armature above the temperature caused by the normal load current, but also cause an undue amount of sparking at the brushes since such currents must circulate from one path to another through the contacts between brushes and commutator. Sparking should, of course, be avoided, because it causes undue burning and wear of the commutator and brushes and, if carried too far, may result

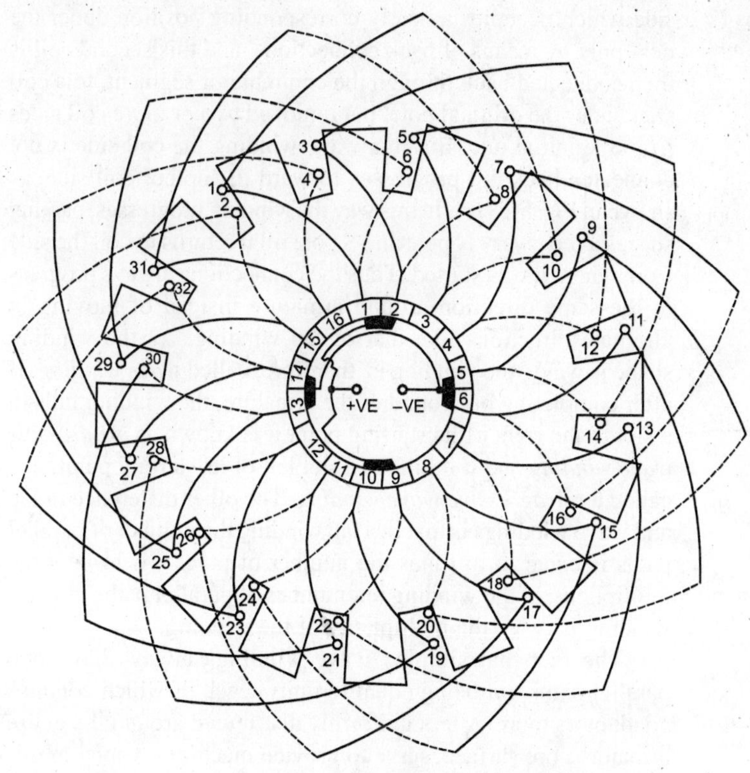

Fig. 5.28 (c) *Circular Duplex Lap Winding Double Re-entrant For 4-poles and 32 Coil Sides*

in *flash-over* from + ve to – ve brushes, a condition representing a short-circuit across the supply lines.

To overcome the detrimental effects resulting from the circulating currents, it is customary to employ equalizer connections in all lap wound armatures. These are low resistance copper wires that connect together points in the armature winding which should, under ideal conditions, be at exactly the same potential at all time but which, because of mechanical and electrical differences, are not. Thus, the equalizer rings relieve the brushes to the circulating current load by causing the latter to be bypassed.

To make these equalizer connections, the number of coils should be multiples of the number of poles and the number of coils per pole should be divisible by a small number 2 or 3. As an example assume a 4-pole generator having 24 coil sides. The number of coil sides per pole will be 6. The winding with equalizer connections is shown in Fig. 5.29. Note that every 3rd coil is connected to an equalizer. The coils that are connected to the same equalizer occupy the same position relative to the poles. This is necessary as such coils, at any instant, should be generating the same emf. Note in Fig. 5.29 that (*i*) one equalizer ring is connected to coil side two pole pitch apart, (*ii*) number of connections to one equalizing ring is equal to the pair of poles and (*iii*) number of rings is equal to the number of coils under one pair of poles *i.e.* equal to

$$\frac{\text{Number of coils}}{\text{Number of pairs of poles}}$$ in case every coil is connected to an equalizer.

Theoretically, every coil should be connected to an equalizer but as this would require an undue number of equalizers, it is sufficient, practically, to connect every third or fourth coil. This is the reason why the number of coils per pole should be divisible by a small number as 2, 3 or 4. In Fig. 5.29 alternate coil has been connected to the equalizer ring and in such a case the winding is said to be 50% equalized. If all the coils were connected to the equalizer rings then the winding would have been 100% equalized.

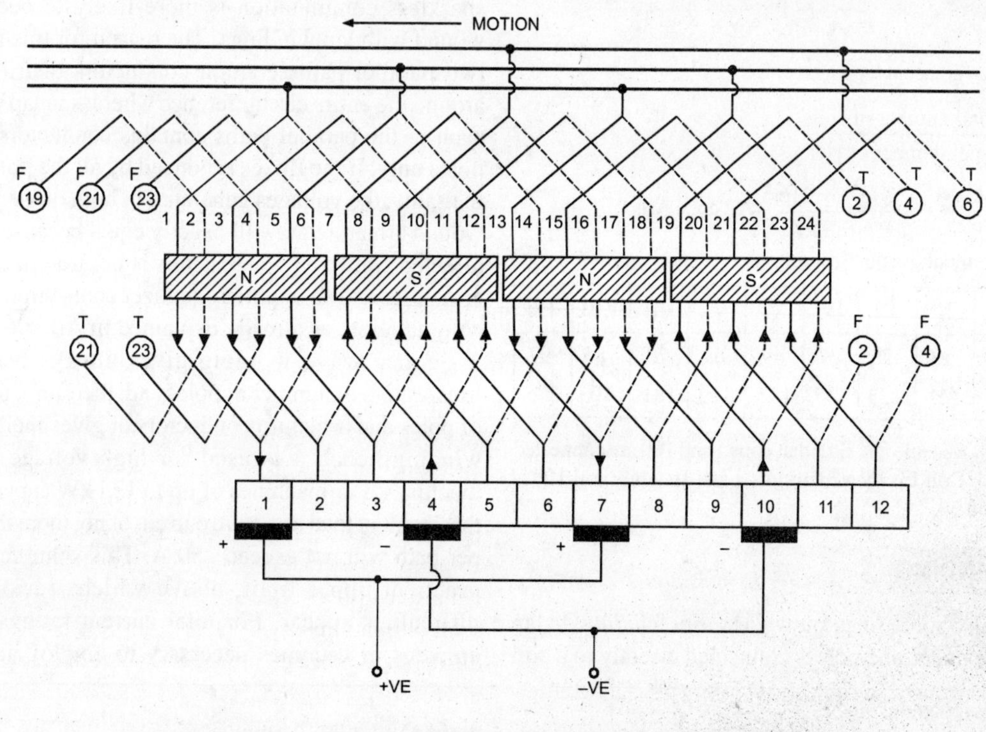

Fig. 5.29 *Simplex Lap Winding With Equalizer Connections*

Example 5.4. A 4-pole dc armature with lap connected coils has 72 slots and 6 coil-sides per slot. Determine the winding pitches and connections to the 9 equalizer rings.

Solution: Number of poles, P = 4

Number of slots = 72

Number of coil sides = Number of slots × number of coil sides per slot

$= 72 \times 6 = 432$

Number of parallel paths, A = P = 4

Number of commutator segments = Number of coils, C

$$= \frac{432}{2} = 216$$

$$\text{Average pitch, } Y_a = \frac{Y_b + Y_f}{2} = \frac{432}{4} = 108 \quad ...(i)$$

$$\text{Back pitch, } Y_b = Y_f \pm 2 \quad ...(ii)$$

Solving Eqs. (*i*) and (*ii*) we have

$$Y_b = 107 \text{ or } 109$$

Thus if we choose $Y_b = 107$, coil side 1 which lies in slot 1 is connected to coil side 108 lying in slot 18, thus the coil span becomes 17 slots. This gives fractional pitch. If we choose back pitch $Y_b = 109$, coil side 1 lying in slot 1 is connected to coil side 110 lying in slot 19. This gives full pitch coils. Hence we decide to choose back pitch

$Y_b = 109$ **Ans.**

Front pitch, $Y_f = Y_b - 2 = 109 - 2 = 107$ **Ans.**

The distance between coils of the same potential say, Y_{eq}* in terms of number of coils

$$Y_{eq} = \frac{\text{Total number of coils}}{\text{Pole-pairs}}$$

$$= \frac{C}{P/2} = \frac{2C}{P} = \frac{2 \times 216}{4} = 108 \text{ coils}$$ **Ans.**

Total number of tappings = Number of equalizer rings × number of pole-pairs

$$= n_r \times \frac{P}{2} = 9 \times \frac{4}{2} = 18$$

Distance between adjacent tappings**,

$$Y_{ph} = \frac{\text{Total number of coils}}{\text{Total number of taps}}$$

$$= \frac{C}{n_r P/2} = \frac{2C}{n_r P} = \frac{2 \times 216}{9 \times 4} = 12 \text{ coils}$$ **Ans.**

Thus the arrangement of equalizer rings is:

Ring No.	I	II	III	IV	V	VI	VII	VIII	IX
Coil No.	1	13	25	37	49	61	73	85	97
	109	121	133	145	157	169	181	193	205

Thus from the above table it is clear that coils 1 and 109 are connected to the equalizer ring I. Coil 1 is located in slots 1 and 19 while coil 109 is located in slots 36 and 54.

5.7 WAVE WINDING

Wave winding is also sometimes known as *series winding*. In lap winding a coil side under one pole is connected directly to a coil side which occupies a nearly corresponding position under the next pole by means of back connections, and this second coil is then connected back through the commutator segment, to a coil side under the original pole, but removed two or more coil sides from the initial one. But in a wave winding the coil side is not connected back but progresses forward to another coil side, as shown in Fig. 5.18 (*e*). In this way the winding progresses, passing successively every N pole and S pole till it returns to coil the side from where it was started. Thus the connections always progress in the same direction around armature instead of moving in alternate directions like that of lap winding. As the winding shape is wavy, the winding is, therefore, called *wave winding*. If after completing one round of the armature, the winding falls in a slot to the right of its starting point it is known as *progressive wave winding* and if it falls in a slot left of its starting point, it is called a *retrogressive wave winding*. The other difference in lap and wave windings is that in a lap winding the number of parallel paths is equal to m times the number of poles where *m* is the multiplicity of the winding but number of parallel paths in wave winding is twice the multiplicity of the winding.

The fact that simplex wave windings always have two parallel paths, (*m* being equal to unity) each of which contains conductors more or less uniformly distributed around the entire armature core surface, serve to provide machines wound in this manner with certain unique advantages. In the first place, wave wound machines need have but two sets of brushes (although as many brush sets as poles are frequently employed), while satisfactory operation of simplex lap wound machines is possible only if there are as many brush sets as there are poles; furthermore, in a wave wound machine equipped with as many brush sets as poles, if one or more of the brush sets develop poor contact with the commutator, satisfactory operation is still possible, this is not true in case of lap wound machines.

A second advantage of wave wound machines is that sparkless commutation is more likely to occur than in those wound with lap windings. The reason for this is that each of the two parallel paths contain conductors distributed completely around the entire circumference whereas in lap wound armatures, each of the parallel paths contains conductors lying under two poles only. If the fluxes produced by all the poles are not exactly the same, the voltages generated in both of the paths of the wave wound armature are still exactly equal because the two paths are affected similarly, whereas this is not true in case of lap wound armatures and so requires equalizer connections to give sparkless commutation, as already explained in Art. 5.6.2.

Since wave winding gives always two parallel paths irrespective of number of poles and, therefore, for a given number of poles and armature conductors, it gives more emf than the lap winding hence it is used for high voltage and low current machines. For machines of up to 125 kW capacity with voltages that result in total armature current of not more than 500 A, current per path will not exceed 250 A. This value represents a sort of practical upper limit, above which serious commutation difficulties appear. For total current ratings larger than 500 amperes, it becomes necessary to employ armature windings

---

* Y_{eq} is termed as equipotential pitch.

** Distance between adjacent tapping Y_{ph} is in terms of number of coils and is called the phase pitch.

giving more than two parallel paths, that is, windings that are not simplex wave. Simplex lap windings, that give as many parallel paths as poles are customarily employed in such cases, but sometimes in order to retain the advantages of the wave type winding over lap type winding and still keep the current per path within the upper limit indicated above, use of multiplex wave winding is made. Such windings, as already told, give the parallel paths twice the multiplicity of the winding in number, regardless the number of poles. Thus duplex, triplex and quadruplex wave windings give four, six and eight paths in parallel. Such windings, however, have limited use because other operating difficulties have developed. In some cases, the arcing at the commutator that these windings were supposed to correct, appeared another way by causing sparking and blackening at every second, third or fourth segment, a condition that resulted when slight differences existed in the electric circuits.

Important Points Regarding Wave Winding. (*a*) *In simplex wave winding*

1. Both the pitches, back pitch and front pitch must be odd numbers.
2. Back and front pitches must be nearly equal to the pole pitch and may be equal or differ by 2 in which case, they will be one more or one less than average pitch.
3. Commutator pitch, Y_c = Average pitch, Y_{av}
4. The average pitch Y_{av} is given by

$$Y_{av} = \frac{Y_b + Y_f}{2} = \frac{Z \pm 2}{P}$$ where Z is the number of conductors or coil sides and P is the number of poles. In order that the wave winding may close itself, the average pitch Y_{av} must be a whole number and agree to the above formulae.

In the above formulae ± 2 has been taken due to the reason that after one round of the armature the winding falls short of two conductors from the starting conductor. If it is not done and the average pitch is taken equal to Z/P then after one round of the armature, the winding will come back to the starting conductor or winding will close itself without including all coil sides.

The + ve sign in the above formulae will give progressive winding and –ve sign a retrogressive winding.

Since average pitch Y_{av} should be a whole number, this winding is not possible with any number of coil sides. For example with 28 conductors in a 4-pole machine, since $Y_{av} = \frac{28 \pm 2}{4} = 7\frac{1}{2}$ or $6\frac{1}{2}$ being fractional number, the wave winding is not possible but for 26 or 30 conductors, this winding can be employed since $Y_{av} = \frac{26 \pm 2}{4} = 7$ or 6 and $Y_{av} = \frac{30 \pm 2}{4} = 8$ or 7.

5. For even number of pairs of poles *i.e.* for 2, 4, 6, or 8 pair pole machine

(*i*) the average pitch, $Y_{av} = \frac{Y_b + Y_f}{2}$ may be odd or even.
(*ii*) the number of coils must be odd.
(*iii*) the number of commutator segments must be odd.

For odd number of pairs of poles such as 3, 5, 7

(*i*) the number of coils may be even or odd.
(*ii*) the number of commutator segments may be even or odd.
(*iii*) if the number of coils is even, average pitch must be odd and if the number of coils is odd, average pitch must be even.

(*b*) *In Multiplex wave Winding.*

1. The relation between the back pitch and the front pitch for multiplex wave winding is also the same as for the multiplex lap winding, so

$$Y_b = Y_f \pm 2m$$

where *m* is the multiplicity of the winding;

+ ve sign indicates progressive winding and –ve sign indicates retrogressive winding.

2. The average pitch for multiplex wave winding is given as $Y_{av} = \frac{Z \pm 2m}{P}$ and it must be an integer.
3. In a duplex winding having an odd number of pairs of poles, whose average pitch is odd, must have an odd number of commutator segments and coils and one whose average pitch is even must have an even number of commutator segments and coils.

If the duplex winding has an even number of pairs of poles, the number of commutator segments and coils must be even irrespective of whether the average pitch is odd or even.

If in a duplex winding average pitch Y_{av} is odd, the winding will be singly re-entrant and if even, doubly re-entrant.

5.7.1. Dummy Coils. The wave winding is possible only with particular number of conductors but sometimes the standard armature punchings available in armature winding shops do not meet with the requirements of the winding as these may accommodate more than the required number of conductors. In such cases *dummy coil* or *coils* are employed. These coils are placed in slots to preserve the balance of machine but are not electrically connected to the rest of the winding.

Say for example it is desired to have a simple wave winding for 16 slots, each slot accommodates two coil sides in a 4-pole machine.

$$Y_{av} = \frac{Z \pm 2}{4} = \frac{32 \pm 2}{4} = 8\frac{1}{2} \text{ or } 7\frac{1}{2}$$

But the average pitch must be an integer, so the winding is not possible with 32 conductors. However, if 30 conductors are used, $Y_{av} = \frac{30 \pm 2}{4} = 8$ or 7. Thus two windings are possible, one progressive and other retrogressive, but in both cases two coil-sides or one coil is inactive and hence is called *dummy coil*. The ends of dummy coil are taped and are not connected to the commutator segments.

5.7.2. Selective Commutation. When more than two brushes are used in simplex wave winding then trouble with the brushes may arise which is known as *"selective commutation"*. When more than two brushes are used in simplex wave winding the number of parallel paths remains the same *i.e.* 2, only current collected from armature is divided between the brushes of same polarity. If brushes fixed are of different contact resistances then the current passing through the individual brush will be different. For example two brushes A and B, shown in Fig. 5.30 (*b*), side by side should each collect half of the load current, but if brush A has more contact resistance than that of B, obviously more current will pass through B and less through A, so brush B will be overloaded. Similarly if brush C has more contact resistance than brush D, brush D will obviously carry more current than C and will be overloaded. So the brushes must be adjusted very carefully when more than two brush sets are used in a wave wound machine. The advantage of using the maximum number of brush sets is that the current per brush set is reduced and so the axial length of the commutator is reduced, which gives a cheaper machine.

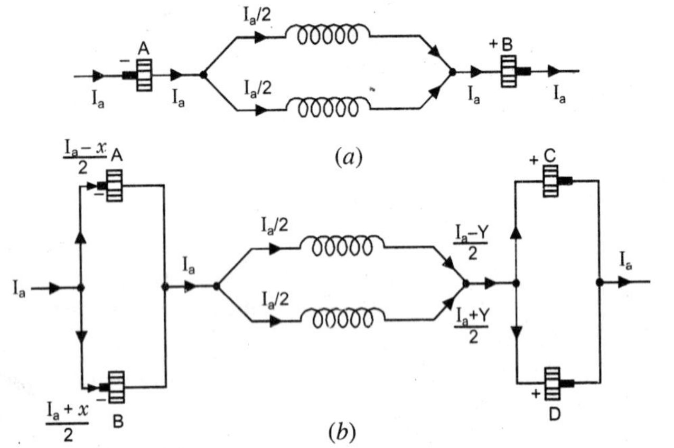

Fig. 5.30 *Selective Commutation*

Example 5.5. Draw up the winding table for a 4 pole, wave-connected armature having 30 coil sides and give a developed diagram of the winding showing the polarity and position of the brushes, the main poles and the direction of motion of the armature for a dc motor.

Solution: Average pitch, $Y_{av} = \frac{Z \pm 2}{P} = \frac{30 \pm 2}{4} = 8$ or 7;

if Y_{av} is taken odd then back pitch Y_b and front pitch Y_f will be equal to Y_{av} *i.e.* 7.

Ist coil side is connected to 8th coil side at the back, which is connected to 15th coil side through commutator segment no. 7 at the front. The 15th coil side is connected to 22nd coil side at the back and 22nd coil side is connected to 29th coil side through commutator segment no 14 at the front and winding diagram is completed with the help of winding table given below. The developed winding diagram is shown in Fig. 5.31.

To determine the brush position the ring winding, as shown in Fig. 5.32, be drawn. On inserting the direction of emfs according to the actual winding, it is clear that the winding divides itself into two halves between M and P and N and R. The point N is the separating point of the emfs and it corresponds to + ve brush. As it is at the back of the winding and not at commutator end, so the +ve brushes may be fixed at points P and Q on the ring winding or on the junctions of coil side nos. 24 and 1; 8 and 15. Point M is the meeting point of the emfs and, therefore, it corresponds to –ve brush. As it is at the back of the winding and not at commutator end so the – ve brushes may be fixed at points R and S on the ring winding or on junctions of coil side nos. 16 and 23 and of coil side nos. 30 and 7.

Since conductors 24, 1, 8, 15, 16, 23, 30, 7 fall in between the poles, so these conductors have got no induced emf and brushes may be fixed at their respective junction points.

Note: *In generator + ve and – ve brushes should be fixed at meeting and separating points respectively but in motor it is vice versa.*

At Back End			*At Front End*	
Coil Side No.	*Connected to Coil Side No.*	*Coil Side No.*	*Connected to Coil Side No.*	*Through Segment No.*
1	8	8	15	7
15	22	22	29	14
29	6	6	13	6
13	20	20	27	13
27	4	4	11	5
11	18	18	25	12
25	2	2	9	4
9	16	16	23	11
23	30	30	7	3
7	14	14	21	10
21	28	28	5	2
5	12	12	19	9
19	26	26	3	1
3	10	10	17	8
17	24	24	1	15

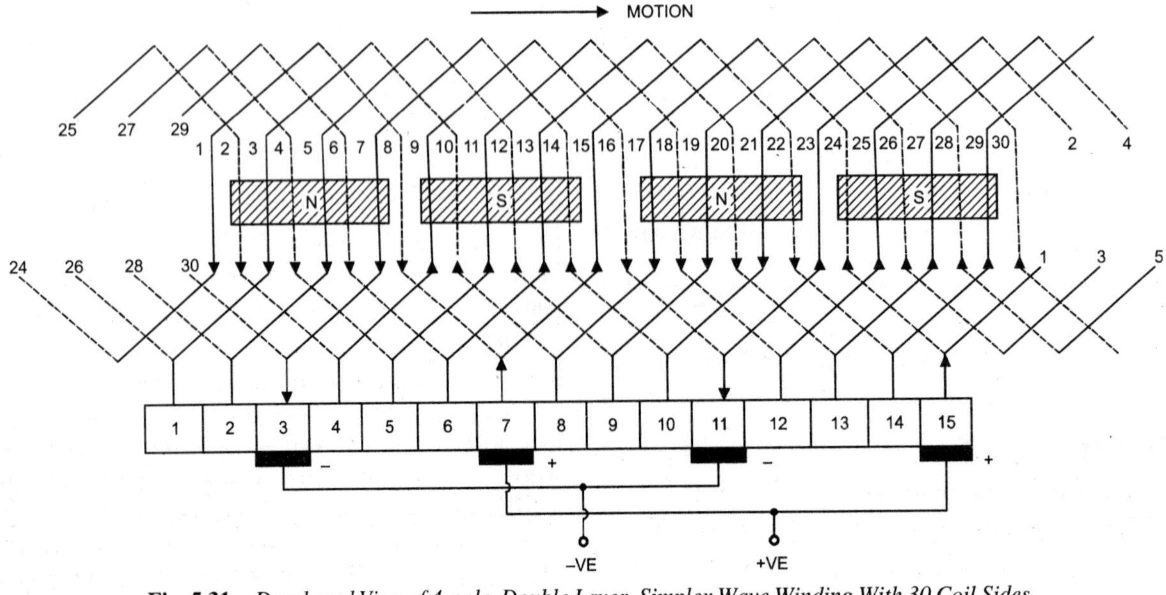

Fig. 5.31 *Developed View of 4-pole, Double Layer, Simplex Wave Winding With 30 Coil Sides*

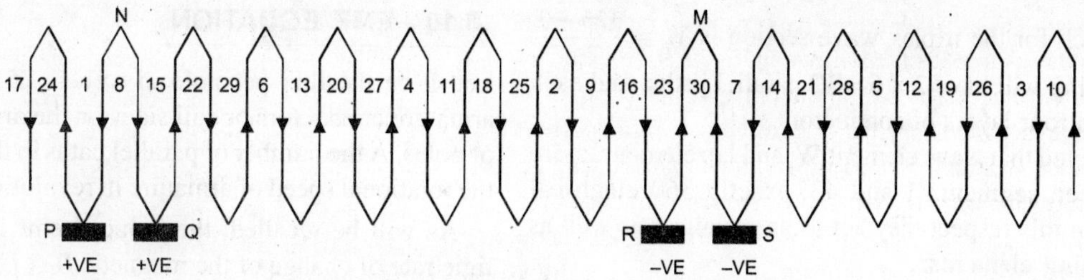

Fig. 5.32 *Equivalent Ring or Spiral Winding*

Direction of Motion: The direction of motion for a motor is determined by Fleming's left hand rule. The direction has been shown in Fig. 5.31.

Characteristics of Simplex Wave Winding. From the above winding diagram, the following facts can be derived for simplex wave winding:

1. Only two brushes are necessary, but parallel brushes may be added to make their number equal to the number of poles.
2. The brushes are connected to the conductors which lie between the poles and have no emf induced in them.
3. The number of parallel paths through the armature winding is two, irrespective of number of poles of the machine.
4. Number of conductors in each path = Z/2.
5. The generated emf is equal to average emf induced in each conductor $\times \frac{Z}{2}$ *i.e.*, $E = e_{av} \times \frac{Z}{2}$
6. Current flowing through each conductor is equal to current per path *i.e.* $I_a / 2$ where I_a is the armature current.
7. The resultant emf round the circuit is zero.

Example 5.6. A 6-pole, dc armature with 36 slots and 2 coil sides/slot is to be wave wound (double layer). Calculate the back pitch, front pitch, and commutator pitch. How many dummy coils, if any required.

[Rajasthan Technical Univ. Electrical Machines-I, 2011]

Solution: Total number of armature conductors

$$Z = \text{Number of slots} \times \text{number of coil sides (conductors)/slot} = 36 \times 2 = 72$$

Number of poles, P = 6

Average pitch is given by

$$Y_{av} = \frac{Z \pm 2}{P} = \frac{72 \pm 2}{6} = \frac{74 \text{ or } 70}{6}$$

Since average pitch must be a whole number, so let average pitch, Y_{av}, be 12 *i.e.* armature conductors used are 70.

Back pitch, $Y_b = Y_{av} + 1 = 12 + 1 = 13$ **Ans.**

Front pitch, $Y_f = Y_{av} - 1 = 12 - 1 = 11$ **Ans.**

Commutator pitch, $Y_c = Y_{av} = 12$ **Ans.**

$$\text{Dummy coil used} = \frac{\text{Total number of armature conductors} - \text{armature conductors actually used}}{2}$$

$$= \frac{72 - 70}{2} = 1 \quad \textbf{Ans.}$$

Example 5.7. Design the winding for 4-pole armature with 2,540 conductors, 127 commutator segments and 32 slots.

[U.P. Technical Univ. Electromechanical Energy Conversion-I, 2006-07]

Solution: Number of conductors, Z = 2,540

Number of slots, S = 32

Number of commutator segments = 127

Number of poles, P = 4

$$\text{Conductors per slot, } Z_S = \frac{Z}{S} = \frac{2{,}540}{32} = 79.375$$

Since Z_S must be a whole number, choose $Z_S = 80$ giving total number of conductors

$$Z = 80 \times 32 = 2{,}560$$

Total number of coils,

C = Number of commutator segments = 127

$$\text{Turns per coil} = \frac{2{,}560/2}{127} = 10.08 \simeq 10$$

Since the coil is a multiturn coil, the winding that can be used is wave winding.

$$\text{Actual number of coils} = \frac{2{,}560/2}{10} = 128$$

Active coils = Number of commutator segments = 127

Dummy coil = Actual number of coils – active coils = 128 – 127 = 1

$$\text{Average pitch, } Y_{av} = \frac{C \pm 1}{P/2} = \frac{127 \pm 1}{4/2} = 64 \text{ or } 63$$

∴ Back pitch, $Y_b = Y_f \pm 2$

i.e. $Y_b = 65$

and $Y_f = 63$ for progressive wave winding and

$Y_b = 63$ and $Y_f = 65$ for retrogressive wave winding.

5.8 FROG-LEG WINDING

Wave windings are, in general, less complicated than equalized-lap windings, easier to install and service, and less expensive but for total current ratings exceeding 500 A, it becomes necessary to use armature windings with more than two parallel paths and, therefore, simplex wave winding is not possible. Very few windings of order higher than duplex have ever been designed and manufactured on commercial basis because of the complicacies involved in proper interconnection between the various armature circuits.

A construction that combines the advantages of both the lap and wave types is the so called *frog-leg winding* developed by the Allis-Chalmers Manufacturing Co. It is composed of a multiplex wave winding in the same slots with a simplex or multiplex lap winding. Both have the same number of parallel paths and are connected to the same commutator. Thus every frog-leg winding has 2 P parallel paths. Moreover, since every series combination of a lap element and a wave element, which is connected to two segments exactly 360 electrical degrees apart, develops zero net voltage, each of these lap-wave combinations acts like an equalizer; all frog-leg windings are, therefore, 100 per cent equalized.

Fig. 5.33 illustrates two frog-leg coils, properly shown in slots and connected to the commutator, for a 6-pole 126-slot and 126-segment armature. The coil pitch $Y_a = \frac{126}{6} = 21$, and the commutator pitch for the triplex wave section is $Y_c = \frac{126-3}{3} = 41$. The winding will have $2 \times 6 = 12$ parallel paths, and each slot will contain four layers of conductors.

It is to be noted that wave element W_1 and lap element L_2 are in series between segments 1 and 43, exactly 360 electrical degrees apart; in this respect they act as an equalizer as well as voltage-generating elements.

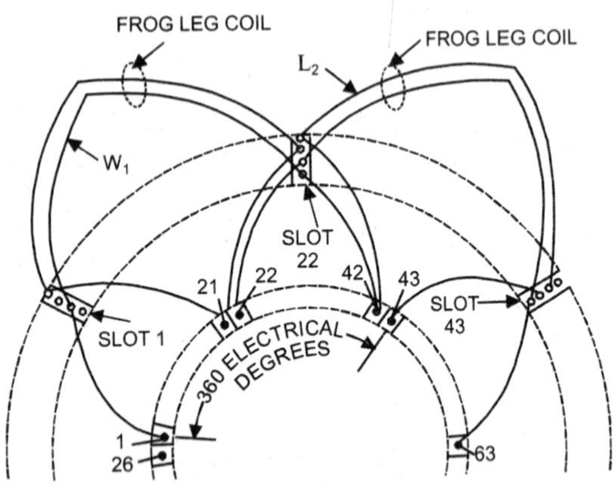

Fig. 5.33 *Two Frog-Leg Coils For a 6-Pole Winding in an Armature Having 126 Slots and 126 Segments*

5.9 COMPARISON BETWEEN SIMPLEX LAP AND SIMPLEX WAVE WINDINGS

Comparison of simplex lap and simplex wave windings is given below in tabular form.

S.No.		*Simplex Lap Winding*	*Simplex Wave Winding*
1.	Number of parallel paths	Number of poles, P	Always 2
2.	Number of brushes	Number of poles, P	2, but in case of larger machines, number of brushes may be equal to P
3.	Dummy coils	Not required	May be required
4.	Equalizer rings	May be required, $= \frac{\text{Number of coils}}{\text{Number of pair of poles}}$	Not required
5.	Conductor current	$I_C = \frac{I_a}{A}$	$I_C = \frac{I_a}{2}$
6.	Induced emf	Independent of number of pole	Dependent upon number of poles
7.	Cost	Lower	Higher
8.	Applications	Low-voltage, high-current machines	High-voltage, low-current machines

5.10 EMF EQUATION

Let Φ be the flux per pole in webers, Z the total number of armature conductors or coil sides on the armature, P the number of poles, A the number of parallel paths in the armature and N be the rotational speed of armature in revolutions per minute (rpm).

As will be recalled, the induced emf is proportional to the time rate of change of the magnetic flux *i.e.*

$$e = \frac{d\Phi}{dt}$$

During one revolution of armature in a P-pole generator each armature conductor cuts the magnetic flux P times, so flux cut by one conductor in one revolution = $P\Phi$ webers.

Since the number of revolutions made by the armature per minute is N so number of revolutions made per second is N/60 and, therefore, flux cut by each conductor per second = Flux cut by one conductor per revolution × number of revolutions of armature/second

$$= \Phi P \times \frac{N}{60} \text{ webers}$$

Consequently the average emf induced in one conductor will be

$$e = \Phi P \frac{N}{60} \text{ volts}$$

The number of conductors in series between a + ve brush and – ve brush is equal to the total number of conductors divided by the number of parallel paths i.e. number of armature conductors per parallel path = Z/A*

∴ The total emf generated between the terminals,

E = Average emf induced in one conductor × number of conductors in each circuit or parallel path

$$= \Phi P \frac{N}{60} \times \frac{Z}{A} \text{ volts} = \Phi Z \frac{N}{60} \times \frac{P}{A} \text{ volts} \qquad ...(5.1)$$

For a given machine the number of poles (P) and number of armature conductors per parallel path (Z/A) are constant, therefore,

$$\text{Generated emf, } E = K\Phi N \quad ...(5.2) \quad \text{where } K = \frac{PZ}{60\,A}$$

$$\text{or } E \propto \Phi N$$

$$\text{or } E \propto \Phi\omega \text{ where } \omega = \frac{2\pi N}{60},$$

the angular velocity in radians/second

Thus we see that the induced emf is directly proportional to flux per pole Φ and speed N. Moreover, the polarity of the induced emf depends upon the direction of the magnetic field and the direction of rotation. If either of the two is reversed, the polarity of the induced emf *i.e* brushes is reversed, but when both are reversed, the polarity remains unchanged.

The induced emf is fundamental phenomenon to all dc machines whether they are operating as generators or motors. However, when the machine is operating as a generator, this induced emf is called the *generated emf,* E_g whereas in case of a machine operating as a motor it is called the *counter* or *back emf,* E_b.

* A = 2 in case of wave winding and = P in case of lap winding.

Example 5.8. A dynamo has a rated armature current at 250 A. What is the current per path of the armature if the armature winding is simplex wave wound or simplex lap wound? The machine has 12 poles.

Solution : Rated armature current, $I_a = 250$ A

Number of poles, $P = 12$

With simplex wave winding

Number of parallel paths, $A = 2$

$$\therefore \text{Current per path, } I_c = \frac{I_a}{A} = \frac{250}{2} = 125 \text{ A } \textbf{Ans.}$$

With simplex lap winding

Number of parallel paths, $A = P = 12$

$$\text{Current per path, } I_c = \frac{I_a}{A} = \frac{250}{12} = 20.833 \text{ A } \textbf{Ans.}$$

Example 5.9. Derive emf equation of a dc generator. What will be change in emf induced if flux is reduced by 20% and the speed is increased by 20%. *[U.P. Technical Univ. Even Semester 2005-06]*

Solution: Since induced emf, $E \propto \Phi N$

$$\text{or } E_2 = E_1 \times \frac{\Phi_2}{\Phi_1} \times \frac{N_2}{N_1} = E_1 \times 0.8 \times 1.2 = 0.96 \, E_1.$$

$$\text{So change in emf} = \frac{E_1 - E_2}{E_1} \times 100$$

$$= \frac{E_1 - 0.96E_1}{E_1} \times 100 = 4\% \text{ decrease } \textbf{Ans.}$$

Example 5.10. A 4-pole lap-connected dc machine has an armature resistance of 0.15 Ω. Find the armature resistance of the machine when rewound for wave connection.

Solution : *When machine is lap wound*

Number of parallel paths, $A = P = 4$

Armature resistance, $R_a = 0.15\ \Omega$

Resistance of each path = Armature resistance × number of parallel paths

$= 0.15 \times 4 = 0.6\ \Omega$

$$\because \quad R_a = \frac{\text{Resistance of each path}}{\text{Number of parallel paths}}$$

When machine is wave wound

Number of parallel paths, $A' = 2$

So resistance of each path = Resistance of each path when lap-connected $\times \frac{A}{A'}$

$$= 0.6 \times \frac{4}{2} = 1.2\ \Omega$$

$$\text{Armature resistance, } R_a = \frac{\text{Resistance of each path}}{A'}$$

$$= \frac{1.2}{2} = 0.6\ \Omega \textbf{ Ans.}$$

Example 5.11. A dc generator has an armature emf of 100 V when the useful flux per pole is 20 mWb, and the speed is 800 rpm. Calculate the generated emf (i) with the same flux and a speed of 1,000 rpm, (ii) with a flux per pole of 24 m Wb and a speed of 900 rpm.

[U.P. Technical Univ. Electrical Engineering, January-2003]

Solution : Generated emf, $E_1 = 100$ V

Flux per pole, $\Phi_1 = 20 \text{ mWb} = 20 \times 10^{-3}$ Wb

Speed, $N_1 = 800$ rpm

Since induced emf is proportional to the product of flux per pole Φ and armature speed N for a given machine i.e.

$$E \propto \Phi N$$

$$\text{or } \frac{E_2}{E_1} = \frac{N_2}{N_1} \times \frac{\Phi_2}{\Phi_1}$$

$$(i) \text{ or } E_2 = \frac{N_2}{N_1} \times \frac{\Phi_2}{\Phi_1} \times E_1 = \frac{1{,}000}{800} \times 1 \times 100 = 125 \text{ V } \textbf{Ans.}$$

$\because N_2 = 1{,}000$ rpm and $\Phi_2 = \Phi_1$

$$(ii) \text{ and } E_3 = \frac{N_3}{N_1} \times \frac{\Phi_3}{\Phi_1} \times E_1$$

$$= \frac{900}{800} \times \frac{24 \times 10^{-3}}{20 \times 10^{-3}} \times 100 = 135 \text{ V } \textbf{Ans.}$$

$\therefore N_3 = 900$ rpm and $\Phi_3 = 24 \times 10^{-3}$ Wb.

Example 5.12. A 6 pole lap wound dc generator has 720 conductors; a flux of 80 mweber/pole is driven at 1,000 rpm. Find the generated emf.

[G.B. Technical Univ. Electrical Engineering Second Semester 2010-11]

Solution: Generated emf, $E = \frac{\Phi Z N}{60} \times \frac{P}{A}$

$$= \frac{80 \times 10^{-3} \times 720 \times 1{,}000}{60} \times \frac{6}{6}$$

$= 960$ V **Ans.**

Example 5.13. A 4-pole dc generator has 51 slots and each contains 20 conductors. Flux per pole is 7 mWb and runs at 1,500 rpm. Find the produced emf of the machine if its armature is wave wound.

[Electrical Machines-I, Rajasthan Univ. 2006; Rajasthan Technical Univ. 2007]

Solution: Flux per pole, $\Phi = 7 \text{ mWb} = 0.007$ Wb

Number of poles, $P = 4$

Speed, $N = 1{,}500$ rpm

Number of armature conductors,

Z = Number of slots × number of conductors $= 51 \times 20 = 1{,}020$

Number of parallel paths, $A = 2$ $\quad \because$ armature is wave wound

$$\text{Generated emf, } E_g = \frac{\Phi Z N}{60} \times \frac{P}{A} \text{ volts}$$

$$= \frac{0.007 \times 1{,}020 \times 1{,}500}{60} \times \frac{4}{2} = 357 \text{ V } \textbf{Ans.}$$

Example 5.14. A 6-pole machine has an armature with 90 slots and 8 conductors per slot and runs at 1,000 rpm, the flux per pole is 0.05 Wb. Determine the induced emf if winding is (*i*) lap connected (*ii*) wave connected.

[U.P. Technical Univ. Electromechanical Energy Conversion-I 2002-03]

Solution : Flux per pole, $\Phi = 0.05$ Wb

Number of poles, $P = 6$

Speed, $N = 1{,}000$ rpm

Number of armature conductors,

Z = Number of slots × number of conductors per slot $= 90 \times 8 = 720$

(*i*) *When machine is lap-connected*

Number of parallel paths, $A = P = 6$

$$\text{Induced emf, } E_g = \frac{\Phi Z N}{60} \times \frac{P}{A} = \frac{0.05 \times 720 \times 1{,}000}{60} \times \frac{6}{6}$$

$= 600$ V **Ans.**

(ii) When machine is wave-connected

Number of parallel paths, A = 2

$$\text{Induced emf, } E_g = \frac{\Phi \times Z \times N}{60} \times \frac{P}{A} = \frac{0.05 \times 720 \times 1{,}000}{60} \times \frac{6}{2}$$

$$= 1{,}800 \text{ V} \quad \textbf{Ans.}$$

Example 5.15. A 4-pole dc machine has lap-connected armature having 60 slots and 8 conductors per slot. The flux per pole is 30 mWb. If the armature is rotated at 1000 rpm, find the emf available across its armature terminals. Also calculate the frequency of emf in the armature coil [*R.G. Technical Univ. Electromechanical Energy Conversion, December-2012*]

Solution: Flux per pole, Φ = 30 mWb = 0.03 Wb

Number of armature conductors

Z = Number of slots × number of conductors per slot

= 60 × 8 = 480

EMF available across armature terminals,

$$E = \frac{\Phi Z N}{60} \times \frac{P}{A}$$

$$= \frac{0.03 \times 480 \times 1{,}000}{60} \times \frac{4}{4} = 240 \text{ V } \textbf{Ans.}$$

Frequency of emf induced in armature coil,

$$f = \frac{PN}{120} \times \frac{4 \times 1{,}000}{120} = 33.33 \text{ Hz } \textbf{Ans.}$$

Example 5.16. The armature of a four-pole dc machine has 100 turns and runs at 600 rpm. The EMF generated on open circuit is 220 V. Find the useful flux per pole when armature is

(i) lap connected (ii) wave connected.

[*G.B. Technical Univ. Electrical Engineering Odd Semester 2010-11*]

Solution: EMF generated in open circuit, E = 220 V

Speed, N = 600 rpm

Number of poles, P = 4

Number of armature conductors, Z = 2 × number of turns

= 2 × 100 = 200

(*i*) When armature is lap connected

Number of parallel paths, A = P = 4

$$\text{Flux per pole, } \Phi = E \times \frac{A}{P} \times \frac{60}{N} \times \frac{1}{Z} \qquad \because E = \frac{\Phi Z N}{60} \times \frac{P}{A}$$

$$= 220 \times \frac{4}{4} \times \frac{60}{600} \times \frac{1}{200} = 0.11 \text{ Wb } \textbf{Ans.}$$

(*ii*) When armature is wave connected,

Number of parallel paths, A′ = 2

$$\text{and Flux per pole, } \Phi' = \frac{220 \times 2}{4} \times \frac{60}{600} \times \frac{1}{200} = 0.055 \text{ Wb } \textbf{Ans.}$$

Example 5.17. An 8 pole dc generator running at 1,200 rpm, and with a flux of 25 mWb/pole generates 440 V. Calculate the number of conductors.

[*M.D. Univ. Electromechanical Energy Conversion, December-2004*]

Solution: Flux per pole, Φ = 25 mWb = 0.025 Wb

Generated emf, E_g = 440 V

(i) When armature is lap-wound

Number of parallel paths, A = P = 8

$$\text{Number of conductors required, } Z = \frac{E_g \times 60}{\Phi \times N} \times \frac{A}{P}$$

$$= \frac{440 \times 60}{0.025 \times 1{,}200} \times \frac{8}{8} = 880 \textbf{ Ans.}$$

$$\because E_g = \frac{\Phi Z N}{60} \times \frac{P}{A} \text{ volts}$$

(ii) When armature is wave-wound

Number of parallel paths, A = 2

$$\text{Number of conductors required, } Z = \frac{E_g \times 60}{\Phi \times N} \times \frac{A}{P}$$

$$= \frac{440 \times 60}{0.025 \times 1{,}200} \times \frac{2}{8} = 220 \quad \textbf{Ans.}$$

Example 5.18. A four pole lap wound dc shunt generator having 70 slots with 12 conductors per slot generates an emf of 420 V at no load, when runs at 980 rpm. How will you obtain a generated open-circuit voltage of 240 V?

[*M.D. Univ. Electromechanical Energy Conversion, December-2010*]

Solution: Number of armature conductors,

Z = 70 × 12 = 840

$$\text{Flux per pole, } \Phi = \frac{E_g \times 60}{Z \times N} \times \frac{A}{P} = \frac{420 \times 60}{840 \times 980} \times \frac{4}{4} = 0.0306 \text{ Wb}$$

Required open-circuit voltage of 240 V can be had either by reducing the flux per pole (keeping speed of 980 rpm constant) or reducing the speed (keeping flux per pole constant).

(*i*) Keeping speed N constant at 980 rpm

$$\text{Required flux per pole, } \Phi' = \Phi \times \frac{E'_g}{E_g}$$

$$= 0.0306 \times \frac{240}{420} = 0.175 \text{ Wb } \textbf{Ans.}$$

(*ii*) Keeping flux constant at 0.0306 Wb

$$\text{Required speed, } N' = N \times \frac{E'_g}{E_g}$$

$$= 980 \times \frac{240}{420} = 560 \text{ rpm } \textbf{Ans.}$$

Example 5.19. An 8-pole dc generator has 500 armature conductors and has a useful flux per pole of 0.065 Wb. What will be the emf generated if it is lap-connected and runs at 1,000 rpm? What must be the speed at which it is to be driven to produce the same emf if it is wave wound?

[*Electrical Machines-I, Rajasthan Technical Univ, 2008, January-2013; J.N. Technological Univ., March-2014*]

Solution: Emf generated when the generator is lap-connected,

$$E_g = \frac{\Phi Z N}{60} \times \frac{P}{A} = \frac{0.065 \times 500 \times 1{,}000}{60} \times \frac{8}{8} = 541.67 \text{ V} \quad \textbf{Ans.}$$

∵ in lap-connected armature, A = P

Speed of generator, when wave-connected, to generate an emf of 541.67 V

$$N' = \frac{E_g \times 60}{\Phi \times Z} \times \frac{A}{P} = \frac{541.67 \times 60}{0.065 \times 500} \times \frac{2}{8} = 250 \text{ rpm} \quad \textbf{Ans.}$$

∵ in wave-connected armature, A = 2

and emf generated is same as in case of lap-connected armature *i.e.* 541.67 V

Example 5.20. A lap-wound dc shunt generator having 80 slots with 10 conductors per slot generates no-load emf of 400 V when running

at 1,000 rpm. At what speed should it be rotated to generate a voltage of 220 V on open circuit?

[*R.G. Technical Univ. Electrical Machine-II, December-2012*]

Solution: No-load emf, $E_1 = 400$ V when speed $N_1 = 1{,}000$ rpm.

Speed of the machine for generating a voltage of 220 V on open circuit,

$$N_2 = \frac{E_2}{E_1} \times N \quad \because E \propto N \text{ assuming } \Phi \text{ to remain constant}$$

$$= \frac{220}{400} \times 1{,}000 = 550 \text{ rpm} \quad \textbf{Ans.}$$

Example 5.21. An 8-pole lap wound armature rotated at 350 rpm is required to generate 260 V. The useful flux per pole is about 0.05 Wb. If the armature has 120 slots, calculate a suitable number of conductors per slot and hence determine the actual value of flux required to generate the same voltage.

Solution : Number of parallel paths, A = P = 8

$\because$ armature is lap wound

Total number of armature conductors required,

$$Z = \frac{E_g \times 60}{\Phi \times N} \times \frac{A}{P} = \frac{260 \times 60}{0.05 \times 350} \times \frac{8}{8} = \frac{260 \times 24}{7}$$

Number of conductors per slot

$$= \frac{Z}{\text{Number of slots}} = \frac{260 \times 24}{7 \times 120} = 8 \quad \text{(an integer number)}$$

Actual value of flux required, Φ

$$= \frac{E_g \times 60}{N \times Z} \times \frac{A}{P} = \frac{260 \times 60}{350 \times 8 \times 120} \times \frac{8}{8} = 0.0464 \text{ Wb} \quad \textbf{Ans.}$$

Example 5.22. Armature of a 2-pole, 200 V generator has 400 conductors and runs at 300 rpm. Calculate the useful flux per pole. If the number of turns in each field coil is 1,200, what is the average value of the emf induced in each coil on breaking the field, if the flux dies away completely in 0.1 second ?

[*J.N. University Hyderabad 2000*]

Solution: Number of parallel paths, A = 2 whether winding may be assumed to be wave or lap as P = 2

$$\text{Flux per pole}, \Phi = \frac{E_g \times 60}{N \times Z} \times \frac{A}{P} = \frac{200 \times 60}{300 \times 400} \times \frac{2}{2} = 0.1 \text{ Wb} \quad \textbf{Ans.}$$

E M F induced in each field coil, e

$$= \text{Number of turns in each field coil} \times \frac{d\Phi}{dt}$$

$$= 1{,}200 \times \frac{0.1 - 0}{0.1} = 1{,}200 \text{ V} \quad \textbf{Ans.}$$

Example 5.23. A 4-pole generator has an induced emf of 262 V when driven at a speed of 400 rpm. The armature is lap wound and has 652 conductors, its resistance being 0.15 Ω. The bore of the pole shoe is 42 cm diameter, the pole subtends an angle of 60° and is 20 cm long. Calculate the flux density in the air gap.

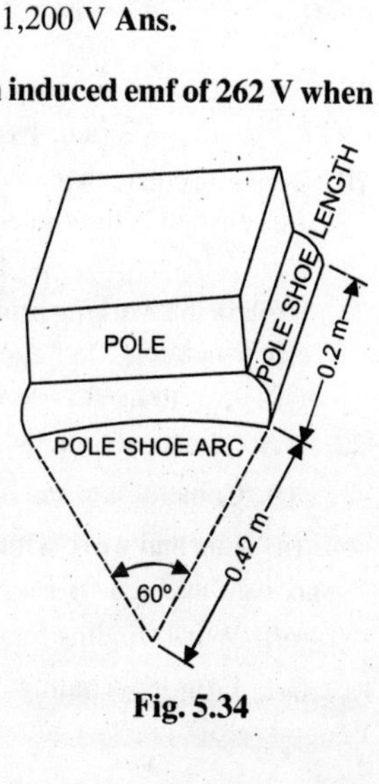

Fig. 5.34

Solution: Number of parallel paths,

A = P = 4

Flux per pole,

$$\Phi = \frac{E_g \times 60}{N \times Z} \times \frac{A}{P}$$

$$= \frac{262 \times 60}{400 \times 652} \times \frac{4}{4} = 0.0603 \text{ Wb}$$

$$\text{Pole shoe arc} = \pi \times D_a \times \frac{\theta}{360}$$

$$= \pi \times 0.42 \times \frac{60}{360} = 0.07\pi$$

Pole shoe length = 20 cm = 0.2 m

Pole shoe area, a = Pole shoe arc × pole shoe length

$= 0.07\pi \times 0.2 = 0.044 \text{ m}^2$

$$\text{Flux density in the air-gap, } B_g = \frac{\Phi}{a} = \frac{0.0603}{0.044} = 1.37 \text{ T} \quad \textbf{Ans.}$$

Example 5.24. A 4-pole dc generator has 1,200 armature conductors and generates 250 V on open circuit when running at a speed of 500 rpm. The diameter of the pole-shoe circle is 0.35 m and the ratio of pole arc to pole pitch is 0.7 while the length of the shoes is 0.2 m. Find the mean flux density in the air gap. Assume lap-connected armature winding.

[*J.N. Technological Univ. Electrical Machines-I, March-2014*]

Solution:

$$\text{Pole pitch} = \frac{\pi \times \text{bore diameter}}{\text{Number of poles}} = \frac{\pi \times 0.35}{4} = 0.275 \text{ m}$$

$$\text{Pole shoe arc} = \text{Pole pitch} \times \frac{\text{pole arc}}{\text{pole pitch}}$$

$= 0.275 \times 0.7 = 0.1925$ m

Axial length of pole shoe = 0.2 m

Area of pole face = Pole shoe arc × axial length

$= 0.1925 \times 0.2 = 0.0385 \text{ m}^2$

$$\text{Flux per pole}, \Phi = \frac{E_g \times 60}{Z \times N} \times \frac{A}{P} = \frac{250 \times 60}{1{,}200 \times 500} \times \frac{4}{4}$$

$= 0.025$ Wb

$$\text{Mean flux density, } B_{av} = \frac{\text{Flux per pole}}{\text{Pole face area}} = \frac{0.025}{0.0385}$$

$= 0.65 \text{ Wb/m}^2$ **Ans.**

Example 5.25. A 6-pole lap wound dc armature has 60 slots with 18 conductors/slot. The ratio of pole arc to pole pitch is 0.68. The diameter of bore of the pole shoe is 0.46 m. The length of the pole shoe is 0.3 m. If the air gap flux density is 0.3 Wb/m² and the emf induced in the armature is 500 V, find the speed at which it runs.

[*J.N. Technological Univ. Electrical Machines-I, March-2014*]

Solution:

$$\text{Pole pitch} = \frac{\pi \times \text{bore diameter}}{\text{Number of poles}} = \frac{\pi \times 0.46}{6} = 0.241 \text{ m}$$

$$\text{Pole shoe arc} = \text{Pole pitch} \times \frac{\text{pole arc}}{\text{pole pitch}}$$

$= 0.241 \times 0.68 = 0.1638$ m

Axial length of pole shoe = 0.3 m

Area of pole face = Pole shoe arc × axial length of pole shoe

$= 0.1638 \times 0.3 = 0.049 \text{ m}^2$

Flux per pole, Φ = Air-gap flux density × area of pole face

$= 0.3 \times 0.049 = 0.0147$ Wb

Number of armature conductors,

$Z = 60 \times 18 = 1{,}080$

Speed of machine to generate emf of 500 V,

$$N = \frac{E_g \times 60}{Z \times \Phi} \times \frac{A}{P}$$

$$= \frac{500 \times 60}{1{,}080 \times 0.0147} \times \frac{6}{6} = 1{,}890 \text{ rpm} \quad \textbf{Ans.}$$

HIGHLIGHTS

1. For dc machines in general, the outstanding advantage lies in their ability and versatility. The principal disadvantage is likely to be the initial investment concerned. Yet the advantages of dc motors are such that they retain a strong competitive position for industrial applications.

   The principal applications of dc generators are to the cases where the primary energy conversion occurs very near the point of consumption and cases where the ability to control output voltage in a prescribed manner is essential. However, the present trend is to generate ac and then convert it into dc by converters (using silicon controlled rectifiers and various other controlled power devices like TRIACs).

2. *DC machine* is commutator type. It is actually an ac machine, but furnished with a special device, called the commutator, which under certain conditions converts ac into dc and vice versa.

3. An *electrical generator* is a machine which converts mechanical energy (or power) into electrical energy (or power). It operates on the principle of electromagnetic induction.

4. In construction a dc machine consists of four parts mainly

   1. Field magnets 2. Armature

   3. Commutator and 4. Brushes and brush gear.

   Field system is stationary part and it is to create a uniform magnetic field, within which the armature rotates.

   Armature is a rotating part, where mechanical energy is converted into electrical energy (in case of generator) or, conversely, electrical energy into mechanical energy (in case of motor).

   Brushes are meant for collecting current from the commutator and supply it to the external load circuit.

5. EMF equation for a generator is given as

$$E_g = \frac{\Phi Z N}{60} \times \frac{P}{A} \text{ volts; } A = 2 \text{ in wave-wound machines}$$

$$A = P \text{ in lap-wound machines.}$$

EXERCISES

1. There is no parallel to dc machine. Justify.
2. Explain the construction and working of dc generator.
   [M.D. Univ. Electromechanical Energy Conversion, May-2009]
3. With the help of neat diagram, explain the working principle of a dc generator. Explain clearly the function of commutator in dc machines.
   [Rajasthan Technical Univ. Electrical Machines-I, 2007]
4. What is the principle of operation of a dc generator? Why is a commutator and brush arrangement necessary for the operation of a dc generator. **[J.N. Technological Univ. Electrical Machines-I, March-2014]**
5. Give constructional features of a dc machine. Describe its working principle explaining how the commutator converts alternating voltage of the armature into direct voltage.
   [G.B. Technical Univ. Electromechanical Energy Conversion-I, 2011-12]
6. Describe the essential parts of dc machines regarding their construction. **[Rajasthan Technical Univ. June/July-2011]**
7. Give the materials and functions of the following parts of a dc machine:
   (*i*) Field poles (*ii*) Yoke (*iii*) Commutator (*iv*) Commutating poles (*v*) Armature
   [U.P. Technical Univ. Electromechanical Energy Conversion-I, 2008-09]
8. Describe the constructional details and working principles of 2-pole and 4-pole dc machines.
   [G.B. Technical Univ. Electromechanical Energy Conversion-I, 2012-13]
9. Draw the neat sketch of a dc generator. State the functions of each part. **[J.N. Technological Univ. Electrical Machines-I, March-2014]**
10. Describe with suitable sketches, the main parts of a dc machine. Explain the main functions of each part making specific reference to the properties of the material used for the construction of each part.
    [M.D. Univ. Electrical Machines-I, May-2009; December-2009]
11. Explain lap winding with suitable example.
    [U.P. Technical Univ. Electromechanical Energy Conversion-I, 2008-09]
12. Illustrate with suitable diagrams the patterns of (*i*) lap winding and (*ii*) wave winding of the armature of a dc machine. Explain the relative merits and the applications of the two types of windings.
13. Define coil pitch, back pitch, front pitch, pole pitch, commutator pitch, resultant pitch, multiplex winding and degree of re-entrant of a winding, with respect to a dc machine.
    [Rajasthan Univ. Electrical Machines-I, 2005]
14. Explain why equalizing connections are used in lap winding and dummy coils are sometimes used in wave winding.
    [Rajasthan Univ. Electrical Machines-I, 2004]
15. What is meant by duplex winding? If a duplex winding does not close after one passage around the armature, is the number of commutator segments even or odd. Explain.
16. Find the expression for the equivalent armature resistance of a P-pole, lap wound armature, when there are total Z conductors, the length of each conductor being l cm. Assume cross-sectional area as a cm^2 and specific resistance of a conductor material as ρ ohm-cm.
17. Derive the emf equation of a dc generator. Discuss the factors affecting the generated emf. **[J.N. Technology Univ. Electrical Machines-I, March-2014]**
18. Derive the expression for generated emf in a dc generator.
    [Electrical Machines-I; Rajasthan Univ. 2004; Rajasthan Technical Univ. 2010, January 2013]
19. Derive the emf equation of a dc generator.
    [R.G. Technical Univ. Electrical Machine-II, December-2012]
20. Derive an equation for induced emf of a dc machine. Which quantities are constant in the emf equation?
    [U.P. Technical Univ. Electrical Machines, 2010-11]
21. Explain the working principle and derive generated emf equation of a dc generator.
    [Gujarat Technological Univ. Electrical Machines-I, Summer-2014]
22. Write short notes on the following:
    (*i*) Construction and function of commutator.
    (*ii*) Lap and wave windings.
    [M.D. Univ. Electrical Machines-I, May-2009; R.G. Technical Univ. December-2013]
    (*iii*) Wave winding merits and demerits over lap winding.
    (*iv*) Frog-leg winding.

SHORT ANSWER TYPE QUESTIONS WITH ANSWERS

Q. 1. On what principle does the generator operate?

Ans. The generator operates on the principle of production of dynamically induced emf.

Q. 2. Draw the sketch and show the path of magnetic flux in a 2-pole dc machine.

Ans. The dotted lines given in the Fig. 1 indicate the mean flux path through the complete magnetic circuit. It will be seen that the flux divides through two paths from each pole through the yoke.

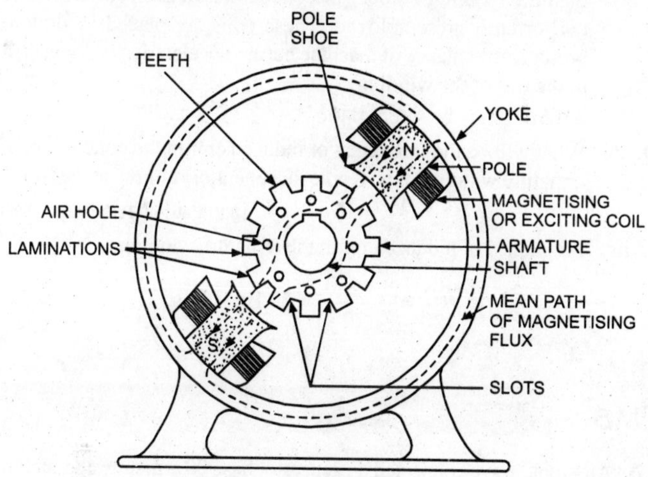

Fig. 1

Q. 3. Why it is necessary to use a large number of coils and commutator segments with the coils evenly distributed around the surface of the armature of a dc machine?

Ans. It is done so as to have practically constant unidirectional voltage at the brushes.

Q. 4 Why do the coils used in commercial generators consist of several turns in series?

Ans. It is done so as to increase the magnitude of generated emf, being in direct proportion to the number of turns in the coil.

Q. 5. Why are electromagnets preferred over permanent magnets for use in large dc machines?

Ans. On account of their greater magnetic effect and field strength regulation.

Q. 6. Why are cast iron yokes used in small dc machines?

Ans. On account of low cost.

Q. 7. Why are fabricated steel yokes used in large machines?

Ans. On account of high permeability thereby resulting in reduced weight.

Q. 8. What for field coils are provided in a dc machine?

Ans. Field coils are provided to give, under various conditions of operation, the number of ampere-turns of excitation required to give proper flux through the armature to induce the emf.

Q. 9. Write the standard value of flux density in each part of the magnetic circuit of a standard machine.

Ans. Standard values of flux density taken for a medium size dc machine is about 0.8 T in the air gap. In the field poles, yoke and armature it is about 1 T and in teeth portion it is taken about 1.5 to 2.0 T.

Q. 10. What are the materials used for the following parts of a dc machine, explain their functions. (*i*) Rotor core (*ii*) Commutator bars (*iii*) Yoke (*iv*) Brushes (*v*) Pole shoes.

[G.G.S.I.P. Univ. Delhi May 2010]

Ans. (*i*) **Rotor core** is made from high permeability silicon steel stampings. Its function is to provide a path of very low reluctance to the magnetic flux.

(*ii*) **Commutator bars** are built up of hard drawn copper or drop forged copper. The main function of commutator are:

1. It provides the electrical connections between the rotating armature coils and the stationary external circuit.
2. As the armature rotates, it performs a switching action reversing the electrical connections between the external circuit and each armature coil in turn so that the armature coil voltages add together and result in a dc output voltage.
3. It also keeps the rotor or armature mmf stationary in space.

(*iii*) **Yoke** is made of cast iron in case of small machines but fabricated steel in case of large machines. It acts as a frame of the machine and carries the magnetic flux produced by poles.

(*iv*) **Brushes** are made of carbon, carbon graphite, metal graphite or copper. The function of brushes is to collect current from the commutator and supply it to the external load circuit.

(*v*) **Pole shoes** are made of thin sheets of steel. The pole shoe acts as a support to the field coils and spreads out the flux over the armature periphery more uniformly and also being of larger cross section reduces the reluctance of the magnetic path.

Q. 11. Why is the pole shoe section of a dc machine made larger than its body?

Ans. The pole shoe acts as a support to the field coils and spreads out the flux over the armature periphery more uniformly and due to larger x-section area reduces the reluctance of the magnetic path.

Q. 12. Why are the pole shoes in a dc machine laminated though the field winding carries a direct current and not an alternating current.

[W.B. Univ. of Technology, 2013]

Ans. The pole faces or pole shoes are always laminated to avoid heating and eddy current losses caused by the fluctuations in the flux distribution on the pole face due to movement of armature slots and teeth.

Q. 13. Why is the armature core of a dc machine laminated?

Ans. To reduce the eddy current loss.

Q. 14. What is the function of armature in a dc generator?

Ans. The function of armature is to rotate the conductors in a uniform magnetic field and provide a path of very low reluctance to the magnetic flux.

Q. 15. Why is the air gap between the pole-pieces and the armature kept very small?

Ans. To reduce the reluctance of the magnetic path.

Q. 16. What is the function of commutator in a dc machine?

[M.D. Univ. Electrical Machines-I, December-2013]

Ans. Refer to Answer of Q. 10.

Q. 17. What is the best-suited material for commutator segments?

Ans. Silvered copper (extruded copper) containing about 0.05% silver.

Q. 18. What for brushes are employed in dc machines?

Ans. The brushes are employed to collect current from the commutator and supply it to the external load circuit.

Q. 19. Why are the graphite or carbon brushes preferred over copper brushes for use in dc machines?

Ans. Due to self lubricating property.

Q. 20. Why are the brushes staggered?

Ans. The brushes are staggered in order to prevent ridge formation on the surface of the commutator.

Q. 21. Why fractional pitch winding is preferred over full pitch winding?

Ans. To affect saving in the copper of end connections and to have improved commutation.

Q. 22. What are the different types of armature windings commonly used in dc machines?

Ans. Lap and wave windings.

Q. 23. For what type of dc machine, wave winding is employed?

Ans. For low current and high voltage rating machine.

Q. 24. For what type of dc machine lap winding is employed?

Ans. For high current and low voltage rating machine.

Q. 25. A particular application employing a dc generator requires high voltage and low current. Should the machine be lap wound or wave wound.

Ans. Wave wound.

Q. 26. In what type of armature winding equalizer connections are used?

Ans. In lap winding.

Q. 27. What is the use of equaliser rings?

Ans. Equaliser rings are used in lap wound machines to avoid unequal distribution of current at the brushes thereby assist in achieving sparkless commutation.

Q. 28. A dummy coil is in the armature winding of a dc machine. Is the machine lap wound or wave wound?

Ans. Wave wound.

Q. 29. Explain the role/purpose of the dummy coils in the dc machine winding. **[R.T. Univ. Electrical Machines-I, January-2013]**

OR

Explain dummy coils in dc machines. **[G.B. Technical Univ., 2012-13]**

Ans. The wave winding is possible only with particular number of conductors but sometimes the standard armature punchings available in armature winding shops do not meet with the requirements of the winding as these may accommodate more than the required number of conductors. In such cases *dummy coil* or *coils* are employed. These coils are placed in slots to preserve the balance of machine but are not electrically connected to the rest of the winding.

Art 5.7.1 may be referred too.

Q. 30. What will be the frequency of induced emf in the conductors of armature winding of an 8 pole dc generator running at 750 rpm? **[B.P. Univ. of Technology, 2008]**

Ans. Frequency of induced emf in the armature winding,

$$f = \frac{PN}{120} = \frac{8 \times 750}{120} = 50 \text{ Hz}. \quad \textbf{Ans.}$$

PROBLEMS

1. Draw the developed diagram of a simple two layer progressive lap winding for a 4-pole generator with 12 slots, each slot containing two coil sides. Indicate the position of brushes. Determine (*i*) the front pitch (*ii*) the back pitch (*iii*) the resultant pitch (*iv*) the number of brushes (*v*) the number of commutator segments for the machine.
 [Ans. (*i*) 5 (*ii*) 7 (*iii*) 2 (*iv*) 4 (*v*) 12**]**

2. Calculate the different pitches for a winding with 21 slots, 21 commutator bars and 4 poles. Winding is simplex wave.
 [Ans. $Y_b = Y_f = 11$; $Y_C = 11$; $Y_P = 11$**]**

3. Calculate the different pitches for a 4-pole lap wound dc machines having 24 slots. Each slot has 2 coil sides.
 [Ans. $Y_f = 11$; $Y_b = 13$; $Y_R = 2$; $Y_P = 12$; $Y_C = 1$**]**

4. Work out the winding details of a 4-pole simplex wave wound dc armature having 21 slots and 4 coil sides per slot. Draw the sequence diagram for the winding.
 [Ans. $Y_f = 21$; $Y_b = 21$; $Y_C = 21$; $Y_P = 21$**]**

5. A separately excited dc generator has an armature emf of 100 V when the flux per pole is 20 mWb and the speed is 800 rpm. Calculate the generated emf with the same flux and the speed at 1500 rpm.
 [Rajasthan Technical Univ. Jan./Feb.-2013]
 [Ans. 187.5 V**]**

6. A 4-pole dc generator has 51 slots and each slot contains 20 conductors. The machine has a useful flux of 0.007 Wb and runs at 1,500 rpm. Find the induced emf, if the machine is (*i*) lap connected (*ii*) wave connected.
 [U.P. Technical Univ. Electromechanical Energy Conversion-I, 2001-02]
 [Ans. (*i*) 178.5 V (*ii*) 357 V**]**

7. An 8-pole wave connected dc generator has 900 armature conductors and a flux per pole of 0.04 Wb. At what speed it must be driven to generate 500 V?
 [R.G. Technical Univ. Jan./Feb. 2007, Nov./Dec.-2007; December-2010]
 [Ans. 208.3 rpm**]**

8. A 4-pole lap wound dc shunt generator having 80 slots with 10 conductors per slot generates at no load an emf of 400 V, when run at 1,000 rpm. How will you obtain a generated open-circuit voltage of 220 V? **[Nagpur University, November-1996]**
 [Ans. 16.5 mWb; 550 rpm**]**

9. The armature of an 8-pole dc generator has 960 conductors and runs at 400 rpm. The flux per pole is 40 mWb. (*i*) Calculate the induced emf, if the armature is lap wound. (*ii*) At what speed should it be driven to generate 400 V, if the armature were wave-connected?
 [Ans. (*i*) 256 V (*ii*) 156.25 rpm**]**

10. A lap wound dc shunt generator having 80 slots with 10 conductors per slot generates at no load an emf of 400 V, when running at 1,000 rpm. Find out the flux per pole. If this generator is required to generate a voltage of 220 V on open-circuit, at what speed it should be rotated?
 [Ans. 30 mWb, 550 rpm**]**

11. Calculate the flux per pole required for a 4-pole generator with 360 conductors generating 250 V at 1,000 rpm, when armature is (*i*) lap-wound (*ii*) wave-wound. **[Ans.** 41.66 mWb; 20.83 mWb**]**

12. The armature of a four pole dc shunt generator is lap wound and generates 216 V when running at 600 rpm. Armature has 144 slots with 6 conductors per slot. If this armature is rewound, wave-connected, find the emf generated with the same flux per pole but running at 500 rpm. **[Ans.** 360 V**]**

❑❑❑

6 CHAPTER

Armature Reaction and Commutation in DC Machines

INSIDE THIS CHAPTER

6.1 Armature Reaction **6.2** Remedies For Field Distortion **6.3** Armature Ampere-Turns **6.4** Commutation **6.5** Reactance Voltage **6.6** Methods of Improving Commutation **6.7** Compensating Windings

6.1 ARMATURE REACTION

The effect of magnetic field set up by the armature current on the distribution of flux under the main poles of a dc machine (a generator or a motor) is known as *armature reaction.*

For illustration of effects of armature current on the distribution of flux under main poles, consider an armature rotating in clockwise direction in the bipolar field (bipolar field has been considered for simplicity and convenience only).

When the field winding is energized and the generator is supplying no load, the field flux is distributed, as shown in Fig. 6.1. The brushes will make contact with the conductors lying in the magnetic neutral plane, (MNP) which coincides with the geometrical neutral plane (GNP). *The plane through the axis, along which no emf is induced in the armature conductors, is known as magnetic neutral plane.* Vector OA represents main mmf both in magnitude as well as in direction, producing the main field flux.

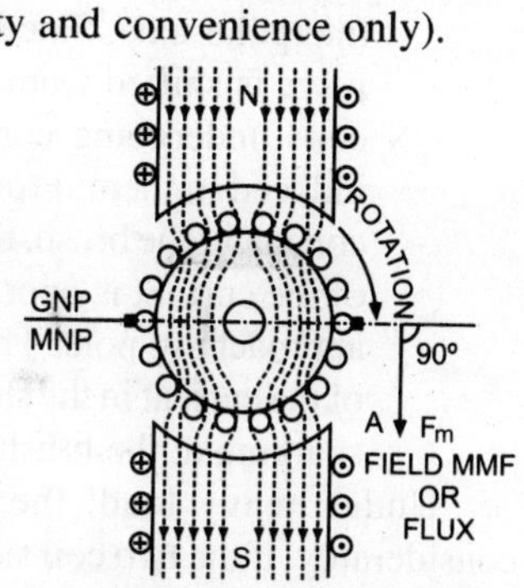

Fig. 6.1 *Distribution of Flux Due To Main Poles When Generator is Supplying No Load*

Current flowing in the armature conductors creates a magnetizing effect or mmf, that acts at right angles to the main field flux. This magnetizing action of the armature current is known as *cross-magnetization.* In Fig. 6.2 the distribution of flux due to current in armature conductors is shown, when the field coils are not energized and the direction of the current in armature conductors is the same as it would actually be when the generator is supplying load, which can be determined by applying Fleming's right hand rule *i.e.* current in all the conductors under N-pole inward (shown by ×) and outward in all those conductors lying under S-pole (shown by •). This distribution of armature current is magnetically equivalent to a coil carrying current in anticlockwise direction, which according to right hand rule or corkscrew rule produces mmf represented by OB, both in direction as well as in magnitude *i.e.* flux produced due to armature current carrying conductors will be perpendicular to the polar axis or parallel to the neutral axis. Cross-magnetization is present only when armature current is flowing and the amount of cross-magnetization so produced is proportional to the amount of current flowing through the armature conductors.

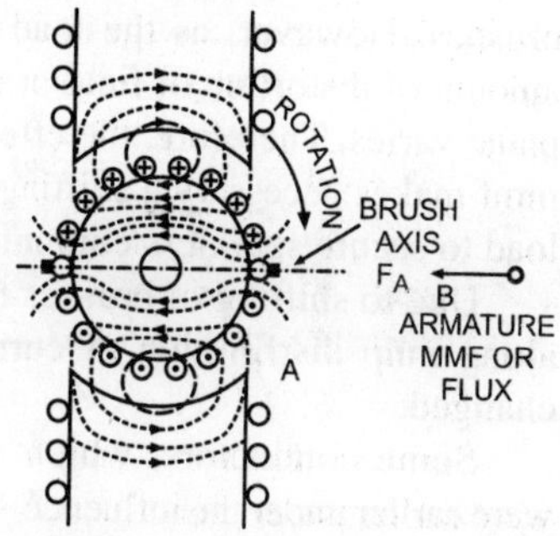

Fig. 6.2 *Distribution of Flux Due To Armature Current Carrying Conductors While Field Coils Carrying No Current*

In Fig. 6.3 distribution of resultant field, formed by the simultaneous action of the main field windings and the field produced by the cross-magnetizing action of the armature current when the armature is rotating in clockwise direction, is shown. When the generator is supplying load, the main mmf is downwards represented by vector OA and mmf produced due to armature current is from right to left represented by vector OB. The resultant mmf is OC. The magnetic neutral plane (mnp) which is always perpendicular to the resultant mmf OC, thus will be shifted round in the direction of rotation. The magnitude of the shift obviously depends upon the length of the vector OB and, therefore, on the magnitude of the armature current. Each of the lines of force produced by the armature crosses the air gap twice and it is observed that the field strength in the gap is weakened under the leading pole tips and strengthened under the trailing pole tips. With no current in the armature the flux density in the air gap is practically uniform, as represented by curve I in Fig. 6.4. With the current flowing in armature conductors the flux density varies as shown by the ordinates of curve II in Fig. 6.4.

The magnetic neutral axis will be shifted along the magnetic neutral plane, the brushes will also be shifted so as to be along

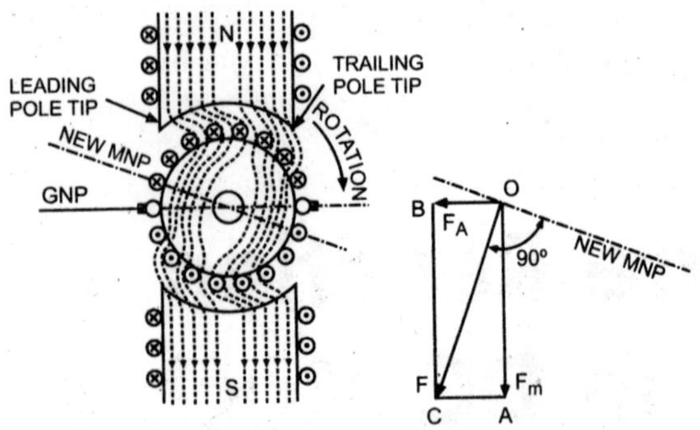

Fig. 6.3 *Distribution of Resultant Flux Established From Simultaneous Action of Field and Armature Current*

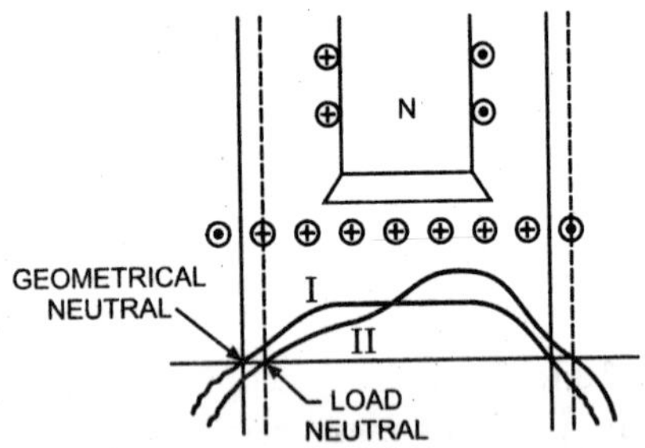

Fig. 6.4

the mnp to have sparkless commutation. If, for instance, the brushes are not shifted to be along mnp and are left along gnp the coils being short circuited by the brushes would be cutting flux and would, therefore, have an emf generated in them causing an arc to be formed as the commutator segments move under the brushes. However, as the load on the generator changes, the amount of distortion of field or shifting of the magnetic neutral plane varies. Therefore, the effect of the action of the armature mmf makes necessary a shifting of the brushes with changes in load to secure sparkless commutation.

Due to shifting of brushes (through an angle θ) so as to be along mnp distribution of current in armature conductors is changed.

Some conductors, which were earlier under the influence of S-pole come under the influence of N-pole and vice versa. The distribution of the current in the armature conductors when placed along mnp is shown in Fig. 6.5. Now the total armature mmf lies along the mnp and is no longer at right angles to the main field. The armature mmf is now no longer all cross-magnetizing in its effect but is partly demagnetizing in its action, since it is partly directed against the main field. The demagnetising component of armature mmf results in a weakened field flux, which in turn results in a lowered generated emf in the generator. The action of the armature current in establishing of mmf that distorts and weakens the field flux is called *armature reaction.*

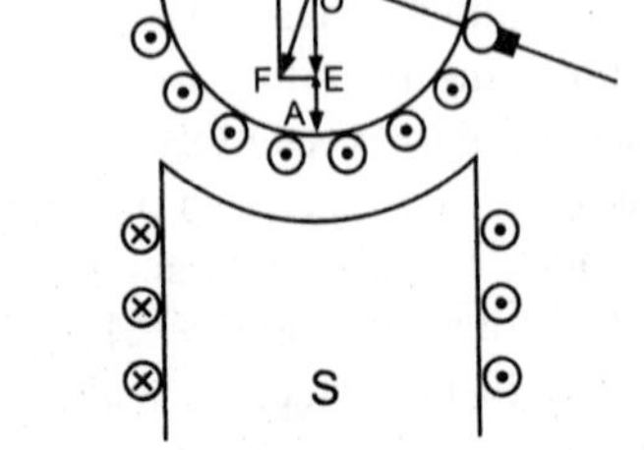

Fig. 6.5

The effects of armature mmf described above, may be summarized as below.

The path of magnetic flux established by armature current is normal to that of main field, therefore, the armature reaction is basically *cross-magnetising*. The effect depends and is directly proportional to the armature mmf and shifts the neutral plane in the same direction as that of rotation of the armature.

1. The *cross-magnetising effect* of armature reaction distorts the field in the air gap. From Figs. 6.3 and 6.4 it is obvious that there are two principal effects of the distortion of the field namely (*i*) the creation of magnetic field in the interpolar region, where the brushes are placed for commutation and (*ii*) weakened field strength in the air gap under leading pole tips and strengthened field under the trailing pole tips.

The above two effects, caused by the distortion of the main field under load, lead to increased iron losses, poor commutation or even sparking at the commutator surface. These are briefly described below:

(*i*) *Iron losses*. The iron losses depend upon the maximum value of flux density B. The increase in iron losses due to increase in flux density under one-half of the pole arc is much more than the decrease in iron losses due to decrease in flux density in the other half. The result is that the iron losses at load are more than those on no load.

(*ii*) *Commutation**. For good commutation, the coils short circuited by the brushes should have zero emf induced in them. The brushes are usually placed along the interpolar axis. Since zero crossing of the flux density wave is shifted from the interpolar axis (Fig. 6.4), the coils undergoing commutation do not have zero emf induced in them. From Fig. 6.4 it is obvious that in the coil under the brush, the emf and, therefore, the direction of the current is opposite to what it would be under the approaching pole. Thus there is a delay in the reversal of the current in the short-circuited coil which may cause sparking at the brushes.

Under heavy load, the flux density wave is distorted considerably. If the two coil sides of a coil are under the points of maximum flux density, a much greater voltage may be induced in this coil. If this voltage between adjacent commutator segments exceeds 30 or 40 V, a spark may occur between these adjacent commutator segments. Sometimes this spark may be too large that it may spread around the commutator in the form of a ring fire.

2. The demagnetising effect of armature reaction reduces the total flux per pole from its no-load value due to magnetic saturation and the reduction in flux depends upon the degree of magnetic saturation. The decrease in main flux on load due to armature reaction may be about 10%. The decrease in flux due to armature reaction on load reduces the magnitude of generated emf in case of a generator and electromagnetic torque developed in case of motor.

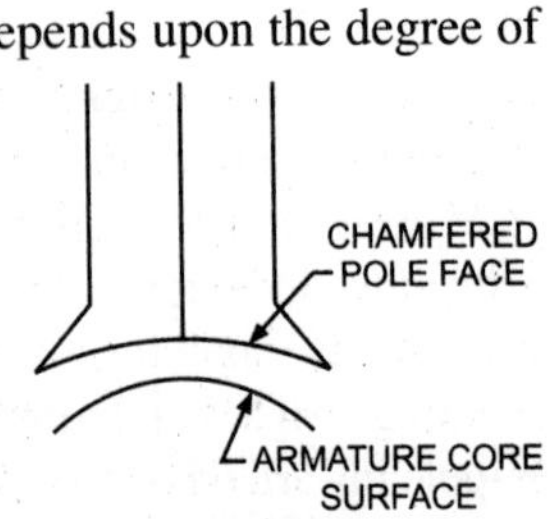

Fig. 6.6

* For details refer to Art 6.4.

3. The effect of armature reaction on the main field flux are of opposite sign for a generator and a motor because for the same direction of armature current under main poles the direction of rotation of armature for a motor is to be opposite to that in case of a generator.

6.2 REMEDIES FOR FIELD DISTORTION

The shift of the magnetic neutral plane in the direction of rotation seriously affects commutation because sparking will occur at the brushes unless they are shifted to the new magnetic neutral plane. Furthermore, the brushes must be shifted back and forth continually as the load changes because the effect of the armature reaction depends upon the value of the armature current. In practice, repeated brush shifting would, of course, be almost as objectionable as the sparking that it attempts to correct. This has led to several corrective methods given below that counter act either in part or completely, the detrimental effects of armature reaction.

1. By flattening of the pole faces slightly so that the air gap is longer at the pole tips than it is at the centre of the pole. This results in increase in reluctance at the pole tips and so in the magnitude of armature cross-flux thereby reducing the distortion of the resultant flux density wave.
2. By making the trailing horn of the pole piece longer than the advancing horn and cutting farther from the surface of the armature so as to equalise the distribution of magnetic flux.
3. The air-gap length may be increased. By doing so the reluctance increases and more ampere-turns are required in the field winding. The field distortion, however, will not be so great as it would be if the magnetic field of the machine were weakened.
4. By reducing the cross section of the pole-pieces, it becomes highly saturated and offers large reluctance to the cross-field.
5. By providing interpoles between the main poles and/or compensating windings if necessary. This will be described in detail later on.

6.3 ARMATURE AMPERE-TURNS

If the brushes lie in the geometric neutral plane, the entire armature reaction effect is cross-magnetising. However, if the brushes are shifted from the gnp, a part of the armature reaction becomes demagnetising and the remaining armature reaction is cross-magnetising.

Let the brush axis be given a forward lead of angle θ so as to lie along the new mnp represented by line AB in Fig. 6.7. If another line CD is drawn making same angle θ with gnp but in opposite direction to that made by brush axis then it is observed that all the conductors lying between angles AOC and BOD carry current in such a direction which causes mmf in opposite direction to that of main mmf. So these conductors cause demagnetising effect and are known as *demagnetising conductors*. Remaining conductors *i.e.* conductors lying between angle AOD and BOC carry current in such a direction which causes mmf in the direction perpendicular to that of main mmf, which results in distortion of main field and these conductors are known as *cross-magnetising conductors.*

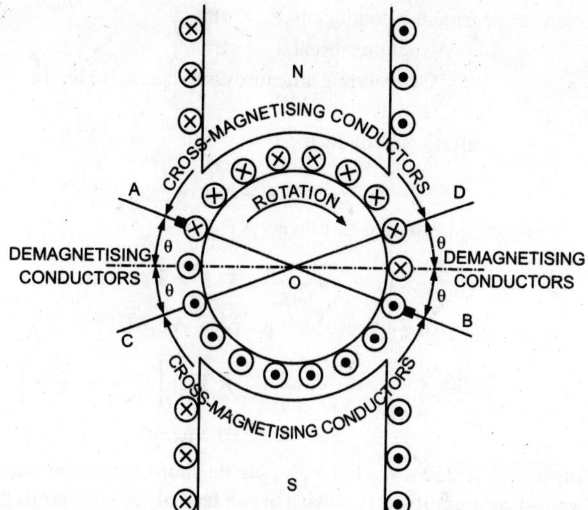

Fig. 6.7

Let total number of armature conductors = Z

Number of parallel paths = A

Angle of lead of brush from gnp = θ* in mechanical degrees

Number of poles = P

Armature current = I_a

Current per conductor, $I_c = \dfrac{I_a}{A}$

Total armature-ampere-turns = $\dfrac{\text{Total armature ampere-conductors}}{2} = \dfrac{ZI_c}{2}$

$$\text{Total armature ampere-turns/pole} = \frac{ZI_c}{2P} \quad \text{...(6.1)}$$

Total demagnetising conductors

= Conductors lying within angles AOC and BOD

$$= \frac{Z}{360}[\angle AOC + \angle BOD] = \frac{Z}{360} \times 4\theta$$

[Since ∠ AOC = ∠ BOD = 2 θ]

Total demagnetising turns

$$= \frac{1}{2} \times \frac{Z}{360} \times 4\theta = \frac{2\theta}{360} \times Z$$

[Since 1 turn = 2 conductors]

Demagnetising ampere-turns/pole, AT_d

$$= \frac{1}{2} \times \frac{2\theta}{360} ZI_c = \frac{Z\theta}{360} I_c \quad \text{...(6.2)}$$

Cross-magnetising ampere-turns/pole, AT_c

$$= AT - AT_d = \frac{ZI_c}{2P} - \frac{\theta}{360} ZI_c$$

$$= ZI_c\left(\frac{1}{2P} - \frac{\theta}{360}\right) \quad \text{...(6.3)}$$

* In a multipolar machine the electrical angle between two consecutive poles, is 180° whereas mechanical angle is 360° divided by number of poles, so $\theta_{elec} = \frac{P}{2}\theta_{mech}$.

Example 6.1. A 4-pole lap wound generator having 480 armature conductors supplies a current of 150 A. If the brushes are given an actual lead of 10°, calculate the demagnetizing and cross-magnetizing ampere-turns per pole.

Solution: Number of poles, $P = 4$

Number of parallel paths, $A = P = 4$

($\because$ generator is lap wound)

Number of armature conductors, $Z = 480$

Armature current, $I_a = 150$ A

(assuming armature current equal to load current)

Current per conductor, $I_c = \frac{I_a}{A} = \frac{150}{4} = 37.5$ A

Brush shift, $\theta = 10°$ (mechanical)

Demagnetizing ampere-turns/pole, AT_d

$$= \frac{Z\theta}{360} \times I_c = \frac{480 \times 10}{360} \times 37.5 = 500 \text{ Ans.}$$

Cross-magnetising ampere turns per pole, AT_c

$$= ZI_c\left(\frac{1}{2P} - \frac{\theta}{360°}\right) = 480 \times 37.5\left(\frac{1}{2\times4} - \frac{10}{360°}\right)$$

$$= 1{,}750 \text{ Ans.}$$

Example 6.2. A 250 kW, 400 V, 6-pole dc shunt generator has 720 lap wound conductors. It is given a brush lead of 2.5 angular degrees (mech.) from the geometric neutral axis. Calculate the cross and demagnetizing ampere-turns. Neglect shunt field current.

[*U.P. Technical Univ. Electromechanical Energy Conversion-I 2005-06*]

Solution: Load current, $I_L = \frac{P}{V} = \frac{250 \times 1{,}000}{400} = 625$ A

Armature current, $I_a = I_L = 625$ A

$\because$ shunt field current is negligible

Current per conductor, $I_c = \frac{I_a}{A} = \frac{625}{6}$ A

Angle of shift, $\theta = 2.5°$ (mechanical)

Demagnetizing AT/pole, $AT_d = ZI_c \times \frac{\theta}{360}$

$$= 720 \times \frac{625}{6} \times \frac{2.5}{360} = 521 \text{ Ans.}$$

Cross-magnetizing AT/pole, $AT_c = ZI_c\left[\frac{1}{2P} - \frac{\theta}{360}\right]$

$$= 720 \times \frac{625}{6}\left[\frac{1}{2\times6} - \frac{2.5}{360}\right]$$

$$= 5{,}729 \text{ Ans.}$$

Example 6.3. Determine (*a*) the demagnetizing ampere-turns per pole and (*b*) cross-magnetising ampere-turns per pole in a 440 V, 4-pole, 25 kW dc generator with 32 slots having 12 conductors per slot in armature winding lap connected when the brushes are given an actual shift of 10°. The generator is not provided with interpoles.

[*W.B. Univ. of Technology Electrical Machines-I, 2010*]

Solution: Load current, $I_L = \frac{P}{V} = \frac{25 \times 1{,}000}{440} = 56.8$ A

Armature current, $I_a = I_L = 56.8$ A neglecting shunt field current

Current per conductor, $I_c = \frac{I_a}{A} = \frac{56.8}{4} = 14.2$ A

Angle of shift, $\theta = 10°$ (mechanical)

Number of armature conductors, $Z = 32 \times 12 = 384$

(*a*) Demagnetising AT/pole,

$$AT_d = ZI_c \times \frac{\theta}{360} = 384 \times 14.2 \times \frac{10}{360} = 152 \text{ Ans.}$$

(*b*) Cross-magnetising AT/pole,

$$AT_c = ZI_c\left[\frac{1}{2P} - \frac{\theta}{360}\right] = 384 \times 14.2\left[\frac{1}{2\times4} - \frac{10}{360}\right] = 530 \text{ Ans.}$$

Example 6.4. A 250 V, 14.9 kW, 8-pole dc motor has single turn coils. The armature is wave wound with 94 commutator segments. If the brushes are shifted by 2 commutator segments at full load, calculate (*i*) total armature reaction ampere-turns (*ii*) demagnetising ampere-turns, (*iii*) cross-magnetising ampere-turns. Assume motor efficiency to be 80%.

[*M.D. Univ. Electrical Machines-I, December-2009*]

Solution: Motor input, $P_{in} = \frac{P_{out}}{\eta} = \frac{14{,}900}{0.8} = 18{,}625$ W

Motor input current, $I_L = \frac{P_{in}}{V} = \frac{18{,}625}{250} = 74.5$ A

Motor armature current, $I_a = I_L = 74.5$ A

assuming motor armature current equal to input current

Current per conductor, $I_c = \frac{I_a}{A} = \frac{74.5}{2} = 37.25$ A

$\because$ motor is wave wound

Number of armature conductors,

Z = Number of commutator segments × number of parallel paths

$= 94 \times 2 = 188$

Brush shift, $\theta = \frac{2}{94} \times 360° = \frac{360°}{47}$ (mechanical)

(*i*) Total armature reaction ampere-turns per pole

$$= \frac{ZI_c}{2P} = \frac{188 \times 37.25}{16} = 438 \text{ Ans.}$$

(*ii*) Demagnetising ampere-turns per pole, AT_d

$$= ZI_c \times \frac{\theta}{360°} = 188 \times 37.25 \times \frac{360}{47} \times \frac{1}{360} = 149$$

(*iii*) Cross-magnetising ampere-turns per pole, AT_c

$$= \frac{ZI_c}{2P} - \frac{ZI_c\theta}{360} = 438 - 149 = 289 \text{ Ans.}$$

Example 6.5. The brushes of a certain lap connected 400 kW, 6-pole generator are given a lead of 18° electrical. From the data given, calculate (*i*) the demagnetising ampere-turns (*ii*) the cross-magnetising ampere-turns (*iii*) series turns required to balance the demagnetising component.

The full load current is 750 A and total number of conductors are 900 and the leakage coefficient is 1.4.

Solution: Armature current, $I_a = 750$ A

Number of parallel paths, $A = P = 6$

$\because$ generator is lap wound

Current per conductor, $I_c = \frac{I_a}{A} = \frac{750}{6} = 125$ A

Angle of lead, $\theta = \frac{\text{Shift in electrical degrees}}{\text{Pair of poles}}$

$$= \frac{18}{6/2} = 6° \text{ (mech)}$$

(*i*) Demagnetising ampere-turns per pole,

$$AT_d = ZI_c \times \frac{\theta}{360} = 900 \times 125 \times \frac{6}{360} = 1{,}875 \text{ Ans.}$$

(*ii*) Cross-magnetising ampere-turns per pole,

$$AT_c = ZI_c\left(\frac{1}{2P} - \frac{\theta}{360}\right) = 900 \times 125\left(\frac{1}{2\times6} - \frac{6}{360}\right) = 7{,}500 \text{ Ans.}$$

(*iii*) *Series-turns required to balance the demagnetising component

$$= \text{Leakage coefficient} \times \frac{AT_d}{I_a} = \frac{1.4 \times 1{,}875}{750} = 3.5 \textbf{ Ans.}$$

Example 6.6. A 220 kW, 4-pole, wave-connected, 440 V shunt generator has 740 conductors and a shunt field current of 10 A. Find the demagnetizing and cross-magnetizing ampere-turns per pole, if the brushes are given a lead of 25 electrical degrees. Find the number of additional shunt field turns to neutralize the demagnetizing effect.

[*J.N. Technological Univ. Electrical Machines-I, March-2014*]

Solution: Load current, $I_L = \frac{P}{V} = \frac{220 \times 1{,}000}{440} = 500$ A

Shunt field current, $I_{sh} = 10$ A

Armature current, $I_a = I_L + I_{sh}$ ...Refer to Eq. (7.10)

$= 500 + 10 = 510$ A

Current per conductor, $I_c = \frac{I_a}{A} = \frac{510}{2} = 255$ A

$$\text{Angle lead, } \theta = \frac{\text{Brush shift in electrical degrees}}{\text{Pair of poles}}$$

$$= \frac{25^\circ}{4/2} = 12.5^\circ \text{ (mechanical)}$$

Demagnetizing ampere-turns per pole,

$$AT_d = Z \times I_c \times \frac{\theta}{360^\circ}$$

$$= \frac{740 \times 255 \times 12.5^\circ}{360^\circ} = 6{,}552 \textbf{ Ans.}$$

Cross-magnetizing ampere-turns per pole,

$$AT_c = ZI_c - \frac{ZI_c\,\theta}{360^\circ}$$

$$= 740 \times 255 - \frac{740 \times 255 \times 12.5^\circ}{360^\circ}$$

$$= 188{,}700 - 6{,}552 = 182{,}148 \textbf{ Ans.}$$

Additional shunt field turns required to neutralize the demagnetizing effect

$$= \frac{AT_d}{I_{sh}} = \frac{6{,}552}{10} = 655 \textbf{ Ans.}$$

Example 6.7. A 4-pole generator supplies a current of 143 A. It has 492 armature conductors (*a*) wave wound (*b*) lap wound. When delivering full load, the brushes are given an actual lead of 10°. Calculate the demagnetising ampere-turns per pole. This field winding is shunt connected and takes 10 A. Find the number of extra shunt field turns to neutralize the demagnetization.

[*Nagpur Univ; 1993; JNTU Hyderabad 2000*]

Solution: Armature current, $I_a = I_L + I_{sh} = 143 + 10 = 153$ A

Brush shift, $\theta = 10^\circ$ (mechanical)

(*a*) **When the generator is wave wound**

Number of parallel paths, $A = 2$

Current per conductor, $I_c = \frac{I_a}{A} = \frac{153}{2} = 76.5$ A

Demagnetising ampere-turns per pole,

$$AT_d = ZI_c \times \frac{\theta}{360^\circ} = 492 \times 76.5 \times \frac{10}{360} = 1{,}045.5 \textbf{ Ans.}$$

Extra shunt field turns required

$$= \frac{AT_d}{I_{sh}} = \frac{1{,}045.5}{10} = 105 \text{ (approximately) } \textbf{Ans.}$$

(*b*) **When the generator is lap wound**

Number of parallel paths, $A = P = 4$

Current per conductor, $I_c = \frac{I_a}{A} = \frac{153}{4} = 38.25$ A

Demagnetising ampere-turns per pole,

$$AT_d = ZI_c \times \frac{\theta}{360^\circ}$$

$$= 492 \times 38.25 \times \frac{10}{360} = 523 \textbf{ Ans.}$$

$$\text{Extra shunt field turns required} = \frac{AT_d}{I_{sh}} = \frac{523}{10}$$

$= 52$ (approximately) **Ans.**

6.4 COMMUTATION

As pointed out earlier, the commutator and its accompanying brushes are very important parts of a dc machine. Two essential actions take place here, namely, the passage of a current from (or to) a moving armature to (or from) external circuit and the commutation process. The commutation process involves the change from a *generated alternating current* to an *externally available direct current.* The transfer of current from the rotating armature to the stationary brushes (and hence to the load) involves a continuously moving contact. Both these actions are to be carefully controlled by the use of suitable materials, good design and proper adjustments, otherwise there would be serious arcing and possible breakdown of the machine.

As one of the armature coil rotates between a positive brush and the succeeding negative brush, current flows in that coil in one direction. Then this coil is short circuited by a brush for an extremely short fraction of a second (usually about $\frac{1}{500}$ second), after which it passes into a region between a negative brush and the succeeding positive brush, where the current reverses. This reversal of current in the armature coil by means of brush and commutator bars, is called the *commutation process* and the period during which the coil remains short circuited is called the *commutation* period (represented by T_c).

Good commutation means no sparking at the brushes and with commutator surface remaining unaffected during continuous

---

* For neutralizing the demagnetising effect of armature reaction, an extra number of turns may be put on each pole determined from the following expressions :

$$\text{Number of extra turns/pole} = \frac{AT_d}{I_{sh}} \text{ for shunt wound generator.}$$

$$\text{Number of extra turns/pole} = \frac{AT_d}{I_a} \text{ for series wound generator.}$$

In case the leakage coefficient is given, multiply each of the above expressions by it.

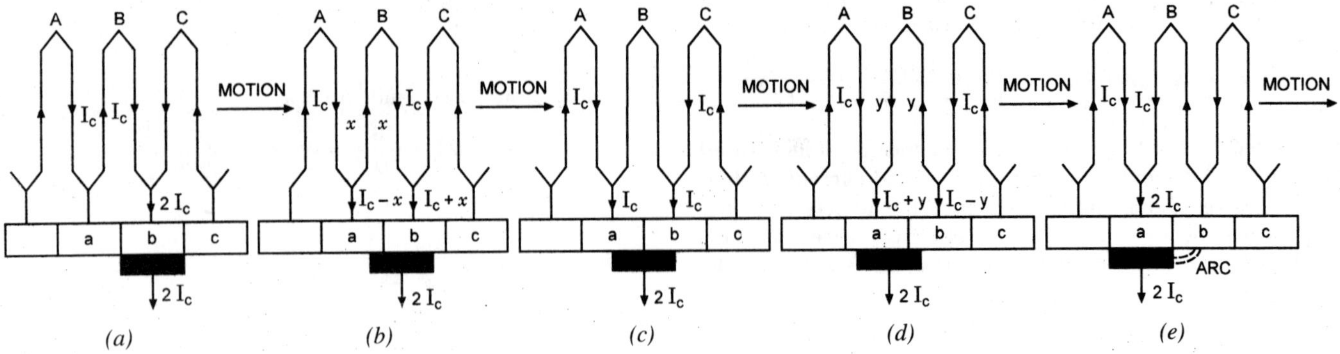

Fig. 6.8 *Commutation Process*

operation of the dc machine. A machine is said to have *poor commutation* if there is sparking at the brushes and the commutator surface gets damaged during the machine operation. Sparking results in overheating at the commutator-brush contact and pitting of the commutator. The effects are cumulative, since any sparking impairs the contact which increases the heating and thus aggravates the situation so that the sparking becomes progressively worse. For satisfactory commutation the current in a coil undergoing commutation must be completely reversed during its commutation period.

The poor commutation may be caused by mechanical or electrical conditions. The mechanical conditions include uneven commutator surface, non-uniform brush pressure, vibrations of brushes in the holders etc. The electrical conditions include an increase in the voltage between the commutator segments, an increase in the current density at the trailing edge of the brush etc.

In order to have physical concept of the commutation process, let us consider that the machine has a ring winding, a part of which is shown in Fig. 6.8. For simplicity it is assumed that the width of the brush is equal to the width of the commutator bar and thickness of the insulation between the commutator bars is negligible. Let the current per conductor be of I_c amperes and the armature is moving from left to right and concentrate on coil B.

In Fig. 6.8 (*a*) the brush is in contact with only bar *b* of the commutator. All the current from the armature path, which is to the left of the brush, passes through coil B, and all of the current from the armature path which is to the right of the brush passes through coil C. Commutator bar *b* conducts the combined current $2\,I_c$ of the two paths (each path carrying current of I_c amperes) to the brush. At instant later the conditions are as illustrated in Fig. 6.8 (*b*). The brush is just starting to be in contact with bar *a*. There are now two paths to the brush for the armature current coming from the armature path, through commutator bars *a* and *b*. As the area of contact with bar *a* increases, current flowing to the brush via bar *a* increases and current via bar b decreases. The total current collected by the brush remains unchanged *i.e.* $2\,I_c$ amperes. If the current distribution between bars *a* and *b* were determined by the areas of contact only, the current through coil B would decrease linearly. When the brush contact areas are equal on bars *a* and *b*, each bar delivers a current of I_c amperes to the brush and coil B carries no current, as illustrated in Fig. 6.8 (*c*). The brush contact area on bar *b* starts to decrease with further rotation of armature and the current delivered by the bar *b* decreases and the coil B carries a current of *y* amperes in counter-clockwise direction (opposite to that of original), as shown in Fig. 6.8 (*d*). When the brush breaks contact with bar *b* and is totally on bar *a*, the short circuit of coil B is removed and coil B carries a current of I_c amperes in counter-clockwise direction. Thus during the period of short circuit, which is also known as *period of commutation,* the current in short-circuited coil should be reversed to full value.

If the current in coil B does not attain its full value in reverse direction by the end of short circuit, the difference between the current through coil B and coil C has to jump from the commutator bar '*b*' to brush in the form of an arc. Thus the cause of sparking at the commutator is the failure of current in the short-circuited coil to reach its full value in the reverse direction by the end of the short-circuit period.

The reversal of current through a coil undergoing commutation may occur in any of the manners illustrated in Fig. 6.9.

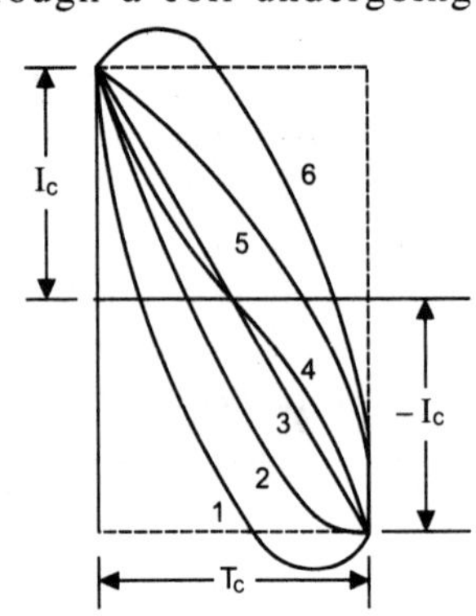

Fig. 6.9

Curve 1 illustrates the *sharply accelerated commutation.* In this case the current may reach its proper final value without a spark but it involves very high localized current densities at the brush contact leading to sparking and excessive heating and deterioration of brush.

Curve 2 illustrates the *accelerated* or *over-commutation* where the current attains its final value with a zero rate of change at the end of commutation period. Usually it results in satisfactory commutation.

Curve 3 illustrates the *linear commutation* where change of current from $+I_c$ to $-I_c$ is uniform. It is an ideal commutation and it gives rise to uniform current density at the brush contact surface and the power loss at the brush contact is minimum.

Curve 4 illustrates *sinusoidal commutation* which generally results in satisfactory commutation.

Curve 5 is a case of *retarded* or *under-commutation.* In this case final rate of change of current is very high. Under such conditions sparking at the trailing edge of the brush is inevitable.

Curve 6 represents *sharply retarded commutation.* This type of commutation also results in excessive current density under the brushes and inevitable sparking.

For *satisfactory commutation* it is essential that the current in a coil undergoing commutation is completely reversed during its commutation period—very short time interval. The main cause which retards or delays the reversal of current through a coil is its inductive property. It may be pointed out that the coil possesses appreciable amount of self-inductance because it contains several turns of wire and lies embedded in the armature which is built up of a material of high magnetic permeability. The inductive property of a coil gives rise to a voltage called the *reactance voltage* which opposes the reversal of current in it. This voltage, even though of a small magnitude, produces a large current through the coil whose resistance is very small due to short circuit. It is noteworthy point that if the brushes are set so that the coils undergoing short circuit are in the magnetic neutral plane, where they are cutting no flux and hence no emf induced in them due to rotation of armature, there will be an emf of self-induction causing severe sparking at the brushes. To offset this reactance voltage it is necessary to produce in each coil, as it passes through its commutating period, a voltage which will help in reversal of current. Such a voltage is called the *commutating emf.* A satisfactory commutation can be produced (*i*) by shifting the brushes in non-interpolar machines, beyond the load magnetic neutral, so that a voltage will be induced in the coil in the direction in which the current is about to flow or (*ii*) strengthen the interpoles more than that is necessary to neutralize the armature reaction flux. For details refer to the forthcoming Arts.

6.5 REACTANCE VOLTAGE

As already mentioned above, the difficulty experienced by the current in attaining its full value in the reverse direction by the end of the short-circuit period is due to the production of self-induced emf, called the *reactance voltage* in the coil undergoing commutation.

Since the current in the coil has to change from $+ I_c$ to $- I_c$ i.e., the total change is $2I_c$. If W_b is the brush width and W_m is the thickness of mica insulation and v_c is the velocity of commutator then the time of commutation or short-circuit is the time required by the commutator in moving a distance equal to circumferential thickness of brush minus the thickness of mica insulation between two commutator segments or

$$T_c = \frac{W_b - W_m}{v_c} \text{ seconds} \quad ...(6.4)$$

If the self-inductance of the coil is L henries, then the value of self-induced emf or reactance voltage

$$= L\frac{di}{dt} = L\frac{2I_c}{T_c} = L\frac{2I_c}{W_b - W_m} v_c \text{ volts}$$

in case of linear commutation ...(6.5)

$$= 1.11\, L \times \frac{2I_c}{T_c} = 2.22\, L \,.\, \frac{I_c}{W_b - W_m} v_c$$

volts in case of sinusoidal commutation ...(6.6)

The self-induction, L is a composite quantity being made up of the true self-induction of the short-circuited coil and the mutual induction of its neighbouring coils. An approximate value for L can be determined by Hobart's formula.

Hobart gave an experimental rule for determination of self-inductance of a coil. He found from experiments that for slots of normal dimensions (depth being 3.5 times the width) a current of 1 amperes flowing through a single turn coil produces a flux of 4 μ*Wb* per metre of embedded length and 0.8 μ*Wb* per metre of free length. So for a coil of free length l_f and embedded length l_e flux per ampere due to the whole coil

$$= (4\, l_e + 0.8\, l_f) \times 10^{-6}\, Wb \text{ for a single turn coil}$$

$$= (4\, l_e + 0.8\, l_f) \times n \times 10^{-6}\, Wb \text{ for a } n\text{-turn coil}$$

Self-induction of coil

$$= (\text{Flux/ampere}) \times \text{number of turns}$$

$$= n^2 (4\, l_e + 0.8\, l_f) \times 10^{-6}\, H$$

Each side of the coil will lie in a slot in which another side at the same instant is undergoing commutation under some other brush, so to take into account the effect of change of flux in the neighbouring coil, the flux due to embedded lengths are doubled.

$$\therefore \text{Total flux/amp} = (8\, l_e + 0.8\, l_f) \times n \times 10^{-6}\, Wb$$

for *n* turn coil ...(6.7)

$$\text{and } L = n^2 (8\, l_e + 0.8\, l_f) \times 10^{-6}\, H \quad ...(6.8)$$

The value of L determined above is true only when the brush spans only one commutator segment *i.e.* only one coil is short circuited at a time. If the brush width is such that it short circuits *m* coils at a time then self inductance L is given by

$$L = mn^2 (8\, l_e + 0.8\, l_f) \times 10^{-6}\, H \quad ...(6.9)$$

$$\text{and reactance voltage} = L\frac{di}{dt}$$

$$= mn^2 (8\, l_e + 0.8\, l_f) \frac{2I}{T_c} \times 10^{-6} \text{ volts} \quad ...(6.10)$$

The above expression shows that the reactance voltage is proportional to the square of the number of turns in an armature coil. So for good commutation single turn coils should be used. This can be done in large machines, where the large number of commutator segments required by such a winding does not make the segments too narrow, but it is impossible with small machines.

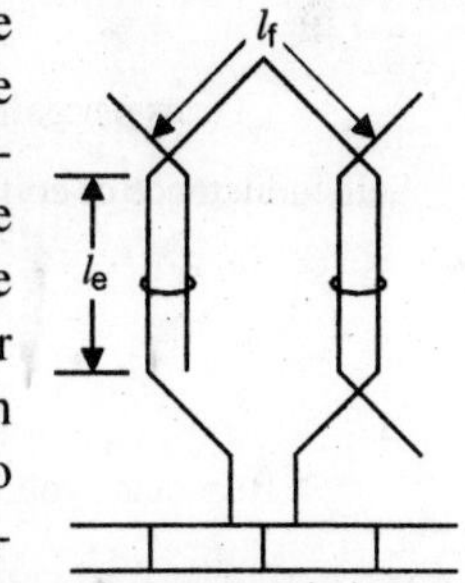

Fig. 6.10

Example 6.8. A 4-pole wave wound dc machine running at 1,500 rpm has a commutator of 30 cm diameter. If armature current is 150 A, thickness of brush 1.25 cm and the self inductance of each armature coil is 0.07 mH. Calculate the average emf induced in each coil during commutation. Assume linear commutation.

[Rajasthan Univ. Electrical Machines-I, 2003]

Solution: Peripheral velocity of commutator,

$$v_c = \frac{\pi DN}{60} = \frac{\pi \times 0.3 \times 1{,}500}{60} = 23.562 \text{ m/s}$$

Neglecting mica insulation, which is not given

$$\text{Commutation time, } T_c = \frac{\text{Brush width, } W_b}{v_c} = \frac{1.25 \times 10^{-2}}{23.562}$$

$$= 5.305 \times 10^{-4} \text{ second}$$

$$\text{Inductance of coil, } L = 0.07 \text{ mH} = 7 \times 10^{-5} \text{ H}$$

$$\text{Current per conductor, } I_c = \frac{\text{Armature current, } I_a}{\text{Number of parallel paths, A}}$$

$$= \frac{150}{2} = 75 \text{ A} \quad \because \text{ machine is wave wound}$$

$$\text{Average induced emf} = L \times \frac{2I_c}{T_c} = \frac{7 \times 10^{-5} \times 2 \times 75}{5.305 \times 10^{-4}}$$

$$= 19.8 \text{ V } \textbf{Ans.}$$

Example 6.9. Calculate the reactance voltage for a 4-pole lap wound generator if speed in rpm = 300; diameter of commutator = 1.12 m; number of commutator segments = 450; brush width = 2.25 cm; length of each conductor = 1m; effective length of core = 0.3 m; turns per commutator segment = 2, full load armature current = 900 A.

Solution: Current per conductor, $I_c = \frac{I_a}{A} = \frac{900}{4} = 225$ A

Peripheral velocity of conductor, $v_c = \frac{\pi DN}{60} = \frac{\pi \times 1.12 \times 300}{60}$

$= 17.6$ m/s

Period of commutation, $T_c = \frac{W_b - W_m}{v_c} = \frac{0.0225}{17.6}$

$= 0.0012784$ second

Free length of conductor $= 1 - 0.3 = 0.7$m

Free length per turn, $l_f = 2 \times 0.7 = 1.4$ m

Embedded length per turn, $l_e = 2 \times 0.3 = 0.6$ m

Segment pitch $= \frac{\pi D}{\text{Number of segments}}$

$= \frac{\pi \times 1.12}{450} = 0.00782$ m

Number of segments short-circuited at a time,

$m = \frac{0.0225}{0.00782} = 3$

Turns per coil, $n = 2$

Self-inductance of coil, $L = mn^2 (8\, l_e + 0.8\, l_f) \times 10^{-6}$ H

$= 3 \times 2^2 (8 \times 0.6 + 0.8 \times 1.4) \times 10^{-6}$

$= 71.04 \times 10^{-6}$ H

Reactance voltage $= L \times \frac{2I_c}{T_c} = \frac{71.04 \times 10^{-6} \times 2 \times 225}{0.0012784}$

$= 25$ volts **Ans.**

Example 6.10. Calculate the reactance voltage for a machine having the following particulars:

Number of commutator segments = 55;
Revolutions per minute = 900
Brush width in commutator segments = 1.74;
Co-efficient of self induction = 153 × 10^{-6} H
Current per coil = 27 A

Solution: Current per coil, $I_c = 27$ A

Self-inductance of coil, $L = 153 \times 10^{-6}$ H

Peripheral velocity of commutator, v_c

$= \text{Number of commutator segments} \times \frac{N}{60}$

$= 55 \times \frac{900}{60} = 825$ segments/second

Commutation period, $T_c = \frac{W_b}{v_c} = \frac{1.74}{825}$ second

Assuming linear commutation

Reactance voltage

$= L \times \frac{2I_c}{T_c} = \frac{153 \times 10^{-6} \times 2 \times 27}{1.74/825} = 3.92$ V **Ans.**

Example 6.11. A 4-pole, lap-wound armature running at 1,500 rpm delivers a current of 150 A and has 64 commutator segments. The brush spans 1.2 segments and inductance of each armature coil 0.05 mH. Calculate the value of reactance voltage assuming (*i*) linear commutation (*ii*) sinusoidal commutation. Neglect mica thickness.

[*Rajasthan Univ. Electrical Machines, 2000*]

Solution: Current per coil, $I_c = \frac{I_a}{P} = \frac{150}{4} = 37.5$ A

Self-inductance of coil, $L = 0.05$ mH $= 5 \times 10^{-5}$ H

Peripheral velocity of commutator,

$v_c = \text{Number of commutator segments} \times \frac{N}{60}$

$= 64 \times \frac{1{,}500}{60} = 1600$ segments/second

Commutation period, $T_c = \frac{W_b}{v_c} = \frac{1.2}{1{,}600} = 0.75$ ms

(*i*) For linear commutation,

Reactance voltage $= L \times \frac{2I_c}{T_c} = \frac{5 \times 10^{-5} \times 2 \times 37.5}{0.75 \times 10^{-3}} = 5$ V **Ans.**

(*ii*) For sinusoidal commutation

Reactance voltage = 1.1 × reactance voltage for linear commutation

$= 1.1 \times 5 = 5.5$ V **Ans.**

6.6 METHODS OF IMPROVING COMMUTATION

There are three methods of improving commutation i.e. of making the current in the short-circuited coil attain its full value in the reverse direction by the end of the short-circuit period.

1. By Use of High Resistance Brushes. The first method of obtaining sparkless commutation is by use of high resistance brushes. When the current from left reaches the commutator segment '*a*' there are two paths open to it. The first path is straight across the bar '*a*' to the brush, the second path is round the short-circuited coil in the clockwise direction and across the bar '*b*' to the brush.

With brushes of low contact resistance, obviously there is no inducement for the current to take second path, but if high contact resistance brushes such as carbon brushes are employed, some of the current will take 2*nd* path through short-circuited coil, then to the bar '*b*', then to the brush, because contact area of the brush with the bar '*a*' is smaller than that with bar '*b*'. As the bar '*b*' passes the brush the contact resistance r_2 increases due to decrease in contact area with bar '*b*' while due to increase in contact area with bar '*a*' contact resistance r_1 decreases and so allows the current to take path through segment '*a*' until the current in the short-circuited coil becomes zero. As the commutator advances further the contact area of bar '*b*' with the brush reduces which causes the current in right hand path to take the path of least resistance *i.e.* through the short-circuited coil, so the direction of the current in short-circuited coil is reversed. Hence the high contact resistance helps the current in attaining its full value in the reverse direction after the period of short circuit, and therefore, in having good commutation. The carbon brushes also lubricate and polish the commutator as it rotates.

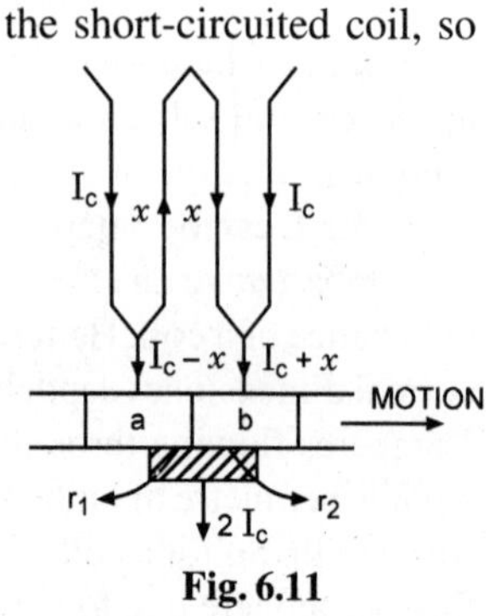

Fig. 6.11

The minor drawbacks of carbon brushes are high voltage drop at brushes due to their high contact resistance and their lower current density. Large voltage drop makes the carbon brushes unsuitable for small machines where this voltage drop forms an appreciable percentage loss. Owing to large voltage drop, the commutator has to be made somewhat larger than with copper brushes in order to dissipate heat efficiently without undue rise of temperature. Owing to lower current density, the brush holders have to be made larger in comparison with those required with copper brushes.

2. By Shifting of Brushes. This method is employed for improving commutation in non-commutating pole machines. The brushes are shifted so that the sides of a coil, as it passes through its commutation period, will cut the sufficient flux from the main poles to develop the necessary commutating emf. The direction of brush shift must be such that the cutting of a portion of the main flux will induce a commutating emf in the coil which will be in the same direction as that of the current through the coil after it has passed through the commutating period. Therefore, for generator operation the brushes must be shifted forward in the direction of rotation of the machine, as illustrated in Fig. 6.12 (*a*). For motor operation the brushes must be shifted backwards with respect to rotation of the machine, as illustrated in Fig. 6.12 (*b*).

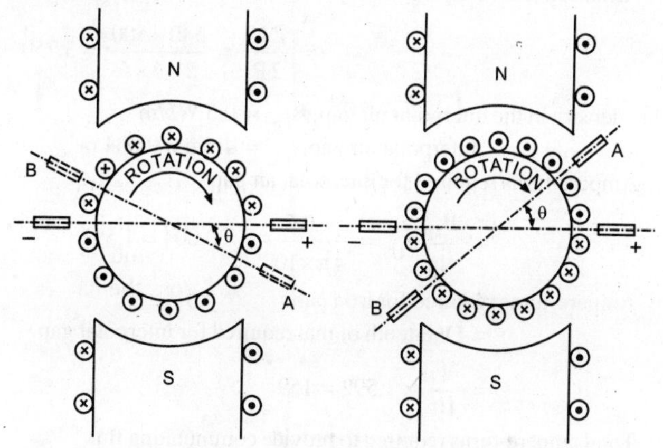

(a) Generator *(b) Motor*

Fig. 6.12 *Brush Shift Method For Satisfactory Commutation in DC Machines*

In general, however, it is desirable to move the brushes a little further forward than the magnetic neutral zone in order that the short-circuited coil may be cutting the fringing flux of the main pole immediately in front and a voltage is induced in the coil in the direction in which the current is about to flow.

The drawback of this method is that, for best commutation, the brushes have to be shifted for every variation of load. Also the larger the forward shift of the brushes of a generator or the backward shift for a motor, the greater are the demagnetizing ampere-turns.

3. By Use of Interpoles or Commutating Poles. In practice, carbon brushes are generally used and there is an appreciable resistance at the brush contacts. The effect of this resistance is to aid commutation but is not sufficient to constitute a solution of the problem.

What is really required, therefore, is to produce in some way in that coil, during the time that it is short circuited, an emf of proper magnitude and direction to overcome the emf of self-induction and so cause complete reversal of the current in the time available (commutating period). This is accomplished in practice by placing a narrow pole, called an *interpole* or a *commutating pole*, over the short-circuited coil, as shown in Fig. 6.13(*a*). When the brushes are located at the mechanical neutral, the sides of a coil are located midway between the main poles at the instant the commutation of the coil takes place. Therefore, the location of the commutating poles must be midway between the main poles. The direction of commutation flux must be such that when the coil cuts across the commutating flux, a voltage will be induced in the coil which will be in the same direction as that of the current through the coil after the coil has passed through the commutating period. Therefore, in case of a generator, the polarity of the commutating pole will be the same as that of the main pole just ahead in the direction of rotation, as shown in Fig. 6.13. For motor operation, the polarity of the commutating pole will be of the main pole just behind.

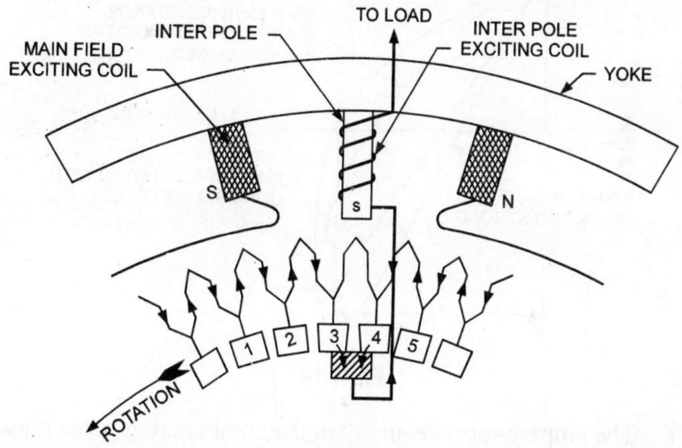

(*a*)

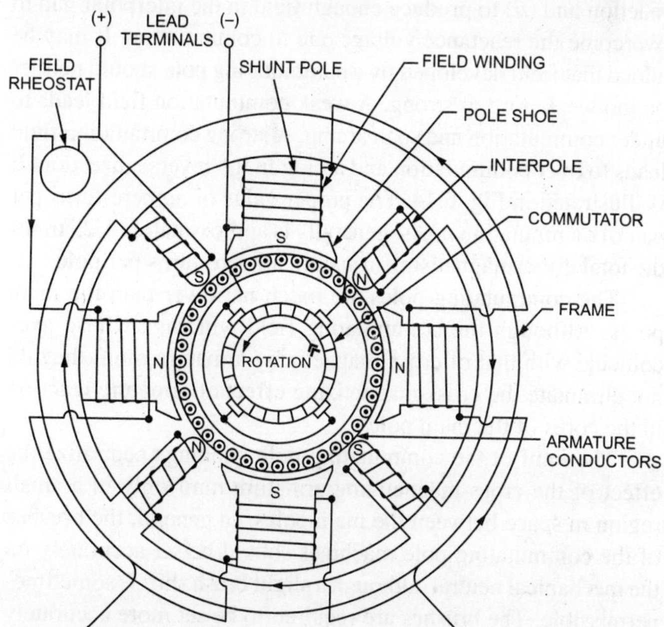

Fig. 6.13 *(b) 4-Pole DC Machine Provided With Interpoles*

Since the amount of emf of self-induction present in the armature coil varies with the amount of armature current flowing, the amount of commutating flux must also, therefore, vary. To make the commutating flux produced by the interpoles proportional to the armature current, interpole windings are connected in series with the armature circuit. The output current is passed around a few turns of heavy copper conductor placed on the interpole and care is taken to see that there is enough iron in it so that it will not become saturated at any armature current that the generator can be expected to carry. The strength of interpole is then directly proportional to the load current. The series connection must be made so that the commutating poles will have the correct polarity, as mentioned in the foregoing para, with respect to the main poles, otherwise they will develop the emf in a coil undergoing commutation, which will oppose the reversal of current instead of aid it. This would result in violent sparking.

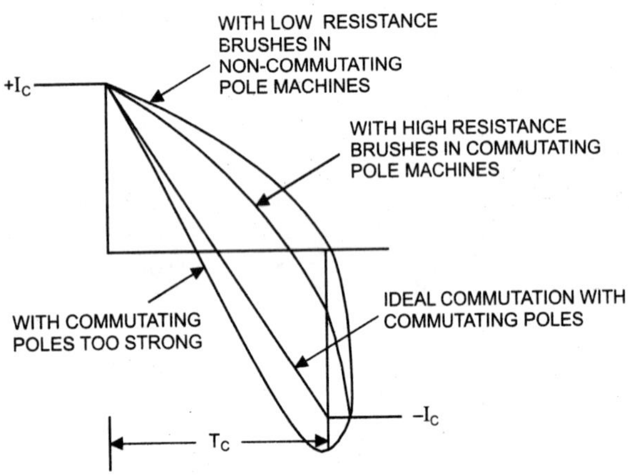

Fig. 6.14

The ampere-turns required on the commutating poles must be sufficient enough (*i*) to neutralize the effect of armature reaction and (*ii*) to produce enough field in the interpolar gap to overcome the reactance voltage due to commutation. It may be added that field developed by a commutating pole should neither be too weak nor too strong. A weak commutation field leads to under commutation and so in arcing. A strong commutation field leads to over commutation and arcing in the reverse direction. It is illustrated in Fig. 6.14. The proper value of ampere-turns per pair of commutating poles, generally is approximately 1.25 times the total cross-magnetising armature ampere-turns per pole.

The commutating poles are much narrower than the main poles. Although the commutating field windings whose axes coincide with that of cross-magnetizing armature mmf, they do not eliminate the cross-magnetizing effect of armature reaction in the cores of the main poles.

The mmf of the commutating-pole windings neutralize the effect of the cross-magnetizing armature mmf only in a small region in space between the main poles. In general, the brushes of the commutating pole machines should be set accurately on the mechanical neutral, although a slight brush shift is sometimes permissible. The brushes are required to be set more accurately than on a non-commutating pole machine; otherwise, there may be instability of speed in case of a motor or difficulty in case of a parallel operation of generators.

As commutating poles carry armature current, therefore, reactance voltage and cross-magnetization are automatically neutralized for all loads because both are produced by the same armature current.

With interpoles, sparkless commutation can be achieved at up to 20-30% overload. In fact, interpoles raise sparking limit of a machine to almost the same value as heating limit. So for a given output an interpole machine can be made smaller and, therefore, cheaper than a non-interpole machine. The use of interpoles is so wide spread that non-interpole machines are rarely employed; if they are employed at all, it is only when the load is known to be constant at all times.

A 4-pole dc machine provided with interpoles is shown in Fig. 6.13(*b*).

Example 6.12. Determine the number of turns on each commutating pole of a 6-pole machine if the flux density in the air gap of the commutating pole is 0.5 Wb/m² at full load and the effective length of the air gap is 4 mm. The full-load current is 500 A and the armature is lap wound with 540 conductors. Assume the number of ampere-turns required for the remainder of the magnetic circuit to be one-tenth of that of the air gap.

Solution: Number of parallel paths, $A = P = 6$

∵ armature is lap wound

Current per conductor, $I_c = \frac{I_a}{A} = \frac{500}{6}$ A

Armature cross-ampere-turns per pole, AT_c

$$= \frac{ZI_c}{2P} = \frac{540 \times 500}{2 \times 6 \times 6} = 3{,}750$$

Flux density in the interpolar air gap, $B_{ip} = 0.5$ Wb/m²

Interpolar air gap, $l_{ip} = 4\text{mm} = 0.004$ m

Ampere-turns required for interpolar air gap

$$= \frac{B_{ip}}{\mu_0} l_{ip} = \frac{0.5}{4\pi \times 10^{-7}} \times 0.004 = 1{,}592$$

Ampere-turns required for iron parts

= One-tenth of that required for interpolar gap

$$= \frac{1}{10} \times 1{,}592 = 159$$

Total ampere-turns required to provide commutating flux

$$= 1{,}592 + 159 = 1{,}751$$

The interpole ampere-turns must be great enough to create commutating flux and also to neutralize the armature across-ampere-turns.

∴ Interpole ampere-turns required, AT_{ip}

$$= 3{,}750 + 1{,}751 = 5{,}501$$

Number of turns on each interpole

$$= \frac{AT_{ip}}{I_a} = \frac{5{,}501}{500} = 11 \textbf{ Ans.}$$

Example 6.13. A 440 V, 4-pole, 25 kW dc generator has a wave-connected armature winding with 846 conductors. The mean flux density in the air gap under the interpoles is 0.5 Wb/m² on full load and the radial gap length is 0.4 cm. Calculate the number of turns required on each interpole.

[*U.P. Technical Univ. Electromechanical Energy Conversion-I, 2004-05*]

Solution: Full-load current, $I_L = \frac{\text{Output in kW} \times 1{,}000}{\text{Supply voltage}}$

$$= \frac{25 \times 1{,}000}{440} = 56.82 \text{ A}$$

Amature current, $I_a = I_L = 56.82$ A

neglecting the shunt field current

Armature cross-ampere-turns per pole,

$$AT_c = \frac{ZI_c}{2P} = \frac{Z}{2P}\times\frac{I_a}{A} = \frac{846}{2\times4}\times\frac{56.82}{2} = 3{,}004$$

Ampere-turns required to provide commutating flux

$$= \frac{B_{ip}}{\mu_0} l_{ip} = \frac{0.5}{4\pi\times10^{-7}}\times0.004 = 1{,}592$$

(neglecting ampere-turns required for iron parts of the commutating pole magnetic circuit)

Total interpole ampere-turns required, AT_{ip}

$$= AT_c + H_{ip}l_{ip} = 3{,}004 + 1{,}592 = 4{,}596$$

Number of turns on each interpole $= \frac{AT_{ip}}{I_a} = \frac{4{,}596}{56.82} =$ **81 Ans.**

6.7 COMPENSATING WINDINGS

When generators are required to supply heavy current with a weak excitation of the magnetic field, as for instance, in reversing rolling mill plants, the provision of commutating poles and the use of carbon brushes do not suffice to afford sparkless commutation. It is, therefore, necessary to provide *compensating winding* in addition for the purpose of neutralizing the effect of armature reaction outside the influence of the interpoles and particularly to maintain a uniform flux distribution under the faces of the main poles. A compensating winding is an auxiliary winding embedded in slots located in the faces of the main poles, as illustrated in Fig. 6.15. The compensating winding is connected in series with armature in a manner so that the direction of current through the compensating conductors in any pole face will be opposite to the direction of current through the adjacent armature conductors, as illustrated in Fig. 6.15. The compensating winding must be designed so that mmf of compensating winding per pole face is equal to mmf produced by the number of armature conductors, which lie under a pole face. In calculating the conductors per pole face required in the compensating winding, it should be noted that the current in the compensating conductors is the armature terminal current I_a, whereas the current in armature conductors is the armature terminal current divided by the number of parallel paths provided by the armature winding between the armature terminals.

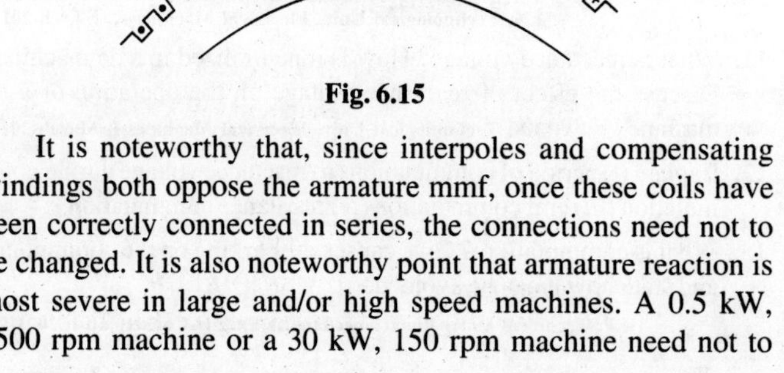

Fig. 6.15

It is noteworthy that, since interpoles and compensating windings both oppose the armature mmf, once these coils have been correctly connected in series, the connections need not to be changed. It is also noteworthy point that armature reaction is most severe in large and/or high speed machines. A 0.5 kW, 1500 rpm machine or a 30 kW, 150 rpm machine need not to have either interpoles or a compensating winding. But a 5 kW, 1500 rpm machine usually has interpoles, and a 100 kW, 1500 rpm machine will probably have both interpoles and compensating windings.

A compensating winding adds considerably to the cost of a machine and doubles the armature copper loss, but it makes the machine able to withstand the most violent fluctuations of load that can be applied to it. Compensating windings are used quite extensively on the generators and motors of Ward Leonard sets particularly in steel mills. They are also used for other types of severe services.

6.7.1. Determination of Compensating Winding Turns. In the actual design of compensating windings, only those conductors in the armature winding, which are directly opposite to the pole faces are neutralized by an equal number of ampere-turns. The commutating poles, which such machines always have, have to take care of the armature reaction effect in the interpolar zones.

Thus, to prevent distortion of flux, under the pole shoe only, the mmf developed per pole by the compensating winding will be $\frac{ZI_c}{2P}\times\frac{\text{Pole arc}}{\text{Pole pitch}}$

So turns per pole for compensating winding

$$= \frac{ZI_c}{2P}\times\frac{\text{Pole arc}}{\text{Pole pitch}}\times\frac{1}{I_a} \qquad ...(6.11)$$

Example 6.14. A 4-pole, 23.75 kW, 250 V lap wound dc shunt generator has 50 slots with 8 conductors per slot and shunt field resistance of 50 Ω. The brushes are given a lead of 8° (mech) when the generator delivers full-load current. Calculate the number of turns on the compensating winding if the pole arc/pole shoe pitch ratio is 0.8.

Solution: Load current, I_L

$$= \frac{\text{Load in kW}\times1{,}000}{V} = \frac{23.75\times1{,}000}{250} = 95 \text{ A}$$

Shunt field current, I_{sh}

$$= \frac{\text{Terminal Voltage}}{\text{Shunt field resistance}} = \frac{250}{50} = 5 \text{ A}$$

Armature current, $I_a = I_L + I_{sh} = 95 + 5 = 100$ A

Current per conductor, $I_c = \frac{I_a}{A} = \frac{100}{4} = 25$ A

Number of armature conductors, $Z = 8 \times 50 = 400$

Ampere-turns per pole for compensating winding

$$= \frac{ZI_c}{2P}\times\frac{\text{Pole arc}}{\text{Pole pitch}} = \frac{400\times25}{2\times4}\times0.8 = 1{,}000$$

Turns per pole for compensating winding

$$= \frac{\text{Armature AT under each pole face}}{I_a} = \frac{1{,}000}{100} = 10 \textbf{ Ans.}$$

Example 6.15. A compensated dc machine has 20,000 armature turns per pole and the ratio of pole arc to pole pitch is 0.7. Interpolar air gap length and the flux density under interpoles are respectively 1 cm and 0.3 T. For the rated armature current of 900 A, calculate the compensating winding conductors per pole and the number of turns on each interpole.

Solution: Compensating winding ampere-turns per pole, AT_c

$$= \frac{\text{Pole arc}}{\text{Pole pitch}}\times AT \text{ per pole} = 0.7 \times 20{,}000 = 14{,}000$$

Compensating winding conductors per pole

$$= \frac{2 \times AT_c}{\text{Armature current, } I_a} = \frac{2 \times 14,000}{900} = 32 \textbf{ Ans.}$$

Flux density in the interpolar air gap, $B_{ip} = 0.3$ T

Interpolar air gap, $l_{ip} = 1$ cm $= 0.01$ m

Ampere-turns required for interpolar air gap

$$= \frac{B_{ip} \times l_{ip}}{\mu_o} = \frac{0.3}{4\pi \times 10^{-7}} \times 0.01 = 2,388$$

Ampere-turns of interpole winding per pole,

AT_i = Armature turns per pole AT_a + ampere-turns required for interpolar air gap – compensating winding ampere turns per pole, AT_c

$= 20,000 + 2,388 - 14,000 = 8,388$

Number of turns on each interpole

$$= \frac{AT_i}{I_a} = \frac{8,388}{900} = 10 \textbf{ Ans.}$$

HIGHLIGHTS

1. The effect of magnetic field set up by the armature current on the distribution of flux under the main poles of a dc machine (a generator or a motor) is known as armature reaction.

   The armature reaction produces demagnetizing and cross-magnetizing effects.

   The *demagnetizing effect of* armature reaction reduces the total flux per pole from its no-load value due to magnetic saturation and the reduction in flux depends upon the degree of magnetic saturation. The decrease in main flux on load due to armature reaction may be about 10%. The decrease in flux reduces the magnitude of generated emf in case of a generator and electromagnetic torque developed in case of a motor.

   The *cross-magnetizing effect* of armature reaction distorts the field in the air gap. Due to this effect increased iron losses, poor commutation or even sparking at the commutator surface occur.

2. Demagnetising ampere-turns per pole are given by equation

$$AT_d = \frac{Z\theta}{360} I_c$$

and cross-magnetizing ampere-turns per pole are given by equation

$$AT_c = ZI_c\left(\frac{1}{2P} - \frac{\theta}{360}\right)$$

where Z is the total number of armature conductors, θ is the angle of lead of brush from geometric neutral plane in mechanical degrees, P is the number of poles and I_c is the current per conductor where

$$I_c = \frac{\text{Armature current, } I_a}{\text{Number of parallel paths, A}}$$

3. The reversal of current in the armature coil by means of brush and commutator bars, is called the *commutation process* and the period during which the coil remains short circuited is called the *commutation period.* Good commutation means no sparking at the brushes and with commutator surface remaining unaffected during continuous operation of the dc machine.

4. Commutation can be improved by (*i*) using high resistance brushes (*ii*) by shifting brushes (*iii*) by using interpoles or commutating poles and (*iv*) by using compensating winding.

   Commutating or interpoles improve commutation by neutralizing the reactance voltage and cross-magnetization while compensating winding neutralize the effect of armature reaction outside the influence of the interpoles and maintains a uniform flux distribution under the faces of the mainpoles.

EXERCISES

1. What is meant by the term 'armature reaction' ? How does it affect the main field flux ? Define GNP and MNP.

   [U.P. Technical Univ. Electromechanical Energy Conversion-I, 2002-03]

2. Draw the field mmf and armature mmf waveforms over one pole pitch in case of a dc generator. Show therein the resultant waveform of these two mmfs. Explain cross-magnetization and demagnetization effects of armature reaction.

   [B.P. Univ. of Technology Electrical Machine-I, 2008]

3. Draw the flux distribution, mmf distribution due to armature conductors carrying current along with the flux distribution due to the main poles of a dc machine.

   [W.B. Univ. of Technology Electrical Machines-I, 2010]

4. What is armature reaction? Describe the effects of armature reaction on the operation of dc machine. How the armature reaction is minimized? **[J.N. Technological Univ. Electrical Machines-I, March-2014; R.G. Technical Univ. Electrical Machine-II, December-2012]**

5. Describe the meaning of "armature reaction." How does it affect the performance of a dc generator? How can the performance be improved with regard to the above? **[U.P.S.C. I.E.S. Electrical Engineering-II, 2007]**

6. What is "armature reaction" in dc machines? What are the effects of armature reaction? How the armature reaction is minimized? Explain with the help of suitable diagrams.

   [U.P. Technical Univ. Electromechanical Energy Conversion-I, 2013-14]

7. What is the effect of armature reaction at leading and trailing pole tips of a dc generator? Explain with the help of neat sketches.

   [J.N. Technological Univ. Electrical Machines-I, March-2014]

8. What do you understand by demagnetizing and cross-magnetizing effects of armature reaction in a dc machine? Derive the expressions for calculations of demagnetizing and cross-magnetizing ampere-turns when brushes are given a shift from geometrical neutral axis.

   [Rajasthan Technical Univ. Electrical Machines-I, 2007]

9. Derive the expressions for calculating the demagnetizing and cross-magnetizing ampere-turns per pole in a dc generator with usual notations. **[J.N. Technological Univ. Electrical Machines-I, March-2014]**

10. What do you understand by linear commutation, under commutation and over commutation in a dc machine? Explain in detail.

    [J.N. Technological Univ. Electrical Machines-I, March-2014]

11. What is reactance voltage? How is it neutralized in a dc machine? Discuss the effect of reactance voltage on the operation of a dc machine. **[J.N. Technological Univ. Electrical Machines-I, March-2014]**

12. Explain (*i*) period of commutation (*ii*) reactance voltage during commutation (*iii*) emf commutation (*iv*) resistance commutation.

13. What is commutation ? Give causes of sparking on the commutator and state how it can be avoided.

    [U.P. Technical Univ. Electromechanical Energy Conversion-I, 2001-02]

14. What is armature reaction in a dc machine? How does it affect commutation? What steps are taken to have effective commutation?
[W.B. Univ. of Technology Electrical Machine-I, 2011]

15. Define commutation. What are the methods for improving the commutation, explain in brief.
[R.G. Technical Univ. Electrical Machine-II, December-2013]

16. What is meant by commutation in dc machines? Describe the process of commutation in dc machines through the reversal of current in a coil. **[G.G.S.I.P. Univ. Electrical Machines, December-2011]**

17. Explain commutation process in dc machiens. What are the causes of poor commutation? Describe the roles of interpoles and compensating winding in dc machines.
[U.P. Technical Univ. Electromechanical Energy Conversion-I, 2006-07]

18. Explain the "commutation process" in dc machines. What are the significance and drawbacks of commutation process in dc machines? Explain in brief.
[G.B. Technical Univ. Electromechanical Energy Conversion-I, 2012-13]

19. Explain briefly the commutation process in a dc machine. Describe how the interpoles and compensating winding help in achieving better commutation. **[B.P. Univ. of Technology Electrical Machine-I, 2008]**

20. Discuss the constructional details and working principle of 2-pole and 4-pole dc machines with interpole and compensating windings. Also mention the types of dc machines and its applications in industry. **[U.P. Technical Univ. Electromechanical Energy Conversion, 2013-14]**

21. Explain the arrangement and action of interpoles in a 2-pole dc machine. Also obtain expression for calculation of ampere-turns needed on interpoles.
[U.P. Technical Univ; Electromechanical Energy Conversion-I, 2005-06]

22. What are the functions of interpoles? How are they wound (*i*) in case of motors, and (*ii*) in case of generators?

23. What are interpoles? Why are the interpoles designed to provide mmf more than the armature mmf in the commutating zone?
[U.P. Technical Univ. Electromechanical Energy Conversion-I, 2008-09]

24. Explain the purpose of compensating winding. Show the section of a dc machine and schematic connection showing compensating winding.
[U.P. Technical Univ. Electromechanical Energy Conversion-I, 2005-06]

25. What is the purpose of compensating windings and interpoles in a dc machine? Explain in detail.
[J.N. Technological Univ. Electrical Machines-I, March-2014]

26. Write short notes on the following:

(*i*) Armature reaction in dc machines. **[G.B. Technical Univ., 2012-13]**

(*ii*) Cross-magnetizing and demagnetizing mmfs and their approximate estimation. **[Rajasthan Univ. 2003]**

(*iii*) Commutation in dc machines. **[Rajasthan Univ. 2001, 2002, 2004]**

(*iv*) Interpoles. **[Rajasthan Univ. 2001, 2002, 2005]**

(*v*) Interpoles and compensating windings in dc machines.
[G.B. Technical Univ., 2012-13]

SHORT ANSWER TYPE QUESTIONS WITH ANSWERS

Q. 1. What is the nature (shape) of the dc armature mmf waveform?
Ans. Triangular.

Q. 2. What is the nature (shape) of air gap flux distribution in space at no load in dc machines?
Ans. Flat topped.

Q. 3. What is armature reaction? **[M.D. Univ., December-2013]**
Ans. Armature reaction is the effect of the magnetic field set up by the armature current on the distribution of flux under the main poles of a dc machine.

Q. 4. What are the two unwanted effects of armature reaction?
Ans. Demagnetization or weakening of main field and cross-magnetization or distortion of field.

Q. 5. Differentiate between geometrical neutral axis and magnetic neutral axis.
Ans. Geometrical neutral axis is the axis between the magnetic poles and magnetic neutral axis is along zero magnetic field.

Q. 6. Define magnetic neutral plane (mnp)?
Ans. The plane through the axis, along which no emf is induced in the armature conductors, is known as magnetic neutral plane.

Q. 7. In a dc machine, along which one of the following axes the armature mmf is directed?
(*a*) Polar axis (*b*) Brush axis (*c*) Interpolar axis.
Ans. Brush axis.

Q. 8. When the brushes in a dc machine are placed at the interpolar axis, is the armature mmf fully demagnetising, fully cross-magnetising or partly demagnetising and partly cross-magnetising?
Ans. Fully cross-magnetising.

Q. 9. How the effect of armature reaction can be minimized in a dc machine? **[G.G.S.I.P. Univ., December-2011]**
Ans. Refer to Art 6.2.

Q. 10. Why in dc machines sometimes brushes are shifted from neutral axis by a small angle?
Ans. So as to lie along mnp to provide sparkless commutation.

Q. 11. What is meant by reactance voltage? **[Anna Univ. May/June-12]**
Ans. The self-induced emf in the coil undergoing commutation is called the reactance voltage.

Q. 12. In a dc machine without interpole to get improved commutation, whether the brush shift should be varied with change in load or brush shift should be fixed?
Ans. Brush shift should be varied with the change in load in a dc machine so as to lie along mnp to provide improved commutation.

Q. 13. Define commutation? **[Anna Univ. May/June-2012]**
Ans. The process of reversal of current in the armature coils by means of brushes and commutator bars, is called the commutation process.

Q. 14. Explain bad commutation. **[R.G.P.V. Bhopal, December-2002]**
Ans. A machine is said to have poor or bad commutation if there is sparking at the brushes and commutator surface. The bad commutation may be caused by mechanical or electrical conditions. The mechanical conditions include uneven commutator surface, non-uniform brush pressure, vibration of brushes in the brush holders etc. The electrical conditions include an increase in voltage between the commutator segments, an increase in the current density at the trailing edge of the brush etc.

Q. 15. What is linear commutation? **[M.D. Univ., December-2013]**
Ans. Linear commutation is an ideal commutation and causes change of current in a coil undergoing commutation from $+ I_c$ to $- I_c$ uniformly.

Q. 16. Name the factor that opposes full reversal of current in a coil undergoing commutation.
Ans. Self-inductance of the coil undergoing commutation.

Q. 17. What is the location of interpole?

Ans. Commutating poles are located midway between the main poles.

Q. 18. Why are the interpoles of a dc machine tapered?

Ans. Interpoles are tapered in order to ensure that there is no saturation at the root of the pole at heavy overloads.

Q. 19. Is the interpole AT required for a compensated dc machine more or less than of an uncompensated one?

Ans. Interpole AT required for a compensated dc machine are less than of an uncompensated one.

Q. 20. Why are commutating poles provided in the construction of a large dc machine?

Ans. In a large dc machine armature reaction effect becomes very large and the commutation problem increases owing to large armature currents. Commutating poles are, therefore, provided to overcome these problems.

Q. 21. What is compensating winding?

Ans. A compensating winding is an auxiliary winding embedded in slots located in the faces of the main poles.

Q. 22. What is the purpose of compensating winding in a dc machine? **[M.D. Univ. December-2013]**

Ans. The purpose of compensating winding is to neutralize the effect of armature reaction outside the influence of the interpoles and particularly to maintain uniform flux distribution under the faces of the main poles.

Q. 23. In what way compensating winding is connected to the armature?

Ans. The compensating winding is connected in series with the armature in a manner so that the direction of current through the compensating winding conductors in any pole face will be opposite to the direction of current through the adjacent armature conductors.

Q. 24. Where are compensating windings placed in a dc machine?

Ans. Compensating windings are placed in slots located in the faces of the main poles.

Q. 25. Air-gap lengths under the main poles of a dc machine are less than those under the interpoles. Why?

Ans. The gaps under the interpoles are kept longer than that under the main poles (1.5 to 1.7 times the main gap) in order to avoid saturation in the interpoles.

PROBLEMS

1. A 4-pole dc generator has wave wound armature with 722 conductors which develops 50 A on full load. Calculate the total ampere-turns per pole. **[Ans.** 2,256.25]
2. A 500 kW, 500 V, 6-pole dc generator has lap wound armature with 468 conductors. Calculate the demagnetising ampere-turns per pole at full load, if the brushes are advanced by 5 mechanical degrees from the geometrical neutral axis. **[Ans.** 1,083]
3. A 250 V, 25 kW, 4-pole dc generator has 328 wave-connected armature conductors. When the machine is delivering full load, the brushes are given a lead of 7.2 electrical degrees. Calculate (*i*) the demagnetising ampere-turns per pole (*ii*) the cross-magnetising ampere-turns per pole. **[Ans.** (*i*) 164 (*ii*) 1,886]
4. A 500 V, wave wound, 750 rpm dc shunt generator supplies a load of 195 A. The armature has 720 conductors and shunt field resistance is 100 Ω. Find the demagnetising ampere-turns/ pole if the brushes are advanced through 3 commutator segments at this load. Also calculate the extra field turns required to neutralize this demagnetization. **[Ans.** 600; 120]
5. A 4-pole dc generator supplies a current of 148 A. It has 492 armature conductors lap connected. The brushes are given a lead of 10° when the machine delivers full load. Calculate the demagnetising armature AT_s per pole. If the shunt field winding takes 6.0 A, determine the number of extra shunt field turns necessary to neutralize this demagnetization. **[Ans.** 526; 88]
6. Determine per pole, the number of (*a*) cross-magnetising ampere-turns (*b*) demagnetising ampere-turns and (*c*) series field turns to balance the back ampere-turns in the case of a dc generator having the following data: 600 conductors, total current 100 A, 6-pole, wave wound, angle of lead 10°, leakage coefficient 1.3. **[Ans.** (*a*) 1,667 (*b*) 833 (*c*) 11]
7. Estimate the number of turns needed on each inter pole of a 6-pole generator delivering 200 kW at 200 V; given number of lap connected armature conductors = 540; interpole air gap = 1.0 cm; flux density in interpolar air gap = 0.3 Wb/m^2. Ignore the effect of iron parts of the circuit and of leakage. **[Ans.** 10]
8. A commutator segment having a diameter of 0.7 metre rotates at 600 rpm. Determine the approximate time of commutation. The brush width is 15 mm. **[Ans.** 0.682 millisecond]
9. Determine the reactance voltage for a dc machine having the datas given below:
   Self inductance of coil = 0.15 mH; current per conductor = 40 A; Brush width = 3 commutator segments; Number of commutator segments = 50; speed = 600 rpm. **[Ans.** 2 V]
10. A 440 V, 1000 ampere lap wound dc machine has 10 poles and 860 armature conductors. Calculate the number of conductors in the pole face to give full compensation if the pole face covers 70% of pole span. **[Ans.** 3,010]

❑❑❑

7 CHAPTER

Operating Characteristics and Applications of DC Generators

INSIDE THIS CHAPTER

7.1 TYPES OF DC GENERATORS

The mmf necessary to establish the flux in the magnetic circuit of a dc generator can be obtained by means of (*i*) permanent magnets (*ii*) field coils excited by some external source and (*iii*) field coils excited by the generator itself.

When permanent magnets are used for establishing the flux in the magnetic circuit, the generator is known as a *permanent magnet generator*. It consists of an armature and one or several permanent magnets encircling the armature. The field developed by the poles of such machines remains fairly constant. Although these machines are very compact but in view of the low power thus generated, permanent magnet generators have not found industrial applications. Such generators are employed only in small sizes like dynamos in motor cycles etc.

7.1.1. Separately-Excited DC Generators. Since the operation of a dc machine depends upon a fixed polarity of the poles which does not vary with time, the field coils need energization from a dc source.

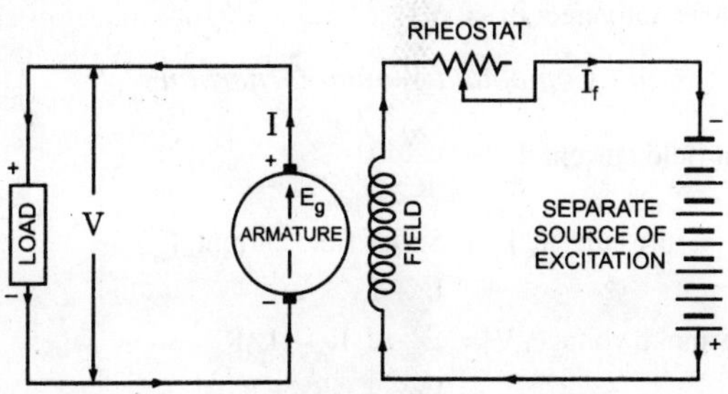

Fig. 7.1

A dc generator whose field winding is excited from an independent external dc source, such as a battery, the generator is called a *separately-excited generator*. The circuit diagram is illustrated in Fig. 7.1. In this case current flowing through the armature, I_a and load, I_L is the same and the terminal voltage (voltage across the load), V is equal to the generated emf, E_g less voltage drop in armature, $I_a R_a$ *i.e.*

$$I_a = I_L = I \text{ (say)} \qquad ...(7.1)$$

$$V = E_g - I R_a \qquad ...(7.2)$$

where R_a is the armature resistance

$$\text{Power developed, } P_g = E_g I \qquad ...(7.3)$$

$$\text{Power delivered to external load, } P_L = V I \qquad ...(7.4)$$

7.1.2. Self-Excited DC Generators. A dc generator whose field winding is excited by the current supplied by the generator itself, is called a *self-excited generator*. In such machines the field coils are interconnected with the armature winding. The field coils may be connected either in *series* with the armature, in parallel with the armature or *partly in series and partly in parallel* with the armature. Accordingly the self-excited generators may be classified as (*i*) *series wound generators* (*ii*) *shunt wound generators* and (*iii*) *compound wound generators*.

Due to residual magnetism, some flux is always present in the poles of such machines. When the armature is rotated, a small voltage is induced in the armature winding owing to residual flux. This induced voltage causes a small current to flow in the field coils and thus increase in flux per pole. The increase in flux causes increase in induced voltage which further increases the field current and so flux per pole. These events take place rapidly and the generator builds up to the rated voltage.

1. Series Wound Generators. In a series wound generator, there is only one field winding (one field coil on the core of each salient pole) which is connected in series with armature winding so that whole current flows through the field winding as well as load. The connection diagram of a series wound generator is given in Fig. 7.2. Since the

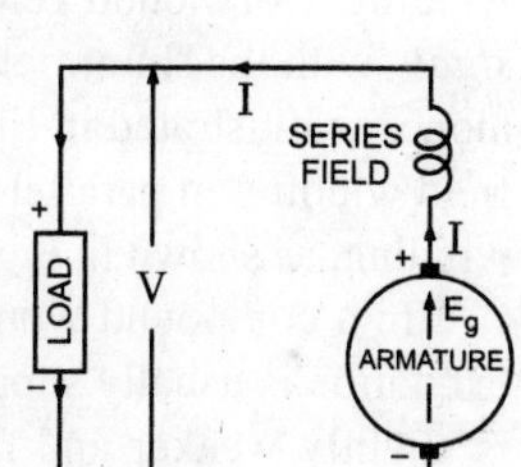

Fig. 7.2

series field winding carries the full-load current, series field winding is designed with a fewer turns of thick wire or strips. The resistance of series field winding R_{se} is, therefore, very low, say of the order of 0.5 Ω. The important relations are given below :

$$I_a = I_{se} = I_L = I \text{ (say)} \qquad ...(7.5)$$

$$V = E_g - I(R_a + R_{se}) \qquad ...(7.6)$$

$$\text{Power developed, } P_g = E_g I \qquad ...(7.7)$$

$$\text{Power delivered, } P_L = VI \qquad ...(7.8)$$

2. Shunt Wound Generators. In a shunt wound generator also there is only one field winding, as in case of series wound generator, but in this case it is connected across the armature circuit forming a parallel or shunt circuit. The voltage across the field winding is, therefore, the same as the terminal voltage of the generator. The connection diagram is illustrated in Fig. 7.3.

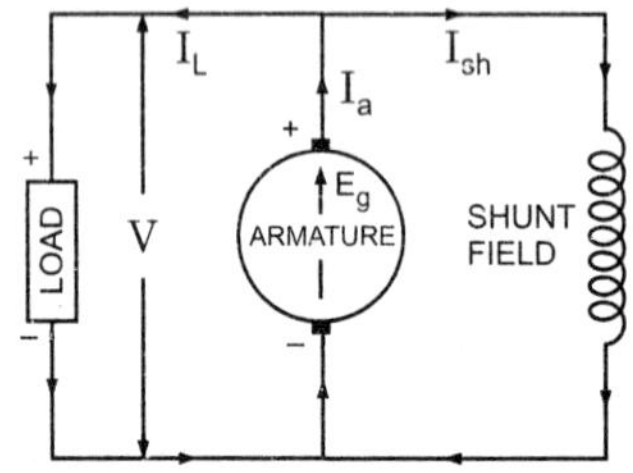

Fig. 7.3

In operation, total armature current I_a flowing from the + ve brush divides between two parallel paths; the external circuit and the field circuit. The load current I_L and the shunt field current I_{sh} close the circuit at the *–ve* brush and their sum is equal to armature current *i.e.*

$$I_a = I_L + I_{sh}$$

The effective power of a generator is proportional to the current delivered to the external load circuit, therefore, it is necessary to keep shunt field current as small as possible. For shunt wound generators it is kept about 2.5 per cent of the rated armature current. Shunt field coils are, therefore, designed to produce the desired mmf by means of a relatively large number of turns of fine wire. The resistance of the shunt field winding R_{sh} will naturally be high, say of the order of 100 Ω. The important relations are given below :

$$\text{Shunt field current, } I_{sh} = \frac{V}{R_{sh}} \qquad ...(7.9)$$

$$\text{Armature current } I_a = I_L + I_{sh} \qquad ...(7.10)$$

$$\text{Terminal voltage, } V = E_g - I_a R_a \qquad ...(7.11)$$

$$\text{Power developed, } P_g = E_g I_a \qquad ...(7.12)$$

$$\text{Power delivered, } P_L = V I_L \qquad ...(7.13)$$

3. Compound Wound Generators. In compound wound generators, there are two field windings (two separate coils on the core of each pole). One of them (having many turns of fine wire) is connected across the armature and the other (having few turns of thick wire) is connected in series with the armature winding. Compound generators may be connected either *short-shunt*, with the shunt field winding in parallel with the armature alone, as illustrated in Fig. 7.4 (*a*) or long shunt, with the shunt field winding in parallel with both the armature and series field winding, as shown in Fig. 7.4 (*b*).

In a compound wound generator, the major portion of the excitation is usually supplied by the shunt field. The shunt field is slightly weaker and the series field is considerably weaker than those of the corresponding machine in which the entire excitation is produced by single shunt or by a single series winding.

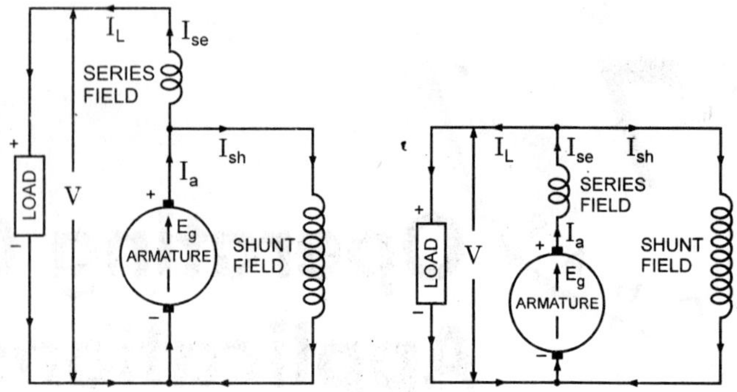

(*a*) *Short-Shunt Compound Wound Generator* (*b*) *Long-Shunt Compound Wound Generator*

Fig. 7.4

Compound generators are also of the two types, known as *cumulative compound wound* and *differential compound wound* generators. In cumulative compound wound generators, series field assists the shunt field, as shown in Fig. 7.5 (*a*) whereas in differential compound wound generators, series field opposes the shunt field, as shown in Fig. 7.5 (*b*).

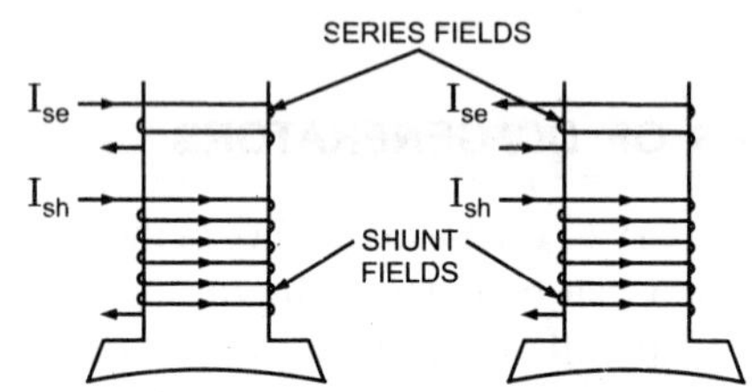

(*a*) *Cumulative Compounding* (*b*) *Differential Compounding*

Fig. 7.5

The important relations for short-shunt and long-shunt compound wound generators are given below :

(a) Short Shunt Compound Wound Generators

$$\text{Series field current, } I_{se} = I_L \qquad ...(7.14)$$

$$\text{Shunt field current, } I_{sh} = \frac{V + I_{se} R_{se}}{R_{sh}} \qquad ...(7.15)$$

$$\text{Armature current, } I_a = I_L + I_{sh} \qquad ...(7.16)$$

$$\text{Terminal voltage, } V = E_g - I_a R_a - I_{se} R_{se} \qquad ...(7.17)$$

$$\text{Power developed, } P_g = E_g I_a \qquad ...(7.18)$$

$$\text{Power delivered, } P_L = V I_L \qquad ...(7.19)$$

(b) Long Shunt Compound Wound Generators

$$\text{Shunt field current, } I_{sh} = \frac{V}{R_{sh}} \qquad ...(7.20)$$

$$\text{Armature current, } I_a = \text{Series field current, } I_{se} \qquad ...(7.21)$$

$$= I_L + I_{sh} \qquad ...(7.22)$$

$$\text{Terminal voltage, } V = E_g - I_a R_a - I_{se} R_{se}$$

$$= E_g - I_a (R_a + R_{se}) \qquad ...(7.23)$$

$$\text{Power developed, } P_g = E_g I_a \qquad ...(7.24)$$

$$\text{Power delivered, } P_L = V I_L \qquad ...(7.25)$$

7.2 ARMATURE RESISTANCE

The resistance offered by the armature circuit is known as *armature resistance* and is represented by R_a. It usually includes (*i*) resistance of armature winding and (*ii*) brush contact resistance.

The armature resistance depends on the type of machine. Except for small machines it is usually of the order of 1 Ω.

7.2.1. Brush Contact Drop.* It is the voltage drop over the brush contact resistance when current flows from the commutator segments to brushes and finally to the external load. It is equal to the product of the values of current and contact resistance.

The voltage drop allowed for all brushes of each polarity shall be 1.0 V for carbon or graphite brushes and 0.3 V for metal carbon brushes, that is, a total drop of 2.0 V for carbon or graphite brushes, and 0.6 V for metal-carbon brushes.

Example 7.1. A four-pole lap-connected shunt generator has 300 armature conductors and a flux/pole of 0.1 Wb. It runs at 1,000 rpm. The armature and field resistances are 0.2 Ω and 125 Ω respectively. Calculate the terminal voltage when it is loaded to take a load current of 90 A. Ignore armature reaction.

Solution: Generator emf,

$$E_g = \frac{\Phi ZN}{60} \times \frac{P}{A} \text{ volts} = \frac{0.1 \times 300 \times 1{,}000}{60} \times \frac{4}{4} = 500 \text{ V}$$

Shunt field current,

$$I_{sh} = \frac{\text{Terminal voltage, V}}{\text{Shunt field resistance, } R_{sh}} \simeq \frac{E_g}{R_{sh}} \simeq \frac{500}{125} \simeq 4 \text{ A}$$

Armature current,

$$I_a = I_L + I_{sh} = 90 + 4 = 94 \text{ A}$$

Terminal voltage,

$$V = E - I_a R_a = 500 - 94 \times 0.2 = 481.2 \text{ V} \quad \textbf{Ans.}$$

Example 7.2. A 30 kW, 300 V dc shunt generator has armature and field resistances of 0.05 Ω and 100 Ω respectively. Calculate the total power developed in the armature when it delivers full-load.

[*M.D. Univ. Electromechanical Energy Conversion, May-2008, December-2012*]

Solution: Generator output, $P_{out} = 30$ kW

Generator terminal voltage, V = 300 V

$$\text{Full-load current, } I_L = \frac{P_{out}}{V} = \frac{30 \times 1{,}000}{300} = 100 \text{ A}$$

$$\text{Shunt field current, } I_{sh} = \frac{V}{R_{sh}} = \frac{300}{100} = 3\text{A}$$

$$\text{Armature current, } I_a = I_L + I_{sh} = 100 + 3 = 103 \text{ A}$$

$$\text{EMF generated, } E_g = V + I_a R_a = 300 + 103 \times 0.05 = 305.15 \text{ V}$$

$$\text{Power developed in the armature, } P_g = \frac{E_g I_a}{1{,}000} = \frac{305.15 \times 103}{1{,}000} = 31.43 \text{ kW} \quad \textbf{Ans.}$$

Example 7.3. A 10 kW, 6-pole dc generator develops an emf of 200 V at 1,500 rpm. The armature has a lap-connected winding. The average flux density over a pole pitch is 0.9 T. The length and diameter of the armature are 0.25 m and 0.2 m respectively. Calculate (*a*) the flux per pole (*b*) the total number of active conductors in the armature and (*c*) the torque developed by the machine when the armature supplies a current of 50 A.

[*W.B. Univ. of Technology Electrical Machine-I, 2013*]

Solution: (*i*) Flux per pole,

$$\Phi = \text{Flux density} \times \text{pole pitch} \times \text{axial length}$$

$$= B \times \frac{\pi \times \text{diameter}}{\text{Number of poles}} \times l = \frac{0.9 \times \pi \times 0.2}{6} \times 0.25$$

$$= 0.0236 \text{ Wb or } 23.6 \text{ mWb} \quad \textbf{Ans.}$$

(*ii*) Total number of active conductors, Z

$$= \frac{E_g \times 60 \times A}{\Phi \times N \times P} = \frac{200 \times 60 \times 6}{0.0236 \times 1{,}500 \times 6} = 340 \quad \textbf{Ans.}$$

Power generated,

$$P_g = E_g \times I_a = 200 \times 50 = 10{,}000 \text{ W}$$

(*iii*) Torque developed,

$$T = \frac{P_g}{\omega} = \frac{10{,}000}{2\pi N/60} = \frac{10{,}000 \times 60}{2\pi \times 1{,}500} = 63.662 \text{ Nm} \quad \textbf{Ans.}$$

Example 7.4. A 4-pole lap-connected armature of a dc shunt generator is required to supply the loads connected in parallel:

(i) 5 kW geyser at 250 V and

(ii) 2.5 kW lighting load also at 250 V.

The generator has an armature resistance of 0.2 Ω and a field resistance of 250 Ω. The armature has 120 conductors in the slots and runs at 1,000 rpm. Allowing 1 V per brush for contact drops find (*i*) flux per pole, (*ii*) armature current per parallel path.

Solution: Total load to be supplied, P = 5 + 2.5 = 7.5 kW

$$\text{Load current, } I_L = \frac{P}{V} = \frac{7.5 \times 1{,}000}{250} = 30 \text{ A}$$

$$\text{Shunt field current, } I_{sh} = \frac{V}{R_{sh}} = \frac{250}{250} = 1 \text{ A}$$

$$\text{Armature current, } I_a = I_L + I_{sh} = 30 + 1 = 31 \text{ A}$$

$$\text{Generated emf, } E_g = V + I_a R_a + \text{brush contact drop} = 250 + 31 \times 0.2 + 2 \times 1 = 258.2 \text{ V}$$

$$(i) \text{ Flux per pole, } \Phi = \frac{E_g \times 60}{Z \times N} \times \frac{A}{P} = \frac{258.2 \times 60}{120 \times 1{,}000} \times \frac{4}{4} = 129.1 \text{ mWb} \quad \textbf{Ans.}$$

$$(ii) \text{ Current per armature path, } I_c = \frac{I_a}{A} = \frac{31}{4} = 7.75 \text{ A} \quad \textbf{Ans.}$$

Example 7.5. A 6-pole, 12 k W, 240 V, dc machine is wave-connected. If the same machine is lap-connected, all other things remaining same, calculate its voltage, current and power ratings.

[*Pb. Technical Univ., December-2002*]

Solution: *When machine is wave-connected*

Rated power, P = 12 k W

Rated voltage, $V_1 = 240$ V

Voltage across each path, V = 240 V

Number of parallel paths, $A_1 = 2$

$$\text{Rated current, } I_1 = \frac{P_1}{V_1} = \frac{12 \times 1{,}000}{240} = 50 \text{ A}$$

$$\text{Current in each parallel path, } I_{c_1} = \frac{I_1}{A_1} = \frac{50}{2} = 25 \text{ A}$$

When machine is lap-connected

Number of parallel paths, $A_2 = 6$

$$\text{Rated voltage, } V_2 = V_1 \times \frac{A_1}{A_2} = 240 \times \frac{2}{6} = 80 \text{ V} \quad \textbf{Ans.}$$

* Refer to Indian Standards IS 4889-1968, March 1969.

Rated current, I_2 = Current in parallel path × number of parallel paths

$= 25 \times 6 = 150$ A **Ans.**

Rated power, $P_2 = V_2 I_2 = \frac{80 \times 150}{1,000} = 12$ kW **Ans.**

Example 7.6. A dc shunt generator has an induced voltage on open-circuit of 127 V. When the machine is on load, the terminal voltage is 120 V. Find the load current if the field circuit resistance is 15 Ω and the armature resistance is 0.02 Ω. Ignore armature reaction.

Solution: Terminal voltage, V = 120 V

Shunt field resistance, $R_{sh} = 15\ \Omega$

Shunt field current, $I_{sh} = \frac{V}{R_{sh}} = \frac{120}{15} = 8$ A

Generated emf, $E_g = 127$ V

Armature current, $I_a = \frac{E_g - V}{R_a}$ neglecting brush drop

$= \frac{127 - 120}{0.02} = 350$ A

Load current, $I_L = I_a - I_{sh} = 350 - 8 = 342$ A **Ans.**

Example 7.7. A 4-pole lap wound dc shunt generator has flux per pole of 0.07 Wb. The armature winding consists of 220 turns, each turn having a resistance of 0.004 Ω. Calculate the terminal voltage when running at 900 rpm if the armature current is 50 A.

[*Gujarat Technological Univ. Electrical Machines-I, Summer-2014*]

Solution: Number of armature conductors,

$Z = 2 \times$ number of turns $= 2 \times 220 = 440$

Generated emf,

$$E_g = \Phi Z \frac{N}{60} \times \frac{P}{A} = 0.07 \times 440 \times \frac{900}{60} \times \frac{4}{4} = 462 \text{ V}$$

Number of turns per path $= \frac{220}{4} = 55$

Armature resistance per path

= Number of turns per path × resistance of each turn

$= 55 \times 0.004 = 0.22\ \Omega$

Armature resistance,

$$R_a = \frac{\text{Armature resistance per path}}{\text{Number of parallel paths}} = \frac{0.22}{4} = 0.055\ \Omega$$

Armature current, $I_a = 50$ A

Terminal voltage,

$V = E_g - I_a R_a = 462 - 50 \times 0.055 = 459.25$ V **Ans.**

Example 7.8. A 4-pole dc shunt generator with lap connected armature has field and armature resistances of 50 Ω and 0.1 Ω respectively. It supplies power to sixty numbers of 100 V, 40 W lamps. Calculate the armature current and the generated emf. Allow a contact drop of 1 V/brush and interpole and compensating winding drops are 0.5 V/pole and 0.25 V/pole respectively.

[*U.P. Technical Univ. Electromechanical Energy Conversion-I, 2013-14*]

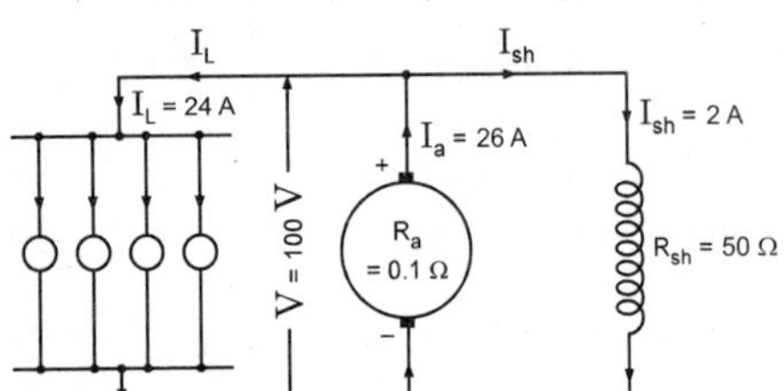

Fig. 7.6

Solution: Total lamp load, P_L = Number of lamps × wattage of each lamp

$= 60 \times 40 = 2,400$ W

Terminal voltage, V = 100 V

Load current, $I_L = \frac{P_L}{V} = \frac{2,400}{100} = 24$ A

Shunt field current, $I_{sh} = \frac{V}{R_{sh}} = \frac{100}{50} = 2$A

Total armature current, $I_a = I_L + I_{sh} = 24 + 2 = 26$ A **Ans.**

Generated emf, $E_g = V + I_a R_a$ + brush drops

$= 100 + 26 \times 0.1 + 2 \times 1 + 2 \times 0.5 + 2 \times 0.25$

$= 106.1$ V **Ans.**

Example 7.9. Estimate the reduction in speed of a generator with constant excitation on bus bars to decrease its load from 500 kW to 250 kW. The resistance between terminals is 0.015 Ω. The busbar voltage is 500 V.

[*R.G. Technical Univ. Electromechanical Energy Conversion, June-2010*]

Solution: Busbar voltage, V = 500 V

Resistance between terminals, R = 0.015 Ω

At load of 500 kW

$$\text{Load current, } I_{L_1} = \frac{\text{Load in kW} \times 1,000}{\text{Busbar voltage}} = \frac{500 \times 1,000}{500} = 1,000 \text{ A}$$

Generated emf, $E_{g_1} = V + I_{L_1} R = 500 + 1,000 \times 0.015 = 515$ V

Speed $= N_1$ rpm (say)

At load of 250 kW

Load current, $I_{L_2} = \frac{250 \times 1,000}{500} = 500$ A

Generated emf, $E_{g_2} = 500 + 500 \times 0.015 = 507.5$ V

$$\text{Speed, } N_2 = \frac{E_{g_2}}{E_{g_1}} \times N_1 = \frac{507.5}{515} \times N_1 = 0.9854\ N_1$$

$\because E_g \propto N$ with constant excitation

$$\text{Reduction in speed} = \frac{N_1 - N_2}{N_1} \times 100$$

$$= \frac{N_1 - 0.9854\ N_1}{N_1} \times 100 = 1.46\%\ \textbf{Ans.}$$

Example 7.10. A 4 pole dc shunt generator with wave wound armature has 40 slots each having 12 conductors. Armature resistance is 1 Ω and shunt field resistance is 200 Ω. The flux per pole is 25 m Wb. If a load of 50 Ω is connected across the armature terminal, calculate the voltage across the load when the generator is driven at 1000 rpm. What will be the load voltage if the generator is lap wound? [*M.D. Univ. Electrical Machines-I, December-2013*]

Solution: *With wave-connected armature*

$$\text{Generated emf, } E_g = \frac{\Phi Z N}{60} \times \frac{P}{A}$$

$$= \frac{25 \times 10^{-3} \times 40 \times 12 \times 1,000}{60} \times \frac{4}{2} = 400 \text{ V}$$

Let the terminal voltage be of V volts

Shunt field current, $I_{sh} = \frac{V}{R_{sh}} = \frac{V}{200}$ A

Load current, $I_L = \frac{V}{R_L} = \frac{V}{50}$ A

Armature current, $I_a = I_L + I_{sh} = \frac{V}{200} + \frac{V}{50} = \frac{V}{40}$ A

Since generated emf, $E_g = V + I_a R_a$ neglecting brush drop

$$400 = V + \frac{V}{40} \times 1 = \frac{41}{40} V \quad ...(i)$$

or Terminal voltage, $V = \frac{400}{41/40} = 390.2$ V **Ans.**

With lap-connected armature

Generated emf, $E_g = \frac{25 \times 10^{-3} \times 40 \times 12 \times 1{,}000}{60} \times \frac{4}{4} = 200$ V

From Eq. (*i*)

Terminal voltage, $V = \frac{200}{41/40} = 195.1$ V **Ans.**

Example 7.11. A 2-pole dc shunt generator charges a 100 V battery of negligible internal resistance. The armature of the machine is made up of 1,000 conductors, each of 2 mΩ resistance. The charging currents are found to be 10 A and 20 A for generator speed of 1055 and 1105 rpm respectively. Find the field circuit resistance and flux per pole of the generator. Neglect armature reaction effects.

[*U.P. Technical Univ. Electromechanical Energy Conversion-I, 2008-09*]

Solution: Number of armature conductors,

$Z = 1{,}000$

Number of poles, $P = 2$

Number of parallel paths, $A = 2$

Armature resistance per path = Number of conductors per path × resistance of each conductor

$$= \frac{1{,}000}{2} \times 2 \times 10^{-3} = 1\ \Omega$$

Armature resistance,

$$R_a = \frac{\text{Armature resistance per path}}{\text{Number of parallel paths}} = \frac{1}{2} = 0.5\ \Omega$$

Let the generated emfs at speeds N_1 of 1,055 rpm and N_2 of 1105 rpm be E_1 and E_2 volts respectively.

$$\text{Then } \frac{E_2}{E_1} = \frac{\Phi Z N_2 \times \frac{P}{A}/60}{\Phi Z N_1 \times \frac{P}{A}/60} = \frac{N_2}{N_1} = \frac{1{,}105}{1{,}055} \quad ...(i)$$

Terminal voltage, $V = 100$

Let the shunt field current be of I_{sh} amperes.

Load current at speed of 1,055 rpm,

$$I_{L_1} = 10 \text{ A}$$

Armature current at speed of 1,055 rpm,

$$I_{a_1} = I_{L_1} + I_{sh} = (10 + I_{sh})$$

Similarly armature current at speed of 1,105 rpm,

$$I_{a_2} = (20 + I_{sh})$$

Now induced emf,

$$E_1 = V + I_{a_1} R_a = 100 + (10 + I_{sh}) \times 0.5 = 105 + 0.5\, I_{sh}$$

and $$E_2 = V + I_{a_2} R_a = 100 + (20 + I_{sh}) \times 0.5 = 110 + 0.5\, I_{sh}$$

and $$\frac{E_2}{E_1} = \frac{110 + 0.5\, I_{sh}}{105 + 0.5\, I_{sh}} \quad ...(ii)$$

Comparing Eqs. (*i*) and (*ii*) we have

$$\frac{1{,}105}{1{,}055} = \frac{110 + 0.5\, I_{sh}}{105 + 0.5\, I_{sh}}$$

or Shunt field current $I_{sh} = 1$ A

Field circuit resistance, $R_{sh} = \frac{V}{I_{sh}} = \frac{100}{1} = 100\ \Omega$ **Ans.**

Substituting $E_1 = 105 + 0.5 \times 1 = 105.5$ V, Z = 1,000 conductors, $N_1 = 1{,}055$ rpm, P = 2, A = 2 in emf equation we have

$$105.5 = \Phi \times 1{,}000 \times \frac{1{,}055}{60} \times \frac{2}{2}$$

or $\Phi = \frac{105.5 \times 60}{1{,}000 \times 1{,}055} = 6$ mWb **Ans.**

Example 7.12. Find the resistance of the load which takes a power of 5 kW from a dc shunt generator whose external characteristic is given by the equation: $V = 250 - 0.5\, I_L$.

[*U.P.S.C. I.E.S. Electrical Engineering-II, 1995*]

Solution: $V = 250 - 0.5\, I_L$ ...(*i*) (given)

and power output, $P = V I_L$

or $V I_L = 5 \times 1{,}000 = 5{,}000$ W ... (*ii*)

Substituting $I_L = \frac{5{,}000}{V}$ from Eq. (*ii*) in Eq. (*i*) we have

$$V = 250 - 0.5 \times \frac{5{,}000}{V}$$

or $V^2 - 250\,V + 2{,}500 = 0$

$$\text{or } V = \frac{250 \pm \sqrt{250^2 - 4 \times 2{,}500}}{2}$$

$= 239.56$ V or 10.44 V

Rejecting lower value 10.44 V of V, as it is not practicable, we have

$V = 239.56$ V

$$I_L = \frac{5{,}000}{239.56} = 20.87 \text{ A}$$

Load resistance, $R_L = \frac{V}{I_L} = \frac{239.56}{20.87} = 11.48\ \Omega$ **Ans.**

Example 7.13. A 4 pole, long shunt lap wound generator supplies 20 kW at a terminal voltage of 440 V. The armature resistance is 0.04 Ω and series field resistance is 0.05 Ω and shunt field resistance is 210 Ω. The brush drop may be taken as 1.0 V. Determine the generated emf. Also calculate number of conductors if the speed is 1,500 rpm and flux per pole is 0.025 weber. Neglect the armature reaction. [*M.D. Univ. Electrical Machines-I, December-2012*]

Solution: Load current, $I_L = \frac{\text{Load supplied in watts}}{\text{Terminal voltage in volts}}$

$$= \frac{20 \times 1{,}000}{440} = 45.45 \text{ A}$$

Shunt field current, $I_{sh} = \frac{V}{R_{sh}} = \frac{440}{210} = 2.095$ A

Armature current, $I_a = I_L + I_{sh} = 45.45 + 2.095 = 47.55$ A

Generated emf, $E_g = V + I_a(R_a + R_{se})$ + brush volt drop

$= 440 + 47.55\,(0.04 + 0.05) + 1$

$= 445.3$ V **Ans.**

Number of conductor required,

$$Z = \frac{E_g \times 60}{\Phi \times N} \times \frac{A}{P}$$

$$= \frac{445.3 \times 60}{0.025 \times 1{,}500} \times \frac{4}{4} = 712 \text{ **Ans.**}$$

Example 7.14. A 12-pole dc generator has a simplex wave wound armature having 144 coils of 10 turns each. The resistance of each turn is 0.011 Ω, the flux per pole is 0.05 Wb and speed is 200 rpm.

(*i*) What is the induced emf?

(*ii*) What is the armature resistance ?

(*iii*) What is the induced counter torque on the shaft of the machine, if a 1,000 Ω resistance is connected to its terminals? Neglect the effect of armature resistance.

[*U.P. Technical Univ. Electromechanical Energy Conversion-I, 2013-14*]

Solution: Number of armature conductors,

$$Z = \text{Number of coils} \times \text{turns in each coil} \times 2$$

$$= 144 \times 10 \times 2 = 2{,}880$$

Number of turns per path

$$= \frac{144 \times 10}{2} = 720$$

Armature resistance per path

= Number of turns per path × resistance of each turn

$$= 720 \times 0.011 = 7.92\ \Omega$$

(*ii*) Armature resistance,

$$R_a = \frac{\text{Armature resistance per path}}{\text{Number of parallel paths}} = \frac{7.92}{2} = 3.96\ \Omega \textbf{ Ans.}$$

(*i*) Induced emf,

$$E_g = \frac{\Phi ZN}{60} \times \frac{P}{A} = \frac{0.05 \times 2{,}880 \times 200}{60} \times \frac{12}{2} = 2{,}880 \text{ V } \textbf{Ans.}$$

(*iii*) Armature current, when a resistance of 1,000 Ω is connected to its terminals,

$$= \frac{E_g}{R_{ext} + R_a} = \frac{2{,}880}{1{,}000 + 3.96} = 2.869 \text{ A}$$

$$\text{Counter torque developed} = \frac{E_g \times I_a}{2\pi \frac{N}{60}} = \frac{2{,}880 \times 2.869}{2\pi \times 200} \times 60$$

$$= 394.5 \text{ N-m } \textbf{Ans.}$$

Example 7.15. A 4-pole, lap-wound, long shunt, dc compound generator has useful flux per pole of 0.07 Wb. The armature consists of 220 turns and resistance per turn is 0.004 Ω. Calculate the terminal voltage if the resistance of shunt and series fields are 100 Ω and 0.02 Ω respectively; when the generator is running at 900 rpm with armature current of 50 A. Also calculate the power output in kW for the generator. [*Nagpur Univ. Elec. Machines, 1993*]

Solution: Generated emf, E_g = 462 V

(Refer to Solution of Example 7.7)

Armature resistance, R_a = 0.055 Ω

(Refer to Solution of Example 7.7)

Armature current, I_a = 50 A

$$\text{Terminal voltage, } V = E_g - I_a (R_a + R_{se})$$

$$= 462 - 50\,(0.055 + 0.02)$$

$$= 458.25 \text{ V } \textbf{Ans.}$$

$$\text{Shunt field current, } I_{sh} = \frac{V}{R_{sh}} = \frac{458.25}{100} = 4.58 \text{ A}$$

$$\text{Output current, } I_L = I_a - I_{sh} = 50 - 4.58 = 45.42 \text{ A}$$

$$\text{Power output, } P_L = \frac{V_L I_L}{1{,}000} = \frac{458.25 \times 45.42}{1{,}000} = 20.814 \text{ kW } \textbf{Ans.}$$

Example 7.16. In a long-shunt compound generator, the terminal voltage is 230 V when it delivers 150 A. Determine: (*i*) induced emf (*ii*) total power generated by armature. The shunt field, series field, diverter, and armature resistances are 92, 0.015, 0.03 and 0.032 ohm respectively.

Solution: Terminal voltage, V = 230 V

$$\text{Shunt current, } I_{sh} = \frac{V}{R_{sh}} = \frac{230}{92} = 2.5 \text{ A}$$

$$\text{Armature current, } I_a = \text{Load current, } I_L + I_{sh}$$

$$= 150 + 2.5 = 152.5 \text{ A}$$

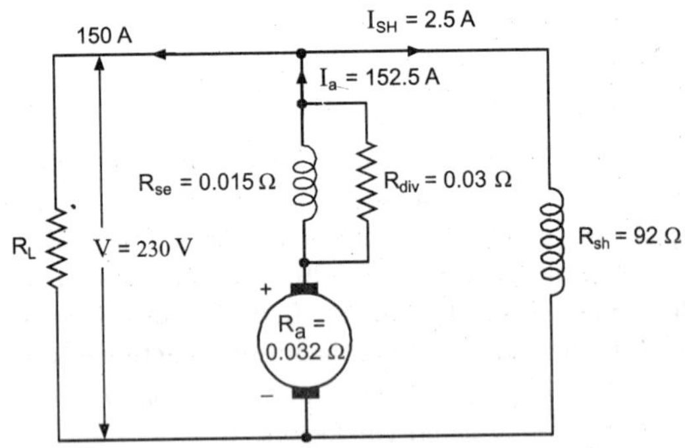

Fig. 7.7

Combined resistance of series field and its diverter,

$$R_{se\ eq} = \frac{R_{se} \times R_{div}}{R_{se} + R_{div}} = \frac{0.015 \times 0.03}{0.015 + 0.03} = 0.01\ \Omega$$

$$\text{(i) Induced emf, } E_g = V + I_a R_a + I_a R_{se\ eq}$$

$$= 230 + 152.5 \times 0.032 + 152.5 \times 0.01$$

$$= 236.4 \text{ V } \textbf{Ans.}$$

$$\text{(ii) Total power generated, } P_g = \frac{E_g \times I_a}{1{,}000} = \frac{236.4 \times 152.5}{1{,}000}$$

$$= 36.051 \text{ kW } \textbf{Ans.}$$

Example 7.17. A short-shunt compound wound dc generator delivers 100 A to a load at 250 V. The generator has shunt field, series field, and armature resistances of 130 Ω, 0.1 Ω, and 0.1 Ω respectively. Calculate the voltage generated in the armature winding. Assume 1 V voltage drop per brush.

[*J.N. Technological Univ. Electrical Machines-I, March-2014*]

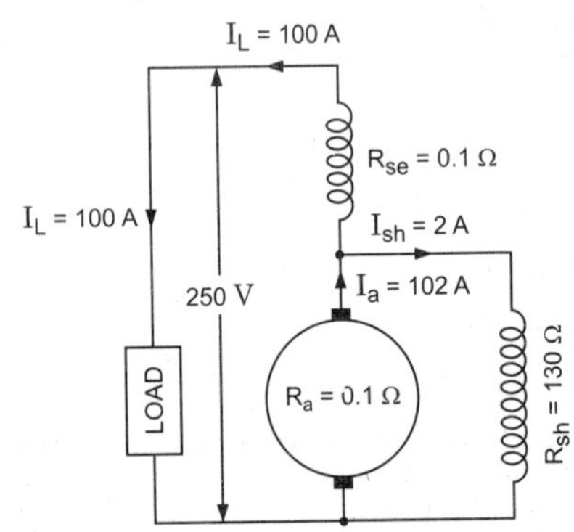

Fig. 7.8

Solution: Load current, I_L = 100 A

Terminal voltage, V_L = 250 V

Series field current, $I_{se} = I_L$ = 100 A

$$\text{Voltage across shunt field, } V_{sh} = V_L + I_{se} R_{se} = 250 + 100 \times 0.1$$

$$= 260 \text{ V}$$

Shunt field current, $I_{sh} = \frac{V_{sh}}{R_{sh}} = \frac{260}{130} = 2$ A

Armature current, $I_a = I_L + I_{sh} = 100 + 2 = 102$ A

Generated voltage, $E_g = V + I_{se} R_{se} + I_a R_a$ + brush contact drop
$= 250 + 100 \times 0.1 + 102 \times 0.1 + 2 \times 1$
$= 250 + 10 + 10.2 + 2 = 272.2$ V **Ans.**

Example 7.18. In a 220 V compound generator, the armature, series and shunt windings have resistances of 0.3 Ω, 0.2 Ω and 60 Ω respectively. The load consists of 80 lamps, each rated at 60 W and 220 V. Find the total emf and armature current when the machine is connected for long shunt and short shunt.

[G.G.S.I.P. Univ. Electromechanical Energy Conversion-I, December-2011]

Solution: Total lamp load, P_L = Number of lamps × wattage of each lamp
$= 80 \times 60 = 4{,}800$ W

Terminal voltage, V = 220 V

Load current, $I_L = \frac{P_L}{V} = \frac{4{,}800}{220} = 21.82$ A

Long Shunt Connection

Shunt field current, $I_{sh} = \frac{V}{R_{sh}} = \frac{220}{60} = 3.67$ A

Armature current, $I_a = I_L + I_{sh} = 21.82 + 3.67$
$= 25.49$ A **Ans.**

Generated emf, $E_g = V + I_a R_a + I_{se} R_{se}$
$= 220 + 25.49 \times 0.3 + 25.49 \times 0.2$
$= 232.745$ V **Ans.**

Short Shunt Connection

Series field current, $I_{se} = I_L = 21.82$ A

Voltage across shunt field, $V_{sh} = V_L + I_{se} R_{se} = 220 + 21.82 \times 0.2$
$= 224.364$ V

Shunt field current, $I_{sh} = \frac{V_{sh}}{R_{sh}} = \frac{224.364}{60} = 3.74$ A

Armature current, $I_a = I_L + I_{sh} = 21.82 + 3.74$ A $= 25.56$ A

Generated emf, $E_g = V + I_{se} R_{se} + I_a R_a$
$= 220 + 21.82 \times 0.2 + 25.56 \times 0.3$
$= 232$ V **Ans.**

Example 7.19. In a 110 V compound generator, the armature, shunt and series winding resistances are 0.06 Ω, 25 Ω, and 0.04 Ω respectively. The load consists of 200 lamps each rated 55 W, 110 V connected in parallel. Find the total emf and armature current, when the machine is connected for (*i*) long shunt (*ii*) short shunt. Ignore the armature reaction and brush drop. (*iii*) If a diverter of resistance 0.1 Ω is connected across the series field winding, find the percentage decrease or increase in series field ampere turns.

[Rajasthan Technical Univ. Electrical Machines-I, 2011]

Solution: Total lamp load, P_L = Number of lamps × wattage of each lamp
$= 200 \times 55 = 11{,}000$ W

Terminal voltage, V = 110 V

Load current, $I_L = \frac{P_L}{V} = \frac{11{,}000}{110} = 100$ A

(*i*) **Long Shunt Connection:**

Shunt field current, $I_{sh} = \frac{V}{R_{sh}} = \frac{110}{25} = 4.4$ A

Armature current, $I_a = I_L + I_{sh} = 100 + 4.4 = 104.4$ A **Ans**

Series field current, $I_{se} = I_a = 104.4$ A **Ans.**

Generated emf, $E_g = V + I_a R_a + I_{se} R_{se}$
$= 110 + 104.4 \times 0.06 + 104.4 \times 0.04$
$= 120.44$ V **Ans.**

(*ii*) **Short Shunt Connection:**

Series field current, $I_{se} = I_L = 100$ A

Voltage across shunt field, $V_{sh} = V_L + I_{se} R_{se} = 110 + 100 \times 0.04$
$= 114$ V

Shunt field current, $I_{sh} = \frac{V_{sh}}{R_{sh}} = \frac{114}{25} = 4.56$ A

Armature current, $I_a = I_L + I_{sh} = 100 + 4.56$
$= 104.56$ A **Ans.**

Generated emf, $E_g = V + I_a R_a + I_{se} R_{se}$
$= 110 + 104.56 \times 0.06 + 100 \times 0.04$
$= 120.3736$ V **Ans.**

(*iii*) Now when a divertor resistance of 0.1 Ω is connected across the series field winding

Current flowing through series field,

$$I'_{se} = I_{se} \times \frac{R_{div}}{R_{se} + R_{div}} = I_{se} \times \frac{0.1}{0.1 + 0.04} = 0.7143 I_{se}$$

Series field ampere-turns
$= 0.7143$ N I_{se}

Percentage decrease in series field ampere-turns

$$= \frac{NI_{se} - 0.7143 NI_{se}}{NI_{se}} \times 100 = 28.57\% \quad \textbf{Ans.}$$

Example 7.20. A 220 V dc compound generator has resistances of armature, shunt, and series windings as 0.1 Ω, 50 Ω and 0.06 Ω respectively. The load current is 100 A at 220 V. Find the induced emf and armature current when the machine is connected as a (*i*) long shunt (*ii*) short shunt (*iii*) how will the ampere-turns of the series winding will change if in (*i*) a diverter resistance of 0.1 Ω be connected in parallel with series winding ? Ignore armature reaction.

[U.P.S.C. I.E.S. Electrical Engineering-I, 2005]

Solution: Terminal voltage, V = 220 V

Load current, I_L = 100 A

(*i*) *When the machine is connected as long shunt*

Shunt field current, $I_{sh} = \frac{V}{R_{sh}} = \frac{220}{50} = 4.4$ A

Armature current, $I_a = I_L + I_{sh}$
$= 100 + 4.4 = 104.4$ A **Ans.**

Induced emf, $E_g = V + I_a (R_a + R_{se})$
$= 220 + 104.4 \times (0.1 + 0.06)$
$= 236.7$ V **Ans.**

(*ii*) *When the machine is connected as short shunt*

Voltage across shunt field, $V_{sh} = V + I_{se} R_{se}$
$= V + I_L R_{se}$
$= 220 + 100 \times 0.06 = 226$ V

Shunt field current, $I_{sh} = \frac{V_{sh}}{R_{sh}} = \frac{226}{50} = 4.52$ A

Armature current, $I_a = I_L + I_{sh} = 100 + 4.52 = 104.52$ A **Ans.**

$$\text{Induced emf, } E_g = V + I_{se}R_{se} + I_aR_a$$
$$= 220 + 100 \times 0.06 + 104.52 \times 0.1$$
$$= 236.452 \text{ V } \textbf{Ans.}$$

When a diverter of 0.1 Ω resistance is connected in parallel with the series field winding, the current through the series field will be reduced and as per current divide rule the current through the series field winding will be given as

$$I'_{se} = \frac{R_{div}}{R_{se} + R_{div}} \times I_{se} = \frac{0.1}{0.06 + 0.1} \times I_{se} = 0.625\ I_{se}$$

Percentage reduction in series field current

$$= \frac{I_{se} - 0.625 I_{se}}{I_{se}} \times 100 = 37.5\%$$

Since number of turns on series field winding remains unchanged, the percentage reduction in ampere-turns of series field winding would be the same as the percentage reduction in series field current *i.e.* 37.5%. **Ans.**

7.3 MAGNETISATION CURVE

According to molecular theory of magnetism the molecules of unmagnetised piece of iron are not arranged in any definite order. When a current is passed through a coil placed around the iron, in order to magnetise it, the molecules become arranged in a definite order. Up to certain stage relatively few ampere-turns of applied mmf are sufficient to arrange the greater part of the molecules in a definite order. In this stage of magnetisation, the amount of flux established in the iron increases almost directly with the increase in ampere-turns applied. However, above this point, which is called the *saturation point,* it becomes increasingly difficult to magnetise the iron further since the unmagnetised molecules become fewer and fewer. Above the saturation point, when much larger increase in ampere-turns are required for corresponding increase in flux in the iron, the iron is said to be saturated. The fact that iron has become saturated does not mean that a further increase in magnetism is impossible; the increase in magnetism merely requires very much larger increase in applied ampere-turns than before the iron become saturated. The curve giving relation between flux per pole and field ampere-turns per pole is known as *magnetisation or saturation curve.*

Saturation curve for a dc generator does not start from zero ordinarily but at some value slightly higher than zero due to *residual magnetism.* As discussed above for lower values of field ampere-turns, the flux is approximately proportional to the field ampere-turns and, therefore, saturation curve is a straight line. At point 'c' the field core begins to become saturated and the curve falls away from the straight line due to varying permeability of iron.

EMF generated in a generator is given by an expression

$$E_g = \Phi Z \frac{N}{60} \times \frac{P}{A} \text{ volts} = \left(\frac{Z}{60} \times \frac{P}{A}\right) \Phi N \text{ volts}$$

The quantity in the bracket is constant for a given machine, let it be K then $E_g = K\Phi N$

i.e. generated emf in a generator is directly proportional to the product of flux per pole and speed. If the generator runs at a constant speed, then generated emf is directly proportional to the flux per pole.

Since the generated emf is proportional to flux per pole if speed is constant and field ampere-turns are proportional to field current I_f, if a curve is drawn between generated emf on no load and field current, when the machine is running at a constant speed, then curve so obtained will be similar to saturation curve. Such curves are known as *magnetic characteristics or open-circuit characteristics* (OCC). Typical OCC is shown in Fig. 7.9 (*b*).

The curve starts from point '*a*' instead of 'O' when the field current is zero which is due to residual magnetism. The curve from point b to point c in Fig. 7.9 (*b*) is practically a straight line and is called the *straight line portion* of the magnetization curve. Point *c*, at which saturation of the magnetic circuit begins is known as the *knee* of the curve. The portion of the curve from point *c* to point d is described as the curve above the knee.

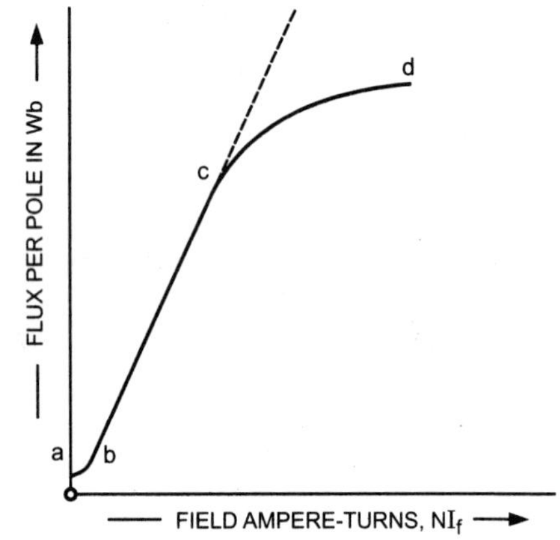

Fig. 7.9 *(a) Saturation or Magnetisation Curve*

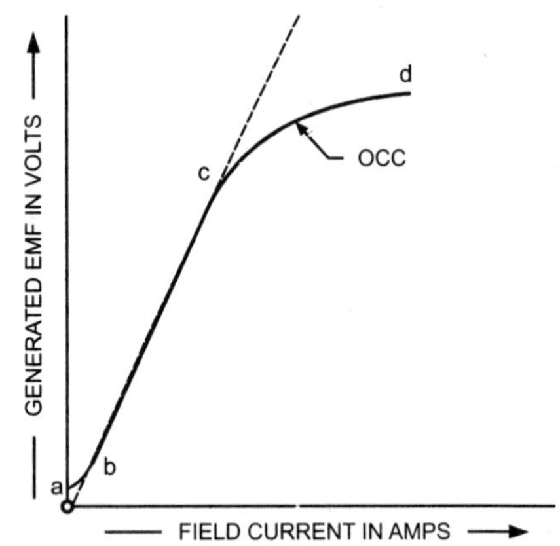

Fig. 7.9 *(b) Magnetic or Open-Circuit Characteristics*

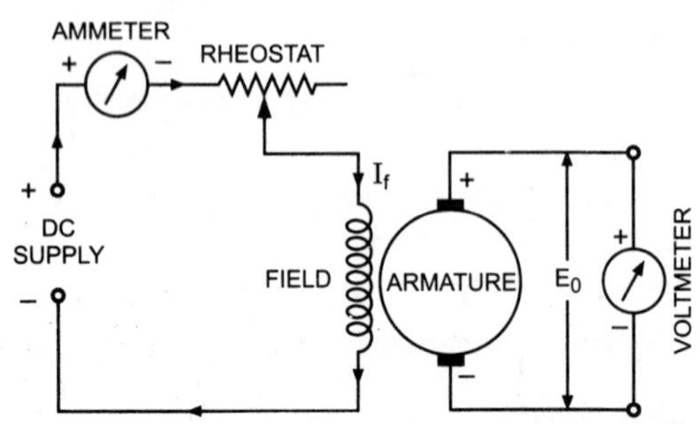

Fig. 7.10 *Experimental Determination of Saturation Curve*

7.3.1. Experimental Determination of Saturation Curve. To determine the saturation curve experimentally, the field winding of the generator (whether series wound or shunt wound) is disconnected from the machine and, instead, is connected in series with an ammeter and rheostat, and put across *dc* supply mains. The field or exciting current is varied with the help of rheostat and its value is read from the ammeter connected in the field circuit for this purpose. The voltmeter across armature terminals measures induced armature emf. As the voltage drop within the armature due to the voltmeter current is negligible so the terminal voltage and the induced emf are identical under these conditions. Keeping the speed of rotation of armature constant, the readings of ammeter and voltmeter are noted for various positions of rheostat connected in the field circuit. On plotting the curve between induced emf (voltmeter reading) and field current (reading of ammeter), *saturation curve* is obtained.

7.4 FIELD RESISTANCE LINE

By Ohm's law, in a simple circuit of constant resistance the current flowing through it is proportional to voltage applied across it *i.e.* $I \propto V$.

If a curve is drawn between V and I, a straight line passing through the origin O is obtained. For example if the resistance of the field circuit is 50 Ω, current will be 3 A when the voltage across it is 150 V; current will be 2 A for voltage of 100 V, 1 A for voltage of 50 V and 0.5 A for voltage of 25 V. If a curve is drawn showing above relations, curve I, shown in Fig. 7.11 is obtained which is known as 50 Ω field resistance line.

Similarly curves No. (II) and (III) give 100 Ω and 150 Ω resistances respectively.

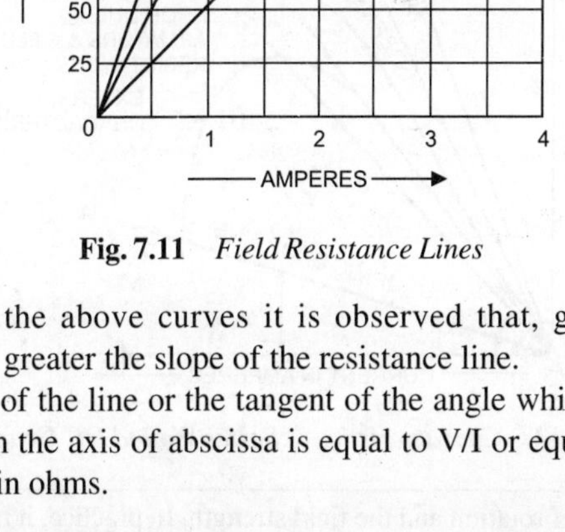

Fig. 7.11 *Field Resistance Lines*

From the above curves it is observed that, greater the resistance, greater the slope of the resistance line.

Slope of the line or the tangent of the angle which the line makes with the axis of abscissa is equal to V/I or equal to field resistance in ohms.

7.5 CHARACTERISTICS OF DC GENERATORS

The curves or graphs giving the relationship between various quantities such as excitation (or field current), generated emf, terminal voltage and the load current etc. are known as *generator characteristics*.

There are three most important characteristics of a dc generator, which are generally studied.

1. Magnetic or Open-Circuit Characteristic. This characteristic, also known as *no-load saturation characteristic*, gives the relation between the generated emf in armature on no load, E_0 and field current, I_f at a given speed. The shape of this characteristic curve is practically the same for all types of generators whether they are separately excited or self-excited.

2. Internal or Total Characteristic. It is a curve which gives the relation between the emf actually generated in the armature, E (less than E_0 due to the demagnetizing effect of armature reaction) and the armature current. This curve, therefore, lies below the OCC.

3. External Characteristic. This curve gives the relation between terminal voltage V and the load current I under the given conditions of speed and excitation. The terminal voltage V is always less than E on account of voltage drop in armature. This curve, therefore, lies below the internal characteristic. This characteristic is of utmost importance for determining the suitability of a generator for a particular service.

7.6 CHARACTERISTICS OF SEPARATELY-EXCITED DC GENERATORS

In a separately-excited dc generator the field coils are energised from an independent source, as already explained in Art 7.1.1.

Since exciting current is independent of load or armature current, if a curve is drawn between flux per pole and load current, keeping field current constant and neglecting armature reaction, a straight line parallel to X-axis is obtained as indicated by curve I in Fig. 7.12. But due to armature reaction, the curve of actual flux is slightly droping as represented by curve II in Fig. 7.12.

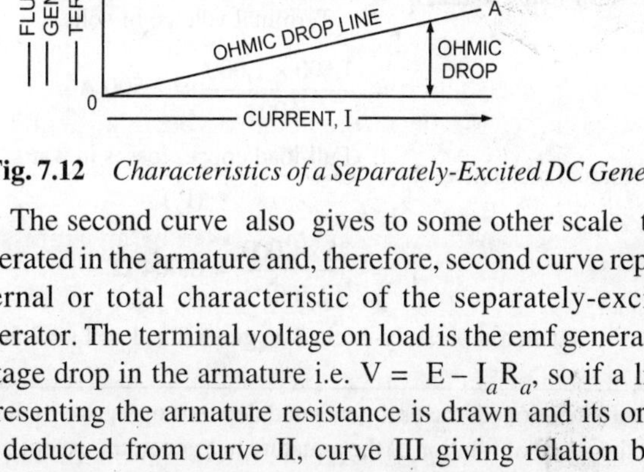

Fig. 7.12 *Characteristics of a Separately-Excited DC Generator*

The second curve also gives to some other scale the emf generated in the armature and, therefore, second curve represents internal or total characteristic of the separately-excited dc generator. The terminal voltage on load is the emf generated less voltage drop in the armature i.e. $V = E - I_a R_a$, so if a line OA representing the armature resistance is drawn and its ordinates are deducted from curve II, curve III giving relation between terminal voltage and load current, known as *external characteristic* is obtained. From curve III, it is obvious that with the increase in load current, terminal voltage decreases slightly.

This decrease can easily be neutralized* by increasing field current slightly and, therefore, constant terminal voltage can be maintained.

The separately-excited dc generator has a decided advantage over the self-excited generators—it operates in a stable condition with any field excitation. Thus a wide range of output voltage may be obtained.The main disadvantage of a separately-excited generator lies in the inconvenience and expense of providing the separate excitation source. For this reason, the use of this type of generator is largely confined to experimental and testing laboratories where such a source is available and a wide variation of output voltage is desirable.

Example 7.21. A separately excited generator, when running at 1,000 rpm supplied 200 A at 125 V. What will be the load current when the speed drops to 800 rpm if I_f is unchanged ? Given that the armature resistance = 0.04 Ω and brush drop = 2 V.

[*Nagpur Univ. Elec. Machines, 1993*]

Solution: Load current, $I_1 = 200$ A

Terminal voltage, $V_1 = 125$ V

Generated emf, $E_{g1} = V_1 + I_a R_a + \text{brush drop}$

$= 125 + 200 \times 0.04 + 2 = 135$ V

Speed, $N_1 = 1{,}000$ rpm

EMF generated at speed N_2 of 800 rpm,

$$E_{g2} = E_{g1} \times \frac{N_2}{N_1} = 135 \times \frac{800}{1{,}000} = 108 \text{ V}$$

$$\text{Load resistance, } R_L = \frac{V_1}{I_1} = \frac{125}{200} = 0.625\ \Omega$$

Let the new load current be I_2 and new terminal voltage V_2

Now $V_2 = E_{g2} - I_2 R_a - \text{brush drop}$

$= 108 - 0.04\, I_2 - 2 = 106 - 0.04\, I_2$ ...(*i*)

and also $V_2 = I_2 R_L = 0.625\, I_2$ ...(*ii*)

From Eqs. (*i*) and (*ii*) we have

$0.625\, I_2 = 106 - 0.04\, I_2$

$$\text{or } I_2 = \frac{106}{0.665} = 159.4 \text{ A } \textbf{Ans.}$$

Example 7.22. A 1500 kW, 600 V, 16-pole separately-excited dc generator runs at 200 rpm. It has 2,500 lap-connected conductors and full-load copper losses are 25 kW. Calculate the useful flux per pole. [*U.P. Technical Univ. Electrical Machines 2004-05*]

$$\textbf{Solution: } \text{Full-load current, } I_L = \frac{\text{Output power in kW} \times 1{,}000}{\text{Terminal voltage in volts}}$$

$$= \frac{1{,}500 \times 1{,}000}{600} = 2{,}500 \text{ A}$$

$$\text{Armature resistance, } R_a = \frac{\text{Full-load copper losses in watts}}{(I_a^2)}$$

$$= \frac{25 \times 1{,}000}{(2{,}500)^2} = 0.004\ \Omega$$

∵ In a separately-excited dc generator, $I_a = I_L = 2{,}500$ A

Generated emf, $E_g = V + I_a R_a$

$= 600 + 2{,}500 \times 0.004 = 610$ V

$$\text{Useful flux per pole, } \Phi = \frac{E_g \times 60 \times A}{Z \times N \times P} = \frac{610 \times 60 \times 16}{2{,}500 \times 200 \times 16}$$

$= 73.2$ mWb **Ans.**

7.7 CHARACTERISTICS OF SERIES WOUND DC GENERATORS

In a series wound dc generator, the armature winding, field winding and external load circuit, all are connected in series with each other, as shown in Fig. 7.2 and, therefore, the same current flows through all parts of the circuit.

1. Magnetic Characteristic. If a curve is drawn between flux per pole and field current, the magnetic characteristic is obtained, as represented by curve I in Fig. 7.13. This curve starts from point F little way up the field axis due to residual magnetism.

2. Internal Characteristic. If there would not have been demagnetising effect of armature reaction, this curve would have given the total flux, when the machine is supplying current to the external load circuit, but due to armature reaction the actual curve lies somewhat below this curve, as represented by curve II. Since the emf generated in the armature is directly proportional to the actual flux at a constant speed, this curve also gives emf generated in the armature to some other scale. This curve is internal or total characteristic because it also gives relation between generated emf E_g and armature current I_a since $I_a = I_f$.

3. External or Load Characteristic. The load current, armature current and field current in series wound generator are the same as already explained. The terminal voltage of the generator is equal to the emf generated E_g less the voltage drop in armature and series field respectively. If R_a and R_{se} are the resistances of the armature and series field respectively and load current is I, then the voltage drop in armature and series field = I

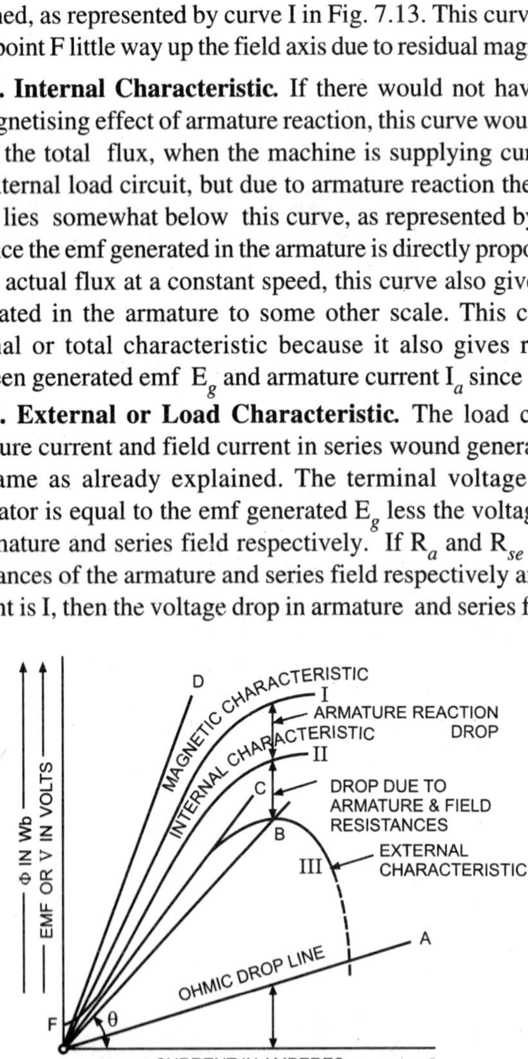

Fig. 7.13 *Characteristics of Series Wound DC Generators*

---

* As already mentioned, the emf generated depends upon two main factors; the speed of rotation and the field strength. In practice, it is not easy to control the speed of rotation of the armature in a generator except by varying the speed of the prime mover (driving engine), and this cannot be done within close limits. In addition, there is usually some convenient speed at which the engine runs most economically, and it is undesirable that it should be varied from this speed merely to adjust the voltage output of the generator. Therefore, it is customary to vary the magnetic flux, and this can be accomplished easily by inserting a variable resistance in the field circuit. This permits a wide range of control of the field current and, therefore, a wide range of the induced emf.

$(R_a + R_{se})$ and terminal voltage, $V = E_g - I\,(R_a + R_{se})$. This voltage drop for different values of load current may be represented by a straight line OA. So if the ordinates of this ohmic drop line OA are deducted from those of internal characteristic, curve III giving the relation between terminal voltage and load current, known as *external characteristic,* is obtained. From external characteristic it is observed that the terminal voltage first increases with the increase in load current, reaches the maximum value and finally decreases. If the load circuit resistance is reduced sufficiently, the terminal voltage may fall to zero. So if the series generator is operated on initial straight portion of the characteristic, it gives voltage approximately proportional to the current and if it is operated on drooping portion of the characteristic, it gives approximately constant current irrespective of the external load circuit resistance.

Critical Load Resistance. If a straight line OB passing through the origin is drawn, then its point of intersection with the external characteristic gives terminal voltage on Y-axis and current on X-axis when the resistance of the external load circuit is equal to the resistance represented by the slope of this line *i.e.* when load resistance = gradient of line OB. Thus if series of such lines are drawn, the points of intersection with the external characteristic will give terminal voltage and load current corresponding to their resistances in the external load circuit. If a line tangential to the external characteristic and passing through origin O, neglecting the initial ordinate due to residual magnetism such as OC, is drawn, the resistance represented by the slope of this line is known as *critical resistance,* since it is the maximum resistance with which the generator will be able to excite. If the resistance in the external load circuit is greater than this, as represented by slope of line OD, the generator will not excite since OD has no point of intersection with the external characteristic, and so it will deliver no current.

Example 7.23. A series generator of total resistance 0.5 Ω is running at 1,000 rpm and delivering 5 kW at a terminal *pd* of 100 V. If the speed is raised to 1,500 rpm and the power is adjusted to 8 kW, find the new current and terminal voltage. Assume that the machine is working on the straight portion of the characteristic such that the flux is proportional to the current and the emf generated is proportional to the product of flux and speed.

[*U.P.S.C. I.E.S. Elec. Engineering-II, 2004*]

Solution: *At 1,000 rpm*

$$\text{Load current, } I_1 = \frac{\text{Output, } P_1}{\text{Terminal voltage, } V_1} = \frac{5 \times 1{,}000}{100} = 50 \text{ A}$$

$$\text{Generated emf, } E_g = V_1 + I\,(R_a + R_{se}) = 100 + 50 \times 0.5 = 125 \text{ V}$$

At 1,500 rpm: Let the current supplied by the generator be I_2 at terminal voltage of V_2 volts.

$$\text{Now } V_2 = \frac{P_2}{I_2} = \frac{8 \times 1{,}000}{I_2} = \frac{8{,}000}{I_2} \quad ...(i)$$

$$\text{Generated emf, } E_{g_2} = V_2 + I_2\,(R_a + R_{se}) = \frac{8{,}000}{I_2} + 0.5\,I_2 \quad ...(ii)$$

Since generated emf is directly proportional to the product of flux and speed,

i.e. $E_{g_2} \propto \Phi N$,

$$\frac{E_{g_2}}{E_{g_1}} = \frac{\Phi_2}{\Phi_1} \times \frac{N_2}{N_1} = \frac{I_2}{I_1} \times \frac{N_2}{N_1}$$

$$\text{or} \quad \frac{\dfrac{8{,}000}{I_2} + 0.5\,I_2}{125} = \frac{I_2}{50} \times \frac{1{,}500}{1{,}000}$$

$$\text{or} \quad \frac{8{,}000}{I_2} + 0.5\,I_2 = 3.75\,I_2$$

$$\text{or} \quad I_2 = \sqrt{\frac{8{,}000}{3.25}} = 49.61 \text{ A} \quad \textbf{Ans.}$$

$$\text{and} \quad V_2 = \frac{8{,}000}{I_2} = \frac{8{,}000}{49.61} = 161.24 \text{ V} \quad \textbf{Ans.}$$

Example 7.24. A dc generator having an external characteristic which is a straight line through zero to 50 V at 200 A, is connected as a booster between a station bus-bar and a feeder of 0.3 Ω resistance. Calculate the voltage between the far end of the feeder and the bus-bar at a current of (*i*) 160 A (*ii*) 50 A.

Solution: The circuit arrangement of the booster is shown in Fig. 7.14.

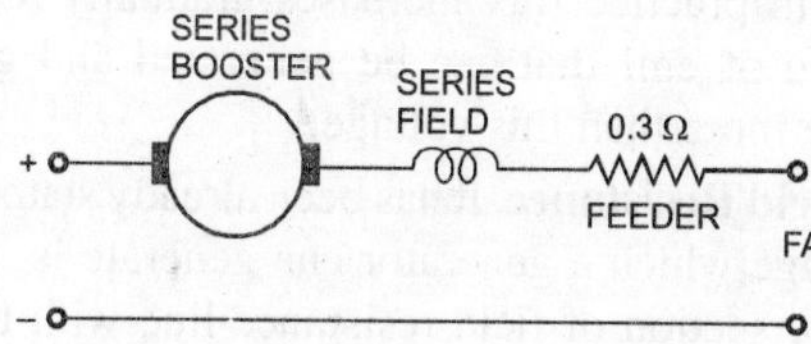

Fig. 7.14

(*i*) Load current, $I_L = 160$ A

$$\text{Voltage rise due to booster} = 50 \times \frac{160}{200} = 40 \text{ V}$$

$$\text{Voltage drop in feeder} = I_L \times \text{feeder resistance} = 160 \times 0.3 = 48 \text{ V}$$

Voltage difference between the station bus-bar and far end of the feeder

$$= 40 - 48 = -8 \text{ V} \quad \textbf{Ans.}$$

(*ii*) Load current, $I_L = 50$ A

$$\text{Voltage rise due to booster} = 50 \times \frac{50}{200} = 12.5 \text{ V}$$

$$\text{Voltage drop in feeder} = 50 \times 0.3 = 15 \text{ V}$$

Voltage difference between the station bus-bar and for end of the feeder

$$= 12.5 - 15 = -2.5 \text{ V} \quad \textbf{Ans.}$$

7.8 CHARACTERISTICS OF SHUNT WOUND DC GENERATORS

Building Up of Voltage of a Shunt Generator at No Load. The curve plotted between the generated emf and shunt field current will be similar to that shown in Fig. 7.15. The generator excites itself due to residual magnetism and develops the voltage as described below :

Line OP represents the shunt field resistance. When the generator is started a small emf (represented by OA in the Fig.) is induced due to residual magnetism. This induced emf causes a flow of current *oa'* in the field circuit. This is obtained by drawing a horizontal line from point A meeting the field resistance line at point '*a*' and then from point '*a*' drawing perpendicular line

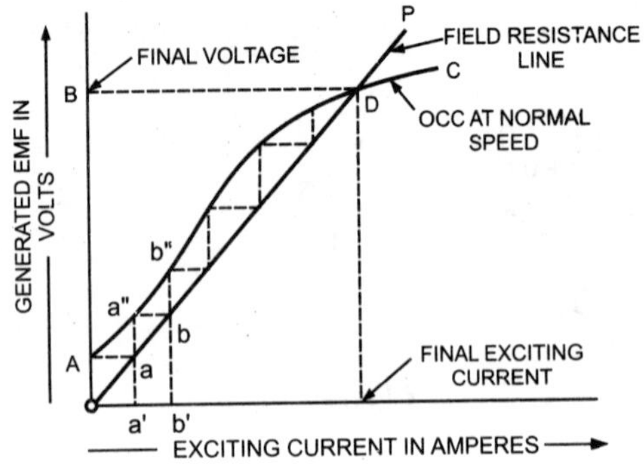

Fig. 7.15 *Building Up of Voltage of a Shunt Generator At No Load*

meeting current axis at a'. When field current is Oa', *the* generated emf. is $a'a''$ which produces a field current Ob', which in turn produces a high voltage $b'b''$. Thus it is observed that effect is cumulative and value of induced emf and field current increases until these reach point D, the point of intersection of the shunt field resistance line and magnetic characteristic obtained. In Fig. 7.15 instantaneous increases in induced emf and field current are shown but in practice flux increases gradually. OB is the maximum value of emf that can be generated and generator cannot generate more than this voltage.

Critical Field Resistance. It has been already stated that the maximum voltage which a generator can generate is given by the point of intersection of field resistance line with the open-circuit characteristic. If the line OA represents the field resistance, then the maximum emf generated is Oa_1, if the field resistance is increased to the value represented by line OB, the generated voltage will be Ob_1 and if it is further increased so that the line representing the field resistance becomes tangent to the curve, as line OC, then emf generated is Oc_1 and the value of resistance given by this line is called the *critical resistance* (Fig 7.16). If field resistance is further increased beyond critical resistance, as represented by line OD the generator will not excite because the line OD has got no point of intersection with the magnetic characteristic (ignoring initial portion of the curve).

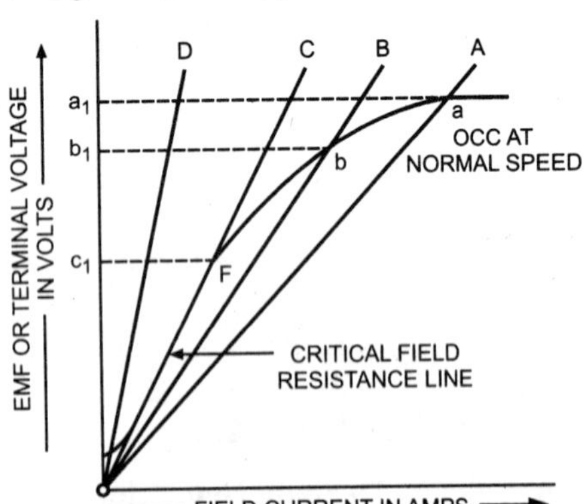

Fig. 7.16

Load Characteristic. It has been observed that the terminal voltage, on no load, of the shunt generator depends upon shunt field resistance and let it be as represented by OA in Fig. 7.17. As the generator is loaded by decreasing the resistance of the external load circuit, the terminal voltage falls. This is due to following three causes.

1. The voltage drop across the armature winding and the contact resistance of the brushes increases with the increase in armature current.
2. As the load is increased, the current in armature conductors increases, so the effect of armature reaction increases and, therefore, field is weakened. Due to weakening of field the induced emf is reduced and, therefore, terminal voltage falls.
3. The decrease in terminal voltage as a result of the first two factors causes decrease in field current. This will in turn cause the emf and, therefore, terminal voltage of the generator to drop too.

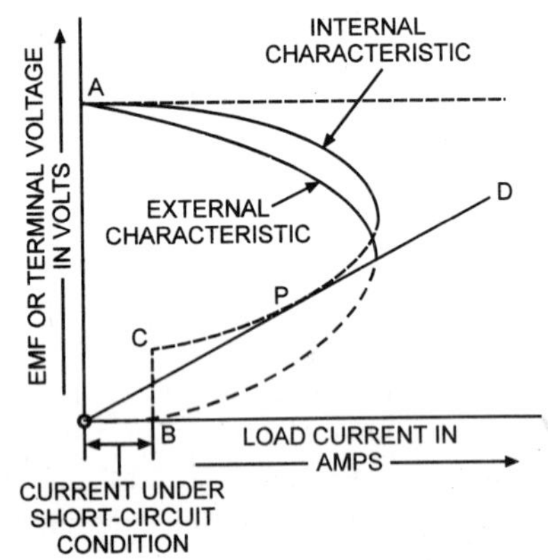

Fig. 7.17 *Load Characteristics of Shunt Wound Generator*

Thus the effects are cumulative and terminal voltage is reduced to a much greater extent. In the beginning the effect of decrease in resistance predominates over the effect of decreased terminal voltage but when the load current reaches a certain value (much higher than the full-load current) the demagnetising effect of armature reaction and voltage drop in the armature become so important that a further decrease in load resistance causes a decrease in current rather than increase and so the characteristic turns back.

When the terminals are short circuited, there is no terminal voltage and thus the shunt winding becomes inactive but a small current is established due to a small voltage induced by the residual magnetism. This external characteristic meets the current axis at point B. *The shunt generators are, therefore, self-protective against accidental short circuit**.

If the armature drop is added to the terminal voltage for all points and plotted against armature current the new curve obtained will give relation between induced emf and armature current.

At the time of short circuit, as already stated, the terminal voltage is zero, the generated emf which is by virtue of residual magnetism is BC and the short-circuit current is OB. Actually generated emf, BC is very small because the residual magnetism will be almost completely neutralized by armature reaction. Sometimes it may even be reversed.

It is for this reason that shunt generators often fail to build up after they have been shut down through a severe type of short circuit.

* If a dc shunt generator is suddenly short-circuited, it will carry a very large current for a moment, because, although the terminal voltage is zero, or nearly zero, the inductance of the field coils tends to maintain the field current. Within a second or two, however, the field current will drop practically to zero, and the armature current will then be merely that produced by the residual magnetism.

Critical Load Resistance: If a line from the origin O is drawn tangential to internal characteristic curve, the gradient of this line will give the minimum value of the external load resistance for which the generator will excite on load. This resistance is known as *critical load resistance.* If the resistance of the external load circuit is less than this value, the generator will fail to build up its voltage.

7.8.1. Determination of Critical Resistance From OCC For Shunt Wound DC Generator. Plot the OCC for the shunt wound generator from given datas. Now if a tangent line is drawn to its initial portion, then the slope of this line will give the value of critical resistance for the generator at the speed, at which OCC datas were obtained.

To determine the slope of the tangent line take any point on this line, read its ordinate in volts and abscissa in amperes. Now the value of critical resistance may be determined from the following relation.

$$\text{Critical resistance in ohms} = \text{Slope of tangent line} = \frac{\text{Ordinate in volts}}{\text{Abscissa in amperes}}$$

7.8.2. Determination of OCC At Different Speed (say N_2) From OCC Drawn For Some Other Speed (Say N_1). If flux is kept constant, generated emf will be directly proportional to speed.

i.e. mathematically $E \propto N$ (for given excitation)

Let from OCC given at speed N_1, the induced emf be E_1 for excitation of I amperes.

Now the induced emf at speed N_2 and for same excitation of I amperes will be given by

$$\frac{E_2}{E_1} = \frac{N_2}{N_1} \quad \text{or} \quad E_2 = E_1 \times \frac{N_2}{N_1}$$

i.e. for given excitation the induced emf at speed N_2 will be $\frac{N_2}{N_1}$ times of induced emf at speed N_1

If any ordinate AD is drawn. The point B, the point of intersection of ordinate AD and given OCC gives the induced emf AB for field current OA. Now the induced emf at speed N_2 for field current OA will be $\frac{N_2}{N_1}$ times of induced emf at speed N_1

$$i.e.\ E_2 = E_1 \times \frac{N_2}{N_1} = AB \times \frac{N_2}{N_1} = AC$$

Thus point C lying on the OCC for speed N_2 is obtained. Similarly other such points can be obtained and new OCC can be drawn passing through such points.

Hence to obtain New OCC

(i) draw given OCC,

(ii) draw any ordinate intersecting the given OCC as AD in Fig. 7.18,

(iii) obtain point C on this ordinate by multiplying the intercept of ordinate between the given OCC and abscissa (as AB in Fig. 7.18) by the ratio of new speed to old speed,

(iv) similarly draw some more ordinates and obtain such more points, and

(v) draw the smooth curve passing through the points obtained.

The curve so drawn will be OCC at new speed.

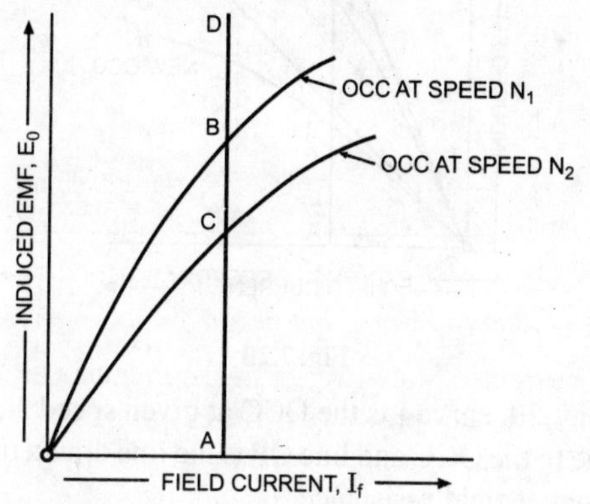

Fig. 7.18

7.8.3. Determination of Speed To Reduce Generated EMF Keeping Shunt Field Resistance Unchanged. In Fig. 7.19, curve I is the OCC at given speed N_1, OA is the line drawn representing the shunt field resistance, point A is the point of intersection of line OA and given OCC. The ordinate of point A represents the induced emf E_1.

Let curve II be the OCC drawn at speed N_2 so that the ordinate of point B, the point of intersection of shunt field resistance line OA and new OCC, may represent the induced emf, E_2.

Now if from point B an ordinate CBD is drawn intersecting given OCC at point C, then new speed will be given by

$$N_2 = \frac{DB}{DC} N_1$$

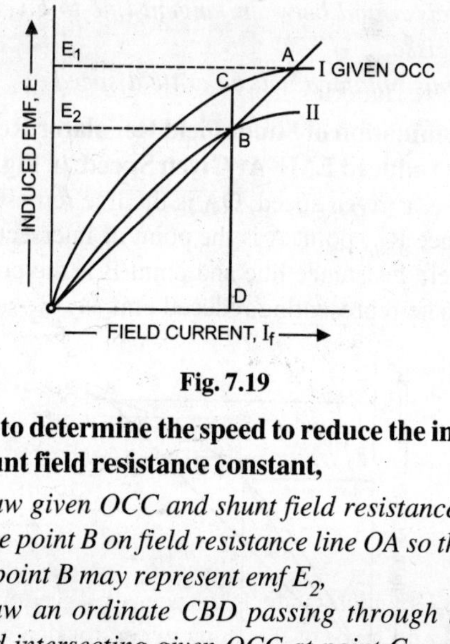

Fig. 7.19

Hence to determine the speed to reduce the induced emf, keeping shunt field resistance constant,

(i) draw given OCC and shunt field resistance line,

(ii) take point B on field resistance line OA so that ordinate of point B may represent emf E_2,

(iii) draw an ordinate CBD passing through the point B and intersecting given OCC at point C, and

(iv) multiply the given speed by the ratio of DB and DC.

Speed thus obtained will be the speed, at which induced emf will be E_2 for given shunt field resistance.

7.8.4. Determination of Critical Speed. *Critical speed of a shunt wound generator is that speed for which the given shunt field resistance will represent critical field resistance.*

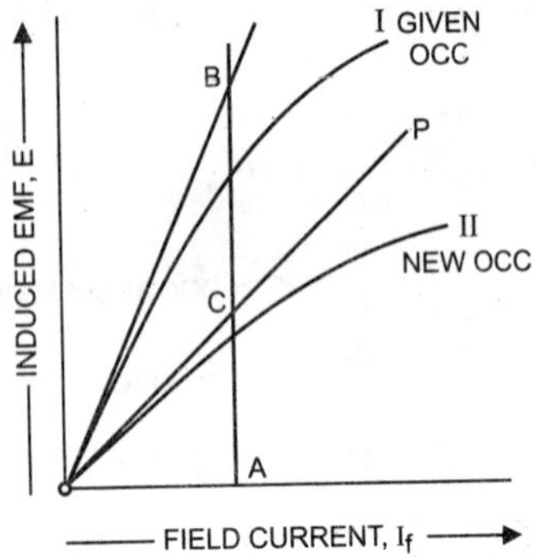

Fig. 7.20

In Fig. 7.20, curve I is the OCC at given speed N, OB is the tangent line to the OCC and line OP is the line drawn to represent the given shunt field resistance.

Let curve II be OCC drawn at the critical speed N_c, to be determined.

By definition of critical speed line OP will be tangent to the new OCC.

If any ordinate AB is drawn intersecting line OP at C then

$$\frac{AC}{AB} = \frac{N_c}{N}$$

$$\text{or } N_c = \frac{AC}{AB} \times N$$

Hence to determine critical speed

(i) draw given OCC, a line tangent to given OCC and line representing the given shunt field resistance,

(ii) draw any ordinate intersecting the field resistance line and tangent line to given OCC and

(iii) multiply the given speed by the ratio of the intercepts of ordinate drawn between field resistance line and abscissa and between tangent line to given OCC and abscissa.

Speed thus obtained will be critical speed.

7.8.5. Determination of Shunt Field Regulator Resistance To Reduce The Induced EMF At Given Speed. In Fig. 7.21, curve shows OCC at a given speed, OA is the line representing shunt field resistance R_{sh}, point A is the point of intersection of OCC and shunt field resistance line and point B is the point on OCC having ordinate representing reduced emf say E_2.

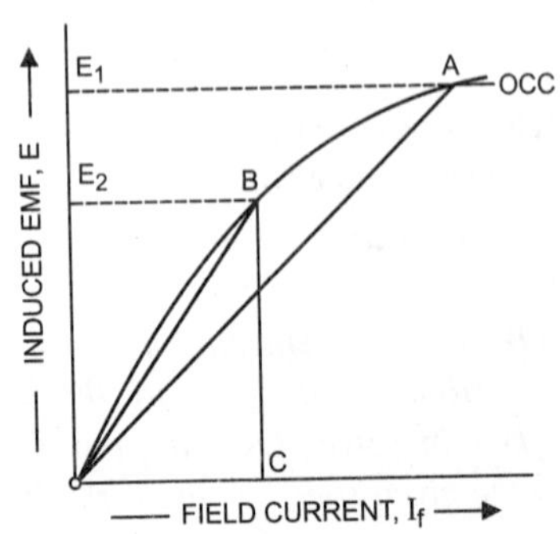

Fig. 7.21

If the line joining point B and point O is drawn, then the slope of this line will represent the total resistance required in the shunt field circuit, say R'_{sh}

$$\text{i.e. } R'_{sh} = \frac{BC}{OC}$$

Hence to determine additional shunt field resistance to reduce the induced emf at given speed

(i) draw OCC, horizontal lines E_1A and E_2B intersecting OCC at points A and B respectively,

(ii) join OA and OB

(iii) determine the shunt field resistance R_{sh} and total resistance in the shunt field, R'_{sh} to induce emf E_2 from the slopes of the lines OA and OB respectively.

Difference of R'_{sh} and R_{sh} gives the additional shunt regulator resistance.

Example 7.25. The open-circuit characteristic of a shunt wound dc generator at 800 rpm gives:

Field current (A)	**0**	**0.5**	**1.0**	**2.0**	**3.0**	**4.0**	**5.0**
Induced emf (V)	**10**	**50.0**	**100.0**	**175.0**	**220.0**	**245.0**	**262.0**

Find graphically the critical resistance of shunt field circuit. If the field circuit resistance is changed to 75 Ω, what will be the critical speed for the machine to build up.

Solution: Draw OCC from given datas. Draw the line OC tangent to OCC at the origin. The slope of the line will give the value of critical resistance. To determine the slope of the line OC, take any point B on this line and from point B draw horizontal and vertical lines meeting induced emf axis and field current axis at point F and A respectively. Critical resistance of the field circuit

$$= \text{Slope of line OC}$$

$$= \frac{\text{OF in volts}}{\text{OA in amperes}} = \frac{200}{2} = 100\ \Omega \textbf{ Ans.}$$

Draw line OD representing resistance of 75 Ω.

Let ordinate drawn from point B intersect line OD at G.

$$\text{Critical speed, } N_c = N \times \frac{AG}{AB} = 800 \times \frac{150}{200} = 600 \text{ rpm } \textbf{Ans.}$$

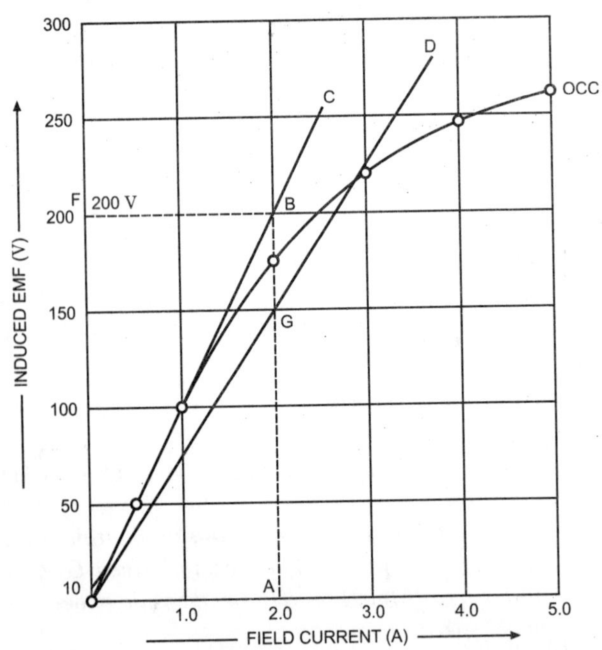

Fig. 7.22

Example 7.26. The magnetisation curve of a dc shunt generator at 1,500 rpm is:

I_f (A)	0	0.4	0.8	1.2	1.6	2.0	2.4	2.8	3.0
E_0 (V)	6	60.0	120.0	172.5	202.5	221.0	231.0	237.0	240.0

For this generator find (*i*) no-load emf for a total shunt field resistance of 100 Ω (*ii*) the critical field resistance at 1,500 rpm and (*iii*) the magnetization curve at 1,200 rpm and therefrom the open-circuit voltage for a field resistance of 100 Ω.

Solution: Draw OCC from given data, as illustrated in Fig. 7.23.
(*i*) Draw 100 Ω (shunt field resistance) line by joining origin (0, 0) and point (1 A, 100 V). The voltage corresponding to point A, the intersecting point of OCC drawn and shunt field resistance line, is 227.5 V. Hence no-load emf is 227.5 V **Ans.**

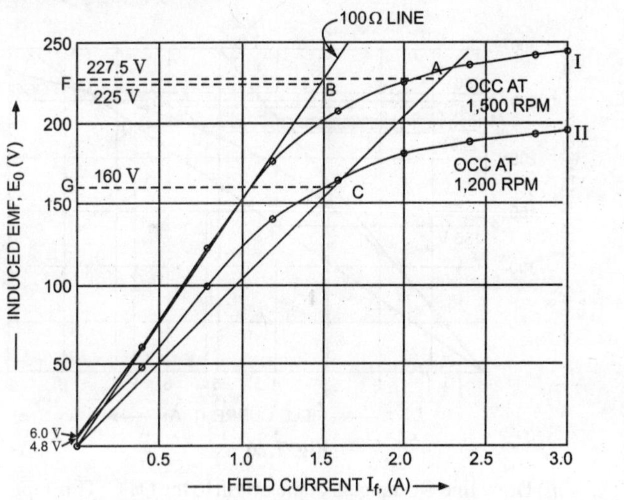

Fig. 7.23

(*ii*) Draw line OB tangent to OCC at the origin. The critical field resistance = Slope of line OB

$$= \frac{\text{Ordinate of B in volts}}{\text{Abscissa of B in amperes}} = \frac{225}{1.5} = 150\ \Omega\ \textbf{Ans.}$$

(*iii*) For 1,200 rpm, the induced emfs for different field currents would be $\left(\frac{1,200}{1,500}\right)$ *i.e.* 0.8 times of those for 1,500 rpm.

The magnetisation curve at 1,200 rpm is given below:

I_f (A)	0	0.4	0.8	1.2	1.6	2.0	2.4	2.8	3.0
E_0 (V)	4.8	48.0	96.0	138.0	162.0	176.8	184.8	189.6	192.0

The voltage corresponding to point C, the intersecting point of OCC drawn at 1,200 rpm and 100 Ω resistance line is 160 V. So no-load induced emf is 160 V **Ans.**

Example 7.27. The open-circuit characteristic of a separately excited dc generator driven at 1,000 rpm is as follows :

Field current :	**0.2**	**0.4**	**0.6**	**0.8**	**1.0**	**1.2**	**1.4**	**1.6**
E M F volts :	**30.0**	**55.0**	**75.0**	**90.0**	**100.0**	**110.0**	**115.0**	**120.0**

If the machine is connected as shunt generator and driven at 1,000 rpm and has a field resistance of 100 Ω, find (*a*) open-circuit voltage and exciting current (*b*) the critical resistance and (*c*) resistance to induce 115 V on open circuit. [*Nagpur Univ. Elec. Machines 1993*]

Solution: Draw OCC from given data, as illustrated in Fig. 7.24.

(*a*) Draw 100 Ω resistance by joining origin (0,0) and point (1 A,100 V). The voltage corresponding to point A the intersecting point of OCC and 100 Ω field resistance line, is 100 V and corresponding current is 1.0 A.

So open- circuit voltage = 100 V **Ans.**

Exciting current = 1.0 A **Ans.**

(*b*) Draw line OB tangent to OCC at the origin.

The critical field resistance = Slope of line OB

$$= \frac{120}{0.8} = 150\ \Omega\ \textbf{Ans.}$$

(*c*) From Fig. 7.24, it is clear that the exciting current corresponding to open- circuit voltage of 115 V is 1.4 A.

So resistance required in field circuit $= \frac{115}{1.4} = 82.14\ \Omega$ **Ans.**

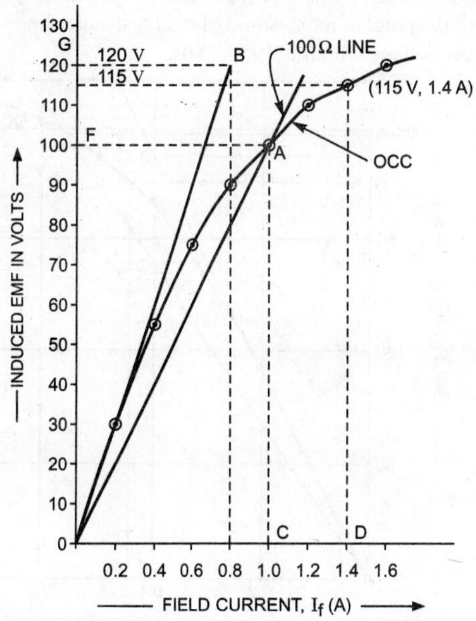

Fig. 7.24

Example 7.28. A shunt generator gave the following results in the OCC test at a speed of 800 rpm.

Field current (A):	**1**	**2**	**3**	**4**	**6**	**8**	**10**
EMF (V):	**90**	**185**	**251**	**290**	**324**	**345**	**360**

The field resistance is adjusted to 50 Ω and the terminal voltage is 300 V. Armature resistance is 0.1 Ω and assuming that flux is reduced by 5% due to armature reaction find the load supplied by the generator.

Solution: Terminal voltage, V = 300 V

Shunt field resistance, R_{sh} = 50 Ω

$$\text{Shunt field current, } I_{sh} = \frac{V}{R_{sh}} = \frac{300}{50} = 6\ A$$

From the given data for OCC, the no-load induced emf corresponding to field current of 6 A is 324 V.

Due to armature reaction, the flux is reduced by 5 per cent, so the induced emf being proportional to flux at a given speed is also reduced in the same ratio (5%).

Hence induced emf on load, $E_g = (1 - 0.05) \times 324 = 307.8$ V

$$\text{Armature current, } I_a = \frac{E_g - V}{R_a} = \frac{307.8 - 300}{0.1} = 78\ A$$

Load current $I_L = I_a - I_{sh} = 78 - 6 = 72$ A

$$\text{Load supplied by generator, } P = \frac{VI_L}{1,000} = \frac{300 \times 72}{1,000} = 21.6\ \text{kW}\ \textbf{Ans.}$$

Example 7.29. The open-circuit characteristic of a dc shunt generator at 300 rpm is:

I_f (A):	0	0.2	0.3	0.4	0.5	0.6	0.7
E_0(V):	7.5	93.0	135.0	165.0	186.0	202.0	215.0

The field resistance is adjusted to 354.5 Ω and the speed is 300 rpm. Determine the following:

(*i*) Graphically the no-load voltage (*ii*) Critical field resistance (*iii*) Critical speed for the given field resistance (*iv*) Additional resistance inserted in the field circuit to reduce the no-load voltage to 175 V.

Solution: (*i*) Plot the OCC from given data as illustrated in Fig. 7.25. Draw field resistance line OA representing resistance of 354.5 Ω. The ordinate of the point of intersection of field resistance line and OCC gives the no-load voltage which is 195 V **Ans.**

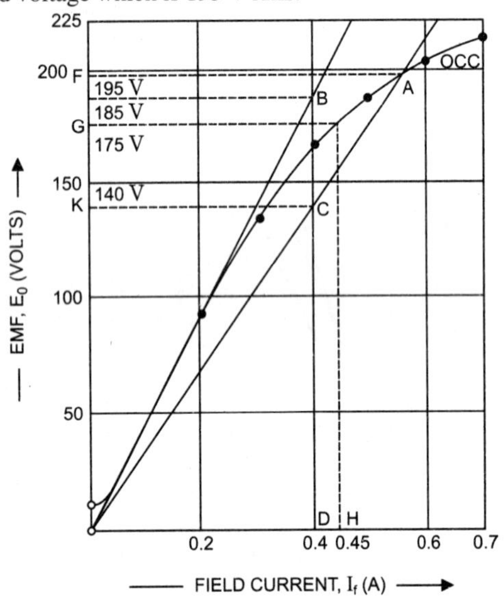

Fig. 7.25

(*ii*) Draw the line OB tangential to OCC. The critical field resistance

$$= \text{Slope of line OB} = \frac{185}{0.4} = 462.5\ \Omega \textbf{ Ans.}$$

(*iii*) Draw ordinate from point B. Let it cut the field resistance line at point C

$$\text{Critical speed } N_c = N \times \frac{DC}{DB} = 300 \times \frac{140}{185} = 227 \text{ rpm}$$

(*iv*) From Fig. 7.25, the field current corresponding to open-circuit voltage of 175 V is 0.45 A.

Field circuit resistance required,

$$R'_{sh} = \frac{175}{0.45} = 388.9\ \Omega$$

Additional resistance required,

$$R = R'_{sh} - R_{sh} = 388.9 - 354.5 = 34.4\ \Omega \textbf{ Ans.}$$

Example 7.30. The OCC of a dc generator driven at 400 rpm is as follows:

Field current (A):	**2**	**3**	**4**	**5**	**6**	**7**	**8**	**9**
Terminal volts:	**110**	**155**	**186**	**212**	**230**	**246**	**260**	**271**

Find (*i*) voltage to which the machine will excite when run as a shunt generator at 400 rpm with shunt field resistance equal to 34 Ω,

(*ii*) resistance of shunt field circuit to reduce the OC voltage to 220 V,

(*iii*) critical value of the shunt field circuit resistance,

(*iv*) the critical speed when the field circuit resistance is 34 Ω and

(*v*) lowest possible speed at which OC voltage of 225 V can be obtained.

Solution: (*i*) Plot OCC from given data, as shown in Fig. 7.26 and draw field resistance line OA representing resistance of 34 Ω. The ordinate of the point of intersection of the field resistance line OA and OCC gives the no-load voltage which is equal to 255 V **Ans.**

(*ii*) From Fig. 7.26 the exciting current corresponding to OC voltage of 220 V is 5.4 A.

Shunt field circuit resistance

$$= \text{Slope of line OB} = \frac{220}{5.4} = 40.74\ \Omega \textbf{ Ans.}$$

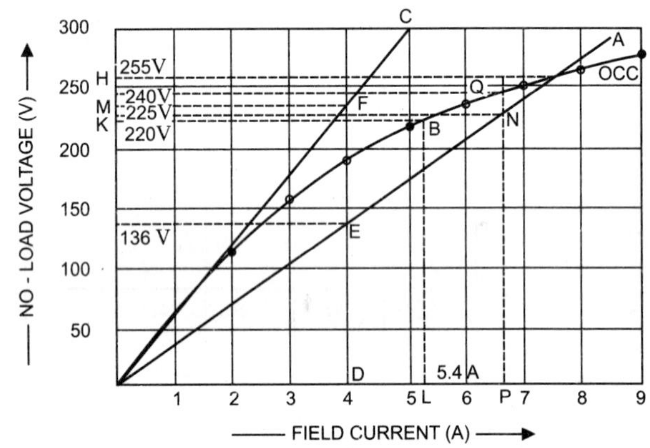

Fig. 7.26

(*iii*) Draw line OC tangent at the origin to the OCC. This represents the critical value of shunt field resistance which is equal to

$$\frac{300}{5} = 60\ \Omega \textbf{ Ans.}$$

(*iv*) Take any convenient point D on field current axis and erect perpendicular intersecting lines OA and OC at points E and F respectively.

$$\text{Critical speed, } N_c = N \times \frac{DE}{DF} = 400 \times \frac{136}{240} = 227 \text{ rpm. } \textbf{Ans.}$$

(*v*) From point M (225 V) draw a horizontal line intersecting line OA at point N. From point N draw a perpendicular line PN intersecting OCC at point Q.

Lowest possible speed for getting 225 V with 34 Ω shunt field circuit resistance,

$$N = N \times \frac{PN}{PQ} = 400 \times \frac{225}{245} = 367 \text{ rpm } \textbf{Ans.}$$

Example 7.31. The OCC of a shunt generator running at 800 rpm is as follows:

Field current (A) :	**1**	**.2**	**3**	**4**	**5**	**6**
Induced emf (V) :	**82.5**	**180**	**225**	**252**	**273**	**282**

If the shunt field resistance is 60 Ω, find (*i*) the voltage to which the machine will build up running at the same speed (*ii*) the value of field regulating resistance if the machine is to build up 120 V when its field coils are grouped in two parallel circuits and generator is run at half the speed. [*Hyderabad Univ. Electrical Technology 1991*]

Solution: (*i*) Draw OCC for speed of 800 rpm from given datas and draw line OA to represent shunt field winding resistance of 60 Ω. From point B, the point of intersection of OCC and line OA, draw horizontal line BD cutting voltage axis at point D. OD represents the terminal voltage on no load, which is found to be 260 volts on measurement.

(*ii*) Draw OCC for speed of 400 rpm from the following data :

Excitation amps	1	2	3	4	5	6
Induced volts when speed is 800 rpm, E_1	82.5	180	225	252	273	282
Induced volts when speed is 400 rpm, $E_2 = E_1 \times \frac{N_2}{N_1} = E_1 \times \frac{400}{800} = \frac{E_1}{2}$	41.25	90	112.5	126	136.5	141

Mark point F on voltage axis so that OF may represent 120 volts. From point F, draw horizontal line cutting OCC drawn for speed of 400 rpm at point C. From point C draw a perpendicular CE to meet excitation axis at point E. OE gives the excitation of 3.4 amperes. It means that the current flowing through each parallel path of shunt field coils is 3.4 A.

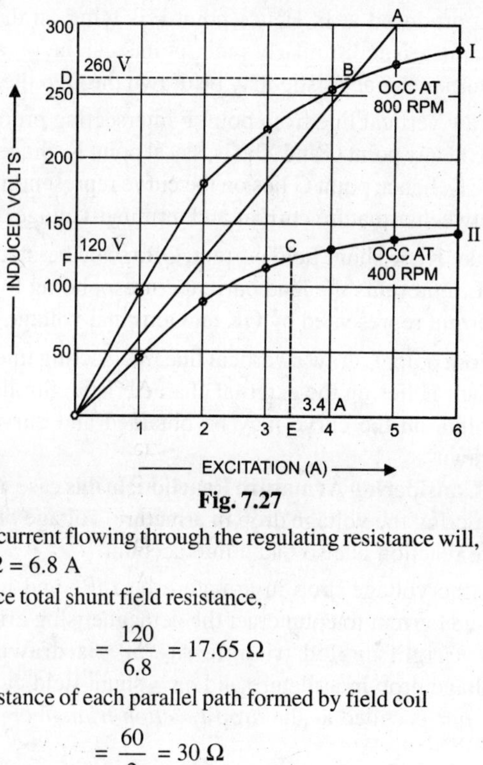

Fig. 7.27

The current flowing through the regulating resistance will, therefore, be 3.4 × 2 = 6.8 A

Hence total shunt field resistance,

$$= \frac{120}{6.8} = 17.65\ \Omega$$

Resistance of each parallel path formed by field coil

$$= \frac{60}{2} = 30\ \Omega$$

Equivalent resistance of two parallel paths formed by field coils

$$= \frac{1}{\frac{1}{30} + \frac{1}{30}} = 15\ \Omega$$

Regulating resistance in series with the field circuit

$$= 17.65 - 15 = 2.65\ \Omega \textbf{ Ans.}$$

7.8.6. Determination of Internal Characteristic From External Characteristic.

Characteristic. In Fig. 7.28. curve I represents external characteristic, line OP represents shunt field resistance and line OQ represents armature resistance.

If point F is considered lying on external characteristic and lines FA and FC are drawn parallel to X-axis and Y-axis respectively intersecting them on points A and C respectively.

For point F

Load current, I_L = OC

Terminal voltage, V = OA

Shunt field current, I_{sh} = AB

Armature current, $I_a = I_L + I_{sh}$

= OC + AB = OC + CD = OD

where CD = AB

Now if from point D, a vertical line DE is drawn, and produced intersecting line AF at H, then corresponding to armature current I_a represented by OD

Terminal voltage, V = OA = DH

Voltage drop in armature = $I_a R_a$ = DE

Generated EMF, E = V + $I_a R_a$

= DH + DE

= DH + HG = DG where HG = DE

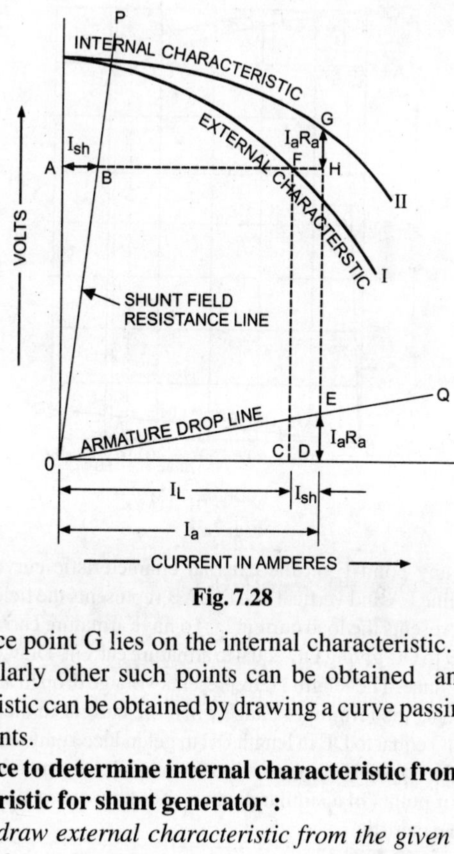

Fig. 7.28

Hence point G lies on the internal characteristic.

Similarly other such points can be obtained and internal characteristic can be obtained by drawing a curve passing through these points.

Hence to determine internal characteristic from external characteristic for shunt generator :

(*i*) *draw external characteristic from the given datas*

(*ii*) *draw shunt field resistance and armature resistance lines*

(*iii*) *take any point say F, on the external characteristic curve*

(*iv*) *from point F draw horizontal and vertical lines intersecting Y and X axis at points A and C respectively. Let these lines be FA and FC respectively*

(*v*) *take point D on X-axis so that CD = AB, representing the shunt field current* I_{sh}

(*vi*) *from point D draw vertical line DE and produce it intersecting line AF produced at H*

(*vii*) *take point G on line DH produced so that HG = DE representing armature drop* $I_a R_a$

(*viii*) *similarly take series of points on external characteristic and find corresponding points lying on internal characteristic and*

(*ix*) *draw curve passing through these points.*

Curve so obtained will be the internal characteristic for the shunt generator.

Example 7.32. The external characteristic of a shunt wound generator is as follows:

Load current in amps	**0**	**10**	**20**	**30**	**40**	**50**	**60**
Terminal voltage in volts	**100**	**98**	**96**	**93.5**	**89**	**83**	**74**

Determine the internal characteristic if the armature and field resistances are 0.1 Ω and 25 Ω respectively.

Solution: Draw the external characteristic curve from the given data, as shown in Fig. 7.29. Draw armature resistance line OQ and field resistance line OP representing the armature resistance of 0.1 Ω and field resistance of 25 Ω respectively.

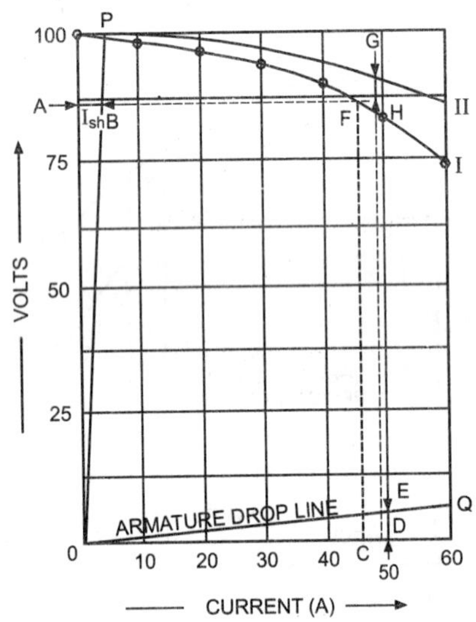

Fig. 7.29

Take any point F on the external characteristic curve and draw horizontal line FA and vertical line FC. AB represents the field current I_{sh} and OC represents the load current I_L. To have armature current, add CD equal to AB to OC giving OD equal to armature current. Draw vertical line DE and produce. The length DE represents voltage drop in armature and DH represents the terminal voltage when the current in the armature is OD. Add HG equal to DE in length DH to get induced emf when armature current is OD. So point G is obtained on the internal characteristic. Determine a number of points in a similar manner and draw curve II, which gives internal characteristic.

7.8.7. Determination of External Characteristic From OCC

(a) Neglecting Armature Reaction

(i) Draw OCC from given data as represented by curve I in Fig. 7.30.

(ii) Draw shunt field resistance line (say ON) intersecting OCC at any point (say A).

(iii) From point A draw horizontal line intersecting Y axis, say at point B. Hence OB is the maximum open-circuit voltage.

(iv) Take any point (say P) on the OCC and draw an ordinate, say PQR intersecting field resistance line at point Q and X axis at point R. Now PR represents the generated emf, QR represents the terminal voltage and PQ represents voltage drop in armature.

(v) From points P and Q draw horizontal lines cutting vertical axis say at points D and E respectively.

(vi) Draw armature resistance line OC.

(vii) From point E, draw line EF parallel to line OC cutting line PD produced at F. Hence point F is lying on the internal characteristic. Similarly other points can be obtained and internal characteristic may be drawn through these points.

(viii) Draw vertical line from point F intersecting produced line QE at any point G and X axis, say at point K. Since FG = PQ = CK, hence point G lies on the curve representing relation between armature current and terminal voltage.

(ix) Take KL = shunt field current, OR (scale being different). OL represents the load current corresponding to armature current represented by OK and terminal voltage OE.

(x) From point L draw a vertical line intersecting line EG at H. Point H lies on the external characteristic. Similarly other points on the curve may be obtained and curve may be drawn.

(b) Considering Armature Reaction. In this case, in addition to considering the voltage drop in armature, voltage drop due to armature reaction is also taken into account.

Let the voltage drop in armature be $I_a R_a$ and increase in shunt field current to counteract the demagnetising effect be I_{sh}. Now if a right angled triangle say *pqr* is drawn, so that *pr* = voltage drop in armature and *qr* = shunt field current. The triangle *pqr* is called as the *drop reaction triangle*.

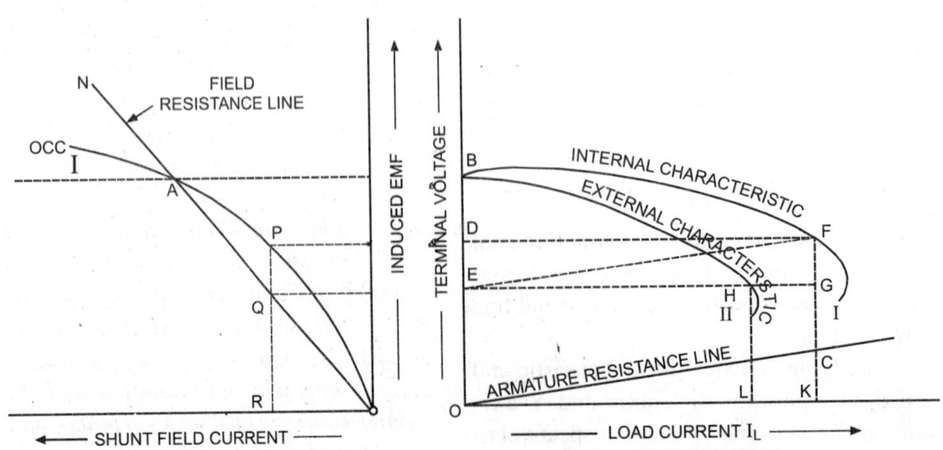

Fig. 7.30

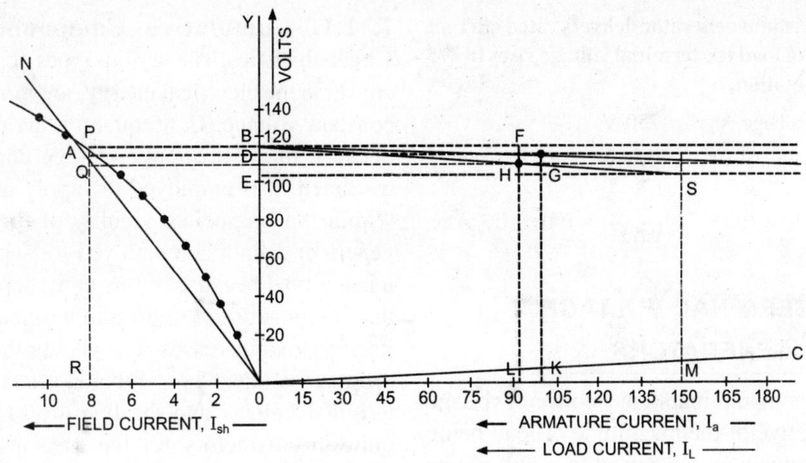

Fig. 7.31

To draw external and internal characteristics, the process in part *(a)* be repeated with the modification of *(iv)* and *(ix)* steps as follows (refer to Fig. 7.32):

(iv) Take any point P on the O C C and draw line PQ parallel to the line *pq* of triangle *pqr* and complete the triangle PQR. Now from points P and Q draw vertical lines cutting X axis at points R' and Q'. Now PR' represents the generated emf, QQ' represents the terminal voltage, PR represents voltage drop in armature due to armature resistance, OR' is the shunt field current to induce an emf represented by PR' and R'Q' is the increase in shunt field current to counteract the demagnetising effect.

(ix) Take KL– shunt field current OQ' (scale being different) OL represents the load current corresponding to armature current represented by OK and terminal voltage QQ.

Example 7.33. A shunt generator has magnetisation curve given by the following figures when driven at constant speed.

Field current Amperes	**1**	**2**	**3**	**4**	**5**	**6**	**7**	**8**	**9**	**10**
Open-circuit voltage	**20**	**38**	**57**	**76**	**91**	**102**	**110**	**116**	**121**	**124**

The resistance of the field shunt circuit is 14 Ω. Find *(a)* the pd given by the dynamo on open-circuit, *(b)* the pd given by it when supplying a load of 150 amperes.

The resistance of the armature is 0.04 Ω and the armature reaction may be neglected.

Solution : *(a)* Draw OCC from the given datas. Draw line ON representing shunt field resistance of 14 Ω. The point of intersection of O C C and shunt field resistance line gives the pd given by the dynamo on open-circuit, which is 118 volts. **Ans.**

(b) Draw internal and external characteristics as discussed in Art 7.8.7 *(a)*. Take point M on load current axis representing 150 A, draw vertical line from point M cutting external characteristic curve at S. From point S draw horizontal line cutting voltage axis at 108.5 volts. Hence pd given by the generator, when supplying a load of 150 A, is 106 volts. **Ans.**

Example 7.34. A shunt generator gave the following OCC at 750 rpm.

Field current in amperes	**0.5**	**1.0**	**1.5**	**2.0**	**2.5**	**3.0**	**3.5**
Generated emf in volts	**54.0**	**107.0**	**153.0**	**185.0**	**210.0**	**230.0**	**245.0**

The armature resistance is 0.15 Ω and shunt field resistance 750 Ω. When armature is short-circuited, a field current of 0.5 ampere circulates 100 A in it. Draw the internal and external characteristics for the generator.

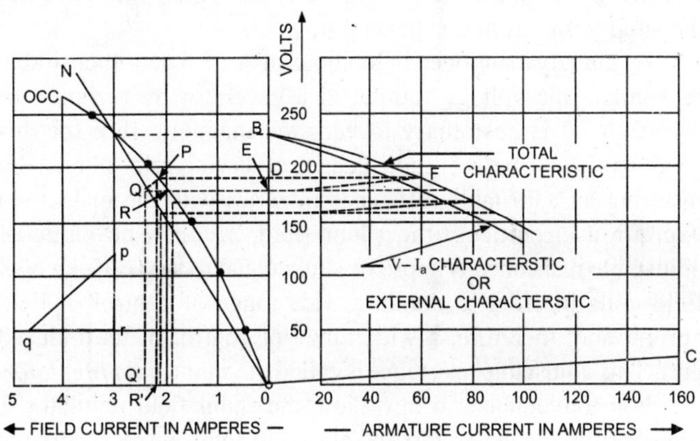

Fig. 7.32

Solution : When armature is short circuited the armature current is 100 A, therefore, voltage drop in armature will be = 100 × 0.15 = 15 volts.

From O C C to induce an emf of 15 volts, field current required is 0.14 A. Hence 0.50 – 0.14 = 0.36 ampere of field current is required to counter effect the demagnetizing effect. Draw drop reaction triangle *pqr* so that *pr* = 15 volts and *qr* has been = 0.36 A on enlarged scale, 5 times of the scale, on which O C C drawn, to achieve accuracy.

Draw internal characteristic and external characteristic as discussed in Art. 7.8.7.

Note: In this case field current is very small as compared to armature current hence line current will be approximately equal to armature current and external characteristic will coincide with $V–I_a$ characteristic.

7.9 VOLTAGE REGULATION

In as much the terminal voltage falls with the increase in load applied to the generator, it becomes desirable to have some means to express the degree of variation in terminal voltage. The term *voltage regulation* is employed for it. By voltage regulation of a generator meant the change in its terminal voltage from full-load to no-load as a percentage of the terminal voltage at full load

$$\text{i.e. percent voltage regulation} = \frac{V_{NL} - V_{FL}}{V_{FL}} \times 100 \qquad ...(7.26)$$

The voltage regulation is a term that may be used for determining the performance of the generator. Lower the voltage regulation better it is. The generator with the smallest per cent regulation will have the least variation in terminal voltage when the load is reduced. Ideally, a shunt generator should maintain a constant voltage *i.e.* zero per cent regulation.

Example 7.35. A 25-kW, 250V shunt generator delivers rated current at rated voltage. On removal of load the terminal voltage rises to 275 V. Determine the voltage regulation.

Solution: Full-load terminal voltage, V_{FL} = 250 V

No-load voltage, V_{NL} = 275 V

$$\text{Percent regulation} = \frac{V_{NL} - V_{FL}}{V_{FL}} \times 100 = \frac{275 - 250}{250} \times 100 = 10\% \textbf{ Ans.}$$

7.10 CONTROL OF TERMINAL VOLTAGE OF DC SHUNT GENERATORS

Ideally, a dc shunt generator should maintain a constant voltage across its terminals irrespective of the magnitude of load being supplied by it so that any equipment connected to it will not be affected by adding or removing other equipment connected to the supply line from this generator.

As already mentioned in footnote of Art 7.6, it is undesirable to regulate the voltage output of a generator by variation of speed and it is customary to vary the magnetic flux for this purpose. It can be accomplished in a very simple manner by inserting an adjustable resistance in series with the main field excitation circuit, *i.e.* the shunt field circuit. The value of adjustable resistance is approximately equal to the resistance of field coils in order to obtain a wide range of control of field current and, therefore, a wide range of control of the induced emf. This adjustable resistance is called the *shunt field regulator.*

For convenience in operation, the shunt field regulator is remote from the generator, and placed in some easily accessible position. The simplest type of regulator is illustrated in Fig. 7.33. It consists of wire spirals mounted in a metal frame; these are suspended vertically to allow them to dissipate the heat developed due to flow of current in them.

In large generator installations, operation of the shunt regulator may be automatic, but in other cases, the variation in resistance is achieved by moving the handle over the studs until the correct value of resistance is found. This will be indicated by a voltmeter connected across generator terminals. Higher the additional resistance in the shunt field circuit, lower the excitation and, therefore, lower induced voltage and vice versa.

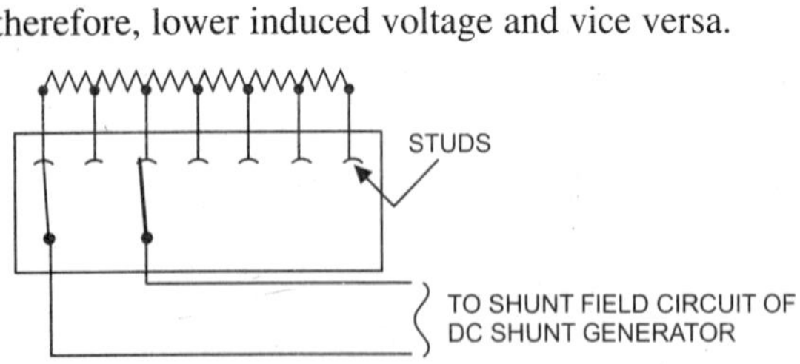

Fig. 7.33

7.11 CHARACTERISTICS OF COMPOUND WOUND DC GENERATORS

Compound wound dc generators are of two types viz *cumulative compound wound* and *differential compound wound generators.*

7.11.1. Cumulative Compound Wound Generator Characteristics. The various types of apparatus which ultimately consume the electrical energy, are mostly designed to operate at constant voltage. Consequently, the use of dc shunt generator, because of its drooping voltage characteristic, is somewhat restricted. It is employed to supply power to apparatus located within the immediate vicinity of the generator. The dc shunt generator cannot be employed for supplying power to remotely located load because in this case there is also a voltage drop in the line in addition to drop in terminal voltage of the generator itself as load increases. The greater the load, the greater the volt drop in the line. The combined effect of reduction in generator terminal voltage and the line drop makes the shunt generator quite unsatisfactory for the transmission of power over long distances.

A cumulative compound wound dc generator, which has an additional winding (having few turns of thick wire, called the *series field winding*) connected in series either with line or with armature, is the solution to the above problem of supplying power to distant loads at constant voltage. The connections are shown in Figs. 7.4 (*a*) and 7.4 (*b*) respectively. In case of short-shunt compound generators, the shunt field is connected directly across the armature terminals and load current flows through the series field (*i.e* series field current is equal to load current) whereas in long shunt compound generator armature current will flow through the series field (*i.e.* series field current equals armature current). The current in series field winding, produces a flux in addition to that produced by shunt field winding, to compensate for fall in terminal voltage due to armature reaction and armature circuit drop. The series field current in the long- shunt generator is greater than that in the short-shunt generator by the amount of shunt field current. Since the increase of ampere-turns of the series field is comparatively small (shunt field current is very small in comparison to armature current), it has little effect on the generated voltage and, therefore, characteristics are almost same in both cases.

The cumulative compound wound generator combines the characteristics of the shunt and series generators. At no load the shunt field winding provides all the field flux since there is no current in the series field winding. With the increase in load on the generator, the series field adds an increasing amount of field flux to that of the shunt field.

If the series excitation is such that the terminal voltage of the generator on full load is the same as that on no load, the generator is said to be level or flat compounded*. If the series excitation becomes more prominent than that of the shunt field, the terminal voltage rises with the increase in load and the generator is said to be *overcompounded.* Similarly when the shunt field excitation plays the prominent part and full-load terminal voltage is less than no-load terminal voltage, the generator is said to be *undercompounded.*

The external characteristics of over-, under- and flat- (or level) compound dc generators, are illustrated in Fig. 7.34.

* Even in case of a flat-compounded generator, the terminal voltage is not constant for all loads from no load to full load. As seen from Fig. 7.34, at half-full load, the terminal voltage is more than the no-load (or full-load) terminal voltage.

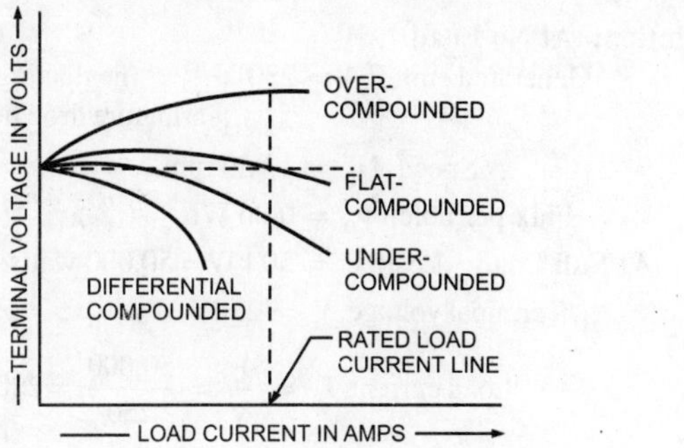

Fig. 7.34 *External Characteristics of DC Compound Generators*

7.11.2. Differential Compound Wound Generator Characteristics. In this type of dc generator, the series field of the compound wound dc generator (may be either short-shunt or long-shunt) is connected so that the flux produced by the series field opposes the flux produced by shunt field. This is accomplished by interchanging the connections of series field winding. In such a generator, the induced voltage on no load would be the same as for the shunt generator because series field does not create any flux. When the generator is loaded, load current (or armature current) flows through the series field windings creating magnetic flux that opposes the flux created by shunt field. This results in decrease of induced emf and, therefore, in terminal voltage. In such generators the terminal voltage drops very rapidly with the increase in load current and, therefore, it is interesting to note that a short-circuit cannot cause any damage to it.

7.11.3. Adjustment of Degree of Compounding. Standard compound wound dc generators, that is, those which are manufactured in comparatively large numbers, are usually constructed with sufficient series field turns to operate as over-compounded. An overcompound dc generator can be modified to give characteristics similar to those of a flat-compound generator or undercompound generator. To accomplish this, it becomes necessary to lower the terminal voltage of the over-compound generator. Since the rise in terminal voltage is a direct result of the series-field ampere turns, reduction in series field ampere-turns becomes imperative to reduce the terminal voltage on load. It is not possible to reduce the number of turns in the series field, so the only possibility is by means of reducing the current flowing through the series field winding. The simplest method of reducing the series field current is to connect a *very low resistance shunt* directly across the series field winding, as shown in Fig. 7.35. Such a *low resistance shunt* is called a *diverter* because it diverts or bypasses, part of the load current through a section of wire that creates no flux. Thus the series field becomes less effective in creating flux to boost the induced voltage to an extent determined by the bypassed current. By adjusting the resistance of the diverter, the current flowing through the series field winding thereby the degree of rise of terminal voltage may be controlled. When the resistance of such a diverter is extremely large, the diverted current I_d will be small and the generator characteristics will be overcompounded. On the other hand, if the resistance of the diverter approaches that of a short-circuit, practically all the load current will be bypassed and the operating external characteristic will resemble with that of a shunt generator. Since the series field resistance R_{se} and the diverter resistance R_d are in parallel, the total line current I_L will divide so that I_{se} and I_d are related to each other by an inverse ratio of the respective resistances and series field current is given by the relation

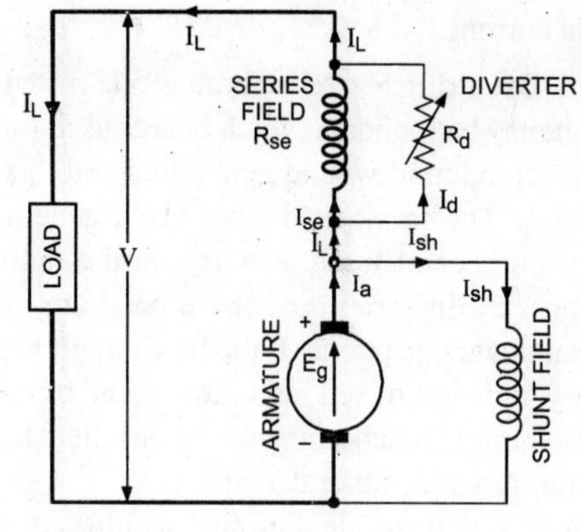

Fig. 7.35

$$\text{Series field current, } I_{se} = I_L \times \frac{R_d}{R_d + R_{se}} \quad \text{...(7.27)}$$

where R_d and R_{se} are the resistances of diverter and series field respectively and I_L is the load current.

7.11.4. Determination of Series Field Ampere-Turns. It is a rather simple task to determine the ampere-turns of the series field required to give any degree of compounding. The degree of compounding required will of course depend upon the desired terminal voltage for a specific value of load current, usually the full-load current. The procedure for obtaining the series-field ampere-turns is as follows:

The generator is operated as a shunt generator with the meters (ammeters and voltmeter) in field and load circuits so as to measure field current, load current and terminal voltage. The field rheostat is adjusted to give desired no-load voltage. Now the load is applied to the generator and adjusted until the ammeter connected in the load circuit indicates the rated value. The terminal voltage of the generator, however, is not the desired value, but is too low. The shunt field current is now increased by reducing the field rheostat resistance and adjusted to raise the terminal voltage to the desired value. Due to increase in terminal voltage, the load current increases. Hence it becomes necessary to increase the load resistance so that the rated value of load current is obtained. The generator is now supplying rated current at the desired terminal voltage. In order to obtain this terminal voltage it was necessary to increase the field ampere-turns to a value greater than that normally obtained as a shunt generator. This increase in the number of ampere-turns is to be provided by the series field, so that when the generator is operating as a cumulative compound generator, the increase in ampere-turns is automatically furnished by the series field.

Let ΔI_{sh} be the increase in shunt field current required to keep terminal voltage at desired value at full load from no load, N_{sh} be the total number of turns per pole on the shunt field winding, N_{se} be the turns per pole on the series field winding and I_{se}* be the current flowing through series field winding.

* I_{se} will be equal to armature current I_a in case of a long-shunt compound dc generator and equal to load current I_L in case of a short-shunt compound generator.

Additional ampere-turns to be provided by series field = $N_{sh} \Delta I_{sh}$

$$\text{or } N_{se} I_{se} = N_{sh} (I_{sh2} - I_{sh1}) \quad ...(7.28)$$

From the above equation, the required series field current is obtained. A diverter may be necessary to give the correct value of series field current.

7.11.5. Field Breaking Switch. There is one further important accessory that may be found on switch boards used for the control of dc shunt or compound wound generators and this is the *field breaking switch.* The shunt field circuit of a generator is made up of a large number of turns of wires on the individual field coils, all connected in series, and these coils are included in a highly efficient magnetic circuit. If the field circuit were suddenly broken, a very high emf of self induction would build up in these field coils and it might reach a magnitude sufficient to breakdown their insulation or cause other damage.

It is essential that the field circuit should never be broken except by means of the special switches designed for the purpose, and it will be imperative that, before the shunt field circuit is broken, it is connected across a non-inductive resistance. Any emf induced in the field coils will be safely dissipated across the resistance, and will not be allowed to build up into a highly dangerous voltage.

Example 7.36. A shunt generator is to be converted into a level compound generator by addition of a series field winding. From a test on the machine with shunt excitation only, it is found that the shunt current is 4.1 A to give 440 V on no load and 5.8 A to give the same voltage when the machine supplies its full load of 200 A. A shunt field winding has 1,200 turns per pole. Find the number of series turns required per pole.

Solution: Ampere-turns required at no load to generate 440 V,

$$AT_0 = 4.1 \times 1{,}200 = 4{,}920$$

Ampere-turns required at full load to generate 440 V,

$$AT_L = 5.8 \times 1{,}200 = 6{,}960$$

Hence, additional ampere-turns required at full load

$$= AT_L - AT_0 = 6{,}960 - 4{,}920 = 2{,}040$$

Full load current, $I = 200$ A

$$\text{Series field turns} = \frac{AT_L - AT_0}{I} = \frac{2{,}040}{200}$$

$$= 10.2 \text{ turns} \simeq 11 \text{ (say) } \textbf{Ans.}$$

Example 7.37. In a 50 kW, 230/250 V (230 V on no load and 250 V on full load) overcompounded dc generator (long shunt) the flux per pole required to produce 230 V on no load at 1,050 rpm is 0.06 Wb. The resistances of the armature and series field are 0.04 Ω and 0.01 Ω respectively and the shunt field resistance is 100 Ω. Calculate the value of the flux per pole at full load, 1000 rpm. Neglect brush drop.

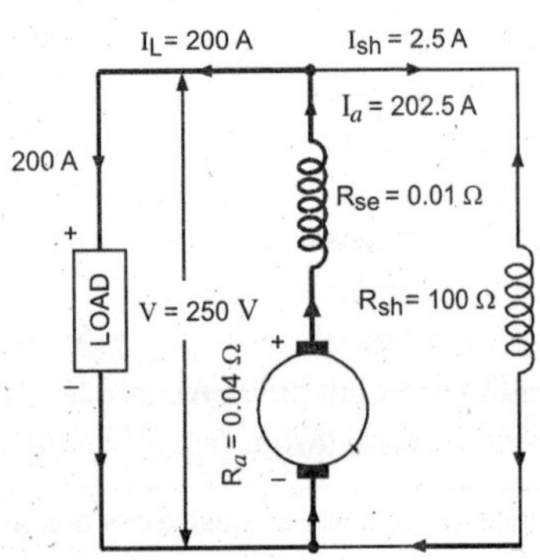

Fig. 7.36

Solution: At No Load

Generated emf, $E_{g0} = 230$ V (neglecting series field and armature drop being very small)

Speed, $N_0 = 1{,}050$ rpm

Flux per pole, $\Phi_0 = 0.06$ Wb

At Full Load Load, P = 50 kW = 50,000 watts

Terminal voltage, V = 250V

$$\text{Load current, } I_L = \frac{P}{V} = \frac{50{,}000}{250} = 200 \text{ A}$$

$$\text{Shunt field current, } I_{sh} = \frac{V}{R_{sh}} = \frac{250}{100} = 2.5 \text{ A}$$

$$\text{Armature current, } I_a = I_L + I_{sh} = 200 + 2.5 = 202.5 \text{ A}$$

$$\text{Generated EMF, } E_{gf} = V + I_a (R_a + R_{se})$$

(Since generator is long shunt)

$$= 250 + 202.5 (0.04 + 0.01)$$

$$= 260.125 \text{ V}$$

$$\frac{E_{gf}}{E_{g0}} = \frac{\Phi_f}{\Phi_0} \times \frac{N_f}{N_0}$$

Since generated emf, $E_g \propto \Phi N$

$$\text{or flux per pole at full load, } \Phi_f = \Phi_0 \times \frac{E_{gf}}{E_{g0}} \times \frac{N_0}{N_f}$$

$$= 0.06 \times \frac{260.125}{230} \times \frac{1{,}050}{1{,}000}$$

$$= 0.0713 \text{ Wb } \textbf{Ans.}$$

Example 7.38. A 500 V, 250 kW, long-shunt compound generator induces an emf of 480 V when running at 1,000 rpm on no load. On full load the speed of the machine drops to 975 rpm, the flux increases by 15% and the terminal voltage rises to 500 V. If the series and shunt field resistances are 0.02 Ω and 100 Ω respectively, calculate the armature resistance. Assume a voltage drop of 1V per brush.

Solution: On no load:

Induced emf, $E_{g0} = 480$V

Speed, $N_0 = 1{,}000$ rpm

On full load:

$$\text{Load current, } I_L = \frac{\text{Rated load in kW} \times 1{,}000}{\text{Rated voltage}}$$

$$= \frac{250 \times 1{,}000}{500} = 500 \text{ A}$$

$$\text{Shunt field current, } I_{sh} = \frac{V}{R_{sh}} = \frac{500}{100} = 5 \text{ A}$$

$$\text{Armature current, } I_a = I_L + I_{sh} = 500 + 5 = 505 \text{ A}$$

$$\text{Induced emf, } E_{gf} = E_{g0} \times \frac{N_f}{N_0} \times \frac{\Phi_f}{\Phi_0}$$

$$= 480 \times \frac{975}{1{,}000} \times 1.15 = 538.2 \text{ V}$$

$$\because N_f = 975 \text{ rpm and } \frac{\Phi_f}{\Phi_0} = 1.15$$

$$\text{Also induced emf} = V + I_a (R_a + R_{se}) + \text{brush drop}$$

$$\text{or } 538.2 = 500 + 505 (R_a + 0.02) + 2$$

$$\text{or Armature resistance, } R_a = \frac{538.2 - 500 - 2}{505} - 0.02 = 0.0517 \ \Omega \ \textbf{Ans.}$$

Example 7.39. A 100 kW, 200 V long shunt (cumulative) compound generator has equivalent armature resistance of 0.03 Ω, a series field resistance of 0.004 Ω. There are 1,200 shunt field turns per pole and 5 series field turns per pole. The data of magnetization curve at armature speed of 1,000 rpm is given below.

I_f (shunt field current) in amp	0	1	2.2	3.3	4.2	5.3	7.1
E_0 (induced emf) in volts	11	33	100	167	200	215	222

Calculate the terminal voltage at the rated current output for the shunt field current of 5 A and a speed of 950 rpm. The armature reaction may be ignored.

Solution: Power rating of generator, P = 100 kW

$$\text{Line current, } I_L = \frac{P \times 1{,}000}{V} = \frac{100 \times 1{,}000}{200} = 500 \text{ A}$$

Series field current, I_{se} = Armature current, $I_a = I_L + I_{sh} = 500 + 5 = 505$ A

Equivalent main field current, I_{feq}

$$= I_{sh} \pm \frac{N_{se}}{N_{sh}} I_{se}$$

where N_{sh} is shunt field turns per pole, N_{se} is series field turns per pole, I_{sh} is shunt field current and I_{se} is series field current, + sign for cumulative compounding and – sign for differential compounding.

$$\text{So } I_{feq} = 5 + \frac{5}{1{,}200} \times 505 = 7.1 \text{ A}$$

From given data of magnetization curve at armature speed of 1,000 rpm

Induced emf corresponding to field current of 7.1 A is 222 V i.e. $E_0 = 222$ V

Induced emf at 950 rpm corresponding to field current of 7.1 A,

$$E'_0 = E_0 \times \frac{N_2}{N_1} = 222 \times \frac{950}{1{,}000} = 210.9 \text{ V}$$

$$\text{Terminal voltage, } V = E'_0 - I_a (R_a + R_{se})$$

$$= 210.9 - 505\,(0.03 + 0.004) = 193.73 \text{ V } \textbf{Ans.}$$

Example 7.40. A long-shunt compound generator has a shunt field winding of 1,000 turns per pole, series field winding of 4 turns per pole and resistance of 0.05 Ω. In order to obtain the rated voltage both at no load and full load for operation as a shunt generator, it is necessary to increase field current by 0.2 A. The full-load armature current of the compound generator is 80 A. Calculate the diverter resistance connected in parallel of series field to obtain flat compound operation.

Solution : Additional ampere-turns required to maintain rated voltage at full load for operation as a dc shunt generator

= Number of turns on shunt field winding × additional shunt field current

= 1,000 × 0.2 = 200

Number of series turns per pole, $N_{se} = 4$

Current necessary to produce 200 AT by the series field

$$= \frac{200}{N_{se}} = \frac{200}{4} = 50 \text{ A}$$

Armature current, I_a = 80 A

Current through diverter, $I_d = I_a - I_{se} = 80 - 50 = 30$ A

If R_d is resistance of diverter then

$$R_d I_d = R_{se} I_{se}$$

$$R_d = \frac{R_{se} I_{se}}{I_d} = \frac{0.05 \times 50}{30} = 0.0833\ \Omega \textbf{ Ans.}$$

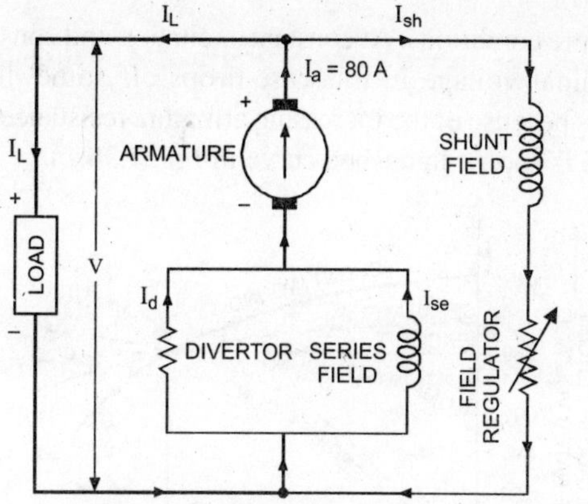

Fig. 7.37

Example 7.41. A 60 kW, 240 V short-shunt compound generator operates as a shunt generator, required an increase in field current of 3 A to provide an overcompensated voltage of 275 V at rated load current of 250 A. The shunt field has 200 turns per pole and the series field 5 turns per pole, with resistance of 240 Ω and 0.005 Ω respectively.

(*i*) Calculate the required diverter resistance.

(*ii*) If the no-load voltage of the compound generator is also 240 V, calculate the air gap mmf/pole at no load and at full load.

Solution: Full load current, $I_L = \frac{P}{V} = \frac{60{,}000}{240} = 250$ A

Additional ampere-turns required to maintain rated voltage at full load for operation as a dc shunt generator

$$= N_{sh} \Delta I_{sh} = 200 \times 3 = 600$$

Number of series turns per pole, $N_{se} = 5$

Current necessary to produce 600 AT by the series field

$$I_{se} = \frac{N_{sh} \Delta I_{sh}}{N_{se}} = \frac{600}{5} = 120 \text{ A}$$

Current through diverter, $I_d = I_L - I_{se} = 250 - 120 = 130$ A

$$\text{Diverter resistance, } R_d = \frac{I_{se} R_{se}}{I_d}$$

$$= \frac{120 \times 0.005}{130} = 0.00462\ \Omega \textbf{ Ans.}$$

(*ii*) Under no-load conditions there is no current in the series field Hence no mmf is provided by the series field

$$\text{Shunt field current, } I_{sh} = \frac{V}{R_{sh}} = \frac{240}{240} = 1\text{A}$$

Air gap mmf per pole on no load = $N_{sh} I_{sh} = 200 \times 1 = 200$ **Ans.**

At full load

$$\text{Shunt field current, } I_{sh} = \frac{V}{R_{sh}} = \frac{V - I_{se} R_{se}}{R_{sh}}$$

$$= \frac{240 - 250 \times 0.005}{240} = 0.9948 \text{ A}$$

MMF provided by shunt field = $N_{sh} I_{sh}$ = 200 × 0.9948 = 198.96

MMF provided by series field = $N_{se} I_{se}$ = 5 × 250 = 1,250

Total mmf per pole = 198.96 + 1,250 = 1,448.96 **Ans.**

7.12 GENERATOR PERFORMANCE CURVES

In a *separately-excited dc generator* the field current is supplied from a separate source and is, therefore, physically independent of armature conditions. At constant excitation and constant speed, the terminal voltage in this case drops off somewhat as load increases because of the increasing armature resistance drop. This decrease is shown in dashed curve in Fig. 7.38.

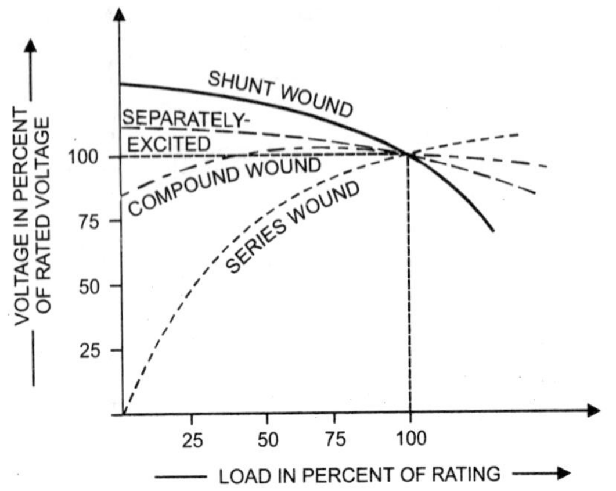

Fig. 7.38

In a *shunt wound dc generator* terminal voltage on no load depends upon shunt field resistance, as already discussed in Art 7.8. The terminal voltage of a dc shunt generator will decrease as load is added because of the increase in armature resistance drop and the decrease in field current accompanying the drop in voltage. The drop in voltage with load is shown in Fig. 7.38.

In a series wound generator, the flux varies widely with load because the load current also flows through the field. The voltage-load characteristic, shown dotted in Fig. 7.38, resembles the magnetization curve but is somewhat lower because of resistance drop.

Buildup of an *unloaded compound dc generator* is the same as for a dc shunt generator since the series field carries negligible current at no load. When the generator is loaded, both the shunt and series field mmfs are present. For a cumulative compound generator, the causes of voltage drop from no load to full load in a dc shunt generator may be partially or completely compensated or overcompensated by the addition of the series field. In a *flat-compounded dc generator*, they are completely compensated, so that no-load and full-load terminal voltages are equal. If the full-load voltage is lower than the no-load voltage, the generator is *undercompounded*; if higher, *overcompounded*. The load characteristic of a flat-compounded dc generator is shown by a chained curve in Fig. 7.38.

7.13 CONDITIONS FOR SELF EXCITATION

The conditions required to be fulfilled before a series or a shunt generator excites itself are given below :

1. There must be some residual magnetism in the field system.
2. The residual magnetism must be in proper direction. The field coils should be connected with the armature in such a way that current flowing through them should increase the emf induced by the residual magnetism.
3. For a series wound generator, the resistance of the external circuit should be less than the critical resistance.
4. For a shunt wound generator, resistance in the field circuit must be less than critical resistance for field circuit and resistance in the load circuit must be greater than critical resistance for the load circuit.

7.14 CAUSES OF FAILURE TO BUILD UP VOLTAGE

There may be one or more of the following reasons due to which a self excited generator may fail to build up voltage.

1. No residual magnetism. The start of the build up process needs some residual magnetism in the magnetic circuit of the generator. If there is little or no residual magnetism, because of inactivity or jarring in shipment, no voltage will be induced that can produce field current.

2. Reversal of Field Connections. The voltage induced owing to residual magnetism acts across the field and results in flow of current in the field coils in such a direction as to produce magnetic flux in the same direction as the residual flux. Reversal of connections of the field winding destroys the residual magnetism which causes the generator failure to build up voltage.

3. In case of dc series wound generators, the resistance in the load circuit may be more than its critical resistance, which may be due to

(*i*) open-circuit
(*ii*) high resistance of load circuit
(*iii*) faulty contact between brushes and commutator and
(*iv*) commutator surface dirty or greasy.

4. In case of shunt wound generator:

(*a*) the resistance of the shunt field circuit may be greater than the critical resistance;
(*b*) the resistance in the load circuit may be lower than the critical resistance;
(*c*) the speed of rotation may not be equal to rated one.

Remedy. In case the generator is started up for the first time, it may be that no voltage will be built up either because the poles have no residual magnetism or the poles have retained some residual magnetism but the field winding connections are reversed so that the magnetism developed by the field winding on start has destroyed the residual magnetism and the machine can not "build up". In both the cases, the field coils must be connected to a dc source (a storage battery) for a short while to magnetise the poles. The application of external source of direct current to the field is called flashing of the field.

7.15 APPLICATIONS OF DC GENERATORS

The choice of a dc generator for a particular service depends upon its performance characteristics. Applications of various types dc generators are given below.

1. Separately-Excited DC Generators. These generators require separate source of supply and are, therefore, usually more expensive than self-excited generators. On this account these are generally used where self-excited generators would be relatively unsatisfactory. These are used in Ward Leonard systems* of speed

* Such systems are used in mine hoists, steel mill drives, paper making machines, diesel-electric locomotives etc.

control, because self excitation would be unstable at lower voltages. Since separate excitation gives a quicker and more precise response to the changes in the resistance of the field circuit and is accordingly used where quick and definite response to control is important.

2. Series Wound DC Generators. These generators have very few applications. They cannot be employed for power supply on account of their rising voltage characteristics. They can be employed as a constant current source by making use of the drooping portion of the voltage-current characteristic.

Their best application is for supplying field current for regenerative braking in dc locomotives. They are also used in series arc lighting. Another important application of these generators is as series boosters.

A *booster* is a series wound dc generator, which is inserted in a circuit to add a certain voltage so that the excessive drop in the feeder is compensated. The series wound generator is coupled to a dc shunt motor, as illustrated in Fig. 7.39. The shunt wound motor draws current from the supply mains and drives the generator coupled to it at a constant speed. Since voltage drop in the feeder is proportional to load, for exact compensation of voltage drop in feeder the voltage injected into the circuit must also be proportional to the load current. The series booster is, therefore, a low voltage and high current dc series generator operating on the linear portion of its voltage-current characteristic. A variable resistor, known as *diverter,* is connected in parallel with the field winding so that the output voltage of the booster may be adjusted so as to match the voltage drop in the line.

3. Shunt Wound DC Generators. Such generators with field regulators are used for light and power supply purposes. These are also used for charging of batteries because they can be made to give constant output voltage.

4. Cumulative Compound Wound DC Generators. Overcompounded type dc generators are generally used for lighting and power services because overcompounding

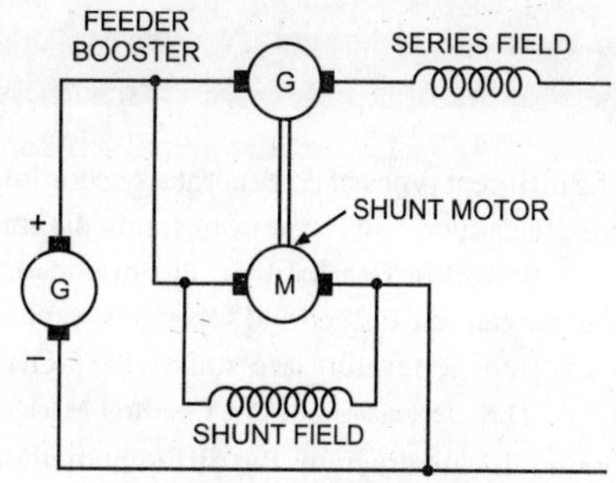

Fig. 7.39 *Series Booster*

compensates voltage drop in the distribution lines and thus the voltage at the consumer terminals remains more or less constant.

5. Differential Compound Wound DC Generator. It is a constant current generator and finds an useful application as an arc welding generator where the generator is practically short circuited every time the electrode touches the metal plates to be welded.

Note: *With the advent of various power electronic devices, there is no doubt that the popularity of dc generators has gone down. At present time bulk of power is generated in the form of alternating current because of technical and economical reasons. Hence the use of dc generators is very limited. They find application mainly in supplying excitation to the small and medium size alternators. For industrial applications of direct current like electrolytic processes, welding processes and variable speed motor drives, the present trend is to generate ac and then convert it into dc by converters. For conversion of ac into dc the dc generator, as a part of motor (ac)-generator (dc) set, has to compete the SCR rectifiers and various other controlled power devices which usually are cheaper in cost, compact in size, relatively noise-free in operation and require minimum maintenance, though they suffer from the disadvantages of having poor power factor, harmonic generation, poor braking etc. However, some of the important applications of dc generators include dynamometers, welding, cross-field generators, for closed-loop control systems, tachogenerators (not covered in this book) etc. Separately-excited dc generators still find use for a wide output-voltage control such as in Ward-Leonard system of speed control.*

HIGHLIGHTS

1. According to method of excitation, dc generators are classified into separately-excited generators and self-excited generators.

   Self-excited generators are further classified in accordance how the field winding is connected w.r.t. armature. Accordingly such generators are (*i*) series wound type (*ii*) shunt wound type and (*iii*) compound wound type. Compound wound generators may be long-shunt type or short-shunt type.

   Compound wound dc generators can also be classified into cumulative compound wound and differential compound wound.

2. There are three main characteristics of a dc generator.

   (*a*) *No-load saturation characteristic*, open-circuit characteristic or magnetisation characteristic — a curve drawn between generated emf and excitation current.

   (*b*) *Internal or total characteristic* giving relation between the emf actually induced in the armature and the armature current I_a.

   (*c*) *External or load characteristic* giving relation between terminal voltage V and the load current I.

   Series wound dc generator has got rising voltage characteristic i.e. its voltage rises with load current.

   Shunt wound dc generator has got slightly drooping characteristic i.e. its terminal voltage falls slightly with the increase in load current.

   Cumulative compound wound generator combines the characteristics of shunt and series wound generators.

3. Among dc generators, separately-excited and cumulatively compounded self- excited machines are the most common. Separately-excited generators have the advantage of permitting a wide range of output voltages, whereas self-excited machines may produce unstable voltages in the lower ranges. Cumulatively compounded dc generators may produce a substantially flat voltage characteristic or one which rises with load, whereas shunt or separately-excited generators produce a drooping voltage characteristic unless regulating means are used.

4. (*a*) The best application of series wound dc generators is for supplying field current for regenerative braking in dc locomotives. They are also used as series boosters and in arc lighting.

   (*b*) Cumulative compound wound dc generators are generally used for lighting and power services.

   (*c*) Differential compound generator finds an useful application as an arc welding generator.

EXERCISES

1. What are the different types of dc generators according to the ways in which fields are excited? Show the connection diagram of each type. **[Rajasthan Technical Univ. Electrical Machines-I, 2008, 2010]**
2. Distinguish between self-excited and separately-excited dc generators. How self-excited generators are classified? Give their circuit diagrams. **[J.N. Technological Univ. Electrical Machines-I, March-2014]**
3. Describe with relevant diagrams the different methods of excitation of dc generators. Explain the meaning of the terms 'differentially compounded', 'cumulatively compounded'. Show the 'long-shunt' and 'short-shunt' connection diagrams. **[B.P. Univ. of Technology Electrical Machines, 2008]**
4. With reference to no-load characteristics of a dc generator, explain the following:
   (*i*) Why is emf not zero when field current is zero?
   (*ii*) Why does the slope of curve change, after a certain value of field current? **[U.P. Technical Univ. Electrical Machines 2005-06]**
5. Explain OCC (no-load characteristic) of a dc generator. How it is obtained for a given generator? **[M.D. Univ. Electromechanical Energy Conversion, May-2008]**
6. Explain the internal and external characteristics of a separately excited dc generator. **[G.G.S.I.P. Univ. Delhi, May-June 2006]**
7. Draw and comment on all the performance characteristics of a separately-excited dc generator. **[R.G. Technical Univ. Electromechanical Energy Conversion-I, June-2011]**
8. Sketch the complete load characteristics of a dc series generator and indicate therein the region of operation of the machine as a voltage booster and as a constant current source. **[J.N. Technological Univ. Electrical Machines-I, March-2014]**
9. Distinguish between external and internal characteristics of a dc generator. How can internal characteristic be derived from external characteristic of a separately-excited generator ? **[U.P. Technical Univ. Electrical Machines 2005-06]**
10. List and explain the factors involved in the voltage build up of shunt generator. **[Rajasthan Technical Univ. Electrical Machines-I, 2011]**
11. Explain the voltage build up phenomenon in dc shunt generator. **[R.G. Technical Univ. Electromechanical Energy Conversion, December-2012]**
12. Explain the internal and external characteristics of dc shunt generator. **[Gujarat Technological Univ. Electrical Machines-I, Summer-2014]**
13. State the reasons for the drop in terminal voltage of the shunt generator when it is loaded. **[J.N. Technological Univ. Electrical Machines-I, March-2014]**
14. Explain why external characteristics of a dc shunt generator is more drooping than that of a separately-excited generator. Discuss their applications. **[J.N. Technological Univ. Electrical Machines-I, March-2014]**
15. With connection diagrams for both shunt and separately-excited dc generators, draw important characteristics of these generators. **[U.P.S.C. I.E.S. Electrical Engineering-II, 2006]**
16. Sketch the load characteristics of: (*i*) dc shunt and (*ii*) dc series generators. Give reasons for the particular shape in each case. **[M.D. Univ. Electromechanical Energy Conversion, December-2006]**
17. Draw and explain the load characteristics for a dc shunt generator. Why the external characteristic of shunt generator turns back as the generator is overloaded. **[G.G.S.I.P. Univ. Delhi Electromechanical Energy Conversion-I, December-2011]**
18. What is the critical field resistance of a dc shunt generator? How is it obtained? What is its significance? **[J.N. Technological Univ. Electrical Machines-I, March-2014]**
19. Define and explain the critical field circuit resistance and critical speed of a dc shunt generator. **[Rajasthan Technical Univ. Electrical Machines-I, 2011]**
20. Define the following:
    (*i*) Critical shunt field resistance.
    (*ii*) Critical speed. **[R.G. Technical Univ. Electromechanical Energy Conversion, December-2012]**
21. Define the terms critical speed and critical field resistance relating to a dc shunt generator. Also explain the voltage build-up process when the generator is driven at constant speed. **[M.D. Univ. Electrical Machines-I, December-2009]**
22. Define the critical speed and critical resistance. Explain how these are determined from OCC of the dc shunt generator and explain the building of voltage. **[J.N. Technological Univ. Electrical Machines-I, March-2014]**
23. Draw external characteristic of DC shunt generator and explain the method to obtain internal characteristic from the same. **[U.P. Technical Univ. Electromechanical Energy Conversion-I, 2009-10]**
24. Draw and explain the drooping characteristics of dc shunt generator. How you obtain the said characteristics in the laboratory? Draw the circuit diagram. **[R.G. Technical Univ. Electrical Machine-II, December-2013]**
25. Draw and explain internal and external characteristics of series and compound generators. **[J.N. Technological Univ. Electrical Machines-I, March-2014]**
26. Draw and explain the external characteristics curve for dc compound generator. **[M.D. Univ. Electromechanical Energy Conversion, December-2011]**
27. Explain the different methods of excitation and characteristics of dc generators with suitable diagrams. **[Anna Univ. Electrical Machines-I, May/June-2012]**
28. What is degree of compounding ? How does it affect the external characteristic of a dc generator ? How may the series field ampere-turns be estimated for a desired degree of compounding ? **[Pb. Univ. Elec. Machines-II, July-1991]**
29. Describe with appropriate circuit configurations the principle of a compound wound dc generator. Show how the terminal voltage of a compound wound generator varies with load current. Indicate when the generator is said to be overcompensated.
30. Draw the external characteristics of shunt, series and compound generators and explain why their shape differ. What is meant by a level compounded dc machine ? **[U.P. Technical Univ. Electromechanical Energy Conversion-I, 2001-02]**
31. What are the various conditions which have to be full filled before a shunt excited dc generator will generate rated voltage across the armature terminals? **[U.P.S.C. I.E.S. Electrical Engineering-II, 2010]**
32. Explain in detail how a dc shunt generator builds up its voltage. Discuss the conditions to be satisfied for the buildup of voltage? What limits the voltage to which it can built up? **[J.N. Technological Univ. Electrical Machines-I, March-2014]**
33. What are the essential conditions for 'building up' of voltage of a dc shunt generator? **[West Bengal Univ. of Technology, Electrical Machine-I, 2013]**
34. List the reasons why a shunt generator fails to build up the voltage. What is critical speed and critical field resistance for a shunt generator? Explain with appropriate diagram. **[R.G. Technical Univ. Electromechanical Energy Conversion-I, June-2011]**
35. The polarity of the terminal voltage on a compound generator is to be reversed. To do so, the machine is stopped. Residual magnetism is reversed by flashing the field, after which the connections are restored as before. The machine is started again.
    (*i*) Would the voltage build up?
    (*ii*) If it does, is the generator cumulative or differentially compound now ? Explain briefly.

SHORT ANSWER TYPE QUESTIONS WITH ANSWERS

Q. 1. In how many ways mmf required to establish the flux in the magnetic circuit of a dc generator can be had?

Ans. The mmf necessary to establish the flux in the magnetic circuit of a dc generator can be obtained by means of (*i*) permanent magnets (*ii*) field coils excited by some external source and (*iii*) field coils excited by the generator itself.

Q. 2. What do you understand by self-excitation mode of dc machine? Name two dc machines working in this mode. **[P.T.U., June-2003]**

Ans. The dc machine in which field winding(s) is/are self excited by the current supplied by the machine itself is called the self-excitation mode of the machine. In such machines the field coils are interconnected with the armature winding.

The self-excited machines may be (*i*) dc series machine (*ii*) dc shunt machine (iii) compound wound.

Q. 3. Why are permanent magnet generators rarely used in industry?

Ans. Because of very small output.

Q. 4. What is the difference between the excitation system of a dc shunt generator and that of a dc series generator?

Ans. In a series wound generator, there is only one field winding which is connected in series with the armature winding so that whole current flows through the field winding as well as load. In a shunt wound generator also there is only one field winding, as in case of a series wound generator, but in this case it is connected across the armature winding forming a parallel or shunt circuit.

Q. 5. Why is it necessary to wind field coils of a medium or large size dc series generators with a small number of turns of thick wire ?

Ans. Since the series field winding has to carry the full-load current, series field winding is wound with a fewer turns of thick wire or strips.

Q. 6. Why is it necessary to wind field coils of a dc shunt generator with a large number of turns of fine wire.

Ans. The field coils of a dc shunt generator are wound with a large number of turns of fine wire so as to have high resistance in order to keep shunt field current as low as possible as well as to provide required mmf.

Q. 7. What is the volt drop allowed for all brushes of each polarity for carbon or graphite brushes?

Ans. 1 V per brush.

Q. 8. What type of current flows in stator winding and rotor winding of a dc shunt generator?

Ans. In stator : dc

In rotor : ac.

Q. 9. Why does saturation curve starts from some value higher than zero ?

Ans. Due to residual magnetism.

Q. 10. In a dc machine flux per pole decreases with increase of load. Justify the statement.

Ans. In a dc machine, flux per pole decreases with the increase of load due to demagnetization effect of armature reaction.

Q. 11. In what respect the separately-excited dc generator is superior to self-excited dc generator?

Ans. The separately-excited dc generator has the advantage of stable operation with any field excitation over self-excited generators.

Q. 12. What is meant by critical resistance of a dc series generator?

Ans. It is the maximum value of resistance with which the dc series generator will be able to excite.

Q. 13. What is the field circuit critical resistance?

[U.P. Technical Univ. 2002-03]

Ans. It is the maximum value of resistance in the field circuit of a dc shunt generator with which the dc shunt generator will be able to excite.

Q. 14. What is the critical load resistance?

Ans. It is the minimum value of load resistance of a dc shunt generator with which it will be able to build up.

Q. 15. What is critical speed in a dc shunt generator?

Ans. It is the speed of the machine for which the given shunt field resistance will represent the critical field resistance.

Q. 16. A dc shunt generator can operate under short-circuit condition but series and cumulative compound generators cannot operate under this condition. Explain the reason.

Ans. In a dc shunt generator the field is connected across the armature and hence under short-circuit condition, some voltage exists across the armature because of residual flux and this keeps the shunt generator to function at low voltage. The shunt generator field also carries small current under these conditions.

In case of series and cumulative compound generators, the series field under the short-circuit condition carries excessive current and hence terminal voltage starts decreasing due to excessive demagnetizing effects of armature reaction. Thus the terminal voltage finally reduces to zero and, therefore, such machines cannot operate under short-circuit condition.

Q. 17. Why it is not advisable to fully load a dc shunt generator until its voltage builds up to the rated value?

Ans. Building up process in a dc shunt generator is a step by step process which means that initially when the machine is started, generated voltage is small circulating small current. In case the generator is loaded fully at this stage, the magnitude of excitation current will become very small and hence resulting into the rate of build up becoming very small. Thus the time taken for the voltage to build up to its rated value will become large. Hence it is advisable not to fully load the dc shunt generator during the process of build-up of voltage.

Q. 18. A dc shunt generator is developing rated terminal voltage at some speed. For this generator, answer the following and give a brief explanation in support of your answers:

(*i*) If only the direction of rotation is reversed, will the generator build up?

(*ii*) If the direction of rotation and also the residual magnetism are reversed, will the generator build up?

Ans. (*i*) The generator will not build up. With the direction of rotation reversed, the polarity of the generated emf will be reversed. The direction of field current will also reverse and hence the residual magnetism present in the poles will be wiped off and the generator will fail to build up.

(*ii*) When the direction of rotation and the residual magnetism both are reversed, then the small emf produced initially will strengthen the field further and hence the generator will build up voltage.

Q. 19. Why does the terminal voltage of a dc shunt generator fall as the current supplied by the machine is increased?

Ans. With the increase in current supplied by the machine (*i*) voltage drop across the armature winding and the contact resistance of the brushes increases (*ii*) field is weakened due to increase in effect of armature reaction. As a result terminal voltage decreases with increase in load current. The decrease in terminal voltage causes further decrease in field current. This will in turn cause the emf and, therefore, terminal voltage to drop too. Thus the effects are cumulative and terminal voltage is reduced to a much greater extent.

Q. 20. Why does the external characteristics of a dc shunt generator turn back as it is overloaded? **[U.P.S.C. I.E.S. 2001]**

Ans. As the generator is loaded by decreasing the resistance of the external load circuit, the terminal voltage falls due to the reasons mentioned in Answer to Q. 19. The effects responsible for fall in terminal voltage are cumulative and the terminal voltage falls to a much greater extent. In the beginning the effect of decrease in resistance predominates over the effect of decreased terminal voltage but when the load current reaches a certain value (much higher than the full-load current) the demagnetising effect of armature reaction and voltage drop in the armature become so important that a further decrease in load resistance causes a decrease in current rather than increase and so the characteristic turns back.

Q. 21. The terminal voltage of a dc generator at full load is more than its open-circuit voltage. What is the mode of excitation of the generator?

Ans. Cumulative compound.

Q. 22. What kind of generator can be expected to have no-load voltage and full-load voltage to be same?

Ans. Flat-compounded dc generator.

Q. 23. A differentially compounded dc generator is delivering full-load current at a terminal voltage of 220 V. If the series winding is short circuited, then its terminal voltage will ......

Ans. Terminal voltage will increase because of series field winding being shorted. This is because series field which opposes the shunt field in a differentially compounded dc generator becomes zero.

Q. 24. A dc generator can run to generate the same no load voltage with different modes of excitation. For which mode of excitation the short-circuit current will be minimum?

Ans. Differential compound wound generator.

Q. 25. Why is a dc series generator not suitable for general electric supply?

Ans. Because of their rising V-I characteristics.

Q. 26. Which type of dc generator is used for increasing the voltage across the feeder carrying current supplied by some other source?

Ans. Series wound dc generator, called the series booster.

Q. 27. Why a dc shunt generator is used for battery charging?

Ans. Because a dc shunt generator can be made to provide constant output voltage.

Q. 28. What type of dc generator is suitable for electric welding ?

Ans. Differentially compounded dc generator.

Q. 29. Why is only a differentially compounded dc generator suitable for arc welding ?

Ans. Because in a differentially compounded dc generator, the terminal voltage drops very rapidly with the increase in load current and, therefore, short-circuit cannot cause any damage to it.

PROBLEMS

1. Determine the power output of a dc armature having 1,152 lap-connected conductors carrying 150 A and rotating at 300 rpm in a 12-pole field. The flux per pole is 60 mWb. **[Ans.** 51.84 kW]

2. Calculate the voltage induced in the armature winding of a 4-pole, lap wound dc machine having 728 active conductors and running at 1,800 rpm. The flux per pole is 30 mWb. If the armature is designed to carry a maximum line current of 100 A, what is the maximum electromagnetic power developed by the armature?
[U.P. Technical Univ. Electrical Engineering Odd Semester 2009-10]
[Ans. 655.2 V, 65.52 kW]

3. A 20 kW, 200 V shunt generator has an armature resistance of 0.05 Ω and a shunt field resistance of 200 Ω. Calculate the power developed in the armature when it delivers rated output. Give sketch.
[U.P. Technical Univ. Electrical Engineering Even Semester 2006-07]
[Ans. 20.71 kW]

4. A 4 pole lap wound dc shunt generator has flux per pole 0.06 wb. The armature winding consists of 200 turns. Each turn having a resistance of 0.002 Ω. Calculate the terminal voltage when running at 1200 rpm if armature current is 50 A.
[M.D. Univ. Electromechanical Energy Conversion, December-2013]
[Ans. 478.75 V]

5. A 4-pole dc shunt generator with lap-connected armature has field and armature resistances of 80 Ω and 0.1 Ω respectively. It supplies power to 50 lamps rated for 100 volts, 60 watts each. Calculate the total armature current and the generated emf by allowing a contact drop of 1 volt per brush.
[U.P. Technical Univ. Electrical Engineering Even Semester 2003-04]
[Ans. 31.25 A, 105.125 V]

6. A 4-pole dc generator with wave-connected armature has 41 slots and 12 conductors per slot. Armature resistance and shunt field resistance are 0.5 Ω and 200 Ω. Flux per pole is 125 mWb. Speed N = 1,000 rpm. Calculate the voltage drop across terminals. The load resistance is 10 Ω. **[U.P. Technical Univ. Odd Semester 2013-14]**
[Ans. 1,947.7 V]

7. A 4-pole dc shunt generator with a shunt field resistance of 100 Ω and an armature resistance of 1 Ω has 378 wave-connected conductors in its armature. The flux per pole is 0.02 Wb. If a load resistance of 10 Ω is connected across the armature terminals and the generator is driven at 1000 rpm, calculate the power absorbed by the load.
[Hyderabad Univ. Elec. Technology, 1991]
[Ans. 5.153 kW]

8. A 4 pole dc shunt generator with wave wound armature has 40 slots each having 12 conductors. Armature resistance is 1 Ω and shunt field resistance is 200 Ω. The flux per pole is 25 mWb. If a load of 50 Ω is connected across the armature terminals, calculate the voltage across the load when the generator is driven at 1,000 rpm. What will be the load voltage if the generator is lap wound ?
[U.P. Technical Univ; Electrical Engineering Odd Semester, 2006-07]
[Ans. 195.1 V]

9. An 8-pole dc shunt generator has 778 wave-connected armature conductors running at 500 rpm, supplies a load of 12.5 Ω resistance at a terminal voltage of 250 V. The armature resistance is 0.24 Ω and the field resistance is 250 Ω. Find out the armature current, the induced emf and the flux per pole.
[Ans. 21 A, 255.04 V, 9.834 mWb]

10. A shunt generator has an induced voltage of 254 V. When the machine is loaded, the terminal voltage drops down to 240 V. Neglecting armature reaction, determine the load current, if the armature resistance is 0.04 Ω, and the field circuit resistance is 24 Ω. **[Ans.** 340 A]

11. A 4-pole lap-wound shunt generator has 300 armature conductors and a flux of 0.1 Wb. It runs at 1,000 rpm. The armature and field resistances are 0.2 Ω, 125 Ω respectively. Calculate the terminal

voltage when it is loaded to take a load current of 90 A. Ignore armature reaction. **[Nagpur Univ., April-1999]**

[Ans. 481.2 V]

12. A 4-pole dc generator has a useful flux per pole of 0.07 Wb. The armature has 400 lap-connected conductors, each of resistance 0.002 Ω and is rotating at a speed of 900 rpm. If the armature current is 50 A, calculate the terminal voltage. **[Ans.** 417.5 V]

13. A 4-pole dc generator has 564 conductors on its armature and is driven at 800 rpm. The flux per pole being 20 mWb and the current in each conductor is 60 A. Calculate (*a*) the total current, (*b*) emf, (*c*) power generated in armature, if the armature is (*i*) wave wound (*ii*) lap wound. **[Ans.** (*i*) (*a*) 120 A (*b*) 300.8 V (*c*) 36.096 kW (*ii*) (*a*) 240 A (*b*) 150.4 V (*c*) 36.096 kW]

14. A shunt generator supplies a load of 10 kW at 200 V through a pair of feeders of total resistance of 0.05 Ω. Armature resistance is 0.1 Ω. Shunt field resistance is 100 Ω. Find the terminal voltage and the generated emf of the generator. **[Ans.** 202.5 V; 207.703 V]

[Hint: Terminal voltage,

V = Load voltage + voltage drop across feeder

$$= 200 + \frac{10 \times 1{,}000}{200} \times 0.05 = 202.5 \text{ V}]$$

15. A short shunt compound generator delivers a load current of 30 A at 220 V and has armature, series and shunt field resistances of 0.05 Ω, 0.03 Ω and 200 Ω respectively. Calculate the induced emf and the armature current. Allow 1.0 V per brush for contact drop. **[U.P Technical Univ. Electrical Machines, 2003–04]**

[Ans. 224.455 V]

16. A long-shunt compound generator delivers a load current of 50 A at 500 V and has armature, series field and shunt field resistances of 0.05 Ω, 0.03 Ω and 250 Ω respectively. Calculate the generated voltage and the armature current. Allow 1 V per brush for contact drop. **[Allahabad Univ. Electrical Science-I 1992]**

[Ans. 506.16 V, 52 A]

17. A short-shunt dc compound generator supplies 200 A at 100 V. The resistance of armature, series field and shunt field windings are 0.04, 0.03, and 60 ohms respectively. Find the emf generated. Also find the emf generated if same machine is connected as a long shunt machine. **[Nagpur Univ., April-1998]**

[Ans. 114.07 V; 114.117 V]

18. In a 110 V compound generator, the armature, shunt and series winding resistances are 0.06 Ω, 25 Ω, and 0.04 Ω respectively. The load consists of 200 lamps each rated 55 W, 110 V connected in parallel. Find the total emf and armature current, when the machine is connected for (*i*) long shunt (*ii*) short shunt. Ignore the armature reaction and brush drop. **[Rajasthan Technical Univ. Electrical Machines-I, January-2013; R.G. Technical Univ. Electromechanical Energy Conversion, December-2012]**

[Ans. (*i*) 104.4 A, 120.44 V (*ii*) 104.56 A (*ii*) 120.3736 V]

19. Series generator having a combined armature and field resistance of 0.4 Ω is running at 1,000 rpm and delivering 5.5 kW at terminal voltage of 110 V. If the speed is raised to 1,500 rpm and load is adjusted to 10 kW, find the new current and terminal voltage. Assume that the machine is working on straight line portion of the magnetisation characteristic. **[Electrical Machines-I; Rajasthan Univ. 2004, Rajasthan Technical Univ. 2007, 2010]**

[Ans. 53.45 A, 187.1 V]

20. The magnetization characteristic of a shunt generator at 1,000 rpm is as follows :

Field Amperes	1.0	2.0	3.0	4.0	5.0	6.0	7.0	8.0
OC Volts	62.5	107.5	155.0	196.5	231.0	256.0	275.0	283.0

Estimate the no-load terminal voltage of the machine when run at 800 rpm with 20 ohms field circuit resistance. **[Ans.** 240 V]

21. A shunt generator has following magnetizing curve:

I_f (amps)	0	0.5	1.0	1.5	2.0
E(volts)	4	42	78	93	100

Calculate the value of critical resistance. Also find the value of open-circuit voltage when field resistance is 60 ohms.

[Ans. 88 ohms; 94 volts]

22. The OCC of a dc generator when separately excited and driven at 1,100 rpm is given by

Field Amperes	0.5	1.0	1.5	2.0	2.5	3.0	3.5
Open-Circuit voltage	61.5	123.0	165.0	198.0	220.0	238.0	253.0

If the generator is shunt connected with a total field circuit resistance of 80 ohms, determine (*i*) the no-load emf and (*ii*) the output resistance when the terminal voltage is 200 V, if the armature resistance is 0.1 ohm. **[Ans.** (*i*) 235 V (*ii*) 1.01266 ohms]

23. The following is the OCC of a dc shunt generator at 300 rpm.

I_f(A)	0	0.2	0.3	0.4	0.5	0.6	0.7
V_{OC} (V)	7.5	93	135	165	186	202	215

The field resistance of the machine is adjusted to 554.5 Ω. Determine

(*i*) No load voltage.

(*ii*) Critical field resistance.

(*iii*) Critical speed for given field resistance.

[U.P. Technical Univ. Electromechanical Energy Conversion-I, 2006-07]

[Ans. (*i*) 195 V (*ii*) 462.5 Ω (*iii*) 227 rpm]

24. The magnetisation curve of a dc shunt generator running at 1000 rpm is as follows:

Field amperes	0.25	0.5	1.0	1.5	2.0	2.5	3.0
EMF volts	36.0	72.0	138.0	188.0	225.0	250.0	270.0

Find (*a*) the value of field resistance to give 240 V on no load (*b*) the speed at which the generator just fails to build up.

[Ans. (*a*)106.67 ohms (*b*) 710 rpm]

25. The magnetisation characteristic of a dc generator is given by:

Field current (amps)	0	0.5	0.75	1.0	1.25	1.5	1.75	2.0
EMF (volts)	40	80.0	112.0	136.0	152.0	160.0	165.0	170.0

(*i*) Determine the value of field resistance to induce 163 volts.

(*ii*) The speed of the generator is increased by 20% and the field resistance is increased by 10%. Determine the induced emf.

[Ans. (*i*) 100 ohms (*ii*) 198 V]

26. A dc shunt generator is to be converted into a level compounded generator by the addition of a series field winding. On no load a current of 4 A in the field winding produces a terminal voltage of 440 V, but on full load when supplying 250 A the shunt field current has to be increased to 6 A for the same terminal voltage at the same speed. The shunt field winding has 1,250 turns per pole. Determine the number of series field turns per pole required for level compounding.

[Ans. 10]

27. A shunt generator is to be converted into a level compounded generator by the addition of a series field winding. From a test on the machine with shunt excitation only, it is found that the shunt field current is 3.1 A to give 400 V on no-load and 4.8 A to give the same voltage when the machine is supplying its full load of 200 A. The shunt field winding has 1,200 turns per pole. Find the number of series turns required per pole. **[Ans.** 10.2]

CHAPTER 8

Parallel Operation of DC Generators

INSIDE THIS CHAPTER

8.1 INTRODUCTION

In most power stations (whether dc or ac) it will be found that power is supplied from several smaller units (generators) operating in parallel rather than from a single large unit capable of taking care of the maximum peak loads. The smaller units can be operated singly or in various parallel combinations, as per actual demand of load. Such practice is considered extremely desirable from the standpoint of efficiency, reliability (continuity of service), maintenance and repair problems, and additions to plant capacity with the growth in load on the power station.

1. Efficiency. The load on an electric power station fluctuates, usually having its peak value sometimes during the day and its minimum value during the night hours. Since generators operate most efficiently when operated on full load (or nearly full load), it is logical to operate a small unit delivering rated output when the load demand is light. Then as the load increases, a larger unit is substituted for the smaller one or another unit is connected to operate in parallel with the one already in operation. This keeps the machines loaded up to their rated capacity and increases *efficiency of operation.*

2. Reliability or Continuity of Service. Continuity of service is one of the utmost requirements of electrical power systems. Several small units are more reliable than a large single unit, since if one unit fails, the continuity of supply can be maintained by remaining units. On the other hand if the power stations consisted only of a single unit, in the event of breakdown of the prime mover or the generator itself there will be complete shut down (failure of supply).

3. Maintenance and Repair. It is considered necessary to carry out regular inspection and maintenance of the machinery so as to avoid the possibility of failure. This is possible only when the unit is out of service which means that the remaining units should be capable to take care of the load. Repairing of a unit is also more convenient and economical if there are several small units in the power station. The cost of standby unit is also small.

4. Additions To Power Plant. The additional unit can be installed as and when required with the growth of load on power station.

Though the generators can be connected in series to operate for supply of power and if it is done all the units will deliver the same current and load on any unit will be proportional to its terminal voltage. So parallel operation is almost universally adopted, in which all the generators operate on the same voltage and load on any unit is proportional to the current delivered by it.

8.2 PARALLEL OPERATION OF DC SERIES GENERATORS

Though the dc series generators are not usually employed for supply of power but dc series motors are frequently arranged to operate as dc series generators during electric braking. The two dc series motors are disconnected from the supply mains and are connected in parallel to the resistors. The motors continue to run due to kinetic energy of train and act as self-excited dc series generators generating emfs. The kinetic energy is converted into electrical energy which is dissipated as heat in the external resistors.

For satisfactory parallel operation of dc series generators, it is necessary to connect the armatures of the two machines through a heavy copper bar, called the *equalizing bar.* It is explained below:

Let two machines of resistance R (the armature and field resistance), induced emfs E_1 and E_2 operate in parallel, as shown in Fig. 8.1.

When E_1 and E_2 are equal they will share the load equally but when one of the induced emf becomes greater than another induced emf, say E_1 becomes slightly greater than E_2, which is quite probable, then a current i will circulate in clockwise direction round the local path, as shown by dotted lines in Fig. 8.1. The magnitude of the circulating current i will be

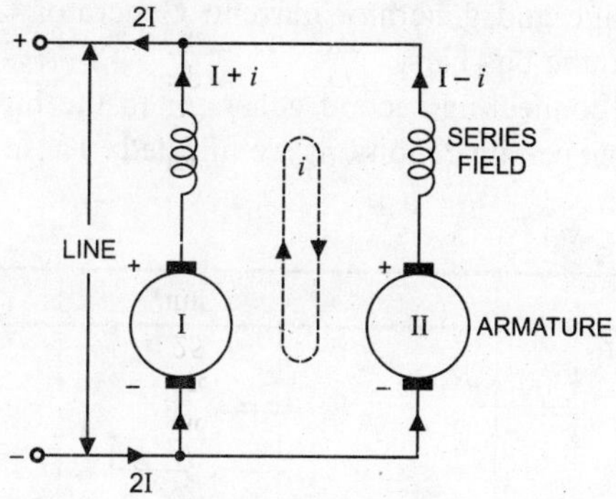

Fig. 8.1 *Series Generators in Parallel Without Equalising Bar Connections*

$\frac{E_1 - E_2}{2R}$. Now total current supplied by machine I will be (I + *i*) and of machine II (I – *i*). Due to increase of field current in machine I and rising characteristic of series generators, the induced emf of machine I will rise and induced emf of machine II will fall. Thus the difference of the two induced emfs E_1 and E_2 will increase, which will cause further increase in circulating current. Thus the effect is cumulative and if there were no fuse or automatic switch in the circuit, the current in machine II will be reversed. This would reverse the direction of induced emf, E_2 and resultant emf in the circuit would be $[E_1 - (-E_2)]$ *i.e.* $(E_1 + E_2)$ and circulating current would be $\frac{E_1 + E_2}{2R}$. Thus the two emfs will then act in series around a circuit of very low resistance and conditions are virtually those of a short circuit on the two machines resulting in damage of the machines.

The possibility of reversal of either machine can be prevented by preventing the flow of circulating current produced due to inequalities of induced emfs of the machines through the series field windings. This object can be achieved by connecting a heavy copper bar of negligible resistance across the two machines, as shown in Fig. 8.2. The circulating current is now confined to the armatures and equalizing bar and the field circuits remain unaffected as long as the total current being supplied does not change. Any tendency to cause increase in armature current of a machine causes drop in its terminal voltage restoring the original condition.

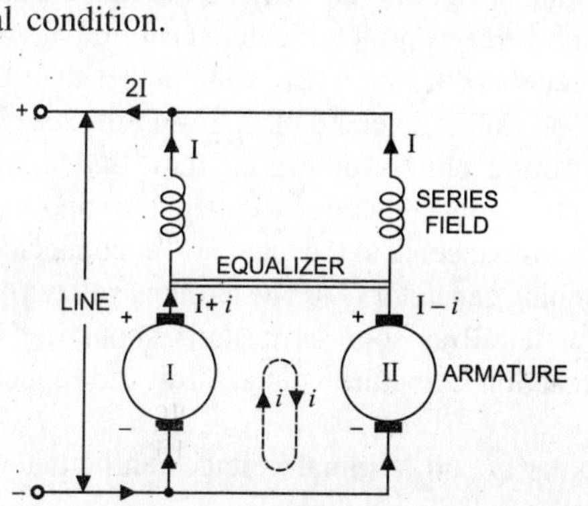

Fig. 8.2 *Series Generators in Parallel With Equalising Bar Connections*

An alternative arrangement for ensuring satisfactory parallel operation of dc series generators is by cross-connecting of field windings, as illustrated in Fig. 8.3. This arrangement is more positive in action since, should the current in one armature tend to rise, its field is weakened due to fall in current in the other armature resulting in more tendency of fall in its terminal voltage. The original condition is, therefore, restored immediately.

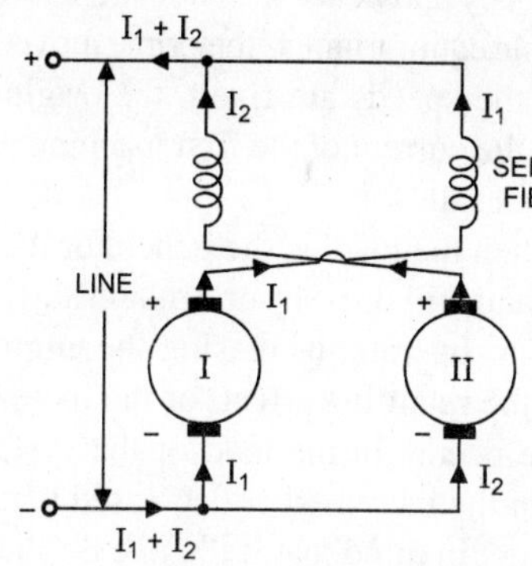

Fig. 8.3 *Series Generators in Parallel With Field Windings Cross-Connected*

8.3 PARALLEL OPERATION OF DC SHUNT GENERATORS

Shunt wound generators due to their drooping characteristics are best suited for parallel operation.

Characteristics of shunt wound generators A and B are shown in Fig. 8.4. If the generators A and B having the current characteristics shown in Fig. 8.4 are connected in parallel, their terminal voltages must be same neglecting drop in connecting leads. Let the terminal voltage of both of the generators be V, then generator A will supply I_A amps and generator B will supply I_B amps. Since generator B has more drooping characteristic to that of generator A, it delivers less current than current delivered by generator A. So we conclude that *the generator with more drooping characteristic carries small load.*

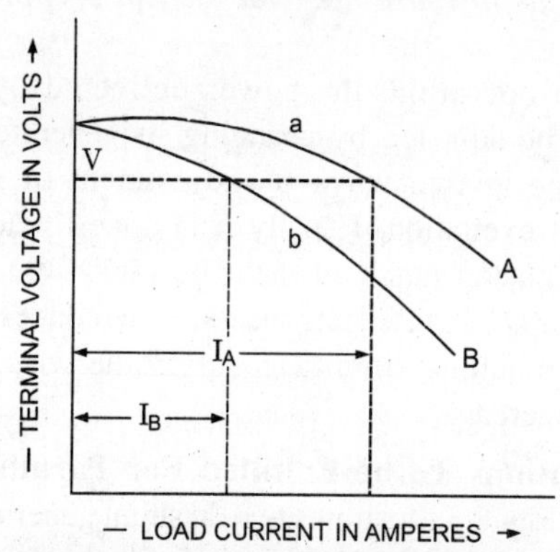

Fig. 8.4 *Load Sharing Characteristics of DC Shunt Generators*

Now let the load on generator B be increased temporarily than its share, which may occur either due to temporary increase in speed of its prime mover or due to momentary change of load on the system. Now when the normal condition restores, if the increased current persists, generator B tends to operate at some point *b* on its characteristic. This results in drop of its terminal voltage, which tends to make it to take less load. Moreover, when generator B takes more load than its own share, generator

A must take less load, because total load is constant. Now generator A operates at some point *a* on its characteristic, which increases its terminal voltage and make it to share more load. Therefore, any tendency of one generator to take more load than its share results in change of terminal voltage in the system opposing this tendency and hence the shunt generators in parallel are said to be in stable equilibrium. If the prime movers are properly governed, that is, the speeds are fixed, the original conditions will be restored as the current of the first machine will rise while that of the other will fall.

In practice, when the load on the generator increases, speed tends to decrease, and will do so by an amount sufficient to cause the governor gear to operate, increasing the engine power and torque to balance the retarding effect of the increased load.

If any change occurs in the load of the system, each unit must share some of the increased or decreased load.

It should be kept in mind that if the two dc shunt generators operating in parallel have

(*i*) equal no-load voltages, the load will be shared in such a ratio that the load current delivered by each machine causes the same voltage drop in the machines.

(*ii*) unequal no-load voltages, the load current shared by the generators will cause sufficient voltage drops in each machine so as to keep their terminal voltages the same.

(*iii*) identical external characteristics, that is, if the voltage changes in both the machines by exactly the same amount for the same per cent of change in load (i.e. identical voltage regulation) then the two machines will share the total load in proportion to their respective power ratings.

(*iv*) different voltage regulation, the generator with higher voltage regulation (i.e. more drooping characteristic) will share smaller load.

For satisfactory operation, it is desirable that all the machines operating in parallel should have same voltage regulation so as to share the total load in proportion to their rated capacity.

While in operation, the power delivered by individual machine can be adjusted by changing its generated emf. This could be done in two ways–by adjustment of speed or by adjustment of excitation. Usually it is done by adjustment of excitation. For a constant load on the power system, if excitation of one machine is increased it is necessary to reduce the excitation of the other machine simultaneously, otherwise, the system voltage will increase.

8.3.1. Conditions To be Fulfilled For Paralleling of DC Generators. Fig. 8.5 illustrates two dc shunt generators and load connected to the bus-bars. Generator G_1 is connected to the bus-bars and supplying load. The polarity of the voltage at the terminals of generator in operation and that of the bus-bar* are indicated in the diagram. Fuses or circuit breakers are employed for protection of equipment in the event of an overload or short circuit and voltmeters and ammeters are included to indicate the output voltage and generator current. Generator G_2 is to be connected to the bus-bars.

Before connecting second generator to the bus-bars it is necessary that certain conditions are fulfilled . These conditions are:

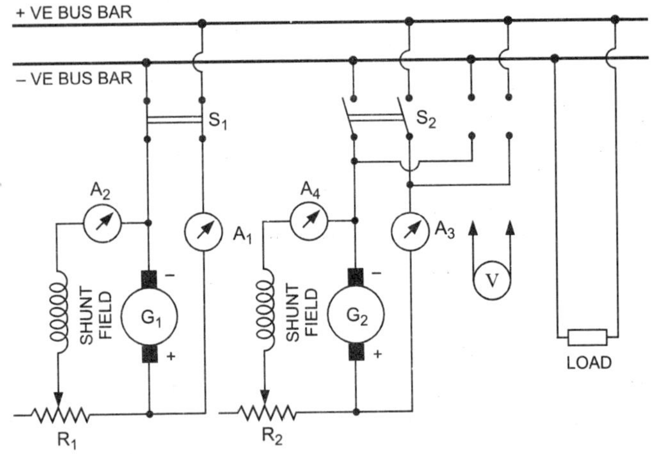

Fig. 8.5 *Two Shunt Generators in Parallel Operation*

1. Same polarity. The polarity of the incoming generator G_2 must be the same as that of the generator G_1, in operation *i.e.* the positive terminals of both the generators are to be connected together and negative terminals together. If the generators are not properly connected, as just indicated, the improper connections will put both generators in series, with only the internal resistance of each generator limiting the magnitude of current. Since internal resistance of the generators is very small, an exceedingly large current will flow, which will trip the circuit breaker or blow fuses.

2. Equal Terminal Voltage. The terminal voltages of both the machines must be equal. If the terminal voltage of generator G_2 is lower than that of generator G_1, and connected in parallel with generator G_1, the generator G_1 with high terminal voltage will supply power not only to load, but also to the second generator G_2 and the machine G_2 will start operating as a motor. The terminal voltage of the generator G_2 should not be too high even when compared with the terminal voltage of generator G_1. If it is too high, it will cause the armature current of generator G_1 to reverse and generator G_1 will be taking power instead of delivering power. It is preferable that the terminal voltage of the incoming generator G_2 be a few volts higher than the terminal voltage of the running generator G_1. With the terminal voltage of the incoming generator higher than that of the running generator, the second generator will begin to supply some power as soon as it is connected to the bus-bars (or connected in parallel with the running generator G_1). The terminal voltage of the second generator will fall as soon as it starts supplying load due to armature reaction, armature voltage drop and reduction in field current.

The polarity and terminal voltage can be determined by a moving coil paralleling voltmeter.

* When a number of generators operate in parallel, positive and negative terminals of all the machines are connected to two separate heavy copper bars placed behind the switch panel. These copper bars, known as bus-bars, may also be regarded as positive and negative terminals for the whole power station.

8.3.2. Starting of Generator and Connecting To Bus-bars. Prime mover of generator G_2 is started and brought up to rated speed keeping the switch S_2 open. The field of generator G_2 is adjusted with the help of shunt regulator R_2 to give its terminal voltage equal or one or two volts more than the bus-bar voltage. The voltages are checked for equality with voltmeters connected across the line and incoming generator or by means of a moving coil paralleling voltmeter V which can be connected either to bus-bars or to incoming generator G_2 by means of plug arrangement, as shown in Fig. 8.5.

The polarity of connections is determined only when a generator is paralleled for the first time. The terminals are then so located as to make a repeated polarity test unnecessary. The polarity test is carried out either with a moving coil voltmeter or incandescent lamps. With the switch of the incoming machine open, any pair of line and generator terminals are short circuited by a strap. Then a voltmeter designed for twice the line voltage is connected to the other pair of terminals. If the connection is correct, the voltmeter will read zero, otherwise it will read twice the line voltage.

After checking the polarity and adjusting the terminal voltage the switch S_2 is closed. This puts generator G_2 in parallel with generator G_1. Under these conditions generator G_2 is not supplying any current and said is to be floating since its generated emf is equal and opposite to the emf of running generator G_1.

Since the speed is kept constant by the engine governor so the induced emf depends upon exciting current only. Therefore, after closing switch S_2 the excitation of the generator G_2 is raised while lowering the excitation of the generator G_1 until the incoming generator takes its share of load. The weakening of the field of generator G_1, while transferring load to incoming generator G_2 by increasing its excitation, is necessary to maintain bus-bar voltage constant.

8.3.3. Taking One Generator Running in Parallel With Another One Out of Service. Now if generator G_1 is required to be taken out of service, the excitation of generator G_1 is gradually reduced and that of generator G_2 is gradually increased until the whole load is transferred to generator G_2. The main switch S_1 of the generator G_1, which is carrying no current at this moment, is opened and supply of steam is cut off.

While the load is shifted from one generator to the other, ammeters A_1 and A_3 giving the load currents should be carefully watched to make sure that overloading does not occur. It is also important that excitation of generator G_1 should not be reduced too much that it may start working as a motor.

To take the generator out of service, its main switch should never be opened suddenly when it is sharing load, because on doing so it will cause violent sparking at the switch contacts and also whole of load will be suddenly transferred to another machine, which is objectionable.

8.4 PARALLEL OPERATION OF DC COMPOUND GENERATORS

The essential connections for the parallel operation of two dc compound generators are shown in Fig. 8.6. Since compound dc generators consist of series field windings and normally have rising external characteristics in operating regions, it is necessary to equalize their field currents. This is usually done by means of an auxiliary bus on the main switch board, called as *equalizing bus-bar* (or bar). A momentary change in load distribution between the generators does not affect the series fields, and so stable operation is obtained. If the machines are provided with interpole windings it is necessary that the interpole windings carry the same current as the armature and the equalising bar should, therefore, be connected at the junction of the interpole winding and series field winding as illustrated in Fig 8.7.

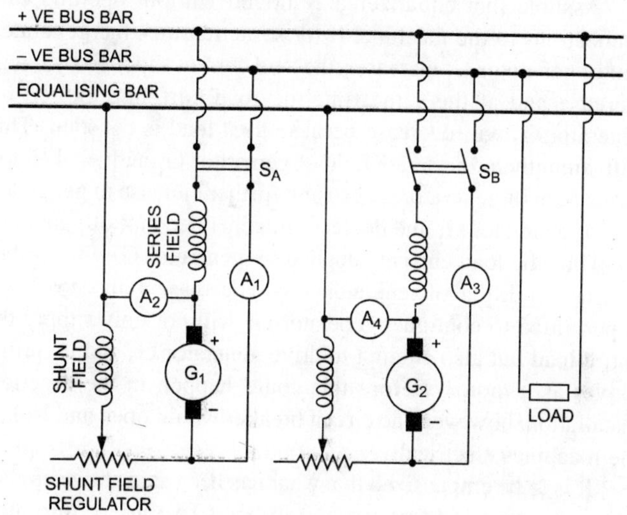

Fig. 8.6 *Two Compound Generators in Parallel Operation*

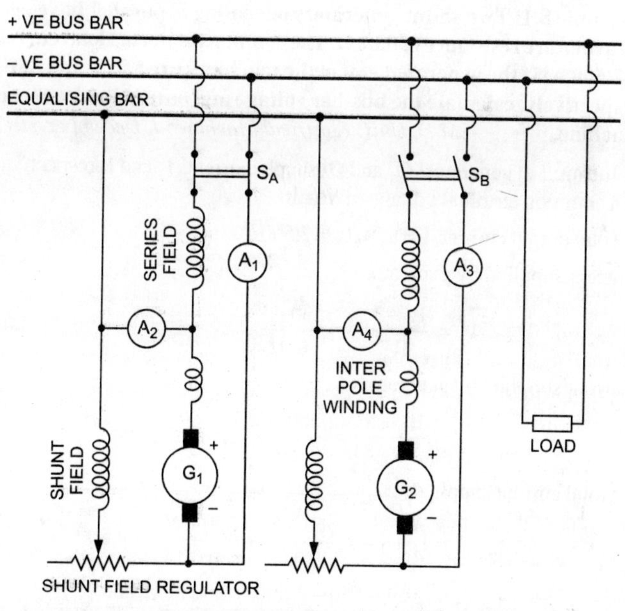

Fig. 8.7

The compound wound dc generators are paralleled to the bus in the same way as dc shunt wound generators. Load transfer or disconnecting one machine from the line, may also be accomplished in the same way. For satisfactory parallel operation it is necessary that the resistance of the equalising bar is very low in comparison to that of series field winding.

For satisfactory parallel operation it is desirable that all the compound dc generators operating in parallel have identical external characteristics so as to share the total load in proportion to their ratings as explained in case of shunt wound generators. Again, since the series field windings are connected in parallel by equalizer connections, the resistances of the series field windings, including their connecting leads must be inversely proportional to the generator's rated currents.

The operator must be careful not to open the equalizer connection on a generator unless at least one of the line connections to that machine has already been removed. Should the equalizer be removed, instability would result if the generators are overcompounded (as usually are). This is explained below.

Assume that equalizer is removed without opening line connections to the machine. If for some reasons, the generated emf of generator G_1 increases, the load current supplied by it will increase and at the same time the load current supplied by generator G_2 will decrease because total load is constant. This will strengthen the series field of generator G_1 and weaken the series field of generator G_2 causing further increase in generated emf of generator G_1 and decrease in generated emf of generator G_2. Thus the load current supplied by generator G_1 will further increase while that of generator G_2 will decrease. If this condition is permitted to continue, generator G_1 will not only supply the entire load but also attempt to drive generator G_2 and its prime mover as a motor. Before this could happen in the practical installation, however, the circuit breaker would open and isolate the machines electrically.

It is to be emphasized that what has been said above applies to overcompound generators and not to undercompound generators which have drooping characteristics.

Example 8.1. Two shunt generators operating is parallel have each an armature resistance of 0.02 Ω. The combined external load current is 2,500 A. If the generated emf of the machines are 560 V and 550 V respectively, calculate the bus-bar voltage and output in kW of each machine. [*M.D. Univ. Electrical Machines-I, December-2012*]

Solution: Let generators G_1 and G_2 supply currents I_1 and I_2 respectively at a common terminal voltage of V volts

Total output current, $I = I_1 + I_2 = 2{,}500$ A ...(*i*)

Current supplied by generator G_1,

$$I_1 = \frac{E_1 - V}{R_a} = \frac{560 - V}{0.02} \quad \text{...(ii)}$$

Current supplied by generator G_2,

$$I_2 = \frac{E_2 - V}{R_a} = \frac{550 - V}{0.02} \quad \text{...(iii)}$$

So total current supplied,

$$I = I_1 + I_2 = \frac{560 - V}{0.02} + \frac{550 - V}{0.02} = 2{,}500$$

$$\text{or } V = \frac{560 + 550 - 2{,}500 \times 0.02}{2} = \frac{1{,}060}{2}\text{ V} = 530 \text{ V}$$

$$\text{Current, } I_1 = \frac{560 - 530}{0.02} = 1{,}500 \text{ A}$$

$$\text{Current, } I_2 = \frac{550 - 530}{0.02} = 1{,}000 \text{ A}$$

$$\text{Output of generator } G_1,\ P_1 = \frac{V \times I_1}{1{,}000} \text{ kW} = \frac{530 \times 1{,}500}{1{,}000}$$

$= 795$ kW **Ans.**

$$\text{Output of generator } G_2,\ P_2 = \frac{V \times I_2}{1{,}000} \text{ kW} = \frac{530 \times 1{,}000}{1{,}000} = 530 \text{ kW}$$

Example 8.2. Two dc generators having rectilinear external characteristics operate in parallel. One machine has the terminal voltage of 270 V on no load and 220 V at the load current of 30 A. The other has a voltage of 280 V at no load and 220 V at the load current of 30 A. Calculate the output current and the bus voltage of each machine when (*i*) the total load current is 50 A (*ii*) the load resistance is 10 Ω. [*U.P.S.C. I.E.S. Electrical Engineering - II 2000*]

Solution: Generator G_1

Voltage drop for 30 A = 270 – 220 = 50 V

$$\text{Voltage drop per ampere of current supplied} = \frac{50}{30} = \frac{5}{3} \text{ V/A}$$

Generator G_2 Voltage drop for 30 A = 280 – 220 = 60 V

$$\text{Voltage drop per ampere of current supplied} = \frac{60}{30} = 2 \text{ V/A}$$

Now let the output currents of generators G_1 and G_2 be I_1 and I_2 respectively and bus-bar voltage be V volts.

Terminal voltage of generator G_1 when it is supplying current I_1 will be equal to bus-bar voltage i.e.

$$V = 270 - \frac{5}{3} I_1 \quad \text{...(i)}$$

Similarly for generator G_2 $V = 280 - 2I_2$ ...(*ii*)

(*i*) Total current supplied, $I = I_1 + I_2 = 50$ A ...(*iii*)

Comparing Eqs. (*i*) and (*ii*) we have

$$270 - \frac{5}{3} I_1 = 280 - 2I_2$$

$$\text{or } 2I_2 - \frac{5}{3} I_1 = 10 \quad \text{...(iv)}$$

Solving Eqs. (*iii*) and (*iv*) we have

Current $I_1 = 24.54$ A **Ans.**

Current $I_2 = 25.46$ A **Ans.**

Terminal voltage $V = 280 - 2I_2$

$= 280 - 2 \times 25.46 = 229.08$ V **Ans.**

$$(ii) \text{ Total current supplied} = I_1 + I_2 = \frac{V}{R_L} = \frac{V}{10} \quad \text{...(v)}$$

Substituting $V = 10 (I_1 + I_2)$ in Eqs. (*i*) and (*ii*) we have

$$10 (I_1 + I_2) = 270 - \frac{5}{3} I_1$$

or $35I_1 + 30I_2 = 810$ ...(*vi*)

and $10 (I_1 + I_2) = 280 - 2I_1$

or $10I_1 + 12I_2 = 280$ ...(*vii*)

Solving Eqs. (*vi*) and (*vii*) we have

$I_1 = 11$ A **Ans.**

$I_2 = 14.17$ A **Ans.**

Terminal voltage, $V = IR_L = (I_1 + I_2)R_L$

$= (11 + 14.17) \times 10 = 251.7$ V **Ans.**

Example 8.3. Two dc shunt generators are operating in parallel. Their no-load voltages are 260 V and characteristics are linear. At 220 V, generator-1 can deliver 310 kW and generator-2 can deliver 600 kW. Find the total load 'P' and load supplied by each at 250 V.

[*J.N. Technological Univ. Electrical Machines-I, March-2014*]

Solution: Load current supplied by generator-1 when supplying 310 kW at 220 V,

$$I_1 = \frac{310 \times 1{,}000}{220} = 1{,}409 \text{ A}$$

Load current supplied by generator-2 when supplying 600 kW at 220 V,

$$I_2 = \frac{600 \times 1{,}000}{220} = 2{,}727 \text{ A}$$

No-load voltage = 260 V in case of both generators-1 and 2

Load voltage = 220 V in case of both generators-1 and 2

Voltage drop, v = 260 – 220

= 40 V in case of both generators 1 and 2

Voltage drop per ampere of current supplied in case of generator-1

$$= \frac{\text{Voltage drop}}{\text{Current supplied}} = \frac{40}{1{,}409} \text{ V/A}$$

Voltage drop per ampere of current supplied in case of generator-2

$$= \frac{40}{2{,}727} \text{ V/A}$$

Current supplied by generator-1, at 250 V,

$$I_1' = \frac{\text{Voltage drop}}{\text{Volt drop per ampere of current}}$$

$$= \frac{260 - 250}{40/1{,}409} = 352.25 \text{ A}$$

Similarly load supplied by generator-2, at 250 V

$$I_2' = \frac{260 - 250}{40/2{,}727} = 681.75 \text{ A}$$

Load supplied by generator-1,

$$P_1 = \frac{\text{Load voltage} \times \text{current supplied } I_1'}{1{,}000} \text{ kW}$$

$$= \frac{250 \times 352.25}{1{,}000} = 88.0625 \text{ kW } \textbf{Ans.}$$

Load supplied by generator-2,

$$P_2 = \frac{V \times I_2'}{1{,}000} = \frac{250 \times 681.75}{1{,}000} = 170.4375 \text{ kW}$$

Total load supplied, P = $P_1 + P_2$ = 88.0625 + 170.4375

= 258.5 kW **Ans.**

Example 8.4. Two shunt generators with straight line characteristics are operated in parallel, their no load voltages being 240 V and 245 V respectively. The rating of the above generators are 500 kW at 230 V and 250 kW at 220V. If the total load supplied is 650 kW, calculate (*i*) the terminal voltage and (*ii*) power supplied by each machine in kW.

Solution: Generator I

Rated load, P_1 = 500 kW

Rated voltage, V_1 = 230 V

$$\text{Rated current, } I_1 = \frac{P_1 \times 1{,}000}{V_1} = \frac{500 \times 1{,}000}{230} = 2{,}173.913 \text{ A}$$

Voltage drop per ampere of current supplied,

$$= \frac{V_{01} - V_1}{I_1} = \frac{240 - 230}{2{,}173.913} = 0.0046 \text{ V/A}$$

Generator II

Rated load, P_1 = 250 kW

Rated voltage V_2 = 220 V

$$\text{Rated current, } I_2 = \frac{P_2 \times 1{,}000}{V_2} = \frac{250 \times 1{,}000}{220} = 1{,}136.3636\text{A}$$

Voltage drop per ampere of current supplied,

$$= \frac{V_{02} - V_2}{I_2} = \frac{245 - 220}{1{,}136.3636} = 0.022 \text{ V/A}$$

Let the currents supplied by the generators be I_1' and I_2' and the terminal voltage be V volts.

Terminal voltage of generator G_1 when it is supplying current I_1', will be equal to load voltage *i.e.*

$$V = 240 - 0.0046\, I_1' \quad ...(i)$$

Similarly for generator G_2, $V = 245 - 0.022\, I_2'$ ...(*ii*)

Total load supplied $V(I_1' + I_2') = 650 \times 1{,}000$ ...(*iii*)

Substituting $I_1' = \dfrac{240 - V}{0.0046}$ from Eq. (*i*) and $I_2' = \dfrac{245 - V}{0.022}$ from Eq. (*ii*) in Eq. (*iii*) we have

$$V\left[\frac{240 - V}{0.0046} + \frac{245 - V}{0.022}\right] = 6{,}50{,}000$$

$$52{,}174\ V - 217.4\ V^2 + 11{,}136.36\ V - 45.45\ V^2 = 6{,}50{,}000$$

$$\text{or } 262.85\ V^2 - 63{,}310.4\ V + 6{,}50{,}000 = 0$$

$$\text{or } V = \frac{63{,}310.4 \pm \sqrt{(63{,}310.4)^2 - 4 \times 6{,}50{,}000 \times 262.85}}{2 \times 262.85}$$

= 230.115 V **Ans.**

Substituting V = 230.115 V in Eqs. (*i*) and (*ii*) we get

$$I_1' = \frac{240 - V}{0.0046} = \frac{240 - 230.115}{0.0046} = 2{,}148.91 \text{ A}$$

$$\text{and } I_2' = \frac{245 - V}{0.022} = \frac{245 - 230.115}{0.022} = 676.59 \text{ A}$$

Power supplied by generator I,

$$P_1 = \frac{VI_1'}{1{,}000} = \frac{230.115 \times 2{,}148.91}{1{,}000} = 494.4 \text{ kW } \textbf{Ans.}$$

Power supplied by generator II,

$$P_2 = \frac{VI_2'}{1{,}000} = \frac{230.115 \times 676.59}{1{,}000} = 155.6 \text{ kW } \textbf{Ans.}$$

Example 8.5. Two similar generators are excited to the same voltage and they carry total load of 400 A. If the resistance of each machine is 0.05 Ω and the terminal voltage is 200 V, find

(*a*) load shared by each machine

(*b*) percentage change in no-load voltage of one so that it does not deliver any current

(*c*) the terminal voltage corresponding to (b).

Solution: Since both of the generators are excited to the same voltage, say E volts, have the same armature resistance and same terminal voltage, each of them shares the load equally *i.e.* half of the total load current of 400 A which is equal to 200 A **Ans.**

Excitation voltage $E = V + I R_a = 200 + 200 \times 0.05 = 210$ V

One of the two generators will stop delivering current when its excitation voltage is reduced and made equal to terminal voltage of the other generator while supplying total load current of 400 A.

Assuming the no-load (excitation) voltage of the machine supplying the total load of 400 A unchanged, its terminal voltage,

$$V_1 = E - I_1 R_a = 210 - 400 \times 0.05 = 190 \text{ V}$$

Generator I will stop supplying current when its no-load voltage is reduced to 190 V.

(*b*) Percentage change in no-load voltage of generator I,

$$= \frac{210 - 190}{210} \times 100 = 9.52\% \textbf{ Ans.}$$

(*c*) Terminal voltage corresponding to (*b*) = V_1 = 190 volts **Ans.**

Example 8.6. The terminal voltage of a shunt generator G_1 falls from 500 V at no load to 470 V when delivering a current of 600 A; for a second generator G_2 the figures are 505 V at no load and 470 V at 400 A. When connected in parallel generators supply a total load of 400 kW. Assuming that the voltage/power characteristics are linear, determine the common bus voltage and the current delivered by each machine.

Solution: Power rating of generator $G_1 = \frac{470 \times 600}{1,000} = 282$ kW

Power rating of generator $G_2 = \frac{470 \times 400}{1,000} = 188$ kW

Let bus-bar voltage = V

Load supplied by generator $G_1 = x_1$ kW

Load supplied by generator $G_2 = x_2$ kW

Since total load to be supplied is 400 kW,

$$x_1 + x_2 = 400 \text{ kW} \qquad ...(i)$$

$$V = 500 - \frac{500 - 470}{282} x_1 = 500 - \frac{20 x_1}{188} \qquad ...(ii)$$

Since voltage/power characteristics are linear

$$\text{and } V = 505 - \frac{505 - 470}{188} x_2 = 505 - \frac{35 x_2}{188} \qquad ...(iii)$$

Subtracting Eq. (*ii*) from Eq. (*iii*) we get

$$0 = 5 + \frac{20 x_1}{188} - \frac{35 x_2}{188}$$

$$\text{or } 7 x_2 - 4x_1 = 188 \qquad ...(iv)$$

Multiplying Eq. (*i*) by 4 and adding it to Eq. (*iv*) we get

$$11 x_2 = 1,788$$

$$\text{or } x_2 = 162.54 \text{ kW}$$

$$\text{and } x_1 = 400 - 162.54 = 237.46 \text{ kW}$$

$$\text{and } V = 505 - \frac{35 \times 162.54}{188} = 474.73 \text{ volts } \textbf{Ans.}$$

Current delivered by generator G_1

$$= \frac{x_1 \times 1,000}{V} = \frac{237.46 \times 1000}{474.73} = 500 \text{ A } \textbf{Ans.}$$

Current delivered by generator G_2

$$= \frac{x_2 \times 1,000}{V} = \frac{162.54 \times 1,000}{474.73} = 342 \text{ A } \textbf{Ans.}$$

Example 8.7. Two shunt wound generators running in parallel have each an armature resistance of 0.02 Ω and a field resistance of 50 Ω. The combined external load current is 5,000 A. The fields are excited so that the emf induced in one machine is 600 V and in the second machine is 610 V. Calculate the bus-bar voltage and the output of each machine.

Solution: Let generators G_1 and G_2 supply currents I_1 and I_2 respectively at a common terminal voltage of V volts.

$$\text{Total output current, } I = I_1 + I_2 = 5,000 \text{ A} \qquad ...(i)$$

$$\text{Armature current of generator, } G_1, I_{a_1} = I_1 + \frac{V}{50} = I_1 + 0.02 \text{ V}$$

$$\text{Armature current of generator } G_2, I_{a2} = I_2 + \frac{V}{50} = I_2 + 0.02 \text{ V}$$

$$\text{Induced emf of generator } G_1, E_1 = V + I_{a1} R_a$$

$$\text{or } 600 = V + (I_1 + 0.02 \text{ V}) \times 0.02 \qquad ...(ii)$$

$$\text{Induced emf of generator } G_2, E_2 = V + I_{a2} R_a$$

$$\text{or } 610 = V + (I_2 + 0.02 \text{ V}) \times 0.02 \qquad ....(iii)$$

Subtracting Eq. (*ii*) from Eq. (*iii*) we get

$$0.02 (I_2 - I_1) = 10$$

$$\text{or } I_2 - I_1 = 500 \qquad ...(iv)$$

Solving Eqs. (*i*) and (*iv*) we get

$$I_1 = 2,250 \text{ A and } I_2 = 2,750 \text{ A}$$

Substituting $I_1 = 2,250$ A in Eq. (*ii*) we get

$$V + (2,250 + 0.02 \text{ V}) \times 0.02 = 600$$

$$\text{or } V = 454.82 \text{ V } \textbf{Ans.}$$

$$\text{Output of generator } G_1 = \frac{454.82 \times 2,250}{1,000} = 1,023.35 \text{ kW } \textbf{Ans.}$$

$$\text{Output of generator } G_2 = \frac{454.82 \times 2,750}{1,000} = 1,250.75 \text{ kW } \textbf{Ans.}$$

Example 8.8. Two shunt generators having full-load regulation of 5% and 4% are rated at 200 kW and 400 V. How will they share a current of 800 A while operating in parallel ? Find also the terminal voltage. If the machines are supplying a load whose resistance is 0.5 Ω, find terminal voltage and their currents?

Solution : Full-load rated voltage, V = 400 V

$$\text{Full-load rated current, } I = \frac{200 \times 1,000}{400} = 500 \text{ A}$$

Let the full-load regulation of generators G_1 and G_2 be 5% and 4% respectively.

$$\text{No-load voltage of generator } G_1, E_1 = V + 0.05 \text{ V} = 400 + 20 = 420 \text{ V}$$

$$\text{No-load voltage of generator } G_2, E_2 = V + 0.04 \text{ V} = 400 + 16 = 416 \text{ V}$$

Voltage drop per ampere output current for generator G_1,

$$= \frac{420 - 400}{500} = 0.04$$

Voltage drop per ampere output current for generator G_2,

$$= \frac{416 - 400}{500} = 0.032$$

Let the currents supplied by generators G_1 and G_2 be I_1 and I_2 amperes respectively and terminal voltage V volts.

$$\text{Total current supplied, } I = I_1 + I_2 = 800 \text{ A} \qquad ....(i)$$

$$\text{Terminal voltage } V = 420 - 0.04 I_1 \qquad ...(ii)$$

$$\text{and also } V = 416 - 0.032 I_2 \qquad ...(iii)$$

Subtracting Eq. (*iii*) from Eq. (*ii*) we get

$$0.04 I_1 - 0.032 I_2 = 4$$

$$\text{or } 5 I_1 - 4 I_2 = 500 \qquad ...(iv)$$

Solving Eqs. (*i*) and (*iv*) we get

$$I_1 = 411 \text{ A}$$

$$\text{and } I_2 = 389 \text{ A } \textbf{Ans.}$$

$$\text{Terminal voltage, } V = 420 - 0.04 \times 411 = 403.56 \text{ V } \textbf{Ans.}$$

Let the common terminal voltage be V volts and currents supplied by generators G_1 and G_2 be I'_1 and I'_2 amperes when a load of 0.5 Ω is being supplied.

$$\text{Total load current } I'_1 + I'_2 = \frac{V'}{0.5} = 2V' \qquad ...(v)$$

$$V' = 420 - 0.04 I'_1 \qquad ...(vi)$$

$$\text{also } V' = 416 - 0.032 I'_2 \qquad ...(vii)$$

Solving Eqs. (*v*), (*vi*) and (*vii*) we get

$$I'_1 = 414.16 \text{ A } \textbf{Ans.}$$

$$I'_2 = 392.7 \text{ A } \textbf{Ans.}$$

$$\text{and } V' = 403.43 \text{ V } \textbf{Ans.}$$

Example 8.9. Two shunt generators operating in parallel deliver a total current of 250 A. One of the generators is rated 50 kW and the other 100 kW. The voltage rating of both machines is 500 V and have regulation of 6 per cent (smaller one) and 4 per cent. Assuming

linear characteristics, determine (*a*) the current delivered by each machine (*b*) terminal voltage.

[*J.N. Technological Univ. Electrical Machines-I, March-2014*]

Solution: 50 kW generator

Full-load current, $I_{f1} = \frac{50 \times 1{,}000}{500} = 100$ A

Full load voltage drop $= 6$ per cent of 500 V $= 0.06 \times 500 = 30$ V

Voltage drop per ampere of current supplied $= \frac{30}{100} = 0.3$ V/A

100 kW generator

Full-load current, $I_{f2} = \frac{100 \times 1{,}000}{500} = 200$ A

Full-load voltage drop $= 0.04 \times 500 = 20$ V

Voltage drop per ampere of current supplied $= \frac{20}{200} = 0.1$ V/A

Let the currents supplied by the 50 kW and 100 kW generators be I_1 and I_2 respectively and terminal voltage be V volts

Terminal voltage of 50 kW generator when it is supplying I_1 amperes

$$V = 500 - 0.3\, I_1 \quad ...(i)$$

Similarly for 100 kW generator

$$V = 500 - 0.1\, I_2 \quad ...(ii)$$

Also $I_1 + I_2 = I_L = 250$ ...(*iii*)

From Eqs. (*i*) and (*ii*)

$$0.3\, I_1 = 0.1\, I_2$$

or $I_2 = 3\, I_1$

Substituting $I_2 = 3\, I_1$ in Eq. (*iii*) we get

$$I_1 + 3\, I_1 = 250$$

or $I_1 = 62.5$ A **Ans.**

and $I_2 = 3\, I_1 = 3 \times 62.5 = 187.5$ A **Ans.**

and $V = 500 - 0.1\, I_2 = 500 - 0.1 \times 187.5 = 481.25$ V **Ans.**

Example 8.10. Two dc shunt generators run in parallel to supply together 2.5 kA. The machines have armature resistances of 0.04 Ω and 0.025 Ω, field resistances of 25 Ω and 20 Ω and induced emfs of 440 V and 420 V respectively. Find the bus-bar voltage and output of each machine.

[*J.N. Technological Univ. Electrical Machines-I, March-2014*]

Solution: Let generators G_1 and G_2 supply currents I_1 and I_2 respectively at a common terminal voltage of V volts.

Total output current, $I_1 + I_2 = 2.5 \times 10^3 = 2{,}500$ A

or $I_2 = (2{,}500 - I_1)$ ...(*i*)

Armature current of generator G_1,

$$I_{a1} = I_1 + \frac{V}{25} = I_1 + 0.04\, V$$

Armature current of generator G_2,

$$I_{a2} = I_2 + \frac{V}{20} = 2{,}500 - I_1 + 0.05\, V$$

Induced emf of generator G_1,

$$E_1 = V + I_{a1} \times 0.04 = V + (I_1 + 0.04\, V) \times 0.04$$

or $440 = 1.0016\, V + 0.04\, I_1$ ...(*ii*)

Induced emf of generator G_2,

$$E_2 = V + I_{a2} \times 0.025$$
$$= V + (2{,}500 - I_1 + 0.05\, V) \times 0.025$$

or $420 = 1.00125\, V + 62.5 - 0.025\, I_1$

or $357.5 = 1.00125\, V - 0.025\, I_1$ ...(*iii*)

Solving Eqs. (*i*), (*ii*) and (*iii*) we get

$I_1 = 1{,}267.138$ A

$I_2 = 1{,}232.862$ A

and $V = 388.7$ V **Ans.**

Output of generator $G_1 = \frac{VI_1}{1{,}000}$

$= \frac{388.7 \times 1{,}267.138}{1{,}000} = 492.54$ kW **Ans.**

Output of generator $G_2 = \frac{VI_2}{1{,}000}$

$= \frac{388.7 \times 1{,}232.862}{1{,}000} = 479.21$ kW **Ans.**

Example 8.11. Two dc shunt generators with emfs of 125 V and 120 V, armature resistances of 0.04 Ω and 0.03 Ω and field resistances of 15 Ω and 20 Ω respectively are in parallel supplying a total load of 25 kW. How do they share the load.

[*J.N. Technological Univ. Electrical Machines-I, March-2014*]

Solution: Let generators G_1 and G_2 supply currents I_1 and I_2 respectively at a common terminal voltage of V volts.

Armature current of generator G_1,

$$I_{a1} = I_1 + \frac{V}{R_{sh1}} = I_1 + \frac{V}{15} = (I_1 + 0.0667\, V) \text{ amperes}$$

Armature current of generator G_2,

$$I_{a2} = I_2 + \frac{V}{R_{sh2}} = I_2 + \frac{V}{20} = (I_2 + 0.05) \text{ amperes}$$

Induced emf of generator G_1,

$$E_1 = V + I_{a1} R_{a1}$$

or $125 = V + (I_1 + 0.0667\, V) \times 0.04 = 1.002667\, V + 0.04\, I_1$ ...(*i*)

Induced emf of generator G_2,

$$E_2 = V + I_{a2} R_{a2}$$

or $120 = V + (I_2 + 0.05) \times 0.03 = 1.00\,15\, V + 0.03\, I_2$ ...(*ii*)

Also $I = I_1 + I_2 = \frac{\text{Load supplied in kW} \times 1{,}000}{V}$

$= \frac{25 \times 1{,}000}{V} = \frac{25{,}000}{V}$ ...(*iii*)

Substituting $I_1 = \frac{25{,}000}{V} - I_2$ in Eq. (*i*) we have

$$125 = 1.002667\, V + 0.04 \left(\frac{25{,}000}{V} - I_2\right) \quad ...(iv)$$

Dividing Eq. (*ii*) by 0.03 and Eq (*iv*) by 0.04 we have

$$4{,}000 = 33.383\, V + I_2 \quad ...(v)$$

and $3{,}125 = 25.067\, V + \frac{25{,}000}{V} - I_2$ ...(*vi*)

Adding Eqs. (*v*) and (*vi*) we have

$$7{,}125 = 58.45\, V + \frac{25{,}000}{V}$$

or $58.45\, V^2 - 7{,}125\, V + 25{,}000 = 0$

or $V = \frac{7{,}125 \pm \sqrt{(7{,}125)^2 - 4 \times 58.45 \times 25{,}000}}{2 \times 58.45}$

$= 118.2$ V or 3.63 V

$= 118.2$ V ∵ figure of 3.63 V is not practicable

Substituting V = 118.2 V in Eq. (*i*) we have

$$I_1 = \frac{125 - 1.002667 \times 118.2}{0.04} = 162 \text{ A}$$

Substituting V = 118.2 V and I_1 = 162 A in Eq. (*iii*) we have

$$I_2 = \frac{25,000}{V} - I_1 = \frac{25,000}{118.2} - 162 = 49.5 \text{ A}$$

Power supplied by generator G_1,

$$P_1 = \frac{V \times I_1}{1,000} \text{ kW} = \frac{118.2 \times 162}{1,000} = 19.149 \text{ kW } \textbf{Ans.}$$

$$P_2 = \frac{V \times I_2}{1,000} \text{ kW} = \frac{118.2 \times 49.5}{1,000} = 5.851 \text{ kW } \textbf{Ans.}$$

Example 8.12. Two shunt generators and a battery are all connected to common bus-bars. The open-circuit voltage of the battery is 247 volts. The open-circuit voltages, armature and field resistances of the generators are 250 V, 0.24 Ω, 100 Ω and 248 V, 0.12 Ω, 100 Ω respectively. Both the generators supply the same current when the load on the bus-bars is 40 A. Calculate the internal resistance of the battery.

Solution: The circuit diagram is shown in Fig. 8.8.

Let current supplied by each generator be I and that from battery I_b.

Since load on the bus-bars is 40 A,

$$I + I + I_b = 40$$

$$\text{or } 2\,I + I_b = 40 \quad ...(i)$$

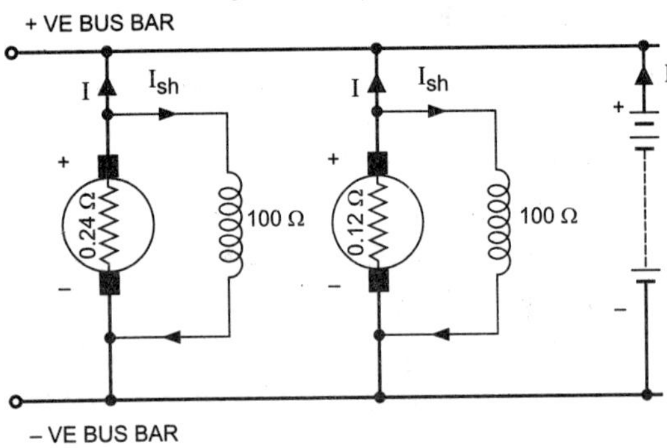

Fig. 8.8

Terminal voltage of generator G_1 is given by

$$V = 250 - \left(I + \frac{V}{100}\right)0.24$$

$$\text{or } 1.0024\,V + 0.24\,I = 250 \quad ...(ii)$$

Terminal voltage of generator G_2 is given by

$$V = 248 - \left(I + \frac{V}{100}\right)0.12$$

$$\text{or } 1.0012\,V + 0.12\,I = 248 \quad ...(iii)$$

Multiplying Eq. (*iii*) by 2 we get

$$2.0024\,V + 0.24\,I = 496 \quad ...(iv)$$

Subtracting Eq. (*ii*) from Eq. (*iv*) we get

$$V = 246 \text{ volts}$$

Substituting V = 246 in Eq. (*ii*) we get

$$1.0024 \times 246 + 0.24\,I = 250$$

$$\text{or } I = 14.266 \text{ A}$$

Substituting I = 14.266 in Eq. (*i*) we get

$$I_b = 40 - 2\,I = 40 - 2 \times 14.266 = 11.468 \text{ amperes}$$

Terminal voltage of battery is given by

$V = E - I_b r$ where r is the internal resistance of battery

$$\text{or } 246 = 247 - 11.468\,r$$

$$\text{or } r = \frac{1}{11.468} = 0.0872\ \Omega \textbf{ Ans.}$$

Example 8.13. In a certain substation there are 5 dc shunt generators in parallel, each having an armature resistance of 0.1 Ω, running at the same speed and excited to give equal induced emfs. Each generator supplies an equal share of a total load of 250 kW at a terminal voltage of 500 V into a load of fixed resistance. If the field current of one generator is raised by 4%, the other remaining unchanged, calculate the power output of each machine and their terminal voltages under these conditions. Assume that the speeds remain constant and flux is proportional to field current.

Solution: The connection diagram for generators is shown in Fig. 8.9.

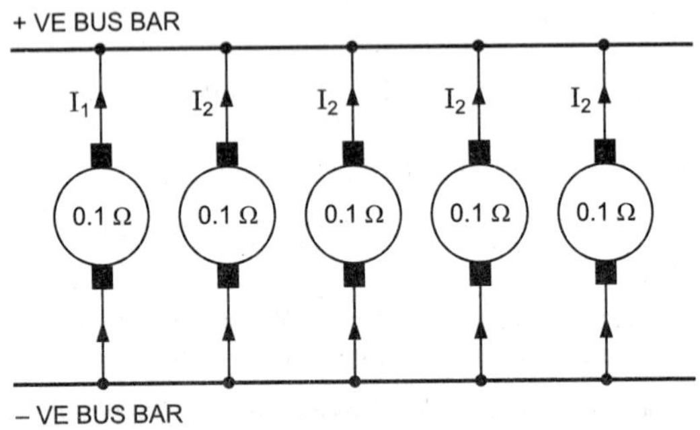

Fig. 8.9

$$\text{Total load supplied} = 250 \text{ kW}$$

$$\text{Load supplied by each generator, } P = \frac{250}{5} = 50 \text{ kW}$$

$$\text{Terminal voltage of each generator, } V = 500 \text{ V}$$

$$\text{Current supplied by each generator, } I = \frac{P \times 1,000}{500} = \frac{50 \times 1,000}{500} = 100 \text{ A}$$

$$\text{EMF generated by each generator, } E = V + IR_a = 500 + 100 \times 0.1 = 510 \text{ V}$$

When field current of one of the generator is increased by 4%, its flux and, therefore, its induced emf is increased by 4% because speed is constant and flux is proportional to field current.

$$\text{So induced emf of one generator} = 510 \times 1.04 = 530.4 \text{ V}$$

Now let the current supplied by the generator, whose excitation has been increased, be I_1 amperes and the current supplied by each of the other four generators be I_2 amperes and V′ be the new terminal voltage.

Terminal voltage of the generator, whose excitation has been increased, is given as

$$V' = 530.4 - 0.1\,I_1 \quad ...(i)$$

Similarly for other generators $V' = 510 - 0.1\,I_2$ ...(*ii*)

$$\text{Fixed resistance of load, } R_L = \frac{V}{5\,I} = \frac{500}{5 \times 100} = 1\ \Omega$$

$$\text{So total current} = I_1 + 4\,I_2$$

$$\text{and } V' = \text{Total current} \times R_L = (I_1 + 4\,I_2) \times 1 = I_1 + 4\,I_2 \quad ...(iii)$$

Subtracting Eq. (*ii*) from Eq. (*i*) we have

$$I_1 - I_2 = 204 \quad ...(iv)$$

Subtracting Eq. (*iii*) from Eq. (*ii*) we have

$$I_1 + 4.1\,I_2 = 510 \quad ...(v)$$

Subtracting Eq. (*iv*) from Eq. (*v*) we have

$$5.1\,I_2 = 306$$

$$\text{or } I_2 = \frac{306}{5.1} = 60 \text{ A}$$

Substituting I_2 = 60 A in Eq. (*iv*) we get

$$I_1 = 204 + 60 = 264 \text{ A}$$

From Eq. (*iii*) $V' = I_1 + 4I_2 = 264 + 4 \times 60 = 504$ V **Ans.**

Output of the machine, whose excitation has been increased,

$$= \frac{504 \times 264}{1{,}000} = 133.056 \text{ kW } \textbf{Ans.}$$

Output of each of other four generators,

$$= \frac{504 \times 60}{1{,}000} = 30.24 \text{ kW } \textbf{Ans.}$$

Example 8.14. Four compound generators operating in parallel supply a load of 1,000 amps. The resistance of series windings and armatures of the generators and their induced emfs are as under:

	A	**B**	**C**	**D**
Series field windings resistances, ohms	**0.004**	**0.004**	**0.006**	**0.006**
Armature resistances, ohms,	**0.01**	**0.01**	**0.02**	**0.02**
EMF generated, Volts	**240**	**235**	**242**	**238**

Determine the current output of each machine, the current and direction in each section of the equalizer bus bar and the bus-bar voltage. Neglect the shunt field currents.

Solution: Let the bus-bar voltage and voltage between equalizer bar and –ve bus-bar be V and v volts respectively, as shown in Fig. 8.10.

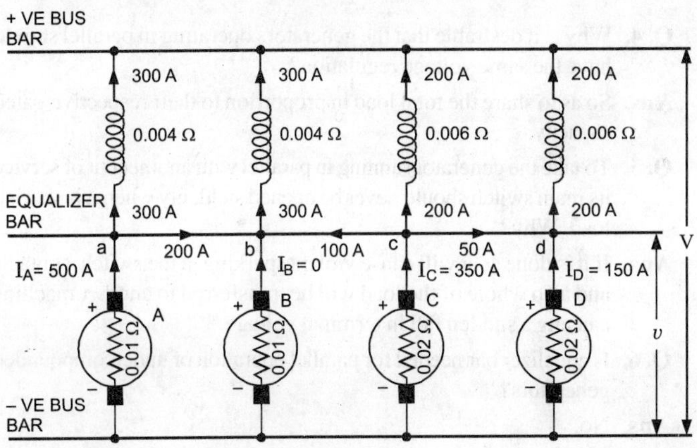

Fig. 8.10

Let the current delivered by 4 machines be I_A, I_B, I_C and I_D respectively

Since $v = E - I_a R_a$

or $I_a = \frac{E - v}{R_a}$,

$$I_A = \frac{240 - v}{0.01} = 24{,}000 - 100\,v \quad ...(i)$$

$$I_B = \frac{235 - v}{0.01} = 23{,}500 - 100\,v \quad ...(ii)$$

$$I_C = \frac{242 - v}{0.02} = 12{,}100 - 50\,v \quad ...(iii)$$

$$I_D = \frac{238 - v}{0.02} = 11{,}900 - 50\,v \quad ...(iv)$$

Adding Eqs. (*i*), (*ii*), (*iii*), and (*iv*) we get

$$I_A + I_B + I_C + I_D = 71{,}500 - 300\,v \quad ...(v)$$

Since $I_A + I_B + I_C + I_D$ = Total current supplied = 1,000 A,

Substituting $I_A + I_B + I_C + I_D = 1{,}000$ in Eq. (v) we get

$$1{,}000 = 71{,}500 - 300\,v$$

$$\text{or } v = \frac{70{,}500}{300} = 235 \text{ volts}$$

Substituting v = 235 volts in Eqs. (*i*), (*ii*), (*iii*) and (*iv*) we get

$I_A = 24{,}000 - 100 \times 235 = 500$ A **Ans.**

$I_B = 23{,}500 - 100 \times 235 = 0$ A **Ans.**

$I_C = 12{,}100 - 50 \times 235 = 350$ A **Ans.**

$I_D = 11{,}900 - 50 \times 235 = 150$ A **Ans.**

Since the pd across all the series fields is the same, currents will divide in the ratio of

$$\frac{1}{0.004} : \frac{1}{0.004} : \frac{1}{0.006} : \frac{1}{0.006}$$

$$\text{or} \quad 3 : 3 : 2 : 2$$

Current in series field of generator A $= \frac{3}{10} \times 1{,}000 = 300$ A

Current in series field of generator B $= \frac{3}{10} \times 1{,}000 = 300$ A

Current in series field of generator C $= \frac{2}{10} \times 1{,}000 = 200$ A

Current in series field of generator D $= \frac{2}{10} \times 1{,}000 = 200$ A

Bus-bar voltage, V = E – voltage drop in any series field

= 235 – 300 × 0.004 = 233.8 volts **Ans.**

Applying Kirchoff's first law at each point *a*, *b*, *c*, and *d* we get

Current from *a* to *b* = 500 – 300 = 200 A **Ans.**

Current from *b* to *c* = 200 + 0 – 300 = –100 A **Ans.**

Current from *c* to *d* = 350 – 200 – 100 = 50 A **Ans.**

HIGHLIGHTS

1. In most power stations (whether dc or ac) it will be found that power is supplied from several units (generators) operating in parallel rather than from a single large unit capable of taking care of the maximum peak loads. Such practice is considered extremely desirable from the standpoint of efficiency, reliability (continuity of service), maintenance and repair problems, and additions to plant capacity with the growth in load on the power station.
2. For satisfactory parallel operation of dc series generators (operation of dc series motors during regenerative braking), it is necessary to connect the armatures of the two machines through a heavy copper bar, called the *equalising bar*. The same purpose is served if field windings of the two machines are cross-connected.
3. DC shunt generators, due to their drooping characteristics, are best suited for parallel operation.

   For satisfactory operation, it is desirable that all the machines operating in parallel should have same voltage regulation so as to share the total load in proportion to their rated capacity.
4. The necessary conditions for paralleling dc generators are (*i*) same polarity (*ii*) equal terminal voltage.

EXERCISES

1. Describe the advantages of using several small dc generators in parallel over the use of single large generator.
2. Why usually parallel operation of series generators is unstable? What remedial measures are taken for its successful operation?
   [J.N. Technological Univ. Electrical Machines-I, March-2014]
3. In what respects do the parallel operation and load sharing of shunt and series generators differ? Explain.
4. List out conditions to be satisfied for running two or more dc shunt generators in parallel.
   [J.N. Technological Univ. Electrical Machines-I, March-2014]
5. Two shunt generators having drooping characteristics are best suited for parallel operation. Explain.
6. Explain the method to bring a second dc shunt generator into parallel operation with one already delivering power. How should their characteristics be related in order that they should satisfactorily share a common load according to their respective ratings.
7. Draw the connection diagram of two shunt generators connected in parallel and discuss their load sharing.
8. What factors determine the load distribution amongst a number of dc shunt generators in parallel? How can load be transferred from one machine to the other?
9. Explain the process of paralleling two dc compound generators.
10. Explain with connection diagrams how two dc compound generators are operated in parallel.
    [J.N. Technological Univ. Electrical Machines-I, March-2014]
11. Explain clearly the importance of an equalizer bar for satisfactory parallel operation of compound generators.
    [J.N. Technological Univ. Electrical Machines-I, March-2014]
12. "If two compounded interpole dc generators are connected in parallel, one equalizing bar has to be used". Explaining the function of equalising bar justify the statement.

SHORT ANSWER TYPE QUESTIONS WITH ANSWERS

Q. 1. Why it is necessary to operate generators in parallel for supply of power?

Ans. For higher efficiency of operation, reliability, convenience and economy in maintenance and repair and possibility of additions to plant capacity with the growth in load on the power station.

Q. 2. Are dc series generators employed for power supply? If not, then what is application of operation of dc series generators in parallel?

Ans. No, dc series motors, employed in locomotives, are frequently arranged to operate as dc series generators during electric braking.

Q. 3. Is parallel operation of two dc series generators inherently unstable? If yes, how is it made stable?

Ans. By connecting the armatures of the two machines operating in parallel through a heavy copper bar or by cross-connecting their field windings.

Q. 4. Why is it desirable that the generators operating in parallel should have the same voltage regulation?

Ans. So as to share the total load in proportion to their respective rated capacity.

Q. 5. To take the generator running in parallel with another out of service its main switch should never be opened suddenly when it is sharing load. Why?

Ans. If it is done so it will cause violent sparking at the switch contacts and also whole of the load will be transferred to another machine causing a sudden fall in terminal voltage.

Q. 6. Is equalizer bar needed for parallel operation of undercompounded generators?

Ans. No.

PROBLEMS

1. Two shunt generators A and B operate in parallel and their load-characteristics may be taken as straight lines. The voltage of generator A falls from 240 V at no load to 220 V at 200 A, while that of B falls from 245 V at no load to 220 V at 150 A. Determine the currents supplied by each machine to a common load of 300 A and the bus-bar voltage. [**Ans.** 168.75 A, 131.25 A, 223.125 V]
2. Two shunt generators operating in parallel give a total output of 603 A. One machine has an armature resistance of 0.02 Ω and generated voltage of 455 V and the other has an armature resistance of 0.025 Ω and a generated voltage of 460 V Calculate the terminal voltage and the kW output of each machine. Neglect field currents.
   [**Ans.** 450.52 V; 100.867 kW; 170.798 kW]
3. Two similar dc shunt generators, each having armature resistance of 0.05 ohm, are connected in parallel to supply 400 A. The no-load voltage of the two generators are 250 V and 260 V respectively. Find the current supplied by each generator and the load terminal voltage.
   [**Ans.** 100 A, 300 A, 245 V]
4. Two dc shunt generators having no-load voltage of 440 V each are operating in parallel. Generator A gives a terminal voltage of 400 V at a load of 100 A while the generator B gives a terminal voltage of 390 V at the same load when operated individually. Assuming the voltage characteristics straight lines find the common terminal voltage and current in each machine when they are sharing a load of 100 A.
   [**Ans.** 55.55 A; 44.45 A; 417.78 V]
5. Two dc shunt generators running in parallel supply a total load current of 125 A. The shunt field and armature resistances of each machine are 50 ohms and 0.4 ohm respectively. The emfs of the two generators at no load are 260 and 250 volts respectively. Determine the current supplied by each generator and the bus-bar voltage.
   [**Ans.** 75 A; 50 A; 228.16 V]
6. Two shunt generators and a battery are working in parallel. The open-circuit voltage, armature and field resistances of the generators are 250 V, 0.24 ohm, 100 ohms and 248 V, 0.12 ohm, 100 ohm respectively. If the generators supply the same current when the load on the bus bars is 40 A, calculate the emf of the battery if its internal resistance is 0.172 ohm. [**Ans.** 248 V]

❑❑❑

9 CHAPTER Operating Characteristics and Applications of DC Motors

INSIDE THIS CHAPTER

9.1 INTRODUCTION

An electric motor is a machine which converts electrical energy into mechanical energy whereas a generator is that machine which converts mechanical energy into electrical one. As regards fundamental principles the dc motors are identical with the dc generators which have the same type of excitation *i.e.* a machine that operates as a motor will also operate satisfactorily as a generator.

The only difference lies, however, in the mode of construction, which is due to the fact that the frame of the generator can, as a rule, be open but those of motors should be either partly or totally enclosed. A generator is usually placed in a suitable position and mechanical protection for the coils and armature may be reduced to minimum. Also the generator is handled by technical persons. Hence there is no risk in having the frame of the generator open, which facilitates cooling, inspection and repair. On the other hand, motors have to work in conditions of dampness, dirt, inflammable gases, chemical fumes and are liable to mechanical damage and, therefore, protection must be adequate and motor frames are made flame-proof, partly enclosed or totally enclosed according to the requirements of service.

The need for motive power has prompted the development of various kinds of electric motors. Since early electric power systems were direct current, the first motors were dc types. With the development of ac power systems, the popularity of dc motor declined, mainly because of its higher cost and its need for more frequent and careful maintenance. However, it never entirely disappeared, partly because the dc motor was inherently the most suitable for smooth, efficient, wide-range speed control and quick reversals and partly because it was the only type that could be employed for automotive and aircraft applications. With electrical utilities transmitting energy on ac, the usual arrangement used to be a motor-generator set (a dc generator driven by an ac motor) supplying power to a dc motor. However, the advent of solid-state diodes and silicon controlled rectifiers (SCRs) eliminated the need for the motor-generator set and made dc motor drives more versatile and more popular than even before.

9.2 WORKING PRINCIPLE OF DC MOTOR

The principle upon which a dc motor works is very simple. If a current carrying conductor is placed in a magnetic field, mechanical force is experienced on the conductor, the direction of which is given by Fleming's left hand rule (also called *motor rule*) and hence the conductor moves in the direction of force. The magnitude of the mechanical force experienced on the conductor is given by

$$F = BI_c\, l_c \text{ newtons}$$

where B is the field strength in teslas (Wb/m^2), I_c is the current flowing through the conductor in amperes and l_c is the length of conductor in metres.

When the motor is connected to the dc supply mains, a direct current passes through the brushes and commutator to the armature winding; while it passes through the commutator it is converted into ac so that the group of conductors under successive field poles carry currents in the opposite directions, as shown in Fig. 9.1. Also the direction of current in the individual conductors reverses as they pass away from the influence of one pole to that of the next.

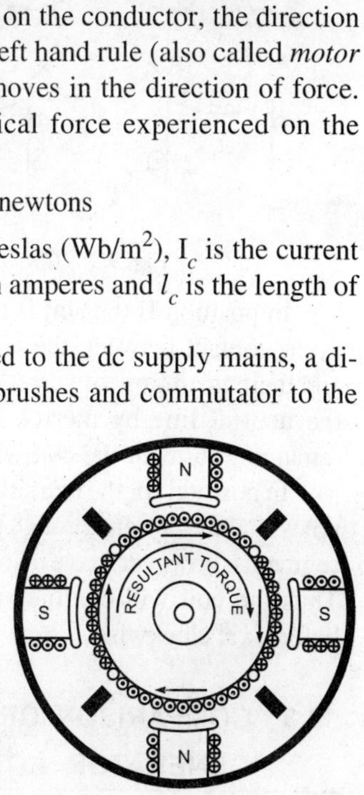

Fig. 9.1

In Fig. 9.1, a 4-pole dc motor is shown when the field and armature circuits are connected across dc supply mains. Let the current in armature conductors be outwards under the N-poles (shown by dots) and inwards under S-poles (shown by crosses). By applying Fleming's left hand rule, the direction of force on each conductor can be determined, which has been illustrated in Fig. 9.1. From Fig. 9.1. it is observed that each conductor experiences a force which tends to rotate the motor armature in clockwise direction. These forces collectively produce a driving torque.

9.2.1. Commutator Action in a DC Motor. In the case of a dc motor, it is necessary that the current through the coils of the armature winding be reversed as a particular coil leaves one pole (say, the north pole), crosses the neutral line and comes under the influence of next pole which is of opposite polarity (*i.e.* the south pole). The operation of the commutator, that serves the above purpose, is given below :

Consider a single turn coil, whose leads are soldered to commutator segments *a* and *b*, each carrying a brush, as illustrated in Fig. 9.2. The positive side of the supply line is connected to left hand brush and negative side to the right hand brush. In position I the line current arrives at the commutator segment a, flows through the bottom side 1 of the coil away from the reader (as shown by cross in the circle) and then through the upper side 2 of the coil towards the reader (as shown by dot in the circle), reaches the commutator segment *b* and flows again into the line through the brush. The coil will tend to rotate in clockwise direction, as determined by Fleming's left hand rule.

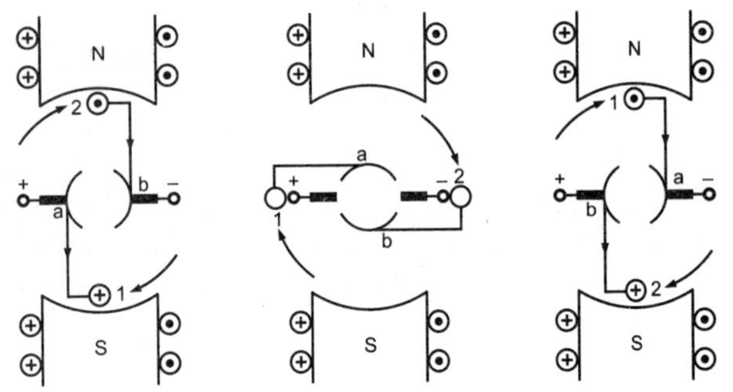

I *II* *III*

Fig. 9.2 *Commutation in DC Motor*

In position II the coil is on the magnetic neutral line; there in no contact between the commutator segments and brushes, and there is no flow of current through the coil. The coil crosses the neutral line by inertia. In case of a multi-turn coil, the remaining turns of the coil will supply the necessary torque.

In position III, the two sides of the coil, 1 and 2, have changed poles, and the current through them has reversed. The commutator segments, however, have also changed contact with the brushes. Thus the coil will continue to rotate in the same direction as before, *i.e.* clockwise.

9.3 COMPARISON OF MOTOR AND GENERATOR ACTION

As mentioned above, dc motor and the dc generator are the same devices at least theoretically. The machine operating as a generator is driven by some external driving force and dc output is obtained from it whereas the machine operating as a motor is supplied by electric current and mechanical rotation is produced. If the excitation of a dc machine (working as a dc shunt generator) operating in parallel with dc line is gradually reduced to an extent that the emf generated by it is less than the line voltage, the machine will start drawing current from the line and working as a motor.

In discussing machine operations the terms, generator action, generation, motor action and motor operation are employed. Generator operation signifies that the machine is actually operating as a generator whereas the term generator action signifies that a voltage is generated by electromagnetic induction. Similarly the term motor operation signifies that the machine is actually operating as a motor and the term motor action signifies the development of mechanical torque by electromagnetic means.

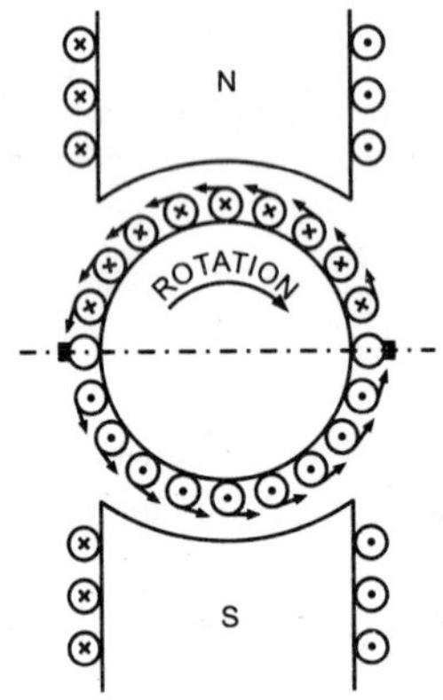

Fig. 9.3

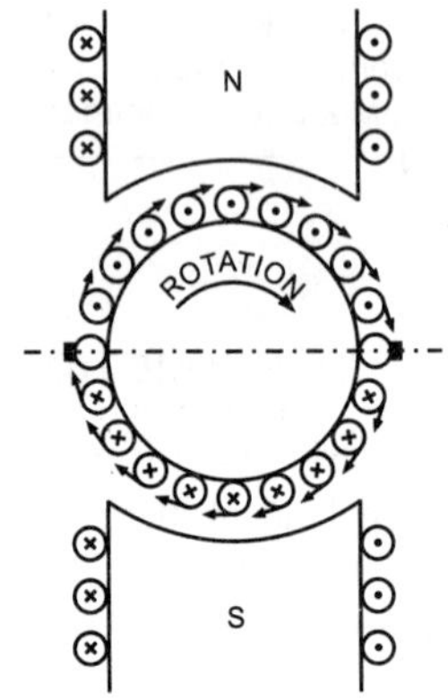

Fig. 9.4

Let us first consider generator operation. In Fig. 9.3 a dc machine driven in a clockwise direction, by its prime mover and supplying direct current to external load circuit is shown. The machine is working as a generator and the direction of the generated emf and current flowing through the armature conductors, as determined by Fleming's right hand rule, are shown in the figure. Since the armature is carrying current and rotating in a magnetic field, electromagnetic forces will be developed between the conductors and field, the direction of the forces so developed will be given by Fleming's left hand rule. These electromagnetic forces acting on the armature conductors will collectively result in a torque acting on the armature in a counter-clockwise direction. This electromagnetic torque, therefore, opposes the outside driving torque, which is causing the rotation of the machine and called the *backward torque* or *magnetic drag* on the conductors. The prime mover has to work against this magnetic drag and the work so done is converted into electrical energy. The larger the output current, more will be the backward torque and, therefore, more mechanical energy will be required to be supplied to the generator.

In the Fig. 9.4 the same machine operating as a motor is shown. This operation takes place when the prime mover is uncoupled from the machine and the machine is connected to the dc supply mains. With the direction of field and armature current shown in the figure, the torque developed by electro-magnetic action will rotate the machine in a clockwise direction (as determined by Fleming's left hand rule). The friction of the machine and the mechanical load that the motor is driving will exert a torque in counter-clockwise direction, opposing the rotation of the motor. Since the armature conductors are revolving in the magnetic field, emf is induced in the armature conductors. The direction of emf so induced, as determined by Fleming's right hand rule, is in direct opposition to the applied voltage. That is why the induced emf in a motor is often called the *counter emf* or *back emf*, E_b. The applied voltage must be large enough to overcome this back emf and to send the current through the resistance of the armature. The electrical energy supplied to overcome this opposition is converted into mechanical energy developed in the armature.

Thus we see that an emf is generated in both generator and motor, therefore, there is a generator action in both motor and generator operations. However, in generator operation, the generated emf produces the armature current, whereas, in motor operation the generated emf opposes the conduction. We also observe that electromagnetic torque is developed in generator as well as motor *i.e.* there is a motor action in both generator and motor operations. However, in motor operation the electromagnetic torque developed causes the armature rotation, whereas in a generator operation the electromagnetic torque produced opposes the armature rotation.

9.4 IMPORTANCE OF BACK EMF

As already explained, when the motor armature continues to rotate due to motor action, the armature conductors cut the magnetic flux and, therefore, emfs are induced in them. The direction of this induced emf, known as *back emf*, is such that it opposes the applied voltage.

Since the back emf is induced due to the generator action, the magnitude of it is, therefore, given by the same expression as that for the generated emf in a generator

$$i.e. \text{ Back emf, } E_b = \frac{\Phi Z N}{60} \times \frac{P}{A} \text{ volts} \quad ...(9.1)$$

the symbols having their usual significance.

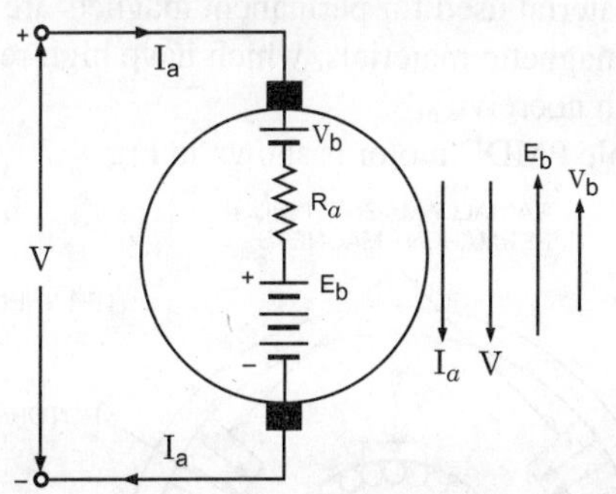

Fig. 9.5 *Equivalent Circuit of a Motor Armature*

The equivalent circuit of a motor is shown in Fig. 9.5. The armature circuit is equivalent to a source of emf, E_b in series with a resistance, R_a and brush contact drop V_b put across a dc supply mains of V volts. It is evident from Fig. 9.5 that the applied voltage V must be large enough to balance the voltage drop in armature resistance brush contact drop and the back emf at all times *i.e.*

$$V = E_b + I_a R_a + V_b \quad ....(9.2)^*$$

where V is the applied voltage across the armature, E_b is the induced emf in the armature by generator action, I_a is the armature current, R_a is the armature resistance and V_b is the brush contact drop.

The Eq. (9.2) may be rewritten as $I_a = \frac{V - E_b}{R_a}$ to give armature current in terms of applied voltage V, induced emf E_b and armature resistance R_a.

As obvious from Eqs. (9.1) and (9.2) the induced emf in the armature of a motor, E_b depends among other factors upon the armature speed and armature current depends upon the back emf E_b for a constant applied voltage and armature resistance. If the armature speed is high, back emf E_b will be large and, therefore, armature current small. If the speed of the armature is low, then back emf E_b will be less and armature current I_a more resulting in development of large torque.

The presence of back emf makes the dc motor a *self-regulating* machine *i.e.* it makes the dc motor to draw as much armature current as is just sufficient to develop the required load torque. This is explained below :

When the motor is operating on no load, small torque is required to overcome the friction and windage losses, therefore, back emf is nearly equal to the applied voltage and armature current is small. When the motor is loaded, the driving torque of the motor is not sufficient to counter the increased retarding torque due to load and the effect is to cause the armature to slow down. With the decrease in the speed of armature back emf falls. The reduced back emf allows a larger current to flow through the armature. The increase in armature current results in higher electromagnetic driving torque. The motor continues to slow down till the electromagnetic torque developed matches the load torque and the steady state conditions are attained. The reverse phenomenon occurs when mechanical load on the motor falls.

When the load on the motor falls, the electromagnetic torque developed is momentarily in excess of the load requirement and, therefore, the motor armature accelerates. With the increase in armature speed, back emf increases causing armature current to decrease. The decrease in armature current causes decrease in electromagnetic torque and the steady-state conditions are attained when the electromagnetic torque developed matches the load torque.

Thus it is evident that back emf E_b acts like a governor *i.e.* it makes a motor self-regulating so that it draws as much current as just required.

9.5 POWER RELATIONSHIP IN MOTOR

The Eq. (9.2) is called the *voltage equation* for a motor.

Multiplying each term of the voltage expression by I_a we get

$$VI_a = E_b I_a + I_a^2 R_a \quad ...(9.3)$$

* Brush contact drop is usually taken into account by including the brush contact resistance in the armature resistance R_a and, therefore, Eq. (9.2) is modified as

$$V = E_b + I_a R_a \quad ...(9.2)$$

The above equation is known as power equation of the dc motor.

The term VI_a represents the power supplied to the motor armature and the term $I_a^2 R_a$ represents the power lost in the armature and, therefore, the term $E_b I_a$ must represent the power developed by the motor armature causing rotation of the armature. The power developed $E_b I_a$ is not all available at the shaft since some of it is used to overcome the mechanical or rotational losses of the motor.

The power lost in the armature, $I_a^2 R_a$ is a small portion (about 5%) of power input to the armature and the remaining portion $E_b I_a$ is converted into mechanical power within the armature.

9.6 MAXIMUM POWER OUTPUT

For a dc motor, maximum torque is developed when the rated voltage is applied to the armature of a motor at standstill and the field is fully excited. Normally the current flowing in the armature under this condition is well beyond the capacity of the armature coils. At this torque and current the whole of the power input to the armature is wasted in the armature resistance (including brushes) and the output is zero. Between the two limiting conditions at which the power output is zero, i.e. the normal no-load condition and the standstill condition, the power output, starting from no-load condition, will increase, steadily as the armature current increases, until at a certain current maximum power output is obtained. Beyond this point of maximum power output, although the current increases, the actual power output falls off owing to marked reduction in the efficiency. This is illustrated in Fig. 9.6.

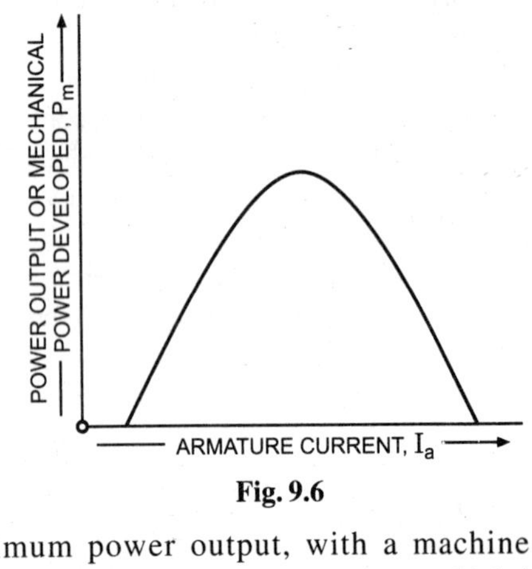

Fig. 9.6

Maximum power output, with a machine of normal proportions, occurs at an armature current which is practically equal to one half of the current obtained at standstill *i.e.* at a load for which armature current, $I_a = \frac{V}{2R_a}$. This is determined as below:

As mentioned in Art. 9.5 the mechanical power developed by the motor is given by the expression

$$P_m = E_b I_a = VI_a - I_a^2 R_a \qquad ...(9.4)$$

Differentiating both sides with respect to armature current I_a we have

$$\frac{dP_m}{dI_a} = V - 2 I_a R_a$$

Mechanical power developed by the motor will be maximum if $\frac{dP_m}{dI_a}$ is zero

$$i.e. \quad V - 2 I_a R_a = 0$$

$$\text{or } I_a R_a = \frac{V}{2}$$

$$\text{and back emf, } E_b = V - I_a R_a = \frac{V}{2} \qquad ...(9.5)$$

Thus the motor develops the maximum power when the armature current is such that the back emf E_b is equal to one half of the applied voltage. This is not attained in practice, since it necessitates a current well beyond the normal operating range. Also an amount of power equal to the mechanical power developed would be wasted in heating the armature, so taking other losses into account, the efficiency would be well below 50%.

9.7 TYPES OF DC MOTORS

Similar to dc generators, the dc motors can also be classified as (1) permanent magnet, (2) separately excited, (3) series wound (4) shunt wound and (5) compound wound dc motors.

9.7.1. Permanent Magnet DC Motor. A permanent magnet dc (PMDC) motor is a dc motor whose poles are made of permanent magnets *i.e.* field flux required in the air gap of a PMDC motor is developed by a set of permanent magnets fixed to the stator. The permanent magnets of a PMDC motor are radially magnetized and mounted on the inner periphery of the cylindrical steel stator. The stator also serves as a return path for the magnetic flux. Field coils are usually not required. However, some of these motors do have coils wound on the poles. If they exist, these coils are intended only for recharging the magnets in the event they lose their strength.

The rotor of this motor is similar to the rotor of a conventional dc motor *i.e.* rotor of a PMDC motor consists of armature core, armature windings and commutator. Stationary carbon brushes are kept pressed on the commutator surface as in a conventional dc motor.

The material used for permanent magnets are ceramics and rare earth magnetic materials, which have high residual flux as well as high coercivity.

A 2-pole PMDC motor is shown in Fig. 9.7.

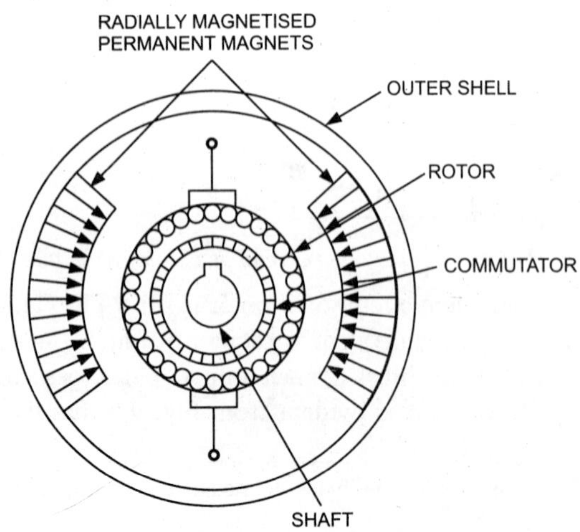

Fig. 9.7 *2-Pole PMDC Motor*

Small fractional and sub-fractional kW motors are now constructed with permanent poles. Most of the PMDC motors operate on 6 V, 12 V or 24 V dc supply obtained from batteries or rectifiers.

The torque is developed by interaction between current carrying rotor conductors and the magnetic flux set up by the permanent magnets.

Equivalent circuit of a PMDC motor is shown in Fig. 9.8. Since the field flux in a PMDC motor is developed by permanent magnets, the field winding is not shown in the equivalent circuit.

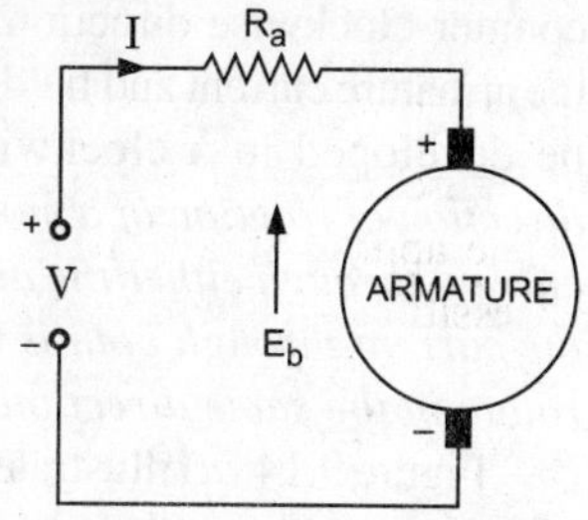

Fig. 9.8 *Equivalent Circuit of a PMDC Motor*

Equations governing the performance of permanent magnet dc (PMDC) motor are basically the same as for dc shunt motor with constant field.

The major advantage of PMDC motor is that they do not have any field winding, so they require no excitation current. Thus there is no continuous loss of energy in the field. Because of absence of field windings, there is saving in space and PMDC motors are smaller in size and cheaper in cost in comparison to corresponding rated conventional dc motors.

The limitation of PMDC motors is that the excessive current in the armature winding may demagnetize the permanent magnets and also the flux density produced in the air gap by permanent magnets is limited. The risk of permanent magnetism getting destroyed by armature reaction (at starting/reversing or heavy overloads) has been greatly reduced by the new permanent magnetic materials like somarium cobalt and neodymium-iron-boron.

These motors offer shunt type characteristic and can only be armature controlled *i.e.* in PMDC motors, speed and torque are controlled through the adjustment of voltage applied to the motor terminals, armature rheostat control and chopper control.

PMDC motors are used extensively in automobiles as starter motors and for windshield wipers and washers, for blowers used in heaters and air-conditioners, to raise and lower windows, in slot cars and electric tooth brushes, in personal computer disc drives etc.

9.7.2. Separately-Excited DC Motors. These motors have field coils similar to those of a shunt wound machine, but the armature and field coils are fed from different supply sources, as illustrated in Fig. 9.9 and may have different voltage ratings.

In a separately excited dc motor,

$$\text{Armature current, } I_a = \text{Line current, } I_L = I \text{ (say)} \quad ...(9.6)$$

$$\text{Back emf developed, } E_b = V - IR_a. \quad ...(9.7)$$

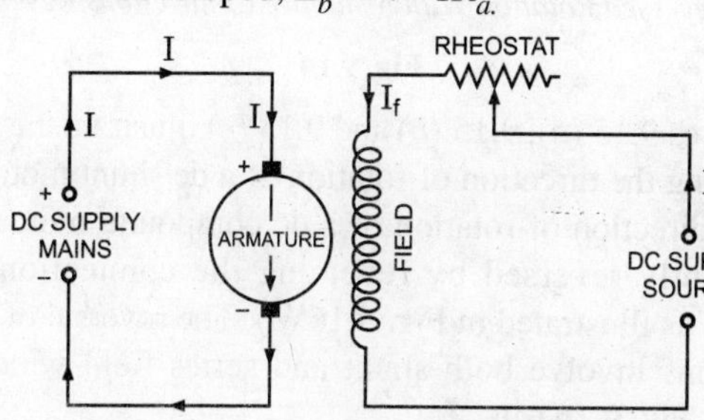

Fig. 9.9 *Separately-Excited DC Motor*

Power drawn from supply mains,

$$P = VI \text{ where V is supply voltage.} \quad ...(9.8)$$

Mechanical power developed,

$$P_m = \text{Power input to armature} - \text{power lost in armature}$$
$$= VI - I^2 R_a = I(V - IR_a) = E_b I \quad ...(9.9)$$

9.7.3. Series Wound DC Motors. As the name implies, the field coils, consisting of few turns of thick wire, are connected in series with the armature, as illustrated in Fig. 9.10. The cross-sectional area of the wire used for field coils has to be fairly large to carry the armature current, but owing to the higher current, the number of turns of wire in them need not be large. In a dc series motor

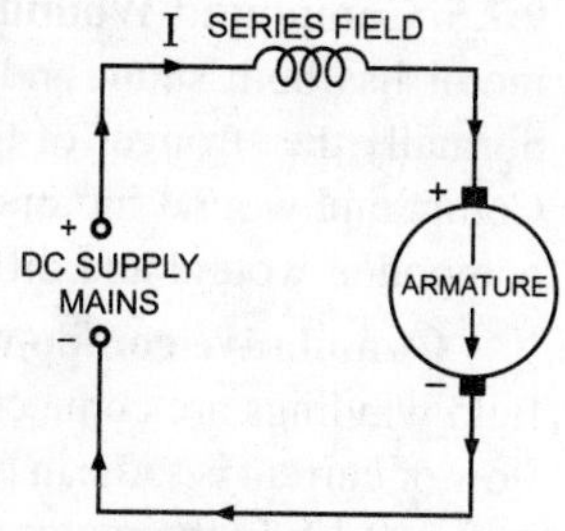

Fig. 9.10 *Series Wound DC Motor*

$$\text{Armature current, } I_a = \text{Series field current, } I_{se}$$
$$= \text{Line current, } I_L = I \text{ (say)} \quad ...(9.10)$$

$$\text{Back emf developed, } E_b = V - I(R_a + R_{se}) \quad ...(9.11)$$

$$\text{Power drawn from supply mains} = VI \quad ...(9.12)$$

Mechanical power developed,

$$P_m = \text{Power input} - \text{losses in armature and field}$$
$$= VI - I^2(R_a + R_{se})$$
$$= I[V - I(R_a + R_{se})] = E_b I \quad ...(9.13)$$

9.7.4. Shunt Wound DC Motors. The word "shunt" means "parallel". These motors are so named because they basically operate with the field coils connected in parallel with the armature. The field winding consists of a large number of turns of comparatively fine wire so as to provide large resistance. The field current is much less than the armature current, sometimes as low as 5%.

The connection diagram is shown in Fig. 9.11. The current supplied to the motor is divided into two paths, one through the field winding and second through the armature *i.e.*,

$$\text{Input line current, } I_L = I_a + I_{sh} \quad ...(9.14)$$

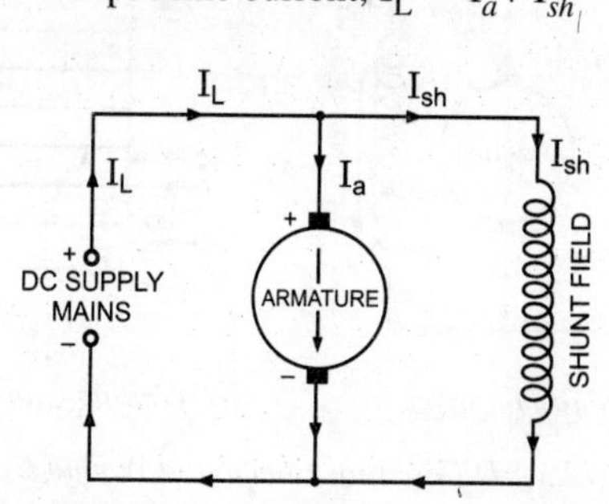

Fig. 9.11 *Shunt Wound DC Motor*

where I_a is the armature current and I_{sh} is the shunt field current and is given by the equation

$$I_{sh} = \frac{V}{R_{sh}} \quad ...(9.15)$$

where V is the supply voltage and R_{sh} is the shunt field resistance.

$$\text{Back emf developed, } E_b = V - I_a R_a \quad ...(9.16)$$

$$\text{Power drawn from supply mains, } P = VI_L \quad ...(9.17)$$

Mechanical power developed,

$$P_m = \text{Power input} - \text{losses in armature and shunt field}$$
$$= VI_L - VI_{sh} - I_a^2 R_a$$
$$= V(I_L - I_{sh}) - I_a^2 R_a$$
$$= V I_a - I_a^2 R_a = I_a (V - I_a R_a) = E_b I_a \qquad ...(9.18)$$

9.7.5. Compound Wound DC Motors. A compound wound dc motor has both shunt and series field coils. The shunt field is normally the stronger of the two (*i.e.*, has more ampere-turns). Compound wound motors are of two types namely cumulative compound wound and differential compound wound motors.

Cumulative compound wound motor is one in which the field windings are connected in such a way that the direction of flow of current is same in both of the field windings, as illustrated in Fig. 9.12. In the motor of this type the flux due to series field winding strengthens the field due to the shunt field winding.

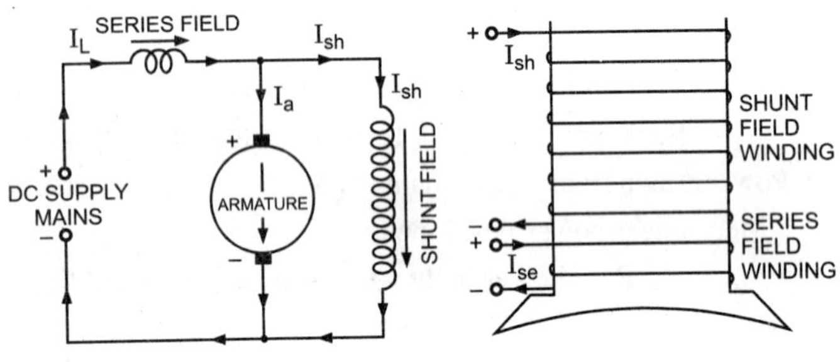

(*a*) Circuit Diagram (*b*) Winding Connection Diagram

Fig. 9.12 *Cumulative Compound Wound DC Motor*

Differential compound wound motor is one in which the field windings are connected in such a way that the direction of flow of current is opposite to each other in the two field windings, as illustrated in Fig. 9.13. In this type of motor the flux due to series field winding weakens the field due to shunt field winding.

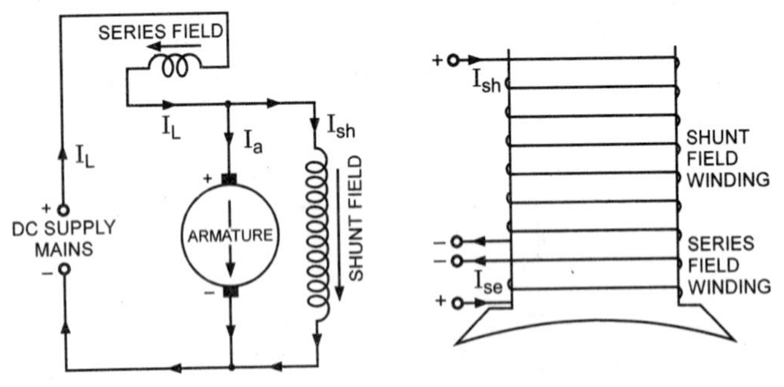

(*a*) Circuit Diagram (*b*) Winding Connection Diagram

Fig. 9.13 *Differential Compound Wound DC Motor*

Compound wound dc motors, like compound wound dc generators, may be either long-shunt connected or short-shunt connected. In long-shunt connected (or long-shunt) compound wound motors, series field and armature are connected in series with each other and in parallel with the shunt field. In short-shunt connected (or short-shunt) compound wound motors, the armature and shunt field are in parallel with each other and the pair is in series with the series field. Long-shunt connection sometimes results in simpler wiring. Changing from long- to short-shunt, or vice versa, has little effect on motor performance.

9.8 DIRECTION OF ROTATION OF DC MOTORS

It is evident from Fig. 9.1 that if the armature current were reversed by reversing the armature leads, but leaving the field polarity the same, torque would be developed in a counter-clockwise direction. Likewise, if the field polarity were reversed leaving the armature current as shown, torque would be developed in a counter-clockwise direction. However, if both, the direction of the armature current and field polarity were reversed, torque would be developed in a clockwise direction as before. *Hence the direction of rotation of a motor can be reversed by reversing the current through either the armature winding or the field coils. If the current through both is reversed, the motor will continue to rotate in the same direction as before.*

Figure 9.14 (*a*) illustrates a dc series motor with the direction of rotation assumed to be clockwise. Figure 9.14 (*b*) shows connections to the field winding reversed giving reversed rotation and Fig. 9.14 (*c*) shows the alternative method when the armature connections at the brushes are interchanged.

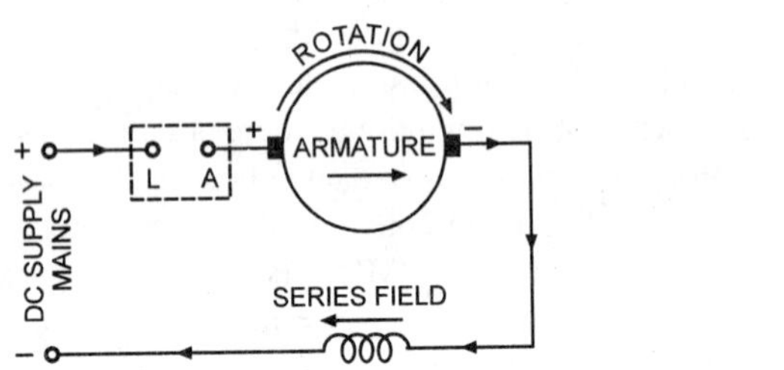

(*a*) Normal Rotation

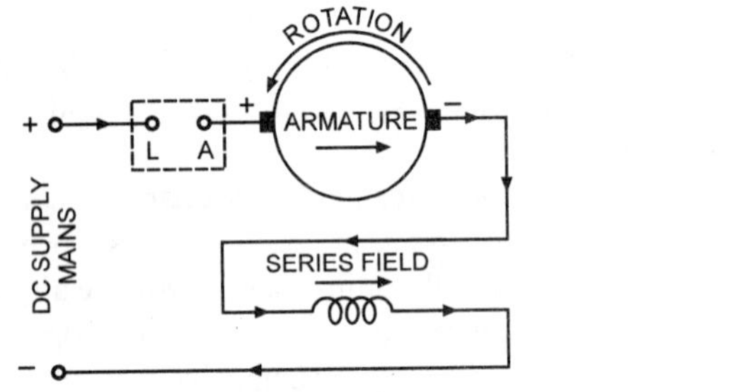

(*b*) Reversed Rotation With Field Connections Reversed

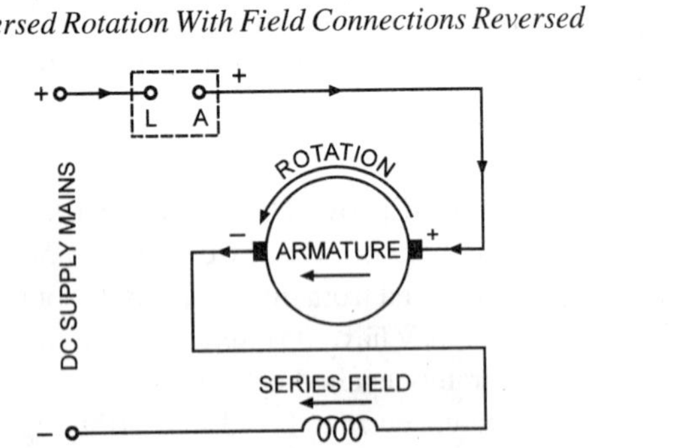

(*c*) Reversed Rotation With Armature Connections Reversed

Fig. 9.14

Figures 9.15 (*a*), 9.15 (*b*) and 9.15 (*c*) illustrate the methods of reversing the direction of rotation of a dc shunt motor.

The direction of rotation of a dc compound motor may be conveniently reversed by reversing the connections to the armature, as illustrated in Fig. 9.16 (*b*). The reversal of the field connections involve both shunt and series field windings, as shown in Fig. 9.16 (*c*).

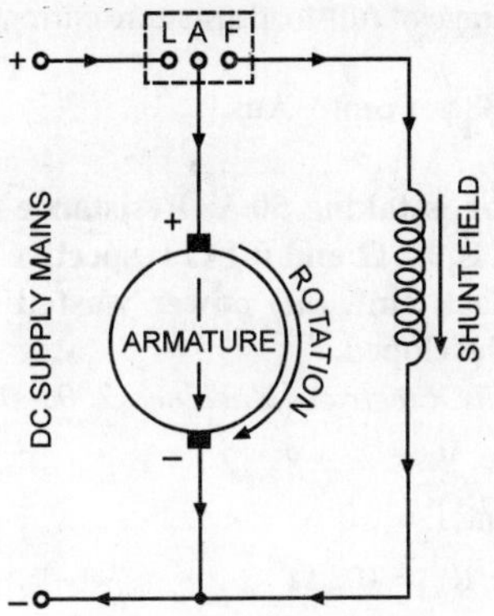

(a) *Normal Rotation*

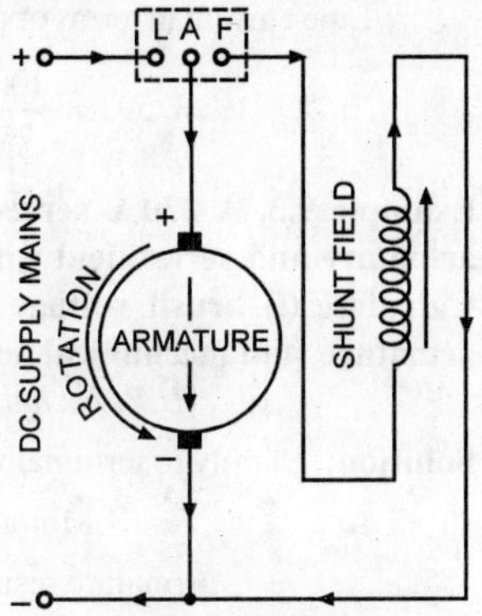

(b) *Reversed Rotation With Field Connections Reversed*

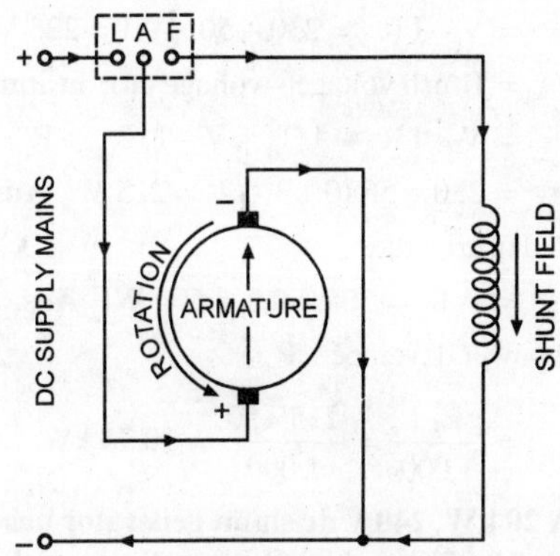

(c) *Reversed Rotation With Armature Connections Reversed*

Fig. 9.15

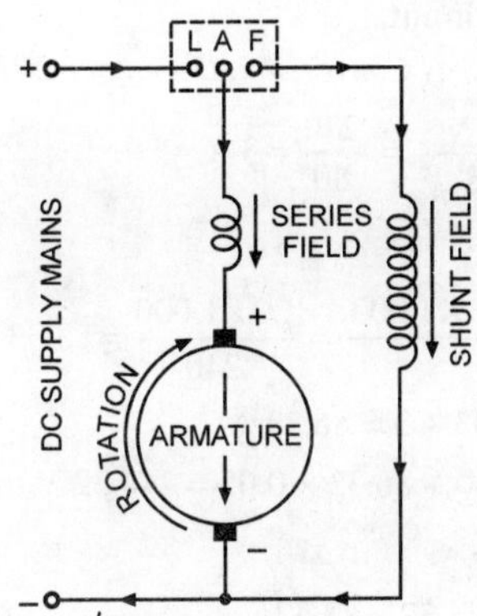

(a) *Normal Rotation*

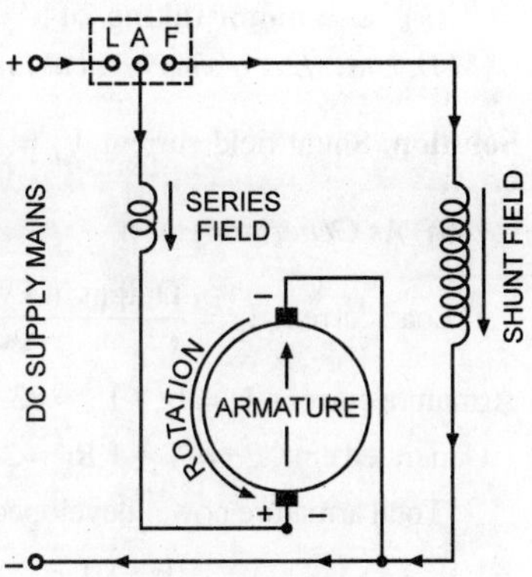

(b) *Reversed Rotation With Armature Connections Reversed*

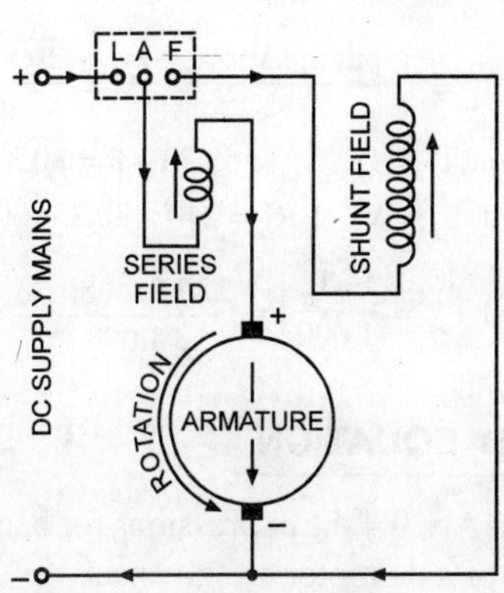

(c) *Reversed Rotation With Field Connections Reversed*

Fig. 9.16

9.9 CIRCUIT MODEL OF DC MACHINES

In a dc machine, all parallel paths in the armature are symmetrical and each path has an induced emf E_a and a resistance R_p. The armature, according to Thevenin's theorem, can be represented by the circuit model with voltage E_a and a series resistance R_a, as depicted in Fig. 9.17, where R_a is equal to R_p divided by number of parallel paths, A i.e. $R_a = R_p/A$. The brush voltage drop is constant and is of the order of 1-2 volts, which is either ignored or its effect is included in armature resistance R_a by linearizing it. The field has a resistance R_f supplied with field current I_f given by $I_f = \frac{V_f}{R_f}$ and is shown at 90° to the brush axis according to the actual arrangement in the machine.* The field provides the flux per pole which induces emf E_a in the armature when it runs at speed N. The power converted (mechanical to electrical or vice versa) is $E_a I_a$.

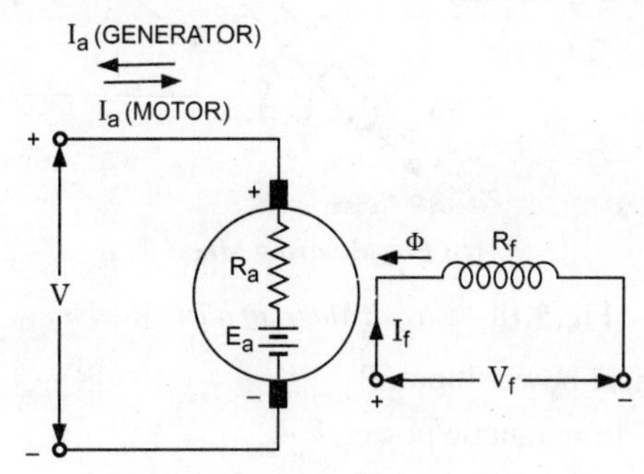

Fig. 9.17 *Circuit Model of a DC Machine*

From circuit point of view it is not necessary to rigidly follow this scheme. Since most of the time steady-state dc behaviour of the machine will be considered, the inductances of the field and armature circuits do not have any effect and are, therefore, not shown in the circuit model.

Generating Mode. The machine operates in generating mode *i.e.* it delivers electrical power when armature current I_a and induced emf E_a are in the same direction, as shown in Fig. 9.18 (*a*). Electromagnetic power is $E_a I_a$ watts and electrical power output is equal to VI watts.

For armature circuit

$$\text{Terminal voltage, } V = E_a - I_a R_a \qquad ...(9.19)$$

Electrical power output, P_{out}

$$= \text{Electromagnetic power, } E_a I_a$$
$$- \text{ ohmic losses including brush contact loss} \qquad ...(9.20)$$

In this mode of operation developed electromagnetic torque T_e is opposite to the direction of rotation of armature i.e. mechanical power is absorbed and a prime mover is required to run the machine (i.e. to provide the necessary mechanical power).

$$\text{Shaft power input} = E_a I_a + \text{rotational losses} \qquad ...(9.21)$$

Motoring Mode. The machine operates in motoring mode when armature current I_a flows in opposition to the generated emf E_a, as depicted in Fig. 9.18 (*b*). For armature circuit

$$\text{Terminal voltage, } V = E_a + I_a R_a \qquad ...(9.22)$$

* In actual machine this angle is 90° electrical.

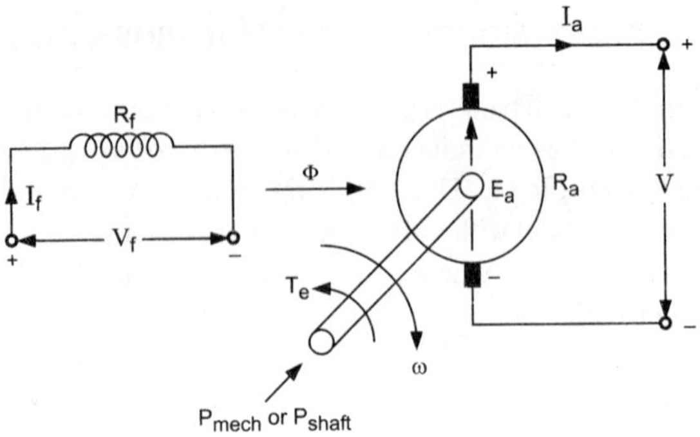

(a) *For Generating Mode*

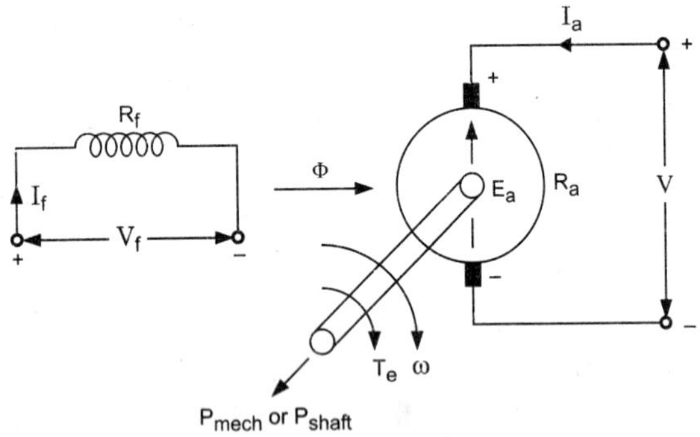

(b) *For Motoring Mode*

Fig. 9.18 *Circuit Model of a DC Machine*

Electrical power input, P_{in}

= Electromagnetic power, $E_a I_a$

+ ohmic losses including brush contact loss ...(9.23)

In this mode of operation developed electromagnetic torque, T_e is in the direction of rotation *i.e.* mechanical power is outputted and absorbed by mechanical load.

Shaft power output = $E_a I_a$ – rotational losses ...(9.24)

Example 9.1. The armature of a dc separately-excited machine has a resistance of 0.1 Ω and is connected to a 230 V supply. Calculate the generated emf when it is running (*i*) as a generator giving 80 A (*ii*) as a motor taking 60 A. [*U.P. Technical Univ. Odd Semester 2004-05*]

Solution: *As a generator:*

Generated emf, $E_G = V + I_a R_a = 230 + 80 \times 0.1 = 238$ V **Ans.**

As a motor:

Generated emf, $E_G = E_b = V - I_a R_a = 230 - 60 \times 0.1 = 224$ V **Ans.**

Example 9.2. A 120 V dc shunt motor has an armature resistance of 0.2 Ω and a brush volt drop of 2 V. The rated full-load armature current is 75 A. Calculate the current at the instant of starting and its value in terms of percentage of full-load armature current. Assume field winding resistance of 120 Ω.

[*U.P. Technical Univ. Electrical Machines, 2004–05*]

Solution: Full load armature current, I_{af} = 75 A

Shunt field current, $I_{sh} = \frac{V}{R_{sh}} = \frac{120}{120} = 1$ A

Back emf at the starting instant, $E_{b_{st}} = 0$

Armature current at the starting instant, I_{a_s}

$$= \frac{\text{V – brush voltage drop}}{\text{Armature resistance}} = \frac{120 - 2}{0.2} = 590 \text{ A}$$

Line current at the starting instant, $I_{L_{st}}$

$= I_{a_{st}} + I_{sh} = 590 + 1 = 591$ A **Ans.**

Line current in terms of percentage of full-load armature current

$$= 591 \times \frac{100}{75} = 788 \text{ per cent} \quad \textbf{Ans.}$$

Example 9.3. A 230 V series motor is taking 50 A. Resistance of armature and series field winding is 0.2 Ω and 0.1 Ω respectively. Calculate (*i*) brush voltage (*ii*) back emf, (*iii*) power wasted in armature, and mechanical power developed.

[*U.P. Technical Univ. Electrical Machines, 2003–04*]

Solution: Supply or terminal voltage, V = 230 V

Motor current, I = 50 A

Armature resistance, R_a = 0.2 Ω

Series field resistance, R_{se} = 0.1 Ω

(*i*) Brush voltage = Terminal voltage – voltage drop in series field

$= V - I R_{se} = 230 - 50 \times 0.1 = 225$ V **Ans.**

(*ii*) Back emf, E_b = Brush voltage – voltage drop in armature

$= V - I R_{se} - I R_a = V - I (R_{se} + R_a)$

$= 230 - 50 (0.1 + 0.2) = 215$ V **Ans.**

(*iii*) Power wasted in armature

$= I^2 R_a = 50^2 \times 0.2 = 500$ W **Ans.**

(*iv*) Mechanical power developed, P_m

$$= \frac{E_b I}{1{,}000} = \frac{215 \times 50}{1{,}000} = 10.75 \text{ kW} \quad \textbf{Ans.}$$

Example 9.4. A 20 kW, 240 V dc shunt generator has armature and field resistance of 0.05 Ω and 80 Ω respectively. Calculate the total armature power developed when working :

(*i*) as a generator delivering 20 kW output.

(*ii*) as a motor taking 20 kW input.

[*M.D. Univ. Electromechanical Energy Conversion, December-2010*]

Solution: Shunt field current, $I_{sh} = \frac{V}{R_{sh}} = \frac{240}{80} = 3$ A

(*i*) *As Generator*

$$\text{Load current, } I_L = \frac{\text{Output in kW} \times 1{,}000}{V} = \frac{20 \times 1{,}000}{240} = 83.33 \text{ A}$$

Armature current, $I_a = I_L + I_{sh} = 83.33 + 3 = 86.33$ A

Generated emf, $E_g = V + I_a R_a = 240 + 86.33 \times 0.05 = 244.32$ V

Total armature power developed,

$$P_g = \frac{E_g \times I_a}{1{,}000} = \frac{244.32 \times 83.33}{1{,}000} = 20.36 \text{ kW} \quad \textbf{Ans.}$$

(*ii*) *As Motor*

$$\text{Line current, } I_L = \frac{\text{Input in kW} \times 1{,}000}{V} = \frac{20 \times 1{,}000}{240} = 83.33 \text{ A}$$

Armature current, $I_a = I_L - I_{sh} = 83.33 - 3 = 80.33$ A

Back emf, $E_b = V - I_a R_a = 240 - 80.33 \times 0.05 = 235.98$ V

$$\text{Power developed, } P_m = \frac{E_b I_a}{1{,}000} = \frac{235.98 \times 80.33}{1{,}000} = 18.96 \text{ kW} \quad \textbf{Ans.}$$

9.10 SPEED EQUATION

As mentioned in Art. 9.4 the expressions for back emf developed in the armature of a dc motor are given as

$$E_b = \frac{\Phi Z N}{60} \times \frac{P}{A} \text{ volts} \quad ...(i)$$

$$\text{and } E_b = V - I_a R_a \text{ volts} \quad ...(ii)$$

Comparing Eqs. (*i*) and (*ii*) we get

$$\frac{\Phi ZN}{60} \times \frac{P}{A} = V - I_a R_a$$

$$\text{or } N = \frac{V - I_a R_a}{\Phi Z} \times \frac{60\,A}{P} = K\frac{V - I_a R_a}{\Phi} = K\frac{E_b}{\Phi}$$

since Z, A and P are constant for a particular machine.

For a dc motor, if initial values of speed, armature current, back emf and flux per pole are N_1, I_{a1}, E_{b1} and Φ_1 respectively and corresponding final values are N_2, I_{a2}, E_{b2} and Φ_2 respectively, then

$$N_1 \propto \frac{E_{b1}}{\Phi_1} \quad \text{where } E_{b1} = V - I_{a1} R_a$$

$$N_2 \propto \frac{E_{b2}}{\Phi_2} \quad \text{where } E_{b2} = V - I_{a2} R_a$$

$$\text{and } \frac{N_2}{N_1} = \frac{E_{b2}/\Phi_2}{E_{b1}/\Phi_1} = \frac{E_{b2}}{E_{b1}} \times \frac{\Phi_1}{\Phi_2} \qquad \text{...(9.25)}$$

In cases of PMDC motor and **dc shunt motor**, flux practically remains constant (*i.e.* $\Phi_2 = \Phi_1$) and

$$\frac{N_2}{N_1} = \frac{E_{b2}}{E_{b1}} = \frac{V - I_{a2} R_a}{V - I_{a1} R_a} \qquad \text{...(9.26)}$$

In above expression since applied voltage V is constant and the voltage drop in armature ($I_a R_a$) is negligible in comparison to supply voltage V, therefore, speed of a dc shunt motor remains almost constant.

In case of a **dc series motor**, prior to saturation,

$$\Phi \propto I_{se} \propto I_a$$

$$\text{or } \frac{\Phi_1}{\Phi_2} = \frac{I_{a1}}{I_{a2}}$$

$$\text{and } \frac{N_2}{N_1} = \frac{E_{b2}}{E_{b1}} \times \frac{I_{a1}}{I_{a2}} \qquad \text{...(9.27)}$$

For a **dc series motor**, after saturation

Flux Φ is independent of field current I_{se} or armature current I_a

and speed $N \propto E_b$

$$\text{or } \frac{N_2}{N_1} = \frac{E_{b2}}{E_{b1}} \qquad \text{...(9.28)}$$

9.11 SPEED REGULATION

The speed regulation of a dc motor is defined as the change in speed when the load on the motor is reduced from full load to zero and is expressed in percentage of rated full-load speed.

$$\therefore \text{ Percentage speed regulation} = \frac{N_0 - N_f}{N_f} \times 100 \qquad \text{...(9.29)}$$

where N_0 and N_f are no-load and full-load speeds respectively. The motor capable of maintaining a nearly constant speed for varying loads is said to have a good speed regulation.

The speed regulation of permanent magnet motors is quite good, about 10 to 15%. This is due to the fact that the voltage drop in armature is quite low. The speed regulation of dc shunt motors is usually less than that for permanent magnet types, somewhat less than 10%. This is due to slight reduction in flux per pole on account of flow of current through armature. The speed regulation of a dc series motor is normally very poor. The no-load speed is usually several times of the rated full-load speed, high enough for centrifugal force to damage the armature of any but the very small machines. The reason for this high no-load speed is that, with the rise in speed, the field flux is weakened causing further increase in speed. The speed regulation for cumulative compound wound dc motors varies from 20% to 50% whereas for differential compound wound dc motors it is about 5%.

Both dc shunt and compound motors have definite no-load speed, while the dc series motor is quite unstable and tends to operate at an extremely high, often runaway, speed when the mechanical load is removed.

Example 9.5. A 6-pole lap wound shunt motor has 500 conductors in the armature. The resistance of armature path is 0.05 Ω. The resistance of shunt field is 25 Ω. Find the speed of the motor when it takes 120 A from dc mains of 100 V supply. Flux per pole is 2×10^{-2} Wb.

[*M.D. Univ. Electromechanical Energy Conversion, December-2006*]

Solution: Shunt field current, $I_{sh} = \frac{V}{R_{sh}} = \frac{100}{25} = 4$ A

Armature current, $I_a = I_L - I_{sh} = 120 - 4 = 116$ A

Back emf, $E_b = V - I_a R_a = 100 - 116 \times 0.05 = 94.2$ V

Speed of motor, $N = \frac{E_b}{\Phi Z} \times \frac{60\,A}{P} = \frac{94.2 \times 60 \times 6}{2 \times 10^{-2} \times 500 \times 6}$

$= 565$ rpm **Ans.**

Example 9.6. The armature resistance of a 200 V shunt motor is 0.4 Ω and no-load current is 2 A, when loaded taking armature current of 50 A, the speed is 1,200 rpm. Determine no-load speed ignoring shunt field current.

[*G.B. Technical Univ. Electromechanical Energy Conversion-I, 2011-12*]

Solution: Supply voltage, V = 200 V

No-load current, $I_{Lo} = 2$ A

Armature current at no load, $I_{ao} = I_{Lo} = 2$A

Back emf on load, $E_{bo} = V - I_{ao} R_a = 200 - 2 \times 0.4 = 199.2$ V

At Load

Armature current, $I_{af} = 50$ A

Back emf, $E_{bf} = 200 - 50 \times 0.4 = 180$ V

Speed, $N_f = 1{,}200$ rpm

No-load speed, $N_o = N_f \times \frac{E_{bo}}{E_f}$

assuming field current remaining unchanged

$= 1{,}200 \times \frac{199.2}{180} = 1{,}328$ rpm **Ans.**

Example 9.7. A 250 V shunt motor has armature and field resistances of 1 Ω and 125 Ω respectively. When running light, it takes a current of 5 A and the speed is 1,500 rpm.

(*i*) Find the motor speed at its full load, the input current being 25 A.

(*ii*) Find also the speed at this load if a resistance of 2.5 Ω is inserted in the armature circuit.

[*U.P. Technical Univ. Electromechanical Energy Conversion-I, 2002-03*]

Solution: Shunt field current, $I_{sh} = \frac{V}{R_{sh}} = \frac{250}{125} = 2$ A

No-load armature current, $I_{a0} = I_{L0} - I_{sh} = 5 - 2 = 3$ A

No-load back emf, $E_{b0} = V - I_{ao}R_a = 250 - 3 \times 1 = 247$ V

No-load speed, $N_0 = 1{,}500$ rpm

(*i*) **At full load:**

Armature current, $I_{af} = I_{lf} - I_{sh} = 25 - 2 = 23$ A

Back emf, $E_{bf} = V - I_{af}R_a = 250 - 23 \times 1 = 227$ V

Assuming flux to remain constant i.e.

$$\Phi_f = \Phi_0$$

$$\text{Speed at full load, } N_f = \frac{E_{bf}}{E_{b0}} \times N_0 = \frac{227}{247} \times 1{,}500$$

$$= 1{,}378 \text{ rpm } \textbf{Ans.}$$

(*ii*) **When a resistance of 2.5 Ω is inserted in the armature circuit**

Resistance in armature circuit,

$$R'_a = R_a + R = 1 + 2.5 = 3.5\ \Omega$$

$$\text{Back emf, } E_{bf_1} = V - I'_{af}R_a = 250 - 23 \times 3.5 = 169.5 \text{ V}$$

$$\text{Speed, } N'_f = \frac{E'_{bf}}{E_{b0}} \times N_0 = \frac{169.5}{247} \times 1{,}500 = 1{,}029 \text{ rpm } \textbf{Ans.}$$

Example 9.8. A 120 V dc shunt motor having an armature circuit resistance of 0.2 Ω and field circuit resistance of 60 Ω, draws a line current of 40 A at full load. The brush voltage drop is 3 V and rated speed (full-load speed) is 1,800 rpm. Calculate (*a*) speed at half load (*b*) speed at 125% of full load.

[Rajasthan Technical Univ. Electrical Machines-I, 2008]

Solution: Shunt field current, $I_{sh} = \frac{V}{R_{sh}} = \frac{120}{60} = 2$ A

At full-load Line current, $I_{lf} = 40$ A

Armature current, $I_{af} = I_{lf} - I_{sh} = 40 - 2 = 38$ A

Back emf, $E_{bf} = V - I_{af}R_a -$ brush contact drop

$= 120 - 38 \times 0.2 - 3 = 109.4$ V

(*a*) **At half-full load**

$$\text{Line current, } I_{lh} = \frac{40}{2} = 20 \text{ A}$$

Armature current, $I_{ah} = I_{lh} - I_{sh} = 20 - 2 = 18$ A

Back emf, $E_{bh} = 120 - 18 \times 0.2 - 3 = 113.4$ V

Since $E_b \propto N\Phi \alpha N$

∵ field current or flux per pole is constant

$$\text{Speed of motor at half-full load, } N_h = \frac{N_f \times E_{bh}}{E_{bf}}$$

$$= \frac{1{,}800 \times 113.4}{109.4} = 1{,}866 \text{ rpm } \textbf{Ans.}$$

(*b*) **At 125% of Full Load**

Line current, $I_l' = I_{lf} \times 1.25 = 40 \times 1.25 = 50$ A

Armature current, $I_a' = I_l' - I_{sh} = 50 - 2 = 48$ A

Back emf, $E_b' = V - I_a'R_a -$ brush drop

$= 120 - 48 \times 0.2 - 3 = 107.4$ V

Speed of motor at 125% of full load,

$$N' = \frac{N_f \times E'_b}{E_{bf}} = \frac{1{,}800 \times 107.4}{109.4} = 1{,}767 \text{ rpm } \textbf{Ans.}$$

Example 9.9. A 230 V dc shunt motor takes an armature current of 3.33 A at rated voltage and at no load speed of 1,000 rpm. The resistance of armature and field windings are 0.3 Ω and 160 Ω respectively. The line current at full load is 40 A. Calculate the full-load speed if the armature reaction weakens no-load flux by 4%.

[U.P.Technical Univ. Energy Conversion 2005-06]

Solution: Shunt field current, $I_{sh} = \frac{V}{R_{sh}} = \frac{230}{160} = 1.44$ A

No-load armature current, $I_{ao} = I_{Lo} - I_{sh} = 3.33 - 1.44 = 1.89$ A

No-load back emf, $E_{bo} = V - I_{ao}R_a$

$= 230 - 1.89 \times 0.3 = 229.43$ V

No-load speed, $N_o = 1{,}000$ rpm

Full-load armature current, $I_{af} = I_{Lf} - I_{sh} = 40 - 1.44 = 38.56$ A

Full-load back emf, $E_{bf} = V - I_{af}R_a$

$= 230 - 38.56 \times 0.3 = 218.43$ V

Full-load flux per pole, $\Phi_f = (1 - 0.04)\Phi_o = 0.96\Phi_o$

$$\text{Full-load speed, } N_f = N_o \times \frac{E_{bf}}{E_{bo}} \times \frac{\Phi_o}{\Phi_f}$$

$$= 1{,}000 \times \frac{218.43}{229.43} \times \frac{1}{0.96}$$

$$= 992 \text{ rpm } \textbf{Ans.}$$

Example 9.10. A 10-hp, 230 V shunt motor has an armature resistance of 0.5 Ω and field circuit resistance of 115 Ω. At no-load and rated (full) voltage the speed is 1,200 rpm and armature current is 2 A. If load is applied, the speed drops to 1,100 rpm. Determine the armature current and the line current.

[U.P. Technical Univ. Electrical Machines, 2002–03]

Solution: At no load Speed, $N_0 = 1{,}200$ rpm

Armature current, $I_{a0} = 2$ A

Back emf, $E_{b0} = V - I_{a0}R_a$

$= 230 - 2 \times 0.5 = 229$ V

Let the armature current at load be of I_{aL} amperes

Back emf, $E_{bL} = V - I_{aL} R_a = 230 - 0.5\ I_{aL}$

$$\text{Since } \frac{\text{Speed at load, } N_L}{\text{Speed at no load, } N_0} = \frac{E_{bL}}{E_{b0}}$$

∵ Flux Φ is same in both cases in case of a dc shunt motor

$$\text{So } E_{bL} = E_{b0} \times \frac{N_L}{N_0} = 229 \times \frac{1{,}100}{1{,}200}$$

$$= 209.917 \text{ V}$$

or $230 - 0.5\ I_{aL} = 209.917$

$$\text{or Armature current at load, } I_{aL} = \frac{230 - 209.917}{0.5} = 40.2 \text{ A } \textbf{Ans.}$$

$$\text{Shunt field current, } I_{sh} = \frac{V}{R_{sh}} = \frac{230}{115} = 2 \text{ A}$$

Line current at load, $I_{LL} = I_{aL} + I_{sh} = 40.2 + 2 = 42.2$ A **Ans.**

Example 9.11. A 230 V dc motor takes no-load armature current of 2.5 A and runs at a speed of 1,000 rpm. If the full-load current is 40 A find the speed on full load and percentage speed regulation. Assume that the flux remains constant. Resistance of armature is 0.25 Ω.

[M.D. Univ. Electromechanical Energy Conversion, December-2013]

Solution: No-load back emf, $E_{b0} = V - I_{a0} R_a$

$= 230 - 2.5 \times 0.25 = 229.375$ V

Full-load back emf, $E_{bf} = V - I_{af} R_a$

$= 230 - 40 \times 0.25 = 220$ V

Full-load flux, $\Phi_f = \Phi_0$ as assumed

$$\text{Full-load speed, } N_f = N_0 \times \frac{E_{bf}}{E_{b0}} \times \frac{\Phi_0}{\Phi_f}$$

$$= 1{,}000 \times \frac{220}{229.375} \times \frac{\Phi_0}{\Phi_0}$$

$$= 959 \text{ rpm } \textbf{Ans.}$$

$$\text{Percentage speed regulation} = \frac{N_0 - N_f}{N_f} \times 100$$

$$= \frac{1{,}000 - 959}{959} \times 100 = 4.275\ \% \textbf{ Ans.}$$

Example 9.12. A 200 V dc series motor draws full-load line current of 38 A at the rated speed of 600 rpm. The motor has armature resistance of 0.4 Ω and the series field resistance is 0.2 Ω. The brush voltage drop irrespective of load is 3.0 volts, find :

(*i*) **The speed of the motor when the load current drops to 19 A.**

(*ii*) **The speed on removal of load when the motor takes only 1 A from supply.**

(*iii*) **The internal horse power developed in each of the above cases. Neglect the effect of armature reaction and saturation.**

Solution: On full load

$$\text{Back emf, } E_{b1} = V - I_1(R_a + R_{se}) - \text{brush voltage drop}$$

$$= 220 - 38\,(0.4 + 0.2) - 3 = 194.2 \text{ V}$$

On load current of 19 A

$$\text{Back emf, } E_{b2} = 220 - 19\,(0.4 + 0.2) - 3 = 205.6 \text{ V}$$

$$\text{On no-load back emf } E_{b0} = 220 - 1\,(0.4 + 0.2) - 3 = 216.4 \text{ V}$$

$$\text{Speed on 19 A load, } N_2 = \text{Rated speed} \times \frac{E_{b2}}{E_{b1}} \times \frac{\Phi_1}{\Phi_2}$$

$$= \frac{600 \times 205.6}{194.2} \times \frac{38}{19} = 1{,}270 \text{ rpm } \textbf{Ans.}$$

assuming unsaturated field *i.e.* $\frac{\Phi_1}{\Phi_2} = \frac{I_1}{I_2}$

Similarly no-load speed, N_0

$$= \frac{600 \times 216.4}{194.2} \times \frac{38}{1} = 25{,}406 \text{ rpm } \textbf{Ans.}$$

$$\text{Internal hp developed on 38 A} = \frac{194.2 \times 38}{735.5} = 10 \textbf{ Ans.}$$

$$\text{Internal hp developed on 19 A} = \frac{205.6 \times 19}{735.5} = 5.31 \textbf{ Ans.}$$

$$\text{Internal hp developed on no-load} = \frac{216.4 \times 1}{735.5} = 0.294 \textbf{ Ans.}$$

Example 9.13. A 230 V dc shunt generator has armature circuit resistance (including brushes) of 0.4 Ω and field circuit resistance of 120 Ω. If the machine run as a motor by connecting it a 230 V dc mains, find the ratio of speed as a generator to the speed as a motor. The line current in each case is 45 A.

[B.P. Univ. of Technology Electrical Machine-I, 2008]

Solution: Shunt field current, $I_{sh} = \frac{V}{R_{sh}} = \frac{230}{120} = 1.9 \text{ A}$

As Generator Line current, $I_{Lg} = 45$ A

$$\text{Armature current, } I_{ag} = I_{Lg} + I_{sh} = 45 + 1.9 = 46.9 \text{ A}$$

$$\text{Generated emf, } E_g = V + I_{ag} R_a$$

$$= 230 + 46.9 \times 0.4 = 248.76 \text{ V}$$

As motor Line current, $I_{Lm} = 45$ A

$$\text{Armature current, } I_{am} = I_{Lm} - I_{sh} = 45 - 1.9 = 43.1 \text{ A}$$

$$\text{Back emf developed, } E_b = V - I_{am} R_a$$

$$= 230 - 43.1 \times 0.4 = 212.76 \text{ V}$$

Ratio of speed as a generator to speed as a motor,

$$\frac{N_g}{N_m} = \frac{E_g}{E_b} = \frac{248.76}{212.76} = 1.169 \textbf{ Ans.}$$

∵ Field current and therefore flux per pole is same in both cases.

Example 9.14. A dc shunt generator delivers 50 kW at 250 V and 400 rpm. The armature and field resistances are 0.02 and 50 Ω respectively. Calculate the speed of the machine running as a shunt motor and taking 50 kW input at 250 V. Allow 1 V per brush for contact drop. *[G.G.S.I.P. Univ. Electrical Machines, December-2011]*

Solution: Shunt field current, $I_{sh} = \frac{V}{R_{sh}} = \frac{250}{50} = 5 \text{ A}$

As Generator

$$\text{Load current, } I_L = \frac{\text{Output in kW} \times 1{,}000}{\text{Terminal voltage, V}} = \frac{50 \times 1{,}000}{250} = 200 \text{ A}$$

$$\text{Armature current, } I_a = I_L + I_{sh} = 200 + 5 = 205 \text{ A}$$

$$\text{Generated emf, } E_g = V + I_a R_a + \text{brush drop}$$

$$= 250 + 205 \times 0.02 + 2 \times 1 = 256.1 \text{ V}$$

At Motor

$$\text{Line current, } I_L = \frac{\text{Power input in kW} \times 1{,}000}{V} = \frac{50 \times 1{,}000}{250} = 200 \text{ A}$$

Field current is same as in case of generator

i.e. $I_{sh} = 5$ A

$$\text{Armature current, } I_a = I_L - I_{sh} = 200 - 5 = 195 \text{ A}$$

$$\text{Back emf, } E_b = V - I_a R_a = 250 - 195 \times 0.02 - 2 \times 1 = 244.1 \text{ V}$$

Since field current is same and armature reaction is ignorable (assumed), flux is same in both cases and speed of the machine as motor,

$$N_m = N_g \times \frac{E_b}{E_g} = 400 \times \frac{244.1}{256.1} = 381.26 \text{ rpm } \textbf{Ans.}$$

Example 9.15. A dc shunt generator driven by a belt from an engine runs at 750 rpm while feeding 100 kW of electric power into 230 V mains. When the belt breaks it continues to run as a motor drawing 9 kW from the mains. At what speed would it run ? Armature resistance 0.08 Ω and field resistance 115 Ω.

[U.P. Technical Univ. Electromechanical Energy Conversion-I, 2005-06]

Solution: As generator

$$\text{Load current, } I_L = \frac{\text{Output in kW} \times 1{,}000}{\text{Terminal voltage, V}}$$

$$= \frac{100 \times 1{,}000}{230} = 434.8 \text{ A}$$

$$\text{Shunt field current, } I_{sh} = \frac{V}{R_{sh}} = \frac{230}{115} = 2 \text{ A}$$

$$\text{Armature current, } I_a = I_L + I_{sh} = 434.8 + 2 = 436.8 \text{ A}$$

$$\text{Generated emf, } E_g = V + I_a R_a$$

$$= 230 + 436.8 \times 0.08 = 264.944 \text{ V}$$

$$\text{Speed, } N_g = 750 \text{ rpm}$$

As motor

$$\text{Line current, } I_L = \frac{\text{Input in kW} \times 1{,}000}{V} = \frac{9 \times 1{,}000}{230}$$

$$= 39.13 \text{ A}$$

Field current, I_f is the same as in case of generator

i.e. $I_f = 2$ A

$$\text{Armature current, } I_a = I_L - I_{sh} = 39.13 - 2 = 37.13 \text{ A}$$

$$\text{Back emf, } E_b = V - I_a R_a$$

$$= 230 - 37.13 \times 0.08 = 227.03 \text{ V}$$

Since field current is same and armature reaction is ignorable (assumed), flux is same in both cases and speed of the machine as motor,

$$N_m = N_g \times \frac{E_b}{E_g} = 750 \times \frac{227.03}{264.944} = 642.7 \text{ rpm} \textbf{ Ans.}$$

Example 9.16. A 220 V, 1.5 kW, 859 rpm, separately-excited dc motor has armature resistance of 2.5 Ω and it draws current of 8 A at rated load condition. If the field current and the armature voltage are fixed at the value of rated load, what will be the no-load speed of the motor ? Assume losses remain constant between no-load and full-load operation.

Solution: At Rated Load

Output, $P_{out} = 1.5 \text{ kW} = 1{,}500 \text{ W}$

Input, $P_{in} = V \times I_L = 220 \times 8 = 1{,}760 \text{ W}$

Losses = Input – output = 1,760 – 1,500 = 260 W

Back emf, $E_{bf} = V - I_{af} R_a = 220 - 8 \times 2.5 = 200 \text{ V}$

At No Load Input = Losses = 260 W

∵ Losses are assumed to remain constant from no load to full load operation.

Armature current, I_a = Line current, I_{L0}

∵ machine is separately excited

$$= \frac{\text{Input}}{V} = \frac{260}{220} = 1.182 \text{ A}$$

Back emf, $E_{b0} = V - I_{a0} R_a = 220 - 1.182 \times 2.5 = 217.045 \text{ V}$

Since voltage at no-load is the same as at rated load

Since field current is same at full load as well as at no load so the flux is same and speed of motor at no load,

$$N_0 = N_f \times \frac{E_{b0}}{E_{bf}} = 859 \times \frac{217.045}{200} = 932.2 \text{ rpm} \textbf{ Ans.}$$

9.12 TORQUE

The measure of causing the rotation of a wheel or the turning or twisting moment of a force about the axis is called the *torque*.

Torque is measured by the product of force and the radius at which this force acts.

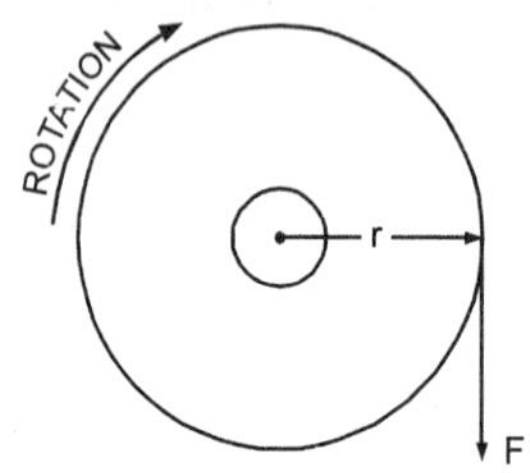

Fig. 9.19

Consider a wheel of radius *r* metres acted by a circumferential force F newtons, as shown in Fig 9.19. Let the force F cause the wheel to rotate at *n* rps.

Torque, $T = F \times r$ newton-metres

Work done per revolution = F × distance moved

$= F \times 2\pi r$ joules

Work done per second $= F \times 2\pi r \times n$

$= F \times r \times 2\pi n$ joules/second or watts

Since $F \times r$ = Torque, T and $2\pi n = \omega$,

angular velocity in radians per second,

Work done per second = $T\omega$ joules per second or watts ...(9.30)

and power developed, $P = T\omega = T \times \frac{2\pi N}{60} = 0.105 \, NT$ watts ...(9.31)

9.13 ARMATURE TORQUE

Let T_e be the electromagnetic torque developed in newton-metres by the motor running at *n* rps.

Power developed = Work done per second

$= T_e\omega = T_e \times 2\pi n$ watts ...(9.32)

Electrical equivalent of mechanical power developed by the armature, as mentioned in Art. 9.5, also

$= E_b I_a$ watts ...(9.33)

Comparing Eqs. (9.32) and (9.33) we have

$$T_e \times 2\pi n = E_b I_a$$

$$\text{or } T_e = \frac{E_b I_a}{2\pi n} = 0.159 \frac{E_b I_a}{n} \text{ N-m} \quad ...(9.34)$$

$$\text{Also } T_e = \frac{E_b I_a}{2\pi \frac{N}{60}} = 9.55 \frac{E_b I_a}{N} \text{ N-m} \quad ...(9.35)$$

where N is speed in rpm

Substituting $E_b = \Phi Z \frac{N}{60} \times \frac{P}{A}$ in Eq. (9.35) we have

$$T_e = 9.55 \times \Phi \times Z \times \frac{N}{60} \times \frac{P}{A} \times \frac{I_a}{N}$$

$$= 0.159 \, \Phi Z P \frac{I_a}{A} \text{ newton-metres} \quad ...(9.36)$$

ALTERNATIVE PROOF. Quantitatively, the force on a straight section of conductor length *l* metres perpendicular to a magnetic field of density B webers per square metre and carrying I_c amperes is given as

$$F_c = B l I_c \text{ newtons}$$

If D is the diameter of the armature, the force F_c acts at a radial distance of D/2 metres and the associated torque per conductor is given as

$$T_c = \frac{B l I_c D}{2}$$

The average flux density, $B_{av} = \frac{\Phi P}{\pi D l}$

So torque developed per conductor,

$$T_c = \frac{\Phi P}{\pi D l} \times \frac{l I_c D}{2} = \frac{1}{2\pi} \Phi P I_c$$

The torque associated with the entire winding is the summation of the torques for the individual conductors or coil sides. Torque developed by such electromagnetic action is called the electromagnetic torque and is given as

$$T = \frac{1}{2\pi} \Phi Z P I_c$$

where Z is the number of armature conductors on the armature surface.

$$= \frac{1}{2\pi} \Phi Z P \frac{I_a}{A} = 0.159 \, \Phi Z P \frac{I_a}{A}$$

∵ Current per conductor, $I_c = \frac{I_a}{A}$

Since Z, P and A are constant for a particular machine,

$$T = K \Phi I_a \qquad \text{where } K = \frac{1}{2\pi} \frac{PZ}{A}$$

$$\text{or } T \propto \Phi I_a$$

Thus it may be concluded that

(*i*) the electromagnetic torque developed by the armature is proportional to the product of flux per pole and armature current.

(*ii*) the direction of electromagnetic torque developed by armature depends upon the direction of flux or magnetic field and the direction of flow of current in armature conductors. If either of the two is reversed the direction of torque developed will be reversed and, therefore, the direction of rotation. When both (the direction of field as well as that of armature current) are reversed the direction of torque (or rotation) will not change. This has been explained in Art. 9.8 also.

In case of a series wound motor flux Φ is directly proportional to armature current I_a (before saturation) because in a series wound motor field winding and armature winding currents are same, therefore,

$$T_e \; \alpha \; I_a^2$$

In case of permanent magnet motors, separately-excited motors and shunt wound motors the field strength i.e. Φ remains practically constant and, therefore,

$$T_e \; \alpha \; I_a$$

On the basis of the same kW output and speed, a dc series motor develops the highest starting torque and the dc shunt motor the least, while the cumulative compound wound dc motor falls somewhere between the first two.

9.14 SHAFT TORQUE

The armature torque is the gross torque, which is developed by the armature. Whole of this torque is not available at the pulley, since certain percentage of torque developed by the armature is lost to overcome the iron and friction losses. *Net torque i.e. gross torque less torque lost in iron and friction losses is known as shaft torque.*

If T_a is the torque developed by armature in N-m, T_f is the torque lost in iron and friction losses, and T_{sh} is the shaft torque or useful torque, then

$$T_a = \frac{E_b I_a}{\frac{2\pi N}{60}} = 9.55 \frac{E_b I_a}{N} \text{ newton-metres}$$

where N is speed of armature in rpm

$$T_f = \frac{\text{Iron and frictional losses in watts}}{2\pi N/60}$$

$$= \frac{9.55}{N} \text{ (iron and frictional losses in watts)}$$

Shaft torque, T_{sh}

$$= T_a - T_f = \frac{E_b I_a - \text{iron and frictional losses}}{2\pi N/60} \text{ newton-metres}$$

$$T_{sh} \text{ also} = \frac{\text{Output in watts}}{2\pi N/60}$$

$$= \frac{9.55 \times \text{output in watts}}{N} \text{ newton-metres} \quad ...(9.37)$$

9.14.1. Brake Horse Power (BHP). In case of electric motors (ac or dc), the mechanical power available at the shaft in horse power is known as *brake horse power* (bhp). If T_{sh} is the shaft torque in newton metres and N is speed in rpm then

$$\text{Output in bhp} = \frac{2\pi N T_{sh}}{60 \times 735.5} \quad ...(9.38)$$

9.15 TORQUE AND SPEED OF A DC MOTOR

From Arts 9.10 and 9.13 we know that

$$N \propto \frac{E_b}{\Phi} \text{ and } T \propto \Phi I_a$$

The above relations reveal that increase in flux would cause increase in torque and decrease in speed. It cannot be so because torque is the cause of rotation and increase in torque should cause increase in speed rather than decrease. The apparent inconsistency between the above two relations may be reconciled as below:

Let the motor be running at a steady speed and the field strength be weakened, which is possible by inserting an additional resistance in the field circuit in case of a dc shunt motor connected across constant voltage supply. The immediate effect is that, owing to the momentary reduction in back emf, the current flowing through armature circuit increases out of all proportion to the reduction in flux. As a result, inspite of the weakened field, the torque increases momentarily tremendously and will exceed considerably the value corresponding to load. The surplus torque, thus available causes the motor to accelerate, and the back emf to increase. The motor will attain steady speed again when the back emf has attained such a value that the current flowing in the armature develops with the slight weakened flux, only just sufficient torque to drive the load.

Consider, for an example, a 200 V motor drawing an armature current of 30 A, the flux per pole being 0.05 Wb, the speed 1,200 rpm and armature resistance 0.2 Ω.

The back emf is then $E_b = 200 - 30 \times 0.2 = 194$ V

Now if the flux is reduced to 0.04 Wb, the back emf will fall momentarily to $194 \times \frac{0.04}{0.05}$ *i.e.* to 155.2 V, the speed remaining the same owing to inertia of heavy armature. The current flowing through the armature will rise up to $\frac{200 - 155.2}{0.2}$ *i.e.* 224 A, if inductance effects are negligible. The torque developed before weakening of field is, therefore, temporarily much exceeded (the torque, being proportional to the product of flux and armature current, increases in the ratio of $\frac{0.04}{0.05} \times \frac{224}{30}$ *i.e.* 6 to 1. The additional torque rapidly accelerates the motor, and if the increase in speed has no effect upon the total value of the load torque to be developed, the speed will increase until the armature current falls to a value I_{a2} such that $I_{a2} \times 0.04 = 30 \times 0.05$ *i.e.* $I_{a2} = 37.5$ A. The ultimate value of the speed then becomes

$$N_2 = N_1 \times \frac{V - I_{a2} R_m}{V - I_{a1} R_m} \times \frac{\Phi_1}{\Phi_2}$$

$$= 1{,}200 \times \frac{200 - 37.5 \times 0.2}{200 - 30 \times 0.2} \times \frac{0.05}{0.04} = 1{,}488.4 \text{ rpm.}$$

For avoiding undue rush of current the field strength should be reduced slowly.

The manner in which the motor slows down on strengthening of the field is also of equally importance. When the field is strengthened, the immediate effect is increase in back emf. Quite a small increase in the flux will normally be sufficient to increase the back emf beyond supply voltage, and in consequence, will not merely reduce the value of armature current, but will cause it to reverse its direction of flow. The machine will, therefore, act temporarily as a generator, and feed energy to the supply system, this energy being available at the expense of the kinetic energy associated with the machine and its load, with the result that a rapid reduction in speed takes place. The decrease in speed will cease when the back emf has fallen to a value such that the motor takes just sufficient current to develop the torque the load requires. By thus strengthening the field the motor speed is reduced by a form of "regenerative braking".

Consider the effect of increasing suddenly the flux from 0.05 Wb to 0.06 Wb in the example just discussed. The back emf raised at once to $194 \times \frac{0.06}{0.05}$ *i.e.* 232.8 V and the current drawn from the supply mains becomes $\frac{200 - 232.8}{0.2} = -164$ A *i.e.* the machine acts as a generator and feeds the current to the supply mains.

Assuming torque to remain constant, the final value of the armature current will become $30 \times \frac{0.05}{0.06} = 25$ A and the speed will attain the value of $1{,}200 \times \frac{200 - 25 \times 0.2}{200 - 30 \times 0.2} \times \frac{0.05}{0.06}$ *i.e.* 1,005 rpm.

From the above discussion it is obvious that torque developed in a motor is certainly a function of flux and armature current and is independent of speed and the speed depends upon the torque.

Example 9.17. A dc shunt motor develops an open-circuit emf of 250 V at 1,500 rpm. Find the developed torque for an armature current of 10 A. [*B.P. Univ. of Technology Electrical Machines, 2008*]

Solution: Open-circuit emf, $E_o = 250$ V

Armature current, $I_a = 10$ A

Assuming back emf equal to open-circuit emf E_o, though it may be slightly less due to armature reaction if field current remains unchanged.

Mechanical power developed,

$$P_{mech} = E_b I_a = 250 \times 10 = 2{,}500 \text{ W}$$

$$\text{Torque developed, } T = \frac{9.55 \times P_{mech}}{N}$$

$$= \frac{9.55 \times 2{,}500}{1{,}500} = 15.92 \text{ N-m } \textbf{Ans.}$$

Example 9.18. 230 V, 4-pole, shunt motor has two circuit armature winding with 500 conductors. The armature resistance is 0.25 Ω, the field resistance is 125 Ω and the flux per pole is 0.02 Wb. Armature reaction is neglected. If the motor draws 14 A from the mains, then find the speed and internal torque developed.

[*West Bengal Univ. of Technology Electrical Machine-I, 2013*]

Solution: Flux per pole, $\Phi = 0.02$ Wb

Number of armature conductors,

$Z = 500$

Number of pole, $P = 4$

Number of parallel paths, $A = 2$

Line current, $I_L = 14$ A

$$\text{Shunt field current, } I_{sh} = \frac{V}{R_{sh}} = \frac{230}{125} = 1.84 \text{ A}$$

$$\text{Armature current, } I_a = I_L - I_{sh} = 14 - 1.84 = 12.16 \text{ A}$$

$$\text{Back emf, } E_b = V - I_a R_a$$

$$= 230 - 12.16 \times 0.25 = 226.96 \text{ V}$$

$$\text{Speed of motor, } N = \frac{E_b}{\Phi Z} \times \frac{60 A}{P} \quad \because E_b = \frac{\Phi ZN}{60} \times \frac{P}{A} \text{ volts}$$

$$= \frac{226.96}{0.02 \times 500} \times \frac{60 \times 2}{4} = 681 \text{ rpm } \textbf{Ans.}$$

$$\text{Internal torque developed, } T = 0.159\, \Phi\, Z\, P \times \frac{I_a}{A} \quad \text{...Refer to Eq. (9.36)}$$

$$= 0.159 \times 0.02 \times 500 \times 4 \times \frac{12.16}{2}$$

$$= 38.67 \text{ N-m } \textbf{Ans.}$$

Example 9.19. A dc shunt motor draws an armature current of 12 A from a 220 V dc supply when developing a gross torque of 20 N-m in the armature running at 1,200 rpm. What will be the induced back emf in the armature?

[*B.P. Univ. of Technology Electrical Machine-I, 2008*]

Solution: Gross torque developed, $T = 20$ N-m

Speed of armature, $N = 1{,}200$ rpm

Armature current, $I_a = 12$ A

$$\text{Back emf, } E_b = T \times \frac{2\pi N}{60} \times \frac{1}{I_a} \quad \because E_b I_a = T \times \frac{2\pi N}{60}$$

$$= 20 \times \frac{2\pi \times 1{,}200}{60} \times \frac{1}{12} = 209.4 \text{ V } \textbf{Ans.}$$

Example 9.20. A dc machine is connected across a 240 V line. It rotates at 1,200 rpm and is generating at 230 V. The armature current is 40 A.

(*a*) Is the machine functioning as a generator or as a motor.
(*b*) Find the resistance of the armature circuit.
(*c*) Determine the electromagnetic torque.
(*d*) If the load is thrown off, what will be the generated voltage and the speed of the machine, assuming that there is negligible effect of armature reaction?

[*U.P.S.C. I.E.S. Electrical Engineering-II, 2009*]

Solution: (*a*) The machine is working as motor because generating voltage of 230 V is less than terminal voltage of the machine (240 V) **Ans.**

$$(b) \text{ Armature resistance, } R_a = \frac{V - E_b}{I_a} \quad \because E_b = V - I_a R_a$$

$$= \frac{240 - 230}{40} = 0.25\ \Omega \textbf{ Ans.}$$

(*c*) Electromagnetic torque developed,

$$T = \frac{E_b I_a}{2\pi N/60} = \frac{230 \times 40}{2\pi \times 1{,}200/60}$$

$$= 73.21 \text{ N-m } \textbf{Ans.}$$

(*d*) When load is thrown off, generated voltage is equal to terminal voltage i.e. 240 V because effect of armature reaction is negligible.

$$\text{Speed of the machine, } N_o = \frac{E_{b_o}}{E_b} \times N$$

Assuming field current constant

$$= \frac{240}{230} \times 1{,}200 = 1{,}252 \text{ rpm } \textbf{Ans.}$$

Example 9.21. Calculate the torque developed by a motor armature carrying a total of 25,000 ampere-conductors, the core diameter being 1.0 m, the length being 0.3 m, the pole arc 70% and flux density in the air gap 0.6 T.

Solution: Force per conductor $= BI_c l_c = 0.6I_c \times 0.3 = 0.18I_c$

Total force acting on armature conductors,

$$F = \text{Force per conductor} \times \text{effective number of conductors}$$

$$= 0.18\, I_c \times \text{pole arc} \times \text{total number of armature conductors}$$

$$= 0.18\, I_c \times 0.7 \times Z = 0.126\, Z\, I_c = 0.126 \times 25{,}000 = 3{,}150 \text{ N}$$

$$\text{Torque developed, } T_e = F \times \frac{\text{core diameter}}{2} = 3{,}150 \times \frac{1.0}{2}$$

$$= 1{,}575 \text{ N-m } \textbf{Ans.}$$

Example 9.22. Calculate the torque developed when a current of 15 A passes through the armature of a motor with the following particulars.

560 conductors, lap wound, 4 poles, pole shoes 15 cm long subtending an angle of 60° at the centre, core radius 14 cm and flux density in the air gap is 1.7 T.

[*J.N. Technological Univ. Electrical Machines-I, March-2014*]

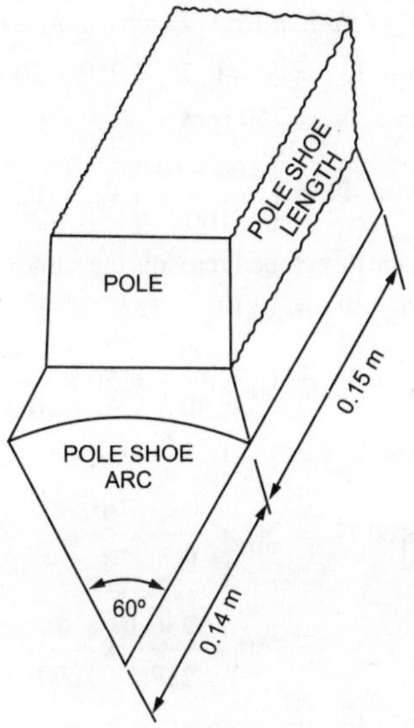

Fig. 9.20

Solution: $\text{Pole shoe arc} = \pi D_a \times \frac{\theta}{360} = \pi \times 0.28 \times \frac{60}{360°}$

$$= 0.1466 \text{ m}$$

$$\text{Pole shoe length} = 0.15 \text{ m}$$

$$\text{Flux per pole, } \Phi = \text{Flux density} \times \text{area of pole face}$$

$$= 1.7 \times 0.1466 \times 0.15 = 0.03738 \text{ Wb}$$

$$\text{Torque developed, } T_a = \frac{1}{2\pi}\, \Phi\, ZP \times \frac{I_a}{A}$$

$$= \frac{1}{2\pi} \times 0.03738 \times 560 \times 4 \times \frac{15}{4}$$

$$= 49.97 \text{ N-m } \textbf{Ans.}$$

Example 9.23. A 440 V, 14-poles lap wound dc shunt motor takes 1,550 A from the supply. There are 280 conductors. The effective diameter of the armature is 1.3 m and active length of each conductor is 0.33 m. The value of average flux density is 0.7 T. Calculate the value of torque developed and speed of the motor.

Solution: Flux per pole Φ

$$= \frac{B_{av} \times \pi \times \text{diameter of armature} \times \text{active length of each conductor}}{\text{Number of poles}}$$

$$= \frac{0.7 \times \pi \times 1.3 \times 0.33}{14} = 0.0674 \text{ Wb}$$

Total number of armature conductors, Z = 280

$$\text{Armature current, } I_a = \text{Current drawn by motor} = 1{,}550 \text{ A}$$

Neglecting shunt field current

$$\text{Back emf developed, } E_b = V = 440 \text{ V}$$

Neglecting voltage drop in armature

$$\text{Developed torque, } T_e = 0.159\Phi ZP\frac{I_a}{A}$$

$$= 0.159 \times 0.0674 \times 280 \times 14 \times \frac{1{,}550}{14}$$

$$= 4{,}651 \text{ N-m } \textbf{Ans.}$$

$$\text{Speed of motor, } N = \frac{E_b \times 60 \times A}{\Phi \times Z \times P} = \frac{440 \times 60 \times 14}{0.0674 \times 280 \times 14}$$

$$= 1{,}400 \text{ rpm } \textbf{Ans.}$$

Example 9.24. A 4-pole dc series motor has wave-connected winding with 600 conductors. Total resistance of motor is 0.8 Ω. When fed from 250 V dc source the motor supplies a load of 10 kW and takes 50 A with a flux per pole of 3 mWb. For these operating conditions, calculate the developed torque and shaft torque.

[*G.G.S.I.P. Univ Delhi Electromechanical Energy Conversion-I, December-2011*]

Solution: Supply voltage, V = 250 V

Line current, I = 50 A

$$\text{Resistance of motor, } R_m = R_a + R_{se} = 0.8\ \Omega$$

$$\text{Back emf, } E_b = V - I(R_a + R_{se}) = 250 - 50 \times 0.8 = 210 \text{ V}$$

$$\text{Speed of motor, } N = \frac{E_b}{\Phi Z} \times \frac{60\, A}{P}$$

$$= \frac{210}{3 \times 10^{-3} \times 600} \times \frac{60 \times 2}{4} = 3{,}500 \text{ rpm}$$

$$\text{Torque developed, } T_e = \frac{E_b I_a}{2\pi N/60} = \frac{210 \times 50}{2\pi \times 3{,}500/60}$$

$$= 28.65 \text{ N-m } \textbf{Ans.}$$

$$\text{Shaft torque} = \frac{\text{Load supplied in kW} \times 1{,}000}{2\pi N/60}$$

$$= \frac{10 \times 1{,}000}{2\pi \times 3{,}500/60} = 27.28 \text{ N-m } \textbf{Ans.}$$

Example 9.25. A 4-pole, lap wound 410 V series motor has the following data: Number of armature conductors is 960, flux per pole is 0.04 Wb, total motor resistance is 0.5 Ω, iron and frictional losses is 2 kW. If the current taken by the motor is 95 A, find (*i*) total torque (*ii*) useful torque at the shaft (*iii*) power output (*iv*) pull at the rim of the pulley of 42 cm diameter connected to the shaft.

[*J.N. Technological Univ. Electrical Machines-I, March-2014*]

Solution: Supply voltage, V = 410 V

Line current, I = 95 A

$$\text{Motor resistance, } R_m = 0.5\ \Omega$$

$$\text{Back emf, } E_b = V - I R_m = 410 - 95 \times 0.5 = 362.5 \text{ V}$$

$$\text{Speed of motor, } N = \frac{E_b}{\Phi Z} \times \frac{60\, A}{P} = \frac{362.5}{0.04 \times 960} \times \frac{60 \times 4}{4}$$

$$= 566.4 \text{ rpm}$$

Mechanical power developed,

$$P_{mech} = \frac{E_b \times I}{1,000} \text{ kW} = \frac{362.5 \times 95}{1,000}$$
$$= 34.4375 \text{ kW}$$

(*i*) Total torque developed, $T_e = \frac{E_b I_a}{2\pi N/60} = \frac{362.5 \times 95}{2\pi \times 566.4/60}$

$$= 580.6 \text{ N-m } \textbf{Ans.}$$

(*iii*) Power output, $P_{out} = P_{mech}$ – Iron and frictional losses

$$= 34.4375 - 2 = 32.4375 \text{ kW}$$

(*ii*) Useful torque at shaft $= \frac{P_{out} \text{ in kW} \times 1,000}{2\pi N/60}$

$$= \frac{32.4375 \times 1,000}{2\pi \times 566.4/60} = 546.88 \text{ N-m } \textbf{Ans.}$$

(*iv*) Pull at rim, $F = \frac{T}{D/2} = \frac{546.88}{0.42/2} \quad \because T = F \times \frac{D}{2}$

$$= 2,604.2 \text{ newtons } \textbf{Ans.}$$

Example 9.26. A 50 kW, 250 V dc shunt motor takes no-load armature current of 4 A at rated voltage and runs at 1,200 rpm. The $r_a = 0.4\ \Omega$ and $r_f = 250\ \Omega$ respectively. At rated load and rated voltage the motor takes 26 A and armature reaction weakens the field flux by 3%. Calculate the full-load speed and corresponding electromagnetic torque of motor.

[*G.G.S.I.P. Univ. Electrical Machines, December-2011*]

Solution: Shunt field current, $I_{sh} = \frac{V}{R_{sh}} = \frac{250}{250} = 1$ A

At No Load

Armature current, $I_{a_0} = 4$ A

No-load back emf, $E_{b_0} = V - I_{a_0} R_a = 250 - 4 \times 0.4 = 248.4$ V

No-load speed, $N_0 = 1,200$ rpm

At Rate Load

Armature current, $I_{af} = I_{Lf} - I_{sh} = 26 - 1 = 25$ A

Back emf, $E_{bf} = V - I_{af} R_a = 250 - 25 \times 0.4 = 240$ V

Flux, $\Phi_f = 0.97\ \Phi_0$ $\because$ Reduction in field flux at load = 3%

Full-load speed, $N_f = N_0 \times \frac{E_{bf}}{E_{b0}} \times \frac{\Phi_0}{\Phi_f}$

$$= 1,200 \times \frac{240}{248.4} \times \frac{\Phi_0}{0.97\Phi_0} = 1,195.3 \text{ rpm } \textbf{Ans.}$$

Electromagnetic torque developed,

$$T_e = \frac{E_{bf} I_{af}}{2\pi N/60} = \frac{240 \times 25 \times 60}{2\pi \times 1,195.3} = 47.934 \text{ N-m } \textbf{Ans.}$$

Example 9.27. A 230 V, dc machine has $R_a = 0.3\ \Omega$, and $R_{sh} = 160\ \Omega$ respectively. It is running as a motor on no load at 1,000 rpm taking an armature current of 3.3 A at rated voltage. When the motor is run on full load at rated voltage, the line current has a value of 40 A. Calculate the speed and torque developed for this condition, assuming that armature reaction weakens the no-load flux by 4%.

[*U.P.S.C. I.E.S. Electrical Engineering-II, 2007*]

Solution: Shunt field current, $I_{sh} = \frac{V}{R_{sh}} = \frac{230}{160} = 1.4375$ A

At No Load Armature current, $I_{a0} = 3.3$ A

No-load back emf, $E_{b0} = V - I_{a0} R_a$

$= 230 - 3.3 \times 0.3 = 229.01$ V

No-load speed, $N_0 = 1,000$ rpm

On Load Armature current, $I_{af} = I_{Lf} - I_{sh}$

$= 40 - 1.4375 = 38.5625$ A

Back emf, $E_{bf} = V - I_{af} R_a = 230 - 38.5625 \times 0.3$

$= 218.43$ V

Flux, $\Phi_f = 0.96\ \Phi_0$

$\because$ Reduction in flux at load = 4%

(*i*) Speed of motor, $N_f = N_0 \times \frac{E_{bf}}{E_{b_0}} \times \frac{\Phi_0}{\Phi_f}$

$$= 1,000 \times \frac{218.43}{229.01} \times \frac{\Phi_0}{0.96\Phi}$$
$$= 993.5 \text{ rpm } \textbf{Ans.}$$

(*ii*) Torque developed, $T_e = \frac{E_{bf} I_{af}}{2\pi N/60} = \frac{218.43 \times 38.5625}{2\pi \times 993.5/60}$

$$= 80.96 \text{ N-m } \textbf{Ans.}$$

Example 9.28. A 250 V dc shunt motor having an armature resistance of 0.25 Ω carries an armature current of 50 A and runs at 750 rpm. If the flux is reduced by 10%, find the speed. Assume that the load torque remains same.

[*Rajasthan Technical Univ. Electrical Machines-I, 2008, 2010*]

Solution: Back emf, $E_{b1} = V - I_{a1} R_a = 250 - 50 \times 0.25 = 237.5$ V

Speed, $N_1 = 750$ rpm

Flux, $\Phi_2 = \left(\frac{100 - 10}{100}\right) \Phi_1 = 0.9\ \Phi_1$

Since total torque developed remains the same,

$$I_{a2}\Phi_2 = I_{a_1}\Phi_1$$

or $I_{a2} = I_{a1} \times \frac{\Phi_1}{\Phi_2} = 50 \times \frac{\Phi_1}{0.9\ \Phi_1} = 55.556$ A

Back emf, $E_{b2} = V - I_{a2}R_a = 250 - 55.556 \times 0.25 = 236.1$ V

Speed $N_2 = N_1 \times \frac{E_{b_2}}{E_{b_1}} \times \frac{\Phi_1}{\Phi_2} \qquad \because N \propto \frac{E_b}{\Phi}$

$$= 750 \times \frac{236.1}{237.5} \times \frac{\Phi_1}{0.9\Phi_1} = 828.4 \text{ rpm } \textbf{Ans.}$$

Example 9.29. A 220 V dc shunt motor has an armature resistance of 0.4 Ω and field resistance of 220 Ω. The no-load input line current of the motor from the 220 V dc mains is 4.5 A while the no-load speed of the motor is 1,500 rpm. The motor takes 22 A from the same 220 V dc mains while driving the full-load torque at 1,470 rpm. Find the induced back emf at no load and at full load. What will be the ratio of full-load torque to the no-load torque developed by the machine? [*B.P. Univ. of Technology Electrical Machine-I, 2008*]

Solution: Shunt field current, $I_{sh} = \frac{V}{R_{sh}} = \frac{220}{220} = 1$ A

At No Load: Line current, $I_{L_0} = 4.5$ A

Armature current, $I_{a_0} = I_{L_0} - I_{sh} = 4.5 - 1 = 3.5$ A

Back emf, $E_{b_0} = V - I_{a_0} R_a$

$= 220 - 3.5 \times 0.4 = 218.6$ V **Ans.**

Speed, $N_0 = 1,500$ rpm

At Full Load: Line current, $I_{L_f} = 22$ A

Armature current, $I_{a_f} = I_{L_f} - I_{sh} = 22 - 1 = 21$ A

$$\text{Back emf, } E_{b_f} = V - I_{a_f} R_a$$
$$= 220 - 21 \times 0.4 = 211.6 \text{ V } \textbf{Ans.}$$

Speed, $N_f = 1{,}470$ rpm

Ratio of full-load torque to no-load torque,

$$\text{i.e. } \frac{T_f}{T_0} = \frac{E_{b_f} I_{a_f}}{E_{b_0} I_{a_0}} \times \frac{N_0}{N_f}$$

$$\because P_{mech} = E_b I_a = T \times \frac{2\pi N}{60} \text{ or } T \propto \frac{E_b I_a}{N}$$

$$= \frac{211.6 \times 21}{218.6 \times 3.5} \times \frac{1{,}500}{1{,}470} = 5.926 \textbf{ Ans.}$$

Example 9.30. A 20 kW, 250 V dc shunt motor has full-load armature current of 85 A at 1,100 rpm. The armature resistance is 0.18 Ω. Determine:

(*i*) The internal torque developed,

(*ii*) The internal torque, if the field current is suddenly reduced to 80% of its original value.

(*iii*) The steady motor speed in part (*ii*) assuming the load torque remaining constant.

[*U.P. Technical Univ. Electromechanical Energy Conversion-I, 2003-04; M.D. Univ. Electrical Machines-I, May-2009*]

Solution: Armature current, $I_{a_1} = 85$ A

$$\text{Back emf, } E_{b_1} = V_1 - I_{a_1} R_a = 250 - 85 \times 0.18 = 234.7 \text{ V}$$

(*i*) Internal developed torque, T_{e_1}

$$= \frac{9.55\, E_{b_1} I_{a_1}}{N_1} = \frac{9.55 \times 234.7 \times 85}{1{,}100}$$
$$= 173.2 \text{ N-m } \textbf{Ans.}$$

(*ii*) Assuming constant power and neglecting magnetic saturation i.e. I_a to be constant and flux proportional to field current

Internal developed torque, T_{e_2}

$$= \frac{T_{e_1} \times I_{a_2} \Phi_2}{I_{a_1} \Phi_1} = 173.2 \times \frac{I_{a_1}}{I_{a_1}} \times 0.8$$
$$= 138.56 \text{ N-m } \textbf{Ans.}$$

(*iii*) $I_{a_2} \Phi_2 = I_{a_1} \Phi_1$ Since torque developed remains constant

$$\text{or Armature current, } I_{a_2} = I_{a_1} \times \frac{\Phi_1}{\Phi_2} = 85 \times \frac{1}{0.8} = 106.25 \text{ A}$$

assuming $\Phi \propto I_f$

$$\text{Back emf, } E_{b_2} = V - I_{a_2} R_a = 250 - 106.25 \times 0.18$$
$$= 230.875 \text{ V}$$

$$\text{Speed, } N_2 = \frac{E_{b_2}}{E_{b_1}} \times \frac{\Phi_1}{\Phi_2} \times N_1 = \frac{230.875}{234.7} \times \frac{1}{0.8} \times 1{,}100$$
$$= 1{,}352.6 \text{ rpm } \textbf{Ans.}$$

Example 9.31. A 200 V dc shunt motor having armature resistance of 0.4 Ω takes armature current of 20 A on full load and runs at 600 rpm. If resistance of 0.5 Ω is placed in the armature circuit, find the speed at half-full load torque. What is the ratio of stalling torque to full-load torque.

Solution: At Full Load

Armature current, $I_{af} = 20$ A

$$\text{Back emf, } E_{bf} = V - I_{af} R_a = 200 - 20 \times 0.4 = 192 \text{ V}$$

$$\text{Torque developed, } T_{ef} = \frac{E_b I_a}{2\pi N/60} = \frac{192 \times 20 \times 60}{2\pi \times 600} = 61.12 \text{ Nm}$$

At Half-full Load Torque

$$\text{Half-full load torque, } T_{ehf} = \frac{1}{2} \times T_{ef} = \frac{1}{2} \times 61.12 = 30.56 \text{ Nm}$$

$$\text{Armature current, } I_{ahf} = \frac{1}{2} I_{af} = \frac{1}{2} \times 20 = 10 \text{ A}$$

$\because T \propto I_a$ for dc shunt motor

$$\text{Back emf, } E_{bhf} = V - I_a(R_a + R)$$
$$= 200 - 10\,(0.4 + 0.5) = 191 \text{ V}$$

$$\text{Speed, } N_{hf} = \frac{E_{bhf} I_{ahf}}{2\pi \times T_{ehf}} \times 60 = \frac{191 \times 10 \times 60}{2\pi \times 30.56}$$
$$= 596.875 \text{ rpm } \textbf{Ans.}$$

$$\because T_e = \frac{E_b I_a \times 60}{2\pi N}$$

Stalling armature current,

$$I_{as} = \frac{V - E_{bs}}{R_a} = \frac{200 - 0}{0.4} = 500 \text{ A}$$

$\because$ Standstill emf = 0

Ratio of stalling torque to full-load torque,

$$\frac{T_{es}}{T_{ef}} = \frac{I_{as}}{I_{af}} = \frac{500}{20} = 25 \textbf{ Ans.}$$

$\because$ for a dc shunt motor $T_e \propto I_a$

Example 9.32. A 220 V dc shunt motor takes 20 A at full load running at 2,500 rpm. The armature resistance is 0.4 Ω and shunt field resistance of 176 Ω. The machine is converted into a long shunt differentially compound motor by augmenting an additional series field winding with 0.1 Ω resistance. There is a 11% change in total flux when compound motor develops the same torque. Determine the speed of compound motor. Assume rotational losses to be constant under all loading conditions.

[*U.P.S.C. I.E.S. Electrical Engineering-II, 2013*]

Solution: Supply voltage, V = 220 V

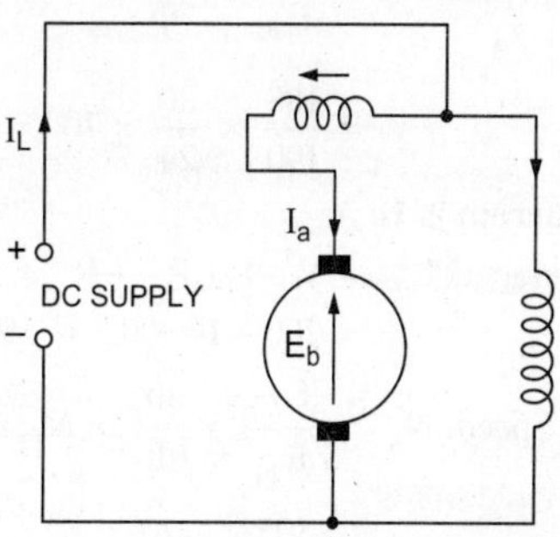

Fig. 9.21 *Differential Compound Motor*

As Shunt Motor

$$\text{Shunt field current, } I_{sh} = \frac{V}{R_{sh}} = \frac{220}{176} = 1.25 \text{ A}$$

Line current, $I_L = 20$ A

$$\text{Armature current, } I_{a_1} = I_L - I_{sh} = 20 - 1.25 = 18.75 \text{ A}$$

$$\text{Back emf, } E_{b_1} = V - I_{a_1} R_a$$
$$= 220 - 18.75 \times 0.4 = 212.5 \text{ V}$$

Speed, $N_1 = 2{,}500$ rpm

As Compound Motor. Since motor is differential-connected, flux will be reduced by 11%.

i.e. $\frac{\Phi_2}{\Phi_1} = 0.89$

Since total torque developed remains the same,

$$I_{a_2}\Phi_2 = I_{a_1}\Phi_1$$

$$\text{or } I_{a_2} = I_{a_1} \times \frac{\Phi_1}{\Phi_2} = \frac{18.75}{0.89} = 21.07 \text{ A}$$

$$\text{Back emf, } E_{b_2} = V - I_{a_2}(R_a + R_{se})$$

$$= 220 - 21.07\,(0.4 + 0.1) = 209.465 \text{ V}$$

$$\text{Speed, } N_2 = \frac{E_{b_2}}{E_{b_1}} \times \frac{\Phi_1}{\Phi_2} \times N_1 \qquad \because E_b \propto N\Phi$$

$$= \frac{209.465}{212.5} \times \frac{1}{0.89} \times 2{,}500 = 2{,}769 \text{ rpm } \textbf{Ans.}$$

Example 9.33. A 200 V dc series motor runs at 700 rpm when operating at its full-load current of 20 A. The motor resistance is 0.5 Ω and the magnetic circuit can be assumed unsaturated. What will be the speed if (i) the load torque is increased by 44% (ii) the motor current is 10 A?

Solution: Supply voltage, V = 200 V

Input current, I_1 = 20 A

$$\text{Back emf developed, } E_{b1} = V - I_1(R_a + R_{se})$$

$$= 200 - 20 \times 0.5 = 190 \text{ V}$$

(i) When the load torque is increased by 44%

$$i.e.\ T_2 = 1.44\,T_1$$

$$\text{or } \Phi_2 I_2 = 1.44\,\Phi_1 I_1 \qquad \because T \propto \Phi I_a$$

$$\text{or } I_2^2 = 1.44\,I_1^2 \qquad \because \Phi \propto I$$

$$\text{or } I_2 = \sqrt{1.44}\,.\,I_1 = 1.2 \times 20 = 24 \text{ A}$$

$$\text{Back emf developed, } E_{b2} = V - I_2(R_a + R_{se})$$

$$= 200 - 24 \times 0.5 = 188 \text{ V}$$

$$\text{Speed } N_2 = \frac{E_{b2}}{E_{b1}} \times \frac{\Phi_1}{\Phi_2} \times N_1$$

$$= \frac{188}{190} \times \frac{20}{24} \times 700 = 577.2 \text{ rpm } \textbf{Ans.}$$

(ii) **When motor current is 10 A**

$$\text{Back emf developed, } E_{b3} = V - I_3(R_a + R_{se})$$

$$= 200 - 10 \times 0.5 = 195 \text{ V}$$

$$\text{Speed, } N_3 = \frac{E_{b3}}{E_{b1}} \times \frac{\Phi_1}{\Phi_3} \times N_1$$

$$= \frac{195}{190} \times \frac{20}{10} \times 700 = 1{,}437 \text{ rpm } \textbf{Ans.}$$

Example 9.34. A 440 V dc shunt motor has a no-load (I_0 = 0) speed of 2,000 rpm. It is running at 1,000 rpm at full-load torque, reduced armature voltage and full field. If load torque is reduced to 50% of rated value with armature voltage and field voltage held constant at previous values, the speed increases to 1,050 rpm. Find the armature voltage drop at full load. Neglect effect of armature reaction.

[*GATE 1994*]

Solution: Supply voltage, V = 440 V

No-load speed, N_o = 2,000 rpm

No-load back emf, E_{b0} = V = 440 V

∵ armature current on no load = 0

Full-load speed, N_f = 1,000 rpm

Half-full load speed, N' = 1,050 rpm

$$\text{Full-load back emf, } E_{bf} = E_{b0} \times \frac{N_f}{N_0} = 440 \times \frac{1{,}000}{2{,}000} = 220 \text{ V}$$

∵ Flux is unchanged

$$\text{Back emf at half full load, } E_b = E_{b0} \times \frac{N'}{N_0} = 440 \times \frac{1{,}050}{2.000} = 231 \text{ V}$$

Let the reduced armature voltage be V_1 volts and armature voltage drop at full load be v volts.

Since flux remains unchanged, when load torque is reduced to 50%, armature current would reduce to 50%, torque T being proportional to the product of flux per pole and armature current. Since voltage drop in armature is proportional to armature current, armature voltage drop will also reduce to 50% *i.e.* become $\frac{v}{2}$ when load torque is reduced to 50%.

$$\text{So } E_{bf} = V - v$$

$$\text{and } E'_b = V - \frac{1}{2}v$$

$$\text{or } 220 = V - v \qquad \ldots(i)$$

$$\text{and } 231 = V - \frac{1}{2}v \qquad \ldots(ii)$$

Solving Eqs. (*i*) and (*ii*) we have

Armature voltage drop at full load, v = 22 V **Ans.**

Example 9.35. The following datas pertain to 250 V dc series motor:

$$\mathbf{Z = 180;\ \frac{P}{A} = 1;\ Flux/Pole = 3.75\ m\ Wb/field\ amp}$$

Total armature circuit resistance = 1 Ω

The motor is coupled to centrifugal pump whose load torque is

$$\mathbf{T_L = 10^{-4}n^2\ N\text{-}m}$$

where n = speed in rpm.

Calculate the current drawn by the motor and the speed at which it will run.

[*G.B. Technical Univ. Electromechanical Energy Conversion-I, 2012-13*]

Solution: Supply voltage, V = 250 V

Number of armature conductors, Z = 180

$$\frac{P}{A} = 1$$

Let the current drawn by motor be of I amperes

In case of dc series motor,

Armature current = Series field current = Line current, I

Flux per pole, Φ = 3.75 mWb/field ampere = 0.00375 I Wb

$$\text{Back emf, } E_b = V - IR = 250 - I \times 1 = (250 - I) \text{ volts} \qquad \ldots(i)$$

$$\text{and } E_{b\ \text{also}} = \frac{\Phi Zn}{60} \times \frac{P}{A} = \frac{0.00375\,I \times 180 \times n}{60} \times 1$$

$$= 0.01125\,nI \qquad \ldots(ii)$$

From Eqs (*i*) and (*ii*) we have

$$n = \frac{250 - I}{I} \times \frac{1}{0.01125} = 88.889 \times \frac{250 - I}{I} \qquad \ldots(iii)$$

Torque developed is given as

$$T_e = 0.159\,\Phi Z \times \frac{P}{A} \times I_a$$

$$= 0.159 \times 0.00375\,I \times 180 \times 1 \times I = 0.107325\,I^2 \qquad \ldots(iv)$$

$$\text{Load torque, } T_L = 10^{-4}\,n^2 \text{ N-m}$$

Neglecting iron, friction and windage losses we have

$$T_e = T_L$$

$$0.107325\ I^2 = 10^{-4}\ n^2$$

$$\text{or } n = \sqrt{0.107325 \times 10^4 \times I^2} = 32.761\ I \quad ...(v)$$

Solving Eqs (*iii*) and (*v*) we have

Current drawn by motor, I = 24.7 **Ans.**

Speed of motor, n = 810 rpm **Ans.**

Example 9.36. A 6-pole, 230 V dc series motor has a flux per pole of 4 mWb/ampere over the working range of the magnetization curve which is assured to be linear. The load torque is proportional to speed squared and its value is 20 N-m at 800 rpm. There are 432 wave-connected conductors and the total resistance of the motor is 1.0 Ω. Determine the motor speed and current when this motor is connected to rated supply voltage.

[U.P. Technical Univ. Electromechanical Energy Conversion-I, 2008-09]

Solution: Supply voltage, V = 230 V

Number of poles, P = 6

Number of parallel paths, A = 2

∵ machine is wave-connected

Number of armature conductors, Z = 432

Motor resistance, $R = R_a + R_{se} = 1\ \Omega$

Let the current drawn by the motor be of I amperes.

In case of series motor, armature current,

I_a = Series field current, I_{se} = Line current, I

Flux per pole, Φ = 0.004 Wb/field ampere = 0.004 I Wb

Back emf is given as

$$E_b = V - IR = (230 - I) \text{ volts} \quad ...(i)$$

Refer to Eq. (9.2)

$$\text{Also } E_b = \frac{\Phi Z N}{60} \times \frac{P}{A}$$

$$= \frac{0.004 I \times 432 \times N}{60} \times \frac{6}{2} = 0.0864\ NI \quad ...(ii)$$

Comparing Eqs. (*i*) and (*ii*) we have

$$0.0864\ NI = 230 - I$$

$$\text{or } I = \frac{230}{1 + 0.0864\ N} \quad ...(iii)$$

Torque developed is given as

$$T = \frac{1}{2\pi} \Phi I_a Z \times \frac{P}{A}$$

$$= \frac{1}{2\pi} \times 0.004\ I \times I \times 432 \times \frac{6}{2} = 0.825\ I^2$$

Substituting $I = \dfrac{230}{1 + 0.0864\ N}$ from Eq. (*iii*) in above equation, we have

$$T = 0.825 \times \left(\frac{230}{1 + 0.0864\ N}\right)^2$$

Given that load torque,

$$T_L \ \alpha\ N^2 = K_L N^2$$

From the given data K_L is evaluated as under

$$20 = K_L \times (800)^2 \quad \because T_L \text{ at 800 rpm is 20 Nm}$$

$$\text{or } K_L = \frac{20}{(800)^2} = \frac{1}{32{,}000}$$

Under steady conditions,

$$T_L = T$$

$$\frac{1}{32{,}000} \times N^2 = 0.825 \times \left(\frac{230}{1 + 0.0864\ N}\right)^2$$

$$\text{or } \frac{N}{\left(\dfrac{230}{1 + 0.0864\ N}\right)} = \sqrt{0.825 \times 32{,}000} = 162.48$$

$$\text{or } N = \frac{230 \times 162.48}{1 + 0.0864\ N}$$

$$\text{or } 0.0864\ N^2 + N - 37{,}370.6 = 0$$

or N = 651.9 rpm **Ans.**

Substituting N = 651.9 rpm in Eq. (*iii*) we have

$$\text{Current drawn by motor, } I = \frac{230}{1 + 0.0864 \times 651.9} = 4 \text{ A} \quad \textbf{Ans.}$$

Example 9.37. A 250 V dc series motor has an armature resistance of 0.4 Ω. From its magnetization curve at 480 rpm, the armature voltage is 125 V at 40 A of field current when running as a motor. Determine the speed at which it will run while drawing a current of 40 A. Determine the torque developed by the motor at this speed.

[G.B. Technical Univ. Electric Drives, 2012-13]

Solution: Line current, I_L = 40 A

Assuming series field resistance to be 0.25 Ω,

$$\text{back emf of motor, } E_{b_1} = V - I_1(R_a + R_{se})$$

$$= 220 - 40\ (0.4 + 0.25) = 194 \text{ V}$$

From magnetization curve

Field current, I_f = 40 A

Back emf, E_{b_2} = Armature voltage = 125 V

Speed, N_2 = 480 rpm

$$\text{Since speed, } N \propto \frac{E_b}{\Phi}$$

$$\text{or } N_1 = N_2 \times \frac{E_{b_1}}{E_{b_2}} \times \frac{\Phi_2}{\Phi_1} = N_2 \times \frac{E_{b_1}}{E_{b_2}}$$

∵ Field current, and therefore flux is same i.e. $\Phi_2 = \Phi_2$

$$= 480 \times \frac{194}{125} = 745 \text{ rpm } \textbf{Ans.}$$

$$\text{Torque developed, } T = \frac{9.55 \times E_b I_a}{N}$$

$$= \frac{9.55 \times 194 \times 40}{745} = 99.5 \text{ N-m } \textbf{Ans.}$$

9.16 ARMATURE REACTION IN DC MOTORS

The effect of the magnetic field developed by the current carrying armature conductors on the distribution of flux under the main poles in dc generators was previously described in chapter 6.

The armature reaction in dc motors is due to the same reasons as in dc generators. Since, however, a dc motor rotates in a direction opposite to that of a dc generator for the same direction of current in the armature winding and for the same polarity of the brushes, the effects of the armature reaction in dc motors will be somewhat different.

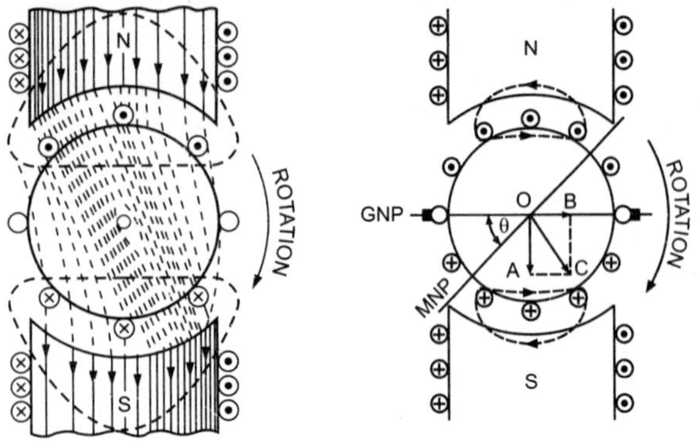

(*a*) *Armature Reaction in a DC Motor* (*b*)

Fig. 9.22

In a generator the effect of armature reaction was to reduce the generated voltage due to reduction in resultant flux but in a motor the reduction of the resultant flux causes an increase in speed.

Fig. 9.22 (*b*) illustrates a motor armature carrying current in outward direction under N pole and in inward direction under S pole corresponding to clockwise rotation. Now main mmf OA acts downward and due to current in the armature conductors, mmf OB is produced in the direction from left to right (at right angles to the polar axis). By adding these two mmfs vectorially the resultant mmf OC is obtained in the direction shown in Fig. 9.22 (*b*). It is observed that the magnetic neutral plane (mnp) which is perpendicular to OC is now shifted backward instead of forward as in the case of a generator. If the direction of the lines of force of the armature cross field is compared with those of main field we see that the main field is strengthened under the leading pole tips and weakened under the trailing pole tips. Hence in a motor, it is necessary to move the brushes backward with the increase in load in order to bring the brushes axis into mnp whereas in generator they are moved forward.

The backward movement of the brushes in the motor is accompanied by a demagnetising and cross-magnetising action of the armature on the field, as shown in Fig. 9.23. In Fig. 9.23 OA is the mmf due to main field, OB is the field due to armature

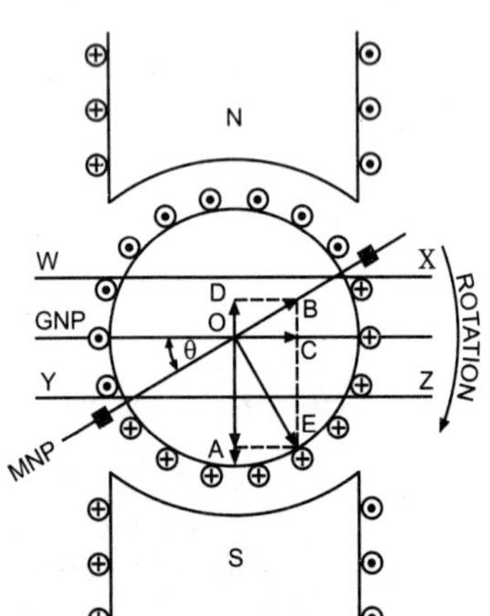

Fig. 9.23

ampere conductors which can be resolved into two components OD and OC, known as *demagnetising* and *cross-magnetising components* respectively. Armature conductors can be divided into two groups, lying above WX and below YZ producing cross-magnetising effect acting from left to right and lying between the two planes producing demagnetising effect. Thus when the brushes in a generator or motor are set up so as to be in mnp a demagnetising action is set up by the armature in both the cases, but the cross-magnetising fields are opposite in direction.

As the load is increased on the motor, the armature reaction tends to increase the motor speed. In fact sometimes motors with short air gaps (producing high armature reaction) have run away when the load was applied. For this reason most shunt and even compound motors are provided with a winding known as *stabilizing winding*, usually consisting of $1\frac{1}{2}$ turns on each pole in series with the armature and aiding the shunt ampere-turns. This prevents weakening of the field through armature reaction which might be caused by an increase in load, thus securing speed stability.

Remedies. 1. The brushes may be given a backward shift in order to bring the brush axis into the magnetic neutral plane in the ordinary motors (the motors which are required to run in one direction only).

2. In cases, where motors are required to be run in both directions without sparking dc motors are commonly provided with interpoles. Interpoles neutralise the cross-magnetising effect of armature and so sparkless commutation is obtained at different loads. In order to oppose field distortion in a motor interpoles must have polarity of the main poles behind them, as shown in Fig. 9.24.
3. In order to increase reluctance of the pole piece in the region where the magnetic flux tends to become most dense, the leading horn of the pole piece may be made longer than the trailing horn and cut farther from the armature surface. Thus the distribution of magnetic flux is equalised.

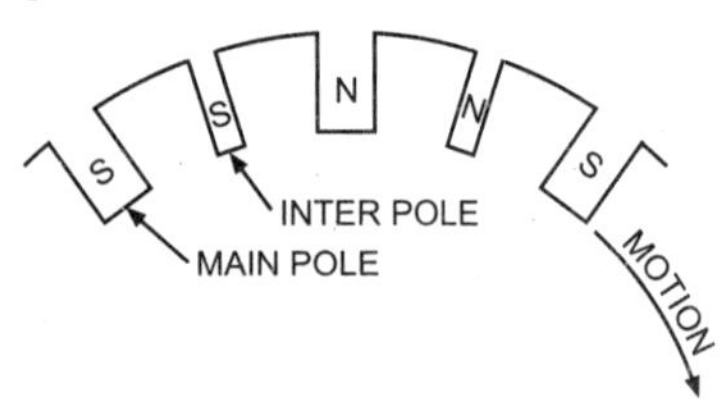

Fig. 9.24

4. In order to increase reluctance, the length of air gap may be increased. It also necessitates more ampere-turns in the field winding. The field distortion, however, will not be so great as it would if the magnetic field of the machine were weaker.
5. Heavy-duty motors, employed in rolling mills, hoists, cranes, mines and traction etc., are also provided with compensating windings which reduce sparking in commutation and prevent field distortion owing to armature reaction.

9.16.1. Connection of Commutating Pole Windings. Practically all new dc machines are provided with commutating poles (or interpoles). The commutating field winding is always connected in series with the armature winding as illustrated in Fig 9.25. The shunt field winding of any commutating-pole machine should not be connected directly across the armature terminals. In connecting the shunt field winding of these machines, the armature winding and the commutating pole winding should be considered as one combined unit.

Comparing the polarity of the commutating poles required for both the motor and generator, it is seen that opposite polarities are necessary. However, while changing the operation of a dc machine (i.e. operating the generator as a motor), it is not necessary to change the internal connections, to the commutating poles. Since the commutating poles are connected in series with the armature the change in direction of current in armature will automatically change the direction of polarity of the commutating poles so that at all times, whether used as generator or motor, the polarity of the commutating poles is correct.

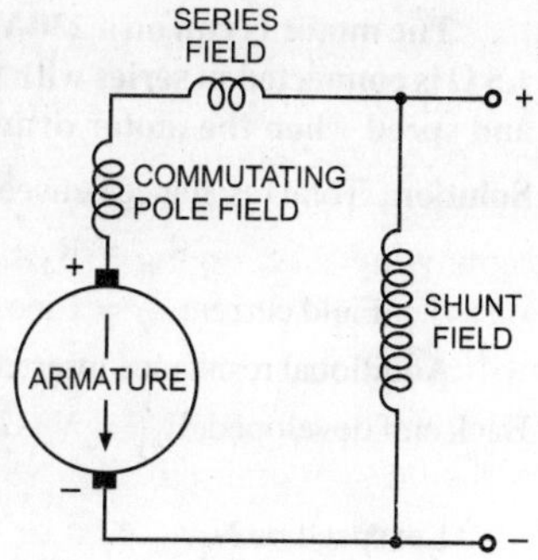

Fig. 9.25

9.17 OPERATING CHARACTERISTICS OF DC MOTORS

The performance and, therefore, suitability of a dc motor is determined from its characteristics, known as *performance characteristics*. The important characteristics of dc motors are:

1. Torque-Armature Current Characteristic. This characteristic curve gives relation between torque developed in the armature, T and armature current I_a. This is also known as *electrical characteristic*.

2. Speed-Armature Current Characteristic. This characteristic curve gives relation between speed N and armature current I_a. This is also known *as speed characteristic*.

3. Speed-Torque Characteristic. This characteristic gives relation between speed N and torque developed in armature, T. This is also known as *mechanical characteristic*. This curve may be derived from the two characteristic curves mentioned above.

The important relations to be kept in mind while discussing motor characteristics are:

(*i*) $I_a = \dfrac{V - E_b}{R_a}$ (*ii*) $N \propto \dfrac{E_b}{\Phi}$ and (*iii*) $T \propto \Phi I_a$.

9.18 OPERATING CHARACTERISTICS OF DC SERIES MOTORS

1. Speed-Current Characteristic. In case of a dc series motor, the mmf due to the exciting coils increases in direct proportion to the line or armature current, so (neglecting armature reaction effects) the value of flux varies with the load current according to the ordinary magnetization curve. Owing to armature reaction, the actual curve representing the useful flux falls below the open-circuit magnetization curve, as illustrated in Fig. 9.26. With larger currents the magnetic circuit gets saturated and flux Φ tends to approach a constant value.

From the speed equation, it is obvious that speed is proportional to back emf E_b and inversely proportional to flux per pole Φ. With the increase in armature current voltage drop in armature circuit and series field $[I (R_a + R_{se})]$ increases and, therefore, back emf E_b decreases, as shown in Fig. 9.26. However, under normal conditions $I (R_a + R_{se})$ drop is quite small and may be neglected. Thus if the applied voltage remains constant, speed N is inversely proportional to flux Φ. If a curve is drawn between speed and input (or line) current I, it will be a rectangular hyperbola before magnetic saturation as up to saturation point the magnetisation curve is a straight line. In this region, the speed decreases abruptly with the increase in input current. After magnetic saturation, the flux Φ tends to become constant and speed-current characteristic becomes a straight line and speed decreases slightly due to voltage drop in armature and series field, as shown in Fig. 9.26. The speed becomes zero when the input current is the normal short-circuit current of the motor *i.e.* equal to applied voltage divided by the motor resistance $(R_a + R_{se})$. Normally this current is many many times full-load value.

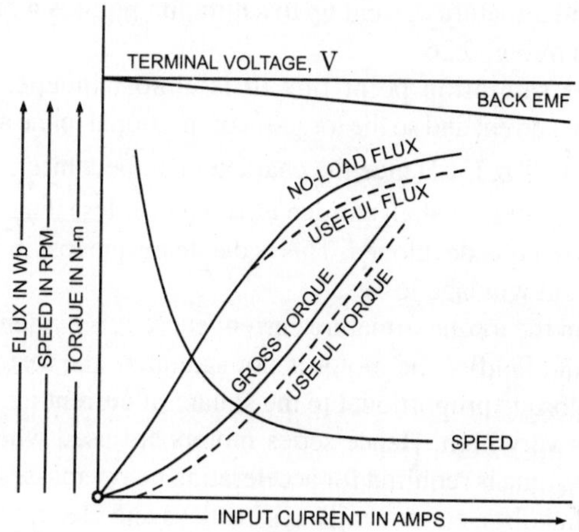

Fig. 9.26

From speed-current characteristic curve it is obvious that the series motor is a variable speed motor *i.e.* speed varies with the variation in line current. With the decrease in load on a dc series motor, the speed increases and may become dangerously high at very light loads.

Since on no load the speed is dangerously high, the machine may get damaged due to heavy centrifugal forces set up in the rotating parts. *This is the reason that series motors are never started on no load, which is explained below.*

When the motor is connected across the supply mains without load, it draws a small current from supply mains flowing through the series field and armature, the speed tends to increase so that back emf may approach the applied voltage in magnitude. The increase in back emf weakens the armature current and hence the field current. This causes again increase in speed so in back emf. Thus the field continues to weaken and speed continues to increase until the armature gets damaged.

To start a dc series motor, mechanical load is first put and then the motor is started.

Since on no load the series motor attains dangerously high speed, which causes heavy centrifugal force resulting in the damage of machine, series motoı s are not suitable for the services:

(*i*) where the load may be entirely removed and

(*ii*) for driving by means of belts because mishap to the belt would cause the motor run on no load.

The motors are suitable for gear drive, because gears provide some load on account of the frictional resistances of the gear teeth in case of sudden release of load.

However, very small series motors may be used with belts, since in case of mishap to belt, the comparatively large frictional resistance would represent an appreciable load on it.

The minimum load on a dc series motor should be large enough (not below 15% of full load) to keep the speed of the motor within limits. In case the speed becomes dangerously high the motor must be disconnected from the supply main.

2. Torque-Current Characteristic. From the expression of mechanical torque T it is obvious that torque is directly proportional to the product of flux per pole Φ and armature current I_a. Up to saturation point flux is proportional to field current and hence to the armature current, because $I_a = I_f$. Therefore, on light load mechanical torque T is proportional to the square of the armature current *i.e.*, $T \alpha I_a^2$ and hence curve drawn between torque and armature current up to saturation point is a parabola, as shown in Fig. 9.26.

After saturation point flux Φ is almost independent of excitation current and so the torque is proportional to the armature current *i.e.* $T \alpha I_a$. Hence the characteristic becomes a straight line. The useful (or shaft) torque is, of course, less than the total (or gross) torque developed. This is due to torque lost in iron and friction and windage losses.

From the torque-armature current curve it is evident that so long as the field of the motor is not saturated, the series motor exerts a torque proportional to the square of current i.e. starting torque is very high. Hence series motors are used where large starting torque is required for accelerating heavy masses quickly such as in hoists, electric railways, trolleys and electric vehicles.

From torque-current characteristic it is evident that series motor develops large starting torque to accelerate the heavy masses and from speed-current characteristic it is also evident that speed falls as the load increases, so series motor is automatically relieved from heavy excessive load, therefore, series motor is best suited for electric traction work.

3. Speed-Torque Characteristic. The speed-torque characteristic can be drawn with the help of above two characteristics, as illustrated in Fig. 9.27. Speed-torque characteristic is also known as mechanical characteristic. Speed sharply falls with the increase in torque for smaller values of load. But at higher loads, the speed drops linearly but slowly with increasing torque.

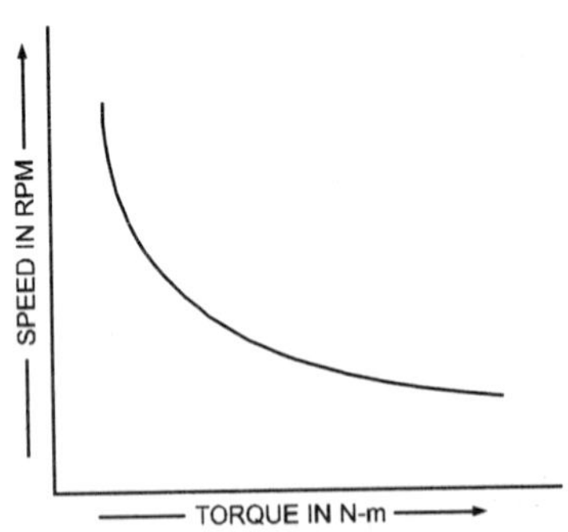

Fig. 9.27 *Speed-Torque Characteristic of DC Series Motor*

Hence series motors are best suited for services where the motor is directly coupled to the load such as fans whose speed falls with the increase in load torque.

Example 9.38. A 230-V dc series motor has an armature resistance of 0.08 Ω and field resistance of 0.05 Ω. The magnetisation characteristic of the machine at 700 rpm is as follows:

Field current	**30**	**60**	**90**	**120**	**150 A**
EMF	**70**	**125**	**180**	**210**	**230 V**

The motor is run on a 230 V supply. An additional resistance of 1.5 Ω is connected in series with the armature. Determine the torque and speed when the motor draws a current of 90 A.

Solution: Total resistance between motor terminals,

$$R_m = R_a + R_{se} = 0.08 + 0.05 = 0.13\ \Omega$$

Field current, I_f = Line current $I_1 = 90$ A

Additional resistance inserted in armature circuit, R = 1.5 Ω

Back emf developed, $E_{b_1} = V - I_1 (R_m + R) = 230 - 90 (0.13 + 1.5) = 83.3$ V

Let speed be N_1

From given datas for magnetisation curve when field current is 90 A

EMF developed, E_2 = 180 V

Speed, N_2 = 700 rpm

Since $N \alpha \frac{E_b}{\Phi}$,

$$\frac{N_1}{N_2} = \frac{E_{b1}}{E_2} \times \frac{\Phi_2}{\Phi_1} = \frac{E_{b1}}{E_2} \qquad \because \Phi_2 = \Phi_1,$$

field current being the same

$$\text{or } N_1 = N_2 \times \frac{E_{b1}}{E_2} = 700 \times \frac{83.3}{180} = 324 \text{ rpm} \quad \textbf{Ans.}$$

$$\text{Torque developed, } T = \frac{9.55 \times E_b I_a}{N} = \frac{9.55 \times 83.3 \times 90}{324} = 221 \text{ N-m } \textbf{Ans.}$$

Example 9.39. A dc series motor running on open circuit at 1,000 rpm gives the following voltage-current relationship. The total resistance of the motor (armature and field) is 0.2 Ω. Neglecting the effects of armature reaction and brush drop, plot the speed-torque curve of the motor when operating on 500 V supply.

Current (A)	**50**	**100**	**150**	**200**	**250**	**300**
Open-circuit voltage (V)	**150**	**295**	**402**	**480**	**530**	**565**

Solution: Voltage applied across the motor, V = 500 V

Resistance of motor, R_m = 0.2 Ω

Speed N_1 = 1,000 rpm

Field current in amps	50	100	150	200	250	300
Back emf E_1 (volts) at speed N_1 (1,000 rpm) = emf generated by armature (given)	150	295	402	480	530	565
Back emf E (volts) at speed N (to be determined) $= V - IR_m$	490	480	470	460	450	440
Speed N (to be determined) in rpm $= \frac{N_1}{E_1} \times (V - IR_m)$	3,267	1,627	1,169	958	849	779
Torque T in N-m $\frac{9.55 (V - IR_m)I}{N}$	71.6	281.75	575.9	917.12	1,265.5	1,618.2

The speed-torque curve now may be drawn from the above datas as shown in Fig. 9.28.

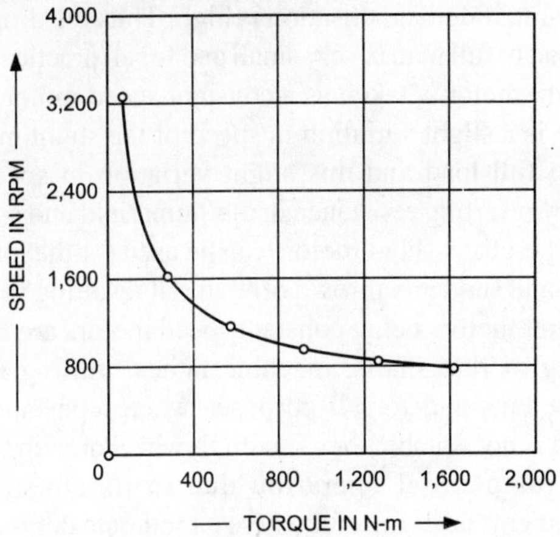

Fig. 9.28

Example 9.40. A series motor takes 20 A at 400 V to drive a fan at 200 rpm. Its resistance is 1 Ω (field and armature). If the torque required to drive the fan varies as the square of the speed, find the necessary applied voltage and current to drive the fan at 300 rpm.

Solution: Let the applied voltage to the motor be V_2 volts and current I_2 amperes for driving fan at 300 rpm.

So $N_2 = 300$ rpm

Speed $N_1 = 200$ rpm

Voltage $V_1 = 400$ V

Current $I_1 = 20$ A

Back emf, $E_1 = V_1 - I_1 R_m = 400 - 20 \times 1.0 = 380$

$$\frac{T_2}{T_1} = \left(\frac{N_2}{N_1}\right)^2 \qquad \text{Since } T \propto N^2$$

$$\text{or } \frac{\Phi_2 I_2}{\Phi_1 I_1} = \left(\frac{300}{200}\right)^2 \qquad \because T \propto \Phi I$$

$$\text{or } \frac{I_2^2}{I_1^2} = \left[\frac{300}{200}\right]^2 \qquad \because \Phi \propto I, \text{ neglecting saturation}$$

or $I_2 = 1.5\, I_1 = 1.5 \times 20 = 30$ A **Ans.**

$$\text{Also } \frac{E_{b2}}{E_{b1}} = \frac{N_2}{N_1} \times \frac{\Phi_2}{\Phi_1} = \frac{N_2}{N_1} \times \frac{I_2}{I_1}$$

$$\text{or } E_{b2} = E_{b1} \times \frac{N_2}{N_1} \times \frac{I_2}{I_1} = 380 \times \frac{300}{200} \times 1.5 = 855 \text{ V}$$

Applied voltage, $V_2 = E_{b2} + I_2 R_m = 855 + 30 \times 1 = 885$ V **Ans.**

Example 9.41. A dc series motor drives a load, the torque of which varies as the square of speed. The motor takes a current of 15 A when the speed is 600 rpm . Calculate the speed and current when the motor field winding is shunted by a diverter of the same resistance as that of field winding. Mention the assumptions made if any.

Solution: Speed, $N_1 = 600$

Series field current, $I_{se1} = I_{a1} = 15$ A

Let the speed be N_2 and armature current I_{a2} when the field winding is shunted by a diverter of the same resistance as that of field winding.

$$\text{Series field current, } I_{se2} = \frac{1}{2} I_{a2}$$

$\because$ current I_{a2} divides equally between series field and diverter

Since torque developed, $T \propto \Phi I_a$

$$\therefore \frac{T_2}{T_1} = \frac{\Phi_2 I_{a2}}{\Phi_1 I_{a1}} = \frac{\frac{1}{2} I_{a2}^2}{I_{a1}^2}$$

$\because \Phi \propto I_{se}$ neglecting magnetic saturation

$$\text{or } \frac{T_2}{T_1} = \frac{I_{a2}^2}{2 I_{a1}^2} \qquad ...(i)$$

$$\text{also } \frac{T_2}{T_1} = \frac{N_2^2}{N_1^2} \qquad ...(ii)$$

Comparing Eqs. (*i*) and (*ii*) we have

$$\frac{N_2^2}{N_1^2} = \frac{I_{a2}^2}{2 I_{a1}^2}$$

$$\text{or } \frac{N_2}{N_1} = \frac{1}{\sqrt{2}} \frac{I_{a2}}{I_{a1}} \qquad ...(iii)$$

Neglecting voltage drop in armature and assuming applied voltage constant

$$N \propto \frac{1}{\Phi}$$

$$\frac{N_2}{N_1} = \frac{\Phi_1}{\Phi_2} = \frac{I_{a1}}{\frac{1}{2} I_{a2}} = \frac{2 I_{a1}}{I_{a2}} \qquad ...(iv)$$

Comparing Eqs. (*iii*) and (*iv*) we have

$$\frac{1}{\sqrt{2}} \frac{I_{a2}}{I_{a1}} = \frac{2 I_{a1}}{I_{a2}}$$

or $I_{a2}^2 = 2\sqrt{2}\, I_{a1}^2 = 2\sqrt{2} \times (15)^2 = 636.396$

or $I_{a2} = 25.227$ A **Ans.**

Substituting $I_{a2} = 25.227$ A in Eq. (*iv*) we have

$$\text{Speed, } N_2 = \frac{2 I_{a1}}{I_{a2}} N_1 = 2 \times \frac{15}{25.227} \times 600 = 713.5 \text{ rpm}$$ **Ans.**

Example 9.42. A 4-pole dc motor runs at 600 rpm on full load and takes 25 A at 450 V. The armature is lap wound with 500 conductors and flux per pole is given by

$$\mathbf{\Phi = 1.7 \times 10^{-2} \times \sqrt{I} \text{ webers}}$$

where I is the motor current. If the supply voltage and torque are both are halved, calculate the speed at which the motor will run. Neglect stray losses.

[*U.P. Technical Univ. Electromechanical Energy Conversion-I, 2010-11*]

Solution: Speed, $N_1 = 600$ rpm

Current, $I_1 = 25$ A

Flux per pole, $\Phi_1 = 1.7 \times 10^{-2} \times I^{0.5}$

$= 1.7 \times 10^{-2} \times 25^{0.5} = 0.085$ Wb

$$\text{Back emf, } E_{b1} = \Phi Z \frac{N}{60} \times \frac{P}{A}$$

$$= 0.085 \times 500 \times \frac{600}{60} \times \frac{4}{4} = 425 \text{ V}$$

$$\text{Armature resistance, } R_a = \frac{V_1 - E_{b1}}{I_1} = \frac{450 - 425}{25} = 1.0\ \Omega$$

Torque developed T $\propto \Phi I$

$$\therefore \frac{T_2}{T_1} = \frac{\Phi_2 I_2}{\Phi_1 I_1} = \frac{1.7\times10^{-2}\, I_2^{0.5}\times I_2}{0.085\times 25} = \frac{I_2^{3/2}}{125}$$

Since $T_2 = \frac{1}{2} T_1$,

$$\frac{I_2^{3/2}}{125} = \frac{T_2}{T_1} = 0.5$$

or $I_2 = (0.5 \times 125)^{2/3} = 15.75$ A **Ans.**

Back emf, $E_{b2} = V_2 - I_2 R_a = \frac{450}{2} - 15.75 \times 1$

$= 209.25$ V $\quad \because V_2 = \frac{V_1}{2}$

$$\frac{N_2}{N_1} = \frac{E_{b2}}{E_{b1}} \times \frac{\Phi_1}{\Phi_2} = \frac{209.25}{425} \times \frac{0.085}{1.7\times10^{-2}\times\sqrt{15.75}} = 0.62$$

Since $N \propto \frac{E_b}{\Phi}$

or speed $N_2 = 0.62\, N_1 = 0.62 \times 600 = 372$ rpm **Ans.**

9.19 OPERATING CHARACTERISTICS OF DC SHUNT MOTORS

1. Speed - Armature Current Characteristic. If applied voltage V is kept constant, the field current will remain constant, hence flux will have maximum value on no load but will slightly decrease due to armature reaction as the load increases but for most purposes the flux is considered to be constant, neglecting armature reaction effect.

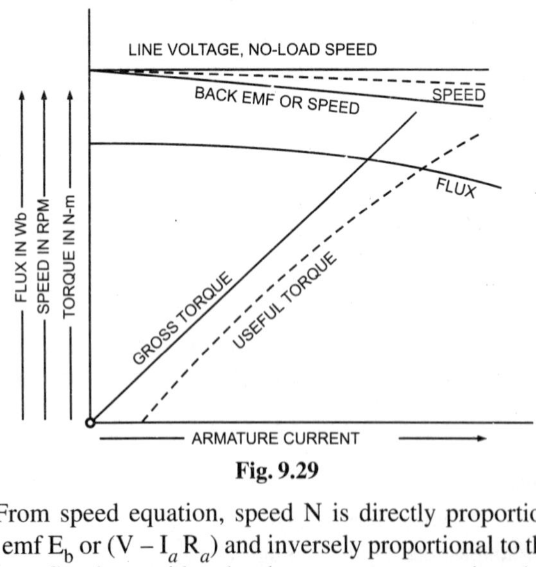

Fig. 9.29

From speed equation, speed N is directly proportional to back emf E_b or $(V - I_a R_a)$ and inversely proportional to the flux Φ. Since flux is considered to be constant as mentioned above, with the increase in armature current the speed slightly falls due to increase in voltage drop in armature and the speed, armature current curve coincides with the back emf-armature current curve. Since voltage drop in armature at full load is very small as compared to applied voltage, drop in speed from no load to full load is very small. If demagnetising effect of armature reaction is considered, the drop in speed due to voltage drop in armature is compensated for up to some extent due to decrease in flux with the increase in armature current and speed-armature current characteristic is less drooping, as shown dotted in Fig. 9.29 or even be rising if demagnetisation is high. Thus the drop in speed from no load to full load is very small and for all practical purposes the dc shunt motor is taken as a constant speed motor.

There is a slight variation in speed of the shunt motor from no load to full load and this slight variation in speed can be made up by inserting resistance in the shunt field and so reducing the flux. Therefore, shunt motors can be used for the loads which are totally and suddenly thrown off without resulting in excessive speed. Shunt motors being constant speed motors are best suited for driving of line shafts, machine lathes, milling machines, conveyors, fans and for all purposes where constant speed is required. It is not suitable for use with flywheel or with fluctuating loads or for parallel operation due to its constant speed characteristic. It is also useful where a moderate degree of speed control is required.

2. Torque-Armature Current Characteristic. From the expression for the torque of a dc motor, torque is directly proportional to the product of flux per pole Φ and armature current I_a. Since in case of dc shunt motor the flux per pole Φ is considered to be constant, torque increases with the increase in load current following linear law, *i.e.* torque-armature current characteristic is a straight line passing through origin O (Fig. 9.29). But the weakening of field due to armature reaction causes the torque line to droop slightly, and the iron and friction losses cause it to be slightly lower than the line representing the electromagnetic developed torque corresponding to

$$T = 0.159\, \Phi Z P \frac{I_a}{A} \text{ Nm.}$$

3. Speed-Torque Characteristic. This characteristic curve can be drawn from the above two characteristics and is shown in Fig. 9.30.

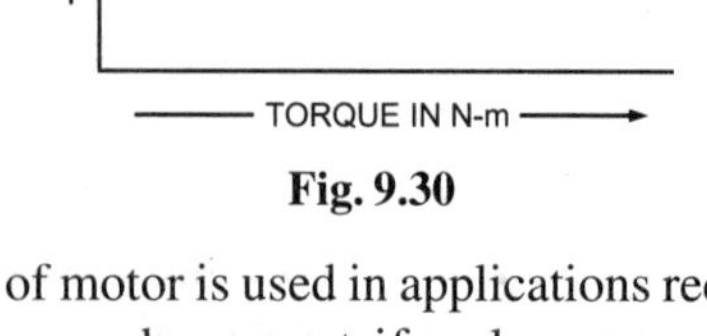

Fig. 9.30

This type of motor is used in applications requiring medium starting torque such as centrifugal pumps, blowers, fans, conveyors, boring mills, shapers, woodworking machines, spinning and weaving machines, printing presses, machine tools etc.

DC shunt motors should never be started on heavy loads because such loads need heavy starting current.

9.19.1. Break in Field Circuit of a DC Motor. It is imperative that the field circuit of a dc motor never be opened while it is running. If the field circuit is opened, the flux will drop to practically zero and, therefore, the speed would increase to an extremely high value. The centrifugal force developed at the high speed may cause the windings to be pulled out of the slots or may cause the commutator segments to buckle, resulting in damage to the motor.

Proper use of overload devices would protect the motor. Back emf would be reduced due to reduction in flux and causes increase in armature current or current drawn from the supply mains by the motor. This excessive current would operate the overload device, disconnect the motor from the supply, and thus protect the motor.

Example 9.43. A 230 V dc shunt motor with constant field drives a load whose torque is proportional to its speed. When running at 750 rpm it takes 30 A. Find the speed at which it will run if a 10 Ω resistance is connected in series with its armature. The resistance of armature may be neglected.

Solution: Armature current, I_a = Line current, $I_L = 30$ A

neglecting shunt field current

Speed, $N_1 = 750$ rpm

Let the speed be reduced to N_2 rpm where $N_2 = KN_1$ after inserting additional resistance of 10 Ω in the armature circuit.

Since with constant field excitation, torque developed is proportional to armature current i.e.

$$T \propto I_a \quad ...(i)$$

Load torque, $T \propto N$ ... (*ii*) given

From Eqs. (*i*) and (*ii*), we have

$$I_a \propto T \propto N$$

$$\text{or } I_{a2} = I_{a1} \times \left(\frac{N_2}{N_1}\right) = K I_{a_1} = 30\text{ K} \quad ...(iii)$$

Since $N \propto [V - I_a(R_a + R)]$, flux Φ remaining the same

$$\frac{N_2}{N_1} = \frac{V - I_{a2}R}{V}$$

∵ armature resistance is negligible

Substituting $\frac{N_2}{N_1} = K$, $I_{a2} = 30$ K from Eq. (*iii*), and R = 10 Ω in above equation, we have

$$K = \frac{230 - 30\text{K} \times 10}{230}$$

$$\text{or } K = \frac{230}{230 + 300} = 0.434$$

New speed, $N_2 = KN_1 = 0.434 \times 750 = 325$ rpm. **Ans.**

9.20 OPERATING CHARACTERISTICS OF SEPARATELY-EXCITED DC MOTORS

The torque-armature current, speed-armature current and speed-torque characteristics of separately-excited dc motors are similar to those of dc shunt motors.

9.21 OPERATING CHARACTERISTICS OF COMPOUND WOUND DC MOTORS

As explained in Art 9.7.5, the compound wound dc motors are of two types viz cumulative compound wound motors and differential compound wound motors.

9.21.1. Operating Characteristics of Cumulative Compound Wound DC Motors. The characteristics of cumulative compound wound motors are combination of shunt and series characteristics and lie between those of shunt and series motors.

As the load is increased, the flux due to series field winding increases and causes the torque greater than it would have been with shunt field winding alone for a given machine and for a given current. The increase in flux due to series field winding on account of increase in load cause the speed to fall more rapidly than it would have done in shunt motor. The characteristics are shown in Fig. 9.31.

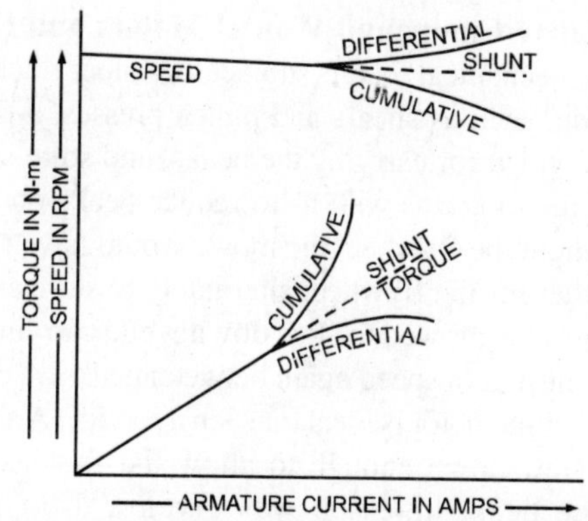

Fig. 9.31 *Characteristics of Compound Wound Motors*

Like the series motor, the cumulative compound wound motor has excellent characteristics for starting heavy loads and yet there is no danger of "running away" at light loads when used without a speed controller.

By varying the proportions of shunt field mmf and series field mmf, compound wound motors may be designed to give speed and torque characteristics anywhere intermediate between those for shunt and series motors. In many cases only enough shunt field is provided to guarantee a safe no-load speed, in which case the motor be called a *stabilised series motor.* Compound wound motors are usually designed so that there will be a drop in speed of about 20% between no load to full load.

Cumulative compound wound motors are used in driving machines which are subject to sudden applications of heavy loads, such as occur in rolling mills, punching and shearing machines, lifts, mine-hoist etc. This type of motor is used also where a large starting torque is required but series motor cannot be employed conveniently such as in cranes and elevators.

Another advantage of the cumulative compound wound motor for suddenly applied loads is that the motor automatically undergoes a substantial drop in speed when the load is applied. Accordingly much of its stored kinetic energy becomes available for supplying a part of the increased load thus reducing the electrical load on the motor as well as the peaks on the power system. The available kinetic energy is frequently increased by the use of a flywheel, particularly with rolling mill motors.

Cumulative compound wound motors are not practical for applications requiring adjustable speed for field control. With a weakened shunt field, the series field becomes dominating hence the changes in load may produce unstable speeds.

Stabilized Shunt Motor. Large dc shunt motors operating at high speeds are subject to large fluctuations in armature current as the line voltage fluctuates, especially if the loads driven by them have high moments of inertia. A sudden drop in line voltage may easily cause the armature current to drop to zero or even to become negative momentarily. Equally violent surges of armature current are caused by sudden increase in line voltage. These fluctuations can be greatly reduced by adding a small amount of series field. Relatively few simple shunt motors, therefore, are now made, and the majority of so-called shunt motors are provided with a relatively weak *series stability winding* connected so as to assist the shunt winding. Such motors are referred to as stabilized shunt motors but they are virtually *cumulative compound motors*.

Cumulative Compound Wound Motors and Flywheels. Whenever a mechanical load is subject to violent fluctuations of short duration, such as shears and punch presses, a flywheel is generally provided for carrying the peaks, and so to smooth out the load on the motor as well as to reduce peaks on the power system. Without the flywheel the motor would have to be much larger. In order for the flywheel alternately to store and give up energy, it must be allowed to slow down while carrying the peak load, and it must gain speed again between peaks. A cumulative compound wound motor is ideal for such a service. A shunt motor would not slow down enough to allow the flywheel to carry much of the peak, unless it was given a drooping speed characteristic by inserting a resistance in series with its armature, and that would involve a substantial I^2R loss.

9.21.2. Operating Characteristics of Differential Compound Wound DC Motors. In differential compound wound motor, since the series field opposes the shunt field, the resultant flux decreases with the increase in load; thus the machine runs at a higher speed than it would do as a shunt motor. The decrease in flux with the increase in load causes the torque to be less than that of a shunt motor as illustrated in Fig. 9.31. Such motors are rarely employed in practice as the differential arrangement causes difficulties during overloads and starting.

Specially designed motors, with weak armature reaction, large air gaps and little magnetic saturation, have been used for constant speed work. Such differential arrangement of the windings is unsafe unless the total weakening of the field due to all causes merely corresponds to the drop in the back emf. For such purposes only a relatively weak series winding is necessary. In this connection if a level compound wound generator is employed as a motor, the connections remaining unaltered, differential action between the shunt and series field will result. For the same polarity of the brushes, the direction of current through shunt field winding will be same for generator and motor operation, but the direction of current through the armature and series field winding will be reversed when the change is made from generator to motor operation.

The differential arrangement of the windings is a disadvantage in very heavy overload as the field may be weakened too much and the motor may tend to race, taking a very large current. This may ultimately result in the series field winding overpowering the shunt causing the motor to reverse. In starting too, if the starter handle be moved too quickly so as to permit the current in the armature, and series field to rise faster than that in high-inductance shunt field, the series field coils will overpower the shunt and the motor starts up in reverse direction. However, this problem can be avoided by keeping the series field shorted during starting.

With a motor intended for constant speed work, however, the series field winding is normally so weak that troubles are not likely to occur.

9.21.3. Speed-Torque Characteristics of Different Types of DC Motors. Speed-torque characteristics of dc series, shunt, separately excited, cumulatively compounded and differentially compounded motors are shown in Fig. 9.32.

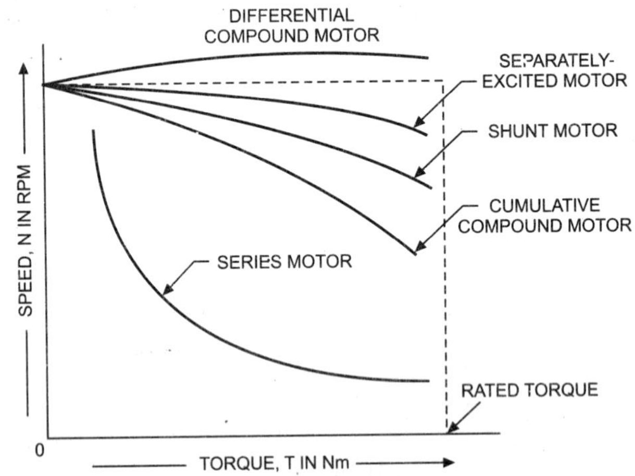

Fig. 9.32 *Speed-Torque Characteristics of Different-Types of DC Motors.*

9.22 SPEED-VOLTAGE CHARACTERISTICS OF DC MOTORS

The speed of a dc motor, according to speed equation, varies as directly proportional to back emf, and inversely proportional to flux per pole, if the armature reaction effect is neglected. If the voltage drop in armature, being very small in comparison to applied voltage, is also neglected then the speed of a dc motor becomes proportional to applied voltage and inversely proportional to flux per pole.

For a dc motor having constant excitation, the speed will vary in direct proportion to the applied voltage to the armature of the motor. This principle is utilized in the Ward-Leonard method of speed control.*

For a dc series motor, the field strength is governed entirely by the magnitude of the load current, therefore, for a given load current the speed of the motor will vary in direct proportion to the applied voltage if voltage drop in armature and series field is negligible.

In case of a dc shunt motor, however, the shunt field current is equally affected by variations in applied voltage to the armature. If the magnetic saturation and armature voltage drop are neglected, the speed of the motor will remain unaffected by variations in applied voltage. As the magnetic circuit approaches saturation, however, the motor speed will tend to vary in direct proportion to applied voltage.

9.23 PERFORMANCE CURVES OF DC MOTORS

9.23.1. Performance Curves of a DC Shunt Motor. In Fig. 9.33, four important characteristics of a dc shunt motor, namely, torque,

* Refer to Art. 10.3.2 (4).

speed, armature current and efficiency, each plotted against the useful output power, are shown. These curves are also known as *performance curves* of a motor.

From the speed-useful output power curve it is seen that the shunt motor has a definite speed at no load. Hence it does not 'run away' when the load is suddenly thrown off provided that the field circuit remains closed. The fall in speed from no load to full load is small, hence, these motors are used where a substantially constant speed is required as in machine shop drives.

From the efficiency-output power curve it is noted that the efficiency of a shunt motor increases rapidly in the beginning, reaches its maximum value and then falls. In shunt wound motors the field, friction and iron losses are approximately constant but at light loads their percentage is high as compared with the load. The armature copper loss varies as the square of the current so this loss, termed as variable loss, increases rapidly with the increase in load. The efficiency of the shunt machine becomes maximum when variable loss is equal to the constant loss.

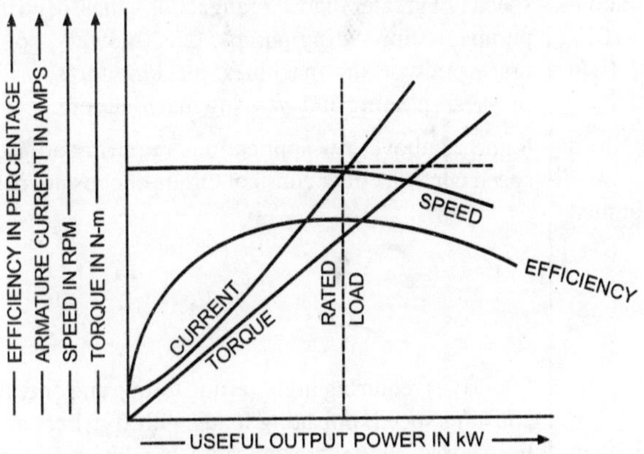

Fig. 9.33 *Performance Curves of a DC Shunt Motor*

The shape of the efficiency curve and the point of maximum efficiency can be varied considerably by the design features of the motor. However, it is advantageous to have an efficiency curve which is fairly flat between 50% and 125% of full load so that there is little change in efficiency and to have the point of maximum efficiency, at about 80 per cent of full load.

From armature current-output power curve it is noted that a certain value of current is required even when the output is zero. The motor input power under no-load condition is consumed in meeting the various losses occuring within the machine.

A shunt wound motor is said to have a low starting torque as compared to other motors but it does not mean that a shunt motor is incapable of starting a heavy load. A shunt motor is capable of starting heavy load also but it would require more excess of current input over normal values than the series and compound motors. For instance, if twice full-load torque is required at start the shunt motor will draw twice full-load current (since $T \alpha I_a$) but a series motor would draw only approximately 1.414 times the full load current (Since $T \alpha I_a^2$). Due to this reason, shunt motors are always started at no load or light load and then the load is applied. When a motor is required to start heavy loads series or compound motor is used.

9.23.2. Performance Curves of a DC Series Motor. In Fig. 9.34, four important characteristics of a dc series motor, namely torque, speed, current and efficiency, each plotted against useful output power, are shown.

From the performance curves for a dc series wound motor it is noted that;

(*i*) The speed of series motor falls rapidly with the increase in load, so a series motor is not suitable for services requiring a substantially constant speed.

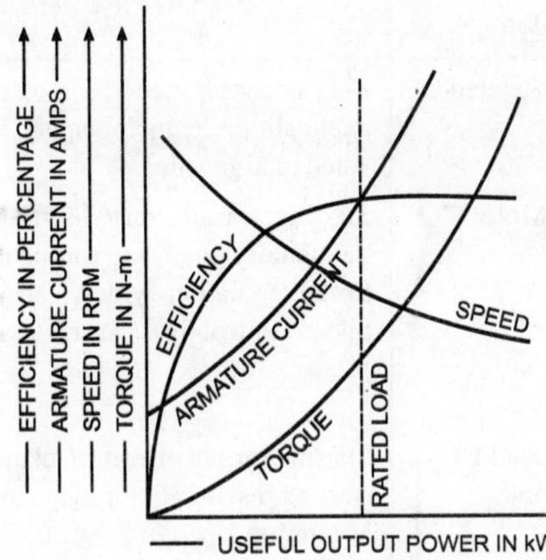

Fig. 9.34 *Performance Curves of a DC Series Motor*

(*ii*) The efficiency increases rapidly in the beginning, reaches its maximum value and then decreases. This is due to the fact that at light loads the friction and iron losses are large compared with the load and the effect to these losses becomes less with the increase in load. The armature and field copper loss varies as the square of the current, so these losses increase rapidly with the increase in load. The efficiency becomes maximum when friction and iron losses are practically equal to the copper loss.

(*iii*) Series motor develops a starting torque comparatively greater than that developed by a shunt motor for a given current. Hence series motors are best suited where huge starting torque is required *i.e.* for streetcars, cranes, hoists and locomotives.

(*iv*) In addition to the large starting torque, there is another unique characteristic of series motors which makes them best suited for traction service. For instance let the streetcar ascend a gradient. In case the shunt wound motor is used for driving the above streetcar, it will maintain the speed of the car at approximately the same value that it had on the level ground and, therefore, will tend to draw an excessive current. A series motor, on the other hand, automatically slows down on such a gradient because of increased current demand and, therefore, it develops more torque at reduced speed. This drop in speed allows the motor to develop a large starting torque with but a moderate increase of power. Hence under the same load conditions, the rating of a series motor would be less than for a shunt motor.

9.24 SPEED AND COST OF A MOTOR

The power that is continuously available from a dc shunt motor without overheating is directly proportional to the speed in case of speed control by varying the voltage applied to the armature. For example, a motor of rating 100 kW at 1,200 rpm will automatically deliver 10 kW if it is operated at 120 rpm. Thus low-speed motors are much more expensive than high- speed motors of the same kW output rating. The use of speed-reduction gears enables high-speed motors to drive low-speed loads and sometimes substantial capital economies may be achieved in this way.

Manufacturers of electric motors build an extensive line of gear motors, in which the motor and the reduction gears are assembled as a single unit in a common housing.

9.25 SUMMARY OF CHARACTERISTICS AND APPLICATIONS OF DC MOTORS

Type of Motor	*Characteristics*	*Applications*
1. Separately-Excited DC Motors	Very accurate speeds can be obtained. Best suited for applications requiring speed variation from very low value to high value.	For paper machines, diesel electric propulsion of ships; in steel rolling mills etc.
2. DC Series Motors	Starting torque very high up to 500%. Maximum momentary operating torque up to 400%. Speed regulation–widely variable, very high at no load. Speed control by series resistance.	For drives requiring very high starting torque and where adjustable varying speed is satisfactory. For hoists, cranes, trolley cars, conveyors, electric locomotives etc. Loads must be positively connected, not belted. To prevent overspeed, lightest load should not be much less than 15 to 20% of full-load torque.
3. Constant Speed DC Shunt Motors	Starting torque–medium, usually limited to 250% by a starting resistor but may be increased. Maximum operating torque usually limited to about 200% by commutation. Speed regulation – 5 to 10%. Speed control–increase up to speed 200% by field control, decrease by armature voltage control.	Essentially for constant-speed applications requiring medium starting torque. May be used for adjustable speed not greater than 2:1 range. For lathes, centrifugal pumps, reciprocating pumps, fans, blowers, conveyors, woodworking machines, machine tools, printing presses, spinning and weaving machines etc.
4. Adjustable Speed DC Shunt Motors	Starting torque–medium, usually limited to 250% by a starting resistance but may be increased. Maximum momentary operating torque–usually limited to about 200% by commutation. Speed regulation–10 to 15%. Speed control–6:1 range by field control, lowered below normal speed by armature voltage control.	Same as above, for applications requiring adjustable speed control, either constant torque or constant output.
5. Cumulative Compound Wound DC Motors	Starting torque–high, up to 450% depending upon the degree of compounding. Maximum momentary operating torque–higher than shunt, up to 350%. Speed regulation–varying depending upon degree of compounding–up to 25 to 30%. Speed control–usually not used but may be up to 125% by field control.	For drives requiring high starting torque and only fairly constant speed, pulsating loads with flywheel action. For shears, conveyors, crushers, bending rolls, punch presses, hoists, elevators, heavy planers, ice making machines, air compressors, rolling mills, printing presses etc.
6. Differential Compound Wound DC Motors With Relatively Weak Series Field	Almost constant torque and constant speed. Tendency towards speed instability with a possibility of motor running away and strong possibility of motor starting in wrong direction.	For experimental and research work.

9.26 SELECTION OF ELECTRIC MOTORS

The proper choice of motor power for a drive is an important condition for economical, efficient and reliable operation of a machine. Installing a motor with a higher rating than is necessary for a given drive leads to extra energy loss during operation of the machine and will occur unnecessary capital expenditure. On the other hand, the installation of a motor of insufficient power reduces the efficiency of the working machine and makes it unreliable; also, the electric motor itself may easily get damaged.

A motor should be chosen for maximum power utilization. During operation it should heat up to the maximum permissible temperature, but should not overheat. In addition, the motor should operate normally under possible temporary overloads and develop the starting torque required by the given working machine. Accordingly, the motor is usually chosen on the basis of heating conditions.

For selection of reliable and efficient motor it is essential that the conditions of service are well known. It is not sufficient to simply specify the output power in kW and the speed but it is also necessary to know the following additional particulars :

(*i*) Torque at the shaft during running, starting and at different loads.

(*ii*) Accelerating torque and braking torque.

(*iii*) Switching frequency.

(*iv*) Efficiency of motor at different loads.

(*v*) Other working requirements.

In studying the behaviour of a motor selected for a particular driven unit, one of the first problems involved is to determine whether the speed-torque characteristic of the motor suits the requirements imposed by the speed-torque characteristic of the driven unit. Drive behaviour during the transient period of a start-up, braking, or speed change-over also depends upon how the speed-torque characteristics of the motor and the driven unit vary with speed. It is, therefore, imperative to study these characteristics in order to be able to select correctly the motor and obtain an economical drive.

HIGHLIGHTS

1. As regards fundamental principles, the dc motors are identical with the dc generators which have the same type of excitation *i.e.*, a machine that operates as a motor will also operate satisfactorily as a generator.
2. An emf is generated in both generator and motor, therefore, there is a generator action in both motor and generator operations. However, in generator operation, the generated emf produces the armature current, whereas, in motor operation the generated emf opposes the conduction. We also observe that electromagnetic torque is developed in generator as well as motor *i.e.* there is a motor action in both generator and motor operations. However, in motor operation the electromagnetic torque developed causes the armature rotation, whereas in a generator operation the electromagnetic torque produced opposes the armature rotation.
3. When a dc motor is connected to dc supply mains, the armature continues to rotate due to motor action, the armature conductors cut the magnetic flux and, therefore, emfs are induced in them. The direction of this induced emf, known as back emf or counter emf, is such that it opposes the applied voltage.

   Back emf in a dc motor, $E_b = \frac{\Phi Z N}{60} \times \frac{P}{A}$ volts
4. Armature current of dc motor is given as
$$I_a = \frac{V - E_b}{R_a}$$
5. Voltage equation of the motor is $V = E_b + I_a R_a$ + brush drop
6. Power balance equation is given as
$$T_e \omega_m = EI \text{ watts}$$
7. Power equation of a dc motor is given as
$$VI_a = E_b I_a + I_a^2 R_a \text{ in case of dc shunt motor}$$
$$\text{and } VI_a = E_b I_a + I_a^2 (R_a + R_{se}) \text{ in case of dc series motor}$$
8. Condition for maximum power developed in armature of a dc motor is
$$E_b = V/2.$$
9. Similar to dc generators, the dc motors can be classified as (*i*) permanent magnet (*ii*) separately-excited (*iii*) series wound (*iv*) shunt wound and (*v*) compound wound dc motors.
10. Direction of rotation of a dc motor can be reversed by reversing the current through either the armature winding or field coils not in both.
11. From emf equation
$$E_b \ \alpha \ \Phi N$$
$$\text{or } N \ \alpha \ \frac{E_b}{\Phi} = K \frac{V - I_a R_a}{\Phi}$$
12. Percentage speed regulation $= \frac{N_0 - N_f}{N_f} \times 100$
13. Torque developed by armature in a dc motor is given as
$$T_a = 0.159 \ \Phi Z P \times \frac{I_a}{A} \text{ newton-metres}$$
$$\text{Also } T_a = 9.55 \frac{E_b I_a}{N} \text{ N-m}$$
   Shaft or useful torque is given as
$$T_{sh} = \frac{\text{Output in watts}}{2\pi N/60} \text{ N-m}$$
14. In a generator the effect of armature reaction was to reduce the generated voltage due to reduction in resultant flux but in a motor the reduction of the resultant flux causes an increase in speed.
15. Among dc motors, the outstanding characteristics are as follows:

    The series motor operates with a decidedly drooping speed with the increase in load on it, the no-load speed usually being prohibitively high; the torque is proportional to almost the square of the current at low saturations and to some power between 1 and 2 as saturation increases. The shunt motor at constant field current operates at a slightly drooping but almost constant speed irrespective of load, the torque being almost proportional to armature current; equally important, however, it is the fact that its speed can be controlled over wide ranges by shunt-field control, armature-voltage control, or a combination of both. Depending on the relative strengths of shunt and series fields, the cumulatively compounded motor is intermediate between the other two and may be given essentially the advantages of one or the other.
16. In a dc series motor, for a given load current the speed of the motor will almost vary in direct proportion to the applied voltage.

    The speed of a dc shunt motor remains unaffected due to variations in supply voltage provided that the magnetic staturation and armature voltage drop are negligible.

EXERCISES

1. Explain construction and working principle of dc motor. **[Rajasthan Technical Univ., February-2010]**
2. Explain the principle of operation of dc motors. **[J.N. Technological Univ. Electrical Machines-I, March-2014]**
3. Explain the motoring and generating actions of a dc machine. **[U.P. Technical Univ., September-2001]**
4. Prove that in a dc generator, generated emf and current in a conductor are in the same direction, whereas in a dc motor generated emf opposes the flow of current in a conductor. **[J.N. Technological Univ. Electrical Machines-I, March-2014]**
5. Explain what is meant by back emf? Explain the principle of torque production in a dc motor. **[R.G. Technical Univ. Electrical Machine-II, December-2012]**
6. What is back emf? Is the back emf greater or lesser than the applied voltage? Why? By what amount do the two voltages differ? Write voltage equation of a motor.
7. What do you mean by back emf in dc motors ? Explain the significance of back emf. Also derive the condition for maximum power in dc motors. **[M.D. Univ. Electromechanical Energy Conversion, December-2010]**
8. Explain how the back emf of a motor causes the development of mechanical power. **[Rajasthan Technical Univ. Electrical Machines-I, 2011]**
9. Describe permanent-magnet dc motors. Discuss their advantages and limitations. Where are these motors used in industry?

   Give the equivalent circuit of a PMDC motor. How does it differ from the equivalent circuit of a conventional dc motor?
10. Derive an expression for the speed of a dc motor in terms of back emf and flux per pole.
11. Define speed regulation and percentage speed regulation of a dc motor.
12. Derive an expression of back emf and torque in dc machines. Also mention the significances of back emf in dc machines. **[U.P. Technical Univ. Electromechanical Energy Conversion-I, 2010-11]**

13. Define torque. Derive the expression for torque developed by a dc motor from fundamentals.

[J.N. Technological Univ. Electrical Machines-I, March-2014]

14. Prove that the armature torque of a dc motor is given by

$$T = \frac{PZ}{2\pi A} \Phi I_a$$

[R.G. Technical Univ. Electromechanical Energy Conversion, June-2010]

15. Write short note on armature reaction in brief.

[M.D. Univ. Electrical Machines-I, December-2012]

16. Explain the armature reaction in a dc motor, indicating the remedies to its adverse effects.

[J.N. Technological Univ. Electrical Machines-I, March-2014]

17. Draw the following characteristics of dc series motors:

(*i*) armature current vs speed and (*ii*) torque vs speed

Also write applications of these motors.

[U.P. Technical Univ. Electromechanical Energy Conversion-I, 2005-06]

18. Explain speed-current, torque-current, speed-torque characteristics of a dc series motor.

[J.N. Technological Univ. Electrical Machines-I, March-2014]

19. Derive the torque equation of a dc series motor. Sketch the speed-torque characteristics of a dc series motor after deriving the necessary relation. [W.B. Univ. of Technology Electrical Machine-I, 2011]

20. Justify that the dc series motor has high starting torque.

[R.G. Technical Univ. Electromechanical Energy Conversion, December-2012]

21. Explain the speed-current, torque-current and speed-torque characteristics of a dc shunt motor.

[U.P.S.C. I.E.S. Elec. Engineering II, 1994]

22. Explain why the speed-torque characteristic of a dc shunt motor is nonlinear in high torque region.

[Rajasthan Technical Univ. Electrical Machines-I, 2011]

23. Derive an expression of torque developed in a dc motor.

Derive necessary expression to draw the speed-torque characteristics of dc series and shunt motor.

Bring out the differences and explain why dc series motor is suitable for traction over shunt motor from the speed-torque characteristics. [W.B. Univ. of Technology Electrical Machines-I, 2010]

24. Compare the speed-current characteristics of various types of dc motors. [U.P.S.C. I.E.S. Electrical Engineering-II, 1997]

25. Derive the expression for the torque developed in a dc motor. Draw and explain the speed-torque characteristic separately in case of shunt and series motors.

[B.P. Univ. of Technology Orissa Electrical Machine-I, 2008]

26. Draw N–I_a and T–I_a curves for DC series and DC shunt motors. From these curves, obtain the N–T curves for both the machines.

[U.P. Technical Univ. Electromechanical Energy Conversion-I, 2009-10]

27. Derive the speed-torque characteristics of dc shunt, series and cumulative compound motors. Sketch these characteristics in one figure on the assumption of same speed at no load.

[G.G.S.I.P. Univ. Electrical Machines, December-2011]

28. Draw the speed-torque characteristics of dc shunt, series and compound motors in same diagram and compare them. Which of the characteristic is more suitable for traction purpose and why?

[G.B. Technical Univ. Electromechanical Energy Conversion-I, 2011-12]

29. Discuss the applications of different types of dc motors.

[J.N. Technological Univ. Electrical Machines-I, March-2014]

30. Explain :

(*i*) A dc shunt motor operates at almost constant speed irrespective of load, why?

(*ii*) A dc series motor develops a high starting torque at starting, why? [U.P. Technical Univ. Electrical Machines, 2005-06]

31. Draw speed-torque characteristics of (*i*) dc shunt motor (*ii*) dc series motor (*iii*) dc cumulatively compounded motor. Mention one industrial application of each of these motors.

[U.P. Technical Univ. Electrical Engineering, February-2001]

32. Which of the following three motors has the poorest speed regulation : Shunt motor, series motor or cumulative compound motor? Explain.

[U.P. Technical Univ. Electrical Machines, 2002-03]

32. List applications of dc shunt, dc series and dc compound motors.

34. A dc compound generator, without any change in machine terminal connections is run as a dc compound motor. Will the nature of compounding change? Explain.

SHORT ANSWER TYPE QUESTIONS WITH ANSWERS

Q. 1. How does a dc motor differ from a dc generator in construction?

Ans. The frame of the generator can, as a rule, be open but those of motors should be either partly or totally enclosed.

Q. 2. What is a commutator in a dc motor?

[R.G.P.V. Bhopal, December-2002]

Ans. The commutator is a form of rotating switch placed between the armature and the supply circuit and so arranged that it will reverse the connections to the supply circuit at the instant of each reversal of current in the armature coils of the motor.

Q. 3. Why is it (commutator) needed? [R.G.P.V. Bhopal, December-2002]

Ans. In case of a dc motor, it is necessary that the current through the coils of the armature winding be reversed as a particular coil leaves one pole (say the N-pole), crosses the neutral line and comes under the influence of next pole which is of opposite polarity (i.e. S-pole). The commutator is required for this purpose.

Q. 4. Explain why the emf generated in the armature of a *dc* motor is called the back emf ? [U.P. Technical Univ. Tutorial Question Bank]

Ans. Since induced emf in a dc motor opposes the voltage applied to the armature of a dc motor, it is called back or counter emf.

Q. 5. Why the torque developed in a dc generator is called the backward torque?

Ans. Since the electromagnetic torque developed in a dc generator opposes the outside driving torque, it is called the backward torque.

Q. 6. When a dc motor produces maximum output power?

Ans. When back emf E_b equals half the applied voltage.

Q. 7. Why the motors are not operated to develop maximum power in practice?

Ans. Because it necessitates a current well beyond the normal operating range. Also an amount of power equal to the mechanical power developed would be wasted in heating the armature, so taking other losses into account, the efficiency would be well below 50%.

Q. 8. What is the significance of back emf in the working of a dc motor. [M.D. Univ. Electrical Technology, December-2011, May-2012, Electromechanical Energy Conversion, December-2012]

Ans. Back emf acts like a governor i.e. it makes the motor self-regulating so that it draws as much current as just required, as explained in Art 9.4

Q. 9. What do you understand by self-excitation mode of dc machine? Name two dc machines working in this mode. **[P.T.U., June-2003]**

Ans. The dc machine in which field winding(s) is/are excited by the current supplied by the machine itself is called the self-excitation mode of the machine. In such machines the field coils are interconnected with the armature winding.

The self-excited machines may be (*i*) dc series machine (*ii*) dc shunt machine.

Q. 10. How can the direction of rotation of a dc motor be determined?

Ans. By applying Fleming's left hand rule.

Q. 11. How can the direction of rotation of a dc shunt motor be reversed?

Ans. Direction of rotation of a dc shunt motor can be reversed by reversing the current flow through either the armature winding or the field winding.

Q. 12. How can the direction of rotation of a dc series motor be changed?

Ans. By reversing the supply terminals to either armature or series field.

Q. 13. A dc shunt motor is running at 1,500 rpm at rated load torque. What happens to the motor operation if the following changes are made:

(*a*) Field winding terminals are reversed.

(*b*) Armature winding terminals are reversed.

(*c*) Supply terminals are reversed.

(*d*) Both field winding and armature winding terminals are reversed. **[P.T.U., June-2003]**

Ans. (*a*) The direction of rotation of the motor will be reversed.

(*b*) The direction of rotation of the motor will be reversed.

(*c*) The motor will keep rotating in the same direction as before the reversal of the line terminals.

(*d*) The direction of rotation of the motor will remain the same.

Q. 14. What is meant by the term speed regulation?

Ans. The speed regulation of a dc motor is defined as the change in speed when the load on the motor is reduced from full load to zero load and is expressed in percentage of rated full-load speed.

Q. 15. Why the shaft torque is always less than that developed by the armature of a motor?

Ans. Because of loss of certain percentage of torque developed by the armature in overcoming the iron, friction and windage losses.

Q. 16. Why the leading horn of pole piece is made longer than the trailing horn and cut farther from the armature surface in case of a dc motor?

Ans. To equalise the distribution of magnetic flux under poles.

Q. 17. Brushes in a dc motor are given a backward shift why?

Ans. In a dc motor, the brushes are given backward shift in order to reduce sparking at the commutator. The backward shift of brushes will induce an emf in the required direction in the short-circuited coils of the armature undergoing commutation and hence improves commutation.

Q. 18. What do you mean by polar axis and brush axis?

Ans. Polar axis is defined as the axis along which the main flux is set up while the axis perpendicular to the polar axis is called the brush axis, along which the brushes are placed.

Q. 19. Why large variable speed dc motors are fitted with compensating windings?

Ans. To reduce sparking in commutation and prevent field distortion owing to armature reaction.

Q. 20. How are interpoles wound in a (*i*) generator (*ii*) motor?

Ans. In case of a motor, interpoles must have polarity of the main poles behind them, while in case of a generator the polarity of the interpoles will be same as that of the main pole just ahead in the direction of rotation.

Q. 21. Out of dc series and dc shunt motor, which motor is called as constant speed motor and why. **[M.D. Univ. December-2010]**

Ans. DC shunt motor is a constant speed motor. From the speed equation, speed N is directly proportional to E_b or $(V - I_aR_a)$ and inversely proportional to flux per pole Φ. In case of a dc shunt motor, the flux per pole Φ is considered to be constant and voltage drop in armature at full load is very small as compared to applied voltage, the drop in speed from no-load to full-load is very small. If the demagnetizing effect of armature reaction is considered, the drop in speed due to voltage drop in armature is compensated up to some extent due to decrease in flux with the increase in armature current. Thus, the dc shunt motor operates at almost constant speed.

But in case of series motor flux per pole Φ varies considerably with the load or armature current.

Q. 22. Why is dc series motor highly variable? **[U.P.S.C. I.E.S. 2004]**

Ans. From the speed equation, it is obvious that speed is proportional to back emf E_b and inversely proportional to flux per pole Φ. With the increase in armature current voltage drop in armature circuit and series field $[I\,(R_a + R_{se})]$ increases and, therefore, back emf E_b decreases, as shown in Fig. 9.26. However, under normal conditions $I\,(R_a + R_{se})$ drop is quite small and may be neglected. Thus if the applied voltage remains constant, speed N is inversely proportional to flux Φ. If a curve is drawn between speed and input (or line) current I, it will be a rectangular hyperbola before magnetic saturation as up to saturation point the magnetisation curve is a straight line. In this region, the speed decreases abruptly with the increase in input current. After magnetic saturation, the flux Φ tends to become constant and speed-current characteristic becomes a straight line and speed decreases slightly due to voltage drop in armature and series field, as shown in Fig. 9.26. The speed becomes zero when the input current is the normal short-circuit current of the motor *i.e.* equal to applied voltage divided by the motor resistance $(R_a + R_{se})$. Normally this current is many many times full-load value.

From speed-current characteristic curve it is obvious that the series motor is a variable speed motor *i.e.* speed varies with the variation in line current. With the decrease in load on a dc series motor, the speed increases and may become dangerously high at very light loads.

Q. 23. Why is the starting torque of a dc series motor more than that of a dc shunt motor of the same rating?

Ans. In case of a dc series motor flux Φ is directly proportional to armature current I_a (before saturation) and, therefore, torque developed $T_e \propto I_a^2$ while in case of a dc shunt motor torque developed $T_e \propto I_a$. Obviously on the basis of the same kW output and speed, a dc series motor develops the starting torque more than that of a shunt motor.

Q. 24. Why dc series motor is never started on no load? **[U.P. Technical Univ. 2013-14]**

Ans. Since on no load the speed is dangerously high, the machine may get damaged due to heavy centrifugal forces set up in the rotating parts. *This is the reason that series motors are never started on no load, which is explained below.*

When the motor is connected across the supply mains without load, it draws a small current from supply mains flowing through the series field and armature, the speed tends to increase so that back emf may approach the applied voltage in magnitude. The increase in back emf weakens the armature current and hence the field current. This causes again increase in speed so in back emf. Thus the field continues to weaken and speed continues to increase until the armature gets damaged.

Q. 25. Why dc series motor is not suitable for belt driven loads? **[Anna Univ. May/June-2012]**

Ans. Because the mishap to the belt would cause the motor to run on no load and on no load the series motor will attain dangerously high speed and the motor may get damaged due to heavy centrifugal forces set up in the rotating parts.

Q. 26. Why series motor is preferred for traction purposes?

Ans. Because dc series motor develops large starting torque and slows down with the increase in load and so gets automatically relieved from heavy excessive load.

Q. 27. Mention with reason the type of dc motor used for loads requiring high starting torque. **[B.P. Univ. of Technology 2008]**

Ans. As explained in Answer to Q. 23, on the basis of the same kW output and speed, a dc series motor develops the torque at start more than that of a shunt motor. So dc series motor is used for loads requiring high starting torque.

Q. 28. What happens if the field winding of a running shunt motor suddenly breaks open? **[M.T.U. First Semester 2011-12]**

Ans. When the field winding of the dc shunt motor is disconnected while in normal operation, the flux will drop to practically zero value, and therefore, the speed will increase to an extremely high value and the motor may get damaged due to heavy centrifugal forces.

Proper use of overload devices should protect the motor. Back emf would be reduced due to reduction in flux and therefore there will be increase in armature current or current drawn from the supply mains by the motor. The excessive current would operate the overload device, disconnect from the supply, and thus protect the motor.

Q. 29. Explain what happens, giving reasons thereof in the following cases:

(*i*) A dc series motor is started at no load.

(*ii*) The field winding of a dc shunt motor gets disconnected while in normal operation.

(*iii*) In a dc shunt motor field terminals are reversed. **[U.P.T.U. 2013-14]**

Ans. (*i*) Refer to Answer of Q. 24.

(*ii*) Refer to Answer of Q. 28.

(*iii*) The direction of rotation of motor will get reversed.

Q. 30. A dc shunt motor runs away at heavy loads. Comment.

Ans. The field may be weakened, due to armature reaction at heavy loads, to the extent that the motor run away.

Q. 31. Why is dc series motor preferred in elevators? **[U.P. Technical Univ. Odd Semester 2013-14]**

Ans. The series motor has a very high starting torque (up to 5 times) of full-load torque. Therefore, dc series motor is preferred in elevators.

Q. 32. For what type of load dc shunt motor is suitable and why?

Ans. DC shunt motors are suitable for loads requiring medium starting torque, approximately constant speed and which are totally and suddenly thrown off because dc shunt motors develop modern starting torque and run almost on constant speed from no load to full load.

Q. 33. On switching a dc motor it is found to rotate in direction opposite to that for which it is designed. Identify the motor.

Ans. Differential compound wound dc motors.

Q. 34. Why cumulative compound wound dc motors are not suitable for applications requiring adjustable speed for field control ?

Ans. Because with a weakened shunt field the series field will overpower the shunt field; hence the changes in load may produce unstable speeds.

Q. 35. What is difference between cumulative compound and differential compound wound motors?

Ans. In cumulative compound wound dc motor, the series field assists the shunt field whereas in differential compound wound motor series field opposes the shunt field.

Q. 36. The speed of a dc shunt motor remains unaffected by variations in applied voltage if magnetic saturation and armature voltage drop are neglected.

Ans. The shunt field current is equally affected by variations in applied voltage and, therefore, upto magnetic saturation, the motor speed remains unaffected due to variations in applied voltage.

Q. 37. Why differentially compounded motor is not used in practice?

Ans. The differential arrangement of the windings is a disadvantage in very heavy overload as the field may be weakened too much and the motor may tend to race, taking a very large current.

Q. 38. A long shunt dc compound motor is cumulatively compounded. Without any change in connection, it is now run as a dc generator. State whether it will now be differentially or cumulatively compounded.

Ans. When a cumulatively compounded long shunt dc motor is run as a generator, the shunt and series field currents produce flux in opposite directions and thus it will operate as a differentially compounded generator.

Q. 39. Identify suitable dc motors for the following applications:

(*a*) Electric traction (*b*) Vacuum cleaners (*c*) Paper making (*d*) Shearing and punching.

Ans. (*a*) DC series motor (*b*) DC series motor (*c*) DC cumulative compound motor (*d*) DC cumulative compound motor.

Q. 40. Why the dc motors are mostly gear motors?

Ans. Because the low-speed motors are much more expensive than high-speed motors of the same kW output rating.

Q. 41. Which of the following three motors has the poorest speed regulation :

Shunt motor, series motor or cumulative compound motor?

Ans. Series motor.

Q. 42. What are the applications of dc series and dc shunt motors? **[U.P. Technical Univ. 2013-14]**

Ans. *DC series motors* are variable speed motors, develop large starting torque but attain dangerously high speed on no load. So these motors cannot be operated on no load. Series motors are best suited for electric traction, elevators, cranes, vacuum cleaners, sewing machines etc.

DC shunt motors are almost constant speed motors, develop medium starting torque and their speeds can be adjusted by using speed control methods. So the shunt motors can be used for the loads which are totally and suddenly thrownoff without resulting in excessive speed. Being constant speed motors, they are best suited for driving of line shafts, machine lathes, milling machines, conveyors, fans and for all purposes where constant speed is required.

PROBLEMS

1. A 25 kW, 250 V dc shunt generator has armature and field resistances of 0.06 Ω and 100 Ω respectively. Determine the total armature power developed when working (i) as generator delivering 25 kW output and (ii) as a motor taking 25 kW input.

[U.P. Technical Univ. Electrical Machines, 2003–04]

[**Ans.** (*i*) 26.26 kW, 23.8 kW]

2. A 4-pole dc shunt motor working on 250 V takes a current of 2A when running at no load at 1,000 rpm. How much back emf is generated ? What will be its back emf, speed and percentage speed drop if the motor takes 51A at a certain load? Armature resistance and shunt field resistance are 0.2 Ω and 250 Ω respectively.

[U.P. Technical Univ. Electrical Machines, 2009-10]

[**Ans.** 249.8 V; 240 V; 961 rpm; 3.9%]

3. A 230 V dc series motor is taking 50 A. Resistances of armature and series field winding are 0.2 Ω and 0.1 Ω respectively. Calculate (*i*) brush voltage (*ii*) back emf.

[U.P. Technical Univ. Even Semester 2013-14]

[**Ans.** (*i*) 225 V (*ii*) 215 V]

4. A 140 V DC shunt motor has an armature resistance of 0.2 ohm and a field resistance 70 ohm. The full-load line current is 40 A and the full-load speed is 1,800 rpm. If the brush contact drop is 3 V, find the speed of the motor at half load.

[G.B. Technical Univ. Electrical Engineering Even Semester 2009-10]

[**Ans.** 1,856 rpm]

5. A dc shunt motor runs at 600 rpm taking 60 A from a 230 V supply. Armature resistance is 0.2 ohm and field resistance is 115 ohms. Find the speed when the current through the armature is 30 A.

[Electrical Engineering U.P. Technical Univ. Even Semester 2007-08; G.B. Technical Univ. Odd Semester 2012-13]

[**Ans.** 616.4 rpm]

6. A 120 V dc shunt motor has an armature resistance of 0.2 Ω and a field resistance of 60 Ω. The full-load line current is 40 A and full-load speed is 1,800 rpm. If the brush contact drop is 3 V, find the speed of the motor at half load.

[U.P. Technical Univ. Electrical Machines and Automatic Control, 2013-14]

[**Ans.** 1,866 rpm]

7. A 250 V dc shunt motor runs at 1,000 rpm and on no-load takes 5 A. Armature and field resistances are 0.2 Ω and 250 Ω respectively. Calculate the speed when loaded and taking current of 50 A. Assume that flux get weakened by 3% due to armature reaction.

[M.D. Univ. Electromechanical Energy Conversion, December-2007]

[**Ans.** 993.7 rpm]

8. A series motor runs at 600 rpm when taking a current of 110 A from a 230 volt supply. The useful flux per pole for 110 A is 24 mWb and that for 50 A is 16 mWb. The armature resistance and series field resistance are 0.12 ohms and 0.03 ohms respectively. Calculate the speed when the current has fallen to 50 A.

[R.G. Technical Univ., June-2014]

[**Ans.** 938 rpm]

9. A dc shunt generator delivers 50 kW at 250 V and 400 rpm. The armature and field resistances are 0.02 Ω and 50 Ω respectively. Calculate the speed of the machine running as shunt motor and taking 50 kW input at 250 V. Allow brush contact drop of 1 V per brush.

[Mahamaya Technical Univ. Electrical Engineering Even Semester 2011-12]

[**Ans.** 381.3 rpm]

10. A dc shunt generator running at 500 rpm delivers 50 kW at 250 V. It has armature resistance of 0.02 Ω and field winding resistance of 50 Ω. Calculate the speed of the machine running as a shunt motor and taking a power of 50 kW at 250 V.

[U.P. Technical Univ. Electrical Machines, 2007-08]

[**Ans.** 484 rpm]

11. A shunt generator delivers 50 kW at 250 V and 400 rpm. The armature and field resistances are 0.02 Ω and 50 W respectively. Calculate the speed of the machine running as shunt motor and taking 50 kW input at 250 V.

[R.G. Technical Univ., June-2013]

[**Ans.** 387.4 rpm]

12. A dc series motor is rated 230 V, 1000 rpm, 80 A (refer to Fig. 1). The series field resistance is 0.11 Ω and the armature resistance is 0.14 Ω. If the flux at an armature current of 20 A is 0.4 times of that under rated condition, calculate the speed at this reduced armature current of 20 A.

[GATE 2001]

[**Ans.** 2,678 rpm]

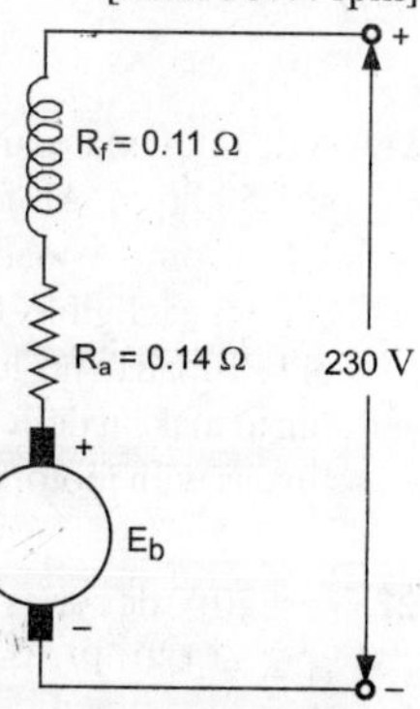

Fig. 1

13. A belt driven dc shunt generator runs at 1,500 rpm delivering 10 kW at 220 V brushes. The belt breaks, following which the machine operates as a motor drawing 2 kW power. What will be its speed as a motor. The armature and field resistances are 0.25 Ω and 55 Ω respectively. Ignore armature reaction and assume the contact drop at each brush to be 1 V.

[GATE 2000]

[**Ans.** 1,387 rpm]

14. A dc shunt generator delivers 60 kW at 240 V and 360 rpm. The armature and field resistances are 0.015 Ω and 60 Ω respectively. Calculate the speed of the machine running as a shunt motor and taking 60 kW input at 240 V. Allow 1 V per brush for contact drop.

[GATE 1998]

[**Ans.** 343 rpm]

15. A dc series motor draws a line current of 100 A from the mains while running at 1,000 rpm. Its armature resistance is 0.15 Ω and the field resistance is 0.1 Ω. Assuming that the flux corresponding to a current of 25 A is 40% of that corresponding to 100 A, find the speed of the motor when it is drawing 25 A from 230 V supply. **[GATE 1996]**

[**Ans.** 2,729 rpm]

16. A 250 V, 4-pole wave wound dc series motor has 782 conductors on its armature. It has armature and series field resistance of 0.75 Ω. The motor takes a current of 40 A. Estimate its speed and gross torque developed if it has a flux per pole of 25 mWb.

[Pune Univ. Elec. Engineering II, 1991]

[**Ans.** 337.6 rpm ; 248.9 N-m.]

17. A 10 hp, 230 V shunt motor has an armature circuit resistance of 0.5 Ω and a field resistance of 115 Ω. At no load and rated voltage the speed is 1,200 rpm and the armature current is 2 A. If the load is applied, the speed drops to 1,100 rpm. Determine the armature current, the line current and torque.

[Pb. Technical Univ. Electrical Engineering, May-2002]

[**Ans.** 40.2 A, 42.2 A, 73.2 N-m]

18. A 5 kW, 230 V, shunt motor has an armature resistance of 0.5 Ω and a field resistance of 230 Ω. At no load the motor runs at a speed of 1,000 rpm and draws a current of 3 *A*. At full load and rated voltage, the current drawn is 23 A and the armature reaction causes a drop of 2% in flux. Determine (*i*) full-load speed (*ii*) full-load torque.

[U.P.S.C. I.E.S. Electrical Engineering-II, 1998]

[**Ans.** (*i*) 976 rpm (*ii*) 47.15 N-m]

19. A 440 V shunt motor takes 105 A as armature current from the supply and runs at 1,000 rpm. The armature resistance is 0.15 Ω. If the total torque developed is unchanged, calculate the speed and armature current, if the magnetic field is reduced to 70% of the initial value.

[Electrical Machines; Rajasthan Univ. 2004; Rajasthan Technical Univ. 2010]

[**Ans.** 1,406 rpm, 150 A]

20. A 220 V dc shunt motor having an armature resistance of 0.25 ohm carries an armature current of 50 A and runs at 600 rpm. If the flux is reduced by 1% by field regulator, find the speed of motor, assuming the load torque to remain the same.

[U.P. Technical Univ. Electrical Machines 2005-06; Electrical Machines and Automatic Control, 2010-11]

[**Ans.** 605.7 rpm]

21. A dc shunt motor draws 10 A at no load from 230 V mains and runs at 1,500 rpm. At full load, armature current is 100 A and speed is 1,470 rpm. Armature resistance of the motor is 0.1 Ω and field current negligible. Find (a) back emf at no load and full load (b) ratio of full-load to no-load torque (c) speed at which armature should be run to make it deliver 100 A at 220 V as generator. Assume the same flux as with motor operation at full load of 100 A. **[Pb. Univ. April-1993]**

[**Ans.** (*a*) 229 V; 220 V; (*b*) 9.8; (*c*) 1,536.8 rpm]

22. A 220 V dc series motor has the following data:

Z = 180, P/A = 1, Flux/pole = 3.75 mWb/field ampere, total armature circuit resistance = 1 Ω

The motor is coupled to a centrifugal pump whose load torque is $T_L = 10^{-4} n^2$ Nm, where n is speed in rpm. Calculate the current drawn by the motor and the speed at which it will run.

[U.P. Technical Univ. Electromechanical Energy Conversion-I, 2004-05]

[**Ans.** 23.1 A, 757 rpm]

23. A 200 V shunt motor with a constant main field drives a load, the torque of which varies as the square of the speed. When running at 600 rpm, it takes 30 A. Find the speed at which it will run and the current it will draw, if a 10 ohm resistor is connected in series with armature. Neglect motor losses.

[U.P. Technical Univ. Electrical Machines, 2006-07]

[**Ans.** 329 rpm]

24. The armature winding of a 200-V, 4-pole, series motor is lap-connected. There are 280 slots and each slot has 4 conductors. The current is 45 A and the flux per pole is 18 mWb. The field resistance is 0.3 Ω; the armature resistance 0.5 Ω and the iron and friction losses total 800 W. The pulley diameter is 0.41 m. Find the pull in newton at the rim of the pulley. [**Ans.** 628.3 newtons]

25. A 15 kW, 230 V, 1150 rpm 4-pole dc shunt motor has a total of 882 armature conductors arranged in four parallel paths and yielding an armature circuit resistance of 0.2 Ω. When it delivers rated power at rated speed, the motor draws an armature current of 73 A at a field current of 1.6 A. Calculate the developed torque. Also find new operating speed if the field flux is reduced to 80% of its original value for the same developed torque. [**Ans.** 130.57 Nm, 1,413 rpm]

26. A series motor takes 20 A at 400 V and runs at 250 rpm. The armature resistance is 0.6 Ω and the field resistance 0.4 Ω. The torque required by the device varies as the square of the speed. Find the applied voltage and current to run the device at 350 rpm.

[**Ans.** 772.8 V; 28 A]

❑❑❑

10 CHAPTER

Speed Control, Starting and Braking of DC Motors

INSIDE THIS CHAPTER

10.1 INTRODUCTION

Speed control means intentional change of the drive speed to a value required for performing the specific work process. This concept of speed control or adjustment should not be taken to include the natural change in speed which occurs due to change in the load on the drive shaft. The desired change in speed is accomplished by acting accordingly on the drive motor or on the transmission connecting it to the unit it serves to drive. This may be done manually by the operator or by means of some automatic control device.

Any given piece of industrial equipment may have its speed changed or adjusted mechanically by means of stepped pulleys, sets of change gears, variable speed friction clutch mechanism, and other mechanical devices. Historically, this proved to be the first step in transition from non-adjustable speed to adjustable speed drive. The electrical speed control has many economical as well as engineering advantages over mechanical speed control.

The nature of the speed control requirement for an industrial drive depends upon its type. Some drives may require continuous variation of speed for the whole of the range from zero to full speed, or over a portion of this range; while the others may require two or more fixed speeds. Some machines may require creeping speed for adjusting or setting up the work. Such a speed is of the order of few rpm. For most of the drives, however, a control of speed within the range of ± 20% may be suitable.

Some of the important terms related to speed are defined below.

Constant Torque Drive. It is a drive in which the motor shaft torque remains constant over a given speed range. In such a drive, shaft power varies as the speed.

Constant Power Drive. It is a drive in which the motor shaft power, given by the product of shaft torque and speed, remains constant over a given speed range. In a constant power drive, higher torques are available at lower speeds and vice versa. The motor size is always determined by the highest torque requirement at the lowest speed.

$$\textbf{Percentage speed regulation} = \frac{N_0 - N_f}{N_f} \times 100 \qquad \text{...(10.1)}$$

where N_0 and N_f are no-load full-load speed respectively.

10.1.1. Speed Control of DC Motors. One of the attractive features the dc motor offers over all other types is the relative ease with which speed control can be achieved and, therefore, dc motors are indispensable for many adjustable speed drives. The various schemes available for speed control can be deduced from the expression of speed for a dc motor which is repeated here with one modification

$$N = K\,\frac{V - I_a\,(R+R_a)}{\Phi}$$

The modification involves the inclusion of an external resistance in the armature circuit. The above expression reveals that the speed can be controlled by adjusting any one of the three factors appearing on the right hand side of the expression: (*i*) applied voltage to the armature terminals, V (*ii*) external resistance in the armature circuit, R and (*iii*) flux per pole, Φ. The first two possibilities involve adjustment affecting the armature circuit, whereas the third involves change in the magnetic field. Therefore, speed control methods are broadly classified as *armature control methods* and *field control methods.* Sometimes a combination of the two methods is employed. With armature control the speed decreases as the voltage applied to the armature terminals is reduced, whereas with field control the speed increases as the flux is reduced.

10.2 SPEED CONTROL OF PERMANENT MAGNET DC (PMDC) MOTORS

It is not possible to control the speed of a PMDC motor by field control because the flux in the case of a PMDC motor remains constant. So it is possible to control the speed and torque of a PMDC motor by armature voltage control, armature resistance control or chopper control. These motors are, therefore, employed only where speeds below base speeds are required. It is not possible to operate these motors above the base speed.

10.3 SPEED CONTROL OF DC SHUNT MOTORS

10.3.1. Field Control Methods. It has previously been pointed out that weakening of field causes increase in the speed of a motor, and strengthening of field reduces the speed. Speed adjustment of dc shunt motors by field control may be obtained by any one of the following methods.

(*i*) Field–rheostat control involving the variation of flux by means of a field rheostat.

(*ii*) Reluctance control involving the variation of the reluctance of the magnetic circuit of the motor.

(*iii*) Field–voltage control involving the variation of voltage applied to the field circuit keeping the voltage applied to the armature terminals constant.

1. Field Rheostat Control. In this method, speed variation is accomplished by means of a variable resistance inserted in series with the shunt field, as illustrated in Fig. 10.1. An increase in controlling resistance reduces the field current with a consequent reduction in flux and an increase in speed. This method of speed control is very simple, convenient and most economical and is, therefore, extensively used in modern electric drives. Since controlling resistance has to carry a small current, it is made up of the slide-wire type of resistor to provide continuously variable speed over the range. The power wasted in the controlling resistance is very little as the field current is very small. This method of speed control is independent of load on the motor and permits remote control of speed.

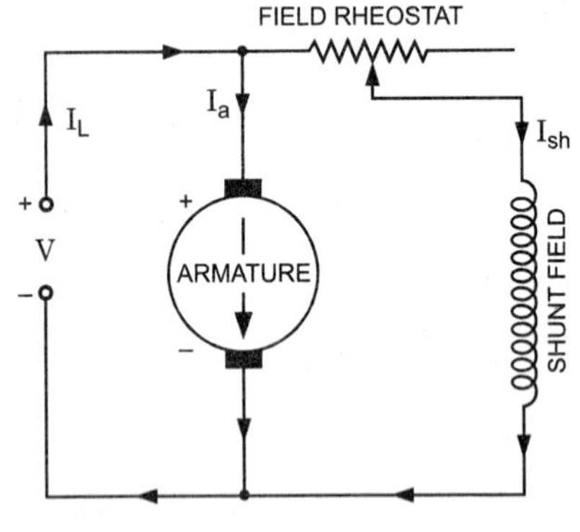

Fig. 10.1 *Field Rheostat Control For a DC Shunt Motor*

The effect of weakening and strengthening of field on the speed of a dc motor has already been discussed in detail in Art. 9.10. The required value of controlling resistance, to provide a desired increase in speed, may be determined from the magnetisation characteristic of the machine. Let initial values of back emf, flux and speed be E_{b1}, Φ_1 and N_1 respectively and let the back emf and flux be E_{b2} and Φ_2 respectively at new speed N_2.

$$\text{Since } N \propto \frac{E_b}{\Phi},$$

$$\frac{N_2}{N_1} = \frac{E_{b2}}{E_{b1}} \times \frac{\Phi_1}{\Phi_2}$$

$$\text{or } \Phi_2 = \frac{E_{b2}}{E_{b1}} \times \frac{N_1}{N_2} \times \Phi_1 \qquad \text{...(10.2)}$$

E_{b1} and E_{b2} can be calculated from the values of supply voltage, armature resistance and armature currents I_{a1} and I_{a2} respectively and, therefore, flux Φ_2 can be determined from above relation. From the magnetisation curve of the machine the value of field current corresponding to the value of Φ_2 may be obtained and from the value of field current, the controlling resistance can be determined.

Since in this method of speed control flux can be only reduced (not increased), the speed only above normal can be obtained. The other limitations and drawbacks of this method of speed control are given below:

1. Creeping speeds cannot be obtained by this method.
2. Top speeds are only obtained at a reduced torque owing to very weak field, so the advantage cannot be taken of the high speeds for increasing the power output of the motor.
3. The speed is maximum at the minimum value of flux, which is governed by the demagnetizing effect of armature reaction on the field as at higher speeds the field is very weak and causes the armature current to increase for the development of certain load torque. This may cause overheating of armature, poor commutation and instability. The high speed limit is also restricted due to mechanical considerations as the centrifugal forces are set up at high speeds.

In non-interpolar machines, the speed can be raised by this method in the ratio of 2 : 1. For overcoming the commutation difficulties interpoles are used and in interpolar machines a ratio of maximum to minimum speeds of 6:1 is fairly common.

To avoid instability caused by relatively powerful effect of armature reaction on high speed, the dc shunt motors are often provided with a relatively weak series field winding, connected so as to assist the shunt field. No danger then arises from momentary heavy loads, as the heavy current through series field winding strengthens the field and tends to reduce the speed.

The speed–torque characteristics obtainable with this method are shown in Fig. 10.2.

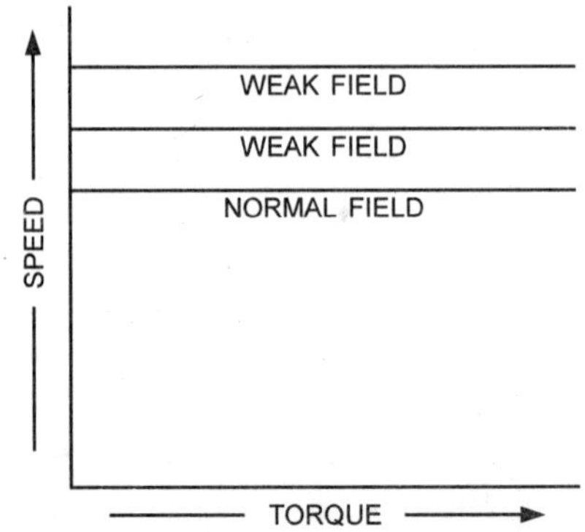

Fig. 10.2 *Speed-Torque Characteristics of DC Shunt Motor With Field Control*

The variable field resistance may be a simple field rheostat or it may be combined with a starting rheostat.

This method of speed control is also employed for dc compound motors, although for any speed setting, there is some variation in the speed with load, in contrast to dc shunt motor, where the speed is but slightly affected by the load.

2. Reluctance Control. This method requires that the motor be constructed with special mechanical features so that the reluctance of the magnetic circuit can be changed. The special features of a motor for reluctance control make the motor more expensive, and, since the same general results can be obtained much more simply by means of a field rheostat, the variable reluctance type of motor is seldom used.

3. Field Voltage Control. This method requires a variable voltage supply for the field circuit which is separate from the main power supply to which the armature is connected. Such a variable supply can be obtained by means of control generator or an adjustable electronic rectifier.

Example 10.1. A 500 V shunt motor runs at its normal speed of 250 rpm when the armature current is 200 A. The resistance of armature is 0.12 Ω. Calculate the speed when a resistance is inserted in the field, reducing the shunt field to 80% of normal value, and the armature current is 100 A.

Solution: Back emf, $E_{b1} = V - I_{a1} R_a = 500 - 200 \times 0.12 = 476$ V

Back emf, $E_{b2} = V - I_{a2} R_a = 500 - 100 \times 0.12 = 488$ V

$\Phi_2 = 80\%$ of $\Phi_1 = 0.8\Phi_1$

Speed, $N_1 = 250$ rpm

$$\text{Speed, } N_2 = N_1 \times \frac{E_{b2}}{E_{b1}} \times \frac{\Phi_1}{\Phi_2}$$

$$= 250 \times \frac{488}{476} \times \frac{\Phi_1}{0.8\Phi_1} = 320.4 \text{ rpm } \textbf{Ans.}$$

Example 10.2. A 250 V dc shunt motor has an armature resistance of 0.5 Ω and a field resistance of 250 Ω. When driving a constant torque load at 600 rpm the motor draws 21 A. What will be the new speed of the motor if an additional 250 Ω resistance is inserted in the field circuit? *[U.P. Technical Univ. Electromechanical Energy Conversion-I, 2007-08; 2010-11]*

Solution: Line voltage, $V = 250$ V

Line current, $I_{L_1} = 21$ A

$$\text{Shunt field current, } I_{sh_1} = \frac{V}{R_{sh}} = \frac{250}{250} = 1 \text{ A}$$

Armature current, $I_{a_1} = I_{L_1} - I_{sh_1} = 21 - 1 = 20$ A

Back emf, $E_{b_1} = V - I_{a_1} R_a = 250 - 20 \times 0.5 = 240$ V

$$\text{New shunt field current, } I_{sh_2} = \frac{V}{R + R_{sh}} = \frac{250}{250 + 250} = 0.5 \text{ A}$$

$$\text{Flux } \Phi_2 = \Phi_1 \times \frac{I_{sh_2}}{I_{sh_1}} = \Phi_1 \times \frac{0.5}{1.0} = 0.5\Phi$$

$\because \Phi \ \alpha \ I_{sh}$

$$\text{Armature current, } I_{a_2} = I_{a_1} \times \frac{\Phi_1}{\Phi_2} = 20 \times \frac{1}{0.5} = 40 \text{ A}$$

$\because T \ \alpha \ \Phi I_a$ is constant

Back emf, $E_{b_2} = V - I_{a_2} R_a = 250 - 40 \times 0.5 = 230$ V

$$\text{New speed, } N_2 = N_1 \times \frac{E_{b_2}}{E_{b_1}} \times \frac{\Phi_1}{\Phi_2}$$

$$= 600 \times \frac{230}{240} \times 2 = 1{,}150 \text{ rpm } \textbf{Ans.}$$

Example 10.3. The armature and field resistances of a 320 V, dc shunt motor are 0.5 Ω and 250 Ω respectively. When driving a load of constant torque at 600 rpm, the armature current is 24 A. If it is required to raise the speed from 600 rpm to 700 rpm, calculate the resistance to be connected in the shunt field circuit. *[J.N. Technological Univ. Electrical Machines-I, March-2014]*

$$\textbf{Solution: } \text{Shunt field current, } I_{sh} = \frac{V}{R_{sh}} = \frac{320}{250} = 1.28 \text{ A}$$

Back emf, $E_{b_1} = V - I_{a_1} R_a = 320 - 24 \times 0.5 = 308$ V

Now the flux and armature current be Φ_2 and I_2 respectively when the speed is to be raised from 600 rpm to 700 rpm.

$$\text{Back emf, } E_{b_2} = V - I_{a_2} R_a = 320 - 0.5 I_{a_2} \quad ...(i)$$

Since load torque is constant,

$$I_{a_2} \Phi_2 = I_{a_1} \Phi_1$$

$$\text{or } I_{a_2} = I_{a_1} \frac{\Phi_1}{\Phi_2} = 24 \frac{\Phi_1}{\Phi_2} \quad ...(ii)$$

$$\text{and } \frac{N_2}{N_1} = \frac{E_{b_2}}{E_{b_1}} \times \frac{\Phi_1}{\Phi_2} \qquad \because N \propto \frac{E_b}{\Phi}$$

$$\text{or } \frac{700}{600} = \frac{320 - 0.5 I_{a_2}}{308} \times \frac{\Phi_1}{\Phi_2}$$

$$\text{or } 359.333 = \left(320 - 0.5 \times 24 \frac{\Phi_1}{\Phi_2}\right) \times \frac{\Phi_1}{\Phi_2}$$

$\because$ From Eq. (*ii*) $I_{a_2} = 24 \frac{\Phi_1}{\Phi_2}$

$$\text{or } 12\left(\frac{\Phi_1}{\Phi_2}\right)^2 - 320\frac{\Phi_1}{\Phi_2} + 359.333 = 0$$

$$\text{or } \frac{\Phi_1}{\Phi_2} = \frac{320 \pm \sqrt{320^2 - 4 \times 12 \times 359.333}}{2 \times 12} = 24.49 \text{ or } 1.1747$$

The figure of 24.49 is not acceptable being impracticable.

Assuming unsaturated magnetic field,

$$\frac{\Phi_1}{\Phi_2} = \frac{I_{sh_1}}{I_{sh_2}} = \frac{V/R_{sh_1}}{V/R_{sh_2}} = \frac{R_{sh_2}}{R_{sh_1}}$$

$$\text{or } R_{sh_2} = R_{sh_1} \times \frac{\Phi_1}{\Phi_2} = 250\ \Omega \times 1.1747 = 293.675\ \Omega$$

Hence resistance to be inserted in the field circuit to raise the motor speed from 600 rpm to 700 rpm,

$$R = R_{sh_2} - R_{sh_1} = 293.675 - 250 = 43.675\ \Omega \ \textbf{Ans.}$$

Example 10.4. Two adjustable speed dc separately-excited motors have maximum speeds of 1,000 and 500 rpm. Speed is controlled by variation of field flux. Motor A drives a load requiring constant developed power over the speed range and motor B drives a load requiring constant torque. All losses, including armature resistance and armature reaction may be neglected and magnetic circuit is assumed linear. If the developed power is equal at 500 rpm and armature currents are each 800 A, what will be the armature currents at 1,000 rpm?

Solution: Motor A. Motor operates at developed constant power and, since all losses are negligible, power input to motor A will also be constant. Assuming supply voltage constant, input current and, therefore, armature current will remain unchanged at 800 A. **Ans.**

Motor B. Armature current, I_{ab} = 800 A

Since motor B operates at constant load torque,

$$\text{Armature current} \propto \frac{1}{\text{Field flux}} \propto N$$

∵ Armature voltage drop is negligible owing to negligible armature resistance

So armature current at, 1,000 rpm,

$$I'_{ab} = I_{aa} \times \frac{N'}{N} = 800 \times \frac{1{,}000}{500} = 1{,}600 \text{ A} \quad \textbf{Ans.}$$

Example 10.5. At 50% of full load, the armature current drawn by a dc shunt motor is 40 A when connected to a 200 V dc mains. By decreasing the field flux, its speed is raised by 20%; this also causes a 10% increase in load torque. Calculate the percentage change in field current. The armature resistance including the brushes is 1 Ω. Neglect saturation and armature reaction. [*GATE 1997*]

Solution: Back emf, $E_{b_1} = V - I_{a_1} R_a = 200 - 40 \times 1.0 = 160$ V

Let the flux and armature current be Φ_2 and I_{a_2} respectively when the speed is raised by 20%.

i.e. $N_2 = 1.2\, N_1$

Since torque increases by 10% i.e.

$$T_2 = 1.1\, T_1$$

$$\text{or} \quad I_{a_2} \Phi_2 = 1.1\, I_{a_1} \Phi_1 = 1.1 \times 40\, \Phi_1 = 44\, \Phi_1$$

$$\text{or} \quad I_{a_2} = 44 \frac{\Phi_1}{\Phi_2}$$

$$\text{Back emf, } E_{b_2} = V - I_{a_2} R_a = 200 - I_{a_2}$$

$$\text{and} \quad \frac{N_2}{N_1} = \frac{E_{b_2}}{E_{b_1}} \times \frac{\Phi_1}{\Phi_2}$$

$$\text{or} \quad 1.2 = \frac{200 - I_{a_2}}{160} \times \frac{\Phi_1}{\Phi_2}$$

$$\text{or} \quad 192 = \left(200 - 44\frac{\Phi_1}{\Phi_2}\right) \times \frac{\Phi_1}{\Phi_2}$$

$$\text{or} \quad 11\left(\frac{\Phi_1}{\Phi_2}\right)^2 - 50\frac{\Phi_1}{\Phi_2} + 48 = 0$$

$$\text{or} \quad \frac{\Phi_1}{\Phi_2} = \frac{50 \pm \sqrt{50^2 - 4 \times 11 \times 48}}{2 \times 11}$$

$$= \frac{50 \pm 19.7}{22} = 3.17 \text{ or } 1.377$$

Since saturation is negligible, field flux is directly proportional to field current i.e.

$$\frac{I_{sh_1}}{I_{sh_2}} = \frac{\Phi_1}{\Phi_2} = 3.17 \text{ or } 1.377$$

Rejecting higher figure which will not give the required increase in speed.

Percentage change in field current

$$= \frac{I_{sh_1} - I_{sh_2}}{I_{sh_1}} \times 100$$

$$= \left(1 - \frac{I_{sh_2}}{I_{sh_1}}\right) \times 100 = \left(1 - \frac{1}{1.377}\right) \times 100 = 27.4\% \quad \textbf{Ans.}$$

Example 10.6. The magnetisation curve for a 4-pole, 220 V shunt motor with 410 wave-connected armature conductors is given below:

Field current (amperes)	**0.5**	**1.0**	**1.5**	**2.0**	**2.5**	**3.0**	**3.5**	**4.0**
Flux per pole (m Wb)	**5.8**	**11.0**	**14.8**	**17.0**	**18.4**	**19.4**	**20.2**	**21.2**

The armature and field resistances of the machine are 1.2 Ω and 60 Ω respectively. With full-load current flowing the machine runs at 750 rpm. Determine the resistance to be inserted in the field circuit to increase the speed from 750 rpm to 1,000 rpm with the torque remaining nearly the same as before. Assume the back emf at higher speed to be the same as before change in speed.

Solution: Shunt field current, $I_{sh1} = \frac{V}{R_{sh1}} = \frac{220}{60} = 3.667$ A

From given magnetisation curve (Fig. 10.3) flux corresponding to field current of 3.667 A

$$\Phi_1 = 20.75 \text{ mWb} = 0.02075 \text{ Wb}$$

$$\text{Speed, } N_1 = 750 \text{ rpm}$$

$$\text{Speed } N_2 = 1{,}000 \text{ rpm}$$

$$\text{Since } N \propto \frac{E_b}{\Phi},$$

$$\frac{N_2}{N_1} = \frac{E_{b2}}{E_{b1}} \times \frac{\Phi_1}{\Phi_2} = \frac{\Phi_1}{\Phi_2} \quad \because E_{b2} = E_{b1} \text{ (given)}$$

$$\text{or } \Phi_2 = \Phi_1 \times \frac{N_1}{N_2} = 0.02075 \times \frac{750}{1{,}000}$$

$$= 0.0156 \text{ Wb} = 15.6 \text{ mWb}$$

From given magnetisation curve field current required to create a flux per pole of 15.6 mWb,

$$I_{sh2} = 1.7 \text{ A}$$

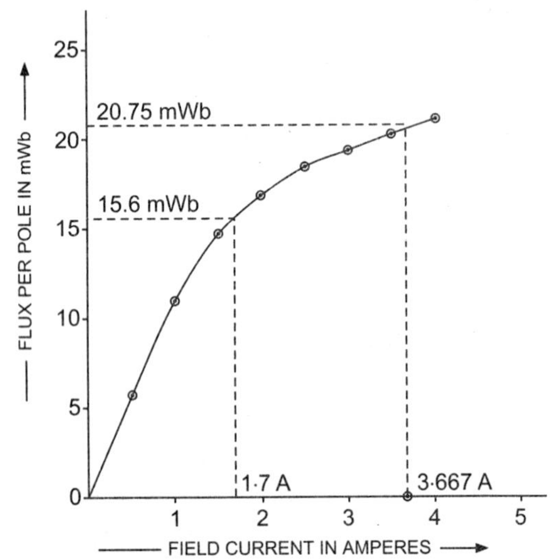

Fig. 10.3

If R is the additional resistance connected in series with the field circuit then

$$R + R_{sh} = \frac{V}{I_{sh2}} = \frac{220}{1.7} = 129.4\ \Omega$$

$$\text{or } R = 129.4 - R_{sh} = 129.4 - 60 = 69.4\ \Omega \quad \textbf{Ans.}$$

10.3.2. Armature Control Methods. Speed variation of shunt motors by armature control requires that voltage applied to the armature terminals shall be changed, without altering the field current. Change in common supply voltage impressed upon both the armature and field will have little effect upon the speed of a shunt motor, unless the voltage is changed by such a large amount that the general performance of the motor would be unsatisfactory.

Speed adjustment of dc shunt motors by armature control may be obtained by any one of the following methods:

1. Armature resistance control involving the variation of voltage applied to the armature terminals by means of a variable resistance connected in series with the armature (Fig. 10.4).
2. Shunted armature control involving the variation of voltage applied to the armature terminals by means of a combination of a variable resistance shunting the armature and a variable resistance in series with the shunted armature (Fig. 10.7).
3. Armature terminal voltage control involving variation of voltage impressed upon the armature circuit.

1. Armature Resistance Control. This method consists simply of a variable resistance connected in series with the armature, as shown in Fig. 10.4. The speed at full load may be reduced to any desired value depending on the amount of resistance. With this method, the voltage across the armature drops as the current passes through the series resistance and the remaining voltage applied to the armature is lower than the line voltage. Thus the speed is reduced in direct proportion to this voltage drop at the armature terminals. Field current will remain unaffected as the shunt field is directly connected across the supply mains. For a constant torque load, the armature current remains the same so input to the motor remains the same but the output decreases in proportion to speed. In case of fans and centrifugal pumps where the load torque decreases with the fall in speed, the losses are considerably low and because of its low initial cost and simplicity this method may be quite convenient and economical for short time or intermittent slowdowns. Wide range of speed (below normal one) can be obtained by this method and at the same time motor will develop any desired torque over its operating range, since the torque depends only upon the armature current, flux remaining unchanged. The main advantage of this method is that speeds below base speed down to creeping speeds of only a few rpm are easily available.

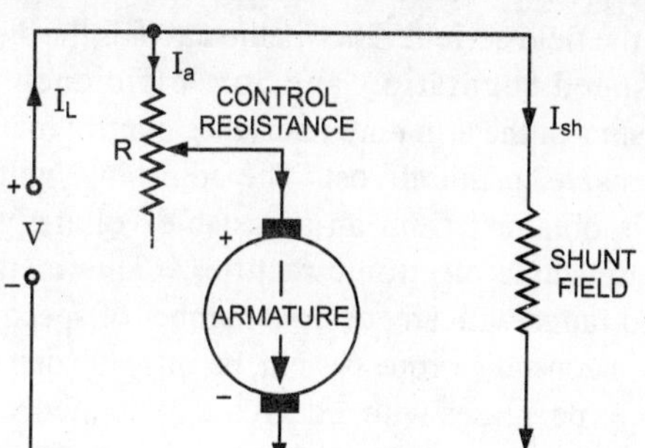

Fig. 10.4 *Armature Resistance Control Method For DC Shunt Motor*

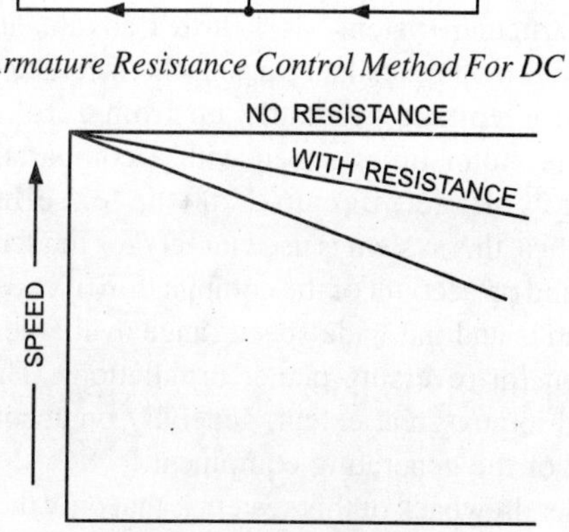

Fig. 10.5 *Speed-Torque Characteristics of DC Shunt Motor With Armature Control*

Let E_{b1} be the back emf at speed N_1, with armature current I_{a1} and no extra resistance in the armature circuit. Let, after insertion of an external resistance R in the armature circuit, the back emf be E_{b2}, speed N_2 and armature current I_{a2} and flux remaining the same.

$$\text{Since } N \propto \frac{E_b}{\Phi}$$

$$\text{or } N \propto E_b \qquad \because \text{ flux } \Phi \text{ is constant}$$

$$\text{or} \quad \frac{N_2}{N_1} = \frac{V - I_{a2}(R_a + R)}{V - I_{a1}R_a} \qquad \text{...(10.3)}$$

Knowing the values of V, I_{a1}, I_{a2}, R_a, N_1 and N_2, the value of additional resistance R to be inserted in series with the armature circuit can be determined. Considering no-load speed, we have

$$\frac{N}{N_0} = \frac{V - I_a(R_a + R)}{V - I_{a0}R}$$

If voltage drop in armature at no load is neglected, the above equation becomes

$$N = N_0\left[1 - \frac{I_a(R_a + R)}{V}\right]$$

From the above equation it is obvious that for a given resistance $(R + R_a)$ in the armature circuit, the speed is a linear function of armature current I_a, as illustrated in Fig. 10.6.

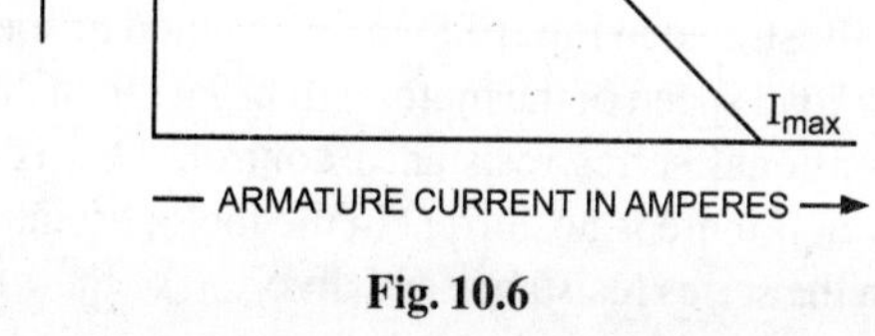

Fig. 10.6

The motor speed will be zero *i.e.* the motor will be stalled when $I_a = \dfrac{V}{R + R_a}$. This is the maximum current and is known as *stalling current.*

The speed-torque characteristics of a dc shunt motor with armature resistance control are shown in Fig. 10.5.

A comparison of speed adjustment of a dc shunt motor by means of field control and by means of resistance in series with the armature shows the superiority of the field control for general use. In field control the adjustment can be obtained by means of a small rheostat and relatively good speed regulation is obtained for all speeds. With the armature control method a bulky rheostat is required, a large amount of power is wasted in the controlling resistance and poor speed regulation results for the lower speeds. The field control method is much more efficient if the rated output is to be delivered by the motor at the different speeds for relatively long periods of time. In armature control method output is reduced in the ratio as the speed. This method of speed control is not desirable for continuous operation.

The armature resistance control method, therefore, is employed where speeds lower than rated one are required for a short period only and that also occasionally as in printing machines, cranes and hoists where the motor is frequently started and stopped. This method of speed control is also employed where the load drops off rapidly with decrease in speed, as in fans and blowers.

2. Shunted Armature Control. Though armature resistance control method is very simple and can be carried out easily but in this method of speed control speed changes with every change in load, because speed variation depends not only upon the controlling resistance but on the load current also. This double dependence makes it impossible to keep the speed sensibly constant on rapidly changing load. A more stable operation can be obtained by using a divertor across the armature in addition to series resistance, as illustrated in Fig. 10.7. Now the changes in armature current (owing to changes in the load torque) will not be so effective in changing the voltage across the armature and hence the speed of motor.

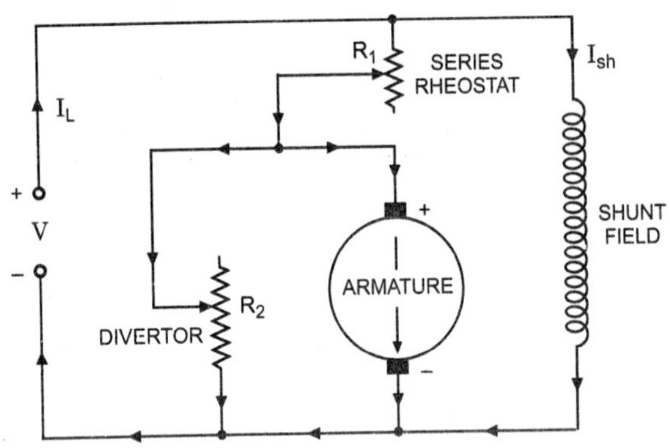

Fig. 10.7 *Shunted Armature Control For DC Shunt Motor*

The speed–torque characteristics for shunted armature control are illustrated in Fig. 10.8. In this method of speed control, the ideal no-load speed of the motor will be less than that obtained with conventional series resistance control. This is due to the fact that even if there is no current in the motor armature, there is a current in the series resistance and divertor, so there is a voltage drop in series resistance and the voltage applied across the armature will be less than supply voltage V, whereas in case of conventional series resistance control this voltage is V.

In Fig. 10.8, natural speed–torque characteristic, speed-torque characteristics with conventional series resistance control and shunted armature control are represented by curves I, II and III respectively.

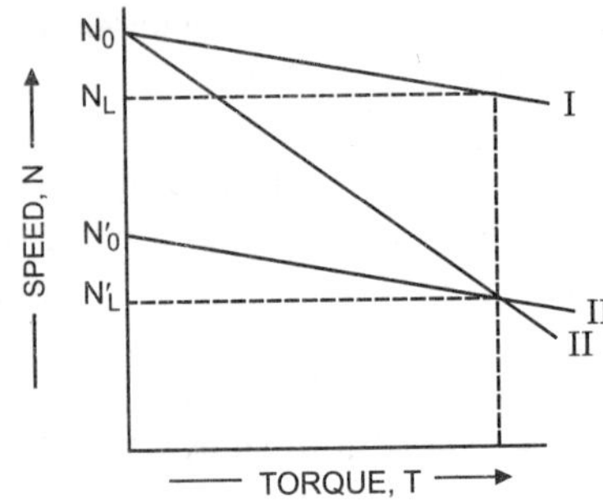

Fig. 10.8

If curves II and III are compared, it is obvious that the shunted armature control has a distinct advantage over the series resistance control. If the speed N'_L is required at a load torque T_L then sufficient resistance can be placed in series with the armature so that the characteristic passes through the required point. However, comparison of this characteristic with that obtained by armature shunting method, reveals that the latter characteristic exhibits greater hardness due to lower slope and will provide more stable operation when the load is liable to fluctuate.

The ideal no-load speed and the hardness of speed-torque characteristic in the shunted armature control depend on the ratio of resistances of divertor and series rheostat. By varying these resistances a suitable characteristic can be had.

The shunted armature control is generally employed in drives of comparatively low power rating. This method is also not economical for continuous operation as there is a relatively large power loss in the series rheostat and divertor. It is usually employed as a means for sharply reducing the speed so as to achieve accurate stopping.

3. Armature Voltage Control. This method of speed control requires a variable source of voltage separate from the source supplying the field current. This method avoids the disadvantages of poor speed regulation and low efficiency which are characteristics of the armature resistance control method but it is more expensive in initial cost. The adjustable voltage for the armature is obtained from an adjustable voltage generator or from an adjustable electronic rectifier.* This method gives a large speed range with any desired number of speed points. It is essentially a constant torque system, because the output delivered by the motor decreases with a decrease in applied voltage and a corresponding decrease in speed.

This particular system has a further advantage that can be employed to provide excellent starting characteristics by bringing the generator voltage gradually up from zero, starting and bringing the motor up to speed with a comparatively slowly increasing voltage. Because of the excellent starting characteristics, this system is used largely for modern high-speed elevators, and on account of the combination of excellent starting characteristics and the wide speed range available, it is used to some extent for reversing planer installations. This method is not applied to any great extent, generally on account of higher initial cost of the generating equipment.

Another drawback of this system is that only the speed below the base speed can be had as the voltage applied to the armature

* For most applications dc motors are fed by rectification of the available ac supply. It is usually necessary to insert a transformer immediately ahead of the rectifier.

can be reduced from its rated value. If the motor is operated with a voltage across armature terminals higher than the rated value for a long duration of time, the armature winding insulation may get damaged.

4. Ward-Leonard Method of Speed Control. The basic adjustable-voltage armature control method of speed control accomplished by means of an adjustable-voltage generator is called the *Ward-Leonard system.* This system consists simply working the motor with a constant excitation and applying a variable voltage to its armature to provide the required speed. The variable voltage supply is obtained from a motor-generator or convertor set. The scheme is illustrated in Fig. 10.9 (*a*). M_1 is the work motor, powered by the generator G, which is driven by a synchronous or induction motor M_2. The excitation current for the work motor M_1 and the generator G is obtained from the exciter E mounted on the same shaft as the generator. The Ward-Leonard set is started by starting the driving motor M_2. The field rheostat R of the generator is gradually brought out of the circuit as the generator picks up the speed and the work motor begins to rotate. The variable voltage across the terminals of the generator or across the motor is obtained by varying the exciting current of the generator G by means of shunt regulator R. The direction of rotation of motor armature can be reversed by reversing the direction of exciting current of the generator G with the help of reversing switch RS. The converter set runs always in the same direction.

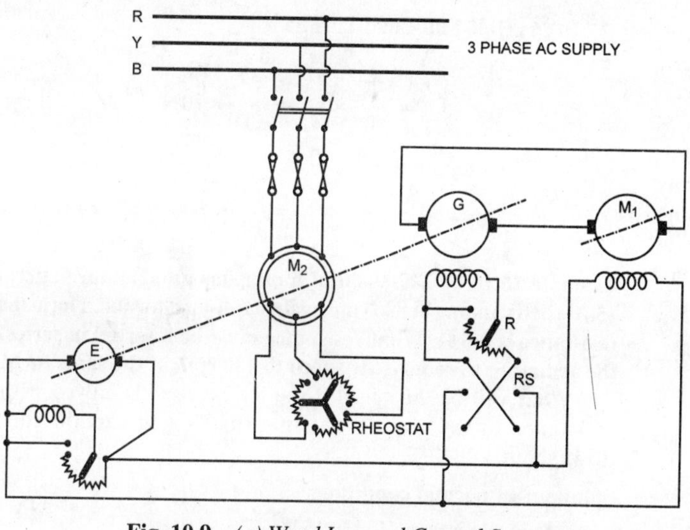

Fig. 10.9 *(a) Ward Leonard Control System*

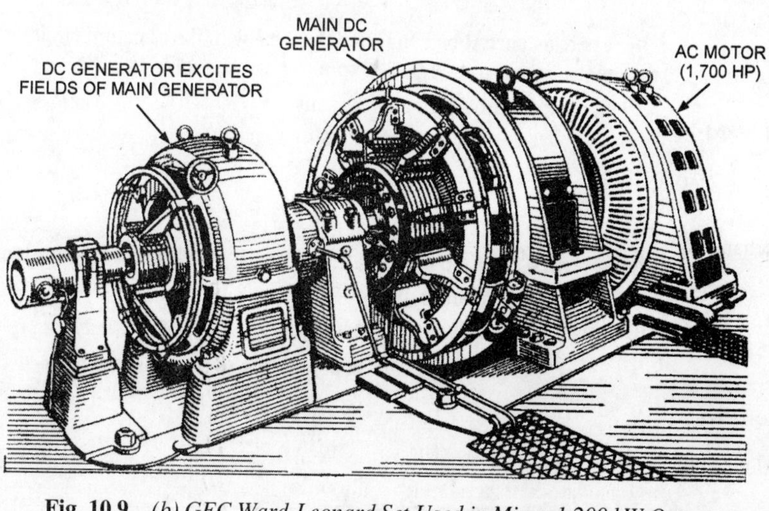

Fig. 10.9 *(b) GEC Ward-Leonard Set Used in Mines 1,200 kW Output Generator Controlling a Large Hoist Motor*

Braking of motor M_1 may be carried out by reducing the generator excitation so that its emf is less than the counter emf of motor M_1. Under these conditions, motor M_1 begins to operate momentarily as a generator, generator G as a motor and ac driving machine M_2 as a generator. As a result kinetic energy of motor M_1 and its load is returned to the supply mains and braking action on the motor M_1 takes place.

Advantages. 1. Very fine speed control over the whole range from zero to normal speed in both directions can be obtained.

2. Uniform acceleration can be obtained.

3. Speed regulation is good.

Disadvantages. 1. Two extra machines are required, so arrangement is costly.

2. Low overall efficiency of the system, especially at light loads.

Applications. This system of speed control is best suited where almost unlimited speed control in either direction of rotation is required such as in steel rolling mills, paper machines, elevators, cranes, mine hoists, diesel-electric propulsion of ships etc.

Ward-Leonard-Ilgner System, which is a modified form of Ward-Leonard system, incorporates a heavy flywheel mounted on the shaft coupling the driving motor M_2 and generator G. In operation with a flywheel, the driving motor has to have a drooping speed–load characteristic *i.e.* its speed must drop with the increase in load on the shaft. The function of the flywheel is to reduce the fluctuations in the power demand from the supply mains as explained below:

An increase in the load on the shaft causes the work motor to draw more current from the generator so more power is required to drive the later and if there were no flywheel, the driving motor would take all the additional power from the supply line, thus causing sharp fluctuations in it. The heavy flywheel, however, stores a large amount of kinetic energy. When an increase in the generator load causes the driving motor to slow down, some of the kinetic energy of the flywheel goes to sustain the peak load on the shaft of the work motor. When the load on the work motor decreases, the driving motor picks up speed and the flywheel stores up kinetic energy.

This system of speed control is employed where the load on the motor shaft sharply varies such as in mine hoists, rolling mills etc.

Example 10.7. A shunt wound motor runs at 500 rpm on a 200 V supply. R_a = 0.5 Ω, the current drawn is 30 A in addition to field current. What resistance must be placed in series with the armature in order that the speed may be reduced to 300 rpm, the current in the armature remaining the same? If the load is changed so that with the inserted resistance, the armature current reduces to 15 A, what will be the speed?

[*U.P.S.C. I.E.S. Electrical Engineering-II, 2004*]

Solution: *With no additional resistance in armature circuit*

Armature current, $I_{a_1} = 30\ A$

$$\text{Back emf, } E_{b_1} = V - I_{a_1}R_a = 200 - 30 \times 0.5 = 185\ V$$

$$\text{Speed, } N_1 = 500 \text{ rpm}$$

With additional resistance (say of R ohms) in armature circuit

Armature current, $I_{a_2} = I_{a_1} = 30\ A$

$$\text{Back emf, } E_{b_2} = V - I_{a_2}(R_a + R)$$
$$= 200 - 30(0.5 + R) = 185 - 30\ R$$

$$\text{Speed, } N_2 = 300 \text{ rpm}$$

Assuming flux to remain constant

$$\frac{E_{b_2}}{E_{b_1}} = \frac{N_2}{N_1}$$

$$\text{or} \quad \frac{185 - 30R}{185} = \frac{300}{500}$$

$$\text{or} \quad R = \frac{185 - 0.6 \times 185}{30} = 2.467\ \Omega \quad \textbf{Ans.}$$

When armature current $I_{a_3} = 15\ A$

$$\text{Back emf, } E_{b_3} = V - I_{a_3}(R + R_a)$$
$$= 200 - 15\ (0.5 + 2.467) = 155.5\ V$$

$$\text{Speed, } N_3 = N_1 \times \frac{E_{b_3}}{E_{b_1}} = 500 \times \frac{155.5}{185} = 420.3 \text{ rpm} \quad \textbf{Ans.}$$

Example 10.8. A 230 V shunt motor is taking a current of 50 A. Resistance of shunt field is 46 Ω and the resistance of the armature is 0.02 Ω. There is a resistance of 0.6 Ω in series with the armature and the speed is 800 rpm. What alteration must be made in the armature circuit to raise the speed to 850 rpm, the torque remaining the same.

Solution: Line current, $I_{L1} = 50\ A$

$$\text{Shunt field current, } I_{sh} = \frac{V}{R_{sh}} = \frac{230}{46} = 5\ A$$

$$\text{Armature current, } I_{a1} = I_L - I_{sh} = 50 - 5 = 45\ A$$

$$\text{Back emf, } E_{b1} = V - I_{a1}(R_a + R)$$
$$= 230 - 45\ (0.02 + 0.6) = 202.1\ V$$

Let the external resistance be reduced from 0.6 Ω to R_1 ohms to raise the speed to 850 rpm.

Since load torque is constant

$$\therefore T_2 = T_1$$

$$\text{or } I_{a2}\Phi_2 = I_{a1}\Phi_1 \qquad \because T \propto \Phi I_a$$

$$\text{or } I_{a2} = I_{a1} \times \frac{\Phi_1}{\Phi_2} = I_{a1} = 45\ A$$

assuming flux unchanged

$$\text{Speed, } N_2 = 850 \text{ rpm}$$

$$\text{Back emf., } E_{b2} = V - I_{a2}(R_a + R_1)$$
$$= 230 - 45\ (0.02 + R_1) = 229.1 - 45R_1$$

$$\text{Also } E_{b2} = E_{b1} \times \frac{N_2}{N_1}$$

$\because$ flux has been assumed to remain constant

$$\text{or } 229.1 - 45R_1 = 202.1 \times \frac{850}{800}$$

$$\text{or } R_1 = 0.3193\ \Omega$$

i.e. Additional resistance of 0.6 Ω will have to be reduced to 0.3193 Ω in order to raise the motor speed from 800 rpm to 850 rpm. **Ans.**

Example 10.9. A 200 V shunt motor having armature resistance of 0.4 Ω takes armature current of 20 A on full load and runs at 600 rpm. If a resistance of 0.5 Ω is placed in the armature circuit, find the speed at (*i*) full-load torque (*ii*) half-full load torque.

[*Electrical Machines-I, Rajasthan Univ. 2006; Rajasthan Technical Univ. 2007*]

Solution: In normal conditions

$$\text{Back emf, } E_{b1} = 200 - 20 \times 0.4 = 192\ V$$

$$\text{Speed, } N_1 = 600 \text{ rpm}$$

(*i*) **At full-load torque**

When an additional resistance of 0.5 Ω is placed in the armature circuit

$$\frac{T_2}{T_1} = \frac{I_{a2}}{I_{a1}} \qquad \because T \propto \Phi I_a \text{ and } \Phi \text{ remains constant}$$

$$\text{or } I_{a2} = I_{a1} = 20\ A \qquad \because T \text{ remains constant}$$

$$\text{Back emf, } E_{b2} = V - I_{a2}(R_a + R) = 200 - 20\ (0.4 + 0.5) = 182\ V$$

$$\text{Speed } N_2 = \frac{E_{b2}}{E_{b1}} \times N_1 = \frac{182}{192} \times 600 = 568.75 \text{ rpm } \textbf{Ans.}$$

$$\because N \propto \frac{E_b}{\Phi} \text{ and } \Phi \text{ is constant}$$

(*ii*) **At Half-Full Load Torque**

$$I_{a3} = \frac{T_3}{T_1} \times I_{a1} = \frac{\frac{1}{2}T_1}{T} \times 20 = 10\ A \qquad \because T_2 = 2\,T_1$$

$$E_{b3} = 200 - 10\ (0.4 + 0.5) = 191\ V$$

$$\text{Speed } N_3 = \frac{E_{b3}}{E_{b1}} \times N_1 = \frac{191}{192} \times 600 = 596.88 \text{ rpm} \quad \textbf{Ans.}$$

Example 10.10. A 240 V shunt motor has an armature current of 15 A when running at 800 rpm against full-load torque. The armature resistance is 0.6 Ω. What resistance must be inserted in series with the armature to reduce the speed to 400 rpm at the same torque.

What will be the speed if the load torque is halved with this resistance in the circuit. Assume the flux to remain constant throughout.

Solution: In normal conditions

$$\text{Armature current, } I_{a1} = 15\ A$$

$$\text{Back emf, } E_{b1} = V - I_{a1}R_a = 240 - 15 \times 0.6 = 231\ V$$

Let resistance R be connected in series with the armature circuit to reduce the speed from $N_1 = 800$ rpm to $N_2 = 400$ rpm.

Since load torque remains the same

$$I_{a2}\Phi_2 = I_{a1}\Phi_1 \qquad \because T \propto \Phi I_a$$

$$\text{or } I_{a2} = \frac{\Phi_1}{\Phi_2} \times I_{a1} = I_{a1} = 15\ A$$

$\because$ flux remains constant

$$\text{Back emf, } E_{b2} = V - I_{a2}(R + R_a)$$
$$= 240 - 15\ (R + 0.6) = 231 - 15\ R$$

$\because E_b \propto N\,\Phi$ and flux Φ remains constant,

$$\frac{E_{b2}}{E_{b1}} = \frac{N_2}{N_1}$$

$$\text{or} \quad \frac{231 - 15\,R}{231} = \frac{400}{800}$$

$$\text{or } R = 7.7\ \Omega \textbf{ Ans.}$$

When load torque is halved

$$i.e.\ T_3 = 0.5\ T_1$$

$$\text{or}\ I_{a3}\ \Phi_3 = 0.5\ I_{a1}\ \Phi_1 \qquad \because T \propto \Phi I_a$$

$$\text{or}\ I_{a3} = 0.5\ I_{a1} \times \frac{\Phi_1}{\Phi_3}$$

$$= 0.5\ I_{a1} = 0.5 \times 15 = 7.5\ A$$

$$\because \Phi_3 = \Phi_2 = \Phi_1$$

$$\text{Back emf, } E_{b3} = V - I_{a3}\ (R + R_a)$$

$$= 240 - 7.5\ (7.7 + 0.6) = 177.75\ V$$

$$\text{and speed, } N_3 = N_1 \times \frac{E_{b3}}{E_{b1}} = 800 \times \frac{177.75}{231} = 614.7 \text{ rpm } \textbf{Ans.}$$

Example 10.11. A 10 kW, 250 V, 1200 rpm dc shunt motor has a full-load efficiency of 80%. The field and armature resistances are 125 and 0.2 ohms respectively. The speed of the motor is to be reduced to 75% with load torque remaining constant.

(*i*) What resistance should be inserted in the armature circuit?

(*ii*) With field current at its normal value, what voltage should be applied to the armature?

Solution: On full-load

$$\text{Total input to the motor} = \frac{\text{Motor output at full load}}{\text{Full-load efficiency}}$$

$$= \frac{10{,}000}{0.8} = 12{,}500 \text{ watts}$$

$$\text{Line current, } I_{L1} = \frac{\text{Motor input}}{\text{Line voltage}} = \frac{12{,}500}{250} = 50\ A$$

$$\text{Shunt field current, } I_{sh} = \frac{V}{R_{sh}} = \frac{250}{125} = 2\ A$$

$$\text{Armature current, } I_{a1} = I_{L1} - I_{sh} = 50 - 2 = 48\ A$$

$$\text{Back emf, } E_{b1} = V - I_{a1} R_a = 250 - 48 \times 0.2 = 240.4\ V$$

(*i*) Let resistance R be connected in series with the armature circuit to reduce the speed to 75% with load torque remaining constant

$$\text{Since } T_2 = T_1,$$

$$I_{a2}\ \Phi_2 = I_{a1} \Phi_1 \qquad \because T \propto \Phi\ I_a$$

$$\text{or } I_{a2} = I_{a1} = 48\ A \qquad \because \Phi_2 = \Phi_1$$

$$\text{Back emf, } E_{b2} = V - I_{a2}\ (R + R_a)$$

$$= 250 - 48\ (R + 0.2) = 240.4 - 48\ R$$

$$\text{Also } \frac{E_{b2}}{E_{b1}} = \frac{N_2}{N_1} \times \frac{\Phi_2}{\Phi_1}$$

$$\text{or } \frac{240.4 - 48\ R}{240.4} = 0.75 \qquad \because N_2 = 0.75\ N_1 \text{ and } \Phi_2 = \Phi_1$$

$$\text{or } R = \frac{240.4 \times 0.75 - 240.4}{-48} = 1.25\ \Omega \textbf{ Ans.}$$

(*ii*) Let the voltage of V_1 volts be applied to the armature circuit to reduce the speed to 75% with normal field current.

Load torque being constant

$$I_{a3} = I_{a1} = 48\ A$$

$$\text{Back emf, } E_{b3} = V_1 - I_{a1}\ R_a = V_1 - 48 \times 0.2$$

$$= (V_1 - 9.6) \text{ volts}$$

$$\text{Also } \frac{E_{b3}}{E_{b1}} = \frac{N_2}{N_1} \times \frac{\Phi_2}{\Phi_1}$$

$$\text{or } \frac{V_1 - 9.6}{240.4} = 0.75 \qquad \because N_2 = 0.75\ N_1 \text{ and } \Phi_2 = \Phi_1$$

$$\text{or } V_1 = 189.9\ V \textbf{ Ans.}$$

Example 10.12. A 200 V dc shunt motor takes 27 A at rated voltage and runs at 800 rpm. Its field resistance is 100 Ω. If an additional resistance of 20 Ω is inserted in the armature circuit, compute the motor speed and line current in case load torque varies as square of speed.

[*M.D. Univ. Electromechanical Energy Conversion, December-2005*]

Solution: Line voltage, $V = 200\ V$

$$\text{Line current, } I_{L1} = 27\ A$$

$$\text{Shunt field current, } I_{sh} = \frac{V_{sh}}{R_{sh}} = \frac{200}{100} = 2\ A$$

$$\text{Armature current, } I_{a1} = I_L - I_{sh} = 27 - 2 = 25\ A$$

$$\text{Back emf, } E_{b1} = V = 200\ V \text{ neglecting armature resistance}$$

After insertion of additional resistance of 20 Ω in the armature circuit, let the armature current and speed be I_{a2} amperes and N_2 rpm.

$$\frac{T_2}{T_1} = \left(\frac{N_2}{N_1}\right)^2 \qquad ...(i) \qquad \because T \propto N^2$$

$$\text{Also } \frac{T_2}{T_1} = \frac{E_{b_2} I_{a_2}}{E_{b_1} I_{a_1}} \times \frac{N_1}{N_2} \qquad ...(ii) \qquad \because T \propto \frac{E_b I_a}{N}$$

From Eqs. (*i*) and (*ii*) we have

$$\left(\frac{N_2}{N_1}\right)^3 = \frac{E_{b_2} I_{a_2}}{E_{b_1} I_{a_1}}$$

$$\left(\frac{E_{b_2}}{E_{b_1}}\right)^3 = \frac{E_{b_2}}{E_{b_1}} \times \frac{I_{a_2}}{I_{a_1}}$$

$$\because N \propto E_b, \text{ flux remaining constant}$$

$$\text{or } \left(\frac{E_{b_2}}{E_{b_1}}\right)^2 = \frac{I_{a_2}}{I_{a_1}}$$

$$\text{or } \left(\frac{200 - 20\ I_{a_2}}{200}\right)^2 = \frac{I_{a_2}}{25} \qquad \because E_2 = 200 - 20\ I_{a_2}$$

$$\text{or } \frac{40{,}000 - 8{,}000\ I_{a_2} + 400\ I_{a_2}^2}{40{,}000} = \frac{I_{a_2}}{25}$$

$$\text{or } I_{a_2}^2 - 24\ I_{a_2} + 100 = 0$$

$$\text{or } I_{a_2} = \frac{24 \pm \sqrt{24^2 - 4 \times 100}}{2} = 5.366\ A \text{ or } 18.63\ A$$

$$\text{Armature current, } I_{a_2} = 5.366\ A$$

ignoring higher value of 18.63 A which is impracticable

$$\text{Line current, } I_{L_2} = I_{a_2} + I_{sh} = 5.366 + 2 = 7.366\ A \textbf{ Ans.}$$

$$\text{Speed, } N_2 = N_1 \times \frac{E_{b_2}}{E_{b_1}} \times \frac{\Phi_1}{\Phi_2} = N_1 \times \frac{E_{b_2}}{E_{b_1}}$$

$$\because \text{flux remains constant}$$

$$= 800 \times \frac{200 - 20 \times 5.366}{200} = 370.7 \text{ rpm } \textbf{Ans.}$$

Example 10.13. A 220 V dc shunt motor takes 22 A at rated voltage and runs at 1,000 rpm. Its field resistance is 100 Ω and armature circuit resistance is 0.1 Ω. Compute the value of additional resistance required in the armature circuit to reduce the speed to 800 rpm when (*i*) the load torque is proportional to speed (*ii*) the load torque varies as the square of the speed.

[*U.P.S.C. I.E.S. Elec. Engineering II, 1993*]

Solution: In normal conditions

Line current, $I_{L1} = 22$ A

Shunt field current, $I_{sh} = \frac{V}{R_{sh}} = \frac{220}{100} = 2.2$ A

Armature current, $I_{a1} = I_{L1} - I_{sh} = 22 - 2.2 = 19.8$ A

Back emf, $E_{b1} = V - I_{a1} R_a = 220 - 19.8 \times 0.1 = 218.02$ V

Speed, $N_1 = 1{,}000$ rpm

(*i*) Let the additional resistance required in the armature circuit be of R ohms to reduce the speed to 800 rpm when the load torque is proportional to speed.

$$T_2 = T_1 \times \frac{N_2}{N_1} = T_1 \times \frac{800}{1{,}000} = 0.8\, T_1$$

$$\text{or } I_{a2}\,\Phi_2 = 0.8\, I_{a1}\,\Phi_1 \qquad \because T \alpha \Phi I_a$$

$$\text{or } I_{a2} = 0.8 \times 19.8 = 15.84 \text{ A} \qquad \because \Phi_2 = \Phi_1$$

$$E_{b2} = V - I_{a2}(R + R_a)$$

$$= 220 - 15.84(R + 0.1) = 218.416 - 15.84\, R$$

$$\text{Since } \frac{E_{b_2}}{E_{b_1}} = \frac{N_2}{N_1}, \qquad \text{flux being constant}$$

$$\frac{218.416 - 15.84R}{218.02} = \frac{800}{1{,}000}$$

or R = 2.778 Ω **Ans.**

(*ii*) Let the additional resistance required in the armature circuit be of R' ohms to reduce the speed to 800 rpm when no load torque varies as the square of the speed

$$T_3 = T_1 \times \left(\frac{N_2}{N_1}\right)^2 = T_1 \times \left(\frac{800}{1{,}000}\right)^2 = 0.64\, T_1$$

$$\text{or } I_{a3}\,\Phi_3 = 0.64\, I_{a1}\,\Phi_1 \qquad \because T \alpha \Phi I_a$$

$$\text{or } I_{a3} = 0.64 \times 19.8 = 12.672 \text{ A} \qquad \because \Phi_3 = \Phi_1$$

$$E_{b3} = V - I_{a3}(R' + R_a) = 220 - 12.672\,(R' + 0.1)$$

$$= 218.7328 - 12.672\, R'$$

$$\text{Since } \frac{E_{b3}}{E_{b_1}} = \frac{N_3}{N_1} \qquad \text{flux being constant}$$

$$\therefore \frac{218.7328 - 12.672\, R'}{218.02} = \frac{800}{1{,}000}$$

or R' = 3.497 Ω **Ans.**

Example 10.14. A separately-excited dc shunt motor is driving a fan whose torque is proportional to the square of the speed. When 100 V are applied to the motor, the current taken by the motor is 8 A, with the speed being 500 rpm. At what applied voltage does the speed reach 750 rpm and then what is the current drawn by the armature? Assume the armature circuit resistance to be 1 Ω. Neglect brush drop and mechanical losses. [*GATE 1999*]

Solution: *With 100 V applied to the armature*

Armature current, $I_{a_1} = 8$ A

Back emf, $E_{b_1} = V - I_{a_1} R_a = 100 - 8 \times 1 = 92$ V

Speed, $N_1 = 500$ rpm

Let the voltage of V volts be applied to the armature to raise the speed to 750 rpm.

$$T_2 = T_1 \times \left(\frac{N_2}{N_1}\right)^2 = T_1 \times \left(\frac{750}{500}\right)^2 = 2.25\, T_1$$

Since $T \propto N^2$

$$\text{or } I_{a_2}\Phi_2 = 2.25\, I_{a_1}\Phi_1 \qquad \because T \propto \Phi I_a$$

Assuming flux to remain constant

$I_{a_2} = 2.25 \times I_{a_1} = 2.25 \times 8 = 18$ A **Ans.**

$$\text{and applied voltage, } V = E_{b_2} + I_{a_2} R_a$$

$$= E_{b_1} \times \frac{N_2}{N_1} + I_{a_2} R_a \qquad \because \Phi_2 = \Phi_1, \text{ as assumed}$$

$$= 92 \times \frac{750}{500} + 18 \times 1 = 156 \text{ V}$$ **Ans.**

10.4 SPEED CONTROL OF DC SERIES MOTORS

Speed control of dc series motors may be accomplished through either armature control, field control or series-parallel control.

10.4.1. Armature Control Methods. Speed adjustment of dc series motors by armature control may be had by any one of the following three methods.

(*i*) Armature resistance control
(*ii*) Shunted armature control and
(*iii*) Armature terminal voltage control.

1. Armature Resistance Control. This is the most common method employed. It is obtained in the same way as for a dc shunt motor with the exception that the control resistance may be connected directly in series with the supply to the complete motor, as shown in Fig. 10.10. Control of armature voltage for the series motor is the same as the voltage applied to the complete motor. The drawbacks of armature resistance control for machines with shunt fields are not as important in the speed control of dc series motors.

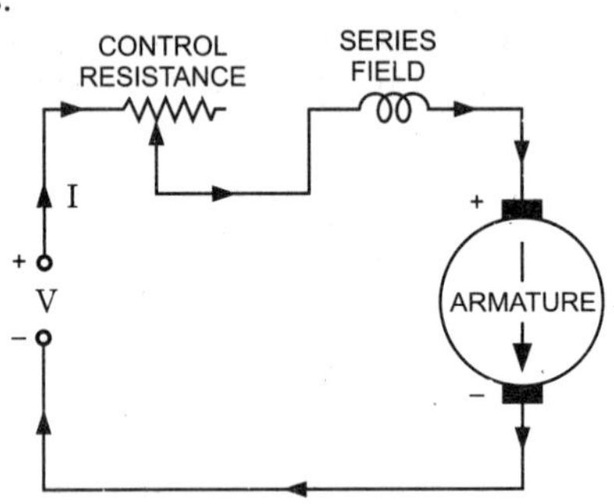

Fig. 10.10 *Armature Resistance Control For DC Series Motors*

The poor speed regulation that is inherent in this method has no significance for the control of dc series motors, since the speed characteristic of a dc series motor is a rapidly drooping curve. The power loss in the control resistance for many applications of dc series motors is not too serious, since in these applications the control is utilized for a large portion of time for reducing the speed under light load conditions and is only employed intermittently when the motor is carrying full load. The speed–torque characteristics of

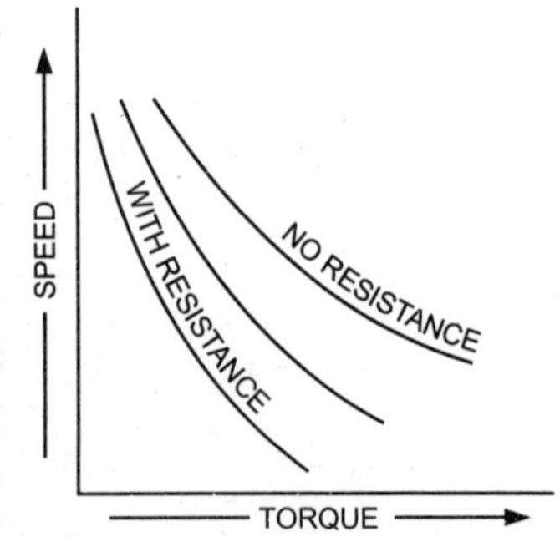

Fig. 10.11

dc series motors with resistance control are shown in Fig. 10.11. The maximum range of speed control of about 3 : 1 will be available depending on the load. This method of speed control is most economical for constant torque drives. This method of speed control is employed chiefly for dc series motors driving cranes, hoists, trains etc. because such drives operate on intermittent duty.

2. Shunted Armature Control. The combination of a rheostat shunting the armature and a rheostat in series with the armature, as shown in Fig. 10.12, is used to give slow speeds at light loads. Such a scheme accomplishes the speed control both by lowering the voltage applied to the motor armature and by varying the flux. The voltage applied to the armature terminals is varied by varying series rheostat R_1. The exciting current can be varied by varying the armature shunting resistance R_2 for the same armature current I_a.

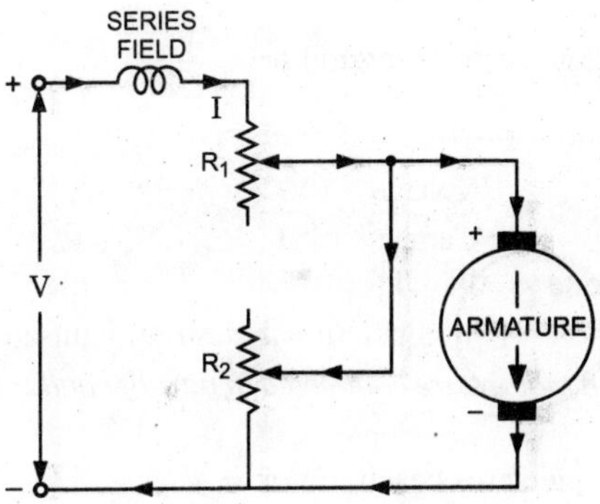

Fig. 10.12 *Shunted Armature Control For DC Series Motors*

For a given constant load torque, if armature current I_a is reduced due to armature divertor, then flux Φ must increase because torque developed by the armature $T \propto \Phi I_a$. This causes increase in current drawn from the supply main, so increase in flux and decrease in speed.

In this method the speed control can be obtained over a wide range but below normal speed. The limit of speed control range depends on the ratio existing between resistances R_1 and R_2 and also on the current at which there is a saturation of the magnetic circuit. No-load speed can effectively be adjusted to any desired low value by this method.

This method of speed control is not economical due to considerable power losses in the speed controlling resistances. Speed control by this method is, therefore, restricted to as short an interval as possible.

3. Armature Terminal Voltage Control. The speed control of dc series motors can be accomplished by supplying the power to the motor from a separate variable voltage supply. This method is seldom used because of high cost of control equipment.

10.4.2. Field Control Methods. The speed of a dc series motor can be controlled by varying the flux in any one of the following manners.

1. Field Divertor Method. The field flux can be reduced by shunting a portion of motor current around the series field, thus reducing the excitation mmf and weakening of field. This method is illustrated in Fig. 10.13. This method gives speeds above normal because flux is reduced by this method. Lesser the divertor resistance, less the field current, less flux and, therefore, more the speed. This method is convenient as well as economical and provides the speed control range usually not exceeding 2 : 1. This method of speed control is used in electric drives in which the speed should rise sharply as soon as the load falls.

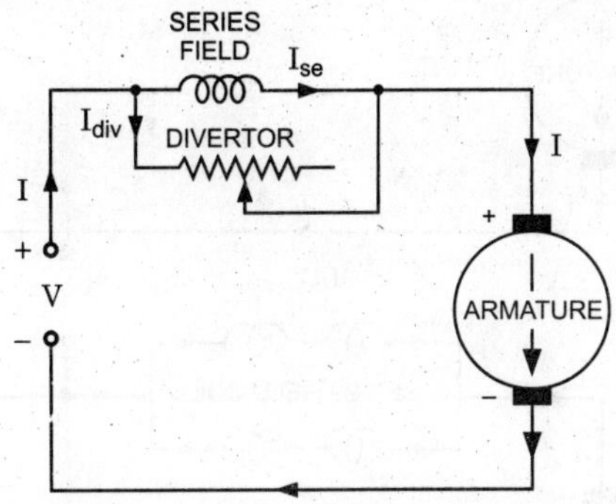

Fig. 10.13 *Field Divertor Method of Speed Control For DC Series Motors*

For satisfactory operation under violently changing loads, the diverting resistance should be highly inductive because it is connected across the series field winding which normally has very low resistance. If the field winding is shunted by means of non-inductive resistance, any sudden change in the load current will not immediately affect the field winding owing to the high inductance of the field circuit impeding any violent change in its current.

2. Tapped Field Control. This is another method of increasing the speed by reducing the flux and it is accomplished by reducing the number of turns of the field winding through which the current flows. In this method of speed control of dc series motors a number of tappings from the field winding are brought outside, as illustrated in Fig. 10.14. A number of series field turns can be taken out according to the requirement. When all field turns are in circuit, the motor runs at the lowest speed and speed increases with cutting out some of the series field turns. This method is usually employed in electric traction.

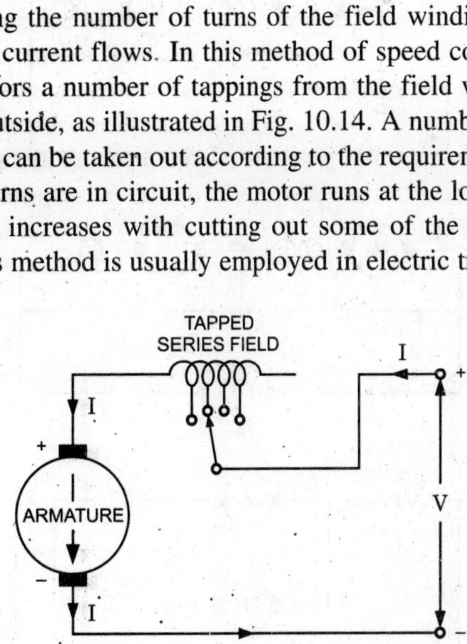

Fig. 10.14 *Tapped Field Control For DC Series Motors*

3. Paralleling Field Coils Method. In this method, used in fan motors, several speeds can be obtained by regrouping of field coils, as illustrated in Fig. 10.15.

10.4.3. Series-Parallel Control. When two or more similar series motors are employed in pairs, as in electric traction, speed control can be obtained by combining series resistance with series and parallel connections, as illustrated in Fig. 10.16.

The machines are started up in series with each other and a starting or control resistance as illustrated in Fig. 10.16 (*a*). The

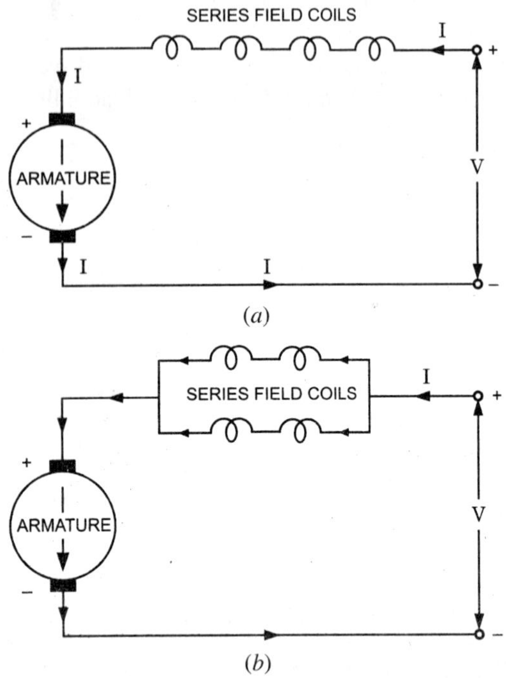

Fig. 10.15

additional resistance is gradually cut out by the controller as the motors attain speed and finally the control resistance is totally removed, then each motor has one half of the line voltage across it, as shown in Fig. 10.16 (*b*). This is the first running position. In this position for any given value of armature current, each motor will run at half of its normal speed.

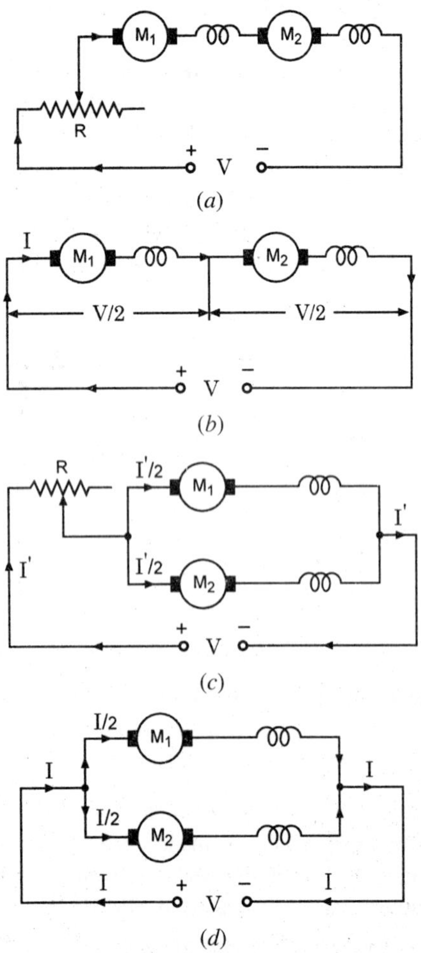

Fig. 10.16

Since there is no external resistance in the circuit, there is no waste of energy and so motors operate at an efficiency nearly equal to that obtainable with full line voltage across the terminal of each motor.

When it is desired to increase the speed of the combination, two motors are connected in parallel and in series with a variable resistance R, as shown in Fig. 10.16 (*c*). This resistance is gradually cut out as the motors attain the speed and finally when this resistance is totally removed from the circuit, as illustrated in Fig. 10.16 (*d*), the second running position is obtained. In this position each motor is connected across the full line voltage.

Let the line voltage be V volts and line current I amperes.

(*i*) *When motors are connected in series and are in running position.*

$$\text{Voltage across each motor} = \frac{V}{2}$$

$$\text{Current through each motor} = I$$

$$\text{Speed} \propto \frac{\text{Voltage}}{\text{Current}} \propto \frac{V/2}{I} \propto \frac{V}{2I} \qquad \text{...(10.4)}$$

$$\text{Torque} \propto \Phi I \propto I^2 \qquad \text{...(10.5)}$$

(Since $\Phi \propto I$, assuming unsaturated field)

(*ii*) *When motors are connected in parallel and are in running position.*

$$\text{Voltage across each motor} = V$$

$$\text{Current through each motor} = \frac{I}{2}$$

$$\text{Speed} \propto \frac{\text{Voltage}}{\text{Current}} \propto \frac{V}{I/2} \propto \frac{2V}{I} \qquad \text{...(10.6)}$$

$$\text{Torque} \propto \text{Flux} \times \text{current} \propto (\text{current})^2 \propto \left(\frac{I}{2}\right)^2 \propto \frac{I^2}{4} \qquad \text{...(10.7)}$$

Comparing Eqs. (10.4) and (10.6) for the speeds of motors we observe that speeds of the motors when running in parallel is

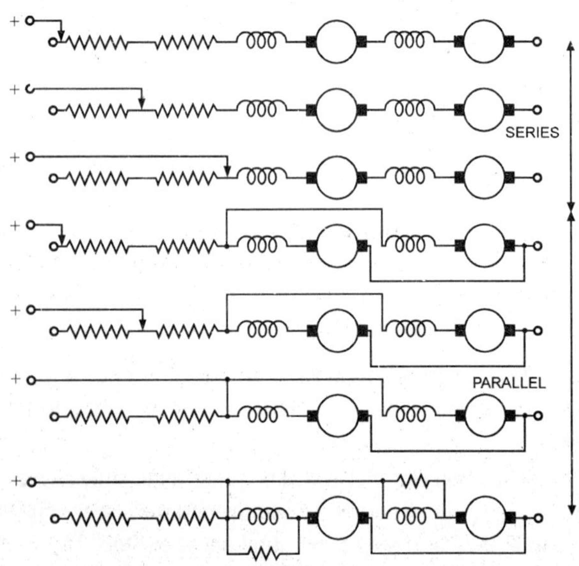

Fig. 10.17 *Combination of Series Parallel and Resistance Control of DC Series Motors*

4 times that while running in series. But from Eqs. (10.5) and (10.7) of torque we observe that torque of the motor when running in series is 4 times that while running in parallel.

There are several advantages of this method of speed control. It is more economical than providing each motor with a separate starting resistance or by connecting the motors permanently in parallel with only single starting resistance. It reduces the starting and braking time by reducing the overall moment of inertia, because in this control system two motors, each rated for half the required load capacity are used. It provides higher reliability of operation. It is more convenient to use two motors of smaller size than one of larger size in view of space limitations. Finally it provides two speeds without wastage of power.

Example 10.15. A series motor of resistance 1.0 Ω between the terminals runs at 1,000 rpm of 220 V with a current of 15 A. Find the speed at which it will run when connected in series with a 4.5 Ω resistance and taking a current of 10 A at the same supply voltage. Assume linear magnetization characteristic.

[U.P.S.C. I.E.S. Electrical Engineering-II, 2003]

Solution: Supply voltage, V = 220 V

Motor resistance, $R_m = 1.0\ \Omega$

Motor input current, $I_1 = 15$ A

Back emf, $E_{b_1} = V - I_1 R_m = 220 - 15 \times 1.0 = 205$ V

Speed, $N_1 = 1{,}000$ rpm

Motor input current, $I_2 = 10$ A

Back emf, $E_{b_2} = V - I_2 (R_m + R) = 220 - 10\,(1.0 + 4.5) = 165$ V

$$\Phi_2 = \Phi_1 \times \frac{I_2}{I_1} \quad \because \text{magnetization characteristic is linear}$$

$$= \Phi_1 \times \frac{10}{15} = \frac{2}{3}\Phi_1$$

$$\text{Speed, } N_2 = N_1 \times \frac{E_{b_2}}{E_{b_1}} \times \frac{\Phi_1}{\Phi_2} \qquad \because N \propto \frac{E_b}{\Phi}$$

$$= 1{,}000 \times \frac{165}{205} \times \frac{\Phi_1}{\frac{2}{3}\Phi_1} = 1{,}207 \text{ rpm} \quad \textbf{Ans.}$$

Example 10.16. A 200 V dc series motor runs at 500 rpm when taking a current of 25 A. The resistance of the armature is 0.5 Ω and that of field is 0.3 Ω. If the current remains constant, calculate the resistance necessary to reduce the speed to 250 rpm.

Solution: Motor input current, $I_1 = 25$ A

Back emf, $E_{b1} = V - I_1 (R_a + R_{se})$

$= 200 - 25\,(0.5 + 0.3) = 180$ V

Speed, $N_1 = 500$ rpm

Let the resistance R be connected in series with the motor to reduce the speed to

$N_2 = 250$ rpm

Motor input current, $I_2 = I_1 = 25$ A

Back emf $E_{b2} = V - I_2 (R + R_a + R_{se})$

$= 200 - 25\,(R + 0.5 + 0.3) = 180 - 25\,R$

$$\text{Since } N \propto \frac{E_b}{\Phi},$$

$$\frac{N_2}{N_1} = \frac{E_{b2}}{E_{b1}} \times \frac{\Phi_1}{\Phi_2} = \frac{E_{b2}}{E_{b1}}$$

$\because \Phi_2 = \Phi_1$ as field current remains the same

$$\text{or} \quad \frac{250}{500} = \frac{180 - 25\,R}{180}$$

or R = 3.6 Ω **Ans.**

Example 10.17. A dc series motor runs at 750 rpm drawing 50 A from 500 V supply. The armature resistance is 0.25 Ω and the series field resistance is 0.2 Ω. Determine the value of external resistance to be added in series with the armature for the motor to run at 550 rpm. The load torque varies as the square of the speed. Assume linear magnetization.

[B.P. Univ. of Technology Electrical Machines, 2008]

Solution: Supply voltage, V = 500 V

Input current, $I_1 = 50$ A

Back emf, $E_{b_1} = V - I_1 (R_a + R_{se})$

$= 500 - 50\,(0.25 + 0.2) = 477.5$ V

Speed, $N_1 = 750$ rpm

Speed, $N_2 = 550$ rpm

Since $T \propto N^2$

$$\frac{T_2}{T_1} = \left(\frac{N_2}{N_1}\right)^2 = \left(\frac{550}{750}\right)^2 = 0.5378$$

$$\text{or} \quad \frac{I_2 \Phi_2}{I_1 \Phi_1} = 0.5378 \qquad \because T \propto \Phi I_a$$

$$\text{or} \quad \left(\frac{I_2}{I_1}\right)^2 = 0.5378 \qquad \because \Phi \propto I \text{ (given)}$$

or $I_2 = 0.7333\ I_1 = 0.7333 \times 50 = 36.667$ A

Let the resistance of R ohms be connected in series with the motor circuit

Back emf, $E_{b_2} = V - I_2 (R + R_a + R_{se})$

$= 500 - 36.667\,(R + 0.25 + 0.2)$

$= 483.5 - 36.667\,R$

$$\text{Since } \frac{E_{b_2}}{E_{b_1}} = \frac{N_2}{N_1} \times \frac{\Phi_2}{\Phi_1} \qquad \because E \propto \Phi N$$

$$\text{or} \quad \frac{483.5 - 36.667\,R}{477.5} = \frac{550}{750} \times \frac{36.667}{50} \qquad \because \frac{\Phi_2}{\Phi_1} = \frac{I_2}{I_1}$$

or R = 6.18 Ω **Ans.**

Example 10.18. A dc series motor with an unsaturated magnetic field and negligible resistance, when running at a certain speed on a given load takes 50 A at 500 V. If the load torque varies as the cube of the speed, calculate the resistance to be inserted to reduce the speed by 50%. *[M.D. Univ. Electrical Machines-I, December 2009]*

Solution: Main input current, $I_1 = 50$ A

Speed, N_1 = N (Say)

Speed, $N_2 = (1 - 0.5)$ N = 0.5 N

Back emf, E_{b1} = V = 500 V $\quad\because$ motor resistance is negligible

$$\frac{T_2}{T_1} = \left(\frac{N_2}{N_1}\right)^3 = 0.5^3 = 0.125 \qquad \because T \propto N^3$$

$$\frac{\Phi_2 I_2}{\Phi_1 I_1} = 0.125 \qquad \because T \propto \Phi I$$

$$\text{or} \quad \left(\frac{I_2}{I_1}\right)^2 = 0.125 \qquad \because \Phi \propto I, \text{ magnetic circuit being unsaturated}$$

$$\text{or} \quad \frac{I_2}{I_1} = \sqrt{0.125} = 0.3536$$

or $I_2 = 0.3536 I_1 = 0.3536 \times 50 = 17.68$ A

Let the resistance of R ohms be connected in series with the motor circuit to reduce the speed by 50%.

$$\text{Back emf, } E_{b2} = V - I_2R = 500 - 17.68\ R$$

$$\text{and } \frac{E_{b2}}{E_{b1}} = \frac{\Phi_2}{\Phi_1} \times \frac{N_2}{N_1} \qquad \because E_b \propto \Phi N$$

$$\text{or } \frac{500 - 17.68R}{500} = \frac{I_2}{I_1} \times \frac{N_2}{N_1} = \frac{17.68}{50} \times 0.5 \qquad \because \Phi \propto I$$

or $R = 23.28\ \Omega$ **Ans.**

Example 10.19. A 250 V dc series motor has a resistance of 0.4 Ω and runs at 2,000 rpm drawing a current of 20 A.

(*i*) What resistance must be added in order to limit the speed to 1,000 rpm when armature current is 30 A? Assume that the flux at 30 A is 20% more than that at 20 A.

(*ii*) At what speed the motor will run if it is connected directly to 250 V supply when the load is such that the current drawn is 30 A?

Solution: Resistance of motor, $R_m = R_a + R_{se} = 0.4\ \Omega$

$$\text{Back emf, } E_{b_1} = V - I_1R_m = 250 - 20 \times 0.4 = 242 \text{ V}$$

(*i*) Let R_1 be the resistance in ohms in series with the motor in order to reduce the speed to 1,000 rpm with armature current of 30 A.

$$\text{Back emf, } E_{b_2} = V - I_2\,(R_1 + R_m)$$

$$= 250 - 30\,(R_1 + 0.4) = 238 - 30\,R_1$$

$$\text{Flux } \Phi_2 = 1.2\,\Phi_1 \qquad \text{...(given)}$$

$$\text{Since } \frac{E_{b_2}}{E_{b_1}} = \frac{N_2}{N_1} \times \frac{\Phi_2}{\Phi_1}$$

$$\text{So } \frac{238 - 30\,R_1}{242} = \frac{1{,}000}{2{,}000} \times 1.2$$

or $R_1 = 3.093\ \Omega$ **Ans.**

(*ii*) When directly connected to 250 V supply

$$I_3 = 30 \text{ A}$$

$$\text{Back emf, } E_{b_3} = V - I_3\,R_m = 250 - 30 \times 0.4 = 238 \text{ V}$$

$$\text{Flux } \frac{\Phi_3}{\Phi_1} = 1.2$$

$$\text{Speed, } N_3 = N_1 \times \frac{E_{b_3}}{E_{b_1}} \times \frac{\Phi_1}{\Phi_3}$$

$$= 2{,}000 \times \frac{238}{242} \times \frac{1}{1.2} = 1{,}639 \text{ rpm } \textbf{Ans.}$$

Example 10.20. A series motor takes 20 A at 400 V and runs at 250 rpm. The armature resistance is 0.6 Ω and field resistance is 0.4 Ω. The torque required by the device varies as the square of the speed. Find the applied voltage and current to run the device at 350 rpm.

[*M.D. Univ. Electrical Machines-I, May 2009*]

Solution: Back emf, $E_{b1} = V_1 - I_1\,(R_a + R_{se})$

$$= 400 - 20\,(0.6 + 0.4) = 380 \text{ V}$$

Let the applied voltage to make the set to run at 350 rpm be V_2 volts

Since $T \propto N^2$,

$$\text{or } \frac{T_2}{T_1} = \left(\frac{N_2}{N_1}\right)^2 = \left(\frac{350}{250}\right)^2 = 1.96$$

$$\text{or } \frac{I_2}{I_1} = \sqrt{1.96} = 1.4 \qquad \text{Assuming field unsaturated}$$

or $I_2 = 1.4I_1 = 1.4 \times 20 = 28$ A **Ans.**

$$\text{Back emf, } E_{b2} = V_2 - I_2\,(R_a + R_{se})$$

$$= V_2 - 28(0.6 + 0.4) = (V_2 - 28) \text{ volts}$$

$$\text{Since } \frac{E_{b2}}{E_{b1}} = \frac{N_2}{N_1} \times \frac{\Phi_2}{\Phi_1} = \frac{N_2}{N_1} \times \frac{I_2}{I_1}$$

$$\text{or } \frac{V_2 - 28}{380} = \frac{350}{250} \times \frac{28}{20} = 1.96$$

or $V_2 = 380 \times 1.96 + 28 = 772.8$ V **Ans.**

Example 10.21. A dc series motor operating at 250 V dc mains and draws 25 A and runs at 1,200 rpm. $R_a = 0.1\ \Omega$ and $R_{se} = 0.3\ \Omega$. A resistance of 25 Ω is placed in parallel with the armature of motor. Determine (*i*) the speed of the motor with the shunted armature connection, if the magnetic circuit remains unsaturated and the load torque remains constant (*ii*) No-load speed of motor.

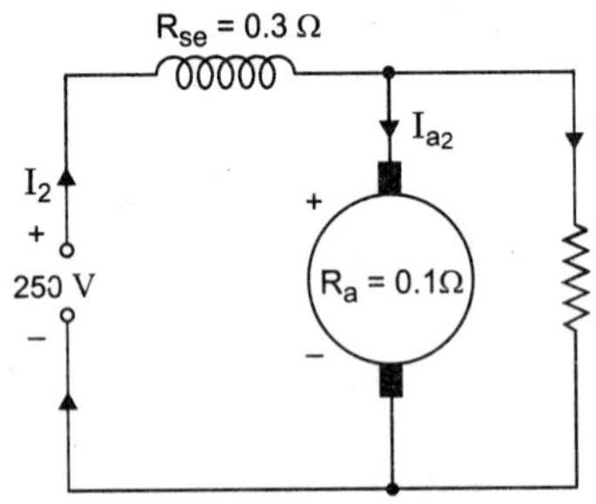

Fig. 10.18

Solution: Back emf, $E_b = V - I_1\,(R_a + R_{se})$

$$= 250 - 25\,(0.1 + 0.3) = 240 \text{ V}$$

When armature is shunted with a resistance of 25 Ω, let the motor input current be of I_2 amperes

$$\text{Potential across the armature} = V - I_2R_{se} = 250 - 0.3\,I_2$$

$$\text{Armature diverter current, } I_{div} = \frac{250 - 0.3\,I_2}{25} = 10 - 0.012\,I_2$$

$$\text{Armature current, } I_{a2} = I_2 - I_{div} = I_2 - (10 - 0.012\,I_2)$$

$$= 1.012\,I_2 - 10$$

$$\text{and } T_2 = T_1 \qquad \text{...(Given)}$$

$$\text{or } I_{a2}\,\Phi_2 = I_{a1}\Phi_1 \qquad \text{...}T \propto \Phi I_a$$

$$\text{or } I_{a2} = I_{a1} \times \frac{\Phi_1}{\Phi_2} = I_1 \times \frac{I_1}{I_2}$$

$$\text{or } 1.012\,I_2 - 10 = \frac{I_1^2}{I_2} = \frac{25^2}{I_2} = \frac{625}{I_2}$$

$$\text{or } 1.012\,I_2^2 - 10\,I_2 - 625 = 0$$

or $I_2 = 30.27$ A ignoring negative value

$$\text{Armature current, } I_{a2} = 1.012\,I_2 - 10 = 1.012 \times 30.27 - 10 = 20.63 \text{ A}$$

$$\text{Back emf, } E_{b2} = V - I_2\,R_{se} - I_{a2}\,R_a$$

$$= 250 - 30.27 \times 0.3 - 20.63 \times 0.1 = 238.856 \text{ V}$$

$$(i) \text{ Speed, } N_2 = N_1 \times \frac{E_{b2}}{E_{b1}} \times \frac{\Phi_1}{\Phi_2}$$

$$= N_1 \times \frac{E_{b2}}{E_{b1}} \times \frac{I_1}{I_2}$$

$$= 1{,}200 \times \frac{238.856}{240} \times \frac{25}{30.27} = 986.4 \text{ rpm } \textbf{Ans.}$$

(ii) **On No Load**

Armature current, $I_{a_0} = 0$

Series field current, $I_{se_0} = \dfrac{V}{R_{div} + R_{se}} = \dfrac{250}{25 + 0.3} = 9.88$ A

Back emf, E_{b_0} = Voltage drop across divertor

$= 25 \times 9.88 = 247$ V

No-load speed of motor, $N_0 = N_1 \times \dfrac{E_{b_0}}{E_{b_1}} \times \dfrac{\Phi_1}{\Phi_2}$

$= 1{,}200 \times \dfrac{247}{250} \times \dfrac{25}{9.88} = 3{,}125$ rpm **Ans.**

$\because \dfrac{\Phi_1}{\Phi_0} = \dfrac{I_{se_1}}{I_{se_0}} = \dfrac{I_1}{I_{se_0}} = \dfrac{25}{9.88}$

Example 10.22. A dc series motor runs at 1,000 rpm and takes 25 A from 250 V mains. The armature current is then reduced to 15 A by inserting a series resistance. Find the new speed and the value of the resistance inserted if the load torque varies as the square of the speed and the field flux is reduced by 15% for above change in the armature current. Assume the combined resistance of the armature and series field to be 1.0 Ω.

Solution: Back emf, $E_{b_1} = V - I_1 (R_a + R_{se}) = 250 - 25 \times 1 = 225$ V

After inserting a resistance of R ohms in series with the motor

Current, $I_2 = 15$ A

Flux, $\Phi_2 = (1 - 0.15)\, \Phi_1 = 0.85\, \Phi_1$

$\dfrac{T_2}{T_1} = \dfrac{I_2 \Phi_2}{I_1 \Phi_1} = \dfrac{15}{25} \times 0.85 = 0.51$ $\because T \propto I$

Since $T \propto N^2$, ... (given)

$N_2 = N_1 \times \sqrt{\dfrac{T_2}{T_1}} = 1{,}000\sqrt{0.51} = 714$ rpm **Ans.**

Back emf, $E_{b2} = V - I_2 (R + R_a + R_{se})$

$= 250 - 15 (R + 1) = 235 - 15$ R

Now since $\dfrac{E_{b2}}{E_{b1}} = \dfrac{N_2}{N_1} \times \dfrac{\Phi_2}{\Phi_1}$,

$\dfrac{235 - 15R}{225} = \dfrac{714}{1{,}000} \times 0.85$

or R = 6.563 Ω **Ans.**

Example. 10.23. The armature resistance of a 25 hp, 250 V series motor is 0.1 Ω, the brush drop is 3 V and the resistance of the series field is 0.05 Ω. When the series motor takes 85 A, the speed is 600 rpm. Calculate the speed when the current is 100 A and a diverter of 0.05 Ω is used. Neglect armature reaction and use linear-portion of saturation curve.

Solution: Back emf, $E_{b1} = V - I_1 (R_a + R_{se})$ – brush drop

$= 250 - 85 (0.1 + 0.05) - 3 = 234.25$ V

When line current is 100 A and a diverter of 0.05 Ω is used

Armature current, $I_{a_2} = 100$ A

Series field current, $I_{se_2} = I_{L_2} \times \dfrac{R_{div}}{R_{se} + R_{div}}$

$= 100 \times \dfrac{0.05}{0.05 + 0.05} = 50$ A

Back emf, $E_{b_2} = V - I_{a_2} R_a - I_{se_2} R_{se}$ – brush drop

$= 250 - 100 \times 0.1 - 50 \times 0.05 - 3$

$= 234.5$ V

Speed, $N_2 = N_1 \times \dfrac{E_{b_2}}{E_{b_1}} \times \dfrac{\Phi_1}{\Phi_2}$

$= 600 \times \dfrac{234.5}{234.25} \times \dfrac{85}{50} = 1{,}021$ rpm **Ans.**

Example 10.24. A dc series motor drives a load, the torque of which varies as the square of speed. The motor takes a current of 30 A, when the speed is 600 rpm. Determine the speed and current when the field winding is shunted by a diverter, the resistance of which is 1.5 times that of the field winding. The losses may be neglected.

[J.N. University Hyderabad, November-2003]

Solution: In normal conditions

Armature current, I_{a_1} = Series field current, I_{se}

= Line current, $I_{L_1} = 30$ A

Back emf, E_{b_1} = Supply voltage = V (say)

After connecting a divertor of resistance of 1.5 R_{se} in parallel with the series field winding, let the speed of motor be N_2 and line current I_{L_2}

Armature current, $I_{a_2} = I_{L_2}$

Series field current, $I_{se_2} = I_{L_2} \dfrac{R_{div}}{R_{se} + R_{div}}$

$= I_{L_2} \times \dfrac{1.5 R_{se}}{R_{se} + 1.5 R_{se}} = 0.6\, I_{L_2}$

$\dfrac{T_2}{T_1} = \left(\dfrac{N_2}{N_1}\right)^2$ $\because T \propto N^2$ (given)

or $\dfrac{I_{a_2}\Phi_2}{I_{a_1}\Phi_1} = \left(\dfrac{N_2}{N_1}\right)^2$

or $\dfrac{I_{L_2} \times 0.6\, I_{L_2}}{I_{L_1} \times I_{L_1}} = \left(\dfrac{N_2}{N_1}\right)^2$ $\because \dfrac{\Phi_2}{\Phi_1} = \dfrac{I_{se_2}}{I_{se_1}} = \dfrac{0.6 I_{L_2}}{I_{L_1}}$

or $\dfrac{0.6\, I_{L_2}^2}{I_{L_1}^2} = \dfrac{N_2^2}{N_1^2}$

or $\dfrac{N_2}{I_{L_2}} = \dfrac{\sqrt{0.6} \times N_1}{I_{L_1}} = \dfrac{\sqrt{0.6} \times 600}{30} = 15.492$ ...*(i)*

Also speed, $N_2 = N_1 \times \dfrac{E_{b_2}}{E_{b_1}} \times \dfrac{\Phi_1}{\Phi_2} = 600 \times \dfrac{V}{V} \times \dfrac{30}{0.6\, I_{L_2}}$

$\because E_{b_2} = V$, losses being negligible

or $N_2 I_{L_2} = \dfrac{600 \times 30}{0.6} = 30{,}000$ ...*(ii)*

Multiplying expressions *(i)* and *(ii)* we have

$N_2^2 = 15.492 \times 30{,}000$

or $N_2 = 681.73$ rpm **Ans.**

From equation *(i)*

$I_{L_2} = \dfrac{30{,}000}{N_2} = \dfrac{30{,}000}{681.73} = 44$ A **Ans.**

Example 10.25. A 500 V dc series motor runs at 500 rpm and takes 60 A. The resistance of the field and the armature are 0.3 and 0.2 ohms respectively. Calculate the value of the resistance to be shunted with the series field winding in order that the speed be increased to 600 rpm, if the torque were to remain constant. Saturation may be neglected. *[J.N. University Hyderabad, November-2003]*

Solution: In normal conditions

Armature current, I_{a1} = Series field current, I_{se1}

= Line current, $I_{L1} = 60$ A

$$E_{b1} = V - I_{L1}(R_a + R_{se})$$

$$= 500 - 60\,(0.2 + 0.3) = 470 \text{ V}$$

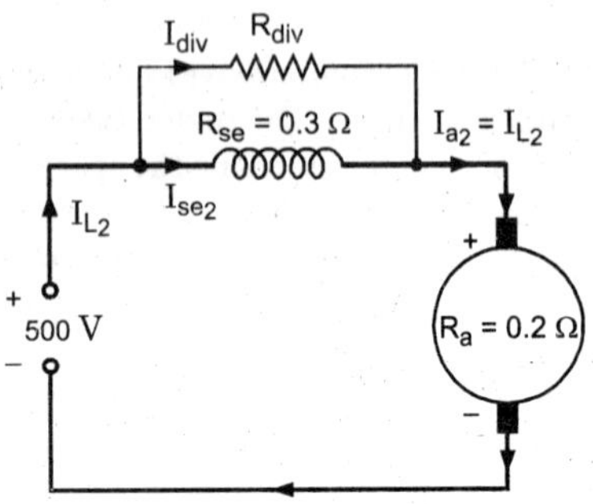

Fig. 10.19

After connecting a diverter across the series field let the line current be I_{L2} and series field current I_{se2}

Armature current, $I_{a2} = I_{L2}$

Since torque remains constant

$$T_2 = T_1$$

or $I_{a2}\,\Phi_2 = I_{a1}\,\Phi_1$ $\qquad\because T \propto \Phi I_a$

or $I_{L2}\,I_{se2} = 60 \times 60$

$\because$ saturation is negligible i.e. $\Phi \propto I_{se}$

or $I_{L2}\,I_{se2} = 3{,}600$ $\qquad$ ...(*i*)

Back emf, $E_{b2} = V - I_{a2}R_a - I_{se2}R_{se}$

$= 500 - 0.2\,I_{L2} - 0.3\,I_{se2}$

Also $E_{b2} = E_{b1} \times \dfrac{N_2}{N_1} \times \dfrac{\Phi_2}{\Phi_1}$

$= 470 \times \dfrac{600}{500} \times \dfrac{I_{se2}}{60} = 9.4\,I_{se2}$

Thus $500 - 0.2\,I_{L2} - 0.3\,I_{se2} = 9.4\,I_{se2}$

or $500 - 0.2\,I_{L2} - 9.7\,I_{se2} = 0$ $\qquad$ ...(*ii*)

Solving Eqs. (*i*) and (*ii*) we have

Line current, $I_{L2} = 71.9$ A

rejecting higher value of 2,428.1 A being impracticable

and $I_{se2} = 50$ A

Since $\dfrac{I_{se2}}{I_{L2}} = \dfrac{R_{div}}{R_{se} + R_{div}}$

or $\dfrac{50}{71.9} = \dfrac{R_{div}}{0.3 + R_{div}}$

or $R_{div} = 0.685\ \Omega$ **Ans.**

Example 10.26. A series motor is required to produce constant full-load torque whilst its speed is raised by 20% above its full-rated value by means of a diverter. Determine the diverter resistance as a percentage of field resistance. Neglect motor losses.

The magnetisation curve of the motor is as follows :

Current	**40**	**60**	**80**	**100%**	**of full-load value**
Flux per pole	**62**	**80**	**92**	**100%**	**of full-load value**

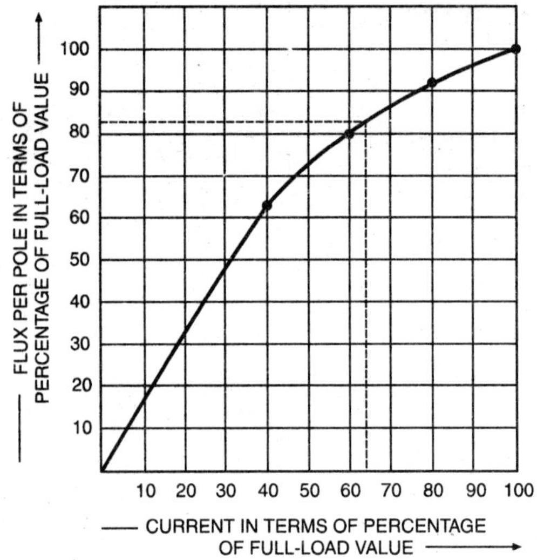

Fig. 10.20

Solution: Let the motor be connected across a supply of V volts and draw current of I amperes while running at normal speed of N rpm and developing full load torque.

Since speed, $N \propto \dfrac{E_b}{\Phi} \propto \dfrac{V - IR}{\Phi} \propto \dfrac{V}{\Phi}$

$\because$ losses are negligible

After connecting a diverter, the speed,

$$N_2 = 1.2\text{ N}$$

and since $\dfrac{N_2}{N_1} = \dfrac{V/\Phi_2}{V/\Phi_1} = \dfrac{\Phi_1}{\Phi_2}$

or $\dfrac{\Phi_1}{\Phi_2} = \dfrac{1.2\text{ N}}{\text{N}} = 1.2$

or $\Phi_2 = \dfrac{\Phi_1}{1.2}$ = 83.3 % of full load value of flux

From saturation curve the field current corresponding to 83.3% of full-load value of flux is 65% of full-load value of field current.

i.e. $I_{se2} = 0.65 I_{se1} = 0.65 I_{a1}$

Since load torque is constant,

$$I_{a2}\,\Phi_2 = I_{a1}\,\Phi_1$$

or $I_{a2} = I_{a1}\dfrac{\Phi_1}{\Phi_2} = 1.2\,I_{a1}$

Since series field current, $I_{se2} = \dfrac{R_{div}}{R_{se} + R_{div}} \times I_{a2}$,

$0.65\,I_{a1} = 1.2\,I_{a1}\dfrac{R_{div}}{R_{se} + R_{div}}$

or $R_{div} = \dfrac{0.65}{0.55} R_{se} = 1.182\,R_{se}$

i.e. Resistance of diverter is 118.2% of series field resistance. **Ans.**

Example 10.27. A 4-pole dc series wound motor takes 20 A from 250 V dc supply to run at 600 rpm when all the four field coils are in series. If the load torque increases as the square of the speed, find the speed and the input current when the field coils are connected in two parallel groups of two coils in series each. Assume the flux is proportional to current and ignore losses.

[*U.P.S.C. I.E.S. Electrical Engineering-II, 2010*]

Solution: Back emf, $E_{b_1} = 250$ V $\qquad\because$ losses are negligible

When the field coils are connected in two parallel groups of 2 in series, the current through each coil is half the new input current I_2. Let the new speed be N_2.

New series field current,

$$I_{se_2} = \frac{1}{2} I_2$$

Back emf, $E_{b_2} = V = 250$ V $\quad \because$ losses are negligible

$$\text{Since } \frac{E_{b_2}}{E_{b_1}} = \frac{N_2}{N_1} \times \frac{\Phi_2}{\Phi_1}$$

$$\text{or } \frac{N_2}{N_1} = \frac{\Phi_1}{\Phi_2} \qquad \because E_{b_1} = E_{b_2} = V = 250 \text{ V}$$

$$= \frac{I_{se_1}}{I_{se_2}} = \frac{I_1}{\frac{1}{2} I_2} = \frac{2I_1}{I_2} = \frac{2 \times 20}{I_2} = \frac{40}{I_2} \qquad \Phi \propto I_{se}$$

$$\frac{T_2}{T_1} = \left(\frac{N_2}{N_1}\right)^2$$

Since load torque varies as the square of speed.

$$\text{or } \frac{I_{a_2}\Phi_2}{I_{a_1}\Phi_1} = \left(\frac{40}{I_2}\right)^2$$

$$\text{or } \frac{I_2}{I_1} \times \frac{I_2/2}{I_1} = \left(\frac{40}{I_2}\right)^2$$

$$\text{or } I_2^2 = \sqrt{2}\, I_1 \times 40 = \sqrt{2} \times 20 \times 40 = 1{,}131.4$$

$$\text{or } I_2 = 33.64 \text{ A}$$

$$\text{Speed, } N_2 = \frac{40}{I_2} \times N_1 = \frac{40}{33.64} \times 600 = 713.524 \text{ rpm } \textbf{Ans.}$$

Example 10.28. Two dc series motors with different air gaps but otherwise identical run at 700 and 750 rpm respectively, when taking 50 A at 500 V. The total resistance of each motor is 0.36 Ω. If the motors are now mechanically coupled and connected in series electrically to 500 V dc supply and together takes 50 A current. Calculate the speed at which the coupled set will run.

[*U.P.S.C. I.E.S. Electrical Engineering-II, 2013*]

Solution: When the first motor is connected to 500 V and takes 50 A at 700 rpm

$$\text{Back emf, } E_{b_1} = V - I_1 R_{m_1} = 500 - 50 \times 0.36 = 482 \text{ V}$$

$$\text{or } E_{b_1} = K\Phi_1 N_1 \qquad \because E_b \propto \Phi N$$

$$\text{or } K\Phi_1 = \frac{E_{b_1}}{N_1} = \frac{482}{700}$$

Similarly when the second motor is connected to 500 V supply and takes 50 A at 750 rpm

$$K_2\Phi_2 = \frac{E_{b_2}}{N_2} = \frac{500 - 50 \times 0.36}{750} = \frac{482}{750}$$

When two motors are mechanically coupled and connected in series to 500 V dc supply they draw a current of 50 A. Let the common speed be N rpm.

Sum of back emf's of two motors

$$E'_{b_1} + E'_{b_2} = 500 - 50\,(0.36 + 0.36) = 464 \text{ V}$$

$$\text{Also } E_{b_1} = K_1\Phi_1 N = \frac{482}{700} N$$

$$\text{and } E_{b_2} = \frac{482}{750} N$$

$$\text{So } \frac{482}{700} N + \frac{482}{750} N = 464$$

$$\text{or } N = 348.55 \text{ rpm } \textbf{Ans.}$$

10.5 NECESSITY OF A STARTER

When the motor is at rest the speed of the motor is zero, therefore, back emf E_b is zero and if a dc motor is connected directly to the supply mains, a heavy current will flow through the armature conductors because from the emf equation for armature circuit ($E_b = V - I_a R_a$)* armature current I_a is given as $\frac{V - E_b}{R_a}$ and armature resistance R_a is very small. For example consider a 400 V, 20 kW dc motor having a total resistance of 0.5 Ω. If switched directly on to the supply, it would draw a current of $\frac{400}{0.5} = 800$ A while the full-load current would probably be about 64 A. The starting current thus would be $\frac{800}{64}$ *i.e.* 12.5 times the full-load current and would give rise to heating effects and mechanical forces 150 times the full-load values. When running, of course, the applied voltage V is opposed by the induced back emf and, therefore, a much smaller current flows.

Heavy inrush of current at the starting instant may cause

(*i*) heavy sparking at the commutator and even flash-overs,

(*ii*) damage to the armature winding, either by the heat developed in the windings, or by the mechanical forces set up by electromagnetic action.

(*iii*) damage to the rotating parts of the motor and load due to development of large starting torque and quick acceleration and

(*iv*) large dip in the supply voltage.

Hence for the protection of the motor against the flow of excessive current during starting period (say 5 to 10 seconds), it is necessary that a high resistance be connected in series with the armature of the motor at the instant of starting and gradually cut in steps as the motor gains speed and develops back emf and ultimately when the motor attains its normal speed, the additional resistance from the armature circuit is totally disconnected. If this additional resistance inserted is left in the armature circuit, it would cause (*i*) additional loss of energy resulting in reduced operating efficiency and (*ii*) reduction in operating speed of the motor.

However, a very small motor (fractional kW motor) may be started simply by closing the switch which connects it to the supply mains. The reasons for it are given below.

1. The resistance and inductance of the armature winding in case of small motors are generally sufficiently large to limit the initial inrush of current to values that are not particularly serious.

* EMF equation for armature circuit for a dc shunt motor is $E_b = V - I_a R_a$ whereas for dc series and compound motors it is $E_b = V - I_a (R_a + R_{se})$

2. The inertia of a small armature is generally so low that it comes up to speed very quickly, thereby minimizing the detrimental effects that might otherwise result from the excessive sustained current.

10.6 STARTERS FOR DC SHUNT AND COMPOUND WOUND MOTORS

Motor starters are generally manufactured in convenient sizes and styles for use as auxiliaries with dc shunt and compound motors. Their primary function is to limit the current in the armature circuit during the starting or accelerating period. They are always rated on the basis of output power and voltage of the motors with which they are to be employed. The simplest type of starter consists of an additional resistance (a rheostat) inserted in series with the armature alone (not with the motor as a whole), as illustrated in Fig. 10.21.

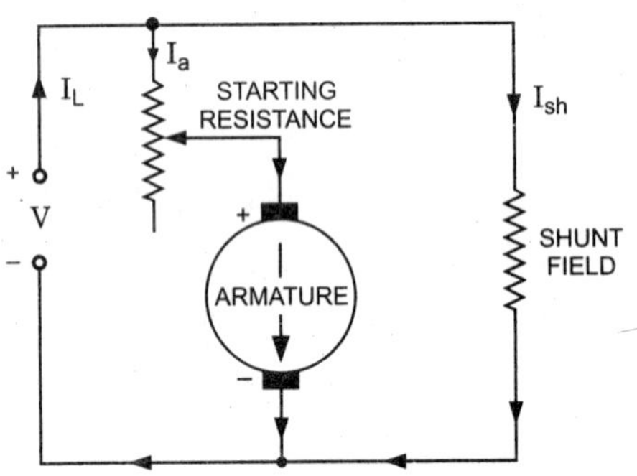

Fig. 10.21

This starter is, however, modified to include a few protective devices, such as no-volt release, overload release etc. There are two standard types of motor starters for dc shunt and compound motors, the three point type and the four point type starters. The four point starter is employed when wide range of speed by shunt field control is required. When no (or little) speed control is required, either type of starter may be used.

In this context it is pertinent to note that while starting dc shunt and compound motors, it is advantageous to keep the field excitation at its maximum value. A large field current, therefore, a higher value of flux will result in a low operating speed and in higher motor torque for a particular value of starting current because motor torque is proportional to the product of flux per pole and armature current. Thus for a given load torque, the motor will accelerate quickly and take less time to reach the lower operating speed from the starting instant. This will result in less heating of the armature during starting. Thus the rheostat, in series with the shunt field winding, should be at zero resistance position at the time of starting of the dc shunt and compound motors.

10.6.1. Three Point Starter. The three point starter with its electrical connections and protective devices (no-volt release and overload release) is illustrated in Fig. 10.22. It consists of a series starting resistance* divided into several sections and connected to brass studs, brass arc by which the connection to shunt field, no-volt release and overload release is made. Since only three terminals (L, F and A) are available from the starter, it is called a *three point starter*. The last stud of the starting resistance is connected to terminal A to which one terminal of the armature is connected. The + ve supply line is connected to the line terminal L through main switch. From line terminal L, supply is connected to the starter arm through overload release. A spiral spring S is placed over the lever to bring the starter arm to the OFF position in case of failure of supply. A soft iron keeper is attached to the starter arm which is pulled by the holding coil or no-volt release under normal running condition. The far end of the brass arc is connected to the terminal F through the no-volt release and to the terminal F one terminal of the field is connected. The –ve supply line is connected directly to the remaining ends of armature and field winding of the dc shunt motor.

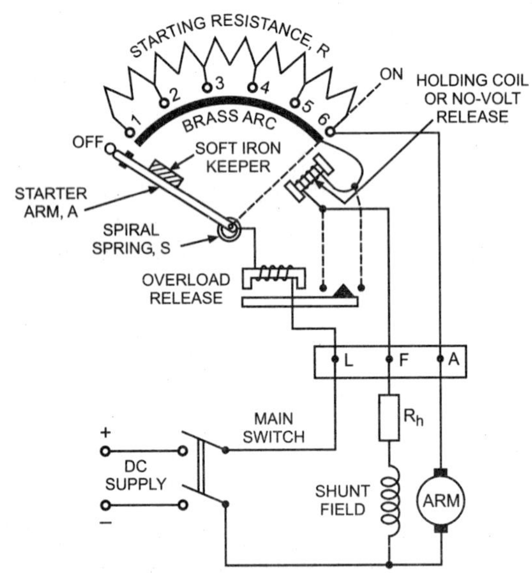

Fig. 10.22 *Three Point Starter*

Operation. When the motor is at rest, the starter arm is in the OFF position due to action of strong spiral spring S. For starting the motor the dc supply is switched on by closing the main switch keeping starter arm in OFF position. The starter arm (or handle) is then turned clockwise to the first stud and brass arc (or strip). As soon as it comes in contact with first stud, whole of the starting resistance R is inserted in series with the armature, the field winding is directly connected across the supply through the brass arc and the holding coil is also energized. As the starter arm is turned further the starting resistance is cut out of the armature circuit in steps and finally entire starting resistance R is cut out of armature circuit. When the starter arm reaches the ON position, it is held against the action of spiral spring S by the force of attraction between holding coil magnet and soft iron keeper attached to the starter arm. The starter arm should not be held for an unduly long time in an intermediate position as it is likely to burn out the starting resistor.

No-Volt Release. When the starter arm reaches "ON" position, the resistance is completely cut off and motor starts running at normal speed. If the supply gets interrupted or

* The resistance is in the armature circuit only during the starting period and, therefore, it need not be wound with such a thick wire, as one which is continuously rated.

disconnected, the starting arm will remain in the same position *i.e.* "ON" position and when the supply is switched on or gets restored, no back emf will be acting in circuit, the armature being directly across the supply mains, and resistance of armature being low, the motor will draw excessive current and will get damaged. Hence for the protection of the motor, some device must be provided, so that starter arm may reach the "OFF" position automatically as soon as the supply is cut off or disconnected or fails and for this purpose *"no-volt"* release coil is provided.

No-volt release coil consists of an electromagnet connected in series with shunt field which holds the arm in the "ON" position. Now when the supply fails or gets disconnected the electromagnet demagnetizes and so releases the starting arm A, which goes back to "OFF" position due to the spring attached to it and gets disconnected from the supply mains.

The other important advantage of connecting the no-volt release in series with the shunt field winding is that it prevents the motor from running away owing to an open shunt field because open-circuited shunt field will demagnetise the electromagnet and release the starter arm A and, thus the starter arm will go back to its OFF position and the supply will be disconnected.

Overload Release Coil. This coil is provided for the protection of the motor against the flow of excessive current due to overload. This coil is connected in series with motor so carries full-load current. When the motor is overloaded, it draws heavy current, which also flows through this coil and magnetises it to such an extent, that it pulls its armature upwards and so short-circuits the no-volt release coil, as shown in Fig. 10.22. The no-volt release coil, being short circuited, demagnetises and releases the starting arm, which goes back to "OFF" position with the action of spring attached to it and the motor is automatically disconnected from the supply mains. Thus the motor is disconnected from the supply and is protected against overloading.

The starters for motors up to 15 kW are provided with overload release to disconnect the motor from the supply mains in the event of an overload. Larger motors are provided with separate automatic circuit breakers.

10.6.2. Four Point Starter. In three point starter (shown in Fig. 10.22) no-volt release coil is connected in series with the shunt field and the field rheostat and, therefore, the current flowing through the field is the same current that flows through the holding coil. Thus if sufficient resistance is *cut in* by the field rheostat so that holding coil current is no longer able to create sufficient electromagnetic pull to overcome the spring tension, the starter arm will fall back to the "OFF" position. It is this undesirable feature of the three point starter that makes it unsuitable for use with speed-controlled motors and that has resulted in the wide spread application of four point starters.

A four point starter with its internal wiring connected to a long shunt compound wound motor is shown in Fig. 10.23. From Fig. 10.23 it is obvious that when the arm touches stud no. 1, line current divides into three parts :

1. One part passes through starting resistance, armature and series field.
2. Second part passes through the shunt field winding.

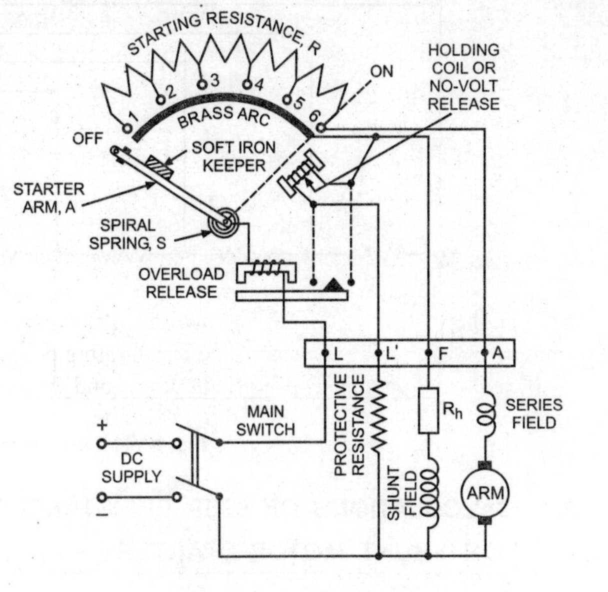

Fig. 10.23 *Four Point Starter*

3. The third part passes through no-volt release coil and protective resistance.

Since in this arrangement, "no-volt release coil" circuit is independent of shunt field circuit, it will not be affected by the change of the current in the shunt field circuit. It means that the electromagnetic pull exerted by the holding coil will always be sufficient and will prevent the spiral spring from restoring the arm to the "OFF" position, no matter how the field rheostat is adjusted.

The possibility of accidently opening the field circuit is quite remote; hence there is greater acceptance of the four point starter over the three point starter.

A motor is started with a four terminal starter in the same way as with a three terminal starter. Any desired speed, above normal, of the motor can be obtained by adjustment of the field rheostat in series with the shunt field.

It is necessary to ensure, before starting a dc motor that the field circuit is closed, the rheostat in series with the shunt field winding is at zero resistance position and starting resistance in series with the armature circuit is at maximum value.

For stopping the motor, the line switch should always be opened rather than throwing back the starting arm. In shunt motors, the line switch can be opened without any appreciable arc, since the motor develops a back emf nearly equal to applied line voltage and the net voltage across the switch contact is small. The electromagnetic energy stored in the field does not appear at the switch but is discharged gradually through the armature. On the other hand if the starting arm is thrown back the field circuit is broken at the last contact button. Owing to the inductive nature of the field, this will cause a hot arc and burn the contact.

Precaution. While stopping the motor, all the resistance in the field rheostat is cut out, so that motor speed falls to its normal value; then the line switch is opened. This procedure ensures that, the next time the motor is started, it will be with a strong field and resultant strong starting torque.

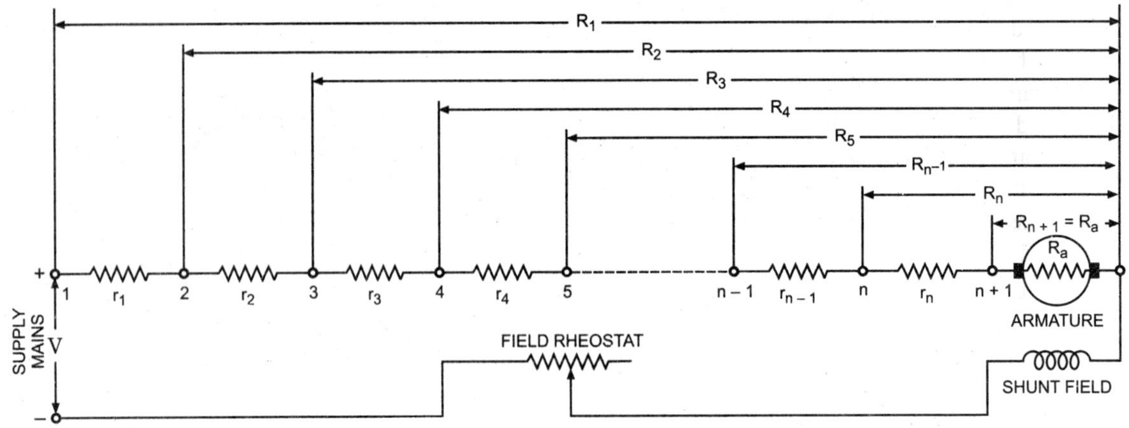

Fig. 10.24 *n-Element Resistance Starter For a DC Shunt Motor*

10.7 CALCULATIONS OF STEP RESISTANCES FOR SHUNT MOTOR STARTER

For starting the motor from standstill to its normal speed, it is always desired that the speed should be increased at constant rate *i.e.* with constant angular acceleration. The angular acceleration depends upon the torque which is proportional to the product of flux Φ and armature current I_a. If supply voltage V remains constant, flux Φ will remain constant and, therefore, for keeping angular acceleration constant armature current I_a should also remain constant.

Since speed, $N \propto \dfrac{V - I_a R}{\Phi}$ where R is the resistance of armature and resistance of starter.

For armature current I_a and flux Φ of constant values and increasing values of speed N it is necessary that R may change satisfying the above expression. For different values of armature current the value of R is given by the equation

$$R = \frac{V - KN}{I_a} \quad \text{where K is a constant}$$

In practice the value of R is changed in steps, and, therefore, armature current will change. The steps of the starter are so designed that the armature current remains within limits which will not change the torque to a greater degree. The variations in motor input current and speed against time are shown in Fig. 10.25.

The fuse or circuit breaker in a dc motor branch circuit is usually not larger than 150% of the motor full-load current. The starter has sufficient resistance to keep the starting current down to about 20%.

Let I_1 and I_2 be the maximum and minimum values of currents during starting and let the starter used for a dc shunt motor have *n*-sections of resistances $r_1, r_2, r_3, r_4, ..., r_n$, as shown in Fig. 10.24. Let the total resistance, when the starting arm is on studs 1, 2, 3, 4, ..., *n* and *n* + 1, be $R_1, R_2, R_3, R_4, ..., R_n$ and R_{n+1} (armature resistance R_a) respectively.

At the instant the starting arm is brought on stud 1, the speed is zero and, therefore, back emf is also zero. Hence maximum value of current, I_1 is given as

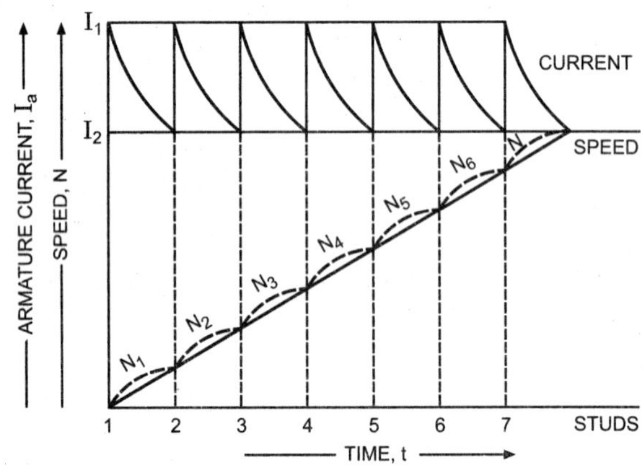

Fig. 10.25

$$I_1 = \frac{V}{R_1} \quad ...(10.8) \text{ where V is supply voltage.}$$

The motor now speeds up and develops a back emf E_b and, therefore, armature current drops to I_2, the arm remaining on the same stud.

$$I_2 = \frac{V - E_b}{R_1}$$

Since back emf is proportional to speed, flux remaining constant, the expression for current I_2 may be written as

$$I_2 = \frac{V - KN_1}{R_1} \quad ...(10.9)$$

where N_1 is the speed attained by the motor and K is a constant.

Now the starter arm is moved to stud 2, the armature current instantaneously rises to value I_1 but the speed has no time to change from N_1

$$\therefore I_1 = \frac{V - KN_1}{R_2} \quad ...(10.10)$$

At the instant of leaving stud 2 the speed has increased to say N_2

$$\therefore I_2 = \frac{V - KN_2}{R_2} \quad ...(10.11)$$

On first making contact with stud 3

$$I_1 = \frac{V - KN_2}{R_3} \quad ...(10.12)$$

Dividing Eq. (10.10) by Eq. (10.9) and Eq. (10.12) by Eq. (10.11) we get

$$\frac{R_1}{R_2} = \frac{R_2}{R_3} = \frac{I_1}{I_2} = \gamma \text{ (say)} \quad ...(10.13)$$

$$\text{or } \frac{R_1}{R_2}\times\frac{R_2}{R_3}\times\frac{R_3}{R_4}\times\frac{R_4}{R_5}\times\cdots\times\frac{R_{n-1}}{R_n}\times\frac{R_n}{R_a} = \gamma^n$$

$$\text{or } \frac{R_1}{R_a} = \gamma^n$$

$$\text{or } \gamma = \sqrt[n]{\frac{R_1}{R_a}} \quad ...(10.14)$$

If the values of armature resistance R_a, limiting currents I_1 and I_2 are known, the values of R_1 and γ can be determined from Eqs. (10.8) and (10.13) respectively and the required number of sections, n can be determined from Eq. (10.14).

Knowing the values of R_1 and γ the values of R_2, R_3, R_4, ...R_n can be determined from Eq. (10.13) and then the step resistances may be calculated as follows :

$$r_1 = R_1 - R_2;\ r_2 = R_2 - R_3;\ r_3 = R_3 - R_4;$$
$$r_{n-1} = R_{n-1} - R_n;\ r_n = R_n - R_a$$

Example 10.29. A 250 V, 25 kW dc shunt motor has an efficiency of 85% when running at 1,000 rpm on full load. The armature resistance is 0.1 Ω and field resistance is 125 Ω. Find the starting resistance required to limit the starting current to 150% of the rated current.

Solution: At full load

$$\text{Input power} = \frac{\text{Full-load power output}}{\text{Efficiency}} = \frac{25}{85/100} = 29.412 \text{ kW}$$

$$\text{Line current, } I_{Lf} = \frac{\text{Input power}}{\text{Supply voltage}} = \frac{29.412 \times 1{,}000}{250} = 117.65 \text{ A}$$

$$\text{Shunt field current, } I_{sh} = \frac{V}{R_{sh}} = \frac{250}{125} = 2 \text{ A}$$

Limiting value of starting current, I_{Ls}

$$= 1.5\, I_{Lf} = 1.5 \times 117.65 = 176.5 \text{ A}$$

Limiting value of armature current, I_{as}

$$= I_{Ls} - I_{sh} = 176.5 - 2 = 174.5 \text{ A}$$

Total resistance in the armature circuit at starting,

$$R_1 = \frac{V}{I_{as}} = \frac{250}{174.5} = 1.433\ \Omega$$

External resistance to be added in the armature circuit

$$= R_1 - R_a = 1.433 - 0.1 = 1.333\ \Omega \quad \textbf{Ans.}$$

Example 10.30. Explain what would happen if the dc motor is directly switched on to the supply, without any starter. A 240 V, 50 A, 1500 rpm dc shunt motor has an armature resistance of 0.2 Ω. Calculate the values of various steps of starter with armature current not exceeding 1.4 pu. Neglect shunt field current.

[*U.P. Technical Univ. Electromechanical Energy Conversion-I, 2005-06*]

Solution: Base voltage, V = 240 V = 1.0 pu

Base armature current, I_a = 50 A = 1.0 pu

$$\text{Base resistance, } R_b = \frac{\text{Base voltage}}{\text{Base current}} = \frac{240}{50} = 4.8\ \Omega$$

$$\therefore \text{ Per unit armature resistance, } R_a = \frac{0.2}{4.8} = 0.041667$$

At the instant the motor is switched on, the total resistance R_1 in the armature circuit

$$R_1 = \frac{V}{I_{max}} = \frac{1}{1.4} = 0.714 \text{ pu or } 0.714 \times 4.8 = 3.43\ \Omega$$

$$\gamma = \frac{I_{max}}{I_{min}} = \frac{1.4}{1} = 1.4 \quad \text{Assuming } I_{min} = \text{Rated current}$$

$$\text{and number of steps required, } n = \frac{\log\frac{R_1}{R_a}}{\log\gamma} \qquad \because \gamma = \left(\frac{R_1}{R_a}\right)^{1/n}$$

$$= \frac{\log\frac{3.43}{0.2}}{\log 1.4} = 8$$

$$\text{Resistance } R_2 = \frac{R_1}{\gamma} = \frac{3.43}{1.4} = 2.45\ \Omega$$

$$\text{Resistance } R_3 = \frac{R_2}{\gamma} = \frac{2.45}{1.4} = 1.75\ \Omega$$

$$\text{Resistance } R_4 = \frac{R_3}{\gamma} = \frac{1.75}{1.4} = 1.25\ \Omega$$

$$\text{Resistance } R_5 = \frac{R_4}{\gamma} = \frac{1.25}{1.4} = 0.89\ \Omega$$

$$\text{Resistance } R_6 = \frac{0.89}{1.4} = 0.636\ \Omega$$

$$\text{Resistance } R_7 = \frac{0.636}{1.4} = 0.454\ \Omega$$

$$\text{Resistance } R_8 = \frac{0.454}{1.4} = 0.324\ \Omega$$

Resistance of first section, $r_1 = R_1 - R_2 = 3.43 - 2.45 = 0.98\ \Omega$

Resistance of second section, $r_2 = R_2 - R_3 = 2.45 - 1.75 = 0.7\ \Omega$

Resistance of third section, $r_3 = R_3 - R_4 = 1.75 - 1.25 = 0.5\ \Omega$

Resistance of fourth section, $r_4 = R_4 - R_5 = 1.25 - 0.89 = 0.36\ \Omega$

Resistance of fifth section, $r_5 = R_5 - R_6 = 0.89 - 0.636 = 0.254\ \Omega$

Resistance of sixth section, $r_6 = R_6 - R_7 = 0.636 - 0.454 = 0.182\ \Omega$

Resistance of seventh section, $r_7 = R_7 - R_8 = 0.454 - 0.324 = 0.130\ \Omega$

Resistance of eighth section, $r_8 = R_8 - R_a = 0.324 - 0.2 = 0.124\ \Omega$

Ans.

Example 10.31. A 230 V shunt motor has an armature resistance of 0.4 Ω. The starting armature current must not exceed 45 A. If the number of sections are 5, calculate the values of resistance steps to be used in the starter.

Solution: Number of sections, n = 5

Resistance in the armature circuit at the starting instant

$$R_1 = \frac{V}{I_{max}} = \frac{230}{45} = 5.11\ \Omega$$

Ratio of maximum current to minimum current during starting

$$\frac{I_{max}}{I_{min}} = \gamma = \left(\frac{R_1}{R_a}\right)^{1/n} = \left(\frac{5.11}{0.4}\right)^{1/5} = 1.6645$$

$$\text{Resistance, } R_2 = \frac{R_1}{\gamma} = \frac{5.11}{1.6645} = 3.07\ \Omega$$

$$\text{Resistance, } R_3 = \frac{R_2}{\gamma} = \frac{3.07}{1.6645} = 1.844\ \Omega$$

$$\text{Resistance, } R_4 = \frac{R_3}{\gamma} = \frac{1.844}{1.6645} = 1.108\ \Omega$$

$$\text{Resistance, } R_5 = \frac{R_4}{\gamma} = \frac{1.108}{1.6645} = 0.666\ \Omega$$

Resistance of 1st section, $r_1 = R_1 - R_2 = 5.11 - 3.07 = 2.04\ \Omega$ **Ans.**

Resistance of 2nd section, $r_2 = R_2 - R_3 = 3.07 - 1.844 = 1.226\ \Omega$ **Ans.**

Resistance of 3rd section, $r_3 = R_3 - R_4 = 1.844 - 1.108 = 0.736\ \Omega$ **Ans.**

Resistance of 4th section, $r_4 = R_4 - R_5 = 1.108 - 0.666 = 0.442\ \Omega$ **Ans.**

Resistance of 5th section, $r_5 = R_5 - R_a = 0.666 - 0.4 = 0.266\ \Omega$ **Ans.**

Example 10.32. A 400 V dc shunt motor taking current of 66.67 A on full load. Calculate number of resistance sections, and resistance of each section, when $R_a = 0.5\ \Omega$ and I_a should not exceed 100 A.

[J.N. Technological Univ. March-2014]

Solution: Armature resistance, $R_a = 0.5\ \Omega$

Ratio of maximum current to minimum current,

$$\gamma = \frac{I_{max}}{I_{min}} = \frac{100}{66.67} = 1.5$$

Resistance in the armature circuit at the starting instant,

$$R_1 = \frac{400}{100} = 4\ \Omega$$

Let the number of resistance sections required be n

$$\text{then } \gamma = \left(\frac{R_1}{R_a}\right)^{1/n}$$

$$\text{or } n = \frac{\log_{10} R_1/R_a}{\log_{10}\gamma} = \frac{\log_{10} 4/0.5}{\log_{10} 1.5} = \frac{0.90309}{0.1761} = 5 \textbf{ Ans.}$$

$$\text{Resistance, } R_2 = \frac{R_1}{\gamma} = \frac{4}{1.5} = 2.667\ \Omega$$

$$\text{Resistance, } R_3 = \frac{R_2}{\gamma} = \frac{2.667}{1.5} = 1.778\ \Omega$$

$$\text{Resistance, } R_4 = \frac{R_3}{\gamma} = \frac{1.778}{1.5} = 1.185\ \Omega$$

$$\text{Resistance, } R_5 = \frac{R_4}{\gamma} = \frac{1.185}{1.5} = 0.79\ \Omega$$

Resistance of 1st section, $r_1 = R_1 - R_2 = 4 - 2.667 = 1.333\ \Omega$ **Ans.**

Resistance of 2nd section, $r_2 = R_2 - R_3 = 2.667 - 1.778 = 0.889\ \Omega$ **Ans.**

Resistance of 3rd section, $r_3 = R_3 - R_4 = 1.778 - 1.185 = 0.593\ \Omega$ **Ans.**

Resistance of 4th section $= R_4 - R_5 = 1.185 - 0.79 = 0.395\ \Omega$ **Ans.**

Resistance of 5th section $= R_5 - R_a = 0.593 - 0.4 = 0.193\ \Omega$ **Ans.**

Example 10.33. The 4-pole, lap wound armature winding of a 500 V, dc shunt motor is housed in a total number of 60 slots, each slot containing 20 conductors. The armature resistance is 1.31 Ω. If during the period of starting the minimum torque is required to be 218 Nm and the maximum torque 1.5 times the minimum torque, find out how many sections the starter should have and calculate the resistances of these sections. Take the useful flux per pole to be 23 mWb.

Solution: Since torque developed is given by the expression

$$T = 0.159\ \Phi\ Z\ I_a \frac{P}{A};$$

$$\text{Minimum current, } I_{min} = \frac{T_{min} \times A}{0.159 \times \Phi \times Z \times P} = \frac{218 \times 4}{0.159 \times 23 \times 10^{-3} \times 60 \times 20 \times 4} = 49.68\ A \simeq 50\ A$$

$$\text{Maximum current, } I_{max} = I_{min} \times \frac{T_{max}}{T_{min}} = 50 \times 1.5 = 75\ A$$

$\because T \propto \Phi I_a$ and Φ is constant

Resistance in the armature circuit at the starting instant,

$$R_1 = \frac{V}{I_{max}} = \frac{500}{75} = 6.67\ \Omega$$

Ratio of maximum current to minimum current, $\gamma = 1.5$

Number of resistance sections required, n

$$= \frac{\log R_1/R_a}{\log\gamma} = \frac{\log\frac{6.67}{1.31}}{\log 1.5} = 4 \textbf{ Ans.}$$

$$\text{Resistance, } R_2 = \frac{R_1}{\gamma} = \frac{6.67}{1.5} = 4.44\ \Omega$$

$$\text{Resistance, } R_3 = \frac{R_2}{\gamma} = \frac{4.44}{1.5} = 2.96\ \Omega$$

$$\text{Resistance, } R_4 = \frac{R_3}{\gamma} = \frac{2.96}{1.5} = 1.98\ \Omega$$

Resistance of 1st section, $r_1 = R_1 - R_2 = 6.67 - 4.44 = 2.23\ \Omega$ **Ans.**

Resistance of 2nd section, $r_2 = R_2 - R_3 = 4.44 - 2.96 = 1.48\ \Omega$ **Ans.**

Resistance of 3rd section, $r_3 = R_3 - R_4 = 2.96 - 1.98 = 0.98\ \Omega$ **Ans.**

Resistance of 4th section, $r_4 = R_4 - R_a = 1.98 - 1.31 = 0.67\ \Omega$ **Ans.**

10.8 SERIES MOTOR STARTER

The series motor starter serves the same purpose as the three and four point starters employed with shunt and compound wound motors. However, series motor starter has different internal and external connections.

A series motor starter with holding coil and overload release is illustrated in Fig. 10.26. Holding coil protects the motor against "no-volt" and "no-load" whereas overload release protects the motor against the flow of excessive current.

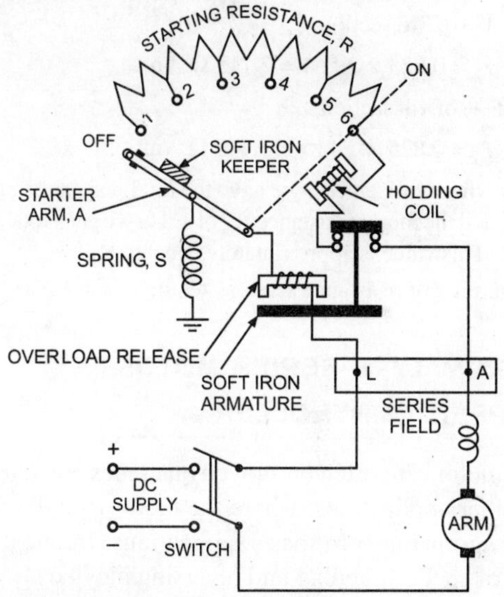

Fig. 10.26 *Face-Plate Starter For DC Series Motor*

The dc supply is switched on and the starter arm is moved to the right for starting the motor. As soon as the starter arm comes in contact with the stud No. 1, the series motor gets connected, through the total starting resistance, to the supply mains and when the starter arm reaches the "ON" position the starting resistance is completely cut off and the motor starts running at normal supply. The arm is held in "ON" position by the holding coil. Should the supply fail while the motor is running, or the motor current falls to too low a value (and consequently the speed becomes excessive), the pull of the holding coil is reduced to an extent that it can no longer hold the arm against the spring action attached to it and the starter arm flies back to the OFF position and thus the motor gets disconnected automatically from the supply mains. Overload release is connected in series with the motor, so carries the total motor current. When the input current to the motor becomes excessive, which also flows through the overload release coil, the soft iron armature gets lifted to break the holding coil circuit. Thus the holding coil gets demagnetised and the starter arm returns back to the OFF position under the action of the spring attached to it and the motor gets disconnected automatically from the supply mains.

For switching off the dc series motor, the line switch should always be opened rather than throwing back the starting arm, as in case of dc shunt and compound motors. If it is done, then heavy sparking is caused at the stud 1, because here the field circuit is broken and the entire energy stored in the magnetic field is dissipated in the form of heavy spark.

However, there is one significant difference. In case of a dc series motor, the flux does not remain constant but varies with the current because the line current, field current and armature current are the same. Thus the back emf at any given speed varies as the current varies between the upper and lower limits. The practical result of this is that a series motor starter has a smaller number of steps than that required for a starter of a dc shunt motor of the same rating with the same current limits. This is because an increase in current causes increase in back emf and thus the current rise tends to be self-limiting. As a result of the fewer steps, the resistance of each section of the starter resistor is greater than for the shunt motor starter.

Since flux varies during the starting operation and its relation to the field current is nonlinear, the determination of the number of steps is rather complicated.

It will be found that the differences between the values of the resistances of the various sections are proportionately less than those for the shunt motor starter, although the number of sections is smaller.

In practice the values are not critical in the series motor starter and it may be possible to use identical values for all the sections. This is due to the self regulating effect, mentioned above.

Example 10.34. A 440 V series motor has a load resistance of 0.8 Ω. It takes 110 A to give the rated output at 800 rpm. Determine the resistance of the resistor to be added to obtain rated torque (*i*) At starting (*ii*) A 600 rpm.

Assume flux is proportional to current.

[*U.P. Technical Univ. Electrical Machines and Automatic Control, 2009-10*]

Solution: Since the torque remains the same in both the cases obviously the current drawn by the motor remains constant at 110 A. This is because $T \propto \Phi I$.

Since current drawn by motor, $I = \dfrac{V - E_b}{R_a + R_{se} + R}$

where R is external resistance inserted in series with motor circuit

$$(i) \text{ or } 110 = \frac{440 - 0}{0.8 + R} \qquad \because \text{ At start } E_b = 0$$

$$\text{or } R = \frac{440}{110} - 0.8 = 3.2\ \Omega \textbf{ Ans.}$$

$$\text{Back emf, } E_{b_1} = V - I(R_a + R_{se})$$

$$= 440 - 110 \times 0.8 = 352 \text{ V}$$

$$(ii) \text{ Back emf, } E_{b_2} = E_{b_1} \times \frac{N_2}{N_1} = 352 \times \frac{600}{800} = 264 \text{ V}$$

$$\text{Since } I = \frac{V - E_b}{R_a + R_{se} + R'}$$

$$\text{or } 110 = \frac{440 - 264}{0.8 + R'}$$

$$\text{or } R' = \frac{176}{110} - 0.8 = 0.8\ \Omega \textbf{ Ans.}$$

Example 10.35. Estimate the number of resistance sections and the resistance of each section for the starter of 7.5 kW, 460 V dc series motor. The starting current varies from 1.5 to 2 times full-load current. The resistance of the machine measured between terminals is 1.8 Ω and the efficiency is 80%.

Assume that the flux density increases by 10% as the current rises from 1.5 to 2 times the rated full-load current.

[*J.N. Technological Univ. Electrical Machines-I, March-2014*]

Solution: Full load current, $I_f = \dfrac{\text{Motor output in watts}}{\text{Supply voltage} \times \text{motor efficiency}}$

$$= \frac{7{,}500}{460 \times 0.8} = 20.38 \text{ A}$$

Maximum allowable current, $I_1 = 2 \times 20.38 = 40.76$ A

Minimum allowable current, $I_2 = 1.5 \times 20.38 = 30.57$ A

Ratio of maximum current to minimum current,

$$\gamma = \frac{I_1}{I_2} = \frac{40.76}{30.57} = 1.333$$

If Φ_1 is the flux for maximum current I_1 and Φ_2 is the flux for the minimum current I_2

$$\beta = \frac{\Phi_1}{\Phi_2} = 1.1$$

$\because$ Φ_1 is 10% more than that of Φ_2

Let the number of resistance sections be n.

When the starter arm is on nth stud, immediately before notching up from nth stud to the $(n + 1)$ th stud

$$I_2 = \frac{V - E_b}{R_n}$$

$$\text{or } E_b = V - I_2 R_n$$

If, now, the starter arm is moved to $(n + 1)$th stud, then

$$E'_b = \frac{\Phi_1}{\Phi_2} E_b = \beta E_b$$

$$\text{and } R_n + 1 = \frac{V - E'_b}{I_1}$$

$$= \frac{V - \beta E_b}{I_1}$$

$$= \frac{V - \beta(V - I_2 R_n)}{I_1} = \frac{V}{I_1}(1 - \beta) + \beta \frac{I_2}{I_1} R_n$$

Now $\frac{V}{I_1} = R_1$, total resistance in the circuit when the starter arm is on the first stud.

$$\text{So } R_{n+1} = R_1 (1 - \beta) + \frac{\beta}{\gamma} R_n$$

$$\text{Similarly } R_n = R_1 (1 - \beta) + \frac{\beta}{\gamma} R_{n-1} \quad ...(i)$$

Resistance of nth section, r_n

$$= R_n - R_{n+1} = \frac{\beta}{\gamma}(R_{n-1} - R_n) = \frac{\beta}{\gamma} r_{n-1}$$

$$\text{or } \frac{r_n}{r_{n-1}} = \frac{\beta}{\gamma} = \frac{\Phi_1}{\Phi_2} \times \frac{I_2}{I_1} = K \text{ Constant} \quad ...(ii)$$

Obviously, the section resistance form a geometric progression series with the ratio K.

From Eq. (i) substituting $n = 2$

$$R_2 = R_1 (1 - \beta) + \frac{\beta}{\gamma} R_1$$

and resistance of first section, r_1

$$= R_1 - R_2 = R_1 - R_1 (1 - \beta) - \frac{\beta}{\gamma} R_1$$

$$\text{or } r_1 = R_1 (1 - 1 + \beta - K) = R_1 (\beta - K) \quad ...(iii)$$

Total resistance in the armature circuit

$$R_1 = \frac{V}{I_1} = \frac{460}{40.76} = 11.3\ \Omega$$

$$K = \frac{\beta}{\gamma} = \frac{1.1}{1.33} = 0.825$$

Resistance of 1st section, r_1

$$= R_1 - R_2 = R_1 (\beta - K) = 11.3 (1.1 - 0.825) = 3.108\ \Omega \textbf{ Ans.}$$

Resistance of 2nd section, r_2

$$= Kr_1 = 0.825 \times 3.108 = 2.564\ \Omega \textbf{ Ans.}$$

Resistance of 3rd section, r_3

$$= Kr_2 = 0.825 \times 2.564 = 2.115\ \Omega \textbf{ Ans.}$$

Resistance of 4th section, r_4

$$= Kr_3 = 0.825 \times 2.115 = 1.745\ \Omega \textbf{ Ans.}$$

Adding these resistances, we have the total resistance of 9.532 Ω. To this if we add the motor resistance R_m of 1.8 Ω we get a total resistance of 11.332 Ω. This value is approximately equal to R_1.

Thus number of resistance sections required, $n = 4$ **Ans.**

10.9 DRUM TYPE SERIES MOTOR SPEED CONTROLLER

DC series motors are often employed on cranes, elevators, tram-cars and other applications, where the motor is under the direct control of an operator. In these applications, frequent starting, variations of speed, stopping and reversing may be necessary. A manually operated controller, more rugged than a starting rheostat, called a *drum controller,* is emp-loyed.

The controller is in the form of rotating drum having segments which makes contact with the fixed points. Speed controller of this type is shown in Fig. 10.27 in which K is an arc braking coil; EM is a braking electromagnet; FLS is a switch for rotation in one direction and RLS is a switch for rotation in reverse direction. The controller has six positions for forward and six positions for reverse rotation of the motor. The working positions of the controller are shown by vertical dotted lines. The electromagnet EM, is connected in parallel with the motor and releases the motor at starting. When the motor is disconnected it is braked mechanically.

Now when the knife blade switch is closed and the controller is set in forward position 1, the connections are along vertical line 1. In this position segments make contact with fixed points 6 and 7, 8 and 9. In this position current flows from + ve bus-bar through armature winding of the motor, arc braking coil K, all the starting resistances 1–6, fixed points 6 and 7 through controller segments, series field winding, the forward limit switch FLS, fixed points 9 and 8 through controller segments and returns back to the negative bus-bar.

In the second forward position, the segments of the controller come in contact with fixed points 5 and 6 thereby bringing some of the starting resistance out of circuit, so speed of the series motor increases. In the subsequent forwarding positions 3rd, 4th and 5th, additional steps of the starting resistances are brought out of the circuit and finally in 6th position all the starting resistances are short circuited and motor attains maximum speed.

In all the six forward positions the direction of current in armature as well as in series field winding is same (from right to left), as shown in Fig. 10.27.

But in first reverse position the current flows from + ve bus-bar through armature winding, arc braking coil K, all the starting resistances 1–6, fixed points 6 and 10 through controller segments, the reverse limit switch RLS, the series field winding, fixed points 7 and 8 through controller segments and returns

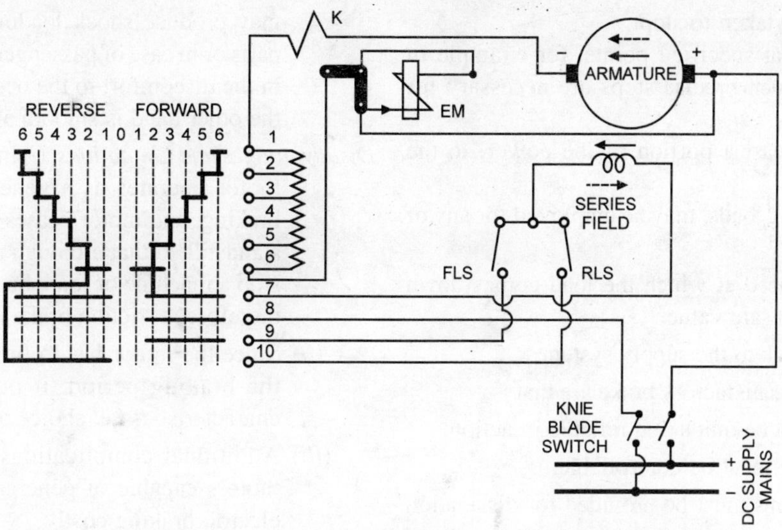

Fig. 10.27 *Drum Type Series Motor Speed Controller*

back to the – ve bus-bar. Hence it is obvious that in reversing positions direction of flow of current in armature winding remains unchanged, while reverses through the series field winding, thereby reversing the direction of rotation of motor.

10.10 AUTOMATIC STARTERS

Push-button type of automatic starters are quite common in use in industry. Even an inexperienced operator, with the help of such starters, called the *auto-starters*, can start and stop the motor without any difficulty.

The operation of such starters depends upon, either the time delay or the counter emf developed across the armature terminals. Here only the basic principle of the counter-emf starter will be discussed.

Counter-EMF Starter. This type of starter depends upon the build up of the back or counter emf to operate the contactors (a heavy duty relay designed to open or close an electrical power circuit), which in turn causes the starting resistors to be shorted. The connections of a counter-emf starter are shown in Fig. 10.28.

When the motor is switched on, the counter emf developed by the motor is zero and the voltage across contactor coil A or B is insufficient to energize the relays. Consequently the contactors

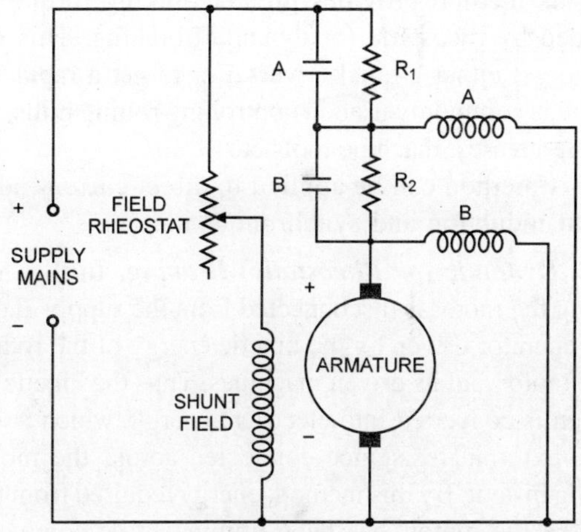

Fig. 10.28

A and B are normally open and, therefore, starting resistances R_1 and R_2 are in series with the armature. As the motor picks up speed and builds up a counter emf the voltage across coils A and B increases until the operating voltage of coil A is reached. Contactor A is then energized, closing the normally open contact A, shorting out the starting resistor R_1. The motor continues to pick up speed building up the counter-emf until the operating voltage of coil B is reached. Contactor B is energized, closing normally open contactor B, which in turn shorts out the starting resistor R_2 and places the armature directly across the supply mains.

The drawback of counter-emf automatic starter is that if the motor fails to start, the counter-emf remains zero, the voltage sensitive relays cannot operate and so the starting resistances may burn out. Such occurrences can, however, be avoided by using definite time-limit starters.

10.11 ELECTRIC BRAKING

If the load is removed from an electric motor and supply to it be disconnected, it will continue to run for some time due to inertia. The time elapsing before it stops will be especially long if the motor is massive and has run at high speed. It is essential, however, in many cases that the motor and its driven machine be stopped quickly (in machine tools, cranes, hoists etc). In fact, quick stopping of a motor is more essential than quick starting. Delay in starting up a motor only causes the machinery to stand idle; a delay in stopping a motor may result in heavy damage to equipment or to the manufactured products and even the loss of human life.

Based on the purpose for which braking is employed, it is of the two types, viz, braking while bringing the drive to rest and braking while lowering loads. In the first type, the device employed for braking absorbs the kinetic energy of the moving parts while in the second one, it absorbs, in addition to the kinetic energy, potential energy, usually gravitational, which can drive the system at an excessively high speed.

Braking, while stopping, may be used for any one of the following objectives:

(*i*) Reducing the time taken to stop.

(*ii*) Stopping exactly at specified points, for example in lifts; sometimes such precise stops are necessary for reasons of safety.

(*iii*) Feeding back, atleast a portion of the power, to the supply system.

Braking, while lowering loads, may be employed for any of the following purposes:

(*i*) Controlling the speed at which the load comes down and limiting it to a safe value.

(*ii*) Feeding power back to the supply system.

The main features of a satisfactory brake are that

(*i*) the braking should be quick and reliable in action,

(*ii*) the braking torque must be controllable,

(*iii*) some suitable means must be provided for dissipation of kinetic energy of the moving parts of the motor and its driven machines and

(*iv*) failure of any part of the braking system must result in application of brakes.

Braking torque can be applied either by mechanical (or friction) brakes or electrodynamically. In mechanical braking, the stored energy of rotating parts is dissipated in the form of heat by a brake shoe or band rubbing on a wheel or a brake drum. In electrical (or electrodynamic) braking the stored energy of rotating parts is converted into electrical energy and dissipated in the resistance in the form of heat or returned to the supply. Here we will deal with electric (or electrodynamic) braking only.

10.11.1. Advantages and Disadvantages of Electric Braking Over Mechanical Braking. Before we deal with electric braking it will be desirable to have the knowledge of advantages and limitations of electric braking over mechanical braking.

Advantages. (*i*) Since mechanical braking, due to excessive wear on the brake blocks or brake lining, needs frequent and costly replacement but no such replacement is required in electric braking, there is a saving on this account.

(*ii*) Frequent adjustments are required to compensate for the wear in mechanical braking thereby making maintenance costly. In electric braking there is no cost by way of maintenance of the brake shoes or lining.

(*iii*) By employing electric braking the capacity of the system can be increased by way of higher speeds and haulage of heavy loads.

(*iv*) In some cases of electric braking, a part of energy is returned to the supply thereby affecting a considerable saving in the operating cost. This is not possible with mechanical braking.

(*v*) In electric braking, heat produced, in no way, is harmful to the braking system whereas in mechanical braking heat is produced at brake blocks or brake lining which may result in failure of brake.

(*vi*) Mechanical braking produces metal dust which, if finds ingress to bearings, will cause overwearing of the bearings. No such dust is produced in electric braking.

(*vii*) Incorrect adjustment of mechanical brakes may result in very harsh and sudden braking being applied which may produce shock loading of ropes and mechanical parts or in case of passenger lifts and trains it may result in the discomfort to the occupants. Electric braking on the other hand is smooth and without shock.

Disadvantages and Limitations. (*i*) In electric braking, the driving motor operates as a generator during the period of braking, and motor ceases to operate as a generator at standstill so that although an electric brake can almost stop a machine or load, but it cannot hold it stationary, therefore, a friction brake is required in addition.

(*ii*) Since the motor has to function as a generator during the braking period, it must have suitable braking characteristics i.e. choice of motor is limited.

(*iii*) Additional complications, high initial cost, special motors capable of generating electrical energy make electric braking costly.

Economic considerations have a considerable bearing on the use of electric braking.

10.11.2. Types of Electric Braking. There are three types of electric braking, all of which are applicable to the usual types of electric motors, viz plugging (or counter-current braking), dynamic (or rheostatic) braking and regenerative braking.

1. *Plugging or Counter-Current Braking.* This is the simplest type of braking. Plugging or counter-current braking occurs when the motor windings are connected for reverse direction of rotation at a time when the armature is still rotating in the forward direction either under the action of an external torque or due to inertia. Plugging involves reconnection of power supply to the motor so that it tends to drive in the opposite direction. It is obvious that, left to itself, the system will come to rest and then accelerate in the reverse direction. In case, it is required to bring the drive system to rest, it is necessary to include a special device to cut off the supply exactly at the instant when the motor stops.

During the braking period, the energy is drawn from the supply. The energy drawn from the supply and stored or kinetic energy of the rotating parts of motor and its driven machines are dissipated in the series current limiting resistor. Thus this method is wasteful of energy. Other drawbacks of this method of electric braking are the shock caused to the motor and equipment and heavy inrush of current at the time of braking.

This method provides greater braking torque than that provided by rheostatic (or dynamic) braking. This method is used to get either a quick reversal or to get a rapid stop. This method is commonly used in controlling rolling mills, elevators, printing presses, machine tools etc.

This method can be applied to direct current, alternating current induction and synchronous motors.

2. *Dynamic (or Rheostatic) Braking.* In this method of braking the motor is disconnected from the supply and operated as a generator driven by the kinetic energy of the rotating parts of the motor and its driven machines. Thus the kinetic energy of rotation is converted into electrical energy, which is dissipated in the external resistance connected across the motor at the braking instant. By this method, energy required from the supply to brake the motor, has been eliminated as compared to the previous method (plugging) and this is marked advantage. This method of braking can be applied to brake the direct current motors, synchronous motors, and induction motors (with a separate source of *dc* excitation during braking).

3. *Regenerative Braking.* In the previous two methods of electric braking namely plugging and rheostatic braking stored energy of the rotating parts of the motor and its driven machine is wasted whilst in plugging extra energy is drawn during the braking period and is wasted. In regenerative braking, mechanical energy is converted into electrical energy, part of which is returned to the supply and the rest of the energy is lost as heat in the windings and the bearings of the electrical machines. Regeneration does not, in general, involve any switching operation, unless it is required to change the speed at which it becomes effective. Most of the electrical machines pass smoothly from motoring to generating operation, when overdriven by load.

For regenerative braking it is necessary for (*i*) supply voltage to drop, (*ii*) the motor to be overexcited or (*iii*) the motor to be running at a speed higher than no-load speed. In all of these cases the armature current is reversed and a retarding torque is produced, slowing down the motor until the back emf of the armature becomes equal to the supply voltage. Regenerative braking will not stop the motor. It is effective only for braking hauling loads. They may occur in a crane motor due to action of the lowering load or in the motor of an electric locomotive moving down grade.

10.12 ELECTRIC BRAKING WITH DC MOTORS

10.12.1. Plugging or Counter-Current Braking. In a dc motor a reversed torque is obtained by reversing the current either in the armature or in the field (not both). Polarity reversal of field winding is rarely used because it results in longer braking time due to relatively large inductance of the field winding in comparison to that can be obtained by polarity reversal of armature winding.

The connections for dc series and shunt motors during normal running and braking conditions are shown in Figs 10.29 and 10.30 respectively. In case of a dc series motor it should be ensured that the direction of flow of current in the field winding remains unchanged when the current flow in the armature winding is reversed.

In the normal running position the back emf is nearly equal to applied voltage and opposite in direction, so that a small voltage acts across the armature circuit to drive the normal current through a small resistance of the motor. In the plugging position, the induced emf acts in the direction of applied voltage, therefore, at the instant of switching the motor to the plugging position, twice of supply voltage acts across the machine circuit and heavy current would flow (about twice the current drawn by the stationary motor on normal rated voltage). Hence to avoid flow of heavy current and limit it to the safer value, it is necessary that switching performing the plugging operation may also re-insert starting resistance and some additional resistance in series with the armature circuit of the motor.

In case the motor is driving a load at the instant of braking, the load will also exert an additional torque and the total braking torque will be that due to the electric braking and the load.

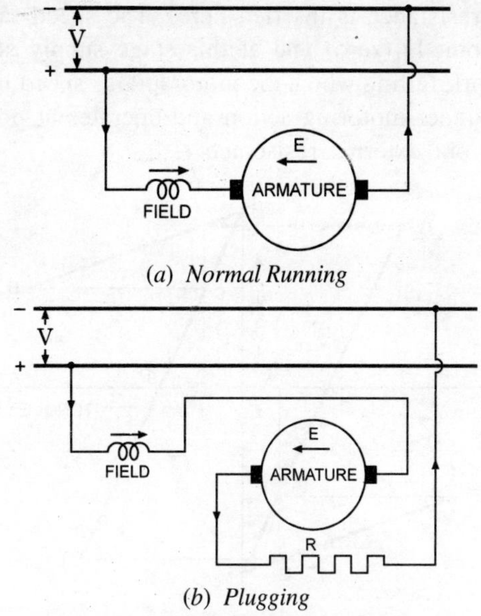

Fig. 10.29 *Plugging With DC Series Motors*

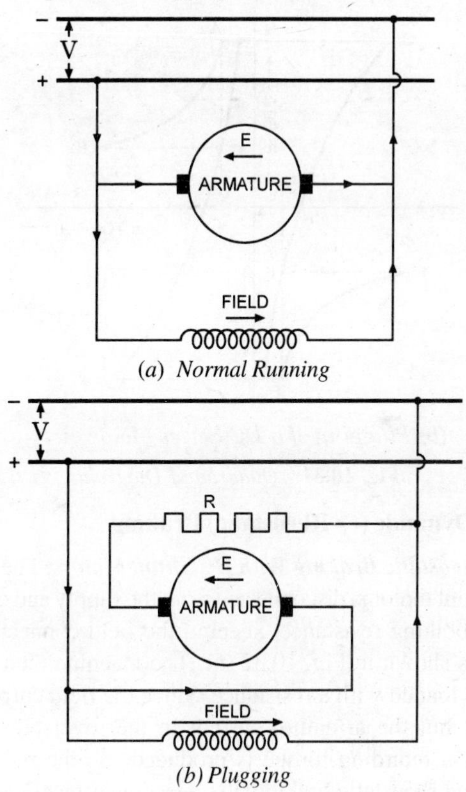

Fig. 10.30 *Plugging With DC Shunt Motors*

Figures 10.31 (*a*) and 10.31 (*b*) explain the plugging operation of the dc shunt and series motors respectively on a *quadrantal diagram.*

AB and EF represent the speed-torque characteristics of the motor in normal and reverse direction of rotation without any external resistance, while CL and KG represent the same with external resistances respectively. A is the initial operating point with load, speed N and load torque T_L and the external resistance zero. When plugging is resorted, the motor continues to run in the normal direction but develops torque in reverse direction and point G is the operating point (on characteristic

KHG, as resistance is inserted here). The speed falls till it reaches point H (zero) and at this stage supply should be switched off, failing which the motor attains speed in reverse direction under motoring action and operates at point E on switching out external resistance.

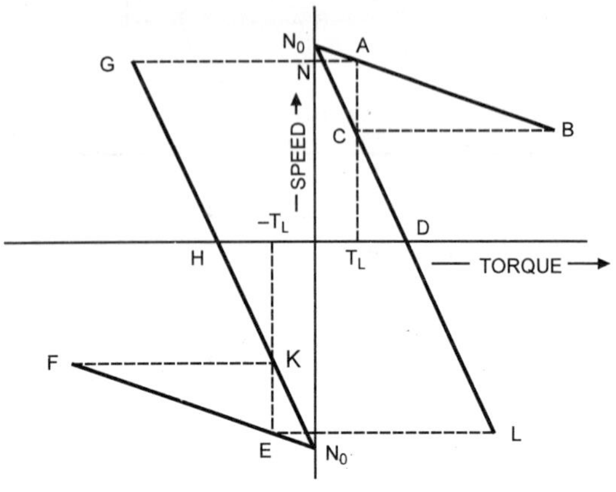

(a) Plugging of a DC Shunt Motor

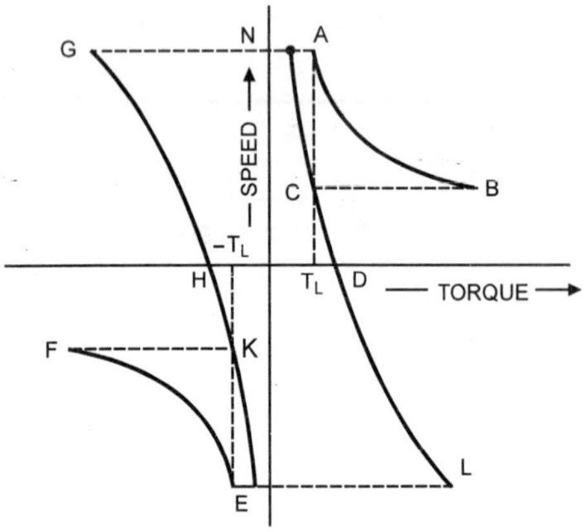

(b) Plugging of a DC Series Motor

Fig. 10.31 *Quadrantal Diagrams*

10.12.2. Dynamic (or Rheostatic) Braking

1. *Rheostatic Braking With DC Shunt Motors.* The armature of a dc shunt motor is disconnected from the supply and connected across a braking resistance, keeping the field connected to the supply, as shown in Fig. 10.32 (*b*). The machine then acts as a generator loaded with a resistance. Since the field current flows as before but the armature current in the reversed direction, therefore, a retarding torque is produced. As the motor speed falls during rheostatic braking, the retarding effect decreases. In order to maintain the retarding torque, some sections of the braking resistance may be then cut out of circuit. If, during rheostatic braking, supply fails, the magnetic flux of the motor will collapse, and there will be no braking effect. This drawback is overcome by providing a series winding in the armature circuit which is connected during the braking period only. The braking effect can be controlled by varying the braking resistance R.

2. Rheostatic Braking With DC Series Motors. In the case of a series wound direct current motor, when it is disconnected from the supply and is connected across a braking resistance, it is essential that the connections of the field winding are changed over. If this is not done the current through the field will also be reversed, and no braking will occur. To avoid this, the field connections are so changed that the current through the field winding will flow in the same direction as before. The resistance inserted in the circuit must be less than the critical resistance, otherwise the generator will not be self-exciting.

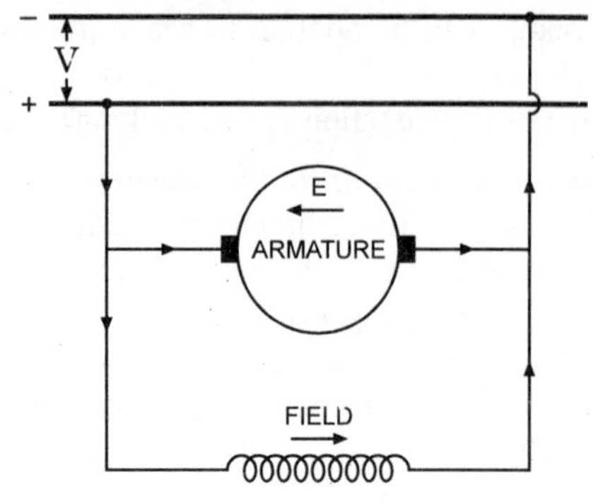

(a) Normal Running

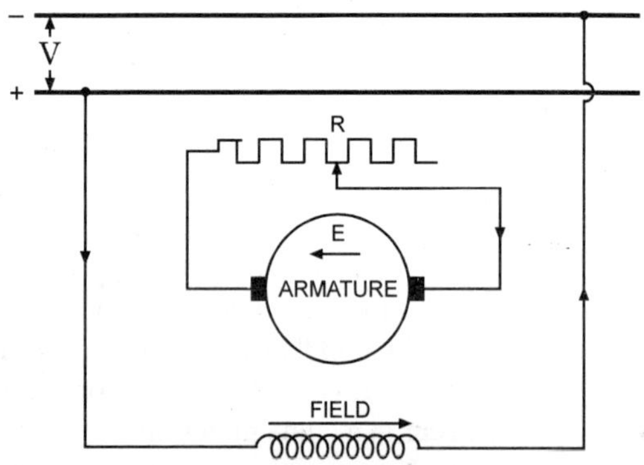

(b) Rheostatic Braking

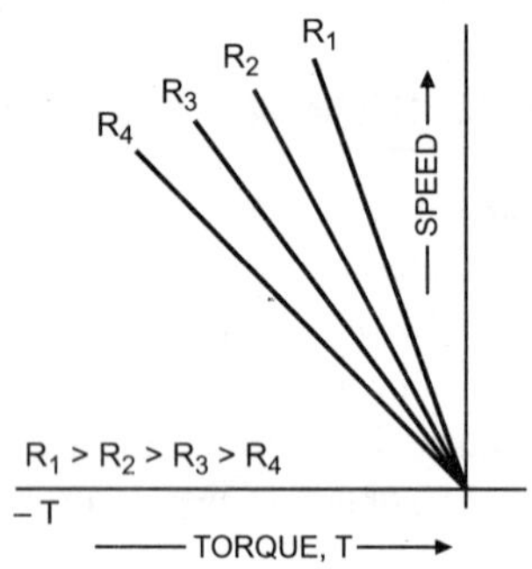

(c) Speed-Torque Characteristics

Fig. 10.32 *Rheostatic Braking With DC Shunt Motors*

Rheostatic braking of a dc series motor can also be affected by feeding its field winding from an independent source and connecting a resistance across the armature terminals i.e. by reconnecting it as a separately-excited machine, as shown in Fig. 10.33 (*c*). An additional resistance is also connected in series with the field winding to limit the field current to a safe value. Since the machine operates as a separately-excited generator, the braking characteristics will be same as those for dc shunt motor during dynamic braking.

Dynamic braking with self excitation is normally not used as the field of the machine may build up suddenly and cause the braking torque to jump to high value and the driven unit may experience sudden jerks.

3. *Rheostatic Braking With DC Compound Motors.* For direct current compound wound motors rheostatic braking is performed in the same way as for dc shunt wound motors.

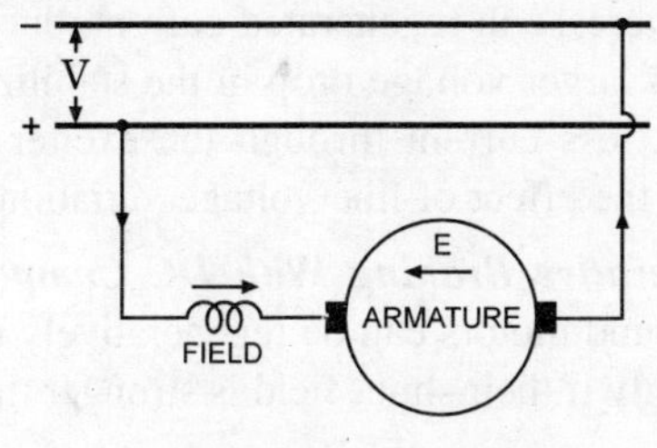

(a) Normal Running

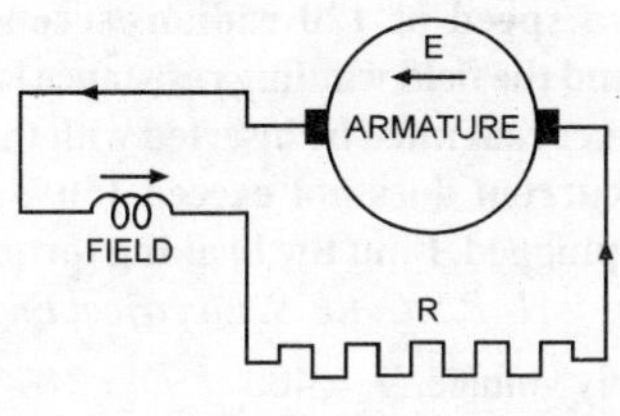

(b) Rheostatic Braking With Self Excitation

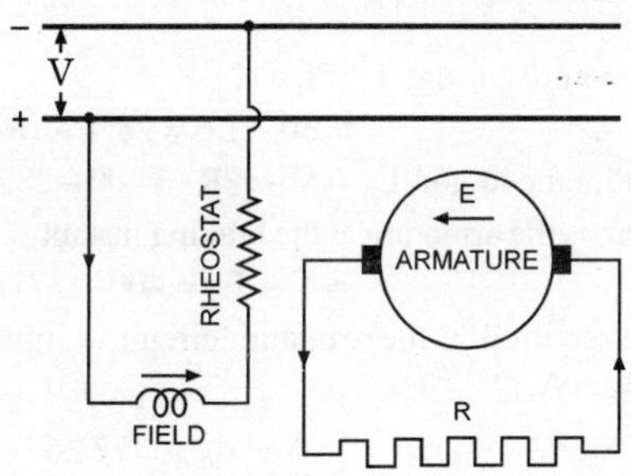

(c) Rheostatic Braking With Separate Excitation

Fig. 10.33 *Rheostatic Braking With DC Series Motors*

The relation between the braking torque and speed for direct current motors can be determined as below.

$$\text{Electric braking torque} = K_1 \Phi I = K_1 \Phi \cdot \frac{E}{R}$$

since braking current, $I = \frac{E}{R}$, where E is the induced emf and R is the total resistance in the motor circuit.

$$\text{Again } E \propto \Phi N$$

$$\text{or } E = K_2 \Phi N$$

$$\therefore \text{Electric braking torque} = \frac{K_1 \Phi}{R} \times K_2 \Phi N = K_3 \Phi^2 N \quad ...(10.15)$$

In case of series wound motors since flux Φ varies with the variation in current the value of torque, therefore, can only be determined graphically from the magnetisation curve.

In case of shunt wound motors since flux remains constant the electric braking torque, therefore, for any given value of R is directly proportional to speed.

In actual practice, in order to have maximum braking torque resistance in the armature circuit is so reduced with speed that armature current and, therefore, braking torque remains constant till external resistance becomes zero. The braking torque falls linearly with the speed and becomes zero at zero speed. In case of dc series motors with self excitation speed-torque curve is not linear because magnetic flux changes with the change in armature current which changes with the variations in speed.

In practice rheostatic braking is preferred for non-reversing drives as this simplifies the drive circuit. For reversing drives, plugging is preferred for braking and starting of the motor in the opposite direction in a short time.

10.12.3. Regenerative Braking.

1. *Regenerative Braking With DC Shunt Motors.* Regenerative braking is an inherent characteristic of shunt wound motors and does not require any changing of connections.

Let us consider a shunt motor running in normal condition. Now if due to, say, hauling load, speed increases above normal, due to increase in speed the induced emf will increase and if it exceeds the line voltage, the machine will start supplying current to the line, thus have tendency to prevent any further increase in speed.

Similarly if the field current is increased so that motor emf exceeds the line voltage and it may start supplying current to the line, the motor will slow down to the speed corresponding to this new value of field current.

Regenerative braking can be easily applied to dc shunt motors, particularly in cases where it is required to hold a load at a certain speed for instance lowering a hoist. It is not, however, possible to obtain regenerative braking down to very low speeds, since a sufficient increase in field current cannot be obtained.

Before bringing the motor to rest, the field excitation is increased to the permissible maximum value, as a result of which the speed of the motor is reduced to the minimum value and the kinetic energy released from the motor is fed back to the supply.

Since the motor is required to operate with a weak field at rated condition, the armature has to be designed to carry a large current to produce the rated torque and hence motor will be larger in size, poor in efficiency and costlier.

2. *Regenerative Braking With DC Series Motors.* The dc series motors cannot be used for regenerative braking in an ordinary way. Since the reversal of armature current necessary to produce regeneration would cause a reversal of field, therefore, series field connections must be reversed. But even if the field connections are reversed at the exact moment, this method would still be useless. Because at the instant of reversal, the emf induced in the motor will be small, so current will flow through the field in wrong direction, which will reverse the field and cause the motor emf to help the supply voltage. This will result in short-circuit of supply. Due to these complications this method is not used for common industrial purposes. Regenerative braking is, however, used with series motors for traction either by modification of windings or by supplying the machines with separate excitation.

One method of obtaining regenerative braking with series motors is the French method. If there is a single series motor as in case of a trolley buses, tramways, it is provided with a main series field winding and auxiliary field windings connected in parallel with the main series field winding as shown in Fig. 10.34 (*a*). During regeneration (braking period) the auxiliary field windings are put in series with each other and are switched across the supply, as shown in Fig. 10.34 (*b*). The machine acts as a compound generator slightly differentially compounded. Such an arrangement is quite stable. Any change in line voltage causes a change in excitation which produces a corresponding change in the induced emf of the machine so that inherent compensation is provided. For example if the line voltage increases beyond the emf of the generator the increased voltage across the generator's field will send a large exciting current through it causing the emf of the generator to rise. The reversal of this will happen when the line voltage decreases.

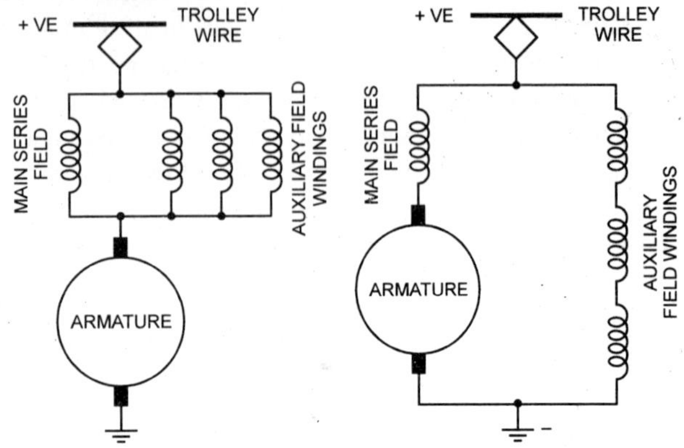

(*a*) *Motoring* (*b*) *Generating*

Fig. 10.34 *French Method of Regenerative Braking*

Another alternative method of regenerative braking, which obviates the losss of power in the resistances in the field circuit, is by using a separate exciter for controlling the excitation of the field windings during regeneration. The exciter may either be axle driven or driven by a motor operated from the auxiliary supply. Many devices have been used to secure compensation against variations in line voltage, most of which make use of some form of differential winding on the exciter. One such method is illustrated in Fig. 10.35 in which the exciter is differentially compounded. The shunt field winding of the exciter is separately excited from an auxiliary supply, the series field winding is connected in the main motor circuit in such a way that it opposes the separately-excited winding during regeneration. It can be seen that a rise in line voltage during regeneration, which will tend to reduce the regenerated current, will strengthen the exciter field and counteract the change. The stabilising resistance R also assists in this action.

Another method used for regenerative braking is depicted in Fig. 10.36. Here the exciter armature along with the field windings of the series motors are connected across the stabilizing resistance in the main circuit. The current through the stabilizing resistance is the sum of the exciter current and the regenerated current. Any increase in regenerated current, due to fall in line voltage, causes larger voltage drop in the stabilizing resistance and, therefore, less current through the exciter circuit, again counter-acting the effect of line voltage variations.

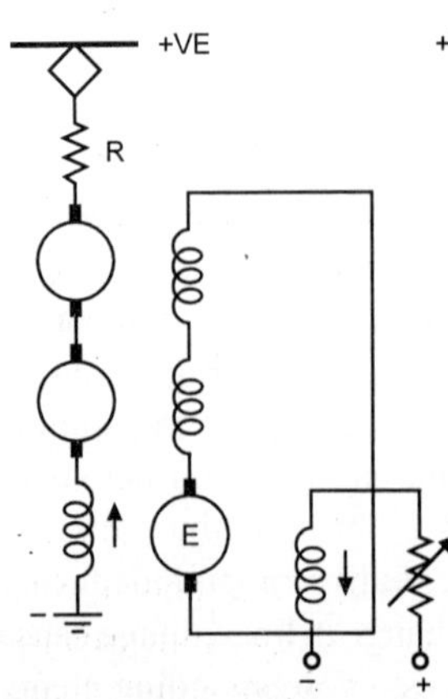

Fig. 10.35 *Differentially Compounded Exciter*

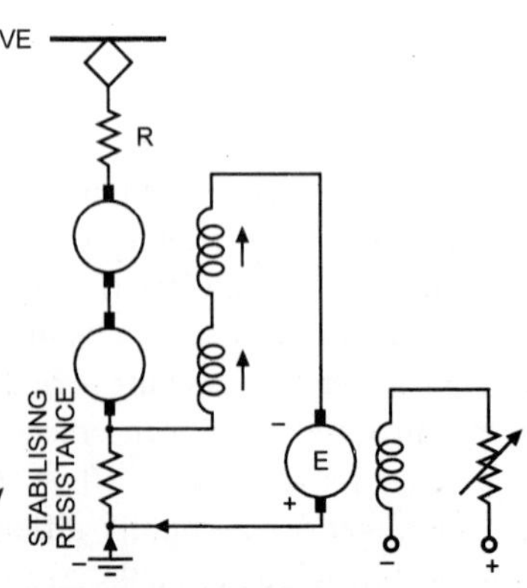

Fig. 10.36 *Separately-Excited Exciter*

3. *Regenerative Braking With DC Compound Motors.* Compound wound motors can be regeneratively retarded by an hauling load only if their shunt field is stronger than their series field.

Example 10.36. A 400 V dc shunt motor draws 30 A while supplying the rated load at a speed of 120 radians/second. The armature resistance is 1.0 Ω and the field winding resistance is 250 Ω. Determine the external resistance that must be inserted with the armature circuit so that armature current does not exceed 150% of its rated value when the motor is plugged. Find the braking torque at the instant of plugging. [*U.P.S.C. I.E.S. Electrical Engineering-II, 2011*]

Solution: Supply voltage, V = 400

$$\text{Shunt field current, } I_{sh} = \frac{V}{R_{sh}} = \frac{400}{250} = 1.6 \text{ A}$$

Full-load armature current, $I_a = I_L - I_{sh}$
$= 30 - 1.6 = 28.4$ A (Rated value)

Induced emf in the motor, $E_b = V - I_a R_a = 400 - 28.4 \times 1 = 371.6$ V

Voltage across the armature at the braking instant
$= V + E_b = 400 + 371.6 = 771.6$ V

Resistance required in the armature circuit to limit the current to 1.5 × 28.4 i.e. 42.6 A,

$$R = \frac{V + E_b}{I_{a\,max}} = \frac{771.6}{42.6} = 18.11\ \Omega$$

External resistance required in the armature circuit
$= R - R_a = 18.11 - 1 = 17.11\ \Omega$ **Ans.**

$$\text{Braking torque, } T_b = \frac{E_b\, I_{a\,max}}{\omega}$$

$$= \frac{371.6 \times 42.6}{120} = 131.92 \text{ N-m } \textbf{Ans.}$$

Example 10.37. A 400 V, 750 rpm, 70 A dc shunt motor has an armature resistance of 0.3 Ω. When running under rated conditions, the motor is to be braked by plugging with armature current limited to 90 A. What external resistance should be connected in series with the armature? Calculate the initial braking torque and its value when the speed has fallen to 300 rpm. Neglect saturation.

[*U.P. Technical Univ. Electric Drives, 2013-14*]

Solution: Supply voltage, V = 400 V

Full-load motor current, $I_f = 70$ A

Full-load armature current, $I_{af} = I_f = 70$ A
neglecting shunt field current

Induced emf in the motor, $E_{bf} = V - I_{af} R_a$
$= 400 - 70 \times 0.3 = 379$ V

Voltage across the armature at the braking instant
$= V + E_{bf} = 400 + 379 = 779$ V

Resistance to be connected in series with the armature to limit the initial braking current to 90 A,

$$R = \frac{V + E_{bf}}{I_b} - R_a = \frac{400 + 379}{90} - 0.3 = 8.3556\ \Omega \textbf{ Ans.}$$

Initial braking torque,

$$T_b = \frac{E_{bf} \times I_{a\,max}}{2\pi N/60} = \frac{379 \times 70}{2\pi \times 750/60} = 377.8 \text{ N-m } \textbf{Ans.}$$

When the speed has reduced to 300 rpm, the induced emf falls in the ratio of speed because the flux is assumed to remain constant.

$$\text{So induced emf, } E'_b = E_{bf} \times \frac{N'}{N} = 379 \times \frac{300}{750} = 151.6 \text{ V}$$

$$\text{Armature braking current, } I'_b = \frac{V + E'_b}{R + R_a} = \frac{400 + 151.6}{8.3556 + 0.3} = 63.73 \text{ A}$$

$$\text{Electric braking torque, } T'_b = \frac{E'_b \times I'_b}{2\pi N'/60} = \frac{151.6 \times 63.73}{2\pi \times 300/60}$$

$$= 307.5 \text{ N-m } \textbf{Ans.}$$

Example 10.38. A 220 V, 17.6 kW dc shunt motor at its rated speed of 1,200 rpm is to be braked by plugging. The armature resistance is 0.1 Ω and the rated efficiency of the motor is 80%. Calculate : (*i*) the resistance to be connected in series with the armature to limit the initial braking current to twice the rated current, (*ii*) the torque when the speed of the motor falls to 400 rpm.

[*U.P. Technical Univ. Electric Drives 2008-09*]

Solution: Full-load (rated) motor current,

$$I_f = \frac{\text{Motor output in kW} \times 1{,}000}{\text{Rated voltage} \times \text{rated efficiency}}$$

$$= \frac{17.6 \times 1{,}000}{220 \times 0.8} = 100 \text{ A}$$

$$\text{Full-load torque, } T_f = \frac{\text{kW output} \times 1{,}000}{2\pi N / 60}$$

$$= \frac{17.6 \times 1{,}000}{2\pi \times 1{,}200 / 60} = 140 \text{ N-m}$$

Full-load armature current,

$$I_{af} = I_f = 100 \text{ A} \quad \text{neglecting shunt field current}$$

Induced emf in motor,

$$E_{bf} = V - I_{af} R_a = 220 - 100 \times 0.1 = 210 \text{ V}$$

Voltage across the armature at the instant of braking

$$= V + E_{bf} = 220 + 210 = 430 \text{ V}$$

(*i*) Resistance required to be connected in series with the armature to limit the initial current to twice the rated current (*i.e.* $I_b = 2I_f = 2 \times 100 = 200$ A),

$$R = \frac{V + E_b}{I_b} - R_a = \frac{430}{200} - 0.1 = 2.05 \ \Omega \ \textbf{Ans.}$$

(*ii*) When the speed has fallen to 400 rpm, the induced emf falls in the ratio of speed because in case of a dc shunt motor flux remains constant,

So, induced emf,

$$E'_b = E_{bf} \times \frac{N'}{N} = 210 \times \frac{400}{1{,}200} = 70 \text{ V}$$

Armature braking current,

$$I'_b = \frac{V + E'_b}{R + R_a} = \frac{220 + 70}{2.05 + 0.1} = 134.9 \text{ A}$$

Electric braking torque,

$$T'_b = T_f \times \frac{I'_b}{I_f} = 140 \times \frac{134.9}{100} = 188.84 \text{ N-m } \textbf{Ans.}$$

Example 10.39. A 220 V, dc shunt motor having an efficiency of 85% drives a hoist having an efficiency of 80%. Calculate the current drawn from the supply to raise a load of 400 kg at 2.5 m/s. What resistance must be added in the armature circuit in order to lower the load at 2.5 m/s using rheostatic braking, assuming that the efficiency of the motor and load will remain the same.

[*U.P. Technical Univ. Electric Drives, 2007-08*]

Solution: Work done per second

$$= \text{Load in kg} \times g \times v = 400 \times 9.81 \times 2.5$$

$$= 9{,}810 \text{ N-m/s or watts}$$

$$\text{Power input} = \frac{\text{Output}}{\eta_m \times \eta_h} = \frac{9{,}810}{0.85 \times 0.8} = 14{,}426.5 \text{ W}$$

Current drawn from supply mains,

$$I = \frac{\text{Power input}}{\text{Rated voltage}} = \frac{14{,}426.5}{220} = 65.6 \text{ A } \textbf{Ans.}$$

When the load is lowered, the machine operates as a generator and the generator output will be dissipated in the resistance connected in the armature circuit.

So, power output of generator,

$$P_{out} = \text{Load in kg} \times g \times v \times \eta_m \times \eta_h$$

$$= 400 \times 9.81 \times 2.5 \times 0.85 \times 0.8 = 6{,}670.8 \text{ watts}$$

If R is the resistance inserted in the armature circuit,

$$\text{Then power dissipated} = \frac{V^2}{R} = 6{,}670.8$$

$$\text{or } R = \frac{V^2}{P_{out}} = \frac{220^2}{6{,}670.8} = 7.255 \ \Omega \ \textbf{Ans.}$$

Example 10.40. The speed of a dc series motor is limited to 250 rpm by application of rheostatic braking system. The total resistance of armature and field of the motor is 1.0 Ω and load exerts a torque of 245 Nm. Calculate the resistance to be connected in the motor circuit, if its magnetisation characteristic at 250 rpm is as follows:

I in Amps	10	20	30	40
E in Volts	90	190	248	295

Solution: Load torque, T = 245 N-m

Input power to machine during rheostatic braking

$$= T \times \frac{2\pi N}{60} = 245 \times 2\pi \times \frac{250}{60}$$

$$= 6{,}414 \text{ N-m/s or watts}$$

During rheostatic braking the machine operates as a dc series generator and the electromagnetic power developed by the generator (neglecting rotational losses), *i.e.*, E I should be 6,414 watts.

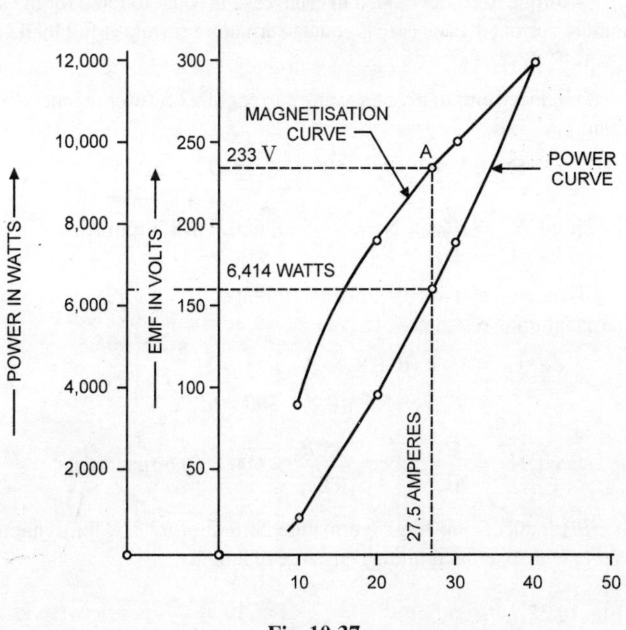

Fig. 10.37

In order to determine E and I separately, plot curves between E and I and power P and I, as shown in Fig. 10.37 from the datas given below:

I in Amps	10	20	30	40
E in Volts	90	190	248	295
Power, P (E I) in watts	900	3,800	7,440	11,800

Current corresponding to power of 6,414 watts is found to be 27.5 A and corresponding induced emf E is 233 V.

Hence total resistance of the circuit $= \frac{E}{I} = \frac{233}{27.5} = 8.47\ \Omega.$

External resistance to be inserted in the motor circuit

$$= 8.47 - 1.0 = 7.47\ \Omega \quad \textbf{Ans.}$$

Example 10.41. A 500 V, 45 kW, 600 rpm dc shunt motor has a full-load efficiency of 90%. The field resistance is 200 Ω and the armature resistance is 0.2 Ω. Find the speed under each of the following conditions at which the machine will develop an electromagnetic torque equal to rated value- (*i*) regenerative braking: no limiting resistance, (ii) plugging: external limiting resistance of 5.5 Ω inserted, (*iii*) dynamic braking: external limiting resistance of 2.6 Ω inserted. The field current is maintained constant and armature reaction and the brush drop may be neglected.

Solution: Rated line current, I_{Lf}

$$= \frac{\text{Output in kW} \times 1{,}000}{\text{Supply voltage} \times \text{full load efficiency}} = \frac{45 \times 1{,}000}{500 \times 0.9} = 100 \text{ A}$$

$$\text{Shunt field current, } I_{sh} = \frac{V}{R_{sh}} = \frac{500}{200} = 2.5 \text{ A}$$

$$\text{Armature current on full load, } I_{af} = I_{Lf} - I_{sh} = 100 - 2.5 = 97.5 \text{ A}$$

$$\text{Induced emf on full load, } E_f = V - I_{af} R_a$$

$$= 500 - 97.5 \times 0.2 = 480.5 \text{ V}$$

$$\text{Rated speed, } N_f = 600 \text{ rpm.}$$

Since field current remains constant and the armature reaction and brush drop are negligible, electromagnetic torque developed is directly proportional to armature current and speed is directly proportional to induced emf.

As torque to be developed in each case is equal to rated torque so armature current in each case is equal to armature current at full load *i.e.* 97.5 A.

(*i*) Induced emf to give armature current of 97.5 A for regenerative braking

$$E_1 = V + I_{af} R_a = 500 + 97.5 \times 0.2 = 519.5 \text{ V}$$

$$\text{and speed, } N_1 = \frac{E_1}{E_f} \times N_f = \frac{519.5}{480.5} \times 600 = 648.7 \text{ rpm } \textbf{Ans.}$$

(*ii*) Induced emf to give armature current of 97.5 A for plugging with external limiting resistance of 5.5 Ω

$$E_2 = I_{af} \times (R + R_a) - V$$

$$= 97.5 \times (5.5 + 0.2) - 500 = 55.75 \text{ V}$$

$$\text{and speed, } N_2 = \frac{E_2}{E_f} \times N_f = \frac{55.75}{480.5} \times 600 = 69.6 \text{ rpm } \textbf{Ans.}$$

(*iii*) Induced emf to give armature current of 97.5 A for dynamic braking with external limiting resistance of 2.6 Ω

$$E_3 = I_{af} \times (R + R_a) = 97.5 \times (2.6 + 0.2) = 273 \text{ V}$$

$$\text{and speed, } N_3 = \frac{E_3}{E_f} \times N_f = \frac{273}{480.5} \times 600 = 340.9 \text{ rpm } \textbf{Ans.}$$

10.13 SINGLE-QUADRANT OPERATION OF MOTORS

Single-quadrant operation means that load and the motor shall be required to run only in one specified direction under the action of developed torque.

When an electric motor is switched to electric supply mains and drives mechanical devices such as line shaft and machine tools, the machine is said to be operating in forward motoring mode or simply motoring mode. In this mode of operation armature current flows in opposition to the emf induced in the armature. The direction of emf induced in the armature is in direct opposition to the applied voltage. Electromagnetic torque developed is in the direction of armature rotation.

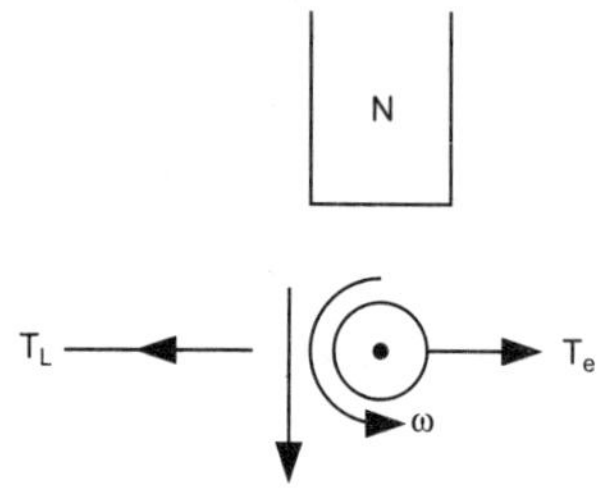

Fig. 10.38 *Single-Quadrant Operation of Motor*

10.14 TWO-QUADRANT OPERATION OF MOTORS

The operation of an electric motor in forward motoring mode and in regenerative braking mode is called the *two-quadrant operation* of an electric motor.

Forward Motoring Mode. When an electric motor is switched to electric supply mains and drives mechanical devices such as line shaft and machine tools, the machine is said to be operating in *forward motoring mode*, explained above.

Regenerative Braking Mode. When an electric motor is to be stopped, brakes are to be applied. In this mode, the motor is made to operate as a generator and the kinetic energy of the motor and load coupled to it is converted into electrical energy, part of which is returned to the supply and the rest of the energy is lost as heat in the windings and bearings of electrical machines. In this mode of operation the armature current and induced emf in the motor are in the same direction but in opposition to supply voltage. In this mode of operation, the electromagnetic torque developed is in direction opposite to that of rotation of armature.

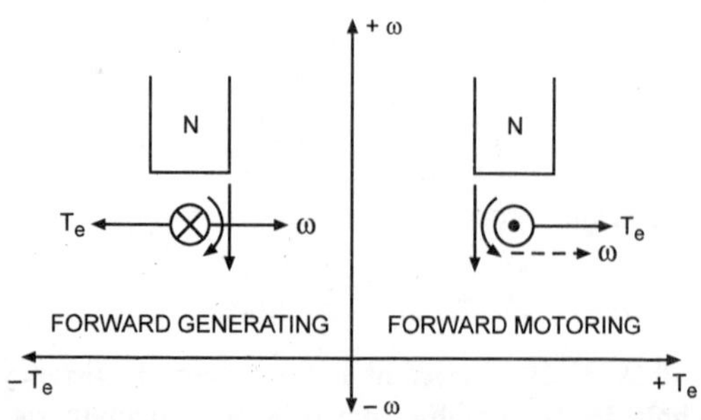

Fig. 10.39 *Two-Quadrant Operation of Motor*

10.15 FOUR-QUADRANT OPERATION OF MOTORS

In view of the fact that both active and passive load torques* can be present in general, in a drive system, the motor driving the load may operate in different regimes - not only as a motor, but for specific periods, also as a generator and as a brake. Further, in many applications, the motor may be required to run in both directions. Therefore, in sketching the speed-torque characteristics of either the load or the motor, it is preferable to make use of all four quadrants of the speed-torque plane for plotting, rather than to confine, the characteristics to the first quadrant alone.

The conventions used for positive and negative values of speed, motor torque and load torque in a diagram of this type are as follows :

The speed is assumed to have a positive sign, if the direction of rotation is counter-clockwise or is in such a way to cause an 'upward' or 'forward' motion of the drive. In case of reversible drives the positive sign for speed may have to be assigned arbitrarily either to counter-clockwise or clockwise direction of rotation. The motor torque is taken to be positive when it causes an increase in speed in the positive sense. The load torque is assigned a positive sign when it acts against the motor torque.

Fig. 10.40 illustrates the four different modes of operation of an electric machine. Although the diagram indicates that the field polarity is maintained and the armature current is reversed to obtain negative torque, the same effect is obtained by reversing the field polarity and maintaining the armature current direction. Field reversal is necessary with some forms of rectifier control. These four quadrants of operation are feasible for any dc or ac rotating machine but the dc machine is much freer to transfer its operation between quadrants and operates satisfactorily at any point within the envelope.

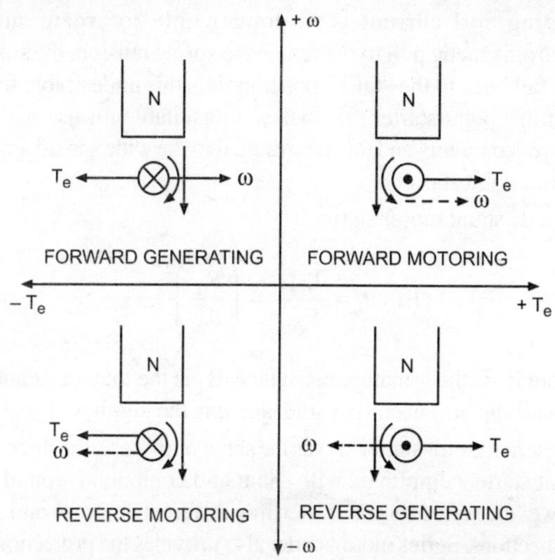

Fig. 10.40 *Four-Quadrant Operation of Motor*

HIGHLIGHTS

1. Speed control means intentional change of the drive speed to a value required for performing the specific work process. Speed of a dc motor can be controlled by adjusting any one of three factors: (*i*) applied voltage to the armature terminals, V (*ii*) external resistance in the armature circuit and (*iii*) flux per pole Φ. The first two possibilities involve adjustment affecting the armature circuit, whereas the third involves change in the magnetic field. Therefore, speed control methods are broadly classified as *armature control methods* and *field control methods.* Sometimes a combination of the two methods is employed. With armature control the speed decreases as the voltage applied to the armature terminals is reduced, whereas with field control the speed increases as the flux is reduced.

2. The speed of a *dc shunt motor* can be controlled by

   (*i*) Field control method (*ii*) Armature resistance control (*iii*) Armature voltage control (*iv*) Ward-Leonard method of speed control.

   Field control method is very simple, convenient and most economical. The power wasted in the controlling resistance is very small and remote control of speed is possible with this method. But creeping speeds cannot be obtained by this method.

   The armature resistance control method is employed where speeds lower than rated one are required for a short period only and that also occasionally as in printing machines, cranes and hoists where the motor is frequently started and stopped. This method of speed control is also employed where the load drops off rapidly with decrease in speed, as in fans and blowers.

   Armature voltage control avoids the disadvantages of poor speed regulation and low efficiency of armature resistance control method but it is more expensive in initial cost.

   Ward-Leonard method of speed control is a armature voltage control method and is best suited where almost unlimited control in either direction of rotation is required.

3. Speed control of dc series motors may be accomplished by any of the following methods:

   1. **Armature Control Methods:** (*i*) Armature resistance control (*ii*) Shunted armature control and (*iii*) Armature terminal voltage control.

   2. **Field Control Methods:** (*i*) Field diverter method (*ii*) Tapped field control (*iii*) Paralleling field coil method (*iv*) Series-parallel control.

   Armature resistance control method is most economical for constant torque drives and is used mainly for dc series motors, driving cranes, hoists, trains etc.

   Shunted armature control method is not economical due to considerable power losses in the speed controlling resistors and its use is, therefore, restricted to as short an interval as possible.

   Armature terminal voltage control is seldom used because of high cost of control equipment.

   Field diverter method is convenient as well as economical, provides the speed control range usually not exceeding 2 : 1 and is used in electric drives in which the speed should rise sharply as soon as load falls.

   Tapped field control and series-parallel control are usually employed in electric traction.

4. Starter is required to be provided with an electric motor in order to limit the initial rush of current to a predetermined value at the time of starting and provide protection against no-voltage and overload.

* Active torques are due to either gravitational force or deformation in elastic bodies and continues to act in one and same direction irrespective of the direction of movement of drive. Passive torques are due to friction or due to shear and deformation in in-elastic bodies and always oppose the motion retarding the rotation of the driven machine.

5. There are two standard types of motor starters for dc shunt and compound motors, the three point type and the four point type starters. The four point starter is employed when wide range of speed by shunt field control is required. When no (or little) speed control is required, either type of starter may be used.
6. In three point starter no-volt release coil is connected in series with the shunt field and the field rheostat and, therefore, the current flowing through the field is the same current that flows through the holding coil. Thus if sufficient resistance is *cut in* by the field rheostat so that holding coil current is no longer able to create sufficient electromagnetic pull to overcome the spring tension, the starter arm will fall back to the "OFF" position. It is this undesirable feature of the three point starter that makes it unsuitable for use with speed-controlled motors and that has resulted in the wide spread application of four point starters.
7. For a dc shunt motor starter

$$\gamma = \frac{I_{max}}{I_{min}} = \left(\frac{R_1}{R_a}\right)^{1/n}$$

where R_a is the armature resistance, R_1 is the total resistance when the starting arm is on first stud, and n is the number of sections.
8. The series motor starter serves the same purpose as the three and four point starters employed with shunt and compound wound motors. However, series motor starter has different internal and external connections. Series motor starter also provides the protection against no-load in addition to no-volt and overload protection.
9. DC series motors are often employed on cranes, elevators, tramcars and other applications, where the motor is under the direct control of an operator. In these applications, frequent starting, variations of speed, stopping and reversing may be necessary. A manually operated controller, more rugged than a starting rheostat, called a *drum controller,* is employed.
10. Push-button type of automatic starters are quite common in use in industry. Even an inexperienced operator, with the help of such starters, called the *auto-starters*, can start and stop the motor without any difficulty.

    The operation of such starters depends upon, either the time delay or the counter emf developed across the armature terminals.
11. Using friction or electric braking, it is possible to bring a motor and its load to rest quickly. In friction braking, it is difficult to get smooth stop. In electric braking it is possible to stop the motor smoothly. There are three kinds of electric braking: (*i*) Plugging (*ii*) Rheostatic or dynamic braking and (*iii*) regenerative braking.

    Plugging or counter-current braking is the simplest type of braking and it occurs when the motor windings are connected for reverse direction of rotation at a time when the armature is still rotating in the forward direction under the action of an external torque or due to inertia. In *dynamic* or *rheostatic braking*, the motor is disconnected from the supply and operated as a generator driven by the kinetic energy of the rotating parts of the motor and its driven machines and electrical energy thus generated is dissipated in the external resistances connected across the motor at the braking instant.

    In the previous two methods of electric braking namely plugging and rheostatic braking, stored energy of the rotating parts of the motor and its driven machine is wasted whilst in plugging extra energy is drawn during the braking period and is wasted. In regenerative braking, mechanical energy is converted into electrical energy, part of which is returned to the supply and the rest of the energy is lost as heat in the windings and the bearings of the electrical machines. Regeneration does not, in general, involve any switching operation, unless it is required to change the speed at which it becomes effective. Most of the electrical machines pass smoothly from motoring to generating operation, when overdriven by load.
12. Single-quadrant operation means that load and the motor shall be required to run only in one specified direction under the action of developed torque.

    In this mode of operation armature current flows in opposition to the emf induced in the armature. The direction of emf induced in the armature is in direct opposition to the applied voltage. Electromagnetic torque developed is in the direction of armature rotation.

    The operation of an electric motor in forward motoring mode and in regenerative braking mode is called the *two-quadrant operation* of an electric motor.

    In view of the fact that both active and passive load torques can be present in general, in a drive system, the motor driving the load may operate in different regimes — not only as a motor, but for specific periods, also as a generator and as a brake. Further, in many applications, the motor may be required to run in both directions. Therefore, in sketching the speed-torque characteristics of either the load or the motor, it is preferable to make use of all four quadrants of the speed-torque plane for plotting, rather than to confine, the characteristics to the first quadrant alone.

EXERCISES

1. Explain the factors which control the speed of dc motors. Explain the flux control method for dc shunt motor.
   [M.D. Univ. Electromechanical Energy Conversion, December-2010]
2. Explain the methods for speed control of dc motor.
   [M.D. Univ. Electromechanical Energy Conversion, December-2013]
3. What are constant torque and constant power operations of an electric drive? **[G.B. Technical Univ. Electric Drives, 2010-11]**
4. What is meant by "constant torque drive" and "constant power drive"? Explain them and mention one application of each such drive. **[U.P. Technical Univ. Electric Drives, 2005-06]**
5. What are the different methods of speed control of a dc motor? Give the advantages and disadvantages of each in brief. Why is it dangerous to uncouple the mechanical load of a dc series motor?
   [U.P. Technical Univ. Electrical Machines, 2003-04]
6. What are the methods of speed control of dc shunt motor? Briefly explain them with the help of neat diagrams.
   [Anna Univ. Electrical Machines-I, May/June-2012]
7. What is meant by speed control ? Explain with circuit diagram a speed control scheme of a dc shunt motor by varying field flux.
   [U.P. Technical Univ. Energy Conversion, 2005-06]
8. What are the advantages of field flux control method over the armature circuit resistance control method employed for speed control of dc motors? **[U.P. Technical Univ. Electromechanical Energy Conversion-I, 2005-06]**
9. State and explain Ward-Leonard method of speed control.
   [W.B. Univ. of Technology Electrical Machine-I, 2011]
10. Explain with neat sketch how speed control of a dc shunt motor is done by Ward-Leonard control system. How the direction of rotation of the motor is usually reversed in this method of speed control?
    [J.N. Technological Univ. Electrical Machines-I, March-2014]
11. Name the various speed control methods of dc motors. Explain speed control of dc series motor by any ***one*** method.
    [M.D. Univ. Electromechanical Energy Conversion, December-2011]
12. Discuss various methods of speed control that can be used for dc series motors. **[J.N. Technological Univ. Electrical Machines-I, March-2014]**

13. Identify the factors that predominantly affect the speed of a dc motor.

Discuss two methods for varying the speed of a dc series motor for constant load torque. List one application of each of these methods.

14. Explain the various methods of speed control of dc series motors including the case when two identical dc series motors are available.
[U.P.S.C. I.E.S. Electrical Engineering-II, 2010]

15. Explain the series-parallel control of two identical series motors. How many speeds are possible? Are the motors run faster in series combination or parallel combination?

16. Why is the starting current high in dc motor? How does the starter reduce the starting current to a safe value?
[U.P. Technical Univ. Electrical Machines and Automatic Control, 2013-14]

17. What is the need of the starter? With a neat diagram explain the construction and working of 3-point starter.
[J.N. Technological Univ. Electrical Machines-I, March-2014]

18. Explain the necessity of starter for dc motor. Discuss three point starter for dc motor.
[Gujarat Technological Univ. Electrical Machines-I, Summer-2014]

19. Describe the working of a three point starter for a dc shunt motor. What are the benefits does a four point starter have over three point starter.
[G.G.S.I.P. Univ. Delhi Electromechanical Energy Conversion-I, December-2011]

20. Explain why a starter is required for starting a dc motor. Describe a 3-point starter, having no-volt and overload protections for a dc shunt motor. What modification is made in 4 point starter? Give relative merits of these two types of starters.
[U.P. Technical Univ. Electromechanical Energy Conversion-I, 2011-12]

21. Describe the working of a 3-point starter for a dc shunt motor with the help of a neat diagram. How does a four point starter differ from this? What are the advantages of the four point starter.
[B.P. Univ. of Technology Electrical Machine-I, 2008]

22. Why is starter used for starting a dc motor?

Sketch a 3-point starter with proper label.

Derive a relation for determining the starter steps.
[W.B. Univ. of Technology Electrical Machine-I, 2011]

23. Draw a neat diagram of dc shunt motor starter. How can the values of resistance elements be calculated? Explain the function of overload release and no-volt release coils.
[U.P. Technical Univ. Electrical Machines, 2010-11]

24. Why is starting current high in a dc motor ? Explain the working of a four point starter for a dc machine.
[U.P.S.C. I.E.S. Electrical Engineering-II, 2002]

25. A DC shunt motor is connected to a 3-point starter. Explain what would happen if:

(*i*) The starter handle is moved rapidly from OFF to the ON position.

(*ii*) The field circuit is open and an attempt is made to start the motor.

(*iii*) The field circuit becomes open circuited with the motor running at no load, with the assumption that the starter is not provided with the no-volt release and the spring.

(*iv*) The field excitation is minimum at the time of starting.

26. What are the advantages and disadvantages of electric braking?
[Pb. Technical Univ. Electric Drives and Utilization, May-2007]

27. Explain the types of braking of dc motors. Which one is usually employed and why? **[G.B. Technical Univ. Electric Drives, 2010-11]**

28. What is meant by braking of dc motors ? Describe briefly, the various methods of braking of dc shunt motors.
[M.D. Univ. Electrical Machines, May-2009; December-2009]

29. Explain the different methods of electrical braking of dc motors and compare their advantages and disadvantages.
[U.P.S.C. I.E.S. Electrical Engineering-II, 2010]

30. Describe the advantages of electric braking. Classify the electric braking. Also explain mathematically why torque developed due to plugging braking is always greater than dynamic braking.
[U.P. Technical Univ. Electric Drives, 2013-14]

31. Explain the various types of braking as used in a dc motor. What necessary modifications are required while braking a dc series motor? **[U.P. Technical Univ. Electric Drives, 2009-10]**

32. Explain the reverse current braking method in dc shunt motors. Also explain the starting of shunt motor with 3-point starter.
[M.D. Univ. Electrical Machines-I, December-2012]

33. How can plugging be applied to a separately-excited dc motor?
[U.P. Technical Univ. Electric Drives, 2008-09]

34. Explain how rheostatic braking is done with dc shunt and series motors. **[J.N. University Hyderabad, November-2003]**

35. Why is series motor not suitable for regenerative braking?

Suggest one method for applying regenerative braking to dc series motor. **[U.P. Technical Univ. Electric Drive, 2005-06]**

36. State the condition under which regenerative braking with dc series motor is possible and with the aid of diagrams of connection, explain the various methods of providing regeneration.
[J.N. University Hyderabad, November-2003]

37. Describe how plugging, rheostatic braking and regenerative braking are employed with dc series motor.
[J.N. University Hyderabad, November-2003]

38. Write short notes on the following:

(*i*) Armature voltage control of dc shunt motors.

(*ii*) Ward-Leonard speed control of dc machines.

(*iii*) Three point starter.
[M.D. Univ. May-2009; W.B. Univ. of Technology, 2010]

(*iv*) Four point starter. **[R.G. Technical Univ., December-2013]**

(*v*) Starter step calculation for dc shunt motor.
[U.P. Technical Univ., 2003-04]

(*vi*) Advantages of electric braking over mechanical braking.

(*vii*) Electric braking of dc motor. **[M.D. Univ., December-2010]**

SHORT ANSWER TYPE QUESTIONS WITH ANSWERS

Q. 1. What are the factors that affect the speed of a dc motor?
[U.P. Technical Univ. Even Semester, 2003-04]

Ans. The factors influencing the speed of a dc motor are :

(*i*) Applied or terminal voltage (*ii*) Flux per pole Φ (*iii*) Resistance of armature circuit and (*iv*) Load or armature current.

Q. 2. Name the methods used for speed control of dc motors.

Ans. (*i*) Field control (*ii*) Armature control.

Q. 3. For a dc motor, the field flux speed control method is called a constant power drive method. Explain.
[U.P. Technical Univ. Electromechanical Energy Conversion-I, 2007-08]

Ans. Constant power drive is a drive in which the motor shaft power, given by the product of shaft torque and speed, remains constant over a given speed range

i.e. $T \times N = \text{Constant}$

or $\Phi I_a \times N = \text{Constant}$

or $\Phi N = \text{Constant}$ because I_a is dependent only on load

Since in field control, speed is approximately inversely proportional to flux Φ *i.e.* ΦN remains almost constant. Thus, field flux speed control method is known as constant power drive method.

Q. 4. What factors do limit the increase of speed of dc shunt motors in field rheostat control?

Ans. The factors limiting the increase of speed of dc shunt motors in field rheostat control are demagnetizing effect of armature reaction on the field, commutation problems at high speed and large centrifugal forces developed at high speed.

Q. 5. Why is field control considered superior to armature resistance control for dc shunt motors?

Ans. Because power wasted in controlling resistance is very small and field control method of speed control is independent of load on motor and permits remote control of speed.

Q. 6. What is the effect of inserting resistance in the field circuit of a dc shunt motor on its speed and torque?

Ans. For a constant supply voltage flux will decrease, therefore, speed will increase and torque will decrease.

Q. 7. Which method we must adopt to control the speed of a dc shunt motor above the base speed?

Ans. Field control method.

Q. 8. Why is it not possible to obtain speed of a dc shunt motor below normal by field rheostat control?

Ans. Because in field rheostat control, the field circuit resistance can be increased (not reduced) to weaken field strength giving increased speed.

Q. 9. Why is it not possible to obtain creeping speeds by field rheostat control of dc shunt motors?

Ans. Field rheostat control in dc shunt motor can provide speed above normal only (not below normal speed).

Q. 10. Why is it not possible to obtain speed of a dc shunt motor above normal by armature resistance control?

Ans. Since in armature resistance control, the voltage across the armature can be reduced (not increased) for a constant supply voltage and constant load torque, speed below normal is obtained.

Q. 11. While controlling the speed of a dc shunt motor what should be done to achieve a constant torque drive?

Ans. Applied voltage should be maintained constant so as to maintain field strength constant.

Q. 12. In the armature circuit of dc motor, an external resistance is introduced. How shall it affect the speed of the motor? **[P.T.U., May-2002]**

Ans. The speed will decrease.

Q. 13. A dc shunt motor is running at rated speed with a resistance R in series with the armature, will the speed of the motor increase or decrease if additional resistance of finite value is connected (*a*) across R and (*b*) in series with R ?

Ans. (*a*) Speed will increase. (*b*) Speed will decrease.

Q. 14. Why is armature control superior to field control scheme in the case of a dc shunt motor? **[U.P.S.C. I.E.S., 2011]**

Ans. Armature control is superior to field control in the control of a dc shunt motor, as explained below:

1. It provides constant torque drive. In case of a dc shunt motor, full-load torque can be obtained at full-load armature current at all speeds by keeping the field current at maximum value.
2. Since the main field ampere-turns are maintained to provide a large value of flux density, distortion caused by armature reaction is restricted.
3. Unlike the field control, speed reversal can be easily implemented in armature control.

Q. 15. In what respect shunted armature control is superior to armature series resistance control?

Ans. More stable operation is obtained.

Q. 16. How is slow speed obtained at light load in case of dc series motors?

Ans. By connecting a control resistance directly in series with the supply to the complete motor.

Q. 17. How the speed above normal can be obtained in case of dc series motors?

Ans. By weakening the field which can be accomplished by

(*i*) shunting a portion of motor current around the series field i.e. field diverter method.

(*ii*) reducing the number of turns of the field winding through which the current flows i.e. tapped field control.

(*iii*) regrouping of field coils i.e. paralleling field coils.

Q. 18. Do the motors run faster in series combination or parallel combination?

Ans. Motors will run faster in parallel combination.

Q. 19. The torque for the series combination of two dc series motors is T_s and that for their parallel combination is T_p. What is the relation between T_s and T_p if the mechanical power is same?

Ans. $T_s = 4T_p$

Q. 20. Why the power loss in control resistance for many applications of dc series motors is not too serious as in case of dc shunt motors?

Ans. Because in case of dc series motors, the control is utilized for a large portion of time for reducing the speed under light load conditions and is only employed intermittently while carrying full load.

Q. 21. What would happen if a dc motor is directly switched on to the supply, without any starter? **[U.P. Technical Univ., 2005-06]**

Ans. If the dc motor is directly switched on to the supply, it will draw heavy current (about 10–12.5 times rated current) and the motor armature winding may get burnt.

Q. 22. Why is the starting current high in a dc motor? **[U.P.S.C. I.E.S., 2002]**

Ans. When the motor is at rest the speed of a dc motor is zero, therefore, back emf is zero and if a dc motor is connected directly to the supply mains a heavy current will be drawn because from emf equation of armature circuit ($E_b = V - I_a R_a$), armature current is given as $\frac{V - E_b}{R_a}$ and armature resistance R_a is very small.

Q. 23. Why is starter used for starting a dc motor? **[W.B. Univ. of Technology, 2011]**

Ans. For protection of motor against the flow of excessive current during starting period, it is necessary to insert high resistance in series with the armature circuit.

Q. 24. Sketch a 3 point starter with proper label. **[W.B. Univ. of Technology, 2011]**

Ans. Refer to Fig. 10.22.

Q. 25. What precautions are taken during starting of dc shunt and series motors and why?

Ans. It is necessary to ensure, before starting a dc motor, that the field circuit is closed, the rheostat in series with the shunt field winding is at zero resistance position and starting resistance in series with the armature circuit is at maximum value. It will restrict armature current to a safer value and develop higher torque at the starting instant.

In case of a dc series motor it is also necessary that there is significant load on it while starting it.

Q. 26. For stopping the motor, the line switch should always be opened rather than throwing back the starting arm. Why?

Ans. If the starter arm is thrown back the field circuit is broken at the last contact button and owing to inductive nature of the field, this will cause a hot arc and burn the contact.

Q. 27. Why is it desirable for the starting lever to fall back to "off" position when the power fails? **[B.P. Univ. of Technology, 2008]**

Ans. If the starter arm remains in the "ON" position on power failure, on restoration of power supply, no back emf will be acting in the armature circuit, the armature being directly across the supply mains will draw excessive current and will get damaged.

Q. 28. A dc shunt motor is connected to a 3-point starter. What would happen if

(*i*) the starter handle is moved rapidly from 'off' to 'on' position;

(*ii*) the field circuit becomes open circuited with the motor running at no load;

Ans. (*i*) When the starter handle is moved rapidly from 'off' to 'on' position, the resistance connected in series with the armature is being cut at a faster rate and the speed of the motor is increased at a slow rate and because of the rates being different, the armature current will increase to a larger value and the ratio of upper and lower values of currents will no more remain constant.

(*ii*) With the opening of field circuit with motor running at no load, the field current will become zero. So flux will be negligibly small, that is due to residual magnetism and the motor speed will suddenly boost up and continue to increase if the supply is not cut off.

Q. 29. What would happen if the external resistance for starting the dc motor, is left in the armature circuit? **[U.P. Technical Univ., 2007-08]**

Ans. It would cause (*i*) additional loss of energy resulting in reduced operating efficiency and (*ii*) reduction in operating speed of motor.

Q. 30. The field circuit is open and attempt is made to start the motor. **[U.P. Technical Univ., 2008-09]**

Ans. Flux produced will be negligibly small, that is due to residual magnetism, and so the motor will fail to pick up speed. The current drawn will also be too large, induced emf being too small, and the motor may get damaged too.

Q. 31. There is a sudden overload of 100%. **[U.P. Technical Univ., 2008-09]**

Ans. When there is a sudden overload of 100%, the overload release coil, connected in series with the motor, carries heavy current and so get magnetised to such an extent, that its armature is pulled upwards and so short circuits the no-volt release coil. The no-volt release coil, being short circuited, demagnetises and release the starting arm, which goes back to OFF position with the action of spring attached to it and the motor is automatically disconnected from the supply mains.

Q. 32. The starter handle is pulled back to stop the motor. **[U.P. Technical Univ., 2008-09]**

Ans. If the starter arm is thrown back the field circuit is broken at the last contact button and owing to inductive nature of the field, this will cause a hot arc and burn the contact.

Q. 33. Explain no-voltage release (NVR) and overload release (OLR) coils provided in a dc motor starter. **[U.P.S.C. I.E.S., 2011]**

Ans. No-volt release coil consists of an electromagnet connected in series with shunt field which holds the arm in the "ON" position. Now when the supply fails or gets disconnected the electromagnet demagnetizes and so releases the starting arm A, which goes back to "OFF" position due to the spring attached to it and gets disconnected from the supply mains.

The other important advantage of connecting the no-volt release in series with the shunt field winding is that it prevents the motor from running away owing to an open shunt field because open-circuited shunt field will demagnetise the electromagnet and release the starter arm A and, thus the starter arm will go back to its OFF position and the supply will be disconnected.

Overload Release Coil. The OLR is a small electromagnet connected in series with the armature circuit. This coil is provided for the protection of the motor against the flow of excessive current due to overload. This coil is connected in series with motor so carries full-load current. When the motor is overloaded, it draws heavy current, which also flows through this coil and magnetises it to such an extent, that it pulls its armature upwards and so short circuits the no-volt release coil, as shown in Fig. 10.22. The no-volt release coil, being short circuited, demagnetises and releases the starting arm, which goes back to "OFF" position with the action of spring attached to it and the motor is automatically disconnected from the supply mains. Thus the motor is disconnected from the supply and is protected against over-loading.

Q. 34. Why do you prefer starter to rheostat in starting dc motors?

Ans. The starter also includes a few protective devices, such as no-volt release, overload release etc.

Q. 35. What is the disadvantage of a 3-point starter as compared to a 4-point starter used to start dc shunt motors? **[B.P. Univ. of Technology, 2008]**

Ans. In three point starter (shown in Fig. 10.22) no-volt release coil is connected in series with the shunt field and the field rheostat and, therefore, the current flowing through the field is the same current that flows through the holding coil. Thus if sufficient resistance is *cut in* by the field rheostat so that holding coil current is no longer able to create sufficient electromagnetic pull to overcome the spring tension, the starter arm will fall back to the "OFF" position. It is this undesirable feature of the three-point starter that makes it unsuitable for use with speed-controlled motors and that has resulted in the wide spread application of four point starters.

Q. 36. How does four point starter differ from three point starter?

Ans. In a four point starter the circuits of no-volt release coil and shunt field are independent and so the operation of no-volt release is not affected due to variations in field current.

Q. 37. Why no-volt release is provided in a dc series motor?

Ans. To protect the dc series motor against no volt and no-load.

Q. 38. Why is it desirable to make an independent circuit of the no-volt release coil?

Ans. To make the operation of no-volt release independent of variations in the field current.

Q. 39. What are the requirements of good electric braking? **[J.N. Univ. April 2003]**

Ans. The requirements of good electric braking are:

(*i*) The braking should be quick and reliable in action.

(*ii*) The braking must be controllable.

(*iii*) Some suitable means must be provided for dissipation of kinetic energy of the moving parts of the motor and its driven machines.

(*iv*) Failure of any part of the braking system must result in application of brakes.

Q. 40. What are various types of electric braking used? **[J.N. Univ., November-2003]**

Ans. There are three types of electric braking, all of which are applicable to the usual types of electric motors, viz plugging (or counter-current braking), dynamic (or rheostatic) braking and regenerative braking.

Q. 41. What are braking systems applicable to a dc shunt motor? **[Anna Univ. Summer, 2003]**

Ans. The braking systems applicable to a shunt motor are plugging, dynamic and regenerative braking.

Q. 42. What is meant by plugging?

Ans. Plugging means stopping of an electric motor by instantaneously reversing its torque till it stops.

Q. 43. How plugging of a dc motor is done?

Ans. Plugging is obtained in a dc motor by reversing the supply terminals to the armature of the motor.

Q. 44. What is dynamic braking? **[Anna Univ. Summer, 2003]**

Ans. In dynamic braking, the motor is disconnected from the supply and operated as a generator driven by the kinetic energy of the motor and its driven machines. The kinetic energy of rotation is converted into electrical energy which is dissipated in the external resistance connected across the motor at the braking instant.

Q. 45. Why rheostatic braking is economical to plugging?

Ans. In rheostatic braking, energy required from the supply to brake the motor is eliminated as compared to the plugging.

Q. 46. What is the most economical method of electric braking?

Ans. Regenerative braking.

Q. 47. What is regenerative braking? **[Anna Univ. Summer, 2003]**

Ans. In regenerative braking, the electric machine operation is changed from motoring to generating, mechanical energy is converted into electrical energy, part of which is returned to the supply and the rest of the energy is lost as heat in the windings and bearings of the electrical machines.

Q. 48. In case of a dc series motor, when it is disconnected from the supply and is connected across a braking resistance, it is essential to change over the field winding connections. Why?

Ans. Otherwise, the current through the field winding will also be reversed, and no braking effect will occur.

PROBLEMS

1. A 230 V dc shunt motor runs at 800 rpm and takes armature current of 50 A. Find the resistance to be added to the field circuit to increase speed to 1,000 rpm at an armature current of 80 A. Assume flux proportional to field current. Armature resistance = 0.15 Ω and field winding resistance = 250 Ω. **[Hyderabad Univ. Electrical Technology, 1991]**
[Ans. 68.95 Ω**]**

2. A 400 V dc shunt motor, with an armature resistance of 0.2 Ω runs at 1,000 rpm and takes an armature current of 60 A. If the field flux is suddenly increased by 20%, obtain at this instant, new values of armature current and the torque in terms of initial torque.
What is the operating mode of dc shunt machine under these conditions? State the assumption made, if any. **[U.P.S.C. I.E.S. Electrical Engineering-II, 2012]**
[Ans. Constant power operation; 60 A, 1.2 times initial torque**]**

3. A 220 V dc shunt motor has an armature resistance of 0.4 Ω and field resistance of 200 Ω. It takes an armature current of 22 A and runs at 600 rpm. It drives a load whose torque is constant. Suggest a suitable method to raise the speed to 900 rpm. Calculate the value of the controllable parameter. **[Nagpur University, April-1998]**
[Ans. 106.8 Ω**]**

4. A 250 V dc shunt motor has an armature resistance of 0.5 Ω and field resistance of 250 Ω. The armature takes 2.5 A while driving a constant load torque at 750 rpm. What resistance must be inserted in the shunt field circuit to raise the speed from 750 rpm to 950 rpm assuming the magnetic curve to be a straight line. **[B.P. Univ. of Technology Electrical Machine-I, 2008]**
[Ans. 71.5 Ω**]**

5. A 230 V dc shunt motor rotates at 1,000 rpm when the armature current is 35 A. The resistance of the armature circuit is 0.3 Ω. Calculate the additional resistance required in the armature circuit to reduce the speed of the motor to 750 rpm, assuming that the armature current is then 25 A. **[G.G.S.I.P. Delhi University, May-June 2006]**
[Ans. 2.315 Ω**]**

6. A 250 V shunt motor with armature resistance of 0.5 Ω runs at 600 rpm on full load and takes an armature current of 20 A. If resistance of 1.0 Ω is placed in the armature circuit, find the speed at (*i*) full-load torque (*ii*) half-full load torque. **[Ans.** (*i*) 550 rpm (*ii*) 587.5 rpm**]**

7. A 240 V shunt motor has an armature current of 15 A when running at 800 rpm against full-load torque. The armature resistance is 0.6 Ω. What resistance must be inserted in series with the armature to reduce the speed to 400 rpm at the same torque?
What will be the speed if the load torque is halved with this resistance in the circuit? Assume flux to remain constant throughout.
[Ans. 7.7 Ω; 614.72 rpm**]**

8. A dc shunt motor drives a centrifugal pump whose torque varies as the square of the speed. The motor is fed from a 200 V supply and takes 50 A when running at 1,000 rpm. What resistance must be inserted in the armature circuit in order to reduce the speed to 800 rpm ? The armature and field resistances of motor are 0.1 Ω and 100 Ω respectively. **[Allahabad Univ. Elec. Machines 1992]**
[Ans. 1.327 Ω**]**

9. The speed of a 15 hp (metric) 400 V dc shunt motor is to be reduced by 25% by the use of a controller. The field current is 2.5 A and the armature resistance is 0.5 Ω. Calculate the resistance of the controller, if the torque remains constant and the efficiency is 82%. **[J.N. Univ. Hyderabad, November-2003]**
[Ans. 3.086 Ω**]**

10. A dc shunt motor takes an armature current of 20 A from a 220 V supply. Armature circuit resistance is 0.5 Ω. For reducing the speed by 50%, calculate the resistance required in the series, with the armature if (*i*) the load torque is constant (*ii*) the load torque is proportional to the square of the speed. **[Sambalpur Univ. 1998]**
[Ans. (*i*) 5.25 Ω (*ii*) 22.5 Ω**]**

11. A dc series motor runs at 500 rpm on 220 V supply drawing a current of 50 A. The total resistance of the machine is 0.15 Ω. Calculate the extra resistance to be connected in series with motor circuit that will reduce the speed to 300 rpm. The torque then being half of the previous value. Assume flux proportional to current. **[G.B. Technical Univ. Electrical Machines and Automatic Control, 2012-13]**
[Ans. 5.95 Ω**]**

12. A series motor takes 20 A at 400 V to drive a load whose torque is proportional to square of the speed at 250 rpm. Determine the necessary applied voltage and current to drive the said load at 350 rpm. Assume straight line magnetising characteristic. **[U.P. Technical Univ. Electrical Machines, 2004-05]**
[Ans. 784 V**]**

13. A dc series motor drives a load, the torque of which varies as the square of speed. The motor takes a current of 15 A when the speed is 600 rpm. Calculate the speed and current when the motor field winding

is shunted by a diverter of equal resistance as that of field winding. Neglect all motor losses and assume the magnetic circuit as unsaturated.

[**Ans.** 713.5 rpm; 25.227 A]

14. A dc series motor with an unsaturated field and negligible resistance, when running at a certain speed on a given load takes 50 A at 460 V. If the load torque varies as the cube of the speed, calculate the resistance required to reduce the speed by 25%.

[Madras University, 1987; Nagpur University, 1999]

[**Ans.** 7.26 Ω]

15. A 4-pole, 250 V dc series motor takes 20 A and runs at 900 rpm. Each field coil has resistance of 0.025 Ω and the resistance of the armature is 0.1 Ω. At what speed will the motor run developing the same torque if (*i*) a diverter of 0.2 Ω is connected in parallel with the series field (*ii*) re-arrranging the field coils in two series and parallel groups. Assume unsaturated magnetic operation.

[Nagpur Univ. Electric Drives & Their Control, 1993]

[**Ans.** (*i*) 1,101.6 rpm (*ii*) 1,275.4 rpm]

16. A dc series motor is driving a fan load where the load torque is proportional to cube of speed. The resistance of the armature and field in series is 1 Ω and the motor takes 10 A and runs at 1000 rpm when operating from 200 V supply. Calculate the value of resistance to be inserted in series with the armature to reduce the operating speed to 800 rpm.

[U.P. Technical Univ. Electromechanical Energy Conversion-I, 2006-07]

[**Ans.** 11.733 Ω]

17. A 3 kW series motor runs normally at 800 rpm on a 200 V supply taking 16 A; when the field coils are all connected in series. Estimate the speed and current taken by the motor if the coils are reconnected in two parallel groups of two each in series. Load torque increases as square of the speed.

[Rajasthan Technical Univ. Electrical Machines-I, 2011, January-2013]

[**Ans.** 951.4 rpm; 26.9 A]

18. Two series motors run at a speed of 500 rpm and 550 rpm respectively when taking 50 A at 500 V. The total resistance of each motor is 0.5 Ω. Calculate the speed of the combination when connected in series and coupled mechanically. The combination is taking 50 A on 500 V supply. [**Ans.** 248.12 rpm]

19. A 120 V dc shunt motor has an armature resistance of 0.2 Ω and a brush volt drop of 2 V. The rated full-load armature current is 75 A. Calculate the current at the instant of starting and its value in terms of percentage of full-load armature current. Assume field winding resistance of 120 Ω. **[U.P. Technical Univ. Electrical Machines, 2004-05]**

[**Ans.** 591 A; 788%]

20. A 440 V, 80 A, 1200 rpm dc shunt motor has an armature resistance of 0.55 Ω. Determine the resistance for each step of the starter for accomplishing starting in 4 steps. The maximum armature current should not exceed 150% of the full-load current. Neglect small amount of shunt field current.

[**Ans.** 1.3847 Ω, 0.8618 Ω, 0.5364 Ω and 0.3338 Ω]

21. A 500 V dc shunt motor is to have a starter with 5 sections. The maximum permissible current is 100 A and the armature resistance is 0.5 Ω. Determine the resistance in each section of the starter.

[**Ans.** 1.845 Ω, 1.165 Ω, 0.734 Ω, 0.464 Ω and 0.292 Ω]

22. The full-load current of a dc motor is 150 A at 600 V. The combined resistance of the armature and interpole winding is 0.25 Ω. Determine the number of steps to be provided in a starter and the resistance value of each step if the ratio of maximum current to full-load current should not exceed 1.5.

[**Ans.** 6 sections, 1.78 Ω, 1.187 Ω, 0.792 Ω, 0.528 Ω, 0.352 Ω and 0.25 Ω]

23. A 220 V separately excited motor running at 1,000 rpm draws 100 A from the source. The motor is braked by plugging. Calculate (*i*) resistance to be inserted in the armature circuit to limit the braking current to twice the full-load current (*ii*) initial braking torque (*iii*) the braking torque when speed has reduced to 500 rpm.

[U.P. Technical Univ. Electric Drives, 2005-06]

[**Ans.** (*i*) 2.2 Ω; (*ii*) Twice full-load torque (*iii*) 1.5 times full-load torque]

24. A 40 kW, 440 V dc shunt motor is braked by plugging. Calculate (*i*) the value of resistance that must be connected in series with the armature circuit to limit the initial braking current to 150 A. (*ii*) the braking torque and (*iii*) the torque when the motor speed falls to 360 rpm. Armature resistance $R_a = 0.1$ Ω, full-load current, $I_a = 100$ A, full-load speed = 600 rpm.

[**Ans.** (*i*) 5.7 Ω; (*ii*) 954.93 Nm; (*iii*) 766 Nm]

25. A 400 V, 25 hp, 450 rpm dc shunt motor is braked by plugging when running on full-load. Determine the braking resistance necessary if the maximum braking current is not to exceed twice the full-load current. Determine also the maximum braking torque and the braking torque when the motor is just reaching zero speed. The efficiency of the motor is 74.6% and the armature resistance is 0.2 Ω.

[**Hint:** Maximum braking torque will be developed initially when motor speed is maximum i.e. at the speed of 450 rpm.]

[Nagpur Univ. Summer, 2004]

[**Ans.** (*i*) 6.1 Ω, 1,028 Nm; 522 N-m]

26. A 250 V dc shunt motor has an armature resistance of 0.1 Ω and with full load has an emf of 240 V at a speed of 1,200 rpm. The motor is driving an overhauling load with a torque of 180 Nm.

Calculate the minimum speed at which the motor can hold the load by means of regenerative braking. [**Ans.** 1,297 rpm]

❑❑❑

11

CHAPTER

Losses, Testing and Efficiency of DC Machines

INSIDE THIS CHAPTER

11.1 POWER LOSSES

The basic function of dynamo is the conversion of energy, either from mechanical to electrical form or from electrical to mechanical form. Whole of the input energy to the dynamo is not converted into useful output energy but a part of the input energy is converted into heat and is lost for the useful output purpose of the machine. The energy converted into heat is called the *energy loss*.

For steady-state operating conditions the average rate of energy conversion is constant, and the energy converted is equal to the product of average power and time duration of steady-state operation.

Therefore, it is more effective to deal with the power losses and power output than to deal directly with the energy values. The basic power equation for a dynamo is

$$P_{input} = P_{output} + P_{losses} \quad ...(11.1)$$

The power losses occurring in a dc machine produce heating in it; therefore, the greater the power losses, as a percentage of the power input, the hotter will the machine tend to become. If such losses attain an excessive value, the temperature rise might be high enough to cause dynamo failure.

There are two general classifications of power losses in a dynamo: (*i*) those that are caused due to flow of electric current in the various parts of a dynamo and (*ii*) those that are developed due to rotation of armature. The former are generally termed *electrical (or copper) losses* and the latter are called the *rotational losses*.

For the purpose of calculation, as well as for understanding what takes place in a dynamo, it is desirable to separate each of the two general losses into its components and then determine what values they have as well as the way in which they vary with load and speed.

1. Electrical (or Copper) Losses. Copper or electrical losses include power lost in heating the shunt field, and also in heating armature circuit, which includes armature winding, brush contacts, series field and interpole field and compensating winding field. These losses are due to power consumed in forcing a current against the resistance of windings and connections. The actual copper losses of the windings will depend upon the effective resistance of the winding for the operating conditions. The effective resistance of winding will vary with the operating conditions of load and excitation, and its exact value even for one particular set of operating conditions is hard to determine. For these reasons it is standard practice to compute losses on the basis of the dc resistance of the winding at 75 °C. These losses are further subdivided into (*a*) armature copper loss and (*b*) field copper loss.

(a) ***Armature Copper Loss.*** This loss is given as $I_a^2 R_a$ where I_a is the armature current and R_a is the armature resistance at operating temperature (assumed to be 75°C). Brush contact loss* is also taken into account by including the brush contact resistance in the armature resistance R_a. This loss is proportional to the square of armature current and is about 30 to 40% of full-load losses.

(b) ***Field Copper Loss.*** This loss is further subdivided into (*i*) shunt field loss, (*ii*) interpole field loss, (*iii*) compensating winding field loss and (*iv*) series field loss. Shunt field loss is given as $I_{sh}^2 R_{sh}$ where I_{sh} is the shunt field current and R_{sh} is the shunt field resistance. The shunt field loss incurs in shunt and compound dc machines and is practically constant if applied

* The brush contact loss is due to brush contact resistance (*i.e.* resistance between the surface of brushes and that of commutator). The voltage drop at a brush is almost independent of armature current I_a and the power loss at the brush contacts may be taken to vary in direct proportion to the armature current. For carbon brushes the voltage drop is about 1V per brush. The power loss due to brush contact resistance is $2V_b I_a$ where V_b is the voltage drop across each brush and I_a is the armature current.

voltage across the machine remains the same. Interpole loss is given as $I_a^2 R_i$ where I_a is the armature current and R_i is the interpole field resistance. Interpole loss incurs only in the machines provided with interpoles. The series field loss, which incurs in series and compound machines, is given as $I_{se}^2 R_{se}$ where I_{se} is series field current and R_{se} is series field resistance. Compensating winding field loss is given as $I_a^2 R_{com}$, where R_{com} is the resistance of compensating winding, and incurs only in the machines provided with compensating winding. With the increase in load on the machine, armature current increases resulting in increase in interpole field, compensating winding field and series field losses. Field copper loss is about 20 to 30% of the full-load losses.

2. Rotational Losses. The rotational losses, obviously vary with the variations in the speed and strength of field of a dc machine. The rotational losses are further subdivided into (*a*) iron or magnetic losses and (*b*) mechanical losses.

(a) *Iron or Magnetic Losses.* These losses are also called *core losses* and include the hysteresis and eddy current losses which are present in any part of the dynamo that is constructed of iron and subjected to variations of magnetic flux.

(*i*) *Hysteresis Loss.* Hysteresis loss incurs in the revolving armature core and teeth of a dc machine since any given part of the armature is subjected to magnetic field reversals as it passes under successive poles. It is directly proportional to the number of reversals per second, that is to speed (speed being equal to $120f/P$). It has been found experimentally that the variation of hysteresis loss with variations in the number of reversals per second (f) and maximum flux density (B_{max}) may be expressed with sufficient accuracy by the equation

$$\text{Hysteresis loss, } P_h = \eta (B_{max})^{1.6} f\,v \text{ joules/second or watts} \quad ...(11.2)$$

where η = Steinmetz hysteresis coefficient, v is volume of core in metres^3, B_{max} is the peak magnitude of the flux density attained in a magnetic cycle and f is the frequency of the magnetic cycles per second and is given as $f = PN/120$, P being the number of field poles and N is the speed of rotation in rpm. The values of Steinmetz hysteresis coefficient for good dynamo sheet-steel and silicon steel are 502.4 and 191 J/m^3 respectively.

Hysteresis loss is unaffected by whether or not the armature core is laminated. In generators, whose speeds are usually constant, and in dc shunt motors that operate at substantially constant speed, the hysteresis loss is fairly constant because the flux density in the core iron changes very little with load. In dc series and compound motors the hysteresis loss does vary somewhat with variations in load because both the speed and flux density are affected by variations in load.

In order to reduce the hysteresis loss in a dynamo, armature core is made of such materials which have a low value of Steinmetz hysteresis coefficient *i.e* silicon steel.

(*ii*) *Eddy Current Loss.* When armature core rotates in the magnetic field of the poles, an emf is induced in it which circulates eddy currents in the armature core. The power loss due to these eddy currents is called the *eddy current loss*. It incurs in armature core and teeth and pole faces. The eddy current loss also depends upon the range of variation of flux density and the frequency of the variation of the flux. In addition it depends upon the thickness of the iron laminations. It can be determined with reasonable accuracy by the following expression:

$$P_e = K_e B_{max}^2 f^2 v t^2 \text{ watts} \quad ...(11.3)$$

where K_e is a constant depending upon the electrical resistance of the magnetic material and the units employed in the expression, B_{max} is the maximum flux density attained throughout the magnetic cycle, f is the frequency of magnetic cycle per second, v is the volume of magnetic material and t is the thickness of the magnetic material laminations.

Eddy currents are electromagnetic in character but are developed only because of the rotation of the armature core; these, too may, therefore, be considered as rotational loss. It is independent of the quality of the magnetic iron used for armature core. This loss is affected in much the same way by the load as hysteresis; if the speed and flux densities remain constant, the eddy current loss does not change; if they vary, as they must in dc series and compound motors, it will change.

It may be noted that eddy current loss varies as the square of thickness of laminations. For this reason the armature core is built of thin laminations insulated from each other by a small layer of varnish and the thickness of laminations is kept as small as possible

The core losses are approximately constant in a machine whose flux and speed remain constant as in case of shunt wound dc machines but are variable in case of series wound and compound wound dc machines. These losses are about 20 to 30% of full-load losses.

(b) *Mechanical Losses.* These losses consist of power losses due to friction of the bearings, air friction or windage, as it is called, caused by the motion of the moving parts through the surrounding medium, and the friction between brushes and commutator or slip rings.

Improved bearings reduce the amount of energy consumed and also bearing wear. For motors having ball or roller bearings, the bearing friction losses are so small compared to the windage losses that in most of the cases they are not computed separately but included in the windage loss. Improper lubrication increases the bearing friction loss. It is roughly proportional to speed.

In all commutator machines, the brush friction loss is quite large. The coefficient of friction of hard brushes is greater than that of soft brushes. Hard brushes, however, will increase wear on the commutator, while soft brushes have less drag and wear faster. The brush friction loss will change with the change in speed.

Windage loss refer to the power required to move the air about the armature. In some machines blades are mounted on the armature to pass large quantities of air over the winding and thus cool the machine. The greater the quantity of air moved, the greater the power consumed. The volume of air moved depends upon the speed of the machine. The windage loss varies as the cube of the speed.

The mechanical losses are constant in a machine operating at approximately constant speed and are independent of load. These losses are about 10 to 20 per cent of full-load losses.

(c) *Stray-Load Loss.* It results from such factors as (*i*) distortion of the flux owing to armature reaction, (*ii*) lack of uniform division of the current in the armature winding through the various paths and through individual conductors of large cross-section area, and (*iii*) short-circuit currents in the coils undergoing commutation. It is difficult to determine the stray-load loss and, therefore, it becomes necessary to assign it a reasonable value arbitrarily. It is usually assumed to be 1 per cent of the output of the machine when the rating is about 150 kW or more; for the smaller ratings, the stray-load loss is generally neglected.

Usually magnetic and mechanical losses collectively termed as *stray power losses* and *shunt field copper losses* are collectively known as *constant losses*.

Those losses which vary with the load are called the *variable losses* and include copper losses occurring in armature winding and series field winding, interpole winding and compensating winding if any provided in the machine.

The total losses occurring in a dc machine are summarized below.

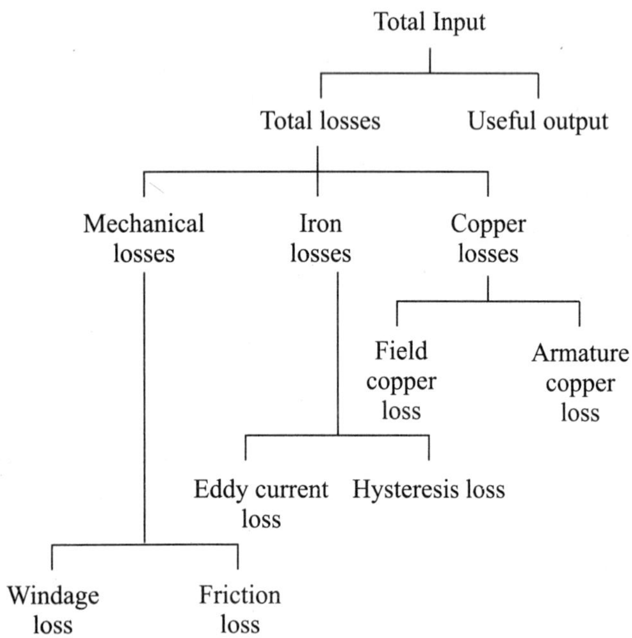

The ratio of useful output to the total input is called the *efficiency* of the machine. It is expressed as

$$\eta = \frac{\text{Output}}{\text{Input}} = \frac{\text{Input} - \text{total losses}}{\text{Input}}$$

$$= \frac{\text{Output}}{\text{Output} + \text{total losses}} \qquad \text{...(11.4)}$$

11.2 POWER FLOW ANALYSIS

Knowledge of sequence of energy conversions is important in the analysis of the operation and characteristics of machines. In the operation of any machine under steady-state conditions, energy is delivered to the machine at some definite rate. Some of this input power is converted into heat through the copper, iron, and mechanical losses. The rest of the input power becomes useful output and is delivered to the load. At some point in the sequence of energy conversion either mechanical energy is converted into electrical, or electrical energy is converted into mechanical. It is important to know at what point this conversion of energy takes place. The criterion for determining the point in the energy conversion sequence at which any particular power loss occurs is as follows:

Determine the energy source which is the cause of the phenomenon that results in conversion of energy into heat. The energy source for the mechanical losses, which are caused by rotation of parts of the machine, must be mechanical one, and the mechanical losses consist of the conversion of energy in mechanical form into heat. Therefore, in the case of generator the mechanical losses must take place before the conversion of mechanical energy into electrical one. On the other hand, in case of a motor the mechanical losses must take place after the conversion of electrical energy into mechanical one. Similarly, the energy source for the iron losses in dc machines, which are caused by rotation of the parts of the machine, must be mechanical one, and the iron losses in these machines consist of the conversion of energy in mechanical form into heat. In a dc generator the iron or magnetic losses must take place before the conversion of mechanical energy into electrical one, whereas in case of a dc motor the iron losses must take place after the conversion of electrical energy into mechanical one. The energy source for copper losses, which are caused due to flow of current through a resistance, must be electrical one and the copper losses consist of the conversion of energy in electrical form into heat. Therefore, in the case of a motor the copper or electrical losses must take place before conversion of electrical energy into mechanical one, whereas for a dc generator the copper losses must take place after the conversion of mechanical energy into electrical one. Power flow diagrams are helpful in depicting graphically the sequence of power conversion. Typical power flow diagrams for a dc generator and a dc motor are shown in Figs. 11.1 and 11.2 respectively.

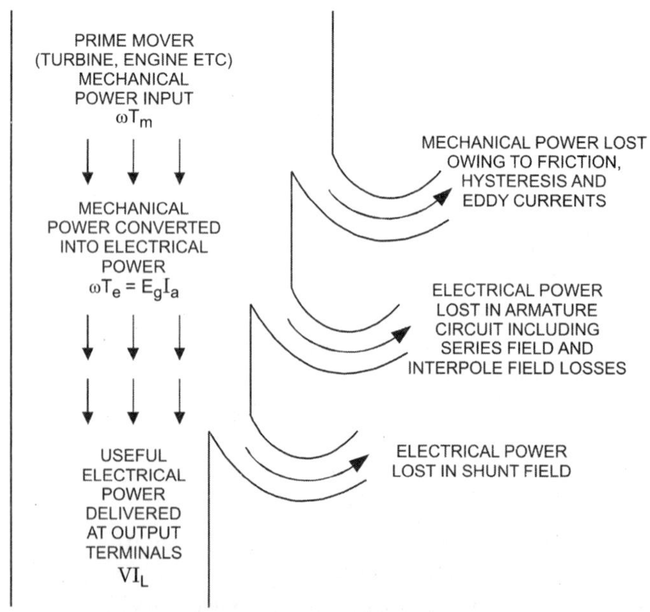

Fig. 11.1 *Power Flow Diagram For DC Generator*

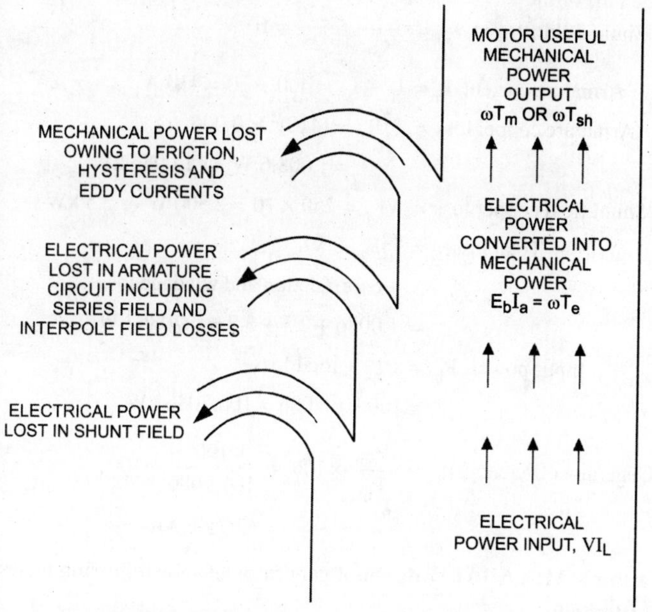

Fig. 11.2 *Power Flow Diagram For DC Motor*

11.3 EFFICIENCY OF DC MACHINES

The various efficiencies of dc machines are expressed as follows:

1. Generator:

(*i*) Mechanical efficiency,

$$\eta_m = \frac{\text{Electrical power developed by armature}}{\text{Total mechanical power input}} = \frac{E_g I_a}{\text{BHP of prime mover} \times 735.5}$$

(*ii*) Electrical efficiency,

$$\eta_e = \frac{\text{Useful electrical power output}}{\text{Electrical power developed}} = \frac{VI_L}{E_g I_a}$$

(*iii*) Commercial or overall efficiency,

$$\eta_G = \eta_m \eta_e = \frac{\text{Useful electrical power output}}{\text{Total mechanical power input}} = \frac{VI_L}{\text{BHP of prime mover} \times 735.5} \quad ...(11.5)$$

2. Motor

(*i*) Electrical efficiency,

$$\eta_e = \frac{\text{Mechanical power developed}}{\text{Total electrical power input}} = \frac{E_b I_a}{VI_L}$$

(*ii*) Mechanical efficiency,

$$\eta_m = \frac{\text{Useful mechanical power output}}{\text{Mechanical power developed}} = \frac{\text{BHP of motor} \times 735.5}{E_b I_a}$$

(*iii*) Commercial or overall efficiency,

$$\eta_M = \eta_m \eta_e = \frac{\text{Useful mechanical power output}}{\text{Total electrical power input}} = \frac{\text{BHP of motor} \times 735.5}{VI_L} \quad ...(11.6)$$

The overall efficiency of dc generator and motor can also be expressed as follows:

Overall efficiency of generator,

$$\eta_G = \frac{\text{Useful power output}}{\text{Total power input}} = \frac{\text{Useful power output}}{\text{Useful power output + total losses}} = \frac{VI_L}{VI_L + \text{total losses}} \quad ...(11.7)$$

Overall efficiency of motor,

$$\eta_M = \frac{\text{Useful power output}}{\text{Total power input}} = \frac{\text{Total power input − total losses}}{\text{Total power input}} = \frac{VI_L - \text{total losses}}{VI_L} \quad ...(11.8)$$

11.4 CONDITION FOR MAXIMUM EFFICIENCY

Condition for maximum efficiency of a dc generator or of a dc motor is the same and for a dc generator it is derived as follows

Generator output $= VI_L$ watts

where V is the terminal voltage and I_L is the output or load current

$$\text{Generator input} = \text{Output + total losses}$$
$$= VI_L + I_a^2 R_a + P_C \text{ (constant losses)}$$
$$= VI_L + (I_L + I_{sh})^2 R_a + P_C$$
$$= VI_L + I_L^2 R_a + P_C$$

∵ Shunt field current I_{sh} is very small in comparison with load current I_L

where R_a is the total resistance of armature circuit (including brush contact resistance and resistance of series, interpole and compensating windings, if any).

$$\text{Generator efficiency, } \eta = \frac{\text{Output}}{\text{Input}} = \frac{VI_L}{VI_L + I_L^2 R_a + P_C} = \frac{1}{1 + \frac{I_L R_a}{V} + \frac{P_C}{VI_L}}$$

Efficiency will be maximum when denominator will be minimum *i.e.* when

$$\frac{d}{dI_L}\left(\frac{I_L R_a}{V} + \frac{P_C}{VI_L}\right) = 0$$

$$\text{or} \quad \frac{R_a}{V} - \frac{P_C}{VI_L^2} = 0$$

$$\text{or } I_L^2 R_a = P_C$$

Hence efficiency will be maximum when variable losses are equal to constant losses.

The load current corresponding to maximum efficiency is given by the relation.

$$I_L = \sqrt{P_C / R_a} \quad ...(11.9)$$

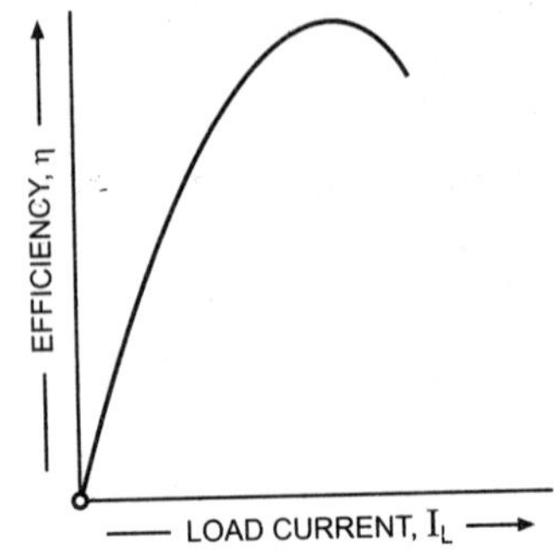

Fig. 11.3

Variation of efficiency with load is illustrated in Fig. 11.3 which indicates that the efficiency increases with the increase in load current, attains a maximum value when the load current equals the value given by Eq. (11.9) and then starts decreasing. The machines are usually designed to give maximum efficiency at or near the rated output of the machine.

Example 11.1. A dc motor drives a 100 kW generator having an efficiency of 87%:

(i) What should be the kW rating of the motor?

(ii) If the overall efficiency of the motor generator set is 74%, what is the efficiency of the motor?

(iii) Also calculate the losses in each machine.

[G.B. Technical Univ., Electromechanical Energy Conversion-I, 2012-13]

Solution: Generator output = 100 kW

Generator efficiency, $\eta_g = 87\%$ or 0.87

(i) kW rating of motor = Motor output in kW = Generator input in kW

$$= \frac{\text{Generator output in kW}}{\eta_g} = \frac{100}{0.87} = 114.94 \text{ kW Ans.}$$

(ii) Motor efficiency,
$$\eta_m = \frac{\text{Overall efficiency}}{\eta_g} \quad \because \eta_{overall} = \eta_m \eta_g$$

$$= \frac{0.74}{0.87} = 0.8506 \text{ or } 85.06\% \textbf{ Ans.}$$

$$\text{Motor input} = \frac{\text{Motor output}}{\eta_m} = \frac{114.94}{0.8506} = 135 \text{ kW}$$

(iii) Losses in generator = Generator input – generator output = 114.94 – 100 = 14.94 kW **Ans.**

Losses in motor = Motor input – motor output = 135 – 114.94 = 20.06 kW **Ans.**

Example 11.2. Calculate the overall efficiency of a 250 V, 100 kW dc shunt generator at full load if the resistance of the armature and shunt field are 0.006 and 25 Ω respectively. The core, friction and windage losses together are 3.2 kW.

[U.P. Technical Univ. Electrical Machines, 2008-09]

Solution: Power output, $P_{out} = 100$ kW

$$\text{Load current, } I_L = \frac{P_{out} \text{ in kW} \times 1{,}000}{V} = \frac{100 \times 1{,}000}{250} = 400 \text{ A}$$

$$\text{Shunt field current, } I_{sh} = \frac{V}{R_{sh}} = \frac{250}{25} = 10 \text{ A}$$

Armature current, $I_a = I_L + I_{sh} = 400 + 10 = 410$ A

Armature copper loss $= I_a^2 R_a = (410)^2 \times 0.006 = 1{,}008.6$ W or 1.0086 kW

Shunt field copper loss $= VI_{sh} = 250 \times 10 = 2{,}500$ W or 2.5 kW

Total losses $= I_a^2 R_a + VI_{sh}$ + core, friction and windage losses $= 1.0086 + 2.5 + 3.2 = 6.7086$ kW

Input power, $P_{in} = P_{out}$ + total losses $= 100 + 6.7086 = 106.7086$ kW

$$\text{Generator efficiency, } \eta_G = \frac{P_{out}}{P_{in}} \times 100 = \frac{100}{106.7086} \times 100 = 93.71\% \textbf{ Ans.}$$

Example 11.3. A 10 kW dc shunt generator has the following losses at full-load.

Mechanical losses = 290 W, Iron loss = 410 W;

Shunt copper loss = 120 W; Armature copper loss = 595 W;

Calculate the efficiency of dc generator at (i) full load (ii) 25% of full load.

[U.P. Technical Univ. Electrical Machines, 2005-06]

Solution: Generator output, P_{out} = 10 kW = 10,000 W

Constant losses, P_c = Mechanical loss + iron loss + shunt field copper loss = 290 + 410 + 120 = 820 W

At Full load

Variable losses = Armature copper loss = 595 W

Total losses = Constant losses + variable losses = 820 + 595 = 1,415 W

$$\text{Generator efficiency, } \eta_1 = \frac{\text{Output}}{\text{Output + total losses}} \times 100 = \frac{10{,}000}{10{,}000 + 1{,}415} \times 100 = 87.6\% \textbf{ Ans.}$$

At 25 per cent of Full Load

$$\text{Output} = \frac{25}{100} \times 10{,}000 = 2{,}500 \text{ W}$$

$$\text{Variable losses} = \left(\frac{25}{100}\right)^2 \times 595 = 37.2 \text{ W}$$

Input = Output + constant losses + variable losses = 2,500 + 820 + 37.2 = 3,357.2 W

$$\text{Generator efficiency, } \eta_2 = \frac{\text{Output}}{\text{Input}} \times 100 = \frac{2{,}500}{3{,}357.2} \times 100 = 74.47\% \textbf{ Ans.}$$

Example 11.4. A shunt generator delivers 195 A at terminal pd o[f] 250 V. The armature resistance and shunt field resistance are 0.02 Ω [and] 50 Ω respectively. The iron and friction losses equal 950 W. Fin[d] (a) emf generated (b) copper losses (c) output of the prime move[r] (d) commercial, mechanical and electrical efficiencies.

[Nagpur Univ. Elec. Machines-I, 199]

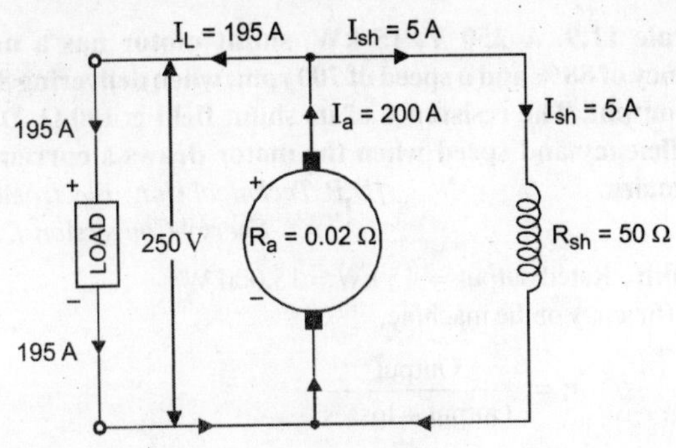

Fig. 11.4

Solution: Output load current, I_L = 195 A

Terminal voltage, V = 250 V

Useful output of generator = $V I_L$

= 250 × 195 = 48,750 watts

Shunt field current, $I_{sh} = \dfrac{V}{R_{sh}} = \dfrac{250}{50} = 5$ A

Armature current, $I_a = I_L + I_{sh} = 195 + 5 = 200$ A

Shunt field copper loss = $V I_{sh} = 250 \times 5 = 1{,}250$ watts

Armature copper loss = $I_a^2 R_a = (200)^2 \times 0.02 = 800$ watts

Iron and friction (or stray) losses, P_S = 950 watts

(*a*) EMF generated, $E_g = V + I_a R_a$

= 250 + 200 × 0.02 = 254 V **Ans.**

(*b*) Total copper losses, $P_{copper} = I_a^2 R_a + V I_{sh}$

= 800 + 1,250 = 2,050 watts **Ans.**

Electrical power developed = Useful power output + total copper loss

= 48,750 + 2,050 = 50,800 watts

Total power input = Electrical power developed + stray losses

= 50,800 + 950 = 51,750 watts

(*c*) Output of prime mover = Total power supplied to generator

= 51,750 W or 51.75 kW **Ans.**

(*d*) Commercial efficiency, $\eta = \dfrac{\text{Useful power output}}{\text{Total power input}} \times 100$

$= \dfrac{48{,}750}{51{,}750} \times 100 = 94.2\%$ **Ans.**

Mechanical efficiency, $\eta_m = \dfrac{\text{Electrical power developed}}{\text{Total power input}} \times 100$

$= \dfrac{50{,}800}{51{,}750} \times 100 = 98.16\%$ **Ans.**

Electrical efficiency, $\eta_e = \dfrac{\text{Useful power output}}{\text{Electrical power developed}} \times 100$

$= \dfrac{48{,}750}{50{,}800} \times 100 = 95.96\%$ **Ans.**

Example 11.5. A shunt generator gives full-load output of 30 kW at a terminal voltage of 200 V. The armature and field resistances are 0.05 Ω and 50 Ω respectively. The iron and friction losses are 1,000 W. Calculate (*a*) generated emf (*b*) copper losses (*c*) efficiency.

[*Rajasthan Technical Univ. Electrical Machines-I, 2008, 2010*]

Solution: Power output, P_{out} = 30 kW = 30,000 W

Terminal voltage, V = 200 V

Output current, $I_L = \dfrac{P_{out}}{V} = \dfrac{30{,}000}{200} = 150$ A

Shunt field current, $I_{sh} = \dfrac{V}{R_{sh}} = \dfrac{200}{50} = 4$ A

Armature current, $I_a = I_L + I_{sh} = 150 + 4 = 154$ A

(*a*) Generated emf, $E_g = V + I_a R_a$

= 200 + 154 × 0.05 = 207.7 V **Ans.**

(*b*) Armature copper loss = $I_a^2 R_a = (154)^2 \times 0.05 = 1{,}185.8$ W

Shunt field copper loss = $V I_{sh} = 200 \times 4 = 800$ W

Total copper losses, $P_{cu} = I_a^2 R_a + V I_{sh}$

= 1,185.8 + 800 = 1,985.8 W **Ans.**

Total losses = Total copper losses + iron and friction losses

= 1,985.8 + 1,000 = 2,985.8 W

Power input, $P_{in} = P_{out}$ + total losses

= 30,000 + 2,985.8 = 32,985.8 W

(*c*) Generator efficiency, $\eta_a = \dfrac{P_{out}}{P_{in}} \times 100 = \dfrac{30{,}000}{32{,}985.8} \times 100$

= 90.95% **Ans.**

Example 11.6. A 10 kW, 200 V, 1200 rpm series dc generator has armature resistance of 0.1 Ω, field winding resistance of 0.3 Ω. The frictional and windage loss of the machine is 200 W and brush contact drop is 1 volt per brush. Find the efficiency of the machine and the load current at which this machine has maximum efficiency.

[*U.P. Technical Univ. Elec. Engineering Even Semester, 2004-05*]

Solution: Output power, P_{out} = 10 kW = 10,000 W

Load current, $I_L = \dfrac{P_{out} \text{ in kW} \times 1{,}000}{V}$

$= \dfrac{10 \times 1{,}000}{200} = 50$ A

Copper power losses, $P_C = I_L^2 (R_a + R_{se}) + I_L \times$ brush drop

$= 50^2 (0.1 + 0.3) + 50 \times 2 = 1{,}100$ W

Friction and windage losses, P_i = 200 W

Power input, $P_{in} = P_{out} + P_C + P_i$

= 10,000 + 1,100 + 200 = 11,300 W

Efficiency of generator, $\eta = \dfrac{P_{out}}{P_{in}} \times 100$

$= \dfrac{10{,}000 \times 100}{11{,}300} = 88.5\%$ **Ans.**

Let the load current for operation with maximum efficiency be I amperes. Maximum efficiency will occur when variable losses $[I^2 (R_a + R_{se}) + 2\,vI]$ are equal to constant losses (friction and windage losses) i.e.

$0.4 I^2 + 2 I = 200$

∵ brush drop per brush, v = 1 V

Solving above equation we have

I = 20 A **Ans.**

Example 11.7. A shunt generator has a full-load current of 195 A at 250 V. The stray losses are 720 W and the shunt field coil resistance is 50 Ω. It has a full-load efficiency of 90%. Find the armature resistance. Also find the current corresponding to maximum efficiency.

Solution: Generator rated output $= V \times I_L = 250 \times 195 = 48{,}750$ W

$$\text{Input power to generator} = \frac{\text{Rated power output}}{\text{Full-load efficiency}} = \frac{48{,}750}{0.9} = 54{,}167 \text{ W}$$

Total losses = Power input – power output
$= 54{,}167 - 48{,}750 = 5{,}417$ W

$$\text{Shunt field current, } I_{sh} = \frac{V}{R_{sh}} = \frac{250}{50} = 5 \text{ A}$$

Shunt field copper loss $= VI_{sh} = 250 \times 5 = 1{,}250$ W

Stray losses = 720 W

Armature current, $I_a = I_L + I_{sh} = 195 + 5 = 200$ A

Armature copper loss, $I_a^2 R_a$ = Total losses – stray losses – shunt field copper loss
$= 5{,}417 - 720 - 1{,}250 = 3{,}447$ W

$$\text{or Armature resistance} = \frac{3{,}447}{I_a^2} = \frac{3{,}447}{(200)^2} = 0.08617\ \Omega \textbf{ Ans.}$$

Constant losses, P_C = Shunt field copper loss + stray losses
$= 1{,}250 + 720 = 1{,}970$ W

Load current for which efficiency is maximum,

$$I_L = \sqrt{\frac{P_C}{R_a}} = \sqrt{\frac{1{,}970}{0.08617}} = 151.2 \text{ A } \textbf{Ans.}$$

Example 11.8. A 250 V shunt generator has full-load current of 200 A. Its armature resistance is 0.06 Ω, field resistance is 100 Ω and the stray losses are 2,000 W. Find the hp of the prime mover when it is delivering full load and find the load for which the efficiency of generator is maximum.

[*M.D. Univ. Electromechanical Energy Conversion, May-2007*]

Solution: Line voltage, V = 250 V

Line current, I_L = 200 A

$$\text{Generator output, } P_{out} = \frac{VI_L}{1{,}000} = \frac{250 \times 200}{1{,}000} = 50 \text{ kW}$$

$$\text{Shunt field current, } I_{sh} = \frac{V}{R_{sh}} = \frac{250}{100} = 2.5 \text{ A}$$

Armature current, $I_a = I_L + I_{sh} = 200 + 2.5 = 202.5$ A

Armature copper loss $= I_a^2 R_a = (202.5)^2 \times 0.06$
$= 2{,}460.375$ W or 2.46 kW

Shunt field copper loss $= I_{sh}^2 R_{sh} = (2.5)^2 \times 100$
$= 625$ W or 0.625 kW

Stray losses, P_s = 2,000 W = 2 kW

Total power input $= P_{out} + I_a^2 R_a + I_{sh}^2 R_{sh} + P_s$
$= 50 + 2.46 + 0.625 + 2 = 55.085$ kW

$$\text{HP of prime mover} = \frac{55.085}{0.7355} = 74.9 \text{ (metric) } \textbf{Ans.}$$

Armature current corresponding to maximum efficiency,

$$I_a = \sqrt{\frac{\text{Constant losses}}{R_a}} = \sqrt{\frac{\text{Shunt field loss + stray loss}}{R_a}}$$

$$= \sqrt{\frac{(0.625 + 2) \times 1{,}000}{0.06}} = 209.2 \text{ A}$$

Load current corresponding to maximum efficiency,

$I_L = I_a - I_{sh} = 209.2 - 2.5 = 206.7$ A **Ans.**

Example 11.9. A 250 V, 15 kW, shunt motor has a maximum efficiency of 88% and a speed of 700 rpm, when delivering 80% of its rated output. The resistance of its shunt field is 100 Ω. Determine the efficiency and speed when the motor draws a current of 78 A from mains. [*U.P. Technical Univ. Electromechanical Energy Conversion-I, 2008-09*]

Solution: Rated output = 15 kW = 15,000 W

Efficiency of the machine,

$$\eta = \frac{\text{Output}}{\text{Output + losses}}$$

$$\text{or Total losses} = \frac{(1-\eta) \times \text{output}}{\eta} = \left(\frac{1}{\eta} - 1\right) \text{output}$$

$$= \left(\frac{1}{0.88} - 1\right) \times 0.8 \times 15{,}000$$

$= 1{,}636.4$ W

At maximum efficiency,

Constant losses (rotational losses + shunt field loss)

= Variable armature circuit losses, $I_a^2 R_a$

$$= \frac{\text{Total losses}}{2} = \frac{1{,}636.4}{2} = 818.2 \text{ W}$$

Input to motor, P_{in} = Motor output, P_{out} + total losses
$= 15{,}000 \times 0.8 + 1{,}636.4 = 13{,}636.4$ W

$$\text{Line current, } I_{L_1} = \frac{P_{in}}{V} = \frac{13{,}636.4}{250} = 54.55 \text{ A}$$

$$\text{Shunt field current, } I_{sh} = \frac{V}{R_{sh}} = \frac{250}{100} = 2.5 \text{ A}$$

Armature current, $I_{a_1} = I_{L_1} - I_{sh} = 54.55 - 2.5 = 52.05$ A

Armature circuit resistance,

$$R_a = \frac{\text{Copper loss}}{I_{a_1}^2} = \frac{818.2}{52.05^2} = 0.302\ \Omega$$

Input of motor, I_{L_2} = 78 A

Armature current, $I_{a_2} = 78 - 2.5 = 75.5$ A

Armature copper loss $= I_{a_2}^2 R_a = (75.5)^2 \times 0.302 = 1{,}721.5$ W

Motor input, $P_{in} = VI = 250 \times 78 = 19{,}500$ W

Motor output, P_{out} = Motor input – constant losses – armature copper loss
$= 19{,}500 - 818.2 - 1{,}721.5 = 16{,}960.3$ W

$$\text{Motor efficiency, } \eta = \frac{P_{out}}{P_{in}} \times 100$$

$$= \frac{16{,}960.3}{19{,}500} \times 100 = 86.98\% \quad \textbf{Ans.}$$

Assuming flux remaining the same,

The speed of the motor,

$$N_2 = N_1 \times \frac{E_{b_2}}{E_{b_1}} = N_1 \times \frac{V \quad I_{a_2} R_a}{V \quad I_{a_1} R_a}$$

$$= 700 \times \frac{250 - 75.5 \times 0.302}{250 - 52.05 \times 0.302}$$

$$= 700 \times \frac{227.2}{234.2809} = 678.8 \text{ rpm} \quad \textbf{Ans.}$$

Example 11.10. A small PMDC motor runs at no-load speed of 10,600 rpm and takes 10 mA when fed from 6 V dc source. Its armature circuit resistance is 6 Ω. Calculate its (*i*) no-load rotational losses (*ii*) stall torque (*iii*) armature current to achieve a shaft power of 1 W.

Solution: Input on no load $= V \times I_{a_o} = 6 \times 10 \times 10^{-3} = 0.06$ W

No-load armature copper loss $= I_{a0}^2 R_a = (10 \times 10^{-3})^2 \times 6$

$= 0.0006$ W

(*i*) No-load losses, $P_C = 0.06 - 0.0006 = 0.0594$ W **Ans.**

(*iii*) Let the armature current be I_{a_f} amperes to provide shaft power of 1 W.

Shaft power, P_S = Power input – armature copper loss – no-load loss

$$\text{or } 1 = 6I_{a_f} - I_{a_f}^2 \times 6 - 0.0594$$

$$\text{or } I_{a_f}^2 - I_{a_f} + 0.176566 = 0$$

$$\text{or } I_{a_f} = 0.771 \text{ A or } 0.2289 \text{ A}$$

Rejecting higher value which is not practicable,

Armature current, $I_{a_f} = 0.2289$ A **Ans.**

$$\text{Speed on load, } N_f = N_0 \times \frac{E_{b_f}}{E_{b_0}}$$

$$= 10{,}600 \times \frac{6 - 0.2289 \times 6}{6 - 0.01 \times 6} = 8{,}256 \text{ rpm}$$

Torque developed on this load, T_f

$$= \frac{E_{b_f} \times I_{a_f} \times 60}{2\pi N_f}$$

$$= \frac{(6 - 0.2289 \times 6) \times 0.2289 \times 60}{2\pi \times 8{,}256}$$

$$= 0.001225 \text{ N-m}$$

Armature current when motor is at rest,

$$I_s = \frac{V - E_{bs}}{R_a} = \frac{6}{6} = 1 \text{ A} \quad \because E_{bs} = 0$$

(*ii*) Torque developed when motor is at rest

$$T_s = \frac{I_s}{I_f} \times T_f = \frac{1}{0.2289} \times 0.001225$$

$= 0.00535$ N-m **Ans.**

Example 11.11. A 550 V, 37 kW, 1000 rpm dc shunt motor has full-load efficiency of 90%. The armature circuit resistance is 0.22 ohm and there is total voltage drop of 2V at brushes. The field current is 1.7 A. Calculate:

(*i*) Full-load line current.

(*ii*) Full-load shaft torque in N-m.

(*iii*) Total resistance in motor starter to limit the starting current to 1.5 times the full-load current.

[*M.D. Univ. Electromechanical Energy Conversion, December-2011*]

Solution: Motor output, $P_{out} = 37$ kW

Full-load efficiency of motor, $\eta = 90\% = 0.9$

$$\text{Input to motor on full load, } P_{in} = \frac{P_{out}}{\eta} = \frac{37}{0.9} = 41.111 \text{ kW}$$

(*i*) Full-load line current,

$$I_{Lf} = \frac{P_{in} \text{ in kW} \times 1{,}000}{V} = \frac{41.111 \times 1{,}000}{550} = 74.75 \text{ A}$$

(*ii*) Full-load line shaft torque,

$$T_{sh} = \frac{\text{Output in watts}}{2\pi N/60} = \frac{37 \times 1{,}000}{2\pi \times 1{,}000/60}$$

$= 353.32$ N-m **Ans.**

(*iii*) Starting current,

$$I_{st} = 1.5\, I_{Lf} = 1.5 \times 74.75 = 112.125 \text{ A}$$

Armature current at starting,

$$I_{as} = 112.125 - 1.7 = 110.425 \text{ A}$$

Total resistance in the motor starter to limit the starting current to 110.425 A

$$= \frac{550 - 2}{110.425} - 0.22 = 4.743\ \Omega \quad \textbf{Ans.}$$

Example 11.12. A 200 V shunt motor has R_a = 0.1 Ω, R_f = 240 Ω and rotational loss 236 W. On full-load, the line current is 9.8 A with the motor running at 1,450 rpm.

Determine :

(*i*) the mechanical power developed,

(*ii*) the power output,

(*iii*) the load torque,

(*iv*) the full-load efficiency.

[*Electromechanical Energy Conversion-I, U.P. Technical Univ. 2005-06; G.B. Technical Univ. 2012-13; R.G. Technical Univ., December-2012*]

Solution: Supply voltage, V = 200 V

Full-load line current, $I_{FL} = 9.8$ A

$$\text{Shunt field current, } I_{sh} = \frac{V}{R_{sh}} = \frac{200}{240} = 0.833 \text{ A}$$

$$\text{Full-load armature current, } I_{af} = I_{FL} - I_{sh} = 9.8 - 0.833 = 8.96667 \text{ A}$$

$$\text{Back emf, } E_{bf} = V - I_a R_a = 200 - 8.96667 \times 0.1 = 199.1 \text{ V}$$

Speed, N = 1,450 rpm

(*i*) Mechanical power developed,

$$P_{mech} = E_{bf} I_{af} \text{ watts} = 199.1 \times 8.96667$$

$= 1{,}785$ W **Ans.**

(*ii*) Power output, $P_{out} = P_{mech}$ – rotational loss

$= 1{,}785 - 236 = 1{,}549$ W **Ans.**

$$\text{(iii) Full-load torque, } T_{FL} = \frac{P_{out}}{2\pi N/60} = \frac{1{,}549 \times 60}{2\pi \times 1{,}450}$$

$= 10.2$ Nm **Ans.**

$$(iv) \text{ Full-load efficiency} = \frac{P_{out}}{P_{in}} \times 100 = \frac{1{,}549}{200 \times 9.8} \times 100$$

$= 79.03\%$ **Ans.**

Example 11.13. A 15 kW, 250 V, 1200 rpm shunt motor has 4 poles, 4 parallel armature paths, and 900 armature conductors; R_a = 0.2 Ω. At rated speed and rated output the armature current is 75 A and I_f = 1.5 A. Calculate (i) flux/pole (ii) the torque developed (iii) rotational losses (iv) η (v) the shaft load (vi) the shaft load found remains fixed, but the field flux is reduced to 7% of its value by field control, determine new operating speed.

[*U.P. Technical Univ. Electromechanical Energy Conversion-I, 2005-06*]

Solution:

Shunt field current, $I_{sh} = I_f = 1.5$ A

Input current, $I_L = I_a + I_{sh} = 75 + 1.5 = 76.5$ A

Back emf, $E_b = V - I_a R_a = 250 - 75 \times 0.2 = 235$ V

(*i*) Flux per pole, $\Phi = \dfrac{E_b \times 60 \times A}{Z \times N \times P} \quad \because E_b = \dfrac{\Phi ZN}{60} \times \dfrac{P}{A}$

$= \dfrac{235 \times 60 \times 4}{900 \times 1{,}200 \times 4}$

$= 0.013$ Wb $= 13$ mWb **Ans.**

(*ii*) Torque developed, $T_e = \dfrac{E_b I_a}{2\pi N/60} = \dfrac{235 \times 75 \times 60}{2\pi \times 1{,}200}$

$= 140.26$ N-m **Ans.**

(*iii*) Rotational losses = Power developed $- P_{out}$

$= E_b I_a - P_{out}$

$= 235 \times 75 - 15{,}000 = 2{,}625$ W **Ans.**

(*iv*) Efficiency, $\eta = \dfrac{\text{Power output}}{\text{Power input}} \times 100$

$= \dfrac{15{,}000}{VI_L} \times 100$

$= \dfrac{15{,}000}{250 \times 76.5} \times 100 = 78.43\%$ **Ans.**

(*v*) Shaft load $= \dfrac{\text{Output in watts}}{2\pi N/60}$

$= \dfrac{15{,}000 \times 60}{2\pi \times 1{,}200} = 119.366$ N-m **Ans.**

(*vi*) When flux is reduced by 7%

Flux, $\Phi' = (1 - 0.07)\,\Phi = 0.93\,\Phi$

Armature current, $I'_a = \dfrac{I_a \Phi}{\Phi'}$

$\because$ load torque remains unchanged

$= \dfrac{75 \times \Phi}{0.93\,\Phi} = 80.645$ A

Back emf $E'_b = V - I'_a R_a$

$= 250 - 80.645 \times 0.2 = 233.87$ V

New operating speed, $N' = N \times \dfrac{E'_b}{E_b} \times \dfrac{\Phi}{\Phi'}$

$= 1{,}200 \times \dfrac{233.87}{235} \times \dfrac{1}{0.93} = 1{,}284$ rpm **Ans.**

Example 11.14. A 200 V shunt motor develops an output of 17.158 kW when taking 20.2 kW. The field resistance is 50 Ω and an armature resistance 0.06 Ω. What is the efficiency and power input when output is 7.46 kW.

[*R.G. Technical Univ. Electrical Machine-II, December-2012*]

Solution: Power input to motor, $P_{in} = 20.2$ kW

Line voltage, $V_L = 200$ V

Line current, $I_L = \dfrac{P_{in} \text{ in kW} \times 1{,}000}{V_L \text{ in volts}}$

$= \dfrac{20.2 \times 1{,}000}{200} = 101$ A

Shunt field current, $I_{sh} = \dfrac{V}{R_{sh}} = \dfrac{200}{50} = 4$ A

Armature current, $I_a = I_L - I_{sh} = 101 - 4 = 97$ A

Armature copper loss $= I_a^2 R_a = 97^2 \times 0.06 = 564.54$ W

Total losses, $P_T = P_{in} - P_{out}$

$= 20.2 - 17.158 = 3.042$ kW or 3,042 W

Constant losses (mechanical loss + iron loss + shunt field copper loss),

P_c = Total losses – variable loss (or armature copper loss)

$= 3{,}042 - 564.54 = 2{,}477.46$ W

When power output is 7.46 kW or 7,460 W

Let the armature current be I'_a amperes

Total losses, P'_T = Constant losses + armature copper loss

$= 2{,}477.46 + (I'_a)^2 \times 0.06$

Power input, $P'_{in} = P'_{out}$ + total losses

$= 7{,}460 + 2{,}477.46 + 0.06\,(I'_a)^2$ ...(*i*)

Also, power input, $P_{in} = VI_L = V\,(I'_a + I_{sh}) = 200\,(I'_a + 4)$ ...(*ii*)

Comparing expressions (*i*) and (*ii*) we have

$7{,}460 + 2{,}477.46 + 0.06\,(I'_a)^2 = 200\,(I'_a + 4)$

or $0.06\,(I'_a)^2 - 200\,I'_a + 9{,}137.46 = 0$

or $I'_a = \dfrac{200 \pm \sqrt{200^2 - 4 \times 0.06 \times 9{,}137.46}}{2 \times 0.06}$

$= 3{,}287$ A or 46.333 A

Rejecting higher value which is not practicable,

New armature current, $I'_a = 46.333$ A

Power input, $P'_{in} = V \times (I'_a + 4)$

$= 200 \times (46.333 + 4)$

$= 10{,}067$ W or 10.067 kW **Ans.**

Efficiency, $\eta = \dfrac{P'_{out}}{P'_{in}} \times 100$

$= \dfrac{7.46}{10.067} \times 100 = 74.1\%$ **Ans.**

Example 11.15. Compute the energy stored in the magnetic field produced by the field winding of a 10 kW, 1440 rpm dc shunt generator under normal operating conditions. Given that the field winding has a time constant of 0.2 s and I^2R loss is 400 W.

[*G.G.S.I.P. Univ. Delhi Electromechanical Energy Conversion-I, December-2011*]

Solution: $I^2R = 400$ W ...(*i*)

Time constant, $\tau = \dfrac{L}{R} = 0.2$ s ...(*ii*)

Multiplying expressions (*i*) and (*ii*) we have

$I^2R \times \dfrac{L}{R} = 400 \times 0.2$ W-s

or $I^2L = 80$ W-s

Energy stored in the magnetic field produced by the field winding

$= \dfrac{1}{2} I^2 L = \dfrac{1}{2} \times 80 = 400$ W-s or joules **Ans.**

Example 11.16. A dc shunt motor rated 250 V, 10 kW, 1,500 rpm has a full-load efficiency of 80%. Its field and armature resistances are 110 Ω and 0.25 Ω respectively. Find the value of the resistance to be inserted in series with the armature and the power lost in the armature circuit to reduce the speed to 1,200 rpm if the load torque is independent of the speed.

[*U.P.S.C. I.E.S. Electrical Engineering-II, 2008*]

Solution: Total power input to motor, $P_{in} = \frac{P_{out}}{\eta} = \frac{10}{0.8} = 12.5$ kW

Line current, $I_{L_1} = \frac{P_{in}}{V} = \frac{12.5 \times 1{,}000}{250} = 50$ A

Shunt field current, $I_{sh} = \frac{V}{R_{sh}} = \frac{250}{110} = 2.27$ A

Armature current, $I_{a_1} = I_{L_1} - I_{sh} = 50 - 2.27 = 47.73$ A

Back emf, $E_{b_1} = V - I_{a_1} R_a$

$= 250 - 47.73 \times 0.25 = 238.0675$ V

Since, load torque is independent of speed, the electromagnetic torque is constant at both the speeds.

Let resistance R be connected in series with the armature circuit to reduce the speed from $N_1 = 1{,}500$ rpm to $N_2 = 1{,}200$ rpm with load torque remaining constant

i.e. $T_2 = T_1$

or $I_{a_2} \Phi_2 = I_{a_1} \Phi_1 \qquad \because T \propto \Phi I_a$

or $I_{a_2} = I_{a_1} = 47.73$ A $\qquad \because \Phi_2 = \Phi_1$

Back emf, $E_{b_2} = V - I_{a_2} (R + R_a)$

$= 250 - 47.73 (R + 0.25)$

$= 238.0675 - 47.73$ R

Also $\frac{E_{b_2}}{E_{b_1}} = \frac{N_2}{N_1} \times \frac{\Phi_2}{\Phi_1} = \frac{N_2}{N_1} \qquad \because \Phi_2 = \Phi_1$

or $\frac{238.0675 - 47.73R}{238.0675} = \frac{1{,}200}{1{,}500}$

or R = 1 Ω **Ans.**

Power lost in armature circuit $= (I_{a_2})^2 (R_a + R)$

$= (47.73)^2 (0.25 + 1) = 2{,}848$ W **Ans.**

Example 11.17. A 20 kW, 500 V dc shunt motor has an efficiency of 90% at full load. The armature copper loss is 40% of full-load losses. The field resistance is 250 Ω. Calculate the resistance values of a 4-section starter for this motor in given two cases (*i*) starting current ≤ 2I$_F$ (*ii*) starting current (min) = 120% I$_F$.

[U.P. Technical Univ. Electromechanical Energy Conversion-I, 2004-05]

Solution: Power output, $P_{out} = 20$ kW $= 20{,}000$ W

Power input, $P_{in} = \frac{P_{out}}{\eta} = \frac{20{,}000}{0.9} = 22{,}222.22$ W

Total losses at full-load = 22,222.22 – 20,000 = 2,222.22 W

Armature copper losses = 40% of total full-load losses

= 0.4 × 2,222.22 = 888.88 W

Full load line current, $I_L = \frac{P_{in}}{V} = \frac{22{,}222.22}{500} = 44.444$ A

Shunt field current, $I_{sh} = \frac{V}{R_{sh}} = \frac{500}{250} = 2$ A

Full-load armature current, $I_a = I_L - I_{sh} = 44.444 - 2 = 42.444$ A

Armature resistance, $R_a = \frac{\text{Full-load armature copper loss}}{(\text{Full-load armature current})^2}$

$= \frac{888.88}{(42.444)^2} = 0.4934\ \Omega$

Starting current (max), $I_{max} \le 2 I_F \le 2 \times 42.444 \le 84.888$ A

So resistance, $R_1 \ge \frac{V}{I_{max}} \ge \frac{500}{84.888} \ge 5.89 = 5.9\ \Omega$ say

$\gamma = \sqrt[n]{\frac{R_1}{R_a}} = \sqrt[4]{\frac{5.9}{0.4934}} = 1.86 \qquad \because n = 4$

Resistance, $R_2 = \frac{5.9}{1.86} = 3.17\ \Omega$

Resistance, $R_3 = \frac{3.172}{1.86} = 1.70\ \Omega$

Resistance, $R_4 = \frac{1.705}{1.86} = 0.92\ \Omega$

Resistance of section, 1, $r_1 = R_1 - R_2 = 5.9 - 3.17 = 2.73\ \Omega$
Resistance of section 2, $r_2 = R_2 - R_3 = 3.17 - 1.70 = 1.47\ \Omega$
Resistance of section 3, $r_3 = R_3 - R_4 = 1.70 - 0.92 = 0.78\ \Omega$
Resistance of section 4, $r_4 = R_4 - R_a = 0.78 - 0.4934 = 0.287\ \Omega$ **Ans.**

(*ii*) Starting current (min), $I_{min} = 1.2\ I_F = 1.2 \times 42.444 = 50.93$ A

$I_{max} = \gamma I_{min} = 50.93\gamma$ A

and $\gamma = \sqrt[n]{\frac{R_1}{R_a}}$ where *n* is the number of sections

$R_1 = \frac{V}{I_{max}} = \frac{500}{50.93\gamma}$

So $\gamma = \sqrt[4]{\frac{500}{50.93\gamma R_a}} = \sqrt[5]{\frac{500}{50.93 \times 0.4934}}$

$= \sqrt[5]{19.9} = 1.82$

Maximum starting current,

$I_{max} = \gamma I_{min} = 1.82 \times 50.93 = 92.7$ A

Resistance, $R_1 = \frac{V}{I_{max}} = \frac{500}{92.7} = 5.4\ \Omega$

Resistance, $R_2 = \frac{R_1}{\gamma} = \frac{5.4}{1.82} = 2.97\ \Omega$

Resistance, $R_3 = \frac{R_2}{\gamma} = \frac{2.97}{1.82} = 1.63\ \Omega$

Resistance, $R_4 = \frac{R_3}{\gamma} = \frac{1.63}{1.82} = 0.89\ \Omega$

Resistance of section 1, $r_1 = R_1 - R_2 = 5.4 - 2.97 = 2.43\ \Omega$
Resistance of section 2, $r_2 = R_2 - R_3 = 2.97 - 1.63 = 1.34\ \Omega$
Resistance of section 3, $r_3 = R_3 - R_4 = 1.63 - 0.89 = 0.74\ \Omega$
Resistance of section 4, $r_4 = R_4 - R_a = 0.89 - 0.4934 = 0.397\ \Omega$ **Ans.**

Example 11.18. The hysteresis and eddy current losses in dc machine running at 1,000 rpm are 250 W and 100 W respectively. If the flux remains constant, at what speed will the total iron losses be halved?

Solution: We know that at same peak flux density

Hysteresis loss, $P_h \propto f \propto N$

and Eddy current loss, $P_e \propto f^2 \propto N^2$

Let $P_h = AN$ and $P_e = BN^2$

where A and B are constants

Substituting $P_h = 250$ watts and $P_e = 100$ watts and N = 1,000 in the above equations for P_h and P_e we get

$$A = \frac{P_h}{N} = \frac{250}{1,000} = 0.25$$

$$\text{and } B = \frac{P_e}{N^2} = \frac{100}{(1,000)^2} = 0.0001$$

Total iron losses at 1,000 rpm, P_i

$$= 250 + 100 = 350 \text{ watts}$$

Let N' be the new speed in rpm at which the total losses will be one-half of the losses at 1,000 rpm *i.e.* $\frac{350}{2} = 175$ watts

Substituting A = 0.25, B = 0.0001 and $P_i = 175$ watts in equation

$$P_i = AN + BN^2 \text{ we get}$$

$$175 = 0.25\,N' + 0.0001\,N'^2$$

$$\text{or } N'^2 + 2,500\,N' - 17,50,000 = 0$$

$$\text{or } N' = \frac{-2,500 \pm \sqrt{(2500)^2 + 4 \times 17,50,000}}{2}$$

$$= 570 \text{ rpm } \textbf{Ans.}$$

11.5 TESTING OF DC MACHINES

The most important performance tests to be conducted on dc machine are:

1. The magnetization or open-circuit test.
2. The load characteristic.
3. The determination of efficiency curve.
4. The temperature rise test.

In addition, when a dc machine is first put on test at the manufacturer's end, it must be given an insulation test and tests must also be made to ensure that commutation is satisfactory. For determination of efficiency of a dc machine it is necessary to have a knowledge of the power input and power output. There are three different methods of determining the efficiency of a dc machine, namely, direct method, indirect method and regenerative method.

1. Direct Method. By this method the efficiency and losses of only small machines can be determined. In this method full load is applied to the machine and output is directly measured. Though this method is very simple in looking but involves complication in the measurement of mechanical power input in the case of generator, and output in the case of a motor. In the case of a generator, the input mechanical power is measured by connecting some form of dynamometer to the prime mover and water resistance load is applied. In the case of small motor, some form of the brake is applied to water pulley for measuring the output. *If brake is to be applied to a series motor, the brake must be tight before the motor is started, otherwise the armature may get damaged and fly to pieces.* In case of large motors the motor is coupled to a generator and generator is loaded by means of a resistance.

This method is wasteful, since all the input is to be wasted and can only be used in case of small machines due to the difficulty of having enough power and suitable brake arrangements in case of larger machines.

2. Indirect Method. By this method, efficiency of shunt and compound dc machines can be determined. This method enables the determination of losses without actually loading the machine. The power is required to supply the losses only, so there is no difficulty in applying this method even to very large machines. Although the efficiency can be calculated with fair accuracy from the results obtained with this method, the disadvantage of this method is that the machine is run light during the test which gives no indication as to the temperature rise on load or to the commutating qualities of the machines.

3. Regenerative Method. It requires two identical machines, one of them operates as a motor and drives the other, which is mechanically coupled to it. The other machine operates as generator and feed back power into the supply. Thus the total power drawn from the supply is only for supplying the internal losses of the two machines. Thus even very large machines may be tested as the power required is small. The machines can also be tested under full-load conditions and for long duration to study their performance regarding commutation, temperature rise etc. conveniently.

11.6 BRAKE TEST

This test is performed with small motors only. The brake is applied to a pulley (either air cooled or water cooled depending on the rating of the motor) mounted on the motor shaft, as illustrated in Fig. 11.5 (*a*). The brake band is fixed with the help of wooden blocks gripping the pulley. One end of the band is fixed to earth through spring S and the other end is connected to suspended weight W_1. The difference of weight W_1 and spring balance reading W_2 gives the effective force on the pulley. The mechanical load is adjusted, by varying suspended weights, to cause a flow of full-load current through the motor circuit. The connection diagram is given in Fig. 11.5 (*a*). The power input to the motor can be measured by measuring the terminal voltage V and the input (or line) current I.

Let the supply voltage measured by voltmeter be V volts, input current measured by ammeter A be I amperes, suspended weight W_1 kg; spring balance reading W_2 kg; speed of rotation of armature N rpm, radius of pulley r metres and thickness of belt t metre.

$$\text{Motor output} = T \times \frac{2\pi N}{60} \text{ kg-m/s}$$

$$= (W_1 - W_2)\left(r + \frac{t}{2}\right)\frac{2\pi N}{60} \text{ kg-m/s}$$

$$\text{since } T = (W_1 - W_2)\left(r + \frac{t}{2}\right)$$

$$= (W_1 - W_2)\, r \times \frac{2\pi N}{60} \text{ kg-m/s}$$

neglecting t, being very small in comparison to r

$$= (W_1 - W_2)\, r \times \frac{2\pi N}{60} \times 9.81 \text{ Nm/s or watts}$$

$$\text{Input to motor} = VI \text{ watts}$$

$$\text{Efficiency of motor, } \eta_m = \frac{\text{Output}}{\text{Input}}$$

$$= \frac{(W_1 - W_2) \times r \times \frac{2\pi N}{60} \times 9.81}{VI} \quad ...(11.10)$$

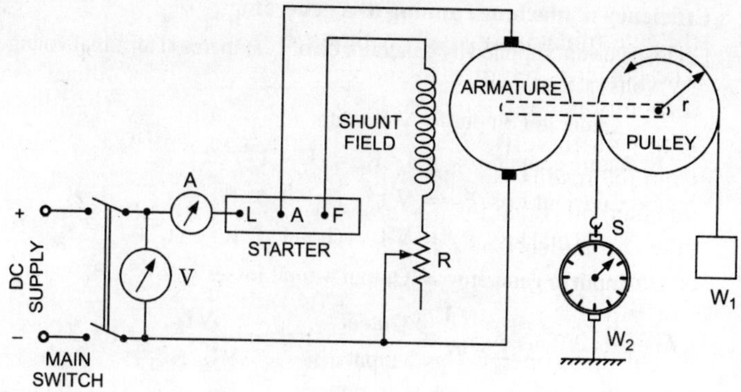

(a) Brake Test

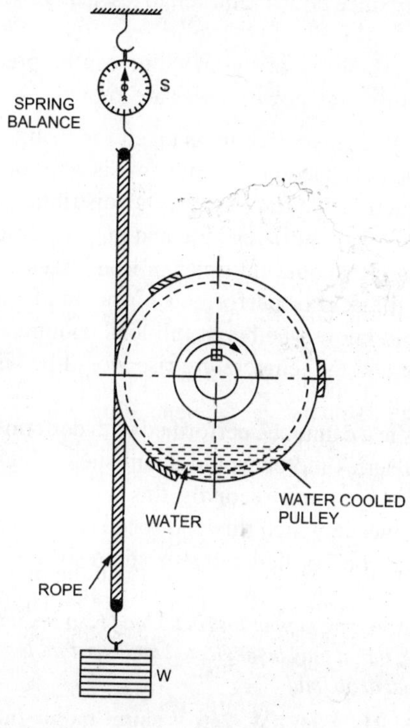

(b)

Fig. 11.5

Alternatively, rope brake is used for testing [Fig. 11.5 (*b*)]. One end of the rope is fixed to a spring balance S attached to a fixed hook while the other end is attached to a hook or pan to which load is added. The mechanical load is adjusted by varying suspended weights to cause a flow of full-load current through the motor circuit.

Let W_1 be the suspended weight in kg, W_2 be the spring balance reading in kg, r be the pulley radius in metres, N be the speed of rotation of armature in rpm, V be the supply voltage in volts and I be the input current in amperes.

Efficiency of the machine is given by Eq. (11.10).

The main drawback of such a method of determining the internal losses or efficiency of a dc machine is that the output of the motor cannot be measured accurately.

The other drawback is that this method cannot be used for determining internal losses and efficiency of large machines, since testing of large machine entails a considerable loss of energy. In fact in the case of very large machines there may be no facility for such a test known as *direct test*.

Precaution: While performing this test with series wound machines care should be taken that brake applied is tight failing which the motor will attain dangerously high speed and get damaged.

Example 11.19. In a brake test on a dc shunt motor, the effective load on the brake pulley was 13 kg, the effective diameter of the pulley was 46 cm, the speed 1,400 rpm, the armature current 23 A, when the supply voltage is 220 V. Calculate the efficiency of the motor at this load when field resistance is 110 Ω.

[*Rajasthan Univ. Electrical Machines-I, 2004*]

Solution: Effective load, F = 13 kg

$$\text{Pulley radius, } r = \frac{46}{2} \text{ cm} = 0.23 \text{ m}$$

$$\text{Torque exerted, } \tau = F \times r$$

$$= 13 \times 0.23 = 2.99 \text{ kg-m}$$

$$= 2.99 \times 9.81 = 29.3319 \text{ N-m}$$

$$\text{Motor output, } P_{out} = T \times \frac{2\pi N}{60}$$

$$= \frac{29.3319 \times 2\pi \times 1,400}{60} = 4,300 \text{ W}$$

$$\text{Armature current, } I_a = 23 \text{ A}$$

$$\text{Shunt field current, } I_{sh} = \frac{V}{R_{sh}} = \frac{220}{110} = 2 \text{ A}$$

$$\text{Input current, } I_L = I_a + I_{sh} = 23 + 2 = 25 \text{ A}$$

$$\text{Input power to motor, } P_{in} = V \times I = 220 \times 25 = 5,500 \text{ W}$$

$$\text{Motor efficiency, } \eta = \frac{P_{out}}{P_{in}} \times 100 = \frac{4,300}{5,500} \times 100 = 78.2\% \textbf{ Ans.}$$

Example 11.20. Brake test was performed on a dc shunt motor and the following readings were obtained. The load on one side of the brake band was 40 kg and on the other side it was 5.5 kg. The motor was taking a current of 80 A from 400 V dc mains while running at 1,470 rpm. The pulley diameter is 110 cm. Neglect the belt thickness and determine the torque, output and efficiency of the motor.

[*B.P. Univ. of Technology Electrical Machine-I, 2008*]

Solution: Torque exerted, T = Effective load × radius of brake pulley

$$= (W_1 - W_2) \times r$$

$$= (40 - 5.5) \times \frac{110}{2 \times 100} = 18.975 \text{ kg-m } \textbf{Ans.}$$

$$\text{Motor output} = T \times \frac{2\pi N}{60} = 18.975 \times 9.81 \times \frac{2\pi \times 1,470}{60}$$

$$= 28,655 \text{ W or } 28.655 \text{ kW } \textbf{Ans.}$$

$$\text{Power input to motor} = V \times I = \frac{80 \quad 400}{1,000} = 32 \text{ kW}$$

$$\text{Motor efficiency, } \eta = \frac{P_{out}}{P_{in}} \times 100 = \frac{28.655}{32} \times 100 = 89.54\% \textbf{ Ans.}$$

11.7 SWINBURNE'S TEST

It is the simplest indirect method of testing dc machines in which the losses are measured separately and efficiency at any desired load is predetermined. The machine is run as a motor on no load, the excitation is adjusted to give normal speed and normal (rated) voltage is applied across the motor terminals. The connection diagram for determining no-load losses of a dc shunt machine is given in Fig. 11.6.

The resistances of the field winding and armature are measured at room temperature (say, at 15 °C) by voltmeter-ammeter method. Since the windings of the machine would be hot under normal operating conditions, the probable hot resistances are computed from the measured cold values for a temperature rise of 40 °C. From the data so obtained, different losses are computed and the efficiency is determined.

Let supply voltage, motor input current and shunt field current measured by voltmeter V, ammeters A_1 and A_2 connected in the circuit be V, I_0 and I_{sh} respectively.

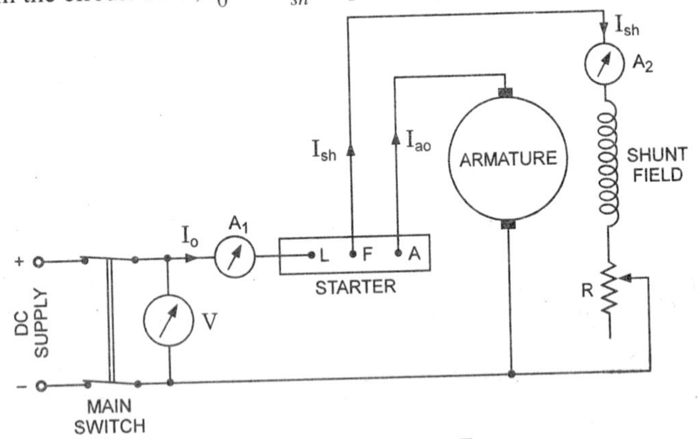

Fig. 11.6 *Swinburne's Test*

Since on no load mechanical output of the machine is zero, whole of the input on no load is used to supply the internal losses in the machine *i.e.* shunt field copper loss, armature copper loss and stray power losses in shunt wound machine and in case of compound wound machine input is also used to supply series field copper loss in addition to shunt field, armature and stray power losses.

$$\text{Now power input} = VI_0 \text{ watts}$$

$$\text{Shunt field copper loss} = V\,I_{sh} \text{ watts}$$

$$\text{Armature copper loss on no load} = I_{a0}^2 R_a = (I_0 - I_{sh})^2 R_a$$

where R_a is the hot resistance of armature.

$$\text{Stray power losses, } P_S = \text{Input on no load} - \text{shunt field loss} - \text{armature copper loss}$$

$$= V\,I_0 - V\,I_{sh} - I_{ao}^2 R_a$$

$$= V\,I_0 - V\,I_{sh} - (I_0 - I_{sh})^2 R_a$$

$$\text{Constant losses, } P_C = P_S + \text{shunt field copper loss}$$

$$= V\,I_0 - V\,I_{sh} - (I_0 - I_{sh})^2 R_a + V\,I_{sh}$$

$$= V\,I_0 - (I_0 - I_{sh})^2 R_a$$

Efficiency of machine running as a motor on load

$$\text{Let the load current} = I_L \text{ amperes}$$

$$\text{Total input} = V\,I_L \text{ watts}$$

$$\text{Constant losses, } P_C = V\,I_0 - (I_0 - I_{sh})^2 R_a$$

[determined from no-load test]

$$\text{Armature copper loss} = (I_L - I_{sh})^2 R_a$$

$$\text{Total losses, } P_T = P_C + (I_L - I_{sh})^2 R_a$$

$$= V\,I_0 - (I_0 - I_{sh})^2 R_a + (I_L - I_{sh})^2 R_a$$

$$\text{Output} = \text{Input} - P_T = V\,I_L - P_T$$

$$\text{Efficiency of motor, } \eta_m = \frac{\text{Output}}{\text{Input}} \times 100 = \frac{V I_L - P_T}{V I_L} \times 100\% \quad \text{...(11.11)}$$

Efficiency of machine running as a generator

Let load current supplied by generator be of I_L amperes at terminal voltage of V volts

$$\text{Generator output} = V\,I_L \text{ watts}$$

$$\text{Armature copper loss} = I_a^2 R_a = (I_L + I_{sh})^2 R_a$$

$$\text{Constant loss, } P_C = V\,I_0 - (I_0 - I_{sh})^2 R_a$$

$$\text{Total loss, } P_T = V\,I_0 - (I_0 - I_{sh})^2 R_a + (I_L + I_{sh})^2 R_a$$

$$\text{Input to generator} = \text{Output} + \text{total losses} = V\,I_L + P_T$$

$$\text{Generator efficiency, } \eta_G = \frac{\text{Output}}{\text{Input}} \times 100 = \frac{V I_L}{V I_L + P_T} \times 100\% \quad \text{...(11.12)}$$

Advantages. 1. It is convenient and economical method of testing dc machines, since power required to test a large machine is very small.

2. The efficiency of the machine can be predetermined at any load, since stray power losses are known.

Disadvantages. 1. No account is taken for change in iron losses from no load to full load on account of distribution of flux due to armature reaction. On full load the flux distribution is very much affected due to armature reaction and in some cases to an extent that iron losses become 1.5 times of iron losses at no load.

2. As the test is performed on no load, therefore, it is impossible to know whether at full load commutation would be satisfactory and the temperature rise would be within specified limits or not.

3. This test cannot be performed with dc series motors, since on no load series motor would attain such a high speed that it will get damaged and secondly this test is only applicable to those machines in which flux and speed remain constant. Hence this test can be applied only to shunt and level compound machines.

Note: *Armature copper loss at no load is so small in comparison to other losses, that it may be neglected and constant losses may be taken equal to input at no load.*

Example 11.21. A 60 kW, 250 V shunt motor takes 16 A when running light at 1,440 rpm. The resistances of armature and field windings are 0.2 ohm and 125 ohm respectively (*i*) Find the efficiency of the motor when taking 152 A. (*ii*) Also estimate the efficiency when working as generator and delivering 152 A at 250 V.

[*Pb. Technical Univ., May-2008*]

Solution: $\text{Shunt field current, } I_{sh} = \frac{V}{R_{sh}} = \frac{250}{125} = 2 \text{ A}$

$$\text{No-load line current, } I_{Lo} = 16 \text{ A}$$

$$\text{No-load input} = VI_{Lo} = 250 \times 16 = 4{,}000 \text{ W}$$

$$\text{No-load armature current, } I_{ao} = I_{Lo} - I_{sh} = 16 - 2 = 14 \text{ A}$$

$$\text{No-load armature copper loss} = I_{a0}^2 R_a = 14^2 \times 0.2 = 39.2 \text{ W}$$

$$\text{Constant losses} = \text{No-load input} - \text{no-load armature copper loss} = 4{,}000 - 39.2 = 3{,}660.8 \text{ W}$$

On full load,

$$\text{Line current, } I_L = 152 \text{ A}$$

$$\text{Input} = VI_L = 250 \times 152 = 38{,}000 \text{ W}$$

$$\text{Armature current, } I_a = I_L - I_{sh} = 152 - 2 = 150 \text{ A}$$

$$\text{Armature copper losses} = I_a^2 R_a = (150)^2 \times 0.2 = 4{,}500 \text{ W}$$

$$\text{Total losses} = \text{Constant losses + armature copper losses}$$
$$= 3{,}660.8 + 4{,}500 = 8{,}160.8 \text{ W}$$
$$\text{Output} = \text{Input} - \text{total losses}$$
$$= 38{,}000 - 8{,}160.8 = 29{,}839.2 \text{ W}$$
$$\text{Efficiency, } \eta = \frac{\text{Output}}{\text{Input}} \times 100 = \frac{29839.2}{38{,}000} \times 100$$
$$= 78.52\% \textbf{ Ans.}$$

As Generator,

$$\text{Line current, } I_L = 152 \text{ A}$$
$$\text{Armature current, } I_a = I_L + I_{sh} = 152 + 2 = 154 \text{ A}$$
$$\text{Armature copper loss} = I_a^2 R_a = 154^2 \times 0.2 = 4{,}743.2 \text{ W}$$
$$\text{Total losses} = \text{Constant losses + armature copper losses}$$
$$= 3{,}660.8 + 4{,}743.2 = 8{,}404 \text{ W}$$
$$\text{Generator output} = VI_L = 250 \times 152 = 38{,}000 \text{ W}$$
$$\text{Generator input} = \text{Output + total losses}$$
$$= 38{,}000 + 8{,}404 = 46{,}404 \text{ W}$$
$$\text{Efficiency, } \eta = \frac{\text{Output}}{\text{Input}} \times 100 = \frac{38{,}000}{46{,}404} \times 100$$
$$= 81.89\% \textbf{ Ans.}$$

Example 11.22. A 220 V dc shunt motor at no load takes a current of 3 A. The resistances of the armature and shunt field are 0.9 Ω and 250 Ω respectively. Estimate the efficiency of the motor when the input current is 18 A. Write the assumptions if any.

[*R.G. Technical Univ. Electrical Machine-II, December-2013*]

Solution: Supply Voltage, V = 220 V

$$\text{Shunt field current, } I_{sh} = \frac{V}{R_{sh}} = \frac{220}{250} = 0.88 \text{ A}$$
$$\text{No-load line current, } I_{L_0} = 3 \text{ A}$$
$$\text{No-load input} = VI_{L_0} = 220 \times 3 = 660 \text{ W}$$
$$\text{No-load armature current, } I_{a_0} = I_{L_0} - I_{sh} = 3 - 0.88 = 2.12 \text{ A}$$
$$\text{No-load armature copper loss} = I_{a_0}^2 R_a = (2.12)^2 \times 0.9 = 4.045 \text{ W}$$
$$\text{Constant losses, } P_c = \text{No-load input} - \text{armature copper loss on no load}$$
$$= 660 - 4.045 = 655.955 \text{ W}$$

On Full Load

$$\text{Line current, } I_L = 18 \text{ A}$$
$$\text{Armature current, } I_a = I_L - I_{sh} = 18 - 0.88 = 17.12 \text{ A}$$
$$\text{Armature copper loss} = I_a^2 R_a = (17.12)^2 \times 0.9 = 263.785 \text{ W}$$
$$\text{Total losses, } P_T = \text{Constant losses + full-load armature copper loss}$$
$$= 655.955 + 263.785 = 919.74 \text{ W}$$
$$\text{Input} = VI_L = 220 \times 18 = 3{,}960 \text{ W}$$
$$\text{Output} = \text{Input} - P_T = 3{,}960 - 919.74 = 3{,}040.26$$
$$\text{Motor efficiency, } \eta_m = \frac{\text{Output}}{\text{Input}} \times 100$$
$$= \frac{3{,}040.26}{3{,}960} \times 100 = 76.77\% \textbf{ Ans.}$$

Example 11.23. A 50 kW, 250 V compound motor takes a current of 9 A while running on no load at rated voltage and speed. The shunt field current is 5 A. The resistances of the windings when hot are: Armature 0.1 Ω, series field 0.07 Ω and interpole 0.03 Ω. The brush drop is 2 V. Determine motor output and the efficiency when the motor intake is 155 A. [*U.P. Technical Univ. Electromechanical Energy Conversion, 2001-02*]

Solution: At No Load

$$\text{Line current, } I_{L_0} = 9 \text{ A}$$
$$\text{Armature current, } I_{a_0} = I_{L_0} - I_{sh} = 9 - 5 = 4 \text{ A}$$

Assuming long-connections

$$\text{Variable losses} = \text{Copper losses in armature, series field, interpole and at brushes}$$
$$= I_{a_0}^2 (R_a + R_{se} + R_i) + I_{a_0} \times \text{brush drop}$$
$$= 4^2(0.1 + 0.07 + 0.03) + 4 \times 2 = 11.2 \text{ W}$$
$$\text{Constant losses, } P_C = \text{Input on no-load} - \text{variable losses}$$
$$= 250 \times 9 - 11.2 = 2{,}238.8 \text{ W}$$

At Full-Load

$$\text{Line current, } I_{L_f} = 155 \text{ A}$$
$$\text{Armature current, } I_{a_f} = I_{L_f} - I_{sh} = 155 - 5 = 150 \text{ A}$$
$$\text{Variable losses} = I_{a_f}^2 (R_a + R_{se} + R_i) + I_{a_f} \times \text{brush drop}$$
$$= 150^2 (0.1 + 0.07 + 0.03) + 150 \times 2$$
$$= 4{,}800 \text{ W}$$
$$\text{Total losses} = P_C + \text{variable losses}$$
$$= 2{,}238.8 + 4{,}800 = 7{,}038.8 \text{ W}$$
$$\text{Power input, } P_{in} = VI_{L_f} = 250 \times 155 = 38{,}750 \text{ W}$$
$$\text{Motor output, } P_{out} = P_{in} - \text{total losses}$$
$$= 38{,}750 - 7{,}038.8$$
$$= 31{,}711.2 \text{ W or } 31.7112 \text{ kW } \textbf{Ans.}$$
$$\text{Motor efficiency, } \eta = \frac{P_{out}}{P_{in}} \times 100 = \frac{31{,}711.2}{38{,}750} \times 100 = 81.84\% \textbf{ Ans.}$$

Example 11.24. A 500 V, 25 hp, dc shunt motor takes 2.4 A while running light. The field and armature resistances are 650 Ω and 0.57 Ω respectively. Calculate full-load efficiency, assuming a brush drop of 2 V. [*U.P.S.C. I.E.S. Electrical Engineering-II, 2002*]

Solution: At No Load

$$\text{Power supplied, } P_{in_0} = VI_{L_0} = 500 \times 2.4 = 1{,}200 \text{ W}$$
$$\text{Shunt field current, } I_{sh} = \frac{V}{R_{sh}} = \frac{500}{650} = 0.77 \text{ A}$$
$$\text{Armature current, } I_{a_0} = I_{L_0} - I_{sh} = 2.4 - 0.77 = 1.63 \text{ A}$$

No-load armature copper loss

$$= I_{a_0}^2 R_a + I_{a_o} \times \text{brush drop}$$
$$= (1.63)^2 \times 0.57 + 1.63 \times 2 = 4.77 \text{ W}$$
$$\text{Constant losses, } P_C = P_{in_0} - \text{no-load armature copper loss}$$
$$= 1{,}200 - 4.77 = 1{,}195.23 \text{ W}$$

At Full-Load

$$\text{Output power, } P_{out} = 25 \times 735.5 = 18{,}387.5 \text{ W}$$

Let the armature current on full load be I_a amperes

$$\text{Power input, } P_{in} = V \times (I_a + I_{sh}) = 500\,(I_a + 0.77)$$
$$\text{Also power input, } P_{in} = P_{out} + \text{armature copper loss + constant losses}$$
$$\text{or } \quad 500\,(I_a + 0.77) = 18{,}387.5 + 0.57\,I_a^2 + 2I_a + 1{,}195.23 \text{ watts}$$
$$\text{or } \quad 0.57\,I_a^2 - 498\,I_a + 19{,}197.8 = 0$$
$$\text{or } \quad I_a = \frac{498 \pm \sqrt{(498)^2 - 4 \times 0.57 \times 19{,}197.8}}{2 \times 0.57}$$
$$= 833.26 \text{ A or } 41.05 \text{ A}$$

Since figure of 833.26 A is impracticable,

Full-load armature current, $I_a = 41.05$ A

$$\text{Motor input} = VI_L = 500 \times (41.05 + 0.77) = 20{,}910 \text{ W}$$

$$\text{Motor efficiency, } \eta = \frac{\text{Motor output}}{\text{Motor input}} \times 100$$

$$= \frac{18{,}387.5}{20{,}910} \times 100 = 87.936\% \textbf{ Ans.}$$

Example 11.25. Calculate the efficiency of a 500 V shunt motor when taking 700 A from the following data recorded when the motor was hot and stationary:

Voltage drop in armature winding 15 V,

Armature current = 510 A

Field current = 9 A

Motor running at normal speed unloaded:

Armature current = 22.5 A

Applied voltage = 550 V

Allow 2 V for brush contact drop and 1% of rated output of 400 kW for stray load losses.

[*Electrical Machines-I; Rajasthan Univ. 2000, Rajasthan Technical Univ. 2010*]

Solution: Armature resistance,

$$R_a = \frac{\text{Voltage drop in armature winding}}{\text{Armature current}}$$

$$= \frac{15}{510} = 0.0294\ \Omega$$

Shunt field resistance,

$$R_{sh} = \frac{\text{Terminal voltage}}{\text{Shunt field current}} = \frac{500}{9} = 55.56\ \Omega$$

$$\text{Rotational losses} = VI_{a0} - I_{a0}^2 R_a - I_{a0} V_b$$

$$= 550 \times 22.5 - 22.5^2 \times 0.0294 - 22.5 \times 2$$

$$= 12{,}315 \text{ W} \quad \text{or} \quad 12.315 \text{ kW}$$

$$\text{Stray losses} = 1\% \text{ of } 400 \text{ kW} = \frac{1}{100} \times 400 = 4 \text{ kW}$$

$$\text{Power input, } P_{in} = \frac{VI_L}{1{,}000} = \frac{500 \times 700}{1{,}000} = 350 \text{ kW}$$

$$\text{Armature current, } I_a = I_L - I_{sh} = 700 - 9 = 691 \text{ A}$$

$$\text{Copper losses} = I_a^2 R_a + V_b I_a + VI_{sh}$$

$$= (691)^2 \times 0.0294 + 2 \times 691 + 500 \times 9$$

$$= 19{,}920 \text{ W} \quad \text{or} \quad 19.92 \text{ kW}$$

$$\text{Total losses} = 12.315 + 4 + 19.92 = 36.235 \text{ kW}$$

$$\text{Motor efficiency, } \eta = \frac{P_{in} - \text{total losses}}{P_{in}}\ 100$$

$$= \frac{350 - 36.235}{350} \times 100 = 89.65\% \quad \textbf{Ans.}$$

11.8 REGENERATIVE OR HOPKINSON'S TEST

This test is also known as *back-to-back test.* It requires two machines, which must be identical. These machines are mechanically coupled and are so adjusted electrically that one of them acts as a motor and the other as a generator. The motor supplies the mechanical power, which is utilized to drive the generator while generator develops electrical power to be utilized in driving the motor in addition to electrical power drawn from the supply mains. The connection diagram for this test is shown in Fig. 11.7.

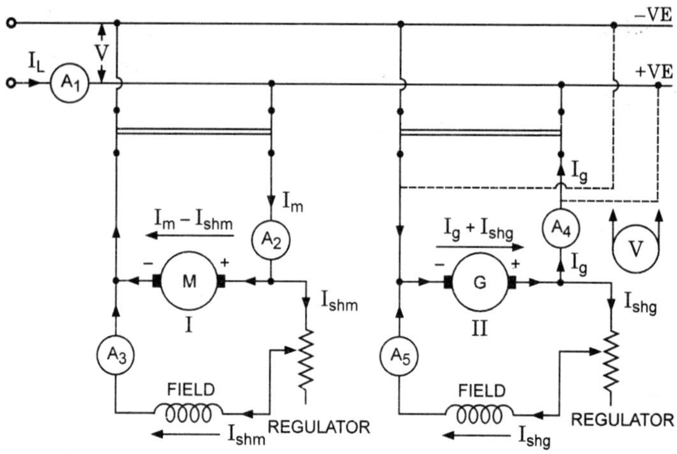

Fig. 11.7 *Hopkinson's Test*

Thus two indentical machines, may be of any size, can be tested under full-load conditions. Under test the power drawn from the supply mains will be that required to overcome the internal losses of the two machines. This method is economical when the large machines are to be tested on full load for longer duration.

The machine acting as a motor, is run up to full speed by means of a starting resistance, the main switch of the generator keeping open. The second machine will, therefore, generate and its voltage is adjusted equal to 1 or 2 V more than bus-bar voltage and of the same polarity. The adjustment is made by means of a paralleling voltmeter V and switched on to the supply.

The machine with smaller excitation acts as a motor and that with larger excitation acts as a generator because back emf of the motor is less than generated emf of generator. Hence to load the machines (both motor and generator) the excitation of the motor is decreased which decreases its back emf so causing the motor to take large current which in turn produces more torque.

During this test the supply voltage is maintained constant at the rated value and the test should be carried out at rated speed.

The armature circuit resistances (*i.e.* armature winding including brush contact, interpole and compensating windings if any) of the two machines are measured at the end of the experiment when the windings have attained their final temperature by passing full-load current through them at standstill and measuring voltage drops across them, keeping field circuit open.

Let ammeters A_1, A_2, A_3, A_4 and A_5 connected in the circuit shown in Fig. 11.7, measure the line current I_L, motor current I_m, motor shunt field current I_{shm}, generator current I_g and generator shunt field current I_{shg} respectively.

According to Kirchhoff's first law

$$\text{Motor current, } I_m = I_L + I_g$$

$$\text{Motor input} = VI_m$$

and generator output $= VI_g$ where V is bus-bar voltage.

If η_m and η_g are the efficiencies of motor and generator respectively,

then motor output $= VI_m \eta_m$

and generator input $= \dfrac{VI_g}{\eta_g}$

But motor output = Generator input

or $V\,I_m\,\eta_m = \dfrac{VI_g}{\eta_g}$

or $\eta_m\,\eta_g = \dfrac{I_g}{I_m}$ ...(11.13)

Assuming armature, field and stray power losses in both of the machines equal then $\eta_m = \eta_g$

and $\eta_m = \eta_g = \sqrt{\dfrac{I_g}{I_m}} = \sqrt{\dfrac{I_g}{I_g + I_L}} = \sqrt{\dfrac{I_m - I_L}{I_m}}$ ...(11.14)

This assumption will not cause any error in case of large machines due to very slight difference in their armature and excitation current, but in case of small machines, the difference between the armature currents and shunt field currents is large and it is not good to make above assumption. To obtain accurate results, armature and shunt field losses are determined separately and stray losses are assumed to be equal in both machines.

Current drawn from supply mains, $I_L = I_m - I_g$

Motor intake current $= I_m$

Motor shunt field current $= I_{shm}$

and motor armature current, $I_{am} = (I_m - I_{shm})$

Generator output current $= I_g$

Generator shunt field current $= I_{shg}$

and generator armature current, $I_{ag} = (I_g + I_{shg})$

Bus-bar voltage = V

Total losses of both machines = Power drawn from supply $= VI_L$

Armature copper loss in motor $= (I_m - I_{shm})^2 R_{am}$, R_{am} being motor armature resistance.

Armature copper loss in generator $= (I_g + I_{shg})^2 R_{ag}$, R_{ag} being generator armature resistance.

Shunt field copper loss in motor $= V\,I_{shm}$

Shunt field copper loss in generator $= V\,I_{shg}$

Total copper losses $= (I_m - I_{shm})^2 R_{am} + (I_g + I_{shg})^2 R_{ag} + V\,I_{shm} + V\,I_{shg}$

Stray power losses of both machines, P_S

$$= V I_L - [(I_m - I_{shm})^2 R_{am} + (I_g + I_{shg})^2 R_{ag} + V\,(I_{shm} + I_{shg})]$$

Stray power losses of each machine $= \dfrac{P_s}{2}$

Efficiency of Motor.

Motor intake $= V\,I_m$

Total losses in motor = Armature loss + field loss + stray power loss.

$$= (I_m - I_{shm})^2 R_{am} + V\,I_{shm} + \frac{P_S}{2}$$

$$\text{Motor output} = VI_m - \left[(I_m - I_{shm})^2 R_{am} + VI_{shm} + \frac{P_S}{2}\right]$$

$$\text{Motor efficiency, } \eta_m = \frac{\text{Output}}{\text{Input}} \times 100$$

Efficiency of Generator.

Generator output $= V\,I_g$

Total losses in generator = Armature loss + field loss + stray loss

$$= (I_g + I_{shg})^2 R_{ag} + V\,I_{shg} + \frac{P_S}{2}$$

$$\text{Generator input} = V\,I_g + (I_g + I_{shg})^2 R_{ag} + V\,I_{shg} + \frac{P_S}{2}$$

$$\text{Generator efficiency, } \eta_g = \frac{\text{Output}}{\text{Input}} \times 100$$

Hence from the above datas determined from the test and relations, the efficiency of motor and generator can be determined.

Advantages of Hopkinson's Test. (*i*) It is economical since power required is small as compared to the full-load power of the two machines.

(*ii*) Since the machines can be tested under full-load conditions for long duration, therefore, the performance of the machines regarding commutation and temperature rise etc. can be conveniently studied.

(*iii*) The efficiency is being determined under load conditions so that the stray load loss is being taken into account.

Disadvantage. The main disadvantage of this test is the necessity of two practically identical machines to be available.

Example 11.26. The Hopkinson test on two identical shunt machines gave the following results:

Input voltage = 250 V
Input current = 10 A
Output current of generator = 60 A
Field current of generator = 4 A
Field current to motor = 3 A

Armature resistance of each machine is 0.06 Ω. Determine efficiency of motor and generator.

[*G.B. Technical Univ. Electromechanical Energy Conversion-I, 2011-12*]

Solution: Refer to Fig. 11.8

Input voltage, V = 250 V

Current drawn from supply main, $I_L = I_m - I_g$

Generator output current, $I_g = 60$ A

Field current of generator, $I_{shg} = 4$ A

Generator armature current, $I_{ag} = I_g + I_{shg} = 60 + 4 = 64$ A

Motor intake current, $I_m = I_L + I_g = 10 + 60 = 70$ A

Motor shunt field current, $I_{shm} = 3$A

Motor armature current, $I_{am} = I_m - I_{shm} = 70 - 3 = 67$ A

Total losses of both machines = Power drawn from supply

$= VI_L = 250 \times 10 = 2{,}500$ W

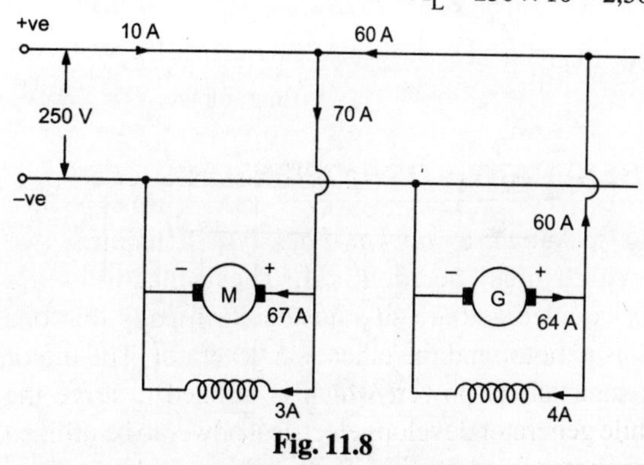

Fig. 11.8

Armature copper loss in motor $= I_{am}^2 R_{am}$

$= 67^2 \times 0.06 = 269.34$ W

Armature copper loss in generator $= I_{ag}^2 R_{ag}$

$= 64^2 \times 0.06 = 245.76$ W

Shunt field copper loss in generator $= VI_{sh\,g} = 250 \times 4 = 1000$ W

Shunt field copper loss in motor $= VI_{sh\,m} = 250 \times 3 = 750$ W

Stray power loss of both machines,

$$P_s = VI_L - (I_{am}^2 R_{am} + I_{ag}^2 R_{ag} + VI_{shm} + VI_{shg})$$
$$= 250 \times 10 - (269.34 + 245.76 + 750 + 1{,}000) = 234.9 \text{ W}$$

Stray power loss of each machine

$$= \frac{P_s}{2} = \frac{234.9}{2} = 117.45 \text{ W} \quad \textbf{Ans.}$$

Efficiency of Motor

Motor intake $= VI_m = 250 \times 70 = 17{,}500$ W

$$\text{Total losses in motor} = I_{am}^2 R_{am} + VI_{sh\,m} + \frac{P_s}{2}$$
$$= 269.34 + 750 + 117.45 = 1{,}136.79 \text{ W}$$

Motor output $= 17{,}500 - 1{,}136.79 = 16{,}363.21$ W

$$\text{Motor efficiency, } \eta_m = \frac{\text{Output}}{\text{Input}} \times 100$$
$$= \frac{16{,}363.21}{17{,}500} \times 100 = 93.5\% \quad \textbf{Ans.}$$

Efficiency of Generator

Generator output $= VI_g = 250 \times 60 = 15{,}000$ W

$$\text{Total losses in generator} = I_{ag}^2 R_{ag} + VI_{sh\,g} + \frac{P_s}{2}$$
$$= 245.76 + 1{,}000 + 117.45 = 1{,}363{,}21 \text{ W}$$

Generator input $= 15{,}000 + 1{,}363.21 = 16{,}363.21$ W

$$\text{Generator efficiency, } \eta_g = \frac{\text{Output}}{\text{Input}} \times 100 = \frac{15{,}000}{16{,}363.21} \times 100$$
$$= 91.67\% \quad \textbf{Ans.}$$

Example 11.27. Two identical dc shunt machines when tested by Hopkinson's method, gave the following data:

Line voltage – 230 V; Line current excluding both the field currents – 30 A; motor armature current – 230 A; field currents 5 A and 4 A.

If the armature resistance of each machine (including brushes) is 0.025 Ω, calculate efficiency of both the machines.

[*U.P. Technical Univ. Electromechanical Energy Conversion-I, 2007-08; W.B. Univ. of Technology Electrical Machine-I, 2011*]

Solution: Line current excluding of field currents, $I'_L = 30$ A

Line voltage, $V = 230$ V

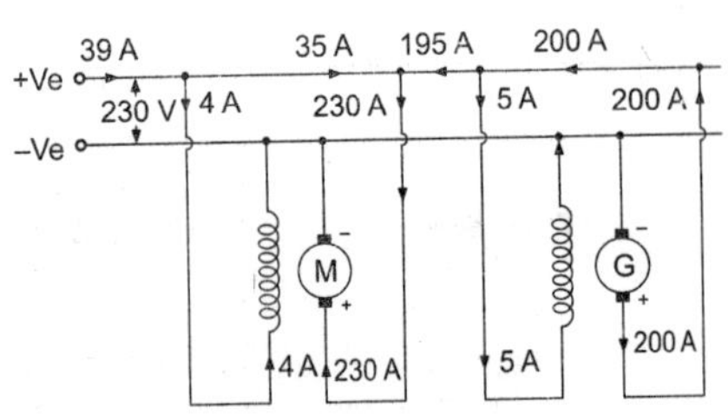

Fig. 11.9

Power input to set excluding field input $= VI'_L$

$= 230 \times 30 = 6{,}900$ W

Motor armature current, $I_{am} = 230$ A

Motor field current, $I_{shm} = 4$ A

Generator field current, $I_{shg} = 5$ A

Generator armature current, $I_{ag} = I_{am} - I'_L$

$= 230 - 30 = 200$ A

$$\text{Motor armature copper loss} = (I_{am})^2 \times R_{am}$$
$$= (230)^2 \times 0.025$$
$$= 1{,}322.5 \text{ W}$$

$$\text{Generator armature copper loss} = (I_{ag})^2 \times R_{ag}$$
$$= (200)^2 \times 0.025 = 1{,}000 \text{ W}$$

$$\text{Total stray losses, } P_s = VI'_L - (I_{am})^2 R_{am} - (I_{ag})^2 R_{ag}$$
$$= 6{,}900 - 1{,}322.5 - 1{,}000 = 4{,}577.5 \text{ W}$$

$$\text{Generator output} = V \times (I_{ag} - I_{shg})$$
$$= 230 \times (200 - 5) = 44{,}850 \text{ W}$$

$$\text{Total losses in generator} = (I_{ag})^2 R_{ag} + VI_{sh} + \frac{P_s}{2}$$
$$= 1{,}000 + 230 \times 5 + \frac{4{,}577.5}{2}$$
$$= 4{,}438.75 \text{ W}$$

Generator input = Generator output + iosses

$= 44{,}850 + 4{,}438.75 = 49{,}288.75$ W

$$\text{Generator efficiency, } \eta_a = \frac{\text{Output}}{\text{Input}} \times 100$$
$$= \frac{44{,}850}{49{,}288.75} \times 100 = 90.99\% \quad \textbf{Ans.}$$

$$\text{Motor input} = V(I_{am} + I_{shm})$$
$$= 230 \times (230 + 4) = 53{,}820 \text{ W}$$

$$\text{Total losses in motor} = (I_{am})^2 R_{am} + VI_{shm} + \frac{P_s}{2}$$
$$= 1{,}322.5 + 230 \times 4 + \frac{4{,}577.5}{2}$$
$$= 4{,}531.25 \text{ W}$$

$$\text{Motor efficiency, } \eta_m = \frac{\text{Input} - \text{total losses}}{\text{Input}} \times 100$$
$$= \frac{53{,}820 - 4{,}531.25}{53{,}820} \times 100 = 91.58\% \quad \textbf{Ans.}$$

Example 11.28. Two Identical dc machines when tested by Hopkinson method gave the following test:

Field currents are 2.5 A and 2 A. Line voltage is 220 V. Line current including both the field currents is 10 A. Motor armature current is 73 A. The armature resistance of each machine is 0.05Ω. Calculate the efficiency of both the machines.

[*U.P.S.C. I.E.S. Elec. Engineering-II, 1993*]

Solution: Supply voltage, $V = 220$ volts

Input line current, $I_L = 10$ A

Motor armature current, $I_{am} = 73$ A

Motor shunt field current, $I_{shm} = 2$ A

$$\text{Motor input current, } I_m = I_{am} + I_{shm} = 73 + 2 = 75 \text{ A}$$
$$\text{Generator output current, } I_g = I_m - I_L = 75 - 10 = 65 \text{ A}$$
$$\text{Generator shunt field current, } I_{shg} = 2.5 \text{ A}$$
$$\text{Generator armature current, } I_{ag} = I_g + I_{shg} = 65 + 2.5 = 67.5$$
$$\text{Input power to set} = V I_L = 220 \times 10 = 2{,}200 \text{ W}$$
$$\text{Motor shunt field copper loss} = V I_{shm} = 220 \times 2 = 440 \text{ W}$$
$$\text{Motor armature copper loss} = I_{am}^2 R_{am} = (73)^2 \times 0.05 = 266.45 \text{ W}$$
$$\text{Generator shunt field copper loss} = V I_{shg} = 220 \times 2.5 = 550 \text{ W}$$
$$\text{Generator armature copper loss} = I_{ag}^2 R_{ag} = (67.5)^2 \times 0.05 = 227.8125 \text{ W}$$
$$\text{Total copper losses} = 440 + 266.45 + 550 + 227.8125 = 1{,}484.2625 \text{ W}$$

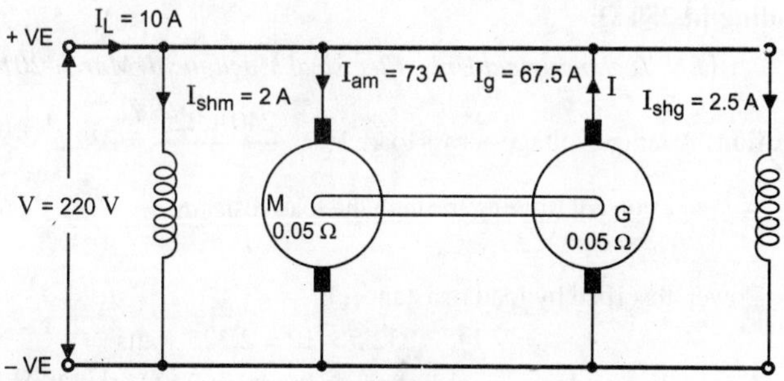

Fig. 11.10

$$\text{Total stray power losses, } P_s = \text{Input to set} - \text{copper losses} = 2{,}200 - 1{,}484.2625 = 715.74 \text{ W}$$

$$\text{Stray loss per machine} = \frac{P_s}{2} = \frac{715.74}{2} = 357.87 \text{ W}$$

Generator efficiency, η_g

$$= \frac{\text{Generator output}}{\text{Generator output + total losses}} \times 100$$

$$= \frac{VI_g}{VI_g + I_{ag}^2 R_{ag} + VI_{shg} + \frac{P_s}{2}} \times 100$$

$$= \frac{220 \times 65}{220 \times 65 + 227.1825 + 550 + 357.87} \times 100 = 92.64\% \textbf{ Ans.}$$

Motor efficiency, η_m

$$= \frac{\text{Motor input} - I_{am}^2 R_{am} - VI_{shm} - \frac{P_s}{2}}{\text{Motor input}} \times 100$$

$$= \frac{220 \times 75 - 266.45 - 440 - 357.87}{220 \times 75} \times 100 = 93.55\% \textbf{ Ans.}$$

11.9 RETARDATION TEST

This test is also known as *running down test*. It is used for finding out the stray losses of shunt wound dc machines.

In this method of testing dc machines, machine under test is speeded up slightly above its normal speed and supply to the armature is cut off. Consequently the armature slows down and its kinetic energy is utilized to meet the rotational losses (stray losses).

$$\text{Kinetic energy of the armature} = \frac{1}{2} J\omega^2$$

where J is the moment of inertia of the armature and ω is the angular speed at this instant.

$$\text{Rotational losses, } P_S = \text{Rate of change of kinetic energy}$$
$$= \frac{d}{dt}\left(\frac{1}{2}J\omega^2\right) = J\omega \frac{d\omega}{dt}$$

Since $\omega = \frac{2\pi N}{60}$, where N is the speed of the armature in rpm at that instant,

$$\text{Rotational losses, } P_S = J\left(\frac{2\pi N}{60}\right)\frac{d}{dt}\left(\frac{2\pi N}{60}\right)$$
$$= \left(\frac{2\pi}{60}\right)^2 JN\frac{dN}{dT} \qquad \text{...(11.15)}$$

Hence to determine stray losses, the values of moment of inertia of armature (J) and rate of change of speed $\frac{dN}{dT}$ must be known.

The value of J and $\frac{dN}{dT}$ are determined as follows.

(i) Determination of $\frac{dN}{dT}$. The machine under test is connected, as shown in Fig. 11.11. The voltmeter V connected across the armature shows the instantaneous back emf of the motor. Since back emf of the motor is directly proportional to speed ($E_b \propto N$), the voltmeter can be suitably calibrated to indicate speed. When the supply to armature is cut off, the speed of the motor decreases.

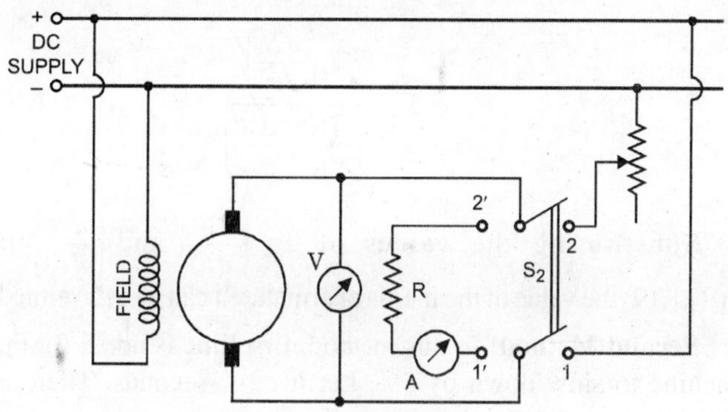

Fig. 11.11

The speed or readings of voltmeter are noted at different intervals of time and a curve is drawn between speed and time, as shown in Fig. 11.12. From any point P lying on speed time curve, tangent is drawn meeting the X-axis and Y-axis at points A and B respectively, then

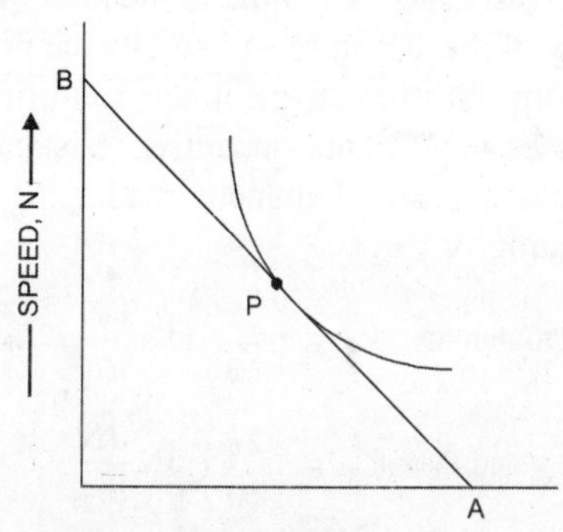

Fig. 11.12

$$\frac{dN}{dT} = \frac{\text{OB in rpm}}{\text{OA in seconds}} \qquad \text{...(11.16)}$$

(ii) Determination of Moment of Inertia. First speed time curve (slowing down curve) is plotted with armature alone as discussed above. Then a flywheel of known Moment of inertia (J_1) is keyed to the shaft and speed-time curve is plotted again. From the above two curves, the value of $\left(\frac{dN}{dt_1}\right)$ and $\left(\frac{dN}{dt_2}\right)$ are determined. Since the moment of inertia in the second case is more and losses in both the cases are almost same, the value of $\frac{dN}{dt_1}$ will be more than the value of $\frac{dN}{dt_2}$.

Hence: In first case, rotational losses,

$$P_S = \left(\frac{2\pi}{60}\right)^2 JN\left(\frac{dN}{dt_1}\right) \quad ...(11.17)$$

In second case, rotational losses,

$$P_S = \left(\frac{2\pi}{60}\right)^2 (J + J_1)N\left(\frac{dN}{dt_2}\right) \quad ...(11.18)$$

Comparing expressions (11.17) and (11.18) we have

$$\left(\frac{2\pi}{60}\right)^2 (J + J_1)N\left(\frac{dN}{dt_2}\right) = \left(\frac{2\pi}{60}\right)^2 JN\left(\frac{dN}{dt_1}\right)$$

$$\text{or } (J + J_1)\left(\frac{dN}{dt_2}\right) = J\left(\frac{dN}{dt_1}\right)$$

$$\text{or } J\left(\frac{dN}{dt_1} - \frac{dN}{dt_2}\right) = J_1\left(\frac{dN}{dt_2}\right)$$

$$\text{or } J = \frac{J_1\left(\frac{dN}{dt_2}\right)}{\left(\frac{dN}{dt_1} - \frac{dN}{dt_2}\right)} \quad ...(11.19)$$

Substituting the values of J_1, $\left(\frac{dN}{dt_1}\right)$ and $\left(\frac{dN}{dt_2}\right)$ in Eq. (11.19) the value of the inertia of armature J can be determined.

Second Method: In this method, first time is noted for the machine to slow down by 5%. Let it be t_1 seconds. Then an additional load known as retarding torque, mechanical or electrical (preferably electrical) is applied and again the time taken by the machine to slow down by 5% in noted, say t_2 seconds. In Fig. 11.11 the additional load (electrical) is applied by throwing the double throw switch S_2 over terminals 1', 2' just after disconnecting the armature from the supply mains. Thus a non-inductive resistance R will be in the motor circuit and now kinetic energy of the armature will supply the power to the load resistance R in addition to meet the rotational losses. The additional losses say P' in non-inductive resistance will be equal to the product of average of ammeter reading, I_{av} and average of voltmeter reading, V.

$$\text{Now rotational loss, } P_S = \left(\frac{2\pi}{60}\right)^2 JN\frac{dN}{dt_1} \quad ...(11.20)$$

$$\text{and } P_S + P'_S = \left(\frac{2\pi}{60}\right)^2 JN\frac{dN}{dt_2} \quad ...(11.21)$$

Dividing Eq. (11.21) by Eq. (11.20) we get

$$\frac{P_S + P'_S}{P_S} = \frac{\frac{dN}{dt_2}}{\frac{dN}{dt_1}}$$

$$\text{or } P_S\left(\frac{dN}{dt_2} - \frac{dN}{dt_1}\right) = P'_S\frac{dN}{dt_1}$$

$$\text{or } P_S = P'_S \times \frac{dN/dt_1}{dN/dt_2 - dN/dt_1} \quad ...(11.22)$$

If change of speed in both the cases is kept same and time taken is noted, let it be t_1 and t_2 seconds respectively, then

$$P_S = P'_S\frac{t_2}{t_1 - t_2} \quad ...(11.23)$$

Example 11.29. A retardation test is made on a separately-excited dc machine as a motor. The induced voltage falls from 240 V to 225 V in 25 seconds on opening the armature circuit and 6 seconds on suddenly changing the armature connection from supply to a load resistance taking 10 A (average). Find the efficiency of the machine when running as a motor and taking a current of 25 A on a supply of 250 V. The resistance of its armature is 0.4 Ω and that of its field winding in 250 Ω.

[*J.N. Technological Univ. Electrical Machines-I, March-2014*]

Solution: Average voltage across load, $V = \frac{240 + 225}{2} = 232.5$ V

Average current flowing through the load resistance,

$$I_{av} = 10 \text{ A}$$

Power absorbed by load resistance, P'

$$= V I_{av} = 232.5 \times 10 = 2{,}325 \text{ watts}$$

Time taken in decreasing the induced voltage 240 V to 225 V on disconnecting supply to armature, $t_1 = 25$ seconds

Time taken in decreasing the induced voltage from 240 V to 225 V on suddenly changing the armature connections from supply mains to load resistance, $t_2 = 6$ seconds.

$$\text{Stray losses, } P_S = P'_S \times \frac{t_2}{t_1 - t_2} = 2{,}325 \times \frac{6}{25 - 6} = 734 \text{ watts}$$

$$\text{Input line current, } I_L = 25 \text{ A}$$

$$\text{Shunt field current, } I_{sh} = \frac{V}{R_{sh}} = \frac{250}{250} = 1 \text{ A}$$

$$\text{Armature current, } I_a = I_L - I_{sh} = 25 - 1 = 24 \text{ A}$$

$$\text{Armature copper loss} = I_a^2 R_a = (24)^2 \times 0.4 = 230.4 \text{ watts}$$

$$\text{Shunt field copper loss} = V I_{sh} = 250 \times 1 = 250 \text{ watts}$$

$$\text{Total losses} = P_S + I_a^2 R_a + V I_{sh}$$

$$= 734 + 230.4 + 250 = 1{,}214.4 \text{ watts}$$

$$\text{Total input to motor} = V I_L = 250 \times 25 = 6{,}250 \text{ watts}$$

$$\text{Motor efficiency, } \eta_m = \frac{\text{Input} - \text{total losses}}{\text{Input}} \times 100$$

$$= \frac{6{,}250 - 1{,}214.4}{6{,}250} \times 100 = 80.6\% \textbf{ Ans.}$$

Example 11.30. A retardation test on a dc motor gave the following results: with field unexcited, the speed fell from 1,530 to 1,470 rpm in 43 seconds; with field normally excited, the same drop in speed occurred in 26 seconds; with an average load of 1.2 kW supplied by armature, the same speed drop occurred in 20 seconds.

Determine the moment of inertia of the rotating parts at 1,500 rpm and the core loss for normal excitation at this speed.

Solution: Let the mechanical losses be P_m and core loss be P_i.

When the field is unexcited, there will be no core loss and stray loss P_S will be equal to mechanical loss P_m

$$\text{So } P_m = \left(\frac{2\pi}{60}\right)^2 JN\left(\frac{dN}{dt_1}\right) = \left(\frac{2\pi}{60}\right)^2 JN\frac{dN}{43} \quad ...(i)$$

With field excited but no load connected across armature

$$P_S = \left(\frac{2\pi}{60}\right)^2 JN\left(\frac{dN}{dt_2}\right) = \left(\frac{2\pi}{60}\right)^2 JN\frac{dN}{26} \quad ...(ii)$$

With field excited and a load of 1.2 kW or 1,200 watts being supplied by the armature

$$P_S + 1{,}200 = \left(\frac{2\pi}{60}\right)^2 JN\left(\frac{dN}{dt_3}\right) = \left(\frac{2\pi}{60}\right)^2 JN\frac{dN}{20} \quad ...(iii)$$

Dividing equation (*iii*) by equation (ii) and dividing equation (*i*) by equation (*ii*) we get

$$\frac{P_S + 1{,}200}{P_S} = \frac{26}{20}$$

$$\text{or } P_S = 4{,}000 \text{ watts}$$

$$\text{and } \frac{P_m}{P_S} = \frac{26}{43}$$

$$\text{or mechanical loss, } P_m = \frac{26}{43} \times P_S = \frac{26}{43} \times 4{,}000 = 2{,}418.6 \text{ watts}$$

$$\text{Core loss, } P_i = \text{Stray power loss} - \text{mechanical loss}$$

$$= 4000 - 2{,}418.6 = 1{,}581.4 \text{ watts } \textbf{Ans.}$$

Substituting P_S = 4,000 watts, N = 1,500 rpm and dN = 1,530 – 1,470 = 60 rpm in equation (*ii*) we get

$$4{,}000 = \left(\frac{2\pi}{60}\right)^2 J \times 1{,}500 \times \frac{60}{26}$$

$$\text{or } J = \frac{4{,}000 \times 26}{1{,}500 \times 60} \times \left(\frac{60}{2\pi}\right)^2 = 105.374 \text{ kg-m}^2 \textbf{ Ans.}$$

11.10 TESTING OF DC SERIES MACHINES

Small series machines can be tested by brake test similar to shunt machines, but the large series machines cannot be tested by Swinburne's test in the same way as shunt machines, because series motors cannot be run on no load due to dangerous high speed. In view of this, the field test* is quite suitable for dc series machines. At the same time, there is no difficulty in having two similar dc series machines, because series motors are used for traction purposes and are, therefore, usually available in pairs. However, the accuracy of this test depends on accuracy with which the motor input and generator output are measured.

In this test two similar dc series machines are required. These two machines are mechanically coupled together and their fields are connected in series, as illustrated in Fig. 11.13 in order to make iron losses of both the machines equal. One of the machines operates as a motor and drives the other machine operating as a separately-excited generator. Load resistance R_L is directly connected to the armature without any switch.

The motor M is started in the usual manner and the output of the generator G is dissipated in the variable resistor load. The voltage V_1 across the motor terminals is kept equal to its rated value. Obviously, the supply voltage V will be equal to V_1 plus the voltage drop across the field winding of the generator. The hot resistances of the various windings are measured by voltmeter-ammeter or other suitable method. Load resistance R_L is varied till the ammeter A_1 shows the full-load current of the motor. Now the readings of various measuring instruments are noted.

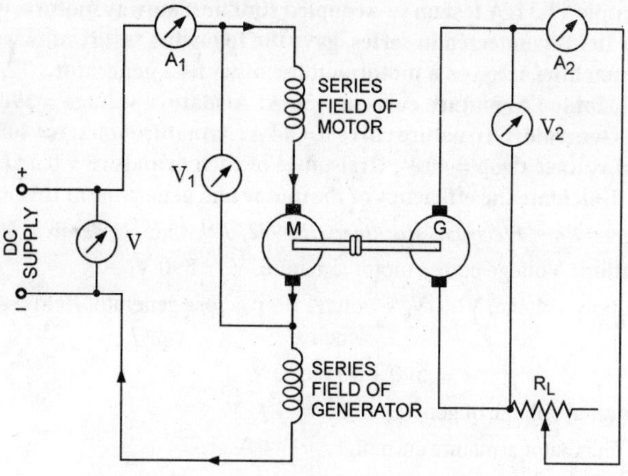

Fig. 11.13 *Field Test On Series Machines*

Let supply voltage = Reading of voltmeter V = V volts

Motor input current = Reading of ammeter $A_1 = I_1$

Terminal voltage of generator = Reading of voltmeter $V_2 = V_2$

Load current of generator = Reading of ammeter $A_2 = I_2$

Armature resistance of each machine = R_a

Series field resistance of each machine = R_{se}

Input to the whole set = $V I_1$

Output = $V_2 I_2$

Total losses of the set, $P_T = V I_1 - V_2 I_2$

Series field and armature copper losses of motor

$$= I_1^2 (R_a + R_{se})$$

Series field and armature copper losses of generator

$$= I_1^2 R_{se} + I_2^2 R_a$$

$$\text{Total copper losses of the set, } P_c = I_1^2 (R_a + 2R_{se}) + I_2^2 R_a$$

$$\text{Stray power losses for the set} = P_T - P_c$$

$$\text{Stray power losses per machine, } P_S = \frac{P_T - P_c}{2}$$

Motor efficiency

$$\text{Motor input} = V_1 I_1$$

$$\text{Motor losses} = I_1^2 (R_a + R_{se}) + P_S$$

$$\text{Motor output} = V_1 I_1 - I_1^2 (R_a + R_{se}) - P_S$$

$$\text{Motor efficiency, } \eta_m = \frac{V_1 I_1 - I_1^2 (R_a + R_{se}) - P_S}{V_1 I_1}$$

Generator Efficiency

$$\text{Generator output} = V_2 I_2$$

$$\text{Generator losses} = I_2^2 R_a + I_1^2 R_{se} + P_S$$

$$\text{Generator input} = V_2 I_2 + I_2^2 R_a + I_1^2 R_{se} + P_S$$

$$\text{Generator efficiency, } \eta_g = \frac{V_2 I_2}{V_2 I_2 + I_2^2 R_a + I_1^2 R_{se} + P_S}$$

Disadvantages. 1. A relatively small error in the measurement of the input to motor or output of generator may cause a relatively large error in the computed efficiency.

2. Total power supplied to the set is wasted.

* Field test is not a regenerative test because output of generator is not fed back to motor but is dissipated in a resistor.

Example 11.31. A test on two coupled similar tramway motors, with their fields connected in series, gave the following test results when one machine acted as a motor and the other as a generator.

Motor : Armature current = 56 A; Armature voltage = 590 V

Generator: Armature current = 44 A; Armature voltage = 400 V; Field voltage drop = 40 V; Resistance of each armature = 0.3 Ω.

Calculate the efficiency of the motor and generator at this load.

[*Nagpur Univ. Electrical Machinery-II 1992; J.N. Univ. Hyderabad 2000*]

Solution: Voltage across motor armature, $V_1 = 590$ V

Supply voltage, V = V_1 + voltage drop across generator field winding

= 590 + 40 = 630 V

Terminal voltage of generator, V_2 = 400 V

Generator armature current, I_2 = 44 A

Motor armature current, I_1 = 56 A

Supply input current, $I = I_1 = 56$ A

Total power input = $VI = 630 \times 56 = 35{,}280$ W

Total power output = $V_2 I_2 = 400 \times 44 = 17{,}600$ W

Total copper losses = $I_1^2 R_a + I_2^2 R_a$

+ I_1 × voltage drop across field windings

= $56^2 \times 0.3 + 44^2 \times 0.3 + 56 \times (40 + 40)$

assuming equal voltage drop across the two field windings

= 940.8 + 580.8 + 4,480 = 6001.6 W

Stray losses of the set = Total power input – power output – total copper losses

= 35,280 – 17,600 – 6001.6 = 11,678.4 W

Stray losses per machine,

$$P_S = \frac{11{,}678.4}{2} = 5{,}839.2 \text{ W}$$

Motor Efficiency

Input power to motor = $V_1 I_1 = 590 \times 56 = 33{,}040$ W

Total losses in motor = $56^2 \times 0.3 + 56 \times 40 + 5{,}839.2 = 9{,}020$ W

Motor output = Input – total losses

= 33,040 – 9,020 = 24,020 W

$$\text{Motor efficiency, } \eta_m = \frac{\text{Motor output}}{\text{Motor input}} \times 100$$

$$= \frac{24{,}020}{33{,}040} \times 100 = 72.7\% \textbf{ Ans.}$$

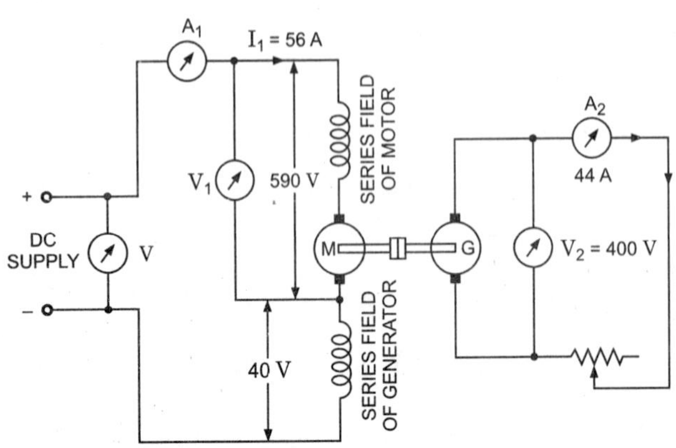

Fig. 11.14

Generator Efficiency

Generator output = $V_2 I_2 = 400 \times 44 = 17{,}600$ W

Total losses in generator = $44^2 \times 0.3 + 56 \times 40 + 5{,}839.2 = 8{,}660$ W

Generator input = Output + total losses

= 17,600 + 8,660 = 26,160 W

$$\text{Generator efficiency, } \eta_G = \frac{\text{Generator output}}{\text{Generator input}} \times 100$$

$$= \frac{17{,}600}{26{,}160} \times 100 = 67.28\% \textbf{ Ans.}$$

11.11 LOAD TEST

This test is performed at rated load and is required for checking the commutation and the temperature rise during operation. For small machines, brake test or testing of a motor coupled to a calibrated generator may be carried out for determination of efficiency directly. For large machines a back-to-back (or Hopkinson test) is performed on the two identical machines for determination of the efficiency of the machines, as already discussed in Art 11.8. In the case of thyristor supply also, two machines are tested back-to-back so that only losses in both the machines are drawn from the test source.

In modern testing laboratories in manufacturer's concerns, the test source is from suitable motor-generator set or from thyristor supply and rectifier power supply. The motor is fitted on the test bed and coupled to a similar machine if possible or another calibrated machine which is feeding the power back to the test supply mains and wasting power only corresponding to the losses in both the machines.

11.12 HEAT-RUN TEST (OR TEMPERATURE-RISE TEST)

The life of insulation of electrical equipment depends upon the maximum temperature attained during operation. The objective of this test is to find out the actual maximum temperature attained while the machine is operating under certain load conditions.

The back-to-back load test is performed for the temperature run also. The normal duration of the hot-run test on large machines is 6 hours. In some cases, to avoid wastage of energy during testing and reduce testing duration (which is normally till the temperature rise attains a steady value), short time tests are carried out and ultimate temperature rise of the machine is determined from the datas obtained from these tests.

The equation for temperature rise is given as $\theta = \theta_m(1 - e^{-t/\tau})$ where θ is the temperature rise at time *t*, θ_m is the maximum temperature rise and τ is the heating time constant depending on the thermal capacity of the insulation and the rate of dissipation of heat.

The above equation for temperature rise is simplified by substituting $e^{-1/\tau}$ by k and we get

$$\theta = \theta_m(1 - k^t)$$

When t = 1, $\theta_1 = \theta_m(1 - k)$ and when t = 2, $\theta_2 = \theta_m(1 - k^2)$

$$\text{Thus } \frac{\theta_2}{\theta_1} = \frac{1 - k^2}{1 - k} = 1 + k$$

$$\text{or } k = \frac{\theta_2 - \theta_1}{\theta_1}$$

and the maximum temperature rise,

$$\theta_m = \frac{\theta_1}{1 - k} \quad \text{...(11.24) if initial temperature rise, } \theta_0 = 0$$

It will thus be possible to determine the maximum temperature rise by taking two or more readings at regular intervals and finding the trend of temperature rise.

The constant k is always given by the ratio of the actual increases in temperatures during successive equal intervals of time *i.e.*,

$$k = \frac{\theta_2 - \theta_1}{\theta_1 - \theta_0} = \frac{\theta_3 - \theta_2}{\theta_2 - \theta_1} = \frac{\theta_4 - \theta_3}{\theta_3 - \theta_2} \text{ and so on} \quad ...(11.25)$$

The interval of time during the test for measurement of temperature may be taken as 30 minutes or even smaller. The application of the short time heat run test is illustrated by worked-out Example 11.32.

11.12.1. Measurement of Temperature Rise. There are three recognized methods of measuring temperature rise of windings and other parts viz (*i*) Thermometer method (*ii*) Resistance Method (*iii*) Embedded temperature-detector method.

1. Thermometer Method. In this method of measurement of temperature, a thermometer is applied to the accessible part of the completed machine. The accessible part is only the surface. Thus by this method we get the temperature of the surface only. When bulb thermometers are used in places where there is a strong varying or moving magnetic field, alcohol thermometers should be used in preference to mercury thermometers.

2. Resistance Method. This method is used for determination of temperature rise of the windings. This method involves the measurement of resistance, cold and hot, and estimation of average temperature rise from the relation

$$R_2 = R_1 (1 + \alpha_1 t_1) \quad ...(11.26)$$

where R_2 is resistance at t_2 °C, R_1 is resistance at t_1 °C and α_1 is temperature coefficient of resistance at t_1 °C

For copper winding the relation is given as

$$\frac{R_2}{R_1} = \frac{t_2 + 235}{t_1 + 235} \quad ...(11.27)$$

For practical purposes, the following alternative formula may be used

$$t_2 - t_a = \frac{R_2 - R_1}{R_1}(235 + t_1) + t_1 - t_a \quad ...(11.28)$$

t_a being the temperature of cooling medium at the end of the test.

When the temperature of a winding is to be determined by resistance, the temperature of the winding before the test, measured by thermometer, shall be practically that of the cooling medium.

3. Embedded Temperature Detector Method. Embedded temperature detectors are resistance thermometers or thermocouples built into the machine during construction at points, which are inaccessible after the machine is completed.

Example 11.32. In a heat run test on a dc machine, three consecutive readings taken at 30 minute time intervals were 35 °C, 45 °C and 50 °C. Determine the final temperature rise likely to be attained.

Solution: Ratio of the temperature rise during successive intervals of time,

$$k = \frac{\theta_2 - \theta_1}{\theta_1 - \theta_0} = \frac{50 - 45}{45 - 35} = 0.5$$

Maximum temperature rise likely to be attained,

$$\theta_m = \theta_0 + \frac{\theta_1 - \theta_0}{1 - k} = 35 + \frac{45 - 35}{1 - 0.5} = 55 \text{ °C } \textbf{Ans.}$$

11.13 MAINTENANCE OF DC MACHINES

The purpose of routine maintenance is to ensure, as far as possible, that a machine is operating, and will continue to operate, in a safe and efficient manner. A systematic inspection routine can result in minor faults being detected and corrected before occurrence of serious trouble. The frequency with which the machine should be inspected is determined by the conditions under which it operates, and the danger and expense which may arise due to breakdown.

The tasks to be normally carried out during a routine inspection/maintenance are discussed briefly below:

1. Mechanical Maintenance.

(*a*) *Bearings.* Satisfactory bearings are vital for a mechanically sound motor. Bearings should be checked for excessive play and correct lubrication; excessive lubrication must be avoided owing to the risk of oil or grease finding its way on to the winding. They should be kept clean as the presence of dirt will cause rapid deterioration, Where ball or roller bearings are used these should be examined periodically for faults or crack. If the bearings of a motor are in good condition and there are no other mechanical faults, the rotor or armature should be able to rotate freely without any jerkiness.

(*b*) *General.* All fixing bolts and driving arrangements should be checked for security, and it should be ensured that all cover plates and guards are correctly in place.

The outside of the machine should be cleaned frequently and the air passages blown out regularly, otherwise dust deposits upon the external surface of a motor will act as a heat insulant and in the air ducts will restrict the flow of cooling air and in consequence the temperature rises. The frequency with which blowing out takes place depends on the type of enclosure of the motor. Where additional provision is required for a motor care should be taken to avoid interference with its ventilation.

2. Electrical Maintenance. Electrical breakdowns in motors may take the form of short-circuit, earth faults or open-circuits. Many short-circuits and earth faults may be prevented by keeping the insulation dry and in good condition. For this the motor is subjected to frequent insulation resistance tests, as described in the suceeding Art. If the insulation resistance is found low, it may normally be improved by drying out either in an open oven or by passing a steady current through the affected windings and then painting with a suitable varnish.

(*a*) *Commutator.* It may give trouble due to dirt deposits, high mica, rough surface or eccentricity.

The commutator should be inspected periodically and cleaned with a dry soft fluff free cloth. If any blackening of the commutator surface is noticed, spring pressure should be checked. Angle of brushes should also be checked. The grooves between the commutator segments should be kept free from carbon dust, etc., as this may lead to short circuiting of the armature coils and excessive sparking. The commutator surface should be smooth, concentric and properly undercut.

(b) Brushes. Brush troubles may arise due to insufficient contact surface, too short a brush, too little spring tension or wrong brush setting.

The brushes should be periodically examined to see that they are making firm contact with the commutator surface and not sticking in their brush-holders owing to accumulation of carbon dust etc. Excessive wear can be prevented by ensuring that brushes for commutators are sufficiently long, fit well into the brush-box and are made of the right grade of carbon. The spring tension should be sufficient to hold the brush firmly against the commutator. New brushes should be carefully bedded into the commutator. The pressure on each brush should be adjusted to 140 g/cm^2 to 180 g/cm^2 pressure. Badly worn brushes should be replaced.

The brush rocker arm should be fixed properly in neutral position.

(c) *Windings.* Insulation resistance of the windings should be periodically checked: both of the armature windings and field winding. If any fault is detected, it should be repaired by isolating the faulty winding.

(d) Insulation Resistance. The insulation resistance of the whole machine should not be less than 1 MΩ. If it is found less than 1 MΩ, then the machine should be cleaned and dried out.

(e) Earthing Conductor. It should be inspected and tested for continuity.

(f) Starter. Before attempting any maintenance operation on a starter it is to be isolated and its outside cleaned to prevent any dirt entering when it is opened. Conduit and cable should be checked and also the starter mounting.

All contacts should be cleaned and any roughness caused due to arcing or mechanical wear should be filed-off. Care should be taken to remove any metallic dust resulting from filing as it can give rise to short-circuits.

Where a carbon deposit has built up as a result of the making and breaking of contacts it must be removed before it gives rise to trackling across insulation.

Any spring-loaded contacts must be seen to move freely but the spring tension must be sufficient to prevent bouncing. Pivot points should be lubricated and sliding contacts may be lightly coated with petroleum jelly.

The operation of *overcurrent* and *no-voltage* devices should be tested at regular intervals, particularly after periods of prolonged idleness or intensive use. The settings of such devices should be kept closely related to load conditions.

11.14 FAULTS IN DC MACHINES

11.14.1. Faults in Field Winding. The faults in the field winding may be an open-circuit, an earth fault or shorting of a coil either completely or some of its turns.

1. Location of Open-Circuit Fault in Field Winding. An open-circuit fault in the field winding of a dc shunt motor will either cause an interruption of supply to the motor owing to operation of overcurrent protection or tremendous increase in speed if it is running light. Such a fault in a dc series motor will cause the motor to stop. In a dc compound motor the effect will depend on whether the break is in series or shunt field winding.

For locating an open-circuit fault an insulation resistance tester or a continuity tester may be used. The effectiveness of the tester is first verified by shorting the test leads together. This should give a very low reading.

Let us take a four-pole dc motor for such a test, as shown in Fig. 11.15. The leads are applied to the opposite ends of the field circuit (*i.e.* at points E and A). An open-circuit will be indicated by a high resistance reading. The test is then repeated between points E and B. Low reading indicates a complete circuit and open-circuit fault in coil 1. If the reading is still high the test is repeated between points E and C. A low reading then shows open-circuit fault in coil 2. If the reading is again high the procedure is to be repeated between points E and D. A low reading will locate the fault in coil 3 and a high reading will indicate open-circuit fault in coil 4. *There is also a possibility of occurrence of open-circuit fault in the leads to the coils rather than the coil itself and this possibility should be investigated before removal of coils for rewinding.*

2. Location of Earth Fault in Field Winding. An earth fault in the field winding of a motor with effective earth leakage protection will cause the disconnection of supply to the motor. The presence of an earth fault can be verified by a test between point E and the motor frame with an insulation resistance tester (Fig. 11.15). Under these circumstances an earth fault will be indicated by zero reading. For location of defective coil, the series connection between coils 2 and 3 is broken at point C. The test between points E and the motor frame is repeated. Zero reading will indicate the fault in coil 3 or 4 whereas high resistance reading will show the fault in coil 1 or 2. Whichever pair of coils is found to be faulty must then be split and each coil tested separately to the frame.

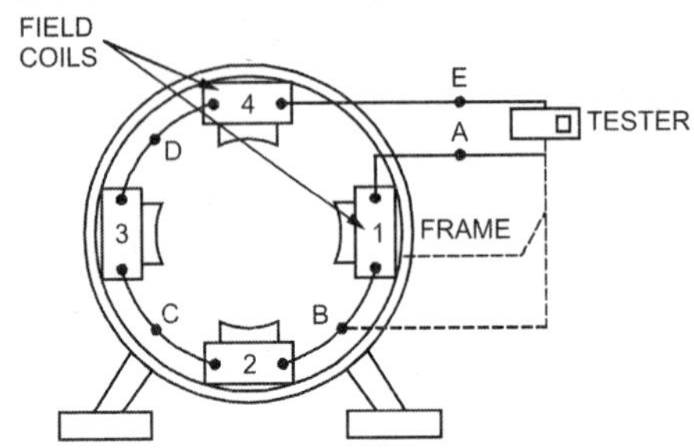

Fig.11.15

3. Location of Shorted Coil in Field Winding. A test circuit is illustrated in Fig. 11.16. The normal supply is connected to the motor through a potentiometer in order to keep test current at a reasonable level. With flow of current through the coils there will be a voltage drop across each coil and it is measured with the voltmeter. If there are short-circuited turns in one of the coils its

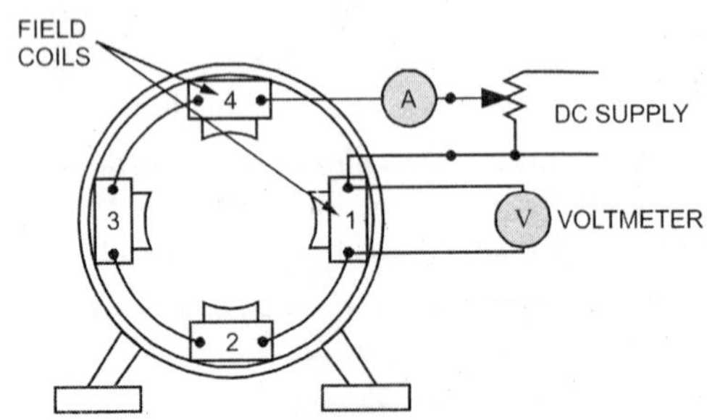

Fig. 11.16

volt drop will be smaller in comparison to those across the other coils. The amount by which the voltage across the faulty coil differs from those of the sound coils will indicate how much of the coil is shorted. A zero reading will indicate a completely shorted coil.

4. Polarity Test. The polarity of the coils can be verified by using the same circuit arrangement and a small compass of the type used for plotting magnetic fields.

11.14.2. Faults in Armature Windings. The possible faults that can be developed in the armature windings are:

1. A short-circuit of a coil or coils in the armature.
2. An open-circuit in one of the armature coils.
3. An "earth" on one or more of the coils of armature.

1. Short-Circuited Coil. If this fault develops when the machine is under operation, there will be intense local heating of the faulty coil, quite independent of the load on the machine, owing to the large local current that will circulate through the faulty coil. The heavy local current, limited solely by the impedance of the short-circuited coil, very often results in the burn out of the coil, and the fault then becomes an open circuit. The short circuit may take place between the risers, or it may take place between adjacent turns of a multi-turn coil. The heavy current is purely local, so it does not overload the brushes.

2. Open-Circuited Coil. The symptoms of this fault are most spectacular when the machine is loaded, but are inappreciable on light load. Half the winding between the brushes having an open-circuited coil carries no current while the other half, having sound coils, carries double the normal current. When the open-circuited coil passes under the influence of the brush, the current in the sound coils changes from double to zero and vice versa. This causes increase in reactance voltage and consequently vicious sparking at the brushes at the instant of breaking of contact of leaving segment connected to the faulty coil with the brush. It is imperative to shut down the machine promptly otherwise the commutator will rapidly be ruined. If the machine is shut down promptly, the commutator will show signs of serious pitting at the segments connected to the faulty coil and those one-pole-pitch distant from it.

3. Earthed Coil. If there is only a single earth, the effect is not serious and noticeable. Serious trouble, however, at once occurs if two separate coils on the armature are earthed. When two separate points on the armature winding become "earthed" the coils between the two faults become short circuited and the symptoms are those of short-circuited coils so far as intense local heating is concerned. If a number of coils are thus short-circuited very violent sparking will occur owing to the marked unbalancing of the armature currents.

Location of Faults in Armature Winding. Armature faults can be located either by using a special test device known as *growler* or by applying a drop test.

The basis of the *growler* is a winding energized by an ac supply. This winding is mounted on a core of ferromagnetic material. When the core is close to the armature winding its alternating magnetic field links with the turns of the armature coils inducing an emf in them. In a sound coil side the induced emf causes a current which circulates throughout the armature. In the case of a short-circuited coil the emf acts in a closed loop and, therefore, gives rise to a much larger current than that in a sound circuit.

For carrying out a drop test the circuit diagram is shown in Fig. 11.17.

This test can be carried out with the armature either in position or removed from its bearings. Current is supplied to the armature from a low-voltage source. A voltmeter is connected by prongs to adjacent commutator segments and the voltage between consecutive commutator segments is measured. If the coil connected between commutator segments is shorted, the voltmeter will give zero reading and if there is a partial fault, the voltmeter indication will be very low in comparison to those between other commutator segment pairs.

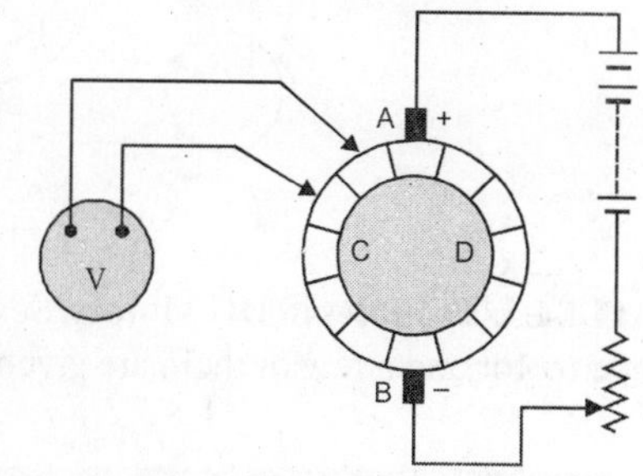

Fig. 11.17

An open-circuited coil prevents the flow of current in its circuit. The voltmeter will then give zero reading while testing the particular circuit containing the fault, until it bridges across the segments joined to the faulty coil when the reading will correspond to the full voltage applied to the brushes. For a lap winding, the faulty coil is thus at once located. For a wave winding the fault will be in one of the coils which are connected in series between the adjacent commutator segments (*i.e.* 2,3 and 4 for a 4-pole, 6-pole or 8-pole machine and so on). For brushes placed one pole-pitch apart with a multipolar wave-wound machine, the potential difference between segments will change sign at the points distant by one or more pole-pitches from the positive brushes. While testing such a machine for open-circuit fault, pairs of segments will be found having no circuit between them. The faulty coil will in every case be in the circuit between these segments and so can be isolated.

While carrying out the drop test, it is always desirable to perform the test twice, moving the commutator round a little between the two tests so that different coils become short circuited by the brushes.

The circuit for testing an 'earthed coil' of the armature is shown in Fig. 11.18. Insulation resistance of the armature is given by the equation

$$R' = r\left(\frac{V}{v} - 1\right)$$

where V is supply voltage, v is the voltmeter reading and r is the voltmeter resistance. The voltmeter reading falls as the contacts are made from the positive brush down to the faulty coil segment. When connected to the faulty coil segment the voltmeter indication will be zero and if the contact is further moved, the voltmeter indication will reverse. For avoiding the phantom earthed point, which is must, the test is repeated by moving the commutator around slightly. This changes the phantom point while the commutator point to which faulty coil is connected remains the same.

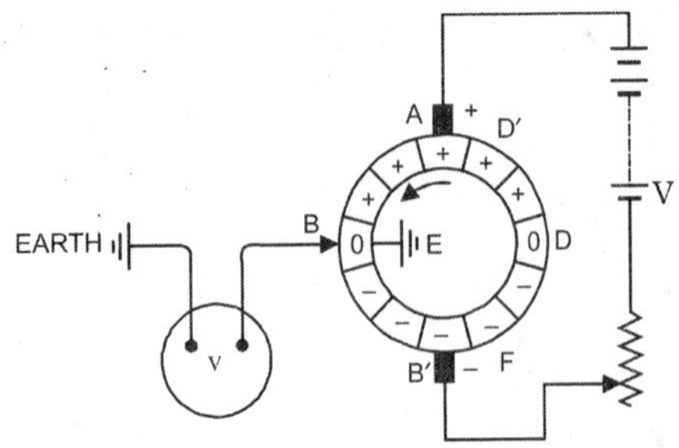

Fig. 11.18

11.14.3. Troubles in DC Motors. Several troubles may arise in a dc motor and a few of them are given below.

1. Sparking at Brushes. It may be due to troubles in brushes, commutator or armature or excessive load.

2. Overheating. It may be due to excessive loading, sparking at brushes, short-circuited field or armature coils, poor ventilation, incorrect voltages or too frequent starts or reversals.

3. Failure To Start. This may be due to open, short-circuit or earth fault, wrong connections, too low supply voltage, excessive load or frozen bearings.

4. Vibration and Pounding Noises. This may be due to loose coupling, loose parts, worn bearings, rotating parts hitting stationary parts, unbalanced armature or misalignment of the machine.

HIGHLIGHTS

1. The total losses occurring in a dc machine are summarized below:

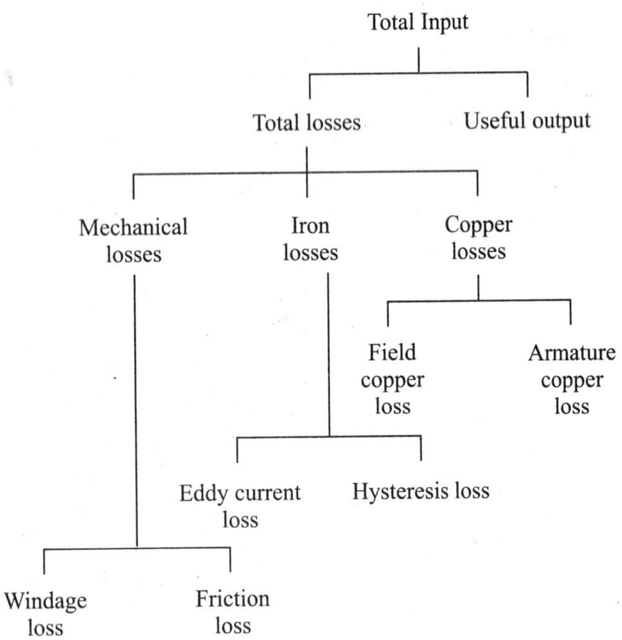

2. Efficiency of electrical machine:

$$\eta = \frac{\text{Power output}}{\text{Power input}} = \frac{\text{Power input} - \text{total losses}}{\text{Power input}}$$

$$= \frac{\text{Power output}}{\text{Power output} + \text{losses}}$$

3. Efficiency of an electrical machine will be maximum when variable losses = Constant losses.
4. Efficiency of a dc machine can be determined by three methods viz. direct method, indirect method, and regenerative method.

   Direct method can be applied to only small machines. Indirect method is used to test shunt and compound machines. Regenerative method requires two identical machines, one to operate as a generator and the other as a motor.
5. **Brake test** is direct method of testing dc machine.

   The main drawback of such a method of determining the internal losses or efficiency of a dc machine is that the output of the motor cannot be measured accurately.

   The other drawback is that this method cannot be used for determining internal losses and efficiency of large machines, since testing of large machine entails a considerable loss of energy. In fact in the case of very large machines there may be no facility for such a test known as *direct test*.
6. **Swinburne's test** is the simplest indirect method of testing of dc machines in which losses are measured separately and efficiency at any desired load is predetermined.

   This test cannot be performed with dc series motors, since on no load series motor would attain such a high speed that it will get damaged and secondly this test is only applicable to those machines in which flux and speed remain constant. Hence this test can be applied only to shunt and level compound machines.
7. **Hopkinson test** is regenerative test. The main disadvantage of this test is the necessity of two practically identical machines.
8. **Retardation test** is known as the running down test and is used for determination of the stray losses of shunt wound dc machines.
9. **Field test** is performed on large capacity dc series machines. It also needs two identical dc series machines. It is not a regenerative test as the generator output is not fed back to motor but is dissipated in an external resistor.
10. **Load test** is performed at rated load and is required to check the commutation and the temperature rise during operation. Brake test is performed in case of small machines whereas in case of large machines a back-to-back test is performed.
11. **Heat-run or temperature-rise test** is performed to find out the actual maximum temperature attained during operation of the machine under certain load conditions. The back-to-back test is performed for the heat run test also.

    There are three recognized methods of measuring temperature rise of windings and other parts viz (*i*) Thermometer method (*ii*) Resistance Method (*iii*) Embedded temperature-detector method.
12. The purpose of routine maintenance is to ensure, as far as possible, that a machine is operating, and will continue to operate, in a safe and efficient manner. A systematic inspection routine can result in minor faults being detected and corrected before occurrence of serious trouble. The frequency with which the machine should be inspected is determined by the conditions under which it operates, and the danger and expense which may arise due to breakdown.
13. The faults in the field winding may be an open circuit, an earth fault or shorting of a coil either completely or some of its turns.

    The possible faults that can be developed in the armature windings are:

    1. A short-circuit of a coil or coils in the armature.
    2. An open-circuit in one of the armature coils.
    3. An "earth" on one or more of the coils of armature.

14. Several troubles may arise in a dc motor and a few of them are given below.

 1. Sparking at Brushes. It may be due to troubles in brushes, commutator or armature or excessive load.

 2. Overheating. It may be due to excessive loading, sparking at brushes, short-circuited field or armature coils, poor ventilation, incorrect voltages or too frequent starts or reversals.

 3. Failure To Start. This may be due to open, short circuit or earth fault, wrong connections, too low supply voltage, excessive load or frozen bearings.

 4. Vibration and Pounding Noises. This may be due to loose coupling, loose parts, worn bearings, rotating parts hitting stationary parts, unbalanced armature or misalignment of the machine.

EXERCISES

1. Identify various losses that occur in the dc shunt machine. **[U.P.S.C. I.E.S. Electrical Engineering-II, 2009]**
2. Describe the different losses in a dc machine. **[W.B. Univ. of Technology Elec. Machine-I, 2011]**
3. What are different losses in dc machines? On what factors they depend? **[U.P.Technical Univ. Even Semester, 2005-06]**
4. What are the various losses occuring in rotating machines ? Mention the methods to reduce them. **[U.P. Technical Univ., September-2001]**
5. Derive efficiency equation of a dc motor. **[U.P. Technical Univ. Electrical Engineering, January-2003]**
6. Explain significance of back emf, state voltage equation, power stages with different losses, and derive condition for maximum power developed for dc motor. **[Gujarat Technological Univ. Electrical Machines-I, Summer-2014]**
7. What are the no-load rotational losses in electrical machines? How can these be determined?

 Which of these losses in rotating electrical machines are influenced by the magnitude of flux, the load and the square of the load? Write a general expression for all the losses occuring in rotating electrical machines and prove therefrom that maximum efficiency occurs when losses proportional to the square of current = constant losses. **[U.P.S.C. I.E.S. Electrical Engineering-II, 2001]**
8. Describe Swinburne's test with the help of a neat diagram to find out the efficiency of a dc machine. **[R.G. Technical Univ. Electrical Machine-II, December-2012]**
9. Describe the Swinburne's test to determine no-load losses of a dc machine. What are limitations of this test? **[J.N. Technological Univ. Elec. Machines-I, March-2014]**
10. Explain Swinburne's test to determine the efficiency of a dc machine with the help of a neat diagram. What are the main advantages and disadvantages of this test? **[R.G. Technical Univ. Electromechanical Energy Conversion, June-2010]**
11. Explain the following:
 (*i*) Brake test
 (*ii*) Swinburne's test
 (*iii*) Hopkinson test.
 Also mention their importances. **[U.P. Technical Univ. Electromechanical Energy Conversion-I, 2013-14]**
12. Explain the Hopkinson's test for dc machine. **[R.G. Technical Univ. Electromechanical Energy Conversion, December-2012]**
13. Explain in brief how efficiency is obtained from Hopkinson's test. **[U.P. Technical Univ. Electromechanical Energy Conversion-I, 2008-09]**
14. With the help of neat circuit diagram, explain Hopkinson's test and derive the relations for efficiency (both for generator and motor). Also state the merits and demerits of this method. **[J.N. Technological Univ. Electrical Machines-I, March-2014]**
15. Explain the Hopkinson's method of testing of the dc machines. Differentiate this method from the Swinburne's method of testing of the dc machines. Can the Swinburne's method of testing of the dc machine be applied on dc series machine? Explain whichever is the answer. **[Rajasthan Technical Univ. Electrical Machines-I, January-2013]**
16. State the advantages of Hopkinson's test over Swinburne's test. **[W.B. Univ. Technology Electrical Machines-I, 2011]**
17. Explain the experimental procedure to conduct retardation test on a dc shunt machine with the help of connection diagram. How the different losses are estimated from the test results? **[J.N. Technological Univ. Electrical Machines-I, March-2014]**
18. Explain the procedure to conduct field's test on series machines in order to determine efficiency. **[J.N. Technological Univ. Electrical Machines-I, March-2014]**
19. Describe methods of measuring temperature in electrical machines. **[Pb. Univ. Electrical Machine Design, Dec.-1995]**
20. What is the purpose of routine maintenance. Discuss the points to be checked during routine inspection, maintenance of a dc machine.
21. List the faults which are likely to occur in the field and armature windings of a dc machine. Describe the tests, with circuit diagrams, to locate the faults in armature winding of a dc machine.
22. Write short notes on the following:
 (*i*) Losses in dc machines
 (*ii*) Brake test
 (*iii*) Swinburne's test
 (*iv*) Hopkinson's test **[R.G. Technical Univ., December-2013]**
 (*v*) Faults in dc machines.

SHORT ANSWER TYPE QUESTIONS WITH ANSWERS

Q. 1. What is meant by "energy loss" in a machine?

Ans. Whole of the input energy to the electrical machine is not converted into useful output energy but a part of the energy is converted into heat and is lost for the useful purpose of the machine. The energy converted into heat is called the energy loss.

Q. 2. What is the basic power equation for a dynamo?

Ans. $P_{input} = P_{output} + P_{losses}$

Q. 3. What are the different types of losses in a dc machine?

Ans. The different types of losses that occur in a dc machine are (*i*) copper or electrical losses (*ii*) iron or magnetic losses and (*iii*) mechanical or friction and windage losses.

Copper or electrical losses include power lost in heating the shunt field and also in heating the armature circuit, which includes armature winding, brush contacts, series field and interpole field.

Iron or magnetic losses include the hysteresis and eddy current losses which are present in any part of the machine that is constructed of iron and subjected to variations of magnetic flux.

Mechanical losses consist of power losses due to friction of the bearings, air friction or windage, as it is called, caused by the motion of the moving parts through the surrounding medium, and the friction between brushes and commutator.

Q. 4. Why do copper losses occur in a dc machine ?

Ans. Copper losses occur in a dc machine due to power consumed in forcing a current against the resistance of windings and connections.

Q. 5. How do the various losses occurring in a dc machine vary with the load ?

Ans. The copper losses occurring in armature winding, series field winding, compensating winding and interpole winding, if provided, almost vary as the square of the load current while that occurring in dc shunt field winding remain practically constant. Rotational losses (iron and mechanical losses) are almost independent of load.

Q. 6. Which of the losses occurring in rotating electrical machines are influenced by the magnitude of the flux, the load and square of the load? **[U.P.S.C. I.E.S. 2001]**

Ans. Core or iron losses are influenced by the magnitude of flux; copper losses (armature, series field, interpoles, compensating windings) are proportional to the square of the load while brush-contact loss is proportional to the load.

Q. 7. How do hysteresis, eddy current, friction and windage losses depend upon the motor speed? **[Rajasthan Technical Univ., January-2013]**

Ans. Hysteresis loss varies as the motor speed, eddy current loss varies as the square of motor speed and the friction and windage losses vary as the cube of the motor speed.

Q. 8. What are the components of iron loss in dc machine and in which part of the machine it occurs?

Ans. Iron loss consists of hysteresis and eddy-current losses caused by changing flux densities in the iron core (stator and rotor cores, poles) of the machine.

Q. 9. Does core loss occur in armature or in the poles of a dc machine?

Ans. Core loss occurs in armature core and pole faces.

Q. 10. On what factors do hysteresis and eddy current losses depend?

Ans. Hysteresis loss depends upon the peak magnitude of the flux density attained in a magnetic cycle, frequency of magnetic cycles, volume of magnetic material and type of magnetic material.

Eddy current loss depends also upon the thickness of magnetic material laminations in addition to factors mentioned above.

Q. 11. How the eddy current losses will change if the thickness of laminations is increased?

Ans. Eddy current losses vary as the square of the thickness of magnetic material laminations.

Q. 12. Why mechanical losses occur in a dc machine?

Ans. Mechanical losses occur in dc machines due to motion of rotating parts through the surrounding medium and the friction between brushes and commutator.

Q. 13. Iron loss in a dc machine increases with increasing load. What is its mode of excitation?

Ans. Series excitation.

Q. 14. Why CROS cannot be used in small rotating machines?

Ans. In small rotating machines it is difficult to correspond the axis of core with the rolling direction.

Q. 15. When does the maximum efficiency of a dc machine take place?

Ans. When variable losses equal constant losses.

Q. 16. What are the different types of tests which can be made for different types of dc machines?

Ans. Direct, indirect and regenerative tests.

Q. 17. Can a brake test be applied to a dc series motor?

Ans. Yes.

Q. 18. What is the main drawback of brake test?

Ans. The output of the motor cannot be measured accurately.

Q. 19. Why brake test is performed with small machines only?

Ans. Testing of a large machine entails a considerable loss of energy, infact in case of very large machines there may be no facility for such a test.

Q. 20. Why is it necessary to keep the brake applied tight while applying brake test to a dc series motor?

Ans. If the brake applied is not tight, the motor will attain dangerously high speed and get damaged.

Q. 21. Why Swinburne's test cannot be performed on dc series machines?

Ans. Because on no load dc series motor will attain such a high speed that it will get damaged and second this test is only applicable to those machines in which flux and speed remain constant.

Q. 22. Why constant losses can be considered equal to input to a machine at no load?

Ans. Because at no load variable losses (armature copper losses and losses in series field winding, interpole winding and compensating winding, if any) are negligible owing to negligible armature current.

Q. 23. Why Swinburne's test is considered convenient and economical method for testing of dc shunt machines?

Ans. Because power required to test a large dc machine by Swinburne's test is quite small.

Q. 24. Why Hopkinson's test is also called the regenerative or back-to-back test?

Ans. The motor supplies mechanical power which is used to drive the generator while generator develops electrical power to be utilized in driving the dc motor in addition to electrical power drawn from the supply mains.

Q. 25. State with reason whether field test on two indentical dc series machines is regenerative method?

Ans. Field test is not regenerative test because output of generator is not fed back to motor but is dissipated in a resistor.

Q. 26. In the Hopkinson's test on two dc machines, machine A has a field current of 1.4 A and machine B has a field current of 1.3 A. Which machine acts as a generator?

Ans. Machine having higher excitation will generate higher emf if the machines are identical and operating at the same speed. So the machine having higher excitation (field current) i.e. machine A is operating as a generator.

Q. 27. At which point of a conductor embedded in a slot does the maximum temperature occur?

Ans. Maximum temperature in a conductor embedded in the slot occurs at its centre.

Q. 28. What is hot spot temperature?

Ans. Hot spot temperature is the temperature of that part of a machine in which heat is produced and it has the highest value.

Q. 29. The outside of the machine should be cleaned frequently and the air passages blown out regularly. Why?

Ans. Otherwise dust deposit upon the external surface of a machine will act as a heat insulant and in air ducts will restrict the flow of cooling air and consequently temperature will rise.

Q. 30. What are the possible causes of sparking at brushes?

Ans. Sparking at brushes may be due to (*i*) trouble in brushes, (*ii*) commutator (*iii*) armature or excessive load.

PROBLEMS

1. A long shunt dynamo running at 1,000 rpm delivers a load of 22 kW at 220 V. Its armature, shunt field and series field resistances are 0.05, 110 and 0.06 ohms respectively. Determine (*i*) copper loss (*ii*) iron and friction losses (*iii*) output of the prime mover and (*iv*) torque exerted by the prime mover. Overall efficiency of the dynamo at full load may be taken 88%.

   [**Ans.** (*i*) 1,584.44 W (*ii*) 1,415.56 W (*iii*) 25 kW (*iv*) 238.73 Nm]

2. A 110 V dc shunt generator has full-load current of 100 A. Its shunt field resistance is 55 Ω and it has constant loss of 500 W. If the full-load efficiency is 88%, find armature resistance. [**Ans.** 0.096 Ω]

3. A 4-pole lap wound dc shunt motor has armature and field resistances as 1 Ω and 200 Ω respectively. The motor is drawing a load current of 20 A from a 220 V dc supply source. Draw electric circuit for the motor and find the value of (*i*) back emf induced in the motor (*ii*) current in each path of the armature and (*iii*) total copper losses.

   [**Ans.** (*i*) 201.1 V (*ii*) 4.725 A (*iii*) 539 watts]

4. A dc shunt generator has a full-load output of 10 kW at a terminal voltage of 240 V. The armature and shunt field winding resistances are 0.6 and 160 ohms respectively. The sum of the mechanical and core losses is 500 W. Calculate the power required, in kW, at the driving shaft at full load, and the corresponding efficiency.

   [Nagpur University, November-1999]

   [**Ans.** 11.978 kW; 83.49%]

5. A 250 V, 20 kW shunt motor running at 1,500 rpm has a maximum efficiency of 85% when delivering 80% of its rated output. The resistance of the shunt field winding is 125 Ω. Determine the efficiency and speed of the motor when it draws 100 A from the mains.

   [**Ans.** 84.25%; 1,458 rpm]

6. A 7.48 kW, 220 V, 900 rpm, shunt motor has a full-load efficiency of 88%, the armature resistance is 0.08 Ω and shunt field current is 2 A. If the speed of the motor is reduced to 450 rpm by inserting a resistance in the armature circuit, find motor output, the armature current, external resistance to be inserted in the armature circuit and overall efficiency. Assume the load torque to remain constant.

   [Nagpur Univ., November-1998]

   [**Ans.** 3.50435 kW; 36.6364 A, 2.962 Ω; 41.23%]

7. A 230 V, 4-pole dc shunt motor draws 36 A from mains when delivering 10 hp. If armature and field circuit resistances are 0.1 Ω and 300 Ω respectively, find the motor current when load is 5 hp. Neglect armature reaction effect.

   [Pb. Univ. Electrical Machines-II, Dec; 1993]

   [**Ans.** 19.627 A]

8. The armature winding of a 200 V, 4-pole series motor is lap connected. There are 280 slots and each slot has 4 conductors. The current is 45 A and the flux per pole is 18 mWb. The field resistance is 0.3 Ω. Armature resistance is 0.5 Ω and the iron and friction losses are 800 W. Find the shaft torque. [**Ans.** 128.734 Nm]

9. A 1500 kW, 550 V, 10-pole dc generator runs at 150 rpm. What must be the useful flux per pole if there are 2,500 lap-connected conductors and the full-load armature copper loss in 25 kW?

   [**Ans.** 89.467 mWb]

10. The armature of a 220 V, 4-pole series motor is lap wound. There are 280 slots and each slot has 4 conductors. The current is 45 A and the flux per pole is 0.018 Wb. The field resistance is 0.3 Ω and armature resistance 0.5 Ω and iron and friction losses total 800 watts. The pulley diameter is 0.41 m. Find the pull at the rim of the pulley.

    [**Ans.** 636 N]

11. Design the resistance section of a seven-stud starter for 36.775 kW, 400 V dc shunt motor. Full-load efficiency is 92% and copper loss is 5% of the input, shunt field resistance is 200 Ω. The lower limit of the current through the armature is to be full load.

    [**Ans.** 0.974, 0.592, 0.359, 0.219, 0.133, 0.08 ohms]

12. The following readings are obtained when doing a load test on a dc motor using a brake drum: Spring balance readings 10 kg and 35 kg; Diameter of the drum 40 cm; Speed of the motor 950 rpm; Applied voltage 200 V; Line current 30 A. Calculate the output power and the efficiency. [**Ans.** 4,880 watts; 81.33%]

13. When running on no load, a 400 V dc shunt motor takes 5 A. Armature resistance is 0.5 Ω and field resistance is 200 Ω. Estimate the power output and efficiency when motor runs on full load and takes 60 A from the line (neglecting contact drop).

    [**Ans.** 20.322 kW; 84.667%]

14. A shunt motor running on no load takes 5 A at 200 V. The resistance of the field is 150 Ω and the armature is 0.1 Ω. Determine the output and efficiency of the motor when the input current is 120 A at 200 V. State the condition if assumed any.

    [M.D. Univ. Electrical Machines-I, December-2012]

    [**Ans.** 21.593 kW; 89.972%]

15. The input to a 220 V dc shunt motor is 11kW. Calculate the efficiency at this load when the following particulars of the motor are given: No-load current = 5A; No-load speed = 1,150 rpm; Armature resistance = 0.6 Ω; Shunt field resistance 110 Ω. [**Ans.** 77.482%]

16. A shunt motor operating on 200 V mains takes 6A on no load. The armature and field resistances are 0.5 Ω and 100 Ω respectively. Calculate the efficiency as (*i*) a generator and (*ii*) motor when the output is 5 kW. [**Ans.** 76.17%; 74.7%]

17. The input to a 200 V dc shunt motor is 11 kW. Calculate (*a*) the torque developed (*b*) the efficiency (*c*) the speed at this load. The particulars of the motor are as follows:

    No-load current = 5A, No-load speed = 1,150 rpm; Armature resistance = 0.5 A; Shunt field resistance = 110 Ω.

    [**Ans.** (*a*) 87.11 N-m (*b*) 79.57% (*c*) 1,031.6 rpm]

18. A 10 kW, 240 V dc shunt motor draws a line current of 5.2 A while running at no-load speed of 1,200 rpm from a 240 V dc supply. It has an armature resistance of 0.25 Ω and a field resistance of 160 Ω. Estimate the efficiency of the motor when it delivers rated load.

    [GATE, 1993]

    [**Ans.** 84.65%]

19. A 10 kW, 250 V dc shunt motor with an armature resistance of 0.8 Ω and a field resistance of 275 Ω takes 3.91 A, when running light at rated voltage and speed. Calculate the efficiency when the machine runs as a generator delivering an output of 10 kW at rated voltage and speed. **[U.P. Technical Univ. Elec. Machines, 2004-05]**

    [**Ans.** 81.24%]

20. When running on no load, a 400-V shunt motor takes 5 A. Armature resistance is 0.5 Ω and field resistance 200 Ω. Find output of the motor and efficiency when running on full-load and taking a current of 50 A. Also find the percentage change in speed from no load to full load. [**Ans.** 16.8525 kW; 84.26%; 5.65%]

21. Hopkinson test on two shunt machines gave the following results for full load:

    Line voltage 250 V, line current excluding field currents 50 A;

    Motor armature current 380 A; Field currents 5 A and 4.2 A.

Calculate the efficiency of each machine. The armature resistance of each machine = 0.02 ohm.

[U.P. Technical Univ. Electromechanical Energy Conversion-I, 2004-05]

[**Ans.** 91.9%; 92.03%]

22. In Hopkinson's test on two identical dc shunt machines, the following readings were obtained:

Line current : 49.4 A, Line voltage : 460 V, Motor armature current : 300 A; Field currents : 5 A and 4.4 A. Armature resistance of each machine : 0.05 Ω. Calculate the efficiency of each machine.

[**Ans.** η_g = 91.47%, η_m = 91.58%]

23. The Hopkinson's Test on two similar shunt machines gave the full-load data : Line voltage = 110 V; Line current = 48 A; Motor armature current = 230 A; Field currents 3 A and 3.5 A; Armature resistance of each machine 0.035 Ω. Calculate the efficiency of each machine assuming a brush contact drop of 1 V per brush.

[Nagpur Univ. Elec. Machines, 1992]

[**Ans.** η_g = 89.76%; η_m = 88.46%]

24. Explain the Hopkinson's test on a pair of shunt motors. In such a test on 250 V machines, the line current was 50 A (not including the field currents of 6 A and 5 A) and the motor current 400 A. The armature resistance of each machine was 0.015 Ω. Calculate the efficiency of each machine. [**Ans.** η_g = 91.04%; η_m = 92.31%]

25. Two similar dc machines, each rated 500 V, 1,000 kW were tested by regenerative method. The test data obtained are as follows:

Output current of generator = 2000 A; Input current from supply mains = 380A; Shunt field current of generator = 22 A; Shunt field current of motor = 17 A; Armature resistance of each machine = 0.01 Ω. Calculate the efficiency of the generator assuming that their iron and mechanical losses are the same. [**Ans.** 91.85%]

26. In a retardation test on a separately excited dc motor, the induced emf in the armature falls from 220 V to 190 V in 30 seconds on disconnecting the armature from the supply. The same fall takes place in 20 seconds if, immediately after, armature is connected to a resistance which takes 10 A (average) during fall. Find stray losses of the machine. [**Ans.** 4,100 watts]

27. In a retardation test on a dc motor, with its field normally excited, the speed fell from 1,525 rpm to 1,475 rpm in 25 seconds. With an average load of 1.0 kW supplied by the armature, the same speed drop occurred in 20 seconds. Find out the moment of inertia of the rotating parts in kg-m^2. [Gujarat Univ. Elec. Machines-III, 1984]

[**Ans.** 121.6 kg-m^2]

28. A test on two coupled similar tramway motors, with their fields connected in series, gave the following results when one machine acted as a motor and the other as a generator.

Motor: Armature current = 56 A; Armature voltage = 590 A; Voltage drop across field winding = 40 V.

Generator: Armature current = 44 A; Armature voltage = 400 V; Field voltage drop = 40 V; Resistance of each armature = 0.3 Ω.

Calculate the efficiency of the motor and generator at this load.

[Nagpur Univ. Machinery-II, 1992]

[**Ans.** η_m = 71.26%; η_g = 64.28%]

❑❑❑

SECTION III

TRANSFORMERS

CHAPTERS COVERED IN THIS SECTION

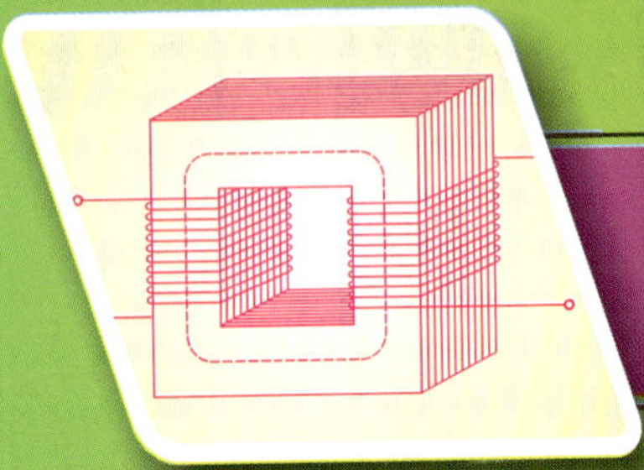

SINGLE-PHASE TRANSFORMERS

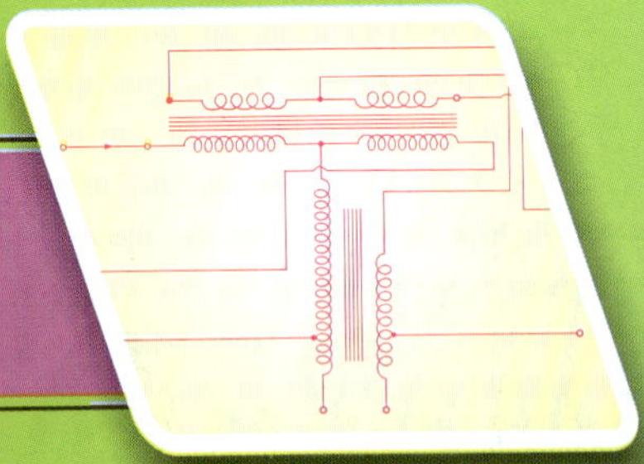

THREE-PHASE TRANSFORMERS

SPECIAL TRANSFORMERS

CHAPTER 12

Single-Phase Transformers

INSIDE THIS CHAPTER

12.1 INTRODUCTION

Electrical energy is generated at places where it is easier to get water head, oil or coal for hydroelectric, diesel or thermal power stations respectively. Then energy is to be transmitted at considerable distances for use in villages, towns and cities located at distant places. As transmission of electrical energy at high voltages is economical, some means are required for stepping up the voltage at generating stations and stepping down the same at the places where it is to be used. Electric machine used for this purpose is "*transformer*". In our country the electrical energy is usually generated at 6.6, 11 or 33 kV, stepped up to 132, 220, 400, or 765 kV with the help of step-up transformers for transmission and then stepped down to 66 kV or 33 kV at grid substations for feeding various substations, which further step down the voltage to 11 kV for feeding distributing transformers stepping down the voltage further to 433/250 volts for the consumer uses.

Transformers are extensively used like the *instrument transformers* for metering purposes. One-to-one ratio transformers (*i.e.* the transformers having equal primary and secondary voltages) are used to electrically isolate the two parts of an electrical circuit. In high-voltage laboratories the transformers are used to provide very high voltages for testing purposes, termed as *testing transformers*. In electric communication circuits, transformers are employed for a variety of purposes e.g., as an impedance transformation device to allow maximum transfer of power from the input circuit to the output device. In radio and television circuits input transformers, interstage transformers and output transformers are widely employed. In power electronics, the transformers find wide use (*i*) in gate pulse triggering and (*ii*) in synchronising the pulse gating signals with the ac supply provided to the main power circuit. The transformers, specifically called the pulse transformers, are widely employed in radar, television and digital computers. Transformers are also employed in telephone circuits, instrumentation circuits and control circuits. Indeed the transformer is a device which plays a vital and essential role in many facets of electrical engineering.

Transformer is an ac machine that (*i*) transfers electrical energy from one electric circuit to another (*ii*) does so without a change of frequency (*iii*) does so by the principle of electromagnetic induction and (*iv*) has electric circuits that are linked by a common magnetic circuit. The energy transfer usually takes place with a change of voltage, although this is not always necessary. When the transformer raises the voltage *i.e.* when the output voltage of a transformer is higher than its input voltage, it is called the *step-up transformer* and when it lowers the voltage it is called the *step-down transformer*.

Since its basic construction requires no moving parts, it is often called the '*static transformer*' and it is very rugged machine requiring the minimum amount of repair and maintenance. Owing to the lack of rotating parts there are no friction or windage losses. Further, the other losses are relatively low, so that the efficiency of a transformer is high. Typical transformer efficiencies at full load lie between 96% and 97% and with extremely large capacity transformers the efficiencies are as high as 99 %. The cost per kVA output of transformers is quite low as compared with other electrical machines. As there are no teeth, slots or rotating parts, and the windings can be immersed in oil, it is not difficult to insulate transformers for very high voltages.

12.2 OPERATING PRINCIPLE

An elementary transformer consists of a soft iron or silicon steel core and two windings placed on it. The windings are insulated from both the core and each other. The core is built up of thin soft iron or silicon steel laminations to provide a path of low reluctance to the magnetic flux. The winding connected to the supply main is called the *primary* and the winding connected to the load circuit is called the *secondary*. The winding connected to higher voltage circuit is called the *high-voltage (hv) winding* while that connected to the lower voltage circuit is called the *low-voltage (lv) winding*. In case of a step-up transformer, low-voltage winding is the primary and high-voltage winding is the secondary while in case of a step-down transformer the high-voltage winding is the primary and low-voltage winding is the secondary.

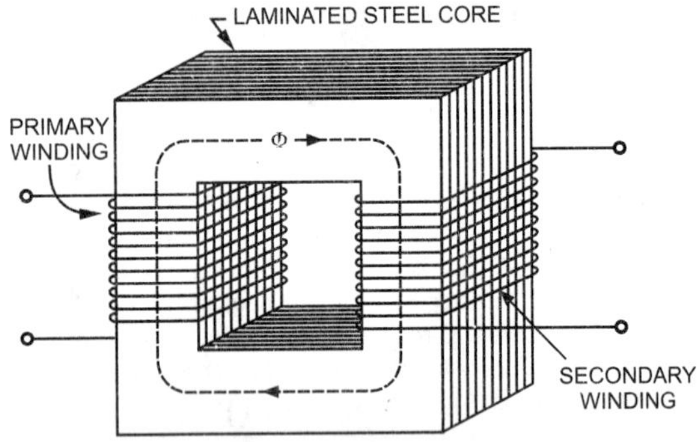

Fig. 12.1 *Simple Transformer*

The action of a transformer is based on the principle that energy may be efficiently transferred by induction from one set of coils to another by means of a varying magnetic flux, provided that both the sets of coils are on a common magnetic circuit. In a transformer, the coils and magnetic circuit are all stationary with respect to one another. The emfs are induced by the variation in the magnitude of flux with time, as illustrated in Fig. 12.3.

Although in the actual construction the two windings are usually wound one over the other, for the sake of simplicity, the figures for analysing transformer theory show the windings on opposite sides of the core as in Fig. 12.1.

When the primary winding is connected to an ac supply mains, a current flows through it. Since this winding links with an iron core, current flowing through this winding produces an alternating flux Φ in the core. Since this flux is alternating and links with the secondary winding also, induces an emf in the secondary winding. The frequency of induced emf in secondary winding is the same as that of the flux or that of the supply voltage. The induced emf in the secondary winding enables it to deliver current to an external load connected across it. Thus the energy is transformed from primary winding to the secondary winding by means of electromagnetic induction without any change in frequency. The flux Φ of the iron core links not only with the secondary winding but also with the primary winding, so produces self-induced emf in the primary winding. This induced emf in the primary winding opposes the applied voltage and, therefore, sometimes it is known as *back emf* of the primary. In fact the induced emf in the primary winding limits the primary current in much the same way as the back emf in a dc motor limits the armature current.

12.3 TRANSFORMER ON DC

A transformer cannot operate on dc supply and never be connected to a dc source. If a rated dc voltage is applied to the primary of a transformer, the flux produced in the transformer core will not vary but remain constant in magnitude and, therefore, no emf will be induced in the secondary winding except at the moment of switching *on*. Thus the transformer is not capable of raising or lowering the dc voltage. Also there will be no self-induced emf in the primary winding, which is only possible with varying flux linkage, to oppose the applied voltage and since the resistance of primary winding is quite low, a heavy current will flow through the primary winding which may result in the burning out of the primary winding. This is reason that *dc is never applied to a transformer*.

12.4 IDEAL TRANSFORMER

For a better understanding and an easier explanation of a practical transformer, certain idealizing assumptions are made which are close approximations for a practical transformer. A transformer having these ideal properties is hypothetical (has no real existence) and is referred to as the *ideal transformer*. It possesses certain essential features of a real transformer but some details of minor significance are ignored which will be introduced step by step while analysing a transformer. The idealizing assumptions made are as follows:

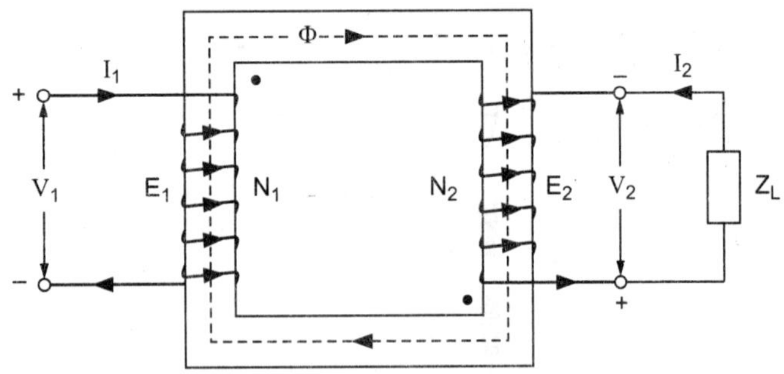

Fig. 12.2 *Ideal Iron-Core Transformer*

(*i*) *No winding resistance i.e.* the primary and secondary windings have zero resistance. It means that there is no ohmic power loss and no resistive voltage drop in an ideal transformer.

(*ii*) *No magnetic leakage i.e.* there is no leakage flux and all the flux set up is confined to the core and links both the windings.

(*iii*) *No iron loss i.e.* hysteresis and eddy current losses in transformer core are zero.

(*iv*) *Zero magnetizing current i.e.* the core has infinite permeability and zero reluctance so that zero magnetizing current is required for establishing the requisite amount of flux in the core.

From the above discussion an ideal transformer is supposed to consists of two pure inductive coils wound on a loss-free core [Fig. 12.2].

12.5 EMF EQUATION

When an alternating (sinusoidal) voltage is applied to the primary winding of a transformer, an alternating (sinusoidal) flux, as shown is Fig. 12.3., is set up in the iron core which links both the windings (primary and secondary windings).

Let Φ_{max} = Maximum value of flux in webers

and f = Supply frequency in hertz.

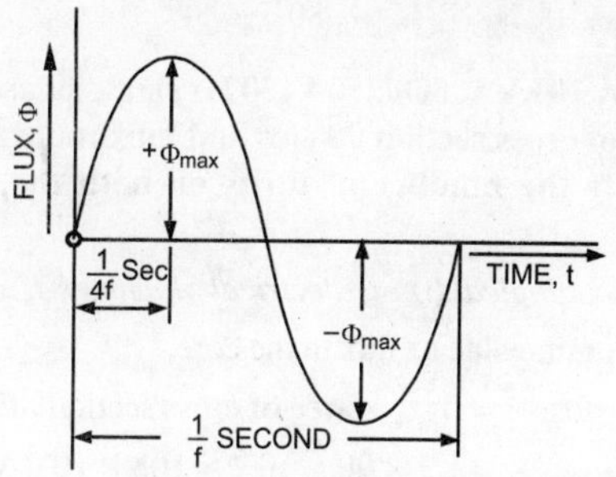

Fig. 12.3 *Sinusoidal Variation of Flux With Time*

As illustrated in Fig. 12.3, the magnetic flux increases from zero to its maximum value Φ_{max} in one-fourth of a cycle *i.e.* in $\frac{1}{4f}$ second.

So average rate of change of flux, $\frac{d\Phi}{dt} = \frac{\Phi_{max}}{1/4f} = 4f\Phi_{max}$

Since average emf induced per turn in volts is equal to the average rate of change of flux,

average emf induced per turn $= 4f\Phi_{max}$ volts

Since flux Φ varies sinusoidally, emf induced will be sinusoidal and form factor for sinusoidal wave is 1.11 *i.e.* the rms or effective value is 1.11 times the average value.

$\therefore$ RMS value of emf induced per turn $= 1.11 \times 4f\Phi_{max}$ volts ...(12.1)

If the number of turns on primary and secondary windings are N_1 and N_2 respectively, then

RMS value of emf induced in primary,

E_1 = EMF induced per turn × number of primary turns

$= 4.44 f\Phi_{max} \times N_1 = 4.44 f N_1 \Phi_{max}$ volts ...(12.2)

Similarly rms value of emf induced in secondary,

$E_2 = 4.44 f\Phi_{max} \times N_2$ volts ...(12.3)

The above relations for emf induced in primary and secondary windings can be derived alternatively as below

The instantaneous value of sinusoidally varying flux may be given as

$\Phi = \Phi_{max} \sin \omega t$

$\therefore$ Instantaneous value of emf induced per turn

$$= \frac{-d\Phi}{dt} \text{ volts} = -\omega\Phi_{max}\cos\omega t = \omega\Phi_{max}\sin\left(\omega t - \frac{\pi}{2}\right) \text{ volts}$$

It is clear from the above equation that the maximum value of emf induced per turn

$= \omega\Phi_{max} = 2\pi f\Phi_{max}$ volts $\quad \because \omega = 2\pi f$

and rms value of emf induced per turn $= \frac{1}{\sqrt{2}} \times 2\pi f\Phi_{max}$

$= 4.44 f\Phi_{max}$ volts

Hence rms value of emf induced in primary,

$E_1 = 4.44 f N_1 \Phi_{max}$ volts

and rms value of emf induced in secondary,

$E_2 = 4.44 f N_2 \Phi_{max}$ volts

In an ideal transformer the voltage drops in primary and secondary windings are negligible, so

EMF induced in primary winding, E = Applied voltage to primary, V_1

and terminal voltage, V_2 = EMF induced in secondary, E_2

Note: If B_{max} is the maximum allowable flux density in Wb/m^2 (or T) and *a* is the area of x-section of iron core in square metres, then in Eqs. (12.1), (12.2) and (12.3), Φ_{max} is given as

$\Phi_{max} = B_{max}\, a$ webers

12.6 VOLTAGE AND CURRENT TRANSFORMATION RATIOS

Referring to Eq. (12.1), it is clear that the *volts per turn* is exactly the same for both the primary and secondary windings *i.e.* in any transformer, the secondary and primary induced emfs are related to each other by the ratio of the number of secondary and primary turns. Thus,

$$\frac{V_2}{V_1} = \frac{E_2}{E_1} = \frac{N_2}{N_1} = K \quad ...(12.4)$$

The same relationship can be derived by dividing Eq. (12.3) by Eq. (12.2).

The constant K in Eq. (12.4) is called the *voltage transformation ratio.*

For step-up transformer, $V_2 > V_1$ or voltage transformation ratio, $K > 1$.

For step-down transformer, $V_2 < V_1$ or voltage transformation ratio, $K < 1$.

In an ideal transformer, the losses are negligible, so the volt-ampere input to the primary and volt-ampere output from secondary can be approximately equated *i.e.*

Output VA = Input VA

or $V_2I_2 = V_1I_1$

$$\text{or } \frac{I_2}{I_1} = \frac{V_1}{V_2} = \frac{E_1}{E_2} = \frac{N_1}{N_2} = \frac{1}{K} \quad ...(12.5)$$

i.e. Primary and secondary currents are inversely proportional to their respective turns.

Note: For a transformer the true ratio of transformation (or turn ratio) is constant, while the voltage ratio $\left(\frac{V_2}{V_1}\right)$ varies about 1 to 8 per cent depending upon the load and its power factor.

Example 12.1. It is desired to have a 4.13 mWb maximum core flux in a transformer at 110 V and 50 Hz. Determine the required number of turns in the primary.

[*Pb. Technical Univ. Electrical Engineering, December-2000*]

Solution: EMF induced in primary, $E_1 = 110$ V

Supply frequency, $f = 50$ Hz

Maximum core flux, $\Phi_{max} = 4.13$ mWb $= 4.13 \times 10^{-3}$ Wb

Required number of turns on primary,

$$N_1 = \frac{E_1}{4.44 f\Phi_{max}} \quad \text{...Refer to Eq. (12.2)}$$

$$= \frac{110}{4.44 \times 50 \times 4.13 \times 10^{-3}} = 120 \textbf{ Ans.}$$

Example 12.2. The emf per turn of a single-phase 10 kVA, 2200/220 V, 50 Hz transformer is 10 V. Calculate (i) the number of primary and secondary turns, (ii) the net cross-sectional area of core for a maximum flux density of 1.5 T.

Solution: EMF per turn = 10 V

Primary induced emf, $E_1 = V_1 = 2{,}200$ V

Secondary induced emf, $E_2 = V_2 = 220$ V

Supply frequency, $f = 50$ Hz

Maximum flux density, $B_{max} = 1.5$ T

(*i*) Number of primary turns, $N_1 = \dfrac{E_1}{\text{EMF per turn}}$

$$= \frac{2{,}200}{10} = 220 \textbf{ Ans.}$$

Number of secondary turns, $N_2 = \dfrac{E_2}{\text{EMF per turn}}$

$$= \frac{220}{10} = 22 \textbf{ Ans.}$$

Maximum value of flux, $\Phi_{max} = \dfrac{\text{EMF per turn}}{4.44\, f}$

$$= \frac{10}{4.44 \times 50} = 0.045 \text{ Wb}$$

...refer to Eq. (12.1)

(*ii*) Net cross-sectional area of core, $a = \dfrac{\Phi_{max}}{B_{max}}$

$$= \frac{0.045}{1.5} = 0.03 \text{ m}^2 \textbf{ Ans.}$$

Example 12.3. A 3,000/200 V, 50 Hz single-phase transformer is built on a core having an effective cross-sectional area of 150 cm^2 and has 80 turns in the low voltage winding. Calculate the maximum flux density in the core.

[*M.D. Univ. Electromechanical Energy Conversion, May-2004*]

Solution: Induced emf in the low-voltage winding $= E_2 = V_2 = 200$ V

Supply frequency, $f = 50$ Hz

Number of turns on *lv* winding $= N_2 = 80$

Maximum voltage of flux, $\Phi_{max} = \dfrac{E_2}{4.44\, f N_2} = \dfrac{200}{4.44 \times 50 \times 80}$

$$= 0.01126 \text{ Wb}$$

Maximum flux density in the core,

$$B_{max} = \frac{\Phi_{max}}{a} = \frac{0.01126}{150 \times 10^{-4}} = 0.75 \text{ T } \textbf{Ans.}$$

Example 12.4. A single-phase, 50 Hz transformer has 80 turns on the primary winding and 280 in the secondary winding. The voltage applied across the primary winding is 240 volts. Calculate :

(*i*) The maximum flux density in the core

(*ii*) Induced emf in the secondary winding.

The net cross-sectional area of the core can be taken 200 cm^2.

[*R.G.T.U. Basic Elec. Engineering Jan./Feb.-2008*]

Solution: Number of primary turns, $N_1 = 80$

Number of secondary turns, $N_2 = 280$

Supply frequency, $f = 50$ Hz

Primary induced emf, $E_1 = V_1 = 240$ V

Area of x-section of core, $a = 200 \text{ cm}^2 = 0.02 \text{ m}^2$

Maximum value of flux, $\Phi_{max} = \dfrac{E_1}{4.44\, f N_1}$

$$= \frac{240}{4.44 \times 50 \times 80} = 0.0135 \text{ Wb}$$

(*i*) The maximum flux density in the core,

$$B_{max} = \frac{\Phi_{max}}{a} = \frac{0.0135}{0.02} = 0.676 \text{ Wb/m}^2 \textbf{ Ans.}$$

(*ii*) Induced emf in secondary winding,

$$E_2 = E_1 \times \frac{N_2}{N_1} = 240 \times \frac{280}{80} = 840 \text{ V } \textbf{Ans.}$$

Example 12.5. A 10 kVA, 500/250 V, 50 Hz single-phase transformer has a net area of cross section 90 cm^2 and maximum flux density is 1.2 T. Calculate the number of turns on both the primary and secondary.

[*J.N. Technological Univ. Electrical Machines-II, January-2014*]

Solution: Maximum value of flux in the core,

$$\Phi_{max} = B_{max} \times \text{area of cross section of core}$$
$$= 1.2 \times 90 \times 10^{-4} = 108 \times 10^{-4} \text{ Wb}$$

Number of turns required on secondary,

$$N_2 = \frac{E_2}{4.44\, f\, \Phi_{max}}$$

$$= \frac{250}{4.44 \times 50 \times 108 \times 10^{-4}} = 106 \textbf{ Ans.}$$

Number of turns required on primary,

$$N_1 = N_2 \times \frac{E_1}{E_2} = 106 \times \frac{500}{250} = 212 \textbf{ Ans.}$$

Example 12.6. A 25 kVA loss-less transformer has 500 turns on the primary and 40 turns on the secondary winding. The primary is connected to 3,000 V, 50 Hz mains determine

(*i*) primary and secondary currents in at full load;

(*ii*) the secondary emf and

(*iii*) the maximum flux in the core. No-load current can be neglected. [*Pb. Technical Univ., May-2012*]

Solution: Primary induced emf, $E_1 = V_1 = 3{,}000$ V

Primary turns, $N_1 = 500$

Secondary turns, $N_2 = 40$

(*ii*) Secondary emf, $E_2 = E_1 \times \dfrac{N_2}{N_1} = 3{,}000 \times \dfrac{40}{500} = 240$ V

(*i*) Primary current, $I_1 = \dfrac{\text{Rated kVA} \times 1{,}000}{E_1}$

$$= \frac{\text{Rated kVA} \times 1{,}000}{V_1}$$

$$= \frac{25 \times 1{,}000}{3{,}000} = 8.333 \text{ A } \textbf{Ans.}$$

Secondary current, $I_2 = \dfrac{25 \times 1{,}000}{240} = 104.167$ A **Ans.**

(*iii*) Maximum flux in the core,

$$\Phi_{max} = \frac{E_1}{4.44 f N_1} = \frac{3{,}000}{4.44 \times 50 \times 500}$$

$$= 0.027 \text{ Wb or } 27 \text{ mWb } \textbf{Ans.}$$

Example 12.7. A 200 kVA, 3300/240 V, 50 Hz single-phase transformer has 80 turns on the secondary winding. Assuming an ideal transformer, calculate (*i*) Primary and secondary currents at full load. (*ii*) The maximum value of flux (*iii*) Number of primary turns. [*J.N. Technological Univ. Electrical Machines-II, January-2014*]

Solution: (*i*) Primary current on full-load

$$I_1 = \frac{\text{Rated kVA} \times 1{,}000}{V_1} = \frac{200 \times 1{,}000}{3{,}300} = 60.6 \text{ A } \textbf{Ans.}$$

Secondary current on full load,

$$I_2 = \frac{\text{Rated kVA} \times 1{,}000}{V_2 \text{ or } (E_2)} = \frac{200 \times 1{,}000}{240} = 833.33 \text{ A } \textbf{Ans.}$$

(*ii*) Maximum value of flux,

$$\Phi_{max} = \frac{E_2}{4.44\, f\, N_2} = \frac{240}{4.44 \times 50 \times 80}$$

$$= 0.0135 \text{ Wb or } 13.5 \text{ mWb } \textbf{Ans.}$$

(*iii*) Number of primary turns,

$$N_1 = \frac{E_1}{E_2} \times N_2 = \frac{3{,}300}{240} \times 80 = 1{,}100 \textbf{ Ans.}$$

Example 12.8. A single-phase, 50 Hz, core type transformer has square cores of 20 cm side, permissible maximum flux density is 1 Wb/m². Calculate the number of turns per limb on the high- and low-voltage sides for a 3,000/220 V ratio.

Solution: Core area = 20 × 20 = 400 cm² = 0.04 m²

Permissible flux density, $B_{max} = 1.0 \text{ Wb/m}^2$

Maximum value of permissible flux,

$$\Phi_{max} = B_{max} \times a = 1 \times 0.04 = 0.04 \text{ Wb}$$

Number of turns on low-voltage (secondary) winding,

$$N_2 = \frac{E_2}{4.44\, f\, \Phi_{max}} = \frac{220}{4.44 \times 50 \times 0.04} = 24.77;\ 26 \text{ (say)}$$

Number of turns on high-voltage (primary) winding,

$$N_1 = \frac{E_1}{E_2} \times N_2 = \frac{3{,}000}{220} \times 26 = 354$$

$$\text{Number of } hv \text{ turns on each limb} = \frac{354}{2} = 177 \quad \textbf{Ans.}$$

$$\text{Number of } lv \text{ turns on each limb} = \frac{26}{2} = 13 \quad \textbf{Ans.}$$

Note: LV turns are calculated first, rounded one figure off to the next higher even number in order that maximum flux density does not exceed the permissible maximum flux density. Then *hv* turns are calculated with corrected number of turns by transformation ratio. Further there are two limbs and each limb accommodates half of the *lv* and half of the *hv* turns from the view point of reducing leakage reactance.

Example 12.9. A single-phase, 50 Hz core type transformer has square cores of 20 cm side. The permissible flux density in the core is 1.0 Wb/m². Calculate the number of turns per limb on the high and low voltage sides for a 3,000/220 V ratio. To allow for insulation of stampings, assume the net iron length to be 0.9 × gross iron length.

Solution: Iron length on one side of the core

= 0.9 × 20 = 18 cm = 0.18 m

Iron length on the other side of the core

= 20 cm or 0.20 m

because only one side of the core is affected by insulation and other side remains unaffected (Fig. 12.4).

Core area, a = 0.18 × 0.20 = 0.036 m²

Permissible flux density, $B_{max} = 1.0 \text{ Wb/m}^2$

Maximum value of flux, $\Phi_{max} = B_{max} \times a$

$= 1.0 \times 0.036 = 0.036$ Wb

Number of turns on *lv* (secondary) winding,

$$N_2 = \frac{E_2}{4.44\, f \Phi_{max}} = \frac{220}{4.44 \times 50 \times 0.036} = 27.53 \simeq 28$$

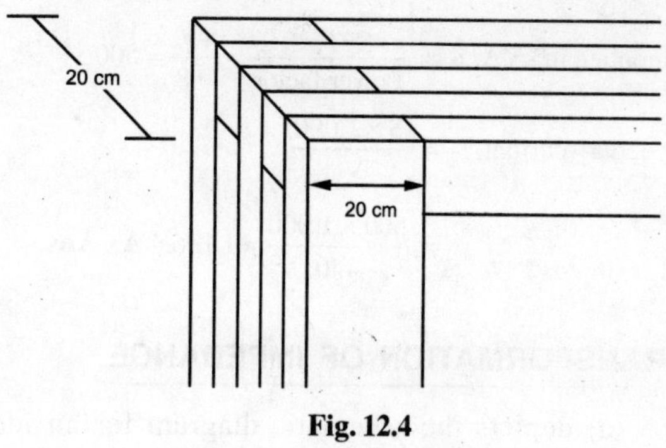

Fig. 12.4

Number of turns on *hv* (primary) winding,

$$N_1 = \frac{E_1}{E_2} \times N_2 = \frac{3{,}000}{220} \times 28 = 382$$

$$\text{Number of } hv \text{ turns on each limb} = \frac{382}{2} = 191 \quad \textbf{Ans.}$$

$$\text{Number of } lv \text{ turns on each limb} = \frac{28}{2} = 14 \quad \textbf{Ans.}$$

Example 12.10. A single-phase transformer has a core whose cross-sectional area is 150 cm²; operates at a maximum flux density of 1.1 Wb/m² from a 50 Hz supply. If the secondary winding has 66 turns, determine the output in kVA when connected to a load of 4 Ω impedance. Neglect any voltage drop in the transformer.

Solution: Peak value of flux density, $B_{max} = 1.1 \text{ Wb/m}^2$

Cross-sectional area of core, a = 150 cm² = 0.015 m²

Maximum value of flux in the core,

$$\Phi_{max} = B_{max} \times a = 1.1 \times 0.015 = 0.0165 \text{ Wb}$$

Number of turns on secondary,

$N_2 = 66$

Supply frequency, f = 50 Hz

RMS value of induced emf in the secondary winding,

$$E_2 = 4.44\, f\, \Phi_{max}\, N_2 = 4.44 \times 50 \times 0.0165 \times 66$$

$$= 240 \text{ volts}$$

Secondary voltage, $V_2 = E_2 = 240$ V

neglecting voltage drop in secondary winding

$$\text{Secondary current, } I_2 = \frac{V_2}{\text{Load impedance}} = \frac{240}{4} = 60 \text{ A}$$

$$\text{Output in kVA} = \frac{V_2 I_2}{1{,}000} = \frac{240 \times 60}{1{,}000} = 14.4 \textbf{ Ans.}$$

Example 12.11. A 6,600/440 V, single-phase 600 kVA transformer has 1,200 primary turns. Find (*i*) Transformation ratio, (*ii*) Secondary turns, (*iii*) Voltage per turn, (*iv*) Secondary current when it supplies a load of 400 kW at 0.8 power factor lagging.

[*R.G.T.U. Basic Elec. Engineering Jan./Feb.-2007*]

Solution: Primary voltage, V_1 = 6,600 V

Secondary voltage, V_2 = 440 V

Primary turns, N_1 = 1,200

Load supplied, P = 400 kW

Power factor of load, cos ϕ = 0.8 (lagging)

$$(i) \text{ Transformation ratio, } K = \frac{E_2}{E_1} = \frac{V_2}{V_1} = \frac{440}{6{,}600} = \frac{1}{15} \text{ Ans.}$$

$$(ii) \text{ Secondary turns, } N_2 = kN_1 = \frac{1}{15} \times 1{,}200 = 80 \text{ Ans.}$$

$$(iii) \text{ Voltage per turn} = \frac{E_1}{N_1} = \frac{6{,}600}{1{,}200} = 5.5 \text{ volts } \textbf{Ans.}$$

(*iv*) Load supplied in kVA, $S = \frac{P}{\text{Power factor}} = \frac{400}{0.8} = 500$

Secondary current, $I_2 = \frac{S \times 1,000}{V_2}$

$= \frac{500 \times 1,000}{440} = 1,136.4$ A **Ans.**

12.7 TRANSFORMATION OF IMPEDANCE

Figure 12.5 (a) depicts the schematic diagram for an ideal transformer with dot points marked on the primary and secondary windings to indicate terminals of the corresponding polarity at any instant. A load impedance Z_2 is connected across the secondary winding.

For the secondary circuit,

$$Z_2 = \frac{V_2}{I_2}$$

For an ideal transformer,

$$\frac{V_2}{V_1} = \frac{N_2}{N_1} = \frac{I_1}{I_2} \quad \text{...Refer to Eq. (12.5)}$$

$$\frac{V_1}{I_1} = \frac{N_1}{N_2} V_2 \times \frac{N_1}{N_2 I_2} \quad \because \text{From above equation}$$

$$V_1 = \frac{N_1}{N_2} V_2 \text{ and } I_1 = \frac{N_2 I_2}{N_1}$$

$$= \left(\frac{N_1}{N_2}\right)^2 Z_2 = a^2 Z_2 = Z'_2 \quad \text{...(12.6)}$$

Equation (12.6) reveals that the impedance on the secondary side reflected on primary side is transformed in the direct ratio of the square of turns i.e. $Z'_2 = \left(\frac{N_1}{N_2}\right)^2 Z_2$. This is shown in Fig. 12.5 (b) and (c).

Thus, the three circuits depicted in Fig. 12.5 are identical as far as their performance viewed from the input terminals *a* and *b* concerned.

Similarly, an impedance Z_1 from the primary side can be referred to the secondary side as

$$Z'_1 = \left(\frac{N_2}{N_1}\right)^2 Z_1 = \frac{1}{a^2} Z_1 \quad \text{...(12.7)}$$

Transferring of an impedance from one side of a transformer to the other side is termed as *referring the impedance* to the other side. Similarly, voltages and currents can be referred to either side of transformer by means of Eq. (12.5).

It may be concluded that for an ideal transformer (*i*) voltages are transformed in the direct turns ratio, (*ii*) currents in the reverse ratio (*ii*) impedances in the direct ratio squared while power and volt-amperes remain unaltered.

Equation (12.6) illustrates the impedance i.e. modifying property of a transformer which may be exploited to match a fixed load impedance to the source impedance for the purpose of transfer of maximum power from source to load. This is achieved by interposing a transformer of suitable turns ratio between the load and source.

Example 12.12. A 5 kVA, 240/120 V, single-phase transformer supplies rated current to a load at 120 V. Determine the magnitude of load impedance as seen from input terminals of transformer. The load impedance is (5 + *j*15) Ω.

[Pb. Technical Univ., December-2002]

Solution: Transformation ratio, $K = \frac{V_2}{V_1} = \frac{120}{240} = 1/2$

Load impedance referred to secondary, $Z_L = (5 + j15)\ \Omega$

Load impedance referred to primary,

$$Z'_L = \frac{Z_L}{K^2} = \frac{5 + j15}{(½)^2} = (20 + j\,60)\ \Omega \textbf{ Ans.}$$

Example 12.13. An audio-frequency ideal transformer is employed to couple a 60 Ω resistance load to an electric source which is represented by a constant voltage of 6 V in series with an internal resistance of 2,400 Ω.

(i) Determine the turn-ratio required to ensure maximum power transfer by matching the load and source impedance.

(ii) Find the load current, voltage and power under the conditions of maximum power transfer.

[U.P. Technical Univ. Electromechanical Energy Conversion-I, 2004-05]

Solution: For maximum power transfer, it is often required to match the soure impedance Z_0 with load impedance Z_L connected to the secondary side, for example in audio circuits of a TV or radio receiver. Since an impedance Z_L connected to the secondary winding is reflected on the primary side as $Z' = \frac{Z_L}{K^2}$ (where K is the transformation ratio), transformer can be employed for impedance transformation. Suitable value for K, the transformation ratio is used to transform Z_L to Z'_L such that $Z'_L = Z_0$

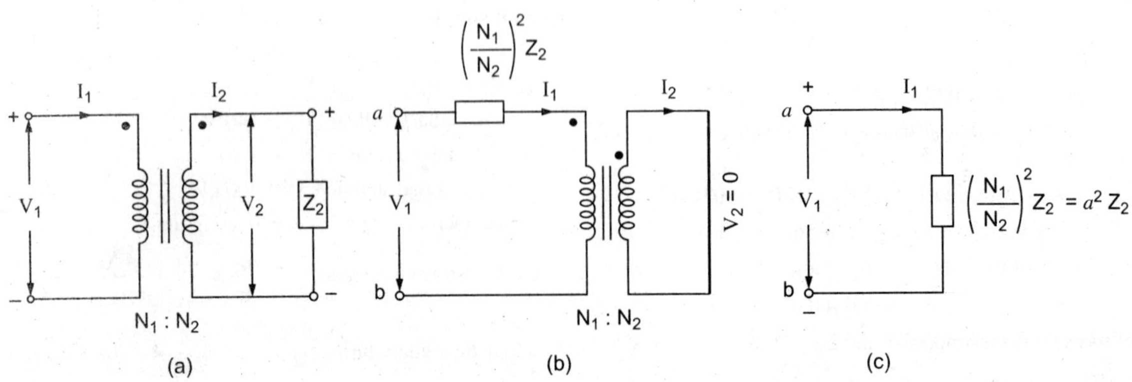

Fig. 12.5 *Impedance Transformation in an Ideal Transformer*

$$Z'_L = \frac{60}{K^2} = 2{,}400$$

$$\text{or } K = \sqrt{\frac{60}{2{,}400}}$$

$$= \frac{1}{\sqrt{40}} = \frac{1}{6.325} \quad \textbf{Ans.}$$

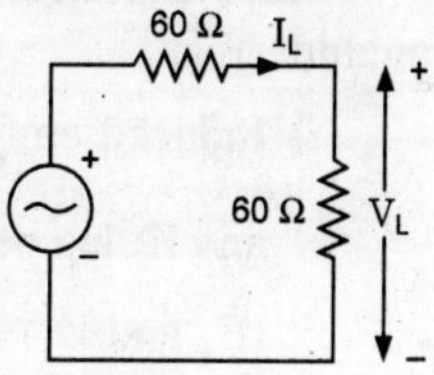

Fig. 12.6

The equivalent circuit, referred to load side is shown in Fig. 12.6.

The source voltage on the load side

$$= \frac{1}{6.325} \times 6 = 0.9487 \text{ V and the source resistance is } 60\ \Omega.$$

$$\text{Load current, } I_L = \frac{0.9487}{60 + 60} = 7.91 \text{ mA} \quad \textbf{Ans.}$$

$$\text{Load voltage, } V_L = I_L \times R_L = 7.91 \times 10^{-3} \times 60 = 0.474 \text{ V} \quad \textbf{Ans.}$$

$$\text{Load power} = I_L^2 R_L = (7.91 \times 10^{-3})^2 \times 60 = 3.754 \text{ mW} \quad \textbf{Ans.}$$

Example 12.14. A 200/50 V, 50 Hz single-phase transformer is connected to a 200 V, 50 Hz supply with secondary winding open. Primary winding has 400 turns.

(*i*) What is the value of maximum flux through the core, if the primary winding has 400 turns?

(*ii*) What is the peak value of flux if the primary voltage is 200V, 25 Hz?

(*iii*) What happens to no-load current ?

[*R.G.P.V. Basic Electrical Engineering, December-2002*]

Solution: (*i*) Maximum value of flux,

$$\Phi_{max} = \frac{E_1}{4.44 f N_1} = \frac{V}{4.44 f N_1}$$

$$= \frac{200}{4.44 \times 50 \times 400}$$

$$= 2.25 \times 10^{-3} \text{ Wb or } 2.25 \text{ m Wb} \quad \textbf{Ans.}$$

(*ii*) When primary voltage is 200 V and supply frequency is 25 Hz

$$\Phi'_{max} = \frac{200}{4.44 \times 25 \times 400} = 4.5 \text{ m Wb} \quad \textbf{Ans.}$$

(*iii*) The no-load primary current I_0, called the exciting current, is very small in comparison to the full-load primary current (2–5% of full-load primary current). This current is made up of a relatively larger quadrature or magnetising component I_m and a comparatively small in-phase or energy component I_e.

When supply frequency is reduced, keeping supply voltage constant, maximum value of flux is increased in inverse ratio of supply frequencies, and therefore, the magnetising component of no-load current and thus the no-load current.

12.8 EFFECTS OF VOLTAGE AND FREQUENCY VARIATIONS

Power transformers are not ordinarily subjected to frequency variations and usually are subject to only modest voltage variations, but it is interesting to consider the effects thereof.

Variations in voltage and/or frequency affects the iron losses (hysteresis and eddy current losses) in a transformer. As long as the flux variations are sinusoidal with respect to line, hysteresis loss (P_h), and eddy current loss (P_e) varies according to the following relations

$$P_h \propto f (\Phi_{max})^x$$

where x lies between 1.5 and 2.5 depending on the grade of iron used in transformer core

$$\text{and } P_e \propto f^2 (\Phi_{max})^2.$$

If the transformer is operated with the frequency and voltage changed in the same proportion, the flux density will remain unchanged as obvious from Eq. (12.2) and apparently the no-load current will also remain unaffected.

The transformer can be operated safely at frequency less than rated one with correspondingly reduced voltage. In this case iron losses will be reduced. But if the transformer is operated with increased voltage and frequency in the same proportion, the core losses may increase to an intolerable level. Increase in frequency with constant supply voltage will cause reduction in hysteresis loss and leave the eddy current losses unaffected. Some increase in voltage could, therefore, be tolerated at higher frequencies, but exactly how much depends on the relative magnitude of the hysteresis and eddy current losses and the grade of iron used in the transformer core.

Example 12.15. A 1 kVA, 220/110 V, 400 Hz transformer is desired to be used at a frequency of 60 Hz. What will be the kVA rating of the transformer at reduced frequency?

Solution: From Eq. (12.2)

$$V_1 = 4.44\, \Phi_{max} N_1 f = 4.44\, B_{max}\, a\, N_1 f$$

Assuming flux density in the core remaining unchanged, we have

$$V_1 \propto f$$

$$\text{or } \frac{V'_1}{V_1} = \frac{f'}{f}$$

$$\text{or } V'_1 = V_1 \times \frac{f'}{f} = 220 \times \frac{60}{400} = 33 \text{ volts}$$

KVA rating, being proportional to the voltage, will change to

$$1 \times \frac{33}{220} \text{ or } 0.15 \text{ kVA or } 150 \text{ VA } \textbf{Ans.}$$

Example 12.16. The core of a 100 kVA, 11000/550 V, 50 Hz, 1-phase, core type transformer has a cross section of 20 cm × 20 cm. Find (*i*) the number of hv and lv turns and (*ii*) the emf per turn if the maximum core density is not to exceed 1.3 T. Assume a stacking factor of 0.9.

What will happen if the primary voltage is increased by 10 % on no load?

Solution: Gross cross-sectional area of core,

$$a_g = 0.20 \times 0.20 = 0.04 \text{ m}^2$$

Net cross-sectional area of core,

$$a = a_g \times \text{stacking factor} = 0.04 \times 0.9 = 0.036 \text{ m}^2$$

Maximum value of flux in the core,

$$\Phi_{max} = B_{max} \times a = 1.3 \times 0.036 = 0.0468 \text{ Wb}$$

$$(ii) \text{ EMF per turn} = 4.44 f \Phi_{max} = 4.44 \times 50 \times 0.0468 = 10.39 \text{ V } \textbf{Ans.}$$

$$(i) \text{ Number of } lv \text{ turns, } N_2 = \frac{E_2}{\text{EMF per turn}} = \frac{550}{10.39} = 54 \textbf{ Ans.}$$

Selecting next higher even integer

$$\text{Number of } hv \text{ turns, } N_1 = \frac{E_1}{E_2} \times N_2 = \frac{11{,}000}{550} \times 54 = 1{,}080 \quad \textbf{Ans.}$$

When the primary voltage of the transformer is increased by 10%, assuming supply frequency remaining the same, flux density will increase by 10%, iron losses will increase by much more than 10% and due to saturation effect the magnetising current will also increase by much more than 10%.

12.9 TRANSFORMER ON NO LOAD

When the primary of a transformer is connected to the source of ac supply and the secondary is open, the transformer is said to be at no load (there is no load on secondary).

Consider an ideal transformer whose secondary side is open and the primary winding is connected to a sinusoidal alternating voltage V_1. The alternating voltage applied to the primary winding will cause flow of alternating current in the primary winding. Since the primary coil is pure inductive and there is no output (secondary being open), the primary draws the magnetising current I_m only. The function of this current is merely to magnetize the core. If the transformer is truely ideal, the magnitude of I_m should be zero by virtue of assumption (*iv*) made in Art 12.4. Since the reluctance of the magnetic circuit is never zero, I_m has definite magnitude. The magnetising current, I_m is small in magnitude and lags behind supply voltage V_1 by 90 °. This magnetising current I_m produces an alternating flux Φ which is, at all times, proportional to the current (assuming permeability of the magnetic circuit to be constant) and, hence, in phase with it.

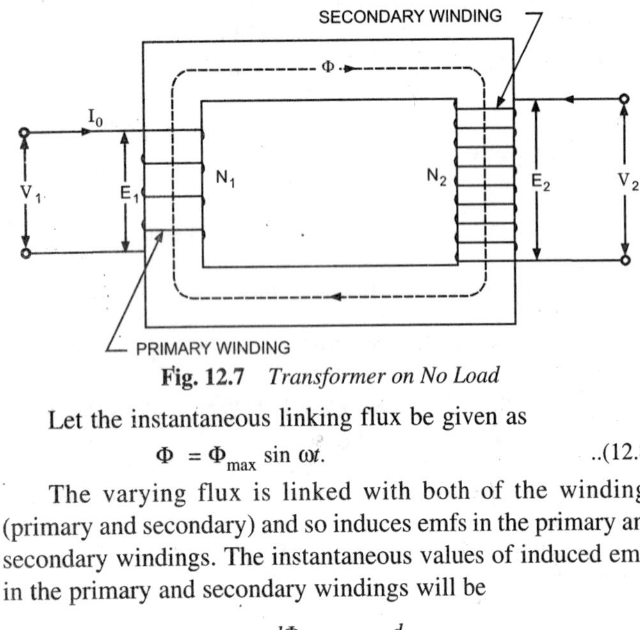

Fig. 12.7 *Transformer on No Load*

Let the instantaneous linking flux be given as

$$\Phi = \Phi_{max} \sin \omega t. \qquad ..(12.8)$$

The varying flux is linked with both of the windings (primary and secondary) and so induces emfs in the primary and secondary windings. The instantaneous values of induced emfs in the primary and secondary windings will be

$$e_1 = -N_1 \frac{d\Phi}{dt} = -N_1 \frac{d}{dt}(\Phi_{max} \sin \omega t)$$

$$= -N_1 \omega \Phi_{max} \cos \omega t$$

$$= N_1 \omega \Phi_{max} \sin\left(\omega t - \frac{\pi}{2}\right) \qquad ...(12.9)$$

$$\text{Similarly, } e_2 = N_2 \omega \Phi_{max} \sin\left(\omega t - \frac{\pi}{2}\right) \qquad ...(12.10)$$

Since primary winding has no ohmic resistance, (as assumed), applied voltage to primary winding is to only oppose the induced emf in the primary winding, hence instantaneous applied voltage to primary will be given by

$$v_1 = -e_1 = -N_1 \omega \Phi_{max} \sin\left(\omega t - \frac{\pi}{2}\right) \qquad ...(12.11)$$

Comparing Eqs. (12.8), (12.9), (12.10), and (12.11) we conclude that

(*i*) Induced emfs in primary and secondary windings, E_1 and E_2 lag behind the main flux Φ by $\frac{\pi}{2}$, so these emfs (E_1 and E_2) are in phase with each other, as shown in Fig. 12.8 vectorially.

(*ii*) Applied voltage to the primary winding leads the main flux by $\frac{\pi}{2}$ and is in phase opposition to the induced emf in the primary winding, as shown in Fig. 12.8 vectorially.

(*iii*) Secondary voltage $V_2 = E_2$ as there is no voltage drop in secondary.

The instantaneous value of applied voltage, induced emfs, flux and magnetising current, in case of an ideal transformer, are illustrated by sinusoidal waves in Fig. 12.9.

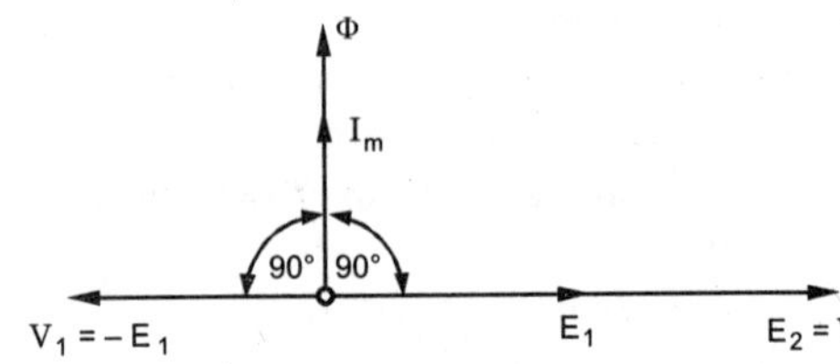

Fig. 12.8 *No-Load Phasor Diagram For an Ideal Transformer*

However, when a varying flux is set up in magnetic material, there will be power loss, called the *iron or core loss.* So the input current to the primary under no-load condition has also to supply the hysteresis and eddy current losses (iron losses) occuring in the core in addition to small amount of copper loss occuring in primary winding (no copper loss occurs in secondary winding on open circuit or on no load). Hence, the no-load primary current I_0 does not lag behind applied voltage V_1 by 90 ° but lags behind V_1 by angle $\phi_0 < 90$ °.

Input power on no load, $P_o = V_1 I_0 \cos \phi_0$ where $\cos \phi_0$ is the primary power factor under no-load condition.

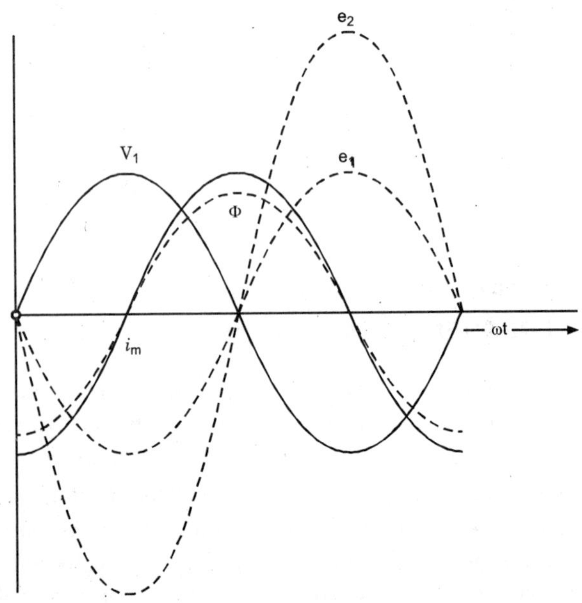

Fig. 12.9

As seen from phasor diagram shown in Fig. 12.10, input current to the primary I_0, called the *exciting current,* has two components (*i*) in-phase, active or energy component, I_e used to meet the iron loss in addition to small amount of copper loss occurring in the primary winding and (*ii*) quadrature component or wattless component called the magnetizing component, I_m used to create the alternating flux in the core.

$$\text{Thus, } I_e = I_0 \cos \phi_0$$
$$\text{and } I_m = I_0 \sin \phi_0$$
$$\text{and } \sqrt{I_e^2 + I_m^2} = I_0$$
$$\text{Angle of lag, } \phi_0 = \text{Tan}^{-1} \frac{I_m}{I_e}$$

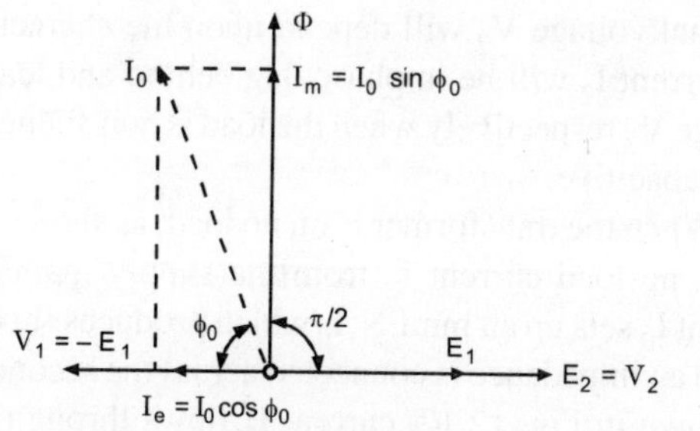

Fig. 12.10

The equivalent circuit of a transformer on no load is shown in Fig. 12.11. The actual transformer is replaced by an ideal transformer with a resistance R_0 and an inductive reactance X_0 in parallel with primary. The two components no-load current I_0 i.e., I_e and I_m are represented by currents drawn by R_0 and X_0 respectively. Both these currents are drawn at induced emf $E_1 = V_1$ for resistanceless, no-leakage primary coil, even otherwise $E_1 = V_1$.

From the equivalent circuit shown in Fig. 12.10.

$$\text{Energy component of no-load current, } I_e = \frac{V_1}{R_0}$$

$$\text{Magnetising component of no-load current, } I_m = \frac{V_1}{X_0}.$$

$$\text{Core or iron loss} = I_e^2 R_0 = \frac{V_1^2}{R_0}$$

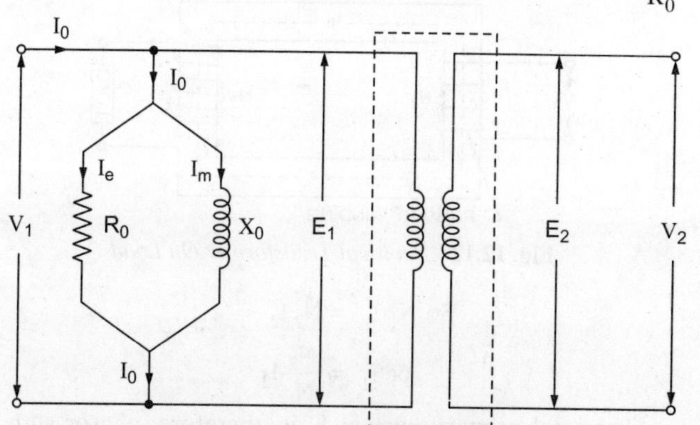

Fig. 12.11 *No-Load Equivalent Circuit For an Actual Transformer*

The noteworthy points are given below:

1. The no-load primary current I_0, called the *exciting current*, is very small in comparison to the full-load primary current. It ranges from 2 to 5 per cent of full-load primary current.
2. The exciting or no-load current I_0 is made up of a relatively large quadrature or magnetizing component I_m and a comparatively small in-phase or energy component I_e, so the power factor of a transformer on no load is very small (usually varies between 0.1 and 0.2 lag). The phase angle between I_0 and V_1 is about 78 ° to 87 °.
3. No-load primary copper loss *i.e* $I_0^2 R_1$ is very small and may be neglected. Thus the no-load primary input power is practically equal to the iron loss occuring in the core of the transformer.
4. Phase angle ϕ_0 is also known as the *hysteresis angle of advance* because it is principally the core loss which is responsible for shift in the current phasor.
5. Since the permeability of the core material varies with the instantaneous value of the exciting current, so the waveform of the exciting or magnetizing current is not truely sinusoidal. As such, it should not be represented by a phasor because only sinusoidally varying quantities are represented by rotating phasors. But, in practice, it makes no appreciable difference.

Example 12.17. A 11 kV/400 V distribution transformer takes a no-load primary current of 1 amp at a power factor of 0.24 lagging. Find (*i*) The core loss current (*ii*) The magnetizing current (*iii*) The iron loss. [*R.G. Technical Univ., June-2012*]

Solution: Exciting or no-load current, $I_0 = 1$ A

Primary voltage, $V_1 = 11$ kV = 11,000 V

(*i*) Core loss current, $I_e = I_0 \cos \phi_0 = 1 \times 0.24 = 0.24$ A **Ans.**

(*ii*) Magnetising current, $I_m = \sqrt{I_0^2 - I_e^2} = \sqrt{1^2 - 0.24^2}$

$= 0.971$ A **Ans.**

(*iii*) Iron loss $= V_1 I_0 \cos \phi_0 = 11{,}000 \times 1 \times 0.24$

$= 2{,}640$ W or 2.64 kW.

Example 12.18. A single-phase transformer of 3,300/220 V, 50 Hz takes a no-load current of 0.8 A and 500 W. Calculate (*i*) the active and magnetising current (*ii*) its pf

[*R.G.T.U. Basic Elec. Engineering, December-2006*]

Solution: Exciting or no-load current, $I_0 = 0.8$ A

Primary voltage, $V_1 = 3{,}300$ V

Iron loss $= V_1 I_0 \cos \phi_0 = 500$ W

(*i*) Iron loss or active component of no-load current,

$$I_e = I_0 \cos \phi_0$$
$$= \frac{500}{V_1} = \frac{500}{3300} = 0.152 \text{ A } \textbf{Ans.}$$

$$\text{Magnetising component, } I_m = \sqrt{I_0^2 - I_e^2}$$
$$= \sqrt{0.8^2 - 0.152^2} = 0.786 \text{ A } \textbf{Ans.}$$

$$(ii) \text{ Power factor, } \cos \phi_0 = \frac{500}{V_1 I_0} = \frac{500}{3{,}300 \times 0.8}$$
$$= 0.1894 \text{ (lagging) } \textbf{Ans.}$$

Example 12.19. A voltage $v = 200 \sin 314\, t$ is applied to the transformer winding in a no-load test. The resulting current is found to be $i = 3 \sin (314\, t - 60°)$

Determine the core loss and the parameters of no-load approximate equivalent circuit.

[U.P. Technical Univ. Electrical Machines 2002-03; Mahamaya Technical Univ. Electrical Engineering Odd Semester 2011-12]

Solution: No-load current, $I_0 = \frac{3}{\sqrt{2}} = 2.12$ A

Core loss = Input at no load

$$= V I_0 \cos \phi_0 = \frac{200}{\sqrt{2}} \times \frac{3}{\sqrt{2}} \cos 60°$$

$$= 150 \text{ W } \textbf{Ans.}$$

Energy component of no-load current,

$$I_e = I_0 \cos \phi_0 = \frac{3}{\sqrt{2}} \cos 60° = 1.06 \text{ A}$$

Magnetising component of no-load current,

$$I_m = \sqrt{I_0^2 - I_e^2} = \sqrt{(2.12)^2 - (1.06)^2} = 1.836 \text{ A}$$

$$\text{No-load resistance, } R_0 = \frac{V}{I_e} = \frac{200/\sqrt{2}}{1.06} = 133.42\ \Omega \textbf{ Ans.}$$

$$\text{No-load reactance, } X_0 = \frac{V}{I_m} = \frac{200/\sqrt{2}}{1.836} = 77\ \Omega \textbf{ Ans.}$$

Example 12.20. In no-load test of single-phase transformer, the following data were obtained:

Primary voltage 230 V, Secondary voltage 115 V, Primary current 0.5 A; Power input 40 W. Find: (*i*) turn ratio (*ii*) magnetising component of no-load current (*iii*) working component (*iv*) the iron losses. Resistance of primary winding is 0.5 Ω.

[Gujarat Technological Univ. Electrical Machines-I, Summer-2014]

Solution: No-load power input to transformer,

$$P_0 = 40 \text{ W}$$

No-load primary current, $I_0 = 0.5$ A

Primary voltage, $V_1 = 230$ V

Secondary voltage, $V_2 = 115$ V

Resistance of primary winding, $R_1 = 0.5\ \Omega$

$$(i) \text{ Turn ratio, } K = \frac{V_2}{V_1} = \frac{115}{230} = 0.5 \textbf{ Ans.}$$

Copper loss in primary winding

$$P_c = I_0^2 R_1 = 0.5^2 \times 0.5 = 0.125 \text{ W}$$

(*iv*) Iron loss, $P_i = P_0 - P_c = 40 - 0.125 = 39.875$ W **Ans.**

(*iii*) Working component of no-load current,

$$I_e = \frac{P_i}{V_1} = \frac{39.875}{230} = 0.173 \text{ A } \textbf{Ans.}$$

(*ii*) Magnetising component of no-load current,

$$I_m = \sqrt{I_0^2 - I_e^2} = \sqrt{0.5^2 - 0.173^2} = 0.469 \text{ A } \textbf{Ans.}$$

Example 12.21. The no-load current of a transformer is 5 A at 0.25 power factor, when supplied at 235 V, 50 Hz. The number of turns on primary side is 200; Calculate : (*i*) The value of maximum flux in the core; (*ii*) The core loss; (*iii*) The components of no-load current.

[M.D. Univ. Electrical Machines-I, May-2009]

Solution: (*i*) Maximum value of flux in the core :

$$\Phi_{max} = \frac{E_1}{4.44 f N_1} = \frac{235}{4.44 \times 50 \times 200}$$

$$= 0.005293 \text{ Wb or } 5.293 \text{ mWb } \textbf{Ans.}$$

(*ii*) Core loss = Input on no load

$$= V_1 I_0 \cos \phi_0 = 235 \times 5 \times 0.25 = 293.75 \text{ W } \textbf{Ans.}$$

(*iii*) Energy component of no-load current,

$$I_e = I_0 \cos \phi_0 = 5 \times 0.25 = 1.25 \text{ A } \textbf{Ans.}$$

Magnetising component of no-load current,

$$I_m = \sqrt{I_0^2 - I_e^2} = \sqrt{5^2 - 1.25^2} = 4.84 \text{ A } \textbf{Ans.}$$

12.10 TRANSFORMER ON LOAD

When the secondary circuit of a transformer is completed through an impedance or load, the transformer is said to be loaded and current flows through the secondary and the load. The magnitude and phase of secondary current I_2 with respect to secondary terminal voltage V_2 will depend upon the characteristic of load *i.e.* current I_2 will be in phase, lag behind and lead the terminal voltage V_2 respectively when the load is non-inductive, inductive and capacitive.

When the transformer is on no load, as shown in Fig. 12.7, it draws no-load current I_0 from the supply mains. The no-load current I_0 sets up an mmf $N_1 I_0$ which produces flux Φ in the core. When an impedance is connected across the secondary terminals, as shown in Fig. 12.12, current I_2 flows through the secondary winding. The secondary current I_2 sets up its own mmf and hence creates a secondary flux Φ_2. The secondary flux Φ_2 opposes the main flux Φ set up by the exciting current I_0 according to Lenz's law. The opposing secondary flux Φ_2 weakens the main flux Φ momentarily, so primary back emf E_1 tends to be reduced. So difference of applied voltage V_1 and back emf E_1 increases, therefore, more current is drawn from the source of supply flowing through the primary winding until the original value of flux Φ is obtained. It again causes increase in back emf E_1 and it adjusts itself as such that there is a balance between applied voltage V_1 and back emf E_1. Let the additional primary current be I_1'. The current I'_1 is in phase opposition with secondary current I_2 and is called the *counterbalancing current.* The additional current I'_1 sets up an mmf $N_1 I'_1$ producing flux Φ_1' in the same direction as that of main flux Φ and cancels the flux Φ_2 produced by secondary mmf $N_2 I_2$ being equal in magnitude.

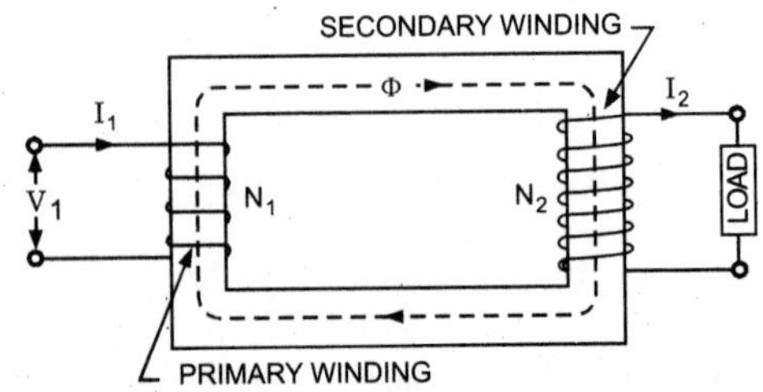

Fig. 12.12 *An Ideal Transformer On Load*

$$\text{So } N_1 I_1' = N_2 I_2$$

$$\text{or } I_1' = \frac{N_2}{N_1} I_2$$

The total primary current I_1 is, therefore, phasor sum of primary counterbalancing current I_1' and no-load current I_0, which will be approximately equal to I'_1 as I_0 is usually very small in comparison to I_1'.

$$\therefore I_1 = I_1' = \frac{N_2}{N_1} I_2$$

or $\frac{I_1}{I_2} = \frac{N_2}{N_1} = K$ (transformation ratio)

Hence primary and secondary currents are inversely proportional to their respective turns.

Since the secondary flux Φ_2 produced by secondary mmf $N_2 I_2$ is neutralized by the flux Φ'_1 produced by mmf $N_1 I_1'$ set up by counterbalancing primary current I'_1, so the flux in the transformer core remains almost constant from no load to full load.

The phasor diagrams for transformer on non-inductive, inductive and capacitive loads are shown in Figs. 12.13 (*a*), (*b*) and (*c*) respectively.

Since the voltage drops in both of the windings of the transformer are assumed to be negligible, therefore

$$V_2 = E_2 \text{ and } V_1 = -E_1$$

The secondary current I_2 is in phase, lags behind and leads the secondary terminal voltage V_2 by an angle ϕ_2 for pure resistive, inductive and capacitive loads respectively.

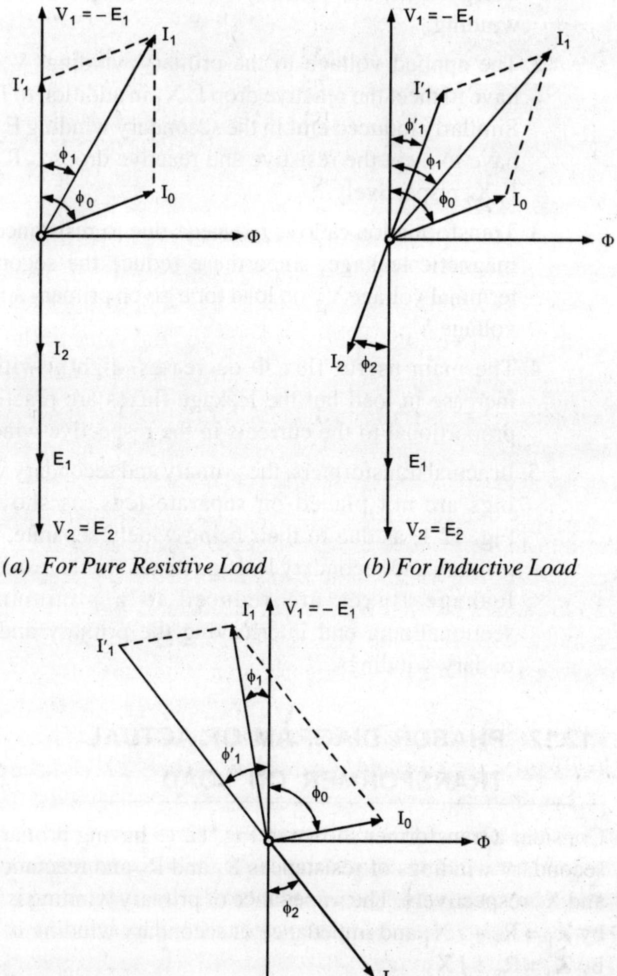

(*a*) *For Pure Resistive Load* (*b*) *For Inductive Load*

(*c*) *For Capacitive Load*

Fig. 12.13 *Phasor Diagram For an Ideal Transformer On Load*

The induced primary current I'_1, also known as *counterbalancing current,* is always in opposition to secondary current I_2 and since no-load current I_0 is very small, the total primary current I_1 is almost opposite in phase to I_2 and K times the secondary current I_2, where K is transformation ratio.

Note. In phasor diagrams shown in Figs. 12.13 (*a*), 12.13 (*b*) and 12.13 (*c*) no-load current has been drawn on exaggerated scale for sake of clarity.

Example 12.22. A single-phase transformer with a ratio of 440/110 V takes a no-load current of 5 A at 0.2 power factor lag. If the secondary supplies a current of 120 A at a pf of 0.8 lagging, calculate the current taken by the primary.

[*Electrical Machines-I; Rajasthan Univ. 2004; M.D. Univ. May-2009; Electromechanical Energy Conversion, M.D. Univ., December-2012*]

Solution: Primary voltage, $V_1 = 440$ V

Secondary voltage, $V_2 = 110$ V

No-load power factor, $\cos \phi_0 = 0.2$ (lag)

Secondary current, $I_2 = 120$ A

Load power factor, $\cos \phi_2 = 0.8$ (lag)

$$\text{Transformation ratio, } K = \frac{V_2}{V_1} = \frac{110}{440} = 0.25$$

Primary counterbalancing current, $I'_1 = K\,I_2 = 0.25 \times 120 = 30$ A lagging behind the supply voltage V_1 by an angie ϕ'_1

where $\phi_1 = \phi_2 = \text{Cos}^{-1}\, 0.8 = 36.87°$

No-load current, $I_0 = 5$ A lagging behind the supply voltage V_1 by an angle $\text{Cos}^{-1}\, 0.2 = 78.46°$

$$\textbf{Primary current, } \mathbf{I_1} = I'_1 + I_0$$
$$= (I'_1 \cos \phi_1 + I_0 \cos \phi_0) + j\,(I_1 \sin \phi_1 + I_0 \sin \phi_0)$$
$$= (30 \times 0.8 + 5 \times 0.2) + j\,[30 \times (-0.6) + 5 \times (-0.98)]$$
$$= (25 - j\,22.9) \text{ A} = 33.9 \angle -42.49° \text{ A } \textbf{Ans.}$$

i.e. 33.9 A lagging behind supply voltage V_1 by 42.49° **Ans.**

12.11 RESISTANCE AND LEAKAGE REACTANCE

In preceding discussions we considered an ideal transformer, which according to our assumptions, has got no resistance in the windings and no leakage flux but in actual practice it is impossible to obtain such an ideal transformer.

In actual transformer both the windings, primary and secondary windings have finite resistances R_1 and R_2 which cause copper losses and voltage drop in them. The result is that:

(*i*) The secondary terminal voltage V_2 is less than the secondary induced emf E_2 and is equal to phasor difference of secondary induced emf E_2 and voltage drop in the secondary winding I_2R_2, if magnetic leakage is negligible, *i.e.*

$$\mathbf{V_2 = E_2 - I_2\,R_2}$$

where I_2 is the secondary current and R_2 is the secondary winding resistance.

(*ii*) Similarly the counter emf of primary, $-E_1$ is equal to phasor difference of voltage applied to the primary winding V_1 and voltage drop in the primary winding $I_1\,R_1$, provided magnetic leakage is negligible, i.e.,

$$\mathbf{-E_1 = V_1 - I_1\,R_1}$$

where I_1 is the primary current and R_1 is the primary resistance.

It was previously assumed that the entire flux Φ, developed by the primary winding, links with and cut every turn of both the primary and secondary windings. In practice, it is impossible to realize this condition. However, part of the flux set up by the primary winding links only the primary turns, as illustrated in Fig. 12.14 by flux Φ_{L1}. Also, some of the flux set up by the secondary winding links only the secondary turns, as illustrated in Fig. 12.14 by Φ_{L2}. These two fluxes Φ_{L1} and Φ_{L2} are known as leakage flux *i.e. that flux which leaks out of the core and does not link both windings.* The flux which does pass completely through the core and links both the windings is known as the *mutual flux* and is illustrated as Φ.

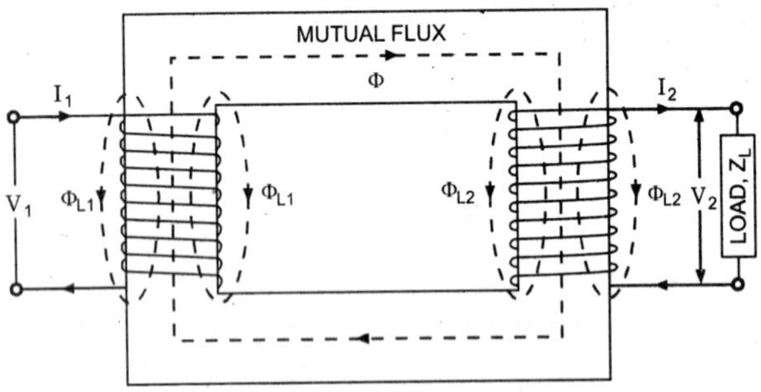

Fig. 12.14 *Magnetic Fluxes in a Transformer*

The primary leakage flux Φ_{L1} linking with the primary winding is produced by primary ampere-turns only, therefore, it is proportional to primary current, number of primary turns being fixed. On no load primary current is so small that leakage flux Φ_{L1} produced by it can be neglected but on load primary current increases resulting in increase in ampere-turns and hence leakage flux Φ_{L1} increases. The primary leakage flux Φ_{L1} is in phase with I_1 and produces self-induced emf E_{L1} given by $E_{L1} = 2\pi f L_1 I_1$ in primary winding, where L_1 is the self-inductance of the primary winding produced by primary leakage flux Φ_{L1}.

The self-induced emf E_{L1}, due to primary leakage flux, in the primary winding must lag leakage flux Φ_{L1} and primary current I by 90°. The emf necessary to balance this counter emf is opposite and equal to it and, therefore, leads the primary current I_1 by 90°. As this emf, induced by the primary leakage flux, is proportional to the current and lags it by 90°, it is nothing more than a reactance voltage and is denoted by $-I_1 X_1$. The component of line voltage that balances this emf is $+I_1 X_1$. The effect of the primary leakage flux, therefore, is to induce an emf that opposes the flow of current to the transformer. The reactance of the primary winding, X_1 can be obtained by dividing self-induced emf E_{L1} by the primary current I_1 *i.e.*

$$X_1 = \frac{E_{L1}}{I_1} = \frac{2\pi f L_1 I_1}{I_1} = 2\pi f L_1$$

Similarly secondary leakage flux Φ_{L2} is set up by secondary ampere-turns and is proportional to secondary current I_2. On no load there is no current in secondary winding and, therefore, no leakage flux exists across the secondary winding on no load. On load leakage flux Φ_{L2} in phase with secondary current I_2 and produces self-induced emf $E_{L2} = 2\pi f L_2 I_2$ in the secondary winding where L_2 is self-inductance of secondary winding due to leakage flux Φ_{L2}. This is also a reactance voltage, and the component that balances it leads the secondary current by 90°. The secondary reactance X_2 opposes the current flowing out of the transformer and can be obtained by dividing self induced emf in secondary winding, E_{L2} by the secondary current I_2 *i.e.*

$$X_2 = \frac{E_{L2}}{I_2} = \frac{2\pi f L_2 I_2}{I_2} = 2\pi f L_2$$

The effect of magnetic leakage is, thus to produce in their respective windings emfs of self-inductance which are proportional to the current, and are, therefore, equivalent in effect to the addition of an inductive coil in series with each winding, the reactance of which is called the leakage reactance.

A transformer with magnetic leakage and winding resistance is equivalent to an ideal transformer (having no resistance and leakage reactance) having inductive and resistive coils connected in series with each winding as shown in Fig. 12.15.

Few important points to be kept in mind are given below:

1. The leakage flux links one or the other winding but not both, hence it in no way contributes to the transfer of energy from the primary winding to the secondary winding.
2. The applied voltage to the primary winding, V_1 will have to meet the reactive drop $I_1 X_1$ in addition to $I_1 R_1$. Similarly induced emf in the secondary winding E_2 will have to meet the resistive and reactive drops $I_2 R_2$ and $I_2 X_2$ respectively.
3. Transformation ratio is reduced* due to resistance and magnetic leakage, since these reduce the secondary terminal voltage V_2 on load for a given primary applied voltage V_1.
4. The main useful flux Φ decreases slightly with the increase in load but the leakage fluxes are practically proportional to the currents in the respective winding.
5. In actual transformers, the primary and secondary windings are not placed on separate legs, as shown in Fig. 12.1, as due to their being widely separate, large primary and secondary leakage fluxes would result. The leakage fluxes are reduced to a minimum by sectionalizing and interleaving the primary and secondary windings.

12.12 PHASOR DIAGRAM OF ACTUAL TRANSFORMER ON LOAD

Consider a transformer shown in Fig. 12.15 having primary and secondary windings of resistances R_1 and R_2 and reactances X_1 and X_2 respectively. The impedance of primary winding is given by $\mathbf{Z}_1 = R_1 + jX_1$ and impedance of secondary winding is given by $\mathbf{Z}_2 = R_2 + jX_2$.

The phasor diagrams of above transformer on (*i*) pure resistive, (*ii*) resistive-inductive, and (*iii*) resistive-capacitive loads are shown in Fig. 12.16 (*a*), (*b*) and (*c*) respectively.

Draw *OA* representing secondary terminal voltage V_2 and OI_2 representing secondary current I_2 in phase as well as magnitude. Since voltage drops due to secondary winding

* Assuming lagging power factor

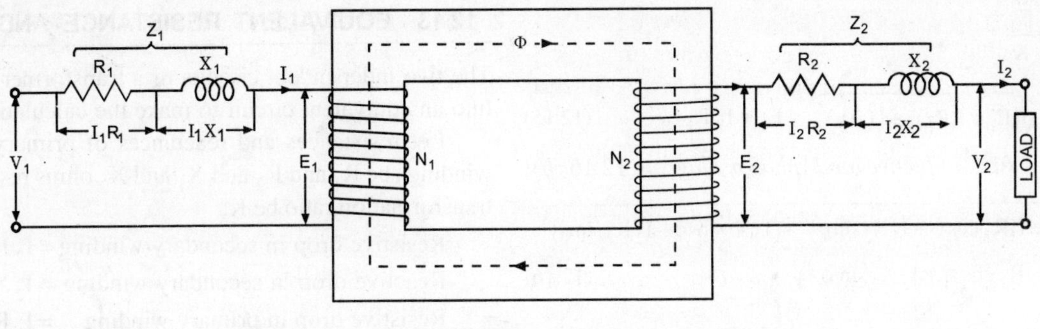

Fig. 12.15 *An Equivalent Diagram of Actual Transformer*

resistance and reactance are $I_2 R_2$ in phase with current I_2 and $I_2 X_2$ leading current I_2 by $\pi/2$ respectively, so draw AB parallel to OI_2 and equal to $I_2 R_2$ in magnitude representing resistive drop in secondary winding and draw BC perpendicular to AB

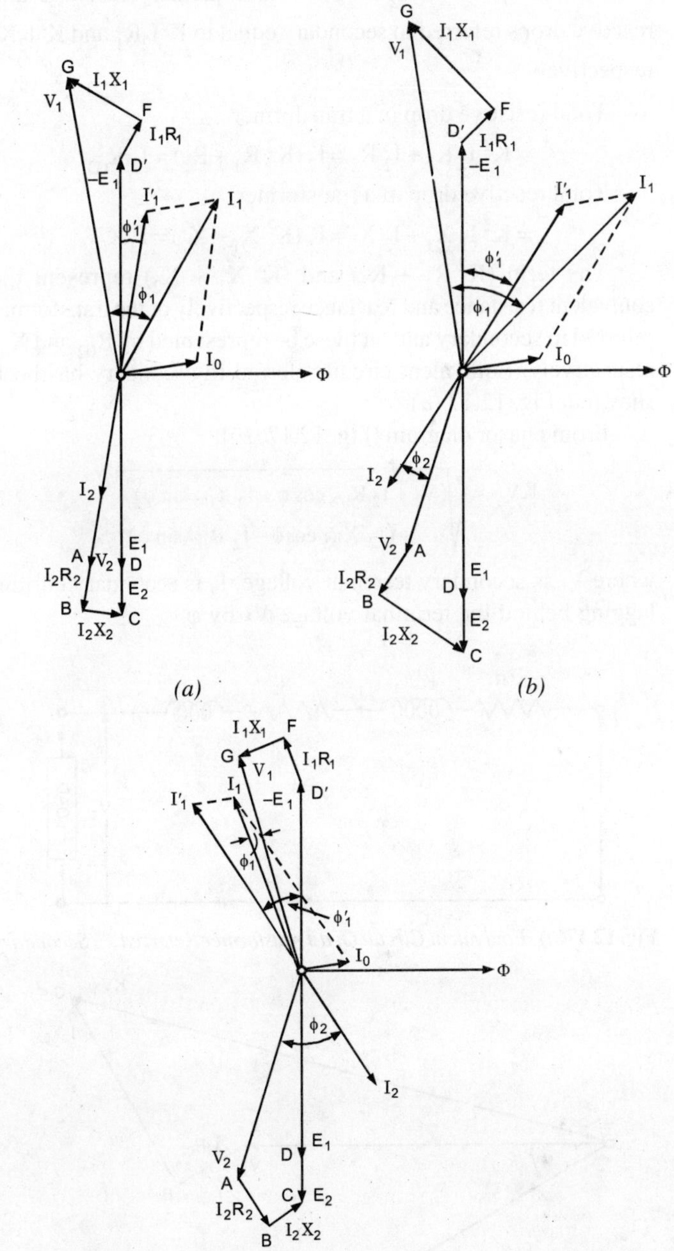

Fig. 12.16 *Phasor Diagram of an Actual Transformer*

and equal to $I_2 X_2$ in magnitude representing reactive drop of secondary winding. Since phasor sum of terminal voltage V_2, secondary resistive drop $I_2 R_2$ and secondary reactive drop $I_2 X_2$ is equal to induced emf E_2 in secondary winding, phasor OC represents secondary induced emf E_2. Hence we have

$$\mathbf{E}_2 = \mathbf{V}_2 + \mathbf{I}_2 (R_2 + j X_2) = \mathbf{V}_2 + \mathbf{I}_2 \mathbf{Z}_2 \quad ...(12.12)$$

The induced emf E_1 in primary winding is in phase with E_2 and equal to $\frac{N_1}{N_2} E_2$ in magnitude, so take $OD = \frac{N_1}{N_2} OC$ representing E_1. Produce DO to D′ taking OD′ = OD hence representing ($-\mathbf{E}_1$).

The induced primary current I_1' is equal to $-I_2 \frac{N_2}{N_1}$ so draw OI_1' equal to $OI_2 \times \frac{N_2}{N_1}$ by producing line I_2 O. Draw line OI_0 representing no-load current in magnitude as well as in phase. The phasor sum of induced primary current I_1' and no-load current I_0 gives primary current represented by phasor OI_1 in Fig. 12.16.

Since voltage drop due to primary winding resistance and reactance are I_1R_1 in phase with primary current I_1 and I_1X_1 leading current I_1 by $\frac{\pi}{2}$ respectively, draw D′F parallel to OI_1 and equal to I_1R_1 in magnitude representing resistive drop in primary winding and draw FG perpendicular to D′F equal to I_1X_1 in magnitude representing reactive drop in primary winding. As the phasor sum of ($-E_1$), primary resistive drop and primary reactive drop gives the applied voltage V_1 to primary winding, hence phasor OG represents the applied voltage V_1 in magnitude as well as in phase.

$$\textit{i.e.}\ \mathbf{V}_1 = -\mathbf{E}_1 + \mathbf{I}_1 (R_1 + j X_1) = -\mathbf{E}_1 + \mathbf{I}_1 \mathbf{Z}_1 \quad ...(12.13)$$

The phase angle ϕ_1 between V_1 and I_1 gives the power factor angle of the transformer.

Since no-load current I_0, resistive drops $I_1 R_1$ and $I_2 R_2$ and reactive drops, $I_1 X_1$ and $I_2 X_2$ are very small, neglecting these we have $\phi_2 = \phi_1' = \phi_1 = \phi$, the phase angle of the load. In Fig. 12.16, no-load current, resistive drops and reactive drops are shown, for clarity, on exaggerated scales.

From phasor diagrams we have

(a) For pure resistive load [phasor diagram 12.16 (a)]

$$E_2 = \sqrt{(V_2 + I_2 R_2)^2 + (I_2 X_2)^2} \approx V_2 + I_2 R_2 \quad ...(12.14)$$

$$E_1 = \frac{E_2}{K}$$

and $$V_1 = \sqrt{(E_1 + I_1R_1)^2 + (I_1X_1)^2} \approx E_1 + I_1 R_1 \quad ...(12.15)$$

(*b*) *For resistive-inductive load* [*phasor diagram* 12.16 (*b*)]

$$E_2 = \sqrt{(V_2 + I_2R_2 \cos\phi + I_2 X_2 \sin\phi)^2 + (I_2 X_2 \cos\phi - I_2 R_2 \sin\phi)^2}$$
$$\approx V_2 + I_2 R_2 \cos\phi + I_2 X_2 \sin\phi \quad ...(12.16)$$

$$E_1 = \frac{E_2}{K} \text{ and}$$

$$V_1 = \sqrt{(E_1 + I_1R_1\cos\phi + I_1 X_1 \sin\phi)^2 + (I_1 X_1 \cos\phi - I_1R_1 \sin\phi)^2}$$
$$\approx E_1 + I_1 R_1 \cos\phi + I_1 X_1 \sin\phi \quad ...(12.17)$$

(*c*) *For resistive-capacitive load* [*phasor diagam* 12.16 (*c*)]

$$E_2 = \sqrt{(V_2 + I_2 R_2 \cos\phi - I_2 X_2 \sin\phi)^2 + (I_2 X_2 \cos\phi + I_2 R_2 \sin\phi)^2}$$
$$\approx V_2 + I_2 R_2 \cos\phi - I_2 X_2 \sin\phi \quad ...(12.18)$$

$$E_1 = \frac{E_2}{K} \text{ and}$$

$$V_1 = \sqrt{(E_1 + I_1 R_1 \cos\phi - I_1 X_1 \sin\phi)^2 + (I_1 X_1 \cos\phi + I_1 R_1 \sin\phi)^2}$$
$$= E_1 + I_1 R_1 \cos\phi - I_1 X_1 \sin\phi \quad ...(12.19)$$

Example 12.23. A 230/460 V transformer has a primary resistance of 0.2 Ω and a reactance of 0.5 Ω and the corresponding values for the secondary are 0.75 Ω and 1.8 Ω respectively. Find the secondary terminal voltage when supplying (*a*) 10 A at 0.8 pf lagging (*b*) 10 A at pf 0.8 leading.

Solution: Transformation ratio, $K = \frac{V_2}{V_1} = \frac{460}{230} = 2$

(*i*) **When Load is 10 A at 0.8 pf lagging** i.e. when

$I_2 = 10$ A and $\cos\phi = 0.8$ lagging.

Primary current $I_1 = K\,I_2 = 2 \times 10 = 20$ A

From Eq. (12.19)

$$\text{Primary induced emf, } E_1 = V_1 - (I_1 R_1 \cos\phi + I_1 X_1 \sin\phi)$$
$$= 230 - (20 \times 0.2 \times 0.8 + 20 \times 0.5 \times 0.6)$$
$$= 220.8 \text{ V}$$

Secondary induced emf, $E_2 = KE_1 = 2 \times 220.8 = 441.6$ V

From Eq. (12.16)

$$\text{Secondary terminal voltage, } V_2 = E_2 - (I_2 R_2 \cos\phi + I_2 X_2 \sin\phi)$$
$$= 441.6 - (10 \times 0.75 \times 0.8 + 10 \times 1.8 \times 0.6)$$
$$= 441.6 - (6 + 10.8) = 424.8 \text{ V}$$ **Ans**

(*ii*) When load is 10 A at 0.8 pf leading

From Eq. (12.19)

$$\text{Primary induced emf, } E_1 = V_1 - (I_1 R_1 \cos\phi - I_1 X_1 \sin\phi)$$
$$= 230 - (20 \times 0.2 \times 0.8 - 20 \times 0.5 \times 0.6)$$
$$= 232.8 \text{ V}$$

and from Eq. (12.18)

Secondary terminal voltage,

$$V_2 = KE_1 - (I_2 R_2 \cos\phi - I_2 X_2 \sin\phi)$$
$$= 2 \times 232.8 - (10 \times 0.75 \times 0.8 - 10 \times 1.8 \times 0.6)$$
$$= 465.6 - (6 - 10.8) = 470.4 \text{ V}$$ **Ans.**

12.13 EQUIVALENT RESISTANCE AND REACTANCE

The two independent circuits of a transformer can be resolved into an equivalent circuit to make the calculations simple.

Let resistances and reactances of primary and secondary windings be R_1 and R_2 and X_1 and X_2 ohms respectively and let transformation ratio be K.

Resistive drop in secondary winding $= I_2 R_2$

Reactive drop in secondary winding $= I_2 X_2$

Resistive drop in primary winding $= I_1 R_1$

Reactive drop in primary winding $= I_1 X_1$

Referred To Secondary Side. Since transformation ratio is K, primary resistive and reactive drops as referred to secondary will be K times, *i.e.* KI_1R_1 and KI_1X_1 respectively. If I_1 is substituted equal to KI_2 then we have primary resistive and reactive drops referred to secondary equal to $K^2I_2R_1$ and $K^2I_2X_1$ respectively.

Total resistive drop in a transformer

$$= K^2 I_2 R_1 + I_2 R_2 = I_2 (K^2 R_1 + R_2) = I_2 R_{02}$$

Total reactive drop in a transformer

$$= K^2 I_2 X_1 + I_2 X_2 = I_2 (K^2 X_1 + X_2) = I_2 X_{02}$$

The term $(K^2 R_1 + R_2)$ and $(K^2 X_1 + X_2)$ represent the equivalent resistance and reactance respectively of the transformer referred to secondary and let these be represented by R_{02} and X_{02} respectively. Equivalent circuit referred to secondary has been shown in Fig. 12.17 (*a*).

From phasor diagram [Fig. 12.17 (*b*)]

$$KV_1 = \sqrt{(V_2 + I_2 R_{02} \cos\phi + I_2 X_{02} \sin\phi)^2 + (I_2 X_{02} \cos\phi - I_2 R_{02} \sin\phi)^2}$$

where V_2 is secondary terminal voltage, I_2 is secondary current lagging behind the terminal voltage V_2 by ϕ.

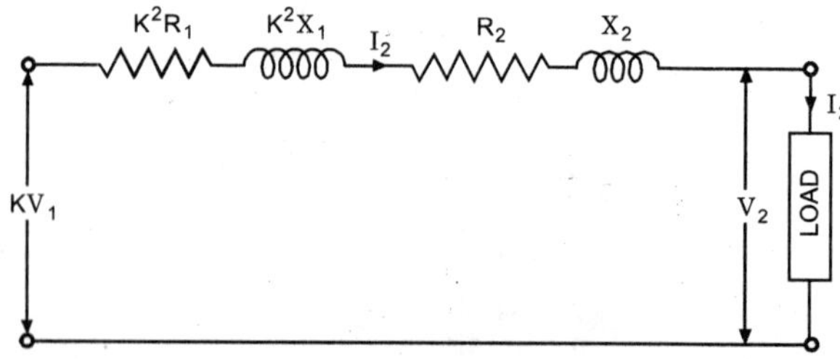

Fig. 12.17(*a*) *Equivalent Circuit Of a Transformer Referred To Secondary*

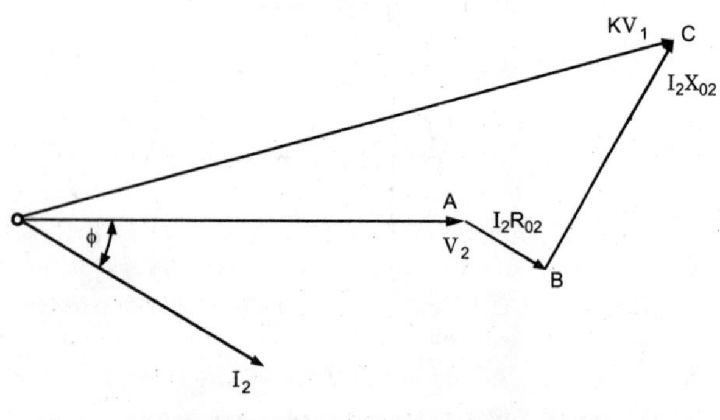

Fig. 12.17(*b*) *Phasor Diagram*

Since term $(I_2 X_{02} \cos\phi - I_2 R_{02} \sin\phi)$ is very small as compared to the term $(V_2 + I_2 R_{02} \cos\phi + I_2 X_{02} \sin\phi)$, neglecting the former we have

$$K V_1 = V_2 + I_2 R_{02} \cos\phi + I_2 X_{02} \sin\phi$$

$$\text{or } V_2 = KV_1 - I_2 R_{02} \cos\phi - I_2 X_{02} \sin\phi \qquad ...(12.20)$$

where V_1 is applied voltage to primary winding.

If load is pure resistive, $\phi = 0$ and $V_2 = KV_1 - I_2 R_{02}$ ...(12.21)

If load is capacitive then ϕ should be taken as –ve hence we have

$$V_2 = KV_1 - I_2 R_{02} \cos\phi + I_2 X_{02} \sin\phi \qquad ...(12.22)$$

Referred To Primary Side. Secondary resistive drop referred to primary

$$= \frac{I_2 R_2}{K} = \frac{I_1 R_2}{K^2} \qquad \text{since } I_2 = \frac{I_1}{K}$$

Secondary reactive drop referred to primary $= \dfrac{I_2 X_2}{K} = \dfrac{I_1 X_2}{K^2}$

Total resistive drop in the transformer referred to primary

$$= I_1 R_1 + \frac{I_1 R_2}{K^2} = I_1\left(R_1 + \frac{R_2}{K^2}\right) = I_1 R_{01}$$

Total reactive drop in the transformer referred to primary

$$= I_1 X_1 + \frac{I_1 X_2}{K^2} = I_1\left(X_1 + \frac{X_2}{K^2}\right) = I_1 X_{01}$$

The terms $\left(R_1 + \dfrac{R_2}{K^2}\right)$ and $\left(X_1 + \dfrac{X_2}{K^2}\right)$ represent the total resistance and reactance of the transformer referred to primary respectively. Let these be represented by R_{01} and X_{01} respectively.

Equivalent circuit referred to primary is shown in Fig. 12.18 (*a*)

From phasor diagram shown in Fig. 12.18 (*b*) we have

$$V_1 = \sqrt{\left(\frac{V_2}{K} + I_1 R_{01} \cos\phi + I_1 X_{01} \sin\phi\right)^2 + (I_1 X_{01} \cos\phi - I_1 R_{01} \sin\phi)^2}$$

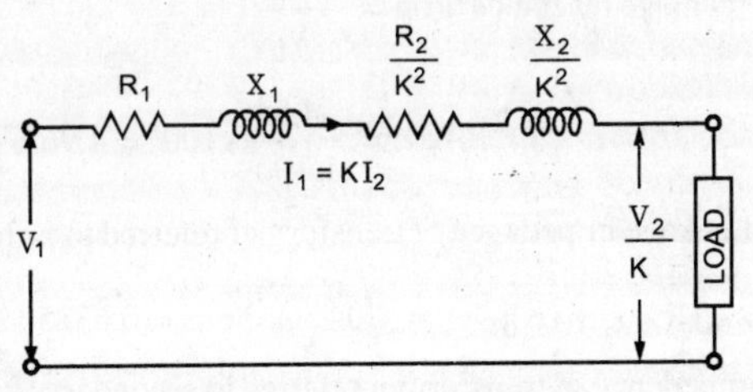

(a) Equivalent Circuit of a Transformer Referred To Primary

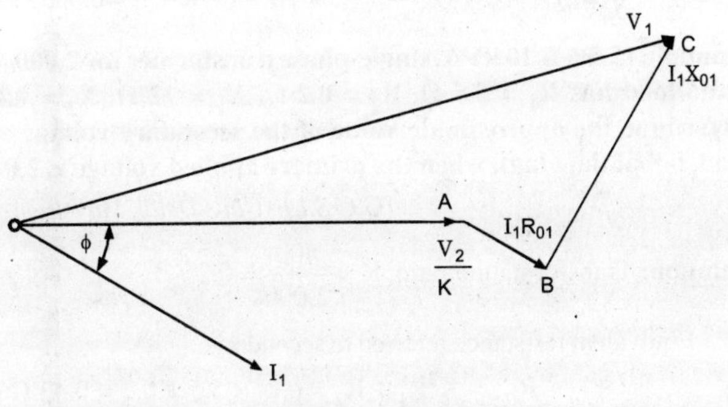

(b) Phasor Diagram

Fig. 12.18

Since term $(I_1 X_{01} \cos\phi - I_1 R_{01} \sin\phi)$ is very small as compared to the term $\left(\dfrac{V_2}{K} + I_1 R_{01} \cos\phi + I_1 X_{01} \sin\phi\right)$, neglecting the former we have

$$V_1 = \frac{V_2}{K} + I_1 R_{01} \cos\phi + I_1 X_{01} \sin\phi \qquad ... (12.23)$$

If load is pure resistive *i.e.* $\phi = 0$ then

$$V_1 = \frac{V_2}{K} + I_1 R_{01} \qquad ... (12.24)$$

If load is capacitive then ϕ should be taken as –ve so we have

$$V_1 = \frac{V_2}{K} + I_1 R_{01} \cos\phi - I_1 X_{01} \sin\phi \qquad ... (12.25)$$

Example 12.24. A 25 kVA, 2,200/220 V, 50-Hz, 1-phase transformer has following parameters:

$R_1 = 1.75\ \Omega$, $R_2 = 0.0045\ \Omega$, $X_1 = 2.6\ \Omega$, $X_2 = 0.0075\ \Omega$.

Calculate: (*i*) Equivalent resistance referred to primary and secondary. (*ii*) Equivalent reactance referred to primary and secondary.

[*U.P. Technical Univ. Electrical Engineering Odd Semester 2013-14*]

Solution: Transformation ratio, $K = \dfrac{V_2}{V_1} = \dfrac{220}{2,200} = 0.1$

(*i*) Equivalent resistance referred to primary,

$$R_{01} = R_1 + \frac{R_2}{K^2} = 1.75 + \frac{0.0045}{(0.1)^2} = 2.2\ \Omega \textbf{ Ans.}$$

Equivalent resistance referred to secondary,

$$R_{02} = K^2 R_1 + R_2 = (0.1)^2 \times 1.75 + 0.0045 = 0.022\ \Omega \textbf{ Ans.}$$

(*ii*) Equivalent reactance referred to primary,

$$X_{01} = X_1 + \frac{X_2}{K^2} = 2.6 + \frac{0.0075}{(0.1)^2} = 3.35\ \Omega \textbf{ Ans.}$$

Equivalent reactance referred to secondary,

$$X_{02} = K^2 X_1 + X_2 = (0.1)^2 \times 2.6 + 0.0075 = 0.0335\ \Omega \textbf{ Ans.}$$

Example 12.25. A transformer has a primary winding of 600 turns and a secondary winding of 150 turns. When the load current on the secondary is 60 A at 0.8 power factor lagging, the primary current is 20 A at 0.707 power factor lagging. Determine the no-load current of the transformer and its phase with respect to the voltage.

[*G.B. Technical Univ. Electrical Engineering Even Semester 2009-10*]

Solution: Turn ratio $K = \dfrac{N_2}{N_1} = \dfrac{150}{600} = \dfrac{1}{4}$

Secondary current, $\mathbf{I}_2 = 60 \angle \cos^{-1} 0.8 = 60 \angle -36.87°$ A

Load current referred to primary,

$$\mathbf{I}'_1 = K\mathbf{I}_2 = \frac{1}{4} \times 60 \angle -36.87° = 15 \angle -36.87° \text{ A}$$

Primary current, $\mathbf{I}_1 = 20 \angle \cos^{-1} 0.707 = 20 \angle -45°$ A

No-load current, $\mathbf{I}_0 = \mathbf{I}_1 - \mathbf{I}'_1 \qquad \because \mathbf{I}_1 = \mathbf{I}_0 + \mathbf{I}'_1$

$$= 20 \angle -45° - 15 \angle -36.87°$$

$$= (14.14 - j\,14.14) - (12 - j\,9)$$

$$= (2.2 - j\,5.14) \text{ A} = 5.59 \angle -66.83° \text{ A} \quad \textbf{Ans.}$$

Example 12.26. A 30 kVA, 2,000/200 V, single-phase, 50 Hz transformer has a primary resistance of 3.5 Ω and reactance of 4.5 Ω. The secondary resistance and reactance are 0.015 Ω and 0.02 Ω respectively. Find (*i*) equivalent resistance, reactance and impedance referred to the primary side (*ii*) total copper loss in the transformer.

[*U.P. Technical Univ. Electrical Engineering Odd Semester 2005-06*]

Solution: Transformation ratio, $K = \frac{V_2}{V_1} = \frac{200}{2,000} = 0.1$

Full-load primary current, $I_1 = \frac{30 \times 1,000}{2,000} = 15$ A

Equivalent resistance referred to primary,

$$R_{01} = R_1 + \frac{R_2}{K^2} = 3.5 + \frac{0.015}{(0.1)^2} = 5\ \Omega \textbf{ Ans.}$$

Equivalent reactance referred to primary,

$$X_{01} = X_1 + \frac{X_2}{K^2} = 4.5 + \frac{0.02}{(0.1)^2} = 6.5\ \Omega \textbf{ Ans.}$$

Equivalent impedance referred to primary,

$$Z_{01} = \sqrt{(R_{01})^2 + (X_{01})^2} = \sqrt{5^2 + 6.5^2} = 8.2\ \Omega \textbf{ Ans.}$$

Total copper loss in transformer

$$= I_1^2 R_{01} = 15^2 \times 5 = 1,125 \text{ watts } \textbf{Ans.}$$

Example 12.27. Given the resistances r_1 and r_2 and currents I_1 and I_2 of the two windings of a transformer, determine the resistance which may be used to represent total ohmic loss in the transformer.

[*M.D. Univ. Electromechanical Energy Conversion, May-2008*]

Solution: Transformation ratio,

$$K = \frac{V_2}{V_1} = \frac{I_1}{I_2} \qquad \text{neglecting no-load current } I_0$$

Equivalent resistance of transformer referred to secondary,

$$R_{02} = K^2 r_1 + r_2 = \left(\frac{I_1}{I_2}\right)^2 r_1 + r_2 \textbf{ Ans.}$$

Equivalent resistance of transformer referred to primary,

$$R_{01} = r_1 + \frac{r_2}{K^2} = r_1 + \frac{r_2}{\left(\frac{I_1}{I_2}\right)^2} = r_1 + r_2 \times \left(\frac{I_2}{I_1}\right)^2 \textbf{ Ans.}$$

Example 12.28. A 100 kVA, 1,100/220 V, 60 Hz transformer has a high-voltage winding resistance of 0.1 Ω and a leakage reactance of 0.3 Ω. The low-voltage winding resistance is 0.004 Ω and the leakage reactance is 0.012 Ω. The source is applied to high-voltage side:

(*i*) Find the equivalent winding resistance and reactance referred to high-voltage side and low-voltage side.

(*ii*) Find the equivalent resistance and reactance drops in volts and in per cent of the rated winding voltages expressed in terms of the primary quantities.

(*iii*) Calculate leakage impedances of the transformer referred to the primary and secondary sides.

[*U.P. Technical Univ. Electrical Engineering Even Semester, 2008-09*]

Solution: Transformation ratio, $K = \frac{V_2}{V_1} = \frac{220}{1,100} = \frac{1}{5}$

(*i*) Equivalent resistance referred to primary (*hv*),

$$R_{01} = R_1 + \frac{R_2}{K^2} = 0.1 + \frac{0.004}{(1/5)^2}$$

$$= 0.1 + 0.004 \times 25 = 0.1 + 0.1 = 0.2\ \Omega \textbf{ Ans.}$$

Equivalent reactance referred to primary (*hv*)

$$X_{01} = X_1 + \frac{X_2}{K^2} = 0.3 + \frac{0.012}{(1/5)^2} = 0.3 + 0.012 \times 25$$

$$= 0.3 + 0.3 = 0.6\ \Omega \textbf{ Ans.}$$

Equivalent resistance referred to secondary (*lv*)

$$R_{02} = K^2 R_1 + R_2 = \left(\frac{1}{5}\right)^2 \times 0.1 + 0.004$$

$$= \frac{0.1}{25} + 0.004 = 0.004 + 0.004 = 0.008\ \Omega \textbf{ Ans.}$$

Equivalent reactance referred to secondary (*lv*)

$$X_{02} = K^2 X_1 + R_2 = \left(\frac{1}{5}\right)^2 \times 0.3 + 0.012$$

$$= \frac{0.3}{25} + 0.012 = 0.012 + 0.012 = 0.024\ \Omega \textbf{ Ans.}$$

(*ii*) Full-load primary current,

$$I = \frac{\text{Output in kVA} \times 1,000}{\text{Primary voltage}} = \frac{100 \times 1,000}{1,100} = \frac{1,000}{11} \text{ A}$$

Primary rated voltage,

$V_1 = 1,100$ V

Equivalent resistance drop referred to primary

$$= I_1 R_{01} = \frac{1,000}{11} \times 0.2 = \frac{200}{11} = 18.18 \text{ V} \textbf{ Ans.}$$

Equivalent reactance drop referred to primary

$$= I_1 X_{01} = \frac{1,000}{11} \times 0.6 = \frac{600}{11} = 54.54 \text{ V} \textbf{ Ans.}$$

$$\text{Percentage resistance drop} = \frac{I_1 R_{01}}{V_1}$$

$$= \frac{18.18}{1,100} \times 100 = 1.653\% \textbf{ Ans.}$$

$$\text{Percentage reactance drop} = \frac{I_1 X_{01}}{V_1}$$

$$= \frac{54.54}{1,100} \times 100 = 4.96\% \textbf{ Ans.}$$

(*iii*) Leakage impedance of transformer referred to primary (*hv*)

$$Z_{01} = \sqrt{(R_{01})^2 + (X_{01})^2} = \sqrt{0.2^2 + 0.6^2} = 0.6325\ \Omega \textbf{ Ans.}$$

Leakage impedance of transformer referred to secondary (*lv*)

$$Z_{02} = \sqrt{(R_{02})^2 + (X_{02})^2} = \sqrt{(0.008)^2 + (0.024)^2} = 0.0253\ \Omega \textbf{ Ans.}$$

Example 12.29. A 10 kVA, single-phase transformer for 2,000/400 V at no load has R_1 = 5.5 Ω, R_2 = 0.2 Ω, X_1 = 12 Ω, X_2 = 0.45 Ω. Determine the approximate value of the secondary voltage at full load, 0.8 pf (lagging), when the primary applied voltage is 2,000 V.

[*G.G.S.I.P. Univ. Delhi, May-June-2007*]

Solution: Transformation ratio, $K = \frac{400}{2,000} = 0.2$

Equivalent resistance referred to secondary,

$$R_{02} = K^2 R_1 + R_2 = 0.2^2 \times 5.5 + 0.2 = 0.42\ \Omega$$

Equivalent reactance referred to secondary,

$$X_{02} = K^2 X_1 + X_2 = 0.2^2 \times 12 + 0.45 = 0.93\ \Omega$$

Secondary full-load current,

$$I_2 = \frac{\text{KVA} \times 1{,}000}{V_2} = \frac{10 \times 1{,}000}{400} = 25 \text{ A}$$

Power factor, $\cos\phi = 0.8$ (lagging)

$$\text{Sin}\,\phi = \text{Sin}\cos^{-1} 0.8 = 0.6$$

Secondary terminal voltage at full load,

$$V_2 = KV_1 - I_2 R_{02} \cos\phi - I_2 X_{02} \sin\phi$$
$$= 0.2 \times 2{,}000 - 25 \times 0.42 \times 0.8 - 25 \times 0.93 \times 0.6$$
$$= 400 - 8.4 - 13.95 = 377.65 \text{ V } \textbf{Ans.}$$

Example 12.30. A single-phase load is supplied through a 34.5 kV feeder and 34.5/2.4 kV transformer. The feeder has an impedance of 50 + *j* 80 ohms and the transformer has an equivalent impedance of 24 + *j* 120 ohms referred to its high-voltage side. The load takes 260 kW at 2.3 kV and 0.866 lagging power factor.

(i) **Find the voltage at the primary side of the transformer.**
(ii) **Determine the voltage at the sending end of the feeder.**
(iii) **Calculate the real and reactive power input at the sending end of the feeder.**

[U.P.S.C. I.E.S. Electrical Engineering-II, 2009]

Solution: The equivalent circuit for the transformer and feeder, referred to *lv* (secondary) side is shown in Fig. 12.19.

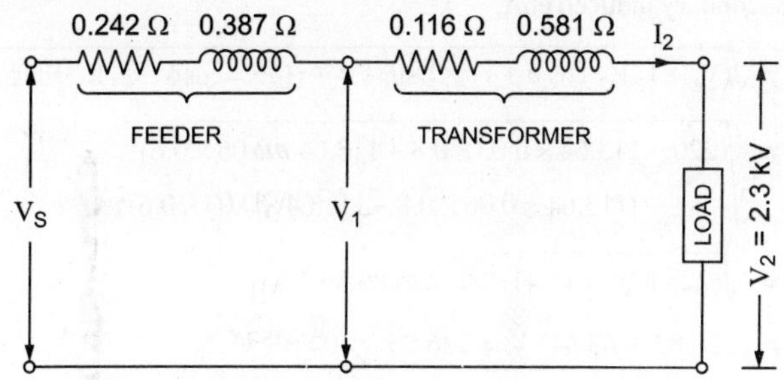

Fig. 12.19 *Equivalent Circuit Referred to LV Side*

$$\text{Transformation ratio, } K = \frac{2.4}{34.5} = \frac{1}{14.375}$$

Feeder impedance referred to *lv* side

$$Z_F = (50 + j\,80) \times \left(\frac{1}{14.375}\right)^2$$
$$= (0.242 + j\,0.387)\ \Omega = 0.4564 \angle 57.98^\circ\ \Omega$$

Equivalent impedance of transformer referred to *lv* side

$$Z_T = (24 + j\,120) \times \frac{1}{(14.375)^2}$$
$$= (0.116 + j\,0.581)\ \Omega = 0.5925 \angle 78.7^\circ\ \Omega$$

Secondary voltage, $V_2 = 2.3 \text{ kV} = 2{,}300 \angle 0^\circ$ V

Taking V_2 as the reference

$$\text{Load current, } I_2 = \frac{\text{Load in kW} \times 1{,}000}{V_2 \times \cos\phi_2}$$
$$= \frac{260 \times 1{,}000}{2{,}300 \times 0.866} = 130.53 \angle -30^\circ \text{ A}$$

(*i*) From Fig. 12.19

$$\mathbf{V_1 = V_2 + I_2 Z_T}$$
$$= 2{,}300 \angle 0^\circ + 130.53 \angle -30^\circ \times 0.5925 \angle 78.7^\circ$$
$$= 2{,}300 + 77.34 \angle 48.7^\circ$$
$$= (2{,}351 + j\,58.1) \text{ V} = 2{,}351 \angle 1.42^\circ \text{ V}$$

(*i*) Voltage on the primary side of the transformer

$$= \frac{V_1}{K} = 2{,}351 \times (14.375) = 33{,}795 \text{ V or } 33.8 \text{ kV } \textbf{Ans.}$$

(*ii*) Voltage at the sending end of the feeder, referred to *lv* side is given by

$$\mathbf{V_S = V_2 + I_2 (Z_F + Z_T)}$$
$$= (2{,}300 + j0) + 130.53 \angle -30^\circ [(0.242 + 0.116) + j\,(0.387 + 0.581)]$$
$$= (2{,}300 + j0) + 130.53 \angle -30^\circ \times 1.032 \angle 69.7^\circ$$
$$= (2{,}300 + j0) + 134.7 \angle 39.7^\circ$$
$$= (2403.64 + j\,86.04) \text{ V} = 2{,}405 \angle 2.05^\circ \text{ V}$$

Voltage at the sending end of the feeder, referred to *hv* side

$$= \frac{V_S}{k} = 2{,}405 \times (14.375) = 34{,}574 \text{ V or } 34.57 \text{ kV } \textbf{Ans.}$$

$$\text{Active power loss} = (130.53)^2 \times (0.242 + 0.116)$$
$$= 6{,}099.6 \text{ W or } 6.1 \text{ kW}$$
$$\text{Reactive power loss} = (130.53)^2 \times (0.387 + 0.581)$$
$$= 16{,}493 \text{ VAR or } 16.493 \text{ kVAR}$$
$$\text{Complex power at load end} = 260 + j\,260 \tan(\cos^{-1} 0.866)$$
$$= (260 + j\,150) \text{ kVA}$$

Complex power at the sending end of the feeder

$$= (260 + j\,150) + (6.1 + j\,16.493)$$
$$= (266.1 + j\,166.493) \text{ kVA } \textbf{Ans.}$$

12.14 EQUIVALENT CIRCUIT OF A TRANSFORMER

The equivalent circuit of any device can be quite helpful in predetermination of the behaviour of the device under various conditions of operation and it can be drawn if the equations describing its behaviour are known. If any electrical device is to be analysed and investigated further for suitable modifications, its appropriate equivalent circuit is necessary. The equivalent circuit for electromagnetic devices consists of a combination of resistances, inductances, capacitances, voltage etc. Such an equivalent circuit (or circuit model) can, therefore, be analysed and studied easily by the direct application of electric circuit theory.

As stated above, equivalent circuit is simply a circuit representation of the equations describing the performance of the device. Eqs. (12.12) and (12.13) describe the behaviour of the transformer under load and are helpful in arriving at the transformer equivalent circuit.

Equivalent circuit of a transformer having transformation ratio $K = E_2/E_1$ is shown in Fig. 12.20.

The induced emf in primary winding E_1 is primary applied voltage V_1 less primary voltage drop. This voltage causes iron loss current I_e and magnetising current I_m and we can, therefore, represent these two components of no-load current by the current drawn by a non-inductive resistance R_0 and pure reactance X_0 having the voltage E_1 or (V_1—primary voltage drop) applied across them, as shown in Fig. 12.20.

$$\text{Secondary current, } I_2 = \frac{I_1'}{K} = \frac{I_1 - I_0}{K}$$

Terminal voltage V_2 across load is induced emf E_2 in secondary winding less voltage drop in secondary winding.

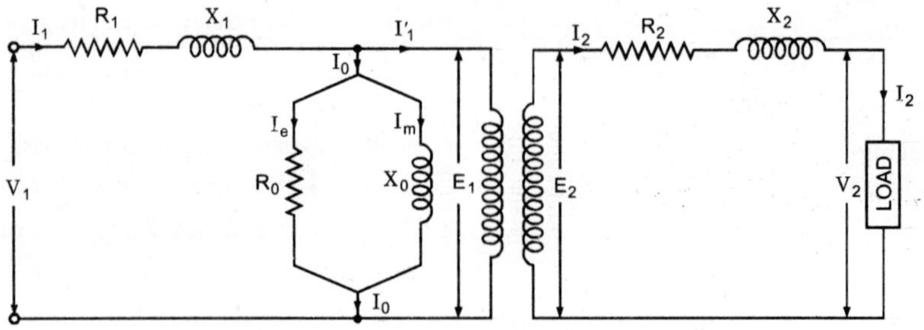

Fig. 12.20 *Equivalent Circuit of a Transformer*

The equivalent circuit can be simplified by transferring the voltage, current and impedance to the primary side. After transferring the secondary voltage, current and impedance to primary side equivalent circuit is reduced to that shown in Fig. 12.21.

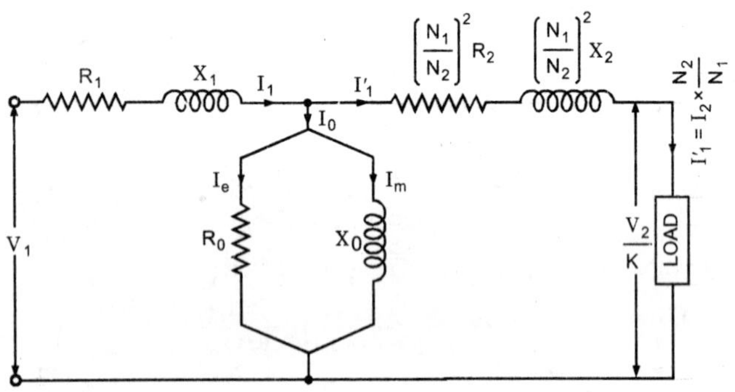

Fig. 12.21 *Equivalent Circuit of a Transformer With All Secondary Impedances Transferred To Primary Side*

The equivalent circuit diagram can further be simplified by transferring the resistance R_0 and reactance X_0 towards left end, as shown in Fig. 12.22. The error introduced by doing so is very small and can be neglected.

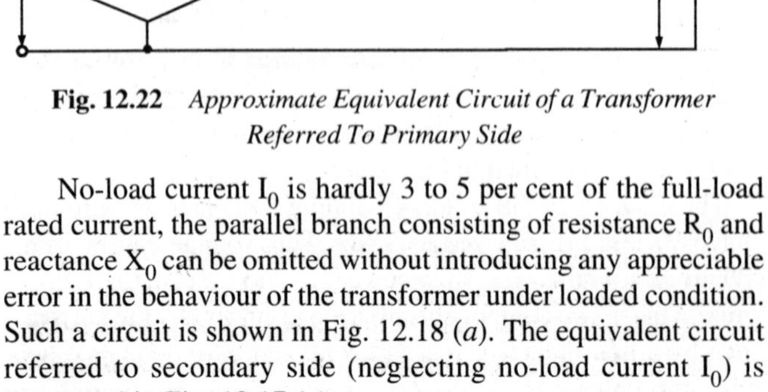

Fig. 12.22 *Approximate Equivalent Circuit of a Transformer Referred To Primary Side*

No-load current I_0 is hardly 3 to 5 per cent of the full-load rated current, the parallel branch consisting of resistance R_0 and reactance X_0 can be omitted without introducing any appreciable error in the behaviour of the transformer under loaded condition. Such a circuit is shown in Fig. 12.18 (*a*). The equivalent circuit referred to secondary side (neglecting no-load current I_0) is illustrated in Fig. 12.17 (*a*).

Example 12.31. A 25 kVA, 2200/220 V, 50 Hz single-phase transformer has the following parameters.

$r_1 = 1.2\ \Omega$; $x_1 = 6.0\ \Omega$; $r_c = 1{,}000\ \Omega$; $x_m = 300\ \Omega$ hv side

$r_2 = 0.03\ \Omega$; $x_2 = 0.06\ \Omega$ (lv side)

Draw the equivalent circuit referred to lv side for a 0.8 pf lagging load being supplied at 220 V.

[*U.P. Technical Univ. Electrical Engineering, June-2001*]

Solution: Rated secondary current, $I_2 = \dfrac{\text{Rated kVA} \times 1{,}000}{\text{Rated secondary voltage}}$

$$= \frac{25 \times 1{,}000}{220} = 113.64 \text{ A}$$

Secondary resistance, $r_2 = 0.03\ \Omega$

Secondary leakage reactance, $x_2 = 0.06\ \Omega$

Power factor $\cos\phi = 0.8$; $\sin\phi = 0.6$

Secondary induced emf,

$$E_2 = \sqrt{(V_2 + I_2 R_2 \cos\phi + I_2 X_2 \sin\phi)^2 + (I_2 X_2 \cos\phi - I_2 R_2 \sin\phi)^2}$$

$$= \sqrt{\begin{aligned}&(220 + 113.64 \times 0.03 \times 0.8 + 113.64 \times 0.06 \times 0.6)^2 \\ &+ (113.64 \times 0.06 \times 0.8 - 113.64 \times 0.03 \times 0.6)^2\end{aligned}}$$

$$= \sqrt{(226.82)^2 + (3.41)^2} = 226.85 \text{ V}$$

$$\mathbf{E_2} = (226.82 + j\,3.41) \text{ V} = 226.85 \angle 0.86^\circ \text{ V}$$

$$\mathbf{I_2} = 113.64\,(0.8 - j\,0.6) \text{ A} = (90.91 - j\,68.2) \text{ A}$$

Transformation ratio,

$$K = \frac{E_2}{E_1} = \frac{220}{2{,}200} = 0.1$$

Shunt branch resistance referred to *lv* side

$$r_c' = r_c \times K^2 = 1{,}000 \times (0.1)^2 = 10\ \Omega$$

Shunt branch reactance referred to *lv* side

$$x_m' = x_m \times K^2 = 300 \times (0.1)^2 = 3\ \Omega$$

Shunt admittance referred to *lv* side,

$$Y_0 = \frac{1}{r_c} - \frac{j}{x_m} = \frac{1}{10} - \frac{j}{3} = (0.1 - j0.333)\mho = 0.348 \angle -73.3^\circ\ \mho$$

Exciting current,

$$I_0 = Y_0 E_2 = 0.348 \angle -73.3^\circ \times 226.85 \angle 0.86^\circ$$

$$= 78.94 \angle -72.44^\circ = (23.82 - j\,75.26) \text{ A}$$

Equivalent secondary current,

$$\mathbf{I_2'} = \mathbf{I_2} + \mathbf{I_0} = (90.91 - j\,68.2) + (23.82 - j\,75.26)$$

$$= (114.73 - j\,143.46) = 183.7 \angle -51.35^\circ \text{ A}$$

$$K\mathbf{V_1} = \mathbf{E_2} + K^2\,\mathbf{I_2'}\,(R_1 + j\,X_1)$$

$$= (226.82 + j\,3.41) + (114.73 - j\,143.46)\,(0.1)^2\,(1.2 + j\,6)$$

$$= 237 \text{ V}$$

Equivalent circuit is shown in Fig. 12.23.

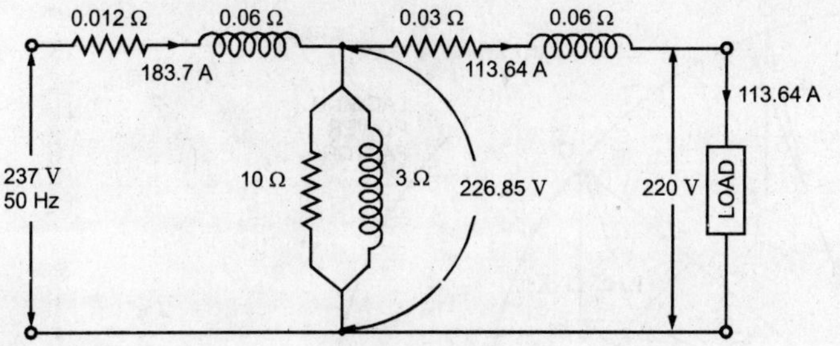

Fig. 12.23

12.15 VOLTAGE REGULATION

The way in which the secondary terminal voltage varies with the load depends upon load current, the internal impedance and the load power factor. The change in secondary terminal voltage from no load to full load, with primary voltage and frequency held constant, is termed the *inherent regulation.*

The figure of merit which determines the voltage characteristic of a transformer is termed as *voltage regulation.* It is usually expressed as a percentage of the rated terminal (secondary) voltage. It is defined as the change in magnitude of terminal (secondary) voltage when full load (rated load) of specified power factor supplied at rated voltage is thrown off (reduced to no load) with primary voltage and frequency held constant, as percentage of the rated terminal voltage.

If V_2 is the secondary (terminal) voltage at full load at specified power factor and E_2 is the secondary (terminal) voltage at no load, then percentage regulation is given as

$$\text{Voltage regulation} = \frac{E_2 - V_2}{\text{Secondary rated voltage}} \times 100$$

As per IS, the secondary rated voltage of a transformer is equal to the secondary voltage at no load *i.e.* E_2

So percentage regulation

$$= \frac{E_2 - V_2}{E_2} \times 100$$

$$= \frac{\text{Voltage drop in transformer at full load}}{\text{No-load rated secondary voltage, } E_2} \times 100$$

$$= \frac{I_2 R_{02} \cos\phi + I_2 X_{02} \sin\phi}{E_2} \times 100 \quad \text{...(12.26)}$$

When the power factor is leading, the percentage regulation is given as

$$\text{Percentage regulation} = \frac{I_2 R_{02} \cos\phi - I_2 X_{02} \sin\phi}{E_2} \times 100 \quad \text{...(12.27)}$$

The percentage regulation can be given as

$$\%\ \text{regulation} = \frac{I_2 R_{02} \cos\phi \pm I_2 X_{02} \sin\phi}{E_2} 100$$

+ ve sign for lagging power factor and

– ve sign for leading power factor

$$= \left(\frac{I_2 R_{02}}{E_2} \cos\phi \pm \frac{I_2 X_{02}}{E_2} \sin\phi\right) \times 100$$

$$= (\text{PU resistance} \times \cos\phi \pm \text{PU reactance} \sin\phi) \times 100 \quad \text{...(12.28)}$$

Voltage regulation of a transformer, on an average, is about 4 per cent.

From the consumer's viewpoint, voltage regulation due to variations in load is undesirable and should be kept smallest possible.

12.15.1. Condition For Zero Regulation.

$$\text{Regulation} = \frac{I_2 R_{02} \cos\phi + I_2 X_{02} \sin\phi}{E_2}$$

Regulation will be zero if the numerator will be equal to zero or $I_2 R_{02} \cos\phi + I_2 X_{02} \sin\phi = 0$

$$\text{or } \tan\phi = \frac{-R_{02}}{X_{02}} \quad \text{... (12.29)}$$

The –ve sign indicates that zero regulation occurs at a leading power factor.

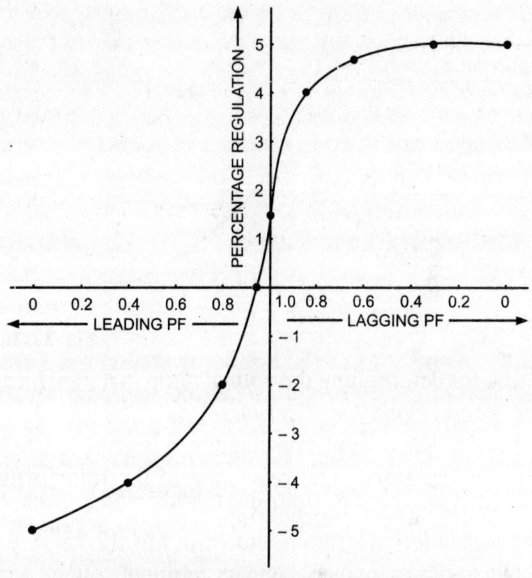

Fig. 12.24 *Percentage Regulation Vs Power Factor (R = 0.02 PU, X = 0.06 PU)*

12.15.2. Condition For Maximum Regulation. Regulation will be maximum if $\frac{d}{d\phi}$ (regulation) = 0

$$\text{or } \frac{d}{d\phi} \frac{I_2 R_{02} \cos\phi + I_2 X_{02} \sin\phi}{E_2} = 0$$

$$\text{or } -\frac{I_2 R_{02}}{E_2} \sin\phi + \frac{I_2 X_{02}}{E_2} \cos\phi = 0$$

$$\text{or } \tan\phi = \frac{X_{02}}{R_{02}} \quad \text{... (12.30)}$$

i.e. Maximum regulation occurs at lagging power factor.

12.16 KAPP REGULATION DIAGRAM

As discussed, the secondary terminal voltage falls with the increase in load on the transformer in case of inductive load (lagging power factor) and may increase with the increase in load in case of capacitive load (leading power factor) i.e. secondary terminal voltage not only depends on the load but on power factor also.

The graphical method of determining the voltage drop in a transformer was proposed by late Dr. Kapp and the graphical

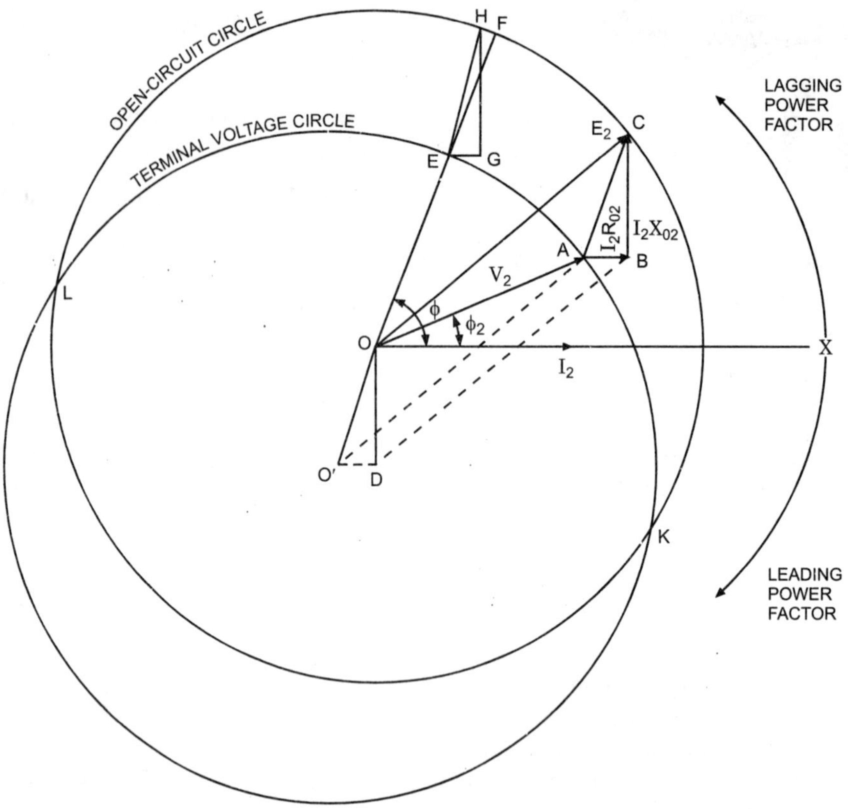

Fig. 12.25 *Kapp Regulation Diagram*

construction for determining the voltage drop in a transformer is called the *Kapp regulation diagram.*

For drawing Kapp regulation diagram, it is necessary to know the equivalent resistance and reactance of the transformer as referred to secondary i.e. R_{02} and X_{02}.

Kapp regulation diagram is shown in Fig. 12.25. A phasor OA is drawn to represent the secondary terminal voltage on load, V_2 and OX is drawn inclined at an angle ϕ_2 to OA, where cos ϕ_2 is the power factor of load. Then OX represents the phase of secondary current. Draw AB parallel to OX and equal to the resistance drop referred to secondary, *i.e* $I_2 R_{02}$; and the perpendicular BC equal to reactance drop referred to the secondary *i.e.* $I_2 X_{02}$. Then CAB is the impedance triangle referred to the secondary, and CA is the total voltage drop. Hence OC is the secondary no-load voltage E_2 and the arithmetic difference (OC – OA) is the secondary drop from no load to full load.

Now if the triangle CAB is transferred to OO′D, as shown, then O′A = OC = E_2. Hence for a given secondary current, the locus of C is a circle with centre O and radius E_2, while the locus of A is a circle of radius E_2 with centre O′. In order to determine the voltage drop on full load at any power factor cos φ, a radius OEF is drawn inclined at an angle φ to OX. If the impedance triangle is drawn in the position HEG, then OH = OF, the secondary no-load voltage, E_2 and OE = secondary voltage on load, V_2. Hence length EF represents the voltage drop. It is obvious that the triangle HEG need not be drawn, but simply the radius OEF. It will be seen that, for all phase angles φ included between ∠ XOK (which is leading) and ∠ LOX (which is lagging), there is a voltage drop on load. For phase angles equal to either of these two limits the voltage on load is equal to that on no-load. For phase angles between ∠LOX and ∠XOK (reckoning in a counter-clockwise direction) there is actually increase of voltage on no load.

The Kapp diagram is very helpful in determining the variation of regulation with power factor, but it has the disadvantage that since the lengths of the sides of the impedance triangle are very small in comparison to the radii of the circles, the diagram has to be drawn on a very large scale for accuracy.

Example 12.32. A single-phase 6300 kVA, 50 Hz, 3300 V/400 V distribution transformer is connected between two 50 Hz supply systems, A and B as shown in Fig. 12.26. The transformer has 12 and 99 turns in the low and high voltage windings respectively. The magnetizing reactance of the transformer referred to the high voltage side is 500 Ω. The leakage reactance of the high and low voltage windings are 1.0 Ω and 0.012 Ω respectively. Neglect the winding resistance and core losses of the transformer. The Thevenin's voltage of system A is 3300 V while that of system B is 400 V. The short-circuit reactance of systems A and B are 0.5 Ω and 0.010 Ω respectively. If no power is transferred between A and B, so that the two system voltages are in phase, find the magnetizing ampere turns of the transformer. [*GATE 2002*]

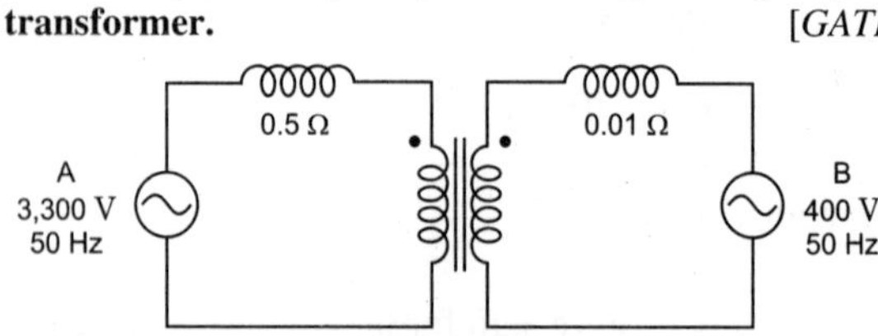

Fig. 12.26

Solution: Equivalent circuit for the Example 12.32 is shown in Fig. 12.27(*a*).

Primary (or hv) voltage, V_1 = 3300 V

Referring to hv side leakage reactance,

$$X_2' = \left(\frac{99}{12}\right)^2 \times 0.012 = 0.8165\ \Omega$$

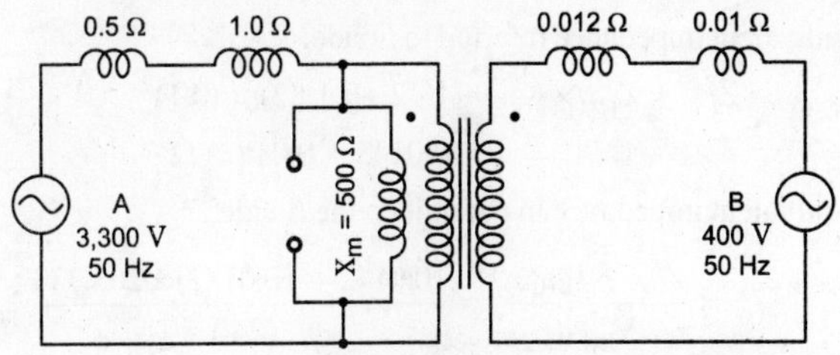

(*a*) *Equivalent*

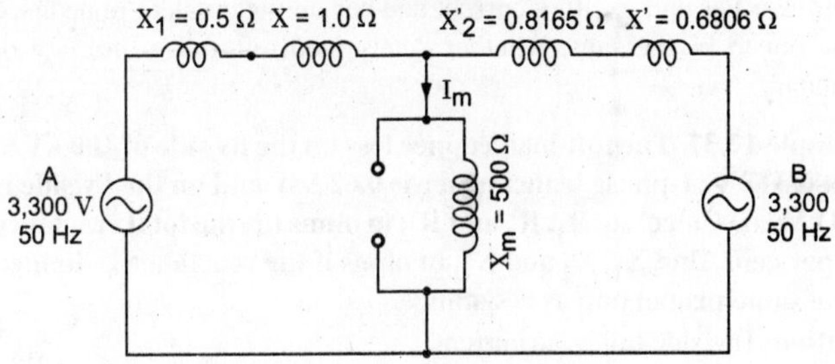

(*b*) *Equivalent Circuit Referred To HV Side*

Fig. 12.27

Short-circuit reactance,

$$X' = \left(\frac{99}{12}\right)^2 \times 0.01 = 0.6806\ \Omega$$

Equivalent circuit, referring all the quantities to *hv* side, is shown in Fig. 12.27(*b*).

Magnetising current,

$$I_m = \frac{3300}{0.5+1.0+500} + \frac{3300}{0.8165+0.6806+500}$$

$$= 6.58 + 6.58 = 13.16\ \text{A}$$

Hence magnetizing ampere-turns = 99 × 13.66 = 1302.84 **Ans.**

Example 12.33. A 20 kVA, 2,500/500 V, 1-ϕ transformer has the parameters $R_1 = 8\ \Omega$, $R_2 = 0.3\ \Omega$, $X_1 = 17\ \Omega$ and $X_2 = 0.7\ \Omega$.

Find the voltage regulation and the secondary terminal voltage for a pf of (*i*) 0.8 lagging (*ii*) 0.8 leading. The primary voltage is held constant at 2,500 V.

[*Rajasthan Univ. Electrical Machines-I, 2001, 2002*]

Solution: Transformation ratio, $K = \frac{E_2}{E_1} = \frac{500}{2,500} = 0.2$

Primary voltage, V_1 = 2,500 V

Secondary voltage on no load, $E_2 = KE_1 \simeq KV_1 = 0.2 \times 2,500 = 500$ V

Equivalent resistance as referred to secondary,

$$R_{02} = K^2R_1 + R_2 = (0.2)^2 \times 8 + 0.3 = 0.62\ \Omega$$

Equivalent reactance as referred to secondary,

$$X_{02} = K^2X_1 + X_2 = (0.2)^2 \times 17 + 0.7 = 1.38\ \Omega$$

Full-load secondary current,

$$I_2 = \frac{\text{KVA} \times 1,000}{V_2} = \frac{20 \times 1,000}{500} = 40\ \text{A}$$

$\text{Cos}\ \phi = 0.8$ lagging

$$\text{Sin}\ \phi = \sqrt{1-\cos^2\phi} = \sqrt{1-0.8^2} = 0.6$$

Voltage drop in transformer windings at full load and 0.8 pf lagging

$$= I_2 R_{02} \cos\phi + I_2 X_{02} \sin\phi$$

$$= 40 \times 0.62 \times 0.8 + 40 \times 1.38 \times 0.6 = 52.96\ \text{V}$$

Secondary terminal voltage, $V_2 = E_2$ – voltage drop

= 500 – 52.96 = 447.04 V **Ans.**

$$\text{Voltage regulation} = \frac{I_2 R_{02} \cos\phi + I_2 X_{02} \sin\phi}{E_2} \times 100$$

$$= \frac{52.96}{500} \times 100 = 10.592\%\quad \textbf{Ans.}$$

Voltage drop in transformer windings at full load and 0.8 pf leading

$$= I_2R_{02} \cos\phi - I_2 X_{02} \sin\phi$$

$$= 40 \times 0.62 \times 0.8 - 40 \times 1.38 \times 0.6 = -13.28\ \text{V}$$

Secondary terminal voltage,

$V_2 = E_2$ – voltage drop

= 500 – (–13.28) = 513.28 V **Ans.**

$$\text{Voltage regulation} = \frac{-13.28}{500} \times 100 = -2.656\%\quad \textbf{Ans.}$$

Example 12.34. A 25 kVA, 230/115 V, 50 Hz, 1-ϕ, transformer has the following data:

R_1	**0.12 Ω**
R_2	**0.04 Ω**
X_1	**0.2 Ω**
X_2	**0.05 Ω**

Find the transformer loading which will make the primary induced emf equal in magnitude to the primary terminal voltage when the transformer is carrying the full-load current. Neglect the magnetizing current.

[*G.B. Technical Univ. Electromechanical Energy Conversion-I, 2012-13*]

Solution: Transformation ratio, $K = \frac{115}{230} = 0.5$

Equivalent resistance referred to primary (*hv*) side,

$$R_{o1} = R_1 + \frac{R_2}{K^2} = 0.12 + \frac{0.04}{(0.5)^2} = 0.28\ \Omega$$

Equivalent reactance referred to primary (*hv*) side,

$$X_{o1} = X_1 + \frac{X_2}{K^2} = 0.2 + \frac{0.05}{(0.5)^2} = 0.4\ \Omega$$

Primary induced emf will be equal to primary terminal voltage when

$I_1 R_{o1} \cos\phi + I_1 X_{o1} \sin\phi = 0$

$$\text{or}\ \ \text{Tan}\ \phi = \frac{-R_{o1}}{X_{o1}} = \frac{-0.28}{0.4} = -0.7$$

or $\phi = \text{Tan}^{-1}\ 0.7 = 35°$ (lead) $\because$ Tan ϕ is negative

and power factor = Cos ϕ = Cos 35° = 0.82 (leading) **Ans.**

Load in kW on full load = kVA × *pf* = 25 × 0.82 = 20.5 kW **Ans.**

Example 12.35. The primary and secondary windings of a 40 kVA, 6,600/250 V single-phase transformer have resistances of 10 Ω and 0.02 Ω respectively. The leakage reactance as referred to primary is 35 Ω. Find (*i*) voltage regulation at full load and power factor 0.8 lagging (*ii*) primary voltage required to circulate full-load current when secondary is short circuited.

[*Rajasthan Univ. Electrical Machines-I, 2005*]

Solution: Transformation ratio, $K = \frac{E_2}{E_1} = \frac{250}{6,600} = \frac{25}{660}$

Primary voltage on no load,

E_1 = 6,600 V

Equivalent resistance as referred to primary,

$$R_{01} = R_1 + \frac{R_2}{K^2} = 10 + \frac{0.02}{(25/660)^2} = 23.94\ \Omega$$

Equivalent reactance as referred to primary,

$$X_{01} = 35\ \Omega$$

Full-load primary current,

$$I_1 = \frac{\text{Rated kVA} \times 1{,}000}{V_1} = \frac{40 \times 1{,}000}{6{,}600} = 6.06\text{ A}$$

$$\text{Cos}\,\phi = 0.8$$

$$\text{Sin}\,\phi = \sqrt{1-0.8^2} = 0.6$$

$$\text{Voltage regulation} = \frac{I_1 R_{01} \cos\phi + I_1 X_{01} \sin\phi}{E_1} \times 100$$

$$= \frac{6.06 \times 23.94 \times 0.8 + 6.06 \times 35 \times 0.6 \times 100}{6{,}600}$$

$$= \frac{243.32}{6{,}600} \times 100 = 3.687\% \quad \textbf{Ans.}$$

When the secondary is short circuited, the load impedance reflected in primary side is zero and total impedance of the transformer referred to primary is

$$Z_{01} = \sqrt{R_{01}^2 + X_{01}^2} = \sqrt{(23.94)^2 + (35^2)} = 42.4\ \Omega$$

Primary voltage required to circulate full-load current,

$$V_{sc} = I_1\, Z_{01} = 6.06 \times 42.4 = 257\text{ V} \quad \textbf{Ans.}$$

Example 12.36. The exciting current measured on the low voltage side of a 50 kVA, 2,400/240 V transformer is 5.41 A. Its equivalent impedance referred to the high voltage is (1.42 + *j* 1.82) ohms. Take the transformer rating as base.

(*i*) Express the exciting current in pu on the LV and HV sides and

(*ii*) Express the equivalent impedance in pu on the high voltage side and also on the low voltage side.

[*U.P. Technical Univ. Electromechanical Energy Conversion-I, 2005-06*]

Solution: Base kVA = Rated kVA = 50

Base LV winding voltage, $V_{B\,(LV)} = 240$ V

Base HV winding voltage, $V_{B(HV)} = 2{,}400$ V

Base LV winding current,

$$I_{B\,(LV)} = \frac{\text{Base kVA} \times 1{,}000}{V_{B(LV)}} = \frac{50 \times 1{,}000}{240} = 208.333\text{ A}$$

$$\text{Base HV winding current, } I_{B\,(HV)} = \frac{50 \times 1{,}000}{2{,}400} = 20.833\text{ A}$$

$$\text{Transformation ratio, } K = \frac{240}{2{,}400} = 0.1$$

No-load current on the low-voltage side,

$$I_0 = 5.41\text{ A}$$

No-load current on the high-voltage side,

$$I_0' = I_0 \times K = 5.41 \times 0.1 = 0.541\text{ A}$$

(*i*) Exciting current on the LV side,

$$I_{0\,(PU)} = \frac{I_0}{I_{B(LV)}} = \frac{5.41}{208.333} = 0.025968 \quad \textbf{Ans.}$$

Exciting current on the HV side,

$$I'_{0\,(PU)} = \frac{I_0'}{I_{B\,(HV)}} = \frac{0.541}{20.833} = 0.025968 \quad \textbf{Ans.}$$

(*ii*) Equivalent impedance in per unit on the *hv* side,

$$Z_{HV\,(PU)} = \frac{Z_{HV} \times I_{B(HV)}}{V_{B(HV)}} = \frac{(1.42 + j1.82) \times 20.833}{2{,}400}$$

$$= (0.01232 + j\,0.0158) \quad \textbf{Ans.}$$

Equivalent impedance referred to *lv* side,

$$Z_{LV} = Z_{HV} \times K^2 = (1.42 + j\,1.82) \times (0.1)^2$$

$$= (0.0142 + 0.0182)\ \Omega$$

Equivalent impedance in per unit on the *lv* side,

$$Z_{LV\,(PU)} = \frac{Z_{LV} \times I_{B(LV)}}{V_{B(LV)}} = \frac{(0.0142 + j0.0182) \times 208.333}{240}$$

$$= (0.01232 + j\,0.0158) \quad \textbf{Ans.}$$

Remarks: Per unit exciting current and per unit equivalent impedance come out to be the same whether determined referred to primary or secondary.

Example 12.37. The full-load copper loss on the hv side of 100 kVA, 11,000/317 V, 1-phase transformer is 0.62 kW and on the *l* v side is 0.48 kW. (*i*) Calculate R_1, R_2 and R'_2 in ohms (*ii*) the total reactance is 4 per cent, find X_1, X_2 and X'_2 in ohms if the reactance is divided in the same proportion as resistance.

Solution: HV side full–load current,

$$I_1 = \frac{\text{Rated KVA} \times 1{,}000}{V_1} = \frac{100 \times 1{,}000}{11{,}000} = 9.09\text{ A}$$

LV side full–load current,

$$I_2 = \frac{100 \times 1{,}000}{317} = 315.46\text{ A}$$

Primary winding resistance,

$$R_1 = \frac{\text{Full-load copper loss in primary winding}}{I_1^2}$$

$$= \frac{0.62 \times 1{,}000}{(9.09)^2} = 7.5\ \Omega\ \textbf{Ans.}$$

Secondary winding resistance,

$$R_2 = \frac{\text{Full-load copper loss in secondary winding}}{I_1^2}$$

$$= \frac{0.48 \times 1{,}000}{(315.46)^2} = 0.00482\ \Omega$$

Transformation ratio,

$$K = \frac{V_2}{V_1} = \frac{317}{11{,}000} = 0.02882$$

Secondary resistance referred to primary,

$$R_2' = \frac{R_2}{K^2} = \frac{0.00482}{(0.02882)^2} = 5.8\ \Omega\ \textbf{Ans.}$$

Percentage reactance referred to hv side

$$= \frac{I_1 X_{01}}{V_1} \times 100 = 4$$

or transformer total reactance referred to primary,

$$X_{01} = \frac{4 \times V_1}{100\ I_1} = \frac{4 \times 11{,}000}{100 \times 9.09} = 48.4\ \Omega$$

$$\text{Given that } \frac{R_1}{R_2'} = \frac{X_1}{X_2'}$$

$$\text{or } \frac{R_1 + R_2'}{R_2'} = \frac{X_1 + X_2}{X_2'}$$

$$\text{or } \frac{7.5 + 5.8}{5.8} = \frac{48.4}{X_2'}$$

or secondary reactance referred to primary

$$X_2' = \frac{48.4 \times 5.8}{13.3} = 21.1\ \Omega\ \textbf{Ans.}$$

Primary winding reactance,

$$X_1 = X_{01} - X_2' = 48.4 - 21.1 = 27.3\ \Omega \textbf{ Ans.}$$

Secondary winding reactance,

$$X_2 = K^2 X_2' = \left(\frac{317}{11,000}\right)^2 \times 21.1 = 0.0175\ \Omega \textbf{ Ans.}$$

Example 12.38. Determine voltage regulation of transformer whose ohmic drop is 1.5% and leakage reactance drop is 5% at 0.8 pf lagging.

[*G.B. Technical Univ. Electromechanical Energy Conversion-I, 2011-12*]

Solution: Percentage regulation

$$= \frac{I_2 R_{02} \cos\phi \times I_2 X_{02} \sin\phi}{E_2} \times 100$$

$$= \frac{I_2 R_{02}}{E_2} \times 100 \cos\phi + \frac{I_2 X_{02}}{E_2} \times 100 \sin\phi$$

$= \%R \cos\phi + \%X \sin\phi$

$= 1.5 \times 0.8 + 5 \times 0.6 = 4.2\%$ **Ans.** $\because \cos\phi = 8$

and $\sin\phi = \sqrt{1 - 0.8^2} = 0.6$

Example 12.39. Calculate the regulation of a transformer in which ohmic losses are 1% of output and reactance drop 5% of the voltage when the pf is (*i*) 0.8 lagging (*ii*) 0.8 leading and (*iii*) unity.

[*M.D. Univ. Electromechanical Energy Conversion, May-2007*]

Solution: Per unit resistance $= \frac{\text{Percentage copper loss}}{100} = \frac{1}{100} = 0.01$

Per unit reactance = Per unit reactance drop $= \frac{5}{100} = 0.05$

(*i*) When pf, $\cos\phi = 0.8$ lagging

$\sin\phi = 0.6$

Percentage voltage regulation = (PU resistance × $\cos\phi$ + PU reactance × $\sin\phi$) × 100

$= (0.01 \times 0.8 + 0.05 \times 0.6) \times 100$

$= 3.8\%$ **Ans.**

(*ii*) When pf, $\cos\phi = 0.8$ leading

$\sin\phi = -0.6$

Percentage voltage regulation $= [0.01 \times 0.8 + 0.05 \times (-0.6)] \times 100$

$= -2.2\%$ **Ans.**

(*iii*) When pf is unity, $\cos\phi = 1.0$

$\sin\phi = 0$

Percentage voltage regulation $= (0.01 \times 1.0 + 0.05 \times 0) \times 100$

$= 1\%$ **Ans.**

Example 12.40. Calculate the percentage voltage drop for a transformer with a percentage resistance of 2.5% and a percentage reactance of 5% of rating 500 kVA when it is delivering 400 kVA at 0.8 pf lagging.

[*J.N. Technological Univ. Electrical Machines-II, January-2014*]

Solution: Percentage resistance = 2.5%

Percentage reactance = 5%

Power factor, $\cos\phi = 0.8$

$$\sin\phi = \sqrt{1 - \cos^2\phi} = \sqrt{1 - 0.8^2} = 0.6$$

Fraction of full-load, $x = \frac{400}{500} = 0.8$

Percentage voltage drop = x (percentage resistance × $\cos\phi$ + percentage reactance × $\sin\phi$)

$= 0.8\,(2.5 \times 0.8 + 5 \times 0.6) = 4\%$ **Ans.**

Example 12.41. Define voltage regulation of transformer. A 100 kVA, 2,200/220 V transformer has leakage reactance drop of 8% and resistance drop of 2%. Find its voltage regulation at full load and 0.8 pf lagging. Find also the power factor at which the regulation will be zero.

[*U.P. Technical Univ. Electromechanical Energy Conversion-I, 2002-03*]

Solution : Regulation = PU resistance × 0.8 + PU reactance × 0.6

$= 0.02 \times 0.8 + 0.08 \times 0.6 = 0.064$ or 6.4% **Ans.**

Regulation will be zero when power factor

$$\cos\phi = \cos\tan^{-1}\frac{-R_{02}}{X_{02}}$$

$$= \cos\tan^{-1}\frac{-I_2 R_{02}/E_2}{I_2 X_{02}/E_2}$$

$$= \cos\tan^{-1}\frac{-0.02}{0.08} = 0.97 \text{ (lead)} \textbf{ Ans.}$$

Example 12.42. A 50 MVA, 76.2 / 33 KV, 1 – phase, 50 Hz, two-winding transformer with tap changer has percentage impedance of 0.5 + *j* 7.0. What tapping must be used to maintain rated voltage at the secondary on (*i*) full load at 0.8 lagging pf and (*ii*) 40 MVA load at 0.6 lagging pf. Assume that the tap changer is provided on the hv side.

Solution : From Eq. (12.26)

(*i*) Percentage regulation

$$= \frac{I_2 R_{02} \cos\phi + I_2 X_{02} \sin\phi}{E_2} \times 100$$

$$= \frac{I_2 R_{02}}{E_2} \times 100 \cos\phi + \frac{I_2 X_{02}}{E_2} \times 100 \sin\phi$$

$= \%R\cos\phi + \%X\sin\phi = 0.5 \times 0.8 + 7 \times 0.6 = 4.6\%$

$\because \cos\phi = 0.8$ and $\sin\phi = \sqrt{1-(0.8)^2} = 0.6$

i.e. voltage to be raised, for maintaining rated voltage = 4.6 %

So tap setting required on the hv side = – 4.6 % or 4.6 % down **Ans.**

(*ii*) At load of 40 MVA *i.e.* 80 % of full load or 0.8 times full load

So voltage regulation $= 0.8\,(\%R\cos\phi + \%X\sin\phi)$

$= 0.8\,(0.5 \times 0.6 + 7 \times 0.8) = 4.72\%$

$\because \cos\phi = 0.6$ and $\sin\phi = \sqrt{1-(0.6)^2} = 0.8$

Tap setting required on the hv side = 4.72 % down **Ans.**

Note: If the tap changer is on the primary side, tap setting will be down and if it is to be provided on secondary side, tap setting will be up for raising the secondary voltage on load.

Example 12.43. A transformer has a reactance drop of 5 % and a resistance drop of 2.5 %. Find the lagging power factor at which the voltage regulation is maximum and the value of this regulation.

Solution : PU resistance $= \frac{2.5}{100} = 0.025$

PU reactance $= \frac{5}{100} = 0.05$

Voltage regulation will be maximum when phase angle

$$\phi = \tan^{-1}\frac{I_2 X_{02}/E_2}{I_2 R_{02}/E_2} = \tan^{-1}\frac{\text{PU reactance}}{\text{PU resistance}} = \tan^{-1}\frac{0.05}{0.025}$$

$= 63.435°$

Power factor $\cos\phi = \cos 63.435° = 0.447$ (lagging) **Ans.**

$\sin 63.435° = 0.8944$

Maximum regulation = PU resistance × cos ϕ + PU reactance × sin ϕ
= 0.025 × 0.447 + 0.05 × 0.8944
= 0.0559 or 5.59% **Ans.**

Example 12.44. A 2,200/220 V, 1-phase transformer has maximum possible voltage regulation of 6% and it occurs at a pf of 0.3. Find the load voltage at full-load at pf of 0.8 lead.

[U.P.S.C. I.E.S. Electrical Engineering-II, 2001]

Solution: Phase angle, $\phi = \text{Cos}^{-1}\, 0.3 = 72.54°$ (lagging)

$\because$ maximum regulation occur at lagging pf

Regulation will be maximum when

$$\tan \phi = \frac{\text{Per unit reactance}}{\text{Per unit resistance}}$$

or PU reactance = Per unit resistance × tan ϕ
= 3.18 PU resistance $\because$ Tan 72.54° = 3.18

Voltage regulation = (PU resistance × cos ϕ + PU reactance sin ϕ) × 100

or 6 = (PU resistance × 0.3 + 3.18 PU resistance × 0.954) × 100

$$\text{or PU resistance} = \frac{6}{100 \times (0.3 + 3.0336)} = 0.018$$

PU reactance = PU resistance × 3.18 = 0.018 × 3.18
= 0.05724

PU voltage regulation = PU resistance cos ϕ + PU reactance × sin ϕ
= 0.018 × 0.8 + 0.05724 × (– 0.6) = – 0.02

$\because$ cos ϕ = 0.8 (leading) and sin ϕ = – 0.6

Load voltage at full load, $V_2 = E_2(1 - \text{PU voltage regulation})$
= 220(1 – (– 0.02)) = 220 × 1.02
= 224.4 V **Ans.**

12.17 TRANSFORMER LOSSES

The transformer is a static machine and, therefore, there are no friction or windage losses. The various power losses occuring in a transformer are enumerated below:

1. Iron or Core Losses. Iron loss is caused by the alternating flux in the core and consists of hysteresis and eddy current losses.

(*a*) Hysteresis Loss. The core of a transformer is subjected to an alternating magnetizing force and for each cycle of emf a hysteresis loop is traced out. The hysteresis loss per second is given by the equation

$$\text{Hysteresis loss, } P_h = \eta'(B_{max})^x f \text{v joules per second or watts} \quad ...(12.31)$$

where f is the supply frequency in Hz, v is the volume of core in cubic metres, η' is the hysteresis coefficient, B_{max} is peak value of flux density in the core and x lies between 1.5 and 2.5 depending upon the material and is often taken as 1.6.

(*b*) Eddy Current Loss. If the magnetic circuit is made up of iron and if the flux in the circuit is variable, currents will be induced by induction in the iron circuit itself. All such currents are known as *eddy currents.*

Eddy currents result in a loss of power, with consequent beating of the material.

The eddy current loss is given by equation

$$P_e = K_e (B_{max})^2 f^2 t^2 \text{ v watts or joules per second} \quad ...(12.32)$$

From the above expression for eddy current loss in a thin sheet it is obvious that eddy current loss varies (*i*) as the square of maximum flux density (*ii*) as the square of the frequency and (*iii*) as the square of thickness of laminations.

The hysteresis and eddy current losses depend upon the maximum flux density in the core and supply frequency. Since it has been determined that the mutual flux varies somewhat with the load (its variation being 1 to 3% from no load to full load), the core losses will vary somewhat with the load and its power factor. It may be emphasized here that *core losses are assumed to remain constant from no load to full load,* the variations in losses from no load to full load being very small and negligible.

These losses are determined from the open-circuit test (refer to Art. 12.18.3).

The input to the transformer with rated voltage applied to the primary and secondary open circuited is equal to the core loss.

These losses are minimized by using steel of high silicon content for the core and by using very thin laminations (0.3 mm to 0.5 mm) insulated from each other either by insulating varnish or by layer of papers.

2. Copper or Ohmic Losses. These losses occur due to ohmic resistance of the transformer windings. If I_1 and I_2 are the primary and secondary currents respectively and R_1 and R_2 are the respective resistances of primary and secondary windings then copper losses occuring in primary and secondary windings will be $I_1^2 R_1$ and $I_2^2 R_2$ respectively. So total copper losses will be $(I_1^2 R_1 + I_2^2 R_2)$. These losses vary as the square of the load current or kVA. For example if the copper losses at full load are P_c then copper losses at one-half or one-third of full load will be

$$\left(\frac{1}{2}\right)^2 P_c \text{ or } \left(\frac{1}{3}\right)^2 P_c \; i.e. \; \frac{P_c}{4} \text{ or } \frac{P_c}{9} \text{ respectively.}$$

Copper losses are determined on the basis of constant equivalent resistance R_{eq} determined from the short-circuit test (refer to Art. 12.18.5) and then corrected to 75 °C (since the standard operating temperature of electrical machines is taken 75 °C).

3. Stray Load Loss. Leakage fields present in a transformer induce eddy currents in the conductors, tank walls, bolts etc. These eddy currents give rise to stray load loss.

4. Dielectric Loss. This loss occurs in the insulating materials, particularly in oil and solid insulations.

The stray load loss and dielectric loss are very small and are, therefore, neglected. Thus the major losses are by far the first two: constant core (or iron) loss, P_i and the variable copper (or ohmic) loss, P_c. Therefore, only these two losses will be considered in futher discussions.

12.17.1. Power Flow Analysis. When the primary of a transformer is connected to an ac supply source, some of the energy supplied is lost due to flow of current through a resistance of the primary circuit and is converted into heat energy and the remainder into a magnetic field energy. Most of the flux of the magnetic field links with both of the primary and secondary circuits. The flux linking with both circuits is called the *mutual flux.* The remainder of the flux links either with primary or secondary (not both). The variation in mutual flux induces an emf in the secondary circuit

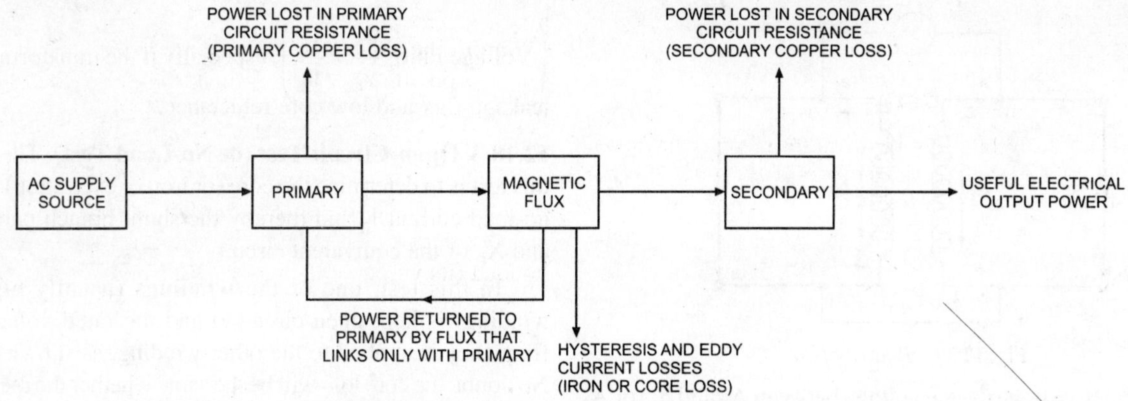

Fig. 12.28 *Power Flow Diagram For a Transformer*

and thereby energy of the magnetic field is converted into electrical energy in the secondary circuit whenever it is closed. All of the energy of the mutual magnetic field is not converted into electrical energy in the secondary circuit but some of the mutual magnetic field energy is converted into heat energy owing to hysteresis and eddy current phenomena in the iron core of the transformer. The energy of the portion of the magnetic field which consists of flux linking only with the primary is returned to the primary circuit as the field collapses. Some of the magnetic field energy converted into electrical energy in secondary circuit is lost due to flow of current in the secondary circuit resistance and the rest is supplied to the load. Power flow diagram for a transformer is given in Fig. 12.28.

Example 12.45. The flux in a magnetic core is alternating sinusoidally at a frequency of 600 Hz. The maximum flux density is 2 T and the eddy current loss is 15 W. Find the eddy current loss if the frequency is raised to 800 Hz and the maximum flux density is reduced to 1.5 T.

[Pb. Univ. Elements of Electrical Engineering Nov; 1994]

Solution: Since from Eq. (12.32)

The eddy current loss,

$$P_e \propto B_{max}^2 f^2$$

$$\therefore \frac{P_{e_2}}{P_{e_1}} = \left(\frac{B_{2\,max}}{B_{1\,max}}\right)^2 \left(\frac{f_2}{f_1}\right)^2$$

$$\text{or } P_{e2} = P_{e_1} \times \left(\frac{B_{2\,max}}{B_{1\,max}}\right)^2 \left(\frac{f_2}{f_1}\right)^2$$

$$= 15 \times \left(\frac{1.5}{2}\right)^2 \times \left(\frac{800}{600}\right)^2 = 15 \text{ watts } \textbf{Ans.}$$

Example 12.46. A transformer is connected to a 1000 V, 50 Hz supply. The total core loss is 1000 W of which 700 W are hysteresis and 300 W are eddy current loss. If the applied voltage is raised to 2000 V and the frequency to 100 Hz, find the new core loss.

[U.P.S.C. I.E.S. Electrical Engineering, 2004]

Solution: At 1,000 V, 50 Hz

Hysteresis loss, $P_{h_1} = 700$ W

Eddy current loss, $P_{e_1} = 300$ W

In a transformer, since induced emf, which is nearly equal to applied voltage is given by

$$E = 4.44\, \Phi_{max} f N$$

When applied voltage (*i.e.* E) and supply frequency, *f* both are doubled (2000 V and 100 Hz respectively), B_{max} will remain unchanged and

$$\text{Hysteresis loss, } P_{h_2} = P_{h_1} \frac{f_2}{f_1} = 700 \times 2 = 1{,}400 \text{ W}$$

$\because P_h \propto f$, B_{max} remaining constant

$$\text{Eddy current loss, } P_{e_2} = P_{e_1} \times \left(\frac{f_2}{f_1}\right)^2 = 300 \times 2^2 = 1{,}200 \text{ W}$$

$\because P_e \propto f^2$, B_{max} remaining constant

$$\text{Total core loss} = P_{h_2} + P_{e_2} = 1{,}400 + 1{,}200$$

$$= 2{,}600 \text{ watts } \textbf{Ans.}$$

12.18 TESTING OF TRANSFORMERS

The main difficulties encountered in testing of large power transformers by direct loading are (*i*) wastage of large amount of energy and (*ii*) a stupendous (impossible for large transformers) task of arranging a load large enough for direct loading. The performance characteristics of a transformer can be conveniently computed from the knowledge of its equivalent circuit parameters which, in turn, may be determined by conducting simple tests called the *open-circuit* or *no-load test* and *short-circuit* or *impedance test* involving very little power cosumption (power needed to supply the losses incurred).

Other tests to be conducted on transformers are *polarity test* and *voltage ratio test.*

12.18.1 Polarity Test. On the primary side of a two winding transformer, one terminal is positive with respect to the other one at any instant. At the same instant, one terminal of the secondary winding is positive with respect to the other one. *Polarity test* is performed to determine the terminals having the same instantaneous polarity (assuming that the terminals are not marked). The relative polarities of the primary and secondary terminals at any instant must be known for connecting windings of the same transformer in parallel, or series, or for interconnecting two or more transformers in parallel, or for connecting single phase transformers for polyphase transformation of voltages.

For determination of relative polarity of the two windings of a transformer the two windings are connected in series across a voltmeter, while one of the windings (preferably hv winding) is excited from a suitable ac voltage source, as illustrated in Fig. 12.29. If the polarities of the windings are as marked on the diagram *i.e.* when the transformer has a subtractive polarity, the voltmeter will read the difference of E_1 and E_2. If the voltmeter reads $E_1 + E_2$, the polarity markings of one of the windings must be interchanged.

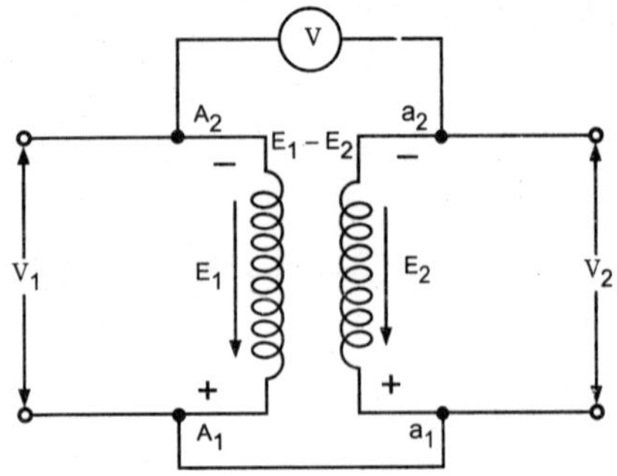

Fig. 12.29 *Polarity Test*

In subtractive polarity, the voltage between A_1 and a_1 (or A_2 and a_2) is reduced. The leads connected to these terminals and the two windings are, therefore, not subjected to high voltage stress. On the other hand in additive polarity the two windings and leads connected to A_1, A_2, a_1 and a_2 are subjected to high voltage stresses. This is the reason that subtractive polarity is preferred over additive polarity.

The above test may not be convenient for determination of relative polarities of the terminals of the two windings of a transformer in the field. Polarity test in the field can be conveniently carried out by using a dc battery, a switch and a dc (permanent mangnet type) voltmeter. The circuit diagram is shown in Fig. 12.30. As the switch on the primary side is closed, the primary current increases and so do the flux linkages of both the windings, inducing emfs in them. The positive polarity of this induced emf in the primary is at the end to which the battery is connected (according to Lenz's law). The end of secondary which simultaneously acquires positive polarity, as determined by the dc voltmeter is the similar polarity end. The reverse happens on opening of the switch *i.e.* the similar polarity end is that end which acquires negative potential.

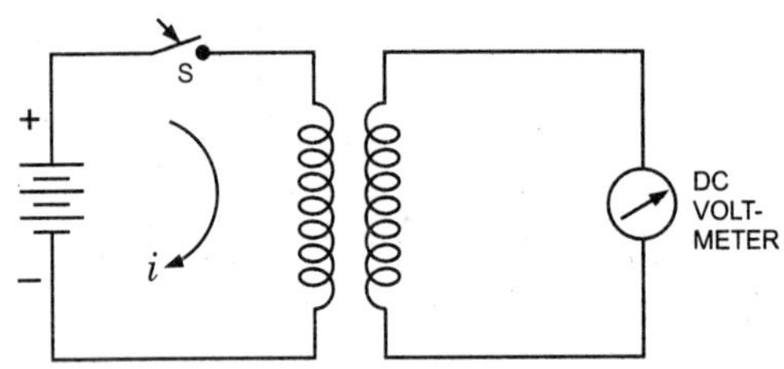

Fig. 12.30

12.18.2. Voltage Ratio Test. The true ratio is based on turn–ratio, as explained in Art. 12.6. If the secondary and primary voltages are measured on no load, their ratio is very nearly equal to the true value. Measurement of primary and secondary currents in short-circuit test also gives fairly accurate results $\left(\text{Voltage ratio } \frac{V_2}{V_1} = \frac{I_1}{I_2}\right)$, especially if the transformer has little leakage flux and low core reluctance.

12.18.3. Open-Circuit Test (or No-Load Test). The purpose of this test is to determine the core (or iron or excitation) loss, P_i and no-load current I_0 and thereby the shunt branch parameters R_0 and X_0 of the equivalent circuit.

In this test, one of the windings (usually high-voltage winding) is kept open circuited and the rated voltage at rated frequency is applied to the other winding, as shown Fig. 12.31. No doubt, the core loss will be the same whether the measurements are made on lv winding or hv winding so long as the rated voltage of that winding is applied to it but in case the measurements are made on hv winding, the voltage required to be applied would often be inconveniently large while the current I_0 would be inconveniently small.

Either an auto-transformer or a voltage divider (VD) is used for varying the voltage applied to the low-voltage winding. Ammeter A and wattmeter W are connected to measure no-load current I_0 and input power W_0. Voltmeter V is connected to measure the applied voltage.

Since no current flows in the open-circuited secondary, the current in the primary will be merely that necessary to magnetize the core at normal voltage. Moreover, this magnetising current is a very small fraction of the full-load current (usually 3 to 10% of full-load current) and may be neglected as far as the copper loss is concerned consequently, the test gives core loss alone practically.

With normal voltage applied to the primary, normal flux will be set up in the core and, therefore, normal iron (or core) loss will occur which are recorded by a wattmeter W.

The open-circuit test gives enough data to compute the equivalent circuit constants R_0, X_0, no–load power factor cos ϕ_0, no-load current I_0 and no load power loss (iron loss) of a transformer.

$$\text{Iron loss, } P_i = \text{Input power on no load}$$
$$= W_0 \text{ watts (say)}$$
$$\text{No-load current} = I_0 \text{ amperes}$$
$$\text{Applied voltage to primary} = V_1 \text{ volts}$$

$$\text{Angle of lag, } \phi_0 = \text{Cos}^{-1} \frac{W_0}{V_1 I_0} \quad \text{...(12.33)}$$

No-load current energy component,

$$I_e = I_0 \cos \phi_0 = \frac{W_0}{V_1} \quad \text{...(12.34)}$$

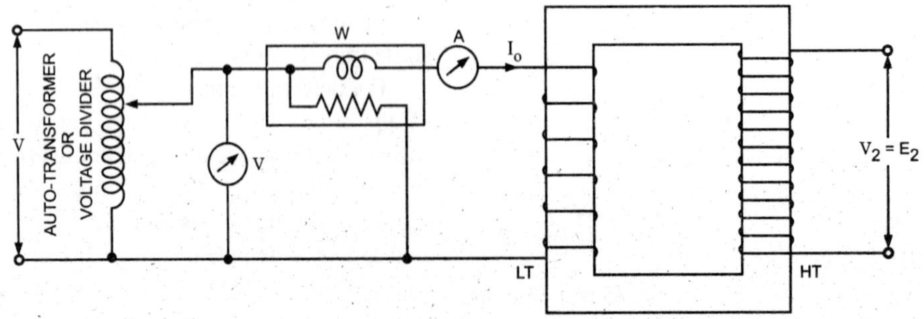

Fig. 12.31 *Circuit Diagram For Open-Circuit Test*

No-load current magnetizing component,

$$I_m = \sqrt{I_0^2 - I_e^2} \qquad ...(12.35)$$

Equivalent circuit parameter, $R_0 = \frac{V_1}{I_e} = \frac{V_1^2}{W_0}$ ...(12.36)

Equivalent circuit parameter, $X_0 = \frac{V_1}{I_m} = \frac{V_1}{\sqrt{I_0^2 - I_e^2}}$ ...(12.37)

Note. 1. Since no-load current I_0 is very small, pressure coils of wattmeter and the voltmeter should be connected such that the currents drawn by them do not flow through the current coils of the wattmeter and ammeter.

2. Since power factor at no load is quite low (in the range of 0.1–0.2 lag), a low power factor wattmeter should be used to ensure accurate measurements.

3. The error due to power loss in ammeter can be eliminated by short circuiting the ammeter while reading wattmeter.

4. Sometimes a high resistance voltmeter is connected across the secondary to indicate the emf induced in the secondary (hv winding). This helps in determination of transformation ratio K.

5. It must, however, be remembered that in making this test, hv side is hot and, therefore, its terminals must be properly insulated.

12.18.4. Separation of Hysteresis and Eddy Current Losses. In Art. 12.17 it has been mentioned that

$$\text{hystersis loss, } P_h \; \alpha \; (B_{max})^{1.6} f$$

$$\text{and eddy current loss, } P_e \; \alpha \; (B_{max})^2 f^2$$

It shows that for constant peak flux density the hysteresis loss varies as the frequency of supply while eddy current loss varies as the square of supply frequency.

For any transformer $\frac{V}{f} \; \alpha \; B_{max}$ and for a particular value of $\frac{V}{f}$ or of peak flux density core loss per cycle are given as

$$\frac{P_i}{f} = A + Bf \qquad ...(12.38)$$

where A is a constant and is equal to $K_h (B_{max})^{1.6}$ and B is another constant having value of $K_e (B_{max})^2$.

The values of constants A and B in Eq. (12.38) can be determined by performing open-circuit test on the transformer. The connections are the same as for open-circuit test (Fig. 12.31) except that the primary of the transformer is connected to variable voltage variable frequency source (a dc motor driven alternator). During this test, the applied voltage V and frequency f are varied together (by adjusting the excitation and speed of the alternator feeding the transformer under test) so as to keep $\frac{V}{f}$ (and, therefore, B_{max}) constant. The wattmeter connected in the circuit indicates the iron loss P_i.

For a series of frequencies, with corresponding changes in applied voltage, the measured input power divided by frequency $\left(\frac{P_i}{f}\right)$ and plotted against frequency will give a straight line, as shown in Fig. 12.32. The intercept on the vertical axis will give the value of constant A while the slope of the line will give the value of constant B. Knowing the values of constants A and B the hysteresis and eddy current loss components for any given frequency can be determined.

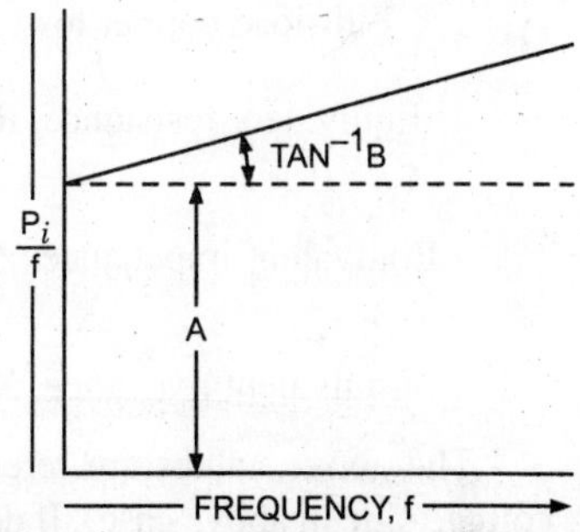

Fig. 12.32

Note: For the separation of iron loss into hysteresis and eddy current loss components for a 3-phase transformer, the readings must be converted to phase values, before plotting $\frac{P_i}{f}$ against f.

12.18.5. Short-Circuit Test (or Impedance Test). The purpose of this test is to determine full-load copper loss and equivalent resistance and equivalent reactance referred to metering side.

In this test, the terminals of secondary winding (usually of low-voltage winding*) are short circuited by a thick wire or strip or through an ammeter (which may serve the additional purpose of indicating secondary rated load current) and variable low voltage is applied to the primary through an auto-transformer or potential divider, as shown in Fig. 12.33. The transformer now becomes equivalent to a coil having an impedance equal to impedance of both the windings.

The applied voltage, V_s to the primary is gradually increased till the ammeter A indicates the full-load (rated) current of the metering side. Since applied voltage is very low (5–8% of the rated voltage), flux linking with the core is very small and, therefore, iron losses are so small that these can be neglected. Thus the power input (reading of wattmeter W) gives total copper

* Voltage required for the short-circuit test is about 5 per cent of the rated value. For a 200 kVA, 2200/220 V transformer, test on high voltage side would need the voltage (to be applied) of $2{,}200 \times \frac{5}{100}$ *i.e.* 110 V (which is standard voltage for instrument coils) and a current of $\frac{200 \times 1{,}000}{2{,}200} = 91$A.

If the test is conducted on low voltage side of the above transformer, the voltage needed would be $220 \times \frac{5}{100} = 11$ V (very small) and the current would be $\frac{200 \times 1{,}000}{220} = 910$ A (very high). At this low voltage, high precision would not be readily obtainable with ordinary instruments.

Thus we see that the if the measurements were made on low-voltage side, the voltage needed would be inconveniently low, while the current would often be inconveniently large.

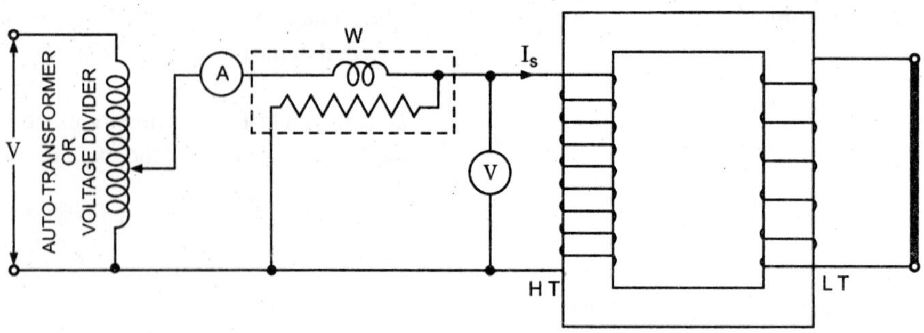

Fig. 12.33 *Circuit Diagram For Short-Circuit Test*

loss at rated load, output being nil. Let the readings of voltmeter, ammeter and wattmeter be V_s, I_s and W_s respectively.

$$\text{Full-load copper loss, } P_c = I_s^2 R_{eq} = W_s \quad ...(12.39)$$

$$\text{Equivalent resistance, } R_{eq} = \frac{W_s}{I_s^2} \quad ...(12.40)$$

$$\text{Equivalent impedance, } Z_{eq} = \frac{V_s}{I_s} \quad ...(12.41)$$

$$\text{Equivalent reactance, } X_{eq} = \sqrt{Z_{eq}^2 - R_{eq}^2} \quad ...(12.42)$$

The above values are referred to the metering side (high-voltage side in above case). If desired, the values could be easily determined referred to the other side, as discussed in Art. 12.13.

12.18.6. Sumpner's Test (or Back To Back Test). Though the efficiency and regulation of the transformer can be determined from open-circuit and short-circuit tests accurately but for determination of temperature rise it is necessary that the transformer be put on full load for a number of hours. Transformers of smaller output can be loaded artificially by means of water load or lamp load but it may be very difficult to arrange suitable loads for loading transformers of large rating. Further there is tremendous wastage of electrical energy.

In Sumpner's test (or back to back test) the two transformers are loaded fully, in similar way as two dc machines in a regenerative test, and the power required from the supply is that necessary for supplying the iron and copper losses of both transformers. For this test two similar transformers are required.

The circuit diagram for performing Sumpner's test is shown in Fig. 12.34.

The primary windings (usually low-voltage windings) of the two transformers are connected in parallel across a single-phase supply with a voltmeter V_1, ammeter A_1 and wattmeter W_1 in the circuit, as shown in the Fig. 12.34. The supply voltage V_1

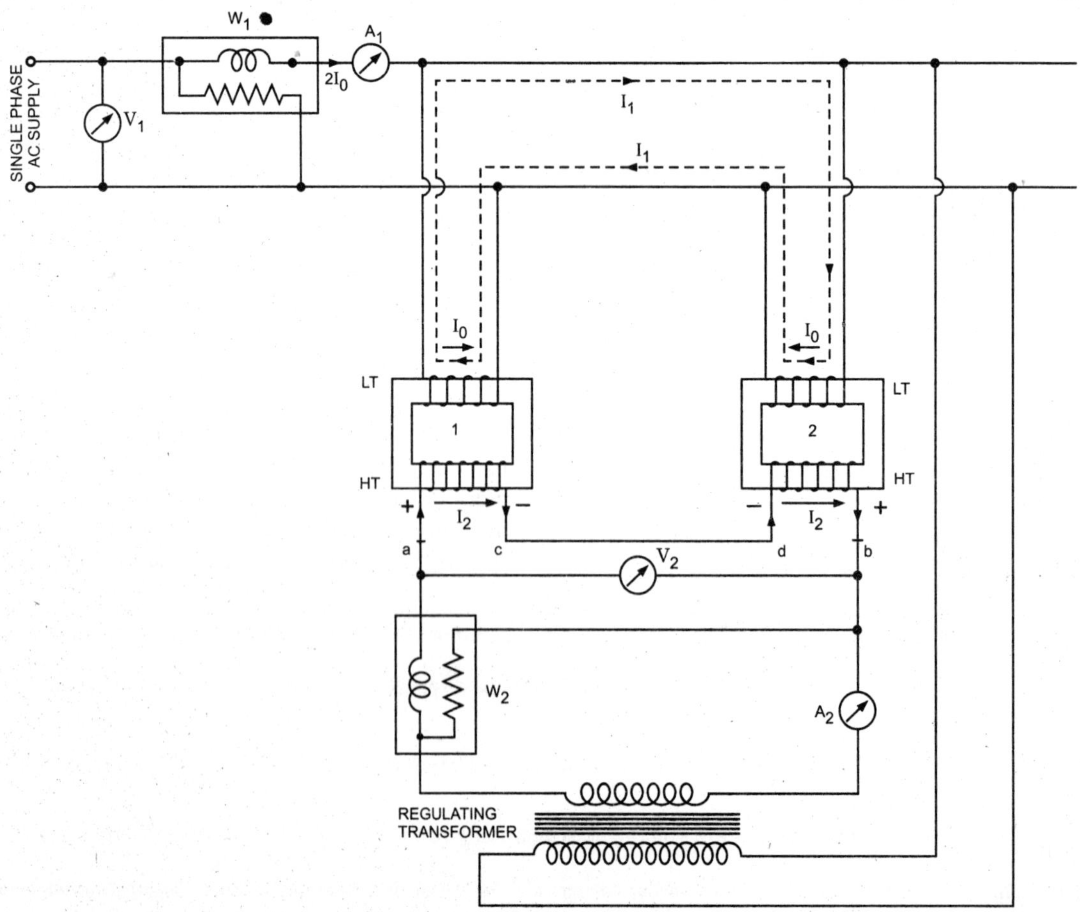

Fig. 12.34 *Circuit Diagram For Sumpner's Test*

must be equal to the rated voltage of the primary windings. The secondary windings of the transformers are then connected together so that their potentials are in opposition to each other. For connecting the two secondaries in opposition, any two terminals of the secondaries are first connected together and the voltage is measured between the other remaining two terminals by a voltmeter V_2 of double the range. If the voltage measured is double that of the secondary voltage of one transformer, it shows that two transformers are not connected in opposition. By interchanging the two connections, the two transformers will be in opposition and under that condition voltmeter V_2 will give zero reading. When the transformers are connected in opposition no current flows through the secondary windings, therefore, there is no secondary flux to oppose the primary flux or the secondary mmf is zero. Hence the transformers act as if their secondaries are open circuited and the wattmeter W_1 measures the iron losses of the two transformers.

Now the two transformers are loaded fully by applying a comparatively low voltage from the regulating transformer to the secondary circuit. The variation in this voltage will cause the change in the circulating current through the secondary windings. This voltage is adjusted to such a value that ammeter A_2 shows full-load secondary currents. The secondary circuit now acts as a pure resistive circuit since combined reactance of the two secondaries, being in opposition, is zero. The secondary current will cause a current to circulate in the two primaries having a path, as shown dotted in Fig. 12.34. The arrows indicate instantaneous direction of current in primaries due to the secondary current. Since the secondaries are connected in opposition, direction of currents in the two primaries are in opposition. Hence the reading of wattmeter W_1 is not affected although the primaries and secondaries of the two transformers carry full-load current. Thus the wattmeter W_2 gives the full-load copper losses of the transformers.

For the assumed directions of no-load currents I_0 and secondary currents I_2 in the two transformers, the primary current of transformer 1 is less (being phasor difference of I_1 and I_0) than that of transformer 2 (being phasor sum of I_1 and I_0) where $2\,I_0$ is the no-load current. Thus the two transformers do not operate under identical conditions—one may have slightly less temperature rise than the other.

For measurement of temperature rise of the two transformers, these are kept under rated load conditions for several hours till maximum steady temperature is attained.

Note: The voltage injected into the secondaries of the two transformers provides the copper losses occuring in them, and may be different from primary supply and may be adjusted to a slightly different frequency from that of the primary supply if this is more convenient.

Example 12.47. A 440 V/220 V, 50 Hz transformer, the total iron loss is 2,500 W, when applied voltage is 220 V at 25 Hz, the corresponding loss is 850 W. Calculate the eddy current loss at normal frequency and voltage.

[*J.N. Technological Univ. Electrical Machines-II, January-2014*]

Solution: When applied voltage and supply frequency are reduced in the same ratio (i.e. halved), B_{max} will remain unaltered.

At the same peak flux density

Hysteresis loss, $P_h \propto f = Af$ (say)

Eddy current loss, $P_e \propto f^2 = Bf^2$ (say)

Substituting $P_{h+e} = 2{,}500$ W where $f = 50$ Hz and $P_{h+e} = 850$ W for $f = 25$ Hz we have

$$2{,}500 = 50\,A + 50^2\,B = 50\,A + 2{,}500\,B \quad ...(i)$$

$$850 = 25\,A + 25^2\,B = 25\,A + 625\,B \quad ...(ii)$$

Solving Eq. (*i*) and (*ii*) we have

$$A = 18 \text{ and } B = 0.64$$

Eddy current loss at normal voltage of 440 V and frequency 50 Hz

$$= Bf^2 = 0.64 \times 50^2 = 1{,}600 \text{ W } \textbf{Ans.}$$

Example 12.48. In a power loss test on a 10 kg specimen of sheet steel laminations, the maximum flux density and waveform factor are maintained constant and the following results are obtained.

Frequency (Hz)	**25**	**40**	**50**	**60**	**80**
Total loss (watts)	**18.5**	**36**	**50**	**66**	**104**

Calculate the eddy current loss per kg at frequency of 50 Hz

Solution: At a given flux density and waveform factor, total iron losses are given as

$$P_{h+e} = P_h + P_e = Af + Bf^2 \text{ or } \frac{P_{h+e}}{f} = A + Bf$$

Total iron loss/cycle *i.e.* P_{h+e}/f for various values of frequency is given below:

f	25	40	50	60	80
P_{h+e}/f	0.74	0.9	1.0	1.1	1.3

A graph is plotted between P_{h+e}/f and f as illustrated in Fig. 12.35. From graph A = 0.5 and B = 0.01

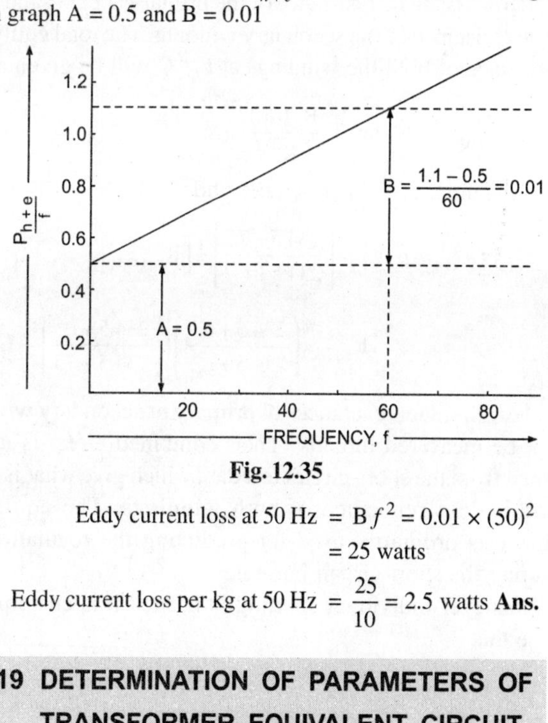

Fig. 12.35

$$\text{Eddy current loss at 50 Hz} = Bf^2 = 0.01 \times (50)^2 = 25 \text{ watts}$$

$$\text{Eddy current loss per kg at 50 Hz} = \frac{25}{10} = 2.5 \text{ watts } \textbf{Ans.}$$

12.19 DETERMINATION OF PARAMETERS OF TRANSFORMER EQUIVALENT CIRCUIT

The parameters for the approximate equivalent circuit can be had readily from open-circuit and short-circuit tests which involve the expenditure of only relatively little power.

The shunt branch parameters R_0 and X_0 of the equivalent circuit of a transformer may be determined, as already discussed in Art 12.18.3.

The values of R_0 and X_0 determined from open-circuit test are referred to the side (usually *lv*) from which the test is conducted and could easily be referred to the other side if so desired by the square of transformation ratio.

The equivalent resistance R_{eq} and equivalent reactance X_{eq} of the two windings referred to the metering side (usually hv side) may be determined from the short-circuit test (Art 12.18.5) and the values of R_{eq} and X_{eq} so determined could easily be referred to the other side, as discussed in Art 12.13.

The equivalent resistance (R_{eq}) of the two windings determined from short-circuit will be somewhat greater than the equivalent resistance of the two windings as determined from dc resistance readings*. The difference is caused by skin and proximity effects and by slight stray losses caused by the leakage flux. The difference between the power supplied under short-circuit condition and the copper loss computed from the equivalent dc resistance is called the *stray-load loss.* Since it is standard practice to base all characteristics of electrical equipment on the basis of operation at 75°C, the values of equivalent resistance determined from short-circuit test and d-c resistance determined from ammeter-voltmeter method (or Wheatstone bridge method) should be corrected to correspond to conditions of operation at 75 °C. However, the dc resistance increases with temperature, whereas the resistance caused by the stray-load loss decreases with temperature. Therefore, it is necessary to determine dc resistance for each of the windings in addition to short-circuit test. R_{eq} at 75 °C may be determined as below.

If R_{eq} (*sc*) is the total equivalent resistance at temperature t_{sc} °C, $R_1(dc)$ is the dc resistance of the primary at t_{dc} °C and $R_{2(dc)}$ is the dc resistance of the secondary winding. The total equivalent dc resistance of both the windings at t_{dc} °C will be given as

$$R_{eq\,(dc)} = R_1(\text{dc}) + \frac{R_2(dc)}{K^2} \qquad \text{...(12.43)}$$

where K is the transformation ratio and

$$R_{eq} \text{ at } 75\,°\text{C} = R_{eq\,(dc)}\left[\frac{234.5+75}{234.5+t_{dc}}\right] + \left[R_{eq\,(sc)} - R_{eq\,(dc)}\left(\frac{234.5+t_{sc}}{234.5+t_{dc}}\right)\right]\left[\frac{234.5+t_{sc}}{234.5+75}\right] \qquad \text{...(12.44)}$$

The individual reactances of primary or secondary windings cannot be measured directly. Their combined effect is usually obtained from the short-circuit test data which give what is called the equivalent reactance of both windings. The equivalent reactance is ordinarily used for predicting the regulation and predicting the short-circuit current.

Whenever individual splitting is needed it is customary to assume that

$$X_1 = \frac{X_2}{K^2} = \frac{X_{01}}{2} \qquad \text{...(12.45)}$$

The above statement is based on the assumption that the transformer is well designed.

Example 12.49. An approximate equivalent circuit of 4 kVA, 200/400 V 1-ϕ transformer, referred to LV side, is shown in Fig. 12.36

(*i*) An open-circuit test is conducted by applying 200 V to the LV side, keeping the HV side open. Calculate the power input, power factor and current drawn by the transformers.

(*ii*) A short-circuit test is conducted by passing full-load current from the HV side keeping the LV side shorted. Calculate the voltage required to be applied to the transformer and the power input and power factor.

[*G.B. Technical Univ. Electromechanical Energy Conversion-I, 2012-13*]

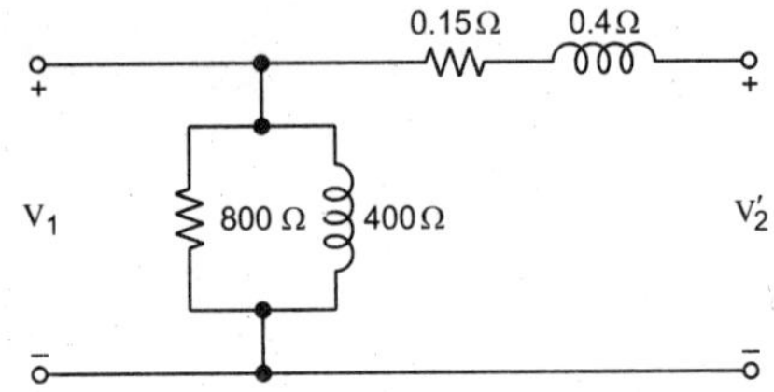

Fig. 12.36

Solution: (*i*) Energy component of no-load current,

$$I_e = \frac{V_1}{R_o} = \frac{200}{800} = 0.25 \text{ A}$$

Magnetizing component of no-load current,

$$I_m = \frac{V_1}{X_o} = \frac{200}{400} = 0.5 \text{ A}$$

Current drawn by transformer,

$$I_o = \sqrt{I_e^2 + I_m^2} = \sqrt{0.25^2 + 0.5^2} = 0.559 \text{ A} \quad \textbf{Ans.}$$

$$\text{Power factor, } \cos\phi_o = \frac{I_e}{I_o} = \frac{0.25}{0.559} = 0.447 \text{ (lagging)} \quad \textbf{Ans.}$$

$$\text{Power input, } P_o = V\,I_e = 200 \times 0.25 = 50 \text{ W} \quad \textbf{Ans.}$$

(*ii*) Transformation ratio,

$$K = \frac{400}{200} = 2$$

Equivalent resistance referred to *hv* (secondary),

$$R_{o2} = K^2\,R_{o1} = 2^2 \times 0.15 = 0.6\ \Omega$$

Equivalent reactance referred to *hv* (secondary),

$$X_{o2} = K^2\,X_{o1} = 2^2 \times 0.4 = 1.6\ \Omega$$

Equivalent impedance referred to *hv* (secondary),

$$Z_{o2} = \sqrt{(R_{o2})^2 + (X_{o2})^2} = \sqrt{0.6^2 + 1.6^2} = 1.71\ \Omega$$

Full-load secondary current,

$$I_2 = \frac{\text{Rated kVA} \times 1{,}000}{V_2} = \frac{4 \times 1{,}000}{400} = 10 \text{ A}$$

Voltage required to be applied to the transformer on *hv* side, keeping *lv* short circuited, to circulate full-load current through secondary,

$$V_s = I_2\,Z_{o2} = 10 \times 1.71 = 17.1 \text{ V} \quad \textbf{Ans.}$$

$$\text{Power factor, } \cos\phi = \frac{R_{o2}}{Z_{o2}} = \frac{0.6}{1.71} = 0.351 \text{ (lagging)} \quad \textbf{Ans.}$$

$$\text{Power input} = V_s\,I_2 \cos\phi = 17.1 \times 10 \times 0.351 = 60 \text{ W} \quad \textbf{Ans.}$$

Example 12.50. The per unit parameters of a 11/0.4 kV, 500 kVA, single phase transformer are $R_1 = R_2 = 0.008$, $X_1 = X_2 = 0.035$, $G_0 = 0.008$, $B_0 = 0.035$. Calculate the values of instrument readings when normal open-circuit and short-circuit tests are applied to this transformer at 50 Hz. [*Pb. Univ. Elec. Machines-I, Dec; 1991*]

Solution: Open-Circuit Test

Open-circuit test is normally conducted on low-voltage side (high-voltage side being open circuited)

Voltmeter reading,

V_1 = Primary (low voltage) rated voltage = 0.4 kV or 400 V **Ans.**

* The recommended methods for measuring resistance of transformer windings are by ammeter-voltmeter method or Wheatstone bridge method.

Primary rated current,

$$I_1 = \frac{kVA \times 1,000}{V_1} = \frac{500 \times 1,000}{400} = 1,250 \text{ A}$$

The shunt branch conductance, $G_0 = 0.008$ pu

The shunt branch susceptance, $B_0 = 0.035$ pu

The shunt branch admittance,

$$Y_0 = \sqrt{G_0^2 + B_0^2} = \sqrt{(0.008)^2 + (0.035)^2} = 0.0359 \text{ pu}$$

Ammeter reading = $I_0 = Y_0 I_1 = 0.0359 \times 1,250 = 44.88$ A **Ans.**

$\because$ Per unit $Y_0 = \dfrac{I_0}{I_1}$

Energy component of no-load current,

$$I_e = \text{Per unit } G_0 \times I_1 = 0.008 \times 1,250 = 10 \text{ A } \textbf{Ans.}$$

Wattmeter reading

$$= V I_e = 400 \times 10 = 4,000 \text{ W or 4 kW } \textbf{Ans.}$$

Short-Circuit Test

Short-circuit test is normally conducted on high-voltage side (low-voltage side being short circuited)

High voltage side rating voltage,

V_2 = 11 kV or 11,000 V

Ammeter reading,

I_s = Rated current of high-voltage side,

$$I_2 = \frac{KVA \times 1,000}{V_2} = \frac{500 \times 1,000}{11,000} = 45.45 \text{ A}$$

Per unit resistance of both windings,

$R = 0.008 + 0.008 = 0.016$

Per unit reactance of both windings,

$X = 0.035 + 0.035 = 0.07$

Per unit impedance of both windings,

$$Z = \sqrt{R^2 + X^2} = \sqrt{(0.016)^2 + (0.07)^2} = 0.0718$$

Voltmeter reading,

V_s = Voltage required to be applied to the high voltage winding to circulate full-load rated current through the windings

$$= \frac{\text{Full-load rated current}}{I_{sc}} \times V_2$$

$$= \frac{\text{Full-load rated current}}{1/\text{per unit impedance}} \times \frac{1}{\text{full-load rated current}} \times V_2$$

$= \text{Per unit impedance} \times V_2$

$= 0.0718 \times 11,000 = 789.8$ V **Ans.**

Wattmeter reading,

$$W_s = I_s^2 \times R_{eq} = (45.45)^2 \times \text{per unit resistance} \times \frac{V_2}{I_2}$$

$$\because \text{per unit resistance} = \frac{I_2 R_{eq}}{V_2}$$

$$= (45.45)^2 \times 0.016 \times \frac{11,000}{45.45} = 8,000 \text{ W or 8 kW } \textbf{Ans.}$$

Example 12.51. A 20 kVA, 250/2,500 V, 50 Hz, 1-ϕ transformer gave the following test results:

OC (LV Side)	250 V	1.4 A	105 W
SC (HV Side)	120V	8A	320W

Find the parameters and draw the equivalent circuit of 1-ϕ transformer as referred to:

(*i*) Low voltage side (*ii*) High voltage side.

[*U.P. Technical Univ. Electromechanical Energy Conversion-I, 2013-14*]

Solution: No-load pf,

$$\cos\phi_0 = \frac{\text{No-load input power, } P_o}{\text{Primary voltage, } V_1 \times \text{No-load current, } I_o}$$

$$= \frac{105}{250 \times 1.4} = 0.3$$

Energy component of no-load current,

$$I_e = I_o \cos\phi_o = 1.4 \times 0.3 = 0.42 \text{ A}$$

Magnetising component of no-load current,

$$I_m = \sqrt{I_o^2 - I_e^2} = \sqrt{1.4^2 - 0.42^2} = 1.3355 \text{ A}$$

(i) Referred to LV side

$$\text{No-load resistance, } R_o = \frac{V_1}{I_e} = \frac{250}{0.42} = 595.2\ \Omega \textbf{ Ans.}$$

$$\text{No-load reactance, } X_o = \frac{V_1}{I_m} = \frac{250}{1.3355} = 187.2\ \Omega \textbf{ Ans.}$$

Short-circuit current, $I_s = 8$ A

Short-circuit voltage, $V_s = 120$ V

Power input on short-circuit, $P_s = 320$ W

Equivalent impedance referred to *hv* (secondary) side,

$$Z_{o2} = \frac{V_s}{I_s} = \frac{120}{8} = 15\ \Omega$$

Equivalent resistance referred to *hv* (secondary) side,

$$R_{o2} = \frac{P_s}{I_s^2} = \frac{320}{8^2} = 5\ \Omega$$

Equivalent reactance referred to *hv* (secondary) side,

$$X_{o2} = \sqrt{(Z_{o2})^2 - (R_{o2})^2} = \sqrt{15^2 - 5^2} = 14.14\ \Omega$$

Transformation ratio,

$$K = \frac{2,500}{250} = 10$$

Equivalent resistance referred to *lv* (primary) side,

$$R_{o1} = \frac{R_{o2}}{K^2} = \frac{5}{10^2} = 0.05\ \Omega \textbf{ Ans.}$$

$$X_{o1} = \frac{X_{o2}}{K^2} = \frac{14.4}{10^2} = 0.144\ \Omega \textbf{ Ans.}$$

Approximate equivalent circuit is shown in Fig. 12.37 (*a*)

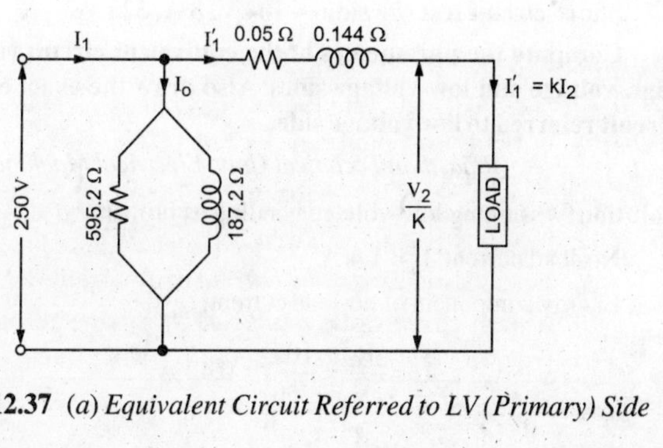

Fig. 12.37 (*a*) *Equivalent Circuit Referred to LV (Primary) Side*

(ii) Referred to HV side

No-load resistance, $R'_o = k^2R_o = 10^2 \times 595.2 = 59{,}520\ \Omega$ **Ans.**

No-load reactance, $X'_o = k^2X_o = 10^2 \times 187.2 = 18{,}720\ \Omega$ **Ans.**

Equivalent resistance referred to hv (secondary) side, as already worked out,

$$R_{o2} = 5\ \Omega \ \textbf{Ans.}$$

Equivalent resistance referred to hv (secondary) side, as already worked out

$$X_{o2} = 14.14\ \Omega \ \textbf{Ans.}$$

Approximate equivalent circuit is shown in Fig. 12.37 (*b*).

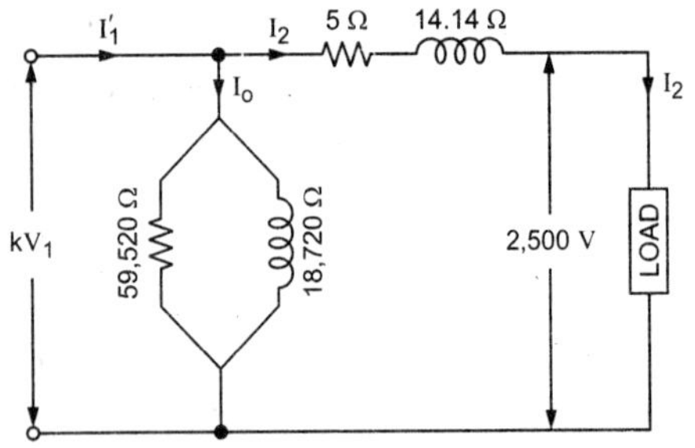

Fig. 12.37 (*b*) *Equivalent Circuit Referred To HV (Secondary) Side*

Example 12.52. The high and low voltage windings of a 6,600/250 V, 50 Hz single-phase transformer have resistances of 0.20 Ω and 3.1×10^{-4} Ω and reactances of 1.0 Ω and 1.69×10^{-3} Ω respectively. Find the input current and power when the high voltage winding is connected to a 200 V, 50 Hz supply and the low voltage winding is short circuited. [*R.G. Technical Univ. Electrical Machine-I, June-2014; Electromechanical Energy Conversion-I, 2011*]

Solution: Transformation ratio, $K = \dfrac{V_2}{V_1} = \dfrac{250}{6{,}600} = \dfrac{1}{26.4}$

Impedance of primary winding; $\mathbf{Z}_1 = (0.2 + j\,1.0)\ \Omega$

Impedance of secondary winding; $\mathbf{Z}_2 = 3.1 \times 10^{-4} + j\,1.69 \times 10^{-3}$

Equivalent impedance referred to primary (*lv*) side,

$$\mathbf{Z}_{01} = \mathbf{Z}_1 + \frac{\mathbf{Z}_2}{K^2}\ (0.2 + j\,1.0) + \frac{3.1\times10^{-4} + j\,1.69\times10^{-3}}{(1/26.4)^2}$$

$$= (0.416 + j\,2.17786)\ \Omega = 2.2172\ \angle 79.18°\ \Omega$$

$$\text{Input current, } I_1 = \frac{\text{Voltage applied to primary}}{Z_{01}} = \frac{200}{2.2172} = 90.2\ \text{A}\ \textbf{Ans.}$$

$$\text{Power input} = I_1^2 R_{01} = (90.2)^2 \times 0.416$$

$$= 3{,}384.9\ \text{W or } 3.385\ \text{kW}\ \textbf{Ans.}$$

Example 12.53. A 20 kVA, 2,500/250 V, 50 Hz, single-phase transformer gave the following test results:

Open-circuit test (*lv* side) – 250 V, 1.4 A, 105 W

Short-circuit test (*hv* side) – 104 V, 8 A, 320 W

Compute the parameters of the equivalent circuit referred to high-voltage and low-voltage sides. Also draw the exact equivalent circuit referred to low voltage side.

[*Rajasthan Technical Univ. Electrical Machines-I, 2010*]

Solution: Assuming low-voltage winding as primary

No-load current, $I_o = 1.4$ A

Energy component of no-load current,

$$I_e = \frac{P_0}{V} = \frac{105}{250} = 0.42\ \text{A}$$

Magnetizing component of no-load current,

$$I_m = \sqrt{I_0^2 - I_e^2} = \sqrt{1.4^2 - 0.42^2} = 1.3355\ \text{A}$$

Equivalent circuit parameter,

$$R_0 = \frac{V_1}{I_e} = \frac{250}{0.42} = 595.24\ \Omega$$

Equivalent circuit parameter,

$$X_0 = \frac{V_1}{I_m} = \frac{250}{1.3355} = 187.2\ \Omega$$

Equivalent impedance referred to *hv* side,

$$Z_{02} = \frac{V_s}{I_s} = \frac{104}{8} = 13\ \Omega$$

Equivalent resistance referred to *hv* side,

$$R_{02} = \frac{W_s}{I_s^2} = \frac{320}{8^2} = 5\ \Omega$$

Equivalent reactance referred to *hv* side,

$$X_{02} = \sqrt{(Z_{02})^2 - (R_{02})^2} = \sqrt{13^2 - 5^2} = 12\ \Omega$$

Transformation ratio,

$$K = \frac{2500}{250} = 10$$

Equivalent resistance referred to *lv* side,

$$R_{01} = \frac{R_{02}}{K^2} = \frac{5}{10^2} = 0.05\ \Omega$$

Equivalent reactance referred to *lv* side,

$$X_{01} = \frac{X_{02}}{K^2} = \frac{12}{10^2} = 0.12\ \Omega$$

Exact equivalent circuit referred to *lv* side is shown in Fig. 12.38.

Equivalent Circuit Parameters referred to *hv* side,

$R'_0 = K^2R_0 = 10^2 \times 595.24 = 59{,}524\ \Omega$ **Ans.**

$X'_0 = K^2X_0 = 10^2 \times 187.2 = 18{,}700\ \Omega$ **Ans.**

$R_{02} = 5\ \Omega$ **Ans.**

$X_{02} = 12\ \Omega$ **Ans.**

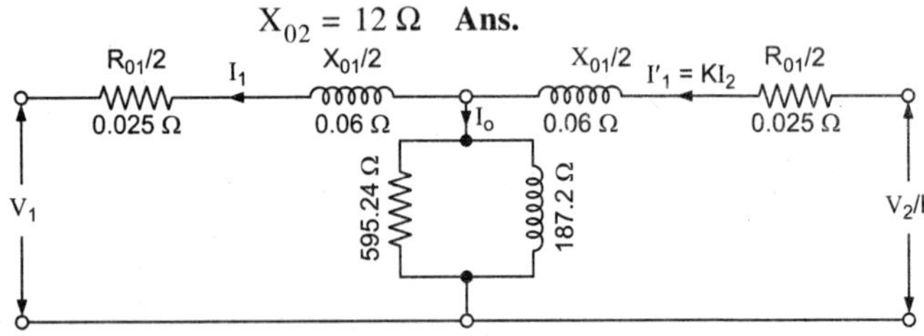

Fig. 12.38 *Exact Equivalent Circuit Referred to LV Side*

Example 12.54. The primary of a single-phase two winding transformer is rated at 12 A and 1,200 V. On open circuit, the readings (referred to hv side) are V_1 = 1,200 V, V_2 = 600 V, I_0 = 0.52 A, P_0 = 150 W. On short circuit the readings on (hv side) are I_{sc} = 12 A, V_{sc} = 144 V and P_{sc} = 450 W. Draw the equivalent circuit for the transformer and determine the parameters. What will be the output voltage across a load impedance of 20 + *j*15 ohms?

[*B.P. Univ. of Technology Electrical Machine-I, 2008*]

Solution: No-load current, $I_0 = 0.52$ A

No-load input power, $P_0 = 150$ W

Primary voltage, $V_1 = 1{,}200$ V

$$\text{No-load power factor, } \cos\phi_0 = \frac{P_0}{V_1 I_0} = \frac{150}{1{,}200 \times 0.52}$$

$$= 0.2404\ \text{(lagging)}$$

Energy component of no-load current, $I_e = I_0 \cos \phi_0$

$$= 0.52 \times 0.2404 = 0.125 \text{ A}$$

Magnetizing component of no-load current,

$$I_m = \sqrt{I_0^2 - I_e^2}$$

$$= \sqrt{0.52^2 - 0.125^2} = 0.5048 \text{ A}$$

$$\text{No-load resistance, } R_0 = \frac{V_1}{I_e} = \frac{1,200}{0.125} = 9,600\ \Omega \text{ or } 9.6\ k\Omega$$

$$\text{No-load reactance, } X_0 = \frac{V_1}{I_m} = \frac{1,200}{0.5048}$$

$$= 2,377\ \Omega \text{ or } 2.377\ k\Omega \textbf{ Ans.}$$

Short-circuit current, $I_{sc} = 12$ A

Short-circuit voltage, $V_{sc} = 144$ V

Power input on short circuit, $P_{sc} = 450$ W

Equivalent impedance referred to *hv* (primary) side,

$$Z_{01} = \frac{V_{sc}}{I_{sc}} = \frac{144}{12} = 12\ \Omega$$

Equivalent resistance referred to *hv* (primary) side,

$$R_{01} = \frac{P_{sc}}{I_{sc}^2} = \frac{450}{12^2} = 3.125\ \Omega$$

Equivalent reactance referred to *hv* (primary) side,

$$X_{01} = \sqrt{Z_{01}^2 - R_{01}^2} = \sqrt{12^2 - 3.125^2} = 11.586\ \Omega$$

The equivalent circuit of the transformer referred to *hv* side is shown in Fig. 12.39.

It is given that on open circuit, voltage on secondary side is 600 V, hence rated voltage of secondary,

$$V_2 = 500 \text{ V}$$

$$\text{Transformation ratio, } K = \frac{V_2}{V_1} = \frac{500}{1,000} = 0.5$$

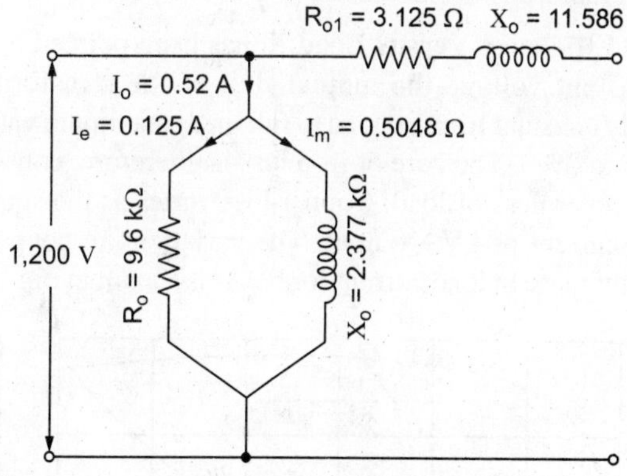

Fig. 12.39

Equivalent impedance referred to secondary

$$\mathbf{Z}_{02} = K^2\mathbf{Z}_{01} = 0.5^2(3.125 + j11.586)$$

$$= (0.78125 + j\,2.8965)\ \Omega$$

Load impedance, $\mathbf{Z}_L = (20 + j\,15)\ \Omega$

Output voltage across load impedance,

$$V_L = \frac{V_2\mathbf{Z}_L}{\mathbf{Z}_L + \mathbf{Z}_{02}} = \frac{600 \times (20 + j15)}{(20 + j\,15) + (0.78125 + j2.8965)}$$

$$= \frac{600 \times 25 \angle 36.87°}{27.425 \angle 40.73°} = 546.95 \angle -3.86° \text{ V } \textbf{Ans.}$$

12.20 DETERMINATION OF REGULATION OF TRANSFORMER FROM OPEN-CIRCUIT AND SHORT-CIRCUIT TESTS

Percentage regulation is given as

$$\%\text{age regulation} = \frac{I_2 R_{02} \cos \phi \pm I_2 X_{02} \sin \phi}{\text{No-load rated secondary voltage}} \times 100$$

...Refer to Eqs. (12.26) and (12.27)

Equivalent resistance R_{02} and reactance X_{02} referred to secondary can be determined from short-circuit test, as explained in Art. 12.18.5. I_2 and $\cos \phi$ are the load current and power factor (lagging or leading) of the load, so known. No-load secondary terminal voltage is equal to emf induced in the secondary E_2.

$$\text{So percentage regulation} = \frac{I_2 R_{02} \cos \phi \pm I_2 X_{02} \sin \phi}{E_2} \times 100$$

Note: Open-circuit test datas are not needed for determination of voltage regulation of transformer.

Example 12.55. Define the voltage regulation of transformer.

A 200 kVA, 1-phase, 3,300/400 V transformer gave the following result in the short-circuit test.

With 200 V applied to primary and the secondary short circuited, the primary current was the full-load value and input power 1,650 W. Calculate the secondary potential drop and percentage regulation when the secondary load is passing 300 A at 0.707 pf lagging with normal primary voltage.

[*M.D. Univ. Electrical Machines-I, December-2012*]

Solution: Full-load primary current,

$$I_1 = \frac{\text{Rated kVA} \times 1,000}{V_1} = \frac{200 \times 1,000}{3,300} = 60.6 \text{ A}$$

Short-circuit current on primary,

$$I_s = I_1 = 60.6 \text{ A}$$

Equivalent impedance referred to hv (or primary) side,

$$Z_{01} = \frac{V_s}{I_s} = \frac{200}{60.6} = 3.3\ \Omega \qquad \because I_s = I_1 = 60.6 \text{ A}$$

Equivalent resistance referred to hv (or primary) side,

$$R_{01} = \frac{W_s}{I_s^2} = \frac{1,650}{60.6^2} = 0.45\ \Omega$$

Equivalent reactance referred to primary side,

$$X_{01} = \sqrt{(Z_{01})^2 - (R_{01})^2} = \sqrt{3.3^2 - 0.45^2} = 3.269\ \Omega$$

Transformation ratio,

$$K = \frac{V_2}{V_1} = \frac{4}{3,300} = \frac{4}{33}$$

Equivalent resistance referred to secondary side,

$$R_{02} = K^2R_{01} = \left(\frac{4}{33}\right)^2 \times 0.45 = 0.0066\ \Omega$$

Equivalent reactance referred to secondary side,

$$X_{02} = K^2X_{01} = \left(\frac{4}{33}\right)^2 \times 3.269 = 0.048\ \Omega$$

Secondary load current, $I_2 = 300$ A

Power factor, $\cos \phi_2 = 0.707$ lagging

$$\text{Sin } \phi_2 = \sqrt{1 - 0.707^2} = 0.707$$

Secondary potential drop $= I_2(R_{02} \cos \phi_2 + X_{02} \sin \phi_2)$

$= 300(0.0066 \times 0.707 + 0.048 \times 0.707)$

$= 11.58$ V **Ans.**

$$\text{Voltage regulation} = \frac{\text{Secondary potential drop}}{V_2} \times 100$$

$$= \frac{11.58}{400} \times 100 = 2.895\% \textbf{ Ans.}$$

Example 12.56. Short-circuit test is conducted on a 5 kVA, 400 V/100 V single-phase transformer with 100 V winding shorted. The input voltage at full load current is 40 V. The wattmeter, on the input reads 250 W. Find the power factor for which regulation at full load is zero. *[GATE 1994]*

Solution: On short-circuit test on primary (400 V) winding

Input power, $W_s = 250$ W

Input voltage, $V_s = 40$ V

Input current, I_s = Full-load primary current, I_1

$$= \frac{\text{Rated kVA} \times 1{,}000}{V_1} = \frac{5 \times 1{,}000}{400} = 12.5 \text{ A}$$

Equivalent resistance as referred to primary,

$$R_{01} = \frac{W_s}{I_s^2} = \frac{250}{(12.5)^2} = 1.6\ \Omega$$

Equivalent impedance referred to primary,

$$Z_{01} = \frac{V_s}{I_s} = \frac{40}{12.5} = 3.2\ \Omega$$

Equivalent reactance referred to primary,

$$X_{01} = \sqrt{(Z_{01})^2 - (R_{01})^2} = \sqrt{(3.2)^2 - (1.6)^2} = 2.77\ \Omega$$

Regulation will be zero when phase angle,

$$\phi = \text{Tan}^{-1} \frac{-R_{02}}{X_{02}} = \text{Tan}^{-1} \frac{-R_{01}}{X_{01}} = \text{Tan}^{-1} \frac{-1.6}{2.77} = -30^\circ$$

or Power factor $= \text{Cos } \phi = \text{Cos } (-30^\circ) = 0.866$ (leading) **Ans.**

12.21 TRANSFORMER EFFICIENCY

The rated capacity of a transformer is defined as the product of rated voltage and full-load (rated) current on the output side. The power output depends upon the power factor of the load.

The efficiency (η) of a transformer, like that of any other apparatus, is defined as the ratio of useful power output to the input power, the two being measured in same units (either in watts or killowatts).

i.e. Transformer efficiency,

$$\eta = \frac{\text{Output}}{\text{Input}} = \frac{\text{Output}}{\text{Output + losses}}$$

$$= \frac{\text{Output}}{\text{Output + iron loss + copper loss}}$$

$$= 1 - \frac{\text{iron loss + copper loss}}{\text{Output + iron loss + copper loss}}$$

Now power output $= V_2 I_2 \cos \phi$

where V_2 is the secondary terminal voltage on load, I_2 is the secondary current at load and $\cos \phi$ is the power factor of the load.

Iron loss, P_i = Hysteresis loss + eddy current loss

Copper loss $= I_1^2 R_1 + I_2^2 R_2 = I_1^2 R_{01} = I_2^2 R_{02}$

12.21.1. Determination of Transformer Efficiency. The ordinary transformer has a very high efficiency (in the range of 96-99%). Hence the transformer efficiency cannot be determined with high precision by direct measurement of output and input, since the losses are of the order of only 1–4%. The difference between the readings of output and input instruments is then so small that an instrument error as low as 0.5% would cause an error of the order of 15% in the losses. Further, it is inconvenient and costly to have the necessary loading devices of the correct current and voltage ratings and power factor to load the transformer. There is also a wastage of large amount of power (equal to that of power output + losses) and no information is available from such a test about the proportion of copper and iron losses.

The best and accurate method of determining of efficiency of a transformer would be to compute losses from open-circuit and short-circuit tests and determine efficiency as follows :

Iron loss, $P_i = W_0$ or P_0, determined from open-circuit test

Copper loss at full load,

$P_c = W_s$ or P_s, determined from short-circuit test

Copper loss at a load x times full load

$= I_2^2 R_{02} = x^2 P_c$

where x is the ratio of load current I_2 to full-load secondary current.

And transformer efficiency,

$$\eta = \frac{V_2 I_2 \cos \phi}{V_2 I_2 \cos \phi + P_i + x^2 P_c} \quad \text{...(12.46)}$$

In Eq. (12.46) the effect of instrument readings is confined to losses only so that overall efficiency obtained from it is far more accurate than that obtained by direct loading. Another great advantage of this method is that it is not necessary for the transformer to be loaded to its full-load rating during testing, and the kW rating of the test plant need be equal to the value of the individual transformer losses.

12.21.2. Efficiency Versus Load. It has been pointed out that with constant voltage the mutual flux of the transformer is practically constant from no load to full load (maximum variation is from 1 to 3%). The core or iron loss is, therefore, considered constant regardless of load. Copper loss varies as the square of the load current or kVA output. The variations in copper loss with the increase in load current (or kVA) is shown in Fig. 12.40.

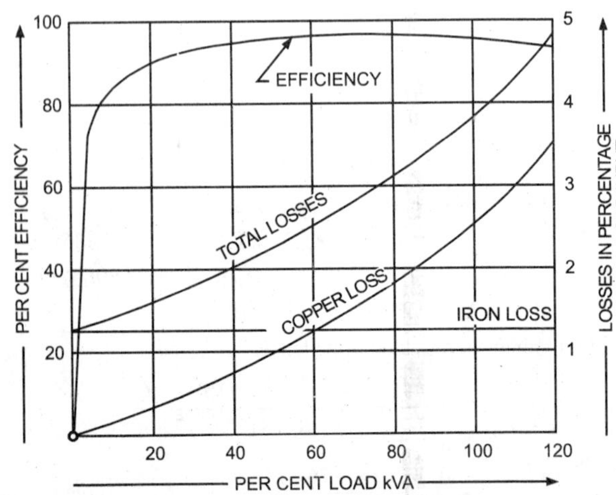

Fig. 12.40

The efficiency vs load curve as deduced from these is also shown in the Fig. 12.40. From the efficiency-load curve shown in Fig. 12.40 it is obvious that the efficiency is very high even at light load, as low as 10 % of rated load. The efficiency is practically constant from about 20% rated load to about 20% overload. At light loads the efficiency is poor because of constant iron loss whereas at high loads the efficiency falls off due to increase in copper loss as the square of load. From Fig. 12.40 it is also obvious that the transformer efficiency is maximum at the point of intersection of copper loss and iron loss curves *i.e.* when copper loss equals iron loss.

12.21.3. Condition For Maximum Efficiency. From Eq. (12.46)

Transformer efficiency,

$$\eta = \frac{V_2 I_2 \cos\phi}{V_2 I_2 \cos\phi + P_i + x^2 P_c} = \frac{x P}{x P + P_i + x^2 P_c} \quad ...(12.47)$$

where P is equal to the full load (rated output in volt-amperers or kVA) × load power factor ($\cos\phi$), P_i is the total iron loss, P_c is the full-load copper loss and x is the fraction of full-load kVA at which efficiency is maximum.

Differentiating both sides of Eq. (12.47) we get

$$\frac{d\eta}{dx} = \frac{(x P + x^2 P_c + P_i) P - x P (P + 2x P_c)}{(x P + P_i + x^2 P_c)^2} = \frac{P(P_i - x^2 P_c)}{(x P + x^2 P_c + P_i)^2}$$

Efficiency η will be maximum if $\frac{d\eta}{dx} = 0$

$$\text{or } P_i - x^2 P_c = 0$$

$$\text{or } x^2 P_c = P_i \quad ...(12.48)$$

Hence the efficiency will be maximum when variable loss (copper loss) is equal to constant loss (iron loss).

Output kVA corresponding to maximum efficiency

$$= x \times \text{Rated kVA} = \text{Rated kVA} \times \sqrt{\frac{P_i}{P_c}} \quad ...(12.49)$$

Since from Eq. (12.48) $x = \sqrt{\frac{P_i}{P_c}}$

Output current I_2 corresponding to maximum efficiency is determined as below:

Copper loss at given load at which efficiency is maximum

$$= I_2^2 R_{02}$$

and since for maximum efficiency to occur it is necessary that copper loss equals P_i,

$$I_2^2 R_{02} = P_i$$

or current corresponding to maximum efficiency,

$$I_2 = \sqrt{\frac{P_i}{R_{02}}} \quad ...(12.50)$$

Power transformers employed for bulk power transmission are operated continuously near about full load and are, therefore, designed to have maximum efficiency at full load. On the other hand, the distribution transformers, which supply load varying over the day through a wide range, are designed to have maximum efficiency at about three-fourths the full load.

It is important to appreciate that, since copper loss depends on current and the iron loss depends on voltage, the total loss in the transformer depends on the volt-ampere product, and not on the phase angle between voltage and current, *i.e.* is independent of the load power factor. The transformers are, therefore, rated in kilovolt amperes (kVA) and not in kilowatts.

12.21.4. Efficiency Versus Power Factor. Transformer efficiency is given as

$$\eta = \frac{\text{Output}}{\text{Output+losses}} = 1 - \frac{\text{losses}}{\text{output + losses}}$$

$$= 1 - \frac{\text{losses}}{V_2 I_2 \cos\phi + \text{losses}}$$

$$= 1 - \frac{\text{losses}/V_2 I_2}{\cos\phi + \text{losses}/V_2 I_2}$$

Substituting $\frac{\text{losses}}{V_2 I_2} = x$ in above equation we get

$$\eta = 1 - \frac{x}{\cos\phi + x} = 1 - \frac{x/\cos\phi}{1 + x/\cos\phi} \quad ...(12.51)$$

The variations of efficiency with power factor at different loadings for a typical transformer are illustrated in Fig. 12.41.

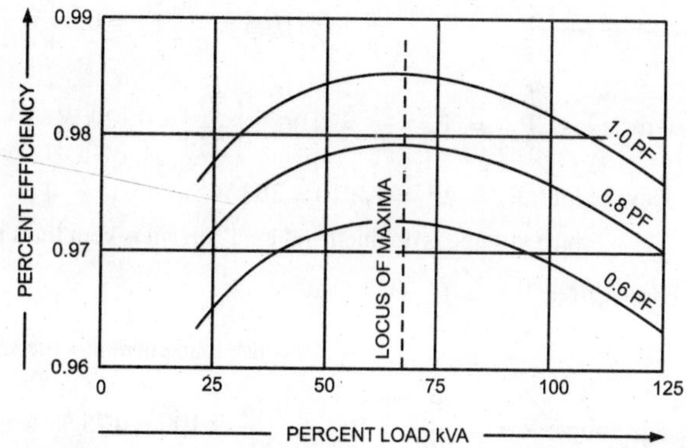

Fig. 12.41

Example 12.57. Calculate the efficiency at half-full load of a 100 kVA transformer for pf of unity and 0.8. The copper loss is 1,000 W at full load and iron loss is 1,000 W.

[Anna Univ. Electrical Machines-I, May/June-2012]

Solution: Copper loss at full load = 1,000 W

$$\text{Copper loss at half full-load} = \left(\frac{1}{2}\right)^2 \times 1{,}000 = 250 \text{ W} = 0.25 \text{ kW}$$

$$\text{Iron loss, } P_i = 1{,}000 \text{ W} = 1 \text{ kW}$$

Transformer output on half-full load and unity pf

$$P_{out_1} = \frac{1}{2} \times 100 \times 1.0 = 50 \text{ kW}$$

$$\text{Transformer efficiency, } \eta_1 = \frac{P_{out_1}}{P_{out_1} + P_c + P_i} \times 100$$

$$= \frac{50}{50 + 0.25 + 1} \times 100 = 97.56\% \textbf{ Ans.}$$

Example 12.58. A 10 kVA, 5,000/440 V, 25 Hz single-phase transformer has copper, eddy current and hysteresis loss of 1.5, 0.5 and 0.6 per cent of output on full load. What will be the percentage losses if the transformer is used on 10 kV, 50 Hz system keeping full load current constant? Assume unity power factor operation. Compare the full-load efficiencies for the two cases.

Solution: When supply voltage,

$$V_1 = 5{,}000 \text{ V and supply frequency, } f_1 = 25 \text{ Hz}$$

Full-load output at unity pf,

$$P_1 = S \times 1.0 = 10 \times 1.0 = 10 \text{ kW}$$

$$\text{Full-load copper loss, } P_{c_1} = \frac{1.5}{100} \times 10 = 0.15 \text{ kW}$$

$$\text{Eddy current loss, } P_{e_1} = \frac{0.5}{100} \times 10 = 0.05 \text{ kW}$$

$$\text{Hysteresis loss, } P_{h_1} = \frac{0.6}{100} \times 10 = 0.06 \text{ kW}$$

Transformer efficiency,

$$\eta = \frac{\text{Output}}{\text{Output} + P_{c_1} + P_{e_1} + P_{h_1}} \times 100$$

$$= \frac{10}{10 + 0.15 + 0.05 + 0.06} \times 100 = 97.466 \text{ \% } \textbf{Ans.}$$

In second case since supply voltage and frequency are increased in the same ratio, the peak flux density remains the same in both cases and, therefore,

$$\text{Eddy current loss, } P_{e_2} = P_{e_1} \times \left(\frac{f_2}{f_1}\right)^2 = 0.05 \times \left(\frac{50}{25}\right)^2 = 0.2 \text{ kW}$$

$$\text{Hysteresis loss, } P_{h_2} = P_{h_1} \times \frac{f_2}{f_1} = 0.06 \times \left(\frac{50}{25}\right) = 0.12 \text{ kW}$$

New output, $P_2 = 2P_1 = 2 \times 10 = 20$ kW

since voltage is doubled and load current is kept the same.

New copper loss, $P_{c_2} = P_{c_1} = 0.15$ kW

since load current is the same.

$$\text{So \% age copper loss} = \frac{P_{c_2}}{P_2} \times 100 = \frac{0.15}{20} \times 100 = 0.75 \text{ \% } \textbf{Ans.}$$

$$\text{\% age eddy current loss} = \frac{P_{e_2}}{P_2} \times 100 = \frac{0.2}{20} \times 100 = 1 \text{ \% } \textbf{Ans.}$$

$$\text{\% age hysteresis loss} = \frac{P_{h_2}}{P_2} \times 100 = \frac{0.12}{20} \times 100 = 0.6 \text{ \% } \textbf{Ans.}$$

Transformer efficiency,

$$\eta_2 = \frac{P_2}{P_2 + P_{c_2} + P_{e_2} + P_{h_2}} \times 100$$

$$= \frac{20}{20 + 0.15 + 0.2 + 0.12} \times 100 = 97.704 \text{ \% } \textbf{Ans.}$$

Example 12.59. The primary and secondary winding of a 500 kVA 1-ph transformer have resistance of 0.4 Ω and 0.0015 Ω respectively. The primary and secondary voltages are 6,000 V and 400 V respectively and the iron loss is 3.2 kW. Calculate the efficiency on

(*i*) Full load and **(*ii*) Half load**

Assuming the pf of the load to be 0.8.

[*M.D. Electrical Machines-I, December-2013*]

Solution: Primary current on full load,

$$I_1 = \frac{\text{Rated kVA} \times 1{,}000}{V_1} = \frac{500 \times 1{,}000}{6{,}000} = 83.33 \text{ A}$$

Secondary current on full load,

$$I_2 = \frac{\text{Rated kVA} \times 1{,}000}{V_2} = \frac{500 \times 1{,}000}{400} = 1{,}250 \text{ A}$$

$$\text{Full-load copper loss} = I_1^2 R_1 + I_2^2 R_2$$

$$= (83.33)^2 \times 0.4 + (1{,}250)^2 \times 0.0015$$

$$= 2{,}778 + 2{,}344 = 5{,}122 \text{ W or } 5.122 \text{ kW}$$

Full-load output,

$$P_{out\,f} = \text{Rated kVA} \times \text{pf} = 500 \times 0.8 = 400 \text{ kW}$$

(*i*) Transformer efficiency at full load,

$$\eta_f = \frac{P_{out\,f}}{P_{out\,f} + \text{copper loss} + \text{iron loss}} \times 100$$

$$= \frac{400}{400 + 5.122 + 3.2} \times 100 = 97.962\% \textbf{ Ans.}$$

(*ii*) Transformer efficiency at half- full load

$$= \frac{\frac{1}{2} \times \text{rated kVA} \times \text{pf}}{\frac{1}{2} \times \text{rated kVA} \times \text{pf} + \left(\frac{1}{2}\right)^2 \text{copper loss} + \text{iron loss}} \times 100$$

$$= \frac{\frac{1}{2} \times 500 \times 0.8}{\frac{1}{2} \times 500 \times 0.8 + \frac{1}{4} \times 5.122 + 3.2} \times 100$$

$$= \frac{200}{200 + 1.2805 + 3.2} \times 100 = 97.81\% \textbf{ Ans.}$$

Example 12.60. A 30 kVA, single-phase transformer has an iron loss of 457 W and copper loss of 125 W when delivering half-full load. At what per cent of the full load will the transformer have maximum efficiency?

[*J.N. Technological Univ. Electrical Machines-II, January-2014*]

Solution: Iron loss = 457 W

Copper loss at half-full load = 125

$$\text{Copper loss at full load} = \frac{\text{Copper loss at half-full load}}{\left(\frac{1}{2}\right)^2}$$

$$= 125 \times 2^2 = 500 \text{ W}$$

Let the transformer maximum efficiency occur at x% of full-load.

$$\text{Then } \frac{x}{100} = \sqrt{\frac{P_i}{P_c}}$$

$$\text{or } x = \sqrt{\frac{P_i}{P_c}} \times 100 = \sqrt{\frac{457}{500}} \times 100 = 95.6\% \textbf{ Ans.}$$

Example 12.61. A 440/110 V transformer has a primary resistance of 0.03 Ω and secondary resistance of 0.02 Ω. Its iron loss at normal input is 150 W. Determine the secondary current at which maximum efficiency will occur and the value of this maximum efficiency at a unity power factor load. [*Rajasthan Univ. Electrical Machines-I, 2005*]

Solution: Transformation ratio, $K = \frac{110}{440} = 0.25$

Total resistance referred to secondary,

$$R_{02} = K^2 R_1 + R_2 = 0.25^2 \times 0.03 + 0.02 = 0.021875 \, \Omega$$

Secondary current corresponding to maximum efficiency,

$$I_2 = \sqrt{\frac{\text{Iron loss}}{R_{02}}} = \sqrt{\frac{150}{0.021875}} = 82.8 \text{ A} \quad \textbf{Ans.}$$

Output corresponding to maximum efficiency and unity pf

$$= V_2 \times I_2 \cos\phi_2 = 110 \times 82.8 \times 1 = 9{,}108 \text{ W}$$

Maximum efficiency,

$$\eta_{max} = \frac{V_2 I_2 \cos\phi_2}{V_2 I_2 \cos\phi_2 + 2 \times \text{iron loss}} \times 100$$

$$= \frac{9{,}108}{9{,}108 + 2 \times 150} \times 100 = 96.8\% \textbf{ Ans.}$$

Example 12.62. A single-phase transformer working at unity pf has an efficiency of 90% on both half load and at full load of 500 kW. Determine (*i*) iron loss (*ii*) full-load copper loss (*iii*) efficiency at 75% of full load (*iv*) maximum efficiency.

[Rajasthan Technical Univ. Electrical Machines-I, 2007, 2013]

Solution: Let the iron loss and full-load copper loss be P_i and P_c respectively:

At full load and unity pf

Power output, $P_{out} = 500$ kW

Transformer efficiency, $\eta = 90\% = 0.9$

$$\text{Power input, } P_{in} = \frac{P_{out}}{\eta} = \frac{500}{0.9} = 555.5 \text{ kW}$$

$$\text{Total loss, } P_i + P_c = P_{in} - P_{out} = 555.5 - 500 = 55.5 \text{ kW} \quad ...(i)$$

A half full load and unity pf

$$\text{Power output, } P'_{out} = \frac{1}{2} \times 500 = 250 \text{ kW}$$

$$\text{Power input, } P'_{in} = \frac{P'_{out}}{\eta} = \frac{250}{0.9} = 277.8 \text{ kW}$$

$$\text{Total loss, } \left(\frac{1}{2}\right)^2 P_c + P_i = P'_{in} - P'_{out} = 277.8 - 250 = 27.8 \text{ kW} \quad ...(ii)$$

Solving Eqs. (*i*) and (*ii*) we have

(*i*) $P_i = 18.6$ kW **Ans.**

(*ii*) $P_c = 36.9$ kW **Ans.**

(*iii*) Output at 75% of full load and unity pf,

$$P_{out} = \frac{3}{4} \times 500 = 375 \text{ kW}$$

$$\text{Total losses} = P_i + \left(\frac{3}{4}\right)^2 P_c = 18.6 + \left(\frac{3}{4}\right)^2 \times 36.9 = 39.36 \text{ kW}$$

$$\eta = \frac{375}{375 + 39.36} \times 100 = 90.5\% \textbf{ Ans.}$$

(*iv*) Output corresponding to maximum efficiency,

$$P_{out} = 500 \times \sqrt{\frac{P_i}{P_c}} = 500 \times \sqrt{\frac{18.6}{36.9}} = 355 \text{ kW}$$

$$\text{Efficiency, } \eta = \frac{355}{355 + 2 \times 18.6} \times 100 = 90.515\% \textbf{ Ans.}$$

Example 12.63. The efficiency of 1,000 kVA, 110/220 V, 50 Hz single-phase transformer is 97.5% at half-full load at 0.8 pf leading and 98% at full load unity pf. Determine iron loss and full-load copper loss. *[Gujarat Technological Univ. Electrical Machines-I, Summer-2014]*

Solution: Let the iron loss and full-load copper loss be P_i and P_c kW respectively.

At full load unity pf

$$\text{Power output, } P_{out_1} = 1{,}000 \times 1.0 = 1{,}000 \text{ kW}$$

Transformer efficiency, $\eta = 98\% = 0.98$

$$\text{Total losses, } P_c + P_i = P_{in_1} - P_{out_1}$$

$$= \frac{1{,}000}{0.98} - 1{,}000$$

$$= 1{,}020.4 - 1{,}000 = 20.4 \text{ kW} \quad ...(i)$$

At half-full load and 0.8 pf leading

$$\text{Power output, } P_{out_2} = \frac{1}{2} \times 1{,}000 \times 0.8 = 400 \text{ kW}$$

Transformer efficiency, $\eta_2 = 97.5\% = 0.975$

$$\text{Total losses} = P_i + \left(\frac{1}{2}\right)^2 P_c$$

$$= \frac{400}{0.975} - 400 = 10.26 \text{ kW} \quad ...(ii)$$

Solving Eqs. (*i*) and (*ii*), we have

Iron loss, $P_i = 6.88$ kW **Ans.**

Full-load copper loss $= P_c = 13.52$ kW **Ans.**

Example 12.64. The maximum efficiency of a 150 kVA, 2,200/220 V, 50 Hz single-phase transformer is 98.5% and occurs at half-full load at unity pf. Determine its full-load efficiency at 0.8 leading pf.

[B.P. Univ. of Technology Electrical Machine-I, 2008]

Solution: Power output corresponding to maximum efficiency,

P_{out} = Fraction of full load × rated kVA × pf

$$= \frac{1}{2} \times 150 \times 1.0 = 75 \text{ kW}$$

$$\text{Since efficiency, } \eta = \frac{\text{Output}}{\text{Output} + P_c + P_i}$$

$$\text{or } P_c + P_i = \frac{75}{0.985} - 75 = 1.142 \text{ kW}$$

$$\text{Iron loss} = \frac{1.142}{2} = 0.571 \text{ kW}$$

∵ For maximum efficiency P_i = copper loss

At half-full load, the copper loss $= P_i = 0.571$ kW

$$\text{Full-load copper loss} = \frac{\text{Copper loss at half-full load}}{\left(\frac{1}{2}\right)^2}$$

$$= \frac{0.571}{\left(\frac{1}{2}\right)^2} = 2.284 \text{ kW}$$

At full load and 0.8 leading pf

Output, $P_{out} = 1 \times 150 \times 0.8 = 120$ kW

$P_i = 0.571$ kW

$P_c = 2.284$ kW

Input, $P_{in} = P_{out} + P_i + P_c$

$= 120 + 0.571 + 2.284 = 122.855$ kW

$$\text{Efficiency, } \eta = \frac{P_{out}}{P_{in}} \times 100 = \frac{120}{122.855} \times 100 = 97.68\% \textbf{ Ans.}$$

Example 12.65. In a 500 kVA, 3,300/500 V, 50 Hz single-phase transformer, the maximum efficiency of 98% occurs at 75% of load at a pf of unity. Calculate the voltage regulation at full load and 0.8 pf lagging if the impedance is 0.1.

[G.G.S.I.P. Univ. Delhi Electromechanical Energy Conversion-I, December-2011]

Solution: At $\frac{3}{4}$ full load and pf of unity

$$\text{Power output, } P_{out} = \frac{3}{4} \times 500 \times 1 = 375 \text{ kW}$$

Maximum efficiency, η_{max} = 98% or 0.98

$$\text{Power input, } P_{in} = \frac{P_{out}}{\eta} = \frac{375}{0.98} = 382.65 \text{ kW}$$

$$\text{Total loss} = \text{Input} - \text{output} = 382.65 - 375 = 7.65 \text{ kW}$$

This loss is equally divided between iron and copper loss, so

$$\left(\frac{3}{4}\right)^2 P_c = P_i = \frac{7.65}{2} = 3.825 \text{ kW}$$

$$\text{or } P_c = \left(\frac{4}{3}\right)^2 \times 3.825 = 6.8 \text{ kW}$$

$$\text{Percentage resistance} = \frac{I_2 R_{02}}{V_2} \times 100 = \frac{I_2^2 R_{02}}{V_2 I_2} \times 100$$

$$= \frac{\text{Copper loss}}{V_2 I_2} \times 100 = \frac{6.8}{500} \times 100 = 1.36\%$$

Percentage impedance = 100 × 0.1 = 10%

$$\text{Percentage reactance} = \sqrt{10^2 - 1.36^2} = 9.907\%$$

$$\text{Percentage regulation} = \frac{I_2 R_{02} \cos\phi_2 + I_2 X_{02} \sin\phi_2}{V_2} \times 100$$

$$= \frac{I_2 R_{02}}{V_2} \times 100 \times \cos\phi_2 + \frac{I_2 X_{02}}{V_2} \times 100 \times \sin\phi_2$$

$$= 1.36 \times 0.8 + 9.907 \times 0.6 = 7.0322\% \textbf{ Ans.}$$

Example 12.66. A single-phase transformer has voltage regulation of 6% and 6.6% for lagging pf of 0.8 and 0.6 respectively. Full-load ohmic loss is equal to iron loss. Calculate (*i*) the lagging pf at which full-load voltage regulation is maximum and (*ii*) the full-load efficiency at unity pf. [*U.P.S.C. I.E.S. Electrical Engineering-II, 2012*]

Solution: Let percentage resistance and reactance drops be v_r and v_x respectively.

$$\text{Percentage regulation} = v_r \cos\phi + v_x \sin\phi$$

where cos ϕ is the load pf.

For 0.8 pf lagging

$$\%\text{age regulation} = v_r \times 0.8 + v_x \times 0.6$$

$$\text{or } 6 = 0.8\, v_r + 0.6\, v_x \quad ...(i)$$

For 0.6 pf lagging

$$\%\text{age regulation} = v_r \times 0.6 + v_x \times 0.8$$

$$\text{or } 6.6 = 0.6\, v_r + 0.8\, v_x \quad ...(ii)$$

Solving Eqs. (*i*) and (*ii*) we have

$$v_r = 3.0 \text{ and } v_x = 6$$

%age copper loss = Percentage resistance drop = 3% of the output

% iron loss = Percentage ohmic loss = 3% of the output

(*i*) Full-load voltage regulation will be maximum for load pf of

$$\cos\phi = \cos\tan^{-1}\frac{X_{02}}{R_{02}}$$

$$= \cos\tan^{-1}\frac{6}{3} = \cos 63.43° = 0.4472 \text{ lag } \textbf{Ans.}$$

$$\text{Full-load efficiency, } \eta = \frac{\text{Output}}{\text{Output + losses}} \times 100$$

$$= \frac{100}{100 + 3 + 3} \times 100 = 94.34\% \textbf{ Ans.}$$

Example 12.67. A 1-phase, 30 kVA, 2300/230 V, 50 Hz transformer gave the following results on tests:

Leakage impedance in the high voltage (HV) winding = 0.55 + *j*0.65 Ω

Leakage impedance in the low voltage (LV) winding = 0.005 + *j*0.006 Ω

Shunt branch admittance as seen from LV side = (0.003 – *j*0.02) S

(*i*) Draw the equivalent circuits of the transformer referred to HV and LV sides.

(*ii*) The transformer is connected to 2,200 volts at the sending end and delivers rated current at 0.8 pf lagging to a load of 0.3 + *j*2 Ω on the LV side. Draw the phasor diagram and hence or otherwise determine the voltage at the load end and the efficiency of the transformer. Assume core losses to be 160 W and ignore voltage drops due to exciting current. [*U.P.S.C. I.E.S. Electrical Engineering-II, 1998*]

Solution: Transformation ratio, $K = \frac{V_2}{V_1} = \frac{230}{2300} = \frac{1}{10}$

Leakage impedance of low-voltage (or secondary) winding referred to primary

$$Z_2' = Z_2 \times \frac{1}{K^2} = (0.005 + j0.006) \times \left(\frac{1}{1/10}\right)^2 = (0.5 + j0.6)\ \Omega$$

Shunt branch admittance as seen from hv side,

$$Y_0 = K^2 Y_0 = \left(\frac{1}{10}\right)^2 \times (0.003 - j0.02) = (0.3 - j2) \times 10^{-4} \text{ S}$$

Transformer equivalent circuit referred to *hv* (primary) side is given in Fig. 12.42(*a*).

Leakage impedance of high voltage (or primary) winding referred to secondary,

$$Z_2' = Z_1 \times K^2 = (0.55 + j0.65) \times \left(\frac{1}{10}\right)^2 = (0.0055 + j0.0065)\ \Omega$$

Transformer equivalent circuit referred to lv (secondary) side is given in Fig. 12.42(*b*).

Rated hv (or primary) current

$$I_1 = \frac{S \times 1{,}000}{V_1} = \frac{30 \times 1{,}000}{2300} = 13 \text{ A}$$

Rated lv (or secondary) current

$$I_2 = \frac{I_1}{K} = \frac{13}{1/10} = 130 \text{ A}$$

$$\cos\phi = 0.8 \text{ (lagging) and } \sin\phi = 0.6$$

Phasor diagram is shown in Fig. 12.42(*c*).

Voltage at load end,

$$V_2 = KV_1 - I_2 R_{02}\cos\phi - I_2 X_{02}\sin\phi$$

$$= \frac{1}{10} \times 2{,}200 - 130 \times (0.0055 + 0.005) \times 0.8 - 130 \times (0.0065 + 0.006) \times 0.6$$

$$= 220 - 1.092 - 0.975 = 217.933 \text{ V} \quad \textbf{Ans.}$$

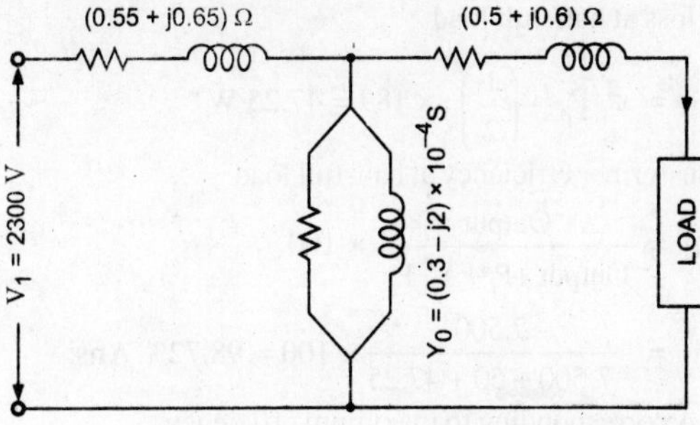

(*a*) *Equivalent Circuit Referred to HV*

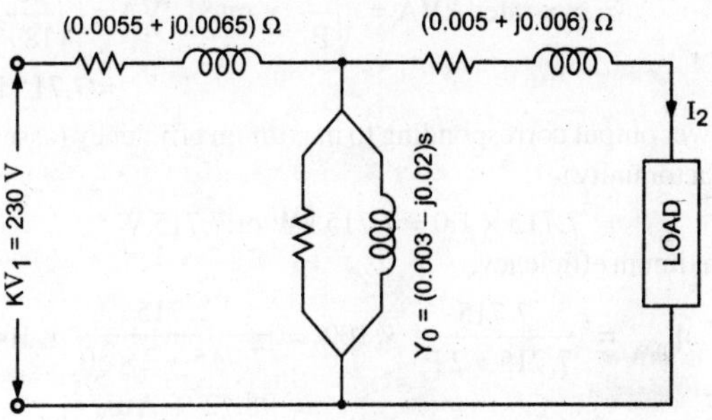

(*b*) *Equivalent Circuit Referred to LV*

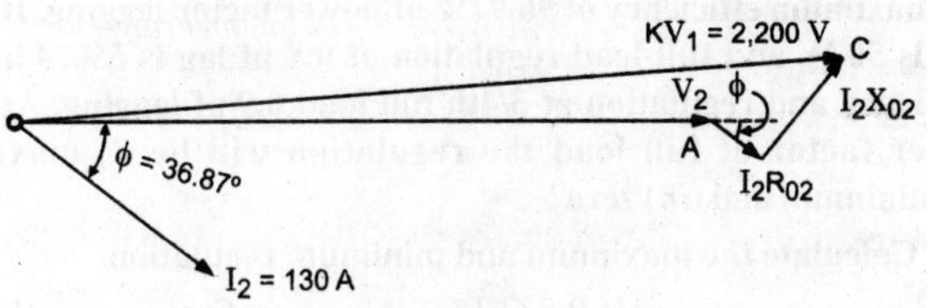

(*c*) *Phasor Diagram*

Fig. 12.42

Iron loss, $P_i = 160$ W (given)

Copper loss, $P_c = I_2^2 R_{02} = (130)^2 \times (0.0055 + 0.005) = 177.45$ W

Transformer output

$$= V_2 I_2 \cos\phi = 217.933 \times 130 \times 0.8 = 22{,}665 \text{ W}$$

Transformer input

$$= V_2 I_2 \cos\phi + P_i + P_c$$

$$= 22{,}665 + 160 + 177.45 = 23002.45 \text{ W}$$

Percentage efficiency,

$$\eta = \frac{\text{Output}}{\text{Input}} \times 100 = \frac{22{,}665}{23002} \times 100 = 98.533\% \quad \textbf{Ans.}$$

Example 12.68. Calculate efficiency, voltage at the secondary terminals and primary input current when supplying full-load secondary current at power factor (*i*) unity (*ii*) 0.8 lag, for a 4 kVA, 230/460 V single-phase transformer.

The following are test results:

Open circuit with 230 V applied to LV side – 0.6 A, 75 W

Short circuit with 20 V applied to HV side – 10 A, 60 W

[U.P.S.C. I.E.S. Electrical Engineering-II, 2005]

Solution: Transformer output at full load, S = 4 kVA

Full-load secondary current,

$$I_2 = \frac{S \times 1{,}000}{V_2} = \frac{4 \times 1000}{460} \text{ A} = 8.6956 \text{ A}$$

Full-load copper loss,

$$P_c = P_s \times \left(\frac{I_2}{I_s}\right)^2 = 60 \times \left(\frac{8.6956}{10}\right)^2 = 45.368 \text{ W}$$

Iron loss, $P_i = 75$ W

(*i*) Transformer efficiency at full load and unity power factor,

$$\eta_1 = \frac{S \times 1{,}000 \times \cos\phi}{S \times 1{,}000 \times \cos\phi + P_c + P_i} \times 100$$

$$= \frac{4 \times 1{,}000 \times 1.0}{4 \times 1{,}000 \times 1.0 + 45.368 + 75} \times 100 = 97.08\% \ \textbf{Ans.}$$

(*ii*) Transformer efficiency at full load and 0.8 pf lagging

$$\eta_2 = \frac{4 \times 1{,}000 \times 0.8}{4 \times 1{,}000 \times 0.8 + 45.368 + 75} \times 100 = 96.37\% \ \textbf{Ans.}$$

Equivalent resistance referred to secondary,

$$R_{02} = \frac{P_s}{I_s^2} = \frac{60}{10^2} = 0.6 \ \Omega$$

Equivalent impedance referred to secondary,

$$Z_{02} = \frac{V_s}{I_s} = \frac{20}{10} = 2 \ \Omega$$

Equivalent reactance referred to secondary,

$$X_{02} = \sqrt{(Z_{02})^2 - (R_{02})^2} = \sqrt{2^2 - 0.6^2} = 1.098 \ \Omega$$

Secondary terminal voltage at unity power factor,

$$V_2 = E_2 - I_2 R_{02} \cos\phi - I_2 X_{02} \sin\phi$$

$$= 460 - 8.6956 \times 0.6 \times 1.0 - 8.6956 \times 1.908 \times 0$$

$$= 454.78 \text{ V} \quad \textbf{Ans.}$$

Secondary terminal voltage at 0.8 lagging power factor,

$$V_2 = 460 - 8.6956 \times 0.6 \times 0.8 - 8.6956 \times 1.908 \times 0.6$$

$$= 460 - 4.174 - 9.955 = 445.87 \text{ V} \quad \textbf{Ans.}$$

$$\text{Primary current, } I_1 = \frac{4 \times 1{,}000}{230} = 17.39 \text{ A} \quad \textbf{Ans.}$$

Example 12.69. A 15 kVA, 2,300/230 V, 50 Hz single-phase transformer gave the following test data:

Open-circuit Test $V_0 = 2{,}300$ V, $I_0 = 0.21$ A, $W_0 = 50$ W

Short-circuit Test $V_s = 47$ V, $I_s = 6.0$ A, $W_s = 160$ W

(*i*) Find the equivalent circuit referred to high-voltage side.

(*ii*) Calculate the full-load voltage regulation at 0.8 pf lagging when the load voltage is held at 220 V.

(*iii*) What is the efficiency at half the rated load at unity power factor.

(*iv*) Find the maximum efficiency and corresponding output power. *[U.P.S.C. I.E.S. Elec. Engineering II, 1997]*

Solution: No-load current $I_0 = 0.21$ A

No-load input power, $P_0 = 50$ W

Primary voltage, $V_1 = 2{,}300$ V

No-load power factor,

$$\cos\phi_0 = \frac{P_0}{V_1 I_0} = \frac{50}{2{,}300 \times 0.21} = 0.1035$$

Energy component of no-load current,

$$I_e = I_0 \cos\phi_0 = 0.21 \times 0.1035 = 0.022 \text{ A}$$

Magnetising component of no–load current,

$$I_m = \sqrt{I_0^2 - I_e^2} = \sqrt{(0.21)^2 - (0.022)^2} = 0.209 \text{ A}$$

No-load resistance,

$$R_0 = \frac{V_1}{I_e} = \frac{2{,}300}{0.022} = 1{,}04{,}500 \ \Omega \text{ or } 104.5 \text{ k}\ \Omega \ \textbf{Ans.}$$

No-load reactance,

$$X_0 = \frac{V_1}{I_m} = \frac{2,300}{0.209} = 11,000\ \Omega \text{ or } 11\ k\ \Omega \textbf{ Ans.}$$

Short-circuit current, I_s = 6.0 A

Short-circuit voltage, V_s = 47 V

Power input on short-circuit, P_s = 160 W

Equivalent impedance referred to primary,

$$Z_{01} = \frac{V_s}{I_s} = \frac{47}{6} = 7.833\ \Omega$$

Equivalent resistance referred to primary,

$$R_{01} = \frac{P_s}{I_s^2} = \frac{160}{6^2} = 4.44\ \Omega$$

Equivalent reactance referred to primary,

$$X_{01} = \sqrt{(Z_{01})^2 - (R_{01})^2} = \sqrt{(7.833)^2 - (4.44)^2} = 6.45\ \Omega$$

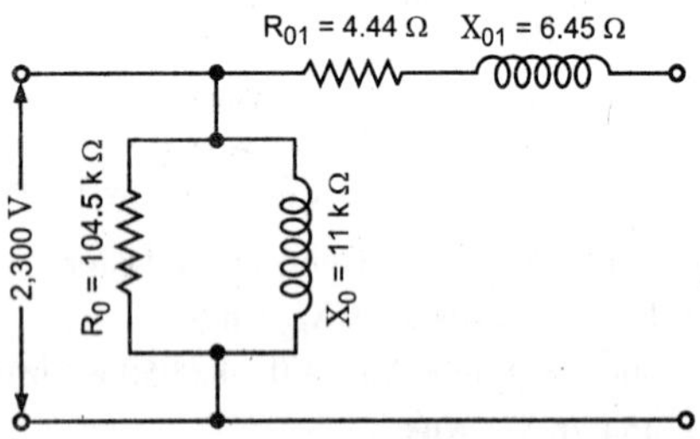

Fig. 12.43

All of the equivalent circuit parameters determined above are referred to primary (*i.e.* high voltage side) and the equivalent circuit of the transformer is shown in Fig. 12.43.

$$\text{Transformation ratio, } K = \frac{V_2}{V_1} = \frac{230}{2,300} = 0.1$$

Equivalent resistance referred to secondary,

$$R_{02} = K^2 R_{01} = (0.1)^2 \times 4.44 = 0.0444\ \Omega$$

Equivalent reactance referred to secondary,

$$X_{02} = K^2 X_{01} = (0.1)^2 \times 6.45 = 0.0645\ \Omega$$

$$\text{Secondary rated current, } I_2 = \frac{15,000}{230} = 65.22 \text{ A}$$

Load voltage, V_2 = 220 V

$$\text{No-load voltage, } E_2 = V_2 + I_2 (R_{02} \cos\phi + X_{02} \sin\phi)$$
$$= 220 + 65.22\,(0.0444 \times 0.8 + 0.0645 \times 0.6)$$
$$= 224.84 \text{ V}$$

(*ii*) Full-load voltage regulation

$$= \frac{I_2 (R_{02} \cos\phi + X_{02} \sin\phi)}{E_2} \times 100$$

$$= \frac{224.84 - 220}{224.84} \times 100 = 2.152\ \% \textbf{ Ans.}$$

Output on half load at 1.0 pf

$$= \text{Rated kVA} \times \frac{1}{2} \times \text{pf} = 15 \times \frac{1}{2} \times 1.0$$

= 7.5 kW or 7,500 W

Iron loss, P_i = 50 W

Full-load copper loss,

$$P_c = \left(\frac{I_1}{I_s}\right)^2 P_s = \left(\frac{15,000/2,300}{6.0}\right)^2 \times 160 = 189 \text{ W}$$

Copper loss at half-full load

$$= x^2 P_c = \left(\frac{1}{2}\right)^2 \times 189 = 47.25 \text{ W}$$

(*iii*) Transformer efficiency at half-full load

$$\eta = \frac{\text{Output}}{\text{Output} + P_i + x^2 P_c} \times 100$$

$$= \frac{7,500}{7,500 + 50 + 47.25} \times 100 = 98.72\% \textbf{ Ans.}$$

(*iv*) KVA corresponding to maximum efficiency

$$= x \times \text{rated kVA} = \sqrt{\frac{P_i}{P_c}} \times \text{rated kVA} = \sqrt{\frac{50}{189}} \times 15$$
$$= 7.715 \text{ kVA}$$

Power output corresponding to maximum efficiency (assuming power factor unity)

$$= 7.715 \times 1.0 = 7.715 \text{ kW or } 7,715 \text{ W}$$

Maximum efficiency,

$$\eta_{max} = \frac{7,715}{7,715 + 2P_i} \times 100 = \frac{7,715}{7,715 + 2 \times 50} \times 100$$
$$= 98.72\ \% \textbf{ Ans.}$$

Example 12.70. A 5 kVA, 220/110 V single-phase transformer has the maximum efficiency of 96.97% at power factor lagging. Its core loss is 50 W and full-load regulation at 0.8 pf lag is 5%. Find the efficiency and regulation at 3/4th full load 0.9 pf lagging. At what power factor at full load the regulation will be (*i*) maximum (*ii*) minimum and (*iii*) zero?

Calculate the maximum and minimum regulation.

[*U.P.S.C. I.E.S. Electrical Engineering-II, 2000*]

Solution: Let power factor corresponding to maximum efficiency be 0.8 (lagging) and corresponding load be x times full load. Now output,

$$P_{out} = xS \text{ (kVA)} \times 1,000 \times \cos\phi$$
$$= x \times 5 \times 1,000 \times 0.8 = 4,000x \text{ watts}$$

Copper loss,

P_c = Iron loss at a load giving maximum efficiency, so

$P_c = P_i = 50$ W

$$\text{Maximum efficiency, } \eta_{max} = \frac{\text{Output, } P_{out}}{P_{out} + P_c + P_i} \times 100$$

$$\text{or} \quad 96.97 = \frac{4,000x}{4,000x + 50 + 50} \times 100$$

or $x = 0.8$

$$\text{Full-load secondary current, } I_2 = \frac{5 \times 1,000}{110} = 45.45 \text{ A}$$

Resistance referred to secondary,

$$R_{02} = \frac{P_c}{(xI_2)^2} = \frac{50}{(0.8 \times 45.45)^2} = 0.0378\ \Omega$$

Full-load regulation as given by $\frac{I(R_{02} \cos\phi + X_{02} \sin\phi)}{V_2} \times 100$ and it is given to be 5%, so

$$\frac{I(R_{02} \cos\phi + X_{02} \sin\phi)}{V_2} \times 100 = 5$$

$$\text{or} \quad \frac{45.45\,(0.0378 \times 0.8 + X_{02} \times 0.6) \times 100}{110} = 5$$

∵ Cos ϕ = 0.8 and sin ϕ = 0.6

or $X_{02} = 0.1513\ \Omega$

Efficiency at $\frac{3}{4}$ th of full-load and at 0.9 pf,

$$\eta = \frac{0.75 \times 5 \times 1{,}000 \times 0.9}{0.75 \times 5 \times 1{,}000 \times 0.9 + \left(\frac{3}{4}\right)^2 \times 50 + 50} \times 100$$

$$= \frac{3{,}375}{3{,}375 + 28.125 + 50} \times 100 = 97.74\% \quad \textbf{Ans.}$$

Regulation at $\frac{3}{4}$ th full-load and 0.9 pf lagging

$$= \frac{0.75 \times 45.45(0.0378 \times 0.9 + 0.1513 \times 0.4359)}{110} \times 100$$

$\because$ Cos ϕ = 0.9 and sin ϕ = 0.4359

= 3.098% **Ans.**

Phase angle corresponding to maximum regulation is given by

$$\phi = \text{Tan}^{-1}\frac{X_{02}}{R_{02}} = \text{Tan}^{-1}\frac{0.1513}{0.0378} = 75.97^\circ$$

Power factor corresponding to maximum regulation

= Cos ϕ = Cos 75.97° = 0.2424 (lagging)

Maximum regulation

$$= \frac{45.45(0.0378 \times 0.2424 + 0.1513 \times 0.97)}{110} \times 100 = 6.44\% \quad \textbf{Ans.}$$

Phase angle corresponding to minimum or zero regulation is given by

$$\phi = \text{Tan}^{-1}\frac{-R_{02}}{X_{02}} = \text{Tan}^{-1}\frac{-0.0378}{0.1513} = -14.03^\circ$$

Power factor corresponding to minimum or zero regulation

= Cos ϕ = Cos (– 14.03°) = 0.97 (leading) **Ans.**

Example 12.71. The efficiency at unity pf, of a 6600/384 V, 200 kVA single-phase transformer is 98%, both at full load and at half full-load. The pf at no-load is 0.2 lagging and the full-load regulation at lagging pf of 0.8 is 4%. Draw the equivalent circuit referred to LV side and insert all values.

[U.P.S.C. I.E.S. Electrical Engineering-II, 2002]

Solution: Let the iron loss and full-load copper loss be P_i and P_c respectively.

At full load and unity power factor

Power output, P_{out} = 200 × 1.0 = 200 kW

Transformer efficiency, η = 98% or 0.98

$$\text{Power input, } P_{in} = \frac{P_{out}}{\eta} = \frac{200}{0.98} = 204.08 \text{ kW}$$

Total losses, $(P_i + P_c) = P_{in} - P_{out}$ = 204.08 – 200 = 4.08 kW ...(*i*)

At half full load and unity power factor

$$\text{Power output, } P_{out} = \frac{1}{2} \times 200 \times 1.0 = 100 \text{ kW}$$

$$\text{Power input, } P_{in} = \frac{P_{out}}{\eta} = \frac{100}{0.98} = 102.04 \text{ kW}$$

$$\text{Total losses } = P_i + \left(\frac{1}{2}\right)^2 P_c = 102.04 - 100 = 2.04 \text{ kW} \quad ...(ii)$$

Solving Eqs. (*i*) and (*ii*) we have

Full-load copper loss, P_c = 2.72 kW

and iron loss, P_i = 1.36 kW

Full-load secondary current,

$$I_2 = \frac{\text{S in kVA} \times 1{,}000}{V_2} = \frac{200 \times 1{,}000}{384} = 520.83 \text{ A}$$

Resistance referred to secondary,

$$R_{02} = \frac{\text{Full-load copper loss}}{I_2^2} = \frac{2.72 \times 1{,}000}{(520.83)^2} = 0.01\ \Omega$$

Full-load regulation at 0.8 pf lagging is given by

$$\frac{I(R_{02}\cos\phi + X_{02}\sin\phi)}{V_2} \times 100 = 4 \quad ...(\text{given})$$

$$\text{or} \quad \frac{520.83(0.01 \times 0.8 + X_{02} \times 0.6)}{384} \times 100 = 4$$

or X_{02} = 0.0358 Ω

$$\text{No load current, } I_0 = \frac{\text{Core loss}}{V_1 \cos\phi_1} = \frac{1.36 \times 1{,}000}{384 \times 0.2} = 17.71 \text{ A}$$

Energy component of no-load current,

$$I_e = I_0 \cos\phi_0 = 17.71 \times 0.2 = 3.54 \text{ A}$$

Magnetising component of no-load current,

$$I_m = \sqrt{I_0^2 - I_e^2} = \sqrt{(17.71)^2 - (3.54)^2} = 17.35 \text{ A}$$

$$\text{Exciting resistance, } R_0 = \frac{V_2}{I_e} = \frac{384}{3.54} = 108.5\ \Omega$$

$$\text{Magnetising reactance, } X_0 = \frac{V_2}{I_m} = \frac{384}{17.35} = 22.13\ \Omega$$

Hence, the equivalent circuit referred to secondary (*lv*) becomes as given in Fig. 12.44.

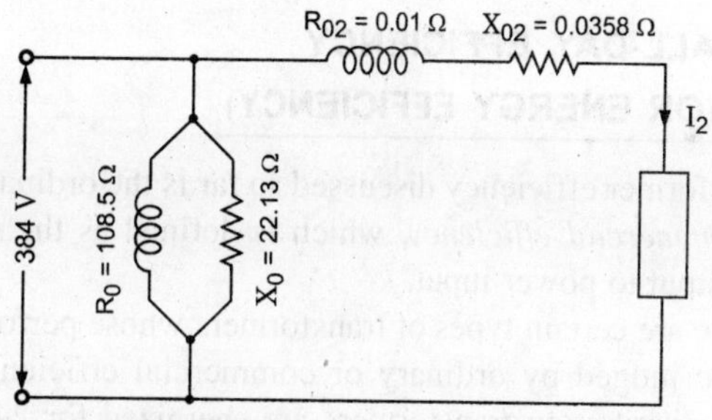

Fig. 12.44 *Equivalent Circuit Referred to LV Side*

Example 12.72. A single-phase transformer is rated at 100 kVA, 2300 / 230 V, 50 Hz. The maximum flux density in the core is 1.2 T and the net x-sectional area is 0.04 m². Determine:

(*i*) The number of primary and secondary turns needed.

(*ii*) Magnetising current if the mean length of the magnetic circuit is 2.5 metres and the relative permeability is 1,200. Neglect the current drawn for the core loss.

(*iii*) The power input on open circuit with rated voltage was 400 watts and power input on short circuit with full-load current flowing was 1,200 watts. Determine the efficiency of the transformer at 75 % of full load with 0.8 pf lag.

(*iv*) What will be the effect on its efficiency if it is connected to a supply of same voltage but double the frequency (*i.e.* 100 Hz).

$$\textbf{Solution:} \text{Transformation ratio, } K = \frac{E_2}{E_1} = \frac{230}{2{,}300} = \frac{1}{10}$$

Supply frequency, f = 50 Hz

Maximum value of flux, $\Phi_{max} = B_{max} \times A$ = 1.2 × 0.04 = 0.048 Wb (*i*) Number of turns on secondary,

$$N_2 = \frac{E_2}{4.44 f \Phi_{max}} = \frac{230}{4.44 \times 50 \times 0.048} = 22 \textbf{ Ans.}$$

Number of turns on primary,

$$N_1 = \frac{N_1}{K} = 22 \times 10 = 220 \textbf{ Ans.}$$

(*ii*) Mean length of magnetic circuit, $l = 2.5$ metres

Relative permeability of core material, $\mu_r = 1{,}200$

Magnetising current,

$$I_m = \frac{B}{\mu_0 \mu_r} \times l \times \frac{1}{N_1} = \frac{1.2 \times 2.5}{4\pi \times 10^{-7} \times 1{,}200 \times 220}$$

$$= 9.04 \text{ A } \textbf{Ans.}$$

(*iii*) Full load copper loss,

$$P_c = 1{,}200 \text{ watts} = 1.2 \text{ kW}$$

$$\text{Iron loss, } P_i = 400 \text{ watts} = 0.4 \text{ kW}$$

Output at 75 % of full load and 0.8 pf

$$= 0.75 \times 100 \times 0.8 = 60 \text{ kW}$$

Input at 75 % of full load and 0.8 pf

$$= \text{Output} + P_i + x^2 P_c = 60 + 0.4 + \left(\frac{3}{4}\right)^2 \times 1.2$$

$$= 61.075 \text{ kW}$$

Efficiency at 75 % of full load and 0.8 pf,

$$\eta = \frac{60}{61.075} \times 100 = 98.24 \% \textbf{ Ans.}$$

(*iv*) When supply frequency is doubled, keeping supply voltage same, the iron losses will increase and efficiency will decrease. **Ans.**

12.22 ALL-DAY EFFICIENCY (OR ENERGY EFFICIENCY)

The transformer efficiency discussed so far is the ordinary, also called *commercial efficiency*, which is defined as the ratio of power output to power input.

There are certain types of transformers whose performance cannot be judged by ordinary or commercial efficiency. For instance, distribution transformers are energized for 24 hours, but they deliver very light loads for major portion of the day. Thus iron or core loss occur for the whole day but the copper loss occurs only when the transformer is loaded. The performance of such a transformer must be judged by its *all-day efficiency*, also called the *energy efficiency* or *operational efficiency* which is computed on the basis of energy consumed during the whole day (24 hours).

The all-day efficiency is defined as the ratio of energy (kWh) output over 24 hours to the energy input over the same period.

$$\textit{i.e. } \text{All-day efficiency} = \frac{\text{Output in kWh}}{\text{Input in kWh}} \quad ...(12.52)$$

Since the distribution transformer does not supply the rated load for the whole day so the all-day efficiency of such a transformer will be lesser than ordinary or commercial efficiency.

For determination of all-day efficiency of a transformer, it is necessary, of course, to know how the load varies from hour to hour during the day.

Higher energy efficiencies are achieved by designing distribution transformers to yield maximum commercial efficiency at less than full load (usually 50–75 per cent of full load). This is achieved by restricting the core flux density to the lower values by using a relatively larger cross section. Thus the ratio of iron loss to copper loss is reduced.

Example 12.73. A 100 kVA transformer has its maximum efficiency of 98% at full load and unity *pf*. During the day it is loaded as follows:

12 hrs — 20 kW at a pf of 0.5 lag
6 hrs — 50 kW at a pf of 0.9 lag
6 hrs — 75 kW at a pf of 0.8 lag

Calculate the "all day efficiency" of the transformer.

Solution: For maximum efficiency

$$\text{Output} = 100 \times 1.0 = 100 \text{ kW}$$

$$\text{Input} = \frac{\text{Output}}{\eta} = \frac{100}{0.98} = 102.04 \text{ kW}$$

$$\text{Total losses} = \text{Input} - \text{output} = 102.04 - 100 = 2.04 \text{ kW}$$

$$\text{Full-load copper loss, } P_c = \text{Iron loss, } P_i$$

$$= \frac{\text{Total losses}}{2} = \frac{2.04}{2} = 1.02 \text{ kW}$$

$$\text{All day output} = 20 \times 12 + 50 \times 6 + 75 \times 6 = 990 \text{ kWh}$$

$$\text{Iron loss for 24 hours} = 24 \times 1.02 = 24.48 \text{ kWh}$$

$$\text{Copper loss for 24 hours} = \left(\frac{20/0.5}{100}\right)^2 \times 1.02 \times 12 + \left(\frac{50/0.9}{100}\right)^2 \times 1.02 \times 6 + \left(\frac{75/0.8}{100}\right)^2 \times 1.02 \times 6$$

$$= 1.9584 + 1.8889 + 5.3789 = 9.226 \text{ kWh}$$

$$\text{All-day efficiency, } \eta = \frac{\text{Output}}{\text{Input}} \times 100$$

$$= \frac{990}{990 + 24.48 + 9.226} \times 100 = 96.7\% \textbf{ Ans.}$$

Example 12.74. Two-100 kW transformers each has a maximum efficiency of 98% but in one the maximum efficiency occurs at full load while in the other it occurs at half load. Each transformer is on full load for 4 hours, on half-load for 6 hours and one-tenth load for 14 hours per day. Determine the efficiency of each transformer.

Solution: Transformer 1

$$\text{Total losses on full load, } P_c + P_i = \frac{\text{Output}}{\eta} - \text{output}$$

$$= \frac{100}{0.98} - 100 = 2.04 \text{ kW}$$

$$\text{Full-load copper loss, } P_c = P_i$$

$$= \frac{P_c + P_i}{2} = \frac{2.04}{2} = 1.02 \text{ kW}$$

Since maximum efficiency occurs at full load and for maximum efficiency $P_c = P_i$

$$\text{All-day output} = 100 \times 4 + \frac{100}{2} \times 6 + \frac{100}{10} \times 14$$

$$= 840 \text{ kWh}$$

$$\text{Iron loss for 24 hours} = 1.02 \times 24 = 24.48 \text{ kWh}$$

Copper loss for 24 hours

$$= 1.02 \times 4 + \left(\frac{1}{2}\right)^2 \times 1.02 \times 6 + \left(\frac{1}{10}\right)^2 \times 1.02 \times 14$$

$$= 5.753 \text{ kWh}$$

All-day efficiency,

$$\eta_1 = \frac{840}{840 + 24.48 + 5.753} \times 100 = 96.53\% \text{ Ans.}$$

Transformer 2

$$\text{Half-load output} = \frac{1}{2} \times 100 = 50 \text{ kW}$$

$$\text{Half-load input} = \frac{50}{0.98} = 51.02 \text{ kW}$$

$$\text{Total losses at half-full load} = 51.02 - 50 = 1.02 \text{ kW}$$

$$\text{Copper loss at half-full load} = \left(\frac{1}{2}\right)^2 \times P_c = \frac{P_c}{4}$$

where P_c is full load copper loss

$$\text{Half-load copper loss, } \frac{P_c}{4} = P_i = \frac{\text{Total loss}}{2} = \frac{1.02}{2} = 0.51 \text{ kW}$$

$$\text{or } P_c = 0.51 \times 4 = 2.04 \text{ kWh}$$

$$\text{All-day output} = 100 \times 4 + \frac{100}{2} \times 6 + \frac{100}{10} \times 14$$

$$= 840 \text{ kWh as above}$$

$$\text{Iron loss for 24 hours} = 24 \times 0.51 = 12.24 \text{ kWh}$$

Copper loss for 24 hours

$$= 2.04 \times 4 + \left(\frac{1}{2}\right)^2 \times 2.04 \times 6 + \left(\frac{1}{10}\right)^2 \times 2.04 \times 14$$

$$= 11.506 \text{ kWh}$$

All-day efficiency,

$$\eta_2 = \frac{840}{840 + 12.24 + 11.506} \times 100 = 97.25\% \text{ Ans.}$$

Example 12.75. The daily variation of load on a 100 kVA transformer is as follows:

8.00 AM to 1.00 PM — 65 kW, 35 kVAR
1.00 PM to 6.00 PM — 80 kW, 50 kVAR
6.00 PM to 1.00 AM — 30 kW, 30 kVAR
1.00 AM to 8.00 AM — No-load

This transformer has no-load core loss of 270 watts and a full-load ohmic loss of 1,200 watts. Determine the all-day efficiency of the transformer.

[*J.N. Technological Univ. Electrical Machines-II, January-2014*]

Solution: All-day output $= 65 \times 5 + 80 \times 5 + 30 \times 7 = 935$ kWh

$$\text{Iron loss for 24 hours} = \frac{270 \times 24}{1,000} = 6.48 \text{ kWh}$$

$$\text{Load in kVA from 8.00 AM to 1.00 PM} = \sqrt{\text{kW}^2 + \text{kVAR}^2}$$

$$= \sqrt{65^2 + 35^2} = 73.824 \text{ kVA}$$

$$\text{Load in kVA from 1.00 PM to 6.00 PM} = \sqrt{80^2 + 50^2} = 94.34 \text{ kVA}$$

$$\text{Load in kVA from 6.00 PM to 1.00 AM} = \sqrt{30^2 + 30^2} = 42.426 \text{ kVA}$$

Copper losses for 24 hours

$$= \left(\frac{73.824}{100}\right)^2 \times 1,200 \times 5 + \left(\frac{94.34}{100}\right)^2 \times 1,200 \times 5 + \left(\frac{42.426}{100}\right)^2 \times 1,200 \times 7$$

$$= 10,122 \text{ Wh or } 10.122 \text{ kWh}$$

All-day efficiency,

$$\eta = \frac{935}{935 + 6.48 + 10.122} \times 100 = 98.26\% \text{ Ans.}$$

12.23 PARALLEL OPERATION OF TRANSFORMERS

12.23.1. Definition. This is an arrangement in which the primaries of two or more transformers are energized from the same source (*i.e.*, the same generator or the same output winding of a bigger transformer or voltage regulator), and the secondaries are connected to supply the same load. In other words the transformers are said to be connected in parallel when their windings are connected to hv and lv bus-bars common to both.

12.23.2. Necessity of Parallel Operation. The operation of transformers in parallel becomes necessary owing to the following reasons.

1. If the amount of power to be transformed is greater than that which can be handled by one transformer, it becomes necessary to employ two or more units in parallel.
2. To meet the additional demand on a power system it becomes necessary either to replace the existing unit by unit of greater capacity or to install additional unit to carry a portion of the load. The cost of making such changes will, in many cases, be greater than if the original transformer is connected in parallel with another to carry a portion of the load.

   Also, sometimes in a new installation, two or more units in parallel, though more expensive, may be preferred over a single unit for reason of reliability—part of the load can be supplied with the remaining healthy units with one unit out.
3. The cost of standby unit is reduced since spare transformer units are invariably required to ensure continuity of service in case of damage or some fault, it is sometimes* found desirable to supply the load through two or more units and thereby to reduce the size of the spare unit.

The main drawback of operating two or more transformers in parallel is that available short-circuit current is increased, sometimes to a level beyond the interrupting capacity of the existing overcurrent devices. A split second bus with a tie breaker interlocked to prevent parallel operation may be necessary in such a case.

12.23.3. Conditions For Satisfactory Operation of Transformers in Parallel. When the transformers are supplied from the same system, of course, have their primaries connected in parallel and have independently loaded secondaries, the operation of individual units is not affected by that of the other so long as the primary voltage is maintained and the line voltage waveform is not affected by either transformer. When the secondaries are also paralleled, however, the transformer characteristics must be properly coordinated for successful operation.

For satisfactory operation of single-phase transformers in parallel, the conditions to be fulfilled are :

(*i*) Polarities of the transformers are the same.

* When spare units are maintained at a central place to serve transformer installations in a certain region, single-unit installations would be preferred.

(*ii*) The voltage ratings of both primaries and secondaries are identical. This obviously implies that the transformation- or turn-ratios are the same.

(*iii*) Per cent (or per unit) impedances of the transformers are the same in magnitude and have the same phase angle (*i.e.* $\frac{X}{R}$ ratio should be the same).

The condition (*i*) is absolutely essential and must be fulfilled and condition (*ii*) should be satisfied to a close degree. There is more latitude as regards condition (*iii*) but the more nearly it is satisfied, the better will be the load sharing.

It is also essential that whenever transformers are connected into parallel banks, disconnecting means must be provided on both the primary and secondary sides of each transformer, and both the associated disconnect switches must be open before any work is done on any unit. If only the primary switch is open, a feedback occurs on the secondary side, and rated voltage will still exist on both the primary and secondary sides of that transformer.

(*i*) *Polarity. The* condition of same polarity is an essential requirement. This can be either right or wrong. If wrong it results in a dead short circuit. So it is essential, before closing of switch on secondary side of incoming transformer, to ensure that polarity of the incoming transformer is correct.

(*ii*) *Voltage Ratio.* An equal voltage ratio (which in cases of high transformation ratio is not necessarily achieved by equality of turn-ratio, unless the designs are identical) is necessary to avoid no-load circulating current. If the voltage ratios of the transformers are not identical, their secondary emfs will be different, and different voltages will appear across the secondary terminals. The difference in emfs induced in secondaries will give rise to a current, called the *circulating current*, flowing in the closed circuit of the secondary windings. Since the leakage impedance is low, even a small voltage difference will give rise to considerable no-load circulating current. This secondary circulating current will cause current, in addition to the magnetizing current, to be drawn from the supply by the primary of each transformer. These additional currents will cause additional copper losses in both primaries and secondaries. When load is connected to the system, the magnitude of secondary currents of the respective transformers will not be in their proper relation to each other. The transformer with higher secondary terminal voltage will tend to carry more than its proper share of the load in order to make the terminal voltages of the same value for both transformers. Thus, one transformer would tend to become overloaded before the other, and the system could not be loaded to the summation of the transformer ratings without overloading one transformer.

If the voltage ratio differs markedly, a fairly large circulating current will flow in the transformer windings, and the parallel operation of the transformers may prove impossible.

(*iii*) *Matched Impedances and Matched Impedance Angles.* For ideal operation, the per cent (or per unit) impedances should be equal in magnitude and have the same phase angle.

If the per unit impedance or the ratios of resistance to reactance are not same, there will be no circulating current but the division of load between the transformers when connected will no longer be proportional to their kVA ratings. If the per unit impedances of transformers are not equal, the one with smaller per unit impedance will carry more load than its share. As the total load increases, this transformer will attain full load, while the other transformer with more per unit impedance will still be operating at part load. The total capacity of two paralleled transformers that have unequal per unit impedances is given by the relation:

$$\text{Total capacity} = S_a + \frac{Z_b}{Z_a} S_b$$

where S_a and Z_a are the respective capacity and per unit impedance of the transformer of larger pu impedance and S_b and Z_b are the respective kVA capacity and pu impedance of transformer of smaller pu impedance. So the maximum permissible kVA load will be less than the sum of the kVA ratings of the units.

If the $\frac{X}{R}$ ratios of the two transformers are different, it results in a divergence of the phase angle of the two currents, so that one transformer will be operating with a higher, and the other with a lower power factor than that of the total load; as a result, the given kW load is not proportionally shared by them.

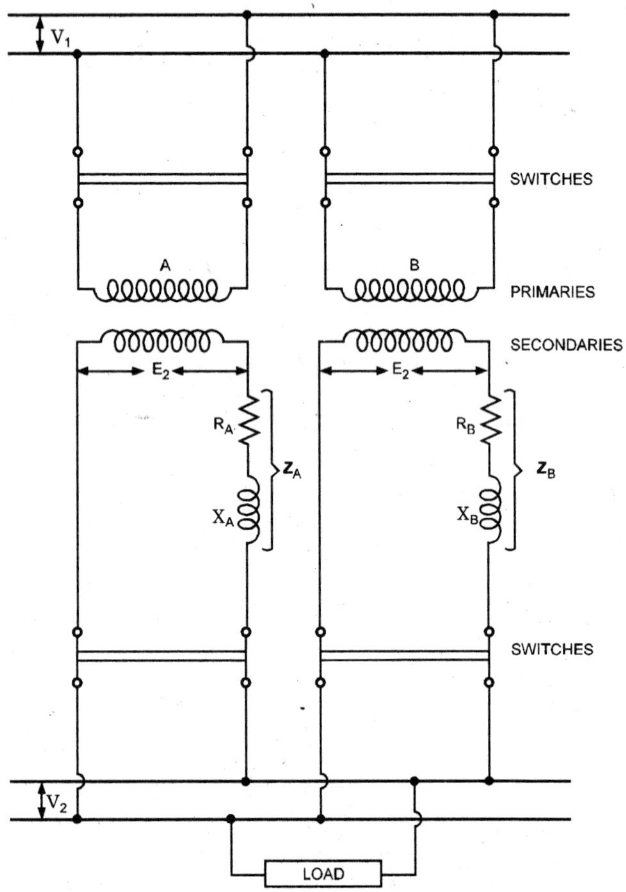

Fig. 12.45

Strictly it is necessary that both pu resistances and pu reactances of paralleled transformers are same for successful parallel operation but generally pu resistance need not be considered with power transformers because it is much smaller than pu reactance. Therefore, satisfactory parallel operation is obtained when the pu impedances are same.

It is not recommended that transformers differing in kVA ratings by more than 3 : 1 should be operated in parallel. In practice, the difference in pu impedances of the transformers to be paralleled should not exceed ± 10 per cent of the average, with kVA rating by no more than 3 times.

Note: It is necessary to distinguish between per unit impedance and ohmic impedance. The currents carried (or kVA loadings) are proportional to their ratings if their ohmic impedances are inversely proportional to their ratings or their pu impedances on their own ratings are equal.

12.23.4. Load Sharing Between Two Transformers Operating in Parallel

Case 1 Equal Voltage Ratio. When the voltage ratios are equal and the voltages are in phase, both sides of a pair of transformers can be connected in parallel and no current will circulate between them on no load. Neglecting magnetizing admittance, the equivalent circuit of the two transformers would then be as shown in Fig. 12.46 (*a*). The phasor diagram is shown in Fig. 12.46 (*b*).

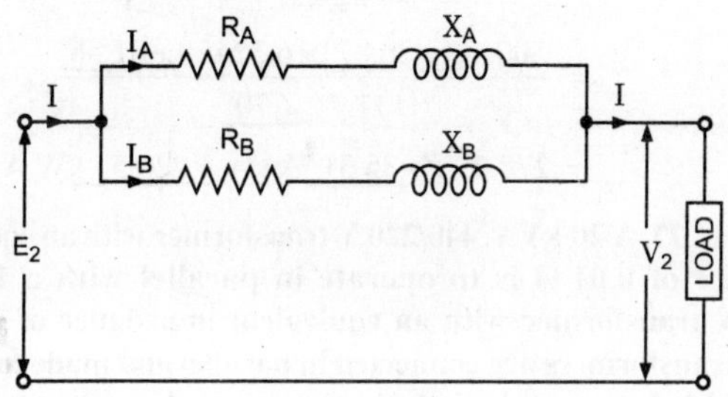

(a) Approximate Equivalent Circuit Diagram

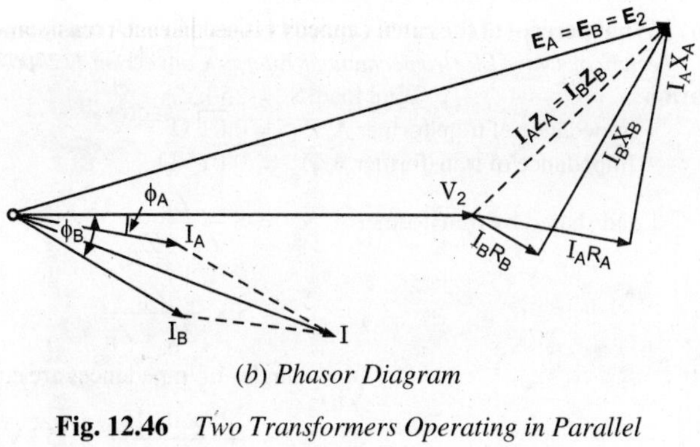

(b) Phasor Diagram

Fig. 12.46 *Two Transformers Operating in Parallel*

From equivalent circuit diagram [Fig. 12.46(*a*)], two voltage equations may be written as (considering all values consistently with reference to secondaries)

$$\mathbf{E}_2 = \mathbf{V}_2 + \mathbf{I}_A \mathbf{Z}_A$$

$$\text{and } \mathbf{E}_2 = \mathbf{V}_2 + \mathbf{I}_B \mathbf{Z}_B$$

$$\text{or } \mathbf{I}_A \mathbf{Z}_A = \mathbf{I}_B \mathbf{Z}_B$$

$$\text{or } \frac{\mathbf{I}_A}{\mathbf{I}_B} = \frac{\mathbf{Z}_B}{\mathbf{Z}_A} = \frac{R_B + jX_B}{R_A + jX_A} \qquad \text{...(12.53)}$$

where R_A, X_A and Z_A are the total equivalent resistance, reactance and impedance respectively of transformer A and R_B, X_B and Z_B are the total equivalent resistance, reactance and impedance respectively of transformer B.

The above equation implies that when two transformers of different kVA ratings are connected in parallel, they divide the total load in proportion to their respective kVA ratings only when their equivalent impedances are inversely proportional to their respective ratings.

$$\text{Also } \mathbf{I}_A \mathbf{Z}_A = \mathbf{I}_B \mathbf{Z}_B = \mathbf{I}\, \mathbf{Z}_{AB}$$

where I is the total current and $\mathbf{Z}_{AB}$ is the combined impedance of $\mathbf{Z}_A$ and $\mathbf{Z}_B$ in parallel.

$$\textit{i.e. } \mathbf{Z}_{AB} = \frac{\mathbf{Z}_A \mathbf{Z}_B}{\mathbf{Z}_A + \mathbf{Z}_B}$$

$$\text{So } \mathbf{I}_A = \mathbf{I}\, \frac{\mathbf{Z}_{AB}}{\mathbf{Z}_A} = \mathbf{I}\, \frac{\mathbf{Z}_A \mathbf{Z}_B}{\mathbf{Z}_A + \mathbf{Z}_B} \times \frac{1}{\mathbf{Z}_A} = \mathbf{I}\, \frac{\mathbf{Z}_B}{\mathbf{Z}_A + \mathbf{Z}_B} \qquad \text{... (12.54)}$$

$$\text{Similarly } \mathbf{I}_B = \mathbf{I}\, \frac{\mathbf{Z}_A}{\mathbf{Z}_A + \mathbf{Z}_B} \qquad \text{... (12.55)}$$

Multiplying both sides of Eqs. (12.54) and (12.55) by common terminal voltage V_2, we have

$$\mathbf{V}_2 \mathbf{I}_A = \mathbf{V}_2 \mathbf{I}\, \frac{\mathbf{Z}_B}{\mathbf{Z}_A + \mathbf{Z}_B} \text{ volt-amperes}$$

or Load shared by transformer A,

$$S_A = \mathbf{V}_2 \mathbf{I}_A \times 10^{-3}\,\text{kVA}$$

$$= \mathbf{V}_2 \mathbf{I}\, \frac{\mathbf{Z}_B}{\mathbf{Z}_A + \mathbf{Z}_B} \times 10^{-3}\,\text{kVA} = \mathbf{S}\, \frac{\mathbf{Z}_B}{\mathbf{Z}_A + \mathbf{Z}_B} \qquad \text{...(12.56)}$$

[where $\mathbf{S} = V_2 I \times 10^{-3}$]

$$\text{and } S_B = \mathbf{V}_2 \mathbf{I}\, \frac{\mathbf{Z}_A}{\mathbf{Z}_A + \mathbf{Z}_B} \times 10^{-3}\,\text{kVA} = \mathbf{S}\, \frac{\mathbf{Z}_A}{\mathbf{Z}_A + \mathbf{Z}_B} \qquad \text{... (12.56 A)}$$

If the equivalent impedances $\mathbf{Z}_A$ and $\mathbf{Z}_B$ are given in pu system or as percentage and the ratings of two transformers are different, then impedances of both transformers should be transferred to common base kVA.

Case 2. Unequal Voltage Ratios. When the turn-ratios are not exactly equal, then with the primaries connected to the same supply, the secondary induced emfs will be unequal and a circulating current will flow, as shown in Fig. 12.47, even at no load. If the transformers are sharing a common load, this circulating current will be superimposed on the load currents.

Let primary supply voltage = $\mathbf{V}_1$ volts

Voltage ratio of transformer A = K_A

Voltage ratio of transformer B = K_B

Equivalent impedance of transformer A referred to secondary,

$$\mathbf{Z}_A = R_A + j\,X_A$$

Equivalent impedance of transformer B referred to secondary,

$$\mathbf{Z}_B = R_B + j\,X_B$$

Output current of transformer A = $\mathbf{I}_A$ amperes

Output current of transformer B = $\mathbf{I}_B$ amperes

Now induced emf in the secondary of transformer A,

$$\mathbf{E}_A = K_A \mathbf{V}_1 = \mathbf{V}_2 + \mathbf{I}_A \mathbf{Z}_A \qquad \text{...(12.57)}$$

Induced emf in the secondary of transformer B,

$$\mathbf{E}_B = K_B \mathbf{V}_1 = \mathbf{V}_2 + \mathbf{I}_B \mathbf{Z}_B \qquad \text{...(12.58)}$$

$$\text{and } \mathbf{V}_2 = \mathbf{I}\, \mathbf{Z}_L = (\mathbf{I}_A + \mathbf{I}_B)\, \mathbf{Z}_L \qquad \text{...(12.59)}$$

where **Z** is the load impedance

$$\text{So } \mathbf{E}_A = \mathbf{I}_A \mathbf{Z}_A + (\mathbf{I}_A + \mathbf{I}_B)\, \mathbf{Z}_L \qquad \text{...(12.60)}$$

$$\text{and } \mathbf{E}_B = \mathbf{I}_B \mathbf{Z}_B + (\mathbf{I}_A + \mathbf{I}_B)\, \mathbf{Z}_L \qquad \text{...(12.61)}$$

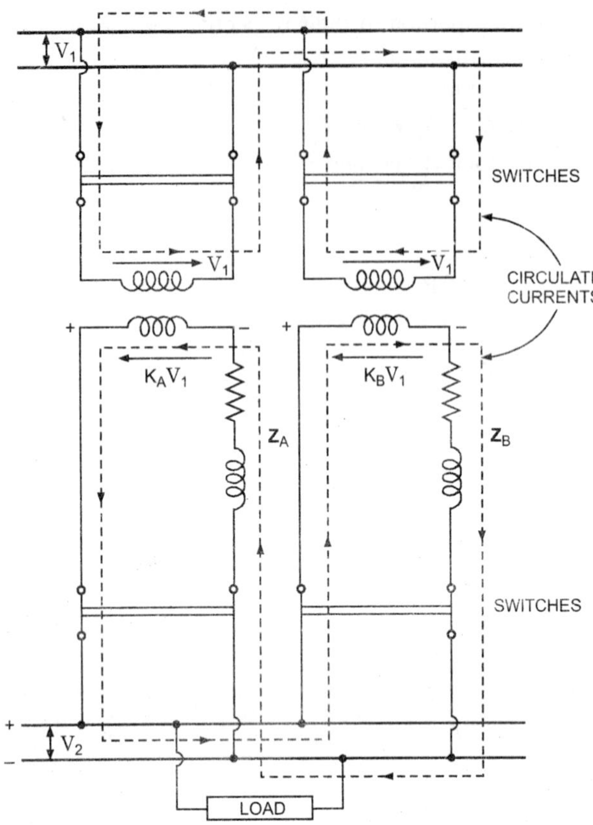

Fig. 12.47

$$\text{or } \mathbf{E}_A - \mathbf{E}_B = \mathbf{I}_A \mathbf{Z}_A - \mathbf{I}_B \mathbf{Z}_B \quad ...(12.62)$$

$$\text{or } \mathbf{I}_A = \frac{(\mathbf{E}_A - \mathbf{E}_B) + \mathbf{I}_B \mathbf{Z}_B}{\mathbf{Z}_A} \quad ...(12.63)$$

Substituting the value of $\mathbf{I}_A$ from Eq. (12.63) in Eq. (12.61), we have

$$\mathbf{E}_B = \mathbf{I}_B \mathbf{Z}_B + \frac{(\mathbf{E}_A - \mathbf{E}_B) + \mathbf{I}_B \mathbf{Z}_B}{\mathbf{Z}_A} \times \mathbf{Z}_L + \mathbf{I}_B \mathbf{Z}_L$$

$$\text{or } \mathbf{I}_B \left(\mathbf{Z}_B + \mathbf{Z}_L + \frac{\mathbf{Z}_B}{\mathbf{Z}_A} \mathbf{Z}_L \right) = \frac{\mathbf{E}_B \mathbf{Z}_A - (\mathbf{E}_A - \mathbf{E}_B) \mathbf{Z}_L}{\mathbf{Z}_A}$$

$$\text{or } \mathbf{I}_B = \frac{\mathbf{E}_B \mathbf{Z}_A - (\mathbf{E}_A - \mathbf{E}_B) \mathbf{Z}_L}{\mathbf{Z}_A \mathbf{Z}_B + \mathbf{Z}_L (\mathbf{Z}_A + \mathbf{Z}_B)} \quad ... (12.64)$$

From the symmetry of the equation we have

$$\mathbf{I}_A = \frac{\mathbf{E}_A \mathbf{Z}_B + (\mathbf{E}_A - \mathbf{E}_B) \mathbf{Z}_L}{\mathbf{Z}_A \mathbf{Z}_B + \mathbf{Z}_L (\mathbf{Z}_A + \mathbf{Z}_B)} \quad ...(12.65)$$

Circulating current,

$$\mathbf{I}_c = \frac{\mathbf{E}_A - \mathbf{Z}_B}{\mathbf{Z}_A + \mathbf{Z}_B} \quad ...(12.66)$$

Example 12.76. Write down the conditions for parallel operation of 3-phase transformers. A load of 500 kVA at 0.8 pf lagging is to be shared by two three phase transformers A and B of equal ratings. If the equivalent impedances as referred to output side are (2 + *j*6) Ω for A and (2 + *j*5) Ω for B, calculate the load shared by each transformer. [*M.D. Univ. Electrical Machines-I, December-2012*]

Solution: Load to be shared,

$$\mathbf{S} = 500 \angle \text{Cos}^{-1}\, 0.8 \text{ lagging} = 500 \angle -36.87^\circ \text{ kVA}$$

Percentage impedance of transformer A,

$$\mathbf{Z}_A = (2 + j6) \text{ or } 6.3245 \angle 71.56^\circ$$

Percentage impedance of transformer B,

$$\mathbf{Z}_B = (2 + j5) \text{ or } 5.385 \angle 68.2^\circ \text{ kVA}$$

Load shared by transformer A

$$= \mathbf{S} \times \frac{\mathbf{Z}_B}{\mathbf{Z}_A + \mathbf{Z}_B}$$

$$= 500 \angle -36.87^\circ \times \frac{2 + j5}{2 + j6 + 2 + j5}$$

$$= \frac{500 \angle -36.87^\circ \times 5.385 \angle 68.2^\circ}{11.7 \angle 70^\circ}$$

$$= 230.128 \angle -38.67^\circ \text{ kVA} = 179.65 \text{ kW} \quad \textbf{Ans.}$$

Load shared by transformer B

$$= \frac{\mathbf{S} \times \mathbf{Z}_A}{\mathbf{Z}_A + \mathbf{Z}_B}$$

$$= \frac{500 \angle -36.87^\circ \times (2 + j6)}{11.7 \angle 70^\circ}$$

$$= \frac{500 \angle -36.87^\circ \times 6.3245 \angle 71.56^\circ}{11.7 \angle 70^\circ}$$

$$= 270.3 \angle -35.31^\circ \text{ kVA} = 220.57 \text{ kW} \quad \textbf{Ans.}$$

Example 12.77. A 20 kVA, 440/220 V transformer with an equivalent impedance of 0.01 Ω is to operate in parallel with a 15 kVA, 440/220 V transformer with an equivalent impedance of 0.015 Ω. The two transformers are connected in parallel and made to carry a load of 25 kVA. Assume both the impedances to have the same angle.

(*i*) Find the individual load currents.

(*ii*) What percent of the rated capacity is used in each transformer?

[*U.P. Technical Univ. Electromechanical Energy Conversion-I, 2005-06*]

Solution: Total load, S = 25 kVA

Impedance of transformer A, Z_A = 0.01 Ω

Impedance of transformer B, Z_B = 0.015 Ω

Load shared by transformer A, $S_A = S \dfrac{\mathbf{Z}_B}{\mathbf{Z}_A + \mathbf{Z}_B}$

$$= S \times \frac{Z_B}{Z_A + Z_B}$$

∵ phase angles of impedances are equal

$$= \frac{25 \times 0.015}{0.01 + 0.015} = 15 \text{ kVA}$$

Load shared by transformer B, $S_B = \dfrac{25 \times 0.01}{0.01 + 0.015} = 10$ kVA

(*i*) Load current of transformer A, $I_A = \dfrac{15 \times 1{,}000}{440} = 34.1$ A on hv side or 68.2 A on lv side **Ans.**

Load current of transformer B, $I_B = \dfrac{10 \times 1{,}000}{440} = 22.73$ A on hv side or 45.46 A on *lv* side **Ans.**

∵ Transformation ratio, $K = \dfrac{1}{2}$

(*ii*) Per cent of the rated capacity used in case of transformer A

$$= \frac{\text{Load shared by transformer A}}{\text{Rated capacity of transformer A}} \times 100 = \frac{15 \text{ kVA}}{20 \text{ kVA}} \times 100$$

$$= 75\% \quad \textbf{Ans.}$$

Per cent of the rated capacity used in case of transformer B

$$= \frac{10 \text{ kVA}}{15 \text{ kVA}} \times 100 = 66.67\% \quad \textbf{Ans.}$$

Example 12.78. A 500 kVA single-phase transformer with 0.012 pu resistance and 0.06 pu reactance, is connected in parallel with a 250 kVA single-phase transformer with 0.014 pu resistance and 0.045 pu reactance to share a load of 600 kVA at 0.8 power factor lagging. Find the kVA and power factor shared by each transformer.

[*U.P.S.C. I.E.S. Electrical Engineering-II, 1999*]

Solution: It should be noted that per unit resistance and reactance given in this problem refer to different ratings. As pointed out in Art. 12.23.4 these should be expressed in terms of the same kVA rating, say 500 kVA.

Per unit impedance of 500 kVA transformer,

$$\mathbf{Z}_1(\text{pu}) = (0.012 + j\,0.06) = 0.0612 \angle 78.69^\circ$$

Per unit impedance of 250 kVA transformer referred to base kVA of 500,

$$\mathbf{Z}_2(\text{pu}) = (0.014 + j0.045) \times \frac{500}{250}$$

$$= (0.028 + j0.09) = 0.09425 \angle 72.72^\circ$$

Total load to be shared, $\mathbf{S} = 600 \angle 36.87^\circ$

$$\because \cos^{-1} 0.8 \text{ (lagging)} = -36.87^\circ$$

Load shared by 500 kVA transformer,

$$\mathbf{S}_1 = \frac{\mathbf{S} \times \mathbf{Z}_{2(pu)}}{\mathbf{Z}_{1(pu)} + Z_{2(pu)}} = \frac{600 \angle -36.87^\circ \times 0.09425 \angle 72.72^\circ}{(0.012 + j0.06) + (0.028 + j0.09)}$$

$$= \frac{600 \angle -36.87^\circ \times 0.09425 \angle 72.72^\circ}{0.15524 \angle 75.069^\circ}$$

$$= 364.27 \angle -39.22^\circ \text{ kVA}$$

or 282.21 kW at a pf of 0.788 (lagging) **Ans.**

Load shared by 250 kVA transformer,

$$\mathbf{S}_2 = \frac{\mathbf{S} \times \mathbf{Z}_{1(pu)}}{\mathbf{Z}_{1(pu)} + \mathbf{Z}_{2(pu)}} = \frac{600 \angle -36.87^\circ \times 0.0612 \angle 78.69^\circ}{0.15524 \angle 75.069^\circ}$$

$$= 236.54 \angle -33.25^\circ \text{ kVA}$$

or 197.82 kW at a pf of 0.836 (lagging) **Ans.**

Example 12.79. Two 6,600/250 V transformers have the following short-circuit characteristics: Applied voltage 200 V, current 30 A, power input 1,200 watts for one of the transformers; the corresponding data for the other transformer being 120 V, 20 A and 1,500 watts. All values are measured on the hv side and with the lv terminals short circuited. Find the approximate current and the power factor of each transformer when working in parallel with each other on the high and low voltage sides and taking a total load of 300 kW at a pf of 0.8 lagging from the high voltage bus-bars.

Solution: Equivalent resistance of transformer 1,

$$R_1 = \frac{1{,}200}{(30)^2} = 1.333\ \Omega$$

Equivalent impedance of transformer 1,

$$Z_1 = \frac{200}{30} = 6.667\ \Omega$$

Equivalent reactance of transformer 1,

$$X_1 = \sqrt{(6.667)^2 - (1.333)^2} = 6.53\ \Omega$$

Equivalent impedance of transformer 1,

$$\mathbf{Z}_1 = (1.333 + j\,6.53) = 6.667 \angle 78.7^\circ\ \Omega$$

Equivalent resistance of transformer 2,

$$R_2 = \frac{1{,}500}{(20)^2} = 3.75\ \Omega$$

Equivalent impedance of transformer 2,

$$Z_2 = \frac{120}{20} = 6\ \Omega$$

Equivalent reactance of transformer 2,

$$X_2 = \sqrt{(6)^2 - (3.75)^2} = 4.684\ \Omega$$

Equivalent impedance of transformer 2,

$$\mathbf{Z}_2 = (3.75 + j\,4.684) = 6 \angle 51.3^\circ\ \Omega$$

$$\mathbf{Z}_1 + \mathbf{Z}_2 = (1.333 + j\,6.53) + (3.75 + j\,4.684)$$

$$= (5.083 + j\,11.214) = 12.31 \angle 65.6^\circ\ \Omega$$

Load to be shared,

$$\mathbf{S} = \frac{300}{0.8} \angle -36.9^\circ = 375 \angle -36.9^\circ \text{ kVA}$$

Total load current,

$$\mathbf{I} = \frac{\mathbf{S} \times 1{,}000}{V_1} = \frac{375 \times 1{,}000}{6{,}600} \angle -36.9^\circ = 56.82 \angle -36.9^\circ \text{ A}$$

Current drawn by transformer 1,

$$\mathbf{I}_1 = \mathbf{I} \times \frac{\mathbf{Z}_2}{\mathbf{Z}_1 + \mathbf{Z}_2} = 56.82 \angle -36.9^\circ \times \frac{6 \angle 51.3^\circ}{12.31 \angle 65.6^\circ}$$

$$= 27.7 \angle 51.2^\circ \text{ A}$$

Hence current drawn by transformer 1

= 27.7 A at pf cos 51.2° *i.e.* 0.627 (lag) **Ans.**

Current drawn by transformer 2,

$$\mathbf{I}_2 = \mathbf{I} \times \frac{\mathbf{Z}_1}{\mathbf{Z}_1 + \mathbf{Z}_2} = 56.82 \angle -36.9^\circ \times \frac{6.667 \angle 78.7^\circ}{12.31 \angle 65.6^\circ}$$

$= 30.8 \angle -23.8^\circ$ or 30.8 A at pf of cos 23.8° *i.e.*, 0.915 (lag) **Ans.**

Example 12.80. A 500 kVA, 500 V single-phase transformer with a reactance drop of 3% and resistance drop of 1% is connected in parallel on the high voltage side with a 250 kVA, 500 V transformer having drops of 6% and 1.5% respectively. The secondary terminal voltage on open circuit is 510 V in the first and 500 V in the second transformer. With the secondary windings connected in parallel calculate (i) the cross-current in the secondaries on no load and (ii) the secondary current in each transformer when the total load is 700 kW at unity pf.

Solution : It is more convenient to work with ohmic impedances than with percentage impedances, we will convert percentage values into numerical values by assuming 480 volts as the terminal voltage (arbitrary value but less than emfs of either transformer).

Full–load current of transformer 1,

$$I_1 = \frac{500 \times 1{,}000}{480} = 1{,}042 \text{ A}$$

Full–load current of transformer 2,

$$I_2 = \frac{250 \times 1{,}000}{480} = 521 \text{ A}$$

OC secondary voltage of transformer 1,

$$E_1 = 510 \text{ V}$$

OC secondary voltage of transformer 2,

$$E_2 = 500 \text{ V}$$

Resistance of 500 kVA transformer,

$$R_1 = \frac{\%R_1 \times E_1}{100 \times I_1} = \frac{1 \times 510}{100 \times 1{,}042} = 0.0049\ \Omega$$

Reactance of 500 kVA transformer,

$$X_1 = \frac{\%X_1 \times E_1}{100 \times I_1} = \frac{3 \times 510}{100 \times 1{,}042} = 0.0147\ \Omega$$

Resistance of 250 kVA transformer,

$$R_2 = \frac{1.5 \times 500}{100 \times 521} = 0.0144\ \Omega$$

Reactance of 250 kVA transformer,

$$X_2 = \frac{6 \times 500}{100 \times 521} = 0.0576\ \Omega$$

$$\therefore\ \mathbf{Z}_1 = 0.0049 + j\,0.0147 = 0.0155\ \angle 71.56^\circ\ \Omega$$

$$\mathbf{Z}_2 = 0.0144 + j\,0.0576 = 0.0594\ \angle 75.96^\circ\ \Omega$$

$$\mathbf{Z}_1 + \mathbf{Z}_2 = (0.0049 + j\,0.0147) + (0.0144 + j\,0.0576)$$

$$= (0.0193 + j\,0.0723)\ \Omega = 0.075\ \angle 75^\circ\ \Omega$$

Load, $\mathbf{S}$ = 700 kW at unity power factor = 700 $\angle 0^\circ$

Load impedance,

$$\mathbf{Z}_L = \frac{V_2^2}{S \times 1{,}000} = \frac{(480)^2}{1{,}000 \times 700 \angle 0^\circ} = 0.329\ \angle 0^\circ\ \Omega$$

$$= (0.329\ \Omega + j\,0)\ \Omega$$

(*i*) Cross-current in the secondaries on no load,

$$I_c = \frac{\mathbf{E}_1 - \mathbf{E}_2}{\mathbf{Z}_1 + \mathbf{Z}_2} = \frac{510 - 500}{0.075 \angle 75^\circ} = 133.3\ \angle -75^\circ \text{ A} \quad \textbf{Ans.}$$

(*ii*) Current supplied by transformer 1,

$$\mathbf{I}_1 = \frac{\mathbf{E}_1 \times \mathbf{Z}_2 - (\mathbf{E}_1 - \mathbf{E}_2)\mathbf{Z}_L}{\mathbf{Z}_1 \times \mathbf{Z}_2 + \mathbf{Z}_L(\mathbf{Z}_1 + \mathbf{Z}_2)}$$

$$= \frac{510 \times 0.0594\ \angle 75.96^\circ + (510 - 500) \times 0.329\ \angle 0^\circ}{0.0155\ \angle 71.56^\circ \times 0.0594\ \angle 75.96^\circ + 0.329\ \angle 0^\circ \times 0.075\ \angle 75^\circ} \text{ A}$$

$$= \frac{10.634 + j\,29.376}{0.00092\ \angle 147.52^\circ + 0.024675\ \angle 75^\circ} \text{ A}$$

$$= \frac{10.634 + j\,29.376}{0.000776 + j\,0.00049 + 0.006386 + j\,0.02383} \text{ A}$$

$$= \frac{10.634 + j\,29.376}{0.007162 + j\,0.02432} = \frac{31.24\ \angle 70^\circ}{0.025\ \angle 73.6^\circ} = 1{,}250\ \angle -3.6^\circ \text{ A } \textbf{Ans.}$$

Similarly,

$$\mathbf{I}_2 = \frac{500 \times 0.0155\ \angle 71.56^\circ (510 - 500) \times 0.329\ \angle 0^\circ}{0.00155\ \angle 71.56^\circ \times 0.0594\ \angle 75.96^\circ + 0.329\ \angle 0^\circ \times 0.075\ \angle 75^\circ} \text{ A}$$

$$= \frac{7.75\ \angle 71.56^\circ - 3.29\ \angle 0^\circ}{0.025\ \angle 73.6^\circ} = \frac{-0.84 + j\,7.35}{0.025\ \angle 73.76^\circ}$$

$$= \frac{7.4 + \angle 76.5^\circ}{0.025\ \angle 73.6^\circ} \text{ A} = 296\ \angle 22.9^\circ \text{ A } \textbf{Ans.}$$

Example 12.81. Two transformers A and B are connected in parallel to a load of (2 + *j* 1.5) Ω. The impedances in secondary are $\mathbf{Z_A = (0.15 + \mathit{j}\,0.5)\ \Omega}$ and $\mathbf{Z_B = (0.1 + \mathit{j}\,0.6)\ \Omega}$. Their no-load terminal voltages are $\mathbf{E_A = 207\ \angle 0^\circ}$ and $\mathbf{E_B = 205\ \angle 0^\circ}$ volts. Find the power output and power factor of each transformer.

[*Pb. Univ. Elec. Machines-I, 1991*]

Solution: Induced emf of transformer A, $\mathbf{E}_A = 207\ \angle 0^\circ = (207 + j\,0)$ volts

Induced emf of transformer B, $\mathbf{E}_B = 205\ \angle 0^\circ = (205 + j\,0)$ volts

Equivalent impedance of transformer A, $\mathbf{Z}_A = (0.15 + j\,0.5)\ \Omega$

Equivalent impedance of transformer B, $\mathbf{Z}_B = (0.1 + j\,0.6)\ \Omega$

Load impedance, $\mathbf{Z}_L = (2 + j\,1.5)\ \Omega$

Let the current supplied by transformers A and B be $\mathbf{I}_A$ and $\mathbf{I}_B$ respectively and terminal voltage V_2.

Now total load current, $\mathbf{I} = \mathbf{I}_A + \mathbf{I}_B$

Voltage across load, $\mathbf{V}_2 = \mathbf{I}\mathbf{Z}_L = (2 + j\,1.5)\,(\mathbf{I}_A + \mathbf{I}_B)$

Since induced emf in transformer A is given as

$$\mathbf{E}_A = \mathbf{V}_2 + \mathbf{I}_A\,\mathbf{Z}_A$$

$$\therefore\ (207 + j\,0) = (2 + j\,1.5)\,(\mathbf{I}_A + \mathbf{I}_B) + (0.15 + j\,0.5)\,\mathbf{I}_A$$

$$\text{or}\ (2.15 + j\,2.0)\,\mathbf{I}_A + (2 + j\,1.5)\,\mathbf{I}_B = 207 + j\,0 \qquad ...(i)$$

Similarly $\mathbf{E}_B = \mathbf{V}_2 + \mathbf{I}_B\ \mathbf{Z}_B$

or $(205 + j\,0) = (2 + j\,1.5)\,(\mathbf{I}_A + \mathbf{I}_B) + (0.1 + j\,0.6)\,\mathbf{I}_B$

$$\text{or}\ (2 + j\,1.5)\,\mathbf{I}_A + (2.1 + j\,2.1)\,\mathbf{I}_B = 205 + j\,0 \qquad ...(ii)$$

Solving Eqs. (*i*) and (*ii*) we get

$$\mathbf{I}_A = \frac{(2.1 + j\,2.1)(207 + j\,0) - (2 + j\,1.5)(205 + j\,0)}{(2.15 + j\,2.0)(2.1 + j\,2.1) - (2 + j\,1.5)(2 + j\,1.5)}$$

$$= \frac{24.7 + j\,127.2}{-1.435 + j\,2.715} = \frac{129.7\ \angle 79^\circ}{3.07\ \angle 117.9^\circ}$$

$$= 42.26\ \angle -38.9^\circ = (32.89 - j\,26.55) \text{ A}$$

$$\text{and } \mathbf{I}_B = \frac{(207 + j\,0)(2 + j\,1.5) - (205 + j\,0)(2.15 + j\,2.0)}{(2 + j\,1.5)(2 + j\,1.5) - (2.1 + j\,2.1)(2.15 + j\,2.0)}$$

$$= \frac{26.75 + j\,99.5}{-1.435 + j\,2.715} = \frac{103\ \angle 75^\circ}{307\ \angle 117.9^\circ} = 33.56\ \angle -42.9^\circ$$

$$= (24.58 - j\,22.84) \text{ A}$$

Terminal voltage,

$$\mathbf{V}_2 = (\mathbf{I}_A + \mathbf{I}_B)(\mathbf{Z}_L) = (32.89 - j\,26.55 + 24.58 - j\,22.84)(2 + j\,1.5)$$

$$= (57.47 - j\,49.39)(2 + j\,1.5)$$

$$= 189 - j\,12.58 = 189.4\ \angle -3.9^\circ \text{ V}$$

Power factor angle of transformer A,

$$\phi_A = -3.9^\circ - (-38.9^\circ) = 35^\circ$$

Power factor of transformer A,

$$\text{Cos}\,\phi_A = \text{Cos}\,35^\circ = 0.8192 \text{ (lag) } \textbf{Ans.}$$

Power factor angle of transformer B,

$$\phi_B = -3.9^\circ - (42.9^\circ) = 39^\circ$$

Power factor of transformer B,

$$\text{Cos}\,\phi_B = \text{Cos}\,39^\circ = 0.7771 \text{ (lag) } \textbf{Ans.}$$

Power output of transformer A,

$$P_A = \frac{V_2 \times I_A \cos\phi_A}{1{,}000} \text{ kW} = \frac{189.4 \times 42.26 \times 0.8192}{1{,}000} = 6.55 \text{ kW } \textbf{Ans.}$$

Power output of transformer B, $P_B = \dfrac{189.4 \times 33.56 \times 0.7771}{1{,}000}$

$$= 4.94 \text{ kW } \textbf{Ans.}$$

Example 12.82. The short circuit tests on two single-phase transformers gave the following results:

200 kVA —3% rated voltage; rated current at 0.25 pf lag

500 kVA —4% rated voltage; rated current at 0.3 pf lag

(*i*) **These two transformers are connected in parallel. How do they share a load of 560 kW at 0.8 pf lag ?**

(*ii*) **If their voltage ratings are 11,000/400 V, calculate the secondary terminal voltage at the load given in part (*i*).**

[*U.P. Technical Univ. Electromechanical Energy Conversion-I, 2003-04*]

Solution: Let $I_{Rated} = I_{Base} = 1.00$

$V_{Rated} = V_{Base} = 1.00$

Under short-circuit condition, $Z_A = \frac{V_{SC}}{I_{SC}} = \frac{0.03 \times 1}{1} = 0.03$

$\because I_{SC} Z_A = V_{SC}$

Short-circuit pf, $\cos \phi_{SC} = 0.25$

and $\sin \phi_{SC} = \sqrt{1 - 0.25^2} = 0.968$

In complex notation, $\mathbf{Z_A} = 0.03\,(0.25 + 0.968)$

$= (0.0075 + j\,0.029)$ pu

Similarly $\mathbf{Z_B} = 0.04\,(0.3 + j\,0.954)$

$= (0.012 + 0.03816)$ pu

Choosing 500 kVA base arbitrarily, we have

$$\mathbf{Z_A} = \frac{500}{200}(0.0075 + j0.029)$$

$$= 0.01875 + j\,0.0725 = 0.0749\ \angle 75.5^\circ$$

$$\mathbf{Z_B} = \frac{500}{200}(0.012 + j\,0.03816) = 0.04\ \angle 72.54^\circ$$

$$\mathbf{Z_A + Z_B} = [(0.01875 + j\,0.0725) + (0.012 + j\,0.03816)]$$

$$= (0.03075 + j\,0.11066)\ \Omega = 0.1149\ \angle 74.47^\circ$$

$$\text{Total kVA, S} = \frac{\text{Total load in kW}}{\text{pf}} = \frac{560}{0.8} = 700 \text{ kVA}$$

$$\mathbf{S} = 700\ \angle \cos^{-1} 0.8 = 700\ \angle -36.87^\circ \text{ kVA}$$

Load shared by transformer A,

$$\mathbf{S_A} = \mathbf{S} \times \frac{\mathbf{Z_B}}{\mathbf{Z_A + Z_B}}$$

$$= \frac{700\ \angle -36.87^\circ \times 0.04\ \angle 72.54^\circ}{0.1149\ \angle 74.47^\circ} \text{ kVA}$$

$= 243.69\ \angle 38.8^\circ$ kVA

$= 243.69 \cos 38.8^\circ$

$= 189.9$ kW at a pf of 0.779 (lag) **Ans.**

Load shared by transformer B,

$$\mathbf{S_B} = \mathbf{S} \times \frac{\mathbf{Z_A}}{\mathbf{Z_A + Z_B}}$$

$$= \frac{700\ \angle -36.87^\circ \times 0.0749\ \angle 75.5^\circ}{0.1149\ \angle 74.47^\circ} \text{ kVA}$$

$= 456.31\ \angle 35.84^\circ$ kVA

$= 369.9$ kW at a pf of 0.8107 (lag) **Ans.**

(*ii*) Current shared by transformer A

$$= \frac{243.69}{200} = 1.2185 \text{ pu}$$

For transformer A, the voltage regulation at 1.2185 pu

$= 1.2185\,(0.0075 \times \cos 38.8^\circ + 0.029 \sin 38.8^\circ)$

$= 1.2185\,(0.005845 + 0.01817) = 0.0292641$

or $\frac{E_2 - V_2}{E_2} = 0.0292641$

or Secondary terminal voltage,

$V_2 = (1 - 0.0292641) \times 400 = 388.3$ V **Ans.**

Example 12.83. Two single-phase transformers rated 1,000 kVA and 500 kVA have per unit leakage impedance of (0.02 + *j* 0.06) and (0.025 + 0.08) respectively. What is largest kVA load that can be delivered by the parallel combination of these two transformers without overloading any one?

[*U.P. Technical Univ. Electromechanical Energy Conversion-I, 2008-09*]

Solution: For transformer A of 1,000 kVA, $\mathbf{Z_A} = (0.02 + j\,0.06)$

For transformer B of 500 kVA, $\mathbf{Z_B} = (0.025 + 0.08)$

Choosing 1,000 kVA base arbitrarily we have

$$\mathbf{Z_A} = (0.02 + j\,0.06) = 0.06325\ \angle 71.565^\circ$$

$$\mathbf{Z_B} = \frac{1,000}{500}(0.025 + j\,0.08) = 0.16763\ \angle 72.646^\circ$$

$$\mathbf{Z_A + Z_B} = (0.02 + 0.06) + (0.05 + j\,0.16)$$

$$= (0.07 + j\,0.22) = 0.2309\ \angle 72.35^\circ$$

$$\text{Total kVA load, } \mathbf{S} = \frac{\mathbf{S_A(Z_A + Z_B)}}{\mathbf{Z_B}}$$

$$= \frac{1,000 \times 0.2309\ \angle 72.35^\circ}{0.16763\ \angle 72.646^\circ} = 1,377.44 \text{ kVA}$$

$$\text{Similarly } \mathbf{S} = \frac{500 \times 0.2309\ \angle 72.35^\circ}{0.06325\ \angle 71.565^\circ} = 1,825.3 \text{ kVA}$$

On per unit basis, $\mathbf{Z_A}$ (0.06325) based on its own kVA rating is less than Z_B (0.0838) based on its kVA. So, as the load on the parallel combination is increased, transformer A attains its load of 1,000 kVA first and, therefore, transformer B shares a load of 1,377.44 – 1,000 = 377.44 kVA

Largest load on transformers, **S** = 1,377.44 kVA **Ans.**

12.24 TRANSFORMER CONSTRUCTION

The transformer is very simple in construction and consists of *magnetic circuit* linking with two windings known as *primary* and *secondary windings*. Besides magnetic circuit and windings it consists of a suitable container for the assembled core and windings, such as a *tank*, a suitable medium for insulating the core and windings from its container such as *transformer oil*, suitable *bushings* (either of porcelain, oil filled or condenser type) for insulating and bringing out terminals of the windings from the container, *temperature gauge* for measurement of temperature of hot oil or hottest spot temperature, and *oil gauge* to indicate the oil level inside the tank. Some transformers are also provided with *conservator tank* in order to slow down deterioration of oil and keep the main tank full of oil, emergency vent to relieve the pressure inside the tank in case the pressure inside the transformer rises to a danger point and *gas operated relay* (*Buchholz relay*) in order to give alarm to indicate the presence of gas in case of some minor fault and take out the transformer out of circuit in case of serious fault.

12.24.1. Core Construction. Since core is a magnetic link between the two systems connected to the transformer and it itself contains a lot of energy, therefore, it is not by any means the passive component as it appears. A transformer core is the steel system which forms the magnetic circuit with all parts pertaining to its construction. Those parts of the magnetic circuit, which carry the transformer windings are called the *limbs* or *legs*, and those parts which connect the legs and serve for closing the magnetic circuit are termed *yokes*.

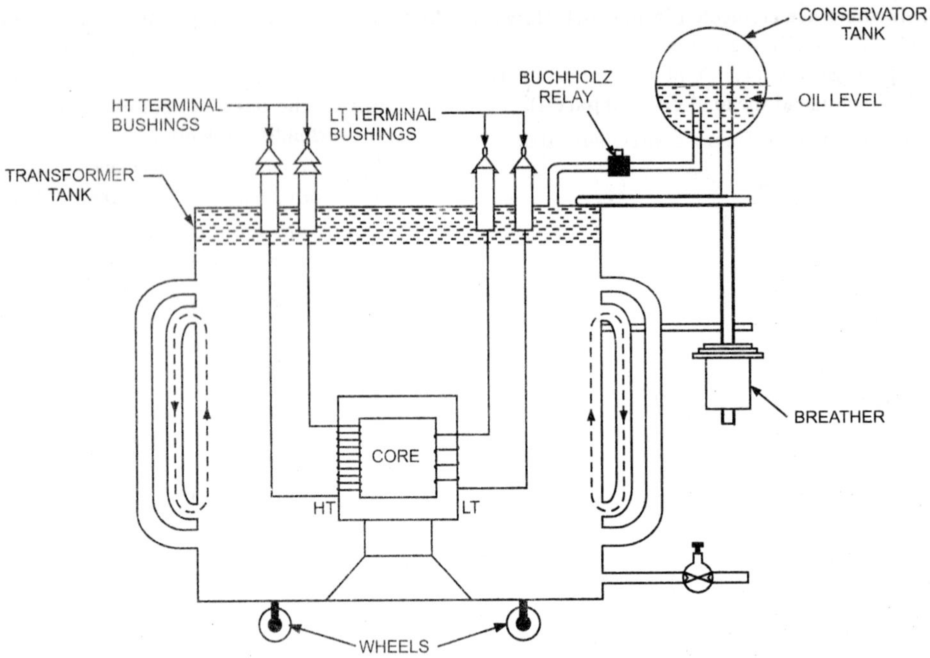

Fig. 12.48

The core material and its construction should be such that the maximum flux is created with minimum magnetizing current and minimum core loss. The use of steel in magnetic circuit introduces iron or core loss but ensures a high permeability of the magnetic circuit. Because of high permeability the magnitude of exciting current necessary to create the required flux in the core, is small. The presence of steel core causes 100 per cent of the magnetic flux created by the primary to be linked with the secondary. The magnetic frame (cores and yokes) of the transformer is built up of laminated electrotechnic steel, called the *transformer grade steel* consisting of 3.5% silicon. The higher content of silicon increases the resistivity of the core, thereby reducing the eddy current core loss. High content silicon steel has a high permeability at low flux densities and narrow hysteresis loop, thereby making magnetizing current and hysteresis core loss small. The silicon content affects the mechanical properties also, *e.g.* it increases tensile strength and impairs ductility. The steel gets brittle, if the silicon content is increased beyond 5%. The steel used for transformer cores may be hot rolled or cold rolled. The hot rolled steel which permitted a maximum flux density of 1.45 T was in use for a considerable length of time. In recent years this type of steel has been completely superseded by cold rolled steel allowing much higher flux densities up to 1.8 T to be used because it has better magnetic properties in the direction of rolling. Although, cold rolled steel is 25–35% more expensive than the hot rolled steel and needs special methods of core assembly, the increase in the value of maximum flux density makes it possible to reduce the amount of core material.

As the flux in the cores is pulsating one, it becomes necessary that the transformer cores are laminated and the laminations should be insulated and made as thin as practicable in order to reduce the eddy current loss to a minimum. However, there is a practical limit beyond which the thickness of laminations cannot be reduced further on account of mechanical considerations. The thickness of laminations or stampings varies from 0.35 mm to 0.5 mm. The thickness should not be made below 0.3 mm because in that case, the laminations become mechanically weak and tend to buckle. As a rule, the cores of high-capacity power transformers (for more than 100 kVA) are assembled of 0.5 mm steel sheets, since such a construction is less labour consuming than with 0.35 mm sheets. The laminations are insulated from each other by a very thin coating of varnish or by using 0.03 mm thick paper. Paper insulation is much cheaper than varnish, but its heat-resisting and heat-conducting properties and mechanical strength are worse. Besides, paper insulation occupies too large a percentage of the stack cross section. For this reason in high capacity power transformers, where these drawbacks are significant, varnish insulation is used.

In all types of transformers, the core is constructed of transformer sheet steel laminations assembled to provide a continuous magnetic path with a minimum of air gap included.

Various types of stampings and laminations employed in the construction of transformers are shown in Figs. 12.49, 12.50 and 12.51 respectively. In Fig. 12.49 (*a*) two L shaped laminations

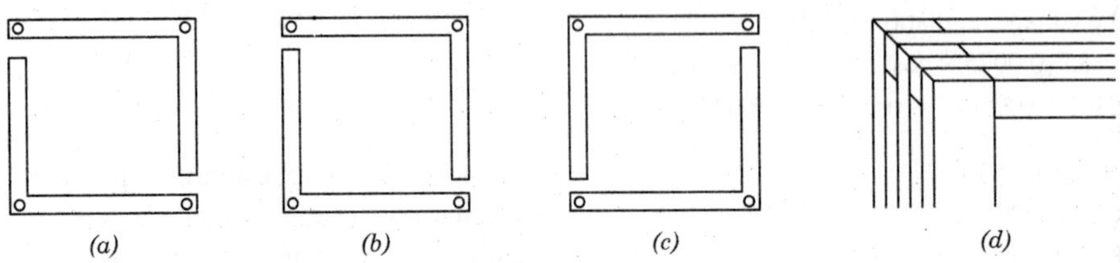

Fig. 12.49

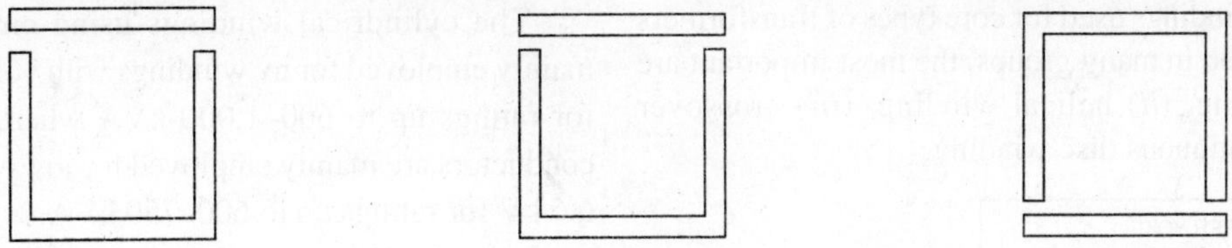

(a) (b) (c)

Fig. 12.50

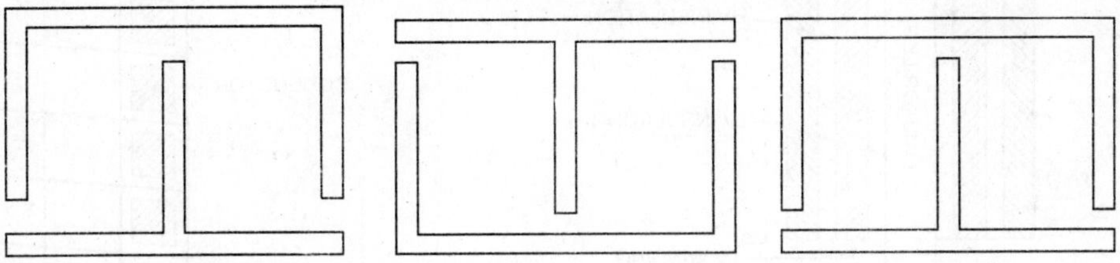

(a) (b) (c)

Fig. 12.51

are shown. They are placed together as illustrated in Figs. 12.49 (*b*) and 12.49 (*c*) to give rectangular laminations. The complete laminated core consists of a number of rectangular laminations placed alternately one over the other, as illustrated in Fig. 12.49 (*d*) so that the joints are staggered. The joints are staggered to avoid continuous gap causing increase in magnetizing current. Also if the joints are not staggered, the core will have less mechanical strength and during operation there would be undue humming noise. After arranging the laminations they are bolted together. In Fig. 12.50 U and I types of laminations are shown and in Fig. 12.51 U and T types of laminations are shown.

12.24.2. Windings. Transformer windings should comply to a number of requirements, the most important of which are: the winding should be economical both as regards initial cost, with a view to the market availability of copper, and the efficiency of the transformer in service; the heating conditions of the windings should meet standard requirements, since departure from these requirements towards allowing higher temperature will drastically shorten the service life of the transformer; the winding should be mechanically stable in respect to the forces appearing when sudden short circuits of the transformer occur; the winding should have the necessary electrical strength in respect to overvoltages.

These requirements are often mutually contradictory. Thus, for example, with a larger current density in the winding less copper is required, but the copper losses increase and, therefore, the efficiency of the transformer decreases. Larger temperature rises allowed in the winding result in reduction in overall dimensions of the transformer, but shorten its service life, etc. Therefore, the winding design of a modern transformer, especially in high-voltage transformer, is of utmost importance.

In a transformer, the high-voltage and low-voltage windings are so designed and are so chosen that the axial short-circuit force is reduced to minimum which allows a much simpler coil support.

Transformer windings are made of solid or stranded copper or aluminium strip conductors. Heavy current capacity needs conductors of large cross section. To reduce eddy current losses in the conductors, several small wires or parallel straps are preferred to one large strap. This gives rise to unequal reactance of the components of the conductor which can be eliminated by transposition of conductors.

Instead of placing primary on one limb and secondary on the other limb (Fig. 12.1) it is a usual practice to wind one-half of each winding on each limb. This ensures tight coupling between the two windings. Consequently leakage flux is considerably reduced. The positioning of the hv and lv windings with respect to the core is also very important from the point of view of insulation requirement. If the hv winding were placed next to the core, it would be necessary to insulate it from both the core and lv winding and two layers of hv insulation would be necessary. By placing the hv winding outside and around the lv winding only one layer of hv insulation that between the high- and low-voltage windings, is necessary.

In, practically, all transformers the coils are wound on a form, dipped in insulating varnish, baked into a rigid mass, and then are assembled with the core. It is easiest to build the core with legs of rectangular cross section, and so it is most economical to wind the coils on a rectangular form. On small transformers rectangular concentric coils are practical, but in a large unit the high repulsion forces generated between the primary and secondary coils under short-circuit conditions tend to "round out" the flat sides of the outer coil; if this happens, the resulting damage to the coil insulation will usually make the transformer unserviceable. *Cylindrical concentric coils* are the obvious answer to this problem of mechanical strength.

According to their arrangement high- and low-voltage windings can be either *concentric* (*i.e.*, the windings, which in each cross section are circles with a common centre) or *sandwich*, in which parts of the hv and lv windings alternate along the height of the leg. Concentric windings are employed in core type transformers as shown in Fig. 12.52 whereas sandwiched windings are almost exclusively used in shell type transformers as shown in Fig. 12.53. On account of the easier insulation facilities, the low-voltage winding is placed nearer to the core in the case of concentric windings and on the outside positions in the case of sandwiched windings. The insulation spaces between low and high-voltage coils also serve to facilitate cooling.

The concentric windings used for core types of transformers can further be classified in many groups, the most important are (*i*) cylindrical winding, (*ii*) helical winding, (*iii*) crossover winding and (iv) continuous disc winding.

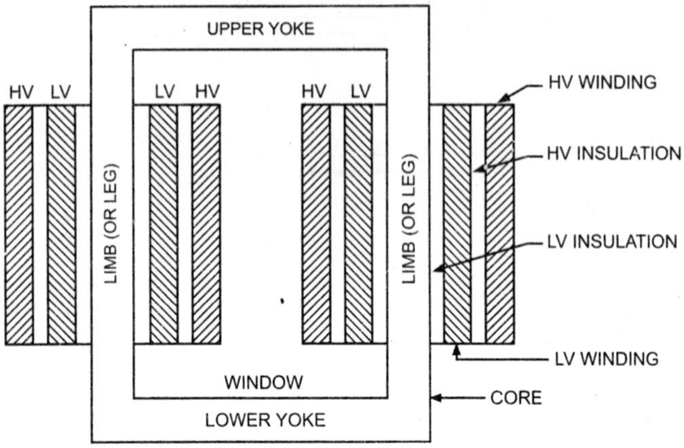

Fig. 12.52 *Concentric Windings*

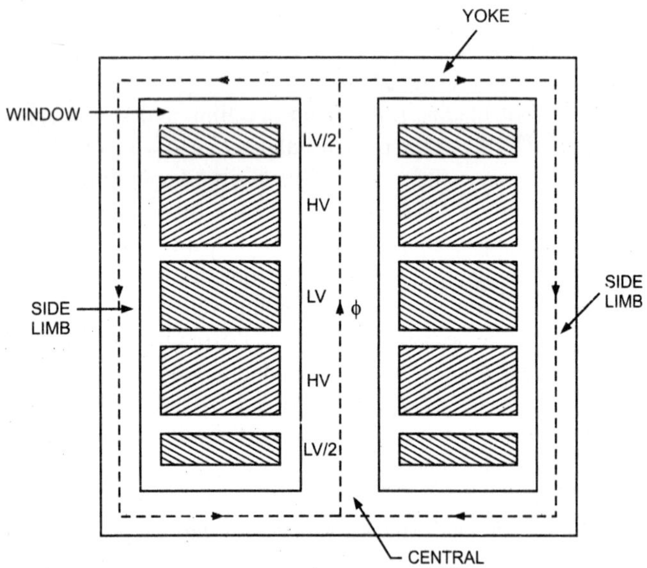

Fig. 12.53 *Sandwich Windings*

1. Cylindrical Windings. These windings are layered type and use either rectangular (or strip) or round conductors. If the cross section of a turn does not exceed 8 to 10 mm^2, the cylindrical winding is made multi-layer of round conductor; for a larger turn cross section the winding is made of rectangular (or strip) conductor, usually double layer. Cylindrical winding with strip conductors is shown in Fig. 12.54. The winding is made up of turns helically wound round the cylinder generatix with the turns close to each other. Hence, the height of the winding is equal to the height of the layer. The rectangular conductor can be wound either *flatwise* or *edgewise*. In the former case the larger size of the conductor is arranged in the axial direction, in the latter case—in the radial direction of the winding. In case the turn cross section exceeds 40-45 mm^2 then a turn is made of several single conductors, which are placed close to each other along the height of the layer, so that they all occupy the same position in relation to the leakage field. For improving the cooling of the winding layers ducts 5 to 8 mm wide are left between them (the greater figure relating to transformers with a higher capacity).

The cylindrical windings using circular conductors are mainly employed for hv windings with voltages 6.6, 11 or 33 kV for ratings up to 600–1,000 kVA whereas those using round conductors are mainly employed for low-voltage windings up to 6.6 kV for ratings up to 600–750 kVA.

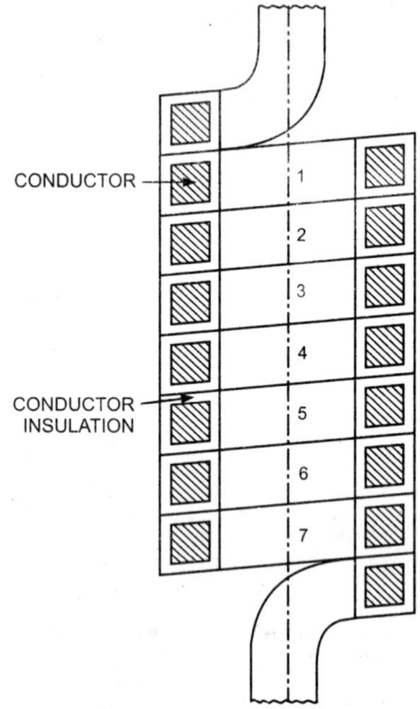

Fig. 12.54 *Cylindrical Winding With Strip Conductors*

2. Helical Windings. Such windings are employed for lv coils of medium and high capacity transformers where the number of coil turns is small but the current is high (as high as 2,000 A). Low-voltage windings of medium and high capacity transformers, therefore, require the use of a conductor consisting of a row of parallel rectangular conductors arranged in one radial direction of the winding flatwise and close to each other. For more uniform distribution of current between the parallel conductors, they are

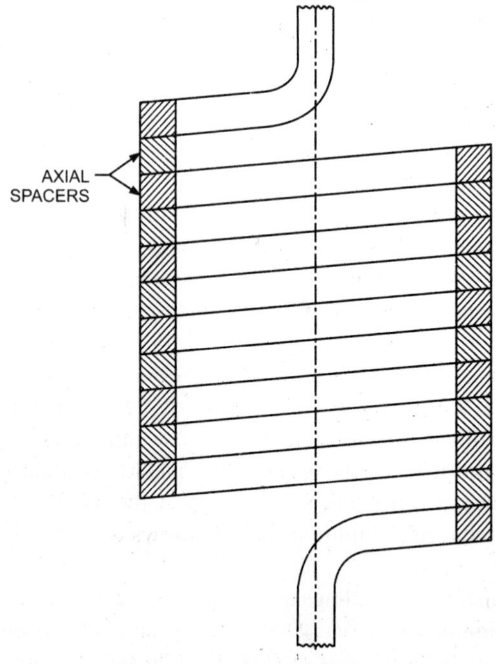

Fig. 12.55 *Helical Winding*

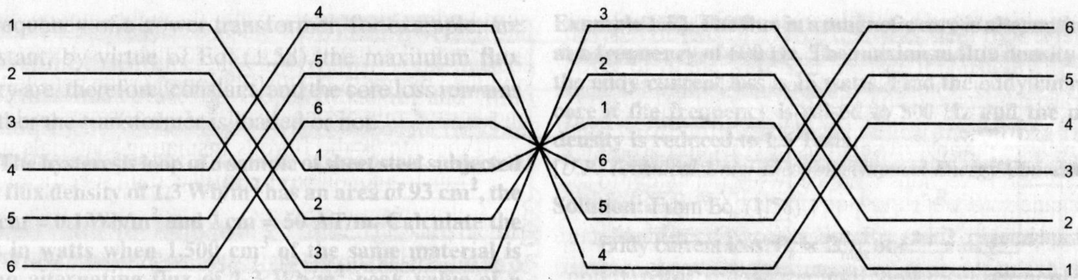

Fig. 12.56

transposed so that each conductor may take each position (Fig. 12.56). The number of parallel conductors in the turn of a helical winding usually varies from 6 to 20.

Basically the helical winding is the same as the cylindrical winding except that in such windings between adjacent turns axial spacers are provided for duct formation to improve oil circulation for better cooling. The ducts are formed by spacers placed all the way round the periphery of the cylinder at regular intervals (Fig. 12.55). The continuous helical winding exhibits high axial mechanical strength and, therefore, finds wide application in lv windings of large size power transformers.

Helical windings are employed in power transformers of rating ranging from 150 kVA to 30 MVA at voltages from 400 V to 11 kV and sometimes up to 33 kV.

3. Cross-over Windings. Such windings are suitable for currents not exceeding 20 A and are very largely employed for hv windings of low rating transformers where number of turns may be large but conductors are of small circular section with double cotton covering or paper covering. The whole winding is divided into a number of coils depending upon the voltage rating. Each coil is wound on a former, usually U pieces, with several layers of several turns per layer. The coil ends, one from inside and one from outside, are joined to other similar coils in series, spaced with blocks of insulating material to allow free circulation of oil. The axial length of each coil and the separation distance between each coil vary according to the voltage rating.

Cross-over windings are employed in the same range of ratings as the cylindrical windings. The cross-over winding has a higher strength than the cylindrical winding under normal operating conditions. However, this winding has a lower impulse strength than the cylindrical winding and also is more labour consuming.

4. Continuous Disc Windings. The disc coils, as their name suggests, consist of a number of flat coils or discs connected in series or parallel. The coils are formed with rectangular strips wound spirally from centre outwards in the radial direction, as illustrated in Fig. 12.57 (*b*). The conductor used is in such lengths as are sufficient for complete winding or section of winding between tappings. The conductor can be a single strip or number of strips in parallel, wound on the flat side. This gives a robust construction for each of the discs. The discs are wound on an insulating cylinder spaced from it by strips along the length of cylinder. The discs are separated from each other with pressboard sectors attached to vertical strips. The vertical and horizontal spacers provide radial and axial ducts for free circulation of oil which comes in contact with every turn. If a winding turn contains several parallel conductors, transposition of them is used as in the case of helical winding.

Continuous disc windings are reliable and strong and, therefore, they are widely employed both as lv and hv windings in large rating transformers. When the continuous winding is used as a hv winding, tappings are made for regulation of the voltage ratio in the range of ± 5 % or 2 (± 2.5 %).

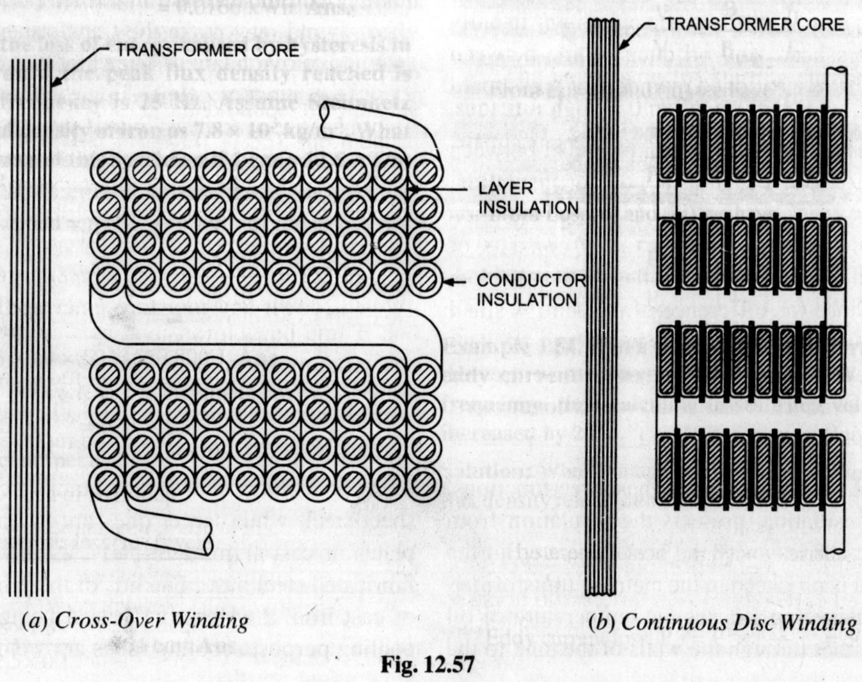

(*a*) *Cross-Over Winding* (*b*) *Continuous Disc Winding*

Fig. 12.57

5. Sandwich Windings. Such windings, as already mentioned, are most commonly employed in shell type transformers and allow easy control over leakage reactance. Leakage reactance can be reduced by subdividing the low- and high-voltage windings into a large number of sections or coils and arranging alternately the hv and lv sections with the lv section nearer to the yoke, as illustrated in Fig. 12.53. The two low voltage coils at the ends have half the turns of a normal low voltage coil. In order to balance the mmf of the adjacent sections, each normal section, whether hv or lv, carries the same number of ampere-turns. The larger the number of sections (or coils), the smaller is the leakage reactance. The sectionalizing of coils not only assists in controlling the leakage reactance but also assists in handling and improving cooling. In comparison to concentric windings they have several drawbacks: they are more labour consuming in manufacture, less stable in respect to short circuits and are more difficult to insulate from each other and from the yoke. This is the reason why the core type transformers with concentric windings are more common.

12.24.3. Insulation. No feature in the construction of a transformer is given more attention and care than the insulating materials because the life of the unit depends, to a very large extent, upon the quality, durability, and handling of these materials. All insulating materials, such as paper, pressboard, cloth, mica, asbestos, and impregnating compounds are selected on the basis of high quality and their ability to preserve this quality after many years of normal service.

Insulation used in a transformer may be classified into two major groups, viz, major insulation and minor insulation.

The insulation provided between windings and earthed parts (insulation provided between lv winding and core and between hv winding and yoke) and the insulation between the windings (insulation between primary and secondary windings) fall into the category of major insulation. For large core type units with concentric windings, major insulation is simply and effectively secured by employing special insulating cylinders and oil ducts extending the full length of the core. In the interleaved construction the major insulation between windings usually consists of sheets of pressboard and oil ducts, while between windings and core the above mentioned cylinders or pressboard sheets may be employed.

Minor insulation includes the insulation provided between the elements of a given winding such as conductor insulation, insulation between turns, layers and coils. The minor insulation between the coils of the same winding usually consists of pressboard sheets and oil ducts. For very small units treated cloth or fibre may be used. Since the difference of pressure is small between the adjacent turns, the insulation need not be very thick. It may consist of one or more wrapping of cotton or treated paper tape around each conductor. For small round conductors, enamel together with the cotton may be used.

12.24.4. Insulating Oil. The insulating oil has three functions. It provides additional insulation, protects the insulation from dirt and moisture and it carries away the heat generated in the cores and coils. The heat is produced in the metal of transformer, passes through the insulation and raises the temperature of oil and is then conducted either through the walls of the tank to the surrounding air or to the water through water cooling tubes. The oil which is extensively used is called the *transformer oil.* It is obtained by fractional distillation of crude petroleum. Vegetable and animal oils are not used in transformers as they form fatty acids that attack the fibrous insulating materials used.

Transformer oil has to fulfill certain specifications, including the following:

1. *High dielectric strength.* As per ISS, the breakdown strength of new transformer oil when treated must be at least 50 kV_{rms} when measured with the help of two spherical electrodes of 12.5 mm diameter and with a gap spacing of 2.5 mm. However, the breakdown strength is greatly reduced due to the presence of impurities like moisture, gas bubbles, solid particles etc.

2. *Low viscosity* to provide good heat transfer. A high viscosity is an obvious disadvantage because of the sluggish flow through the many small orifices in the windings.

3. *Purity.* The oil must not contain impurities such as acid, alkali and sulphur or its compounds to prevent corrosion of metal parts and insulation. Sulphur compounds, if present, accelerate the formation of sludge.

4. *High flash point.* The temperature at which oil vapour ignites spontaneously is called the *flash point.* The flash point of transformer oil should not be less than 135 ° C.

5. *Free from sludging under normal operating conditions.* Sludging means slow formation of semi-solid hydrocarbon owing to heating and oxidation. The sludge deposits itself on windings, tank walls and in cooling ducts. Sludges, being bad conductor of heat, greatly reduce the heat transfer from the windings to the oil and so increases the temperature of windings. Moreover, sludges block the cooling tubes and cause further increase in the temperature of the transformer and may make the transformer unserviceable because of overheating. The chief cause of sludge formation is contact of oxygen in air with oil under hot conditions. The contact of air with oil is minimized by using conservator and breather. Sometimes for preventing sludging certain chemicals called the *inhibitors* are added to the transformer oil.

6. *Good resistance to emulsion* so that the oil may throw down any moisture entering the apparatus instead of holding it in suspense.

If the oil is found to contain moisture or suspended contaminants it should be filtered and if thistreatment is considered to be inadequate, the oil should be replaced by fresh one.

12.24.5. Tank. Small capacity tanks are fabricated from welded sheet steel, while larger ones are assembled from plain boiler plates or cast-aluminium parts, usually mounted on a shallow fabricated steel base. The lids of these transformer tanks can be of cast iron, a waterproof gasket being used at the joints. For cooling purpose, cooling tubes are welded with the tank, but in case of radiators, separate radiators are individually welded and then bolted-on to the transformer tank afterwards. A tank must be capable to withstand the stresses developed by jacking and lifting and shall be not larger than necessary to accommodate the core, windings and internal connections with appropriate electrical clearance.

12.24.6. Conservator or Expansion Tank. The satisfactory operation of a transformer depends so largely on the condition of the oil and, therefore, devices and methods for keeping the oil clean and dry are of utmost importance.

The oil level of a transformer changes with changes in temperature of oil which in turn depends upon the load on the transformer. The oil expands with the increase in load and contracts when the load decreases. Large power transformers are also liable to overloads which may overheat the oil and consequently there is a sludge formation if air is present. Occasionally large power transformers also suffer short circuits and temperature rise becomes excessively high. This causes vapourisation of a part of the oil. The oil vapours form explosive mixture with air that ignites and may cause a considerable damage. For these reasons it is necessary to prevent the oil from having contact with air as well as moisture. For this purpose *conservators* are employed.

Small transformers are sometimes not provided with conservator or expansion tank. So the oil level has to be kept some distance below the top cover to provide space for oil expansion under temperature rise. When the oil expands, the air is expelled out and air is drawn inside under contraction of oil. This process is termed *breathing*. Thus the oil is in contact with air. Exposure to air accelerates the deterioration of oil due to increased sludge and acid formation through moisture absorption and oxidation.

Conservator is a small auxiliary oil tank (an airtight cylindrical metal drum) that may be mounted above the transformer and connected to the main tank by a pipe. Its function is to keep the main tank of the transformer completely filled with oil in all circumstances despite expansion or contraction of oil with the changes in temperature, conservator is always partly filled with oil and absorbs the expansion and contraction of oil and keeps the main tank full of oil. It also reduces the rate of oxidation of oil, partly because less oil surface is exposed to air and partly because of the reduced temperature of the oil exposed to air. Thus the sludge formation is considerably reduced and whatever sludge is formed settles to the bottom of the expansion chamber into a sludge pan from where it is periodically withdrawn by means of a drain tap. Normally the capacity of conservator should be approximately 10–12 % of the oil volume of the main tank.

The expansion tank is usually installed on the low-voltage side of the transformer tank above the level of the transformer cover on a supporting frame. The position of the installation is governed by voltage clearance to leads and bushings. A small pipe connection between the gas space in the expansion tank and the cover of the transformer tank permits the gas above the oil in the transformer to pass into the expansion tank, so that the transformer main tank will be completely filled with oil.

12.24.7. Elbow Relief Device. An elbow relief device is also provided on transformers equipped with expansion tanks. This device consists of a large diameter steel pipe on the transformer cover or in the wall near the top. It is usually at an angle to the vertical, but may be vertical. The pipe, which has an elbow at its end, is of sufficient length so that the oil can rise in it to the maximum level in the expansion tank without splitting over the elbow portion.

A thin glass relief diaphragm is placed at the top of the device, above the maximum oil level. A fault in the transformer (say a breakdown or a short circuit in the transformer winding) may produce a considerable pressure inside the tank, and under these circumstances the diaphragm should burst, thus protecting the tank. A protective hood extends past the diaphragm and prevents moisture from entering the expansion tank in case the diaphragm is accidently ruptured.

12.24.8. Temperature Gauge. Every transformer is provided with a temperature gauge to indicate hot oil or hottest spot temperature. It is a self-contained weatherproof unit made with alarm contacts. It is of dial type operated by Bourdon gauge connected to a thermometer bulb located in the region of hottest oil.

12.24.9. Oil Gauge. Every transformer is provided with an oil gauge to indicate the oil level so that it can be read by a person standing on the floor. The oil gauge may be provided with the alarm contacts which give an alarm to the switchboard when the oil level has dropped beyond permissible level due to oil leak, damage of cooler and due to any other reason.

12.24.10. Breather. When the transformer becomes warm, the oil and gas expand. The gas at the top of the oil is expelled out. When the transformer cools, air is drawn into the transformer. Unless preventive measures are taken, moisture is drawn during this process, called the *breathing*. This moisture is readily absorbed by the oil, and the dielectric properties of the oil are correspondingly reduced. The air entering the transformer is made moisture free by letting it pass through an apparatus called the *breather*. A breather consists of a small container connected to the vent pipe and contains a dehydrating material like silica gel crystals impregnated with cobalt chloride. The material is blue when dry and a whitish pink when damp. The colour can be observed through a glass window provided in front of container.

12.24.11. Gas Operated Relay-Buchholz Relay. It is a gas and oil actuated protective device and it is practically universally used on all oil-immersed transformers having rating more than 750 kVA. It is installed in the pipe joining the main tank of the transformer to the conservator and is used to give alarm in case of minor fault and to disconnect the transformer from supply mains in case of severe internal fault. The use of such a relay is possible only with transformers having conservators.

The construction of Buchholz relay is illustrated in Fig. 12.58. It consists of two elements mounted in a small chamber located in the pipe connection between the conservator and the transformer tank. When minor fault occurs, heat is produced due to current leakage, some of the oil in the transformer tank evaporates and some vapour collects in the top of the chamber while passing to the conservator. When a predetermined amount of vapour accumulates in the top of the chamber, the oil level falls, the mercury type switch attached to the float is tilted and closes the alarm circuit to ring the bell. A release cock is provided

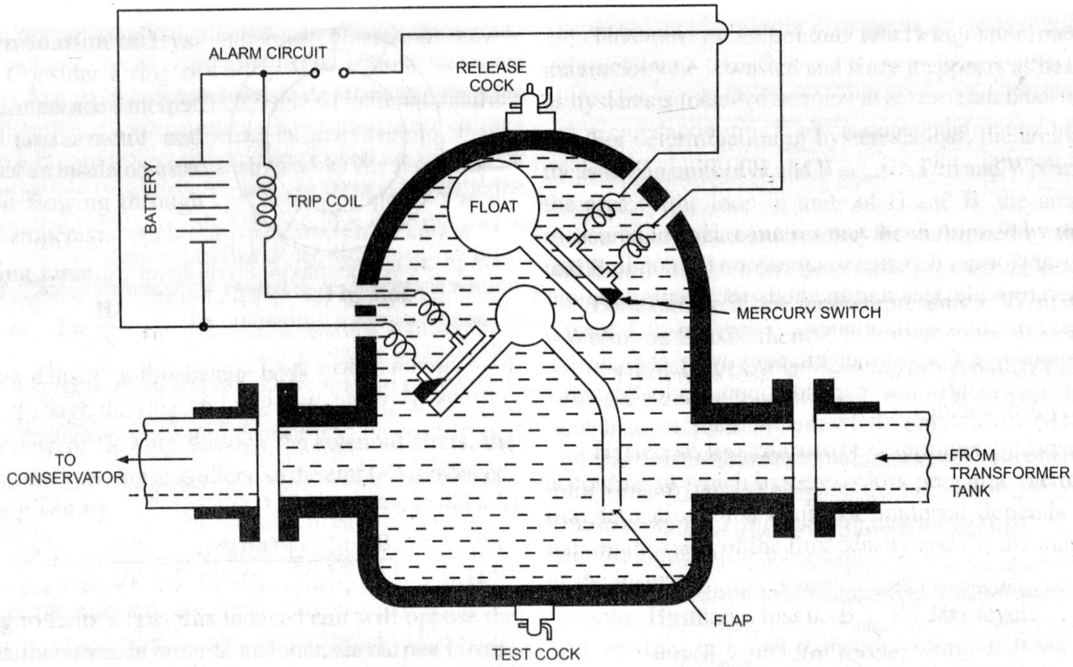

Fig. 12.58 *Gas Operated Buchholz Relay*

at the top of the chamber so that after operation the pressure in the chamber can be released and gas emitted to allow the chamber to refill the oil. When severe fault occurs large volume of gas is evolved so that the lower element containing a mercury switch mounted on a hinged type flap is tilted and the trip coil is energized. A test cock is provided at the bottom of the chamber to allow air to be pumped into the chamber for test purposes.

12.24.12. Leads and Terminals. The connections to the windings are by copper rods or bars, insulated wholly or in part, and taken to the bus-bars directly in the case of air cooled transformers, or to the insulator bushings mounted on the top of the transformer tank in case of oil cooled transformers. The shape and size of leads are very important in hv transformers, not due to current carrying capacity, but due to dielectric stresses and corona, etc., at sharp bends and corners.

12.24.13. Bushings. Transformers are connected to hv lines, and, therefore, care is to be taken to prevent flash-over from the high voltage connections to the earthed tank. Connections from cables are made in cable boxes, but overhead connections are to be brought through *bushings* specially designed for different classes of voltages.

The bushing consists of a current carrying element in the form of a conducting rod, and a porcelain cylinder installed in the hole of the cover of the transformer and employed to isolate the current carrying element. Up to 33 kV, ordinary porcelain insulators can be used. Above this voltage oil-filled or condenser type bushings are used.

The *oil-filled bushing* consists of a hollow porcelain cylinder with a conductor, (usually a hollow tube) through its axis. The space between the conductor and the inner surface of the porcelain is filled with oil. Oil is filled into the bushing from the top, where there is a glass cylinder to indicate the oil level and to act as an expansion chamber to accommodate the variation of oil bulk due to variations in operating temperature. An hv oil-filled bushing is shown in Fig. 12.59 (*a*).

The capacitor type bushing is constructed of thick layers of backelized paper or synthetic resin bonded paper interleaved with thin graded layers of tin foils. This results into a series of capacitors formed by the conductors and first tin foil layer, the first and second tin foil layers within synthetic resin bonded paper cylinders in between and so on. The lengths and thicknesses are so arranged as to give an approximate uniform electric stress distribution throughout the radial depth of the bushing.

For outdoor application, a bushing is covered by porcelain rainshed, corrugated circumferentially to provide a longer and partially protected leakage path.

12.24.14. Tappings. The transformers are usually provided with few tappings so that output voltage can be varied over a small range for constant input voltage, Most industrial transformers are provided with four tappings on the hv winding, two on each side of the 'normal' voltage, so that the tap changer has five positions. In special cases there may be tappings on the lv winding, usually in addition to those provided on hv winding. Though the tappings can be changed while the transformer is supplying load by making some suitable arrangement but in most of the transformers the tappings are to be changed after disconnecting them from the supply mains.

12.25 TYPES OF TRANSFORMERS

Depending upon the type of construction used, the transformers are classified into two categories viz (*i*) *core type* and (*ii*) *shell type.* Another recent development is *spiral core* or *wound core type*, the trade name being *spiracore* transformer.

Depending upon the type of service, in the field of power system, they are classified as (*i*) *power transformers* (*ii*) *distribution transformers*.

12.25.1. Core Type and Shell Type Transformers. In core type **construction**, as shown in Fig. 12.60, the coils are wound around

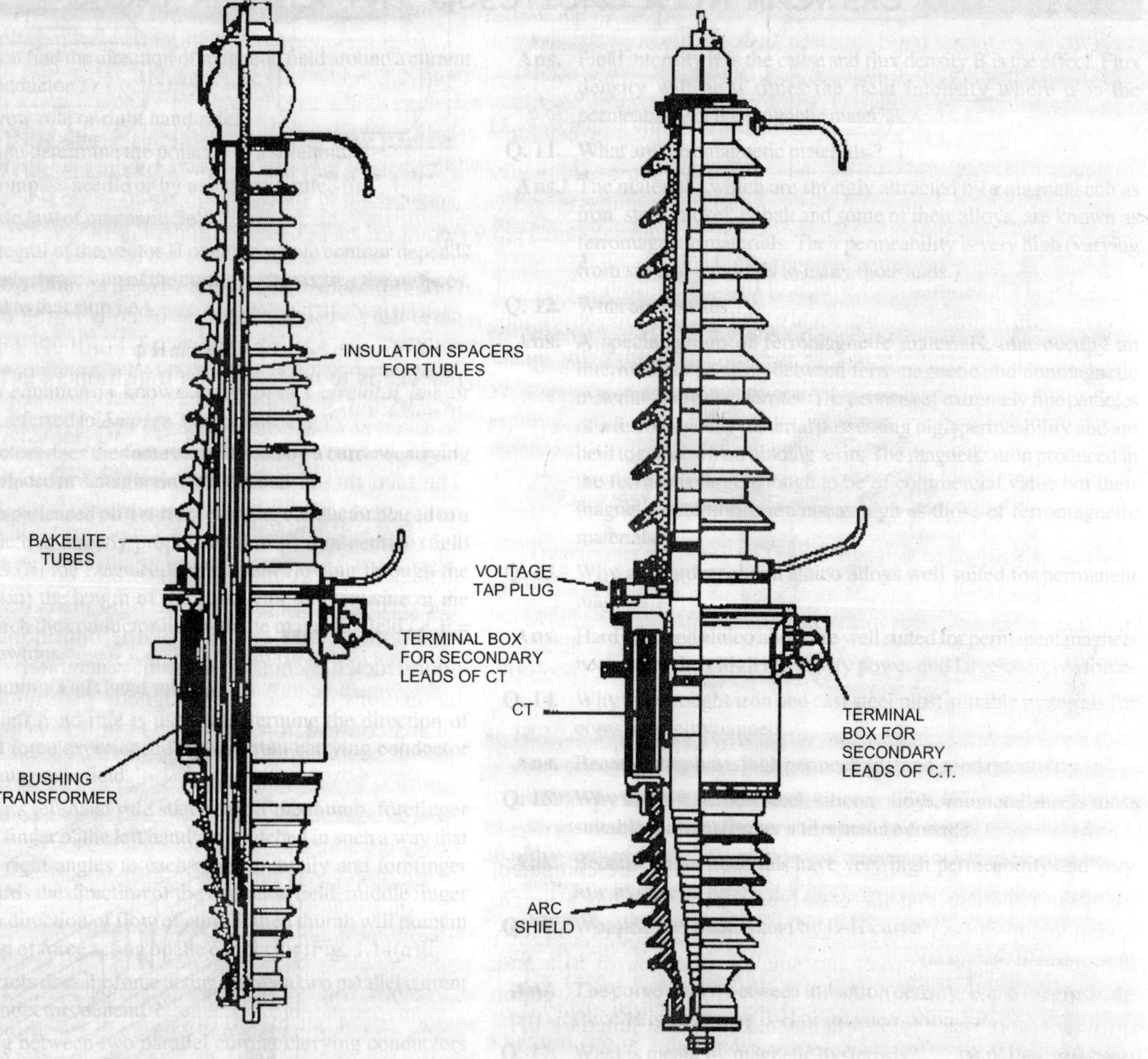

(a) Hv Oil-filled Bushing *(b) Hv Capacitor Type Bushing*

Fig. 12.59

the two limbs of a rectangular magnetic core. Each limb carries one half of the primary winding and one half of the secondary winding so as to reduce the leakage reactance to the minimum possible. The lv winding is wound on the inside nearer to the core while the hv winding is wound over the lv winding away from the core in order to reduce the amount of insulation materials required.

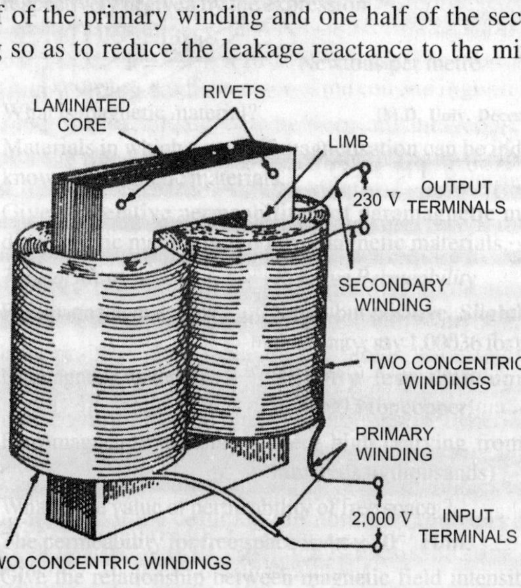

Fig. 12.60 *Core Type Transformer*

Small transformers may have cores of rectangular or square x-section with rectangular or circular coils but it is wasteful in case of large capacity transformers. In case of large sized transformers stepped cruciform core with circular cylindrical coils is employed, as illustrated in Fig. 12.61. Such a core employs laminations of different sizes. Though the cost of manufacturing of such a cruciform core is much greater, but the circular coils that are used are easier to wind and provide more mechanical strength, especially when short circuit occurs. Other advantages of using cruciform core are high space factor and reduced mean length of turns resulting in reduced copper loss.

In shell type construction, as shown in Fig. 12.62, the coils are wound on the central limb of a three limb core. The entire flux passes through the central limb and divides into two parts going to side limbs, as shown in Fig. 12.53. Consequently, the

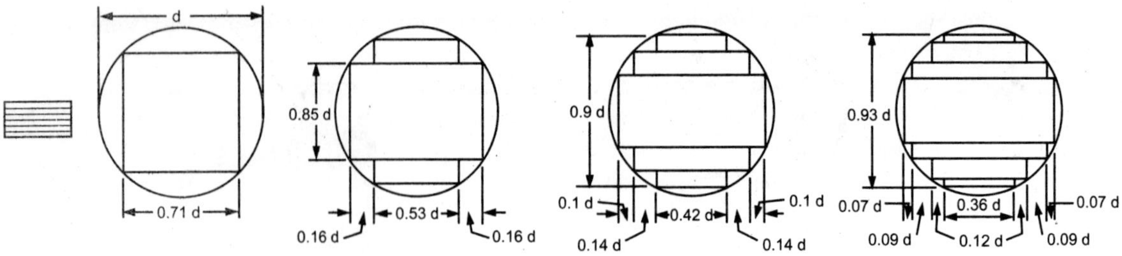

Fig. 12.61

x-sectional area (and hence width) of the central limb is twice of that of each of the side limbs. Sandwich type winding, as already described in Art. 12.24.2, is used in such a construction.

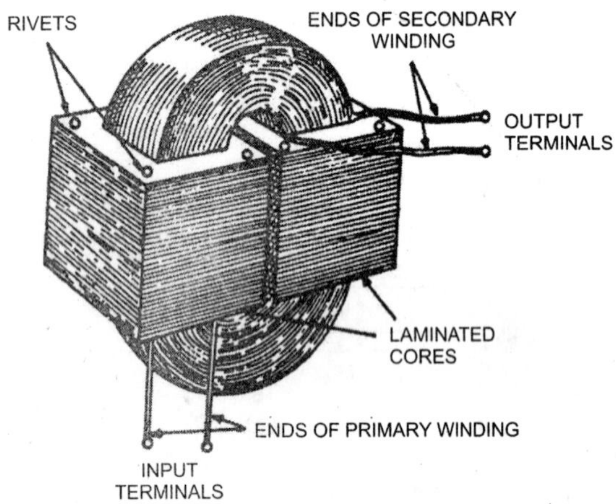

Fig. 12.62 *Shell Type Transformer*

If the core type and shell type transformers are compared, then we conclude that the core type transformer has a longer mean length of iron core and a shorter mean length of coil turn. The core type transformer has smaller x-section of iron and, therefore, a greater number of turns. Core type construction provides more space for insulation making it best suited for EHV requirements.

In shell type transformers advantage is gained through the core being used to protect the windings from mechanical damage. The shell type construction gives better support against electromagnetic forces between current carrying conductors. These forces are of considerable magnitude under short-circuit conditions. The shell type construction is commonly used for small transformers where a square or rectangular core cross section is suitable for economic consideration. The shell type construction needs more specialized fabrication facilities than core type,while the latter offers an additional advantage of permitting visual inspection of coils in the case of a fault and ease of repair at substation site. For these reasons, the present practice is to use core type transformers in large high voltage installations.

12.25.2. Spiracore Transformer

The fact, that with flux in the direction of grain orientation of iron, permeability is increased while core losses are reduced, has led to the development of the *wound core,* wherein spirals of strip steel are wound through the window of a preformed coil. The resultant structure is more rigid than cores made of stacked laminations, and because of the continuous magnetic path unobstructed by butt or lap joints encountered where individual laminations are used, the magnetic path reluctance is reduced. Such a construction is illustrated in Fig. 12.63.

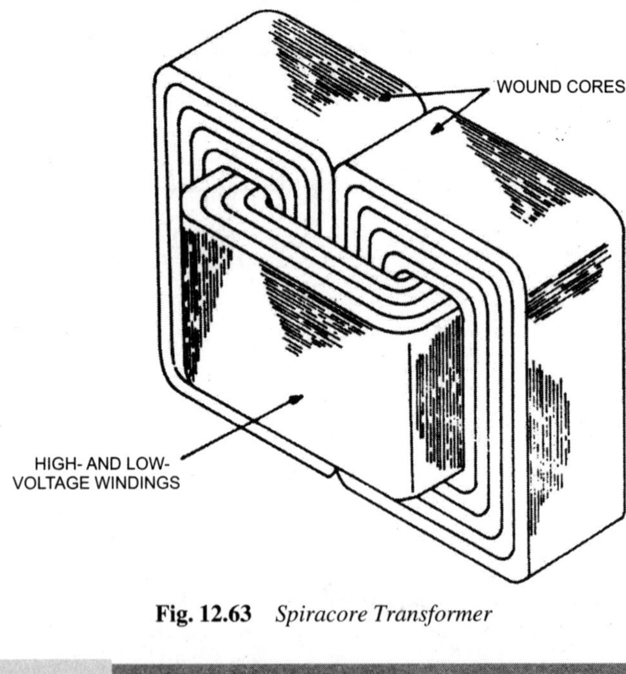

Fig. 12.63 *Spiracore Transformer*

HIGHLIGHTS

1. *Transformer* is an ac machine that (*i*) transfers electrical energy from one electric circuit to another (*ii*) does so without a change of frequency (*iii*) does so by the principle of electromagnetic induction and (*iv*) has electric circuits that are linked by a common magnetic circuit. The energy transfer usually takes place with a change of voltage, although this is not always necessary. It is a static machine and has very high efficiency (in the range of 96-99%).

   A transformer is never operated on dc. If dc supply is connected to the transformer primary or secondary, it may burn out since there is no self-induced emf and heavy current will flow through it.

2. In an *ideal transformer,* winding resistance, magnetic leakage, and iron losses are neglected and core is assumed to have infinite permeability and zero reluctance.

3. The emfs induced in the transformer windings are given as

$$E_1 = 4.44 f N_1 \Phi_{max} \text{ volts}$$
$$\text{and } E_2 = 4.44 f N_2 \Phi_{max} \text{ volts}$$

4. Voltage transformation ratio,

$$K = \frac{E_2}{E_1} = \frac{N_2}{N_1} = \frac{V_2}{V_1} = \frac{I_1}{I_2}$$

5. The impedance on the secondary side reflected on the primary side is transformed in the direct ratio of the square of the turns *i.e.*

$$Z_2' = \left(\frac{N_1}{N_2}\right)^2 Z_2 = a^2 Z_2.$$

Similarly an impedance, Z_1 from the primary side can be referred to secondary side as

$$Z_1' = \left(\frac{N_2}{N_1}\right)^2 Z_1 = \frac{1}{a^2} Z_1$$

6. If the transformer is operated with the frequency and voltage changed in the same proportion, the flux density will remain unchanged and so the no-load current.

7. The current drawn by transformer on no load is called the no-load current or exciting current. It has two components (*i*) in-phase, active or *energy component* I_e ($I_0 \cos \phi_0$) used to meet the iron loss and small amount of copper loss occurring in primary winding and (*ii*) quadrature or wattless component, called the *magnetizing component*, I_m ($I_0 \sin \phi_0$) used to create alternating flux in the core.

No-load phase angle, $\phi_0 = \text{Tan}^{-1} \frac{I_m}{I_e}$.

8. When a transformer is loaded, the secondary ampere-turns set up a magnetic field that opposes the main field. To neutralize this effect, transformer draws extra primary current from the mains. Flux remains almost constant from no load to full load.

The primary current, $\mathbf{I_1} = \mathbf{I'_1} + \mathbf{I_0} = \mathbf{K\, I_2} + \mathbf{I_0}$

9. When a practical transformer is loaded

Induced emf in secondary winding,

$$\mathbf{E_2} = \mathbf{V_2} + \mathbf{I_2}\,(\mathbf{R_2} + j\,\mathbf{X_2}) = \mathbf{V_2} + \mathbf{I_2\,Z_2}$$

Induced emf in primary,

$$E_1 = E_2/K$$

and applied voltage to primary winding,

$$\mathbf{V_1} = -\mathbf{E_1} + \mathbf{I_1}\,(\mathbf{R_1} + j\,\mathbf{X_1}) = -\mathbf{E_1} + \mathbf{I_1\,Z_1}$$

10. The total resistance and reactance of the transformer referred to primary are given as

$$R_{01} = R_1 + \frac{R_2}{K^2},\; X_{01} = X_1 + \frac{X_2}{K^2}$$

$$\text{and } Z_{01} = \sqrt{R_{01}^2 + X_{01}^2}$$

11. The total resistance and reactance of the transformer referred to secondary are given as

$$R_{02} = (K^2 R_1 + R_2),\; X_{02} = (K^2 X_1 + X_2)$$

$$\text{and } Z_{02} = \sqrt{R_0^2 + X_{02}^2}$$

12. Approximate voltage drop in the transformer referred to secondary $= I_2 (R_{02} \cos \phi \pm X_{02} \sin \phi)$ + ve sign for lagging power factor and – ve sign for leading power factor.

13. The condition for zero regulation is

Load phase angle, $\phi = \text{Tan}^{-1}\left(\frac{-R_{02}}{X_{02}}\right)$ – ve sign indicates that zero regulation occurs at a leading power factor.

14. The condition for maximum regulation is

Load phase angle, $\phi = \text{Tan}^{-1} \frac{X_{02}}{R_{02}}$

Maximum regulation always occur at lagging power factor.

15. Percentage voltage regulation

$$= \frac{E_2 - V_2}{E_2} \times 100$$

$$= \frac{I_2\,(R_{02} \cos \phi \pm X_{02} \sin \phi)}{\text{No-load secondary voltage, } E_2} \times 100$$

16. Transformer losses are of two types namely iron loss and copper loss. Iron losses are almost constant from no load to full load while copper losses vary as the square of load in kVA.

17. Transformer efficiency is given as

$$\eta = \frac{V_2\, I_2 \cos \phi}{V_2\, I_2 \cos \phi + P_i + x^2\, P_c}$$

Transformer efficiency will be maximum when copper losses equal iron or core losses.

Output kVA corresponding to maximum efficiency

$$= \text{Rated kVA}\sqrt{\frac{\text{Iron losses}}{\text{Full-load copper losses}}}$$

18. *Open-circuit lest* is performed to determine the core or iron losses, P_i, and no-load current I_0 and thereby the shunt branch parameters R_0 and X_0 of the equivalent circuit.

19. *Short-circuit test* is performed to determine the copper losses at rated load and equivalent resistance and reactance referred to metering side.

20. Sumpner's test, also called the *back to back test*, is the most important test for determination of rise of maximum temperature of a transformer. For this test two identical transformers are required which are loaded fully in this test but the power required from the supply is that necessary for supplying the iron and copper losses of both of the transformers.

21. The performance of distribution transformers is judged by its all-day efficiency, also called the *energy efficiency* or *operational efficiency* which is computed on the basis of energy consumed during the whole day (24 hours) and is given as

$$\text{Transformer all-day efficiency, } \eta = \frac{\text{kWh output in 24 hours}}{\text{kWh input in 24 hours}}$$

22. For satisfactory operation of single-phase transformers in parallel, the conditions to be fulfilled are :

(*i*) Polarities of the transformers are the same.

(*ii*) The voltage ratings of both primaries and secondaries are identical. This obviously implies that the transformation- or turn-ratios are the same.

(*iii*) Per cent (or per unit) impedances of the transformers are the same in magnitude and have the same phase angle (*i.e.* $\frac{X}{R}$ ratio should be the same).

EXERCISES

1. What is a transformer? Differentiate between step-up and step-down transformers? **[J.N. Technological Univ. Electrical Machines-II, January-2014]**

2. Discuss in brief about the construction and basic principle of transformer. **[R.G. Technical Univ., December-2010]**

3. Describe the constructional features and principle of operation of single phase transformer. **[Rajasthan Univ. Electrical Machines-I, 2006]**
4. What and why shall be the consequences when the hv winding of a 50 Hz, 1 kVA, 230/115 V transformer is connected to a 10 V dc supply ? **[M.D. Univ. Electrical Machines-I, December 2009]**
5. Explain the assumptions taken for ideal transformer. **[M.D. Univ. Electrical Machines-I, December-2012]**
6. Derive an expression for the emf induced in a transformer winding. Show that the emf induced per turn in primary is equal to the emf per turn in secondary. **[M.D. Univ. Electromechanical Energy Conversion, May-2007]**
7. Derive an expression for the emf of an ideal transformer. **[Anna Univ. Electrical Machines-I, May/June-2012]**
8. Derive the electromotive force (emf) equation of a single phase transformer. **[B.P. Univ. of Technology Electrical Machines, 2008]**
9. Draw general schematic of a single phase transformer. Describe its working principle and deduce the expression for emf in secondary winding. **[J.N. Technological Univ., Electrical Machines-II, January-2014]**
10. What are the assumptions made for an ideal transformer. Draw the equivalent circuit and phasor diagram of an ideal transformer. **[R.G. Technical Univ., December-2012]**
11. Draw and explain the phasor diagram of a single-phase transformer at lagging pf. **[G.G.S.I.P. Electromechanical Energy Conversion-I, 2011]**
12. Draw and explain the phasor diagram of a transformer on load. How it affects the power factor of the loaded transformer. **[M.D. Univ. Electrical Machines-I, December-2013]**
13. Draw and explain the no-load phasor diagram of a single phase transformer. Discuss how primary leakage flux is accounted for in the phasor diagram. **[M.D. Univ. Electrical Machines-I, December-2009]**
14. Why does the primary current in a transformer increase when a load is connected across the secondary ? Explain the mechanism through which primary draws more current on loading the secondary. Is the ratio of primary current to secondary current exactly equal to the ratio of secondary turns to primary turns? Explain. **[R.G.P.V. Bhopal Basic Electrical Engineering, December-2004]**
15. Explain the meaning of referred quantities. Obtain expressions for resistance, reactance and impedance of transformer equivalent circuit when seen from primary side. **[U.P. Technical Univ. Electromechanical Energy Conversion-I, 2005-06]**
16. Draw the equivalent circuit of a transformer and explain the significance of each term in it. **[G.G.S.I.P. Electromechanical Energy Conversion-I, December-2011]**
17. In what way a practical transformer differs from an ideal transformer? Develop an equivalent circuit for the practical transformer. **[Rajasthan Technical Univ., Electrical Machines-I, January-2013]**
18. Develop step by step the exact equivalent circuit of a single-phase two-winding transformer. **[B.P. Univ. of Technology Electrical Machines-I, 2008]**
19. Draw the approximate equivalent circuit of a transformer referred to the primary side and indicate how it differs from the exact equivalent circuit. **[J.N. Technological Univ., Electrical Machines-II, January-2014]**
20. What is regulation? How it can be obtained from equivalent circuit parameters? **[J.N. Technological Univ. Electrical Machines-II, January-2014]**
21. What is voltage regulation? What is its importance? For which type of load, the voltage regulation is negative? Derive an expression for voltage regulation under lagging pf load. **[R.G. Technical Univ., Electromechanical Energy Conversion-I, June-2010]**
22. The "voltage regulation" of a transformer will be zero for a leading pf load. Justify this statement. **[B.P. Univ. of Technology Electrical Machines, 2008]**
23. Define voltage regulation of transformer.

Derive conditions for (*i*) zero regulation (*ii*) maximum regulation. **[Rajasthan Technical Univ. Electrical Machines-I, 2008]**
24. Derive an expression for computing the per unit voltage regulation of a transformer for lagging and leading power factors. **[R.G.P.V. Bhopal, June-2005]**
25. Derive the condition for zero voltage regulation. Also show that the magnitude of maximum voltage regulation equals the pu (per unit) value of equivalent leakage impedance. **[U.P. Technical Univ. Electromechanical Energy Conversion-I, 2004-05]**
26. Draw the Kapp's regulation diagram and explain how the regulation of the transformer is determined from this diagram.
27. Draw equivalent circuit and the phasor diagram of a transformer on load and therefrom do the power flow analysis. **[U.P. Technical Univ. Electromechanical Energy Conversion-I, 2001-02]**
28. State the various losses which take place in the transformer. On what factors do they depend? Explain the steps taken to minimise these losses. **[J.N. Technology Univ., Electrical Machines-II, January-2014]**
29. What are the various losses in a transformer? Where do they occur and how do they vary with load? **[M.D. Univ. Electromechanical Energy Conversion, May-2007]**
30. Explain, why the hysteresis and eddy-current losses occur in a transformer and how they can be reduced. Also write the factors on which these losses depend. **[U.P. Technical Univ. Electrical Engineering Second Semester, 2002-03]**
31. What are the different losses that occur in a transformer on load? How can these losses be determined experimentally? **[J.N. Technological Univ., Electrical Machines-II, January-2014]**
32. What is meant by similar polarity ends of the two windings of a single phase transformer? How are these ends identified? **[U.P.S.C. I.E.S. Electrical Engineering-II, 1997]**
33. What is 'polarity test'? Why and how this should be performed in a single-phase two-winding transformer? Explain clearly with the help of connection diagram. **[B.P. Univ. of Technology, Electrical Machines, 2008]**
34. Explain polarity test of a transformer and mention its importance. **[U.P. Technical Univ. Electrical Machines, 2009-10]**
35. Discuss how to conduct open-circuit and short-circuit tests on a single-phase transformer in the laboratory. From the test results how the efficiency and regulation of the transformer is determined? **[J.N. Technological Univ., Electrical Machines-II, January-2014]**
36. Explain indirect test method for determination of voltage regulation and efficiency of a single-phase transformer with necessary diagram. **[Gujarat Technological Univ., Electrical Machines-I, Summer-2014]**
37. Give the laboratory circuit diagram to perform open- and short-circuit tests of single phase transformer with brief explanation and then draw the equivalent circuit diagram using data obtained from the OC and SC tests. How will you determine the losses from these tests? **[R.G. Technical Univ., Electrical Machine-I, June-2013]**
38. Describe Sumpner's (back to back) test used for transformers. Indicate its advantages. **[U.P.S.C. I.E.S. Electrical Engineering-II, 1998]**
39. In a back-to-back test, show that one transformer may have less temperature rise than the other. In Sumpner's test, the frequency of voltage injected in the secondary circuit may not be equal to the rated frequency. Explain.
40. Describe the back-to-back test for determining the regulation and efficiency of a pair of similar transformers. What are the limitations of this test? **[U.P. Technical Univ. Electrical Machines, 2005-06]**
41. Explain the process of finding the efficiency of a transformer by Sumpner's test. **[Electrical Machines-I; Rajasthan Univ. 2003, 2005, 2006; Rajasthan Technical Univ., 2010]**

42. What are the losses in transformer? Write an expression for calculating the efficiency of a transformer. Develop the condition for maximum efficiency.

[U.P. Technical Univ. Electrical Machines and Automatic Control, 2013-14]

43. Define the efficiency of a transformer and find the condition for obtaining maximum efficiency.

[M.D. Univ. Electromechanical Energy Conversion, December-2006]

44. Derive the condition for maximum efficiency of a transformer.

[U.P.S.C. I.E.S. Electrical Engineering-II, 2004]

45. What is all day efficiency and what is its importance? For what type of transformer it is used?

[U.P. Technical Univ. Electromechanical Energy Conversion-I, 2005-06]

46. In what way is all day efficiency different from conventional efficiency? How is all day efficiency calculated?

[U.P. Technical Univ. Electrical Machines, 2009-10]

47. Explain the conditions in detail, that must be fulfilled for the satisfactory parallel operation of two single-phase transformers.

[U.P.S.C. I.E.S. Elec. Engineering II, 1994]

48. Discuss conditions for parallel operation of two single-phase transformers.

[Gujarat Technological Univ., Electrical Machines-I, Summer-2014]

49. What are the conditions for satisfactory parallel operation of single-phase transformer? Deduce expressions for the load shared by two transformers in parallel when no-load voltages of these transformers are not equal. **[M.D. Univ., Electrical Machines-I, December-2010]**

50. Discuss the effect of circulating current at no load, in two single-phase transformers operating in parallel.

Two transformers operating in parallel have different reactance to resistance ratios. Show that one transformer operates at a better pf than the other.

Show that if two transformers have same pu impedances, they will share the load in proportion to their kVA capacities.

[J.N. Technological Univ. Electrical Machines-II, January-2014]

51. Define transformer. Explain important factors that affect the selection of transformer core construction.

[U.P. Technical Univ. Electromechanical Energy Conversion-I, 2005-06]

52. What do you mean by "Voltage Regulation" and "Efficiency" of 1-ϕ transformers?

Determine an expression for voltage regulation and efficiency in terms of equivalent circuit parameters of 1-ϕ transformers.

[U.P. Technical Univ. Electromechanical Energy Conversion-I, 2010-11]

SHORT ANSWER TYPE QUESTIONS WITH ANSWERS

Q. 1. What is the basic function of transformer? **[Pb. Technical Univ., May-2013]**

Ans. The basic function of a transformer is to transform energy from one voltage level to another (may be from low voltage to high voltage or vice versa).

Q. 2. What do you understand by step-up and step-down transformers?

Ans. When transformer raises the voltage i.e. when the output voltage is higher than its input voltage, it is called the step-up transformer. When the transformer lowers the voltage, it is called the step-down transformer.

Q. 3. If the primary current of a transformer is twice as much as the secondary current, is this a step-up or step-down transformer? **[R.G. Technical Univ., June-2011]**

Ans. Step-up transformer, since primary and secondary currents are inversely proportional to their respective turns and primary and secondary voltages are directly proportional to their respective turns.

Q. 4. What is transformation ratio in a transformer?

Ans. The ratio of secondary and primary induced emfs is known as transformation ratio, *i.e.*,

$$\text{Transformation ratio, } K = \frac{E_2}{E_1} = \frac{V_2}{V_1} = \frac{N_2}{N_1}$$

Q. 5. Why does a transformer has iron core? **[P.T.U., December-2000]**

Ans. The use of iron or steel for transformer core ensures a high permeability of the magnetic circuit and because of high permeability the magnitude of exciting current necessary to create the required flux in the core is small. The presence of steel core also causes 100% of magnetic flux created by primary to be linked with secondary.

Q. 6. Why is the efficiency of a transformer maximum among electrical equipments? Explain. **[P.T.U. 1999]**

Ans. Owing to lack of rotating parts in a transformer, there are no friction and windage losses. The other losses such as iron losses are also comparatively small because of better magnetic material and less handling of the material. So the efficiency of transformer is maximum among electrical equipments.

Q. 7. What is normally the efficiency of a transformer ?

Ans. Typical transformer efficiencies at full load lie between 96 % and 97 % and with extremely large transformers they may be as high as 99 %.

Q. 8. Prove that power on both primary as well as secondary remains same. **[Pb. Technical Univ., May-2012]**

Ans. Since power losses are negligible, the power on both primary as well as secondary remains same.

Q. 9. What name is given to the coils through which current flows from the source?

Ans. Primary winding.

Q. 10. What name is given to the coils across which load is connected?

Ans. Secondary winding.

Q. 11. Why the transformer is also called the static transformer?

Ans. Since basic construction of transformer does not require any moving part, it is also called the static transformer.

Q. 12. What is one-to-one transformer ? What is its application ?

[R.G.P.V. Bhopal, June-2002]

Ans. The transformer having equal turns in primary and secondary windings is called the one-to-one transformer. Such transformers are used for isolating the load from supply.

Q. 13. What happens if dc supply is applied to the transformer?

[Anna Univ., May/June-2012]

Ans. If a rated dc voltage is applied to the primary of the transformer, the flux produced in the transformer core will not vary but remain constant in magnitude and, therefore, there will be no self-induced emf in the primary winding (which is only possible with varying flux linkage) to oppose the applied voltage and since the resistance of the primary winding is quite low, a heavy current will flow through the primary winding that may cause burning out of the primary winding.

Q.14. What do you understand by the term "ideal transformer"?

[U.P. Technical Univ. Second Semester, 2013-14]

Ans. The properties of an ideal transformer are (*i*) no winding resistance (*ii*) no magnetic leakage (*iii*) no iron loss and (*iv*) zero magnetizing current i.e. an ideal transformer is assumed to consist of two purely inductive coils wound on a loss-free core.

Q. 15. What is the difference between an ideal and practical transformer? **[P.T.U., January-2000]**

Ans. An *ideal transformer* is an imaginary transformer that has no winding resistance, no magnetic leakage, no iron loss and zero reluctance while a *practical transformer* is an actual transformer which operates at a particular frequency and has winding resistance, magnetic leakage, iron loss, copper loss, and magnetic reluctance.

Q. 16. The primary of a single-phase transformer having a turn ratio 1:2 is connected to a 10 V dc source while its secondary is connected to a 10 Ω resistor. What is the value of the steady state current in the resistor ?

Ans. Zero.

Q. 17. Is there a definite relationship between the number of turns and voltages in transformers ?

Ans. Yes; the voltage varies in exact proportion to the number of turns connected in series in each winding.

Q. 18. What is transformer and transformation ratio? **[M.D. Univ., July-2001]**

Ans. The transformer is an ac machine that (*i*) transfers electrical energy from one electric circuit to another (*ii*) does so without a change of frequency (*iii*) does so by the principle of electromagnetic induction and (*iv*) has electric circuits that are linked by a common magnetic circuit. Voltage transformation ratio is defined as the ratio of secondary induced emf to primary induced emf.

Q. 19. Does the flux linking both the windings of a transformer remain same ?

Ans. Yes; the flux linking the primary and secondary windings is the same.

Q. 20. Does the transformer draw any current when its secondary is open ?

Ans. Yes; no-load primary current.

Q. 21. What is meant by magnetising current ?

Ans. Magnetising current is quadrature or wattless component of exciting current of transformer and is used to create alternating flux in the core.

Q. 22. What do you know about the no-load current of a transformer ?

Ans. The current drawn by primary winding of a transformer on no load is called the no-load or exciting current. It is very small in magnitude in comparison to the full-load primary current (2 to 5 per cent of full-load current). It has two components—energy component (comparatively very small) to meet the iron loss in addition to small amount of copper loss occuring in the primary winding and magnetising component relatively large to create the alternating flux in the core.

Q. 23. Does the magnetising current of a transformer lie in phase with the applied voltage ?

Ans. No; the magnetising current of a transformer lags behind the applied voltage by 90°.

Q. 24. Ordinarily, what is the phase relationship between the primary and secondary voltages of a transformer ?

Ans. 180 ° out of phase.

Q. 25. Explain how in a transformer, as the secondary current increases, primary current also increases. **[P.T.U., May-2002]**

Ans. When the secondary current of a transformer increases, the secondary ampere-turns (N_2I_2) creates more secondary flux. This increased secondary flux opposes the main flux Φ set up by the exciting current I_0. Thus main flux Φ is weakened momentarily by the opposing increased secondary flux Φ_2 and, therefore, difference of applied voltage V_1 and back emf E_1 increases, consequently, more current is drawn from the source of supply flowing through the primary winding until the original value of flux Φ restores. The extra current drawn from the supply source increases the primary ampere-turns (N_1I_1') and compensates the extra secondary ampere-turns i.e. $N_1I_1' = N_2I_2$

Thus in a transformer when the secondary current increases, primary current also increases.

Q. 26. Why the main flux in a transformer remains practically constant from no load to full load ?

Ans. When the transformer is on no load, it draws no-load or exciting current I_0 from the supply mains. The no-load current I_0 sets up mmf N_1I_0 which produces flux Φ in the core. When an impedance is connected across the secondary terminals, current I_2 flows through the secondary. The secondary current I_2 sets up its own mmf and hence creates a secondary flux Φ_2 opposing the main flux Φ. In other words, the secondary ampere-turns weaken the main flux in the core. Momentarily, self-induced emf E_1 decreases due to which primary draws extra current from the supply source. This extra current drawn from the supply source increases the primary ampere-turns (N_1I_1') and compensates the secondary ampere-turns N_2I_2.

Since the secondary flux Φ_2 produced by secondary mmf N_2I_2 is neutralized by the flux Φ_1' produced by mmf N_1I' set up by counterbalancing primary current I_1', the flux in the transformer core remains constant from no load to full load.

Q. 27. How does the leakage flux occur in a transformer? **[P.T.U., December-2000]**

Ans. In actual practice the total flux created does not link the primary and secondary windings but is divided into three parts namely the main flux Φ linking both the primary and secondary windings, primary leakage flux Φ_{L_1}, linking with primary winding only and secondary leakage flux Φ_{L_2}, linking with secondary winding only.

The flux set up by the primary or secondary which is set up in the core linking with its own turns and not linking with the other is known as *leakage* flux.

Q. 28. What is leakage reactance of transformers? **[P.T.U., December-2003]**

Ans. Leakage fluxes of primary and secondary produce self-inductances L_1 given as $\frac{N_1\Phi_{L1}}{I_1}$ and L_2 given as $\frac{N_2\Phi_{L2}}{I_2}$ in the respective windings i.e. primary and secondary. These inductances, in turn, produce inductive reactances X_1 given as $2\pi f L_1$ and X_2 given as $2\pi f L_2$ which cause voltage drop in respective windings lagging behind the currents I_1 and I_2 respectively by 90°. The reactances produced by leakage fluxes are called the leakage reactances.

Q. 29. What is the effect of leakage flux in a transformer ? What is meant by magnetising current ? **[P.T.U., June-2000]**

Ans. The effect of leakage flux is to develop into their respective windings emfs of self-inductance which are proportional to the current, and are, therefore, equivalent in effect to the addition of an inductive coil in series with each winding. The component of no-load current used to create the flux in the transformer core is known as magnetising current.

Q. 30. Explain with reasons what happens when the primary terminals of a single phase transformer are connected to the rated ac supply while secondary terminals are short circuited.

Ans. The transformer with secondary short circuited becomes equivalent to a coil having an impedance equal to impedance of both windings, which is quite small. When the primary terminals of the transformer are connected to the rated ac supply, heavy current will flow through primary and secondary windings that may cause burning out of these windings.

Q. 31. Define voltage regulation of a transformer.

[UP Technical Univ. Second Semester 2003-04; G.G.S.I.P. Univ. Delhi, May-2005]

Ans. Voltage regulation of a transformer is defined as the change in magnitude of terminal or secondary voltage, when full load (rated load) of specified power factor supplied at rated voltage is thrown off (reduced to no load) with primary voltage and frequency held constant, as percentage of the rated terminal voltage.

Q. 32. What is voltage regulation in case of a transformer?

For which type of load power factor will this regulation be zero?

[B.P. Univ. of Technology, 2008]

Ans. Refer to Answer of Q. 31.

Regulation will be zero in case of leading pf load.

Q. 33. Is voltage regulation better or worse at 0.6 *pf* lagging than at 0.8 *pf* lagging for the same kVA output explain. **[P.T.U., May-2001]**

Ans. The voltage regulation will be the maximum and worst when power factor becomes equal to $\cos \tan^{-1} \dfrac{X_{02}}{R_{02}}$ (lagging).

Since equivalent reactance of transformer is usually more than double its equivalent resistance, the voltage regulation will increase (i.e. become more and more worse) till power factor falls even below $\cos \tan^{-1} 2$ i.e. 0.45 (lagging), the voltage regulation will worsen at 0.6 pf lagging than at 0.8 pf lagging.

Q. 34. What is meant by regulation of a transformer and how is it affected by the power factor of the load supplied by it?

Ans. The change in secondary terminal from no load to full load is termed the regulation. It is usually expressed as a percentage or a fraction of the rated no-load terminal voltage.

The voltage regulation increases with the fall in power factor of the load, if the load is inductive, and becomes maximum when the power factor of the inductive load becomes equal to

$$\cos\left(\tan^{-1}\frac{X_{02}}{R_{02}}\right).$$

The voltage regulation drops with the fall in power factor of the load, if the load is capacitive, and becomes zero, when the power factor of the capacitive load becomes equal to $\cos\left(\tan^{-1}\dfrac{-R_{02}}{X_{02}}\right)$, and then increases with the fall in power factor of load in the negative direction.

Q. 35. Why sometimes the voltage regulation of a transformer is positive and sometimes negative? **[G.G.S.I.P. Univ. Delhi, May-2010]**

Ans. Voltage regulation depends upon the load current and pf of load. For lagging power factor, the voltage regulation will be positive and for leading power factor, the voltage regulation may be positive or negative.

Q. 36. Is the regulation at rated load of a transformer same at 0.8 pf lag and 0.8 pf lead?

Ans. No; the regulation obtained is different at leading and lagging power factors.

Q. 37. What type of load should be connected to a transformer for getting maximum voltage regulation ?

Ans. Inductive load having phase angle of $\phi = \tan^{-1} \dfrac{X_{eq}}{R_{eq}}$.

Q. 38. What type of load should be connected to a transformer for getting negative voltage regulation ?

Ans. Highly capacitive load.

Q. 39. Is the percentage impedance of a transformer different on the primary or on the secondary side ?

Ans. No; the percentage impedance of a transformer remains the same on both sides.

Q. 40. Will the voltage regulation of a transformer based on its primary referred equivalent circuit be the same as that when the equivalent circuit is referred to the secondary ?

Ans. Yes; voltage regulation will be the same if magnetising component of current is neglected.

Q. 41. On what factors does the hysteresis loss depend ? Why can this loss be assumed to be constant ? **[P.T.U., December-2000]**

Ans. Hysteresis loss depends upon the type of material, volume of the material, frequency of reversal of magnetic field and maximum value of flux density in the core of the material i.e.

$$P_h = \eta\, (B_{max})^{1.6} f\, v \text{ watts}$$

Since from no load to full load the flux linking with the core and supply frequency remains constant, hysteresis loss is assumed to remain constant.

Q. 42. Write an expression of hysteresis loss in a transformer. Why it decreases at higher frequencies with constant V?

[U.P. Technical Univ. Odd Semester, 2013-14]

Ans. The core of a transformer is subjected to an alternating magnetizing force and for each cycle of emf a hysteresis loop is traced out. The hysteresis loss per second is given by the equation

Hysteresis loss, $P_h = \eta'(B_{max})^x f\, v$ joules per second or watts

where f is the supply frequency in Hz, v is the volume of core in cubic metres, η' is the hysteresis coefficient, B_{max} is peak value of flux density in the core and x lies between 1.5 and 2.5 depending upon the material and is often taken as 1.6.

Hysteresis loss is proportional to $(B_{max})^x f$ and for a transformer with constant V, the product of maximum flux density B_{max} and frequency f is constant [Refer to Eq. (12.1)]. So at higher frequencies product $(B_{max})^{1.6} f$ and therefore hysteresis loss decreases, x being more than unity.

Q. 43. What are eddy currents?

[M.D. Univ., December-2009; May-2010; May-2011]

Ans. We have seen that whenever the flux linking with a closed electric circuit changes, an emf is induced in the circuit and a current flows, the value of which depends on the emf around the circuit and the resistance of the circuit. It is not necessary that the circuit be a wire and that the flux passes entirely through it. If a solid block of metal is traversed by a varying flux, metallic circuits in the block itself, which are linked by the flux, will carry current. If the magnetic circuit is made up of iron and if the flux in the circuit is variable, currents will be induced by induction in the iron circuit itself. All such currents are known as *eddy currents.*

Q. 44. On what factors does the eddy current loss depend ? Why this loss can be assumed to be constant ? **[P.T.U., June-2000]**

Ans. The eddy current loss depends on the maximum value of flux density in the core and supply frequency for a given core (i.e nature of magnetic material, volume of the core, and thickness of laminations remaining unchanged). Since from no load to full load the flux linking with the core and supply frequency remain constant, the eddy current loss remains constant.

Q. 45. How can eddy current loss be minimised ?

Ans. By using laminated iron core.

Q. 46. List out various losses that take place in a transformer. Which loss is independent of load ? **[P.T.U. June 2003; May-2010]**

Ans. Iron or core losses consisting of hysteresis and eddy current losses. These losses occur in the core and yoke due to alternating flux.

Copper losses occuring in primary and secondary windings due to their resistances when the transformer is loaded.

Iron losses are independent of load, and therefore, are called the constant losses.

Q. 47. What are the different losses in a transformer ? How do they depend on voltage and frequency ?

[G.G.S.I.P. Univ. Delhi Electrical Science, May-2008]

Ans. The various power losses occurring in a transformer are iron or core losses (hysteresis loss and eddy current loss) and copper or ohmic losses. Hysteresis loss and eddy current loss depend upon supply frequency *f* (hysteresis loss varies directly as the supply frequency and eddy current loss varies as the square of the supply frequency) and peak value of flux density (hysteresis loss varies as $(B_{max})^{1.6}$ and eddy current loss varies as B^2_{max}) which in turn depends on supply voltage and frequency.

Ohmic loss depend upon the load current supplied by transformer and are independent of supply voltage and frequency.

Q. 48. Classify the losses in power transformer.

[G.B.T.U. First Semester 2011-12; U.P. Technical Univ. Even Semester, 2013-14]

Ans. Transformer losses may be classified into two categories namely constant losses and variable losses.

Iron losses are independent of load, and therefore, are called the constant losses. These losses consist of hysteresis and eddy current losses and occur in the core and yoke of the transformer due to alternating flux.

Copper losses vary as the square of load current or kVA, and therefore, they are called variable losses. These losses occur in primary and secondary windings of the transformer due to their resistances when the transformer is loaded.

Q. 49. Name the constant losses taking place in a transformer.

[G.B.T.U. Even Semester, 2011-12]

Ans. Iron losses are constant losses that take place in a transformer. Eddy current loss and hysteresis loss constitute the iron losses.

Q. 50. Why eddy current loss of a transformer depends on applied voltage. **[M.D. Univ., December-2013]**

Ans. If the transformer is operated with increased voltage but at the same frequency the eddy current loss increases due to increase in value of peak value of flux density and if applied voltage is increased with increase in frequency, the proportion of increase being the same in both, the eddy current loss increases due to increase in supply frequency.

Q. 51. If the supply frequency is doubled, which loss component in transformer will double ?

Ans. Due to doubling of supply frequency, the hysteresis loss component will be double.

Q. 52. What is the purpose of the markings on transformer leads ?

Ans. They are there for standardization, so that the transformer polarities may be recognizable for any type of use.

Q. 53. Under what conditions, open-circuit test is performed on transformer ? What are the results obtained ? **[P.T.U., June-2003]**

Ans. Open-circuit test is performed to determine the core or iron or excitation loss, P_i and no-load current I_0 and thereby the shunt branch parameters R_0 and X_0 of the equivalent circuit.

In this test, one of the windings (usually *hv* winding) is kept open circuited and the rated voltage at rated frequency is applied to the other winding. Either an auto-transformer or voltage divider (VD) is used for varying the voltage applied to the *lv* winding. Ammeter A and wattmeter W are connected to measure no-load current I_0 and input power at no load, W_0 respectively. Voltmeter is connected to measure the applied voltage.

Q. 54. What do we mean by an equivalent circuit of an electrical machine/ device? Why is it necessary ?

[G.G.S.I.P. Univ. Delhi Electrical Science, May-2008]

Ans. An electrical circuit that can represent behaviour of the electrical machine/device to an approximation is called its equivalent circuit.

The equivalent circuit of any device can be quite helpful in predetermination of the behaviour of the device under various conditions of operation and it can be drawn if the equations describing its behaviour are known. If any electrical device is to be analysed and investigated further for suitable modifications, its appropriate equivalent circuit is necessary. The equivalent circuit for electromagnetic devices consists of a combination of resistances, inductances, capacitances, voltage etc. Such an equivalent circuit (or circuit model) can, therefore, be analysed and studied easily by the direct application of electric circuit theory.

Q. 55. What is the advantage of predetermining the performance of electrical machines through open-circuit and short-circuit test data ?

[P.T.U., January-2000]

Ans. The advantage of predetermining the performance of electrical machines through open-circuit and short-circuit test data is that we can know the behaviour and characteristics of the machine that will show during actual operating conditions and on this basis we can select the machine meeting our requirements.

Q. 56. What are the performance parameters of single-phase transformers?

[U.P. Technical Univ., 2013-14]

Ans. The performance parameters of single phase transformers are the shunt branch parameters R_0 and X_0 and equivalent resistance and reactance R_{eq} and X_{eq} of transformer primary and secondary windings.

Q. 57. How can copper loss be measured in a transformer ?

Ans. Copper loss can be determined from short-circuit test. The power supplied during short-circuit test is equal to copper loss corresponding to short-circuit current.

Q. 58. What are the various tests that give the complete parameters of the equivalent circuit of the transformer ?

Ans. Open-circuit and short-circuit tests.

Q. 59. What are typical informations that can be had by performing open- and short-circuit tests on transformer ?

Ans. Iron and copper losses and the parameters of the equivalent circuit of the transformer can be determined from the open-circuit and short-circuit tests and so the voltage regulation and efficiency of the transformer can be determined.

Q. 60. Why core loss is neglected in short-circuit test and copper loss in open-circuit test ?

Ans. Since in short-circuit test the voltage applied to the primary of transformer is very low (5 to 8% of rated voltage), the flux produced in the core is very small and, therefore, iron losses are very small and can be neglected.

Since in open-circuit test no current flows in the open-circuited secondary and the current in the primary will also be very small (usually 3 to 10% of full-load current), the copper losses in the open-circuit test will be so small that these can be neglected.

Q. 61. Why are iron or core losses assumed to remain constant from no load to full load in a power transformer ?

Ans. Since iron or core losses depend on the flux density and supply frequency for a given transformer core and flux density and supply frequency remain the same from no load to full load, iron or core losses are assumed to remain constant from no load to full load.

Q. 62. Why the *lv* winding is normally short circuited during short-circuit test ?

Ans. Voltage required for the short-circuit test is very low (about 5 per cent of the rated value) and if the measurements are made, by short circuiting *hv* winding, on *lv* side, the voltage required would be inconveniently low, while the current would often be inconveniently large.

Q. 63. Why is the iron loss in a transformer substantially independent of the load current ? Explain.

[U.P. Technical Univ. Electrical Machines, 2001-02]

Ans. Iron or core loss is caused by the alternating flux in the core and consists of hyteresis and eddy current losses which depend upon the maximum flux density in the core and supply frequency. Since it has been determined that the mutual flux varies somewhat with the load (its variation being 1 to 3 per cent from no load to full load), the core loss will vary somewhat with the load and it's power factor. It may be emphasized here that core loss is assumed to remain constant from no load to full load, the variations in losses from no load to full load being very small and negligible.

Q. 64. Explain how a short-circuit test is performed on a single-phase transformer and why it gives the copper losses only and not the iron losses? **[U.P. Technical Univ. Electrical Machines, 2001-02]**

Ans. Short-circuit test is performed by short circuiting the secondary (low-voltage) winding by a thick wire or strip and applying variable low voltage to the primary (*hv* winding) through an auto-transformer or potential divider. The applied voltage to the primary is gradually increased till the ammeter indicates the full-load current of metering side.

Since applied voltage is very low (5 – 8% of the rated voltage), flux linking with the core is very small and, therefore, iron losses are so small that these can be neglected and thus the power input gives the copper losses only and not the iron losses.

Q. 65. Explain why in testing large transformers the open-circuit test is carried out with the high-voltage winding open and the short circuit test with the low-voltage winding shorted. Write the parameters obtained from these tests.

[U.P. Technical Univ. Electromechanical Energy Conversion-I, 2007-08]

Ans. In open-circuit test on large transformers, one of the windings (usually high voltage winding) is kept open circuited and the rated voltage at rated frequency is applied to the other winding, as shown in Fig. 12.31. No doubt, the core loss will be the same whether the measurements are made on *lv* winding so long as the rated voltage of that winding is applied to it but in case the measurements are made on hv winding, the voltage required to be applied would often be inconveniently large while the current I_0 would be inconveniently small.

Voltage required for the short-circuit test is about 5 per cent of the rated value. For a 200 kVA, 2200/220 V transformer, test on high voltage side would need the voltage (to be applied) of $2{,}200 \times \frac{5}{100}$ *i.e.* 110 V (which is standard voltage for instrument coils) and a current of $\frac{200 \times 1{,}000}{2{,}000} = 91$ A.

If the test is conducted on low voltage side of the above transformer, the voltage needed would be $220 \times \frac{5}{100} = 11$ V and the current would be $\frac{200 \times 1{,}000}{220} = 910$ A (very high). At this low voltage, high precision would not be readily obtainable with ordinary instruments.

Thus we see that the if the measurements were made on low voltage side, the voltage needed would be inconveniently low, while the current would often be inconveniently large.

Q. 66. Define efficiency of a transformer. **[M.D. Univ., December-2012]**

Ans. The efficiency (η) of a transformer, like that of any other apparatus, is defined as the ratio of useful power output to the input power, the two being measured in same units (either in watts or killowatts).

Q. 67. Write the condition for efficiency in transformer to be maximum.

[M.T.U. First Semester, 2011-12]

Ans. Efficiency of transformer will be maximum when variable loss (copper loss) is equal to constant loss (iron loss)

$$i.e., \quad x^2P_c = P_i$$

where P_i is the iron loss, P_c is the copper loss on full load and x is the fraction of full-load kVA at which efficiency is maximum.

Q. 68. Is the efficiency of a transfromer same at the same load at 0.8 *pf* lag and 0.8 *pf* lead ?

Ans. Yes, the efficiency of a transformer remains the same for a given load at a given power factor irrespective of the fact whether the *pf* is lagging or leading.

Q. 69. What is meant by all-day efficiency of a power transformer and why is it lower than commercial efficiency ?

Ans. The all-day efficiency, also known as energy efficiency or operational efficiency, is defined as the ratio of energy (kWh) output over 24 hours to the energy input over the same period

$$\text{i.e.} \quad \text{All-day efficiency} = \frac{\text{Output in kWh}}{\text{Input in kWh}}$$

There are certain types of transformers, such as distribution transformers, which remain energized for 24 hours but they supply very light loads for major portion of the day. Thus iron or core loss occurs for the whole day but the copper loss occurs only when the transformer is loaded and so the all-day efficiency of transformer is lower than its commercial efficiency.

Q. 70. Why all day efficiency is lower than commercial efficiency.

[Anna Univ., May/June-2012]

Ans. Refer to Ans. of Q. 69.

Q. 71. Two single-phase transformers A and B of identical voltage and turn-ratio operate in parallel. If the impedances of the two transformers are equal while resistance /reactance ratio of A is higher than that of B, what would be the magnitudes of I_A and I_B and their phase position?

Ans. Magnitudes of I_A and I_B are equal but phase angle of I_A is less than that of I_B.

Q. 72. If transformers with different electrical characteristics are connected in parallel, what will happen?

Ans. They will not share the load equally.

Q. 73. Name four materials used in construction of transformer.

[G.G.S.I.P. Univ. Delhi, May/June-2006]

Ans. Four materials used in construction of transformer are:

(*i*) Core : Soft iron or silicon steel

(*ii*) HV and LV windings: Copper or aluminium wire

(*iii*) Insulating materials: Paper, pressboard, cloth, mica, asbestos and impregnating compounds according to needs.

(*iv*) Container : Steel tank

Q. 74. Why is laminated core used in the transformer?

[M.D. Univ., May-2003]

Ans. Laminated core is used in the transformer to reduce the eddy current loss.

Q. 75. Why is an iron-silicon alloy core used in a transformer?

[P.T.U., December-2005]

Ans. Iron-silicon alloy core is used in a transformer to ensure a high permeability of the magnetic circuit and keep the iron loss the minimum possible.

Q. 76. Why silicon content in electrical sheet steel is limited to 4.5 to 5 %?

Ans. Silicon content exceeding 5 % makes the sheet steel brittle and so causes problem in punchings.

Q. 77. Why is the area of yoke of a transformer usually kept 15–20 % more than that of core ?

Ans. By keeping yoke area 15–20% more than the core, flux density in the yoke is reduced, resulting into reduction in iron loss for the yoke.

Q. 78. Explain why the cross section of the limbs in a core type transformer are stepped instead of being circular.

[B.P. Univ. of Technology, 2008]

Ans. Small transformers may have cores of rectangular or square x-section with rectangular or circular coils but it is wasteful in case of large capacity transformers. In case of large sized transformers stepped cruciform core with circular cylindrical coils is employed, as illustrated in Fig. 12.61. Such a core employs laminations of different sizes. Though the cost of manufacturing of such a cruciform core is much greater, but the circular coils that are used are easier to wind and provide more mechanical strength, especially when short circuit occurs. Other advantages of using cruciform core are high space factor and reduced mean length of turns resulting in reduced copper loss.

Q. 79. What is the cause of noise in transformers ?

Ans. In a transformer, noise occur mainly due to loosening of stampings and mechanical forces developed during operation.

Q. 80. Why mittred core is preferred in large transformers ?

Ans. The iron loss and the exciting current are reduced with mittred joints because flux lines are made along the direction of grain orientation.

Q. 81. Why lv winding is placed first on the core and then hv winding in case of a core type transformer ?

Ans. Placing of lv winding near the core and hv winding around the lv winding in a core type transformer reduces the amount of insulation materials required.

Q. 82. Is it possible for any voltage winding to serve as primary ?

Ans. Any voltage winding can serve as primary or secondary winding provided its current and voltage ratings are not exceeded.

Q. 83. Which kind of core is preferred in the medium power transformers and why ? [Pb. Technical Univ., June-2013]

Ans. Medium power transformers may have cores of rectangular or square cross section with rectangular or circular coil. This is because of lower cost of manufacturing.

Q. 84. Why circular coils are always preferred over rectangular coils for winding a transformer ?

Ans. Circular coils are preferred for winding a transformer as they can be easily wound on machines, conductors can easily be bent and winding does not bulge out due to radial forces developed during operation.

Q. 85. In what type of transformers sandwich coils are used ? What advantages are gained by the use of sandwich coils ?

Ans. Sandwich coils are used in shell type transformers for lv and hv windings. The advantage gained by the use of sandwich coils is very low value of leakage reactance resulting in low voltage drops and so voltage regulation.

Q. 86. Why excessive insulation is harmful to a coil. ?

Ans. Excessive insulation makes heat transfer poor.

Q. 87. What are the functions of transformer oil ?

Ans. It provides additional insulation, protects the insulation from dirt and moisture and it carries away the heat generated in the cores and coils.

Q. 88. Why tappings are provided in the hv windings of a transformer ?

Ans. The hv winding has more turns hence it is convenient to provide tappings on this winding to give better regulation.

PROBLEMS

1. The maximum flux density in the core of 250/3,000 V, 50 Hz single phase transformer is 1.2 Wb/m^2. If the emf per turn is 8 V, determine (*i*) primary and secondary turns (*ii*) area of the core.

   [**Ans** (*i*) 32; 376 (*ii*) 0.03 m^2]

2. A single-phase transformer has 400 primary and 1,000 secondary turns. The net cross-sectional area of the core is 60 cm^2. If the primary winding be connected to a 50-Hz supply at 520 V, calculate (*i*) the peak value of flux density in the core (*ii*) the voltage induced in the secondary winding. [**Ans** 0.976 T; 1,300 V]

3. A single phase transformer has 350 primary and 1,050 secondary turns. The net cross-sectional area of the core is 55 cm^2. If the primary winding be connected to a 400 V, 50 Hz single phase supply, calculate (*i*) maximum value of the flux density in the core and (*iii*) the voltage induced in the secondary winding.

   [RGPV 2001; G.G.S.I.P. Delhi Univ., May/June-2006]

   [**Ans.** (*i*) 0.936 T (*ii*) 1,200 V]

4. A single-phase 2,200/220 V, 50 Hz transformer has core area 3,600 mm^2, and the maximum flux density 1.6 T. Determine the number of turns of primary and secondary windings.

   [J.N. Technological Univ. Electrical Machines-II, January-2014]

   [**Ans.** 172, 1,720]

5. A single-phase transformer, 50 Hz, 220/300 V, transformer has a net cross-sectional area of core is 400 cm^2. If the peak value of flux density in the core is 1.24 Wb/m^2, calculate the suitable values for the number of turns on the primary and secondary windings.

   [R.G.T.U., July-2008]

   [**Ans.** 20, 28]

6. A 200 kVA, 6600/400 V, 50 Hz single-phase transformer has 80 turns on the secondary. Calculate (*i*) the approximate values of the primary and secondary currents (*ii*) the approximate number of primary turns and (*iii*) the maximum value of flux.

   [Pb. Univ. Elec. Engineering-I, 1994]

   [**Ans.** (*i*) 30.3 A, 500 A; (*ii*) 1,320 V (*iii*) 22.5 mWb]

7. The no-load current of a transformer is 4.0 A at 0.25 pf when supplied at 250 V, 50 Hz. The number of turns on the primary winding is 200. Calculate

   (*i*) flux in the core (*ii*) the core loss, and (*iii*) magnetising current.

   [U.P. Technical Univ. Electrical Machines, 2003-04]

   [**Ans.** (*i*) 5.63 mWb (*ii*) 250 W (*iii*) 3.873 A]

8. A single-phase transformer is rated 25 kVA, 600/200 V, 50 Hz. Calculate the impedance of load in ohms to fully load the transformer when connected to (a) 600 V side (b) 240 V side.

   [Mahamaya Technical Univ. Even Semester, 2011-12]

   [**Ans.** 14.4 Ω, 1.6 Ω]

9. The no-load current of a transformer is 5 A at 0.25 pf when supplied at 235 V, 50 Hz. The number of turns on the primary winding is 200. Calculate (*i*) the maximum value of flux in the core, (*ii*) the core loss (*iii*) the magnetizing component. [Nagpur Univ. Elec. Machines-I, 1993]

   [**Ans.** (*i*) 5.293 mWb (*ii*) 293.75 W (*iii*) 4.84 A]

10. A single-phase transformer with a ratio of 440/110 V takes a no-load current of 5 A at 0.2 power factor lagging. If the secondary supplies a current of 120 A at a pf of 0.8 lagging, estimate the current taken by the primary. [**Ans.** (*i*) 33.9 A, 42.49° (lag)]

11. A 230/115 V, single-phase transformer is supplying a load of 5A, at pf 0.866 lagging. The no-load current is 0.2 A at pf of 0.208 lagging. Calculate the primary current and primary power factor.

    [Nagpur Univ. Summer, 2000]

    [**Ans.** 2.64 A; 33.26° (lagging)]

12. Draw and explain phasor diagram of transformer under loaded condition. A single phase 440/110 V transformer takes a no-load current of 4 A at 0.2 power factor. If the secondary supplies a current of 100 A at a power factor of 0.8 lagging determine (i) the current taken by the primary winding (ii) the magnetising reactance and resistance representing core losses.

[U.P. Technical Univ. Electrical Engineering, February-2002]

[**Ans.** (*i*) $28.12 \angle 42.29^\circ$ A ; 550 Ω, 112.25 Ω]

13. A 10 kVA, 1-phase transformer with 2,000/400 V at no load, has resistance and leakage reactance of primary winding 5.5 Ω and 12 Ω respectively, the corresponding values of secondary winding being 0.5 Ω and 0.45 Ω. Determine the value of secondary voltage at full load, 0.8 pf lagging, when the primary applied voltage is 2,000 V.

[U.P. Technical Univ., July-2002]

[**Ans.** 377.65 V]

14. A 50 kVA, 4400/220 V transformer has $R_1 = 3.45\ \Omega$, $R_2 = 0.009\ \Omega$. The values of reactances are $X_1 = 5.2\ \Omega$ and $X_2 = 0.015\ \Omega$. Calculate for the transformer (*i*) equivalent resistance as referred to primary (*ii*) equivalent resistance referred to secondary (*iii*) equivalent reactance as referred to both primary and secondary (*iv*) equivalent impedance as referred to both primary and secondary (*v*) total copper loss, first using individual resistances of the two windings and secondly using equivalent resistances as referred to each side.

[Nagpur Univ. Elec. Engineering-I, 1993]

[**Ans.** (*i*) 7.05 Ω (*ii*) 0.017625 Ω (*iii*) 11.2 Ω, 0.028 Ω (*iv*) $13.23 \angle 57.81^\circ$ Ω ; $0.0331 \angle 57.81^\circ$ Ω (*v*) 910.4 W]

15. A single phase load is fed through a 66 kV feeder whose impedance is $(120 + j400)$ ohm and a 66/6.6 kV transformer whose equivalent impedance (referred to LV) is $(0.4 + j1.5)$ ohm. The load is 250 kW at 0.8 leading power factor and 6 kV.

(*i*) Calculate the voltage at the sending-end of the feeder.

(*ii*) Calculate the voltage at primary terminals of the transformer.

(*iii*) Calculate the complex power input at the sending-end of the feeder. **[U.P. Technical Univ. Electromechanical Energy Conversion-I, 2004-05]**

[**Ans.** (*i*) 58.95 kV (*ii*) 59.698 kV (*iii*) $(254.3264 - j\ 172.628$ kVA)]

16. The ohmic values of the circuit parameters of a transformer, having a turns ratio of 5, are $R_1 = 0.5\ \Omega$, $R_2 = 0.021\Omega$, $X_1 = 3.2\ \Omega$, $X_2 = 0.12\ \Omega$, $R_c = 350\ \Omega$, referred to the primary and $X_m = 98\ \Omega$ referred to primary. Draw the approximate equivalent circuit of the transformer referred to secondary. Show the numerical values of the circuit parameters.

[U.P. Technical Univ. Electrical Engineering Odd Semester, 2008-09]

[**Ans.**

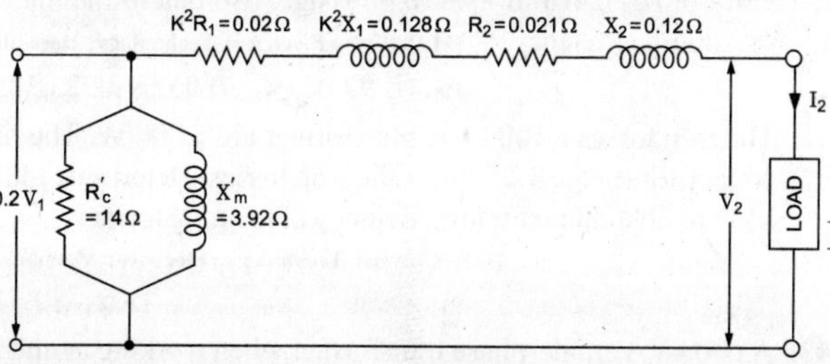

]

17. A 3300/230 V, single-phase transformer has an equivalent resistance of 0.015 pu and an equivalent leakage reactance of 0.04 pu. Find the secondary terminal voltage at full load at 0.8 pf lag, for a primary voltage of 3,300 V. **[Pb. Univ. Elec. Machines–I, Dec.; 1992]**

[**Ans.** 2,229.28 V; 221.72 V]

18. Calculate the percentage voltage drop for a transformer with a percentage resistance of 2.5 % and a percentage reactance of 5 % of rating 500 kVA when it is delivering 400 kVA at 0.8 pf lagging.

[**Ans.** 4%]

19. Calculate the regulation of a transformer in which the ohmic loss is 1% of the output and reactance drop of 5% of the voltage, when the power factor is (*i*) 0.8 lag (*ii*) unity (*iii*) 0.8 leading.

[Madras Univ., 1997]

[**Ans.** (*i*) 3.8% (*ii*) 1% (*iii*) – 2.2%]

20. A transformer has copper loss of 1.5% and reactance drop of 3.5% when tested at full load. Calculate its full-load regulation at (*i*) *u* pf (*ii*) 0.8 pf lagging and (*iii*) 0.8 pf leading.

[Bharathidhasan Univ., April-1997]

[**Ans.** (*i*) 1.5%, (*ii*) 3.33%, (*iii*) – 0.9%]

21. Resistances and leakage reactances of 10 kVA, 50 Hz, 2300/230 V single phase distribution transformer are $r_1 = 3.96\ \Omega$, $r_2 = 0.0396\ \Omega$, $x_1 = 15.8\ \Omega$, $x_2 = 0.158\ \Omega$. Subscript 1 refers to hv and 2 to lv winding. (*a*) The transformer delivers rated kVA at 0.8 pf lagging to a load on the lv side. Find the hv side voltage necessary to main 230 V across load terminals. Also find percentage voltage regulation. (*b*) Find the pf of the rated load current at which the regulation will be zero, hence find the hv side voltage.

[Nagpur Univ., November-1997]

[**Ans.** (*a*) 2409.6 V (*b*) 0.97(leading), 2300 V]

22. If in a transformer core, the hysteresis and eddy current losses are 80 W and 50 W at normal voltage and frequency, calculate the losses when voltage and frequency are increased by 20%.

[U.P. Technical Univ. Elec. Engineering First Semester, 2004-05]

[**Ans.** 96 W, 72 W]

23. A transformer is connected to 2200 V, 40 Hz supply. The core loss is 800 watts out of which 600 watts are due to hysteresis and remaining eddy current losses. Determine the core loss if the supply voltage and frequency are 3300 V and 60 Hz respectively.

[Bharathiyar Univ., Nov.-1997]

[**Ans.** 1,350 W]

24. A transformer with normal voltage impressed has a flux density of 1.4 Wb/m^2 and a core loss comprising of 1,000 W eddy current loss and 3,000 W hysteresis loss. What do these losses become under the following conditions:

(*a*) increasing the applied voltage by 10 % at rated frequency.

(*b*) reducing the frequency by 10 % with normal voltage impressed.

(*c*) increasing both impressed voltage and frequency by 10 %.

[J.N. Technological Univ. Electrical Machines-II, January-2014]

[**Ans.** (*a*) 3,494 W; 1,210 W (*b*) 3,196 W; 1,000 W (*c*) 3,300 W; 1,210 W]

25. In a transformer, the core loss is found to be 52 W at 40 Hz and 90 W at 60 Hz measured at the same peak flux density. Calculate the hysteresis and eddy current losses at 50 Hz.

[Nagpur Univ. Electrical Machines, 1993]

[**Ans.** 45 W, 25 W]

26. A 50 Hz, single phase transformer has turn ratio of 6. The resistances and reactances of the high voltage and low voltage windings are:

$r_1 = 0.95\ \Omega$, $x_1 = 5\ \Omega$, $r_2 = 0.03\ \Omega$, $x_2 = 0.13\ \Omega$

Determine the voltage to be applied to the high voltage side so that full-load current of 200 A flows in low-voltage winding on short circuit. **[M.D. Univ. Electromechanical Energy Conversion, May-2008]**

[**Ans.** 329.7 V]

27. The following data were obtained on a 20 kVA, 50 Hz, 2,000/200 V distribution transformer:

	Voltage (V)	Current (A)	Power (W)
OC test with HV open circuited:	200	4	120
SC test with LV short circuited:	60	10	300

Draw the approximate equivalent circuits of transformer referred to the *hv* and *lv* side respectively.

[Rajasthan Univ. Electrical Machines-I, 2003]

[Ans.

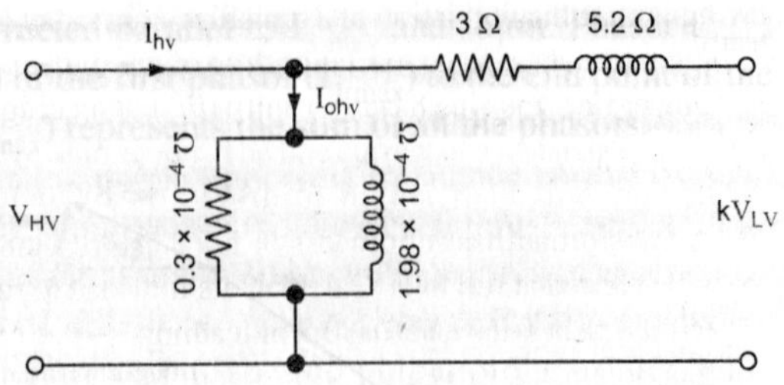

Equivalent Circuit Referred to HV side

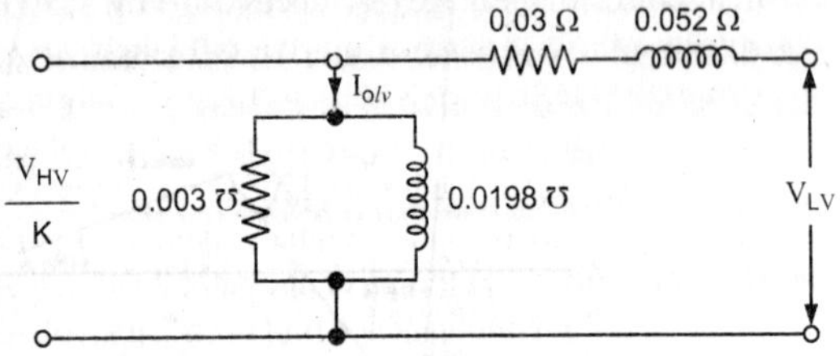

Equivalent Circuit Referred to LV side]

28. Define efficiency of transformer. A 20 kVA, 2500/250 V, 50 Hz, 1-phase transformer has the following test results :

OC Test (LV Side) : 250 V, 1.4 A, 105 W

SC Test (HV Side) : 104 V, 8 A, 320 W

Compute the parameters of the approximate equivalent circuit referred to low voltage side and also draw the circuit.

[M.D. Univ. Electrical Machines-II, December-2012]

[**Ans.** R_0 = 595.24 Ω, X_0 = 187.2 Ω, R_{01} = 0.05 Ω, X_{01} = 0.12 Ω]

29. Given below are the results conducted on a 50 kVA, 2200/220 V transformer.

Open-Circuit (LV Side) Test: 405 W, 5 A, 220 V

Short-Circuit (HV Side) Test: 805 W, 20.2 A, 95 V

Calculate the parameters of the equivalent circuit.

[Pb. Technical Univ., December-2003]

[**Ans.** 11.95 kΩ, 4.73 kΩ, 1.973 Ω, 4.266 Ω]

30. The following data were obtained on a 50 kVA, 2400/120 V transformer.

Open-circuit test, instruments on *lv* side:

Wattmeter reading = 396 W; Ammeter reading = 9.65 A;

Voltmeter reading = 120 V

Short-circuit test, instruments on *hv* side:

Wattmeter reading = 810 W; Ammeter reading = 20.8 A;

Voltmeter reading = 92 V

Find the equivalent circuit parameters referred to high voltage side.

[Pb. Technical Univ., May-2002]

[**Ans.** 14.5 kΩ, 5.3 kΩ, 1.872 Ω, 4 Ω]

31. A short-circuit test when performed on the hv side of a 10 kVA, 2,000/400 V single-phase transformer, gave the following data: 60 V, 4 A, 100 W

If the lv side is delivering full load current at 0.8 pf lag and at 400 V, find the applied voltage to hv side.

[Nagpur Univ. Elec. Machines-I, 1993]

[**Ans.** 2,065.9 V]

32. Define voltage regulation of a transformer. The following test data were taken on 25 kVA, 2,400/240 V, 50 Hz transformer with low-voltage winding shorted and measurement made on hv side :

V = 41.5 V, I = 21.7 A, P = 350 W.

Calculate the voltage regulation of this transformer when it supplies full load at 0.8 pf lagging. Neglect the magnetising current.

[U.P. Technical Univ. Electrical Machines, 2002-03]

[**Ans.** 0.72%]

33. In a 25 kVA, 2000/200 V transformer, the iron and copper losses are 350 W and 400 W respectively. Calculate the efficiency of the transformer at unity power factor at (*i*) full load (*ii*) half load.

[G.G.S.I.P. Univ. Delhi, May-2011]

[**Ans.** (*i*) 97.087% (*ii*) 96.53%]

34. A 1 kVA, single phase transformer has an iron loss of 12 W, and full load copper loss of 18 W. Find the full-load efficiency assuming the power factor be 0.9. **[R.G.T.U., April-2009]**

[**Ans.** 96.77%]

35. The full-load copper and iron losses of 15 kVA 1-phase transformer are 320 W and 200 W respectively. Calculate the efficiency of the transformer for (*a*) full load (*b*) half load, when the load pf is 0.8 lagging in each case. **[M.D. Univ. Electrical Machines-I, December-2010]**

[**Ans.** (*i*) 95.85% (*ii*) 95.55%]

36. A 25 kVA, 1-phase transformer, 2200 V to 220 V has a primary resistance of 1.0 Ω and a secondary resistance of 0.01 Ω. Find the equivalent secondary resistance and the full-load efficiency at 0.8 pf if the iron loss of the transformer is 80% of the full-load copper loss.

[Utkal Univ. Electrical Technology, 1998]

[**Ans.** 0.02 Ω, 97.93%]

37. The no-load transformation ratio in a 200 kVA, 50 Hz single-phase transformer is 4,000/250 V. For a maximum core flux of 0.06 Wb, find

(*i*) the efficiency at half the rated kVA and at unity power factor,

(*ii*) the efficiency at full load, 0.8 power factor lagging and

(*iii*) kVA load for maximum efficiency, given that the full-load copper losses are 1.8 kW and core losses are 1.5 kW.

[U.P.Technical Univ. Electrical Machines, 2001-2002]

[**Ans.** (*i*) 98.087% (*ii*) 97.98% (*iii*) 182.574 kVA]

38. A 11,000 / 230 V, 150 kVA, 1-phase, 50 Hz transformer has core loss of 1.4 kW and full-load copper loss of 1.6 kW. Determine (*i*) kVA load for maximum efficiency and value of maximum efficiency at unity pf. (*ii*) the efficiency at half-full load 0.8 pf leading.

[Nagpur Univ. Basic Elec. Machines, 1993]

[**Ans.** 140.0311 kVA; 98.04 % (*ii*) 97.09 %]

39. A 20 kVA, 50 Hz, 2000/200 V, single phase transformer has iron loss of 120 W and full-load copper loss of 300 W. Low-voltage side of the transformer is loaded at 0.8 lagging power factor. Calculate maximum efficiency of transformer.

[U.P. Technical Univ. Electrical Machines, 2006-07]

[**Ans.** 97.68%]

40. In a 25 kVA transformer the iron and full-load copper losses are 350 W and 400 W respectively. Calculate efficiency at (*i*) full load unity pf (*ii*) half full load, 0.8 pf lag. Also determine the load for maximum efficiency. **[M.D. Univ. Electrical Technology, December-2008]**

[**Ans.** (*i*) 97.087% (*ii*) 95.69%, 23.385 kVA]

41. The iron losses a 100 kVA transformer are 1,000 W. The full-load copper losses are 1.25 times the iron losses. Determine the load in kVA to obtain maximum efficiency of the transformer.

[B.P. Univ. of Technology, Electrical Machines-I, 2008]

[**Ans.** 89.44 kVA]

42. A 600 kVA single-phase transformer when working at unity pf has an efficiency of 92% at full load and also at half-full load. Determine the efficiency when it operates at unity pf and 60% of full load.

[U.P. Technical Univ., Electrical Machines, 2003-04; R.G. Technical Univ., Electromechanical Energy Conversion-I, June-2010]

[**Ans.** 92.328%]

43. A 100 kVA, 50 Hz, 440/11,000 V, 1-phase transformer has an efficiency of 98.5% when supplying full-load current at 0.8 pf lagging, and efficiency of 99% when supplying half of full-load current at unity power factor. Find the core losses and the copper losses corresponding to full-load current. At what value of load current will the efficiency be maximum.

[Rajasthan Technical Univ. Electrical Machines-I, 2008]

[Ans. 267.4 W, 950.6 W, 4.82 A]

44. The efficiency of a 400 kVA, 1-phase transformer is 98.77% when delivering full-load at 0.8 pf, and 99.13% at half-load and unity power factor. Calculate (*i*) the iron loss (*ii*) full-load copper loss.

[M.D. Univ. Electromechanical Energy Conversion, December-2005, December-2006]

[Ans. 1.013 kW, 2.967 kW]

45. A 500 kVA transformer has 90% efficiency at full load and at 70% of full load both at upf.

(*i*) Separate out the transformer losses.

(*ii*) Determine the transformer efficiency at 80% of full load, upf.

[R.G. Technical Univ., June-2014]

[Ans. (*i*) 32.68 kW, 22.876 kW (*ii*) 90.13%]

46. The efficiency of a 400 kVA, single phase transformer is 98.77% at full load 0.8 power factor and 99.13% at half full load unity power factor. Find:

(*i*) Iron losses at full and half full-loads

(*ii*) Cu losses at full and half full loads.

[U.P. Technical Univ. Electrical Engineering Second Semester, 2007-08]

[Ans. (*i*) 1.0117 kW each (*ii*) 2.9733 kW, 0.743325 kW]

47. The maximum efficiency of a 100 kVA, 1100/440 V 50 Hz transformer is 96%. This occurs at 75% of full load at 0.8 pf lagging. Find the efficiency of transformer at $\frac{3}{4}$ FL at 0.6 pf leading.

[G.B. Technical Univ. Electrical Engineering Second Semester, 2011-12]

[Ans. 94.74%]

48. In a 50 kVA, 11 kV/400 V transformer, the iron and copper losses are 500 W and 600 W respectively under rated conditions. Calculate the efficiency on unity power factor at full load. Find the load for maximum efficiency and the iron and copper losses corresponding to this load.

[GATE, 1998]

[Ans. 97.85%, 45.645 kVA, 500 W]

49. A transformer rated 200 V/50 V, 10 kVA has a core loss of 100 watts. What is the maximum efficiency of the transformer at 0.8 lagging power factor ? Assume full-load copper loss as 200 watts. At what load this maximum efficiency is obtained? How and why would your answer be different if the power factor were unity?

[R.G.P.V. Bhopal Basic Electrical Engineering, December, 2004]

[Ans. 7.07 kVA, 96.585%, 97.249%]

50. A 200 kVA transformer has an efficiency of 98 % at full load. If the maximum efficiency occurs at three quarters of full load, calculate the efficiency at half load. Assume negligible magnetising current and pf of 0.8 at all loads.

[Pb Univ. Electrical Technology Jan.-1991]

[Ans. 97.92%]

51. The maximum efficiency of a 500 kVA, 3,300 / 500 V, 50 Hz, single phase transformer is 97 % and occurs at 3/4 full load, unity power factor. If the impedance is 10 %, calculate the regulation at full load, power factor 0.8 lagging.

[Pb. Univ. Elec. Machines-I June 1993; U.P.T.U. Elec. Machines 2005-06; Madurai Kamraj Univ., November-1997]

[Ans. 7.52%]

52. A 25 kVA, 2,000/200 V, 50 Hz transformer offers its maximum efficiency at 80% of full-load. Its per unit resistance and impedance are 0.012 and 0.05 respectively. The no-load current is 4% of its full-load current at 0.2 power factor. Determine the efficiency and voltage regulation at half of full-load and 0.8 power factor lagging.

[M.D. Univ. Electrical Machines-I, December-2009]

[Ans. 97.294%, 1.98%]

53. The OC and SC tests on a 5 kVA, 230/110 V, 50 Hz transformer gave the following data:

OC test (HV side)	230 V,	0.6 A,	80 W
SC test (LV side)	6 V,	15 A,	20 W

Calculate percentage efficiency and regulation of a transformer on full load at 0.8 pf lagging.

[R.G. Technical Univ., December-2011]

[Ans. 93.8%, 12.6%]

54. The maximum efficiency of a 100 kVA, single phase transformer is 98 % and occurs at 80 % of full load. If the leakage impedance of the transformer is 5 %, find the voltage regulation at rated load of 0.8 pf lagging.

[Nagpur Univ. Elec. Machines -I, 1993]

[Ans. 3.74 %]

55. A 25 kVA, 50 Hz, 1–phase, 24,000 / 240 V distribution transformer has the following parameters referred to the low voltage side:

Total series impedance $(0.1 + j\,1.0)\ \Omega$; Shunt conductance 0.012 ℧; Shunt susceptance = 0.09 ℧.

(*a*) When the transformer is operated at 240 V and is supplying a unity power factor load, determine the maximum efficiency and the load at which it occurs.

(*b*) If the load is changed to zero pf load, calculate the kVA at which both the primary supply voltage and secondary output voltage are at their nominal values.

[GATE, 1996]

[Ans. 93.52%; 19.95 kW]

56. A 100 kVA, 1,000/220 V, 1-phase transformer gave the following test results:

OC Test (*hv*): 1 kV, 0.8 kW, 2A

SC Test (*lv*): 15 V, 1.5 kW, 455 A

Determine parameters of equivalent circuit referred to hv side and efficiency at full load 0.8 pf lagging.

[U.P. Technical Univ. Electromechanical Energy Conversion-I, 2011-12]

[Ans. 1.250 kΩ, 0.5455 kΩ, 0.1497 Ω, 0.6645 Ω, 97.2%]

57. Consider a 4 kVA, 200/400 V single phase transformer supplying full load current at 0.8 lagging power factor. The OC / SC test results are as follows:

OC Test :	200 V	0.8 A	70 W (lv side)
SC Test :	20 V	10 A	60 W (hv side)

Calculate efficiency, secondary voltage and current into primary at the above load.

Calculate the load at unity power factor corresponding to maximum efficiency.

[Nagpur Univ. Elec. Machines, 1993]

[Ans. 96.1%, 383.75 V, 4.32 kVA]

58. A single phase transformer has percentage regulation of 4 and 4.4 for lagging power factors of 0.8 and 0.6 respectively, The full load copper loss is equal to iron loss. Calculate (*i*) the lagging power factor at which full-load regulation is maximum (*ii*) The full-load efficiency at unity power factor.

[U.P.S.C. I.E.S. Elec. Engineering II, 1996]

[Ans. (*i*) 0.4472 lagging; 96.15%]

59. The open-circuit and short-circuit tests on a 10 kVA, 500/250 V, 50 Hz, 1-phase transformer gave the following results:

Open-circuit : 500 V, 2 A, 100 W (hv side)

Short circuit : 25 V, 20 A, 90 W (hv side)

Determine:

(*i*) components of the no load current,

(*ii*) approximate equivalent circuit referred to the primary side, and

(*iii*) regulation and efficiency at full load and 0.8 pf lagging.

[U.P. Technical Univ. Electromechanical Energy Conversion-I, 2001-02]

[Ans. (*i*) 0.2 A, 1.99 A (*iii*) 3.72%, 97.68%]

(*ii*)

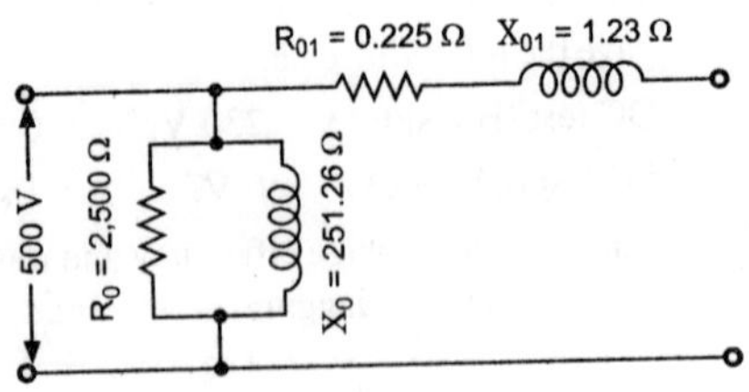

60. Find all-day efficiency of a transformer having maximum efficiency of 98 % at 15 kVA at unity power factor and loaded as follows:

12 hours — 2 kW at 0.5 pf lag:

6 hours — 12 kW at 0.8 pf lag

6 hours — at no-load. **[Nagpur Univ. Elec. Machines, 1993]**

[**Ans.** 95.31%]

61. What do you mean by all-day Efficiency ? A 5-kVA transformer has 34 W core loss and 40 W copper loss at full load. It operates at rated kVA and 0.8 power factor lagging for 6 hours, one-half rated kVA and 0.5 power factor lagging for 12 hours and no load for 6 hours. What is its all day efficiency ?

[Chhattisgarh Vivekanand Technical Univ., 2006-07]

[**Ans.** 97.07%]

62. A 100 KVA transformer has its maximum efficiency of 0.98 at full-load at unity power factor. During the day it is loaded as follows:

12 hours : 20 kW at power factor 0.5

6 hours : 45 kW at power factor 0.9

6 hours : 80 kW at power factor 0.8

Calculate all-day efficiency of the transformer.

[U.P. Technical Univ. Electrical Machines, 2008-09]

[**Ans.** 96.67%]

63. Define 'All day efficiency'. A transformer has its maximum efficiency of 0.98 at 15 kVA unity power factor load. Find the all day efficiency for the following loading cycle : 2 kW at 0.5 pf for 12 hours, 12 kW at 0.8 pf for 6 hours and 18 kW at 0.9 pf for 6 hours.

[U.P. Technical Univ. Electromechanical Energy Conversion-I, 2006-07]

[**Ans.** 96.98%]

64. Two 1-phase transformers with equal turns have impedances of $(0.5 + j\,3)$ ohms and $(0.6 + j\,10)$ ohms with respect to the secondary. If they operate in parallel, determine how they will share total load of 100 kW at pf 0.8 lagging.? **[Madras Univ. Electrical Technology, 1987]**

[**Ans.** $96\angle{-35.46^\circ}$ kVA or 78.2 kW

$29.128\angle{-41.91^\circ}$ kVA or 21.8 kW]

65. Two transformers each of 800 kVA are connected in parallel. One has a resistance and reactance of 1% and 4% respectively and the other has 1.5% and 6% respectively. Calculate the load shared by each transformer and the corresponding pf when the total load shared by the two transformers is 1,000 kVA at 0.8 pf lagging.

[U.P. Technical Univ. Electromechanical Energy Conversion-I, 2001-02]

[**Ans.** $600\angle{-36.87^\circ}$ kVA or 480 kW at 0.8 pf lag

$400\angle{-36.87^\circ}$ kVA or 320 kW at 0.8 pf lag]

66. Two transformers have following characteristics

No. 1 — % IR = 1.0% and % I X = 5.0%

No. 2 — %IR = 1.5% and % I X = 4.0%

How they will share a load of 100 kVA at 0.8 pf lagging

[Pb Univ. Elec. Machines-I, June-1992]

[**Ans.** $45.74\angle{-41.91^\circ}$ kVA; $54.59\angle{-32.66^\circ}$ kVA]

67. Two single-phase transformers A and B have the following parameters:

Transformer A : 10 kVA, 400 V/200 V, percentage resistance and percentage reactance 3% and 4% respectively;

Transformer B : 5 kVA, 400 V/200 V, percentage resistance and percentage reactance are 4% and 3% respectively.

These two transformers are connected in parallel and they share a common load of 12 kW at a power factor of 0.8 lagging. Determine the active and reactive power delivered by transformer A.

[GATE, 1999]

[**Ans.** 7.466 kW; 6.787 kVAR (inductive)]

68. Two single phase transformers rated at 500 kVA and 400 kVA respectively are connected in parallel to supply a load of 1,000 kVA at 0.8 lagging power factor. The resistance and reactance of the first transformer are 2.5% and 6% respectively, and of the second transformer 1.6% and 7% respectively. Calculate the kVA loading and the power factor at which each transformer operates.

[**Ans.** 582 kVA at pf 0.84 (lag); 421.5 kVA at pf 0.737 (lag)]

69. Two single phase transformers, one of 1,000 kVA and the other of 500 kVA are connected in parallel to the same bus-bars on the primary side; their no-load secondary voltages being 500 V and 510 V respectively. The impedance voltage of the first transformer is 3 % and that of the second 5 %. Assuming the ratio of resistance to reactance the same and equal to 0.4 in each, what will be the cross-current when the secondaries are connected in parallel.

[**Ans.** 315.46 A]

70. Two single-phase transformers, rated 1,000 kVA and 500 kVA respectively, are connected in parallel on both HV and LV sides. They have equal voltage ratings of 11 kV/400 V and their per unit impedances are $(0.02 + j\,0.07)$ and $(0.025 + j\,0.0875)$ respectively. What is the largest value of the unity power factor load that can be delivered by the parallel combination at the rated voltage?

[U.P. Technical Univ. Electromechanical Energy Conversion-I, 2004-05]

[**Ans.** 1,400 kW at unity pf]

❑❑❑

13 CHAPTER

Three-Phase Transformers

13.1 THREE-PHASE TRANSFORMERS

The voltage transformation in a 3-phase network can be had by means of either a bank* of three single-phase transformers or a single 3-phase transformer. A single unit 3-phase transformer consists essentially of three single-phase transformers with their three cores united into a single-core assembly. Like single-phase transformers, the three-phase transformers are also of the *core type or shell type.*

The unique arrangement of the windings and the core in a 3-phase transformer makes it possible to save a great deal of iron by interlinking the magnetic circuits so that the same iron is used by the three phases simultaneously (a 3-phase transformer used for the same purpose requires one third as much iron for the magnetic circuit as a bank of three single-phase transformers). Proper flux densities are maintained because the three phase currents are displaced 120 electrical degrees with respect to each other.

Moreover, since the entire three phase assembly is reduced in size and is placed in a single tank, additional savings result because less transformer oil is used, fewer transformer bushings are required (all polyphase connections are made inside the tank), and the complete transformer occupies less floor space than a bank of three single-phase transformers.

A 3-phase transformer is cheaper (about 15%) than a bank of three single-phase transformers. It is due to saving in cost of the iron core, of the tank and oil, of the bushings and of the auxiliary apparatus.

A 3-phase transformer is slightly more efficient than a bank of three single-phase transformers. This is due to the fact that it has shorter magnetic path and consequently the core volume and hence the core loss is smaller.

Installation and operational costs are also smaller in case of 3-phase units because in case of three-phase transformers there is only one unit to install and operate.

The three-phase transformer, however, does present some disadvantages. It is larger and heavier than any one single-phase transformer, and is more difficult to handle, ship and set in place for operation. In underground applications (mines) a bank of single-phase units maybe preferred on account of ease of transport. Secondly three-phase transformers are much more difficult and costly to repair than are single-phase units. Moreover, in case of breakdown in any coil, the entire unit must be removed from service, rather than just one transformer, as is the case when single-phase transformers are employed. Thus the cost of spare units and of repairs is greater with the use of three-phase transformer. The bank also offers the advantage of a derated open-delta operation when one single-phase unit becomes inoperative (In case of delta-connected banks, the faulty unit is simply disconnected and continuity of supply is maintained by operating the bank in open delta at 57.7% of the original power capacity till the faulty single-phase unit is replaced.)

The advantages of three-phase transformer (such as lesser weight, lower initial cost, lesser space requirement) overweighs its disadvantages and so it is a common practice to use a three-phase transformer unit.

13.1.1 Three-Phase Core Type Transformer. A 3-phase core type transformer is shown in Fig. 13.1. The core consists of three legs with the magnetic circuit completed through two yokes (one at the top and another at the bottom.) A primary and

* *A transformer bank* used in 3-ϕ system consists of three independent 1-phase transformers with their primary and secondary windings connected either in star or in delta.

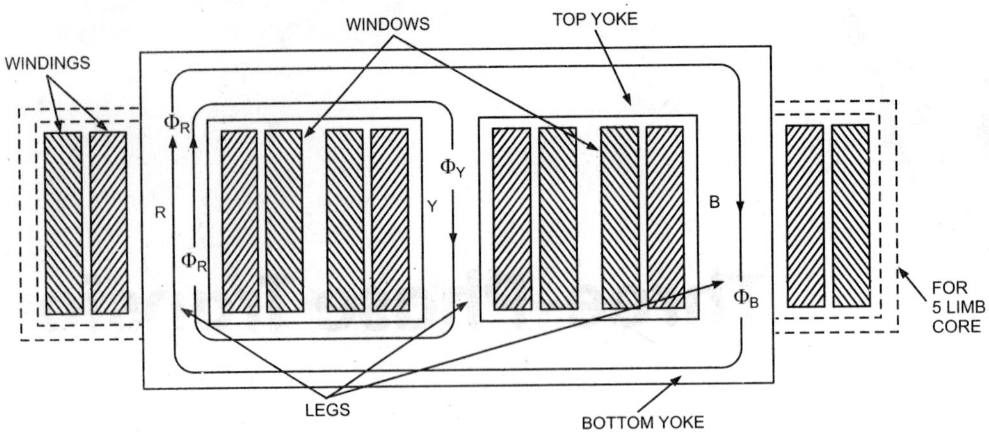

Fig 13.1 *Three-Phase Core Type Transformer*

secondary of one phase is wound on each limb (on each limb the lv winding is placed next to the core and hv winding is placed over the lv winding with suitable insulation between the core and lv winding as well as between lv and hv windings). It is possible to build core with only 3 legs since the fluxes established by three windings are 120° apart in time phase. Two core legs act as the return for the flux in the third leg. Flux flows up each leg in turn and down the other two legs in general, so that the magnetic circuits for different phases are is series and, therefore, interdependent. The addition of two half-section outer limbs as indicated makes a five limb core, the windings being confined to the middle three limbs. This is done to reduce the height of the yoke and hence overall height of the transformer. The magnetic circuits for each phase are then virtually independent. In either case, the magnetic circuits of the three phases are somewhat unbalanced, the middle phase having less reluctance than the outer two. This makes the magnetising current of the middle phase slightly less than that of the two outer phases, but the magnetizing currents are so small that this has no noticeable effect on the operation of the transformer.

It should be noted that the transformer has only two windows and each window contains two primary and two secondary windings.

13.1.2. Three-Phase Shell Type Transformer. A shell type transformer, which is not commonly used, is shown in Fig. 13.2 (*a*). It appears like three single-phase shell type cores built on top of one another. Owing to the joint use of the magnetic paths between the coils, there is less iron in this type of transformer than in three equivalent single-phase units. Some economy* in the common yoke sections can be further affected by reversing the middle coils as illustrated so that the flux carried corresponds to the phasor sum of the time phasors $\Phi_{R/2}$ and $\Phi_{Y/2}$ instead of their difference. The three phases, in this case, are more independent than they are in the core type transformers, because each phase has a magnet circuit independent of the other, if saturation effects in the common paths are neglected.

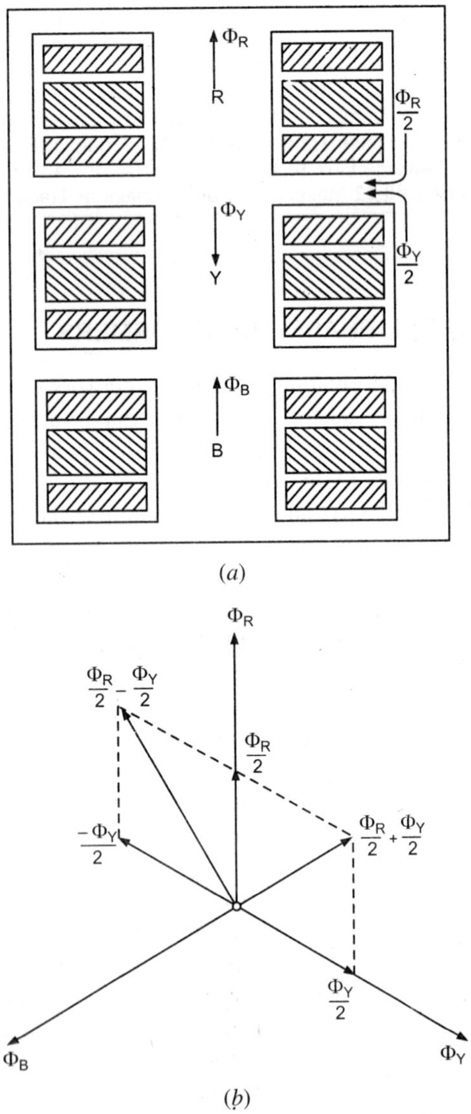

Fig 13.2 *Three-Phase Shell Type Transformer*

* With three phase coils wound in the same direction, the flux in the common area between adjacent phases, being the phasor difference of the two phase fluxes, is $\sqrt{\frac{3}{2}}\,\Phi$ or 0.866 of the flux in the middle leg and for maintaining uniform flux density less iron will be required in the common legs. With middle winding, phase Y, reversed, the flux in the common leg, being the phasor sum of the two phase fluxes, is 0.5 Φ [Fig. 13.2 (*b*)]. Thus there is a considerable saving in the amount of core material.

One advantage of the shell type transformer over the core type transformer is that the former one can be operated in open delta should one of the windings be damaged; operation can proceed at 57.7 per cent of the transformer rating if the primary and secondary of the damaged phase are disconnected from the circuit and are then short circuited.

13.2 PHASING OUT OF PRIMARY AND SECONDARY WINDINGS WITH RESPECT TO EACH OTHER

In manufacture, where all connections are traceable, the phases on primary and secondary sides may readily be grouped. Otherwise all phases are short circuited except a primary and a supposedly secondary. A small direct current is circulated in the primary and a voltmeter is connected across the secondary. A momentary deflection of the voltmeter on making and breaking of primary current confirms that the secondary is corresponding to the primary chosen. The test is repeated for other windings.

For this test, all the phase ends should be separate and be available. Difficulty may be experienced with internally connected zig-zag (interconnected star) arrangements in three-phase windings.

13.3 THREE-PHASE TRANSFORMER CONNECTIONS

The standard lettering for phase is A B C. Capital letters A, B, C are used for hv winding and lower case letters *a*, *b*, *c* for lv winding with (A), (B), (C) for a tertiary winding, where provided. Neutral terminals precede line terminals. Each winding has two ends designated by the subscript numbers 1, 2, or if there are intermediary tappings (or there are sections of the same winding) these are numbered in order of their separation from end 1, as illustrated in Fig. 13.3.

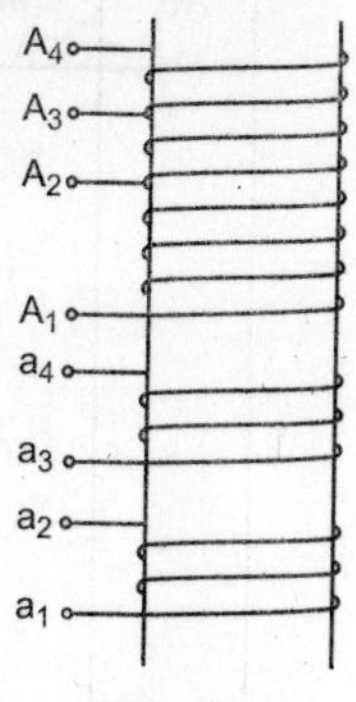

Fig 13.3

The interconnection of the phase windings to give a three phase, three- or four-wire supply provides three alternative modes of connection (*i*) mesh or delta (*ii*) star (*iii*) zig-zag, and each of these can be achieved in two ways. For example a star-connection may be had by joining together A_1, B_1, and C_1, to form the neutral and using A_2, B_2 and C_2 as line terminals. Alternatively A_2, B_2, and C_2 may be joined to give neutral and A_1, B_1, and C_1 may be used as line terminals.

Now since primary and secondary can be treated in either of these ways, at least twelve methods of connection are possible.

The twelve methods are illustrated in Fig. 13.4. These are arranged in four main groups, as given below, according to the

Main Group No.	Phase Displace-ment	Phasor Group	Symbol	Phase Diagrams of Induced Voltages (HV, LV) and Winding Connections and Relative Position of Terminals (HV, LV)
1	0°	1_1	Yy0	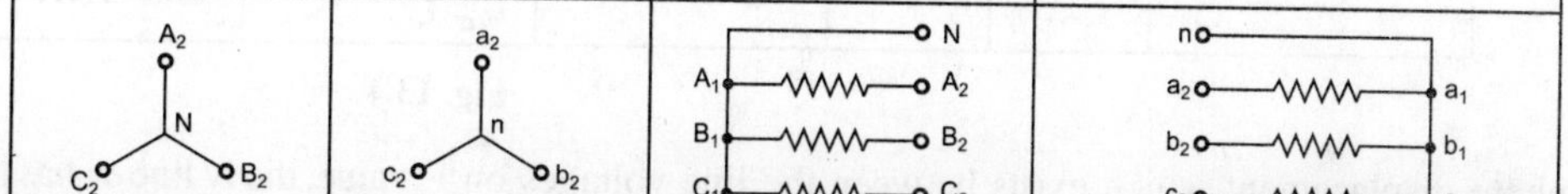
		1_2	Dd0	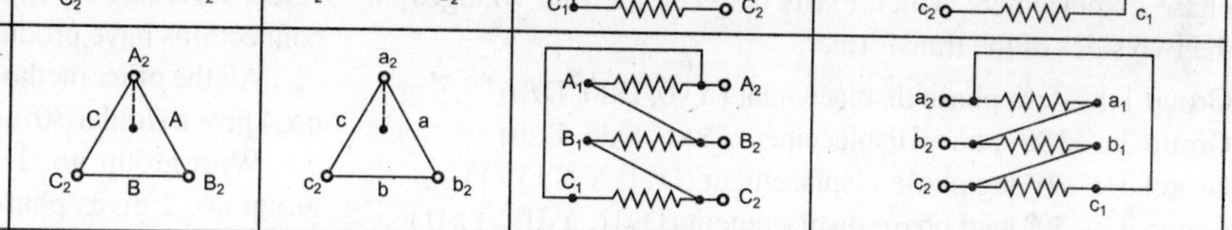
		1_3	Dz0	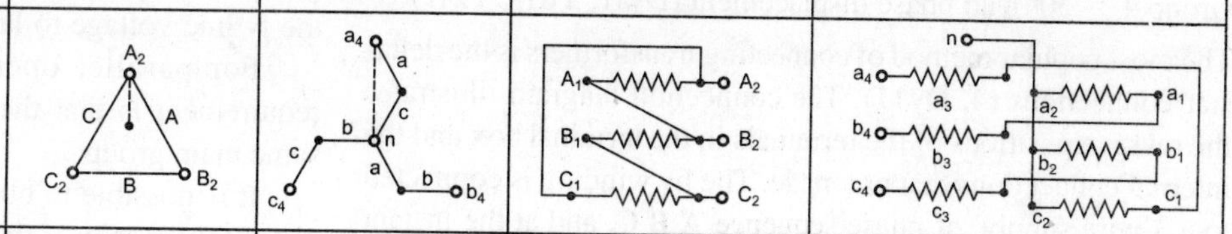
2	180°	2_1	Yy6	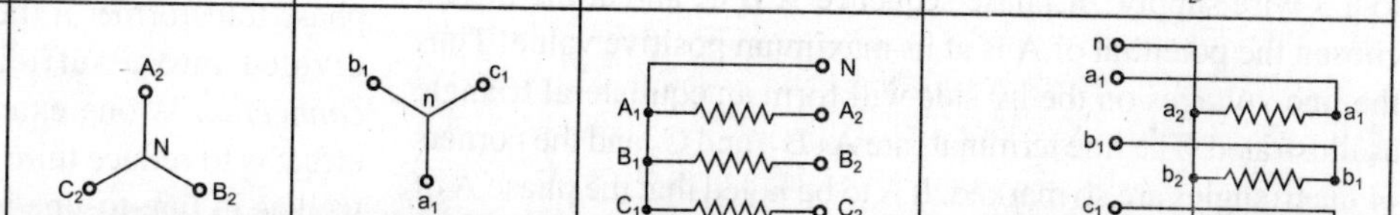
		2_2	Dd6	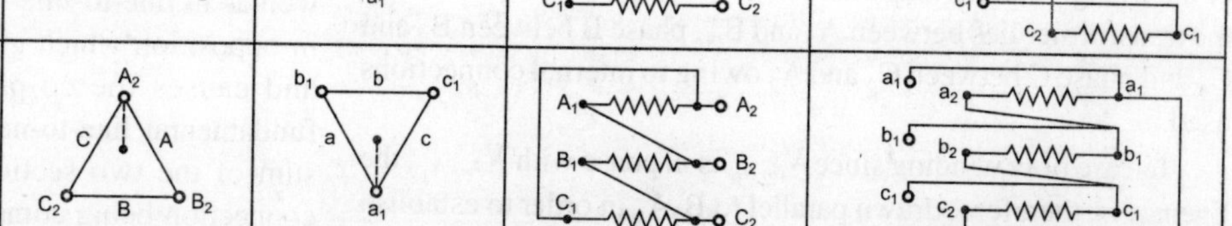
		2_3	Dz6	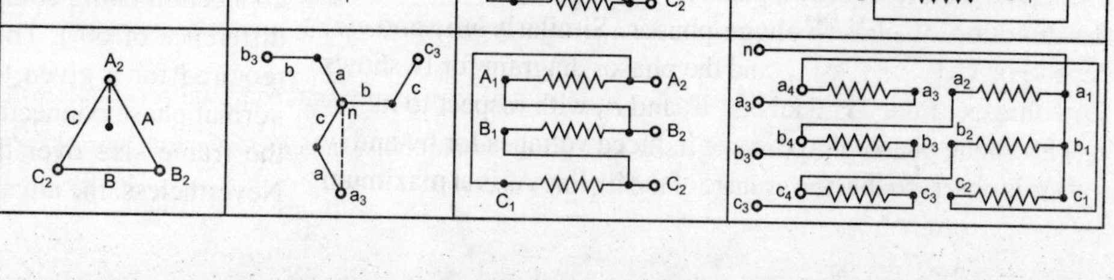

Main Group No.	Phase Displacement	Phasor Group	Symbol	Phase Diagrams of Induced Voltages		Winding Connections and Relative Position of Terminals	
				HV	LV	HV	LV
3	–30°	3_1	Dy1	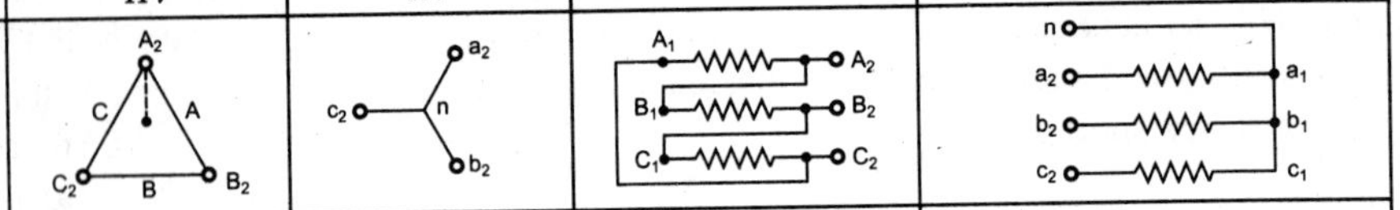			
		3_2	Yd1	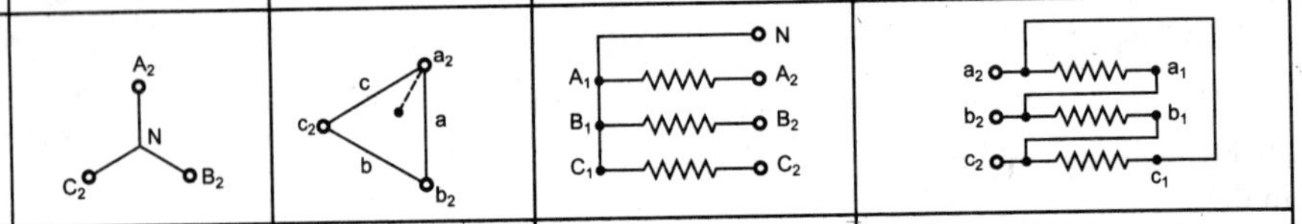			
		3_3	Yz1	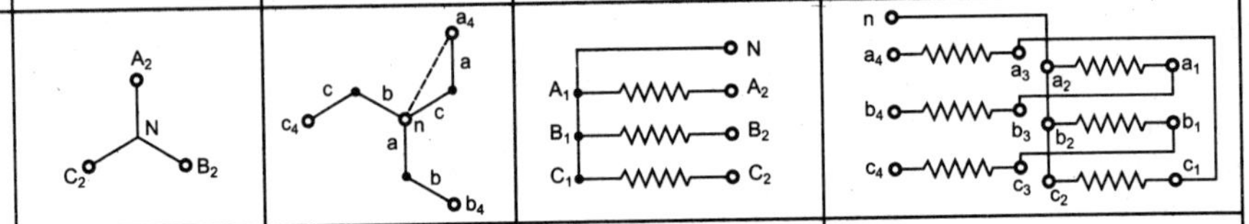			
4	+30°	4_1	Dy11	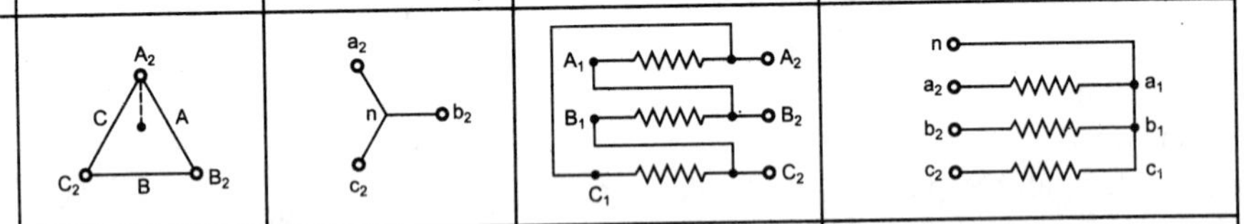			
		4_2	Yd11	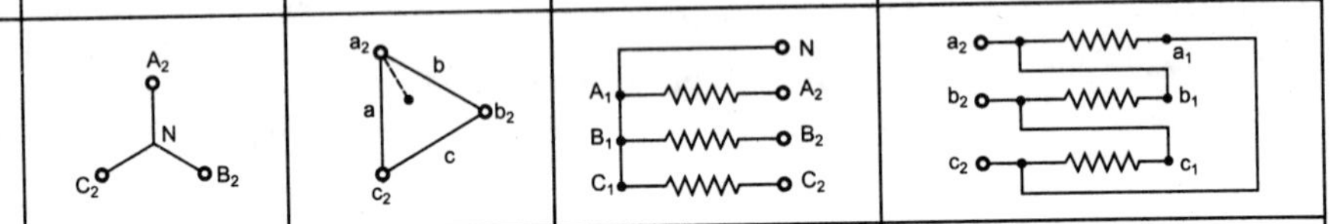			
		4_3	Yz11	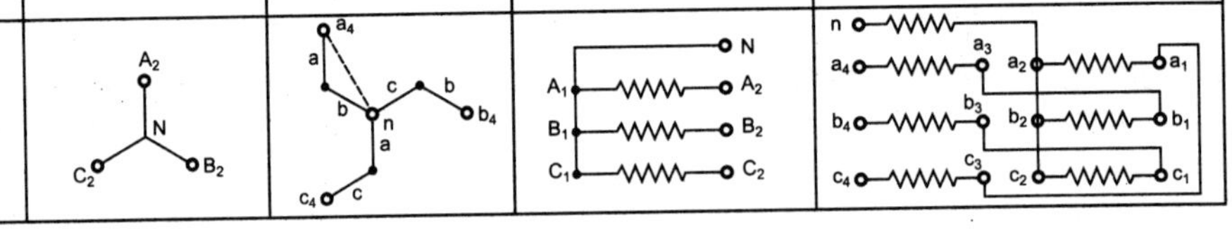			

Fig. 13.4

phase displacement which exists between the line voltages on the two sides of the transformer.

Group 1 : Zero phase displacement (Yy0, Dd0, Dz0)
Group 2 : 180° phase displacement (Yy6, Dd6, Dz6)
Group 3 : 30° lag phase displacement (Dy1, Yd1, Yz1)
Group 4 : 30° lead phase displacement (Dy11, Yd11, Yz11)

The most popular method of connecting transformers is the delta-star connections (4, Dy11). The connection diagram illustrates the relative positions of the terminals in the terminal box and the internal connections that are made. The hv winding is connected to a 3-wire supply, of phase sequence A B C, and at the instant chosen the potential of A is at its maximum positive value. Thus the line voltages on the hv side will form an equilateral triangle as illustrated. The line terminals are A_2 B_2, and C_2 and the corners of the triangles are so marked. It is to be noted that the phase A of the transformer lies between A_2 and B_2, phase B between B_2 and C_2 and phase C between C_2 and A_2 owing to internal connections used.

In case of lv winding since $V_{a_2 a_1}$ is in phase with $V_{A_2 A_1}$, the line na_2 is, therefore, drawn parallel to B_2A_2 in order to establish the direction of phase-voltage phasor. Similarly nb_2 and nc_2 depend on C_2B_2 and A_2C_2, and the phasor diagram for lv shows the voltages of the terminals a_2, b_2 and c_2 with respect to n.

From the phasor diagrams of induced voltages for hv and lv sides it is observed that the voltage of the hv line A_2 is at maximum value, the lv line a_2 has been advanced 30°. In fact the transformer connections have produced this displacement of 30° for all lines.

All the other methods of connections in the same main group no. 4 give a similar 30° advance which is the basis of classification.

With group no. 1 there is no phase displacement, but the group no. 2 gives phase displacement of 180°. Group 3 causes the lv line voltage to lag by 30°.

For parallel operation of transformers the essential requirement is that the transformer connections belong to the same main group.

It is possible to obtain any desired phase shift with a three phase transformer if the primary and/or secondary windings are divided into a sufficient number of sections. The *zig-zag connection* is one example of a sectionalized winding and its effect is to reduce third harmonics in line-to-neutral voltages, as well as in line-to-line voltages. The half sections are connected in opposition which gives the larger total fundamental voltage and causes the co-phasal third harmonics to cancel. The fundamental line-to-neutral voltage is 0.866 times the arithmetic sum of the two section voltages (phase voltages in zig-zag, connection being composed of two half-voltages with a phase difference of 60°). This means that 15 per cent more turns are required for a given total voltage per phase compared with a normal phase connection, which may necessitate an increase in the frame size over that normally used for the given rating. Nevertheless, the advantages of zig-zag arrangement may offset the cost; unbalanced loads on the secondary side are distributed better on the primary side. The zig-zag/star connection has been employed where delta connections were mechanically weak (on account of large number of turns and small copper sections) in hv transformers; also for rectifiers.

13.3.1. Choice Between Star and Delta Connections. In star-connection with earthed neutral the maximum voltage of the phase winding to ground is $1/\sqrt{3}$ (or 57.7%) of the line voltage, while in delta connection this is equal to the line voltage in case of earthing of one of the line during a fault. Therefore, phase voltage, turns and insulation are less in star connection than those in delta connection and for very high voltage transformers the star-connection on hv side is about 10% cheaper than delta connection. A delta-connected primary is necessary for a star-connected lv secondary feeding mixed three-phase and single-phase loads. In case of unbalanced 1-phase loads on secondary, delta-connected primary is necessitated to permit the flow of zero-sequence current

$$\mathbf{I}_0 = \frac{\mathbf{I}_n}{3} = \frac{1}{3}(\mathbf{I}_A + \mathbf{I}_B + \mathbf{I}_C) \text{ so that } \mathbf{I}_A + \mathbf{I}_B + \mathbf{I}_C = 0$$

This means that only *positive* and *negative sequence* currents flow in the lines on the delta side.

This could also be achieved by star-connected primary provided the primary and secondary star points are earthed, but this is not recommended because of flow of ground current for unbalanced secondary loads.

13.3.2. Choice of Transformer Connections.

1. Star-Star (Yy0 or Yy6) Connections. This type of transformer is most economical for small current and high voltage transformers. The advantages and disadvantages of such an arrangement are given below :

Advantages. (*i*) The number of turns per phase and the amount of insulation is minimum because phase voltage = $1/\sqrt{3}$ of line voltage.

(*ii*) There is no phase displacement between the primary and secondary voltages.

(*iii*) With star points available on both sides it is possible to provide a neutral connection.

Disadvantages. (*i*) If the load on the secondary side of the transformer is unbalanced, the phase voltages of load side change unless the load star point is earthed. The difficulty of shifting neutral can be overcome by connecting the primary star point to the star point of the generator.

(*ii*) The primary of the transformer draws a magnetizing current which has third and fifth harmonics. If neutral of primary winding is not connected to neutral of generator, the third and fifth harmonics current cannot flow hence the flux in the core cannot be of sinuosoidal wave and, therefore, the voltages will be distorted. By connecting primary neutral to the generator neutral, the path for return of these third and fifth harmonics current is provided and, therefore, the trouble of distortion of voltages is overcome.

(*iii*) Even if neutral point of primary is connected to neutral of generator or earthed, still if third harmonic is present in the alternator voltage form, it will appear on secondary side. Though the secondary line voltages do not contain third harmonic voltage; but the 3rd harmonic voltages are additive in the neutral and causes the current in the neutral of tripple frequency (3rd harmonic) which will cause interference to the nearby communication system.

Third harmonic phase voltages may be high in shell type 3-phase transformers.

The star-star connection is rarely used* owing to the difficulties associated with the exciting current.

2. Delta-Delta (Dd0 or Dd6) Connections. This arrangement is generally used in systems which carry large currents on low voltages and especially when continuity of service must be maintained even though one of the phase develops fault. As will be pointed out subsequently, when one of the phases is removed in Δ-Δ connected transformers, operation continues on what is known as *open delta*. By being operated in this way, the transformer delivers three phase currents and voltages in their correct phase relationship but the capacity of the transformer is reduced to 57.7% of what it was with all three phase windings sound.

Advantages. (*i*) There is no phase displacement between the primary and secondary voltages.

(*ii*) There is no distortion of flux, because the third harmonic component of magnetising current can flow in the delta-connected primary winding without flowing in the line wires.

(*iii*) The cross section of conductor is reduced because the phase current is $1/\sqrt{3}$ times of line current.

(*iv*) No difficulty is experienced due to unbalancing of loads on secondary side.

Disadvantages. (*i*) More insulation is required in comparison to Y-Y connections.

(*ii*) The absence of star point may be disadvantageous. If one line gets earthed due to fault, maximum voltage between windings and core will be full line voltage.

3. Star-Delta (Yd1 or Yd11) Connections. This type of transformer connection is used principally where the voltage is to be stepped down as, for example, at the end of a transmission line. In this type of connection of transformers the neutral of the primary winding is earthed. In this system line voltage ratio is $1/\sqrt{3}$ times of transformer turn-ratio and secondary line voltages have a phase shift of ± 30° with respect to primary line voltages. On the hv side of the transformer, insulation is stressed only to the extent of 57.7% of the line voltage and, therefore, there is some saving in the cost of insulation.

4. Delta-Star (Dy1 or Dy11) Connections. This scheme of connection is generally used where it is necessary to step up the voltage as for example, at the beginning of a ht transmission system. On the high voltage side of the transformer, insulation is

---

* Now there is a favourable trend for reasons of economy.

stressed only to the extent of 57.7% of line to line voltage. Another advantage is that the neutral point is stable and will not "float" when load is unbalanced. There is no distortion of flux because the existence of a Δ connection allows a path for the 3rd harmonic contents.

In this scheme of connection the line voltage ratio is $\sqrt{3}$ times of transformer turn-ratio and the secondary line voltages have a phase shift of ± 30° with respect to primary line voltages.

The neutral of the secondary is grounded to provide 3-phase, 4-wire system and this scheme of connections is widely used in distribution systems because it can be used to serve both the three-phase power equipment and single-phase lighting circuits. In such case the 11,000/433 V transformer connections provide a four-wire secondary, with neutral supplying the 4th wire. The 3-phase equipment is connected to the line wires to operate at 415 V while the lighting equipment is connected between one of the line wires and neutral to operate at 240 volts.

Star-delta or delta-star connected transformers cannot be operated in parallel with star-star or delta-delta connected transformers even though the voltage ratio are correctly adjusted as there will be a 30° phase difference between corresponding voltages on the secondary side.

5. Delta-Zigzag Star Connection. The winding of each phase on the star side is divided into two equal halves with labelling as shown in Fig. 13.5. Each leg of the star connection is formed by using halves from two different phases, as shown in Fig. 13.6. The phasor diagram for the connection (0° phase shift) is given in Fig. 13.7. Reversal of connection on the star side provides the 180° phase shift connection. If voltage of each half of the transformer secondary is V_2, the phase voltage of the

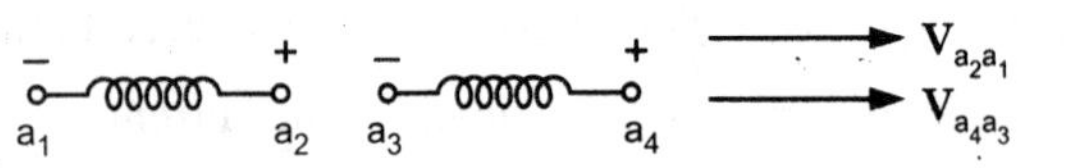

Fig. 13.5

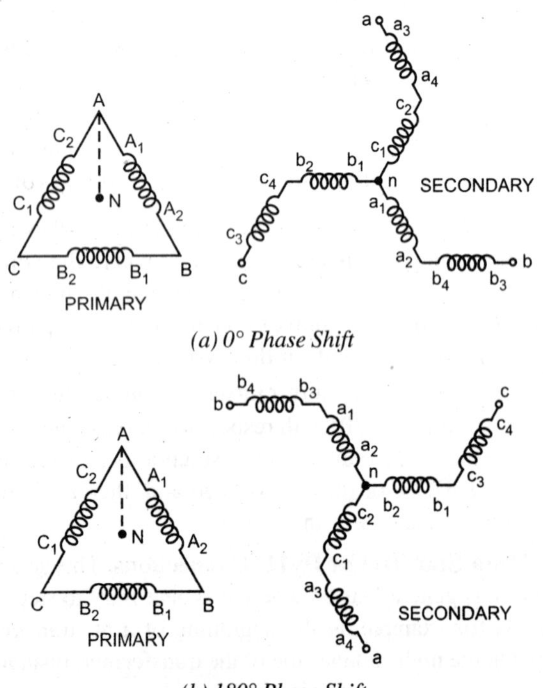

(a) 0° Phase Shift

(b) 180° Phase Shift

Fig. 13.6 *Delta-Zigzag Star Connection of Transformer*

secondary system is $\sqrt{3}V_2$ and line voltage is $3V_2$. If I_2 be the rated current of each secondary winding, the volt-ampere provided by the three phase transformer and banks of single phase transformers is $3\sqrt{3}V_2I_2$ while the sum of the individual units is $6V_2I_2$. Thus only $\frac{\sqrt{3}}{2}$ *i.e.* 0.866 of the combined individual capacity is available.

There is no phase difference between phase voltages of the primary and secondary systems in the connection shown in Fig. 13.6 (*a*) while there is a phase shift of 180° in the connection shown in Fig. 13.6 (*b*). If the transformer is polyphase and the delta is on the hv side which is usually the case, the connection shown in Fig. 13.6 (*a*) is termed DZ0 and that shown in Fig. 13.6 (*b*) is termed DZ6. Delta-zigzag connection is useful for supply of power to rectifiers.

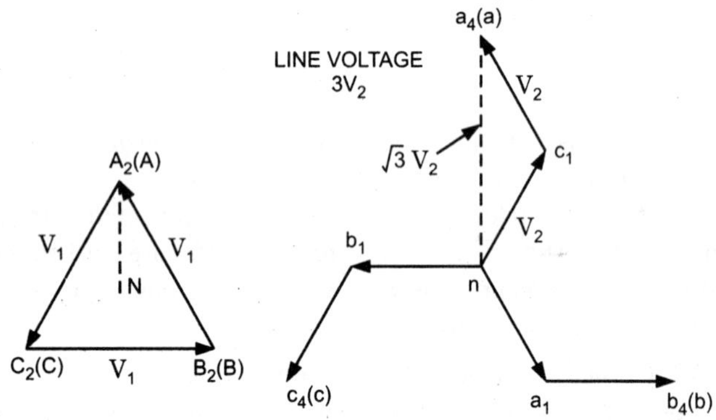

Fig. 13.7 *Phasor Diagram For Delta Zigzag Star 0°-Connection*

6. Star-Zigzag Star Connection. Star-Zigzag star connection is shown in Fig. 13.8. Zigzag connection has already been explained above. In Fig. 13.8 (*a*) the phase voltage of

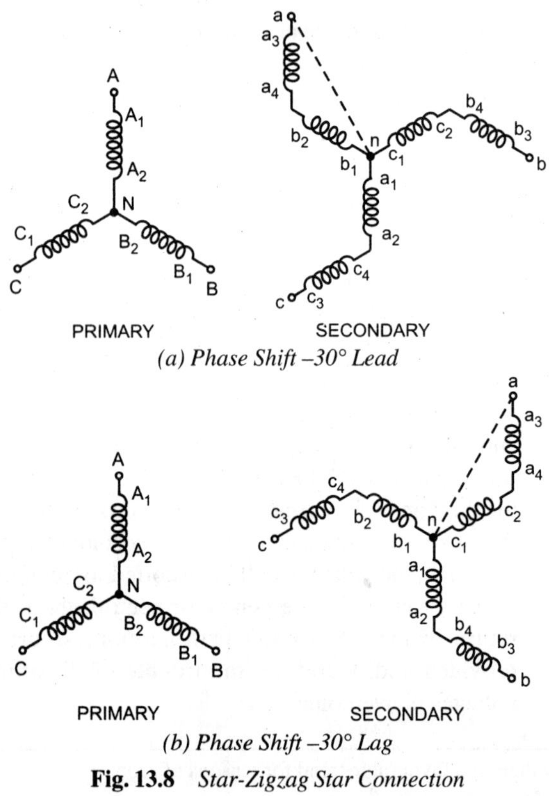

(a) Phase Shift –30° Lead

(b) Phase Shift –30° Lag

Fig. 13.8 *Star-Zigzag Star Connection*

Table 13.1

Type of Connection	*Primary Side*			*Secondary Side*			
	Line Voltage	*Phase Voltage, V_P*	*Phase Current I_P*	*Phase Voltage*	*Phase Current*	*Line Voltage*	*Line Curernt*
Star-Star	V_L	$V_L/\sqrt{3}$	I_L	$\frac{N_2}{N_1}\times\frac{V_L}{\sqrt{3}}$	$\frac{N_1}{N_2}I_L$	$\frac{N_2}{N_1}V_L$	$\frac{N_1}{N_2}I_L$
Delta-Delta	V_L	V_L	$\frac{I_L}{\sqrt{3}}$	$\frac{N_2}{N_1}\times V_L$	$\frac{N_1}{N_2}\frac{I_L}{\sqrt{3}}$	$\frac{N_2}{N_1}\times V_L$	$\frac{N_1}{N_2}I_L$
Star-Delta	V_L	$\frac{V_L}{\sqrt{3}}$	I_L	$\frac{N_2}{N_1}\times\frac{V_L}{\sqrt{3}}$	$\frac{N_1}{N_2}I_L$	$\frac{N_2}{N_1}\times\frac{V_L}{\sqrt{3}}$	$\sqrt{3}\times\frac{N_1}{N_2}I_L$
Delta-Star	V_L	V_L	$\frac{I_L}{\sqrt{3}}$	$\frac{N_2}{N_1}\times V_L$	$\frac{N_1}{N_2}\times\frac{I_L}{\sqrt{3}}$	$\sqrt{3}\,\frac{N_2}{N_1}\times V_L$	$\frac{N_1}{N_2}\times\frac{I_L}{\sqrt{3}}$

secondary system leads that of primary by 30° while in Fig. 13.8 (*b*) they lag by 30°. If the zigzag is on low voltage side, which is normally the case and the transformer is polyphase, then connection shown in Fig. 13.8 (*a*) is designated as YZ11 and that shown in Fig. 13.8 (*b*) as YZ1. There is similar reduction in available capacity as delta-zigzag star connection.

The zigzag connection in this case affects a reduction of third harmonic voltage and simultaneously permits unbalanaced loading even though the primary is star-connected with isolated neutral.

13.3.3. Comparison of Voltage and Current Relationships For Different Types of Transformer Connections. Table 13.1 provides the comparison of voltage and current relationships for different types of transformer connections.

Example 13.1. A 3-phase 50 Hz transformer has an iron cross section of 400 cm^2 (gross). If the flux density be limited to 1.2 Wb/m^2, find the number of turns per phase on high and low-voltage windings. The voltage ratio is 2,200/110 V, the hv side being connected in star and low voltage in mesh.

Solution: Net iron cross section, $A_i = 400 \times 0.9 = 360$ cm^2 or 0.036 m^2

assuming stacking factor, the ratio of net iron cross section to gross iron cross section to be 0.9

Peak value of flux, $\Phi_{max} = B_{max} \times A_i = 1.2 \times 0.036 = 0.0432$ Wb

HV side phase voltage, $E_{P1} = \frac{2,200}{\sqrt{3}} = 1,270$ V **Ans.**

LV side phase voltage, $E_{P2} = 110$ V

Turn per phase on low voltage winding,

$$N_2 = \frac{E_{P2}}{4.44\Phi_{max}\times f} = \frac{110}{4.44\times 0.0432\times 50} = 11.47 \text{ say } 12 \textbf{ Ans.}$$

Turn per phase on high voltage winding,

$$N_1 = N_2\times\frac{E_{P1}}{E_{P2}} = 12\times\frac{1,270}{110} = 138$$

Example 13.2. A 3-phase, delta/star connected 11 kV/400 V, 50 Hz transformer takes a line current of 5 amp, when secondary load of 0.8 lagging pf is connected. Determine each coil current and output of transformer. [*U.P. Technical Univ. Electrical Machines, 2003-04*]

Solution: Primary line voltage, $V_{L1} = 11,000$ V

Primary line current, $I_{L1} = 5$ A

Power factor of load, $\cos\phi = 0.8$

Coil current on primary side $= I_{P1} = \frac{I_{L1}}{\sqrt{3}} = \frac{5}{\sqrt{3}} = 2.887$ A **Ans.**

Secondary line voltage, $V_{L2} = 400$ V

$$\text{Turn-ratio, } \frac{N_2}{N_1} = \frac{1}{\sqrt{3}}\frac{V_{L2}}{V_{L1}} = \frac{1}{\sqrt{3}}\times\frac{400}{11,000} = 0.021$$

Coil current on secondary side

$$= I_{P2} = I_{L2} = I_{L1}\times\frac{N_1}{\sqrt{3}N_2}$$

$$= 5\times\frac{1}{\sqrt{3}\times 0.021} = 137.5 \text{ A } \textbf{Ans.}$$

$$\text{Transformer output} = \frac{\sqrt{3}\,V_{L2}\,I_{L2}\cos\phi}{1,000}\text{ kW}$$

$$= \frac{\sqrt{3}\times 400\times 137.5\times 0.8}{1,000} = 76.21 \text{ kW } \textbf{Ans.}$$

Example 13.3. An ideal 3-phase delta/star step-down transformer delivers power to a balanced 3-ϕ load of 120 kVA at 0.8 pf. The input line voltage is 11 kV and the turns ratio of the transformer, phase to phase is 10. Determine the line voltage, line currents, phase voltages and phase currents on both the primary and secondary sides.

[*Rajasthan Technical Univ. Electrical Machines-I, 2011*]

Solution: Power output, $P_{out} = 120$ kVA

Load power factor, $\cos\phi = 0.8$

Primary line voltage, $V_{L1} = 11$ kV $= 11,000$ V **Ans.**

Primary phase voltage, $V_{P1} = V_{L1} = 11,000$ **Ans.**

∵ primary side is delta-connected

Secondary phase voltage, $V_{P2} = \frac{V_{P1}}{\text{Turn-ratio}} = \frac{11,000}{10} = 1,100$ V **Ans.**

Secondary line voltage, $V_{L2} = \sqrt{3}\,V_{P2} = \sqrt{3}\times 1,100 = 1,905$ V

∵ Secondary is delta-connected

$$\text{Secondary line current, } I_{L2} = \frac{\text{Power output in kVA}\times 1,000}{\sqrt{3}\,V_{L2}}$$

$$= \frac{120\times 1,000}{\sqrt{3}\times 1,905} = 36.37 \text{ A } \textbf{Ans.}$$

Secondary phase current, $I_{P2} = I_{L2} = 36.37$ A **Ans.**

Primary phase current, $I_{P1} = \frac{I_{P2}}{10} = \frac{36.37}{10} = 3.637$ A **Ans.**

Primary line current, $I_{L1} = \sqrt{3}\,I_{P1} = \sqrt{3}\times 3.637 = 6.3$ A **Ans.**

Example 13.4. The HV terminals of a 3-phase bank of three single-phase transformers are connected to a 3-wire, 3-phase, 11 kV (LL) system. The LV terminals are connected to a 3-wire, 3-phase load rated of 1,000 kVA and 2200 V line-to-line (LL). Specify the voltage, current and kVA ratings of each transformer (both HV and LV windings) for the following connections

(*i*) Y-Y (*ii*) Δ-Δ (*iii*) Y-Δ (*iv*) Δ-Y.

[*U.P. Technical Univ. Electrical Machines 2006-07*]

Solution: Primary line voltage, $V_{L_1} = 11$ kV

Output = 1,000 kVA

Secondary line voltage, $V_{L_2} = 2.2$ kV

$$\text{Secondary line current, } I_{L_2} = \frac{1{,}000}{\sqrt{3} \times 2.2} = 262.43 \text{ A}$$

$$\text{Primary line current, } I_{L_1} = \frac{1{,}000}{\sqrt{3} \times 11} = 52.486 \text{ A,}$$

neglecting losses

(*i*) **Y-Y Connections :**

HV voltage rating of each transformer

$$= V_{P_1} = \frac{V_{L1}}{\sqrt{3}} = \frac{11}{\sqrt{3}} = 6.35 \text{ kV} \textbf{ Ans.}$$

HV current rating of each transformer

$$= I_{P_1} = I_{L_1} = 52.486 \text{ A}$$

LV voltage rating of each transformer

$$= V_{P_2} = \frac{V_{L_2}}{\sqrt{3}} = \frac{2.2}{\sqrt{3}} = 1.27 \text{ kV} \textbf{ Ans.}$$

LV current rating of each transformer

$$= I_{P_2} = 262.43 \text{ A} \textbf{ Ans.}$$

KVA rating of each transformer

$$= V_{P_1} \times I_{P_1} = 6.35 \text{ kV} \times 52.486 \text{ A} = 333.33 \text{ kVA} \textbf{ Ans.}$$

(*ii*) **Δ-Δ Connections :**

HV voltage rating of each transformer

$$= V_{P_1} = V_{L_1} = 11 \text{ kV} \textbf{ Ans.}$$

HV current rating of each transformer

$$= I_{P_1} = \frac{I_{L1}}{\sqrt{3}} = \frac{52.486}{\sqrt{3}} = 30.3 \text{ A} \textbf{ Ans.}$$

LV voltage rating of each transformer

$$= V_{P_2} = V_{L_2} = 2.2 \text{ kV} \textbf{ Ans.}$$

LV current rating of each transformer

$$= I_{P_2} = \frac{I_{L_2}}{\sqrt{3}} = \frac{262.43}{\sqrt{3}} = 151.5 \text{ A} \textbf{ Ans.}$$

kVA rating of each transformer will remain the same in each case and will be equal to one-third of total rating of load *i.e.* $\frac{1{,}000}{3} = 333.33$ kVA **Ans.**

(*iii*) **Y-Δ Connections :**

HV voltage rating of each transformer

$$= V_{P_1} = \frac{V_{L_1}}{\sqrt{3}} = \frac{11}{\sqrt{3}} = 6.35 \text{ kV} \textbf{ Ans.}$$

HV current rating of each transformer $= I_{P_1} = I_{L_1} = 52.486$ A **Ans.**

LV voltage rating of each transformer $= V_{P_2} = V_{L_2} = 2.2$ kV **Ans.**

LV current rating of each transformer

$$= I_{P_2} = \frac{I_{L_2}}{\sqrt{3}} = \frac{262.43}{\sqrt{3}} = 151.5 \text{ A} \textbf{ Ans.}$$

(*iv*) **Δ-Y Connections :**

HV voltage rating of each transformer

$$= V_{P_1} = V_{L_1} = 11 \text{ kV} \textbf{ Ans.}$$

HV current rating of each transformer

$$= I_{P_1} = \frac{I_{L_1}}{\sqrt{3}} = \frac{52.486}{\sqrt{3}} = 30.3 \text{ A} \textbf{ Ans.}$$

LV voltage rating of each transformer

$$= V_{P_2} = \frac{V_{L_2}}{\sqrt{3}} = \frac{2.2}{\sqrt{3}} = 1.27 \text{ kV} \textbf{ Ans.}$$

LV current rating of each transformer

$$= I_{P_2} = I_{L_2} = 262.43 \text{ A} \textbf{ Ans.}$$

Example 13.5. A three-phase transformer bank, containing three 1-phase transformers, is used to step down the voltage of a 3-phase, 6,600 V transmission line. If the primary line current is 10 A, calculate the secondary line voltage, line current and output kVA for the following connections:

(*i*) Y/Δ and (*ii*) Δ/Y.

The turn ratio is 12. Neglect losses.

[*U.P. Technical Univ. Electromechanical Energy Conversion-I, 2003-04; Electrical Machines 2006-07; Rajasthan Technical Univ. Electrical Machines-I, 2010*]

Solution: Primary line voltage, $V_{L1} = 6{,}600$ V

Primary line current, $I_{L1} = 10$ A

$$\text{Turn-ratio} = \frac{N_1}{N_2} = 12$$

(*i*) *Y/Δ Connections*

$$\text{Primary phase voltage, } V_{P1} = \frac{V_{L1}}{\sqrt{3}} = \frac{6{,}600}{\sqrt{3}} \text{ V}$$

$$\text{Secondary line voltage, } V_{L2} = V_{P2} = V_{P1} \times \frac{N_2}{N_1}$$

$$= \frac{6{,}600}{\sqrt{3}} \times \frac{1}{12} = 317.5 \text{ V} \textbf{ Ans.}$$

Primary phase current, $I_{P1} = I_{L1} = 10$ A

Secondary line current,

$$I_{L2} = \sqrt{3}\, I_{P2} = \sqrt{3} \times I_{P1} \times \frac{N_1}{N_2}$$

$$= \sqrt{3} \times 10 \times 12 = 207.85 \text{ A} \textbf{ Ans.}$$

(*ii*) *Δ/Y Connections*

Primary phase voltage, $V_{P1} = V_{L1} = 6{,}600$ V

Secondary line voltage,

$$V_{L2} = \sqrt{3}\, V_{P2} = \sqrt{3} \times V_{P1} \times \frac{N_1}{N_2}$$

$$= \sqrt{3} \times \frac{6{,}600}{12} = 952.6 \text{ V} \textbf{ Ans.}$$

Primary phase current,

$$I_{P1} = \frac{I_{L1}}{\sqrt{3}} = \frac{10}{\sqrt{3}} \text{ A}$$

Secondary line current,

$$I_{L2} = I_{P2} = I_{P1} \times \frac{N_1}{N_2} = \frac{10}{\sqrt{3}} \times 12 = 69.28 \text{ A } \textbf{Ans.}$$

$$\text{Output} = \frac{\sqrt{3} \times V_L \times I_L}{1{,}000} = \frac{\sqrt{3} \times 6{,}600 \times 10}{1{,}000} = 114.315 \text{ kVA } \textbf{Ans.}$$

Example 13.6. A 500 kVA, 11/0.43 kV, 3-phase delta/star connected transformer has on rated load an HV copper loss of 2.5 kW and an LV loss of 2 kW. The total leakage reactance of 0.06 per unit. Find the ohmic values of the equivalent resistance and leakage reactance on the delta side.

[U.P. Technical Univ. Electromechanical Energy Conversion-I, 2004-05]

Solution: Primary phase voltage, $V_{P_1} = V_{L_1} = 11$ kV

∵ primary is delta-connected

$$\text{Secondary phase voltage, } V_{P_2} = \frac{V_{L_2}}{\sqrt{3}} = \frac{0.43}{\sqrt{3}} = 0.248 \text{ kV}$$

∵ secondary is star connected

$$\text{Turn-ratio, } K = \frac{V_{P_2}}{V_{P_1}} = \frac{0.248}{11} = 0.02257$$

$$\text{Primary phase current, } I_{P_1} = \frac{\text{Load in kVA/3}}{\text{Primary phase voltage in kV}}$$

$$= \frac{500}{3 \times 11} = 15.152 \text{ A}$$

$$\text{Secondary phase current, } I_{P_2} = \frac{500}{3 \times 0.248} = 672 \text{ A}$$

Copper loss in primary winding per phase,

$$P_{C_1} = \frac{2.5 \times 1{,}000}{3} = 833.33 \text{ W}$$

Resistance of each primary winding,

$$R_1 = \frac{P_{C_1}}{I_{P_1}^2} = \frac{833.33}{(15.152)^2} = 3.63 \ \Omega$$

Copper loss in secondary winding per phase,

$$P_{C_2} = \frac{2 \times 1{,}000}{3} = 666.66 \text{ W}$$

Resistance of each secondary winding,

$$R_2 = \frac{P_{C_2}}{I_{P_2}^2} = \frac{666.66}{(672)^2} = 0.001476 \ \Omega$$

Equivalent resistance per phase referred to primary,

$$R_{01} = R_1 + \frac{R_2}{K^2} = 3.63 + \frac{0.001476}{(0.02257)^2}$$

$$= 3.63 + 2.9 = 6.53 \ \Omega \ \textbf{Ans.}$$

Equivalent reactance per phase referred to primary,

$$X_{01} = \frac{\text{Total leakage reactance per unit} \times V_{P_1}}{I_{P_1}}$$

$$= \frac{0.06 \times 11{,}000}{15.152} = 43.56 \ \Omega \ \textbf{Ans.}$$

Example 13.7. A 3-phase transformer has its primary connected in delta and secondary in star. It has an equivalent resistance of 1% and equivalent reactance of 6%. The primary applied voltage is 6,600 V. What must be the ratio of transformation in order that it will deliver 4,800 V at full-load current and 0.8 power factor lag.

[J.N. Technological Univ. Electrical Machines-II, January-2014]

Solution: Percentage regulation is given by the expression

$$\%\text{age regulation} = v_r \cos\phi + v_x \sin\phi$$

where v_r is the percentage resistive drop or percentage equivalent resistance and v_x is the percentage reactive drop or percentage equivalent reactance

Since $v_r = 1\%$; $v_x = 6\%$; $\cos\phi = 0.8$ and $\sin\phi = 0.6$

$$\text{Percentage regulation} = 1 \times 0.8 + 6 \times 0.6 = 4.4\%$$

Secondary induced emf (line to line)

$$= 4{,}800 + 4.4\ \% \text{ of } 4{,}800 = 5{,}011.2 \text{ V}$$

Secondary induced emf per phase

$$= \frac{5{,}011.2}{\sqrt{3}} = 2{,}893.22 \text{ V}$$

$$\text{Transformation ratio, } K = \frac{\text{Secondary phase emf}}{\text{Primary phase emf}}$$

$$= \frac{2{,}893.22}{6{,}600} = 0.438 \ \textbf{Ans.}$$

Example 13.8. A 3-phase transformer rated at 1,000 kVA, 11/3.3 kV has its primary windings star-connected and its secondary windings delta-connected. The actual resistances per phase of these windings are : primary 0.375 Ω, secondary 0.095 Ω; and the leakage reactances per phase are : primary 9.5 Ω, secondary 2 Ω. Calculate the voltage at normal frequency which must be applied to primary terminals in order to obtain full-load current in the windings when the secondary terminals are short circuited. Also calculate the power input under these conditions.

Solution: Primary phase voltage, $E_{P1} = \dfrac{11}{\sqrt{3}}$ kV

Secondary phase voltage, $E_{P2} = 3.3$ kV

$$\text{Turn-ratio, } K = \frac{E_{P2}}{E_{P1}} = \frac{3.3}{11/\sqrt{3}} = 0.5196$$

Equivalent resistance per phase referred to primary,

$$R_{01} = R_1 + \frac{R_2}{K^2} = 0.375 + \frac{0.095}{(0.5196)^2} = 0.727 \ \Omega$$

Equivalent reactance per phase referred to primary,

$$X_{01} = X_1 + \frac{X_2}{K^2} = 9.5 + \frac{2}{(0.5196)^2} = 16.91 \ \Omega$$

Equivalent impedance per phase referred to primary,

$$Z_{01} = \sqrt{(R_{01})^2 + (X_{01})^2} = \sqrt{(0.727)^2 + (16.91)^2}$$

$$= 16.926 \ \Omega$$

Full-load primary current,

$$I_1 = \frac{1{,}000 \times 1{,}000}{\sqrt{3} \times 11{,}000} = 52.5 \text{ A}$$

Applied potential difference per phase required to circulate this current with the secondary terminals short circuited

$$V_S = I_1 Z_{01} = 52.5 \times 16.926 = 888.4 \text{ V}$$

Line voltage to be applied to the primary terminals

$$V_{LS} = \sqrt{3}\ V_S = \sqrt{3} \times 888.4 = 1{,}539 \text{ V or } 1.54 \text{ kV } \textbf{Ans.}$$

∵ since primary windings are star-connected

Power input under these conditions

$$= 3 I_{P1}^2 R_{01} = 3 \times 52.5^2 \times 0.727 = 6{,}000 \text{ W or } 6 \text{ kW } \textbf{Ans.}$$

Example 13.9. A 120 kVA, 6,000/400 V, Y/Y, 3-ϕ, 50 Hz transformer has an iron loss of 1,800 W. The maximum efficiency occur at 3/4 full load. Find the efficiency of the transformer at *(i)* full load and 0.8 power factor and *(ii)* the maximum efficiency at unity power factor.

Solution: Iron loss, $P_i = 1{,}800$ watts

Since maximum efficiency occurs at 3/4 full load

$\therefore$ Copper loss at 3/4 full load = Iron loss, P_i = 1,800 W = 1.8 kW

$$\text{Copper loss at full load, } P_c = 1{,}800 \times \left(\frac{4}{3}\right)^2 = 3{,}200 \text{ W} = 3.2 \text{ kW}$$

(*i*) Full-load output at 0.8 pf = 120 × 0.8 = 96 kW

Full-load input at 0.8 pf = 96 + 1.8 + 3.2 = 101 kW

$$\text{Efficiency, } \eta = \frac{96}{101} \times 100 = 95\% \textbf{ Ans.}$$

(*ii*) Maximum efficiency occurs at 3/4 full load

$\therefore$ Output corresponding to maximum efficiency

$$= 120 \times \frac{3}{4} \times 1 = 90 \text{ kW}$$

$$\text{Input} = 90 + 1.8 + 1.8 = 93.6 \text{ kW}$$

$$\text{Efficiency, } \eta = \frac{90}{93.6} \times 100 = 96.15\% \textbf{ Ans.}$$

Example 13.10. A 3-phase transformer, ratio 33/6.6 kV, Δ/Y, 2 MVA has a primary resistance of 8 Ω per phase and a secondary resistance of 0.08 Ω per phase. The percentage impedance is 7%. Calculate the secondary voltage with rated primary voltage and hence the regulation for full-load 0.75 pf lagging conditions.

[*M.D. Univ. Electrical Machines, December 2009*]

Solution: Full-load secondary current,

$$I_2 = \frac{\text{MVA} \times 10^6}{\sqrt{3}\, V_{L_2}} = \frac{2 \times 10^6}{\sqrt{3} \times 6.6 \times 10^3} = 175 \text{ A}$$

$$\text{Turn-ratio, } K = \frac{V_{P2}}{V_{P1}} = \frac{6.6/\sqrt{3}}{33} = 0.1155$$

Equivalent resistance per phase referred to secondary,

$$R_{02} = R_2 + K^2 R_1 = 0.08 + (0.1155)^2 \times 8 = 0.1867 \ \Omega$$

Secondary impedance drop per phase,

$$I_2 Z_{02} = \frac{7}{100} \times \frac{6{,}600}{\sqrt{3}} = 266.7 \text{ V}$$

So, equivalent impedance per phase referred to secondary,

$$Z_{02} = \frac{266.7}{175} = 1.524 \ \Omega$$

Equivalent reactance per phase referred to secondary,

$$X_{02} = \sqrt{(Z_{02})^2 - (R_{02})^2} = \sqrt{(1.524)^2 - (0.1867)^2} = 1.513 \ \Omega$$

Secondary phase voltage,

$$E_{P2} = \frac{6{,}600}{\sqrt{3}} = 3{,}810 \text{ V}$$

Secondary voltage on full load,

$$V_{P2} = E_{P2} - I_2 (R_{02} \cos\phi + X_{02} \sin\phi)$$
$$= 3{,}810 - 175\,(0.1867 \times 0.75 + 1.513 \times 0.6614)$$
$$= 3{,}610 \text{ V} \qquad \because \cos\phi = 0.75 \text{ and } \sin\phi = 0.6614$$

$$\text{Secondary line voltage} = \sqrt{3} \times 3{,}610 = 6{,}250 \text{ V } \textbf{Ans.}$$

$$\text{Percentage regulation} = \frac{I_2 (R_{02}\cos\phi + X_{02}\sin\phi)}{E_{P2}} \times 100$$

$$= \frac{175\,(0.1867 \times 0.75 + 1.513 \times 0.6614)}{3{,}810} \times 100$$

$$= 5.25\% \textbf{ Ans.}$$

Example 13.11. The maximum efficiency of a 3-phase 11,000/400 V, 500 kVA transformer is 98.8% and occurs at 80 % full load, unity power factor. Its percentage impedance is 4.5%. Load power factor is now varied while the load current and the supply voltage are held constant at their rated values. Determine the load power factor at which the secondary terminal voltage is minimum and find the value of the latter.

[*GATE, 1992*]

Solution: The output at 80% full load and unity pf

$$= 500 \times 0.8 \times 1.0 = 400 \text{ kW}$$

Input at 80% full load and unity pf

$$= \frac{400}{\eta} = \frac{400}{0.988} = 404.8583 \text{ kW}$$

$\because$ at 80% full load, efficiency is maximum and is equal to 98.8%.

Total losses at 80% full load = Input – output

= 404.8583 – 400 = 4.8583 kW

So, copper losses at 80% full load

$$= \frac{4.8583}{2} = 2.42915 \text{ kW}$$

$$\text{Full-load copper losses, } P_c = \frac{2.42915}{0.8^2} = 3.7955 \text{ kW}$$

Percentage resistance = Percentage copper loss of full load output

$$= \frac{3.7955}{500} \times 100 = 0.7591\%$$

Percentage impedance = 4.5%

$$\text{Percentage reactance} = \sqrt{(4.5)^2 - (0.7591)^2} = 4.4355\%$$

For terminal voltage to be minimum, the drop or percentage regulation will be maximum *i.e.* $v_r \cos\phi + v_x \sin\phi$ will be maximum.

$$\text{or } \frac{d}{d\phi}(v_r \cos\phi + v_x \sin\phi) = 0$$

$$\text{or } -v_r \sin\phi + v_x \cos\phi = 0$$

$$\text{or } \phi = \text{Tan}^{-1}\frac{v_x}{v_r} = \text{Tan}^{-1}\frac{4.4355}{0.7591} = 80.3°$$

Thus load power factor for minimum voltage,

$$\cos\phi = \text{Cos } 80.3° = 0.1687 \text{ (lag) } \textbf{Ans.}$$

Example 13.12. A 100 kVA, 3-phase, 50 Hz 3,300/400 V transformer is delta-connected on the hv side and star-connected on the lv side. The resistance of the hv winding is 3.5 Ω per phase and that of the lv winding 0.02 Ω per phase. Calculate the iron losses of the transformer at normal voltage and frequency if its full-load efficiency be 95.8% at 0.8 pf (lag).

Solution: Output at full-load and 0.8 pf

$$= \text{kVA} \times \text{pf} = 100 \times 0.8 = 80 \text{ kW}$$

Full-load efficiency = 0.958

$$\text{Input at full-load and 0.8 pf} = \frac{\text{Output}}{\eta} = \frac{80}{0.958} = 83.5 \text{ kW}$$

Total losses at full load = 83.5 – 80 = 3.5 kW

$$\text{Turn-ratio} = \frac{(V_2)_p}{(V_1)_p} = \frac{400/\sqrt{3}}{3{,}300} = \frac{4}{33\sqrt{3}}$$

The equivalent resistance referred to secondary,

$$R_{02} = R_2 + K^2 R_1$$

$$= 0.02 + \left(\frac{4}{33\sqrt{3}}\right)^2 \times 3.5 = 0.037 \ \Omega$$

The secondary phase current at full load, $(I_2)_p$

$$= \frac{\text{Rated kVA} \times 1{,}000}{3 \times (V_2)_p} = \frac{100 \times 1{,}000}{3 \times \frac{400}{\sqrt{3}}} = 144.1 \text{ A}$$

Total copper loss $= 3\,(I_2)^2_p \times R_{02}$

$$= 3 \times (144.1)^2 \times 0.037 = 2{,}300 \text{ watts or } 2.3 \text{ kW}$$

Iron loss, P_i = Total loss – total copper loss

$$= 3.5 - 2.3 = 1.2 \text{ kW } \textbf{Ans.}$$

Example 13.13. A 300 kVA, 11 kV/440 V, 50 Hz, 3-phase Δ/Y transformer gave the following test results:

At rated voltage on no load: 440 V, 21.1 A, 1.3 kW (lv side)

At rated current on short circuit: 630 V, 15.7 A, 3.08 kW (hv side)

Evaluate the rated loss, impedance voltage, per unit resistance and leakage reactance, efficiency and regulation on full load at 0.8 pf lagging. [*U.P.S.C. I.E.S. Electrical Engineering-II, 2008*]

Solution: Iron loss, P_i = No-load input on rated voltage = 1.3 kW

Full-load copper loss, P_c = Short-circuit input at rated current = 3.08 kW

Rated losses $= P_i + P_c = 1.3 + 3.08 = 4.38$ kW **Ans.**

$$\text{Impedance voltage, } v_z = \frac{V_s}{V_1} = \frac{630}{11{,}000} = 0.0573 \text{ per unit } \textbf{Ans.}$$

Per unit resistance, v_r = Unit copper loss

$$= \frac{3.08}{300} = 0.01027 \text{ pu } \textbf{Ans.}$$

$$\text{Per unit reactance, } v_x = \sqrt{v_z^2 - v_r^2}$$

$$= \sqrt{0.0573^2 - 0.01027^2} = 0.0564 \textbf{ Ans.}$$

Full-load output, P_{out} = Rated kVA × pf = 300 × 0.8 = 240 kW

Efficiency on full-load at 0.8 pf lagging

$$\eta = \frac{\text{Rated output at 0.8 pf lag, } P_{out}}{P_{out} + \text{rated losses}} \times 100$$

$$= \frac{300 \times 0.8}{300 \times 0.8 + 4.38} \times 100$$

$$= 98.21\% \textbf{ Ans.}$$

Voltage regulation on full load at 0.8 pf lagging

$$= (v_r \cos\phi + v_x \sin\phi) \times 100$$

$$= (0.01027 \times 0.8 + 0.0564 \times 0.6) \times 100$$

$$\because \sin\phi = \sin(\cos^{-1} 0.8) = 0.6$$

$$= 4.206\% \textbf{ Ans.}$$

Example 13.14. A 1,500 kVA, 6.6/1.1 kV, 3 phase delta/star connected transformer has the following test result:

S.C. test: (Instruments placed on hv side): 300 V, 131.21 A, 30 kW.

Calculate its percentage resistance, percentage reactance drops, percentage efficiency and percentage regulation on full load at 0.8 power factor lagging. The iron loss during OC test with rated voltage supplied is 25 kW. [*U.P.S.C. I.E.S. Electrical Engineering-II, 2006*]

Solution: Phase voltage, $V_{Ps} = V_{Ls} = 300$ V

$$\text{Phase current, } I_{Ps} = \frac{I_{Ls}}{\sqrt{3}} = \frac{131.21}{\sqrt{3}} = 75.75 \text{ A}$$

This is because short-circuit test has been conducted on the hv (primary) side.

Equivalent impedance referred to primary,

$$Z_{01} = \frac{V_{Ps}}{I_{Ps}} = \frac{300}{75.75} = 3.96\ \Omega$$

Equivalent resistance referred to primary,

$$R_{01} = \frac{P_s}{3I_s^2} = \frac{30 \times 1{,}000}{3 \times (75.75)^2} = 1.743\ \Omega$$

Equivalent reactance referred to primary,

$$X_{01} = \sqrt{(Z_{01})^2 - (R_{01})^2}$$

$$= \sqrt{(3.96)^2 - (1.743)^2} = 3.56\ \Omega$$

$$\text{Base impedance, } Z_{\text{base (hv side)}} = \frac{(KV_B)^2 \times 1{,}000}{KVA_B}$$

$$= \frac{(6.6)^2 \times 1{,}000}{1{,}500} = 29.04\ \Omega$$

Converting R_{01} and X_{01} into star equivalent we have

$$R'_{01} = \frac{R_{01}}{3} = \frac{1.743}{3} = 0.581\ \Omega$$

$$X'_{01} = \frac{X_{01}}{3} = \frac{3.56}{3} = 1.187\ \Omega$$

Percentage resistance drop

$$= \frac{R'_{01}}{Z_{base}} \times 100 = \frac{0.581}{29.04} \times 100 = 2\% \textbf{ Ans.}$$

Percentage reactance drop

$$= \frac{X'_{01}}{Z_{base}} \times 100 = \frac{1.187}{29.04} \times 100 = 4.09\% \textbf{ Ans.}$$

Power factor, $\cos\phi = 0.8$ (lagging)

$$\sin\phi = \sqrt{1 - 0.8^2} = 0.6$$

Percentage regulation = Percentage resistance × cos φ + Percentage reactance × sin φ

$$= 2 \times 0.8 + 4.09 \times 0.6 = 4.054\% \textbf{ Ans.}$$

So, full-load copper loss, P_c = 30 kW

Iron loss, P_i = 25 kW

Full-load output at 0.8 pf,

$$P_{out} = 1{,}500 \times 0.8 = 1{,}200 \text{ kW}$$

$$\text{Transformer efficiency, } \eta = \frac{P_{out}}{P_{out} + P_c + P_i} \times 100$$

$$= \frac{1{,}200}{1{,}200 + 30 + 25} \times 100 = 95.62\% \textbf{ Ans.}$$

13.4 THREE-PHASE TRANSFORMATION WITH TWO TRANSFORMERS

It is possible to transform three-phase power by means of only two single-phase transformers. There are two methods of doing so, namely (*i*) Open-delta connection method and (*ii*) T-connection method. Both of these methods result in slightly unbalanced output voltage under load because of the unsymmetrical relations. This is not serious in commercial transformers, as their regulation is seldom poorer than 2 or 3%.

13.4.1. Open Δ (or V-V) Connection. In a Δ-connection the voltage of any one phase at any instant is equal to the sum of the instantaneous voltages in the other two phases, so that if one transformer A of a bank of Δ-connected transformers is disconnected, the difference of potentials between the line terminals *a* and *b* remains unchanged and it is maintained by transformers B and C in series. Thus the three-phase power can still be obtained from the lines *a, b* and *c*. Such type of connection is called the *open-delta* or *V-V connection.*

Such an arrangement is employed when

1. The three-phase load is comparatively small, so that installation does not warrant a three-transformer bank.
2. One of the transformers in a Δ-Δ bank fails, so that service may be maintained (although at reduced capacity) until the faulty transformer is repaired or a new one is substituted.
3. It is anticipated that the future load will increase to warrant the closing of the open Δ at some later date.

In Δ-Δ connection (Fig. 13.9), if the rated full-load secondary current of each transformer is I amperes, then full-load line current $I_L = \sqrt{3}$ I amperes, being the phasor difference of two currents which are equal to I and have a phase difference of 120°.

The kVA rating of each transformer

$$= \frac{VI}{1,000} \text{ where V is the secondary voltage of each transformer}$$

Total power supplied by the line

$$= \sqrt{3}\, VI_L \cos\phi$$

But $I_L = \sqrt{3}\, I$

∴ kVA rating of the bank of three transformers

$$= \frac{\sqrt{3} VI_L}{1,000}$$

$$= \frac{3VI}{1,000} = 3 \text{ times the rating of one transformer.}$$

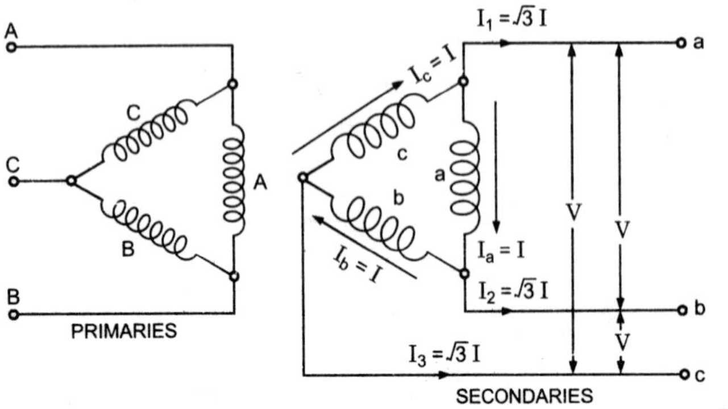

Fig. 13.9 *Δ-Δ Connection*

Now if one transformer is removed from the bank, each remaining transformer is still able to carry its rated full-load current I, but the line current is now the current from a single transformer winding. Thus removal of one transformer from the bank reduces

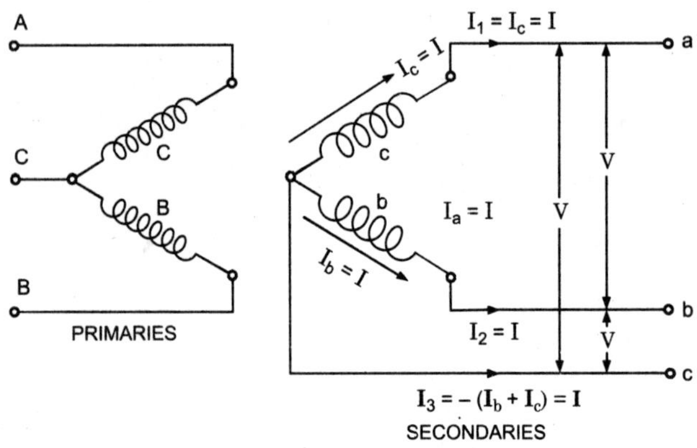

Fig. 13.10 *Open Δ-Connection*

the full-load line current from $\sqrt{3}$ I to I amperes, which is reduction of 42.3% instead of 33.33% as might be expected.

For example, if three single-phase transformers, each of 10 kVA rating, are connected to form a closed Δ bank, the rating of the bank is 30 kVA; but when one of the transformers is removed, the rating of the resulting open-Δ bank is not 20 kVA (as expected) but 30 × 0.577 = 17.31 kVA. Thus when transformers are connected in open Δ, the rating of the bank is approximately 13.4% less than the sum of the individual ratings of the transformers. The reason for the loss of capacity is explained below:

On removal of transformer A, the remaining transformers B and C must carry the current of the missing transformer A in addition to their own currents. Since I_A is not in phase with either I_B or I_C so I_B and I_C are shifted in phase by 30° – one in one direction and the other in other direction. The result is that if the load power factor is unity, the transformers B and C are operating at a power factor of 0.866 (cos 30°). Thus we see that the reduction in rating comes about because the average power factor becomes 0.866 times the balanced load power factor.

A significant point in this scheme of connections is that, except for a balanced unity pf load, the two transformers in the V-V bank operate at different power factors. When an open delta bank of two transformers delivers a balanced 3-ϕ load, whose power factor is cos ϕ, one of the unit will operate at a pf of cos (30° + ϕ) while the other at a pf of cos (30° – ϕ). This means for example that :

1. When the power factor of the load is unity, the transformers themselves will be operating at pf of cos 30° *i.e.* at 0.866.
2. When the power factor of the load is 0.866, one transformer will be operating at unity pf and the other at 0·5 power factor.
3. When the pf of the load is 0.5, one transformer will operate at zero power factor and deliver no power, while the other operating at a pf of 0.866 will handle the entire load.

The main disadvantage of this type of connection is that the secondary terminal voltage tend to become unbalanced to a greater extent with the increase in load. This is true even when the 3-phase load is balanced.

However, it is noteworthy that if three single-phase transformers are delivering their rated load and one transformer is removed, the two remaining transformers become overloaded. The load line current, which is $\sqrt{3}$ times the transformer current in closed Δ, is the same as the phase current in open Δ. Each transformer supplies the load $\sqrt{3}$ (or 1.732) times the rated load *i.e.* 73.2% additional load. This overload may be carried temporarily, but some provision should be made to transfer or remove part of the load if overheating and consequent breakdown of the remaining two transformers is to be avoided.

Another noteworthy point is that if the two transformers are operating in open-delta system and supplying their rated load (in the above example 17.32 kVA), the addition of a third transformer increases the total capacity of $\sqrt{3}$ or 1.732 times (*i.e.* 30 kVA). It means the rating of bank is increased by 73% with an investment increase of approximately 50%.

13.4.2. T-T Connection. This is another method of transformation of 3-phase power from one voltage to another by using two transformers. Since it was first proposed by C.F. Scott, it is frequently called the *Scott connection*. This scheme of connection is widely used for 3-phase to 2-phase , and vice versa transformation, as explained in Art. 13.5.

In this arrangement two transformers are required. One of the two transformers, the so called *main* transformer, must have at least two primary and two secondary coils so that a centre tap may be brought out from each side. The other transformer, known as *teaser* transformer must have primary and secondary windings, the number of turns of which are 0.866 times of the respective turns on the main transformer. The current ratings of the two transformers should be the same. As a rule, it is more economical to manufacture both transformers alike with 86.6 per cent taps. The wiring diagram is shown in Fig. 13.11. One end of both the primary and secondary of the teaser transformer is connected to the centre taps on both primary and secondary of the main transformer respectively, as shown in Fig. 13.11. The two ends of main transformer and 86.6% tapping point on teaser transformer are connected to the 3-phase supply on the primary side and to the load on secondary side.

If V_{AB}, V_{BC} and V_{CA} are the applied voltages across the line terminals as shown in Fig. 13.11, then

$$V_{CD} = V_{CA} + V_{AD} = V\angle 120^\circ + \frac{1}{2}V\angle 0^\circ$$
$$= V(-0.5 + j\,0.866) + 0.5\,V$$
$$= j\,0.866\,V = 0.866\,V\angle 90^\circ$$

The voltage applied across the teaser is 0.866 times of that across the main and has a phase difference of 90°. If both the primary and secondary of the teaser have 0.866 times of the respective turns on the main transformer, then the induced voltages in the secondary circuit will have same phase and magnitude relationship as have the applied voltages on the primary. Thus the voltage induced in the teaser is 0.866 times of that induced in the main transformer and has a phase difference of 90°. Consequently, a balanced 3-ϕ system of voltages across points *a*, *b* and *c* is available.

If V is the line voltage and I is current rating of the winding, which is also the line current, the combined volt-ampere rating of the two transformers is VI + 0.866 VI *i.e.* 1.866 VI. The volt-amperes supplied are $\sqrt{3}$ VI, therefore, utility factor is $\sqrt{3}/1.866$ *i.e.* 0.928. Thus the T-T connection is more economical than open-Δ connection with utility factor of 0.866. The two transformers used, however, are usually identical, as already mentioned. This permits interchangeability, whether either transformer may be used as the main or teaser but 13.4 per cent of the winding will be idle. With identical transformers, the total volt-ampere capacity of the two transformers is 2VI and the utility factor is 0.866, the same as in case of open-Δ system. This implies, therefore, that each transformer in a T-T connection must have a rating 15% greater than each of the transformers in a three transformer bank.

For any balanced load of power factor cos ϕ (lag), though the teaser operates at a pf of cos ϕ (lag), one of the two halves of the main transformer operates at a pf of cos (30° + ϕ) and the other at cos (30° – ϕ). This is similar to the condition as in open Δ. Mainly because of this (*i.e.* different loading effects) the voltages on the secondary side tend to become unbalanced to a greater extent with the increase in load.

It may be noted that this arrangement provides the 3-phase, 4-wire system. Three phase power loads may be connected between lines a, b and c while lighting load may be connected between *ad* and *bd*. There is a further advantage

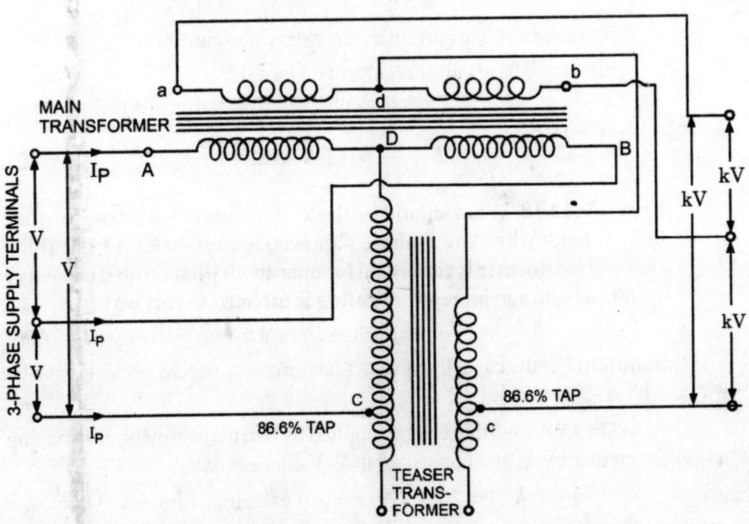

Fig. 13.11 *Connection Diagram For T-T Connection*

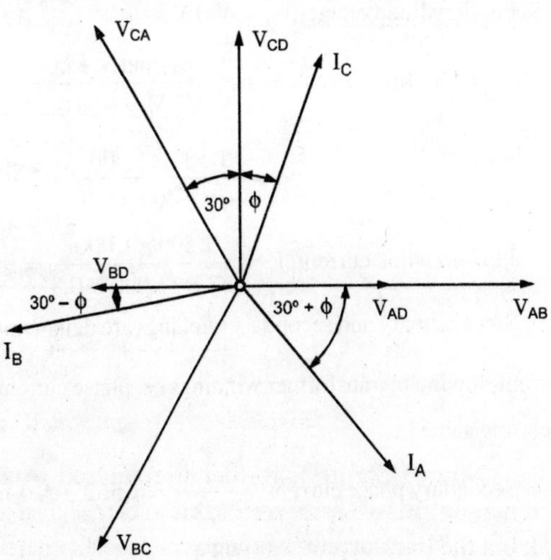

Fig. 13.12 *Phasor Diagram For T-T Connection*

of availability of neutral on the teaser transformer. This permits a true three phase four-wire system with the use of two transformers, which could not be had in open-Δ system.

Example 13.15. Two-transformers are connected in open delta and supply a balanced 3-phase load of 240 kW at 400 volts and a pf of 0.866, determine:

(a) the secondary line current,

(b) the kVA load on each transformer,

(c) the power delivered by the individual transformers.

[*M.D. Univ. Electrical Machines-I, December-2013*]

Solution: Load supplied = 240 kW.

Secondary voltage, V = 400 volts

Load pf, $\cos\phi$ = 0.866

(*a*) Secondary line current $= \dfrac{\text{kW output} \times 1{,}000}{\sqrt{3} \times V \times \cos\phi}$

$$= \frac{2{,}40{,}000}{\sqrt{3} \times 400 \times 0.866} = 400 \text{ A } \textbf{Ans.}$$

(*b*) kVA load on each transformer

$$= \frac{V \times I}{1{,}000} = \frac{400 \times 400}{1{,}000} = 160 \textbf{ Ans.}$$

(*c*) For a power factor of 0.866, $\phi = 30°$

Power delivered by one transformer,

P_1 = kVA cos (30° – ϕ) = 160 cos 0° = 160 kW **Ans.**

Power delivered by the other transformer,

P_2 = kVA cos (30° + ϕ) = 160 cos 60° = 80 kW **Ans.**

Example 13.16. Three single-phase transformers, connected in delta-delta, supply a balanced 3-phase load of 1,500 kW at 400 V at 0.8 pf lagging. The transformers are supplied from 3-phase mains at 11,000 V. Find the current in the windings of each transformer. If one of the transformer is removed because of a sudden fault, and the supply is maintained in V-V connection, determine the currents in the windings and the power supplied by each of the transformers.

[*W.B. Univ. of Technology, Electrical Machine-I, 2013*]

Solution: When the three transformers are in service

Load, P = 1,500 kW

Power factor, $\cos\phi$ = 0.8 lagging

Primary line voltage, V_1 = 11,000 V

Secondary line voltage, V_2 = 400 V

Secondary line current, $I_2 = \dfrac{\text{kW output} \times 1{,}000}{\sqrt{3}\, V_2 \times \cos\phi}$

$$= \frac{1{,}500 \times 1{,}000}{\sqrt{3} \times 400 \times 0.8} = 2{,}706 \text{ A } \textbf{Ans.}$$

Primary line current, $I_1 = \dfrac{1{,}500 \times 1{,}000}{\sqrt{3} \times 11{,}000 \times 0.8} = 98.4$ A

Since primary and secondary windings are delta-connected therefore, current flowing in transformer windings i.e. phase currents will $\dfrac{1}{\sqrt{3}}$ times their line value i.e.,

Secondary phase current $= \dfrac{2{,}706}{\sqrt{3}} = 1{,}562.3$ A **Ans.**

Primary phase current $= \dfrac{98.4}{\sqrt{3}} = 56.8$ A **Ans.**

When one transformer, being faulty removed i.e. when two transformers, connected in open delta, are in service

Secondary line current $= \dfrac{\text{kW output}}{\sqrt{3} \times V \times \cos\phi}$

$$= \frac{1{,}500 \times 1{,}000}{\sqrt{3} \times 400 \times 0.8} = 2{,}706 \text{ A}$$

Secondary phase current = Secondary line current = 2,706 A **Ans.**

Primary phase current = Secondary phase current × transformation ratio

$$= 2{,}706 \times \frac{400}{11{,}000} = 98.4 \text{ A}$$

kVA load on each transformer $= \dfrac{400 \times 2{,}706}{1{,}000} = 1{,}082.4$ kVA

Load phase angle, $\cos\phi = \text{Cos}^{-1}\, 0.8 = 36.87°$ (lagging)

Power delivered by one transformer,

P_1 = kVA cos (30° – ϕ)

= 1,082.4 cos (30° – 36.87°) = 1,074.7 kW **Ans.**

Power delivered by the other transformer,

P_2 = kVA cos (30° + ϕ)

= 1,082.4 cos (30° + 36.87°) = 425.3 kW **Ans.**

Example 13.17. Two transformers, each rated 250 kVA, 11/2 kV and 50 Hz, are connected in open-delta on both the primary and secondary sides.

(i) Find the kVA that can be supplied from this transformer connection.

(ii) A delta-connected three phase load 250 kVA, 0.8 pf, 2 kV is connected to the lv terminals of this open-delta transformer. Determine the transformer currents on the 11 kV side of the connections.

[*U.P. Technical Univ. Electromechanical Energy Conversion-I, 2005-06; Rajasthan Technical Univ. Electrical Machines-I, 2011*]

Solution : (*i*) The kVA load that can be supplied by two transformers, each having rating of 250 kVA

= 2 × kVA rating of each transformer × 0.866

= 433 kVA **Ans.**

Secondary line current,

$$I = \frac{\text{kVA output}}{\sqrt{3} \times V} = \frac{250 \times 1{,}000}{\sqrt{3} \times 2{,}000} = 72 \text{ A}$$

Secondary phase current = Secondary line current = 72 A

Primary phase current (current on hv side)

= Secondary phase current × transformation ratio

$$= 72 \times \frac{2}{11} = 13 \text{ A } \textbf{Ans.}$$

Example 13.18. What should be the kVA rating of each transformer in V-V bank when the 3-phase balanced load is 40 kVA? If a third similar transformer is connected for operation, what is rated capacity? What percentage increase in rating is affected in this way?

[*Rajasthan Univ. Electrical Machines-I, 2006*]

Solution: Let the capacity of each transformer connected in V-V bank be *x* kVA.

The kVA load that can be supplied by two transformers, each having rating of *x* kVA, and connected in V-V connections

= 2 × kVA rating of each transformer × 0.866

= 2*x* × 0.866 = 1.732 *x*

Load supplied = 40 kV

So $1.732\,x = 40$

or $x = \frac{40}{1.732} = 23$ kVA **Ans.**

When a third similar transformer is connected for operation, the rated capacity becomes 3 times the rated capacity of each transformer *i.e.* $3 \times 23 = 69$ kVA **Ans.**

Percentage increase in capacity $= \frac{69 - 40}{40} \times 100 = 72.5\%$ **Ans.**

Example 13.19. Two T-connected transformers are used to supply a balanced load of 100 kVA at 400 V from a balanced 11 kV 3-ϕ supply. Determine (*i*) current and voltage rating of each transformer coil (*ii*) kVA rating of the main and teaser transformers.

[*J.N. Technological Univ. Electrical Machines-II, January-2014*]

Solution: Voltage rating of primary of main transformer

$=$ Primary line voltage, $V_1 = 11{,}000$ V **Ans.**

Voltage rating of primary of teaser transformer

$= 0.866\ V_1 = 0.866 \times 11{,}000 = 9{,}526$ V **Ans.**

Current ratings of primaries of both of the transformers (main and teaser)

$=$ Primary line current $= \frac{100 \times 1{,}000}{\sqrt{3} \times 11{,}000} = 5.25$ A **Ans.**

Voltage rating of secondary of main transformer

$=$ Secondary line voltage, $V_2 = 400$ V **Ans.**

Voltage rating of secondary of teaser transformer

$= 0.866 \times 400 = 346.4$ V **Ans.**

Current ratings of secondaries of both of the transformers

$=$ Primary current × transformation ratio

$= 5.25 \times \frac{11{,}000}{400} = 144.34$ A **Ans.**

Rating of main transformer

$= \frac{V_1 I_1}{1{,}000} = \frac{11{,}000 \times 5.25}{1{,}000} = 57.75$ kVA **Ans.**

Rating of teaser transformer

$= 0.866 \times 57.75 = 50$ kVA **Ans.**

Example 13.20. Two T-connected transformers are used to supply a 440 V, 30 kVA balanced load from a balanced 3-phase supply of 3000 V. Calculate

(*i*) voltage and current rating of each coil.

(*ii*) kVA rating of main and teaser transformer.

[*M.D. Univ. Electrical Machines-I, December-2012*]

Solution: (*i*) Voltage rating of primary of main transformer

$=$ Primary line voltage, $V_1 = 3{,}000$ V **Ans.**

Voltage rating of primary of teaser transformer

$= 0.866 V_1 = 0.866 \times 3{,}000 = 2{,}598$ V **Ans.**

Voltage rating of secondary of main transformer

$=$ Secondary line voltage, $V_2 = 440$ V **Ans.**

Voltage rating of secondary of teaser transformer

$= 0.866 \times V_2 = 0.866 \times 440 = 381$ V **Ans.**

Current ratings of primary of both the transformers (main and teaser)

$=$ Primary line current $= \frac{30 \times 1{,}000}{\sqrt{3} \times 3{,}000} = 5.774$ A **Ans.**

Current ratings of secondary of both the transformers (main and teaser)

$=$ Primary current × transformation ratio

$= 5.774 \times \frac{3{,}000}{440} = 39.368$ A **Ans.**

(*ii*) Rating of main transformer

$= \frac{V_1 \times I_1}{1{,}000} = \frac{3{,}000 \times 5.774}{1{,}000} = 17.322$ kVA **Ans.**

Rating of teaser transformer

$= 0.866 \times 17.322 = 15$ kVA **Ans.**

13.5 3-PHASE TO 2-PHASE CONVERSION AND VICE VERSA

In some cases such as for electric furnaces, it is desired to work with two-phase current. Since nowadays 3-phase ac supply is available, it is necessary to convert from 3-phase to 2-phase. This is achieved by means of Scott connection. It employs two identical transformers, one with primary centre tap, known as *main transformer*, and the other with primary 0.866 tap, known as *teaser* transformer. The main transformer primary centre tap point D is connected with one end of the teaser primary, as shown in Fig. 13.13. Now ends A, B and C are connected to 3-phase ac supply and true two-phase four wire supply is obtained, as shown in the figure. The two-phase side makes use of all the turns on teaser winding (not 86.6% turns) as well as all those on the main

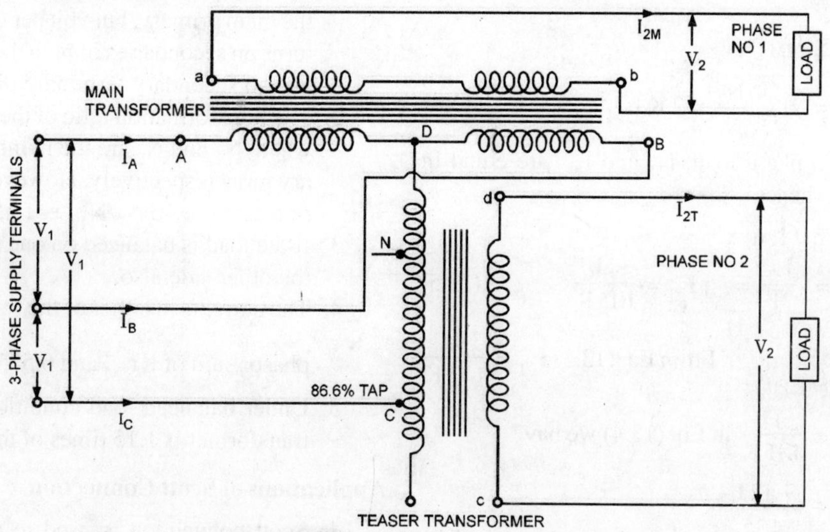

Fig. 13.13 *Scott-Connected Transformers*

winding. This connection will cause the voltage per turn of the two transformers to be equal.

Since point D is located midway on A B, V_{CD} leads V_{AB} by 90° *i.e.* voltages across the primaries of the transformer are 90° apart. It follows that the secondary voltages are 90° apart. With equal fluxes in the core, the secondary windings require an equal number of turns (say, N_2) to provide equal secondary voltages. Hence the two transformers have unequal ratios of transformation.

Transformation ratio of main transformer,

$$K = \frac{N_2}{N_1} \qquad ...(13.1)$$

Transformation ratio of teaser transformer

$$= \frac{N_2}{0.866\ N_1} = 1.15 \qquad ...(13.2)$$

It is to be noted that point D is not the neutral point of the primary system, as its voltage with respect to line terminals A, B and C are not equal to $\frac{V_1}{\sqrt{3}}$. Let N be such a point, the voltages from point N to points A, B and C are equal. Point N, therefore, is the neutral of the primary system.

The neutral point on the three phase side can be located on the teaser transformer. The neutral must have a voltage of $\frac{V}{\sqrt{3}}$ *i.e.* 0.577 to C and since the voltage C to D is 0.866 V, the neutral point N will be 0.866 V – 0.577 V from D or 0.289 V *i.e.* a number of turns below D equivalent to 28.9 percent of the primary turns in the first or main transformer. Since 0.289 is one-third of 0.866, the neutral point is one third the way down the teaser transformer winding from D to C or point N divides the teaser winding in the ratio of 1 : 2.

For determination of primary currents let us neglect the magnetizing current and consider load component currents I_{2M} and I_{2T} only.

For the teaser transformer,

$$I_C \times 0.866\ N_1 = I_{2T}\ N_2$$

$$\text{or } I_C = I_{2T} \times \frac{N_2}{0.866\ N_1} = 1.15\ K\ I_{2T} \qquad ...(13.3)$$

For the main transformer

$$\frac{I_A N_1}{2} - \frac{I_B N_1}{2} = I_{2M}\ N_2$$

$$\text{or } I_A - I_B = 2I_{2M}\frac{N_2}{N_1} = 2\ K\ I_{2M} \qquad ...(13.4)$$

With a balanced two phase load, I_{2T} and I_{2M} are equal in magnitude but 90° apart, hence

$$\mathbf{I}_{2T} = j\ \mathbf{I}_{2M}$$

$$\text{or } \mathbf{I}_{2M} = \frac{\mathbf{I}_{2T}}{j} = -j\ \mathbf{I}_{2T} = \frac{-j\ \mathbf{I}_C}{1.15\ K} \qquad ...(13.5)$$

$$\because \text{From Eq. (13.3) } \mathbf{I}_{2T} = \frac{\mathbf{I}_C}{1.15\ K}$$

Substituting $\mathbf{I}_{2M} = \frac{-j\ \mathbf{I}_C}{1.15\ K}$ in Eq. (13.4) we have

$$\mathbf{I}_A - \mathbf{I}_B = \frac{-j\,2\,\mathbf{I}_C}{1.15}$$

$$\text{or } \mathbf{I}_A - \mathbf{I}_B + j\sqrt{3}\ \mathbf{I}_C = 0 \qquad ...(13.6)$$

Since the phasor sum of three line currents is zero,

$$\therefore\ \mathbf{I}_A + \mathbf{I}_B + \mathbf{I}_C = 0 \qquad ...(13.7)$$

Solving Eqs. (13.6) and (13.7) we have

$$\mathbf{I}_A = \mathbf{I}_B\,(-0.5 + j\,0.866) = \mathbf{I}_B\ \angle 120^\circ$$

$$\text{and } \mathbf{I}_C = \mathbf{I}_B\ (-0.5 - j\,0.866) = \mathbf{I}_B\ \angle -120^\circ$$

Thus I_A, I_B and I_C are equal in magnitude but are 120° apart from each other and so they give a balanced system.

Phasor diagram for voltages and currents, neglecting the transformer impedance and magnetizing current, is shown in Fig. 13.14. The load currents are lagging behind their respective secondary voltages by an angle φ. One half of the main transformer primary winding AB, has an angle of lag of (30° – φ) while the other has angle of lag of (30° + φ) where teaser primary has an angle of lag of φ. This is the reason why the volt-ampere capacity of the primary winding of main transformer must be greater than that of its secondary.

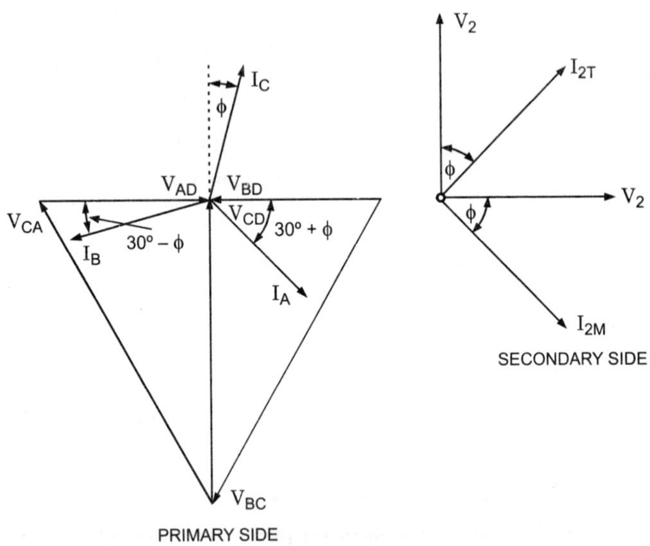

Fig. 13.14 *Phasor Diagram*

In Scott-connected transformers we can summarize as follows:

1. Primary of teaser transformer has 0.866 times the turns of the main primary, but volt per turn is the same. Number of turns on secondaries of both the transformers are the same and so secondary terminal voltages are equal.
2. The transformation ratio of the main transformer is N_2/N_1 where N_1 and N_2 are the number of primary and secondary turns respectively. However, the transformation ratio of teaser is 1.15 N_2/N_1 i.e. 1.15 K.
3. If the load is balanced on one side, it will be balanced on the other side also.
4. The currents in either of the two halves of the primary are phasor sum of KI_{2M} and 0.575 KI_{2T} $\left(\text{or } \frac{1}{2} I_1 T\right)$.
5. Under balanced load conditions, the rating of the main transformer is 1.15 times of that of the teaser.

Applications of Scott Connection

1. Scott connection is used to convert 3-phase supply to 2-phase one for supply to electric furnaces.

2. For supply of single-phase loads such as electric trains which are so scheduled as to keep the load on the 3-phase system almost balanced.
3. To provide linkage between 3-phase and two-phase systems with the flow of power in either direction.

   Though the Scott connection permits conversion of 2-phase supply to 3-phase one, but because of non-availability of 2-phase supply the conversion from 2-phase to 3-phase is not used in practice.

Example 13.21. A Scott-connected transformer is fed from 11 kV 3-phase supply mains and delivers 2-phase power at 400 V. Each phase on the secondary side has 100 turns. Calculate the number of turns on the primary of both the transformers and position of neutral.

[G.G.S.I.P. Univ. Delhi Electromechanical Energy Conversion-I, December-2011]

Solution: Primary voltage, $V_1 = 11{,}000$ V

Secondary voltage, $V_2 = 400$ V

$$\text{Turn ratio, } \frac{N_2}{N_1} = \frac{V_2}{V_1} = \frac{400}{11{,}000} = \frac{1}{27.5}$$

Secondary turns, $N_2 = 100$

Number of primary turns on main transformer,

$$N_1 = \frac{N_2}{K} = \frac{100}{1/27.5} = 2{,}750 \textbf{ Ans.}$$

Number of primary turns on teaser transformer

$$= 0.866 \times \frac{100}{1/27.5} = 2{,}382 \textbf{ Ans.}$$

Number of turns below point D

= 28.9 per cent of the primary turns on the main transformer

$$= \frac{28.9}{100} \times 2{,}750 = 795 \textbf{ Ans.}$$

Example 13.22. Two single phase furnaces A and B are supplied at 100 V by means of Scott-connected transformers from a 3-phase, 6 kV system. Furnace A is supplied from the teaser transformer. Calculate the line currents on the three phase side when (*i*) each furnace takes 600 kW at pf 0.8 (lag), (*ii*) furnace A takes 500 kW at unity power factor and furnace B 600 kW at 0.8 pf (lag).

Solution: Secondary voltage, $V_2 = 100$ V

Primary main voltage, $V_1 = 6 \text{ kV} = 6{,}000$ V

$$\text{Transformation ratio, } K = \frac{V_2}{V_1} = \frac{100}{6{,}000} = \frac{1}{60}$$

$$(i) \text{ Secondary currents, } I_{2T} = I_{2M} = \frac{600 \times 1{,}000}{100 \times 0.8} = 7{,}500 \text{ A}$$

$$\text{Teaser primary current, } I_{1T} = 1.15\, K\, I_{2T} = 1.15 \times \frac{1}{60} \times 7{,}500 = 144 \text{ A}$$

Hence current in line C, I_C = 144 A

$$\text{Main primary current, } I_1 = K\, I_{2M} = \frac{1}{60} \times 7{,}500 = 125 \text{ A}$$

The current in each of the lines A and B consists of the main primary current K I_{2M} and one half of the teaser primary current I_C vectorially combined. As the power factor is the same on both secondaries, the line currents are in quadrature.

Line currents,

$$I_A = I_B = \sqrt{I_1^2 + \left(\frac{1}{2} I_C\right)^2} = \sqrt{(125)^2 + \left(\frac{144}{2}\right)^2} = 144 \text{ A } \textbf{Ans.}$$

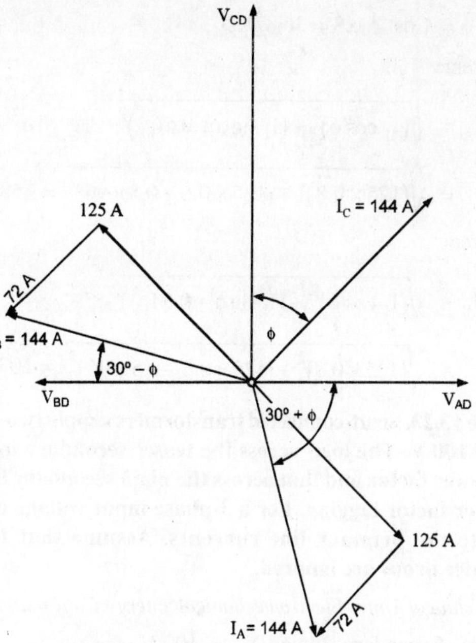

Fig. 13.15 *Phasor Diagram*

Phase angle between phasor I_C and I_A

$$= 90° + \tan^{-1}\frac{72}{125} = 120°$$

Similar phase difference exists between phasors I_C and I_B and phasors I_A and I_B.

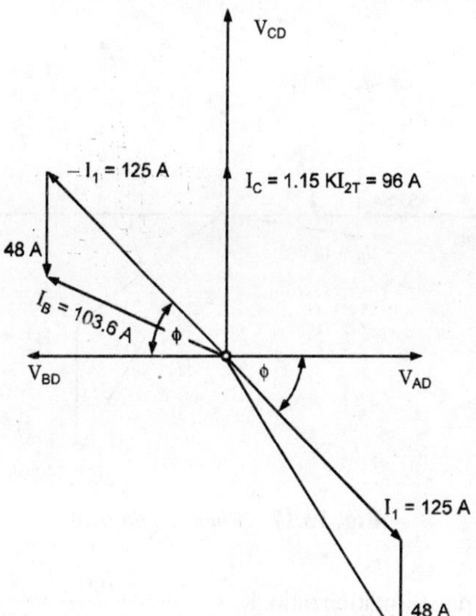

Fig. 13.16 *Phasor Diagram*

$$(ii) \text{ Secondary current, } I_{2T} = \frac{500 \times 1{,}000}{100 \times 1.0} = 5{,}000 \text{ A}$$

$$\text{Teaser primary current, } I_{1T} = \frac{k\, I_{2T}}{0.866} = \frac{5{,}000}{60 \times 0.866} = 96 \text{ A}$$

Hence current in line C = 96 A

$$\text{Secondary current, } I_{2M} = \frac{600 \times 1{,}000}{100 \times 0.8} = 7{,}500 \text{ A}$$

$$\text{Main primary current, } I_1 = k\, I_{2M} = \frac{7{,}500}{60} = 125 \text{ A}$$

$$\phi = \text{Cos}^{-1}\ 0.8 = 36.87°$$

Line current,

$$I_A = \sqrt{(I_1 \cos\phi)^2 + (I_1 \sin\phi + 0.5 I_{1T})^2}$$

$$= \sqrt{(125 \times 0.8)^2 + (125 \times 0.6 + 0.5 \times 96)^2} = 158.5 \text{ A } \textbf{Ans.}$$

Line current,

$$I_B = \sqrt{(I_1 \cos\phi)^2 + (I_1 \sin\phi - 0.5 I_{1T})^2}$$

$$= \sqrt{(125 \times 0.8)^2 + (125 \times 0.6 - 0.5 \times 96)^2} = 103.6 \text{ A } \textbf{Ans.}$$

Example 13.23. Scott-connected transformers supply two single phase loads at 100 V. The load across the teaser secondary in 300 kW at unity power factor and that across the main secondary is 200 kW at 0.8 power factor lagging. For a 3-phase input voltage of 11,000 V, calculate the primary line currents. Assume that the leakage impedance drops are ignored.

[U.P. Technical Univ. Electromechanical Energy Conversion-I, 2008-09]

Solution: Secondary voltage, V_2 = 100 V

Primary main voltage, V_1 = 11,000 V

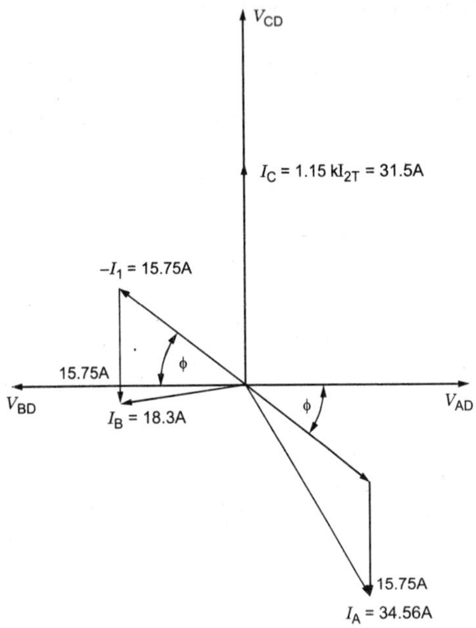

Fig. 13.17 *Phasor Diagram*

$$\text{Transformation ratio, K} = \frac{V_2}{V_1} = \frac{100}{11,000} = \frac{1}{110}$$

$$\text{Teaser secondary current, } I_{2T} = \frac{300 \times 1,000}{100 \times 1.0} = 3,000 \text{ A}$$

$$\text{Teaser primary current, } I_{1T} = \frac{k I_{2T}}{0.866} = \frac{3,000}{110 \times 0.866} = 31.5 \text{ A}$$

Hence current in line C = 31.5 A

$$\text{Secondary main current, } I_{2M} = \frac{200 \times 1,000}{100 \times 0.8} = 2,500 \text{ A}$$

$$\text{Main primary current, } I_1 = k I_2 M = \frac{2,500}{110} = 22.73 \text{ A}$$

$$\phi = \text{Cos}^{-1}\ 0.8 = 36.87°$$

$$\text{Line current, } I_A = \sqrt{(I_1 \cos\phi)^2 + (I_1 \sin\phi + 0.5\, I_{1T})^2}$$

$$= \sqrt{(22.73 \times 0.8)^2 + (22.73 \times 0.6 + 0.5 \times 31.5)^2}$$

$$= \sqrt{(18.184)^2 + (13.638 + 15.75)^2} = 34.56 \text{ A } \textbf{Ans.}$$

$$\text{Line current, } I_B = \sqrt{(I_1 \cos\phi)^2 + (I_1 \sin\phi - 0.5\, I_{1T})^2}$$

$$= \sqrt{(22.73 \times 0.8)^2 + (22.73 \times 0.6 - 0.5 \times 31.5)^2}$$

$$= 18.3 \text{ A } \textbf{Ans.}$$

Example 13.24. It is desired to transform 2,400 V, 5,000 kVA three phase power to 2-phase power at 600 V by Scott-connected transformers. Determine the voltage and current ratings of both primary and secondary of each transformer. Neglect the transformer no-load currents. [*U.P. Technical Univ. Electromechanical Energy Conversion-I, 2007-08; Rajasthan Univ. Electrical Machines-I, 2004*]

Solution: Primary volts, V_1 = 2,400 V

Secondary volts, V_2 = 600 V

$$\text{Transformation ratio, K} = \frac{V_2}{V_1} = \frac{600}{2,400} = \frac{1}{4}$$

Teaser Transformer

Voltage rating of secondary = V_2 = 600 V

Voltage rating of primary = 0.866 V_1 = 0.866 × 2,400 = 2,080 V **Ans.**

Current rating of secondary,

$$I_{2T} = \frac{\text{Load on teaser transformer} \times 1,000}{V_2}$$

$$= \frac{5,000 \times 1,000}{2 \times 600} = 4,167 \text{ A } \textbf{Ans.}$$

Current rating of primary,

$$I_{1T} = 1.15\, K\, I_{2T} = 1.15 \times \frac{1}{4} \times 4,167 = 1,200 \text{ A } \textbf{Ans.}$$

Main Transformer

Voltage rating of secondary = V_2 = 600 V **Ans.**

Voltage rating of primary = V_1 = 2,400 V **Ans.**

Current rating of secondary = $I_{2M} = I_{2T}$ = 4,167 A

∵ load is same on both transformers

$$\text{Main primary current, } I_1 = K\, I_{2M} = \frac{1}{4} \times 4,167 = 1,042 \text{ A}$$

Current rating of primary,

$$I_A \text{ or } I_B = \sqrt{I_1^2 + \left(\frac{1}{2} I_{1T}\right)^2} \quad \because \text{pf is same on both secondaries}$$

$$= \sqrt{(1,042)^2 + \left(\frac{1}{2} \times 1,200\right)^2} = 1,200 \text{ A } \textbf{Ans.}$$

Example 13.25. A balanced 3-phase, 100 kW load at 400 V and 0.8 pf lagging is to be obtained from a balanced two phase 1,100 V lines. Determine the kVA rating of each unit of the Scott-connected transformer.

[*J.N. Technological Univ. Electrical Machines-II, January-2014*]

Solution: Secondary line voltage, V_2 = 400 V

Primary voltage, V_1 = 1,100 V

Voltage rating of main transformer secondary

$$= \text{Supply line voltage} = 400 \text{ V}$$

Voltage rating of teaser transformer secondary

$$= 0.866 \text{ V}_2 = 0.866 \times 400 = 346.4 \text{ V}$$

Voltage rating of primary of each of transformer

$$= \text{V}_1 = 1{,}100 \text{ V}$$

Current rating of primary of each of transformers,

$$\text{I}_1 = \frac{\text{Load in kW} \times 1{,}000}{2 \times \text{V}_1 \times \text{pf}} = \frac{100 \times 1{,}000}{2 \times 1{,}100 \times 0.8} = 56.82 \text{ A}$$

Current rating of secondary of each transformer,

$$\text{I}_2 = \frac{\text{Load in kW} \times 1{,}000}{\sqrt{3}\,\text{V}_2 \times \cos\phi} = \frac{100 \times 1{,}000}{\sqrt{3} \times 400 \times 0.8} = 180.42 \text{ A}$$

The volt-ampere rating of teaser primary as well as secondary is the same i.e. $1{,}100 \times 56.82 \times 10^{-3} = 62.5$ kVA **Ans.**

The volt-ampere rating of secondary main transformer

$$= \frac{400 \times 180.42}{1{,}000} = 72.17 \text{ kVA } \textbf{Ans.}$$

The volt-ampere rating of primary of main transformer

$$= \frac{1{,}100 \times 56.82}{1{,}000} = 62.5 \text{ kVA } \textbf{Ans.}$$

Incidentally, if two identical transformers are used to provide interchangeability, then both must of rating 72.17 kVA **Ans.**

Example 13.26. Two 110 V single-phase furnaces take loads of 500 kW and 800 kW respectively at a power factor of 0.71 lagging and are supplied from 6,600 V, 3-ϕ mains through a Scott-connected transformer combination. Calculate the currents in 3-ϕ lines, neglecting transformer losses. Draw the phasor diagram.

[*G.B. Technical Univ. Electromechanical Energy Conversion-I, 2012-13; R.G. Technical Univ. Electrical Machines-I, June-2014*]

Solution: Transformation ratio, $\text{K} = \frac{110}{6{,}600} = \frac{1}{60}$

Main secondary current, $\text{I}_{2\text{M}} = \frac{800 \times 1{,}000}{110 \times 0.71} = 10{,}243$ A

Main primary current, $\text{I}_1 = \text{K I}_{2\text{M}} = \frac{10{,}243}{60} = 170.7$ A

Teaser secondary current, $\text{I}_{2\text{T}} = \frac{500 \times 1{,}000}{110 \times 0.71} = 6{,}402$ A

Teaser primary current, $\text{I}_{1\text{T}} = \frac{\text{K I}_{2\text{T}}}{0.866} = \frac{1}{60} \times \frac{6{,}402}{0.866} = 123.2$ A

Hence current in line C, $\text{I}_\text{C} = \text{I}_{1\text{T}} = 123.2$ A

$$\text{Line currents, } \text{I}_\text{A} = \text{I}_\text{B} = \sqrt{\text{I}_1^2 + \left(\frac{1}{2}\text{I}_\text{C}\right)^2}$$

$$= \sqrt{(170.7)^2 + \left(\frac{123.2}{2}\right)^2}$$

$$= 181.5 \text{ A } \textbf{Ans.}$$

Phase angle, $\phi = \cos^{-1} 0.71 = 44.765°$ (lagging)

Phasor diagrams are shown in Fig. 13.18.

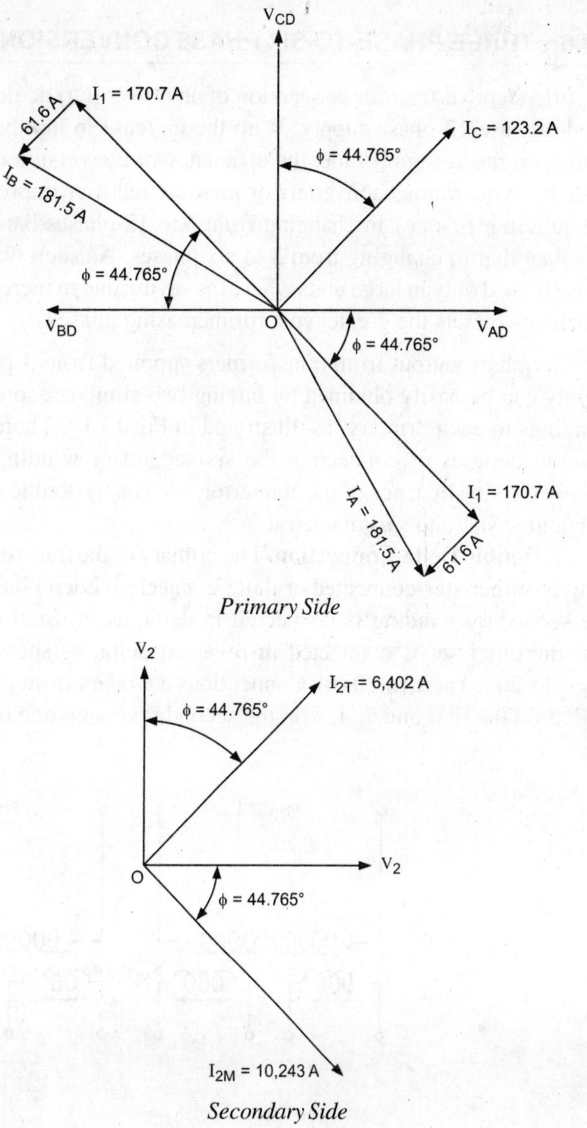

Fig. 13.18 *Phase Diagram*

Example 13.27. Two single-phase Scott-connected transformers supply a 3-phase 4-wire 50 Hz distribution system with 400 V between lines. The hv windings are connected to a 2-phase 6,000 V (per phase) system. The core area is 250 cm^2, while the maximum allowable flux density is 1.2 T. Determine the number of turns on each winding and the point to be tapped for the neutral wire on the 3-phase side.

Solution: Maximum allowable flux,

$$\Phi_{max} = \text{B}_{max} \times \text{core area} = 1.2 \times 0.025 = 0.03 \text{ Wb}$$

Voltage on primary (hv) side,

$$\text{V}_1 = 6{,}000 \text{ V}$$

Number of turns of hv side of both of the transformers,

$$\text{N}_1 = \frac{\text{V}_1}{4.44 f \Phi_{max}} = \frac{6{,}000}{4.44 \times 50 \times 0.03} = 900 \textbf{ Ans.}$$

Number of turns of lv side of main transformer,

$$\text{N}_2 = \frac{\text{V}_2}{4.44 f \Phi_{max}} = \frac{400}{4.44 \times 50 \times 0.03} = 60 \textbf{ Ans.}$$

Number of turns on lv side of teaser transformer

$$= 0.866 \text{ N}_2 = 0.866 \times 60 = 52 \textbf{ Ans.}$$

Number of turns CN (Fig. 13.13)

$$= \frac{2}{3} \times 52 = 35 \textbf{ Ans.}$$

13.6 THREE PHASE-TO-SIX PHASE CONVERSION

Rectifiers/converters, for conversion of ac power into dc power need 6-phase/12-phase supply. With the increase in number of phases on the ac supply side, the efficiency of conversion of ac to dc by synchronous converters or mercury rectifier improves, the gain in efficiency in changing from 6 to 12-phases is much less than that in changing from 3 to six phases. As such twelve phase is used only in large units where the saving due to increased efficiency offsets the greater cost for increasing phases.

Six phase output from transformers supplied from 3-phase supply can be easily obtained by having two similar secondary windings to each primary, as illustrated in Fig 13.19. There are several methods of connecting the six secondary windings to supply the six slip rings of the convertor, such as (*i*) double delta (*ii*) double star and (*iii*) diametral.

1. Double Delta Connection. The primary of the transformer may be either star-connected or delta-connected. Each phase of the secondary winding is connected in delta, as in usual form and the other set is connected in reversed delta, as shown in Figs. 13.20 (*a*) and 13.20 (*b*). Connections are taken from points 1, 3, 5 on the first and 2, 4, 6 on the second reversing one of the two triangles with respect to each other. Since both windings are connected to a symmetrical armature, therefore, this fixes the relative position of two voltage triangles, hence length of six phasors obtained by joining their corners are equal to another and represent ac voltage applied to slip-rings. Since there is a phase difference of 60° between the two consecutive phasors, so they give true 6-phase supply.

2. Double Star Connection. In this case one set of secondary is connected in star as in usual form and the other set of secondary is connected in reversed star, as shown in Fig. 13.21. The two star points coincide due to symmetry of the armature winding. The length of six phasors obtained by joining the star ends is equal and represents ac voltage applied to the slip-rings of a 6-phase convertor. This connection is often used for 3-wire dc supply because neutral point of star supplies neutral wire for the arrangement.

3. Diametral Connection. In this case, each phase of secondary has one winding. The six ends of three secondary windings are connected to diametral points on the armature winding and the middle points are connected together, as shown in Fig. 13.22. This method is very simple and most commonly

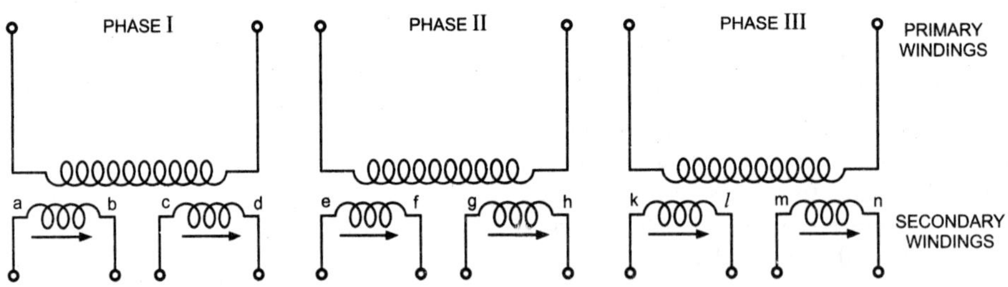

Fig. 13.19

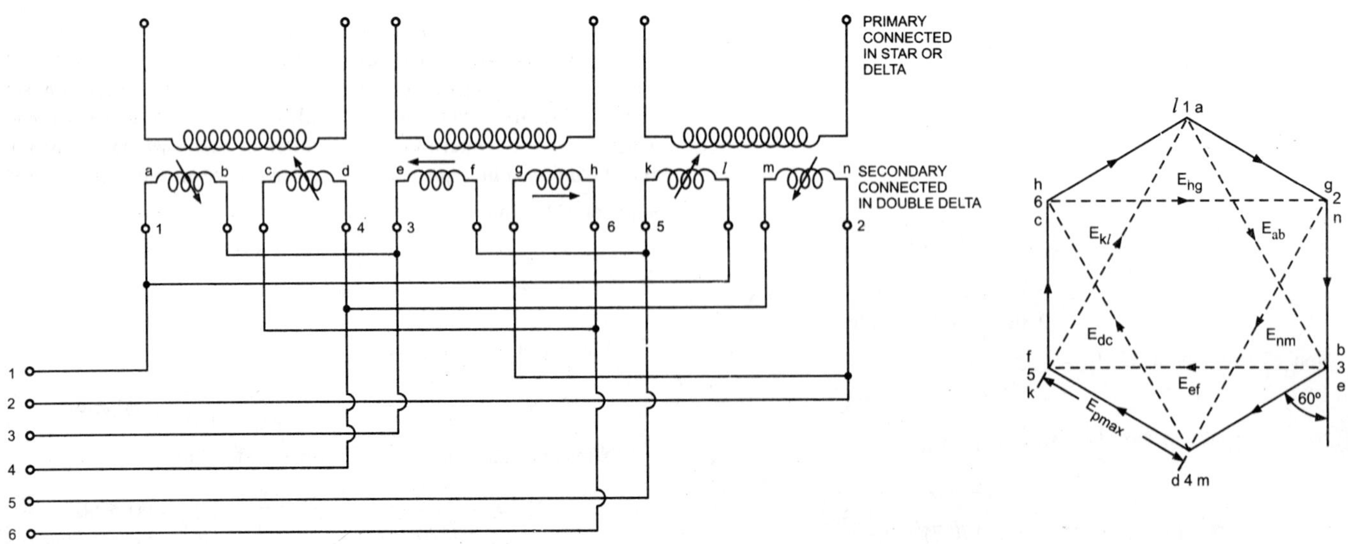

(*a*) (*b*)

Fig. 13.20

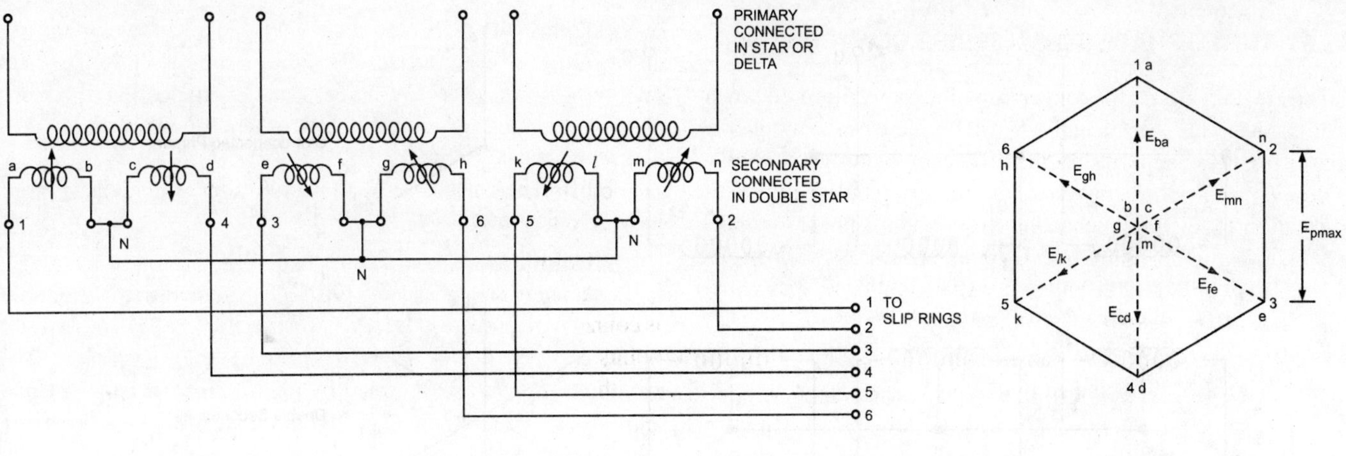

(a) (b)

Fig. 13.21

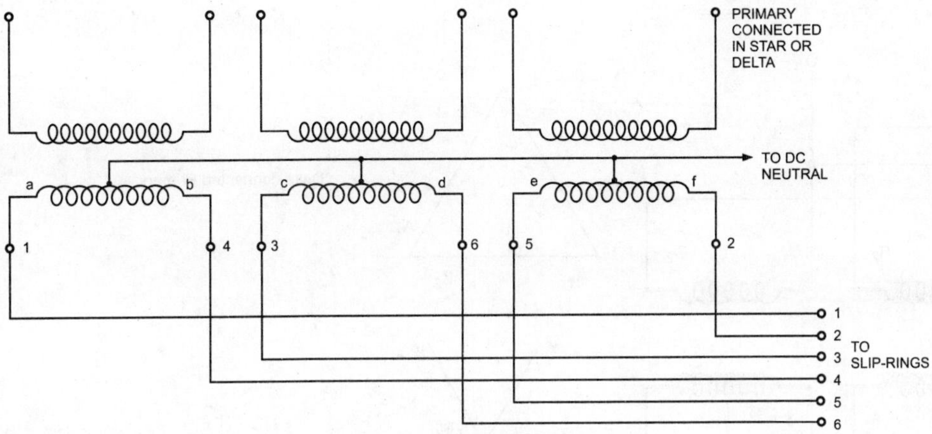

Fig. 13.22(a)

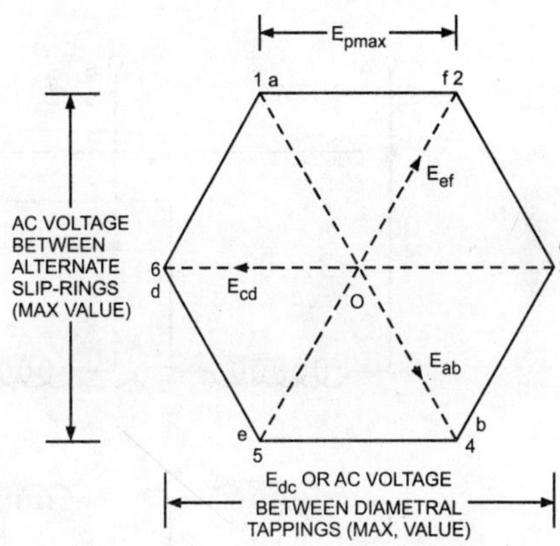

Fig. 13.22(b)

used. The rotary convertor when supplying a 3-wire dc system on dc side, the middle wire is connected to this star point, otherwise star point needs not to be formed, because potential of the middle points is fixed by the armature itself. In this case maximum value of emf induced in each phase of the secondary in represented by the diameter of the circle, whereas in double star method is given by the radius.

13.7 THREE PHASE-TO-TWELVE PHASE CONVERSION

There are the following commonly used schemes for three-phase to twelve-phase conversion.

1. Star-delta double star
2. Double chord
3. Double star zig-zag.

1. Star-Delta Double Star Connections. This arrangement needs two banks of three single-phase transformers or two three-phase transformers. The connections are shown in Fig. 13.23. The secondary windings are arranged in the form of a pair of double-star connections. The primaries of one set of transformers or one of three phase transformers is star-connected while that of the other is delta connected. There is then a phase shift of 30° between the secondary star voltages of the two six phase systems resulting in a balanced twelve phase system.

Each phase winding of the star-connected primary is designed for $\frac{V_1}{\sqrt{3}}$, V_1 being the input line voltage, while each phase winding of delta-connected primary is designed for V_1 volts. If V_2 is the line-to-line voltage of the twelve-phase output *i.e.* V_2 = length of phasors $b'f$ or fc or cd', etc.

$$\text{then} \quad \frac{b'f/2}{ne'} = \text{Sin}\ \frac{30°}{2}$$

$$\text{or} \quad \frac{V/2}{ne'} = \text{Sin } 15°$$

$$\text{or } n\,e' = \frac{V_2}{2 \sin 15°} = 1.932\ V_2$$

Thus line-to-neutral voltage on the secondary side is 1.932 V_2 and the secondary with a mid-tap should be designed for twice the line-to-neutral voltage on the secondary side *i.e.* 2 × 1.932 V_2 = 3.864 V_2.

The transformer with primaries connected in delta must have a tertiary winding connected in delta so that a path for third harmonic currents be provided.

2. Double-Chord Connections. In this arrangement three single-phase transformers or one three-phase transformer is required. The secondary of each phase has two equal sections.

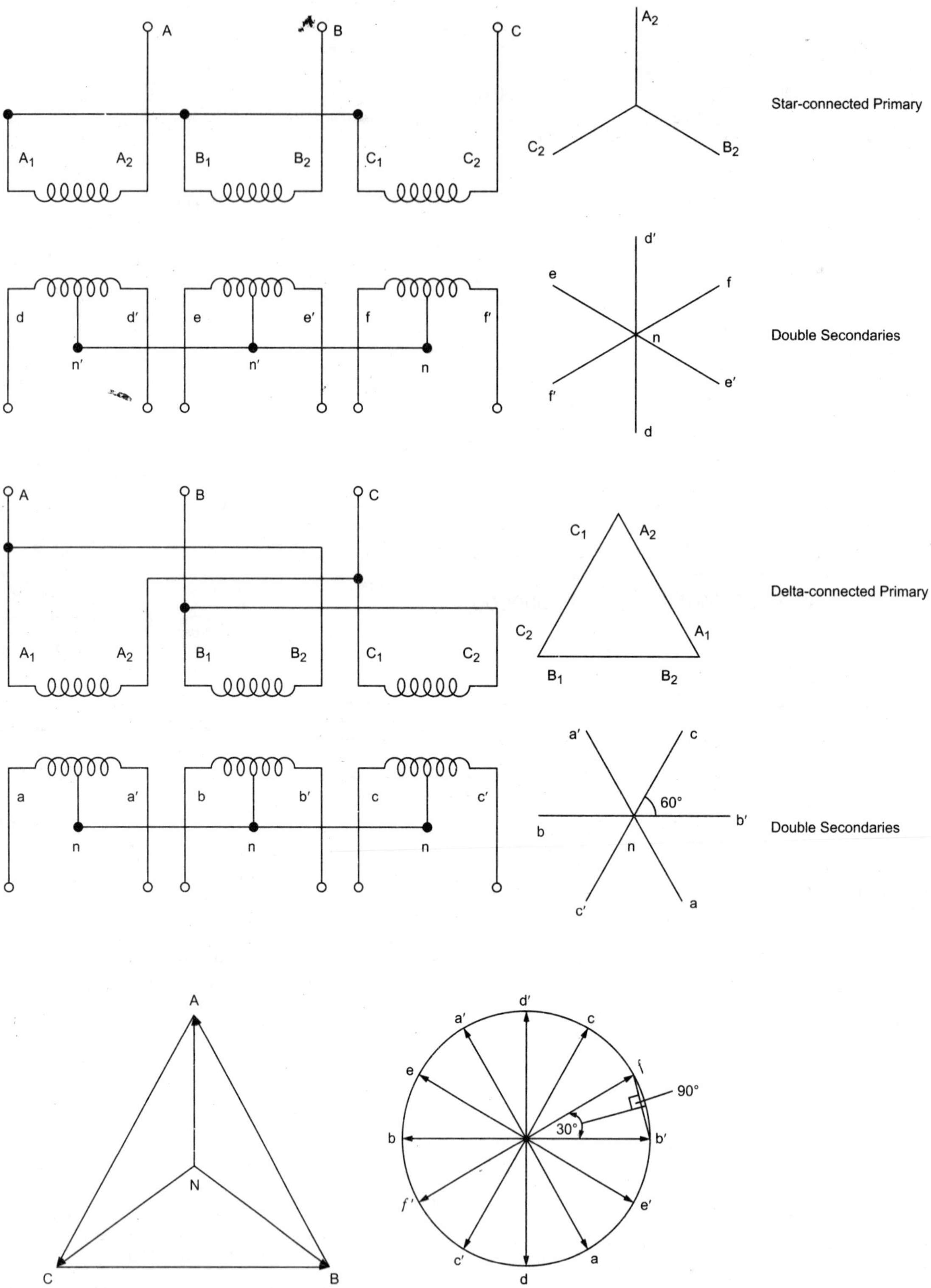

Fig. 13.23 *Three-To-Twelve Phase Conversion (Star-Delta Double-Star-Connection)*

The connections are shown in Fig. 13.24. When secondary terminals are connected to the corresponding taps of rotary converter or motor converter, the line-to-line voltages will form a balanced 12-phase system.

3. Double Star-Zig-zag Connections. For this arrangement, three single-phase transformers or one three phase transformers is required, but each must have 5 secondaries. The primary is delta-connected. One of the secondaries is the main secondary with centre-taps and provides voltages *na*, *na′* etc. Four other secondaries have equal but less number of turns so as to provide the voltages *a*1, *b*5 etc. The number of turns of different windings should be such as to provide balanced 12-phase voltages. Connections are shown in Fig. 13.25.

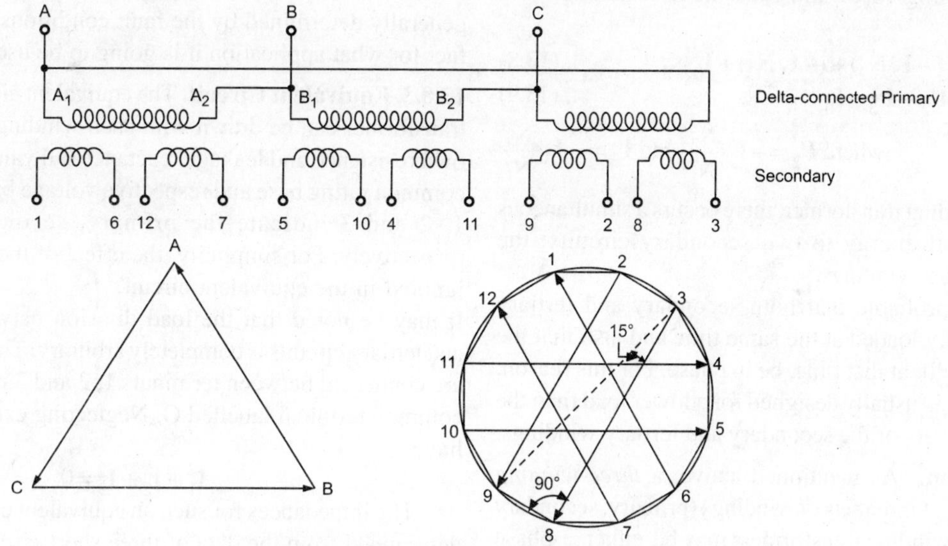

Fig. 13.24 *Double-Chord Connections For Three-Phase-To-Twelve-Phase Conversion*

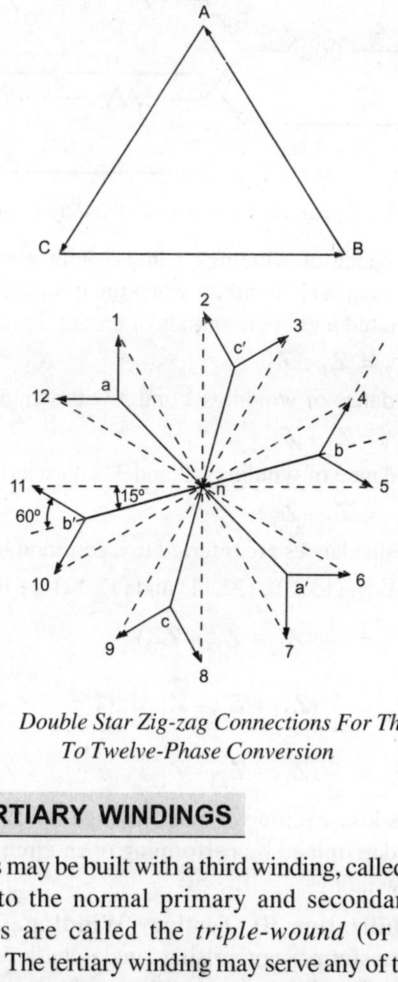

Fig. 13.25 *Double Star Zig-zag Connections For Three-Phase To Twelve-Phase Conversion*

13.8 TERTIARY WINDINGS

Transformers may be built with a third winding, called the *tertiary*, in addition to the normal primary and secondary and such transformers are called the *triple-wound* (or *3-winding*) transformers. The tertiary winding may serve any of the following purposes:

(*i*) To supply the substation auxiliaries at a voltage different from those of the primary and secondary windings.

(*ii*) To supply phase compensating devices, such as condensers* operated at a voltage which is different from both primary and secondary voltage.

(*iii*) To interconnect three supply systems operating at different voltages.

(*iv*) To load large split-winding generators.

(*v*) To measure voltage of an HV testing transformer.

(*vi*) In star/star connected transformers, to allow sufficient earth fault current to flow for operation of protective equipment, to suppress harmonic voltages, and to limit voltage unbalance when the main load is unsymmetrical.

The tertiary winding is called the *auxiliary winding* when it is employed for supplying an additional small load at a different voltage. On the other hand, it is called the *stabilising winding* when it is employed to limit the short-circuit current as in (*vi*).

Tertiary windings are normally delta-connected so that when faults and short circuits occur on the primary or secondary sides (particularly between lines and earth), the considerable unbalance produced in phase voltages may be compensated by the circulating currents flowing in the closed Δ. The reactance of the windings should be large enough to limit the circulating currents in order that there is no overheating of the windings.

13.8.1. Principle of Operation. The *operating principle* of a 3-winding (or triple wound) transformer is essentially the same as for the two-winding type. The primary winding of a three winding transformer acts as the magnetizing winding, and its current produces the main magnetic flux. The flux links the secondary and tertiary windings and induces emfs in them in proportion to the number of turns in the windings. When loads are connected across the secondary and tertiary windings, currents I_2 and I_3 flow in them. The demagnetizing effect of these currents will be determined by the geometrical sum of the magnetizing forces of the secondary and tertiary windings. So the magnetizing force of the primary winding must balance this combined

* In practically all extensive high voltage systems, regulation of voltage at certain points is provided by means of synchronous condensers. These condensers are connected to the high voltage system through transformers and from economic considerations, in many cases the condenser connections are made through a third winding in the transformers, tying two sections of the system together.

secondary magnetizing force and must have a magnetizing component.

$$i.e. \quad \mathbf{I}_1 N_1 = (-\mathbf{I}_2 N_2) + (-\mathbf{I}_3 N_3) + \mathbf{I}_0 N_1 \qquad ...(13.8)$$

$$\text{or } \mathbf{I}_1 = \mathbf{I'}_2 + \mathbf{I'}_3 + \mathbf{I}_0 \qquad ...(13.9)$$

$$\text{where } \mathbf{I'}_2 = -\mathbf{I}_2 \frac{N_2}{N_1} \text{ and } \mathbf{I'}_3 = -\mathbf{I}_3 \frac{N_3}{N_1}$$

Thus in a 3- winding transformer, there occurs a simultaneous transfer of electrical energy to two secondary circuits: the secondary and tertiary windings.

It is highly improbable that both secondary and tertiary windings will be fully loaded at the same time and also that the currents $\mathbf{I'}_2$ and $\mathbf{I'}_3$ will, at that time, be in phase. For this reason, the primary winding is usually designed for a lower load than the sum of the rated powers of the secondary and tertiary windings.

13.8.2. Construction. As mentioned above, a *three-winding transformer* consists of three sets of windings (primary, secondary and tertiary). Three winding transformers may be either 3-phase units or 1-phase units connected in a 3-phase bank. Such transformers have a large kVA rating. Since the three windings are operated at three different voltages, they may be termed as high voltage, medium voltage and low voltage windings.

Two alternative arrangements of windings of a 3-winding transformer are illustrated in Figs. 13.26 (*a*) and 13.26 (*b*) respectively. The hv winding in both cases is the outermost winding because of insulation consideration. The lv and mv windings are placed adjacent to each other with either of the two nearer to the core. The lv and mv windings are kept near to each other in order to reduce the leakage reactance between this pair of windings.

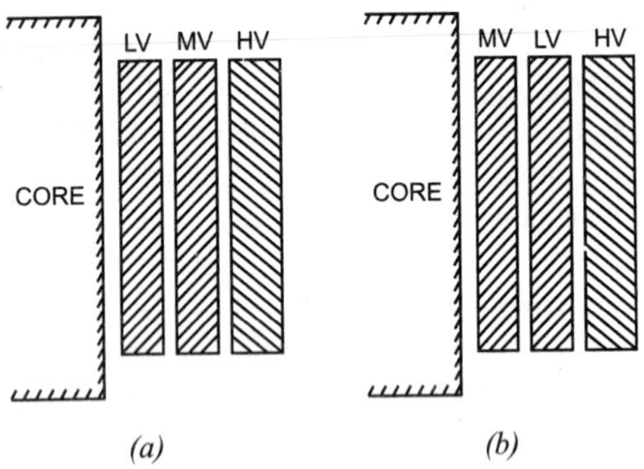

Fig. 13.26 *Winding Arrangement of a Triple- Wound Transformer*

Thus the two possible winding arrangements for windings starting from core outwards are : (*i*) lv, mv and hv and (*ii*) mv, lv and hv.

In two-winding transformers, the kVA ratings of both the primary and the secondary windings are the same but it is not so in case of 3-winding transformers. The kVA rating of the 3-winding transformer is considered to be equal to the largest kVA rating of any of its windings.

The rating of tertiary winding depends upon the intended application. If it is provided for supply of an additional load, the winding is designed and calculated on the same basis as the primary and secondary. When employed for balancing of loads and controlling short-circuit currents, the tertiary winding carries current only for short periods, and its rating depends mainly on its heat capacity. In practice, the x-section of the winding wire is generally determined by the fault conditions irrespective of the fact for what application it is going to be used.

13.8.3. Equivalent Circuit. The equivalent circuit of a 3-winding transformer can be drawn with each winding represented by its own resistance and leakage reactance (all values are reduced to a common rating base and respective voltage basis. The subscripts 1, 2 and 3 indicate the primary, secondary and tertiary respectively. For simplicity, the effect of the exciting current is ignored in the equivalent circuit.

It may be noted that the load division between the secondary and tertiary circuits is completely arbitrary. Three external circuits are connected between terminals 1, 2 and 3 respectively and the common terminal labelled O. Neglecting exciting current I_0 we have

$$\mathbf{I}_1 + \mathbf{I}_2 + \mathbf{I}_3 = 0.$$

The impedances for such an equivalent circuit can be readily determined from the data of three short-circuit tests as follows. For circuit shown in Fig. 13.27.

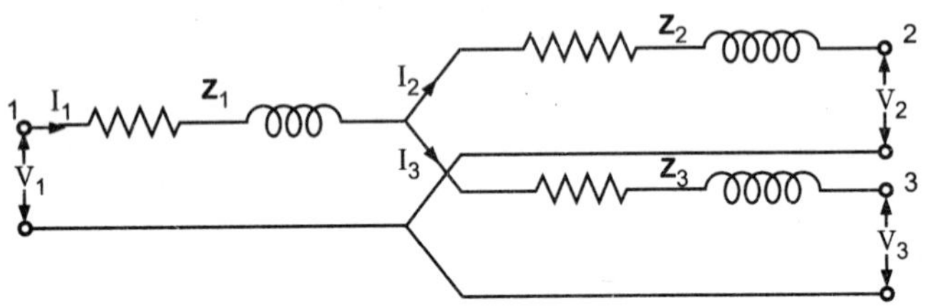

Fig. 13.27 *Equivalent Circuit of a 3-Winding Transformer*

SC impedance of windings 1 and 2 with winding 3 open (impedance of equivalent circuit when the terminals of circuit 2 are short circuited and the terminals of circuit 3 are open),

$$\mathbf{Z}_{12} = \mathbf{Z}_1 + \mathbf{Z}_2 \qquad ...(13.10)$$

SC impedance of windings 1 and 3 with winding 2 open

$$\mathbf{Z}_{13} = \mathbf{Z}_1 + \mathbf{Z}_3 \qquad ...(13.11)$$

and SC impedance of windings 2 and 3 with winding 1 open,

$$\mathbf{Z}_{23} = \mathbf{Z}_2 + \mathbf{Z}_3 \qquad ...(13.12)$$

All the impedances are referred to a common base.

Solving Eqs. (13.10), (13.11) and (13.12) we have

$$\mathbf{Z}_1 = \frac{1}{2}(\mathbf{Z}_{12} + \mathbf{Z}_{13} - \mathbf{Z}_{23}) \qquad ...(13.13)$$

$$\mathbf{Z}_2 = \frac{1}{2}(\mathbf{Z}_{23} + \mathbf{Z}_{12} - \mathbf{Z}_{13}) \qquad ...(13.14)$$

$$\text{and } \mathbf{Z}_3 = \frac{1}{2}(\mathbf{Z}_{13} + \mathbf{Z}_{23} - \mathbf{Z}_{12}) \qquad ...(13.15)$$

The core-loss, exciting or magnetizing impedance and turn-ratio can be determined by performing open-circuit test on any of the three windings.

13.8.4. Stabilization By Tertiary Winding. The star/star connected transformer comprising single-phase units, or three phase shell- or 5 limb core type units have the drawbacks (*i*) that they cannot readily supply unbalanced loads between line and neutral and (*ii*) that their phase emfs may get distorted by third harmonics emfs. The reason is that magnetic flux set up in the iron core due to third harmonic and zero sequence currents have an iron path. The zero phase sequence impedance is, therefore, high (0.5 pu). The flow of earth fault is severely restricted, and may be insufficient to operate the protective gear. The zero sequence impedance to out-of-balance or earth-fault currents include the very high impedance between the secondary winding of the loaded or faulty phase and the primary windings of the other two phases. Abnormal voltages, therefore, occur when zero phase sequence current flows. The output load or fault current flows through one phase only, whereas the corresponding input load current, in the absence of a primary neutral, has to return through the other two phases B and C (Fig. 13.28) which have very high reactance owing to absence of their secondary currents. The voltages of phases B and C rise and become nearly equal to the primary line voltage and the voltage of the faulty or loaded phase A reduces to very low value. Thus serious voltage unbalance is developed. By use of delta-connected tertiary winding zero sequence currents are provided low reactance paths and voltage unbalance on unbalanced loads or ground faults is prevented. This is illustrated in Fig. 13.29. It has been assumed that each winding has equal turns per phase. This action is in brief, to interlink magnetically the separate magnetic circuits.

The mesh-connected tertiary provides a path for the third harmonic currents required for magnetizing to the comparatively high saturation flux densities now employed. The circulation of the third harmonic currents then provides, with the primary current, the excitation necessary to secure a substantially sinusoidal flux and emf.

The two drawbacks, mentioned above, are reduced to some extent in three limb core type transformers, because zero-sequence flux is forced out of the core limbs to high-reluctance air and tank (Fig. 13.29) leading to reduction in flux and zero-sequence impedance thereby reducing voltage unbalanced considerably. The third-harmonic flux also has similarly high reluctance path and so magnitude of the third harmonic flux is small. The distortion in voltage waveshape is small and the transformer has more or less sinusoidal flux and emf when the input voltage is sinusoidal. It is nevertheless usual to provide a delta-connected tertiary even to the three limb core type transformer if it is star/star connected.

Example 13.28. A small 10 VA, 115 V primary transformer has two secondary windings of 6.3 V and 5 V respectively with resistive impedances of 0.2 Ω and 0.15 Ω respectively. Find the circulating current in secondary when low voltage windings are connected in parallel.

Solution: When the two low voltage windings are connected in parallel, the circulating current will be given as

$$I = \frac{E_2 - E_1}{Z_1 + Z_2} = \frac{6.3 - 5}{0.2 + 0.15} = 3.714 \text{ A } \textbf{Ans.}$$

Example 13.29. A delta/star 6.6/0.4 kV bank of three identical single-phase transformers supplies a load of 80 kW at unity power factor between the line R and neutral and a balanced load of 600 kW at 0.8 lagging power factor. Determine the current magnitude in each primary winding and each input line. State assumptions. Draw the relevant diagrams.

Solution: Secondary phase voltage, $V_{P2} = \dfrac{0.4 \times 1{,}000}{\sqrt{3}} = 231$ V

Primary phase voltage, $V_{P1} = 6.6 \times 1{,}000 = 6{,}600$ V

Turn-ratio, $K = \dfrac{V_{P2}}{V_{P1}} = \dfrac{231}{6{,}600}$

Current in red phase of secondary due to single phase load of 80 kW at unity.

$$= \frac{80 \times 1{,}000}{231} = 346.32 \text{ A}$$

Current in all phases due to a balanced load of 600 kW at 0.8 pf lagging

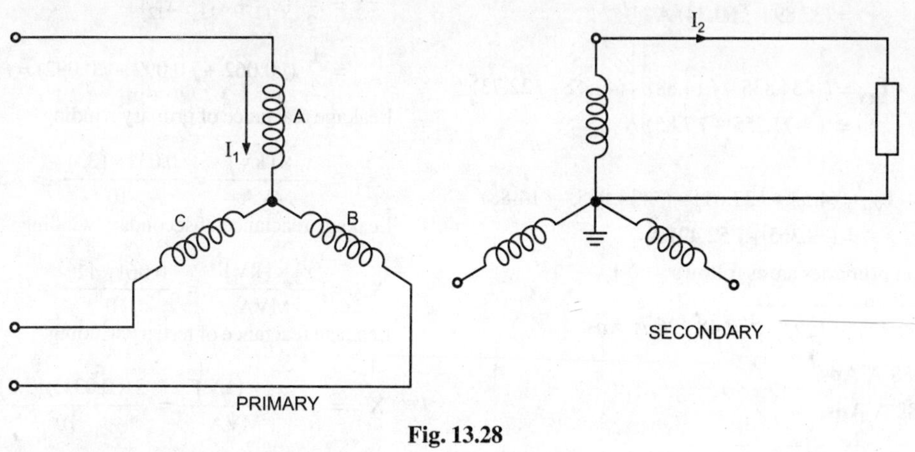

Fig. 13.28

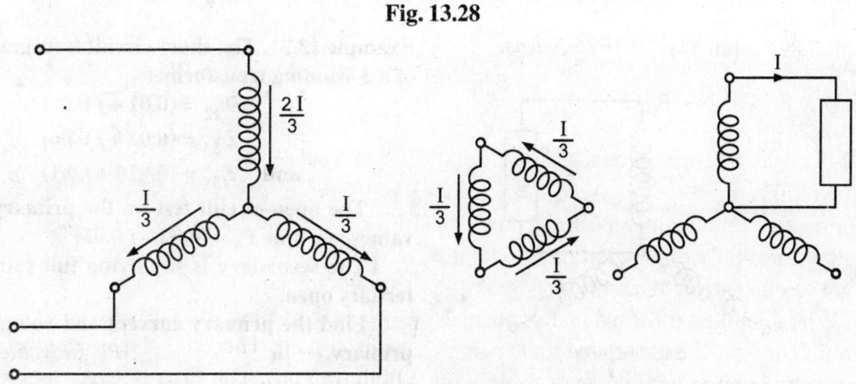

Fig. 13.29

$$= \frac{600 \times 1{,}000}{3 \times 231 \times 0.8} = 1{,}082.53 \text{ A}$$

lagging behind their respective phase voltages by an angle $\cos^{-1} 0.8$ (or 36.87°)

Assuming V_{RY} as the reference phasor for primary

$$V_{RY} = 6{,}600 \angle 0° \text{ V};$$
$$V_{YB} = 6{,}600 \angle -120°$$
$$\text{and } V_{BR} = 6{,}600 \angle 120°$$

Current in phase RY due to single phase load of 80 kW at unity pf,

$$= 346.32 \times \frac{231}{6{,}600} = 12.12 \text{ A in phase with voltage } V_{RY}$$

Current in phase RY due to balanced load of 600 kW at 0.8 pf lagging

$$= 1{,}082.53 \times \frac{231}{6{,}600} = 37.88 \text{ A} \angle -36.87°$$

Current in phase RY,

$$\mathbf{I}_{RY} = 12.12 \angle 0° + 37.88 \angle -36.87°$$
$$= 12.12 + (30.3 - j\,22.73) = (42.42 - j\,22.73) \text{ A}$$

Current in phase YB,

$$\mathbf{I}_{YB} = 37.88 \angle -120° - 36.87°$$
$$= 37.88 \angle -156.87° = (-34.835 - j\,14.88) \text{ A}$$

Current in phase BR,

$$\mathbf{I}_{BR} = 37.88 \angle -120° - 36.87°$$
$$= 37.88 \angle 83.13° = (4.53 + j\,37.61) \text{ A}$$

Line current,

$$\mathbf{I}_R = \mathbf{I}_{RY} - \mathbf{I}_{BR} = (42.42 - j\,22.73) - (4.53 + j\,37.61)$$
$$= (37.89 - j\,60.34) \text{ A}$$

Line current,

$$\mathbf{I}_Y = \mathbf{I}_{YB} - \mathbf{I}_{RY} = (-34.835 - j\,14.88) - (42.42 - j\,22.73)$$
$$= (-77.255 + j\,7.85) \text{ A}$$

Line current,

$$\mathbf{I}_B = \mathbf{I}_{BR} - \mathbf{I}_{YB} = (4.53 + j\,37.61) - (-34.835 - j\,14.88)$$
$$= (39.365 + j\,52.49) \text{ A}$$

Currents in different primaries are as follows

$$I_{RY} = \sqrt{(42.42)^2 + (22.73)^2} = 48.126 \text{ A } \textbf{Ans.}$$
$$I_{YB} = 37.88 \text{ A } \textbf{Ans.}$$
$$I_{BR} = 37.88 \text{ A } \textbf{Ans.}$$

$$\text{Current in line R, } I_R = \sqrt{(37.89)^2 + (60.34)^2} = 71.25 \text{ A } \textbf{Ans.}$$

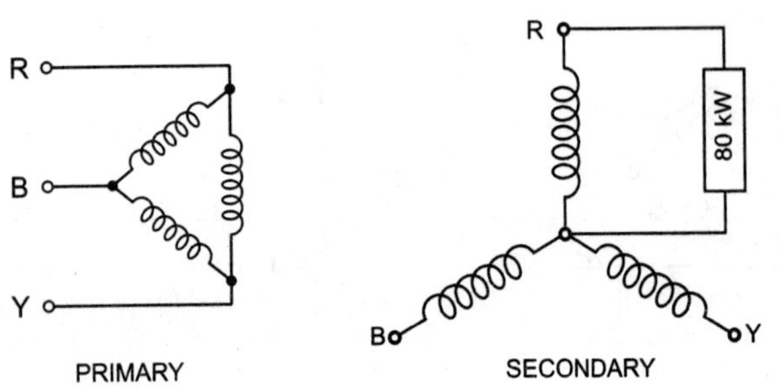

Fig. 13.30

$$\text{Current in line Y, } I_Y = \sqrt{(77.255)^2 + (7.85)^2} = 77.65 \text{ A } \textbf{Ans.}$$

$$\text{Current in line B, } I_B = \sqrt{(39.365)^2 + (52.49)^2} = 65.61 \text{ A } \textbf{Ans.}$$

Example 13.30. The following is the data pertaining to a 3-winding transformer:

HV in star ; 10 MVA, 33 kV
MV in star ; 7.5 MVA, 11kV
LV in delta ; 7.5 MVA, 3.3 kV
The per unit leakage reactances are :

$$\left.\begin{aligned} x_{12} &= j\,0.092 \\ x_{13} &= j\,0.062 \end{aligned}\right\} \textbf{measured on } \textbf{\textit{hv}} \textbf{ side}$$

$$\textbf{and } x_{23} = j\,\textbf{0.069–measured on mv side}$$

For the Y-equivalent circuit, find the coil leakage reactance in pu and also in ohms. [*Pb Univ. Elec. Machines-I, June 1993*]

Solution: $x_{12} = j\,0.092$ measured on hv side

$x_{13} = j\,0.062$ measured on hv side

$x_{23} = j\,0.069$ measured on mv side

$x_{23} = j\,0.069$ measured on mv side *i.e.* referred to MVA of 7.5

$$= j\,0.069 \times \frac{10}{7.5} = j\,0.092$$

measured on hv side *i.e.* referred to base MVA of 10

Coil leakage reactance referred to common base MVA of 10

$$x_1 = \frac{1}{2}(x_{12} + x_{13} - x_{23})$$
$$= \frac{1}{2}(j\,0.092 + j\,0.062 - j\,0.092) = j\,0.031 \textbf{ Ans.}$$
$$x_2 = \frac{1}{2}(x_{23} + x_{12} - x_{13})$$
$$= \frac{1}{2}(j\,0.092 + j\,0.092 - j\,0.062) = j\,0.061 \textbf{ Ans.}$$
$$x_3 = \frac{1}{2}(x_{13} + x_{23} - x_{12})$$
$$= \frac{1}{2}(j\,0.062 + j\,0.092 - j\,0.092) = j\,0.031 \textbf{ Ans.}$$

Leakage reactance of primary winding,

$$X_1 = \frac{x_1 \times (\text{kV})^2}{\text{MVA}} = \frac{0.031 \times (33)^2}{10} = 3.3759\ \Omega \textbf{ Ans.}$$

Leakage reactance of secondary winding,

$$X_2 = \frac{x_2 \times (\text{kV})^2}{\text{MVA}} = \frac{0.061 \times 11^2}{10} = 0.7381\ \Omega \textbf{ Ans.}$$

Leakage reactance of tertiary winding,

$$X_3 = 3 \times \frac{x_3 \times (\text{kV})^2}{\text{MVA}} = \frac{3 \times 0.031 \times (3.3)^2}{10} = 0.1013\ \Omega \textbf{ Ans.}$$

Example 13.31. The short-circuit tests gave the following pu values of a 3-winding transformer

$$\mathbf{Z}_{12} = \mathbf{(0.01 + j\,0.06)}$$
$$\mathbf{Z}_{23} = \mathbf{(0.02 + j\,0.06)}$$
$$\textbf{and } \mathbf{Z}_{31} = \mathbf{(0.016 + j\,0.1)}$$

The open-circuit test on the primary side gave the following values per unit $Y_0 = (0.02 - j\,0.05)$

The secondary is supplying full rating at 0.8 pf lagging and tertiary open.

Find the primary current and voltage at the terminals of the primary. [*Pb. Univ. Elec. Machines- I, June-1991*]

Solution: $\mathbf{Z}_{12} = 0.01 + j\,0.06$

$$\mathbf{Z}_{23} = 0.02 + j\,0.06$$
$$\mathbf{Z}_{31} = 0.016 + j\,0.1$$
$$\mathbf{Z}_1 = \frac{1}{2}(\mathbf{Z}_{12} + \mathbf{Z}_{13} - \mathbf{Z}_{23})$$
$$= \frac{1}{2}[(0.01 + j\,0.06) + (0.016 + j\,0.1) - (0.02 + j\,0.06)]$$
$$= 0.003 + j\,0.05$$
$$\mathbf{Z}_2 = \frac{1}{2}(\mathbf{Z}_{23} + \mathbf{Z}_{12} - \mathbf{Z}_{13})$$
$$= \frac{1}{2}[(0.02 + j\,0.06) + (0.01 + j\,0.06) - (0.016 + j\,0.1)]$$
$$= 0.007 + j\,0.01$$
$$\mathbf{Z}_3 = \frac{1}{2}(\mathbf{Z}_{13} + \mathbf{Z}_{23} - \mathbf{Z}_{12})$$
$$= \frac{1}{2}[(0.016 + j\,0.1) + (0.02 + j\,0.06) - (0.01 + j\,0.06)]$$
$$= 0.013 + j\,0.05$$

Secondary load current,

$$\mathbf{I}_2 = 1\,(0.8 - j\,0.6)$$

No-load current,

$$\mathbf{I}_0 = 1\,(0.02 - j\,0.05)$$

Primary current,

$$\mathbf{I}_1 = \mathbf{I}_2 + \mathbf{I}_0 = (0.8 - j\,0.6) + (0.02 - j\,0.05)$$
$$= (0.82 - j\,0.65) = 1.0464 \textbf{ Ans.}$$

Voltage at the terminals of the primary

$$= 1 + \mathbf{I}_2\,\mathbf{Z}_2 + \mathbf{I}_1\,\mathbf{Z}_1$$
$$= 1 + (0.8 - j\,0.6)\,(0.007 + j\,0.01 + (0.82 - j\,0.65)\,(0.003 + j\,0.05)$$
$$= 1.04656 + j\,0.006 = 1.04658 \textbf{ Ans.}$$

Example 13.32. In a single-phase, three-winding transformer, the turns ratio for primary; secondary; tertiary winding is 20 : 4 : 1. With lagging currents of 50A at a power factor of 0.8 in the secondary, and 60 A at a pf of 0.6 in the tertiary winding, find the primary current and power factor. *[GATE, 2000]*

Solution: Secondary winding current, $\mathbf{I}_2 = 50\,\angle -36.87°$ A

$\because$ Phase angle $\phi = \text{Cos}^{-1}\,0.8$ (lagging) $= -36.87°$

Tertiary winding current, $\mathbf{I}_3 = 60\,\angle -53.13°$ A

$\because$ $\phi = \text{Cos}^{-1}\,0.6$ (lagging) $= -53.13°$

Secondary winding current referred to primary side,

$$\mathbf{I'}_2 = \mathbf{I}_2 \frac{N_2}{N_1} = 50 \times \frac{4}{20}\,\angle -36.87° \text{ A}$$
$$= 10\,\angle -36.87° \text{ A} = (8 - j6) \text{ A}$$

Tertiary winding current referred to primary side,

$$\mathbf{I'}_3 = \mathbf{I}_3 \times \frac{N_3}{N_1} = 60 \times \frac{1}{20}\,\angle -53.13° \text{ A}$$
$$= 3\,\angle -53.13° = (1.8 - j2.4) \text{ A}$$

Total primary current,

$$\mathbf{I}_1 = \mathbf{I'}_2 + \mathbf{I'}_3 = (8 - j6) + (1.8 - j2.4) \text{ A}$$
$$= (9.8 - j8.4) \text{ A} = 12.9\,\angle -40.6° \text{ A } \textbf{Ans.}$$

i.e. Primary current, $I_1 = 12.9$ A

Primary power factor $= \text{Cos}\,\phi_1 = \text{Cos}\,(-40.6°) = 0.759$ (lagging) **Ans.**

Example 13.33. A 6,600 V/400 V/110 V star/star/mesh connected three-phase transformer has a magnetizing current of 5.5 A and the balanced three-phase loads of 1,000 kVA at 0.8 pf lag on secondary and 200 kVA, 0.5 leading pf load on the tertiary. Find the primary current and power factor. Neglect loss.

[U.P.S.C. I.E.S., Electrical Engineering-II, 2011]

Solution: Turn ratio $= N_1 : N_2 : N_3 = \frac{6,600}{\sqrt{3}} : \frac{400}{\sqrt{3}} : 110$

Let the currents in primary, secondary and tertiary windings be I_1, I_2 and I_3 respectively.

Tertiary current per phase,

$$I_3 = \frac{\text{Power supplied from tertiary winding}}{3 \times \text{tertiary voltage per phase}}$$
$$= \frac{200 \times 1,000}{3 \times 110} = 606.06\,\angle 60° \text{ A}$$

$\because$ Phase angle $= \text{Cos}^{-1}\,0.5$ lead $= 60°$

Tertiary winding current referred to primary side,

$$I'_3 = \left(\frac{N_3}{N_1}\right) I_3 = \frac{110}{6,600/\sqrt{3}} \times 606.06\,\angle 60° = 17.495\,\angle 60° \text{ A}$$

Secondary current per phase,

$$I_2 = \frac{\text{Power supplied from secondary winding}}{3 \times \text{secondary voltage per phase}}$$
$$= \frac{1,000 \times 1,000}{3 \times 400/\sqrt{3}}\,\angle -36.87°$$

$\because$ phase angle $= \text{Cos}^{-1}\,0.8$ lag $= -36.87°$

$$= 1,443.37\,\angle -36.87° \text{ A}$$

Secondary current referred to primary,

$$I'_2 = \left(\frac{N_2}{N_1}\right) I_2 = \frac{400/\sqrt{3}}{6,600/\sqrt{3}} \times 1,443.37\,\angle -36.87° \text{ A}$$
$$= 87.477\,\angle -36.87° \text{ A}$$

Magnetising current lags behind the voltage by 90° as losses are negligible, so

$$I_m = I_0 = 5.5\,\angle -90° \text{ A}$$

Primary current, $I_1 = I'_2 + I'_3 + I_0$

$$= 87.477\,\angle -36.87° + 17.495\,\angle 60° + 5.5\,\angle -90°$$
$$= (69.98 - j\,52.486) + (8.748 + j\,15.15) - j\,5.5$$
$$= (78.728 - j\,42.836) \text{ A} = 89.627\,\angle -28.55° \text{ A}$$

So magnitude of primary current,

$$I_1 = 89.627 \text{ A } \textbf{Ans.}$$

and primary pf $= \text{Cos}\,28.55° = 0.878$ lagging **Ans.**

13.9 TRANSFORMER NOISE

Under no-load conditions, the ''hum'' caused by energized power transformers originates in the ferromagnetic core, where the laminations tend to vibrate by magnetic forces. The noise is transmitted through the oil to the tank side and thence to the surroundings. The essential factors for noise production in transformers are (*i*) magnetostriction;* (*ii*) the mechanical vibrations caused by the laminations, depending upon the tightness of clamping, size, gauge, associated structural parts, etc; (*iii*) the mechanical vibration of tank walls, and (*iv*) the damping.

* The permeability of ferromagnetic materials changes when they are subjected to varying magnetization as B-H characteristic curves of ferromagnetic materials are nonlinear. As a result dimensional changes occur both parallel to, and perpendicular to, the direction of magnetization. This phenomenon is called the *magnetostriction*.

In general, the total noise emission may be reduced by (*i*) preventing vibration of core plate, which needs the use of a lower flux density and attention to constructional features such as clamping bolts, proportions, and dimensions of the "steps" in plate width, tightness of clamping and uniformity of plates; (*ii*) sound insulating the transformer from the tank by cushions, padding, or oil-barriers; (*iii*) preventing tank wall vibration by suitable design of tank and stiffeners; and (*iv*) sound insulating the tank from the ground or surrounding air. There is no complete solution to the noise problem.

13.10 NO-LOAD CURRENT WAVESHAPE

With a sinusoidal voltage V_1 applied to the primary of a transformer, the induced emf in the primary E_1 which balances the applied voltage V_1 (neglecting ohmic drop in the winding) has to be a sinusoid, and since E_1 is proportional to the rate of change of flux Φ_1 the latter must vary consinusoidally with respect to time (*e.g.*, a 90° phase difference must exist between Φ and E_1). But the no-load current wave will be distorted (non-sinusoidal) owing to the hysteresis loop.

Assume that the steady state operation has been attained so that the hysteresis loop shown in Fig. 13.31 (*a*) is being repeated in successive cycles of the applied voltage. At instant O in Fig. 13.31 (*b*), the flux is zero and the magnetising current is OA. At instant B, the flux is OB [Fig. 13.31 (*a*)] and the current is BC. When the flux is at its maximum value OF, the current is also at its maximum value FG. Thus, by projecting from Fig. 13.31 (*a*) to the flux curve in Fig. 13.31 (*b*) and erecting ordinates representing the corresponding value of the current, the waveform of the magnetising current can be derived. It can be shown that this waveform contains a sinusoidal component having the same frequency as that of the supply voltage and referred to as the *fundamental* component, together with other sinusoidal components having frequencies that are odd multiples of the supply voltage, the strongest being the third harmonic which may be as large as 40% of the fundamental.

Such a wave represents the magnetising component I_m and the hysteresis component of no-load current. It reaches its maximum value at the same time as the flux wave, but the two waves do not go through the zero simultaneously. The hysteresis component supplies the hysteresis loss and will be in phase with $-E_1$ (or V_1). Thus the current lags behind the applied voltage V_1 by angle slightly less than 90°.

The above current is required to be modified to account for the eddy current loss in the core and the modified current is called the *no-load current*, I_0. The corresponding current component apart from being in phase with applied voltage V_1 is sinusoidal in nature as it balances the effect of sinusoidal eddy currents caused by sinusoidal core flux. It is, therefore, seen that eddy currents do not introduce any harmonic in the exciting current. However, eddy current loss reduces the angle between the applied voltage V_1 and the no-load current, I_0.

When a transformer supplies current to a linear load, the load current is sinusoidal and being much larger than the excitation current would 'swamp out' the non-sinusoidalness in the resultant primary current; as a consequence the primary current on load is sinusoidal for all practical purposes.

In certain 3-phase transformer connections, third-harmonic current cannot flow, as will be discussed later on, as a result the magnetising current I_m is almost sinusoidal. For satisfaction of BH curve, the core flux Φ must then be nonsinusoidal; it is a flat-topped wave, as shown in Fig. 13.32 (*b*). This can be verified by assuming a sinusoidal magnetising current and then finding out the flux waveshape from the flux-magnetising current relationship, the normal magnetising curve, neglecting hysteresis component which is in phase with V_1. Since the flux is flat-topped, the emf which is its derivative will now be peaky with a strong third harmonic component. The various waveforms are illustrated in Fig. 13.32.

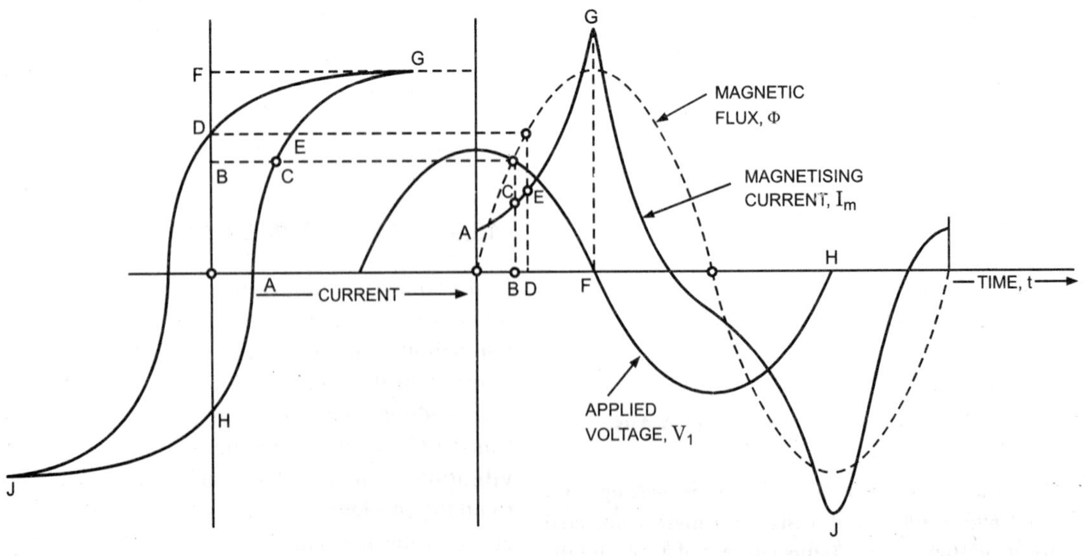

(*a*) Hysteresis Loop (*b*) Waveforms

Fig. 13.31

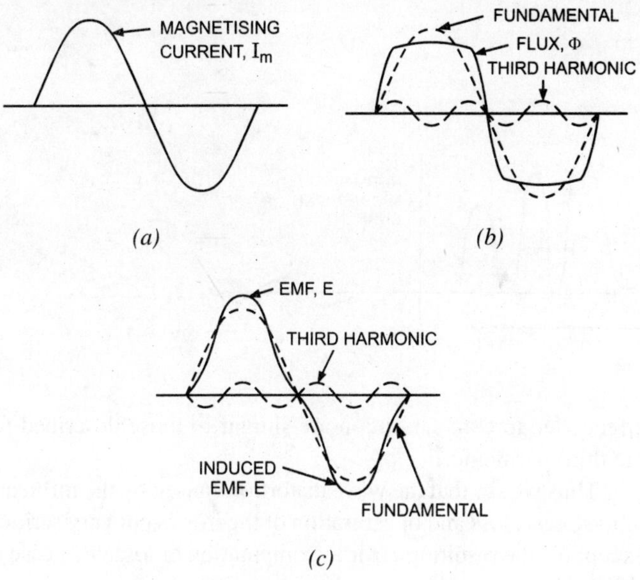

Fig. 13.32

13.11 INITIAL RUSH OF CURRENT (OR SWITCHING-IN TRANSIENTS)

When the primary side of a transformer is switched on to normal voltage with secondary open-circuited, a transient rush of current may take place. This rush depends upon the point in the cycle at which the voltage wave is switched on, the value and direction of the residual core flux, the shape of the saturation curve, and the normal flux density used. The worst conditions may occur when the applied voltage has zero value at the instant of switching.

The approximate value of the magnetizing current-rush may be determined as follows, neglecting the limiting effects of primary impedance drop and the effects of saturation.

Let the applied voltage be $v = V_{max} \cos(\omega t + \alpha)$ while the primary is switched on at time $t = 0$.

At every instant the voltage applied must be balanced by the emf induced by the flux created in the core by the magnetizing current (neglecting the small drops in resistance and leakage reactance).

$$\text{So } e = -v = -V_{max} \cos(\omega t + \alpha) \qquad \text{...(13.16)}$$

$$\text{Also } e = -N_1 \frac{d\Phi}{dt} \qquad \text{...(13.17)}$$

where N_1 is number of primary turns and Φ is the flux in the core. Comparing Eqs. (13.16) and (13.17) we have

$$\frac{d\Phi}{dt} = \frac{V_{max}}{N_1} \cos(\omega t + \alpha)$$

$$\text{or } \Phi = \frac{V_{max}}{N_1} \int \cos(\omega t + \alpha)\, dt$$

$$= \frac{V_{max}}{\omega N_1} \sin(\omega t + \alpha) + C = \Phi_{max} \sin(\omega t + \alpha) + C \qquad \text{...(13.18)}$$

where $\Phi_{max} = \dfrac{V_{max}}{\omega N_1}$ and C is asymmetrical component of core flux.

Now the core is initially demagnetized and it may contain a small residual flux Φ_{res}. So $\Phi = \Phi_{res}$ at $t = 0$. Thus asymmetrical component of core flux

$C = \Phi_{res} - \Phi_{max} \sin\alpha$ and the Eq. (13.18) may be rewritten as

$$\Phi = \Phi_{max} \sin(\omega t + \alpha) + \Phi_{res} - \Phi_{max} \sin\alpha \qquad \text{...(13.19)}$$

If the transformer is switched on at the instant that the applied voltage passes through a peak value *i.e.* when $\alpha = 0$, a flux corresponding to steady state operation is immediately established in the transformer and also does the magnetizing current. Such conditions are those of normal no-load operation and there is no transient.

The worst conditions for connection are when $\alpha = \dfrac{-\pi}{2}$ *i.e.* when the switch is closed at the instant the applied voltage passes through a zero value. In this case

$$\Phi = -\Phi_{max} \cos\omega t + \Phi_{res} + \Phi_{max}$$

i.e. the flux attains nearly double the normal flux after half a cycle, as illustrated in Fig. 13.33. This is referred to as the *doubling effect*. The corresponding magnetizing current will be very large as the core goes into deep saturation region of magnetization; it may indeed be as large as 100 times the normal magnetizing current. If the normal magnetizing current is 5 per cent of the normal rated current, the doubling effect may raise it initially to several times rated current. The double flux density will not actually be reached owing to considerable resistance and leakage reactance drops with large currents. Further, the current gradually falls to the normal magnetizing currents. Because of the low time constant of the transformer circuit, the distortion effects of the transients may last several seconds. The transformer switching transient is referred to as the *inrush current*. Nevertheless the high initial current may cause intense mechanical stress in the transformer windings. This is why the windings of large power transformers are strongly braced. A typical oscillogram of the inrush current is illustrated is Fig. 13.34.

The inrush current does not present a direct danger to the transformer but it may lead to its being tripped out of the power circuit. Therefore, the protection apparatus must be so designed as to avoid the undue tripping of transformer out of power circuit.

The above discussion pertains to a single-phase transformer. When a three-phase transformer is switched on a more or less considerable current inrush should be expected, since there will always be a phase whose voltage at the switching instant is nearly zero.

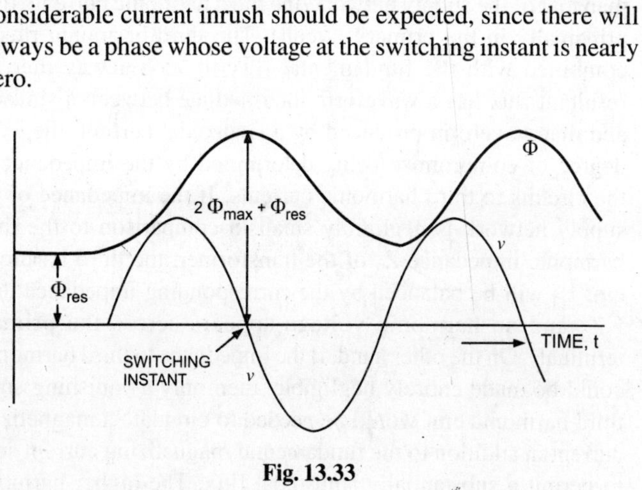

Fig. 13.33

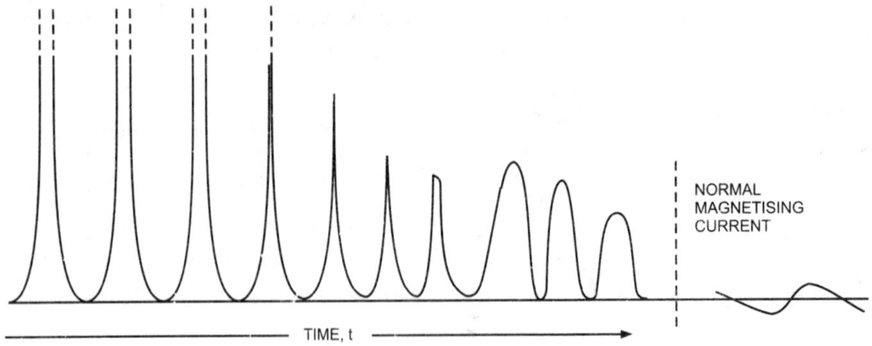

Fig. 13.34

13.12 HARMONICS IN TRANSFORMERS (OR EFFECTS OF THIRD HARMONIC COMPONENTS IN MAGNETIZING CURRENTS)

The use of high flux densities in the cores of power transformers, imposed by the requirements of an economical design and the reduction in size, results in high saturation level and the departure from rectilinearity of the flux-current relation or B-H (or B-AT) curve. Due to saturation effects, a sinusoidal flux and emf necessitate a pronounced third and less pronounced higher-order harmonic components in the magnetizing current. If for any reason the third harmonic current is not permitted to flow *i.e.* when the magnetizing current is sinusoidal, the flux is flat-topped containing ''depressing'' third harmonic and as a consequence third harmonic voltages are present in the induced emf.

The above problem in single-phase and three-phase transformers is discussed below.

1. Single-Phase Transformers. A sinusoidal supply voltage cannot itself give rise to any third harmonic current; it supplies fundamental reactive power, part of which is converted to the harmonic power in the transformer core by reason of its nonlinear B-H characteristic. The flow of a third harmonic current implies the presence of a third harmonic emf. and, therefore, a corresponding harmonic in the flux waveform. The harmonic currents flow through the effective third harmonic impedances of the primary winding and its supply network, and also to a limited extent in the secondary winding and load on the secondary side (since the load impedance is usually much larger than that of the supply network, the third harmonic currents flow principally in the primary circuit). The third harmonic flux is combined with the fundamental flux in such a way that the resultant flux has a waveform intermediate between a sinusoid and that waveform produced by a sinusoidal current, the exact degree of compromise being determined by the impedance of the circuits to third harmonic currents. If the impedance of the supply network is negligibly small in comparison to the third harmonic impedance Z_3 of the transformer, the third harmonic emf E_3 will be balanced by the corresponding impedance drop I_3Z_3 and no harmonic voltage appears across the primary terminals. On the other hand, if the impedance to third harmonics could be made entirely negligible, then only a vanishing small third harmonic emf would be needed to circulate a magnetizing current in addition to the fundamental magnetizing current so as to permit a substantially sinusoidal flux. The higher harmonic effects due to core saturation are similar to those described for the third harmonic.

Thus we see that the wave distortion caused by the influence of hysteresis loss and of saturation of the iron is not very serious, except for the resulting error in computation of losses, in case of single-phase transformers.

2. Three Phase Bank of Single-Phase Transformers. While considering three-phase transformers it is essential to distinguish between the cases where the phases are magnetically separate and where they are magnetically as well as electrically inter-linked. In case of three phase banks of single-phase transformers the magnetic circuits are obviously separate, and each core must itself produce the flux demanded by the conditions of its electrical connections, which then determine the flow of harmonic currents. In the following description both supply and load are considered to be balanced and star-connected.

(*a*) *Delta/delta (Dd) Connection.* It is to be observed that the phase difference in third harmonic currents and voltages on a 3-phase system in $3 \times 120° = 360°$ or 0° which means that these are *cophasal*. Therefore, third harmonic (in general of the order $3n$ called triplen) currents and voltages cannot be present on lines of a 3-phase system as these do not add up to zero.

Delta connection provides a closed path to triplen emfs, which circulate currents in the closed mesh and are absorbed by the triplen leakage impedance drop : there is no resultant triplen voltage at the line terminals. The impedance to harmonic currents is usually small (although it may be greater than fundamental impedance of the transformer), and very small emfs are sufficient to circulate considerable harmonic currents.

The action in delta/delta connection is as follows :

The supply voltage, being sinusoidal, provides only sinusoidal magnetizing current. Such a current produces a nonsinusoidal flat-topped flux wave, since there is no peaky current to overcome the saturation at high flux densities. This flat-topped flux wave induces a peaky emf with a strong third harmonic component, which does not balance the applied sinusoidal voltage, but leaves as resultant, third harmonic components (and also higher harmonics). The third harmonic components now circulate a triplen current in the mesh, which is in addition to the sinusoidal supply current, and restores the flux to almost sinusoidal. If the triplen mesh impedance is low, a very small departure of the flux from a perfect sine is sufficient to give a large third harmonic magnetizing current. Thus the conditions are established whereby the primary voltage and primary current are sinusoidal, the flux is very nearly a sinusoidal, and the necessary triplen magnetizing current component is circulated round the closed mesh. Inspite of the presence of a third harmonic in the magnetizing current of each phase, the line currents do not have any third harmonic component, since the combination of the two third harmonic components of the phase currents connected from any line cancel each other. In a Δ-Δ connection the primaries carry all of the fundamental component of the exciting current while the third harmonic component of the exciting current can divide so that part of it exists in the primaries and the remaining portion in the secondaries.

(b) Star/Delta or Delta/Star (Yd or Dy) Connection Without Neutral. So long as there is no neutral wire, either of these connections operates substantially as in delta-delta connections but since there is a mesh connection on only one side, the triplen mesh impedance is now larger, and the compensation of the magnetizing current will necessitate a greater divergence of the flux wave from a true sinusoid. In star-delta (Y-Δ) connection the primaries carry the fundamental component of the exciting current and the secondaries carry the third harmonic component whereas in delta-star (Δ-Y) connection the secondary voltages do not contain any harmonic component.

(c) Star/Star (Yy) Connection With Neutrals. As in Δ connection of the primaries, each primary is directly connected to the respective phase of the supply. Again, there is nothing to interfere with the existence of a third harmonic in the magnetizing current, and the voltage induced in each transformer will be nearly sinusoidal. The line currents in this case will contain the third harmonic component of the magnetizing current. Under balanced conditions the third harmonic components of the three phases will be in phase with each other and of the same magnitude. The current in the neutral wire of the supply system, therefore, will not be zero but will be equal to three times the third harmonic component of one phase.

(d) Star/Star (Yy) Connection Without Neutral. The triplen emfs are all directed simultaneously away from or towards the star point and, therefore, cancel between any pair of lines. Consequently no triplen currents can flow, and (apart from the fifth, seventh, ... harmonic components) the input magnetizing currents are sinusoidal. The flux waveform is flattened, its triplen harmonics providing the triplen emf. The balance between applied voltage and induced emf between line terminals is maintained since the triplen emf balance out, but the effect on the star-point voltage is to make it oscillate. The complexor conditions are illustrated in Fig. 13.35. The voltages, $e_{r_1}, e_{y_1}, e_{b_1}$ represent the fundamental terms, mutually displaced by 120°. There is a third harmonic emf, e_3 co-phasal in all legs, additional to the fundamental, and the resultant voltages across the legs are shown dotted. The diagram illustrates four successive instants displaced by one-twelfth period. It is to be noted that the third harmonic phasor e_3 rotates thrice as fast as the fundamental system. All the line-to-star point voltages fluctuate, causing ''neutral oscillation''. The non-triplen harmonics form balanced systems with a 120° phase displacement and appear across the lines to cause a self suppressing circulating current.

The phenomenon of *oscillating neutral* is highly undesirable and because of this star-star connection with isolated neutrals is not used in practice.

(e) Other Star/Delta (Yd) Connections. All the variants of the possible connections will fall under one or other of the cases (a) to (d) already dealt with. A mesh connection always provides a closed path for triplen currents, while a neutral conductor on the supply side allows triplen currents to flow into the supply

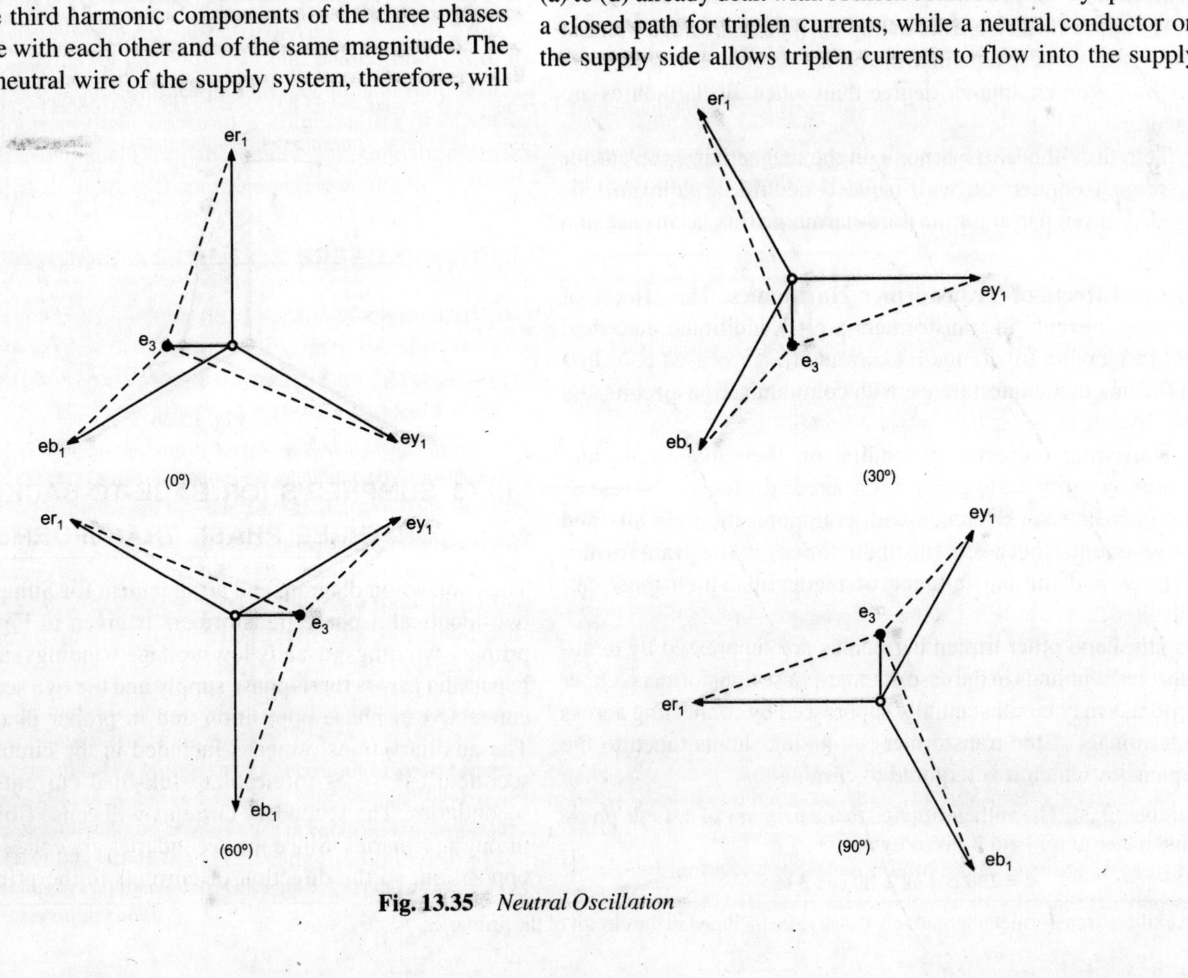

Fig. 13.35 *Neutral Oscillation*

network. A neutral conductor is not operative on the secondary side on open circuit.

3. Three-Phase Transformer Units. In the 3-phase shell type transformer, the magnetic circuits are separate and so not interact, and the discussion made above for the bank of single-phase transformers will apply. With the very common three-limbed core type transformer, however, the phases are magnetically interlinked. Any triplen flux harmonic that exists will be directed either all upwards or all downwards in the limbs together at any instant. They must, therefore, find return paths outside the iron circuit (through the air or oil and walls of the tank, if any), as shown in Fig. 13.36. The magnetic reluctance of these paths is high and the third harmonic flux components are reduced to negligible proportions. Occasionally the third harmonic fluxes have been found to cause losses in the tank walls.

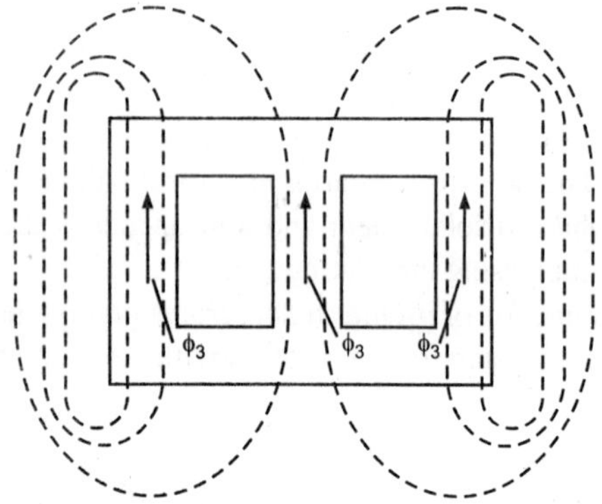

Fig. 13.36 *Third Harmonic Flux Paths*

Harmonics which are not multiples of three are not co-phasal in the limbs, so these remain to distort the flux and emf waveform though to a much smaller degree than when all harmonics are present.

To reduce the fifth harmonic in the magnetizing current for the star/star connection with isolated neutral, a path must be provided through iron for the third-harmonic flux, as in case of a five limb core.

13.12.1. Effects of Transformer Harmonics. The effects of harmonic currents in transformers are (*i*) additional electrical (I^2R) loss owing to circulating currents (*ii*) increased core loss and (*iii*) magnetic interference with communication circuits and protective gear.

Harmonic voltages, depending on their magnitude and frequency may cause (*i*) increased dielectric stresses (*ii*) electrostatic interference with communication circuits and (*iii*) resonance between the inductance of the transformer windings and the capacitance of feeder to which these are connected.

Third and other triplen harmonics are suppressed by delta-connected windings in the case of three-phase transformers. Other harmonics may be substantially suppressed by connecting across the terminals of the transformer resonating shunts tuned to the frequencies which it is required to eliminate.

Example 13.34. The voltage applied to the primary of a single phase transformer at no-load is given by

$$v = 200 \cos \omega t + 80 \cos 3 \omega t$$

If the primary has 300 turns, then calculate the maximum value of the flux in the core for a fundamental frequency of 50 Hz.

[*Pb. Univ. Elec. Machines I, Dec. 1992*]

Solution : RMS value of fundamental component of emf,

$$E_1 = \frac{200}{\sqrt{2}} = 100\sqrt{2} \text{ or } 141.42 \text{ V}$$

RMS value of third harmonic component of emf,

$$E_3 = \frac{80}{\sqrt{2}} = 40\sqrt{2} \text{ or } 56.57 \text{ V}$$

Peak value of fundamental component of flux,

$$\Phi_{1\,max} = \frac{E_1}{4.44\, f N_1} = \frac{141.42}{4.44 \times 50 \times 300} = 2.12 \text{ mWb}$$

Peak value of third harmonic component of flux,

$$\Phi_{3\,max} = \frac{E_3}{3 \times 4.44\, f N_1} = \frac{56.57}{3 \times 4.44 \times 50 \times 300} = 0.283 \text{ mWb}$$

Maximum value of flux in the core,

$$\Phi_{max} = \Phi_{1max} - \Phi_{3max}, \text{ as obvious from Fig. 13.38.}$$
$$= 2.12 - 0.283 = 1.837 \text{ mWb } \textbf{Ans.}$$

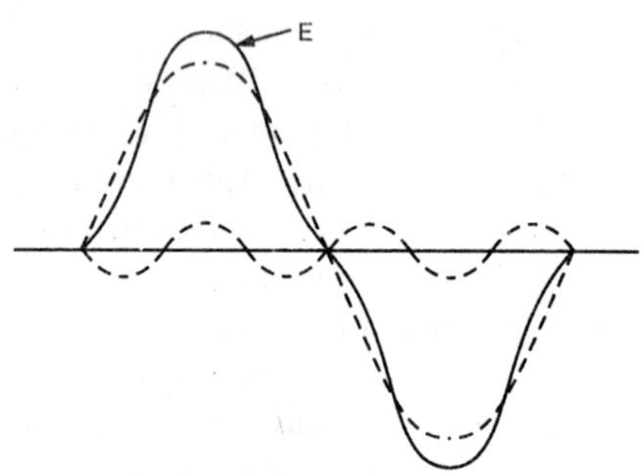

Fig. 13.37

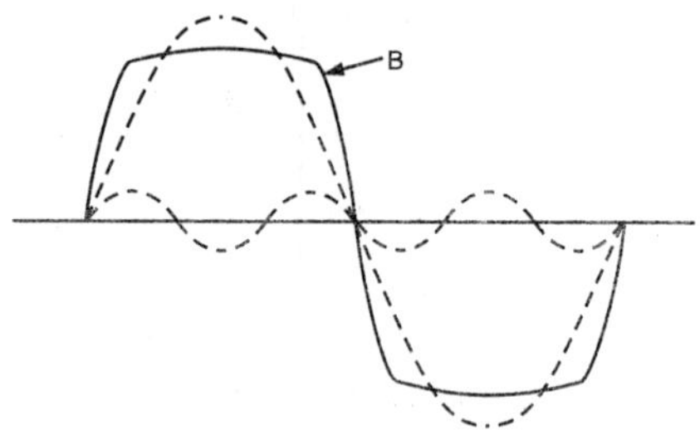

Fig. 13.38

13.13 SUMPNER'S (OR BACK-TO-BACK) TEST ON THREE PHASE TRANSFORMERS

The connection diagram and arrangement for Sumpner's test on two identical 3-phase transformers is given in Fig. 13.39. The primary windings (usually low-voltage windings) are connected in parallel across three-phase supply and the two secondaries are connected in phase opposition and in proper phase sequence. The auxiliary transformer is included in the circuit of the two secondaries* so as to circulate full-load current through the secondaries. The secondary currents will cause flow of currents through primaries. Since the secondaries are connected in phase opposition, so the direction of currents in the primaries of the

* Auxiliary transformer alternatively can also be included in the circuit of the primaries.

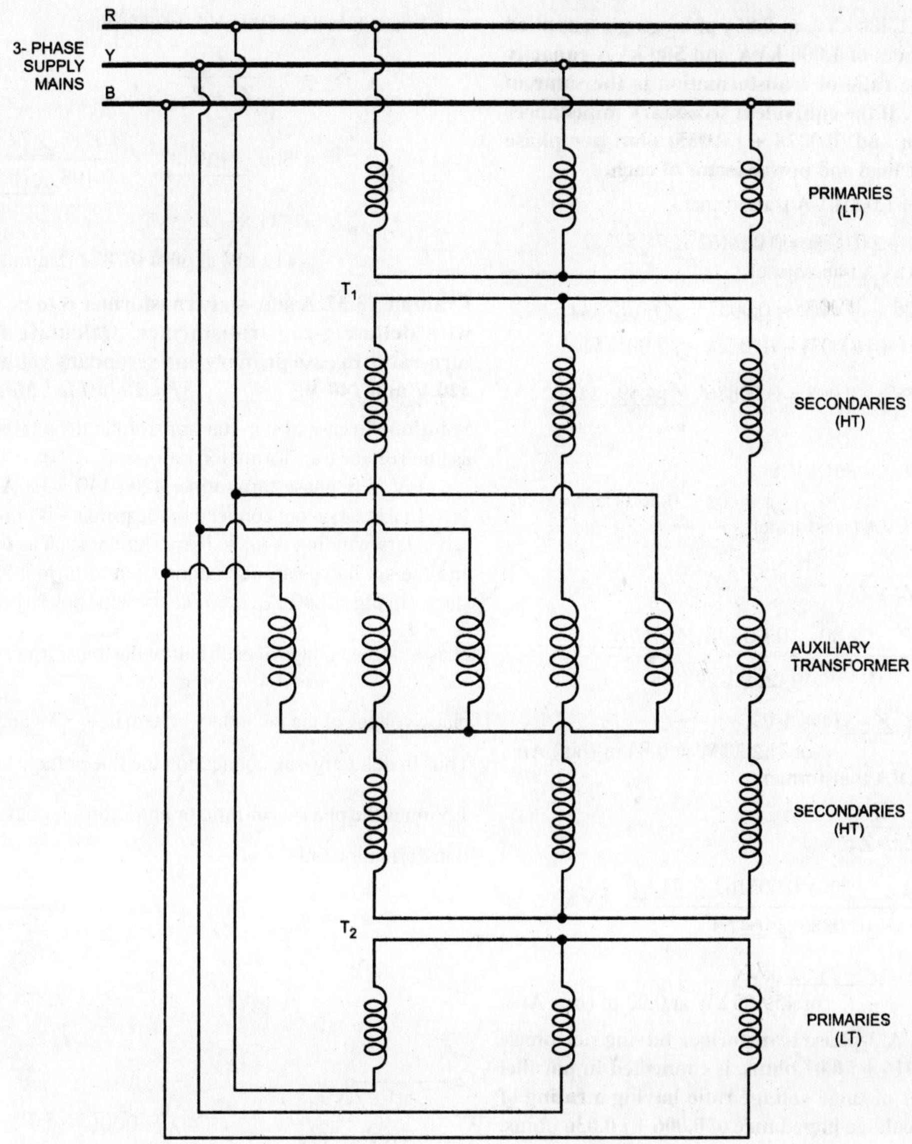

Fig. 13.39 *Circuit Diagram For Sumpner's Test On Two Identical 3-Phase Transformers*

two transformers will be in opposition. Thus the power required from the supply mains is that necessary for supplying the iron and copper losses of both transformers. Thus the heat run or temperature rise test can be performed.

13.14 PARALLEL OPERATION OF 3-PHASE TRANSFORMERS

Parallel operation of three-phase transformers must fulfil for each phase the conditions discussed for satisfactory parallel operation of single-phase transformers (Art 12.23.3). In addition the condition of same phase sequence and same relative phase displacement are to be fulfilled.

1. Phase sequence must be the same. Phase sequence refers to the order in which the terminal voltages attain their maximum values. In paralleling of two three-phase transformers those terminals whose voltages attain their maximum values simultaneously are paired.
2. The 3-phase transformers to be operated in parallel must have the same phase displacement between primary and secondary line voltages; *i.e.* they must belong to the same group number. Only a few of the possible connections (Art. 13.3) can be worked in parallel without excessive circulating current on small loads; for example, the secondary voltages of star/star and delta/star transformers have a phase difference of –30º, making parallel connection inadmissible. The various connections produce various magnitudes and phase displacements; magnitudes can be adjusted by tappings, but phase divergence cannot be compensated. The following are typical of the connections for which, from the view point of sequence and phase divergence, transformers can be connected in parallel.

Transformer 1	:	Yy	Dd	Yy	Yd	Yd
Transformer 2	:	Yy	Dd	Dd	Dy	Yz

Thus all transformers in the same group can be operated in parallel. Further, transformers with $a + 30º$ and $– 30º$ angle can be paralleled by reversing the primary and secondary phase sequence of one.

Example 13.35. A load of 1,400 kVA at 0.866 pf lagging is supplied by two 3-phase transformers of 1,000 kVA and 500 kVA capacity operating in parallel. The ratio of transformation is the same in both: 6,600/400 delta-star. If the equivalent secondary impedances are (0.001 + j 0.003) ohm and (0.0028 + j 0.005) ohm per phase respectively, calculate the load and power factor of each.

Solution: The impedance of 1,000 kVA transformer,

$$Z_1 = (0.001 + j\,0.003) = 0.003162 \angle 71.57° \,\Omega$$

The impedance of 500 kVA transformer,

$$Z_2 = (0.0028 + j\,0.005) = 0.00573 \angle 60.75° \,\Omega$$

$$Z_1 + Z_2 = (0.001 + j\,0.003) + (0.0028 + j\,0.005)\,\Omega$$

$$= 0.0038 + j\,0.008 = 0.00886 \angle 64.59° \,\Omega$$

Total load to be shared,

$$S = 1{,}400 \angle -30° \text{ kVA}$$

$$\because \phi = \cos^{-1} 0.866 \text{ (lag)} = -30°$$

Load shared by 1,000 kVA transformer,

$$S_1 = S \times \frac{Z_2}{Z_1 + Z_2}$$

$$= \frac{1{,}400 \angle -30° \times 0.00573 \angle 60.75°}{0.00886 \angle 64.59°} \text{ kVA}$$

$$= 905.8 \angle -33.84° \text{ kVA}$$

or 752.3 kW at 0.83 pf (lag) **Ans.**

Load shared by, 500 kVA transformer,

$$S_2 = S \times \frac{Z_1}{Z_1 + Z_2}$$

$$= \frac{1{,}400 \angle -30° \times 0.003162 \angle 71.57°}{0.00886 \angle 64.59°} \text{ kVA}$$

$$= 499.64 \angle -23.02° \text{ kVA}$$

or 459.85 kW at 0.92 pf (lag) **Ans.**

Example 13.36. A 300 kVA 3-phase transformer having per phase leakage impedance of 0.016 + *j* 0.07 ohms, is connected in parallel with another transformer of same voltage ratio having a rating of 500 kVA and per phase leakage impedance of 0.006 + *j* 0.036 ohms. Find the load shared by each transformer and their operating power factors for a total load of 800 kVA at 0.8 pf lagging.

[G.G.S.I.P. Univ., Electromechanical Energy Conversion-I, December-2011]

Solution: Impedance of 300 kVA transformer,

$$Z_1 = (0.016 + j\,0.07) = 0.0718 \angle 77.125° \,\Omega$$

Impedance of 500 kVA transformer

$$Z_2 = (0.006 + j\,0.036) = 0.0365 \angle 80.54° \,\Omega$$

$$Z_1 + Z_2 = (0.016 + j\,0.07) + (0.006 + j\,0.036)$$

$$= (0.022 + j\,0.106)\,\Omega = 0.108 \angle 78.27° \,\Omega$$

Total load to be supplied, $S = 800 \angle -36.87°$ kVA

Load shared by 300 kVA transformer,

$$S_1 = S \times \frac{Z_2}{Z_1 + Z_2}$$

$$= 800 \angle -36.87° \times \frac{0.0365 \angle 80.54°}{0.108 \angle 78.27°}$$

$$= 270.37 \angle -34.6°$$

or 221 kW at pf of 0.823 (lagging) **Ans.**

Load shared by 500 kVA transformer

$$S_2 = S \times \frac{Z_1}{Z_1 + Z_2}$$

$$= 800 \angle -36.87° \times \frac{0.0718 \angle 77.125°}{0.108 \angle 78.27°}$$

$$= 531.85 \angle -38°$$

= 419 kW at pf of 0.7878 (lagging) **Ans.**

Example 13.37. A star-star transformer is to be operated in parallel with delta-zig-zag transformer. Calculate the hv to lv phase turn-ratio, in case primary and secondary voltages are respectively 420 V and 140 V. *[Pb. Univ. Elec. Machines- I, June-1993]*

Solution : In case of star-star transformer the phase turn ratio is the same as line voltage transformation ratio, so

HV to lv phase-turn ratio = 420 : 140 = 3:1 **Ans.**

In delta-zig-zag connection the primary is connected in delta. Each secondary winding is made in two halves such as a_1-a_2, a_3-a_4, b_1-b_2 etc. and the six halves are then connected to form a zig-zag connection as shown in Fig. 13.40. A_1, a_1, a_3 are of same polarity and similarly for other phases. If the voltage of each half of the transformer secondary is $\frac{V_2}{2}$, the phase voltage of the secondary system is $\frac{\sqrt{3}}{2}V_2$ and line voltage is $\frac{3}{2}V_2$. Thus in delta-zig-zag connection the line voltage transformation ratio is 1.5 times the phase turn-ratio or phase-turn ratio is $\frac{2}{3}$ times line voltage transformation ratio.

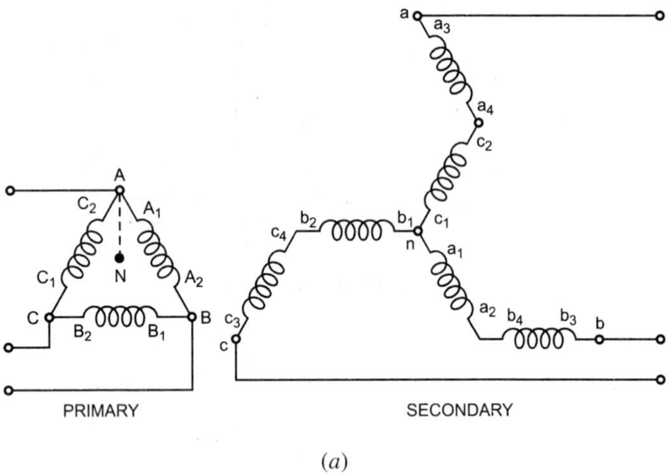

(a)

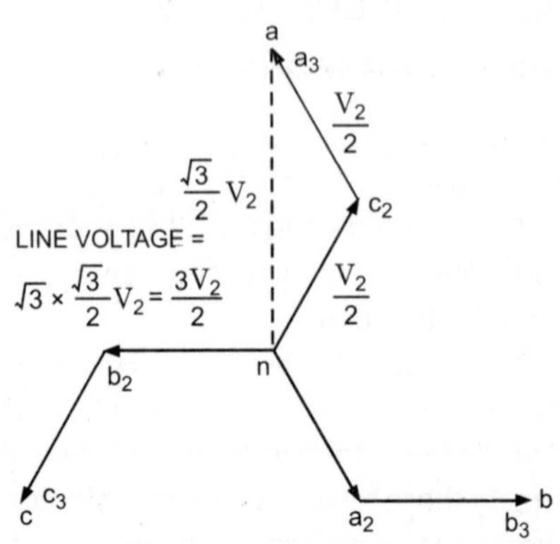

(b) Phasor Diagram For Zig-zag Connected Secondary

Fig. 13.40

Thus in delta-zig-zag transformer

Phase turn-ratio (secondary to primary or hv to lv) = $\frac{2}{3}$ times line voltage transformation ratio = $3 : 1 \times \frac{2}{3} = 9 : 2$ **Ans.**

13.15 FREQUENCY RESPONSE OF TRANSFORMER

Transformers are widely used in the output stage of electronic amplifiers where frequency varies over a wide range. So the transformer frequency response is very significant in variable frequency applications. Frequency response of a transformer is the variation it causes, if any, in the transformation ratio i.e. $\frac{V_2}{V_1}$ when the supply frequency varies. A flat frequency response is desirable over the frequency band of interest.

Variation in output voltage w.r.t. the source or input voltage V_1 is usually accompanied by a variation in the phase angle of the output voltage w.r.t. input voltage. The variation of the phase angle of output or load voltage V_2 w.r.t. source or input voltage V_1 is termed as the phase response. A small phase angle is desirable.

The performance of a transformer depends to a considerable extent upon its frequency response. In the design of transformers used in electronic amplifiers, appropriate steps are taken to ensure that the transformation ratio is essentially uniform over some specified frequency response. The frequency response of a transformer is shown in Fig. 13.41. It is found that the transformation remains constant over a middle range of frequencies and drop at low and high frequencies.

At very low frequencies, the magnetising reactance is small enough to be comparable with or even less than the load impedance so that this tends to be shorted out and the secondary terminal voltage is largely reduced. The output voltage is of course zero for dc ($f = 0$). In the intermediate or middle frequency range magnetising reactance X_m is large enough and does not effect its voltage ratio behaviour and leakage reactances of the windings are small enough so that the transformation ratio $\frac{V_2}{V_1}$ remains substantially constant as shown in Fig. 13.41. With the further increase of frequency, however the leakage reactance and voltage drops tends to cause the output voltage to fall appreciably again. Capacitances between turns and between the windings and earth can no longer be ignored since these are shunted across the circuit in a distributed fashion causing behaviour similar to leading pf load. The combined effects of increasing series inductive reactance and decreasing shunt capacitive reactance with increasing frequency cause the output voltage to drop sharply.

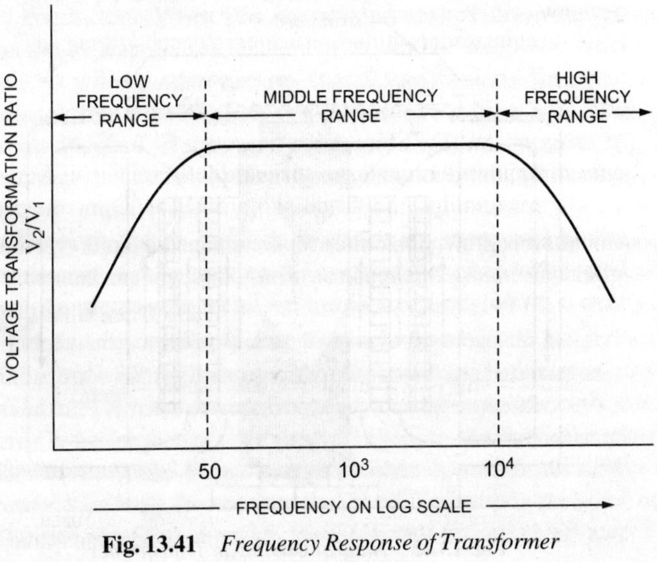

Fig. 13.41 *Frequency Response of Transformer*

13.16 POWER AND DISTRIBUTION TRANSFORMERS

13.16.1. Power Transformers. The term is used to include all transformers of large size (250 kVA and above) used in generating stations and substations for transforming the voltage at each end of a power transmission line. They may be single-phase or 3-phase; 3-wire delta/delta or delta/ star-connected; and of voltage rating 220/11 kV or in high voltage range. They are put in operation during load hours and disconnected during light load hours *i.e* they are usually operated on approximately full load. This is possible because they are arranged in banks and can be thrown in parallel with other units or disconnected at will. So power transformers are designed to have maximum efficiency at or near full load (i.e. with iron loss to full load copper loss ratio of 1 : 1). Power transformers are designed to have considerable greater leakage reactance than is permissible in distribution transformers because in power transformers inherent voltage regulation is not as much important as current limiting effect of the higher leakage reactance. Power transformers usually make use of flux density of 1.5 to 1.7 T, have percentage impedance ranging from 6–18 % and regulation of 6–10 per cent. They may be self oil cooled, forced air cooled or forced water cooled.

13.16.2. Distribution Transformers. Transformers of rating up to 200 kVA, used to step down the distribution voltage to a standard service voltage are known as *distribution transformers.* They are kept in operation all the 24 hours a day whether they are carrying any load or not. In such transformers iron loss occurs for all the time where copper loss occurs only when they are loaded. Therefore, distribution transformers should be designed with iron loss smaller in comparison to full-load copper loss (say with iron loss to full-load copper loss ratio of 1 : 3). In other words, they are designed for good 'all-day efficiency' and not for highest efficiency on full load. They are of the self cooling type and are almost invariably oil-immersed. The distribution transformers are usually 3-phase, 4-wire 11 kV/433 V delta/star connected. Such transformers are designed to have an inherently good voltage regulation which is possible by arranging the coils in such a way as to have minimum leakage reactance. Such transformers have percentage impedance of 4–5 % and voltage regulation of 4–8 %. They make use of cold rolled steel with flux density of 1.7 T.

13.17 METHODS OF REDUCING LEAKAGE FLUX

The leakage flux can be practically eliminated by winding the primary and secondary, one over the other, uniformly around a laminated iron ring of uniform x-section. But such an arrangement is not commercially practicable except in very small sizes, owing to the cost of threading a large number of turns through the ring.

The various methods used for reducing leakage flux are:

1. By Increasing Height of Window. The area of window is determined by the x-sectional area of primary and secondary windings. For a given area of the window the reluctance of leakage flux path is increased by increasing the height and reducing the width of window. Usually the ratio of height to width is kept up to 4 because if window is made very narrow, an excessive proportion of the width is occupied by the insulation.

2. By Arranging Primary and Secondary Windings Concentrically. Each winding is equally divided between the two limbs, the general practice being to place lv winding next to the core and hv winding on the outside, as illustrated in Fig. 12.52. With concentric windings, the leakage flux is approximately a half of that with the windings on separate limbs. Furthermore, the leakage flux is linked with only a part of primary or secondary winding, so that the leakage flux is considerably reduced.

3. By Sandwiching Primary and Secondary Windings. The aim of sandwiching is again to arrange the primary and secondary windings as close to each other as possible, so that leakage flux is reduced. The leakage flux can be further reduced by using more and more sandwiched sections for both primary and secondary windings; but this arrangement has the disadvantage that adjacent sections must be well insulated from each other. So with a larger number of sections an excessive amount of space is taken by insulation.

4. By Adopting Shell Type Construction. In shell type transformers the low-voltage and high-voltage windings are sandwiched, as illustrated in Fig. 12.53. This arrangement of using sandwiched coils results in more closeness of the primary and secondary coils and so the leakage flux is considerably reduced. In shell type construction, since iron surrounds both the windings, so the fluxes created by primary and secondary windings tend to follow the least reluctance iron path around them rather than the high reluctance air path. Thus leakage flux is further reduced.

13.18 METHODS OF COOLING

Since there is no rotating part in the transformer which could produce a draught, cooling of transformers is more difficult than other electric machines with rotating parts.

In transformation of energy from one voltage level to another voltage level some energy is lost in between, in the core and in the windings. All the energy lost in a transformer must be dissipated as heat. Although the energy loss is a small proportion of the total energy undergoing transformation, but it becomes quite large in magnitude in transformers of higher kVA rating. The greater the rating of a transformer, the more difficult it becomes to dissipate the heat, for the kVA rating of the transformer increases much more rapidly than the heat-dissipating surface (losses increase in proportion to the weight of transformers i.e. as the cube of their linear dimensions whereas cooling surface area increases only as the square of their linear dimensions).

There are a number of methods used for cooling of transformers. The choice of method depends upon the size, type of application and the type of conditions available at the site where the transformer is installed.

In so far as cooling methods are concerned, the transformers are either of dry type or oil-immersed type. The cooling methods used for dry type transformers are:

1. Air Natural (AN) Cooling. For smaller output (5–10 kVA) the external surface is sufficient to dissipate the heat produced by losses. Such transformers are cooled by the combination of air convection currents within the enclosing case and by natural radiation from the case itself. Instrument transformers, except for those for the highest voltages, are ordinarily of dry type. Dry type transformers are frequently used indoor since it is not necessary to install them in fireproof vaults.

2. Air Blast Cooling. This type of cooling is also used in dry type transformers. In this method of cooling, heat dissipation is improved by a continuous blast of air forced through the core and windings. The air blast is produced by means of external fans or blowers. This type of cooling is limited to transformers of voltage rating not exceeding 25 kV. It is used in situations located in thickly populated places where oil is considered a fire hazard.

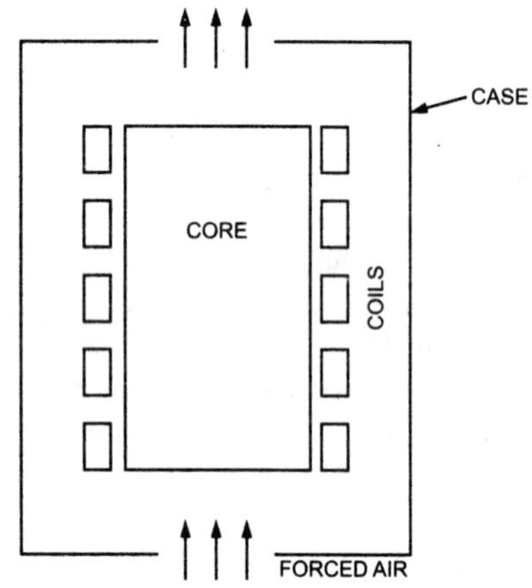

Fig. 13.42 *Air Blast Cooling*

Usually transformers are placed over an air chamber in which the air is maintained under pressure by blowers. The air is forced through core and windings and is discharged through the top of case. The main disadvantage of this process is that increase in insulation strength, due to immersion in oil, is not obtained. A filter to remove the dust from the cooling air, thus preventing its deposits in the transformer ducts, is desirable.

The cooling methods used in oil-immersed type transformers are:

1. Oil Natural (ON) Cooling. This is the most usual method of cooling. Here the assembly of core and windings is immersed in

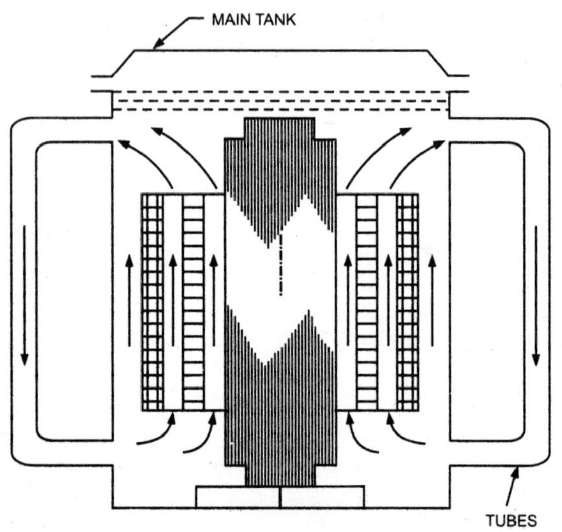

Fig. 13.43 *Natural Cooling in Transformers*

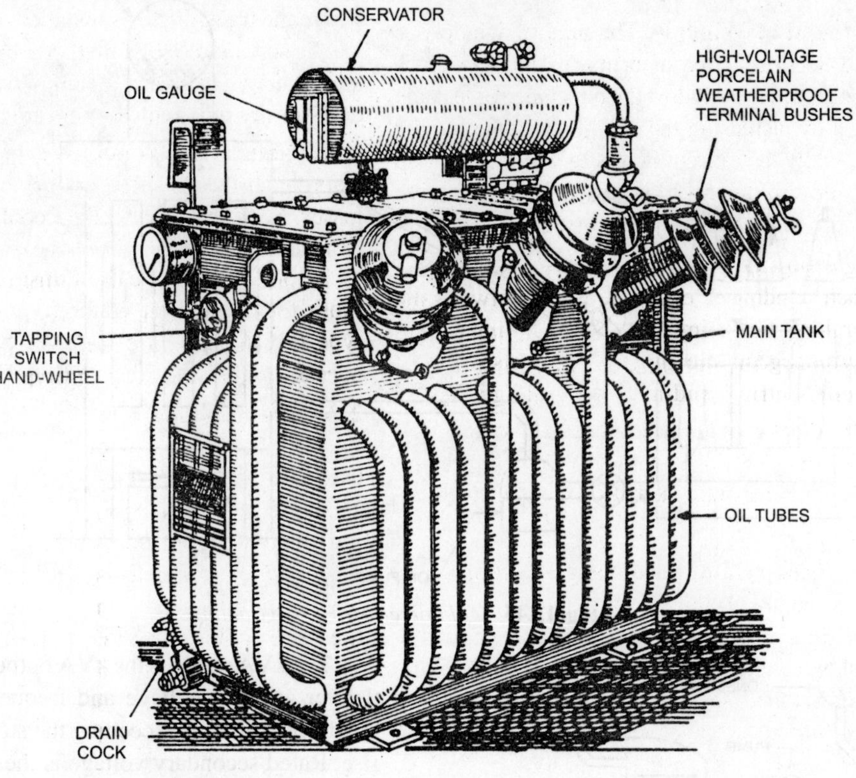

Fig. 13.44 *50 KVA, 11/0.4 KV Oil Immersed Natural Cooled Transformer*

an insulating oil contained in an iron tank. The heat produced in core and windings is passed on to oil by conduction. Oil in contact with heated parts rises and its place is taken by cool oil from the bottom. The heated oil transfers its heat to the tank surface which dissipates it to the surroundings. The advantages of this method of cooling are (*i*) freedom from dust (*ii*) freedom from atmospheric moisture which would have affected the insulation (*iii*) easy extraction of heat from innermost parts as oil is better conductor of heat (*iv*) good amount of natural convection currents since oil has high coefficient of volume expansion.

In many cases smooth tank walls are not sufficient to dissipate the heat, therefore, the tank dissipating surface area is increased by providing ribs, also called the *fins*, and corrugations. A corrugated tank has about 50 % more radiating surface than a smooth tank. For further increasing the heat dissipating capacity radiators or tubes are provided on all the four walls of the transformer.

2 Oil Natural Air Forced (OB) Cooling. Here again the assembly of core and windings is immersed in an insulating oil and cooling is improved by forced air over the cooling surfaces. The air is forced over external surfaces such as the case, tubes, and radiators, usually by means of fans mounted external to the transformer. Medium to large capacity transformers are cooled by this method.

3. Oil Natural Water Forced (OW) Cooling. For such a cooling, the core and windings are immersed in an insulating oil and the cooling is increased by circulation of water through copper cooling coils mounted above the transformer core but below the oil surface. The heated water is cooled in a spray pond or cooling tower. This method will be cheaper where a natural water head is already available.

4. Oil Forced Air Natural (OFN) Cooling. In this method of cooling, oil is circulated through the transformer with the help of pump and cooled in a heat exchanger by natural circulation of air. This method proves very useful where the coolers have to be well removed from the transformer. This method is not very common.

5. Oil Forced Air Forced (OFB) Cooling. In this method of cooling, as shown in Fig. 13.45, the oil is cooled in external heat exchangers using air blast produced by fans. It is not necessary that the oil pump and fans may be used at all times. At light loads, say up to 50% of rated load (when the losses are small) natural circulation of oil (ON cooling) may be sufficient to cool the transformer. At higher loads, the pump and fans may be switched on by temperature sensing elements. This arrangement results in higher efficiency of the system.

6. Oil Forced Water Forced (OFW) Cooling. In this method of cooling, the heated oil is pumped out from the main tank to the radiator (heat exchanger) where the oil is cooled by the water passing through copper tubes. The pressure of oil is kept higher than that of water and, therefore, any leakage that occurs is from oil to water (not from water to oil). Also there is no condensation problem in this method.

Power transformers with a capacity of 10 MVA have a cooling radiator system with natural cooling. The oil forced air forced (OFB) cooling is usual one for power transformers of rating 30 MVA and higher. Oil forced water forced (OFW) cooling is used for transformers designed for hydroelectric power plants.

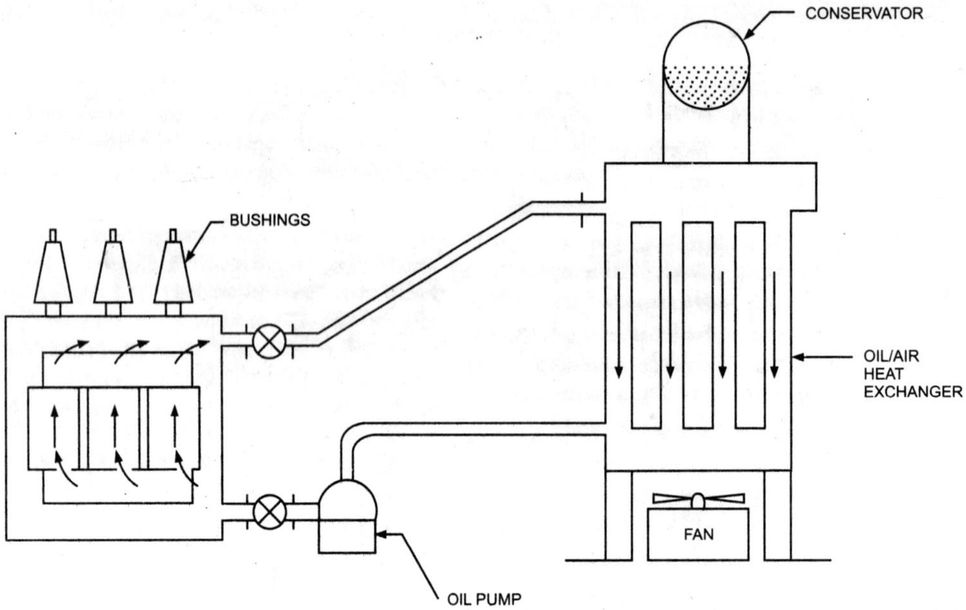

Fig. 13.45 *Oil Forced Air Forced Cooling*

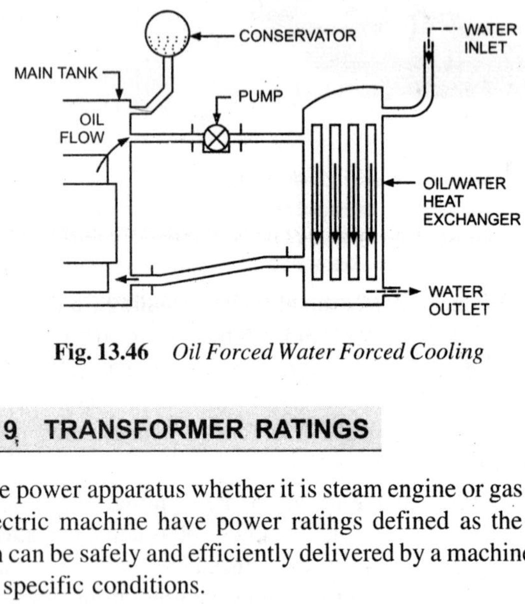

Fig. 13.46 *Oil Forced Water Forced Cooling*

13.19 TRANSFORMER RATINGS

All the power apparatus whether it is steam engine or gas engine or electric machine have power ratings defined as the power which can be safely and efficiently delivered by a machine under some specific conditions.

Electrical apparatus or machine is usually rated at the load, which it can carry without overheating and damage to insulation *i.e.*, rating of electric machine is governed by the temperature rise caused by the internal losses of the machine.

The copper loss in the windings depends upon the magnitude of currents flowing through the primary and secondary windings and core loss depends on the voltage and thus losses are independent of pf.

The output in kW is proportional to power factor for the transformer of a given kVA. For example, output of 1,000 kVA transformer on full load will be 200, 500, 800 and 1,000 kW at power factor 0.2, 0.5, 0.8 and unity respectively, but copper losses in windings will remain same regardless of power factor.

For the foregoing reasons transformers are rated in kVA, not in kW.

A transformer is specified by its kVA (or MVA), voltage and frequency.

The kVA rating is the kVA output which the transformer can deliver at rated voltage and frequency under usual operating conditions without exceeding the standard limits of temperature rise. Rated secondary voltage is the voltage that appears across the secondary terminals when rated current flows. The rated primary voltage is equal to the rated secondary voltage multiplied by the turn-ratio between primary and secondary i.e.

$$V_1\,(\text{rated}) = V_2\,(\text{rated}) \times \frac{N_1}{N_2}$$

For example a typical transformer carries a name plate bearing the information : 2 MVA; 33,000/11,000 V. It means that the rated secondary voltage V_2 (the voltage that appears across the load when delivering rated current) is 11,000 V, the primary winding voltage rating is 33,000 V, which is obtained by multiplying the rated secondary voltage by the turn-ratio between the primary and secondary, as mentioned above. The impressed voltage to the primary would be somewhat different from 33,000 V when the transformer delivers rated load of 2 MVA at 11,000 V. Finally, the kVA figures always refer to the *output kVA* appearing at the secondary load terminals. Of course, the input kVA is slightly higher because of internal losses (core and winding losses).

13.20 TRANSFORMER SPECIFICATIONS

The transformer specifications give the rating and performance expectations of the transformer. These are broadly as below:

1. kVA rating. 2. Rated voltage. 3. Number of phases (1-phase or 3-phase). 4. Rated frequency. 5. Connections (Δ or $\curlywedge$ for 3-phase transformer). 6. Tappings if any. 7. Type of core (core or shell). 8. Type (power or distribution). 9. Ambient temperature (generally average 40 °C). 10. Type of cooling [(a) cooling medium—air, oil or water (b) circulation type—natural or forced (c) simple or mixed cooling]. 11. Temperature rise above ambient in °C depending upon the class of winding insulation. 12. Voltage

regulation [(a) Per cent or pu at full load at 75 ºC unity pf or 0.8 pf lag (b) Impedance-per cent or pu (c) Reactance—per cent or per unit]. 13. No-load current in amperes or per cent of rated current at rated voltage and rated frequency. 14. Efficiency—in per cent or per unit at full load, ½ load, ¾ load at unity pf or 0.8 pf.

The specification items and limitations are laid down by the following IS specifications:

IS : 1180-1964 : Specifications for outdoor type three-phase distribution transformers up to and including 100 kVA and 11 kV.

IS : 2026 -(Part I, II, III, IV 1962 : Specifications for power transformers beyond 100 kVA.

1. Outdoor Type Distribution Transformers (IS : 1180-1964)

According to this the **standard ratings** for distribution type transformers are 16, 25, 40, 50, 63, 80 and 100 kVA.

The **no-load voltage ratios** are 3,300 /433 V, 6,600/433 V and 11,000/433 V.

The **tappings** shall be provided on hv side and shall be in 5 steps.

The ranges shall be ± 2.5% and ± 5%. Off-load tap changers are used.

Connections : Δ/Y (Dy11) with neutral brought out to a separate insulated terminal.

Cooling is by low viscosity transformer oil

Conservator tank is provided on transformer of rating 50 kVA or above.

Limits of Temperature Rise: The following temperature rises shall be permitted over the ambient temperature of 45 ºC.

(*i*) Temperature rise in winding measured by resistance method — 55 ºC

(*ii*) Temperature rise in oil measured by thermometer in the top oil — 45 ºC

The above temperature rises are for ON, OB and OW type cooling.

Impedance: The percentage impedance at 75 ºC is 4.5% subject to the tolerance limits of ± 10%.

2. Power Transformers (IS 2026-1962)

The **standard kVA ratings for 3-phase transformers are** 25, 40, 63, 100, 125, 160, 200, 250, 315, 400, 500, 630, 800, 1000, 1250, 1600, 2000, 2500, 3125, 4000, 6300, 8000, 10000, 12500, 16000, 20000, 25000, 31500, 40000, 50000, 63000, and 80000 kVA

The **standard ratings for 1-phase transformers are** 1, 2, 5, 10, 16 and 25 kVA.

Above 25 kVA, the standard rating for single-phase transformers shall be one-third of the value given for 3-phase transformers.

Tappings : The standard tapping ranges are ± 2½% and ± 5%

Tap changing is carried out by means of an externally operated off-circuit switch capable of being locked in positions.

If required, the transformer may be equipped with on-load tap-changer.

Temperature Rise Limits. The temperature rise of transformer windings, oil and core shall not exceed the limits prescribed as given below.

1. Windings, measured by resistance
   — ON, OB, OW cooling — 55 ºC
   — OFN, OFB cooling — 60 ºC
   — OFW cooling — 65 ºC.
2. Oil, measured by thermometer in top oil—all types — 45 ºC.
3. Cores — core shall be designed so that the temperature rise on any part of the external surface does not exceed that of the winding.

Impedance. The per cent impedances at 75 ºC are 4.5% for transformers of rating up to and including 100 kVA, 11kV and 4.75% for transformers of rating above 100 kVA including 1000 kVA, 11kV.

13.21 HEATING AND COOLING OF TRANSFORMERS

It is the same as for other rotating electrical machines explained in Art 4.18. The expression for temperature rise is given below:

$$\theta = \theta_F (1 - e^{-t/\tau}) \qquad \ldots (13.20)$$

where θ is the temperature rise, θ_F is the final temperature rise and τ is the heating time constant.

During cooling, the expression for temperature rise is given as

$$\theta = \theta_F e^{-t/\tau'} \qquad \ldots (13.21)$$

where τ′ is the cooling time constant.

The heating and cooling time curves for continuous and constant loading of a transformer are determined from test.

Example 13.38. A transformer has a full-load copper loss equal to twice its core loss. In a temperature-rise test at full load the following rises were obtained — 12° C after 1 hour, 21° C after 2 hours. Find (*a*) the final steady temperature rise and time constant (*b*) the steady temperature rise after 1 hour at 50 percent overload from cold.

Solution: From Eq. (13.20), $\theta = \theta_F (1 - e^{-t/\tau})$ ... (*i*)

Now from given data when t = 1 hour, θ = 12° C and when t = 2 hours, θ = 21° C

Substituting the above values of t and θ in equation (*i*) we get

$$12 = \theta_F (1 - e^{-t/\tau}) \qquad \ldots (ii)$$

$$\text{and } 21 = \theta_F (1 - e^{-2/\tau}) \qquad \ldots (iii)$$

Dividing equation (*iii*) by equation (*ii*) we get

$$1.75 = \frac{\theta_F\left(1 - e^{-2/\tau}\right)}{\theta_F\left(1 - e^{-1/\tau}\right)} = 1 + e^{-1/\tau}$$

(*a*) or time constant,

$$\tau = \frac{-1}{\log_e 0.75} = 3.476 \text{ hours } \textbf{Ans.}$$

Substituting τ = 3.476 in equation (*i*) we get

Final steady temperature rise,

$$\theta_F = \frac{12}{1 - e^{-1/3.476}} = 48° \text{ C } \textbf{Ans.}$$

Total loss on full-load

$$= P_c + P_i = 2\,P_i + P_i = 3\,P_i$$

∵ full-load copper loss, $P_c = 2 \times$ iron loss $= 2 \times P_i$

Total loss on 50% over-load

$$= \left(\frac{150}{100}\right)^2 \times 2\,P_i + P_i = 5.5\,P_i$$

$$\frac{\text{Total loss on 50\% over-load}}{\text{Total loss on full-load}} = \frac{5.5\,P_i}{3\,P_i} = \frac{11}{6}$$

Hence, final temperature rise on 50% overload

$$\theta'_F = \text{Final temperature rise on full load} \times \frac{\text{Total loss on 50\% overload}}{\text{Total loss on full load}} = 48 \times \frac{11}{6} = 88°\,C$$

(*b*) Temperature rise after 1 hour on, 50% over-load,

$$\theta = \theta'_F\,(1 - e^{-t/\tau}) = 88\,(1 - e^{-1/3.476}) = 22°\,C \textbf{ Ans.}$$

Example 13.39. The full-load temperature rise of an oil immersed transformer is 15° C after 1 hour and 23° C after 2 hours : (*a*) find the final steady temperature rise on full-load, (*b*) estimate the 1-hour rating for the same temperature rise if the full-load copper loss is twice the iron loss.

Solution: From Eq. (13.20), $\theta = \theta_F\,(1 - e^{-t/\tau})$ ... (*i*)

Now from given data when t = 1 hour, θ = 15° C and when t = 2 hours, θ = 23° C

Substituting the above values of t and θ in equation (i) we get

$$15 = \theta_F\,(1 - e^{-1/\tau}) \quad ...(ii)$$

$$\text{and } 23 = \theta_F\,(1 - e^{-2/\tau}) \quad ...(iii)$$

Dividing equation (*iii*) by equation (*ii*) we get

$$\frac{23}{15} = \frac{\theta_F\left(1-e^{-2/\tau}\right)}{\theta_F\left(1-e^{-1/\tau}\right)} = 1 + e^{-1/\tau}$$

$$\text{or} \quad e^{-1/\tau} = \frac{23}{15} - 1 = \frac{8}{15}$$

Substituting $e^{-1/\tau} = \frac{8}{15}$ in equation (*ii*) we have

(*a*) Final steady temperature rise,

$$\theta_F = \frac{15}{1-8/15} = 32.14°\,C \textbf{ Ans.}$$

If, on overload, the temperature rise was 32.14° C, after one hour, the final temperature rise would be given by

$$32.14 = \theta'_F\,(1 - e^{-1/\tau})$$

$$\text{or } \theta'_F = \frac{32.14}{1-8/15} = 68.87\,°C$$

Let 1-hour rating be x times full-load

$$\text{Total loss on full-load} = P_c + P_i = 2\,P_i + P_i = 3\,P_i$$

$$\because \text{ Full-load copper loss, } P_c = 2 \times \text{iron loss} = 2\,P_i$$

$$\text{Total loss on } x \text{ times full-load} = x^2 \times P_c + P_i = x^2 \times 2\,P_i + P_i = (1 + 2\,x^2)$$

$$\therefore \frac{\text{Total loss on } x \text{ times full-load}}{\text{Total loss on full load}} = \frac{\text{Final temperature rise on } x \text{ times full-load}}{\text{Final temperature rise on full load}}$$

$$\text{or} \quad \frac{P_i\left(1+2x^2\right)}{3P_i} = \frac{68.87}{32.14}$$

$$\text{or } x = \sqrt{\frac{\frac{68.87}{32.14}\times 3 - 1}{2}} = 1.6475$$

So 1-hour rating will be 64.75 per cent overload **Ans.**

13.22 MAINTENANCE AND REPAIRS OF TRANSFORMERS

13.22.1. Maintenance Requirements

1. Deterioration. The normal life of a well operated transformer may be taken as 25-35 years. It may serve longer if careful installation, rigid inspection and preventive maintenance is carried out and may become unserviceable earlier when operated with negligence.

The following points need consideration.

(*a*) *Overheating*. Overheating damages the insulation. The paper deteriorates by becoming brittle, the amount of deterioration depending on the degree of overheating and length of time duration it is allowed to continue. When brittle, the paper still provides reasonable insulation, but it gets very easily damaged by mechanical means such as vibration or movement of the transformer. At higher temperature the paper begins to char, and its insulation properties are much reduced. In extreme cases, of course, the transformer may burst into flames. The oil, when overheated, forms sludge, which is deposited on the windings and tends to clog the cooling ducts. This causes further increase in temperature and speeds up the deterioration, which again depends on both the temperature and period of overheating.

(*b*) *Impurities*. Impurities in the oil reduces its insulating properties and speeds up sludging. The moisture is absorbed by the paper from the atmosphere through the oil. Dampness is the enemy of most types of insulation.

(*c*) *Mechanical Damage*. The connections may become loose or dirty, including the connections on tapping switch or link board if either is provided. The porcelain of bushings may get cracked or chipped. The steel tank is liable to corrode, particularly if provided with tubes.

2. Frequency of Maintenance. Routine maintenance is carried out with the object of finding any defects at any early stage and taking necessary steps for their rectification. It is impossible to provide a rigid schedule appropriate in all circumstances, so it is necessary to check the various parts at comparatively short intervals when the transformer is first installed and then build up a routine from the experience thus gained. Maintenance schedule on a typical transformer installation is given below.

S.No.	*Items*	*After installation*	*Subsquently*
1.	Breather	monthly	3 monthly
	Free air or oil seal	weekly	monthly
	Dehydrating		
2.	Oil level	twice weekly	weekly
3.	Temperature	often	monthly when lightly loaded; weekly when heavily overloaded.
4.	Oil		
	(*i*) Crackle and dielectric tests	3 monthly	annually
	(*ii*) Acidity test	half yearly	every two years

S.No.	Items	After installation	Subsequently
5.	Bushings	3 monthly	3 monthly
6.	External connections	half yearly	annually
7.	Paint work	half yearly	annually
8.	Internal above oil level (non-conservator)	half yearly	annually
9.	Internal including core and windings	half yearly	every two years

(*a*) ***Breather.*** Change in weather or in the load cycle may increase the intake of moisture considerably, and therefore, it becomes necessary to attend the breather more frequently than given in the schedule.

(*b*) ***Oil.*** The oil in high voltage transformers needs testing more frequently and must be maintained in better condition than is necessary for low voltage equipment.

(*c*) ***Temperature.*** When the transformer is installed and put in service for the first time, the oil temperature must be taken at interval of about 2 hours for the first two days to find the hours during which the transformer is likely to attain maximum temperature. The temperature of oil may then be checked for further two or three days during these hours to see whether the same temperature is reached every day. If the temperature remains below 70° C, temperature may be checked once a month or so unless the load increases or weather gets considerably hot. In case the temperature exceeds 70° C, more frequent checking of temperature is required.

(*d*) ***Complete Overhaul.*** From time to time the transformer should be taken out of service for a complete examination and overhaul, which should include lifting of core and windings from the tank for inspection.

If a transformer has been short circuited it is well worth checking that the coils have not been loosened, and in such circumstances where the transformer is put to repeated short-circuits (or near short circuits such as in case of arc furnace and high-voltage testing transformers) an annual examination is a very good safeguard.

Where a breakdown is less important it is unnecessary to disturb the windings annually, and they can be left for 2 or 3 years. In such cases it is sufficient with non-conservator transformers to remove the cover every year and examine the inside of the tank for condensed moisture or rust. The tapping switch can be checked at the same time.

Typical maintenance schedule is given below. The intervals may be varied for individual installation on the basis of experience.

S. No.	Items	Period between inspection
1.	Oil level, temperature (if high)	week
2.	Dehydrating breather, temperature (if low)	month
3.	Free air or oil-seal breather, bushings	3 months
4.	Oil (crackle and dielectric tests), paint work, external connections, internal above oil level (non-conservator)	year
5.	Complete examination including acidity test of oil and inspection of core and windings	2 or 3 years

13.22.2. Maintenance Procedure. The noteworthy point is that the transformer oil is not injurious to the human skin in normal circumstances and can be washed out of overalls, but it soaks into concrete floors making them unsightly and dangerous.

1. Safety Precautions. Better it will be if breather, oil level and temperature are inspected first safely when the transformer is energized. In case of small units breather, oil level indicator and thermometer (dial type) are visible from outside. On larger units the three items concerned may be mounted some distance away from live metal, in which case it may be possible to enter the enclosure (if there is any one) and inspect them in reasonable safety whilst the transformer is energised. The rest of maintenance is carried out with the transformer isolated from the supply.

2. Breather. In case of a plain breather, its ends must be kept clean and ventilation holes free from dust. If an oil seal is provided the oil should be wiped out and replaced to the correct value. Silica-gel dehydrating breathers are fitted with a sight glass so that colour of the crystals can be noted. The colour changes from *blue to pink* as the crystals absorb moisture, and when they are predominantly pink they are approaching saturation and need reactivation. The crystals are baked at about 200ºC until the whole mass of the material is at this temperature and the blue colour restored. The breather is cleaned, the crystals are replaced and the oil of the sealing cup at the bottom is renewed. It is preferable to have a spare charge of silicagel so that the breather can be replaced immediately. The crystals just removed should be dried as mentioned above and then sealed up with adhesive tap until required.

3. Oil Level. The oil level varies with its temperature. The oil level indicator usually indicates the level when the oil is cold (at a temperature of 15° C). This means that the oil level can be checked accurately only when the transformer has been off for some hours. In some cases the oil level indicator is marked for higher temperature.

The transformer oil should be topped up as necessary with clean transformer oil, but only small amounts should be required at frequent intervals. If the oil level drops appreciably over a short period, the tank should be checked for leaks, and repair.

A leaking gasket is often remedied by retightening the bolts. Gasket may be replaced if necessary.

In case the seal on the tapping switch shaft leaks, the manufacturer may be consulted.

A leaking weld can be repaired satisfactorily only by re-welding.

4. Gaskets. It is preferable to have in stock a set of spare gaskets obtained from the manufacturer.

The large gasket for the tank cover is sometimes made up of strips of gasket material supplied in rolls. Lengths of material are measured off against the tank flange, overlapping the strips by at least 5 mm to allow for the joint. Each joint should preferably be close to a bolt hole so that the bolt does not go through the joint but supplies a good pressure to it.

The first strip is placed along the flange, either with the sealing compound or by tacking it in place with a small quantity of oil-proof adhesive.

The length is trimmed and the joint is cut as shown in Fig. 13.47, with a sharp knife, each surface of the joint is smeared

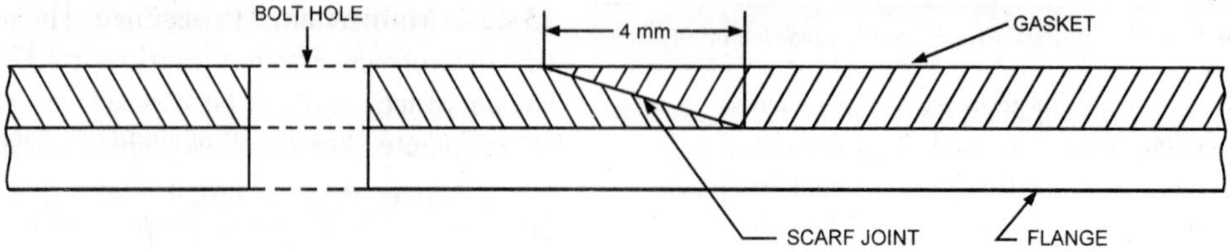

Fig. 13.47

with sealing compound and the second strip is stuck to the flange. This is carried out all round the tank and then the cover is placed carefully in place, lining up the bolt holes accurately. The gasket is pierced for the bolts and the bolts are tightened evenly all round until they are firmly tight without compressing the gasket.

5. Welding. In case the transformer is fitted with removable radiators and one develops a leak, the valves are closed to the faulty unit, the oil is drained out and then it is removed for repair. When the radiator is refitted care is taken that the valves are opened.

To avoid trapping air in the transformer, after refitting of radiator, air-vent plug from the top of the radiator is removed and bottom valve is slightly opened so that the radiator is filled with oil. When oil appears at the air-vent, the bottom valve is closed, the vent plug is replaced, both bottom and top valves are opened and the oil is added to the main tank to correct the level.

If the radiator is not fitted with an air-vent, the oil level in the tank is lowered below the level of the top opening to the radiator, both top and bottom valves are opened and then the tank is refilled with oil to the correct level.

If the main tank needs rewelding, the core and the windings are removed and the oil is drained out.

The presence of oil fumes during welding can be dangerous, so the item to be repaired must be thoroughly ventilated before hand and the welding should not be done in an enclosed space.

6. Temperature. The transformer oil at the top of tank should not normally exceed 90° C, since at higher temperatures it deteriorates rapidly. Overheating is usually caused by an excessive load, and the remedy is to reduce the load current. Overheating may also be caused by a fault in the transformer. If it is cooled by means of radiators or water coolers they are often fitted with valves. It should be checked and ensured that valves are open.

A possible source of trouble is the formation of oil sludge in the cooling paths, but if the transformer is new, or is properly maintained, this is unlikely. The only remedy is to overhaul the transformer completely.

A partial short circuit in the winding may cause overheating. If this is suspected, the transformer should be taken out of service immediately. In this case the load is removed from the transformer and no-load primary current is determined and if it exceeds the normal value of magnetizing current considerably, certainly the winding is faulty. It may be worth lifting the winding out of the tank to see if the fault is visible, but usually it is preferable for faults in the windings to be dealt by a repair organisation.

The surrounding air temperature affects the transformer temperature and may cause overheating.

7. Bushings. Porcelain bushings are cleaned and examined for cracks and chips. Very slight chips may be ignored but any serious damage will require replacement. In some cases the porcelain only needs change, in other cases the conducting rod is cemented into the porcelain and the complete bushing must be renewed.

If the bushing is below oil level, oil is lowered until it is below the bushing hole.

If only the porcelain is to be replaced, it may not be necessary to remove the internal bushing connection, for in some cases the bushing stems are joined by an insulated bar to prevent them turning when the nuts are unscrewed . After removing all the nuts from the top of the bushing the old porcelain is lifted straight up over the central stem and the new porcelain is slided down over the stem and nuts are screwed.

If the insulated bar between the bushing stems is not provided, the internal connections are removed and the whole bushing is removed before the porcelain is changed.

If the complete bushing is to be replaced, the internal connections to the bushings must be removed. If the replacement bushing has a socket at the bottom end, the method is usually to unclamp the old bushing and withdraw it from the tank, internal lead being flexible and long enough to permit this. This lead can now be unplugged from the old bushing and plugged into the new one, which is then lowered into the hole in the tank and reclamped firmly but not too tightly.

If the replacement bushing shows that the internal connection is secured by a nut, it is necessary to have access to the tank. This may mean removing the cover or handholes may be provided adjacent to the bushing. With the cover or handhole plate removed the internal connection and then the bushing clamps are removed. The old bushing is taken out and new one is fitted. The clamps are replaced and then internal connections are made, making sure that the latter are tight. The cover is then placed.

New gaskets may be required, but in any case it must be ensured that all the necessary gaskets are in place before topping up the oil.

8. Oil Sampling. Samples of oil are taken from the bottom of the tank through the sampling valve or drain valve if provided otherwise oil is trickled out from sampling device, usually provided in drain plug in the form of small hole pointing downwards, by unscrewing the plug partly.

Samples must be true representative of the oil in the transformer, bad sampling can make the oil appear much worse than is actually the case.

The utensils required for this purpose are (*i*) one or more stoppered glass bottles of capacity 3 litres and (*ii*) a wide-mouth container which will fit under the sampling valve or drain valve.

These utensill should be cleaned initially by being washed thoroughly in warm water and detergent, swilled out with clean (preferably distilled) water, drained and then thoroughly dried in a warm oven. Wiping dry with cloth will almost certainly leave loose fibres which will affect the tests. Alternatively, the containers can be washed with petrol, drained and then left to dry in an clean ventilated place.

If the utensils have been used previously for the same purpose, they need not be cleaned again unless the previous sample proved verty dirty, or they have become dirty in the mean time.

In addition to the above items, a bucket, a lint-free cleaning cloth and some thin, smooth paper (toilet paper) are also required.

The samples should be taken when the transformer is warm, and if it is outdoor, the weather should be reasonably dry. The floor may be protected by placing a metal (or plastic) sheet under the valve. An extra precaution is to make a ring of saw dust on the sheet to contain split oil:

The following procedure is adopted for taking oil samples :

1. Loose material, if any, is removed from the valve and tank round it, and the valve is cleaned with a rag and then wiped out round the outlet with paper for removing the fibre, if any.
2. Sufficient oil is run off to flush the bore of the valve and is allowed to run straight into bucket, if possible, as this oil is not required.
3. Some oil is run into the jug, some clean paper is dipped into it and the outlet valve is again wiped out. The jug is rinsed out into the bucket.
4. Sufficient oil is drawn off into the jug to rinse out all the sampling bottles. The sampling bottles are thoroughly rinsed into the bucket.
5. The oil is drawn into the jug and then poured into the sample bottles, putting the stopper into each bottle as it is filled. About 3 litres of oil is collected for sample.
6. The valve is tightly closed and the transformer is filled up with fresh oil.

Small transformers are often fitted with a drain plug which is not intended for sampling. Since the quantity of oil is small the tank can be easily drained and refilled. The drain plug and surrounding tank is cleaned and the plug is slackened off until the oil seeps out fairly fast and about half a litre of oil is collected in a clean and dry container. Plug is completely removed and a new gasket is fitted before refilling.

Since free water in the oil collects at the bottom of the tank, a crackle test on this sample will show if an undue amount of moisture is being drawn into the tank.

After use, the sampling utensils should be rinsed out with clean oil, the stoppers replaced in the bottles, and then stored in a dust free box in a clean and dry place.

9. Oil Testing.

(*a*) *Crackle test* for free water is performed as per IS: 335-1953. This test is only qualitative.

(*b*) *Dielectric test* is performed as per IS: 335-1953. The oil shall withstand a test voltage of at least 30 kV for one minute without breakdown.

(*c*) The *acidity of oil* is determined as per IS: 1866-1961. The limits of permissible acidity are :

(*i*) Acidity below 0.5 mg KOH/g: No action need be taken if oil is satisfactory in all other respects.

(*ii*) Acidity between 0.5 mg and 1.0 mg (KOH)/g: Oil is to be kept under observation.

(*iii*) Acidity exceeds 10 mg (KOH)/g : oil should be treated or discarded.

(*d*) *Sludge Test* The traces of solid matter in oil samples may be examined as per IS: 1866-1961.

Appearance. The oil should be a clear light colour with a faint characteristic smell. Cloudiness in oil may be due to suspended moisture or suspended solid matter. Dark brown colour may indicate dissolved asphalt, green colour dissolved compounds; acrid small indicates the presence of volatile acids.

Test Results. If the oil looks satisfactory but fails on the crackle and dielectric tests it is probably contaminated with moisture. Oil which fails the dielectric test but is satisfactory otherwise probably contains fibrous matter.

Oil which has been contaminated by a fault in a transformer is unsuitable for use.

10. Oil Treatment . Oil, which is unsatisfactory, can be treated either by drying out process in which small amounts of moisture is removed or by filteration process in which large amounts of water and solid impurities are removed. The latter process is usually preferred. Drying out process is carried out by stoving, internal heaters or by short circuiting.

11. External Connections. Connections must be tight. If they appear blackened or corroded, the connections are removed and cleaned down to bright metal with emery paper. The connections are then made and a heavy coating of grease is provided. It is particularly important that heavy current connections should be properly maintained. If the metal has the characteristic bluish tinge which indicates that it has been hot, then in most cases the connections cannot be considered satisfactory. Either it has become loose or dirty, or the conductor is not suitable for the current being carried.

12. Paint Work. Loose paint and rust, if any, is removed down to bright metal and touched up with red lead or zinc-chromate primer. One coating of good quality paint either locally or overall is then provided for the sake of appearance.

The insulation resistance is measured by megger along the temperature. This is because the insulation resistance gets reduced to nearly half for every 10°C temperature rise. Insulation resistance should not be less than 2 MΩ for each 1,000 V of operating voltage.

13.22.3. Internal Inspection. The core and windings should be lifted in suitable conditions. The surroundings should be such that the air is not damp (i.e. they should be weather proof, with no water standing about and no dripping of condensation from the roof) and atmosphere is not dusty. It is advisable to keep it well away from the shops in which metal is being worked, to avoid metal filings and swarf. The transformer should be several degrees warmer than the surrounding air in order to avoid condensation of dampness on the coils. Sufficient head room is necessary, and a hoist is usually required. The floor should be swept and the dust settled before the core and coils are exposed.

Insulation resistance must be measured before disturbing the transformer.

1. Opening of Transformer. The tank contains oil with oil-vapour above it, and care should be taken to prevent a fire. Naked lights must be kept well away whilst the tank is open. If an inspection lamp is required, a low voltage electric lamp is the ideal choice. The spanner used must be clean, particularly of metal filings. It is advisable to tie a loop of cotton tap or string to the spanner and hang it round the wrist, otherwise the spanner may end up at the bottom of the tank. It needs care to see that loose nuts and bolts do not do the same. If any thing does fall into the oil it must be fishedout again of course, which may be a very tiresome business.

It is sometimes useful to start by removing the hand hole cover plate and have a look inside, but it should be ensured that the oil level is below the cover plate. The cover being removed should be clean and dry so that the dust and water do not fall into the tank.

2. Removal of Cover. The method of removing the cover varies according to the type of construction. If the core and windings are suspended from the tank cover substantial lifting plugs are usually provided on the cover. These are sometimes in the form of threaded eyes which screw on the projecting studs and are removed when not in use.

Larger transformers have the core and windings separate from the cover. In such case, if the bushings are mounted in tank walls, the cover is simply lifted. If the bushings are mounted on the cover, bushings are removed before removal of cover as mentioned in Art 13.22.2 (7).

3. Cores and Coils. With the cover removed it is easy to see what is to be done. If the bushings are in the tank walls, the connections are opened and removed if they project into the tank. The core and coils are usually held in the tank by a bolt at each end near the top which must be removed. The mechanical connections to the tapping switch handle, if provided, are removed. Lifting lugs are usually provided on the core and coils but it is to ensured that the slings do not foul against the tapping switch or bushing connections as they pull tight. To avoid this a spreader between the slings may be necessary in some cases.

The core and coils are allowed to drain into the tank for a few minutes before lowering them to the floor.

4. Inspection. Nothing should be disturbed unnecessarily; particularly leads should not be pulled out of place unless they are definitely unsatisfactory.

1. There are horizontal bolts or studs at each end of the top and bottom clamps, and these should be tight. They will not be tightened up rigidly, otherwise clamps will be bent. There may be similar bolts right through the core at the centre of each clamp, tightened to about the same degree as the end bolts .
2. Bolted connections including the connections on the tapping board or switch should be checked for tightness, and these should be firm.
3. Bolts used in general assembly should also be checked.
4. The coils should be clean and free from sludge. Slight traces of sludge may be washed off with clean transformer oil, but if the deposit is heavy the internal ducts may be blocked and the matter may have to be referred to the manufacturer.
5. The windings should be clamped firmly with no movement anywhere and no loose spacers or end blocks. The methods used for tightening are:
   (*i*) Vertical tie bars joining the top and bottom clamps are adjusted. These are used to put pressure on the coil stacks.
   (*ii*) Special coil-adjustment bolts provided at the top of the leg are tightened.
   (*iii*) Wedges of insulation board (presspahn) are tapped into place at the top and bottom of each coil stock.

   It is preferable to consult the manufacturer before attempting to tighten coil stacks (particularly by using wedges). Overclamping may cause damage.
6. If a tapping switch is provided it should be examined as follows:

   The position of the moving contact should be marked on the tapping switch by making a light scratch on the two contacts it joins. The switch is moved over the whole range making sure that the moving arm provides a good contact pressure on the fixed pins. Finally, the moving arm is brought back to the original position so that it mates up correctly with the handle on the tank.
7. If a tapping link board is fitted it should be ensured that the nuts are tight. It should be also ensured that the permanent connections do not slaken off when the links are changed. Sometimes a plug-and-socket type tap-changer is provided. In such a case there should be reasonable contact pressure when the moving part is pushed. Each pin is examined separately because the socket fits over two pins at the same time and good contact pressure on one pin disguise the lack of pressure on the other.
8. The megger tests are repeated on the core and coils out of tank, connecting the earth terminals to the core, if the results of megger tests carried out before untanking were too low. If the results of the megger tests carried out after untanking are satisfactory, the trouble obviously lies elsewhere.
9. The inside of the tank and cover should show no sign of rust. If they do this must be removed, of course, and treated as a sign that the insulation is damp.
10. Sludge at the bottom of the tank can be removed with transformer oil.

5. Retanking. When every thing is satisfactory, the core and windings are brought back to the tank and bushings and tapping switch handle are connected. It is advisable to fit new gaskets. While tightening up gasketted joints the bolts should be tightened until they are firm but not overtightened, as this only compresses the gasket and does not improve the seal.

The bushings are fitted, if they have been removed, and insulation resistance is measured as a check that the bushing leads are not touching the tank.

HIGHLIGHTS

1. A three-phase transformer operates on similar principles and behaves the same way as a single-phase transformer. A single three-phase transformer has the advantages of lesser weight, lower initial cost, lesser space requirement. Three phase transformers are of two types viz. core type and shell type. Core type transformers are commonly used.
2. The primary and secondary windings of three-phase transformers can be connected in several ways like star and delta and the voltage can be raised or lowered according to needs. The most commonly used transformer connections are star-star, delta-delta, star-delta and delta-star.
3. It is possible to transform three-phase power by means of only two single-phase transformers. There are two methods of doing so, namely (*i*) Open-delta connection method and (*ii*) T-connection method. Both of these methods result in slightly unbalanced output voltage under load because of the unsymmetrical relations. This is not serious in commercial transformers, as their regulation is seldom poorer than 2 or 3%.
4. In some cases such as for electric furnaces, it is desired to work with two-phase current. Since nowadays 3-phase ac supply is available, it is necessary to convert from 3-phase to 2-phase. This is achieved by means of Scott-connection. It employs two identical transformers, one with primary centre tap, known as *main transformer*, and the other with primary 0.866 tap, known as *teaser* transformer.
5. Transformers may be built with a third winding, called the *tertiary*, in addition to the normal primary and secondary; and such transformers are called the *triple-wound* (or 3-*winding*) transformers.

   The tertiary winding is called the *auxiliary winding* when it is employed for supplying an additional small load at a different voltage. On the other hand, it is called the *stabilising winding* when it is employed to limit the short-circuit current.

   Tertiary windings are normally delta-connected so that when faults and short circuits occur on the primary or secondary sides (particularly between lines and earth), the considerable unbalance produced in phase voltages may be compensated by the circulating currents flowing in the closed Δ. The reactance of the windings should be large enough to limit the circulating currents in order that there is no overheating of the windings.
6. Parallel operation of three-phase transformers must fulfil for each phase the conditions discussed for satisfactory parallel operation of single-phase transformers (Art 12.23.3). In addition the condition of same phase sequence and same relative phase displacement are to be fulfilled.

EXERCISES

1. What is difference between a 3-phase transformer bank and a 3-phase transformer unit? What are the advantages of a 3-phase unit transformer over three single-phase transformer of the same kVA rating. **[Rajasthan Technical Univ. Electrical Machines-I, 2007, 2008]**
2. Compare bank of three single-phase transformers and three-phase core type transformer. **[Rajasthan Technical Univ., Electrical Machines-I, January-2013]**
3. Explain the basic principle of working of three-phase transformer. **[M.D. Univ., Electromechanical Energy Conversion, May-2009]**
4. Discuss unbalanced operation of 3-phase transformers. Also mention its advantages and disadvantages. **[U.P. Technical Univ., Electromechanical Energy Conversion-I, 2013-14]**
5. Explain the constructional details of a 3-phase transformer. **[M.D. Univ. Electrical Machines-I, December-2013]**
6. Explain how the 3-phase core type transformer was evolved from three single-phase core type transformers having both their windings on one leg. **[U.P. Technical Univ. Electromechanical Energy Conversion-I, 2008-09]**
7. In 3-phase shell type transformer, a considerable economy is achieved in core material if the middle phase winding is wound is reverse direction as compared with the outer two phase windings. Explain. **[U.P. Technical Univ. Electromechanical Energy Conversion-I, 2008-09]**
8. Describe in detail the four phasor groups pertaining to three-phase transformers. Draw the phasor diagrams and connection schemes for each of these groups. **[G.G.S.I.P. Electromechanical Energy Conversion-I, December-2011]**
9. Draw the physical connection and phasor diagram of the following transformer connections.
   (*a*) Yd0 (*b*) Dy0 (*c*) Yd6 **[R.G. Technical Univ., Electrical Machine-I, June-2014]**
10. Draw the phasor diagram and connection diagram of the following three-phase transformer groups:
    (*a*) Dy1 (*b*) Dz6 (*c*) Yz11 (*d*) Dy6 (*e*) Dd6 **[W.B. Univ. of Technology Electrical Machine-I, 2011]**
11. (*i*) Discuss three-phase transformer phasor groups. How the displacement is expressed as the clock hour number?
    (*ii*) Show the terminal connections of a 3-phase transformer with corresponding phasor diagram having the vector groups: D_{y1} and Y_{d11}. **[G.B. Technical Univ. Electromechanical Energy Conversion-I, 2011-12]**
12. What are distinguishing features of Δ-Y and Δ-Δ 3-phase connections. **[U.P. Technical Univ. Electrical Machines, 2009-10]**
13. Explain the delta/delta, star/star, delta/star and star/delta connections of 3 single-phase transformer bank. What are their distinguishing features? State their applications. **[Rajasthan Technical Univ. Electrical Machines-I, 2007]**
14. Draw time phase relationship between terminal voltages of hv and lv side of the following three phase transformer configurations. (*i*) Y/Δ; (*ii*) Δ/Y (*iii*) Y/Y (*iv*) Δ/Δ **[Rajasthan Univ. Electrical Machines-I, 2000, 2003]**
15. Three single-phase transformers are connected in star-delta with hv winding in star connection. Draw the connection diagram and phasor diagram for both the hv and lv windings. Label the phasors clearly. **[B.P. Univ. of Technology Electrical Machines, 2008]**
16. What are the distinguishing features of Y-Y, Y-Δ, Δ-Y and Δ-Δ 3-phase connections. Compare their advantages and disadvantages. **[U.P. Technical Univ. Electrical Machines, 2010-11]**
17. Explain a simple zig-zag connection and state the group. **[M.D. Univ. Electrical Machines-I, December-2009]**
18. In open-delta transformers, show that the secondary line voltages form a balanced 3-phase system of voltages, in case the supply voltages are balanced. **[U.P. Technical Univ. Electromechanical Energy Conversion-I, 2008-09]**
19. Explain open-delta connection of a 3-phase transformer. What are the applications of this system? Discuss. **[U.P. Technical Univ. Electromechanical Energy Conversion-I, 2006-07]**
20. Explain the following connections by which 3-phase power is transformed using 2 single-phase transformers
    (*i*) Scott connection (*ii*) Open-delta connection. **[R.G. Technical Univ. Electromechanical Energy Conversion-I, June-2010]**

21. Explain Scott connection for 3-phase to 2-phase conversion and show that the primary neutral divides the teaser primary winding in the ratio 2 : 1.

[Rajasthan Technical Univ. Electrical Machines-I, January-2013]

22. Draw the Scott connection of transformers and also mark the terminals and turn ratio. What are the applications of Scott connection.

[U.P. Technical Univ. Electrical Machines, 2010-11]

23. Explain with necessary diagrams how two three-phase transformers can be used to convert 3-phase supply to 2-phase one. If the load is balanced on one side, show that it will be balanced on the other side.

[J.N. Technological Univ. Electrical Machines-II, January-2014]

24. Describe in brief Scott connection of two single-phase transformers for conversion of a balanced 3-phase to a balanced 2-phase supply. Draw circuit diagram and phasor diagram?

[R.G. Technical Univ. Electrical Machine-I, June-2014]

25. What are the various methods of three-to-six phase conversion? Explain any one in detail. [M.D. Univ. Electrical Machines-I, December-2009]

26. With the help of circuit and phasor diagrams explain the 3-phase to 6-phase transformation using double star transformation method. What are their relative merits and demerits?

[Electrical Machines-I, Rajasthan Univ. 2005; Rajasthan Technical Univ., 2010; 2013]

27. State the reasons of using tertiary windings in a transformer. [U.P.S.C. I.E.S. Electrical Engineering-II, 1997]

28. Draw and explain circuit connection and single phase equivalent circuit of a 3-winding transformer. Also mention its applications.

[U.P. Technical Univ. Electromechanical Energy Conversion-I, 2005-06]

29. Explain the use of tertiary winding in a star-star transformer.

[W.B. Univ. of Technology Electrical Machine-I, 2011]

30. Discuss the construction and working principle of a three winding transformer. [J.N. Technological Univ. Electrical Machines-II, January-2014]

31. Explain the working principle and constructional details of three winding transformers. Also mention the importances of third winding in three winding transformers.

[U.P. Technical Univ. Electromechanical Energy Conversion-I, 2010-11]

32. Explain excitation phenomenon in transformers.

[U.P. Technical Univ. Electromechanical Energy Conversion-I, 2005-06]

33. Explain the basic difference between the magnetizing current of a large capacity transformer and the magnetizing inrush current.

[U.P.S.C. I.E.S. Electrical Engineering-II, 2005]

34. What is inrush phenomena in transformer? Discuss qualitatively this phenomena if single-phase transformer is switched on at the instant applied voltage is maximum positive. Show that doubling effect is not so dangerous as it appears. [Pb. Univ. Elec. Machines-I, Dec; 1992]

35. Show that when the magnetising current of a transformer is sinusoidal, the flux produced is nonsinusoidal and when the magnetising current is nonsinusoidal, the flux produced is sinusoidal.

Show that third harmonic current and its multiples are co-phasal. Show that when the flux is nonsinusoidal, the effect of harmonics is more pronounced on voltage induced.

[W.B. Univ. of Technology Electrical Machine-I, 2011]

36. Why is the waveshape of magnetising current of a transformer nonsinusoidal? Explain the phenomenon of inrush of magnetizing current. What factors contribute to the magnitude of inrush current?

[M.D. Univ. Electrical Machines-I, December-2010]

37. What do you mean by "inrush current" in 3-ϕ transformers? What is the importance of inrush current in 3-ϕ transformers? Also mention its drawbacks.

[G.B. Technical Univ. Electromechanical Energy Conversion-I, 2012-13]

38. What are the disadvantages of current and voltage harmonics in transformer? Explain how these harmonics can be eliminated.

[U.P. Technical Univ. Electromechanical Energy Conversion-I, 2006-07; Electrical Machines, 2010-11]

39. Explain with suitable diagrams how harmonics are produced in transformers even when the supply voltage be purely sinusoidal. Which order of harmonics are usually prominent? What is done to neutralize the effect of third harmonic voltages in high voltage star-star connected transformers?

[U.P. Technical Univ. Electromechanical Energy Conversion-I, 2011-12]

40. Summarize the conditions for proper parallel operation of

(*i*) single-phase transformers, and

(*ii*) 3-phase transformers. [U.P.S.C. I.E.S. Electrical Engineering-II, 2005]

41. Discuss briefly the essential and desirable conditions that must be fulfilled for operating two three-phase transformers in parallel.

[G.G.S.I.P. Univ. Electromechanical Energy Conversion-I, December-2011]

42. State and explain the conditions of parallel operation of two three-phase transformers. [W.B. Univ. of Technology, Electrical Machine-I, 2011]

43. What do you mean by "PARALLEL OPERATION of 3-ϕ TRANSFORMERS"? Also discuss the advantages/disadvantages of parallel operation of 3-ϕ transformers.

Also discuss the excitation phenomenon and harmonics in 3-ϕ transformers.

[U.P. Technical Univ. Electromechanical Energy Conversion-I, 2010-11]

44. What do you mean by "load sharing" in 3-ϕ transformers? Also mention its basic role and limitations in 3-ϕ transformers.

[U.P. Technical Univ. Electromechanical Energy Conversion-I, 2013-14]

45. State the necessary conditions for satisfactory operation of two transformers in parallel. Derive expression for load shared by two transformers in parallel, when the voltage ratios are equal.

[M.D. Univ. Electrical Machines-I, May-2009]

46. Explain various methods of reducing magnetic flux in transformers.

47. Discuss the different methods of cooling used for transformers. Draw neat sketches to illustrate your answer. What is the function of oil in transformer?

48. What is meant by rating of transformer ? Explain what is referred from 100 kVA, 11/0.4 kV rating of a transformer.

49. List important specifications of transformer as per ISI.

[Pb. Univ. Elec. Machine Design, May 1995]

50. Write short notes on the following:

(*i*) Three-phase transformer connections.

[U.P. Technical Univ., 2003-04]

(*ii*) Open-delta connection. [Rajasthan Technical Univ., 2008, 2013]

(*iii*) Scott connection. [R.G. Technical Univ., June-2013]

(*iv*) Tertiary windings. [U.P. Technical Univ., 2002-03]

(*v*) Transformer noise.

(*vi*) Inrush phenomenon in transformer. [M.D. Univ., May-2009]

(*vii*) Harmonics in 3-phase transformers. [Rajasthan Univ., 2001, 2002]

(*viii*) Parallel operation of 3-phase transformers.

[U.P. Technical Univ., 2002-03]

(*ix*) Power and distribution transformers.

[R.G. Technical Univ., June-2013]

(*x*) Transformer specifications.

SHORT ANSWER TYPE QUESTIONS WITH ANSWERS

Q. 1. What is a 3-ϕ transformer?

Ans. A 3-ϕ transformer is that transformer which is equivalent to three single-phase transformers but wound on one core and enclosed within one common case.

Q. 2. To which vector group(*s*) will a star (*hv* side) and delta (lv side) connected transformer belong?

Ans. 3_2 and 4_2

Q. 3. What advantages has the star connection over the delta connection? **[M.D. Univ., December-2013]**

Ans. Each star-connected transformer is wound for only 57.7% of line voltage. In hv transmission, this admits of much smaller transformers being built for high voltage than possible with the delta connection, because of less insulation.

Q. 4. What advantage is obtained with the delta-connection?

Ans. When three transformers are connected in delta, one may be removed and the two remaining units will carry 57.7% of the original three-phase load and thus maintain the continuity of supply.

Q. 5. Compare the behaviour of star-star transformer with star-zig-zag transformer. **[U.P. Technical Univ., 2007-08]**

Ans. Zig-zag connections can be achieved in two ways, known as 30° lag phase displacement (YZ1) and 30° lead phase displacement YZ11. There is a reduction in available capacity (only 0.866 times the combined individual capacity) in zig-zag connection.

In case of star-star connection, there is no phase displacement between primary and secondaries but in case of star zig-zag connection, there is a phase displacement of 30°.

The zig-zag connection affects a reduction of third harmonic voltage and at the same time permits unbalanced loading even though the primary is star-connected with isolated neutral.

Q. 6. In Yy0 transformer, the polarities of phase winding B are reversed. Explain with phasor diagrams, its effect on the secondary output voltages. **[U.P. Technical Univ., 2007-08]**

Ans. Connection and phasor diagrams with polarities of phase winding B reversed are shown in Figs. 1(*a*) and 1(*b*) respectively. Secondary output voltages are shown in Fig. 1.

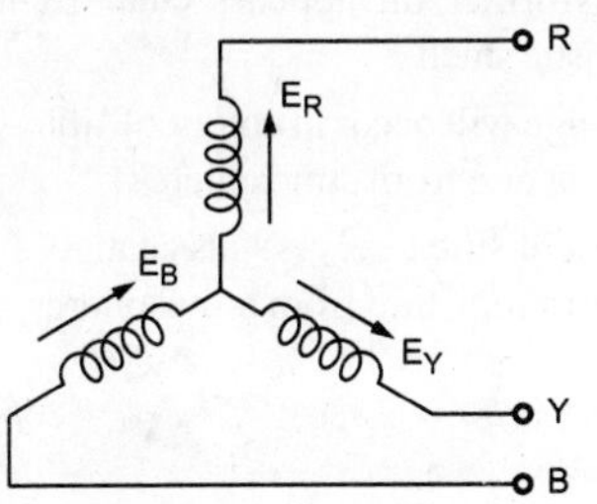

(a) Connection Diagram

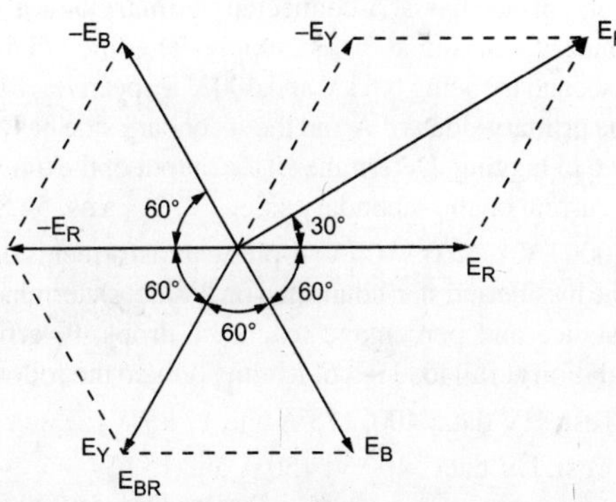

(b) Phasor Diagram of Line and Phase Voltages

Fig. 1

Line voltage E_{RY} leads phase voltage E_R by 30° and is equal to $\sqrt{3}$ times the phase value in magnitude.

Line voltage E_{YB} is in phase opposition to phase voltage E_R and is equal to phase value in magnitude.

Q. 7. Explain the clock method of angle designation for representating 3-phase transformers. **[U.P. Technical Univ. 2007-08]**

Ans. The method of denoting group connections is based on comparison of the relative positions of vectors of high and low line voltages with the positions of the minute and hour hands of a clock. In order to perform the comparison, the minute hand is made to point at 12 and to coincide with the vector of high line voltage, and the hour hand in a clockwise direction the position of the hour hand determines the type of group connection. This is known as clock method of angle designation for representing 3-phase transformers.

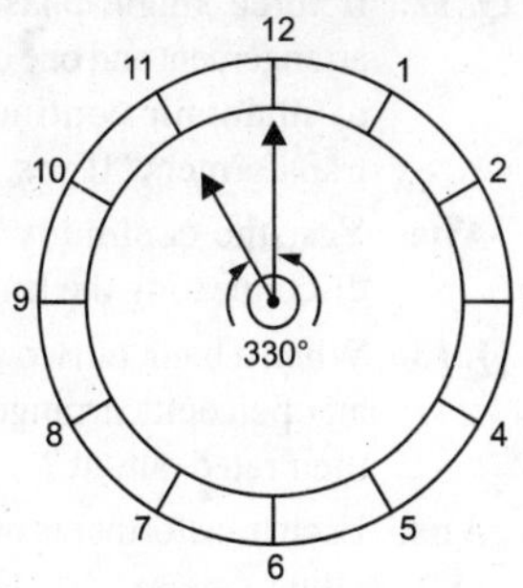

Fig. 2

Q. 8. What is an open-delta system? **[U.P. Technical Univ. 2005-06]**

Ans. If one of the transformers of delta-delta connection is inoperative while a 3-ϕ supply is connected to primary, three equal supplies will be available at the secondary terminals on no load and supply is still possible but with reduced capacity. Such a system is known as an open-delta system.

Q. 9. If the polarity of one transformer is reversed on a delta bank of single-phase transformers that are connected for 3-phase operation, what would be the result ?

Ans. Instead of zero voltage on the tie-point there would be a voltage that is twice the proper value.

Q. 10. One supply line feeding the primaries of Dy transformer gets disconnected. For a phase turn ratio equal to unit, determine the line to line and line-to-neutral output voltages. **[U.P. Technical Univ. 2007-08]**

Ans. Let the winding C be get disconnected. Phasor diagram is shown in Fig. 3.

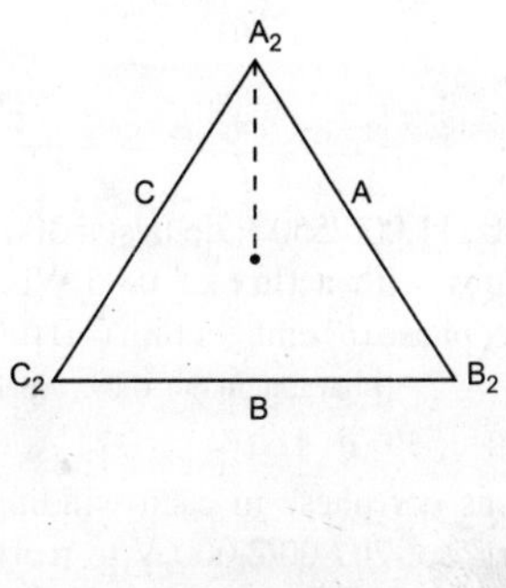

(a) Primaries

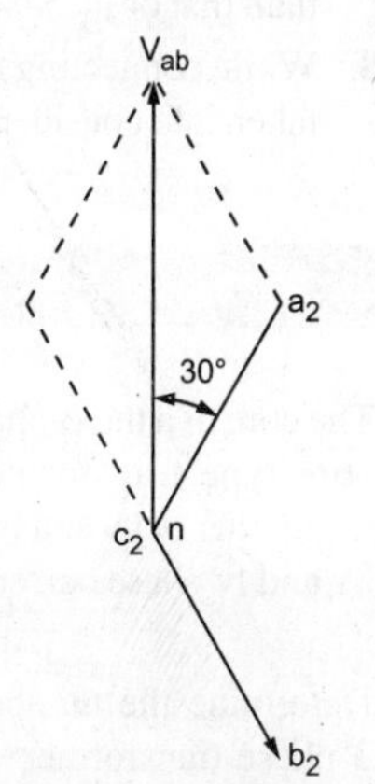

(b) Secondaries

Fig. 3

Line to neutral voltages

V_a = Phase value, say V volts

V_b = Phase value, V volts

$V_c = 0$

Line values

$$V_{ab} = \sqrt{3}\ V \text{ leading } V_a \text{ by } 30^\circ$$
$$V_{bc} = V \text{ in phase with } V_b$$
$$V_{ca} = V \text{ in phase with } V_a$$

Q. 11. Is it possible to connect two single-phase transformers to give a 3-phase output from a 3-phase input?

Ans. Yes, they would have to be connected in an open delta.

Q. 12. If three single-phase transformers are connected in a delta arrangement and one of the transformers get damaged, is it possible to maintain continuity of supply with the remaining two transformers? If yes, how?

Ans. Yes, the continuity of supply can be maintained by merely disconnecting the leads to the disabled transformer.

Q. 13. When a bank of two single-phase transformers are connected in an open delta arrangement to supply a 3-ϕ load, do they supply their rated output ?

Ans. Each transformer is only capable of supplying 86.6% of its output rating.

Q. 14. When a bank of two single-phase transformers, each with a rating of 100 kVA, are connected in open-delta to supply 3-phase load, what would be the output capacity of the arrangement?

Ans. 173.2 kVA.

Q. 15. In a star-star connected transformer, if a delta tertiary is provided, what is the major reason?

Ans. In a star-star connected transformer, a delta tertiary is provided to allow sufficient earth fault current to flow for operation of protective equipment, to suppress harmonic voltages, and to limit voltage unbalance when the main load is unsymmetrical.

Q. 16. Is it possible to connect phasor group I and phasor group II transformers in parallel?

Ans. Yes.

Q. 17. Two single-phase transformers A and B of identical voltage and turn-ratio operate in parallel. If the impedances of the two transformers are equal while resistance /reactance ratio of A is higher than that of B, what would be the magnitudes of I_A and I_B and their phase position?

Ans. Magnitudes of I_A and I_B are equal but phase angle of I_A is less than that of I_B.

Q. 18. While connecting transformers in parallel, what factors must be taken into consideration ?

Ans. Their electrical characteristics such as voltage ratio, per cent impedance, phase displacement, phase sequence, polarity.

Q. 19. If transformers with different electrical characteristics are connected in parallel, what will happen?

Ans. They will not share the load equally.

Q. 20. In back to back test, show that one transformer may have slightly less temperature rise than other. **[U.P. Technical Univ., 2007-08]**

Ans. For the assumed directions of no-load currents I_0 and secondary currents I_2 in the two transformers, the primary current of transformer 1 is less (being phasor difference of I_1 and I_0) than that of transformer 2 (being phasor sum of I_1 and I_0) where 2 I_0 is the no-load current. Thus the two transformers do not operate under identical conditions—one may have slightly less temperature rise than the other.

Q. 21. Why the rating of transformer is express in kVA rather than in kW? **[G.G.S.I.P. Univ., December-2011]**

Ans. All the power apparatus whether it is steam engine or gas engine or electric machine have power ratings defined as the power which can be safely and efficiently delivered by a machine under some specific conditions.

Electrical apparatus or machine is usually rated at the load, which it can carry without overheating and damage to insulation *i.e.*, rating of electric machine is governed by the temperature rise caused by the internal losses of the machine.

The copper loss in the windings depends upon the magnitude of currents flowing through the primary and secondary windings and core loss depends on the voltage and thus losses are independent of pf.

The output in kW is proportional to power factor for the transformer of a given kVA. For example, output of 1,000 kVA transformer on full load will be 200, 500, 800 and 1,000 kW at power factor 0.2, 0.5, 0.8 and unity respectively, but copper losses in windings will remain same regardless of power factor.

For the foregoing reasons transformers are rated in kVA, not in kW.

Q. 22. How does transformer oil appear when it is new?

Ans. The transformer oil appears clear light colour with a faint characteristic smell.

Q. 23. What change will occur in colour of silica-gel crystals when they absorb moisture from atmosphere?

Ans. The colour of silica-gel crystals changes from blue to pink when they absorb moisture from atmospheres.

PROBLEMS

1. The core of a three-phase, 50 Hz, 11,000/550 V delta/star 300 kVA, core type transformer operates with a flux of 0.05 Wb. Find (*i*) number of hv and lv turns per phase (*ii*) emf per turn (*iii*) full-load hv and lv phase currents. **[Bharathidhasan Univ., April-1997]**
[**Ans.** (*i*) 1,040; 30 (*ii*) 11.1 V (*iii*) 9.1 A, 315 A]

2. Determine the number of turns per phase in each winding of a 3-phase transformer with a ratio of 20,000/2,000 V to work on a 50 Hz network. The hv winding is delta-connected and the low voltage winding is star-connected. Each core has a net section of 504 cm^2. Assume a flux density of 1.2 Wb/m^2. [**Ans.** 1,490; 86]

3. A 3-phase transformer has 400 turns on the primary and 40 turns on the secondary. The supply voltage is 3,300 V. Find the secondary voltage on no-load when the windings are connected in (*a*) star-delta (*b*) delta-star. [**Ans.** (*i*) 190.5 V (*ii*) 571.6 V]

4. A transformer has a Δ-connected primary and a star-connected secondary working at 50 Hz supply. The line voltages of primary and secondary being 6.6 kV and 400 V respectively. The line current of the primary side is 8 A and the secondary side has a balanced load of 0.6 pf lagging. Determine (*i*) the output of the transformer (*ii*) the line current on the secondary side. [**Ans.** 54.87 kW; 132 A]

5. A 2,000 kVA, 6,600/400 V, 3-phase transformer is delta-connected on the hv side and star-connected on lv side. Determine its percentage resistance and percentage reactance drops, % efficiency and % regulation at full load 0.8 pf leading. Given the following data.
SC Test; HV data: 400, 175 A and 17 kW
OC Test; LV data : 400 V, 150 A and 15 kW
[Nagpur Univ. Basic Elec. Machines, 1993]
[**Ans.** 0.85%, 6 %, 98.04 %, – 2.92%]

6. (*a*) Two identical single-phase transformers are connected in open-delta across 3-phase mains and deliver a balanced load of 3,000 kW at 11 kV and 0.8 pf lagging. Calculate the line and phase currents and the power factors at which the two transformers are working.

(*b*) If one more identical unit is added and the open-delta is converted to delta, calculate the additional load of the same power factor that can now be supplied for the same temperature rise. Also calculate the phase and line currents.

[**Ans.** (*a*) 196.8 A,196.8 A, 0.993 (lag); 0.393 (lag) (*b*) 2,196 kW; 341A; 196.8 A]

7. Three 1,100/110 V transformers connected Δ/Δ supply a lighting load of 100 kW. One of these transformers is damaged and hence removed for repairs. What currents will be flowing in each transformer when (*i*) the three transformers were in service (*ii*) the two transformers are in service. [**Ans.** 303 A, 30.3 A, 525 A, 52.5 A]

8. Two 100 V single-phase furnaces are supplied through Scott-connected transformers from 600 V, 3-phase mains. Calculate the current in each of 3-phase lines when the power taken by each furnace is 600 kW at pf of 0.71 (lag). [**Ans.** 1625 A, 1625 A, 1625 A]

9. In a Scott-connection, calculate the vaiues of line currents on 3-phase side if the loads on the 2-phase side are 300 kW and 450 kW both at 100 V and 0.707 pf (lag) and the 3-phase line voltage is 3,300 V. The 300 kW load is on the leading phase on the 2-phase side. Neglect transformer losses. [**Ans.** 148.5 A, 206.7 A; 206.7A]

10. Two furnaces are supplied with 1-phase current at 50 V from a 3-phase 4.6 kV system by means of two single phase Scott-connected transformers with similar secondary windings. When the load on the main transformer is 350 kW and that on the other transformer is 200 kW at 0.8 pf lagging, what will be the current in each 3-phase line? Neglect phase displacement and losses in transformers.

[**Ans.** 62.76 A; 100 A; 100 A]

❑❑❑

CHAPTER 14

Special Transformers

INSIDE THIS CHAPTER

14.1 INTRODUCTION

As already explained, transformers are extensively used for transmission and distribution of power and transform energy from one voltage level to another (may be from low voltage to high voltage or vice versa). But there are many applications to which conventional transformers do not suit and for every specific purpose some modifications are required to be made according to needs. Specific purpose transformers include auto-transformers, variac transformers, booster transformers, induction voltage regulators, tap-changing transformers, isolation transformers, phase shifting transformers, constant current transformers, low-voltage transformers, high leakage transformers, audio-frequency transformers, high-frequency transformers, pulse transformers, grounding transformers, welding transformers, traction transformers, instrument transformers etc.

14.2 AUTO-TRANSFORMER

The operating principle and general construction of an auto-transformer is the same as that of conventional two-winding transformer. The auto-transformer differs from a conventional two-winding transformer in a way in which the primary and secondary are interrelated. In a conventional two-winding transformer, the primary and secondary windings are completely insulated from each other but are magnetically linked by a common core. In an auto-transformer, the two windings, primary and secondary, are connected electrically as well as magnetically, in fact, a part of the single continuous winding is common to primary and secondary. The single winding is wound on a laminated silicon steel core and, therefore, both primary and secondary sections of this one winding are on the same magnetic circuit.

The auto-transformers are of two types in construction. In one type of auto-transformers, there is a continuous winding with taps brought out at convenient points determined by the designed secondary voltages and in the other type of auto-transformers, there are two or more distinct coils which are electrically connected to form a continuous winding. In either case, the same laws governing conventional two-winding transformers apply equally well to auto-transformer.

In an auto-transformer shown in Fig. 14.1, the primary winding is AB consisting of whole turns between terminals A and B, let it be N_1. Voltage V_1 is applied across primary terminals A and B, as illustrated in Fig. 14.1. To obtain the secondary winding the winding AB is tapped at C and, therefore, secondary winding consists of turns between terminals B and C, let it be N_2, and secondary voltage is available across terminals B and C. If the internal impedance drop is neglected, then the voltage per turn is $\frac{V_1}{N_1}$ and, therefore, the voltage across BC is $\frac{V_1 N_2}{N_1}$.

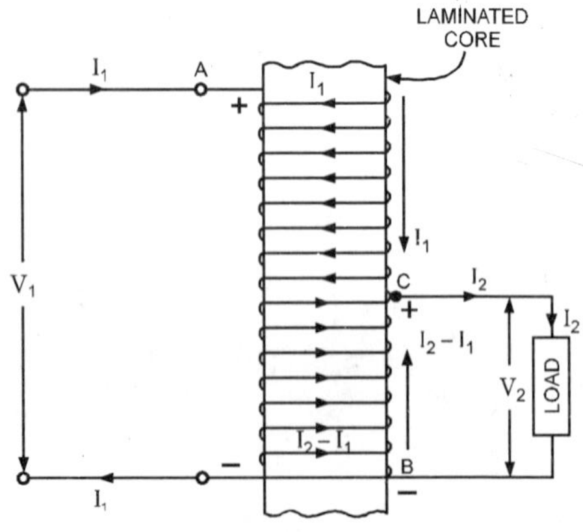

Fig. 14.1 *Step-Down Auto-Transformer*

Although diagrammatically the auto-transformer has the appearance of a resistance type potential divider, its operation is quite different, as described below:

A resistive potential divider cannot step up voltage, whereas an auto-transformer can do. In a potential divider the input current is always more than its output current but it is not so in an auto-transformer. In a step-down auto-transformer, the output (load) current is more than the input current. In a potential divider, almost entire power to the load flows by conduction, whereas in auto-transformer, a part of the power is conducted and the rest is transferred to the load by transformer action. Thus in a potential divider, there is a great loss of power in the different resistances whereas it is not so in case of an auto-transformer. Voltage regulation is also poor in case of a potential divider.

14.2.1. Volt-Ampere Relations. Consider the case where V_1 is applied voltage and voltage V_2 is available across the secondary terminals, as illustrated in Fig. 14.1.

Neglecting losses

$$V_1 I_1 \cos\phi_1 = V_2 I_2 \cos\phi_2$$

If internal leakage impedance drops and losses are negligible, then

$$\cos\phi_1 = \cos\phi_2$$

$$\text{and } V_1 I_1 = V_2 I_2$$

$$\text{or } \frac{V_2}{V_1} = \frac{I_1}{I_2} = \frac{N_2}{N_1} = K,$$

transformation ratio (as in case of two-winding transformers) ...(14.1)

Note: Because the exciting current and the leakage reactance are small in case of auto-transformers so it is an accepted practice to assume the input power and output power to be the same.

In an auto-transformer, only a part of the power input is transferred from the primary to the secondary side by transformer action, while the remainder is transferred directly from the primary to the secondary side as mentioned above. The relative amounts of power transformed (or inductively transferred) and power conductively transferred depend upon the ratio of transformation, as worked out below;

Under ideal conditions, the mmf due to primary current and secondary current will balance exactly, leaving the mutual flux unchanged by load currents. I_2 will, therefore, oppose I_1 in the common section. The current flowing in section AC of winding is primary current I_1 flowing from A to C, current supplied to load is I_2 and current in section BC will be $(I_2 - I_1)$ flowing from B to C, as illustrated in Fig. 14.1. Neglecting the exciting current $(I_2 - I_1)$ may be determined algebraically.

Power delivered to load = $V_2 I_2$ volt-amperes

Power in winding A C = $E_{AC} I_1 = (V_1 - V_2) I_1$ volt-amperes

Power transformed = Power in winding BC

$$= V_2 (I_2 - I_1)$$

$$= V_2 I_2 \left(1 - \frac{I_1}{I_2}\right) = V_2 I_2 (1 - K) \quad ...(14.2)$$

$$\because \text{From Eq. (14.1) } \frac{I_1}{I_2} = K$$

Ratio of power transformed to total power delivered

$$= \frac{V_2 I_2 (1-K)}{V_2 I_2} = (1 - K) \quad ...(14.3)$$

Power conducted directly

= Power delivered to load

– power transferred by transformer action

$= V_2 I_2 - V_2 I_2 (1 - K) = KV_2 I_2 = K \times$ power output ...(14.4)

Equation (14.3) may be rewritten as

$$\frac{\text{Inductively transferred power}}{\text{Total power}} = 1 - K = 1 - \frac{V_2}{V_1} = \frac{V_1 - V_2}{V_1}$$

$$= \frac{\text{High voltage } - \text{low voltage}}{\text{High voltage}}$$

14.2.2. Phasor Diagram. The phasor diagram of an auto-transformer is derived as follows:

Let the positive direction of applied voltage V_1 be downwards *i.e.* from A to B, then the positive direction of current I_1 round the primary circuit is clockwise, and that of current I_2 round the load circuit is also clockwise. The positive direction of the induced emfs are upwards i.e. from B to C and from C to A. The applied voltage V_1 is balanced by induced emfs in these two sections, plus voltage drop in them. For drawing phasor diagram start from load side. Draw phasor OA downwards to represent the voltage across load, V_2 and phasor OB lagging behind phasor OA by an angle ϕ_2 to represent load current I_2 where $\phi_2 = \cos^{-1}$ (power factor of the load).

The current in section BC of winding is given by

$$I_{BC} = I_2 - I_1 = I_2 \left(1 - \frac{I_1}{I_2}\right) = I_2 (1 - K)$$

where K is transformation ratio

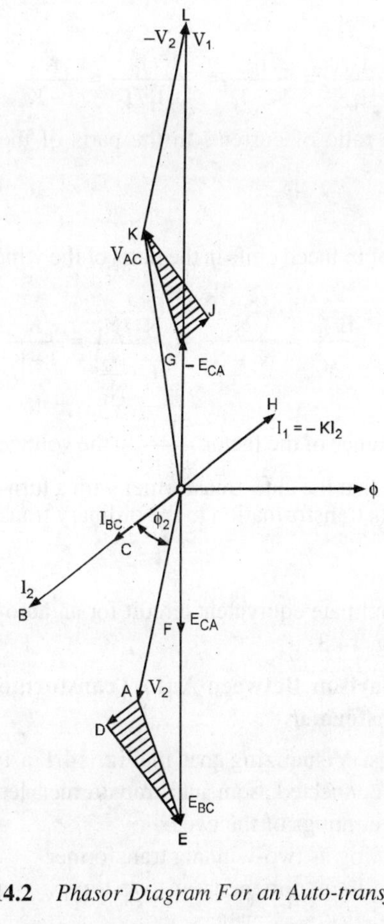

Fig. 14.2 *Phasor Diagram For an Auto-transformer*

The current in winding section BC is represented by phasor OC where OC = (1 – K) OB. Draw phasor AD parallel to current phasor OC and equal to I_2 (1 – K) R_2 in magnitude to represent resistive drop in section BC of winding. Now draw phasor DE perpendicular to current phasor OC and equal to I_2 (1 – K) X_2 in magnitude to represent reactive drop in section BC of winding. The phasor OE, therefore, represents the induced emf E_{BC} in section BC.

The induced emf in section CA, $E_{CA} = E_{BC}\left(\frac{1-K}{K}\right)$ in magnitude and in phase with E_{BC}, so take OF = OE $\left(\frac{1-K}{K}\right)$ representing E_{CA}. Produce FO to OG taking OG = OF hence representing – E_{CA}.

The current I_1 is in phase opposition to I_2 and is given by $I_1 = - K\, I_2$ so draw OH equal to K × OB by producing line BO. Now draw impedance voltage triangle GJK for the section CA, where G J = I_1R_1 and J K = I_1X_1.

The potential difference V_{AC} across section AC is, therefore, the phasor sum of the emf OG and impedance drop GK and it is, therefore, represented by OK. Finally the applied voltage V_1 is the phasor sum of V_{AC} and V_{CB}

i.e. $\mathbf{V_1} = \mathbf{V_{AC}} + \mathbf{V_{CB}} = \mathbf{V_{AC}} - \mathbf{V_2}$, and is represented by OL.

14.2.3. Equivalent Circuit. The auto–transformer may be considered as a two winding transformer when the part AC of winding (Fig. 14.1) can be taken as primary with impressed voltage V_{AC} and primary current as I_1 and CB as the secondary with the output voltage V_2 and current $I_2 - I_1$. The current ratio is, therefore, given as

$$\frac{I_{AC}}{I_{BC}} = \frac{I_1}{I_2 - I_1} = \frac{I_1/I_2}{1 - I_1/I_2} = \frac{K}{1-K}$$

Hence the ratio of currents in the parts of the winding is $\left(\frac{K}{1-K}\right)$.

The ratio of induced emfs in the parts of the winding CB and AC *i.e.*

$$\frac{E_{CB}}{E_{AC}} = \frac{N_2}{N_1 - N_2} = \frac{N_2/N_1}{1-\frac{N_2}{N_1}} = \frac{K}{1-K}$$

The occurrence of the factor $\frac{K}{1-K}$ in the voltage and current ratios signifies that the auto-transformer with a turn-ratio of K is equivalent in its transformation to the ordinary transformer with a ratio of $\frac{K}{1-K}$.

The approximate equivalent circuit for an auto-transformer is shown in Fig. 14.3.

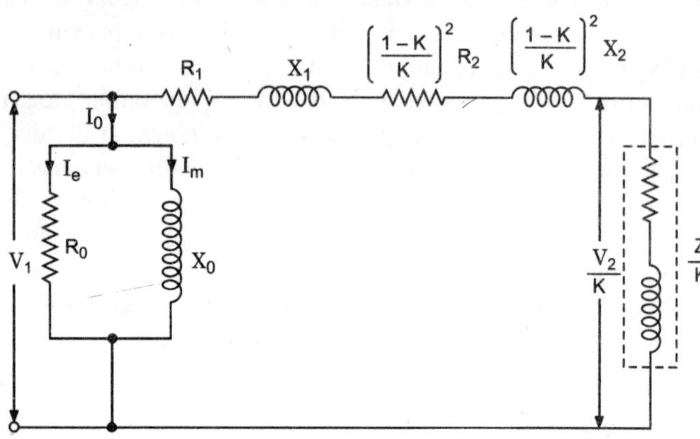

Fig. 14.3 *Approximate Equivalent Circuit Diagram For an Auto-transformer*

14.2.4. Comparison Between Auto-Transformer and Two Winding Transformer.

1. VA Ratings. Visualizing that in Fig. 14.1 a two-winding transformer is connected as an auto-transformer, let us compare the volt-ampere ratings of the two.

Voltage rating as two-winding transformer

$$= (V_1 - V_2)\, I_1 = (I_2 - I_1)\, V_2 \qquad \text{...(14.5)}$$

When used as an auto-transformer,

Voltage rating as an auto-transformer

$$= V_1I_1 = V_2I_2 \qquad \text{... (14.6)}$$

$$\frac{\text{Volt-ampere rating as an auto-transformer}}{\text{Volt-ampere rating as two winding transformer}}$$

$$= \frac{V_1I_1}{(V_1 - V_2)I_1} = \frac{V_2I_2}{V_2(I_2 - I_1)} = \frac{1}{1-K} \qquad \text{...(14.7)}$$

From Eq. (14.7) it is obvious that a two-winding transformer of a given volt-ampere rating when connected as an auto-transformer can handle high volt-amperes. This is because in the auto-transformer connection part of VA is conducted directly from the primary to the secondary.

2. Conductor Material Requirements. The cross section of conductor (copper or aluminium) is proportional to the current to be carried and length of the conductor in a winding is proportional to the number of turns. Hence the weight of the conductor material in the winding being proportional to product of cross-sectional area and length of the conductor is proportional to the product of number of turns and current to be carried.

In an ordinary transformer, the total weight of conductor material required is proportional to

$$(N_1I_1 + N_2I_2)\, \alpha\, 2N_1I_1 \text{ because } N_2I_2 = N_1I_1$$

In an auto-transformer (Fig. 14.1),

The top section AC has $(N_1 - N_2)$ turns and carries current I_1 and bottom section BC has turns N_2 and carries current $I_2 - I_1$.

So weight of conductor material in an auto-transformer having the same input (V_1 I_1), the same output (V_2 I_2) and the same voltage ratio $\left(\frac{V_2}{V_1}\right)$ as that of the two-winding transformer is proportional to $[(N_1 - N_2)I_1 + N_2(I_2 - I_1)]$ or $[(N_1 - N_2)I_1 + N_2I_2 - N_2I_1]$ or $[(N_1 - N_2)I_1 + N_1I_1 - N_2I_1]$ or $2\,(N_1 - N_2)\,I_1$

So $\frac{\text{Weight of conductor in an auto-transformer}}{\text{Weight fof conductor in a two winding transformer}}$

$$= \frac{2\,(N_1 - N_2)\,I_1}{2\,N_1I_1} = 1 - \frac{N_2}{N_1} = 1 - K \qquad \text{...(14.8)}$$

Hence saving in conductor material required for winding affected by using an auto-transformer

$= K \times$ weight of conductor in a two winding transformer ...(14.9)

From Eq. (14.9) it is obvious that nearer the ratio of transformation is to unity, the greater is the economy in conductor material requirement. If $V_2 = V_1$ *i.e.* when K is equal to unity, the auto-transformer, according to Eq. (14.8), needs no conductor material, for actually no transformer is required. If voltage is to be transformed from very high voltage to very low voltage *i.e.* when K is to be very small, there will be a little saving in the conductor material and in such cases it is preferable to use an ordinary two-winding transformer.

3. Core Size. The voltage per turn, and, therefore, the flux, is the same so the x-section of the core is not affected. However, the reduction in conductor material means a smaller window area and, therefore, reduced core length. Thus for the same core area the size and weight of the core of the auto-transformer is reduced. Accordingly the size and weight of the auto-transformer is reduced.

4. Cost. Saving in both conductor and core materials is affected by using an auto-transformer, as explained above in (2) and (3). The saving in material and cost, is, of course, less than that indicated, on account of the fact that the core is nearly as large as in case of the two-winding transformer. For transformation ratio of $\frac{1}{2}$, approximately 50 per cent saving in conductor material would be affected, and the cost might be 65 to 70 per cent of that of a two-winding transformer of the same rating.

5. Losses and Efficiency. Owing to the reduction in conductor and core materials, the ohmic losses in conductor and the core loss are lowered. Thus an auto-transformer has lower losses and, therefore, higher efficiency than those of the two-winding transformer of the same output.

6. Leakage Impedance and Voltage Regulation. Reduction in conductor material means lower value of ohmic resistance. A part of the winding being common, leakage flux and, therefore, leakage reactance is less. In other words, an auto-transformer has the lower value of leakage impedance and has better voltage regulation than a two-winding transformer of the same volt-amperes output.

Besides above, variable output voltages can be obtained from an auto-transformer. When used as a variable ratio auto-transformer, it is known as *variac, autostat, dimmerstat* etc.

The series connection of windings in an auto-transformer, however, involves the necessity for insulating a winding to a higher value than that of the emf induced in it.

There is another **drawback** of an auto-transformer over a conventional two-winding transformer which **prohibits** its use and it is due to the direct electrical connection between the primary and secondary circuits. If an open circuit develops in common winding CB, the supply voltage would be impressed directly upon the external load circuit in series with the AC portion of the winding. With a step-down auto-transformer this could result in burned out secondary loads and/or a serious shock hazard, particularly if the voltage ratio $\frac{V_2}{V_1}$ is quite low. Hence if an auto–transformer is used to supply a low voltage system from an high voltage system, it becomes essential to earth the low voltage side, for example terminal B in Fig. 14.1. In general, however, an auto-transformer is never used for interconnecting hv and lv systems.

14.2.5. Applications. As said earlier, auto-transformers are used when transformation ratio K is nearly equal to unity and where there is no objection to direct electrical connection between primary and secondary. Hence auto-transformers are extensively used for ratio ranging from at 0.4 to 1.0 for the following applications:

(*i*) To obtain a neutral in a 3-wire a c distribution system in the same way as a balancer set is used in a 3-wire dc distribution system. The connection of an auto-transformer used in this way is shown in Fig. 14.4. If the load on the upper half of system is greater than on the lower half, the winding BC acts as the primary and winding AC acts as secondary to supply the extra load required by upper half of the system. The auto-transformer used in this manner is known as a *balance coil*. The balance coil is cheaper and more efficient than balancer set.

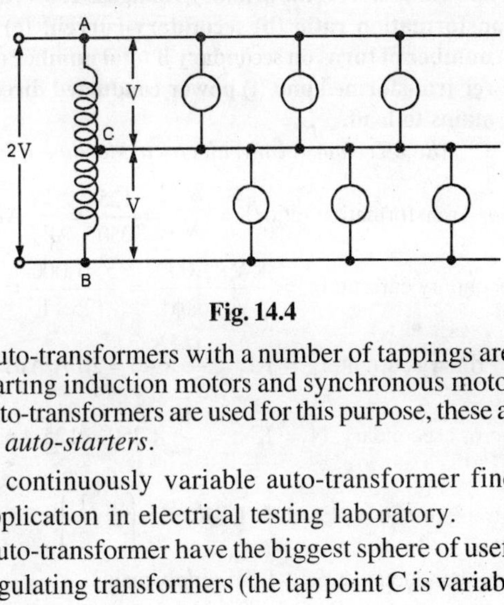

Fig. 14.4

(*ii*) Auto-transformers with a number of tappings are used for starting induction motors and synchronous motors. When auto-transformers are used for this purpose, these are known as *auto-starters*.

(*iii*) A continuously variable auto-transformer finds useful application in electrical testing laboratory.

(*iv*) Auto-transformer have the biggest sphere of usefulness as regulating transformers (the tap point C is variable in such cases).

(*v*) Auto-transformers are also used as boosters to raise the voltage in an ac feeder.

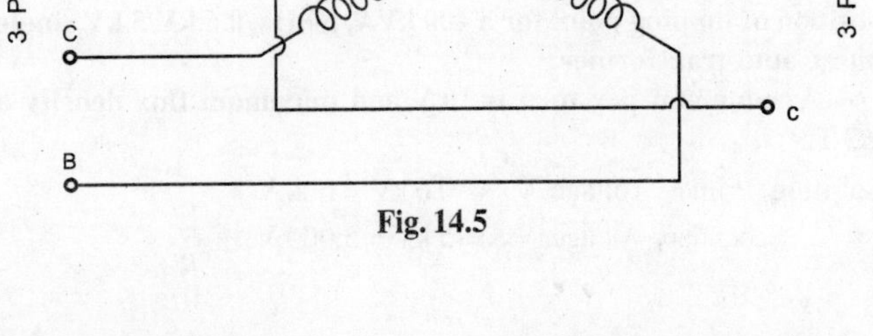

Fig. 14.5

(*vi*) As furnace transformers for getting a convenient supply to suit the furnace winding from normal 230 V ac supply.

(*vii*) Auto-transformers are mainly used for interconnecting systems that are operating at roughly the same voltage. The interconnection of EHV system (*e.g.* 220 kV and 132 kV) by the auto-transformers results in considerable saving of bulk and cost as compared to the conventional two winding transformers.

Example 14.1. An auto-transformer having 1,250 turns is connected across a 250 V ac supply. What secondary voltage will be obtained if a tap is taken at 800 th turn.

Solution: Supply voltage, $V_1 = 250$ V

Primary turns, $N_1 = 1{,}250$

Secondary turns, $N_2 = 800$

$$\text{Secondary voltage, } V_2 = V_1 \times \frac{N_2}{N_1} = 250 \times \frac{800}{1{,}250} = 160 \text{ V } \textbf{Ans.}$$

It may be worked out by an alternative method as follows:

$$\text{EMF per turn} = \frac{V_2}{N_1} = \frac{250}{1{,}250} = 0.2 \text{ V}$$

and secondary voltage, V_2 = EMF per turn × secondary turns $= 0.2 \times 800 = 160$ V **Ans.**

Example 14.2. An auto-transformer supplies a load of 5 kW at 125 V and at unity power factor. If the primary voltage is 250 V, determine (a) transformation ratio (b) secondary current (c) primary current (d) number of turns on secondary if total number of turns is 250. (e) power transformed and (f) power conducted directly from the supply mains to load.

[*R.G. Technical Univ. Electrical Machine-I, June-2014*]

Solution: (*a*) Transformation ratio, $K = \frac{V_2}{V_1} = \frac{125}{250} = \frac{1}{2}$ **Ans.**

(*b*) Secondary current, $I_2 = \frac{\text{kW} \times 1{,}000}{V_2 \cos\phi} = \frac{5 \times 1{,}000}{125 \times 1} = 40$ A **Ans.**

(*c*) Primary current, $I_1 = KI_2 = \frac{1}{2} \times 40 = 20$ A **Ans.**

(*d*) Turns across secondary, $N_2 = KN_1 = \frac{1}{2} \times 250 = 125$ **Ans.**

(*e*) Power transformed = Load × (I – K) $= 5\left(1 - \frac{1}{2}\right) = 2.5$ kW **Ans.**

(*f*) Power conducted directly from supply mains $= 5 - 2.5 = 2.5$ kW **Ans.**

Example 14.3. An auto-transformer is used to step down primary voltage of 220 V to 200 V. Determine the amount of copper saved to equivalent two winding transformer required to transform 220 V to 200 V. [*Pb. Univ. Elec. Machines-I, January-1991*]

Solution: Transformation ratio, $K = \frac{200}{220} = 0.909$

Economy or amount of copper saved to equivalent two winding transformer

$= K = 0.909$ or 90.9% **Ans.**

Example 14.4. Determine the core area, the number of turns and the position of tapping point for a 400 kVA, 50 Hz, 0.6 kV/5 kV single-phase auto-transformer.

Assume emf per turn is 10 V and maximum flux density at 1.2 T.

Solution: Primary voltage, $V_1 = 0.6 \text{ kV} = 600$ V

Secondary voltage, $V_2 = 5 \text{ kV} = 5{,}000$ V

Number of turns on secondary,

$$N_2 = \frac{V_2}{\text{EMF per turn}} = \frac{5{,}000}{10} = 500 \textbf{ Ans.}$$

$$\text{Number of turns on primary, } N_1 = \frac{V_1}{\text{EMF per turn}} = \frac{600}{10} = 60$$

$$\text{Position of tapping} = \frac{60}{500} \times 100$$

= 12% from the bottom of the hv winding. **Ans.**

$$\text{Core flux, } \Phi_{max} = \frac{\text{EMF per turn}}{4.44 f} = \frac{10}{4.44 \times 50} = 0.045 \text{ Wb}$$

$$\text{Core area, } a = \frac{\Phi_{max}}{B_{max}} = \frac{0.045}{1.2} = 0.0375 \text{ m}^2 = 375 \text{ cm}^2 \textbf{ Ans.}$$

Example 14.5. An auto-transformer is used to stepdown voltage level from 230 V to 200 V while the load is 20 kW at upf. Neglecting losses and magnetising current find the current in different sections of the winding.

Solution: Transformation ratio, $K = \frac{V_2}{V_1} = \frac{200}{230} = \frac{20}{23}$

$$\text{Secondary current, } I_2 = \frac{\text{Load in kW} \times 1{,}000}{V_2 \times pf} = \frac{20 \times 1{,}000}{200 \times 1} = 100 \text{ A}$$

$$\text{Primary current, } I_1 = KI_2 = \frac{20}{23} \times 100 = 86.96 \text{ A}$$

Current distribution is shown in Fig. 14.6.

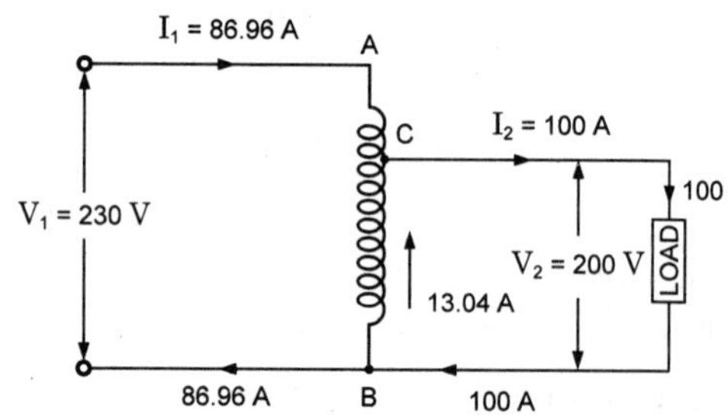

Fig. 14.6

Example 14.6. A 11,000/2,200 V, single-phase transformer is rated at 1,000 kVA. If the two windings are connected in series to form an auto transformer, determine its rated voltage and power.

[*U.P. Technical Univ. Electrical Machines, 2008-09*]

Solution: When the two windings are connected in series, the total voltage available is (11,000 + 2,200) *i.e.* 13,200 V. The other side can have the voltage of 11,000 V or 2,200 V. So the voltage rating can be either 13,200/ 11,000 V or 13,200/2,200 V **Ans.**

$$\text{In the first case, transformation ratio, } K_1 = \frac{11{,}000}{13{,}200} = 0.8333$$

$$\text{Rated output} = \frac{\text{Rating as a two-winding transformer}}{1 - K} = \frac{1{,}000}{1 - 0.8333} = 6{,}000 \text{ kVA } \textbf{Ans.}$$

In the second case, transformation ratio,

$$K_2 = \frac{2{,}200}{13{,}200} = 0.1667$$

$$\text{Rated output} = \frac{1{,}000}{1 - 0.1667} = 1{,}200 \text{ kVA } \textbf{Ans.}$$

Example 14.7. A 200/100, 1-phase transformer is rated 10 kVA. If the transformer is connected as an auto-transformer across a 50 V supply, what is the maximum output voltage that can be obtained from the transformer? If the load current is now 8 A, determine the kVA output and the input current.

[*M.D. Univ. Electromechanical Energy Conversion, December-2005*]

Solution: Maximum output voltage will be obtained when the windings are connected with additive polarity, as shown in Fig. 14.7.

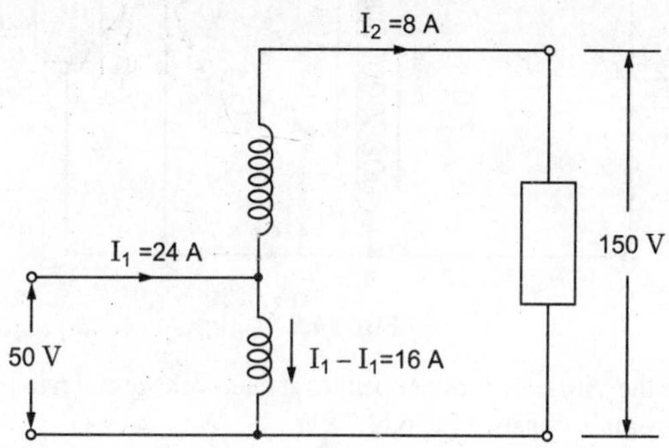

Fig. 14.7

Turn-ratio of 2-winding transformer, K = 2

Turn-ratio of auto-transformer with two windings connected with additive polarity

$$K' = K + 1 = 3$$

Voltage connected to primary, V_1 = 50 V

Output voltage, $V_2 = KV_1 = 3 \times 50 = 150$ V **Ans.**

Load current, I_2 = 8 A

$$\text{KVA output} = \frac{V_2 I_2}{1{,}000} = \frac{150 \times 8}{1{,}000} = 1.2 \textbf{ Ans.}$$

$$\text{Input current, } I_1 = \frac{V_2 I_1}{V_1} = \frac{150 \times 8}{50} = 24 \text{ A } \textbf{Ans.}$$

Example 14.8. A 5 kVA, 2,300/230 V two-winding transformer is connected as an auto-transformer and supplied by a 2,530 V source. The transformer is loaded so that the rated currents of the windings are not exceeded. Calculate: currents in different sections of the auto-transformer, kVA output and kVA transferred inductively and conductively.

[*G.B. Technical Univ. Electromechanical Energy Conversion-I, 2011-12*]

Solution: When the two windings are connected in series so as to supply from 2,530 V, the other side can have the voltage of 230 V or 2,300 V. Thus the voltage rating can be either 2,530/230 V or 2,530/2,300 V.

In first case [Fig. 14.8(*a*)]

$$\text{Transformation ratio} = \frac{230}{2{,}530} = \frac{1}{11}$$

$$\text{Rated output} = \frac{\text{Rating as two-winding transformer}}{1 - K}$$

$$= \frac{5}{1 - \frac{1}{11}} = 5.5 \text{ kVA } \textbf{Ans.}$$

KVA transformed inductively = (1 – K) × total power delivered

$$= \left(1 - \frac{1}{11}\right) \times 5.5 = 5 \text{ kVA } \textbf{Ans.}$$

KVA transformed conductively = K × power output

$$= \frac{1}{11} \times 5.5 = 0.5 \text{ kVA } \textbf{Ans.}$$

$$\text{Load current, } I_2 = \frac{\text{KVA output} \times 1{,}000}{V_2}$$

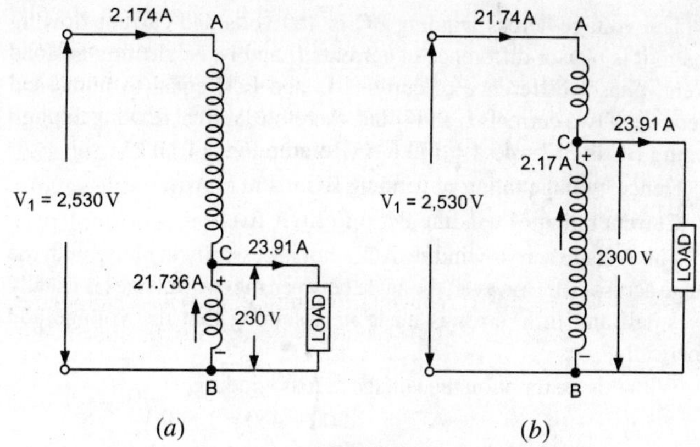

(*a*) (*b*)

Fig. 14.8

$$= \frac{5.5 \times 1{,}000}{230} = 23.91 \text{ A}$$

$$\text{Supply current, } I_1 = \frac{\text{KVA output} \times 1{,}000}{V_1}$$

$$= \frac{5.5 \times 1{,}000}{2{,}530} = 2.174 \text{ A}$$

Thus current flowing through winding AC = I_1 = 2.174 A **Ans.**

The current flowing through winding BC = $I_1 - I_2$ = 23.91 – 2.174 = 21.736 A **Ans.**

In second case [Fig. 14.8(*b*)]

$$\text{Transformation ratio, K} = \frac{2{,}300}{2{,}530} = \frac{10}{11}$$

$$\text{Rated output} = \frac{5}{1 - \frac{10}{11}} = 55 \text{ kVA } \textbf{Ans.}$$

KVA transformed inductively = (1 – K) × total power delivered

$$= \left(1 - \frac{10}{11}\right) \times 55 = 5 \text{ kVA } \textbf{Ans.}$$

KVA transformed conductively = K × total power delivered

$$= \frac{10}{11} \times 55 = 50 \text{ kVA } \textbf{Ans.}$$

$$\text{Load current, } I_2 = \frac{55 \times 1{,}000}{2{,}300} = 23.91 \text{ A}$$

$$\text{Supply current, } I_1 = \frac{55 \times 1{,}000}{2{,}530} = 21.74 \text{ A}$$

Thus current flowing through winding AC = 23.91 – 21.74 = 2.17 A

Example 14.9. A load of 100 kVA is to be supplied at 460 volts from 2,300 V supply mains by an auto-transformer. Determine the current and voltage rating for each of the two windings. What would be the kVA rating of the transformer, if it were used as a two-winding transformer.

Solution: Load to be supplied = 100 kVA

Secondary voltage, V_2 = 460 V

Primary voltage, V_1 = 2,300 V

$$\text{Load current, } I_2 = \frac{\text{KVA} \times 1{,}000}{V_2} = \frac{100 \times 1{,}000}{460} = 217 \text{ A } \textbf{Ans.}$$

Assuming input equal to output (*i.e.* neglecting losses)

$$\text{Input current, } I_1 = \frac{\text{KVA} \times 1{,}000}{V_1} = \frac{100 \times 1{,}000}{2{,}300} = 43.4 \text{ A } \textbf{Ans.}$$

The voltage across winding BC is 460 volts and current flowing through it is phasor difference of currents I_1 and I_2. Neglecting no-load current, phasor difference of currents I_1 and I_2 is equal to numerical difference of two currents I_1 and I_2 and, therefore, current flowing through winding BC is 217 – 43.4 = 173.6 A (Refer to Fig. 14.1)

Hence voltage rating of winding BC = 460 V **Ans.**

Current rating of winding BC = 173.6 A **Ans.**

The voltage across winding AC is not necessarily in phase with the voltage across CB. However, the angle between the two voltages is usually very small and little error is made in assuming that the voltage add numerically.

With this assumption the voltage across winding,

$$AC = 2{,}300 - 460 = 1{,}840 \text{ V}$$

Current flowing through the winding AC is I_1 *i.e.* 43.4 A

Hence voltage rating of winding

$$AC = 1{,}840 \text{ volts } \textbf{Ans.}$$

Current rating of winding AC = 43.4 A **Ans.**

The auto-transformer, when connected as two winding transformer;

The permissible voltage input = 1,840 V

Permissible voltage output = 460 V

The permissible current output = 173.6 A

The rating of transformer as a two winding transformer

$$= \frac{\text{Permissible voltage output} \times \text{permissible current output}}{1{,}000}$$

$$= \frac{460 \times 173.6}{1{,}000} = 80 \text{ kVA } \textbf{Ans.}$$

Example 14.10. A 200 kVA, 2300/460 V, 50 Hz, 2-winding transformer is to be used as an auto-transformer to step-up the voltage of 2,300 V to 2,760 V. If the transformer has an efficiency of 96% at 0.8 pf lagging, impedance of 4% and regulation of 3%, determine (i) voltage and current ratings of each side (ii) kVA rating (iii) efficiency at unity power factor (iv) percentage impedance (v) regulation, (vi) short-circuit current of each side (vii) kVA transformed and kVA conducted at full load, while it is used as an auto-transformer.

Solution : (*i*) Rated voltage of the auto-transformer on lv side is 2,300 V and on hv side is

$$2{,}300 + 460 = 2{,}760 \text{ V } \textbf{Ans.}$$

The rated currents of hv and lv windings are respectively

$$I_{hv} = \frac{200 \times 1{,}000}{2{,}300} = 86.96 \text{ A}$$

$$I_{lv} = \frac{200 \times 1{,}000}{460} = 434.8 \text{ A}$$

Rated current of hv side of the auto-transformer

= Rated current of 460 V side of the two-winding transformer

= 434.8 A **Ans.**

Rated current of 434.8 A in the lv winding needs, by transformer action, a rated current of 86.96 A in the 2,300 V winding acting as the primary

Applying Kirchhoff's current law at point C, we get line current

$$I_1 = 434.8 + 86.96 = 521.76 \text{ A } \textbf{Ans.}$$

(*ii*) kVA rating of the auto-transformer

$$= \frac{2{,}760 \times 434.8}{1{,}000} = 1{,}200 \text{ kVA } \textbf{Ans.}$$

(*iii*) Power output of two-winding transformer at 0.8 pf lagging

$$= 200 \times 0.8 = 160 \text{ kW}$$

$$\text{Full-load losses} = \frac{\text{Output}}{\text{Efficiency}} - \text{output} = \frac{160}{0.96} - 160 = 6.667 \text{ kW}$$

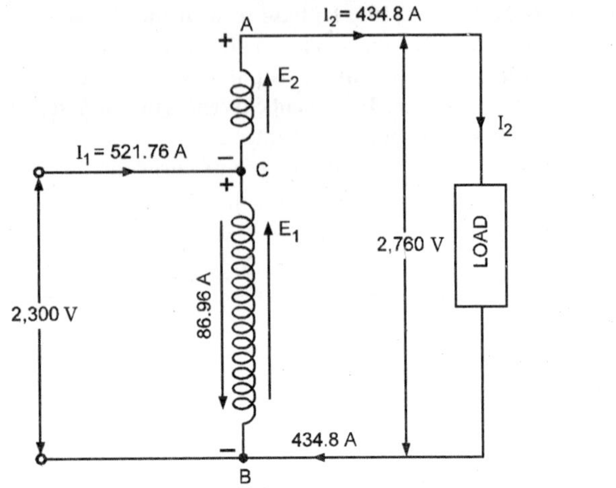

Fig. 14.9

Since the auto-transformer operates at rated voltage and rated currents, the losses remain constant i.e. 6.667 kW

Efficiency of auto-transformer for an output of 1,200 × 1.0 kW is

$$\eta = \frac{\text{Output}}{\text{Output + losses}} \times 100$$

$$= \frac{1{,}200}{1{,}200 + 6.667} \times 100 = 99.45\% \textbf{ Ans.}$$

It may be worked out by an alternative method as follows :

Per cent full-load losses at unity pf output as a two-winding transformer

$$= \frac{6.667}{200} \times 100 = 3.333\%$$

Transformation ratio,

$$K = \frac{\text{LV side voltage}}{\text{HV side voltage}} = \frac{2{,}300}{2{,}760} = 0.833$$

Per cent full-load losses as an auto-transformer

= (1 – K) × per cent full-load losses as a 2-winding transformer

= (1 – 0.833) × 3.333 = 0.55 %

and auto-transformer full-load efficiency

= 100 – per cent full-load losses

= 100 – 0.55 = 99.45%, the same as before,

(*iv*) Per cent impedance of two-winding transformer = 4%

The ohmic drop at full load is the same in both cases. If the impedance drop is referred to hv side of the auto-transformer (Fig. 14.1), the pu or pc drop in the auto-transformer is with respect to V_1 while that in the two-winding transformer it is with respect to

$$V_1 - V_2 = V_2 (1 - K).$$

If the impedance drop is referred to lv side of the auto-transformer (Fig. 14.1), then the winding current is

$$I_2 - I_1 = I_2 (1 - K),$$

so the impedance drop with respect to I_2 is to be reduced (1 – K) times in the case of an auto-transformer.

Therefore pu or pc impedance drop in an auto-transformer

= (1 – K) × pu or pc impedance drop as a two-winding transformer

= (1 – 0.833) × 4 = 0.667 % **Ans.**

(*v*) As regulation is proportional to pu or pc impedance drop,

So regulation as an auto-transformer

= (1 – K) × regulation as a two-winding transformer

= (1 – 0.833) × 3 = 0.5% **Ans.**

(*vi*) Short-circuit current as an auto-transformer

$$= \frac{1}{\text{pu impedance of the auto-transformer}}$$

$$= \frac{1}{0.00667} = 150 \text{ pu}$$

$$\text{So short-circuit current on lv side} = \frac{150 \times 434.8}{1{,}000}$$

$$= 65.22 \text{ KA } \textbf{Ans.}$$

$$\text{Short–circuit current on lv side} = \frac{150 \times 521.76}{1{,}000}$$

$$= 78.264 \text{ KA } \textbf{Ans.}$$

(*vi*) Here winding BC acts as the primary and the winding AC as the secondary

$$\text{So kVA transformed} = \frac{86.96 \times 2{,}300}{1{,}000}$$

$$\text{or } \frac{434.8 \times 460}{1{,}000} = 200 \text{ kVA } \textbf{Ans.}$$

$$\text{kVA conducted} = 1{,}200 - 200 = 1{,}000 \text{ kVA } \textbf{Ans.}$$

Example 14.11. A 100/400 V, 2 kVA single-phase transformer has 5% impedance. If it is connected as a 500/400 V auto-transformer, deduce from fundamentals its kVA rating and percentage impedance as an auto-transformer.

[*U.P. Technical Univ. Electromechanical Energy Conversion-I, 2006-07*]

Solution: Transformation ratio,

$$K = \frac{400}{500} = 0.8$$

$$\text{Rated output} = \frac{\text{Rating at a two-winding transformer}}{1 - K}$$

$$= \frac{2}{1 - 0.8} = 10 \text{ kVA } \quad \textbf{Ans.}$$

Percentage impedance

= Percentage impedance of two-winding transformer × (1 – K)

= 5 × (1 – 0.8)% = 1% **Ans.**

Example 14.12. A 400/100 V, 10 kVA two-winding transformer is to be employed as an auto-transformer to supply a 400 V circuit from a 500 V source. When tested as a 2-winding transformer at rated load, 0.85 pf lagging, its efficiency is 0.97. (*i*) Determine its kVA rating as an auto-transformer (*ii*) Find its efficiency as an auto-transformer. [*Rajasthan Technical Univ. Electrical Machines-I, 2011*]

Solution: Rated voltage of auto-transformer on lv side is 400 V and hv side is (400 + 100) *i.e.* 500 V

The rated currents of hv and lv windings are respectively

$$I_{hv} = \frac{10 \times 1{,}000}{400} = 25 \text{ A}$$

$$I_{lv} = \frac{10 \times 1{,}000}{100} = 100 \text{ A}$$

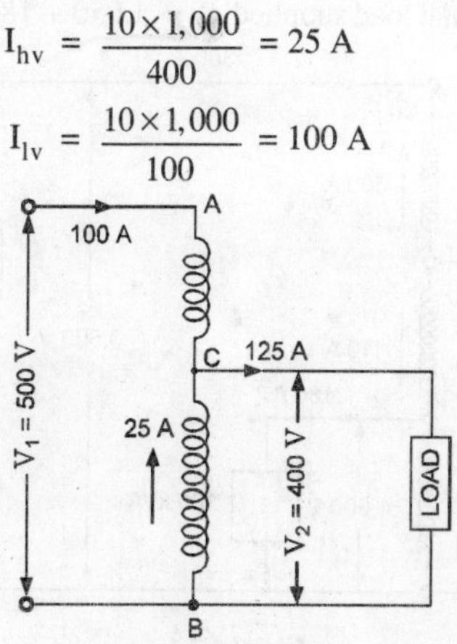

Fig. 14.10

Rated current of hv side of auto-transformer

= Rated current of 100 V side of two-winding transformer

= 100 A

Rated current of 100 A in lv winding needs, by transformer action, a rated current of $100 \times \frac{100}{400} = 25$ A in the 400 V winding acting in the secondary winding.

Applying KCL at point C, we get load current,

$$I_2 = 100 + 25 = 125 \text{ A}$$

(*i*) KVA rating as an auto-transformer

$$= \frac{400 \times 125}{1{,}000} = 50 \text{ kVA}$$

Transformation ratio,

$$K = \frac{\text{LV side voltage}}{\text{HV side voltage}} = \frac{400}{500} = 0.8$$

Power output of two-winding transformer at 0.85 pf lagging

= 10 × 0.85 = 8.5 kW

Full-load losses

$$= \frac{\text{Output}}{\eta} - \text{output} = \frac{8.5}{0.97} - 8.5 = 0.2629 \text{ kW}$$

Since auto-tranformer operates at rated voltage and rated current, the losses remains constant *i.e.* 0.2629 kW.

Efficiency as an auto-transformer for an output of 50 × 1.0 = 50 kW is

$$\eta = \frac{\text{Output}}{\text{Output + losses}} \times 100 = \frac{50}{50 + 0.2629} \times 100$$

$$= 99.77\% \textbf{ Ans.}$$

Example. 14.13. Two resistive loads R_1 and R_2 each of 200 Ω are to be supplied with voltage of 100 V and 300 V respectively, from the available 400 V supply. To this end, one auto-transformer connected as shown in Fig. 14.11 is designed.

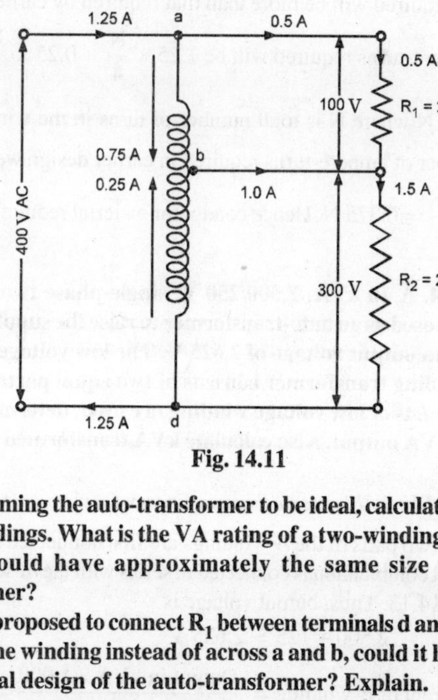

Fig. 14.11

Assuming the auto-transformer to be ideal, calculate the currents in its windings. What is the VA rating of a two-winding transformer which would have approximately the same size as the auto-transformer?

It is proposed to connect R_1 between terminals d and one suitable tap c on the winding instead of across a and b, could it lead to a more economical design of the auto-transformer? Explain.

$$\textbf{Solution: } \text{The current through the load, } R_1 = \frac{100}{200} = 0.5 \text{ A}$$

$$\text{The current through the load, } R_2 = \frac{300}{200} = 1.5 \text{ A}$$

$$\text{The total load supplied} = 100 \times 0.5 + 300 \times 1.5 = 500 \text{ VA}$$

The total current drawn from the supply mains

$$= \frac{500}{400} = 1.25 \text{ A}$$

Applying Krichhoff 's current law at the appropriate points we get the distribution of currents in all parts of the circuit, as shown in Fig. 14.11.

VA rating of the two winding transformer of the same size as that of the auto-transformer

$$= 300 \times 0.25 \text{ or } 100 \times 0.75 \textit{ i.e. } 75 \text{ VA } \textbf{Ans.}$$

When the load R_1 is connected between terminals *d* and *c*, so that voltage across terminal *c* and *d* is 100 V (Fig. 14.12).

The total load supplied and, therefore, the total current drawn from the supply mains remain the same as before *i.e.* 1.25 A.

Applying Kirchhoff 's current law at the appropriate points we get the distribution of currents in various parts of the circuit, as shown in Fig. 14.12.

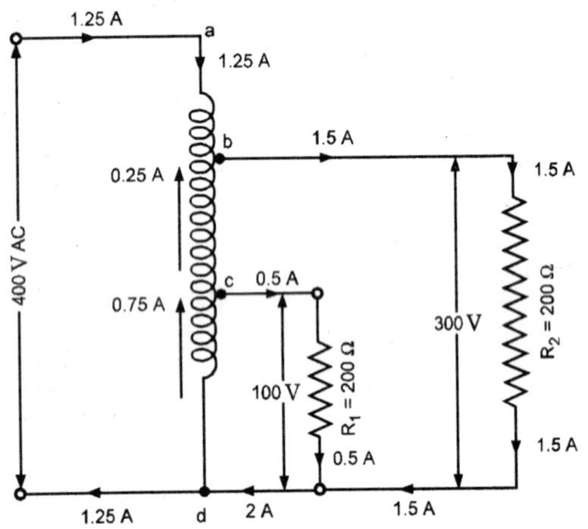

Fig. 14.12

No, the proposed design will not be economical, because in this ampere-turns required will be more than that required by earlier design.

Now total ampere-turns required will be $1.25 \times \frac{N}{4} + 0.25 \times \frac{N}{2} + 0.75 \times \frac{N}{4}$ *i.e.* 0.625 N where N is total number of turns in the winding and while the number of ampere-turns required in earlier design were $0.75 \times \frac{N}{4} + 0.25 \times \frac{3N}{4} = 0.375$ N. Hence conductor material requirement will increase.

Example 14.14. A 10 kVA, 2,500/250 V, single-phase two-winding transformer is used as an auto-transformer to raise the supply voltage of 2,500 V to an output voltage of 2,625 V. The low voltage winding of the two winding transformer consists of two equal parts of 125 V each. If both parts of low voltage winding are used, determine auto-transformer kVA output. Also calculate kVA transformed and kVA conducted.

[U.P. Technical Univ. Electromechanical Energy Conversion-I, 2007-08]

Solution: The two parts of the *lv* windings are first connected in parallel and this parallel combination is connected in series with the *hv* winding, as shown in Fig. 14.13. Thus, output voltage is

$$2{,}500 + 125 = 2{,}625 \text{ V}$$

$$\text{The rated current of } lv \text{ winding} = \frac{\text{Output in kVA} \times 1{,}000}{V_2}$$

$$= \frac{10 \times 1{,}000}{250} = 40 \text{ A}$$

Thus, total output current, $I_2 = 2 \times 40 = 80$ A

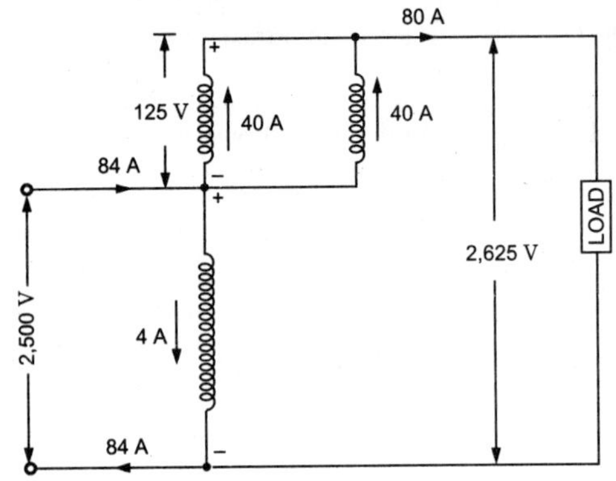

Fig. 14.13

$$\text{kVA output of auto-transformer} = \frac{V_2 I_2}{1{,}000}$$

$$= \frac{2{,}625 \times 80}{1{,}000} = 210 \text{ kVA } \textbf{Ans.}$$

$$\text{The rated current of } hv \text{ winding} = \frac{10 \times 1{,}000}{2{,}500} = 4 \text{ A}$$

Thus, current drawn from supply mains = 80 + 4 = 84 A

$$\text{kVA transformed} = \frac{80 \times 125}{1{,}000} = 10 \text{ kVA} \quad \textbf{Ans.}$$

and kVA conducted = 210 − 10 = 200 kVA **Ans.**

Example 14.15. A 2,000/1,000/500 V, single-phase three-winding transformer is to be used as an auto-transformer, with supply voltage of 3,000 V. Two loads, one of 1,050 kVA at 3,500 V and the other of 180 kVA at 1,000 V, are to be energised from this auto-transformer output. Draw a suitable diagram of connections and find the currents in various parts of the circuit. Assume the loads to have the same power factor.

[U.P. Technical Univ. Electromechanical Energy Conversion-I, 2008-09]

Solution: The required arrangement is shown in Fig. 14.14.

The current drawn by the load of 1,050 kVA at 3,500 V,

$$I_2 = \frac{1{,}050 \times 1{,}000}{3{,}500} = 300 \text{ A}$$

The current drawn by load of 180 kVA at 1,000 V

$$I_3 = \frac{180 \times 1{,}000}{1{,}000} = 180 \text{ A}$$

Total load supplied, S = 1,050 + 180 = 1,230 kVA

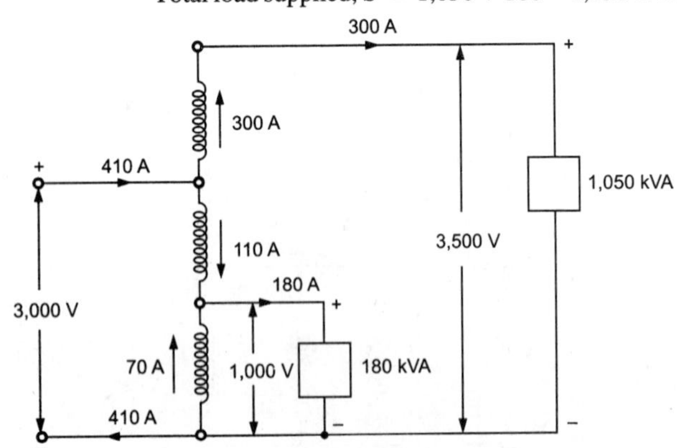

Fig. 14.14

Neglecting, losses, the current drawn from supply mains

$$I_1 = \frac{1,230 \times 1,000}{3,000} = 410 \text{ A}$$

By applying Kirchhoff's current law at the appropriate points currents in different sections of the circuit are obtained, as shown in Fig. 14.14.

Example 14.16. A 3-phase star-connected auto-transformer, supplies a balanced 3-phase load of 100 kW at 350 V at 0.8 pf lagging from 400 V ac supply mains. The efficiency of auto-transformer is 99 per cent. Determine the current in various branches of the auto-transformer.

Solution: Load, P = 100 kW

Power factor, cos ϕ = 0.8 (lag)

Secondary voltage, V_{L_2} = 350 V

Secondary or load current,

$$I_2 = \frac{P \times 1,000}{\sqrt{3} V_{L_2} \cos\phi} = \frac{100 \times 1,000}{\sqrt{3} \times 350 \times 0.8} = 206.2 \text{ A}$$

Since efficiency of transformation is 99 per cent

So per phase VA input

$$= \frac{\text{Per phase VA output}}{\eta} = \frac{\frac{350}{\sqrt{3}} \times 206.2}{0.99} = 42,088$$

So input current

$$I_1 = \frac{\text{Input VA per phase}}{400/\sqrt{3}} = \frac{42,088}{400/\sqrt{3}} = 182.24 \text{ A}$$

$\therefore$ The current flowing from neutral N to the tapping points a, b, c is 206.2 – 182.24 i.e. 23.96 A.

The magnitude and direction of the currents in the various branches of the 3-phase auto-transformer is shown in Fig. 14.15.

Example 14.17. A 3-phase load of 40 kW, 380 V having a pf of 0.8 (lagging) receives its power through a 3-phase auto-transformer from a 3-phase 440 V supply. The auto-transformer is connected in star. The neutral is not grounded. Neglect exciting current and impedance of transformer winding.

(*a*) Draw the diagram depicting the above situation.

(*b*) Calculate the currents entering the load from the auto-transformer.

(*c*) Calculate the currents entering the transformer from the supply side.

(*d*) Show the currents flowing in the transformer winding along with directions.

[*U.P.S.C. I.E.S. Electrical Engineering-II, 2007*]

Solution: Load, P = 40 kW

Power factor, cos ϕ = 0.8 (lag)

Secondary voltage, V_{L_2} = 380 V

(*b*) Currents entering the load from the auto-transformer,

$$I_2 = \frac{P \times 1,000}{\sqrt{3} V_{L_2} \cos\phi_2} = \frac{40 \times 1,000}{\sqrt{3} \times 380 \times 0.8} = 75.97 \text{ A } \textbf{Ans.}$$

(*c*) Currents entering the transformer from the supply side,

$$I_1 = \frac{P \times 1,000}{\sqrt{3} V_{L_1} \cos\phi}$$ $\because$ Exciting currents impedance of transformer windings are negligible

$$= \frac{40 \times 1,000}{\sqrt{3} \times 440 \times 0.8} = 65.6 \text{ A}$$

(*a*), (*d*) Circuit diagram showing the currents flowing in the transformer windings along with direction is depicted in Fig. 14.16.

14.3 VARIAC TRANSFORMER

Small transformers are sometimes provided with tap changing facilities without the complicated control involved with on-load tap-changer. Such devices are sold under the trade names of Variac, Regavolt, Regulac, Powerstat, Auto-stat etc. Such devices make use of various circuits. One commonly used arrangement is shown in Fig. 14.17. The core used is laminated and circular in shape. The winding is placed on the core with usual precaution taken for insulation, and a small portion of each turn is left exposed for the various contact points. A sliding contact with winding is made with a thin carbon brush that covers at least two segments as shown in Fig. 14.17. The position of sliding contact along the exposed surface of the winding may be varied by a handwheel to give the desired output. The output voltage can be varied between 0 and 120 per cent of input voltage. The short-circuit current in the coil must pass through the brush, the resistance of which is sufficient to limit the current to a safer value. Schematic diagram for a single-phase variac transformer is shown in Fig. 14.17 (b). Schematic diagram for a 3-phase variac transformer is given in Fig. 14.5.

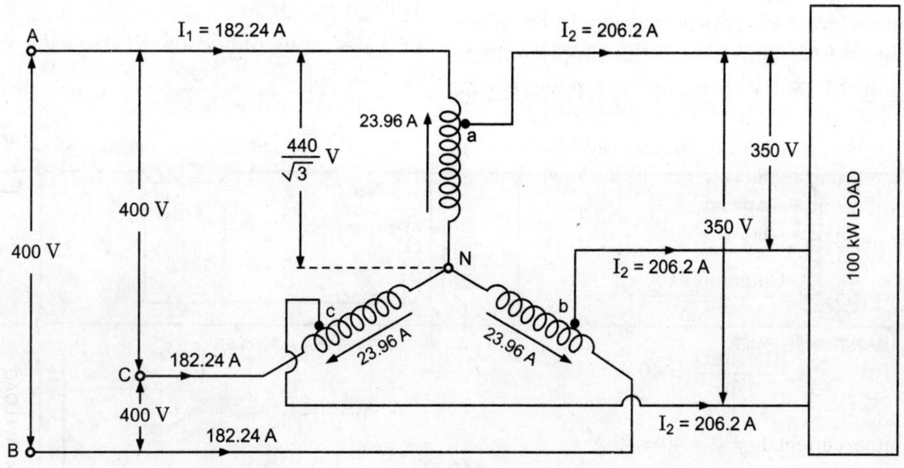

Fig. 14.15

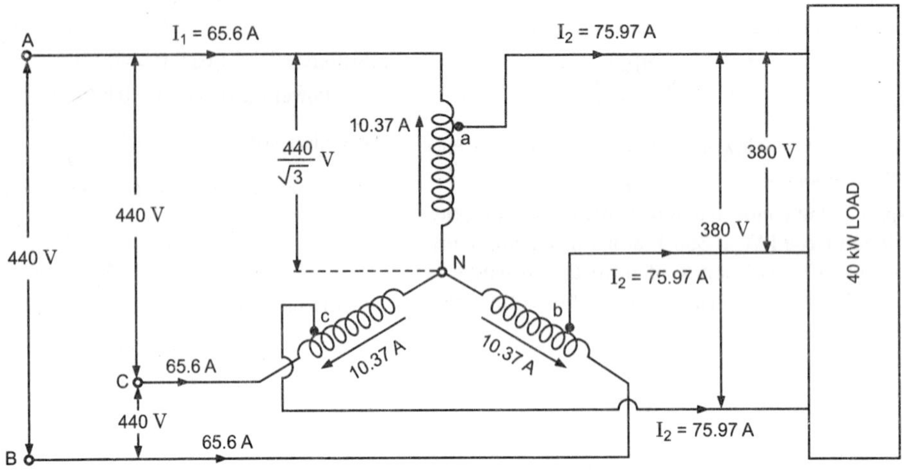

Fig. 14.16

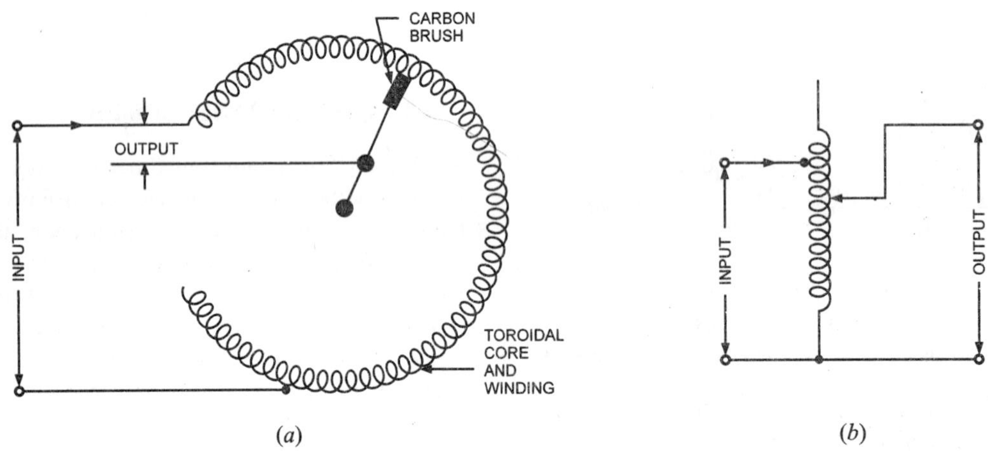

Fig. 14.17 *Variac Transformer*

14.4 BOOSTER TRANSFORMER

In some cases it is not convenient to have tap changing gear in the main transformer, and sometimes it may be required to control the voltage of a feeder at a point far away from the main transformer. Booster transformer may be employed on such occasions. The secondary of the booster transformer is connected in series with the line and booster transformer primary is supplied from the secondary of the regulating transformer fitted with on load tap changing gear similar to that mentioned in Art 14.6.2. The output (secondary) winding of the regulating transformer is so connected to the primary of the booster transformer that the voltage injected in the line, V_B is in phase with supply voltage V_S, as shown in the Fig. 14.18. By changing the tapping on the regulating transformer, magnitude of V_B can be changed and thus feeder voltage V_F can be regulated.

The rating of such regulating transformer is only the fraction (equal to a percentage boost) of that of the main transformer. It is given by the expressions $\frac{\text{Line current} \times \text{boost voltage}}{1{,}000}$ kVA for single-phase boosters

and $\frac{\sqrt{3} \times \text{Line current} \times \text{boost voltage}}{1{,}000}$ kVA for 3-phase voltage boosters.

or kVA rating of booster = Load kVA × percentage boost in voltage.

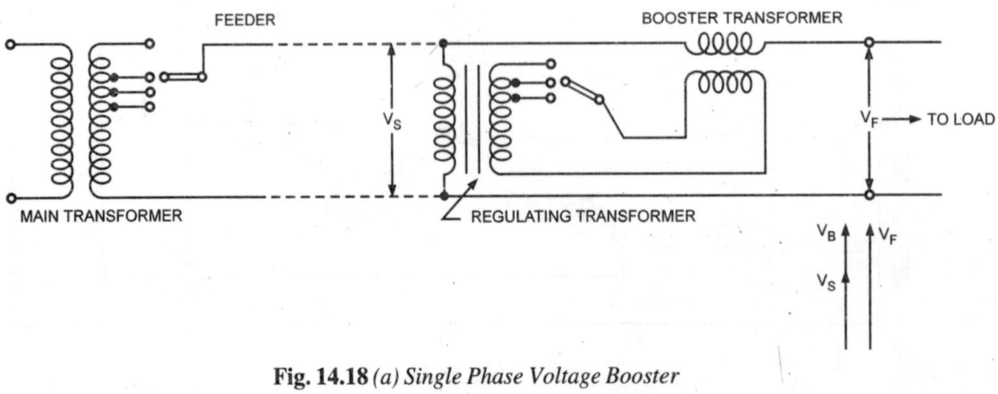

Fig. 14.18 *(a) Single Phase Voltage Booster*

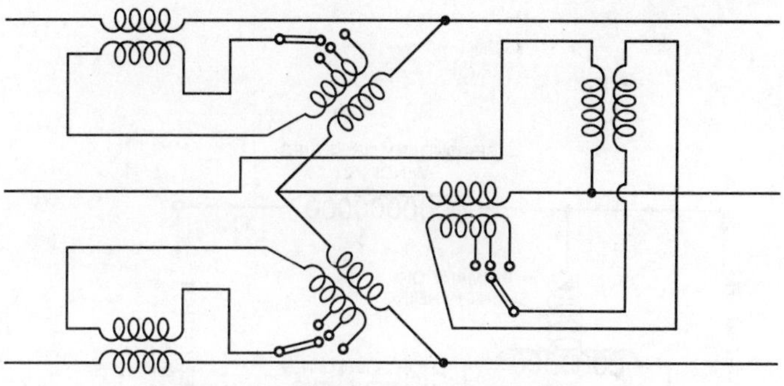

Fig. 14.18 *(b) Three-Phase Voltage Booster*

The advantages of the above system are that the regulating equipment is independent of the main transformer so that a failure in the former will not throw the latter out of service for any length of time; and that it is much cheaper method when there is no main transformer at the point where regulation is desired. When the regulation is required at a point where a main transformer is to be placed, this system is costlier, requires more floor space, and increases the losses.

14.5 INDUCTION VOLTAGE REGULATOR

The fundamental principle of transformer action requires that primary and secondary windings be coupled electromagnetically. When the flux created by the primary passes through the secondary winding, maximum emf is induced in the latter. However, if the secondary is rotated and brought into such a position that the flux passes across (not through) it, no emf will be induced in it. The idea of rotating the primary with respect to secondary is applied in the induction voltage regulator.

The induction voltage regulator is essentially a step-down transformer, but is so constructed that the secondary voltage may be varied from zero to a certain maximum value, dependent upon the ratio of turns in the two windings, merely by altering the position of the primary-coil axis with respect to the secondary-coil axis. The primary winding is connected across the circuit to be regulated and the secondary is connected in series with the circuit. An induction voltage regulator may be of single phase or three phase type and comprises of a stator and a rotor.

14.5.1. Single-Phase Induction Voltage Regulator. Schematic diagram of a single-phase induction voltage regulator is shown in Fig. 14.19. The primary winding is connected across the single-phase supply and the secondary winding is connected in series with the outgoing lines. The primary current I_1 consists of a no-load magnetizing current together with a load component, the magnitude of which is determined by the main circuit current I_2 and the ratio of primary to secondary turns on the induction regulator. The phase of current I_2 with respect to output voltage V_2 depends, of course, on the nature of load connected on the output side.

The flux created is alternating and when the axis of the two windings coincide, the flux set up by the primary winding links with the whole of the secondary winding and maximum voltage is induced in the secondary. As the rotor is rotated through 90°, none of the flux set up by the primary links with the secondary and hence no emf is induced in secondary. And if rotor is further rotated, the direction of induced emf changes i.e. becomes negative. Thus the regulator either adds or subtracts from the circuit voltage depending upon the relative positions of the two windings of the regulator. Single-phase induction regulator does not cause any phase shift. Means are provided for raising or lowering the circuit voltage throughout the range of the regulator in an infinite number of steps in the induction voltage regulator. The voltage ratio is usually 10 : 1, and since primary may be rotated through 180°, the load voltage may be as much as 10% above or below that across the primary. To provide a voltage regulation of 20% (or ± 10%) the induction regulator of rating 10% of the kVA rating of the circuit is required.

The actual single-phase induction regulator slightly differs from that shown in Fig. 14.19. The primary winding is placed in slots in the surface of the laminated cylindrical core since it has to carry smaller current and has small conductor area. No slip rings are required to feed the primary winding, since the rotor movement is restricted from 0° to 180°. Flexible leads are sufficient to feed the primary winding. The rotor also carries, in addition to the primary winding, a short-circuited compensating winding (also called the tertiary winding), whose magnetic axis is always 90° away from that of the primary winding, to neutralize the detrimental series reactance effect of the secondary. The rotor structure is mounted on a shaft so that it can be rotated on its axis and thereby change the relative position between the primary and secondary windings. Since the rotor is not required to rotate like that of a motor, the shaft is generally arranged vertically for convenience. The secondary winding connected in series with the outgoing lines is housed in the stator slots, due to larger

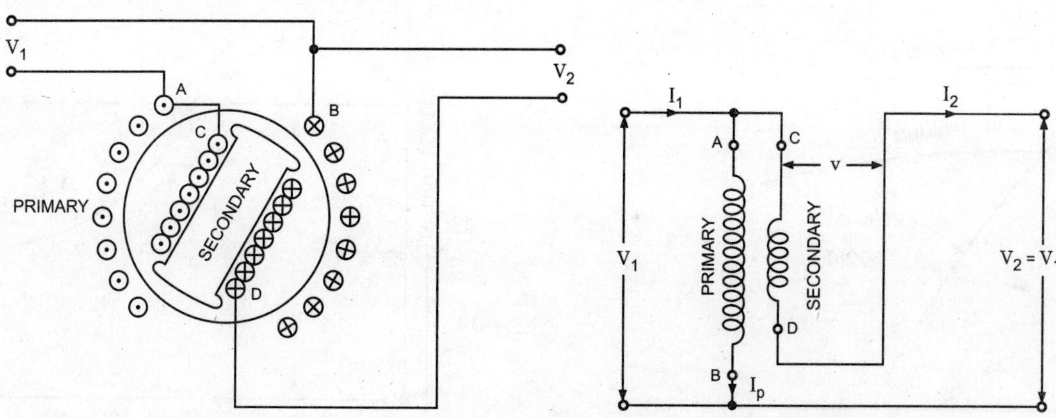

Fig. 14.19 *Schematic Diagram of a Single-Phase Induction Regulator*

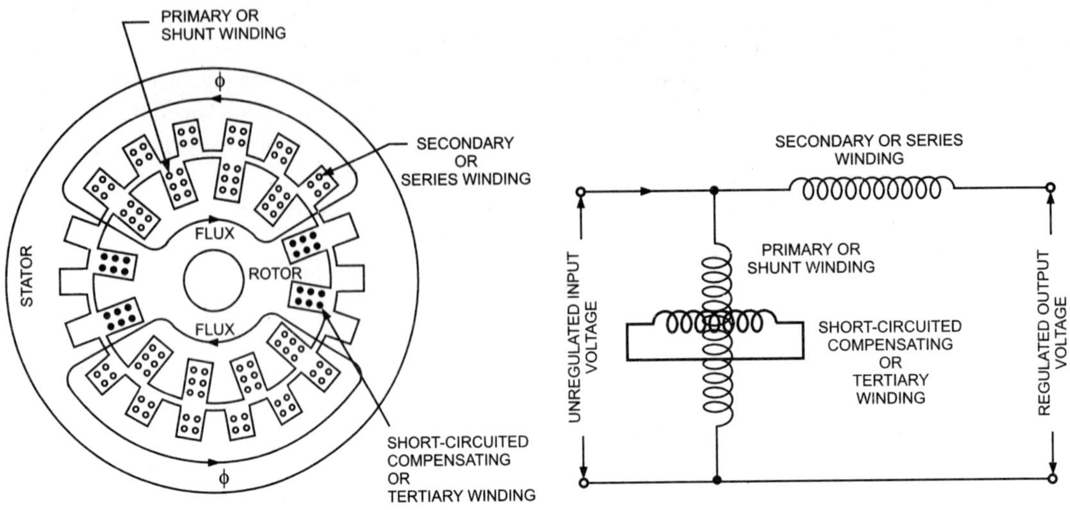

(a) Cross Section *(b) Connection Diagram*

Fig. 14.20 *Single-Phase Induction Regulator*

conductor area. Both of the primary and secondary windings are usually designed to provide 2-poles or 4-poles because the smaller the number of poles, the larger the mechanical angle corresponding to one pole pitch and the easier the voltage adjustment. The spread over of both windings is about 120°. Fig. 14.20 shows diagrammatically the x-section of a bipolar single-phase induction regulator and its connections.

If the regulator were not provided with the short-circuited compensating winding, its operation would not be satisfactory (voltage regulation would be very poor with adjustments for intermediate voltages), as obvious from induction voltage regulator output characteristics shown in Fig. 14.21.

When the axes of primary and secondary windings coincide, there is maximum magnetic coupling between them and zero magnetic coupling with compensating winding. Consequently the voltage v induced in the secondary winding is maximum (V_{max}) and the output voltage is $V_1 + V_{max}$. If the primary is rotated through 180°, the secondary induced voltage is again maximum, but of reversed polarity. The output voltage is, therefore, $V_1 - V_{max}$. This induced voltage in secondary is always either in phase with or in phase opposition to supply voltage V_1. This is due to the fact that only alternating (or pulsating) flux is developed by the single-phase primary winding. However, the magnitude of induced voltage in secondary varies with magnetic coupling between the primary and secondary windings, which depends upon the relative positions of the axes of primary and secondary windings. In general, the secondary induced voltage v is given as $V_{max} \cos \theta$ and output voltage $V_2 = V_1 + v = V_1 + V_{max} \cos \theta$ where θ is the angle measured from the maximum boost position or an angle between the axes of primary and secondary windings. The curve showing the variations in output voltage with θ measured from the maximum boost position is given in Fig. 14.22. The noteworthy point is that the compensating winding is not at all affected by the primary winding, since their axes are fixed in quadrature.

The short-circuited compensating winding (or tertiary winding) reduces the secondary leakage reactance by air-gap flux in the axis perpendicular to the primary winding because it is displaced from primary winding by 90°. Consider two extreme cases. When the axes of the two windings (primary and secondary) are 90° apart, little or no voltage is induced in the secondary because of minimum flux linkage between the stator and rotor. However, the secondary winding carries the load current I_2. In

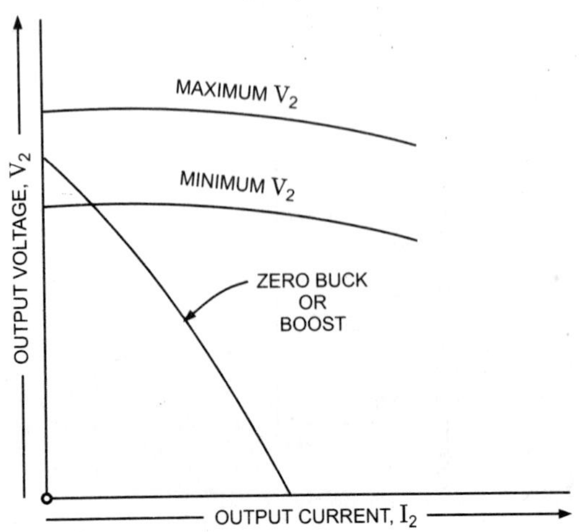

Fig. 14.21

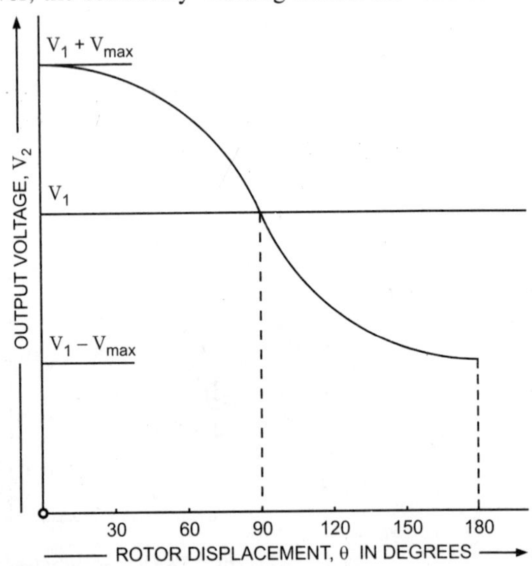

Fig. 14.22

the absence of compensating or tertiary winding, the secondary mmf I_2N_2 would not be neutralized by any mmf. As a result of it, secondary mmf I_2N_2 would saturate the core and the leakage reactance offered by the secondary winding would be quite large. Therefore, a large voltage drop would occur across the secondary winding resulting in a considerable drop in the output voltage. The compensating winding, however, does not permit this to happen. When the rotor is in or near the position of zero boost i.e. when $\theta = 90°$, the magnetic coupling (or flux linkage) between the secondary and the compensating winding becomes maximum and the secondary winding current I_2, therefore, induces maximum current in the compensating winding by transformer action. Thus the secondary and tertiary act together as a short-circuited transformer in which the secondary of the regulator acts as the primary and compensating or tertiary as the short-circuited secondary. The secondary mmf N_2I_2 is almost completely neutralized by the compensating winding mmf and the reactance offered by the secondary winding becomes negligibly small. When the regulator is adjusted to either of the positions resulting in maximum secondary voltage (i.e when $\theta = 0°$ or $180°$), the axis of the compensating winding is 90° to that of the secondary, and there is no flux linkage between the tertiary and the secondary. When the position of the regulator is adjusted for any intermediate position, a certain portion of the secondary mmf reacts with the primary to produce straight transformer action in the secondary, and the remainder of the secondary mmf reacts as the primary of a transformer with a short-circuited secondary which consists of compensating or tertiary winding. The three windings of the single-phase induction regulator, therefore, become equivalent to the circuit shown in Fig. 14.23. The actual secondary winding is replaced by two windings with turns equal to $N_2 \cos\theta$ and $N_2 \sin\theta$ respectively. The regulator then becomes equivalent to the two transformers I and II, as shown in Fig. 14.23.

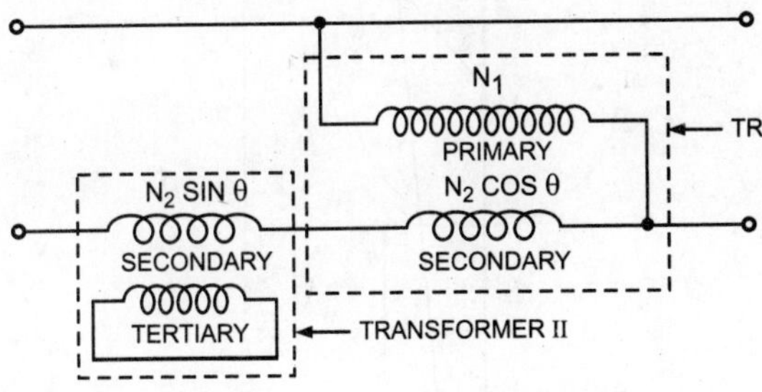

Fig. 14.23 *Equivalent Circuit of a Single-Phase Induction Regulator With Compensating or Tertiary Regulator*

The compensating winding tends to keep the copper or I^2R losses constant for a given load current and various degrees of boost or buck. The variation of primary and short-circuit mmf or ampere-turns with angular displacement of the rotor is shown in Fig. 14.24. Since the copper losses vary as the sum of the squares of the currents, the I^2R losses remain almost, constant for any rotor angle, assuming that the primary and the compensating windings have the same effective resistance (which usually happens to be in practice). The typical curves giving per cent leakage reactance and I^2R loss in relation to the rotor displacement are shown in Fig. 14.25.

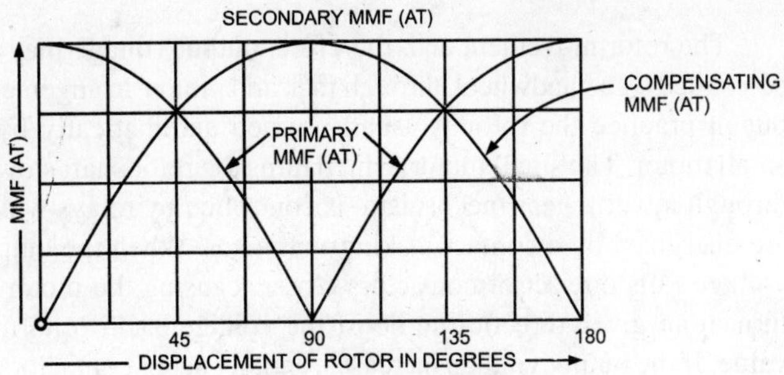

Fig. 14.24

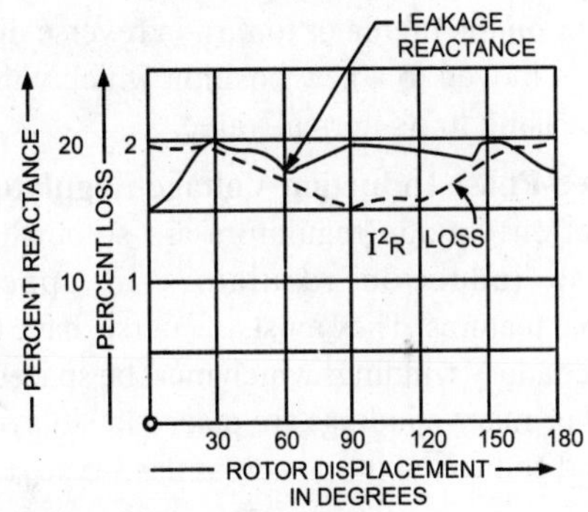

Fig. 14.25

The phasor diagram for currents in a single-phase induction regulator is shown in Fig. 14.26.

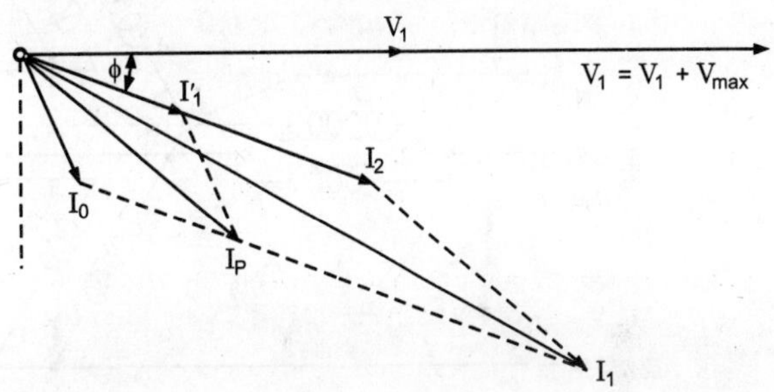

Fig. 14.26 *Single-Phase Induction Voltage Regulator*

If N_1 and N_2 are the number of turns on the primary and secondary windings respectively, then for any rotor displacement angle θ

$$I'_1 N_1 = I_2 N_2 \cos\theta$$

$$\text{or } I'_1 = I_2 \frac{N_2}{N_1} \cos\theta$$

where I'_1 is the load component of primary current I_1.

The secondary mmf component $I_2 N_2 \sin\theta$ is balanced by the tertiary or compensating winding. The total primary current I_p is the phasor sum of I'_1 and the primary exciting current, I_0. The supply current I_1 is the phasor sum of I_p and I_2. The primary winding current I_p is much smaller than secondary current I_2.

The single-phase induction voltage regulator can transfer power in either direction when connected, as in Fig. 14.19. The single-phase induction voltage regulator provided with compensating winding cannot be employed to transform power from its secondary to the primary winding when the secondary is connected to an alternating voltage supply. Under such conditions of operation the compensating winding is directly coupled to the secondary for the intermediate positions and would provide a short circuit on the transformer.

The rotor movement and, therefore, output voltage may be controlled by a handwheel, through rack and pinion arrangement, but in practice the rotor is usually turned automatically by a small motor. The small motor, which turns the rotor shaft slowly through a worm gear mechanism, is controlled by relays which are energized by a contact making voltmeter. When the output voltage falls one set of contactors closes, causing the motor to turn in an given direction to boost the voltage back to normal value. If the output voltage increases, another set of contactors is activated by the contact making voltmeter causing the small driving motor on the regulator to turn in reverse direction, thus, the primary is moved to a new position which will result in the voltage decreasing to its normal value.

14.5.2. Three-Phase Induction Voltage Regulator. The three-phase induction voltage regulators are somewhat similar to single-phase induction regulators in appearance and constructional features. They must, of course, have three primary and three secondary windings which must be spaced 120 degrees apart. Three primary windings are placed in slots of a laminated rotor core and connected in star across the 3-ϕ ac supply because of the few turns per phase as compared to those required in delta connection. Three secondary windings are placed in slots of a laminated stator core and are connected in series with the load as shown in Fig. 14.27. No compensating or tertiary winding is required in case of a 3-phase induction voltage regulator because, in this case, for every position of setting of the regulator each secondary winding is magnetically coupled to one or more of the primary windings. No slip rings are required and the primary windings are energized through flexible leads only as in case of single-phase induction voltage regulators. Since in this case rotating magnetic field of constant magnitude is produced, so induced voltages in the secondary are of constant magnitude but of varying phase, which changes with the change in position of the rotor with respect to stator. Rotor movement is obtained as in the case of a single phase induction regulator.

The voltage phasor diagram for any particular phase is shown in Fig. 14.27 (*b*) where V_1 is the supply voltage, V_r is the induced voltage in the secondary and V_2 is the output voltage, each per phase. The output voltage V_2 is obtained as the phasor sum of V_1 and V_r for any rotor displacement angle θ. The locus of V_2 is

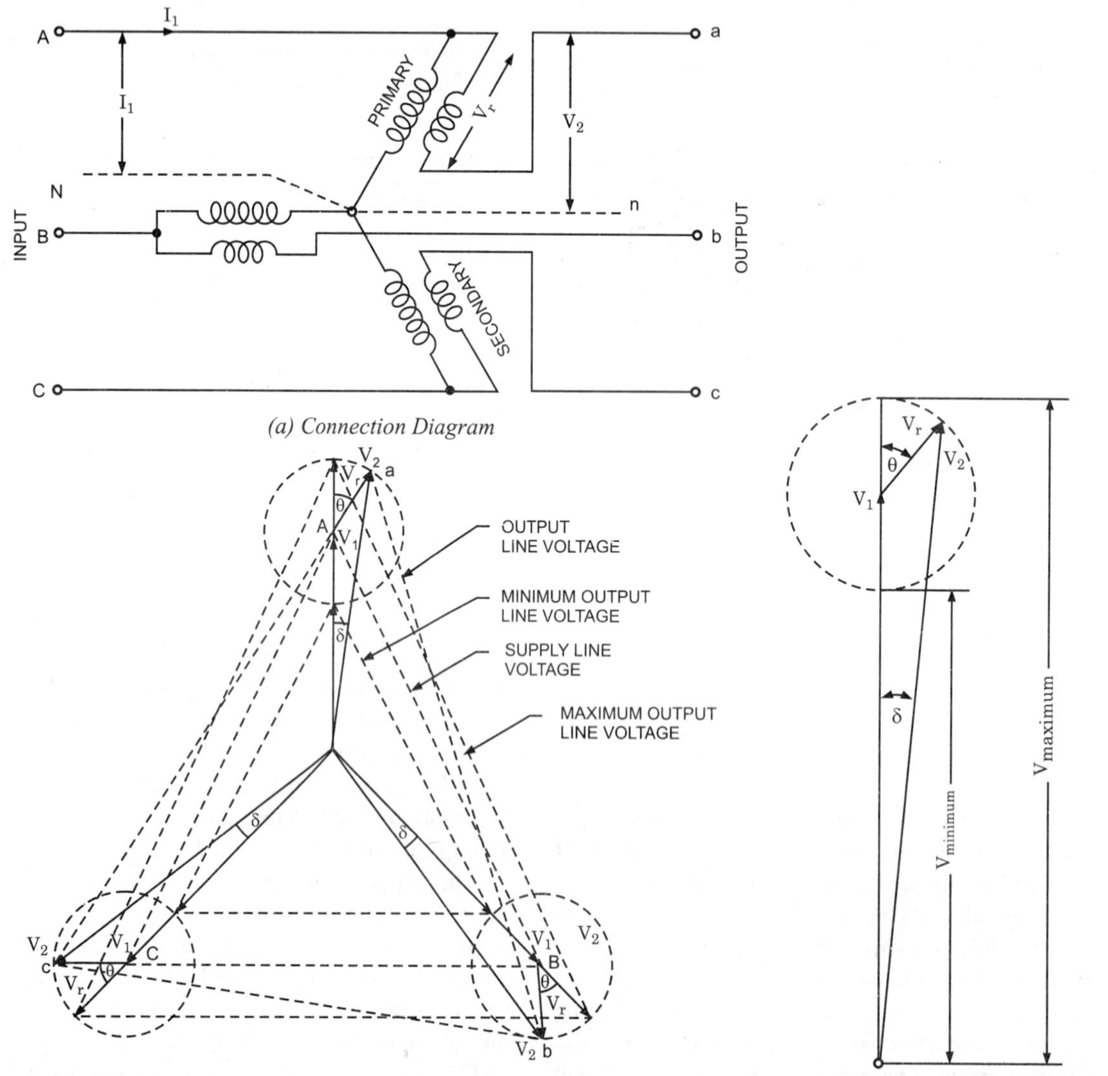

(a) Connection Diagram

(c) Complete Phasor Diagram

(b) Voltage Phasor Diagram For One Phase

Fig. 14.27 *Three Phase Induction Voltage Regulator*

consequently a circle drawn with centre on the extremity of V_1 and of radius V_r. The maximum output voltage $V_2 = V_1 + V_r$ is obtained when V_r is in phase with V_1 and minimum output voltage $V_1 = V_1 - V_r$ is obtained when V_r is in phase opposition to V_1. Thus output voltage varies from $(V_1 - V_r)$ to $(V_1 + V_r)$. For intermediate values of output voltage V_2, there is a phase displacement between V_1 and V_2. It may be seen from phasor diagram shown in Fig. 14.27 (*b*) that

$$V_2 = \sqrt{(V_1 + V_r \cos\theta)^2 + (V_r \sin\theta)^2}$$

and the phase displacement angle θ is given as

$$= \text{Tan}^{-1} \frac{V_r \sin\theta}{V_1 + V_r \cos\theta}$$

The complete phasor diagram for the three phases is shown in Fig. 14.27 (*c*) where ABC and abc are the input and output line voltages respectively. The supply and output line voltages are in phase only at maximum boost and maximum buck positions and for all other positions there is a phase displacement between the supply line and output voltages, which is given as $\tan^{-1} \frac{V_r \sin\theta}{V_1 + V_r \cos\theta}$. This may be objectionable in certain cases of interconnection between systems where such a phase displacement would give rise to large circulating current. Such a phase displacement, however, may be avoided by using either a *twin three-phase regulator* or a bank of three single-phase induction regulators.

The 3-phase induction regulator is unable to balance the line voltages when supplying unbalanced loads because it can supply an identical correction only between any two lines. Two or three single-phase regulators are often used on 3-phase lines. The corrective voltage of the respective singe-phase regulators can be controlled entirely independently of each other.

14.5.3. Twin Three-Phase Voltage Regulator. It comprises two coupled three-phase regulators each of half of the total required boost kVA. The secondary windings are connected in series and the rotors have opposite phase rotation, in order to give resultant output voltage V_2 exactly in phase with the supply voltage V_1. The primary windings are connected in parallel, as shown in Fig. 14.28.

A connection diagram and a voltage phasor diagram for one phase of a twin 3-phase voltage regulator are given in Fig. 14.28. For the line voltages exceeding 11 kV, it is customary to use transformers to supply the regulator primary and to inject secondary induced voltage into the main circuit.

Induction voltage regulators like transformers are available in air-cooled or noninflammable liquid cooled designs for indoor use. Oil cooled regulators are ordinarily used outdoors.

Rating. An induction voltage regulator is rated by the voltage (either raise or lower) it provides, and the load current passing through the series winding. In other words, it is designed according to regulator capacity and not circuit capacity. In all cases, the induction regulator rating is much less than the circuit rating (10 or 15% of circuit rating).

Since an induction regulator inherently introduces into the circuit the same amount of voltage, whether boosting or bucking, select the higher of the two as the required range of regulation.

In single-phase circuits, the kVA rating of induction regulator is given by the expression

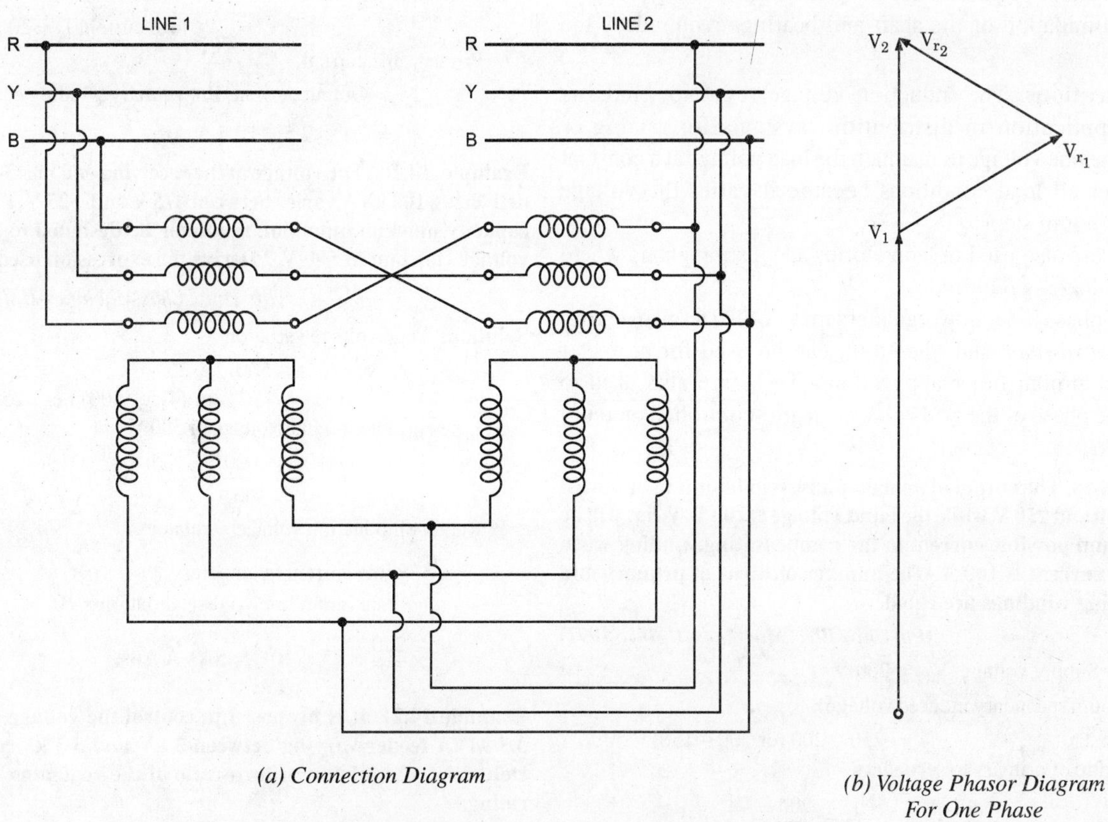

(*a*) Connection Diagram

(*b*) Voltage Phasor Diagram For One Phase

Fig. 14.28 *Twin 3-Phase Voltage Regulator*

$$\frac{\text{Percentage raise or lower} \times \text{circuit rated voltage}}{100} \times \frac{\text{maximum load in amps.}}{1,000}$$

In three-phase circuits, the kVA rating of induction regulator is given by the expression

$$\frac{\sqrt{3} \times \%\text{ age raise or lower} \times \text{circuit rated voltage}}{100 \times 1,000} \times \text{maximum load in amps}$$

A 3-phase induction regulator may be designed as one unit or may be formed by three single-phase induction regulators connected either in star or delta. When three single-phase induction regulators are connected in delta, the range of regulation is 50% greater than the regulation of the single unit. Single-phase regulators for ± 10 % of voltage regulation will provide voltage regulation of ± 15% when connected in delta in a 3-phase circuit.

Advantages (*i*) It is simple and rugged in construction.

(*ii*) It can withstand overloads well.

(*iii*) It is reliable in operation because of the absence of any part subject to wear.

(*iv*) Its operation is not affected by load and power factor variations.

(*v*) It provides stepless voltage variation without any arcing or short–circuiting of turns as in the case of transformer with tap-changing mechanism.

Disadvantages. The induction voltage regulator is more expensive than the transformer with tap-changing mechanism. The other advantages of tap-changing transformers over the induction voltage regulators are the small magnetizing currents, and the elimination of the shaft and bearings required by the regulator.

Applications. The induction voltage regulator finds its greatest application in distribution systems for raising or lowering the line voltage to maintain the load voltage at a constant value under all load conditions because it varies the voltage uniformly, not in steps.

These are also used in laboratories and testing plants where variable voltage is required.

The 3-phase induction regulators may be also termed *phase-shifting transformer* and, therefore, can be used for ac meter testing and similar other applications. These are also used to advance the phase of the grid voltages in grid-controlled mercury arc rectifiers.

Example 14.18. The output of a single-phase regulator is to be varied between 150 and 250 V while the input voltage is 200 V. What will be the maximum possible current in the compensating winding when the output current is 100 A. The number of turns of primary and compensating windings are equal.

[*Pb. Univ. Elec. Machines–II, July-1991*]

Solution: Supply voltage, V_1 = 200 V

Maximum secondary induced voltage,

$$V_{max} = 250 - 200 \text{ (or } 200 - 150) = 50 \text{ V}$$

Turn ratio of primary to secondary

$$= \frac{N_1}{N_2} = \frac{200}{50} = \frac{4}{1}$$

Since number of turns of primary and compensating windings are equal so, the turn ratio of compensating winding to secondary winding i.e.

$$\frac{N_c}{N_2} = 4$$

Secondary current, I_2 = 100 A

Maximum possible current in the compensating winding will be for rotor displacement θ to be equal to 90° from the position of maximum boost and is given as

$$I_c = \frac{N_2 I_2}{N_c} = \frac{N_2}{N_c} \times I_2 = \frac{1}{4} \times 100 = 25 \text{ A } \textbf{Ans.}$$

Example. 14.19. A single-phase induction regulator has an effective turn-ratio of 4 : 1. If the supply voltage is 400 V, determine the maximum and minimum value of output voltages.

If the load current is 20 A and the secondary winding is set at 60° away from the maximum boost position, calculate the primary current and the primary line current. Effect of no-load current may be ignored. [*Pb. Univ. Elec. Machines–II, May-1995*]

Solution: Supply voltage, V = 400 V

Effective turn ratio, $\frac{N_1}{N_2} = 4$

Maximum value of voltage induced in secondary,

$$V_{max} = \frac{N_2}{N_1} \times V_1 = \frac{1}{4} \times 400 = 100 \text{ V}$$

So output voltage limits = (400 ± 100) V giving maximum and minimum values of output voltages 500 V and 300 V respectively.

Neglecting no-load current, primary winding current,

$$I_p = I_2 \frac{N_2}{N_1} \cos\theta$$

$$= 20 \times \frac{1}{4} \times \cos 60° = 20 \times \frac{1}{4} \times \frac{1}{2} = 2.5 \text{ A } \textbf{Ans.}$$

∵ load current I_2 = 20 A and θ = 60°

Primary line current,

$$I_1 = \text{Output current, } I_2 + \text{primary winding current, } I_p$$

$$= 20 + 2.5 = 22.5 \text{ A } \textbf{Ans.}$$

Example. 14.20. The voltage at the receiving end of a 3-phase feeder delivering 100 kVA varies between 475 V and 525 V. For what kVA capacity must an induction regulator be designed to maintain the voltage constant at 500 V ? Derive the expression used.

[*Pb. Univ. Elec. Machines-II, January-1991*]

Solution: Total voltage variation

= (525 – 500) i.e. 25 V

or (475 – 500) i.e. – 25 V i.e. ± 25 V

Line current with line voltage of 500 V

$$= \frac{100 \times 1,000}{\sqrt{3} \times 500} = \frac{200}{\sqrt{3}} \text{ A}$$

Rating of induction voltage regulator

$$= \sqrt{3} \times \text{line current in amperes} \times \text{maximum line voltage variation} \times 10^{-3}$$

$$= \sqrt{3} \times \frac{200}{\sqrt{3}} \times 25 \times 10^{-3} = 5 \text{ kVA } \textbf{Ans.}$$

Example 14.21. It is proposed to control the voltage of a 3-phase, 3.3 MVA feeder varying between 3 kV and 3.5 kV about 3.3 kV. Determine the minimum turn-ratio of the regulator. Find also its rating.

Solution: Rated voltage of the circuit = 3.3 kV = 3,300 V

$$\text{Rated current of the circuit} = \frac{3.3\times10^6}{\sqrt{3}\times3,300} = \frac{1,000}{\sqrt{3}}\text{ A}$$

The regulator must increase the line voltage by 3.5 – 3.3 = 0.2 kV or 200 V and reduce it by 3.3 – 3 = 0.3 kV or 300 volts. It will, therefore, be rated on the larger voltage variation of 300 volts.

$$\therefore \text{Rating of regulator} = \sqrt{3}\times\frac{1,000}{\sqrt{3}}\times300\text{ VA} = 300\text{ kVA } \textbf{Ans.}$$

$$\text{Turn-ratio, } \frac{N_1}{N_2} = \frac{V_1}{V_r} = \frac{3,300}{300} = 11:1 \textbf{ Ans.}$$

14.6 TAP CHANGING TRANSFORMERS

The modern electrical equipments are normally designed to operate satisfactorily at a fixed voltage level. On the other hand voltage variation in power system is a normal phenomenon owing to the rapid growth in industries and distribution network. It is, therefore, very important to keep the consumer's terminal voltage within the statutory limits. The transformer output voltage and, so, the consumer's terminal voltage may be controlled by providing tappings on the transformer windings (either on the primary or on the secondary).

The principle of regulating the secondary voltage is based on changing the number of turns on the primary or secondary *i.e.* on changing the ratio of transformation. Decrease in primary turns causes increase in emf per turn, and so in secondary output voltage. Secondary output voltage can also be increased by increasing secondary turns and keeping primary turns fixed. In other words, decrease in primary turns has the same effect as that of increase in secondary turns.

The tappings used in a transformer are shown in Fig. 14.29.

There is always a tapping on the hv winding which when connected to the rated voltage gives rated voltage on the lv side. This tapping is called the *principal tapping*.

The number of turns included at other taps may be either larger or lesser than the number of turns used at principal tapping. The tapping at which the turns included are more than that of principal tapping is called the *positive tapping*, while a tapping at which the number of turns included is less than the number of turns included at principal tapping is called the *negative tapping*.

Standard distribution transformers have taps arranged in 2½% steps so that rated secondary voltage can be obtained when the primary supply voltage is 0, 2½, 5, 7½% below the nominal primary voltage rating. Transformers used in unit substations often have taps arranged in 2½ % steps and may include some

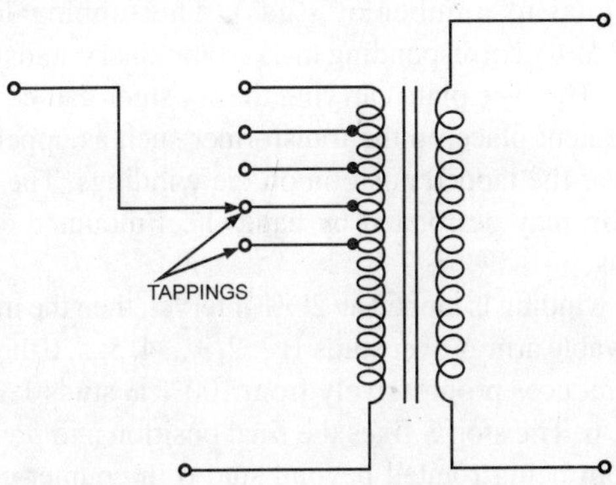

Fig. 14.29 *Transformer Tappings*

overvoltage taps to accommodate primary voltages above the nominal rating. Taps in 2¼% steps may be used on transformers with automatic tap changing equipment. Most of the transformers carry rated kVA on any overvoltage tap and/or rated current on a reduced voltage tap.

The tappings which cause the change in turn ratio, may be placed either on the primary or on the secondary side. The choice between the two sides should be based on maintaining the emf per turn constant, as far as possible. Decrease in primary turns (or increase in emf per turn) causes increase in core flux, and therefore, saturation of the core, increase in core loss, increase in magnetizing current and pronounced third harmonic. On the other hand decrease in emf per turn results in reduction in core flux, and therefore, reduction in core losses and poor utilization of the core.

In transformers at the generating stations, the primary voltage can be kept almost constant, so the tappings should be provided on the secondary side while in a transformer installed at the end of a transmission line i.e. where primary voltage varies, the tappings should be provided on the primary side.

Tappings are usually provided on the high voltage windings due to the following reasons :

1. The number of the turns on the hv winding is large, and therefore, a fine and smoother control of the output voltage is obtained.
2. It is not possible to tap other than an integral number of turns, and this may cause difficulty with *lv* windings. For example 260 V phase winding with 13 V per turn cannot be taped closer than 5%.
3. It is difficult to tap *lv* winding, since it is placed next to the core, due to insulation consideration. On the other hand the hv winding, which is placed outside the lv winding, is easily accessible and can, therefore, be tapped easily.
4. Tap changers on hv side will have to handle low currents though it will need more insulation.

   There is an additional advantage of providing tappings on hv winding in case of step-down transformers. On light load, for decreasing the voltage on the secondary side, the number of turns included on the hv side is increased resulting in reduced emf per turn and so in the flux density and the core loss.

Another important point about the tappings is that whether these should be provided at the phase ends, at the neutral point, or in the middle of the windings. The choice is largely a constructional question. The advantage of having the tapping at the phase ends is that the number of bushing insulators required is reduced, which may be useful where cover space is limited. Some transformers have reinforced insulation at the phase ends, it is essential that in such cases either the tappings should not be provided or the reinforcement should be carried beyond the lowest tap. The insulation between various phase parts, when the tappings are made at the neutral ends, is small. This arrangement is economical especially in the case of high voltage transformers. For achieving large voltage variations, tappings should be placed near the centres of the phase windings in order to reduce magnetic asymmetry. However, this arrangement is not possible with *lv* windings placed next to the core (as in core type transformer) because of accessibility and insulation considerations.

An effect of great importance is that unbalanced axial forces are developed under faulty conditions when tappings have produced magnetic asymmetry between the primary and secondary windings, as discussed below.

The currents flowing in primary and secondary windings interact with the flux in between the two windings and develop a radial force tending to compress the lv winding on to the core and force the hv winding away from the core, as shown in Fig. 14.30 (*a*). This is because the currents in the primary and secondary windings are in opposite directions.

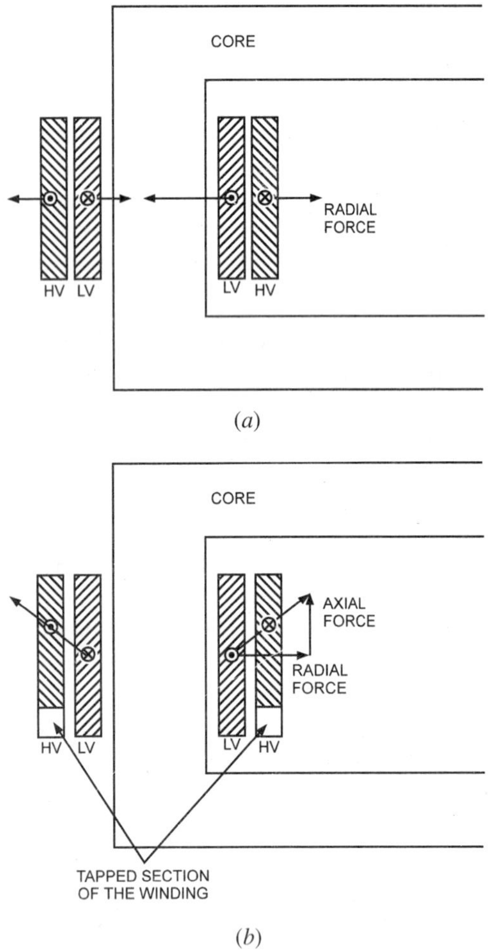

Fig. 14.30

Now in case tappings are provided at one end and some of the turns are cut out by the tap changer as shown in Fig. 14.30 (*b*); axial forces in addition to radial forces are also developed. Under short-circuit conditions, the axial forces become very large and may cause irrepairable damage to the transformer windings.

It is, therefore, necessary to arrange the windings to limit the amount of unbalance. Various methods of providing the tappings on transformer windings are shown in Fig. 14.31.

In case of small transformers the tappings are provided at one end of the windings, as shown in Fig. 14.31 (*a*). In case of large transformers they are usually placed centrally, both for (b) delta and (c) for star-connected windings. Axial mmf unbalance is mitigated by 'thinning' the *lv* windings, as shown in Fig. 14.31 (*d*), or by balancing parts of the winding more symmetrically, as

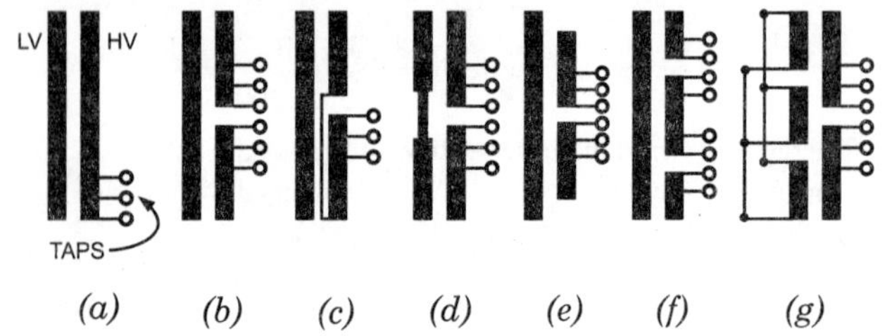

Fig. 14.31 *Methods of Tapping*

shown in Fig. 14.31 (*e*) and (*f*). The axial mmf balance may also be achieved by splitting the untapped winding into a number of parts and then connecting these parts in parallel, as shown in Fig. 14.31 (*g*).

Tap changing involves changes in leakage reactance, copper loss and core loss and possibly some difficulty in operating transformers in parallel.

The tappings may be changed with the transformer isolated from the supply on both sides. This is called *off-load tap changing*. This arrangement is used for occasional adjustments, as in distribution transformers which are provided with ± 5% and ± 2% taps. In some transformers special arrangements are made to avoid interruption of supply so that tap-changing is affected with the transformer energized or on load. This is called the *on-load tap changing*. Daily and short-time adjustments are made with the help of *on-load tap-changing* gear.

14.6.1. Off-Load Tap Changing. This is the cheapest method of changing the turn ratio of a transformer. As the name indicates, it needs disconnection of transformer from the supply before changing the tap. This adjustment is made by tapping the respective windings as required and taking connections of tappings to some position near the top of the transformer. Tap changing is affected manually through hand holes provided in the core. Alternatively the reconnection can be made by carrying the tapping leads through the cover for changing either by hand or by manually operated switches. The commonly used switches are (i) vertical tapping switches and (ii) face-plate switches. *Vertical tapping switch* comprises an insulated bar, one for each phase, carrying knife contacts, to be raised or lowered by hand-wheel and gearing to make connections to the tapping contacts. In a *face-plate switch* the spindles of movable contact arms are geared to a shaft with a handwheel external to the tank. Such switches may be mounted on the upper yoke, or located nearer to the tapped positions on the windings.

A simple off-load tap changer is illustrated in Fig. 14.32. It has six brass or copper studs. The winding is also tapped at six points (equal to number of studs). The tapping leads are connected to six corresponding marked stationary studs arranged in a circle. The face plate carrying the six studs can be mounted at a convenient place on the transformer such as upper yoke or located near the tapped position on the windings. The movable contact arm may be rotated by handwheel mounted externally on the tank.

If the winding is tapped at 2½% interval, then the movement of the movable arm A over studs 1, 2; 2, 3 ... 4, 5; 5, 6 the winding in circuit reduces progressively from 100% at studs 1, 2 to 90% at studs 5, 6. The stop S fixes the final position and prevents the arm A from being rotated beyond stud 6 in counter-clockwise direction or from stud 1 in clockwise direction so as to avoid

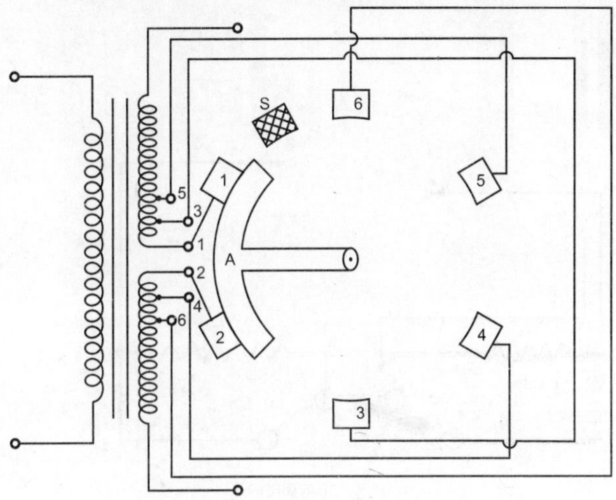

Fig. 14.32 *Off-Load Tap Changer*

bridging of studs 1 and 6 by the arm A. The tap changing is affected only after the transformer is disconnected from the supply. In Fig. 14.32 the arm A is bridging the studs 1 and 2 and the whole winding is in series. Now for bringing arm A at the studs 2 and 3, the transformer is first disconnected and then the arm A is rotated to bridge the studs 2 and 3. After this the transformer is switched on to the supply. Now 97.5% of the winding is in the circuit.

A mechanical lock is provided to prevent unauthorised operation of the off-load tap changer. Further, for preventing inadvertent operation, an electromagnetic latching device or microswitch is provided to operate the circuit breaker so as to disconnect the transformer from the supply as soon as the movable arm is moved; well before the contact of the arm with the stud, with which it was in contact, opens.

14.6.2. On-Load Tap Changing. Such an arrangement is employed for changing the turn ratio of the transformer to regulate the system voltage while the transformer is delivering load. Nowadays almost all the large power transformers are provided with on-load tap clanger. It is invariable practice to provide the tappings at the neutral end of hv windings of a generator transformer. The tappings on the windings are brought out through a terminal board to a separate oil-filled compartment in which the on-load tap changer switch is housed. The tap-changer is in the form of a selector switch. The changer is operated by a motor operated driving mechanism by local or remote control and a handle is also fitted for manual operation in case of an emergency.

The essential features of an on-load tap-changer are that during its operation the main circuit should not be opened to prevent sparking and no part of the tapped winding should get short-circuited. All types of on-load tap-changing circuits are provided with an impedance, which is introduced to limit the short-circuit current during the operation of tap-changing. The impedance may be either a resistor or centre-tapped reactor on the basis of which the on-load tap-changers can, in general, be classified as *resistor* or *reactor type*. In modern designs the current limiting is almost invariably carried out by a pair of resistors.

On-load tap-changing gear with resistor transition, in which one winding tap is changed over for each operating position, is illustrated in Fig. 14.33. The sequence of operations during the transition from one tap to the next (adjoining) (in this case from tap 2 to tap 3). Backup main contactors are provided which short-circuit the resistors for normal operation.

In order to limit the loss of energy, it is essential that the resistors are in the circuit for as minimum of time durations as possible. The resistors are designed for short-time rating for economical consideration and therefore, it is desirable to reduce to minimum their duty time duration. So it becomes essential to incorporate some form of energy storage in the driving mechanism to ensure that a tap change, once initiated, shall be completed irrespective of failure of auxiliary control supply. All modern on-load tap–changers use springs energy as storage elements. They reduce the time that a resistor is in the circuit to a few seconds. The current break is eased by the fact that the short-circuit resistor current has unity power factor.

On-load tap-changer control gear may be from simple push-button initiation to complex automatic control of several transformers operating in parallel. The aim of control is to maintain a given voltage level within a specified tolerance, or to raise it with load to compensate for the voltage drop in the transmission line. The main components are an automatic voltage regulator, a time-delay relay, and compounding elements. The time-delay prevents initiation of a tap-changer due to minor transient voltage fluctuation. It may be set for a delay up to 60 seconds.

Another form of on-load tap-changer, provided with a centre-tapped reactor is illustrated in Fig. 14.34. The function of the reactor is again to prevent the short-circuiting of the tapped windings. During normal operation the short-circuiting switch, S remains closed. The reactor used is suitably designed to prevent large values of current flowing in any section of the primary winding when two tapping switches are closed simultaneously. For understanding applications of this type of on-load tap-changer assume that the tapping switch 1 is closed and the output voltage is minimum. For raising the output voltage, the short-circuiting switch S is opened, the second tapping switch is closed, the first tapping switch is opened and finally the short-circuiting switch S is closed. When the short-circuiting switch S is in open position and the two tapping switches are in closed position, a part of the transformer winding between the two tappings is shunted by the reactor, but large circulating current is not established on account of its high reactance. The line current remains unaffected by this reactance, as the current equally divides and flows in opposite directions in the two halves of the reactor. When only one tapping switch is closed the reactor carries the full primary current. Instead of switches sliding contacts may be used. Sliding contacts are so attached to the ends of reactor that one makes before the other breaks and in normal operating conditions both contacts touch the same tapping stud. If the reactor is designed to carry the full-load current continuously then a working position is possible with two consecutive tapping switches closed. This provides twice as many turn ratios as there are taps, minus one. Usually, the tappings are located midway between the end-turns of the winding in order to prevent the surge voltages from getting into the load-ratio control elements.

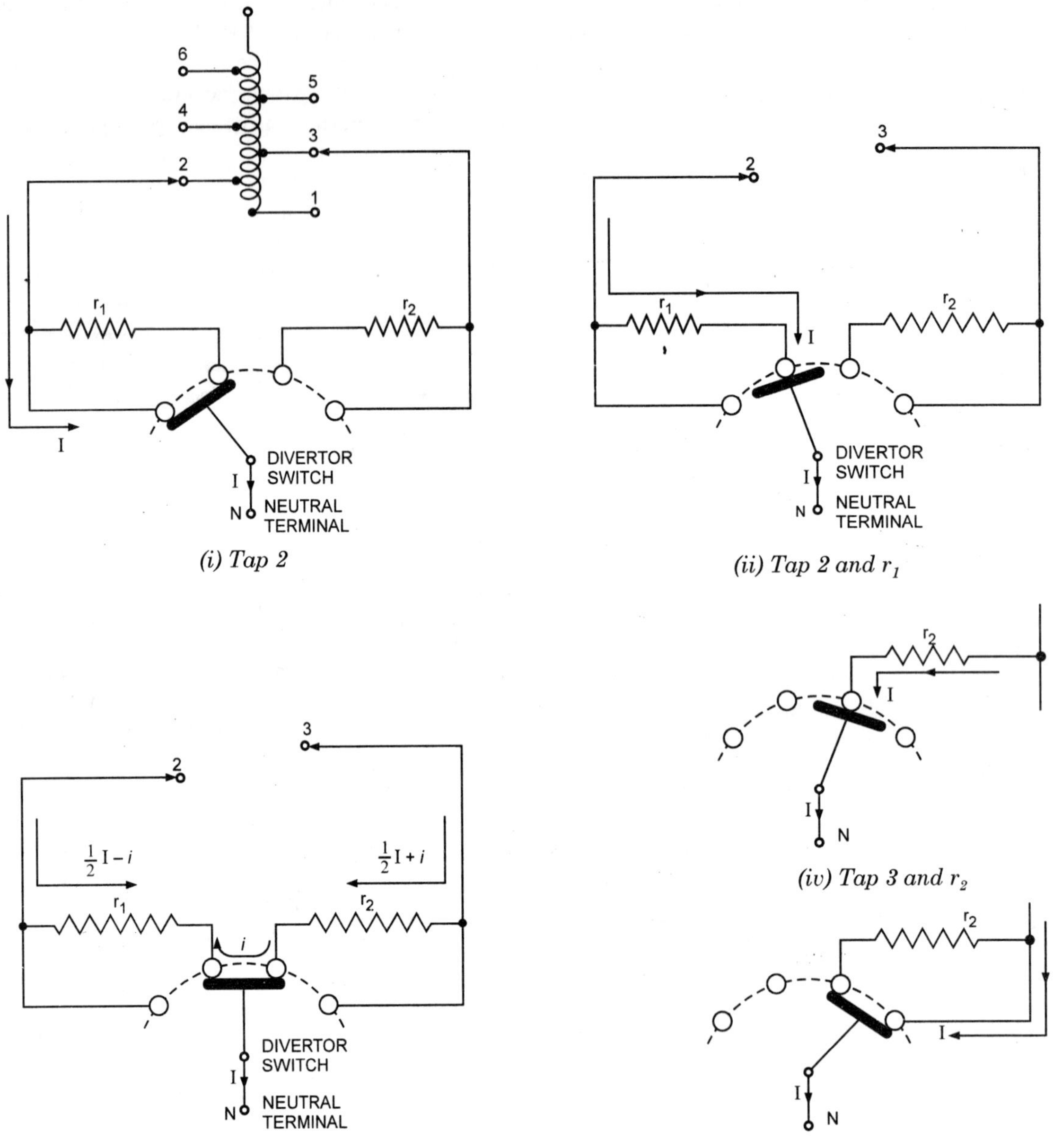

(i) Tap 2

(ii) Tap 2 and r_1

(iii) Tap 2 and 3

(iv) Tap 3 and r_2

(v) Tap 3

Fig. 14.33 *Switching Sequence For On-Load Tap Changing*

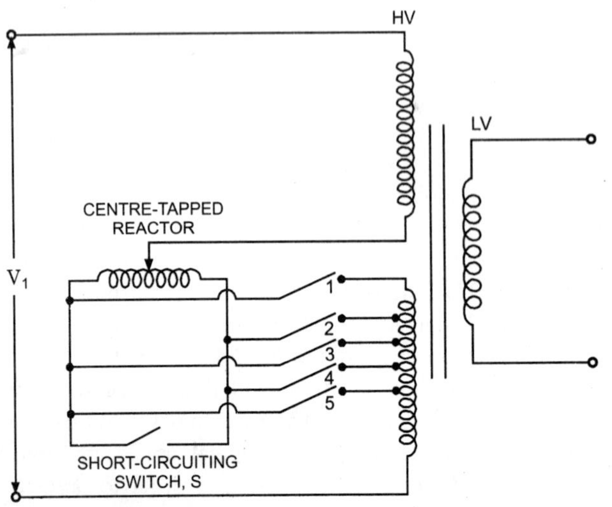

Fig. 14.34 *On-Load Tap Changer*

14.7 PHASE-SHIFTING TRANSFORMER

Existing transmission systems are usually operated and stressed to the limit of their performance capability of their original design so as to maximize utilisation of the assets. In order to maintain the economical, reliable and secure operation of the grid in such conditions, the requirement for various aspects of power flow management within the power systems is becoming obvious. Phase-shifting transformers assist in control of the real power flow in transmission lines and all systems interties. They allow better utilization of existing networks concerning load growths.

A phase-shifting transformer (Fig. 14.35) is a special type of system intertie transformers which control the flow of power through specific lines in a complex power transmission network by providing the possibility to inject a voltage with an arbitrary phase angle in a power system.

Fig. 14.35 *600 MVA Phase-Shifting Transformer*

Phase-shifting transformers are mainly used in hv systems (up to 420 kV) and the throughout power rating is quite large (up to 1,630 MVA).

The phase-shifting transformers are employed for (*i*) controlling the flow of power between two large independent power systems (*ii*) for balancing the loading when power systems are connected together at more than one point so that loops exist and the impedances in parallel paths results in undesired distribution of power flow in the paths (*iii*) for changing the effective phase displacement between the input voltage and the output voltage of a transmission line, thus controlling the amount of active power that can flow in the line and (*iv*) for protection from thermal overload and for improving transmission system stability.

Operation. Consider two power systems with designation as A and B connected by a transmission line TL with impedance Z. The active power will be transmitted from system A to system B, if only the voltage in the left of the transmission line, V_A leads the voltage in the right of transmission line, V_B, this is accomplished by using a phase-shifting transformer which inject extra voltage ΔV into the left side of the TL. The ΔV is used to balance the voltage drop in the TL as follows.

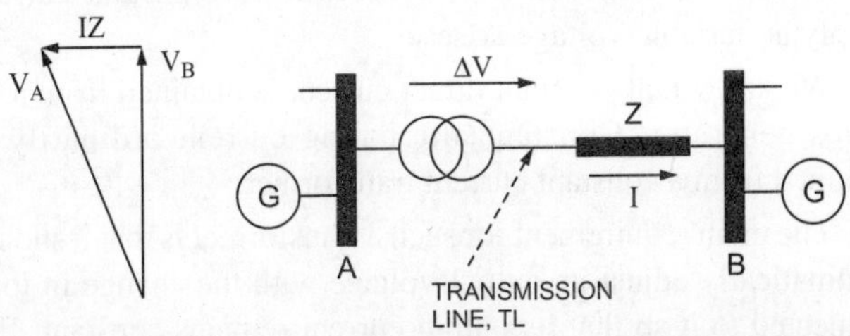

Fig. 14.36

$\mathbf{V_A = V_B + IZ}$ $\quad$ V_A will lead V_B by a certain angle

If the two power systems A and B are connected through two transmission lines having impedances Z_A and Z_B, the power flow in the two transmission lines will be controlled by the magnitudes of Z_A and Z_B and it will result in undesirable distribution of power flow.

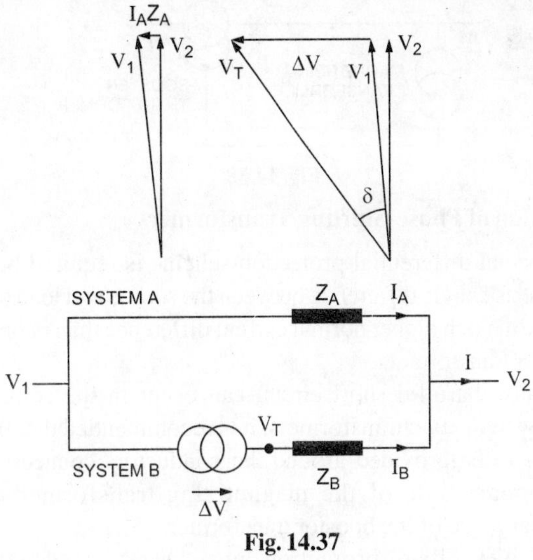

Fig. 14.37

$$\mathbf{V_1 = V_2 + I_A Z_A}$$

$$\text{or } \mathbf{V_1 = V_2 + I_B Z_B}$$

In order to control power flow in this case, a phase-shift transformer is inserted in one of the two systems, as shown in Fig. 14.37, thus inserting a voltage ΔV causing increase in current I_B and decreasing current I_A. Assuming unity pf load

$$\mathbf{V_1 = V_2 + I_A Z_A}$$

$$\mathbf{V_T = V_2 + I_B Z_B}$$

$$\mathbf{V_T = V_1 + \Delta V = V_2 + I_A Z_A + \Delta V}$$

Then the phase shifting transformer must have a secondary voltage V_T which is higher than the primary voltage V_1 which must be displaced in the right direction by an angle δ.

The basic way of providing variable angle displacement (phase shifting) between the primary and secondary voltages in a transformer is to connect the winding of one limb in a three-phase transformer in series with a winding situated on one of the other limbs. Also, the angle displacement can be had by means of a single core three-phase transformer in the same way as that of 3-core, three-phase transformers.

Important Points.

1. The pf of load has a considerable effect in determination of the shift angle.
2. The phase-shifting transformer impedance will add to the impedance of the system and may require a large angle displacement.
3. The phase-shifting transformer has an impact on the exchange of both active and reactive power between systems.
4. The phase-shifting transformer must be able to change the displacement angle in appropriate steps within a certain range.

Phase-Shifting Transformer Tap Changers. In order to fit voltage and current to available tap changers, it may be necessary to use two separate transformers—one magnetizing transformer and one boosting transformer [Fig. 14.38]. In circuit where the tap changer is located, the voltage can be chosen independently of the voltage of the power system. The tap changer is typically located on the secondary side of the magnetizing transformer.

Note: The two transformers are quite large and because of transport limitations they are usually placed in separate tanks.

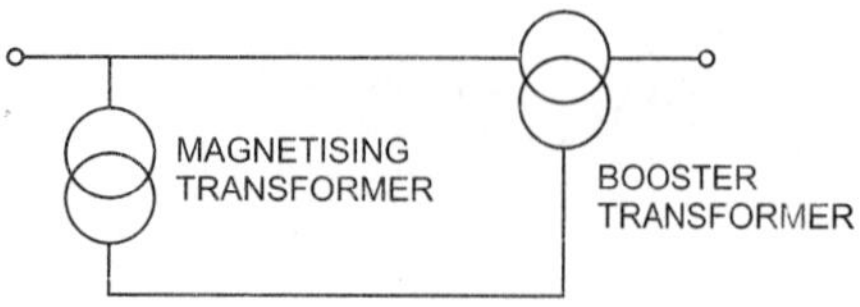

Fig. 14.38

Protection of Phase-Shifting Transformers

- A special differential protection scheme is required because the phase angle difference between the source and load current cause a much higher normal current difference than in ordinary power transformers.
- Because harmful short circuit can occur in the connection between the two transformers, it is recommended that the bus ducts to be provided around the conductors connecting the secondary side of the magnetizing transformer to the primary side of the booster transformer.
- The overvoltage protection must be reviewed with the manufacturers to minimize the effects of transient voltages which can occur at the transformers windings resulting in high voltages and high frequencies oscillations.

14.8 ISOLATION TRANSFORMERS

The surges riding on supply lines may not have any harmful effect on major power equipments but can cause maloperation, loss of function or heavy damage to electronic equipments. Also, the single-phase power which is usually obtained from one of the line terminals and the supply neutral which may be floating because of oscillations of neutral potential. Even a shifting of a neutral potential by 20 to 30V rms may cause damages to MOSFET devices base electronic equipments like computers. Hence isolation transformers are required to provide voltage of the same rms value.

Isolation transformers are two-winding transformers with centre-tap on the primary winding grounded via a capacitor and secondary winding may be grounded solidly to the ground if required. Also, a thick copper sheet is wound between the two windings (primary and secondary) but it does not form a closed turn. This shield is solidly grounded and does not permit the shifting of secondary winding potential through capacitive coupling. Metal oxide varistors are also employed at the input terminals so as to absorb the surge energy.

14.9 CONSTANT CURRENT TRANSFORMER

The transformers considered so far are *constant voltage transformers;* There are services in which constant current is desired such as in street and airport runway lighting. Constant voltage parallel system is somewhat objectionable as voltage drop becomes a serious problem owing to use of large lengths of conductor. Also, the conductors to the lamps nearest the distribution centre may carry the currents of all the lamps and must be of much larger cross section that the conductors to the lamps at the end of the line. In between the conductor size must, of course, vary between the two extremes. With the lamps connected in series, the conductor size remains the same throughout, and all lamps carry the same current. Furthermore, the total conductor length is less, since only one line is required to each lamp.

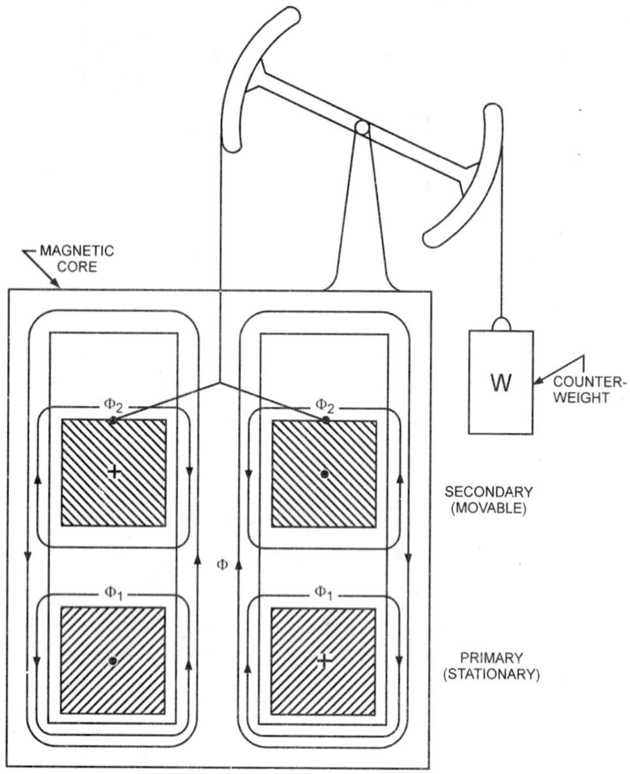

Fig. 14.39 *Constant Current Transformer*

However, there is one major drawback of series circuit lighting — if one lamp burns out, all the lamps in the circuit will go out. To avoid break in series lighting circuits owing to blowing off one or more lamps, provision is made at each lamp for establishing a short circuit across its terminals upon its failure. This is accomplished by connecting a film cut-out in parallel with each lamp. Under normal operation the film of the cut-out acts as an insulator but upon failure of the lamp, the entire voltage of the series circuit is momentarily applied across the cut-out, puncturing the insulation and thus shorting out the lamp.

There is another drawback of series lighting — when any one or more of the lamps connected in the circuit burn out, the total impedance of the series circuit drops and if a constant voltage were maintained the current could increase and cause other lamps to burn out. So the necessity of constant current supply at variable voltage arises.

We know that constant direct current is obtained from a dc series generator. Constant alternating current ordinarily is obtained from a constant current transformer.

The main requirement for such a transformer is that it should automatically adjust its output voltage with the change in load connected to it so that its output current remains constant. If a transformer is designed with a large leakage reactance which can be automatically controlled in inverse relation to the load, it will be possible to maintain a constant output current since secondary voltage of a transformer supplying a given load depends to a large extent upon its leakage reactance.

In such a transformer, the primary coil is stationary at the base while the secondary coil is free to move between vertical guides and is provided with an adjustable counterweight, the setting of which governs the value of output current—increase in counterweight will cause the transformer to operate at lower output current and vice versa. The stationary primary winding receives the power at constant voltage source and the movable secondary winding delivers a constant current to the load circuit (lighting circuit). Both the stationary primary and the movable secondary winding surround the centre leg of the silicon steel laminated core; the core is quite long and operates at a higher flux density. The transformer is of shell type. The weight of the secondary winding is always slightly greater than that of the counterweight. The secondary winding takes its lowest position at full load and have its maximum output voltage in this position. In order to avoid mechanical oscillations resulting from change in load or in the input voltage to the primary oil dash pots are used.

The operation of this transformer is based upon the fact that primary and secondary currents are opposite in direction and, therefore, develop opposing fields, and the repulsion between the two windings resulting from the force developed by the action of the primary leakage flux on the secondary current tends to lift the secondary winding. The secondary winding adjusts its position automatically so as to establish equilibrium between the forces of the winding weight, of repulsion between the two windings, and of magnitude of counterweight. The leakage reactance of the transformer varies with the position of the movable winding, the greater the distance between the two windings, the greater will be leakage reactance.

If the mechanical system is in equilibrium, a reduction of load, which would result if one or more lamps were short circuited, would cause an increase in the seconary current and a corresponding increase in primary current along with an increase in the primary leakage flux Φ_1. This would result in an increase in repulsive force which would lift the secondary to a new position where it would be subjected to the same flux density of the primary leakage flux in which it was supported prior to the change in load. With the upward movement of secondary, the secondary current reduces until it returns to its original value at the point of mechanical equilibrium. As a result of the change in position the leakage fluxes of both windings have increased and then, since leakage reactances are large, the output voltage is reduced. With a sensitive mechanical suspension on the secondary, the secondary current will adjust itself to a given value for a given counterweight.

The input voltage remains constant. The primary current also remains fairly constant over the entire operating range; in practice, this is usually for loads that vary between the rated number of lamps to about half as many. The slight variation in input current with variation in load is because of variation in magnetizing current. The power factor of the supply, however, varies considerably owing to variation in magnetizing current. The overall power factor is high only when the coils are close together but falls with the increase in distance between the coils. This is one of the main objection to its use.

Typical characteristics of a constant current transformer are shown in Fig. 14.40.

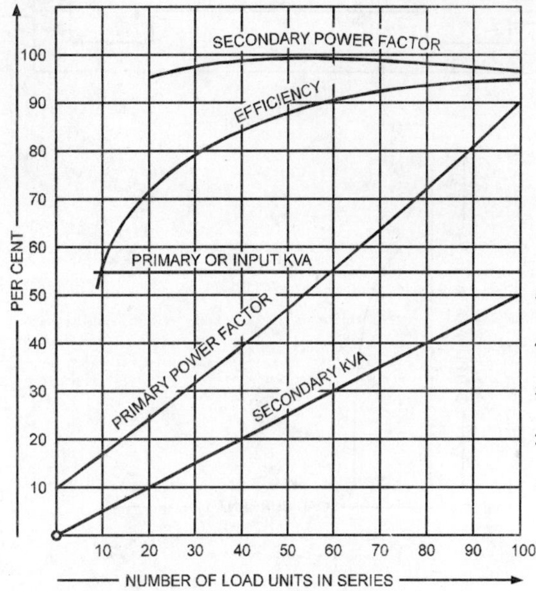

Fig. 14.40

14.10 LOW-VOLTAGE TRANSFORMERS

Large currents at variable low voltages are required for electric arc furnaces and electrolytic processes. A shell type construction is employed so that the low-voltage windings can be sectionalized, and therefore, uniform leakage impedance for parallel connection can be had. A 3-phase furnace with a single turn per secondary phase may, however, be of core type construction and make use of sheet type conductors. The phase terminal arrangements are designed for minimum magnetic leakage, as even a small inductance may cause relatively large voltage drop. Voltage regulation is provided by taps on the hv side in such a way as to develop minimum unbalanced mechanical forces in the event of a short circuit, which is a frequent occurrence.

14.11 HIGH-LEAKAGE REACTANCE TRANSFORMERS

In a conventional transformer, efforts are made to give output voltage almost constant from no load to full load. Such transformers are designed to have the leakage reactance minimum possible. On the other hand, in some special applications (such as furnace ignition and gaseous discharge lamps) it is desirable to make use of transformers that give very high no-load voltages but small short-circuit currents (i.e. the transformers giving a drooping output voltage characteristic). Such a drooping-output voltage characteristic can be had by designing the transformer with high leakage reactance. Gaseous discharge lamps require at start a voltage considerably higher than that required for normal operation so as to breakdown the gas and cause it to conduct. The high leakage-reactance transformers can be designed to take care automatically of these voltage requirements. High leakage transformers can be designed to give open-circuit voltage ranging between 20 kV and 30 kV and the short-circuit current of the order of 1 or 2 mA.

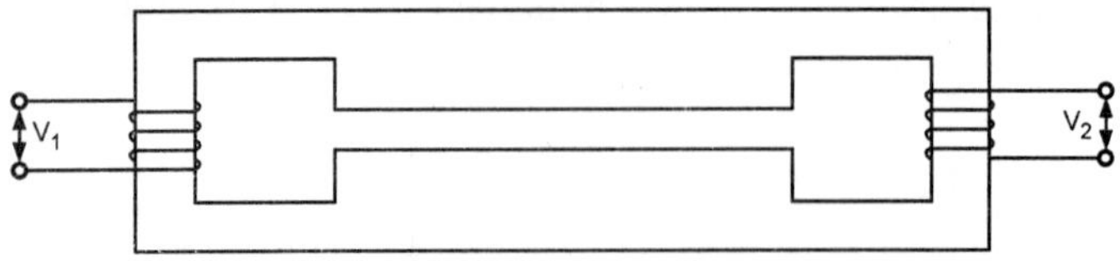

(*a*)

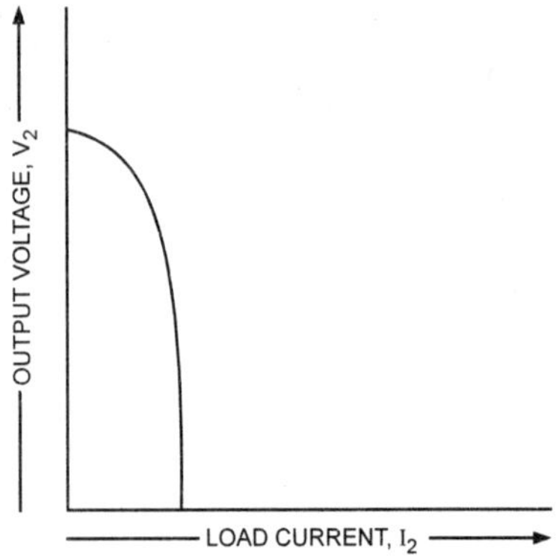

(*b*)

Fig. 14.41 *High-Leakage Reactance Transformer*

Such transformers do not differ much from conventional two-winding transformers, except that they have small volt-ampere rating and design proportions quite different. A structure, that can be designed with a high-voltage secondary winding, is shown in Fig. 14.41 (*a*). The windings are spaced well apart and a shunt magnetic circuit is provided between the windings. The shunt magnetic circuit has an air gap that prevents the mutual flux from being passed at no load, but on load, leakage flux of large value is forced through the shunt circuit, causing a large voltage drop. The output voltage characteristics of a high leakage reactance transformer are shown in Fig. 14.41 (*b*).

14.12 AUDIO-FREQUENCY TRANSFORMERS

Small iron-core transformers employed in the AF range of 20 Hz to 20 kHz are called *audio-frequency* or simply *audio transformers.* Audio transformers are the transformers specifically designed for use in AF circuits to carry audio signals. They can be employed to block radio frequency interference or the dc component of an audio signal, to split or combine audio signals, or to provide impedance matching between high and low impedance circuits, such as between a high impedance amplifier output and a low impedance loudspeaker, or between a high impedance instrument output and the low impedance input of a mixing console. AF transformers that operate with loudspeaker voltages and currents are larger than those which operate at microphone or line level, carrying much less power.

Being magnetic devices, AF transformers are susceptible to external magnetic fields such as those produced by ac current-carrying conductors. **Hum** is a term commonly used to describe unwanted signals originating from the mains power supply (typically 50 or 60 Hz). AF transformers used for low-level signals, such as those from microphones, usually include magnetic shielding to protect against extraneous magnetically coupled signals.

Fig. 14.42 *Two Speaker Level AF Transformers in a Tube Amplifier (on Left) and The Power Supply Toroidal Transformer (on Right)*

Figure 14.43 shows five AF transformers for different line level purposes. The two black boxes on the left contain 1 : 1 transformers for splitting signals, balancing unbalanced signals, or isolating two different ac ground systems to eliminate buzz and hum. The two cylindrical metal case fit into octal sockets each one containing a 1 : 1 transformer, the first is rated at 600 Ω, the second is rated at 15,000 Ω. On the far right is a **DI unit**; its 12 : 1 transformer converts a high impedance unbalanced input to a low impedance balanced output.

Fig. 14.43 *Five AF Transformers For Different Line Level Purposes*

Audio transformers were originally designed to connect different telephone systems to one another while keeping their respective power supplies isolated, and are still commonly used to interconnect **professional audio** systems or system components, to eliminate buzz and hum. Such transformers typically have a 1 : 1 ratio between the primary and the secondary. These can also be used for splitting signals, **balancing unbalanced signals**, or feeding a balanced signal to unbalanced equipment. Transformers are also used in **DI boxes** to convert high-impedance instrument signals (e.g. bass guitar) to low impedance signals to enable them to be connected to a microphone input on the **mixing console**.

A particularly critical component is the output transformer of a **tube amplifier**. Tube circuits for quality reproduction have long been produced with no other (interstage) audio transformers, but an output transformer is required to couple the relatively high impedance (up to a few hundred ohms depending upon configuration) of the output tube(s) to the low impedance of a loudspeaker. (The tubes can deliver a low current at a high voltage; the speakers require high current at low voltage.) Most solid-state power amplifiers require no output transformer at all.

Audio transformers affect the sound quality because they are nonlinear. Harmonic distortion is added to the original signal, especially odd-order harmonics with an emphasis on third-order harmonics. When the incoming signal amplitude is very low there is not enough level to energize the magnetic core. When the incoming signal amplitude is very high the transformer saturates and adds ringing harmonics. Another nonlinearity comes from limited frequency response. For good low-frequency response a relatively large magnetic core is required; high power handling increases the required core size. Good high-frequency response requires carefully designed and implemented windings without excessive leakage inductance or stray capacitance. All this makes for an expensive component.

Early transistor audio power amplifiers often had output transformers, but they were eliminated as advances in semiconductors allowed the design of amplifiers with sufficiently low output impedance to drive a loudspeaker directly.

14.13 RADIO-FREQUENCY (RF) TRANSFORMERS

Beyond the normal power frequency range at frequencies exceeding 50-60 Hz, the thickness of the laminations have to be reduced for controlling the iron losses to an acceptable levels. In case the operating frequency exceeds beyond a few kHz, it becomes necessary to use iron-dust cores or the ferrites, special oxides which exhibit reasonable good magnetic properties with very small eddy current loss.

There are several types of transformers used in RF work. Steel laminations do not suit to RF transformers.

1. Air-Core Transformers. These are used for HF work. The lack of a core means very low inductance. Such transformers may be nothing more than a few turns of wire soldered onto a printed circuit board (PCB).

2. Ferrite-Core Transformers. These transformers are extensively employed in intermediate frequency (IF) stages in superheterodyne radio receivers. They are mostly tuned transformers, containing a threaded ferrite slug that is screwed in or out to adjust IF tuning. These transformers are usually canned (shielded) for stability and to reduce interference.

3. Transmission-Line Transformers. For radio frequency (RF) use, transformers are sometimes made from configurations of transmission line, sometimes bifilar or coaxial cable, wound around ferrite or other types of core. This style of transformer provides an extremely wide bandwidth but only a limited number of ratios (such as 1 : 9, 1 : 4 or 1 : 2) can be achieved with this technique.

The core material increases the inductance dramatically, thereby raising its Q factor. The cores of such transformers help improve performance at the lower frequency end of the band. RF transformers sometimes used a third coil (called a tickler winding) to inject feedback into an earlier (detector) stage in antique regenerative radio receivers.

In RF and microwave systems, a **quarter-wave impedance transformer** provides a way of matching impedances between circuits over a limited range of frequencies, using only a length of transmission line. The line may be coaxial cable, waveguide, stripline or microstripline.

14.14 VARIABLE FREQUENCY TRANSFORMERS

Variable frequency transformers provide transmission solutions for a smarter grid, enabling transmission system operators to control power flows between power grids with high reliability, speed and efficiency, while offering flexibility in how utilities provide power to meet customer's demand.

Fig. 14.44 *Variable Frequency Transformer (VFT)*

A **variable frequency transformer (VFT)** is used to transmit electricity between two (asynchronous or synchronous) alternating current domains. The VFT is a relatively recent development (first deployed in 2004). Most asynchronous grid interties use high-voltage direct current converters, while synchronous grid interties are connected by lines and "ordinary" transformers, but without the ability to control power flow between the systems.

Five small variable frequency transformers with a total power rate of 25 MVA were in use at Neuhof Substation, Bad Sachsa, Germany for coupling power grids of former East and West Germany between 1985 and 1990.

Smaller VFTs are used in large land-based wind turbines, so that the turbine rotation speed can vary while connected to a distribution grid.

The VFT system is based on a combination of hydro generator and transformer technologies. It consists of a rotary transformer for continuously controllable phase shift, together with a drive system and control that adjust the angle and speed of the rotary transformer to regulate the power flow through the VFT.

The VFT system provides a means to control power between two grids. The grids need not be synchronous. A common situation is where two grids of the same nominal frequency cannot economically be directly connected with ac lines. The VFT allows controlled power exchange between the grids, while retaining many of the inherent virtues of an ac interconnection.

The simple design of the VFT components, based on established and widely used rotating machinery, ensures its long-term maintainability.

Construction and Operation.

A variable frequency transformer (VFT) a doubly fed electric machine resembling a vertical shaft hydroelectric generator with a three-phase wound rotor, connected by slip rings to one external power circuit. The stator is connected to the other. With no applied torque, the shaft rotates due to the difference in frequency between the networks connected to the rotor and stator. A direct-current torque motor is mounted on the same shaft; changing the direction of torque applied to the shaft changes the direction of power flow.

VFT behaves as a continuously adjustable phase-shifting transformer. It allows control of the power flow between two networks. Unlike power electronics solutions such as back-to-back hvdc, the variable frequency transformer does not demand harmonic filters and reactive power compensation. Limitations of the concept are the current-carrying capacity of the slip rings for the rotor winding. A **phase angle regulating transformer, phase angle regulator (PAR,** American usage), **phase-shifting transformer, phase shifter** (West coast American usage), or **quadrature booster (quad booster,** British usage), is a specialised form of transformer used to control the flow of real power on three-phase electricity transmission networks.

For an alternating current transmission line, power flow through the line is proportional to the sine of the difference in the phase angle of the voltage between the transmitting end and the receiving end of the line. Where parallel circuits with different capacity exist between two points in a transmission grid (for example, an overhead line and an underground cable), direct manipulation of the phase angle allows control of the division of power flow between the paths, preventing overload. Quadrature boosters thus provide a means of relieving overloads on heavily laden circuits and re-routing power via more favourable paths.

Alternately, where an interchange partner is intentionally causing significant "inadvertent energy" to flow through an unwilling interchange partner's system, the unwilling partner may threaten to install a phase shifter to prevent such "inadvertent energy", with the unwilling partner's tactical objective being the improvement of his system's stability at the expense of the other system's stability. As power system reliability is really a regional or national strategic objective, the threat to install a phase shifter is usually sufficient to cause the other system to implement the required changes to his system to reduce or eliminate the "inadvertent energy".

The capital cost of a quadrature booster can be very high, however, the utility to transmission system operators in flexibility and speed of operation, and particularly savings in permitting more economical dispatch of generation, can soon recover the cost of ownership.

By means of a voltage derived from the supply that is first phase shifted by 90° (hence is in quadrature), and then reapplied to it, a phase angle is developed across the quadrature booster. It is this induced phase angle that affects the flow of power through specified circuits.

Arrangement. A quadrature booster typically consists of two separate transformers: a shunt unit and a series unit. The shunt unit has its windings connected across the phases, so it produces output voltages shifted by 90° with respect to the supply. Its output is then applied as input to the series unit, which, because its secondary winding is in series with the main circuit, adds the phase-shifted component. The overall output voltage is hence the phasor sum of the supply voltage and the 90° quadrature component.

Tap connections on the shunt unit allow the magnitude of the quadrature component to be controlled, and thus the magnitude of the phase shift across the quadrature booster. The flow on the circuit containing the quadrature booster may be increased (*boost tapping*) or reduced (*buck tapping*). Subject to system conditions, the flow may even be bucked enough to completely reverse from its neutral-tap direction.

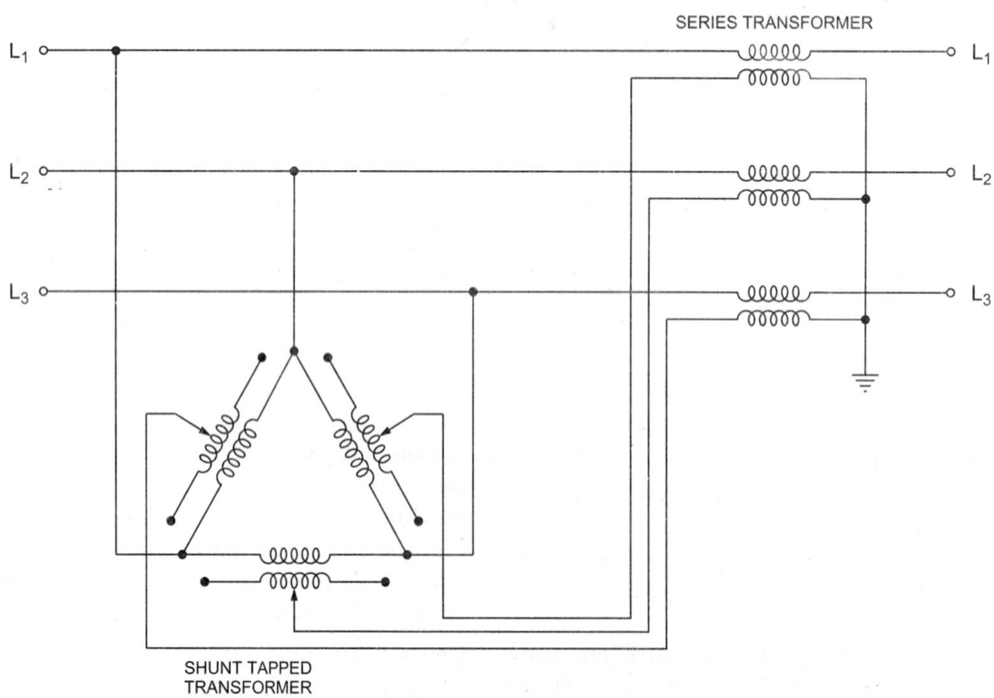

Fig. 14.45 *Simplified Circuit Diagram of a Three-Phase Quadrature Booster.*
[Arrows shown on shunt transformer secondary windings are movable taps; the windings have floating ends shown, and grounded centre taps (not shown)]

14.15 PULSE TRANSFORMERS

There are certain applications where a dc pulse of extremely short duration is to be applied on to a circuit at high potential from a source at ground level. The faithfulness of the transmission is the main requirement. This is achieved through specially designed transformers, called the *pulse transformers*.

Figure 14.46(*a*) shows ideal rectangular pulse input to the pulse transformer which couples a source of pulses of electrical energy to the load. Pulse transformers are quite small in size and have few turns on both primary and secondary windings. The interwinding capacitance and leakage inductances of these transformers are quite small. Transformer core is made of ferrites or wound strips of high permeability alloys like permalloy in order that it has maximum magnetizing inductance. A high voltage insulation is used between windings and ground. Fig. 14.46(*b*) shows practical rectangular input pulse to such transformers.

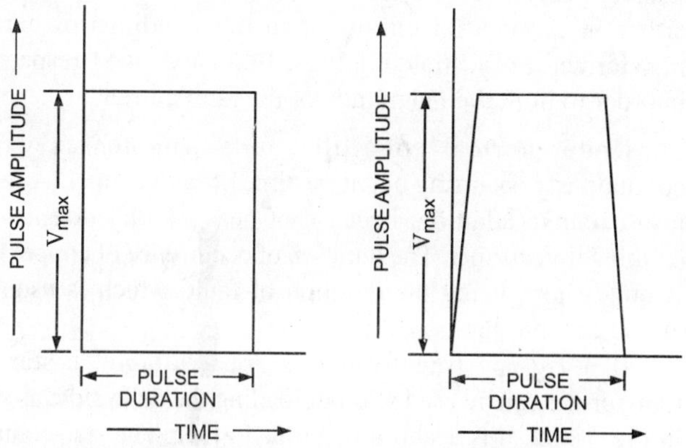

(*a*) *Ideal Rectangular Pulse Input*

(*b*) *Practical Rectangular Input Pulse*

Fig. 14.46

A *pulse transformer* is a transformer that is optimised for transmission of rectangular electrical pulses (i.e. pulses with fast rise and fall times and a relatively constant amplitude, as shown in Fig. 14.46). In power transformers, magnetic fluxes are alternating in positive and negative values, but in case of pulse transformers, this phenomenon is not applicable. The basic construction of pulse transformer differs slightly from that of power transformer. The types, construction and operating principle of a pulse transformer is discussed as follows:

14.15.1. Types of Pulse Transformers. In general *pulse transformers* are classified in two categories, namely *signal type pulse transformer* and *power type pulse transformer*. The classification of pulse transformers are done with the help of a determining factor 'Voltage Time Integral'. Each pulse has its peak voltage (that is the maximum amplitude) and a certain time period (duration of the pulse). The product of those is termed as *voltage time integral*. Large voltage time integral means the pulse peak voltages are high.

- **Signal Type Pulse Transformer:** This type of pulse transformers are relatively smaller in size and used for low power applications. Therefore, the transmitted pulses are smaller in amplitude. Suppose, a sophisticated device requires a pulse series of low volts. In such cases, signal type pulse transformers are used efficiently. This type of pulse transformers are widely used in telecommunication services, as the large part of telecommunication system deals with signal transmission and modulation. Special decoration with lights also needs the pulse transformers. *Signal type pulse transformers* are also used in control of digital logics like gate controlling, SCR turn ON process etc. As they are used in digital communication system, so they are designed to subpress the adverse effects of signal distortion.
- **Power Type Pulse Transformer:** The Voltage time integral factor is high in case of *power type pulse transformer*. Therefore the size of this type of transformer is larger than signal type. As the working voltage peaks are high, so sufficient insulation is provided with this type of transformer. Medium sized power pulse transformers are widely used in several types of flashing devices, like camera flashes etc. But, large sized power pulse transformers are used in power transmission and distribution sectors. They are also used to combine low voltage circuits and high voltage semiconductor gates. By increasing the capacity, high voltage pulse transformers are manufactured which are capable to deliver high amplitude power pulses.

14.15.2. Construction of Pulse Transformers. The general *construction of pulse transformer* is quite similar to that of the power transformers, but some special improvement or modification is provided. As they are used in sophisticated electronics applications, therefore the design of such transformers should be carried out in such a way, that they can minimise the effects of signal distortion. The pulse transformers are designed on shell type core construction. The main reason is, shell type core construction provides reduced susceptibility for eddy currents as well it can withstand with high voltage level (specially in case of power pulse transformer). Additionally, this offers sufficient mechanical support for the core. In order to suppress the effects of signal distortion, the number of turns is lesser in pulse transformer. This type of construction offers lesser leakage inductance. As those transformers are based on shell type core construction, therefore the capacitance effect in between the primary and secondary winding is reduced.

Fig. 14.47

The rising and falling process is very rapid in pulse series, thus the core is subjected to rapid magnetisation and demagnetisation. To withstand with this rapid change in magnetisation, the core (frame) is usually made with high permeability material. Instead of silicon steel, they are constructed with permalloy or ferrite materials as they offer more permeability and inductance. Sufficient insulation is provided for the core and copper windings, which ensures the electrical isolation for high voltage pulse transformers.

14.15.3. Operating Principle of Pulse Transformer. The conventional transformers deals with sinusoidal waveforms, but pulse transformers deals with square pulse waveforms. Signal processing is the vital part of such transformers. Initially, input signal waves are provided in the primary side of pulse transformer, it may be distorted sinusoidal or square pulse signal. But, this distorted signal usually randomize the flux distribution in primary coil. Then the high reluctance path (core) confines those fluxes into it and reduces the leakage fluxes. The distortion of pulses are also reduced as the core is made with high permeability material.

Then the flux linkages in secondary winding leads to develop another pulse series, which is nothing but the *dV/dt* of the input signal. Amplitude of the pulses are dependent upon the turns ratio. So, it provides a distortionless and modified pulse series as the output signal.

14.15.4. Difference Between Conventional Transformer and Pulse Transformer

- In traditional power transformers, the operating signal is sinusoidal in nature, whereas, in pulse transformers, the operating signal is a series of square or rectangular pulses.
- Unlike traditional transformers, pulse transformers operate on the transient i.e. they provide response for fast signal pulses.
- Conventional transformers are used to transfer electric power, from one circuit to another, but pulse transformers are used to transmit pulses from one circuit to another.

14.15.5. Applications of Pulse Transformers. Small sized pulse transformers, i.e. signal transformers, are used in low-voltage electronics applications like camera flasher, and they are also used in controlling of digital logic circuits.

The generated pulses from such transformers are used for gate turn on process. They are also used in telecommunication system. High voltage-pulse transformers, i.e. power pulse transformers, are used in radars, particle acceleration purposes.

Fig. 14.48

14.16 GROUNDING TRANSFORMER

The neutral or star point is usually available at every voltage level from generator or transformer neutral. However, if no such point is available due to delta connections or if neutral point is desired on bus-bars, a zig-zag transformer is most commonly used. Such a transformer has no secondary. It is a core type transformer having three limbs built up in the same fashion, as illustrated in Fig. 14.49. The connections are also shown in Fig. 14.50. It is seen that the currents in the two halves of the winding on each limb are in opposite directions. As a result, there will not be any undesirable harmonic prevailing in the circuit and so stresses on the insulation of the transformer are considerably reduced.

The fault current is quite high as the impedance of earthing transformer is quite low. The resistance or resistances are inserted either in the neutral circuit or in the windings of earthing transformer, as illustrated in Figs. 14.50 and 14.51 respectively in order to limit the magnitude of the fault current.

Under normal condition, only iron losses will be continuously occurring but at the time of fault copper losses will also occur in addition, because of heavy fault current, in the earthing transformer. The duration of occurrence of copper losses is quite short, being the duration of fault, which is usually of 30–60 second duration.

If a zig-zag transformer is not available, a star-delta transformer can be used without loading the delta side, as shown in Fig. 14.52. It is a step-down transformer. The star-connected primaries are connected to the bus-bars and its neutral is grounded. The secondaries are delta-connected and generally do not supply any load but provide a closed path for triple harmonic currents to circulate in them. Under normal operation the current in the grounding transformer is only its own exciting current. However, in the event of a single line-to-ground fault

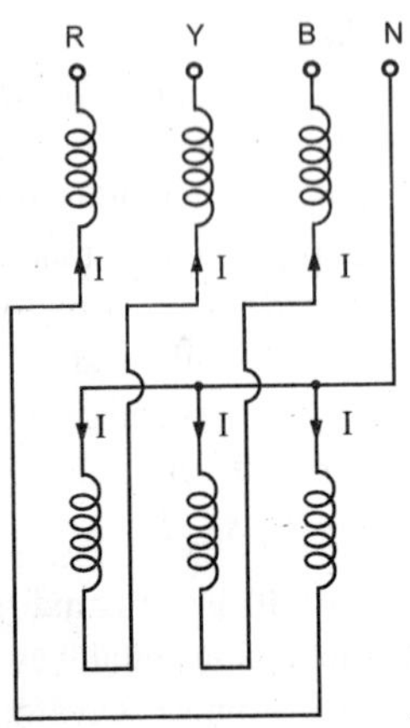

Fig. 14.49 *Zig-Zag Transformer*

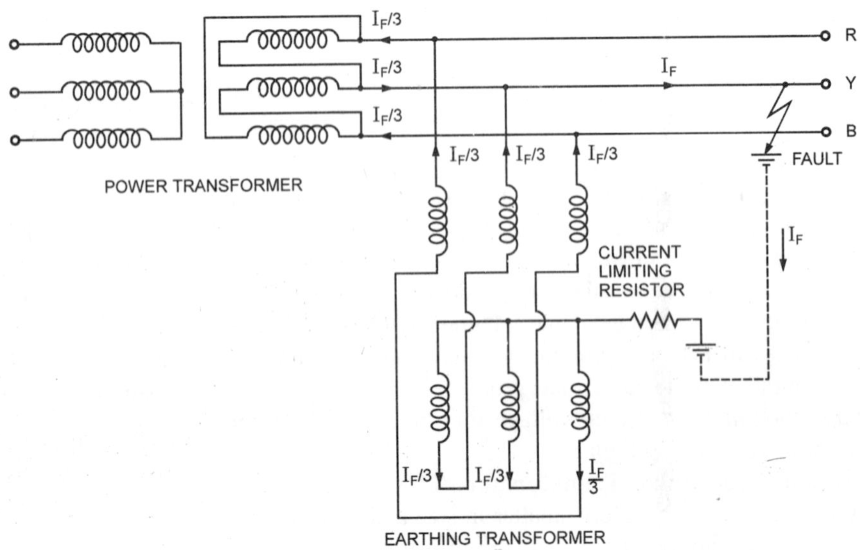

Fig. 14.50

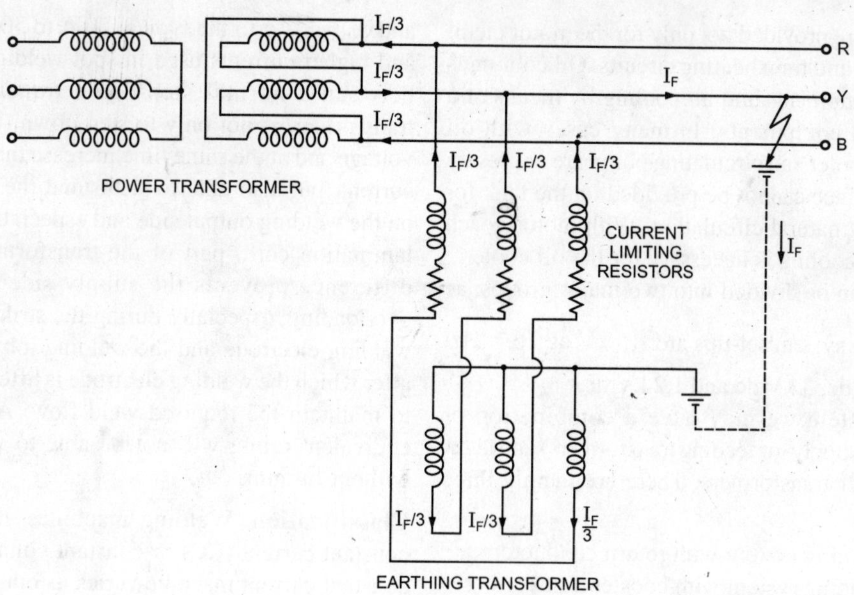

Fig. 14.51

condition, large current may flow. So it should be of sufficient rating to handle the fault current.

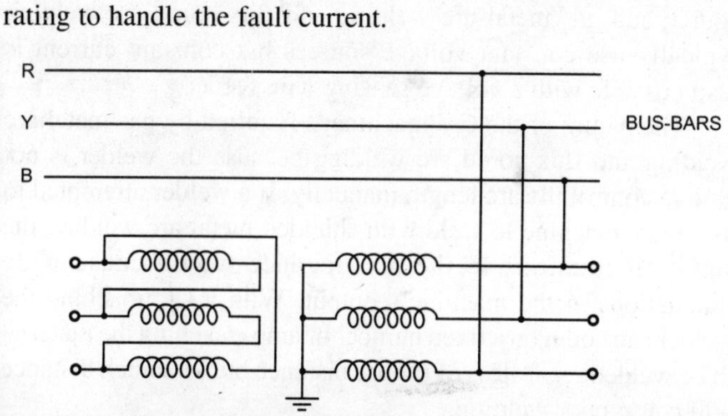

Fig. 14.52. ***Star-Delta Transformer Grounding***

As compared to the power transformer the rating of a zig-zag transformer is quite different. The *rated voltage* of an earthing or grounding transformer is the line-to-line voltage on which it is intended to be used. *Current rating* of this transformer is the maximum neutral electric current in amperes that the transformer is designed to carry in fault condition for a specific time. Generally the time interval, for which transformer designed to carry the maximum fault current through it safely, is taken as 30 or 60 seconds. A 60 second grounding transformer will be more expensive as compared to 30 second one.

14.17 TRACTION TRANSFORMERS

Traction transformers are always designed and manufactured according to a specific railway system and based on detailed parameters. The single-winding auto type transformers are usually

(*a*) (*b*)

Fig. 14.53 *Grounding Transformers*

employed and tappings are provided not only for the motor cicuit but also for the auxiliary and train heating circuits. Oil cooling is usual with European equipments and air cooling by means of a blower, with American equipments. In many cases with oil cooling an external oil cooler and circulating pump are necessary as sufficient cooling surface cannot be provided on the tank for the cooling of the oil by natural circulation. With motor-coach equipments an air blast cooling is necessary for the oil cooler.

Traction systems can be divided into two major groups, ac and dc systems. Typical system set-ups are: 15 kV ac $16\frac{2}{3}$ Hz; 25 kV ac 50 Hz; 1.5 kV dc, 3 kV dc and 1.2 kV dc.

In the traction system we may have a combination of stationary transformers (supply- or feeding transformers) and auto-transformers and booster transformers. There are mainly three different systems in use:

- Single-phase feeding system with return conductor.
- Single-phase feeding system with booster transformer.
- Two-phase feeding system with auto transformer.

Features Specifications of Traction Transformer. Railway balance traction transformer is the main transformer in the traction substation of railway electrification.

It can suit for traction electrification performance characteristics, such as short-time serious overload, 3-phase unbalanced load, frequent close to earth short-circuit, etc. It has the distinguishing features of low loss, high mechanical strength, simple construction, convenient manufacture.

Because of traction transformer's long working in no-load condition, the proportion of no-load loss to load loss is selected reasonable in order to be in transformer economic performance during designing.

Each of these systems have their advantages that should be taken into account when choosing the traction system.

14.18 WELDING TRANSFORMERS

A *welding transformer* is a transformer used in welding power supply.

It pulls relatively low current drawn from the mains power (typically limited to 15 A to avoid tripping the circuit breaker) and converts it to the typical 50 A to 500 A used in arc welding and higher currents used in spot welding. The main difference between a normal step-down transformer and a welding transformer, is not only to step down (lower) the outlet supply voltage and at the same time increase the available output circuit current, but also able to withstand the short-circuit conditions on the welding output side and especially for the magnetic (iron lamination core) part of the transformer. This magnetic path difference prevents the supply side electrical circuit from overloading, especially during the striking of the arc, when the welding electrode and the welding job touch to initiate an arc, after which the welding electrode is lifted slightly by experience, to maintain the required weld flow. A normal transformer of equivalent rating will not be able to withstand this operation without burning out.

Classification. Welding machines are usually classified as constant current (CC) or constant voltage (CV) transformers; a constant current machine varies its output voltage to maintain a steady current while a constant voltage machine will fluctuate its output current to maintain a set voltage. Shielded metal arc welding and gas tungsten arc welding will use a constant current source and gas metal arc welding and flux-cored arc welding typically use constant voltage sources but constant current is also possible with a voltage sensing wire feeder.

The nature of the CV machine is required by gas metal arc welding and flux-cored arc welding because the welder is not able to control the arc length manually. If a welder attempted to use a CV machine to weld with shielded metal arc welding the small fluctuations in the arc distance would cause wide fluctuations in the machine's output. With a CC machine the welder can count on a fixed number of amps reaching the material to be welded regardless of the arc distance but too much distance will cause poor welding.

Transformer. A *transformer*-style welding power supply converts the moderate voltage and moderate current electricity from the *utility mains* (typically 115 or 230 volts) into a high current and low-voltage supply, typically between 17 to 45 open-circuit volts and 55 to 590 amperes. A rectifier converts ac into dc on more expensive machines.

Fig. 14.54 *40 MVA 220 kV Transformer*

1000 A Welding Transformer

3-Phase Aluminium Foiled 100 kVA Welding Transformer

Fig. 14.55

For obtaining the requisite external characteristic welding transformers are provided with a relatively large reactance either in the transformer proper or in a separate reactor (Fig. 14.56). The former is called the *high reactance type welding transformer* while the latter is known as the *external reactor type welding transformer.*

14.18.1. High Reactance Type Welding Transformer. The windings, primary and secondary both, of the welding transformer should have high reactance in order that it may give steeply drooping V-I characteristic. This is attained by placing the primary and secondary windings either on separate limbs or on the same limbs but spaced some distance apart. With the increase in current, the leakage flux increases so does the emf of the self-induction. This is why an increase in primary or secondary current entails an increase in reactance voltage drop across the respective windings.

Control of current in high reactance welding transformers can be affected by three methods. One of them involves a moving primary coil. As it is moved, the spacing between the windings varies, and so does the reactance and hence the output welding current. A second method is based on the use of tapped windings either on the primary or the secondary side and the variation of the transformation ratio by bringing in or out of circuit the requisite number of turns. The third method utilises movable magnetic shunts. The position of these shunts placed in the path of the leakage flux controls the output welding current through control of reactance.

14.18. 2. The External Reactor Type Welding Transformer. This type of welding transformer consists of a normal reactance single phase step-down transformer and a separate reactor (choke). The inductive reactance and resistance of the windings in such a transformer are small, so that its secondary voltage varies but little with the welding current. The requisite drooping (negative) volt-ampere characteristic is obtained by placing the reactor in the secondary circuit. The reactor consists of a steel core and windings wound with a wire designed to carry the maximum welding current. The arc voltage decreases with the increasing current or increasing voltage drop across the reactor. This provides drooping (negative) volt-ampere characteristic.

Control of welding current can be affected by two methods either by varying the reluctance of the reactor (the moving core reactor) or by varying the number of turns of the reactor winding brought in circuit (the tapped reactor). The core of the moving core-reactor [(Fig. 14.56) (*a*)] consists of a fixed portion carrying the winding, and a moving limb which can be shifted away from and towards the fixed core by a suitable arrangement, thus varying the air gap between them. An increase in air gap adds to the reluctance of the magnetic circuit of the reactor while its self-induction and inductive reactance drops so that the welding current increases. When the air gap is reduced, the reluctance of the magnetic circuit is also reduced, the magnetic flux increases, as does the inductive reactance of the coil and the welding current drops. In this way the welding current can be controlled very accurately and continuously.

In the tapped reactor, the core is made solid, but the coil is divided into several sections, each section having a tap brought out to the regulator cover [(Fig. 14.56 (*b*)]. Moving of a contact across the taps will vary the number of turns brought in circuit and with this, the magnitude of welding current is controlled in steps.

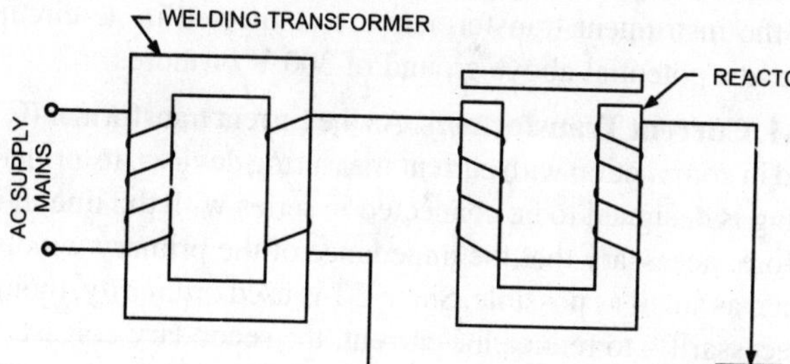

(*a*) *Welding Transformer With Moving Core Reactor*

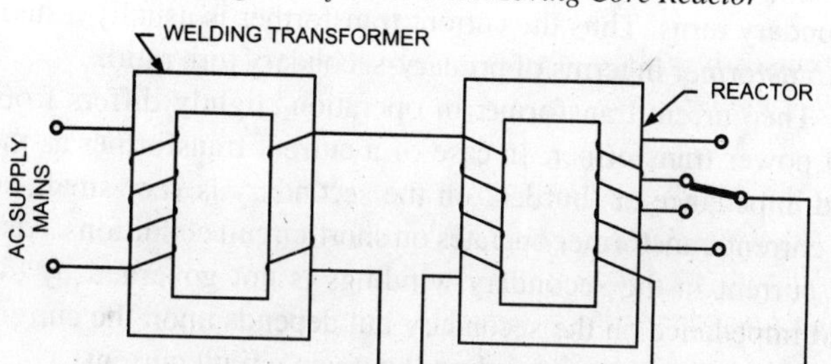

(*b*) *Welding Transformer With Moving Tapped Reactor*

Fig. 14.56

It is usual to use capacitor for correction of inherently low power factor.

The trade against being the least expensive is that pure transformer designs are often bulky and massive because they operate at the utility mains frequency of 50 or 60 Hz. Such low-frequency transformers must have a high magnetizing inductance to avoid wasteful shunt currents. The transformer may also have significant leakage inductance for short-circuit protection in the event of a welding rod becoming stuck to the workpiece. The leakage inductance may be variable so the operator can set the output current.

14.19 INSTRUMENT TRANSFORMERS

When it is necessary to measure comparatively high values of current or voltage, it is particularly desirable to use standard low range instruments in conjunction with specially constructed *accurate-ratio transformers*. The latter are called the *instrument transformers* and are of two types : *current (or series) transformers* and *potential (or parallel) transformers.* The general principles of their operation are the same as for power transformers, the main difference being that they are required to transform relatively small amounts (almost negligible) of power because their only loads, called *burdens,* constitute the delicate moving elements of instruments (ammeters, voltmeters, wattmeters, energy meters etc.) and protective relays.

Current transformer is used whenever the current of an ac circuit exceeds the safe current of the measuring instrument. Potential transformer is used wherever and whenever the voltage of an ac circuit exceeds the voltage of 750 V as it is not easy to provide adequate insulation on measuring instruments for voltage exceeding 750 V.

Note that the secondary circuit (one side of a secondary winding) is earthed so that the instrument potential never rise above that of the secondary voltage. This is general safety practice when the instrument transformers are connected in ac circuits that have a potential above ground of 300 V or more.

14.19.1. Current Transformers. As the current transformer (CT) is used in conjunction with current measuring devices, its primary winding is designed to be connected in series with the line. It is, therefore, necessary that the impedance of the primary winding be made as small as possible. Since CT is used ordinarily, though not necessarily, to reduce the current, the secondary consists of more turns than the primary, the ratio of primary to secondary current being inversely proportional to the ratio of primary to secondary turns. Thus the current transformer is usually a step-up transformer in terms of primary-secondary turn ratio.

The current transformer in operation slightly differs from that power transformer. In case of a current transformer as the load impedance or 'burden' on the secondary is very small, so the current transformer operates on short-circuit conditions. Also the current in the secondary windings is not governed by the load impedance on the secondary but depends upon the current flowing in the primary winding (i.e. main circuit current).

The amount of power which a CT handles is small. The product of voltage and current on the secondary side when the CT is supplying the instrument with its maximum rated value of current, is known as *rated burden* and is measured in volt-amperes. The usual rating of a CT is very small (e.g. 15 VA and in no case exceeds 30 VA).

A current transformer comprises a laminated iron (usually high permeability steel) core carrying a large number of turns of fine insulated wire for the secondary. The secondary is usually designed to carry a current of 5 A. Over the secondary winding are wound two or three turns of thick insulated wire or strip of copper. The cross section area of this wire depends upon the maximum current it has to carry, which is the total current to be measured in the main circuit.

An ideal transformer, that is, one having an overall efficiency of 100 per cent would provide in secondary an exact fraction of the main circuit current. For 100 per cent efficiency, i.e. for constant ratio of transformation it is necessary that the impedance of the magnetizing path is infinite and that the impedance is zero. In practice, this is not possible and allowance has to be made for this inaccuracy.

In general, a current transformer has a high overall operating efficiency when:

(*i*) A high grade core material is used (the better the grade of core material the lower the loss).

(*ii*) The flux density in the core is low and there are a large number of ampere-turns in the windings. It is obvious that when the primary consists only of one or two turns, high efficiency is not possible. On the other hand, it is not generally practicable to have more than three turns owing to size of large copper strip relative to the fine secondary winding and the small yoke. Efficiency is, therefore, to some extent sacrificed for compactness.

(*iii*) The load on the secondary is quite low. When the current drawn by the instrument is small, the secondary can consist of a large number of turns and a high accuracy is achieved.

In CTs, the current transformation ratio is not exactly equal to the ratio of secondary to primary turns, mainly because of the magnetizing current. The percentage ratio error in modern CTs seldom exceeds 0.5 per cent. Ratio error may, however, be compensated for making the actual ratio of turns slightly different from the nominal current ratio : for example, a CT having nominal current ratio of 100 to 5 has an actual turn ratio of 4.98 to 100.

When the primary winding has a large current rating it may consist of a straight conductor which passes through the centre of a hollow metal core. The secondary winding is wound on the

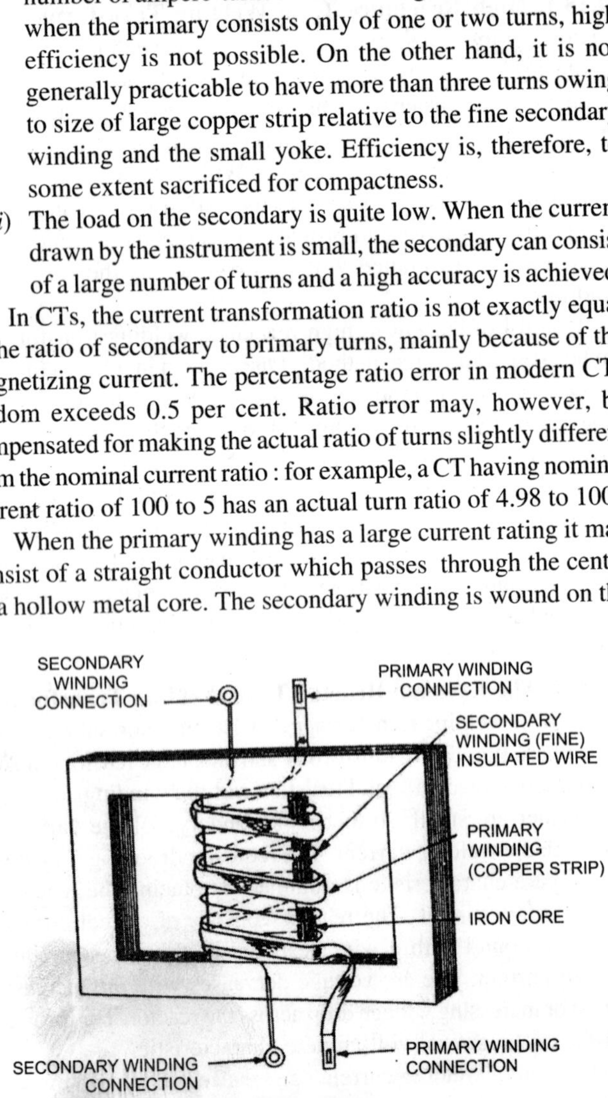

Fig. 14.57

core. This is called a *bar-type* CT, so named from the construction of the primary which consists of a copper busbar. All standard CTs of 100 A or larger ratings are of this type.

Some CTs of rating 100 A or more do not have any primary. The secondary is wound over a ring shaped core which encircles the cable comprising part of the circuit of distribution feeder. The cable itself then serves as a primary winding having single turn.

It should be kept is mind that the voltage across the primary of a CT has nothing to do with the voltage rating of the circuit in which it is placed. The voltage between the winding and the earth, however, will be the phase-to-earth voltage of the system and may be very high. Since the framework and core of the CT are connected to earth, they must be adequately insulated from the conductors. Insulation must be of a high order when the voltage is more than 700 V.

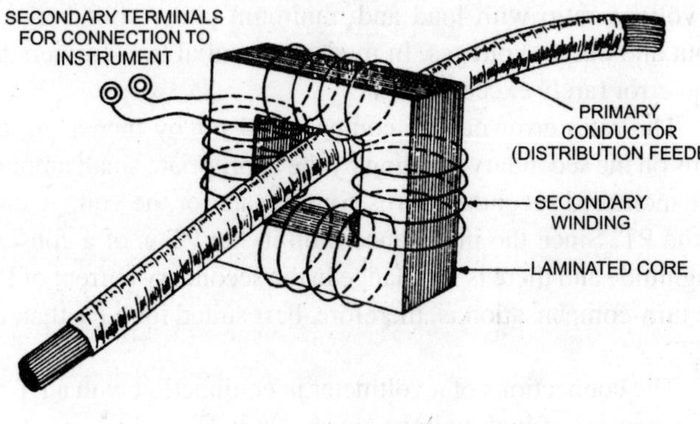

Fig. 14.58 *Current Transformer Without a Wound Primary*

The current transformers are air cooled, but when they are for use in high voltage lines, it is necessary to provide oil cooling. For systems above 12 kV, special transformers are available. The principle of construction is similar to that of the lower voltage types, except that extra percautions are taken against the leakage or flash-over.

An exceedingly important precaution that is to be taken while using a CT is to maintain at all times a closed circuit on the secondary side. If the instrument is to be taken out of the circuit, then the secondary terminals must first be short circuited by a low resistance not exceeding 0.5 Ω. The reason for the precaution is given below :

Under normal operating conditions the secondary ampere-turns of a CT almost balances the primary ampere-turns and the flux set up is only due to small number of ampere-turns I_0T_P. If by chance the secondary is open circuited and there is no change in the primary current, then primary ampere-turns remain the same and opposing ampere-turns set up by the secondary become zero, consequently the entire primary ampere-turns T_PI_P are used in creating the flux. This causes an extraordinary large flux and flux density in the iron core. The large flux leads to dangerously high emfs for both the operator and the transformer insulation. The high flux density causes an increase of iron loss in the core.

If the secondary winding of a CT is accidentally opened while in use and the winding is not damaged or direct current is passed through the winding then the CT must be demagnetized before being used again. The CT can be demagnetized by connecting a high resistance across the secondary terminals and full rated primary current flowing, reducing the resistance gradually to zero. CTs are sometimes fitted with a short-circuiting switch across the secondary terminals, in order that the secondary circuit may be closed while instruments are connected or disconnected.

The connections of an ammeter in conjunction with a CT for measurement of large current are shown in Fig. 14.59.

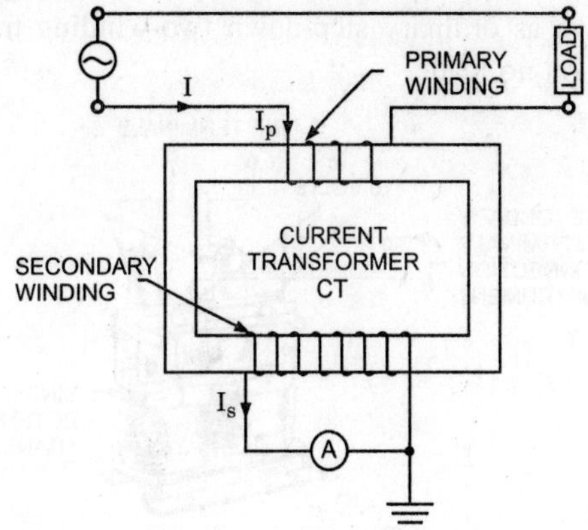

Fig. 14.59 *Connections of CT*

14.19.2. Potential Transformers. The potential transformers are used for measurement of high voltages by means of low-range voltmeters or for energizing the potential coils of wattmeters and energy meters. These are also used for energizing relays and other protective devices.

The potential transformer (PT) operates on the same principle as a power transformer. Both primary and secondary windings are wound on a high grade steel—low-voltage winding is kept next to the earth core and the high voltage winding is on the outside. The potential transformers are always step-down transformers, since they are meant for reducing the voltage to a reasonable operating value.

As a rule, the potential transformers are of the shell type because such a construction develops a high degree of accuracy.

The input voltage may be as high as 138,000 V or higher even, and in this case the pressure exists between the primary terminals as they are connected across the supply system and not in series with it as in the case of a CT. The secondary circuit is extremely well insulated from the high voltage primary and that, in addition, it is grounded from the point of view of safety. Such grounding serves to protect the operator from the high voltage hazard.

Up to voltage of 5,000 V, PTs are usually of dry type, between 5,000 V and 138,000 V, they may be either dry type or oil immersed type, although for voltages exceeding 138,000 V they are always oil immersed type.

The output voltage of potential transformers has been standardized at 110 V irrespective of the high input voltage. This comparatively low voltage has been chosen for reasons of safety and reduces the necessary insulation of the windings and terminals of the instrument. Further, as the voltage is standardized, the design of the voltage coil portion of the instrument may also be standardized.

In a step-down power transformer, the primary or high voltage winding usually has a high voltage but of small current, while the secondary has heavy current at low voltage. The potential transformers, however, differ in that currents in both windings are comparatively small. This is because the load on a potential transformer is an instrument having a high impedance (mostly resistance). Since only a very small current is required to operate the potential coil of the instruments, so it is not necessary to design potential transformers for heavy currents. In other words, PTs behave as ordinary step-down two-winding transformers operating on no load.

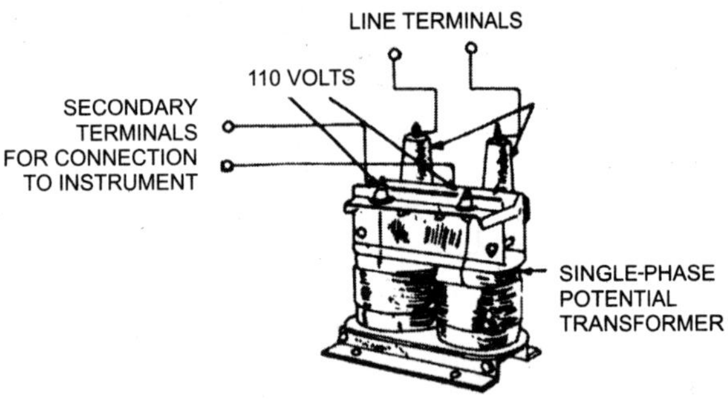

Fig. 14.60 *Potential Transformer*

The very high input voltages necessitate a special arrangement of coils. For example, assuming that the high voltage was connected to 33,000 V supply, and the winding consisted of two layers of turns, the first layer running from the left to the right and the second layer from right to left, the voltage between first and the last turn would be 33,000 V. This pressure between adjacent turns would need very high insulation. For reducing the insulation and the risk of breakdown, high voltage windings are divided into sections as illustrated in Fig. 14.61 (b). The voltage between two turns where there are two layers of turns, cannot exceed one-eighth of 33,000 V, as there are eight such coils. The greater the number of layers per section, the lower the voltage between adjacent turns. Heavy insulation in the form of oil ducts is provided between the coils, between sections, and between layers.

Both windings are protected from overload by fuses. The fuses for low-voltage winding are usually of the cartridge type and are mounted either on the top of the chamber or in the terminal box, so they are easily accessible. Owing to the very small current in the high voltage winding, little reliance could be placed on the use of ordinary fuse and fuse of special types are used. These fuses are spring loaded and the fuse element consists of a short piece of silver wire. When a fuse operates, the spring is released and prevents the setting of an arc. The fuses are located in hv bushings and are easily removed for replacement.

The small currents in the secondary permit the use of very fine wire and many turns. According to basic emf a large number of turns permit the use of a low flux density in the core and also a comparatively small cross section of the iron. Thus a potential transformer is much smaller and lighter than the power transformer with the same voltage rating. The rating of a potential transformer rarely exceeds 500 VA.

No doubt there are ratio and phase angle errors in potential transformers owing to magnetizing current and impedance drops, but careful design (high grade steel operated at very low flux density not exceeding 0.6 to 0.8 T) ensures minimum variation of voltage ratio with load and minimum phase shift between input and output voltages. In modern potential transformers the ratio error rarely exceeds 0.5%.

The ratio error can be compensated for by increasing the turns on the secondary winding by the appropriate small amount. The increased secondary turns compensate for the voltage drop in the PT. Since the impedance drop in the PT is of a constant magnitude and there is no change in the secondary current of PT, the turn-compensation is, therefore, best suited for a PT than for a CT.

The connections of a voltmeter in conjunction with a PT for measurement of high voltage are shown in Fig. 14.62.

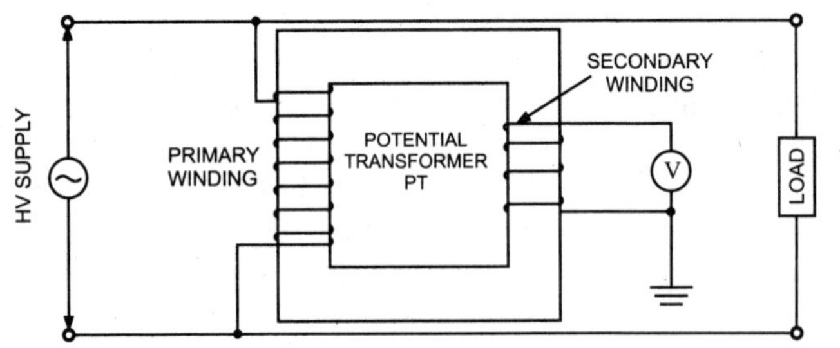

Fig. 14.62 *Connection Diagram of a PT*

Example 14.22. Power of a single-phase 6.6 kV load drawing a current of about 50 A is required to be measured by means of a wattmeter having volt terminals marked as 110 V and current

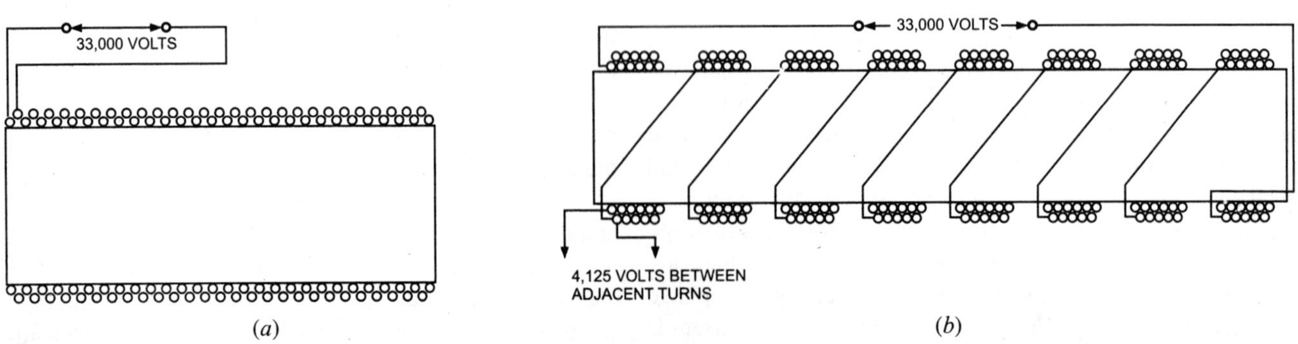

Fig. 14.61

terminals as 5 A. Draw the circuit diagram showing the connection of the wattmeter using the instrument transformers. Write suitable transformation ratio against each instrument transformer shown in the diagram.

Solution: Primary voltage, $V_1 = 6.6\text{ kV} = 6{,}600$ volts

Secondary voltage, $V_2 = 110$ volts

Hence transformation ratio of PT $= \dfrac{V_1}{V_2} = \dfrac{6{,}600}{110} = 60$

Primary current, $I_1 = 50$ A

Secondary current, $I_2 = 5$ A

Transformation ratio of CT $= \dfrac{I_2}{I_1} = \dfrac{50}{5} = 10$ **Ans.**

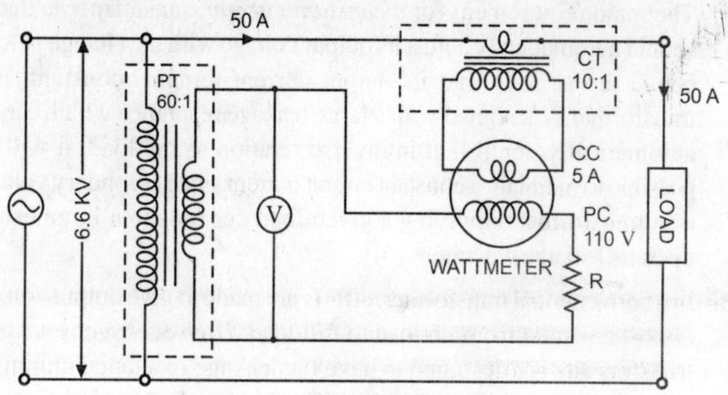

Fig. 14.63

HIGHLIGHTS

1. The operating principle and general construction of an auto-transformer is the same as that of conventional two winding transformer. The auto-transformer differs from a conventional two winding transformer in a way in which the primary and secondary are interrelated. In a conventional two-winding transformer, the primary and secondary windings are completely insulated from each other but are magnetically linked by a common core. In the auto-transformer, the two windings, primary and secondary, are connected electrically as well as magnetically, in fact, a part of the single continuous winding is common to primary and secondary. The single winding is wound on a laminated silicon steel core and, therefore, both primary and secondary sections of this one winding are on the same magnetic circuit.

   In an auto-transformer:

   Ratio of power transformed to total power delivered = (1 – K)

   Ratio of power conducted directly to power output = K where K is transformation ratio

   $$\frac{\text{Volt-ampere rating as an auto-transformer}}{\text{Volt-ampere rating as two winding transformer}} = \frac{1}{1-K} \quad ...(i)$$

   $$\frac{\text{Weight of conductor in an auto-transformer}}{\text{Weight of conductor in a two winding transformer}} = 1 - K \quad ...(ii)$$

   Hence saving in conductor material required for winding affected by using an auto-transformer

   = K × weight of conductor in a two winding transformer ...(*iii*)

2. In some cases it is not convenient to have tap changing gear in the main transformer, and sometimes it may be required to control the voltage of a feeder at a point far away from the main transformer. Booster transformer may be employed on such occasions.

   The advantages of the booster transformers are that the regulating equipment is independent of the main transformer so that a failure in the former will not throw the latter out of service for any length of time; and that it is much cheaper method when there is no main transformer at the point where regulation is desired. When the regulation is required at a point where a main transformer is to be placed, this system is costlier, requires more floor space, and increases the losses.

3. The induction voltage regulator is essentially a step-down transformer, but is so constructed that the secondary voltage may be varied from zero to a certain maximum value, dependent upon the ratio of turns in the two windings, merely by altering the position of the primary-coil axis with respect to the secondary-coil axis. The primary winding is connected across the circuit to be regulated and the secondary is connected in series with the circuit. An induction voltage regulator may be of single phase or three phase type and comprises of a stator and a rotor.

   The 3-phase induction regulators may be also termed *phase-shifting transformer* and, therefore, can be used for ac meter testing and similar other applications. These are also used to advance the phase of the grid voltages in grid-controlled mercury arc rectifiers.

4. The modern electrical equipments are normally designed to operate satisfactorily at a fixed voltage level. On the other hand voltage variation in power system is a normal phenomenon owing to the rapid growth in industries and distribution network. It is, therefore, very important to keep the consumer's terminal voltage within the statutory limits. The transformer output voltage and, so, the consumer's terminal voltage may be controlled by providing tappings on the transformer windings (either on the primary or on the secondary).

   Tappings are usually provided on the hv winding.

   The tappings may be changed with the transformer isolated from the supply on both sides. This is called *off-load tap changing.* This arrangement is used for occasional adjustments, as in distribution transformers which are provided with ± 5% and ± 2% taps. In some transformers special arrangements are made to avoid interruption of supply so that tap-changing is affected with the transformer energized or on load. This is called the *on-load tap changing.* Daily and short-time adjustments are made with the help of *on-load tap-changing* gear.

5. A phase-shifting transformer (Fig. 14.35) is a special type of system intertie transformers which control the flow of power through specific lines in a complex power transmission network by providing the possibility to inject a voltage with an arbitrary phase angle in a power system.

   Phase-shifting transformers are mainly used in hv systems (up to 420 kV) and the throughout power rating is quite large (up to 1,630 MVA).

6. Isolation transformers are two-winding transformers with centre-tap on the primary winding grounded via a capacitor and secondary winding may be grounded solidly to the ground if required. Also, a thick copper sheet is wound between the two windings (primary and secondary) but it does not form a closed turn. This shield is solidly grounded and does not permit the shifting of secondary winding potential through capacitive coupling. Metal oxide varistors are also employed at the input terminals so as to absorb the surge energy.

7. There are services in which constant current is desired such as in street and airport runway lighting.

The main requirement for *a constant current* transformer is that it should automatically adjust its output voltage with the change in load connected to it so that its output current remains constant. If a transformer is designed with a large leakage reactance which can be automatically controlled in inverse relation to the load, it will be possible to maintain a constant output current since secondary voltage of a transformer supplying a given load depends to a large extent upon its leakage reactance.

8. In a conventional transformer, efforts are made to give output voltage almost constant from no load to full load. *High-leakage reactance* transformers are designed to have the leakage reactance minimum possible. In some special applications (such as furnace ignition and gaseous discharge lamps) it is desirable to make use of transformers that give very high no-load voltages but small short-circuit currents (i.e. the transformers giving a drooping output voltage characteristic). Such a drooping-output voltage characteristic can be had by designing the transformer with high leakage reactance. Gaseous discharge lamps require at start a voltage considerably higher than that required for normal operation so as to breakdown the gas and cause it to conduct. The high leakage-reactance transformers can be designed to take care automatically of these voltage requirements. High leakage transformers can be designed to give open-circuit voltage ranging between 20 kV and 30 kV and the short-circuit current of the order of 1 or 2 mA.

9. Small iron-core transformers employed in the AF range of 20 Hz to 20 kHz are called *audio-frequency* or simply *audio transformers*. Audio transformers are the transformers specifically designed for use in AF circuits to carry audio signals. They can be employed to block radio frequency interference or the dc component of an audio signal, to split or combine audio signals, or to provide impedance matching between high and low impedance circuits, such as between a high impedance amplifier output and a low impedance loudspeaker, or between a high impedance instrument output and the low impedance input of a mixing console. AF transformers that operate with loudspeaker voltages and currents are larger than those which operate at microphone or line level, carrying much less power.

10. Beyond the normal power frequency range at frequencies exceeding 50-60 Hz, the thickness of the laminations have to be reduced for controlling the iron losses to an acceptable levels. In case the operating frequency exceeds beyond a few kHz, it becomes necessary to use iron-dust cores or the ferrites, special oxides which exhibit reasonable good magnetic properties with very small eddy current loss. There are several types of transformers, such as air-core transformers, ferrite-core transformer and transmission line transformer, used in RF work.

11. Variable frequency transformers provide transmission solutions for a smarter grid, enabling transmission system operators to control power flows between power grids with high reliability, speed and efficiency, while offering flexibility in how utilities provide power to meet customers' demand.

    A variable frequency transformer (VFT) a doubly fed electric machine resembling a vertical shaft hydroelectric generator with a three-phase wound rotor, connected by slip rings to one external power circuit. The stator is connected to the other. With no applied torque, the shaft rotates due to the difference in frequency between the networks connected to the rotor and stator. A direct-current torque motor is mounted on the same shaft; changing the direction of torque applied to the shaft changes the direction of power flow. VFT behaves as a continuously adjustable phase-shifting transformer.

12. A *pulse transformer* is a transformer that is optimised for transmission of rectangular electrical pulses (i.e. pulses with fast rise and fall times and a relatively constant amplitude, as shown in Fig. 14.46). In power transformers, magnetic fluxes are alternating in positive and negative values, but in case of pulse transformers, this phenomenon is not applicable. The basic construction of pulse transformer differs slightly from that of power transformer.

    The general *construction of pulse transformer* is quite similar to that of the power transformers, but some special improvement or modification is provided. As they are used in sophisticated electronics applications, therefore the design of such transformers should be carried out in such a way, that they can minimise the effects of signal distortion. The pulse transformers are designed on shell type core construction.

    The rising and falling process is very rapid in pulse series, thus the core is subjected to rapid magnetisation and demagnetisation. To withstand with this rapid change in magnetisation, the core (frame) is usually made with high permeability material. Instead of silicon steel, they are constructed with permalloy or ferrite materials as they offer more permeability and inductance. Sufficient insulation is provided for the core and copper windings, which ensures the electrical isolation for high voltage pulse transformers.

13. Traction transformers are always designed and manufactured according to a specific railway system and based on detailed parameters. The single-winding auto type transformers are usually employed and tappings are provided not only for the motor cicuit but also for the auxiliary and train heating circuits. Oil cooling is usual with European equipments and air cooling by means of a blower, with American equipments. In many cases with oil cooling an external oil cooler and circulating pump are necessary as sufficient cooling surface cannot be provided on the tank for the cooling of the oil by natural circulation. With motor-coach equipments an air blast cooling is necessary for the oil cooler.

14. A welding transformer, the same as a dc welding machine, must have a no-load voltage of about 60-70 volts sufficient to fire the arc and also a sharply drooping external characteristic. For obtaining the requisite external characteristic welding transformers are provided with a relatively large reactance either in the transformer proper or in a separate reactor (Fig. 14.56). The former is called the *high reactance type welding transformer* while the latter is known as the *external reactor type welding transformer*.

15. When it is necessary to measure comparatively high values of current or voltage, it is particularly desirable to use standard low-range instruments in conjunction with specially constructed *accurate-ratio transformers*. The latter are called the *instrument transformers* and are of two types : *current (or series) transformers* and *potential (or parallel) transformers*. The general principles of their operation are the same as for power transformers, the main difference being that they are required to transform relatively small amounts (almost negligible) of power because their only loads, called *burdens*, constitute the delicate moving elements of instruments (ammeters, voltmeters, wattmeters, energy meters etc.) and protective relays.

    Current transformer is used whenever the current of an ac circuit exceeds the safe current of the measuring instrument. Potential transformer is used wherever and whenever the voltage of an ac circuit exceeds the voltage of 750 V as it is not easy to provide adequate insulation on measuring instruments for voltage exceeding 750 V.

EXERCISES

1. Define an auto-transformer. Distinguish clearly the difference between a resistive potential divider and auto-transformer. **[U.P. Technical Univ. Electrical Machines, 2010-11]**
2. What is an auto-transformer? State its merits and demerits over the two winding transformer. Give the constructional features and explain the working principle of a 1-phase auto-transformer. **[M.D. Univ. Electrical Machines-I, December-2010]**
3. Show that in case of an auto-transformer
$$\frac{\text{Inductively transferred power}}{\text{Total power}} = \frac{\text{High voltage} - \text{low voltage}}{\text{High voltage}}$$
[U.P. Technical Univ. Electromechanical Energy Conversion-I, 2007-08]
4. Derive the expression of saving of copper in auto-transformer as compared to an equivalent two winding transformer. **[R.G. Technical Univ. Electrical Machine-I, June-2014]**
5. What is an auto-transformer? State its merits and demerits over the two winding transformer. **[M.D. Univ. Electrical Machines-I, December-2013]**
6. State the applications of auto-transformers. What are the reasons of higher efficiency of auto-transformers as compared to conventional transformers. **[U.P. Technical Univ. Electrical Machines, 2009-10]**
7. Give the construction, working and applications of variac transformers.
8. What is a booster transformer and what special precautions are required to be taken when using a booster transformer.
9. Explain the operation of a single-phase induction regulator. Why is a compensating winding is used in a single-phase induction regulator and why is it not used in 3-phase? **[W.B. Univ. of Technology Electrical Machine, 2011]**
10. What is the essential difference between the principles of operation of single-phase and three-phase induction regulators? Explain the operation and construction of a single-phase regulator. **[Pb. Univ. Elec. Machines II, December-1991]**
11. Explain the working of a single-phase induction regulator. How does the compensating tertiary winding help is balancing ampere-turns?
Find the rating of a 3-phase induction regulator in terms of applied voltage and load kW to give ± 10% variation in output voltage. **[Pb. Univ. Elec. Machines-II, May-1994]**
12. Discuss the working of a 3-phase induction regulator with the help of phasor diagram. Explain the advantage of twin regulator. **[Pb. Univ. Elec. Machines-II, Dec.-1995]**
13. Explain the principle of a 3-phase induction regulator. Under what conditions there is no phase displacement between incoming and outgoing voltages? **[Pb. Univ. Elec. Machines-II, Dec.-1993]**
14. What do you understand by tap-changing transformers? **[R.G. Technical Univ. Electrical Machine-I, June-2014]**
15. Explain with diagram off-line tap changing transformer. **[M.D. Univ. Electrical Machines-I, December-2010]**
16. Why are tap-changing transformers required? Explain the operation of no-load tap-changing transformer. **[J.N. Technological Univ. Electrical Machines-II, January-2014]**
17. Explain the function and principle of operation of no-load and on-load tap-changing transformers.
18. What is isolation transformer and for what it is used?
19. Explain construction and principle of operation of a constant current transformer. Give its typical characteristics and applications.
20. How does an high-leakage reactance transformer differ from a conventional two-winding transformer, explain. Give its output characteristic. Where does it find application?
21. What is an RF transformer? Give various types of RF transformers with their applications.
22. Give the construction, operation and applications of variable frequency transformers.
23. What is pulse transformer and where is it used? Give the construction and operating principle of pulse transformer.
24. What is an earthing transformer. How does it differ from conventional power transformer and where is it used?
25. Give features specifications of traction transformer.
26. Explain construction of welding transformer. How does it differ from power transformer?
27. Describe the working principle of current transformer and potential transformer. **[U.P. Technical Univ. Electrical Machines, 2005-06]**
28. Explain the constructional difference between a current transformer and potential transformer. **[U.P. Technical Univ. Electrical Machines, 2005-06]**
29. What is need of potential transformers? How it works? **[M.D. Univ. Electromechanical Energy Conversion, May-2009]**
30. Explain construction of welding transformer. How does it differ from power transformer?
31. Explain the following in brief and write the field of applications.
    (*i*) Tap-changers
    (*ii*) Pulse and high-frequency transformers. **[R.G. Technical Univ. Electrical Machine-I, June-2013]**
32. Write short notes on the following:
    (*i*) Auto-transformer. **[R.G. Technical Univ., June-2013]**
    (*ii*) Advantages of auto-transformer over two-winding transformer. **[W.B. Univ. of Technology, 2010]**
    (*iii*) Variac transformer.
    (*iv*) Booster transformer.
    (*v*) Single-phase and three-phase induction regulators. **[W.B. Univ. of Technology, 2010]**
    (*vi*) Tap-changing transformer. **[U.P. Technical Univ., 2002-03]**
    (*vii*) On-load tap-changer. **[W.B. Univ. of Technology, 2009-10]**
    (*viii*) Phase-shifting transformer. **[M.D. Univ., December-2010]**
    (*ix*) Low voltage transformers.
    (*x*) Audio transformer.
    (*xi*) RF transformer.
    (*xii*) Grounding transformer.
    (*xiii*) Traction transformer.
    (*xiv*) Pulse transformer. **[U.P. Technical Univ. 2002-03; 2003-04]**
    (*xv*) Welding transformer. **[U.P. Technical Univ., 2002-03]**
    (*xvi*) Current transformer. **[M.D. Univ., December-2012]**
    (*xvii*) Potential transformer. **[M.D. Univ., December-2012]**

SHORT ANSWER TYPE QUESTIONS WITH ANSWERS

Q. 1. How does auto-transformer differ from potential divider?

Ans. Although diagrammatically the auto-transformer has the appearance of a resistance type potential divider, its operation is quite different, as described below:

A resistive potential divider cannot step up voltage, whereas an auto-transformer can do. In a potential divider the input current is always more than its output current but it is not so in an auto-transformer. In a step-down auto-transformer, the output (load) current is more than the input current. In a potential divider, almost entire power to the load flows by conduction, whereas in auto-transformer, a part of the power is conducted and the rest is transferred to the load by transformer action. Thus in a potential divider, there is a great loss of power in the different resistances whereas it is not so in case of an auto-transformer. Voltage regulation is also poor in case of a potential divider.

Q. 2. What is relation between rating as an auto-transformer and an equivalent two-winding transformer?

Ans. The relation between rating of an auto-transformer and an equivalent two-winding transformer is given below:

$$\frac{\text{VA rating as an auto-transformer}}{\text{VA rating as two winding transformer}} = \frac{1}{1-K}$$

Q. 3. Give the saving in conductor material required for winding affected by using as an auto-transformer.

Ans. Saving in conductor material required for winding effects by using an auto-transformer is equal to the K times the weight of conductor in a two-winding transformer.

Q. 4. Why auto-transformers have higher efficiency than those of the two-winding transformer of the same output.

Ans. Owing to the reduction in conductor and core materials, the ohmic losses in conductor and the core loss are lowered. Thus an auto-transformer has lower losses and, therefore, higher efficiency than those of the two-winding transformer of the same output.

Q. 5. Where may auto-transformers be used?

Ans. Auto-transformers are used when transformation ratio is nearly equal to unity and where there is no objection to direct electrical connection between primary and secondary.

Q. 6. A 2-winding transformer is used to step down the voltage from 220 V to 110 V. If an auto-transformer is used for the same purpose, what would be the ratio of copper weights in them?

Ans. 0.5.

Q. 7. What is booster transformer? **[M.D. Univ., December-2013]**

Ans. Booster transformer is one which is often used toward the end of a power line in order to raise the voltage to the desired value. The secondary of the booster transformer is connected in series with the line and booster transformer primary is supplied from the secondary of the regulating transformer with on load tap changing gear.

Q. 8. May an ordinary transformer be used as a booster transformer ?

Ans. Yes.

Q. 9. When connecting an ordinary transformer as a booster transformer, what important factors must be considered ?

Ans. The hv side of the transformer must be able to withstand the voltage of the line in which it is to be connected, the lv side must have a voltage of approximately the value by which the line voltage is to be boosted up, and the low-voltage side must also have a current carrying capacity that is sufficient to carry the line current.

Q. 10. What special precaution must be taken when using a booster transformer?

Ans. There must be no fuse on the hv side, or primary; since the booster transformer is similar to a current transformer, if the fuse should blow, an extremely high voltage could be built up on the secondary side.

Q. 11. What is an induction regulator?

Ans. Induction regulator is similar to a booster transformer. It has a primary and secondary windings which are wound on separate cores. The primary can be moved in either direction; this is usually done by an electric motor. The primary bucks or boosts the line voltage according to the position of primary with respect to secondary.

Q. 12. Why is a compensating winding used in a single-phase induction regulator and why is not used in 3-phase?

[W.B. Univ. of Technology, 2011]

Ans. The short-circuited compensating winding reduces the secondary leakage reactance by air-gap flux in the axis perpendicular to the primary winding and improves the operation and voltage regulation. The compensating winding tends to keep the copper losses constant for a given load current and various degrees of boost or buck.

No compensating winding is required in case of a 3-phase induction regulator because, in this case, for every position of setting of the regulator each secondary winding is magnetically coupled to one or more of the primary windings.

Q. 13. What is a phase-shifting transformer?

Ans. A phase-shifting transformer is a special type of system intertie transformers which control the flow of power through specific lines in a complex power transmission network by providing the possibility of injecting voltage with an arbitrary phase angle in a power system.

Q. 14. Give important points regarding phase-shifting transformer.

Ans.
1. The pf of load has a considerable effect in determination of the shift angle.
2. The phase-shifting transformer impedance will add to the impedance of the system and may require a large angle displacement.
3. The phase-shifting transformer has an impact on the exchange of both active and reactive power between systems.
4. The phase-shifting transformer must be able to charge the displacement angle in appropriate steps within a certain range.

Q. 15. What is an isolating transformer?

Ans. Isolation transformers are two-winding transformers with centre-tap on the primary winding grounded via a capacitor and secondary winding may be grounded solidly to the ground if required. Also, a thick copper sheet is wound between the two windings (primary and secondary) but it does not form a closed turn. This shield is solidly grounded and does not permit the shifting of secondary winding potential through capacitive coupling. Metal oxide varistors are also employed at the input terminals so as to absorb the surge energy.

Q. 16. How does pulse transformer differ from power transformer in construction?

Ans. The general *construction of pulse transformer* is quite similar to that of the power transformers, but some special improvement or modification is provided. As they are used in sophisticated electronics applications, therefore the design of such transformers should be carried out in such a way, that they can minimise the effects of signal distortion. The pulse transformers are designed on shell type core construction. The main reason is, shell type core construction provides reduced susceptibility for eddy current as well it can withstand with high voltage level (specially in case of power pulse transformer). Additionally, this offers sufficient mechanical support for the core. In order to suppress the effects of signal distortion, the number of turns is lesser in pulse transformer. This type of construction offers lesser leakage inductance. As those transformers are based on shell type core construction, therefore the capacitance effect in between the primary and secondary winding is reduced.

Q. 17. Give the applications of pulse transformer.

Ans. Small sized pulse transformers, i.e. signal transformers, are used in low-voltage electronics applications like camera flasher, and they are also used in controlling of digital logic circuits.

The generated pulses from such transformers are used for gate turn on process. They are also used in telecommunication system. High voltage-pulse transformers, i.e. power pulse transformers, are used in radars, particle acceleration purposes.

Q. 18. What is an earthing transformer? [U.P.T.U. 2006-07]

or

What is the use of earthing transformer ? [U.P.T.U. 2007-08]

or

What is the role of earthing transformer in neutral grounding ?

Ans. In cases where the neutral points of a three-phase system are not accessible or where the transformers or generators are delta connected, an artificial neutral point is required to be created. This is accomplished by means of a zig-zag transformer, called the earthing transformer.

As compared to the power transformer the rating of a zig-zag transformer is quite different. A grounding transformer is usually specified by the single-phase fault current that it handles. The two most common intervals specified for the fault current are 30 and 60 s. A 60 s grounding transformer will be more expensive as compared to a 30 one.

Q. 19. Why an additional central limb with an air gap is sometimes provided in welding transformers?

Ans. To increase the value of leakage flux so that its leakage reactance increases.

Q. 20. How are CT*s* rated?

Ans. CT*s* are rated as 100/5 or 50/5 A. The first number is the total current that the CT is supposed to handle, and the second number is the current on the secondary when the full-load current is flowing through the primary.

Q. 21. What is general safety practice in use of instrument transformers?

Ans. The secondary circuit (one side of secondary winding) should be earthed so that the potential of the instrument never rise above that of the secondary voltage.

Q. 22. What is meant by rated burden of an instrument transformer?

Ans. The amount of power in volt-amperes which an instrument transformer handles at maximum loading is called the rated burden of the instrument transformer.

Q. 23. What is the usual secondary voltage rating of a PT?

Ans. 110 V.

Q. 24. How ratio error is compensated in case of a PT?

Ans. Ratio error in case of a PT can be compensated for by increasing the turns on the secondary by appropriate small amount.

PROBLEMS

1. A single-phase auto-transformer is to deliver a power of 200 kW at unity pf at 240 V with an input voltage of 300 V. Determine the kVA rating of each section of the auto-transformer winding.

   [B.P. Univ. of Technology Electrical Machine-I, 2008]

   [**Ans.** 40 kVA]

2. A single-phase auto-transformer is to deliver a power of 350 kW at unity pf at 350 V with an input voltage of 230 V. Determine the kVA rating of each section of the auto-transformer winding.

   [B.P. Univ. of Technology Electrical Machines, 2008]

   [**Ans.** 120 kVA]

3. A 10,000/2,500 V transformer is rated at 100 kVA as a two winding transformer. It is desired to connect the two windings of the transformer in series to a 12,500 V supply and operate the same as an auto-transformer. Explain the connections for achieving the same. What are the possible voltage ratio and the corresponding maximum output? [**Ans.** 4/5; 500 kVA and 1/5; 125 kVA]

4. A transformer has a primary voltage rating of 11,500 V and secondary voltage rating of 2,300 V. Two windings are connected in series and the primary is connected to a supply of 11,500 V, to act as a step-up auto-transformer. Determine the output voltage of the transformer.

   [**Ans.** 13,800 V]

5. A 11,500/2300 V transformer is rated at 100 kVA as a two-winding transformer. If the two windings are connected in series to form an auto-transformer, what will be the possible voltage ratio?

   [**Ans.** 13,800/2300 V, 13,800/11,500 V]

6. A two-winding transformer is rated at 2,400/240 V, 50 kVA. It is reconnected as a step-up auto-transformer with 2,400 V input. Calculate the rating of auto-transformer and the inductively and conductively transferred powers while delivering the rated output at unity power factor. [**Ans.** 550 kW, 50 kW, 500 kW]

7. A 400/100 V, 10 kVA, 2 winding transformer is to be employed as an autotransformer to supply a 400 V circuit from 500 V source. When tested as a 2-winding transformer at rated load, 0.8 pf lagging, its efficiency is 0.97. Determine its kVA rating and efficiency as an auto-transformer.

   [U.P. Technical Univ. Electromechanical Energy Conversion-I, 2004-05]

   [**Ans.** 50 kVA, 99.5%]

8. A 400/100 V, 5 kVA, single-phase two-winding transformer is to be used as an auto-transformer to supply 400 V from a 500 V source. When tested as a two-winding transformer at rated load and 0.8 pf lagging, its efficiency was found to be 0.95. Determine its kVA rating as an auto-transformer. Also calculate the transformed kVA and conducted kVA. Find its efficiency as an auto-transformer at rated load and 0.8 pf lagging.

   [M.D. Univ. Electromechanical Energy Conversion. May-2006]

   [**Ans.** 25 kVA, 5 kVA, 20 kVA, 98.96%]

9. The voltage at the receiving end of a 3-phase feeder delivering 200 kVA varies between 480 V and 520 V. For what kVA capacity must an induction regulator be designed to maintain the output voltage constant at 500 V? [**Ans.** 8 kVA]

10. A single phase induction voltage regulator has an effective turn-ratio 5 %. If the supply voltage is 400 V, determine the maximum and minimum values of output voltages. If the load current is 25 A and the secondary winding is set at 30° away from the maximum boost position, calculate the primary current and primary line current. State the assumption if made. [**Ans.** 480 V and 320 V; 4.33 A; 29.33 A]

□□□

SECTION IV

SYNCHRONOUS MACHINES

CHAPTERS COVERED IN THIS SECTION

SYNCHRONOUS GENERATORS OR ALTERNATORS

SYNCHRONOUS MOTORS

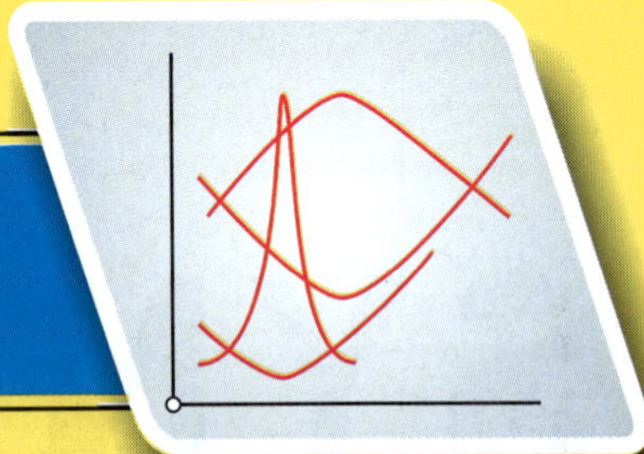

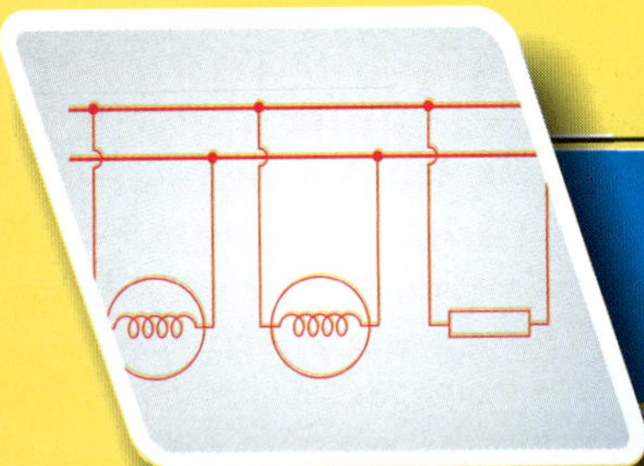

PARALLEL OPERATION OF ALTERNATORS

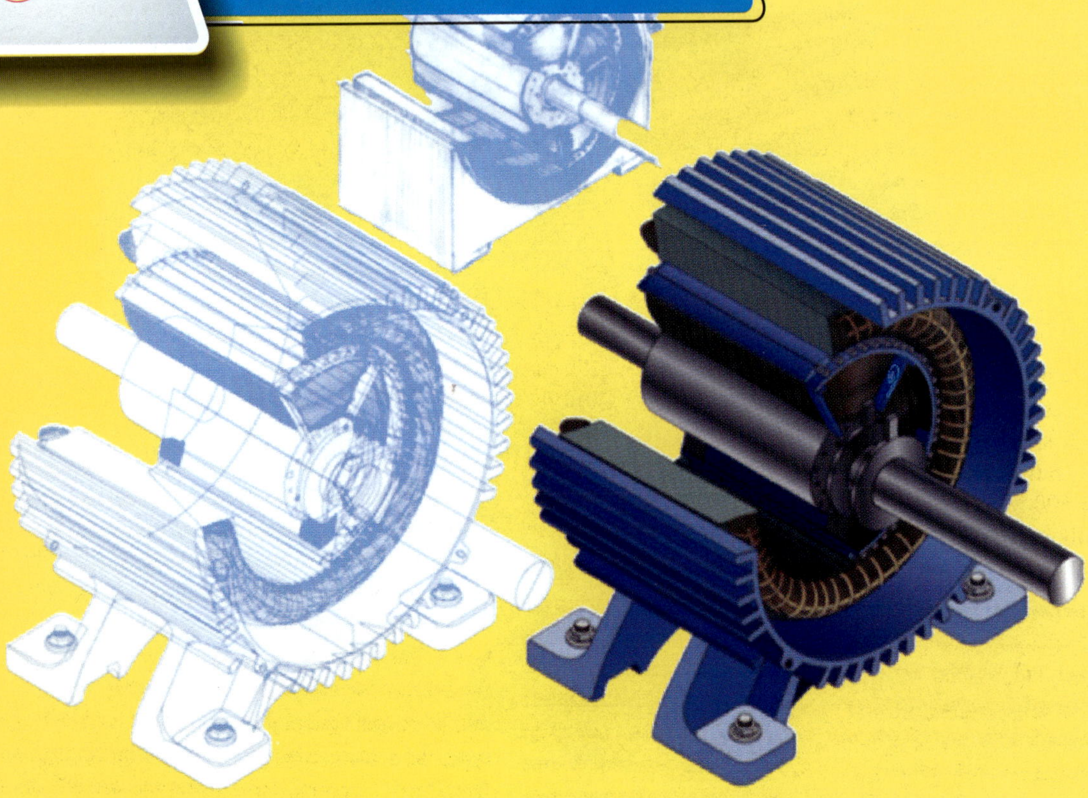

15 CHAPTER

Synchronous Generators or Alternators

INSIDE THIS CHAPTER

15.1 INTRODUCTION

A synchronous machine is an ac machine in which the rotor moves at a speed which bears a constant relationship to the frequency of currents, in the armature winding. A synchronous machine is one of the important types of electric machines. Large ac networks operating at constant frequency of 50 Hz (or 60 Hz) rely almost exclusively on *synchronous generators,* also called the *alternators*, for the supply of electrical energy, and may have *synchronous compensators* at key points for control of reactive power. Private, stand-by and peak load plants with diesel or gas-turbine prime movers also have synchronous generators. Non-land based synchronous plants (with a generating capacity of 50 MVA or more) are found on oil rigs, on large aircrafts (with hydraulically driven generators working at 400 Hz), and on ships for variable frequency supply to synchronous propeller motors. Synchronous motors provide constant speed industrial drives with the possibility of power factor correction.

Synchronous machines are generally constructed in larger sizes. Small size alternators are not economical. The modern trend is to build alternators of very large sizes capable of generating 500 MVA or even more. The synchronous motor is rarely built in small sizes owing to superior performance characteristics and economical construction of induction motors.

15.1.1. Operating Principle. Synchronous machines operate on the same fundamental principle of electromagnetic induction as dc machines, but, unlike the latter, in the synchronous machine there is no need to rectify the time varying emf which is induced in the armature winding. Consequently a synchronous machine does not require a commutator. It is, in fact, quite possible to use a dc generator as an alternator by placing a set of collector rings on the shaft and connecting these rings to the proper points on the armature winding; brushes riding on the rings can then be connected to the load. But unlike dc generator, they are to be driven at a very definite constant speed as the frequency of generated emf is determined by that speed. The latter is usually referred to as the *synchronous speed,* for which reason these machines are frequently called the synchronous generators.

Synchronous generators, because of absence of commutator, are comparatively simple and possess several important advantages over the dc generators.

15.1.2. Classification of Synchronous Machines. Synchronous machines, according to their applications, may be *synchronous generators, synchronous motors* or *synchronous compensators.* A synchronous generator is a synchronous machine which receives mechanical energy from a prime mover (steam turbine, hydraulic turbine or diesel engine) to which it is mechanically coupled and delivers electrical energy. A synchronous motor receives electrical energy from ac supply main and drives mechanical load. Synchronous compensator (or phase modifier) is a synchronous machine designed to operate on no load with its shaft connected neither to a prime mover nor to a mechanical load and is used to control reactive power in power supply networks. Synchronous compensators are designed for ratings up to 100 MVAR and speeds up to 3,000 rpm.

The synchronous machines may be *single-, two-* or *3-phase* types. For the single-phase type, the armature has all the coils connected for addition of individual voltages; this provides a pair of output terminals and a single ac output. For the two-phase type, the armature has two sets of windings placed so the outputs from three terminals (one common) are 90° out of phase. For the three-phase type, the armature has three sets of windings placed so that three outputs with a mutual phase difference of 120° are available. Three windings may be connected either in star or delta. Large ac machines are invariably 3-phase type.

Based on the construction of the machines, the synchronous machines may be classified as (*i*) rotating armature type and (*ii*) rotating field type.

Since it is immaterial for generation of an induced emf whether a conductor moves across a magnetic field or vice versa, synchronous generators may be constructed with either the armature or the field structure as the revolving member.

Rotating armature type alternator looks very much like a dc generator except that there are 3 slip rings in place of commutator (or 4 slip rings if it is desired to provide a connection to the generator neutral). In such generators the required magnetic field is produced by dc electromagnets placed on the stationary member called the *stator*, and the current generated is collected by means of brushes and slip rings on the revolving member, called the *rotor*. Such an arrangement is economical for the small low-voltage generators. Rotating armature type alternators are built only in small ratings up to about 200 or 250 kVA because the voltage generated is comparatively low and the current to be collected by the brushes small, no difficulty being experienced in collecting such a current. Such machines are suitable for small power plants, isolated lighting plants, where medium or small size machines are required.

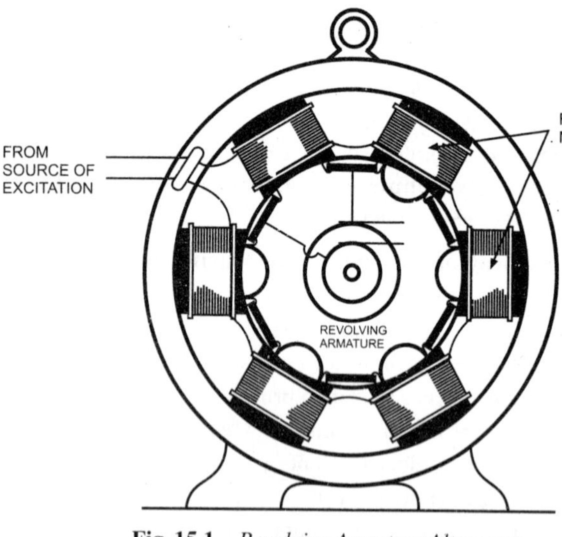

Fig. 15.1 *Revolving Armature Alternator*

Practically all medium and large sized machines are always constructed with revolving field. The advantages of stationary armature and revolving field system are given below:

1. It is easier to insulate stationary armature winding for very high voltage e.g. as high as 33,000 volts because insulation of stationary armature is not subjected to mechanical stresses due to centrifugal action and more space is available on the stator for providing more insulation as the stator is outside the rotor.
2. The load circuit can be connected directly with the fixed terminals of the stator (or armature windings) without passing through slip rings and brushes. In the case of rotating armature the load circuit is required to be connected with the armature winding by means of slip rings and brushes in which case a difficulty is experienced in collection of heavy currents at very high voltages.
3. The armature winding can be more easily braced in a rigid frame to prevent any deformation which could be developed by the mechanical stresses set up due to short-circuit currents and the high centrifugal forces brought into play.
4. The armature winding is cooled more readily because the stator core can be made large enough with many air passages or cooling ducts for forced air circulation.
5. Only two slip rings are required for the supply of direct current to the rotor and since the exciting current is to the supplied at low voltage of 125 or 250 V, there is no difficulty in insulating them.
6. Since the exciting current is relatively small, therefore, the slip rings and the brush gear need be of only light construction.
7. Due to simple, light and robust construction of the rotor, higher speed of rotating dc field is possible. This results in increased output from the machine of given dimensions.

Since in dc machines, the commutator makes necessary that either the armature should be rotating one, or brushes should revolve with the field, it is convenient to have armature as a rotating member in dc machines.

Revolving field alternator with essential parts is shown in Fig. 15.2.

The synchronous machines may be classified as (*i*) salient pole machines and (*ii*) cylindrical rotor machines depending upon the type of construction used for the rotor. The salient pole construction is used for generators and motors of all ranges of output and up to all but the higher speeds. Medium and large sized generators for the highest speeds are of the cylindrical rotor type.

The synchronous generators, based on the type of prime movers to which they are mechanically coupled, may be classified

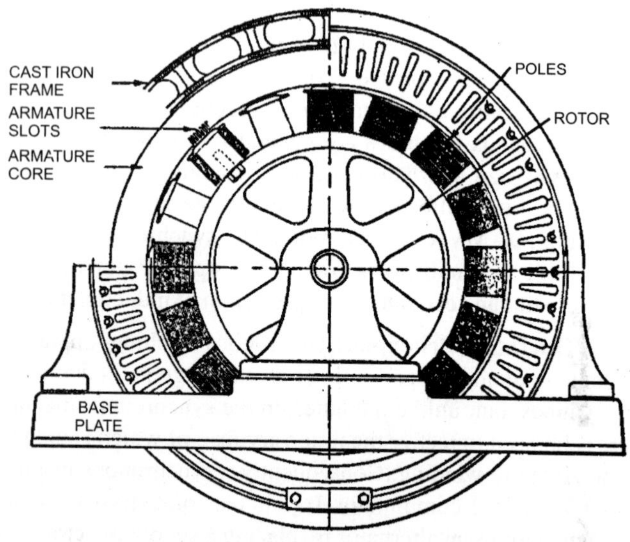

Fig. 15.2 *Revolving Field Salient Pole Type Alternator*

as (*i*) turbo-generators, (*ii*) hydrogenerators and (*iii*) engine driven generators.

Turbo-generators are driven by steam turbines. The efficiency of steam turbines is high at large speeds, and therefore, synchronous machines driven by steam turbines (*i.e.* turbo-generators) are high speed machines. The maximum speed of turbo-generators is 3,000 rpm corresponding to 2 poles and 50 Hz. These have small diameters (limited to about 5.2 m) and horizontal configurations. Cylindrical rotor construction is used as the salient pole construction is impractical owing to large mechanical forces. Two-pole construction with a speed of 3,000 rpm for 50 Hz is commonly used for turbo-generators. The 4-pole construction, with a synchronous speed of 1,500 rpm is now obsolete. Turbo-alternators have ratings up to 1,000 MVA.

The synchronous generators driven by water turbines are called the *waterwheel generators* or *hydrogenerators*. Such generators have to match the specific speed requirements of the turbines driving them. These generators are low speed machines with speeds between 1,500 and 100 rpm (*i.e.* number of poles ranging from 4 to 60) and of rating up to 750 MVA. Hydroelectric power plants usually have vertical shaft.

Diesel-engine driven synchronous generators are of small ratings from 100 kVA or less to 1,000 kVA. The speed is generally between 375 to 1,500 rpm, *i.e.* the number of poles is 16 to 4. Generally 4 to 6 pole alternators are common for diesel driven generators. The diesel engines are manufactured as horizontal type, and therefore, both the diesel engine and the synchronous generator are mounted horizontally and are connected by a horizontal shaft. Such generators employ salient pole type construction. The torque of the diesel engine is not uniform and this makes the synchronous generator sensitive to torque variations.

15.2 CONSTRUCTION

Synchronous machine consists essentially of two parts namely the armature (or stator) and the field magnet system.

1. Armature (or Stator). The armature is an iron ring, formed of laminations of special magnetic iron or steel alloy (silicon steel*) having slots on its periphery to accommodate armature conductors and is known as *stator*. The whole structure is held in a frame** which may be of cast iron or welded steel plate. Since the field rotates in between the stator, so that flux of the rotating field cuts the core of the stator continuously and causes eddy current loss in the stator core. To minimize the eddy current loss, the stator core is laminated. The laminations (usually of thickness 0.5 mm or less) are stamped out in complete rings (for smaller machines) or in segments (for larger machines) and insulated from each other with paper or varnish. The stampings also have openings which make axial and radial ventilating ducts to provide efficient cooling. A general view of stator and frame is shown in Fig. 15.3.

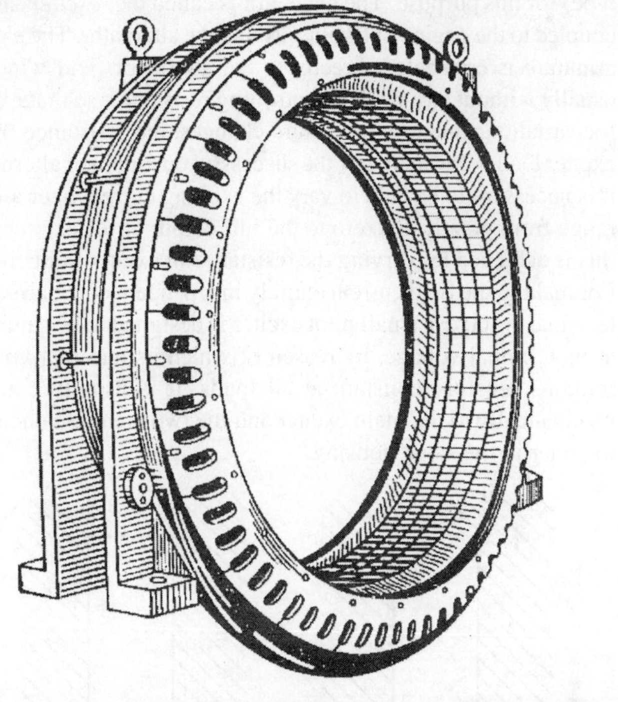

Fig. 15.3 *Stator of a Synchronous Generator*

The open slots are more commonly used because the coils can be form wound and insulated prior to being placed in the slots giving least expenditure and more satisfactory winding method. This type of slots also facilitate in removal and replacement of defective coils. But this type of slots have disadvantage of distributing the air-gap flux into branches or tufts which tends to produce ripples in the emf wave. The semi-closed type slots are better in this respect but do not permit the use of form-wound coils. Totally closed slots are rarely used.

Number of slots per pole per phase should be large to give sinusoidal generated emf. In general, the number of slots per pole per phase is 3 to 4 for small machines and more than 5 for large machines. Small machines may have integral slot winding with enough chording. For large machines fractional slot winding is used and so the number of slots per pole per phase may not be an integer.

The minimum tooth width should be such that the maximum flux density in it does not exceed the limit of flux density in teeth permissible to avoid saturation. The depth of the slot and shape is such as to accommodate the required number of conductors per slot for the rated voltage to be produced at the rated speed. Generally depth is 4 to 6 times the width. If the depth is increased, synchronous reactance in slot increases, core losses in teeth are larger and more excitation will be required for the machine.

2. Field Magnet System or Rotor. The magnetic field required for the generation of alternating voltage is provided by field magnets similar to those of a dc machine and these need to be supplied with direct current. This may be had from some external source, but so as to make the alternator self-contained and independent of other source of supply, it is more usual to install a small dc generator (usually a flat compound wound

* Although grain oriented steels give lower core loss, the orientation is not everywhere favourable, and cold-rolled non-oriented steels containing 3 per cent silicon are usually used.

** In a dc machine the outer frame (or yoke) serves to carry the magnetic flux but in an alternator, it is merely used for holding the armature stampings and coils in position.

type) for this purpose. The generator is called the '*exciter*' and is coupled to the engine or turbine driving the alternator. The exciter armature is connected directly to the alternator field winding, usually without any kind of controlling resistance, so that except for variations due to temperature changes, the resistance of the exciter load is constant. For the successful operation of alternator, it is necessary to be able to vary the exciting current over a wide range from practically zero to the full output of the exciter, and this is achieved by varying the resistance in exciter field circuit. For making the field current entirely independent of the armature terminal voltage, a small pilot exciter is designed to be saturated at its terminal voltage, by reason of which its terminal voltage remains sensibly constant at all loads. It is usually coupled mechanically to the main exciter and the two are often enclosed together in the same housing.

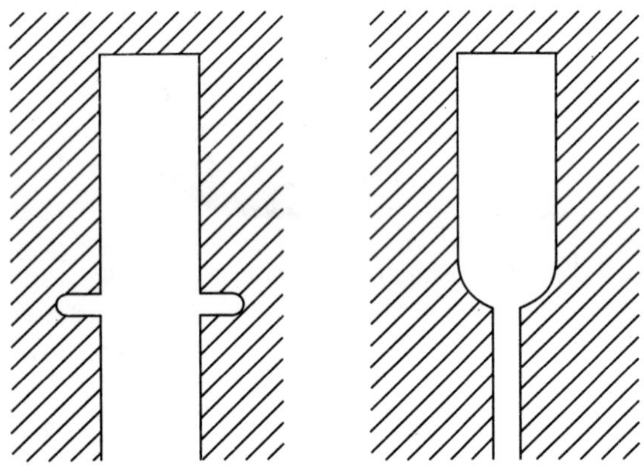

(a) Open Slot *(b) Semi-Closed Slot*

Fig. 15.4

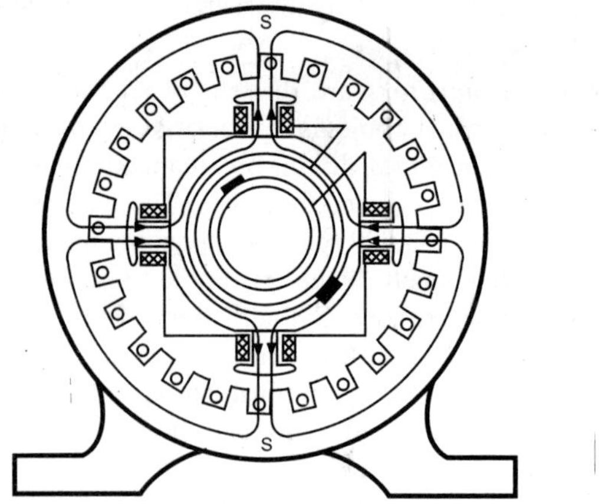

(a) Salient Pole Type Field Structure

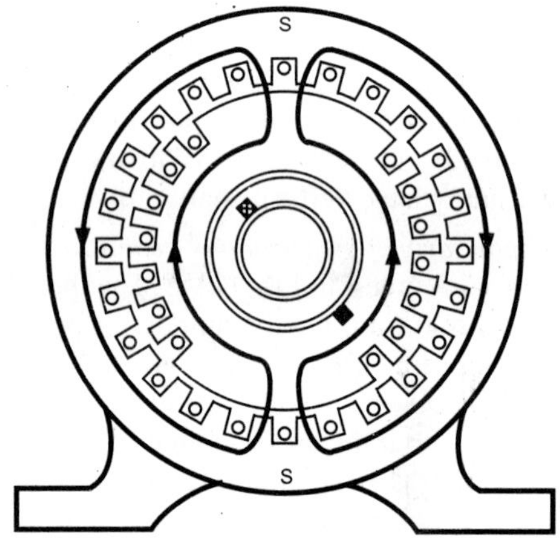

(b) Non-Salient Pole Type Field Structure

Fig. 15.5

The field system of the alternator is rotated within the armature ring and is known as *rotor*. The exciting current is supplied to the rotor through two slip rings and brushes. The polarities of the field produced is alternately north (N) and south (S). The power rating of the exciter is ordinarily 0.5 to 1% of power rating of the synchronous generator. The voltage rating of the exciter is usually between 125 and 250 V.

Rotors, as already mentioned in Art. 15.1.2, are of two types namely (*i*) salient pole type and (*ii*) smooth cylindrical type or non-salient pole type. Both types are shown in Fig. 15.5.

(a) Salient Pole Type Field Structure. The field structure of this type is used almost entirely for slow and moderate speed alternators, since it is least expensive and provide ample space for the field ampere-turns. Salient pole machines are cheaper than cylindrical rotor machines when speed does not exceed 1,000 rpm. Salient pole type field structure cannot be employed in high speed machines on account of high peripheral speed (100 to 170 m/s) and the difficulty of obtaining sufficient mechanical strength. The salient field poles would also cause an excessive windage loss and produce noise if driven at high speed.

The rotor of a low-speed machine resembles a flywheel with electromagnets, similar to the field magnets of a dc machine, bolted on all round the periphery. It is usually made of cast iron or cast steel and has a heavy central hub, a number of radial arms, and a heavy rim which forms the yoke of the magnet system. The flywheel type of rotor is constructed to give adequate mechanical strength to withstand centrifugal and driving forces and at the same time to provide the necessary paths for the magnetic fluxes. The poles may be either solid or laminated but in any case, the pole shoes are laminated because the armature teeth between the slots cause concentrations of flux which generate emfs in the pole shoes. If they were not laminated the resulting eddy currents would cause excessive loss and heating. The pole faces are so shaped that the radial air-gap length increases from the pole centre to the pole tips, as shown in Fig. 15.6, so that the flux distribution over the armature surface is sinusoidal and waveform of generated emf is sinusoidal.

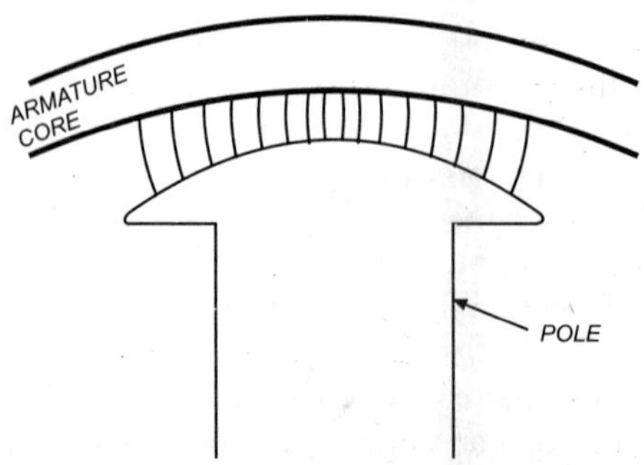

Fig. 15.6

The poles have concentrated windings and are wound with ordinary wire coils for the smaller sizes, but in larger machines

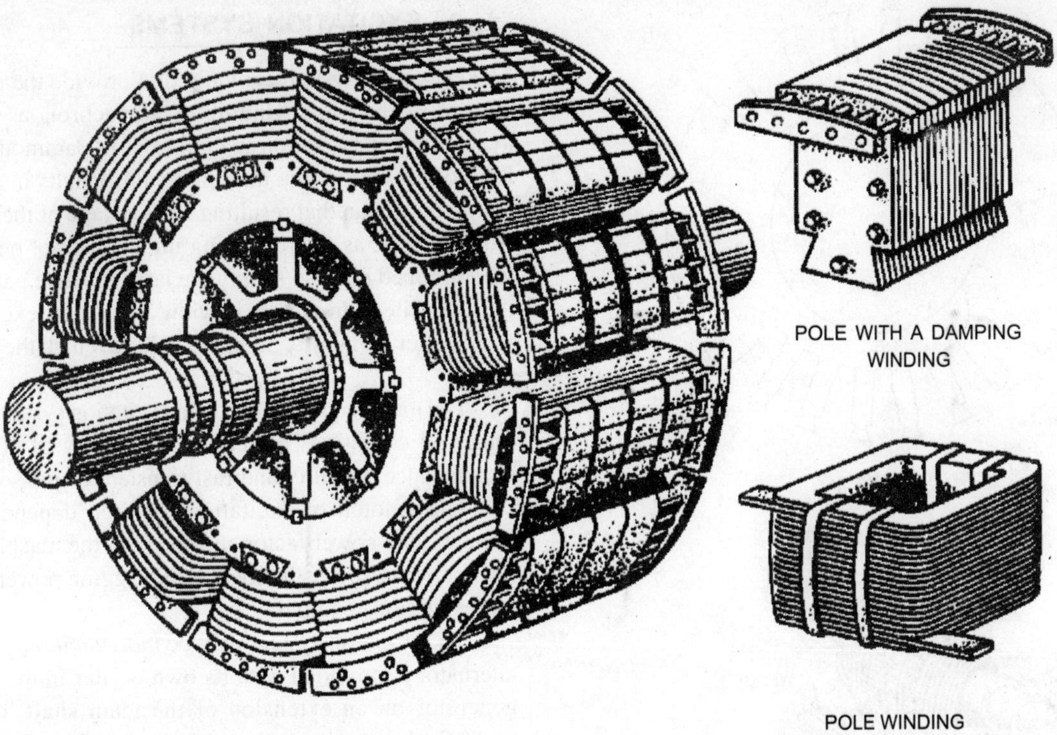

SALIENT POLE FIELD STRUCTURE

Fig. 15.7

the field windings are made of rectangular copper strip wound on edge. The latter arrangement provides good mechanical strength and saves wastage of space.

Salient pole machines are frequently provided with a damper winding on the rotor to damp rotor oscillations during transient conditions and to facilitate operation under unbalanced load conditions. The damper winding consists of copper bars laid in slots on the faces of the pole pieces and connected to each other at the ends by means of short-circuiting segments. The segments on individual poles are generally joined together to form common rings resulting in a short-circuited squirrel cage winding similar to that used in induction machines with squirrel cage rotors.

It should be clearly understood that under normal operating conditions, damper winding does not carry any current because rotor runs at synchronous speed. The damper winding also tends to maintain balanced 3-phase voltage under unbalanced load conditions.

Since the speed of salient pole machines is quite low, so to give required frequency, such machines have large number of poles (4 to 60). To accommodate such a large number of poles, these machines have large diameter and small length, because the output of a machine is roughly proportional to its volume.

The salient pole field structure has the following special features :

(*i*) They have large diameter and short axial length.

(*ii*) The pole shoes cover about 2/3 of pole pitch.

(*iii*) Poles are laminated in order to reduce eddy current losses.

(*iv*) These are employed with hydraulic turbines or diesel engines. The speed is 100 to 375 rpm. A general view of salient pole field structure is shown in Fig. 15.7.

(*b*) *Smooth Cylindrical or Non-salient Pole Type Field Structure.* The field structure of this type is used in very high speed alternators (turbo-alternators). To reduce the peripheral velocity the diameter of the rotor is reduced and axial length is increased. The ratio of diameter to axial length ranges from 1/3 to 1/2. A small number of poles is required to give the required frequency.

Due to high peripheral speed, the rotating part of the turbo-generator is subjected to high mechanical stresses. As a result the rotor of a large turbo-generator is normally built from solid steel forging. Chromium-nickel-steel or special chrome-nickel-molybdenum steel is used for the rotors of turbo-alternators. The forging has radial slots in which the field copper, usually in strip form, is placed. The coils are held in place by steel or bronze wedges and the coil ends are fastened by metal rings. Normally two-thirds of the rotor is slotted for the field winding and one-third is left without slots so as to form the pole faces, as shown in Fig. 15.8. Rectangular slots with tapered teeth are milled out in the rotor so that rectangular conductors can be used for the field winding. Each slot is provided with a ventilation hole at the bottom. To reduce harmful tooth ripples, either stator slots or poles are skewed. General view of an assembled smooth cylindrical rotor is shown in Fig. 15.9.

The non-salient field structure has the following special features.

(*i*) They are of smaller diameter and of very long axial length.

(*ii*) Robust construction and noiseless operation.

(*iii*) Less windage (air-resistance) loss.

(*iv*) Better in dynamic balancing.

(*v*) High operating speed (3,000 rpm).

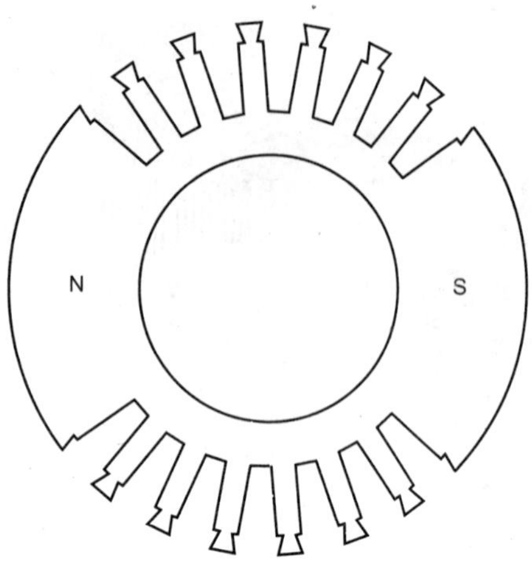

Fig. 15.8

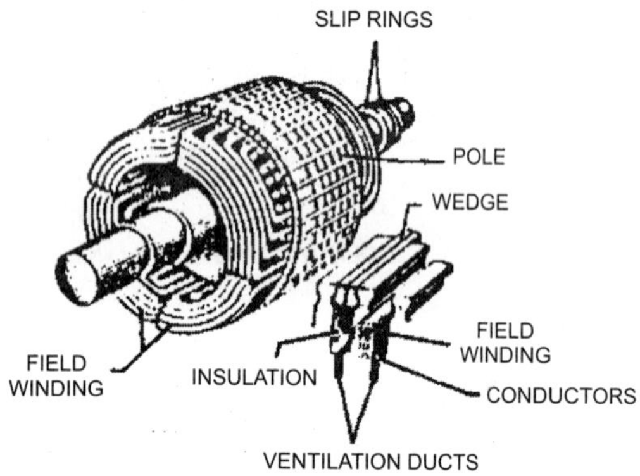

Fig. 15.9

(*vi*) Nearly sinusoidal flux distribution around the periphery, and therefore, gives a better emf waveform than that obtainable with salient pole field structure.

(*vii*) There is no need of providing damper windings (except in special cases to assist in synchronising) because the solid field poles themselves act as efficient dampers.

Air gap. A very small air gap increases the stray-load loss; also the eccentricity in the air gap can result from mechanical difficulties. This also increases the synchronous reactance X_d. A large air gap needs larger excitation current. A compromise has to be made. Generally the ratio of air gap to the pole pitch is between 0.008 to 0.02.

15.3 EXCITATION SYSTEMS

The excitation system is required to provide the necessary field current to the rotor winding of a synchronous machine. The availability of excitation at all times is of paramount importance. Loss of excitation of a unit on the bus results in a more serious disturbance than that resulting from outage of the generator unit from the bus, as the remaining units must not only pick up the load dropped but also supply the large reactive current drawn by the unexcited alternator. In view of this an excitation system with better reliability is preferable, even if the initial cost is more.

The main requirements of an excitation system are reliability under all conditions of service, simplicity of control, ease of maintenance, stability and fast transient response.

The amount of excitation required depends on the load current, load power factor and speed of the machine. Larger the load currents, lower the speeds and lagging power factors, more the excitation required.

An excitation system may be *individual* one, in which each alternator is provided with its own exciter in the form of small generator on an extension of the main shaft, or *centralised excitation system* having two or more exciters feeding a bus-bar to which field systems of all the alternators in the power plant are connected. Though centralised excitation system is a cheaper arrangement but a fault in this system adversely affects all the alternators in the power plant. As such individual excitation system is widely used.

The excitation systems can be broadly classified as: 1. DC excitation system, 2. AC excitation system, and 3. Static excitation system.

15.3.1. DC Excitation System. In dc *excitation system*, the system has two exciters—the main exciter (a separately-excited dc generator providing the field current to the alternator) and a pilot exciter (a compound wound self-excited dc generator providing the field current to the main exciter). The exciter output is regulated by an automatic voltage regulator (AVR) for controlling the output terminal voltage of the alternator. The current transformer input to the AVR ensures limiting of alternator current during fault. With the opening of field breaker, the field discharge resistor is connected across the field winding so as to

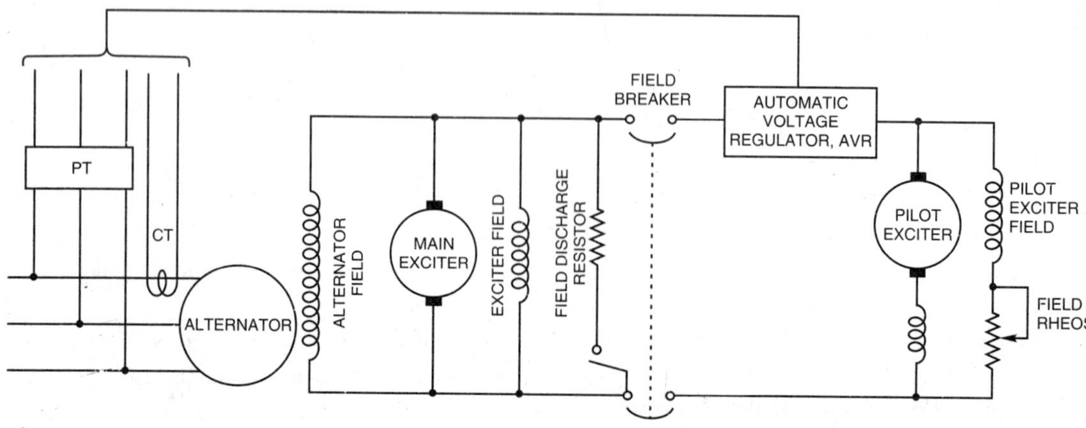

Fig. 15.10 *DC Excitation System*

dissipate the stored energy in the field winding which is highly inductive. The main and pilot exciters can be either driven by the main shaft (directly or through gearing) or separately driven by a motor. Direct driven exciters are usually preferred as these preserve the unit system of operation and the excitation is not affected by external disturbances. Exciters for small sized turbo-alternators are usually directly coupled to the shaft whereas for medium and large size turbo-alternators, the exciters are coupled to the main shaft through gear and are usually driven at 1,000 rpm. The voltage rating of the main exciter is about 400 V and its capacity is about 0.5% of the capacity of the alternator. Troubles in the exciters of turbo-alternators are quite frequent because of their high speeds and as such separate motor driven exciters are provided as standby exciters. In some cases the dc excitation system is equipped with amplidyne controller. The amplidyne is a high response cross-field dc generator, which has a very high amplification factor and requires very small control power. It is supplied from an automatic voltage regulator (AVR). The main shortcomings of dc excitation system are large time constant (around 3 seconds) and commutation difficulties. As such the dc excitation systems have been superseded by ac excitation systems and static systems.

15.3.2. AC Excitation System. This *system* consists of an alternator and thyristor rectifier bridge directly connected to the main alternator shaft. The main exciter may either be self excited or separately excited. A *rotating thyristor excitation system* employs self excited main exciter whereas the *brushless excitation system* employs a separately excited main exciter.

1. Rotating Thyristor Excitation System. Block diagram of a rotating thyristor excitation system is depicted in Fig. 15.11, the rotating portion being shown enclosed by dashed line rectangle. This system comprises of an ac exciter having a stationary field and a rotating armature. The exciter output, rectified by a full-wave thyristor bridge rectifier circuit, is supplied to the main alternator field winding. The alternator field winding is also supplied from the exciter output through another rectifier circuit. Exciter voltage can be built up using its residual flux. For convenience and to shorten the build up time, stationary battery supply is also incorporated.

The power supply and rectifier control, incorporated as a part of rotating portion, generates the required phase controlled triggering signals in response to a dc control level fed by the voltage regulator.

An automatic voltage regulator (AVR) is also incorporated in the system. The alternator voltage signal (from the PT) is averaged and compared directly with the operators voltage adjustment (voltage reference) in the auto mode of operation. In the manual mode of operation the excitation current of the alternator is compared with a separate manual voltage adjustment (reference) and supplied to the rotating thyristor bridge through separate regulating elements.

For maintaining the terminal voltage of the exciter at the desired level, the exciter voltage is averaged, compared with the exciter voltage reference and applied to the rectifier control and rectifier in a suitable manner.

The system also incorporates additional features like overvoltage protection, overcurrent protection in the exciter field control, lead-lag compensation for stabilization of voltage control, watt and VAR signals for regulating the voltage and field discharge resistor, which are not shown in the block diagram.

2. Brushless Excitation System. With the increase in rating of alternators, the problem of brush maintenance gets more and more complicated, so the brushless excitation systems are becoming more and more popular. The advent of silicon diodes and thyristors made it possible to have compact rectified system converting ac to dc at higher power levels.

This system is depicted in Fig. 15.12, the rotating portion being shown enclosed by a dashed line rectangle. The excitation system comprises of an alternator rectifier main exciter and a permanent magnet generator (PMG) pilot exciter. Both of the main and pilot exciter are driven directly from the main shaft. The main exciter has a stationary field and a rotating armature directly connected, through silicon rectifiers, to the field of the main alternator. The pilot exciter is a shaft driven permanent magnet generator having rotating permanent magnets attached to the shaft and a 3-phase stationary armature, which feeds the main exciter's field through 3-phase full-wave phase controlled thyristor bridges. This system eliminates the use of commutator,

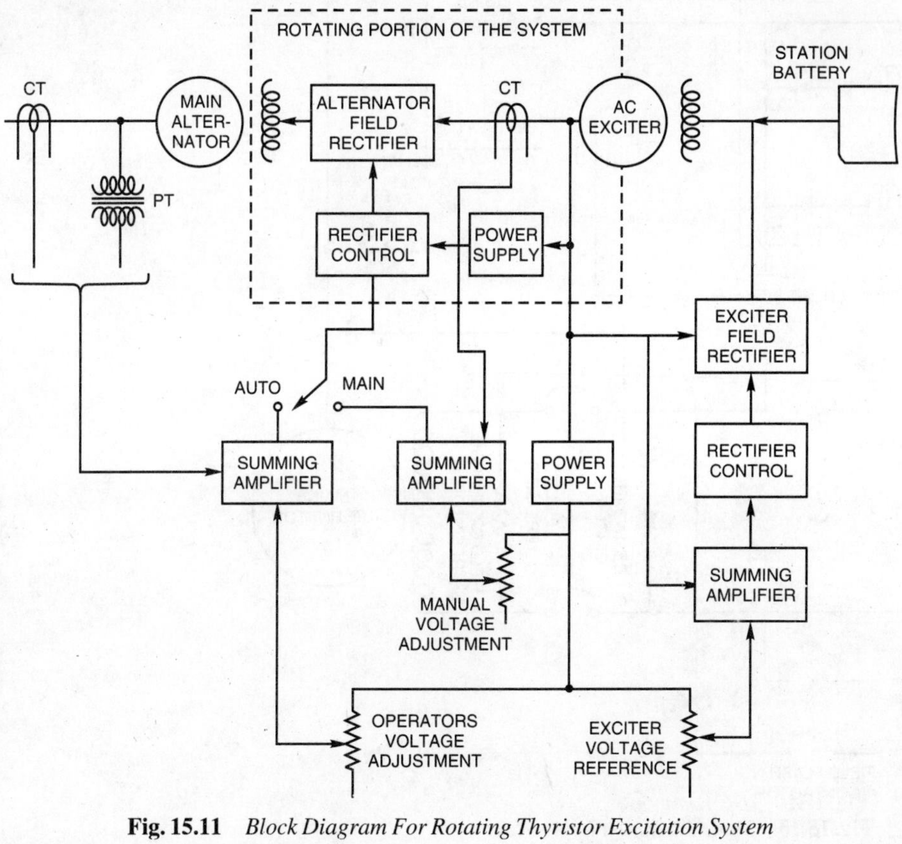

Fig. 15.11 *Block Diagram For Rotating Thyristor Excitation System*

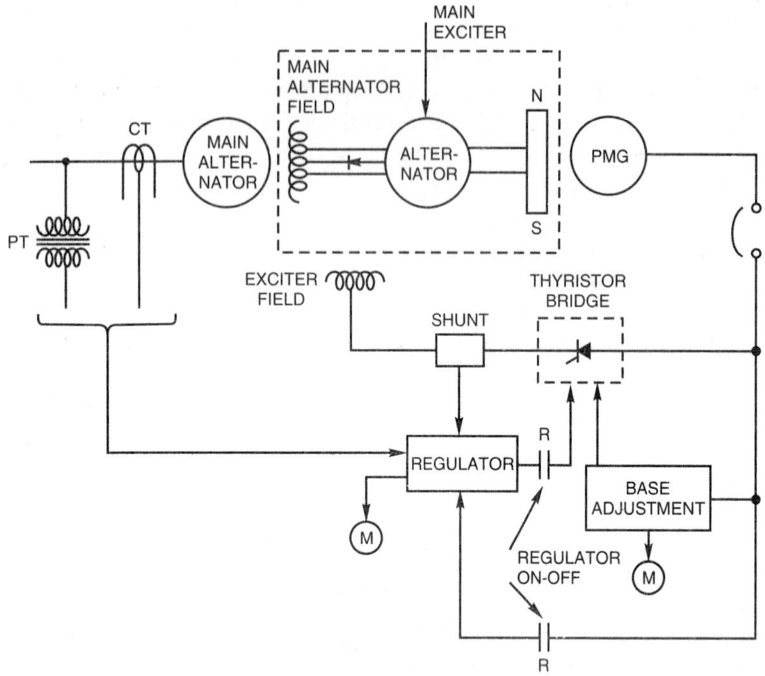

Fig. 15.12 *Brushless Excitation System*

collectors and brushes and has a short time constant and a response time of less than 0.1 second. The short time constant has the advantage in improved small signal dynamic performance and facilitates the application of supplementary power system stabilizing signals.

3. Static Excitation System. Figure 15.13 depicts the static excitation system using SCRs.

In *static excitation system*, the excitation supply is taken from the alternator itself through a 3-phase star/delta connected, oil immersed, forced air cooled, indoor type step-down transformer and a rectifier system employing mercury-arc rectifiers or silicon controlled rectifiers. The star-connected primary is connected to the alternator bus, the delta-connected secondary supplies power to the rectifier system and the delta-connected tertiary feeds power to grid control circuits and other auxiliary equipment. The rectifiers are connected in parallel to give sufficient current carrying capacity. Each leg of rectifier is protected with series fuse, surge protection and fault indicating light. The rectifiers are forced air cooled. This system has a very

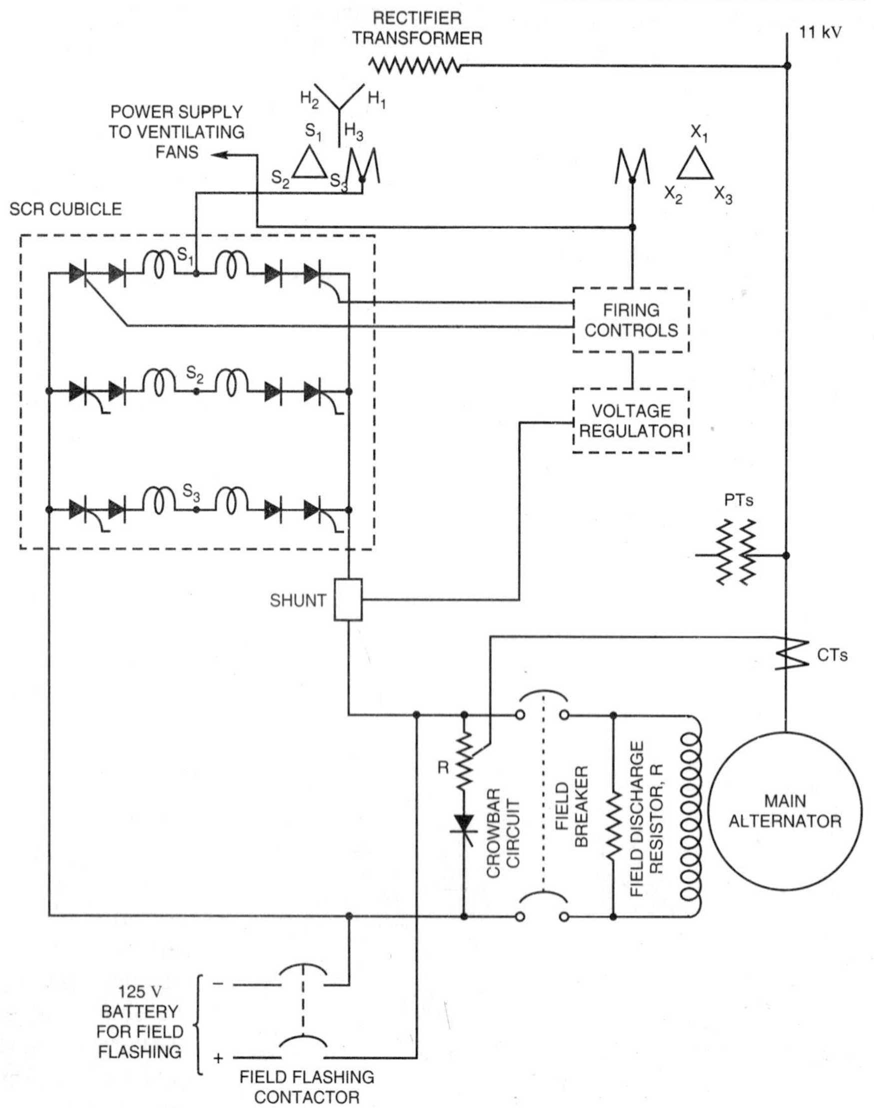

Fig. 15.13 *Static Excitation Using SCRs*

small response time (about 20 milliseconds) and provides excellent dynamic performance. SCRs are ideally suited for a static excitation system because they have high speed of response, high power gain and can be easily paralleled. The advantages of the static excitation system are elimination of exciter windage loss and commutator wearing and winding maintenance resulting in reduced operating costs and *electronic speed response.* The fact that in static excitation the voltage is proportional to the speed, affords a major advantage in load rejection.

15.4 PRODUCTION OF SINUSOIDAL ALTERNATING EMF

When the rotor is rotated by means of some prime mover, the armature conductors cut the magnetic flux, therefore, an emf is induced in the armature conductors, due to electromagnetic induction effect.

When the conductor is opposite the neutral planes, as at *A, C* and *E,* induced emf in it is minimum because flux density is minimum there. When the conductor is opposite the middle of the poles, as at *B,D* and *F,* induced emf in it is maximum, the direction of induced emf depends upon the name of the pole influencing the conductor at any given instant. Thus an alternating emf is induced in the conductors which goes through one complete cycle in an angular distance equal to twice of the pole pitch. Though the shape of the wave of alternating induced emf is not exactly the sinusoidal but taken as sinusoidal.

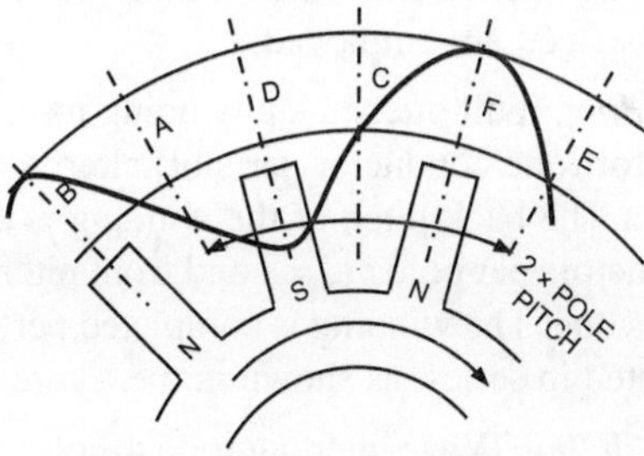

Fig. 15.14

15.5 FREQUENCY OF INDUCED EMF

As mentioned above, an emf induced in conductor goes through one complete cycle in an angular distance equal to twice the pole pitch.

If the number of poles on rotor of an alternator is *P*, then *P*/2 cycles of emf are completed in one revolution. The number of cycles per second, *known as frequency f,* will be equal to the product of number of cycles of emf per revolution and number of revolutions made per second by the rotor.

i.e. Frequency, $f = \frac{P}{2} \times n$

where *n* is the number of revolutions made per second by rotor

$$\text{or } f = \frac{P}{2} \times \frac{N}{60} = \frac{PN}{120} \qquad ...(15.1)$$

where N is the number of revolutions made per minute by the rotor.

Hence frequency of induced emf or current induced in stator conductors depends upon the number of poles and speed of rotor.

Example 15.1. A 4-pole alternator rotates at 1,500 rpm. What is the frequency of the generated voltage.

[*U.P. Technical Univ. Electrical Machines 2008-09*]

Solution: Number of poles, P = 4

Alternator speed, N = 1,500 rpm

Frequency of generated voltage,

$$f = \frac{P \times N}{120} = \frac{4 \times 1{,}500}{120} = 50 \text{ Hz} \quad \textbf{Ans.}$$

Example 15.2. A 6-pole ac generator is running and producing the frequency of 60 Hz. Calculate the revolutions per minute of the generator. If the frequency is decreased to 20 Hz, how many number of poles will be required if the generator is to be run at the same speed.

Solution: Frequency generated, f = 60 Hz

Number of poles, P = 6

$$\text{Speed, N} = \frac{120\, f}{P} = \frac{120 \times 60}{6} = 1{,}200 \textbf{ Ans.}$$

When frequency, f' = 20 Hz

$$\text{Number of poles required, P}' = \frac{120\, f'}{N} = \frac{120 \times 20}{1{,}200} = 2 \textbf{ Ans.}$$

15.6 ARMATURE WINDINGS

The armature windings of dc machines are usually closed circuit windings but alternators windings may be either closed giving *delta connections* or open giving *star connections.*

The general principles governing windings for dc machines also hold good for alternator windings e.g.

(*i*) The span of each coil must be equal to pole pitch *i.e.* the two sides of any coil must be under adjacent poles.

(*ii*) The coils must be so connected that emfs induced in them help each other.

(*iii*) The winding can be either single layer or double layer. The single layer winding will have one coil side per slot, while double layer winding will have two coil sides/ slot.

(*iv*) In addition to above, winding must be so designed that it may approximately give induced emf of sinusoidal form.

15.6.1. Classification of Armature Windings. The classification may be made according to several points of view e.g.

(*a*) Single-phase windings and polyphase windings.
(*b*) Concentrated windings and distributed windings.
(*c*) Half coiled (or hemitropic) windings and whole-coiled windings.
(*d*) Single layer windings and double layer windings.
(*e*) Lap windings, wave windings and concentric (or spiral) windings.
(*f*) Full pitched coil windings and fractional pitched coil windings.
(*g*) Integral slot winding and fractional slot winding.

1. *Single-Phase Armature Windings.* There are two types of single-phase armature windings, namely

(*a*) concentrated and (*b*) distributed windings.

(*a*) *Concentrated Winding*: If one slot per pole or slots equal to number of poles are employed, then *concentrated winding* is obtained. Such windings give maximum induced emfs for a given number of conductors but the waveform of induced emf is not exactly of sinusoidal form.

Simplest possible winding known as "*skelton wave winding*" is shown in Fig. 15.15 (*a*). In this winding the number of conductors or coil sides is equal to the number of poles. The direction of induced emf in the conductor under north pole is shown upward while in the conductor under south pole downward. The conductor no. 1 under north pole is connected to conductor no. 2 under south pole, at the back and conductor no. 2 is connected to conductor no. 3 at the front and so on. It is single layer, full pitched coil winding.

Single layer winding requires considerable space for the end connections of the coils and thus it is rarely used in practice.

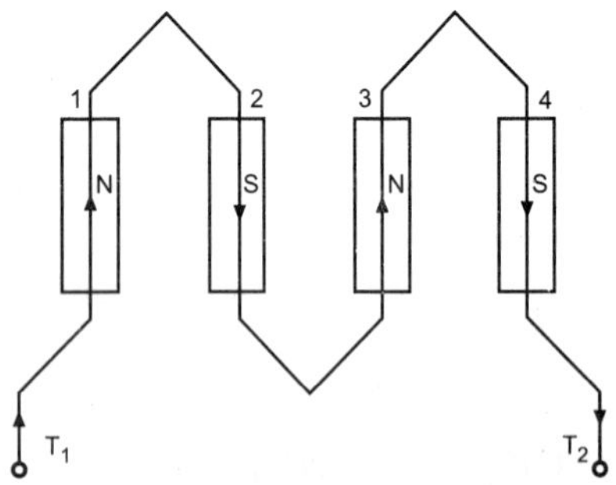

(a) Skelton Wave Winding

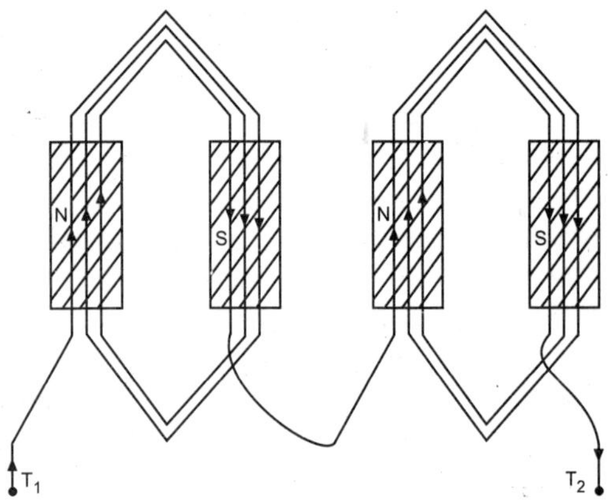

(b) Half Coiled (or Hemitropic) Winding

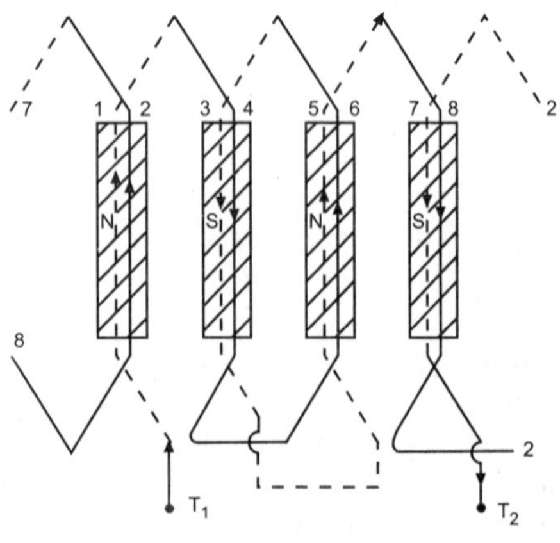

(c) Whole Coiled Winding

Fig. 15.15

If single turn coils of skelton wave winding are replaced by multi-turns coils in order to obtain higher emf, the multi-turn half coiled (*hemitropic*) winding is obtained. Since coils only cover one half of the armature periphery, therefore, this winding is known as *half coiled or hemitropic* winding.

If the coils are distributed over the whole of the periphery, the winding so obtained is called *whole coiled winding.*

In this winding the number of coils is equal to the number of poles. One side of each coil lying on the top is shown by continuous (or solid) line and the other side lying in the bottom of the slot is shown by dotted line in Fig.15.15 (c). It is double layer winding.

(*b*) *Distributed Winding.* If the conductors are placed in several slots under one pole, the winding is known as *distributed winding*. The winding may be partially or completely distributed accordingly as the slots are spread over only a portion of it or over the entire armature surface. Though it reduces the induced emf it yet is most commonly employed due to its manifold advantages.

Advantages. (*i*) The harmonic emfs are reduced and so the waveform is improved.

(*ii*) Certain distorting harmonics can be altogether eliminated.

(*iii*) The distributed winding diminishes armature reaction and armature reactance.

(*iv*) Higher current density in the copper can be adopted due to even distribution of copper over the surface of armature resulting in even distribution of copper loss and, therefore, in efficient cooling.

(*v*) The core is better utilised as a number of small slots evenly spaced are employed.

(*i*) *Lap Winding.* Full pitched lap winding for a 4-pole, 12-slot, 12–conductor (one conductor per slot) alternator is shown in Fig. 15.16 (*a*). The back pitch of the winding is equal to the number of conductors per pole *i.e.* = 3 and front pitch is equal to back pitch minus one. The winding is completed per pair of pole and then connected in series, as shown in the figure.

(*ii*) *Wave Winding.* Wave winding for a 4-pole, 12-slot, 12–conductor (one conductor per slot), alternator is shown in Fig. 15.16 (*b*). The back pitch and front pitch, each is equal to number of conductors per pole.

(*iii*) *Concentric or Spiral Winding*. Concentric or spiral winding for a 4-pole, 12-slot, 12– conductor (one conductor per slot) alternator is shown in Fig. 15.16 (*c*).

In this winding the coils are of different pitches. The outer coil pitch is 5, the middle coil pitch is 3 and inner coil pitch is one.

Lap windings have their widest fields of application in the stators of most high-speed synchronous machines, wave windings are ideal for the rotors of wound rotor induction type motors and concentric or spiral windings are used most frequently in slow speed, large diameter alternators :

2. ***Polyphase Armature Winding.*** Terms used frequently in connection with polyphase windings are:

Coil Group. It is the product of number of phases and number of poles i.e. coil group = Number of poles × number of phases.

Balanced Winding. If the number of coils per coil group is a whole number, the winding is known as *balanced winding.* Each

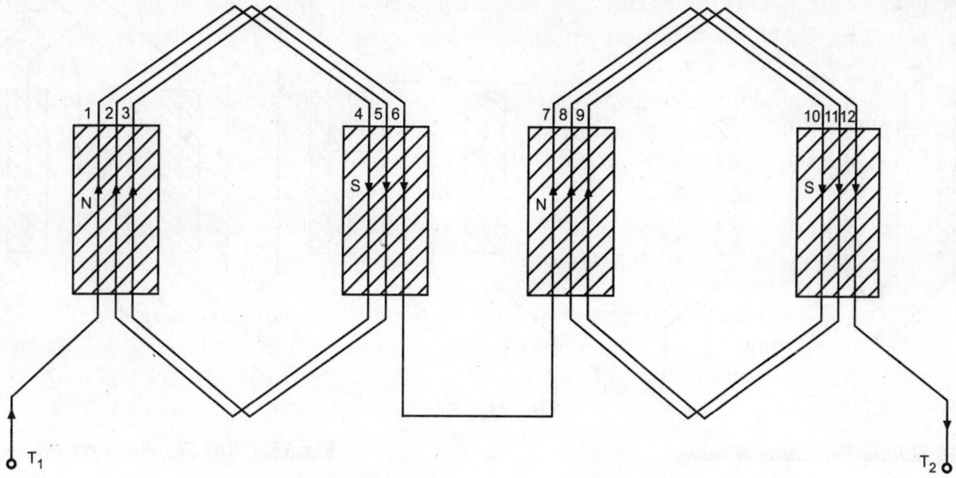

Fig. 15.16 (*a*) *Single Layer Lap Winding*

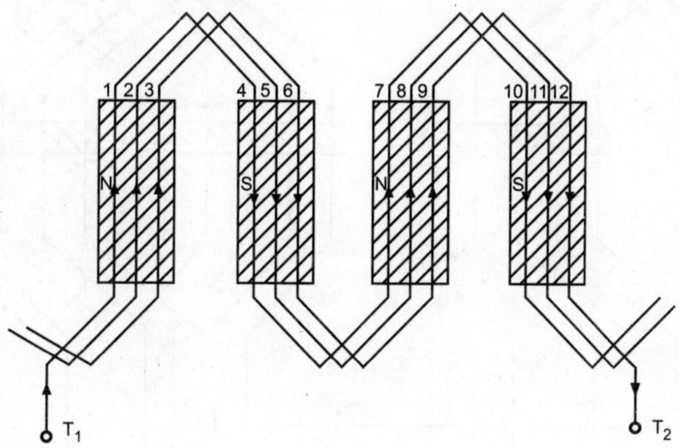

Fig. 15.16 (*b*) *Single Layer Wave Winding*

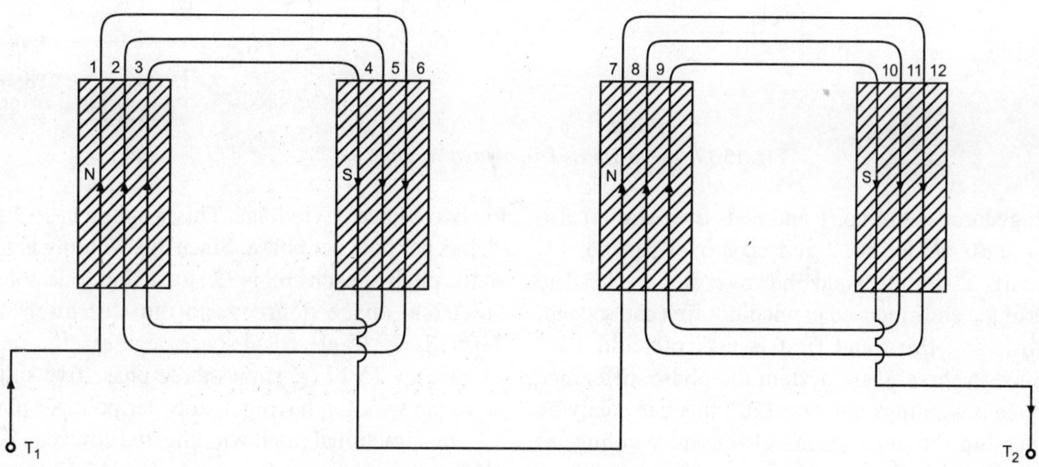

Fig. 15.16 (*c*) *Single Layer Spiral Winding*

pole contains equal number of coils of different phases in such a case.

Unbalanced Winding. If the number of coils per coil group is not a whole number, the winding is known as *unbalanced winding*. In such a case each pole contains unequal number of coils of different phases.

It is important to note that there must be equal number of coils in each phase, either the winding is balanced or unbalanced.

The polyphase windings are arranged like single-phase windings, the only difference is that in two-phase alternators there are two separate single-phase windings placed 90° (electrical degrees) apart and in a 3-phase alternator there are three separate windings placed 60 electrical degrees apart for convenience.

Figure 15.17 (*a*) represents a skelton 2-phase, 4-pole winding, two slots per pole. The electrical phase difference between adjacent slots $= \frac{180°}{2} = 90°$. Each of the winding occupies 4 slots. T_1 and T_2 are the starts of phase windings no. I and no. II respectively. T_1' and T_2' are the finish of phase windings no. I and no. II respectively.

Figure 15.17 (*b*) represents a skelton 3-phase, 4-pole winding, 3 slots per pole. The electrical phase difference between adjacent slots $= \frac{180°}{3} = 60°$. Each of the winding occupies 4 slots, T_1, T_2, and T_3 are the starting terminals and T_1', T_2' and T_3' are the finishing terminals of the three phases.

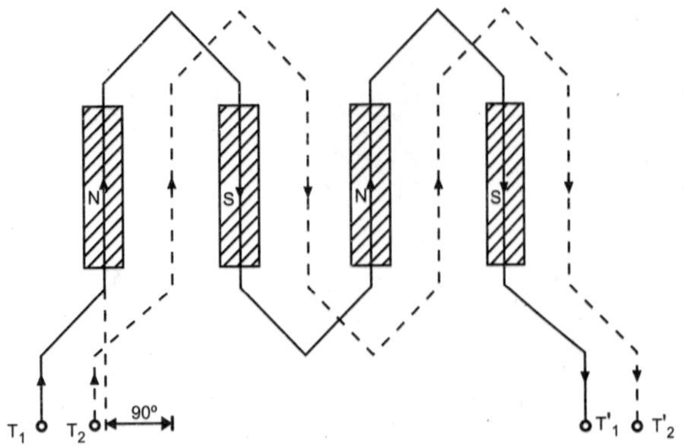

Fig. 15.17 *(a) Skelton Two Phase Winding*

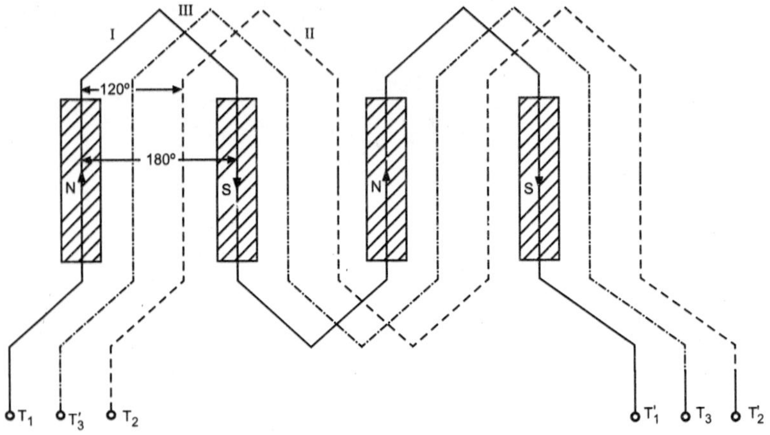

Fig. 15.17 *(b) Skelton Three Phase Winding*

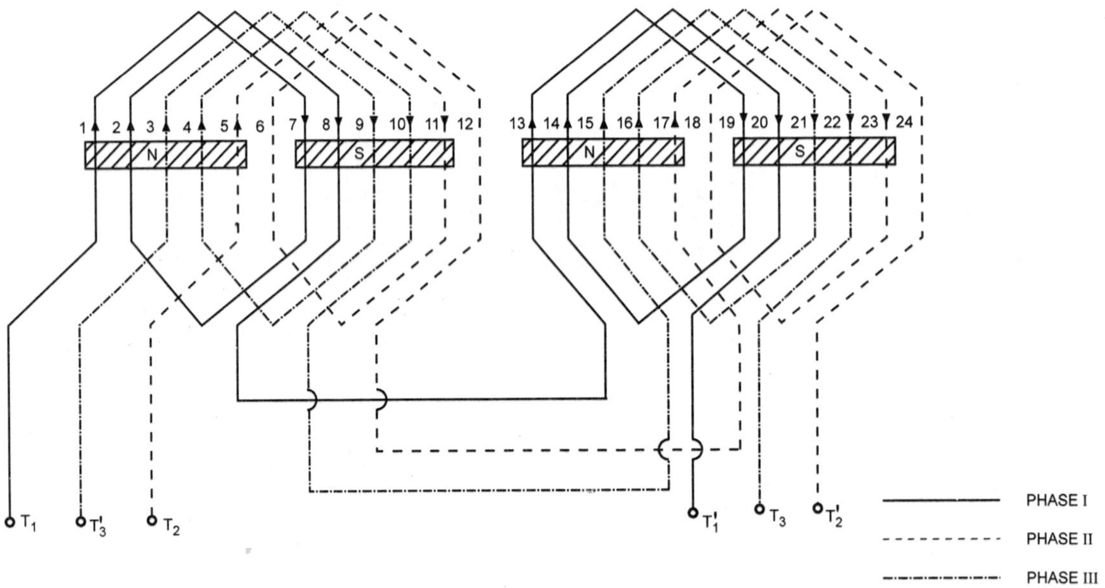

Fig. 15.17 *(c) 3-Phase Distributed Winding*

First winding starts at slot no. 1 and ends over slot no. 10. Second winding starts at slot no. 2 and ends over slot no. 11. Third winding starts at slot no. 3 and ends over slot no. 12. The phase difference of the emfs induced in windings first and second, second and third, and third and first is 60°, 60° and 120° respectively. Since in three phase system the phase difference between the adjacent windings must be 120°; this can easily be obtained by reversing the connections of second winding, as shown in Fig. 15.17 *(b)*.

Figure 15.17 *(c)* represents 4-pole, 24-slot, single layer, full-pitched 3-phase distributed winding.

$$\text{No. of slots per pole per phase} = \frac{\text{No. of slots}}{\text{No. of poles} \times \text{No. of phases}}$$

$$= \frac{24}{4 \times 3} = 2$$

$$\text{Phase angle between adjacent slots} = \frac{180^\circ}{6} = 30^\circ$$

Hence slots No. 1, 2, 7, 8, 13, 14, 19 and 20 for phase I
slots No. 5, 6, 11, 12, 17, 18, 23 and 24 for phase II
and slots No. 3, 4, 9, 10, 15, 16, 21 and 22 for phase III

Figure 15.17 *(d)* shows three-phase full-pitched, double layer lap winding. Each winding is spaced 120 electrical degrees from the two adjacent windings. This winding has 12 slots per pole or 4 slots per pole per phase. Since this winding is full-pitched coil, so the pitch of each coil is 12 slots. Since one pole represents 180 electrical space degrees, so the slot pitch corresponds to 180°/12, i.e. 15 electrical space degrees.

Figure 15.17 *(e)* shows three-phase five-sixth pitch, double layer lap winding having 4 slots per pole per phase.

In a fractional pitch winding, the coil span is made less than 180° electrical space degrees. In Fig. 15.17 *(e)* a coil instead of having a pitch of 12 slots, now has a pitch of 10 slots, so that its spread is no longer equal to pole pitch. Except pitch, the winding is similar to the winding shown in Fig. 15.17 *(d)* in every respect.

This type of winding improves the waveform of the induced emf, saves copper in the connections of coil ends and reduces the inductance of the windings. Such winding induces slightly less emf than full pitch winding under the same conditions.

Full Pitch and Short Pitch Windings. When the two coil sides forming a complete coil of a winding are 180 electrical space degrees apart, the winding is known as *full pitch winding*, as shown in Fig. 15.17 *(d)*. In this type of winding when one side of the coil is under N pole, the other side is in corresponding position under S pole. The induced emfs differ by 180° in phase,

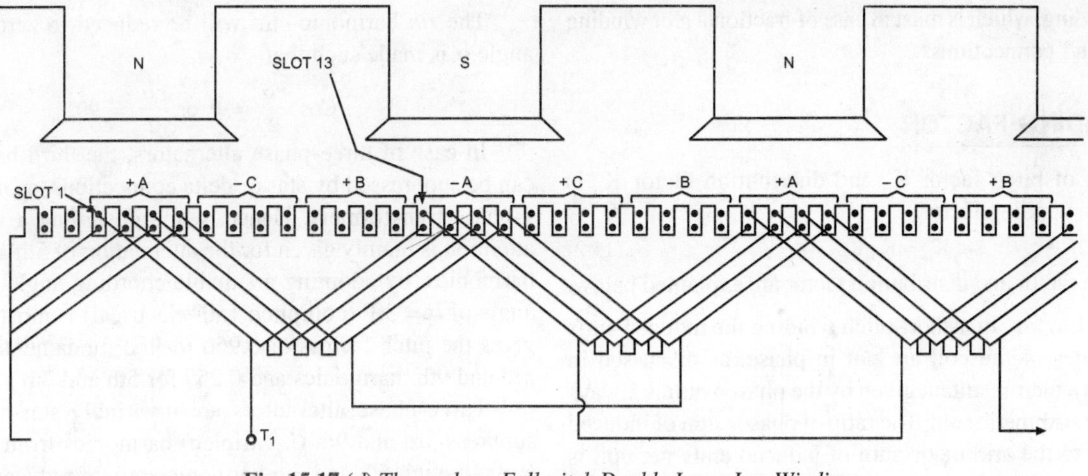

Fig. 15.17 *(d) Three-phase Full-pitch Double Layer Lap Winding*

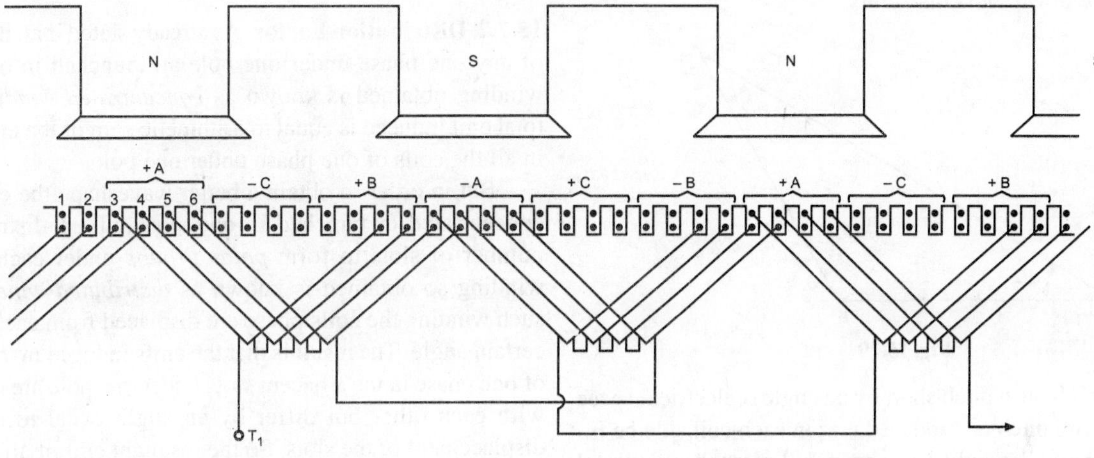

Fig. 15.17 *(e) Three-phase Five-sixth-pitch Double Layer Lap Winding*

but the coil is connected in such a way that emfs add giving resultant emf E, as shown in Fig. 15.18.

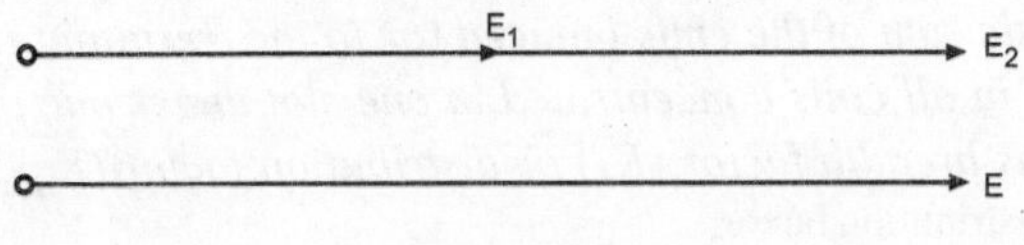

Fig. 15.18

When the coil span of the winding is less than 180 electrical space degrees, i.e. the two coil sides forming a complete coil of the winding are less than 180 electrical space degrees apart, the winding is known as *fractional pitch winding.* In this type of winding, the induced emf in each coil side is not in phase, so the resultant emf is the phasor sum of induced emfs in the coil sides, which is slightly less than their arithmetic sum, hence induced emf in short pitch winding is less than that induced in full pitch winding under the same conditions.

The advantages of short-pitch windings are :

(*i*) Waveform of the induced emf is improved and the distorting harmonic can be reduced or totally eliminated.

(*ii*) Copper is saved in the coil ends due to less span.

(*iii*) The inductance of the winding is reduced due to lesser length of coil ends as well as lesser mutual inductance between those conductors which lie in slots containing also conductors of either of the other two phases.

(*iv*) Fractional number of slots per pole can be used which in turn reduces the tooth ripples.

(*v*) Mechanical strength of the coil is increased.

Integral Slot and Fractional Slot Windings. When the number of slots per pole per phase, m is an integer, the winding is called the *integer slot* winding but when the number of slots per pole per phase, m is a fractional number the winding is referred to as *fractional slot winding*. Of course in both cases the number of slots per phase must be whole number so that each phase has the same number of coils.

Fractional slot winding is practicable only with double layer arrangement. It limits the number of parallel circuits available, because phase group under several poles must be connected in series before a unit is formed and the winding repeats the pattern to give a second unit that can be put in parallel with the first.

Though the fractional slot winding may appear to be little complicated, it has certain technical advantages as well as ease in manufacturing. Since the number of armature slots chosen need not necessarily be an integral multiple of the number of poles, one may choose a particular number of slots for which the lamination notching gear is available. This results in saving in cost of drawings and equipment. The flexibility can be effectively employed where the number of poles varies over a wide range as in the case of synchronous machines. Further mmf harmonics, leakage reactance of windings and cost of copper is reduced owing to chording, which is must in case of fractional slot winding and shorter end connections.

15.7 WINDING FACTOR

The product of pitch factor K_p and distribution factor K_d is referred to the winding factor

$$i.e.\ K_w = K_p K_d \qquad ..(15.2)$$

The pitch factor and distribution factor are explained below.

15.7.1. Pitch Factor. In a short-pitch winding the induced emfs in the two sides of the coil are not in phase, as discussed in Art 15.6, hence their resultant given by the phasor sum is always less than their arithmetic sum. The ratio of phasor sum of induced emfs per coil to the arithmetic sum of induced emfs per coil is known as *coil pitch factor, coil-span factor* or simply *pitch factor* (K_c or K_p). It is always less than unity.

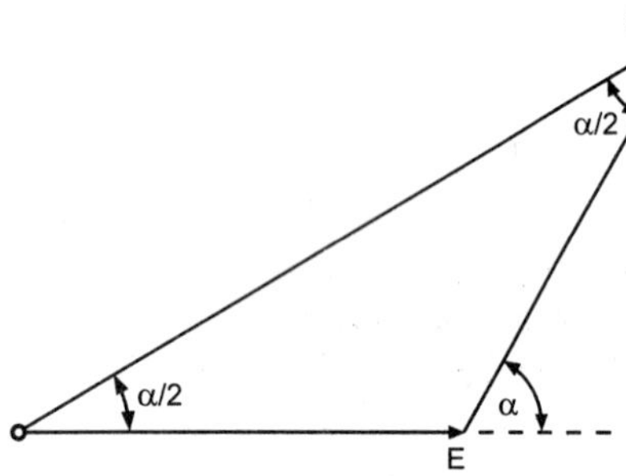

Fig. 15.19

Let the coil have a pitch short by an angle α electrical space degrees from full pitch and induced emf in each coil side be E.

Now if the coil would have been full-pitched, then total induced emf in the coil would have been 2 E.

If the coil is short pitched by α electrical space degrees the resultant induced emf, E_R in the coil is phasor sum of two voltages α° (electrical) apart

$$i.e.\ E_R = 2\,E \cos\frac{\alpha}{2}$$

$$\text{Pitch factor, } K_p = \frac{\text{Phasor sum of coil side emfs}}{\text{Arithmetic sum of coil side emfs}}$$

$$= \frac{2E\cos\frac{\alpha}{2}}{2E} = \text{Cos}\,\frac{\alpha}{2} \qquad ...(15.3)$$

If the coil span is reduced by one slot, then the phase angle α between the induced emfs in the two sides of the coil is given as

$$\alpha = \frac{180^\circ}{n} \quad \text{where } n \text{ is the number of slots per pole.}$$

The pitch factor given by Eq. (15.3) is for the fundamental component of emf. The flux wave may consist of space-field harmonics also, which give rise to the corresponding time harmonics in the generated emf wave. A 3rd harmonic component of the flux wave, may be imagined as produced by 3 poles as compared to one pole for the fundamental component. In view of this, the chording angle for the *rth* harmonic becomes *r* times the chording angle for the fundamental component and the pitch factor for the *rth* harmonic is given as

$$\text{Pitch factor, } K_{pr} = \text{Cos}\,\frac{r\alpha}{2} \qquad ...(15.4)$$

The *rth* harmonic emf will be reduced to zero if chording angle α is made such that

$$\text{Cos}\,\frac{r\alpha}{2} = 0 \text{ or } \frac{r\alpha}{2} = 90^\circ$$

In case of three-phase alternators, the third harmonic emf can be suppressed by star or delta connections, as in the case of 3-phase transformers. Hence, while designing a winding, the attention is mainly given for the attenuation of 5th and 7th order harmonics, by adopting a suitable chording angle. A chording angle of α = 30° (coil pitch 150° electrical) is most useful, as it gives the pitch factors of 0.966 for the fundamental, 0.707 for 3rd and 9th harmonics and 0.259 for 5th and 7th harmonics.

Three-phase alternators are invariably star-connected to suppress 3rd and 9th (*i.e.* triplen) harmonics from the induced emfs; though 5th and 7th harmonic emfs of reduced magnitude are present in the lines.

15.7.2. Distribution Factor. As already stated if all the coil sides of any one phase under one pole are bunched in one slot, the winding obtained is known as *concentrated winding* and the total emf induced is equal to arithmetic sum of the emfs induced in all the coils of one phase under one pole.

But in order to obtain a better waveshape the coils are not concentrated or bunched in one slot, but are distributed in a number of slots to form *polar groups* under each pole. The winding so obtained is known as *distributed winding* and in such winding the coils/phase are displaced from each other by a certain angle. The result is that the emfs induced in the coil sides of one phase in the adjacent slots under one pole are not in phase with each other but differ by an angle equal to the angular displacement of the slots. So the resultant emf of all the coils of one phase under one pole would be the phasor sum of the individual coil emfs and will be less than that if all the coil sides of one phase under one pole were bunched in one slot.

The ratio of the phasor sum of the emfs induced in all the coils distributed in a number of slots under one pole to the arithmetic sum of the emfs induced (or to the resultant of emfs induced in all coils concentrated in one slot under one pole) is known as breadth factor (K_b) or distribution factor (K_d)

or distribution factor,

$$K_d = \frac{\text{EMF induced in distributed winding}}{\text{EMF induced if the winding would have been concentrated}}$$

$$= \frac{\text{Phasor sum of component emfs}}{\text{Arithmetic sum of component emfs}}$$

The distribution (or breadth) factor is always less than unity.

Let number of slots per pole = n

and number of slots per pole per phase = m

Induced emf in each coil side = E_c

$$\text{Angular displacement between the slots, } \beta = \frac{180^\circ}{n}$$

The emfs induced in different coils of one phase under one pole are represented by side AC, CD, DE, EF, ...which are equal in magnitude (say each equal to E) and differ in phase (say by β°) from each other (Fig. 15.20).

If bisectors are drawn on AC, CD, DE, EF... they would meet at common point (say at O). The point O would be the circum centre of the circle having AC, CD, DE, EF...as the chords and representing the emfs induced in the coils in different slots.

$$\text{EMF induced in each coil side, } E = AC = 2OA \sin \frac{\beta}{2}$$

$$\text{Arithmetic sum } = m \times 2 \times OA \sin \frac{\beta}{2}$$

The resultant emf induced in one polar group of one phase would be the phasor sum as represented by the phasor AB shown in Fig. 15.20.

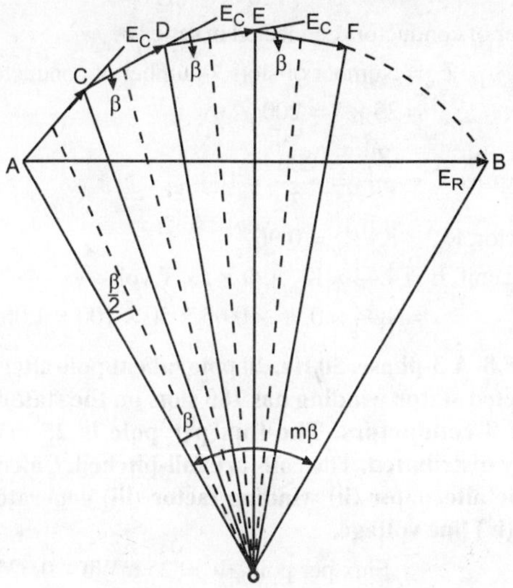

Fig. 15.20

∴ The resultant emf,

$$E_R = AB = 2 \times OA \sin \frac{AOB}{2} = 2 \times OA \sin \frac{m\beta}{2}$$

and distribution factor,

$$K_d = \frac{\text{Phasor sum of components emfs}}{\text{Arithmetic sum of components emfs}} = \frac{\sin \frac{m\beta}{2}}{m \sin \frac{\beta}{2}} \quad ...(15.5)$$

$m\beta$ is also known as the *phase spread* and is spread in electrical degrees or radians.

The distribution factor K_d given by Eq. (15.5) is for the fundamental component of emf. If the flux distribution contains space-harmonics, the slot angular pitch β on the fundamental scale, would become r β for the *r th* harmonic component and thus the distribution factor for the *r th* harmonic would be

$$K_{dr} = \frac{\sin \frac{rm\beta}{2}}{m \sin \frac{r\beta}{2}} \quad ...(15.6)$$

Example 15.3. An alternator has 9 slots per pole. The coil span is 8 slots. Find the pitch factor for fundamental frequency.

[*Rajasthan Technical Univ. Electrical Machines-II, 2011*]

Solution: Number of slots per pole, $n = 9$

Angular displacement between the slots,

$$\beta = \frac{180°}{n} = \frac{180°}{9} = 20° \text{ (electrical)}$$

$$\text{Coil span} = \frac{180° \times \text{coil span in terms of slots}}{\text{Number of slots per pole}} = \frac{180 \times 8}{9} = 160°$$

Chording angle, $\alpha = 180° - \text{coil span} = 180° - 160° = 20°$

Pitch factor for the fundamental frequency

$$k_p = \text{Cos}\frac{\alpha}{2} = \text{Cos}\frac{20°}{2} = 0.9848 \textbf{ Ans.}$$

Example 15.4. Calculate the distribution factor for a 3-phase distributed single layer winding of the armature of an alternator. The alternator has 2 poles and a total of 18 slots.

[*G.G.S.I.P. Univ. Electromechanical Energy Conversion-II, May-2011*]

Solution: Number of slots per pole, $n = \frac{18}{2} = 9$

Number of slots per pole per phase,

$$m = \frac{n}{\text{Number of phases}} = \frac{9}{3} = 3$$

Angular displacement between the slots,

$$\beta = \frac{180°}{n} = \frac{180°}{9} = 20° \text{ (electrical)}$$

Distribution factor,

$$k_d = \frac{\text{Sin} \frac{m\beta}{2}}{m \sin \frac{\beta}{2}} = \frac{\text{Sin} \frac{3 \times 20°}{2}}{3 \sin \frac{20°}{2}} = \frac{1}{3} \frac{\sin 30°}{\sin 10°} = 0.96 \textbf{ Ans.}$$

Example 15.5. An 8-pole, 3-phase, 60° spread, double layer winding has 72 coils in 72 slots. The coils are short-pitched by two slots. Calculate the winding factor for the fundamental and third harmonic.

[*U.P. Technical Univ. Electromechanical Energy Conversion-II, 2010-11*]

Solution: Number of slots per pole, $n = \frac{72}{8} = 9$

Number of slots per pole per phase,

$$m = n \times \frac{\text{phase spread}}{180} = 9 \times \frac{60}{180} = 3$$

Angular displacement between the slots,

$$\beta = \frac{180°}{n} = \frac{180°}{9} = 20° \text{ (electrical)}$$

$$\text{Coil span} = \frac{180° \times \text{coil span in terms of slots}}{\text{Number of slots per pole}}$$

$$= \frac{180\,(9-2)}{9} = 140°$$

Chording angle, $\alpha = 180° - \text{coil span} = 180° - 140° = 40°$

For the fundamental component

$$\text{Distribution factor, } K_d = \frac{\text{Sin} \frac{m\beta}{2}}{m \sin \frac{\beta}{2}} = \frac{\text{Sin} \frac{3 \times 20°}{2}}{3 \sin \frac{20°}{2}} = 0.96$$

$$\text{Pitch factor, } K_p = \text{Cos} \frac{\alpha}{2} = \text{Cos} \frac{40°}{2} = 0.94$$

Winding factor, $K_w = K_d \times K_p = 0.96 \times 0.94 = 0.9$ **Ans.**

For the third harmonic component (i.e. $r = 3$)

$$\text{Distribution factor, } K_{d_3} = \frac{\text{Sin} \frac{rm\beta}{2}}{m \sin \frac{r\beta}{2}} = \frac{\text{Sin} \frac{3 \times 3 \times 20}{2}}{3 \sin \frac{3 \times 20°}{2}} = 0.666$$

$$\text{Pitch factor, } K_{p_3} = \text{Cos} \frac{r\alpha}{2} = \text{Cos} \frac{3 \times 40°}{2} = 0.5$$

Winding factor, $K_{w_3} = K_{d_3} \times K_{p_3} = 0.666 \times 0.5 = 0.333$ **Ans.**

15.8 EMF EQUATION

Let Z_p = Number of conductors in series per phase

P = Number of poles

Φ = Useful flux per pole in webers

N = Rotational speed in rpm

and f = Frequency in hertz and is equal to $\frac{NP}{120}$

The flux cut by any conductor while passing from the centre of one interpolar gap to the centre of the next is Φ webers and since during the movement, the emf wave completes half cycle *i.e.* the time taken is $\frac{1}{2f}$ seconds, therefore, the average rate of cutting the flux,

$$\frac{d\Phi}{dt} = \frac{\Phi}{1/2f} = 2f\Phi \text{ Wb/s}$$

Hence average emf induced in each conductor = $2f\Phi$ volts

Average emf per phase,

E_{av}/phase = Number of conductors in series/phase arranged in one slot/pole × average emf induced per conductor

$= Z_p \times 2f\Phi = 2\,T \times 2f\Phi = 4\,\Phi f T$ volts

$\because Z_p = 2$ T (turns)

For distributed winding the average value of emf per phase will be K_d times the above value

i.e. E_{av} / phase = $4K_d \Phi f T$ volts

For short pitch winding the true average value of emf per phase will be K_p times the above value

E_{av}/phase = $4\,K_d\,K_p\,\Phi f T$ volts

and E_{rms}/phase = Form factor, $K_f \times E_{av}$/phase

$= 4\,K_f K_d\,K_p\,\Phi f T$ volts

For sinusoidal wave of emf, K_f = 1.11

$$\therefore E_{rms} / \text{phase} = 4.44\,K_d\,K_p\,\Phi f T \text{ volts} \quad ...(15.7)$$

If the alternator is star-connected, as is usually the case, the line voltage is $\sqrt{3}$ times the phase value.

So line induced emf, $E_L = \sqrt{3} \times 4.44\,K_d\,K_p\,\Phi f T$ volts ...(15.8)

For full-pitched and concentrated windings, $K_p = 1$ and $K_d = 1$

Example 15.6. The armature of a 1-phase alternator is completely wound with 100 single-turn coils distributed uniformly. The induced emf in each turn is 2 V. What is the emf induced in the whole winding with all 100 coils connected in series?

Solution: In case of single-phase winding, the number of slots per pole per phase,

m = number of slots per pole, n

and distribution factor is given as

$$K_d = \frac{\text{Sin}\frac{m\beta}{2}}{m\sin\frac{\beta}{2}} = \frac{\text{Sin}\, n \times \frac{180°/n}{2}}{n \times \sin\frac{180°/n}{2}} \quad \because m = n \text{ and } \beta = \frac{180°}{n}$$

$$= \frac{\text{Sin } 90°}{n \times \frac{\pi}{2n}} = \frac{1}{\pi/2} = \frac{2}{\pi}$$

$\because$ For very small angle sin θ = θ in radians

EMF induced in the whole winding

= Number of turns × induced emf in each turn × K_d

$= 100 \times 2 \times \frac{2}{\pi} = \frac{400}{\pi}$ or 127.32 V **Ans.**

Example 15.7. A 4-pole alternator has an armature with 25 slots and 8 conductors per slot and rotates at 1,500 rpm and flux per pole is 0.05 Wb. Calculate the emf generated if winding factor is 0.96 and all conductors are in series.

[*Anna Univ. Electrical Machines-II, November/December-2012*]

Solution: Flux per pole, Φ = 0.05 Wb

$$\text{Frequency}, f = \frac{PN}{120} = \frac{4 \times 1,500}{120} = 50 \text{ Hz}$$

Number of conductors connected in series,

Z_p = Number of slots × number of conductors per slot

= 25 × 8 = 200

Number of turns, $T = \frac{Z_p}{2} = \frac{200}{2} = 100$

Winding factor, $K_w = K_d\,K_p = 0.96$

Generated emf, $E = 4.44 \times K_w \times \Phi \times f \times T$ volts

$= 4.44 \times 0.96 \times 0.05 \times 50 \times 100 = 1,065.6$ V **Ans.**

Example 15.8. A 3-phase, 50 Hz, 20 pole salient pole alternator with star-connected stator winding has 180 slots on the stator. Each slot consists of 8 conductors. The flux per pole is 25 mWb and is sinusoidally distributed. The coils are full-pitched. Calculate (i) the speed of the alternator (ii) winding factor (iii) generated emf per phase and (iv) line voltage.

Solution: Flux per pole, Φ = 25 mWb = 0.025 Wb

Frequency, f = 50 Hz

Number of armature conductors, Z = 180 × 8 = 1,440

Number of armature conductors per phase = $\frac{1,440}{3}$ = 480

Number of turns per phase, T = $\frac{480}{2}$ = 240

Number of poles, P = 20

(*i*) Speed, $N = \frac{120 f}{P} = \frac{120 \times 50}{20} = 300$ rpm **Ans.**

(*ii*) Number of slots per pole, $n = \frac{180}{20} = 9$

Number of slots per pole per phase,

$$m = \frac{n}{\text{Number of phases}} = \frac{9}{3} = 3$$

Angular displacement between the slots,

$$\beta = \frac{180°}{n} = \frac{180°}{9} = 20° \text{ (elec.)}$$

$$\text{Distribution factor}, K_d = \frac{\text{Sin}\frac{m\beta}{2}}{m\sin\frac{\beta}{2}} = \frac{\text{Sin}\frac{3 \times 20°}{2}}{3\sin\frac{20°}{2}} = \frac{\text{Sin } 30°}{3 \sin 10°} = 0.96$$

Pitch factor, K_p = 1 $\because$ Coils are full-pitched

(*ii*) Winding factor, $K_W = K_d\,K_p = 0.96 \times 1 = 0.96$ **Ans.**

(*iii*) Generated emf per phase = $4.44\,K_d\,K_p\,\Phi f T$ volts

$= 4.44 \times 0.96 \times 1 \times 0.025 \times 50 \times 240$

= 1.280 V **Ans.**

(*iv*) Line voltage, $V_L = \sqrt{3} \times 1,280 = 2,215$ V **Ans.**

Example 15.9. A 3-phase, 4-pole, star-connected synchronous generator runs at 1,500 rpm. The stator has 80 slots and 18 conductors per slot. The flux in the stator yoke is 0.006 Wb. Determine generated phase and line voltages, if the winding factor is 0.96.

[*G.B. Technical Univ. Electromechanical Energy Conversion-II, 2012-13*]

Solution: Frequency, $f = \frac{PN}{120} = \frac{4 \times 1{,}500}{120} = 50$ Hz

Winding factor, $k_w = k_d k_p = 0.96$

Flux per pole, $\Phi = 0.006$ Wb

Number of turns,

$$T = \frac{\text{Number of slots} \times \text{number of conductors per slot}}{2 \times \text{number of phases}}$$

$$= \frac{80 \times 18}{2 \times 3} = 240$$

Generated phase voltage,

$$E_p = 4.44\, k_w \Phi f T = 4.44 \times 0.96 \times 0.006 \times 50 \times 240 = 306.9 \text{ V } \textbf{Ans.}$$

Generated line voltage,

$$E_L = \sqrt{3} \times E_p = \sqrt{3} \times 306.9 \text{ V} = 531.5 \text{ V} \quad \textbf{Ans.}$$

Example 15.10. Find the no-load phase and line voltages of a star-connected 3-phase, 6-pole alternator which runs at 1,200 rpm. Having flux per pole of 0.1 Wb sinusoidally distributed, its stator has 54 slots having double layer winding. Each coil has 8 turns and each coil is short pitched by 1 slot.

[Gujarat Technological Univ. Electrical Machines-I, Summer-2014]

Solution: Flux per pole, $\Phi = 0.1$ Wb

$$\text{Frequency}, f = \frac{PN}{120} = \frac{6 \times 1{,}200}{120} = 60 \text{ Hz}$$

Number of armature conductors,

$$Z = 54 \times 8 \times 2 = 864$$

$$\text{Number of turns per phase, } T = \frac{864}{3 \times 2} = 144$$

$$\text{Number of slots per pole, } n = \frac{54}{6} = 9$$

$$\text{Phase angle, } \alpha = \frac{180°}{n} = \frac{180°}{9} = 20°$$

$$\text{Pitch factor, } K_p = \text{Cos}\frac{\alpha}{2} = \text{Cos}\frac{20°}{2} = 0.9848$$

Number of slots per pole per phase,

$$m = \frac{n}{3} = \frac{9}{3} = 3$$

Angular displacement between slots,

$$\beta = \frac{180°}{n} = \frac{180°}{9} = 20° \text{ (electrical)}$$

$$\text{Distribution factor, } K_d = \frac{\text{Sin}\frac{m\beta}{2}}{m \sin\frac{\beta}{2}} = \frac{\text{Sin}\frac{3 \times 20°}{2}}{3 \sin\frac{20°}{2}} = 0.96$$

$$\text{No-load phase voltage, } E_p = 4.44\, K_d\, K_p\, \Phi f T$$

$$= 4.44 \times 0.96 \times 0.9848 \times 0.1 \times 60 \times 144$$

$$= 3{,}627 \text{ V } \textbf{Ans.}$$

$$\text{No-load line voltage, } E_L = \sqrt{3}\, E_p = \sqrt{3} \times 3{,}627 = 6{,}282 \text{ V } \textbf{Ans.}$$

Example 15.11. A 3-phase, 16-pole alternator has a star-connected winding with 144 slots and 12 conductors per slot. The flux per pole is 0.03 Wb and is sinusoidally distributed. The speed of the machine is 375 rpm. Find the frequency, the phase and line electromotive forces. *[J.N. Technological Univ. Electrical Machines-III, May-2013]*

Solution: Number of slots per pole, $n = \frac{144}{16} = 9$

$$\text{Number of slots per pole per phase, } m = \frac{9}{3} = 3$$

Number of conductors connected in series per phase,

$$Z_p = \frac{144 \times 12}{3} = 576$$

Number of series turns per phase,

$$T = \frac{Z_p}{2} = \frac{576}{2} = 288$$

Angular displacement between the slots,

$$\beta = \frac{180°}{9} = 20°$$

$$\text{Frequency}, f = \frac{PN}{120} = \frac{16 \times 375}{120} = 50 \text{ Hz } \textbf{Ans.}$$

Distribution factor,

$$K_d = \frac{\text{Sin}\, m \times \frac{\beta}{2}}{m \sin\frac{\beta}{2}} = \frac{\text{Sin}\, 3 \times \frac{20°}{2}}{3 \sin\frac{20°}{2}} = 0.9598 \textbf{ Ans.}$$

Assuming pitch factor K_p to be unity

$$\text{Phase emf, } E_p = 4.44 K_d K_p \Phi f T \text{ volts}$$

$$= 4.44 \times 0.9598 \times 1.0 \times 0.03 \times 50 \times 288$$

$$= 1{,}841 \text{ V } \textbf{Ans.}$$

$$\text{Line emf, } E_L = \sqrt{3}\, E_p = \sqrt{3} \times 1{,}841 = 3{,}189 \text{ V } \textbf{Ans.}$$

Example 15.12. Calculate the rms value of the induced emf per phase of a 10 pole, three phase, 50 Hz alternator with 2 slots per pole per phase and 4 conductors per slot in two layers. The coil span is 150°. The flux per pole is 0.12 Wb.

[U.P.S.C. I.E.S. Electrical Engineering-II, 2006]

Solution: Number of slots per pole per phase, $m = 2$

Number of slots per pole, $n = m \times$ number of phases $= 2 \times 3 = 6$

Number of slots per phase $= m \times$ number of poles $= 2 \times 10 = 20$

Number of conductors connected in series per phase,

$Z_p =$ Number of conductors per slot $\times$ number of slots per phase $= 4 \times 20 = 80$

$$\text{Number of turns per phase, } T = \frac{Z_p}{2} = \frac{80}{2} = 40$$

Flux per pole, $\Phi = 0.12$ Wb

Supply frequency, $f = 50$ Hz

Angular displacement between slots,

$$\beta = \frac{180°}{n} = \frac{180°}{6} = 30°$$

$$\text{Distribution factor, } K_d = \frac{\text{Sin}\frac{m\beta}{2}}{m \sin\frac{\beta}{2}} = \frac{\text{Sin}\frac{2 \times 30°}{2}}{2 \sin\frac{30°}{2}} = 0.966$$

$$\text{Chording angle, } \alpha = 180° - \text{coil span} = 180° - 150° = 30°$$

$$\text{Pitch factor, } K_p = \text{Cos}\frac{\alpha}{2} = \text{Cos}\frac{30°}{2} = 0.966$$

RMS value of induced emf per phase,

$$E_{p\,(\text{rms})} = 4.44 K_d K_f \Phi f T$$

$$= 4.44 \times 0.966 \times 0.966 \times 0.12 \times 50 \times 40$$

$$= 994.37 \text{ V } \textbf{Ans.}$$

Example 15.13. A 3-phase, star-connected alternator has the following data:

Voltage required to be generated on open circuit = 4,000 V at 50 Hz; Speed = 500 rpm; slots per pole per phase = 3; Conductors per slot = 12. Calculate number of poles and (*ii*) useful flux per pole.

Assume all conductors per phase to be connected in series and coils to be full pitch.

[J.N. Technological Univ. Electrical Machines-III, May-2013]

Solution: Open-circuit line voltage, $E_L = 4{,}000$ V

Number of slots per pole per phase,

$$m = 3$$

Number of slots per pole, $n = 3 \times m = 3 \times 3 = 9$

(*i*) Number of poles, $P = \frac{120f}{N} = \frac{120 \times 50}{500} = 12$ **Ans.**

Number of conductors connected in series per phase

$$Z_p = 3 \times 12 = 36$$

Number of series turns per phase,

$$T = \frac{Z_p}{2} = \frac{36}{2} = 18$$

Angular displacements between slots,

$$\beta = \frac{180°}{n} = \frac{180}{9} = 20°$$

$$\text{Distribution factor, } K_d = \frac{\text{Sin } \frac{m\beta}{2}}{m \sin \frac{\beta}{2}} = \frac{\text{Sin} \times \frac{3 \times 20°}{2}}{3 \sin \frac{20°}{2}} = 0.9598$$

Coil span factor, $K_p = 1$ ∵ Coils are full-pitched

$$(ii) \text{ Flux per pole, } \Phi = \frac{E_L}{\sqrt{3} \times 4.44 \times K_d \times K_p \times f \times T}$$

$$\because E_L = \sqrt{3}\ 4.44 \times K_d \times K_p \times \Phi \times f \times T$$

$$= \frac{4{,}000}{\sqrt{3} \times 4.44 \times 0.9598 \times 1 \times 50 \times 18}$$

$= 0.602$ Wb **Ans.**

Example 15.14. A 3.75 MVA, 10 kV, 3-phase, 50 Hz, 10 pole low-speed hydrogenerator has 144 slots containing a two-layer diamond winding with 5 conductors per coil side in each slot. The coil span is 12 slot pitches. The flux per pole is 0.116 Wb. Calculate the phase emf. [*U.P.S.C. I.E.S. Electrical Engineering, 2008*]

Solution: Flux per pole, $\Phi = 0.116$ Wb

Supply frequency, $f = 50$ Hz

$$\text{Slots per pole, } n = \frac{144}{10} = 14.4$$

$$\text{Slots per pole per phase, } m = \frac{14.4}{3} = 4.8$$

Angular displacements between slots,

$$\beta = \frac{180°}{n} = \frac{180°}{14.4} = 12.5°$$

Coil span $= 12 \times 12.5 = 150°$

Chording angle, $\alpha = 180° - 150° = 30°$

$$\text{Pitch factor, } K_p = \text{Cos}\frac{\alpha}{2} = \text{Cos}\frac{30°}{2} = 0.9659$$

$$\text{Distribution factor, } K_d = \frac{\text{Sin } \frac{m\beta}{2}}{m \sin \frac{\beta}{2}} = \frac{\text{Sin } \frac{4.8 \times 12.5°}{2}}{4.8 \sin \frac{12.5°}{2}} = 0.9568$$

Number of turns per phase,

$$T = \frac{144 \times 5 \times 2}{2 \times 3} = 240$$

Phase voltage, $E_p = 4.44\ K_d\ K_p\ \Phi f T$

$= 4.44 \times 0.9568 \times 0.9659 \times 0.116 \times 50 \times 240$

$= 5{,}712$ V **Ans.**

Example 15.15. Calculate the rms value of the induced emf per phase of a 10-pole, 3-ϕ, 50 Hz alternator with 2 slots per pole per phase and 4 conductors per slot in two layers. The coil span is 150°. The flux per pole has a fundamental component of 0.12 Wb and a 20 % third harmonic component.

[*J.N. Technological Univ. Electrical Machines-III, May-2013*]

Solution: Number of slots / pole / phase, $m = 2$

Number of slots / pole, $n = 2 \times 3 = 6$

Number of slots / phase $= 2 \times 10 = 20$

Number of conductors connected in series, $Z_p = 20 \times 4 = 80$

Number of series turns/phase, $T = \frac{Z_p}{2} = \frac{80}{2} = 40$

Angular displacement between adjacent slots,

$$\beta = \frac{180°}{n} = \frac{180°}{6} = 30°$$

$$\text{Distribution factor, } K_d = \frac{\text{Sin } m\frac{\beta}{2}}{m \sin \frac{\beta}{2}} = \frac{\text{Sin } \frac{2 \times 30°}{2}}{\text{Sin } \frac{30°}{2}} = 0.966$$

$$\text{Coil span factor, } K_p = \text{Cos } \frac{\alpha}{2} = \text{Cos } \frac{180° - 150°}{2}$$

$$= \text{Cos } 15° = 0.966$$

Induced emf per phase (fundamental component),

$E_1 = 4.44\ K_d\ K_p\ \Phi f T$

$= 4.44 \times 0.966 \times 0.966 \times 0.12 \times 50 \times 40$

$= 994.4$ V.

For third harmonic component of flux

$$\text{Distribution factor, } K_{d_3} = \frac{\text{Sin } \frac{mr\beta}{2}}{m \sin \frac{r\beta}{2}} = \frac{\text{Sin } \frac{2 \times 3 \times 30°}{2}}{2 \sin \frac{3 \times 30°}{2}} = 0.707$$

$$\text{Coil span factor, } K_{p_3} = \text{Cos } 3 \times \frac{180° - 150°}{2} = \text{Cos } 45° = 0.707$$

Frequency, $\Phi_3 = 3 \times f = 3 \times 50 = 150$

$$\text{Flux per pole, } \Phi_3 = \frac{1}{3} \times 0.12 \times \frac{20}{100} = 0.008 \text{ Wb}$$

Induced emf per phase (third harmonic component)

$E_3 = 4.44\ K_d\ K_p\ \Phi_3 f_3\ T$

$= 4.44 \times 0.707 \times 0.707 \times 0.008 \times 150 \times 40 = 106.56$ V

Induced emf per phase,

$$E_p = \sqrt{E_1^2 + E_3^2} = \sqrt{(994.4)^2 + (106.56)^2} = 1{,}000 \text{ V } \textbf{Ans.}$$

Example 15.16. The flux distribution curve of a smooth-core, 50 Hz generator is B = 10 sin θ + 2 sin 3θ + 2 sin 5θ + 2 sin 7 θ kilo-gauss when θ is measured from the neutral axis. The pole pitch is 35 cm, the core length 32 cm, and the stator coil span four-fifths of the pole pitch. Determine the equation for the emf induced in one turn, and its rms value.

Solution: Area of pole pitch $= 0.35 \times 0.32 = 0.112\ m^2$

Fundamental component of flux per pole,

Φ_1 = Average flux density × area

$$= \frac{2}{\pi} \times 10 \times 1{,}000 \times 10^{-4} \times 0.112 = 0.0713 \text{ Wb}$$

∵ Average value of flux density $= \frac{2}{\pi}$ peak value of flux density

and 1 kilo-gauss $= 10^3 \times 10^{-4}$ Wb/m^2

RMS value of fundamental voltage per conductor

$$= 1.11 \times 2 f \times \Phi_1 = 1.11 \times 2 \times 50 \times 0.0713$$

$$= 7.914 \text{ V}$$

Peak value of fundamental voltage per conductor

$$= \sqrt{2} \times 7.914 = 11.193 \text{ V}$$

Since harmonic conductor voltages are in proportion to their flux-densities, so

Third harmonic voltage (maximum value)

$$= 11.193 \times \frac{2}{10} = 2.24 \text{ V}$$

Fifth harmonic voltage (maximum value)

$$= 11.193 \times \frac{2}{10} = 2.24 \text{ V}$$

Seventh harmonic voltage (maximum value)

$$= 11.193 \times \frac{2}{10} = 2.24 \text{ V}$$

$$\text{Coil span} = \frac{4}{5} \times 180° = 144°$$

So chording angle, $\alpha = 180° - 144° = 36°$

Pitch factor for fundamental component,

$$K_{p_1} = \text{Cos } \frac{\alpha}{2} = \text{Cos } \frac{36°}{2} = 0.951$$

Pitch factor for third harmonic component,

$$K_{p_3} = \text{Cos } \frac{3\alpha}{2} = \text{Cos } \frac{3 \times 36°}{2} = 0.588$$

Pitch factor for fifth harmonic component,

$$K_{p_5} = \frac{\text{Cos } 5\alpha}{2} = \frac{\text{Cos } 5 \times 36°}{2} = 0$$

Pitch factor for seventh harmonic component,

$$K_{p_7} = \frac{\text{Cos } 7\alpha}{2} = \frac{\text{Cos} \times 7 \times 36°}{2} = -0.588$$

Since peak values of induced emfs in one turn

$$= 2 \times K_{p_r} \times \text{peak value of induced emf per conductor}$$

So peak values of induced emfs in one turn

(*i*) for fundamental component,

$$E_{1\max} = 2 \times 0.951 \times 11.193 = 21.3 \text{ V}$$

(*ii*) for third harmonic component,

$$E_{3\max} = 2 \times 0.588 \times 2.24 = 2.64 \text{ V}$$

(*iii*) for 5th harmonic component,

$$E_{5\max} = 2 \times 0 \times 2.24 = \text{zero}$$

(*iv*) for 7th harmonic component,

$$E_{7\max} = 2 \times (-0.588) \times 2.24 = -2.64 \text{ V}$$

Hence the equation for the emf induced in one turn is given as

$$e = 21.3 \sin\theta + 2.64 \sin 3\theta - 2.64 \sin 7\theta \text{ volts } \textbf{Ans.}$$

RMS value of emf induced in one turn,

$$E_{rms} = \sqrt{\left(\frac{21.3}{\sqrt{2}}\right)^2 + \left(\frac{2.64}{\sqrt{2}}\right)^2 + \left(\frac{2.64}{\sqrt{2}}\right)^2} = 15.29 \text{ V } \textbf{Ans.}$$

15.9 WAVESHAPE

The flux distribution curve in the air gap of an alternator (Fig. 15.21) is not usually so rectangular as in a dc machine, but it is never perfectly sinusoidal unless the machine has a salient pole rotor with specially shaped pole shoes graduating the air gap (Fig. 15.6), or a cylindrical rotor with a sinusoidally distributed field winding. The induced emf wave per conductor is similar to flux wave since induced emf per conductor is directly proportional to flux density at constant speed. (e being equal to Blv volts, where l, the length of conductor is constant and at constant speed v is also constant).

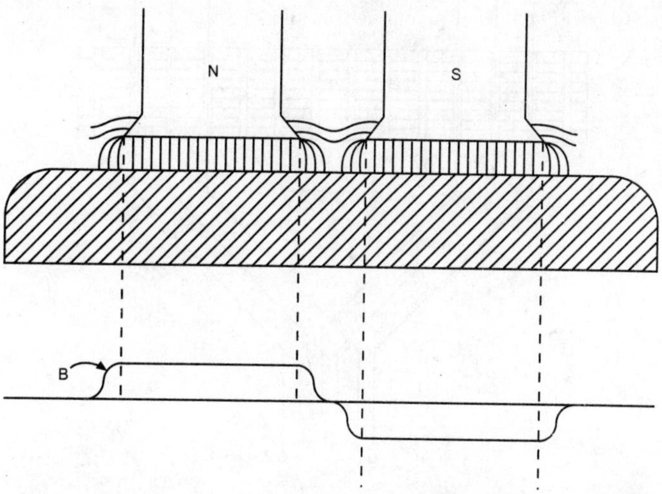

Fig. 15.21

If the winding is full-pitched one, the emfs, induced in two coil sides of each coil will be in phase and of the same magnitude because at any instant both the sides of the coil lie under corresponding positions of opposite poles. Hence for full-pitched winding, the induced emf wave in each coil will have the same shape as the emf induced in each coil side. If concentrated winding is employed, the resulting emf wave will be of the same shape as the flux density curve, shown in Fig. 15.21, which is flat topped.

In Fig. 15.22, a phase belt consisting of 3 coils of a 3 phase alternator having 3 slots per pole per phase is shown. The shape of emf wave for each of the 3 full-pitch coils giving one phase of the winding will be of the shape, as shown in Fig. 15.23 by curve I, II and III respectively. The displacement angle, β between the adjacent slots is $\frac{180°}{9} = 20$ electrical space degrees. Hence three emfs in the three successive coils are 20 electrical space degrees apart, as shown in Fig. 15.23. The resultant emf per phase would be the phasor sum of the induced emfs in the three coils, which may be obtained by adding the ordinates of the three emfs waves. The resultant emf per phase E_p wave is not flat topped but is almost sine wave. Hence the distributed winding is employed to give emf wave nearly sinusoidal. By using distributed winding, breadth factor for harmonics is very much less than for the fundamental as explained in Art. 15.7.2, hence harmonics are reduced in the resulant emf wave.

For a fractional pitch winding, the emf in each coil may be obtained first by adding the emfs of each coil side graphically and then resultant emf per phase may be obtained by adding the coil emfs, as shown in Fig. 15.23. The resultant emf wave for a fractional pitch winding will be more nearly sinusoidal than for full pitch winding. Pitch factor or coil span factor for harmonics is very much less than for the fundamental component, as already explained in Art. 15.7.1.

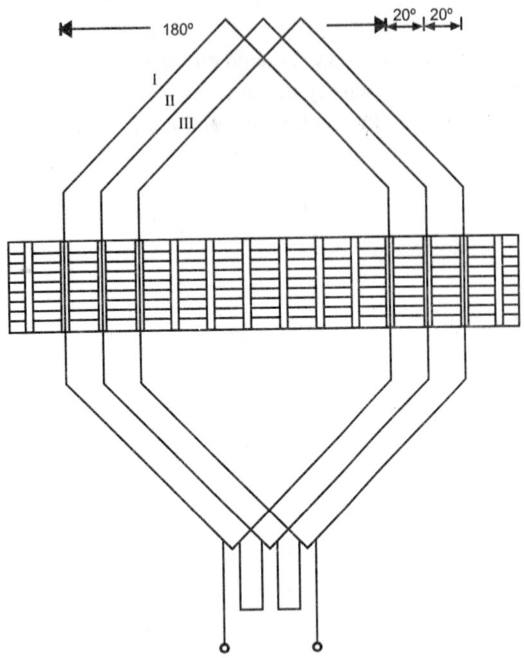

Fig. 15.22

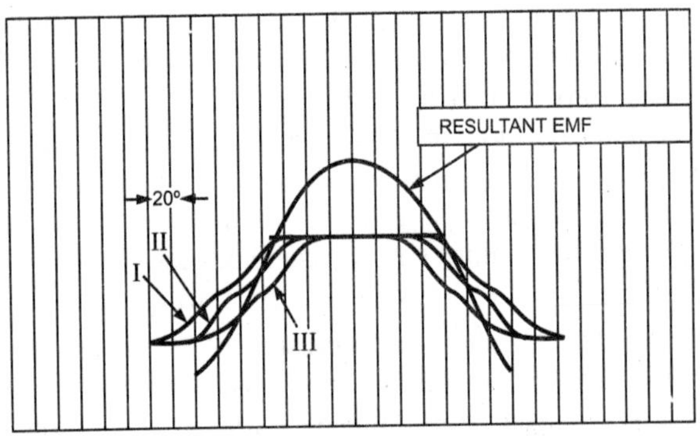

Fig. 15.23

15.10 HARMONICS IN VOLTAGE WAVEFORM

The sources of harmonics in the output voltage waveform are the (*i*) nonsinusoidal waveform of the field flux, as already explained in the foregoing article, and (*ii*) variation in reluctance of the air gap due to the slotting of the stator core.

15.10.1. Suppression of Harmonics. In case of salient pole machines if the air gap is made to vary sinusoidally around the machine, the field produced will be sinusoidal. However, it suffices to increase the air gap gradually from centre toward the pole tips, as shown in Fig. 15.6. An approximate of sinusoidal field form can be obtained by skewing the pole faces.

In cylindrical rotor machines, the length of air gap is uniform throughout, therefore, to have field of sinusoidal form, the mmf of the field winding is made to vary as nearly sinusoidal as possible. This is achieved by distributing the winding in slots.

It may be mentioned here that on account of saturation in iron parts, the sinusoidal field form cannot be obtained in salient pole machines even if the air-gap length is varied sinusoidally or in cylindrical rotor machines even if the mmf distribution in space is made sinusoidal. An ideal sinusoidal field form is very difficult to achieve, and therefore, harmonic emfs will be induced in the winding. The redeeming feature is that the harmonics can be easily eliminated from the output voltage waveform by properly designing the windings. The various methods for elimination of harmonics from the output voltage are:

(*i*) *Distributing* the armature winding as already explained in the foregoing Arts.

(*ii*) *Short chording* the armature winding by making the coil span less than a full pole-pitch, as already explained in the foregoing Arts.

(*iii*) *Skewing the poles* usually through one slot-pitch.

(*iv*) *Fractional slot winding* : Higher order harmonics can be drastically reduced by using fractional slot winding. The flux pulsations can be reduced by having number of slots per pole arc as an integer plus 1/2.

(*v*) *Larger air-gap length* causes the increase in reluctance which will reduce the harmonics.

15.11 RATING OF ALTERNATORS

All the power apparatus whether it is steam engine or gas engine or electric machine have power ratings defined as the power which can be safely and efficiently delivered by a machine under some specific conditions.

Electrical apparatus or machine is usually rated at the load, which it can carry without overheating and damage to insulation i.e. rating of electric machine is governed by the temperature rise caused by the internal losses of the machine. The copper loss in the armature (I^2R) depends upon the strength of the armature current and core loss on voltage and thus these losses are independent of power factor.

The output in kW is proportional to power factor for the alternator of a given kVA. For example output of 1,000 kVA alternator on full load will be 200, 500, 800 and 1,000 kW at power factor 0.2, 0.5, 0.8 and unity respectively, but copper losses in armature will remain same regardless of power factor.

For the foregoing reasons alternators are rated in kVA or kW at specified power factor. Other name plate datas include voltage, current, frequency, speed, number of phases, field ampere and voltage and maximum temperature rise.

The prime movers which drive the alternators, have a rating independent of power factor. It depends on kilowatt output. Thus a turbine, which can be used to drive a 25,000 kVA alternator operating at 0.8 pf, can also drive another alternator of rating 30,000 kVA but operating at 0.66 pf. However, it cannot drive an alternator of 22,500 kVA capacity operating at unity power factor.

In addition to mentioning of kVA/MVA rating of alternator, it is also necessary to mention the pf (usually from 0.8 to 0.9 lagging) for which they are designed to operate under steady-state conditions. This is because an alternator designed to operate 0.8 pf (lagging) at rated load would need more field current when operated at pf lower than the designed value. Higher field current would cause overheating of the field system, which is quite objectionable. Since for constant terminal voltage, field current required for operation at lagging pf is higher than that in case of operation at leading pf, a lagging pf puts a restriction on exciter output and current in the excitation coils. So, in case pf given on the name plate does not bear, whether the pf is leading or lagging one, then it should be assumed as lagging one. Under rated voltage and power conditions, reactive power flow handled by an alternator is restricted by its armature heating, for operating power factors near the rated value. In case of operating pf deviated (usually lower) from its designed value, the flow of reactive power is restricted both by heating field system as well as armature.

Example 15.17. A 3-ϕ, hydroelectric synchronous generator is read to be 110 MW, 0.8 pf lagging, 6 kV, Y-connected, 50 Hz, 100 rpm. Determine the following:

(*i*) the number of poles

(*ii*) the kVA rating

(*iii*) the prime-mover rating if the full-load generator efficiency is 97.1% (leave out field loss).

[*U.P. Technical Univ. Electromechanical Energy Conversion-II, 2013-14*]

Solution: (*i*) Number of poles, $P = \frac{120f}{N} = \frac{120 \times 50}{100} = 60$ **Ans.**

(*ii*) kVA rating $= \frac{\text{Power output in MW} \times 1{,}000}{\text{pf}}$

$= \frac{110 \times 1{,}000}{0.8} = 137{,}500$ **Ans.**

(*iii*) Prime-mover rating $= \frac{\text{Power output}}{\eta} = \frac{110}{0.971} = 113.28$ **Ans.**

15.12 FACTORS CONTROLLING SIZE OF ALTERNATOR

With the increase in kVA rating of alternator operating efficiency improves. For instance, an alternator of rating 10 kVA may have efficiency as low as 50% while one of 10 MVA may have efficiency of about 90% for the given pf. Another advantage of large size machines is increase in power output/kg with the increase in machine rating i.e. larger machines weigh relatively less than smaller ones and are, consequently cheaper.

However, with large-sized machines, ventilation problem becomes acute. Large sized machines tend to overheat because of production of higher power loss per unit surface area (W/m^2). So far keeping the temperature rise within acceptable limits, efficient cooling systems which becomes even more elaborate with the increase in power output, is required. For cooling of machines of rating up to 50 MVA, circulating cold-air system is adequate, but for large sized machines (say 500 MVA), hydrogen cooling is required. For very large sized machines, in the range of 1,000 MVA, we have to restort to hollow water-cooled conductors. Ultimately, a point is reached where increase in cost of cooling system exceeds the saving made elsewhere and this fixes the upper limit of the machine size.

So far as the speed is concerned, low speed machines are always larger than high-speed machines of the same output rating and therefore cooling problem is simplified in low speed machines.

15.13 LEAKAGE REACTANCE

When current flows through the stator conductors, the flux is set up, a portion of this flux does not cross the air gap, but completes its path in the stator, as shown in Fig. 15.24. Such a flux is known as the *leakage flux*. This leakage flux is proportional to stator (or armature) current, since the magnetic path it covers is not normally saturated. It also depends on the phase angle between the stator current and voltage applied across the stator. The leakage flux sets up an emf of self-inductance leading the load current I by $\pi/2$ and proportional to load current I in magnitude. Hence armature winding is assumed to possess leakage reactance X_L such that the voltage drop due to it, IX_L is equal to an emf set up by leakage flux. A part of the generated emf is used up to overcome this leakage reactace drop in addition to armature resistance drop.

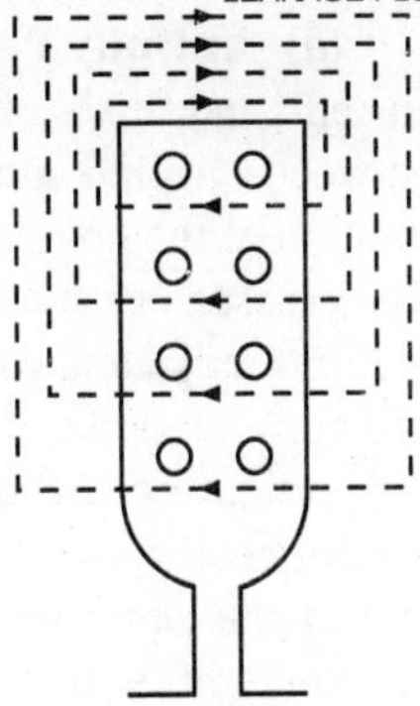

Fig. 15.24

i.e. Generated emf

= Phasor sum of terminal voltage, armature resistance drop and leakage reactance drop.

$$i.e.\ \mathbf{E} = \mathbf{V} + \mathbf{I}R_e + j\,\mathbf{I}\,X_L = \mathbf{V} + \mathbf{I}\,(R_e + j\,X_L) \quad ...(15.9)$$

Phasor diagram illustrating this fact is shown in Fig. 15.25.

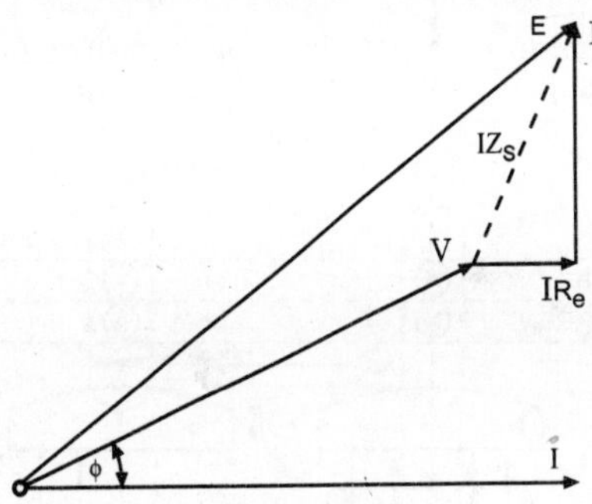

Fig. 15.25

15.14 ARMATURE REACTION

In a dc machine, the armature mmf (ampere-turns) act on the magnetic circuit of the machine in such a way as to distort the air-gap flux and to change its magnitude. For a given armature current, the direction and magnitude of this armature reaction depend on the position of brushes, as already discussed in Chapter 6 of Electrical Machines-I. In an alternator, somewhat similar conditions exist. For a given armature current, the magnitude and direction of the armature reaction cannot depend on the brush position, as in case of a dc machine, but depends on the phase displacement of the stator current I with respect to the emf E induced in the stator winding by the field winding flux.

In an alternator the phase displacement or phase angle ϕ between current I and the emf E can be within the limits of $\frac{-\pi}{2} \le \phi \le \frac{+\pi}{2}$.

We will consider three cases (*i*) when $\phi = 0$ *i.e.* when the power factor of load is unity (*ii*) when $\phi = \frac{+\pi}{2}$ *i.e.* when the power factor is zero and lagging and (*iii*) when $\phi = \frac{-\pi}{2}$ *i.e.* when the power factor of the load is zero and leading.

(*a*) At Unity Power Factor. The distribution of the stator currents and fluxes of an alternator with zero phase displacement between current I and emf E (*i.e.* when pf is unity), assuming the rotation of the poles to be clockwise, is shown in Fig. 15.26 (*a*). The crosses (+) and dots (•) appearing in the conductor cross sections indicate the instantaneous direction of the emfs induced and currents flowing—cross indicating the inward direction and dot indicating the outward direction. The maximum of the fundamental wave of the field will be opposite to the pole centres, and at the same points the conductors have their maximum induced emf; with $\phi = 0$ the conductors carrying the maximum current will also be at the same points, as shown in the figure. The armature reaction mmf is directed perpendicular to the main field mmf, as in case of a dc machine with brushes on the neutral axis. This causes the distortion of the flux due to main field and asymmetrical distribution of the flux density under the pole shoe. The flux density under the trailing pole tips increases somewhat, while under the leading pole tips it decreases. The axis of the resultant field is displaced under the action of armature reaction mmf in a direction opposite to that of rotation of the rotor.

Hence armature reaction at unity power factor has got distorting effect.

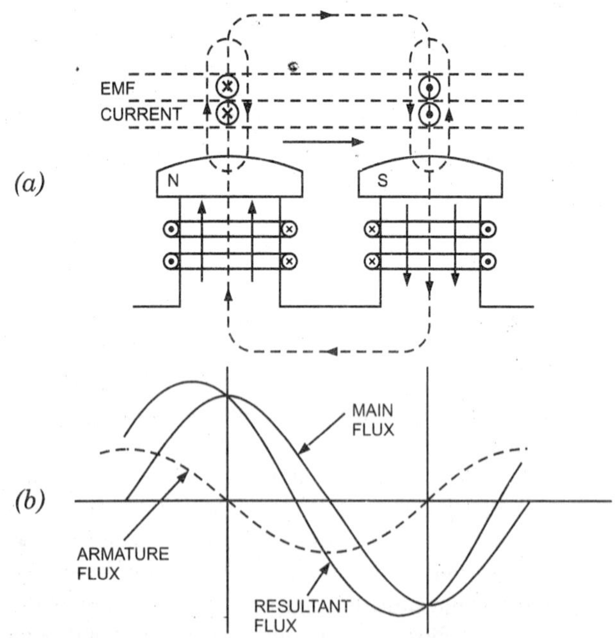

Fig. 15.26 *Armature Reaction At Unity Power Factor*

(*b*) At Lagging Zero Power Factor. The distribution of currents and fluxes with phase angle $\phi = +\frac{\pi}{2}$ radians *i.e.*, for a purely inductive load is shown in Fig. 15.27 (*a*). The current maximum will be shifted in space by an angle $\frac{\pi}{2}$ from the emf maximum, which coincides with the centres of the poles. This shift will be opposite to the direction of rotation, because the fundamental armature reaction wave rotates in step with the field poles, while when $\phi = +\frac{\pi}{2}$, current wave lags behind the emf wave by an angle $\frac{\pi}{2}$. The field created by the armature reaction mmf will be in opposite to the main field flux and will, therefore, have a *wholly demagnetising effect.*

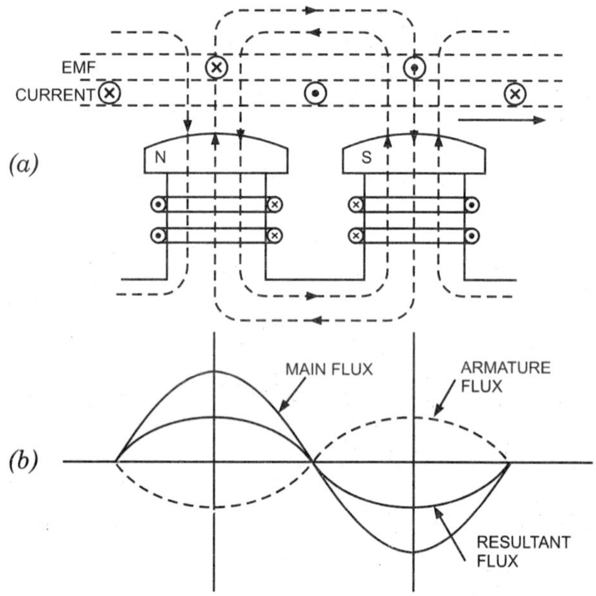

Fig. 15.27 *Armature Reaction At Zero Power Factor (Lagging)*

(*c*) At Leading Zero Power Factor. The distribution of currents and fluxes with phase angle $\phi = -\frac{\pi}{2}$ *i.e.* for a purely capacitive load is shown in Fig. 15.28 (*a*). The current maximum will be shifted to the right from the emf maximum, which remains as before under the pole centres, and the armature reaction will, therefore, have a wholly magnetizing effect on the main field.

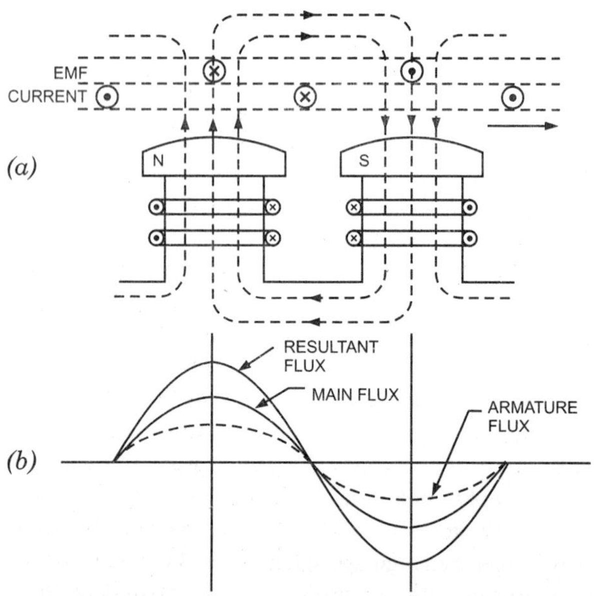

Fig. 15.28 *Armature Reaction At Zero Power Factor (Leading)*

For any power factor cos ϕ of load, the armature reaction has cross-magnetizing component proportional to I cos ϕ and demagnetizing component proportional to I sin ϕ, and is taken + ve for lagging power and – ve for leading power factor.

The phasor sum of the fundamental armature winding and field winding mmf waves constitutes, in a synchronous machine, the mmf creating the resultant magnetic flux.

15.15 SYNCHRONOUS IMPEDANCE

15.15.1. Synchronous Reactance. The emf set up due to armature reaction mmf F_a is always in quadrature with the load current I and is proportional to it. Thus it is equivalent to an emf induced in an inductive coil and the effect of armature reaction can, therefore, be considered equivalent to reactance drop IX_a where X_a is the fictitious reactance which takes care of the armature reaction effect. The armature winding possesses a certain leakage reactance X_L, as mentioned in Art. 15.13. The sum of leakage reactance X_L and fictitious reactance X_a is called the *synchronous reactance*.

$$i.e.\ X_s = X_L + X_a \qquad ...(15.10)$$

15.15.2. Effective Resistance. The effective resistance of the armature winding is somewhat greater than the conductor resistance, called the *dc resistance,* as measured by direct current. This is due to additional loss, over the purely I^2R loss, inside and sometimes outside the conductor, owing to alternating current. The main sources of this additional loss are (*i*) eddy currents in the surrounding material; (*ii*) magnetic hysteresis in the surrounding material; and (*iii*) eddy currents or unequal current distribution in the conductor itself.

In many cases it is sufficiently accurate to measure the armature resistance by direct current and increase it to a fictitious value, called the *effective resistance*, R_e which is large enough to take care of these additional losses. Effective resistance, R_e can vary widely from 1.25 to 1.75 or more times the dc resistance, depending upon design.

15.15.3. Synchronous Impedance. When the synchronous reactance X_s is combined with the armature effective resistance R_e, the quantity obtained is called the *synchronous impedance.*

$$i.e.\ \mathbf{Z}_s = R_e + j\,X_s \qquad ...(15.11)$$

Armature winding effective resistance R_e in alternators is usually very small in comparison to synchronous reactance X_s, and therefore, synchronous impedance Z_s may be assumed equal to the synchronous reactance for many purposes.

15.16 ALTERNATOR ON LOAD

If the excitation of an alternator is adjusted to give normal voltage at no load and then a load is applied, the terminal voltage of the alternator changes, even though the speed is kept constant. The voltage usually falls, but in certain circumstances it may rise even.

The change in terminal voltage of an alternator with the change in load supplied by it is due to the following three reasons:

(*i*) Voltage drop on account of armature effective resistance.

(*ii*) Voltage drop on account of armature leakage reactance.

(*iii*) Voltage drop on account of armature reaction.

15.16.1. Phasor Diagram. Let the effective armature resistance of an alternator be R_e, leakage reactance X_L and voltage drop due to armature reaction be equal to voltage drop on account of fictitious reactance X_a. All the values are referred to phase values.

The phasor diagrams for inductive, purely resistive and capacitive loads are shown in Fig. 15.29 (*a*), 15.29 (*b*) and 15.29 (*c*) respectively. All these phasor diagrams apply to one phase of a 3-phase machine.

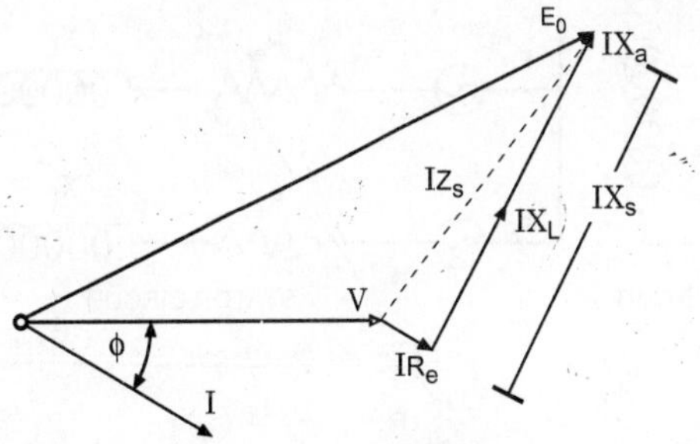

(*a*) *At Lagging PF*

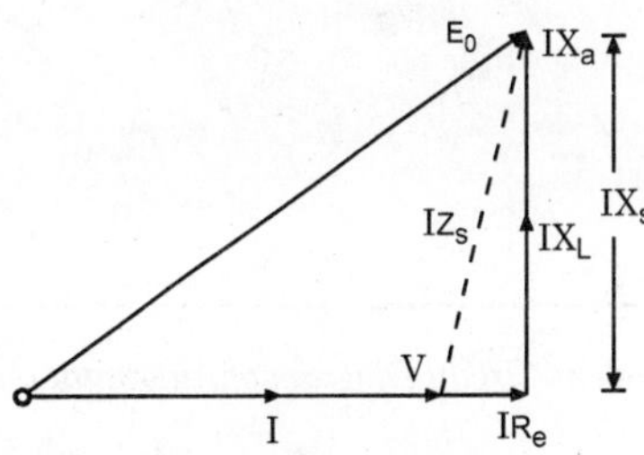

(*b*) *At Unity PF*

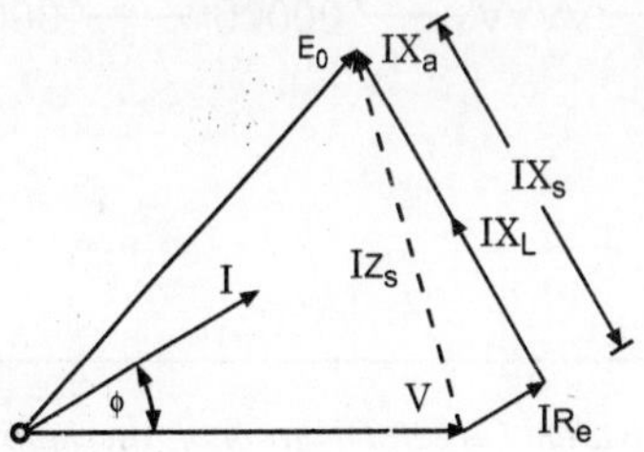

(*c*) *At Leading PF*

Fig. 15.29 *Phasor Diagrams For a Loaded Alternator*

Let the terminal voltage per phase be V volts and load current per phase be I amperes and power factor cos ϕ. The voltage drop due to armature effective resistance, I R_e will be in phase with current I, the voltage drop due to leakage reactance, I X_L and voltage drop due to armature reaction I X_a will be in quadrature with the load current I. The terminal voltage V differs from the terminal voltage on no load, also called the *excitation voltage,* E_0 by the phasor sum of above voltage drops *i.e.* $\mathbf{IZ}_s$, where $\mathbf{Z}_s$ is the synchronous impedance.

$$\text{Total voltage drop on load} = \mathbf{I}R_e + j(\mathbf{I}X_L + \mathbf{I}X_a)$$
$$= \mathbf{I}[R_e + j(X_L + X_a)]$$
$$= \mathbf{I}\,\mathbf{Z}_s \qquad ...(15.12)$$

$$\text{and voltage on no load } \mathbf{E}_0 = \mathbf{V} + \mathbf{I}\,\mathbf{Z}_s \qquad ...(15.13)$$

15.16.2. Equivalent Circuit (or Circuit Model). When the alternator is operating on no load (*i.e.* the rotor is rotating and energized and stator is open circuited) its circuit diagram is shown in Fig. 15.30 (*a*). Each phase generates an emf E_0 per phase. The equivalent circuit of an alternator for one phase is shown in Fig. 15.30 (*b*). At no load, the terminal voltage V per phase is equal to excitation or generated emf E_0 per phase *i.e.* $V = E_0$.

The per phase equivalent circuit of an alternator on load may be represented as given in Fig. 15.30 (*c*).

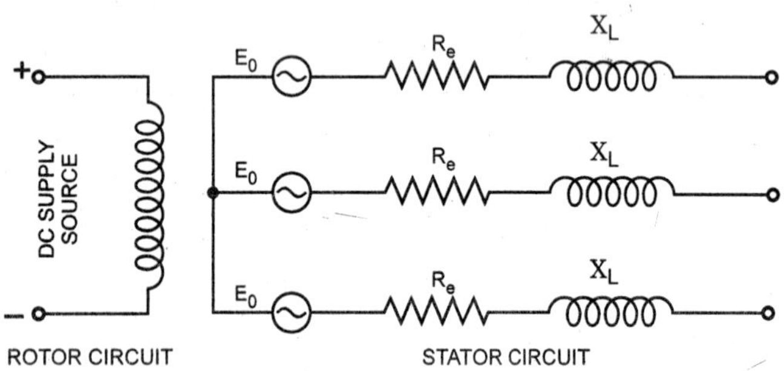

(a) Circuit Diagram

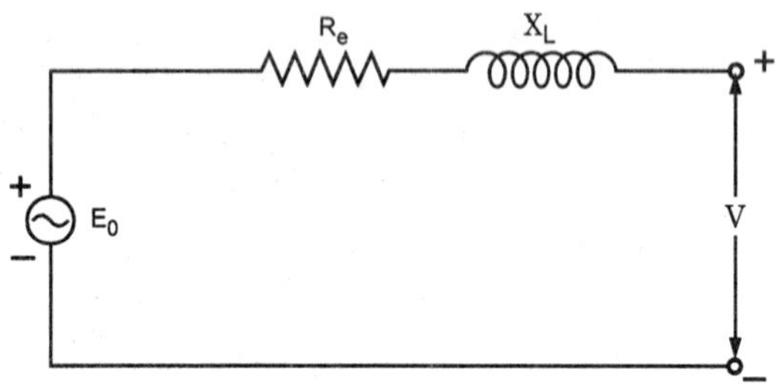

(b) Equivalent Circuit Diagram of Alternator on No Load

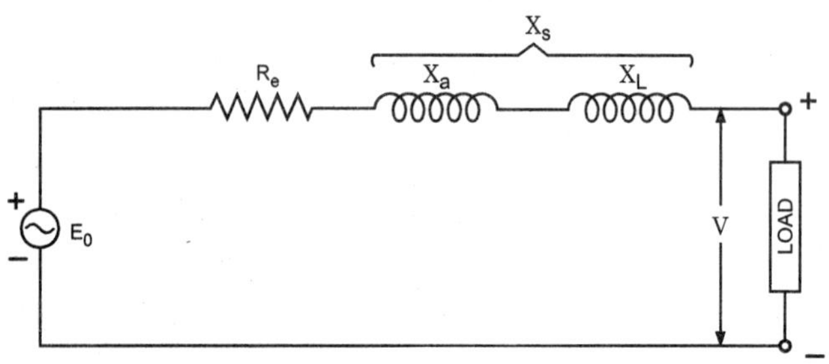

(c) Equivalent Circuit Diagram of Alternator on Load

Fig. 15.30

15.16.3. Load Characteristic. While the exciting current and the speed remain constant, the terminal voltage of an alternator changes with the change in the load or armature current, as already mentioned above. The relationship between the terminal voltage, V and load current, I of an alternator is known as its *load characteristic.* The curves showing variation of terminal voltage V with load current I for constant excitation (or no-load voltage E_0) for three different power factor loads are given in Fig. 15.31. The curves are plotted in terms of percentage values because general operating conditions can be illustrated better in this way.

Normally the terminal voltage falls with the increase in load current, but when the power factor is leading one, the load characteristic curve may rise at first. Each curve is nearly straight at the beginning but tends to droop because, with the increase in load current, the angle of lag between current and emf, owing to the original field, increases.

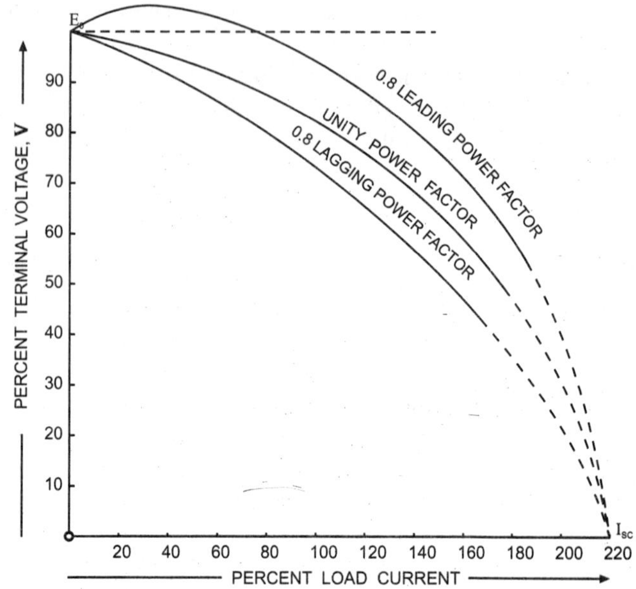

Fig. 15.31

The highest current is obtained when the alternator terminals are short circuited, the value being given as $I_{sc} = \dfrac{E_0}{Z_s}$ amperes where E_0 is no-load terminal voltage and Z_s is the synchronous reactance. All curves meet at the short-circuit point.

In modern alternators, the steady short-circuit current is not much greater than full-load rated current. This is purposely arranged to prevent excessive current in the event of a short circuit.

The synchronous impedance can be determined from open- and short-circuit tests, as will be discussed in the following article (Art. 15.17).

15.16.4. Effect of Variation of Power Factor On Terminal Voltage (Load Current, Excitation and Speed Remaining Unchanged). On changing the power factor from leading to lagging one, if the load current, excitation and speed are kept constant, the terminal voltage falls. Since when the power factor is leading (highly capacitive load) the effect of armature flux is to help the main flux, hence to generate more emf but when power factor is lagging (highly inductive load) the effect of armature flux is to oppose the main flux, hence to generate less emf. Thus at lagging power factor the terminal voltage falls from that on leading power factor mainly owing to decrease in generated emf.

15.17 LABORATORY METHOD OF DETERMINATION OF SYNCHRONOUS REACTANCE (TESTING OF SYNCHRONOUS GENERATOR)

As already defined, the sum of leakage reactance X_L and fictitious reactance X_a, to account for the voltage drop due to armature reaction is known as synchronous reactance X_s and it is given as

$$X_s = \sqrt{Z_s^2 - R_e^2}$$

where Z_s is the synchronous impedance and R_e is the effective resistance.

Determination of Effective Resistance of Armature. Armature resistance per phase can be measured directly by voltmeter-ammeter method or by using Wheatstone bridge. The connection diagram for measuring armature resistance by ammeter-voltmeter method is shown in Fig. 15.32.

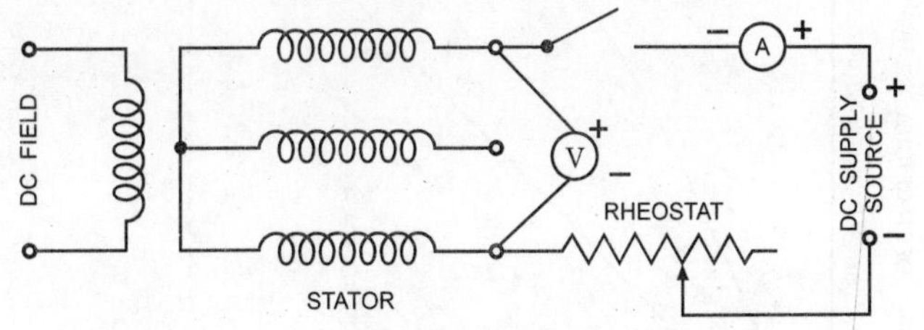

Fig. 15.32

DC resistance between pair of terminals, with the dc field winding open, is measured. The voltmeter reading divided by ammeter gives the value of resistance of two phases connected in series. So resistance of each phase, R_a will be equal to half the resistance measured. Since the resistance under measurement is very low so it will be necessary to use a high current rheostat to adjust the current to about rated value, a dc ammeter of proper range, and a low-range voltmeter. The alternator should be at rest. For operating conditions this measured value of armature resistance R_a is increased by about 60% or so to allow the skin effect and give effective value of armature resistance, R_e.

Synchronous impedance is determined from the open-circuit and short-circuit tests, as explained below :

1. Open-Circuit Test. The connections for the open-circuit test are shown in Fig. 15.33. With the armature- winding circuit open, the machine is run at rated speed. The field is connected to a dc source in series with a field rheostat and an ammeter so that the field current be adjusted and noted. The field current is raised in suitable steps until the voltage between any pair of terminals of the armature winding is somewhat above rated emf and corresponding values of voltmeter V indicating no-load line voltage are noted. The open-circuit voltages per phase, E_0 are obtained by dividing the voltmeter readings by $\sqrt{3}$. Now the curve is drawn between E_0 and field current I_f, as shown in Fig. 15.34. The curve so drawn is known as *open-circuit characteristic*. The initial straight part of the curve yields the air-gap characteristic since at that low excitation the iron is unsaturated and its reluctance is negligible.

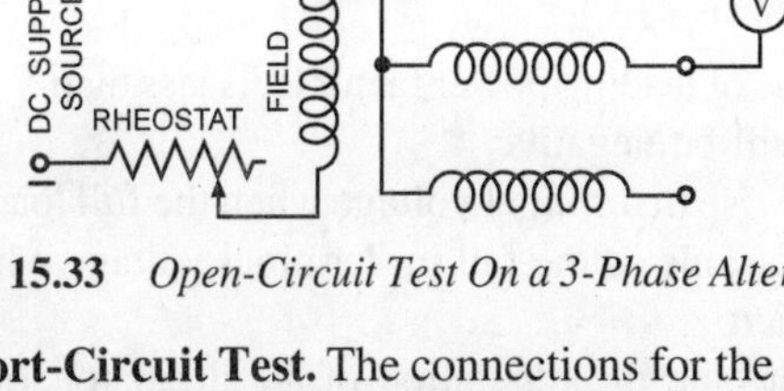

Fig. 15.33 *Open-Circuit Test On a 3-Phase Alternator*

2. Short-Circuit Test. The connections for the short-circuit test are shown in Fig. 15.35. All the three phases are shorted, as shown in Fig. 15.35. Since the three currents will be equal, only one of them need be measured and so one ammeter is connected in any one of the three phases. Rheostat of sufficiently high ohmic value is inserted in the dc field circuit to keep the current in the circuit very low. The machine is run at synchronous speed,

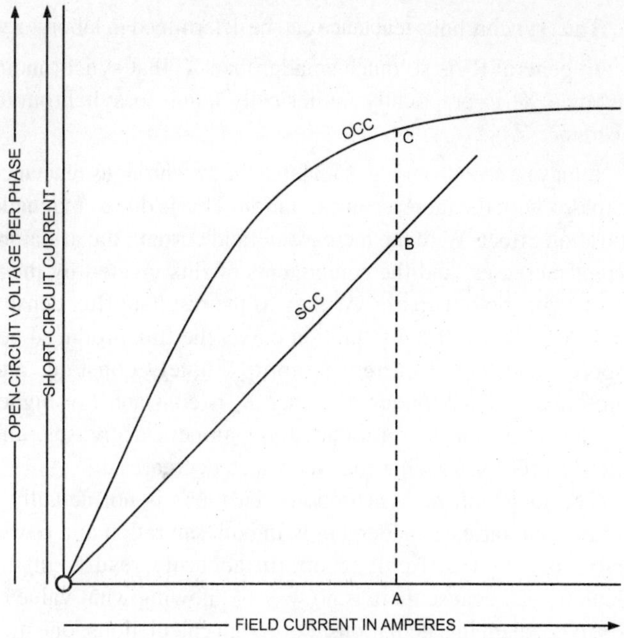

Fig. 15.34

the field excitation is raised from zero to an amount sufficient to circulate 150 to 200 % of full-load current. The readings of the ammeters connected in the field circuit and armature circuit A_1 and A_2 are noted. The short-circuit characteristic is drawn by plotting a curve between short-circuit current I_{sc} and field current I_f, as shown in Fig. 15.34. The short-circuit characteristic is normally a straight line through the origin because the net excitation is so small that there is no saturation in the magnetic circuit. Further, the short-circuit characteristic is not affected due to slight deviation in speed from the rated.

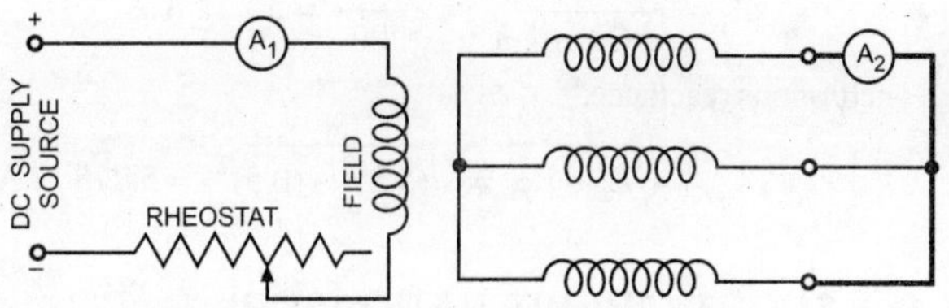

Fig. 15.35 *Short-Circuit Test On a 3 Phase Alternator*

The synchronous impedance per phase Z_s can be determined by knowing the value of short-circuit curent, I_{sc} corresponding to a field current that gives rated terminal voltage per phase. The open-circuit voltage divided by the short-circuit current, corresponding to field current which gives rated emf per phase, gives the value of synchronous impedance per phase Z_s.

If OA is the field current that gives the rated emf per phase represented by AC and AB gives the short-circuit current I_{sc} corresponding to field current OA. Since on short circuit the terminal voltage is zero and whole of the induced emf is utilized to create a short circuit current I_{sc} through the synchronous impedance Z_s . That is

$$E_0 = I_{sc} Z_s$$

$$\text{or synchronous impedance per phase, } Z_s = \frac{E_0}{I_{sc}} = \frac{\text{AC (in volts)}}{\text{AB (in amperes)}}$$

$$\text{and synchronous reactance, } X_s = \sqrt{Z_s^2 - R_e^2}$$

Thus synchronous reactance can be determined in laboratory.

In general R_e is so much smaller than X_s that synchronous reactance X_s is practically numerically equal to synchronous impedance Z_s.

It may be seen from Fig. 15.34 that the synchronous reactance decreases with the increase in excitation. This is due to magnetic saturation effect. With the increase in field current, the armature current increases, and the components of flux created by these two currents both increase. As long as the resultant flux density is below the knee of the saturation curve, the flux produced per ampere of armature current is approximately constant, and therefore, the synchronous reactance X_s is constant. For higher flux densities, the flux produced per ampere decreases, and consequently the synchronous reactance decreases.

The fact that the synchronous reactance is not actually a constant but varies considerably with both saturation and power factor makes it difficult to obtain accurate results in the calculations, because there is no way of knowing what value of X_s is to be taken in calculations. For rough calculations, one may use the value of X_s corresponding to rated no-load voltage.

Example 15.18. The effective resistance of a 2200 V, 50 Hz, 440 kVA, 1-phase alternator is 0.5 Ω, on short circuit a field current of 40 A gives the full load current of 200 A. The emf on open circuit with the same excitation is 1,160 V. Calculate the synchronous impedance and reactance.

Solution: For a field current of 40 A

Short-circuit current, I_{sc} = 200 A

Open-circuit voltage, E_0 = 1,160 V

Synchronous impedance,

$$Z_s = \frac{\text{OC voltage}}{\text{SC current}} = \frac{1,160}{200} = 5.8\ \Omega \textbf{ Ans.}$$

Synchronous reactance,

$$X_s = \sqrt{Z_s^2 - R_e^2} = \sqrt{(5.8)^2 - (0.5)^2} = 5.778\ \Omega \textbf{ Ans.}$$

15.18 SHORT-CIRCUIT RATIO (SCR)

The short-circuit ratio (SCR) of a synchronous machine is defined as the ratio of field current to produce rated voltage on open circuit to field current required to circulate rated current on short circuit while the machine is mechanically driven at synchronous speed.

From OCC and SCC shown in Fig. 15.36, according to definition

Short-circuit ratio,

$$\text{SCR} = \frac{I_{f1}}{I_{f2}} = \frac{OA}{OD} = \frac{AB}{DE} = \frac{AB}{AC} = \frac{1}{AC/AB}$$

$$\because \frac{AC}{AB} = \frac{\text{PU voltage on open-circuit}}{\text{Corresponding PU current on short-circuit}} = X_s$$

$$\text{So SCR} = \frac{1}{X_s} \qquad ...(15.14)$$

Thus SCR is reciprocal of per unit synchronous reactance X_s. The value of synchronous reactance depends upon saturated conditions of the machine but SCR is specific and defined at rated voltage.

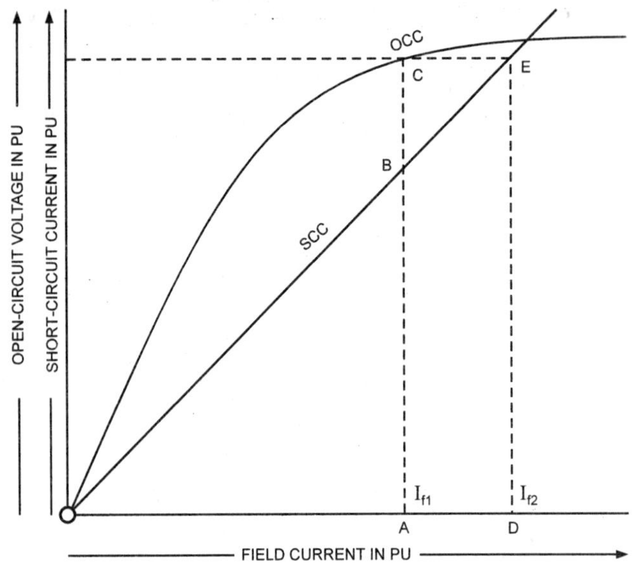

Fig. 15.36

A small value of SCR indicates a smaller value of current under short-circuit conditions owing to large value of synchronous reactance. The SCR of a high speed turbo-alternator usually lies between 0.5 and 0.75 while that of low speed salient pole generators lies between 1.0 and 1.5.

With high value of SCR, stability limit increased, voltage regulation improved and machines with a low value of SCR have difficulty during parallel operation owing to small value of synchronising power.

15.19 VOLTAGE REGULATION

From the previous discussions it is obvious that the terminal voltage of an alternator changes from no load to full load. The magnitude of this change in terminal voltage depends not only on load current but also on the power factor of the load.

Voltage regulation of an alternator is defined as the increase in terminal voltage, expressed as percentage (or per unit) of the rated terminal voltage, when the load at a given power factor is thrown off, with speed and field current remaining the same.

$$\textit{i.e.} \text{ voltage regulation} = \frac{E_0 - V}{V} \text{ per unit} \qquad ...(15.15)$$

$$= \frac{E_0 - V}{V} \times 100 \text{ in percentage} \qquad ...(15.16)$$

where E_0 is no-load terminal voltage and V is the full-load terminal voltage.

Note: In case of leading power factor E_0 is less than V, so voltage regulation will be negative.

Voltage rise in terminal voltage when the full load is thrown off is not the same as the fall in terminal voltage when the full load is applied.

To determine the regulation of an alternator, open-circuit and short-circuit tests are performed, as discussed in Art. 15.17, which give open-circuit characteristic and short-circuit characteristic.

From short-circuit and open-circuit tests, the voltage regulation can be determined by any one of the two methods given below.

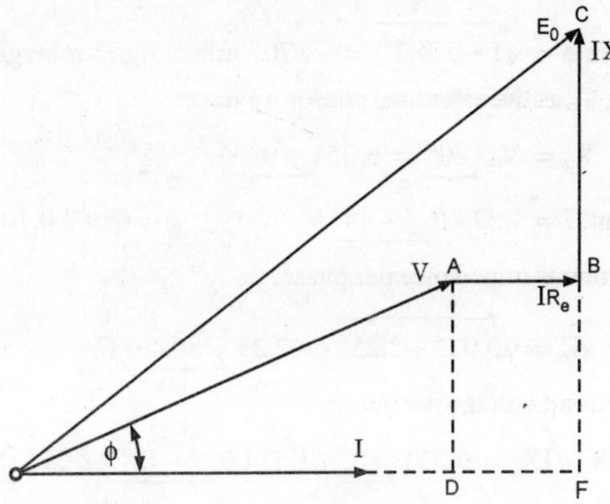

Fig. 15.37

1. Synchronous Impedance Method or EMF Method. After determining R_e and X_s from open-circuit and short-circuit tests, the phasor diagram for any load of any power factor can be drawn, as shown in Fig. 15.37.

Let load current be I lagging behind the terminal phase voltage V by ϕ

No-load terminal voltage per phase, E_0 = OC

Full-load terminal voltage, V = OA

Armature resistance drop, $I R_e$ = AB

Synchronous reactance drop $I X_s$ = BC

$$\text{Now } OC^2 = OF^2 + FC^2 = (OD + DF)^2 + (FB + BC)^2$$

$$\text{or } E_0^2 = (V \cos \phi + I R_e)^2 + (V \sin \phi + IX_s)^2$$

[Since BF = AD = V sin ϕ]

$$\text{or } E_0 = \sqrt{(V \cos \phi + IR_e)^2 + (V \sin \phi + IX_s)^2}$$

$$\text{And \% regulation} = \frac{E_0 - V}{V} \times 100$$

Notes. 1. For leading pf taking ϕ as negative and for unity pf taking $\phi = 0$ voltage regulation can be determined at leading pf and unity pf respectively.

2. Of various methods for determination of regulation, this one yields results most likely to be inexact. The regulation obtained by this method is always higher than actual value and, therefore, this is called the *pessimistic* method. The results are, however, more likely to be 'on safe side'. However, this method is theoretically accurate for non-salient pole machines with distributed field windings when saturation is not considered. The reason of error is that in this method synchronous impedance or reactance is assumed to remain constant while actually it is not. The value of synchronous impedance (or reactance) varies with the saturation. At low saturation its value is larger because the effect of armature reaction is greater than that at high saturation. Now under short-circuit conditions, saturation is very low and, therefore, the value of synchronous impedance measured is higher than that in actual working conditions.

2. *Ampere-Turn (or MMF) Method.* In this method, the datas determined from open-circuit tests are utilized.

From the above two tests the open-circuit and short-circuit characteristics are drawn, as shown in Fig. 15.34.

From the above characteristic field current I_{f1} is determined to give rated voltage V on no load, neglecting armature resistance drop, and field current I_{f2} is determined to cause short-circuit current, equal to full-load current, on short circuit.

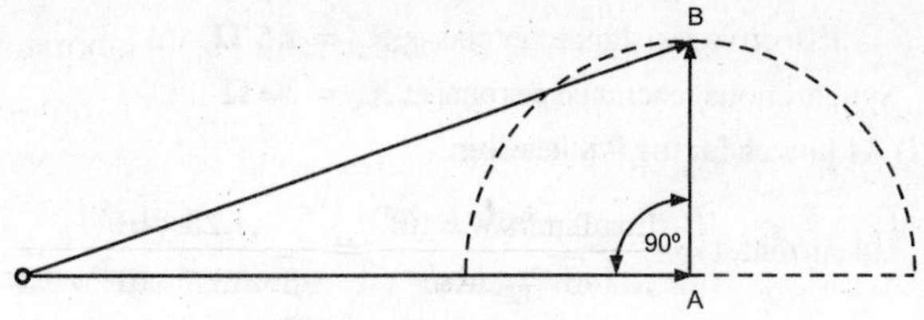

(a) For Unity P F

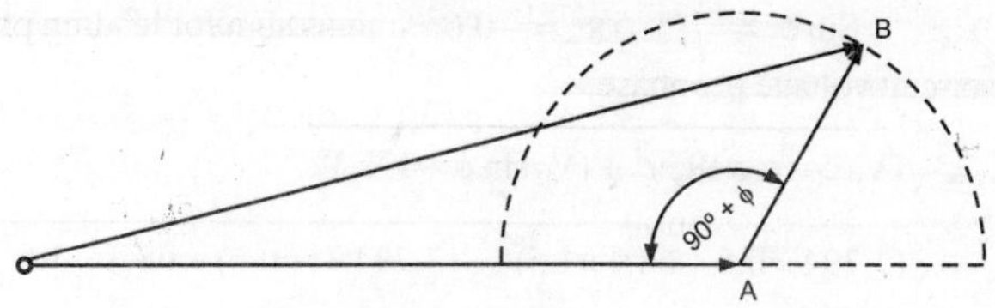

(b) For Lagging P F

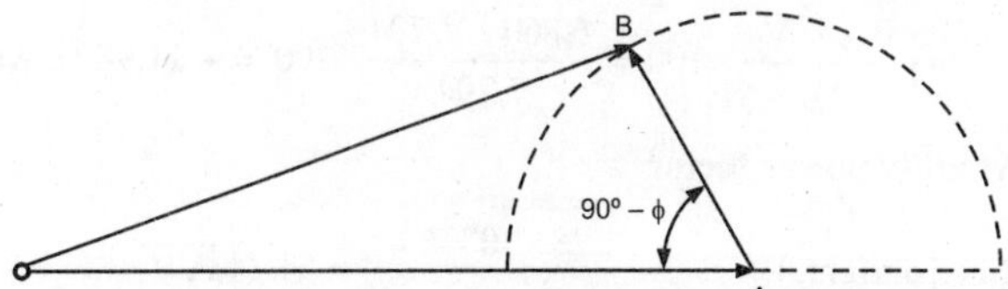

(c) For Leading P F

Fig. 15.38

On short circuit, the field excitation I_{f2} balances the impedance drop in addition to armature reaction on full load. But since R_e is usually very small and X_L is also small for low voltage on short circuit, so impedance drop can be neglected. Hence pf on short circuit is almost zero lagging and the field amp-turns are used entirely to overcome the armature reaction. Therefore, I_{f2} gives demagnetising amp-turns at full load.

Now let the alternator supply full load current at a pf of cos ϕ. Draw OA representing I_{f1} to give full load rated voltage, V [or more exactly $V + I R_e \cos \phi$] then draw AB at an angle (90º ± ϕ) representing I_{f2} to give full load current on short circuit; + ve sign for lagging pf and –ve sign for leading pf, as shown in Fig. 15.38. Now find field current I_f, measuring OB, which will give open-circuit EMF E_0, which can be determined from OCC.

Then percentage regulation can be determined from the relation,

$$\text{\% Regulation} = \frac{E_0 - V}{V} \times 100.$$

This method of determination of synchronous impedance is known as *optimistic method* since it gives values lower than actual values. The reason of it is that the excitation to overcome armature reaction is determined on unsaturated part of the saturation curve.

Example 15.19. A 3-phase star-connected alternator is rated at 1,600 kVA, 13.5 kV. The per phase armature effective resistance and synchronous reactance are 1.5 Ω and 30 Ω respectively. Calculate voltage regulation for a load of 1.280 MW at power factors of *(i)* 0.8 leading *(ii)* unity and *(iii)* 0.8 lagging.

[U.P. Technical Univ. Electromechanical Energy Conversion-II, 2005-06]

Solution: Load, P = 1.28 MW

$$\text{Phase voltage, } V_P = \frac{V_L}{\sqrt{3}} = \frac{13.5 \times 1,000}{\sqrt{3}} = 7,794 \text{ V}$$

Effective resistance per phase, $R_e = 1.5\ \Omega$

Synchronous reactance per phase, $X_S = 30\ \Omega$

***(i)* At power factor 0.8 leading.**

$$\text{Load current, } I = \frac{\text{Load in MW} \times 10^6}{\sqrt{3}\ V_L \cos\phi} = \frac{1.28 \times 10^6}{\sqrt{3} \times 13.5 \times 10^3 \times 0.8} = 68.4 \text{ A}$$

Power factor, $\cos\phi = 0.8$

$$\sin\phi = \sqrt{1-0.8^2} = -0.6 \quad \text{minus sign for leading pf}$$

Open-circuit voltage per phase,

$$E_{OP} = \sqrt{(V_P \cos\phi + IR_e)^2 + (V_P \sin\phi + IX_S)^2}$$

$$= \sqrt{(7{,}794 \times 0.8 + 68.4 \times 1.5)^2 + [7{,}794 \times (-0.6) + 68.4 \times 30]^2}$$

$$= 6{,}860 \text{ V}$$

Percentage regulation

$$= \frac{E_{OP} - V_P}{V_P} \times 100 = \frac{6{,}860 - 7{,}794}{7{,}794} \times 100 = -11.98\ \% \textbf{ Ans.}$$

***(ii)* At unity power factor**

$$\text{Load current, } I = \frac{1.28 \times 10^6}{\sqrt{3} \times 13.5 \times 10^3 \times 1.0} = 54.74 \text{ A}$$

$\cos\phi = 1.0$ and $\sin\phi = 0$

Open-circuit voltage per phase,

$$E_{OP} = \sqrt{(7{,}794 \times 1.0 + 54.74 \times 1.5)^2 + (7{,}794 \times 0 + 54.74 \times 30)^2}$$

$$= \sqrt{(7{,}794 + 82.11)^2 + (0 + 1{,}642.2)^2} = 8{,}045.5 \text{ V}$$

$$\text{Percentage regulation} = \frac{8{,}045.5 - 7{,}794}{7{,}794} \times 100 = 3.227\ \% \textbf{ Ans.}$$

***(iii)* At power factor 0.8 lagging**

Load current, $I = 68.4$ A, same as in case (*i*)

$\cos\phi = 0.8$ and $\sin\phi = 0.6$

Open-circuit voltage per phase,

$$E_{OP} = \sqrt{(7{,}794 \times 0.8 + 68.4 \times 1.5)^2 + (7{,}794 \times 0.6 + 68.4 \times 30)^2}$$

$$= 9{,}243 \text{ V}$$

$$\text{Percentage regulation} = \frac{9{,}243 - 7{,}794}{7{,}794} \times 100 = 18.6\ \% \textbf{ Ans.}$$

Example 15.20. A 30 MVA, 11 kV, 50 Hz, star-connected three-phase alternator has an effective armature resistance per phase = 0.02 Ω and synchronous reactance per phase = 2.25 Ω. Find the voltage regulation for a pf of 0.707 lagging. What is the load angle of the alternator under this condition?

[B.P. Univ. of Technology Electrical Machines, 2008]

Solution: Load, $S = 30$ MVA

$$\text{Phase voltage, } V_P = \frac{V_L}{\sqrt{3}} = \frac{11 \times 1{,}000}{\sqrt{3}} = 6{,}351 \text{ V}$$

Effective resistance per phase,

$R_e = 0.02\ \Omega$

Synchronous reactance per phase,

$X_S = 2.25\ \Omega$

Phase current, I = Line current

$$= \frac{\text{Load MVA} \times 10^6}{\sqrt{3} \times V_L} = \frac{300 \times 10^6}{\sqrt{3} \times 11{,}000} = 1{,}574.6 \text{ A}$$

$$\sin\phi = \sqrt{1-0.707^2} = -0.707 \text{ minus sign for lagging pf}$$

Taking V_P as the reference phasor we have

$$\mathbf{V_P} = V_P \angle 0^\circ = 6{,}351 \angle 0^\circ \text{ V}$$

Current, $\mathbf{I} = 1{,}574.6 \angle{-45^\circ}$ V $\quad \because \phi = \cos^{-1} 0.707 = -45^\circ$

Synchronous impedance per phase,

$$\mathbf{Z_S} = \sqrt{0.02^2 + 2.25^2} = 2.25 \angle 89.49^\circ\ \Omega$$

Open-circuit voltage per phase,

$$\mathbf{E_{OP}} = \mathbf{V} + \mathbf{IZ_S} = 6{,}351 \angle 0^\circ + 1{,}574.6 \angle{-45^\circ} \times 2.25 \angle 89.49^\circ$$

$$= (6{,}351 + j\,0) + 3{,}542.8 \angle 44.49^\circ$$

$$= (6{,}351 + j\,0) + (2{,}527 + j\,2{,}501)$$

$$= (8{,}878 + j\,2{,}501) = 9{,}223.5 \text{ V} \angle 15.73^\circ \text{ V}$$

So open-circuit voltage per phase,

$$E_{OP} = 9{,}223.5 \text{ V}$$

$$\text{Percentage voltage regulation} = \frac{E_{OP} - V}{V_P} \times 100$$

$$= \frac{9{,}223.5 - 6{,}351}{6{,}351} \times 100 = 45.23\% \textbf{ Ans.}$$

Load angle, $\delta = 15.73^\circ$ **Ans.**

Example 15.21. A full-load current of 100 A requires an excitation current of 3 amperes when a single-phase 1,200 V alternator is short-circuited by an ammeter of resistance 0.01 Ω. On open-circuit the same excitation produces 350 V. The resistance of the armature is 0.5 Ω. Calculate the regulation of the alternator at 0.8 power factor leading. *[U.P. Technical Univ. Electrical Machines, 2001-02]*

Solution: Rated voltage, $V = 1{,}200$ V

Rated current, $I = 100$ A

Effective resistance of armature, $R_e = 0.5\ \Omega$

If X_S is the synchronous reactance of the alternator, then

$$\sqrt{(R_e + \text{ammeter resistance})^2 + X_S^2} = \frac{\text{OC voltage}}{\text{SC current}}$$

$$\text{or } \sqrt{(0.5 + 0.01)^2 + X_S^2} = \frac{350}{100} = 3.5\ \Omega$$

$$\text{or } X_S = \sqrt{3.5^2 - 0.51^2} = 3.4626\ \Omega$$

At 0.8 leading pf, $\cos\phi = 0.8$ and $\sin\phi = -0.6$

Open-circuit voltage,

$$E_0 = \sqrt{(V\cos\phi + IR_e)^2 + (V\sin\phi + IX_S)^2}$$

$$= \sqrt{(1{,}200 \times 0.8 + 100 \times 0.5)^2 + (1200 \times (-0.6) + 100 \times 3.4626)^2}$$

$$= 1{,}077 \text{ V}$$

$$\text{Voltage regulation} = \frac{1{,}077 - 1{,}200}{1{,}200} \times 100 = -10.25\% \textbf{ Ans.}$$

Example 15.22. Find the synchronous impedance and reactance of a single-phase alternator in which a given field current produces an armature current of 250 A, on short circuit and a generated emf of 1500 V on open circuit. The armature resistance is 2.0 ohms. Calculate the terminal potential difference when a load of 250 A at 6.6 kV at a lagging pf of 0.8 is switched off.

[U.P.S.C. I.E.S. Electrical Engineering-II, 2003]

Solution: Effective armature resistance, $R_e = 2.0\ \Omega$

Synchronous impedance,

$$Z_S = \frac{\text{OC voltage}}{\text{SC current}} = \frac{1{,}500}{250} = 6\ \Omega \quad \textbf{Ans.}$$

Synchronous reactance,

$$X_S = \sqrt{Z_S^2 - R_e^2} = \sqrt{6^2 - 2^2} = 5.657\ \Omega \quad \textbf{Ans.}$$

Load current, I = 250 A

Terminal voltage, V = 6.6 kV = 6,600 V

Power factor, $\cos\phi$ = 0.8 (lagging) and $\sin\phi = 0.6$

Open-circuit voltage,

$$E_0 = \sqrt{(V\cos\phi + I R_e)^2 + (V\sin\phi + IX_S)^2}$$

$$= \sqrt{(6{,}600 \times 0.8 + 250 \times 2)^2 + (6{,}600 \times 0.6 + 250 \times 5.657)^2}$$

= 7,892.3 V **Ans.**

Example 15.23. An alternator has an armature resistance of 0.3 Ω and leakage reactance of 1.22 Ω. The alternator supplies 100 A current to a feeder of resistance 1.5 Ω and reactance 2.0 Ω. The voltage at the far end of the feeder is 3,000 V. The load current has a pf of 0.8 lagging with respect to this voltage. Find the terminal voltage of the alternator and the emf generated.

[U.P.S.C. I.E.S. Electrical Engineering-II, 2004]

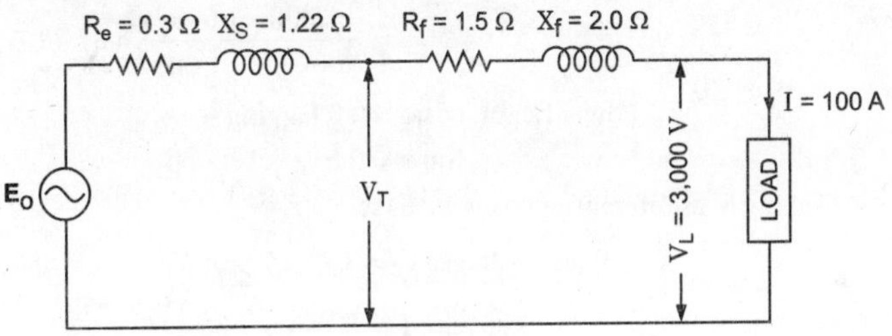

Fig. 15.39

Solution: Load voltage,

$$V_L = 3{,}000 \angle 0^\circ \text{ Taking } V_L \text{ as reference phasor}$$

Load current, $I_L = 100 \angle -36.87^\circ$ A

$\because \phi = \cos^{-1} 0.8 \text{ (lagging)} = -36.87^\circ$

Feeder impedance,

$$Z_f = (R_f + jX_f) = (1.5 + j\,2.0)\ \Omega = 2.5 \angle 53.13^\circ \Omega$$

Synchronous impedance of alternator,

$$Z_S = (R_e + jX_S) = (0.3 + j\,1.22)\ \Omega = 1.256 \angle 76.18^\circ\ \Omega$$

Terminal voltage of alternator

V_T = Load voltage + Voltage drop in feeder

$= 3{,}000 \angle 0^\circ + 100 \angle -36.87^\circ \times 2.5 \angle 53.13^\circ$

$= 3{,}000 \angle 0^\circ + 250 \angle 16.26^\circ$

$= (3{,}000 + j0) + (240 + j70)$

$= (3{,}240 + j70) = 3{,}240.76 \angle 1.24^\circ$ V **Ans.**

Generated emf,

$E_0 = V_T + I_L Z_S$

$= (3{,}240 + j70) + 100 \angle -36.87^\circ \times 1.256 \angle 76.18^\circ$

$= (3{,}240 + j70) + 125.6 \angle 39.31^\circ$

$= (3{,}240 + j70) + (97.18 + j79.57)$

$= (3{,}337.18 + j149.57) = 3{,}340.5 \angle 2.57^\circ$ V **Ans.**

Example 15.24. A 3-phase, star-connected synchronous generator rated at 10 kVA and 230 V has a synchronous reactance of 1.2 Ω per phase and armature resistance of 0.5 Ω per phase. Calculate the following: (*a*) The % voltage regulation at full load and 0.8 lagging power factor. (*b*) The power factor of load such that the voltage regulation is zero on full load.

[U.P.S.C. I.E.S. Electrical Engineering, 2009]

Solution: Phase voltage, $V_P = \frac{230}{\sqrt{3}} = 132.79$ V

Load, S = 10 kVA

Load current, $I = \frac{10 \times 1{,}000}{\sqrt{3} \times 230} = 25.1$ A

Effective resistance per phase, $R_e = 0.5\ \Omega$

Synchronous reactance per phase, $X_S = 1.2\ \Omega$

Power factor, $\cos\phi = 0.8$

$\sin\phi = \sqrt{1^2 - 0.8^2} = 0.6$

Open-circuit voltage per phase,

$$E_{0P} = \sqrt{(V_P\cos\phi + IR_e)^2 + (V_P\sin\phi + IX_S)^2}$$

$$= \sqrt{(132.79 \times 0.8 + 25.1 \times 0.5)^2 + (132.79 \times 0.6 + 25.1 \times 1.2)^2}$$

= 161.75 V

$$\text{Percentage regulation} = \frac{E_{0P} - V_P}{V_P} \times 100 = \frac{161.75 - 132.79}{132.79} \times 100$$

= 21.8% **Ans.**

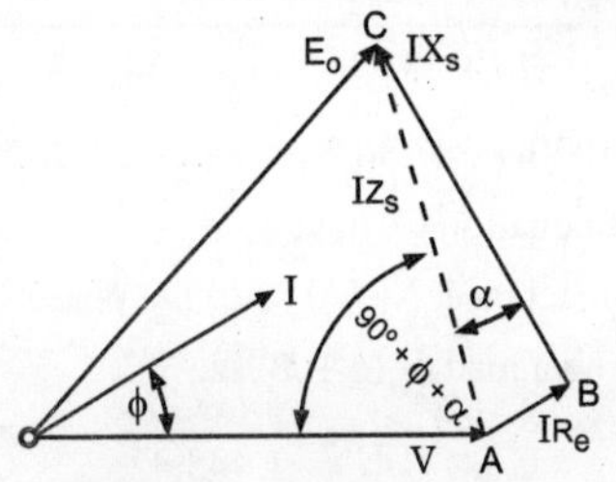

Fig. 15.40 *Phasor Diagram*

In phasor diagram shown in Fig. 15.40.

$$\alpha = \tan^{-1}\frac{IR_e}{IX_S} = \tan^{-1}\frac{0.5}{1.2} = 22.62^\circ$$

$OC = E_0$

$OA = V$

$$AC = \sqrt{(IR_e)^2 + (IX_S)^2}$$

$$= \sqrt{(25.1 \times 0.5)^2 + (25.1 \times 1.2)^2} = 32.63 \text{ V}$$

Now $OC^2 = OA^2 + AC^2 + 2O\,A \times AC \times \sin(\phi + \alpha)$

or $\sin(\phi + \alpha) = \frac{-AC^2}{2 \times OA \times AC}$

$\because$ For zero regulation $E_0 = V$ or $OC = OA$

$= \frac{-AC}{2 \times OA} = -\frac{32.63}{2 \times 132.79} = -0.12286$

or $\phi + \alpha = \sin^{-1}(-0.12286) = -7.057^\circ$

or $\phi = -7.057^\circ - \alpha = -7.057^\circ - 22.62^\circ = -29.68^\circ$

Power factor $= \cos\phi = \cos(-29.68^\circ) = 0.8688$ leading **Ans.**

Example 15.25. A star-connected 3-phase alternator delivers a 3-phase star-connected load at a power factor of 0.8 lagging. A wire connects the load and the alternator. The terminal voltage at no load is 2,500 V, and at full load of 1,460 kW it is 2,200 V. Determine the terminal voltage when it delivers a 3 phase star-connected load having a resistance of 6 Ω and reactance 8 Ω per phase respectively. Assume constant current and field excitation.

[U.P.S.C. I.E.S. Electrical Engineering II, 2006]

Solution: Load supplied, $P = 1{,}460$ kW

Load voltage, $V_L = 2{,}200$ V

$$\text{Load current, } I_L = \frac{P \times 1{,}000}{\sqrt{3}\, V \times \cos\phi} = \frac{1{,}460 \times 1{,}000}{\sqrt{3} \times 2{,}200 \times 0.8} = 478.94 \text{ A}$$

$$\text{Load voltage per phase, } V_P = \frac{V_L}{\sqrt{3}} = \frac{2{,}200}{\sqrt{3}} = 1{,}270 \text{ V}$$

$$\text{No-load voltage per phase, } E_{OP} = \frac{E_{0L}}{\sqrt{3}} = \frac{2{,}500}{\sqrt{3}} = 1{,}443.4 \text{ V}$$

No-load emf = Terminal voltage + synchronous reactance drop neglecting armature resistance

$$= V_P + j I_L X_S$$

$$\text{or } \mathbf{E_{OP}} = (1{,}270 + j0) + j478.94 \angle{-36.87^\circ} \times X_S$$

$$\text{or } \mathbf{E_{OP}} = (1270 + j0) + j383.15X_S + 287.36X_S$$

$$= (1270 + 287.36X_S) + j383.15X_S$$

Magnitude of E_{OP} is given as

$$E_{OP} = \sqrt{(1270 + 287.36X_S)^2 + (383.15X_S)^2}$$

$$\text{or } (1{,}443.4)^2 = (1270 + 287.36\, X_S)^2 + (383.15\, X_S)^2$$

Solving above equation we have

Synchronous reactance, $X_S = 0.55$ Ω per phase

When supplying a load of $(6 + j8)$ Ω,

$$\text{Load current, } \mathbf{I_L} = \frac{\mathbf{E_{OP}}}{\mathbf{X_S} + \mathbf{Z_L}} = \frac{1{,}443.4}{j\,0.55 + (6 + j8)}$$

$$= \frac{1{,}443.4}{6 + j\,8.55} = \frac{1{,}443.4}{10.445 \angle 54.94}$$

$$= 138.2 \angle{-54.94^\circ}$$

$$\text{Load impedance, } \mathbf{Z_L} = (6 + j\,8) = 10 \angle 53.13^\circ \,\Omega$$

$$\text{Terminal voltage (phase value)} = I_L\, Z_L = 138.2 \times 10 = 1{,}382 \text{ V}$$

$$\text{Terminal voltage (line value)} = \sqrt{3} \times 1{,}382 = 2{,}393.7 \text{ V}$$ **Ans.**

Example 15.26. A 5 MVA, 11 kV, 3-phase star-connected alternator is synchronized to the bus bars and is operating with an induced EMF of 125% of the rated voltage. If the load current is 500 A, what is the power factor of operation? The machine has a synchronous reactance of 5 Ω and negligible resistance per phase.

Solution: Load current, $\mathbf{I_L} = 500$ A

Power factor angle = ϕ(lagging) or $-\phi$(say)

$$\text{Terminal voltage per phase, } V = \frac{11{,}000}{\sqrt{3}} = 6{,}350.85 \text{ V}$$

$$\text{Induced emf per phase, } E_0 = \frac{11{,}000}{\sqrt{3}} \times 1.25 = 7{,}938.57 \text{ V}$$

Synchronous reactance per phase, $X_S = 5$ Ω

Now induced emf per phase,

$$\mathbf{E_0} = \mathbf{V} + j\mathbf{I_L X_S} \quad \text{neglecting resistance}$$

$$\text{or } E_0 = 6{,}350.85 + j500 \times (\cos\phi - j\sin\phi) \times 5$$

Taking V as reference phasor

$$= (6{,}350.85 + 2{,}500 \sin\phi) + j\,2{,}500 \cos\phi$$

$$\text{or } (E_0)^2 = (6{,}350.85 + 2{,}500 \sin\phi)^2 + (2{,}500 \cos\phi)^2$$

$$\text{or } (7{,}938.57)^2 = (6{,}350.85 + 2{,}500 \sin\phi)^2 + (2{,}500 \cos\phi)^2$$

$$\text{or } \sin\phi = \frac{(7{,}938.57)^2 - (6{,}350.85)^2 - (2{,}500)^2}{2 \times 6{,}350.85 \times 2{,}500} = 0.5176$$

$$\text{Phase angle, } \phi = \sin^{-1} 0.5176 = 31.174^\circ$$

Power factor = $\cos\phi = \cos 31.174^\circ = 0.855$ (lagging) **Ans.**

Example 15.27. A 1,000 kVA, 6.6 kV, 3-phase, star-connected synchronous generator has a synchronous reactance of 2.5 Ω/phase. It supplied full-load current at 0.8 lag pf and rated terminal voltage. Compute the terminal voltage for the same excitation when generator supplies full-load current at 0.8 pf leading.

Solution: Rated voltage per phase, $V = \dfrac{6.6 \times 1{,}000}{\sqrt{3}} = 3{,}810.5$ V

$$\text{Full-load current, } I = \frac{\text{Rated kVA} \times 1{,}000}{\sqrt{3} \times V_L}$$

$$= \frac{1{,}000 \times 1{,}000}{\sqrt{3} \times 6.6 \times 1{,}000} = 87.48 \text{ A}$$

Power factor, $\cos\phi = 0.8$ lagging

$\sin\phi = 0.6$

Taking V as reference phasor we have

Phase voltage, $\mathbf{V} = 3{,}810.5 \angle 0^\circ$ V

Current, $\mathbf{I} = 87.48 \angle{-36.87^\circ}$

$\because \cos^{-1} 0.8 = -36.87^\circ$ minus sign for lagging pf

Open-circuit voltage per phase

$$\mathbf{E_{OP}} = \mathbf{V} + \mathbf{IZ_S} = 3{,}810.5 \angle 0^\circ + 87.48 \angle{-36.87^\circ} \times 2.5 \angle 90^\circ$$

$$= (3{,}810.5 + j\,0) + (131.22 + j\,174.96)$$

$$= 3{,}941.72 + j\,174.96 = 3{,}945.6 \text{ V}$$

When supplying the same load current (i.e. 87.48 A) at 0.8 pf leading for the same excitation.

Let V′ be the terminal voltage per phase in volts and reference phasor.

$$\mathbf{V'} = \mathbf{V'} \angle 0^\circ \text{ V}$$

$$\mathbf{I} = 87.48 \angle 36.87^\circ \text{ A}$$

$$\mathbf{Z_S} = 2.5 \angle 90^\circ$$

Open-circuit voltage,

$$\mathbf{E_{OP}} = V' \angle 0^\circ + \mathbf{IZ_S}$$

$$= (V' + j\,0) + 87.48 \angle 36.87^\circ \times 2.5 \angle 90^\circ$$

$$= (V' + j\,0) + 218.7 \angle 126.87^\circ$$

$$\text{or } \mathbf{E_{OP}} = (V' - 131.22) + j\,174.96$$

$$\text{or } E_{OP} = \sqrt{(V' - 131.22)^2 + (174.96)^2}$$

$$\text{or } 3{,}945.6 = \sqrt{(V' - 131.22) + (174.96)^2} \qquad \because E_{OP} = 3{,}863 \text{ V}$$

$$\text{or } V' = \sqrt{(3{,}945.6)^2 - (174.96)^2} + 131.22$$

$$= 3{,}941.72 + 131.22 = 4{,}072.94 \text{ V}$$

$$\text{Terminal voltage (line-to-line)} = \sqrt{3} \times 4{,}072.94$$

$$= 7{,}054 \text{ V or } 7.054 \text{ kV}$$ **Ans.**

Example 15.28. A 2000 kVA, 11 kV, 3-phase star-connected alternator has a resistance of 0.3 ohm and reactance of 5 ohms per phase. It delivers full-load current at a pf of 0.8 lagging and normal rated voltage. Compute the terminal voltage for the same excitation and load current at a 0.8 pf leading.

[U.P.S.C. I.E.S. Electrical Engineering, 2000; U.P. Technical Univ. Electromechanical Energy Conversion, 2006-07]

Solution: Rated voltage per phase, $V = \frac{11 \times 1,000}{\sqrt{3}} = 6,350.85$ V

Full-load current,

$$I = \frac{\text{Rated kVA} \times 1,000}{\sqrt{3} \times V_L} = \frac{2,000 \times 1,000}{\sqrt{3} \times 11,000} = 104.97 \text{ A}$$

Power factor, $\cos\phi = 0.8$ (lagging) and $\sin\phi = 0.6$

Open-circuit voltage per phase,

$$E_0 = \sqrt{(V\cos\phi + IR_e)^2 + (V\sin\phi + IX_S)^2}$$

$$= \sqrt{\begin{matrix}(6,350.85 \times 0.8 + 104.97 \times 0.3)^2 \\ + (6,350.85 \times 0.6 + 104.97 \times 5)^2\end{matrix}} = 6,703 \text{ V}$$

When supplying same load current at 0.8 pf (leading) for the same excitation

$$E_0 = \sqrt{(V\cos\phi + IR_e)^2 + (V\sin\phi + IX_S)^2}$$

or $6,703 = \sqrt{(V \times 0.8 + 104.97 \times 0.3)^2 + [V \times (-0.6) + 104.97 \times 5]^2}$

or $V = 6,978$ V

Terminal voltage (line-to-line)

$= \sqrt{3} \times 6,978 = 12,086$ V or 12.086 kV **Ans.**

Example 15.29. A synchronous generator supplies rated power at 0.8 pf lagging. Its resistance and synchronous reactance are 0.01 and 1.0 pu. Calculate the terminal voltage if the open-circuit voltage is 2.1 p.u. What is the minimum allowable excitation voltage?

[U.P.S.C. I.E.S. Electrical Engineering-II, 2005]

Solution: Open-circuit voltage, $E_0 = 2.1$ pu

Effective armature resistance, $R = 0.01$ pu

Synchronous reactance, $X_S = 1.0$ pu

Load current, $I = 1.0$ pu

Power factor, $\cos\phi = 0.8$ (lagging)

$\sin\phi = 0.6$

From phasor diagram given in Fig. 15.41 we have

$$E_0^2 = (V\cos\phi + IR_e)^2 + (V\sin\phi + IX_S)^2$$

or $(2.1)^2 = (V \times 0.8 + 1 \times 0.01)^2 + (V \times 0.6 + 1 \times 1.0)^2$

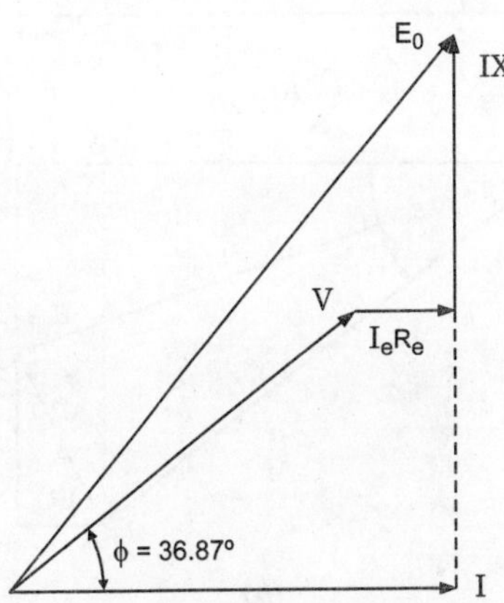

Fig. 15.41 *Phasor Diagram*

or $4.41 = (0.8V + 0.01)^2 + (0.6V + 1)^2$

or $V^2 + 1.216\,V - 3.4099 = 0$

or $V = \frac{-1.216 \pm \sqrt{1.478656 + 13.6396}}{2} = 1.336$ pu **Ans.**

For minimum excitation, power factor,

$\cos\phi = 0$

$\phi = 90°$ (leading)

and $\sin\phi = -1$

and $E_{0\,min}^2 = (V\cos\phi + IR_e)^2 + (V\sin\phi + IX_S)^2$

$= (1.336 \times 0 + 1 \times 0.01)^2 + (-1.336 + 1 \times 1)^2$

$= (0.01)^2 + (-0.336)^2$

or $E_{min} = \sqrt{(0.01)^2 + (-0.336)^2} = 0.33615$ pu **Ans.**

Example 15.30. In 50 kVA, star-connected, 440 V, 3-phase, 50 Hz alternator, the effective armature resistance is 0.25 Ω per phase. The synchronous reactance is 3·2 Ω /phase and the leakage reactance is 0.5 Ω / phase. Determine at rated load and unity power factor:

***(a)* internal emf E_a *(b)* no-load emf E_0 *(c)* percentage regulation on full load *(d)* value of synchronous reactance which replaces armature reaction.**

Solution: Rated load = 50 kVA = 50,000 volt-amperes

Line voltage, $V_L = 440$ V

Phase voltage $V_P = \frac{440}{\sqrt{3}} = 254$ V

Full load current, $I = \frac{50,000}{\sqrt{3} \times 440} = 65.6$ A

Effective armature resistance per phase, $R_e = 0.25\ \Omega$

Leakage reactance per phase, $X_L = 0.5\ \Omega$

Synchronous reactance per phase, $X_S = 3.2\ \Omega$

Power factor, $\cos\phi = 1$

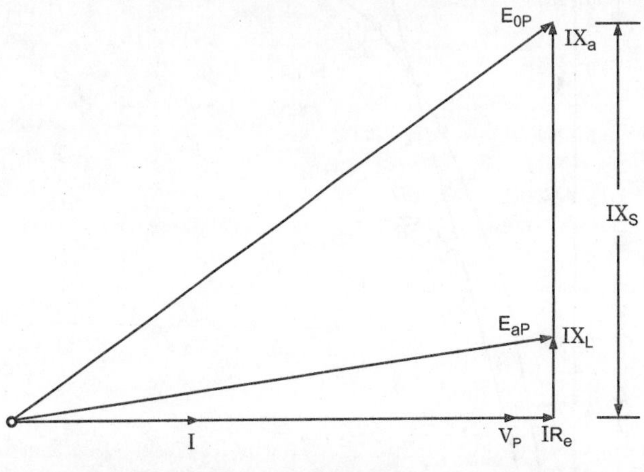

Fig. 15.42

From phasor diagram (Fig. 15.42)

Internal emf per phase,

$$E_{aP} = \sqrt{(V_P + IR_e)^2 + (IX_L)^2}$$

$$= \sqrt{(254 + 65.6 \times 0.25)^2 + (65.6 \times 0.5)^2} = 272 \text{ V}$$

(*a*) Line value of internal emf,

$E_a = \sqrt{3} \times 272 = 472$ V **Ans.**

No-load emf per phase,

$$E_{0P} = \sqrt{(V_p + IR_e)^2 + (IX_S)^2}$$

$$= \sqrt{(254 + 65.6 \times 0.25)^2 + (65.6 \times 3.2)^2} = 342 \text{ V}$$

(*b*) Line value of no-load emf,

$$E_{0L} = \sqrt{3} \times 342 = 593 \text{ volts } \textbf{Ans.}$$

(*c*) Percentage regulation

$$= \frac{593 - 440}{440} \times 100 = 34.6\ \% \ \textbf{Ans.}$$

(*d*) Fictitious reactance/phase,

$$= X_a = X_S - X_L = 3.2 - 0.5 = 2.7\ \Omega \ \textbf{Ans.}$$

Example 15.31. **(*a*) A 3-phase, star-connected, 5 kVA, 400 V, 50 Hz, 4-pole alternator has the following test data at rated speed :**

(*i*) Exciting current (in amperes)

0.5	1.0	1.5	2.0	2.5	3.0	3.5	4.0	4.5	5.0	6.0	8.0

Per phase OC volts

75	140	173	202	224	240	250	257	260	263	266	271

(*ii*) Exciting current (amperes)	**1.0**	**2.0**	**3.0**
SC line current	**3.6**	**7.2**	**10.8**

(*iii*) Armature resistance per phase 2.0 Ω

Draw OC and SC characteristics on a squared graph paper and then determine unsaturated value of synchronous reactance per phase in ohms and in per unit.

(*b*) For the synchronous machine (*a*) determine per cent voltage regulation at rated load at 0.8 pf lag and lead by synchronous impedance method under unsaturated condition. Draw relevant phasor diagrams.

Solution : OCC and SCC of the machine are shown in Fig. 15.43 alongwith the appropriate scales.

Unsaturated value of the synchronous reactance is the value of the reactance which is obtained for a linear portion of OCC.

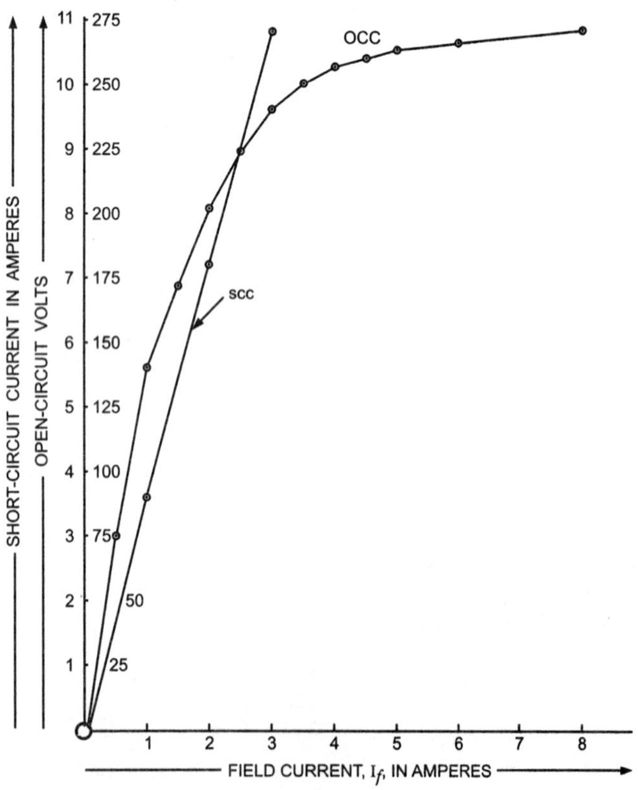

Fig. 15.43

For a field current of 1 A

Open–circuit line emf = 140 V

Short-circuit current, I_{sc} = 3.6 A

Synchronous impedance per phase,

$$Z_S = \frac{\text{OC voltage per phase}}{\text{SC phase current}} = \frac{140}{3.6} = 38.88\ \Omega$$

Armature resistance per phase, $R_e = 2.0\ \Omega$

Synchronous reactance per phase,

$$X_S = \sqrt{(Z_S)^2 - R_e^2} = \sqrt{(38.88)^2 - (2)^2} = 38.84\ \Omega \ \textbf{Ans.}$$

Rated phase voltage,

$$V_P = \frac{V_L}{\sqrt{3}} = \frac{400}{\sqrt{3}} = 231 \text{ V}$$

Phase current,

$$I = \text{Line current, } I_L = \frac{5 \times 1{,}000}{\sqrt{3} \times 400} = 7.217 \text{ A}$$

Per unit value of synchronous reactance,

$$X_S = \frac{IX_S}{V_P} = \frac{7.217 \times 38.84}{231} = 1.213 \text{ pu } \textbf{Ans.}$$

For 0.8 lagging power factor

$\cos\phi = 0.8$ and $\sin\phi = 0.6$

Open-circuit voltage per phase,

$$E_{0P} = \sqrt{(V_P \cos\phi + IR_e)^2 + (V_P \sin\phi + IX_S)^2}$$

$$= \sqrt{(231 \times 0.8 + 7.217 \times 2)^2 + (231 \times 0.6 + 7.217 \times 38.84)^2} = 463.88 \text{ V}$$

Percentage regulation

$$= \frac{E_{0P} - V_P}{V_P} \times 100 = \frac{463.88 - 231}{231} \times 100$$

$$= 100.81\% \ \textbf{Ans.}$$

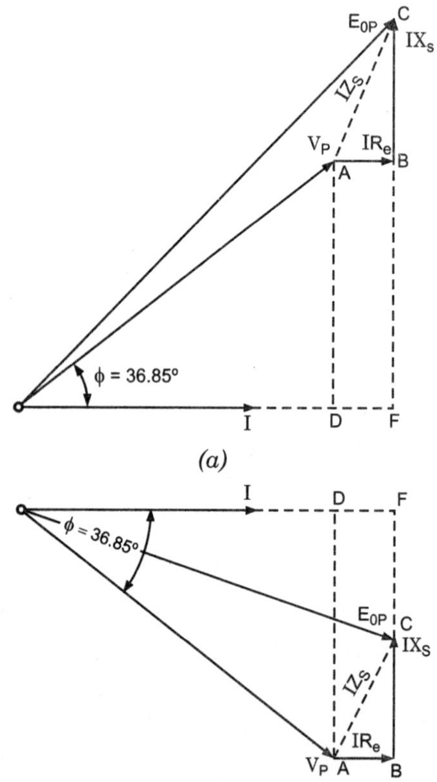

Fig. 15.44

For 0.8 leading power factor

$\cos\phi = 0.8$ and $\sin\phi = -0.6$

Open-circuit voltage per phase

$$E_{OP} = \sqrt{(231\times0.8+7.217\times2)^2 + [231\times(-0.6)+7.217\times38.84]^2} = 244.5 \text{ V}$$

Voltage regulation

$$= \frac{E_{OP}-V_P}{V_P}\times100 = \frac{244.5-231}{231}\times100 = 5.844\ \% \ \textbf{Ans.}$$

Phasor diagrams for 0.8 power factor lagging and leading are shown in Fig. 15.44 (*a*) and 15.44 (*b*) respectively.

Example 15.32. The no-load excitation of an alternator required to give rated voltage is 100 A. In a short-circuit test with full-load current flowing in the armature, the field excitation was 80 A. Determine the approximate excitation that will be required to give full load current at 0.8 pf lag and at the rated terminal voltage.

Solution: Field current required to give full-load rated voltage on open circuit,

$$I_{f1} = 100 \text{ A}$$

Field current required to give full-load current on short circuit in armature,

$$I_{f2} = 80 \text{ A}$$

To find the field current, required to give full-load current at 0.8 pf lag and at the rated terminal voltage draw OA equal to 100 A, then draw AB equal to 80 A making an angle $(90° + \phi)$ with OA where $\phi = \cos^{-1} 0.8$. The required field current will be OB, the phasor sum of OA and AB.

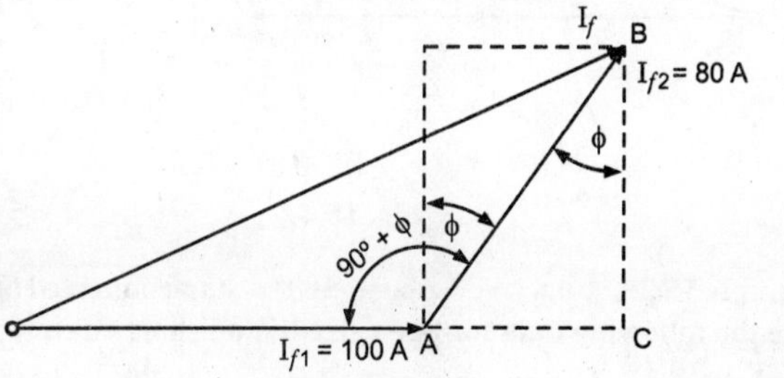

Fig. 15.45

The required field current I_f to give full–load current at 0.8 pf lag and at the rated terminal voltage

$$= \text{OB} = \sqrt{(\text{OA}+\text{AC})^2+(\text{CB})^2}$$

$$= \sqrt{(\text{OA}+\text{AB}\sin\phi)^2+(\text{AB}\cos\phi)^2}$$

$$= \sqrt{(100+80\times0.6)^2+(80\times0.8)^2} = 161.25 \text{ A } \textbf{Ans.}$$

Example 15.33. A 3.5 MVA Y-connected synchronous generator rated 4,160 V at 50 Hz has open-circuit characteristics given by the following data:—

Field current (A)	**50**	**100**	**150**	**200**	**250**	**300**	**350**	**400**
Line EMF (V)	**1620**	**3150**	**4160**	**4750**	**5130**	**5370**	**5550**	**5650**

A field current of 200 A is found necessary to circulate full-load current on short circuit of synchronous generator. Calculate by ampere-turn method, the full-load voltage regulation at 0.8 pf lagging. Neglect R_α.

[U.P. Technical Univ. Electromechanical Energy Conversion II, 2003-04]

Solution: From OCC [Fig. 15.46 (*a*)] drawn from given data, the field current required to give normal emf of 4,160 V on open circuit is 150 A. Field current required to give full-load current on short-circuit is 200 A.

To determine the field current required to give terminal voltage of 4,160 V on full load, determine the phasor sum of the field excitation of 200 A and normal value of 150 A [Fig. 15.46 (*b*)]

$$I_f = \sqrt{(150)^2+(200)^2-2\times150\times200\times\cos(90°+36.87°)} = 313.85 \text{ A}$$

Generated emf, E_0 corresponding to this excitation of 313.85 A, as found from OCC [Fig. 15.46 (*a*)] is 5,440 V

$$\text{Percentage regulation} = \frac{E_0 - V}{V}\times100$$

$$= \frac{5,440-4,160}{4,160}\times100 = 30.77\ \% \ \textbf{Ans.}$$

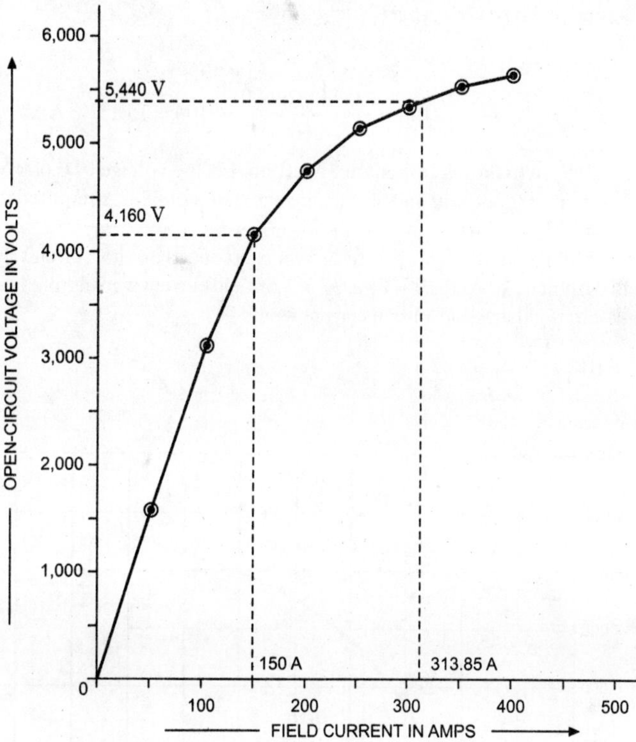

(*a*)

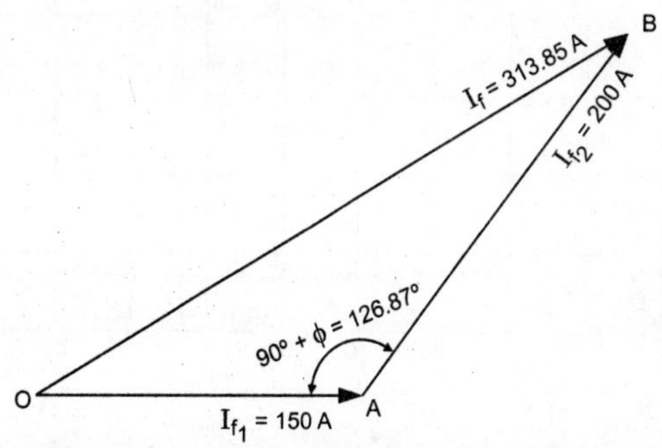

(*b*)

Fig. 15.46

Example 15.34. The following test results are obtained on a 6,600 V alternator.

Open-circuit voltage	**3,100**	**4,900**	**6,600**	**7,500**	**8,300**
Field current amperes	**16**	**25**	**37.5**	**50**	**70**

A field current of 20 A is found necessary to circulate full-load current on short circuit of the armature. Calculate by (*i*) the ampere-turn method and (*ii*) the synchronous impedance method the full-

load regulation at 0.8 pf (lag). Neglect resistance and leakage reactance. State the drawbacks of each of these methods.

Solution: ***(i)*** **Ampere-turn Method.** From OCC (Fig. 15.47) drawn from given data the field current required to give normal emf of 6,600 volts on open circuit is 37.5 A.

Field current required to give full load current on short circuit is 20 A.

To determine the field current required to give terminal pd of 6,600 V on full load, determine the phasor sum of the field excitation of 20 A and normal value of 37.5 A (Fig. 15.48)

$$I_f = \sqrt{(37.5)^2 + (20)^2 - 2 \times 37.5 \times 20 \times (-0.6)} = 52 \text{ A}$$

The generated emf E_0, corresponding to this excitation of 52 A, as found from OCC, (Fig. 12.43) is 7,600 V.

Percentage regulation

$$= \frac{E_0 - V}{V} \times 100 = \frac{7{,}600 - 6{,}600}{6{,}600} \times 100 = 15.15 \text{ \% } \textbf{Ans.}$$

(ii) **Synchronous Impedance Method.** Let the voltage of 6,600 V be taken as 100 per cent and also let the excitation of 37.5 A required to give 6,600 V on open-circuit be 100%.

Full load or 100 % armature current is produced on short circuit by field current of 20 A. If 100 % *i.e.* 37.5 A of field current were applied on short-circuit, then short-circuit current would be

$$100 \times \frac{37.5}{20} = 187.5 \text{ \%}.$$

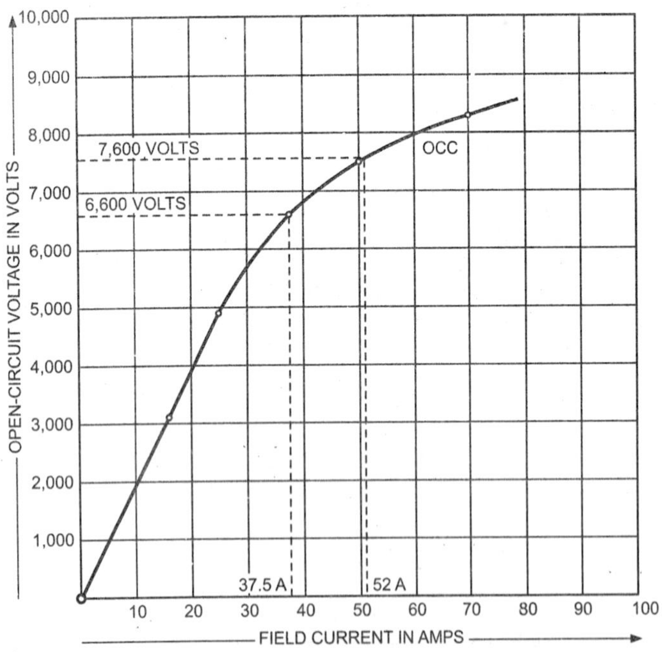

Fig. 15.47

Percentage impedance,

$$Z_s = \frac{\text{OC voltage}}{\text{SC current}} \text{ for given excitation } = \frac{100}{187.5} \times 100 = 53.3 \text{ \%}$$

Since armature resistance is negligible, therefore, impedance drop I Z_s = 53.3 % of the normal voltage will lead the load current by 90° or $\angle OAB = 90° + \phi$.

The phasor sum of normal voltage and impedance drop will give open-circuit voltage, E_0

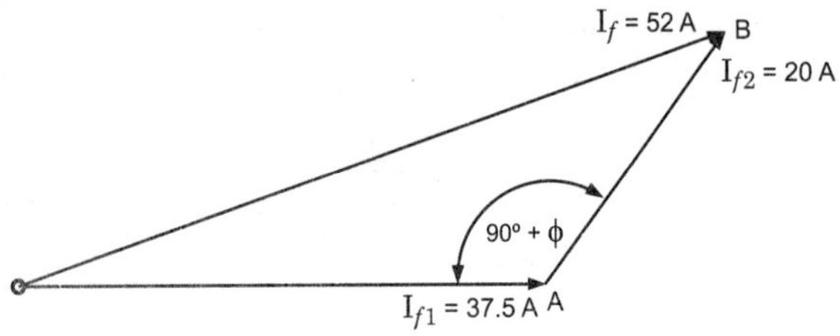

Fig. 15.48

$$i.e.\ E_0 = \sqrt{(100)^2 + (53.3)^2 + 2 \times 100 \times 53.3 \times (90° + \phi)}$$

$$= \sqrt{(100)^2 + (53.3)^2 - 2 \times 100 \times 53.3 \cos(-0.6)}$$

$$= 138.7 \text{ \%}$$

$$\text{Percentage regulation} = \frac{138.7 - 100}{100} \times 100 = 38.7 \text{ \% } \textbf{Ans.}$$

The two values of regulation determined by the two methods are found to differ widely from each other. The first method gives a little lower value as compared to the actual value while the other gives somewhat higher value. However, the first value is more likely to be nearer to the actual value because the second method employs synchronous impedance Z_s which is not constant but depends upon field excitation.

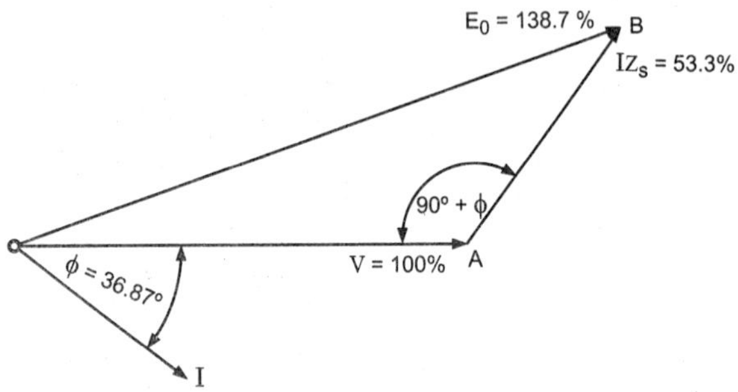

Fig. 15.49

Example 15.35. A 6.6 kV, 3-phase, 50 Hz, star-connected alternator gave the following data for open circuit and short circuit.

I_f **(A)**	**3.2**	**5.0**	**7.50**	**10.00**	**14.00**
E_0 **(kV)**	**3.10**	**4.90**	**6.60**	**7.50**	**8.24**
I_{sc} **(A)**	**500**	**778**	**1170**		

Compute the synchronous impedance under saturated and unsaturated conditions and voltage regulation of alternator at full load and 0.8 pf lagging. The dc resistance per phase of armature winding is 0.02 Ω.

[*M.D. Univ. Electromechanical Energy Conversion, December-2005*]

Solution: The OCC and SCC are drawn as per data and is shown in Fig. 15.50. The air-gap line, drawn by extending the straight-line part of the OCC, is indeed the OCC, if iron did not get saturated.

Corresponding to E_0 = 6.6 kV on the air-gap line, short-circuit current I_{sc} is 1,050 A for the same field current.

∴ Unsaturated synchronous impedance,

$$Z_s = \frac{\text{OC Voltage (phase value)}}{I_{sc}}$$

$$= \frac{6.6 \times 1{,}000 / \sqrt{3}}{1{,}050} = 3.629\ \Omega \textbf{ Ans.}$$

Corresponding to E_0 = 6.6 kV on the OCC;

I_{sc} = 1,170 A for the same field current

∴ Saturated synchronous impedance,

$$Z_s = \frac{6.6 \times 1{,}000 / \sqrt{3}}{1{,}170} = 3.257\ \Omega \textbf{ Ans.}$$

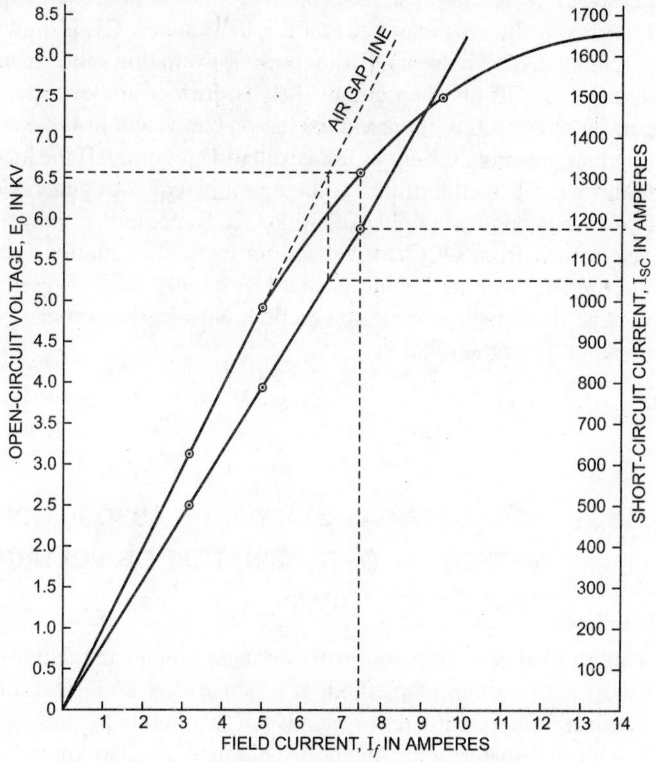

Fig. 15.50

Since neither rated output nor rated current is given for the generator, short-circuit current of 1,170 A corresponding to V_{OC} = 6.6 kV for the same field current of 7.5 A may be taken as rated current.

DC resistance per phase of armature winding,

$$R_{dc} = 0.02\ \Omega$$

Effective resistance per phase of armature winding,

$$R_e = 1.5\ R_{dc} = 1.5 \times 0.02 = 0.03\ \Omega$$

$\because$ Effective resistance is roughly 1.25 to 1.75 times R_{dc}

Synchronous reactance,

$$X_s = \sqrt{Z_s^2 - R_e^2} = \sqrt{(3.257)^2 - (0.03)^2} = 3.257\ \Omega$$

$$\text{Terminal voltage, } V = \frac{6.6 \times 1{,}000}{\sqrt{3}} = 3{,}810.5 \text{ V}$$

Power factor of load, $\cos\phi$ = 0.8 lagging

$$\sin\phi = \sqrt{1 - 0.8^2} = 0.6$$

Open-circuit voltage,

$$E_0 = \sqrt{(V\cos\phi + IR_e)^2 + (V\sin\phi + IX_s)^2}$$

$$= \sqrt{(3{,}810.5 \times 0.8 + 1{,}170 \times 0.03)^2 + (3{,}810.5 \times 0.6 + 1{,}170 \times 3.257)^2}$$

$$= 6{,}831 \text{ V}$$

$$\text{Open-circuit voltage, } E_{0L} = \sqrt{3} \times 6{,}831 = 11{,}832 \text{ V}$$

$$\text{Percentage regulation} = \frac{E_{0L} - V_L}{V_L} \times 100$$

$$= \frac{11{,}832 - 6{,}600}{6{,}600} \times 100 = 79.27\%$$ **Ans.**

15.20 ZERO POWER FACTOR OR POTIER METHOD OF DETERMINATION OF VOLTAGE REGULATION OF AN ALTERNATOR

The regulation obtained by mmf and emf methods is based on the total synchronous reactance (the sum of reactance due to armature leakage flux and due to armature reaction effect). This method is based on the separation of reactances due to leakage flux and that due to armature reaction flux, therefore, it is more accurate.

For determination of voltage regulation by this method, the data required are (*i*) effective resistance of armature winding (*ii*) open-circuit characteristic (*iii*) field current to circulate full-load current in the stator. and (*iv*) zero power factor full-load voltage characteristic—*a curve between terminal voltage and excitation while the machine is being run on synchronous speed and delivering full-load current at zero power factor.*

The first three have already been discussed in Art. 15.17. The zero power factor characteristic curve can be obtained as follows.

The machine is run at rated synchronous speed by a prime mover. A purely inductive load (variable load reactors) is connected across the armature terminals and the excitation or field current is raised so as to cause flow of full-load armature current. The value of the reactance is then increased step by step in such a way that the excitation current is adjusted to a value that causes full-load rated armature current to flow. The armature terminal voltages are varied from 125% to 25% of the rated voltage in steps, maintaining the speed and rated armature current constant throughout the test. The armature terminal voltages and excitation currents are noted at each step. The curve drawn between terminal voltage and excitation current, as shown in Fig. 15.51, gives the *zero power factor (lagging) characteristic.*

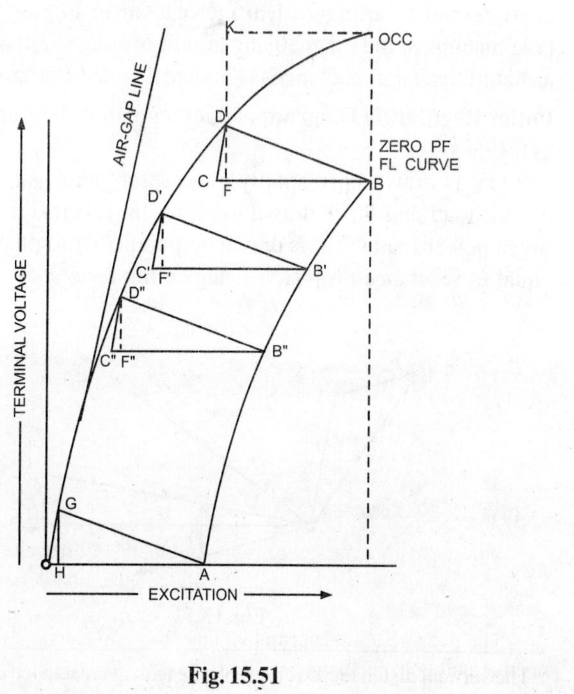

Fig. 15.51

When the zero power factor (lagging) characteristic is to be used only for obtaining the Potier reactance, it is sufficient to determine the point representing rated armature current and rated voltage. This is indicated by point B in Fig. 15.51.

The alternator can be loaded alternatively by an underexcited synchronous motor.

There is a definite relationship between the zero power factor (lagging) characteristic and an open-circuit characteristic of an alternator. The zero power factor characteristic curve is of exactly the same shape, as the OCC but it is shifted vertically downward by leakage reactance drop IX_L and horizontally by the armature reaction mmf.

Zero power factor full-load voltage-excitation characteristic can be drawn by knowing two points A and B. Point A is obtained from a short-circuit test with full-load armature current. Hence OA represents excitation (field current) required to overcome demagnetising effect of armature reaction and to balance leakage reactance drop at full load. Point B is obtained when full-load current flows through the armature and wattmeter reads zero, as discussed above.

From point B, line BC is drawn equal and parallel to AO. Then a line is drawn through C parallel to initial straight part of OCC (parallel to OG), intersecting the OCC at D. BD is joined and a perpendicular DF is dropped on BC. The triangle BFD is imposed at various points OCC to obtain corresponding points on the zero factor curve.

In Δ BDF the length BF represents armature reaction excitation and the length DF represents leakage reactance drop ($I X_L$)*. This is known as Potier reactance voltage drop and the triangle is known as *Potier triangle*. The potier reactance is given as

$$X_p = \frac{\text{DF (voltage drop per phase)}}{\text{Zero power factor current per phase}}$$

In case of cylindrical rotor machines, Potier reactance is nearly equal to armature leakage reactance. In case of salient pole machines, the magnetising circuit is more saturated and the armature leakage reactance is smaller than the Potier reactance.

Potier Regulation Diagram : Potier regulation diagram is drawn as follows.

OV is drawn horizontally to represent terminal voltage, V on full load and OI is drawn to represent full-load current at a given power factor. VE is drawn perpendicular to phasor OI and equal to reactance drop (IX_L), neglecting resistance drop. Now phasor OE represents generated emf E. From OCC field excitation I_1 corresponding to generated emf E is determined, OI_1 is drawn perpendicular to phasor OE to represent excitation required to induce emf OE on open circuit. $I_1 I_2$ is drawn parallel to load current phasor OI to represent excitation equivalent to full-load armature reaction. OI_2 gives total excitation required. If the load is thrown off, then terminal voltage will be equal to generated emf corresponding to field excitation OI_2. Hence emf E_0 may be determined from OCC corresponding to field excitation OI_2. Phasor OE_0 will lag behind phasor OI_2 by 90°. EE_0 represents voltage drop due to armature reaction. Now regulation can be obtained from the relation

$$\% \text{ Regulation} = \frac{E_0 - V}{V} \times 100$$

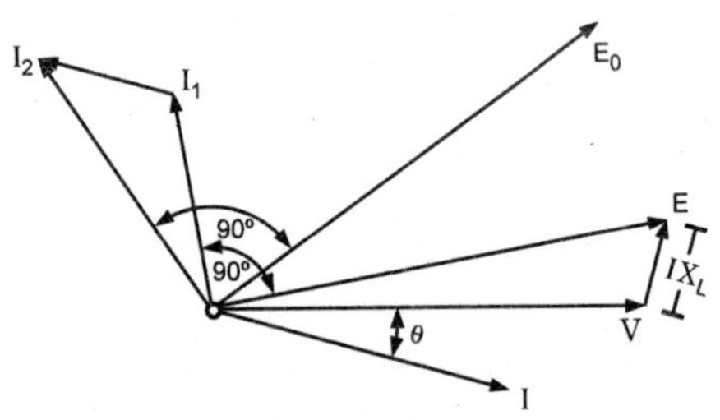

Fig. 15.52

15.21 ASA (AMERICAN STANDARDS ASSOCIATION) METHOD OF DETERMINATION OF VOLTAGE REGULATION OF AN ALTERNATOR

This method of determination of alternator voltage regulation is modification of mmf method. It is somewhat empirical and involves an approximation which is not required in the leakage-reactance-armature mmf method. This method, also known as A.I.E.E. method, is now official in USA and provides reliable results for the magnitude of open-circuit voltage and regulation for both salient pole and nonsalient pole machines.

The procedure usually adopted for determination of open-circuit emf of an alternator is as follows:

First of all, the field current corresponding to terminal voltage V is determined from air-gap line. This current I_1 is the excitation current that would be required if the machine had no armature impedance or armature reaction and if there was no saturation of the magnetic circuit.

Then field current required to circulate full-load current on short circuit of alternator is determined from the short-circuit characteristic [I_2 in Fig. 15.53 (*a*)].

Line *ab* representing field current I_2 is drawn making an angle 90° ± ϕ from line *oa* representing field current I_1 where + sign is for lagging pf generation operation and for leading pf motor operation while – sign is for leading pf generator operation and lagging pf motor operation. The field current *ob* (i.e. I_3) (the phasor sum of current I_1 and I_2) is the field current that would be required for the considered operating condition provided the magnetic circuit remained unsaturated.

Now E_G is determined for the operating condition that is being considered using the X_a determined by the Potier triangle method. Line parallel to field axis is drawn from E_G to meet OCC at point *c* and intersecting air-gap line at point *b*. The horizontal distance in Fig. 15.53 (*a*) from point *b* to *c* i.e. *bc* represents the additional excitation current that is required because of saturation of the magnetic circuit. The arithmetic sum of excitation currents I_3 and I_4 [excitation current represented by distance *bc* in Fig. 15.53 (*a*)] gives the excitation current I_{exc} that is required for the operating conditions under consideration.

* The vertical distance FD represents the leakage reactance drop and DK represents the voltage drop due to armature reaction.

Finally open-circuit emf E_o is determined from OCC corresponding to field excitation represented by OC in Fig. 15.53(*b*).

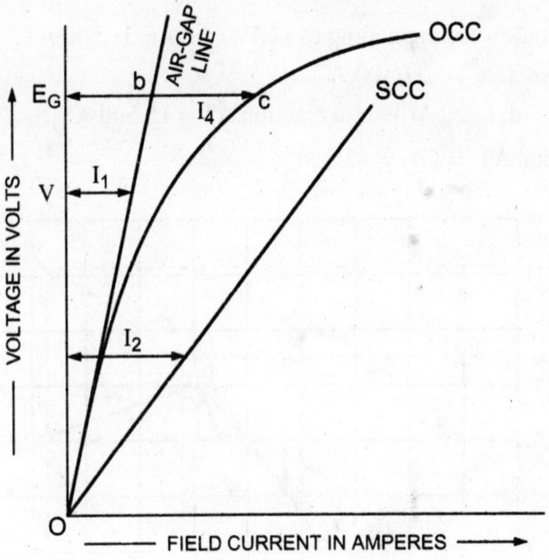

(*a*)

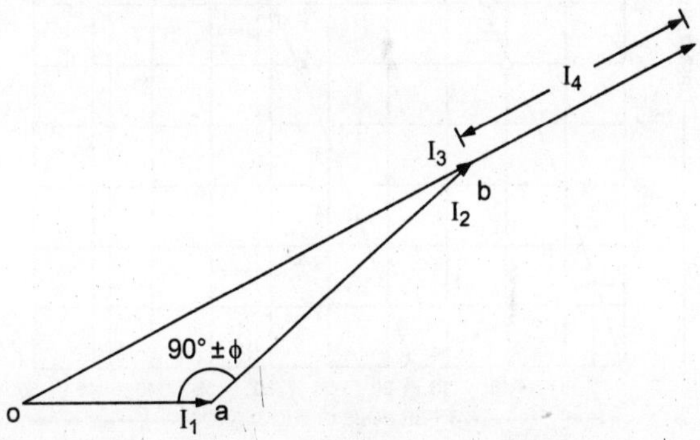

(*b*)

Fig. 15.53 *ASA Method For Determination of Voltage Regulation of Alternator*

Now regulation can be had from the relation

$$\% \text{ Regulation} = \frac{E_o - V}{V} \times 100$$

Example 15.36. A 10 kVA, 440 V, 50 Hz, three-phase alternator has the following OCC:

Field current (Amp.)	Terminal voltage (Volts)
1.0	100
3.0	300
5.0	440
8.0	550
11.0	600
15.0	635

With full load zero power factor load applied, an excitation of 14 A produced a terminal voltage of 500 V. On short circuit, 4 A excitation was required to circulate the full load current. Using mmf method, determine the full load percentage regulation for 0.6 pf lagging and 0.6 pf leading. *[U.P.S.C. I.E.S. Electrical Engineering-II, 1999]*

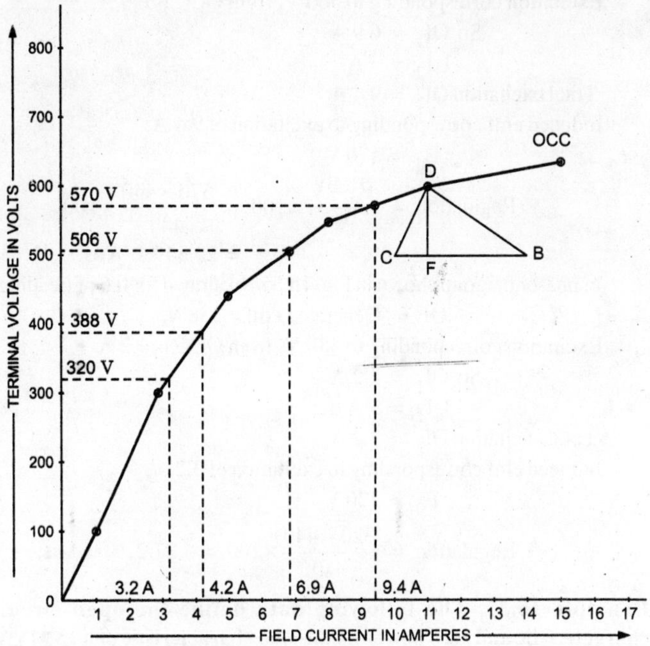

Fig. 15.54

Solution: From Potier triangle BDF

Field current required to overcome demagnetizing effect of armature reaction on full load BF = 3 A

Full-load reactance drop, $\sqrt{3} I X_L$ = DF = 100 V

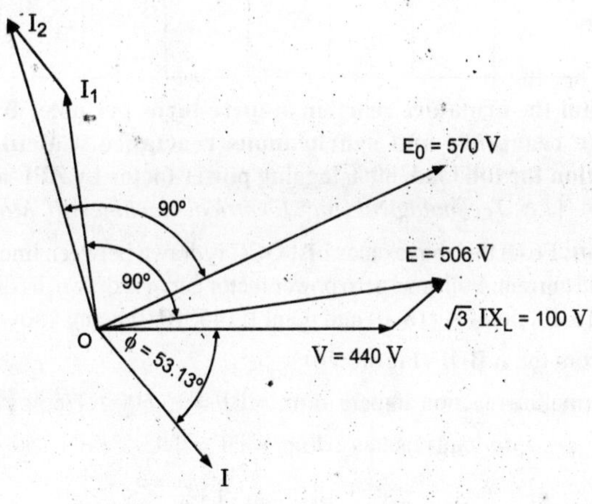

(*a*) *Potier Regulation Diagram For Lagging PF*

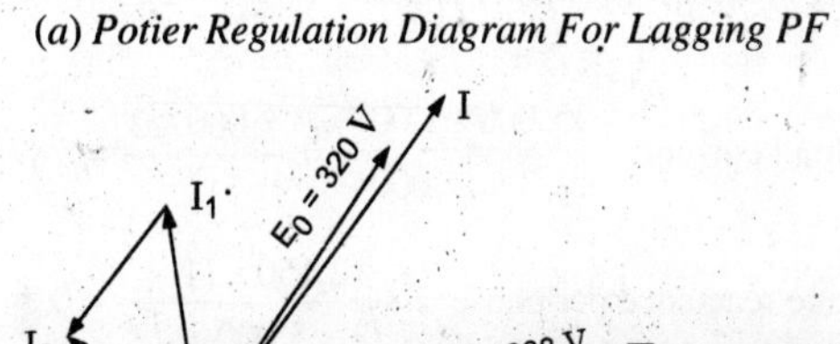

(*b*) *Potier Regulation Diagram For Leading PF*

Fig. 15.55

In phasor diagram shown in Fig. 15.55 (*a*), drawn for 0.6 pf (lagging)

OV = 440 V

VE = $\sqrt{3} I X_L$ = 100 V

OE = Generated emf = 506 V

Excitation corresponding to 506 V, from OCC is 6.9 A

So $OI_1 = 6.9$ A

$I_1I_2 = 3$ A

Total excitation $OI_2 = 9.4$ A

Induced emf corresponding to excitation of 9.4 A,

$E_0 = 570$ V

$$\% \text{ Regulation} = \frac{E_0 - V}{V} \times 100 = \frac{570 - 440}{440} \times 100 = 29.54\% \quad \textbf{Ans.}$$

In phasor diagram shown in Fig. 15.55 (*b*), drawn for 0.6 pf (leading)

OE = Generated emf = 388 V

Excitation corresponding to 388 V, from OCC is 4.2 A

So $OI_1 = 4.2$ A

$I_1I_2 = 3$ A

Total excitation $OI_2 = 3.2$ A

Induced emf corresponding to excitation of 3.2 A,

$E_0 = 320$ V

$$\% \text{ Regulation} = \frac{320 - 440}{440} \times 100 = -27.27\% \quad \textbf{Ans.}$$

Example 15.37. The following data define the open-circuit characteristic and zero power factor load characteristic of a 15 MVA, 11 kV, 3-phase, 50 Hz, Y-connected turbo-generator.

Field Ampere-Turns	**10,000**	**18,000**	**25,000**	**30,000**	**40,000**	**45,000**	**50,000**
V_{OC} line (kV)	**5.0**	**8.4**	**10.0**	**11.2**	**12.8**	**13.3**	**13.65**
Full-load, zero power factor line voltage in kV	**–**	**0**	**–**			**10**	

Find the armature reaction ampere-turns per pole, armature leakage reactance and synchronous reactance. Calculate the regulation for full load, 80% lagging power factor by ZPF method.
[*J.N. Technological Univ. Electrical Machines-III, May-2013*]

Solution: From the above datas first OCC is drawn between line voltage and field current. Full-load zero power factor curve is drawn, as discussed in Art 15.20. point A (18, 0) and point B (45, 10.0) being known.

From the Δ BDF (Fig. 15.56)

Armature reaction ampere turns = BF = 15,500 AT/pole **Ans.**

Full-load reactance drop = DF = 1.15 kV = 1,150 volts

$$\text{Leakage reactance drop per phase} = \frac{1,150}{\sqrt{3}} = 664 \text{ volts}$$

$$\text{Full-load current} = \frac{15,000 \times 1,000}{\sqrt{3} \times 11,000} = \frac{5,000\sqrt{3}}{11} = 787.3 \text{ A}$$

$$\text{Leakage reactance per phase} = \frac{1,150 \times 11}{\sqrt{3} \times 5,000 \times \sqrt{3}} = 0.843\ \Omega \ \textbf{Ans.}$$

From OCC, we find that 18,000 AT corresponds to OC voltage of

$$\frac{8.4 \times 10^3}{\sqrt{3}} = 4,850 \text{ V}$$

$$\text{Synchronous impedance, } Z_s = \frac{\text{OC voltage}}{\text{SC current}} = \frac{4,850}{787.3} = 6.16\ \Omega$$

Since R_e is negligible, hence

$X_s = Z_s = 6.16\ \Omega$ **Ans.**

In phasor diagram

OV = Terminal line voltage = 11,000 volts

VE = 1,150 volts representing reactance drop

OE = 11,700 volts

Excitation corresponding to 11,700 volts = 31,500 AT.

∴ $OI_1 = 31,500$ AT

I_1I_2 = Armature reaction AT = 15,500 AT.

Total AT = $OI_2 = 43,500$

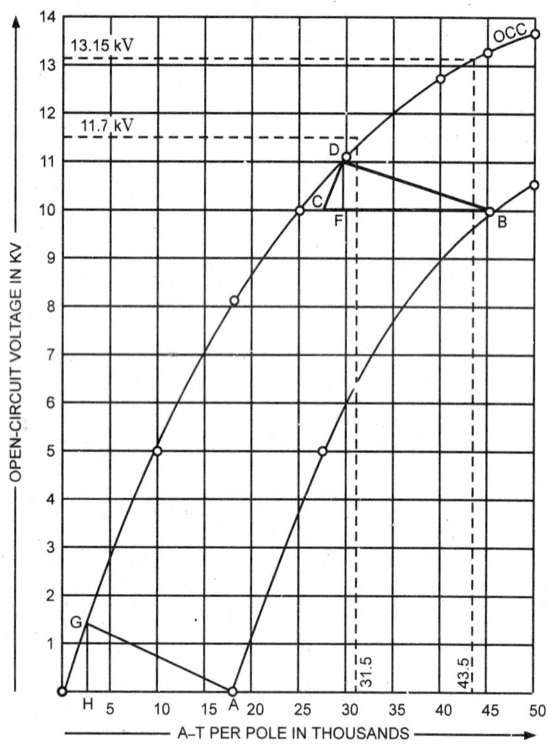

Fig. 15.56

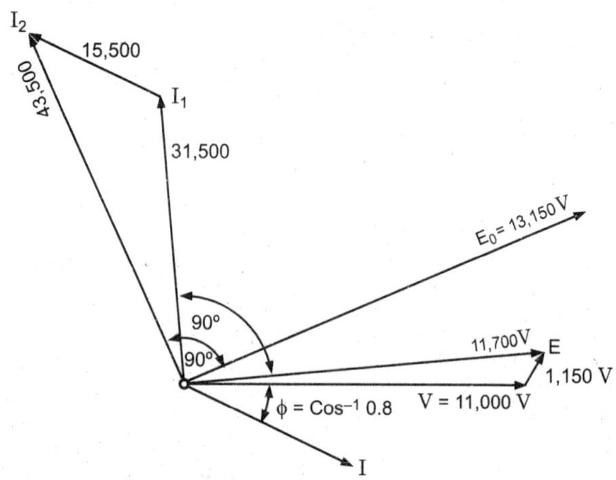

Fig. 15.57

Induced emf corresponding to 43,500 AT,

$E_0 = 13,150$ volts

$$\% \text{ Regulation} = \frac{13,150 - 11,000}{11,000} \times 100 = 19.55\ \% \ \textbf{Ans.}$$

Example 15.38. Find the regulation by the zero-power factor method of 5,000 kVA, 6,600 V, 3-phase, 50 Hz star-connected alternator at full load, unity power factor having the following test data:

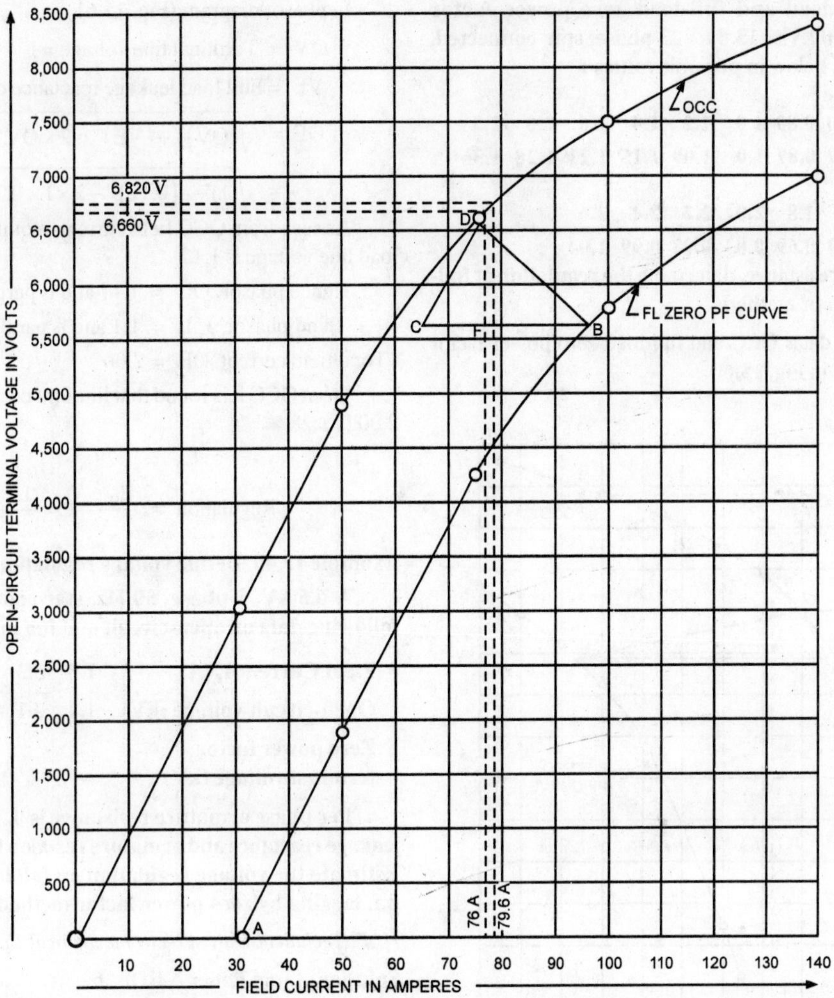

Fig. 15.58

Field current in amperes	**32**	**50**	**75**	**100**	**140**
Open-circuit terminal voltage in volts	**3,100**	**4,900**	**6,600**	**7,500**	**8,300**
Full-load current zero pf tests Line pd in volts	**0**	**1,850**	**4,250**	**5,800**	**7,000**

Neglect armature resistance.

Solution: From the above datas OCC and full-load zero power factor curves are drawn, as shown in Fig. 15.58.

From the Potier ΔBDF

Field current required to overcome demagnetising effect of armature reaction on full load, BF = 23 amperes

Full load reactance drop, DF = 900 volts

In phasor diagram,

OV = Terminal line voltage of 6,600 volts

VE = 900 volts representing line reactance drop

$$\therefore\ OE = \sqrt{(OV)^2 + (VE)^2 - 2 \times OV \times VE \cos(90° + \phi)}$$

$$= \sqrt{(6{,}600)^2 + (900)^2 - 2 \times 6{,}600 \times 900 \cos 90°} = 6{,}660\text{ V}$$

As seen from OCC field current required to induce 6,660 line volts is 76 amperes. Hence phasor $OI_1 = 76$ A and is perpendicular to phasor OE and phasor $I_1 I_2 = 23$ A and is parallel to phasor OI.

Total field current, $OI_2 = 79.6$ A

From OCC it is found that line emf corresponding to this field current of 79.6 A is 6,820 volts.

$$\therefore E_0 = 6{,}820 \text{ volts}$$

$$\text{Percentage regulation} = \frac{E_0 - V}{V} \times 100 = \frac{6{,}820 - 6{,}600}{6{,}600} \times 100$$

$= 3.33$ % **Ans.**

Fig. 15.59

Example 15.39. The no-load and full-load zero power factor characteristics for a 23,500 kVA, 13.8 kV, 3-phase, star-connected, turbo-generator are given below in per unit values :

No-load characteristic :

I_f	0.10	0.20	0.40	0.60	0.80	1.0	1.2	1.4	1.6	2.0	2.5
V	0.13	0.23	0.45	0.69	0.87	1.0	1.09	1.15	1.21	1.28	1.36

Zero pf characteristic:

I_f	1.2	1.3	1.4	1.6	1.8	2.0	2.2	2.4	2.6
V	0.015	0.13	0.25	0.49	0.69	0.83	0.92	0.99	1.04

Neglecting armature resistance, determine the regulation at full-load, 0.8 pf lag by the zero pf method.

Solution : From the above datas OCC and full-load zero power factor curves are drawn, as shown in Fig. 15.60.

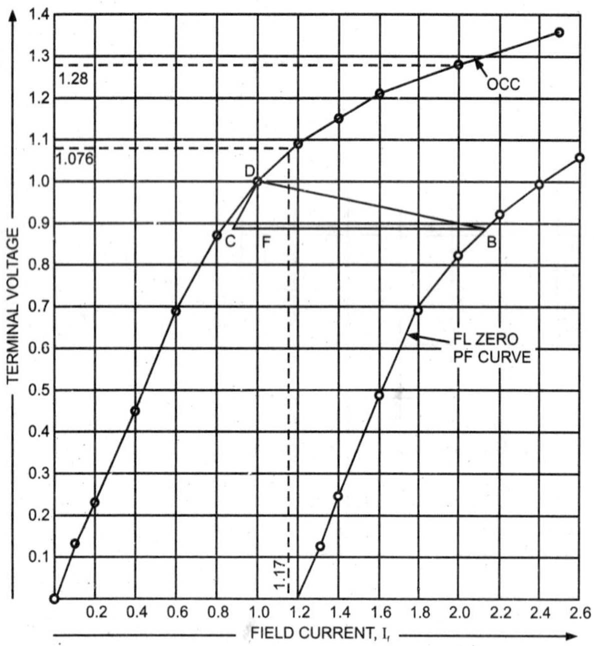

Fig. 15.60

From the Potier Δ BDF

Field current required to overcome demagnetising effect of armature reaction on full load, BF = 1.10

Full load leakage reactance drop, DF = 0.12

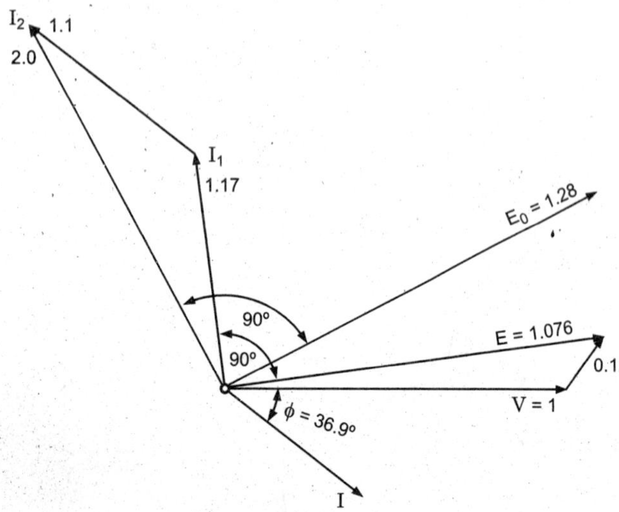

Fig. 15.61

In phasor diagram (Fig. 15.61)

OV = Terminal line voltage = 1

VE = Full load leakage reactance drop = 0.12

$$OE = \sqrt{(OV)^2 + (VE)^2 - 2 \times OV \times VE \cos(90° + \phi)}$$

$$= \sqrt{(1)^2 + (0.12)^2 - 2 \times 1 \times 0.12 \times (-0.6)} = 1.076$$

As seen from OCC field current, required to induce 1.076 times full load line voltage is 1.17.

Hence phasor, OI_1 = 1.17 and is perpendicular to phasor OE

and phasor, $I_1 I_2$ = 1.1 and is parallel to phasor OI

Total field current, OI_2 = 2.00

From OCC it is found that line emf corresponding to field current of 2.00 is 1.28

$$\therefore E_0 = 1.28$$

$$\text{Regulation} = \frac{E_0 - V}{V} = \frac{1.28 - 1.00}{1.00} = 0.28 \textbf{ Ans.}$$

Example 15.40. Define voltage regulation.

A 6.6 kV, 3-phase, 50 Hz, star-connected alternator gave the following data on open-circuit and full load, zero power factor tests:

Field Current I_f (A)	1.5	3.3	5.0	7.5	10.0	14.0
Open-circuit voltage (kV)	1.4	3.1	4.9	6.6	7.5	8.24
Zero power factor terminal voltage (kV)	—	0	1.85	4.24	5.78	7.0

Per phase armature resistance is 0.2 Ω. Determine the value of leakage reactance and armature reaction by Potier's triangle method. Estimate the voltage regulation on full load current of 500 A at 0.8 p.f. lagging by zero power factor method.

[U.P. Technical Univ. Electromechanical Energy Conversion II, 2002-03]

Solution: From Potier Δ BDF

Field current required to overcome demagnetising effect of armature reaction on full-load, BF = 2.4 A **Ans.**

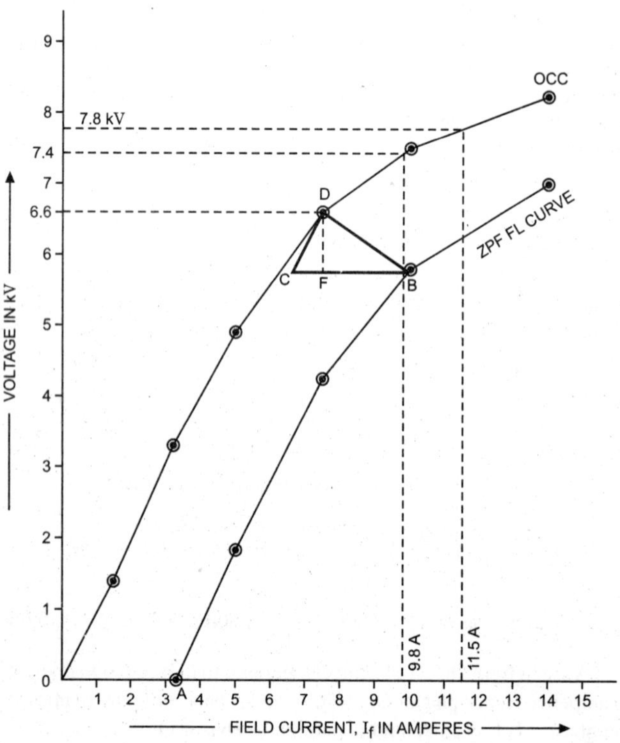

Fig. 15.62

Full-load reactance drop, DF = 900 V

Full-load current, I = 500 A

$$\text{Leakage reactance per phase} = \frac{900}{\sqrt{3}\times 500} = 1.04\ \Omega\ \textbf{Ans.}$$

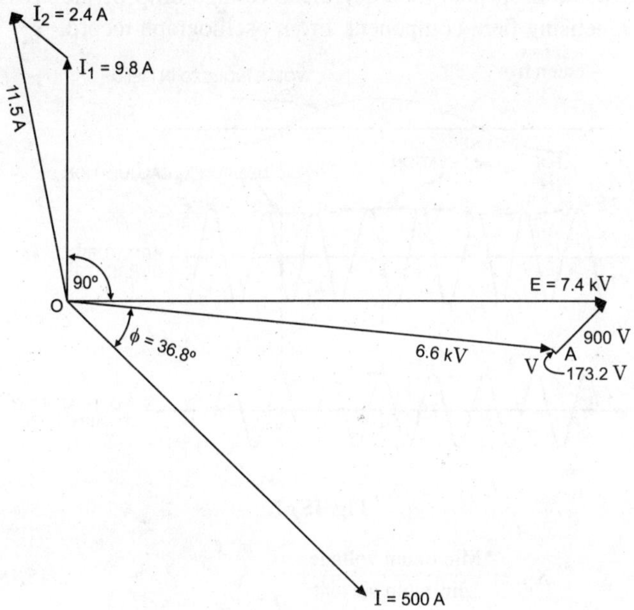

Fig. 15.63

In phasor diagram shown in Fig. 15.63,

OV = V_L = 6.6 kV

VA = Armature resistance drop

$= \sqrt{3}\times 500\times 0.2 = 173.2$ V

AE = Reactance drop = 900 V

OE = Generated emf = 7.4 kV

Excitation corresponding to 7.4 kV = 9.8 A

So, OI_1 = 9.8 A

I_1I_2 = Armature reaction amperes = 2.4 A

Total excitation = OI_2 = 11.5 A

Induced emf corresponding to excitation of 11.5 A, E_0 = 7.8 kV

$$\%\ \text{Regulation} = \frac{E_0 - V}{V}\times 100$$

$$= \frac{7.8-6.6}{6.6}\times 100 = 18.182\ \%\ \textbf{Ans.}$$

15.22 TWO REACTANCE THEORY FOR SALIENT POLE SYNCHRONOUS MACHINES

A multipolar machine with cylindrical rotor has a uniform air gap and therefore, its reactance remains the same, irrespective of the spatial positon of rotor. The effect of armature reaction, fluxes and voltages induced can, therefore, be treated in a simple way with concept of a synchronous reactance and taking it as constant for all positions of field poles with respect to the armature. However, a salient pole synchronous machine has non-uniform air gap due to which its reactance varies with the rotor position. Thus the salient pole machine possesses two axes of geometric symmetry (*i*) field pole axis, called *direct axis* or *d-axis* and (*ii*) axis passing through the centres of the interpolar space, called the *quadrature axis* or *q-axis*, as shown in Fig. 15.64 whereas in case of a cylindrical rotor machine there is only one axis of symmetry (pole axis or direct axis).

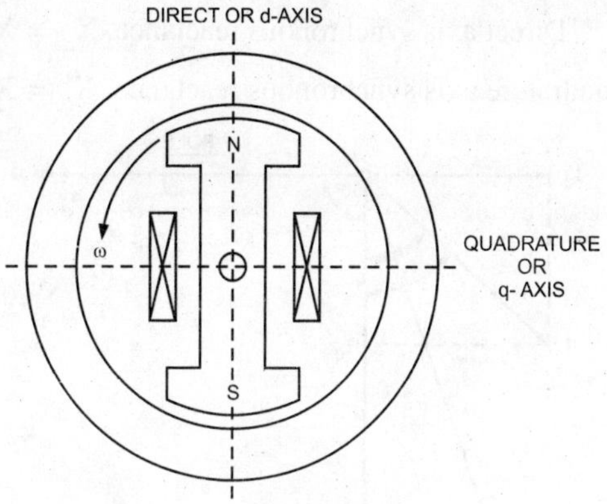

Fig. 15.64

In case of salient pole machine, the reluctance of the magnetic paths on which the emf acts are different along the direct axis and quadrature axis. The reluctance of the direct axis magnetic circuit is due to yoke and teeth of the stator, air gap, and pole and core of the rotor. In quadrature axis, the reluctance is mainly due to large air gap in the interpolar space.

Because of non-uniformity of the reluctance of the magnetic paths, the mmf of the armature is divided into two components namely (*i*) a direct acting component and (*ii*) a quadrature (or cross) component. It has been explained in Art. 15.14 that when armature current is in phase with the excitation voltage, the entire mmf of the armature acts at right angles to the axis of the salient poles and, therefore, all the armature mmf is in quadrature. On the other hand, if the armature current is in quadrature with the excitation voltage E_0, the entire mmf of the armature acts directly upon the magnetic paths through the salient poles and all of the armature mmf is direct acting, either directly opposing or directly aiding the mmf of the salient pole field windings. When the phase difference between armature current and excitation voltage is of some angle in between 0 and 90°, the armature mmf will have both a direct acting and a quadrature component. The direct-acting component is proportional to the sine of the phase angle between the armature current and excitation voltage whereas the quadrature (or cross) component is proportional to the cosine of the phase angle between the armature current and the excitation voltage. The two reactance concept is similar to the synchronous impedance concept in that the effect of armature reaction is taken into account by means of equivalent armature reactance voltage. However, owing to the difference in the reluctance of the magnetic paths upon which the two components of the armature mmf act, the value of the equivalent reactance for the direct component of armature mmf is greater than the value of the equivalent reactance for the quadrature component of the armature mmf. Thus the two reactance concept for salient-pole machines replaces the effect of armature reaction by two fictitious voltages. These reactance voltages are respectively $I_d X_{ad}$ and $I_q X_{aq}$, where I_d and I_q are the components of the armature current along direct and quadrature axis respectively. Each of these components of armature current also, produce a leakage reactance voltage caused by the armature leakage flux. The armature leakage reactance is assumed to have the same value X_L for both components of the armature current, Therefore, there is synchronous reactance for each component of the armature mmf as follows :

$$\text{Direct axis synchronous reactance, } X_d = X_{ad} + X_L$$

$$\text{Quadrature axis synchronous reactance, } X_q = X_{aq} + X_L$$

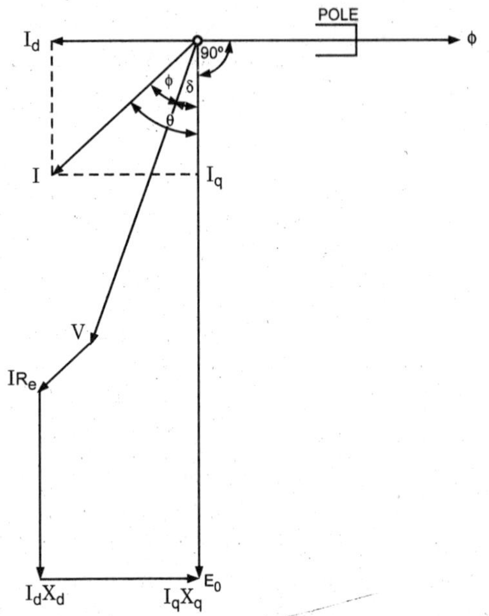

Fig. 15.65

The voltage equation for each phase of the armature based on the two-reactance concept is given as:

$$V = E_0 - I R_e - I_d X_d - I_q X_q \quad \text{...(15.17)}$$

In salient pole synchronous machines X_q = 0.6 to 0.7 times X_d whereas in cylindrical rotor machines $X_q = X_d$

15.22.1. Determination of X_d and X_q By Low Slip Test. The values of X_d and X_q are determined by applying a balanced reduced external voltage (say, V volts) to an unexcited machine at a speed a little less than the synchronous speed (the slip being less than 1%). Connection diagram is shown in Fig. 15.66. Applied voltage to the armature, armature current and the voltage induced in the field winding are measured by oscillographs. Due to voltage V applied to the stator terminals a current I will flow causing a stator mmf. This stator mmf moves slowly relative to the poles and induces an emf in the field circuit in a similar fashion to that of rotor in an induction motor at slip frequency. The effect will be that the stator mmf will move slowly relative to the poles. The physical poles and the armature reaction mmf are alternately in phase and out, the change occuring at slip frequency. When the axis of the poles and the axis of the armature reaction mmf wave coincide, the armature mmf acts through the field magnetic circuit. The voltage applied to the armature is then equal to drop caused by the direct component of armature reaction and leakage reactance. When the armature reaction mmf is in quadrature with the field poles, the applied voltage is equal to the leakage reactance drop plus the equivalent voltage drop of the cross-magnetising field component. From oscillograph record.

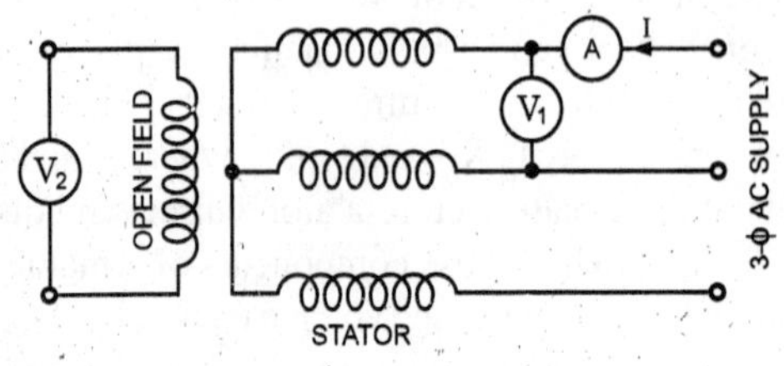

Fig. 15.66.

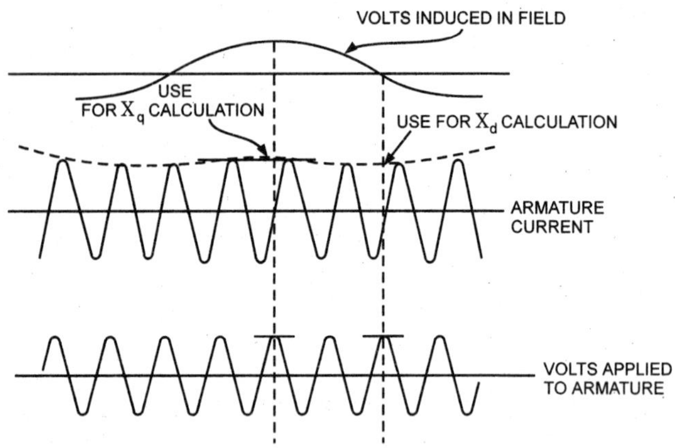

Fig. 15.67

$$X_d = \frac{\text{Maximum voltage}}{\text{Minimum current}} \quad \text{...(15.18)}$$

$$X_q = \frac{\text{Minimum voltage}}{\text{Maximum current}} \quad \text{...(15.19)}$$

Example 15.41. A 3-phase star-connected salient pole synchronous generator is driven at a speed near synchronous with the field circuit open, and the stator is supplied from a balanced 3-phase supply. Voltmeter connected across the line gave minimum and maximum readings of 2,800 and 2,820 volts. The line current fluctuated between 360 and 275 amperes. Find the direct and quadrature axis synchronous reactances per phase. Neglect armature resistance.

Solution: Maximum voltage = 2,820 volts

Minimum voltage = 2,800 volts

Maximum current = 360 A

Minimum current = 275 A

Direct-axis synchronous reactance between lines

$$= \frac{\text{Maximum voltage}}{\text{Minimum current}} = \frac{2,820}{275} = 10.25\ \Omega$$

Direct-axis synchronous reactance per phase,

$$X_d = \frac{10.25}{\sqrt{3}} = 5.92\ \Omega. \textbf{ Ans.}$$

Quadrature synchronous reactance between lines

$$= \frac{\text{Minimum voltage}}{\text{Maximum current}} = \frac{2,800}{360} = 7.778\ \Omega$$

Quadrature axis synchronous reactance per phase,

$$X_q = \frac{7.778}{\sqrt{3}} = 4.5\ \Omega \textbf{ Ans.}$$

15.23 CONSTRUCTION OF TWO REACTION DIAGRAM

For construction of two reaction diagram, the values of X_d and X_q must be known.

From the load specified the voltage phasor and current phasor at the load power factor angle are drawn as shown in Fig. 15.68.

From the extremity of the voltage phasor a line parallel to current phasor OI and equal to I R_e is drawn. Now from the extremity of the phasor I R_e, a line perpendicular to current phasor OI and equal to IX_q is drawn then a line from the origin O and passing through the extremity of IX_q phasor is drawn. This line gives the direction of excitation voltage E_0. Knowing the direction of E_0 the values of I_d and I_q are determined. By adding vectorially $I_d X_d$ and $I_q X_q$ to extremity of resistive drop phasor IR_e, the magnitude of excitation voltage E_0 is determined.

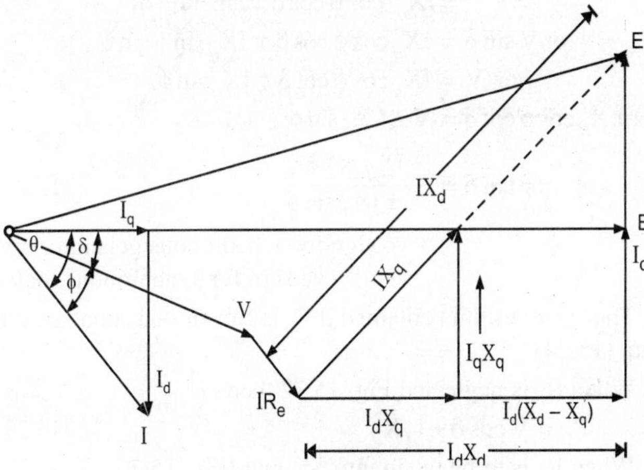

Fig. 15.68

It should be noted that the triangle involving the voltages IX_q and $I_q X_q$ is similar to the triangle of the currents. Therefore, the left hand portion of the phasor $I_d X_d$ must be equal to $I_d X_q$. The portion of the components of $I_d X_q$ shown provides the ratio of X_q and X_d.

If the perpendicular drawn from the extremity of the phasor I R_e is produced so that its length is equal to $IX_d - IX_q$; the phasor E'_0, which is normally obtained from the synchronous impedance method, can be determined. In Fig. 15.68 the extremity of E'_0 is shown on the line perpendicular to $I_d X_d$ and it is slightly greater than E_0, making the regulation as obtained by the synchronous impedance method greater than that obtained by the two-reaction method. The difference depends upon the saliency (ratio of pole pitch to pole arc) and increases with the increase in the value of saliency. One of the greatest difference between the two methods is the magnitude of the torque angle, which is much larger when determined from the synchronous impedance phasors than when found from the two-reaction diagram.

The angle θ between E_0 and I is called the *internal power factor angle* whereas δ, the angle between E_0 and V is called the *load* or *power angle.*

The phasor diagram is redrawn in Fig. 15.69 taking E_0 in horizontal direction.

From Fig. 15.69

$$I_d = I \sin \theta$$

$$I_q = I \cos \theta$$

In ΔABC, $\angle \underline{ACB} = \theta$ as angle between the two lines is the same as between their perpendiculars. Line AC is perpendicular to phasor I and line CB perpendicular to phasor E_0.

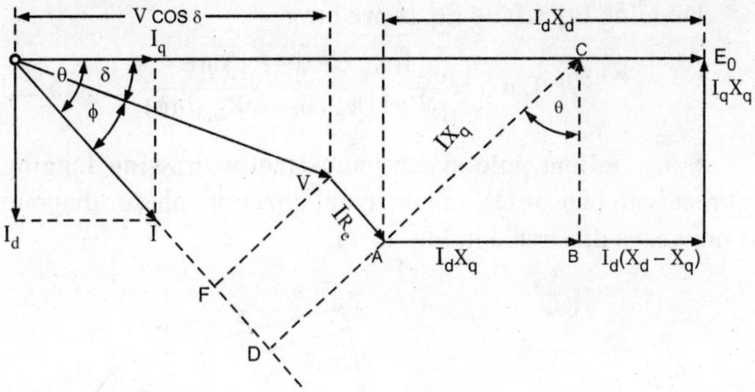

Fig. 15.69

From ΔODC

$$\text{Tan } \theta = \frac{DC}{OD} = \frac{DA + AC}{OF + FD}$$

$$= \frac{V \sin \phi + IX_q}{V \cos \phi + IR_e} \quad \text{...(15.20) generating}$$

$$= \frac{V \sin \phi - IX_q}{V \cos \phi - IR_e} \quad \text{...(15.21) motoring}$$

The angle θ can be determined from the above equations.

Load angle δ = θ – φ in case of generating and

δ = φ – θ in case of motoring.

Angle φ is taken + ve for lagging power factor and will be taken negative for leading power factor.

Power angle δ can be determined from the phasor diagram, of a salient pole synchronous generator supplying lagging current, drawn with phasor OV in horizontal direction (Fig. 15.70).

From phasor diagram shown in Fig. 15.70 we have

$$\mathbf{E} = \mathbf{V} + \mathbf{I}\,(R_e + j\,X_q)$$

$$\text{or } E\,(\cos \delta + j \sin \delta) = V\,(1 + j\,0) + I\,(\cos \phi - j \sin \phi)(R_e + j\,X_q)$$

$$\text{or } E \cos \delta + j\,E \sin \delta = V + I\,(R_e \cos \phi + X_q \sin \phi) + j\,I\,(X_q \cos \phi - R_e \sin \phi)$$

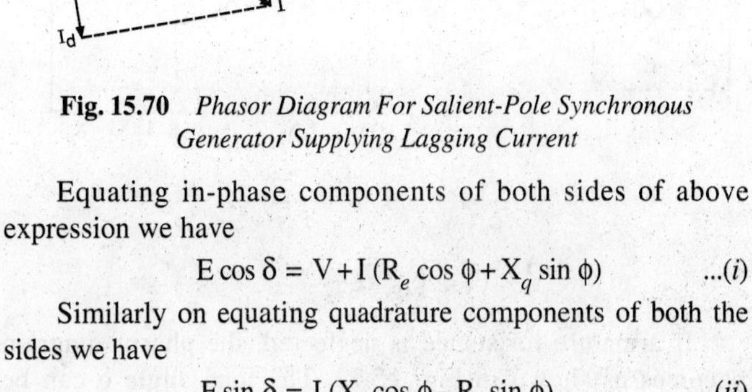

Fig. 15.70 *Phasor Diagram For Salient-Pole Synchronous Generator Supplying Lagging Current*

Equating in-phase components of both sides of above expression we have

$$E \cos \delta = V + I\,(R_e \cos \phi + X_q \sin \phi) \quad ...(i)$$

Similarly on equating quadrature components of both the sides we have

$$E \sin \delta = I\,(X_q \cos \phi - R_e \sin \phi) \quad ...(ii)$$

Dividing Eq. (*ii*) by Eq. (*i*) we have

$$\text{Tan}\,\delta = \frac{I(X_q \cos\phi - R_e \sin\phi)}{V + I(R_e \cos\phi + X_q \sin\phi)} \quad ...(15.22)$$

For a salient-pole synchronous motor drawing lagging current with phasor OV in horizontal direction, phasor diagram is drawn, as illustrated in Fig. 15.71.

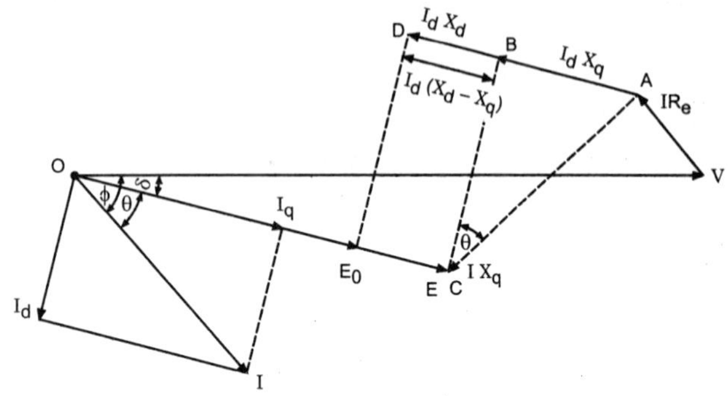

Fig. 15.71 *Phasor Diagram For Salient-Pole Synchronous Motor Drawing Lagging Current*

From phasor diagram shown in Fig. 15.71, we have

$$\mathbf{E} = \mathbf{V} - \mathbf{I}\,(R_e + j\,X_q)$$

$$\text{or } E\,(\cos\delta - j\sin\delta) = V\,(1 + j\,0) - I\,(\cos\phi - j\sin\phi)\,(R_e + j\,X_q)$$

$$\text{or } E\,(\cos\delta - j\sin\delta) = V - I\,(R_e\cos\phi + X_q\sin\phi) - j\,I\,(X_q\cos\phi - R_e\sin\phi)$$

From above equation we have

$$E\cos\delta = V - I\,(R_e\cos\phi + X_q\sin\phi) \quad ...(i),$$

considering in-phase components

$$\text{and } E\sin\delta = I\,(X_q\cos\phi - R_e\sin\phi) \quad ...(ii),$$

considering quadrature components

Dividing expression (*ii*) by expression (*i*) we have

$$\text{Tan}\,\delta = \frac{I(X_q \cos\phi - R_e \sin\phi)}{V - I(R_e \cos\phi + X_q \sin\phi)} \quad ...(15.23)$$

Thus power angle δ for a salient pole synchronous machine can be given by expression.

$$\text{Tan}\,\delta = \frac{I(X_q \cos\phi - R_e \sin\phi)}{V \pm I(R_e \cos\phi + X_q \sin\phi)} \quad ...(15.24)$$

\+ *ve* sign for synchronous generator and
\– *ve* sign for synchronous motor

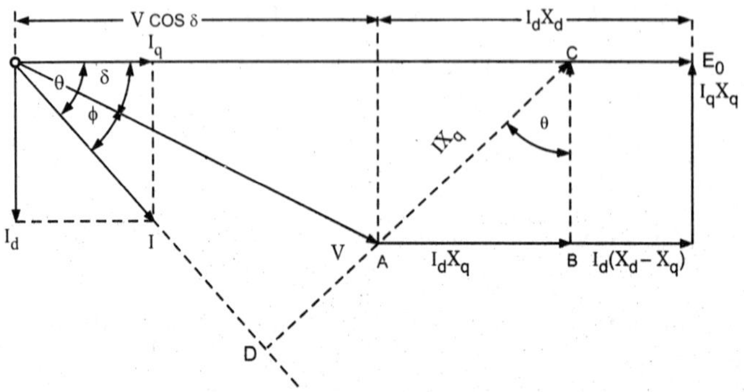

Fig. 15.72

If armature resistance is neglected, the phasor diagram becomes as shown in Fig. 15.72. The load angle δ can be determined directly as below :

$$\delta = \theta - \phi \text{ in case of generating}$$
$$\text{and } \delta = \phi - \theta \text{ in case of motoring}$$
$$\text{or } \theta = \phi \pm \delta \text{ in general} \quad ...(15.25)$$

Direct-axis component,

$$I_d = I\sin\theta = I\sin(\phi \pm \delta) \quad ...(15.26)$$

Quadrature axis component,

$$I_q = I\cos\theta = I\cos(\phi \pm \delta) \quad ...(15.27)$$

$$V\sin\delta = I_q X_q = I\cos\theta\, X_q = I\,X_q\cos(\phi \pm \delta)$$
$$= IX_q\,(\cos\phi\cos\delta \mp \sin\phi\sin\delta)$$
$$\text{or } V\sin\delta = IX_q\cos\phi\cos\delta \mp IX_q\sin\phi\sin\delta$$
$$\text{or } V = IX_q\cos\phi\cot\delta \mp IX_q\sin\phi$$
$$\text{or } I\,X_q\cos\phi\cot\delta = V \pm I\,X_q\sin\phi$$

$$\text{or } \tan\delta = \frac{IX_q\cos\phi}{V \pm IX_q\sin\phi} \quad ...(15.28)$$

\+ ve sign for synchronous generator and
\– ve sign for synchronous motor

The same result is obtained if R_e is substituted equal to zero in Eq. (15.24).

When R_e is neglected Fig. 15.72 then

$$E_0 = V\cos\delta \pm I_d X_d \quad ...(15.29)$$

When R_e is to be taken into account (Fig. 15.70)

$$E_0 = V\cos\delta + I\,R_e\cos\theta + I_d X_d$$
$$= V\cos\delta + I_q R_e + I_d X_d \quad ...(15.30) \text{ ...generating}$$
$$\text{and } E_0 = V\cos\delta - I_q R_e - I_d X_d \quad ...(15.31) \text{ ...motoring}$$

15.24 POWER DEVELOPED BY A SYNCHRONOUS GENERATOR

Power developed per phase,

$$P_d = \text{Power output } (P_{out}) \text{ per phase}$$
$$= VI\cos\phi \quad \text{provided R or copper loss is negligible}$$

Now from Fig. 15.72

$$I_q X_q = V\sin\delta \quad ...(i)$$
$$I_d X_d = E_0 - V\cos\delta \quad ...(ii)$$

Also $I\cos\phi = I_d\sin\delta + I_q\cos\delta$

So power developed per phase,

$$P = VI_d\sin\delta + V\,I_q\cos\delta$$

Substituting values of I_d and I_q from Eqs. (*ii*) and (*i*) we get

$$P = V\cdot\frac{E_0 - V\cos\delta}{X_d}\sin\delta + V\cdot\frac{V\sin\delta}{X_q}\cos\delta$$

$$= \frac{E_0 V}{X_d}\sin\delta - \frac{V^2}{X_d}\sin\delta\cos\delta + \frac{V^2\sin\delta\cos\delta}{X_q}$$

$$= \frac{E_0 V}{X_d}\sin\delta + V^2\left[\frac{1}{X_q} - \frac{1}{X_d}\right]\sin\delta\cos\delta$$

$$= \frac{E_0 V}{X_d}\sin\delta + \frac{V^2}{2}\left[\frac{1}{X_q} - \frac{1}{X_d}\right]\sin 2\delta \quad ...(15.32)$$

Total power developed will be three times of the above *i.e* total power developed by a 3-phase synchronous generator

$$= \frac{3\,E_0 V}{X_d}\sin\delta + \frac{3V^2}{2}\left[\frac{1}{X_q} - \frac{1}{X_d}\right]\sin 2\delta \quad ...(15.33)$$

Above expression for power developed consists of two terms, the first term representing power due to field excitation and the second term gives the reluctance power *i.e.* power due to saliency. In case of a cylindrical rotor machine $X_d = X_q$ and so the second term becomes zero and the power is given by the first term only. In case there is no field exicitation *i.e.* E_0, the first term becomes zero and the power developed is given by the second term.

It may be noted that value of δ is positive in case of an alternator and negative for a motor.

Example 15.42. A salient pole, 3-phase, star-connected alternator at 5 MVA, 6.6 kV, has a resistance of 1.5 % and leakage reactance of 10 %. The short-circuit armature reaction for full load current is equivalent to 60 A field current. The armature cross reaction per armature turn is half of the direct reaction. The open-circuit characteristic is as follows:

Field current in amps	**32**	**50**	**75**	**100**	**140**
Terminal pd in volts	**3,100**	**4,700**	**6,600**	**7,500**	**8,300**

Find the percentage regulation on full load at pf 0.8 (lagging).

Solution: The OCC is drawn as shown in Fig. 15.73 (*a*)

The phasor diagram for full load 0.8 power factor (lagging) is drawn to represent terminal voltage (6,600 V) taken as 100 % [Fig. 15.73 (*b*)].

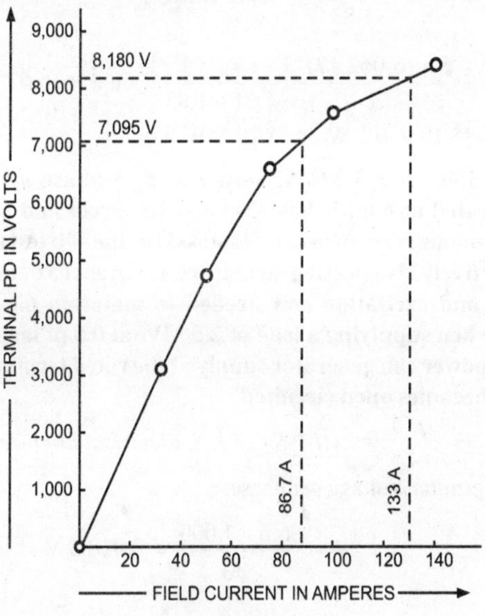

Fig. 15.73 (*a*)

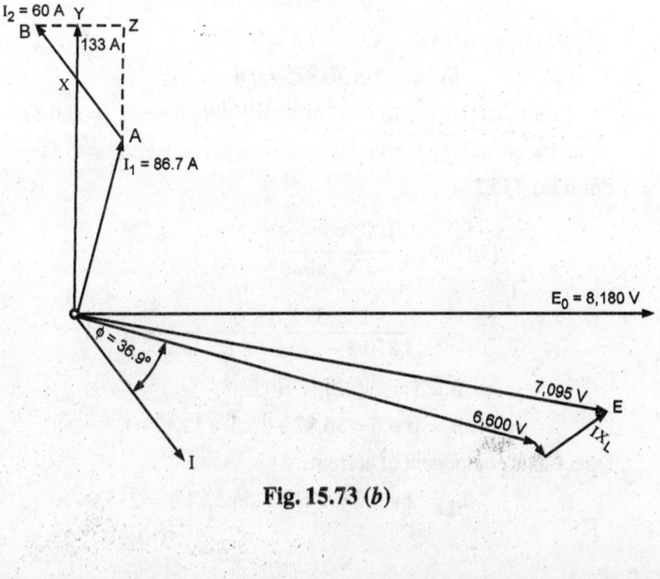

Fig. 15.73 (*b*)

Then to the same scale the resistance and leakage reactance drops of 1.5 % and 10 % are added in phase and in quadrature respectively with the current to give generated emf OE,

$$OE = 107.5\ \% = 7{,}095\ V$$

From OCC the field excitation corresponding to 7,095 volts, I = 86.7 **A.** Now OI_1 is drawn perpendicular to phasor OE to represent excitation required to induce emf OE on open circuit. $I_1 I_2 = 60$ A is drawn parallel to load current phasor OI to represent excitation equivalent to full-load armature reaction on short circuit. I_1I_2 is now divided at X such that $I_1X = KI_1 I_2$ where K is the ratio of the cross-reaction to the direct reaction per ampere-turn. In this case K = 0.5 so that X bisects AB.

A line is now drawn through OX and produced; a perpendicular from I_2 is drawn to this line meeting it at Y. Then OY is the excitation required and is made up of the phasor sum of OI_1, I_1Z (to balance the direct reaction) and ZY (to balance the cross-reaction).

$$\text{By measurement OY} = 133\ A$$

The emf generated corresponding to excitation of 133 amperes = 8,180 volts

$$\text{Percentage regulation} = \frac{8{,}180 - 6{,}600}{6{,}600} \times 100 = 23.94\ \%\ \textbf{Ans.}$$

Example 15.43. A 3-phase star-connected synchronous generator supply current of 10 A having phase angle of 20° lagging at 400 V. Find the load angle and the components of armature current I_d and I_q if X_d = 10 Ω and X_q = 6.5 Ω. Assume armature resistance to be negligible. *[Rajasthan Technical Univ. Electrical Machines-II, 2010]*

Solution: Direct axis synchronous reactance per phase, $X_d = 10\ \Omega$

Quadrature-axis synchronous reactance per phase, $X_q = 6.5\ \Omega$

Armature current, I = 10 A

Power factor angle, $\phi = 20°$ (lagging)

Terminal voltage per phase,

$$V = \frac{400}{\sqrt{3}} = 230.94\ V$$

$$\text{Tan}\ \delta = \frac{IX_q \cos\phi}{V + IX_q \sin\phi} = \frac{10 \times 6.5 \times \cos 20°}{230.94 + 10 \times 6.5 \times \sin 20°}$$

$$= \frac{61.08}{253.17} = 0.24126$$

Load angle, $\delta = \text{Tan}^{-1}\ 0.24126 = 13.564°$

Angle, $\theta = \delta + \phi = 13.564° + 20° = 33.564°$

Direct axis component of armature current,

$I_d = I \sin\theta = 10 \sin 33.564° = 5.53$ A **Ans.**

Quadrature axis component of armature current,

$I_q = I \cos\theta = 10 \cos 33.564° = 8.33$ A **Ans.**

Example 15.44. A synchronous generator has a direct-axis synchronous reactance 'X_d' of 0.8 pu and a quadrature-axis synchronous reactance 'X_q' of 0.5 pu, Determine the pu open-circuit voltage for full-load at a lagging pf of 0.8. Neglect saturation.

Solution: Direct-axis synchronous reactance, X_d = 0.8 pu

Quadrature-axis synchronous reactance, X_q = 0.5 pu

Power factor, cos φ = 0.8 lagging

Phase angle, $\phi = \text{Cos}^{-1}\ 0.8 = 36.87°$

Sin φ = Sin 36.87° = 0.6

Terminal voltage, V = 1.0 pu

Armature current, I = 1.0 pu

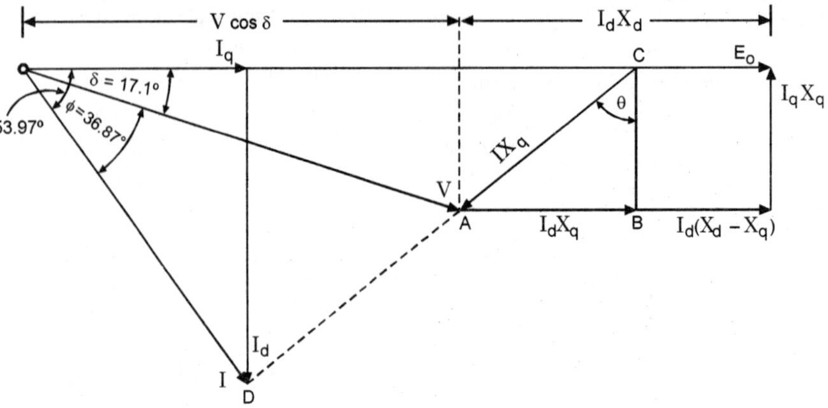

Fig. 15.74

$$\text{Tan }\theta = \frac{DC}{OD} = \frac{DA + AC}{OD} = \frac{V\sin\phi + IX_q}{V\cos\phi}$$

$$= \frac{1\times 0.6 + 1\times 0.5}{1\times 0.8} = 1.375$$

$$\text{Angle }\theta = \text{Tan}^{-1}\,1.375 = 53.97°$$

$$\text{Load angle, }\delta = \theta - \phi = 53.97° - 36.87° = 17.1°$$

Direct-axis component of current,

$$I_d = I\sin\theta = 1 \times \sin 53.97° = 0.8087 \text{ pu}$$

$$\text{No-load emf, } E_0 = V\cos\delta + I_d X_d$$

$$= 1.0\cos 17.1° + 0.8087 \times 0.8 = 1.60278 \text{ pu}$$ **Ans.**

Example 15.45. A 44 MVA, 10.5 kV, 50 Hz, star-connected three-phase salient pole synchronous generator has X_d = 1.83 Ω and X_q = 1.21 Ω. It delivers rated load at 0.8 pf lagging. The armature resistance is negligible. Determine the power developed by the generator and the %age voltage regulation.

[*U.P.S.C. I.E.S. Electrical Engineering-II, 2013*]

Solution: Terminal voltage per phase,

$$V = \frac{10.5\times 1{,}000}{\sqrt{3}} = 6{,}062.17 \text{ V}$$

$$\text{Armature current, } I = \frac{\text{Rated MVA}\times 10}{\sqrt{3}\,V} = \frac{44\times 10^6}{\sqrt{3}\times 10.5\times 1{,}000}$$

$$= 2{,}419.37 \text{ A}$$

$$\text{Load phase angle, } \phi = \text{Cos}^{-1}\,0.8 = 36.87°$$

$$\text{Sin }\phi = \text{Sin }36.87° = 0.6$$

Direct-axis synchronous reactance per phase, X_d = 1.83 Ω

Quadrature-axis synchronous reactance per phase X_q = 1.21 Ω

From Eq. (15.28)

$$\text{Tan }\delta = \frac{IX_q\cos\phi}{V + IX_q\sin\phi} = \frac{2{,}419.37\times 1.21\times 0.8}{6{,}062.17 + 2{,}419.37\times 1.21\times 0.6}$$

$$= 0.2995$$

$$\text{Load angle, } \delta = \text{Tan}^{-1}\,0.2995 = 16.67°$$

$$\text{Angle, } \theta = \phi + \delta = 36.87° + 16.67° = 53.545°$$

Direct-axis component of current,

$$I_d = I\sin\theta = 2{,}419.37\sin 53.545° = 1{,}945.95 \text{ A}$$

Excitation voltage per phase,

$$E_o = V\cos\delta + I_d X_d$$

$$= 6{,}062.17\cos 16.67° + 1{,}945.95 \times 1.83 = 9{,}368.49 \text{ V}$$

$$\%\text{ regulation} = \frac{E_o - V}{V}\times 100 = \frac{9{,}368.49 - 6{,}062.17}{6{,}062.17}\times 100$$

$$= 54.54\%$$ **Ans.**

Power developed by generator

$$= \frac{3E_oV}{X_d}\sin\delta + \frac{3V^2}{2}\left[\frac{1}{X_q} - \frac{1}{X_d}\right]\sin 2\delta$$

$$= \frac{3\times 9{,}368.49\times 6{,}062.17}{1.83}\sin 16.67°$$

$$+ \frac{3\times(6{,}062.17)^2}{2}\left[\frac{1}{1.21} - \frac{1}{1.83}\right]\sin 2\times 16.67°$$

$$= 35.19\times 10^6 \text{ W or } 35.19 \text{ MW}$$ **Ans.**

Example 15.46. A 3.5 MVA, slow speed, 3-phase synchronous generator rated at 6.6 kV has 32 poles. Its direct and quadrature axis synchronous reactances as measured by the slip test are 9.6 and 6 Ω respectively. Neglecting armature resistance, determine the regulation and excitation emf needed to maintain 6.6 kV at the terminals when supplying a load of 2.5 MW at 0.8 pf lagging. What maximum power can generator supply at the rated terminal voltage, if the field becomes open circuited?

[*U.P.S.C. I.E.S. Electrical Engineering, 1994*]

Solution: Terminal voltage per phase,

$$V = \frac{6.6\times 1{,}000}{\sqrt{3}} = 3{,}810.5 \text{ V}$$

$$\text{Armature current, } I = \frac{2.5\times 10^6}{\sqrt{3}\times 6{,}600\times 0.8} = 273.37 \text{ A}$$

$$\text{Load phase angle, } \phi = \text{Cos}^{-1}\,0.8 = 36.87°$$

$$\text{Sin }\phi = \text{Sin }36.87° = 0.6$$

Direct-axis synchronous reactance per phase, X_d = 9.6 Ω

Quadrature axis synchronous reactance per phase, X_q = 6 Ω

From Eq. (15.28)

$$\text{Tan }\delta = \frac{IX_q\cos\phi}{V + IX_q\sin\phi}$$

$$= \frac{273.37\times 6\times 0.8}{3{,}810.5 + 273.37\times 6\times 0.6} = 0.2737$$

$$\text{or } \delta = \text{Tan}^{-1}\,0.2737 = 15.3°$$

$$\text{Angle }\theta = \phi + \delta = 36.87 + 15.3 = 52.17°$$

Direct-axis component of current,

$$I_d = I\sin\theta = 273.31\sin 52.17° = 215.94 \text{ A}$$

Excitation voltage per phase,

$$E_0 = V \cos \delta + I_d X_d$$
$$= 3{,}810.5 \times \cos 15.3° + 215.94 \times 9.6 = 5{,}748 \text{ V}$$

Excitation voltage (line-to-line)

$$= \sqrt{3} \times 5{,}748 = 9{,}956 \text{ V} \textbf{ Ans.}$$

$$\text{Percentage regulation} = \frac{9{,}956 - 6{,}600}{6{,}600} \times 100 = 50.85\ \% \textbf{ Ans.}$$

When the field get open circuited, the power developed

$$= \frac{V_L^2}{2}\left[\frac{1}{X_q} - \frac{1}{X_d}\right] \sin 2\delta \qquad \text{Refer to Eq. (15.33)}$$

The power developed will be maximum for sin 2δ = 1 and so the maximum power, that the generator can supply at the rated terminal voltage, with field open-circuited

$$= \frac{V_L^2}{2}\left[\frac{1}{X_q} - \frac{1}{X_d}\right] = \frac{(6{,}600)^2}{2}\left[\frac{1}{6} - \frac{1}{9.6}\right]$$
$$= 1{,}361{,}250 \text{ W or } 1.361 \text{ MW} \textbf{ Ans.}$$

Example 15.47. A 10 kVA, 380 V, 50 Hz, 3-phase, star-connected salient pole alternator has direct-axis and quadrature-axis reactances of 12 Ω and 8 Ω respectively. The armature has a resistance of 1 Ω per phase. The generator delivers rated load at 0.8 pf lagging with the terminal voltage being maintained at rated value. If the load angle is 16.15°, determine (*i*) the direct axis and quadrature axis components of armature current, (*b*) excitation voltage of the generator. *[GATE 1993]*

Solution: Armature resistance, $R_e = 1\ \Omega$

Direct-axis synchronous reactance, $X_d = 12\ \Omega$

Quadrature-axis synchronous reactance, $X_q = 8\ \Omega$

Power factor, $\cos \phi = 0.8$

$\phi = \text{Cos}^{-1}\ 0.8 = 36.87°$

Load angle, $\delta = 16.15°$

$\theta = \phi + \delta = 36.87° + 16.15° = 53.02°$

$$\text{Terminal voltage per phase, } V = \frac{380}{\sqrt{3}} = 219.4 \text{ V}$$

$$\text{Armature current, } I = \frac{\text{kVA} \times 1{,}000}{\sqrt{3}\ V_L} = \frac{10 \times 1{,}000}{\sqrt{3} \times 380} = 15.2 \text{ A}$$

Phasor diagram is shown in Fig. 15.75.

Angle θ can also be determined as follows

$$\text{Tan}\ \theta = \frac{DC}{OD} = \frac{DA + AC}{OF + FD} = \frac{V \sin \phi + IX_q}{V \cos \phi + IR_e}$$
$$= \frac{219.4 \times 0.6 + 15.2 \times 8}{219.4 \times 0.8 + 15.2 \times 1} = 1.3278$$

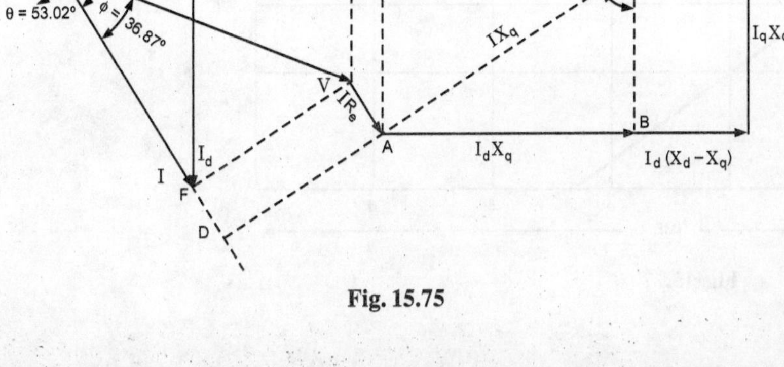

Fig. 15.75

or $\theta = \text{Tan}^{-1}\ 1.3278 = 53.02°$ The same as determined above.

Direct-axis component of armature current,

$$I_d = I \sin \theta = 15.2 \sin 53.02° = 12.14 \text{ A} \textbf{ Ans.}$$

Quadrature-axis component of armature current,

$$I_q = I \cos \theta = 15.2 \cos 53.02° = 9.14 \text{ A} \textbf{ Ans.}$$

$$\text{Excitation voltage, } E_0 = V \cos \delta + I_q R_e + I_d X_d$$
$$= 219.4 \cos 16.15° + 9.14 \times 1 + 12.14 \times 12$$
$$= 210.74 + 9.14 + 145.68 = 365.56 \text{ V}$$

Excitation voltage (line-to-line)

$$= \sqrt{3} \times 365.56 = 633 \text{ V} \textbf{ Ans.}$$

15.25 TRANSIENT CONDITIONS OF ALTERNATORS

In normal operation, the alternator may be subjected to transient conditions which cannot be explained from steady-state theory, discussed so far. These transients may occur from (*i*) switching, (*ii*) sudden changes of load, (*iii*) sudden short circuit (either line-to-line or line to neutral) or from a symmetrical short circuit (short circuit on all output terminals). These short circuits may develop severe mechanical stresses on the armature coils and their end turns or large torques, which may damage the alternator or its prime mover. The analysis of the synchronous machine under conditions of such transients is useful in predicting the possible conditions that may result from abnormal operation resulting from short circuits.

The complete analysis of the transient conditions that may occur from short circuits is extensive and far beyond the scope of this book. Here the discussions will be limited to the phenomena that determine the short-circuit transient currents and the reactances that limit these currents.

15.25.1. Constant Flux Linkage Concept. Effects of short-circuit currents can be conveniently determined by using the concept of *constant flux-linkage,* which may be stated as below:

It is impossible to change the flux linkages in a closed electrical circuit that contains no resistance or capacitance (i.e. in a purely inductive circuit).

In case of a synchronous machine, the armature and field windings can be assumed almost purely inductive because they do not contain any capacitance and their resistances are almost negligible in comparison to their inductive reactances. So the flux linkages in the armature circuit and the field circuit cannot be changed suddenly by the application of short circuit to the armature winding. For maintaining these flux linkages constant, large changes of current may take place in both windings when the short circuit occurs in order to keep their respective flux linkages constant.

15.25.2. Transient Reactances. The characteristics of short-circuit currents of a 3-φ synchronous generator are shown in Fig. 15.76. These curves indicate that the short circuit occurred at the instant when the voltage of phase R was maximum (assuming that the machine was not delivering any current to the load prior to fault). The transient reactances can be obtained from any one of the three oscillograms by graphical methods. In case of using Fig. 15.76 (*b*) or 15.76 (*c*), it is necessary to

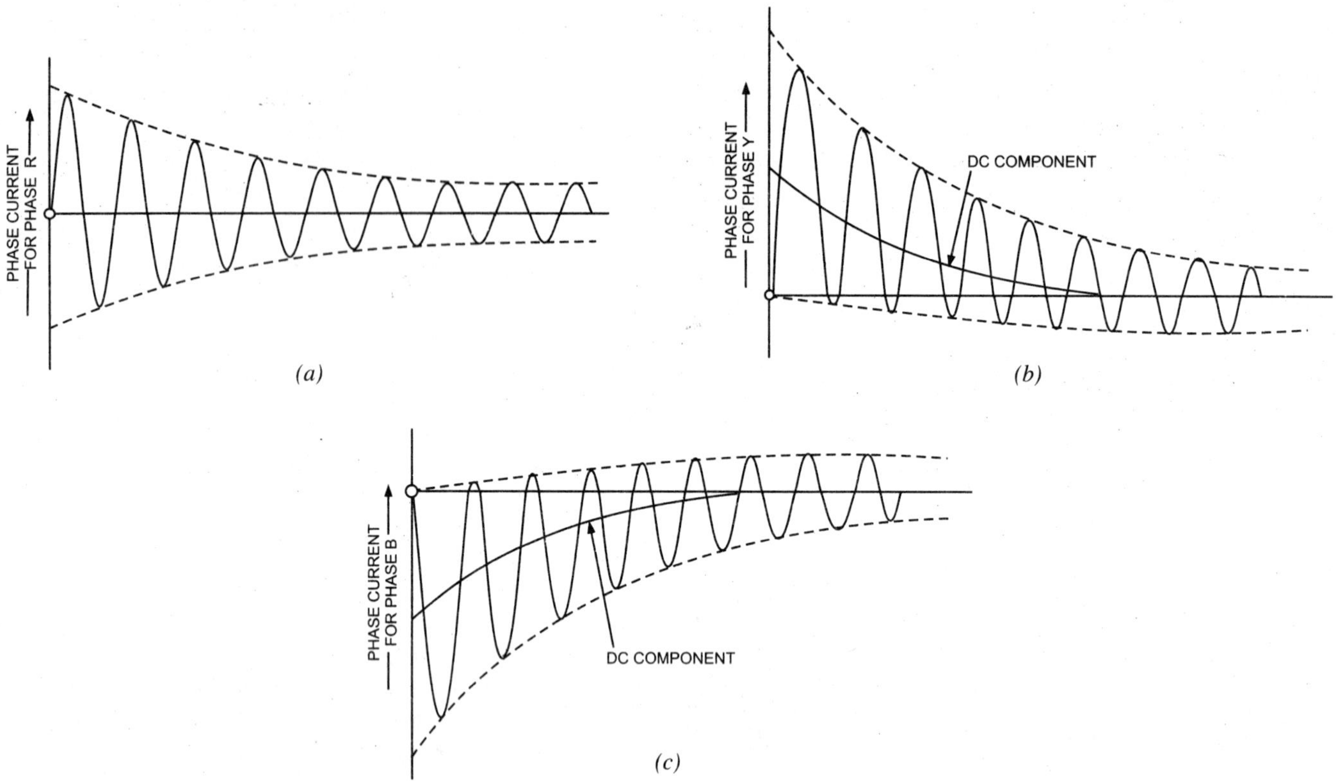

Fig. 15.76 *Short-Circuit Current Waveforms in Three Phases of a Synchronous Generator*

eliminate the dc components and plot the envelopes of the ac components only. Now they will also appear as in Fig. 15.76 (*a*). The transient reactance may be determined as below:

The peak values of each positive wave of Fig. 15.76 (*a*) are plotted as shown in Fig. 15.77.

Envelope of symmetrical ac component is constructed as represented by curve I in Fig. 15.77. Sustained peak value of the short-circuit current is determined and curve II is drawn in terms of peak values. Now curve III is obtained by subtracting the ordinates of curve II from those of curve I. The left portions of

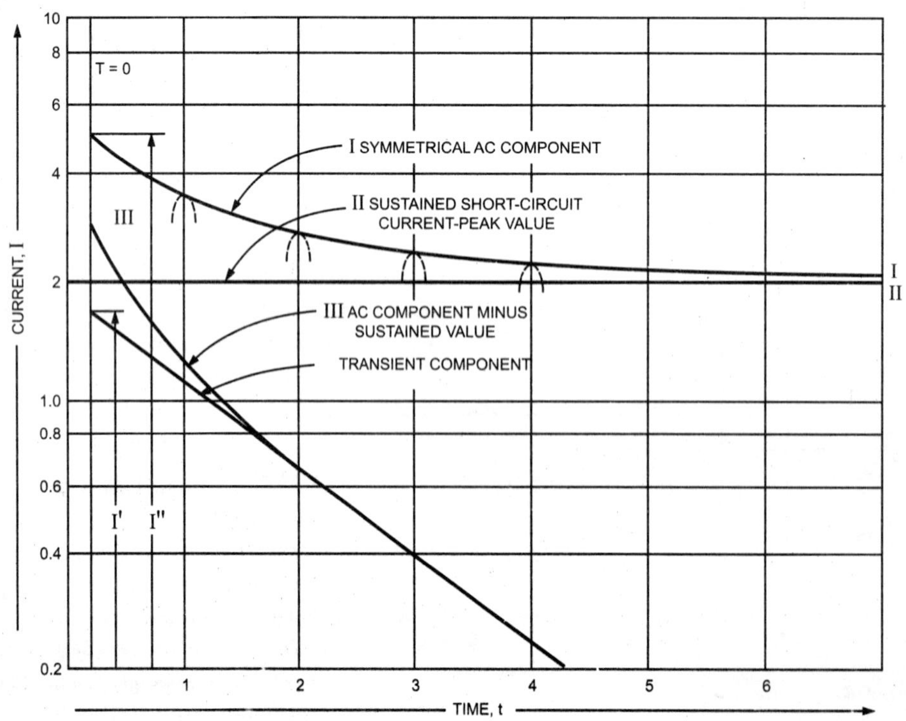

Fig. 15.77

curves I, II and III are extended, if necessary, to intersect the zero -time ordinate (T = 0). A tangent to the lower part of curve III is drawn and its intersection with zero time ordinate is determined.

If the terminal voltage of the machine before the fault taking place is E_0, the rated voltage is V and the rated rms value of the armature current is I, the direct transient reactance is given as

$$X'_d = \frac{E_0/V}{I'/1.414\,I} \text{ pu or } \frac{\sqrt{2}\,E_0}{I'} \text{ ohms} \quad ...(15.34)$$

The direct-axis subtransient reactance is given as

$$X''_d = \frac{E_0/V}{I''/1.414I} \text{ pu or } \frac{\sqrt{2}\,E_0}{I''} \text{ ohms} \quad ...(15.35)$$

Direct-axis subtransient reactance X_d'' almost equals the leakage reactance of the machine.

The short-circuit produces an armature reaction tending to reduce the main field flux. As the flux of armature reaction is being established, a voltage is induced in the field circuit tending to increase the field current momentarily. This transient component of the field current is eventually dampened by the field resistance and the air-gap flux is gradually reduced at the rate, determined by the time constant of the field circuit. The rate of decay of this transient d-c current in the field and likewise the transient component of armature current is illustrated in Fig. 15.77. It is shown as a straight line tangent to curve III.

The initial rate of decay of ac component less the sustained value (curve III) is larger than that produced by the field time constant. The higher rate of decay is due to the action of damper winding provided for this purpose. The time constant of the damper winding is lower than that of the main field winding and, therefore, transient effects on the armature current are of short duration. The actions of damper winding under short-circuit condition are similar to those of field circuit in that it tends to keep its flux linkages constant momentarily. The damper winding current decays rapidly. Its action alongwith that of the field winding determines the initial rate of decay of the armature current at the instant that the short circuit occurs.

Example 15.48. A 3-ϕ, star-connected, 11 kV, 165 MVA alternator has a synchronous reactance of 1 Ω and armature resistance of 0.1 Ω per phase. Calculate the short-circuit current at the instant of short-circuit and sustained short-circuit current.

Solution: The magnitude of the short-circuit current at the instant of fault depends upon the magnitude of voltage at that instant and the value of sub-transient reactance. Since nothing is mentioned, it is assumed that the sub-transient reactance is the same as the synchronous reactance.

Synchronous impedance,

$$Z_s = \sqrt{R_e^2 + X_s^2} = \sqrt{1^2 + 0.1^2} = 1.00\,5\ \Omega$$

Short-circuit current at the instant of fault

$$= 2\sqrt{2}\,\frac{\text{Phase voltage}}{Z_s} = 2\sqrt{2}\,\frac{11{,}000/\sqrt{3}}{1.005} = 17{,}874 \text{ A } \textbf{Ans.}$$

Sustained short-circuit current

$$= \frac{\text{Phase voltage}}{Z_s} = \frac{11{,}000/\sqrt{3}}{1.005} = 6{,}319 \text{ A } \textbf{Ans.}$$

15.26 LOSSES AND EFFICIENCY

Just as in case of generators, the input to the alternators is not readily measured. The direct measurement of efficiency by actually loading is accompanied by the difficulties of providing the necessary power and finding suitable load. Efficiency is, therefore, determined by measuring the losses of the machine.

The losses in an alternator are :

1. Electrical losses that include field loss, armature winding loss and brush contact loss.

The copper loss in the field circuit is obtained by adding the $I_f^2 R_f$ loss of field winding and the electrical loss of the field rheostat or, more simply, by multiplying the excitation voltage by the field current. This loss is constant for any given voltage and power factor, but varies with these two. The maintenance of rated voltage at low lagging pf needs comparatively large field current and gives the maximum field (or excitation) loss.

The armature copper loss in the stator windings is determined on the basis of the armature resistance. Whether the ohmic or the effective resistance should be used, depends upon the treatment of the stray load loss.

The electrical loss at the brush contacts, between the brushes and slip rings, is usually quite small and is often neglected.

2. Core loss is due to eddy currents and hysteresis caused by the flux resulting from the combined rotor and stator fields. Such loss occurs in pole faces and armature teeth and core. This loss is assumed to be independent of load, but varies with the excitation, if the excitation is to be varied with load and with the change in power factor, then this loss changes also. It is equal to the difference of power required to drive the alternator with and without the field excited.

3. Friction and windage loss includes ventilation loss (power required to circulate the cooling air), and loss due to bearing and brush friction. Since the speed is constant, so this loss for a particular machine remains constant. It is equal to the mechanical power required to drive the alternator at rated speed with no excitation.

4. Load or stray power loss is due to armature leakage flux which causes eddy currents and hysteresis in the iron surrounding the armature conductors in addition to the additional core loss caused by distortion of the magnetic field under load conditions. It may, however, be included in the efficiency calculation by using the effective value of the armature resistance instead of the dc resistance (refer to Art. 15.15).

After determining the foregoing losses, the efficiency may be determined from the following relation :

Percent efficiency,

$$\eta = \left[1 - \frac{\text{kW losses}}{\text{kVA output} \times \text{pf} + \text{kW losses}}\right] \times 100 \quad ...(15.36)$$

15.26.1. Determination of Losses. A simple and quite accurate method of determining the losses is by using a calibrated dc motor for driving the alternator. The motor need be large enough to supply only the losses of the alternator. The motor is calibrated in the sense that all its losses have been determined in different conditions of operation, so that its output is readily determined. If possible, dc motor and alternator should be direct coupled; if belted, correction is necessary to account for the belt loss.

1. The generator is run at rated speed without excitation. The power input to the alternator or power output of the dc motor gives friction and windage loss (bearing, brush and windage loss).
2. Now the generator is run at rated speed with field excitation. The additional power output of the motor is a measure of core loss *i.e.* the core loss is equal to the difference of motor output required to drive the alternator with and without the field excited. The excitation should be that at which the alternator normally operates.

The armature winding of the alternator is short circuited through ammeters, and the alternator is operated at rated speed with reduced excitation (adjusted to give rated line current). The motor output less friction and windage loss (determined in step 1) gives the armature copper loss plus stray power loss. The flux density under short-circuit condition is so low as to make the core loss negligible. Thus if the calculated armature copper loss and the friction and windage loss are subtracted from the motor output, the stray power loss is obtained.

The armature copper loss and excitation loss are determined as mentioned in the foregoing paragraphs.

An alternative method is to operate the test machine as an unloaded synchronous motor from an ac supply source. In this method of determination of losses, the effective resistance of the armature of the machine is determined, as mentioned in Art. 15.17. The field copper loss is determined as before by dc measurements. Finally the alternator is operated as a synchronous motor without load. The field current is adjusted to normal value (as mentioned on the name plate). Both power input and the armature current are measured. Power input less armature copper loss gives the sum of the frictional and core losses. Unless the alternator is designed for a wide frequency range, the latter sum of the losses may be taken as a constant loss, and the efficiency at any load may be determined. This method does not give results as accurate as of the previous method, but fairly close to them. The advantage of this method is that auxiliary calibrated motor is not required.

When a pair of identical machines is available they may be operated back-to-back with rotors mechanically coupled and given a suitable angular displacement.

Regardless of which test method is adopted, efficiency is calculated from the Eq. (15.36).

Example 15.49. A 500 kVA, 11 kV, 3-ϕ, star-connected alternators has the following data:

Friction and windage loss	**= 1,500 W**
Open-circuit core loss	**= 2,500 W**
Effective armature resistance/phase	**= 4 Ω**
Field copper loss	**= 1,000 W**

Find the following regarding above synchronous alternators:

(*i*) Alternator efficiency at half-full load and at 0.85 power factor lagging.

(*ii*) Maximum efficiency of the alternator.

[U.P. Technical Univ. Electromechanical Energy Conversion-II, 2011-12]

Solution: (*i*) Full-load armature current per phase,

$$I = \frac{\text{KVA} \times 1{,}000}{\sqrt{3} \times \text{line voltage}} = \frac{500 \times 1{,}000}{\sqrt{3} \times 11{,}000} = 26.243 \text{ A}$$

Short-circuit load loss at half full-load

$$= 3 \times \left(\frac{1}{2}I\right)^2 \times R_e + \text{stray-load loss}$$

$$= 3 \times \left(\frac{26.243}{2}\right)^2 \times 4 = 2{,}066 \text{ W} \quad \text{or} \quad 2.066 \text{ kW}$$

$\because$ stray-load loss is zero here

Total loss at half full-load = Friction and windage loss + open-circuit core loss + field copper loss + short-circuit load loss at half full-load

$= 1{,}500 + 2{,}500 + 1{,}000 + 2{,}066$

$= 7{,}066$ W or 7.066 kW

$$\text{Output at half full-load} = 500 \times \frac{1}{2} \times 0.8 = 200 \text{ kW}$$

$$\text{Efficiency at half full-load} = \frac{\text{Output}}{\text{Input}} \times 100$$

$$= \frac{\text{Output}}{\text{Output + total loss at half full-load}} \times 100$$

$$= \frac{200}{200 + 7.066} \times 100 = 96.588\% \quad \textbf{Ans.}$$

For maximum efficiency,

Variable losses = Fixed losses

$3\,I^2R_a$ = Rotational loss + field circuit loss

or $3\,I^2 \times 4 = 1{,}500 + 2{,}500 + 1{,}000 = 5{,}000$ W

or The current at which maximum efficiency occurs is given as

$$I_m = \sqrt{\frac{5{,}000}{12}} = 20.4124 \text{ A}$$

Output corresponding to maximum efficiency

$$= \sqrt{3} \times V_L \times I_m \cos\phi = \sqrt{3} \times 11{,}000 \times 20.4124 \times 0.85$$

$$= 330{,}572 \text{ W}$$

Total loss at maximum efficiency

= 2 × fixed losses = 2 × 5,000 = 10,000 W

Maximum efficiency

$$= \frac{\text{Output corresponding to maximum efficiency}}{\text{Output corresponding to maximum efficiency + losses}} \times 100$$

$$= \frac{330{,}572}{330{,}572 + 10{,}000} \times 100 = 97.06\% \quad \textbf{Ans.}$$

Example 15.50. A 10,000 kVA, 3-phase, star-connected 11,000 V, 2-pole turbo-generator has a synchronous impedance of (0.0145 + *j* 05) ohms per phase. The various losses in this generator are as follows:

Open-circuit core loss at 11,000 V	**90 kW**
Windage and friction loss	**50 kW**
Short-circuit load loss at 525 A	**220 kW**
Field winding resistance	**3 Ω**
Field current	**175 A**

Ignoring the change in field current, compute the efficiency at (*a*) rated load 0.8 pf leading (*b*) half rated load 0.9 pf lagging.

[GATE 1996]

Solution: Phase current, $I_L = \dfrac{10{,}000 \times 1{,}000}{\sqrt{3} \times 11{,}000} = 525$ A

Friction and windage loss $= 50$ kW

Core loss $= 90$ kW

Armature copper loss $= 220$ kW being given at full-load current of 525 A

$$\text{Field copper loss} = \frac{I_f^2 R_f}{1,000} = \frac{(175)^2 \times 3}{1,000} = 91.875 \text{ kW}$$

(*a*) At full load 0.8 pf (lead)

Output = Rated kVA × pf = 10,000 × 0.8 = 8,000 kW

Total losses = 50 + 90 + 220 + 91.875 = 451.875 kW

$$\text{Efficiency, } \eta = \frac{\text{Output}}{\text{Output + losses}} \times 100$$

$$= \frac{8,000}{8,000 + 451.875} \times 100 = 94.65\ \% \textbf{ Ans.}$$

(*b*) At half load 0.9 pf (lag)

$$\text{Output} = \frac{1}{2} \times 10,000 \times 0.9 = 4,500 \text{ kW}$$

$$\text{Armature copper loss} = \left(\frac{1}{2}\right)^2 \times 220 = 55 \text{ kW}$$

Total losses = 50 + 90 + 55 + 91.875 = 286.875 kW

$$\text{Efficiency, } \eta = \frac{4,500}{4,500 + 286.875} \times 100 = 94\ \% \textbf{ Ans.}$$

Example 15.51. A 9 kVA, 208 V, 1,200 rpm 3-phase, 60 Hz, star-connected generator has a field winding of 4.5 Ω. The armature impedance is (0.3 + *j* 5) Ω per phase. When the generator operates at full load and 0.8 pf lagging, the field winding current is 5 A. Its rotational loss is 500 W. Determine (*i*) voltage regulation (*ii*) efficiency of alternator and (*iii*) torque applied by the prime mover.

[*W.B. Technological Univ. Electrical Machines-II, 2012-13*]

Solution:

$$\text{Phase voltage, } V_P = \frac{208}{\sqrt{3}} = 120 \text{ V}$$

$$\text{Load current, } I = \frac{\text{Load in kVA} \times 1,000}{\sqrt{3}\, V_L}$$

$$= \frac{9 \times 1,000}{\sqrt{3} \times 208} = 25 \text{ A}$$

Power factor, $\cos\phi = 0.8$ (lagging)

$$\sin\phi = \sqrt{1 - 0.8^2} = 0.6$$

Open-circuit voltage per phase,

$$E_{op} = \sqrt{(V_p \cos\phi + IR_e)^2 + (V_p \sin\phi + IX_s)^2}$$

$$= \sqrt{(120 \times 0.8 + 25 \times 0.3)^2 + (120 \times 0.6 + 25 \times 5)^2} = 222.5$$

$$(i) \text{ Percentage regulation} = \frac{E_{op} - V_p}{V_p} \times 100$$

$$= \frac{222.5 - 120}{120} \times 100 = 85.44\% \textbf{ Ans.}$$

Full-load output at 0.8 pf lagging = 9 × 0.8 = 7.2 kW or 7,200 W

Rotational loss = 500 W

Field copper loss = $I_f^2 R_f = 5^2 \times 4.5 = 112.5$ W

Armature copper loss = $3I_a^2 R_a = 3 \times 25^2 \times 0.3 = 562.5$ W

Full-load input at 0.8 pf lagging

= Full-load output at 0.8 pf
+ armature copper loss + field copper loss
+ rotational loss + core loss
= 7,200 + 562.5 + 112.5 + 500 + 600 = 8,975 watts

Assuming core loss to be 600 W

$$\text{Efficiency of alternator} = \frac{\text{Input} - \text{output}}{\text{Input}} \times 100$$

$$= \frac{8,975 - 7,200}{8,975} \times 100 = 19.78\%$$

(*iii*) Torque applied by the prime mover

$$= \frac{\text{Output of prime mover}}{2\pi N/60}$$

$$= \frac{\text{Input to alternator}}{2\pi N/60} = \frac{8,975}{2\pi \times 1,200/60} = 71.42 \text{ N-m } \textbf{Ans.}$$

Example 15.52. A 1.0 MVA, 11 kV, 3-phase star-connected alternator has an effective resistance of 2 Ω per phase. The characteristics on open circuit with full-load current at zero pf and core losses (on open circuit) are given below:

Field Current (A)	Open-Circuit Voltage, V (Line Value)	Core Loss kW	Saturation Curve Zero PF; V
40	–	–	0
50	7,000	7.5	–
110	12,500	16.6	8,500
140	13,750	22.4	10,500
180	15,000	33.5	12,400

Deduce by the zero power factor method

(*i*) the percentage regulation for full-load at a pf of 0.8 lag and

(*ii*) the efficiency at this load, given that the mechanical and additional losses add up to 10 kW. Draw the phasor diagram.

[*W.B. Technological Univ. Electrical Machines-II, 2011-12*]

Solution: From the given data, first OCC, full-load zero power factor curve and core-loss curves are drawn, as shown in Fig. 15.78.

$$\text{Full load current, } I = \frac{1.0 \times 10^6}{\sqrt{3} \times 11,000} = 52.486 \text{ A}$$

From the data of z.p.f. characteristic, $I_f = 40$ A produces full-load current on short circuit. Note that the short-circuit characteristic current has a pf of almost zero (lagging). Therefore short-circuit current, $I_{sc} = 52.486$ A and $I_f = 40$ A is a point on the SCC, so the straight line joining this point with origin provides the SCC, as shown in Fig. 15.78.

From point B on full-load ZPF curve, line BC is drawn parallel to line OA and equal to OA. From point C, a line parallel to the air-gap line intersecting OCC at point D is drawn. Line DF is drawn parallel to ordinate intersecting line BC at F.

From ΔBDF (Fig. 15.78)

Armature reaction mmf = BF = 25 A

Full-load reactance drop = DF = 2,400 V

$$\text{Leakage reactance drop per phase} = \frac{2,400}{\sqrt{3}} = 1,386 \text{ V}$$

$$\text{Leakage reactance per phase} = \frac{1,386}{52.486} = 26.4\ \Omega$$

In phasor diagram (Fig. 15.79)

OV = Terminal line voltage = 11,000 V

VE = 2,400 volts representing reactance drop

OE = 12,700 V

Excitation corresponding to 12,700 V = 114 A

$\therefore$ $OI_1 = 114$ A

I_1I_2 = Armature-reaction mmf = 25 A

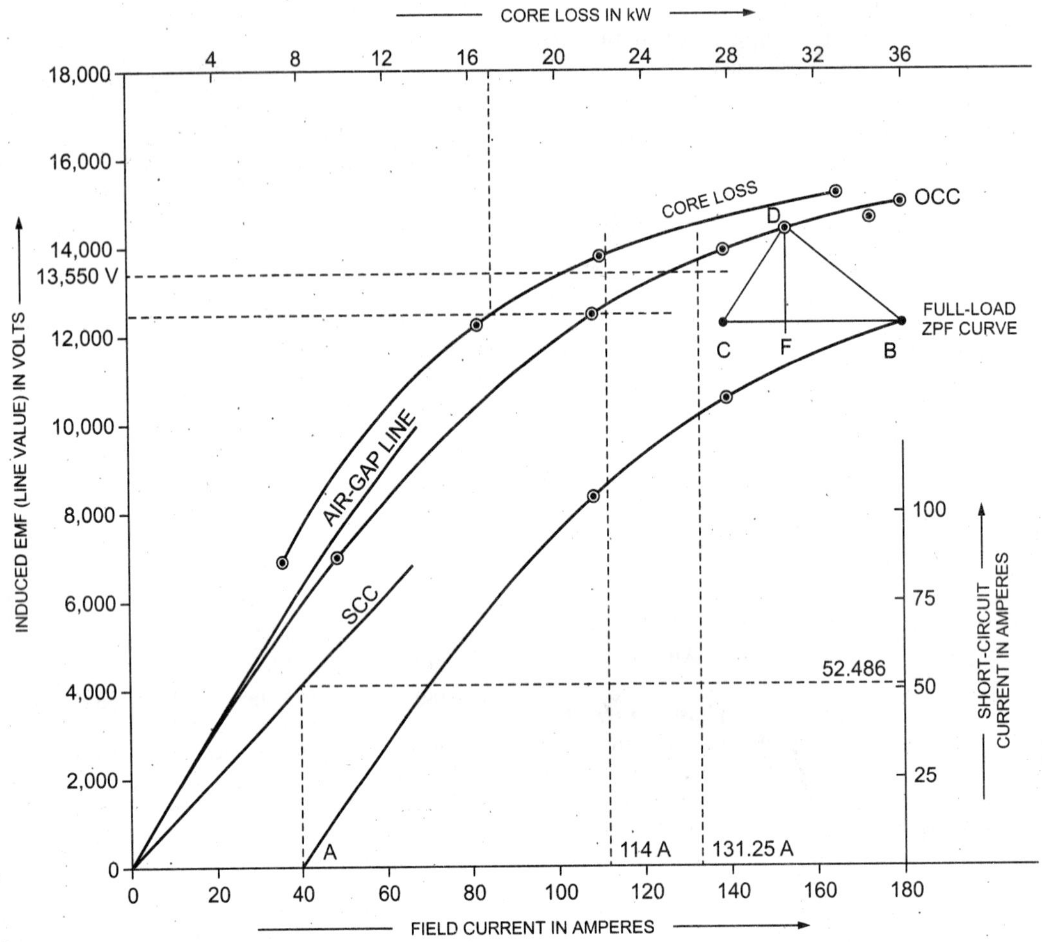

Fig. 15.78

Total Ampere-turns = OI_2 = 131.25 A

Induced emf corresponding to 131.25 A,

$$E_0 = 13,550 \text{ V}$$

(*i*) Percentage voltage regulation

$$= \frac{13,550 - 11,000}{11,000} \times 100 = 23.18\% \textbf{ Ans.}$$

Core loss corresponding to induced emf (on open circuit),

E_0 = 12,700 V due to air-gap flux is found out from core-loss curve (Fig. 15.78), which gives 17.6 kW

$$\text{Armature copper loss} = 3I^2R_a$$
$$= 3 \times (52.486)^2 \times 2$$
$$= 16,529 \text{ W or } 16.529 \text{ kW}$$

Mechanical and additional losses

$$= 10 \text{ kW}$$

$$\text{Total loss} = 17.6 + 16.529 + 10 = 44.129 \text{ kW}$$

$$\text{Output} = 1.0 \times 1,000 \times 0.8 = 800 \text{ kW}$$

$$(ii) \text{ Efficiency, } \eta = \frac{\text{Output}}{\text{Output + losses}} \times 100$$

$$= \frac{800}{800 + 44.129} \times 100 = 94.77\% \textbf{ Ans.}$$

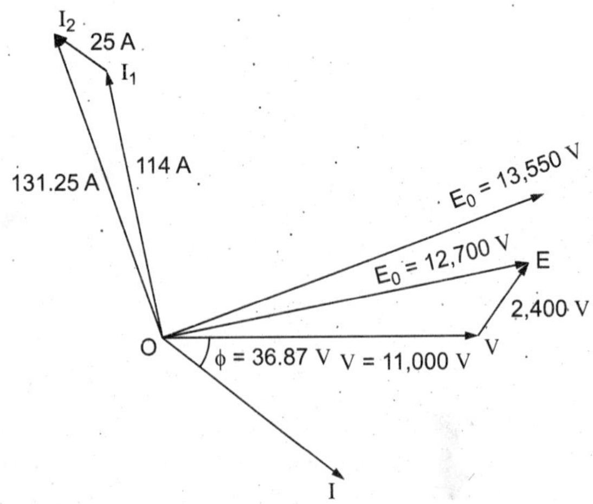

Fig. 15.79

Example 15.53. A 1,500 kVA, 11 kV, 50 Hz, star-connected synchronous generator is tested by operating it as a synchronous motor on no-load. The results are as follows :

1. **When operated as a synchronous motor on no-load, the power input to the motor is 55.8 kW and the armature line current is 70.0 A.**
2. **The excitation of 38.6 A at 250 V dc gives the normal open-circuit voltage.**

3. DC resistance test : DC line voltage 17.5 V and dc line current 70 A.

Take effective resistance 1.6 times the dc resistance.

Determine the efficiency of the machine on 80 % of full-load and 0.75 pf (lag).

Solution : From synchronous motor test, armature phase current,

$$I_p = I_L = 70.0 \text{ A}$$

From DC resistance test dc resistance of armature per phase,

$$R_a = \frac{1}{2} \times \frac{17.5}{70} = 0.125 \ \Omega$$

Effective resistance of armature,

$$R_e = 1.6 R_a = 1.6 \times 0.125 = 0.2 \ \Omega/\text{phase}$$

$$\text{Armature copper loss} = 3 \times (70.0)^2 \times 0.2 = 2{,}940 \text{ watts}$$

Constant losses (core and friction and windage losses)

$$= \text{Input to motor} - \text{armature copper loss}$$

$$= 55.8 \times 1{,}000 - 2{,}940 = 52{,}860 \text{ watts}$$

Rated armature phase current,

$$I_a = \frac{1{,}500 \times 1{,}000}{\sqrt{3} \times 11 \times 1{,}000} = 78.73 \text{ A}$$

Armature current at 80 % of full load

$$= 0.8 \times 78.73 = 63 \text{ A}$$

At 80 % of full load and 0.75 pf

$$\text{Output of the machine, } P_{out} = 1{,}500 \times 0.8 \times 0.75 = 900 \text{ kW}$$

$$\text{Armature copper loss} = 3 \times (63)^2 \times 0.2$$

$$= 2{,}381 \text{ watts or } 2.381 \text{ kW}$$

$$\text{Total field copper loss} = \frac{250 \times 38.6}{1{,}000} = 9.65 \text{ kW}$$

$$\text{Total losses} = 52.86 + 2.381 + 9.65 = 64.891 \text{ kW}$$

$$\text{Efficiency, } \eta = \frac{\text{Output}}{\text{Output+losses}} \times 100$$

$$= \frac{900}{900 + 64.891} \times 100 = 93.28 \ \% \ \textbf{Ans.}$$

15.27 POWER FLOW IN AN ALTERNATOR

For generator operation the rotor of the machine is supplied with mechanical energy from its prime mover. A part of the mechanical energy is utilized or lost to overcome friction and windage, and hysteresis and eddy current losses (all losses except armature copper loss) and remainder of the energy supplied is converted into electrical energy in the armature circuit. Some of the electrical energy developed in the armature circuit is lost in overcoming armature copper loss. The remainder becomes the useful electrical output of the generator. The power flow diagram in case of a synchronous generator is given in Fig. 15.80.

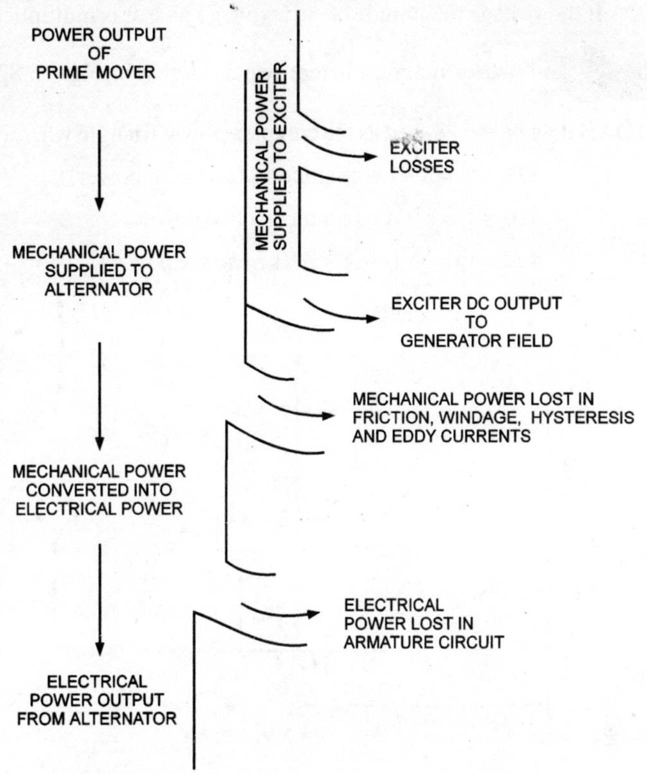

Fig. 15.80 *Power Flow Diagram For An Alternator*

15.28 CAPABILITY CURVES OF ALTERNATOR

The limits within which the alternator operate safely is called *capability curves*, also sometimes known as *operating capability charts*. The safe region of operation is restricted by the following considerations.

1. The MVA loading should not exceed the alternator rating. This limit is determined by the stator (armature) heating owing to armature current.
2. The MW load should not exceed the rating of the prime mover.
3. The excitation (or field) current should not exceed a specified value determined by the field heating.
4. For steady-state stable operation it is necessary that the load angle δ be kept less than 90°. The theoretical limit of stability occurs when δ is 90°.

Let us consider a 3-phase cylindrical rotor synchronous generator having synchronous reactance X_s ohms per phase. Let it be connected to the bus-bar having phase voltage of V volts and supply current I amperes at pf cos ϕ (lagging). For simplicity, the armature resistance and saturation are assumed to be negligible. Let the phase angle between induced emf and bus-bar voltage V be δ.

The phasor diagram is drawn with bus-bar voltage phasor as reference phasor (Fig. 15.81).

In Fig. 15.81

$$OA = I X_s$$

$$OB = I X_s \sin \phi$$

$$AB = I X_s \cos \phi$$

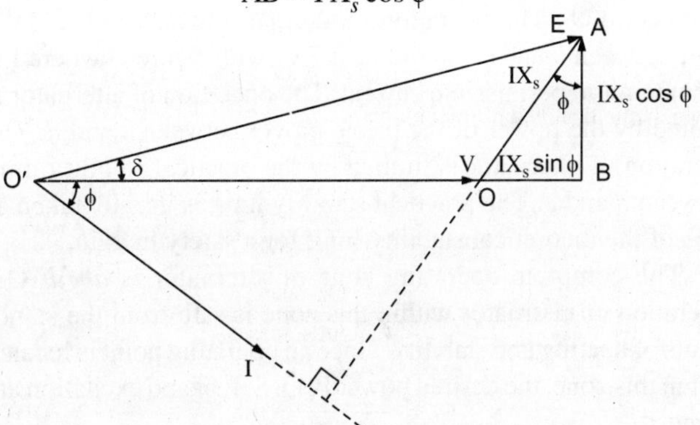

Fig. 15.81

If the voltage magnitude of each voltage phasor is multiplied by $\frac{3V}{X_s}$ and phasor diagram is redrawn as shown in Fig. 15.82.

ΔOAB may be recognised as the complex power triangle wherein

$$OA = S = 3\ VI \text{ volt-amperes (apparent power)}$$

$$AB = P = 3VI \cos\phi \text{ watts (active power)}$$

$$OB = Q = 3VI \sin\phi \text{ VARs (reactive power)}$$

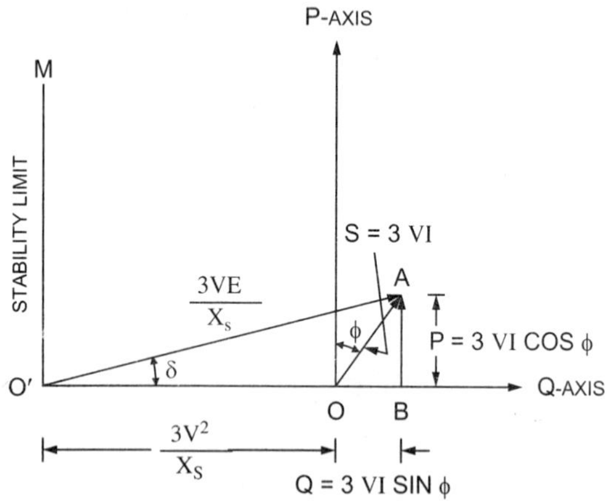

Fig. 15.82

By convention Q is positive for lagging pf.

A mere change of scale will convert these values to the units of kVA, kW and kVAR (or MVA, MW and MVAR).

For constant S operation, the locus is a circle with centre O having radius OA (i.e. 3VI). For constant excitation operation, the locus is a circle with centre at O′ and radius O′A $\left(i.e. \frac{3VE}{X_s}\right)$. The radial straight lines through point O are constant pf lines. Constant P operation lies on a line parallel to Q-axis. OO′ represents short-circuit current at rated voltage for E = 0. The theoretical stability limit is the straight line O′M which is at right angle to OO′ and $\delta = 90°$.

A typical capability curve for a cylindrical rotor alternator is shown in Fig. 15.83. It is plotted on S-plane where P is vertical axis and Q is horizontal axis.

In Fig. 15.83, the operation of alternator is limited by the maximum excitation current between *a* and *b* and radius is $\frac{3VE}{X_s}$ with centre O′. The operation of alternator is limited by MVA limit between *b* and *c* and radius is 3VI with centre O where I is the maximum permissible current. The operation of alternator is limited by the power of the prime mover between *c* and *d*. The operation of alternator is limited by the practical stability limit between *d* and *e*. The practical stability limit is usually taken as 90% of the theoretical stability limit for a safety margin.

The complete operating zone of alternator is *abcdf* O*a*. Operation of alternator within this zone is safe from the stand-points of heating and stability. Once an operating point is located within this zone, the desired power P, Q, S, I, pf and excitation are found.

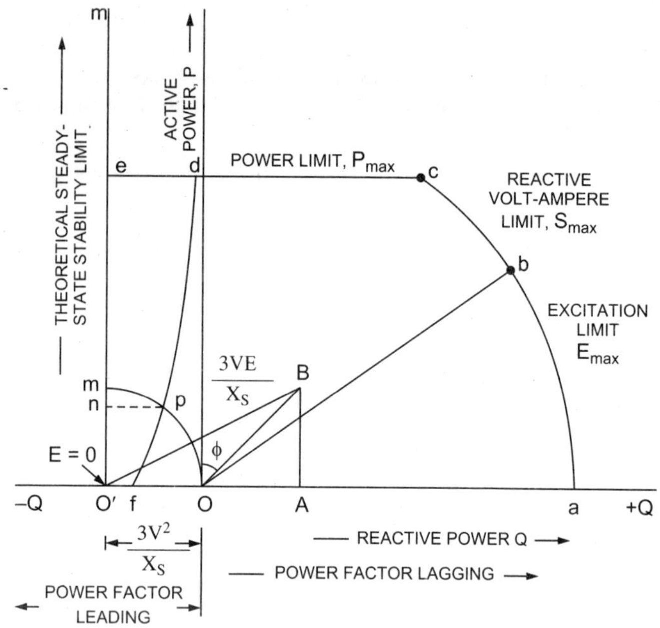

Fig. 15.83 *Capability Curve for a Cylindrical Rotor Alternator*

15.29 VENTILATION

In the operation of **salient field pole alternators,** that have small axial length and large diameter, there is no particular problem in obtaining sufficient ventilation for the armature core. In addition to the large area available for direct radiation, there is the fanning action of the rotating field which may be increased by providing fan blades. Since the axial length of the core is small the heat produced in the embedded portion of the armature winding will be conducted to the end connections and dissipated from there if the fanning action is sufficient. However, armature core is usually subdivided by providing ventilating ducts at intervals of about 6–8 cm to allow the flow of cooling air radially.

However, there are ventilation problems with the **turbo-generators,** which have axial length many times the diameter. The smooth cylindrical rotor provides little fan effect in circulating air. Further, the surfaces of such alternators, which are exposed to the surrounding air, are small.

If open-circuit ventilation is adopted for such machines, large volume of air will be required. In such a system of ventilation it is necessary to pass the incoming air through the suitable filters to remove any dust before it is moved round the machine so as to prevent clogging of dust. The air brought in must also be moisture free. Thus the operating cost of handling several tons of air, depending upon the rating of the machine will be very heavy. Further it cannot ensure dust free entry of air.

The usual procedure is to have the machine completely enclosed with a cooling system using either air or hydrogen. The larger machines also have hollow conductors through which, water, oil or hydrogen are circulated as coolant.

In closed circuit air cooling system, used in machines of rating upto 60 MW, the air is repeatedly circulated round the machine. After the ventilating air has been heated by the various parts of the generator, it is cooled by passing it through a water cooled heat exchanger. For machines of higher ratings hydrogen gas is employed as a cooling medium. Figure 15.84 shows the

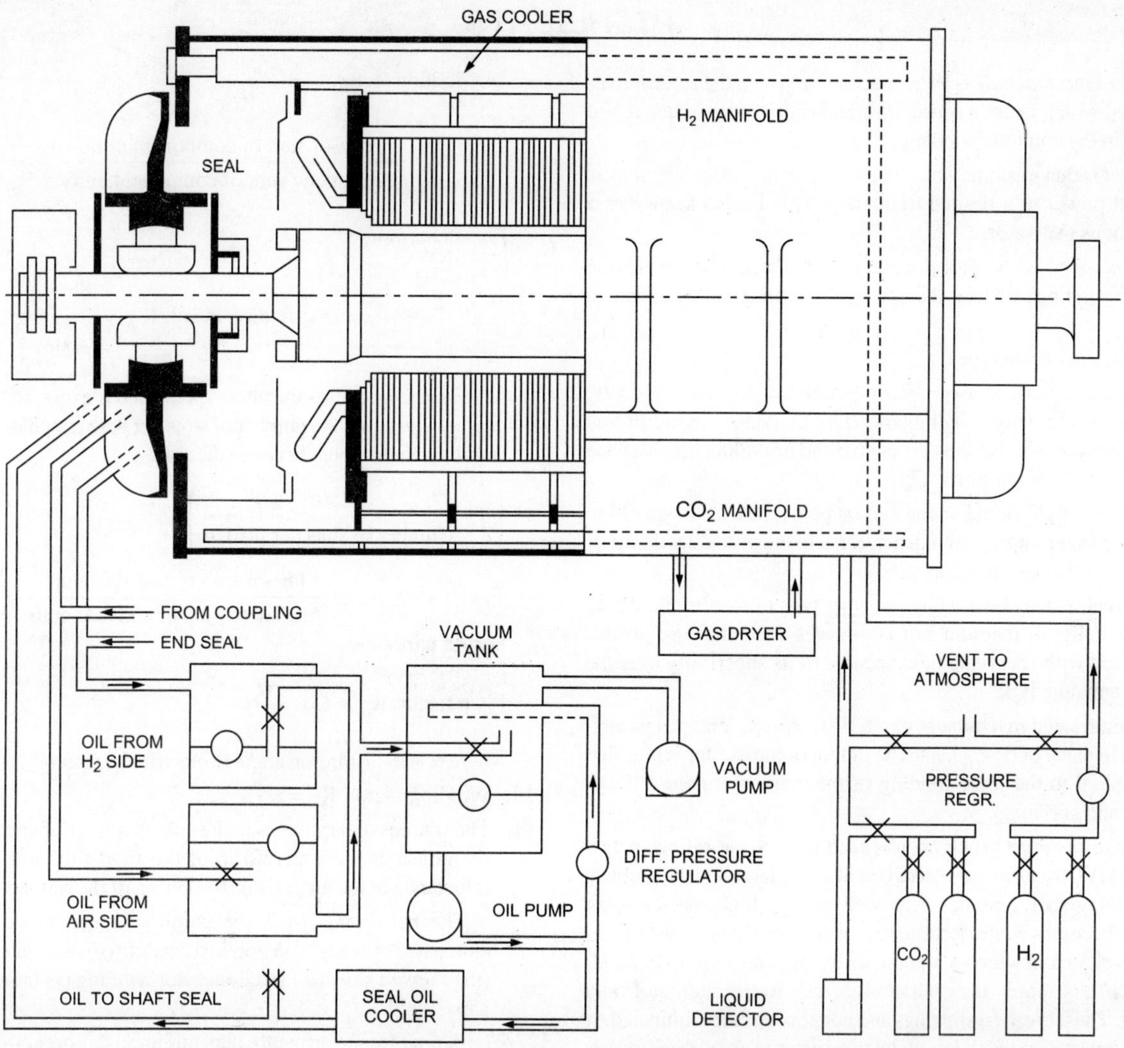

Fig. 15.84 *Hydrogen-Cooling System*

hydrogen cooling system. There are several advantages of using hydrogen as coolant in place of air in closed circuit ventilation system such as increased efficiency, increased output rating, increased life, elimination of fire hazard, less noise, smaller size of coolers etc.

1. The density of hydrogen is 1/14 that of air, the power required to circulate hydrogen is about 1/14th of the power required for an equivalent quantity of air. Hence the *losses are reduced* and the *efficiency is improved* by about 1 per cent at full load.
2. Hydrogen has specific heat 14 times, a heat transfer coefficient 1.5 times and thermal conductivity 7 times that of air so it takes away heat more effectively. So an increased output can be taken from a given frame size. A generator using hydrogen as a coolant may be *rated about* 20 *per cent higher* than of the same physical size using air.
3. The life of a machine is mainly the life of the winding insulation and air pockets in insulation can be source of such high local temperature that there is always the risk of insulation breakdown. The thermal conductivity of hydrogen is nearly 7 times that of air and is of the same order as the winding insulation. Therefore, if there are pockets filled with hydrogen, heat conduction through them will be as good as through winding insulation and thus local high temperatures are avoided and *life of insulation is increased.*

   Insulation corona which is present when air is used as a coolant in high voltage machines is greatly reduced under hydrogen, thus again increasing the insulation life.
4. *Fire hazards are eliminated* as hydrogen does not support combustion so long as the hydrogen / air mixture exceeds 3/1.
5. *Smaller size of heat exchanger / cooler* is required to cool the heated hydrogen.
6. *Less noise* as the rotor moves in a medium of lower density.
7. Hydrogen as a coolant does not oxidize the insulation and paint work.

Because of the explosive nature and the cost of replacement of hydrogen, it is necessary to have a gas-tight sealing system for the alternator frame. The hydrogen is circulated by a blower system through ducts in the turbo-alternator and is cooled by a water cooling system.

Hydrogen is used as a coolant in the machines of rating upto 200 MW. For machines of higher rating direct water cooling method is used.

HIGHLIGHTS

1. A *synchronous machine* is an ac machine in which the rotor moves at a speed which bears a constant relationship to the frequency of currents in the armature winding.
2. *Alternator* is that machine which converts mechanical power into ac electrical power at a desired frequency. This is also known as a synchronous generator.
3. An *alternator* consists of a stator and a rotor. The stator provides the armature windings whereas the rotor provides the rotating magnetic field.

   The rotors are of two types :

   (*i*) Salient pole type used almost entirely for low and moderate speed alternators (hydrogenerators or diesel-engine driven generators) being least expensive and providing ample space for field ampere-turns.

   (*ii*) Smooth cylindrical or non-salient pole type used in very high speed alternators (turbo-generators), being of smaller diameter and robust construction.
4. The alternators may be rotating armature type or rotating field type but practically all medium and large sized machines are always constructed with revolving field because of its superiority over the rotating armature type.
5. For medium sized machines ac exciters (three-phase ac generators) are used in place of dc exciters. The output of the exciter is rectified and supplied to the rotor winding of the main alternator through brushes and slip rings.

   In case of large synchronous machines, of the rating of few hundred MW, the excitation needs become very large and the problem of supplying such amounts of power through high-speed sliding contacts becomes acute. Recently, brushless excitation system has been developed in which a 3-phase ac exciter and a group of solid-state rectifiers supply the excitation directly to the alternator field winding. Thus, brushes, slip rings and commutator are eliminated. A brushless excitation needs less maintenance due to absence of brushes and slip rings.
6. In a multipolar machine one pair of poles or one cycle of flux distribution equals 360 electrical degrees or 2π electrical radians. Since there P/2 complete wavelengths or cycles in one complete revolution, it follows that

$$\theta_{elec} = \frac{P}{2}\,\theta_{mech}$$

7. The frequency of induced emf in a coil in a multipolar machine is given by the expression

$$f = \frac{PN}{120}\ \text{Hz}$$

   where P is number of poles and N is the mechanical speed of rotation in rpm.
8. The radial frequency ω_{elec} of the wave of emf induced in P-pole machine is given as

$$\omega_{elec} = \frac{P}{2}\,\omega_{mech}$$

9. The expression for induced emf in a three phase alternator is given by expression

$$E_{rms}/\text{phase} = 4.44\, K_d\, K_p\, \Phi f T \text{ volts}$$

   where Φ is the useful flux per pole in webers, f is frequency in Hz and T is the number of series turns per phase and K_d and K_p are respectively the distribution factor and pitch factor and are given below:

   Distribution factor,

$$K_d = \frac{\text{Phasor sum of component emfs}}{\text{Arithmetic sum of component emfs}} = \frac{\sin \frac{m\beta}{2}}{m \sin \frac{\beta}{2}}$$

   For *r*th harmonic

$$\text{Distribution factor, } K_{dr} = \frac{\sin r\frac{m\beta}{2}}{m \sin \frac{r\beta}{2}}$$

   where $m\beta$ is known as the phase spread and is expressed in electrical degrees – m being the number of slots per pole per phase and β is the angular displacement between the slots

$$\left(\beta = \frac{180°}{\text{Number of slots per pole, } n}\right)$$

$$\text{Pitch factor, } K_p = \frac{\text{Phasor sum of coil side emfs}}{\text{Arithmetic sum of coil side emfs}} = \cos \frac{\alpha}{2}$$

   For *r*th harmonic

$$\text{Pitch factor, } K_{pr} = \cos \frac{r\alpha}{2}$$

   where α is the chording angle in electrical degrees.
10. Winding factor, $K_w = K_d\, K_p$.
11. The sources of harmonics in the output voltage of an alternator are the nonsinusoidal waveform of the field flux and variation in reluctance of the air gap due to slotting of the stator core.

    The harmonics from the output voltage, however, can be eliminated by using (*i*) distributed winding (*ii*) short-pitched winding (iii) skewed slots (*iv*) fractional slot winding (*v*) larger length air gap.
12. The copper loss in armature depends upon the strength of armature current and the core loss on the voltage and thus the power losses, responsible for temperature rise are independent of pf. So alternators are rated in kVA/MVA.
13. With the increase in kVA rating of alternator, operating efficiency is improved, output/kg increases and cost is reduced consequently. However, with large-sized machines, ventilation problem become acute.
14. The effect of armature flux on the main field flux is known as *armature reaction*. Armature reaction has distorting effect on unity power factor, wholly demagnetising at zero *pf* lagging and wholly magnetizing at zero power factor leading.
15. The sum of leakage reactance X_L and fictitious reactance X_a (representing armature reaction effect) is known as synchronous reactance X_s *i.e.*,

$$X_s = X_L + X_a$$

16. The phasor sum of effective resistance of armature R_e and synchronous reactance X_s is known as synchronous impedance Z_s *i.e.*,

$$\mathbf{Z}_s = R_e + j\,X_s$$

    Armature winding effective resistance R_e in alternators is usually very small in comparison to synchronous reactance X_s, and therefore, synchronous impedance may be assumed equal to the synchronous reactance for many purposes.
17. With the variation in load on an alternator, voltage drops due to (*i*) armature resistance (*ii*) armature leakage reactance and (*iii*) on account of armature reaction vary, so its terminal voltage varies.

18. For constant load current, excitation and speed, the terminal voltage of an alternator decreases when power factor is changed from leading to lagging.

19. Voltage regulation of an alternator is given by the expression

$$\text{Voltage regulation} = \frac{E_0 - V}{V} \text{ per unit}$$

or $\frac{E_0 - V}{V} \times 100$ in percentage

where E_0 is no-load terminal voltage and V is the full-load terminal voltage.

The voltage regulation of an alternator is positive and increases with the decrease in lagging power factor. It may become negative for leading power factor.

20. Voltage regulation may be determined by any of the following methods:
    1. Synchronous impedance (or emf) method – pessimistic method.
    2. Ampere-turn (or mmf) method – optimistic method.
    3. Potier (or zero power factor) method.
    4. A.S.A. method.

21. Zero power factor characteristic of a synchronous machine is used for determination of (*i*) leakage reactance drop (*ii*) voltage drop due to armature reaction and (*iii*) armature reaction ampere-turns.

22. A multipolar machine with cylindrical rotor has a uniform air gap and, therefore, its reactance remains the same, irrespective of the spatial position of rotor. But in case of salient pole machines, the air gap is not uniform and its reactance varies with the rotor position. Because of non-uniformity of the reluctance of the magnetic paths, the mmf of the armature is divided into two components viz. (*i*) a direct-acting component along the field pole axis, called the direct-axis and (*ii*) quadrature-component along the axis passing through centre of the interpolar space, called the quadrature-axis. These facts form the basis of the two-reactance theory applied to the salient pole machines.

23. Power developed by a 3-phase salient pole synchronous machine is given by

$$\frac{3E_0V}{X_d}\sin\delta + \frac{3V^2}{2}\left[\frac{1}{X_q} - \frac{1}{X_d}\right]\sin 2\delta$$ where E_0 and V are phase values.

Above expression for power developed consists of two terms, the first term representing power due to field excitation and the second term gives the reluctance power *i.e.* power due to saliency. In case of a cylindrical rotor machine $X_d = X_q$ and so the second term becomes zero and the power is given by the first term only. In case there is no field excitation *i.e.* E_0, the first term becomes zero and the power developed is given by the second term.

It may be noted that value of δ is positive in case of an alternator and negative for a motor.

24. Electrical losses (including armature winding loss, brush contact loss and field loss), core loss, friction and windage loss and stray *power loss occur in an alternator.*

Efficiency of a synchronous generator is given by

$$\eta = 1 - \frac{\text{kW losses}}{\text{kVA output} \times \text{pf} + \text{kW losses}} \times 100$$

25. The short-circuit ratio (SCR) of a synchronous machine is defined as the ratio of field current to produce rated voltage on open circuit to field current required to circulate rated current on short circuit while the machine is mechanically driven at synchronous speed.

$$\text{SCR} = \frac{1}{X_s}$$

26. In large-sized machines hydrogen gas is employed as a cooling medium. There are several advantages of using hydrogen as coolant in place of air in closed circuit ventilation system such as increased efficiency, increased output rating, increased life, elimination of fire hazard, less noise, smaller size of coolers etc.

EXERCISES

1. Why is rotating field system preferred against a stationary field in a synchronous machine?
[G.B. Technical Univ. Electromechanical Energy Conversion-II, 2012-13]

2. Explain why in case of 3-ϕ synchronous machines, the armature windings put on stator and field windings put on rotor whereas in case of dc machines, the armature windings put on rotor and field windings put on stator poles? Explain in brief.
[G.B. Technical Univ. Electromechanical Energy Conversion-II, 2011-12]

3. Explain the construction and principle of operation of synchronous machines. **[Anna Univ. Electrical Machines-I, May/June-2012]**

4. By means of a neat diagram, describe the main parts of an alternator. Describe the difference in construction of rotors of alternators used in hydroelectric plants and steam plants.
[U.P. Technical Univ. Electrical Machines 2009-10]

5. Differentiate between different rotor constructions of synchronous machines with neat sketch of each.
[Pb. Technical Univ. Synchronous Machines, May-2013]

6. With neat sketch describe the construction and principle of operation of salient pole alternator. **[Anna Univ. Electrical Machines-II, December-2012]**

7. Explain the essential difference between cylindrical and salient pole rotors used in large alternators. Also suggest various methods of excitation system for large alternators.
[U.P. Technical Univ. Electromechanical Energy Conversion II, 2005-06]

8. Make neat, labelled sketches of salient pole and cylindrical synchronous machines and compare them with reference to their (*i*) number of poles, (*ii*) speed (*iii*) axial length (*iv*) applications.
[Rajasthan Univ. Electrical Machines-II, 2005]

9. Describe various schemes used for exciting large synchronous machines. Which method is being preferred nowadays for exciting very large turbo-generators and why?
[Rajasthan Technical Univ. Electrical Machines-II, 2011]

10. Describe the various schemes used for exciting large synchronous generators. Explain any one with the help of neat sketch.
[Rajasthan Technical Univ. Electrical Machines-II, 2012]

11. Explain construction, excitation and principle of synchronous generator. **[Rajasthan Technical Univ. Electrical Machines-II, 2013]**

12. By help of neat diagram explain concentric and distributed type windings in an electrical machine.
[Rajasthan Technical Univ. Electrical Machines-II, June/July-2013]

13. What are the advantages of using distributed winding and short pitch coil ? Derive, from the fundamental, the expressions for winding factor and pitch factor.
[U.P. Technical Univ. Electromechanical Energy Conversion II, 2003-04]

14. Explain the importance of distributed armature winding in a synchronous generator. Derive an expression for distribution factor for such a winding.
[J.N. Technological Univ. Electrical Machines-III, May-2013]

15. How do the winding factors affect the waveform of the generated emf of synchronous generator? What steps are taken to eliminate third and fifth harmonic components of the generated emf?

[J.N. Technological Univ. Electrical Machines-III, May-2013]

16. Explain the effects of distributed winding and use of short-pitched coils on the magnitude of the generated voltage of alternator.

[J.N. Technological Univ. Electrical Machines-III, May-2013]

17. Explain the following in brief.

(*a*) Chording (*ii*) Skewing (*iii*) Breadth factor (*iv*) Pitch factor.

[Rajasthan Technical Univ. Electrical Machines-II, June/July-2011]

18. Describe basic concept of alternator. Define and explain (*i*) Distribution factor (*ii*) Pitch factor.

[Gujarat Technological Univ. Electrical Machines-I, Summer, 2014]

19. Derive the expressions for distribution factor and pitch factor of a synchronous machine.

[W.B. Technological Univ. Electrical Machines-II, 2012-13]

20. What is the distribution factor ? What is its effect ? Derive an expression for distribution factor of a winding having β slots per pole per phase and a slot angle β. [M.D. Univ. Electrical Machines-II, December-2010]

21. Derive the expression for induced emf in one "N-turn" full-pitch coil. Show that the emf generated in a short pitch coil is reduced by a factor cos $\alpha/2$ where α is the "chording angle".

[Rajasthan Technical Univ. Electrical Machines-II, 2012]

22. Derive an expression for the winding factor of an alternator.

[U.P. Technical Univ. Electromechanical Energy Conversion-II, 2010-11]

23. Derive emf equation for an alternator. Explain the effect of distribution of winding and use of short-pitch coil on the magnitude of the generated voltage of an alternator.

[U.P. Technical Univ. Electromechanical Energy Conversion-II, 2007-08]

24. Derive, from the fundamental, generalized equation of the EMF generated in the non-salient pole synchronous generator, taking into account the effect of 5th and 7th harmonic components.

[U.P. Technical Univ. Electromechanical Energy Conversion II, 2003-04]

25. Derive an expression for the emf generated in the armature winding of a synchronous machine. Also explain how harmonics in the emf wave are eliminated by short-pitch winding.

[U.P.Technical Univ. Energy Conversion, 2005-06]

26. Derive the general equation of induced emf. What are the effects of the harmonics in generated emf?

[Rajasthan Technical Univ. Electrical Machines-II, June/July-2011]

27. Describe the harmonics in generated voltage and explain how they are suppressed. [J.N. Technological Univ. Electrical Machines-III, May-2013]

28. Define and explain armature reaction, its bad effects on synchronous generator. [Rajasthan Technical Univ. Electrical Machines-II, 2010]

29. Discuss the nature of armature reaction of an alternator for zero pf lagging, zero pf leading and unity pf load.

[W.B. Technological Univ. Electrical Machines-II, 2012-13]

30. Draw the phasor diagram of a loaded alternator for the following conditions :

(*i*) lagging pf (*ii*) leading pf and (*iii*) unity pf.

[U.P. Technical Univ. Electrical Machines, 2008-09]

31. Draw and explain the equivalent circuit diagram and phasor diagram of a synchronous machine.

[U.P. Technical Univ. Electromechanical Energy Conversion-II, 2009-10]

32. Sketch and explain the open-circuit and short-circuit characteristics of a synchronous machine. [U.P.S.C. I.E.S. Electrical Engineering 1997; U.P. Technical Univ. Electromechanical Energy Conversion, 2010-11]

33. Discuss about the 'open-circuit' and 'short-circuit' characteristics of a synchronous generator. Draw the phasor diagram under short-circuit condition. What do you understand by the term 'short-circuit ratio'? Discuss how the short-circuit ratio can be calculated from the two characteristic curves.

[U.P.S.C. I.E.S. Electrical Engineering-II, 2000; U.P. Technical University Electromechanical Energy Conversion-II, 2006-07]

34. Draw the open-circuit characteristic and short-circuit characteristic of synchronous machine and show the variation of synchronous impedance Z_s with the field current I_f. Explain the nature of variation of Z_s. [U.P.S.C. I.E.S. Electrical Engineering-II, 2009]

35. What is short-circuit ratio? Explain the method to determine the short-circuit ratio of a synchronous machine using OCC and SCC.

[Pb. Technical Univ. Synchronous Machines, May-2013]

36. What do you mean by "OCC" and "SCC" in synchronous machines? Determine the value of synchronous reactance and short-circuit ratio from OCC and SCC.

[U.P. Technical Univ. Electromechanical Energy Conversion-II, 2013-14]

37. Define regulation of an alternator. Name the different methods of determination of regulation of an alternator. Which of these methods is known 'optimistic' and why? Briefly describe the method only.

[W.B. Technological Univ. Electrical Machines-II, 2011-12]

38. Explain the synchronous impedance method of finding the voltage regulation of a three-phase alternator. Give the circuits to perform tests required for the same.

[G.G.S.I.P. Univ. Electromechanical Energy Conversion-II, May-2011]

39. Explain the emf method of determining the regulation of an alternator.

[Anna Univ. Electrical Machines-II, Nov./Dec; 2012]

40. Explain the mmf method of determining the voltage regulation of alternator. [M.D. Univ. Electrical Machines-II, May-2009]

41. Explain clearly how stator leakage reactance and armature reaction are estimated by Potier triangle method.

[G.B. Technical Univ. Electromechanical Energy Conversion-II, 2012-13]

42. Give the circuit diagram used in laboratory to determine the voltage regulation of alternator by ZPF method and explain the procedure in brief. [R.G. Technical Univ. Electrical Machine-II, December-2013]

43. Describe with all relevant circuit diagrams and graphs, the method of determining the regulation of a 3-phase alternator using ZPF method. Compare the results of this method with emf and mmf methods.

[U.P.S.C. I.E.S. Electrical Engineering-II, 2007]

44. List the methods of determining voltage regulation of an alternator. Compare the merits and demerits of various methods. Explain how to obtain (*i*) leakage reactance (ii) synchronous reactance (*iii*) armature reaction mmf using Potier triangle method.

[Rajasthan Univ. Electrical Machines-II, 2005, 2006]

45. Define voltage regulation of an alternator.

Explain how the "Potier Triangle" can be drawn with the help of open-circuit characteristics and any two points on zero power factor characteristics. [Rajasthan Technical Univ. Electrical Machines-II, 2012]

46. Define voltage regulation of an alternator. Discuss ASA method of finding the voltage regulation.

[Rajasthan Technical Univ. Electrical Machines-II, June/July-2011]

47. Discuss the effect of saliency and explain the two reaction theory for salient pole synchronous machine.

[G.G.S.I.P. Univ. Delhi Electromechanical Energy Conversion-II, May-2011]

48. Discuss the two reactance theory of salient pole machines and draw its phasor diagram. [Pb. Technical Univ. Synchronous Machines; May-2013]

49. Discuss Blondel two reaction theory of salient pole synchronous machines.

Describe the slip test method for measurement of X_d and X_q of synchronous machines.

[R.G. Technical Univ. Electrical Machines-II, December-2012]

50. Develop phasor diagram for a salient pole alternator supplying a leading pf load and 2-reaction theory.

[J.N. Technological Univ. Electrical Machines-III, May-2013]

51. Derive the equation for developed power in a salient pole synchronous generator. What is reluctance power in a salient pole synchronous generator? **[W.B. Technological Univ. Electrical Machines-II, 2012-13]**

52. What is two reaction theory of salient pole synchronous machines? Draw and explain the phasor diagram of salient pole alternator supplying full-load lagging power current. Show that the power output per phase is given by

$$P = \frac{EV}{X_d}\sin\delta + \frac{V^2}{2}\left[\frac{1}{X_q} - \frac{1}{X_d}\right]\sin 2\delta$$

where X_d, X_q are direct-axis and quadrature-axis synchronous reactance, E, V are no-load voltage and terminal voltage and δ is load angle. **[Rajasthan Univ. Electrical Machines-II, 2002, January 2003, 2004]**

53. Give the analysis of short-circuit oscillogram of an alternator to determine its various parameters. **[R.G. Technical Univ. Electrical Machines-II, December-2013]**

54. Discuss the phenomenon of sudden 3-phase short circuit at the armature terminals of an alternator. Draw a typical waveshape of current and mark the different regions. Write an expression for current. **[M.D. Univ. Electrical Machines-II, December-2010]**

55. A 3-phase synchronous generator is excited by dc field to give rated voltage at its terminals on no load when a sudden short (symmetric) circuit occurs at the terminals.

Show the typical plots of three armature phase and field current and also show the typical plot of the ac component of short-circuit current in both actual and semilogarithmic plot identifying different operational conditions. **[U.P.S.C. I.E.S. Electrical Engineering-II, 2010]**

56. Describe different methods of cooling of alternators. What are the advantages of hydrogen gas as a cooling medium as compared to air? What special precautions should be taken for hydrogen cooled alternators? **[U.P. Technical Univ. Electrical Machines, 2010-11]**

57. Write short notes on the following:

(*i*) Construction of synchronous machines.

(*ii*) Excitation systems for synchronous machines.

(*iii*) Armature windings. **[Rajasthan Technical Univ. June/July-2011]**

(*iv*) Coil span factor. **[R.G. Technical Univ., December-2012]**

(*v*) Distribution factor. **[R.G. Technical Univ., December-2012]**

(*vi*) Armature reaction in alternator. **[M.D. Univ., May-2009]**

SHORT ANSWER TYPE QUESTIONS WITH ANSWERS

Q. 1. What are the essential elements for generating emf in alternators? **[Bharathiyar Univ., April-1997]**

Ans. The essential elements for generating emf in alternators are prime mover, stator, rotor and exciter.

Q. 2. Why is it possible to design alternators to generate much higher voltages than dc generators ?

Ans. In case of an alternator, field system is moving and armature is stationary and it is easier to insulate stationary armature winding for very high voltage because insulation of stationary armature is not subjected to mechanical stresses due to centrifugal action and more space is available on the armature for providing more insulation as the stator is outside the rotor. In case of dc generators field is stationary and armature is moving.

Q. 3. Why rotating field system is preferred over the stationary field system in synchronous generator ? **[U.P. Technical Univ. 2008-09]**

Ans. Refer to Art. 15.1.2.

Q. 4. Explain the choice of synchronous machine on the basis of the generating station. **[U.P.S.C. I.E.S. 2007]**

Ans. Synchronous machines are of two types: (*i*) Cylindrical rotor type (*ii*) Salient pole type.

A cylindrical rotor synchronous machine is characterised by long core length and small diameter and so the centrifugal forces developed at high speed are limited. These are employed in steam generating stations.

The salient pole synchronous machine is characterised by small core length and large diameter. These are very suitable for multipolar slow-speed water-turbine generators.

Most of the synchronous motors are of the salient pole type.

Q. 5. What type of rotor is adopted for high speed alternators ?

Ans. Cylindrical or non-salient pole rotor construction is used for high speed alternators.

Q. 6. What type of alternators will be used for (*i*) hydraulic turbine (*ii*) steam turbines as prime movers? **[G.G.S.I.P. Univ., December-2011]**

Ans. Salient pole type alternators will be use for hydraulic turbine whereas smooth cylindrical rotor type alternators will be used for steam turbines.

Q. 7. Name the two types of alternators depending on the rotor construction. **[Bharathiyar Univ., April-1996]**

Ans. Salient pole alternators and cylindrical rotor alternators.

Q. 8. Name the types of alternators depending on the type of prime mover. **[Bharathiyar Univ. Nov; 1995]**

Ans. (*i*) Turbo-generators (*ii*) hydro-generators (or waterwheel generators) and (iii) diesel-engine driven generators.

Q. 9. What will be the number of poles of a 50 Hz alternator if it runs at its greatest speed ? **[M.K. Univ., Nov.-1996]**

Ans. Two.

Q. 10. What is the maximum speed of a 50 Hz alternator?

Ans. 3,000 rpm.

Q. 11. Why are salient pole alternators more suitable for low speed and non-salient pole for high speed operation? **[Madras Univ., April-1996]**

Ans. The salient pole type field structure is used almost entirely for low speed alternators, since it is least expensive and provides ample space for the field ampere-turns but it cannot be used for high speed alternators on account of high peripheral speed and the difficulty of obtaining sufficient mechanical strength. So non-salient pole field structure is used for high speed operation.

Q. 12. Which one of the cylindrical and salient pole alternators is more stable ?

Ans. Salient pole alternator is more stable.

Q. 13. Do we use laminations for all the iron parts of an electrical machine?

Ans. All the iron parts of an electrical machine are laminated except the end covers and outer stator frame. The laminations are used to reduce the eddy current loss.

Q. 14. Why are the poles and pole shoes laminated ? **[Bharathiyar Univ., November-1996]**

Ans. For reducing eddy current loss.

Q. 15. Why machines having long core length are provided with radial ventilating ducts while machines having deep cores are provided with axial cooling.

Ans. Radial ventilating ducts provided with long core machines take away heat at faster rate. Similarly in case of deep cores axial ventilating ducts do. Otherwise the central portion of long core and the inner portion of deep core may attain a temperature higher than the permissible value.

Q. 16. What is the role of damper winding in (*a*) synchronous generator and (*b*) synchronous motor ?

Ans. (*a*) In a synchronous generator damper winding is provided to suppress the negative sequence field and to damp the oscillations when the generator starts hunting.

(*b*) In a synchronous motor, damper winding is provided to develop starting torque so as to make the motor self starting and to develop damping torque when the machine starts hunting.

Q. 17. Why rotating machines with aluminium armature coils have increased leakage reactance?

Ans. Rotating machines with aluminium armature coils need more space to accommodate conductors and so large size of stator slots resulting into increased leakage reactance.

Q. 18. At what voltage is the field of an alternator usually excited?

Ans. 125 or 250 V dc.

Q. 19. What is an exciter?

Ans. An exciter is a small dc generator (usually flat compound wound type) to supply dc power to the field magnet system or rotor of the alternator.

Q. 20. Why are distributed windings preferred over concentrated windings? **[M.S. Univ., April-1996]**

Ans. Distributed windings are preferred over concentrated windings as they (*i*) improve waveform of induced emf (*ii*) provide better utilization of core (*iii*) diminish armature reaction and (*iv*) facilitate cooling.

Q. 21. Why fractional pitched winding is preferred over full-pitched winding? **[Anna Univ., May/June-2012]**

Ans. Fractional pitched winding is preferred over full-pitched winding owing to following advantages.

(*i*) Waveform of the induced emf is improved and the distorting harmonic can be reduced or totally eliminated.

(*ii*) Copper is saved in the coil ends due to less span.

(*iii*) The inductance of the winding is reduced due to lesser length of coil ends as well as lesser mutual inductance between those conductors which lie in slots containing also conductors of either of the other two phases.

(*iv*) Fractional number of slots per pole can be used which in turn reduces the tooth ripples.

(*v*) Mechanical strength of the coil is increased.

Q. 22. A full pitch coil in an alternator has a span of 18 slots. What should be the span if the third harmonic component of the emf is to be eliminated ?

Ans. 12 slots.

Q. 23. Define pitch factor and distribution factor. **[Pb. Technical Univ., May-2013]**

Ans. In a short-pitch winding the induced emfs in the two sides of the coil are not in phase, hence their resultant given by the phasor sum is always less than their arithmetic sum. The ratio of phasor sum of induced emfs per coil to the arithmetic sum of induced emfs per coil is known as *coil pitch factor, coil-span factor* or simply *pitch factor* (K_c or K_p). It is always less than unity.

The ratio of the phasor sum of the emfs induced in all the coils distributed in a number of slots under one pole to the arithmetic sum of the emfs induced (or to the resultant of emfs induced in all coils concentrated in one slot under one pole) is known as breadth factor (K_b) or distribution factor (K_d)

or distribution factor,

$$K_d = \frac{\text{EMF induced in distributed winding}}{\text{EMF induced if the winding would have been concentrated}}$$

$$= \frac{\text{Phasor sum of component emfs}}{\text{Arithmetic sum of component emfs}}$$

The distribution (or breadth) factor is always less than unity.

Q. 24. What is the other name for distribution factor ? **[M.S. Univ., April-1996]**

Ans. Breadth factor.

Q. 25. Is it possible to make a balanced 3 – ϕ, 6-pole winding with 48 slots?

Ans. Number of slots per pole per phase, $m = \frac{48}{3 \times 6} = \frac{8}{3}$. As the denominator of m is not divisible by 3, it is not possible to have balanced winding?

Q. 26. What is the value of distribution factor for a uniformly distributed 3-phase winding?

Ans. Distribution factor for a uniformly distributed 3-phase winding with 60° phase spread is 0.955.

Q. 27. Define winding factor. **[M.D. Univ., December-2011]**

Ans. The product of pitch factor k_p and distribution factor k_d is referred to as the winding factor.

Q. 28. What is meant by "armature reaction" of a synchronous machine? **[Madras Univ., April-1997]**

Ans. The effect of armature flux on the main field flux is known as armature reaction. Armature reaction has distorting effect on unity power factor, wholly demagnetising at zero power factor lagging and wholly magnetising at zero power factor leading.

Q. 29. What parameter of load influences the armature reaction of an alternator?

Ans. Power factor of load.

Q. 30. Does the armature mmf of an alternator delivering a capacitive load magnetise or demagnetise the main field.

Ans. Magnetises.

Q. 31. What are the effects of armature reaction in an alternator at (*a*) zero leading pf (*b*) unity pf loads. **[G.G.S.I.P. Univ, May-2011]**

Ans. Armature reaction effect is wholly magnetizing at zero pf leading and wholly distorting at unity pf loads.

Q. 32. Explain the concept of replacing armature reaction by a reactance. **[J.N. Technological Univ., May-2013]**

Ans. The emf set up due to armature reaction mmf F_a is always in quadrature with the load current I and is proportional to it. Thus it is equivalent to an emf induced in an inductive coil and the effect of armature reaction can, therefore, be considered equivalent to reactance drop IX_a where X_a is the fictitious reactance which takes care of the armature reaction effect.

Q. 33. Define synchronous reactance. **[Pb. Technical Univ., May-2013]**

Ans. The emf set up due to armature reaction mmf F_a is always in quadrature with the load current I and is proportional to it. Thus it is equivalent to an emf induced in an inductive coil and the effect of armature reaction can, therefore, be considered equivalent to reactance drop IX_a where X_a is the fictitious reactance which takes care of the armature reaction effect. The armature winding possesses a certain leakage reactance X_L, as mentioned in Art 15.13. The sum of leakage reactance X_L and fictitious reactance X_a is called the *synchronous reactance*.

$$i.e.\ X_s = X_L + X_a$$

Q. 34. Explain synchronous impedance of an alternator. **[U.P.S.C. I.E.S., 2003]**

Ans. In the short-circuit test, entire induced emf E_o is used in circulating the short-circuit current I_{sc} through the synchronous impedance Z_s, therefore synchronous impedance Z_s is given as

$$Z_s = \frac{\text{Open-circuit terminal voltage for certain field current, } E_o}{\text{Short-circuit current for the same field current, } I_{sc}}$$

The phasor sum of effective resistance of armature R_e and synchronous reactance X_s is known as synchronous impedance Z_s i.e.,

$$Z_s = R_e + j\,X_s$$

Armature winding effective resistance R_e in alternators is usually very small in comparison to synchronous reactance X_s, and therefore, synchronous impedance may be assumed equal to the synchronous reactance for many purposes.

It may be seen from Fig. 15.34 that the synchronous reactance decreases with the increase in excitation. This is due to magnetic saturation effect. With the increase in field current, the armature current increases, and the components of flux created by these two currents both increase. As long as the resultant flux density is below the knee of the saturation curve, the flux produced per ampere of armature current is approximately constant, and therefore, the synchronous reactance X_s is constant. For higher flux densities, the flux produced per ampere decreases, and consequently the synchronous reactance decreases.

Synchronous impedance, being phasor sum of effective resistance (almost remains constant) and synchronous reactance, remains constant where there is no saturation but for higher flux densities synchronous impedance decreases as shown in the figure.

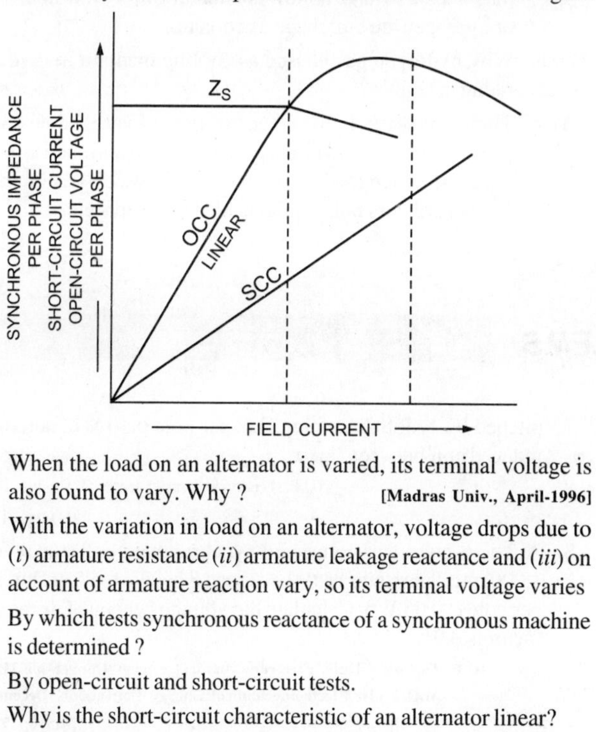

Q. 35. When the load on an alternator is varied, its terminal voltage is also found to vary. Why ? **[Madras Univ., April-1996]**

Ans. With the variation in load on an alternator, voltage drops due to (*i*) armature resistance (*ii*) armature leakage reactance and (*iii*) on account of armature reaction vary, so its terminal voltage varies

Q. 36. By which tests synchronous reactance of a synchronous machine is determined ?

Ans. By open-circuit and short-circuit tests.

Q. 37. Why is the short-circuit characteristic of an alternator linear?

Ans. The short-circuit characteristic is normally a straight line through the origin because the net excitation is so small that there is no saturation in the magnetic circuit.

Q. 38. What do you mean by the short-circuit characteristic of an alternator? **[M.S. Univ., April-1996]**

Ans. The short-circuit characteristic of an alternator is a curve (usually a straight line passing through origin) plotted between short-circuit current and field current while the alternator being driven at its rated speed.

Q. 39. Define voltage regulation of an alternator. **[Rajasthan Technical Univ., 2012]**

Ans. Voltage regulation of an alternator is defined as the increase in terminal voltage, expressed as percentage (or per unit) of the rated terminal voltage, when the load at a given power factor is thrown off, with speed and field current remaining the same.

i.e. Voltage regulation $= \dfrac{E_0 - V}{V}$ per unit

$= \dfrac{E_0 - V}{V} \times 100$ in percentage

where E_0 is no-load terminal voltage and V is the full-load terminal voltage.

Note: In case of leading power factor E_0 is less than V, so voltage regulation will be negative.

Voltage rise in terminal voltage when the full-load is thrown off is not the same as the fall in terminal voltage when the full load is applied.

Q. 40. What are the effects on voltage regulation of an alternator of power factor (*a*) at lagging power factor and (*b*) leading power factor. **[Madras Univ., April-1997]**

Ans. The regulation of an alternator is positive and increases with the decrease in lagging power factor. The regulation of an alternator improves first with the decrease in leading power factor, becomes unity and further decrease in leading power factor makes the regulation negative and increasing.

Q. 41. Why voltage regulation of an alternator is negative for leading power factor ? **[Madras Univ., April-1996]**

Ans. When the power factor is leading, the effect of armature flux is to assist the main flux, hence to generate more emf and so to increase the terminal voltage when the alternator is loaded. Thus the terminal voltage of an alternator decreases when the load of leading pf (i.e capacitive load) is thrown off and voltage regulation is –ve.

Q. 42. Why does synchronous impedance method give a poorer voltage regulation ?

Ans. In synchronous impedance method of determination of voltage regulation synchronous reactance is assumed to be constant while actually it varies with the saturation (at low saturation its value is higher because the effect of armature reaction is larger than that at high saturation). Now under short-circuit conditions, saturation is very low and, therefore, the value of synchronous impedance measured is higher than that in actual operating conditions and the regulation determined is higher than actual one.

Q. 43. Which method of regulation is called pessimistic and why? **[Bharathiyar Univ., April-1997]**

Ans. Synchronous impedance or emf method of determination of voltage regulation of an alternator is pessimistic method. For reasons see answer to above question.

Q. 44. What is the use of Potier triangle?

Ans. Potier triangle is used in determination of voltage regulation of an alternator.

Q. 45. Which quantities are usually determined using the zero-power factor characteristic of a synchronous machine?

Ans. Leakage reactance drop, voltage drop due to armature reaction and armature reaction ampere-turns.

Q. 46. What are the characteristics required for Potier method of computing voltage regulation? **[Bharathidhasan Univ., Nov.-1995]**

Ans. OCC, SCC and zero-power-factor full load voltage characteristic.

Q. 47. What is an air-gap line? **[M.S. Univ., April-1996]**

Ans. Tangent to OCC is called the air-gap line.

Q. 48. Two reactance theory is applied only to salient pole machines. State the reason. **[M.S. Univ., Nov.-1996]**

Ans. A multipolar machine with cylindrical rotor has a uniform air gap and, therefore, its reactance remains the same, irrespective of the spatial position of rotor. But in case of salient pole machines, the air gap is not uniform and its reactance varies with the rotor position. Because of non-uniformity of the reluctance of the magnetic paths, the mmf of the armature is divided into two components viz. (*i*) a direct-acting component along the field pole axis, called the direct-axis and (*ii*) quadrature-component along the axis passing through centre of the interpolar space, called the quadrature-axis. These facts form the basis of the two-reactance theory applied to the salient pole machines.

Q. 49. State the use of slip test on an alternator.

[Bharathiyar Univ., April-1997]

Ans. The slip test is performed on an alternator to determine X_d and X_q.

Q. 50. Upon what factor does the power angle 'δ' depend ?

[Bharathiyar Univ., April-1996]

Ans. Power angle δ depends upon supply voltage, armature current, power factor of load and quadrature component of synchronous reactance.

Q. 51. Why an alternator with low value of SCR has lower limit stability ?

Ans. An alternator with low value of SCR has high value of X_d $\left(X_d = \frac{1}{SCR}\right)$ and, therefore, has lower stability limit because maximum power output of a machine is inversely proportional to X_d.

Q. 52. What are the losses that take place in an alternator.

Ans. 1. Electrical losses including armature winding loss, brush contact loss and field loss, 2. core loss 3. Friction and windage loss and 4. Stray power loss.

Q. 53. There is no difficulty in obtaining sufficient ventilation in case of operation of salient pole alternators?

Ans. In the operation of **salient field pole alternators,** that have small axial length and large diameter, there is no particular problem in obtaining sufficient ventilation for the armature core. In addition to the large area available for direct radiation, there is the fanning action of the rotating field which may be increased by providing fan blades. Since the axial length of the core is small the heat produced in the embedded portion of the armature winding will be conducted to the end connections and dissipated from there if the fanning action is sufficient. However, armature core is usually subdivided by providing ventilating ducts at intervals of about 6–8 cm to allow the flow of cooling air radially.

Q. 54. Why open-circuit ventilation cannot be adopted for cooling of turbo-generator?

Ans. Ventilation problems in case of turbo-generators are very acute as they have axial length many times diameter. The smooth cylindrical rotor provides little fan effect in circulating air. Further, the surfaces of such alternators, which are exposed to the surrounding air, are small.

If open-circuit ventilation is adopted for such machines, large volume of air will be required. In such a system of ventilation it is necessary to pass the incoming air through the suitable filters to remove any dust before it is moved round the machine so as to prevent clogging of dust. The air brought in must also be moisture free. Thus the operating cost of handling several tons of air, depending upon the rating of the machine will be very heavy. Further it cannot ensure dust free entry of air.

The usual procedure is to have the machine completely enclosed with a cooling system using either air or hydrogen. The larger machines also have hollow conductors through which, water, oil or hydrogen are circulated as coolant.

Q. 55. Why hydrogen gas is used as cooling medium in case of large rating machines?

Ans. There are several advantages of using hydrogen as coolant in place of air in closed circuit ventilation system such as increased efficiency, increased output rating, increased life, elimination of fire hazard, less noise, smaller size of coolers etc.

PROBLEMS

1. A two-pole alternator rotates at 3,000 rpm. What is the frequency of generated emf?
[G.B. Technical Univ. Electromechanical Energy Conversion-II, 2012-13]
[**Ans.** 50 Hz]

2. An 8-pole ac generator is running at 750 rpm. What is the frequency. At what speed must the generator be run so that frequency shall be 25 Hz ? [**Ans.** 50 Hz; 375 rpm]

3. Calculate the winding factor of a 3-phase balanced winding having a coil span of 9 slots with 54 stator slots and 6 poles.
[B.P. Univ. of Technology Electrical Machine-I, 2008]
[**Ans.** 0.96]

4. Calculate the 'pitch factor' and 'distribution factor' of a 3-phase balanced winding having a coil span of 8 slots with 72 stator slots and 8 poles. [B.P. Univ. of Technology Electrical Machines, 2008]
[**Ans.** 0.9848; 0.96]

5. Calculate the fundamental, third and fifth harmonic distribution factor for a stator with 36 solts wound for 3-phase 4-poles.
[U.P. Technical Univ. Energy Conversion, 2005-06]
[**Ans.** 0.96, 0.6666, 0.2176]

6. The armature of a l-phase alternator is completely wound with T single turn coils distributed uniformly. The induced emf in each turn is 1.6 V. Determine the emf induced for the whole winding with all T coils in series. [**Ans.** 3.2 T/π volts]

7. A 3-phase, 8-pole, 900 rpm star-connected alternator has 72 slots on the armature and each slot has 10 conductors. The winding is short pitched by 1 slot. Assuming flux per pole 0.10 Wb, determine the induced emf between lines.
[U.P. Technical Univ. Electrical Machines, 2006-07]
[**Ans.** 4,995 V]

8. A 3-ϕ, 6-pole, star-connected synchronous generator revolves at 1,000 rpm. The stator has 90 slots and 8 conductors per slot. The flux per pole is 0.05 Wb. Calculate the voltage generated, if the winding factor is 0.96.
[U.P. Technical Univ. Electromechanical Energy Conversion II, 2003-04; M.D. Univ. Electromechanical Energy Conversion, December-2012]
[**Ans.** 2,215 V]

9. A 3-ϕ, 6-pole star-connected alternator revolves at 1,000 rpm. The stator has 90 slots and 8 conductors per slot. The flux per pole is 0.05 Wb (sinusoidally distributed). Calculate the voltage generated/phase by the machine if winding factor is 0.96.
[U.P. Technical Univ. Electrical Machines, 2008-09]
[**Ans.** 1,278.7 V]

10. Derive the emf equation for synchronous generator and calculate no-load phase and line voltage of a star-connected 3-phase, 4-pole alternator which runs at 1,500 rpm, having flux per pole of 0.12 wb sinusoidally distributed. Its stator has 52 slots having double layer winding. Each coil has 8 turns and the coil is chorded by 1 slot.
[M.D. Univ. Electromechanical Energy Conversion, December-2011]
[**Ans.** 3,511 V, 6082 V]

11. A 3-phase, 16-pole, star-connected alternator has 144 slots and 6 conductors per slot. The flux per pole is 0.03 Wb sinusoidally distributed and the speed is 375 rpm. If the coil span is 160°, determine: (*i*) Frequency (*ii*) Pitch factor (*iii*) Distribution factor (*iv*) Phase and line emfs.

[U.P. Technical Univ. Electromechanical Energy Conversion II, 2004-05]

[**Ans.** 50 Hz, 0.9848, 0.9598, 906.5 V, 1,570 V]

12. Calculate the no-load terminal voltage of 3-phase, 6-pole, star-connected alternator running at 1,000 rpm having the following data:

Sinusoidally distributed flux per pole = 60 mWb

Total number of slots on the stator inner periphery = 54

Number of conductors per slot = 10 and coil span 8 slots

[B.P. Univ. of Technology Electrical Machine-I, 2008]

[**Ans.** 5,888 V]

13. A 1100 V, 3-phase alternator is running at 300 rpm and has 24 poles. Find the number of conductors in stator winding, if the magnetic flux is 5×10^{-2} Wb/pole by assuming the breadth factor equal to 0.96 and coil to be full-pitched. **[U.P. Technical Univ. Electrical Machines, 2001–02]**

[**Ans.** 300]

14. Find the no-load phase and line voltage of star-connected 3-phase, 6-pole alternator which runs at 1,200 rpm, having flux per pole of 0.1 Wb sinusoidally distributed. Its stator has 54 slots having double layer winding. Each coil has 8 turns and the coil is chorded by 1 slot.

[**Ans.** 1,813 V; 3,141 V]

15. A 3-phase, 4-pole, 50-Hz, star-connected alternator has 60 slots with 2 conductors per slot and having armature winding of the double layer type. Coils are short-pitched, that is, if one coil side lies in slot number 1, the other coil side lies in slot number 13. Find the useful flux per pole required to induce a line voltage of 6.6 kV.

[Bharathiyar Univ. Elec. Machines II, April-1997]

[**Ans.** 0.943 Wb]

16. A 3-phase, 50 Hz star-connected alternator runs at 600 rpm. It has 60 stator slots with 4 conductors per slot. The flux per pole is 0.06 Wb per pole and sinusoidally distributed. Determine the phase and line emfs. **[J.N. Technological Univ. Electrical Machines-III, May-2013]**

[**Ans.** 514.6 V; 891.4 V]

17. Calculate the rms value of induced emf per phase of a 10-pole, 3-phase, 50 Hz alternator with 2 slots/pole/phase and 4 conductors per slot. Assume coil span of 150°. Flux per pole has a fundamental component of 0.12 Wb and a third harmonic component of 20 % of the fundamental. What is the value of line-to-line emf?

[**Ans.** 1,000; 1,722 V]

[**Hint:** Since phase emfs induced by the 3rd, 9th and 15th harmonics are eliminated from the line voltages, the line voltage for star connection = $994.4 \times \sqrt{3}$ = 1,722 V]

18. For a 3-phase winding with 3 slots per pole per phase and coil span of 8 slots, compute the following:

(*i*) Breadth and pitch factors.

(*ii*) Percentage increase in per phase emf (rms) due to third harmonic where flux density in the air gap of the machine is found to have 30% third harmonic.

[Rajasthan Technical Univ. Electrical Machines-II, 2012]

[**Ans.** 0.9598; 0.9848; 1.66%]

19. A 4-pole, 50 Hz, 3-phase Y-connected alternator has a single layer full pitched winding with 21 slots per pole and two conductors per slot. The fundamental flux is 0.6 Wb and air-gap flux contains a third harmonic of 5% amplitude. Find the rms values of phase emf due to fundamental and third harmonic flux and the total induced emf.

[M.D. Univ. Electrical Machines-II, December-2011]

[**Ans.** 3,564.8 V; 39.9 V; 3,565 V]

20. A 500 kVA, 1,100 V, 50 Hz star-connected 3-phase alternator has armature resistance per phase of 0.1 Ω and synchronous reactance per phase 1.5 Ω. Find its voltage for (*a*) unity pf (*b*) 0.9 lagging and (*c*) 0.8 leading pf. Also, calculate voltage regulation in each case.

[Allahabad Univ; Elec. Machines, 1993]

[**Ans.** (*a*) 1,333 V; 21.187% (*b*) 1,556 V; 41.447% (*c*) 925.7 V; – 15.845%]

21. Find the voltage regulation at full load 0.8 power factor lagging for a three phase, 1500 kVA, 6600 volt, star connected alternator having an armature resistance of 0.093 ohm per phase and a synchronous reactance of 8.5 ohm/phase.

[U.P. Technical Univ. Electrical Machines, 2005-06]

[**Ans.** 20%]

22. A 100 kVA, 3,000 V, 50 Hz 3-phase star connected alternator has effective armature resistance of 0.25 Ω. The field current of 38 A produces short-circuit current of 195 A and an open-circuit emf of 1025 V as line voltage. Calculate the full-load voltage regulation at 0.85 pf lagging and 0.8 pf leading. Also draw the phasor diagram for both the cases.

[M.D. Univ. Electromechanical Energy Conversion, December-2012]

[**Ans.** 2.04%, –1.753%]

23. A 60 kVA, 220 V, 50 Hz, single phase alternator has effective armature resistance of 0.012 ohm and armature leakage reactance of 0.06 ohm. Compute the voltage induced in armature when alternator is delivering rated current at a load power factor of :

(*i*) unity, (*ii*) 0.8 lagging, (*iii*) 0.8 leading.

[M.D. Univ. Electromechanical Energy Conversion, December-2010]

[**Ans.** 224 V, 232.7 V, 213.34 V]

24. A 3-phase star-connected cylindrical rotor synchronous generator rated at 10 kVA and 230 V has a synchronous reactance of 1.2 Ω/phase and an armature resistance of 0.5 Ω/phase. Calculate the percentage voltage regulation at full load with 0.8 lagging pf.

[G.G.S.I.P. Univ. Electromechanical Energy Conversion-II, May-2011]

[**Ans.** 21.59%]

25. A 550 V, 55 kVA, single-phase synchronous generator has an effective resistance of 0.2 Ω. A field current of 10 amps produced an armature current of 200 A on short-circuit and an emf of 450 V on open circuit. Calculate synchronous impedance and voltage regulation at full load with pf 0.8 (lagging).

[U.P. Technical Univ. Electromechanical Energy Conversion II, 2003-04]

[**Ans.** 30.91%]

26. A 100 kVA, 3000 V, 50 Hz 3-phase star-connected alternator has effective armature resistance of 0.25 Ω. The field current of 35 A produces short-circuit current of 200 A and open-circuit emf of 1,040 V (line value). Calculate full-load voltage regulation at 0.8 pf lagging and 0.8 leading. Also draw the phasor diagrams for both cases. **[M.D. Univ. Electromechanical Energy Conversion, December-2011]**

[**Ans.** 2.25%, 1.732%]

27. Determine the voltage regulation of a 2,000 V, 1-phase alternator giving current of 100 A at 0.8 pf lagging. Use the test data given below :

Full-load current of 100 A is produced on short circuit by field excitation of 2.5 A; an emf of 500 V is generated on open circuit by same excitation. The armature resistance is 0.8 Ω.

[U.P. Technical Univ. Electrical Machines, 2002-03]

[**Ans.** 19.25%]

28. A 3-phase, 1500 kVA, star-connected, 50 Hz, 2300 V alternator has a resistance between each pair of terminals as measured by direct current is 0.16 Ω. Assume that the effective resistance is 1.5 times the ohmic resistance. A field current of 70A produces a short-circuit current equal to full-load current of 376 A in each line. The same field current produces an emf of 700 V on open circuit. Determine the synchronous reactance of the machine and its full-load regulation at 0.8 power factor lagging. [U.P. Technical Univ. Electrical Machines, 2009-10]

[**Ans.** 1.068 Ω; 22.875%]

29. A full-load current of 100 A requires an excitation current of 3 amperes when a single phase 1,200 V alternator is short circuited by an ammeter of resistance 0.01 Ω. On open circuit the same excitation produces 350 V. The resistance of the armature is 0.5 Ω. Calculate the regulation of the alternator at 0.8 power factor leading.

[U.P. Technical Univ. Electrical Machines, 2001-02]

[**Ans.** –10.25%]

30. A 100 kVA, 3,000 V, 50 Hz, 3-phase star-connected alternator has effective armature resistance of 0.25 Ω. The field current of 38 A produces a short-circuit current of 195 A and an open-circuit emf of 1,025 V as line voltage. Calculate the full-load voltage regulation at 0.85 pf lagging and 0.8 pf leading. Also draw the phasor diagram for both the cases. [M.D. Univ. Electrical Machines-II, December-2011]

[**Ans.** 1.097%; –0.736%]

31. A 150 kVA, 3,300 V, 50 Hz star-connected, three-phase alternator has an effective stator resistance 0.125 Ω/phase. During open-circuit test of the machine, for a field current of 8.5 A, the terminal voltage is measured to be 600 V line-to-line voltage. While performing the short-circuit test on the same machine, the short-circuit current is 120 A for the same field current of 8.5 A. Find the terminal voltage per phase which rated full load at 0.8 leading pf is switched off from the generator without changing the excitation and speed. What is the load angle in this case? [B.P. Univ. of Technology Electrical Machines-I, 2008]

[**Ans.** –2.19%]

32. A 3-phase star-connected synchronous generator is rated at 1.5 MVA, 11 KV. The armature effective resistance and synchronous reactance are 1.2 Ω and 25 Ω respectively per phase. Calculate the percentage voltage regulation for a load of 1.4375 MVA at (*i*) 0.8 pf lagging and (*ii*) 0.8 pf leading. Also find out the pf at which the regulation becomes zero. [U.P.S.C. I.E.S. Elec. Engineering II, 1992]

[**Ans.** 21.6%, –13.1%; 0.981 leading]

33. Find the voltage regulation at full-load 0.8 power factor lagging for a three phase, 1500 kVA, 6600 volt, star connected alternator having an armature resistance of 0.093 ohm per phase and a synchronous reactance of 8.5 ohm/phase.

[U.P. Technical Univ. Electrical Machines 2005-06]

[**Ans.** 20%]

34. A 3-phase, star connected alternator is rated at 1600 kVA, 13500 V. The armature effective resistance and synchronous reactance are 1.5 Ω and 30 Ω respectively per phase. Calculate the percentage regulation for a load of 1,280 kW at power factor of (*i*) 0.8 lagging and (*ii*) 0.8 leading. [U.P. Technical Univ. Electrical Machines 2007-08]

[**Ans.** 18.59%, –11.99%]

35. A 15 kVA, 400 V, 50 Hz, 3-phase star-connected alternator gave the following data.

Field current (Amp)	2.0	2.5	3.0	3.5	4	4.5	5.0
V_{dc} line (volts)	266	334	377	422	450	484	508

An excitation of 2 A gave full-load current on short circuit. Armature resistance per phase is 1 Ω. Calculate full-load regulation at 0.8 pf leading using MMF method.

[J.N. Technological Univ. Electrical Machines-III, May-2013]

[**Ans.** –15%]

36. The OC and SC test readings for a 3-phase, 50 Hz, 1,000 kVA, 2,000 V, star-connected alternator are as follows:

Field current (A)	10	20	25	30	40	50
OC terminal voltage (V)	800	1,500	1,760	2,000	2,350	2,600
SC armature current (A)	–	200	250	300	–	–

The effective armature resistance per phase is 0.2 Ω. Estimate the full-load percentage regulation at 0.8 pf lagging and 0.6 pf leading. Use mmf method.

[Manonmanian Sundaranar Univ. Elec. Machines-II Nov.-1996]

[**Ans.** 34.6%; – 6.4%]

37. An 11 kV, 3-phase alternator has X_d and X_q respectively as 1 pu and 0.6 pu. Armature resistance is negligible. Compute the generated emf when the machine delivers rated kVA at 0.8 pf lag.

[**Ans.** 19.525 kV]

38. The direct and quadrature axis reactances of a salient pole synchronous generator are 1.0 pu and 0.6 pu respectively. The armature resistance is negligible. Calculate the excitation voltage in pu when the generator delivers rated kVA at 0.8 pf lagging and rated voltage.

[**Ans.** 1.775 pu]

39. A 3-phase alternator has a direct-axis synchronous reactance of 0.8 pu and a quadrature-axis synchronous reactance of 0.45 pu. Draw the phasor diagram for full load of 0.8 pf lagging and obtain therefrom (*i*) the load angle and (*ii*) the open-circuit voltage in pu.

[**Ans** (*i*) 15.826° (*iii*) 1.5984 pu]

40. A 5 kVA, 240 V, star-connected, 3-phase salient pole alternator with direct -axis and quadrature-axis synchronous reactance of 12 Ω and 7 Ω respectively delivers full load current at unity pf. Calculate the excitation voltage and reluctance power.

[**Ans.** 334.93 V; 1,521.5 watts]

41. A salient pole synchronous generator has the following parameters: $X_d = 1.2$, $X_q = 0.8$, $r_a = 0.025$. Compute the excitation voltage E on a per unit basis when the alternator is delivering at rated voltage for 0.8 pf lagging. [**Ans.** 1 9696 pu]

42. A 20 kVA, 220 V, 50 Hz, star-connected, 3-phase salient pole synchronous generator supplies load at a lagging power factor angle 45°. The constants of generators are X_d = 4.0 Ω and X_q = 2.0 Ω, R_a = 0.5 Ω. Calculate power angle and voltage regulation under the given load conditions. [U.P. Technical Univ. Electrical Machines, 2003-04]

[**Ans.** 14.22°, 149.52%]

43. A synchronous generator has a direct-axis synchronous reactance 'X_d' of 0.8 pu and a quadrature-axis synchronous reactance 'X_q' of 0.5 pu. Determine the pu open-circuit voltage for full-load at a lagging pf of 0.8. Neglect saturation.

[U.P. Technical Univ. Electromechanical Energy Conversion II, 2003-04]

[**Ans.** 1.60278 pu]

44. A 3.5 MVA, 3-phase alternator rated at 6.6 kV has 32 poles. Its direct and quadrature-axis synchronous reactance are 9.6 Ω and 6 Ω respectively. Neglecting armature resistance, determine the regulation and excitation emf need to maintain 6.6 kV at the terminals when supplying a load of 2.5 MW at 0.8 pf lagging. What maximum power can the generator supply at rated terminal voltage, if the field becomes open-circuited? [J.N. Technological Univ. Electrical Machines-III, May-2013]

[**Ans.** 1.361 MW]

❑❑❑

16 CHAPTER

Synchronous Motors

INSIDE THIS CHAPTER

16.1 INTRODUCTION

A synchronous motor has the same relationship to an alternator as a dc motor has to a dc generator *i.e.* if an alternator is supplied ac power it is capable of rotating as a motor and doing mechanical work. If the mechanical power supplied to a rotating alternator is removed while the dc field remains energized, and an ac supply is then connected across the armature terminals, torque will be developed and the alternator will continue to rotate at a speed determined by the ac supply frequency and number of poles on the synchronous machine. Changes in mechanical load within the machine's rating will not cause a change in speed.

In case of a dc motor the field and armature, both, are supplied from dc supply mains, since both require direct current. But in case of a synchronous motor the field structure is to be energized by direct current, as in case of an alternator, whereas the armature winding is connected to a 3-phase ac supply mains. Such a machine, therefore, requires two sources of supply, viz., an ac source to supply the power for driving the armature and a dc source to excite the field. DC excitation may be provided either from an exciting plant dc system or, if no direct current is available, from separate dc generators. Direct-connected exciters are frequently found on high-speed motors. These are overhung from specially constructed motor brackets with the exciter armature mounted on an extension of the motor shaft. On low-speed machines, belted exciters are sometimes employed, notably on air compressor drives. Separate motor-generator sets are widely used for excitation of low-speed synchronous motors. Where a group of synchronous motors is installed, it is sometimes economical to use a single motor-generator set for exciting the entire group, although greater flexibility is obtained when each motor is provided with its own exciter. Synchronous motor with a directly connected exciter is shown in Fig. 16.1.

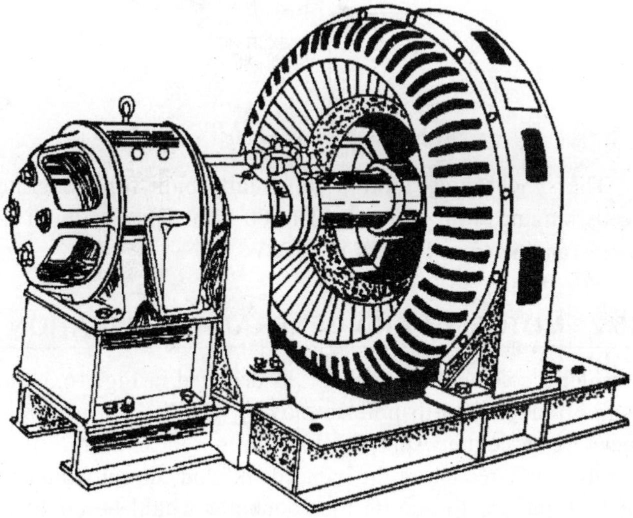

Fig. 16.1 *Synchronous Motor With Directly Connected Exciter*

Essentially, the construction of a polyphase synchronous motor is the same as that of an alternator. Certain features of its design may be varied to suit the particular requirements. A synchronous motor may have either a revolving armature or a revolving field, but the revolving field construction is the more common.

The essential parts of a three-phase synchronous motor are (*i*) laminated stator core with 3-phase armature winding (*ii*) rotating field structure complete with damper winding and slip rings (*iii*) brushes and brush holders and (*iv*) two end shields to house the bearings that support the rotor shaft.

The stator core and windings of a synchronous motor are similar to those of a 3-phase alternator. The leads from the stator windings terminate in a terminal box usually mounted on the side of the motor frame. The rotor of the synchronous motor usually has salient field poles connected to give alternate polarity. The number of rotor field poles must be the same as that of stator field poles. The field circuit leads are brought out to two slip rings mounted on the shaft. Carbon brushes mounted in brush holders make contact with the two slip rings. The terminals of the field circuit are brought out from the brush holders to a second terminal box mounted on the motor frame.

In the most common type of field construction, the poles are somewhat different from those of the alternator. In addition to the exciting winding, a synchronous motor often is provided with a squirrel cage winding on the rotor. The slots of this winding are in the pole faces and are parallel to the shaft. The ends of the copper or bronze bars embedded in the slots are short circuited at both ends by conducting rings. This short-circuited winding on the rotor is known as the *cage* winding, the *damper* winding, the *amortisseur* winding, or sometimes, to describe its function, the *starting* winding. A typical rotor for a polyphase synchronous motor is shown in Fig. 16.2.

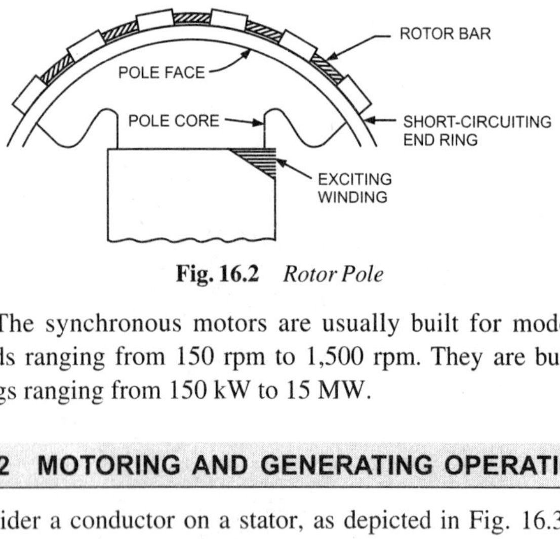

Fig. 16.2 *Rotor Pole*

The synchronous motors are usually built for moderate speeds ranging from 150 rpm to 1,500 rpm. They are built in ratings ranging from 150 kW to 15 MW.

16.2 MOTORING AND GENERATING OPERATION

Consider a conductor on a stator, as depicted in Fig. 16.3 and assume that the north pole is moving past it in a counter-clockwise direction. Applying Fleming's right hand rule it is seen that the direction of the induced emf and current would be outward and the torque on the conductor would be counter-clockwise. Consequently for this condition to be sustained, a counter-clockwise mechanical torque would have to be applied to the rotor shaft to counterbalance the electromagnetic

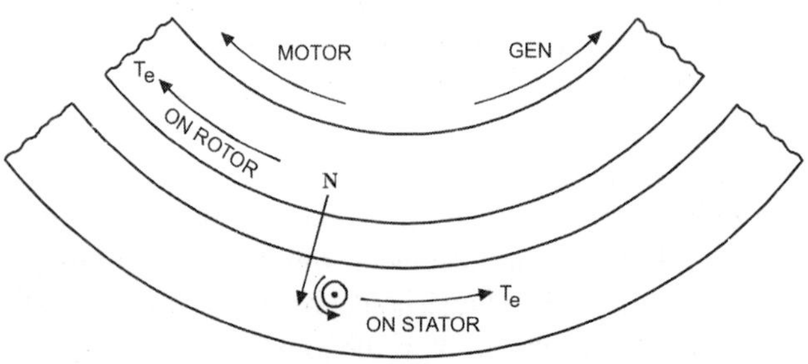

Fig. 16.3 *Motoring and Generating Operation*

developed torque T_e. There would be a conversion of mechanical power to electrical power; the *generating condition*. Conversely, if the conductor current was sustained in the direction shown by an external electrical source, electromagnetic torque T_e would be available for driving a coupled mechanical load in the clockwise direction; the motoring condition. Because of the reversal in rotation, the induced emf would be in the opposite sense to the direction of current shown, another indication of motoring mode.

For a synchronous generator, also called an alternator, the frequency of the conductor currents would vary directly as speed N, being equal to $f = \frac{PN}{120}$. The rotating armature mmf wave due to currents at this frequency would move in space at speed $N_s = \frac{120f}{P} = \frac{PN}{120} \times \frac{120}{P}$, which, as might be expected, is the same speed as that of the rotor. For a motor, the rotating armature- field fixes the motion of rotor at the synchronous speed.

16.3 PRINCIPLE OF OPERATION

The action of the motor may be followed by referring to Fig. 16.4, which illustrates a 3-phase, 2-pole synchronous machine. The rotor has two poles and the stator has two poles per phase. In practice it is usual to have salient poles on the rotor, but the armature winding is housed in slots in the concave periphery of the stator. In the figure, salient type stator poles have been shown for convenience only.

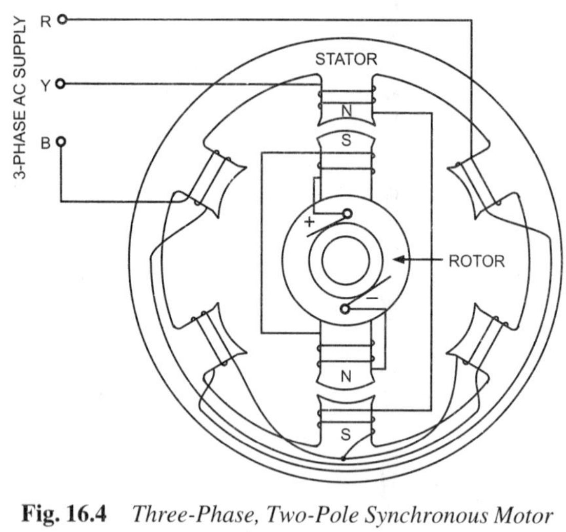

Fig. 16.4 *Three-Phase, Two-Pole Synchronous Motor*

As the rotor is excited from dc supply, so the poles of the rotor retain the same polarity throughout but the polarity of the stator poles changes as it is connected to an ac supply mains.

First of all let us consider the rotor as stationary and in the position shown in Fig. 16.5 (*a*), At this instant the rotor S-pole is attracted to the stator N-pole and, therefore, the rotor tends to rotate in clockwise direction. After half a period $\left(i.e. \text{ after } \frac{T}{2} \text{ second whereas } T = \frac{1}{f}\right)$ the polarity of stator poles is reversed but the polarity of the rotor poles remains the same [Fig. 16.5 (*b*)]. At this instant, the rotor S-pole is repelled by the stator S-pole, being similar in nature and, therefore, rotor tends to rotate in counter-clockwise direction. Thus the torque acting on the rotor of a synchronous motor is not unidirectional but pulsating one and due to inertia of the rotor, it will not move in any direction. So the *synchronous motor has got no self-starting torque*.

Now let us consider the rotor rotating in clockwise direction, which is possible by external means, and in the position shown in Fig. 16.5 (*a*). At this instant, as mentioned above, the rotor S-pole is attracted to stator N-pole and so the torque acts on the rotor in clockwise direction. After half a period the stator polarity is reversed *i.e.* stator S-pole becomes N-pole and N-pole becomes S-pole but if the rotor is rotated at such a speed by some external means at the starting moment that rotor S-pole advances by a pole pitch so that it is again under the influence of stator N-pole, as shown in Fig. 16.5 (*c*), the torque acting on the rotor will be again clockwise. Hence a continuous (unidirectional) torque will be obtained and the two fields will get locked into each other. Now if the external means is removed the rotor will continue to rotate in clockwise direction under the influence of clockwise continuous torque acting on the rotor.

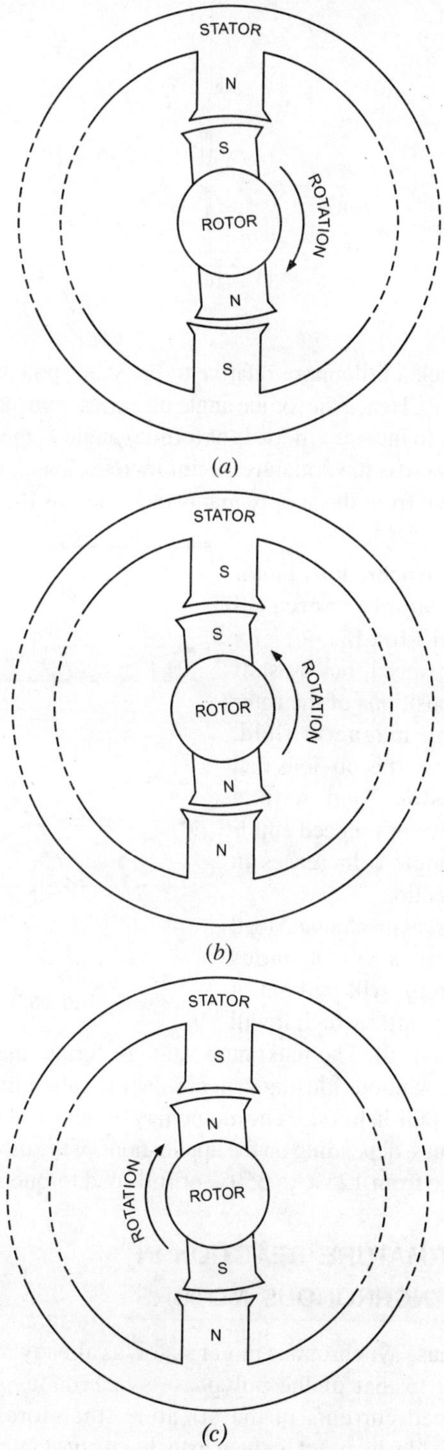

(*a*) (*b*) (*c*)

Fig. 16.5

Hence to obtain a continuous torque it is necessary that the rotor rotates at such a speed that it moves through the distance equal to pole pitch in half the period *i.e.* in $\frac{T}{2}$ or $\frac{1}{2f}$ second where f is the supply frequency. If there are P poles then to complete one revolution it should take $\frac{P}{2f}$ second or the rotor should rotate at a speed of $\frac{2f}{P} \times 60$ revolutions per minute.

Hence to obtain a continuous torque the rotor should rotate at synchronous speed given by the expression $N_s = \frac{120f}{P}$.

16.4 EFFECT OF LOAD ON A SYNCHRONOUS MOTOR

All the electric motors, whether dc or ac act as generators at the same time they are motoring. This is true because an emf is always induced opposing the applied voltage when the machine is motoring. In a dc machine, the rotating armature conductors cut magnetic flux under the stationary field poles; in an ac machine, as in an induction motor or a synchronous motor, the rotating magnetic field cuts the stationary armature conductors of the stator. In fact, it is this very opposing induced emf that limits the armature current to a value that is just enough to take care of the power requirements of the motor for a particular load.

When the load on a dc shunt motor is increased, speed of motor falls slightly. This reduces the induced emf, called the back emf, so that resultant voltage across the armature $(V - E_b)$ increases, which in turn increases the armature current, since armature current is given by the expression $I_a = \left(\frac{V - E_b}{R_a}\right)$, to supply extra torque and power required by the increased load. In the case of synchronous motor, however, the speed, as we have already discussed, remains constant irrespective of magnitude of load, therefore, method of drawing increased current from supply mains on synchronous motor cannot be same as that of a dc shunt motor.

The behaviour of a synchronous motor on load is discussed below:

Since the synchronous motor has got no self-starting torque, first it is driven by some external means as an alternator and is synchronised with the supply. At the instant of synchronising the induced emf across the stator circuit of the motor (also called the back or counter emf or excitation voltage) is equal to the applied line voltage V but opposite in direction, as shown in Fig. 16.7 (*a*) vectorially. The relative positions of a stator pole

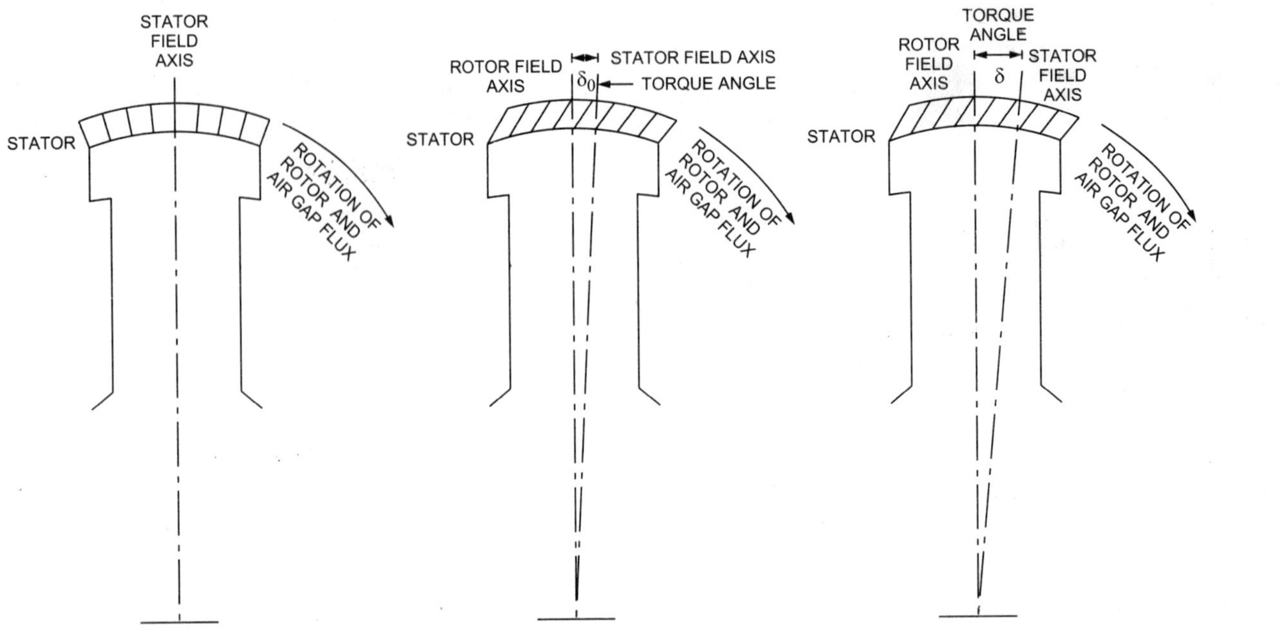

(*a*) (*b*) (*c*)

Fig. 16.6

and dc field pole of a synchronous motor at the synchronising instant are shown in Fig. 16.6 (*a*). The centre lines of the two poles coincide. Since V = E, hence the resultant voltage is zero and so is the armature current. Motor intake is zero, as there is neither load nor losses to be met by it. Under these conditions the motor is said to be *floating*.

After synchronising, the external means is removed and the motor tends to slow down. There is no change in speed; there is merely a shift in relative positions of the two poles, both of which continue to rotate at synchronous speed. The rotor pole falls back a little relative to the stator pole, as shown in Fig. 16.6 (*b*). The angular displacement between the rotor and stator poles δ is called the *torque angle*. Now the applied voltage and induced emf are no longer in direct opposition since induced emf at this instant lags behind the induced emf at the synchronising instant, say by an angle δ_0, as shown in Fig. 16.7 (*b*). The resultant voltage E_R (the resultant of applied voltage **V** and induced emf **E**) causes a current $I_0 = \frac{E_R}{Z_s}$ to flow in the stator winding; where Z_s is the synchronous impedance per phase of the armature circuit. I_0 lags E_R by an angle θ, called the *internal angle*, where $\theta = \tan^{-1}\frac{X_s}{R_e}$; X_s and R_e being the stator synchronous reactance and effective resistance respectively. Hence on no load motor draws power equal to $VI_0 \cos \phi_0$ (for one phase) from the supply mains which is sufficient to make the motor running continuously at synchronous speed.

At no load the torque angle δ_0 is very small, so the resultant voltage E_R acting across the armature circuit is small, due to which a small current is drawn from the supply mains. At no load, the torque angle δ_0 is so small that it can be assumed zero (neglecting the motor internal losses).

Now when the motor is loaded, it slows down momentarily to adjust itself to the change in loading conditions, so the rotor pole falls back a little more relative to the stator pole, as shown in Fig. 16.6 (*c*). Hence the torque angle increases with the increase in load. Due to increase in load (or torque) angle δ, the resultant voltage E_R across the armature circuit increases, and, therefore, current drawn from the supply mains increases, as illustrated in Fig. 16.7 (*c*).

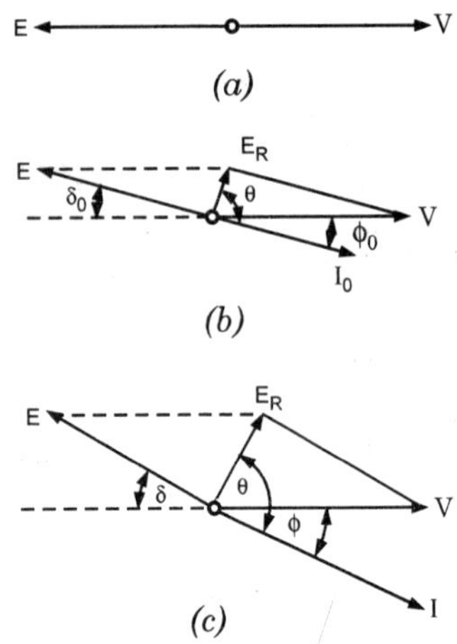

Fig. 16.7

Thus a synchronous motor is able to supply increased mechanical load, not by reduction in speed, but by shift in relative positions of the rotor and rotating magnetic field. From Fig. 16.7 it is obvious that for increasing load with a constant value of induced emf E the phase angle ϕ increases in lagging direction.

If too great mechanical load is applied to a synchronous motor, the rotor will pull out of synchronism, after which it will come to standstill. The maximum value of torque that a motor can develop without loosing synchronism is called the *pull-out torque*, like pull-in torque. The motor may be designed for widely varying values, depending on the applications of the motor. These values range from 125% to 350% of full-load torque values.

16.5 ARMATURE REACTION IN SYNCHRONOUS MOTORS

The polyphase synchronous motor stator, as already mentioned, is identical to that of the polyphase synchronous generator. The balanced currents in the armature, therefore, set up a magnetic field which, for a given load, has a constant magnitude and always rotates at synchronous speed. Moreover, in case of an alternator, the armature reaction flux induces a voltage in a given phase which always lags behind the armature current by 90° in phase. This is true regardless of the polarity or the position of the rotor. It depends only on the fact that the armature is wound for 3-phase and carries balanced currents.

In a synchronous motor too, the flux created by the armature reaction lags the respective armature currents by 90°. But now since the generated or excitation voltage E is more than 90° out of phase with the armature current, armature reaction flux has a different effect on the rotor flux than it has in the case of an alternator, except when the armature current I and excitation voltage E are exactly 180° apart as illustrated in Fig. 16.8 (*a*). Under this condition, the armature reaction is cross-magnetizing, as it is when the generated or excitation voltage of an alternator is in phase with the armature current. When the motor operates at lagging power factor such that the armature flux leads the excitation voltage by less than 90°, as shown in Fig. 16.8 (*b*), the armature reaction has a magnetising component as well as cross-magnetising one. With a leading power factor on the motor, the armature reaction flux has a demagnetising as well as cross-magnetising component, as shown in Fig. 16.8 (*c*). In Fig. 16.8, Φ_{AR} represents the armature reaction flux. The component along the axis of the rotor poles is shown along the line of excitation voltage E, assisting the rotor flux for the lagging power factor and opposing it in the case of the leading power factor.

It is thus seen that the armature reaction is one which opposes the action of the main field flux. When the dc field current is reduced so as to draw the current from the ac supply mains at lagging power factor, the armature current drawn from the ac supply mains acts to strengthen the field. On the other hand, if the field current is increased so as to draw a current at leading

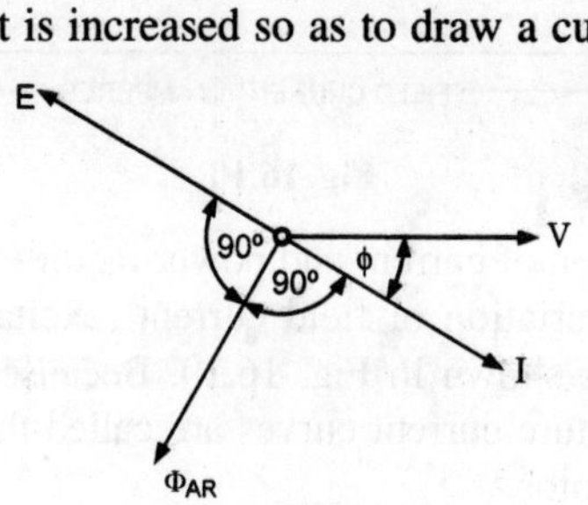

(*a*) *Cross-Magnetising*

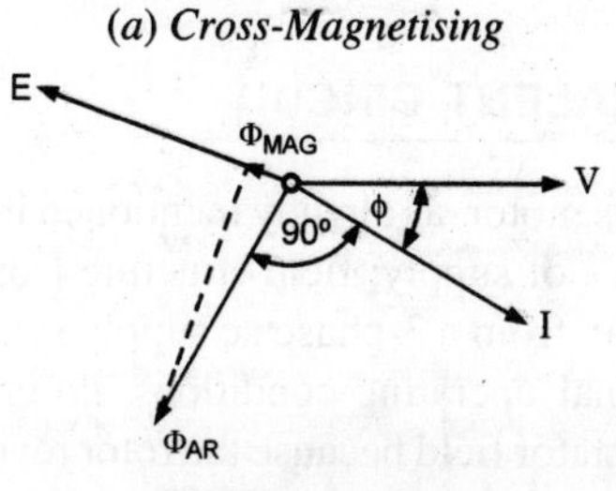

(*b*) *Magnetising*

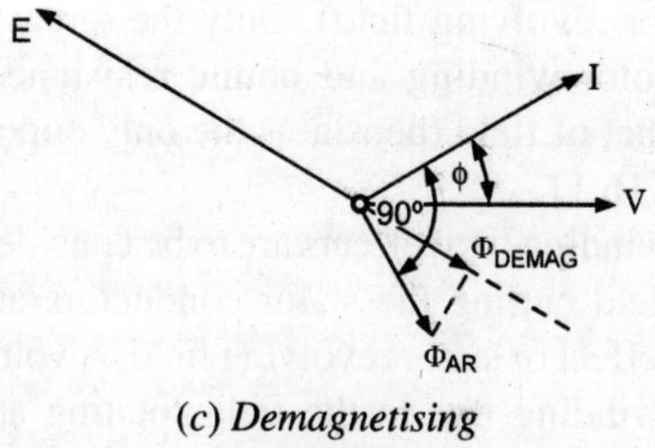

(*c*) *Demagnetising*

Fig. 16.8

power factor, the armature current now sets up a flux tending to reduce the flux created by the main field. In any case, the motor tends to operate at a nearly constant flux or generated voltage, as does the transformer.

16.6 EFFECTS OF VARYING EXCITATION ON ARMATURE CURRENT AND POWER FACTOR

The change in field excitation neither affects the speed of the motor nor the output of the motor but does affect the power factor and consequently armature current for constant supply voltage and constant input power.

The operation of a synchronous motor at a constant supply voltage demands nearly constant flux, as mentioned in the foregoing section. Both the dc source and ac source cooperate in establishing this resultant flux. When the dc excitation is weak, the flux produced by it is weak and needs stengthening, so the stator draws the current lagging behind the supply voltage by 90° so as to magnetise the air gap to the extent needed to provide the necessary resultant flux. The magnetising component of current being a large part of the total input current results in low lagging power factor. For constant applied voltage and constant input power, the active component of current drawn from the supply mains should be constant and so because of large wattless (magnetising) current; armature current I is much more. With the increase in excitation to 100%, the magnetising current drawn from the 3-phase ac supply mains is reduced *i.e.* the magnetising component becomes a smaller part of the total input current and so the power factor improves. This results in reduction of armature current I drawn from 3-ϕ ac supply mains. If the excitation is further increased such that power factor becomes unity, the stator draws only energy or active current, and the dc field circuit supplies all the current necessary to magnetize the rotor field. For unity power factor, the armature current drawn will be minimum. On increasing the dc excitation further, the field produced by it becomes stronger than it is required and needs demagnetization. So the motor draws the leading current so as to demagnetise the main field. This makes the power factor leading and lesser than unity and so the armature current increases.

The effect of variation in field excitation on armature current and power factor can also be explained as below.

Consider a synchronous motor loaded with a constant mechanical load and operating at a constant supply voltage. When the excitation is 100% *i.e.* when E = V, armature current I lags behind the resultant voltage E_R (phasor sum of V and E) by a fixed angle θ, given as $\theta = \text{Tan}^{-1}\dfrac{X_s}{R_e}$ and lags behind the applied voltage by an angle ϕ, power factor angle [Fig. 16.9 (*a*)] where X_s is the synchronous reactance per phase, R_e is the effective resistance of armature per phase, θ is the *internal angle* and δ is the load or torque angle. The power drawn per phase from the 3-ϕ ac supply mains is VI cos ϕ. Since applied voltage and input power or load is constant so the active component of the current drawn from the supply mains, I cos ϕ remains constant for constant mechanical load (neglecting internal losses). As the speed of the motor remains constant so increase in dc excitation causes increase in excitation voltage E and decrease in excitation causes reduction in excitation voltage E.

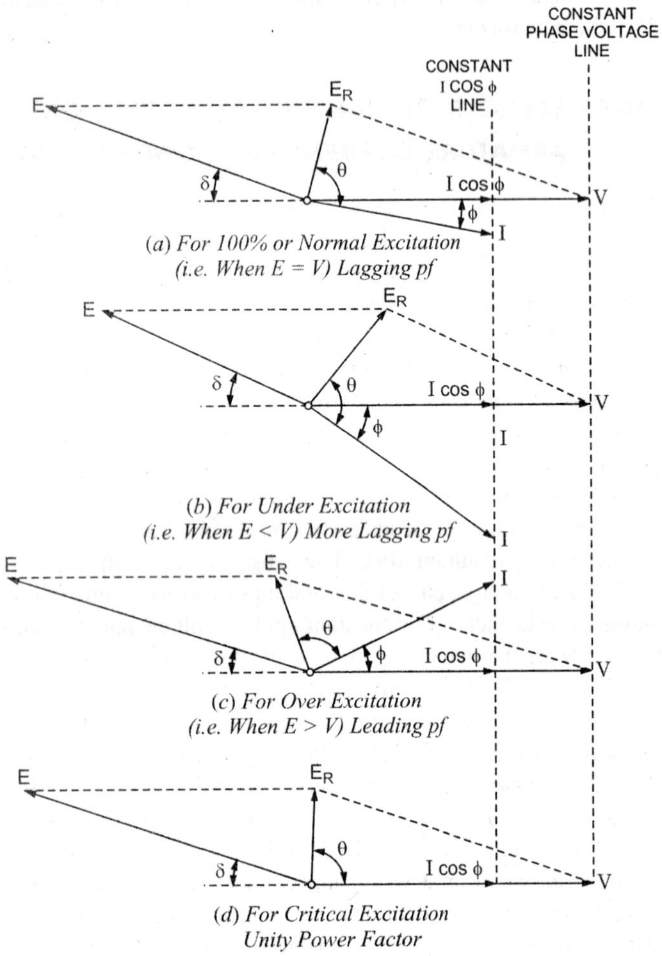

(*a*) *For 100% or Normal Excitation (i.e. When E = V) Lagging pf*

(*b*) *For Under Excitation (i.e. When E < V) More Lagging pf*

(*c*) *For Over Excitation (i.e. When E > V) Leading pf*

(*d*) *For Critical Excitation Unity Power Factor*

Fig. 16.9

When the excitation is reduced the excitation voltage E decreases in magnitude to the value shown in Fig. 16.9 (*b*) which shifts the resultant voltage E_R in clockwise direction and since the phase angle θ between the resultant voltage E_R and current I, being equal to $\tan^{-1} \frac{X_s}{R_e}$, is fixed, so the armature current phasor is also shifted in clockwise direction. Thus power factor angle ϕ increases, power factor decreases and armature current increases because I cos ϕ is constant. *Hence with reduction in dc excitation synchronous motor draws more current from the supply mains at lower lagging power factor.*

When the excitation is increased, the excitation voltage E increases in magnitude to a value shown in Fig. 16.9 (*c*) which shifts the resultant voltage E_R in counter-clockwise direction. This results in change in position of phasor of current I with respect to terminal voltage so that the motor then operates at a leading power factor. For a certain value of excitation phase angle ϕ between voltage V and current I is zero *i.e.* the current I will be in phase with voltage V and at this instant the value of current drawn from supply mains, I will be minimum, as shown in Fig. 16.9 (*d*). Now if the excitation is further increased, it draws current with leading power factor, as shown in Fig. 16.9 (*c*).

Hence when the excitation of a synchronous motor is increased above 100 per cent, first power factor improves until it becomes unity; at this instant the current drawn from the supply mains is minimum and is in phase with supply voltage. With further increase in field current the power factor becomes leading one and decreases and current drawn from supply main increases. The details are given in Table 16.1.

Table 16.1 Comparison of Various Excitations

Type of Excitation	*Comparison of E and V*	*Nature of Power Factor*	*Armature Current I*
Normal Excitation	E = V	Lagging	Increased I
Under Excitation	E < V	Lagging	Increased I
Over Excitation	E > V	Leading	Increased I
Critical Excitation	E > V	Unity	Minimum I

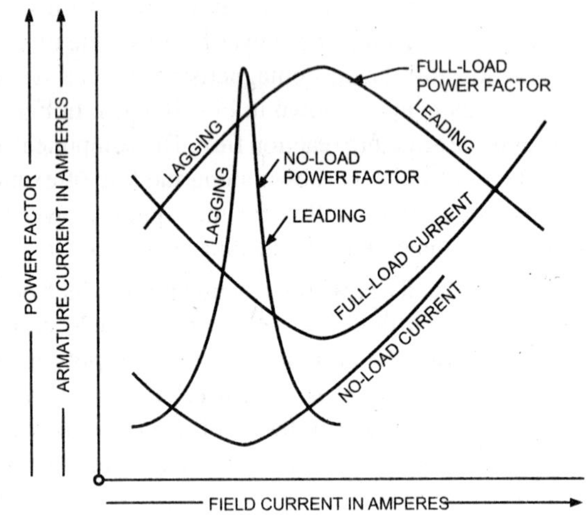

Fig. 16.10

The variation of current and power factor of a synchronous motor with a variation of field current (excitation) and for a constant load is shown in Fig. 16.10. Because of their shape, excitation-armature current curves are called the *V-curves* of a synchronous motor.

16.7 EQUIVALENT CIRCUIT

The synchronous motor, as already mentioned in Art. 16.1, is fed from two sources of supply; field structure from dc source and armature winding from a 3-phase ac supply source.

Under normal operating conditions, no emf is induced in the rotor due to stator field because the rotor revolves at the same speed as the stator field (*i.e.* there is no relative motion between rotor and stator revolving field). Only the direct current flows through the rotor winding and ohmic resistance of the field winding and that of field rheostat is the only opposition to it, as shown in Fig. 16.11.

In stator winding, two effects are to be considered; the effect of the rotor field cutting the stator conductors at synchronous speed and the effect of stator revolving field. A voltage is induced in the stator winding due to the field rotating at synchronous speed. This induced emf E is known as *excitation voltage* or

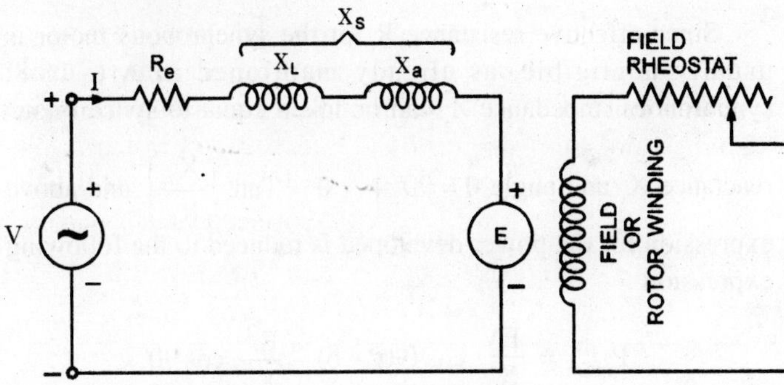

Fig. 16.11

back emf or *counter emf* and opposes the applied voltage, V, to the stator. The magnitude of this induced emf depends upon the rotor flux per pole or strength of excitation current I_{dc} because the motor operates at a constant speed.

The effect of stator revolving field on the stator (or armature) conductors is accounted for by including an inductive reactance in the armature (or stator) winding. This is called the *synchronous reactance* X_s. An effective resistance R_e is considered to be in series with this reactance to account for copper loss in the armature winding and stray power loss. This resistance R_e when combined with synchronous reactance X_s gives the *synchronous impedance* Z_s of the machine.

The equivalent circuit for one phase of a star-connected synchronous motor is shown in Fig. 16.11. From equivalent circuit diagram shown in Fig. 16.11 we have

Excitation voltage per phase(reversed) *i.e.*

$$-\mathbf{E} = \mathbf{V} - \mathbf{I}\,\mathbf{Z}_s$$

or armature current per phase,

$$\mathbf{I} = \frac{\mathbf{V}+\mathbf{E}}{\mathbf{Z}_s} = \frac{\mathbf{E}_R}{\mathbf{Z}_s} \quad ...(16.1)$$

where $\mathbf{Z}_s = R_e + j\,X_s$ and is equal to $\sqrt{R_e^2 + X_s^2}$ in magnitude.

16.8 PHASOR DIAGRAM

The phasor diagram of a synchronous motor is similar to that of an alternator in many respects. The effect of armature reaction can be substituted by an equivalent reactance X_a which when combined with the leakage reactance of the armature gives the synchronous reactance X_s, as illustrated in equivalent circuit diagrams (Fig. 16.11). The synchronous reactance X_s combined with armature effective resistance* R_e gives the synchronous impedance, Z_s.

The phasor diagram of a synchronous motor differs from that of a synchronous generator in two respects. First one is the voltage equation for the stator circuit. In case of synchronous generator the excitation voltage E plays the role of a source voltage and is equal to the phasor sum of terminal voltage V and the synchronous impedance drop $\mathbf{I}\,\mathbf{Z}_s$. But in the synchronous motor supply voltage V is the source voltage applied to the stator winding of the synchronous motor and E is a reaction or counter emf which is internally generated. The supply voltage V has to overcome the synchronous impedance drop $\mathbf{I}\,\mathbf{Z}_s$ and counter emf E. Therefore, the voltage equation for a synchronous motor is given as

$$\mathbf{V} = -\mathbf{E} + \mathbf{I}\,\mathbf{Z}_s$$

where V is the applied voltage per phase, E is the counter emf or excitation voltage per phase, I is the armature current per phase and Z_s is the synchronous impedance per phase.

The second point of view in which the phasor diagram of the synchronous motor differs from that of a synchronous generator involves the angle δ, called *power* or *torque angle*. In a synchronous generator excitation voltage E is ahead of terminal voltage V in time because of the driving action of the prime mover. The angle δ represents the displacement between the centre lines of the stator and rotor poles and the centre line of the rotor poles is ahead of the centre line of the stator poles by an electrical angle δ in case of a synchronous generator. But in case of a synchronous motor the situation is reverse of that in a synchronous generator *i.e.* the rotor falls behind the stator and this is because of load on the motor shaft. Thus in a synchronous motor the phasor of applied voltage V leads the phasor of excitation voltage E.

On the basis of the modifications mentioned above the phasor diagrams for a synchronous motor with armature current (*a*) lagging behind (*b*) in phase with and (*c*) leading the applied voltage are shown in Figs. 16.12 (*a*), (*b*) and (*c*) respectively. In

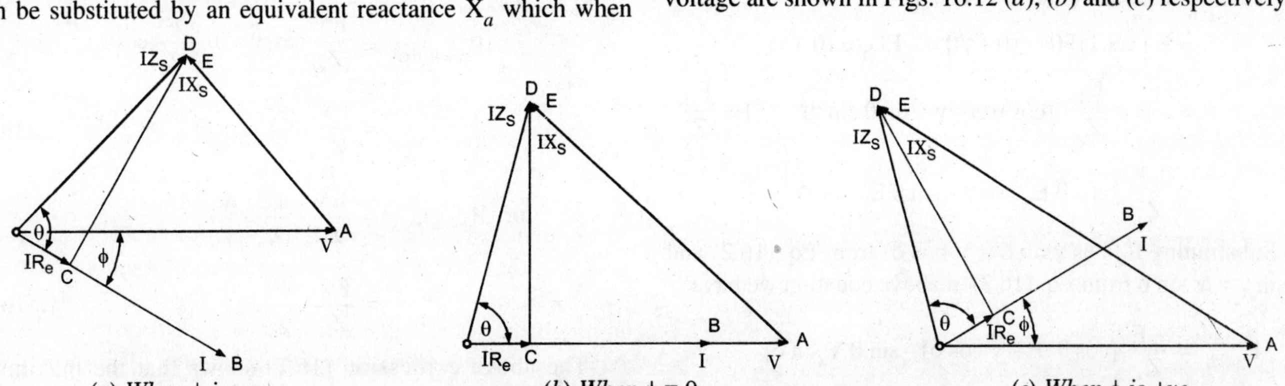

(*a*) When ϕ is –ve (*b*) When ϕ = 0 (*c*) When ϕ is +ve

Fig. 16.12

* In synchronous machines the armature effective resistance is usually very small in comparision to synchronous reactance $\left(R_e \text{ is usually } \frac{1}{100} \text{ to } \frac{1}{10}\text{th of } X_s\right)$

Fig. 16.12. phasors OA, OB, OC, CD, OD and AD represent the applied voltage V, load current I, resistive drop IR_e, reactive drop IX_s, resultant voltage across the stator winding $\mathbf{E_R = I Z_s}$ and excitation voltage E (all phase values) respectively.

16.9 POWER DEVELOPED IN A SYNCHRONOUS MOTOR

Let us consider a 3-phase synchronous motor having effective resistance R_e ohms/phase and synchronous reactance X_s ohms/phase. Let the motor be connected across 3-phase ac supply having voltage per phase equal to V volts and have induced emf (or excitation voltage) per phase of E volts and load angle δ electrical degrees.

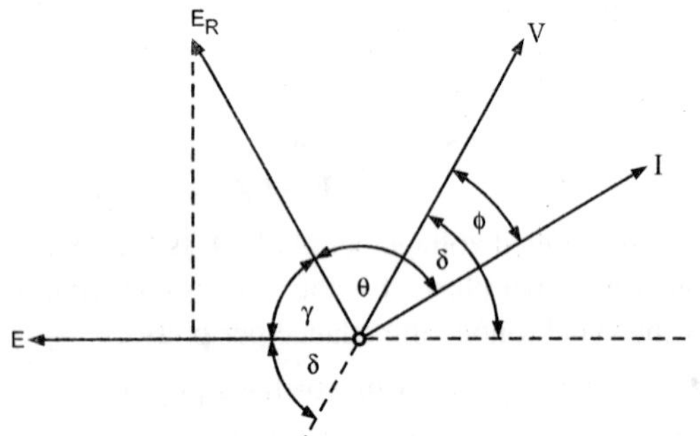

Fig. 16.13

Taking induced voltage E horizontally, resolving V and E_R horizontally and vertically and equating the algebraic sum of their horizontal and vertical components separately with the horizontal and vertical components of resultant voltage E_R respectively we have

$$E_R \cos\gamma = E - V\cos\delta \quad ...(16.2)$$

$$\text{and } E_R \sin\gamma = V\sin\delta \quad ...(16.3)$$

Mechanical power developed per phase,

P_{mech} = EMF induced/phase × armature current per phase × cosine of the angle between the two (*i.e.* angle between I and E reversed)

$$= E\,I\cos[180° - (\theta+\gamma)] = -EI\cos(\theta+\gamma)$$

$$= -E\times\frac{E_R}{Z_s}(\cos\theta\cos\gamma - \sin\theta\sin\gamma) \quad \because I = \frac{E_R}{Z_s}$$

$$= \frac{-E}{Z_s}(\cos\theta\, E_R\cos\gamma - \sin\theta\, E_R\sin\gamma)$$

Substituting $E_R \cos\gamma = (E - V\cos\delta)$ from Eq. (16.2) and $E_R \sin\gamma = V\sin\delta$ from Eq. (16.3) in above equation we have

$$P_{mech} = \frac{-E}{Z_s}[\cos\theta(E - V\cos\delta) - \sin\theta\, V\sin\delta]$$

$$= \frac{EV}{Z_s}(\cos\delta\cos\theta + \sin\theta\sin\delta) - \frac{E^2}{Z_s}\cos\theta$$

$$= \frac{EV}{Z_s}\cos(\theta-\delta) - \frac{E^2}{Z_s}\cos\theta \quad ...(16.4)$$

Since effective resistance R_e of the synchronous motor is usually negligible, as already mentioned in Art. 16.8, synchronous impedance Z_s can be taken equal to synchronous reactance X_s and angle $\theta = 90° \left(\because \theta = \text{Tan}^{-1}\frac{X_s}{R_e}\right)$ and above expression for the power developed is reduced to the following expression

$$P_{mech} = \frac{EV}{X_s}\cos(90° - \delta) - \frac{E^2}{X_s}\cos 90°$$

$$= \frac{EV}{X_s}\sin\delta \quad ...(16.5)$$

16.9.1. Conditions For Maximum Power Developed. Equation (16.4) is for the mechanical power developed in terms of the load angle δ and internal angle θ of the motor for a constant supply voltage V and induced emf E (or excitation as the induced emf depends upon the excitation).

Now following two cases may be considered

(*i*) *Fixed V, E, and θ* $\left(\text{or } R_e \text{ and } X_s \text{ because } \theta = \text{Tan}^{-1}\frac{X_s}{R_e}\right)$.

Under these conditions the power developed P_{mech} will vary with the load angle δ and the maximum power developed can be determined by differentiating Eq. (16.4) with respect to load angle δ and then equating it to zero.

$$\therefore \frac{dP_{mech}}{d\delta} = \frac{EV}{Z_s}\sin(\theta-\delta) = 0$$

$$\text{or } \sin(\theta-\delta) = 0$$

$$\text{or } \delta = \theta = \text{Tan}^{-1}\frac{X_s}{R_e} \quad ...(16.6)$$

i.e. mechanical power developed for constant supply voltage and constant excitation will be maximum when load angle δ will be equal to internal angle θ (or equal to $\tan^{-1}\frac{X_s}{R_e}$) and the value of maximum power developed is determined by substituting δ = θ in Eq. (16.4) so we get

$$(P_{mech})_{max} = \frac{EV}{Z_s}\cos(\theta-\theta) - \frac{E^2}{Z_s}\cos\theta$$

$$= \frac{EV}{Z_s} - \frac{E^2}{Z_s}\cos\theta \quad ...(16.7)$$

$$\text{or } (P_{mech})_{max} = \frac{EV}{Z_s} - \frac{E^2}{Z_s}\times\frac{R_e}{Z_s}$$

$$= \frac{EV}{Z_s} - \frac{E^2}{Z_s^2}R_e \quad ...(16.8)$$

The above expression (16.7) shows that the maximum mechanical power developed or torque (since speed is constant) depends upon the applied voltage V, induced emf E (or excitation) and internal angle, θ. Maximum value of θ (and hence δ) is 90° and, therefore, maximum mechanical power developed

$$(P_{mech})_{max} = \frac{EV}{Z_s} \quad ...(16.9)$$

$(P_{mech})_{max}$ is the value of power at which the motor falls out of step.

Solving for E from Eq. (16.8), we get

Excitation voltage,

$$E = \frac{Z_s}{2R_e}\left[V \pm \sqrt{V^2 \; 4R_e (P_{mech})_{max}}\right] \quad ...(16.10)$$

The two values of E so obtained give the excitation limits for any load.

(ii) Fixed V and θ (or R_e and X_s). For determining the value of excitation voltage E to give maximum power developed, differentiate Eq. (16.8) with respect to E and equate to zero

$$\frac{d(P_{mech})_{max}}{dE} = \frac{V}{Z_s} - \frac{2E}{Z_s^2}R_e = 0$$

$$\text{or } E = \frac{VZ_s}{2R_e}$$

Substituting $E = \dfrac{VZ_s}{2R_e}$ in Eq. (16.8) we have

$$(P_{mech})_{max} = \frac{V^2}{2R_e} - \frac{V^2}{4R_e} = \frac{V^2}{4R_e} \quad ...(16.11)$$

The value of excitation voltage to develop maximum power, given by $\dfrac{VZ_s}{2R_e}$ would be very large, because effective resistance of the stator winding is very small. Usually it is physically impossible to achieve these maximum conditions because of saturation and practical limits on excitation. Thus this equation is of theoretical interest only because the assumptions on which it is based are not in sufficient agreement with the physical facts.

16.10 POWER FLOW IN A SYNCHRONOUS MOTOR

In motor operation, electrical energy from some source outside the machine is supplied to the armature circuit. Some of this supplied energy is utilised to overcome the armature copper loss $I^2 R_e$ and remainder of it is converted into mechanical energy through the electromagnetic motor action. Some of this energy which is converted into mechanical energy is lost through friction, windage, and hysteresis and eddy current phenomena. The remainder of the mechanical energy is available for useful work. The power flow diagram is shown in Fig. 16.14.

Power supplied to stator of the synchronous motor

$$= \sqrt{3}\, V_L I_L \cos\phi \text{ watts}$$

where V_L is the line voltage across the stator terminals, I_L is the line current supplied to the stator and cos ϕ is the power factor (cos of the phase angle between phase voltage V and phase current I).

Armature copper losses for 3-phases $= 3 I^2 R_e$ watts

Mechanical power developed $= \sqrt{3}\, V_L I_L \cos\phi - 3 I^2 R_e$

The mechanical power developed per phase also

= EMF induced/phase × armature current per phase × cosine of the angle between E and I

$= -EI \cos(\theta + \gamma)$ Refer to Fig. 16.13.

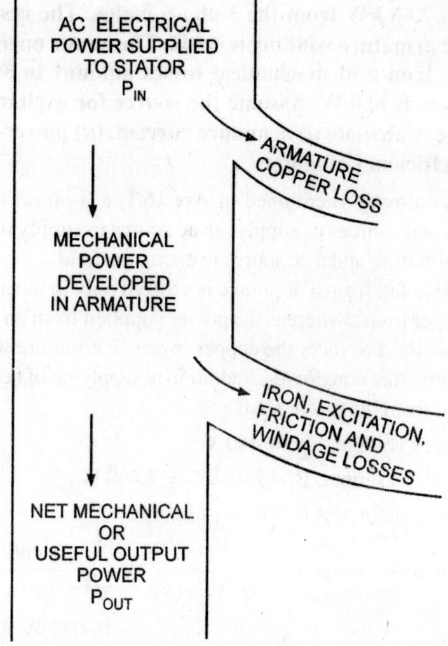

Fig. 16.14 *Power Flow Diagram For a Synchronous Motor*

16.11 DIFFERENT TORQUES OF A SYNCHRONOUS MOTOR

The torque required to operate the driven machine at every moment between the initial breakaway and the final shutdown is important in determination of the motor characteristics. The various torques associated with synchronous motors are termed starting torque, running torque, pull-in torque, and pull-out torque.

1. Starting Torque. It pertains to the ability of the motor to accelerate the load. The starting torque, sometimes also called the *breakaway torque*, required by the driven machine may be as low as 10%, as in case of centrifugal pumps, and as high as 200 or 250% of full-load torque, as in case of loaded reciprocating two-cylinder compressors.

The synchronous motor has got no self-starting torque, but in modern synchronous motors, almost any reasonable torque can be had by proper design of the damper windings (by changes in the resistance and size of the damper winding).

2. Running Torque. It is the torque developed by the motor under running conditions. It is determined by the output power and speed of the driven machine. The peak output power determines the maximum torque that would be required by the driven machine. The motor must have a breakdown or a maximum running torque greater than this value so as to avoid stalling of the machine.

3. Pull-in Torque. It refers to the ability of the motor to pull-into synchronism when changing from induction to synchronous motor operation.

4. Pull-out Torque. It pertains to the ability of the motor to remain in synchronism under rated load conditions.

The maximum torque which the motor will develop without pulling out of step (or synchronism) is called the *pull-out torque*. Its value varies from 1.25 to 3.5 times the full-load torque.

Example 16.1. A 3-phase, 440 V, 50 Hz star-connected synchronous motor takes 7.46 kW from the 3-phase mains. The resistance per phase of the armature winding is 0.5 Ω. The motor operates at a pf of 0.75 lag. Iron and mechanical losses amount to 500 W. The excitation loss is 650 W. Assume the source for excitation to be a separate one. Calculate (*i*) armature current, (*ii*) power supplied to motor, (*iii*) efficiency of motor.

Solution: As already mentioned in Art 16.1, a 3-phase synchronous motor needs two sources of supply; an ac source to supply the power for driving the armature and a dc source to excite the field.

The power fed from a dc source is utilized only in meeting the field winding copper losses whereas the power supplied from an ac source to armature is utilised to meet the copper losses in armature winding, and friction and armature core losses in addition to supplying of net mechanical power output from the motor shaft.

Supply voltage, V_L = 440 V

Power input, P = 7.46 kW = 7,460 W

Power factor, cos ϕ = 0.75 (lagging)

$$(i)\ \text{Armature current, } I = \frac{P}{\sqrt{3}\,V_L \cos\phi} = \frac{7,460}{\sqrt{3} \times 440 \times 0.75}$$

$$= 13.052 \text{ A} \quad \textbf{Ans.}$$

Armature winding copper loss

$$= 3 \times I^2 R_a = 3 \times (13.052)^2 \times 0.5 = 255.5 \text{ W}$$

(*ii*) Total power supplied to motor,

P_{in} = Power supplied from 3-phase mains + excitation loss

= 7,460 + 650 = 8,110 W **Ans.**

Iron and mechanical losses = 500 W

Output power = Power input to motor armature – armature winding copper loss – iron and friction losses

= 7,460 – 255.5 – 500 = 6,704.5 W

(*iii*) Efficiency of motor,

$$\eta = \frac{\text{Output power}}{\text{Input power}} \times 100 = \frac{6,704.5}{8,110} \times 100 = 82.67\% \quad \textbf{Ans.}$$

Example 16.2. A 2300 V, 3-phase synchronous motor driving a pump is provided with a line ammeter and a field rheostat. When the rheostat is adjusted such that the ac line current is minimum, the ammeter reads 8.8 A. What is the power being delivered to the pump, neglecting losses ? How should the rheostat be adjusted so that the motor operates at 0.8 leading power factor? How many kVARs is the motor supplying to the system at this new power-factor ? *[GATE, 2000]*

Solution: Line voltage, V_L = 2,300 V

Line current, I_L = 8.8 A

Power factor, cos ϕ = 1.0

because line current will be minimum only at unity power factor.

Power delivered to pump (neglecting losses)

$$= \sqrt{3}\, V_L I_L \cos\phi = \sqrt{3} \times 2,300 \times 8.8 \times 1.0$$

= 35,057 Watts or 35.057 kW

For operating motor at 0.8 leading power factor, excitation is to be increased by reducing field rheostat resistance. With change in excitation, active power remains same and, therefore,

Armature current = Line current, I'_L

$$= \frac{I_L}{\text{New pf}} = \frac{8.8}{0.8} = 11 \text{ A}$$

$$\text{KVAR supplied} = \frac{\sqrt{3} \times V_L \times I'_L \times \sin\phi}{1,000}$$

$$= \frac{\sqrt{3} \times 2,300 \times 11 \times 0.6}{1,000} = 26.29 \quad \textbf{Ans.}$$

$$\because \sin\phi = \sqrt{1 - \cos^2\phi} = \sqrt{1 - 0.8^2} = 0.6$$

Example 16.3. A 50 kW synchronous motor is tested by driving it by another motor. When the excitation is not switched on, the driving motor takes 800 W. When the armature is short circuited and the rated armature current of 10 A is passed through it, the driving motor requires 2,500 W. On open circuiting the armature with rated excitation, the driving motor takes 1,800 W. Calculate the efficiency of the synchronous motor at 50% load. Neglect the losses in the driving motor. *[GATE, 2001]*

Solution: Power input to a synchronous machine without excitation gives friction and windage loss (refer to Art. 15.26.1), therefore,

Friction and windage loss, W_f = 800 W

When the armature winding of a synchronous machine is short circuited, and the machine is operated at rated speed with reduced excitation (adjusted to give rated line current), the power input equals the armature winding copper loss plus stray loss in addition to friction and windage losses. So

Armature winding loss plus stray loss

W_1 = Short-circuit loss - friction and windage loss

= 2,500 – 800 = 1,700 W

When the synchronous machine is driven at rated speed with rated excitation but with armature open circuited, the power input equals the open-circuit core loss plus friction and windage losses. So,

Open-circuit core loss, W_2 = 1,800 – 800 = 1,000 W

Total loss = $W_f + W_1 + W_2$ = 800 + 1,700 + 1,000

= 3,500 W or 3.5 kW

At 50% load

Motor output, P_{out} = 0.5 × 50 = 25 kW

$$\text{Total losses} = \left(\frac{1}{2}\right)^2 W_1 + W_2 + W_f$$

$$= \left(\frac{1}{2}\right)^2 \times 1,700 + 1,000 + 800$$

= 2,225 W or 2.225 kW

Motor input, P_{in} = P_{out} + total losses

= 25 + 2.225 = 27.225 kW

Efficiency of synchronous motor,

$$\eta = \frac{P_{out}}{P_{in}} \times 100 = \frac{25}{27.225} \times 100 = 91.83\% \quad \textbf{Ans.}$$

Example 16.4. A 3,300 V delta-connected motor has synchronous reactance per phase (delta) of 18 Ω. It operates at a leading pf of 0.707 when drawing 800 kW from the mains. Calculate its excitation emf. *[Pb. Technical Univ. Synchronous Machines, May-2013; U.P. Technical Univ. Electrical Machines, 2010-11]*

Solution: Phase voltage, $V_P = V_L$ = 3,300 V

Impedance per phase, Z_s = 18 Ω

Power factor, cos ϕ = 0.707 (leading)

Phase angle, $\phi = \text{Cos}^{-1}\, 0.707 = 45°$ (leading)

Internal angle, $\theta = 90°$ $\because$ armature resistance is negligible.

$\text{Cos}(\phi + \theta) = \text{Cos}(45° + 90°) = -0.707$

$$\text{Phase current, } I_P = \frac{\text{Power input in watts}}{3 V_P \cos\phi}$$

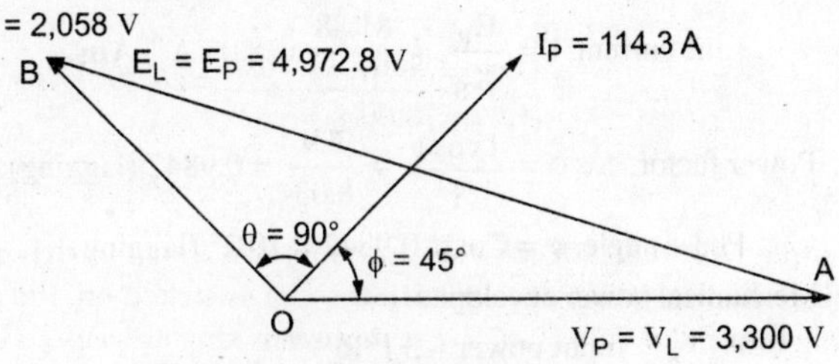

Fig. 16.15 *Phasor Diagram*

$$= \frac{800 \times 1{,}000}{3 \times 3{,}300 \times 0.707} = 114.3 \text{ A}$$

$$E_R = I_P Z_s = 114.3 \times 18 = 2{,}058 \text{ V}$$

From phasor diagram shown in Fig. 16.15 we have

Excitation emf,

$$E_L = E_P = \sqrt{V_P^2 + E_R^2 - 2V_P E_R \cos(\phi + \theta)}$$

$$= \sqrt{(3{,}300)^2 + (2{,}058)^2 - 2 \times 3{,}300 \times 2{,}058 \times (-0.707)}$$

$= 4{,}972.8$ V **Ans.**

Example 16.5. A 3-phase, 400 V, 50 Hz, 37.3 kW, star-connected synchronous motor has a full-load efficiency of 88%. The synchronous impedance of motor is (0.2 + *j*1.6) Ω/phase. If the excitation of the motor is adjusted to give a leading pf of 0.9, calculate for full load (*i*) the induced emf (*ii*) total mechanical power developed.

[*J.N. Technological Univ. Electrical Machines-III, May-2013*]

Solution: Supply voltage per phase, $V = \frac{V_L}{\sqrt{3}} = \frac{400}{\sqrt{3}} = 230.94$ V

$$\text{Power input, } P_{in} = \frac{\text{Output power, } P_{out}}{\eta} = \frac{37.3}{0.88} = 42.386 \text{ kW}$$

$$\text{Full-load current, } I = \frac{P_{in} \text{ in watts}}{\sqrt{3} \times V_L \times \cos\phi} = \frac{42.386 \times 1{,}000}{\sqrt{3} \times 400 \times 0.9} = 67.98 \text{ A}$$

$$\text{Impedance per phase, } Z_s = \sqrt{0.2^2 + 1.6^2} = 1.6125\ \Omega$$

$$\text{Impedance drop per phase, } E_R = IZ_s = 67.98 \times 1.6125 = 109.6 \text{ V}$$

$$\text{Phase angle, } \phi = \text{Cos}^{-1}\ 0.9 = 25.84°$$

$$\text{Internal angle, } \theta = \text{Tan}^{-1}\frac{X_s}{R_s} = \text{Tan}^{-1}\frac{1.6}{0.2} = 82.87°$$

$$(\theta + \phi) = 82.87° + 25.84° = 108.71°$$

$$\text{Cos}(\theta + \phi) = \text{Cos } 108.71° = -0.321$$

From phasor diagram shown in Fig. 16.16, we have

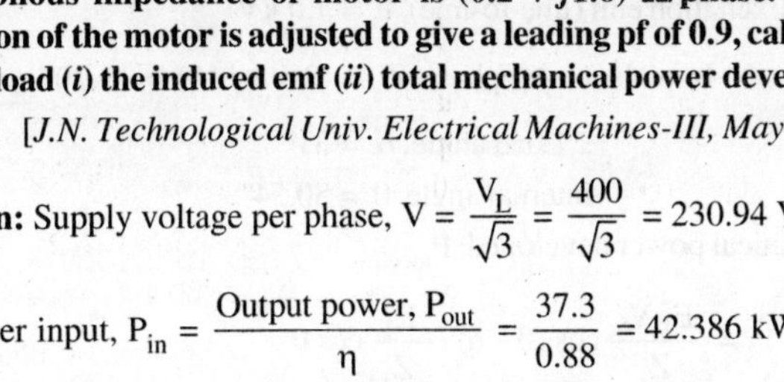

Fig. 16.16 *Phasor Diagram*

Excitation emf per phase,

$$E = \sqrt{V^2 + E_R^2 - 2VE_R \cos(\theta + \phi)}$$

$$= \sqrt{230.94^2 + 109.6^2 - 2 \times 230.94 \times 109.6 \times (-0.321)}$$

$= 285.79$ V

(*i*) Excitation emf (line value) $= \sqrt{3} \times 285.79 = 495$ V **Ans.**

$$\text{Total copper loss} = 3 \times I^2R_a = 3 \times (67.98)^2 \times 0.2 = 2{,}773 \text{ W or } 2.773 \text{ kW}$$

(*ii*) Total mechanical power developed,

$$P_{mech} = \text{Power input, } P_{in} - \text{total copper loss}$$

$= 42.386 - 2.773 = 39.613$ kW **Ans.**

Example 16.6. The full-load current of a 3.3 kV, star-connected synchronous motor is 160 A at 0.8 pf lagging. The resistance and synchronous reactance of the motor are 0.8 Ω and 5.5 Ω per phase respectively. Calculate the excitation emf, torque angle, efficiency and shaft output of the motor. Assume the mechanical stray load loss to be 30 kW.

[*U.P. Technical Univ. Electromechanical Energy Conversion-II, 2009-10*]

Solution: Full-load current, $I = 160\ \angle -36.87°$ A

Synchronous impedance per phase,

$$\mathbf{Z}_s = (0.8 + j\,5.5) = 5.558\ \angle 81.72°\ \Omega$$

Excitation emf per phase = Terminal voltage per phase, $V - I\mathbf{Z}_s$

$$= \frac{3{,}300}{\sqrt{3}} - 160\ \angle -36.87° \times 5.558\ \angle 81.72° \text{ V}$$

$$= 1{,}905.26 - 889.28\ \angle 44.85° \text{ V}$$

$$= 1{,}905.26 - (630.46 + j627.1675) \text{ V}$$

$$= (1{,}274.8 - j627.1675) = 1{,}420.8\ \angle -26.2° \text{ V}$$

Excitation emf (line-to-line)

$= \sqrt{3} \times 1{,}420.8 = 2{,}461$ V or 2.461 kV **Ans.**

Torque angle,

$\delta = -26.2°$ **Ans.**

Mechanical power developed,

$$P_{mech} = 3 \times 1{,}420.8 \times 160 \cos(\phi - \delta)$$

$$= 3 \times 1{,}420.8 \times 160 \cos[-36.87° - (-26.2°)] \times 10^{-3} \text{ kW}$$

$$= 670.2 \text{ kW}$$

Mechanical power developed can also be computed from the relation

$$P_{mech} = \text{Power supplied to motor, P} - \text{total copper losses}$$

$$= \left[\sqrt{3}\ V_L I_L \cos\phi - 3I^2R_e\right] \times 10^{-3} \text{ kW}$$

$$= \left[\sqrt{3} \times 3{,}300 \times 160 \times 0.8 - 3 \times 160^2 \times 0.8\right] \times 10^{-3} \text{ kW}$$

$= 670.2$ kW, the same as computed before

Shaft output,

$P_{out} = P_{mech}$ – stray load loss $= 670.2 - 30 = 640.2$ kW **Ans.**

Power input,

$$P_{in} = \sqrt{3}\ V_L I_L \cos\phi = \sqrt{3} \times 3{,}300 \times 160 \times 0.8 \times 10^{-3}$$

$$= 731.62 \text{ kW}$$

Motor efficiency,

$$\eta = \frac{P_{out}}{P_{in}} \times 100 = \frac{640.2}{731.62} \times 100 = 87.5\% \textbf{ Ans.}$$

Example 16.7. A 500 V, 1-ϕ synchronous motor gives a net output mechanical power of 7.46 kW and operates at 0.9 pf lagging. Its effective resistance is 0.8 Ω. If the iron and mechanical losses are 500 W and excitation losses are 800 W, estimate the armature current. Calculate the commercial efficiency.

Solution : Let the input armature current be I amperes

$$\text{Power input} = VI\cos\phi = 500 \times I \times 0.9 = 450\text{ I watts}$$

$$\text{Power output} = \text{Power input} - \text{mechanical losses} - \text{copper losses}$$

$$\text{or } 7{,}460 = 450\,I - 500 - I^2 \times 0.8$$

$$\text{or } 0.8\,I^2 - 450\,I + 7{,}960 = 0$$

$$\text{or } I = \frac{450 \pm \sqrt{(450)^2 - 4\times 0.8\times 7{,}960}}{2\times 0.8} = 18.3 \text{ A } \textbf{Ans.}$$

Power input inlcuding excitation losses

$= (500 \times 18.3 \times 0.9 + 800)$ watts $= 9{,}035$ W or 9.035 kW

Commercial efficiency,

$$\eta = \frac{\text{Output}}{\text{Input}} \times 100 = \frac{7.46}{9.035} \times 100 = 82.57\% \textbf{ Ans.}$$

Example 16.8. A 3-phase star-connected 400 V synchronous motor takes a power input of 5,472 W at rated voltage. Its synchronous reactance is 10.2 Ω per phase and resistance is negligible. If its excitation voltage is adjusted equal to rated voltage of 400 V. Compute the load angle, power factor and armature current.

[Rajasthan Technical Univ. Electrical Machines-II, 2012]

Solution: Supply voltage per phase, $V_1 = \frac{400}{\sqrt{3}} = 230.94$ V

Induced emf per phase, E = V = 230.94 V

Reactance per phase, $X_s = 10.2\ \Omega$

Since induced emf is not greater than supply voltage, the motor must operate with lagging power factor

$$\text{Since power input} = \sqrt{3}\, V_L\, I_L \cos\phi = \sqrt{3}\, V_L\, I \cos\phi$$

∵ in star-connected synchronous motor, phase current, I = Line current, I_L

$$\text{or } I\cos\phi = \frac{\text{Power input}}{\sqrt{3}\,V_L} = \frac{5{,}472}{\sqrt{3}\times 400} = 7.9 \text{ A}$$

Since resistance is negligible, synchronous impedance is equal to synchronous reactance, i.e., 10.2 Ω and internal angle, θ = 90°.

Impedance drop per phase, $E_R = I\,X_S = 10.2\,I$

In ΔABC of phasor diagram shown in Fig. 16.17 we have

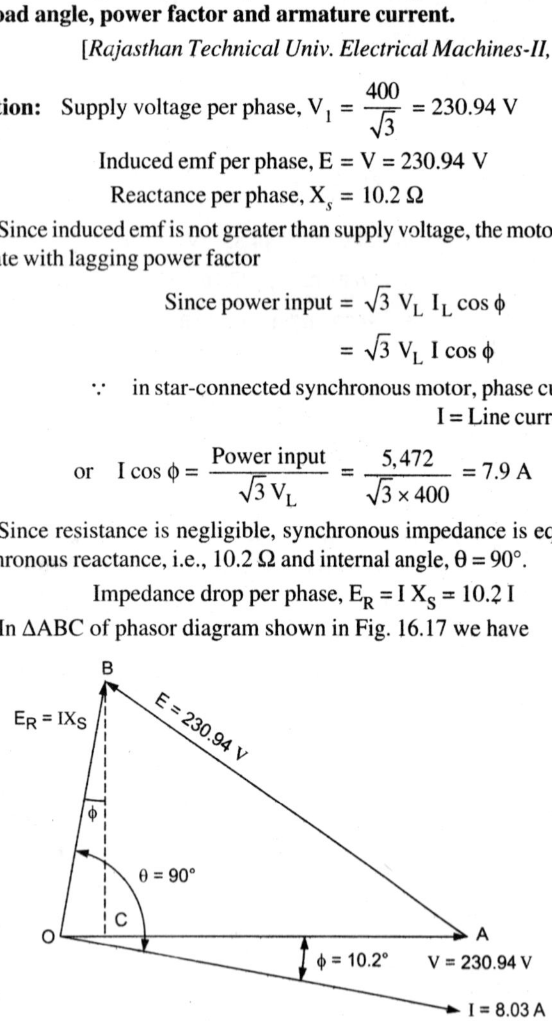

Fig. 16.17 *Phasor Diagram*

$$AB^2 = BC^2 + AC^2$$

$$\text{or } E^2 = (E_R\cos\phi)^2 + AC^2$$

$$\text{or } AC = \sqrt{E^2 - (E_R\cos\phi)^2}$$

$$= \sqrt{(230.94)^2 - (10.2\,I\cos\phi)^2}$$

$$= \sqrt{(230.94)^2 - (10.2\times 7.9)^2} = 216.43 \text{ V}$$

$$OC = AO - AC = 230.94 - 216.43 = 14.51 \text{ V}$$

$$\text{Now } E_R = \sqrt{BC^2 + OC^2}$$

$$= \sqrt{(10.2\times 7.9)^2 + (14.51)^2} = 81.88 \text{ V}$$

$$\text{Line current, } I = \frac{E_R}{X_S} = \frac{81.88}{10.2} = 8.03 \text{ A } \textbf{Ans.}$$

$$\text{Power factor, } \cos\phi = \frac{I\cos\phi}{I} = \frac{7.9}{8.03} = 0.9842 \text{ (lagging) } \textbf{Ans.}$$

Phase angle, $\phi = \text{Cos}^{-1}\, 0.9842 = 10.2°$ (lagging)

Mechanical power developed,

$$P_{mech} = \text{Input power} - 3\,I^2 R_e$$

$= 5{,}472$ ∵ armature resistance is negligible

$$\text{Load angle, } \delta = \text{Sin}^{-1}\frac{P_{mech}\times X_s}{V_L\,E_L} \qquad \because P_{mech} = \frac{E_L V_L}{X_s}\sin\delta$$

$$= \text{Sin}^{-1}\frac{5{,}472\times 10.2}{400\times 400} = 20.4° \textbf{ Ans.}$$

Example 16.9. A 3-phase star-connected non-salient pole synchronous motor connected to a 6.6 kV mains has an armature impedance of (2.5 + j 15) Ω/phase. The excitation of the machine gives a generated emf of 7.0 kV. The iron and friction losses amount to 10 kW. Determine the output of the motor when operating at a load angle of 31° (electrical).

[J.N. Technological Univ. Electrical Machines-III, May-2013]

Solution: Supply voltage (line-to-line), $V_L = 6.6$ kV

Excitation emf (line-to-line), E = 7.0 kV

Impedance per phase, $\mathbf{Z}_s = (2.5 + j\,15) = 15.207\ \angle 80.54°$

Load angle, α = 31°

Internal angle, θ = 80.54°

Mechanical power developed, P_{mech}

$$= \frac{E_L V_L}{Z_s}\cos(\theta - \delta) - \frac{E_L^2}{Z_s}\cos\theta$$

$$= \frac{7{,}000\times 6{,}600}{15.207}\cos(80.54° - 31°) - \frac{7{,}000^2}{15.207}\cos 80.54°$$

= 1,971,458.3 − 529,597.6 = 1,441,860.6 W or 1,441.86 kW

Motor output = Mechanical power developed − iron and friction losses

= 1,441.86 − 10 = 1,431.86 kW **Ans.**

Example 16.10. A 2,300 V, three-phase, 60 Hz, star-connected cylindrical rotor synchronous motor has a synchronous reactance of 11 Ω/phase. When it delivers 200 hp the efficiency is found to be 90% excluding the field loss. The power angle is 15 electrical degrees. Neglecting the ohmic resistance, determine (*i*) the induced excitation voltage per phase and the line current.

[G.G.S.I.P. Univ. Electromechanical Energy Conversion-II, May-2011]

Solution: Output power, $P_{out} = 200 \times 735.5 = 147{,}100$ W

$$\text{Power input to motor, } P_{in} = \frac{P_{out}}{\eta} = \frac{147{,}100}{0.9} = 163{,}444 \text{ W}$$

Since ohmic resistance is negligible, mechanical power developed,

$$P_{mech} = P_{in} = 163{,}444 \text{ W}$$

$$\text{Also } P_{mech} = \frac{E_L V_L}{Z_s}\cos(\theta - \delta) - \frac{E_L^2}{Z_s}\cos\theta$$

$$= \frac{E_L V_L}{Z_s}\cos(90° - \delta) - \frac{E_L^2}{Z_s}\cos 90°$$

∵ θ = 90°, R_e being negligible.

$$= \frac{E_L V_L}{Z_s}\sin\delta$$

or Induced excitation voltage (line value)

$$E_L = \frac{P_{mech} \times Z_s}{V_L \sin \delta} = \frac{163,444 \times 11}{2,300 \times \sin 15°} = 3,020 \text{ V}$$

(*i*) Induced excitation voltage per phase,

$$E = \frac{E_L}{\sqrt{3}} = 1,743.72 \textbf{ Ans.}$$

Supply voltage per phase,

$$V = \frac{V_L}{\sqrt{3}} = \frac{2,300}{\sqrt{3}} = 1,327.9 \text{ V}$$

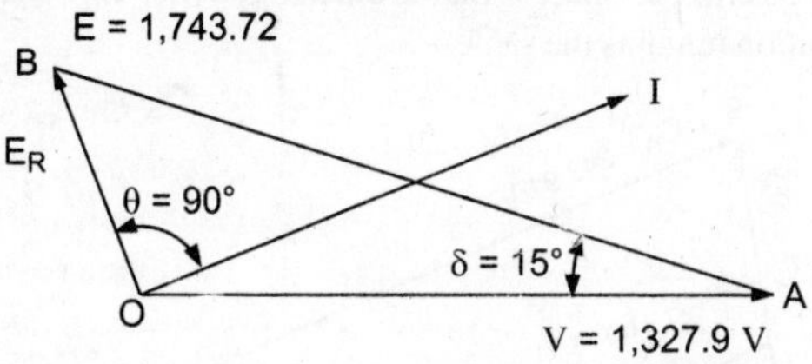

Fig. 16.18

From phasor diagram shown in Fig. 16.18, we have

$$E_R = \sqrt{V^2 + E^2 - 2VE \cos \delta}$$

$$= \sqrt{1327.9^2 + 1,743.72^2 - 2 \times 1,327.9 \times 1,743.72 \times \cos 15°}$$

$$= 575 \text{ V}$$

Line current,

$$I_L = \text{Phase current, } I = \frac{E_R}{Z} = \frac{575}{11} = 52.3 \text{ A}$$

Example 16.11. A 3-phase, 415 V, 6-pole, 50 Hz star-connected synchronous motor has emf of 520 V (L – L). The stator winding has a synchronous reactance of 2 Ω per phase, and the motor develops a torque of 220 Nm. The motor is operating at 415 V, 50 Hz bus. (*a*) Calculate the current drawn from the supply and its power factor (*b*) draw the phasor diagram showing all the relevant quantities.

[GATE 1992]

Solution: Supply voltage per phase, $V = \frac{415}{\sqrt{3}} = 239.6$ V

Induced emf per phase, $E = \frac{520}{\sqrt{3}} = 300$ V

Synchronous speed of motor,

$$N_s = \frac{120 f}{P} = \frac{120 \times 50}{6} = 1,000 \text{ rpm}$$

Torque developed, T = 220 Nm

Total power developed,

$$3\, P_{mech} = \frac{T \times 2\pi N_s}{60} = \frac{220 \times 2\pi \times 1,000}{60} = 23,038 \text{ watts}$$

Power developed per phase

$$= \frac{23,038}{3} = 7,679 \text{ watts}$$

Synchronous impedance/phase,

$$Z_s = 2 \angle 90° \ \Omega$$

Power developed per phase is given as

$$P_{mech} = \frac{EV}{Z_s} \cos(\theta - \delta) - \frac{E^2}{Z_s} \cos \theta$$

$$= \frac{EV}{Z_s} \sin \delta \qquad \theta = 90°$$

So $\frac{EV}{Z_s} \sin \delta = 7,679$

$$\text{or } \sin \delta = \frac{7,679 \times Z_s}{EV} = \frac{7,679 \times 2}{300 \times 239.6} = 0.2135$$

$$\text{or } \delta = \sin^{-1} 0.2135 = 12.33°$$

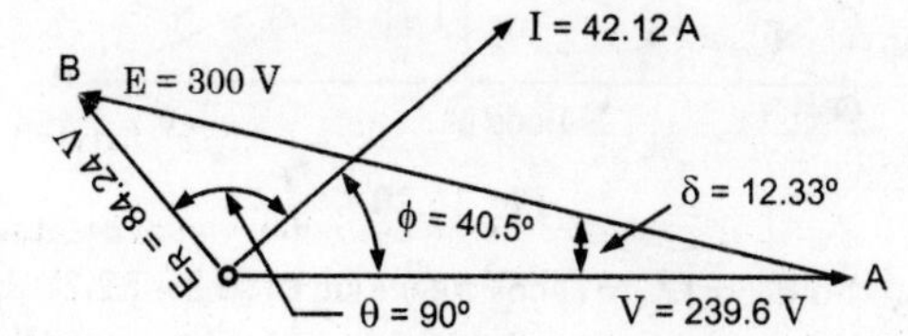

Fig. 16.19

From phasor diagram shown in Fig 16.19.

$$E_R = \sqrt{V^2 + E^2 - 2VE \cos\delta}$$

$$= \sqrt{(239.6)^2 + (300)^2 - 2 \times 239.6 \times 300 \times \cos 12.33°}$$

$$= 84.24 \text{ V}$$

Current drawn per phase,

$$I = \frac{E_R}{Z_s} = \frac{84.24}{2} = 42.12 \text{ A } \textbf{Ans.}$$

Again from phasor diagram

$$\sin(\phi + \theta) = \frac{E}{E_R} \sin \delta = \frac{300}{84.24} \sin 12.33° = 0.76$$

$$\text{or } \phi + \theta = \sin^{-1} 0.76 = 130.5°$$

$$\text{or } \phi = 130.5° - 90° = 40.5°$$

Power factor $= \cos \phi = \cos 40.5° = 0.76$ (leading) **Ans.**

Example 16.12. A 2 kV, 50 Hz, 3-phase, star-connected synchronous motor has an effective resistance and synchronous reactance of 0.2 Ω, 2.2 Ω per phase respectively. The input is 800 kW at rated load and the induced emf (line value) is 2,500 V. Evaluate (*i*) line current (*ii*) the power factor (*iii*) the motor speed (if it is a 6 pole machine) and (*iv*) the readings of the two wattmeters if the power is measured by 2-wattmeter methods.

Draw the per-phase phasor diagram at this condition.

[W.B. Technological Univ. Electrical Machines-II, 2011-12]

Solution: Since the induced emf is greater than the applied or terminal voltage, the motor must be operating at a leading pf.

Active component of motor phase current,

$$I \cos \phi = \frac{\text{Input power}}{\sqrt{3} \times V_L} = \frac{800 \times 1,000}{\sqrt{3} \times 2,000}$$

$$= 230.94 \text{ A}$$

Impedance per phase, $Z_s = \sqrt{0.2^2 + 2.2^2} = 2.209 \ \Omega$

Internal angle, $\theta = \tan^{-1} \frac{X_s}{R_e} = \tan^{-1} \frac{2.2}{0.2} = 84.8°$

Applied voltage per phase, $V = \frac{2 \times 1,000}{\sqrt{3}} = 1,154.7$ V

Induced emf per phase, $E = \frac{2,500}{\sqrt{3}} = 1,443.4$ V

In phasor diagram shown in Fig. 16.20.

$$AB = 1,443.4 \text{ V}$$

OI leads terminal voltage V by an angle ϕ, so motor phase current may be expressed as $(I \cos \phi + j\, I \sin \phi)$ if V is taken as reference phasor.

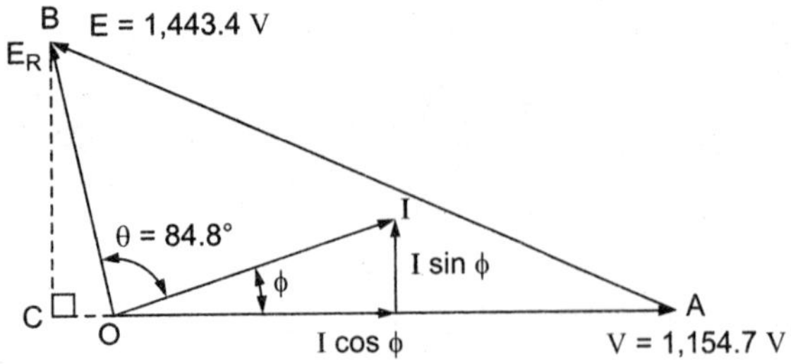

Fig. 16.20

$$\text{So } \mathbf{E}_R = \mathbf{IZ}_S = (\text{I} \cos \phi + j\, \text{I} \sin \phi)\,(0.2 + j\,2.2)$$
$$= (230.94 + j\, \text{I} \sin \phi)\,(0.2 + j\,2.2)$$
$$= (46.2 - 2.2\, \text{I} \sin \phi) + j\,(508 + 0.2\, \text{I} \sin \phi)$$

Obviously OC = 46.2 – 2.2 I sin ϕ

and BC = 508 + 0.2 I sin ϕ

From the right-angled ΔABC, we have

$$AB^2 = BC^2 + AC^2 = BC^2 + (AO - OC)^2$$

∵ OC is on negative side of O

$$\text{or } 1{,}443.4^2 = (508 + 0.2\, \text{I} \sin \phi)^2 + [1{,}154.7 - (46.2 - 2.2\, \text{I} \sin \phi)]^2$$

Solving the above equation we have

$$\text{I} \sin \phi = 106.5 \text{ A}$$

(*i*) Line current, I_L = Phase current, I

$$= \sqrt{(\text{I} \cos \phi)^2 + (\text{I} \sin \phi)^2}$$
$$= \sqrt{(230.94)^2 + (106.5)^2} = 254.3 \text{ A } \textbf{Ans.}$$

(*ii*) Power factor, $\cos \phi = \dfrac{\text{I} \cos \phi}{\text{I}} = \dfrac{230.94}{254.3} = 0.908$ (leading) **Ans.**

(*iii*) Motor speed, $N = \dfrac{120 f}{P} = \dfrac{120 \times 50}{6} = 1{,}000$ rpm

Phase angle, $\phi = \text{Cos}^{-1} \text{pf} = \text{Cos}^{-1}\, 0.908 = 24.77°$

(*iv*) Reading on wattmeter $W_1 = V_L I_L \cos (30° + \phi)$

$$= 2{,}000 \times 254.3 \cos (30° + 24.77°)$$
$$= 293{,}510 \text{ W or } 293.5 \text{ kW}$$

Reading on wattmeter, $W_2 = V_L I_L \cos (30° - \phi)$

$$= 2{,}000 \times 254.3 \cos (30° - 24.77°)$$
$$= 506{,}482 \text{ W or } 506.5 \text{ kW}$$

Example 16.13. A 415 V, 2 pole, 3 phase, 50 Hz, star-connected, non-salient pole synchronous motor has synchronous reactance of 2 Ω per phase and negligible stator resistance. At a particular field excitation, it draws 20 A at unity power factor from a 415 V, 3 phase, 50 Hz supply. The mechanical load on the motor is now increased till the stator current is equal to 50 A. The field excitation remains unchanged. Determine:

(*a*) **the per phase open-circuit voltage E_0.**

(*b*) **the developed power for the new operating condition and corresponding power factor.** *[GATE, 2002]*

Solution: At unity power factor operation, the phasor diagram can be drawn as shown in Fig. 16.21.

Supply voltage per phase, $V = \dfrac{415}{\sqrt{3}}$ V = 239.6 V

Reactance dropper phase, $E_R = IX_S = 20 \times 2 = 40$ V

(*a*) Induced emf per phase, $E_0 = \sqrt{V^2 + E_R^2}$

$$= \sqrt{(239.6)^2 + 40^2} = 242.91 \text{ V } \textbf{Ans.}$$

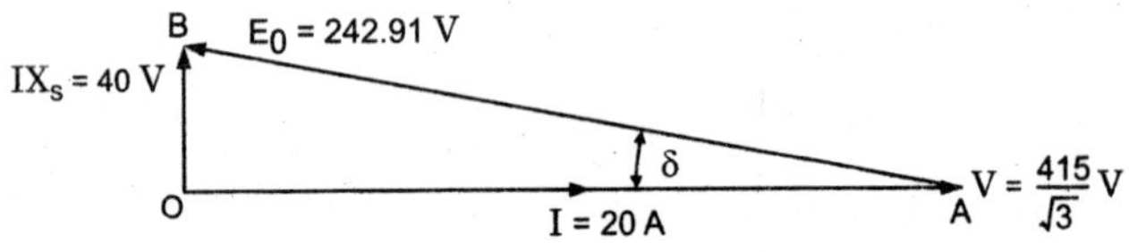

Fig. 16.21 *Phasor Diagram For UPF Operation*

(*b*) With the increase in load,

Load current, I′ = 50 A

Reactance drop per phase, $E'_R = I'X_S = 50 \times 2 = 100$ V

Induced emf per phase remains unchanged (i.e. $E'_0 = 242.91$ V), as field excitation remains the same.

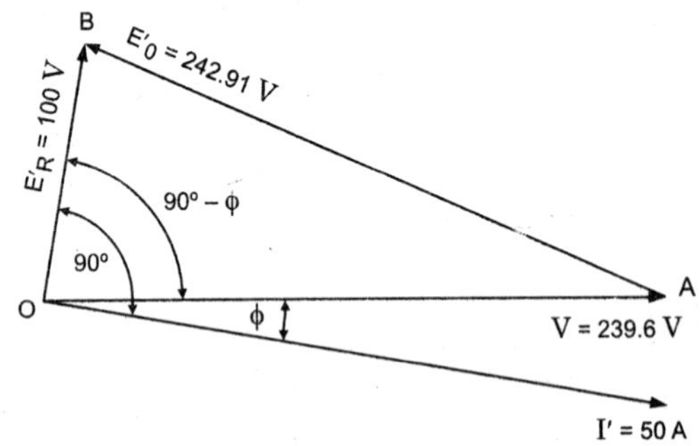

Fig. 16.22

Assuming armature current lagging behind the supply voltage by angle ϕ, as shown in phasor diagram (Fig. 16.22).

In triangle AOB we have

$$E_0'^2 = V^2 + E_R'^2 - 2VE'_R \cos (90° - \phi)$$

$$\text{or } (242.91)^2 = (239.6)^2 + (100)^2 - 2 \times (239.6) \times 100 \times \sin \phi$$

$$\text{or } \text{Sin}\, \phi = \frac{(239.6)^2 + (100)^2 - (242.91)^2}{2 \times 239.6 \times 100} = 0.17535$$

or Phase angle, $\phi = \text{Sin}^{-1}\, 0.17535 = 10.1°$

Power factor = Cos ϕ = Cos 10.1° = 0.9845 (lagging) **Ans.**

Power developed = Power input – copper losses

$$= 3 \times V \times I' \times \cos \phi - I^2 R_a$$
$$= 3 \times 239.6 \times 50 \times 0.9845 - 0$$

∵ Stator resistance is negligible

$$= 35{,}383 \text{ W or } 35.383 \text{ kW } \textbf{Ans.}$$

Example 16.14. A 2,000 V, 3-phase, 4-pole, Y-connected synchronous motor runs at 1,500 rpm. The excitation is constant and corresponds to an open-circuit voltage of 2000 V. The resistance is negligible as compared to synchronous reactance of 3 Ω per phase. Determine power input, power factor and torque developed for an armature current of 150 A.

[Rajasthan Technical Univ. Electrical Machines-II, 2012]

Solution: Supply voltage per phase, $V = \dfrac{2{,}000}{\sqrt{3}} = 1{,}154.7$ V

Induced emf per phase, $E = \dfrac{2{,}000}{\sqrt{3}} = 1{,}154.7$ V

Input phase current, I = Input line current = 150 A

∵ motor is star-connected

Impedance drop per phase, $E_R = I\,Z_S = 150 \times 3 = 450$ V

Internal angle, θ = 90° ∵ resistance is negligible

Assuming armature current lagging behind the supply voltage by an angle ϕ, as shown in phasor diagram (Fig. 16.23).

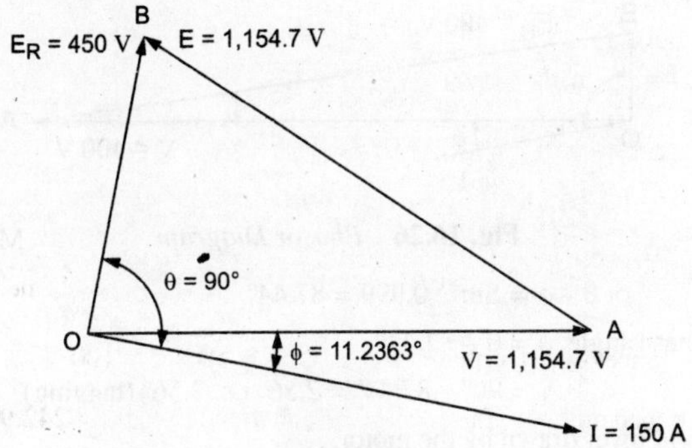

Fig. 16.23

In ΔAOB we have

$$E^2 = V^2 + E_R^2 - 2VE_R \cos(90° - \phi)$$

$$1{,}154.7^2 = 1{,}154.7^2 + 450^2 - 2 \times 1{,}154.7 \times 450 \times \sin\phi$$

$$\text{or} \quad \text{Sin}\,\phi = \frac{450}{2 \times 1{,}154.7} = 0.1948$$

Power factor angle, φ = 11.2363° **Ans.**

Power factor = Cos 11.2363° = 0.981 (lagging)

$$\text{Power input} = \sqrt{3}\, V_L\, I_L \cos\phi$$

$$= \sqrt{3} \times 2{,}000 \times 150 \times 0.981$$

= 509,655 W or 509.655 kW **Ans.**

Copper losses in armature is zero ∵ armature resistance is negligible

$$\text{Torque developed, T} = \frac{\text{Power output}}{2\pi N_s/60}$$

$$= \frac{\text{Power input} - \text{copper losses in armature}}{2\pi N_s/60}$$

$$= \frac{509{,}655 - 0}{\frac{2\pi \times 1{,}500}{60}} = 3{,}245 \text{ Nm} \quad \textbf{Ans.}$$

Example 16.15. A 20 MVA, 11 kV, 3-phase, delta-connected synchronous motor has a synchronous impedance of 15 ohm/phase. Windage, friction and iron losses amount to 1200 kW.

(*i*) Find the value of unity power factor current drawn by the motor at a shaft load of 15 MW. What is the excitation emf under this condition?

(*ii*) If the excitation emf is adjusted to 15.5 kV (line) and the shaft load is adjusted so that the motor draws unity power factor current. Find the net motor output.

[U.P. Technical Univ. Electromechanical Energy Conversion-II, 2006-07]

Solution:(*i*) Shaft load = 15 MW = 15,000 kW

Mechanical power developed,

$$P_{mech} = \text{Shaft load} + \text{windage friction and iron losses}$$

$$= 15{,}000 + 1{,}200 = 16{,}200 \text{ kW}$$

Power input to motor,

$$P = P_{mech} = 16{,}200 \text{ kW} \quad \text{neglecting armature resistance}$$

$$\text{Line current, } I_L = \frac{P \text{ in watts}}{\sqrt{3}\, V_L \cos\phi} = \frac{16{,}200 \times 1{,}000}{\sqrt{3} \times 11{,}000} = 850.28 \text{ A} \quad \textbf{Ans.}$$

$$\text{Phase current, } I_P = \frac{I_L}{\sqrt{3}} = \frac{850.28}{\sqrt{3}} = 490.9 \text{ A}$$

Impedance drop per phase, $E_R = I_P \times Z_S$ = 490.9 × 15 = 7,363.7 V

Internal angle, θ = 90° Assuming negligible armature resistance

From phasor diagram shown in Fig. 16.24.

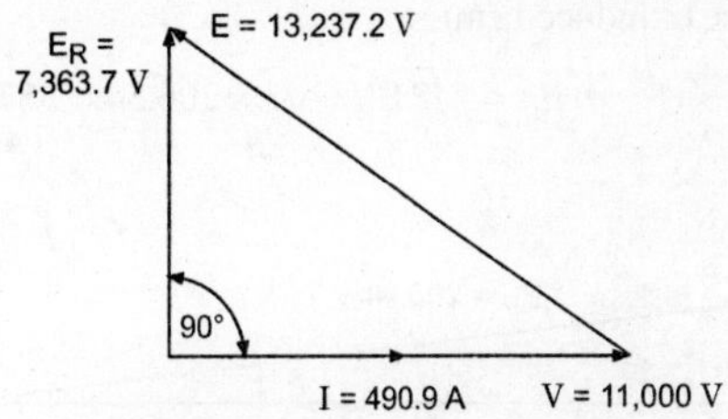

Fig. 16.24 *Phasor Diagram*

$$\text{Excitation emf, E} = \sqrt{V^2 + E_R^2}$$

$$= \sqrt{(11{,}000)^2 + (7{,}363.7)^2} = 13{,}237.2 \text{ V} \quad \textbf{Ans.}$$

(*ii*) Impedance drop per phase,

$$E_R = \sqrt{E^2 - V^2} \because \text{phase angle } \phi = 0° \text{ and } \theta = 90°$$

$$= \sqrt{(15{,}500)^2 - (11{,}000)^2} = 10{,}920 \text{ V}$$

Load current per phase,

$$I_P = \frac{E_R}{Z_S} = \frac{10{,}920}{15} = 728 \text{ A}$$

$$\text{Line current, } I_L = \sqrt{3}\, I_P = \sqrt{3} \times 728 = 1{,}261 \text{ A}$$

$$\text{Power input to motor, P} = \sqrt{3}\, V_L\, I_L \cos\phi$$

$$= \sqrt{3} \times 11{,}000 \times 1{,}261 \times 1.0$$

= 24006224 W = 24,006.22 kW

Net motor output = 24,006.22 – 1,200

= 22,806.22 kW or 22.80622 MW **Ans.**

Example 16.16. A 20 kW, 400 V, 3-phase, star-connected synchronous motor has per phase impedance of (0.15 + *j* 0.90) Ω. Determine the induced emf, torque angle and mechanical power developed for full load at 0.8 pf lagging. Assume 92% efficiency of the motor. Draw phasor diagram.

[G.B. Technical Univ. Electromechanical Energy Conversion-II, 2012-13]

Solution:

$$\text{Motor input} = \frac{\text{Motor output}}{\eta}$$

$$= \frac{20}{0.92} = 21.739 \text{ kW or } 21{,}739 \text{ W}$$

$$\text{Armature current, I} = \frac{\text{Motor input}}{\sqrt{3}\, V_L \cos\phi} = \frac{21{,}739}{\sqrt{3} \times 400 \times 0.8}$$

= 39.222 A

Supply voltage per phase,

$$V = \frac{400}{\sqrt{3}} = 230.94 \text{ V}$$

$$\text{Resultant voltage, } E_R = I \times Z_S = 39.222 \times \sqrt{0.15^2 + 0.90^2}$$

$$= 39.222 \times 0.9124 = 35.787 \text{ V}$$

$$\text{Internal angle, } \theta = \text{Tan}^{-1}\frac{X_S}{R_e} = \text{Tan}^{-1}\frac{0.90}{0.15} = 80.5377°$$

$$\text{Induced emf per phase, } E_P = \sqrt{V^2 + E_R^2 - 2\,VE_R \cos(\theta - \phi)}$$

$$= \sqrt{(230.94)^2 + (35.787)^2 - 2 \times 230.94 \times 35.787 \times \cos(80.5377° - 36.87°)}$$

$$= \sqrt{53{,}333.3 + 1{,}280.7 - 11{,}956.6}$$

= 206.54 V **Ans.**

Line value of induced emf,

$$E_L = \sqrt{3}\,E_P = \sqrt{3} \times 206.54 = 357.73 \text{ V} \quad \textbf{Ans.}$$

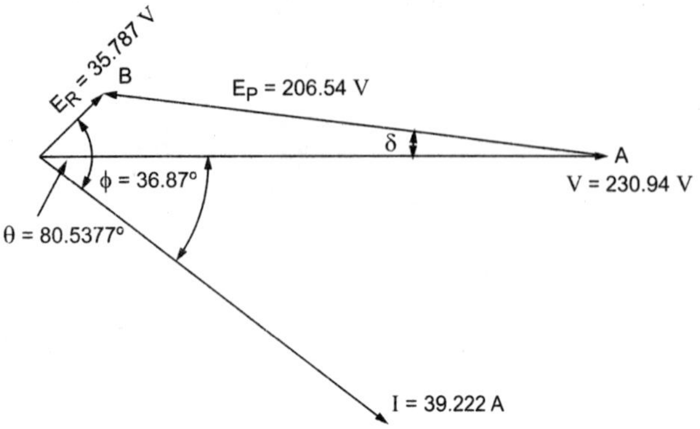

Fig. 16.25 *Phasor Diagram*

Mechanical power developed = Power input – $3I^2R_e$

$= 21{,}739 - 3 \times (39.222)^2 \times 0.15$

$= 21{,}047$ W or 21.047 kW **Ans.**

Also $\dfrac{E_R}{\sin\delta} = \dfrac{E_P}{\sin(\theta - \phi)}$

or $\dfrac{35.787}{\sin\delta} = \dfrac{206.54}{\sin(80.5377° - 36.87°)}$

$$\sin\delta = \frac{35.787 \sin 43.6677°}{206.54} = 0.11964$$

Torque angle, $\delta = \sin^{-1} 0.11964 = 6.87°$ **Ans.**

Example 16.17. A 20 pole, 693 V, 50 Hz, 3-φ, Δ-connected synchronous motor is operated at no load with normal excitation. It has armature reactance per phase of 10 Ω and negligible resistance. If rotor is retarded by 0.5° (mechanical) from its synchronous position, compute (*i*) Rotor displacement in electrical degrees (*ii*) armature emf per phase (*iii*) armature current per phase (*iv*) power drawn by the motor (*v*) power developed by the armature.

[*J.N. Technological Univ. Electrical Machines-III, May-2013*]

Solution: Supply voltage per phase, V = 693 V

(*ii*) Induced emf per phase, E = V = 693 V **Ans.**

(*i*) Rotor displacement in electrical degrees,

$$\delta_{(elec)} = \frac{P}{2}\,\delta_{(mech)} = \frac{20}{2} \times 0.5 = 5° \text{ (elec.)} \quad \textbf{Ans.}$$

Now from ΔAOB of phasor diagram shown in Fig. 16.26 we have

$$E_R = \sqrt{V^2 + E^2 - 2VE\cos\delta}$$

$$= \sqrt{693^2 + 693^2 - 2 \times 693 \times 693 \times \cos 5°} = 60.456 \text{ V}$$

(*iii*) Armature current per phase,

$$I = \frac{E_R}{Z} = \frac{60.456}{10} = 6.0456 \text{ A} \quad \textbf{Ans.}$$

From ΔAOB in Fig. 16.26

$$\frac{AB}{\sin \angle AOB} = \frac{OB}{\sin \angle OAB}$$

$$\frac{693}{\sin(\theta - \phi)} = \frac{60.456}{\sin 5°}$$

or $\sin(\theta - \phi) = \dfrac{693 \sin 5°}{60.456} = 0.999$

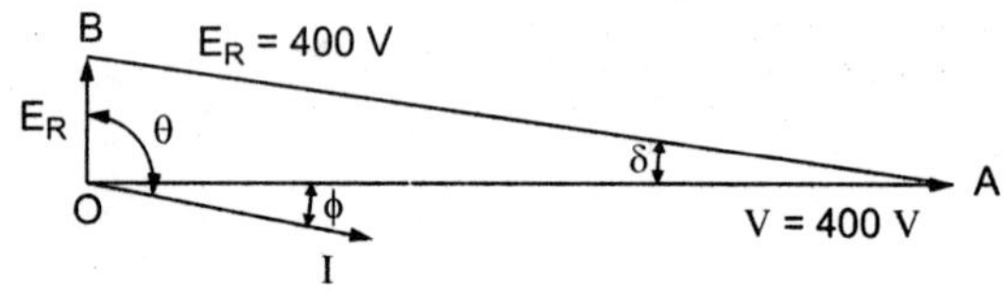

Fig. 16.26 *Phasor Diagram*

or $\theta - \phi = \sin^{-1} 0.999 = 87.44°$

Phase angle, $\phi = \theta - 87.44°$

$= 90° - 87.44° = 2.56°$ i.e. 2.56° (lagging)

(*iv*) Power drawn by the motor,

$$P_{in} = \sqrt{3}\,V_L I_L \cos\phi = \sqrt{3} \times 693 \times 6.0456 \times \cos 2.56°$$

$= 7{,}249$ W **Ans.**

(*v*) Power developed by the motor,

$P_{mech} = P_{in} = 7{,}249$ W **Ans.** $\quad\because R_e$ is negligible.

Example 16.18. A 50 Hz, 4-pole, 3-phase, star-connected synchronous motor has a synchronous reactance of 12.0 Ω/phase and negligible armature resistance. The excitation is such as to give an open-circuit voltage of 13.2 kV. The motor is connected to 11.5 kV, 50 Hz supply.

What maximum load can the motor supply before loosing synchronism? What is the corresponding motor torque, line current and power factor?

Solution: Supply voltage/phase, $V = \dfrac{11.5 \times 1{,}000}{\sqrt{3}} = 6{,}640$ V

Induced emf/phase, $E = \dfrac{13.2 \times 1{,}000}{\sqrt{3}} = 7{,}621$ V

Internal angle, θ = 90°

∵ armature resistance is negligible

Synchronous impedance/phase,

$$Z_s = X_s = 12.0\ \Omega$$

Power developed will be maximum when

load angle, δ = θ = 90°

Impedance drop per phase,

$$E_R = V^2 + E^2 - 2\,VE\cos\delta$$

$$= \sqrt{(6{,}640)^2 + (7{,}621)^2 - 2 \times 6{,}640 \times 7{,}621 \times \cos 90°}$$

$= 10{,}108$ V

Line current,

$$I_L = \text{Phase current, } I = \frac{E_R}{Z_s} = \frac{10{,}108}{12} = 842.3 \text{ A} \quad \textbf{Ans.}$$

The maximum load that motor can deliver,

$$(P_{mech})_{max} = 3\left[\frac{EV}{Z_s} - \frac{E^2}{Z_s}\cos\theta\right]$$

$$= 3\left[\frac{6{,}640 \times 7{,}621}{12} - \frac{(7{,}621)^2}{12}\cos 90°\right]$$

= 12,650,000 watts or 12.65 MW **Ans.**

Power supplied to motor,

$P_{IN} = P_{mech}$ + armature copper loss $= P_{mech} = 12.65$ MW

∵ armature resistance is negligible

So $\sqrt{3}\,V_L I_L \cos\phi = 12.65 \times 10^6$

or Power factor, $\cos\phi = \dfrac{12.65\times10^6}{\sqrt{3}\times11{,}500\times842.3} = 0.754$ **Ans.**

Synchronous speed, $N_s = \dfrac{120f}{P} = \dfrac{120\times50}{4} = 1{,}500$ rpm

Torque developed corresponding to maximum power developed,

$$(T_{mech})_{max} = \frac{(P_{mech})_{max}}{2\pi N_s/60}$$

$$= \frac{12.65\times10^6}{2\pi\times1{,}500/60} = 80{,}532 \text{ N-m } \textbf{Ans.}$$

Example 16.19. The excitation of a 415 V, 3-phase, mesh-connected synchronous motor is such that the induced emf is 520 V. The impedance per phase is (0.5 + *j* 4.0) ohms. If the friction and iron losses are constant at 1,000 watts, calculate the power output, line current, power factor and efficiency for maximum power output.

Solution: Supply voltage/phase, V = 415 V

∵ motor is mesh connected

Induced emf/phase, E = 520 V

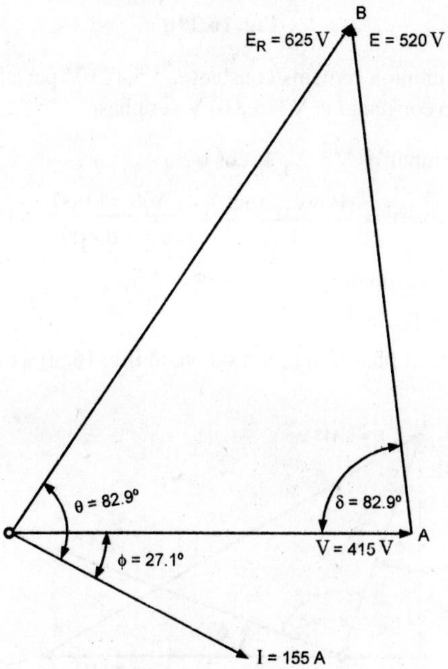

Fig. 16.27

Synchronous impedance/phase, $Z_s = \sqrt{(0.5)^2 + 4^2} = 4.03\ \Omega$

Internal angle, $\theta = \text{Tan}^{-1}\dfrac{X_s}{R_e} = \text{Tan}^{-1}\dfrac{4}{0.5} = 82.9°$

For a constant value of supply voltage, excitation (or induced emf), effective resistance R_e and synchronous reactance X_s maximum power will be developed when load angle

$$\delta = \theta = 82.9° \qquad \text{...(Art. 16.9.1)}$$

Now from phasor diagram ΔOAB (Fig. 16.27) we have

$$E_R = \sqrt{V^2 + E^2 - 2VE\cos\delta}$$

$$= \sqrt{(415)^2 + (520)^2 - 2\times415\times520\cos 82.9°}$$

$= 625$ V/phase

Load current, $I = \dfrac{E_R}{Z_s} = \dfrac{625}{4.03} = 155$ A/phase

Line current, $I_L = \sqrt{3}\,I = \sqrt{3}\times155 = 268.2$ A **Ans.**

Maximum power developed per phase,

$$(P_{mech})_{max} = \frac{EV - E^2\cos\theta}{Z_s}$$

$$= \frac{520\times415 - 520^2\times\cos 82.9°}{4.03} = 45{,}255 \text{ watts}$$

Maximum power for 3-phase,

$= 3\times45{,}255 = 1{,}35{,}765$ watts

Output power = Power developed – iron and friction losses

$= 1{,}35{,}765 - 1{,}000$

$= 1{,}34{,}765$ watts or 134.765 kW **Ans.**

Total copper losses $= 3I^2R_e = 3\times(155)^2\times0.5 = 36{,}000$ watts

Input to motor = Power developed + copper losses

$= 1{,}35{,}765 + 36{,}000 = 1{,}71{,}765$ watts

Power factor, $\cos\phi = \dfrac{\text{Input to motor in watts}}{\sqrt{3}V_L I_L}$

$= \dfrac{1{,}71{,}765}{\sqrt{3}\times415\times268.2} = 0.89$ **Ans.**

Efficiency, $\eta = \dfrac{\text{Output}}{\text{Input}}\times100 = \dfrac{1{,}34{,}765}{1{,}71{,}765}\times100$

$= 78.46\%$ **Ans.**

Example 16.20. A 400 V, 6-pole, 3-ϕ, 50 Hz, star-connected synchronous motor has a resistance and synchronous impedance of 0.5 Ω and 4 Ω per phase respectively. It takes a current of 15 A at unity power factor when operating with a certain field current. If the torque load is increased until the line current is increased to 60 A, the field current remaining the same determine the gross torque developed and the new power factor.

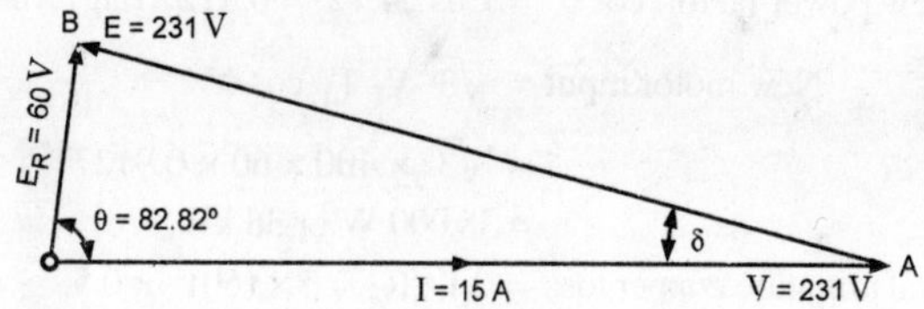

Fig. 16.28 (*a*)

Solution: Supply voltage per phase, $V = \dfrac{400}{\sqrt{3}} = 231$ V

Effective resistance/phase, $R_e = 0.5\ \Omega$

Synchronous impedance/phase, $Z_s = 4\ \Omega$

Synchronous reactance/phase, $X_s = \sqrt{(4)^2 - (0.5)^2} = 3.968\ \Omega.$

Internal angle, $\theta = \text{Cos}^{-1}\dfrac{R_e}{Z_s} = \text{Cos}^{-1}\dfrac{0.5}{4} = 82.82°$

When input current, I = 15 A

and cos ϕ = unity

Impedance drop, $E_R = IZ_s = 15\times4 = 60$ V

Induced emf/phase,

$$E = \sqrt{V^2 + E_R^2 - 2VE_R\cos\theta}$$

$$= \sqrt{(231)^2 + (60)^2 - 2\times231\times60\cos 82.82°} = 231 \text{ V}$$

Now when the load on the motor is increased, the angle of retardation δ increases. The phasor diagram is shown in Fig. 16.28 (*b*).

Input current, $I' = 60$ A

Voltage per phase, $V = 231$ V, as before

Induced emf per phase, $E = 231$ V, as before

Since excitation is constant

Impedance drop/phase, $E'_R = I'Z_s = 60 \times 4 = 240$ V

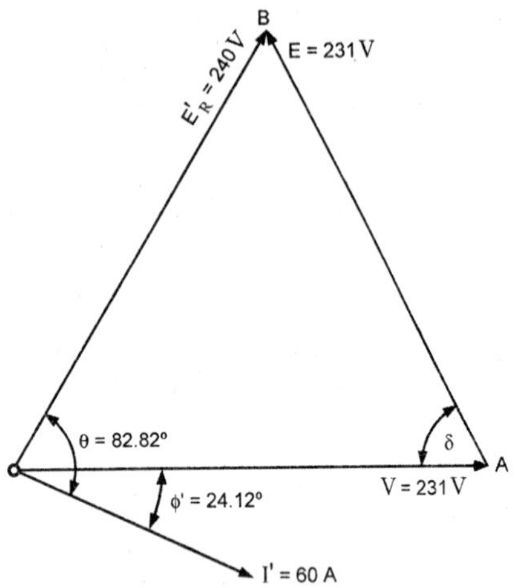

Fig. 16.28 (b)

Now in Δ OAB of phasor diagram [Fig. 16.28 (b)] we have

$$E^2 = V^2 + E'^2_R - 2\,V\,E'_R \cos \angle AOB$$

$$\text{or } 231^2 = 231^2 + 240^2 - 2 \times 231 \times 240 \cos \angle AOB$$

$$\text{or } \angle AOB = \text{Cos}^{-1}\, 0.5195 = 58.7°$$

Internal angle, $\theta = 82.82°$

∴ Power factor angle,

$$\phi' = \theta - \angle AOB = 82.82° - 58.7° = 24.12° \text{ (lag)}$$

New power factor, $\cos\phi' = \text{Cos } 24.12° = 0.9127$ (lag) **Ans.**

$$\text{New motor input} = \sqrt{3}\; V_L I'_L \cos\phi'$$

$$= \sqrt{3} \times 400 \times 60 \times 0.9127$$

$$= 38{,}000 \text{ W or } 38 \text{ kW}$$

$$\text{Total armature copper loss} = 3\, I'^2 R_e = 3 \times (60)^2 \times 0.5$$

$$= 5{,}400 \text{ W or } 5.4 \text{ kW}$$

$$\text{Mechanical power developed} = 38 - 5.4 = 32.6 \text{ kW}$$

$$\text{Synchronous speed, } N_s = \frac{120f}{P} = \frac{120 \times 50}{6} = 1{,}000 \text{ rpm}$$

$$\text{Gross torque developed, } T_g = \frac{\text{Power developed in watts}}{2\pi N_s/60}$$

$$= \frac{32.6 \times 1{,}000}{2\pi \times \frac{1{,}000}{60}} = 311 \text{ N-m}$$ **Ans.**

Example 16.21. A 6.6 kV star-connected, 3-phase synchronous motor works at constant voltage and constant excitation. Its synchronous reactance is 20 Ω per phase. When the input is 1,000 kW, the power factor is 0.8 leading. Find the power factor when the input is increased to 1,500 kW. Neglect armature resistance.

Solution: Supply voltage per phase,

$$V = \frac{6.6 \times 1{,}000}{\sqrt{3}} = 3{,}810 \text{ V}$$

Synchronous impedance per phase,

$Z_s = X_s = 20\ \Omega$ ∵ armature resistance is negligible

Internal angle, $\theta = 90°$

When the input is 1,000 kW at 0.8 pf leading

$$\text{Armature current/phase, } I = \frac{1{,}000 \times 1{,}000}{\sqrt{3} \times 6{,}600 \times 0.8} = 109.35 \text{ A}$$

$$\text{Power factor angle, } \phi = \text{Cos}^{-1}\, 0.8 = 36.87°$$

$$\text{Impedance drop per phase, } E_R = IZ_s = 109.35 \times 20 = 2{,}187 \text{ V}$$

Induced emf per phase,

$$E = \sqrt{V^2 + E_R^2 - 2\,VE_R \cos(\theta + \phi)}$$

$$= \sqrt{(3{,}810)^2 + (2{,}187)^2 - 2 \times 3{,}810 \times 2{,}187 \cos(90° + 36.87°)}$$

$$= 5{,}413 \text{ V.}$$

Now when the load on the machine is increased, the angle of retardation δ will increase. The phasor diagram is shown in Fig. 16.30.

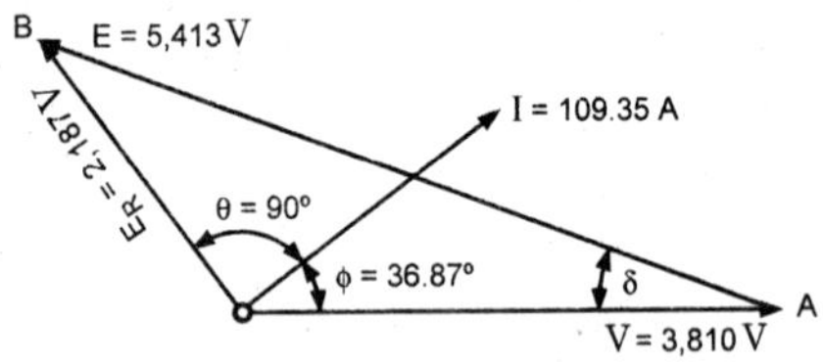

Fig. 16.29

Since excitation remains constant *i.e.* 5,413 V per phase. Supply voltage is also constant *i.e.* V = 3,810 V per phase.

$$\text{Since power input} = \sqrt{3}\; V_L I_L \cos\phi$$

$$\text{So } I'\cos\phi' = \frac{\text{Power input}}{\sqrt{3}\,V_L} = \frac{1{,}500 \times 1{,}000}{\sqrt{3} \times 6{,}600} = 131.2 \text{ A}$$

Impedance drop per phase,

$$E'_R = I'X = 20\,I'$$

In Δ ABC of phasor diagram shown in Fig. 16.30 we have

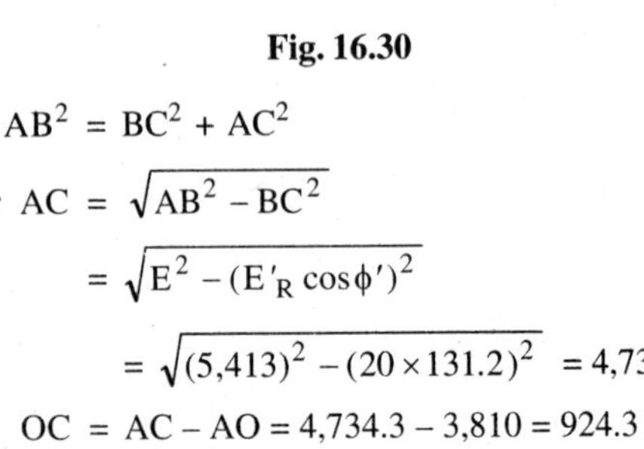

Fig. 16.30

$$AB^2 = BC^2 + AC^2$$

$$\text{or } AC = \sqrt{AB^2 - BC^2}$$

$$= \sqrt{E^2 - (E'_R \cos\phi')^2}$$

$$= \sqrt{(5{,}413)^2 - (20 \times 131.2)^2} = 4{,}734.3 \text{ V}$$

$$OC = AC - AO = 4{,}734.3 - 3{,}810 = 924.3 \text{ V}$$

$$E'_R = \sqrt{BC^2 + OC^2}$$

$$= \sqrt{(20 \times 131.2)^2 + (924.3)^2} = 2{,}782 \text{ V}$$

$$\text{Line current, } I' = \frac{E'_R}{Z_s} = \frac{2{,}782}{20} = 139.1 \text{ A}$$

$$\text{Power factor, } \cos\phi' = \frac{I'\cos\phi'}{I'} = \frac{131.2}{139.1} = 0.9432 \text{ (leading)}$$ **Ans.**

Example 16.22. A 3-phase, Y-connected synchronous motor takes 48 kW at 693 V (line), the pf being 0.8 lag. The induced emf is now increased by 30%, the power input being the same. Find the new current and pf. Z_s equals (0 + *j* 2) ohm/phase.

Solution: Supply voltage per phase, $V = \dfrac{693}{\sqrt{3}} = 400$ V

Synchronous impedance per phase, $Z_s = X_s = 2\ \Omega \qquad \because R_e = 0$

Internal angle $\theta = 90°$

At normal excitation

Power factor, $\cos\phi = 0.8$ (lag)

Power factor angle, $\phi = \cos^{-1} 0.8 = -36.87°$

Armature current/phase,

$$I = \frac{\text{kW input} \times 1{,}000}{\sqrt{3}\, V_L \cos\phi} = \frac{48 \times 1{,}000}{\sqrt{3} \times 693 \times 0.8} = 50 \text{ A}$$

Impedance drop per phase,

$E_R = I Z_s = 50 \times 2 = 100$ V

Induced emf per phase,

$$E = \sqrt{V^2 + E_R^2 - 2 V E_R \cos(\theta + \phi)}$$

$$= \sqrt{400^2 + 100^2 - 2 \times 400 \times 100 \cos[90° + (-36.87°)]}$$

$$= 349.3 \text{ V}$$

When the excitation is increased by 30% the induced emf for phase E′ will become 1.3 × 349.3 = 454 V.

Since power input and supply voltage is constant, active component of current drawn remains unchanged *i.e.*

$$I' \cos\phi' = 50 \times 0.8 = 40 \text{ A}$$

The phasor diagram is shown in Fig. 16.31 assuming that the current drawn from supply mains is leading one now.

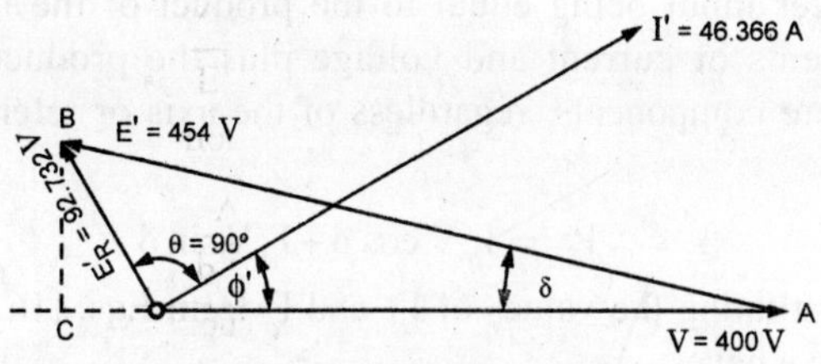

Fig. 16.31

In Δ ABC of the phasor diagram shown in Fig. 16.31

$$AC = \sqrt{AB^2 - BC^2} = \sqrt{E'^2 - (E'_R \cos\phi')^2}$$

$$= \sqrt{454^2 - (I' Z_s \cos\phi')^2}$$

$$= \sqrt{454^2 - (40 \times 2)^2} = 446.9 \text{ V}$$

$$OC = AC - OA = 446.9 - 400 = 46.9 \text{ V}$$

$$E'_R = \sqrt{BC^2 + OC^2} = \sqrt{(40 \times 2)^2 + (46.9)^2} = 92.732 \text{ V}$$

$$\text{Current, } I' = \frac{E'_R}{Z_s} = \frac{92.732}{2} = 46.366 \text{ A } \textbf{Ans.}$$

$$\cos\phi' = \frac{I' \cos\phi'}{I'} = \frac{40}{46.366} = 0.863 \text{ (lead) } \textbf{Ans.}$$

Example 16.23. A 20 hp, 3-phase, 400 V synchronous motor operating on full load from infinite bus-bar, has its excitation so adjusted that power factor is 0.87 lagging. Load being kept constant, excitation is now increased by 20%. Find the new power factor. Synchronous reactance may be taken as 1.0 per unit.

Solution: Assuming synchronous motor star-connected

Supply voltage per phase, $V = \dfrac{400}{\sqrt{3}} = 231$ volts

Synchronous reactance = 1.0 per unit

Since per unit reactance $= \dfrac{I X_s}{V}$

$\therefore$ Synchronous reactance drop, $I X_s = V \times$ per unit reactance $= V \times 1 = 231$ volts

Neglecting effective resistance

Impedance drop per phase, $E_R = I Z_s = I X_s = 231$ V

and internal angle $\theta = 90°$

Power factor, $\cos\phi = 0.87$ (lagging)

$\phi = \text{Cos}^{-1}\ 0.87 = 29.54°$

From phasor diagram shown in Fig. 16.32

$$E = \sqrt{V^2 + E_R^2 - 2 V E_R \cos(\theta - \phi)}$$

$$= \sqrt{(231)^2 + (231)^2 - 2 \times 231 \times 231 \cos(90° - 29.54°)}$$

$$= 232.6 \text{ V}$$

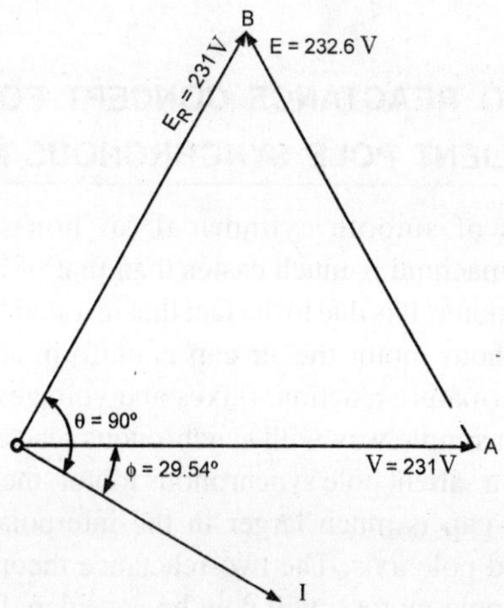

Fig. 16.32

The phasor diagram for increased emf is shown in Fig. 16.33. Now induced emf, E′ = 1.2 E = 1.2 × 232.6 = 279 V. Let the angle between phase voltage V and current I be ϕ′. Let the new current be I′. As load remains the same so neglecting losses input power can be assumed to remain the same and since supply voltage is same so active component of current drawn from the 3-ϕ supply main would remain the same.

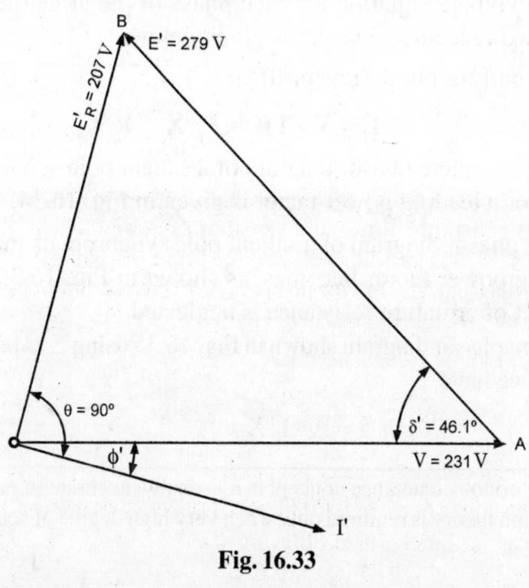

Fig. 16.33

$$\therefore I' \cos \phi' = I \cos \phi = 0.87 I$$

$$\text{and impedance drop/phase} = E'_R = I' Z_s = I' X_s$$

$$\text{and } E'_R \cos \phi' = I' X_s \cos \phi' = I \cos \phi \times X_s$$

$$= 0.87 I X_s$$

$$= 0.87 \times 231 = 201 \text{ V}$$

From Δ AOB of phasor diagram shown in Fig. 16.33.

$$\frac{E'}{\sin(90° - \phi')} = \frac{E'_R}{\sin \delta'}$$

$$\text{or } \sin \delta' = \frac{E'_R \cos\phi'}{E'} = \frac{201}{279}$$

$$\text{or } \delta' = 46.1°$$

Again from ΔAOB, we get

$$E'_R = \sqrt{V^2 + (E')^2 - 2 V E' \cos\delta'}$$

$$= \sqrt{(231)^2 + (279)^2 - 2 \times 231 \times 279 \cos 46.1°} = 207 \text{ volts}$$

New power factor,

$$\cos \phi' = \frac{E'_R \cos\phi'}{E'_R} = \frac{201}{279} = 0.971 \text{ (lagging)} \textbf{ Ans.}$$

16.12 TWO REACTANCE CONCEPT FOR SALIENT POLE SYNCHRONOUS MOTORS

The analysis of smooth cylindrical (or non-salient pole) synchronous machine is much easier than that of a salient pole synchronous motor. It is due to the fact that in a smooth cylindrical rotor synchronous motor the air gap is uniform and, therefore, the effect of armature reaction, fluxes and voltages induced can be treated in a simple way with synchronous reactance concept. But in case of a salient pole synchronous motor, the air gap is not uniform—air gap is much larger in the interpolar space than along the field pole axis. The two-reactance theory, applicable to the salient pole motor*, will now be considered.

The salient pole synchronous motor has two axis—field pole axis, called the *direct axis* or *d-axis* and the axis passing through the centre of interpolar space, called the *quadrature axis* or *q-axis*, as in case of an alternator. The armature current of the motor can also be resolved along *d*-axis and *q*-axis called I_d and I_q components as in case of an alternator.

The voltage equation for each phase of the armature based on the two reactance concept is given below:

Induced emf for phase (reversed) *i.e.*

$$-\mathbf{E} = \mathbf{V} - \mathbf{I} \mathbf{R}_e - \mathbf{I}_d \mathbf{X}_d - \mathbf{I}_q \mathbf{X}_q \qquad ...(16.12)$$

The complete phasor diagram of a salient pole synchronous motor, for a leading power factor is given in Fig. 16.34.

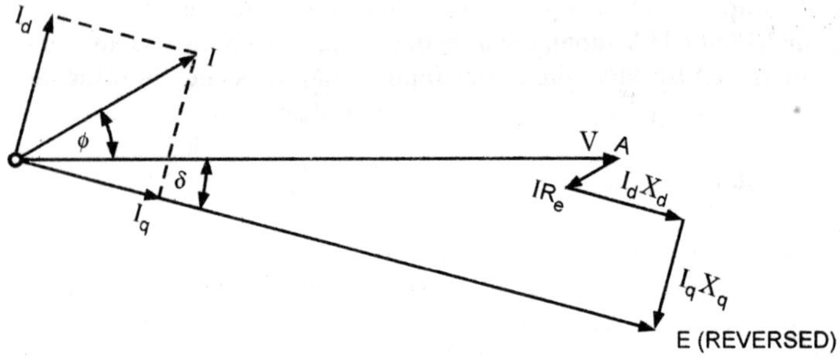

Fig. 16.34 *Phasor Diagram With Leading Current For Synchronous Motor*

The phasor diagram of a salient pole synchronous motor for a leading power factor becomes, as shown in Fig. 16.35, when the effect of armature resistance is neglected.

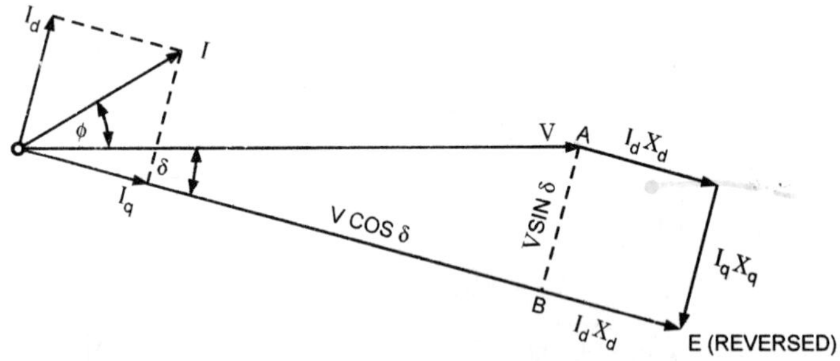

Fig. 16.35 *Phasor Diagram With Leading Current For Synchronous Motor (Neglecting Effect of Armature Resistance)*

From phasor diagram shown in Fig. 16.35 using E as reference phasor, we have

$$V \cos \delta = E - I_d X_d \qquad ...(16.13)$$

$$\text{and } V \sin \delta = I_q X_q \qquad ...(16.14)$$

$$\text{So } I_d = \frac{E - V \cos\delta}{X_d} \qquad ...(16.15)$$

$$\text{and } I_q = \frac{V \sin\delta}{X_q} \qquad ...(16.16)$$

Power input being equal to the product of the in-phase components of current and voltage plus the product of the quadrature components, regardless of the axis of reference, is given as

$$P_{in} = I_q V \cos \delta + I_d V \sin \delta$$

Substituting the values of I_d and I_q from Eqs. (16.15) and (16.16) we have

$$P_{in} = \frac{V^2}{X_q} \cos \delta \sin \delta + \frac{E - V\cos\delta}{X_d} V \sin \delta$$

$$= \frac{EV}{X_d} \sin \delta + \frac{V^2}{2}\left[\frac{1}{X_q} - \frac{1}{X_d}\right] \sin 2\delta \qquad ...(16.17)$$

Total power input will be 3 times of the above

i.e. Total power input

$$= \frac{3EV}{X_d} \sin\delta + \frac{3V^2}{2}\left(\frac{1}{X_q} - \frac{1}{X_d}\right) \sin 2\delta$$

$$= \frac{E_L V_L}{X_d} \sin\delta + \frac{V_L^2}{2}\left(\frac{1}{X_q} - \frac{1}{X_d}\right) \sin 2\delta \qquad ...(16.18)$$

For determination of power developed copper losses will be deducted from the power input determined from the above expression.

* Synchronous reactance concept is reasonably accurate in predicting the steady-state performance of salient pole synchronous motors and two-reaction theory is required only when very high degree of accuracy is required or when problems concerning transients or power stability are to be handled.

16.13 STABILITY AND MAXIMUM LOAD ANGLE

Differentiating the Eq. (16.17) with respect to load angle δ we have

$$\frac{dP_{in}}{d\delta} = \frac{EV}{X_d}\cos\delta + V^2\left[\frac{1}{X_q} - \frac{1}{X_d}\right]\cos 2\delta \quad ...(16.19)$$

The above expression gives the rate of change of power as a function of the load angle δ. This is called the *stability factor*, *rigidity factor*, or simply the *stiffness of coupling*. This is approximately because it is based on power input rather than power developed.

In case of smooth cylindrical rotor motor the corresponding expression is obtained by differentiating Eq. (16.5) w.r.t. load angle δ so

$$\frac{dP_{in}}{d\delta} = \frac{EV}{X_S}\cos\delta$$

Equation (16.19) contains a second term, implying greater stiffness of the coupling for the salient pole motor.

Maximum load angle can be determined by equating, the RHS of the Eq. (16.19) equal to zero

$$i.e. \quad \frac{EV}{X_d}\cos\delta + V^2\left[\frac{1}{X_q} - \frac{1}{X_d}\right]\cos 2\delta = 0$$

Solving above equation for cos δ we have

$$\text{Cos } \delta \simeq \frac{-E}{4V\frac{X_d - X_q}{X_q}} + \sqrt{\frac{1}{2} + \left[\frac{EX_q}{4V(X_d - X_q)}\right]^2} \quad ...(16.20)$$

This is the cosine of the maximum load angle when the effect of armature resistance is neglected.

Example 16.24. A salient pole synchronous motor has $X_d = 0.85$ pu and $X_q = 0.55$ pu

It is connected to bus-bars of 1.0 pu voltage, while its excitation is adjusted to 1.2 pu. Calculate the maximum power output, the motor can supply without loss of synchronism.

Compute the minimum pu excitation that is necessary for the machine to stay in synchronism while supplying the full-load torque (i.e., 1.0 pu power).

[U.P. Technical Univ. Electromechanical Energy Conversion-II, 2013-14]

Solution: For maximum power to be delivered by a salient pole synchronous motor, according to Eq. (16.20)

$$\text{Cos } \delta = \frac{-EX_q}{4V(X_d - X_q)} + \sqrt{\frac{1}{2} + \left[\frac{EX_q}{4V(X_d - X_q)}\right]^2}$$

$$= \frac{-1.2\times 0.55}{4\times 1(0.85-0.55)} + \sqrt{\frac{1}{2} + \left[\frac{1.2\times 0.55}{4\times 1(0.85-0.55)}\right]^2}$$

$$= -0.55 + 0.896 = 0.346$$

or Load angle $\delta_{(max)} = \text{Cos}^{-1}\, 0.346 = 69.8°$

Now maximum power is given by Eq. (16.18)

$$i.e. \quad P_{max} = \frac{EV}{X_d}\sin\delta + \frac{V^2}{2}\left[\frac{1}{X_q} - \frac{1}{X_d}\right]\sin 2\delta$$

$$= \frac{1.2\times 1}{0.85}\sin 69.8° + \frac{1}{2}\left[\frac{1}{0.55} - \frac{1}{0.85}\right]\sin(2\times 69.8°)$$

$= 1.533$ pu **Ans.**

The power delivered due to excitation is given as

$$P = \frac{EV}{X_d}\sin\delta$$

Substituting $V = 1.0$, $X_d = 0.85$, $\delta = 69.8°$ and $P = 1.533$ in the above equation we have

$$1.533 = \frac{E\times 1.0}{0.85}\sin 69.8°$$

$$\text{or } E = \frac{1.533\times 0.85}{\sin 69.8°} = 1.388 \text{ pu } \textbf{Ans.}$$

i.e., the minimum pu excitation required for the machine to remain in synchronous while delivering full-load torque = 1.388 pu **Ans.**

Example 16.25. A 6.6 kV, 3-phase, star-connected synchronous motor is running in parallel with an infinite bus. Its direct and quadrature-axis synchronous reactances are 10 Ω and 5 Ω, respectively. If the field current is reduced to zero, find the maximum load that can be put on the synchronous motor. Also calculate the armature current and the maximum power developed. Neglect armature resistance.

[U.P. Technical Univ. Electromechanical Energy Conversion-II, 2007-08]

Solution: Line voltage, $V_L = 6.6$ kV = 6,600 V

Direct-axis synchronous reactance, $X_d = 10\ \Omega$

Quadrature-axis synchronous reactance, $X_q = 5\ \Omega$

Field current, $I_f = 0$, so induced emf $E = 0$

Power input is given by

$$P = \frac{E_L V_L}{X_d}\sin\delta + \frac{V_L^2}{2}\left(\frac{1}{X_q} - \frac{1}{X_d}\right)\sin 2\delta \quad \text{...Refer to Eq. (16.18)}$$

Substituting $E_L = 0$ in above equation we have

$$P = \frac{V_L^2}{2}\left[\frac{1}{X_q} - \frac{1}{X_d}\right]\sin 2\delta$$

For maximum reluctance power $\sin 2\delta = 1$ or $\delta = 45°$

$$\text{and } P_{max} = \frac{V_L^2}{2}\left[\frac{1}{X_q} - \frac{1}{X_d}\right] = \frac{(6{,}600)^2}{2}\left[\frac{1}{5} - \frac{1}{10}\right]$$

$= 2{,}178{,}000$ W = 2,178 kW **Ans.**

For maximum power $\delta = 45°$

$$\text{and } I_d = \frac{V_p\cos\delta}{X_d} = \frac{6.6\times 1{,}000/\sqrt{3}}{10}\cos 45° = 269.45 \text{ A}$$

$$\text{and } I_q = \frac{V_p\sin\delta}{X_q} = \frac{6.6\times 1{,}000/\sqrt{3}}{5}\sin 45° = 538.9 \text{ A}$$

Armature current for maximum power,

$$I_a = \sqrt{I_d^2 + I_q^2} = \sqrt{(269.45)^2 + (538.9)^2} = 602.5 \text{ A} \quad \textbf{Ans.}$$

16.14 V- AND INVERTED V-CURVES

It has already been discussed how the variation in field current affects the power factor at which a synchronous motor operates. In case of a synchronous motor driving a constant mechanical load, variation in field or dc excitation current will not only affect the power factor but also the current drawn by the motor.

Except for the change in copper losses due to variation in armature current and a slight change in core loss due to variation in flux, the power input to the motor is almost constant for a constant load.

The power drawn by a 3-phase synchronous motor is given by, $P = \sqrt{3}\ V_L\ I_L \cos\phi$ where V_L is the line voltage, I_L is the armature current and $\cos\phi$ is the power factor. Since input power P and supply voltage V is constant, so decrease in power factor causes increase in armature current and vice versa.

Hence variation in excitation or in field current causes the variation in armature current and curves drawn between armature current and field current for different constant loads are known as "V-curves" due to their shape similar to English letter V. *The V-curves of a synchronous motor give relation between armature current and field current for different power inputs.*

Similarly the variation of power factor with a variation in field current (dc excitation) for a constant load gives inverted V curve, shown in Fig. 16.10.

The connections for determining the V-curves experimentally are shown in Fig. 16.36.

The synchronous motor is coupled to a shunt wound dc machine which is used as a dc motor for driving the synchronous machine up to the time of synchronising. After synchronising the dc supply to the shunt machine is interrupted by throwing double pole double throw (DPDT) switch S_1 to the terminals 2-2′. Now synchronous machine operates as a motor and drives the dc shunt machine, now working as a generator and supplying power to load resistance R_L.

First of all synchronous motor is started by the dc shunt motor and brought to its rated speed. Resistances R_1 and R_f are employed respectively for the purpose of starting and controlling the field current to dc shunt machine. During starting field control resistance R_f is kept zero.

When the synchronous motor attains nearly the synchronous speed, its field is energised and voltage of the machine is built up to the bus-bar voltage (ac). Now the synchronous motor is synchronised with the bus-bars, and dc supply to the dc shunt machine is interrupted by opening the switch S_1. Now the synchronous motor is running on no load and it can drive the dc shunt machine working as a dc shunt generator. DC shunt machine is made to operate as a dc shunt generator by making resistance R_1 in the armature circuit as zero and exciting it and then loading it.

(*i*) Keeping the synchronous motor on no load, take the readings of ammeter, voltmeter and wattmeters on ac side for different values of excitation. The excitation is varied with the help of rheostat provided in the field circuit of the synchronous motor.

Now the dc generator is loaded so that the motor is loaded to about one-fourth of its full load. Keeping this load constant again note the readings of ammeter, voltmeter and wattmeters on ac side for different excitations. Repeat this process for one-half full load, full load and 1.25 times full load on synchronous motor.

The V-curves for no load, half load, full load and 1.25 times full load are shown in Fig. 16.37.

By the aid of V-curves, the power factor for any value of armature current I and given input power can be determined from the relation power factor, $\cos\phi = \frac{I_{min}}{I}$ where I_{min} is the minimum current drawn by the motor at given load.

From V–curves for different power inputs it is observed that with low value of field current, the armature current is large and lagging. As the field current is increased the power factor increases and armature current decreases until it reaches its minimum value. When the armature current is minimum, the power factor is unity and the corresponding field current is known as *normal field current or excitation* of the motor for that particular load. The point at which unity power factor occurs is shown by dotted curve. The lines drawn through points of equal power factor at different loads are known as *compounding curves*. The region in which field current is less than its normal value is known as

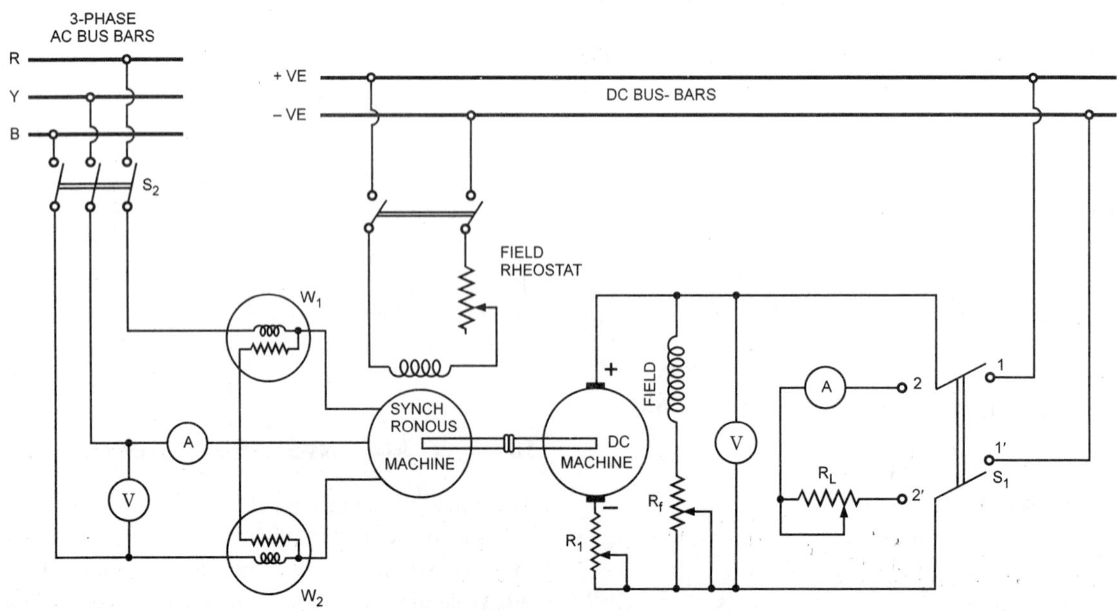

Fig. 16.36 *Connections For Obtaining V-Curves of a Synchronous Motor*

region of underexcitation or region of lag. If the field current is further increased, the power factor becomes leading and begins to decrease, so armature current begins to increase. This region, in which field current is more than normal value of field current or armature current leads, is known as *region of overexcitation or region of lead*.

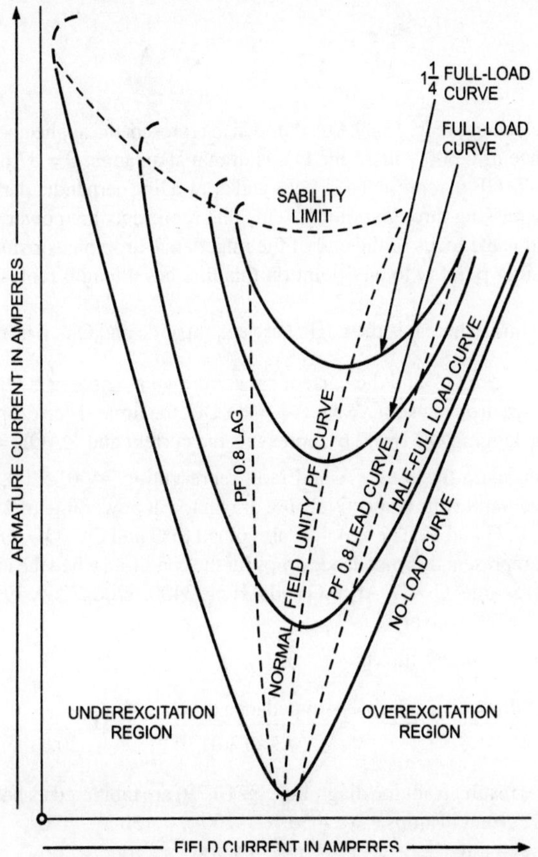

Fig. 16.37 *V-Curves of a Synchronous Motor*

From V–curves it is also to be noted that the no-load armature current rises sharply, with the change in field current, on either side of its unity power factor point. A synchronous motor when operated on no load with an unexcited field, but at synchronous speed, may draw as large as 150% of the rated current or even more. A large change of excitation makes relatively less difference in the armature current at full load. With the reduction in excitation, the limit of stability (for the given load) is reached; at this point the motor pulls out of synchronism. Under light load, say, up to 30% of rated, the excitation may be reduced to zero in most of the motors without causing them to loose synchronism.

In operation, the excitation current of the motor is usually adjusted at the rated value (value of dc excitation mentioned on the name plate), and is kept at this value at all loads. Thus by keeping the excitation at rated value, full pull-out torque is maintained and at the same time maximum amount of reactive kVA are available for power factor correction. If the motor is to be operated for long duration of time at a considerably reduced load, it may be desirable, from the point of view of efficiency, to reduce the excitation to a value below that marked on the name plate to give power factor mentioned on the name plate for the actual reduced load point of operation.

16.15 CONSTANT POWER LINES

Let OA represent applied voltage per phase of the motor, and AB represent the motor induced emf/phase (Fig. 16.38). Then OB, the resultant of OA and AB is the resultant emf acting in the armature circuit. Let it be E_R.

The armature current, $I = \frac{E_R}{Z_s}$ will lag induced emf E_R by θ

where $\theta = \text{Tan}^{-1} \frac{X_s}{R_e}$.

Draw phasor OI representing the current in phase as well as in magnitude. Then ∠AOI = ϕ, the phase of the current with respect to applied voltage V. Draw OY inclined at θ to OA (or at ϕ to OB). Draw BC ⊥ to OX which is at right angle to OY. Since OB is proportional to the armature current, the phasor OB, when referred to OY, represents the current in both magnitude and phase.

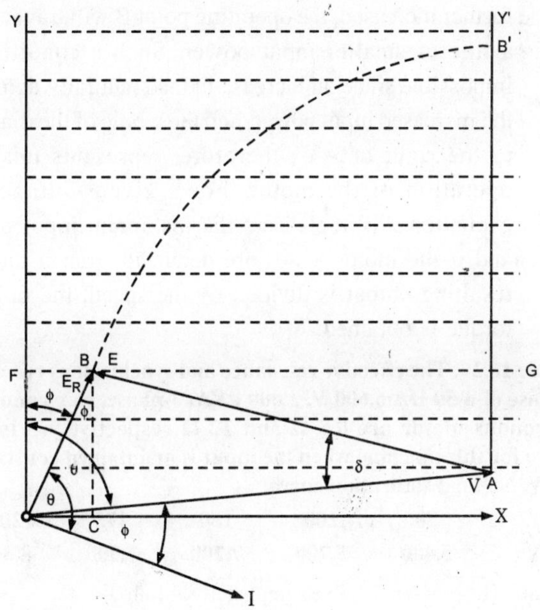

Fig. 16.38

The power input/phase to the motor

$$= V I \cos\phi = V \times BC.$$

As V is constant, power input is proportional to BC. Thus if the motor is operating with constant input power, then locus of the working point B is a straight line passing through point B and parallel to OX (or perpendicular to OY) since this line is at a constant distance from OX. Hence FG represents a constant power input line for a given voltage but varying excitation (or varying induced emf). Similarly, a series of such parallel lines can be drawn, each representing a definite input power of the motor.

The important points regarding constant power lines are:

(*i*) For equal increments of input power, the power lines are equally spaced.

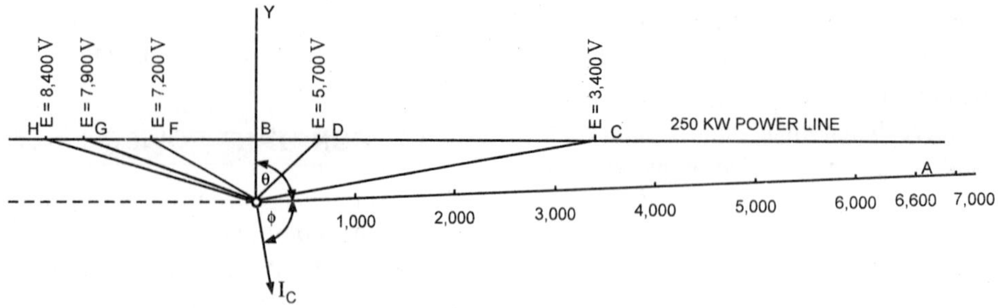

Fig. 16.39

(*ii*) The perpendicular distance from working point B to line OX represents the motor input.

(*iii*) The line OX represents zero power line.

(*iv*) If the excitation is constant, the magnitude of induced emf, E (or AB) is constant and the locus of the working point B is a circle shown dotted in Fig. 16.38 with A as centre and AB as radius. With the increase in load, the load angle δ increases, and so operating point B goes on to line of higher power point till B′ is reached at which the input power is obviously a maximum. If the load is further increased, the operating point B will travel on to a line of smaller input power. Such a condition is impossible since an increase in load naturally demands the increased input power, and the whole of the diagram to the right of AY′, therefore, represents unstable operation of the motor. For a given voltage and excitation AB' represents the maximum input power and if the motor losses are deducted from it and the resulting output is divided by the speed, the pull-out torque is obtained.

Example 16.26. The effective resistance and synchronous reactance per phase of a 50 Hz, 6,600 V, 1,000 kVA, 3-phase, star-connected synchronous motor are 0.5 Ω and 12 Ω respectively. Plot the V-curve for this machine when the input is maintained constant at 250 kW. No-load saturation curve:

I_f	50	100	150	175	200
V	**3,400**	**5,700**	**7,200**	**7,900**	**8,400**

Solution: The problem is solved graphcially as follows:

Motor input = 250 kW = 250,000 watts

Supply line voltage, V_L = 6,600 V

$$\text{Supply current, } I = \frac{250{,}000}{\sqrt{3}\, V_L \cos\phi} = \frac{250{,}000}{\sqrt{3}\times 6{,}600\,\cos\phi} = \frac{21{,}9\,\text{A}}{\cos\phi}$$

$$\text{or } I\cos\phi = 21.9 \text{ A}$$

$$\text{Impedance, } Z_s = \sqrt{(0.5)^2 + (12)^2} = 12.01\ \Omega$$

$$\text{Internal angle, } \theta = \text{Tan}^{-1}\frac{X_s}{R_e} = \text{Tan}^{-1}\frac{12}{0.5} = 87.6°$$

$$\text{Line internal drop, } E_R = \sqrt{3}\ I Z_s$$

$$= \sqrt{3} \times \frac{21.9}{\cos\phi} \times 12.01 = \frac{455}{\cos\phi} \text{ volts}$$

Referring to Fig. 16.39, OA is drawn to represent the applied potential difference of 6,600 volts. Line OY is drawn at an angle θ = 87.6° from OA. Make OB to represent 455 volts and draw a line perpendicular to line OY and passing through point B. This line represents the power of 250 kW, and is the locus of the end of the impedance drop phasor since any line joining point O to any point on this line has a length representing $\frac{455}{\cos\phi}$ volts, where φ is the angle between this line and OY. φ is also the angle between OA and the current phasor drawn at angle of 87.6° from impedance drop phasor *e.g.* when φ = ∠YOC, the impedance drop = OC then OI_c lagging OC by 87.6° represents the current and $\angle AOI_c = \phi$.

Now with the centre A and radii representing 3400, 5700, 7200, 7900 and 8400 volts respectively are drawn to cut power line of 250 kW at C, D, F, G and H. These points are joined to O and OC, OD, OF, OG and OH represent the impedance drops for the condition when the induced (line) emfs are AC, AD, AF, AG and AH *i.e.* 3400, 5700, 7200, 7900 and 8400 volts respectively.

Now armature current,

$$I = \frac{\text{Impedance drop}}{\sqrt{3}\,Z_s} = \frac{\text{Impedance drop}}{\sqrt{3}\times 12.01} = \frac{\text{Impedance drop}}{20.8}$$

The results from the diagram (Fig. 16.39) are tabulated as below

Field current in amps	50	100	150	175	200
Open-circuit emf (line) in volts	3,400	5,700	7,200	7,900	8,400
Impedance drop (line) in volts	3,400	1,100	700	1,300	1,750
Armature current, I in amperes	164	53	33.7	62.5	84.3

From the above table V-curve is drawn, as shown in Fig. 16.40.

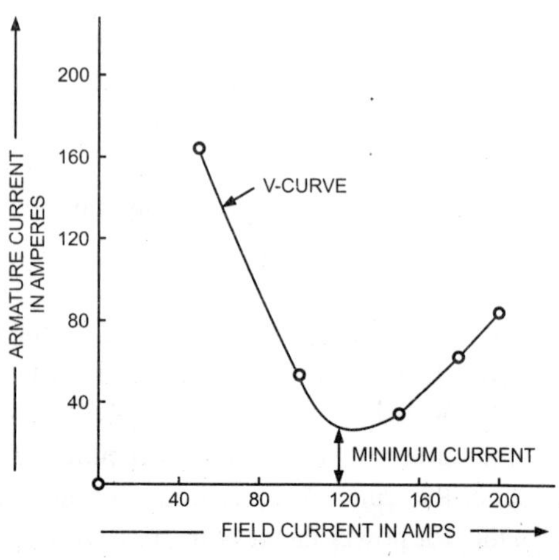

Fig. 16.40

16.16 CIRCLE DIAGRAMS

The circle diagram is an extension of the synchronous impedance diagram and is most accurate when applied to the smooth cylindrical (non-salient pole) motor. The complete diagram contains two groups of circles viz the excitation circles and the power circles.

16.16.1. Excitation Circles (Constant Excitation Circles). The voltage equation for the synchronous motor, as mentioned in Art 16.8, is given as

$$\mathbf{V} = -\mathbf{E} + \mathbf{I}\,\mathbf{Z}_s$$

or

$$\mathbf{I} = \frac{\mathbf{V}}{\mathbf{Z}_s} + \frac{\mathbf{E}}{\mathbf{Z}_s} \quad ...(16.21)$$

Taking V as the reference, the phasor E, although constant in magnitude for the given condition of constant excitation, varies in phase with respect to V in accordance with the necessary variations in load angle δ. Since constant E traces out a circular locus (Fig. 16.38), the same must be true of $\frac{E}{Z_s}$.

The right-hand components of the Eq. (16.21) gives values of two current phasors and their phasor sum gives armature current I. The first component of armature current is produced by applied voltage V and its limiting impedance is Z_s. This component of current $\frac{V}{Z_s}$ must lag V by an angle θ where $\theta = \text{Tan}^{-1}\frac{X_s}{R_e}$. The second component of armature current produced by induced emf E must likewise lag E by an angle θ. The applied voltage V is assumed fixed in position and so the current $\frac{V}{Z_s}$ will be constant and fixed in position w.r.t. V. The second component of the armature current is constant in magnitude for a given excitation, but its phase position is always θ degrees behind E. Since E varies in position with the load, the current $\frac{E}{Z_s}$ must vary in position with load. When these two voltages V and E are in phase opposition, the armature current I is the arithmetic difference of the two components, $\frac{V}{Z_s}$ and $\frac{E}{Z_s}$. For any load angle δ the current $\frac{E}{Z_s}$ is retarded in time phase from the no-load position and the value of armature current is determined by the phasor sum of the two components, as expressed by Eq. (16.21).

From the relations mentioned above the extremity of the phasor $\frac{E}{Z_s}$ and likewise the extremity of the phasor I must follow the path of a circle as the load is varied on the motor. The locus is known as an *excitation circle.* It defines the magnitude and power factor of the armature current for the fixed excitation and varying load. The centre of this circle regardless of the excitation will be on a line θ degrees behind the voltage V phasor and at a distance from $\frac{V}{Z_s}$ amperes. For 100% excitation the no-load current is zero. For other excitations the no-load current is determined by the upper intersection of the particular excitation circle. Group of excitation circles for different degrees of excitation is given in Fig. 16.41.

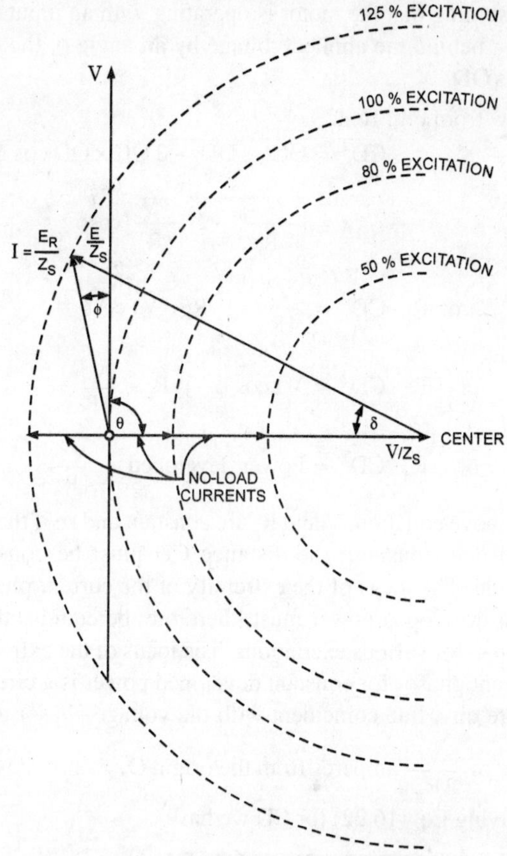

Fig. 16.41 *Constant Excitation Circle Diagram*

16.16.2. Constant Power Circle. Consider a reproduced portion diagram of Fig. 16.41, as illustrated in Fig. 16.42. The distance OA represents the current component $\frac{V}{Z_s}$. Draw a perpendicular to phasor OA from point A intersecting voltage line at point B, and

$$OB = \frac{OA}{\cos\theta} = \frac{V/Z_s}{R_e/Z_s} = \frac{V}{R_e} \text{ amperes}$$

Making $OC = \frac{1}{2}$ OB we have

$$OC = \frac{V}{2R_e}$$

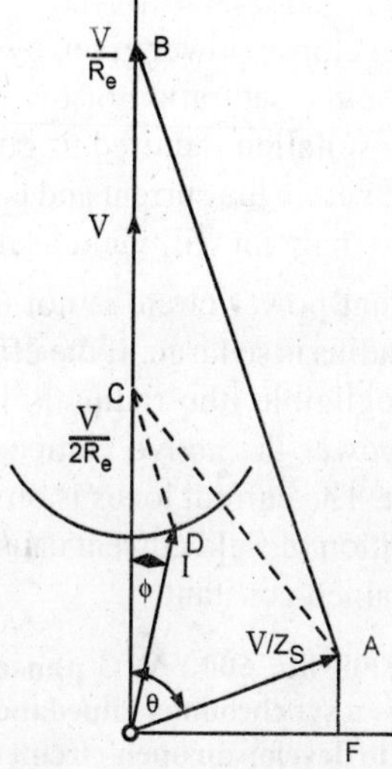

Fig. 16.42 *Constant Power Circle Diagram*

Now assuming that the motor is operating with an input current I lagging behind the applied voltage by an angle ϕ, the current phasor is OD.

Now from Fig. 16.42

$$CD^2 = OC^2 + OD^2 - 2\,OC \times OD \cos\phi$$

$$= \left(\frac{V}{2R_e}\right)^2 + I^2 - \frac{V}{R_e} I \cos\phi$$

or $$R_e \cdot CD^2 = \frac{V^2}{4R_e} + I^2 R_e - VI\cos\phi$$

or $$-R_e \cdot CD^2 = VI\cos\phi - I^2 R_e - \frac{V^2}{4R_e}$$

or $$-R_e \cdot CD^2 = \text{Power developed} - \frac{V^2}{4R_e} \quad ...(16.22)$$

In above equation, V and R_e are constant and so if the power developed is constant, the distance CD must be constant in magnitude. The locus of the extremity of the current phasor for constant developed power must, therefore, be equidistant from the point C for various excitations. The locus of the extremity of the current phasor for constant developed power is a circle with its centre on a line coincident with the voltage phasor and at a distance of $\frac{V}{2R_e}$ amperes from the origin O.

Solving Eq. (16.22) for CD we have

$$CD = \sqrt{\frac{V^2}{4R_e^2} - \frac{\text{power developed}}{R_e}} \text{ amperes} \quad ...(16.23)$$

From above expression it is evident that for zero developed power CD = $\frac{V}{2R_e}$ and since CA = OC, the no-load circle will pass through O and A. Circles of smaller radii give greater power. The maximum possible developed power will coincide points D and C and for this condition

$$\text{Power developed} = \frac{V^2}{4R_e^2} \times R_e = \frac{V^2}{4R_e} \quad ...(16.24)$$

The power supplied to the armature under these conditions will be equal to $\frac{V}{2R_e} \times V$ *i.e.* $\frac{V^2}{2R_e}$ watts.

The maximum developed power given by Eq. (16.24) is an imaginary one because the saturation of the machine will not permit the degree of excitation required to give this particular adjustment. However, within the current and power factor range of the motor the circle diagram will yield useful results.

Note. The constant power circle is not of great practical interest because the radius is so large. If the effective resistance of the armature is negligible, the radius is infinite since for constant developed power the active component of current I cos ϕ will be constant. The current locus is obviously a straight line in this case, variation of field current causing the kVAR to vary, with kW maintained constant.

Example 16.27. A 1,000 hp, 6000 V, 3-phase, star-connected synchronous motor has a synchronous impedance of (1.5 + *j* 16) Ω per phase. It is excited to develop an open-circuit emf of 5 kV. Draw the locus diagram of the current for loads up to 1,250 hp with constant excitation. Determine the maximum value of the power factor.

Solution: Supply voltage per phase, $V = \frac{6,000}{\sqrt{3}} = 3,464$ V

Induced emf per phase, $E = \frac{5 \times 1,000}{\sqrt{3}} = 2,887$ V

Impedance, $Z_s = \sqrt{(1.5)^2 + 16^2} = 16.07\ \Omega$

Internal angle, $\theta = \text{Tan}^{-1}\frac{X_s}{R_e} = \text{Tan}^{-1}\frac{16}{1.5} = 84.6°$

Current due to supply voltage,

$$\frac{V}{Z_s} = \frac{3,464}{16.07} = 215.6 \text{ A}$$

Current due to induced emf,

$$\frac{E}{Z_s} = \frac{2,887}{16.07} = 179.6 \text{ A}$$

Since power input = $\sqrt{3}\ V_L I \cos\phi$

So $$I\cos\phi = \frac{\text{Power input}}{\sqrt{3}\,V_L} = \frac{1,250 \times 735.5}{\sqrt{3} \times 6,000} = 88.5 \text{ A}$$

Locii of induced emf of 5 kV and current for loads up to 1,250 hp (neglecting losses) are shown in Fig. 16.43.

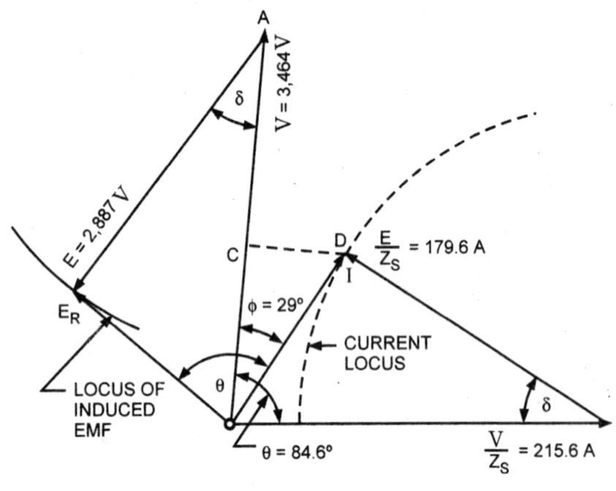

Fig. 16.43

Figure 16.43 is drawn with OA = 3,464 V as reference phasor. Locus of E is an arc with centre A and radius 2,887 V; OB = 215.6 A making an angle θ = 84.6°; CD a line perpendicular to phasor V with OC = I cos φ = 88.5 A. The arc of the circle with centre B and radius of current 179.6 A between origin O and point D is the locus of the stator current. Maximum value of power factor = Cos ∠COD = Cos 29° = 0.875 (lag) **Ans.**

16.17 MAIN CHARACTERISTICS OF SYNCHRONOUS MOTORS

1. Synchronous motor operates only at one speed, known as *synchronous speed*, given by the expression $N_s = \frac{120f}{P}$. The speed is independent of load and can be varied either by varying the supply frequency or by changing the number of poles.

2. It is, inherently, not a self-starting motor and, therefore, before connecting it to an ac supply it is run up to the synchronous speed by some external means.
3. It can be operated under a wide range of power factor both leading and lagging by controlling its excitation current.
4. On no load synchronous motor draws very little current from the supply mains to meet the internal losses. With fixed excitation the input current increases with the increase in load. After the input current reaches maximum no further increase in load is possible. If the motor is further loaded, the motor will stop.
5. If field excitation is small, the stator inductance will be more so the motor draws lagging current and, therefore, power factor is low. If the excitation is increased the stator flux is neutralized by the rotor flux and, therefore, power factor becomes unity. In this position, the current and voltage are in phase and motor current will be minimum.

If excitation is further increased rotor flux increases. It neutralizes the stator flux and causes leading current. This property makes the motor extremely useful as a power factor corrector. By taking leading current it compensates the lagging current drawn by other equipment from the supply mains and thus brings the total current drawn from the supply mains more nearly in phase with the voltage.

16.18 SYNCHRONOUS CONDENSERS

An overexcited synchronous motor running on no load is called the *synchronous condenser* or *synchronous phase advancer* and behaves like a capacitor, the capacitive reactance of which depends upon the motor excitation. Power factor can be improved by using synchronous condensers like shunt capacitors connected across the supply.

In phasor diagram (Fig. 16.44), phasor I_L represents the current drawn by the industrial load, lagging behind the applied voltage V by a large angle ϕ_L and phasor I_M represents the current drawn by the synchronous condenser leading the applied voltage V by the angle ϕ_M. The resultant current I is the phasor sum of I_L and I_M and now angle of lag ϕ is much smaller than ϕ_L. Thus overall power factor is improved from cos ϕ_L to cos ϕ by the use of the synchronous condenser. In this way the power factor can be made unity even.

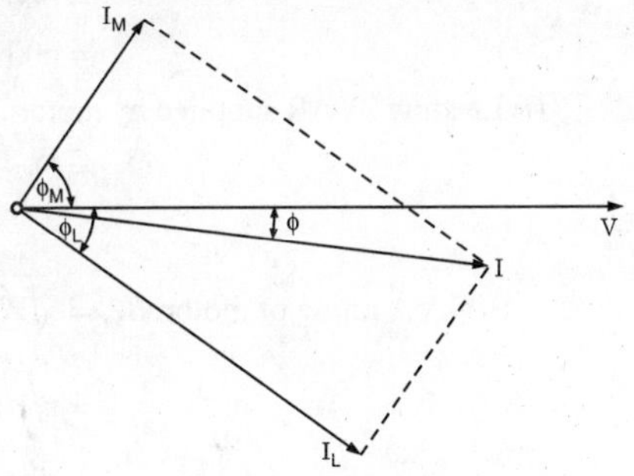

Fig. 16.44

In large industrial plants, which have a low lagging power-factor load, it is often found economical to install a synchronous motor, even though the motor is not required to drive a load. The motor is operated overexcited at no load so that the current drawn by it leads the voltage by nearly 90°. A synchronous motor used in this way is said to "float"on the line because it has no mechanical output. Since the motor operating at no load acts in the same manner as a static capacitor and when operated in this manner it is called the *synchronous condenser* or a *synchronous capacitor*.

When a synchronous motor is operated on no load, it still takes power for its own rotational losses (*i.e.* friction, windage and iron losses). Although the inphase component of armature current is usually comparatively small under no-load operation, the total current may be rather large, as it depends upon dc excitation. When a synchronous condenser is carrying full-rated leading kVA, the current leads the supply voltage by 88 to 89°, depending on the size of the machine. The angles correspond to power factors of 3.5 and 1.7 per cent respectively.

The synchronous condenser is especially designed so that practically all its rated kVA are available for power factor correction. The design of a synchronous capacitor is characterized by its large synchronous reactance and field windings, because its function is to develop zero-power factor currents with the least expenditure of power in losses. In comparison to a synchronous motor with equal armature voltage and current ratings, a synchronous condenser needs more copper in the field winding to carry the increased field current. The synchronous condenser does not require so large shaft and bearings as the synchronous motor because no shaft torque is required.

Synchronous condensers are usually built in large units and are employed where a large quantity of corrective kVAR (say 5,000 kVAR or more) is required.

The advantages of synchronous condensers over static capacitors as a power factor correction devices are (*i*) a finer control can be obtained by variation of field excitation; (*ii*) inherent characteristic of synchronous condensers of stabilizing variations in the line voltage and thereby automatically aid in regulation, (*iii*) possibility of overloading a synchronous condenser for short periods, and (*iv*) improvement in the system stability and reduction of the effect of sudden changes in load owing to inertia of synchronous condenser. By use of synchronous condensers at intermediate stations, the voltage of the line can be kept constant at various points along its length, thereby, increasing the current carrying capacity of the line and improvement of power factor.

The disadvantages of synchronous condensers over static capacitors as power factor correcting devices are : (*i*) except in size above about 5,000 kVAR, the cost is higher than that of static capacitors of the same rating ; (*ii*) comparatively higher maintenance and operating costs ; (*iii*) comparatively lower efficiency (say 97%) due to losses in rotating parts and heat losses, (*iv*) noise is produced in operation (*v*) an auxiliary equipment is required for starting synchronous condensers ; (*vi*) possibility of synchronous condensers falling out of synchronism causing in interruption of supply ; and (*vii*) increase of short-circuit currents when the fault occurs near the synchronous condenser.

Synchronous condensers are largely employed by utilities at large substations for improving the power factor and voltage regulation. Machines up to 100 MVAR rating or even higher have been used. The excitation current is regulated automatically to give a desired voltage level.

Example 16.28. A 500 V, 50 Hz, 3-phase circuit takes 20 A at a lagging power factor of 0.8. A synchronous motor is used to raise the power factor to unity. Calculate the kVA input to the motor, and its

power factor when driving a mechanical load of 7.5 kW. The motor has an efficiency of 85%.

Solution: kVAR drawn by the 3-phase circuit,

$$Q = \sqrt{3}\, V_L I_L \sin\phi \times 10^{-3}$$

$$= \sqrt{3} \times 500 \times 20 \times 0.6 \times 10^{-3} = 10.3923 \text{ kVAR}$$

Power supplied by the motor,

$$P = \frac{\text{Output in kW}}{\eta} = \frac{7.5}{0.85} = 8.8235 \text{ kW}$$

Power factor will be raised to unity when kVAR (leading) drawn by a 3-phase synchronous motor will become equal to the kVAR drawn by 3-ϕ ac circuit.

i.e. kVAR drawn by synchronous motor, Q = 10.3923 (leading)

KW drawn synchronous motor, P = 8.8235

KVA input to the motor,

$$S = \sqrt{P^2 + Q^2} = \sqrt{(8.8235)^2 + (10.3923)^2}$$

$$= 13.63 \text{ kVA } \textbf{Ans.}$$

$$\text{Power factor, } \cos\phi = \frac{P}{S} = \frac{8.8235}{13.63} = 0.6472 \text{ (leading) } \textbf{Ans.}$$

Example 16.29. An alternator supplying 500 kW at 0.7 power factor lagging has its power factor raised to unity by means of an overexcited synchronous motor. At a constant armature current how much input power is required for the synchronous motor? Find the power factor of the syncrhonous motor.

Solution: Load, P = 500 kW

$$\text{Load kVA, } S = \frac{P}{\cos\phi} = \frac{500}{0.7} = 714.3$$

$$\text{Power factor angle, } \phi = \cos^{-1} 0.7 = 45.57°$$

$$\text{Load kVAR} = S \times \sin\phi$$

$$= 714.3 \sin 45.57° = 714.3 \times 0.714 = 510$$

Thus a synchronous motor when connected to improve power factor under overexcited condition will operate as a synchronous condenser. The power required to be supplied by synchronous condenser, therefore, is 510 kVAR and the power factor of the synchronous motor will be zero (leading).

Example 16.30. A substation operating at its full load of 1,000 kVA supplies a load of power factor 0.71 lagging. Calculate the permissible additional load at this power factor and the rating of synchronous condenser to raise the substation power factor to 0.87 lagging.

Solution: Rating of power plant, S = 1,000 kVA

Load supplied at 0.71 pf = 0.71 S = 0.71 × 1,000 = 710 kW

Load supplied at 0.87 pf = 0.87 S = 0.87 × 1,000 = 870 kW

Increase in load supplied = 870 – 710 = 160 kW

Additional load supplied at 0.71 pf (lagging)

$$= \frac{160}{0.71} = 225.35 \text{ kVA } \textbf{Ans.}$$

Reactive power drawn by load

$$= \text{New load in kW} \times \tan\phi_1$$

$$= 870 \tan \cos^{-1} 0.71 = 863 \text{ kVAR}$$

Reactive power supplied by the plant

$$= S \sin\phi_2 = S \sin \cos^{-1} 0.87$$

$$= 1{,}000 \times \sin 29.54° = 493 \text{ kVAR}$$

Rating of ideal synchronous condenser

= Reactive power drawn by load – reactive power supplied by the plant

= 863 – 493 = 370 kVAR **Ans.**

Example 16.31. A synchronous motor improves the power factor of a load of 500 kW from 0.707 lag to 0.95 lag. Simultaneously motor carries a load of 100 kW. Find (*i*) the leading kVA supplied by the motor. (*ii*) kVA rating of motor (*iii*) power factor at which the motor operates. [*G.G.S.I.P Univ. Electrical Machines, December-2011*]

Solution: Load supplied, P_L = 500 kW

Load power factor, $\cos\phi_L = 0.707$

$$\text{Load kVAR, } Q_L = P_L \tan\phi_L$$

$$= P_L \tan(\cos^{-1} 0.707)$$

$$= 500 \tan 45° = 500 \text{ kVAR}$$

Synchronous motor load, P_M = 100 kW

Total load, $P = P_L + P_M = 500 + 100 = 600$ kW

Combined pf, $\cos\phi = 0.95$ (lagging)

Phase angle, $\phi = \cos^{-1} 0.95 = 18.195°$

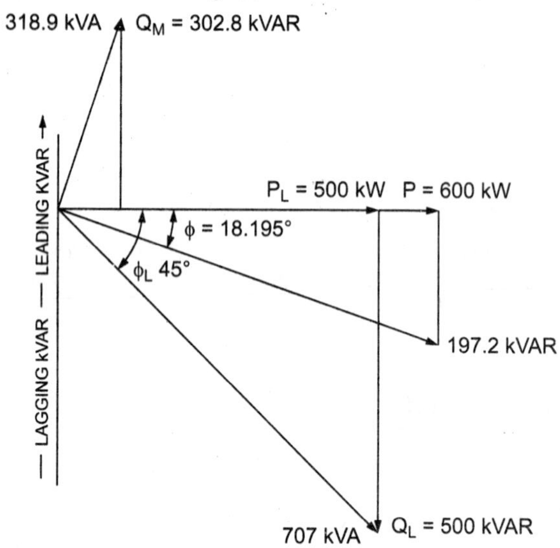

Fig. 16.45

$$\text{Combined kVAR} = P \tan\phi$$

$$= 600 \tan 18.195°$$

$$= 600 \times 0.328684 = 197.21 \text{ kVAR}$$

(*i*) Leading kVAR supplied by motor,

$$Q_M = Q_L - Q = 500 - 197.2$$

$$= 302.8 \textbf{ Ans.}$$

$$(ii) \text{ kVA rating of motor, } S_M = \sqrt{P_M^2 + Q_M^2}$$

$$= \sqrt{100^2 + 302.8^2} = 318.9 \textbf{ Ans.}$$

$$\text{Power factor of motor, } \cos\phi_M = \frac{P_M}{S_M} = \frac{100}{318.9}$$

$$= 0.3136 \text{ (leading) } \textbf{Ans.}$$

Example 16.32. An industrial load of 4,000 kW is supplied at 11 kV, the pf being 0.8 lagging. A synchronous motor is required to meet an additional load of 1,500 hp (1,103.25 kW) and at the same time to raise the resultant power factor to 0.95 (lagging). Determine the kVA capacity of the motor and the power factor at which it must operate. Take the efficiency of the motor as 80%.

[*U.P.S.C. I.E.S. Electrical Engineering-II, 2011*]

Solution: Industrial load, P_L = 4,000 kW

Power factor of load, $\cos\phi_L = 0.8$ lagging

$$\text{Load kVAR, } Q_L = P_L \tan\cos\phi_L$$

$$= 4{,}000 \tan(\cos^{-1} 0.8) = 3{,}000 \text{ (lagging)}$$

$$\text{Input to synchronous motor, } P_M = \frac{\text{Additional load in kW}}{\text{Motor efficiency}}$$

$$= \frac{1,103.25}{0.8} = 1,379.0625 \text{ kW}$$

$$\text{Total load, } P = P_L + P_M = 4,000 + 1,379.0625$$

$$= 5,379.0625 \text{ kW}$$

Power factor of combined load,

$$\cos\phi = 0.95 \text{ (lagging)}$$

$$\text{Phase angle, } \phi = \text{Cos}^{-1}\, 0.95 = 18.195^\circ$$

$$\text{Combined kVAR, } Q = P\tan\phi = 5,379.0625 \tan 18.195^\circ$$

$$= 1,768 \text{ (leading)}$$

kVAR supplied by the synchronous motor,

$$Q_M = Q_L - Q = 3,000 - 1,768 = 1,232 \text{ (leading)}$$

kVA capacity of the motor,

$$S_M = \sqrt{P_M^2 + Q_M^2} = \sqrt{(1,379.0625)^2 + (1,232)^2}$$

$$= 1,849.23 \text{ kVA } \textbf{Ans.}$$

Power factor of synchronous motor,

$$\cos, \phi_M = \frac{P_M}{S_M} = \frac{1,379.0625}{1,849.23} = 0.746 \text{ (leading) } \textbf{Ans.}$$

16.19 SYNCHRONOUS PHASE MODIFIERS OR COMPENSATORS

It is well known that a synchronous motor can be made to take either a lagging or leading current from the line by altering its excitation. Idle-running synchronous motors were first employed in connection with power plants to correct for low power factor of the load, and thus reduce the current and power losses in the feeders and generators. The synchronous machines used for power factor improvement are usually referred to as '*synchronous condensers*' as they are always required to take a leading current.

However, an idle-running synchronous motor can be used for voltage regulating purposes by connecting it in parallel with the load at the receiving end of a line, as shown in Fig. 16.46. Since the machines used for voltage regulation are required to run part of the time with leading current and part of the time with lagging current depending upon the load conditions, therefore, it is more appropriate to refer to these machines as '*synchronous phase modifiers*'. They are usually of salient pole design with 6 or 8 poles with ratings up to 60 MVA, 11 kV and connected to high-voltage system through transformers.

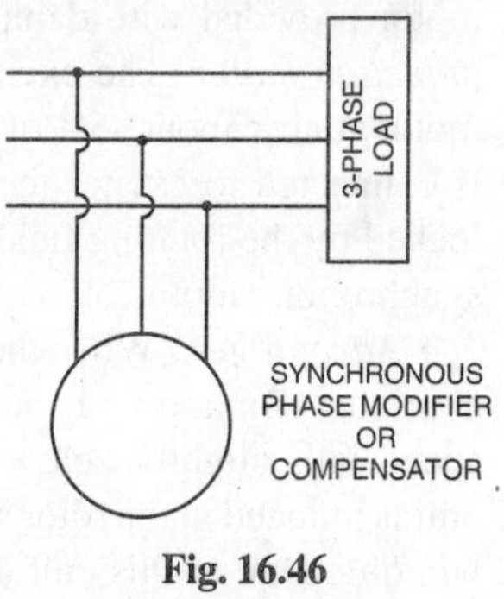

Fig. 16.46

Synchronous phase modifiers differ from the ordinary synchronous motors in as much as they are built for the highest economical speeds, and provided with smaller shafts and bearings and special attention is paid for securing a high overall efficiency. Standard machines for this purpose are designed to give their full-load output at leading power factor, and can carry about 50% of their rated capacity at lagging power factor. Machines can, however, be designed to operate at full rating on both leading and lagging power factors but they are larger in size, have poor efficiency and are more expensive than standard machines.

A transmission line needs lagging VARs (reactive volt-amperes) at the receiving end during peak load conditions to avoid excessive voltage drop and these lagging VARs are supplied by the synchronous phase modifier when it is overexcited as it absorbs leading VARs when overexcited. During off-peak hours the transmission line needs leading VARs at the receiving end to prevent any voltage rise. These leading VARs are supplied by the synchronous phase modifier when it is underexcited as it absorbs lagging VARs when underexcited. Thus synchronous phase modifier or compensator is a very flexible source of supply of reactive VARs (leading or lagging as per need) and is well suited for voltage regulation (or compensation).

16.20 HUNTING AND DAMPING

We have seen that when a synchronous motor is loaded the rotor poles fall back by a certain angle, called the load angle and represented by δ behind the poles of the forward rotating magnetic field. If the load is suddenly thrown off, the rotor poles are pulled into almost exact position to the poles of the forward field, but due to inertia of rotor the rotor poles travel too far. They are then pulled back again, and so on. Thus an oscillation is set up about the equilibrium position corresponding to new load. The oscillation of the rotor about its equilibrium position is called the *hunting*. The hunting is caused by change in load, change in excitation, or change in some other condition, such as explained in Art. 16.20.1, of the system. In some cases, when the mass of the rotating parts has an oscillating period that is the same as, or some multiple of, the hunting period, mechanical resonance is set up and the amplitude of the swing of the rotor poles relative to the poles of rotating field becomes too great that either the motor circuit gets opened by the circuit protective devices, or the motor falls out of synchronism and stall. Hunting is, of course, an *objectionable characteristic* of all synchronous motors. The most significant adverse effect of hunting is increase in probability of loosing synchronism. Hunting also produces severe mechanical stresses as well as large variations in current and power drawn by the motor. Other effects of hunting are increase in losses resulting in increase of temperature and increase in possibility of resonance. In case the frequency of the torque component becomes equal to that of the transient oscillation of the machine, resonance may occur.

Hunting is suppressed by providing a damper winding, also known as the *squirrel cage* winding or *damper grids*. The damper winding consists of short-circuited copper or bronze bars embedded in the faces of the field poles of the synchronous motors (Fig. 16.2). When the machine is operating at exactly synchronous speed, there is no relative motion between the rotor and air-gap flux and so no voltage is induced in this damper or squirrel cage winding. Thus at synchronous speed of the motor no current flows through this additional winding and operation of the machine is not affected. But when the speed of the machine deviates from the synchronous speed or hunting takes place, there is a relative motion between the rotor (or damper winding) and the air-gap flux and, therefore, eddy currents are set up in this winding in such a way as to suppress the oscillations. The effectiveness of the damping winding in suppressing the rotor oscillations, however, depends upon the resistance of this winding; the lower the resistance, the stronger is the damping action. Fortunately this winding also serves to make the synchronous motor self-starting and a high squirrel cage resistance is necessary for developing good starting torque, it is customary to use a compromise value for the cage. The term *amortisseur* is generally used for the squirrel cage in connection with its damping action.

The dampers cannot prevent hunting completely because their operation depends upon the presence of oscillating motion.

Other methods of suppressing hunting by using flywheels or by designing synchronous machines with suitable synchronising power coefficients.

16.20.1. Causes of Hunting

1. Sudden change in load or excitation current, as already mentioned in Art 16.20.
2. Occurrence of fault in the system that is supplied by the alternator.
3. Cyclic variations in load torque.

16.21 STARTING METHODS OF SYNCHRONOUS MOTORS

As already stated the synchronous motor has got no self-starting torque and the motor must first be brought up to synchronous speed by some external means before it is capable of operating with load angle δ in the manner previously described. Various methods of starting of synchronous motors are described as follows:

1. From DC Source. If the dc supply and dc compound motor is available, the synchronous motor is coupled and started by means of a dc compound motor. The speed of dc motor is adjusted by the speed regulator. The synchronous motor is then excited and synchronised with ac supply mains. At the moment of synchronising the synchronous motor is switched on with the ac mains, and either the dc motor is disconnected from the dc supply mains or the field of the dc machine is strengthened until it begins to function as a generator. Now the synchronous machine is operating, as a motor, from ac supply mains and dc machine acts as load on it.

The synchronous motor can also be started by the exciter mounted on an overhung synchronous motor bracket and shaft extension. Here again, an available dc source operates the exciter as a motor during the starting period; then, after the synchronous machine is brought up to speed and synchronised, the exciter assumes its normal function.

2. By Means of AC Motor. A small direct-coupled induction motor, called the *pony motor*, may be used for starting the synchronous motor unless the motor is required to start against full-load torque. The induction motor frequently has two poles less than the synchronous motor and so is capable of raising the speed of the latter to synchronous speed. Before switching the ac supply to the synchronous motor, it must be synchronised with the bus-bars. After normal operation is established, the pony motor is sometimes uncoupled from the synchronous motor. Since

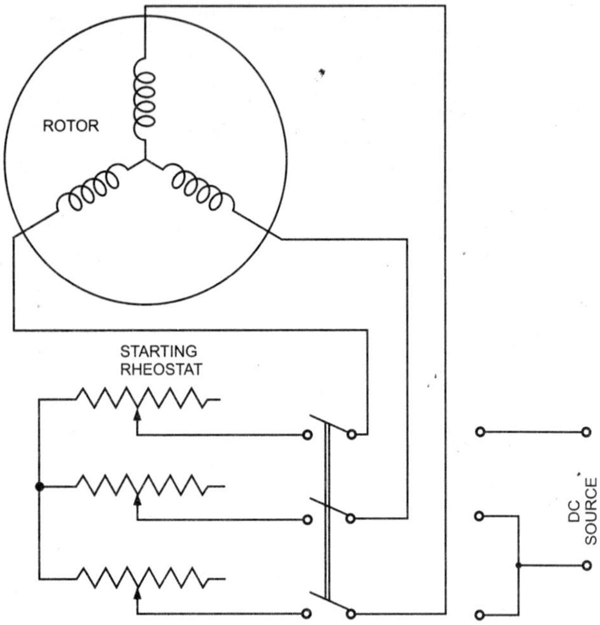

Fig. 16.47

motor operates on no load prior to synchronising, the starting motor has to overcome the inertia of the synchronous motor at no load and therefore, the rating of the starting motor is much smaller than the rating of the synchronous motor. This method is not very satisfactory and not suited to industrial needs. Modern machines are usually of the self-starting type and are arranged to start as induction motors.

3. By Means of Damper Grids in The Pole Faces. The synchronous motor is made self-starting by providing a special winding on the rotor poles, known as *damper winding* or *squirrel cage winding*. The damper winding consists of short-circuited copper bars embedded in the face of the field poles. AC supply given to the stator produces a rotating magnetic field which causes the rotor to rotate, therefore, in the beginning synchronous motor provided with damper winding starts as a *squirrel cage induction motor*. The exciter moves along the rotor. When the motor attains about 95% of synchronous speed, the rotor winding is connected to exciter terminals and the rotor is magnetically locked by the rotating field of the stator and the motor runs as a synchronous motor.

Advantages. When the motor is overloaded it does not stop, continues running as a squirrel cage induction motor. When the speed falls slightly below synchronism due to overloading an emf is induced in the rotor short-circuited winding called *damper winding*. Due to this emf a torque is produced which keeps the motor running even on overload.

Disadvantages. (*i*) Since damper winding resistance is low, at start it takes large current from supply mains.

To avoid large starting current drawn from the supply mains the synchronous motor is started usually as slip-ring induction motor. The rotor winding is connected in series with triplex rheostat at the time of starting by throw-over switch. Rotor circuit is completed and an emf is induced in the rotor circuit as in the rotor of a slip-ring induction motor.

As the motor attains synchronous speed, the throw-over switch is put on the side of the exciter terminals and the dc supply passes through the rotor winding. While running, the rotor acts like rotor of synchronous motor. Hence the motor runs as a synchronous motor. If the starting and excitation is done automatically the motor is known as *auto-synchronous motor*.

(*ii*) Field windings meant for 110 or 250 volts, the exciting voltage generally being this much, should be highly insulated for withstanding high induced emf due to induction at starting.

Precautions. Under normal running conditions, there is no relative motion between the rotating magnetic field and the field winding of a synchronous motor and, therefore, no voltage is induced in the field winding by the air-gap flux. However, if the voltage is applied to the stator winding and the motor is at standstill with the field winding unexcited, the revolving field produced by the polyphase armature (or stator) currents will cut across the field windings at synchronous speed and thereby induce a high voltage in the field winding. This induced voltage would be dangerous and often would result in breakdown of the insulation of the field winding. *Therefore, in starting a synchronous motor the field winding must be shorted through a suitable resistance so that the induced voltage is distributed throughout the whole winding and no part can be subjected to the high voltage induced in the entire winding*.

Another method of eliminating the hazard of breakdown of insulation of the field winding due to induced high voltage in it is by means of dividing the field winding into several sections during the starting period. While emf is still induced in the field coils, the number of coils in series is reduced, so that the magnitude of emf induced in any section is kept within safe limits. Sectionalizing of field winding may be accomplished by a centrifugally operated switch that keeps the field circuit sectionalized until the rotor revolves at near synchronous speed.

16.21.1. Starting of Synchronous Motor Under Load. Synchronous motor with the usual low resistance type of damper has a comparatively low starting torque. The starting torque of a synchronous motor may be increased by increasing the resistance of the rotor winding. Since in a synchronous motor the squirrel cage windings are not effective during normal operation, (as the high resistance starting winding gives at running speed high slip and low efficiency in case of induction motors) therefore, the resistance of squirrel cage can be made sufficiently high to give high starting torque and the motor can be started under considerable load. The high resistance of the squirrel cage may produce such a large slip that locking of rotor with the synchronously rotating stator poles may become impossible. This difficulty is overcome by short-circuiting the field winding near the synchronous speed of the motor.

For the purpose, where the motor is required to be started under considerable loads, synchronous motors with phase wound dampers are employed. These motors consist of usual salient poles but the damper bars, instead of being connected to end rings are phase connected and brought out to the resistors through slip rings. Thus such motors have five slip rings, two for conducting current to dc field winding and other three to be connected to the terminals of the 3-phase rotor winding. The resistors are put in the circuit at start and are taken out as the motor attains the nearly synchronous speed, just as done with slip-ring induction motors. When the three slip rings are short circuited, the winding acts as a damper winding.

This type of motor is used in heavy duty machinery such as conveyors, ball mills and crushers etc.

16.22 SPEED CONTROL OF SYNCHRONOUS MOTORS

The speed of a synchronous motor depends upon two factors viz number of poles, P and supply frequency, f as already mentioned in Art. 16.3. The construction of the rotor of a synchronous motor being fixed, the number of poles on the rotor is likewise fixed and cannot be changed during operation. However, when a single synchronous motor is supplied power from an alternator, as in case of ship propulsion, the speed of the motor can be changed by changing the speed of the alternator–the speed of the motor changes exactly in the same proportion as that of the alternator supplying power to it. It is to be noted here that the voltage and frequency are directly proportional to the speed at which the alternator is driven. So no special control is required for maintaining V/f ratio constant. Field excitation control, however, is required as voltage applied to the terminals of stator is changed. If not, power factor is greatly affected. Assuming the motor is operating at near unity power factor and normal voltage, increase in speed causes increase in applied voltage across the motor stator terminals and the motor is badly underexcited (*i.e.* low lagging power factor) for fixed dc excitation. On the other hand, lowering of speed causes overexcitation (*i.e.*, low leading power factor). Thus field excitation control is essential.

16.23 ELECTRIC BRAKING OF SYNCHRONOUS MOTORS

In principle, a synchronous motor may be stopped by either plugging or rheostatic braking. But the former method has certain limitations, as explained below.

If the direction of dc excitation is reversed, dc and ac fields will rotate in opposite directions and no braking torque will be developed in case of plain synchronous motors (*i.e.*, synchronous motors provided with no damper windings). The braking effect will be produced, however, due to eddy currents induced in damper windings, if provided for starting purposes. Plugging as applied to synchronous motors provided with damper windings, does not prove satisfactory. Because of usually large size of such motors it results in heavy surges of stator current causing line disturbances beyond permissible limits and also it calls for additional control equipment to switch off the motor when the speed falls to zero. Also, since the plugging torque arises from the damper windings and unless it is specially designed for the purpose, the braking torque developed during plugging are of smaller magnitudes. Hence, *rheostatic or dynamic braking* is almost exclusively used for braking synchronous motors.

As already mentioned above, synchronous motors are usually braked with rheostatic braking. In rheostatic braking, the stator of synchronous motor is disconnected from the ac supply and is short circuited through a set of balanced 3-phase resistors, keeping the field excitation unchanged. In case field excitation is increased, greater braking torques will be developed. During dynamic braking, the machine acts as an alternator and dissipates the stored energy in the resistors. The terminal voltage and frequency drop with the fall in speed and both become zero at standstill. The maximum torque occurs at approximately that speed for which the machine reactance becomes equal to the braking resistance. The appearance of high torque immediately prior to stopping has a very positive effect on the braking process.

16.24 SYNCHRONOUS MOTOR RATINGS

The synchronous motor rating includes horse power or kW output, as expected, voltage, frequency, rotational speed in rpm, power factor, full-load current rating and normal dc excitation current.

The voltage rating and frequency are essential because operating flux is determined by these factors. Frequency determines the speed also for a given machine. For normal industrial use the synchronous motors are rated at unity power factor, 90% lead or 80% lead. For special applications synchronous motors may be rated at other power factors. Unity-power-factor motors may be operated at leading power factors but with load sufficiently reduced in order to reduce the stator current to the rated value. When a synchronous motor is rated at 0.8 or 0.9 leading power factor, it will have a larger current carrying capacity for a given kW output.

16.25 MERITS AND DEMERITS OF SYNCHRONOUS MOTORS

Merits.

(*i*) The ease with which the power factor can be controlled. An overexcited synchronous motor having a leading power factor can be operated in parallel with induction motors and other power apparatus operating at lagging power factors, thereby improving the power factor of the supply system.

(*ii*) The speed is constant and independent of load. This characteristic is mainly of use when the motor is required to drive another alternator to generate a supply at a different frequency as in frequency changers.

(*iii*) Electromagnetic power varies linearly with the voltage.

(*iv*) These motors can be constructed with wider air gaps than induction motors, which make them better mechanically.

(*v*) These motors usually operate at higher efficiencies, especially in the low-speed unity-pf ranges.

Demerits

(*i*) The cost per kW output is generaly higher than that of an induction motor.

(*ii*) It requires dc excitation which must be supplied from external source.

(*iii*) The synchronous motor is inherently not self-starting motor and needs some arrangement for its starting and synchronizing.

(*iv*) It cannot be used for variable speed jobs as there is no possibility of speed adjustment.

(*v*) It cannot be started under load. Its starting torque is zero.

(*vi*) It has a tendency to hunt.

(*vii*) It may fall out of synchronism and stop when overloaded.

(*viii*) Collector rings and brushes are required.

(*ix*) For some purposes synchronous motors are not desirable as for driving shafts in small workshops having no other power available for starting and in cases where frequent starting or strong starting torque is required.

16.26 APPLICATIONS OF SYNCHRONOUS MOTORS

Synchronous motors are rarely used below 40 kW output in the medium speed range because of their much higher initial cost in comparison to that of induction motors. In addition they need a dc excitation source, and the starting and control devices are usually more expensive—especially where automatic operation is required. However, there is a kW output and speed range where the disadvantage of higher initial cost vanishes, even to the point of putting the synchronous motors to the advantage. Where low speeds and high kW outputs are involved the induction motors are no longer cheaper because they need large amounts of iron in order to restrict the air-gap flux density to 0.7 T. In the synchronous machines, on the other hand, a value twice this figure is permissible because of the separate excitation.

The various classes of service for which synchronous motors are employed may be classified as (*i*) power factor correction (*ii*) voltage regulation and (*iii*) constant-speed, constant-load drives.

1. They are used in powerhouses and substations in parallel to the bus-bars to improve the power factor. For this purpose they are run without mechanical load on them and overexcited.
2. In factories having a large number of induction motors, or other power apparatus operating at lagging power, they are employed to improve the power factor.
3. Such motors are also used to regulate the voltage at the end of transmission lines, as discussed in Art 16.19.
4. Because of the higher efficiency possible with synchronous motors, they can be employed advantageously for the loads where constant speed is required. Typical applications of high-speed synchronous motors (above 500 rpm) are such drives as fans, blowers, dc generators, line shafts, centrifugal pumps and compressors, reciprocating pumps and compressors, constant speed frequency changers, rubber and paper mills etc.

The fields of applications of low-speed synchronous motors (below 500 rpm) are drives such as reciprocating compressors when started unloaded, dc generators, centrifugal and screw type pumps, vacuum pumps, electroplating generators, line shafts, rubber and band mills, ball and tube mills, chippers, metal rolling mills etc. Flywheel is used for pulsating loads.

High power electronic convertors generating very low frequencies made the operation of synchronous motors at ultra low speeds. Synchronous motors in very large sizes (as high as 10 MW size) operating at ultra low speeds are employed to drive crushers, rotary kilns and variable speed ball mills.

16.27 TROUBLES IN SYNCHRONOUS MOTORS

The possible problems alongwith their symptoms, possible cause and remedies are given below in tabular form :

Problem	*Symptoms and possible causes*	*Possible remedies*
1. Starting (Failure to Start)	(*i*) Faulty connection	Inspect for open or poor connections.
	(*ii*) Wrong connections causing reverse direction of rotation	Reverse any two leads to supply mains.
	(*iii*) One phase open circuit or short circuit	Test, locate and remove the fault.
	(*iv*) Supply voltage too low	Reduce the external impedance connected for starting purpose.
	(*v*) High friction	Check alignment, load friction, belt tightness and bearing tightness and ensure proper lubrication of bearings.
	(*vi*) Field excited	Ensure that the field applying contactor is open and field discharge contactor is closed through discharge resistance.
	(*vii*) Large load	Remove part of the load.
	(*viii*) Some fault in automatic field relay	Check connections, contactor tips and power supply to solenoid.
2. Failure to attain normal (synchronous) speed	(*i*) Low voltage	Increase voltage across the motor stator terminals
	(*ii*) Excessive load	Remove part of load.
	(*iii*) Field excited	As in 1 (*vi*).
3. Failure to pull into synchronism	(*i*) Excessive load	Remove part of the load.
	(*ii*) No field excitation	Check circuit connections (ensure operation of field applying contactor), field rheostat, exciter output and contacts of switches. Check for open circuit in the field or exciter and setting of field rheostat to give rated field current when field is applied.
4. Pulling out of synchronism or Tripping of circuit Breaker	(*i*) Power failure	Restore power supply when available.
	(*ii*) Line voltage too low	Increase dc excitation.
	(*iii*) Low exciter voltage	Increase dc excitation.
	(*iv*) Short circuit in field	Check with low-voltage and polarity indicator and remove the fault.
	(*v*) Open circuit in field and exciter circuit	Test with magneto and remove the break.
	(*vi*) Reversed field	Check with low-voltage and polarity indicator and reverse the leads.
	(*vii*) Excessive torque peak	Check driven machine for bad adjustment or consult motor manufacturer.
	(*viii*) Wide load fluctuations	As given below.
5. Hunting	Wide load fluctuations	Correct excessive torque peak at driven machine or consult motor manufacturer. If the driven machine is compressor, check valve operations. Increase or decrease flywheel size. Try dc field current adjustment.
6. Stator overheat in spots	(*i*) Unbalanced currents	Check connections and ensure tight and proper connections.
	(*ii*) Open-phase	Check connections and set them right.
	(*iii*) Rotor not centered	Realign and shim stator or bearings.
7. Over heating of one or more coils	Short circuit	Replace faulty coil.
8. Over heating of field	(*i*) Excessive field current	Reduce excitation to give stator current of rated value.
	(*ii*) Short circuit in a field coil	Replace or repair the faulty field coil.
9. Overheating of all parts of motor	(*i*) Improper voltage	Check if supply voltage as per rated value.
	(*ii*) Overload	Reduce load.
	(*iii*) No field excitation	Check circuit and exciter.
	(*iv*) Excessive or poor excitation	Adjust the dc excitation to the rated value.
	(*v*) Reversed field coil	Check polarity and, if wrong, reverse leads.
	(*vi*) Excessive room temperature	Supply cool air.
	(*vii*) Improper ventilation	Remove any obstruction and clean out dirt.

HIGHLIGHTS

1. *Synchronous motor* is identical in construction with the alternator and converts ac electrical power into mechanical power.
2. When 3-phase supply is given to the stator of a 3-phase synchronous motor, a revolving field is produced. This field tries to drag the rotor with it but could not do so owing to rotor inertia. Hence starting torque is not developed. Thus a synchronous motor is inherently not self-starting and, therefore, it is necessary to provide some external means to start it and bring its speed near synchronous speed.
3. Synchronous motor operates only at one speed known as synchronous speed given as $N_s = 120f/P$. It can be operated under a wide range of power factor both leading and lagging by controlling its excitation current.
4. A synchronous motor is able to supply increased mechanical load, not by reduction in speed, but by shift in relative positions of the rotor and rotating magnetic field. For increased load with constant excitation, phase angle ϕ increases in lagging direction.
5. In synchronous motors, the armature reaction is one which opposes the action of the main field flux. When the dc field current is reduced so as to draw the current from the ac supply mains at lagging power factor, the armature current drawn from the ac supply mains acts to strengthen the field. On the other hand, if the field current is increased so as to draw a current at leading power factor, the armature current now sets up a flux tending to reduce the flux created by the main field. In any case, the motor tends to operate at a nearly constant flux or generated voltage, as does the transformer.
6. Power developed in a synchronous motor is given by the expression
$$P_{mech} = \frac{EV}{X_s} \sin \delta$$
where E is excitation or induced emf, V is terminal voltage, X_s is synchronous reactance (all phase values) and δ is the load angle in electrical degrees.
7. Maximum value of mechanical power developed is given by expression
$$(P_{mech})_{max} = \frac{EV}{X_s}$$
$\because$ Load angle corresponding to maximum power, $\delta = \theta = 90°$
8. Mechanical power developed is given by expression.
$$\text{Mechanical power developed} = \sqrt{3}\, V_L I_L \cos \phi - 3 I^2 R_e$$
Mechanical power developed per phase is also given by expression
$$P_{mech} = -E\, I \cos (\theta + \gamma)$$
where E is induced emf per phase, I is armature current per phase and $(\theta + \gamma)$ is the angle between E and I.
9. Total power input in a 3-phase salient synchronous motor is given by
$$P_{in} = \frac{E_L V_L}{X_d} \sin \delta + \frac{V_L^2}{2}\left(\frac{1}{X_q} - \frac{1}{X_d}\right) \sin 2\delta$$
Power developed will be equal to power input less copper losses.
10. The maximum value of load angle δ, assuming negligible, stator resistance is given by
$$\text{Cos}\,\delta = \frac{-E}{4V \dfrac{X_d - X_q}{X_q}} + \sqrt{\frac{1}{2} + \left[\frac{EX_q}{4V(X_d - X_q)}\right]^2}$$
11. The V-curves of a synchronous motor give relation between armature current and field current for different power inputs.

    Similarly variation of power factor with a variation in field current (dc excitation) for a constant load gives inverted V-curve.
12. Synchronous condenser is a specially designed synchronous motor for operation without mechanical load with its field overexcited so as to draw leading armature current. It is used for power factor correction.
13. Synchronous condensers are sometimes operated at power factors ranging from lagging through unity to leading for voltage control. When operated in this manner a synchronous condenser is called a synchronous *phase modifier* or *reactor.*
14. The oscillation of synchronous motor rotor about its equilibrium position is known as *hunting*. It is minimised by providing damper winding on rotor.
15. Synchronous motors are rarely used in small ratings (below 40 kW output) in the medium speed range because of being expensive in comparison to induction motors. Where low speeds and high kW outputs are involved the synchronous motors prove economical.

    The various services for which synchronous motors are used may be classified as (*i*) power factor correction (*ii*) voltage regulation and (*iii*) constant speed constant load drives.

EXERCISES

1. Explain the principle of operation of a 3-phase synchronous motor. Why is starting torque not produced in this motor?
[U.P. Technical Univ., Electrical Engineering First Semester, 2005-06]
2. Explain with neat sketch the principle of operation of a 3-phase synchronous motor. Also explain why it will not run at other than synchronous speed. **[U.P. Technical Univ. Electrical Machines, 2010-11]**
3. Explain why 3-ϕ synchronous machines are always run at synchronous speed $\left(N_s = \frac{120\,f}{P}\right)$? The symbols having their usual meanings. **[Electromechanical Energy Conversion-II, G.B. Technical U.P. 2011-12; U.P. Technical Univ., 2013-14]**
4. Describe in detail about the effect of load change on load angle and power factor of a 3-ϕ synchronous motor operating on infinite bus bar and constant excitation.
[Anna Univ. Electrical Machines-II, Nov./Dec.; 2012]
5. Explain the mechanism of torque production in a synchronous motor.
[R.G.P.V. Bhopal, December-2002]
6. Describe briefly the effect of varying excitation upon armature current and power factor of synchronous motor when the input power to the motor is maintained constant.
[G.B. Technical Univ. Electrical Machines and Automatic Control, 2011-12]
7. Explain the effect of excitation on armature current and power factor in a synchronous motor.
[Rajasthan Technical Univ. Electrical Machines-II, 2010]
8. Explain the effect of varying excitation on armature current and power factor in a synchronous motor. Draw V curves and state their significance. **[Rajasthan Univ. Electrical Machines-II, 2006, 2007]**
9. Is it possible to vary the speed of synchronous motor by varying the field excitation or by any other method ? Explain what happens when the field current is increased or decreased.
[U.P. Technical Univ. Electrical Machines, 2006-07]

10. Draw a curve between field current and armature current of a synchronous motor, when power input to the motor and supply voltage are constant. What is the shape of the curve ? Show the region of lagging and leading pf. Why is power input at unity pf minimum ?

[U.P. Technical Univ. Electromechanical Energy Conversion II, 2003-04]

11. A synchronous motor is running at a certain load. When its field current is increased its armature current decreases. Before this change, explain with the help of phasor diagram whether the motor was delivering or absorbing the lagging reactive kVA.

[G.G.S.I.P. Univ. Electrical Machines, December-2011]

12. Draw the equivalent circuit of a synchronous motor and explain how in a synchronous motor by changing the field excitation, it is possible to operate the motor from lagging to leading power factor while keeping the load constant.

[M.D. Univ. Electromechanical Energy Conversion, December-2013]

13. What are the types of synchronous motors? Explain working principle and equivalent circuit of the synchronous motor.

[Rajasthan Technical Univ. Electrical Machines-II, 2011]

14. Describe with the help of diagrams, the principle, construction and phasor diagram of a synchronous motor.

[Rajasthan Technical Univ. Electrical Machines-II, 2013]

15. Draw the equivalent circuit and phasor diagram and derive the expression of electromagnetic power developed by a cylindrical rotor in synchronous motor. **[Rajasthan Univ. Electrical Machines-II, 2005]**

16. Derive an expression for power developed in a synchronous motor.

[J.N. Technological Univ. Electrical Machines-III, May-2013]

17. From the phasor diagram of the salient pole synchronous machine, show that:

$$\text{Tan}\ \delta = \frac{I_a X_q \cos\phi - I_a R_q \sin\phi}{V_t - I_a X_q \sin\phi - I_a R_q \cos\phi}$$

where the symbols having their usual meanings.

[U.P. Technical Univ. Electromechanical Energy Conversion-II, 2013-14]

18. A 3-phase salient pole synchronous motor is connected to an infinite bus. Derive an expression for the electromechanical power developed. What will happen if the excitation is reduced to zero ? Also comment on the stability of the machine as compared to that of a cylindrical rotor machine.

19. Define the V-curves and inverted V-curves at different loading conditions of synchronous motors.

[G.B. Technical Univ. Electromechanical Energy Conversion-II, 2011-12]

20. Discuss in detail how V curves are obtained for a synchronous motor. **[Anna Univ. Electrical Machines-II, Nov./Dec.-2012]**

21. Draw neatly and explain the family of "V-curves" showing the variation of stator input current with respect to the field exciting current of a three-phase synchronous motor operating at 25%, 50%, 75% and 100% of full load. Why more field-ampere turns are required to maintain the unity power factor at higher loads.

[B.P. Univ. of Technology Electrical Machines, 2008]

22. Draw suitable phasor diagrams and explain the mode of operation of a synchronous motor under:

(*i*) Constant load, varying excitation

(*ii*) Constant excitation, varying load.

[Rajasthan Univ. Electrical Machines-II, 2005]

23. Discuss in detail about the excitation circles.

[J.N. Technological Univ. Electrical Machines-II, May-2013]

24. Develop the excitation circle for a cylindrical rotor synchronous motor. How are the circles helpful in studying the steady-state behaviour of synchronous motors.

[J.N. Technological Univ. Electrical Machines-III, May-2013]

25. What is meant by constant power circle for synchronous motor ? How is it derived ?

26. Show that the locus of stator current for a constant output of 3-ϕ synchronous motor connected to a constant voltage, constant frequency bus-bars is a circle.

27. What are the main characteristics of synchronous motor? Explain *V*-curves of synchronous motor.

[Rajasthan Technical Univ. Electrical Machines-II, 2011]

28. What are the salient features of a synchronous motor?

[J.N. Technological Univ. Electrical Machines-III, May-2013]

29. What is a synchronous condenser? Explain with the help of phasor diagram its operation. What are its applications?

[M.D. Univ. Electrical Machines-II, December-2010]

30. What is a synchronous condenser? Explain its operation and utility with phasor diagram.

[W.B. Technological Univ. Electrical Machines-II, 2011-12]

31. Why does a static condenser differ from synchronous condenser ? Discuss the basic philosophy of synchronous condenser with the help of phasor diagram.

[U.P. Technical Univ. Electromechanical Energy Conversion II, 2003-04]

32. What do you understand by a synchronous condenser ? Explain with the help of phasor diagrams its operation and application.

[U.P.S.C. I.E.S. Elec. Engineering II, 1995]

33. What is synchronous condenser? What are the advantages of installing a synchronous condenser in an electrical system?

[Rajasthan Technical Univ. Electrical Machines-II, 2012]

34. What is synchronous condenser? How does it improves the efficiency and regulation of a system?

[Rajasthan Technical Univ. Electrical Machines-II, 2011]

35. Briefly describe the phenomenon of 'hunting' in a synchronous machine. How is it remedied ? **[U.P.S.C. I.E.S. Elec. Engineering II, 1996]**

36. Explain hunting phenomenon in synchronous motors. How can it be reduced, describe in detail.

[G.G.S.I.P. Univ. Electromechanical Energy Conversion-II, May-2011]

37. What is hunting? Discuss various methods to suppress the hunting.

[J.N. Technological Univ. Electrical Machines-III, May-2013]

38. What are the causes of "HUNTING" in synchronous machines? How is it minimised? What are the advantages and disadvantages of "HUNTING" in synchronous machines?

[U.P. Technical Univ. Electromechanical Energy Conversion-II, 2013-14]

39. Explain function of damper winding in a synchronous motor.

[Rajasthan Technical Univ. Electrical Machines-II, 2012]

40. What is a damper winding and in which machine it is found? Where is it placed within the machine? Briefly explain its working.

[W.B. Technological Univ. Electrical Machines-II, 2011-12]

41. What are the methods of starting of synchronous motors?

[U.P.S.C. I.E.S. Electrical Engineering-II, 2006]

42. Explain why a synchronous motor is not self-starting. Describe any one method of starting a synchronous motor.

[J.N. Technological Univ. Electrical Machines-III, May-2013]

43. Explain various starting methods for a synchronous motor.

[Pb. Technical Univ. Synchronous Machines, May-2013]

44. Discuss the advantages and disadvantages of synchronous motors.

[U.P.Technical Univ. Elec. Engineering First Semester 2004-05]

45. Discuss various applications of the synchronous motor in brief.

[Rajasthan Technical Univ. Electrical Machines-II, 2011]

46. Write short notes on the following:

(*i*) V-curve and its importance. **[M.D. Univ., December-2012]**

(*ii*) V-curves and inverted V-curves.

[R.G. Technical Univ., December-2013; Pb. Technical Univ., May-2013]

(*iii*) Salient pole synchronous machine and its particular applicability in synchronous motors. [W.B. Technological Univ., 2011-12]

(*iv*) Power factor control of synchronous motor. [Rajasthan Technical Univ., 2013]

(*v*) Synchronous condenser. [M.D. Univ., May-2009; December-2012; G.B. Technical Univ., 2012-13]

(*vi*) Power factor improvement using synchronous condensers. [Rajasthan Univ., 2002, 2005]

(*vii*) Modes of operations of synchronous motors. [G.B. Technical Univ., 2011-12]

(*viii*) Hunting phenomenon in 3-ϕ synchronous motors. [G.B. Technical Univ., 2011-12]

(*ix*) Hunting and damping. [G.B. Technical Univ., 2012-13]

(*x*) Hunting and damper winding. [R.G. Technical Univ., December-2013]

(*xi*) Starting of synchronous motor. [G.B. Technical Univ., 2012-13; Rajasthan Technical Univ., 2013]

SHORT ANSWER TYPE QUESTIONS WITH ANSWERS

Q. 1. Why synchronous motors having no starting torque of their own? [U.P.S.C. I.E.S. Electrical Engineering-II, 2002]

Ans. In a 3-phase synchronous motor, armature winding is supplied power from a 3-phase supply mains while field winding is excited from a dc source. For production of torque it is essential that the relative velocity between the fields developed by armature and field windings is zero but at starting instant of a 3-phase synchronous motor, the field developed by field winding is stationary while the field produced by armature winding is a rotating magnetic field rotating at a synchronous speed of N_s given by $120f/P$. Hence synchronous motors have no starting torque of their own.

Q. 2. Why a synchronous motor runs only at synchronous speed? [Pb. Technical Univ., May-2013]

Ans. In a 3-phase synchronous motor, the armature winding is energized from a 3-phase supply mains while the field winding is excited from a dc source. For development of continuous torque, it is necessary that the two fields remain stationary with respect to each other. This is possible only when the rotor also rotates at the speed at which the stator field rotates i.e. synchronous speed N_s. When the rotor revolves at synchronous speed and is operating with a constant load, the axis of the mmf of the stator is fixed with respect to the axis of the mmf of the field. It means that the synchronous motor runs only at synchronous speed.

Q. 3. What factors determine the speed of a synchronous motor?

Ans. The speed of a synchronous motor is dependent on (*i*) the supply frequency and (*ii*) the number of poles for which the motor has been designed.

Q. 4. What factors determine the number of poles of a synchronous motor ?

Ans. The number of poles of a synchronous motor is determined by supply frequency and speed.

Q. 5. A synchronous motor runs at N_S rpm at full load. What will be its speed if the load is reduced to half the full-load ?

Ans. N_s rpm.

Q. 6. A synchronous motor designed to operate in USA is brought in India. If in USA it runs at 1,800 rpm, at what speed it will run in India? [G.G.S.I.P. Univ., May-2011]

Ans. The supply frequency in USA is 60 Hz while in India it is 50 Hz and speed of the given synchronous motor is directly proportional to the supply frequency, so the speed of motor in India will be

$$1{,}800 \times \frac{50}{60} \text{ i.e. } 1{,}500 \text{ rpm.}$$

Q. 7. In a 3-phase synchronous motor field mmf lags the air-gap flux, then what is the position of air-gap flux vis-a-vis armature mmf ?

Ans. It also lags.

Q. 8. Will the air-gap flux in a 3-phase synchronous motor lag or lead the armature mmf ?

Ans. Air-gap flux will lag behind the armature mmf.

Q. 9. How the power factor of a synchronous motor can be controlled. [Pb. Technical Univ., May-2013]

Ans. The power factor of a synchronous motor supplying a constant load can be controlled by varying its excitation.

Q. 10. Does the change in excitation affect the synchronous motor speed and power factor ?

Ans. Change in excitation will affect the power factor (not the speed) of the synchronous motor.

Q. 11. What is the effect of increase in excitation of a synchronous motor ?

Ans. When the excitation of a synchronous motor is increased, first power factor improves until it becomes unity and with the further increase in excitation the power factor becomes leading one and decreases.

Q. 12. A three-phase cylindrical rotor synchronous motor, fed from an infinite bus, is operating at unity pf while driving its rated full load. What will be the nature of its operating power factor if its excitation is increased to 120% of its normal value without changing its shaft load? [B.P. Univ. of Technology, 2008]

Ans. Leading. When the excitation of a synchronous motor is increased above 100%, first pf improves until it becomes unity and with the further increase in excitation, the pf becomes leading and decreases.

Q. 13. If V and E_0 be respectively the terminal voltage/phase and open-circuit emf/phase and θ is the internal angle of a 3-phase synchronous motor, what would be the relationship among them if the pf is to be lagging ?

Ans. $E_0^2 = V^2 + E_R^2 - 2VE_R\cos(\theta + \phi)$

Q. 14. Define the term coupling angle in synchronous motor. [M.D. Univ., December-2011]

Ans. The angular displacement between the axes of the stator mmf and rotor mmf is called the torque or coupling angle δ.

Q. 15. What should be the relationship between power angle δ and internal angle θ for stable operation of a 3-phase synchronous motor working with constant excitation. If excitation is increased, what will happen to the power angle ?

Ans. Power angle δ should not exceed the internal angle θ for stable operation of a 3-phase synchronous motor.

With the increase in excitation power angle will decrease.

Q. 16. A 3-phase synchronous motor driving a constant load torque draws power from infinite bus at leading power factor. How power angle and power factor will change if the excitation is increased ?

Ans. Both power angle and power factor will decrease.

Q. 17. What are pull-in and pull-out torques? [Pb. Technical Univ., May-2013]

Ans. **Pull-in Torque.** It refers to the ability of the motor to pull-into synchronism when changing from induction to synchronous motor operation.

Pull-out Torque. It pertains to the ability of the motor to remain in synchronism under rated load conditions.

The maximum torque which the motor will develop without pulling out of step (or synchronism) is called the *pull-out torque*.

Its value varies from 1.25 to 3.5 times the full-load torque.

Q. 18. A synchronous motor develops some mechanical power, even if the field is unexcited. Is it cylindrical or salient pole machine ?

Ans. Salient pole.

Q. 19. Compare the salient pole rotor and cylindrical pole rotor synchronous motor on stability point of view.

Ans. Salient pole rotor machine is more stable than cylindrical pole rotor machine.

Q. 20. What is meant by V curves of synchronous motor ?

Ans. The curves drawn between armature current and field current for different constant loads are known as V-curves.

Q. 21. What is synchronous condenser ? **[Anna Univ. Nov./December-2012]**

Ans. Synchronous condenser is specially designed synchronous motor for operation without mechanical load with its field overexcited so as to draw leading armature current.

Q. 22. In what operating condition is a synchronous motor referred to as synchronous condenser ?

Ans. A synchronous motor operated without mechanical load with its field overexcited is referred to as a synchronous condenser.

Q. 23. What is the use of synchronous condenser ?

Ans. Power factor correction.

Q. 24. An alternator can be used as a synchronous condenser. Under what conditions it should operate ?

Ans. Overexcited synchronous motor.

Q. 25. What is hunting in synchronous machines?

[Pb. Technical Univ., May-2013]

Ans. When a synchronous motor is loaded the rotor poles fall back by a certain angle, called the load angle and represented by δ behind the poles of the forward rotating magnetic field. If the load is suddenly thrown off, the rotor poles are pulled into almost exact position to the poles of the forward field, but due to inertia of rotor the rotor poles travel too far. They are then pulled back again, and so on. Thus an oscillation is set up about the equilibrium position corresponding to new load. The oscillation of the rotor about its equilibrium position is called the *hunting*. The hunting is caused by change in load, change in excitation, or change in some other condition such as explained in Art 16.20.1 of the system.

Q. 26. How will you minimise hunting?

Ans. Hunting is minimised by providing a damper winding, also known as the squirrel cage winding or damper grids (short-circuited copper or bronze bar embedded in the faces of the field poles). Other methods of reducing hunting are by using flywheels or by designing synchronous machines with suitable synchronising power coefficients.

Q. 27. What is the purpose of damping winding?

[Pb. Technical Univ., May-2013]

Ans. The purpose of damper windings is to minimise hunting and make the synchronous motor self-starting.

Q. 28. What is the common starting method used for synchronous motor ?

Ans. By means of damper grids in the pole shoes.

Q. 29. How is the speed of a synchronous motor varied ?

Ans. By varying the frequency of supply source.

Q. 30. What are the advantages of synchronous motor ?

Ans. The advantages of synchronous motor are (*i*) constant operation speed, (*ii*) higher efficiency, especially in the low-speed unity pf ranges and (*iii*) capability of being operated under a wide range of power factor both lagging and leading.

Q. 31. A synchronous motor is connected to an infinite bus and supplying constant torque and operates at unity power factor. If the field current is increased, explain what happens to the following:

(*a*) The magnitude of resultant flux wave.

(*b*) The magnitude of armature current and its power factor.

(*c*) Rotor current away from resultant mmf wave or towards it.

(*d*) Space angle between armature mmf and resultant air-gap mmf.

Ans. (*a*) The motor will become overexcited, the field mmf will increase but increase in field mmf will be neutralised by the increase in counter armature mmf and the magnitude of result flux wave will remain almost constant.

(*b*) The armature current will increase and power factor will drop.

(*c*) Rotor current will move towards the resultant mmf because the resultant flux wave will move towards the leading power factor side causing reduction in angle between rotor current and mmf wave.

(*d*) Space angle between armature mmf and resultant air gap mmf will also decrease because resultant flux wave will move towards the air-gap mmf.

Q. 32. A synchronous motor start as usual but fails to develop its full torque. What could be the reasons? **[M.D. Univ., December-2011]**

Ans. The reasons may be (*i*) too low exciter voltage (*ii*) open circuit or short-circuit in the field (*iii*) reversing of field spool.

Q. 33. Will the synchronous motor start with field excited?

Ans. No.

Q. 34. Under which conditions a synchronous motor will fail to pull-up into step?

Ans. A synchronous motor may fail to pull-up into step in the condition(s) of (*i*) no field excitation (*ii*) excessive load (*iii*) excessive load inertia.

PROBLEMS

1. A 400 V, 50 Hz, 3-phase, 37.3 kW star-connected synchronous motor has a full-load efficiency of 80%. The synchronous impedance of the motor is $(0.2 + j1.6)$ ohm per phase. If the motor is adjusted to give a leading power factor of 0.9, calculate the induced emf for full-load. **[U.P. Technical Univ. Electrical Machines, 2006-07]**

[Ans. 292.8 V**]**

2. A 6,000 V, 3-phase, star-connected synchronous motor has an input current of 40 A at 0.8 pf lagging. The effective resistance and leakage reactance per phase are 0.5Ω and 10 Ω. respectively. Determine: (*i*) Power input to motor (*ii*) Induced emf (*iii*) Internal mechanical power developed.

[U.P. Technical Univ. Electromechanical Energy Conversion II, 2002-03]

[Ans. (*i*) 332.55 kW (*ii*) 5,582 V (*iii*) 330.15 kW**]**

3. Define the term coupling angle in synchronous motor.

The input to an 11 kV, 3-phase, Y-connected synchronous motor is 60 A. The effective resistance and synchronous reactance per phase are 1 Ω and 30 Ω respectively. Find (*i*) power supplied to the motor and (*ii*) induced emf for a pf of 0.8 leading.

[M.D. Univ. Electrical Machines-II, December-2011]

[Ans. 904.2 kW; 13.042 kV**]**

4. A 3-phase, 400 V synchronous motor takes 52.5 A at a pf of 0.8 lead. Calculate the power supplied and the induced emf. The motor impedance per phase is $(0.25 + j3.2)$ Ω.

[Rajasthan Univ. Electrical Machines-II, 2003]

[Ans. 29.1 kW, 555.56 V**]**

5. A 2.3 kV, 3-ϕ, star-connected synchronous motor has $Z_s = (0.2 + j\,2.2)$ ohms per phase. The motor is operating at 0.5 power factor leading with a line current of 200 A. Determine the generated emf per phase. **[Rajasthan Univ. Electrical Machines-II, 2003]**

[Ans. 1,708 V]

6. The input to a 1.1 kV, three-phase star-connected synchronous motor is 50 A. The effective resistance and synchronous reactance per phase are 0.95 Ω and 29 Ω respectively. Calculate the power supplied to the motor and the induced emf for a pf of (*i*) 0.8 lagging (*ii*) 0.8 leading.

[U.P. Technical Univ. Electrical Machines and Automatic Control, 2009-10]

[Ans. (*i*) 76.21 kW; 2,016 V (*ii*) 76.21 kW; 3,270 V]

7. A 6600 V, 3-phase, star-connected synchronous motor draws a full-load current of 80 A at 0.8 power factor leading. The per phase armature resistance is 2.2 Ω and synchronous reactance is 22 Ω. If the stray losses of the machine are 3,200 W, calculate the induced emf, output power and efficiency.

[Rajasthan Technical Univ. Electrical Machines-II, 2002, 2006; U.P. Technical Univ. Electromechanical Energy Conversion-II, 2005-06]

[Ans. 7,276 V, 686.18 kW, 93.79%]

8. A 1,000 kVA, 11,000 V, 3-ϕ, star-connected synchronous motor has an armature resistance and reactance per phase of 3.5 Ω and 40 Ω respectively. Determine the induced emf and angular retardation of the motor when fully loaded at (*a*) unity pf (*b*) 0.8 pf lagging (*c*) 0.8 pf leading. **[Bangalore Univ. Elec. Engg. II, 1992]**

[Ans. (*a*) 11,284 V; 18.8° (*b*) 8,985 V; 17.6° (*c*) 13,295 V; 13.48°]

9. A 2.3 kV, 3-ϕ, star-connected synchronous motor has $Z_s = (0.2 + j\,2.2)$ ohms per phase. The motor is operating at 0.5 power factor leading with a line current of 200 A. Determine the generated emf per phase. **[Bharathidhasan Univ. Elec. Machines II, Nov; 1996]**

[Ans. 1,708 V]

10. A 10 hp (metric), 400 V, 3-phase, star-connected synchronous motor has Z_s equal to $0.35 + j\,2.8$ ohm per phase. Find the angle of retard and the voltage to which it must be excited to give a full-load output at 0.866 leading pf. Assume an efficiency 0.88.

[Mananmanian Sundaranar Univ. Elec. Machines II, April-1996]

[Ans. 8.36°, 431 V]

11. A synchronous motor takes 25 kW from 400 V supply mains. The synchronous reactance of the motor is 4 Ω. Calculate the power factor at which the motor would operate when the field excitation is so adjusted that the generated emf is 500 V. **[R.G.P.V. Bhopal, 2000]**

[Ans. 0.99 (leading)]

12. A 400 V, 3-ϕ, star-connected synchronous motor has a synchronous reactance of 1Ω per phase. The input to motor is 100 kW at normal voltage and induced line emf is 600 V. Calculate line current and power factor. Neglect armature resistance.

[Mananmanian Sundaranar Univ. Elec. Machines II Nov; 1996]

[Ans. 167A; 0.8645 (lead)]

13. A 200 V, three-phase, star-connected synchronous motor has an effective resistance and synchronous reactance of 0.2 Ω and 2.2 Ω per phase respectively. The input is 800 kW at normal voltage and induced line emf is 2,500 V. Calculate the line current and power factor. **[Anna Univ. Electrical Machines-I, May-June-2012]**

[Ans. 254.3 A, 0.908 (leading)]

14. A 3,300 V, star-connected synchronous motor has synchronous impedance of $(0.4 + j\,5)$ Ω per phase for excitation emf of 4,000 V and motor input power of 1,000 W at rated voltage. Calculate the line current and power factor.

[Rajasthan Technical Univ. Electrical Machines-II, June/July-2011]

[Ans. 184.5 A; 0.9485 (leading)]

15. The synchronous reactance per phase of a 3-phase star-connected 6,600 V synchronous motor is 10 Ω. For a certain load the input is 900 kW and the induced line emf is 8,900 V (line to line). Evaluate the line current. Neglect resistance. **[Nagpur Univ. Elec. Machines; 1993]**

[Ans. 149.23 A]

16. Consider a 3,300 V delta-connected synchronous motor having a synchronous reactance per phase of 18 Ω. It operates at a leading pf of 0.707 when drawing 800 kW from mains. Calculate the excitation emf and the rotor angle (δ), explaining the latter term.

[Nagpur Univ. Elec. Machines 1993]

[Ans. 4,972 V; 17°]

17. A 2,000 V, 3-phase, 4-pole, Y-connected synchronous motor runs at 1,500 rpm. The excitation is constant and corresponds to an open-circuit voltage of 2,000 V. The resistance is negligible as compared to synchronous reactance of 3 Ω/phase. Determine the power input, power factor and torque developed for an armature current of 200 A.

[Nagpur Univ. Elec. Engg-I, 1993]

[Ans. 669.03 kW; 0.966 (lag) 4,259 N-m]

18. A 3-phase synchronous motor of 8,000 W at 1,100 V has synchronous reactance of 8 Ω per phase. Find the minimum current and the corresponding induced emf for full-load condition. Neglect armature resistance. The efficiency of the machine is 0.8.

[Rajasthan Univ. Electrical Machines-II, 1999]

[Ans. 5.25 A, 1102.4 V]

19. A 2200 V, 400 kW, 3-phase star-connected synchronous motor has a synchronous impedance of $0.3 + j\,3$Ω/ph. Determine the excitation emf (back emf) and the power angle if the motor works at rated load at 0.8 pf leading with an efficiency of 94 %.

[U.P. Technical Univ. Electromechanical Energy Conversion II 2004-05]

[Ans. 2,640 V, 13.66°]

20. A 75 kW, 400 V, 4-pole, 3-phase, star-connected synchronous motor has a resistance and synchronous reactance per phase of 0.04 Ω and 0.4 Ω respectively. Compute for full load 0.8 pf lead the open-circuit emf per phase and gross mechanical power developed. Assume an efficiency of 92.5%. **[Rajasthan Univ. Electrical Machines-II, 2003]**

[Ans. 266.18 V; 78.512 kW]

21. A 700 kW, 11 kV, 50 Hz, 3-phase, star-connected synchronous motor has a full-load efficiency of 92%. The synchronous impedance of the motor is $(0.15 + j\,15)$ ohms per phase. The excitation of the motor is adjusted to give an operating pf of 0.85 leading. Under this operating pf and full-load conditions, compute (*i*) the per phase induced emf (*ii*) the load angle.

[B.P. Univ. of Technology Electrical Machines, 2008]

[Ans. (*i*) 6,743 V, 5.12°]

22. A 10 kW, 400 V, 3-phase star-connected synchronous motor has synchronous impedance per phase of $(0.35 + j\,2.8)$ Ω. Find the torque angle and voltage to which the motor must be excited to give a full-load output at 0.856 pf load. Assume efficiency to be 88 per cent. **[J.N. Technological Univ. Electrical Machines-III, May-2013]**

[Ans. 11.5° and 429V]

23. A 20-pole, 30 kW, 660 V, 50 Hz, 3-ϕ, star-connected synchronous motor is operating with its generated voltage per phase exactly equal to the phase voltage applied to its armature. At loaded condition the motor is retarded by 5 mechanical degrees from its synchronous position. The synchronous reactance and armature resistance are 10 Ω and 1 Ω per phase. Calculate (*i*) armature current per phase (*ii*) the power per phase and (*iii*) the total power drawn by the motor from the bus and (*iv*) the developed power.

[Ans. (*i*) 32 A (*ii*) 11.5 kW (*iii*) 34.5 kW (*iv*) 31.43 kW]

24. A 3-phase, 8-pole, 3,300 V, 50-Hz, star-connected synchronous motor has armature resistance of 0.5 Ω / phase and synchronous reactance of 5 Ω / phase. While running on no-load, the excitation has been adjusted so as to make the emf numerically equal to and anti-phase with terminal voltage. With a certain load torque applied the rotor gets retarded by 3 mechanical degrees, calculate the armature current and power factor of the motor. **[Ans.** 79.6A; 0.9992 (lag)]

25. A 400 V, 6-pole, 3-ϕ, 50 Hz, star-connected synchronous motor has negligible armature resistance and 4Ω synchronous reactance per phase. It takes a current of 15 A at unity power factor when operating with a certain field current. If the load torque is gradually increased until the line current is 60 A, the field current remaining unchanged, find the gross torque developed and the new power factor.

[Ans. 346.64 N-m; 0.8732 (lag)]

26. A 400 V, 50-Hz, 6-pole, 3-ϕ, star-connected synchronous motor has a synchronous reactance of 4 Ω / phase and resistance of 0.5 Ω / phase. On full load, the excitation is adjusted so that machine takes an armature current of 60 A at 0.866 pf leading. Keeping the excitation unchanged, find the maximum power output. Excitation, friction, windage and iron-losses total 2 kW. **[Ans.** 51.42 kW]

27. A 3-ϕ, 3,300 V, Y-connected synchronous motor has an effective resistance and synchronous reactance of 2.0 and 18.0 Ω per phase respectively. If the open-circuit generated emf is 3,800 V between lines, calculate (*i*) the maximum total mechanical power that the motor can develop and (*ii*) the current and pf at the maximum mechanical power. **[Ans.** 604.35 kW; 151.4 A; 0.8572 (lead)]

28. Construct curves showing the relation (*a*) between armature and field current, (*b*) between pf and field current, for a star-connected synchronous motor with a synchronous reactance of 8.25 Ω/phase and negligible resistance. The machine takes a constant power input of 800 kW at 6.6 kV. The open-circuit characteristic is as follows:

Terminal voltage, kV	3	4	5	6	6.6	7	7.5	8
Field current, Amps	16	23	31	41	50	56	69	85

29. A synchronous motor absorbing 50 kW is connected in parallel with a factory load of 200 kW having a lagging pf of 0.8. If the combination has a pf of 0.9 lagging, how many leading kVAR are to be supplied by the motor ? What is its power factor ?

[Bharathidhasan Univ. Elec. Machines II Nov; 1996]

[Ans. 29 kVAR (leading); 0.865 (leading)]

30. The load in a factory is 800 kVA at 0.85 pf lagging and in addition, there is a 3-phase synchronous motor having an input of 200 kW. Determine the input to the synchronous motor in kVA and the power factor at which it must operate in order that the power factor of the combined load comes to be 0.95 lagging.

[Ans. 239.7 kVA; 0.834 (leading)]

31. A manufacturing plant takes 360 kW, at a lagging power factor of 0.6. Calculate (*a*) the rating of the synchronous motor which would deliver 140 kW, and at the same time make it possible for the plant to operate without overloading its 600 kVA supply transformer (*b*) the resultant pf of the system and (*c*) the power factor of the synchronous motor. **[Ans.** (*a*) 204 kVA (*b*) 0.8333 (lag) (*c*) 0.687 (lead)]

32. A 3-phase synchronous motor absorbing 60 kW is connected in parallel with a factory load of 240 kW having a lagging power factor of 0.8. If the combined load has a pf of 0.9, what is the value of the leading kVAR supplied by the motor and at what power factor is it working? **[Madras Univ. Elec. Machines II, April-1997]**

[Ans. 34.7 kVAR, 0.866 (leading)]

33. A small industrial load of 500 kW at 0.6 pf lagging is supplied from a 3300 V, 3-phase, 50 Hz system. It is desired to raise the pf of the entire system to 0.8 lagging by means of a synchronous motor, which is also drawing a pump so that the synchronous motor takes 100 kW from the line. Determine the rating of the synchronous motor. **[U.P. Technical Univ. Electromechanical Energy Conversion II, 2002-03]**

[Ans. 238.63 kVA]

34. A manufacturing plant takes 200 kW at 0.6 pf from a 600 V, 60 c/s, 3-phase system. It is desired to raise the power factor of the entire system to 0.9 by means of a synchronous motor, which at the same time is driving a load requiring the synchronous motor take 80 kW from the line. What should be the rating of the synchronous motor in volts and amperes? **[Rajasthan Univ. Electrical Machines-II, 2000]**

[Ans. 600 V, 147.7 A]

❑❑❑

17 CHAPTER

Parallel Operation of Alternators

INSIDE THIS CHAPTER

17.1 PARALLEL OPERATION OF ALTERNATORS

In most power stations (whether dc or ac), it will be found that the power is supplied from several smaller units (generators) operating in parallel rather than from a single large unit capable of taking care of the maximum peak loads. There are a number of good reasons for this practice, even though in some instances it might be cheaper to install only one or two large units to take care of the system load (as a rule, large machines and their auxiliary equipment cost less per kW output than small machines). The demand for electrical energy continues to grow at a steady space, electrical utilities and others find it necessary to increase generating capacity at regular intervals. Since it is not economical to discard serviceable alternator (or alternators) in favour of larger ones, it becomes necessary to either operate alternators in parallel or else subdivide the load and feed it from multiple isolated systems. Of these two alternatives, parallel operation is much preferred because it does not require major changes in the existing system and provides greater reliability, improves efficiency, facilitates repairs and maintenance and makes the supply of load possible when it (load) exceeds the capacity of the largest single unit available.

1. Efficiency. The load on an electric power station fluctuates, usually having its peak value sometimes during the day and its minimum value during the night hours. Since generators operate most efficiently when operated on full load (or nearly full load), it is logical to operate a small unit delivering rated output when the load demand is light. Then as the load increases, a larger unit is substituted for the smaller one or another unit is connected to operate in parallel with the one already in operation. This keeps the machines loaded up to their rated capacity and increases the *efficiency of operation.*

2. Reliability or Continuity of Service. Continuity of service is one of the uttermost requirements of electrical power systems. Several small units are more reliable than a large single unit, since if one unit fails, the continuity of supply can be maintained by the remaining units. On the other hand if the power station consisted only of a single unit, in the event of breakdown of the prime mover or the generator itself, there will be a complete shutdown (failure of supply).

3. Maintenance and Repair. It is considered necessary to carry out regular inspection and maintenance of the machinery so as to avoid the possibility of failure. This is possible only when the unit is out of service which means that the remaining units should be capable to take care of the entire load on the system. Repairing of a unit is also more convenient and economical if there are several small units in the power station. The cost of stand-by unit is also small.

4. Physical Size. No doubt that alternators and their associated equipment are built in sizes much larger than those of dc generators, but there are instances where the load on the system exceeds the sizes to which a single unit (alternator) can be built. Hence it becomes necessary to operate two or more units in parallel.

17.2 REQUIREMENTS FOR PARALLEL OPERATION

The alternators to be operated in parallel should meet some requirements so as to make their operation proper. These requirements are as follows:

1. They must have the same output voltage rating.
2. The rated speeds of the machines should be such as to give the same frequency (f being equal to $\frac{PN}{120}$).
3. The alternators should be of the same type so as to generate voltages of the same waveform. They may differ in their kVA ratings.

4. The prime movers of the alternators should have same speed-load characteristics, which of course must be drooping ones, so as to load the generators in proportion to their output ratings.
5. The alternators should have reactances in their armatures, otherwise they will not operate in parallel successfully.

17.3 SYNCHRONIZATION OF ALTERNATORS

It is imperative that changing from one unit to another or connecting an additional unit in parallel with the others (already operating in parallel) is carried out with the minimum possible disturbance to the existing system. Exactly how an alternator is taken off or put on a line without disturbing the voltage and frequency of the system will be discussed here.

The process of connecting an alternator in parallel with another alternator or with the common bus bars (the bus bars to which a number of alternators are connected) is called *synchronization*.

17.3.1. Conditions For Proper Synchronization. It will be logical to assume that only one alternator is to be placed in parallel with one or more alternators already in operation to meet an additional load. Those alternators, which are in operation and sharing the load, are called the *running machines*, while that which is to be introduced in the system is known as the *incoming machine*. Before the *incoming machine* is connected in parallel with the running machine or machines it is necessary that it meets certain conditions, which are given below.

1. The terminal voltage of the incoming machine must be exactly equal to that of the others, or of the bus bars connecting them.
2. The speed of the incoming machine must be such that its frequency $\left(\text{being equal to } \frac{PN}{120}\right)$ equals bus bar frequency.
3. The phase of the incoming machine voltage must be the same as that of the bus bar voltage relative to the load *i.e.* the phase voltages of the incoming machine and the bus bar should be in phase opposition. This implies that there will be no circulating current between the windings of the alternators already in operation (the bus bars) and the incoming machine.

In case of 3-phase alternators, an additional requirement to be met is that the phase sequence of the incoming machine is the same as that of the bus bars. Mismatch between two phase sequences out of three phase sequences will result in short-circuit current which is ultimately restricted by the combined synchronous impedances of the windings involved. This short-circuit current will be even excessive at the moment of closure of switch. This is due to the fact that the armature reaction cross-magnetisation requires time for establishment.

17.3.2. Synchronization of Single-Phase Alternators. Before connecting an alternator, to the bus bars or in parallel with another alternator already supplying the load, the three conditions mentioned in Art. 17.3.1. must be fulfilled.

The equality of the terminal voltages* can be determined readily by connecting a voltmeter first to the running alternator or the bus bars and then to the incoming alternator. The remaining two conditions *i.e.* equality of frequency and phasing out can be determined with the help of two lamps connected across the ends of the double-pole paralleling switch and known as *synchronising lamps*.

Let alternator B is to be connected to bus bars, to which alternator A is already connected. The method to be followed for doing so is given below.

The prime mover of alternator B (incoming alternator) is started and brought up close to the rated speed. The alternator is then excited and its voltage is raised, by increasing its excitation, to that of bus bars or the running alternator. The synchronising lamps L_1 and L_2 are connected across the double pole paralleling switch of the incoming alternator B, as shown in Fig. 17.1. If the frequencies of the alternators A and B are exactly the same and their terminal voltages, with respect to local series circuit, are in exact phase opposition, no resultant voltage will act across the lamps L_1 and L_2 and, therefore, these lamps will remain dark. If the frequencies of the alternators A and B are not equal, the phase angle between their voltages will be continually changing, and therefore, the current through the synchronising lamps and through the local series circuit (shown dotted in the figure) will be changing, resulting in flickering of lamps. The frequency of flickering of lamps will be equal to the difference of frequencies of the two alternators *i.e.* greater the difference in frequencies, greater will be frequency of flickering and lesser the difference in frequencies, lesser will be the frequency of flickering. In the middle of the dark period the two voltages will be in phase opposition with respect to the local series circuit. Synchronising is done in the middle of the dark period. This is known as *dark lamp method*.

Hence at the time of synchronising of alternators the speed of the incoming machine is adjusted until the lamps go in and out very slowly, the terminal voltage of the incoming alternator is made equal to bus-bar voltage by adjusting the excitation of the incoming alternator and then switch of incoming alternator is closed in the middle of the dark period.

Since it is somewhat easier to judge the middle of bright than middle of dark period, therefore, some engineers prefer to synchronise in the middle of the bright period. This method is known as *bright lamp method* and necessitates the crossing over of the lamp connections, as in the case of alternaor C in Fig. 17.1.

The maximum voltage across individual lamp in dark lamp method will occur when, during synchronising process alternator B is in phase with alternator A with respect to local series circuit. In case of bright lamp method the maximum voltage across individual lamp will occur when alternator C is in phase opposition with alternator A. The magnitude of maximum voltage that can exist across a lamp is double that of the voltage of one machine.

As it is obviously impossible to use lamps on any other except low voltages, small instrument transformers, known as potential transformers, are, therefore, necessary for reducing the voltage applied to the synchronising lamps.

* The equality of terminal voltages can also be determined by taking the readings of the voltmeters connected to the running alternator or the bus bars and to the incoming alternator.

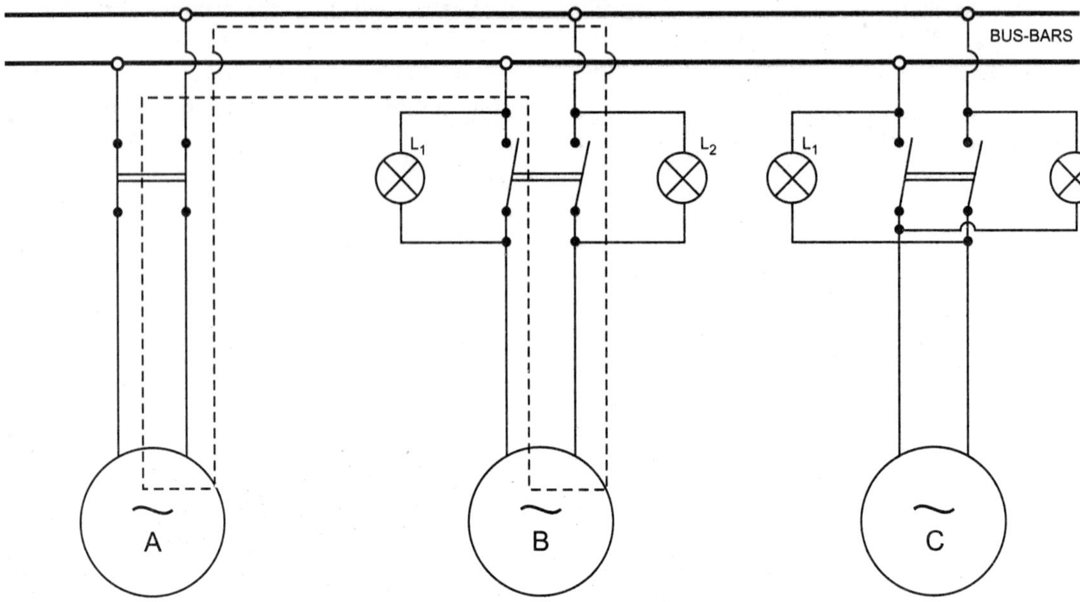

Fig. 17.1 *Synchronising of Single Phase Alternators*

After alternator B or C has been synchronised and is operating in parallel with alternator A, the incoming alternator may be made to deliver the load in part or in full, assuming of course, that its rating is sufficient to do so. The load is shifted from the running alternator to the incoming alternator by increasing the mechanical power input to the prime mover of the incoming alternator and simultaneously reducing the mechanical power input to the prime mover of existing alternator A. This may be readily done by admitting more steam (opening the steam valve) to the prime mover of the incoming alternator (if it is a steam machine) and simultaneously closing the steam valve of alternator A. When the incoming alternator carries the entire load, as indicated by the ammeters and wattmeters in the line circuits, the line circuit breaker and field breaker of alternator A are opened, the later unit is then shut down if desired.

The most important point, to be emphasized, is that load cannot be shifted from one machine to another by adjustment of excitation. The adjustment of excitation changes the voltage and power factor of the alternator.

17.3.3. Synchronization of 3-Phase Alternators. The same two methods of synchronising, as already discussed for synchronising of single-phase alternators, in Art. 17.3.2, but with slight modifications are used for three-phase alternators.

The additional condition to be met in case of synchronising of 3-phase alternators is that the phase sequence of the incoming machine should match with that of the system to which it is going to be connected. The process of checking the phase sequence and getting it correct is known as *'phasing out the alternator.'* The different methods of phasing out of alternators are given below:

I. Three Dark Lamp Method. Figure 17.2 shows an alternator A supplying a load, with alternator B all wired up and ready for start-up. If these are low voltage alternators, the alternator B can be phased out by connecting three lamps across the ends of the triple-pole paralleling switch as shown in Fig. 17.2. The prime mover of the incoming alternator B is started and brought up close to the rated speed. The alternator is then excited and its voltage is raised by increasing the excitation to that of the running alternator or the bus bars and the lamps are watched closely. If the incoming alternator B is properly connected all the three lamps should become bright and dim together. If they bright and dim in sequence, it shows that incoming alternator B is not properly connected with the bus bars, and the phase sequence of the incoming alternator B, therefore, must be reversed relative to the system. The usual way to do this is to interchange any two of the leads on either the alternator side or the line side of the switch. It is possible to achieve the same effect by reversing the direction of rotation of the alternator, but this is very seldom convenient.

Once the alternator B has been properly phased out, it will always be correct unless either the wiring or the rotation is altered so establishing the proper phase sequence is not a continual part of the synchronising process of the alternators.

The speed of the prime mover of the incoming machine is further adjusted until the lamps flicker at a very low rate (usually less than one dark period per second) and the voltage of the incoming machine is again adjusted and made equal to that of the running generator (or bus bars) and the paralleling switch is closed at the instant all the three lamps are dark.

The disadvantage of this method is that lamps go dark at somewhat less than half their rated voltage, and so the paralleling switch might be closed when there is a considerable phase difference between the machines, with a high circulating current resulting in possible damage to the machines. Another disadvantage is that the lamp filaments might burn out. For this reason it is desirable to have the two frequencies nearly equal, so that the synchronising is not dependent on a constant dark period. A third disadvantage is that the flicker of lamps does not indicate whether the incoming alternator is slow or fast.

The above difficulties are eliminated in *two bright and one dark lamp method* in which two of the lamps are cross-connected, as shown in case of incoming alternator C in Fig. 17.2.

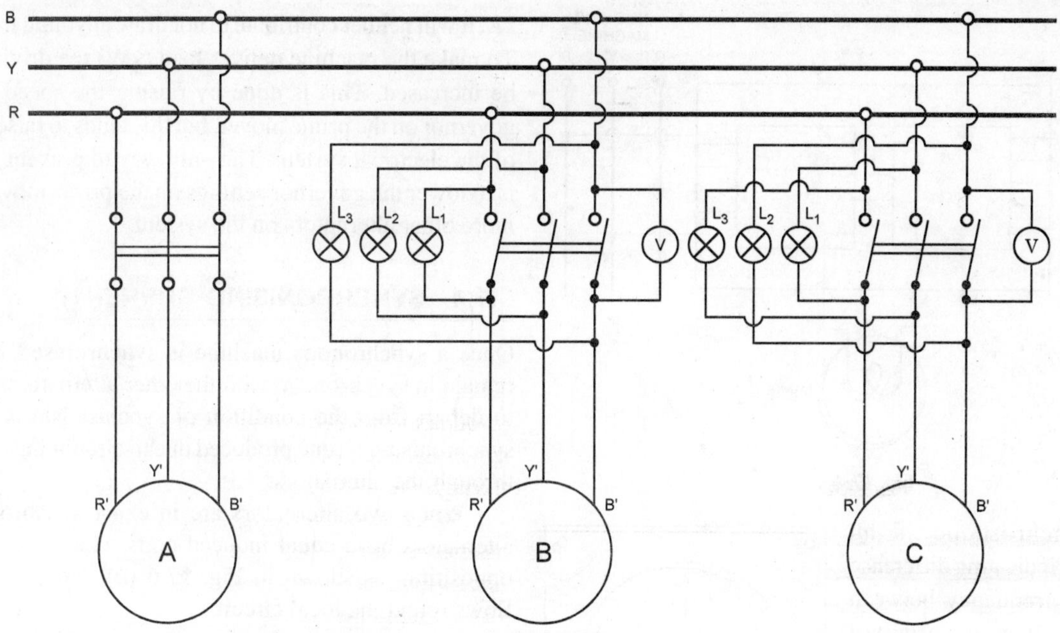

Fig. 17.2 *Synchronising of 3-Phase Alternators*

II. Two Bright and One Dark Lamp Method. When the incoming alternator is in synchronism, with alternator A or bus bars, lamps L_1 and L_3 are bright and L_2 is dark.

Since near the point of synchronism, the brightness of one lamp increases and of another decreases so the instant at which the incoming machine is in synchronism with the bus bars can be accurately determined and the paralleling switch should be closed at this instant.

Let the voltage star RYB refer to bus bars and star shown in dotted, the R'Y'B' to incoming machine C. The instantaneous voltage across the three lamps are given by the phasors RY', R'Y and BB'. When the frequency of incoming alternator is the same as that of the bus bar, both phasor diagrams rotate in space with the same angular velocity, so the potential difference across each lamp remains constant.

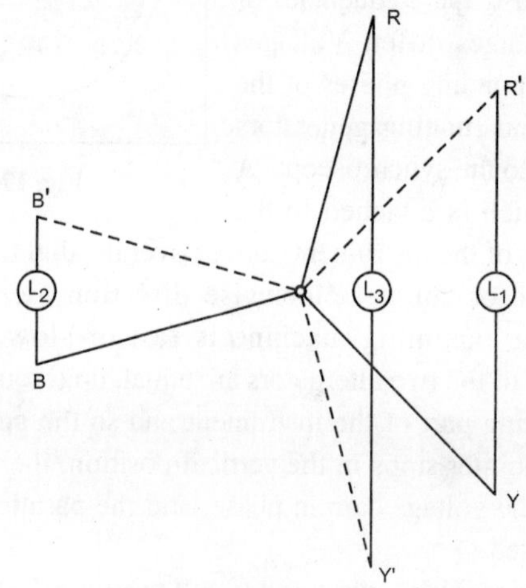

Fig. 17.3

If the incoming machine is too fast, then phasor diagram R′Y′B′ will rotate faster than RYB so the voltage across lamp L_3 represented by RY' decreases, the voltage across lamp L_1 represented by R'Y increases and voltage across lamp L_2 represented by BB' decreases. And if the incoming machine is too slow then diagram R'Y'B' will rotate slower than RYB so at that instant voltage across lamp L_3 represented by RY' increases, voltage across lamp L_1 represented by R'Y decreases and voltage across lamp L_2 represented by BB' increases.

Hence when the three lamps are placed in a ring, a light wave travelling in counter-clockwise direction indicates that the incoming machine is slow and light wave travelling in clockwise direction indicates that the incoming machine is fast, so by observing the sequences of brightness of the lamps, it can be determined whether the incoming machine is fast or slow.

The paralleling switch is closed when the changes in light are very slow and at the instant the lamp L_2 connected directly across one phase is dark.

In case of high-voltage generators, which is the usual case, direct connection of lamps is not practical. In such cases the lamps may be connected through potential transformers (the potential transformers directly across the switch and the lamps into the secondary circuits of the potential transformers). However, in most installations there will be potential transformers connected across the line on both sides of the paralleling switch.

Alternatively, the synchronising can be done with the help of one three limbed transformer and a lamp, the two primary windings on the outer limbs being connected to the bus bars and the incoming alternator respectively, and the secondary winding on the central limb to a lamp (Fig. 17.4). The correct moment for closing the paralleling switch is the middle of the dark period if the primaries are each connected the same way round, and the middle of the bright period if the connections of one winding are reversed.

If the phase sequence is known to be correct, the best method of synchronising alternators is by means of a single phase device, called the *synchroscope*, which provides a more accurate indication of synchronism than do lamps. The dial markings of the synchroscope are shown in Fig. 17.5.

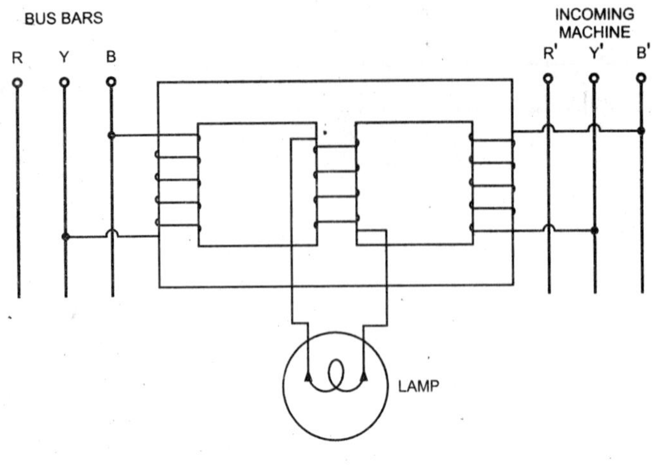

Fig. 17.4

The synchroscope is an instrument for indicating difference of phase and frequency between two voltages. It is essentially a split-phase motor in which torque is developed if the frequencies of the two voltages differ. Voltages from corresponding phases of the incoming and running generators are applied to the synchroscope. A pointer, which is attached to the rotating part of the instrument, moves over the dial face in either a clockwise or counter-clockwise direction, depending on whether the incoming machine is fast or slow, when the frequencies of the two alternators are equal, no torque is exerted on the rotating part of the instrument and so the pointer stops. When the pointer stops in the vertical position, the frequencies are equal, the voltages are in phase, and the paralleling switch may be closed.

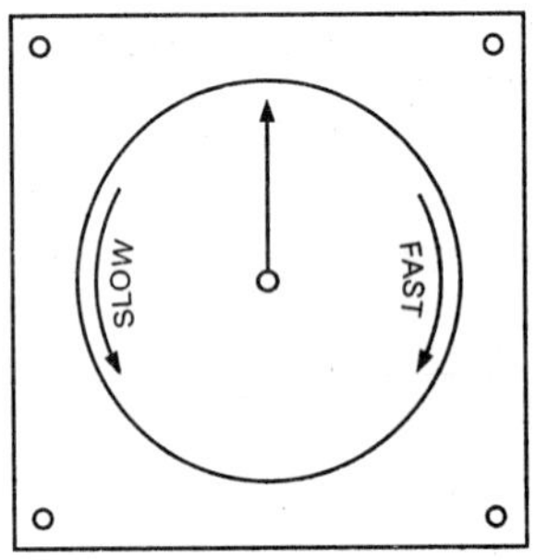

Fig. 17.5

In practice it is sometimes difficult to adjust the speed of the incoming machine closely enough to stop the synchroscope pointer in the vertical position. In such a case, the speed of the incoming machine should be adjusted as closely as possible and the paralleling switch closed just before the pointer reaches the vertical position when travelling in the fast direction. This makes the incoming machine to take a small amount of load immediately after closure of switch and results in stable operation.

It is very important that the synchronising operation be done as accurately as possible. If the switch is closed when the alternator is out of phase with the system, a high circulating current will momentarily flow through the machine and pull it ahead (or back) to its correct position. However, this creates a voltage disturbance on the system, and sometimes the magnetic forces developed on the alternator windings will damage the coils.

If the synchronising operation has been correctly accomplished, the machine will now be "floating on the line," *i.e.*, it will neither contribute to nor draw anything from the system. To make the machine deliver load (kW) the driving torque is to be increased. This is done by raising the speed setting of the governor on the prime mover, but this tends to raise the frequency of the electrical system. The only way to prevent frequency rise is to lower the governor settings on the prime mover (s) of one or more other alternators on the system.

17.4 SYNCHRONISING CURRENT

Once a synchronous machine is synchronised, it will tend to remain in synchronism with the other alternators. Any tendency to depart from the condition of synchronism is opposed by a synchronising torque produced due to circulating current flowing through the alternators.

When two alternators are in exact synchronism, the two alternators have equal induced emfs which are in exact phase opposition, as shown in Fig. 17.6 (*a*). No circulating current flows round the local circuit.

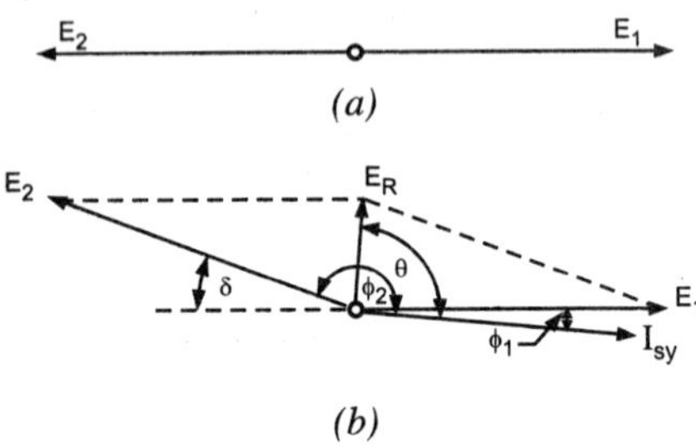

Fig. 17.6

When the induced emfs of the two alternators are equal in magnitude but not in exact phase opposition, as shown in Fig. 17.6 (*b*), their resultant emf acts round the local circuit and causes flow of current called the *synchronising current*, I_{sy}.

If any alternator (say, the second alternator) due to some disturbance tends to retard, E_2 falls back by a phase angle δ electrical degrees, as shown in Fig. 17.6 (*b*). Now though their induced emfs E_1 and E_2 are equal in magnitude but have a phase difference of 180° – δ. Let each of the induced emfs E_1 and E_2 be equal to E.

$$\text{Resultant emf, } E_R = 2\,E\cos\left[\frac{180^\circ-\delta}{2}\right]$$

$$= 2\,E\cos\left(90^\circ-\frac{\delta}{2}\right) = 2\,E\sin\frac{\delta}{2} \qquad \text{...(17.1)}$$

$$= 2\,E\times\frac{\delta}{2} = E\,\delta \qquad \because \delta \text{ is very small}$$

$$\text{Synchronising current, } I_{sy} = \frac{E_R}{Z_s} = \frac{E\delta}{Z_s} \qquad \text{...(17.2)}$$

where Z_s is the combined synchronous impedance per phase of the two alternators (or of one alternator only if it is connected to infinite bus bars*).

* A supply system with a large number of synchronous machines connected in parallel is referred to as *infinite bus bars*. Any additional machine, whether to operate as a generator or as a motor is connected in parallel with the bus bars.

Infinite bus bars represent a system of large capacity whose frequency and the phase and magnitude of voltage are not affected even if there is a variation of excitation or power of a synchronous machine connected to it.

The synchronising current I_{sy} lags behind the resultant emf E_R by an angle θ given by $\theta = \tan^{-1}\frac{X_s}{R_e}$ where X_s is the combined synchronous reactance and R_e is the effective resistance of the two alternators (or of one alternator only if it is connected to infinite bus bars).

If resistance R_e is very small as compared to synchronous reactance X_s then

Synchronising current $I_{sy} = \frac{E\delta}{X_s}$ and lags behind E_R by 90°

i.e. almost in phase with E_1.

Thus I_{sy} is generating current with respect to machine no. 1 and motoring current with respect to machine no. 2 (in generator action the current flows in the direction of the induced emf while in motor action current flows in a direction opposite to that of the induced emf). This current I_{sy} sets up a synchronising torque T_{sy} which tends to accelerate the motoring machine (*i.e.* machine no. 2) and deaccelerate the generating machine (*i.e.* machine no. 1).

Similarly if the machine no 2 tends to speed up, induced emf E_2 will tend to advance in phase and the synchronosing current I_{sy} will be generating current with respect to machine no 2 and motoring current with respect to machine no 1. The torque developed due to synchronising current I_{sy} will now tend to retard machine no 2 and accelerate machine no 1.

Thus, any departure from synchronism results in development of a synchronising torque T_{sy} which tends to keep the machines in synchronism.

17.4.1. Synchronising Power. In above case (Art 17.4), machine no. 1 supplies power equal to $E_1 I_{sy} \cos \phi_1$ and the machine no. 2 receives power equal to $E_2 I_{sy} \cos (180° - \phi_2)$.

Power supplied by machine no. 1 = Power supplied to machine no. 2 + copper losses.

The power supplied by machine no. 1 is called the *synchronising power* and is given by the expression

$$P_{sy} = E_1 I_{sy} \cos \phi_1 = E_1 I_{sy} = E \times \frac{E\delta}{X_s} = \frac{\delta E^2}{X_s} \quad ...(17.3)$$

$\because E_1 = E$ and ϕ_1 is very small

Total synchronising power for 3-phases

$$= 3 P_{sy} = \frac{3\,\delta E^2}{X_s} \quad ...(17.4)$$

17.4.2. Synchronising Torque. If T_{sy} be the synchronising torque in Nm, then total synchronising power

$$3 P_{sy} = T_{sy} \times \frac{2\pi N_s}{60}$$

$$\text{or Synchronising torque, } T_{sy} = \frac{3P_{sy} \times 60}{2\pi N_s} = \frac{3\delta E^2}{2\pi N_s X_s} \quad ...(17.5)$$

where N_s is the synchronous speed in rpm and is equal to $\frac{120 f}{P}$.

17.5 EFFECT OF REACTANCE

Since with respect to local circuit, the emf of an alternator is in phase opposition to the emf of another alternator with which it is working in parallel, the machines run as synchronous motor with respect to another. Hence if due to certain reasons, the input to machine no. 2 is cut off, which is quite probable, it must receive wattful motor current from the other.

Consider two machines having resistance but negligible reactance. Their emfs E_1 and E_2 will be practically in phase opposition, as shown in Fig. 17.8 (*a*) and their resultant emf E_R will be almost in quadrature with E_1 and E_2. The synchronising current I_{sy} given by the relation $I_{sy} = \frac{E_R}{R_1 + R_2}$ will be in phase with E_R and, therefore, in quadrature with E_1 and E_2. Hence the synchronising current I_{sy} will be wattless current and convey no power from machine no. 1 to machine no. 2, which needs help.

Now consider the two machines having the synchronous reactance, but no resistance. The synchronising current I_{sy} will be in quadrature with resultant emf E_R, therefore, almost in phase with one of the emfs, say in phase with E_1, as shown in Fig 16.8 (*b*). Thus machine no. 1 will supply power to machine no. 2 and keep the machine no. 2 running.

From the above discussions it is obvious that without reactance the machines could not develop motor and generator power, respectively to restore synchronism; parallel operation would not be possible.

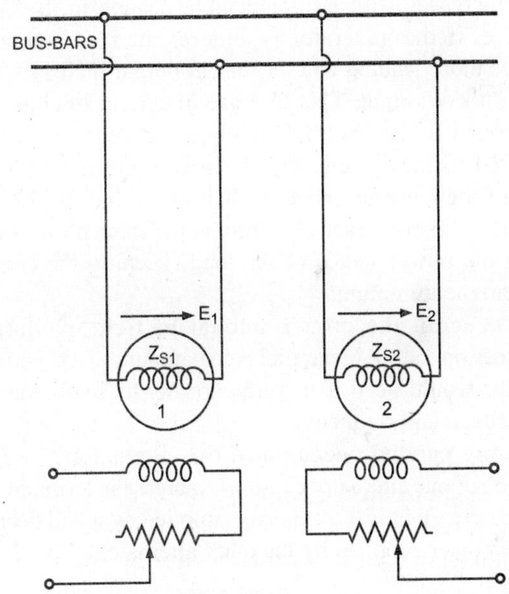

Fig. 17.7

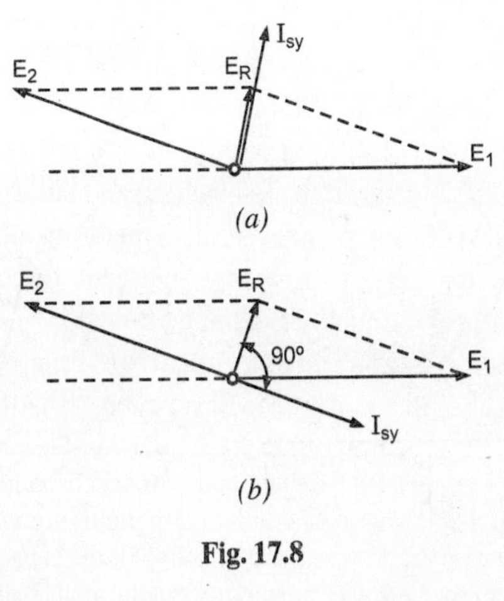

Fig. 17.8

17.6 EFFECT OF INCREASING EXCITATION OF ONE OF THE ALTERNATORS

For simplicity, let us consider two identical alternators sharing equally a load whose power factor is cos ϕ. If both machines have exactly the same excitation it will be found that their currents I_1 and I_2 are equal in magnitude (say I each) and in phase, since the conditions are identical for both machines. The phasor diagram for the total load, for one phase is given in Fig. 17.9 (*a*). Now, if the excitation of one of the alternators is increased, it will cause flow of synchronising current I_{sy} almost in quadrature with supply voltage V. Therefore, the load current of alternator 1, whose excitation has been increased, will be I_1, the phasor sum of I_{sy} and I and that of alternator 2 will be I_2, the phasor difference of I_{sy} and I. Hence power factor cos ϕ_1 of alternator 1 decreases and that of the other improves. Because synchronising current I_{sy} is in quadrature with V, therefore, it does not change wattful (active) components but changes wattless (reactive) components. Hence by changing the excitation, the power factors of the alternators are changed.

If the excitation of an alternator operating in parallel with other alternators is increased above its normal value of excitation, its power factor changes in the lagging direction and its current output increases with no appreciable change in its kW load. Likewise, if the generator is underexcited its power factor becomes more leading and its current output increases with no change in kW output. This increase in current in either case is not supplied to the load but circulates between the alternators connected to the system, thereby increasing their losses and reducing their useful capacity. It is desirable in most cases, therefore, to operate each alternator at the same power factor (at or near the power factor of the load) keeping the circulating current to the minimum.

In general, the proper amount of field excitation for alternators operating in parallel is the amount of excitation each alternator would need if it were carrying its load alone at the same voltage and frequency.

During parallel operation of two alternators, increase in excitation of one alternator causes (*i*) increase in terminal voltage and (*ii*) increase in reactive power supplied by it and decrease in reactive power supplied by the other alternator.

17.6.1. Maintaining Voltage. Now when excitation of one machine is increased, it becomes necessary to reduce the excitation, by the same amount, of the other machine operating in parallel with it so that the normal voltage is maintained at the original value V. The fact that the voltage remains unchanged means that in neither machine has the effective field strength altered, despite the variations in exciting current.

What actually happens is that the machine with the increased excitation automatically produces a lagging wattless current I_q just sufficient to offset the excess excitation, through the action of armature reaction. It will be remembered that a lagging current has a demagnetising effect on the main field.

On the other hand, the machine with reduced excitation produces a leading wattless current I'_q just sufficient, through its magnetising effect, to maintain the field strength unchanged. The conditions are illustrated in Fig. 17.9 (*c*), where cos ϕ_1 and cos ϕ_2 are the power factors of the loads carried by the individual machines. Thus the variation in relative excitation results in variations of wattless current and, therefore, of reactive kVA, without affecting the sharing of the kW load or power.

The voltage of the system in which several alternators are operating in parallel may be controlled (raised or lowered) by varying the excitation of all the machines, up or down as required; this being done automatically by the voltage regulators.

17.7 EFFECT OF INCREASING DRIVING TORQUE OF ONE OF THE ALTERNATORS

When the driving torque of one of the alternator is increased (*e.g.* by increasing the steam supply in case of steam turbine driven alternator), it immediately starts to accelerate. The rotor of the alternator 1 (whose driving torque has been increased) takes lead in relation of rotor of alternator no. 2, and Fig. 17.10 shows the condition when the rotor of alternator 1 has taken a lead of δ degrees (electrical). This causes a phase difference between E_1 and E_2, the magnitude of induced emfs remaining the same, resulting in flow of synchronising current I_{sy}. Thus the power delivered by the alternator, whose driving torque has been increased, will increase by an amount equal to $E_1 I_{sy} \cos \phi_1$. Hence by increasing the driving torque of one of the alternators, it is further loaded and other is relieved of its load. If the output of the

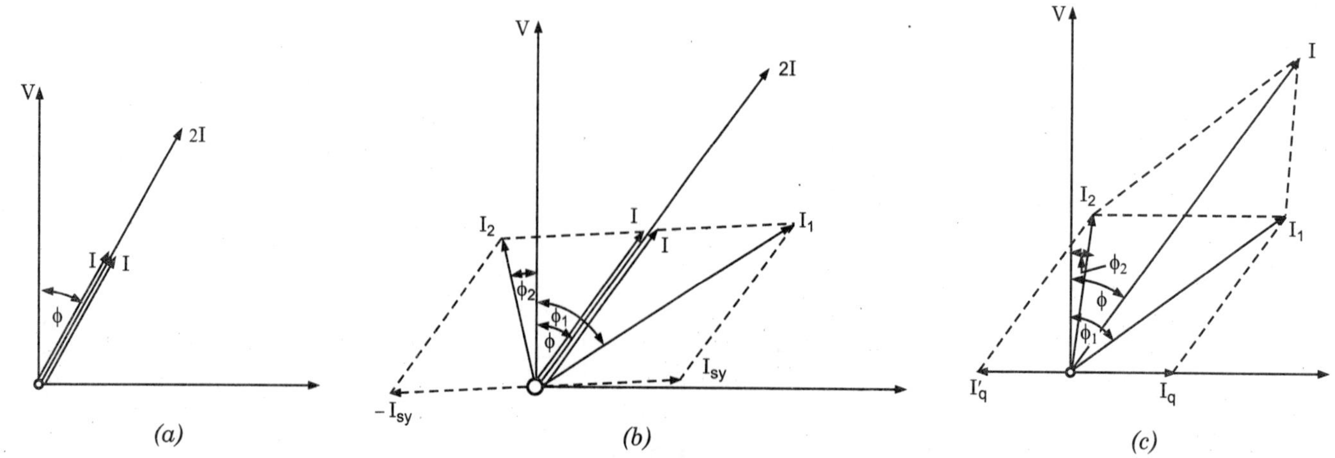

Fig. 17.9

alternator, whose driving torque has been increased, becomes more than the total load supplied, then the other alternator will run as a synchronous motor.

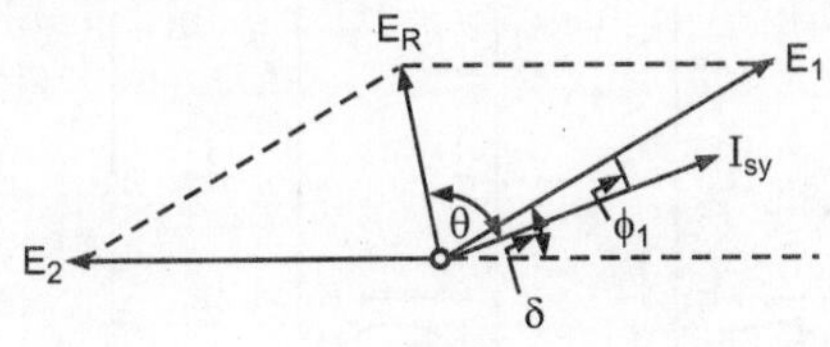

Fig. 17.10

The kW load division between alternators is made by adjusting the governor controls*. One prime mover governor is opened while the other is closed simultaneously. Thus the system frequency is maintained constant while the load is shifted from one machine to the other. Governor control switches are mounted on the switchboard so that the operator is able to watch the switchboard instruments while making adjustments of load division.

17.8 EFFECT OF CHANGE IN SPEED OF ONE OF THE ALTERNATORS

When the two alternators are in exact synchronism the two alternators have equal potential differences and are in exact phase opposition, as shown in Fig. 17.6 (*a*). No circulating current flows round the local circuit. But when the speed of machine 2 is reduced, E_2 falls back by a phase angle δ electrical degrees, as shown in Fig. 17.11, the magnitude of induced emfs E_1 and E_2 remaining the same. It will cause resultant emf E_R acting in the local circuit, which will cause a flow of synchronising current I_{sy} in the local circuit. The synchronising current I_{sy} will lag behind E_R by θ where $\theta = \tan^{-1}\dfrac{X_s}{R_e}$. Since R_e is negligible as compared to X_s, therefore, θ = 90°. The synchronising current I_{sy} will be generating current with respect to machine 1. This synchronising current sets up a synchronising torque tending to retard the generating machine 1 and accelerating the motoring machine *i.e.* machine 2. Thus the two machines come again in synchronism.

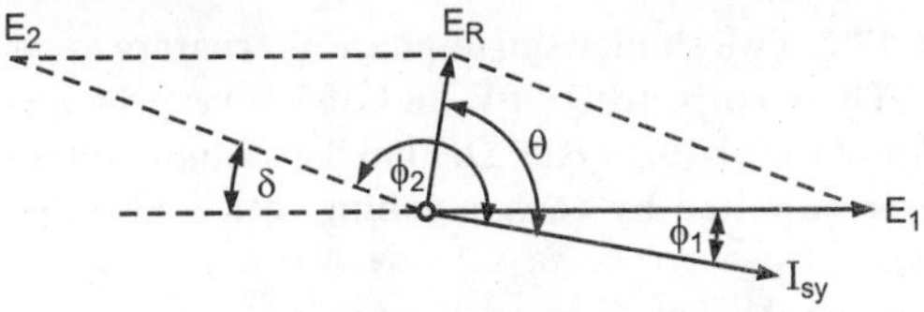

Fig. 17.11

The frequency of the system can be raised or lowered by increasing or reducing the speed of all the machines simultaneously.

17.9 EFFECT OF UNEQUAL VOLTAGES

Let us consider two alternators having their emfs E_1 and E_2 exactly in phase (relative to external load circuit) but of different magnitudes ($E_1 > E_2$). The resultant emf, E_R being equal to $E_1 - E_2$ acts in the local circuit and causes synchronising current I_{sy} around the local circuit. This synchronising current I_{sy} lags behind E_R or E_1 by 90° and leads E_2 by 90° as shown in Fig. 17.12. Thus synchronising current I_{sy} produces demagnetising effect on machine no. 1, resulting thereby reduction in E_1 and magnetising effect on machine no. 2 resulting thereby increase in E_2. Thus difference of E_1 and E_2 is reduced and stable condition is established.

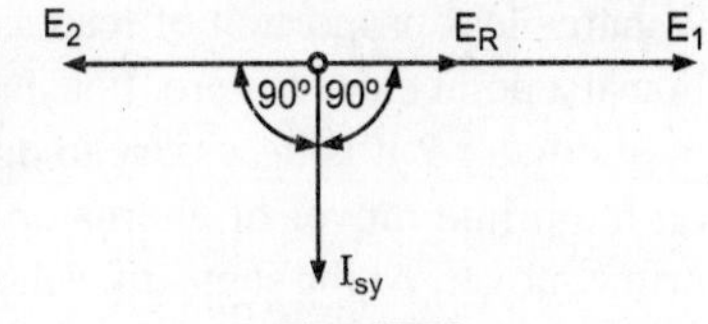

Fig. 17.12

17.10 GOVERNOR CHARACTERISTICS

Unlike dc machines in parallel, disturbing of balance between the emfs does not change the power (kW) of either machine, only the reactive kVA. Furthermore, the speeds must be identical, because of the synchronising action which ties together the speed-torque characteristics. Fig. 17.13 explains how the speed-load characteristics determine the power supplied by different machines operating in parallel. Variation in load sharing necessitates a change in mechanical characteristic, and for a mechanical type of drive this involves an alteration to the governor setting.

The speed-load characteristic of the prime mover of the alternator must be a drooping curve, as shown in Fig. 17.13 so that its operation is stable. Otherwise any increase in load would be accompanied by an increase in speed which means resumption of still more load and this process continues until it takes abnormally large load and run away. The governors usually employed with the generator prime movers act on the centrifugal principle *i.e.* the momentary drop in speed caused by increase in load makes the valve mechanism to operate. The speed-load characteristics of such governors are usually linearly drooping, as illsutrated in Fig. 17.13.

Let us consider that alternator 1 is supplying the system load and the alternator 2 has been just synchronised. Now both of the alternators are operating at the same frequency with major portion of load on alternator 1, as shown in Fig. 17.13. Alternator

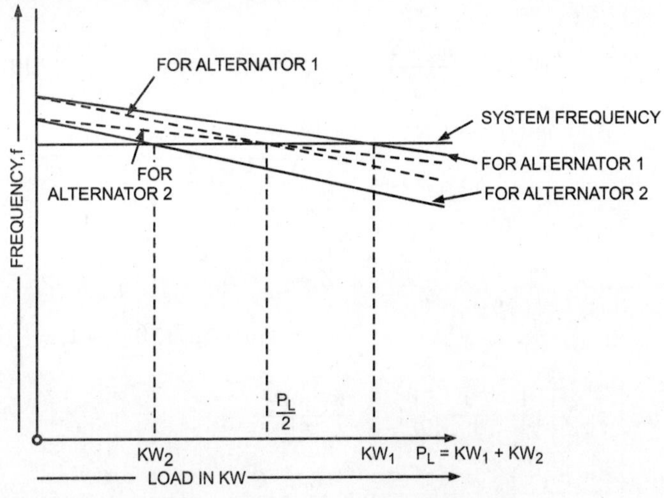

Fig. 17.13 *Speed-Load Characteristics*

* The input to the prime movers (steam turbines, water turbines and diesel units) of the alternators is controlled by sensitive governors.

2 shares less proportion of load as it cuts the system frequency line at a point close to zero. For shifting of load from alternator 1 to alternator 2 it is necessary to open very slightly the governor on the prime mover of alternator 2 to increase the input to the prime mover. At the same time, the governor on the prime mover of alternator 1 is closed very slightly, reducing the input to the prime mover of alternator 1. Thus the input to the prime mover of alternator 1 is reduced resulting in reduced power output of alternator 1 and the input to the prime mover of alternator 2 is increased resulting in increased power output of alternator 2. By careful adjustment of the governors of both prime movers, it is possible to load both of the alternators equally as illustrated in Fig. 17.13.

For satisfactory operation it is necessary that all the machines operating in parallel should have the same speed regulation so as to share the total load in proportion to their rated capacity.

During parallel operation of two alternators, an increase in governor set causes (*i*) increase in system frequency and (*ii*) increase in power supplied by one alternator and reduction in power supplied by the other alternator.

17.11 LOAD SHARING BETWEEN TWO ALTERNATORS

Consider two machines with identical speed-load characteristics running in parallel with a common terminal voltage of V volts and load impedance Z.

Let the generated emfs of the two machines 1 and 2 operating in parallel be E_1 and E_2 respectively and synchronous impedances per phase be Z_{s_1} and Z_{s_2} respectively.

Terminal voltage of machine 1,

$$V = E_1 - I_1 Z_{s_1} \quad ...(17.6)$$

Similarly terminal voltage of machine 2,

$$V = E_2 - I_2 Z_{s_2} \quad ...(17.7)$$

Also $V = IZ = (I_1 + I_2) Z \quad ...(17.8)$

From Eqs. (17.6) and (17.7) we have

$$I_1 = \frac{E_1 - V}{Z_{s_1}} \quad ...(17.9)$$

and $$I_2 = \frac{E_2 - V}{Z_{s_2}} \quad ...(17.10)$$

Adding Eqs. (17.9) and (17.10) we have

$$I_1 + I_2 = \frac{E_1 - V}{Z_{s_1}} + \frac{E_2 - V}{Z_{s_2}}$$

or $$\frac{V}{Z} = \frac{E_1 - V}{Z_{s_1}} + \frac{E_2 - V}{Z_{s_2}}$$

$\because$ From Eq. (17.8) $I_1 + I_2 = \frac{V}{Z}$

or $$V\left(\frac{1}{Z_{s_1}} + \frac{1}{Z_{s_2}} + \frac{1}{Z}\right) = \frac{E_1}{Z_{s_1}} + \frac{E_2}{Z_{s_2}}$$

or $$V = \frac{\frac{E_1}{Z_{s_1}} + \frac{E_2}{Z_{s_2}}}{\frac{1}{Z_{s_1}} + \frac{1}{Z_{s_2}} + \frac{1}{Z}} \quad ...(17.11)$$

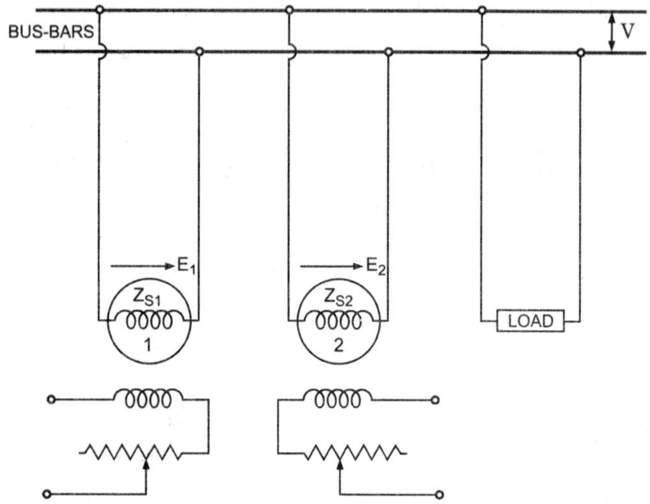

Fig. 17.14

Example 17.1. Synchronising lamps are connected across single-phase alternators having emfs of 228 and 230 V and running at 50 Hz and 50.8 Hz respectively. Before closure of switch, determine

(*i*) maximum and minimum voltage across each lamp
(*ii*) frequency of voltage across the lamps
(*iii*) how often will the lamps light up in 1 minute
(*iv*) peak value of voltage across each lamp
(*v*) phase relations at the instants of occurrence of maximum and minimum voltages.

Solution: Maximum voltage across each lamp,

$$E_{max} = \frac{E_1 + E_2}{2} = \frac{228 + 230}{2} = 229 \text{ V } \textbf{Ans.}$$

Minimum voltage across each lamp,

$$E_{min} = \frac{E_1 \sim E_2}{2} = \frac{228 \sim 230}{2} = 1 \text{ V } \textbf{Ans.}$$

(*ii*) Frequency of voltage across the lamps,

$$f = (f_1 \sim f_2) = (50 \sim 50.8) = 0.8 \text{ Hz } \textbf{Ans.}$$

(*iii*) Lamps light up in one minute

$$= 0.8 \times 60 \text{ i.e. } 48 \text{ times}$$

(*iv*) Peak value of voltage across each lamp,

$$E_{peak} = 229 \times \sqrt{2} = 323.85 \text{ V } \textbf{Ans.}$$

(*v*) In-phase and in phase opposition respectively in the local circuit.

Example 17.2. Two similar single-phase alternators are running in parallel. Their emfs are 100 V and 105 V respectively and the impedance of each is (0.2 + *j*1.0) Ω. Find the terminal voltage, current and power supplied by each machine to a load impedance of (2 + *j*3) Ω.

[*G.B. Technical Univ. Electromechanical Energy Conversion-II, 2012-13*]

Solution: Load admittance,

$$Y = \frac{1}{Z} = \frac{1}{2 + j3} = \frac{2 - j3}{2^2 + 3^2} = \frac{2 - j3}{13} = (0.154 - j0.2308) \text{ S}$$

Admittance of each machine,

$$Y_1 = Y_2 = \frac{1}{0.2 + j1.0} = \frac{0.2 - j1.0}{(0.2)^2 + 1^2} = (0.1923 - j0.9615) \text{ S}$$

Terminal voltage,

$$V = \frac{E_1 Y_1 + E_2 Y_2}{Y_1 + Y_2 + Y}$$

$$= \frac{100\,(0.1923 - j0.9615) + 105\,(0.1923 - j0.9615)}{(0.1923 - j0.9615) + (0.1923 - j0.9615) + 0.154 - j0.2308}$$

$$= \frac{39.4215 - j197.1075}{0.5386 - j2.1538} = \frac{201.011 \angle -78.69^\circ}{2.22 \angle -75.96^\circ}$$

$$= 90.545 \angle -2.73^\circ$$

$$= (90.442 - j4.313) \text{ V } \textbf{Ans.}$$

Current supplied by machine 1,

$$\mathbf{I}_1 = \frac{\mathbf{E}_1 - \mathbf{V}}{\mathbf{Z}_1} = (\mathbf{E}_1 - \mathbf{V})\,\mathbf{Y}_1$$

$$= [100 - (90.442 - j\,4.313)] \times (0.1923 - j\,0.9615)$$

$$= (9.558 + j\,4.313)] \times (0.1923 - j\,0.9615)$$

$$= 10.486 \angle 24.287^\circ \times 0.9805 \angle -78.69^\circ$$

$$= 10.282 \angle -54.403^\circ \text{ A } \quad \textbf{Ans.}$$

Current supplied by machine 2,

$$\mathbf{I}_2 = (\mathbf{E}_2 - \mathbf{V})\,\mathbf{Y}_2 = [105 - (90.442 - j\,4.313)] \times (0.1923 - j\,0.9615)$$

$$= (14.558 + j\,4.313)] \times (0.1923 - j\,0.9615)$$

$$= 15.183 \angle 16.5^\circ \times 0.9805 \angle -78.69^\circ$$

$$= 14.887 \angle -62.19^\circ \text{ A } \quad \textbf{Ans.}$$

Power supplied by machine 1,

$$P_1 = \mathbf{V}_C\,\mathbf{I}_1 = 90.545 \angle 2.73^\circ \times 10.282 \angle -54.403^\circ$$

$$= 930.99 \angle 51.673^\circ \text{ VA} = 577.35 \text{ watts } \quad \textbf{Ans.}$$

Power supplied by machine 2,

$$P_2 = \mathbf{V}_C\,\mathbf{I}_2 = 90.545 \angle 2.73^\circ \times 14.887 \angle -62.19^\circ$$

$$= 1{,}347.94 \angle -59.46^\circ \text{ VA} = 684.9 \text{ watts } \quad \textbf{Ans.}$$

Example 17.3. Two star-connected synchronous generators connected in parallel have an emf of 1,200 V per phase share a common star-connected load of impedance (2 + *j* 1.0) Ω/phase. The synchronous impedances of the machines are $\mathbf{Z}_{S_1}$ = (0.1 + *j*2) Ω/phase and $\mathbf{Z}_{S_2}$ = (0.2 + *j* 3) Ω/phase respectively. Determine the common terminal voltage, line current, power outputs and no-load circulating current when two machines internal emfs have a phase divergence of 5° electrical. [*J.N. Technological Univ. Electrical Machines-III, May-2013*]

Solution: Load impedance, $\mathbf{Z} = (2 + j\,1.0) = 2.236 \angle 26.565^\circ\ \Omega$ /phase

Synchronous impedance of generator 1,

$$\mathbf{Z}_{S_1} = (0.1 + j\,2) = 2.0025 \angle 87.138^\circ \ \Omega/\text{phase}$$

Synchronous impedance of generator 2,

$$\mathbf{Z}_{S_2} = (0.2 + j\,3) = 3.0067 \angle 86.186^\circ \ \Omega/\text{phase}$$

Let emf of generator 1 be,

$$\mathbf{E}_1 = 1{,}200 \angle 0^\circ \text{ V} = (1{,}200 + j\,0) \text{ V/phase}$$

EMF of generator 2 be,

$$\mathbf{E}_2 = 1{,}200 \angle 5^\circ \text{ V} = (1{,}195.4 + j\,104.6) \text{ V/phase}$$

Common terminal voltage,

$$\mathbf{V} = \frac{\dfrac{\mathbf{E}_1}{\mathbf{Z}_{S_1}} + \dfrac{\mathbf{E}_2}{\mathbf{Z}_{S_2}}}{\dfrac{1}{\mathbf{Z}_{S_1}} + \dfrac{1}{\mathbf{Z}_{S_2}} + \dfrac{1}{\mathbf{Z}}}$$

$$= \frac{\dfrac{1{,}200 \angle 0^\circ}{2.0025 \angle 87.138^\circ} + \dfrac{1{,}200 \angle 5^\circ}{3.0067 \angle 86.186^\circ}}{\dfrac{1}{2.0025 \angle 87.138^\circ} + \dfrac{1}{3.0067 \angle 86.180^\circ} + \dfrac{1}{2.236 \angle 26.565^\circ}}$$

$$= \frac{599.25 \angle -87.138^\circ + 399.11 \angle -81.186^\circ}{0.4994 \angle -87.138^\circ + 0.3326 \angle -86.186^\circ + 0.4472 \angle -26.565^\circ}$$

$$= \frac{(29.925 - j\,598.5) + (61.155 - j\,394.397)}{(0.0249 - j\,0.4488) + (0.0221 - j\,0.3319) + (0.4 - j\,0.2)}$$

$$= \frac{91.08 - j\,992.897}{0.447 - j\,0.9807} = \frac{997.066 \angle -84.76^\circ}{1.07777 \angle -65.497^\circ}$$

$$= 925.12 \angle -19.263^\circ = (873.326 - j\,305.2) \text{ V}$$

Line terminal voltage,

$$V_L = \sqrt{3} \times 925.12 \text{ V} = 1{,}602.4 \text{ V } \textbf{Ans.}$$

Line current = Phase current $\mathbf{I}_1$

$$= \frac{\mathbf{E}_1 - \mathbf{V}}{\mathbf{Z}_{S_1}} = \frac{(1{,}200 + j0) - (873.326 - j\,305.2)}{2.0025 \angle 87.138^\circ}$$

$$= \frac{326.674 + j\,305.2}{2.0025 \angle 87.138^\circ}$$

$$= \frac{447.06 \angle 43.054}{2.0025 \angle 87.138^\circ}$$

$$= 223.25 \angle -44.1^\circ \text{ A } \textbf{Ans.}$$

Line current = Phase current, $\mathbf{I}_2$

$$= \frac{\mathbf{E}_2 - \mathbf{V}}{\mathbf{Z}_{S_2}} = \frac{(1{,}195.4 + j\,104.6) - (873.326 - j305.2)}{3.0067 \angle 86.186^\circ}$$

$$= \frac{322.074 + j\,409.8}{3.0067 \angle 86.186^\circ}$$

$$= \frac{521.2175 \angle 51.835^\circ}{3.0067 \angle 86.186^\circ}$$

$$= 173.35 \angle -34.351^\circ \text{ A } \textbf{Ans.}$$

Power output of alternator 1

$$= 3\mathbf{V}_C\,\mathbf{I}_1 = 3 \times 925.12 \angle 19.263^\circ \times 223.25 \angle -44.1^\circ$$

$$= 619{,}599 \angle -24.837^\circ \text{ VA} = 562.3 \text{ kW } \textbf{Ans.}$$

Power output of alternator 2

$$= 3\mathbf{V}_C\,\mathbf{I}_2 = 3 \times 925.12 \angle 19.263^\circ \times 173.35 \angle -34.351^\circ$$

$$= 481{,}108 \angle -15.088 \text{ VA} = 464.5 \text{ kW } \textbf{Ans.}$$

Circulating current on no load

$$= \frac{\mathbf{E}_1 - \mathbf{E}_2}{\mathbf{Z}_{S_1} + \mathbf{Z}_{S_2}} = \frac{(1{,}200 + j\,0) - (1{,}195.4 + j\,104.6)}{(0.1 + j\,2) + (0.2 + j\,3)}$$

$$= \frac{4.6 - j\,104.6}{(0.3 + j\,5)} = \frac{104.7 \angle -87.482^\circ}{5.009 \angle 86.566^\circ}$$

$$= 20.9 \angle -174^\circ \text{ A } \textbf{Ans.}$$

Example 17.4. A 12.5 MVA 3-ϕ star-connected 11 kV, 4-pole, 50 Hz synchronous motor has a reactance of 0.15 pu and negligible armature resistance. Calculate the synchronizing power per mechanical degree, when it supplies full load at 11 kV and 0.8 pf leading.

[*U.P. Technical Univ. Electromechanical Energy Conversion-II, 2003-04*]

Solution: Voltage per phase, $V = \dfrac{11 \times 1{,}000}{\sqrt{3}} = 6{,}350$ V

$$\text{Full-load current, } I = \frac{12.5 \times 10^6}{\sqrt{3} \times 11{,}000} = 656 \text{ A}$$

Now IX = 15 % of 6,350 V = 952.5 V

Synchronous reactance, $X_s = \frac{952.5}{656} = 1.452\ \Omega$

Rotor displacement in electrical degrees

$= \text{Rotor displacement in mechanical degrees} \times \frac{P}{2}$

$= 1 \times \frac{4}{2} = 2^\circ$ or $\frac{2}{180}\pi$ ie 0.0349 radian

Power factor, $\cos\phi = 0.8$ (lag)

$\phi = \text{Cos}^{-1}\,0.8 = -36.87^\circ$

$\text{Sin}\,\phi = \text{Sin}(-36.87^\circ) = -0.6$

Induced emf, $\mathbf{E} = \mathbf{V} + j\mathbf{I}\mathbf{X}$

$= 6{,}350 + j656\,(0.8 + j\,0.6) \times 1.452$

$= 5{,}778.5 + j\,762 = 5{,}828.5\ \angle 7.51^\circ$ V

Synchronising power for 3-phases

$$= \frac{3\delta \mathbf{E}\mathbf{V}}{X_s}$$

$$= \frac{3 \times 0.0349 \times 5{,}828.5 \cos 7.51^\circ \times 6{,}350}{1.452}$$

= 2,645,874 W or 2,645.874 kW **Ans.**

Example 17.5. A 2-pole, 50 Hz, 3-ϕ, turbo-alternator is excited to generate the bus-bar voltage of 11 kV on no load. Calculate the synchronising power per degree of mechanical displacement of the rotor and the corresponding synchronising torque. The machine is star-connected and the short-circuit current for this excitation is 1,200 amperes.

Solution: EMF per phase, $E = \frac{11{,}000}{\sqrt{3}} = 6{,}350$ V

Rotor displacement,

$$\delta = 1 \times \frac{P}{2} = 1 \times \frac{2}{2} = 1 \text{ electrical degree or } \frac{\pi}{180} \text{ elec. radian}$$

Synchronous speed,

$$N_s = \frac{120f}{P} = \frac{120 \times 50}{2} = 3{,}000 \text{ rpm}$$

Synchronous reactance,

$$X_s = \frac{E}{I_{sc}} = \frac{6{,}350}{1{,}200} = 5.292\ \Omega \text{ (neglecting resistance)}$$

Total synchronising power for 3-phases,

$$3\,P_{sy} = \frac{3\delta E^2}{X_s} = \frac{3 \times \pi \times (6.350)^2}{180 \times 5.592}$$

= 3,77,655 W or 377.655 kW **Ans.**

Synchronising torque,

$$T_{sy} = \frac{3P_{sy} \times 60}{2\pi N_s} = \frac{3{,}77{,}655 \times 60}{2\pi \times 3{,}000} = 1{,}202 \text{ N-m } \textbf{Ans.}$$

Example 17.6. A 5,000 kVA, 10,000 V, 1,500 rpm, 50 Hz 3-phase alternator runs in parallel with other machines. Its synchronous reactance is 20%. Find for (*a*) no load (*b*) full load at power factor 0.8 lagging, synchronising power, per unit mechanical angle of phase displacement and calculate the synchronising torque if the mechanical displacement is 0.5°.

Solution: Voltage per phase, $V = \frac{10{,}000}{\sqrt{3}} = 5{,}774$ V

assuming alternator star-connected

Full-load current, $I = \frac{5{,}000 \times 1{,}000}{\sqrt{3} \times 10{,}000} = 288.7$ A

Synchronous reactance, $X_s = \frac{20}{100} \times \frac{5{,}774}{288.7} = 4\ \Omega$

Number of poles, $P = \frac{120f}{N_s} = \frac{120 \times 50}{1{,}500} = 4$

Rotor displacement, $\delta = 1 \times \frac{P}{2} = 1 \times \frac{4}{2} = 2$ electrical degrees

$= 2 \times \frac{\pi}{180} = \frac{\pi}{90}$ elec. radian

(*a*) At no load

Total synchronising power for 3-phases,

$$3\,P_{sy} = \frac{3\delta E^2}{X_s} = 3 \times \frac{\pi}{90} \times \frac{(5{,}774)^2}{4}$$

= 8,72,665 W or 872.665 kW **Ans.**

Synchronising torque per degree mechanical displacement,

$$T_{sy} = \frac{3P_{sy}}{2\pi N_s/60} = \frac{8{,}72{,}665}{2\pi \times 1{,}500/60} = 5{,}555.6 \text{ N-m } \textbf{Ans.}$$

Synchronising torque for 0.5 degree mechanical displacement

= 0.5 × 5,555.6 = 2,778 Nm **Ans.**

(*b*) At full-load 0.8 pf lagging

Phasor diagram is illustrated in Fig. 17.15

Induced emf, $\mathbf{E} = \mathbf{V} + j\,\mathbf{I}\,X_s$

$= (5{,}774 + j\,0) + j\,288.7\,(\cos\phi - j\sin\phi) \times X_s$

$= (5{,}774 + j\,0) + j\,288.7\,(0.8 - j\,0.6) \times 4$

$= 6{,}467 + j\,924 = 6{,}533\ \angle 8.1^\circ$ V

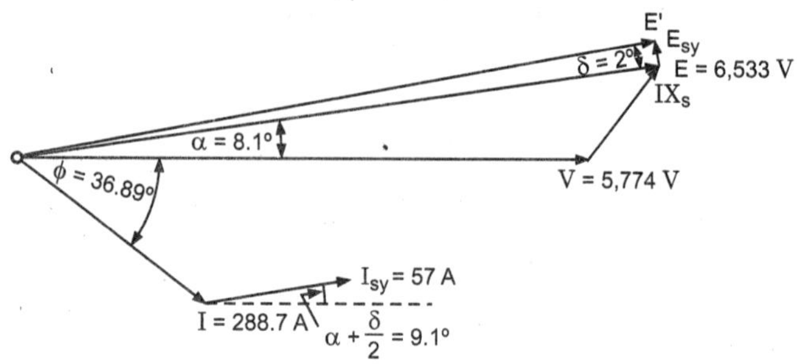

Fig. 17.15

As a result of mechanical displacement an additional internal emf E_{sy} is created which causes synchronising current I_{sy} to flow lagging behind E_{sy} by 90°

$$E_{sy} = 2\,E\sin\frac{\delta}{2} = 2 \times 6{,}533 \times \sin 1^\circ = 228 \text{ V}$$

$$I_{sy} = \frac{E_{sy}}{X_s} = \frac{228}{4} = 57 \text{ A and leads V by}$$

$$\left(\alpha + \frac{\delta}{2}\right) \textit{ i.e. } \left(8.1 + \frac{2}{2}\right) \text{ or } 9.1^\circ$$

Synchronising power for 3-phases

$$= \frac{3VI_{sy}\cos\left(\alpha + \frac{\delta}{2}\right)}{1{,}000} \text{ kW}$$

$$= \frac{3 \times 5{,}774 \times 57 \times \cos 9.1^\circ}{1{,}000} = 975 \text{ kW } \textbf{Ans.}$$

Synchronising torque per degree mechanical displacement

$$T_{sy} = \frac{975 \times 1{,}000}{2\pi \times 1{,}500/60} = 6{,}207 \text{ Nm } \textbf{Ans.}$$

Synchronising torque for 0.5 degree mechanical displacement

$= 0.5 \times 6{,}207 = 3{,}103.5$ N-m **Ans.**

Example 17.7. A 750 kVA, 11 kV, 4-pole, 3-phase star-connected alternator has a resistance of 1.4 Ω per phase and a synchronous reactance of 30 Ω per phase. Calculate the synchronous power per mechanical degree displacement for full load for 0.8 power factor lagging, the terminal voltage being 11 kV.

Solution: Voltage per phase, $V = \dfrac{11 \times 1{,}000}{\sqrt{3}} = 6{,}351$ V

Full-load current, $I = \dfrac{750 \times 1{,}000}{\sqrt{3} \times 11{,}000} = 39.36$ A

Synchronous impedance per phase,

$$\mathbf{Z}_s = (1.4 + j\,30)\ \Omega$$

Rotor displacement,

$$\delta = 1 \times \frac{P}{2} = 1 \times \frac{4}{2} = 2 \text{ electrical degrees}$$

$$= 2 \times \frac{\pi}{180} = \frac{\pi}{90} \text{ elec. radian}$$

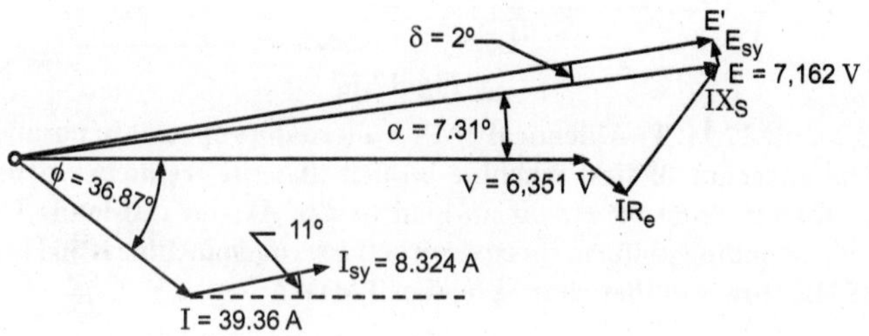

Fig. 17.16

Power factor, $\cos\phi = 0.8$

Phase angle, $\phi = \text{Cos}^{-1}\, 0.8 = 36.87°$

$\text{Sin}\,\phi = \text{Sin}\, 36.87° = 0.6$

Induced emf per phase,

$$\mathbf{E} = \mathbf{V} + \mathbf{I}\,\mathbf{Z}_s$$

$$= (6{,}351 + j\,0) + 39.36\,(0.8 - j\,0.6)\,(1.4 + j\,30)$$

$$= 7{,}103.56 + j\,911.58 = 7{,}162\ \angle 7.31°\ \text{V}$$

Synchronising emf,

$$E_{sy} = 2\,E \sin\frac{\delta}{2} = 2 \times 7{,}162 \sin 1° = 250 \text{ V}$$

$$\mathbf{I}_{sy} = \frac{\mathbf{E}_{sy}}{\mathbf{Z}_s} = \frac{250}{(1.4 + j30)} = 8.324\ \angle 87.33°\ \text{A}$$

I_{sy} lags E_{sy} by 87.33° or leads V by $\left[\alpha + \frac{\delta}{2} + (90° - 87.33°)\right]$ *i.e.* $\left(7.31° + \frac{2}{2} + 2.67°\right)$ or 11°

Synchronising power for 3-phases,

$$3\,P_{sy} = \frac{3\,V\,I_{sy} \cos 11°}{1{,}000} \text{ kW}$$

$$= \frac{3 \times 6{,}351 \times 8.324 \cos 11°}{1{,}000} = 155.68 \text{ kW } \textbf{Ans.}$$

Example 17.8. Two alternators working in parallel supply the following loads (1) lighting load of 500 kW (2) 1,000 kW at 0.9 pf lag (3) 500 kW at 0.9 pf lead (4) 800 kW at 0.8 pf lag. One alternator is supplying 1,500 kW at 0.95 pf lag. Calculate the load on the other machine.

[*U.P. Technical Univ. Electromechanical Energy Conversion-II, 2010-11*]

Solution: Phase angle of first load, $\phi_1 = \text{Cos}^{-1}\, 1.0 = 0°$

Phase angle of second load, $\phi_2 = \text{Cos}^{-1}\, 0.9 = 25.84°$ (lag)

Phase angle of third load, $\phi_3 = \text{Cos}^{-1}\, 0.9 = 25.84°$ (lead)

Phase angle of fourth load, $\phi_4 = \text{Cos}^{-1}\, 0.8 = 36.87°$ (lag)

Total load, $P = P_1 + P_2 + P_3 + P_4$

$= 500 + 1{,}000 + 500 + 800 = 2{,}800$ kW

Reactive power of first load,

$$Q_1 = P_1 \tan\phi_1 = 500 \tan 0° = 0$$

Reactive power of second load,

$$Q_2 = P_2 \tan\phi_2 = 1{,}000 \tan 25.84° = 484.32 \text{ kVAR (lag)}$$

Reactive power of third load,

$$Q_3 = P_3 \tan\phi_3 = 500 \tan 25.84° = 242.16 \text{ kVAR (lead)}$$

Reactive power of fourth load,

$$Q_4 = P_4 \tan\phi_4 = 800 \tan 36.87° = 600 \text{ kVAR (lag)}$$

Total reactive power to be supplied

$$Q = Q_1 + Q_2 + Q_3 + Q_4$$

$$= 484.32 + (-242.16) + 600 = 842.16 \text{ kVAR (lag)}$$

Load supplied by first alternator, $P_A = 1{,}500$ kW

Power factor, $\cos\phi_A = 0.95$ (lag)

Phase angle, $\phi_A = \text{Cos}^{-1}\, 0.95 = 18.2°$ (lag)

Reactive power supplied by first alternator,

$$Q_A = P_A \tan\phi_A = 1{,}500 \tan 18.2° = 493 \text{ kVAR (lag)}$$

Reactive power supplied by second alternator,

$$Q_B = Q - Q_A = 842.16 - 493 = 349.16 \text{ kVAR (lag)}$$

Power supplied by second alternator,

$$P_B = P - P_A = 2{,}800 - 1{,}500 = 1{,}300 \text{ kW } \textbf{Ans.}$$

Phase angle of second alternator,

$$\phi_B = \text{Tan}^{-1}\frac{Q_B}{P_B} = \text{Tan}^{-1}\frac{349.16}{1{,}300} = 15°$$

Power factor of second alternator

$= \text{Cos}\,\phi_B = \cos 15° = 0.9658$ (lag) **Ans.**

Example 17.9. Two alternators working in parallel supply a lighting load of 3,000 kW and a motor load aggerating to 5,000 kW at a pf of 0.72. One machine is loaded up to 5,000 kW at 0.8 pf lagging. What is the load and pf of the other machine?

[*U.P.S.C. I.E.S. Electrical Engineering-II, 2006*]

Solution: Lighting load, $P_1 = 3{,}000$ kW

$\text{Cos}\,\phi_1 = 1.0$

∵ PF of lighting load is assumed to be unity

Reactive component of lighting load,

$$Q_1 = 0$$

Motor load, $P_2 = 5{,}000$ kW

$\text{Cos}\,\phi_2 = 0.72$ (lagging)

∵ Motor load is assumed to be lagging pf load

Reactive component of motor load,

$$Q_2 = P_2 \tan\phi_2 = 5{,}000 \tan(\cos^{-1} 0.72)$$

$$= 4{,}819 \text{ kVAR (lagging)}$$

Total power supplied, $P = P_1 + P_2 = 3{,}000 + 5{,}000 = 8{,}000$ kW

Total reactive power supplied,

$$Q = Q_1 + Q_2 = 0 + 4{,}819 = 4{,}819 \text{ kVAR (lagging)}$$

Load supplied by alternator A,

$$P_A = 5{,}000 \text{ kW}$$

Reactive power supplied by alternator A

$$Q_A = P_A \tan \phi_A = 5{,}000 \tan (\cos^{-1} 0.8)$$
$$= 3{,}750 \text{ kVAR (lagging)}$$

Power supplied by alternator B,

$$P_B = P - P_A = 8{,}000 - 5{,}000 = 3{,}000 \text{ kW}$$

Reactive power supplied by alternator B,

$$Q_B = Q - Q_A = 4{,}819 - 3{,}750 = 1{,}069 \text{ kVAR (lagging)}$$

$$\text{Phase angle, } \phi_B = \text{Tan}^{-1} \frac{Q_B}{P_B} = \text{Tan}^{-1} \frac{1{,}069}{3{,}000} = 19.61^\circ$$

Power factor of alternator B

$$= \text{Cos } \phi_B = \text{Cos } 19.61^\circ = 0.942 \text{ (lagging)}$$ **Ans.**

Example 17.10. Two similar alternators display the following data: 3-phase, 6600 volts, 50 Hz, 1,000 kVA, resistance per phase of each of the star-connected alternator = 1.02 Ω.

Saturation Curve:

Field current	**150**	**200**	**250**	**300**	**350**	**400**	**500**
Terminal voltage	**5,600**	**6,490**	**7,000**	**7,400**	**7,750**	**8,000**	**8,500**

Synchronous reactance per phase = 4 ohms

When operating in parallel with a terminal voltage of 6,600 V the first alternator supplies 85 amps at a pf of 0.85 lagging. If the load power factor is 0.78 lagging and total load is 1,500 kW, determine the excitation of the second alternator.

Solution: Phase angle of the load, $\phi = \text{Cos}^{-1} 0.78 = 38°44'$

$$\text{Reactive power of the load} = \text{Load in kW} \times \tan \phi$$
$$= 1{,}500 \times \tan 38°44'$$
$$= 1{,}500 \times 0.8021 = 1{,}203.15 \text{ kVAR}$$

Power supplied by first alternator

$$= \frac{\sqrt{3} V_L I_1 \cos\phi_1}{1{,}000} = \frac{\sqrt{3} \times 6{,}600 \times 85 \times 0.85}{1{,}000} = 825 \text{ kW}$$

Reactive power supplied by first alternator

$$= \text{Load in kW} \times \tan \phi_1 = 825 \times \tan [\cos^{-1} 0.85]$$
$$= 825 \times 0.62 = 511.5 \text{ kVAR}$$

Power supplied by other alternator

$$= 1{,}500 - 825 = 675 \text{ kW}$$

Reactive power supplied by other alternator

$$= 1{,}203.15 - 511.50 = 691.65 \text{ kVAR}$$

$$\text{Since power supplied} = \frac{\sqrt{3} V_L I_2 \cos\phi_2}{1{,}000}$$

$$\therefore I_2 \cos \phi_2 = \frac{\text{Power supplied in kW} \times 1{,}000}{\sqrt{3} V_L} = \frac{675 \times 1{,}000}{\sqrt{3} \times 6{,}600}$$
$$= 59 \text{ A}$$

$$\text{Since reactive power supplied} = \frac{\sqrt{3} V_L I_2 \sin\phi_2}{1{,}000}$$

$$\therefore I_2 \sin \phi_2 = \frac{691.65 \times 1{,}000}{\sqrt{3} \times 6{,}600} = 60.4 \text{ A}$$

$$\text{Phase voltage} = \frac{6{,}600}{\sqrt{3}} = 3{,}810.4 \text{ volts}$$

$$(E)^2 = (V + I_2 R \cos \phi_2 + I_2 X \sin \phi_2)^2 + (I_2 X \cos \phi_2 - I_2 R \sin \phi_2)^2$$
$$= (3810.4 + 59 \times 1.02 + 60.4 \times 4)^2 + (59 \times 4 - 60.4 \times 1.02)^2$$
$$= 1{,}69{,}40{,}639$$

$$E = \sqrt{1{,}69{,}40{,}639} = 4{,}115 \text{ volts}$$

$$E_L = \sqrt{3} \times 4{,}115 = 7{,}130 \text{ volts}$$

From saturation curve excitation required to induce an emf of 7,130 volts = 265 amperes. **Ans.**

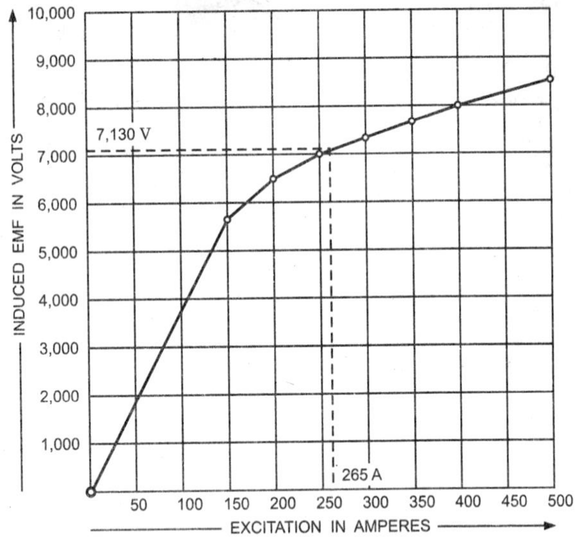

Fig. 17.17

Example 17.11. Two identical 2 MVA alternators operate in parallel. The governor of first machine is such that the frequency drops uniformly from 50 Hz on no load to 47.5 Hz on full-load. The corresponding uniform speed drop of the second machine is 50 Hz to 48 Hz. How will they share a load of 3 MW ?

Solution: Speed-load characteristics are drawn for driving turbines of machines 1 and 2 in Fig. 17.18. CA represents the speed-load characteristic for the turbine driving machine 1 and CB represents the speed-load characteristic of machine 2.

Since the machines are operating in parallel, therefore, their frequency must be same. Let DF be drawn through G, so that the total load supplied is 3 MW

Load supplied by machine 1

$$= DG = OA - AH = 2 - \frac{DH}{CO} \times OA$$
$$= 2 - \frac{OG}{50 - 47.5} \times 2 = 2 - 0.8\, OG$$

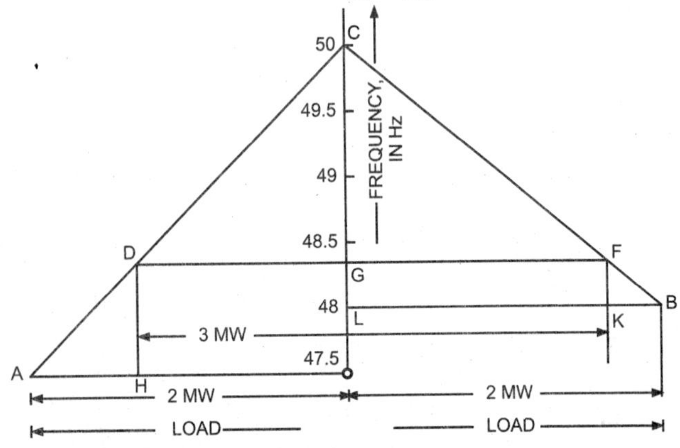

Fig. 17.18

Load supplied by machine 2

$$= GF = LB - BK = 2 - \frac{FK}{CL} \times BL$$
$$= 2 - \frac{OG - 0.5}{50 - 48} \times 2 = 2.5 - OG$$

Total load supplied $= DG + GF = 2 - 0.8\ OG + 2.5 - OG = 3$

$$\text{or } OG = \frac{1.5}{1.8} = \frac{5}{6}$$

So load supplied by machine 1

$$= 2 - 0.8 \times OG = 2 - 0.8 \times \frac{5}{6}$$

$= 1.333$ MW (assuming unity pf) **Ans.**

And load supplied by machine 2

$$= 2.5 - OG = 2.5 - \frac{5}{6}$$

$= 1.667$ MW (assuming unity pf) **Ans.**

Example 17.12. The speed regulations of two 800 kW alternators A and B running in parallel are 100% to 104% and 100 to 105% from full load to no load respectively. How will the two alternators share a load of 1,000 kW?

[R.G. Technical Univ. Electrical Machines-II, December-2013]

Solution: Speed-load characteristics for alternator A and B are represented by lines CA and DB respectively as shown in Fig. 17.19.

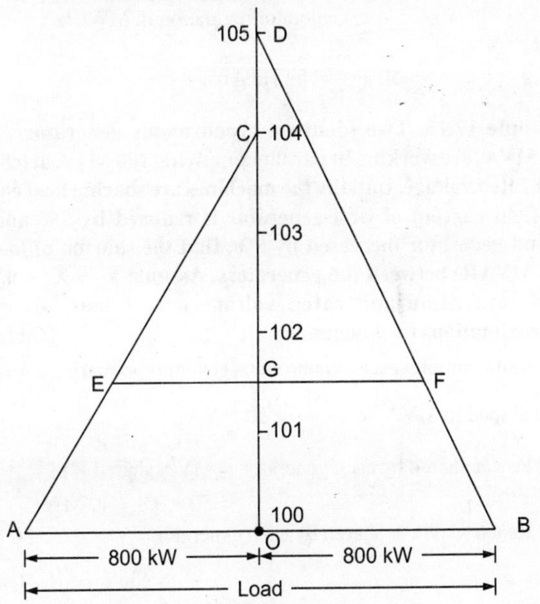

Fig. 17.19

Since the alternators are operating in parallel, their frequency or percentage speed must be the same. Let EF be drawn through G so that the total load is 1,000 kW.

Load supplied by alternator A,

$$EG = AO \times \frac{CG}{CO} = 800 \times \frac{4 - OG}{4}$$

$$= 800 - 200\ OG$$

Load supplied by alternator B,

$$GF = BO \times \frac{DG}{DO} = 800 \times \frac{5 - OG}{5}$$

$$= 800 - 160\ OG$$

Since EG + GF = EF = Total load supplied

$$800 - 200\ OG + 800 - 160\ OG = 1{,}000$$

$$\text{or } OG = \frac{600}{360} = \frac{5}{3}$$

So load supplied by alternator A,

$$P_A = 800 - 200\ OG$$

$$= 800 - 200 \times \frac{5}{3} = 466.67 \text{ kW } \textbf{Ans.}$$

Load supplied by alternator B,

$$P_B = 800 - 160 \times \frac{5}{3} = 533.33 \text{ kW } \textbf{Ans.}$$

Example 17.13. The following two synchronous machines are operating in parallel

Machine A	**50 MW**	**6% speed regulation**
Machine B	**50 MW**	**3% speed regulation**

(*i*) Determine the load taken by each machine for a total load of 80 MW when the speed changers are set to give rated speed at 100% rated output.

(*ii*) The speed changer of machine A is so adjusted that 80 MW is equally shared. Find the output of machine A for the rated speed and also its percentage speed at no-load.

[U.P.S.C. I.E.S. Elec. Engineering II, 1992]

Solution: Speed-load characteristics for machines A and B are represented by lines CA and DB respectively in Fig. 17.20 (*a*).

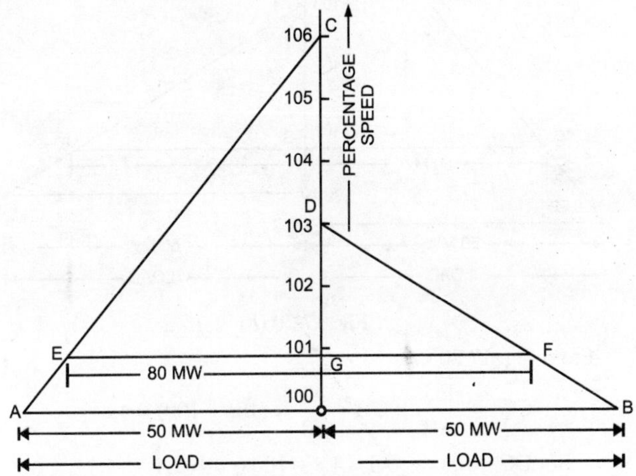

Fig. 17.20 (*a*)

(*i*) Since the machines are operating in parallel, therefore, their frequency or percentage speed must be the same. Let EF be drawn through G so that the total load supplied is 80 MW

Load supplied by machine A,

$$EG = AO \times \frac{CG}{CO} = 50 \times \frac{CO - OG}{CO}$$

$$= 50 \times \frac{6 - OG}{6} = \frac{25}{3}(6 - OG)$$

Load supplied by machine B,

$$GF = BO \times \frac{DG}{DO} = 50 \times \frac{3 - OG}{3} = \frac{50}{3}(3 - OG)$$

Since EG + GF = EF = Total load supplied

$$\frac{25}{3}(6 - OG) + \frac{50}{3}(3 - OG) = 80$$

$$\text{or } OG = \frac{60}{75} = 0.8$$

So load supplied by machine A,

$$P_A = \frac{25}{3}(6 - OG) = \frac{25}{3}(6 - 0.8) = 43.333 \text{ MW}$$

Load supplied by machine B,

$$P_B = \frac{50}{3}(3 - OG) = \frac{50}{3}(3 - 0.8) = 36.667 \text{ MW}$$

(*ii*) Now speed changer of machine A is so adjusted that a total load of 80 MW is shared equally i.e. load shared by machine B is $\frac{1}{2} \times 80$ MW = 40 MW

$$\text{So GF} = \frac{50}{3}(3 - \text{OG}) = 40$$

$$\text{or OG} = 3 - \frac{12}{5} = 0.6$$

i.e. speed of operation will be 100.6%

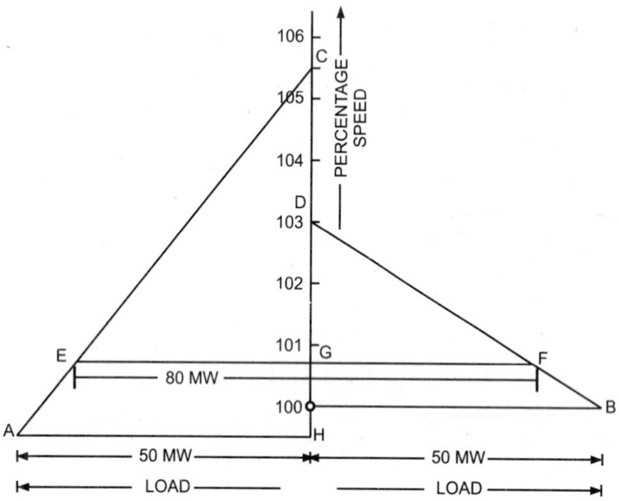

Fig. 17.20 (*b*)

From Fig. 17.20 (*b*)

$$\text{CG} = \frac{\text{EG}}{\text{AH}} \times \text{CH} = \frac{40}{50} \times (106 - 100) = 4.8$$

$$\text{So OC} = \text{CG} + \text{OG} = 4.8 + 100.6 = 105.4$$

So percentage no-load speed of machine A is 105.4% **Ans.**

Load at rated speed

$$= \text{AH} \times \frac{\text{CO}}{\text{CH}} = 50 \times \frac{105.4 - 100}{106 - 100} = 45 \text{ MW } \textbf{Ans.}$$

Example 17.14. Two synchronous generators are supplying a common load. Generator 1 has no-load frequency of 51.5 Hz and regulation of 1 MW/Hz. Generator 2 has no-load frequency of 51.0 Hz and regulation of 1 MW/Hz. The total load is 2.5 MW at 0.8 pf lagging.

(*i*) At what frequency, are the generators supplying this load and how much power is supplied by each generator ?

(*ii*) An additional load of 1 MW is attached to this system. What will be the new frequency.

(*iii*) How much is governor set–point of generator 2 to be adjusted to bring the system frequency at 50 Hz for 3.5 MW system load.

[U.P.S.C. I.E.S. Elec. Engineering II, 1997]

Solution: (*i*) Let the frequency of supply be x Hz at a load of 2.5 MW

Load supplied by generator 1

$$= (51.5 - x) \times 1 = (51.5 - x) \text{ MW}$$

Load supplied by generator 2

$$= (51 - x) \times 1 = (51 - x) \text{ MW}$$

Total load supplied

$$= (51.5 - x) + (51 - x) = 2.5$$

$$\text{or } x = 50 \text{ Hz } \textbf{Ans.}$$

Load supplied by generator 1

$$= 51.5 - 50 = 1.5 \text{ MW } \textbf{Ans.}$$

Load supplied by generator 2

$$= 51 - 50 = 1.0 \text{ MW } \textbf{Ans.}$$

(*ii*) Let the new frequency of supply be x_1 Hz

Now total load supplied

$$= (51.5 - x_1) \times 1 + (51 - x) \times 1 = 3.5$$

$$\text{or } x = 49.5 \text{ Hz } \textbf{Ans.}$$

Load supplied by generator 1

$$= (51.5 - 49.5) \times 1 = 2 \text{ MW } \textbf{Ans.}$$

Load supplied by generator 2

$$= (51.0 - 49.5) \times 1 = 1.5 \text{ MW } \textbf{Ans.}$$

(*iii*) Load supplied by generator 1 at system frequency of 50 Hz

$$= (51.5 - 50) \times 1 = 1.5 \text{ MW}$$

Load to be supplied by generator 2 at system frequency of 50 Hz

$$= 3.5 - 1.5 = 2 \text{ MW}$$

No-load frequency of generator 2

$$= 50 + \frac{\text{Load to be supplied in MW}}{\text{Generator regulation in MW/Hz}}$$

$$= 50 + \frac{2}{1} = 52 \text{ Hz } \textbf{Ans.}$$

Example 17.15. Two identical synchronous generators, each of 100 MVA, are working in parallel supplying 100 MVA at 0.8 lagging pf at rated voltage. Initially the machines are sharing load equally. If the field current of first generator is reduced by 5% and of the second generator increased by 5%, find the sharing of load (MW and MVAR) between the generators. Assume $X_d = X_q = 0.8$ pu no field saturation and rated voltage across load. Reasonable approximations may be made. *[GATE, 2001]*

Solution: Initially each synchronous generator is sharing a load of half of total load in MVA i.e. $\frac{100}{2}$ or 50 MVA.

$$\text{Power shared by each generator} = \text{MVA shared} \times \text{pf} = 50 \times 0.8 = 40 \text{ MW}$$

$$\text{Reactive power shared by each generator} = 50 \times 0.6 = 30 \text{ MVAR}$$

$$\because \sin\phi = \sqrt{1 - 0.8^2} = 0.6$$

When the field of first generator is reduced by 5%, its induced emf becomes (1 – 0.05) times of initially induced emf E_{A_1}

i.e. $E_{A_2} = 0.95\, E_{A_1}$

When the field of second generator is increased by 5%, its induced emf becomes (1 + 0.05) times of initially induced emf E_{B_1}

i.e. $E_{B_2} = 1.05\, E_{B_1}$

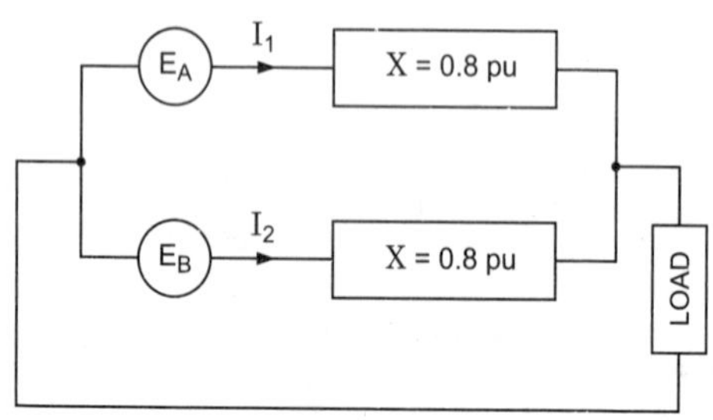

Fig. 17.21

Let the base MVA be 50 MVA

$$\text{Terminal voltage, } \mathbf{V} = 1\angle 0^\circ \text{ pu}$$

$$\text{Load current, } \mathbf{I} = 1\angle -36.87^\circ \text{ pu}$$

Initially induced emfs

$$\mathbf{E}_{A_1} = \mathbf{E}_{B_1} = \mathbf{V} + j\mathbf{IX}$$
$$= (1 + j \times 1 \angle -36.87^\circ \times 0.8)\text{ pu}$$
$$= (1 + 0.8 \angle -36.87^\circ + 90^\circ)\text{ pu}$$
$$= 1 + 0.48 + j\,0.64 = (1.48 + j0.64)\text{ pu}$$
$$\mathbf{E}_{A_2} = (1.48 + j\,0.64) \times 0.95 = (1.406 + j\,0.608)\text{ pu}$$
$$\mathbf{E}_{B_2} = (1.48 + j0.64) \times 1.05 = (1.554 + j0.672)\text{ pu}$$

$$\text{Load current, } \mathbf{I}_1 = \frac{\mathbf{E}_{A_2} - \mathbf{V}}{jX} = \frac{(1.406 + j0.608) - 1}{j0.8}\text{ pu}$$
$$= (0.76 - j0.5075)$$
$$= 0.91387 \angle -33.73^\circ \text{ pu}$$

$$\text{Load current, } \mathbf{I}_2 = \frac{\mathbf{E}_{B_2} - \mathbf{V}}{jX} = \frac{1.554 + j0.672 - 1}{j0.8}$$
$$= (0.84 - j\,0.6925)$$
$$= 1.0886 \angle -33.73^\circ \text{ pu}$$

MVA shared by generator A

$$= 50 \times 0.91387 \angle -33.73^\circ = 45.69 \angle -33.73^\circ$$

Power shared by generator A

$= 45.69 \cos 33.73^\circ = 38$ MW **Ans.**

Reactive power shared by generator A

$= 45.69 \sin 33.73^\circ = 25.37$ MVAR **Ans.**

MVA shared by generator B

$$= 50 \times 1.0886 \angle -39.5^\circ = 54.43 \angle -39.5^\circ \text{ MVA}$$

Power shared by generator B

$= 54.43 \cos 39.5^\circ = 42$ MW **Ans.**

Reactive power shared by generator B

$= 54.43 \sin 39.5^\circ = 34.63$ MVAR **Ans.**

Example 17.16. Two identical 3-phase star-connected alternators running in parallel supply total demand of 1,500 kW at 11 kV and pf 0.867 (lag). Each alternator supplies half the demand and has a reactance (syn) of 50 Ω/ph and resistance of 4 Ω/ph. The field excitation of the first alternator is so adjusted that its armature current is 50 A (lag). Determine the armature current of the second alternator and the generated voltage of the first alternator.

Solution: Total demand, $P = 1{,}500$ kW

Line voltage, $V_L = 11\text{ kV} = 11{,}000$ V

Power factor, $\cos\phi = 0.867$

$$\sin\phi = \sqrt{1 - (\cos\phi)^2} = \sqrt{1 - (0.867)^2} = 0.4983$$

$$\text{Load current, } I_L = \frac{P}{\sqrt{3} V_L \cos\phi} = \frac{1{,}500 \times 1{,}000}{\sqrt{3} \times 11{,}000 \times 0.867} = 90.8 \text{ A}$$

Wattfull component of load current = 90.8 × 0.867 = 78.72 A

Wattless component of load current = 90.8 × 0.4983 = 45.24 A

In identical conditions the current supplied by each machine $= \dfrac{90.8}{2} = 45.4$ A

Assuming power supplied constant, the wattfull component of both machines is the same as $\dfrac{78.72}{2} = 39.36$ A. But due to variation in excitation of first machine the wattless components of load current would be redivided.

The armature current of first machine after changing the excitation is 50 A (lag). The active or wattfull component of current of first machine after changing the excitation = 39.36 A as before.

Wattless component of current of first machine after changing the excitation

$$= \sqrt{50^2 - (39.36)^2} = 30.83 \text{ A}$$

Wattless component of current of second machine after changing the excitation

$$= 45.24 - 30.83 = 14.41 \text{ A}$$

Armature current of second alternator after changing the excitation

$$= \sqrt{(39.36)^2 + (14.41)^2} = 41.91 \text{ A } \textbf{Ans.}$$

Power factor of machine,

$$\cos\phi_1 = \frac{39.36}{50} = 0.7872$$

$$\sin\phi_1 = \sqrt{1 - (\cos\phi_1)^2} = \sqrt{1 - (0.7872)^2} = 0.6166$$

Generated voltage of first machine

$$= \sqrt{(V\cos\phi_1 + IR_e)^2 + (V\sin\phi_1 + IX_s)^2}$$

$$= \sqrt{\left(\frac{11{,}000}{\sqrt{3}} \times 0.7872 + 50 \times 4\right)^2 + \left(\frac{11{,}000}{\sqrt{3}} \times 0.6166 + 50 \times 50\right)^2}$$

$= 8{,}258.6$ V

Line generated voltage of first machine $= \sqrt{3} \times 8{,}258.6 = 14{,}304$ V or 14.3 kV **Ans.**

Example 17.17. Two identical, 3-phase, star-connected generators operating in parallel, share equally a total load of 750 kW at 6,000 V and power factor 0.8 lagging. The synchronous reactance and resistance of each machine are respectively 50 Ω and 2.5 Ω per phase. The field of the first generator is excited so that the armature current is 40 A (lagging). Find (*i*) the armature current of the second alternator (*ii*) the pf of each machine.

[*J.N. Technological Univ. Electrical Machines-III, May-2013*]

Solution: Total load supplied by the two generators,

$$I_L = \frac{P}{\sqrt{3}\, V_L \cos\phi} = \frac{750 \times 1{,}000}{\sqrt{3} \times 6{,}000 \times 0.8} = 90.21 \text{ A}$$

$$\mathbf{I}_L = I_L \angle -\cos^{-1} 0.8 = 90.21 \angle -36.87^\circ = (72.168 - j\,54.126) \text{ A}$$

$$\text{Load supplied by generator } G_1 = \frac{P}{2} = \frac{750}{2} = 375 \text{ kW}$$

Power factor of load supplied by generator G_1,

$$\cos\phi_1 = \frac{P/2}{\sqrt{3} V_L I_{a1}} = \frac{375 \times 1{,}000}{\sqrt{3} \times 6{,}000 \times 40} = 0.90211 \text{ (lagging)}$$

Thus armature current,

$$\mathbf{I}_{a_1} = I_{a_1} \angle -\cos^{-1} 0.90211 = 40 \angle -25.56^\circ \text{ A}$$
$$= (36.084 - j\,17.258) \text{ A}$$

Armature current of generator G_2,

$$\mathbf{I}_{a_2} = \mathbf{I}_L - \mathbf{I}_{a_1} = (72.168 - j\,54.126) - (36.084 - j\,17.258)$$
$$= (36.084 - j\,36.868) = 51.588 \angle -45.62^\circ \text{ A}$$

(*i*) Armature current of the second generator, $I_{a_2} = 51.588$ A

(*ii*) Generator G_1 operates at pf of 0.90211 (lagging) **Ans.**

Generator G_2 operates at pf of cos (–45.62°) = 0.6995 (lagging) **Ans.**

17.12 SYNCHRONOUS MACHINES ON INFINITE BUS BARS

A supply system with a large number of synchronous machines connected in parallel is referred to as *infinite bus bars*. Any additional machine, whether to operate as a generator or as a motor is connected in parallel with the bus bars.

Infinite bus bars represent a system of large capacity whose frequency and the phase and magnitude of voltage are not affected even if there is a variation of excitation or power of a synchronous machine connected to it. Such system behaves like a large generator having virtually zero impedance and infinite rotational inertia. An infinite bus has the following characteristics: (*i*) Constant terminal voltage (*ii*) Constant frequency and (*iii*) Negligible impedance.

Here we will investigate the effect of varying the excitation and torque of the prime mover of a synchronous machine connected to infinite bus bars.

17.12.1. Effect of Varying The Excitation of a Synchronous Machine Connected To Infinite Bus bars. Consider an alternator, which has been just paralleled to infinite bus bars and supplying no load (the torque supplied to its shaft being sufficient to meet with internal losses only). At this instant the open-circuit voltage of the incoming machine, E is equal and opposite to bus bar voltage V, as shown in Fig. 17.22 (*a*) and there is no current through the alternator circuit.

Let the excitation of the incoming machine be increased resulting increase in open-circuit voltage of the machine. The phasor sum of V and E is no longer zero and their phasor sum E_R will cause a synchronising current I_{sy} to flow in the alternator circuit making it to act as a generator supplying power to the bus bars. This can be seen from the projections of I_{sy} on the phasors E and V of Fig. 17.22 (*b*). But the torque supplied to the shaft of the incoming machine were such that it was supplying the friction, windage and iron losses of the machine only, and the condition of Fig. 17.22 (*b*) shows an electrical output of $VI_{sy} \cos \theta$. This indicates that at this particular instant, the prime mover is not supplying enough energy to make up the internal losses and electrical output, and therefore, the incoming generator will tend to slow down. Actually it will not slow down but will get retarded momentarily to change the time phase of E to the position shown in Fig. 17.22 (*c*). The delay in the time phase of E will cause a shift in E_R in a clockwise direction and likewise a clockwise shift of I_{sy}.

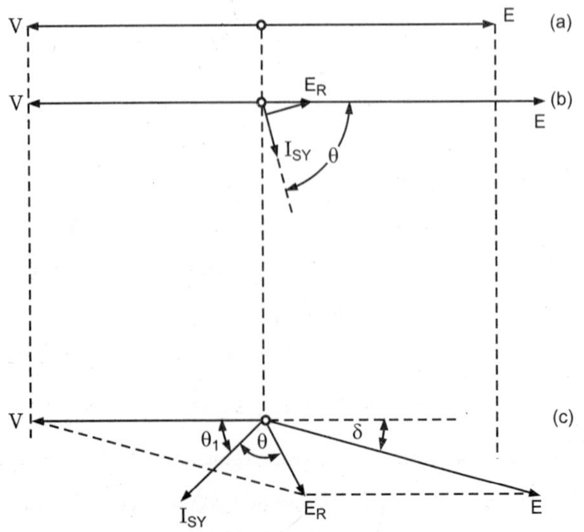

Fig. 17.22

Due to shifting of I_{sy} from the position of Fig. 17.22 (*b*) to that of 17.22 (*c*) the alternator synchronising current I_{sy} shifts the direction of its power component and it will now receive energy from the bus bars ($VI_{sy} \cos \theta_1$) sufficient to meet the copper losses of the alternator. The amount of power transferred as a result of this change in excitation is determined by the change in the copper losses only. The magnitude of the synchronising current I_{sy} for this condition of operation is

$$I_{sy} = \frac{E - V}{R + jX_s} = \frac{E_R}{R + jX_s} \quad ...(17.12)$$

If there were no armature resistance, the synchronising current I_{sy} would always be at 90° with respect to E and V. Thus variation in excitation will cause a large change in the reactive current supplied by the alternator, and because of armature resistance, there will be a slight change in the power that it receives or delivers to the bus. *For all practical purposes it is assumed that variation in excitation causes change in power factor.*

Fig. 17.22 (*c*) shows stable operation of the machine on the bus bars. A tendency of E to advance in time would tend to make it supply power to the bus bars and thereby experiencing a retarding torque. A tendency for E to delay in time would tend to convert it into a synchronous motor, and it would be driven by energy drawn from the bus bars.

It may be noted that when the alternator is overexcited, it operates at lagging power factor and when underexcited it operates at leading power factor.

When the excitation of an alternator connected to infinite bus bars and supplying load at lagging pf is increased, the alternator falls back (*i.e.* load angle decreases), power factor further drops and armature current increases due to reduced power factor; active component being constant.

17.12.2. Effect of Varying the Torque of The Prime Mover. When the torque of the prime mover of the alternator connected to the infinite bus bars is increased (by increasing steam supply in the case of steam drive) the alternator connected to it will swing ahead in phase with respect to other alternator or alternators connected to the common bus bars. It will cause a phase difference between V and E, the magnitude of E remaining unchanged. The resultant of these two voltages, E_R will cause a synchronising current I_{sy}. Thus the power delivered by the alternator, whose input power has been increased will increase by an amount equal to $E I_{sy} \cos \theta_1$. Hence by increasing the torque of the prime mover of one alternator, it is further loaded and an equivalent load is removed from the other unit or units with which the machine is paralleled. If the output of the alternator, whose prime mover torque has been increased, become more than the total load being supplied, then the other machine or machines will operate as synchronous motor(s).

Thus active and reactive power loading of an alternator operating on infinite bus is controlled by controlling the input power to it and excitation respectively.

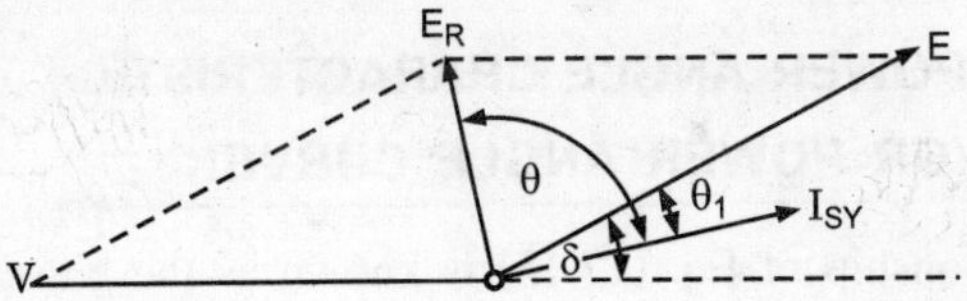

Fig. 17.23

17.13 POWER FLOW TRANSFER EQUATIONS FOR A SYNCHRONOUS GENERATOR

17.13.1. Output Power Equation. Let us consider a 3-phase star-connected synchronous generator having effective resistance R_e ohms per phase and synchronous reactance X_s ohms per phase. Let it be connected to the bus bars having phase voltage of V volts and supply current I amperes at power factor cos ϕ. Let the phase angle between induced emf E and bus-bar voltage V be δ.

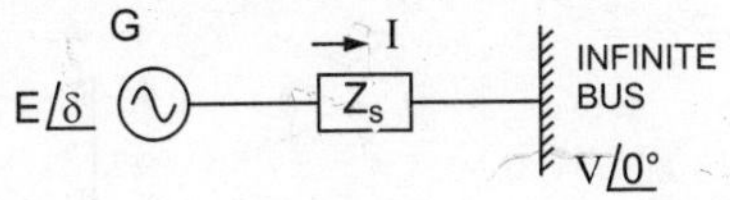

Fig. 17.24

Taking bus-bar voltage as reference phasor we have

$$\mathbf{V} = V\angle 0°$$

$$\mathbf{E} = E\angle\delta$$

$$\text{Impedance per phase, } \mathbf{Z}_s = (R + jX_s) = Z_s\angle\theta$$

$$\text{where } \theta = \text{Tan}^{-1}\frac{X_s}{R_e}$$

$$\text{Armature current, } \mathbf{I} = \frac{E\angle\delta - V\angle 0°}{Z_s\angle\theta}$$

$$\text{Conjugate of armature current, } I_C = \frac{E\angle-\delta - V\angle 0°}{Z_s\angle-\theta}$$

Complex output power is given as

$$S_{out} = P_{out} + jQ_{out} = VI_C = V\angle 0°\left[\frac{E\angle-\delta - V\angle 0°}{Z_s\angle-\theta}\right]$$

$$\text{or } P_{out} + jQ_{out} = \frac{EV}{Z_s}\angle\theta-\delta - \frac{V^2}{Z_s}\angle\theta$$

$$= \frac{EV}{Z_s}\cos(\theta-\delta) - \frac{V^2}{Z_s}\cos\theta$$

$$+ j\left[\frac{EV}{Z_s}\sin(\theta-\delta) - \frac{V^2}{Z_s}\sin\theta\right] \quad ...(17.13)$$

Equating real and imaginary parts expressions for true output power and reactive power can be had.

So true output power per phase is given as

$$P_{out} = \frac{EV}{Z_s}\cos(\theta-\delta) - \frac{V^2}{Z_s}\cos\theta \quad ...(17.14)$$

Since effective resistance R_e is generally negligible, synchronous impedance Z_s can be taken equal to synchronous reactance X_s and internal angle θ = 90° and the expression for power output is reduced to the following expression

$$P_{out} = \frac{EV}{X_s}\cos(90° - \delta) - \frac{V^2}{X_s}\cos 90° = \frac{EV}{X_s}\sin\delta \quad ...(17.15)$$

The Eq. (17.14) is for electrical output in terms of constant bus-bar voltage V, induced emf E, (or excitation as the induced emf depends upon the excitation), synchronous impedance Z_s, internal angle θ being equal to $\tan^{-1}\frac{X_s}{R_e}$ and load angle δ.

Condition for maximum power output can be determined by differentiating the Eq. (17.14) with respect to δ and equating it to zero.

$$\frac{dP_{out}}{d\delta} = \frac{EV}{Z_s}\sin(\theta-\delta) = 0$$

$$\text{or } \sin(\theta-\delta) = 0$$

$$\text{or } \delta = \theta$$

i.e. Power output for constant bus-bar voltage V and fixed excitation will be maximum when δ = θ and the expression for maximum power output per phase will become

$$P_{out\,max} = \frac{EV}{Z_s}\cos(\theta-\theta) - \frac{V^2}{Z_s}\cos\theta$$

$$= \frac{EV}{Z_s} - \frac{V^2}{Z_s}\cos\theta \quad ...(17.16)$$

Now from Δ AOB of phasor diagram shown in Fig. 17.25

$$IZ_s = \sqrt{V^2 + E^2 - 2VE\cos\delta}$$

$$\text{or } I = \frac{\sqrt{V^2 + E^2 - 2VE\cos\delta}}{Z_s}$$

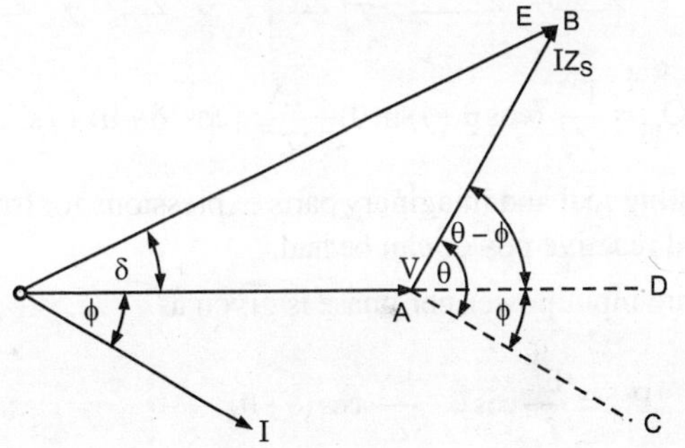

Fig. 17.25

Current corresponding to maximum power output *i.e.* when δ=θ

$$I_{max} = \frac{\sqrt{V^2 + E^2 - 2VE\cos\theta}}{Z_s}$$

and corresponding power factor,

$$\cos\phi = \frac{P_{out\,max}}{VI_{max}} = \frac{\frac{EV}{Z_s} - \frac{V^2}{Z_s}\cos\theta}{\frac{V}{Z_s}\sqrt{V^2 + E^2 - 2VE\cos\theta}}$$

$$= \frac{E - V\cos\theta}{\sqrt{V^2 + E^2 - 2VE\cos\theta}} \quad ...(17.17)$$

Reactive power output per phase is given as

$$Q_{out} = \frac{EV}{Z_S}\sin(\theta-\delta) - \frac{V^2}{Z_S}\sin\theta \quad ...(17.18)$$

Since effective resistance R_e is generally negligible, therefore, synchronous impedance can be taken equal to synchronous reactance and internal angle θ = 90°, and the expression for reactive power output becomes

$$Q_{out} = \frac{EV}{X_S}\sin(90° - \delta) - \frac{V^2}{X_S}\sin 90°$$

$$= -\frac{V^2}{X_S} + \frac{EV}{X_S}\cos\delta \quad ...(17.19)$$

Noteworthy points are:

1. When E cos δ exceeds V, the generator is overexcited and it delivers reactive power to the infinite bus. Under this condition machine operates on lagging pf.
2. When E cos δ equals V, the generator is normally excited and it neither absorbs reactive power from the infinite bus bar nor delivers reactive power to infinite bus bar. Under this condition the machine operates at unity pf.
3. When E cos δ is lesser than V, the generator is underexcited and it absorbs reactive power from the infinite bus bar. Under this condition the machine operates at leading power factor.

17.13.2. Input Power Equation. Complex input power is given as

$$S_{in} = P_{in} + jQ_{in} = \mathbf{EI}_C = E\angle\delta\left[\frac{E\angle{-\delta} - V\angle 0°}{Z_s\angle{-\theta}}\right]$$

$$= E\angle\delta\left[\frac{E}{Z_s}\angle\theta-\delta - \frac{V}{Z_s}\angle\theta\right] = \frac{E^2}{Z_s}\angle\theta - \frac{EV}{Z_s}\angle\delta+\theta$$

$$\text{or } P_{in} + jQ_{in} = \frac{E^2}{Z_s}(\cos\theta + j\sin\theta) - \frac{EV}{Z_s}[\cos(\delta+\theta) + j\sin(\delta+\theta)]$$

Equating real and imaginary parts expressions for true input power and reactive power can be had.

So true input power per phase is given as

$$P_{in} = \frac{E^2}{Z_s}\cos\theta - \frac{EV}{Z_s}\cos(\delta+\theta) \quad ...(17.20)$$

and reactive input power per phase is given as

$$Q_{in} = \frac{E^2}{Z_s}\sin\theta - \frac{EV}{Z_s}\sin(\delta+\theta) \quad ...(17.21)$$

Since effective resistance is generally negligible, synchronous impedance Z_s can be taken equal to synchronous reactance X_s and internal angle θ = 90° and expressions for P_{in} and Q_{in} are reduced to the following expressions

$$P_{in} = \frac{EV}{X_s}\sin\delta \quad ...(17.22)$$

$$\text{and } Q_{in} = \frac{E^2}{X_s} - \frac{EV}{X_s}\cos\delta \quad ...(17.23)$$

Condition for maximum input power can be determined by differentiating Eq. (17.20) w.r.t. δ and equating it to zero.

$$\text{So } \frac{dP_{in}}{d\delta} = \frac{-EV}{Z_s}\sin(\delta+\theta) = 0$$

$$\text{or } \delta = 180° - \theta$$

$$\text{So } P_{in\,max} = \frac{E^2}{Z_s}\cos\theta + \frac{EV}{Z_s} \quad ...(17.24)$$

17.14 POWER ANGLE CHARACTERISTIC (OR POWER ANGLE CURVE)

The relationship of Eq. (17.15) is known as the *power angle characteristic* of the synchronous machine and is plotted in Fig. 17.26 for given E and V. The maximum power occurs at power angle δ = 90° beyond which the machine falls out of synchronism. The machine can be taken up to P_{max} only by gradually increasing the load. This is known as the *steady-state stability limit* of the machine.

The machine is normally operated at power angle δ much less than 90° in order to avoid the machine from going into an unstable region, where it will fall out of step during transient power swings.

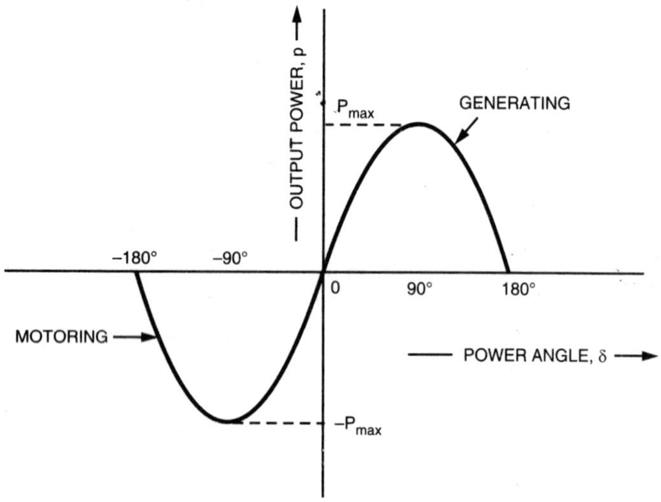

Fig. 17.26 *Power Angle Characteristic*

Example 17.18. A generator has synchronous reactance of 1.7241 pu and is connected to a very large system. The terminal voltage of the generator is 1.0∠0° pu and the generator is supplying to the system a current of 0.8 pu at 0.9 pf lagging. Neglecting resistance, calculate:

(*i*) internal voltage

(*ii*) active and reactive power output of the generator

(*iii*) the power angle and reactive power output of the generator if the excitation of the generator is increased by 20%, keeping active power constant.

[U.P.S.C. I.E.S. Electrical Engineering-II, 1998]

Solution: Current, I = 0.8 pu

Terminal voltage, V = 1.0

Synchronous reactance, X_S = 1.7241 pu

Power factor, cos φ = 0.9 pf (lagging)

φ = 25.842°

Sin φ = 0.4359

(*i*) Internal voltage,

$$\mathbf{E}_0 = \mathbf{V} + j\mathbf{I}X_S$$

$$= 1.0\angle 0° + j0.8\angle{-25.842°} \times 1.7241$$

$$= (1.0 + j0) + 1.37928\angle 64.158°$$

$$= (1.0 + j0) + (0.6012 + j1.2413)$$

$$= (1.6012 + j1.2413) = 2.026\angle 37.78° \text{ pu} \quad \textbf{Ans.}$$

(*ii*) Load angle, $\delta = 37.78°$

$$\text{Active power} = \frac{E_0 V}{X_S} \sin\delta = \frac{2.026 \times 1.0}{1.7241} \sin 37.78° = 0.72 \text{ pu } \textbf{Ans.}$$

$$\text{Reactive power} = \frac{E_0 V}{X_S} \cos\delta - \frac{V^2}{X_S}$$

$$= \frac{2.026 \times 1.0}{1.7241} \cos 37.78° - \frac{1.0^2}{1.7241}$$

$$= 0.92876 - 0.58 = 0.3487 \text{ pu} \quad \textbf{Ans.}$$

(*iii*) When excitation is increased by 20%,

Internal voltage,

$$E'_0 = 1.2\, E_0 = 1.2 \times 2.026 = 2.4312 \text{ pu}$$

Neglecting saturation

Let the new load angle be δ'

$$\text{Active power} = \frac{E'_0 V \sin\delta'}{X_S} = 0.72$$

$\because$ active power remains constant

$$\text{or } \sin\delta' = \frac{0.72 \times X_S}{E'_0 \times V} = \frac{0.72 \times 1.7241}{2.4312 \times 1.0} = 0.51$$

$$\text{or Power angle } \delta' = \text{Sin}^{-1}\, 0.51 = 30.7° \quad \textbf{Ans.}$$

$$\text{Reactive power} = \frac{E'_0 V}{X_S} \cos\delta' - \frac{V^2}{X_S}$$

$$= \frac{2.4312 \times 1.0}{1.7241} \cos 30.70° - \frac{1.0^2}{1.7241} = 0.6325 \text{ pu } \textbf{Ans.}$$

Example 17.19. A 2,200 V, 50 Hz, 3-phase, star-connected alternator has an effective resistance of 0.5 Ω per phase. A field current of 30 A produced the full-load current of 200 A on short-circuit and a line to line emf of 1,100 V on open circuit. Determine (*i*) the power angle of the alternator when it delivers full-load at 0.8 pf (lag) (*ii*) the SCR of the alternator.

[*U.P. Technical Univ. Electromechanical Energy Conversion-II, 2010-11*]

Solution:

$$\text{Phase voltage, } V = \frac{V_L}{\sqrt{3}} = \frac{2{,}200}{\sqrt{3}} = 1{,}270.2 \text{ V}$$

Full-load current, I = 200 A

Effective resistance per phase, R_e = 0.5 Ω

Synchronous impedance per phase,

$$Z_S = \frac{\text{OC phase voltage}}{\text{SC current per phase}} \text{ for the same excitation}$$

$$= \frac{1{,}100/\sqrt{3}}{200} = 3.175\ \Omega$$

Synchronous reactance per phase,

$$X_S = \sqrt{Z_S^2 - R_e^2} = \sqrt{(3.175)^2 - (0.5)^2} = 3.136\ \Omega$$

Open–circuit voltage per phase,

$$E = \sqrt{(V\cos\phi + IR_e)^2 + (V\sin\phi + IX_S)^2}$$

$$= \sqrt{(1{,}270.2 \times 0.8 + 200 \times 0.5)^2 + (1270.2 \times 0.6 + 200 \times 3.136)^2}$$

= 1,782 volts

Power output per phase,

$$P = V_p I_p \cos\phi = \frac{2{,}200}{\sqrt{3}} \times 200 \times 0.8 = 2{,}03{,}227 \text{ watts}$$

From Eq. (15.32)

Power developed per phase due to field excitation

$$P = \frac{E_0\, V}{X_S} \sin\delta \text{ neglecting losses}$$

$$\text{So } \sin\delta = \frac{P X_S}{E_0\, V} = \frac{2{,}03{,}227 \times 3.136}{1{,}782 \times 1{,}270.2} = 0.2816$$

(*i*) Power angle of alternator

$$\delta = \text{Sin}^{-1} 0.2816 = 16.35° \textbf{ Ans.}$$

(*ii*) SCR of alternator

$$= \frac{1}{X_S} = \frac{1}{3.136} = 0.319 \textbf{ Ans.}$$

Example 17.20. A 3-phase, 11,000 V, star-connected turbo-alternator having a synchronous reactance of 6 Ω per phase and negligible resistance has an armature current of 200 A at unity power factor when operating on constant frequency, constant voltage bus bars. If the steam admission remains the same and the emf is raised by 25%, determine the new values of current and power factor.

Solution: Terminal voltage per phase, $V = \dfrac{11{,}000}{\sqrt{3}} = 6{,}350$ V

Reactance per phase, X_S = 6 Ω

Armature current, I = 200 A

Power factor, $\cos\phi$ = 1.0

Phase angle, $\phi = \text{Cos}^{-1}\, 1.0 = 0°$

At unity power factor the reactance drop IX_S

$$= I_R X_S = 200 \times 6 = 1{,}200 \text{ V}$$

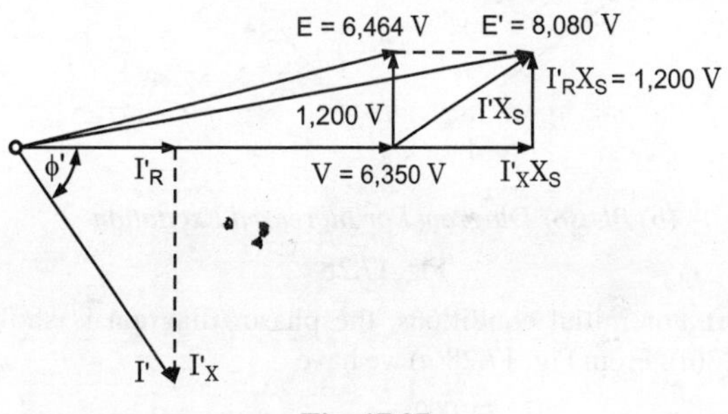

Fig. 17.27

The phasor diagram for this condition is shown in Fig. 17.27

The emf generated per phase,

$$E = \sqrt{(6{,}350)^2 + (1{,}200)^2} = 6{,}464 \text{ V}$$

If the excitation is increased keeping the steam supply or power input same, the power output will not change. The result is to give the armature current a lagging reactive component which exerts a demagnetising effects on the main field and neutralise the increases in excitation. The phasor diagram is given in Fig. 17.27, which shows that the locus of the emf phasor is the dotted line perpendicular to $I'_R X_S$. I'_R is the power or active component of new armature current = 200 A as before and I'_X is the reactive component of new armature current.

With increased excitation the emf generated per phase

$$E' = 1.25\, E = 1.25 \times 6{,}464 = 8{,}080 \text{ V}$$

From phasor diagram shown in Fig. 17.27

$$(E')^2 = (V + I'_X X_S)^2 + (I'_R X_S)^2$$

$$\text{or } (8{,}080)^2 = (6{,}350 + 6\, I'_X)^2 + (200 \times 6)^2$$

$$\text{or } 6{,}350 + 6\, I'_X = \sqrt{(8{,}080)^2 - (1200)^2} = 7{,}990 \text{ V}$$

$$\text{or } I'_X = \frac{7{,}990 - 6350}{6} = 273.33 \text{ A}$$

New value of armature current,

$$I' = \sqrt{(I'_R)^2 + (I'_X)^2} = \sqrt{(200)^2 + (273.33)^2}$$

$$= 338.7 \text{ A } \textbf{Ans.}$$

New power factor, $\cos\phi' = \dfrac{I'_R}{I'} = \dfrac{200}{338.7} = 0.59$ (lag) **Ans.**

Example 17.21. A star-connected alternator is synchronized with an infinite bus of 11 kV; its steam input is then increased till its output power is 15 MW. Now, when its excitation emf is increased to 130%, the synchronous machine starts operating at a pf of 0.8 lagging. Compute the synchronous reactance of the machine. Neglect armature resistance.

Determine the power factor, load angle and armature current of the machine before the excitation emf is increased.

[U.P.S.C. I.E.S. Electrical Engineering-II, 2001]

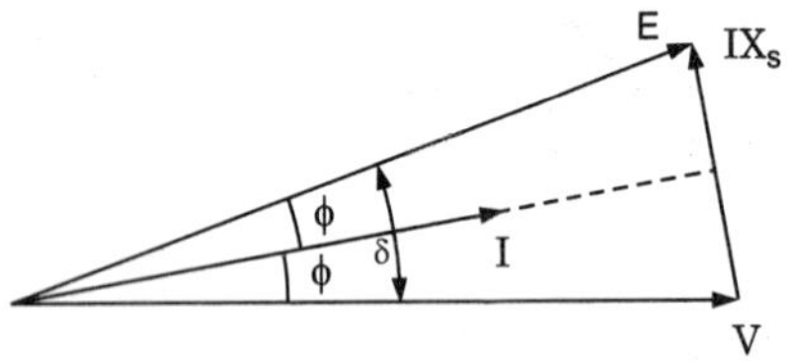

(*a*) *Phasor Diagram For Initial Condition*

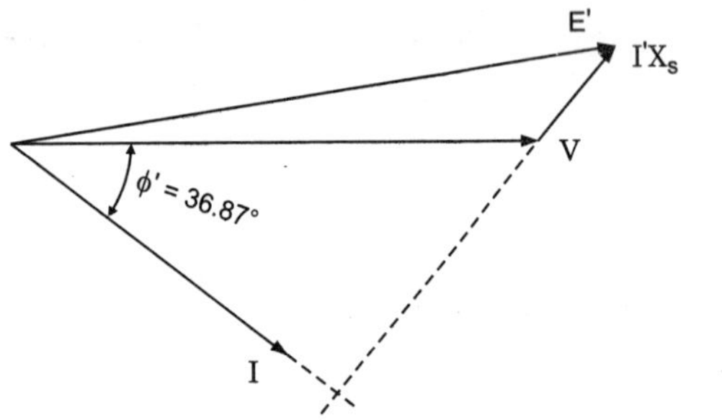

(*b*) *Phasor Diagram For Increased Excitation*

Fig. 17.28

Solution: For initial conditions, the phasor diagram is shown in Fig. 17.28(*a*). From Fig. 17.28(*a*) we have

$$E = V = \frac{11,000}{\sqrt{3}} = 6,350.85 \text{ V}$$

Load angle, $\delta = 2 \times$ power factor angle $= 2\phi$

$$\text{Power output, } P = \frac{EV}{X_S}\sin\delta$$

$$\text{or} \quad \frac{15\times10^6}{3} = \frac{\dfrac{11,000}{\sqrt{3}} \times \dfrac{11,000}{\sqrt{3}} \times \sin\delta}{X_S}$$

$\because$ Power developed per phase $= \dfrac{15}{3}$ MW

$$\text{or} \quad \frac{\sin\delta}{X_S} = \frac{15}{121} = 0.124 \qquad ...(i)$$

Now when excitation is increased to 130%, the phasor diagram is shown in Fig. 17.28 (*b*).

$$E' = 1.3E = \frac{1.3\times11,000}{\sqrt{3}} = 8,256 \text{ V}$$

Phase angle, $\phi = \cos^{-1} 0.8 = 36.87°$ (lagging)

$$\text{Load current, } I' = \frac{\text{Power input}}{\sqrt{3}\,V\cos\phi'} = \frac{15\times10^6}{\sqrt{3}\times11,000\times0.8} = 984.12 \text{ A}$$

$\because$ For negligible armature resistance power input = Power developed

From phasor diagram shown in Fig. 17.28 (*b*) we have

$$(E')^2 = (V\cos\phi')^2 + (V\sin\phi + I'X_S)^2$$

$$\text{or } (8,256)^2 = (6,350.85\times0.8)^2 + (6,350\times0.6 + 984.12X_S)^2$$

$$\text{or } (8,256)^2 = (5,080.68)^2 + (3810.51 + 984.12X_S)^2$$

$$\text{or } 984.12X_S = \sqrt{(8,256)^2 - (5,080.68)^2} - 3,810.51 = 2,697$$

$$\text{or} \quad X_S = \frac{2,697}{984.12} = 2.74$$

Before increasing of excitation

From Eq. (*i*)

$$\sin\delta = 0.124X_S = 0.124\times2.74 = 0.34$$

or Load angle $\delta = \sin^{-1} 0.34 = 19.86°$

$$\text{Initial power factor} = \cos\frac{\delta}{2} = \cos\frac{19.86°}{2} = 0.985 \text{ (leading)}$$

$$\text{Armature current, } I = \frac{15\times10^6}{\sqrt{3}\times11,000\times0.985} = 800 \text{ A} \quad \textbf{Ans.}$$

Example 17.22. A 3-phase, 7 MVA, 11 kV, star-connected alternator is synchronised with an infinite bus at rated voltage. Now the steam input to the machine set is increased till the synchronous machine begins to operate with a load angle of 40°. The synchronous machine has $Z_S = (0 + j\,12)\ \Omega$. Calculate the pf, armature current and the active and reactive powers delivered to infinite bus under these conditions.

Without any change in steam input, how can this alternator be made to deliver zero reactive power to the bus? Calculate armature current, excitation voltage and load angle under these conditions. Draw phasor diagrams at the time of synchronising, when (*i*) load angle is 40° and (*ii*) no reactive power flows.

[U.P.S.C. I.E.S. Electrical Engineering-II, 2004]

Solution: Phasor diagram at the time of synchronising, when load angle $\delta = 40°$, is shown in Fig. 17.29.

Induced emf per phase, E = Bus bar voltage per phase, V

$$= \frac{11,000}{\sqrt{3}} = 6,350.8 \text{ V}$$

Phase impedance per phase,

$$Z_S = (0 + j\,12) = 12\angle 90°\ \Omega \text{ i.e. } X_S = 12\ \Omega$$

$$\text{Resultant emf, } E_R = IX_S = 2V\sin\frac{\delta}{2} = 2\times6,350.8\sin\frac{40°}{2} = 4,344 \text{ V}$$

$$\text{Armature current, } I = \frac{IX_S}{X_S} = \frac{4,344}{12} = 362 \text{ A}$$

$$\text{Phase angle, } \phi = \frac{\delta}{2} = \frac{40°}{2} = 20°$$

Power factor, $\cos\phi = \cos 20° = 0.9397$ (leading) **Ans.**

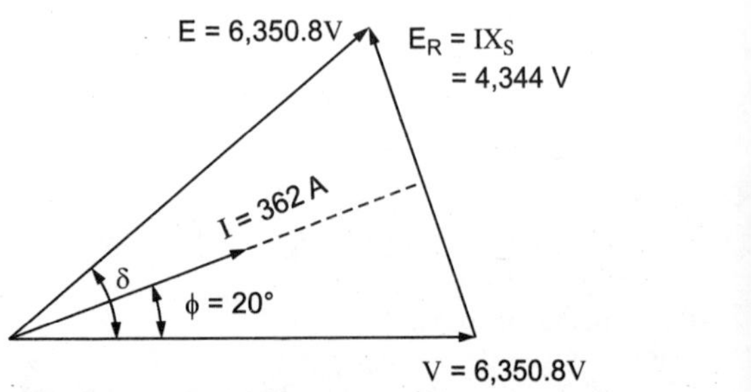

Fig. 17.29 *Phasor Diagram For Initial Conditions*

$$\text{Active power, } P = \frac{\sqrt{3}\,V_L I_L \cos\phi}{1{,}000} = \frac{\sqrt{3}\times 11{,}000\times 362\cos 20^\circ}{1{,}000}$$

$$= 6{,}481 \text{ kW}$$

$$\text{Reactive power, } Q = \frac{\sqrt{3}\,V_L I_L \sin\phi}{1{,}000} = \frac{\sqrt{3}\times 11{,}000\times 362\times\sin 20^\circ}{1{,}000}$$

$$= 2{,}358.9 \text{ kVAR}$$

(*ii*) When the alternator delivers zero reactive power,

Power factor, $\cos\phi' = 1.0$

Supply current, I' = Active component of supply current determined above i.e.

$$I' = I\cos\phi = 362\cos 20^\circ = 340.17 \text{ A}$$

Resultant emf, $E'_R = I'X_S = 340.17 \times 12 = 4{,}082$ V

From phasor diagram shown in Fig. 17.30.

$$\text{Induced emf, } E = \sqrt{V^2 + E_R^2}$$

$$= \sqrt{(6{,}350.8)^2 + (4{,}082)^2} = 7{,}549.6 \text{ V}$$

Induced emf (line value), $E'_L = \sqrt{3}\times 7{,}549.6 = 13{,}076$ V **Ans.**

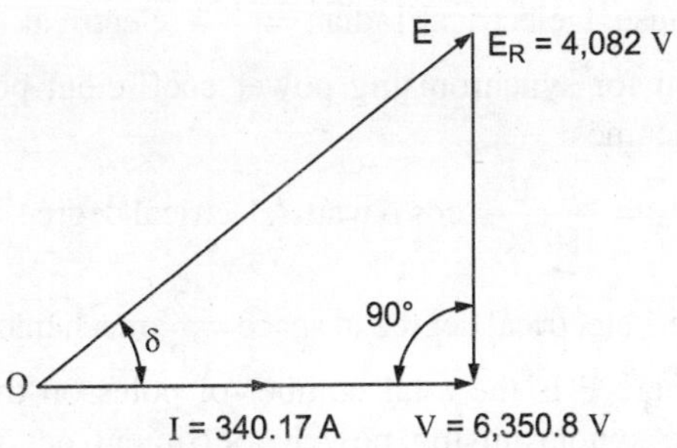

Fig. 17.30 *Phasor Diagram For No Reactive Power Flow*

The alternator can be made to deliver zero reactive power by increasing excitation so that induced emf is 13,076 V.

$$\text{Load angle, } \delta = \cos^{-1}\frac{V}{E} = \cos^{-1}\frac{6{,}350.8}{7{,}549.6} = 32.73^\circ \textbf{ Ans.}$$

Armature current, $I' = 340.17$ A **Ans.**

Excitation voltage, $E'_L = 13{,}076$ V **Ans.**

Example 17.23. A 3-phase, Y-connected alternator connected to 6.0 kV bus bar has a synchronous impedance of 10 Ω per phase and a resistance of 2 Ω per phase. The excitation corresponds to an open-circuit emf of 6.6 kV. What is the maximum load the machine can supply to an external circuit before dropping out of synchronism and what are the corresponding values of armature current and pf?

[*J.N. Technological Univ. Electrical Machines-III, May-2013*]

Solution: Bus-bar voltage per phase, $V = \frac{6{,}000}{\sqrt{3}} = 3{,}464.1$ V

Open-circuit voltage per phase, $E = \frac{6{,}600}{\sqrt{3}} = 3{,}810.5$ V

Synchronous impedance per phase, $Z_S = 10\ \Omega$

Resistance per phase, $R_e = 2\ \Omega$

$$\text{Internal angle, } \theta = \cos^{-1}\frac{R_e}{Z_S}$$

$$= \cos^{-1}\frac{2}{10} = 78.463^\circ$$

Maximum power output per phase,

$$= \frac{EV}{Z_S} - \frac{V^2}{Z_S}\cos\theta$$

$$= \frac{3{,}810.5\times 3{,}464.1}{10} - \frac{(3{,}464)^2}{10}\times 0.2$$

$$= 1{,}320{,}000 - 240{,}000 = 1{,}080{,}000 \text{ W or } 1.08 \text{ MW}$$

Total maximum power that can be supplied by the machine

$$= 3 \times 1.08 = 3.24 \text{ MW } \textbf{Ans.}$$

Current corresponding to maximum output power,

$$I_{max} = \frac{\sqrt{V^2 + E^2 - 2VE\cos\theta}}{Z_S}$$

$$= \frac{\sqrt{(3{,}464.1)^2 + (3{,}810.5)^2 - 2\times 3464.1\times 3{,}810.5\times 0.2}}{10}$$

$$= 460.87 \text{ A}$$

Corresponding power factor,

$$\cos\phi = \frac{P_{max}/\text{phase}}{I_{max}\times V} = \frac{1{,}080{,}000}{460.87\times 3{,}464.1} = 0.6765 \textbf{ Ans.}$$

Example 17.24. A cylindrical rotor hydrogenerator is feeding an active power of 0.25 pu into a large network bus which is held at 1.0 pu voltage. The generator is overexcited with an induced voltage of 1.5 pu. The synchronous impedance of the generator and the connecting link are *j*0.725 pu/phase and *j* 0.11 pu/phase respectively.

Calculate the percent change in reactive power output measured at the network bus in each of the following cases:

(*i*) If the turbine torque is increased by 100%, keeping the excitation of the generator constant.

(*ii*) If the turbine torque is held constant at the initial value, but the excitation is increased by 20% (neglect saturation).

[*U.P.S.C. I.E.S. Electrical Engineering-II, 2002*]

Solution: Synchronous impedance per phase,

$$Z_S = j\,0.725 \text{ pu}$$

Connecting link impedance, $Z_l = j\,0.11$ pu

Total impedance per phase, $Z = j\,0.725 + j\,0.11 = j\,0.835$ pu

It means that synchronous generator and connecting link have total reactance of 0.835 pu and negligible resistance.

If δ is the load angle, then

$$\sin\delta = \frac{\text{Power developed per phase}}{V\times E}\times X$$

$$= \frac{0.25}{1\times 1.5}\times 0.835 = 0.1392$$

Reactive power output per phase,

$$Q = \frac{-V^2}{X} + \frac{VE}{X}\cos\delta \qquad ...(i)$$

(*i*) When the turbine torque is increased by 100%

$$P' = \frac{VE\sin\delta'}{X} = 2P = \frac{2VE\sin\delta}{X}$$

∵ V and E remain unchanged

or New load angle, $\delta' = \sin^{-1}(2\sin\delta) = \sin^{-1}(2\times 0.1392) = 16.16^\circ$

Change in reactive power output

$$= \frac{Q'-Q}{Q}\times 100 = \frac{\frac{VE}{X}(\cos\delta' - \cos\delta)}{-\frac{V^2}{X} + \frac{VE}{X}\cos\delta}\times 100$$

$$= \frac{\text{Cos}\,\delta' - \text{Cos}\,\delta}{\text{Cos}\,\delta - \frac{V}{E}} \times 100$$

$$= \frac{\text{Cos}\,16.16° - \text{Cos}\,8.08°}{\text{Cos}\,16.16° - \frac{1}{1.5}} \times 100 = -10.07\%$$ **Ans.**

(*ii*) When turbine torque is held constant but excitation is increased by 20%.

New load angle $\delta' = \delta = \text{Sin}^{-1}\,0.1392 = 8°$

New induced emf $E' = 1.2E$

Change in reactive power output

$$= \frac{E'\cos\delta - E\cos\delta}{E\cos\delta - V}$$

$$= \frac{(E' - E)\cos\delta}{E\cos\delta - 1}$$

$$= \frac{(1.20 - 1) \times 1.5 \times 0.99}{1.5 \times 0.99 - 1} = 0.6124$$

Percentage change in reactive power output

$= 0.6124 \times 100 = 61.24\%$ **Ans.**

17.15 SYNCHRONISING POWER AND SYNCHRONISING TORQUE COEFFICIENTS

A synchronous machine (whether a generator or a motor), after properly synchronising to infinite bus bars, has an inherent tendency to remain in synchronism. Consider an alternator transferring power P to infinite bus bars at a steady-state power angle δ_0 and there is a sudden transient disturbance due to increase in load causing an increase in load angle $d\delta$. Thus, the operating point of the machine shifts to a new constant-power line and the load on the machine becomes $P + \delta P$. Since the steady-state power input remains unaltered, the speed of the machine decreases due to increase in load and the machine returns back to steady-state position. Similarly, if due to a transient disturbance, the rotor of the machine decelerate, the load angle will decrease and the operating point of the machine will shift to a new constant-power line and the load on the machine will fall to $P - \delta P$. Since the steady-state power input remains constant, the rotor will accelerate and consequently the machine will return back to steady-state position. Thus, it may be concluded that the effectiveness of this connecting action is dependent upon the change in power transfer for a given change in load. A measure of effectiveness is given by *synchronising power coefficient,* also known as *stiffness of coupling, rigidity factor* or *stability factor.*

It is denoted by P_{sy} and is defined as the rate at which the synchronous power P varies with load angle.

For a cylindrical rotor machine, the power output per phase is given as

$$P = \frac{V}{Z_S}[E\cos(\theta - \delta) - V\cos\theta] \quad ...(17.25)$$

and the synchronising power coefficient is given as

$$P_{sy} = \frac{dP}{d\delta} = \frac{EV}{Z_S}\sin(\theta - \delta) \quad ...(17.26)$$

and the synchronising torque coefficient is defined by

$$\tau_{sy} = \frac{d\tau}{d\delta} = \frac{1}{2\pi n_s}\frac{dP}{d\delta} = \frac{EV}{2\pi n_s Z_S}\sin(\theta - \delta) \quad ...(17.27)$$

Since effective resistance is generally negligible, synchronous impedance can be taken equal to synchronous reactance X_S and internal angle $\theta = 90°$ and the expression for synchronising power coefficient and synchronising torque coefficient are reduced to

$$P_{sy} = \frac{EV}{X_S}\cos\delta \text{ watts per electrical radian} \quad ...(17.28)$$

$$\text{and } \tau_{sy} = \frac{EV}{2\pi n_s X_S}\cos\delta \text{ N-m per electrical radian} \quad ...(17.29)$$

For a salient pole machines

$$P = \frac{EV}{X_S}\sin\delta + \frac{1}{2}V^2\left[\frac{1}{X_q} - \frac{1}{X_d}\right]\sin 2\delta$$

$$P_{sy} = \frac{dP}{d\delta} = \frac{EV}{X_S}\cos\delta + V^2\left[\frac{1}{X_q} - \frac{1}{X_d}\right]\cos 2\delta \quad ...(17.30)$$

The units of synchronising power coefficient are watts per electrical radian. For conversion of unit of synchronising power in watts/electrical degree, the expression for P_{sy} is multiplied by $\frac{\pi}{180}$ because 1 electrical radian $= \frac{180}{\pi}$ electrical degree. The expression for synchronising power coefficient per electrical degree becomes

$$P_{sy} = \frac{\pi}{180}\frac{EV}{X_S}\cos\delta \text{ watts/electrical degree} \quad ...(17.31)$$

Since 1 electrical degree in space $= \frac{2}{P}$ mechanical degree in space, where P is the total number of poles on the machine. Therefore synchronising power coefficient per mechanical degree is given as

$$P_{sy} = \frac{\pi P}{360}\frac{EV}{X_S}\cos\delta \text{ watts per mechanical degree} \quad ...(17.32)$$

Synchronising power coefficient gives rise to synchronising torque coefficient at synchronous speed. It means that the synchronising torque is the torque which results in synchronising power at synchronous speed. The synchronising torque is given by the expression

$$\tau_{sy} = \frac{1}{\omega_s} m \frac{dP}{d\delta}\frac{\pi P}{360} \text{ N-m per mechanical degree} \quad ...(17.33)$$

where m = number of phases and $\omega_s = 2\pi n_s$, n_s being the synchronous speed of the machine in rps.

$$\text{So } \tau_{sy} = \frac{P_{syn}}{2\pi n_s} \quad ...(17.34)$$

17.15.1. Significance of Synchronising Power Coefficient. The synchronising power coefficient P_{sy} is a measure of the stiffness (or rigidity) of the electromagnetic coupling between the stator and rotor fields. Very large value of synchronising power coefficient means that the coupling is stiff or rigid and too large stiffness of coupling means that the motor has tendency of following closely the variation of speed caused by the disturbance in electric power supply. In case of absence of power supply disturbance, too much rigidity of coupling would result in practically constant speed operation of the machine regardless of the fluctuations in mechanical load. It means that for a too large stiff coupling, with change of load or power supply the machine will be subjected to undue mechanical shocks which may damage the rotor or the windings.

Equations (17.26) and (17.30) indicates that the synchronising power coefficient P_{sy} is inversely proportional to synchronous reactance and is directly proportional to the excitation voltage. Machines with larger air gaps have relatively smaller reactance and are therefore more stiffer than the machines with smaller air gap. Since P_{sy} is directly proportional to excitation voltage, an overexcited synchronous machine is more stiffer than an underexcited synchronous machine.

Equations (17.26) and (17.30) also reveal that the restoring action is the strongest when $\delta = 0$, that is, at no load. The restoring action is zero when $\delta = \pm 90°$. At these values of load angle δ, the machine will operate at the steady-state limit of the stability and in the condition of unstable equilibrium. At the steady-state limit of stability, the ability of the machine to resist small changes of load is zero, therefore, it is not possible to operate a machine at the steady-state limit of stability. However, the machine provided with special fast-acting excitation system may be operated at the steady-state limit of stability.

Example 17.25. A 5 MVA, 11 kV, 50 Hz, 4-pole, star-connected synchronous generator with synchronous reactance of 0.7 pu is connected to an infinite bus. Find synchronizing power and the corresponding torque per unit of mechanical angle displacement: (*i*) at no load and (*ii*) at full load of 0.8 pf lagging.

[*G.B. Technical Univ. Electromechanical Energy Conversion-II, 2011-12*]

Solution: Voltage per phase, $V = \dfrac{11,000}{\sqrt{3}} = 6,350$ V

Full-load current per phase, $I = \dfrac{5 \times 10^6}{\sqrt{3} \times 11,000} = 262.4$ A

Now $IX_S = 0.7$ times voltage per phase $= 0.7 \times 6,350 = 4,445$ V

Synchronous reactance,

$$X_S = \frac{IX_S}{I} = \frac{4,445}{262.4} = 16.94\ \Omega$$

Synchronous speed of the machine,

$$N_s = \frac{120 f}{P} = \frac{120 \times 50}{4} = 1,500 \text{ rpm}$$

(*i*) *At no load*

Torque angle, $\delta \cong 0$ and $E = V = 6,350$ V

Synchronising power per mechanical degree,

$$P_{sy} = \frac{3E^2}{Z_S} \times \frac{\pi P}{360} \times \cos\delta = \frac{3 \times (6,350)^2}{16.94} \times \frac{\pi \times 4}{360} \times 1$$

$$= 249,266 \text{ W or } 249.266 \text{ kW} \quad \textbf{Ans.}$$

The corresponding synchronising torque per mechanical degree,

$$\tau_{sy} = \frac{60}{2\pi N_s} \times P_{sy} = \frac{60}{2\pi \times 1,500} \times 249,266 = 1,587 \text{ Nm} \quad \textbf{Ans.}$$

(*ii*) Induced emf per phase,

$$\mathbf{E} = \mathbf{V} + j\mathbf{I}X_S$$

$$= 6,350 + j\, 262.4\, (0.8 - j\, 0.6) \times 16.94$$

$$= 6,350 + 2,667 + j\, 3,566$$

$$= 9,017 + j\, 3,566 = 9,678 \angle 21.3° \text{ V}$$

Synchronising power per mechanical degree

$$P_{sy} = \frac{3\pi P}{360} \times \frac{EV}{X_S} \cos\delta = \frac{3\pi \times 4}{360} \times \frac{9,678 \times 6,350}{16.94} \cos 21.3°$$

$$= 353,954 \text{ W} \quad \text{or} \quad 353.954 \text{ kW} \quad \textbf{Ans.}$$

The corresponding synchronising torque per mechanical degree,

$$\tau_{sy} = \frac{60}{2\pi N_s} \times P_{sy} = \frac{60}{2\pi \times 1,500} \times 353,954 = 2,253 \text{ Nm} \quad \textbf{Ans.}$$

Example 17.26. A 3-phase, 400 kVA, 6.6 kV, 1,500 rpm, 50 Hz, star-connected alternator is running in parallel with infinite bus bars. Its synchronous reactance drop is 25%. Calculate the synchronising power and torque per unit mechanical angle of phase displacement when the alternator operates at full load 0.8 pf lagging.

[*J.N. Technological Univ. Electrical Machines-III, May-2013*]

Solution: Voltage per phase, $V = \dfrac{6,600}{\sqrt{3}} = 3,810.5$ V

Full-load current per phase, $I = \dfrac{400 \times 10^3}{\sqrt{3} \times 6,600} = 35$ A

Reactance drop, $IX_S = 25\%$ of terminal voltage per phase

$$= \frac{25}{100} \times 3,810.5 = 952.625 \text{ V}$$

Synchronous reactance, $X_S = \dfrac{IX_S}{I} = \dfrac{952.625}{35} = 27.218\ \Omega$

Synchronous speed of the machine,

$$n_s = \frac{N_s}{60} = \frac{1,500}{60} = 25 \text{ rps}$$

Induced emf per phase,

$$\mathbf{E} = \mathbf{V} + j\, \mathbf{I}X_S$$

$$= 3,810.5 + j\, 35\, (0.8 - j\, 0.6) \times 27.218$$

$$= (4,382 + j\, 762.1) \text{ V} = 4,447.8 \angle 9.866° \text{ V}$$

Number of poles, $P = \dfrac{120 f}{N} = \dfrac{120 \times 50}{1,500} = 4$

Synchronising power per mechanical degree,

$$P_{sy} = \frac{3\pi P}{360} \times \frac{EV}{X_S} \cos\delta$$

$$= \frac{3\pi \times 4}{360} \times \frac{3,810.5 \times 4,447.8}{27.218} \cos 9.866°$$

$$= 64,243.5 \text{ W or } 64.2435 \text{ kW } \textbf{Ans.}$$

The corresponding synchronising torque per mechanical degree

$$\tau_{sy} = \frac{1}{2\pi n_s} \times P_{sy}$$

$$= \frac{64,243.5}{2\pi \times 25} = 408.99 \text{ N-m } \textbf{Ans.}$$

17.16 HUNTING

Any alternator that is running at a constant speed, is working under a condition of dynamic equilibrium between the mechanical driving torque of its prime mover and the electro-magnetic retarding torque of its electrical load. If it is connected to the bus bars to which other alternators and synchronous motors are connected, then any change in either its mechanical driving torque or its electric load upsets the balance and causes it to swing away from the position of perfect synchronism. Circulating currents immediately flow between this machine and the other synchronous machines connected to the common bus bars to pull it back into synchronism. But the change to correct phase position cannot be accomplished without some "overshooting" on the part of the rotors, accompanied by a retardation, with a repetition of the entire cycle. That is, the alternators hunt, their actions being exactly equivalent to those of synchronous motors under similar conditions. The period of the swing agrees with the natural oscillating period of the rotor as a torsional pendulum.

In some cases, alternators employ prime movers which have a pulsating torque output such as a diesel engine. This causes the alternator rotor periodically moving slightly faster, then slower. This pulsating or oscillating effect, called 'hunting,' causes current to surge back and forth between alternators operating in parallel, which is quite objectionable. If the pulsation has a "forced" frequency within 20% of that of the natural oscillating frequency, the oscillation following any load change will be cumulative and the alternators may be thrown out of synchronism.

It is highly desirable that these oscillations should die out quickly before they are reinforced by other disturbances, as otherwise there is a possibility of the alternator being thrown out of synchronism. The tendency of diesel engines to hunt is reduced by using a heavy flywheel to reduce torque variations. Damper windings, often called *amortisseur* windings, are placed on the surface of some generator rotors to reduce the tendency to hunt. There is no need of providing damper windings (except in some special cases to assist in synchronising) because in such machines, solid rotor cores provide path for eddy currents, especially in the quadrature axis, where the iron may form an equivalent rotor circuit, thus producing the same effects as those damper bars. Dash-pots are also placed on the engine governors to reduce tendency to hunt. Generators driven by steam turbines generally do not have a tendency to hunt since the torque produced does not pulsate.

17.17 TIME PERIOD OF OSCILLATION

Many causes, such as variation in load or variation in mechanical driving torque creates phase swinging of the machine. If the time period of these oscillations coincides with that of the natural time period, the amplitude of these oscillations may become so high that the machine may fall out of synchronism. The expression for the natural time period of oscillations for a synchronous machine is derived below.

Let the torque per mechanical radian displacement of rotor be T N-m

Moment of inertia of the rotor Σmr^2 be J kg-m^2

If the rotor is displaced by δ electrical radians the total synchronising power developed by all the three phases $= \dfrac{3\delta E^2}{Z_S}$

where E is generated voltage per phase

Synchronising power per electrical radian displacement

$$= \frac{3E^2}{Z_S}$$

Since displacement in electrical radians

$$= \frac{P}{2} \times \text{displacement in mechanical radians}$$

Synchronising power per mechanical radian displacement

$$P_{sy} = 3\,\frac{E^2}{Z_S} \times \frac{P}{2} = \frac{3E^2P}{2Z_S}$$

Synchronising torque,

$$T_{sy} = \frac{\text{Synchronising power in watts}}{2\pi n_s}$$

where n_s is synchronous speed in rps

$$= \frac{3E^2P}{2Z_S \times 2\pi n_s} = \frac{3EP}{4\pi n_s} \times \frac{E}{Z_S} = \frac{3PEI_{sc}}{4\pi n_s}$$

$\because \dfrac{E}{Z_S} = I_{sc}$, the short-circuit current

$$= \frac{3EI_{sc}}{4\pi} \times \frac{2f}{(n_s)^2} = 0.4775\,\frac{EI_{sc}\,f}{(n_s)^2} \qquad \text{...(17.35)}$$

$$\because P = \frac{2f}{n_s}$$

Natural time period,
$$T_n = 2\pi\sqrt{\frac{J}{T_{sy}}} = 2\pi\sqrt{\frac{J(n_s)^2}{0.4775\,EI_{sc}\,f}}$$

$$= 9.1\,n_s\sqrt{\frac{J}{EI_{sc}\,f}} \qquad \text{...(17.36)}$$

Natural frequency of oscillation,

$$f_n = \frac{1}{T_n} = \frac{1}{9.1 n_s}\sqrt{\frac{EI_{sc}\,f}{J}} \qquad \text{...(17.37)}$$

Example 17.27. A 10 MVA, 10 kV, 3-ϕ, 50 Hz, 1,500 rpm alternator is paralleled with others of much greater capacity. The moment of inertia of the rotor is 2 × 10^5 kg-m^2 and the synchronous reactance of the machine is 40%. Calculate the frequency of oscillation of the rotor.

Solution: Moment of inertia of the rotating system,

$$J = 2 \times 10^5 \text{ kg-m}^2$$

Phase voltage, $E = \dfrac{10{,}000}{\sqrt{3}}$ V

Percentage synchronous reactance, X_S = 40 %

Full-load current, $I = \dfrac{\text{Rated MVA} \times 10^6}{\sqrt{3} \times \text{rated line voltage}}$

$$= \frac{10 \times 10^6}{\sqrt{3} \times 10{,}000} = \frac{1{,}000}{\sqrt{3}} \text{ A}$$

Ratio of short-circuit current to full-load current,

$$\frac{I_{sc}}{I} = \frac{100}{\text{Percentage } X_S} = \frac{100}{40} = 2.5$$

So short-circuit current,

$$I_{sc} = 2.5\,I = 2.5 \times \frac{1{,}000}{\sqrt{3}} = \frac{2{,}500}{\sqrt{3}} \text{ A}$$

Synchronous speed,

$$n_s = \frac{N_s}{60} = \frac{1{,}500}{60} = 25 \text{ rps}$$

Natural frequency of oscillation

$$= \frac{1}{9.1\,n_s}\sqrt{\frac{E\,I_{sc}\,f}{J}}$$

$$= \frac{1}{9.1 \times 25}\sqrt{\frac{\frac{10{,}000}{\sqrt{3}} \times \frac{2{,}500}{\sqrt{3}} \times 50}{2 \times 10^5}} = 0.2 \text{ Hz } \textbf{Ans.}$$

Example 17.28. A 10,000 kVA, 4-pole, 6,600 V, 50 Hz, 3-phase star-connected alternator has a synchronous reactance of 25% and operates on constant voltage, constant-frequency bus bars. If the natural period of oscillation while operating at full load and unity power factor is to be limited to 1.5 seconds, calculate the moment of inertia of the rotating system.

[Andhra Univ. Electric Machinery II 1990]

Solution: Phase voltage, $E = \frac{6,600}{\sqrt{3}} = 2,200\sqrt{3}$ volts

Frequency, $f = 50$ Hz

Synchronous speed, $n_s = \frac{2f}{P} = \frac{2 \times 50}{4} = 25$ rps

Full-load current, $I = \frac{10,000 \times 1,000}{\sqrt{3} \times 6,600} = \frac{10,000}{\sqrt{3} \times 6.6}$ A

Short-circuit current,

$$I_{sc} = I \times \frac{100}{25} = \frac{4 \times 10,000}{\sqrt{3} \times 6.6} = \frac{40,000}{6.6\sqrt{3}} \text{ A}$$

Natural period of oscillation,

$T = 1.5$ seconds

Since $T = 9.1\, n_s \sqrt{\frac{J}{EI_{sc} f}}$

so moment of inertia of rotating system,

$$J = \frac{T^2 \times E \times I_{sc} \times f}{(9.1)^2\, n_s^2} = \frac{(1.5)^2 \times 2,200\sqrt{3} \times 40,000 \times 50}{6.6\sqrt{3} \times (9.1)^2 \times (25)^2}$$

$= 2.9 \times 10^4$ kg-m^2 **Ans.**

Example 17.29. A 100 MVA synchronous generator operates on full load at frequency of 50 Hz. The load is suddenly reduced to 50 MW. Due to time lag in governor system, the steam valve begins to close after 0.4 second. Determine the change in frequency that occurs in this time. Given: H = 5 kW-s/kVA of generating capacity.

[U.P.S.C. I.E.S. Elec. Engineering II, 1995]

Solution: Generating capacity, G = 100 MVA = 100,000 kVA

Inertia constant, H = 5 kW-s/kVA

Kinetic energy stored in rotating parts of the synchronous generator and prime-mover at synchronous speed of 50 Hz

$= H \times G = 5 \times 100,000 = 5 \times 10^5$ kW-seconds

Excess power input to the generator shaft before the steam valve begins to close = 50 MW

Excess energy transferred to the rotating parts in 0.4 second

$= 50 \times 1,000 \times 0.4 = 20,000$ kW-seconds

Since kinetic energy stored is proportional to the square of speed

So $\frac{5 \times 10^5 + 0.2 \times 10^5}{5 \times 10^5} = \left(\frac{f'}{f}\right)^2$

or frequency at the end of 0.4 *s*,

$$f' = f\sqrt{\frac{5.2 \times 10^5}{5 \times 10^5}} = 50\sqrt{1.04} = 51 \text{ Hz}$$ **Ans.**

17.18 PARALLEL OPERATION OF ALTERNATORS THROUGH TRANSMISSION LINES

The impedance of a line connecting the two alternators adds directly to the sum of the synchronous impedances of the machines. This increase in impedance acts to reduce the synchronising current and power and thus it makes them less strongly held in synchronism. They become more sensitive to the disturbances (voltage and torque) of large value, beyond the ability of the softer coupling that results. The reduction in synchronising power is more likely to lead to disconnection of individual units of apparatus and shutdowns from the operation of automatically reclosing type circuit breakers. The voltage drop in a transmission line reduces the voltage and, therefore, the high line impedances may result in instability of operation; the synchronising power being directly proportional to the square of voltage.

HIGHLIGHTS

1. Requirements of parallel operation:
   (*i*) Same output voltage rating.
   (*ii*) The rated speeds of the machines should be such as to give the same frequency.
   (*iii*) The alternators should be of the same type so as to generate voltages of the same waveform.
   (*iv*) The prime movers of the alternators should have same speed-load characteristics, which of course must be drooping ones.
   (*v*) The alternators should have reactances in their armatures.
2. Conditions for proper synchronizing:
   (*i*) Terminal voltage of incoming machine must be exactly equal to that of other machines or bus bar.
   (*ii*) The frequency of the incoming machine must be equal to bus-bar frequency.
   (*iii*) The phase of the incoming machine must be same as that of the bus-bar voltage.
   (*iv*) In case of 3-phase synchronous generators, phase sequence of the incoming machine and that of the bus bar must be same.
3. Synchronising current is given by

   $$I_{sy} = \frac{E\delta}{Z_s}$$

   where E is induced emf per phase, δ is load angle and Z_s is synchronous impedance per phase.
4. Total synchronising power for 3-phases is given by

   $$P_{sy} = \frac{3\delta E^2}{X_s}$$

5. Synchronising torque is given by

   $$T_{sy} = \frac{3P_{sy} \times 60}{2\pi N_s}$$

6. Without reactance in armature circuits of machines, the machines could not develop motor and generator power, respectively to restore synchronism, parallel operation would not be possible.
7. In general, the proper amount of field excitation for alternators operating in parallel is the amount of excitation each alternator would need if it were carrying its load alone at the same voltage and frequency.

8. The voltage of the system in which several alternators are operating in parallel may be controlled (raised or lowered) by varying the excitation of all the machines, up or down as required; this being done automatically by the voltage regulators.
9. The kW load division between alternators is made by adjusting the governor controls.
10. The frequency of the system can be raised or lowered by increasing or reducing the speed of all the machines simultaneously.
11. For satisfactory operation it is necessary that all the machines operating in parallel should have the same speed regulation so as to share the total load in proportion to their rated capacity.
12. Terminal voltage of two alternators having respectively induced emfs E_1 and E_2, synchronous impedances Z_{S_1}, Z_{S_2} and supplying load Z is given by

$$\mathbf{V} = \frac{\dfrac{\mathbf{E}_1}{\mathbf{Z}_{S_1}} + \dfrac{\mathbf{E}_2}{\mathbf{Z}_{S_2}}}{\dfrac{1}{\mathbf{Z}_{S_1}} + \dfrac{1}{\mathbf{Z}_{S_2}} + \dfrac{1}{\mathbf{Z}}}$$

13. When the excitation of an alternator connected to infinite bus bars and supplying load at lagging pf is increased, the alternator falls back (*i.e.* load angle decreases), power factor further drops and armature current increases due to reduced power factor; active component being constant.
14. Active and reactive power loading of an alternator operating on infinite bus is controlled by controlling the input power to it and excitation respectively.
15. Power developed per phase by a 3-phase non-salient synchronous machine is given by

$$P = \frac{EV}{Z_S}\cos(\theta-\delta) - \frac{V^2}{Z_S}\cos\theta = \frac{EV}{X_S}\sin\delta$$

if stator resistance is negligible.

16. Reactive power developed per phase by a 3-phase non-salient synchronous generator is given by

$$Q = \frac{EV}{Z_S}\sin(\theta-\delta) - \frac{V^2}{Z_S}\sin\theta$$

$$= \frac{EV}{X_S}\cos\delta - \frac{V^2}{X_S}$$

if stator resistance is negligible.

17. Input power per phase to a 3-phase non-salient synchronous generator is given as

$$P_{in} = \frac{E^2}{Z_s}\cos\theta - \frac{EV}{Z_s}\cos(\delta+\theta)$$

Input reactive power per phase to a 3-phase non-salient synchronous generator is given as

$$Q_{in} = \frac{E^2}{Z_s}\sin\theta - \frac{EV}{Z_s}\sin(\delta+\theta)$$

18. A measure of effectiveness is given by synchronising power coefficient, also known as stiffness of coupling, rigidity factor or stability factor. Synchronising power coefficient and synchronising torque coefficient for a non-salient synchronous machine are given as

$$P_{syn} = \frac{EV}{Z_s}\sin(\theta-\delta) \text{ watts/electrical radian}$$

$$\tau_{syn} = \frac{1}{2\pi n_s}\frac{dP}{d\delta} = \frac{EV}{2\pi n_s Z_s}\sin(\theta-\delta) \text{ watts per electrical radian.}$$

19. Machines with larger air gaps have relatively smaller reactance and, are therefore, more stiff than the machines with smaller air gaps.

Since synchronising power coefficient P_{syn} is directly proportional to excitation voltage, an overexcited synchronous machine is more stiffer than an underexcited machine.

Restoring action is strongest when load angle $\delta = 0$, i.e. on no load and is zero when $\delta = \pm 90°$.

20. Synchronosing power per mechanical radian displacement

$$3P_{sy} = \frac{3E^2P}{2Z_S}$$

$$\text{Synchronising torque, } T_{sy} = 0.4775\frac{EI_{se}f}{(n_s)^2}$$

$$\text{Natural time period, } T_n = 9.1 n_s\sqrt{\frac{J}{EI_{sc}f}}$$

$$\text{Natural frequency of oscillation, } f_n = \frac{1}{T_n} = \frac{\sqrt{EI_{sc}f}}{9.1 n_s\sqrt{J}}$$

EXERCISES

1. What is the necessity of parallel operation of alternators? State the conditions necessary for paralleling alternators.
[M.D. Univ. Electrical Machines-II, December-2010]
2. State and explain the conditions for parallel operation of alternators.
[Anna Univ. Electrical Machines-II; November/December-2012]
3. Explain synchronisation of alternators with necessary conditions. Give any one method in detail.
[Gujarat Technological Univ. Electrical Machines-I, Summer-2014]
4. Explain what is meant by synchronising of alternator. What are the various methods of synchronising? Explain any one method with the help of neat sketch. **[Rajasthan Technical Univ. Electrical Machines-II, 2012]**
5. What are the conditions of parallel operation of synchronous generators ? Write, in sequence, the steps needed to connect a synchronous generator with infinite bus by dark bright lamp method.
[U.P. Technical Univ. Electromechanical Energy Conversion II, 2003-04]
6. With appropriate circuit and connection diagrams, explain the "2 bright and 1 dark" lamp method of synchronising a 3-phase alternator to the grid or to another existing alternator. Is a "1 bright and 2 dark" lamp method possible? Justify your answer in brief.
[W.B. Technological Univ. Electrical Machines, 2011-12]
7. Explain how an alternator is synchronised to the bus bars.
[J.N. Technological Univ. Electrical Machines-III, May-2013]
8. What is meant by infinite bus bars? State the conditions required to be satisfied by connecting a synchronous generator to an infinite bus bar. Explain how the instant for synchronizing can be determined.
[U.P.S.C. I.E.S. Electrical Engineering-II, 1998]
9. Explain the parallel operation of synchronous generators with drooping characteristics. Also draw the phasor diagram under the operating conditions for parallel operation and elaborate the same.
[U.P. Technical Univ. Electromechanical Energy Conversion-II, 2006-07]

10. The synchronous alternators are having slightly unequal open-circuit emfs and slightly unequal load-frequency. Suggest suitable measures and method of synchronising two alternators in parallel. Explain using schematic diagram.

[U.P. Technical Univ. Electromechanical Energy Conversion-II, 2006-07]

11. Discuss the use of synchroscope in the parallel operation of alternators.

[Bharathiyar Univ. Elec. Machines II, April-1996]

12. Derive expressions for synchronising power and synchronising torque when two alternators are connected in parallel.

13. With the help of a phasor diagram, explain the effect of excitation control of an alternator operating with another alternator supplying a common load. **[J.N. Technological Univ. Electrical Machines-III, May-2013]**

14. Describe the effect of change of excitation and mechanical input of an alternator. **[W.B. Technological Univ. Electrical Machines, 2012-13]**

15. How does the parallel operation of two alternators differ from the condition when an alternator run with an infinite bus bar ? How is the power sharing affected in parallel operation when (*i*) Excitation of one of them is changed and (*ii*) Steam input to one of them is changed ?

[U.P. Technical Univ. Electromechanical Energy Conversion-II, 2009-10]

16. What is an Infinite Bus? An alternator connected to infinite bus is operating at rated load and unity power factor. How its operation shall be affected by increasing the field current in respect of induced emf, torque angle and power factor ?

[G.B. Technical Univ. Electromechanical Energy Conversion-II, 2012-13]

17. Explain the effect of varying of excitation of a synchronous generator connected to infinite busbar on the power factor, armature current and load angle.

[U.P. Technical Univ. Electromechanical Energy Conversion, 2010-11]

18. Deduce the expressions for output real and reactive powers from the equivalent circuit (per phase) in case of a cylindrical rotor synchronous generator. **[B.P. Univ. of Technology Electrical Machine-I, 2008]**

19. What do you understand by 'power angle of an alternator ? Will it increase or decrease if the input to the prime mover of the alternator connected to an infinite bus is increased ? Explain.

20. Draw the power angle characteristic and get the condition for maximum power in synchronous machines.

[Pb. Technical Univ. Synchronous Machines, May-2013]

21. Explain power angle characteristic of cylindrical rotor alternator.

[U.P. Technical Univ. Electromechanical Energy Conversion-II, 2009-10]

22. Derive the condition for maximum power output of a synchronous generator connected to infinite bus bars and operating at constant excitation.

23. What is synchronising power? Explain its significance with parallel operation of alternators.

[J.N. Technological Univ. Electrical Machines-III, May-2013]

24. What is hunting in an alternator? Discuss the measures to be taken to minimise the hunting.

[W.B. Technological Univ. Electrical Machines-II, 2012-13]

25. What are the causes of hunting in an alternator operating in parallel with other machines ? What are its bad effects and how they can be minimised ?

26. What is hunting in synchronous machines and what are its effects? Does it occur in synchronous generators only or in synchronous motors also? How is it reduced in salient pole synchronous generators and in turbo-alternators? Formulate analytically the dynamic of hunting in case of round rotor synchronous generator and identify when hunting may lead to operational instability?

[U.P.S.C. I.E.S. Electrical Engineering-II, 2010]

27. Obtain an expression for the natural frequency of undamped oscillation of the rotor of a synchronous motor.

28. Write short notes on the following:

(*i*) Parallel operation of alternators. **[U.P. Technical Univ., 2010-11]**

(*ii*) Process of synchronisation. **[U.P. Technical Univ., 2010-11]**

(*iii*) Power flow equations of cylindrical and salient pole machines.

[G.B. Technical Univ., 2011-12]

(*iv*) Hunting in synchronous generators.

SHORT ANSWER TYPE QUESTIONS WITH ANSWERS

Q. 1. Why are alternators put in parallel ?

Ans. For high efficiency of operation, reliability, convenience and economy in maintenance and repair and possibility of additions to plant capacity with the growth in load on the power station.

Q. 2. What is meant by synchronising of alternators ?

Ans. The process of connecting an alternator in parallel with another alternator or with the common bus bars is called the synchronising.

Q. 3. What are the conditions for parallel operation of alternators ?

Ans. Terminal voltage, frequency and the phase of the incoming machine must be the same as that of the bus bars (or other alternators already operating in parallel).

In case of 3-phase alternators, an additional requirement to be met is that the phase sequence of the incoming machine must be the same as that of the bus bars.

Q. 4. What are the two methods by which two alternators are put in parallel ?

Ans. Synchronising lamp method and synchroscope method.

Q. 5. A 3-phase, 60-pole alternator is to be synchronised to an infinite bus bar using two bright one dark lamp method. If the lamps glow 60 times per minute in a particular sequence, determine the speed of the alternator.

Ans. 98 rpm or 102 rpm.

Q. 6. A machine with a large air gap has a higher synchronising power. Why ?

Ans. A large air gap offers a large reluctance to the path of the flux produced by the armature mmf and thus reduces the effect of armature reaction. This results in a small value of synchronous reactance and, therefore, a higher synchronising power because synchronising power is inversely proportional to the synchronous reactance.

Q. 7. What is the effect of increasing the steam admission to the prime mover connected to an alternator ?

Ans. By increasing the steam admission to the prime mover of an alternator will load it further.

Q. 8. How can the distribution of load between two alternators operating in parallel be varied ?

Ans. Distribution of load between two alternators operating in parallel can be varied by adjusting the governor control. One prime mover governor is opened while the other is closed simultaneously so as to maintain the system frequency constant.

Q. 9. What is the effect of varying excitation of an alternator running in parallel with another alternator ?

Ans. Variation in excitation of an alternator running in parallel with another alternator causes variation in wattless component and, therefore, of reactive kVA, without affecting the sharing of the kW load or power.

Q. 10. What is meant by infinite bus bars ?

Ans. Infinite bus bars represent a system of large capacity whose frequency and the phase and magnitude of the voltage are not affected even if there is a variation of excitation or power of a synchronous machine connected to it.

Q. 11. Will an increase in the excitation of an alternator connected to an infinite bus increase its real power generation ?

Ans. No.

Q. 12. How will the power angle and power factor of a 3-phase alternator supplying power to an infinite bus at constant excitation, change if the steam input is reduced ?

Ans. Both power angle and power factor will decrease if the alternator was supplying load to the infinite bus at lagging power factor.

Q. 13. At what power angle a synchronous generator will develop maximum power ?

Ans. Synchronous generator will develop maximum power when power angle $\delta = \theta$, internal angle where $\theta = \text{Tan}^{-1} \dfrac{X_s}{R_e}$.

Q. 14. A synchronous machine has been synchronised with an infinite bus. Now the machine is made to deliver active power to the bus, keeping the excitation constant. Will it, at the same time, generate or absorb reactive power.

Ans. For proper synchronisation, E = V

and reactive power is given as

$$Q = \frac{EV}{X_s}\cos\delta - \frac{V^2}{X_s} = \frac{V^2}{X_s}\cos\delta - \frac{V^2}{X_s} = \frac{V^2}{X_s}(\cos\delta - 1)$$

When the active power is delivered to the bus, the load angle can't be zero and therefore cos δ is lesser than unity. This means negative reactive power, which indicates that the synchronous generator is absorbing reactive power.

Q. 15. What is hunting in synchronous machines and what are its effects? Does it occur in synchronous generators only or in synchronous motors also? [U.P.S.C. I.E.S. Electrical Engineering-II, 2010]

Ans. The oscillation of synchronous machine about its equilibrium position is called the *hunting*. The effects of hunting are as follows:

1. It causes variations of the supply voltage.
2. It increases the chances of loosing synchronism.
3. It causes great surges in current and power flow.
4. It increases the possibility of resonance.
5. It may develop huge mechanical stresses in the rotor shaft.
6. It results in increase in machine losses and thus the rise in temperature of the machine.

Hunting occurs in both synchronous generators and motors.

Q. 16. Do the generators driven by the steam turbines have a tendency to hunt ?

Ans. No, because the solid field poles act as efficient dampers.

Q. 17. How hunting is reduced in alternators ?

Ans. The hunting in alternators is reduced by employing heavy flywheels, putting dash-pots on the engine governors, and to use amortisseur or squirrel cage windings on the surface of the rotors.

PROBLEMS

1. Synchronising lamps are connected across 440 V alternators running at 50 Hz and 50.6 Hz respectively. How often will the lamps light up in 1 minute and what is the periodicity of the high frequency component ? [**Ans.** 36 times a minute]
2. Two single-phase generators operate in parallel and supply a load impedance of $(3 + j\,4)$ ohms. Determine the terminal voltage, power factor and kW output of each machine if the impedance of each machine is $(0.2 + j\,2)\ \Omega$ and emfs are $(200 + j\,0)$ and $(220 + j\,0)$ volts respectively. [**Ans.** 178 $\angle{-5.06°}$ V; 0.7193 (lag); 1.747 kW; 0.518 (lag); 2.081 kW]
3. Two single-phase generators operate in parallel on a load impedance of Z ohms. Their emfs are E_1 and E_2 and their synchronous impedances Z_1 and Z_2. Deduce the terminal voltage in terms of these emfs and admittances Y, Y_1 and Y_2.
   Determine the terminal voltage and kW output of each machine if $E_1 = 100$, $E_2 = 110$ volts, $Z = 3 + j\,4$ and $Z_1 = Z_2 = 0.2 + j\,1$ ohms. [**Ans.** 96.08 V, 443 W, 665 W]
4. A 3 MVA, 6-pole alternator runs at 1,000 rpm in parallel with other machines on 3.3 kV bus bars. The synchronous reactance is 25%. Calculate the synchronising power for one mechanical degree of mechanical displacement and the corresponding synchronising torque. [**Ans.** 628.4 kW; 6,000 N-m]
5. Calculate the value of synchronising power in kW for one mechanical degree of displacement for a 3-phase, 2,000 kVA, 6,600 V, 50 Hz, 12-pole machine having a synchronous reactance of 25% and negligible resistance. [**Ans.** 952.8 kW]
6. A 3 MVA, 6-pole alternator runs at 1,000 rpm on 3.3 kV bus bars. The synchronous reactance is 25%. Calculate the synchronising power and torque per mechanical degree of displacement when the alternator is supplying full-load at 0.8 pf lag. [**Ans.** 722.563 kW; 6,900 Nm]
7. A 2,000 kVA, 3-phase, 8-pole alternator runs at 750 rpm in parallel with other machines on 6,000 V bus bars. Find synchronising power on full load 0.8 pf lagging per mechanical degree of displacement and the corresponding synchronising torque. The synchronous reactance is 6 ohms per phase. [**Ans.** 502.6 kW; 6,400 Nm]
8. Two similar 400 V, 3-phase alternators share equal kW power delivered to a balanced 3-φ, 50 kW, 0.8 pf lag load. If the power factor of one machine is 0.95 lag, find the power factor and the current supplied by the other machine. [Manamanian Sundaranar Univ. Elec. Machines II, April-1996] [**Ans.** 0.6493 lag; 55.6 A]
9. Two 6,600 V, star-connected alternators in parallel supply the following loads :
   400 kW at unity power factor lagging,
   400 kW at 0.85 power factor lagging,
   300 kW at 0.8 power factor lagging,
   800 kW at 0.7 power factor lagging.
   The armature current of one machine is 100 A at 0.9 pf lagging. Determine the armature current, the output and the power factor of the other machine. [**Ans.** 103 A, 872 kW, 0.74 (lagging)]
10. An alternator of 1,500 kVA runs in parallel with one of 1,000 kVA capacity. What load should each supply and at what power factor should each operate in order that the currents and power outputs shall be proportional to their ratings if the combined load is 2,000 kW at 0.8 pf ? [**Ans.** 1,200 kW at 0.8 pf and 800 kW at 0.8 pf]

11. The speed regulation of two 1,000 kW alternators running in parallel are 100% to 104% and 100% to 105% from full load to the no load respectively. How will the two alternators share a load of 1,200 kW and also find the load at which one machine ceases to supply any portion of the load. [**Ans.** 555.55 kW and 644.45 kW; 200 kW]

12. Two 75 MW, 3-ϕ alternators are operating in parallel and are sharing equally a load of 150 MW corresponding to full load of each machine. The settings of the governors are such that the rise in speed from full-load to no-load is 2% in the first machine and 3% in the other machine. The speed-regulation characteristics may be assumed to be straight line in both the cases. Determine the load on each machine when the total load is 100 MW. [**Ans.** 45 MW, 55 MW]

13. Two alternators of ratings 600 kVA and 500 kVA operate in parallel sharing a total load of 800 kW. The frequency of first alternator drops from 51 Hz to 49 Hz when fully loaded at unity power factor and that of second alternator drops from 51.5 Hz to 48.5 Hz on full loading at unity power factor. Find out the load shared by each alternator and the frequency at which they will operate in parallel.

[U.P. Technical Univ. Electromechanical Energy Conversion II, 2002-03]

[**Ans.** 460.7 kW, 339.3 kW; 49.464 Hz]

14. Two identical, 3-phase alternators operating in parallel, share equally a load of 1,000 kW at 6,600 V and 0.8 lagging power factor. The field excitation of the first machine is adjusted so that the armature current is 50 A at lagging pf. Determine

(*i*) the armature current of the second alternator

(*ii*) the power factor at which each machine operates.

[U.P. Technical Univ. Electromechanical Energy Conversion-II, 2007-08]

[**Ans.** (*i*) 60.22 A (*ii*) 0.87477 (lagging); 0.7263 (lagging)]

15. Two similar 6.6 kV alternators supply a total load of 200 kW at 0.8 power factor lagging so that the power supplied by each machine is same. If the excitation of one of the alternator is decreased so that its load current reduces to 100 A, determine the load current of the second alternator and the power factor at which each alternator is working. [**Ans.** 119.8 A; 0.87256 (lag); 0.7286 (lag)]

16. A 3-phase, star-connected turbo-alternator having a synchronous reactance of 10 Ω per phase and negligible armature resistance has an armature current of 20 A at unity pf. The supply voltage is constant at 11 kV at constant frequency. If the steam admission is unchanged and the emf is raised by 25%, determine the current and power factor.

If the higher value of excitation is maintained and the steam supply is slowly increased, at what power output will the alternator break away from synchronism ?

Draw the phasor diagram under maximum power condition.

[**Ans.** 288.6 A; 0.762 (lagging); 16,200 kW]

17. A 3-ϕ, 11-kV, 5 MVA, star-connected alternator has a synchronous impedance of $(1 + j\,10)$ ohms per phase, and the excitation is such that the generated line emf is 14 kV. If the alternator is connected to infinite bus bars, determine the maximum power output at the given excitation. [**Ans.** 14,130 kW]

18. A 3-phase star-connected alternator is operated at a constant voltage of 6.6 kV and its excitation voltage is 6.4 kV. Find the maximum output power and pf at this power assuming $X_s = (1 + j\,10)\ \Omega$ per phase. **[J.N. Technological Univ. Electrical Machines-III, May-2013]**

[**Ans.** 3.793 MW; 0.658]

19. A 12.5 MVA 3-ϕ star-connected 11 kV, 4-pole, 50 Hz synchronous motor has a reactance of 0.15 pu and negligible armature resistance. Calculate the synchronizing power per mechanical degree, when it supplies full load at 11 kV and 0.8 pf leading.

[U.P. Technical Univ. Electromechanical Energy Conversion-II, 2003-04]

[**Ans.** 2.646 MW]

20. A 150 kVA, 3-phase star-connected, 6.6 kV, 8-pole, 50 Hz synchronous generator has a reactance of 0.6 pu and negligible resistance. Calculate the synchronising power per mechanical degree at full load and 0.8 pf lagging.

[U.P. Technical Univ. Electrical Machines, 2010-11]

[**Ans.** 23.734 kW]

21. A 3-phase star-connected alternator is delivering 20 MW and 8 MVAR to an infinite bus at 11 kV. Synchronous impedance is $(0 + j\,3)\ \Omega$. Find the load angle and excitation emf of the alternator.

[J.N. Technological Univ. Electrical Machines-III, May-2013]

[**Ans.** 22.48°, 14.266 kV]

22. A 5000 kVA, 3-ϕ, 10,000 V, 50 Hz alternator runs at 1,500 rpm connected to constant frequency, constant voltage bus bars. If the moment of inertia of the entire rotating system is 1.5×10^4 kg-m^2 and the steady short-circuit current is 5 times the normal full load current, find the natural time period of oscillations. [**Ans.** 1.364 seconds]

❑❑❑

SECTION V

THREE-PHASE INDUCTION MOTORS FRACTIONAL KILOWATT MOTORS AND SPECIAL ELECTRICAL MACHINES

CHAPTERS COVERED IN THIS SECTION

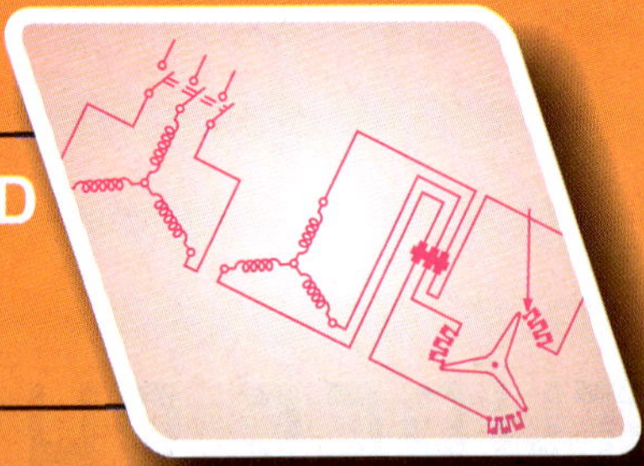

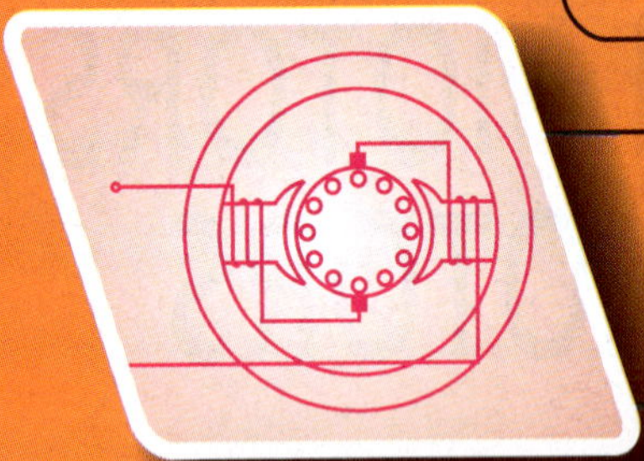

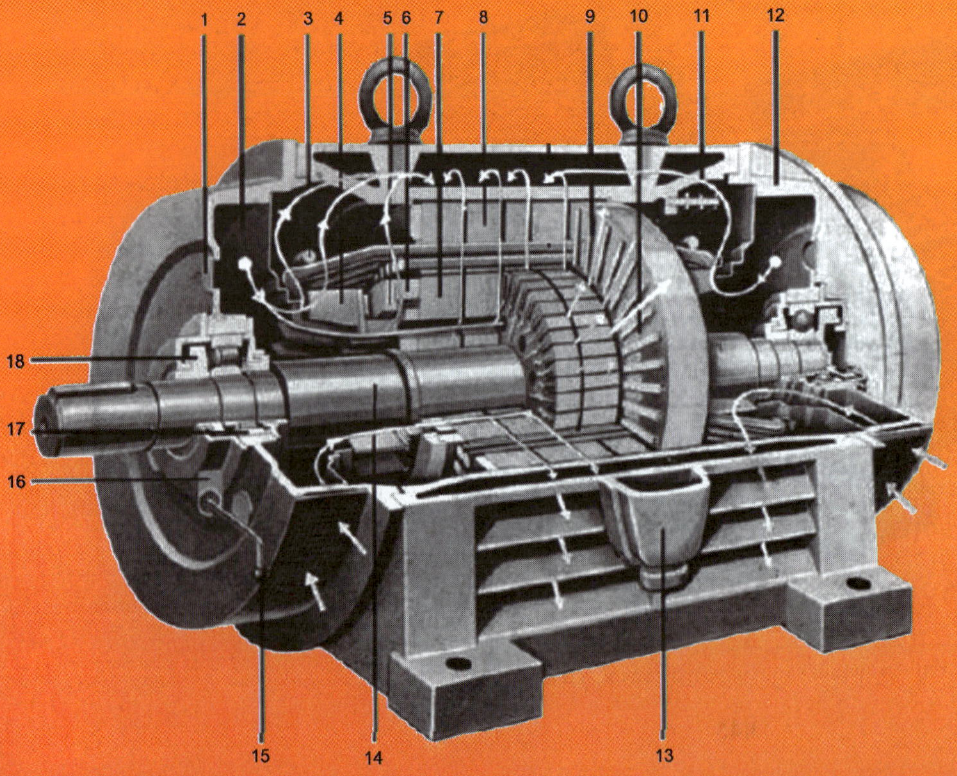

1. Access for checking air gap
2. Air-deflecting baffle
3. Coil bracing ring
4. Fan
5. Rotor end ring
6. Rotor base
7. Stator core
8. Fully-formed coils of the two layer stator winding
9. Core duct separator
10. Preformed coil in section
11. End winding connections
12. Bearing endshield
13. Terminal box with bolted on cable sealing end
14. Shaft
15. Grease ejector handle
16. Grease collector
17. Anti-friction bearing with grease regulator
18. Grease impeller

CHAPTER 18

Construction, Theory and Operation of Three-Phase Induction Motors

INSIDE THIS CHAPTER

18.1 INTRODUCTION

The polyphase induction motor is, by a very considerable margin, the most widely used ac motor (almost more than 90 per cent of the mechanical power used in industry is provided by 3-phase induction motors). The reasons are its low cost, simple and rugged construction, absence of commutator, good operating characteristics (reasonably good power factor, sufficiently high efficiency and good speed regulation). An induction motor of a medium size may have an efficiency as high as 90 per cent and a power factor of 0.89 It is substantially a constant speed motor with a shunt characteristic; a few per cent speed drop from no load to full load. The physical size of such a motor for a given output rating is relatively small as compared with other types of motors.

The distinguishing feature of such a motor is that it is a singly-excited machine, although such machines are equipped with both field and armature windings. In such a machine the field* (or stator) winding is connected to an ac supply and there is no electrical connection from the armature* (or rotor) to any source of supply. Currents are made to flow in the armature (or rotor) conductors by induction which interact with the field produced by the field (or stator) winding and thereby produce a net unidirectional torque. Such motors are also called *asynchronous motors* as they run at a speed other than the synchronous speed of the rotating field developed by the stator currents.

Like other electrical machines, the asynchronous machine is reversible, *i.e.*, it can operate as both a motor and a generator. The mode of operation of the machine is determined by the speed of the rotating field in relation to the rotor.

An asynchronous machine may be considered to be a transformer in the sense that the power is transferred from the stator (primary) to the rotor (secondary) winding only by mutual induction. For this reason such a machine is often called the *induction* machine.

There is no simple and cheaper method of controlling the speed of the induction motor as is possible in case of a dc shunt motor and the torque developed by a polyphase induction motor is also not so strong as that developed by a dc shunt motor. It still finds stiff competition from the dc shunt motor in industrial applications.

18.2 CONSTRUCTION

The 3-phase induction motor is very simple in construction compared to a dc motor or a synchronous motor. The essential features of a polyphase induction motor are: a laminated stator core carrying a polyphase winding; a laminated *rotor* core carrying either a cage or polyphase winding, the latter with shaft-

* The terminology of the field and armature windings seldom is used for polyphase induction motors.

mounted slip rings; a stiff *shaft* to preserve the very short air gap; a *frame* to form the stator housing and carry the end covers, bearings, and terminal box. It is not possible, as in the dc machines, to use the frame as part of the magnetic circuit. The end covers receive the ball- or roller-bearings with their clamping plates. Non-salient pole construction is used for all polyphase induction motors.

18.2.1. Frame. It is the outer body of the motor. Its functions are to support the stator core and winding, to protect the inner parts of the machine and serve as a ventilating housing or means of guiding the coolant into effective channels. The frame may be die cast or fabricated. Machines up to about 50 kW rating may have frame die cast in a strong silicon-aluminium alloy, sometimes with the stator core cast in. The frames for medium and large machines are almost exclusively fabricated. The process of die casting has the advantage that it facilitates the use of thicker cross section of frame where greater mechanical strength is required whereas the advantage of fabrication is its adaptability to new designs and modifications.

For large motors of rating 250 kW to 10,000 kW, the frames are generally fabricated in the form of box type enclosures having detachable top cover. Frames are provided with feet by which they are fixed to the base plate.

The frame should be strong and rigid as air-gap length in an induction motor is very small. If the frame is not rigid, the rotor will not remain concentric with stator giving rise to unbalanced magnetic pull.

18.2.2. Stator Core. The stator of an induction motor is quite similar in construction to that of a 3-phase synchronous alternator or motor. The stator core is to carry the alternating flux which produces hysteresis and eddy current losses. In order to reduce eddy currents and hysteresis loss in the stator core it is assembled of high grade, low-electrical loss, silicon steel punchings. The thickness of punchings varies from 0.35 mm to 0.65 mm. In small motors, or if low core loss is not so important, slightly thicker laminations may be used. In the smaller sizes, the stator plates form complete annular rings, as shown in Fig. 18.2, but for larger machines they are prepared of segmental laminations to avoid wastage of steel. The punchings are insulated from one another by coating of varnish or oxide produced by heat treatment. The stampings are assembled under hydraulic pressure and are keyed to the frame. Careful alignment is required to prevent the rotor from rubbing against the stator, the air gap between the stator and rotor being very small. The radial ventilating ducts are provided along the length of the stator core, spaced every 5 or 7 cm. These are provided by the use of spacers placed between the laminations. The air gap between the stator and rotor is made as small as practicable (0.3 to 0.35 mm in small machines and 1.0 to 1.5 mm in high power machines) so as to make air-gap reluctance minimum. The slots are punched out on the inner periphery of the stator laminations, as shown in Fig. 18.2, to accommodate the stator winding, also sometimes known as *primary winding*. Some motors, especially of large ratings, have open type slots so that form wound coils can be used but usually the motors are provided with semi-closed type slots to make the air-gap reluctance minimum and to prevent excessive tooth loss. In large, high-speed motors it is not essential to employ semi-closed slots to reduce the reluctance. The reason of it is that in large, high-speed motors, pole pitch is large, therefore, the number of conductors per pole is large, so the desired flux density in the air gap can be attained without an excessive magnetizing current, even if open type slots are employed.

Fig. 18.1 *Cutaway View Through Stator of Induction Motor*

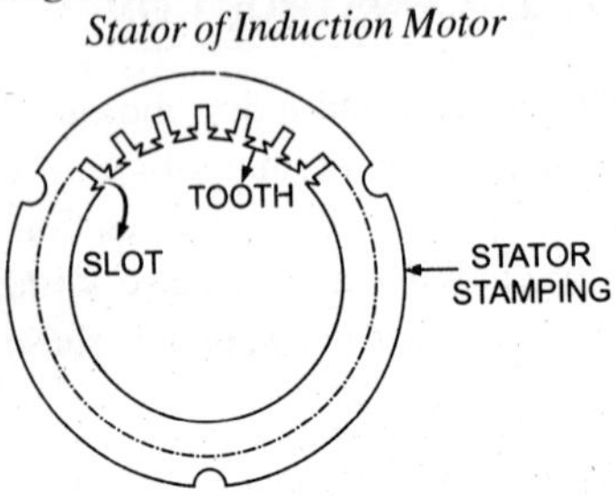

Fig. 18.2

18.2.3. Stator or Primary or Field Winding. Motors up to 250 kW operate usually at voltages in the range 380-440 V. For 3.3 kV the minimum economical rating is 250-300 kW, and for 11 kV about 750 kW.

In a polyphase induction motor the stator winding is usually 3-phase winding which is usually supplied from a 3-phase supply mains. The three phases of the winding can be connected in either star or delta depending upon the methods of starting used. The squirrel cage motors are usually started by star-delta starters and, therefore, their stators are designed for delta connections and six terminals (two from each phase) are brought out to be connected to the starter. The wound rotor motors are started by inserting resistance in the rotor circuit and, therefore, the stator winding can be connected in either star or delta as desired. It is wound for a definite number of poles as per requirement of speed. Greater the number of poles, lesser the speed or vice versa for a supply of given frequency.

In general, the same windings can be used on the stator of induction motor as were suitable for synchronous generators or motors. The double layer winding is mostly used in medium size motors because of its greater ease of manufacture, assembly, and repair. Moreover, stator windings are almost always short pitched because of reduced copper weight and winding resistance as well as reduced leakage reactance and harmonic torque disturbances that result. Small motors, operating at ordinary supply voltages, with a small number of slots having a large number of turns per phase may use single layer mush winding. For large motors, or high-voltage motors the stator windings may be formed by single layer concentric coils. Double layer coils are placed in open slots so that form wound coils can be used while mush winding is housed in semi-closed slots.

Stator of a 3-phase induction motor and 3-phase, 4-pole, star-connected stator winding are illustrated in Figs. 18.3 (*a*) and 18.3 (*b*) respectively.

18.2.4. Rotor. The rotor comprises a cylindrical laminated iron core, much the same as that of a dc machine, with slots, around the core, carrying the rotor conductors. In general the same sheet steel laminations are employed for the rotor core as for the stator, but owing to the lower frequencies of the rotor flux, thicker laminations can be employed without excessive iron loss. The rotor stamping is shown in Fig. 18.4. Like stator, the rotor

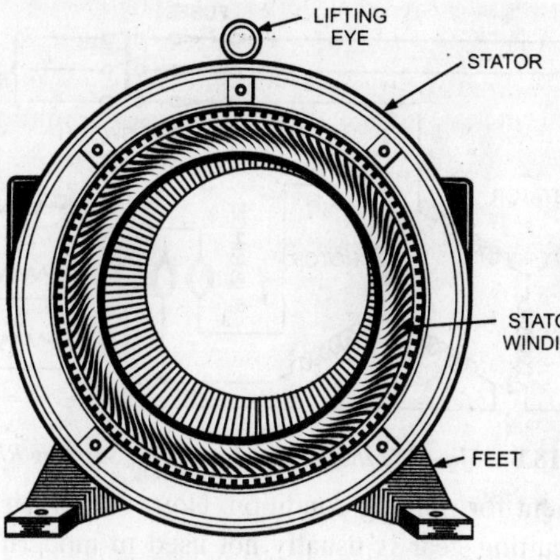

Fig. 18.3 (*a*) *Stator of a 3-Phase Induction Motor*

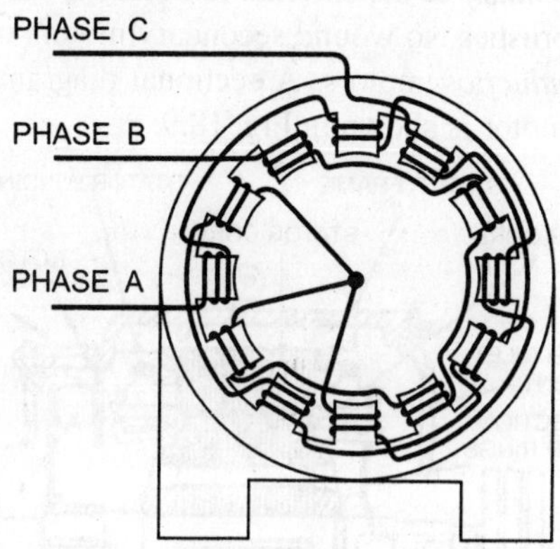

Fig. 18.3 (*b*) *3-Phase, 4-Pole, Star-Connected Stator Winding*

laminations are punched in one piece for small machines; in larger machines the laminations are segmented and dovetailed to a fabricated spider. If there are ventilating ducts on the stator core, an equal number of such ducts is provided on the rotor core also. Fan blades are usually employed on the ends of the rotor core to force circulating air through the motor for cooling.

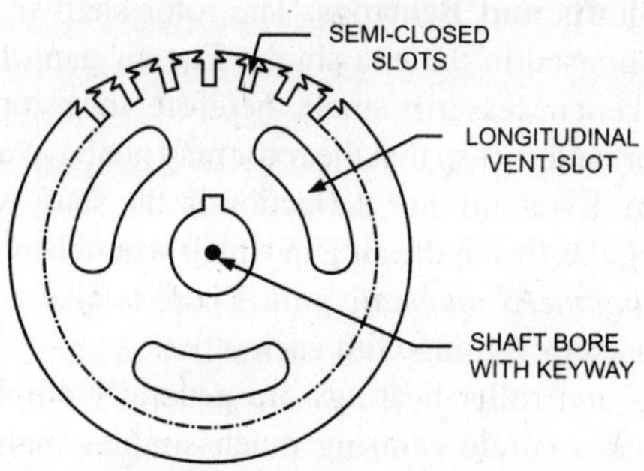

Fig. 18.4 *Rotor Stamping*

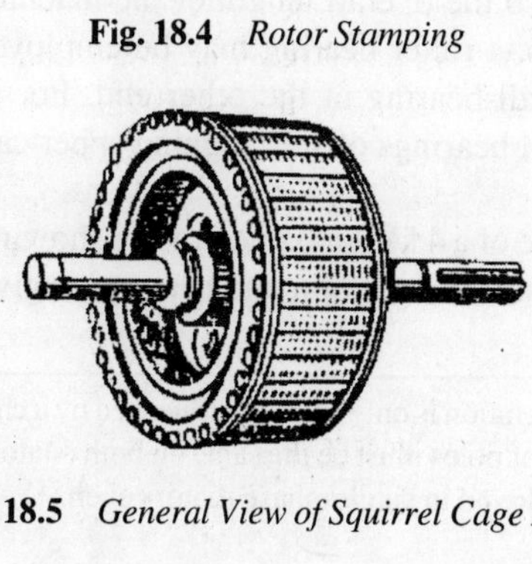

Fig. 18.5 *General View of Squirrel Cage Rotor*

The rotor has a smaller number of slots than the stator and must be a non-integral multiple of stator slots so as to prevent magnetic locking of rotor and stator teeth at the starting instant.

The rotors employed in 3-phase induction motors, according to the type of windings used, are of two types, viz; squirrel cage rotor and the wound rotor.

1. *Squirrel Cage Rotor. Almost* 90 per cent of induction motors are provided with squirrel cage rotor because of its very simple, robust and almost instructible construction. In cage construction, copper, brass, or aluminium bars are placed, as the rotor conductors, parallel or approximately parallel to the shaft (one bar in each slot) and close to the rotor surface. The conductors are not insulated from the core, since the rotor currents naturally follow the path of least resistance, *i.e.*, the rotor conductors. At both ends of the rotor, the rotor conductors are all short circuited by the continuous end rings of similar material to that of the rotor conductors. The rotor conductors and their end rings form a complete closed circuit in itself, resembling a squirrel cage, thus explaining the name. In motors with ratings up to 100 kW, the squirrel cage structure is formed by aluminium cast (under pressure) into the slots of the rotor (Fig. 18.6). In large motors the rotor bars, instead of being cast, are wedged into the rotor slots and are then welded securely to the end rings. The slots on the rotor are always not parallel to the motor shaft but are usually skewed in order to obtain a uniform torque, reduce the magnetic locking of the stator and rotor and reduce the magnetic humming noise while running.

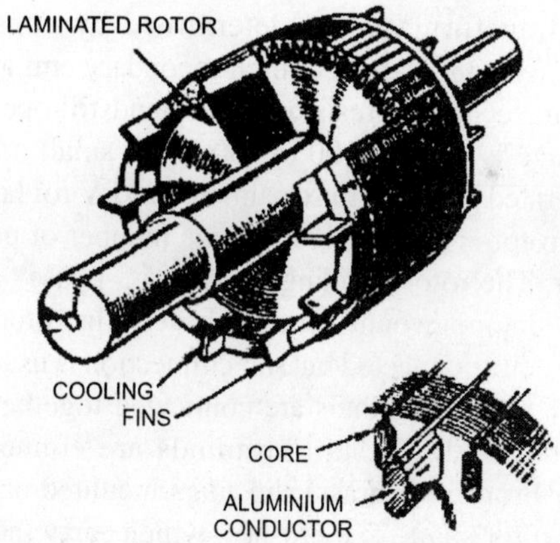

Fig. 18.6 *Cutaway View of Squirrel Cage Rotor*

The slots on rotor are either of semi-closed type or of totally closed type, because there is little difficulty in inserting the rotor bars in such slots. The advantage of semi-closed and totally closed slots is that the effective cross-sectional area of the air gap is increased, therefore, magnetizing current is reduced. Such slots also reduce the pulsations of flux in the individual teeth, therefore, tooth loss is reduced. The disadvantage of such slots is that these give higher slot inductance than the open slots, which lowers the power factor and reduces the starting and breakdown torques of the motor.

The squirrel cage rotor windings are perfectly symmetrical and have the advantage of being adaptable to any number of pole pairs. The distribution of current due to electromagnetic induction in the rotor bars varies from bar to bar sinusoidally and depends upon the position and time, assuming sinusoidal distribution of radial flux density in space and also the applied voltage to be varying sinusoidally with time.

Since the rotor winding is permanently short circuited in cage construction, so there is no possibility of adding any external resistance in the rotor circuit.

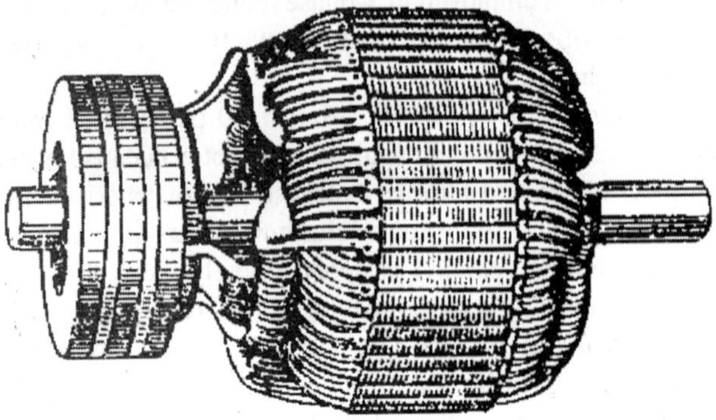

Fig. 18.7 *Rotor With Slip Rings*

2. *Wound Rotor.* As the name implies, such a rotor is wound with an insulated winding similar to that of the stator except that the number of slots is smaller and fewer turns per phase of a heavier conductor are used. Bar, strap, or wire is used for rotor windings, the last being used where many turns are desired. A large number of rotor turns increases the secondary voltage and reduces the current that flows through the slip rings. The secondary voltage determines the insulation that must be provided; furthermore, the voltage and current influence the value of the resistance to be employed across the slip rings. The motor operation is not influenced by the number of rotor turns, but the ratio of transformation is determined by consideration of secondary current, danger of high secondary emf at starting, and distance to secondary resistors. The standstill open-circuit slip-ring voltage is usually 100 to 400 V for small machines using hand-operated gear and maximum up to 1 kV for large machines.

The rotor is wound for the same number of poles as that of the stator. The rotor winding is always 3-phase* winding even when the stator is wound for two phases. The rotor winding may be star or delta connected but star connection is usually preferred. The three finish terminals are connected together to form star point and the three start terminals are connected to three phosphor-bronze (or brass) slip rings mounted on but insulated from the rotor shaft. The brushes, which carry the current from and to the rotor windings, are held in box type holders mounted on insulated steel rods and securely bolted to the end shield. Each brush is fed forward by a lever held in tension by an adjustable spring. These brushes are further externally connected to a 3-phase star-connected rheostat for the purpose of starting and speed control (Fig. 18.8). At the time of starting, the entire resistance is included in the rotor circuit and this resistance is gradually cut out as the rotor picks up the speed. For the normal running condition the entire external resistance is cut out and the rotor windings are short circuited automatically through the slip rings by means of a metal collar which is pushed along the shaft and connects all the rings together. Some machines are provided with brush lifting and slip-ring short-circuiting

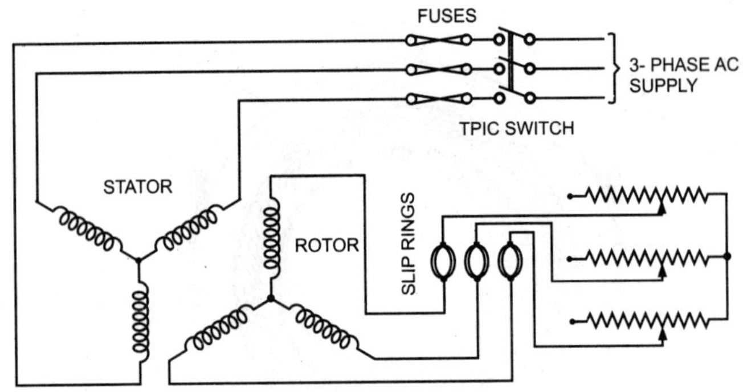

Fig. 18.8 *Slip-ring Induction Motor With Starting Rheostat*

arrangement for running condition. However, brush lifting and short-circuiting gear is usually not used in modern machines. The rotor is skewed in this case also. Since the connection of the wound secondary to the external terminals is made through slip rings and brushes, so wound secondary motors often are called *slip-ring induction* motors. A sectional diagram of a slip-ring induction motor is shown in Fig. 18.9.

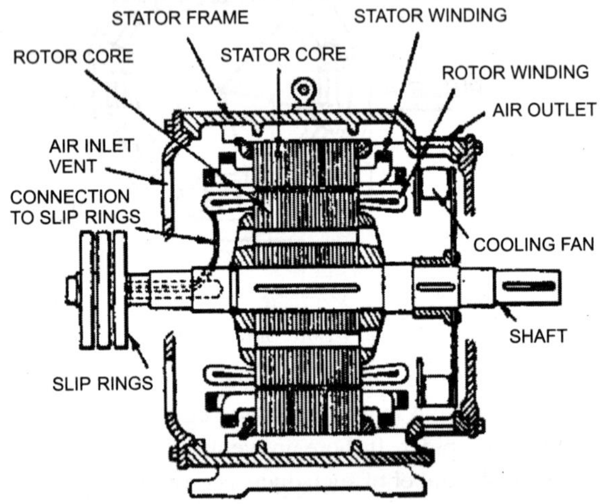

Fig. 18.9 *Section of a Slip-ring Induction Motor*

18.2.5. Shafts and Bearings. The rotor shaft is supported by bearings housed in the end shield. The air gap of an induction motor in kept necessarily small, therefore, the rotor shaft must be made short and stiff so that the rotor may not have any significant deflection. Even a minor deflection in the shaft would develop large irregularities in the air gap which would lead to production of an *unbalanced magnetic pull.* There is also a possibility of rotor and stator fouling with each other.

Ball- and roller-bearings are generally employed as their use makes accurate centring much simpler than with journal bearings. Also the overall length of the machine is reduced. In small motors, a roller bearing may be employed at the driving end and a ball-bearing at the other end. For large and heavy rotors journal bearings of self aligning spherical seated type are useful.

Structure of a 4 kW induction motor showing stator core and windings, stator frame, rotor and bearings is given in Fig. 18.10.

* The flux set up by the stator windings is a rotating flux with a definite number of poles; operation is only slightly influenced by a change in the number of phases provided on the rotor winding so long as it is greater than one, but the number of poles must be the same on both (stator and rotor). Three-phase rotors are standard for both three- and two-phase motors and are sometimes employed in single-phase motors even.

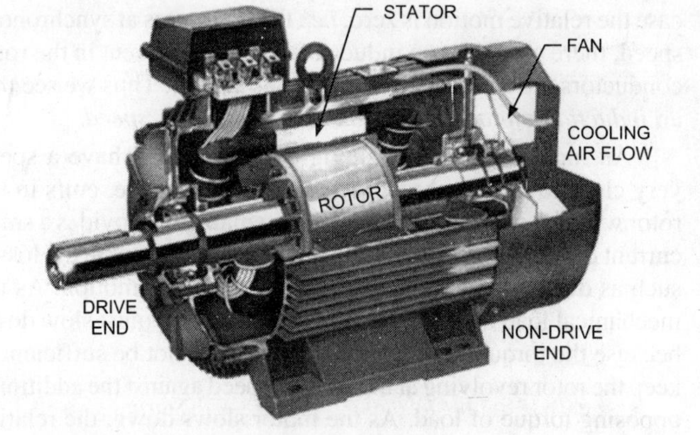

Fig. 18.10 *Structure of a 4 kW Induction Motor Showing Stator Core and Windings, Stator Frame, Rotor and Bearings*

18.3 TYPES OF 3-PHASE INDUCTION MOTORS

The 3-phase induction motors are of two types namely (*i*) squirrel cage and (*ii*) wound rotor or slip-ring induction motors. Both motors operate on the same principle and the same stator construction but differ in rotor construction. The rotor of the squirrel cage induction motor is of squirrel cage type while slip-ring induction motor employs wound rotor. The slip-ring induction motors are less extensively used than squirrel cage type because of their higher first cost and greater maintenance cost. The slip-ring induction motors are employed only when speed control or high starting torque is required.

Induction motor with squirrel cage rotor and that with wound rotor, in disassembled form are shown in Fig. 18.11 (*a*) and 18.11 (*b*) respectively.

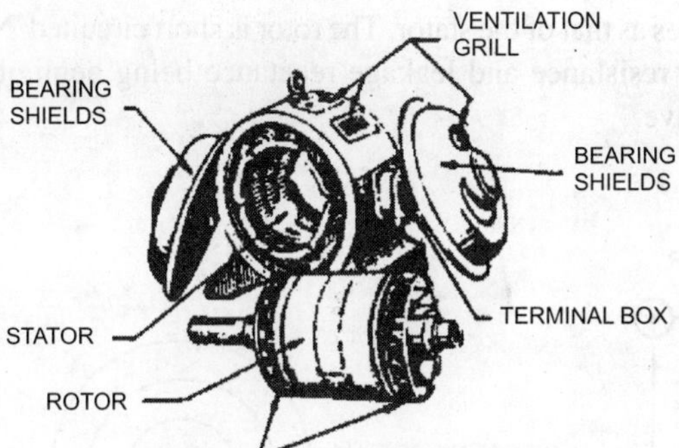

(*a*) *Induction Motor With Squirrel Cage Rotor (Disassembled)*

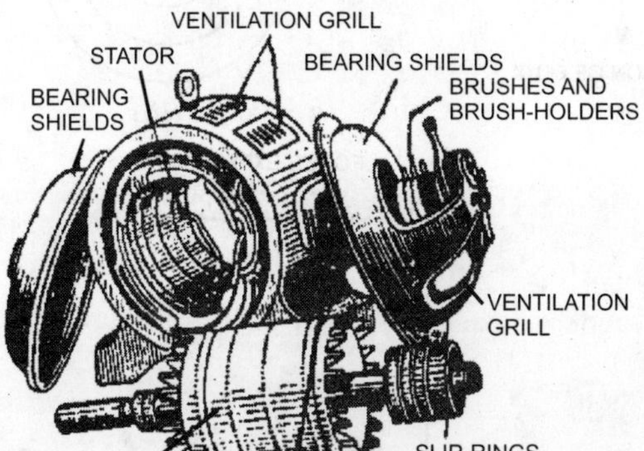

(*b*) *Induction Motor With Wound Rotor (Disassembled)*

Fig. 18.11

18.4 PRINCIPLE OF OPERATION

In a direct current motor, current is drawn from the supply and conducted into the armature conductors through the brushes and commutator. When the armature conductors carry current in the magnetic field established by the field, a force is exerted on the conductors which tends to move them at right angles to the field.

Though in an induction motor, there is no electrical connection to the rotor, but currents are induced in the rotor circuit, and therefore, the same condition exists as in the dc motors *i.e.*, the rotor conductors carry current in the stator magnetic field and thereby have a force exerted upon them tending to move them at right angles to the field.

When the stator or primary winding of a 3-phase induction motor is connected to a 3-phase ac supply, a rotating magnetic field is established which rotates at synchronous speed. The direction of revolution of this field will depend upon the phase sequence of the primary currents and, therefore, will depend upon the order of connection of the primary terminals to the supply. The direction of rotation of the field can be reversed by interchanging the connection to the supply of any two leads of a 3-phase induction motor. The number of magnetic poles of the revolving field will be the same as the number of poles for which each phase of the primary or stator winding is wound. The speed at which the field produced by the primary currents will revolve is called the *synchronous speed* of the motor and is given by an expression, $N_s = \frac{120f}{P}$ where f is supply frequency and P is the number of poles on stator.

As the rotating magnetic field produced by the primary currents sweeps across the rotor conductors, an emf is induced in these conductor just as an emf is induced in the secondary winding of a transformer by the flux set up by the primary current. Since the rotor winding is either directly shorted or closed through some external resistance, the emf induced in the secondary by the revolving field causes a current to flow in the rotor conductors.

The setting up of the torque for causing the rotor to rotate is explained below:

A section of an induction motor stator and rotor, with the magnetic field assumed to be rotating in a clockwise direction and with the rotor stationary, as at starting, is shown in Fig. 18.12 (*a*). The relative motion of the rotor with respect to the stator field is counter-clockwise. By applying right hand rule, the direction of induced emf or current in the rotor conductor shown is found outward. Hence the direction of the flux due to rotor current alone is anticlockwise, as shown in the figure. Now,

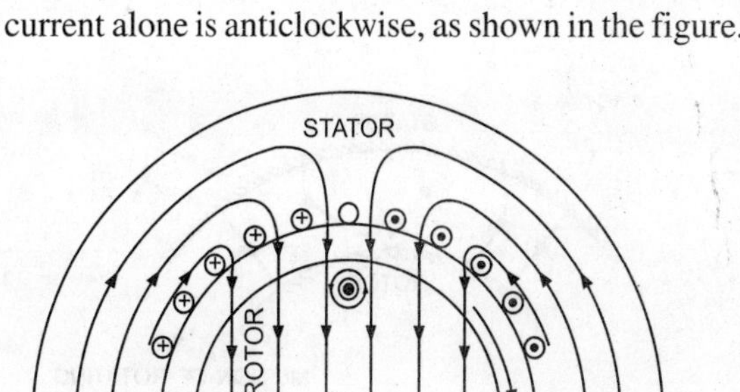

Fig. 18.12 (*a*)

by applying the left hand rule, or by the effect of combined field, it is clear that the rotor conductor shown experiences a force tending to move the conductor to the right. For simplicity, only one rotor conductor is shown. However, other adjacent rotor conductors in the stator field likewise carry current in the same direction as the conductor shown and also have a force exerted upon them tending to move them towards the right. One half cycle later, the stator field direction will have reversed, but the rotor currents will have also reversed, so that the force on the rotor is still the same. Likewise rotor conductors under other stator field poles will have a force exerted upon them all tending to turn the rotor in the clockwise direction. If the developed torque is great enough to overcome the resisting torque of the load, the rotor will accelerate in the clockwise direction or in the same direction as the rotation of the stator field.

The above facts have been illustrated diagrammatically in Fig. 18.12 (*b*).

This may also be explained in terms of Lenz's law. According to Lenz's law we know that the direction of induced emf would be in such a direction that it would try to oppose the very cause for which it is due. Because the cause producing the induced currents is the relative speed between the rotating magnetic field and stationary rotor conductors, therefore, they circulate in such a way that a torque is produced in the rotor tending to cause it to follow the rotating magnetic field and thus reducing the relative speed.

When the rotor is stationary and about to start, the frequency of induced emf in the rotor is equal to that of the supply fed to the stator, because the relative motion is at synchronous speed. As the motor picks up the speed, the relative motion between the rotor and the synchronously rotating magnetic field becomes less and the frequency of emf induced in the rotor decreases. The magnitude of rotor induced emf, induced rotor current and so that of torque developed depends upon the relative motion. In case the relative motion is zero, *i.e.*, the rotor runs at synchronous speed, there would be no induced emf and no current in the rotor conductors, no rotor field and hence no torque. Thus we see that an *induction motor cannot run at synchronous speed.*

An induction motor running at no load will have a speed very close to the synchronous speed and, therefore, emfs in the rotor winding will be very small. This small emf provides a small current developing a torque just sufficient to overcome the losses such as due to friction and maintain the rotor in motion. As the mechanical load is applied on the motor shaft, it must slow down because the torque developed at no load will not be sufficient to keep the rotor revolving at the no load speed against the additional opposing torque of load. As the motor slows down, the relative motion between the magnetic field and the rotor is increased. This results in greater rotor currents and greater developed torque. Thus, as the load is increased, the motor slows down until the relative motion between the rotor and the rotating magnetic field is just sufficient to result in the development of the torque necessary for that particular load. The decrease in speed from no load to full load is usually 4 to 5 per cent in case of small and medium size induction motors and in case of large size motors it varies from 2 to 2.25 per cent. In respect of speed-load characteristics, a 3-phase induction motor is quite similar to a dc shunt motor.

18.5 FLUX AND MMF WAVES IN INDUCTION MOTORS

The cross-sectional view of a three-phase induction motor is depicted in Fig. 18.13, the stator being supplied from a 3-phase supply. The rotor is wound three phase and for the same number of poles as that of the stator. The rotor is short circuited. Neglecting stator resistance and leakage reactance being negligibly small, we have

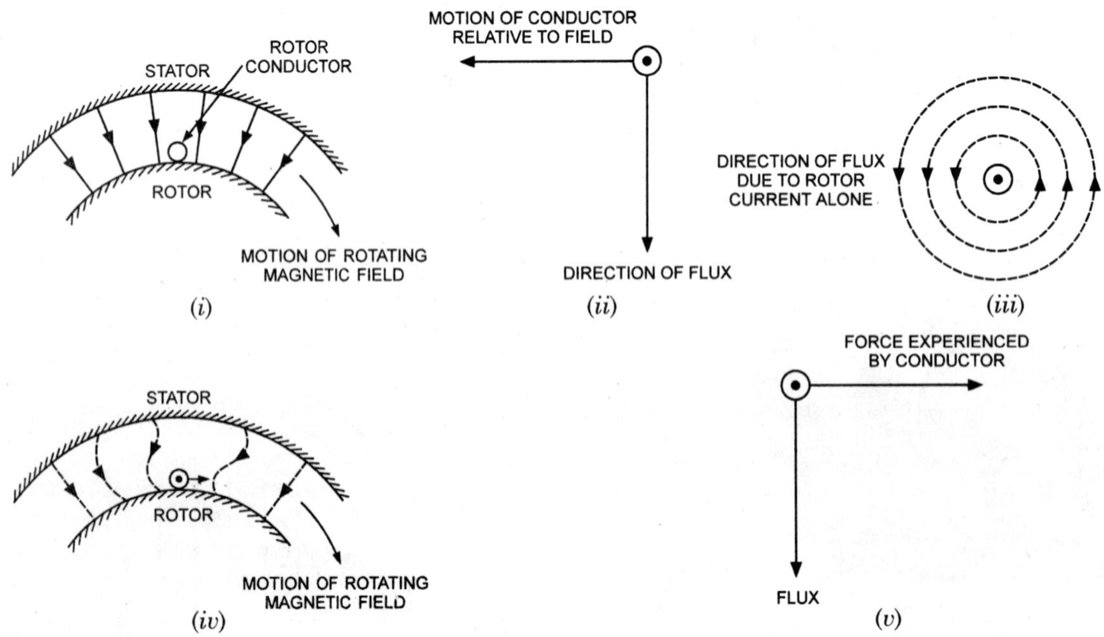

(*b*) *Production of Torque*

Fig. 18.12

Terminal voltage per phase,

$$V = \text{Stator induced emf per phase,}$$

$$E_1 = 4.44\, K_{w1}\, N_1\, \Phi_r f \qquad ...(18.1)$$

where K_{w1} is the stator winding factor, N_1 is the number of series turns per phase, Φ_r is the resultant air-gap flux per pole and f is supply frequency in Hz.

Equation (18.1) reveals that irrespective of load conditions existing on the rotor, resultant air-gap flux per pole Φ_r is constant, related to the supply voltage V in view of the assumption made. The mmf phasor $\mathbf{F}_r$ with associated flux density phasor $\mathbf{B}_r$, which is the cause for generation of Φ_r, rotates at synchronous speed because it is associated with balanced 3-phase current supplied to the stator. Due to the relative speed between $\mathbf{B}_r$ and the rotor, an emf is induced in the rotor windings which causes a current to flow in the rotor conductors. The torque is developed due to interaction between $\mathbf{B}_r$ and the rotor currents and the torque so developed, according to Lenz's law, tend to turn the rotor in the direction of rotation of $\mathbf{B}_r$, so as to reduce the relative speed. Thus, the motor is *self-starting* and the rotor attains a steady speed $N < N_s$ depending upon the load coupled to it. In case the rotor runs at synchronous speed N_s, there would be no induced emf and no current in the rotor conductors, no rotor field and hence no torque because in this case relative speed between $\mathbf{B}_r$ and rotor becomes zero.

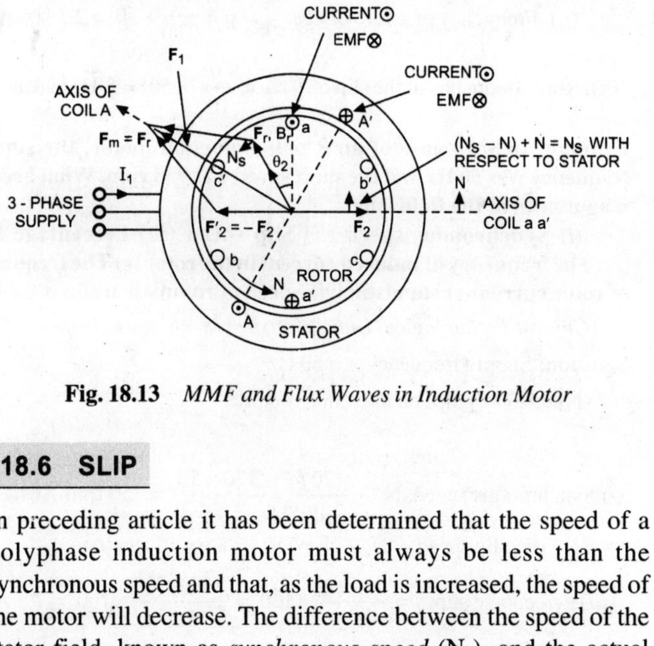

Fig. 18.13 *MMF and Flux Waves in Induction Motor*

18.6 SLIP

In preceding article it has been determined that the speed of a polyphase induction motor must always be less than the synchronous speed and that, as the load is increased, the speed of the motor will decrease. The difference between the speed of the stator field, known as *synchronous speed* (N_s), and the actual speed of the rotor (N) is known as the *slip* and is denoted by *s*. Though the slip can be expressed in rpm or in radians per second, but usually it is expressed as a fraction or percentage of synchronous speed.

$$\text{Fraction slip, } s = \frac{\text{Synchronous speed} - \text{rotor speed}}{\text{Synchronous speed}}$$

$$= \frac{N_s - N}{N_s} \qquad ...(18.2)$$

$$\text{and percentage slip} = \frac{N_s - N}{N_s} \times 100 \qquad ...(18.3)$$

At normal load the slip of an induction motor is usually between 2 and 5 per cent. At no load the slip is as small as 0.5 per cent and for many purposes the machine may then be considered to be running at synchronous speed. As the load is applied, the natural effect of the load torque is to cause the machine to slow down. As it does so, slip increases, and with it current and torque until the driving torque of the machine balances the retarding torque of the load. This determines the speed at which the machine runs on load.

The induction motor is thus a motor with substantially constant speed and plays the same role as a dc shunt motor.

Example 18.1. An induction motor has a rated speed of 715 rpm. How many poles have its rotating magnetic field?

[*G.B. Technical Univ. Electrical Engineering Even Semester, 2012-13*]

Solution: The number of poles, even in number and to provide a synchronous speed nearest higher than actual speed of motor (*i.e.* 715 rpm), that the motor must have is 8 **Ans.**

$$\because P_m = \frac{120 f}{N_m} \simeq \frac{120 \times 50}{715} \simeq 8 \text{ assuming } f = 50 \text{ Hz}$$

Example 18.2. A 60 Hz induction motor has 2 poles and runs at 3,510 rpm. Calculate (*a*) the synchronous speed and (*b*) the per cent slip.

[*U.P. Technical Univ. Electrical Engineering Even Semester, 2008-09*]

Solution: (*a*) Synchronous speed,

$$N_S = \frac{120 \times f}{P} = \frac{120 \times 60}{2} = 3{,}600 \text{ rpm} \quad \textbf{Ans.}$$

$$(b) \text{ Per cent slip} = \frac{N_S - N}{N_S} \times 100 = \frac{3{,}600 - 3{,}510}{3{,}600} \times 100$$

$$= 2.5\% \textbf{ Ans.}$$

Example 18.3. Calculate the speed of a 4-pole, 50 Hz, 400 V, 3-phase induction motor when it is operating at a slip of 1.5 per cent.

[*G.G.S.I.P. Univ. Delhi, May-2010*]

Solution: Synchronous speed,

$$N_s = \frac{120 f}{P} = \frac{120 \times 50}{4} = 1{,}500 \text{ rpm}$$

$$\text{Slip, } s = 1.5\% = 0.015$$

$$\text{Operating speed, } N = N_s(1 - s)$$

$$= 1{,}500\,(1 - 0.015) = 1{,}477.5 \text{ rpm } \textbf{Ans.}$$

Example 18.4. A-12 pole alternator is coupled to an engine running at 500 rpm. It supplies a 3-phase induction motor having full-load speed at 1,440 rpm. Find percentage slip and number of poles of the motor. [*Mahamaya Technical Univ. Electrical Engineering Even Semester, 2011-12*]

Solution: Frequency of generated emf, $f = \frac{P_a \times N_a}{120} = \frac{12 \times 500}{120} = 50$ Hz

The number of poles, even in number and to give a synchronous speed nearest higher than actual speed of motor (i.e. 1,440 rpm), that the motor must have is 4. **Ans.**

$$\because P_m = \frac{120 f}{N_m} = \frac{120 \times 50}{1{,}450} \simeq 4$$

Synchronous speed of motor,

$$N_s = \frac{120 f}{P} = \frac{120 \times 50}{4} = 1{,}500 \text{ rpm}$$

$$\text{Slip, } s = \frac{N_s - N}{N_s} = \frac{1{,}500 - 1{,}440}{1{,}500} = 0.04 \text{ or } 4\ \% \textbf{ Ans.}$$

Example 18.5. A motor-generator set used for providing variable frequency ac supply consists of a 3-phase, 10-pole synchronous motor and a 24-pole, 3-phase synchronous generator. The motor-generator set is fed from a 25 Hz, 3-phase ac supply. A 6-pole, 3-phase induction motor is electrically connected to the terminals of the synchronous generator and runs at a slip of 5%. Determine (i) the frequency of the generated voltage of the synchronous generator, (ii) the speed at which the induction motor is running.

[*U.P. Technical Univ., February-2001*]

Solution: Speed of motor-generator set,

$$N_s = \frac{120 \times \text{supply frequency}, f_1}{\text{Number of poles on synchronous motor}}$$

$$= \frac{120 \times 25}{10} = 300 \text{ rpm}$$

(*i*) Frequency of generated voltage,

$$f_2 = \frac{\text{Speed of motor-generator set} \times \text{number of poles on synchronous generator}}{120}$$

$$= \frac{300 \times 24}{120} = 60 \text{ Hz} \quad \textbf{Ans.}$$

(*ii*) Speed of induction motor,

$$N_m = N_s(1-s) = \frac{120 f_2}{P_m}(1-s) = \frac{120 \times 60}{6}(1-0.05)$$

$$= 1{,}140 \text{ rpm} \quad \textbf{Ans.}$$

18.7 FREQUENCY OF ROTOR CURRENT (OR EMF)

It has been shown in Art. 18.4 that the rotor runs in the direction of rotating magnetic field. When the rotor is stationary, rotor conductors are being cut by the rotating flux at synchronous speed N_s $\left(N_s \text{ being equal to } \frac{120f}{P}\right)$, therefore, the frequency of rotor current (or emf) is the same as that of supply frequency. But when the rotor starts revolving, the rate at which the rotor conductors are being cut by the rotating flux depends upon the relative speed between the rotor and the stator revolving magnetic field, called the *slip speed.* The frequency of rotor current (or emf) induced by the relative motion between rotor conductors and the stator revolving magnetic field is given by the equation:

$$\text{Rotor emf frequency}, f' = \frac{\text{Relative speed in rpm}}{120/P}$$

$$= \frac{N_s - N}{120/P} \quad ...(18.4)$$

$$\text{Since slip}, s = \frac{N_s - N}{N_s}$$

$$\text{or } N_s - N = s N_s = s \times \frac{120f}{P}$$

Substituting $N_s - N = s N_s = s \times \frac{120f}{P}$ in Eq. (18.4) we have

$$\text{Rotor emf frequency}, f' = s \times \frac{120f}{P} \times \frac{P}{120} = sf \quad ...(18.5)$$

Thus the frequency of rotor current (or emf) in an induction motor is given by the product of slip *s* and the supply frequency *f*. That is why, it is also called the *slip frequency.*

Example 18.6. A 50 Hz, 4-pole, 3-phase induction motor has rotor current frequency of 2 Hz. Obtain slip and speed of motor.

[*Pb. Technical Univ., May-2005; May-2010*]

Solution: $\text{Slip}, s = \frac{\text{Rotor current frequency}}{\text{Supply frequency}}$

$$= \frac{f'}{f} = \frac{2}{50} = 0.04 \text{ or } 4\% \quad \textbf{Ans.}$$

$$\text{Speed of motor, } N = N_s(1-s) = \frac{120f}{P}(1-s) = \frac{120 \times 50}{4}(1-0.04)$$

$$= 1{,}440 \text{ rpm} \quad \textbf{Ans.}$$

Example 18.7. A 6-pole, 50 Hz, 3-phase induction motor runs at 950 rpm at full load. Calculate the following parameters.

(*i*) **What is the percentage slip at full load?**
(*ii*) **What is frequency of rotor voltage?**
(*iii*) **Rotor frequency at the slip of 10%.**

[*Rajasthan Technical Univ. Electrical Machines-II, 2012*]

Solution: Supply frequency, $f = 50$ Hz

$$\text{Synchronous speed}, N_S = \frac{120f}{P} = \frac{120 \times 50}{6} = 1{,}000 \text{ rpm}$$

Full-load sped, $N_f = 950$ rpm

$$(i) \text{ Percentage slip}, s = \frac{N_s - N}{N_s} = \frac{1{,}000 - 950}{1{,}000} \times 100 = 5\%$$

$$(ii) \text{ Frequency of rotor voltage}, f' = sf = \frac{5}{100} \times 50 = 2.5 \text{ Hz} \quad \textbf{Ans.}$$

$$(iii) \text{ Rotor frequency at the slip of } 10\% = \frac{10}{100} \times 50 = 5 \text{ Hz} \quad \textbf{Ans.}$$

Example 18.8. In case of an 8 pole induction motor, the supply frequency was 50 Hz and the shaft speed was 735 rpm. What are the magnitudes of the following:

(*i*) Synchronous speed (*ii*) Slip speed (*iii*) Percentage slip (*iv*) The frequency of induced current in the rotor (*v*) The frequency of rotor current at standstill (*vi*) Speed of rotor when slip is 0.04.

[*Gujarat Technological Univ. Electrical Machines-I, Summer-2014*]

Solution: Supply frequency, $f = 50$ Hz

Number of poles on induction motor,

$$P = 8$$

$$(i) \text{ Synchronous speed}, N_s = \frac{120f}{P} = \frac{120 \times 50}{8} = 750 \text{ rpm} \quad \textbf{Ans.}$$

$$(ii) \text{ Slip speed} = N_s - N = 750 - 735 = 15 \text{ rpm} \quad \textbf{Ans.}$$

$$(iii) \text{ Percentage slip}, s = \frac{N_s - N}{N_s} \times 100 = \frac{750 - 735}{750} \times 100 = 2\% \quad \textbf{Ans.}$$

(*iv*) Frequency of induced current in the rotor,

$$f' = sf = 0.02 \times 50 = 1 \text{ Hz} \quad \textbf{Ans.}$$

(*v*) Frequency of rotor current at standstill.

$$f_1' = s_1 f = 1.0 \times 50 = 50 \text{ Hz} \quad \textbf{Ans.}$$

$\because$ Slip at standstill, $s = 1.0$

(*vi*) Speed of rotor when slip is 0.04,

$$N = N_s(1-s) = 750(1-0.04) = 720 \text{ rpm} \quad \textbf{Ans.}$$

18.8 SPEED OF ROTOR FIELD (OR MMF)

The rotating field set up by stator currents rotates at synchronous speed N_s revolutions per minute, relative to the stator surface.

Similarly the currents having a frequency of f' (f' being equal to sf) when flowing through the rotor winding set up the rotor magnetic field which rotates at a speed of $s\,N_s$ rpm, relative to the rotor surface, in the direction of rotation of the rotor. However, the rotor itself is running at a speed of N rpm with respect to stator surface.

So the speed of rotor magnetic field with respect to stator surface is equal to the sum of actual speed of rotor N and rotor field speed sN_s, with respect to rotor surface. Thus the speed of rotor field (or mmf) with respect to stator surface is given as

$$N + s\,N_s = N_s\,(1 - s) + s\,N_s = N_s \text{ rpm}$$

Thus the rotor magnetic field also rotates, in space, at the same speed and in the same direction as that of the stator field. This concludes that both the rotor and stator fields rotate synchronously, which means that they are stationary w.r.t. each other at all possible rotor speeds. These two synchronously rotating magnetic fields, in fact, superimpose on each other and give rise to the actually existing rotating field, which corresponds to the magnetizing current of the stator winding.

Since the two fields set up by stator (primary) and rotor (secondary) currents are stationary relative to each other, so the polyphase induction motor can be considered as being equivalent to a transformer having an air gap separating the iron portions of the magnetic circuit carrying the primary and secondary windings. Owing to this air gap, the magnetizing current and the magnetic leakage for an induction motor are large compared with the corresponding values for a transformer of the same kVA rating. Also, the friction and windage losses contribute towards making the efficiency of the induction motor less than that of the corresponding transformer. On the other hand, the stator mmf has to balance the rotor mmf and also provide the magnetizing and no-load loss components of the stator current, as in a transformer. Hence, an increase in slip due to increase in load is accompanied by an increase in rotor currents and, therefore, by a corresponding increase in the stator currents. Because of distributed windings in induction motors the ratio of stator and rotor currents is not equal to the ratio of turns per phase in the rotor and stator winding.

Example 18.9. A 3-phase, 400 V, 100 kW, 50 Hz, 6-pole induction motor operates at 3% slip at full load. Determine the following:

(a) The speed of the motor and its direction relative to the rotating magnetic field.
(b) The rotor frequency.
(c) The speed of the stator field.
(d) The speed of the air-gap field.
(e) The speed of the rotor field relative to stator rotating field.

[U.P.S.C. I.E.S. Electrical Engineering-II, 2009]

Solution: Synchronous speed of motor,

$$N_s = \frac{120f}{P} = \frac{120 \times 50}{6} = 1{,}000 \text{ rpm}$$

(*a*) Motor field, $N = (1 - s)\,N_s$
$= (1 - 0.03) \times 1{,}000$
$= 970$ rpm in the direction of the rotating field **Ans.**

(*b*) The rotor frequency, $f' = sf = 0.03 \times 50 = 1.5$ Hz **Ans.**

(*c*) Speed of stator field $= N_s = 1{,}000$ rpm **Ans.**

(*d*) The speed of air-gap field $= N_s = 1{,}000$ rpm **Ans.**

(*e*) The speed of rotor field relative to the stator rotating field
= zero **Ans.**

Example 18.10. A 3-phase, 50 Hz induction motor has full-load speed of 1,440 rpm. Calculate the following:

(i) Number of poles.
(ii) Full-load slip and rotor frequency.
(iii) Speed of stator field with respect to stator structure and rotor structure.
(iv) Speed of rotor field with respect to rotor structure, stator structure and stator field.

[J.N. Technological Univ. Electrical Machines-II, January-2014]

Solution: (*i*) The number of poles, even in number and provide a synchronous speed nearest higher than rotor speed of 1,440 rpm is 4 **Ans.**

$$\because P \simeq \frac{120f}{N_f} = \frac{120 \times 50}{1{,}440} \simeq 4$$

(*i*) Synchronous speed, $N_s = \dfrac{120f}{P} = \dfrac{120 \times 50}{4} = 1{,}500$ rpm

Full-load slip, $s_f = \dfrac{N_s - N_f}{N_s} = \dfrac{1{,}500 - 1{,}440}{1{,}500}$
$= 0.04$ or 4% **Ans.**

Full-load rotor frequency, $f_f' = s_f f = 0.04 \times 50 = 2$ Hz **Ans.**

(*iii*) Speed of stator field with respect to stator structure
$= N_s = 1{,}500$ rpm

Speed of stator field with respect to rotor structure
$= N_s - N = 1{,}500 - 1{,}440 = 60$ rpm **Ans.**

Speed of rotor field with respect to rotor structure
$= N_s - N = 1{,}500 - 1{,}440 = 60$ rpm **Ans.**

Speed of rotor field with respect to stator structure
$= N_s = 1{,}500$ rpm **Ans.**

Speed of rotor field with respect to stator field is zero because both of the stator field and rotor field revolve at the same speed of 1,500 rpm **Ans.**

18.9 ROTOR EMF

When the rotor is at standstill, the motor is equivalent to a 3-phase transformer with secondary short circuited. So induced emf per phase E_2 in the rotor, when it is at standstill *i.e.* at the instant of starting is given by

$$E_2 = E_1 \times \frac{N_2}{N_1} \quad \text{...(18.6)}$$

where E_1 is applied voltage per phase to primary *i.e* stator winding, N_2 and N_1 are the numbers of turns per phase on rotor and stator respectively.

When the rotor starts running, the relative speed of the rotor with respect to stator flux drops in direct proportion with the relative speed or slip *s* and induced emf in the rotor is given by sE_2.

Hence for slip *s*, the induced emf in the rotor is *s* times the induced emf in the rotor at standstill.

18.10 ROTOR CURRENT AND POWER FACTOR

The circuit diagram of an induction motor rotor is given in Fig. 18.14.

Let the resistance and inductance per phase of rotor be R_2 and L_2 respectively and induced emf per phase in rotor at standstill be E_2.

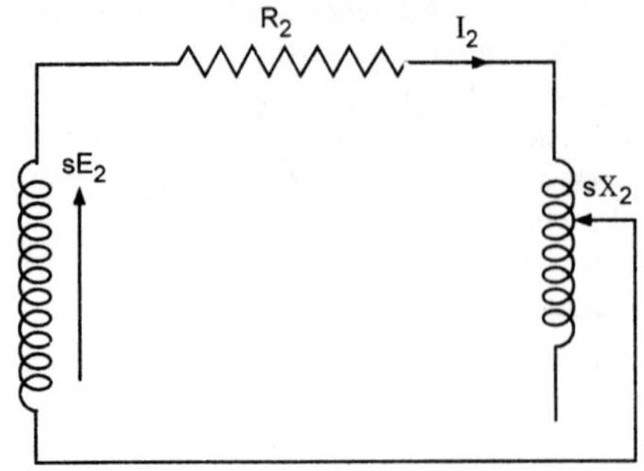

Fig. 18.14

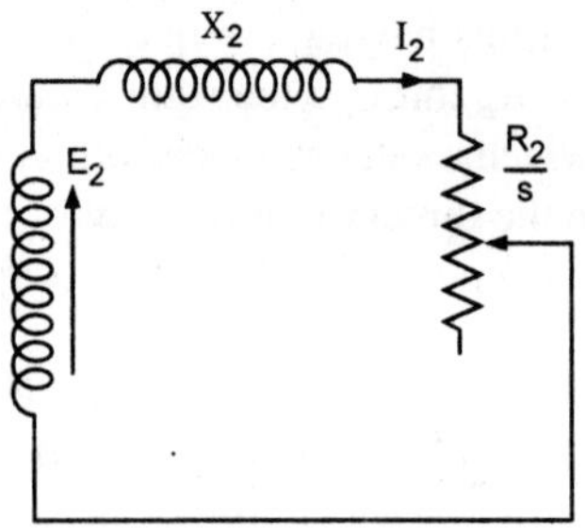

Fig. 18.15 *Equivalent Circuit of Rotor*

At Standstill

Induced emf per phase in rotor = E_2

Rotor winding resistance / phase = R_2

Rotor winding reactance / phase, $X_2 = 2\pi f L_2$

where f is supply frequency

$$\text{Rotor winding impedance / phase } Z_2 = \sqrt{R_2^2 + X_2^2} \quad ...(18.7)$$

$$\text{Rotor current, } I_2 = \frac{E_2}{Z_2} = \frac{E_2}{\sqrt{R_2^2 + X_2^2}} \quad ...(18.8)$$

$$\text{Power factor of rotor current, } \cos\phi_2 = \frac{R_2}{\sqrt{R_2^2 + X_2^2}} \quad ...(18.9)$$

At Slip *s*

Induced emf per phase in rotor winding = sE_2

Rotor winding resistance / phase = R_2

Rotor winding reactance / phase

$$= 2\pi f' L_2 = 2\pi s f L_2 = s(2\pi f L_2) = sX_2$$

Rotor winding impedance / phase

$$= \sqrt{R_2^2 + (sX_2)^2} \quad ...(18.10)$$

Rotor current per phase,

$$I_2 = \frac{sE_2}{\sqrt{R_2^2 + (sX_2)^2}} = \frac{sE_2}{\sqrt{R_2^2 + s^2X_2^2}} \quad ...(18.11)$$

Power factor of rotor current,

$$\cos\phi_2 = \frac{R_2}{\sqrt{R_2^2 + s^2X_2^2}} \quad ...(18.12)$$

The Eq. (18.11) for rotor current I_2 may be rewritten as

$$I_2 = \frac{E_2}{\sqrt{\left(\frac{R_2}{s}\right)^2 + X_2^2}} \quad ...(18.13)$$

The above equation gives convenient form of rotor equivalent circuit, as shown in Fig. 18.15.

The resistance $\frac{R_2}{s}$ is a function of slip and can be splitted into two parts R_2 and $R_2 \frac{(1-s)}{s}$ where R_2 is rotor resistance and $R_2 \frac{(1-s)}{s}$ is the fictitious resistance representing electrical load on the rotor. The final simplified equivalent circuit of an induction motor rotor is given in Fig. 18.16.

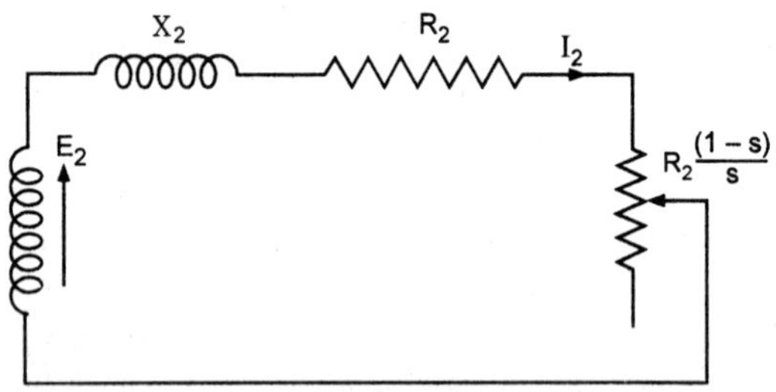

Fig. 18.16

The power consumed by the fictitious resistance representing the rotor electrical load is given as

$I_2^2 R_2 \frac{(1-s)}{s}$ and is converted into mechanical power.

Example 18.11. A three phase 440 volt, 50 hp, 50 Hz induction motor delivers rated output power at 1,440 rpm. Find

(*a*) Number of poles of machine.

(*b*) Synchronous speed.

(*c*) Slip.

(*d*) Slip rpm.

(*e*) Rotor speed w.r.t. (*i*) Rotor structure (*ii*) Stator (*iii*) Stator rotating mmf.

(*f*) Rotor emf at operating speed if stator to rotor turn ratio is 1:0.5. Assume winding factor is unity.

[*R.G. Technical Univ., December-2012*]

Solution: Supply frequency, f = 50 Hz

Full-load speed, N_f = 1,440 rpm

(*a*) The number of poles, even in number to give a synchronous speed nearest higher than the actual speed of motor (*i.e.*, 1,440 rpm) that the motor must have is 4. **Ans.**

$$\because \; P = \frac{120 \times f}{N_f} = \frac{120 \times 50}{1{,}440} = 4$$

(*b*) Synchronous speed, $N_s = \frac{120 \times f}{P} = \frac{120 \times 50}{4} = 1{,}500$ rpm Ans.

(*c*) Slip, $s = \frac{N_s - N_f}{N_s} \times 100 = \frac{1{,}500 - 1{,}440}{1{,}500} = 0.04$ or 4% Ans.

(*d*) Slip rpm = $s \times N_s = 0.04 \times 1{,}500 = 60$ rpm **Ans.**

(*e*) Rotor speed w.r.t. rotor structure = 0

Rotor speed w.r.t. stator = N_f = 1,440 rpm **Ans.**

Rotor speed w.r.t. stator rotating mmf = Slip rpm = 60 rpm **Ans.**

(*f*) Rotor emf at operating speed

= Slip × turn ratio × V_1 × winding factor

= 0.04 × 0.5 × 440 = 8.8 V **Ans.**

Example 18.12. A three-phase delta-connected 440 volts, 50 Hz, 4-pole induction motor has a rotor standstill emf per phase of 130 volts. If the motor is running at 1440 rpm, calculate slip, frequency of rotor induced emf, the value of the rotor induced emf per phase and stator to rotor turn ratio for same speed.

[Mahamaya Technical Univ. Electrical Engineering Odd Semester, 2011-12]

Solution: Synchronous speed of induction motor,

$$N_s = \frac{120 \times \text{supply frequency}}{\text{Number of poles}} = \frac{120 \times 50}{4} = 1{,}500 \text{ rpm}$$

Motor speed, N = 1,440 rpm

(*i*) Slip, $s = \frac{N_s - N}{N_s} = \frac{1{,}500 - 1{,}440}{1{,}500} = 0.04$ or 4% **Ans.**

(*ii*) Frequency of rotor induced emf

= Slip × supply frequency = sf = 0.04 × 50 = 2 Hz **Ans.**

(*iii*) Rotor induced emf per phase at the speed of 1,440 rpm or at slip of 0.04

= Slip × induced emf per phase in rotor winding at standstill

$= s \times E_2 = 0.04 \times 130 = 5.2$ V **Ans.**

(*iv*) Stator-to-rotor turn-ratio

$$= \frac{\text{Induced emf per phase in stator}}{\text{Induced emf per phase in rotor at standstill}}$$

$$= \frac{E_{P_1}}{E_{P_2}} = \frac{440}{130} = \frac{44}{13} \textbf{ Ans.}$$

Example 18.13. A 1200 V, 50 Hz star-connected induction motor has a star-connected slip-ring rotor with a transformation ratio of 3.75. The rotor resistance per phase is 0.016 ohm and leakage inductance of 0.8 mH/phase. Neglect stator impedance. Find:

(*i*) rotor starting current per phase with slip rings short circuited,

(*ii*) the rotor pf at starting,

(*iii*) the external resistance per phase required to obtain a starting current of 125 A in the stator,

(*iv*) the rotor current at 4% slip and

(*v*) the rotor pf at 4% slip.

[U.P.S.C. I.E.S. Electrical Engineering-II, 2003]

Solution: Voltage applied to stator per phase,

$$E_1 = \frac{1{,}200}{\sqrt{3}} = 692.82 \text{ V}$$

Emf induced per phase in rotor at standstill,

$$E_2 = KE_1 = \frac{692.82}{3.75} = 184.752 \text{ V}$$

Rotor resistance per phase, $R_2 = 0.016\ \Omega$

Rotor standstill reactance per phase,

$$X_2 = 2\pi f L_2 = 2\pi \times 50 \times 0.8 \times 10^{-3} = 0.2513\ \Omega$$

At standstill (with slip rings shorted)

Rotor impedance per phase,

$$\mathbf{Z}_2 = R_2 + jX_2 = (0.016 + j0.2513) = 0.2518\ \angle 86.36^\circ\ \Omega$$

(*i*) Rotor starting current, $I_2 = \frac{E_2}{Z_2} = \frac{184.752}{0.2518} = 733.72$ A

(*ii*) Rotor pf at starting,

$\text{Cos}\ \phi_2 = \text{Cos}\ 86.36^\circ = 0.0635$ (lagging) **Ans.**

(*iii*) Rotor current per phase corresponding to stator line current of 125 A

$$I_2 = \frac{I_1}{K} = \frac{125}{1/3.75} = 3.75 \times 125 = 468.75 \text{ A}$$

Rotor impedance per phase,

$$Z'_2 = \frac{E_2}{I_2} = \frac{184.752}{468.75} = 0.394\ \Omega$$

Required rotor resistance per phase,

$$R'_2 = \sqrt{(Z'_2)^2 - (X_2)^2} = \sqrt{(0.394)^2 - (0.2513)^2} = 0.30345\ \Omega$$

External resistance required per phase,

$$R = R'_2 - R_2 = 0.30345 - 0.016 = 0.28745\ \Omega \quad \textbf{Ans.}$$

(*iv*) **At 4% slip (with slip rings shorted)**

Rotor impedance per phase,

$$\mathbf{Z}_2 = R_2 + jsX_2 = 0.016 + j \times 0.04 \times 0.2513$$

$$= (0.016 + j0.01)\ \Omega = 0.01887\ \angle 32^\circ\ \Omega$$

Rotor current, $I_2 = \frac{sE_2}{Z_2} = \frac{0.04 \times 184.752}{0.01887} = 391.67$ A **Ans.**

(*v*) Rotor power factor, $\cos \phi_2 = \text{Cos}\ 32^\circ = 0.848$ (lagging) **Ans.**

18.11 ROTOR TORQUE

The torque of an induction motor, being due to interaction of the rotor and stator fields, depends on the strength of those fields and the phase relation between them. Thus the torque developed by the rotor of an induction motor is directly proportional to (*i*) rotor current, I_2 (*ii*) stator flux per pole, Φ and (*iii*) power factor of the rotor circuit, $\cos \phi_2$

i.e. $T \propto \Phi I_2 \cos \phi_2$

Since rotor emf per phase at standstill, $E \propto \Phi$

$\therefore$ Torque developed, $T \propto E_2 I_2 \cos \phi_2$

$$\text{or } T = K E_2 I_2 \cos \phi_2 \qquad \text{...(18.14)}$$

where K is any constant

18.11.1. Torque Under Running Conditions

Substituting $I_2 = \frac{sE_2}{\sqrt{R_2^2 + s^2 X_2^2}}$ from Eq. (18.11) and $\cos \phi_2 = \frac{R_2}{\sqrt{R_2^2 + s^2 X_2^2}}$ from Eq. (18.12) in Eq. (18.14) we have torque developed by an induction motor rotor,

$$T = KE_2 \times \frac{sE_2}{\sqrt{R_2^2 + s^2 X_2^2}} \times \frac{R_2}{\sqrt{R_2^2 + s^2 X_2^2}}$$

$$= \frac{K s R_2 E_2^2}{R_2^2 + s^2 X_2^2} \qquad \text{...(18.15)}$$

It may be seen from Eq. (18.15) that the running torque is proportional to the square of supply voltage because rotor induced emf E_2 is proportional to supply voltage.

18.11.2. Starting Torque. At start the rotor is stationary and so the slip $s = 1.0$ and the starting torque may be obtained by substituting $s = 1.0$ in Eq. (18.15).

$$\text{Starting torque, } T_{st} = \frac{KR_2 E_2^2}{R_2^2 + X_2^2} \qquad \text{...(18.16)}$$

Generally, the stator supply voltage V is constant, so the rotating flux Φ set up by the stator currents and rotor induced emf E_2 are also constant.

$$\therefore T_{st} = \frac{K_1 R_2}{R_2^2 + X_2^2} = \frac{K_1 R_2}{Z_2^2} \quad ...(18.17)$$

where K_1 is another constant.

18.11.3. Condition For Maximum Running Torque. Torque under running condition given by Eq. (18.15) is reproduced here

$$T = \frac{K R_2 s E_2^2}{R_2^2 + s^2 X_2^2}.$$

Torque T for fixed supply voltage will be maximum when

$$\frac{sR_2}{R_2^2 + s^2 X_2^2} \text{ or } \frac{R_2}{\frac{R_2^2}{s} + sX_2^2} \text{ or } \frac{R_2}{\left(\frac{R_2}{\sqrt{s}} - X_2\sqrt{s}\right)^2 + 2R_2X_2} \text{ is}$$

maximum viz. $\frac{R_2}{\sqrt{s}} - X_2\sqrt{s} = 0$ or $s = \frac{R_2}{X_2}$

Substituting $s = \frac{R_2}{X_2}$ in Eq. (18.15) we have

$$T_{max} = \frac{KR_2^2 E_2^2/X_2}{R_2^2 + R_2^2} = \frac{KE_2^2}{2X_2} \quad ...(18.18)$$

From Eq. (18.18) it is obvious that

(*i*) Maximum torque is independent of rotor circuit resistance.

(*ii*) The slip at which maximum torque occurs is governed by the rotor resistance. As seen from above, the torque becomes maximum when rotor reactance equals the rotor resistance. So by varying the rotor circuit resistance, which is possible with wound rotors, maximum torque can be made to occur at any desired slip or motor speed.

(*iii*) Maximum torque varies inversely as standstill reactance of the rotor, hence to have maximum torque, standstill reactance *i.e.* inductance of the rotor should be kept as small as possible. This is achieved by placing the rotor conductors very close to the surface of the rotor and reducing the air gap between stator and rotor to the smallest possible value.

(*iv*) Maximum torque varies directly as the square of supply voltage.

18.11.4. Condition For Maximum Starting Torque. Starting torque T_{st} will be maximum when $\frac{R_2}{X_2} = s = 1$ or $R_2 = X_2$

Since the rotor resistance is not more than 1 or 2 per cent of leakage reactance otherwise efficiency will be low, so to increase the starting torque, extra resistance is required to be inserted in the rotor circuit at start and cut out gradually as the motor picks up speed. This is possible in case of wound rotor motors. This is the reason that wound rotor motors are used where heavy loads are to be accelerated at start such as cranes, lifts, elevators etc.

18.12 RELATION BETWEEN ROTOR TORQUE AND ROTOR POWER FACTOR

As in the case of a dc motor the armature torque T_a varies as the product of flux per pole and armature current i.e. $T_a \propto \Phi I_a$ (refer to Art. 9.13), similarly, in the case of an induction motor, the rotor torque varies as the product of flux per stator pole and the rotor current. In addition, rotor power factor is also to be considered in case of an induction motor i.e.

$$T \propto \Phi I_2 \cos\phi_2 \quad \text{(refer to Art 18.11)}$$

$$\text{and } T = K E_2 I_2 \cos\phi_2 \quad \text{[refer to Eq. (18.14)]}$$

Above equation reveals that with the increase in rotor phase angle ϕ_2, power factor $\cos\phi_2$ decreases causing rotor torque to reduce and vice versa.

The noteworthy point is that the stator flux distribution is assumed to be sinusoidal. The magnitude of emf induced in the rotor, owing to revolving flux depends upon the flux density, in which the conductor lies at the instant consider ($e = Blv$ volt). Hence, the induced emf in the rotor is also sinusoidal.

The effect of rotor power factor $\cos\phi_2$ on the rotor torque is illustrated in Fig. 18.17 (*a*) and (*b*).

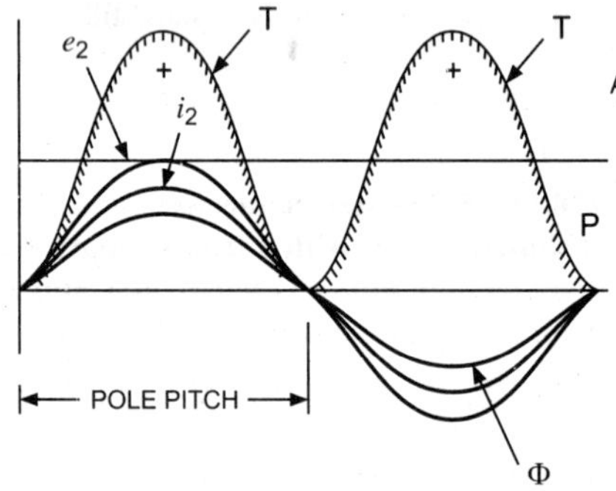

(*a*)

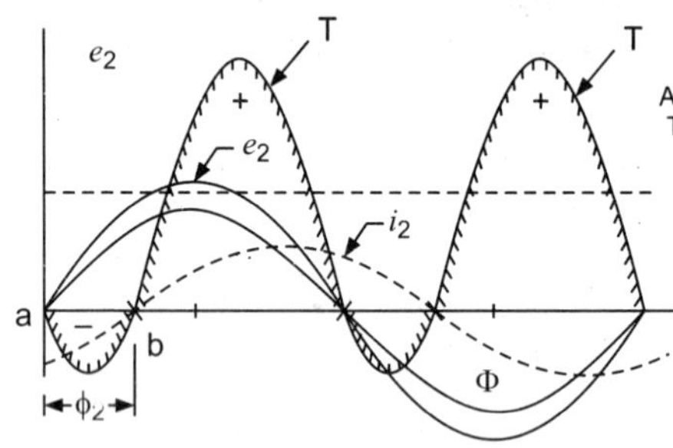

(*b*)

Fig. 18.17

If the rotor circuit is assumed to be non-inductive (i.e. phase angle between rotor emf and rotor current, $\phi_2 = 0$), the instantaneous value of the torque experienced by each conductor is given by the product of instantaneous values of flux and the rotor current ($\because F \propto BI_2l$). Hence, torque can be had by plotting the product of flux Φ and rotor current I_2. As seen from Fig. 18.17 (*a*), the rotor torque is always positive i.e. unidirectional.

Now consider that the rotor circuit is inductive, as shown in Fig. 18.17 (*b*). In this case rotor current I_2 lags behind induced voltage in rotor circuit, E_2 by an angle ϕ_2 where phase angle

$\phi_2 = \text{Tan}^{-1}\dfrac{X_2}{R_2}$ and R_2 and X_2 are rotor resistance/phase and standstill reactance per phase respectively.

From Fig. 18.17 (*b*), it is seen that for a portion of the pole pitch, the rotor toque is negative i.e. backward. Hence, the net torque, being the difference of forward torque and backward torque, is considerably reduced. In case the rotor circuit is pure inductive i.e. phase angle ϕ_2 is 90º, then the net torque is zero because in such a case the forward and backward torques become equal and opposite.

18.13 TORQUE-SLIP AND TORQUE-SPEED CURVES

Torque developed by an induction motor rotor is given by the equation $T = \dfrac{KsR_2E_2^2}{R_2^2 + s^2X_2^2}$.

From the above equation it is revealed that

(*i*) When speed is synchronous *i.e.* when slip *s* is zero, the torque T is zero so the torque-slip curve starts from origin O.

(*ii*) When the speed is very near to synchronous speed N_s *i.e.* when slip s is very low the value of term sX_2 is very small and is negligible in comparison with rotor resistance R_2,therefore, torque T is approximately proportional to slip s, if rotor resistance R_2 is constant *i.e.* at speeds near synchronous speed the torque-speed and torque-slip curves are approximately straight lines, as shown in Fig. 18.18 (*a*) and 18.18 (*b*) respectively.

(*iii*) As the slip increases , *i.e.* as the speed drops with the increase in load, torque increases, reaches its maximum value when slip $s = \dfrac{R_2}{X_2}$. The maximum torque is also known as *breakdown or pull-out torque*. The slip corrresponding to breakdown torque is called the breakdown slip s_b.

(*iv*) With the further increase in slip or drop in speed due to increase in load beyond the point of maximum torque the torque begins to decrease. The result is that the motor slows down and eventually stops. The motor operates for the value of slip between zero and that corresponding to breakdown or pull-out torque.

With higher values of slip, R_2 becomes negligible as compared to $s\,X_2$ and the torque varies according to the relation

$T \,\alpha\, \dfrac{s}{s^2X_2^2} \,\alpha\, \dfrac{1}{s}$ if standstill reactance X_2 is constant. It means that torque-speed curves are rectangular hyperbola with the speed or slip beyond that corresponding to breakdown or pull-out torque.

Torque-speed and torque-slip curves are shown in Figs. 18.18 (*a*) and 18.18 (*b*) respectively for different values of rotor resistance R_2.

It is seen that although maximum torque is independent of rotor resistance R_2, yet the exact location of T_{max} is dependent on it. Greater the R_2, greater is the value of slip at which the maximum torque occurs.

Induction machine can be operated in three different modes i.e. motoring mode, generating mode and braking mode as depicted in Fig. 18.19.

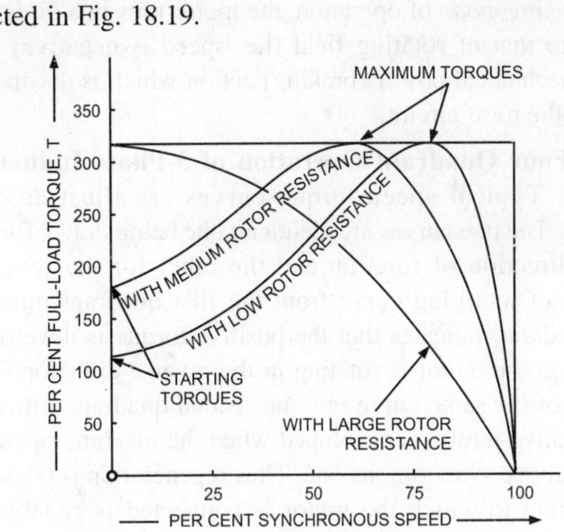

(a) Torque-Speed Curves

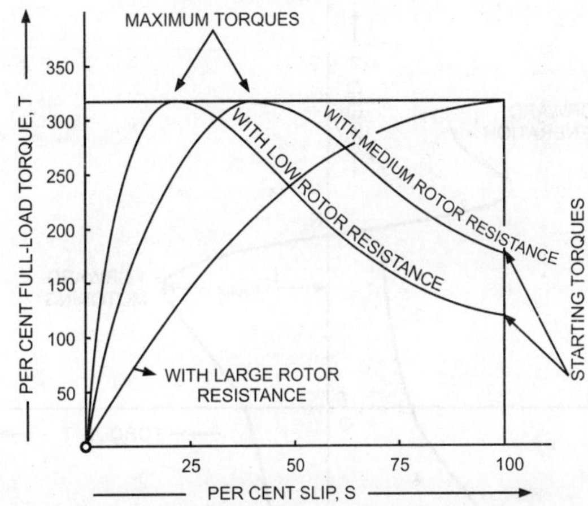

(b) Torque-Slip Curves

Fig. 18.18

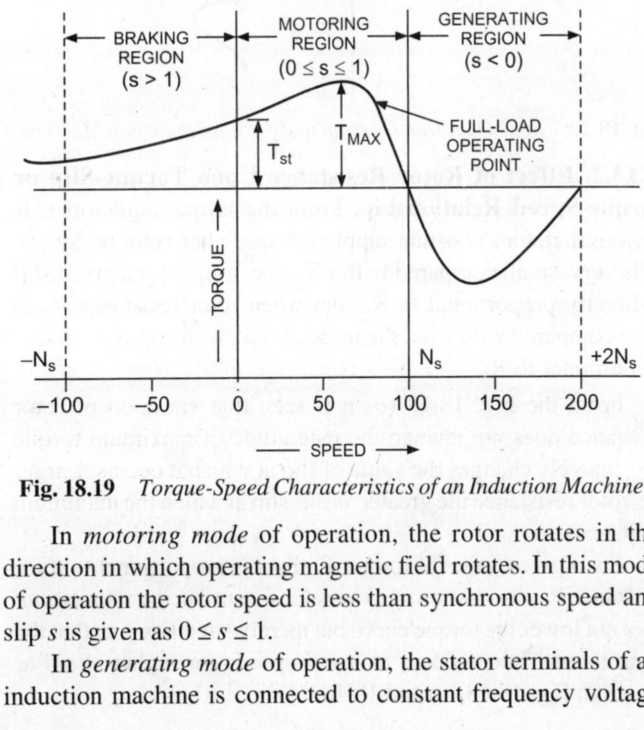

Fig. 18.19 *Torque-Speed Characteristics of an Induction Machine*

In *motoring mode* of operation, the rotor rotates in the direction in which operating magnetic field rotates. In this mode of operation the rotor speed is less than synchronous speed and slip *s* is given as $0 \le s \le 1$.

In *generating mode* of operation, the stator terminals of an induction machine is connected to constant frequency voltage source and the rotor is driven above the synchronous speed by the prime mover. In this mode of operation slip s is negative.

In *braking mode* of operation, the motor runs in a direction opposite to that of rotating field (i.e. speed is negative) and absorbs mechanical power (braking) action which is dissipated as heat in the rotor circuit.

18.13.1. Four Quadrant Operation of 3-Phase Induction Machine. Typical speed-torque curves are illustrated in Fig. 18.20. The two curves are identical; one being drawn for the forward direction of rotation and the other for the reverse. Extension of motoring curve from the first quadrant into the fourth quadrant indicates that the positive torque is developed even though the motor is rotating in the reverse direction. The extension of the same curve into the second quadrant indicates that a negative torque is developed when the machine operates at speeds above synchronous one. Thus regeneration is possible if the system to which the motor is connected is capable of supplying required reactive power for excitation.

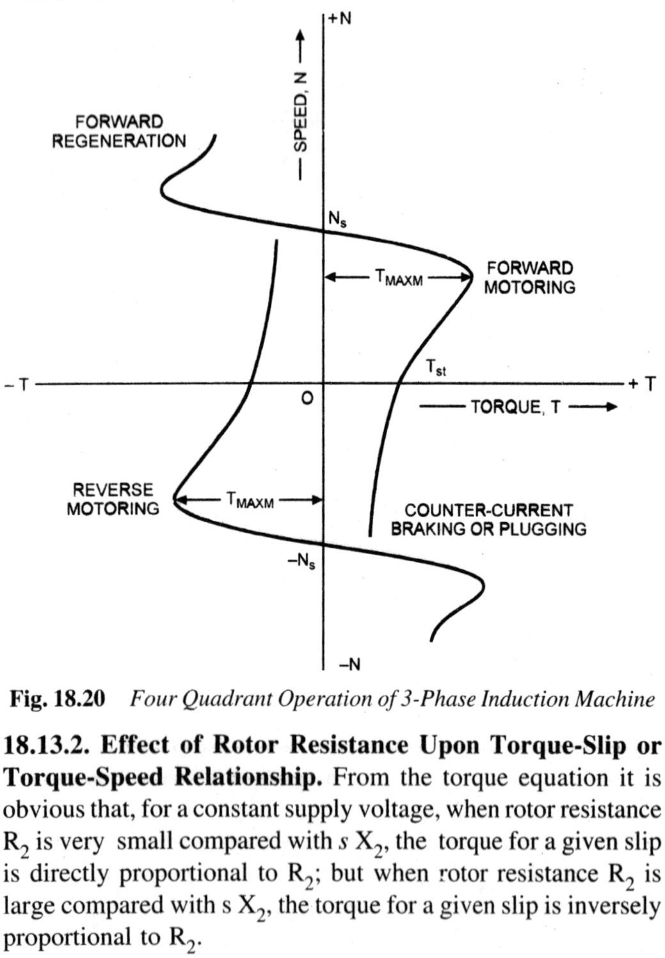

Fig. 18.20 *Four Quadrant Operation of 3-Phase Induction Machine*

18.13.2. Effect of Rotor Resistance Upon Torque-Slip or Torque-Speed Relationship. From the torque equation it is obvious that, for a constant supply voltage, when rotor resistance R_2 is very small compared with $s\,X_2$, the torque for a given slip is directly proportional to R_2; but when rotor resistance R_2 is large compared with s X_2, the torque for a given slip is inversely proportional to R_2.

From the Fig. 18.18 (*b*) it is seen that variation of rotor resistance does not change the magnitude of maximum torque T_{max}; merely changes the value of slip at which it occurs. Larger the rotor resistance the greater is the slip at which the maximum torque occurs.

For a given value of torque T, slip s is proportional to rotor resistance R_2, so addition of external resistance in the rotor circuit does not lower the torque curve but merely stretches it so that the same torque values occur at lower speeds (or higher slips) as shown in Fig. 18.18 (*a*) and 18.18 (*b*).

18.13.3. Variation of Starting Torque With Rotor Resistance. If the rotor circuit resistance R_2 were zero, then the rotor current at standstill would lag 90° behind the rotor induced emf and the induction motor would, therefore, develop no torque at start. On the other hand, if the rotor resistance were infinitely large, there would be no rotor current, and, therefore, no torque. Since the starting torque is zero when $R_2 = 0$, and also when R_2 is infinitely large, it follows that there must be some intermediate value of rotor resistance that will give a maximum value of starting torque. The variation of starting current and starting torque with rotor resistance, as plotted from Eqs. (18.8) and (18.16) are shown in Fig. 18.21.

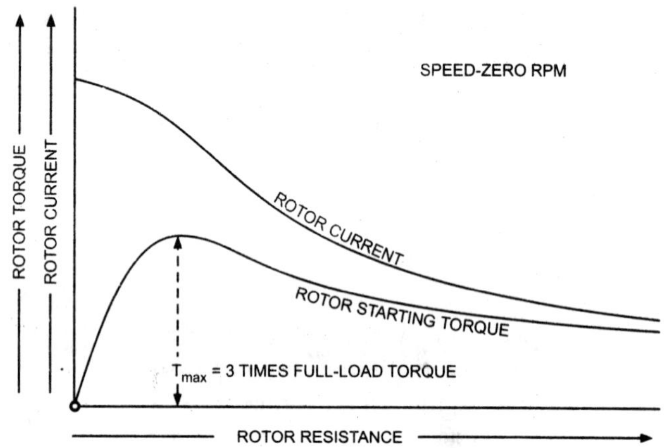

Fig. 18.21

As the rotor resistance R_2 is increased from zero, power factor is improved but rotor impedance is increased and so the rotor current is reduced. At first, the effect of improved power factor predominates over the effect of decrease in rotor current due to increase in impedance and the starting torque, therefore, increases. But after a certain value of rotor resistance, the effect of increased impedance predominates over the effect of improved power factor and so the torque starts decreasing.

Torque will be maximum when $R_2 = sX_2$, as determined in Art. 18.11.3, So for having maximum starting torque, and also maximum acceleration throughout the accelerating period, the motor should be started with a rotor resistance equal to rotor reactance at standstill, and rotor resistance R_2 should be reduced continuously, as the motor gains speed, always keeping the rotor resistance equal to the rotor reactance.

18.14 EFFECT OF CHANGE IN SUPPLY VOLTAGE ON STARTING TORQUE

From Eq. (18.16),

$$\text{Starting torque, } T_{st} = \frac{KR_2E_2^2}{R_2^2 + X_2^2}$$

Since emf induced in rotor at standstill, $E_2 \propto \Phi \propto V$, the supply voltage

$$\text{So starting torque, } T_{st} = \frac{K'R_2V^2}{R_2^2 + X_2^2}$$

where K′ is another constant or $T_{st} \propto V^2$, Rotor resistance R_2 and standstill reactance X_2 being fixed for a given machine *i.e. starting torque is proportional to the square of the applied voltage.* It means that the starting torque is very sensitive to any change in the applied voltage. For example, a change of 10 per cent in supply voltage will cause a change of 19 % in the rotor torque. This fact is very important in connection with starting methods of squirrel cage induction motors.

18.15 EFFECT OF CHANGE IN SUPPLY VOLTAGE ON TORQUE AND SLIP

When the motor is running with slip *s*, then torque acting on the rotor is given by

$$T = \frac{KsR_2E_2^2}{R_2^2 + s^2X_2^2} \quad \text{...Refer to Eq. (18.15)}$$

Since emf induced in the rotor at standstill, $E_2 \propto V$

So torque developed $T = \dfrac{K' s R_2 V^2}{R_2^2 + s^2 X_2^2}$

where K′ is another constant

Since at full load, slip *s* is very low, therefore, $s^2 X_2^2$ can be neglected in comparison with R_2^2 and equation for torque may be

written as $T = \dfrac{K' s R_2 V^2}{R_2^2} = \dfrac{K' s V^2}{R_2}$ or $T \propto sV^2$

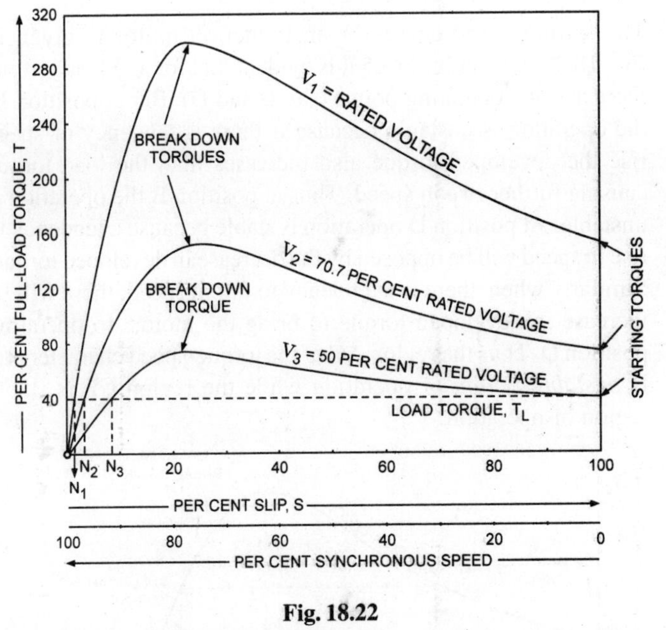

Fig. 18.22

If the supply voltage is changed, it changes the torque under running conditions also. With the decrease in supply voltage, torque under running condition decreases. As the motor now develops a reduced torque, the motor slows down (*i.e.* slip increases) so as to cause increase in torque to meet the load torque and the motor may come to a standstill (if the load, torque exceeds the pull-out torque of the motor corresponding to reduced supply voltage) unless the load is removed. Figure 18.22 shows the variation of torque with slip for three different values of line voltage.

18.16 EFFECTS OF VARYING SUPPLY VOLTAGE AND SUPPLY FREQUENCY TOGETHER ON TORQUE AND SPEED

This kind of variation can occur for example, if the generator supplying the power to the induction motor (or motors) is subject to speed variations, either due to momentary overloads or because of a noticeable speed regulation. Both output voltage and frequency will vary as the speed if no automatic correction is provided for it. On some large marine drives, the propellor motors are of induction type and are speed-controlled from such a local supply which is provided by synchronous generators coupled to variable speed turbines.

Since emf induced in stator, E_1 is proportional to the product of stator flux Φ and supply frequency f (*i.e.* $E_1 \propto \Phi f$), a change of frequency will give rise to a change of flux level unless the induced emf is changed in the same ratio. An imbalance in this ratio will cause either an excessive flux and saturation, or diminished flux and reduced torque per ampere of current. Constant flux per pole is approximated if the supply volts per Hz (V/f)) is maintained constant because E_1 is approximately equal to V, the supply voltage. However, an idealized variable-frequency control scheme might aim to ensure that any particular torque should be obtained at the same flux.

With this criterion of controlling $\dfrac{E_1}{N_s}$ to a constant value for a particular torque, the speed-difference $(N_s - N)$ is also fixed for the same torque, since $(N_s - N)$ in turn determines the slip frequency f', the rotor emf E_2, impedance Z_2 and current I_2, which remain unchanged. From this it follows that as the frequency is reduced, the shape of the torque-slip (or torque-speed) curves do not change, but displaced horizontally, as illustrated in Fig. 18.23.

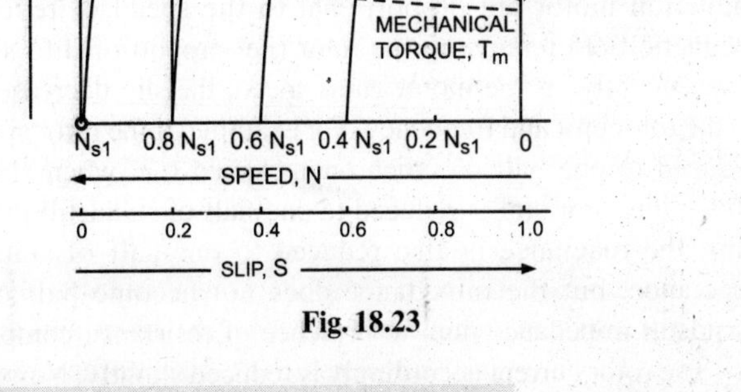

Fig. 18.23

18.17 FULL-LOAD TORQUE AND MAXIMUM TORQUE

Let the full-load slip of the motor be s_f.

Since full-load torque T_f and maximum torque T_{max} are given by equation

$$T_f = \frac{K s_f R_2 E_2^2}{R_2^2 + s_f^2 X_2^2} \quad \text{...Refer to Eq. (18.15)}$$

$$\text{and } T_{max} = \frac{K E_2^2}{2 X_2} \quad \text{...Refer to Eq. (18.18)}$$

$$\text{so } \frac{T_f}{T_{max}} = \frac{\dfrac{K s_f R_2 E_2^2}{R_2^2 + s_f^2 X_2^2}}{\dfrac{K E_2^2}{2X_2}}$$

$$= \frac{2 s_f R_2 X_2}{R_2^2 + s_f^2 X_2^2}$$

$$= \frac{2 s_f \dfrac{R_2 X_2}{X_2^2}}{\dfrac{R_2^2 + s_f^2 X_2^2}{X_2^2}} = \frac{2 s_f \dfrac{R_2}{X_2}}{\dfrac{R_2^2}{X_2^2} + s_f^2} = \frac{2 a s_f}{a^2 + s_f^2} \quad \text{...(18.19)}$$

where a is the ratio of rotor resistance to rotor standstill reactance. In fact $a = s_{max\,T}$, the slip corresponding to maximum torque.

$$\text{So } \frac{T_f}{T_{max}} = \frac{2 s_{max\,T} s_f}{s_{max\,T}^2 + s_f^2} = \frac{2}{\dfrac{s_{max\,T}}{s_f} + \dfrac{s_f}{s_{max\,T}}} \quad \text{...(18.20)}$$

18.18 STARTING TORQUE AND MAXIMUM TORQUE

Substituing $s_f = 1$ in Eq. (18.19) we have

$$\frac{T_{st}}{T_{max}} = \frac{2a}{a^2 + 1} \quad \text{where } a = \frac{R_2}{X_2} \quad \text{... (18.21)}$$

18.19 CURRENT-SPEED CHARACTERISTIC

The current-speed characteristic of an induction motor may be analysed from Eq. (18.11). Since the machine can operate as a motor between no-load speed and zero speed, the characteristic is studied between zero speed to synchronous speed.

The magnitude and frequency of emf induced in rotor of an induction motor are proportional to the speed of revolving magnetic field relative to the rotor (i.e. proportional to slip s). Consequently, as the motor gains speed, the slip decreases and so the rotor emf and frequency. For example, if the rotor speed is reduced to one half of synchronous speed (*i.e.* when slip s is 50%); the rotor emf is reduced to one-half of standstill induced emf, the reactance is also reduced to one-half of standstill reactance, but the impedance does not become half of the standstill impedance due to presence of resistance component R_2. The rotor current accordingly is reduced slightly. Now when the rotor attains 75% of synchronous speed, the rotor emf is reduced to one-fourth of standstill emf; the reactance is also reduced to one-fourth of standstill reactance but rotor current is not reduced to one fourth of standstill value. This is obviously due to fixed value of rotor resistance R_2. The rotor current has reduced about 30 per cent, as illustrated in the figure. With the further increase in speed, the rotor current drops rapidly (almost linearly) until at synchronous speed the rotor emf and rotor current are both zero. This is because at high speeds, the slip s is low and so the rotor reactance (sX_2) becomes small as compared to rotor resistance R_2 and the rotor current becomes proportional to rotor induced emf or proportional to slip (∵ rotor induced emf is proportional to slip). The variation of rotor current with speed for a low impedance squirrel cage motor (class A motor) are shown in Fig. 18.24.

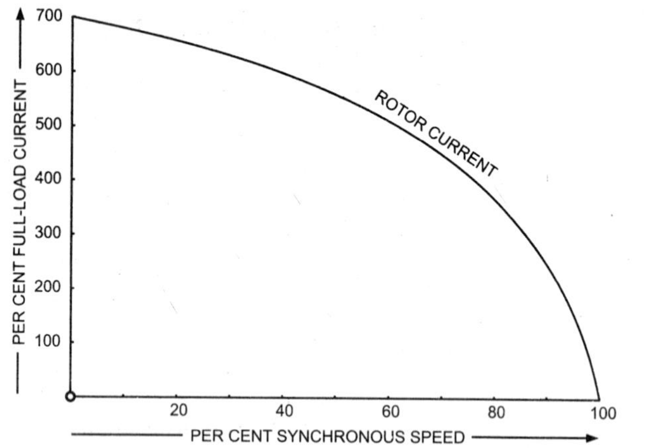

Fig. 18.24 *Current-Speed Characteristic of Low Impedance Type Squirrel Cage Induction Motor*

18.20 TORQUE-SPEED CURVE AND OPERATING REGION

The torque-speed curve for an induction motor is given in Fig. 18.25. From Fig. 18.25 it is evident that for any load torque, there are two operating points (say B and D). But at position B, the operation is unstable because if there is tendency of speed rise the developed torque also increases than the load torque causing further rise in speed. Thus at position B the operation is unstable. At position D operation is stable because a tendency to rise in speed will be opposed by the decrease in developed torque. Similarly when there is a tendency to fall in speed, there will be increase in developed torque to bring the motor to operating position D. Thus the region AC of the torque-speed characteristic is *unstable region* of operation while the region CF is stable region of operation.

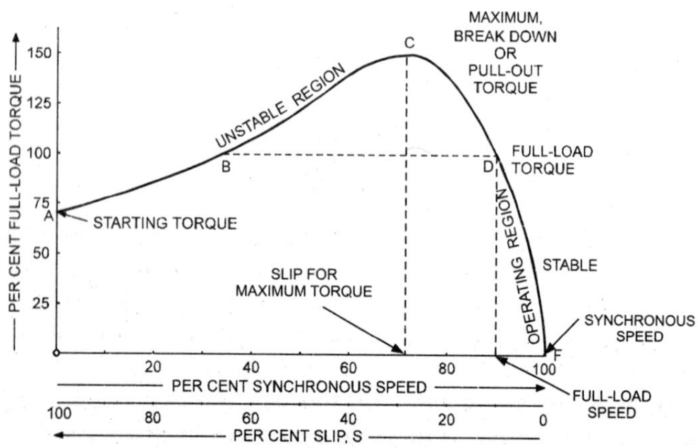

Fig. 18.25

At full load, the motor runs at a speed of N rpm. If the mechanical load applied to the motor is increased, motor speed drops till the torque developed by the motor equals the load torque. As long as the two torques are in balance, the motor will run at constant speed. However, if the load torque exceeds the breakdown torque of the motor, the motor will immediately stop.

The value of breakdown torque varies with the design of the motor but ranges from 200 to 300 per cent of full-load torque in standard squirrel cage motors.

The torque-speed characteristics of an induction motor in the operating region are quite similar to those of a dc shunt motor. Even at full load, the drop in speed from no load does not exceed 5 per cent.

Example 18.14. A 4-pole, 50 Hz, 3-phase induction motor has a rotor resistance of 0.02 Ω per phase and standstill reactance of 0.5 Ω per phase. Determine the speed at which the maximum torque is developed. [*M.D. Univ. Electrical Machine-II, December-2010*]

Solution: Rotor resistance per phase, $R_2 = 0.02\ \Omega$

Rotor standstill reactance per phase, $X_2 = 0.5\ \Omega$

Since the torque under running is maximum at that value of slip, which makes rotor reactance per phase equal to the rotor resistance per phase

∴ Slip corresponding to maximum torque,

$$s_{\max T} = \frac{R_2}{X_2} = \frac{0.02}{5} = 0.04$$

Speed corresponding to maximum torque,

$$N_{\max T} = \frac{120f}{P}(1 - s_{\max T})$$

$$= \frac{120 \times 50}{4}(1 - 0.04) = 1{,}440 \text{ rpm} \quad \textbf{Ans.}$$

Example 18.15. A 12-pole, 3-phase, 50-Hz induction motor runs on full load with a slip of 4.5% having rotor resistance and standstill reactance per phase 0.016 Ω and 0.078 Ω respectively. Find

(*a*) the available maximum torque in terms of full-load torque.

(*b*) the speed at which the maximum torque occurs.

[*Pb. Technical Univ. Asynchronous Machines, May-2013*]

Solution: Rotor resistance per phase, $R_2 = 0.016\ \Omega$

Rotor standstill reactance per phase,

$X_2 = 0.078\ \Omega$

$$\text{Synchronous speed, } N_s = \frac{120f}{P} = \frac{120 \times 50}{12} = 500 \text{ rpm}$$

Full-load slip, $s_f = 4.5\%$ or 0.045

Ratio of rotor resistance to standstill reactance,

$$a = \frac{R_2}{X_2} = \frac{0.016}{0.078} = 0.20513$$

(*i*) Ratio of maximum torque to full-load torque,

$$\frac{T_{\max}}{T_f} = \frac{a^2 + s_f^2}{2\,as_f} = \frac{0.20513^2 + 0.045^2}{2 \times 0.20513 \times 0.045} = 2.389$$

So maximum torque, $T_{\max}$ = 2.389 times full-load torque **Ans.**

(*ii*) Slip corresponding to maximum torque,

$s_{\max T} = 0.20513$

The speed at which the maximum torque occurs,

$$N_{\max T} = N_s(1 - s_{\max T})$$

$$= 500\,(1 - 0.20513) = 397.4 \text{ rpm} \quad \textbf{Ans.}$$

Example 18.16. A 6.6 kV, 10-pole 50 Hz, 3-phase, star-connected three-phase induction motor has a slip-ring rotor of resistance 0.015 Ω/phase and standstill reactance of 0.35 Ω/phase. The motor has a speed of 560 rpm while delivering it rated full-load torque. Compute the slip at maximum torque and the ratio of maximum torque to full-load torque. Neglect the stator impedance drop.

[*B.P. Univ. of Technology Electrical Machines, 2008*]

Solution: $$\text{Synchronous speed, } N_s = \frac{120f}{P} = \frac{120 \times 50}{10} = 600 \text{ rpm}$$

Full-load speed, $N_f = 560$ rpm

$$\text{Full-load slip, } s_f = \frac{N_s - N_f}{N_s} = \frac{600 - 560}{600} = 0.06667$$

$$\text{Slip at maximum torque, } s_{\max T} = \frac{R_2}{X_2} = \frac{0.015}{0.35} = 0.04286 \ \textbf{Ans.}$$

Ratio of maximum torque to full-load torque,

$$\frac{T_{\max}}{T_f} = \frac{\dfrac{s_{\max T}}{s_f} + \dfrac{s_f}{s_{\max T}}}{2} = \frac{\dfrac{0.04286}{0.06667} + \dfrac{0.06667}{0.04286}}{2} = 1.099 \ \textbf{Ans.}$$

Example 18.17. A 4-pole, 50 Hz, 3-phase induction motor develops a maximum torque of 200 N-m at 1,400 rpm. Find the torque at a slip of 5%. Neglect stator impedance.

[*G.G.S.I.P. Univ. Delhi Electromechanical Energy Conversion-II; May-2011*]

Solution: $$\text{Synchronous speed, } N_s = \frac{120f}{P} = \frac{120 \times 50}{4} = 1{,}500 \text{ rpm}$$

Speed corresponding to maximum torque,

$N_{\max T} = 1{,}400$ rpm

Slip corresponding to maximum torque,

$$s_{\max T} = \frac{N_s - N_{\max T}}{N_s} \times 100 = \frac{1{,}500 - 1{,}400}{1{,}500} = 0.06667$$

Given slip, $s = 0.05$

Torque at a slip of 5 per cent

$$= \frac{2}{\dfrac{s_{\max T}}{s} + \dfrac{s}{s_{\max T}}} \times T_{\max} = \frac{2}{\dfrac{0.06667}{0.05} + \dfrac{0.05}{0.06667}} \times 200$$

$$= 192 \text{ N-m} \ \textbf{Ans.}$$

Example 18.18. A 4-pole, 50-Hz, 3-phase induction motor develops a maximum torque of 110 N-m at 1,360 rpm. The resistance of the star-connected rotor is 0.25 Ω/phase. Calculate the value of resistance that must be inserted in series, with each rotor phase to produce a starting torque equal to half the maximum torque.

[*U.P. Technical Univ. Electrical Machines, 2010-11*]

Solution: $$\text{Synchronous speed, } N_s = \frac{120f}{P} = \frac{120 \times 50}{4} = 1{,}500 \text{ rpm}$$

Speed corresponding to maximum torque,

$N_{\max T} = 1{,}360$ rpm

Slip corresponding to maximum torque,

$$s_{\max T} = \frac{N_s - N_{\max T}}{N_s} = \frac{1{,}500 - 1{,}360}{1{,}500} = 0.09333$$

Rotor standstill reactance per phase,

$$X_2 = \frac{R_2}{s_{\max T}} = \frac{0.25}{0.09333} = 2.679\ \Omega$$

According to Eq. (18.21), ratio of starting torque to maximum torque,

$$\frac{T_{st}}{T_{max}} = \frac{2a}{a^2+1} = \frac{1}{2}$$

or $a^2 - 4a + 1 = 0$

$$\text{or } a = \frac{4 \pm \sqrt{16-4}}{2} = 0.268 \text{ rejecting higher value}$$

If r is the additional resistance introduced per phase in the rotor circuit to provide the starting torque equal to half the maximum torque then

$$\frac{R_2 + r}{X_2} = 0.268$$

$$\text{or } r = 0.268\, X_2 - R_2$$

$$= 0.268 \times 2.679 - 0.25 = 0.468\ \Omega \textbf{ Ans.}$$

Example 18.19. A 4 pole, 50 Hz, 7.4 kW motor, at rated voltage and frequency has a starting torque of 150% and a maximum torque of 200% of full-load torque. Calculate

(*i*) Full-load speed.

(*ii*) Speed at maximum torque.

[*M.D. Univ. Electromechanical Energy Conversion, December-2011*]

Solution: Starting torque, T_{st} = 150% of T_f or equal to 1.5 T_f

Maximum torque, T_{max} = 200% of T_f or equal to 2 T_f

$$\text{or } \frac{T_{st}}{T_{max}} = \frac{2a}{a^2+1} \quad \text{where } a = \frac{R_2}{X_2}$$

$$\text{or } \frac{1.5T_f}{2\,T_f} = \frac{2a}{a^2+1}$$

or $1.5a^2 - 4a + 1.5 = 0$

$$\text{or } a = \frac{4 \pm \sqrt{16 - 4 \times 1.5 \times 1.5}}{2 \times 1.5}$$

$$= 0.4514 \text{ neglecting higher value}$$

Hence slip corresponding to maximum torque,

$$S_{max\,T} = 0.4514$$

From Eq. (18.19), we have

$$\frac{T_f}{T_{max}} = \frac{2aS_f}{a^2 + S_f^2}$$

$$\text{or } \frac{1}{2} = \frac{2 \times 0.4514 \times S_f}{(0.4514)^2 + S_f^2}$$

or $S_f^2 - 1.8056S_f + 0.203762 = 0$

$$\text{or Full-load slip, } S_f = \frac{1.8056 \pm \sqrt{(1.8056)^2 - 4 \times 0.203762}}{2}$$

$$= 0.121 \text{ neglecting higher value}$$

$$\text{Synchronous speed, } N_s = \frac{120f}{P} = \frac{120 \times 50}{4} = 1{,}500 \text{ rpm}$$

(*i*) Full-load speed, $N_f = N_s(1 - S_f)$

$= 1{,}500(1 - 0.121) = 1{,}318.5$ rpm **Ans.**

(*ii*) Speed at maximum torque, $N_{max\,T}$

$= N_s(1 - S_{max\,T})$

$= 1{,}500(1 - 0.4514) = 822.9$ rpm **Ans.**

Example 18.20. A 6-pole, 50 Hz, 3-phase slip-ring induction motor has resistance and reactance of 0.5 Ω and 5 Ω per phase respectively. Calculate (*i*) at what speed the torque is maximum? (*ii*) the ratio of maximum torque/starting torque. What must be the external resistance per phase so that the starting torque is half of the maximum torque.

[*U.P. Technical Univ. Electromechanical Energy Conversion-II, 2010-11; J.N. Technological Univ. Electrical Machines-II, January-2014*]

Solution: Rotor resistance per phase, $R_2 = 0.5\ \Omega$

Rotor standstill reactance per phase, $X_2 = 5\ \Omega$

Since the torque under running condition is maximum at that value of slip, which makes rotor reactance per phase equal to the rotor resistance per phase.

∴ Slip corresponding to maximum torque,

$$s_{max\,T} = \frac{R_2}{X_2} = \frac{0.5}{5} = 0.1$$

Speed corresponding to maximum torque,

$$N_{max\,T} = N_s(1 - s_{max\,T})$$

$$= \frac{120f}{P}(1 - s_{max\,T}) = \frac{120 \times 50}{6}(1 - 0.1) = 900 \text{ rpm } \textbf{Ans.}$$

(*ii*) Ratio of rotor resistance to rotor standstill reactance,

$$a = \frac{R_2}{X_2} = \frac{0.5}{5} = 0.1$$

$$(ii) \text{ Ratio of } \frac{\text{maximum torque}}{\text{starting torque}} = \frac{a^2+1}{2a} = \frac{(0.1)^2+1}{2 \times 0.1} = 5.05 \textbf{ Ans.}$$

...Refer to Eq. (18.21)

(*iii*) Ratio of starting torque T_{st} and maximum torque T_{max}, according to Eq. (18.21) is given as

$$\frac{T_{st}}{T_{max}} = \frac{2a}{a^2+1} = 0.5 \quad \because \text{ It is given that } T_{st} = \frac{1}{2}T_{max}$$

or $a^2 - 4a + 1 = 0$

$$\text{or } a = \frac{4 \pm \sqrt{16-4}}{2} = 0.268 \text{ rejecting higher value.}$$

If r is the additional resistance introduced per phase in the rotor circuit to give the starting torque equal to half of the maximum torque then

$$\frac{R_2 + r}{X_2} = 0.268$$

or $r = 0.268\, X_2 - R_2 = 0.268 \times 5 - 0.5 = 0.84\ \Omega$ **Ans.**

Example 18.21. A 746 kW, 3-phase, 50 Hz, 16-pole induction motor has a rotor impedance of (0.02 + *j* 0.15) Ω at standstill. Full-load torque is obtained at 360 rpm. Calculate (*i*) the ratio of maximum to full-load torque (*ii*) the speed for maximum torque and (*iii*) the rotor resistance to be added to get maximum starting torque.

[*Electrical Machines-II, Rajasthan Technical Univ. Electrical, 2013; J.N. Technological Univ., January-2014*]

Solution: Synchronous speed, $N_s = \frac{120f}{P} = \frac{120 \times 50}{16} = 375$ rpm

Full-load speed, $N_f = 360$ rpm

$$\text{Full-load slip, } s_f = \frac{N_s - N_f}{N_s} = \frac{375 - 360}{375} = 0.04$$

Slip corresponding to maximum torque,

$$s_{max\,T} = \frac{R_2}{X_2} = \frac{0.02}{0.15} = 0.133$$

(*i*) From Eq. (18.20) the ratio of maximum to full-load torque,

$$\frac{T_{max}}{T_f} = \frac{s_{max\,T}/s_f + s_f/s_{max\,T}}{2} = \frac{\frac{0.133}{0.04} + \frac{0.04}{0.133}}{2} = 1.817 \textbf{ Ans.}$$

Speed corresponding to maximum torque,

$N_{max\,T} = N_s(1 - s_{max\,T}) = 375\,(1 - 0.133) = 325$ rpm **Ans.**

(*iii*) For obtaining maximum starting torque,

$R_2 + r = X_2$

or $r = X_2 - R_2 = 0.15 - 0.02 = 0.13\ \Omega$ **Ans.**

Example 18.22. The starting and maximum torques of a 3-phase induction motor are 1.5 times and 2.5 times its full-load torque. Determine the percentage change in rotor circuit resistance to obtain a full-load slip of 0.03. Neglect stator impedance.

[*Rajasthan Technical Univ. Electrical Machines-II, 2013*]

Solution: Starting torque, $T_{st} = 1.5\,T_f$

Maximum torque, $T_{max} = 2.5\,T_f$

$$\frac{T_{st}}{T_{max}} = \frac{2a}{a^2+1} = \frac{1.5\,T_f}{2.5\,T_f} = \frac{1.5}{2.5} = 0.6$$

...Refer to Eq. (18.21)

or $3a^2 - 10a + 3 = 0$

or $(3a - 1)(a - 3) = 0$

or $a = \frac{1}{3}$ rejecting higher value

or $\frac{R_2}{X_2} = a = \frac{1}{3}$

or rotor resistance, $R_2 = \frac{1}{3} X_2$

From Eq. (18.19)

$$\frac{T_f}{T_{max}} = \frac{2as_f}{a^2 + s_f^2}$$

or $\frac{1}{2.5} = \frac{2a \times 0.03}{a^2 + 0.03^2}$ $\because T_{max} = 2.5\,T_f$ and $s_f = 0.03$

or $a^2 - 0.15\,a + 0.0009 = 0$

$$\text{or } a = \frac{0.15 \pm \sqrt{0.15^2 - 4 \times 0.0009}}{2}$$

$= 0.1437$, other value is not feasible.

So percentage reduction in rotor resistance

$$= \frac{R_2 - R_2'}{R_2} \times 100 = \frac{\frac{X_2}{3} - 0.1437\,X_2}{\frac{X_2}{3}} \times 100$$

$= 56.9\,\%$ **Ans.**

Example 18.23. A squirrel cage induction motor has a slip of 4% at full load. Its starting current is 5 times the full-load current. The stator impedance and magnetising current may be neglected; the rotor resistance is assumed constant.

(*i*) Calculate the maximum torque and the slip at which it would occur

(*ii*) Calculate the starting torque. Express torque in pu of the full-load torque [*U.P.S.C. I.E.S. Elec. Engineering II, 1995*]

Solution: From Eq. (18.8) rotor starting current,

$$I_{st} = \frac{E_2}{\sqrt{R_2^2 + X_2^2}}$$

and from Eq. (18.13) rotor current at full-load,

$$I_f = \frac{E_2}{\sqrt{\left(\frac{R_2}{s_f}\right)^2 + X_2^2}}$$

$$\text{so } \frac{I_{st}}{I_f} = \sqrt{\frac{(R_2/s_f)^2 + X_2^2}{R_2^2 + X_2^2}} = \sqrt{\frac{\left(\frac{R_2/X_2}{s_f}\right)^2 + 1}{\left(\frac{R_2}{X_2}\right)^2 + 1}} = \sqrt{\frac{\left(\frac{a}{s_f}\right)^2 + 1}{a^2 + 1}}$$

where $a = \frac{R_2}{X_2}$

$$\text{or } 5 = \sqrt{\frac{\left(\frac{a}{0.04}\right)^2 + 1}{a^2 + 1}} \quad \because I_{st} = 5\,I_f \text{ and } s_f = 0.04$$

or $a = 0.2$

(*i*) The slip at which maximum torque occur,

$s_{max\,T} = a = 0.2$ **Ans.**

From Eq. (18.20)

$$T_{max} = \frac{1}{2}\left(\frac{s_{max\,T}}{s_f} + \frac{s_f}{s_{max\,T}}\right) \times T_f$$

$$= \frac{1}{2}\left(\frac{0.2}{0.04} + \frac{0.04}{0.2}\right) \times T_f = 2.6 \text{ pu of full-laod torque}$$ **Ans.**

(*ii*) From Eq. (18.21) starting torque

$$T_{st} = \frac{2a}{a^2+1} T_{max}$$

$$= \frac{2 \times 0.2}{0.2^2 + 1} T_{max} = 0.3846\,T_{max} = 0.3846 \times 2.6\,T_f$$

$= T_f$ or 1.0 pu of full-load torque **Ans.**

Example 18.24. Explain constant (V/*f*) method of speed control of 3-phase induction motor. A 50 Hz, 3-phase induction motor has a rated voltage V_1. The motor's breakdown torque at rated voltage and frequency occurs at slip of 0.2. The motor is run from a 60 Hz supply of voltage V_2. The stator impedance can be neglected. Find the ratio of currents and torques at starting if $V_2 = V_1$.

[*U.P. Technical Univ. Electromechanical Energy Conversion-II, 2009-10*]

Solution: $s_{max\,T} = \frac{R_2'}{X_2'} = 0.2$

where R_2' and X_2' are the rotor resistance and standstill reactance at 50 Hz supply frequency referred to stator.

$$\text{Starting primary current, } I_{st(1)} = I'_{2(1)} = \frac{V_1}{\sqrt{(R_2')^2 + (X_2')^2}} \quad ...(i)$$

$$\text{Starting torque, } T_{st(1)} = \frac{K V_1^2 R_2'}{(R_2')^2 + (X_2')^2} \quad ...(ii)$$

At voltage V_2 and frequency $f_2 = 60$ Hz

Starting primary current,

$$I_{st(2)} = I'_{2(2)} = \frac{V_2}{\sqrt{(R_2')^2 + \left(\frac{6}{5}X_2'\right)^2}}$$

$$= \frac{V_2}{\sqrt{(R_2')^2 + (1.2\,X_2')^2}} \quad ...(iii)$$

$$\text{Starting torque, } T_{st(2)} = \frac{K V_2^2 R_2'}{(R_2')^2 + (1.2\,X_2')^2} \quad ...(iv)$$

Dividing Eqs. (*iii*) and (*iv*) respectively by Eqs. (*i*) and (*ii*) we have

$$\frac{I_{st(2)}}{I_{st(1)}} = \frac{V_2}{V_1}\sqrt{\frac{(R_2')^2+(X_2')^2}{(R_2')^2+(1.2)^2(X_2')^2}}$$

$$= \frac{V_2}{V_1}\sqrt{\frac{(s_{\max T})^2+1^2}{(s_{\max T})^2+1.2^2}} = 1\times\sqrt{\frac{0.2^2+1}{0.2^2+1.2^2}} = 0.838 \textbf{ Ans.}$$

$$\frac{T_{st(2)}}{T_{st(1)}} = \frac{V_2^2}{V_1^2}\times\frac{(s_{\max})^2+1^2}{(s_{\max})^2+1.2^2} = 1\times\frac{0.2^2+1^2}{0.2+1.2^2} = 0.7027 \textbf{ Ans.}$$

Example 18.25. An induction motor runs at a slip frequency of 2 Hz when supplied from a 3-phase, 400 V, 50 Hz supply. For the same developed torque, find the slip frequency at which it will run when supplied from a 3-phase, 340 V, 40 Hz system. Slip at which the machine develops maximum torque using 50 Hz supply is 0.1. Neglect stator impedance and assume linear torque-slip characteristic between zero torque and maximum torque in the working region.

[*GATE 1992*]

Solution: At 400 V, 50 Hz supply

$$\text{Slip, } s_1 = \frac{f'}{f} = \frac{2}{50} = 0.04$$

Ratio of rotor resistance to standstill reactance,

$$\frac{R_2}{X_2} = s_{\max T} = 0.1$$

$$\text{or } R_2 = 0.1\, X_2$$

Torque developed at slip $s_1 = 0.04$,

$$T_1 = \frac{K s_1 E_2^2}{R_2^2+s_1^2X_1^2} = \frac{K_1 s_1 V_1^2}{R_2^2+s_1^2X_1^2} \qquad \because E_2 \alpha V$$

$$= \frac{K_1\times 0.04\times(400)^2}{(0.1X_2)^2+(0.04X_2)^2} = \frac{6{,}400\,K_1}{0.0116\,X_2^2}$$

At 340 V, 40 Hz supply

Let the machine develop torque equal to T_1 at slip s_2

$$\text{So } \frac{K_1 s_2 V_2^2}{R_2^2+s_2^2X_2^2} = T_1$$

$$\text{or } \frac{K_1 s_2\times 340^2}{(0.1X_2)^2+s_2^2X_2^2} = \frac{6{,}400\,K_1}{0.0116\,X_2^2}$$

$$\text{or } \frac{1{,}15{,}600\, s_2}{0.01+s_2^2} = \frac{6{,}400}{0.0116}$$

$$\text{or } 6{,}400\, s_2^2 - 1{,}340.906\, s_2 + 64 = 0$$

$$\text{or } s_2^2 - 0.2095\, s_2 + 0.01 = 0$$

$$\text{or } s_2 = \frac{0.2095 \pm \sqrt{(0.2095)^2-0.04}}{2}$$

$$= 0.07356 \text{ or } 0.1359$$

Rejecting higher value,

$$\text{slip } s_2 = 0.07356$$

Slip frequency $f' = s_2 f = 0.07356 \times 40 = 2.94$ Hz **Ans.**

18.21 POWER LOSSES AND EFFICIENCY

The power losses in an induction motor fall in the same general classifications as in any dynamo. There are electrical losses due to heat produced by the current in the stator and rotor windings (usually called the *copper losses*), magnetic losses in the iron due to hysteresis and eddy currents (usually called the *iron* or *core losses*), mechanical losses due to friction in the bearings and ventilation (usually called the *friction* and *windage losses*) and stray load losses. There are copper losses and core losses in the stator, and copper losses and mechanical losses in the rotor. Actually there is some core loss in the rotor but under normal operating conditions, the rotor frequency is so low that rotor core loss may be neglected.

The iron and mechanical losses are considered to be constant and are based on no-load conditions.

Friction losses actually vary slightly with the load, especially if the load is belted to the motor. These losses also vary with the speed. But since the speed varies slightly over the operating range, these losses can be considered to remain constant without involving noticeable error.

The core losses are due to both the main and leakage fluxes. The magnitude and distribution of such losses vary with the load. With the increase in load, the leakage fluxes increase due to increase in currents and the mutual flux decrease. Thus the loss due to mutual flux is reduced and that due to leakage fluxes is increased with the increase in load. Because of this partially neutralization effect; it is quite accurate for practical purposes to assume the core losses as constant at all loads and speeds within the operating range of the motor under constant voltage conditions*. Except for the rotor tooth losses, practically all the core losses occur in the stator, owing to the low frequency of the rotor flux.

The copper losses, of course, vary with the load. The stator and rotor copper losses vary with the square of their respective currents.

The stray load losses include rotor iron loss, additional fundamental frequency and high frequency losses in the iron, strand and circulating current losses in the stator winding and harmonic losses in rotor conductors under load. Such losses are approximately 0.5 per cent of power supplied to the motor.

The standard definition of efficiency (output /input) applies to the induction motors and it is given as

$$\eta_m = \frac{\text{Mechanical power output in watts (or kW)}}{\begin{matrix}\text{Mechanical power output in watts (or kW)}\\ \text{+ fixed losses in watts (or kW)}\\ \text{+ copper losses in watts (or kW)}\end{matrix}}\times 100$$

$$\text{or } \eta_m = \frac{P_{mech}}{P_{mech}+P_{const}+P_{cu}}\times 100 \qquad \text{... (18.22)}$$

18.22 POWER FLOW DIAGRAM

Figure 18.26 gives a power flow diagram for a typical induction motor and clearly indicates how the electrical power supplied to the stator winding is converted into useful mechanical power available at the shaft.

Power input to the stator winding $= 3V_1 I_1 \cos\phi_1$ where V_1 is phase supply voltage, I_1 is stator current per phase and $\cos\phi_1$

* For small changes in supply voltage, such losses are assumed to vary in direct proportion of supply voltage.

is the stator power factor,

Some of the power supplied to the stator winding is lost as heat in the iron and in the copper coils. Stator copper loss is about 3.5% and stator iron loss is about 2.5% of power input to the motor. The remaining power (about 94% of power input to the motor) is transferred magnetically across the air gap to the rotor as to a transformer secondary. Part of this rotor power received (about 3.5% of power input to motor) is lost as heat in the rotor conductors and the rest is converted into mechanical power called the mechanical power developed in the rotor. This is about 90.5% of power input to the motor. Then there are friction and windage losses (about 2% of power input to the motor), as usual and the remaining power (about 88.5 per cent of power input to the motor) is available at the motor shaft, called the mechanical power output.

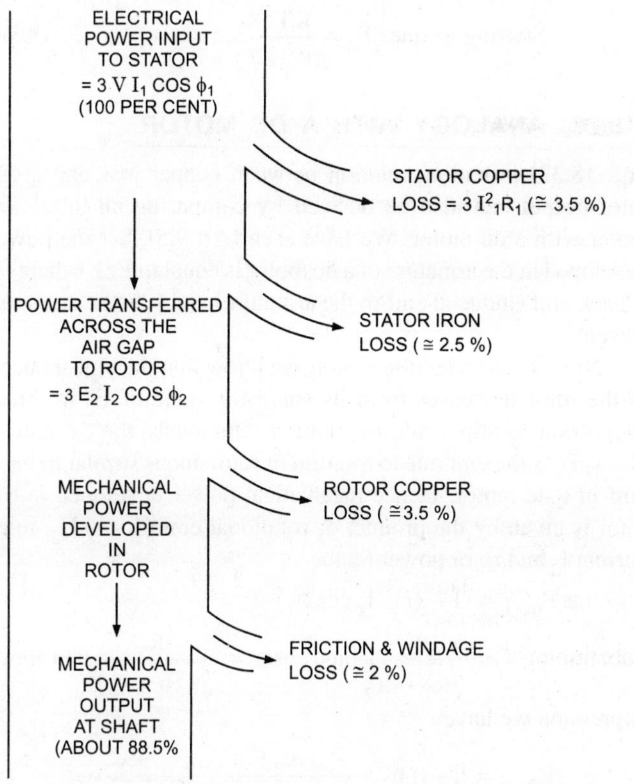

Fig. 18.26 *Power Flow Diagram For an Induction Motor*

18.23 POWER STAGES IN AN INDUCTION MOTOR

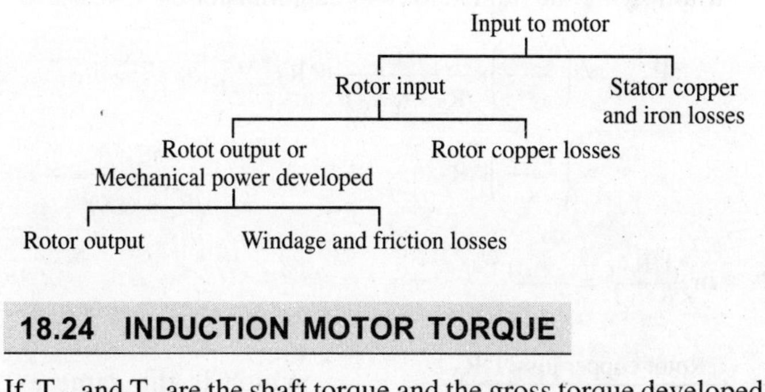

18.24 INDUCTION MOTOR TORQUE

If T_{sh} and T_g are the shaft torque and the gross torque developed in newton metres and N is the rotational speed in rpm, then

$$\text{Power output, } P_{out} = T_{sh} \times \frac{2\pi N}{60} \text{ watts} \quad ...(18.23)$$

and the mechanical power developed,

$$P_{mech} = T_g \times \frac{2\pi N}{60} \text{ watts} \quad ...(18.24)$$

$$\text{or } T_{sh} = \frac{P_{out}}{\frac{2\pi N}{60}} = 9.55 \frac{P_{out}}{N} \text{ N-m} \quad ...(18.25)$$

$$\text{and } T_g = \frac{P_{mech}}{2\pi N/60} = 9.55 \frac{P_{mech}}{N} \text{ N-m} \quad ...(18.26)$$

The difference between gross torque developed, T_g and shaft torque T_{sh} is equal to the torque lost due to friction and windage losses in the motor.

18.25 ROTOR OUTPUT

When a 3-phase induction motor is operating on a certain load, the power transferred from the stator to the rotor (*i.e.* rotor input) can be expressed in terms of the torque exerted on the rotor by the rotating flux and the synchronous speed at which this flux is rotating.

Power transferred from stator to rotor or input power to rotor or air–gap power,

$$P_2 = \frac{2\pi N_s}{60} T \text{ watts} \quad ...(18.27)$$

where T is the torque in newton-metres, exerted on the rotor by the rotating flux and N_s is the synchronous speed of rotating magnetic field in rpm.

This power transferred from the stator to the rotor (or power input to the rotor) is often referred to as the torque in synchronous watts.

Total mechanical power developed by rotor,

$$P_{mech} = \frac{2\pi N}{60} T \text{ watts} \quad ...(18.28)$$

where N is the speed of rotor, at full load, in rpm.

Rotor copper loss = Power input to rotor – output power of rotor (or mechanical power developed by rotor)

$$= \frac{2\pi N_s T}{60} - \frac{2\pi N T}{60}$$

$$= \frac{2\pi T}{60}(N_s - N) = \frac{2\pi T}{60} s N_s$$

$$= s \times \frac{2\pi N_s}{60} \times T = s \times \text{power input to rotor}$$

So total copper loss in rotor

$$= \text{Slip} \times \text{power input to rotor} \quad ...(18.29)$$

$$\text{or Power input to rotor} = \frac{\text{Total copper loss in rotor}}{s} \quad ...(18.30)$$

or Mechanical power developed by rotor,

$$P_{mech} = \text{Power input to rotor} - \text{rotor copper loss}$$

$$= \text{Power input to rotor} - \text{slip} \times \text{power input to rotor}$$

$$= (1 - s) \times \text{power input to rotor} \quad ...(18.31)$$

From Eqs. (18.29) and (18.31) we have

$$\frac{\text{Total copper loss in rotor}}{\text{Gross rotor output, } P_{mech}} = \frac{s}{1-s} \quad ...(18.32)$$

Hence from above equations

Power input to rotor : Mechanical power developed by rotor (gross rotor output power) : rotor copper loss

i.e. $P_2 : P_{mech} : P_{cu} :: 1 : (1-s) : s$ ...(18.33)

18.26 SYNCHRONOUS WATT

It is a new unit of torque and is defined as the torque which develops a power of 1 watt at the synchronous speed of the motor. When we say that the motor is developing a torque of 100 synchronous watts, we mean that the rotor input is 100 watts and that the torque is such that the power developed would be 100 watts provided the rotor is running synchronously and developing the same torque.

$$\text{Rotor input} = T_{sy} \times 2\pi N_s$$

$$\text{or } T_{sy} = \frac{\text{Rotor input in watts}}{2\pi N_s} \quad ...(18.34)$$

18.27 INDUCTION MOTOR TORQUE EQUATION

When a 3-phase induction motor is operating, on a certain load, with slip s, then

Induced emf per phase in rotor

$= sE_2$ volts where E_2 is standstill rotor induced emf

Rotor impedance per phase,

$$\mathbf{Z}_2 = R_2 + sX_2$$

where R_2 and X_2 are rotor resistance and standstill reactance per phase

Rotor current per phase,

$$I_2 = \frac{sE_2}{Z_2} = \frac{sE_2}{\sqrt{R_2^2 + s^2 X_2^2}}$$

Rotor circuit power factor,

$$\cos\phi_2 = \frac{R_2}{\sqrt{R_2^2 + s^2 X_2^2}}$$

Total rotor copper loss = Number of phases × rotor induced emf per phase × rotor current per phase × rotor circuit power factor

$$= 3 \times sE_2 \times \frac{sE_2}{\sqrt{R_2^2 + s^2 X_2^2}} \times \frac{R_2}{\sqrt{R_2^2 + s^2 X_2^2}}$$

$$= \frac{3R_2 s^2 E_2^2}{R_2^2 + s^2 X_2^2} \quad ...(18.35)$$

Also from Eq. (18.29)

Total rotor copper loss

$$= s \times \text{power input to rotor} = \frac{s \times T \times 2\pi N_s}{60}$$

$$\text{So } s \times T\, \frac{2\pi N_s}{60} = \frac{3R_2 s^2 E_2^2}{R_2^2 + s^2 X_2^2}$$

or Torque developed by rotor,

$$T = \frac{3 \times 60}{2\pi N_s} \cdot \frac{R_2 s E_2^2}{R_2^2 + s^2 X_2^2} \quad ...(18.36)$$

Substituting $E_2 = kE_1$ in above equation we have

Torque developed by rotor,

$$T = \frac{3 \times 60}{2\pi N_s} \times \frac{sK^2 E_1^2 R_2}{R_2^2 + (sX_2)^2} \quad ...(18.37)$$

In fact $\frac{3K^2}{2\pi N_s/60} = K'$, is called the constant of the given machine.

Hence, the torque equation [Eq. (18.37)] may be simplified to

$$T = K' \frac{sE_1^2 R_2}{R_2^2 + (sX_2)^2} \quad ...(18.38)$$

Starting torque can be had by substituting $s = 1$ in Eq. (18.38), and therefore,

$$\text{Starting torque, } T_{st} = \frac{K' E_1^2 R_2}{R_2^2 + X_2^2} \quad ...(18.39)$$

18.28 ANALOGY WITH A DC MOTOR

Eq. (18.32) providing relation between copper loss and gross rotor output can also be derived by comparing an induction motor with a dc motor. We have seen (Art 9.5) that the power developed in the armature of a dc motor is equal to $E_b I_a$ where E_b is back emf (induced emf in the armature) and I_a is the armature current.

Now, in an induction motor, we know that the emf induced in the rotor decreases from its standstill value of E_2 in direct proportion to slip while in rotation. Obviously the difference $(1-s)E_2$ is the emf due to rotation of rotor and is similar to back emf in a dc motor. Hence mechanical power developed in the rotor is given by the product of rotational emf $(1-s)E_2$, rotor current I_2 and rotor power factor

i.e. $P_{mech} = (1-s) E_2 I_2 \cos\phi_2$

Substituting $I_2 = \frac{sE_2}{\sqrt{R_2^2 + sX_2^2}}$ and $\cos\phi_2 = \frac{R_2}{\sqrt{R_2^2 + (sX_2)^2}}$ in above expression we have

$$P_{mech} = (1-s) E_2 \times \frac{sE_2}{\sqrt{R_2^2 + (sX_2)^2}} \times \frac{R_2}{\sqrt{R_2^2 + (sX_2)^2}}$$

$$= \frac{s(1-s) E_2^2 R_2}{R_2^2 + (sX_2)^2}$$

Multiplying the numerator and denominator by s, we have

$$P_{mech} = \left(\frac{1-s}{s}\right) \times \frac{s^2 E_2^2}{R_2^2 + (sX_2)^2} \times R_2$$

$$= \left(\frac{1-s}{s}\right) I_2^2 R_2 \qquad \because \frac{sE_2}{\sqrt{R_2^2 + (sX)^2}} = I_2$$

$$\text{or } \frac{I_2^2 R_2}{P_{mech}} = \frac{s}{1-s}$$

i.e. $\frac{\text{Rotor copper loss, } I_2^2 R_2}{\text{Gross rotor output, } P_{mech}} = \frac{s}{1-s}$, which is the same as Eq. (18.32).

Example 18.26. A squirrel-cage induction motor is rated 25 kW, 440 V, 3-ϕ, 50 Hz. On full load it draws 28.7 kW with line current 50 A and runs at 720 rpm. Calculate:

(*i*) the slip(s)

(*ii*) the power factor, and

(*iii*) the efficiency (η).

[*U.P. Technical Univ. Electromechanical Energy Conversion-II, 2013-14*]

Solution: The number of poles, even in number, to provide a synchronous speed nearest higher than actual speed of motor (720 rpm) that the motor must have is 8.

$$\because \quad P_m = \frac{120f}{N_m} \simeq \frac{120 \times 50}{720} \simeq 8$$

$$\text{Synchronous speed, } N_s = \frac{120f}{P} = \frac{120 \times 50}{8} = 750 \text{ rpm}$$

$$(i) \text{ The slip, } s = \frac{N_s - N}{N_s}$$

$$= \frac{750 - 720}{750} = 0.04 \text{ or } 4\% \textbf{ Ans.}$$

$$(ii) \text{ Power factor} = \frac{\text{Power drawn in kW} \times 1{,}000}{\sqrt{3}\, V_L I_L}$$

$$= \frac{28.7 \times 1{,}000}{\sqrt{3} \times 440 \times 50} = 0.753 \text{ (lagging)}$$

$$(iii) \text{ Efficiency, } \eta = \frac{\text{Power output}}{\text{Power input}} \times 100 = \frac{25}{28.7} \times 100 = 87.1\% \textbf{ Ans.}$$

Example 18.27. The power input to rotor of 440 V, 50 Hz, 6 pole, 3-phase induction motor is 75 kW. The rotor electromotive force is observed to make 110 complete alterations per minute. Calculate :

(*i*) slip

(*ii*) rotor speed

(*iii*) rotor copper loss per phase.

[*M.D. Univ. Electromechanical Energy Conversion, December-2010*]

Solution: Rotor frequency, $f' = \frac{110}{60} = 1.8333$ cycles per second or Hz

$$(i) \text{ Slip, } s = \frac{f'}{f} = \frac{1.8333}{50} = 0.03667 \quad \textbf{Ans.}$$

$$\text{Synchronous speed, } N_s = \frac{120f}{P} = \frac{120 \times 50}{6} = 1{,}000 \text{ rpm}$$

$$(ii) \text{ Rotor speed, } N = N_s(1 - s) = 1{,}000(1 - 0.03667)$$

$$= 963.3 \text{ rpm } \textbf{Ans.}$$

(*iii*) Rotor copper loss per phase = Slip × power input to rotor per phase

$$= 0.03667 \times \frac{75}{3}$$

$$= 0.912 \text{ kW or } 912 \text{ W } \textbf{Ans.}$$

Example 18.28. The power input to the 3-phase induction motor is 50 kW and the corresponding stator losses are 2 kW. Calculate the total mechanical power developed and the rotor I²R loss when the slip is 3%. [*U.P. Technical Univ. Energy Conversion, 2008-09*]

Solution: Power input to motor, P_1 = 50 kW

Stator losses = 2 kW

Power input to rotor, P_2 = Power output from stator

= Input to motor – stator losses

= 50 – 2 = 48 kW

Slip, s = 3% or 0.03

Total mechanical power developed,

$$P_{mech} = P_2(1 - s) = 48(1 - 0.03) = 46.56 \text{ kW } \textbf{Ans.}$$

Rotor copper loss = $s \times P_2 = 0.03 \times 48 = 1.44$ kW **Ans.**

Example 18.29. A 3-ϕ, 400 V, 6-pole, 50 Hz, IM develops mechanical power of 20 kW at 985 rpm. Calculate:

(*i*) the rotor copper loss,

(*ii*) the total input power, and

(*iii*) rotor frequency (f_2).

The stator losses are equal to 1,800 W. Neglect mechanical loss.

[*U.P. Technical Univ. Electromechanical Energy Conversion-II, 2013-14*]

Solution: Synchronous speed of motor,

$$N_s = \frac{120f}{P} = \frac{120 \times 50}{6} = 1{,}000 \text{ rpm}$$

$$\text{Slip, } s = \frac{N_s - N}{N_s} = \frac{1{,}000 - 985}{1{,}000} = 0.015$$

Mechanical power developed,

$$P_{mech} = 20 \text{ kW}$$

$$(i) \text{ Rotor copper loss} = \frac{sP_{mech}}{1 - s} = \frac{0.015 \times 20}{1 - 0.015}$$

$$= 0.3047 \text{ kW or } 304.7 \text{ W } \textbf{Ans.}$$

$$\text{Power input to rotor, } P_2 = \frac{P_{mech}}{1 - s} = \frac{20}{1 - 0.015} = 20.3047 \text{ kW}$$

(*ii*) Total power input, $P_1 = P_2$ + stator copper loss

= 20.3047 + 1.8 = 22.1047 kW **Ans.**

(*ii*) Rotor frequency, $f_2 = sf = 0.015 \times 50 = 0.75$ Hz **Ans.**

Example 18.30. The power input to the rotor of a 3-phase, 50 Hz, 6-pole, slip-ring induction motor is 40 kW and the motor runs at 960 rpm. The rotor resistance per phase is 0.25 Ω. Determine the rotor current per phase.

[*M.G. Univ. Electrical Machines-III, November-2005*]

Solution: Power input to rotor, P_2 = 40 kW

$$\text{Synchronous speed, } N_s = \frac{120f}{P} = \frac{120 \times 50}{6} = 1{,}000 \text{ rpm}$$

Motor speed, N = 960 rpm

$$\text{Slip, } s = \frac{N_s - N}{N_s} = \frac{1{,}000 - 960}{1{,}000} = 0.04$$

Rotor copper loss = $s \times P_2 = 0.04 \times 40 \times 1{,}000 = 1{,}600$ W

$$\text{Rotor current per phase, } I_2 = \sqrt{\frac{\text{Total rotor copper loss}}{3 \times \text{rotor resistance/phase}}}$$

$$\because \text{ total rotor copper loss} = 3I_2^2 R$$

$$\text{or } I_2 = \sqrt{\frac{1{,}600}{3 \times 0.25}} = 46.2 \text{ A } \textbf{Ans.}$$

Example 18.31. The input power to a 6-pole, 3-phase, 50 Hz induction motor is 42 kW, the speed is 970 rpm. The stator losses are 1.2 kW and friction and windage losses 1.8 kW. Find (*i*) the rotor copper losses and (*ii*) efficiency of motor.

[*M.D. Univ. Electromechanical Energy Conversion; December-2012; R.G. Technical Univ. Electromechanical Energy Conversion-I, June-2011; Electrical Machine-I, June-2014*]

Solution: Synchronous speed, $N_s = \frac{120f}{P} = \frac{120 \times 50}{6} = 1{,}000$ rpm

Motor speed, N = 970 rpm

$$\text{Slip, } s = \frac{N_s - N}{N_s} = \frac{1{,}000 - 970}{1{,}000} = 0.03$$

Power input to stator, $P_1 = 42$ kW

Stator losses = 1.2 kW

Power input to rotor, $P_2 = P_1$ – stator losses
$= 42 - 1.2 = 40.8$ kW

Rotor copper losses $= sP_2 = 0.03 \times 40.8 = 1.224$ kW **Ans.**

Power output, $P_{out} = P_2$ – (rotor copper losses + friction and windage losses)
$= 40.8 - (1.224 + 1.8) = 37.776$ kW

$$\text{Motor efficiency, } \eta = \frac{P_{out}}{P_1} \times 100 = \frac{37.776}{42} \times 100 = 89.94\% \text{ **Ans.**}$$

Example 18.32. The core loss in 3-phase induction motor is 100 W and equals the mechanical loss, the stator copper loss is 120 W when developing 1,500 W at the shaft output. What is efficiency of machine? Assume the slip as 4%.

[*M.D. Univ. Electromechanical Energy Conversion, December-2013*]

Solution: Shaft output, $P_{sh} = 1{,}500$ W

Mechanical power developed, $P_{mech} = P_{sh}$ + mechanical losses
$= 1{,}500 + 100 = 1{,}600$ W

$$\text{Power input to rotor, } P_2 = \frac{P_{mech}}{1 - s} = \frac{1{,}600}{1 - 0.04} = 1{,}666.67 \text{ W}$$

Power input to motor, P_{in} = Power input to stator
$= P_2$ + stator losses
$= 1{,}666.67 + 100 + 120$
$= 1{,}886.67$ W

$$\text{Efficiency of machine, } \eta = \frac{P_{sh}}{P_{in}} \times 100 = \frac{1{,}500 \times 100}{1{,}886.67} = 79.505\% \text{ **Ans.**}$$

Example 18.33. The power input to the rotor of a 440 V, 50 Hz, 3-phase, 6-pole induction motor is 50 kW. The rotor emf makes 120 cycles per minute. Friction and windage losses are 2 kW. Calculate (*i*) slip (*ii*) rotor speed (*iii*) rotor copper losses (iv) mechanical power developed (*v*) output power (*vi*) output torque.

[*U.P. Technical Univ. Electromechanical Energy Conversion-II, 2006-07*]

Solution: Supply frequency, $f = 50$ Hz

$$\text{Rotor emf frequency, } f' = \frac{120}{60} = 2 \text{ Hz}$$

$$(i) \text{ Slip, } s = \frac{f}{f'} = \frac{2}{50} = 0.04 \quad \text{**Ans.**}$$

(*ii*) Rotor speed, $N = N_s(1 - s)$

$$= \frac{120 \times f}{p}(1 - s) = \frac{120 \times 50}{6}(1 - 0.04)$$

$= 960$ rpm **Ans.**

Power input to rotor, $P_2 = 50$ kW

(*iii*) Rotor copper losses $= sP_2 = 0.04 \times 50 = 2$ kW

(*iv*) Mechanical power developed,

$P_{mech} = P_2(1 - s) = 50(1 - 0.04) = 48$ kW **Ans.**

(*v*) Output power, $P_{out} = P_{mech}$ – friction and windage losses
$= 48 - 2 = 46$ kW

$$(vi) \text{ Output torque, } T = \frac{P_{out} \times 60}{2\pi N} = \frac{46 \times 1{,}000 \times 60}{2\pi \times 960}$$

$= 457.57$ N-m **Ans.**

Example 18.34. A 3-phase, 500 V, 50-Hz induction motor with 6-poles gives an output of 20 kW at 950 rpm with a pf of 0.8. The mechanical losses are equal to 1 kW. Calculate for this load (*i*) slip (*ii*) rotor copper loss (*iii*) input if the stator losses are 1,500 W (*iv*) line current. [*R.G. Technical Univ. Electrical Machine-I, June-2014*]

Solution: Synchronous speed of motor,

$$N_s = \frac{120 f}{P} = \frac{120 \times 50}{6} = 1{,}000 \text{ rpm}$$

$$(i) \text{ Slip, } s = \frac{N_s - N}{N_s} = \frac{1{,}000 - 950}{1{,}000} = 0.05 \text{ or } 5\% \text{ **Ans.**}$$

Motor output, $P_{out} = 20$ kW

Rotor output, $P_{mech} = P_{out}$ + mechanical losses $= 20 + 1 = 21$ kW

$$\text{Power input to rotor, } P_2 = \frac{P_{mech}}{1 - s} = \frac{21}{1 - 0.05} = 22.1 \text{ kW}$$

(*ii*) Rotor copper loss $= P_2 - P_{mech} = 22.1 - 21 = 1.1$ kW **Ans.**

Stator losses = 1,500 W = 1.5 kW

(*iii*) Power input to stator, $P_1 = P_2$ + stator losses
$= 22.1 + 1.5 = 23.6$ kW **Ans.**

$$(iv) \text{ Line current, } I_L = \frac{P_1}{\sqrt{3}\, V_L \cos\phi} = \frac{23.6 \times 1{,}000}{\sqrt{3} \times 500 \times 0.8}$$

$= 34$ A **Ans.**

Example 18.35. An induction motor has an efficiency of 0.9 when the shaft load is 45 kW. At this load, stator ohmic loss and rotor ohmic loss each equal to the iron loss. The mechanical loss is one-third of the no-load losses. Neglect ohmic losses at no load. Calculate the slip.

[*J.N. Technological Univ. Electrical Machines-II, January-2014*]

Solution: Motor output, P_{out} = Shaft load = 45 kW

$$\text{Motor input, } P_{in} = \frac{P_{out}}{\eta} = \frac{45}{0.9} = 50 \text{ kW}$$

Total losses, that include (*i*) stator copper and iron losses (*ii*) rotor copper loss and (*iii*) rotor mechanical losses (rotor iron loss being negligible),

$= 50 - 45 = 5$ kW

No-load loss of an induction motor consists of (*i*) stator iron loss and (*ii*) mechanical losses, (stator copper loss at no load being negligible). Moreover, these two losses are independent of load.

Let each of the stator copper loss, rotor copper loss and stator iron loss be equal to P_i and mechanical loss be P_{wf}.

Then no-load loss $= P_i + P_{wf}$

So mechanical losses, P_{wf} one-third of no-load loss

$$= \frac{1}{3}(P_i + P_{wf})$$

$$\text{or } P_{wf} = \frac{P_i}{2}$$

Total losses = Stator copper losses + stator iron losses + rotor copper losses + mechanical losses

$$\text{or } 5 = 3 P_i + \frac{P_i}{2}$$

$$\text{or } P_i = \frac{5}{3.5} = 1.4286 \text{ kW}$$

Now rotor output = Motor output + mechanical losses

$$= 45 + \frac{1.4286}{2} = 45.7143 \text{ kW}$$

$$\text{Rotor input} = \text{Rotor output} + \text{rotor copper losses}$$

$$= 45.7143 + 1.4286 = 47.1429 \text{ kW}$$

$$\text{Slip, } s = \frac{\text{Rotor copper losses}}{\text{Rotor input}}$$

$$= \frac{1.4286}{47.1429} = 0.03 \text{ or } 3\ \% \textbf{ Ans.}$$

Example 18.36. A 50 hp, 6-pole, 50 Hz, slip-ring induction motor runs at 960 rpm on full-load with a rotor current of 40 A. Allowing 300 W for copper loss in the short-circuiting gear and 1,200 W for mechanical losses, find the resistance R_2 per phase of the 3-phase rotor winding. *[GATE 1998]*

Solution: Motor Output,

$$P_{out} = 50 \times 735.5 = 36{,}775 \text{ watts}$$

$$\text{Rotor output} = \text{Motor output} + \text{short-circuiting gear loss} + \text{mechanical losses}$$

$$= 36{,}775 + 300 + 1{,}200 = 38{,}275 \text{ watts}$$

$$\text{Synchronous speed, } N_s = \frac{120\,f}{P} = \frac{120 \times 50}{6} = 1{,}000 \text{ rpm}$$

$$\text{Slip, } s = \frac{N_s - N}{N_s} = \frac{1{,}000 - 960}{1{,}000} = 0.04$$

$$\text{Rotor copper losses} = 3I^2 R_2 = \frac{\text{Rotor output}}{1 - s} \times s$$

$$= \frac{38{,}275}{1 - 0.04} \times 0.04 = 1{,}594.8 \text{ W}$$

or Rotor resistance per phase,

$$R_2 = \frac{1{,}594.8}{3 \times I^2} = \frac{1{,}594.8}{3 \times 40^2} = 0.3322\ \Omega \textbf{ Ans.}$$

Example 18.37. The power input to a 6-pole, 50 Hz, 3-phase induction motor is 700 W at no load and 10 kW at full load. The no-load copper losses may be assumed negligible while the full-load stator and rotor copper losses are 295 W and 310 W respectively. Find the full-load speed, shaft torque and efficiency of the motor assuming rotational and core losses to be equal.

[Pb. Univ. Elec. Machines II, February, 1993]

Solution: Full-load input, P_{in} = 10 kW or 10,000 watts

Power input on no load = Core loss + rotational losses

$\because$ no-load copper losses are negligible

or 700 W = 2 × core loss (or rotational loss)

$\because$ both the rotational loss and core loss are equal,

$$\text{or Rotational loss} = \text{Core loss} = \frac{700}{2} = 350 \text{ W}$$

Stator core loss = Core loss = 350 W

assuming rotor core loss to be negligible

$$\text{Power input to rotor, } P_2 = P_{in} - \text{stator core loss} - \text{stator copper loss}$$

$$= 10{,}000 - 350 - 295 = 9{,}355 \text{ watts}$$

$$\text{Rotor output power} = P_2 - \text{rotor copper loss}$$

$$= 9{,}355 - 310 = 9{,}045 \text{ W}$$

$$\text{Full-load speed, } N = N_s \times \frac{\text{Rotor output power}}{\text{Power input to rotor}}$$

$$= \frac{120 \times 50}{6} \times \frac{9{,}045}{9{,}355} = 967 \text{ rpm } \textbf{Ans.}$$

$$\text{Motor output power} = \text{Rotor output power} - \text{mechanical loss}$$

$$= 9{,}045 - 350 = 8{,}695 \text{ watts}$$

$$\text{Shaft torque, } T_{sh} = \frac{\text{Motor output power in watts}}{\frac{2\pi N}{60}}$$

$$= \frac{8{,}695}{2\pi \times \frac{967}{60}} = 85.865 \text{ Nm } \textbf{Ans.}$$

$$\text{Motor efficiency, } \eta = \frac{\text{Output}}{\text{Input}} \times 100 = \frac{8{,}695}{10{,}000} \times 100$$

$$= 86.95\ \% \textbf{ Ans.}$$

Example 18.38. An 8-pole, 3-phase, 50-Hz induction motor running with a slip of 4% is taking 20 kW. Stator losses amount to 500 W. If the mechanical torque lost in friction is 16.25 N-m, find *(i)* rotor copper loss *(ii)* bhp *(iii)* efficiency.

[R.G. Technical Univ. Electromechanical Energy Conversion-I, June-2010]

Solution: Power input to stator, P_1 = Power drawn by motor, P_{in} = 20 kW

$$\text{Synchronous speed, } N_s = \frac{120f}{P} = \frac{120 \times 50}{8} = 750 \text{ rpm}$$

$$\text{Motor speed, } N = N_s(1 - s)$$

$$= 750\,(1 - 0.04) = 720 \text{ rpm}$$

Stator losses = 500 W = 0.5 kW

$$\text{Power input to rotor, } P_2 = P_1 - \text{stator losses}$$

$$= 20 - 0.5 = 19.5 \text{ kW}$$

$$(i) \text{ Rotor copper loss} = sP_2 = 0.04 \times 19.5 = 0.78 \text{ kW}$$

$$\text{Mechanical power developed, } P_{mech} = P_2 - \text{rotor copper loss}$$

$$= 19.5 - 0.78 = 18.72 \text{ kW}$$

$$\text{Mechanical losses} = \text{Mechanical torque lost in friction} \times \frac{2\pi N}{60}$$

$$= 16.25 \times \frac{2\pi \times 720}{60} = 1{,}225 \text{ W or } 1.225 \text{ kW}$$

$$\text{Power output, } P_{out} = P_{mech} - \text{mechanical losses}$$

$$= 18.72 - 1.225 = 17.495 \text{ kW}$$

$$(ii) \text{ BHP} = \frac{P_{out}}{0.7355} = \frac{17.495}{0.7355} = 23.8 \textbf{ Ans.}$$

$\because$ 1 bhp = 0.7355 kW

$$(iii) \text{ Efficiency, } \eta = \frac{P_{out}}{P_1} \times 100 = \frac{17.495}{20} \times 100 = 87.475\% \textbf{ Ans.}$$

Example 18.39. A 6-pole, 50-Hz, 3-phase induction motor running on full load and with a 4% slip, develops a torque of 15.21 kg-m at its pulley rim. The friction and windage losses are 200 W and stator copper and iron losses are 1,620 watts. Calculate *(i)* rotor copper loss and the efficiency at full load. *[R.G. Technical Univ. Electromechanical Energy Conversion-I, December-2012]*

Solution: Synchronous speed, $N_s = \dfrac{120f}{P} = \dfrac{120 \times 50}{6} = 1{,}000$ rpm

$$\text{Motor speed, } N = N_s(1 - s)$$

$$= 1{,}000\,(1 - 0.04) = 960 \text{ rpm}$$

Power output at the shaft = Torque at pulley in kg-m × 9.81 × $\dfrac{2\pi N}{60}$ watts

$$= \frac{15.21 \times 9.81 \times 2\pi \times 960}{60} = 15{,}000 \text{ W}$$

Mechanical power developed,

$$P_{mech} = \text{Shaft power} + \text{friction and windage losses}$$

$$= 15{,}000 + 200 = 15{,}200 \text{ W}$$

Power input to rotor, $P_2 = \frac{P_{mech}}{1-s} = \frac{15,200}{1-0.04} = 15,833$ W

Power input to stator, $P_1 = P_2$ + stator losses

$= 15,833 + 1,620 = 17,453$ W

(*i*) Rotor copper loss $= P_2 - P_{mech} = 15,833 - 15,200 = 633$ W **Ans.**

(*ii*) Efficiency at full load, $\eta_f = \frac{\text{Shaft power}}{P_1} \times 100$

$$= \frac{15,000}{17,453} \times 100 = 85.94\% \text{ **Ans.**}$$

Example 18.40. A 4-pole, 50-Hz 3-phase induction motor delivers a shaft torque of 110 N-m at full load running at 950 rpm. Calculate (*i*) rotor copper loss (*ii*) Power input to rotor. The mechanical losses account for 100 W. [*U.P.S.C. I.E.S. Electrical Engineering-II, 2011*]

Solution: Synchronous speed, $N_s = \frac{120f}{P} = \frac{120 \times 50}{4} = 1,500$ rpm

Motor speed, N = 950 rpm

Slip, $s = \frac{N_s - N}{N_s} = \frac{1,500 - 950}{1,500} = 0.3667$

Net mechanical output or shaft power

$$= \text{Shaft torque} \times \frac{2\pi N}{60} = \frac{110 \times 2\pi \times 950}{60} = 10,943 \text{ W}$$

Gross mechanical output of rotor,

P_{mech} = Shaft power + mechanical loss

$= 10,943 + 100 = 11,043$ W or 11.043 kW

(*i*) Rotor copper loss $= \frac{P_{mech}}{1-s} \times s = \frac{11.043}{1-0.3667} \times 0.3667$

$= 6.393$ kW **Ans.**

(*ii*) Power input to rotor $= \frac{P_{mech}}{1-s} = \frac{11.043}{1-0.3667} = 17.436$ kW **Ans.**

Example 18.41. A 440 V, 6-pole, 50 Hz, 3-phase delta-connected induction motor running on full load develops a useful torque of 160 N-m when the rotor emf makes 120 complete cycles/min. Determine the shaft power output. If the mechanical torque lost in friction and for core loss is 10 N-m, estimate:

(*i*) Copper loss in rotor windings

(*ii*) Input to the motor

(*iii*) Efficiency

The total stator loss is 1 kW.

[*U.P. Technical Univ. Electromechanical Energy Conversion-II 2004-05*]

Solution: Rotor frequency,

f' = Number of complete cycles per second

$= \frac{120}{60} = 2$ cycles per second or 2 Hz

Slip, $s = \frac{f'}{f} = \frac{2}{50} = 0.04$

Synchronous speed, $N_s = \frac{120f}{P} = \frac{120 \times 50}{6} = 1,000$ rpm

Useful torque, $T_{sh} = 160$ N-m

Mechanical torque lost in friction and core loss,

$T_f = 10$ N-m

Torque exerted on the rotor by the rotating flux

$T = T_{sh} + T_f = 160 + 10 = 170$ N-m

Power transferred from the stator to rotor,

$$P_2 = T \times \frac{2\pi N_s}{60} = 170 \times \frac{2\pi}{60} \times 1,000$$

$= 17,800$ W or 17.8 kW

(*i*) Copper loss in rotor windings

$= s \times P_2 = 0.04 \times 17.8 = 0.712$ kW **Ans.**

(*ii*) Input to motor, $P_{in} = P_2$ + stator losses

$= 17.8 + 1.0 = 18.8$ kW **Ans.**

Motor output or shaft power output

$$= T_{sh} \times \frac{2\pi N}{60}$$

$$= 160 \times \frac{2\pi}{60} \times 1,000\,(1 - 0.04)$$

$= 16,085$ W or 16.085 kW **Ans.**

(*iii*) Motor efficiency, $\eta = \frac{\text{Motor output}}{\text{Motor input}} \times 100$

$$= \frac{16.085}{18.8} \times 100 = 85.56\ \% \text{ **Ans.**}$$

Example 18.42. A 10 kW, 400 V, 3-phase, 4-pole, 50 Hz slip-ring induction motor develops rated output (i.e. 10 kW) at rated voltage and frequency and with its slip rings short circuited. The maximum torque equal to twice the full-load torque, occurs at a slip of 10% with zero external resistance in the rotor circuit. Stator resistance and rotational losses are neglected. Determine (*i*) slip and rotor speed at full-load torque (*ii*) rotor ohmic loss at full-load torque (*iii*) starting torque at rated voltage and frequency.

[*W.B. Univ. of Technology Electrical Machines-I, 2010*]

Solution: Slip corresponding to maximum torque, $s_{max\,T} = 10\%$ or 0.1

Ratio of maximum torque to full-load torque, $\frac{T_{max}}{T_f} = 2$

Let the full-load slip be s_f

From Eq. (18.20) we have

$$\frac{(s_{max\,T})^2 + (s_f)^2}{2\,s_{max\,T}\,s_f} = \frac{T_{max}}{T_f}$$

$$\text{so } \frac{0.1^2 + s_f^2}{2 \times 0.1\,s_f} = 2$$

or $s_f^2 - 0.4\,s_f + 0.01 = 0$

$$(i) \text{ or } s_f = \frac{0.4 \pm \sqrt{0.4^2 - 4 \times 1 \times 0.01}}{2}$$

$= 0.0268$ or 2.68% rejecting high value

Synchronous speed, $N_s = \frac{120f}{P} = \frac{120 \times 50}{4} = 1,500$ rpm

Rotor speed, $N = N_s(1 - s) = 1,500\,(1 - 0.0268)$

$= 1,460$ rpm

(*ii*) Mechanical power developed, P_{mech}

= Shaft power = 10 kW = 10,000 W

∵ rotational losses are negligible.

Rotor ohmic loss at full-load torque

$$= \frac{P_{mech}}{1-s} \times s = \frac{10,000 \times 0.0268}{1 - 0.0268} = 275.4 \text{ W **Ans.**}$$

(*iii*) Full-load torque,

$$T_f = \frac{\text{Air-gap power}}{2\pi N_s/60} = \frac{10{,}000/(1-0.0268)}{2\pi\times1{,}500/60} = 65.42 \text{ N-m}$$

and $$\frac{T_{st}}{T_{max}} = \frac{2}{\frac{s_{max\,T}}{1}+\frac{1}{s_{max\,T}}}$$

or Starting torque,

$$T_{st} = \frac{2}{0.1+\frac{1}{0.1}}\times T_{max} = \frac{2}{10.1}\times 2T_f = \frac{4}{10.1}\times 65.42$$

$$= 25.91 \text{ N-m } \textbf{Ans.}$$

Example 18.43. A 40 kW, 3-phase slip-ring induction motor of negligible stator impedance runs at a speed of 0.96 times synchronous speed at rated torque. The slip at maximum torque is 4 times the full-load value. If the rotor resistance of the motor is increased by 5 times, determine :

(*a*) The speed, power output and rotor copper loss at rated torque

(*b*) The speed corresponding to maximum torque.

[*GATE 1993*]

Solution: Motor output, P_{out} = 40 kW

Speed at rated torque, N = 0.96 N_s

$$\text{Full-load slip, } s_f = \frac{N_s - N}{N_s} = \frac{N_s - 0.96\,N_s}{N_s} = 0.04$$

Slip corresponding to maximum torque,

$$s_{max\,T} = 4\,s_f = 4\times 0.04 = 0.16$$

From Eq. (15.20)

$$= \frac{\text{Full-load torque}}{\text{Maximum torque}} = \frac{T_f}{T_{max}}$$

$$= \frac{2\,s_{max\,T}\,s_f}{(s_{max\,T})^2+s_f^2} = \frac{2\times0.04\times0.16}{(0.16)^2+(0.04)^2} = 0.4706$$

When the rotor circuit resistance is increased 5 times, the magnitude of maximum torque will remain unchanged because it is independent of load but the slip corresponding to the maximum torque will change. Let the new slip corresponding to maximum torque be $s'_{max\,T}$.

Since slip corresponding to maximum torque is proportional to rotor resistance provided its standstill reactance is fixed so

$$s'_{max\,T} = s_{max\,T}\times\frac{R'}{R} = s_{max\,T}\times 5 = 0.16\times5 = 0.8$$

Now $$\frac{T_f}{T_{max}} = \frac{2\times s'_f\,s'_{max\,T}}{(s'_f)^2+(s'_{max\,T})^2}$$

where s'_f is the new full-load slip

or $$0.4706 = \frac{2\times s'_f\times0.8}{(s'_f)^2+0.8^2}$$

or $$(s'_f)^2 - \frac{1.6}{0.4706}s'_f + 0.64 = 0$$

or $$(s'_f)^2 - 3.4\,s'_f + 0.64 = 0$$

or $$s'_f = \frac{3.4\pm\sqrt{(3.4)^2-4\times0.64}}{2} = 0.2$$

So new speed at full load,

$N' = (1 - s'_f)\,N_s = (1 - 0.2)\,N_s = 0.8\,N_s$ **Ans.**

Gross torque at full-load

$$= \frac{40\times1{,}000}{2\pi\times\frac{N}{60}} = \frac{40\times1{,}000}{2\pi\times0.96\,N_s/60}$$

New power output at full load

$$= \text{Gross torque at full-load}\times\frac{2\pi N'}{60}$$

$$= \frac{40\times1{,}000}{2\pi\times0.96\,N_s/60}\times\frac{2\pi N'}{60}$$

$$= \frac{40{,}000\times0.8\,N_s}{0.96\,N_s} = 33{,}333 \text{ W or } 33.333 \text{ kW } \textbf{Ans.}$$

Rotor copper loss at rated torque

$$= \frac{\text{Power output}}{1-s}\times s = \frac{33.333}{0.8}\times0.2 = 8.33 \text{ kW } \textbf{Ans.}$$

Speed corresponding to maximum torque

$= N_s(1 - s'_{max\,T}) = N_s(1 - 0.8) = 0.2\,N_s$ **Ans.**

Example 18.44. A 4 pole, 3-phase, 50 Hz, 230 V induction motor has a delta connected stator and a star connected rotor. Each phase of rotor winding has one-fourth the number of turns of each stator. The full-load speed is 1,455 rpm. The rotor resistance is 0.3 Ω and rotor standstill reactance is 1.0 Ω per phase. The rotor and stator windings are similar. Stator losses are equal to 50 watts. Friction and windage losses are equal to 30 W. Calculate :

(*i*) blocked rotor voltage per phase.

(*ii*) rotor current per phase under full-load running condition.

(*iii*) total rotor power input at full-load.

(*iv*) rotor gross loss at full-load.

(*v*) total mechanical power developed.

[*U.P. Technical Univ. Electrical Machines 2007-08*]

Solution: Stator emf per phase, $E_1 = V = 230$ V

∵ Motor is delta connected on stator side

(*i*) Blocked rotor voltage per phase

= Induced emf in rotor at standstill

$$= E_2 = \frac{E_1}{\text{Turn-ratio}} = \frac{230}{4} = 57.5 \text{ V } \textbf{Ans.}$$

$$\text{Synchronous speed, } N_s = \frac{120f}{P} = \frac{120\times50}{4} = 1{,}500 \text{ rpm}$$

$$\text{Full-load slip, } s = \frac{N_s - N}{N_s} = \frac{1{,}500-1{,}455}{1{,}500} = 0.03$$

(*ii*) Rotor current per phase under full-load running condition,

$$I_2 = \frac{sE_2}{\sqrt{R_2^2+s^2X_2^2}} = \frac{0.03\times57.5}{\sqrt{0.3^2+0.03^2\times1^2}} = 5.72 \text{ A } \textbf{Ans.}$$

(*iii*) Total rotor power input at full-load,

$$P_2 = \frac{3R_2 sE_2^2}{R_2^2+s^2X_2^2} = \frac{3\times0.3\times0.03\times57.5^2}{0.3^2+0.03^2\times1^2} = 982 \text{ W } \textbf{Ans.}$$

(*iv*) Rotor copper loss at full-load = sP_2 = 0.03 × 982 = 29.5 W

Rotor gross loss = sP_2 + friction and windage losses

= 29.5 + 30 = 59.5 W **Ans.**

(*v*) Total mechanical power developed = 982 – 59.5 = 922.5 W **Ans.**

Example 18.45. A 440 V, 3-phase, 50 Hz, 4-pole, star-connected induction motor has a full-load speed of 1,420 rpm. The rotor has an impedance of (0.3 + *j* 4.5) ohm and rotor/startor turn ratio of 0.8. Calculate

(*i*) Full-load torque.

(*ii*) Starting current.

(*iii*) Starting torque.

[*M.D. Univ. Electromechanical Energy Conversion, December-2011*]

Solution: Synchronous speed of motor,

$$N_s = \frac{120f}{P} = \frac{120 \times 50}{4} = 1{,}500 \text{ rpm}$$

Full-load speed, N_f = 1,420 rpm

$$\text{Full-load slip, } s_f = \frac{N_s - N_f}{N_s} = \frac{1{,}500 - 1{,}420}{1{,}500} = 0.05333$$

Voltage applied to stator per phase,

$$E_1 = \frac{440}{\sqrt{3}} = 254 \text{ V}$$

Induced emf per phase in rotor at standstill,

$$E_2 = KE_1 = 0.8 \times 254 = 203.2 \text{ V}$$

$$(i) \text{ Full-load torque, } T_f = \frac{3 \times 60}{2\pi N_s} \times \frac{R_2 s_f E_2^2}{R_2^2 + s_f^2 X_2^2}$$

$$= \frac{3 \times 60}{2\pi \times 1{,}500} \times \frac{0.3 \times 0.05333 \times (203.2)^2}{0.3^2 + (0.05333 \times 4.5)^2}$$

$$= \frac{0.06}{\pi} \times \frac{660.64}{0.1476} = 85.483 \text{ N-m} \quad \textbf{Ans.}$$

$$(ii) \text{ Starting rotor current, } I_{2st} = \frac{E_2}{Z_2} = \frac{203.2}{\sqrt{0.3^2 + 4.5^2}} = 45 \text{ A}$$

Starting current (stator current at starting)

$$= \frac{I_{2st}}{K} = \frac{45}{0.8} = 56.25 \text{ A} \quad \textbf{Ans.}$$

$$(iii) \text{ Starting torque, } T_{st} = \frac{3 \times 60}{2\pi N_s} \times \frac{R_2 \times E_2^2}{R_2^2 + X_2^2} \quad \because \text{ slip at start} = 1.0$$

$$= \frac{3 \times 60}{2\pi \times 1{,}500} \times \frac{0.3 \times (203.2)^2}{0.3^2 + 4.5^2}$$

$$= \frac{0.06}{\pi} \times \frac{12{,}387}{20.34} = 11.63 \text{ N-m} \quad \textbf{Ans.}$$

Example 18.46. A 3-phase induction motor has a 4-pole, star-connected stator winding. The motor runs on a 50 Hz supply with 200 V between lines. The rotor resistance and standstill rotor reactance per phase are 0.1 Ω and 0.9 Ω respectively. The ratio of rotor to stator turns in 0.67. Calculate:

(*i*) total torque at 4% slip

(*ii*) maximum torque developed

(*iii*) speed at maximum torque

(*iv*) maximum mechanical power.

Neglect stator impedance.

[*U.P. Technical Univ. Electromechanical Energy Conversion-II, 2007-08*]

Solution: Voltage applied per phase to stator, $E_1 = \frac{200}{\sqrt{3}} = 115.47$ V

EMF induced per phase in the rotor at standstill,

$$E_2 = K E_1 = 0.67 \times 115.47 = 77.36 \text{ V}$$

$$\text{Synchronous speed, } N_s = \frac{120f}{P} = \frac{120 \times 50}{4} = 1{,}500 \text{ rpm}$$

(*i*) Torque at 4% slip,

$$T = \frac{3 \times 60}{2\pi N_S} \times \frac{R_2 \times s \times E_2^2}{R_2^2 + s^2 X_2^2}$$

$$= \frac{3 \times 60}{2\pi \times 1{,}500} \times \frac{0.1 \times 0.04 \times (77.36)^2}{0.1^2 + 0.04^2 \times 0.9^2} = 40.475 \text{ N-m} \quad \textbf{Ans.}$$

(*ii*) Torque will be maximum when $s = \frac{R_2}{X_2} = \frac{0.1}{0.9} = 0.1111$

Substituting s = 0.1111 in equation

$$T = \frac{3 \times 60}{2\pi N_s} \times \frac{R_2 s E_2^2}{R_2^2 + s^2 X_2^2} \quad \text{we have}$$

$$T_{max} = \frac{3 \times 60}{2\pi N_s} \times \frac{R_2 \times \frac{R_2}{X_2} E_2^2}{R_2^2 + R_2^2} = \frac{3 \times 60}{2\pi N_s} \times \frac{E_2^2}{2X_2}$$

$$= \frac{3 \times 60}{2\pi \times 1{,}500} \times \frac{(77.36)^2}{2 \times 0.9} = 63.5 \text{ N-m} \quad \textbf{Ans.}$$

(*iii*) Slip corresponding to maximum torque,

$$s_{max\,T} = \frac{R_2}{X_2} = \frac{0.1}{0.9} = 0.1111$$

Speed corresponding to maximum torque,

$$N_{max\,T} = N_s(1 - s_{max\,T})$$

$$= 1{,}500(1 - 0.1111) = 1{,}333.3 \text{ rpm} \quad \textbf{Ans.}$$

(*iv*) Maximum mechanical power,

$$P_{mech\,max} = \frac{2\pi N_{max\,T}}{60} \times T_{max}$$

$$= \frac{2\pi \times 1{,}333.3}{60} \times 63.5 = 8{,}866 \text{ watts or } 8.866 \text{ kW} \quad \textbf{Ans.}$$

Example 18.47. A 4-pole, 400 V, 3 phase induction motor has a standstill reactance of 0.5 ohm per phase, standstill rotor emf of 100 V per phase and rotor resistance of 50 milliohm per phase. Calculate the maximum torque and the slip at which it occurs. Neglect the stator impedance.

[*U.P. Technical Univ. Electromechanical Energy Conversion-II, 2009-10*]

Solution: Rotor induced emf at standstill, E_2 = 100 V

Slip at standstill, s = 1

Rotor resistance, R_2 = 0.05 Ω per phase

Rotor standstill reactance, X_2 = 0.5 Ω per phase

$$\text{Synchronous speed, } N_s = \frac{120f}{P} = \frac{120 \times 50}{4} = 1{,}500 \text{ rpm}$$

Substituting the above values for s, R_2, X_2, N_s and E_2 in Eq. (5.35) we have starting torque

$$T_{st} = \frac{3 \times 60}{2\pi N_s} \times \frac{R_2 s E_2^2}{R_2^2 + s^2 X_2^2}$$

$$= \frac{3 \times 60}{2\pi \times 1{,}500} \times \frac{0.05 \times 1 \times 100^2}{0.05^2 + 1^2 \times 0.5^2} = 47.16 \text{ N-m}$$

Slip corresponding to maximum torque,

$$s_{max\,T} = \frac{R_2}{X_2} = \frac{0.05}{0.5} = 0.1 \text{ Ans.}$$

Ratio of rotor resistance to rotor standstill reactance,

$$a = \frac{R_2}{X_2} = \frac{0.05}{0.5} = 0.1$$

$$\text{Maximum torque, } T_{max} = T_{st} \times \frac{a^2 + 1}{2a} \quad \text{(Refer to Eq. (5.20))}$$

$$= 47.16 \times \frac{0.1^2 + 1}{2 \times 0.1} = \frac{47.16 \times 1.01}{0.2}$$

$$= 238.16 \text{ N-m} \quad \textbf{Ans.}$$

Example 18.48. A 4-pole, 50 Hz, 3-ϕ, IM has a rotor resistance of 4.5 Ω/phase and standstill reactance of 8.5 Ω/phase. With no external resistance in the rotor circuit, the starting torque of the motor is 85 N-m.

(i) What is the rotor voltage at standstill?

(ii) What would be the starting torque if 3Ω resistance were added in each rotor phase?

(iii) Neglecting stator voltage drop, what would be the induced rotor voltage and the torque at a slip of 0.03?

[*U.P. Technical Univ. Electromechanical Energy Conversion-II, 2013-14*]

Solution: Synchronous speed, $N_s = \frac{120f}{P} = \frac{120 \times 50}{4} = 1{,}500$ rpm.

Since torque developed by 3-ϕ induction motor at standstill is given as

$$T_s = \frac{3 \times 60}{2\pi N_s} \cdot \frac{R_2 E_2^2}{R_2^2 + X_2^2} = 85$$

or Rotor voltage at standstill,

$$E_2 = \sqrt{\frac{2\pi N_s \times 85}{3 \times 60} \times \frac{R_2^2 + X_2^2}{R_2}}$$

$$= \sqrt{\frac{2\pi \times 1{,}500}{3 \times 60} \times 85 \times \frac{4.5^2 + 8.5^2}{4.5^2}} = 302.46 \text{ V/phase} \quad \textbf{Ans.}$$

(*ii*) When 3Ω resistance is inserted in each rotor phase, R_2 becomes 4.5 + 3, i.e., 7.5Ω and starting torque would be

$$T_s = \frac{3 \times 60}{2\pi N_s} \cdot \frac{R_2 E_2^2}{R_2^2 + X_2^2} = \frac{3 \times 60}{2\pi \times 1{,}500} \times \frac{7.5 \times 302.46^2}{7.5^2 + 8.5^2}$$

$$= 101.98 \text{ N-m} \quad \textbf{Ans.}$$

(*iii*) Induced rotor voltage at a slip of 0.03

$$= sE_2 = 0.03 \times 302.46 = 9.07 \text{ V per phase} \quad \textbf{Ans.}$$

Torque developed at a slip of $s = 0.03$

$$= \frac{3 \times 60}{2\pi N_s} \times \frac{R_2 s E_2^2}{R_2^2 + s^2 X_2^2}$$

$$= \frac{3 \times 60}{2\pi \times 1{,}500} \times \frac{4.5 \times 0.03 \times 302.46^2}{4.5^2 + (0.03 \times 8.5)^2} = 11.16 \text{ N-m} \quad \textbf{Ans.}$$

18.29 EQUIVALENT CIRCUIT OF AN INDUCTION MOTOR

The energy is transferred from primary (stator) winding to secondary (rotor) winding entirely by induction, therefore, induction motor is essentially a transformer. At standstill, the induction motor is actually a static transformer having its secondary (rotor) winding short circuited.

When the motor operates at slip *s*, the frequency of rotor currents is *s* times the frequency of the stator currents, therefore, the revolving field produced by the rotor currents revolves with respect to rotor itself at speed

$$\frac{120sf}{P} = \frac{120f}{P} \times s = N_s \times s,$$

Mechanical speed of the motor, $N = N_s(1 - s)$. The speed of the revolving field of the rotor with respect to stator or space is obtained by combining the rotational speed of rotor field with respect to rotor with mechanical speed of the rotor. Hence speed of rotor revolving field with respect to stationary stator or space $= s N_s + N_s(1 - s) = N_s$.

Hence from the point of view of stator, the induction motor still can be considered as static transformer, even when its rotor is rotating, and it is possible to represent the performance of an induction motor by a transformer phasor diagram. Actually the rotor field does not exist alone but combines with the revolving field of stator to produce a resultant field, just as in the transformer, the resultant field is produced by the combination of primary and secondary ampere-turns.

In the transformer the load on secondary is electrical and in an induction motor the load is mechanical which can be replaced by an equivalent electrical load of resistance R'_L given by

$$R'_L = \frac{R_2}{K^2}\left(\frac{1}{s} - 1\right) \quad \text{...(Refer to Art. 18.31),}$$

where R_2 is the resistance per phase of rotor and K is the turn-ratio of the secondary (rotor) to primary (stator).

The equivalent circuit for a polyphase induction motor is shown in Fig. 18.27 where V is applied voltage per phase; R_1 and X_1 are stator resistance and leakage reactance per phase respectively; R_2 and X_2 are rotor resistance and standstill leakage reactance per phase respectively; K is the turn-ratio of secondary to primary ; R_0 and X_0 are the no-load resistance and reactance per phase respectively, with the rotor being driven at synchronous speed.

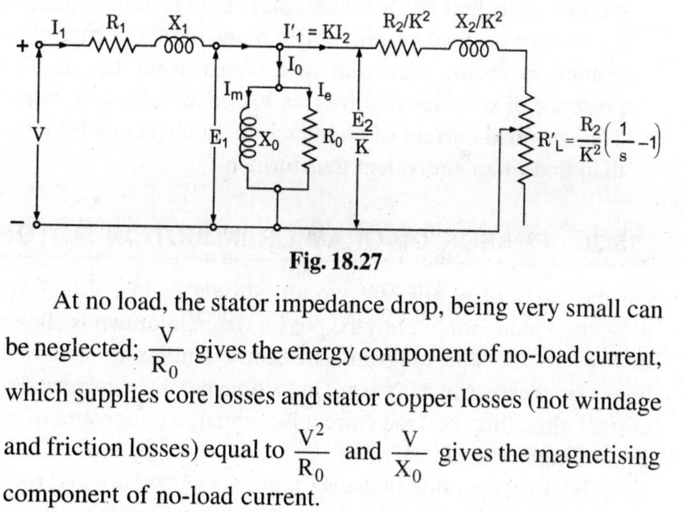

Fig. 18.27

At no load, the stator impedance drop, being very small can be neglected; $\frac{V}{R_0}$ gives the energy component of no-load current, which supplies core losses and stator copper losses (not windage and friction losses) equal to $\frac{V^2}{R_0}$ and $\frac{V}{X_0}$ gives the magnetising component of no-load current.

E_1 is the emf induced per phase in the stator, and is equal to applied voltage per phase minus impedance drop in primary (stator) winding. E_2 is the emf induced in the rotor, if it is assumed that the mechanical load is replaced by equivalent electrical load, as shown in Fig. 18.27.

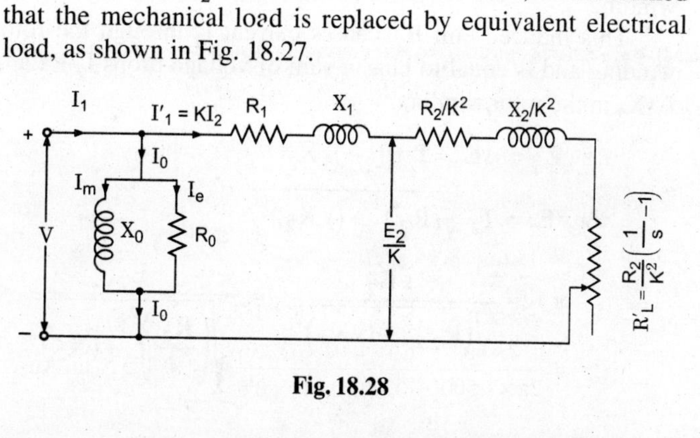

Fig. 18.28

The equivalent circuit can be simplified by transferring no-load current components to the supply terminals, as shown in Fig. 18.28.

18.29.1. Comparison Between 3-Phase Induction Motor and Transformer. The important differences between a 3-phase induction motor and transformer are as given below:

1. Because of distributed windings in an induction motor, the ratio of stator and rotor currents is not equal to the ratio of turns per phase in the rotor and stator windings.
2. In an induction motor the magnetic leakage and leakage reactances of rotor and stator are higher than those in a transformer.
3. The friction and windage losses are also present in induction motor, so the efficiency is lower than that of a transformer.
4. No-load current in an induction motor varies from about 30 to 50 per cent of full-load current whereas in transformer it varies from 2 to 5 per cent of full-load primary current. This is because of the fact that the mutual flux in a transformer completes its path through low-reluctance core, whereas the mutual flux in an induction motor has to cross the air gap between stator and rotor. Thus the reluctance offered to mutual flux in an induction motor is much higher than that in a transformer and, therefore, more magnetizing force (mmf) is required to develop the required mutual flux in case of induction motor than that in a transformer. It means that the magnetising current is large for the induction motor as compared to that for a transformer. Energy component of exciting or no-load current is also much higher in case of an induction motor than that in a transformer because of presence of windage and friction losses in induction motor. Thus no-load current of an induction motor is much higher than that of an equivalent transformer.

18.30 PHASOR DIAGRAM OF INDUCTION MOTOR

On the basis of equivalent circuit, shown in Fig. 18.27 for polyphase induction motor, the phasor diagram drawn is shown in Fig. 18.29. The diagram is exactly similar to that of a transformer except that in case of induction motor the magnetizing current, therefore, no-load current is very large on account of air gap (of high reluctance).

The revolving flux in the air gap links with stator and rotor windings and induces emfs in them. At any load, when the motor is operating at slip s, the induced emf in the rotor winding $E_2' = sE_2$ where E_2 is the standstill induced emf per phase in rotor winding.

This induced emf E_2' causes current I_2 through the rotor winding and is equal to phasor sum of voltage drops $I_2 R_2$ and $I_2 sX_2$ in the rotor winding.

$$i.e. \quad \mathbf{E'}_2 = s\mathbf{E}_2 = \mathbf{I}_2 (R_2 + js X_2)$$

$$\text{or } s E_2 = I_2 \sqrt{(R_2)^2 + (s X_2)^2}$$

$$\text{or } I_2 = \frac{s E_2}{\sqrt{(R_2)^2 + (sX_2)^2}} = \frac{E_2}{\sqrt{\left(\frac{R_2}{s}\right)^2 + (X_2)^2}}$$

From the above equation, the rotor having true resistance R_2, variable leakage reactance sX_2 and variable induced emf sE_2 appears as a rotor having fixed reactance X_2, constant induced emf E_2 and variable resistance $\frac{R_2}{s}$.

$$\text{Total power supplied to rotor} = 3 I_2^2 \frac{R_2}{s}$$

$$\text{Power lost in rotor} = 3 I_2^2 R_2$$

$$\text{Mechanical power developed} = 3 I_2^2 \left(\frac{R_2}{s} - R_2\right)$$

$$= 3 I_2^2 R_2 \left(\frac{1}{s} - 1\right) = 3 I_2^2 R_L$$

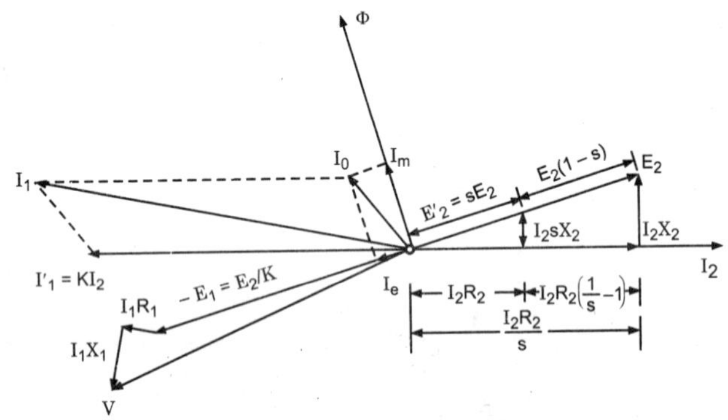

Fig. 18.29 *Phasor Diagram For a Polyphase Induction Motor*

Hence mechanical power developed can be replaced by a pure resistive load of equivalent resistance R_L given by the relation $R_L = R_2\left(\frac{1}{s} - 1\right)$ *and induction motor may be considered* just as static transformer, in which stator winding acts as primary, rotor winding acts as secondary and load resistance $R_L = R_2\left(\frac{1}{s} - 1\right)$ equivalent to mechanical load acts as an electrical load connected across the terminals of the secondary winding.

In phasor diagram shown in Fig. 18.29, $\frac{I_2 R_2}{s}$ is the total resistive drop, I_2X_2 is the total leakage reactance drop where X_2 is the rotor standstill reactance per phase. The phasor sum of voltage drops in rotor *i.e.* of $\frac{I_2 R_2}{s}$ and $I_2 X_2$ gives emf E_2 induced in rotor. The difference of emf E_2 and actual emf sE_2 induced in the rotor *i.e.* $E_2 (1 - s)$ is due to slip and is known as *speed emf*, which would be induced in the rotor by its cutting the air-gap flux at actual rotor speed. Although under running conditions no such emf exists, but it is necessary to consider it in the operation of replacing the mechanical load by electrical one.

As in transformer, an emf E_1 equal to $\frac{E_2}{K}$ is induced in the stator. The air-gap flux Φ leads the induced emfs E_1 and E_2 by 90°. I_m and I_e are the magnetising and energy components of no-load current I_0. Stator current I_1 is the phasor sum of I_1' and I_0 where I_1' is the counterbalancing secondary (rotor) current and is equal to $K I_2$. I_1R_1 and I_1X_1 are the resistive and leakage reactance drops in the stator winding, V is the applied voltage per phase and is equal to the phasor sum of reversed induced emf E_1 and impedance drop in stator winding.

18.31 MAXIMUM POWER OUTPUT

Equivalent circuit of the induction motor shown in Fig. 18.28 is further simplified as shown in Fig. 18.30.

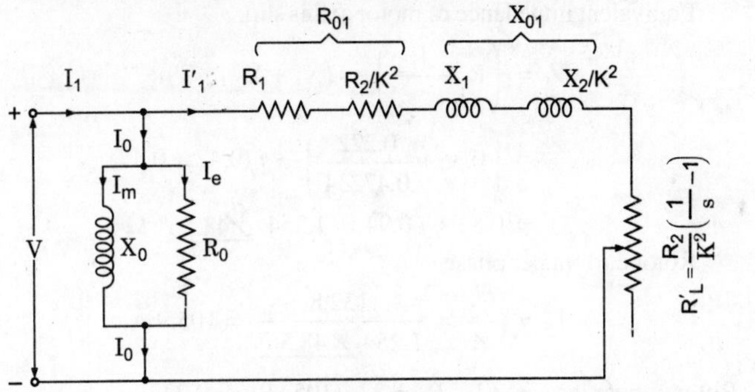

Fig 18.30 *Approximate Equivalent Circuit of an Induction Motor*

where $R_{01} = R_1 + R_2/K^2$ and $X_{01} = X_1 + X_2/K^2$

$$\text{Load resistance, } R'_L = \frac{R_2}{K^2}\left(\frac{1}{s}-1\right)$$

$$\text{Gross mechanical power output, } P_g = 3\, I_1'^2 R'_L$$

$$\text{But, } I'_1 = \frac{V}{\sqrt{(R_{01}+R'_L)^2+(X_{01})^2}}$$

$$\therefore P_g = \frac{3V^2 R'_L}{(R_{01}+R'_L)^2+(X_{01})^2} \quad \text{... (18.40)}$$

The condition for maximum power output can be determined by differentiating the above equation and by equating the first derivative to zero. If it is done, it will be found that

$$R'_L = \sqrt{(R_{01})^2+(X_{01})^2} = Z_{01}$$

Hence the gross mechanical power output of the motor is maximum when the equivalent load resistance R'_L is equal to the standstill leakage impedance of the motor Z_{01}.

Substituting $R'_L = Z_{01}$ in Eq. (18.40) we get

$$P_{g\,max} = \frac{3V^2 Z_{01}}{(R_{01}+Z_{01})^2+(X_{01})^2}$$

$$= \frac{3V^2 Z_{01}}{(R_{01})^2+(X_{01})^2+(Z_{01})^2+2R_{01}Z_{01}}$$

$$= \frac{3V^2}{2(R_{01}+Z_{01})} \quad \text{... (18.41)}$$

The slip corresponding to maximum gross mechanical power output will be given by the expression

$$Z_{01} = R'_L = \frac{R_2}{K^2}\left(\frac{1}{s}-1\right)$$

or slip corresponding to maximum gross mechanical power output,

$$s = \frac{\frac{R_2}{K^2}}{\frac{R_2}{K^2}+Z_{01}} \quad \text{... (18.42)}$$

18.32 MAXIMUM TORQUE

Equivalent circuit of an induction motor shown in Fig. 18.30 may be redrawn as shown in Fig. 18.31.

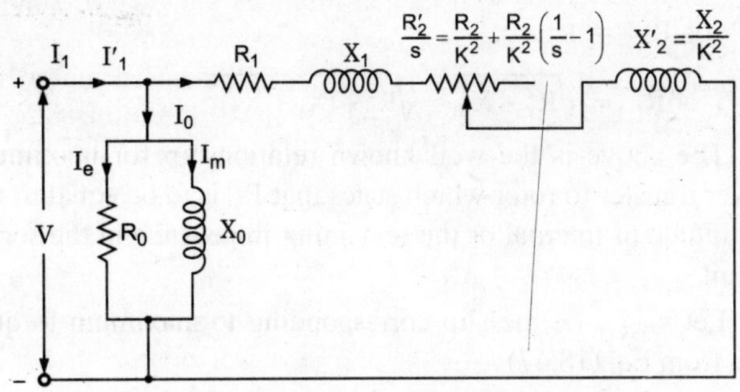

Fig. 18.31 *Modified Equivalent Circuit of a 3-Phase Induction Motor*

From the circuit shown in Fig. 18.31.

$$\text{Current, } I'_1 = \frac{V}{\sqrt{\left(R_1+\frac{R'_2}{s}\right)^2+(X_1+X'_2)^2}}$$

Power input to rotor

$$= 3(I'_1)^2\frac{R'_2}{s} = \frac{3V^2\frac{R'_2}{s}}{\left(R_1+\frac{R'_2}{s}\right)^2+(X_1+X'_2)^2} \quad \text{...(18.43)}$$

$$\text{or } T = \frac{3\times 60}{2\pi N_s}\times\frac{V^2\frac{R'_2}{s}}{\left(R_1+\frac{R'_2}{s}\right)^2+(X_1+X'_2)^2} \quad \text{...(18.44)}$$

At small slips $\frac{R'_2}{s} > R_1, X_1$ and X'_2 and therefore, neglecting R_1, X_1 and X'_2 compared to $\frac{R'_2}{s}$ at small slips, Eq. (18.44) may be written as

$$T = \frac{3\times 60}{2\pi N_s}\times\frac{V^2\frac{R'_2}{s}}{\left(\frac{R'_2}{s}\right)^2} = \frac{3\times 60}{2\pi N_s}\times\frac{V^2 s}{R'_2} \quad \text{...(18.45)}$$

Equation (18.44) may be written as

$$T = \frac{\frac{3V^2R'_2}{s}}{(R_1+R_L)^2+X_{01}^2}$$

$$= \frac{3V^2R_L}{(R_1+R_L)^2+X_{01}^2} \text{ synchronous watts} \quad \text{...(18.46)}$$

where $R_L = R'_2/s$ and $X_{01} = X_1 + X'_2$

Again the torque may be considered in terms of the variable resistance $\frac{R'_2}{s} = R_L$ of Fig. 18.31. For maximum torque or maximum power transfer to rotor or R_L we have to equate

$$\frac{dT}{dR_L} = 0.$$

From Eq. (18.46)

$$\frac{dT}{dR_L} = \frac{3V^2\{(R_1+R_L)^2+X_{01}^2-2R_L(R_1+R_L)\}}{\{(R_1+R_L)^2+X_{01}^2\}^2} = 0.$$

$$\text{or } R_L^2 = R_1^2 + X_{01}^2$$

$$\text{or } R_L = \sqrt{R_1^2+X_{01}^2} = \sqrt{R_1^2+(X_1+X_2')^2} \quad ...(18.47)$$

The above is the well known relationship for maximum power transfer to rotor which states that R_L is to be equal to the magnitude of internal or the remaining impedance in the series circuit.

Let $s_{max\,T}$ be the slip corresponding to maximum torque, then from Eq. (18.47)

$$R_L = \frac{R_2'}{s_{max\,T}} = \sqrt{R_1^2+(X_1+X_2')^2}$$

$$\text{or } s_{max\,T} = \frac{R_2'}{\sqrt{R_1^2+(X_1+X_2')^2}} \quad ...(18.48)$$

Substituting the value of R_L from Eq. (18.47) in Eq. (18.46) we have

Maximum torque,

$$T_{max} = \frac{3V^2\sqrt{R_1^2+X_{01}^2}}{2\left[R_1\sqrt{R_1^2+X_{01}^2}+(R_1^2+X_{01}^2)\right]}$$

$$\text{or } T_{max} = \frac{3V^2}{2(R_1+\sqrt{R_1^2+X_{01}^2})}$$

$$= \frac{3V^2}{2[R_1+\sqrt{R_1^2+(X_1+X_2')^2}]} \text{ synchronous watts}$$

$$\text{or } T_{max} = \frac{3\times 60}{2\pi N_S}\times\frac{V^2}{2}\times\frac{1}{R_1+\sqrt{R_1^2+(X_1+X_2')^2}} \quad ...(18.49)$$

Mechanical power developed

$$P_{mech} = \frac{3V^2(1-s)\frac{R_2'}{s}}{\left(R_1+\frac{R_2'}{s}\right)^2+(X_1+X_2')^2} \text{ watts} \quad ...(18.50)$$

The power at the slip corresponding to maximum torque is given by

$$P_{max\,T} = T_{max}(1-s_{max\,T})$$

$$= \frac{3V^2[\sqrt{R_1^2+(X_1+X_2')^2}-R_2']}{2\sqrt{R_1^2+(X_1+X_2')^2}[\sqrt{R_1^2+(X_1+X_2')^2}+R_1]} \text{ watts}$$

...(18.51)

However, this is not the maximum power. The maximum power occurs at a different slip as discussed in Art. 18.31.

Example 18.49. A 7.5 kW, 230 V, 3-phase, star-connected, 50 Hz, 4-pole, squirrel cage induction motor has its full-load internal torque at a slip of 0.04. The parameters of the motor are:

R_1 = 0.36 Ω /phase; R_2 = 0.222 Ω/phase; $X_1 = X_2$ = 0.47 Ω/phase; X_m = 15.5 Ω/phase. Assume that the shunt branch is connected across the supply terminals. Find (*a*) maximum internal torque at rated voltage and frequency (*b*) slip at maximum torque (*c*) internal starting torque at rated voltage and frequency. [*GATE 1996*]

Solution : It is assumed that there is unity turn -ratio between stator and rotor.

$$\text{Phase voltage, } V = \frac{230}{\sqrt{3}} = 132.8 \text{ V}$$

At maximum torque

(*b*) Slip, $s_{max\,T} = \frac{R_2}{X_2} = \frac{0.222}{0.47} = 0.47234$ **Ans.**

Equivalent impedance of motor at this slip,

$$\mathbf{Z} = \left(R_1+\frac{R_2}{s}\right)+j(X_1+X_2)$$

$$= \left(0.36+\frac{0.222}{0.47234}\right)+j(0.47+0.47)$$

$$= 0.83 + j\,0.94 = 1.254\angle 48.56^\circ\ \Omega$$

Rotor current per phase,

$$I_2 = \frac{E_2}{Z} = \frac{132.8}{1.254\angle 48.56^\circ} = 105.9 \text{ A}$$

$$\text{Rotor copper loss} = 3I_2^2R_2 = 3\times(105.9)^2\times 0.222 = 7{,}469 \text{ watts}$$

Power input to rotor,

$$P_2 = \frac{\text{Rotor copper loss}}{s_{max\,T}} = \frac{7{,}469}{0.47234} = 15{,}813 \text{ watts}$$

Synchronous speed,

$$N_s = \frac{120f}{P} = \frac{120\times 50}{4} = 1{,}500 \text{ rpm}$$

Maximum torque,

$$T_s = \frac{9.55\times P_2}{N_s} = \frac{9.55\times 15{,}813}{1{,}500} = 100.7 \text{ N-m } \textbf{Ans.}$$

At standstill

Slip, s = 1.0

Equivalent motor impedance,

$$\mathbf{Z} = (R_1+R_2)+j(X_1+X_2)$$

$$= (0.36+0.222)+j(0.47+0.47)$$

$$= 0.582 + j\,0.94 = 1.106\angle 58.24^\circ\ \Omega$$

$$\text{Rotor current, } I_2 = \frac{E_2}{Z} = \frac{132.8}{1.106} = 120.1 \text{ A}$$

Power input to rotor,

$$P_2 = \text{Total rotor copper losses} = 3\times(120.1)^2\times 0.222 = 9{,}608 \text{ watts}$$

$$\text{Starting torque, } T_{st} = \frac{9.55\times P_2}{N_s} = \frac{9.55\times 9{,}608}{1{,}500} = 61.2 \text{ N-m } \textbf{Ans.}$$

Example 18.50. A 25 hp, 400 V, 50 Hz, 4-pole, star-connected induction motor has the following impedances per phase in ohms referred to the stator side:

R_s = 0.641, R_r = 0.332; X_s = 1.106, X_r = 0.464 and X_{mag} = 26.30

Rotational losses are assumed constant and are 1.1 kW and the core losses are assumed negligible. If the slip is 2.2 % at rated voltage and frequency, find; (*i*) speed (*ii*) stator current (*iii*) power factor (*iv*) output and input power (*v*) efficiency of motor.

[*U.P.S.C. I.E.S. Elec. Engineering II, 1997*]

Solution: Stator impedance per phase,

$$\mathbf{Z}_1 = (0.641 + j\,1.106)\ \Omega$$

Equivalent rotor impedance per phase,

$$\frac{\mathbf{Z_2}}{K^2} = (0.332 + j\,0.464)\ \Omega$$

Equivalent load resistance per phase,

$$R'_L = \frac{R_2}{K^2}\left(\frac{1}{s} - 1\right) = 0.332\left(\frac{1}{0.022} - 1\right) = 14.759\ \Omega$$

Voltage applied per phase,

$$V = \frac{400}{\sqrt{3}} = 231\ V$$

Effective impedance per phase,

$$\mathbf{Z} = \mathbf{Z_1} + \frac{\mathbf{Z_2}}{K^2} + R'_L$$

$$= (0.641 + j\,1.106) + (0.332 + j\,0.464) + (14.759 + j\,0)$$

$$= (15.732 + j\,1.57) \text{ or } 15.81\ \angle 5.7^\circ\ \Omega$$

Counterbalancing rotor current per phase,

$$\mathbf{I'_1} = \frac{231}{15.81\ \angle 5.7^\circ} = 14.6\ \angle{-5.7^\circ}\ A = (14.53 - j\,1.45)\ A$$

Magnetising component of no–load current per phase,

$$I_m = \frac{V}{X_{mag}} = \frac{231}{26.30} = 8.78\ A$$

Energy component of no-load current,

$I_e = 0$ $\qquad\because$ core losses are negligible

Hence no-load current per phase,

$$\mathbf{I_0} = (0 - j\,8.78)\ A$$

Stator current per phase,

$$\mathbf{I_1} = \mathbf{I'_1} + \mathbf{I_0} = (14.53 - j\,1.45) + (0 - j\,8.78)$$

$$= (14.53 - j\,10.23) \text{ or } 17.77\ \angle{-35.15^\circ}\ A$$

(*i*) Speed, $N = N_s(1 - s) = \frac{120\,f}{P}(1 - s)$

$$= \frac{120 \times 50}{4}(1 - 0.022) = 1{,}467 \text{ rpm } \textbf{Ans.}$$

(*ii*) Stator current, $I_1 = 17.77$ A **Ans.**

(*iii*) Power factor,

$$\cos\phi_1 = \text{Cos}(-35.15^\circ) = 0.8176 \text{ (lag) } \textbf{Ans.}$$

(*iv*) Output power,

$$P_{out} = 3\,(I'_1)^2\,R'_L - \text{rotational losses}$$

$$= [3 \times (14.6)^2 \times 14.759] - 1{,}100$$

$$= 8{,}338 \text{ watts } \textbf{Ans.}$$

Power input, $P_{in} = 3\,V\,I_1\cos\phi_1 = 3 \times 231 \times 17.77 \times 0.8176$

$$= 10{,}066 \text{ watts } \textbf{Ans.}$$

(*v*) Motor efficiency,

$$\eta = \frac{P_{out}}{P_{in}} \times 100 = \frac{8{,}338}{10{,}066} \times 100 = 82.83\ \% \textbf{ Ans.}$$

Example 18.51. A 400 V, 3-phase, star-connected induction motor has a stator impedance of (0.06 + *j* 0.2) ohm and an equivalent rotor impedance of (0.06 + *j* 0.22) ohm. Determine the maximum gross power output and the slip at which it occurs. Exciting current may be neglected.

Solution: The equivalent motor resistance referred to stator,

$$R_{01} = R_1 + \frac{R_2}{K_2} = 0.06 + 0.06 = 0.12\ \Omega$$

The equivalent motor reactance referred to stator,

$$X_{01} = X_1 + \frac{X_2}{K^2} = 0.2 + 0.22 = 0.42\ \Omega$$

The equivalent motor impedance referred to stator,

$$Z_{01} = \sqrt{(R_{01})^2 + (X_{01})^2} = \sqrt{0.12^2 + 0.42^2}$$

$$= 0.437\ \Omega$$

As per Art 18.31, slip corresponding to maximum gross power output,

$$s = \frac{\frac{R^2}{K^2}}{\frac{R^2}{K^2} + Z_{01}} = \frac{0.06}{0.06 + 0.437} = 0.12 \text{ or } 12\ \% \textbf{ Ans.}$$

Maximum gross power output

$$= \frac{3\,V^2}{2\,(R_{01} + Z_{01})} = \frac{3 \times \left(\frac{400}{\sqrt{3}}\right)^2}{2\,(0.12 + 0.437)}$$

$$= 143{,}626 \text{ W or } 143.63 \text{ kW } \textbf{Ans.}$$

18.33 INDUCTION MOTOR TESTING

Various tests, which are performed on a polyphase induction motor, for determination of its efficiency and other operating characteristics are (*i*) stator resistance test (*ii*) no-load test (*iii*) blocked rotor test and (*iv*) heat run test.

As the circuit model of an induction motor is similar to that of a transformer, the parameters of the model can be had by means of no-load test as in the case of a transformer—*no-load test* (corresponding to the open-circuit test on the transformer) and the blocked rotor test (corresponding to the short-circuit test on the transformer.)

The stator core loss and friction and windage losses of an induction motor can be determined from no-load test, stator resistance can be determined from stator resistance test and rotor resistance can be determined from blocked rotor test and so the stator and rotor copper losses can be computed for any load and so the efficiency.

Heat run test is performed to determine the temperature rise of different parts of the motor while it is running at rated speed and that is why, it is sometimes called the *temperature rise test.*

If the motor is a 3-phase induction motor, it is customary to assume that the stator winding is star-connected*.

1. Stator Resistance Test. The stator winding resistance is measured between any two terminals, using direct current. This gives the resistance of two phases in series, which must be divided by 2 to obtain stator winding resistance per phase. For greater accuracy, the resistance between three pairs of stator terminals is measured. The average of these three values divided by 2 gives the resistance per phase. Since the effective ac resistance is higher than the dc resistance, the latter is usually multiplied by a figure of about 1.25 to obtain the former.

* This may not be actual case, but the result would be exactly the same, if delta connections were assumed.

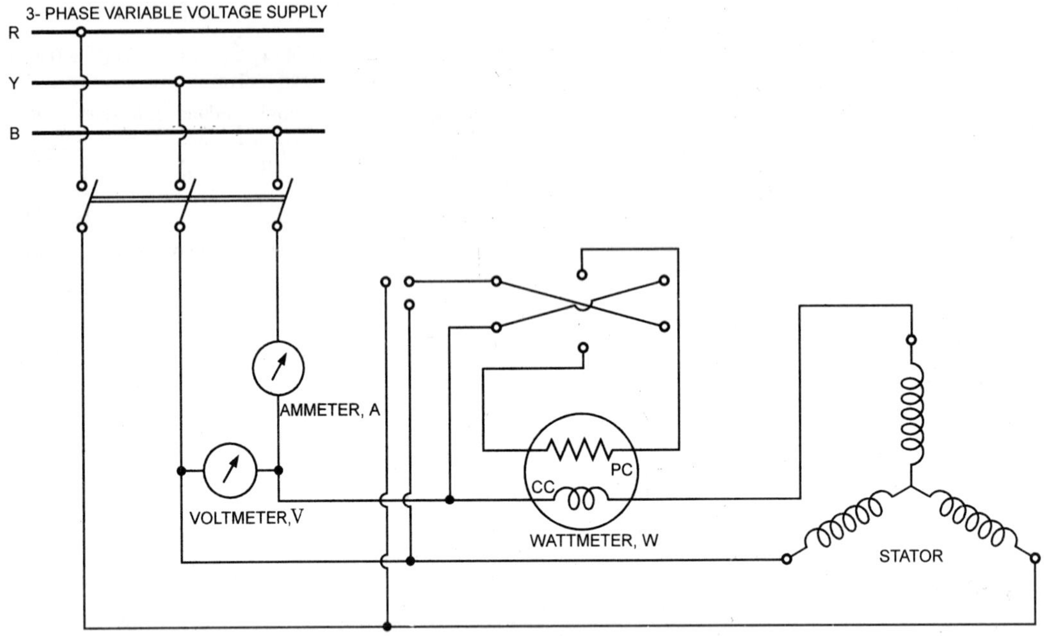

Fig. 18.32 *Circuit Diagram For No-Load Test on a 3-Phase Induction Motor*

2. No-Load Test. This test is performed to determine no-load current I_0, no-load power factor cos ϕ_0, windage and friction losses P_{wf}, no-load core loss P_i, no-load power input P_o, and no-load resistance R_0 and reactance X_0. Moreover, it also reveals any mechanical fault, noise etc.

The no-load test is performed with different values of applied voltage below and above rated voltage, while the motor is running light (without load).

The power input for all three phases P_0, voltage V_0 (line to line value) and current I_0 are measured by wattmeter W, voltmeter V and ammeter A connected in the circuit Fig. (18.32). If the motor is of wound rotor type, the external resistor should be cut out completely. Since motor is running without load, the power factor would be less than 0.5 (about 0.05 to 0.15), hence total power input will be equal to the difference of the two readings of the wattmeter.

At no load, power input is equal to the core loss P_i, stator copper loss P_{c1} and windage and frictional losses P_{wf}. Core loss occurs only in the stator as the slip is extremely small (of the order of 0.001) and so the frequency of rotor current is as low as 0.05 Hz. The magnitude of no-load current in an induction motor is about 30 to 40 % of full-load current because of the air gap. So the stator copper loss at no load needs to be accounted for. This can be estimated by determining stator resistance, as mentioned above.

The mechanical power developed corresponds to windage and friction losses P_{wf} only and the rotor circuit can, therefore, be considered to be open. This is also evident from the magnitude of rotor circuit resistance $\frac{R_2/K^2}{s}$. This resistance becomes very high because the slip at no load is extremely small and so the rotor circuit is practically open at no load. Thus the equivalent circuit of the motor at no load will be as illustrated in Fig. 18.33.

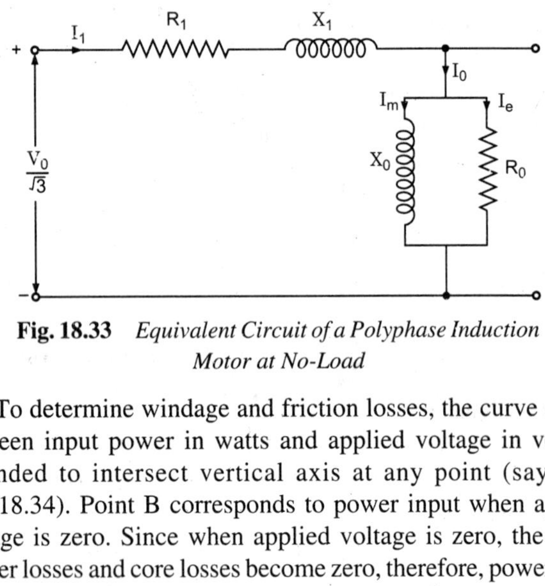

Fig. 18.33 *Equivalent Circuit of a Polyphase Induction Motor at No-Load*

To determine windage and friction losses, the curve drawn between input power in watts and applied voltage in volts is extended to intersect vertical axis at any point (say B in Fig. 18.34). Point B corresponds to power input when applied voltage is zero. Since when applied voltage is zero, the stator copper losses and core losses become zero, therefore, power input at no load and zero voltage given by OB represents the windage and friction losses.

If windage and friction losses P_{wf}, so determined, and stator copper losses (3 I_0^2 R_1) are subtracted from power input at no

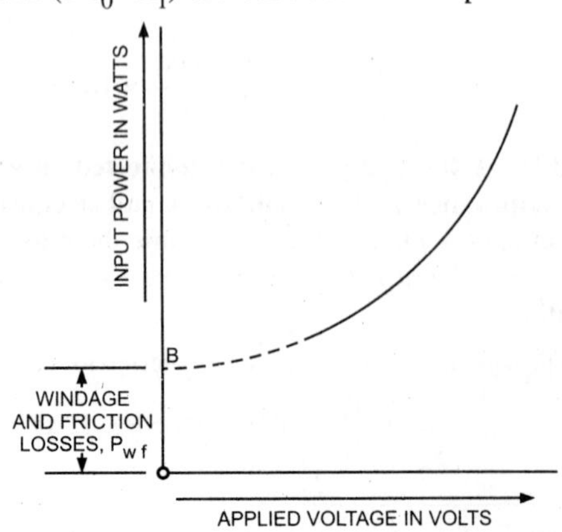

Fig. 18.34

load, P_0, core losses P_i can be determined. Knowing the total core losses P_i, no-load current I_0 and applied voltage V_0 (line-to-line value), the values of magnetising component and energy component of no-load current (I_m and I_e respectively), no-load resistance R_0 and no-load reactance X_0 can be determined from the following relations.

$$\text{No-load power factor, } \cos\phi_0 = \frac{P_i}{\sqrt{3}V_0 I_0} \quad ...(18.52)$$

Energy component of no-load current,

$$I_e = I_0 \cos\phi_0 = I_0 \times \frac{P_i}{\sqrt{3}\,V_0\,I_0} = \frac{P_i}{\sqrt{3}\,V_0} \quad ...(18.53)$$

Magnetising component of no-load current,

$$I_m = \sqrt{I_0^2 - I_e^2} \quad ...(18.54)$$

$$\text{No-load impedance, } Z_0 = \frac{V_0/\sqrt{3}}{I_0} \quad ...(18.55)$$

$$\text{No-load resistance, } R_0 = \frac{V_0/\sqrt{3}}{I_e} = \frac{V_0/\sqrt{3}}{P_i/\sqrt{3}\,V_0} = \frac{V_0^2}{P_i} \quad ...(18.56)$$

$$\text{No-load reactance, } X_0 = \sqrt{Z_0^2 - R_0^2} \quad ...(18.57)$$

3. Voltage-Ratio Test. This test can only be performed on a wound rotor motor by exciting the stator winding at rated voltage and frequency (the rotor circuit being kept open circuited and standstill). The ratio of rotor to stator voltage can be measured by means of a voltmeter. It is to be noted that emf appearing at the slip rings (induced emf in rotor) is of the supply frequency because the rotor is at standstill.

4. Blocked Rotor Test. This test is performed to determine the short-circuit current I_{sc} with normal applied voltage to stator; power factor on short circuit, total equivalent resistance and reactance of the motor as referred to stator (R_{01} and X_{01}).

In this test the rotor is held firmly (rotor windings are short circuited at slip rings in case of wound rotor motor) and stator is connected across supply of variable voltage. This test is just equivalent to short-circuit test on transformer. The connection diagram remains the same as in case of no-load test. Instruments must usually be changed to record the high power and current indications.

Though this test does not reveal any mechanical defect but is of importance because the short-circuit current I_{sc} and power factor $\cos\phi_s$ alongwith the no-load current I_0 and power factor $\cos\phi_0$ enables the circle diagram to be drawn, (as will be discussed in Art. 18.35). In addition, the copper losses computed from this test are necessary for determination of efficiency.

Starting with zero voltage across the stator, the applied voltage is gradually increased in steps till the full-load current flows in the stator. The readings of voltmeter, ammeter and wattmeter are noted. In this test, the wattmeter readings are additive because the power factor is much greater than 0.5. To avoid overheating of the motor, the readings must be taken as quick as possible. For greater accuracy, the temperature of the stator winding should also be noted during the test so that the appropriate corrections can be made to the recommended temperature of 75° C.

The curve drawn between stator current and applied voltage, if drawn, will be a straight line.

The windage and friction losses are absent because the rotor is at standstill and core losses being very small (main flux being considerably reduced under short-circuit condition at the low short-circuit test voltage) can also be ignored for practical purposes. Thus power input during this test can be considered to be entirely stator and rotor copper losses and no-load resistance R_0 of the equivalent circuit representing core losses can also be omitted. But the magnetizing reactance X_0, being much lower in an induction motor compared to a transformer cannot be ignored. The equivalent circuit of an induction motor under blocked rotor condition will, therefore, be as illustrated in Fig. 18.35.

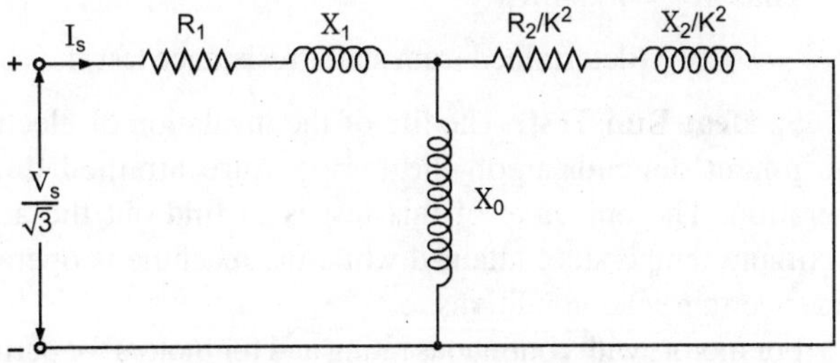

Fig. 18.35 *Equivalent Circuit of a Polyphase Induction Motor Under Blocked Rotor Condition*

If V_s is the applied voltage (line-to-line value) and causes current I_s in the stator winding and P_s is the total input power at short circuit then short-circuit current I_{sc} with normal voltage V (line-to-line) applied across the motor is given by

$$I_{sc} = I_s \frac{V}{V_s} \quad ...(18.58)$$

$$\text{and power factor, } \cos\phi_s = \frac{P_s}{\sqrt{3}V_s I_s} \quad ...(18.59)$$

Since input on short circuit meets with stator and rotor copper losses, as justified above, so

Input on short circuit, $P_s = 3 I_s^2 R_{01}$

or motor equivalent resistance per phase, as referred to stator,

$$R_{01} = \frac{P_s}{3 I_s^2} \quad ...(18.60)$$

Motor equivalent impedance per phase, as referred to stator,

$$Z_{01} = \frac{V_s/\sqrt{3}}{I_s} \quad ...(18.61)$$

Motor equivalent reactance per phase, as referred to stator,

$$X_{01} = \sqrt{Z_{01}^2 - R_{01}^2} \quad ...(18.62)$$

These values constitute the series equivalent of the blocked rotor test (Fig. 18.36)

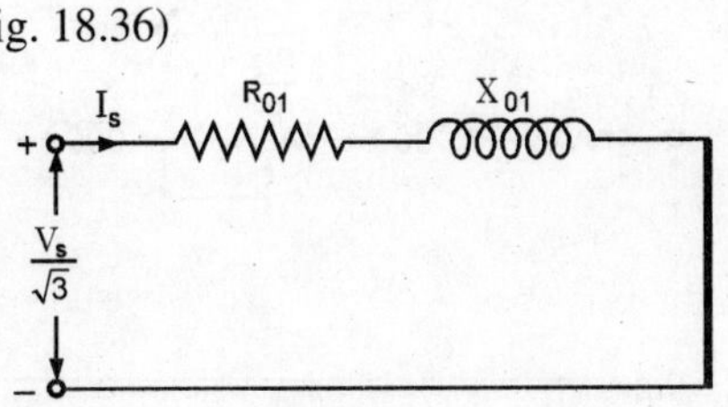

Fig. 18.36

Usually the stator resistance per phase, X_1 is assumed to be equal to rotor reactance per phase, as referred to stator

$$X'_2 = \left(\frac{X_2}{K^2}\right) \text{ and so}$$

$$X_1 = X'_2 = \frac{X_{01}}{2} \qquad ...(18.63)$$

In case of wound rotor motors, the stator and rotor resistances are separated by dividing R_{01} in the ratio of dc resistances of the stator and rotor windings. In case of squirrel cage induction motors, the rotor resistance per phase, as referred to stator, can be determined by subtracting R_1 from R_{01}.

Thus $R'_2 = R_{01} - R_1$

(determined from stator resistance test) ...(18.64)

5. Heat Run Test. The life of the insulation of electrical equipment depends upon the temperature attained during operation. The objective of this test is to find out the actual maximum temperature attained while the machine is operating under certain load conditions.

For motors with continuous rating and for motors for periodic duty and for continuous duty with intermittent load, this test is continued till equilibrium is reached. In the case of intermittent duty cycle motors, the duration of one cycle is kept 10 minutes.

The temperature is measured both while the motor is operating and after it is shut down. The motor should be stopped as soon as the motor attains the thermal equilibrium.

Example 18.52. A 440 V, 50 Hz, 4-pole, 3-phase delta connected induction motor has a leakage impedance of (0.3 + *j* 5.5 + 0.25/s) ohm/phase (delta phase) referred to the stator. The stator to rotor voltage ratio is 2.5. Determine the external resistance to be inserted in each star-phase of the rotor winding such that the motor develops a gross torque of 150 N-m at a speed of 1250 rpm.

[*U.P. Technical Univ. Electromechanical Energy Conversion-II, 2006-07*]

Solution: Synchronous speed,

$$N_s = \frac{120f}{P} = \frac{120 \times 50}{4} = 1{,}500 \text{ rpm}$$

$$\text{Slip, } s = \frac{N_s - N}{N_s} = \frac{1{,}500 - 1{,}250}{1{,}500} = 0.1667$$

Total leakage impedance referred to stator (equivalent star basis)

$$= \frac{1}{3}\left(0.3 + \frac{0.25}{s} + j5.5\right)$$

$$= \left(0.1 + \frac{0.0833}{s} + j1.833\right)\Omega\Big/\text{phase (star)}$$

From the equivalent circuit of a 3-phase induction motor shown in Fig. 18.37.

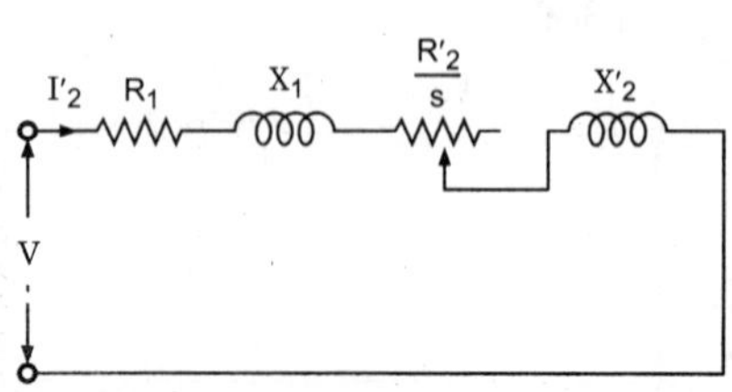

Fig. 18.37 *Equivalent Circuit of a 3-phase Induction Motor*

$$\text{Current, } I'_2 = \frac{V}{\sqrt{\left(R_1 + \frac{R'_2}{s}\right)^2 + (X_1 + X'_2)^2}}$$

$$\text{Power input to rotor} = 3\,(I'_2)^2 \times \frac{R'_2}{s}$$

$$= \frac{3V^2 \frac{R'_2}{s}}{\left(R_1 + \frac{R'_2}{s}\right)^2 + (X_1 + X'_2)^2}$$

$$\text{or Gross torque, } T = \frac{3}{2\pi N_s/60} \times \frac{V^2 \frac{R'_2}{s}}{\left(R_1 + \frac{R'_2}{s}\right)^2 + (X_1 + X'_2)^2}$$

Substituting values we have

$$150 = \frac{3}{2\pi \times 1500/60} \times \frac{\left(440/\sqrt{3}\right)^2 \times \frac{R'_2}{0.1667}}{\left(0.1 + \frac{R'_2}{0.1667}\right)^2 + (1.833)^2}$$

or $36\,R_2'^2 - 48.1\,R'_2 + 3.37 = 0$

or $R'_2 = 1.262\ \Omega;\ 0.0742\ \Omega$

Lower value is less than 0.0833 Ω, hence ignored.

External resistance to be inserted, $R'_{ext} = 1.262 - 0.0833 = 1.1787\ \Omega$

$$R_{ex}\text{ (rotor)} = \frac{R'_{ext}}{(\text{Stator to rotor voltage ratio})^2}$$

$$= \frac{1.1787}{(2.5)^2} = 0.1886\ \Omega \textbf{ Ans.}$$

Example 18.53. Consider the equivalent circuit diagram of 3-ϕ, Y-connected, 440 V, 50 Hz, 4-pole, IM shown in Fig. 18.38:

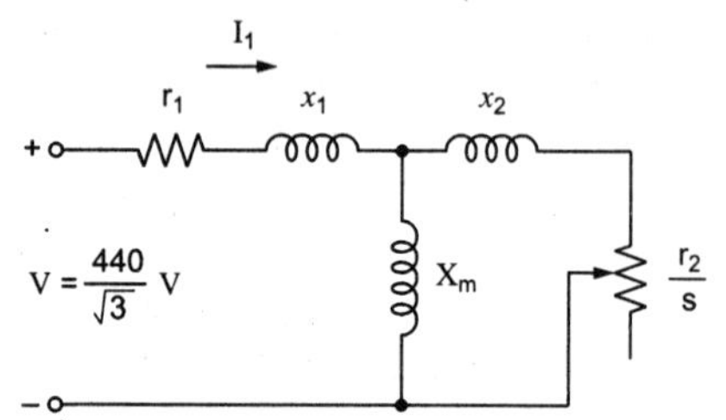

Fig. 18.38 *Equivalent circuit diagram of IM.*

where r_1 = 0.294 Ω; x_1 = 0.503 Ω; X_m = 13.25 Ω; r_2 = 0.209 Ω; x_2 = 0.144 Ω.

(*i*) **Determine the stator current and power factor (motor runs at N_r = 1,460 RPM)**

(*ii*) **Determine the air-gap power and rotor copper losses (motor runs at N_r = 1,450 RPM).**

(*iii*) **What is the value of motor speed at which it takes 30 ampere current at 0.8 pf lagging from supply mains?**

(*iv*) **What is the value of slip when motor runs at N_r = 1,480 RPM?**

[*G.B. Technical Univ. Electromechanical Energy Conversion-II, 2011-12*]

Solution: Equivalent impedance of the circuit given in Fig. 18.38.

$$\mathbf{Z} = r_1 + jx_1 + \left[\left(\frac{r_2}{s} + jx_2\right) \| \, jX_m\right]$$

$$= r_1 + jx_1 + \frac{\left(\frac{r_2}{s} + jx_2\right)(jX_m)}{\frac{r_2}{s} + j(x_2 + X_m)} \quad ...(i)$$

$$= r_1 + jx_1 + R_f + jX_f$$

where Z_f is the per-phase impedance offered to the stator by the rotating air-gap field. It includes the reflected effect of rotor leakage impedance $\frac{r_2}{s} + jx_2$ and the magnetizing reactance X_m.

(*i*) Slip of the motor, $s = \frac{N_s - N}{N_s} = \frac{\frac{120 \times 50}{4} - 1,460}{\frac{120 \times 50}{4}} = \frac{40}{1,500} = 0.02667$

Substituting the values of r_1, x_1, r_2, x_2, X_M and s in Eq. (*i*) we have

$$\mathbf{Z} = r_1 + jx_1 + R_f + jX_f$$

$$= 0.294 + j0.503 + \frac{\left(\frac{0.209}{0.02667} + j0.144\right)(j13.25)}{\frac{0.209}{0.02667} + j(0.144 + 13.25)}$$

$$= 0.294 + j\,0.503 + 5.7136 + j\,3.4858$$

$$= 6 + j\,3.9888 = 7.2 \angle 33.62° \ \Omega \quad \textbf{Ans.}$$

$$\text{Stator current, } \mathbf{I}_1 = \frac{440/\sqrt{3}}{7.2 \angle 33.62°} = 35.28 \angle -33.62° \text{ A } \textbf{Ans.}$$

Stator power factor = Cos (– 33.62°) = 0.8327 (lagging) **Ans.**

(*ii*) Slip of motor, $s = \frac{N_s - N}{N_s} = \frac{\frac{120 \times 50}{4} - 1,450}{\frac{120 \times 50}{4}} = \frac{50}{1,500} = 0.03333$

Equivalent circuit impedance

$$\mathbf{Z} = r_1 + jx_1 + R_f + jX_f$$

$$= 0.294 + j0.503 + \frac{\left(\frac{0.209}{0.3333} + j\,0.144\right)(j13.25)}{\frac{0.209}{0.3333} + j\,13.394}$$

$$= 0.294 + j\,0.503 + 5.03 + j\,2.5 \quad ...(ii)$$

$$= 5.324 + j\,3.003 = 6.113 \angle 29.425° \ \Omega$$

From Eq. (*ii*) $R_f = 5.03\ \Omega$

$$\text{Stator current, } \mathbf{I}_1 = \frac{440/\sqrt{3}}{6.113 \angle 29.425°} = 41.56 \angle -29.425° \text{ A}$$

$$\text{Total air-gap power, } P_g = 3I_1^2 R_f = 3 \times (41.56)^2 \times 5.03$$

$$= 26,060 \text{ W} \quad \textbf{Ans.}$$

$$\text{Rotor copper losses} = s \times P_g = 0.03333 \times 26,060$$

$$= 868.65 \text{ W} \quad \textbf{Ans.}$$

(*iii*) Stator current, $I_1 = 30$ A

Stator power factor, $\cos\phi_1 = 0.8$ (lagging)

$$\phi_1 = \text{Cos}^{-1}\,0.8 = 36.87° \text{ (lagging)}$$

So stator current, $\mathbf{I}_1 = 30 \angle -36.87°$

$$\text{Stator impedance } \mathbf{Z} = \frac{440/\sqrt{3}}{30 \angle -36.87°}$$

$$= 8.4678 \angle 36.87° = (6.774 + j\,5.081)\ \Omega \quad ...(iii)$$

Also, $\mathbf{Z} = r_1 + x_1 + \frac{\left(\frac{r_2}{s} + jx_2\right)(jX_m)}{\frac{r_2}{s} + jx_2 + jX_m}$ where s is the slip

$$= 0.294 + 0.503 + \frac{\left(\frac{0.209}{s} + j\,0.144\right)(j\,13.25)}{\frac{0.209}{s} + j\,13.394} \quad ...(iv)$$

Equating expressions (*i*) and (*ii*) we have

$$6.774 + j\,5.081 = 0.294 + 0.503 + \frac{-1.908 + j\frac{2.76925}{s}}{\frac{0.209}{s} + j\,13.394}$$

$$\text{or} \quad 6.48 + j\,4.578 = \frac{-1.908 + \frac{j2.76926}{s}}{\frac{0.209}{s} + j\,13.394}$$

$$\text{or} \quad (6.48 + j\,4.578)\left(\frac{0.209}{s} + j\,13.394\right) = -1.908 + j\frac{2.76926}{s}$$

Solving above equation we have

slip, $s = 0.022$

$$\text{Motor speed, } N = N_s(1 - s) = \frac{120 \times 50}{4} \times (1 - 0.022)$$

$$= 1,467 \text{ rpm } \textbf{Ans.}$$

(*iv*) Rotor speed, N = 1,480 rpm

$$\text{Slip, } s = \frac{N_s - N}{N_s} = \frac{1,500 - 1,480}{1,500}$$

$$= 0.01333 \quad \text{or} \quad 1.333\% \quad \textbf{Ans.}$$

Example 18.54. A 3-phase, star-connected, 400 V, 50-Hz, 4-pole induction motor has the following per phase parameters referred to the stator:

$\mathbf{R_1 = 0.15\ \Omega,\ X_1 = 0.45\ \Omega\ R'_2 = 0.12\ \Omega,\ X'_2 = 45\ \Omega}$ **and** $\mathbf{X_m = 28.5\ \Omega}$

Calculate stator current, and power factor when the motor is operated at rated voltage and frequency with slip $s = 0.04$**.**

[*R.G. Technical Univ. Electromechanical Energy Conversion-I, June-2011*]

Solution: Motor equivalent impedance per phase, at the slip of 0.04,

$$\mathbf{Z} = \mathbf{Z}_1 + \frac{\left(\frac{R'_2}{s} + jX'_2\right) \times jX_m}{\frac{R'_2}{s} + jX'_2 + jX_m}$$

$$= 0.15 + j0.45 + \frac{\left(\frac{0.12}{0.04} + j0.45\right) \times j28.5}{\frac{0.12}{0.04} + j0.45 + j28.5}$$

$$= 0.15 + j\,0.45 + 2.81 + j\,0.724$$

$$= 2.96 + j\,1.174 = 3.1843 \angle 21.63° \ \Omega$$

$$\text{Stator current, } \mathbf{I}_1 = \frac{\mathbf{V}_L/\sqrt{3}}{\mathbf{Z}} = \frac{400 \angle 0°}{\sqrt{3} \times 3.1843 \angle 21.63°}$$

$$= 72.52 \angle -21.63° \text{ A}.$$

Taking phase voltage as reference phasor

So stator current, $I_1 = 72.52$ A

and power factor, $\cos\phi_1 = \cos(-21.63°) = 0.9296$ (lagging) **Ans.**

Example 18.55. A 3-phase, star-connected, 400 V, 50-Hz, 4-pole induction motor has the following per phase constants in ohms referred to the stator.

$$r_1 = 0.15, x_1 = 0.45, r_2' = 0.12, x_2' = 0.45, x_m = 28.5$$

Fixed losses (core, friction and windage) are 400 watts.

Calculate the stator current, rotor speed, output torque of the motor when it is operated at rated voltage, frequency and at 4% slip.

[W.B. Univ. of Technology, Electrical Machine-I, 2011]

Solution: Synchronous speed, $N_s = \frac{120 \times f}{P} = \frac{120 \times 50}{4} = 1{,}500$ rpm

Slip, s = 4% or 0.04

Fixed losses = 400 watts

Stator current, I_1 = 72.52 A as worked out in Example 18.54

Rotor speed, $N = N_s(1 - s) = 1{,}500\,(1 - 0.04) = 1{,}440$ rpm

Air-gap power, $P_2 = 3I_1^2 R = 3 \times (72.52)^2 \times 2.96 = 46{,}700$ watts

∵ R = 2.96 as worked out in Example 18.54

Mechanical power developed,

$$P_{mech} = (1 - s)P_2 = (1 - 0.04) \times 46{,}700 = 44{,}832 \text{ watts}$$

Shaft power, $P_{sh} = P_{mech}$ – fixed losses $= 44{,}832 - 400 = 44{,}432$ watts

Output torque of motor, $T_{sh} = \frac{P_{sh}}{2\pi N_s/60} = \frac{44{,}432 \times 60}{2\pi \times 1{,}440}$

= 294.65 N-m **Ans.**

Example 18.56. Determine motor efficiency of an induction motor with parameters given in Example 18.55.

Solution: Stator copper loss $= 3I_1^2 \times$ stator resistance

$= 3 \times (72.52)^2 \times 0.15 = 2{,}367$ W

Rotor copper losses $= s \times P_2 = 0.04 \times 46{,}700 = 1{,}868$ W

Total losses = Stator copper losses + rotor copper losses + fixed losses

$= 2{,}367 + 1{,}868 + 400 = 4{,}635$ watts

Power input, $P_{in} = P_{sh}$ + total losses $= 44{,}432 + 4{,}635 = 49{,}067$ watts

Motor efficiency, $\eta_m = \frac{P_{sh}}{P_{in}} \times 100 = \frac{44{,}432}{49{,}067} \times 100$

= 90.55% **Ans.**

Example 18.57. A 10 kW, 400 V, 50 Hz, 4-pole, Y-connected squirrel cage IM gave the following test results:

No-load Test	400 V	8 A	250 Watts
Blocked rotor Test	90 V	35 A	1,350 watts.

The dc resistance of the stator winding per phase measured immediately after the blocked rotor test is 0.6 Ω.

Calculate the following:

(*i*) **Equivalent circuit parameters of 3-ϕ SCIM.**

(*ii*) **Rotational losses.**

[G.B. Technical Univ. Electromechanical Energy Conversion-II, 2011-12]

Solution: (*i*) Applied voltage per phase, V = 400 V

No-load line current, $I_{L0} = 8$ A

No-load phase current, $I_0 = \frac{8}{\sqrt{3}}$ A

Effective resistance of stator winding,

$$R_e = 1.2\, R_{dc} = 1.2 \times 0.6 = 0.72\ \Omega$$

∵ Effective stator resistance per phase is taken equal to 1.2 times its dc value

No-load impedance, $Z_0 = \frac{V_0}{I_0} = \frac{400}{8\sqrt{3}} = 86.6\ \Omega$

No-load resistance, $R_0 = \frac{P_0}{3 \times I_0^2} = \frac{250}{3 \times \left(\frac{8}{\sqrt{3}}\right)^2} = 3.906\ \Omega$

No-load reactance, $X_0 = \sqrt{Z_0^2 - R_0^2} = \sqrt{86.6^2 - 3.906^2} = 86.51\ \Omega$ **Ans.**

From Block Rotor Test

Applied voltage per phase, $V_s = 90$ V

Line current, $I_{Ls} = 35$ A

Stator current per phase, $I_s = \frac{35}{\sqrt{3}}$ A

Input, $P_s = 1{,}350$ W

Equivalent impedance per phase, $Z_{01} = \frac{V_s}{I_s} = \frac{90}{35\sqrt{3}} = 4.454\ \Omega$

Equivalent resistance per phase, $R_{01} = \frac{P_s}{3I_s^2} = \frac{1{,}350}{3 \times (35/\sqrt{3})^2} = 1.102\ \Omega$

Equivalent reactance per phase $X_{01} = \sqrt{Z_{01}^2 - R_{01}^2}$

$= \sqrt{(4.454)^2 - (1.102)^2} = 4.316\ \Omega$

Stator resistance per phase, $X_1 = X_2' = \frac{X_{01}}{2} = \frac{4.316}{2} = 2.158\ \Omega$

Magnetising reactance $X_m = X_0 - X_1 = 86.51 - 2.158 = 84.352\ \Omega$

Rotor reactance per phase, $X_2 = X_m + X_2' = 84.352 + 2.158 = 86.51\ \Omega$

Rotor resistance per phase, $R_2 = (R_{01} - R_e) \times \left(\frac{X_2}{X_m}\right)^2$

$= (1.102 - 0.72) \times \left(\frac{86.51}{84.352}\right)^2 = 0.402\ \Omega$

Thus the parameters of the induction motor equivalent circuit are:

$R_1 = 0.72\ \Omega$, $R_2 = 0.402\ \Omega$; $X_1 = X_2' = 2.158\ \Omega$; $X_m = 84.352\ \Omega$ **Ans.**

(*ii*) No-load rotational losses

= Input at no load – no-load stator copper loss

$= 250 - 3\left(\frac{8}{\sqrt{3}}\right)^2 \times 0.72 = 203.92$ W **Ans.**

Example 18.58. For the short-circuit test on a 50 hp, 440 V, 50 Hz, 6-pole star-connected, 3-phase induction motor, the following data were obtained.

V = 230 V, I = 140 A, cos ϕ = 0.35 (line values). The motor drives a load having a constant torque of 300 N-m. Estimate the percentage reduction in the supply voltage possible before the motor stalls. Assume that the copper losses are equally divided between the stator and the rotor. Neglect magnetising current.

[U.P.S.C. I.E.S. Electrical Engineering-II, 2005]

Solution: From the short-circuit test

Voltage applied, V_s = 230 V (line-to-line value)

$= \frac{230}{\sqrt{3}} = 132.79$ V (phase value)

Short-circuit current per phase, $I_s = I_{L_s} = 140$ A

Equivalent impedance referred to stator per phase

$$Z_{01} = \frac{V_s}{I_s} = \frac{132.79}{140} = 0.9485\ \Omega$$

Short-circuit power factor, $\cos \phi_s = 0.35$

$$\sin \phi_s = \sqrt{1 - 0.35^2} = 0.9367$$

$$\text{So } R_1 + R_2' = Z_{01} \cos \phi_s = 0.9485 \times 0.35 = 0.332\ \Omega$$

$$\text{and } X_1 + X_2' = Z_{01} \sin \phi_s = 0.9485 \times 0.9367 = 0.8885\ \Omega$$

Since the copper losses are equally divided between the stator and rotor.

$$R_1 = R_2' = \frac{0.332}{2} = 0.166\ \Omega$$

$$\text{Synchronous speed, } N_s = \frac{120 \times f}{P} = \frac{120 \times 50}{6} = 1{,}000 \text{ rpm}$$

Line minimum voltage per phase possible is V′. At this voltage maximum torque developed is given by

$$T_{max} = \frac{3 \times 60}{2\pi N_s} \times \frac{(V')^2}{2} \times \frac{1}{R_1 + \sqrt{R_1^2 + (X_1 + X_2')^2}}$$

$$= \frac{3 \times 60}{2\pi \times 1{,}000} \times \frac{(V')^2}{2} \times \frac{1}{0.166 + \sqrt{0.166^2 + 0.8885^2}}$$

The motor will stall if

$$\frac{3 \times 60}{2\pi \times 1{,}000} \times \frac{(V')^2}{2} \times \frac{1}{0.166 + \sqrt{0.166^2 + 0.8885^2}} = 300$$

$$\text{or } V' = 149.7 \text{ V}$$

$$\text{Normal voltage per phase} = \frac{440}{\sqrt{3}} = 254 \text{ V}$$

∴ Percentage reduction in voltage possible

$$= \frac{V - V'}{V} \times 100 = \frac{254 - 149.7}{254} \times 100 = 41.08\%$$ **Ans.**

Example 18.59. A 3 kW, 400 V/200 V, Delta/Star, 50 Hz, three phase, 6-pole induction motor is found to draw a line current of 25 A at a power factor of 0.4, when a blocked rotor test is conducted at the rated voltage. Determine the stator and rotor winding resistances in ohms per phase, if the torque developed by the motor under the above conditions is 25 N-m. *[GATE 1999]*

Solution: Voltage applied (phase values),

V_s = Line value = 400 V

∵ motor is delta connected on primary

Short circuit current per phase,

$$I_s = \frac{I_{Ls}}{\sqrt{3}} = \frac{25}{\sqrt{3}} = 14.434 \text{ A}$$

Equivalent impedance referred to stator per phase

$$Z_{01} = \frac{V_s}{I_s} = \frac{400}{14.434} = 27.713\ \Omega$$

Equivalent resistance referred to stator per phase

$$R_{01} = R_1 + R_2' = Z_{01} \times \cos \phi_s = 27.713 \times 0.4 = 11.085\ \Omega$$

$$\text{Turns-ratio, } K = \frac{200/\sqrt{3}}{400} = 0.2887$$

$$\text{Synchronous speed, } N_s = \frac{120 f}{P} = \frac{120 \times 50}{6} = 1{,}000 \text{ rpm}$$

Assuming, f = 50 Hz

$$\text{Rotor current, } I_2 = \frac{I_1'}{K} = \frac{14.434}{0.2887} = 50 \text{ A}$$

Since rotor copper loss = s × power input to rotor

$$\text{or } 3I_2^2 R_2 = s \times \frac{2\pi N_s}{60} T = \frac{2\pi N_s}{60} \times T$$

∵ Under blocked condition $s = 1$

or Rotor resistance,

$$R_2 = \frac{T \times 2\pi N_s}{60} \times \frac{1}{3I_2^2} = \frac{25 \times 2\pi \times 1{,}000}{60} \times \frac{1}{3 \times 50^2}$$

$= 0.349\ \Omega$/phase **Ans.**

Rotor resistance referred to stator,

$$R_2' = \frac{R_2}{K^2} = \frac{0.349}{(0.2887)^2} = 4.19\ \Omega/\text{phase}$$

Stator resistance per phase,

$$R_1 = R_{01} - R_2' = 11.085 - 4.19 = 6.897\ \Omega$$ **Ans.**

Example 18.60. A 400 V, 6 pole, 50 Hz, three-phase star-connected induction motor running at rated voltage takes 7.5 A with a power input of 700 W. With the rotor blocked and 150 V applied to the stator, the input current is 35 A and the power input is 5,000 W. The rotor and the stator copper losses are equal under the blocked rotor condition. The standstill leakage reactance of the stator and rotor as seen from the stator are estimated to be in the ratio of 1 : 0.5. Obtain the equivalent circuit of the induction motor. Calculate the net mechanical power output, torque, input power and the efficiency at a slip of 4%. *[U.P.S.C. I.E.S. Electrical Engineering-II, 1999]*

Solution: Blocked Rotor Test

$$\text{Applied voltage per phase, } V = \frac{150}{\sqrt{3}} = 50\sqrt{3} \text{ V}$$

Short-circuit stator phase current, $I_s = I_{Ls} = 35$ A

Input on short-circuit current, P_s = 5,000 W

Power factor on short circuit,

$$\cos \phi_s = \frac{P_s}{3VI_s} = \frac{5{,}000}{3 \times 50\sqrt{3} \times 35} = 0.54986 \text{ (lagging)}$$

Equivalent impedance referred to stator,

$$Z_{01} = \frac{V_s}{I_s} = \frac{50\sqrt{3}}{35} = 2.474\ \Omega$$

Equivalent resistance referred to stator,

$$R_{01} = Z_{01} \cos \phi_s = 2.474 \times 0.54986 = 1.36\ \Omega$$

Equivalent reactance referred to stator,

$$X_{01} = Z_{01} \sin \phi_s = 2.474 \times 0.835 = 2.067\ \Omega$$

Stator resistance,

R_1 = Rotor resistance referred to stator, R_2'

$$= \frac{R_2}{K^2} = \frac{1.36}{2} = 0.68\ \Omega$$

∵ Stator and rotor losses are equal under blocked rotor condition

Stator reactance,

$$X_1 = \frac{X_{01} \times 1}{1 + 0.5} = \frac{2.067}{1.5} \times 1 = 1.378\ \Omega$$

Rotor reactance referred to stator,

$$X_2' = \frac{X_2}{K^2} = \frac{2.067}{1 + 0.5} \times 0.5 = 0.689\ \Omega$$

No-load Test:

$$\text{Applied voltage per phase, } V_0 = \frac{400}{\sqrt{3}} \text{ V}$$

$$\text{No-load stator phase current, } I_0 = I_{L0} = 7.5 \text{ A}$$

$$\text{Input on no load, } P_0 = 700 \text{ W}$$

Stator copper loss,

$$P_{c1} = 3I_0^2R_1 = 3 \times 7.5^2 \times 0.68 = 114.75 \text{ W}$$

Total stator core loss,

$$P_i = P_0 - P_{c1} \quad \text{neglecting windage and friction losses}$$
$$= 700 - 114.75 = 585.25 \text{ W}$$

No-load power factor,

$$\cos\phi_0 = \frac{P_i}{3VI_0} = \frac{585.25}{3 \times \frac{400}{\sqrt{3}} \times 7.5} = 0.11263 \text{ (lagging)}$$

Energy component of no-load phase current,

$$I_e = I_0 \cos\phi_0 = 7.5 \times 0.11263 = 0.8447 \text{ A}$$

Magnetising component of no-load phase current,

$$I_m = I_0 \sin\phi_0 = 7.5 \times 0.9936 = 7.4523 \text{ A}$$

$$\text{No-load leakage resistance, } R_0 = \frac{V}{I_e} = \frac{400/\sqrt{3}}{0.8447} = 273.4\ \Omega$$

$$\text{No-load leakage reactance, } X_0 = \frac{V}{I_m} = \frac{400/\sqrt{3}}{7.4523} = 30.989\ \Omega$$

$$\text{Slip, } s = 0.04$$

Equivalent circuit of the induction motor is given in Fig. 18.39.

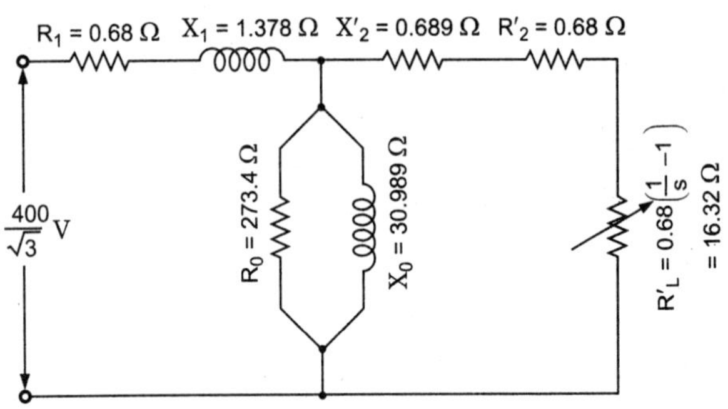

Fig. 18.39 *Equivalent Circuit Diagram*

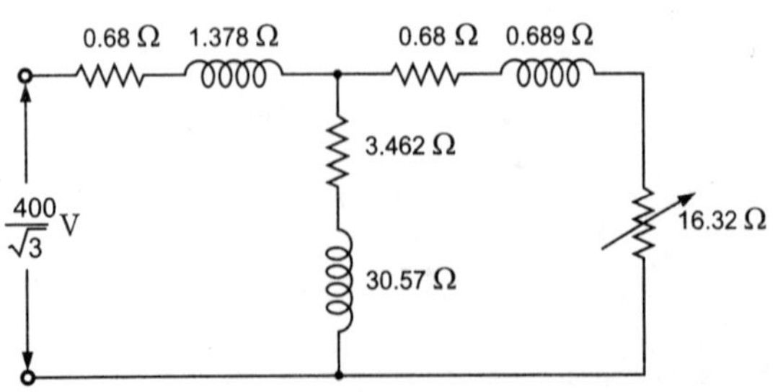

Fig. 18.40

Series equivalent of shunt branch,

$$\mathbf{Z}_{se} = \frac{1}{\mathbf{Y}_p} = \frac{1}{\frac{1}{R_0} - \frac{j}{X_0}} = \frac{1}{\frac{1}{273.4} - \frac{j}{30.989}}$$

$$= \frac{1}{0.00366 - j0.0323}\ \Omega$$

$$= \frac{1}{0.0325 \angle -83.54^\circ}\ \Omega$$

$$\text{or } \mathbf{Z}_{se} = 30.77 \angle 83.54^\circ\ \Omega = (3.462 + j30.57)\ \Omega$$

Modified equivalent circuit is shown in Fig. 18.40.

Thevenin voltage,

$$V_{Th} = \frac{\mathbf{Z}_{se}}{\mathbf{Z}_1 + \mathbf{Z}_{se}} \times V$$

$$= \frac{30.77 \angle 83.54^\circ \times \frac{400}{\sqrt{3}} \angle 0^\circ}{(0.68 + j1.378) + (3.462 + j30.57)}$$

$$= \frac{30.77 \angle 83.54^\circ \times \frac{400}{\sqrt{3}} \angle 0^\circ}{4.142 + j31.948}$$

$$= \frac{30.77 \angle 83.54^\circ \times \frac{400}{\sqrt{3}} \angle 0^\circ}{32.215 \angle 82.61^\circ} = 220.58 \angle 0.93^\circ \text{ V}$$

Thevenin's impedance,

$$\mathbf{Z}_{Th} = \frac{\mathbf{Z}_{se}\mathbf{Z}_1}{\mathbf{Z}_{se} + \mathbf{Z}_1}$$

$$= \frac{30.77 \angle 83.54^\circ \times 1.5366 \angle 63.735^\circ}{32.215 \angle 82.61^\circ}$$

$$= 1.4677 \angle 64.665^\circ = (0.628 + j\,1.327)\ \Omega$$

Rotor current with reference to stator,

$$\mathbf{I}'_2 = \frac{\mathbf{V}_{Th}}{\mathbf{Z}_{Th} + \mathbf{Z}'_2 + R'_L}$$

$$= \frac{220.58 \angle 0.93^\circ}{(0.628 + j1.327) + (0.68 + 16.32 + j0.689)}$$

$$= \frac{220.58 \angle 0.93^\circ}{17.628 + j2.016}$$

$$= \frac{220.58 \angle 0.93^\circ}{17.743 \angle 6.53^\circ} = 12.432 \angle -5.6^\circ \text{ A}$$

Net mechanical power output,

$$P_{mech} = 3 \times (I'_2)^2 \times R'_L = 3 \times (12.432)^2 \times 16.32$$
$$= 7{,}567 \text{ W or } 7.567 \text{ kW} \quad \textbf{Ans.}$$

Speed of motor,

$$N = N_s(1 - s) = \frac{120 \times 50}{6}(1 - 0.04) = 960 \text{ rpm}$$

$$\text{Torque, } T = \frac{P_{mech}}{2\pi N/60} = \frac{7{,}567 \times 60}{2\pi \times 960} = 75.27 \text{ N-m} \quad \textbf{Ans.}$$

Impedance of the equivalent circuit shown in Fig. 18.40.

$$Z_{in} = 0.68 + j1.378 + (3.462 + j30.57) \,||\, (0.68 + 16.32 + j0.689)$$

$$= (0.68 + j1.378) + \frac{(3.462 + j30.57)(17 + j0.689)}{(3.462 + j30.57) + (17 + j0.689)}$$

$$= 0.68 + j1.378 + \frac{30.765 \angle 83.54^\circ \times 17.014 \angle 2.32^\circ}{37.36 \angle 56.79^\circ}$$

$$= 0.68 + j1.378 + 14 \angle 29.07^\circ$$

$$= 0.68 + j1.378 + 12.236 + j6.8$$

$$= 12.916 + j8.178 = 15.287 \angle 32.34^\circ \ \Omega$$

$$\text{Input current, } I_1 = \frac{400/\sqrt{3}}{15.287 \angle 32.34^\circ} = 15.107 \angle -32.34^\circ \text{ A}$$

$$\text{Input power, } P_{in} = \sqrt{3}\, V_L\, I_1 \cos \phi$$

$$= \sqrt{3} \times 400 \times 15.107 \cos 32.34^\circ = 8{,}843 \text{ watts}$$

$$\text{Motor efficiency, } \eta = \frac{P_{mech}}{P_{in}} \times 100 = \frac{7{,}567}{8{,}843} \times 100 = 85.57\% \ \textbf{Ans.}$$

18.34 MEASUREMENT OF SLIP

The various methods of measurement of slip are given below:

1. Actual Measurement of Rotor Speed. In this method of determination of slip, the actual speed of the rotor is measured and then subtracted from the synchronous speed to obtain slip

$$i.e. \text{ slip, } s = N_s - N$$

This method of determination of slip is not accurate, since slip is very small as compared to synchronous or actual rotor speed and is the difference of two nearly equal quantities and, therefore, small percentage error in measurement of rotor speed will cause large percentage error in slip.

2. By Measure of Rotor Frequency. Since the rotor frequency f', being the product of slip and supply frequency, is very low, it can be accurately measured by means of a low value centre-reading dc moving coil voltmeter connected across the slip rings in case of wound rotor motor. The current in the voltmeter follows the variations of rotor current and so the pointer oscillates about its mean zero position. The number of complete cycles (movement from zero to a maximum to right, back to zero and on a maximum to the left and then back to zero) made by the pointer can be easily counted. The slip can be determined from the following relation

$$s = \frac{\text{Rotor frequency}}{\text{Supply frequency}} = \frac{f'}{f}$$

$$= \frac{\text{Number of oscillations, counted over a period of } t \text{ seconds}}{50 \times t} \qquad \text{...(18.65)}$$

With a squirrel cage motor the above method cannot be applied. If, however, a large flat search coil having many turns is placed centrally against the end plate on the non-driving end of the motor, it is sometimes possible to pick up sufficient voltage by induction from the leakage fluxes to obtain a reading. A larger 50 Hz voltage will also be induced, but this is too rapid to affect the instrument. Commercial slip indicators have been developed employing such a search coil in conjunction with low-pass filter-amplifier to eliminate the fundamental frequency and then comparing f' and f by means of a bridge circuit or digital counter.

It is sometimes possible to obtain an indication by connecting a dc millivoltmeter between the ends of the shaft.

3. Stroboscopic Method. This is another method for measurement of slip. In this method a disc known as *stroboscopic disc*, with the geometrical pattern on it, is rigidly attached to the shaft of the motor, whose slip is to be determined, as shown in Fig. 18.41. The number of black sectors as well as white sectors, each are equal to the number of poles for which the motor is wound. For example for a 6-pole motor there will be 12 sectors, 6 black and 6 white. A neon glow lamp supplied from the motor line either through a transformer or directly is arranged to illuminate the stroboscopic disc. The neon lamp employed has the characteristics that it ceases to glow when voltage across it falls below a certain limit *i.e.* such a lamp glows twice in a cycle. If the disc were to rotate at synchronous speed, it would appear to be stationary. Since, in actual practice, its speed is slightly less than the synchronous speed, it appears to rotate slowly backwards. This fact is explained as follows.

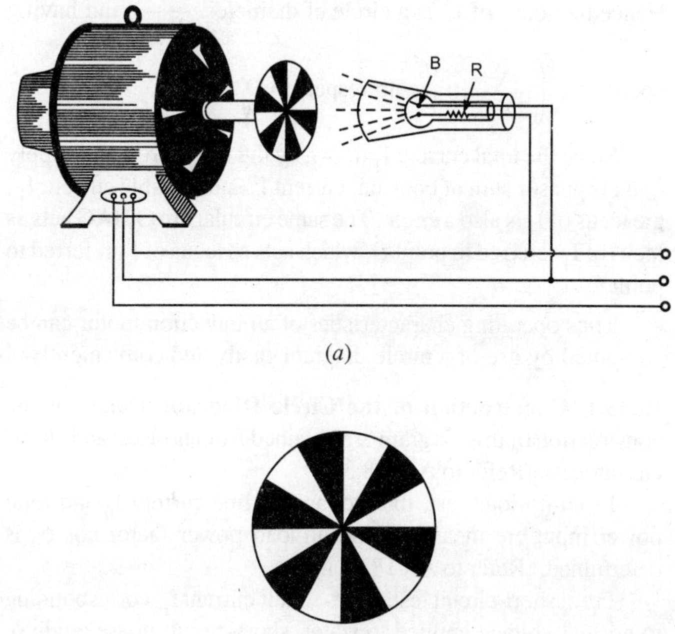

(*a*)

(*b*)

Fig. 18.41

In one half cycle the rotor of the motor would advance one pole if there were no slip and during this period each black sector (or white sector) would advance to the position just occupied by the adjacent black sector (or white sector) that preceded it. During the period of advancement the sectors are but faintly visible, for the illumination from the lamp is practically zero. Hence the disc would appear stationary. Owing to the fact that each conductor on the rotor does not advance one pole each half cycle, the sectors will not reach the position of the next adjacent sector of same colour but will fall short of this distance, owing to the slip. The sectors on the disc will not appear stationary but will seem to be rotating slowly backward.

The complete apparent revolutions of the disc are counted in a given time and slip is determined from the following relation.

$$\text{Slip, } s = \frac{\text{Apparent revolutions counted in the given time}}{\text{Time in seconds}} \times \frac{\text{pairs of poles}}{\text{supply frequency}} \qquad \text{...(18.66)}$$

18.35 CIRCLE DIAGRAM

It is known that in a series circuit consisting of constant reactance but variable resistance, the locus of the current is a circle of diameter $\frac{V}{X}$.

In a circuit shown in Fig. 18.30 the current I_0 to the shunt circuit is constant, in the circuit to the right $X_{01} = X_1 + \frac{X_2}{K^2}$, the equivalent reactance referred to primary (stator) is constant; $R_{01} = R_1 + \frac{R_2}{K^2}$, the equivalent resistance referred to primary is also constant but load resistance R'_L varies with the load on the motor, Hence the locus of I_1' is a circle of diameter $\frac{V}{X_{01}}$ and having coordinate $\left(\frac{V}{2X_{01}}, 0\right)$ with respect to O'.

Since the total current I_1 drawn by the motor from the supply main is phasor sum of constant current I_0 and variable current I_1', the locus of I_1 is also a circle. The same circular arc O'HAG acts as locus of I_1 referred to point O, which acts as locus of I'_1 referred to point O'.

Thus operating characteristics of an induction motor can be computed by use of a circle diagram easily and conveniently.

18.35.1. Construction of the Circle Diagram. Datas for the construction of this diagram are obtained from no-load and short-circuit tests. [Refer to Art. 18.33]

From no-load test, line voltage V, line current I_0 and total power input are measured and no-load power factor cos ϕ_0 is determined. [Refer to Art 18.33 (2)].

From short-circuit test, short-circuit current I_{sc} corresponding to normal voltage applied to stator, short-circuit phase angle ϕ_s are determined.

The voltage phasor V is taken along Y-axis, *i.e.* in phasor diagram (Fig. 18.42) OV represents the normal voltage applied to the stator. Phasor OO ' is drawn at an angle ϕ_0 from OV (lagging) representing no-load line current I_0 in magnitude as well as in direction. Line OF is drawn perpendicular to voltage phasor OV.

Phasor OA be drawn at an angle ϕ_s from phasor OV (lagging) and equal to I_{sc} in magnitude. Thus phasor OA represents the line current that would exist were the rated line voltage V impressed across the stator when motor blocked and phasor O'A represents the corresponding rotor current I_1' as referred to primary (stator).

Obviously the points O' and A lie on the circle. To determine the centre of the required circle, line O'G is drawn parallel to OF, O'A is joined and line BC is drawn perpendicular to line O'A bisecting the line O'A at point B and intersecting line O'G at C. Now with C as center and O'C as radius the semicircle O'HP'AG is drawn.

Since voltage phasor is drawn vertically, therefore, all vertical distances represent the active components of currents and power to different scales. For example O'J represents the active component of no-load current, which supplies core loss, friction and windage loss and small amount of stator copper losses. O'J also represents power input at no-load to some other scale. Similarly vertical component AF of OA represents the active component of short-circuit current I_{sc} and also represents power input on short circuit to some other scale.

Thus AF represents the power input when rotor is blocked. The power input when rotor is blocked meets core losses, and stator and rotor copper losses. Line AE represents the stator and rotor copper losses when EF = O'J is subtracted from AF.

Now the line AE is divided by D into two segments such that $\frac{AD}{DE} = \frac{\text{Rotor copper losses}}{\text{Stator copper losses}}$. Line joining points O' and D is drawn. Line O' D is known as *torque line or rotor input line.*

In case of wound rotor motor, the line AE is divided directly into two segments such that

$$\frac{AD}{DE} = \frac{R_2 / K^2}{R_1} = \frac{R_2'}{R_1}$$

In case of a squirrel cage motor, stator resistance R_1 is determined and DE is made equal to stator copper losses $[3I_{sc}'^2 R_1]$, where I_{sc}' is the short-circuit current per phase.

In circle diagram (Fig. 18.42) line O'A represents the output line or mechanical power developed line.

If a line N'P'F' is drawn parallel to line NPF and tangential to the semicircle, then the vertical intercept P'P between this tangent line N'P'F' and horizontal line NPF represents the maximum power input to the motor.

If a line L'Q'D' is drawn parallel to torque line LQD and tangential to the semicircle, then the vertical intercept Q'Q between the tangential line L'Q'D' and torque line LQD represents the maximum torque.

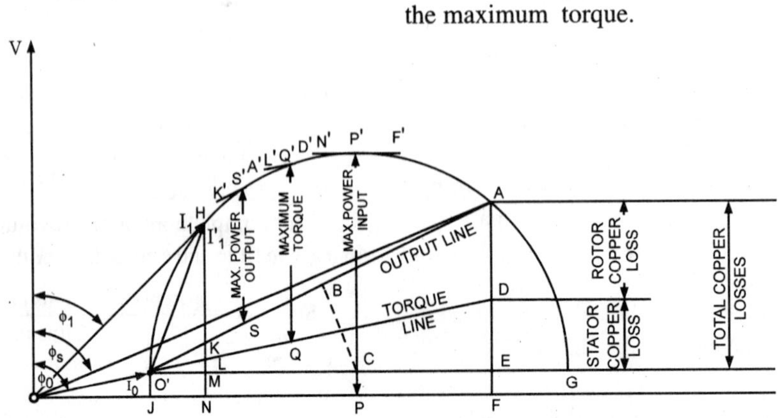

Fig. 18.42 *Circle Diagram For An Induction Motor*

If a line K'S'A' is drawn parallel to output line O'A and tangential to the semicircle, then the vertical intercept S'S between the tangential line K'S'A' and output line O'A represents the maximum power output.

At any load current I_1 (represented by OH), the secondary current referred to primary is I_1' (represented by O'H) being equal to phasor difference of I_1 and I_0 *i.e.* $(I_1 - I_0)$, HN represents the energy component of load current I_1. Now

$$\text{Motor input} = \sqrt{3}\ V \cdot NH$$

$$\text{Constant losses} = \sqrt{3}\ V \cdot NM$$

$$\text{Stator copper losses} = \sqrt{3}\ V \cdot ML$$

$$\text{Rotor input power} = \sqrt{3}\ V \cdot LH$$

$$\text{Rotor copper losses} = \sqrt{3}\ V \cdot LK$$

$$\text{Motor output} = \sqrt{3}\ V \cdot HK$$

$$\text{Efficiency} = \frac{\text{Output}}{\text{input}} = \frac{\sqrt{3}\ V \cdot KH}{\sqrt{3}\ V \cdot NH} = \frac{KH}{NH}$$

$$\text{Slip, } s = \frac{\text{Rotor copper losses}}{\text{Rotor input}} = \frac{\sqrt{3}\ V \cdot LK}{\sqrt{3}\ V \cdot LH} = \frac{LK}{LH}$$

$$\text{Power factor, } \cos\phi_1 = \frac{NH}{OH}$$

Example 18.61. Make use of the circle diagram or otherwise, prove that the maximum output of an induction motor per phase $= \dfrac{V_1 (I_{sc} - I_0)}{2(1+\cos\phi_{sc})}$ **where V_1 = source voltage per phase, I_{sc} = source current during blocked rotor test, I_0 = no-load currrent and ϕ_{sc} = angle of lag during blocked rotor test between source voltage and current.**

Solution: Maximum power output of an induction motor per phase, according to Eq. (18.41) is given as

$$P_{gmax} = \frac{V_1^2}{2(R_{01} + Z_{01})} \text{ watts}$$

where V_1 is the source voltage per phase and R_{01} and Z_{01} are the motor equivalent resistance and reactance respectively, in ohms, as referred to the stator.

Neglecting the phase angle difference caused by the exciting current during blocked rotor test

$$\text{Cos}\ \phi_{sc} = \frac{R_{01}}{Z_{01}}$$

$$\text{and } I_{sc} - I_0 = \frac{V_1}{Z_{01}}$$

$$\text{So } P_{gmax} = \frac{V_1 \times \dfrac{V_1}{Z_{01}}}{2\left[\dfrac{R_{01}}{Z_{01}} + 1\right]} = \frac{V_1 (I_{sc} - I_0)}{2[\cos\phi_{sc} + 1]} \text{ proved.}$$

Example 18.62. Draw the circle diagram of a 20 hp, 400 V, 50 Hz, 3-phase star-connected induction motor from the following test data (line values)

No-load	**400 V**	**9 A**	**pf 0.2**
Blocked rotor	**200 V**	**50 A**	**pf 0.4**

From the circle diagram find (*i*) line current (*ii*) power factor (*iii*) slip (*iv*) efficiency at full load and (*v*) maximum power output.

Solution: No-load current, $I_0 = 9$ A

No-load power factor, $\cos\phi_0 = 0.2$

or no-load phase angle, $\phi_0 = \cos^{-1} 0.2 = 78.5°$

Short-circuit current with normal voltage applied across the stator,

$$I_{sc} = 50 \times \frac{400}{200} = 100 \text{ A}$$

Short-circuit power factor, $\cos\phi_s = 0.4$

or $\phi_s = \cos^{-1} 0.4 = 66.5°$

From the above datas, the circle diagram (Fig. 18.43) is constructed as follows:

The current scale is assumed 10.0 A per cm.

OV is drawn vertical to represent the input voltage. OO' is drawn lagging behind phasor OV by an angle 78.5° to represent no-load current of 9 A. Phasor OA is drawn lagging behind phasor OV by an angle 66.5° representing short-circuit I_{sc} with full applied voltage (i.e. 100 A).

A horizontal line O'G is drawn through O' and O'A are joined. The line O'A is bisected at right angle by line BC. Then point C on line O'G is the centre of the required semicircle. This semicircle drawn through points O' and A with centre C is the locus of the current phasor for all loading conditions from no load to short circuit.

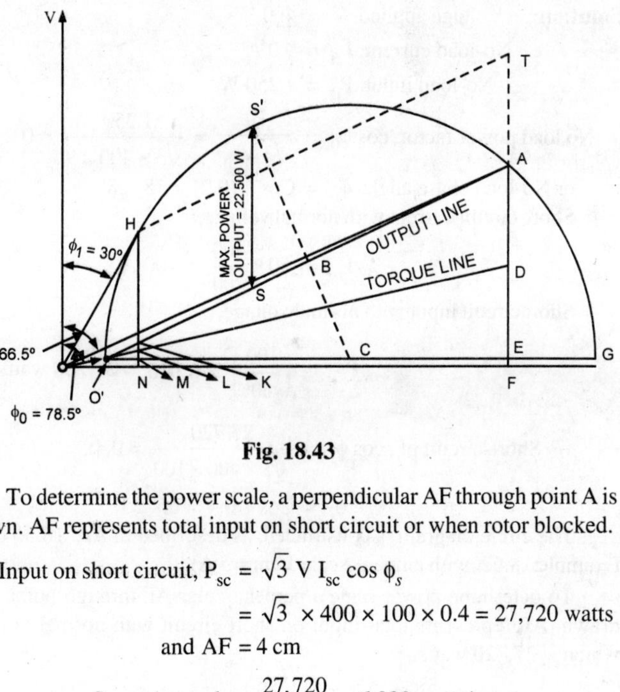

Fig. 18.43

To determine the power scale, a perpendicular AF through point A is drawn. AF represents total input on short circuit or when rotor blocked.

$$\text{Input on short circuit, } P_{sc} = \sqrt{3}\ V I_{sc} \cos\phi_s$$

$$= \sqrt{3} \times 400 \times 100 \times 0.4 = 27{,}720 \text{ watts}$$

and AF = 4 cm

$$\text{So power scale} = \frac{27{,}720}{4} = 6{,}930 \text{ watts/cm}$$

AE represents total copper loss in the rotor and stator. Assuming copper losses being equally divided between rotor and stator losses, AE is bisected at point D.

Now motor output = 20 hp = 20 × 735.5 = 14,710 watts, which will be represented by $\dfrac{14{,}710}{6{,}930} = 2.12$ cm. Line FA is extended to point T so that AT = 2.12 cm. From point T line TH is drawn parallel to output line O' A cutting the circle at point H. Point H is joined with origin O and perpendicular HN is drawn.

Now from circle diagram

(*i*) Line current on full load = OH = 3 cm = 3 × 10 = 30 A **Ans.**

(*ii*) Full-load power factor, $\cos\phi_1 = \dfrac{NH}{OH} = \dfrac{2.6}{3.0} = 0.866$ **Ans.**

(*iii*) Slip $= \dfrac{KL}{LH} = \dfrac{0.13}{2.3} = 0.056$ or 5.6% **Ans.**

(*iv*) Efficiency, $\eta = \frac{KH}{NH} \times 100 = \frac{2.12}{2.6} \times 100$

$= 81.54\%$ **Ans.**

(*v*) Maximum power output is obtained by producing CB intersecting semicircle at S′ and then dropping perpendicular from point S′ on horizontal line OF meeting output line at S. Now SS′ represents the maximum power output,

Maximum power output = SS′ × power scale

= 3.25 × 6,930 = 22,500 watts **Ans.**

Example 18.63. Draw the circle diagram for a 3-phase, 6-pole, 50 Hz, 400-V, star-connected induction motor from the following datas (line values) :

No-load test	**400 V: 9A;**	**1,250 watts**
Short circuit test	**200 V; 50 A;**	**6,930 watts**

The stator loss at standstill is 55% of total copper losses and full-load current is 32.0 A. From circle diagram determine (i) pf, slip, output, efficiency, speed and torque at full load (ii) starting torque (iii) maximum power factor (iv) maximum power output (v) maximum power input (vi) maximum torque in synchronous watts and slip for maximum torque.

Solution: Voltage applied, V = 400 V

No-load current, I_0 = 9.0 A

No-load input, P_0 = 1,250 W

$$\text{No load power factor, } \cos\phi_0 = \frac{P_0}{\sqrt{3}\,VI_0} = \frac{1{,}250}{\sqrt{3} \times 400 \times 9} = 0.2$$

or No-load phase angle, $\phi_0 = \text{Cos}^{-1}(0.2) = 78°\ 28'$

Short-circuit current with normal voltage,

$$I_{sc} = 50 \times \frac{400}{200} = 100 \text{ A}$$

Short circuit input with normal voltage,

$$P_{sc} = \left(\frac{100}{50}\right)^2 \times 6{,}930 = 27{,}720 \text{ watts}$$

$$\text{Short-circuit pf, } \cos\phi_s = \frac{27{,}720}{\sqrt{3} \times 400 \times 100} = 0.4;$$

$$\phi_s = \text{Cos}^{-1}\,0.4 = 66°\ 25'$$

The circle diagram is constructed, as described in Art. 18.35 or in Example 18.62, with ampere-scale 1 cm = 5 A.

To determine power scale a perpendicular AF through point A is drawn. AF represents total input on short circuit with normal voltage which is 27,720 watts.

Since AF = 8.1 cm

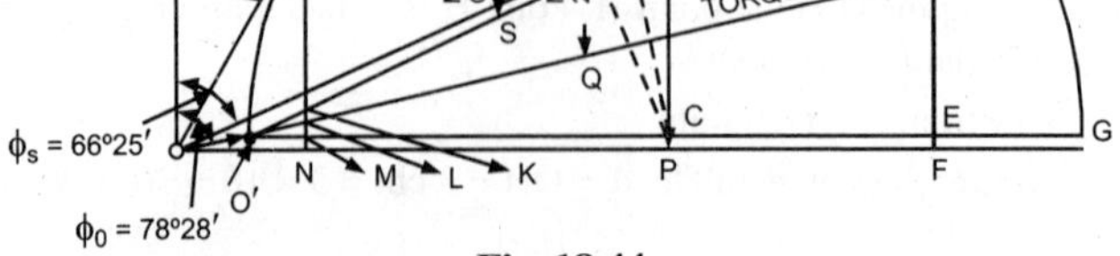

Fig. 18.44

Hence power scale is 1 cm $= \frac{27{,}720}{8\,1} = 3{,}420$ watts

Since stator copper loss is 55% of total copper loss

∴ Torque line O′ D is obtained by taking ED = 0.55 EA

(*i*) Since full-load current is 32.0 A, hence it is represented by a length of 32/5 = 6.4 cm.

With O as centre and 6.4 cm as radius an arc is drawn intersecting semicircle at point H. This point represents full-load conditions. Point O and H are joined. From point H, perpendicular HN is drawn to line OF intersecting lines O′ A, O′ D, O′ E and O′ F at points K, L, M and N respectively.

At full load

(*a*) pf $= \frac{HN}{OH} = \frac{5.55}{6.4} = 0.866$ **Ans.**

also pf = Cos ∠VOH = Cos 30° = 0.866 **Ans.**

(*b*) Slip $= \frac{KL}{LH} = 0.0625$ or 6.25% **Ans.**

(*c*) Output = KH × power scale

= 4.5 × 3,420 = 15,390 watts or 15.39 kW **Ans.**

(*d*) Efficiency $= \frac{KH}{NH} \times 100 = \frac{4.5}{5.55} \times 100 = 81\%$ **Ans.**

(*e*) Speed at full load $= N_s(1 - s) = \frac{120 f}{P}(1 - s)$

$= \frac{120 \times 50}{6}(1 - 0.625) = 937.5$ rpm **Ans.**

Torque at full load T_f = LH × power scale

= 4.8 × 3,420 =16,416 synchronous watts **Ans.**

(*ii*) Starting torque T_{st} = AD × Power scale

= 3.50 × 3,420 =11,970 synchronous watts **Ans.**

(*iii*) Maximum power factor is obtained by drawing a line from point O and tangential to semicircle. In this problem full-load current line OH is tangential (which is not necessary in every case) to the semicircle, therefore, maximum power factor,

$\text{Cos}\,\phi_{max} = \text{Cos}\,30° = 0.866$ **Ans.**

(*iv*) Maximum power output is obtained by producing CB intersecting semicircle at S′ and then dropping perpendicular from point S′ on horizontal line OF meeting output line at S. Now SS′ represents maximum power output.

Maximum power output = SS′ × power scale

= 6.5 × 3,420

= 22, 230 watts or 22.23 kW **Ans.**

(*v*) Maximum power input is obtained by drawing vertical line PP′ from point P. PP′ represents maximum input power.

∴ Maximum input power = PP′ × power scale

= 10.25 × 3,420

= 35,000 watts or 35 kW **Ans.**

(*vi*) Maximum torque is obtained by drawing line CQ′ perpendicular to O′D and dropping a vertical line from Q′ intersecting torque line O′D at point Q. Now Q′Q represents maximum torque.

∴ Maximum torque = Q′Q × power scale = 7.2 × 3,420

= 27,000 synchronous watts **Ans.**

Slip for maximum torque $= \frac{1.65}{7.9} = 0.209$ or 20.9% **Ans.**

Note: In circle diagrams (For Examples 18.62 ... 18.69), actual length are different from the values mentioned, owing to reduction in making figures.

Example 18.64. A 3-phase, 6-pole, 50 Hz, 400 V, star-connected induction motor has the following test results:

No-load test :	**400 V,**	**9 A,**	**1,250 watts**
Short-circuit test:	**200 V,**	**50 A,**	**6,930 watts**

Determine the power scale using circle diagram.

[*Gujarat Technological Univ. Electrical Machines-II, Summer-2014*]

Solution: Refer to Solution of Example 18.63.

Example 18.65. Draw the circle diagram for a 5 hp, 200 V, 50 Hz, 4-pole, 3-ϕ, star-connected induction motor from the following data: (i) 200 V, 5 A, 350 watts (ii) 100 V, 26 A, 1700 watts (iii) Rotor copper loss at standstill = half of the total copper loss.

Estimate therefrom the full load current, power factor, speed and torque.

Solution: No-load power factor,

$$\text{Cos}\,\phi_0 = \frac{P_0}{\sqrt{3}\,V\,I_0} = \frac{350}{\sqrt{3}\times 200\times 5} = 0.202$$

or No-load phase angle,

$$\phi_0 = \text{Cos}^{-1}\,0.202 = 78.35°$$

Short circuit current with normal voltage,

$$I_{sc} = 26\times\frac{200}{100} = 52\text{ A}$$

Short-circuit input with normal voltage,

$$P_{sc} = 1{,}700\times\left(\frac{200}{100}\right)^2 = 6{,}800\text{ watts}$$

Short-circuit power factor,

$$\text{Cos}\,\phi_s = \frac{6{,}800}{\sqrt{3}\times 200\times 52} = 0.3775$$

$$\text{or } \phi_s = \text{Cos}^{-1}\,0.3775 = 67.82°$$

The circle diagram is constructed, as described in Art. 18.35 or Example 18.62, with ampere scale 1 cm = 5A.

To determine the power scale a perpendicular AF through point A on O′G is drawn. AF represents total input on short circuit with normal voltage applied, which is 6,800 watts.

Since AF = 3.9 cm

$$\text{Hence power scale, 1 cm} = \frac{6{,}800}{3.9} = 1{,}744\text{ watts}$$

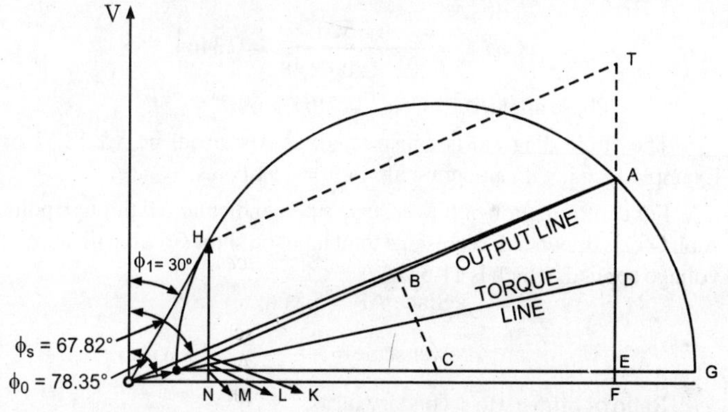

Fig. 18.45

Since rotor copper losses at standstill = half of the total copper losses, line AE is bisected at point D to separate the stator and rotor copper losses.

Now power output = 5 hp = 5 × 735.5 = 3677.5 watts which will be represented by $\frac{3{,}677.5}{1{,}744}$ = 2.1 cm on the circle diagram. Line FA is extended to point T so that AT = 2.1 cm. From point T line TH is drawn parallel to output line O′A cutting the circle at point H. Point H is joined with origin O and perpendicular HN is drawn.

Now from circle diagram (Fig. 18.45)

$$\text{Full load current, } I_f = OH = 3\text{ cm} = 3\times 5 = 15\text{ A} \textbf{ Ans.}$$

$$\text{Full load power factor, } \cos\phi_f = \frac{NH}{OH} = \frac{2.6}{3} = 0.866 \textbf{ Ans.}$$

$$\text{Full load speed, } N_f = \frac{KH}{LH}\times N_s$$

$$= \frac{2.2}{2.3}\times\frac{120\times 50}{4} = 1{,}435\text{ rpm} \textbf{ Ans.}$$

$$\text{Full load torque} = LH\times\text{power scale}$$

$$= 2.3\times 1{,}744$$

$$= 4{,}010\text{ synchronous watts } \textbf{Ans.}$$

Example 18.66. A 3-phase, 400 V, induction motor gave the following test readings:

No-load	**400 V**	**1,250 W**	**9 A**
Short-circuit	**150 V**	**4 kW**	**38 A**

Draw the circle diagram. If the normal rating is 20.27 hp (metric), find the circle diagram, the full-load values of current, power factor and slip.

[*J.N. Technological Univ. Electrical Machines-II, January-2014*]

Solution: No-load power factor,

$$\cos\phi_0 = \frac{P_0}{\sqrt{3}\,V\,I_0} = \frac{1{,}250}{\sqrt{3}\times 400\times 9} = 0.2005$$

No-load phase angle,

$$\phi_0 = \cos^{-1}\,0.2005 = 78.5°$$

Short-circuit current with normal voltage,

$$I_{sc} = 38\times\frac{400}{150} = 101.33\text{ A}$$

Short-circuit input power with normal voltage,

$$P_{sc} = 4{,}000\times\left(\frac{400}{150}\right)^2 = 28{,}444\text{ Watts}$$

Short-circuit pf,

$$\text{Cos}\,\phi_s = \frac{28{,}444}{\sqrt{3}\times 400\times 101.33} = 0.405$$

$$\text{or } \phi_s = \cos^{-1}\,0.405 = 66.1°$$

The circle diagram is constructed, as described in Art 18.35 or Example 18.62, with ampere scale 1 cm = 5 A.

For determination of power scale a perpendicular AF through point A on O′G is drawn. AF represents total input on short circuit with normal voltage applied which is 28,444 watts

Since AF = 8.1 cm

$$\text{Hence power scale, 1 cm} = \frac{28{,}444}{8.1} = 3{,}510\text{ watts}$$

Assuming rotor copper losses at standstill as half of the total copper losses, line AE is bisected at point D to separate the stator and copper losses.

Now power output = 20.27 hp = 20.27 × 735.5 = 14,910 W

which will be represented by $\frac{14{,}910}{3{,}510}$ = 4.25 cm on the circle diagram.

Line FA is extended to point T so that AT = 4.25 cm. From point T, line TH is drawn parallel to line O′A cutting the circle at point H. Point H is joined with origin O and perpendicular HN is drawn.

Now from circle diagram (Fig. 18.46).

Full-load current, I_f = OH = 6 cm = 6 × 5 = 30 A **Ans.**

$$\text{Full-load power factor, } \cos\phi_f = \frac{NH}{OH} = \frac{5.2}{6} = 0.866 \textbf{ Ans.}$$

$$\text{Slip, } s = \frac{LK}{LH} = \frac{0.3}{4.5} = 0.0667 \text{ or } 6.67\% \textbf{ Ans.}$$

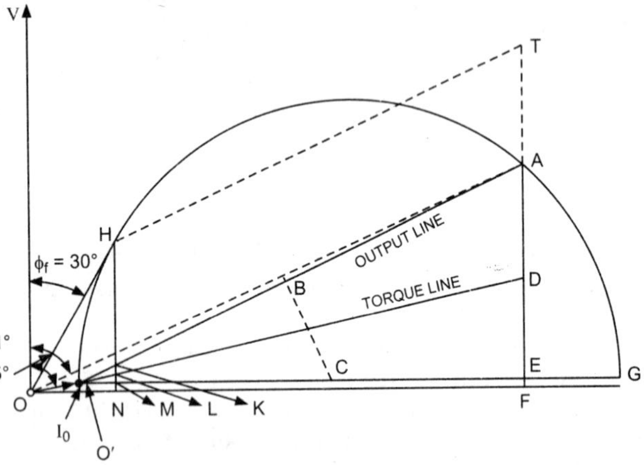

Fig. 18.46

Example 18.67. A 400 V, 3-phase, 50 Hz, star-connected induction motor has the following test results :

No-load test	**400 V**	**8.5 A**	**1,100 W**
Blocked rotor test	**180 V**	**45 A**	**5,799 W**

Calculate the line current and power factor when operating at 4% slip. The stator resistance per phase is 0.5 Ω.

Solution: No- load power factor,

$$\cos\phi_0 = \frac{1,100}{\sqrt{3}\times 400\times 8.5} = 0.187$$

$$\text{and } \phi_0 = \cos^{-1} 0.187 = 79.2°$$

Short-circuit current with normal voltage,

$$I_{sc} = \frac{45\times 400}{180} = 100 \text{ A}$$

Short-circuit input with normal voltage,

$$P_{sc} = 5,700\times\left(\frac{400}{180}\right)^2 = 28,150 \text{ W}$$

Short-circuit power factor,

$$\cos\phi_s = \frac{5,700}{\sqrt{3}\times 180\times 45} = 0.4063$$

$$\text{and } \phi_s = \cos^{-1} 0.4063 = 66°$$

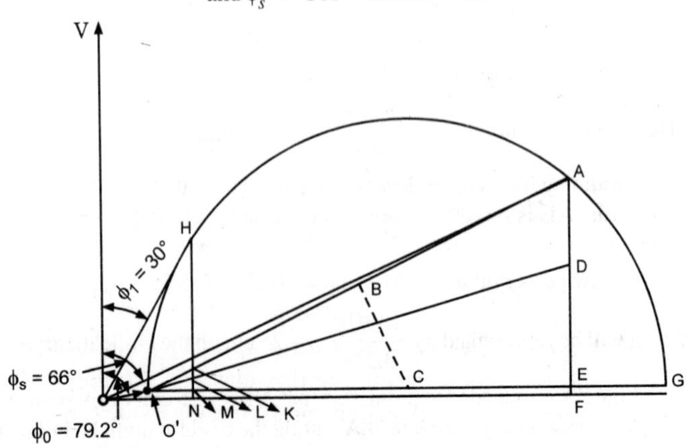

Fig. 18.47

The circle diagram is constructed, as described in Art. 18.35 or in Example 18.62, with ampere scale 1 cm = 10 A. Stator copper losses on short-circuit = $3\, I_{sc}^2\, R_1 = 3 \times 100^2 \times 0.5$ = 15,000 W

To determine power scale a perpendicular AF through point A on O′G is drawn. AF represents total input on short-circuit with normal voltage applied, which is 28,150 W. Since AF = 4.05 cm

$$\text{Hence power scale, 1 cm} = \frac{28,150}{4.05} = 6,950 \text{ watts}$$

Stator losses of 15,000 W are represented by ED

$$\text{So ED} = \frac{15,000}{6,950} = 2.16 \text{ cm}$$

Thus point D is located on line AF to separate out the stator and rotor copper losses.

$$\text{Slip, } s = \frac{\text{Rotor copper loss}}{\text{Rotor input}} = \frac{KL}{LH} = 0.04$$

Hence by trial and error point H is so located on the circle that $\frac{LH}{KL}$ = 25.

Now line current, I_L = OH = 10 × 2.42 = 24.2 A **Ans.**

Power factor, $\cos\phi_1 = \cos 30° = 0.866$ **Ans.**

Example 18.68. Draw the circle diagram for a 5.6 kW, 400 V, three-phase four-pole, 50 Hz, slip-ring induction motor from the following data:

No-load Readings	**:**	**400 V**	**6 A**	**$\cos\phi_0 = 0.087$**
Short-circuit Readings:		**100 V**	**12 A**	**720 watts**

The ratio of primary to secondary turns = 2.62

Stator resistance per phase is 0.67 Ω and the rotor is 0.185 Ω.

Calculate (*i*) Full-load current (*ii*) Full-load slip (*iii*) Full-load power factor (*iv*) Maximum power.

[*R.G. Technical Univ. Electrical Machine-I, June-2013*]

Solution: No-load phase angle, $\phi_0 = \cos^{-1} 0.087 = 85°$

Short-circuit current with normal voltage,

$$I_{sc} = 12\times\frac{400}{100} = 48 \text{ A}$$

Short-circuit input power with normal voltage,

$$P_{sc} = 720\times\left(\frac{400}{100}\right)^2 = 11,520 \text{ watts or } 11.52 \text{ kW}$$

Short-circuit power factor,

$$\cos\phi_s = \frac{11,520}{\sqrt{3}\times 400\times 48} = 0.3464$$

$$\text{Phase angle, } \phi_s = \cos^{-1} 0.3464 = 69.7°$$

The circle diagram is constructed, as described, in Art 18.35 or Example 18.62, with ampere scale 1 cm = 3 amperes.

For determination of power scale a perpendicular AF through point A on O′G is drawn. AF represents total input on short circuit with normal voltage applied, which is 11.52 kW

Since AF = 5.5 cm

$$\text{Power scale} = \frac{11.52}{5.5} = 2.1 \text{ kW/cm}$$

Ratio of primary to secondary turns,

K = 2.62

Stator resistance per phase, $R_1 = 0.67\ \Omega$

Rotor resistance per phase, $R_2 = 0.185\ \Omega$

$$\frac{\text{Rotor copper loss}}{\text{Stator copper loss}} = K^2\times\frac{R_2}{R_1} = 2.62\times\frac{0.185}{0.67} = 1.9$$

$$\text{or } \frac{\text{Rotor copper loss}}{\text{Total copper loss}} = \frac{1.9}{1.9+1.0} = \frac{1.9}{2.9} = 0.655$$

Line AE is divided at point D in the ratio 1.9 : 1.

Full-load output = 5.6 kW which will be represented by $\frac{5.6}{2.1} = 2.7$ cm on the circle diagram. Line FA is extended to point T so that AT = 2.7 cm. From point T line TH is drawn parallel to output line O′A cutting the circle at point H. Point H is joined with origin O and perpendicular HN is drawn.

Now from circle diagram (Fig. 18.48)

(*i*) Full-load current, I_f = OH = 3.8 cm = 3.8 × 3 = 11.4 A **Ans.**

(*ii*) Full-load slip, $s = \frac{LK}{LH} = \frac{0.13}{2.8} = 0.0464$ or 4.64% **Ans.**

(*iii*) Full-load power factor,

$$\cos\phi_f = \frac{NH}{OH} = \frac{3.1}{3.8} = 0.81 \textbf{ Ans.}$$

(*iv*) Maximum power = SS′ = 5.2 cm = 5.2 × 2.1 = 10.9 kW **Ans.**

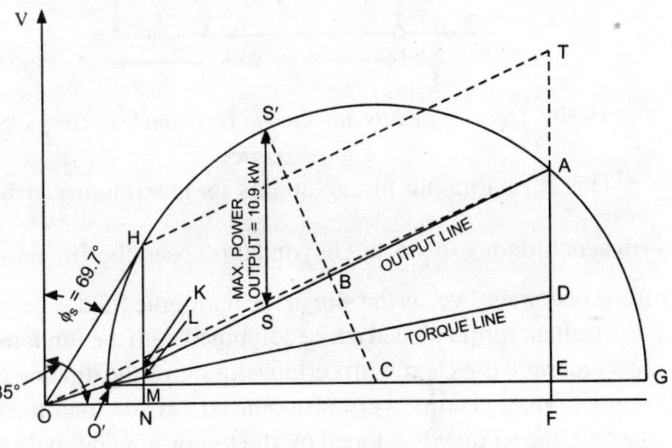

Fig. 18.48

Example 18.69. Draw the circle diagram of a 3-phase mesh-connected, 30 hp, 500 V, 4-pole, 50 Hz cage-type induction motor. The figures given below give the measurements of line current and voltage and readings of two wattmeters connected to measure the input power.

No-load	**500 V**	**8.3 A**	**+ 2.85 kW**	**– 1.35 kW**
Short-circuit (blocked rotor)	**100 V**	**32 A**	**– 0.75 kW**	**+ 2.35 kW**

Find from the diagram for full-load (*i*) line current (*ii*) power factor (*iii*) efficiency and (*iv*) maximum output.

Solution: No–load power input,

$$P_0 = 2.85 - 1.35 = 1.5 \text{ kW or } 1{,}500 \text{ W}$$

No-load power factor, $\cos\phi_0 = \frac{P_0}{\sqrt{3}\,V\,I_0}$

$$= \frac{1{,}500}{\sqrt{3}\times 500\times 8.3} = 0.2087$$

and $\phi_0 = \text{Cos}^{-1}\, 0.2087 = 78°$

Short-circuit line current with normal voltage applied,

$$I_{sc} = I_s \times \frac{V}{V_s} = 32 \times \frac{500}{100} = 160 \text{ A}$$

Short-circuit input power with normal voltage,

$$P_{sc} = P_s \times \left(\frac{V}{V_s}\right)^2$$

$$= (2.35 - 0.75) \times \left(\frac{500}{100}\right)^2 = 40 \text{ kW}$$

Short-circuit power factor, $\cos\phi_s = \frac{P_s}{\sqrt{3}\,VI_s}$

$$= \frac{4\times 1{,}000}{\sqrt{3}\times 500\times 160} = 0.28868$$

and $\phi_s = \text{Cos}^{-1}\, 0.28868 = 73.22°$

The circle diagram is constructed, as described in Art 18.35 or in Example 18.62, with ampere scale 1 cm = 16 A (line value).

To determine power scale a perpendicular AF through point A on O′G is drawn. AF represents total power input on short circuit with normal voltage applied which is 40 kW or 40,000 W.

$$\text{Hence power scale} = \frac{40{,}000}{AF} = \frac{40{,}000}{2.89} = 13{,}800 \text{ watts}$$

Alternatively the power scale = $\sqrt{3}$ × line voltage × ampere per cm

$$= \sqrt{3} \times 500 \times 16 = 13{,}850 \text{ watts/cm}$$

$$\text{Motor output} = 30 \times 735.5 = 22{,}065 \text{ watts,}$$

which will be represented by $\frac{22{,}065}{13{,}850} = 1.6$ cm. Line FA is extended to point T so that AT = 1.6 cm. From point T draw line TH parallel to output line O′A cutting the circle at H. Point H is joined with origin O and perpendicular HN is drawn.

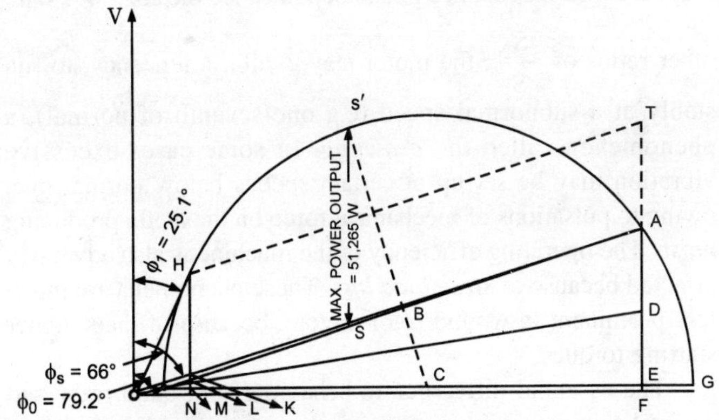

Fig. 18.49

Now from circle diagram (Fig. 18.49)

(*i*) Line current on full load = OH = 2 × 16 = 32 A **Ans.**

(*ii*) Full-load power factor, $\cos\phi_1 = \frac{NH}{OH} = \frac{1.81}{2} = 0.905$ **Ans.**

(*iii*) Motor efficiency, $\eta = \frac{KH}{NH} \times 100 = \frac{1.75}{2} \times 100$

$= 87.5\%$ **Ans.**

(*iv*) Maximum power is obtained by producing CB intersecting circle at S′ and then dropping perpendicular from point S′ on the horizontal line OF meeting output line at S. Now SS′ represents the maximum power output.

So maximum power output = 3.7 × 13,850 = 51,245 watts **Ans.**

18.36 COGGING AND CRAWLING

In the discussions of the induction motor made so far, it has been assumed that the flux in the air gap was distributed sinusoidally in space i.e. the rotating flux has a sinusoidal distribution and rotates at synchronous speed. Owing to presence of harmonics, actually it is not so, as assumed. The harmonics may be either time harmonics or space harmonics.

If the voltage applied to a polyphase induction motor contains harmonics, the air-gap flux may have components that rotate at speeds other than that corresponding to the fundamental frequency. For example, a fifth *time harmonic* would tend to rotate the rotor at 5 times the speed of the fundamental. Such harmonics are of little significance because the torques developed by such harmonics are usually very small throughout the operating range of the motor. The time harmonic voltage is usually small in proportion to the fundamental voltage, the motor reactance to the time harmonics is high and consequently harmonic currents and, therefore, harmonic torques developed are very small.

Space harmonic fields are developed by the windings, slotting, magnetic saturation, gap-length irregularity. These harmonic fields induce emfs and circulate harmonic currents in the rotor windings. These harmonic currents in the rotor interact with the harmonic fields and develop harmonic torques, vibration and noise.

With certain relationships between the number of poles and of stator and rotor slots in cage rotor motors, peculiar behaviour may be observed when the motor is started. For an example, with the number of stator slots S_1 equal to rotor slots S_2, the motor may refuse to start at all, a phenomenon called the *cogging*. With other ratios of $\frac{S_2}{S_1}$, the motor may exhibit a tendency, to run stably at a subnormal speed (e.g one-seventh of normal), a phenomenon called the *crawling*. In some cases excessive vibration may be set up at certain speeds below normal one, owing to pulsations of mechanical force on the teeth, producing *noise*. The operating efficiency of the machine is also adversely affected because of *stray load loss*. These phenomena are much less prominent in wound rotor motors, because of their higher starting torques.

The essential difference in behaviour of wound rotor and squirrel cage rotor in this context is that a cage winding, being in effect a multi-phase winding, can circulate a current under any harmonic emf developed by the gap flux, except those having wavelength equal to the pitch of the rotor bars. Wound rotors, on the other hand, have a tendency to reduce the effect of most harmonics of pole-pitch different from the coil pitch (i.e. the effect of most of the harmonics other than fundamental).

A 3-phase winding carrying sinusoidal currents produces harmonics of the order of $n = 6m \pm 1$, where m is an integer. The movement of harmonics is with or against the direction of rotation for positive and negative sign respectively. The number of poles for the nth harmonic is n times the number of poles of the fundamental and, therefore, the synchronous speed of the nth harmonic is $\frac{1}{n}$ th of the synchronous speed of the fundamental.

Thus, for $m = 1$, a 3-phase winding will produce a forward rotating 7th harmonic rotating at a speed of $\frac{1}{7}$th of synchronous speed and a backward rotating 5th harmonic rotating at a speed of $\frac{1}{5}$th of synchronous speed. The 5th and 7th harmonic fluxes may be deemed as set up by sets of additional poles superimposed on the fundamental poles. They generate rotor emfs, currents and torque of the same general torque-speed shape as that of the fundamental but with synchronous speeds equal to $\frac{1}{5}$ (backward) and $\frac{1}{7}$ (forward) of that of the fundamental, as illustrated in Fig. 18.50.

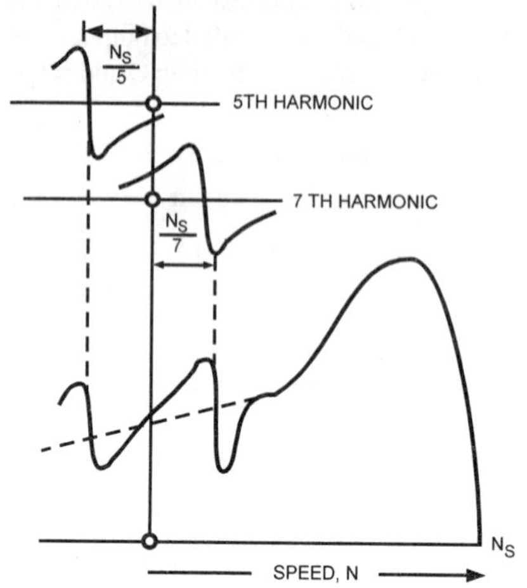

Fig. 18.50 *Dips Caused By 5th and 7th Harmonics in Torque-Speed Characteristic*

The 7th harmonic torque attains its maximum just before $\frac{1}{7}$th synchronous speed, but beyond this speed the 7th harmonic torque becomes – ve, as the slip in the harmonic field is negative. The resultant torque-speed curve combined with the fundamental shows marked dips and with certain slot ratios the dip due to 7th harmonic may become very pronounced. If the load torque is constant, the torque developed by the motor may fall below this load torque and, when this occurs, the motor cannot accelerate to its full speed but continues to run at a speed a little lower than the $\frac{1}{7}$th synchronous speed. The motor is now said to be crawling.

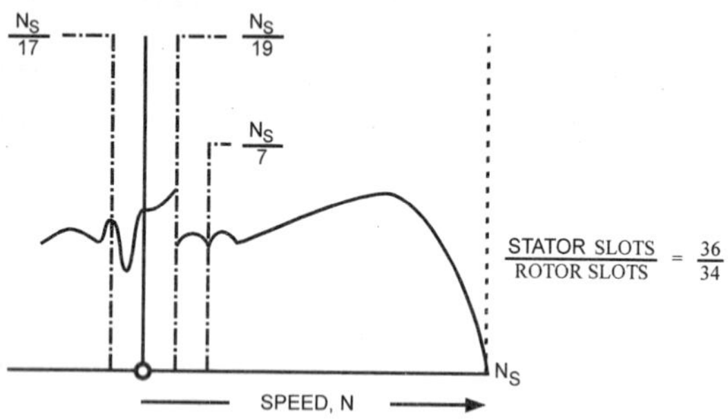

Fig. 18.51 *Dips Caused by Slot Harmonics in Torque-Speed Characteristic*

18.36.1. Harmonic Induction Torques. In a slotted m-phase winding, the emf distribution factors of harmonics of order $n = 6Ag' \pm 1$, where A is an integer and g' is the number of slots per pole per phase, are the same as those for the fundamental. A 4-pole motor with 36 stator slots ($S_1 = 36, g' = 3$) will thus encourage 19 (18 + 1)th and 17 (18 – 1)th harmonics due to slotting. The 19th harmonic field rotates forward and 17th harmonic field rotates backward. Thus the dips in the torque-speed characteristic would be produced at $+\frac{1}{19}$ and $-\frac{1}{17}$ of synchronous speed, as shown in Fig. 18.51. The effect of production of dips may be intensified by rotor slotting. For a 4-pole, 36 slot stator motor, if the rotor has 76 slots, there would be one rotor bar corresponding to every 19 th harmonic pole. Thus the 19 th harmonic torques would be very large and the rotor would vibrate considerably as illustrated in Fig. 18.52.

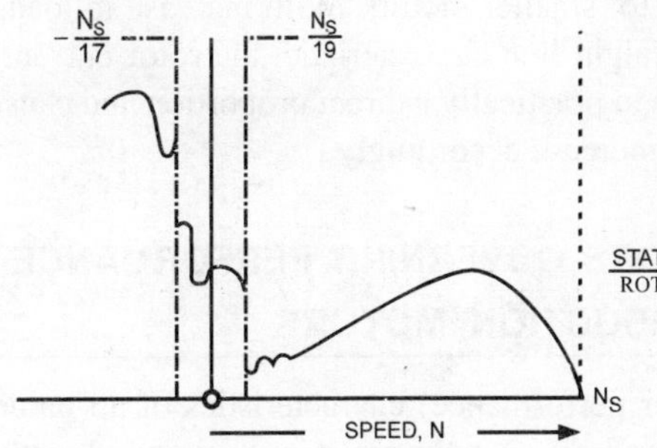

Fig. 18.52 *Torque Intensification Produced Due to Slot Harmonics Owing To Improper Slot Combination*

18.36.2. Harmonic Synchronous Torques. The rotor slotting produces space harmonics of the order $\frac{S_2}{P} \pm 1$ (plus sign for movement in the direction of rotation of the rotor). For an example, in an induction motor having 24 stator slots and 18 rotor slots (*i.e.* $S_1 = 24$ and $S_2 = 18$), the slot combination causes the stator *mmf* to have a reversed 11th and a forward 13th harmonic while the rotor has a reversed 13th and forward 15th harmonic. For a fundamental synchronous speed N_s, the stator 13th harmonic mmf rotates at speed $\frac{N_s}{13}$ with respect to stator surface and the rotor *mmf* of 13th harmonic rotates at speed $-(N_s - N)/13$ with respect to rotor when the rotor is running at speed N. These two mmfs lock into each to develop a synchronous torque when

$$\frac{N_s}{13} = N - \left(\frac{N_s - N}{13}\right)$$

$$\text{or } N = \frac{N_s}{7}$$

Thus there is a torque discontinuity at $\frac{N_s}{7}$ in the torque-speed characteristic produced not by the 7th but by the 13th harmonics. This is illustrated in Fig. 18.53.

The induction harmonic torque cannot be avoided, but can be reduced by making a proper choice of coil span and by *skewing* rotor slots. The synchronous harmonic torques can be avoided totally by a proper combination of stator and rotor slots.

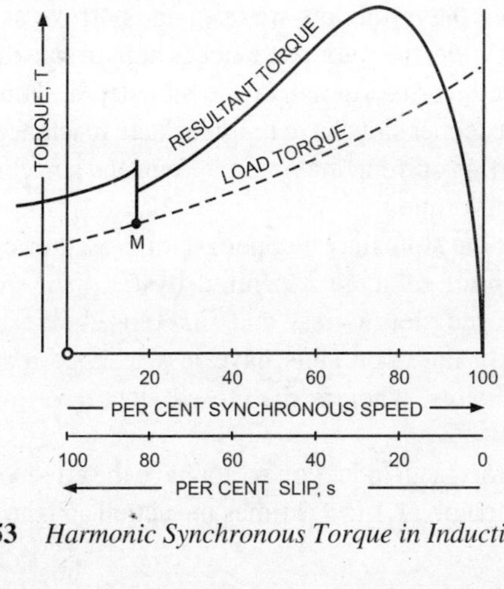

Fig. 18.53 *Harmonic Synchronous Torque in Induction Motor*

Example 18.70. Explain the terms, cogging and crawling.

A 3-phase, 400 V, 4-pole, 50-Hz cage rotor induction motor has 24 stator slots and 44 rotor slots and has a crawling speed of 115 rpm. Explain whether it is due to induction crawling or synchronous crawling.

[*U.P. Technical Univ. Electromechanical Energy Conversion II, 2002-03*]

Solution: In a slotted m-phase stator winding, the emf distribution factors of the harmonics of order $n' = 6\,Ag' \pm 1$, where A is an integer and g' is the number of stator slots per pole per phase, are the same as those of the fundamentals. A 4-pole motor with 24 stator slots ($s_1 = 24$, $g' = 2$) will thus encourage eleventh and thirteenth harmonics due to slotting. The thirteenth harmonic field rotates forward and eleventh harmonic field rotates backward. Thus the dips in the torque-speed characteristics would be produced at + 1/13 and – 1/11 of synchronous speed. For a motor having 24 stator slots and 44 rotor slots, the eleventh harmonic will have 44 half-waves in the gap, each corresponding to one rotor bar. Not only will the eleventh harmonic torque be large, but it will vibrate strongly.

The crawling speed of 115 rpm is less than $\frac{N_s}{11}\left(\text{i.e. } \frac{120 \times 50}{4 \times 11} = 136.4 \text{ rpm}\right)$ and, therefore, it is due to induction crawling.

Example 18.71. Explain crawling and cogging. A 3-phase, 430 V, 50 Hz, 4-pole squirrel cage induction motor has 24 stator slots and 28 rotor slots. What are possible crawling speeds?

[*G.B. Technical Univ. Electromechanical Energy Conversion-II, 2012-13*]

Solution: Synchronous speed of machine,

$$N_s = \frac{120\,f}{P} = \frac{120 \times 50}{4} = 1{,}500 \text{ rpm.}$$

In a slotted *m*-phase stator winding, the emf distribution factors of the harmonics of order $n' = 6\,Ag' \pm 1$ where A is an integer and g' is the number of stator slots per pole per phase, are the same as those of the fundamentals. A 4-pole motor with 24 stator slots ($s_1 = 24$, $g' = 2$) will thus encourage eleventh and thirteen harmonics due to slotting. The thirteenth harmonic field rotates forward and 11th harmonic field rotates backward.

The thirteen harmonic field produced by stator winding revolves at a speed of $\frac{N_s}{13}$ with respect to stator.

Order of backward rotating field produced by rotor slotting.

$$N = \frac{2 \times 28}{4} - 1 = 13$$

Its speed with respect to rotor is $-\frac{(N_s - N)}{13}$ where N is the speed of rotor. The rotor revolves forward at a speed N and therefore it revolves its 13th harmonic at a speed:

$$\frac{(N_s - N)}{13} + N \text{ with respect to stator}$$

Hence in order that the two fields may revolve synchronously with respect to each other we have

$$\frac{N_s}{13} = N - \frac{N_s - N}{13}$$

$$\text{or Speed of rotor, } N = \frac{N_s}{7} = \frac{1{,}500}{7} = 214.3 \text{ rpm} \quad \textbf{Ans.}$$

18.37 PERFORMANCE CURVES OF INDUCTION MOTORS

Steady-state operating characteristics of an induction motor, give graphically the variation of speed (or slip), power factor, stator current and efficiency with the variation in power output from no-load to full load. The performance curves of a 75 kW, 2200 V, 3-phase, 50 Hz squirrel cage induction motor are shown in Fig. 18.54.

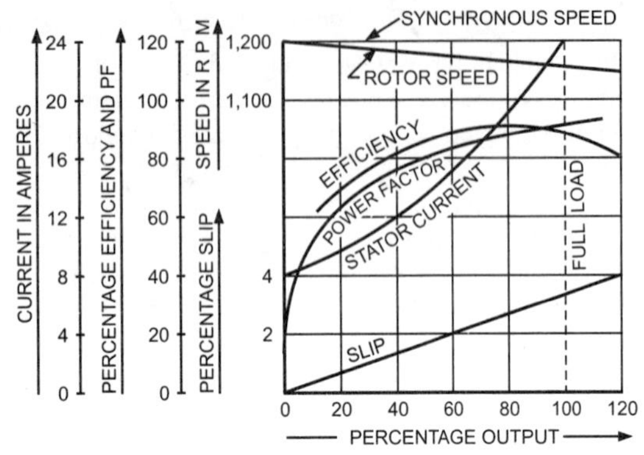

Fig. 18.54 *Performance Curves of a 75 kW, 2,200 V, 3-Phase, 50 Hz Squirrel Cage Motor*

1. Speed Characteristic. The three-phase, squirrel cage induction motor like a dc shunt motor, operates at substantially constant speed from no load to full load. As the rotor cannot attain the speed of the rotating magnetic field (i.e. synchronous speed), it must at all times operate with a certain amount of slip. Because of the extremely low impedance of the rotor, only a slight decrease in speed is enough to cause a large increase in rotor current to develop the torque necessary to drive the load. The per cent slip at no load is less than 1 per cent while at full load it is usually 2 to 5 per cent. This small percentage change in slip (or speed) from no load to full load likewise indicates why a squirrel cage induction motor is considered a constant speed motor.

2. Power Factor Characteristic. Because of the air gap between the stator and rotor of an induction motor, the reluctance of the magnetic circuit is high. As a result an induction motor draws a large magnetising current (I_m) to produce the required flux in the air gap.

The current drawn by an induction motor operating at no load is largely a magnetising current, so no-load current lags behind the applied voltage by a large angle and thus the power factor of an induction motor at no load is quite low (about 0.1 lagging). As the load is increased, the active or power component of the current increases, but the magnetising component remains practically the same. Consequently, the power factor improves. However, because of the large value of magnetising current, which is present regardless of load, the power factor of an induction motor, even at full load, seldom exceeds 90 per cent.

3. Efficiency Characteristic. At all loads, there are certain fixed losses (such as core loss, friction and windage loss) in addition to variable losses that include stator and rotor copper losses varying nearly as the square of the load. At light loads, the efficiency is quite low because the fixed losses are a relatively large part of the input. With the increase in load, the efficiency, at first, increases rapidly and attains a maximum value, the fixed and variable losses being equal at this point. Beyond this point, the copper losses (I^2 R losses) become relatively large, causing the efficiency to decrease.

The maximum efficiency occurs at about 80 to 95% of rated output, the higher values being applicable to larger motors.

4. Stator Current Characteristic. The no-load current is about 30 to 40 per cent of the rated current, the higher values being applicable to smaller motors. With increase in load, slip increases in a straight line characteristic, the rotor current will likewise increase in practically a direct proportion and cause the stator current to increase accordingly.

18.38 FACTORS GOVERNING PERFORMANCE OF INDUCTION MOTORS

The operating (or performance) characteristics of an induction motor are governed by its rotor resistance, air-gap length and shape of both stator and rotor slots and teeth.

One of the most important of the points given above is **rotor resistance.** Its value is determined primarily by the material used in construction and the area of cross section of the end-rings in case of a squirrel cage induction motor. The two most widely used materials are copper and aluminium, the latter having a resistivity that is about 1.65 times as much as that of the former. An *increase in rotor resistance* affects the motor performance as below :

With the increase in starting resistance the starting torque increases until it attains the maximum value, after which it diminishes (Art 18.13.3). An increase in rotor resistance reduces the starting current and operating efficiency and causes increase in per cent regulation. The maximum torque remains unaffected because it depends upon the rotor reactance only (Art 18.13.2).

The **air gap** is one of the two main sources of the low power factor at which the induction motor operates. In induction motors, for constant supply voltage, the air-gap flux remains almost constant. Increase in air-gap length causes increase in magnetizing current necessary to set up the air-gap flux. This reduces the no-load power factor as well as full-load power factor of the induction motor. So in order to reduce the magnetizing current it is obvious to have air gap between stator and rotor as small as practicable (it cannot be too small because extremely small value of air gap presents mechanical difficulties). Too much reduction of the air gap may cause increase in noise, pulsation losses and magnetic pull and reduce the overload capacity.

The shapes of the rotor teeth and slots affect the reactance, which, in turn, affects the starting current and the maximum torque. When the rotor slots are deep and narrow, as opposed to shallow and wide, the rotor reactance is high because slot leakage flux is directly proportional to the slot depth. Thus induction motors with deeper slots have more leakage reactance and so less value of starting current, maximum torque and slip corresponding to maximum torque.

Further the slots may be open, semi-closed or closed. For a given slot mmf reluctance offered by (*i*) open slots is more (*ii*) semi-closed slots is less and (*iii*) closed slots is still less. Consequently the open slots have less leakage reactance than semi-closed slots, whereas the closed slots have more leakage reactance.

Open slots in an induction motor have the effect of increasing the air-gap length (1.1 to 1.2 times the actual air-gap length) as a result of which more magnetizing current is required and the operating power factor of the motor becomes poor. But an induction motor with open slots has less leakage reactance and, therefore, more starting torque and more maximum torque. Likewise an induction motor with semi-closed or closed slots needs less magnetizing current and, therefore, better power factor but reduced starting torque and maximum torque due to more leakage reactance.

The slots in the rotor are not always made parallel to the shaft, but are given a slight twist or *skew*. The motor noise and vibration, cogging and synchronous cusps can be reduced or even entirely eliminated by skewing either the stator or rotor slots. The practice usually followed in India is to skew the rotor slots. In order to eliminate the effect of any harmonic, the rotor bars should be skewed through an angle so that the bar lies under alternate harmonic poles of the same polarity (i.e. bars must be skewed through two pitches). The skewing of rotor slots also results in a smoother torque curve for different positions of the rotor and causes increase in effective transformation ratio between stator and rotor, increase in rotor resistance due to increased length of bars, increase in motor impedance at a given slip and also increase in slip for a given torque.

With certain relationship between the number of poles and of stator and rotor slots in cage rotor motors, peculiar behaviour may be observed when the motor is started*. The motor may refuse to start or crawl at subnormal speed. So to avoid cogging and crawling of induction motor at start (*i*) the number of rotor slots should never be equal to stator slots but must either be larger or smaller (*ii*) the difference between the stator slots and rotor slots should not be equal to P, 2P, or 5 P to avoid synchronous cusps (*iii*) the difference between the number of stator slots and rotor slots should not be equal to 1, 2, (P ± 1) or (P ± 2) to avoid noise and vibration and (*iv*) the difference between the number of stator slots and rotor slots should not be equal to 3 P for 3-phase machines in order to avoid magnetic locking.

Harmonic torques can also be reduced by chording, integral slot winding and increasing air-gap length in addition to by skewing of rotor slots. Chorded windings with integral number of slots per pole per phase weakens the stator winding mmf harmonics and so eliminates the harmonic induction torques. Windings with fractional number of slots per pole per phase create asymmetrical mmf distribution around the air gap and produces noise in the motor. So fractional slot windings are not employed for induction motor stators and only integral slot windings are employed. Though by increasing the air-gap length harmonic torques can be reduced but it will cause increase in no-load current and so in poor power factor.

18.39 HIGH TORQUE CAGE MOTORS (DEEP BAR AND DOUBLE CAGE MOTORS)

The great simplicity and robustness of the cage motor are very valuable characteristics of the motor from the standpoint of cost and maintenance. The main drawback of a squirrel cage motor is its poor starting torque, because of its low rotor resistance. If the rotor were made of high resistance (to obtain a good starting torque with reasonable starting current) its rotor copper loss will increase and, therefore, efficiency will be reduced and full-load slip (speed regulation) will increase. This also is not desirable. The difficulty with a cage rotor is that its cage is permanently short circuited, so there is no possibility of introducing external resistance temporarily in the rotor circuit at the start. Many efforts have been made to build squirrel cage motors with improved starting characteristics without sacrificing significantly its excellent running characteristics. Increased starting torque with short-circuited rotors can be obtained by using two types of rotors (either deep bar rotors or double cage rotors) which employ change of rotor circuit impedance with change in rotor frequency to have high starting torque and also high efficiency under normal running conditions.

18.39.1. Deep Bar Cage Rotor Motor. In deep bar motors use is made of skin effect with rotor bars caused by the slot leakage fluxes.

Let us first consider the phenomena at starting. At starting the rotor frequency is equal to the supply frequency. The slot leakage flux paths of the rotor for these conditions are depicted in Fig. 18.55(*a*). A glance at this figure shows that bar portions of different heights are linked by different number of leakage flux lines; the lower parts with greatest number of lines, the upper parts with the least number. For this reason the maximum leakage emf is induced in the lower parts of the bar, the minimum emf in the upper parts of the bar. The result is that the current in the rotor bars is forced to flow in the top layers of the bars. The current density is distributed along the conductor height as illustrated by curve 1 in Fig. 18.55(*b*). This is called the *skin effect*. With the usual conductor height of 10-12 mm, the skin effect is almost unnoticeable. In deep bar motors, where the bars are 20-25 mm high (the depth-to-width ratio being 10 or 12 to 1), the effect is very strong and appreciably changes the rotor parameters.

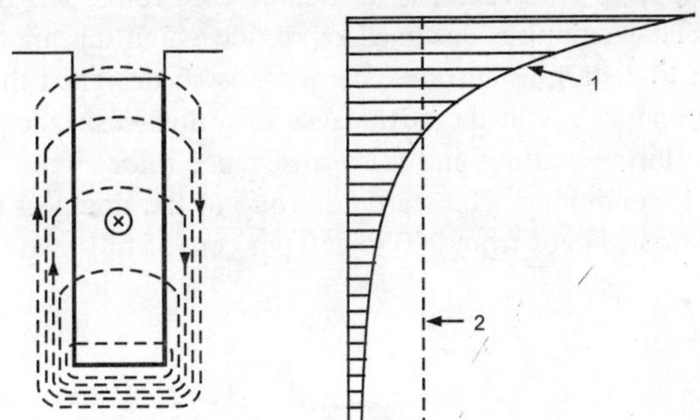

Fig. 18.55

Non-uniform current distribution causes greater ohmic loss meaning thereby that the effective bar resistance (ac resistance) becomes much more than its dc resistance. As a result, the starting torque increases and the starting current decreases. As the motor accelerates, the frequency of rotor currents decreases to values corresponding to steady state operating conditions. The skin effect gradually diminishes, and the current becomes more uniformly distributed over the conductor cross section, as shown by dotted line 2 in Fig. 18.55(*b*). The resistance of the rotor gradually decreases, and the inductive reactance related to the

* For details refer to Art. 18.36.

circuit frequency increases. At normal speed the deep bar motor is a plain squirrel-cage motor with the usual resistance, but with a somewhat higher inductive react-ance, owing to which the power factor, efficiency and the overload capacity of the motor are somewhat reduced.

Fig. 18.56 *Cutaway View of Deep Bar Rotor*

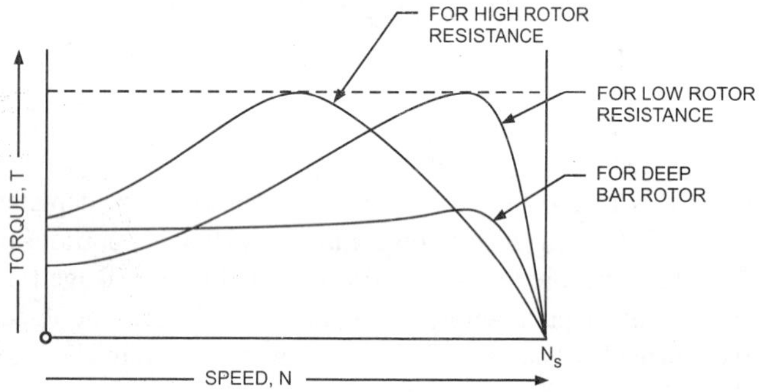

Fig. 18.57

Cutaway view of deep bar rotor is given in Fig. 18.56. The torque-speed characteristic of a deep bar rotor compared to a normal-bar design with low and high rotor resistances is illustrated in Fig. 18.57.

18.39.2. Double Cage Rotor Motor. A rotor design, which though expensive gives still better starting and running characteristics than the deep bar design, is the double cage rotor. The double squirrel cage motor is designed to provide a high starting torque with a low starting current. The rotor is so designed that the motor operates with the advantages of a high-resistance rotor circuit during starting and a low-resistance rotor circuit under running conditions. The starting torque of the double squirrel cage motor ranges from 200 to 250 per cent of full-load torque with a starting current of 400 to 600 per cent of full-load value. It is classed as a high-torque low starting current motor.

The rotor of a double squirrel cage motor carries two squirrel cage windings embedded in two rows of slots. The outer slots contain a high resistance and low leakage reactance winding and the inner slots contain a low resistance and high leakage reactance winding.

The outer winding bars are usually of manganese brass while the inner winding bars are of red copper. The inner winding has greater cross section than that of outer winding in order to have resistance one fourth or one fifth of outer winding. The end rings may be separate or combined.

At starting, the rotor current has the same frequency as the supply current, therefore, leakage reactance of the inner winding is high. As a result (*i*) the impedance of the outer winding is less than that of the inner winding and so bulk of the current flows through the outer high resistance winding and (*ii*) the phase difference between the induced emf and current in the inner winding will be large and the torque will be small. Because of high resistance and low leakage reactance of the outer winding, the phase difference between the induced emf and current in it will be small and, therefore, high starting torque will be developed.

At normal speed the leakage reactance of the squirrel cage windings is of minor importance and the current divides between them almost inversely as their resistances. Because of low resistance inner winding carries bulk of the current and provides greater proportion of torque.

Thus, the outer cage has high resistance and low reactance to resistance ratio, whereas the inner cage has low resistance and large reactance-to-resistance ratio. Thus the outer cage produces high starting and accelerating torque, while the inner winding provides the running torque at good efficiency.

The torque-speed characteristics of individual cages as well as resultant are shown in Fig. 18.59. It may be assumed to a first approximation, the two cages develop two separate torques, as shown in the figure. The resultant torque-speed characteristic can be modified and any desired characteristic can be obtained by modifying the individual cage resistances and leakage reactances. The resistances can be changed by changing the areas

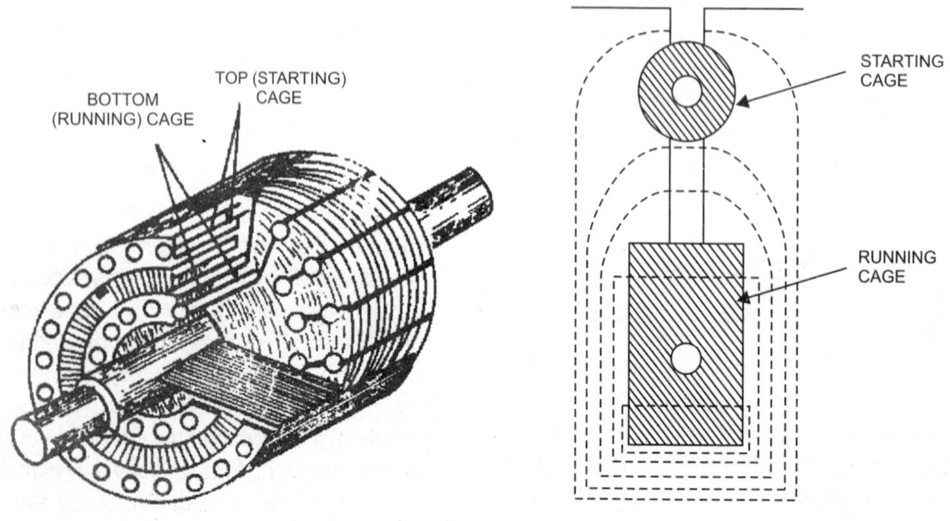

Fig. 18.58

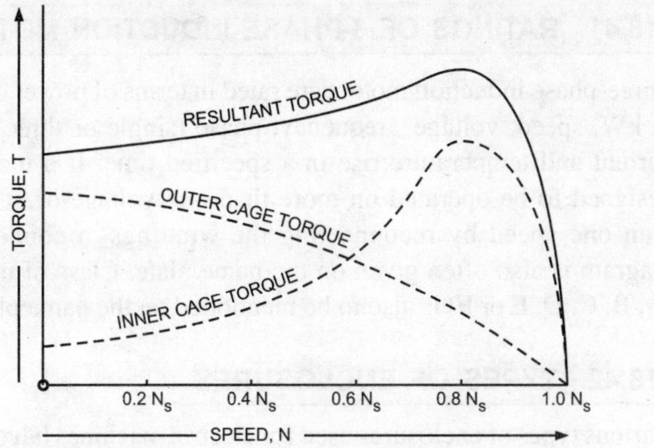

Fig. 18.59

of cross section of bars while the leakage reactances can be changed by changing the width of the slot openings and the depth of the inner cage.

The pull-out torque of a double cage motor is smaller than that of a plain squirrel cage motor because two cages produce their maximum torques at two different speeds. Also, the additional reactance of the inner cage reduces the full-load power factor while the high resistance of the outer cage increases the full-load copper loss thereby reducing the operating efficiency.

The cost of a double cage motor is 20 to 30 per cent higher than that of plain squirrel cage motor.

A deep bar rotor construction is employed for higher starting torque applications and double cage construction for still higher starting requirements. For large size motors with stringent starting torque requirements, the most expensive wound rotor construction is employed.

18.40 EQUIVALENT CIRCUIT OF DOUBLE CAGE MOTOR

Let the resistance and reactance per phase of stator winding be R_1 and X_1 respectively, also let the phase resistance and reactance of outer and inner cages referred to the stator be $\frac{R_{2o}}{K^2}$ and $\frac{X_{2o}}{K^2}$ and $\frac{R_{2i}}{K^2}$ and $\frac{X_{2i}}{K^2}$ respectively. Since both cages completely link the main flux so the impedance of the two cages can be considered in parallel. The equivalent circuit for the slip *s* for one of the phases and referred to primary (or stator) is shown in Fig. 18.60.

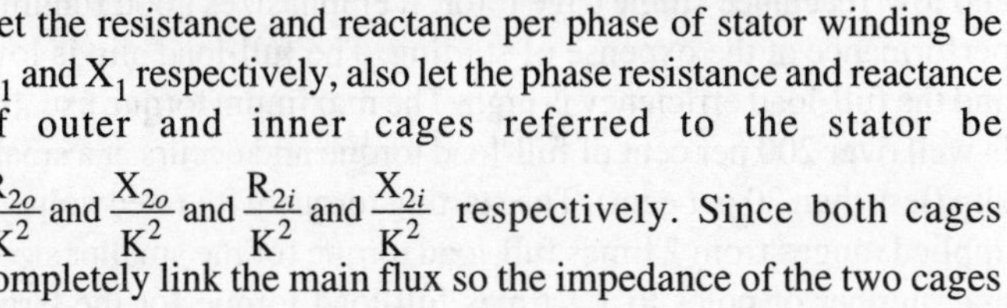

Fig. 18.60 *Equivalent Circuit of a Double Cage Motor*

Example 18.72. A double cage motor has standstill impedance of (1 + *j*1) Ω and (0.3 + *j*5) Ω for its two cages. Compare the relative torques of the two cages (*i*) at standstill and (*ii*) at a slip of 0.05.

[U.P.S.C. I.E.S. Electrical Engineering-II, 2001]

Solution: The equivalent circuit diagram of the double cage motor is shown in Fig. 18.61.

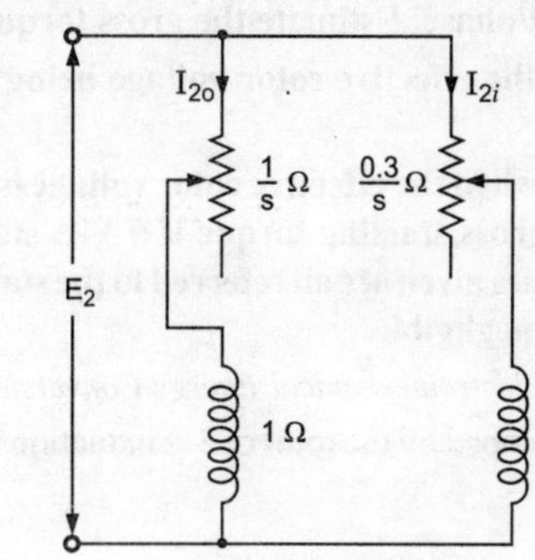

Fig. 18.61

At Standstill

(*i*) The slip, $s = 1$

Outer cage impedance, $\mathbf{Z}_{2o} = \frac{1}{1} + j1 = (1 + j1)\,\Omega = \sqrt{2}\;\angle 45°\;\Omega$

$$\text{Outer cage current, } I_{2o} = \frac{E_2}{Z_{2o}} = \frac{E_2}{\sqrt{2}}\text{ A}$$

Power input to outer cage,

$$P_{2o} = I_{2o}^2 R_{2o} = \left(\frac{E_2}{\sqrt{2}}\right)^2 \times 1 = \frac{E_2^2}{2} \text{ watts}$$

Inner cage impedance,

$$\mathbf{Z}_{2i} = \frac{0.3}{1} + j5 = (0.3 + j5) = 5.009\;\angle 86.57°\,\Omega$$

$$\text{Inner cage current, } I_{2i} = \frac{E_2}{Z_{2i}} = \frac{E_2}{5.009} = 0.1996\,E_2$$

Power input to inner cage,

$$P_{2i} = I_{2i}^2 \times R_{2i} = (0.1996 E_2)^2 \times 0.3 = 0.011957 E_2^2$$

$$\frac{\text{Torque of outer cage, } T_o}{\text{Torque of inner cages, } T_i} = \frac{P_{2o}}{P_{2i}} = \frac{E_2^2/2}{0.011957 E^2} = 41.82 \quad \textbf{Ans.}$$

***(ii)* At a slip of 0.05**

Outer cage impedance,

$$\mathbf{Z}_{2o} = \left(\frac{1}{0.05} + j1\right) = (20 + j1) = 20.025\;\angle 2.8°\,\Omega$$

Outer cage current,

$$\mathbf{I}_{2o} = \frac{E_2}{Z_{2o}} = \frac{E_2}{20.025} = 0.0499\,E_2$$

Power input to outer cage,

$$P_{2o} = I_{2o}^2 R_{2o} = (0.0499\,E_2)^2 \times 20 = 0.0499\,E_2^2$$

Inner cage impedance,

$$\mathbf{Z}_{2o} = \left(\frac{0.3}{0.05} + j5\right) = (6 + j5) = 7.81\;\angle 39.8°\,\Omega$$

Inner cage current,

$$I_{2i} = \frac{E_2}{Z_{2i}} = \frac{E_2}{7.81} = 0.128\,E_2$$

Power input to inner cage,

$$P_{2i} = I_{2i}^2 R_{2i} = (0.128 E_2)^2 \times 6 = 0.09836 E_2^2$$

$$\frac{\text{Torque of outer cage, } T_o}{\text{Torque of inner cage, } T_i} = \frac{P_{2o}}{P_{2i}} = \frac{0.0499 E_2^2}{0.09836 E_2^2} = 0.5 \quad \textbf{Ans.}$$

Example 18.73. The two-cages of a 3-ϕ, 50 Hz, 4-pole, delta-connected induction motor have respective standstill leakage impedances of (2 + *j*8) and (9 + *j*2)Ω/phase. Estimate the gross torque developed:

(*i*) at standstill, the effective rotor voltage being 230 V/phase, and

(*ii*) at 1,450 rpm when the effective rotor voltage is 400 V/phase. What is the gross starting torque if a Y-Δ starter is used? Rotor quantities given are all referred to the stator; the stator impedance is negligible.

[*U.P. Technical Univ. Electromechanical Energy Conversion-II, 2013-14*]

Solution: Torque developed by the rotor of 3-ϕ induction motor is given as

$$T = \frac{3 \times 60}{2\pi N_s} \cdot \frac{R_2 \, sE_2^2}{R_2^2 + s^2 X_2^2} \qquad \text{Refer to Eq. (18.36)}$$

$$\text{Torque developed at standstill} = \frac{3 \times 60}{2\pi N_s} \cdot \frac{R_2 E_2^2}{R_2^2 + X_2^2} \qquad \because \text{At standstill } s = 1.0$$

Torque developed by inner cage at standstill,

$$T_i = \frac{3 \times 60}{2\pi N_s} \times \frac{R_{2i} E_2^2}{R_{2i}^2 + X_{2i}^2} \qquad ...(i)$$

Torque developed by outer cage at standstill,

$$T_o = \frac{3 \times 60}{2\pi N_s} \times \frac{R_{2o} E_2^2}{R_{2o}^2 + X_{2o}^2} \qquad ...(ii)$$

Total torque developed by double cage induction motor,

$$T = T_i + T_o = \frac{3 \times 60}{2\pi N_s} \times \frac{R_{2i} E_2^2}{R_{2i}^2 + X_{2i}^2} + \frac{3 \times 60}{2\pi N_s} \times \frac{R_{2o} E_2^2}{R_{2o}^2 + X_{2o}^2}$$

$$= \frac{3 \times 60}{2\pi \times N_s} E_2^2 \left[\frac{R_{2i}}{R_{2i}^2 + X_{2i}^2} + \frac{R_{2o}}{R_{2o}^2 + X_{2o}^2} \right] \qquad ...(iii)$$

Synchronous speed of motor,

$$N_s = \frac{120 f}{P} = \frac{120 \times 50}{4} = 1{,}500 \text{ rpm}$$

(*i*) Substituting values of N_s, R_{2i}, X_{2i}, R_{2o}, X_{2o} and E_2 in Eq. (*iii*) we have

$$T = \frac{3 \times 60}{2\pi \times 1{,}500} \times 230^2 \left[\frac{2}{2^2 + 8^2} + \frac{9}{9^2 + 2^2} \right]$$

$= 136.7$ N-m **Ans.**

$$(ii) \text{ Slip, } s = \frac{N_s - N}{N_s} = \frac{1{,}500 - 1{,}450}{1{,}500} = \frac{1}{30}$$

Voltage induced in rotor while running is 400 V/phase

i.e., $sE_2 = 400$

or Voltage induced in rotor at standstill,

$$E_2 = \frac{400}{s} = \frac{400}{1/30} = 12{,}000 \text{ V}$$

Substituting again values of E_2, N_s, R_{2i}, X_{2i}, R_{2o} and X_{2o} in Eq. (*iii*) we have

$$T = \frac{3 \times 60}{2\pi \times 1{,}500} \times 12{,}000^2 \left[\frac{2}{2^2 + 8^2} + \frac{9}{9^2 + 2^2} \right]$$

$= 372{,}114$ N-m

Gross torque developed in star-delta starting

$= \frac{1}{3} \times$ torque developed in direct staring

$= \frac{1}{3} \times 372{,}114 = 124{,}038$ N-m **Ans.**

18.41 RATINGS OF 3-PHASE INDUCTION MOTORS

Three-phase induction motors are rated in terms of power output in kW, speed, voltage, frequency, phase (single or three), line current and temperature rise in a specified time. If a motor is designed to be operated on more than one voltage or at more than one speed by reconnecting the windings, a connection diagram is also often given on the name plate. Class of motors (A, B, C, D, E or F) is also to be mentioned on the name plate.

18.42 TYPES OF ENCLOSURES

Various types of enclosures used for electric machines have been described in Art 4.22.

18.43 STANDARD TYPES OF SQUIRREL CAGE MOTORS

The three-phase squirrel cage induction motors are available in the market in a range of standard ratings up to 150 kW at various standard frequencies, voltages and speeds in order to meet the usual requirements of industry. Large motors are generally regarded as a special-purpose rather than general-purpose motors. The motor manufacturers of the United States through the National Electrical Manufacturers Association (NEMA) have standardized the types of 3-phase squirrel cage induction motors according to their electrical characteristics into six types designated as class A, B, C, D, E and F respectively. The original commercial squirrel cage induction motors, which were of the shallow slot type, are designated as class A. For this reason class A motors often are used as a reference and referred to as normal starting torque, normal starting current, normal slip motors.

1. *Class A Motors (Low Impedance Squirrel Cage Rotor Motors).* The motors of this class usually have a low-resistance and low-reactance single cage rotor. It emphasizes good running performance at the expense of starting. The full-load slip is low and the full-load efficiency is high. The maximum torque usually is well over 200 per cent of full-load torque and occurs at a small slip (less than 20 per cent). The starting torque with rated voltage applied ranges from 2 times full-load torque for the smaller sizes and number of poles to 1.1 times full-load torque for the large sizes and number of poles. The high starting current (5 times to 8 times full-load current when started at rated voltage) is the main drawback of this class. The motors of this type and of rating above 4 kW require reduced voltage for starting. Reduced voltage starting involves extra auxiliary equipment and results in decrease in starting torque. Such motors are used for machine tools, centrifugal pumps, motor-generator sets, fans, blowers, compressors, and other applications requiring low-starting torque.

The general configuration of slot construction of such motors is shown in Fig. 18.62. As seen the rotor bars are placed close to the surface of rotor, resulting in relatively low reactance.

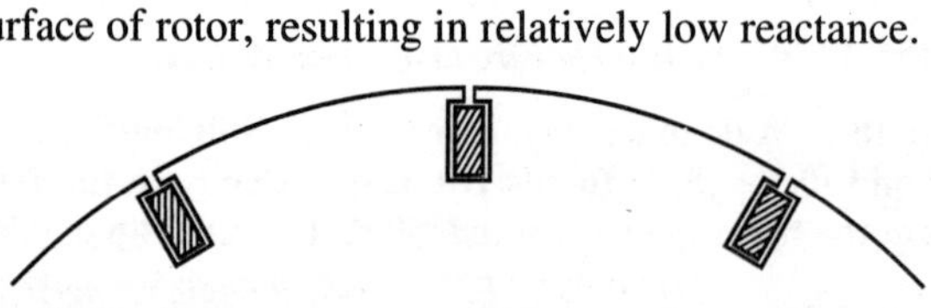

Fig. 18.62 *Rotor Construction of Class A Squirrel Cage Motor*

2. *Class B Motors (High Reactance Squirrel Cage Rotor Motors).* This type of motor has largely replaced the class A motors for new installations as its running characteristics are quite similar to those of the class A but on account of its limited starting current the voltage need not be reduced for starting. The starting current with rated voltage applied is usually about 5 to $5\frac{1}{2}$ times the full-load current. The starting torque, full-load slip and efficiency are good — about the same as for class A motors. However, the use of high reactance slightly decreases the power factor and decidedly lowers the maximum torque. In motors of this class high reactance is obtained by employing totally enclosed, deep narrow slots.

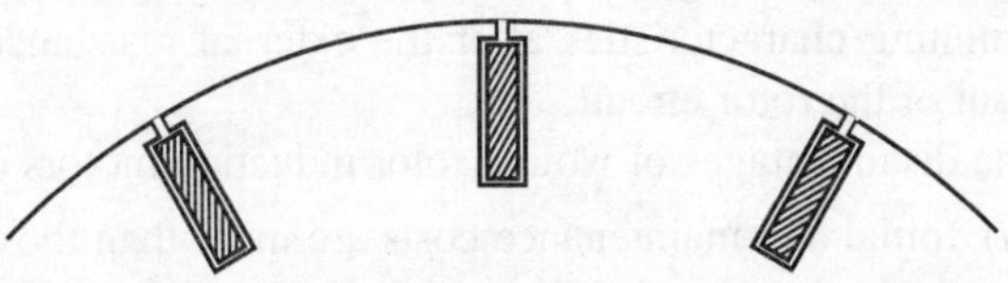

Fig. 18.63 *Rotor Construction of Class B Squirrel Cage Motor*

The design is the commonest in the 5-150 kW range of sizes. Its applications are constant-speed drives where starting-torque requirements are not severe, such as in driving fans, blowers, pumps and machine tools.

3. *Class C Motors (High Starting Torque, Low Starting Current Double Squirrel Cage Rotor Motors).* These motors are of double squirrel cage type (Fig. 18.64) and give high starting torque with low starting current. The starting current and slip with rated voltage applied are in the same range as for class B motors. The starting torque with normal voltage applied is 2.75 times the full-load torque. However, operating efficiency is lower and the slip is higher as compared to those of class A and class B motors. Typical applications are in driving air compressors, conveyors, reciprocating pumps, crushers, mixers, large refrigerating machines and other equipments requiring high starting torque (say starting torque of more than 1.5 times full-load torque).

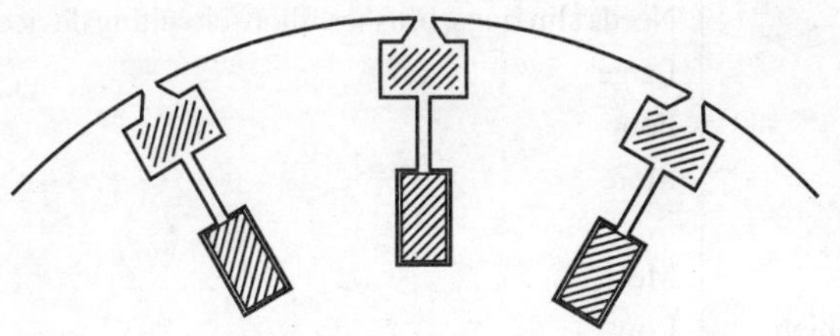

Fig. 18.64 *Rotor Construction of Class C Squirrel Cage Motor*

4. *Class D Motors. (High Resistance Squirrel Cage Rotor Motors).* Such motors are provided with a high resistance squirrel cage rotors and, therefore, give high starting torque with low starting current. The starting current with normal voltage applied are of the same order as for class C motors. The full-load slip varies from 5% to 20% depending upon the application for which the motor is designed. The operating efficiency of motors of this type is also low. Class D motors are used only for driving intermittent loads involving high accelerating duty and for driving high impact loads such as punch presses, shears, bulldozers, small hoists, flywheel machinery and metal drawing equipment. When driving high impact loads, the motor is generally provided with a flywheel, which reduces the pulsations in power drawn from the supply system.

For obtaining high starting torque with low starting current, thin rotor bars are used which make the leakage flux in the rotor low and useful flux high.

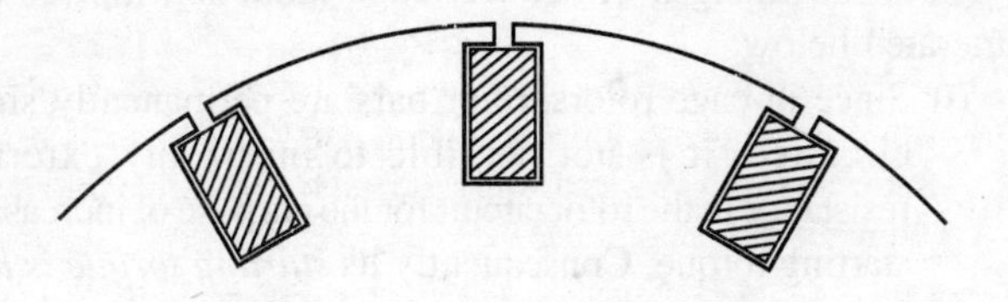

Fig. 18.65 *Rotor Construction of Class D Squirrel Cage Motor*

5. *Class E Motors.* These motors have a relatively low starting torque, normal starting current and have a relatively low slip at rated load. The class E motors of rating above 5 kW require compensators or resistance starters.

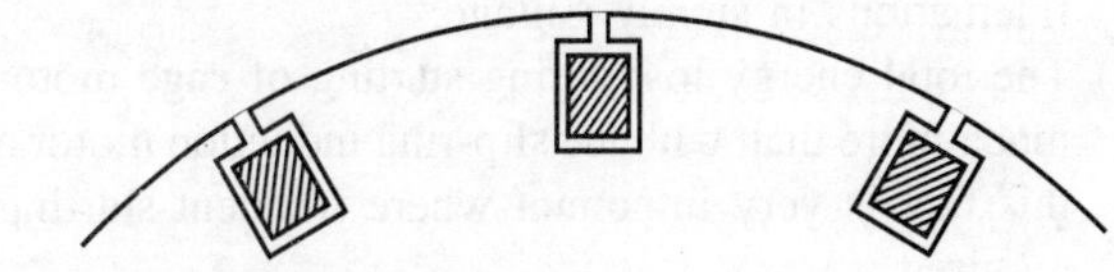

Fig. 18.66 *Rotor Construction of Class E Squirrel Cage Motor*

6. *Class F Motors.* The class F motors have a relatively low starting torque, low starting current and normal slip. The starting current with normal voltage applied and the full-load slip are of the same order as for class B and C motors. The starting torque with rated voltage applied is about 1.25 times of full-load torque. Its starting current is low due to the design of the rotor which has high reactance during starting.

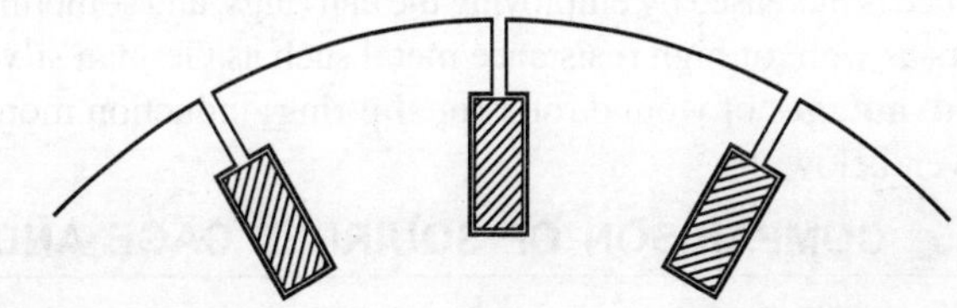

Fig. 18.67 *Rotor Construction of Class F Squirrel Cage Motor*

18.44 ADVANTAGES, DISADVANTAGES AND APPLICATIONS OF INDUCTION MOTORS

Squirrel cage induction motors are comparatively *cheaper* than wound rotor induction motors and are *simple* and *rugged* in construction.

A cage rotor needs considerably less conductor material than a wound rotor, so copper loss in cage rotor is low and *efficiency* of a cage motor is a *little more* than that of a wound rotor motor.

Cage motor is more rugged in construction and needs no slip rings, brushes etc., therefore, its *maintenance cost is low.*

Cage motor is *explosion-proof,* because the absence of slip rings and brushes eliminates risk of sparking.

Cage motor can be *cooled better* because of its bare end rings. Also there is more space for rotor fans.

A squirrel cage induction motor has very small length of overhang, so it has low rotor overhang leakage flux. This causes standstill reactance X_2 low for a cage rotor than wound rotor. So a cage motor has *more pull-out torque, greater maximum power output* and *better operating power factor* as compared to a wound rotor motor.

Other advantages of squirrel cage motors are (*i*) nearly constant speed, high overload capacity and simple starting arrangement which may be made automatic.

The disadvantages of squirrel cage induction motors are enumerated below:

(*i*) Since in cage rotors, rotor bars are permanently short circuited, it is not possible to insert any external resistance in the rotor circuit for the purpose of increasing starting torque. Consequently its *starting torque is low* with *large starting current* : — starting torque varying between 1.5 and 2 times the full-load torque with starting current 5 to 9 times the full-load current. *Power factor* is also *low* at start in case of cage motors.

(*ii*) No possibility of speed control.

(*iii*) Squirrel cage induction motors are very sensitive to fluctuations in supply voltage.

(*iv*) The total energy lost during starting of cage motor is much more than with the slip-ring induction motor and this fact is very important where frequent starting is required.

Squirrel cage induction motor is suitable for constant speed industrial drives of small power where speed control is not required and where starting torque requirements are of medium or low value, such as for printing machinery, flour mills and other shaft drives of small power.

When the squirrel cage motors are used for crane or hoist work where large starting torque is of more importance than high efficiency owing to the intermittent nature of load, the rotor resistance is increased by employing the end rings, and sometimes the bars as well, of high resistance metal such as German silver.

Advantages of wound rotor (or slip-ring) induction motors are given below :

(*i*) *High starting torque* with low starting current. By inserting an external resistance of proper value (so that rotor resistance R_2 becomes equal to its standstill reactance X_2) in each phase of the rotor circuit, maximum torque can be obtained at start.

(*ii*) *Adjustable speed.* Speed can be controlled easily in case of a wound rotor motor simply by adding external resistors in the rotor circuit. This is the reason that wound rotor induction motors are considered to be variable speed motors.

Other advantages are high overload capacity, nearly constant speed, smooth acceleration under heavy loads, no abnormal heating during starting, adoptability to automatic starting and good running characteristics after the external resistances are taken out of the rotor circuit.

The disadvantages of wound rotor induction motors are :

(*i*) Initial and maintenance costs are more than those of a squirrel cage motor, because a slip-ring induction motor needs slip rings, brushes, short-circuiting devices etc.

(*ii*) Speed regulation is poor when operated with external resistances in the rotor circuit.

(*iii*) Lower efficiency and power factor as compared with cage rotor motor.

(*iv*) Low power factor at light loads.

(*v*) Sensitivity to fluctuations in supply voltage.

Wound rotor (or slip-ring) induction motors are used for loads requiring severe starting conditions or for loads requiring speed control such as for driving line shafts, lifts, pumps, generators, winding machines, cranes, hoists, elevators, compressors, small electric excavators, printing presses, turn tables, strokers, large ventilating fans, crushers etc.

18.45 COMPARISON OF SQUIRREL CAGE AND SLIP-RING (PHASE WOUND) INDUCTION MOTORS

S.No.	*Characteristics*	*Squirrel Cage Induction Motor*	*Slip-Ring Induction Motor*
1.	Construction	Simple and rugged	Needs slip rings, brushes, short-circuiting device etc.
2.	Overhang	Less	Large
3.	Space factor in slots	Better	Poor
4.	Cost (initial as well as maintenance)	Less	More
5.	Copper losses	Small	More
6.	Efficiency	High (only for machines, not designed for high starting torque)	Low
7.	Power factor	Better	Poor
8.	Cooling	Cooling better because of its bare end rings and availability of more space for rotor fans	Not quite efficient
9.	Speed regulation	Better	Poor when operated with external resistances in the rotor circuit
10.	Starting	Simple	The motor needs slip rings, brushgear, short-circuiting device and starting resistors etc.
11.	Starting torque	Starting torque is low with large starting current	Possibility of increasing starting torque by insertion of external resistances in the rotor circuit
12.	Power factor at start	Poor	Can be improved
13.	Speed control	No possibility	Possible by insertion of external resistor in the rotor circuit
14.	Protection against explosion	Explosion proof	Not explosion proof

18.46 COMPARISON BETWEEN SYNCHRONOUS AND INDUCTION MOTORS

S. No.	Synchronous Motors	Induction Motors
1.	It has got no self-starting torque and some external means is required for its starting.	It has got self-starting torque and no special means is required for its starting.
2.	Its average speed is constant and independent of load.	Its speed falls with the increase in load and is always less than synchronous speed.
3.	It can be operated under a wide range of power factor, both lagging and leading.	It operates at only lagging power factor, which becomes very poor at light loads.
4.	It requires dc excitation so it is a doubly excited machine.	It requires no dc excitation so it is a singly excited machine.
5.	No speed control is possible.	Speed can be controlled but to small extent.
6.	It is used for supplying mechanical load as well as for power factor improvement.	It is used for supplying mechanical load only.
7.	Its torque is less sensitive to change in supply voltage.	Its torque is more sensitive to change in supply voltage.
8.	Breakdown torque is proportional to the supply voltage.	Breakdown torque is proportional to the square of the supply voltage.
9.	It is more complicated and costs more comparatively. However, synchronous motors with speeds below 500 rpm and ratings exceeding about 40 kW or with medium speeds from 500 to 1,000 rpm and ratings exceeding about 500 kW cost less than induction motors.	It is more simple and costs less comparatively.

18.47 EFFECTS OF OPERATING CONDITIONS

(i) *Effect of Loading.* With the increase in load, speed falls and consequently torque increases. A point will reach when the maximum torque will be developed. A further increase in load causes further drop in speed, consequently the driving torque decreases and the rotor stops ultimately.

(ii) *Effect of Unbalance Supply Voltage.* If the 3-phase supply to the motor is not balanced, the rotating magnetic field will move at a non-uniform rate and in a non-uniform strength. This will cause more unbalanced currents in the stator windings and, therefore, produce an unequal heating.

(iii) *Effect of Break in One Phase.* If one phase of a polyphase induction motor breaks, may be due to any reason, when the motor is in operaton, the motor will continue to operate single phase, provided that the load does not exceed 57.7% of the normal rating, with about the same temperature rise as when carrying rated load as a 3-phase induction motor. It will not, however, start, single phase but if brought up to speed by some external means, such as pulling on the belt, it will continue to operate satisfactorily on load below 57.7% of rated one.

A break in one phase of the rotor of a wound rotor induction motor may prevent it from starting but if it is brought up to speed before the circuit is opened, it will operate under reduced load but usually with considerable vibrations.

(iv) *Effects of Variation in Line Frequency and Line Voltage.* It is usual for manufacturers to guarantee satisfactory operation of motors with only slight altered characteristics with voltage or frequency variations amounting to not more 10% above or below normal, provided, however, that when both voltage and frequency vary at the same time the combined variation shall not exceed 10% above or below.

Since the speed of an induction motor depends upon the speed of the revolving field, which in turn depends upon the line frequency, hence any variation in line frequency will cause corresponding changes in motor speed. But with modern power systems there is seldom any marked change in frequency, hence no trouble need be anticipated on this account. It should be understood, however, that the motors designed for a certain commercial frequency should not be operated upon a system of decidedly different frequency, otherwise aside from the resulting change in speed there would be changes in the torque which the motor would develop and in the operating temperature of its various parts and thus motor operation will not be satisfactory.

With voltages below normal the torque developed is reduced in proportion to the square of the voltage; therefore, with a below-normal supply voltage the motor may not be able to start its customary load or may become stalled on a small overload. If the motor is operated with its rated kW load at reduced voltage, the strength of the stator field will be weaker comparatively, resulting in decrease in speed as well as in core losses but increase in stator and rotor currents and the operating temperature may become abnormal owing to increased copper losses. If, on the other hand, the motor is operated on line voltage above normal, the stator field strength will be increased resulting in increase in speed as well as core losses but decrease in input current and, therefore, reduction in copper losses. The tendency in either case is to overheat some part of the machine even though the temperature as a whole does not become abnormal.

The variation in supply frequency and supply voltage will not affect only the speed and torque of the motor but also the starting current, full-load current, starting torque, maximum torque (as given below) and operating temperature of various parts of the machine and thus make their operation unsatisfactory.

Variation in	*Synchronous speed*	*%age slip*	*Full-load speed*	*Starting current*	*Full-load current*	*Starting torque and maximum running torque*
Supply voltage, V	Constant	$\alpha \frac{1}{V^2}$	Decreases with the decrease in V	α V	Increases with decrease in V	α V^2
Supply frequency, f	αf	Constant	αf	$\alpha \frac{1}{f}$	Increases slightly with decrease in f	$\alpha \frac{1}{f^2}$

HIGHLIGHTS

1. An induction motor is simply a transformer whose magnetic circuit is separated by an air gap into two relatively movable portions, one carrying the primary winding and the other the secondary winding.

   It essentially consists of (*i*) a stator and (*ii*) a rotor.

   Rotors are of two types (*i*) squirrel-cage rotor and (*ii*) phase-wound or wound rotor.

2. In a polyphase induction motor, slip-frequency currents are induced in the rotor windings as the rotor slips past the synchronously rotating stator flux wave. These rotor currents in turn produce a flux wave which rotates in synchronism with the stator flux wave, the torque is produced from the interaction of these two flux waves. For increased load on the motor, the rotor speed falls, resulting in larger slip and increased induced rotor currents, and greater torque.

   Examination of the flux-mmf interactions in a polyphase induction motor shows that, electrically, the machine is a generalized transformer. The synchronously rotating air-gap flux wave in the induction machine is the counterpart of the mutual core flux in the transformer. The rotating field induces emfs of stator frequency in stator windings and of slip frequency in the rotor windings for all rotor speeds other than synchronous speed. Thus, the induction machine transforms voltages and at the same time changes frequency. When viewed from the stator, all rotor electrical and magnetic phenomena are transformed to stator frequency. The rotor mmf reacts on the stator windings in the same way that the mmf of secondary current in a transformer reacts on the primary.

3. *Working principle.* When a 3-ϕ supply is provided to a 3-phase wound stator of an induction motor, a rotating magnetic field is established rotating at synchronous speed. At start, stationary rotor conductors cut across the revolving magnetic field and an emf is induced in them by the electromagnetic phenomenon. Current flows through the rotor conductors, as they form a closed path, and so rotor field is developed. By interaction of stator magnetic field and rotor field, torque develops and causes the rotation of rotor in the same direction as that of revolving magnetic field.

4. The direction of rotation of a 3-phase induction motor can be reversed by interchanging the connections to the supply of any two terminals.

5. *Synchronous speed,* $N_s = \frac{120\,f}{P}$, f is supply frequency and P is the number of poles on stator.

6. **Slip.** The rotor never attains the synchronous speed because at that speed there would be no relative speed between rotor conductors and stator revolving field and induction phenomenon is not possible. Its speed is always less than synchronous speed.

   The difference between the speed of the stator field, known as synchronous speed (N_s), and the actual speed of the rotor (N) is known as the slip and is denoted by s

$$s = \frac{N_s - N}{N_s}.$$

7. Both the stator and rotor fields rotate synchronously, which means that they are stationary w.r.t. each other at all possible rotor speeds.

8. Frequency of rotor current or emf, $f' = sf$.

9. Rotor emf at start, $E_2 = E_1 \times \frac{N_2}{N_1}$.

10. Rotor current at start, $I_2 = \frac{E_2}{\sqrt{R_2^2 + X_2^2}}$.

    Rotor current at slip s, $I_2' = \frac{sE_2}{\sqrt{R_2^2 + (sX_2)^2}}$.

11. Starting torque, $T_{st} = \frac{KR_2E^2}{R_2^2 + X_2^2} = \frac{K_1R_2}{R_2^2 + X_2^2}$, if applied voltage is constant.

    Torque under running condition,

    $T = \frac{K_1 sR_2E_2^2}{R_2^2 + s^2X_2^2} = \frac{K_2 sR_2}{R_2^2 + s^2X_2^2}$, if applied voltage is constant.

12. Condition for maximum torque under running condition

$$R_2 = s\,X_2.$$

13. Slip corresponding to maximum torque, $s_{\max T} = \frac{R_2}{X_2}$.

14. Condition for maximum starting torque, $R_2 = X_2$.

15. Maximum torque, $T_{\max} \;\alpha\; \frac{1}{X_2}$.

16. One of the salient facts affecting induction motor applications is that the slip at which maximum torque occurs can be controlled by varying the rotor resistance. A high rotor resistance gives optimum starting conditions but poor running performance. A low rotor resistance, on the other hand, may result in unsatisfactory starting conditions.

17. Starting torque of an induction motor is proportional to the square of applied voltage.

18. When the induction motor is running with slip s, the torque acting on the rotor,

$$T \propto s \times (\text{supply voltage})^2 \propto s\,V^2.$$

19. A change of frequency will give rise to a change of flux level unless the induced emf is changed in the same ratio. An unbalance in this ratio will cause either an excessive flux and saturation, or diminished flux and reduced torque per ampere of current.

With the lowering of frequency, keeping $\frac{V}{f}$ constant, the shape of torque-slip (or torque-speed) curves do not change, but displaced horizontally.

20. $$\frac{T_f}{T_{max}} = \frac{2}{\frac{s_{max\,T}}{s_f} + \frac{s_f}{s_{max\,T}}}.$$

21. $$\frac{T_{st}}{T_{max}} = \frac{2R_2/X_2}{\left(\frac{R_2}{X_2}\right)^2 + 1} = \frac{2a}{a^2+1}.$$

22. The rotor current of a 3-phase induction motor reduces slightly with the slight increase in speed but with large increase in speed the current drops rapidly (almost nearly) until at synchronous speed the rotor emf and rotor current are both zero. This is because at high speed, the slip s is low and so the rotor reactance (sX_2) becomes small as compared to rotor resistance R_2 and the rotor current becomes proportional to rotor induced emf which is proportional to emf.

23. The value of breakdown torque varies with the design of motor but ranges from 200 to 300% of full-load torque in standard squirrel cage motors.

24. Gross torque, $T_g = \frac{P_{mech}}{2\pi N/60} = 9.55\frac{P_{mech}}{N}$ Nm

Shaft torque, $T_{sh} = \frac{P_{output}}{2\pi N/60} = 9.55\frac{P_{output}}{N}$ Nm

where $P_{output} = P_{mech}$ – copper losses

Power Relations

Rotor copper loss $= s \times$ power input to rotor

$$\frac{\text{Rotor copper loss}}{\text{Gross rotor output}} = \frac{s}{1-s}$$

$$\frac{\text{Gross rotor output}}{\text{Rotor input}} = \frac{N}{N_s} = 1 - s$$

Power input to rotor : gross rotor output power : rotor copper loss

i.e. $P_2 : P_{mech} : P_{cu} :: 1 : (1-s) : s$

25. Torque developed by rotor is given by

$$T = \frac{3\times 60}{2\pi N_s}\cdot\frac{R_2 s E_2^2}{R_2^2 + s^2X_2^2}.$$

26. Equivalent electrical load of resistance R'_L is given as

$$R'_L = \frac{R_2}{K^2}\left(\frac{1}{s} - 1\right).$$

27. Maximum gross mechanical power output of a 3-phase induction motor is given by

$$P_{g\,max} = \frac{3V^2}{2(R_{01} + Z_{01})}.$$

and slip corresponding to maximum gross mechanical power output is given by

$$s = \frac{R_2/K^2}{\frac{R_2}{K^2} + Z_{01}}.$$

28. The expression for maximum torque in a 3-phase induction motor is

$$T_{max} = \frac{3\times 60}{2\pi N_s}\times\frac{V^2}{2}\times\frac{1}{R_1 + \sqrt{R_1^2 + \left(X_1 + \frac{X_2}{K^2}\right)^2}}.$$

and the expression for slip corresponding to maximum torque is

$$s_{max\,T} = \frac{R_2/K^2}{\sqrt{R_1^2 + \left(X_1 + \frac{X_2}{K^2}\right)^2}}.$$

29. No-load test is performed to determine no-load current I_0, no-load pf $\cos\phi_0$, windage and friction losses $P_{\omega f}$, no-load core loss P_i, no-load power input P_0, and no-load resistance R_0 and reactance X_0.

Blocked rotor test is performed for determination of short-circuit current I_{sc} with normal voltage applied to stator, power factor on short-circuit, total equivalent resistance and reactance of the motor as referred to stator (R_{01} and X_{01}).

30. The operating characteristics of an induction motor can be computed by using a circle diagram easily and conveniently. The locus of stator load current I′ is a circle of diameter $\frac{V}{X_{01}}$ and having coordinate $\left(\frac{V}{2X_{01}}, 0\right)$ with respect to O′.

31. Increased starting torque with short-circuited rotors can be obtained by using two types of rotors (either deep bar rotors or double cage rotors).

In deep-bar rotors use is made of skin effect with rotor bars caused by the slot leakage fluxes.

The rotor of a double cage motor carries two squirrel cage windings embedded in two rows of slots. The outer slots contain a high resistance and low leakage reactance winding and the inner slots contain a low resistance and high leakage reactance winding.

Double cage rotor motor is classed as a high-torque, low-starting current motor and costs approximately 20 to 30 per cent higher than that of plain squirrel cage motor.

32. For applications requiring a substantially constant speed without excessive severe starting conditions, the squirrel cage induction motor usually is unrivaled because of its ruggedness, simplicity, and relatively low cost. Its only disadvantage is its relatively low pf (about 0.85 to 0.90 at full load and considerably lower at light loads and for lower-speed motors). A wound rotor motor can be used for very severe starting conditions or when speed control by rotor resistance is required. A wound rotor motor is more expensive than a squirrel cage motor.

EXERCISES

1. Give classification of ac motors. Describe construction of three-phase squirrel cage induction motor with diagram.

   [Gujarat Technological Univ. Electrical Machines-I, Summer-2014]

2. Explain with neat diagram, the constructional features and working principle of a 3-ϕ induction motor.

   [Anna Univ. Electrical Machines-II, Nov./Dec.-2012]

3. Bring out clearly, with the help of neat sketches the differences between the 3-phase slip-ring induction motor and 3-phase squirrel cage induction motor.

   [J.N. Technological Univ. Electrical Machines-II, January-2014]

4. Explain how a synchronously rotating magnetic field with constant magnitude is produced in a 3-phase induction motor when a balanced 3-phase supply is applied to its balanced 3-phase stator winding.

   [W.B. University of Technology; Electrical Machine-I, 2011]

5. Explain how torque is produced in a polyphase induction motor. **[M.G. Univ. Electrical Machines-III, 2005]**

6. Why should the rotor of a 3-phase induction motor rotate in the same direction as that of rotating magnetic field? How can the direction of rotation of 3-phase induction motors be reversed. **[G.G.S.I.P. Univ. Delhi Electrical Machines; December-2011]**

7. Discuss the principle of operation of 3-phase induction motor. How induction motor acts as transformer. **[Rajasthan Technical Univ. Electrical Machines-II, 2011]**

8. Explain why 3-ϕ induction motor cannot run at synchronous speed $\left(N_s = \frac{120f}{P}\right)$, where symbols having their usual meaning. **[U.P. Technical Univ. Electrical Engineering Odd Semester 2013-14]**

9. Explain the principle of operation of a 3-phase induction motor. What is slip? Explain.

   Derive the expression for the rotor frequency. **[U.P. Technical Univ. Electrical Machines, 2010-11]**

10. Derive an expression for the frequency of rotor current is an induction motor. **[M.D. Univ. Electrical Machines-II, December-2010]**

11. Explain the terms slip, slip frequency, wound rotor and cage rotor in induction motor.

12. Discuss the theory of operation of a 3-phase induction motor. What is meant by slip of an induction motor ? Explain the importance of slip in operation and performance of induction motor. **[U.P. Technical Univ. Electrical Machines, 2009-10]**

13. Develop an expression to show that the torque developed in a 3-ϕ induction motor depends upon the supply voltage, rotor reactance and slip. Also obtain the conditions of maximum torque. **[U.P. Technical Univ. Electrical Machines, 2001-02]**

14. Show how increase in rotor circuit resistance will increase the starting torque of an induction motor. State the maximum limit of such increased resistances. **[U.P. Technical Univ. Electromechanical Energy Conversion-II, 2005-06]**

15. For an induction motor, derive a relationship between the starting torque and maximum torque explaining clearly the symbols used. **[J.N. Technological Univ. Electrical Machines-II, January-2014]**

16. Derive the condition for maximum torque of a 3-phase induction motor under running condition. **[Rajasthan Technical Univ. Electrical Machines-II, 2012]**

17. Show that slip at which maximum torque of a polyphase induction motor occurs is directly proportional to the rotor resistance r_2 but the maximum torque T_{em} is independent of r_2. **[W.B. Univ. of Technology Electrical Machines-I, 2010]**

18. If stator impedance of a three-phase induction motor is neglected show from its equivalent circuit that maximum torque T_{em} per phase is given by

$$T_{em} = \frac{1}{\omega_s}\frac{V_1^2}{2X_2}$$

and hence show that

$$\frac{T_e}{T_{em}} = \frac{2}{\dfrac{s_{mT}}{s} + \dfrac{s}{s_{mT}}}$$

where s is any slip and s_{mT} = slip at which maximum torque occurs. **[U.P.S.C. I.E.S. Electrical Engineering-II, 2012]**

19. In the case of three phase induction motor derive the ratio of maximum torque to full-load torque as

$$\frac{T_m}{T_{FL}} = \frac{1}{2}\frac{\beta^2 + s_{FL}^2}{\beta\, s_{FL}} \text{ where } \beta = \frac{R_2}{X_2}$$

[U.P.S.C. I.E.S. Elec. Engg. II, 1992]

20. Any variation in the rotor resistance of a three-phase induction motor does not affect the magnitude of its maximum torque whereas it does affect the speed at which the maximum torque occurs. Justify the answer with proof. **[B.P. Univ. of Technology Electrical Machines, 2008]**

21. Draw and explain torque-slip characteristics of the three-phase induction motor. **[R.G. Technical Univ. Electrical Machine-I, June-2013]**

22. Draw and explain the torque-slip characteristics of an induction motor. Mark the starting and maximum torque on the diagram. Explain the effect of rotor resistance on starting and maximum torques. **[Rajasthan Technical Univ. Electrical Machines-II, 2012]**

23. Obtain an expression for the condition of maximum torque of an induction motor. Sketch and explain torque-slip curves for several values of rotor resistance and indicate the condition for maximum torque to be obtained at starting. **[Gujarat Technological Univ. Electrical Machines-I, Summer-2014]**

24. Explain the Torque-Slip characteristics of a 3-phase induction motor. Starting with the expression for torque as a function slip, show that the value of maximum torque is independent of rotor resistance. **[U.P.S.C. I.E.S. Electrical Engineering-II, 1998]**

25. Derive the expression for developed torque for a 3-phase induction motor and obtain the condition for maximum torque. Obtain the ratio of full-load torque to maximum torque. Also draw torque-slip curves and discuss the effect of rotor resistance. **[G.B. Technical Univ. Electromechanical Energy Conversion-II, 2012-13]**

26. Sketch and explain the torque-slip characteristics of the 3-ϕ cage and slip-ring induction motors. Show the stable region in the graph. **[Anna Univ. Electrical Machines-II, Now./Dec.-2012]**

27. A squirrel cage induction motor is supplied from a variable voltage, variable frequency source, the ratio of voltage to frequency being maintained constant. Give a sketch showing the approximate nature of torque versus speed characteristics at 25 per cent, 50 per cent, 75 per cent and 100 per cent rated voltage. Neglect effect of harmonics.

28. Explain the power flow diagram of three-phase induction motor. **[R.G. Technical Univ. Electrical Machine-I, June-2013]**

29. Show through power flow diagram, how electrical input is converted into mechanical power output to an induction motor. **[R.G. Technical Univ. Electromechanical Energy Conversion-I, June-2010]**

30. What are the terms air-gap power, internal mechanical power developed and shaft power? How are these terms related with each other? **[U.P.S.C. I.E.S. Electrical Engineering-II, 2001]**

31. Enumerate various components of power loss in an induction motor and name the parts wherein these occur.

    Give typical values of power factor and efficiency under full load condition for a 100 kW, 1480 rpm motor.

32. Explain the terms air-gap power P_g, internal mechanical power developed P_m and derive the relation

    $P_g : P_{\text{rotor Cu loss}} : P_m = 1 : s : 1 - s$ where s is the slip. **[R.G. Technical Univ. Electrical Machine-I, June-2014]**

33. Derive the equation for torque under running condition for 3-phase induction motor. **[M.D. Univ. Electrical Machines-II, December-2011]**

34. Draw the exact equivalent circuit of an induction motor. What are the various parameters present? **[Rajasthan Technical Univ. Electrical Machines-II, 2012]**

35. Deduce and discuss the equivalent circuit of 3-ϕ induction motor. **[Anna Univ. Electrical Machines-II, Nov./Dec.-2012]**

36. Develop an equivalent circuit for a 3-phase induction motor and how the mechanical load is considered in the equivalent circuit of an induction motor. **[G.B. Technical Univ. Electrical Machines and Automatic Control, 2012-13]**

37. Give the procedure of conducting no-load and blocked rotor tests on 3-phase induction motor. How are the parameters of the equivalent

circuit determined from the test results? Draw the equivalent circuit of the 3-phase induction motor.

[R.G. Technical Univ. Electrical Machine-I, June-2014]

38. Write four differences between transformer and 3-phase induction motor. [R.G. Technical Univ. Electrical Machine-I; June-2014]

39. Explain the similarities between the transformer and induction motor equivalent circuits.

[M.D. Univ. Electromechanical Energy Conversion, December-2014]

40. Explain why the no-load test on 3-ϕ induction motor is conducted at rated voltage whereas only a small voltage is applied to stator winding during blocked rotor test.

[U.P. Technical Univ. Electromechanical Energy Conversion II, 2005-06]

41. Draw the equivalent circuit and phasor diagram of induction motor and explain blocked rotor test on induction motor.

[M.D. Univ. Electromechanical Energy Conversion, December-2013]

42. How would you determine circle diagram of a 3-phase induction motor experimentally?

[R.G. Technical Univ. Electromechanical Energy Conversion-I, December-2012]

43. Explain the procedure of drawing circle diagram of an induction motor. What information can be drawn from the circle diagram ? Discuss. [U.P. Technical Univ. Electromechanical Energy Conversion-II, 2005-06; R.G. Technical Univ. Electromechanical Energy Conversion-I, June-2010]

44. For the rotor circuit of a 3-phase induction motor, show that the locus of the tip of rotor current phasor I_2 is a circle with diameter equal to $\frac{E_2}{X_2}$. Symbols have their usual meaning.

[U.P.S.C. I.E.S. Electrical Engineering-II, 2012]

45. With the help of a neat diagram, explain the steps for the construction of circle diagram of a 3-phase induction motor on the basis of data for its no-load and blocked rotor tests. Show the output and torque lines assuming equal copper losses in the stator and rotor. Further, identify the starting torque and indicate how (*a*) power factor (*b*) slip (*c*) efficiency can be computed for a given point of operation on the circle diagram. [U.P.S.C. I.E.S. Electrical Engineering-II, 2008]

46. What do you understand by time and space harmonics in a 3-phase induction motor? [R.G. Technical Univ. Electrical Machine-I, June-2014]

47. Discuss significance of space harmonics and their effect on induction motor performance. [Pb. Technical Univ. Asynchronous Machines, May-2013]

48. Explain the phenomena cogging and crawling of a 3-phase squirrel cage induction motor. [W.B. Univ. of Technology Electrical Machine-I, 2011]

49. What do you mean by crawling? Does it depend on space harmonics or time harmonics?

[U.P. Technical Univ. Electromechanical Energy Conversion-II, 2006-07]

50. What is meant by leakage reactance? How does it affect the performance of three-phase induction motor?

[Pb. Technical Univ. Asynchronous Machines, May-2013]

51. For high starting torque application, which type of rotor constructions are preferably adopted in induction motor? Explain above rotor constructions with its suitable torque-slip characteristic.

[U.P. Technical Univ. Electromechanical Energy Conversion-II, 2006-07]

52. Write a note on the construction of deep bar induction motor.

[G.B. Technical Univ. Special Electrical Machines, 2012-13]

53. Discuss the constructional details of double squirrel cage induction motor. [J.N. Technological Univ. Electrical Machines-II, January-2014]

54. Why do you use double cage induction motor? Explain disadvantages if any. [U.P.S.C. I.E.S. Electrical Engineering-II, 2007]

55. Explain how the desirable features of high-starting torque and low-operating slip are obtained in a double-cage polyphase induction motor. [U.P.S.C. I.E.S. Electrical Engineering-II, 2001]

56. Discuss the relative merits and demerits of single cage and double cage induction motors. [U.P.S.C. I.E.S. Electrical Engineering-II, 2003]

57. How is it possible to obtain a high starting torque and good running performance with a double cage induction motor ? Explain with the help of torque-slip curves. Compare a double cage induction motor with a deep bar induction motor.

[U.P. Technical Univ. Electromechanical Energy Conversion II, 2002-03]

58. Write differences between double cage and deep bar rotor of a 3-phase induction motor.

[R.G. Technical Univ. Electrical Machine-I, June-2014]

59. Compare deep bar induction motor with a double cage induction motor.

[U.P. Technical Univ. Electromechanical Energy Conversion-II, 2004-05; 2013-14]

60. Explain the importance and construction of double squirrel cage induction motor.

[Gujarat Technological Univ. Electrical Machines-II, Summer-2014]

61. Discuss the advantages of slip-ring induction motor over squirrel cage motor. [R.G. Technical Univ. Electrical Machine-I, June-2014]

62. Distinguish between 3-phase slip-ring and squirrel cage induction motors, from the points of view of starting torque, pull-out torque, efficiency and power factor. [Madras Univ. Elec. Machines-II, April-1996]

63. Clearly differentiate between a synchronous motor and an induction motor and state the advantages of having rotating field system rather than a rotating armature system in a synchronous machine.

[U.P. Technical Univ. Elec. Engineering Even Semester, 2003-04]

64. Explain briefly various effects of unbalanced power supply on performance of 3-phase induction motor.

[R.G. Technical Univ. Electromechanical Energy Conversion-I, June-2014]

65. Justify the following statements:

(*i*) The air gap in a 3-phase induction motor is kept as short as possible.

(*ii*) Power factor of a 3-phase induction motor is low at no load.

(*iii*) The rotor of a squirrel cage motor is skewed.

(*iv*) Double squirrel cage rotor improves the starting torque.

(*v*) The maximum torque of a SCIM is called pull-out torque.

[R.G. Technical Univ. Electromechanical Energy Conversion-I, June-2011]

66. Write short notes on the following:

(*i*) Circle diagram of 3-phase induction motor. [M.D. Univ., May-2009]

(*ii*) Crawling and cogging. [R.G. Technical Univ., December-2012]

(*iii*) Deep bar motors. [M.D. Univ. December-2010]

(*iv*) Double cage rotor in 3-ϕ induction motors.

[U.P. Technical Univ., 2013-14]

(*v*) Double cage and deep bar motors. [M.D. Univ., May-2009]

(*vi*) Standard types of squirrel cage induction motors.

(*vii*) Comparison between synchronous and induction motors.

SHORT ANSWER TYPE QUESTIONS WITH ANSWERS

Q. 1. What is meant by an induction motor ? [M.K. Univ. Nov; 1995]

Ans. An induction motor derives its name from the fact that the current in the rotor conductors is induced by the motion of rotor conductors relative to the magnetic field developed by the stator currents.

Q. 2. What are the two types of 3-phase induction motors ? Which type is generally preferred ? [M.S. Univ. Nov.-1996]

Ans. Two types of 3-phase induction motors are (*i*) squirrel cage and (*ii*) wound rotor or slip-ring induction motors. Squirrel cage induction motor is generally preferred.

Q. 3. Why is an induction motor called asynchronous.

[R.G. Technical Univ., June-2013]

Ans. Because the rotor does not turn in synchronism with the rotating field developed by the stator currents.

Q. 4. What do you mean by space phase difference? **[R.G. Technical Univ., June-2013]**

Ans. Space displacement is known as the space phase difference.

Q. 5. Why closed type slots are often used for small induction motors ?

Ans. Use of closed slots makes the reluctance more uniform, reduces exciting current and gives better power factor.

Q. 6. Why the air gap between stator core and rotor of an induction motor is made very small ?

Ans. The air gap is made as small as possible so as to produce required flux with a minimum exciting current and give leakage reactance as small as possible. This results in improved power factor.

Q. 7. What are the drawbacks of wider air gap in 3-phase induction motor? **[Pb. Technical Univ., May-2013]**

Ans. Wider air gap in 3-phase induction motor causes increase in magnetizing current necessary to set up air-gap flux. This reduces no-load power factor as well as full-load power factor of the induction motor. Increase in air-gap length also results in increase in leakage reactance. Increase in leakage reactance makes the pull-out torque small.

Q. 8. Why the slots on the rotor of an induction motor are usually skewed ?

Ans. The slots on the rotor of an induction motor are usually skewed in order to obtain a uniform torque, reduce the magnetic locking of the stator and rotor, and reduce the magnetic humming noise while running.

Q. 9. Explain why the number of poles of the stator and of the rotor of an ac motor should be equal. **[U.P. Technical Univ. Electrical Machines, 2001-02]**

Ans. The flux set up by the stator windings is a rotating flux with a definite number of poles; operation is only possible when the number of poles is same on both (stator and rotor).

Q. 10. Which type of induction motor develops higher starting torque?

Ans. Slip-ring induction motor.

Q. 11. What is the necessity of short-circuited rotor conductors in a squirrel cage induction motor ?

Ans. If the rotor conductors in a squirrel cage motor are not short circuited, the rotor winding will be open circuited and when the stator is connected to 3-ϕ ac supply, the rotating magnetic field will be set up in the air gap and an emf will be induced in the rotor but no current will flow through the rotor conductors and, therefore, no torque will be developed and the rotor will remain stationary.

Q. 12. When is a rotating magnetic field produced ? What is its importance in electrical machinery ? **[P.T.U., January-2000]**

Ans. When a 3-phase wound machine is energized by 3-phase supply, a rotating magnetic field is developed that tends the rotor of the machine to move in its direction of rotation. Energy transfer phenomenon occurs due to such rotating magnetic field.

Q. 13. Explain what is rotating magnetic field and how this is made use of in electrical machinery ? **[P.T.U., December-2001]**

Ans. A rotating magnetic field is that which is constant in magnitude but whose axis of direction rotates in space as field system of a dc machine. When a 3-phase winding is energized by 3-phase supply, a rotating magnetic field is developed that tends the rotor of the machine to move in the direction of rotation.

The operation of ac rotating machines (such as synchronous machines and induction machines) is based on the production of rotating magnetic field.

Q. 14. How can the direction of rotation of the 3-phase induction motor be reversed ? **[Madras Univ., April-1996]**

Ans. The direction of rotation of a 3-phase induction motor may be reversed by interchanging the connection to the supply of any two leads of the motor.

Q. 15. Give the expression of speed in terms of poles and frequency of supply. **[U.P. Technical Univ., 2002-03]**

Ans. $\text{Speed, N} = \dfrac{120 \times \text{frequency of supply}}{\text{Number of poles}}$.

Q. 16. Define slip in 3-phase induction motor. What is its value at starting and at synchronous speed? **[U.P.Technical Univ. Elec. Engineering Second Semester, 2005-06]**

Ans. The difference between the speed of the stator field and the actual speed of the rotor expressed as a fraction of synchronous speed is known as slip of an induction motor.

i.e. Fraction slip, $s = \dfrac{N_s - N}{N_s}$

At starting, slip, $s = 1.0$ **Ans.**

At synchronous speed, $s = 0$ **Ans.**

Q. 17. Define terms : Slip and slip frequency in case of 3-phase induction motors. **[P.T.U. B.E.E.E., Dec.-2006]**

Ans. The difference between the speed of the stator field and the actual speed of the rotor expressed as a fraction of synchronous speed is known as slip of an induction motor.

The frequency of rotor current (or emf) in an induction motor is given by the product of slip s and the supply frequency f known as slip frequency.

Q. 18. Define 'slip' of an induction motor. **[Anna Univ., Nov;/Dec;-2012]**

Ans. Refer to Answer of Q. 16.

Q. 19. What is meant by rotor frequency in 3-phase induction motor? **[Pb. Technical Univ., May-2013]**

Ans. The frequency of induced current/emf in rotor is known as the rotor frequency.

Q. 20. What is frequency of rotor currents of an induction motor? **[R.G. Technical Univ., June-2013]**

Ans. Frequency of rotor currents of an induction motor, known as slip frequency, is slip times the supply frequency i.e. sf.

Q. 21. An induction motor is running at 50 Hz and at a speed of 1,440 rpm. Motor has 4 poles. At what speed in space will the field produced by rotor rotate? **[G.G.S.I.P. Univ., May-2011]**

Ans. Field produced by rotor will rotate at synchronous speed given by the expression $N_s = \dfrac{120f}{P}$.

Q. 22. Define slip speed.

Ans. When the rotor of an induction motor rotates, the rate at which the rotor conductors are cut by the rotating flux, dependent upon the relative speed between the rotor and stator revolving field, is called the slip speed. In other words, the difference between the synchronous speed and the actual rotor speed is known as the slip speed.

Q. 23. State the operating conditions of the motor when the slip is either positive greater than one or negative. **[U.P. Technical Univ., 2005-06; 2008-09]**

Ans. When the slip is positive greater than 1, the induction motor acts as a brake and it absorbs kinetic energy from the still revolving field causing its speed to fall. When the slip is negative (i.e. when the speed is more than synchronous speed), the induction machine operates as an induction generator and feeds power back to the supply line.

Q. 24. What is the speed of rotor mmf of a 3-phase induction motor w.r.t. its stator mmf ?

Ans. zero.

Q. 25. In case of a 3-phase induction motor why is slip always positive ? **[P.T.U., 1999]**

Ans. The induction motor cannot run at synchronous speed because at synchronous speed, there will not be relative speed between the revolving field and the rotor conductors and so the basic phenomenon will not occur to develop torque. It means that a 3-phase induction motor always run at a speed lesser than that of the field. The slip being the difference of synchronous speed, N_s and actual speed N ($N_s > N$) is, therefore, always positive.

Q. 26. Explain why rotor of an induction motor can never attain synchronous speed. **[P.T.U., June-2003]**

Ans. An induction motor cannot run at synchronous speed. If it were possible, by some means, for the rotor to attain synchronous speed, the rotor would then be standstill with respect to rotating flux. Then no emf would be induced in the rotor, no rotor current would flow, and therefore, there would be no torque developed.

Q. 27. Show that in induction motor, the rotor always run in the direction of stator field. **[P.T.U., December-2002]**

Ans. According to Lenz's law we know that the direction of induced emf would be in such a direction that it would try to oppose the very cause for which it is due. Because the cause producing the induced currents in the rotor conductors is the relative speed between the rotating magnetic field and stationary rotor conductors, therefore, they circulate in such a way that a torque is produced in the rotor tending to cause it to follow the rotating magnetic field and thus reducing the relative speed.

Q. 28. How does the slip vary with load ? **[P.T.U., May-2001]**

Ans. With the increase in load, the motor slows down until the relative motion between the rotor and the rotating magnetic field is just sufficient to result in the development of torque required for a particular load. Thus slip increases with the increase in load.

Q. 29. If the full-load speed of an induction motor is 960 rpm, what will be the synchronous speed ? Give an intuitive answer. **[M.S. Univ., April-1996]**

Ans. 1,000 rpm.

Q. 30. At what slip the torque developed in an induction motor will be maximum ? **[M.S. Univ. Nov.-1996]**

Ans. Torque developed in an induction motor will be maximum at a slip given by the ratio of rotor resistance to rotor standstill reactance.

i.e. $s = \dfrac{R_2}{X_2}$.

Q. 31. What is the condition for maximum torque at starting in a 3-phase induction motor ? **[Bharathidhasan Univ., Nov.-1996]**

Ans. Torque developed in an induction motor at start will be maximum when rotor resistance equals rotor standstill reactance i.e. $R_2 = X_2$.

Q. 32. Is the maximum torque of a 3-phase induction motor dependent on the rotor resistance ?

Ans. No, it is independent of rotor resistance.

Q. 33. What is the relationship of developed torque of a 3-phase induction motor with the supply voltage?

Ans. $T \alpha sV^2$.

Q. 34. When the applied rated voltage per phase is reduced to one-half, what will be the starting torque of a squirrel cage induction motor in terms of its starting torque with full voltage ? **[M.K. Univ. Nov.-1996]**

Ans. One-fourth of starting torque with full rated voltage.

Q. 35. What is the effect of increasing the rotor resistance on starting current and torque? **[Anna Univ., Nov;/Dec;-2012]**

Ans. Refer to Art 18.13.3.

Q. 36. How the starting torque and maximum torque vary with rotor circuit resistance ? **[U.P. Technical Univ., 2002-03]**

Ans. As the rotor resistance R_2 is increased from zero, power factor is improved but rotor impedance is increased and so the rotor current is reduced. At first, the effect of improved power factor predominates over the effect of decrease in rotor current due to increase in impedance and the starting torque, therefore, increases. But after a certain value of rotor resistance, the effect of increased impedance predominates over the effect of improved power factor and so the torque starts decreasing.

Maximum torque is independent of rotor circuit resistance.

Q. 37. Why is that the V/*f* ratio is kept constant while controlling the speed of a 3- phase induction motor by varying the supply frequency ?

Ans. In order to maintain flux level constant.

Q. 38. Speed of a 3-phase induction motor is varied by varying the supply frequency while V/*f* is kept constant. How will the maximum torque vary ?

Ans. The magnitude of maximum torque does not change but the speed at which it occurs changes in the same ratio as the supply frequency.

Q. 39. What are the different losses taking place in a 3-phase induction motor ? **[G.G.S.I.P. Delhi Univ., May/June-2006]**

Ans. Stator copper loss; stator iron loss; rotor copper loss, friction and windage losses and stray load loss.

Q. 40. Why are the iron losses in the rotor of an induction motor normally neglected ?

Ans. Because of very low frequency of emf induced in rotor.

Q. 41. What are the no-load rotational losses in electrical machines? **[I.E.S., 2001]**

Ans. The mechanical loss due to friction in bearings and ventilation (usually called the friction and windage losses) and open-circuit core loss combined together are called the no-load rotational losses.

Q. 42. Which of the losses in rotating electrical machines are influenced by the magnitude of the flux, the load, and the square of load? **[I.E.S. 2001]**

Ans. No-load rotational losses (open-circuit core loss and friction and windage losses) are dependent on the magnitude of flux.

Brush contact loss occurs when current is to be conducted to, or, from, the rotating part of the electrical machine through sliding contacts and is proportional to load.

Copper loss in different windings of the machine and stray load loss are proportional to the square of load.

Q. 43. What are the terms air-gap power, internal mechanical power developed and shaft power? How are these terms related with each other? **[I.E.S. 2001]**

Ans. The power transferred from the stator to the rotor via the air-gap magnetic field is known as the *air-gap power*, P_g.

Air-gap power, P_g less the rotor copper loss is known as the internal mechanical power developed.

Shaft power is the power available at the shaft and is equal to internal mechanical power developed less the mechanical losses.

Q. 44. How is the mechanical power output represented in the circuit model of an induction motor ? **[Bharathiyar Univ., Nov.-1996]**

Ans. Mechanical power output per phase is represented by a pure resistive load of equivalent resistance R_L given by the relation

$R_L = R_2 \left(\dfrac{1}{s} - 1\right)$ where R_2 is the rotor circuit resistance per phase and s is the slip.

Q. 45. The induction motor can be called a generalised transformer. Explain why ? **[P.T.U., May-2001]**

Ans. Since the two fields set up by stator (primary) and rotor (secondary) are stationary relative to each other, so the polyphase induction motor can be considered equivalent to a transformer having an air gap, separating the iron portions of the magnetic circuit carrying the primary and secondary windings. Thus the induction motor can be called a generalized transformer.

Q. 46. The reactance of the rotor of the induction motor varies greatly between starting and running conditions. Explain why ? **[P.T.U., June-2000]**

Ans. The reactance of the rotor of an induction motor is given as

$$X_2' = 2\pi f' L_2$$

where f' is the frequency of rotor current and is equal to slip multiplied by supply frequency f

i.e $f' = sf$

So $X_2' = 2\pi s f L_2 = sX_2$ where X_2 is standstill reactance of the rotor.

Since the slip s varies greatly from starting to normal running condition, rotor reactance varies greatly between starting and running conditions.

Q. 47. An induction motor can be represented by an equivalent transformer. Are their pu magnetising currents of the same order ?

Ans. No.

Q. 48. List out the tests necessary to draw the circle diagram of a 3-phase induction motor. **[Bharathidhasan Univ., Nov.-1996]**

Ans. No-load and blocked rotor tests.

Q. 49. What parameter of the equivalent circuit should change the diameter of the circle diagram and have it effect the pull-out torque ?

Ans. Total standstill leakage reactance per phase referred to stator changes the diameter of the circle diagram and increase in standstill leakage reactance reduces the pull-out torque.

Q. 50. What measure can be taken for minimizing the effect of crawling ?

Ans. Effect of crawling can be minimized by chording, skewing and integral slot winding.

Q. 51. How the locking tendency of the rotor of 3-phase induction motor can be reduced. **[Pb. Technical Univ., May-2013]**

Ans. With certain relationships between the number of poles and of stator and rotor slots in cage rotor motors, peculiar behaviour may be observed when the motor is started. For an example, with the number of stator slots S_1 equal to rotor slots S_2, the motor may refuse to start at all, a phenomenon called the *cogging*.

The teeth of rotor and stator appear to be locked magnetically, and this exerts a reverse torque when the rotor teeth are moving away from those on the stator. To reduce this problem, the number of rotor slots is made prime to the number of slots on the stator. Many induction motors are built with skewed or twisted rotor slots. Skewing reduces cogging and other torque defects and tends to eliminate noise.

Q. 52. Stator of an induction motor has full pitch coil span of 18 slots. What must be short pitch angle in degrees if 5th harmonic component of the emf is to be eliminated ?

Ans. 36°.

Q. 53. Explain the reasons for lower power factor of low-speed, 3-phase induction motors as compared to that of high-speed motors.

Ans. Because of larger number of poles in low-speed 3-phase induction motors as compared to high-speed motors, magnetising current is more due to increase in leakages, which increases with the increase in number of poles. So the power factor of low-speed induction motors is poor in comparison to that for high-speed induction motors.

Q. 54. Why the power factor of a squirrel cage induction motor is low at light loads ?

Ans. The current drawn by an induction motor operating at light load is largely a magnetising current so current at light loads lags behind the applied voltage by a large angle and thus the power factor of an induction motor at light loads is quite low.

Q. 55. The magnetising current in a transformer is 2 to 5 % whereas in 3-phase induction motor, it is about 25-40 % of rated current. Mention the reason. **[U.P. Technical Univ., 2002-03]**

Ans. Owing to the air gap in an induction motor, the magnetising current is large for the induction motor as compared to that for a transformer. The friction and windage losses are also present in the induction motor. So no-load current of an induction motor is much higher than that of an equivalent transformer.

Q. 56. Why the power factor of an induction motor is low at starting ?

Ans. The rotor frequency and rotor reactance are high under starting conditions and, therefore, rotor currents lag the rotor emf by a large angle. This results in low power factor at starting.

Q. 57. Why the efficiency of a three phase induction motor is less than that of a transformer ? **[M.S. Univ., Nov.-1996]**

Ans. The large magnetising current, due to air gap and friction and windage losses in the induction motor are the factors which make the efficiency of the induction motor less than that of the transformer.

Q. 58. The starting torque of a squirrel cage induction motor cannot be altered, when the applied voltage is constant, why ? **[M.K. Univ., Nov.-1996]**

Ans. The starting torque of a squirrel cage induction motor cannot be increased as there is no provision for inserting resistance in the rotor circuit.

Q. 59. How will you improve the starting torque of a 3-phase induction motor ? **[Madras Univ. Elec. Machines-II, October-1995]**

Ans. The starting torque of a 3-phase induction motor can be improved by inserting external resistance in each phase of the rotor circuit, which is possible only in case of wound rotor motors.

Q. 60. What are the advantages of double squirrel cage induction motor? **[Anna Univ. Nov;/Dec;-2012]**

Ans. The double squirrel cage induction motor provides a high starting torque with a low starting current, and therefore, it is quite suitable for direct-on-line starting.

By a proper choice of resistances and reactances of the outer and inner cages, a wider range of torque-slip characteristics can be had with double cage induction motors. This is not possible with a single cage motor.

Q. 61. What are the main advantages of a cage motor ? **[M. S. Univ., April-1996]**

Ans. The main advantages of a cage motor are simple and rugged construction, low initial as well as maintenance cost, better efficiency and power factor, better cooling, more pull-out torque, greater maximum power output, nearly constant speed and simple starting arrangement.

Q. 62. What are the main disadvantages of a cage motor ? **[M. S. Univ., April-1996]**

Ans. The main drawbacks of a cage motor are low starting torque with large starting current and no possibility of a speed control.

Q. 63. What are the advantages and disadvantages of wound rotor over squirrel cage rotor in case of induction motor ? **[P.T.U., December-2002]**

Ans. In wound rotor external resistance can be inserted and, therefore,

(*i*) high starting torque can be obtained by inserting an external resistance of proper value in each phase of the rotor circuit and (*ii*) adjustable speed can be obtained simply by adding external resistors in the rotor circuit.

The drawback of wound rotor over squirrel cage rotor are initial and maintenance costs are more because wound rotor needs slip rings, brushes, short-circuiting devices etc.

Q. 64. What is meant by single phasing ?

[Madras Univ. Elec. Machines-II, Nov.-1996]

Ans. The term 'single phasing' refers to operation with one stator terminal accidently disconnected (as by a blown fuse or a faulty contact on the run side of a star-delta switch).

Q. 65. What may happen if one of the three lines of a 3-phase cage rotor induction motor suddenly gets disconnected?

[W.B. Univ. of Technology, 2011-12]

Ans. *Effect of Break in One Phase*. If one phase of a polyphase induction motor breaks, may be due to any reason, when the motor is in operaton, the motor will continue to operate single phase, provided that the load does not exceed 57.7% of the normal rating, with about the same temperature rise as when carrying rated load as a 3-phase induction motor. It will not, however, start, single phase but if brought up to speed by some external means, such as pulling on the belt, it will continue to operate satisfactorily on load below 57.7% of rated one.

A break in one phase of the rotor of a wound rotor induction motor may prevent it from starting but if it is brought up to speed before the circuit is opened, it will operate under reduced load but usually with considerable vibrations.

Q. 66. What is the usual proportion of 3-phase induction motors, 3-phase synchronous motors, dc motors and fractional horse power motors in a typical distribution system ?

[P.T.U., January-2000]

Ans. About 90 per cent 3-phase induction motors, the rest 10 per cent fractional horse power motors. 3-phase synchronous motors and dc motors both type negligible in a typical power distribution system.

Q. 67. What is the place of application of cage and wound rotor induction motors. **[P.T.U., January-2000]**

Ans. **Applications of 3-Phase Squirrel Cage Induction Motor:** Squirrel cage induction motor is suitable for constant speed industrial drives of small power where speed control is not required and where starting torque requirements are of medium or low value as for printing machinery, flour mills, saw mills, pumps, prime movers for small generators and other shaft drives of small power.

Applications of 3-Phase Wound Rotor Motor: Wound rotor induction motor is used for loads requiring severe starting conditions or for loads requiring speed control such as for driving line shafts, lifts, pumps, generators, winding machines, cranes, hoists, elevators, compressors, small electric excavators, printing machines, turn-tables, strokers, large ventilating fans, crushers etc.

PROBLEMS

1. The synchronous speed of induction motor is 1,500 rpm and rotor speed is 1,440 rpm. Find slip. **[Pb. Technical Univ., May-2008]**

   [Ans. 4%]

2. Calculate the 'slip frequency' and 'slip speed' of a 6 pole induction motor drawing power from a 3-phase, 110 V, 60 Hz source and running at 1,140 rpm. **[B.P. Univ. of Technology Electrical Machines, 2008]**

   [Ans. 60 rpm; 3 Hz]

3. A 3-phase, 4-pole induction motor is supplied from 3-phase, 50 Hz ac supply. Calculate:

   (*i*) the synchronous speed.

   (*ii*) the rotor speed when slip is 4%.

   (*iii*) the rotor frequency when rotor runs at 600 rpm.

   [U.P. Technical Univ. Electrical Engineering Second Semester, 2005-06; Basic Electrical Engineering Second Semester, 2013-14]

   [Ans. 1,440 rpm; 3 Hz]

4. A 4-pole, 1 HP, 3-phase induction motor has a speed of 1,440 rpm at a particular load when connected to 400 V, 50 Hz supply. Calculate (*i*) the slip and (*ii*) rotor frequency of the motor.

   [G.G.S.I.P. Univ. Delhi Electrical Science, May-2008]

   [Ans. 4%, 2 Hz]

5. A 3-phase, 6-pole, 50 Hz induction motor has a slip of 1% at no-load and 3% at full-load. Find (*i*) synchronous speed, (*ii*) no-load speed (*iii*) full-load speed (*iv*) frequency of rotor current at standstill (*v*) Frequency of rotor current at full-load.

   [M.D. Univ. Electromechanical Energy Conversion, May-2005]

   [Ans. (*i*) 1,000 rpm, (*ii*) 990 rpm, (*iii*) 970 rpm, (*iv*) 50 Hz, (*v*) 1.5 Hz]

6. A 4-pole, 50 Hz induction motor runs with a slip of 0.01 pu on full load. Calculate the frequency of rotor current (*a*) at standstill (*b*) on full load. **[M.D. Univ. Electrical Machines-II, May-2009]**

   [Ans. (*a*) 50 Hz (*b*) 0.5 Hz]

7. A 3-phase, 50 Hz induction motor has a full-load speed of 960 rpm. Calculate the speed of rotor field with respect to rotor structure, with respect to stator structure and with respect to stator field.

   [Pb. Univ. Elec. Machines II, May-1995]

   [Ans. 40 rpm, 1,000 rpm; zero]

8. A 3-phase 60 Hz induction motor having 6 poles runs at a full-load speed of 1,140 rpm. What will be the speed of rotation of its rotor field with respect to the stator structure and also with respect to the stator field. **[B.P. Univ. of Technology Electrical Machines, 2008]**

   [Ans. 1,200 rpm; zero]

9. A 3-phase, 50 Hz induction motor has 6 poles and operates with a slip of 5% at a certain load. Determine

   (*i*) the speed of the rotor with respect to the stator.

   (*ii*) the frequency of rotor current.

   (*iii*) the speed of the rotor magnetic field with respect to rotor.

   (*iv*) the speed of the rotor magnetic field with respect to stator and

   (*v*) the speed of the rotor magnetic field with respect to the stator magnetic field.

   [U.P. Technical Univ. Electrical Engineering February 2002; Odd Semester 2013-14]

   [Ans. (*i*) 950 rpm, (*ii*) 2.5 Hz, (*iii*) 50 rpm, (*iv*) 1,000 rpm, (*v*) zero]

10. A 3-phase, 50 Hz induction motor has a full-load speed of 1,440 rpm. For this motor calculate:

    (*i*) Number of poles

    (*ii*) Full load slip

    (*iii*) Rotor frequency

    (*vi*) Speed of stator field with reference to stator structure.

    [Rajasthan Technical Univ. Electrical Machines-II, 2011]

    [Ans. (*i*) 4, (*ii*) 4%, (*iii*) 2 Hz, (*iv*) 1,500 rpm]

11. A 3-phase induction motor runs at a speed of 1,485 rpm at no load and at 1,350 rpm at full load when supplied from a 50 Hz, 3-phase line:

(i) How many poles does the motor have?

(ii) What is the slip at no-load and at full-load?

(iii) What is the frequency of rotor voltages at no-load and at full-load?

(iv) What is the speed at both no-load and full-load of, the rotor field with respect to rotor conductors, the rotor field with respect to the stator and the rotor field with respect to the stator field.

[U.P. Technical Univ. Electromechanical Energy Conversion-II, 2006-07]

[**Ans.** (i) 4, (ii) 1%, 10%, (iii) 0.5 Hz, 5 Hz, (iv) 15 rpm, 150 rpm, 1,500 rpm, 0]

12. A 5 hp, 230 V, 50 Hz induction motor has a rated full-load speed of 950 rpm. The induced voltage per phase of rotor at standstill is 100 V. Calculate: (i) Number of poles and % full-load slip. (ii) Rotor induced voltage and its frequency at full-load.

[U.P. Technical Univ. Electrical Engineering Odd Semester, 2013-14]

[**Ans.** (i) 6, 5%, (ii) 5 V, 2.5 Hz]

13. The star-connected rotor of an induction motor has a standstill impedance of (0.4 + j4) ohm per phase and the rheostat impedance per phase is (6 + j2) ohm.

The motor has an induced emf of 80 V between slip rings at standstill when connected to its normal supply voltage. Find (i) rotor current at standstill with the rheostat in the circuit (ii) when the slip rings are short circuited and motor is running with a slip of 3%.

[Nagpur Univ. Elec. Engg. I, 1993]

[**Ans.** (i) 5.265 A lagging behind induced emf by an angle 43.15°
(ii) 3.32 A lagging behind induced emf by an angle of 16.7°]

14. A 4-pole, 50 Hz, 3-phase induction motor has rotor resistance of 0.21 Ω and standstill reactance of 0.7 Ω per phase. Calculate the speed at which maximum torque is developed.

[M.D. Univ. Electromechanical Energy Conversion, December-2007]

[**Ans.** 1,050 rpm]

15. Three-phase, 50 Hz, 8-pole induction motor has full-load slip of 2%, the rotor resistance and standstill rotor reactance per phase are 0.001 and 0.005 ohms respectively. Find the ratio of maximum to full-load torque and the speed at which maximum torque occurs.

[Rajasthan Technical Univ. Electrical Machines-II, 2010]

[**Ans.** 5.005, 600 rpm]

16. A 4-pole, 50 Hz induction motor has full-load slip of 5%. Each rotor phase has a resistance of 0.3 Ω and standstill reactance of 1.2 Ω. Find the ratio of maximum torque to full-load torque and the speed at which maximum torque occurs.

[M.D. Univ. Electrical Machines-II, December-2011]

[**Ans.** 2.6; 1,125 rpm]

17. A 3-phase, 50 Hz, 8-pole induction motor has a full-load slip of 4%. The rotor resistance is 0.001 Ω per phase and standstill reactance 0.005 Ω per phase. Find the ratio of maximum to full-load torque and the speed at which the maximum torque occurs.

[J.N. Technological Univ. Electrical Machines-II, January-2014]

[**Ans.** 2.6; 600 rpm]

18. A 3.3 kV, 20-pole, 50-Hz, 3-phase star-connected induction motor has a slip-ring rotor of resistance 0.025 Ω and standstill reactance 0.28 Ω per phase. The motor has a speed of 294 rpm when full-load torque is applied. Compute slip at maximum torque and the ratio of maximum to full-load torque. Neglect stator impedance.

[U.P. Technical Univ. Electromechanical Energy Conversion-II, 2006-07]

[**Ans.** 0.0893, 2.344]

19. A 3-phase induction motor has a synchronous speed of 250 rpm and 4% slip at full load. The rotor has a resistance of 0.02 Ω per phase and a standstill leakage reactance of 0.15 Ω per phase. Calculate :

(i) Speed at which maximum torque is developed.

(ii) Ratio of maximum to full-load torque.

(iii) Ratio of maximum to starting torque.

[M.D. Univ. Electrical Machine-II, May-2009]

[**Ans.** (i) 216.7 rpm, (ii) 1.8167, (iii) 3.817]

20. The slip-ring rotor resistance and standstill rotor reactance per phase of a 3300 V, 24-pole, 50 Hz, 3-phase star-connected induction motor are 0.02 Ω and 0.3 Ω respectively. Calculate:

(i) The speed at maximum torque (ii) Ratio of full-load torque to maximum torque if the full-load torque is obtained at 246 rpm.

[Rajasthan Univ. Electrical Machines-II, 2005]

[**Ans.** 233.33; 0.454]

21. A 6-pole, 50 Hz, three-phase induction motor has rotor resistance of 0.4 Ω/phase, maximum torque is 200 N-m at 850 rpm. Find (i) torque at 4% slip (ii) additional rotor resistance to get $\frac{2}{3}$ rd of maximum torque at starting. [U.P. Technical Univ. Electrical Machines, 2005-06]

[**Ans.** (i) 99.6 N-m, (ii) 0.6187 Ω]

22. The rotor resistance and reactance per phase of a 4-pole, 50 Hz, 3-phase induction motor are 0.025 Ω and 0.12 Ω respectively. Make simplifying assumption, state them and find (i) speed at maximum torque (ii) find value of additional rotor resistance per phase required to give three-fourth of maximum torque at starting.

[Nagpur Univ. Elec. Machines 1993]

[**Ans.** 1,187.5 rpm; 0.0292 Ω]

23. A 8-pole, 50 Hz, 3 phase induction motor develops maximum torque of 150 N-m at 700 rpm. The rotor resistance is 0.8 Ω/phase. Find the torque at 4% slip. Neglect stator impedance.

[M.D. Univ. Electromechanical Energy Conversion, December-2013]

[**Ans.** 132.35 N-m]

24. A 4 pole, 50 Hz, 3 phase induction motor develops a maximum torque of 160 N-m at 1,360 rpm. The resistance of star-connected rotor is 0.25 Ω per phase. Calculate value of resistance that must be inserted in series with each rotor phase to produce a starting torque equal to the half of maximum torque.

[M.D. Univ. Electromechanical Energy Conversion, December-2010]

[**Ans.** 0.468 Ω]

25. An 8-pole, 50 Hz, 3-phase induction motor develops a maximum torque of 150 N-m at 650 rpm. The rotor resistance is 0.6 Ω/phase. Find torque at 4% slip. Neglect stator impedance.

[U.P. Technical Univ. Electromechanical Energy Conversion II, 2005-06]

[**Ans.** 82.6 N-m]

26. A 3-phase induction motor has a starting torque of 100% and maximum torque of 200% of the full-load torque. Determine (i) slip at which maximum torque occurs (ii) full-load slip.

[U.P. Technical Univ. Electromechanical Energy Conversion II, 2005-06]

[**Ans.** (i) 0.268 (ii) 0.072]

27. A 3-phase induction motor has starting torque of 100 % and a maximum torque of 200 % of the full-load torque. Find slip at maximum torque. [U.P.S.C. I.E.S. Elec. Engineering II, 1994]

[**Ans.** 26.8%]

28. The starting and maximum torques of a 3-phase induction moor are 1.5 times and 2.5 times its full-load torque. Determine the percentage change in rotor circuit resistance to obtain a full-load slip of 0.03. Neglect stator impedance. [Pb. Univ. Elec. Machines II, May-1994]

[**Ans.** 56.9%]

29. A delta-connected, 6-pole, 50 Hz, 3-phase induction motor has a rotor resistance of 0.15 Ω per phase and exerts maximum torque at 880 rpm. Calculate the percentage maximum torque that would be exerted (i) at standstill and (ii) at 940 rpm.

[**Ans.** (i) 23.66% maximum torque, (ii) 80% maximum torque]

30. The power input to the rotor of a 400 V, 50 Hz, 6-pole, 3-phase. Induction motor is 80 kW. The rotor emf is observed to make 100 complete alternations per minute. Calculate slip, mechanical power developed and rotor copper loss.

[U.P. Technical Univ. Electrical Machines, 2002-03]

[**Ans.** 0.0166; 78.667 kW, 1.333 kW]

31. Draw the slip-torque curve of 3-ϕ induction motor and show the region of stable operation. The stator loss of 3-ϕ induction motor is 2 kW. When the input is 90 kW, what will be rotor mechanical power developed and rotor copper loss, if the motor is running at slip of 4 % ? [U.P. Technical Univ. Electromechanical Energy Conversion-II, 2003-04]

[**Ans.** 84.48 kW; 3.52 kW]

32. The power input to a 3-phase, 50 Hz induction motor is 50 kW. The total stator loss is 800 W. Find the total mechanical power developed, if the rotor emf makes 90 complete cycles per minute.

[U.P. Technical Univ. Electromechanical Energy Conversion-II, 2006-07]

[**Ans.** 47.724 kW]

33. The power supplied to a 3-phase, 4-pole, 50 Hz induction motor is 40 kW and the corresponding stator losses are 1.5 kW. While motor is operating with 4% slip, the mechanical losses are 0.8 kW. Calculate efficiency of the motor and shaft torque.

[U.P. Technical Univ. Electrical Machines, 2006-07]

[**Ans.** 90.4%, 239.8 N-m]

34. A 3-phase, 400 V, 50 Hz, induction motor takes a power input of 35 kW at its full-load speed of 980 rpm. The total stator losses are 1 kW and the friction and windage losses are 1.5 kW. Calculate slip, rotor ohmic losses, shaft power, shaft torque and efficiency.

[M.D. Univ. Electromechanical Energy Conversion, May-2006]

[**Ans.** 2%, 680 W, 31.82 kW; 310 N-m; 90.91%]

35. The power input to the rotor of a 440 V, 50 c/s, 6-pole, 3-phase induction motor is 80 kW. The rotor emf is observed to make 100 complete alternations per minute. Calculate (*a*) the slip (*b*) the rotor speed (*c*) the mechanical power developed (*d*) the rotor copper loss per phase (*e*) the rotor resistance per phase if the rotor current is 65 A. [Rajasthan Univ. Electrical Machines-II, 2001]

[**Ans.** (*a*) 0.0333, (*b*) 966.7 rpm, (*c*) 77.336 kW, (*d*) 888 W, (*e*) 0.21 Ω]

36. A 6-pole, 3-phase, 60 Hz induction motor takes 48 kW input power at 1,140 rpm. The stator copper loss is 1.4 kW, stator core loss 1.6 kW and rotor mechanical losses are 1 kW. Find efficiency.

[G.G.S.I.P. Univ. Delhi Electromechanical Energy Conversion-III, May-2011]

[**Ans.** 86.98%]

37. A 4-pole 3-ϕ induction motor delivers 37 hp at the shaft at a speed of 1,425 rpm on 500 V, 50 Hz supply. The mechanical losses total 3 hp and power factor is 0.9. Calculate for this load : (*i*) The slip, (*ii*) The copper losses, (*iii*) The total power input if the stator losses are 2,500 W, (*iv*) The efficiency, (*v*) The line current.

[U.P. Technical Univ. Electrical Machines, 2008-09]

[**Ans.** (*i*) 5%, (*ii*) 1.548 kW, (*iii*) 33.468 kW, (*iv*) 81.3%, (*v*) 42.94 A]

38. A 4-pole, 400 V, 3-phase, 50 Hz squirrel cage induction motor runs at 1,450 rpm at 0.85 pf lagging developing 11 kW power. The stator losses are 1,100 W and mechanical losses are 400 W. Determine (*i*) Slip (*ii*) Rotor copper loss (*iii*) Rotor frequency (*iv*) Line current (*v*) Efficiency.

[U.P. Technical Univ. Electromechanical Energy Conversion II, 2002-03]

[**Ans.** (*i*) 3.33%, (*ii*) 393 W, (*iii*) 1.665 Hz, (*iv*) 21.89 A, (*v*) 85.32%]

39. An induction motor has an efficiency of 0.9 when delivering an output of 37 kW. At this load, the stator copper loss and rotor copper loss each equals the iron loss. The mechanical losses are one-third of the no-load loss. Calculate the slip.

[Bharathidhasan Univ. Elec. Machines II, Nov.-1996]

[**Ans.** 3%]

40. A 25 hp, 6-pole, 50 c/s, 3-phase slip-ring induction motor runs at 960 rpm on full load with a rotor current per phase of 35 A. Allowing 250 W for the copper loss in the short-circuiting gear and 1,000 W for mechanical losses, find the resistance per phase of the 3-phase rotor winding. [Rajasthan Univ. Electrical Machines-II, 2000]

[**Ans.** 0.223 Ω]

41. A 6-pole, 50 Hz, 3-ϕ induction motor running on full load develops a useful torque of 150 N-m at a rotor frequency of 1.5 Hz. Calculate the shaft power output. If the mechanical torque lost in friction be 10 N-m, determine:

(*i*) rotor copper loss

(*ii*) the input to the motor

(*iii*) the efficiency.

[U.P. Technical Univ. Electromechanical Energy Conversion-II, 2008-09]

[**Ans.** (*i*) 502.65 W, (*ii*) 17.755 kW, (*iii*) 85.82%]

42. A 6-pole, 50 Hz, 3 phase induction motor running on full load develops a useful torque of 160 N-m when the rotor emf makes 120 complete cycles per minute. Calculate the shaft power output.

[U.P. Technical Univ. Electromechanical Energy Conversion-II, 2009-10]

[**Ans.** 16.085 kW]

43. A 1,000 V, 24 pole, 50 Hz, 3-phase star connected induction motor has a slip ring rotor having a resistance of 0.02 Ω and stand still reactance of 0.3 Ω per phase. The motor develops full-load torque at a speed of 245 rpm. Find :

(*i*) Full load torque

(*ii*) Speed at maximum torque and

(*iii*) Maximum torque.

Neglect stator impedance. The ratio of stator to rotor turns is 2.

[U.P. Technical Univ. Electrical Machines, 2009-10]

[**Ans.** (*i*) 8,761 N-m, (*ii*) 233.3 rpm, (*iii*) 15,915 N-m]

44. A 3-phase, star-connected , 220 V (line-to-line), 50 Hz, 4-pole induction motor has the following constants in ohm per phase referred to stator :

$r_1 = 0.29$, $r_2 = 0.14$, $x_1 = 0.5$, $x_2 = 0.21$ and magnetising reactance $x_{mag} = 13.25$ ohms. The core loss may be assumed to be constant at 400 watts. For a slip of 2% compute (*i*) speed (*ii*) output torque and (*iii*) stator current of the motor. Neglect frictional and windage losses.

[**Ans.** (*i*) 1,470 rpm (*ii*) 40 Nm (*iii*) 21.5 A]

45. The following data refer to a 12-pole, 420 V, 50 Hz, 3-phase, mesh connected induction motor: R_1 = 2.95; X_1 = 6.82; R_2 = 2.08, X'_2 = 4.11 ohms per phase. On no-load the line value of magnetising current is 6.7 A and total core loss is 269 W. Determine power factor, input current, equivalent rotor current and torque at a slip of 3% from the approximate equivalent circuit.

[Manomanian Sundaranar Univ. Elec. Machines II, April-1996]

[**Ans.** 0.78 (lag); 13.1 A; 9.95A; 393.3 N-m]

46. A 10 kW, 400 V, 4-pole, delta connected squirrel cage induction motor gave the following test results:

No-load test : 400 V, 8.1 A, 750 W

Blocked rotor test: 90 V, 34 A, 1350 W

Calculate equivalent circuit parameters.

[Rajasthan Technical Univ. Electrical Machines-II, 2012]

[**Ans.** 640 Ω, 86.3 Ω; 4.585 Ω, 1.1678 Ω, 4.434 Ω, 2.217 Ω; 0.5678 Ω; all are phase values]

47. Draw the circle diagram for 5 hp (3.73 kW), 200 V, 50 Hz, 4-pole, 3-ϕ star-connected induction motor from the following test data :

No-load	Blocked rotor
Line voltage 200 V, line current 5 A, total input 350 W.	Line voltage 100 V, line current 26 A, total input 1,700 W.

Estimate from the diagram for full load condition, the line current, power factor and also the maximum torque in terms of the full-load torque. The rotor copper loss at standstill is half the total copper loss.

[**Ans.** 15 A, 0.83, 180%]

48. Draw the circle diagram of a 10 hp, 200 V, 50 Hz, 3-ϕ slip ring induction motor with a star-connected stator and rotor, a winding ratio of unity, a stator resistance of 0.38 Ω/phase and a rotor resistance of 0.24 Ω per phase. The following are test readings:

No-load	:	200 V,	7.7 A,	pf 0.195
Short-circuit	:	100V,	47.6 A,	pf 0.454

Find (*a*) the starting torque (*b*) maximum torque in synchronous watts (*c*) the maximum power factor (*d*) the slip for maximum torque (*e*) the maximum output.

[**Ans.** (*a*) 5,600 synchronous watts (*b*) 12,500 watts (*c*) 0.88 (*d*) 0.195 (*e*) 10.4 kW]

49. An induction motor has a double cage rotor with equivalent impedance at standstill of $(1.0 + j\,1.0)$ and $(0.2 + j\,4.0)$ ohms. Find the relative values of torque given by each cage (*a*) at starting and (*b*) at slip of 5%.

[**Ans.** (*a*) 40 : 1 (*b*) 0.4 : 1]

50. The impedances at standstill of the inner and outer cages of a double cage rotor are $(0.01 + j\,0.5)$ Ω and $(0.05 + j\,0.1)$ Ω respectively. The stator impedance may be assumed to be negligible.

Calculate the ratio of the torques due to the two cages:

(*i*) at starting, and

(*ii*) when running with a slip of 5%.

[Electromechanical Energy Conversion-II; U.P. Technical Univ. 2009-10; G.B. Technical Univ., 2011-12]

[**Ans.** (*i*) 100, (*ii*) 1.4355]

❑❑❑

CHAPTER 19

Starting, Speed Control and Braking of Three-Phase Induction Motors

INSIDE THIS CHAPTER

19.1 STARTING OF SQUIRREL CAGE INDUCTION MOTORS

As a general rule, a motor is subjected to the most severe service during its starting and accelerating period; it is during this interval of time that an extremely large current tends to flow unless steps are taken to keep its value down to a reasonable level. In the case of an induction motor, at start when the rotor is at standstill, the squirrel cage rotor is just like a short-circuited secondary. Therefore, the current in the rotor circuit will be very high and consequently the stator also will draw a large current from the supply lines if full-line voltage were applied at start. Of course, this high current is momentary and falls off rapidly as the motor picks up speed. The magnitude of this current depends on the electrical design of the motor and is independent of the mechanical load. The duration of starting current, however, depends upon the time required for acceleration, which in turn depends on the nature of the driven load.

It is true that the starting of a squirrel cage induction motor because of its rugged construction does not impose electrical stresses that are as severe as those experienced by commutator type motors; nevertheless, it is important that the starting current be minimised to such an extent that the line voltage drop does not affect the operation of other equipment connected to the same distribution network, Modern well-designed induction motors will usually draw about 5 to 7 times the rated full-load current at the starting instant if rated voltage is applied to it. One obvious way to reduce the starting current is to impress the lower voltage across the stator winding at starting instant.

With the reduction in voltage applied to the stator of an induction motor the short-circuit current will be reduced in the same proportion. The main flux, which is determined by the applied voltage, and, therefore, the magnetising current will fall in the same proportion if the magnetic circuit is not saturated. Further, since the core losses are approximately proportional to the square of the flux density and, therefore, the voltage, the energy component of the no-load current will be reduced in proportion to the reduced voltage.

One serious objection to the practice of employing reduced voltage for starting of squirrel cage induction motor is the large reduction in starting torque, because starting torque is proportional to the square of impressed voltage [refer to Eq. (18.16)]. Whether or not such a large reduction in starting torque is permissible will depend, of course, upon the application.

There are several methods of starting squirrel cage induction motors. The most common methods are described below.

1. Direct-On-Line Starting or Full-Voltage Starting. Although there is no limitation on the size of the motor that may be started by this method, it should be understood that objectionable line voltage drops will usually occur, especially if large motors are started frequently. Whether or not normal voltage is employed will, therefore, depend upon the following factors :

(*i*) the size and design of the motor
(*ii*) the kind of application
(*iii*) the location of the motor in the distribution system and
(*iv*) the capacity of the power system and the rules governing such installations as established by power supply companies.

Squirrel cage motors of capacity up to 1.5 kW, double cage motors and squirrel cage motors of large capacity having a large rotor resistance are started by this method. This is the most economical method of starting of induction motors.

As the name suggests, this method involves direct switching of polyphase squirrel cage induction motor to the supply mains, as illustrated in Fig. 19.1. In this method of starting of cage motors the rate of temperature rise is high and the motor may get damaged if the starting period is large, which may be due to (*i*) excessive load (*ii*) insufficient rotor resistance or (*iii*) excessive voltage drop in the supply lines. The starting period usually lasts a few seconds because starting torque for induction motor is about twice the full-load torque.

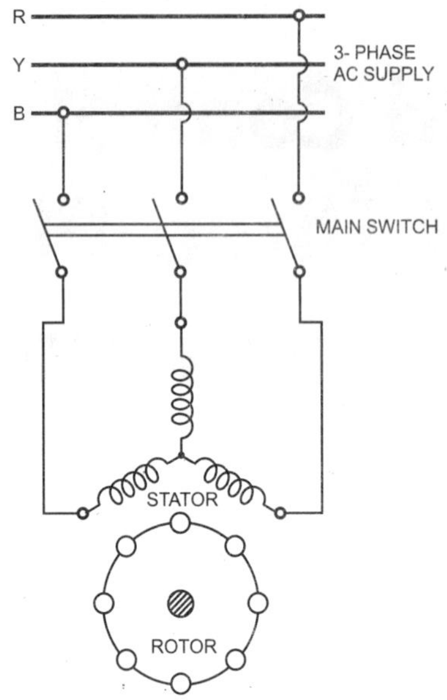

Fig. 19.1

The push button type direct-on-line starter, which is very common in use, is shown in Fig. 19.2. It is simple, inexpensive and easy to install and maintain. It consists of a set of 'start' and 'stop' push buttons, a contactor (an electromagnet) with its associated contacts and usually an overload and undervoltage protection devices. The start button is a momentary contact switch that is held normally open by a spring. The stop button is held normally closed by a spring. When the start button is pressed the operating coil (or main contactor) gets energized through the overload relay contacts OL (normally closed). This closes the three main contacts M that connects the motor to the supply. At the same time a set of auxiliary or maintaining contacts MC are closed. When the maintaining contacts MC closed, a new circuit is established through stop button, maintaining contacts MC and operating coil. Since operating coil circuit is now maintained by the auxiliary contacts MC, the start button may be released. When the stop button is pressed, the operating coil gets de-energized, thereby opening all main contacts and auxiliary contacts.

If the supply fails or line voltage drops below a certain value, the main contacts and the maintaining contacts are both opened. Upon return of the supply, the contactor cannot close until the start button is again closed. Because a contactor that is controlled by a three-wire control circuit maintains the interruption of the circuit even after the supply is restored, it is said to provide *undervoltage protection* for the motor. This protection is employed when it is desired to prevent the unexpected starting of a motor.

Overload protection is employed for protecting the motor and control apparatus from excessive heating due to overloads on motor. Thermal overload relays are commonly used for motor overload protection. Thermal overload relays are of two types. Both types are operated from the heat generated in a thermal element through which the motor current flows. In one type, the heat bends a bimetallic strip, and in the other the heat melts a film of solder. Both act to open the motor control circuit and, therefore, to disconnect the motor from the source of supply.

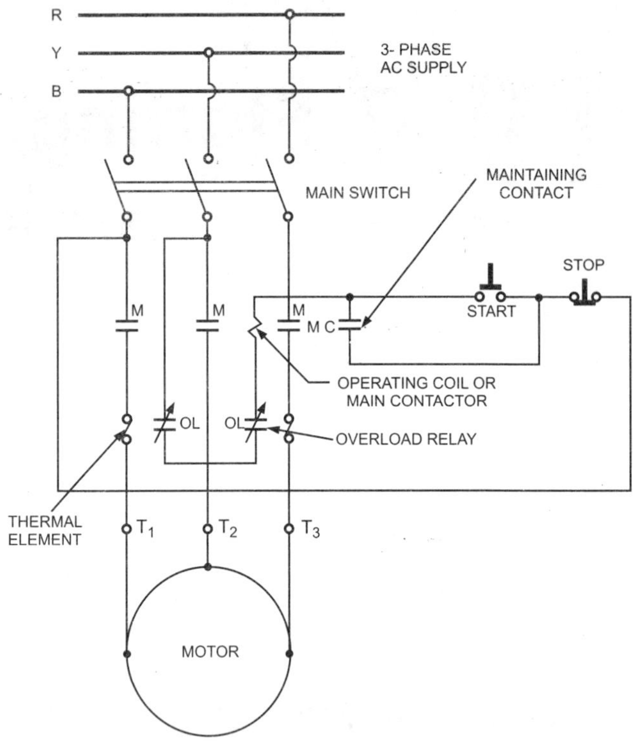

Fig. 19.2 *Wiring Diagram of a Direct-On-Line Starter With Protective Devices*

A bimetallic strip is made of two different metals whose surface has been welded together. One of the metal expands rapidly as compared to the other, when heated. When heat is applied, the strip is caused to deflect or curl up due to unequal expansion of the two metals. When the motor current attains a predetermined value, the heat generated deflects the strip far enough to trip a latch that opens the motor control circuit. When the strip cools down sufficiently, the relay may be reset and the motor re-started.

In the solder film type of overload relay the heat generated at a given overload current melts the film, which releases a latch arrangement and opens the motor control circuit. When the solder has cooled enough to hold the latch, the relay may be reset.

In either type of the thermal relay, the time required to operate is determined by the magnitude of current flowing in the heater. Thus, the relay operates slowly for light overloads but disconnects the motor in a short time for dangerously heavy overloads. Relays in which the time of operation is inversely proportional to the amount of current flowing are known as the *inverse time relays.*

Torque Developed On Starting of Induction Motor By Direct Switching.

$$\text{Input power to rotor} = T\omega$$

$$\text{Rotor copper loss} = s \times \text{input power to rotor}$$

$$\text{or } 3 I_2^2 R_2 = s T \omega$$

where I_2 is the rotor current per phase, R_2 is rotor resistance per phase, T is the torque developed, s is the slip and ω is the supply angular velocity.

$$\text{or Torque developed, } T = \frac{3 I_2^2 R_2}{s\omega}$$

$$i.e.\ T \propto \frac{I_2^2}{s} \text{ if rotor resistance is constant}$$

Since rotor current is proportional to stator current I_1, so

$$T \propto \frac{I_1^2}{s}$$

$$\text{or } T = K\frac{I_1^2}{s} \text{ where K is any constant}$$

At starting slip, $s = 1$

So starting torque, $T_{st} = K(I_{st})^2$

where I_{st} is the starting current

$$\text{Full-load torque, } T_f = \frac{K I_f^2}{s_f}$$

where I_f is the full-load motor current and s_f is the full-load slip

$$\text{So } \frac{T_{st}}{T_f} = \frac{K I_{st}^2}{K I_f^2/s_f} = \left(\frac{I_{st}}{I_f}\right)^2 s_f$$

$$\text{or } T_{st} = T_f \times \left(\frac{I_{st}}{I_f}\right)^2 s_f$$

When the motor is connected directly across the supply mains, the starting current is equal to the short-circuit current. I_{sc}

$$\therefore \text{ Starting torque, } T_{st} = T_f\left(\frac{I_{sc}}{I_f}\right)^2 s_f \quad \text{...(19.1)}$$

2. Primary Resistor (or Reactor) Starting. This method of starting of a polyphase cage motor is very simple and provides smooth acceleration of the motor. In this method of starting of a 3-phase squirrel cage induction motor, reduced voltage is obtained by means of resistors (or reactors) that are connected in series with each stator lead, as shown in Fig. 19.3, during the starting period. The voltage drop in resistors (or reactors) causes a reduced voltage across the motor terminals. As the motor picks up the speed, the resistors (or reactors) are cut out in steps and finally short circuited when the motor attains the operating speed.

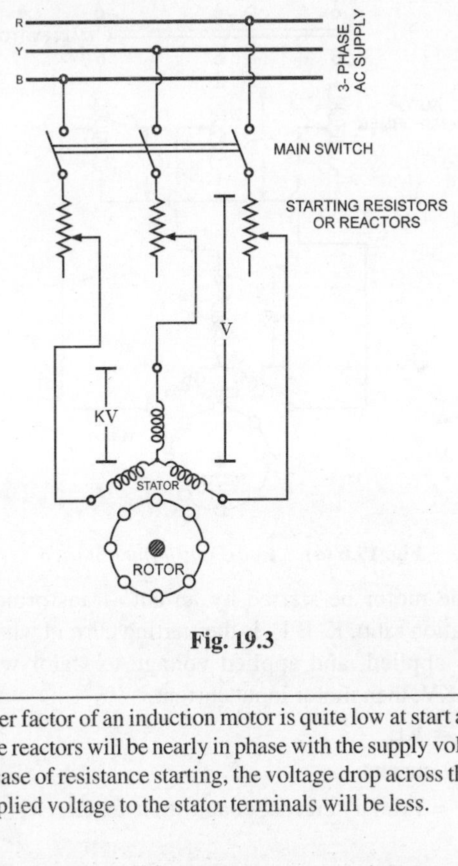

Fig. 19.3

The primary resistor method is quite similar to the dc shunt motor starter with its series armature resistor, because both provide a voltage drop between the voltage supply and the motor terminals. There are, however, some very important differences. In a dc shunt motor the starting torque is directly proportional to the starting armature current whereas in case of a polyphase induction motor the torque varies as the square of the motor terminal voltage i.e., there is a far greater reduction of starting torque for an equivalent reduction of starting current.

Although the initial cost of reactors is high in comparison to that of resistors, reactor starting is preferred because this method incurs small power losses and is more effective* in reducing the voltage applied to the stator at starting.

The advantages and disadvantages of this method of starting of cage motors are given below:

Advantages. 1. Smooth acceleration.
2. High power factor during start.
3. Less expensive than auto-transformer starter in lower output ratings.
4. Closed transition starting.
5. Available with as many as 7 accelerating points.

Disadvantages 1. Resistors give off heat.
2. Low torque efficiency.
3. Starting duration usually exceeds 5 seconds, so needs expensive resistors.
4. Starting voltage is difficult to adjust to meet varying conditions.

Circuit diagram of a magnetically operated primary resistance starter is illustrated in Fig. 19.4. It consists of three line resistors, two sets of three line contacts, two contactors, overload protection, a set of start and stop buttons and several auxiliary contacts. When the start button is pressed, the main contactor is energized through the overload relay contacts OL (normally closed), which in turn closes the three main contacts M and one auxiliary contact MC. Thus the motor is connected to the line through the starting resistors. At the same time, a mechanical escapement timer is started. After a definite time has elapsed during which the motor has speeded up, the time delay contacts TC close. This energizes the accelerating contactor A, which in turn closes the three contacts A and short circuiting the starting resistors applies full voltage to the motor.

Overload and undervoltage protections are provided in the same manner as in the direct-on-line starter.

In the starter described above, the starting resistors are cut out in one step. Starters are also available, in which arrangement is to cut out starting resistances in several steps and thus to give smooth acceleration with less line disturbance.

If the normal supply phase voltage is V volts and by using line resistance starter the voltage is reduced to KV volts, then

* The power factor of an induction motor is quite low at start and if reactors are used for reducing voltage across the motor terminals, the voltage drop across the reactors will be nearly in phase with the supply voltage and so the net reduction in the applied voltage to the stator will be more. On the other hand in case of resistance starting, the voltage drop across the resistors will be nearly in quadrature with the applied voltage and so the net reduction in the applied voltage to the stator terminals will be less.

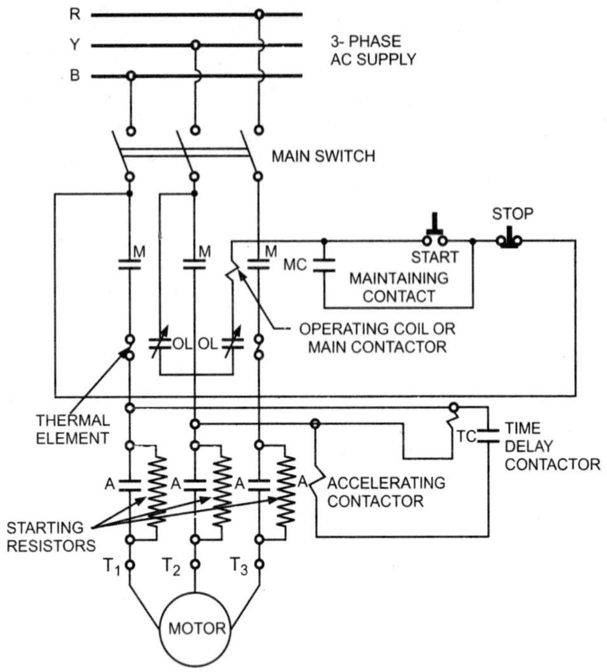

Fig. 19.4 *Wiring Diagram of a Primary Resistance Starter*

starting current is also reduced in the same ratio *i.e.* starting current $I_{st} = KI_{sc}$ where I_{sc} is the short-circuit current.

Starting torque,

$$T_{st} = T_f \times \left(\frac{I_{st}}{I_f}\right)^2 s_f = T_f \times K^2 \left(\frac{I_{sc}}{I_f}\right)^2 s_f \quad ...(19.2)$$

$= K^2 \times$ Torque obtained by switching the motor directly.

This method finds its application in some cases where only very low starting torques are required. A curve plotted between the starting-current/full-load current and the starting-torque/full-load torque in case of stator resistance method of starting of induction motor is depicted in Fig. 19.5.

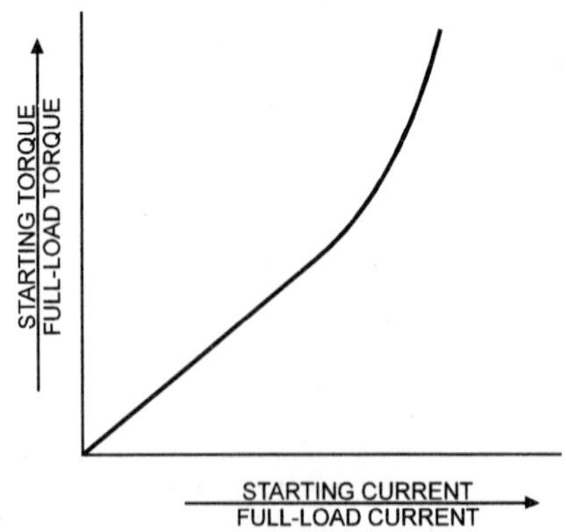

Fig. 19.5

3. Auto-Transformer Starting. In auto-transformer starting method the reduced voltage is obtained by taking tappings at suitable points from a three phase auto-transformer, as shown in Fig. 19.6 (*a*). The auto-transformers are generally tapped at the 50, 60 and 80 per cent points, so that adjustment at these voltages may be made for proper starting torque requirements. Since the contacts frequently break large values of current, arcing is sometimes quenched effectively by having them assembled to operate in an oil bath.

Auto-transformer starters may be either manually or magnetically operated. The wiring diagram of an auto-transformer starter is shown in Fig. 19.6 (*b*).

The manual auto-transformer starter is essentially a multi-pole double-throw switch. It consists of three sets of contacts starting, running (both stationary) and the movable contacts (the contacts attached to the operating handle). The operating handle is spring loaded lever mounted on the outside of a steel cabinet to stand vertically in the "*off*" position. When the operating handle is moved to the start position, the movable contacts are moved against the starting contacts. This energizes the star-connected auto-transformer and impresses reduced voltage (50 or 60 or 80 per cent according to the setting on auto-transformer) across the motor. After the motor has accelerated to about full speed, the operating handle is moved to the run position. This instantly opens the starting contacts (or disconnects the auto-transformer from the line) and connects the motor directly to the line through the running contacts. A latch, fixed to the mechanism, is made to drop into a notch so that the operator is prevented from throwing the handle accidently to the run position first, however, when the handle is quickly pushed from the *start* to the '*run*' position, the latch is kicked up to make the lever free for its forward motion. The operating handle is held in the run position by an electromagnet or undervoltage (latch) coil until the stop button is pressed. If the supply fails or supply voltage drops to low value, the electromagnet will release and trip the holding mechanism. Overload protection is provided by thermal overload relays.

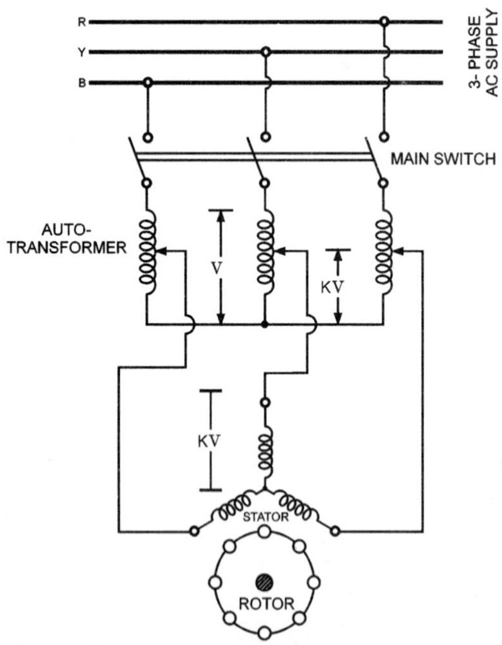

Fig. 19.6 (*a*) *Auto-Transformer Starter*

Let the motor be started by an auto-transformer having transformation ratio, K. If I_{sc} is the starting current when normal voltage is applied, and applied voltage to stator winding at starting = KV, then motor input current,

$$I_{st} = KI_{sc} \quad ....(19.3)$$

Supply current

= Primary current of auto-transformer

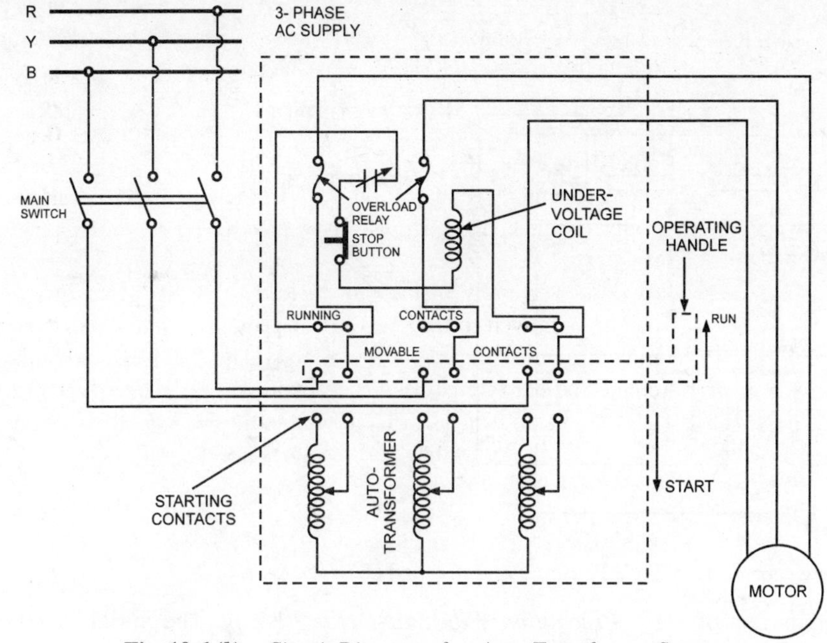

Fig. 19.6 (*b*) *Circuit Diagram of an Auto-Transformer Starter*

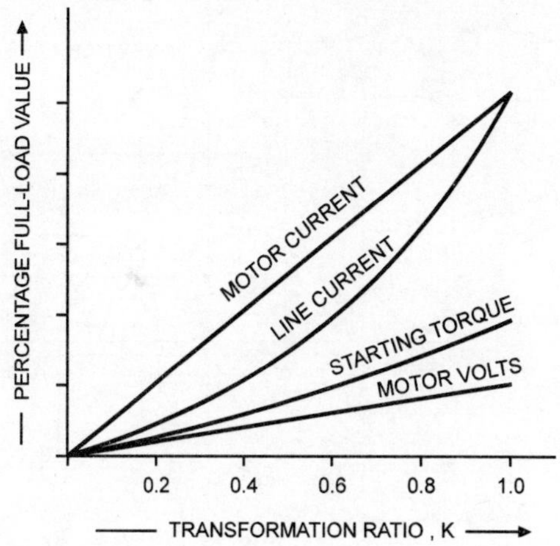

Fig. 19.6(*c*)

$= K \times$ secondary current of auto-transformer

$= K^2 I_{sc}$

Starting torque,

$$T_{st} = T_f \left(\frac{I_{st}}{I_f}\right)^2 \times s_f = T_f K^2 \left(\frac{I_{sc}}{I_f}\right)^2 \times s_f \qquad ...(19.4)$$

$= K^2 \times$ Torque obtained by direct switching.

Hence the line current and the starting torque are reduced in the square ratio.

The advantages of this method of starting of cage induction motors lies in the fact that the voltage is reduced by transformation and not by dropping the voltage in the resistors (or reactors) and, therefore, the current and power drawn from the supply mains are also reduced in comparison to primary resistor (or reactor) starting. The internal losses of the starter itself are small during long starting periods. Other advantages of this method are (*i*) availability of highest torque per ampere of supply current (*ii*) adjustment of starting voltage by selection of proper tap on the auto-transformer (*iii*) suitability for long starting periods (*iv*) closed transition starting and (*v*) motor current larger than supply current.

The drawbacks of such starters are low power factor and higher cost in case of lower output rating motors.

This method can be employed for starting of star-connected as well as delta-connected motors.

For starting of large cage motors (of output rating exceeding 20 kW) this method of starting is often used.

The values of starting torque available, together with the corresponding values of the voltage applied to the stator and the motor and line currents expressed as percentages of the normal full-load values, for different tap positions are shown in Fig. 19.6 (*c*).

4. Star-Delta Starting. This method of starting of cage induction motors is based upon the principle that with 3 windings connected in star, the voltage across each winding is $1/\sqrt{3}$ *i.e.* 57.7 % of the line to line voltage whereas the same winding connected in delta will have full line-to-line voltage across each.

The star-delta starter connects the three stator windings in star across the rated supply voltage at the starting instant. After the motor attains speed the same windings, through a change-over switch, are reconnected in delta across the same supply voltage.

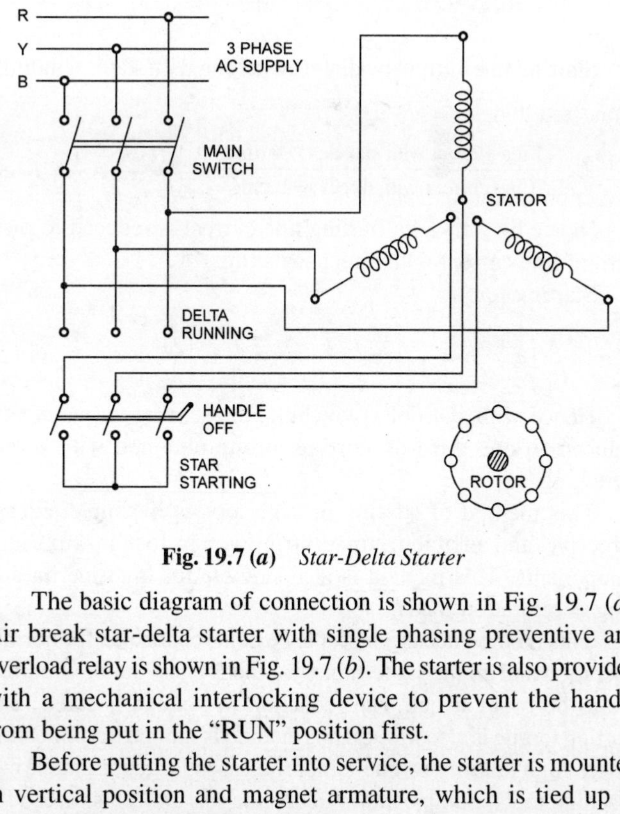

Fig. 19.7 (*a*) *Star-Delta Starter*

The basic diagram of connection is shown in Fig. 19.7 (*a*). Air break star-delta starter with single phasing preventive and overload relay is shown in Fig. 19.7 (*b*). The starter is also provided with a mechanical interlocking device to prevent the handle from being put in the "RUN" position first.

Before putting the starter into service, the starter is mounted in vertical position and magnet armature, which is tied up to avoid damage in transit, is released.

For starting the handle is moved to START position and quickly to RUN position after motor has accelerated.

For stopping of motor, STOP push button is used.

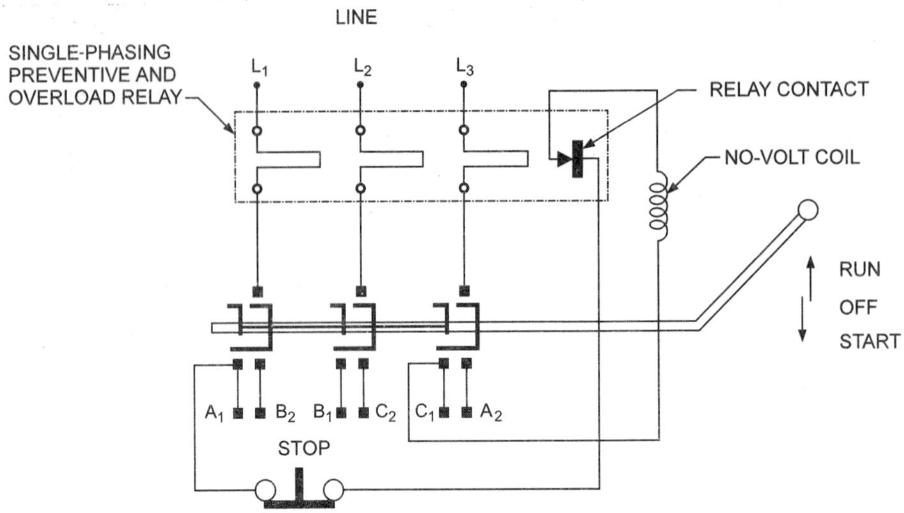

Fig. 19.7 (*b*) *Air Break Star-Delta Starter With Single Phasing Preventive and Overload Relay*

The load indicator may be set at full load rated current of motor when thermal overload trips will be operated at approximately 20 to 25 % overload.

Since at starting instant, the stator windings are connected in star, voltage across each phase winding is reduced to $\frac{1}{\sqrt{3}}$ of line voltage and, therefore, starting current per phase becomes equal to $I_{sc}/\sqrt{3}$.

Starting line current by connecting the stator windings in star at the starting instant = starting motor current per phase $= I_{sc}/\sqrt{3}$.

Starting line current by direct switching with stator windings connected in delta = $\sqrt{3}\ I_{sc}$.

$$\therefore \quad \frac{\text{Line current with star-delta starting}}{\text{Line current with direct switching}} = \frac{I_{sc}/\sqrt{3}}{\sqrt{3}\, I_{sc}} = \frac{1}{3}$$

Hence by star-delta starting line current is reduced to one-third of line current with direct switching.

Starting torque,

$$T_{st} = T_f\left(\frac{I_{st}}{I_f}\right)^2 s_f = T_f\left[\frac{I_{sc}/\sqrt{3}}{I_f}\right]^2 s_f = \frac{1}{3}T_f\left(\frac{I_{sc}}{I_f}\right)^2 s_f \quad ...(19.5)$$

Hence with star-delta switching, the starting torque is also reduced to one-third of starting torque obtained with direct switching.

This method of starting of cage motors is simple, cheap, effective and efficient since no power is lost in auxiliary components. This method is also suitable for high inertia and long acceleration loads.

This method needs a motor to be delta-connected for normal operation and all the six terminals of the 3-phase stator windings are to be brought out. The reduction in voltage is fixed and starting torque is also low. So this method is limited to application where high starting torque is not the essential requirement e.g. machine tools, pumps, motor generator sets etc. This method is unsuitable for line voltage exceeding 3,000 V, because of excess number of stator turns required for delta connection.

Such starters are employed for starting 3-phase squirrel cage induction motors of rating between 4 and 20 kW.

Precaution With Star-Delta Starting. The initial current flowing when the motor is started in star is 57.7 % of the short-circuit current in delta together with a transient in each phase. The transient currents decay rapidly but the steady state is not reached until the motor has attained 70 per cent of its synchronous speed. The change-over from star to delta connection should not be made until the motor attains about 90 per cent of synchronous speed, otherwise there will be a current surge considerably greater than full-load current which may even be greater than the standstill current with star connection.

Example 19.1. Calculate the reduction in starting current and starting torque when the supply voltage to a cage rotor motor is 80 per cent instead of 100 per cent.

[Rajasthan Technical Univ. Electrical Machines-II, 2013]

Solution: Starting current with normal voltage = I_{sc}

Starting current with 80 % of normal supply voltage = $0.8\ I_{sc}$

$$\text{Reduction in starting current} = \frac{I_{sc} - 0.8\, I_{sc}}{I_{sc}} \times 100 = 20\ \%\ \textbf{Ans.}$$

$$\text{Starting torque with normal voltage} = T_f\left(\frac{I_{sc}}{I_f}\right)^2 s_f$$

where I_f, s_f, T_f, and I_{sc} are the full-load current, full-load slip, full-load torque and short-circuit current with normal voltage applied to the stator respectively.

Starting torque with 80 % of normal supply voltage applied to the stator

$$= T_f\left(\frac{0.8\, I_{sc}}{I_f}\right)^2 s_f$$

= 0.64 times of starting torque with normal supply voltage.

Reduction in starting torque = (1 – 0.64) × 100 = 36 % **Ans.**

Example 19.2. A 3-phase induction motor has a ratio of maximum torque to full-load torque as 2.5 : 1. Determine the ratio of actual starting torque to full-load torque for star-delta starting. Given $R_2 = 0.4\ \Omega$ and $X_2 = 4\ \Omega$.

Solution: Ratio of maximum torque to full-load torque, according to Eq. (18.19), is given as

$$\frac{T_{max}}{T_f} = \frac{R_2^2 + s_f^2 X_2^2}{2\, s_f R_2 X_2}$$

where R_2 is rotor resistance per phase, X_2 is standstill rotor reactance per phase and s_f is the full-load slip.

$$\text{or } 2.5 = \frac{0.4^2 + 4^2 s_f^2}{2 s_f \times 0.4 \times 4}$$

or $s_f^2 - 0.5 s f + 0.01 = 0$

or $s_f = 0.02087$ rejecting higher value

Full-load rotor current per phase,

$$I_f = \frac{s_f E_2}{\sqrt{R_2^2 + s_f^2 X_2^2}} \quad \text{[Refer to Eq. (18.11)]}$$

Short-circuit rotor- current per phase,

$$I_{sc} = \frac{E_2}{\sqrt{R_2^2 + X_2^2}} \quad \text{at standstill slip } s \text{ is } 1$$

$$\text{So } \frac{I_{sc}}{I_f} = \frac{\sqrt{R_2^2 + s_f^2 X_2^2}}{s_f \sqrt{R_2^2 + X_2^2}} = \frac{\sqrt{0.4^2 + 0.02087^2 \times 4^2}}{0.02087 \sqrt{0.4^2 + 4^2}} = 4.87$$

Starting torque with star-delta starter,

$$T_{st} = \frac{1}{3} T_f \left(\frac{I_{sc}}{I_f} \right)^2 s_f$$

$$= \frac{1}{3} \times T_f \times (4.87)^2 \times 0.02087 = 0.165 \, T_f$$

So the ratio of actual starting torque to full-load torque for star-delta starting = 0.165 **Ans.**

Example 19.3. A 3-ϕ squirrel cage IM (SCIM) has maximum torque equal to twice the full-load torque. Determine the ratio of motor torque to its full-load torque, if it is started by:

(*i*) DOL Starter

(*ii*) Auto-transformer starter with 70% tapping.

(*iii*) Star-delta Starter.

The per phase rotor resistance and per phase standstill reactance referred to stator are 0.2 Ω and 2 Ω respectively. Neglect stator impedance.

[*G.B. Technical Univ. Electromechanical Energy Conversion-II, 2011-12*]

Solution: Slip corresponding to maximum toque,

$$s_{\max T} = \frac{R_2}{X_2} = \frac{0.2}{2} = 0.1$$

$$\frac{\text{Starting torque, } T_{st}}{\text{Maximum torque, } T_{\max}} = \frac{2a}{a^2 + 1} \quad \text{where } a = \frac{R_2}{X_2} = 0.1$$

... Refer to Eq. (18.21)

$$= \frac{2 \times 0.1}{(0.1)^2 + 1} = \frac{0.2}{1.01}$$

(*i*) Starting torque with DOL starting,

$$T_{st} = \frac{0.2}{1.01} \times T_{\max} = \frac{0.2}{1.01} \times 2T_f \qquad \because T_{\max} = 2T_f$$

$$= 0.396 \, T_f \quad \textbf{Ans.}$$

(*ii*) Starting torque with auto-transformer starter with 70% tapping

$$= 0.7^2 \times 0.396 \, T_f = 0.194 \, T_f \quad \textbf{Ans.}$$

(*iii*) Starting torque with star-delta starter

$$= \frac{1}{3} \times \text{starting torque with DOL starting}$$

$$= \frac{1}{3} \times 0.396 \, T_f = 0.132 \, T_f \quad \textbf{Ans.}$$

Example 19.4. A 12 kW, 3–ϕ, 6-pole, 50 Hz, 400 V, delta-connected induction motor runs at 960 rpm on full load. If it takes 85 A on direct starting, find the ratio of the starting torque to full-load torque with a star-delta starter. Full-load efficiency and power factor are 88 % and 0.85 respectively.

Solution: Full-load line current drawn by a 3-phase induction motor (delta-connected)

$$= \frac{\text{Output in kW} \times 1{,}000}{\sqrt{3} \times \text{line voltage} \times \text{full-load efficiency} \times \text{full-load power factor}}$$

$$= \frac{12 \times 1{,}000}{\sqrt{3} \times 400 \times 0.88 \times 0.85} = 23.16 \text{ A}$$

Full-load current per phase,

$$I_f = \frac{23.16}{\sqrt{3}} = 13.37 \text{ A}$$

Short-circuit current per phase,

$$I_{sc} = \frac{85}{\sqrt{3}} \text{ A}$$

$$\text{Synchronous speed, } N_s = \frac{120 f}{P} = \frac{120 \times 50}{6} = 1{,}000 \text{ rpm}$$

Full-load speed *f*, N_f = 960 rpm

$$\text{Full-load slip, } s_f = \frac{N_s - N_f}{N_s} = \frac{1{,}000 - 960}{1{,}000} = 0.04$$

$$\frac{\text{Starting torque}}{\text{Full-load torque}} = \frac{T_{st}}{T_f} = \frac{1}{3} \left(\frac{I_{sc}}{I_f} \right)^2 \times s_f$$

$$= \frac{1}{3} \left(\frac{85/\sqrt{3}}{13.37} \right)^2 \times 0.04 = 0.18 \textbf{ Ans.}$$

Example 19.5. A 3-phase delta-connected cage-type induction motor when connected directly to 400 V, 50 Hz supply takes a starting current of 100 A in each stator phase. Calculate:

(*i*) The line current for direct-on-line starting

(*ii*) line and phase starting currents for star-delta starting

(*iii*) line and phase starting currents for a 70% tapping on auto-transformer starting.

[*U.P. Technical Univ. Electromechanical Energy Conversion-II, 2008-09; Rajasthan Univ. Electrical Machines-II, 2006*]

Solution: (*i*) Starting current (phase value), I_{SC} = 100 A

Line current for direct-on-line starting,

$$I_{st} = \sqrt{3} \times 100 = 173.2 \text{ A } \textbf{Ans.}$$

(*ii*) Line current for Y-Δ starting

= One-third of line current with direct switching

$$= \frac{1}{3} \times 173.2 = 57.73 \text{ A } \textbf{Ans.}$$

Phase current for Y-Δ starting

= Line current for Y-Δ starting = 57.73 A **Ans.**

(*iii*) **Auto-Transformer Starting:** With 70 per cent tapping on auto-transformer, the line voltage across the delta-connected motor = 0.7 × 400 = 280 V. For delta-connection phase voltage = Line voltage = 280 V

$$\text{Motor phase current} = \frac{280}{400} \times 100 = 70 \text{ A} \quad \textbf{Ans.}$$

$$\text{Motor line current} = \sqrt{3} \times 70 = 121.24 \text{ A} \quad \textbf{Ans.}$$

Supply line current = K × motor line current

= 0.7 × 121.24 = 84.87 A **Ans.**

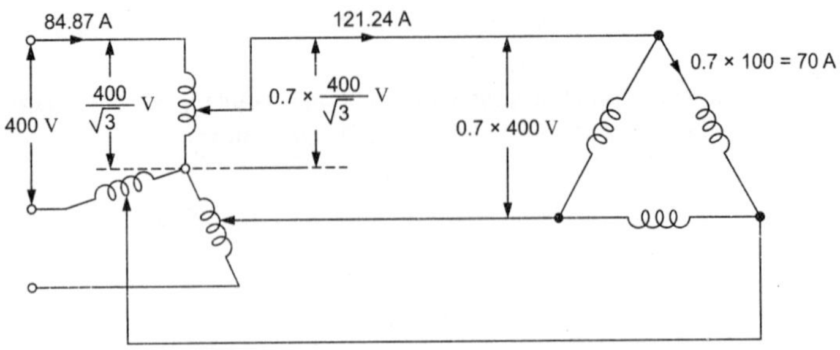

Fig. 19.8 *Auto-Transformer Starting*

Example 19.6. A squirrel cage induction motor takes 5 times full-load current and develops 2.5 times full-load torque when switched directly on the line. Determine the torque developed and the internal value of the current drawn from the supply mains when started through an auto-transformer with 60% tapping.

[*J.N. Technological Univ. Electrical Machines-II, January-2014*]

Solution: Starting current with direct switching = $5\ I_f$

Starting torque with direct switching = $2.5\ T_f$

Auto-transformer transformation ratio, K = 60% or 0.6

Torque developed at start,

$T_s = K^2 \times$ starting torque obtained with direct switching

$= (0.6)^2 \times 2.5\ T_f = 0.9\ T_f$ or 90% of full-load torque **Ans.**

Internal value of current drawn from the supply main at start,

= K × starting current with direct switching

$= 0.6 \times 5I_f = 3I_f$ or 300% of full-load current

Example 19.7. A squirrel-cage induction motor has a starting current of six times the full-load current at a slip of 0.04. Calculate the line current and starting torque in pu of full-load values for the following methods of starting:

(*i*) Direct switching

(*ii*) Auto-transformer starting with motor current limited to 2.0 pu.

(*iii*) Star-Delta starting.

[*U.P.S.C. I.E.S. Electrical Engineering-II, 2002*]

Solution: Short circuit current, $I_{sc} = 6I_f$

Full-load slip, $s_f = 0.04$

(*i*) Direct Switching:

$$\text{Starting current, } I_{st} = 6I_f = 6\frac{I_f}{I_f}\text{ pu} = 6 \text{ pu} \quad \textbf{Ans.}$$

$$\text{Starting torque, } T_{st} = T_f \times \left(\frac{I_{sc}}{I_f}\right)^2 \times s_f$$

$$= T_f \times 6^2 \times 0.04 = 1.44\ T_f$$

$$= \frac{1.44 T_f}{T_f} = 1.44 \text{ pu} \quad \textbf{Ans.}$$

(*ii*) Auto-Transformer Starting:

Starting current, $I_{st} = 2.0 \text{ pu} = 2I_f$

$$\text{Transformation ratio, } K = \frac{I_{st}}{I_{sc}} = \frac{2I_f}{6I_f} = \frac{1}{3}$$

Motor starting current, I_{st} (motor) = 2.0 pu

Line starting current, I_{st} (line) = KI_{st} (motor)

$$= \frac{1}{3} \times 2.0 = 0.667 \text{ pu} \quad \textbf{Ans.}$$

Starting torque,

$T_{st} = K^2 \times$ torque obtained by direct switching

$$= \left(\frac{1}{3}\right)^2 \times 1.44 \text{ pu} = 0.16 \text{ pu} \quad \textbf{Ans.}$$

(*iii*) Star-Delta Starting

$$\text{Starting current, } I_{st} = \frac{1}{3} \times \text{Starting current with direct switching}$$

$$= \frac{1}{3} \times 6 \text{ pu} = 2 \text{ pu} \quad \textbf{Ans.}$$

Starting torque,

$$T_{st} = \frac{1}{3} \times \text{starting torque obtained with direct switching}$$

$$= \frac{1}{3} \times 1.44 \text{ pu} = 0.48 \text{ pu} \quad \textbf{Ans.}$$

Example 19.8. A 50 kVA , 400 V, 3-phase, 50 Hz squirrel cage induction motor has full-load slip of 5 %. Its standstill impedance is 0.866 Ω/ phase. It is started using a tapped auto-transformer. If the maximum allowable supply current at the time of starting is 100 A, calculate the tap position and the ratio of starting torque to full-load.

[*GATE 1997*]

Solution: Full-load current drawn by a 3-phase induction motor,

$$I_f = \frac{\text{Output in kVA} \times 1{,}000}{\sqrt{3} \times \text{line voltage}} \quad \text{neglecting losses}$$

$$= \frac{50 \times 1{,}000}{\sqrt{3} \times 400} = 72.2 \text{ A}$$

$$\text{Short-circuit current, } I_{sc} = \frac{\text{Supply voltage (line value)}/\sqrt{3}}{\text{Standstill impedance per phase}}$$

$$= \frac{400/\sqrt{3}}{0.866} = 266.7 \text{ A}$$

Supply current at start, I_{st} = 100 A

Tap position of auto-transformer,

$$K = \sqrt{\frac{I_{st}}{I_{sc}}} = \sqrt{\frac{100}{266.7}} = 0.6124 \text{ or } 61.24\ \% \ \textbf{Ans.}$$

Full-load slip, s = 5 % or 0.05

Ratio of starting torque to full-load torque,

$$\frac{T_{st}}{T_f} = K^2\left[\frac{I_{sc}}{I_f}\right]^2 s_f = 0.6124^2\left[\frac{266.7}{72.2}\right]^2 \times 0.05$$

$$= 0.256 \ \textbf{Ans.}$$

Note: 3-phase induction motor has been assumed to be star-connected.

Example 19.9. A cage induction motor when started by means of a star-delta starter takes 180% of full-load line current and develops 35% of full-load torque at starting. Calculate the starting torque and current in terms of full-load values, if an auto-transformer with 75% tapping were employed.

[*U.P. Technical Univ. Electromechanical Energy Conversion-II, 2007-08*]

Solution: Supply line current at start with direct-on-line starting,

I_{sc} = 3 times the current drawn with star-delta starter

= 3 × 180% of full load current or $5.4\ I_f$

Starting torque with star-delta starter,

$$T_{st} = \frac{1}{3}\left[\frac{I_{sc}}{I_f}\right]^2 s_f = 35\% \text{ of } T_f = 0.35\ T_f$$

$$\text{or} \quad \frac{T_{st}}{T_f} = \frac{1}{3}\left[\frac{I_{sc}}{I_f}\right]^2 s_f = 0.35$$

or $\frac{1}{3} \times (5.4)^2 \times s_f = 0.35$

or Full-load slip, $s_f = \dfrac{0.35 \times 3}{(5.4)^2} = 0.036$

Auto-transformer ratio, K = 75% = 0.75

Starting current (line value) = $K^2 I_{sc} = 0.75^2 \times 5.4\, I_f = 3.0375\, I_f$ **Ans.**

$$\text{Starting torque, } T_{st} = K^2 \left(\frac{I_{sc}}{I_f}\right)^2 s_f\, T_f$$

$$= (0.75)^2 \times (5.4)^2 \times 0.036\, T_f$$

$$= 0.59\, T_f \text{ or } 59\% \text{ of full-load torque } \textbf{Ans.}$$

Example 19.10. A squirrel cage induction motor has a full-load slip of 0.05. The motor starting current at rated voltage is 5.5 times of full-load current. Find tapping of auto-transformer starter which should give full-load torque at starting and also find line current at starting. [*Rajasthan Technical Univ. Electrical Machines-II, 2011*]

Solution: $\dfrac{\text{Starting torque}}{\text{Full-load torque}} = K^2 \left(\dfrac{I_{sc}}{I_f}\right)^2 s_f$

where K is the auto-transformer tapping

or $K = \sqrt{\left(\dfrac{I_f}{I_{sc}}\right)^2 \times \dfrac{1}{s_f} \times \dfrac{T_{st}}{T_f}}$

$= \sqrt{\left(\dfrac{1}{5.5}\right)^2 \times \dfrac{1}{0.05} \times 1} = 0.813$ or 81.3% tapping **Ans.**

Starting line current, $I_{st} = K^2 I_{sc} = (0.813)^2 \times 5.5\, I_f = 3.636\, I_f$ **Ans.**

Example 19.11. The full-load slip of a 400-volts, 3-phase cage induction motor is 3.5 % and with locked rotor, full load current is circulated when 92 volts is applied between lines. Find the necessary tapping on an auto-transformer to limit the starting current to twice the full-load current of the motor. Determine also the starting torque in terms of the full-load torque.

Solution: Short-circuit current with full voltage (rated voltage) applied,

$$I_{sc} = \frac{400}{92} I_f = \frac{100}{23} I_f \text{ amperes}$$

Limiting line current, $I_{st} = 2\, I_f$ ...(given)

Tapping required on auto transformer,

$$K = \sqrt{\frac{I_{st}}{I_{sc}}} = \sqrt{\frac{2\, I_f}{\frac{100}{23} I_f}}$$

$$= \sqrt{0.46} = 0.678 \text{ or } 67.8\ \% \textbf{ Ans.}$$

$$\text{Starting torque, } T_{st} = K^2 \left(\frac{I_{sc}}{I_f}\right)^2 s_f\, T_f$$

$$= 0.46 \times \left(\frac{100}{23}\right)^2 \times 0.035 \times T_f = 0.304\, T_f$$

i.e. 30.4 % of full-load torque **Ans.**

Example 19.12. A 10 kW, 400 V, three-phase induction motor has full-load efficiency of 0.87, and power factor of 0.85. At standstill and at rated voltage motor draws 5 times its full-load current and develops a starting torque of 1.5 times its full-load torque. An auto-transformer is installed to reduce the starting current and to give full load torque at starting. Neglecting exciting current of auto-transformer, determine at the time of starting the voltage applied to the motor terminals.

[*Rajasthan Technical Univ. Electrical Machines-II, 2010*]

Solution: Motor full-load current,

$$I_f = \frac{10 \times 1{,}000}{\sqrt{3} \times 400 \times 0.87 \times 0.85} = 19.52 \text{ A}$$

Short-circuit current,

$$I_{sc} = 5 \times I_f = 5 \times 19.52 = 97.6 \text{ A}$$

Starting torque developed with direct switching (400 V applied to the stator)

$$T_{st} = 1.5\, T_f$$

Let the voltage applied to develop starting torque equal to full-load torque be V′ volts

Because starting torque ∝ (applied voltage)^2

$$\therefore \left(\frac{V'}{V}\right)^2 = \frac{T'_{st}}{T_{st}} = \frac{T_f}{1.5 T_f} = \frac{1}{1.5}$$

$$\text{or } V' = V \times \sqrt{\frac{1}{1.5}} = \frac{400}{\sqrt{1.5}} = 326.6 \text{ V } \textbf{Ans.}$$

Example 19.13. A 20 hp, 400 V, 950 rpm, three phase, 50 Hz, 6-pole cage induction motor draws 6 times the full-load current at standstill from a 400 V mains and develops 1.8 times full-load running torque. The full-load current is 30 A.

(*i*) What voltage must be applied to produce full-load torque at starting ?

(*ii*) What current will this voltage produce ?

(*iii*) If this voltage is obtained by an auto-transformer, what will be the line current ?

(*iv*) If the starting current in the line is limited to full-load current by means of an auto-transformer, what will be the starting torque as a percentage full-load torque?

Assume full-load slip as 5% and neglect magnetising current and the stator impedance drops.

Solution: Short-circuit current, $I_{sc} = 6\, I_f = 6 \times 30 = 180$ A

Starting torque developed with direct switching (400 V applied to the stator),

$$T_{st} = 1.8\, T_f$$

(*i*) Let the voltage applied to develop starting torque T_{st} equal to full-load torque be V′ volts.

Starting torque α (applied voltage)^2

$$\therefore \left(\frac{V'}{V}\right)^2 = \frac{T'_{st}}{T_{st}} = \frac{T_f}{1.8\, T_f} = \frac{1}{1.8}$$

$$\text{or } V' = \sqrt{\frac{1}{1.8}} \times V = \frac{400}{\sqrt{1.8}} = 298.14 \text{ V } \textbf{Ans.}$$

(*ii*) Current produced at start by V′ (298.14 V)

$$= I_{st} \times \frac{V'}{V} = 180 \times \frac{298.14}{400} = 134.163 \text{ A } \textbf{Ans.}$$

∵ starting current drawn by motor is proportional to applied voltage

$$\text{Transformation ratio, } K = \frac{V'}{V} = \frac{298.14}{400} = 0.7454$$

(*iii*) Line current with auto-transformer of ratio K of 0.7454 = $K^2 I_{sc}$ = $(0.7454)^2 \times 180 = 100$ A **Ans.**

(*iv*) Auto transformer transformation ratio to produce starting line current equal to full-load current

$$K = \sqrt{\frac{I_{st}}{I_{sc}}} = \sqrt{\frac{I_f}{6\,I_f}} = 0.408.$$

$$\text{Starting torque} = K^2 \left[\frac{I_{sc}}{I_f}\right]^2 s_f T_f = 0.408^2\,[6]^2 \times 0.05\,T_f = 0.3\,T_f$$

i.e. 30 per cent of full-load torque **Ans.**

19.2 STARTING OF SLIP-RING INDUCTION MOTORS

Though the methods described in the preceding section can also be employed for starting of 3-phase slip-ring induction motors but this is not done in general because then the advantages of wound rotor cannot be realized fully. *Rotor resistance starter* cannot be employed in case of squirrel cage type motors because external resistance cannot be introduced in the cage rotor.

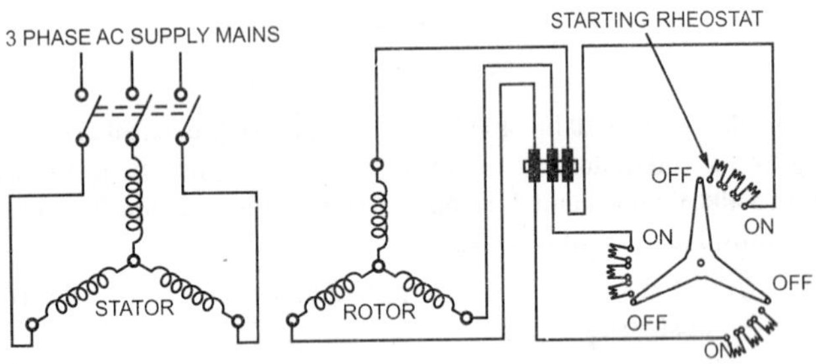

Fig. 19.9 *Slip-Ring Induction Motor With Starting Rheostat*

Slip-ring induction motors are usually started with full line voltage across the stator terminals and by introducing variable resistance in each phase of the rotor circuit. The external resistance introduced in each phase of the rotor circuit not only reduces the current at the starting instant but increases the starting torque also due to improvement in power factor. The external resistance at the starting instant is introduced in the rotor circuit by taking the rotor winding terminals out to the slip rings mounted on the shaft with brushes resting on them, as shown in Fig. 19.9. As the motor accelerates, the external resistance is cut out in steps so that the available electromagnetic torque remains maximum during the accelerating period. Ultimately when the motor attains the normal speed, the rotor winding is short circuited automatically through the slip rings by means of a metal collar which is pushed along the shaft under the slip rings touching them all there by eliminating the resistance of brush contacts on the slip rings. The brushes are then lifted to reduce the wear and frictional losses.

For small and medium size motors face-plate starters are employed but for large size motors liquid starters are employed.

A starter is also provided with no-voltage (or undervoltage) and overcurrent protective devices. It is also provided with some type of interlocking to ensure proper sequential operation of the line contactor and starter. This interlocking arrangement prevents the closing of stator contactor unless the starter is 'all in'.

Since slip-ring induction motor develops high starting torque due to introduction of external resistance in the rotor circuit at the starting instant with reasonable starting current, such motors can be started under load also.

19.2.1. Calculations of Resistance Steps For 3-phase Slip-ring Induction Motor Starters. Consider one phase of a slip-ring rotor with resistance R_2 and standstill reactance X_2. Let $r_1, r_2 \ldots r_{n-1}$ be the resistances of the different steps in each phase and $R'_1, R'_2, R'_3 \ldots R'_{n-1}$ and R'_n be the total resistance per phase in the rotor circuit on the 1*st*, 2*nd*, 3*rd*...(*n* – 1)*th* and *nth* stud respectively (Fig. 19.10).

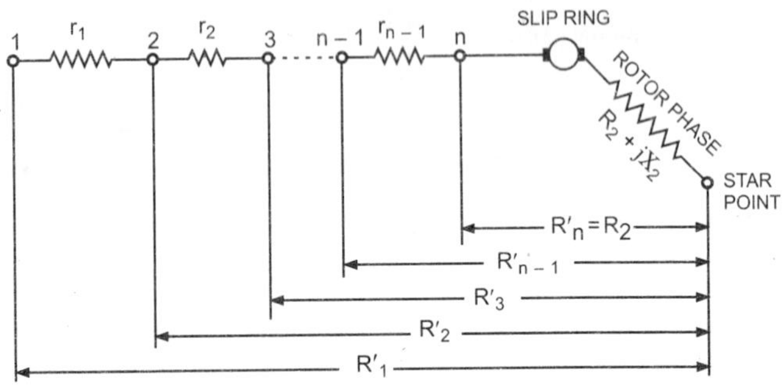

Fig. 19.10

Let us further assume that the motor starts against a constant torque and the rotor current fluctuates between two fixed values of $I_{2\,max}$ and $I_{2\,min}$ respectively.

At the commencement of each step, the current is $I_{2\,max}$ and the resistance in the rotor circuit is R'_1 for slip $s_1 = 1$, R_2' for slip s_2 and so on

$$I_{2max} = \frac{E_2}{\sqrt{\left(\frac{R_1'}{s_1}\right)^2 + (X_2)^2}} = \frac{E_2}{\sqrt{\left(\frac{R_2'}{s_2}\right)^2 + (X_2)^2}}$$

$$= \frac{E_2}{\sqrt{\left(\frac{R_3'}{s_3}\right)^2 + (X_2)^2}} = \ldots$$

where E_2 is the emf induced per phase in the rotor at standstill

$$\text{or} \quad \frac{R_1'}{s_1} = \frac{R_2'}{s_2} = \frac{R_3'}{s_3} = \ldots = \frac{R'_{n-1}}{s_{n-1}} = \frac{R_2}{s_{max}} \qquad \ldots(19.6)$$

where s_{max} is the slip under normal operating conditions when the rotor current per phase is $I_{2\,max}$.

When the motor is switched on, it is on 1st stud, the resistance in the rotor circuit per phase is R_1', the slip s_1 is unity and rotor current per phase is the maximum permissible value $I_{2\,max}$. As the motor gains speed, the current decreases till it becomes $I_{2\,min}$ at slip s_2,

$$\text{Hence } I_{2\,min} = \frac{E_2}{\sqrt{\left(\frac{R_1'}{s_2}\right)^2 + (X_2)^2}}$$

Similarly for other studs also

$$I_{2\,min} = \frac{E_2}{\sqrt{\left(\frac{R_2'}{s_3}\right)^2 + (X_2)^2}} = \frac{E_2}{\sqrt{\left(\frac{R_3'}{s_4}\right)^2 + (X_2)^2}} = \ldots$$

$$\text{So that } \frac{R_1'}{s_2} = \frac{R_2'}{s_3} = \frac{R_3'}{s_4} = \ldots \frac{R_{n-1}'}{s_{max}} \qquad \ldots(19.7)$$

Combining Eqs. (19.6) and (19.7) we get

$$\frac{s_2}{s_1} = \frac{s_3}{s_2} = \frac{s_4}{s_3} = ... \frac{s_{max}}{s_{n-1}} = \frac{R'_2}{R'_1} = \frac{R'_3}{R'_2} = \frac{R'_4}{R'_3} = ... \frac{R_2}{R'_{n-1}} = K \text{ (say)}$$

From Eq. (19.6) total resistance in the rotor circuit per phase on first stud is

$$R'_1 = R_2 \times \frac{s_1}{s_{max}} = \frac{R_2}{s_{max}} \quad ...(19.8) \quad \text{since } s_1 = 1$$

while thereafter $R'_2 = KR'_1;\ R'_3 = KR'_2 = K^2R'_1$

$$\text{and } R_2 = KR'_{n-1} = K^{n-1}\ R'_1 = K^{n-1}\frac{R_2}{s_{max}}$$

$$\text{or } K = (s_{max})^{\frac{1}{n-1}} \quad ...(19.9)$$

where $(n-1)$ is number of sections in the starter

The resistances of various sections are determined as follows:

$$r_1 = R'_1 - R'_2 = R'_1(1-K)$$

$$r_2 = R'_2 - R'_3 = R'_1(K - K^2) = Kr_1$$

$$r_3 = R'_3 - R'_4 = R'_1(K^2 - K^3) = Kr_2 = K^2r_1 \quad \text{etc.}$$

Example 19.14. Design the five sections of a 6-stud rotor starter for a 3-phase wound rotor induction motor. The slip at full-load is 2 per cent and the starting current is 1.5 times the full-load current. The resistance of rotor is 0.02 Ω per phase.

Solution: Full-load slip, $s_f = 2\%$ or 0.02

Rotor current at full-load is given as

$$I_{2f} = \frac{s_f E_2}{\sqrt{R_2^2 + s_f^2 X}} \simeq \frac{s_f E_2}{R_2}$$

∵ when slip s is very low, the value of term sX_2 is very small and can be neglected *i.e.* slip s directly varies with rotor current

So slip at 1.5 times full-load current,

$$s_{max} = 1.5 \times 0.02 = 0.03$$

The total resistance in the rotor circuit per phase on first stud,

$$R'_1 = \frac{R_2}{s_{max}} = \frac{0.02}{0.03} = 0.6667\ \Omega$$

$$\text{and } K = (s_{max})^{\frac{1}{n-1}} = (0.03)^{\frac{1}{5}} = 0.496$$

The resistances of the various sections of the starter are, therefore,

$$r_1 = (1-K)R'_1 = (1-0.496) \times 0.6667 = 0.336\ \Omega$$

$$r_2 = K\,r_1 = 0.496 \times 0.336 = 0.1667\ \Omega$$

$$r_3 = K^2\,r_1 = (0.496)^2 \times 0.336 = 0.0826\ \Omega$$

$$r_4 = K^3\,r_1 = (0.496)^3 \times 0.336 = 0.041\ \Omega$$

$$\text{and } r_5 = K^4\,r_1 = (0.496)^4 \times 0.336 = 0.0203\ \Omega$$

19.3 SPEED CONTROL OF INDUCTION MOTORS

The problem of speed control of electrical motors in general and of induction motors in particular is of great practical importance.

In a number of industries motors must satisfy very strict speed characteristic requirements, both with respect to the range and smoothness of control and also with respect to economical operation. From the view point of speed control characteristics, induction motors are inferior to dc motors. The speed of a dc shunt motor can be adjusted between wide range with good efficiency and speed regulation, but in induction motors speed cannot be varied without loosing efficiency and good speed regulation.

The speed of an induction motor is given by the expression $N = \frac{120 f}{P}(1-s)$. Thus there are three factors viz, supply frequency f, number of poles P and slip s on which the speed of an induction motor depends. Hence to change the speed of an induction motor it is essential to change at least one of the above three factors.

Methods of speed control are distinguished according to the main action on the motor : (*i*) from the stator side, and (*ii*) from the rotor side.

Various methods of speed control from stator side are (*a*) variation of supply frequency (*b*) variation of applied voltage and (c) by changing the number of poles. From the rotor side the speed may be controlled (*a*) by changing the resistance in the rotor circuit and (*b*) by introducing into the rotor circuit an additional emf of the same frequency as the fundamental emf of the rotor.

For the latter method of speed control of induction motors, an additional electrical machine or several such machines are required. A set consisting of a regulated induction motor and one or more additional electrical machines connected to it electrically or mechanically is called *cascade.* Commutator machines are commonly used as the additional machines.

Speed Control By Variation of Supply Frequency. This method of speed control provides wide speed-control range with gradual variation of the speed throughout this range. The major difficulty with this method is how to get the variable frequency supply. The auxiliary equipment required for this purpose results in a high first cost, increased maintenance and lowering of the overall efficiency. That is why, this method is not employed for general purpose speed control applications. Inspite of the fact that this scheme is complicated, there are certain applications in which its wide, continuously variable, speed range and good speed regulation makes its use highly desirable.

If an induction motor is to be operated at different frequencies with practically constant values of efficiency, power factor, overload capacity and a constant absolute slip, then, with the iron unsaturated, it is essential that the supply voltage V be varied with the change in frequency according to the equation

$$\frac{V}{V_0} = \frac{f}{f_0}\sqrt{\frac{T}{T_0}} \quad ... (19.10)$$

where V and T are the voltage and torque corresponding to the frequency f and V_0 and T_0 to the frequency f_0.

For constant torque *i.e.* $T = T_0$, we have

$$\frac{V}{V_0} = \frac{f}{f_0} \quad ...(19.11)$$

i.e., the voltage applied to the stator must vary in proportion to the frequency.

This kind of variation can occur for example, if the supply generator is subjected to the speed variations, either due to momentary overloads or because of a noticeable speed regulation. Both output voltage and frequency vary as the speed if no automatic correction is provided. On some large marine drives, the propellor motors are induction type and are speed controlled from such a local supply which is provided by synchronous generators coupled to variable speed turbines. Even in such cases the range of speed variation is limited, because the efficiency of the prime movers falls rapidly with the change in speed from that for what they are designed.

From the emf equation, it can be understood that a change of frequency will result in a change of flux level unless the induced emf is changed in the same ratio. An imbalance in this ratio will cause either an excessive flux and saturation, or reduced flux and reduced torque per ampere of current. Excessive flux created due to reduction in frequency made for the purpose of reducing speed at constant voltage will cause increase in iron losses and increased iron losses alongwith reduced ventilation caused by the lower speed can result in excessive heating.

Constant flux per pole is approximated if the applied voltage per hertz $\frac{V}{f}$ is maintained constant, because $V \simeq E_1$, induced emf in stator, However, an idealized variable-frequency control scheme might aim to ensure that any particular torque should be obtained at the same flux $\frac{E_1}{f}$, as when operating at normal voltage and frequency.

19.3.1. Torque-Slip Characteristics With Variation of Supply Frequency $\left(\frac{V}{f}\text{ Ratio Remaining Constant}\right)$

Let nominal supply voltage per phase and nominal frequency be V_0 and f_0 respectively and X'_{20} be the nominal rotor standstill reactance referred to stator at nominal frequency f_0.

Then at any frequency f

$$V = \frac{f}{f_0} V_0 \qquad \text{...(19.12)}$$

$$X'_2 = \frac{f}{f_0} \times X'_{20} \qquad \text{...(19.13)}$$

and $$N_s = \left(\frac{f}{f_0}\right) N_{s0} \qquad \text{...(19.14)}$$

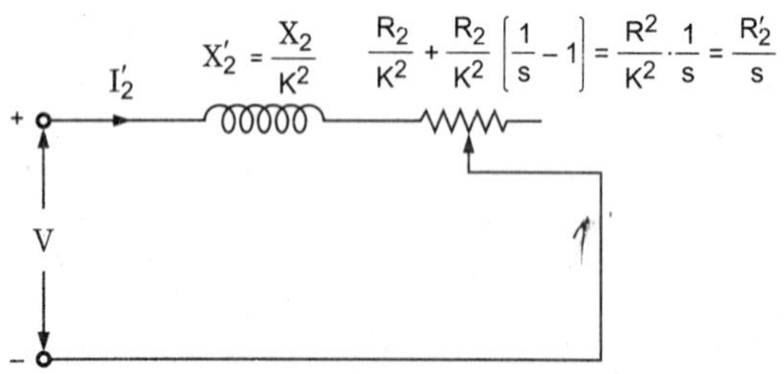

Fig. 19.11 *Equivalent Circuit of Rotor Referred to Stator.*

The equivalent circuit of an induction motor referred to stator, neglecting stator impedance, can be redrawn as shown in Fig. 19.11.

Rotor current referred to stator

$$I'_2 = \frac{V}{\sqrt{\left(\frac{R'_2}{s}\right)^2 + (X'_2)^2}} \qquad \text{...(19.15)}$$

$$\text{Power input to rotor} = 3(I'_2)^2 \frac{R'_2}{s} = \frac{3V^2 R'_2/s}{\left(\frac{R'_2}{s}\right)^2 + (X'_2)^2}$$

or $$\text{Torque, } T = \frac{60}{2\pi N_S} \times \frac{3V^2 R'_2/s}{\left(\frac{R'_2}{s}\right)^2 + (X'_2)^2} \qquad \text{...(19.16)}$$

Slip corresponding to maximum torque,

$$S_{\max T} = \frac{R'_2}{X'_2} = \frac{R_2}{X_2}$$

$$= \frac{\text{Rotor resistance}}{\text{Standstill rotor reactance}} \qquad \text{...(19.17)}$$

$$T_{\max} = \frac{60}{2\pi N_s} \times \frac{3V^2}{2X'_2} \qquad \text{...(19.18)}$$

and starting torque,

$$T_{st} = \frac{60}{2\pi N_s} \times \frac{3V^2 R'_2}{(R'_2)^2 + (X'_2)^2} \qquad \text{...(19.19)} \because s = 1$$

The maximum starting torque is obtained when $R'_2 = X'_2$, so

$$T_{st\,\max} = \frac{60}{2\pi N_s} \times \frac{3V^2}{2X'_2} = T_{\max} \qquad \text{...(19.20)}$$

Near the rated (full-load) speed, the slip of the induction motor is so small that

$$R'_2/s >> X'_2$$

and, therefore, X'_2 can be altogether neglected and Eqs. (19.15) and (19.16) become

$$I'_2 = \frac{sV}{R'_2} \qquad \text{...(19.21)}$$

and $$T = \frac{60}{2\pi N_s} \times \frac{3sV^2}{R'_2} \qquad \text{...(19.22)}$$

Now from Eqs. (19.17) and (19.13) we have

$$s_{\max T} = \frac{R'_2}{X'_2} = \left(\frac{f_0}{f}\right)\frac{R'_2}{X'_{20}} = \frac{f_0}{f} s_{\max T_0} \qquad \text{...(19.23)}$$

From Eqs. (19.18), (19.12), (19.13) and (19.14) we have

$$T_{\max} = \frac{60}{2\pi N_s} \times \frac{3V^2}{2X'_2}$$

$$= \frac{60}{2\pi\left(\frac{f}{f_0}\right)N_{s0}} \times \frac{3\left(\frac{f}{f_0}\right)^2 V_0^2}{2\left(\frac{f}{f_0}\right)X'_{20}} = \frac{60}{2\pi N_{s0}} \times \frac{3V_0^2}{2X'_{20}}$$

$$= \text{Constant independent of } f \qquad \text{...(19.24)}$$

and $$T_{st} = \frac{60}{2\pi N_s} \times \frac{3V^2 R'_2}{(R'_2)^2 + (X'_2)^2}$$

$$= \frac{60}{2\pi\left(\frac{f}{f_0}\right)N_{s0}} \times \frac{3\left(\frac{f}{f_0}\right)^2 V_0^2 R'_2}{(R'_2)^2 + \left(\frac{f}{f_0}\right)^2 X'^2_{20}} \qquad \text{...(19.25)}$$

From Eqs. (19.23), (19.24) and (19.25) we conclude that

(*i*) Slip corresponding to maximum torque decreases with increasing f.

(*ii*) Maximum torque is independent of frequency and

(*iii*) Starting torque decreases with increasing f.

For low values of slip, the approximate expressions of Eqs. (19.21) and (19.22) we may write

$$I_2' = \frac{sV}{R_2'} = \frac{s\frac{f}{f_0}V_0}{R_2'} \quad ...(19.26)$$

and
$$T = \frac{60}{2\pi N_s} \times \frac{3s\left(\frac{f}{f_0}\right)^2 V_0^2}{R_2'}$$

$$= \frac{60}{2\pi N_s} \times 3\left(\frac{f}{f_0}\right)^2 \left(\frac{sV_0^2}{R_2'}\right) \quad ...(19.27)$$

Above Eqs. (19.26) and (19.27) indicate that current and torque both increase with f at a given slip.

From the above conclusions the torque-slip characteristics of motor can be drawn as shown in Fig. 19.12. From Fig. 19.12 it is obvious that as the frequency is reduced from a value corresponding to N_s, say, the torque-speed characteristics curves are of the same shape but displaced horizontally. Under these conditions the maximum developed torque remains approximately constant and output power varies approximately in direct proportion to synchronous speed.

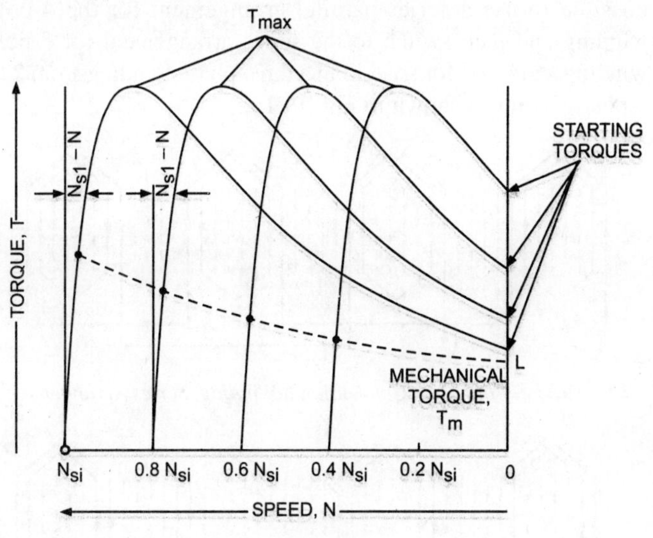

Fig. 19.12

Since supply frequency f can be given as $f = \frac{PN}{120} + f'$ (rotor frequency) and f' is known for a particular value of torque required. The voltage required to be applied to the stator of the motor can be determined from the equivalent circuit.

With rapid growth made in power electronics and circuits combined with ruggedness and higher efficiency of an induction motor, motors with this type of drive are replacing dc motors in several applications. Variable frequency supply may be had from various types of frequency conversion equipments such as adjustable frequency generators, rotating frequency changers etc. However, dc link invertors are presently most popular in practice for providing variable frequency variable voltage supply to induction motors for smooth and wide range speed control.

19.3.2. Speed Control By Variation of Supply Voltage. This is a slip-control method with constant frequency variable supply voltage. In this method of speed control of induction motors, the voltage applied to the stator is varied for varying the speed.

By reference to Eq. (18.15) it is seen that, for a given slip and constant internal parameters (rotor resistance and standstill reactance), the torque developed by an induction motor is directly proportional to the square of induced emf in rotor, E_2, which is approximately proportional to the impressed voltage, V. Thus torque developed is proportional to the square of supply voltage, V. By reference to Art 18.11.3, it is seen that slip at maximum torque is independent of supply voltage. The variation in supply voltage does not cause any change in synchronous speed. Torque-speed characteristics for a particular induction motor for rated impressed voltage, $\frac{1}{\sqrt{2}}$ rated voltage and one-half the rated value are given in Fig. 18.22. It readily can be seen from the torque-speed characteristics that for a given load, the speed of motor cannot be varied over a wide range:

From Eq. (18.13), rotor current is given as

$$I_2 = \frac{E_2}{\sqrt{\left(\frac{R_2}{s}\right)^2 + (X_2)^2}}$$

and rotor current with reference to stator is given as

$$I_2' = \frac{V}{\sqrt{\left(\frac{R_2'}{s}\right)^2 + (X_2')^2}}$$

where R_2' and X_2' are the rotor resistance and standstill reactance per phase as referred to stator, s is slip and V is supply voltage per phase.

Around the rated speed, slip of the induction motor is so small that $\frac{R_2'}{s}$ is much greater than X_2' and so X_2' can be neglected in comparison to $\frac{R_2'}{s}$ and rotor current referred to stator is given

$$I_2' = \frac{sV}{R_2'} \quad ...(19.28)$$

In Art 18.15, we have seen that torque developed by an induction motor, $T \propto sV^2$

From the above relations it is revealed that for an induction motor operating at full-load slip, if the slip is to be doubled for constant load torque, then the supply voltage must be reduced by a factor of $\frac{1}{\sqrt{2}}$ and consequently the rotor current I_2' will become $\sqrt{2}$ times of the full-load value. The motor, therefore, tends to get overheated. It is, therefore, necessary to take care, while using this method of speed control, that the motor is not damaged due to excessive current at the reduced speeds.

This method of speed control is simple, low in first cost and has low maintenance cost but it has limited use because (*i*) the operation at voltages exceeding rated voltage is restricted by magnetic saturation, (*ii*) a large change in voltage is required for a relatively small change in speed (*iii*) the developed torque reduces greatly with the reduction in supply voltage and the motor will come to standstill if the load torque exceeds the pull-out torque of the motor corresponding to reduced supply voltage and (*iv*) the range of speed control is very limited in the downward direction *i.e.*, from rated speed to lower speeds.

In general, this method of speed control is used only on loads where the torque required drops off considerably as the speed is reduced such as with small squirrel cage motors driving fans.

The variable voltage may be obtained by means of either saturable reactors, variac, or tap-changing transformers.

19.3.3. Speed Control By Changing The Number of Poles. This method is easily applicable to squirrel cage motors because a cage winding automatically reacts to create the same number of poles as the stator. This method of speed control is generally not practicable with wound rotor motors as in such machines this method would involve considerable complications of design and switching, since the interconnections of both primary and secondary would have to be changed simultaneously in a manner to produce the same number of poles in both windings. Otherwise, negative torque will be developed by some of the rotor conductor belts.

The number of pole pairs in the stator can be changed as follows:

(*a*) by using multiple stator windings—by placing two or more independent windings on the stator, each producing a different number of poles;

(*b*) by using consequent pole technique — by placing one or two independent windings on the stator and changing the number of poles by changing the interconnections of primary coils;

(c) by using pole amplitude modulation technique — a suppressed– carrier modulation technique.

(*a*) By Using Multiple Stator Windings. In this method of speed control two or more completely independent windings, each wound for different number of poles, are placed in the same stator slots. The number of stator winding poles are in this case in no way interrelated and can be arbitrarily chosen depending on the operating conditions of this motor. For example a two speed motor may have two stator windings, one wound for 4 poles and another for 6 poles which will give synchronous speeds of 1,500 rpm and 1,000 rpm with a supply frequency of 50 Hz. Motors with four independent stator windings are also used sometimes and they give four different synchronous (and hence running) speeds. Of course, one winding is used at a time, the others being entirely inoperative. Change over from one speed to the other may be done by a mechanical switch or by contactors. With such an arrangement, the winding or windings which is/are not being used must be kept open circuited by the switch, or at least left in star. Otherwise, because of transformer action, the winding which is connected to the supply would induce voltages in the idle winding(s) and cause overheating due to the subsequent circulating currents.

The regulation itself amounts to changing the motor speed in steps as one or another stator winding is connected to the supply mains. At each change, the motor finds itself under conditions essentially analogous to the starting conditions *i.e.*, a large current and reduced torque. Because of inactive conductors, this method of pole changing requires a large stator than a single-speed motor of the same rating. This method has been used for elevator motors, traction motors and also for small motors driving machine tools.

(b) Consequent Pole Method. The method of pole changing for speed control of 3-phase induction motor is described above in (*a*) *i.e.* multiple stator winding method is employed only in small motors. In large motors, it is more expedient to employ *consequent pole technique* which eliminates the necessity of having multiple stator windings for changing the number of stator poles. This method of speed control makes use of the entire winding for each speed. Normally, for a single speed motor, successive coil groups are wound so as to create poles of opposite polarity in succession. This is illustrated in Fig. 19.13, where one phase of a four-pole winding is developed. It should be noted that all the coils are in series but that the end connections are made to alternate groups, *i.e.* group 1 to group 3, and group 4 to group 2. Now the end connections of the groups, with the same coils, can be changed so as to have successive coils of the same polarity, as illustrated in Fig. 19.14. By paralleling the series connections of groups 2-4 with that groups 1-3, the polarity of group 2 and 4 has changed, so that there are now four successive north poles. As a consequence of this south poles are created between each of the north poles, and the motor now has an 8-pole winding. Thus for a 50 Hz supply, the series arrangement gives a synchronous speed of 1,500 rpm while the series-parallel arrangement gives a synchronous speed of 750 rpm. It is also possible to use a series-parallel arrangement for the 4-pole winding and then switch to the series arrangement for 8-pole winding. A circuit for switching a three phase winding from 2 to 4 pole winding is shown in Fig. 19.15.

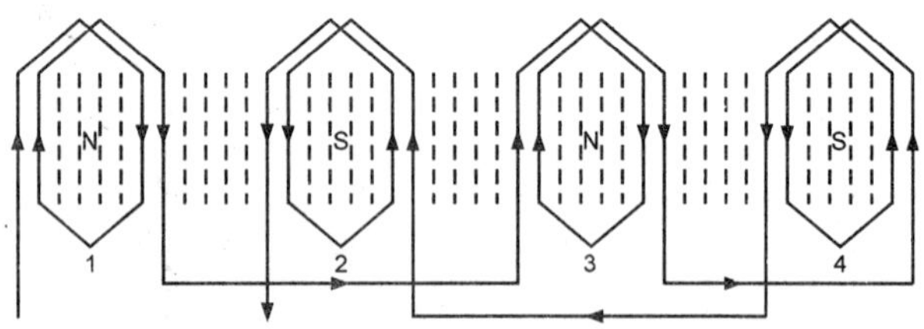

Fig. 19.13 *Four-Pole Conventional Winding, Series Arrangement*

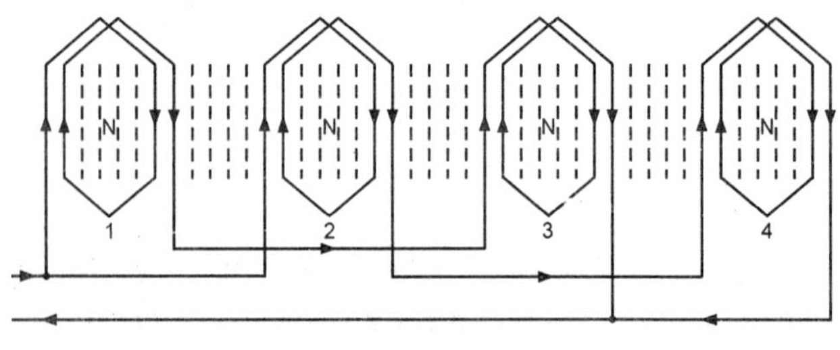

Fig. 19.14 *Consequent-Pole Method For 8-pole Winding, Series-Parallel Arrangement*

By reconnecting all the coil groups for the same polarity, the stator acts as though it has twice as many poles as pole groups. The additional poles so created are called the *consequent poles*, and the winding is known as a *consequent pole winding*. By an unique switching arrangement, it is possible to have the standard (or conventional) connection on one side of a double-throw switch and the consequent pole connection on the other side of the switch. Thus two speeds can be obtained, the higher speed with the conventional connection and half speed with consequent-pole connection.

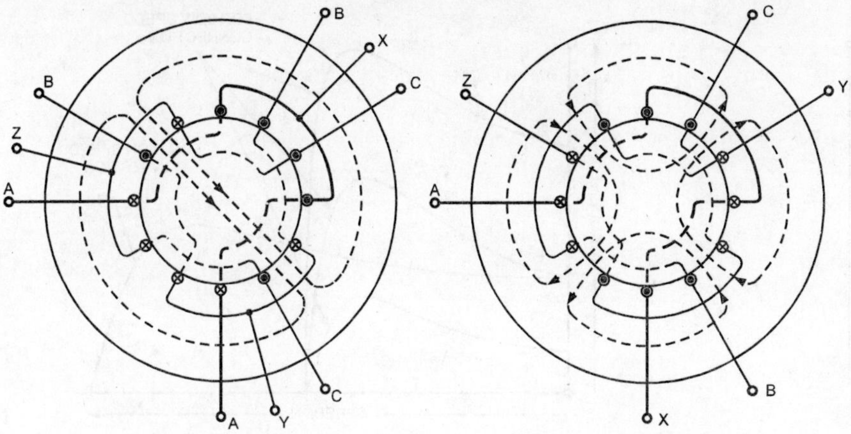

Fig. 19.15 *Switching Stator Winding From Two To Four Poles*

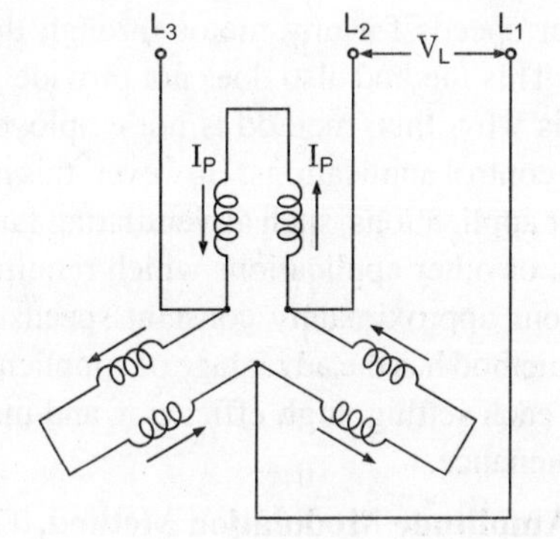

Low Speed Series Delta

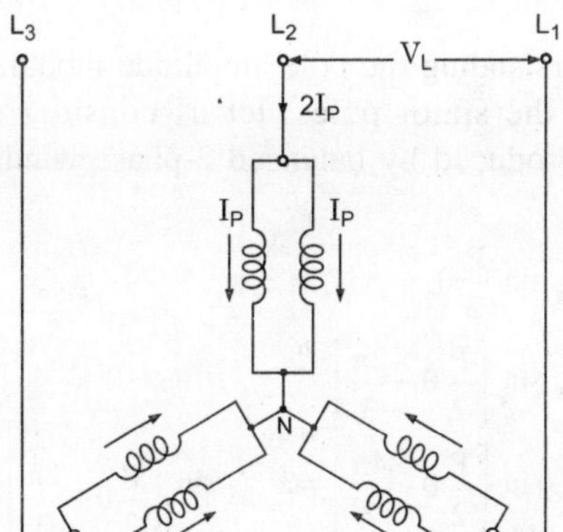

High Speed Parallel Star

Fig. 19.16 *Constant Torque Operation*

In the motors using consequent pole technique for speed control, the best possible design is not usually obtainable at both speeds. That is, desirable characteristics, such as high power factor, are sacrificed at one speed in order that a reasonably good power factor be obtained at the other speed. Sometimes the stator connections are changed from delta to star at the same time that the pole connections are changed. This changes the voltage and makes possible a better motor at each speed.

Two speed motors are usually made with one winding on the stator, the number of poles being changed in the ratio 1:2. Three-and four-speed motors are provided with two windings on the stators, one or both of which are made with changing the number of poles. For example if it is desired to obtain a motor for four synchronous speeds; 1500, 1,000, 750 and 500 rpm, two windings should be placed on the stator, one of which giving 4 and 8 poles and the other giving 6 and 12 poles.

Employing a series-parallel connection of phase groups of individual phases, the phases can themselves be connected in star/delta resulting in two speed operation with the three types of torque-speed characteristics viz *constant torque, constant power output* and *variable torque.* The connections are given in Figs. 19.16, 19.17 and 19.18 respectively. Let V_L be the line voltage and I_P be the nominal current of each phase winding.

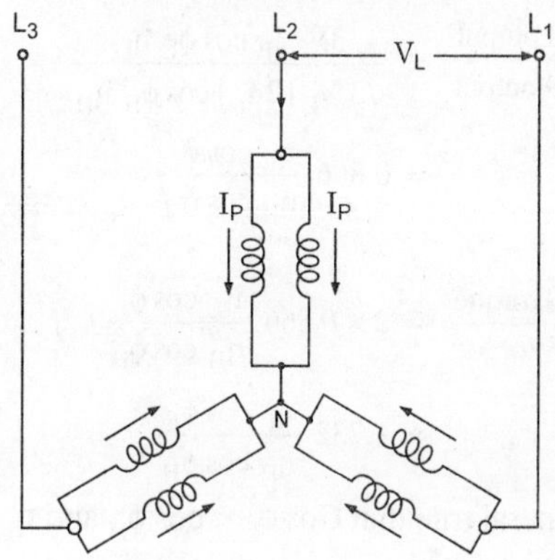

Low Speed Parallel Star

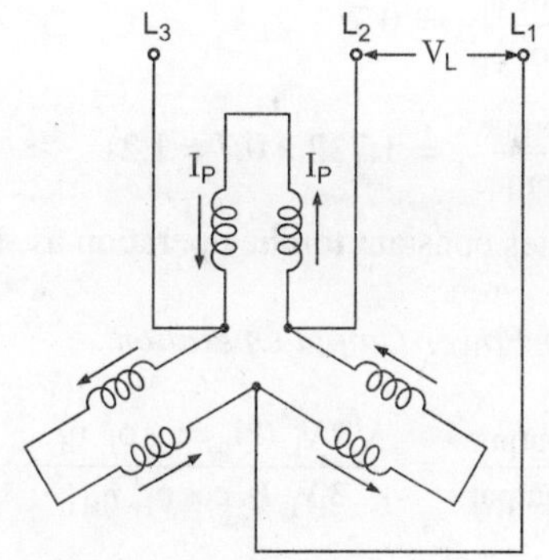

High Speed Series Delta

Fig. 19.17 *Constant Power Output Operation*

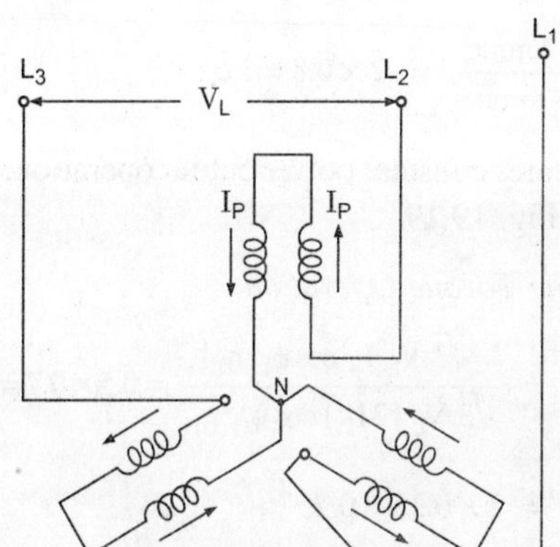

Low Speed Series Star

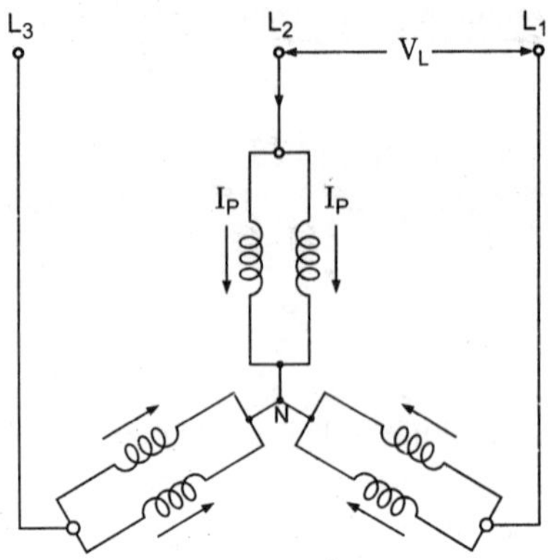

High Speed Parallel Star

Fig. 19.18 *Variable Torque Operation*

1. *Constant Torque Operation.*

$$\frac{\text{LS output}}{\text{HS output}} = \frac{3V_L\, I_P \cos\phi_L\, \eta_L}{\sqrt{3}\, V_L\, (2\, I_p) \cos\phi_H\, \eta_H}$$

$$= 0.866\, \frac{\eta_L \cos\phi_L}{\eta_H \cos\phi_H} \quad \text{... (19.29)}$$

and, therefore,

$$\frac{\text{LS torque}}{\text{HS torque}} = 2 \times 0.866\, \frac{\eta_L \cos\phi_L}{\eta_H \cos\phi_H}$$

$$= 1.732\, \frac{\eta_L \cos\phi_L}{\eta_H \cos\phi_H} \quad \text{...(19.30)}$$

It is in general true that Cos $\phi_L < \cos\phi_H$ and $\eta_L < \eta_H$

Their typical values are such that

$$\frac{\eta_L \cos\phi_L}{\eta_H \cos\phi_H} = 0.7$$

So $$\frac{\text{LS torque}}{\text{HS torque}} = 1.732 \times 0.7 = 1.21 \quad \text{... (19.31)}$$

This indicates constant torque operation as shown by curve 1 in Fig. 19.19.

2. *Constant Power Output Operation.*

$$\frac{\text{LS output}}{\text{HS output}} = \frac{\sqrt{3}\, V_L\, (2I_p) \cos\phi_L\, \eta_L}{3\, V_L\, I_P \cos\phi_H\, \eta_H}$$

$$= 1.15\, \frac{\eta_L \cos\phi_L}{\eta_H \cos\phi_H}$$

$$= 1.15 \times 0.7 = 0.805 \simeq 0.8 \quad \text{... (19.32)}$$

and $$\frac{\text{LS torque}}{\text{HS torque}} = 2 \times 0.8 = 1.6 \quad \text{...(19.33)}$$

This indicates constant power output operation, as illustrated by curve 2 in Fig. 19.19.

3. *Variable Torque Operation*

$$\frac{\text{LS output}}{\text{HS output}} = \frac{\sqrt{3}\, V_L\, I_p \cos\phi_L\, \eta_L}{\sqrt{3}\, V_L\, (2I_P) \cos\phi_H\, \eta_H} = 0.5 \times 0.7 = 0.35 \quad \text{...(19.34)}$$

and $$\frac{\text{LS torque}}{\text{HS torque}} = 2 \times 0.35 = 0.7 \quad \text{...(19.35)}$$

This indicates the variable torque operation as illustrated by curve 3 in Fig. 19.19.

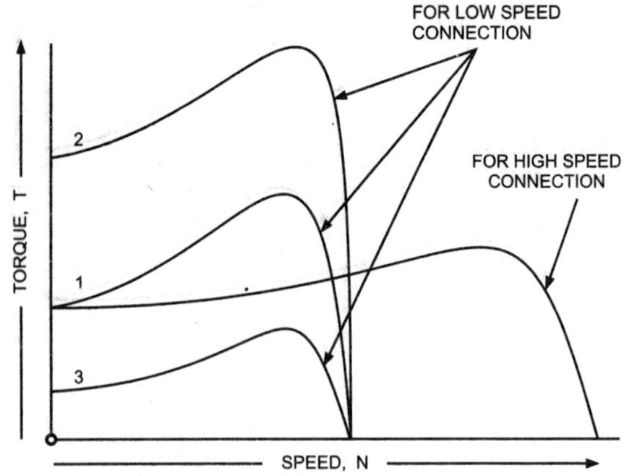

Fig. 19.19

Owing to complications in the design and switching of interconnection of the stator windings, it is not feasible to obtain more than four speeds for any motor through this method of speed control. This method also does not provide gradual speed control. That is why, this method is not employed for general purpose speed control applications. However, this method is very satisfactory for applications, such as ventilating fans, conveyors, machine tools, or other applications which require operation at only two or four approximately constant speeds which it will provide. This method has the advantage of simplicity, good speed regulation for each setting, high efficiency and moderate initial cost and maintenance.

***(c)* Pole Amplitude Modulation Method.** This method of pole changing has the advantage that the ratio of the two speeds obtainable need not be necessarily 2:1 as in the consequent pole method.

For understanding the pole amplitude modulation method of changing the stator poles, let us consider spatial mmf distribution produced by balanced 3-phase winding which is given as

$$\left.\begin{aligned} F_R &= F_{max} \sin \frac{P}{2}\theta \\ F_Y &= F_{max} \sin\left[\frac{P}{2}\theta - \frac{2\pi}{3}\right] \\ \text{and } F_B &= F_{max} \sin\left[\frac{P}{2}\theta - \frac{4\pi}{3}\right] = F_{max} \sin\left[\frac{P}{2}\theta + \frac{2\pi}{3}\right] \end{aligned}\right\} \quad \text{...(19.36)}$$

where P is the number of stator poles and θ is the mechanical angle.

Let the above mmf be modulated by another sinusoidal mmf given, for three phases, as

$$\left.\begin{aligned} F_{MR} &= F \sin \frac{n}{2}\,\theta \\ F_{MY} &= F \sin\left[\frac{n}{2}\theta - \frac{2\pi}{3}\right] \\ \text{and } F_{MB} &= F\sin\left[\frac{n\theta}{2} - \frac{4\pi}{3}\right] = F \sin\left[\frac{n\theta}{2} + \frac{2\pi}{3}\right] \end{aligned}\right\} \quad \text{...(19.37)}$$

where n is any integer.

Now the modulated mmfs are given as

$$F'_R = F_R \cdot F_{MR} = F_{max} \sin\frac{P}{2}\theta \cdot F \sin\frac{n\theta}{2}$$

$$= F'_{max}\left[\cos\left(\frac{P-n}{2}\right)\theta - \cos\left(\frac{P+n}{2}\right)\theta\right]$$

$$\left.\begin{aligned} F'_Y &= F_Y F_{MY} \\ &= F'_{max}\left[\cos\left(\frac{P-n}{2}\right)\theta - \cos\left\{\left(\frac{P+n}{2}\right)\theta + \frac{2\pi}{3}\right\}\right] \\ \text{and}\quad F'_B &= F_B F_{MB} \\ &= F'_{max}\left[\cos\left(\frac{P-n}{2}\right)\theta - \cos\left\{\left(\frac{P+n}{2}\right)\theta - \frac{2\pi}{3}\right\}\right] \end{aligned}\right\} \quad ...(19.38)$$

where $F'_{max} = F_{max}\, F$

From above Eq. (19.38) it is found that the modulated mmf wave has (P + *n*) number of poles and rotate in a direction opposite to that of the unmodulated field because the first terms are co-phasal in space and so the net mmf of pole number (P – *n*) will be zero.

Again, if the modulating mmfs are given as

$$\left.\begin{aligned} F_{MR} &= F\sin\frac{n}{2}\theta \\ F_{MY} &= F\sin\left[\frac{n\theta}{2}+\frac{2\pi}{3}\right] \\ \text{and } F_{MB} &= F\sin\left[\frac{n\theta}{2}-\frac{2\pi}{3}\right] \end{aligned}\right\} \quad ...(19.39)$$

Then modulated mmfs will be given as

$$\left.\begin{aligned} F_R' &= F'_{max}\left[\cos\left(\frac{P-n}{2}\right)\theta - \cos\left(\frac{P+n}{2}\right)\theta\right] \\ F_Y' &= F'_{max}\left[\cos\left\{\left(\frac{P-n}{2}\right)\theta + \frac{2\pi}{3}\right\} - \cos\left(\frac{P+n}{2}\right)\theta\right] \\ F_B' &= F'_{max}\left[\cos\left\{\left(\frac{P-n}{2}\right)\theta - \frac{2\pi}{3}\right\} - \cos\left(\frac{P+n}{2}\right)\theta\right] \end{aligned}\right\} \quad ...(19.40)$$

From above Eq. (19.40) it is found that the modulated mmf wave has (P – *n*) number of poles and rotate in the direction opposite to that of the unmodulated field.

From the above analyses it is concluded that if an unmodulated 8-pole motor is modulated with an mmf for which $n = 2$, it will result in either a 10-pole or 6-pole motor.

By this method of pole changing wide range of speed control is possible. It also allows great reduction in size and cost of the motor.

19.3.4. Speed Control By Variation of Rotor Resistance (or Rotor Resistance Control).

As the name implies, this type of control is only possible with wound rotor induction motors *i.e.* this method cannot be applied to squirrel cage motor.

Wound rotor motors are usually started by connecting starting resistances in the secondary circuit (refer to Art 19.2), which are shorted out as the motor speeds up. If the ohmic values of these resistors are properly chosen and if these resistors are designed for continuous operation, they can serve dual purpose — starting and speed control. This method of speed control has characteristics similar to those of dc shunt motor speed control by means of resistance in series with the armature.

From Eq. (18.15) torque developed by an induction motor is given as $T = \dfrac{Ks\, R_2\, E_2^2}{R_2^2 + s^2 X_2^2}$

When the speed is very near to synchronous speed N_s *i.e.*, when slip *s* is very low the value of the term sX_2 is very small and can be neglected as compared to R_2 and torque developed becomes proportional to $\dfrac{s}{R_2}$ $\left(i.e.\ T \propto \dfrac{s}{R_2}\right)$. So it is obvious that for a given torque, slip s can be increased (or speed can be reduced) by increasing the rotor resistance. It is also obvious that by this method speeds only below rated speed can be had. This method of speed control is stepped one. However, the larger the number of steps of the external resistances, the smoother is the speed control.

Secondary resistance control is simple and low in initial as well as in maintenance cost and it is possible to have a large starting torque with low starting current and large pull-out torques at small values of slip, as obvious from Fig. 18.18 (b). The circuit arrangement is the same as shown in Fig. 19.9.

However, this method of speed control is associated with very large power losses in the external resistances, especially at low speeds.

From Eqs. (18.31) and (18.32) we have

Mechanical power developed by rotor,

$$P_{mech} = \text{Power input to rotor } (1 - s)$$

and

$$\frac{\text{Total copper loss in rotor}}{\text{Gross rotor output, } P_{mech}} = \frac{s}{1-s} = \frac{(N_s - N)/N_s}{1-s}$$

$$= \frac{N_s - N}{N_s(1-s)} = \frac{N_s - N}{N} \quad ...(19.41)$$

From above discussion it is concluded that the mechanical power developed is reduced in direct proportion to the reduction in the rotational speed of the rotor and at the same time power losses in the rotor circuit resistances are correspondingly increased. As a result of these power losses, the efficiency is largely reduced at low speeds. For instance, at 50 per cent of synchronous speed the efficiency of rotor is only 50 per cent and the efficiency of the motor would be somewhat less. The poor speed regulation at low speeds can be appreciated from a study of torque-speed characteristics of an induction motor [Fig. 18.18 (*a*)]:

This method of speed control of polyphase induction motor has the following disadvantages.

(*i*) Reduction in speed is accompanied by reduction in efficiency.

(*ii*) Double dependence of speed not only on rotor resistance R_2 but on load as well— with a large resistance in the rotor circuit, the speed varies considerably with variation in torque [Fig. 18.18 (*a*)].

(*iii*) The external rotor resistors are comparatively bulky and expensive as they have to dissipate a good deal of power without getting overheated.

This method of speed control as such is, therefore, not suitable for controlling speed at constant torque. But this method is widely used for loads where the torque required drops off considerably as the speed is reduced such as fan loads for which the power input drops noticeably as the speed is reduced which in turn reduces the rotor copper losses. Moreover, this method is not adopted for continuous speed control but is preferred for intermittent (short-time) operation. Such a method of speed control is widely used in practice, for low-power motors and in overhead cranes. It is sometimes used, however, for speed regulation of rolling mills especially where they are provided with flywheels for reducing the load peaks in the circuit. Here the rheostat, called the *speed regulator,* is automatically switched on when the load increases, and as a result speed decreases and part of the load is compensated at the expense of kinetic energy of the flywheel. Conversely, when the load decreases, the rotor circuit resistance is reduced; the speed increases and the flywheel begins to store kinetic energy.

19.3.5. Secondary Foreign Voltage Control. In this method, the speed of an induction motor is controlled by injecting a voltage of slip frequency in the secondary circuit. If injected voltage is in phase with the induced emf in the secondary circuit, it is equivalent to decreasing of resistance in the secondary circuit, therefore, slip decreases or speed increases. It is possible to achieve super-synchronous speeds by injecting a voltage of slip frequency in phase with the induced emf in the rotor circuit. If injected voltage is in phase opposition to the induced emf in the secondary circuit, it is equivalent to increasing of resistance in the secondary circuit, hence resulting in increase in slip or reduction in speed. Thus a large range of speed control is possible by this method by including frequency-converting equipment in the rotor circuit.

In this method the disadvantages of lower efficiency and poor speed regulation are overcome, but, since it requires one and usually more auxiliary machines to inject emf of slip frequency in the secondary circuit, therefore, it is costlier and used with motors of very large rating, such as for motors in steel mills.

Several methods have been developed to control speed of induction motors by injecting a voltage of slip frequency in the secondary circuit. The different methods vary in their requirements for auxiliary equipment; from the incorporation of the auxiliary equipment in the design of induction motor itself to a fairly complicated system of auxiliary rotating machines and adjustable ratio transformers. The basic creteria that such a system must ful fill are : (*i*) that the variable foreign voltage have a slip frequency *i.e.* frequency equal to *sf* and (*ii*) that the energy associated with the foreign voltage be recoverable. A comprehensive discussion of this method of speed control is beyond the scope of this book. The most common systems based on this principle viz the Kramer system, Leblanc system and Scherbius system of speed control of polyphase induction motors will be discussed in this very chapter in the following articles and the Schrage brush-shifting induction motor employing this principle of speed control will be discussed in chapter 21 (Art 21.3).

19.3.6. Speed Control By Concatenation. In this method of speed control, two motors are required, at least one of which must have a wound rotor. The speed is changed by changing the slip of one motor, which changes the frequency supplied to the other motor. In this method of speed control, instead of dissipating energy in the control rheostat, this energy can be supplied to another induction motor connected to the first one as a cascade, and converted into mechanical energy. Hence the power drawn from the power circuit can be utilised more effectively.

For normal operation of the cascade, motor I and II (Fig. 19.20) must be joined both electrically and mechanically. An electrical connection alone does not suffice, since motor I begins to operate from a power circuit with normal voltage and frequency, and runs at normal speed, whereas motor II receives a very low voltage with a very low frequency from the slip rings of motor I. Hence the motor does not develop any torque and is only an additional resistance introduced into the rotor circuit of motor I.

Motors I and II should be so selected that their characteristics correspond to each other. Thus, for instance, if both motors are identical in design (conditions prevailing in railway traction), the stators of both the motors being high-voltage circuits and the rotor, low-voltage circuits, it is necessary to interconnect the rotors of both motors; to connect the rotor of motor I to the stator of motor II would not be rational. In special installations motor II may have a squirrel cage rotor; here the rotor of motor I is connected to the stator of motor II, as shown in Fig. 19.20 (*b*), but the characteristics of both circuits should be matched. Motor I should have a one-to-one ratio of transformation between stator and rotor so that at standstill with the external circuit of rotor open, the voltage across the rotor slip rings should be equal to line voltage. When two rotors are connected together, as shown in Fig. 19.20 (*a*), it is possible to dispense with slip rings, but they are generally retained to permit the use of either motor alone. The variable resistances are employed to start the unit and are short circuited during running condition. The speed of the combination will be determined by the sum or difference of the number of poles in the two machines. In practice it is customary, to connect the two machines in such a way that the phase rotation of the stator fields of both motors is in the same direction. The synchronous speed of the cascade set, in this case, is given as

$$N_{s1} = \frac{120\,f}{P_1 + P_2}$$

where f is supply frequency and P_1 and P_2 are the number of poles on machines I and II respectively.

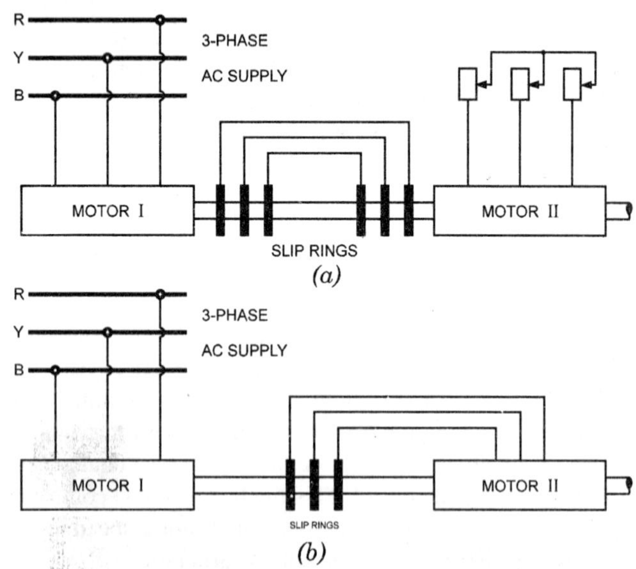

Fig. 19.20 *Cascade Connections of Induction Motors*

The above relation may be proved as below:

Let the supply frequency be f and the number of poles on machines I and II be P_1 and P_2 respectively.

The synchronous speed of motor I, $N_{s1} = \dfrac{120\,f}{P_1}$

The speed of motor I at slip s_1, $N_1 = N_{s1}(1 - s_1) = \dfrac{120\,f}{P_1}\,(1 - s_1)$

The impressed frequency on the second motor,

$$f_2 = s_1 f$$

Synchronous speed of motor II, $N_{s2} = \frac{120\, f_2}{P_2} = \frac{120\, s_1 f}{P_2}$

Speed of motor II at slip s_2, $N_2 = \frac{120\, s_1 f}{P_2}(1 - s_2)$

When the motor shafts are rigidly coupled, the speeds N_1 and N_2 must be the same

$$\text{So } \frac{120\, f}{P_1}(1 - s_1) = \frac{120\, s_1 f}{P_2}(1 - s_2) \quad \text{... (19.42)}$$

$$\text{or } \frac{1 - s_1}{P_1} = \frac{s_1 - s_1 s_2}{P_2}$$

But the product of two slips s_1 and s_2 is usually very small compared with the other term and so can be neglected and we get

$$\frac{1 - s_1}{P_1} = \frac{s_1}{P_2}$$

$$\text{or } s_1 = \frac{P_2}{P_1 + P_2} \quad \text{...(19.43)}$$

and concatenated speed of the set,

$$N = \frac{120\, f}{P_1}(1 - s_1) = \frac{120\, f}{P_1}\left(1 - \frac{P_2}{P_1 + P_2}\right)$$

$$= \frac{120 f}{P_1 + P_2} \quad \text{...(19.44)}$$

Now if the two motors are so connected electrically that the torques of the two motors are in opposite directions, then Eq. (19.42) becomes

$$\frac{120\, f}{P_1}(1 - s_1) = -\frac{120\, s_1 f}{P_2}(1 - s_2)$$

$$\text{or } \frac{1 - s_1}{P_1} = \frac{s_1 s_2 - s_1}{P_2}$$

Again neglecting $s_1 s_2$ we have

$$s_1 = \frac{P_2}{P_2 - P_1} \quad \text{...(19.45)}$$

and concatenated speed of the set,

$$N = \frac{120\, f}{P_1}(1 - s_1)$$

$$= \frac{120\, f}{P_1}\left(1 - \frac{P_2}{P_2 - P_1}\right) = \frac{120\, f}{P_1 - P_2} \quad \text{...(19.46)}$$

If the two machines are operated individually, then

Synchronous speed of machine I,

$$N_{s1} = \frac{120\, f}{P_1} \quad \text{...(19.47)}$$

and synchronous speed of machine II,

$$N_{s2} = \frac{120\, f}{P_2} \quad \text{...(19.48)}$$

Thus four different speeds are possible by cascade connections of two induction motors. Since the shaft torque is largely reduced in differential cascade [Eq. (19.46)], so it is not employed in practice.

If the two motors have the same number of poles and are connected in cumulative cascade [Eq. (19.44)], they will operate at practically half speed.

The power distribution in a cascade is the same as in a normal motor. The power input to rotor of machine I is equal to the power supplied to the stator of motor I minus the copper and core losses. The power input to the rotor of machine I is further divided into two parts; one part proportional to the speed of the set (*i.e.*, $1 - s_1$) and is converted into mechanical power and the other proportional to s_1 *is* developed as electrical power at the slip frequency and supplied to machine II, which uses it for producing mechanical power and losses. Thus the mechanical outputs of the two machines are approximately in the ratio of $(1 - s_1) : s_1$ or $\left(1 - \frac{P_2}{P_1 + P_2}\right) : \frac{P_2}{P_1 + P_2}$ or $P_1 : P_2$ *i.e.* in the ratio of the number of poles on the machines.

The set will not start, if connected in concatenation, because rotors will tend to rotate in opposite directions. Hence it must first be brought up to speed by an auxiliary motor or one motor alone be started and then second motor be connected.

It will be recognized that the concatenation method of speed control is similar to the series-parallel control of dc motors. In concatenation, at starting and for intermediate speeds, resistance is introduced in the rotor circuit of the second motor. When the motors are connected in parallel across the line, resistance is introduced in each rotor circuit and is gradually cut out. Full torque is available only when the motors are cascade connected.

A cascade of two induction motors has substantial disadvantages. Indeed, the magnetising current necessary to produce the fluxes in both motors is drawn from the supply mains through motor I. Therefore, no-load current of the set is appreciably greater than no-load current of one motor. On the other hand motor II is connected is series with the motor I and, therefore, the short-circuit current of the set is lesser than that of motor I, alone. The cascade characteristics are much inferior to the corresponding characteristics of a motor operating alone. A cascade of two induction motors are, therefore, used only in special cases, mainly on railways with 3-phase current.

Example 19.15. A 50 Hz, 440 V, 3-phase, 4-pole induction motor develops half the rated torque at 1,490 rpm. With the applied voltage magnitude remaining at the rated value, what should be its frequency if the motor has to develop the same torque at 1,600 rpm ? Neglect stator and rotor winding resistances and leakage reactances and iron losses. [*GATE, 1995*]

Solution: Synchronous speed at supply frequency of 50 Hz,

$$N_s = \frac{120\, f}{P} = \frac{120 \times 50}{4} = 1{,}500 \text{ rpm}$$

Slip at a speed of 1,490 rpm (*i.e.* N = 1,490 rpm),

$$s = \frac{N_s - N}{N_s} = \frac{1{,}500 - 1{,}490}{1{,}500} = 0.00667$$

Since torque developed by an induction motor, $T \propto sV^2$, slip s for constant torque and constant applied voltage remains unchanged.

New speed, $N' = 1{,}600$ rpm

Slip, $s = 0.00667$

New synchronous speed,

$$N_s' = \frac{N'}{1 - s} = \frac{1{,}600}{1 - 0.00667} = 1{,}610.7 \text{ rpm}$$

$$\text{New frequency, } f' = \frac{P \times N_s'}{120} = \frac{4 \times 1{,}610.7}{120} = 53.7 \text{ Hz } \textbf{Ans.}$$

Example 19.16. A star-connected cage induction motor has the following data :

440 V, 50 Hz, 4-pole, 1460 rpm, $R_s = 2\ \Omega$, $R_r' = 2\ \Omega$, $X_s = X_r' = 3.0\ \Omega$.

Calculate starting torque and starting current of this motor at 50 Hz and 10 Hz for V/f control.

Solution : At 50 Hz supply

Supply voltage per phase, $V = \frac{440}{\sqrt{3}} = 254$ V

Equivalent resistance of motor as referred to stator,

$$R_{01} = R_s + R'_r = 2 + 2 = 4\ \Omega,$$

Equivalaent reactance of motor as referred to stator,

$$X_{01} = X_s + X'_r = 3 + 3 = 6\ \Omega$$

$$\text{Starting current, } I_{st} = \frac{V}{\sqrt{(R_{01})^2 + (X_{01})^2}}$$

$$= \frac{254}{\sqrt{4^2 + 6^2}} = 35.23 \text{ A } \textbf{Ans.}$$

$$\text{Total copper losses in rotor} = 3\, I_{st}^2 \times R'_r = 3 \times (35.23)^2 \times 2$$

$$= 7{,}446 \text{ watts}$$

$$\text{Rotor power input at standstill, } P_2 = \text{Total copper losses}$$

$$= 7{,}446 \text{ watts} \qquad \because s = 1$$

$$\text{Synchronous speed, } N_s = \frac{120\, f}{P} = \frac{120 \times 50}{4}$$

$$= 1{,}500 \text{ rpm}$$

$$\text{Starting torque, } T_{st} = \frac{9.55\, P_2}{N_s} = \frac{9.55 \times 7{,}446}{1{,}500}$$

$$= 47.41 \text{ N-m } \textbf{Ans.}$$

At 10 Hz supply

Supply voltage per phase,

$$V' = V \times \frac{f'}{f} = 254 \times \frac{10}{50} = 50.8 \text{ V}$$

Equivalent resistance of motor referred to stator,

$$R'_{01} = 4\ \Omega, \text{ the same as before}$$

Equivalent reactance of motor as referred to stator,

$$X'_{01} = X_{01} \times \frac{f'}{f} = 6 \times \frac{10}{50} = 1.2\ \Omega$$

$$\text{Starting current, } I'_{st} = \frac{V'}{\sqrt{(R'_{01})^2 + (X_{01})^2}} = \frac{50.8}{\sqrt{4^2 + 1.2^2}}$$

$$= 12.166 \text{ A } \textbf{Ans.}$$

Power input to rotor at standstill,

$$P'_2 = \text{Total copper losses}$$

$$= 3\,(I'_{st})^2\, R'_r = 3 \times (12.166)^2 \times 2$$

$$= 888.1 \text{ watts}$$

$$\text{Synchronous speed, } N'_s = \frac{120\, f'}{P} = \frac{120 \times 10}{4} = 300 \text{ rpm}$$

$$\text{Starting torque, } T'_{st} = \frac{P'_2 \times 9.55}{N'_s} = \frac{888.1}{300} \times 9.55 = 28.27 \text{ Nm } \textbf{Ans.}$$

Example 19.17. A star-connected squirrel cage induction motor has the following parameters and rating:

$R_s = 2\ \Omega$, $R'_r = 3\ \Omega$, $X_s = X'_r = 3.5\ \Omega$, V = 400, 50 Hz, 1,380 rpm

It is driven by a voltage source inverter which gives constant $\frac{V}{f}$ control to IM. If the inverter frequency is variable from 10% of rated frequency to base frequency, calculate:

(*a*) frequency for a speed of 1,000 rpm and full-load torque

(*b*) torque for a frequency of 35 Hz and speed of 950 rpm.

[*U.P.S.C. I.E.S. Electrical Engineering-II, 2004*]

Solution: Supply voltage per phase, $V = \frac{400}{\sqrt{3}} = 231$ V

Supply frequency = 50 Hz

Rotor speed, N = 1,380 rpm

Number of poles, even in number, to provide a synchronous speed nearest higher than rotor speed (i.e. 1,380 rpm) that the motor must have is 4.

$$\text{Synchronous speed, } N_s = \frac{120 f}{P} = \frac{120 \times 50}{4} = 1{,}500 \text{ rpm}$$

$$\text{Full-load slip, } s_f = \frac{N_s - N}{N_s} = \frac{1{,}500 - 1{,}380}{1{,}500} = 0.08 \text{ or } 8\%$$

$$\omega_s = \frac{2\pi N_s}{60} = \frac{2\pi \times 1{,}500}{60} = 50\pi \text{ radians/second}$$

$$\text{Full-load torque, } T_f = \frac{3}{\omega_s} \times \frac{V^2 R'_r / s_f}{\left(R_s + \frac{R'_r}{s_f}\right)^2 + X_{01}^2}$$

$$= \frac{3}{50\pi} \times \frac{231^2 \times 3/0.08}{\left(2 + \frac{3}{0.08}\right)^2 + (3.5 + 3.5)^2}$$

$$= 23.736 \text{ N-m}$$

Frequency range 10 of rated frequency to 50 Hz i.e. 5–50 Hz

$$\frac{V}{f} = \frac{400/\sqrt{3}}{50} = 4.62$$

(*a*) Rotor speed, N = 1,000 rpm

Full-load torque, $T_f = 23.736$ N-m

$$\omega_s = \frac{2\pi N_s}{60} = \frac{2\pi}{60} \times \frac{120 f}{P} = \frac{4\pi f}{4} = \pi f$$

where f is new supply frequency

$$X_{01} = 2\pi f L = \frac{f}{50} \times X_{50\text{ Hz}} \qquad \because X \propto f$$

$$\text{or } X_{01} = \frac{f}{50} \times 7 = 0.14 f \qquad \because X_{50\text{ Hz}} = 3.5 + 3.5 = 7\ \Omega$$

$$\text{Full-load slip, } s_f = \frac{\omega_s - \omega_r}{\omega_s} = 1 - \frac{\omega_r}{\omega_s} = 1 - \frac{\frac{2\pi \times 1{,}000}{60}}{\pi f} = 1 - \frac{33.33}{f}$$

Full-load torque is given as

$$T_f = \frac{3}{\omega_s} \frac{V^2 R'_2 / s_f}{\left(R_s + \frac{R'_r}{s_f}\right)^2 + X_{01}^2}$$

and for low-slip region i.e. when full-load slip s_f is very small

$$\left(R_s + \frac{R'_r}{s_f}\right)^2 + X_{01}^2 \cong \left(\frac{R'_r}{s_f}\right)^2$$

$\because$ R_s and X_{01} are negligibly small as compared to $\frac{R'_r}{s_f}$

$$\text{So } T_f = \frac{3}{\omega_s} \times \frac{V^2 R'_r}{(R'_r)^2 / s_f}$$

$$\text{or } 23.736 = \frac{3}{\pi f} \times \frac{(4.62\, f)^2 \times 3}{3^2 \Big/ \left(1 - \frac{33.33}{f}\right)} \qquad \because \frac{V}{f} = 4.62$$

$$= \frac{3}{\pi} \times (4.62)^2 \left(\frac{f - 33.33}{3}\right)$$

$$\text{or } f = \frac{23.736 \times \pi}{(4.62)^2} + 33.33 = 36.82 \text{ Hz } \textbf{Ans.}$$

(*b*) Frequency, $f = 35$ Hz

$$\text{Synchronous speed, } N_s = \frac{120 f}{P} = \frac{120 \times 35}{4} = 1{,}050 \text{ rpm}$$

Rotor speed, N = 950 rpm

$$\text{Slip, } s_{f_1} = \frac{N_s - N}{N_s} = \frac{1{,}050 - 950}{1{,}050} = 0.095$$

$$X_{01} = \frac{35}{50} \times 7 = 4.9\ \Omega$$

$$V = 4.62 f = 4.62 \times 35 = 161.7 \text{ V}$$

$$\omega_s = \frac{2\pi N_s}{60} = \frac{2\pi \times 1{,}050}{60} = 35\pi \text{ radians/speed}$$

$$\text{Torque, } T = \frac{3}{\omega_s} \times \frac{V^2}{\left(R_s + \frac{R'_r}{s_{f_1}}\right)^2 + X_{01}^2} \times \frac{R'_r}{s_{f_1}}$$

$$= \frac{3}{35\pi} \times \frac{(161.7)^2 \times 3/0.095}{\left(2 + \frac{3}{0.095}\right)^2 + (4.9)^2} = 19.563 \text{ N-m } \textbf{Ans.}$$

Example 19.18. A 4-pole, 400 V, 10 kW slip-ring 3-phase induction motor has rotor resistance = 0.16 Ω/phase and standstill reactance = 0.27 Ω/phase. The voltage across the slip rings at standstill is 141 V per phase. The motor develops a torque of 62.4 N-m at a slip of *s* = 0.035 and rotor current of 31.1 A. Find the magnitude and phase of emf to be injected to the rotor circuit so that the motor runs at *s* = 0.02 and develops the same torque.

[*U.P.S.C. I.E.S. Electrical Engineering-II, 2008*]

Solution: Rotor resistance per phase, $R_2 = 0.16\ \Omega$

Standstill reactance per phase, $X_2 = 0.27\ \Omega$

Slip, $s = 0.035$

$$\text{Rotor current, } I_2 = \frac{E_2}{\sqrt{\left(\frac{R_2}{s}\right)^2 + X_2^2}}$$

$$\text{or } 31.1 = \frac{E_2}{\sqrt{\left(\frac{0.16}{0.035}\right)^2 + 0.27^2}}$$

$$\text{or } E_2 = 31.1\sqrt{\left(\frac{0.16}{0.035}\right)^2 + 0.27^2} = 142.42 \text{ V}$$

$$\theta_2 = \text{Tan}^{-1}\frac{sX_2}{R_2} = \text{Tan}^{-1}\frac{0.035 \times 0.27}{0.16} = 3.38°$$

$$\text{Developed torque, } T = \frac{1}{\omega_s} \times \frac{3I_2^2 R_2}{s}$$

$$\text{or } T \propto \frac{I_2^2}{s}$$

$$\text{So } \frac{I_2'^2}{0.02} = \frac{(31.1)^2}{0.035}$$

$$\text{or } I_2 = \sqrt{\frac{0.02}{0.035} \times (31.1)^2} = 23.51 \text{ A}$$

For current of 23.51 A induced emf E'_2 is given as

$$23.51 = \frac{E'_2}{\sqrt{\left(\frac{0.16}{0.02}\right)^2 + (0.27)^2}}$$

$$\text{or } E'_2 = 23.51 \times \sqrt{64 + (0.27)^2} = 188.187 \text{ V}$$

$$\theta'_2 = \text{Tan}^{-1}\frac{0.02 \times 0.27}{0.16} = 1.93°$$

$$\text{Injected emf} = \mathbf{E'_2} - \mathbf{E_2} = 188.187 \angle 1.93° - 142.42 \angle 3.38°$$

$$= (188.08 + j\,6.3378) - (142.178 + j\,8.397)$$

$$= (45.902 - j\,2.0592) = 45.948 \angle -2.57° \text{ V } \textbf{Ans.}$$

Example 19.19. A 3-phase, 440 V, 1,000 rpm slip-ring induction motor is operating with 2 % slip and taking a stator current of 50 A. Speed of the motor is reduced at constant torque to 500 rpm using stator voltage control. Calculate the new value of stator current.

Solution: Supply voltage, V = 440 V

Slip, s = 2 % or 0.02

Stator current, I_1 = 50 A

Synchronous speed, N_s = 1,000 rpm

Reduced speed, N′ = 500 rpm

$$\text{Slip at reduced speed, } s' = \frac{N_s - N'}{N_s} = \frac{1{,}000 - 500}{1{,}000} = 0.5$$

Torque developed by an induction motor

$T \propto sV^2$, so for constant torque

$$V \propto \sqrt{\frac{1}{s}}$$

$$\text{or } V' = V \times \sqrt{\frac{s}{s'}} = 440\sqrt{\frac{0.02}{0.5}} = 88 \text{ V}$$

Stator current $I_1 \propto sV$

$$\text{So new stator current, } I_1' = I_1 \times \frac{s'V'}{sV} = \frac{50 \times 0.5 \times 88}{0.02 \times 440}$$

$$= 250 \text{ A } \textbf{Ans.}$$

Example 19.20. A 75 kW, 415 V, 4-pole, 3-ϕ, 50 Hz induction motor is delivering rated torque at 5 % slip driving a pump load. The motor delivers maximum torque at 10 % slip. A 3-phase ac regulator controls the rms supply voltage to effect limited speed control. Assume the torque-speed characteristics of the motor in the stable region to be linear. Calculate the rms voltage at which the motor runs at 1,350 rpm.

Solution: Full-load slip, s_f = 5 % or 0.05

$$\text{Full-load speed, } N = N_s(1 - s) = \frac{120 f}{P}(1 - s)$$

$$= \frac{120 \times 50}{4}(1 - 0.05)$$

$$= 1{,}425 \text{ rpm}$$

Slip corresponding to maximum torque,

$$s_{\max T} = 10\ \% \text{ or } 0.1$$

Ratio of rotor resistance to standstill reactance,

$$\frac{R_2}{X_2} = s_{\max T} = 0.1$$

$$\text{or } X_2 = 10\,R_2$$

When speed, $N' = 1{,}350$ rpm

$$\text{Full-load slip, } s_f' = \frac{\frac{120\,f}{P} - N'}{\frac{120\,f}{P}} = \frac{\frac{120\times 50}{4} - 1{,}350}{\frac{120\times 50}{4}} = 0.1$$

For pump load assuming that the torque varies as the speed so torque corresponding to 1,350 rpm,

$$T' = T \times \frac{1{,}350}{1{,}425}$$

Since from Eq. (18.15),

$$\text{Torque developed} = \frac{K\,R_2\,s\,E_2^2}{R_2^2 + s^2 X_2^2},$$

$$\frac{T'}{T} = \frac{s_f'\left(E_2'\right)^2 \Big/ \left[R_2^2 + \left(s_f'\,X_2\right)^2\right]}{s_f\,E_2^2 / R_2^2 + s_f^2\,X_2^2}$$

$$= \frac{s_f'\,(V')^2\left[R_2^2 + s_f^2\,X_2^2\right]}{s_f\,V^2\left[R_2^2 + \left(s_f'\,X_2\right)^2\right]} \qquad \because\ E_2 \propto V, \text{ the supply}$$

$$\text{So } V' = V \times \sqrt{\frac{T'}{T} \times \frac{s_f}{s_f'} \times \frac{R_2^2 + \left(s_f'\,X_2\right)^2}{R^2 + \left(s_f\,X_2\right)^2}}$$

$$= 415 \times \sqrt{\frac{1{,}350}{1{,}425} \times \frac{0.05}{0.1} \times \frac{R_2^2 + (0.1 \times 10\,R_2)^2}{R_2^2 + (0.05 \times 10 \times R_2)^2}}$$

$$= 415 \times \sqrt{\frac{1{,}350}{1{,}425} \times 0.5 \times \frac{8}{5}} = 361.3 \text{ V} \quad \textbf{Ans.}$$

Example 19.21. A 230 V, 20 hp, 60 Hz, 6-pole, 3-phase induction motor driving a constant torque load at rated frequency, rated voltage and rated horse-power, has a speed of 1,175 rpm and an efficiency of 92.1%. Determine the new operating speed if a system disturbance causes 10% drop in voltage and 6% drop in frequency. Assume that friction, windage and stray power losses remain constant.

[*GATE, 2000*]

Solution: Synchronous speed, $N_{s1} = \dfrac{120 f}{P} = \dfrac{120 \times 60}{6}$

$$= 1{,}200 \text{ rpm}$$

$$\text{Slip, } s_1 = \frac{1{,}200 - 1{,}175}{1{,}200} = \frac{1}{48}$$

$$\text{New frequency, } f_2 = 60\left(1 - \frac{6}{100}\right) = 56.4 \text{ Hz}$$

$$\text{Synchronous speed, } N_{s2} = \frac{120 \times 56.4}{6} = 1{,}128 \text{ rpm}$$

$$\text{New supply voltage, } V_2 = \left(1 - \frac{10}{100}\right)V_1 = 0.9\,V_1$$

Since load torque remains constant, the product $\dfrac{sV^2}{f}$ remains constant (*T* being proportional to $\dfrac{sV^2}{f}$),

$$\frac{s_2 V_2^2}{f_2} = \frac{s_1 V_1^2}{f_1}$$

$$\text{or } s_2 = s_1 \times \left(\frac{V_1}{V_2}\right)^2 \times \frac{f_2}{f_1} = \frac{1}{48} \times \left(\frac{1}{0.9}\right)^2 \times \frac{56.4}{60} = 0.02418$$

New operating speed,

$$N_2 = N_{s_2}(1 - s_2) = 1{,}128(1 - 0.2418) = 1{,}100.7 \text{ rpm} \quad \textbf{Ans.}$$

Example 19.22. A 50 Hz, 3-phase induction motor designed for voltage V_1 is switched on to 40 Hz supply of voltage V_2. Find

(*i*) the ratio of starting torques, starting currents and maximum torques when $V_2 = 1.5\,V_1$.

(*ii*) the ratio of V_1 to V_2 to give equal values of starting torques both at 40 Hz and 50 Hz. Ignore stator, impedance and neglect rotor resistance.

[*U.P.S.C. I.E.S. Electrical Engineering-II, 2003*]

Solution: $\dfrac{V_2}{V_1} = 1.5$

Ratio of standstill reactance,

$$\frac{X_{22}}{X_{21}} = \frac{2\pi f_2 L_2}{2\pi f_1 L_2} = \frac{f_2}{f_1} = \frac{40}{50} = 0.8$$

From Eq. (18.16), if rotor resistance is negligibly small compared to its standstill reactance, then

Starting torque,

$$T_{st} \propto \left(\frac{V}{X_2}\right)^2$$

(*i*) So ratio of starting torques,

$$\frac{T_{st_2}}{T_{st_1}} = \left(\frac{V_2}{V_1}\right)^2 \left(\frac{X_{21}}{X_{22}}\right)^2 = 1.5^2 \times \frac{1}{0.8^2} = 3.52 \quad \textbf{Ans.}$$

Ratio of starting currents

$$\frac{I_{st_2}}{I_{st_1}} = \frac{V_2}{V_1} \times \frac{X_{21}}{X_{22}} = 1.5 \times \frac{1}{0.8} = 1.875 \quad \textbf{Ans.}$$

Ratio of maximum torques,

$$\frac{T_{\max_2}}{T_{\max_1}} = \left(\frac{V_2}{V_1}\right)^2 \times \frac{X_{21}}{X_{22}} \qquad T_{\max} \propto \frac{V_2^2}{X_2}$$

$$= (1.5)^2 \times \frac{1}{0.8} = 2.8125 \quad \textbf{Ans.}$$

(*ii*) Since starting torque is proportional to $\left(\dfrac{V}{X_2}\right)^2$.

$$\left(\frac{V_2}{X_{22}}\right)^2 = \left(\frac{V_1}{X_{21}}\right)^2$$

$$\text{or Ratio } \frac{V_1}{V_2} = \frac{X_{21}}{X_{22}} = \frac{1}{0.8} = 1.25 \quad \textbf{Ans.}$$

Example 19.23. An 8-pole, 50 Hz, 3-phase induction motor is running at 4 % slip when delivering full load torque. It has standstill rotor resistance of 0.1 Ω and reactance of 0.6 Ω per phase. Calculate the speed of the motor if an additional resistance of 0.5 Ω per phase is inserted in the rotor circuit. The full-load torque remains constant.

Solution: Synchronous speed of motor,

$$N_s = \frac{120\,f}{P} = \frac{120 \times 50}{8} = 750 \text{ rpm}$$

Full load slip, $s_1 = 0.04$

The motor torque Eq. (18.15) is given by the expression

$$T = \frac{K\, sR_2\, E_2^2}{R_2^2 + s^2\, X_2^2}$$

Let the new slip at full load be s_2

$$\text{Now } T_1 = \frac{K s_1\, R_2\, E_2^2}{R_2^2 + s_1^2\, X_2^2} = \frac{K \times 0.04 \times 0.1 \times E_2^2}{(0.1)^2 + (0.04 \times 0.6)^2} = \frac{K\, E_2^2}{2.644}$$

$$\text{and } T_2 = \frac{K\, s_2\, (R_2 + R)\, E_2^2}{(R_2 + R)^2 + (s_2 X_2)^2}$$

$$= \frac{K \times s_2 \times (0.1 + 0.5)\, E_2^2}{(0.1 + 0.5)^2 + (0.6\, s_2)^2} = \frac{0.6\, s_2\, K\, E_2^2}{0.36 + 0.36\, s_2^2}$$

Since torque remains the same *i.e.* $T_1 = T_2$

$$\therefore \frac{0.6\, s_2\, KE_2^2}{0.36 + 0.36\, s_2^2} = \frac{KE_2^2}{2.644}$$

or $0.6\,(1 + s_2^2) = 2.644\, s_2$

or $s_2^2 - 4.407\, s_2 + 1 = 0$

or $s_2 = 0.24$ neglecting higher value, being impractical

Speed of motor, $N_2 = N_s\,(1 - s) = 750\,(1 - 0.24) = 570$ rpm **Ans.**

Example 19.24. The rotor of a 4-pole, 50 Hz, slip-ring induction motor has a resistance of 0.25 Ω per phase and runs at 1,440 rpm at full-load. Calculate the external resistance per phase which must be added to lower the speed to 1,200 rpm, the torque being the same as before.

[*U.P. Technical Univ. Electrical Machines 2003-04; Electromechanical Energy Conversion-II, 2010-11; Bharathiyar Univ. Elec. Machines II Nov. 1996; Madurai Kamarajar Univ. Elec. Machines II Nov. 1996*]

Solution: Synchronous speed of motor,

$$N_s = \frac{120\, f}{P} = \frac{120 \times 50}{4} = 1{,}500 \text{ rpm}$$

Full-load speed, N = 1,440 rpm

$$\text{Full-load slip, } s_1 = \frac{N_s - N}{N_s} = \frac{1{,}500 - 1{,}440}{1{,}500} = 0.04$$

After inserting resistance (say R ohm per phase) in the rotor circuit,

Motor speed, N_2 = 1,200 rpm

$$\text{Slip, } s_2 = \frac{1{,}500 - 1{,}200}{1{,}500} = 0.2$$

$$\text{Since slip, } s = \frac{\text{Rotor copper loss}}{\text{Input power to rotor}} = \frac{3 I_2^2\, R_2}{\text{Input power to rotor}}$$

For constant power input to rotor and rotor current $s \propto R_2$

∵ for constant load torque power input to rotor and rotor current remains the same

$$\text{So } \frac{s_2}{s_1} = \frac{R_2 + R}{R_2}$$

$$\text{or } \frac{0.20}{0.04} = \frac{0.25 + R}{0.25}$$

or R = 1 Ω **Ans.**

Example 19.25. A 3-phase, 440 V, 50 Hz, 110 kW, 24-pole, 245 rpm slip-ring induction motor has both its stator and rotor windings connected in star. The ratio of stator turns to rotor turns is 1.25. The resistance measured between each pair of slip rings is 0.04 Ω. This motor drives a fan which requires 110 kW at the full speed of the motor. The torque required to drive the fan varies as the square of speed. What resistance should be connected in series with each slip-ring so that the fan will run at 175 rpm. Neglect stator resistance and leakage reactance and rotational losses.

Solution: Synchronous speed of motor,

$$N_s = \frac{120 \times 6}{P} = \frac{120 \times 50}{24} = 250 \text{ rpm}$$

$$\text{Full-load slip, } s_f = \frac{N_s - N}{N_s} = \frac{250 - 245}{250} = 0.02$$

$$\text{Torque at full-load, } T_f = \frac{\text{Output in kW} \times 1{,}000}{2\pi N/60}$$

$$= \frac{110 \times 1{,}000 \times 60}{2\pi \times 245} = 4{,}287.44 \text{ N-m}$$

Rotor resistance per phase,

$$R_2 = \frac{\text{Resistance between a pair of slip rings}}{2}$$

$$= \frac{0.04}{2} = 0.02\ \Omega$$

Rotor resistance referred to stator,

$$R'_2 = R_2 \times K^2 = 0.02 \times (1.25)^2 = 0.03125\ \Omega$$

Torque developed by rotor is given as

$$T_f = \frac{3}{2\pi N_s/60} \times \frac{V_1^2 R'_2\, s_f}{(R'_2)^2 + s_f^2 (X'_2)^2} \qquad \text{Refer to Eq. (18.44)}$$

Substituting $V_1 = \frac{440}{\sqrt{3}}$ V (stator being connected in star), $s_f = 0.02$, $R'_2 = 0.03125\ \Omega$, $T_f = 4{,}287.44$ N-m, and $N_s = 250$ rpm in above equation, we have

$$4{,}287.44 = \frac{3 \times (440)^2 \times 0.03125 \times 0.02 \times 60}{3 \times 2\pi \times 250\left[(0.03125)^2 + (0.02\, X'_2)^2\right]}$$

$$\text{or } (X'_2)^2 = \left[\frac{440^2 \times 0.03125 \times 1.2}{4{,}287.44 \times 500\, \pi} - (0.03125)^2\right] \times \left(\frac{1}{0.02}\right)^2$$

$$= 0.2536$$

$$\text{or } X'_2 = 0.5036\ \Omega$$

$$\text{Slip at 175 rpm, } s = \frac{250 - 175}{250} = 0.3$$

Let the rotor resistance referred to stator per phase corresponding to speed of 175 rpm be R''_2 and torque developed be T

$$\text{So } \frac{T}{T_f} = \left[\frac{N}{N_f}\right]^2 = \left(\frac{175}{245}\right)^2$$

or Torque at 175 rpm,

$$T = T_f \times \left(\frac{175}{245}\right)^2 = 4{,}287.44 \times \left(\frac{175}{245}\right)^2$$

$$= 2{,}187.47 \text{ Nm.}$$

Substituting T = 2,187.47 Nm, $V_1 = \frac{440}{\sqrt{3}}$ V, $s = 0.3$ and $X_2 = 0.5036\ \Omega$ in torque equation, we have

$$2{,}187.47 = \frac{(440)^2 \times R''_2 \times 0.3 \times 60}{2\pi \times 250[(R''_2)^2 + (0.3 \times 0.5036)^2]}$$

or $(R''_2)^2 + 0.022825 = 1.0142 R''_2$

or $R''_2 = 0.023\ \Omega$ or $0.991\ \Omega$

The first value, being lesser than original value, is rejected. So resistance to be added to each slip ring

$$= \frac{0.991 - 0.03125}{(1.25)^2} = 0.6142\ \Omega \quad \textbf{Ans.}$$

Example 19.26. The open-circuit voltage across the slip rings of a 100 hp induction motor is 273 V at standstill. What resistance in rotor circuit will reduce its full-load speed by 25 %. The full-load slip is 2 % with no additional rotor resistance. Assume rotor to be star-connected. [*Pb. Univ. Elec. Machines II Dec; 1994*]

Solution: Mechanical power developed by rotor,

P_{mech} = Motor output in watts neglecting windage and frictional losses

$= 100 \times 735.5 = 73{,}550$ watts

Full-load slip, $s_1 = 2\%$ or 0.02

Standstill induced emf per phase in rotor,

$$E_2 = \frac{273}{\sqrt{3}} = 157.62 \text{ V}$$

Let the rotor current per phase be I_2 amperes, rotor resistance per phase be R_2 and rotor reactance negligible

$$\text{So } I_2 = \frac{s_1 E_2}{R_2} = \frac{0.02 \times 157.62}{R_2} \quad \ldots(i)$$

Mechanical power developed by rotor

$$= \frac{\text{Total rotor copper loss}}{s}(1-s)$$

$$\text{or } 73{,}550 = \frac{3I_2^2 R_2}{0.02} \times (1-0.02) = 147\, I_2^2 R_2$$

Substituting $I_2 = \dfrac{0.02 \times 157.62}{R_2} = \dfrac{3.1524}{R_2}$ in above equation we get

$$147 \times \left(\frac{3.1524}{R_2}\right)^2 R_2 = 73{,}550$$

$$\text{or } R_2 = \frac{147 \times (3.1524)^2}{73{,}550} = 0.02\ \Omega$$

New speed $N_2 = N_s(1 - 0.02) \times (1 - 0.25) = 0.735\ N_s$

$$\text{So slip, } s_2 = \frac{N_s - N_2}{N_s} = \frac{N_s - 0.735 N_s}{N_s} = 0.265$$

Assuming load torque to be constant

$s \;\alpha$ rotor resistance

$$\text{So } \frac{s_2}{s_1} = \frac{R_2 + R}{R_2}$$

$$\text{or } \frac{0.265}{0.02} = \frac{0.02 + R}{0.02}$$

or $R = 0.245\ \Omega$ **Ans.**

Example 19.27. Two 50 Hz, 3-phase induction motors having six and four poles respectively are cumulatively cascaded, the 6-pole motor being connected to the main supply. Determine the frequency of the rotor currents and slips referred to each stator field if the set has a slip of 2 %.

Solution: Synchronous speed of the set,

$$N_{sc} = \frac{120\,f}{P_1 + P_2} = \frac{120 \times 50}{6+4} = 600 \text{ rpm}$$

Actual speed of the set,

$$N = N_{sc}(1 - s') = 600\,(1 - 0.02) = 588 \text{ rpm}$$

Synchronous speed of 6-pole motor stator field,

$$N_s = \frac{120 \times f}{P_1} = \frac{120 \times 50}{6} = 1{,}000 \text{ rpm}$$

Slip referred to the stator field,

$$s = \frac{N_s - N}{N_s} = \frac{1{,}000 - 588}{1{,}000} = 0.412 \text{ or } 41.2\,\% \textbf{ Ans.}$$

The 6-pole motor rotor current frequency or the impressed frequency on 4-pole motor stator field,

$$f' = s\,f = 0.412 \times 50 = 20.6 \text{ Hz } \textbf{Ans.}$$

The synchronous speed of 4-pole induction motor with frequency f',

$$N_s' = \frac{120\,f'}{P_2} = \frac{120 \times 20.6}{4} = 618 \text{ rpm}$$

The slip as referred to the 4-pole induction motor,

$$s'' = \frac{N_s' - N}{N'_s} = \frac{618 - 588}{618} = 0.0485 \text{ or } 4.85\,\% \textbf{ Ans.}$$

Frequency of rotor current of 4-pole induction motor,

$$f'' = s''f' = 0.0485 \times 20.6 = 1.0 \text{ Hz } \textbf{Ans.}$$

Example 19.28. A 4-pole induction motor and 6-pole induction motor are connected in cumulative cascade at 50 Hz supply. The frequency in the secondary circuit of the 6-pole motor is observed to be 1.0 Hz. Determine the slip in each machine and combined speed of the set.

[*J.N. Technological Univ. Electrical Machines-II, January-2014*]

Solution: Synchronous speed of the set,

$$N_{sc} = \frac{120f}{P_1 + P_2} = \frac{120 \times 50}{6+4} = 600 \text{ rpm}$$

Frequency of rotor current of 6-pole induction motor,

$f'' = 1.0$ Hz

$$\text{Slip of the set, } s' = \frac{f''}{f} = \frac{1.0}{50} = 0.02 \textbf{ Ans.} \qquad \because f = 50 \text{ Hz}$$

Actual speed of the set,

$$N = N_{sc}(1 - s') = 600\,(1 - 0.02) = 588 \text{ rpm}$$

Synchronous speed of 4-pole motor stator field,

$$N_s = \frac{120f}{P_2} = \frac{120 \times 50}{4} = 1{,}500 \text{ rpm}$$

Slip referred to stator field,

$$s = \frac{N_s - N}{N_s} = \frac{1{,}500 - 588}{1{,}500} = 0.608 \text{ or } 60.8\% \textbf{ Ans.}$$

The 4-pole motor current frequency or the impressed frequency on 6-pole motor stator field

$$f' = sf = 0.608 \times 50 = 30.4 \text{ Hz}$$

Synchronous speed of 6-pole induction motor with frequency f'

$$N_s' = \frac{120 \times 30.4}{6} = 608 \text{ rpm}$$

Slip referred to the 6-pole induction motor,

$$s'' = \frac{N_s' - N}{N_s} = \frac{608 - 588}{608} = 0.033 \text{ or } 3.3\% \textbf{ Ans.}$$

Example 19.29. A cascade set consists of two motors A and B with 4 poles and 6 poles respectively. The motor is connected to a 50 Hz supply. Find (i) the speed of the set (ii) the electric power transferred to motor B when the input to motor A is 25 kW. Neglect losses.

Solution: Synchronous speed of the set,

$$N_{sc} = \frac{120\,f}{P_A + P_B} = \frac{120 \times 50}{4+6} = 600 \text{ rpm}$$

Because the outputs of the two motors are proportional to the number of their poles, so power transferred to motor B with 6 poles

= Power output of motor B neglecting losses

$$= P \times \frac{P_2}{P_1 + P_2} = 25 \times \frac{6}{4+6} = 15 \text{ kW } \textbf{Ans.}$$

Example 19.30. A 10 hp, 4-pole, 50 Hz, 220 V induction motor is connected in cascade with another 15 hp, 6-pole, 50 Hz, 220 V motor. What are the possible speeds obtainable with this combination and what is the maximum load in hp which may be delivered without

over loading either machine ? Also determine the ratio of mechanical power outputs of the two machines at this load.

Solution: Synchronous speed of machine I (6-pole machine),

$$N_{s1} = \frac{120\,f}{P_1} = \frac{120 \times 50}{6} = 1{,}000 \text{ rpm}$$

Synchronous speed of machine II (4-pole machine),

$$N_{s2} = \frac{120\,f}{P_2} = \frac{120 \times 50}{4} = 1{,}500 \text{ rpm}$$

Concantenated speed of the set when cumulatively connected,

$$N_{sc1} = \frac{120\,f}{P_1 + P_2} = \frac{120 \times 50}{6+4} = 600 \text{ rpm}$$

Concantenated speed of the set when differentially connected,

$$N_{sc2} = \frac{120\,f}{P_1 - P_2} = \frac{120 \times 50}{6-4} = 3{,}000 \text{ rpm}$$

So speeds available are 1,000, 1,500, 600 and 3,000 rpm **Ans.**

In cascade connection the rating of the main motor should be such that it covers the power output of the auxiliary motor also. So the maximum load which can be delivered is 15 hp **Ans.**

The ratio of the mechanical power outputs of the main and auxiliary induction motors are

$P_1 : P_2$ or 6 : 4 or 1.5 : 1 **Ans.**

19.4 KRAMER SYSTEM OF SPEED CONTROL

This system is based on the extraction of power from the rotor circuit of a polyphase wound rotor induction motor via the slip rings. This power at slip frequency is supplied to a synchronous convertor, and the resulting direct current is used to supply the armature of a separately-excited dc motor directly coupled to the induction motor. By varying the excitation of the dc motor, its emf and hence the induction motor terminal voltage, are varied, thereby changing the motor speed. Variation in excitation of the synchronous convertor controls the power factor of the main induction motor.

Constant output power Kramer drive is illustrated in Fig. 19.21. It develops no torque at synchronous speed. The main induction motor M is mechanically coupled to the auxiliary dc machine. Power from the rotor circuit of the main induction motor M is supplied to the slip rings of the synchronous convertor C and dc output of the convertor is supplied to the dc machine D, which develops mechanical torque. Power flow in reverse direction is also possible. When power is extracted from the rotor of the main induction motor and supplied to the synchronous convertor which supplies dc power to the auxiliary dc machine, auxiliary dc machine will now operate as a motor and develop an additional torque. The developed torque is, therefore, the sum of the torques developed by motors M and D. Thus we get high torque at low speed. Separate excitations are used for dc machine and the synchronous convertor. Change in excitation of the synchronous convertor will regulate the power factor of the ac input. This in turn will change the power factor of the main induction motor.

If the dc excitation of the auxiliary dc machine is increased, it will tend to reduce its speed. Main induction motor will, therefore, be driving the auxiliary dc machine which will now operate as a dc generator. The dc power generated will now be converted to power at slip frequency by the synchronous convertor and fed into the rotor circuit of the main induction motor. Thus the speed of the induction motor will increase. Some of the torque developed by the induction motor will be consumed by the dc machine operating as a generator now. Hence the cascaded set with excessive excitation of auxiliary dc machine will run at high speed and with low torque. Speed of the main motor is thus controlled by the field regulator of the auxiliary dc machine. This is, sometimes, called *constant output power cascading.*

If instead of auxiliary dc machine being mechanically coupled to the main induction motor, it were to drive an ac

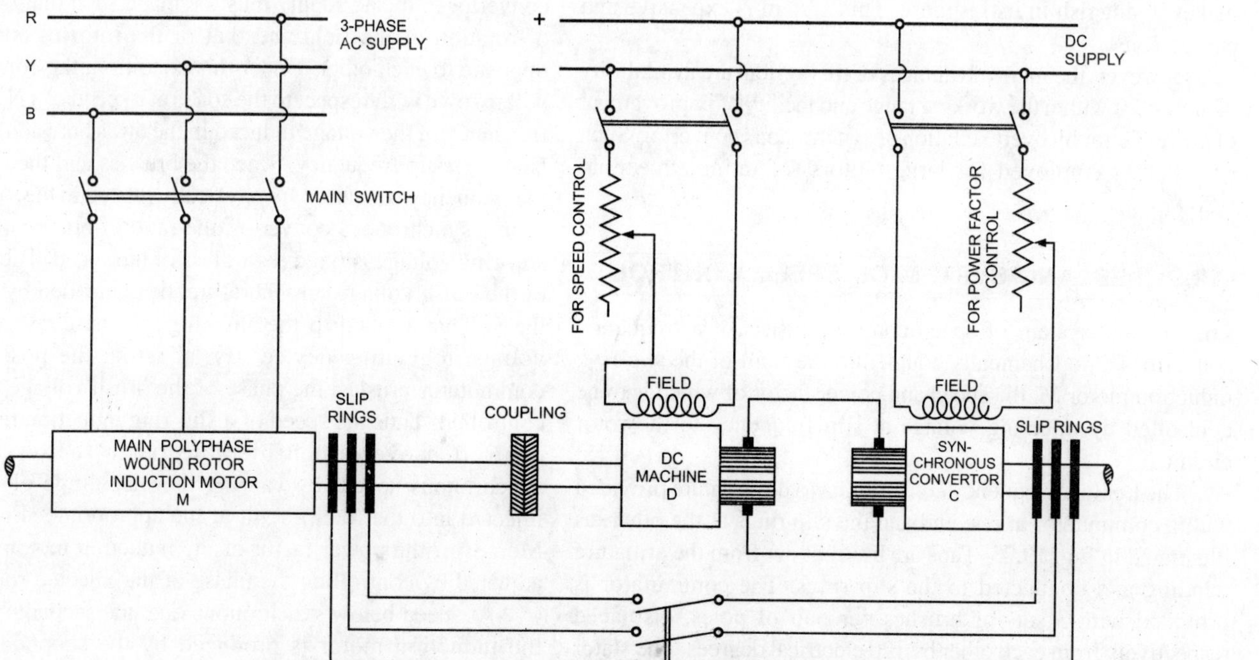

Fig. 19.21 *Constant Output Power Kramer Drive*

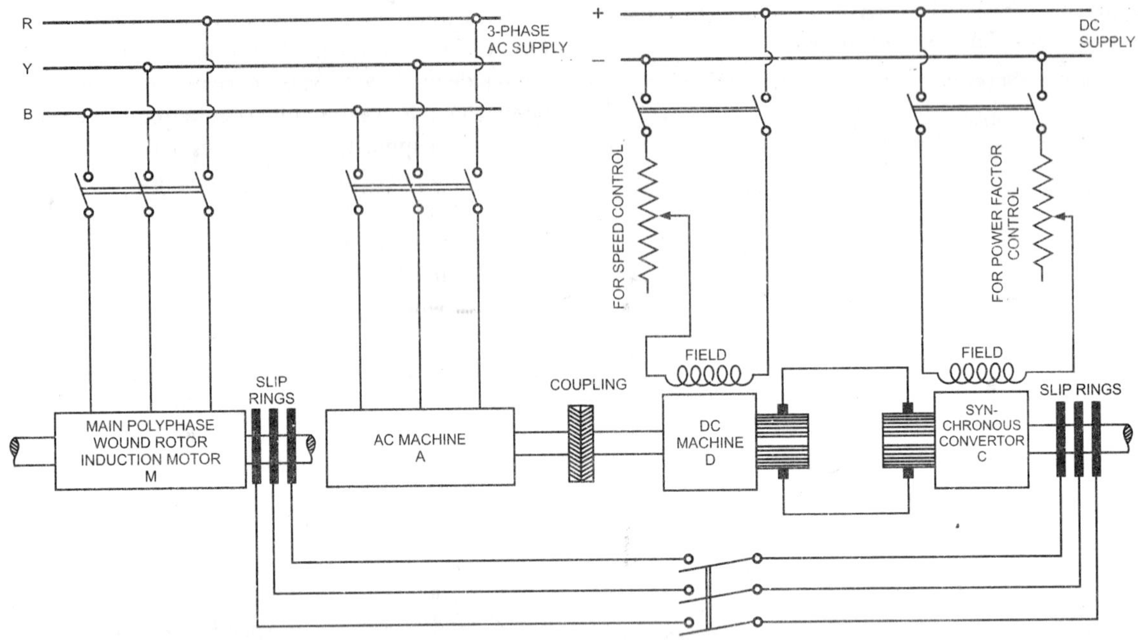

Fig. 19.22 *Constant Torque Kramer Drive*

generator which would feed back to the supply line, as shown in Fig. 19.22, this cascaded arrangement would provide constant torque. For speeds below synchronous, power is drawn from the rotor circuit of the main induction motor M and supplied to the synchronous convertor C, dc output of the convertor C to the dc machine D and back to supply through ac generator A. For speeds above synchronous the ac generator becomes an ac motor supplying power to the rotor circuit of the main motor through the dc generator D and convertor C.

In both of the Kramer drives the convertor changes the speed with speed adjustments of the main induction motor. It always operates at a speed corresponding to the slip frequency of the main motor. This feature adds inertia to the moving system which makes it sluggish in its response. This system is expensive and cumbersome.

However, the main advantages of this system are availability of any speed within the working range and facility of improvement of power factor by overexcitation of synchronous convertor. Such a system is employed for large motors (of rating exceeding 4,000 kW).

19.5 LEBLANC SYSTEM OF SPEED CONTROL

The Leblanc system of speed control consists of a frequency convertor FC mechanically coupled to the shaft of the slip-ring induction motor M, the speed and power factor of which is to be controlled by injecting voltage at slip frequency to the rotor circuit.

The Leblanc frequency convertor has a dc armature provided with a commutator at one end and the slip rings at the other, as illustrated in Fig. 19.23. Taps are brought out from the armature winding and connected to the slip rings. The commutator is provided with 3 sets of brushes per pair of poles, displaced respectively from each other by 120 electrical degrees. The stator frame of frequency convertor FC is in the form of a cylindrical ring and is made of laminated steel and is not provided with any sort of winding. The purpose of the stator frame of frequency convertor is to provide a low and uniform reluctance path to the magnetic flux of the machine. The commutator brushes of the convertor are connected to the slip rings of the main induction motor M, as illustrated in Fig. 19.23. The slip-ring terminals of the frequency convertor FC are connected, through a voltage regulating device, to the same ac supply mains to which the main induction motor M is connected.

The power supply through the slip rings of the frequency convertor FC develops a magnetic field in the convertor which revolves at synchronous speed N_s with respect to the rotor (armature). The rotor of the frequency convertor FC is driven by the main motor at a speed N_s $(1 - s)$. The connections of the convertor to the ac supply must be made such that the direction of rotation of the field and that of the rotor of convertor are opposite to each other. Under this condition, the convertor field will revolve with respect to the space at a speed of $s\,N_s$. Hence the frequency of the voltage induced in the turns between the brushes is always slip frequency. Since the brushes and the slip rings of the frequency convertor are physically related in the same manner as in a synchronous convertor, the ratio of brush voltage to the slip-ring voltage is fixed regardless of the speed. The magnitude of the brush voltage can, therefore, be controlled by controlling the voltage applied to the slip rings of the convertor through voltage regulating device. By adjusting the position of the commutator brushes the phase of the brush voltage can also be controlled. Thus the speed of a slip-ring induction motor can be controlled over a considerable range (below and above synchronous speed) by varying the magnitude of the voltage injected into the rotor circuit at the appropriate slip frequency. Moreover, the power factor of the induction motor can also be adjusted by controlling the phase of the injected rotor voltage.

At speed below synchronous one, the secondary current of the induction motor is produced by the secondary induced voltage of the induction motor. Power is supplied by the rotor circuit to the convertor, and the armature of the convertor returns

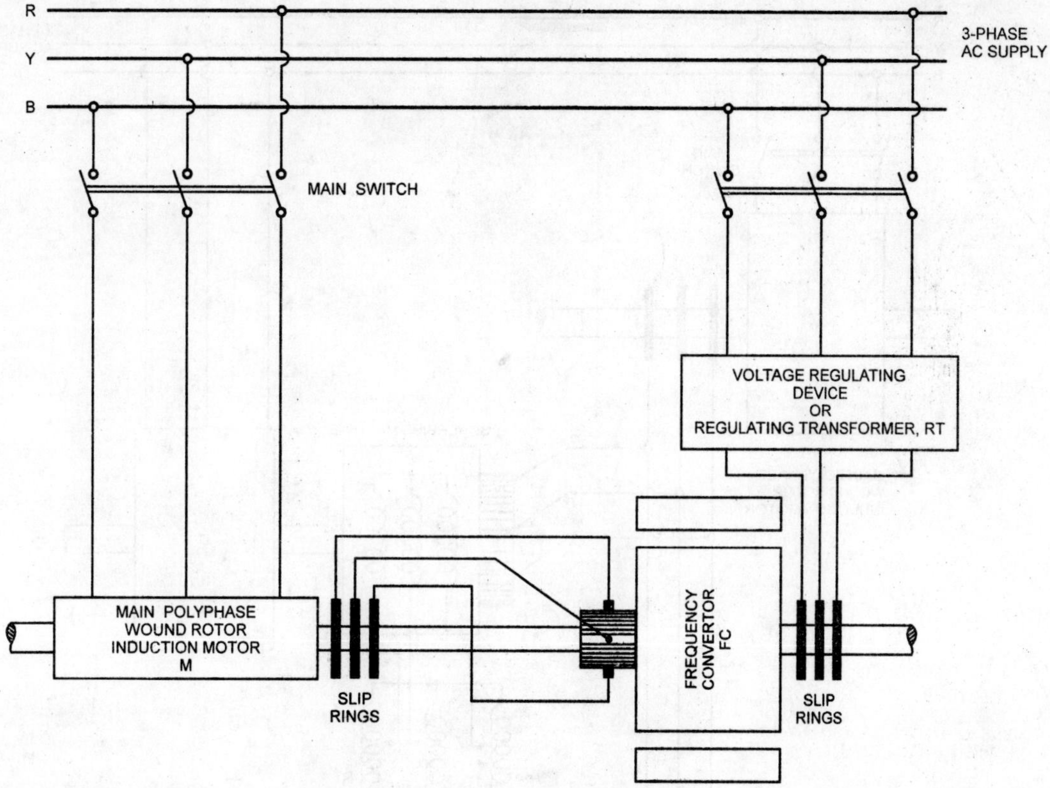

Fig. 19.23 *Leblanc Speed Control System*

its major portion to the ac supply mains. On the other hand at a speed above synchronous one, the secondary current of the induction motor is produced by the combination of the secondary induced voltage and the foreign voltage. Power is drawn from the ac supply mains by the convertor and passed on by it to the secondary of the induction motor. The mechanical output of the induction motor is obtained from a combination of electrical power supplied to its stator from the supply mains and the electrical power supplied to its rotor from the convertor.

19.6 SCHERBIUS SYSTEM OF SPEED CONTROL

Scherbius system of speed control of a 3-phase wound rotor induction motor is shown in Fig. 19.24. Here the slip energy is not converted into dc and fed to a dc motor, rather it is fed directly to a special 3-phase commutator motor, called a *Scherbius machine*. The Scherbius machine has rotor similar to that of a dc machine and stator similar to that of three-phase induction motor. The power from the slip rings of the main induction motor is supplied to the commutator of the Scherbius machine directly and to the stator through regulating transformer RT. The Scherbius machine may be coupled to the main induction motor shaft to give a constant power characteristic, as illustrated in the Fig. 19.24. The Scherbius machine is a variable speed motor and its speed (and so that of main motor M) is controlled by either varying the tappings on RT or by adjusting the brush position on Scherbius machine.

In this system, torque is developed by the Scherbius machine S as well as the main induction motor M. High torques can be obtained from this system. Its speed range is limited to speeds below synchronous. The size of the regulating machine S is proportional to the speed reduction below synchronism that is provided in the design. The power factor of the system is inherently low.

A Scherbius system for providing a constant-torque drive is shown in Fig. 19.25. Here the Scherbius motor is coupled to separate induction machine D, acting as an induction generator to feed the power back to the ac supply mains. For light loads the induction machine may act as a motor to supply rotational losses of the motor-generator set SD.

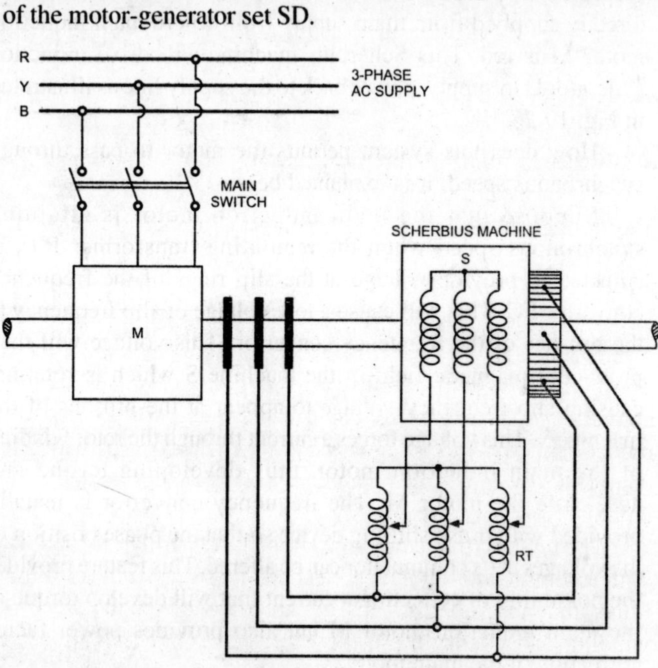

Fig. 19.24 *Scherbius Constant Output Power Drive*

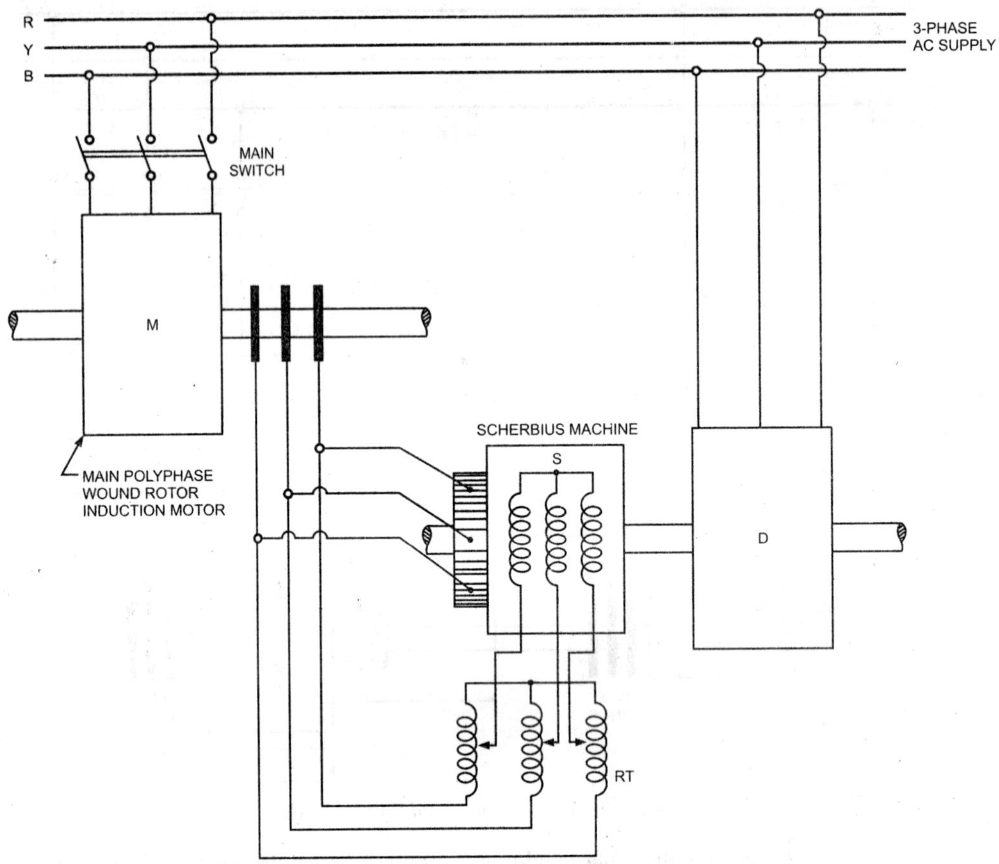

Fig. 19.25 *Scherbius Constant Torque Drive*

In case of Scherbius machine S whose stator is fed from the slip rings of the main induction motor M, it is not possible to obtain speed above synchronous. This is because as soon as synchronous speed is approached there is no rotor power to excite the Scherbius machine S. To overcome this, stator is fed from separate frequency convertor FC, coupled to the main motor and directly supplied from main supply from where main induction motor M is fed. This Scherbius machine will drive induction generator D to supply power back to the supply line as illustrated in Fig. 19.26.

How does this system permits the motor to pass through synchronous speed, it is explained below.

Suppose that the main induction motor is attaining synchronous speed when the regulating transformer RT_2 is adjusted to provide voltage at the slip rings of the frequency convertor FC. This will cause a low voltage of slip frequency at the brushes of the frequency convertor. This voltage will then produce a magnetic field in the machine S which is rotating, causing slip frequency voltage to appear at the brushes of the machine S. This voltage forces a current through the rotor winding of the main induction motor, thus developing torque and accelerate the motor M. The frequency convertor is usually provided with brush shifting device so that the phase position of the voltages at its commutator can be altered. This feature provides the possibility of collecting a current that will develop torque in the main induction motor M but also provides power-factor correction of the main motor.

Though the Scherbius drive is costlier than Kramer system but it is superior to the Kramer drive when fine speed control is required near synchronous speed.

19.7 ELECTRICAL BRAKING OF POLYPHASE INDUCTION MOTORS

The simplest method of stopping of an induction motor, or any other type of motor, is to disconnect the motor from the supply mains. Torque is no longer developed, and the combined effect of the rotor and the external load brings the motor to standstill. When rapid and more positive action is required, mechanical or electrical braking may be employed, but the latter has many advantages, particularly where precise control of the stopping moment and smoothness of operation are required.

The motor is said to operate on electrical braking when the direction of developed torque is opposite to that of rotation. During electrical braking the motor operates on torque-speed characteristic corresponding to the method of braking employed. There are three methods of electrical braking of induction motors: plugging (or counter-current braking), dynamic (or rheostatic braking) and regenerative braking.

19.7.1. Plugging (or Counter-Current Braking). Plugging can be achieved in an induction motor merely by reversing any two of the three phases which causes a reversal of the direction of the rotating magnetic field. At the instant of switching the motor to the plugging position the motor runs in the opposite direction to that of the field and the relative speed is approximately twice $[(2 - s)$ times$]$ of synchronous speed *i.e.* the slip is very nearly equal to two, being equal to $(2 - s)$. So voltage induced in the rotor will be twice of normally induced voltage at standstill and the winding must be provided with the additional insulation to withstand this much voltage.

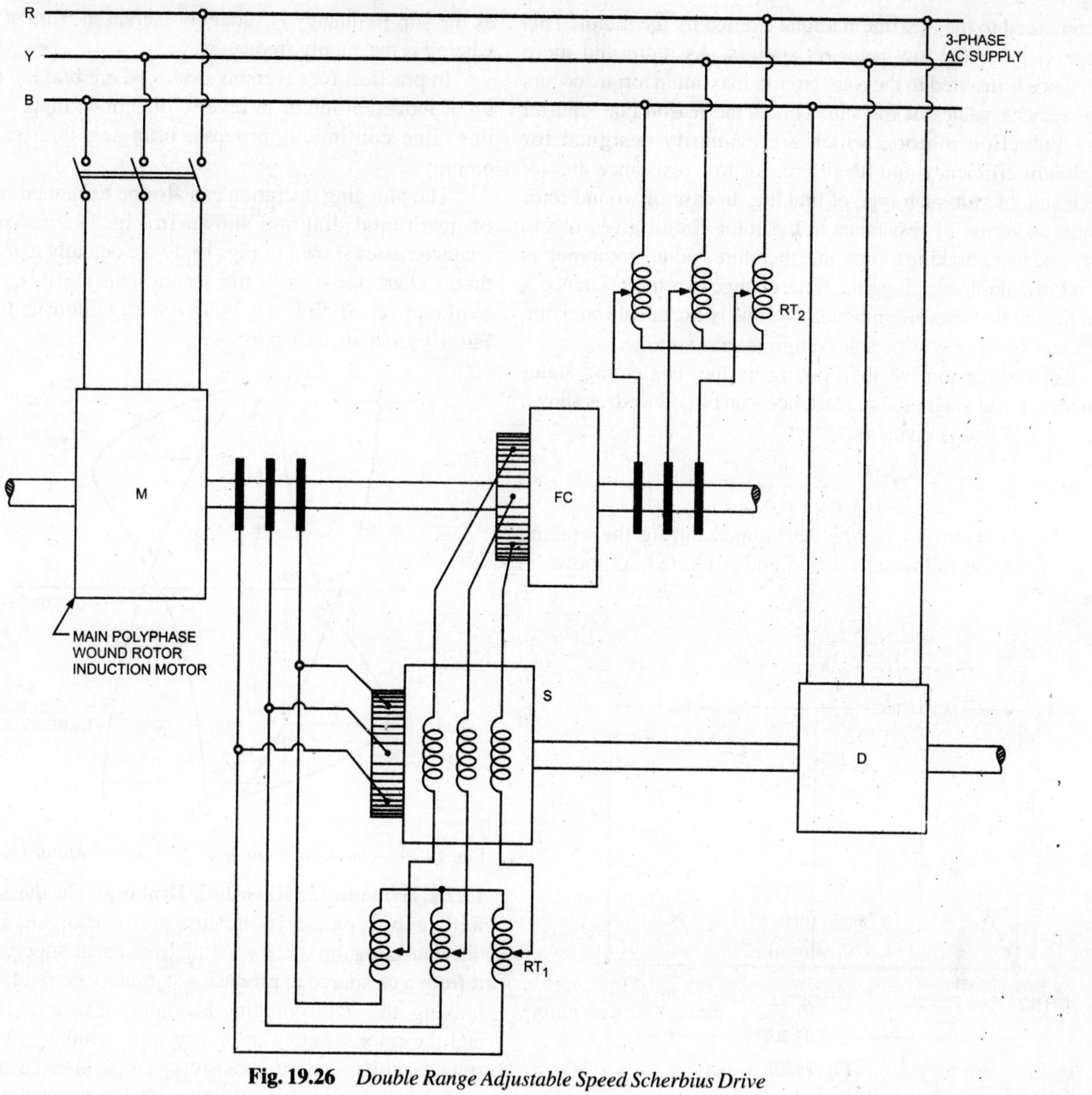

Fig. 19.26 *Double Range Adjustable Speed Scherbius Drive*

During plugging period, the motor acts as a brake and it absorbs kinetic energy from the still revolving load causing its speed to fall. The associated power P_{mech} is dissipated as heat in the motor. At the same time, the rotor also continues to draw power from the stator which is also dissipated as heat. Due to increased rotor frequency there are additional iron losses in the rotor. Thus, the heat developed in the rotor during braking period are about three times the heat developed during starting period (or under blocked condition of rotor).

The size of the induction motor where plugging is to be applied, is therefore, decided not by loading conditions alone but by braking condition also.

In case of a squirrel cage motor, energy is dissipated wholly within the machine; whereas in case of wound rotor motor this energy is dissipated also in the external resistance added in the rotor circuit for this purpose.

The conditions for braking an induction motor can be studied by considering the torque-slip curves of the motor when extended beyond the point of 100 % slip, as shown in Fig. 19.27. The ordinate at the point B represents the torque at the instant of plugging and we can see that the torque increases gradually as the motor approaches the standstill, after which, of course, if the supply line is not disconnected it will start up again in the reverse direction in accordance with the ordinary torque-speed curves from A to O. Thus we see that for bringing the motor to rest, it is essential to disconnect the supply to the motor at zero speed.

From the torque-slip curves shown in Fig. 19.27, it is observed that magnitude of the braking torque is very much low

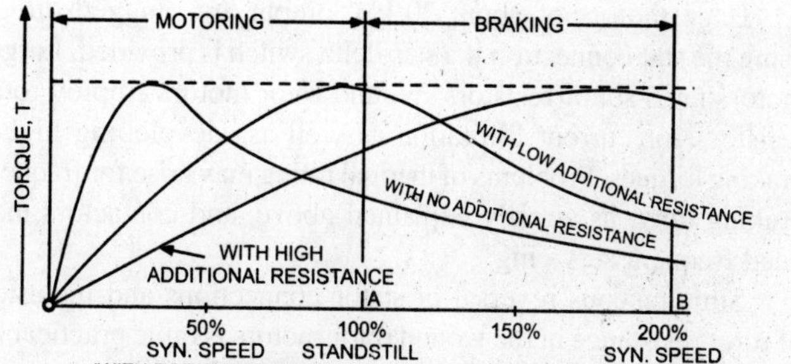

Fig. 19.27

as compared to the maximum torque exerted by the motor. This is specially so for low rotor resistances. As more and more resistance is inserted in the rotor circuit, maximum torque occurs at increasing values of the slip. This is the reason that squirrel cage induction motors, which are primarily designed for maximum efficiency and, therefore, for low resistance are not much suitable for such type of braking. In case of wound rotor motors, addition of resistance in the rotor circuit gives rise to more and more braking torque and, therefore, wound rotor motors are more suitable for plugging. If the change in rotor resistance is so adjusted that maximum braking torque is exerted throughout, optimum braking will be achieved in minimum time.

The expression for the braking torque, neglecting stator impedance and magnetizing reactance, can be deduced, as shown in Art 18.27 and is given as

$$T_b = \frac{3 \times 60}{2\pi N_s} \cdot \frac{R_2\, sE_2^2}{R_2^2 + s^2 X_2^2} \qquad ...(19.49)$$

The rotor current can be determined during the braking period from the following relation and is plotted, as shown in Fig. 19.28.

$$I_2 = \frac{s\,E_2}{\sqrt{R_2^2 + s^2 X_2^2}} \qquad ...(19.50)$$

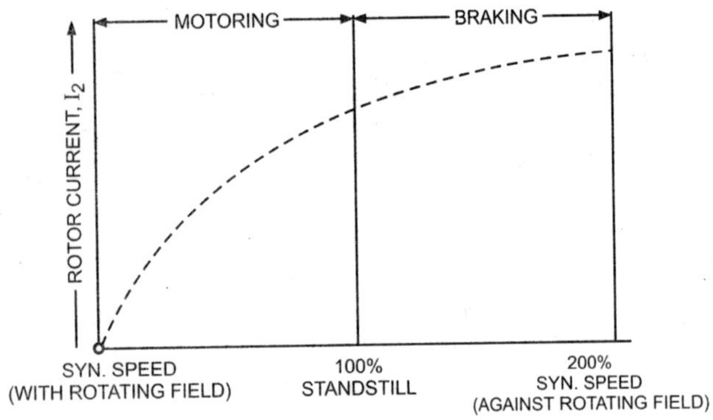

Fig. 19.28

where E_2 is the emf induced in each phase of the rotor at standstill, R_2 is the rotor resistance per phase and X_2 is the rotor standstill reactance per phase and s is the slip.

During plugging period, since the induced emf in the rotor is very high, as already stated above, so the rotor current and thereby stator current are very high. However, braking current can be reduced by inserting external resistance in the rotor circuit. From this point also, the wound rotor motors are beneficial as compared to squirrel cage rotor motors.

Cage motors of about 20 kW output are plugged direct, using the star connection if a star-delta switch is provided. Larger motors need stator resistors. Wound rotor motors employ rotor resistors for current limitation as well as developing higher braking torques. Problems of thermal rating may arise for frequent braking duty, as already explained above, and contactors may need frequent servicing.

Simultaneous reversal of stator connections and inclusion of rotor resistance in the wound rotor motors is quite practicable, but the same effect can be achieved by including saturistors in the rotor windings. Alternatively a deep bar rotor cage may give rise in effective rotor resistance following the plugging operation as the slip frequency is suddenly increased from sf to $(2 - s) f$ where f is the supply frequency.

In practice, for reversing drives where braking and starting up of induction motor in reverse direction comprises stages of the same continuous process, plugging is advantageously employed.

The plugging operation can also be explained with the help of quadrantal diagram shown in Fig. 19.29. Speed-torque characteristics shown in Fig. 19.29 are actually extension of the motor characteristics in the second and fourth quadrant. The explanation of Fig. 19.29 is exactly identical to one for Fig. 10.31 (*a*) or 10.31 (*b*).

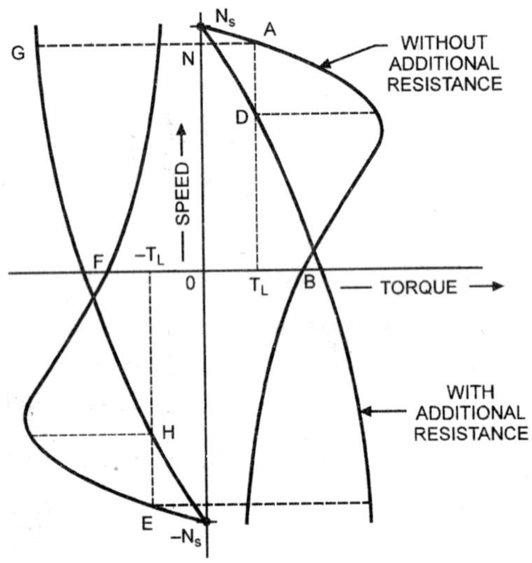

Fig. 19.29 *Quadrantal Diagram of Induction Motor During Plugging*

19.7.2. Dynamic (or Rheostatic Braking). The rheostatic braking with a polyphase induction motor can be obtained by disconnecting the stator winding from the ac supply and exciting it from a dc source to produce a stationary dc field. In rheostatic braking, the stator winding is employed as a dc field winding and the rotor winding as an armature winding. With a wound rotor machine, external resistors can be inserted into the rotor circuit to provide a load. With squirrel cage machines, however, the rotor winding itself has to form the load.

The source of excitation may be provided either by an independent dc source or from the ac mains through a transformer-rectifier set, as shown in Fig. 19.31.

Various methods of connecting the stator winding to a dc source are shown in Fig. 19.30. Though connections shown in Figs. 19.30 (*c*) and 19.30 (*f*) give uniform current loading of all the three phases but are complicated in switching operation. Being more simple in this respect, the connections shown in Figs. 19.30 (*a*) and (*b*) are usually employed.

A typical connection diagram is given in Fig. 19.31; the machine operates as motor with contactors L closed, while with L open and B closed a direct current is supplied through two stator phases, the third being left open circuited. The resistance R′ is inserted in the stator circuit to limit its current. In case of wound rotor motors additional resistance R is inserted in the rotor circuit to control the braking effect.

While the machine is operating normally as a motor, its stator magnetic field is rotating at a synchronous speed in the same direction as that of the rotor, but slightly faster than the

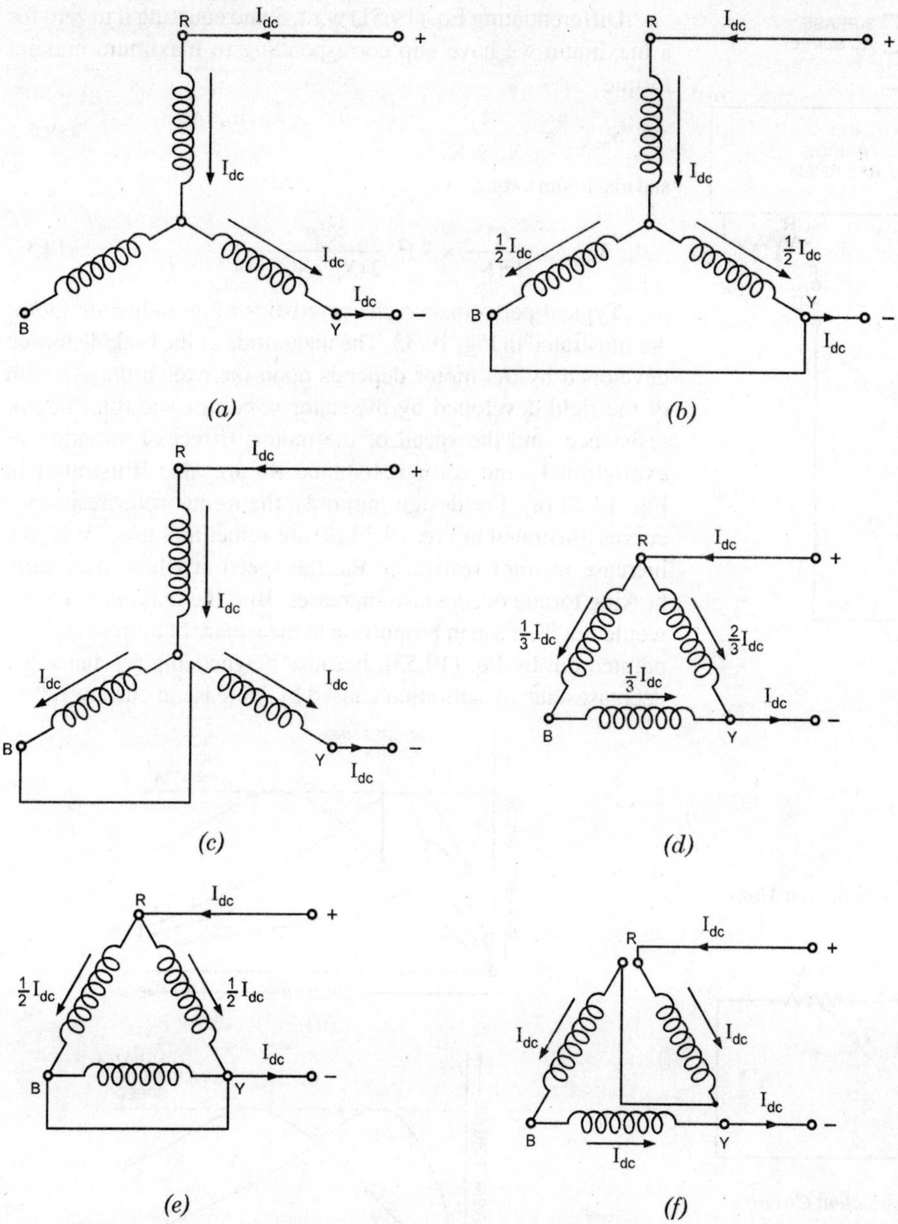

Fig. 19.30 *Connection of Stator Windings of 3-Phase Induction Motors For Dynamic Braking*

rotor conductors. When the stator windings are disconnected from the ac supply and excited with dc, the magnetic field produced will be stationary in space, thus making the rotor conductors move past the field with a speed $(1 - s)\,N_s$ or $S\,N_s$. The currents induced in the rotor conductors will be opposite in direction to that corresponding to motoring operation, producing a braking torque.

Although the air-gap flux is stationary, there will be the same number of poles as when the machine is excited with ac and the rotor currents will, therefore, have a frequency which is near to the frequency corresponding to synchronous speed (*i.e.* N_s) initially, but which decreases with rotor speed and becomes zero at standstill *i.e.* rotor current frequency can be expressed as $S\,f$, where f is the frequency of ac supply. Likewise, the emf induced in the rotor decreases from a maximum when the rotor is running, to zero at standstill. In other words, the emf induced in the rotor will be given by SE_1 where E_1 is the magnitude of the emf induced in the rotor when it rotates past the field at speed N_s. It is, thus seen that the conditions in rotor during dc rheostatic braking with the speed falling from synchronous to standstill are very much the same as when the motor accelerates in the normal manner. Hence the equivalent circuit of the rotor can be represented as shown in Fig. 19.32 (*a*). Dividing the voltage and the impedance by S, we have the circuit shown in Fig. 19.32 (*b*) carrying the same current I_2.

Since the stator winding carries only direct current, the inductance of the stator has no effect under steady state operation. The dc voltage applied across the stator winding is fixed only by the stator winding resistance. There will be no iron loss in the stator core. But the rotor core loss will be of considerable magnitude and hence the rotor resistance value referred to stator, which determines the torque developed during braking, must be corrected to take this into account. The emf induced in the rotor, at synchronous speed, E_1 will be given by $I_m X_m$ where I_m represents the magnetizing current passing through the magnetizing reactance X_m. The circuit shown in Fig. 19.32 (*b*) can be modified into a circuit shown in Fig. 19.32 (*c*). The phasor diagram corresponding to this modified circuit is shown in Fig. 19.32 (*d*).

It may be noted that the rotor current is alternating in nature although it is produced by a constant flux, caused by a magnetizing mmf $I_m N_1$, which remains stationary in space. However, when viewed from the rotor, both the magnetizing mmf and the resultant flux in the stator will appear as alternating. Also, since the rotor ampere-turns $I_2 N_2 (= I'_2 N_1)$ must be balanced by the stator ampere-turns so as to maintain the magnetizing ampere turns $I_m N_1$, the stator ampere-turns $I_1 N_1$ as seen from the rotor, represent the phasor sum of $I_m N_1$ and $I'_2 N_1$. The effective alternating current I_1 depends on the magnitude of direct current carried by the stator and the nature of the stator winding connections, as explained above.

From above discussion, it can be concluded that the operation of the motor during dc rheostatic braking can be analysed by using the equivalent circuit (the type of which we are already familiar) shown in Fig. 19.32 (*c*).

The braking torque can be determined from the relation

$$T_b = \frac{60}{2\pi N_s} \cdot 3(I'_2)^2 \frac{R'_2}{S} \text{ N-m}$$

$$= \frac{60}{2\pi N_s} \cdot 3I_1^2 \frac{X_m^2 \left(\dfrac{R'_2}{S}\right)}{\left(\dfrac{R'_2}{S}\right)^2 + (X_m + X'_2)^2} \text{ N-m} \qquad \text{...(19.51)}$$

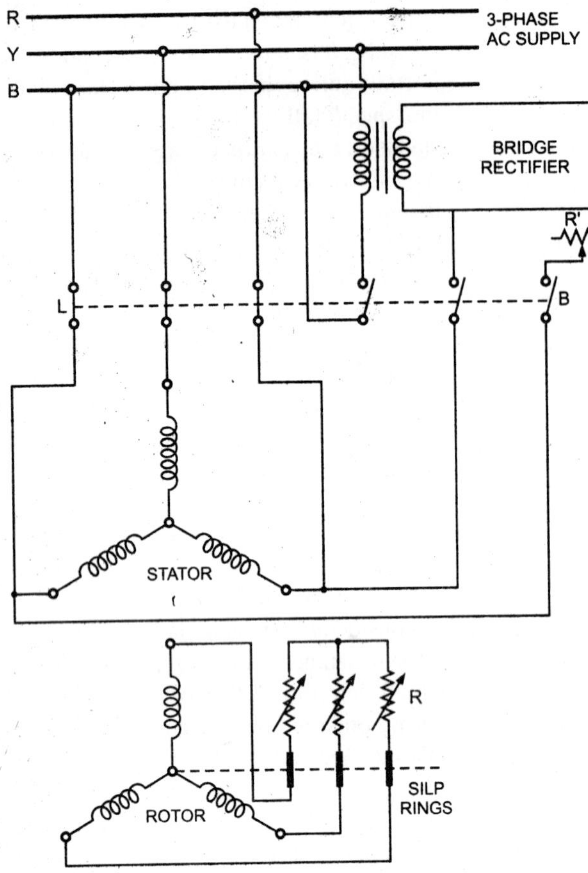

Fig. 19.31 *DC Rheostatic Braking of a 3-phase Induction Motor*

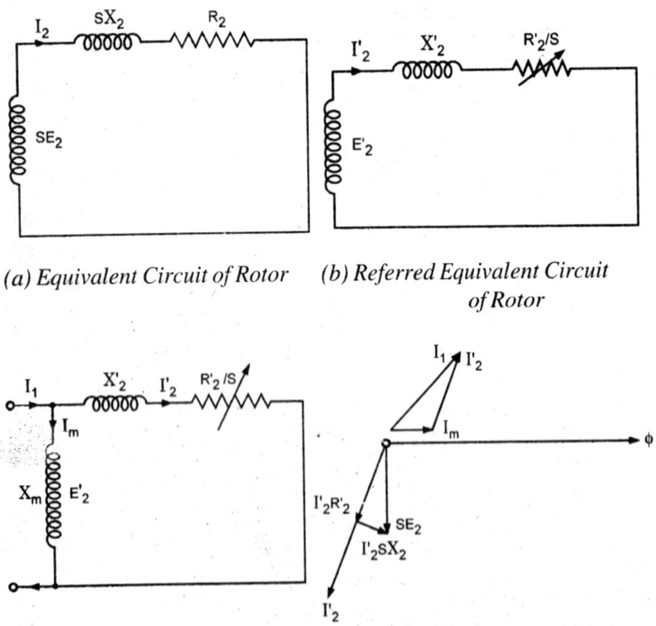

(a) Equivalent Circuit of Rotor *(b) Referred Equivalent Circuit of Rotor*

(c) Referred Equivalent Circuit For Braking Operation *(d) Phasor Diagram*

Fig. 19.32 *DC Rheostatic Braking of Polyphase Induction Motor*

The above equation enables us to determine the braking torque at any speed of a given induction motor, when excited by direct current, I_1 corresponding to an equivalent ac, which would have produced the same mmf as that by the actual direct current carried by the stator windings.

Differentiating Eq. (19.51) w.r.t. S and equating it to zero for a maximum we have slip corresponding to maximum braking torque

$$S_{\max T} = \frac{R'_2}{X_m + X'_2} \quad \text{... (19.52)}$$

and maximum torque,

$$T_{\max} = \frac{60}{2\pi N_s} \times 3 I_1^2 \frac{X_m^2}{2(X_m + X'_2)} \quad \text{... (19.53)}$$

Typical performance characteristics of an induction motor are illustrated in Fig. 19.33. The magnitude of the braking torque developed by the motor depends upon the excitation (strength of the field developed by the stator winding), the rotor circuit resistance, and the speed of the motor. Effect of variation of excitation I_1 and rotor resistance R_2 are also illustrated in Fig. 19.33 (*a*). For design purposes, the torque-rotor resistance curves illustrated in Fig. 19.33 (*b*) are sometimes used. With the increase in rotor resistance R_2, the speed at which maximum braking torque occurs also increases. But, the maximum torque would not increase in proportion to the square of current (I_1^2), as pointed out by Eq. (19.53), because magnetising reactance X_m decreases due to saturation caused by increase in current I_1.

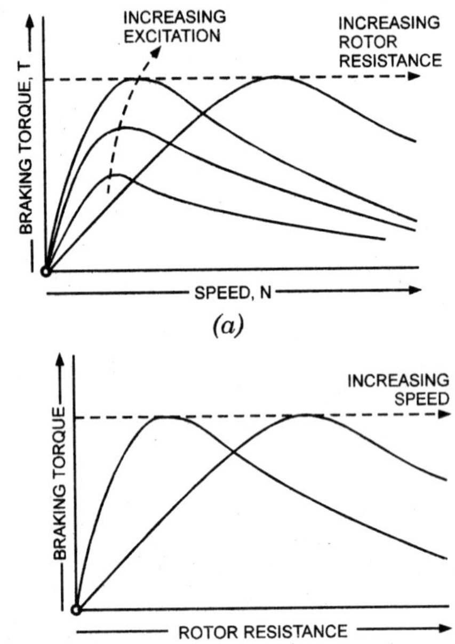

Fig. 19.33 *DC Rheostatic Braking Characteristics*

Stable braking operation is on the steeper portion of torque-speed curve [*i.e.* on the left hand side of the maximum torque position in Fig. 19.33 (*a*)]. If inadvertently the speed rises above that at which the torque is a maximum, further rise of speed is caused by a decrease in braking torque, which may lead to a further rise in speed in the case of an overhauling load. At the start of applying braking, therefore, we require high resistance in the rotor circuit, which has to be gradually reduced with decrease in speed. Braking torque can be controlled by any or both of the following methods: (*i*) By controlling the dc excitation (*ii*) By varying the rotor resistance.

Determination of Equivalent AC Excitation. Equivalent ac excitation I_1 can be determined by equating the mmfs produced by dc and ac.

Let us first consider the connection shown in Fig. 19.30 (*a*). Phases R and Y carry equal direct currents, but in opposite directions. Phase B does not carry any current. Assuming that the space mmfs developed by the direct current flowing in phases R and Y are sinusoidal in nature (or neglecting space harmonic mmfs), the resultant mmf can be determined as shown in Fig. 19.34.

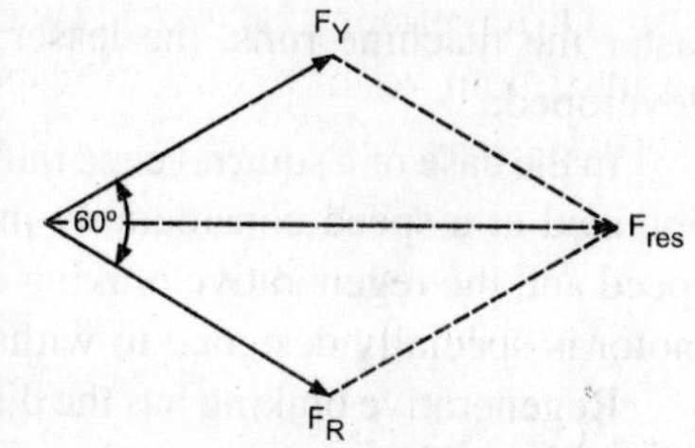

Fig. 19.34

Resultant mmf, $F_{res} = 2\,I_{dc}\,N_{eff}\cos 30° = \sqrt{3}\,I_{dc}\,N_{eff}$ ...(19.54)

where N_{eff} is the effective turns per phases

MMF due to I_{ac} (rms) flowing in all the three phases

$$= \frac{3}{2}\left(\sqrt{2}\,I_{ac}\right)N_{eff} \quad ...(19.55)$$

Equating both mmfs, we have

$$\sqrt{3}\,I_{dc}\,N_{eff} = \frac{3}{\sqrt{2}}\,I_{ac}\,N_{eff}$$

$$\text{or } I_{ac} = \sqrt{\frac{2}{3}}\,I_{dc} \quad ...(19.56)$$

The magnitude of dc voltage required to supply current I_{dc} is given by

$$V_{dc} = 2\,I_{dc}\,R_1$$

where R_1 is the stator resistance per phase

So excitation power,

$$P_{exc} = V_{dc}\,I_{dc} = 2\,I_{dc}^2\,R_1 = 3\,I_{ac}^2\,R_1 \quad ...(19.57)$$

The corresponding values for different methods of connection are given below in tabular form.

In all of the above schemes, for the same value of the equivalent alternating current I_{ac}, same magnetic fields and, therefore, the same magnitude of braking torques are obtained and excitation power for all the connections is also the same. From the table given below it is also clear that the scheme shown in 19.30 (*a*) is the most preferred one, because it needs the least current at maximum voltage.

AC dynamic braking may be obtained with self excitation by connecting the stator windings to a bank of static capacitors as shown in Fig. 19.35. In this case the induction machine operates as an induction generator receiving its ac excitation from the static capacitors. The generated electrical energy is dissipated as heat in the rotor circuit. AC dynamic braking is not popular owing to relatively high cost of capacitor banks.

DC dynamic braking takes little power from the supply and provides smooth braking torque, useful for mine winders and high inertia loads. Its advantages over plugging are the absence of the reverse-rotation air-gap field [and, therefore, no tendency for the machine to run backwards) and a lower rotor I^2R loss. By increasing rotor resistance, the braking torque can be made more effective at higher speeds. This method is, therefore, very much suitable for retarding overhauling loads. In case of a squirrel cage motor, since rotor resistance is small, it will apply braking torque at very low speeds and, therefore, useful only for bringing the load to rest.

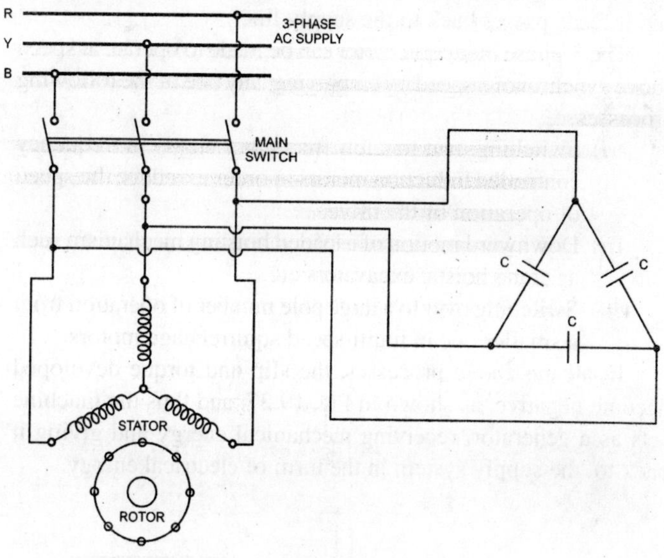

Fig. 19.35

The advent of automatic control of dynamic braking of induction motors employing closed loop systems, has made the induction motors more popular than dc motors, especially for the drives employing in mine hoists.

Figure 19.36 shows the quadrantal characteristics of induction motor under dynamic braking condition.

Connection shown in Fig. 19.30	*Equivalent I_{ac} rms*	*DC Excitation I_{dc}*	*DC Excitation Voltage V_{dc}*	*Excitation Power*	
				DC	*AC*
(a)	$\sqrt{\frac{2}{3}}\,I_{dc}$	$\sqrt{1.5}\,I_{ac}$	$2\,I_{dc}\,R_1$	$2\,I_{dc}^2\,R_1$	$3\,I_{ac}^2\,R_1$
(b)	$\frac{1}{\sqrt{2}}\,I_{dc}$	$\sqrt{2}\,I_{ac}$	$1.5\,I_{dc}\,R_1$	$1.5\,I_{dc}^2\,R_1$	$3\,I_{ac}^2\,R_1$
(d)	$\sqrt{\frac{2}{3}}\,I_{dc}$	$\frac{3}{\sqrt{2}}\,I_{ac}$	$\frac{2}{3}\,I_{dc}\,R_1$	$\frac{2}{3}\,I_{dc}^2\,R_1$	$3\,I_{ac}^2\,R_1$
(e)	$\frac{1}{\sqrt{6}}\,I_{dc}$	$\sqrt{6}\,I_{ac}$	$\frac{1}{2}\,I_{dc}\,R_1$	$\frac{1}{2}\,I_{dc}^2\,R_1$	$3\,I_{ac}^2\,R_1$

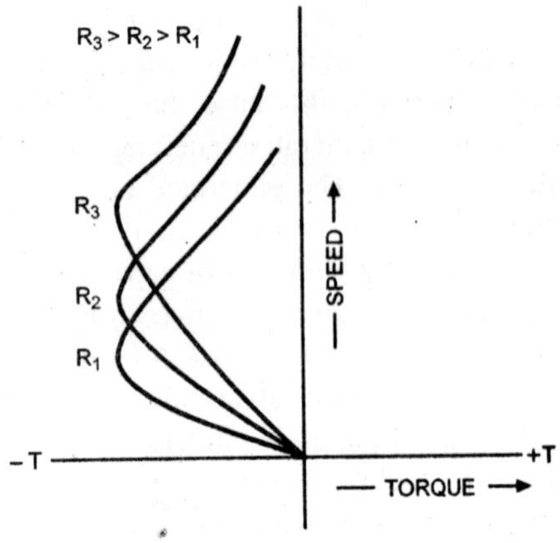

Fig. 19.36 *Quadrantal Diagram of Induction Motor During Dynamic Braking*

19.7.3. Regenerative Braking. Regenerative braking is an inherent characteristic of an induction motor, since it operates as an induction generator when it runs at speed above synchronous and it feeds power back to the supply line.

The 3-phase induction motor can be made to operate at speed above synchronous speed by employing any one of the following processes.

(*i*) Switching over to a low frequency supply in frequency controlled induction motors in order to reduce the speed of operation of the drive.

(*ii*) Downward motion of a loaded hoisting mechanism such as crane hoists, excavators etc.

(*iii*) Switching over to a large pole number of operation from a smaller one in multi-speed squirrel cage motors.

In all the above processes, the slip and torque developed become negative, as shown in Fig. 19.37, and thus the machine acts as a generator, receiving mechanical energy and giving it back to the supply system in the form of electrical energy.

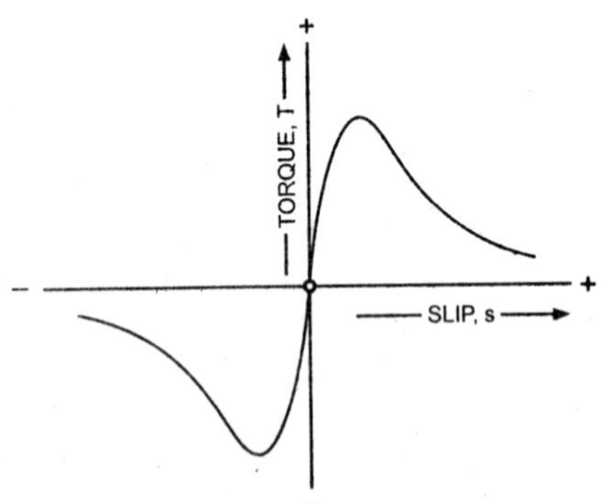

Fig. 19.37

If the load drives the motor above synchronous speed, no switching operation is required. Once the machine is driven above synchronous speed, the braking operation automatically starts. The operating point will depend upon the magnitude of load torque and the nature of torque-speed characteristic of the machine during generating operation. By varying the resistance in the rotor circuit, it is possible to operate at any speed above synchronous speed during braking. In case the driving torque of the load exceeds the maximum braking torque, of which the machine is capable, the system will become unstable and the speed will rise further, probably to a disastrous value, since, the faster the machine runs, the lesser will be the braking torque developed.

In the case of a squirrel cage induction motor, stable speed is obtained at a speed considerably in excess of the synchronous speed and the regenerative braking cannot be applied unless the motor is specially designed to withstand the excessive speed.

Regenerative braking has the disadvantage of the possibility of braking only at super-synchronous speeds and, therefore, is seldome used for braking. This method can be used only in hoisting type of mechanism or with a multi-speed squirrel cage motor. It is advantageous on mountain railways too. It returns about 20 % of the total energy on certain railway runs and saves a great deal of brake shoe wear.

Example 19.31. A 30 kW, 400 V, 3-phase, 4-pole, 50 Hz induction motor has full-load slip of 5%. If the ratio of standstill reactance to resistance per rotor phase is 4, estimate the plugging torque at full speed.

[*Pb. Technical Univ. Electric Drives and Utilization December 2008*]

Solution: Synchronous speed, $N_s = \dfrac{120f}{P} = \dfrac{120 \times 50}{4} = 1{,}500$ rpm

Full-load slip, $s_f = 5\%$ or 0.05

Full-load speed, $N_f = N_s(1 - s_f) = 1{,}500\,(1 - 0.05) = 1{,}425$ rpm

$$\text{Full-load torque, } T_f = \frac{\text{Power developed in kW} \times 1{,}000}{2\pi N/60} = \frac{30 \times 1{,}000}{2\pi \times 1{,}425/60} = 201.04 \text{ N-m}$$

Since torque developed,

$$T \;\alpha\; \frac{sR_2E_2^2}{R_2^2 + (sX_2)^2}$$

$$\text{So} \quad \frac{T_2}{T_1} = \frac{s_2R_2E_2^2/(R_2^2 + s_2^2X_2^2)}{s_1R_2E_2^2/(R_2^2 + s_1^2X_2^2)}$$

$$= \frac{s_2(R_2^2 + s_1^2X_2^2)}{s_1(R_2^2 + s_2^2X_2^2)}$$

$$= \frac{s_2}{s_1}\,\frac{1 + s_1^2\left(\dfrac{X_2}{R_2}\right)^2}{1 + s_2^2\left(\dfrac{X_2}{R_2}\right)^2} = \frac{s_2}{s_1}\,\frac{1 + 16s_1^2}{1 + 16s_2^2} \qquad \frac{X_2}{R_2} = 4$$

Slip, $s_1 = s_f = 0.05$

Slip, $s_2 =$ Slip at plugging $= 2 - s_f = 2 - 0.05 = 1.95$

$$\text{So plugging torque, } T_P = \frac{1.95}{0.05} \times \frac{1 + 16(0.05)^2}{1 + 16(1.95)^2} \times T_f$$

$$= 39 \times \frac{1.04}{61.84} \times 201.04 = 131.86 \text{ N-m } \textbf{Ans.}$$

Example 19.32. The following test results were obtained on a 7.5 kW, 400 V, 4-pole, 50 Hz, delta connected induction motor with a stator resistance of 2.1 ohm/phase.

No-load	**:**	**400 V**	**5.5 A**	**410 W**
Rotor blocked	**:**	**140 V**	**20 A**	**1550 W**

Obtain the approximate equivalent circuit model. Also estimate the breaking torque developed when the motor running with a slip of 0.05 has two of its terminals suddenly interchanged.

[*U.P. Technical Univ. Electromechanical Energy Conversion-II, 2006-07*]

Solution: No-Load Test

Applied voltage per phase, V = 400 V

No-load line current, I_{L0} = 5.5 A

No-load phase current, $I_0 = \frac{I_{L0}}{\sqrt{3}} = \frac{5.5}{\sqrt{3}} = 3.175$ A

Input on no-load, P_0 = 410 W

Stator copper loss, $P_{c_1} = 3\, I_0^2\, R_1$

$= 3 \times (3.175)^2 \times 2.1 = 63.525$ W

Total stator core loss, $P_i = P_0 - P_{c1} - P_{wf}$

$= 410 - 63.525$ neglecting friction and windage losses

$= 346.475$ W

No-load power factor, $\cos \phi_0 = \frac{P_i}{3\, V I_0} = \frac{346.475}{3 \times 400 \times 3.175} = 0.091$

Energy component of no-load phase current,

$$I_e = I_0 \cos \phi = 3.175 \times 0.091 = 0.2887 \text{ A}$$

No-load resistance, $R_0 = \frac{V}{I_e} = \frac{400}{0.2887} = 1{,}385.4\ \Omega$

Magnetising component of no-load current,

$$I_m = \sqrt{I_0^2 - I_e^2} = \sqrt{(3.175)^2 - (0.2887)^2}$$

$= 3.162$ A

No-load leakage reactance, $X_0 = \frac{V}{I_m} = \frac{400}{3.162} = 126.5\ \Omega$ **Ans.**

Short-circuit Test:

Applied voltage per phase, V_s = 140 V

Stator line current, I_{L_s} = 20 A

Stator phase current, $I_s = \frac{20}{\sqrt{3}} = 11.547$ A

Input, P_s = 1,550 W

Power factor, $\cos \phi_s = \frac{P_s}{3\, V I_s} = \frac{1{,}550}{3 \times 140 \times 11.547} = 0.32$

Equivalent impedance, $Z_{01} = \frac{V_s}{I_s} = \frac{140}{11.547} = 12.12\ \Omega$

Equivalent reactance, $R_{01} = \frac{P_s}{3 I_s^2} = \frac{1{,}550}{3 \times (11.547)^2} = 3.875\ \Omega$ **Ans.**

Equivalent reactance, $X_{01} = \sqrt{(Z_{01})^2 - (R_{01})^2}$

$= \sqrt{(12.12)^2 - (3.875)^2} = 11.484\ \Omega$ **Ans.**

Approximate equivalent circuit is shown in Fig. 18.30.

During plugging

Slip, $s = 2 - s_f = 2 - 0.05 = 1.95$

Synchronous speed, $N_s = \frac{120 f}{P} = \frac{120 \times 50}{4} = 1{,}500$ rpm

Initial rotor current during plugging

$$I'_2 = \frac{V}{\sqrt{\left(R_1 + \frac{R'_2 + R'}{s}\right)^2 + (X_1 + X'_2)^2}}$$

$$= \frac{400}{\sqrt{\left(2.1 + \frac{3.875 - 2.1}{1.95}\right)^2 + (11.484)^2}} = 33.69 \text{ A}$$

$\because R'_2 = R_{01} - R_1 = 3.875 - 2.1$ and additional resistance inserted, $R' = 0$

Initial braking torque,

$$T_b = \frac{3 \times 60}{2\pi N_s} (I'_2)^2 \times \frac{R'_2 + R'}{s}$$

$$= \frac{3 \times 60}{2\pi \times 1{,}500} \times (33.69)^2 \times \frac{1.775}{1.95} = 19.73 \text{ N-m } \textbf{Ans.}$$

Example 19.33. A 3-phase, 400 V, 50 Hz, 8-pole, star-connected induction motor has the following parameters:

$R_1 = 0.1\ \Omega$, $R_2 = 0.5\ \Omega$, $X_1 = 0.6\ \Omega$, $X_2 = 2.4\ \Omega$, $s_f = 0.05$ and the ratio of effective stator to rotor turns $\frac{1}{2}$. The motor is to be braked at rated speed and an external resistance of 2 Ω per phase (referred to stator) has been inserted into the rotor circuit. Determine the initial braking torque in (i) plugging and (ii) dc dynamic braking.

Solution: Rotor resistance referred to stator,

$$R'_2 = 0.5 \times \left(\frac{1}{2}\right)^2 = 0.125\ \Omega$$

Rotor leakage reactance referred to stator,

$$X'_2 = 2.4 \times \left(\frac{1}{2}\right)^2 = 0.6\ \Omega$$

(*i*) During plugging

Slip, $s = 2 - s_f = 2 - 0.05 = 1.95$

Synchronous speed, $N_s = \frac{120 f}{P} = \frac{120 \times 50}{8} = 750$ rpm

Initial rotor current during plugging,

$$I'_2 = \frac{V}{\sqrt{\left[R_1 + \frac{(R'_2 + R')}{s}\right]^2 + (X_1 + X'_2)^2}}$$

$$= \frac{400/\sqrt{3}}{\sqrt{\left(0.1 + \frac{0.125 + 2}{1.95}\right)^2 + (0.6 + 0.6)^2}} = 136.7 \text{ A}$$

Initial braking torque,

$$T_b = \frac{3 \times 60}{2\pi N_s} \times (I'_2)^2 \times \frac{R'_2 + R'}{s}$$

$$= \frac{3 \times 60}{2\pi \times 750} \times (136.7)^2 \times \frac{2.125}{1.95} = 777 \text{ N-m } \textbf{Ans.}$$

(*ii*) During dc dynamic braking

The equivalent circuit of the motor corresponding to slip of 0.05 and the equivalent circuit of the machine during dynamic braking at rated speed are given in Figs. 19.38 (*a*) and 19.38 (*b*) respectively.

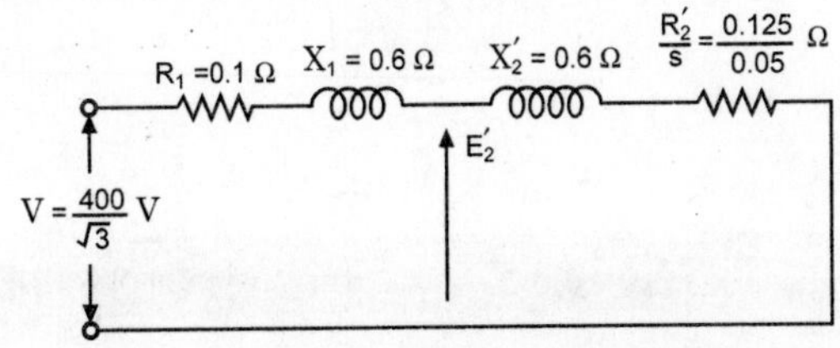

(*a*)

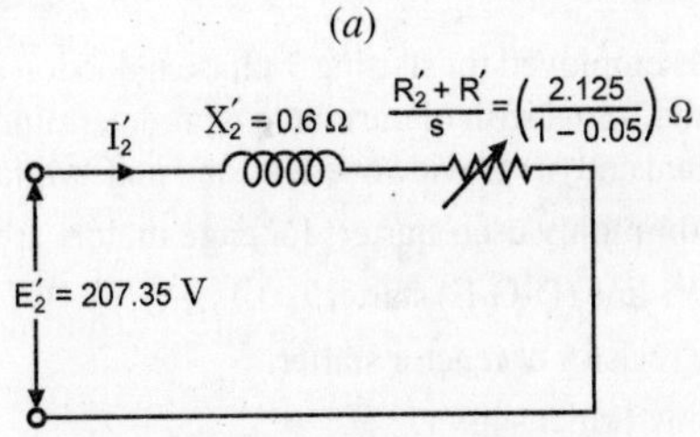

(*b*)

Fig. 19.38

$$E_1 = \frac{\sqrt{\left(\frac{0.125}{0.05}\right)^2 + 0.6^2}}{\sqrt{\left(\frac{0.125}{0.05} + 0.1\right)^2 + (1.2)^2}} \times \frac{400}{\sqrt{3}} \text{ V} = 207.35 \text{ V}$$

During dc dynamic braking,

$$S = 1 - 0.05 = 0.95$$

$$I'_2 = \frac{207.35}{\sqrt{0.6^2 + \left(\frac{2.125}{0.95}\right)^2}} = 89.53 \text{ A}$$

Initial braking torque,

$$T_b = \frac{3 \times 60}{2\pi N_s} \times (I'_2)^2 \times \frac{R'_2 + R'}{S}$$

$$= \frac{3 \times 60}{2\pi \times 750} \times (89.53)^2 \times \frac{2.125}{0.95} = 684.9 \text{ N-m } \textbf{Ans.}$$

Example 19.34. A 400 V, Y-connected, 3-phase, 6-pole, 50 Hz wound rotor induction motor has the following parameters referred to stator:

$R_1 = R'_2 = 1\ \Omega$; $X_1 = X'_2 = 2\ \Omega$

Stator to rotor turns ratio = 2.3

(*i*) For regenerative breaking at what speed the motor will hold an overhauling torque of 100 N-m.

(*ii*) For plugging operation the initial full-load speed is 950 rpm. The motor is braked by inserting additional resistance in the rotor circuit during plugging so as to reduce the braking current to 1.5 times its full-load current. Compute the external resistance and initial breaking torque. [*U.P.S.C. I.E.S. Electrical Engineering-II, 2013*]

Solution: Synchronous speed, $N_s = \frac{120f}{P} = \frac{120 \times 50}{6} = 1{,}000$ rpm

(*i*) **For regenerative braking**

Torque developed is given as

$$T_b = \frac{3 \times 60}{2\pi N_s} \times \frac{V^2}{\left(R_1 + \frac{R'_2}{s}\right)^2 + (X_1 + X'_2)^2} \times \frac{R'_2}{s}$$

$$\therefore\ 100 = \frac{3 \times 60}{2\pi \times 1{,}000} \times \frac{\left(400/\sqrt{3}\right)^2 \times \frac{1}{s}}{\left(1 + \frac{1}{s}\right)^2 + (2+2)^2}$$

$$\text{or } \frac{s}{(s+1)^2 + 16s^2} = \frac{100 \times 2\pi \times 1{,}000}{3 \times 60 \times \left(\frac{400}{\sqrt{3}}\right)^2} = 0.06545$$

or $17s^2 - 13.28s + 1 = 0$

$$\text{or } s = \frac{13.28 \pm \sqrt{(13.28)^2 - 4 \times 17}}{2 \times 17} = 0.0844 \text{ or } 0.697$$

Larger value is discarded as the motor is operating near full-load condition,

$$s = 0.0844$$

and so operating speed $= 1{,}000(1 - 0.0844) = 915.6$ rpm **Ans.**

(*ii*) Full-load speed, $N_f = 950$ rpm

$$\text{Full-load slip, } s_f = \frac{1{,}000 - 950}{1{,}000} = 0.05$$

$$\text{Full-load current, } I_f = \frac{V/\sqrt{3}}{\sqrt{\left(R_1 + \frac{R'_2}{s_f}\right)^2 \times (X_1 + X'_2)^2}}$$

$$= \frac{400/\sqrt{3}}{\sqrt{\left(1 + \frac{1}{0.05}\right)^2 + (2+2)^2}} = 10.8 \text{ A}$$

Initial rotor current during plugging

$$I'_2 = 1.5 \times 10.8 = 16.2 \text{ A}$$

$$\text{Slip, } s = 2 - s_f = 2 - 0.05 = 1.95$$

Since rotor current is given as

$$I'_2 = \frac{V/\sqrt{3}}{\left(R_1 + \frac{R'_2 + R'_{ex}}{s}\right)^2 + (X_1 + X'_2)^2}$$

$$\text{so } 16.2 = \frac{400/\sqrt{3}}{\sqrt{\left(1 + \frac{1 + R'_{ex}}{1.95}\right)^2 + (2+2)^2}}$$

$$\text{or } \frac{2.95 + R'_{ex}}{1.95} = \sqrt{203.221 - 16} = 13.683$$

$$\text{or } R'_{ex} = (13.683 \times 1.95) - 2.95 = 23.73\ \Omega$$

So addition resistance in the rotor circuit,

$$R_{ex} = \frac{R'_{ex}}{(\text{stator to rotor turn ratio})^2}$$

$$= \frac{23.73}{(2.3)^2} = 4.486\ \Omega \textbf{ Ans.}$$

$$\text{Initial braking torque, } T_b = \frac{3 \times 60}{2\pi N_s} \times (I'_2)^2 \times \frac{R_2 + R'_{ex}}{s}$$

$$= \frac{3 \times 60}{2\pi \times 1{,}000} \times (16.2)^2 \times \frac{1 + 23.73}{1.95} \text{ N-m}$$

$= 95.35$ N-m **Ans.**

HIGHLIGHTS

1. The starter is employed for starting 3-phase induction motor in order to (*i*) limit the initial rush of current to a predetermined value at the starting instant and (*ii*) provide overload and undervoltage protection.
2. The most commonly used starters for cage motors are :
   (*a*) Direct-on-line (D-O-L) starter.
   (*b*) Primary resistor or reactor starter.
   (*c*) Auto-transformer starter.
   (*d*) Star-delta starter.

(*a*) **Direct-on-line starting** is the cheapest method of starting of a 3-phase induction motor. Squirrel cage motors of capacity up to 1.5 kW capacity can be started by this method.

In D-O-L starting

Starting current, $I_{st} = I_{sc}$

and starting torque, $T_{st} = T_f \left(\frac{I_{sc}}{I_f}\right)^2 s_f$

(*b*) **Primary resistor or reactor starting method** is very simple and provides smooth acceleration of the motor. In this method reduced voltage is obtained by means of resistors (or reactors) connected in series with each stator lead. The drawbacks of this method are requirement of expensive resistors and low torque efficiency.

$$\text{Starting current, } I_{st} = KI_{sc}$$

$$\text{Starting torque, } T_{st} = K^2 T_f \left(\frac{I_{sc}}{I_f}\right)^2 s_f = K^2 T_f$$

(*c*) **Auto-transformer starting** is often employed for starting of large cage motors (of output rating exceeding 20 kW). In this method, the reduced voltage is obtained by taking tappings at suitable points from a 3-phase auto-transformer. The advantages of auto-transformer starting are (*i*) availability of highest torque per ampere of supply current (*ii*) suitability for long starting periods (*iii*) closed transition starting and (*iv*) motor current larger than supply current. The drawbacks of this method are poor power factor and higher cost in case of lower output rating motors.

$$\text{Starting current, } I_{st} = K^2 I_{sc}$$

$$\text{Starting torque, } T_{st} = K^2 T_f \left(\frac{I_{sc}}{I_f}\right)^2 \times s_f$$

$$= K^2 \times \text{Torque obtained by direct switching}$$

(*d*) **Star-delta starting method** is simple, cheap, effective and efficient method of starting cage motors as no power is lost in auxiliary equipment. The star-delta starter connects the three stator windings in star across the rated supply voltage at the starting instant. After attainment of normal speed, the same windings are re-connected in delta.

Star-delta starter reduces line current at starting and starting torque to one-third of those with direct switching.

3. Slip-ring induction motors are usually started with full-line voltage across the stator terminals and by introducing variable resistor in each phase of the rotor circuit. The external resistances introduced in each phase of the rotor circuit not only reduces the current at the starting instant but increases the starting torque also due to improvement in *pf*.

4. The problem of speed control of electrical motors in general and of induction motors in particular is of great practical importance.

The speed of a 3-phase induction motor can be controlled by any of the following methods.

1. **From stator side :** (*i*) By variation of supply frequency (*ii*) By variation of applied voltage and (*iii*) by changing the number of poles.
2. **From rotor side :** (*i*) By changing the resistance in the rotor circuit and (*ii*) By introducing into the rotor circuit an additional emf of the same frequency as the fundamental emf of the rotor.

(*i*) **Speed control by variation of supply frequency.** This method of speed control provides wide speed-control range with gradual variation of the speed throughout this range. The major difficulty with this method is how to get the variable frequency supply.

(*ii*) **Speed control by variation of supply voltage.** This method of speed control is simple, low in first cost and has low maintenance cost but it has limited use because (*i*) the operation at voltages exceeding rated voltage is restricted by magnetic saturation, (*ii*) a large change in voltage is required for a relatively small change in speed (*iii*) the developed torque reduces greatly with the reduction in supply voltage and the motor will come to standstill if the load torque exceeds the pull-out torque of the motor corresponding to reduced supply voltage and (*iv*) the range of speed control is very limited in the downward direction *i.e.*, from rated speed to lower speeds.

In general, this method of speed control is used only on loads where the torque required drops off considerably as the speed is reduced such as with small squirrel cage motors driving fans.

(*iii*) **Speed control by variation of rotor resistance.** This type of control is only possible with wound rotor induction motors. This method of speed control is simple and low in initial as well as maintenance cost and it is possible to have a large starting torque with low starting current and large pull-out torques at small values of slip.

(*iv*) **In secondary foreign voltage control.** The speed of an induction motor is controlled by injecting a voltage of slip frequency in the secondary circuit. It is costlier method and is used with motors of very large rating, such as for motors in steel mills.

(*v*) **Speed control by concatenation.** In this method of speed control, two motors are required, at least one of which must have a wound rotor.

The synchronous speed of the cascade set is given as

$$N_s = \frac{120 f}{P_1 \pm P_2}$$

+ sign for connections providing torques of the two motors in the same direction.

– sign for connections providing torques of the two motors in the opposite directions.

5. There are three methods of electric braking of induction motors viz. plugging (or counter-current braking), dynamic (or rheostatic) braking, and regenerative braking.

Plugging can be achieved in an induction motor merely by reversing any two of the three phases which causes a reversal of the direction of the rotating magnetic field.

Rheostatic braking with a 3-phase induction motor can be obtained by disconnecting the stator winding from the ac supply and exciting it from a dc source to produce a stationary dc field. Thus stator winding is used as a dc field winding and the rotor winding as an armature winding.

Regenerative braking is an inherent characteristic of an induction motor since it operates as an induction generator when it runs at speed above synchronous speed and it feeds power back to the supply line.

EXERCISES

1. Draw and explain the equivalent circuit of a 3-phase induction motor. Also explain with the help of mathematical equations, the requirement of starter in the starting of 3-phase induction motor.
[M.D. Univ. Electrical Machines-II, December-2011]
2. Draw and discuss direct-on-line starter for a 3-phase squirrel induction motor. **[M.G. Univ. Electrical Machines-III, November-2005]**
3. Write different starters used for 3-phase induction motor and explain any one of them.
[Gujarat Technological Univ. Electrical Machines-II, Summer-2014]
4. Name various methods of starting the polyphase induction motors and describe auto-transformer method of starting, in detail.
[U.P. Technical Univ. Electromechanical Energy Conversion II, 2002-03]
5. Why starters are necessary for starting 3-phase induction motors? What are the various types of starters? Explain star-delta type starter in detail. **[Anna Univ. Electrical Machines-II, Nov./Dec.-2012]**
6. Why "Starter" is required to start a 3-phase induction motor? Name different starting methods and describe star-delta method of starting a squirrel cage induction motor. Determine the ratio of starting to full-load torque.
[G.B. Technical Univ. Electromechanical Energy Conversion-II, 2012-13]
7. What is the need of starters for induction motors? Briefly describe various techniques used for the starting of 3-phase induction motors.
[U.P.S.C. I.E.S. Electrical Engineering-II, 1998]
8. Describe with the help of an appropriate diagram, the star-delta method of starting a three-phase induction motor.
[U.P.S.C. I.E.S. Electrical Engineering-II, 1999]
9. Compare the star/delta starter and auto-transformer method of starting of 3-phase squirrel cage induction motor on the basis of (*i*) starting torque and starting current (*ii*) adoptability to motor rating, terminal arrangements and (*iii*) starting conditions. Discuss the limitation of these methods.
[J.N. Technological Univ. Electrical Machines-II, January-2014]
10. Explain the different methods of starting squirrel cage induction motors. Compare them.
[U.P. Technical Univ. Electromechanical Energy Conversion, 2006-07]
11. Explain the principle of rotor-resistance starting of slip-ring type induction motors. **[W.B. Univ. of Technology Electrical Machines-I, 2010]**
12. What are different methods of starting 3-ϕ induction motor ? Describe, with the help of neat diagram, the method of starting slip-ring induction motor and squirrel cage induction motor.
[U.P. Technical Univ. Electromechanical Energy Conversion II, 2003-04]
13. Explain various methods of starting large induction motors and compare their values of starting torques.
[U.P.S.C. I.E.S. Electrical Engineering-II, 2010]
14. Describe briefly various speed control techniques for induction motors (both squirrel cage and wound rotor types).
[U.P.S.C. I.E.S. Electrical Engineering-II, 2013]
15. Name the chief methods of speed control of a 3-phase induction motor. Explain briefly one of the most modern methods of speed control of an induction motor namely V/*f* control.
[U.P.S.C. I.E.S. Electrical Engineering-II, 2005]
16. Sketch a typical torque-speed curve of a polyphase induction motor under rated supply voltage V and frequency *f*. On the same diagram, sketch the torque-speed curve for the following voltages and frequencies but with rotor resistance unchanged.

(*i*) $\frac{V}{2}$ and f (*ii*) $\frac{V}{2}$ and $\frac{f}{2}$ (*iii*) V and $\frac{f}{2}$.
[Pb. Univ. Elec. Machines-II, May-1995]
17. How is the speed of a 3-phase induction motor controlled by its stator voltage control?
[J.N. Technological Univ. Electrical Machines-II, January-2014]
18. How is the speed of a 3-phase induction motor controlled by its stator voltage control ? Show that the region of speed control by this method is limited by the slip at which maximum torque develops.
[Pb. Univ. Elec. Machines II, Dec.-1994]
19. Explain the basic principle of induction motor speed control by pole-changing.

Discuss how the three different torque-speed characteristics are obtained by pole-changing and by suitable connections amongst the three phases. Sketch these characteristics on the same diagram.
[Pb. Univ. Elec. Machines II, May-1995]
20. Discuss pole-changing method of speed control for a 3-phase induction motor.
[U.P. Technical Univ. Electromechanical Energy Conversion II, 2005-06]
21. Explain the pole-changing technique adopted for the control of 3-phase induction motor.
[M.G. Univ. Electrical Machines-III, November-2005]
22. Describe the various methods of speed control of three-phase slip-ring induction motors.
[G.G.S.I.P. Univ. Electromechanical Energy Conversion-II, May-2011]
23. Explain principle of speed control of a 3-phase induction motor by (*i*) adding resistance (*ii*) injecting voltage. Draw the corresponding torque-speed characteristics and discuss the applications and limitations of these methods.
[J.N. Technological Univ. Electrical Machines-II, January-2014]
24. Explain the principle of slip control by slip frequency emf injection in rotor circuit with neat sketch.
[G.B. Technical Univ. Special Electrical Machines, 2012-13]
25. Explain by help of neat diagram about cascade connection in induction motor. **[Rajasthan Technical Univ. Electrical Machines-II, 2013]**
26. Explain cascade arrangement for controlling speed of three-phase induction motor. **[Rajasthan Technical Univ. Electrical Machines-II, 2012]**
27. Give salient features of Kramer scheme of speed control of induction motor. **[Pb. Technical Univ. Asynchronous Machines, May-2013]**
28. Explain Scherbius scheme for speed control of induction motor.
[Pb. Technical Univ. Asynchronous Machines, May-2013]
29. State different types of electric braking applied to induction motor. Explain each in detail. **[G.B. Technical Univ. Electric Drives, 2012-13]**
30. Explain the plugging of a 3-phase induction motor.
[M.D. Univ. Electrical Machines-II, December-2011]
31. A 3-phase induction motor is operating in steady state at slip s. By suddenly interchanging two stator leads, the motor is plugged to a quick stop. Derive an expression for the braking torque of the motor.
32. Mention two disadvantages of counter-current braking of a 3-phase induction motor. How does dynamic braking overcome the above disadvantages ? Draw the torque-speed characteristics of the motor under dynamic braking condition.
33. Explain dc dynamic braking of three phase induction motor. Discuss effects of excitation and rotor resistance on the torque-slip characteristics of the motor. **[U.P. Technical Univ. Electric Drive, 2005-06]**
34. Explain "regenerative braking" as applied to 3-phase induction motor. What are the conditions for regenerative braking of an induction motor to be possible. **[Rajasthan Technical Univ. Electrical machines-II, 2012]**
35. Write short notes on the following:
    (*i*) Starting methods for 3-phase induction motor.
    [R.G. Technical Univ., December-2012]
    (*ii*) Speed control of 3-phase induction motor.
    [W.B. Univ. of Technology, 2010]
    (*iii*) Plugging as applied to 3-phase induction motor.
    (*iv*) Regenerative braking as applied to 3-phase induction motor.

SHORT ANSWER TYPE QUESTIONS WITH ANSWERS

Q. 1. Why are starters used for starting of induction motors (3-phase)? **[U.P. Technical Univ., December-2003]**

or

What is the necessity of starter in 3-phase induction motor? **[G.G.S.I.P. Delhi Univ. May/June-2006]**

Ans. At the starting instant, a 3-phase induction motor acts as a polyphase transformer having an air gap with a short-circuited secondary. If the motor of higher capacity (larger than 2 kW) is connected directly to the supply mains, it will draw heavy current (about 5-8 times the rated full-load current) at the starting instant which is objectionable because it will cause large voltage drop in the distribution network and thus affect adversely the operation of other electrical equipment connected to the same distribution network. Hence, to control the starting current, use of starter is essential for starting of 3-phase induction motors of larger size (rating exceeding 2 kW).

Q. 2. Give methods for starting of 3-phase squirrel cage induction motors? **[U.P.S.C. I.E.S., 2009]**

Ans. Starters used for starting of a squirrel cage induction motor are

(*a*) Direct-on-line starter
(*b*) Line resistance starter
(*c*) Star-delta starter and
(*d*) Auto-transformer starter

Q. 3. Why do you require starters for the starting of 3-phase induction motors. Name any three starters used for starting 3-phase induction motors. **[G.G.S.I.P. Univ., December-2011]**

Ans. Refer to Answers of Qs. 1 and 2.

Q. 4. Which is the cheapest method of starting a 3-phase induction motor.

Ans. Direct-on-line starting.

Q. 5. How many terminals do you expect to find on the terminal box of a squirrel cage induction motor to be started by star-delta starter ?

Ans Six.

Q. 6. In what ratio line current and starting torque are reduced with star-delta starting ?

Ans. Line current and starting torque are reduced to one-third with star-delta starting.

Q. 7. What is the serious objection to the practice of employing reduced voltage for the starting of a squirrel cage induction motor ?

Ans. Large reduction in staring torque because the starting torque varies as the square of voltage applied to the stator.

Q. 8. What type of protection is provided in the starter meant for 3-phase induction motors ?

Ans. Overload and undervoltage protection.

Q. 9. A 3-phase induction motor is to be started first using an auto-transformer with 80 per cent tapping and next direct-on-line. What will be the ratio of the starting torques ?

Ans. Starting torque with auto-transformer with 80 % tapping/starting torque on direct on line = 0.64.

Q. 10. What is the parameter other than the ratio of rotor current limits which decides the number of resistance sections of a slip-ring induction motor starter ?

Ans. Slip at full load.

Q. 11. Is it possible to add an external resistance in the rotor circuit of a 3-phase cage induction motor ? Give reason for your answer. **[Madras Univ., April-1997]**

Ans. No; because the rotor winding is permanently short circuited.

Q. 12. Why is the rotor rheostat starter unsuited for a cage motor ?

Ans. Because there is no provision for inserting resistance in the rotor circuit of a cage motor.

Q. 13. Rotor resistance starting is preferred to reduced voltage starting in a wound rotor induction motor. Why ?

Ans. Because rotor resistance starting of a wound rotor induction motor not only reduces the starting current at the starting instant but increases the starting torque also due to improvement in power factor whereas the reduced voltage starting reduces only the starting current at the starting instant.

Q. 14. For which type of loads, induction motor with voltage controller is suitable and why ?

Ans. Induction motor with voltage controller is suitable only for loads where the torque required drops off considerably with the drop in speed such as for driving fans because a large change in voltage is required for a relatively small change in speed and developed torque reduces greatly with the reduction in supply voltage.

Q. 15. On what factors does the speed of an induction motor depend ?

Ans. The speed of an induction motor depends on supply frequency, slip and number of poles on stator.

Q. 16. List out the methods of speed control of cage type 3-ϕ induction motors. **[Anna Univ., Nov./Dec.-2012]**

Ans. Various methods of speed control of cage type 3-ϕ induction motors are:

(*i*) Variation of supply frequency (*ii*) Variation of supply voltage (*iii*) Changing of number of poles.

The number of pole pairs in the stator can be changed as follows:

(*a*) by using multiple stator windings—by placing two or more independent windings on the stator, each producing a different number of poles;
(*b*) by using consequent pole technique — by placing one or two independent windings on the stator and changing the number of poles by changing the interconnections of primary coils;
(*c*) by using pole amplitude modulation technique — a suppressed– carrier modulation technique.

Q. 17. If frequency variation is used to control the speed of an induction motor, what else must be changed ?

Ans. The voltage applied to the stator must be varied in direct proportion to the frequency so as to maintain flux level constant.

Q. 18. If the number of poles on a motor is increased to lower the speed, how will the power factor be affected ?

Ans. Power factor will be reduced.

Q. 19. What are the advantages of introducing resistance in the rotor circuit of a 3-phase wound rotor induction motor? **[G.G.S.I.P. Univ., May-2011]**

Ans. By introducing resistance in the rotor circuit it is possible to have a large starting torque with low starting current and large pull-out torques at small values of slip.

Q. 20. While controlling induction motor speed, how super-synchronous speed is achieved ?

Ans. Super-synchronous speed can be achieved, while controlling the speed of an induction motor, by injecting a slip frequency voltage in phase with emf induced in the rotor.

Q. 21. What will be the effect on the torque developed by an induction motor when the applied voltage is reduced to half, supply frequency remaining unchanged ?

Ans. Torque developed by an induction motor with half-rated voltage, supply frequency remaining unchanged, will be reduced to one-fourth because $T \alpha V^2$ (*i.e.* square of the supply voltage).

Q. 22. What will be the effect on the torque developed by an induction motor when both the applied voltage and frequency are reduced to half ?

Ans. The torque developed by an induction motor will be reduced to one half of normal value when both the applied voltage and frequency are reduced to half because $T \propto \frac{V^2}{f}$.

Q. 23. What are the different ways of controlling slip?
[G.B. Technical Univ., 2012-13]

Ans. Slip can be controlled by (*i*) variation of supply voltage (*ii*) introducing resistance in rotor circuit, (*iii*) injecting a voltage of slip frequency in the rotor circuit and (*iv*) in concatenation method of speed control, by changing the slip of one motor.

Q. 24. What is meant by plugging ?

Ans. Plugging means stopping of an electric motor by instantaneously reversing its torque till it stops.

Q. 25. Electric braking is accomplished in an induction motor by interchanging two stator terminals. Name the method.

Ans. Plugging or counter-current braking.

Q. 26. Explain, the conditions under which rotor frequency is smaller/ equal/greater than stator current frequency in an induction motor.

Ans. The frequency of rotor currents is equal to slip times the frequency of the stator current. During motoring mode the slip is very small (much less than unity) and so the rotor current frequency is small than that of stator current. At standstill the slip is unity and so the rotor current frequency is equal to stator current frequency. During counter-current braking (or plugging) the slip is larger $(2 - s)$ and so the frequency of rotor currents is much larger than that of stator currents.

Q. 27. When it is required to control the induction motor speed during deceleration, what kind of braking is preferable to employ ?

Ans. DC dynamic braking is highly suitable for controlling the speed of the induction motor during deceleration by controlling the dc excitation.

Q. 28. What are the conditions for regenerative braking of an induction motor to be possible ?

Ans. The driven load should drive the motor above synchronous speed but the driving torque of the load should not exceed the maximum torque, which the machine is capable to develop during regeneration.

Q. 29. What is the standard direction of rotation of an induction motor ?

Ans. Counter-clockwise when looking from non-drive end of the motor.

Q. 30. Why regenerative braking cannot be applied to a squirrel cage induction motor? **[U.P. Technical Univ., 2006-07]**

Ans. In the case of a squirrel cage induction motor, stable speed is obtained at a speed considerably in excess of the synchronous speed and the regenerative braking cannot be applied unless the motor is specially designed to withstand the excessive speed.

PROBLEMS

1. Determine approximately the starting torque of an induction motor in terms of its full-load torque when started by means of (*i*) star-delta starter and (*ii*) by an auto-transformer starter with 50 % taps. The motor draws 5 times full-load current when switched directly on line and has a full-load slip of 4%.
   [Ans. One-third and one-fourth of full-load torque]
2. Find the ratio of starting to full-load current in a 10 kW (output), 400 V, 3 phase induction motor with star/delta starter, given that full-load power factor is 0.85, the full-load efficiency is 0.88 and the blocked rotor current at 200 V is 40 A. Ignore magnetising current.
   [Ans. 1.38]
3. A 3-phase squirrel cage induction motor has a ratio of maximum torque to full-load torque as 2.5 :1. Determine the ratio of actual starting torque to full-load torque for auto-transformer starting with a tapping of 75 percent. The rotor resistance = 0.4 Ω and $X_2 = 4\ \Omega$.
   [Ans. 0.278]
4. The full-load efficiency and power factor of a 16 hp, 440 V, 3-phase induction motor are 85 % and 0.8 respectively. The blocked rotor line current is 45 A at 220 V. Calculate the ratio of starting to full-load current if the motor is provided with a star-delta starter. Neglect magnetising current. **[Ans.** 2.288]
5. A 3-phase squirrel cage type induction motor, when started by means of a star-delta starter, takes 200 % of full-load current (line) and develops 44 % of full-load torque at starting. Calculate the starting torque and current if an autotransformer with 75 per cent tapping was employed. **[U.P. Technical Univ. Electrical Machines, 2002-03]**
   [Ans. 74.2% of T_f; 337.5% of I_f]
6. A 3-phase induction motor takes a starting current which is 5 times full-load current at normal voltage. Its full-load slip is 4 per cent. What auto-transformer ratio would enable the motor to be started with not more than twice the full-load current drawn from the supply? What would be the starting torque under this condition?
   [Ans. 40% of full-load torque]
7. A 3-ϕ, Δ-connected squirrel cage induction motor, when connected directly to 400 V, 3-ϕ, 50 Hz supply takes a starting current of 100 A in each stator phase. Calculate line current for direct on-line starting, and line and phase currents for Y-Δ starting.
   [U.P. Technical Univ. Electromechanical Energy Conversion, 2003-04]
   [Ans. 173.2 A, 57.73 A, 57.73 A]
8. Determine approximately the starting torque of a 3-phase induction motor in terms of full-load torque when started by (*i*) star-delta starter (*ii*) an auto-transformer starter with 50 per cent tapping. The short-circuit current of motor is 5 times the full-load current and the full-load slip is 5 percent.
   [Bharathiyar Univ. Elec. Machines-II, April-1997]
   [Ans. 41.67% of full-load torque, 31.25% of full-load torque]
9. Why starter is needed for operation of 3-phase induction motor ? List different starting methods of slip ring type and squirrel cage type induction motors. An induction motor has a starting current of six times the full-load current and a full-load slip of 0.05. Find in PU of full-load values the line current and starting torque with
   (*i*) Direct starting. (*ii*) Star-delta starting.
   [U.P. Technical Univ. Electromechanical Energy Conversion-II, 2009-10]
   [Ans. (*i*) 6 pu, 1.8 pu, (*ii*) 2 pu, 0.6 pu]
10. Determine approximately the starting torque of an induction motor in terms of full-load torque when started by means of (*i*) star-delta switch (*ii*) an auto-transformer with 70.7% tappings.
   The short-circuit current of the motor at normal voltage is 6 times the full-load current and full-load slip is 4%.
   [R.G. Technical Univ. Electrical Machine-I, June-2013]
   [Ans. (*i*) 48% of full-load torque (*ii*) 72% of full-load torque]

11. A cage type induction motor when started by means of a star-delta starter takes 180 % of full-load line current and develops 35 % of full-load torque at starting. Calculate the starting torque and current in terms of full-load values, if an auto-transformer with 80 % tappings were employed.

[U.P. Technical Univ. Electromechanical Energy Conversion II, 2005-06]

[Ans. 345.6% of full-load current, 67.2% of full-load torque**]**

12. A 3-phase squirrel cage induction motor has a starting current 175% of full-load line current and develops 35% of full-load when operated by a star-delta switch. What should be the starting torque and current if an auto-transformer with 80% tapping is employed.

[M.G. Univ. Electrical Machines-III, November-2003]

[Ans. 67.2% of full-load torque and 3.36 times of full-load current**]**

13. Calculate the steps in a 5-section rotor starter for a 3-phase induction motor. The maximum starting current = full-load current. Full-load slip = 1.8 % with rings shorted. Rotor resistance per phase = 0.015 Ω.

[Ans. r_1 = 0.46 Ω, r_2 = 0.206 Ω, r_3 = 0.0923 Ω; r_4 = 0.0413 Ω and r_5 = 0.0185 Ω**]**

14. The rotor of a 6 pole, 50 Hz slip-ring induction motor has a resistance of 0.2 Ω per phase and runs at 960 rpm on full load. Calculate the approximate resistance per phase of a rotor rheostat such that speed is reduced to 750 rpm for full-load torque.

[U.P. Technical Univ. Electrical Machines, 2006-07]

[Ans. 1.05 Ω/phase**]**

15. The voltage across two slip rings of a stationary star-connected 100 hp, 415 V induction motor is 273 V with full voltage across stator. If full-load slip is 2 per cent, find the resistance to be added in series with each slip-ring to reduce the speed by 20 %. Assume load torque to be constant and rotor reactance negligible.

[Pb. Univ. Elec. Machines II Dec.-1993]

[Ans. 0.196 Ω**]**

16. A cascade set consists of two motors A and B with 4 poles and 6 poles respectively. The motor A is connected to 50 Hz supply. Determine (*i*) the speed of the set (*ii*) the electric power transferred to motor B when the input to motor A is 25 kW. Losses may be neglected.

[Ans. (*i*) 600 rpm (*ii*) 10 kW**]**

17. Two 3-phase slip-ring induction motors each having 4-poles operate in cascade connection. The supply is 440 V, 50 Hz, what are the obtainable speeds? Give suitable reasons for speed during differential cascade.

[U.P. Technical Univ. Electromechanical Energy Conversion II, 2006-07]

[Ans. 1,500 rpm and 750 rpm; Motors with equal number of poles cannot be operated in differential cascade**]**

18. A 3-phase 440 V, 50 Hz, 10-pole star-connected induction motor has $\mathbf{Z}_1 = (0.15 + j\,0.6)$ Ω and $\mathbf{Z}_2 = (0.45 + j\,1.8)$ Ω. Full-load slip is 5 percent and ratio of effective stator to rotor turns $\frac{1}{\sqrt{3}}$. The motor is to be braked at rated speed and external resistance of 1.75 Ω per phase (referred to stator) has been inserted into the rotor circuit. Determine the initial braking torque in case of (*i*) plugging and (*ii*) dc dynamic braking.

[Ans. (*i*) 1,182 N-m (*ii*) 1,226 N-m**]**

19. A 25 kW, 400 V, 3-ϕ, 4-pole, 50 Hz induction motor has full-load slip of 5%. If the ratio of standstill reactance to resistance per phase is 4, estimate the plugging torque at full speed. **[Ans.** 109.86 N-m**]**

❑❑❑

C H A P T E R

20 Fractional Kilowatt Motors

INSIDE THIS CHAPTER

20.1 INTRODUCTION

There are two basic reasons for the use of single-phase rather than 3-phase motors.

First, for reasons of economy, most houses, offices and also rural areas are supplied with single-phase ac, as power requirements of individual load items are rather small.

The second factor is the economics of the motor and its branch circuit. Fixed loads requiring not more than 0.5 kW can generally be served most economically with single-phase power and a single-phase motor. Single-phase motors are simple in construction, reliable, easy to repair and comparatively cheaper in cost and, therefore, find wide use in fans, refrigerators, vacuum cleaners, washing machines, other kitchen equipment, tools, blowers, centrifugal pumps, small farming appliances etc. To avoid the need for special branch circuits, portable tools are also usually single-phase.

Because of above reasons motors of comparatively small ratings (mostly in fractional kW ratings) are manufactured in large number to operate on single-phase ac at standard frequencies. An indication of the number of such motors can be had from the fact that the sum of total of all fractional kilowatt motors in use today far exceeds the total of integral kilowatt motors of all types. Single-phase motors perform a great variety of useful services in the home, the office, the factory, in business establishments, on the farm, and many other places where electricity is available. They find interesting applications in automatic control devices of various types. Since the requirements of the numerous applications differ so widely, the motor manufacture industry has developed several types of such machines, each type having operating characteristics that meet definite demands. For example, one type operates satisfactorily on dc or any frequency up to 50 Hz; another rotates at absolutely constant speed, regardless of the load; another develops considerable starting torque; and still another, although not capable of developing much starting torque is nevertheless extremely cheap to make and very rugged.

Single-phase ac motors may be divided in three general classes namely (*i*) induction motors (*ii*) commutator motors and (*iii*) synchronous motors.

Induction motors are further classified as split-phase motors, shaded-pole motors and repulsion– start induction motors according to the method of producing starting torque. The commutator motors are the series motor, universal (ac/dc) motor, the repulsion-induction motor, with various modifications and combinations of these types being in use.

Single-phase induction motors in very small sizes $\left(\frac{1}{400}\text{ kW to }\frac{1}{25}\text{ kW}\right)$ are used is toys, hairdryers, vending machines etc. The universal motor is widely used in portable tools, vacuum cleaners and kitchen equipment.

The main drawbacks of single-phase motors are low overload capacity, low efficiency, low power factor and low output as compared to that of a 3-phase motor of the given frame size.

20.2 SINGLE-PHASE INDUCTION MOTORS

If one line of a three-phase induction motor is opened while the motor is running with moderate or light load, it is found that the motor continues to run though at slightly lower speed. This condition is known as *single-phase operation*. Opening of one line (phase) leaves the equivalent of a single phase connected to the stator. In other words a 3-phase induction motor has become a single-phase motor.

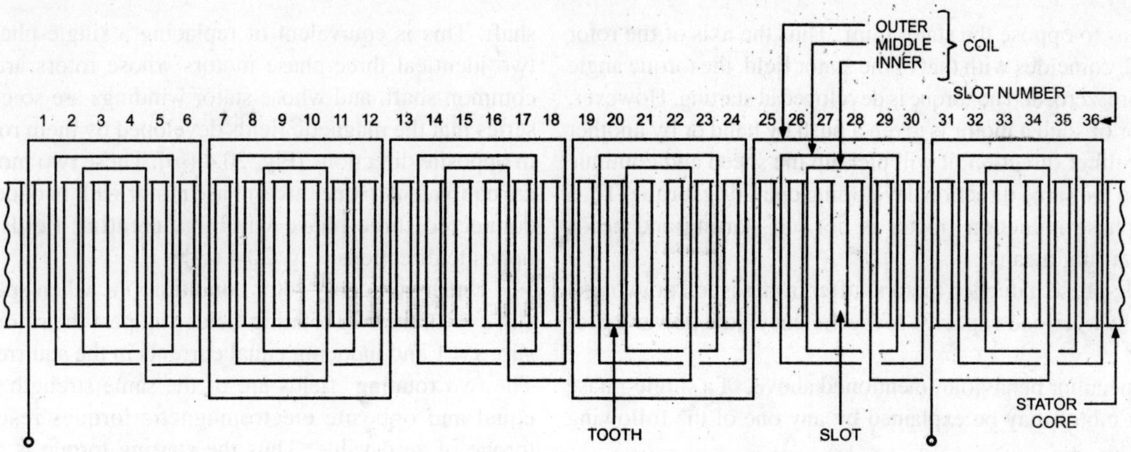

Fig. 20.1 *Winding Diagram For a Single-Phase 6-Pole Motor With 36 Slots (Main Winding Only)*

20.2.1. Construction. A single-phase induction motor is similar to a 3-phase squirrel cage induction motor in physical appearance. The rotor of a single-phase squirrel cage motor is essentially the same as that employed in 3-phase induction motors and needs no further description. There is uniform air gap between stator and rotor but no electrical connection between them (stator and rotor). Except for shaded-pole types, the stator core is also very similar. A single-phase motor can be wound for any even number of poles, two, four, and six being most common. Like three-phase machines, adjacent poles have opposite magnetic polarity and synchronous speed equation ($N_s = \frac{120f}{P}$) also applies.

The stator windings, however, are distinctively different in two aspects. First, single-phase motors are usually provided with concentric coils, as illustrated in Fig. 20.1. In this diagram the coils appear to have only one turn of wire, but in practice each coil has many turns. With concentric coils, the number of turns per coil can be adjusted to provide an approximately sinusoidal distribution of mmf along the air gap and this is one of their advantages. Second, single-phase squirrel cage motors normally have two stator windings. Except for shaded-pole motors, the two windings are similar to that illustrated in Fig. 20.1, but one of them usually have few turns of much thin wire, and the two windings are in space quadrature with respect to each other. In motors that operate with both windings energized, the winding with the heaviest wire is known as the *main winding* and the other is called the *auxiliary winding*. If the motor runs with auxiliary winding open, these windings are usually referred as running and starting windings respectively. A motor that runs with auxiliary winding open, uses a centrifugal switch to cut out the starting (or auxiliary) winding after starting. In most of the motors the main winding is placed at the bottom of the slots and the starting winding on top but shifted 90° from the running winding.

If the motor has four or more poles, the main winding of a shaded-pole motor is similar to the windings that have already been described, except that it generally has only one coil per pole. A 2-pole shaded-pole motor may be constructed similarly to the magnetic circuit given in Fig. 20.2.

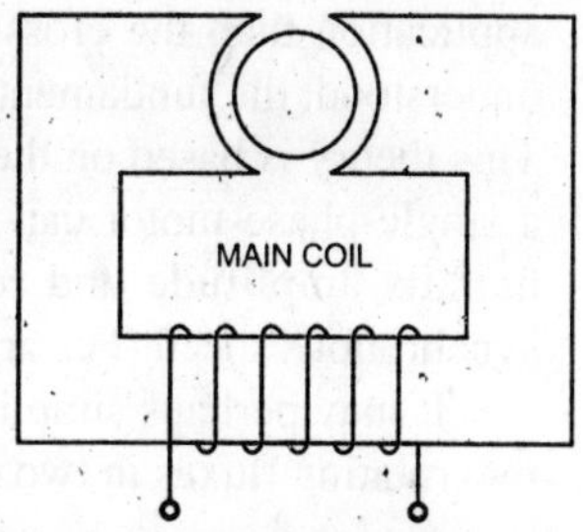

Fig. 20.2

An elementary single-phase induction motor is shown in Fig. 20.3.

20.2.2. Action. When the stator winding of a single-phase induction motor is connected to single-phase ac supply, a magnetic field is developed, whose axis is always along the axis of stator coils. With alternating current in the fixed stator coil the mmf wave is stationary in space but pulsates in magnitude and varies sinusoidally with time. Currents are induced in the rotor conductors by transformer action, these currents being in such a

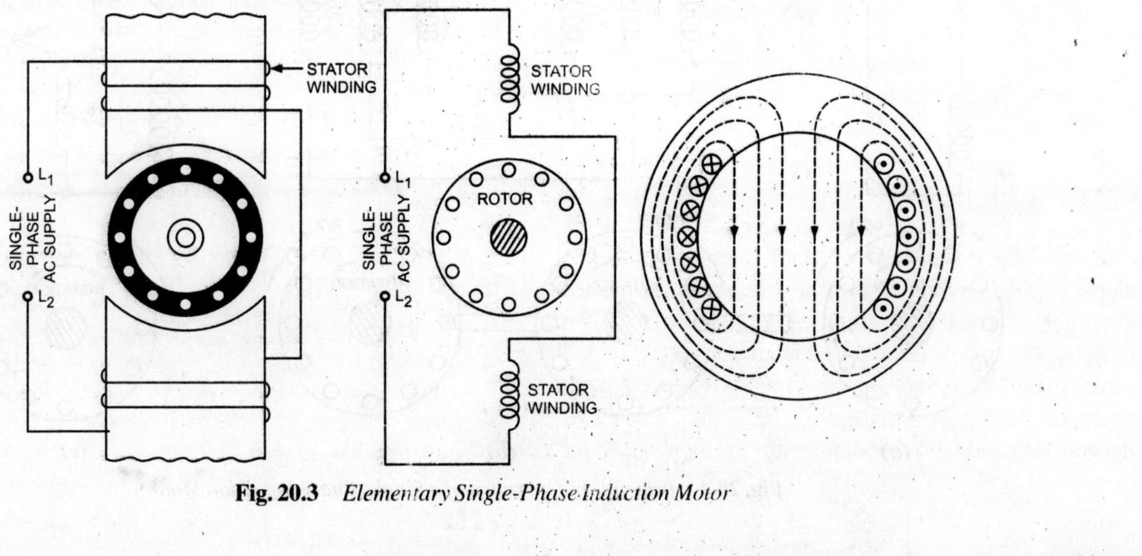

Fig. 20.3 *Elementary Single-Phase Induction Motor*

direction as to oppose the stator mmf. Thus the axis of the rotor mmf wave coincides with that of the stator field, the torque angle is, therefore, zero, and no torque is developed at starting. However, if the rotor of such a motor is given a push by hand or by another means in either direction, it will pick up the speed and continue to rotate in the same direction developing operating torque. Thus a single-phase induction motor is not self-starting and needs special starting means.

Single-phase induction motors suffer from several drawbacks such as low overload capacity, low efficiency, low power factor etc.

The peculiar behaviour, mentioned above, of a single-phase induction motor may be explained by any one of the following two theories

1. Double revolving field theory
2. Cross-field theory.

Both theories have their own advantages and fields of application. They ultimately lead to the result that are approximately same.

20.3 DOUBLE REVOLVING FIELD THEORY

Though double revolving field theory is not simpler in application than the cross-field theory, but once this theory is understood, the fundamental concepts are more readily grasped. This theory is based on the idea that pulsating field produced in a single-phase motor can be resolved into two components of half its amplitude and rotating in opposite directions with synchronous speed (Ferraris principle—Refer to Art. 4.11.1).

It may perhaps simplify the visual picture by placing these two rotating fluxes in two separate motors mounted on the same shaft. This is equivalent of replacing a single-phase motor by two identical three-phase motors whose rotors are fitted on a common shaft, and whose stator windings are so connected in series that the magnetic fields developed by them rotate in space in opposite directions [Fig. 20.4 (*b*)]. These two motors, in turn, are equivalent to one three-phase motor with two identical series connected three-phase windings creating fields rotating in opposite directions [Fig. 20.4 (*c*)].

Under stationary rotor condition (*i.e.* when speed N = 0 or slip $s = 1$), the two rotating fields slip past the rotor at the same slip, $s = 1$ and inducing equal currents in the squirrel cage rotor. The two rotating fields are of the same strength and develop equal and opposite electromagnetic torques resulting in net torque of zero value. Thus the starting torque is zero and the single-phase induction motor is non-self starting. Further, the two rotating fields induce a resultant emf in the stator which balances the applied voltage assuming low leakage impedance of the stator winding.

If, however, the rotor is made to run at speed N by some external means in any direction, say in the direction of forward field, the two slips are now s and $(2 - s)$, as shown below:

The slip of the rotor with respect to the forward rotating field F_f,

$$s_f = \frac{N_s - N}{N_s} = s \qquad \text{...(20.1)}$$

The slip of the rotor with respect to the backward rotating field F_b,

$$s_b = \frac{N_s - (-N)}{N_s} = \frac{2N_s - (N_s - N)}{N_s} = (2 - s) \qquad \text{...(20.2)}$$

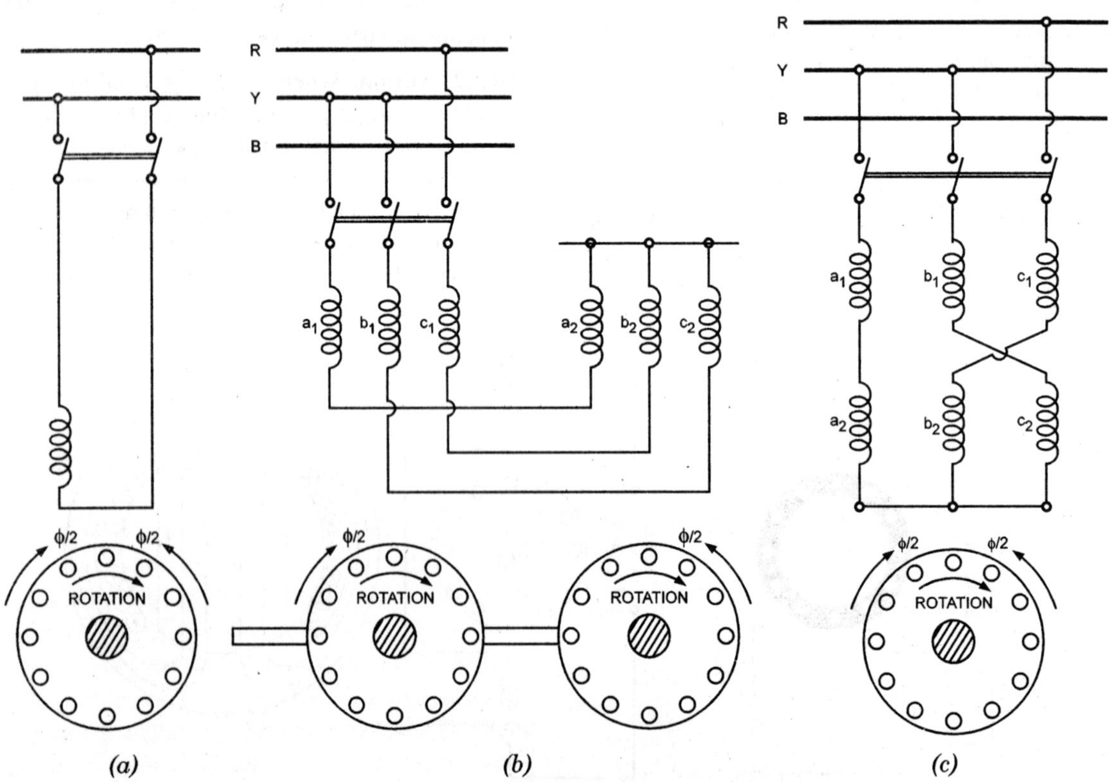

Fig. 20.4 *Principle of Operation of a Single-Phase Induction Motor*

For normal operation $(2 - s) >> s$ and as a consequence the backward field rotor currents are much larger than at standstill and have a low power factor. The corresponding opposing rotor mmf, owing to stator impedance, causes the backward field to be greatly reduced in strength. On the other hand, the low-slip forward rotating field induces smaller currents of a high power factor in the rotor than at standstill. This leads to greatly strengthening of the forward field. Thus weakening of backward field and strengthening of forward field depends upon the slip (or speed of rotor) and the difference increases with the decrease in slip with respect to the forward field or with the increase in rotor speed in forward direction. In fact, at near about the synchronous speed, the forward field may be several times the backward field. As a result there is a net running torque.

Let us assume that the torque developed by forward field in the direction of rotor rotation is positive. Then the torque developed by the backward field will be negative and, evidently, be a braking torque.

The torque-slip curve of actual motor may be obtained by applying the principle of superposition to the hypothetical constituent motors. Their individual torque-slip curves will be of the form shown in Fig. 20.5 by dotted line curves. The net or resultant torque at different slips is the algebraic sum of individual torques and is shown by the full-line curve in Fig. 20.5.

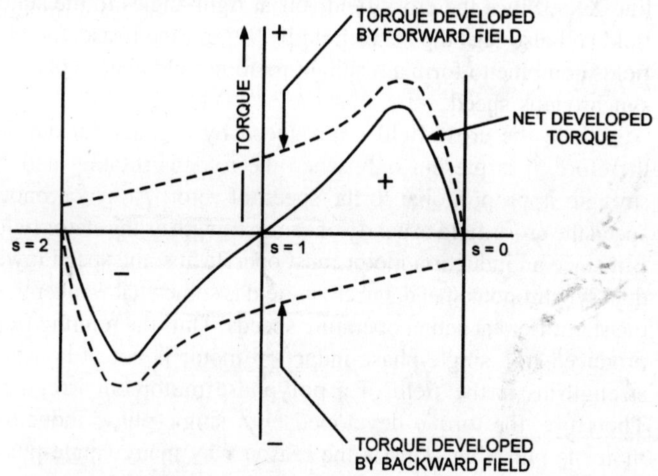

Fig. 20.5 *Torque-Slip Curves For a Single-Phase Induction Motor*

From Fig. 20.5 it is noted that the curve of the resultant torque passes through zero at a slip of unity (or at standstill), thus showing that no starting torque is developed and the motor cannot start rotating independently. However, for other values of slip, the motor develops a net torque in the direction of rotation and the operating conditions with the rotor running in either direction at a certain speed are identical. The breakdown or maximum torque is also less than that what is expected in a 3-phase induction motor. A further point is that the resultant developed torque drops to zero at a speed slightly below synchronous speed whereas in a polyphase induction motor the torque does not drop to zero until synchronous speed is attained. The fact is that the torque developed at synchronous speed is negative. This is due to the torque developed by the backward field.

The effect of increasing rotor resistance is rather different in a single-phase induction motor than in a 3-phase induction motor. In the case of a 3-phase induction motor the value of the maximum torque is not affected by the change in rotor resistance, the only change being the value of slip at which maximum torque occurs. In a single-phase machine, the increase in rotor resistance increases the effectiveness of the backward field, which reduces the breakdown torque, lowers the efficiency and increases the slip corresponding to maximum torque. Double squirrel cages and deep rotor bars are also impractical for the same reasons, and some other means must be adopted to obtain high starting torque.

The performance characteristics of a single-phase induction motor, in general, are of the same nature as those for polyphase induction motor. However, the performance of a single-phase induction motor, in every respect, is somewhat inferior to that of a 3-phase induction motor for the given frame size. This is due to the presence of backward rotating field. It has a lower breakdown torque at larger slip and increased power losses. Further, it has greater power (true as well as apparent) input because of their consumption in the backward rotating field. Even the copper losses occuring in the stator winding of a single-phase induction motor are higher because all the current flows through a single winding. As a consequence, efficiency is lowered and temperature rise is increased. Further, the speed regulation of a single-phase induction motor tends to be poorer than that for a polyphase motor. The normal slip of a single-phase induction motor under load conditions is rather greater than that of the corresponding 3-phase motor and, therefore, power factor of a single-phase motor tends to be lower. For a given power and speed rating, a single-phase motor must therefore, have a larger frame size than that of a 3-phase motor.

The forward field and the rotor's backward reaction field and also the backward field and the rotor's forward reaction field move in opposite directions with relative speeds of double the synchronous speed and develop second harmonic pulsating torques with zero average value. As a consequence single-phase motors tend to be somewhat noisier than three-phase ones, which have no such pulsating torque.

20.4 CROSS-FIELD THEORY

Although the double revolving field theory of a single-phase induction motor gives some idea of its method of operation, it does not give a conception of the reaction, that actually takes place in the machine. The mechanism of torque production is as follows:

Consider standstill conditions with the stator winding connected to a single-phase ac supply. The stator current establishes a field along the vertical axis YY', similar to that shown in Fig. 20.6. During the half cycle that the stator current is flowing in the direction shown, north pole is established on the stator surface at A and south pole at C. During the next half cycle, the stator poles are reversed. Although the stator field strength is varying and reversing its polarity periodically, its action line is always along the vertical line YY'. Thus with alternating current in the stator winding, the stator mmf wave is stationary in space but pulsates in magnitude, the stator field strength alternating in polarity and varying sinusoidally with time.

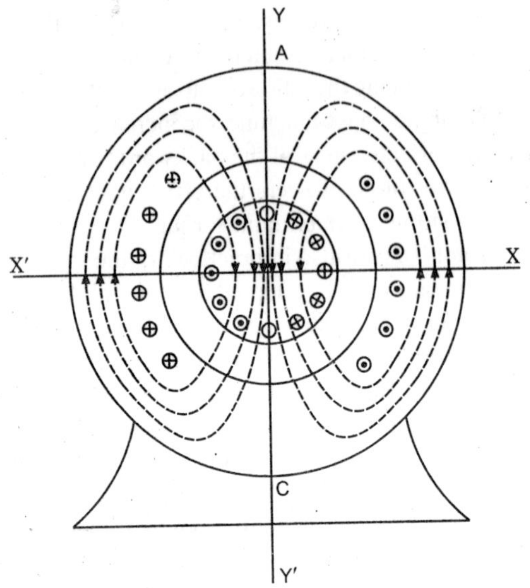

Fig. 20.6 *Action of a Single-Phase Motor at Standstill*

The rotor winding acts like a short-circuited secondary winding of a transformer and, therefore, carry induced currents. The directions of currents which would exist throughout one half cycle are shown in Fig. 20.6. Throughout the other half cycle the currents would be in respective opposite directions. The rotor currents establish poles on the rotor surface, but since these poles are always in direct line (along the vertical axis YY') with the stator poles no torque is developed in either direction. Therefore, no starting torque is developed to start the machine.

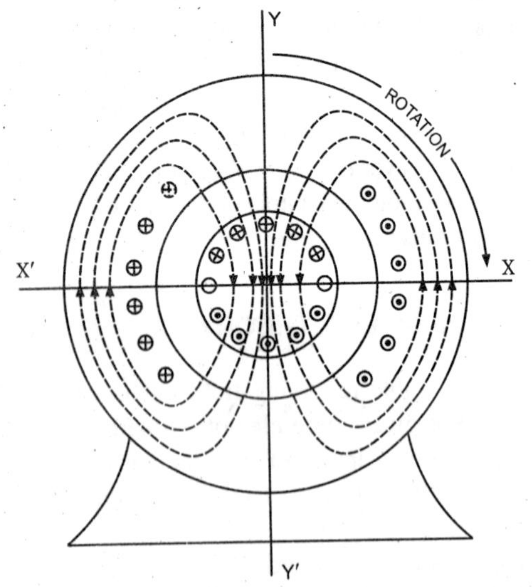

Fig. 20.7 *Action of a Single-Phase Motor Running at Synchronous Speed*

When, however, the single-phase stator winding is connected to single-phase ac supply and the rotor is made to rotate by some external means, the rotor conductors cut across the stator field, causing an emf known as *speed emf*, to be generated in them. In Fig. 20.7 for example, the stator field is downward and the rotor is being turned in clockwise direction, the direction of the generated emfs as determined by Fleming's right hand rule, will be outward in the lower half of the rotor and inward in the upper half of the rotor as indicated by the dots and crosses. During the next half-cycle, the direction of the generated emfs will be reversed. The generated rotor emfs vary in phase with the stator current and flux. The rotor current due to these emfs lags by nearly 90° owing to low resistance and high inductance of the rotor winding. The phase relations of the stator current and flux, the rotor emf and the rotor current and flux are shown in Fig. 20.8.

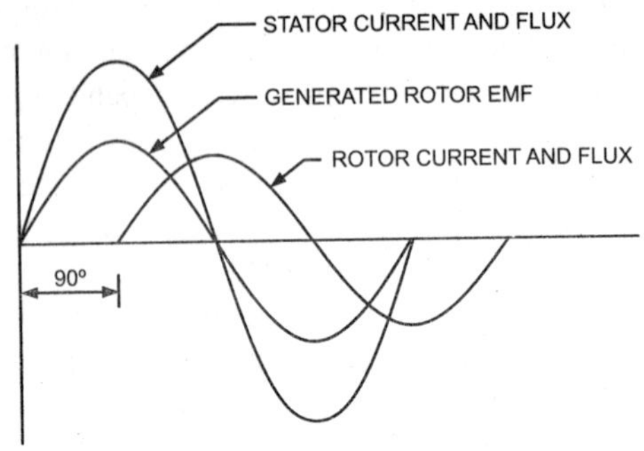

Fig. 20.8

The field created by the rotor currents, as shown in Fig. 20.8, becomes maximum nearly one-fourth cycle after the generated emf has reached its maximum value. Since the field created by rotor currents is at right angles to the field by the stator currents, it is known as *cross–field.* As the rotor currents are caused to flow by an alternating emf, the field resulting from these currents also alternates and its action is always along the line XX'. Since the cross-field acts at right angles to the stator field and also lags the stator field by 90° in time phase, the two fields combine to form a resultant rotating field that revolves at synchronous speed.

Since the cross-field is produced by a generator action, therefore, it is present only when the rotor is rotating and its strength is proportional to the speed of rotor. At synchronous speed the cross-field is nearly of same strength as the stator field. But since an induction motor must operate at some speed lower than synchronous speed, the cross-field is somewhat weaker than the stator field at actual operating speeds. Thus the rotating field produced in a single-phase induction motor is not of constant strength as is the field of a polyphase motor but irregular. Therefore, the torque developed by a single-phase induction motor is pulsating. This is the reason why many single-phase motors are set on rubber or spring mounts to reduce the vibrations and noise which are inherent in such motors.

From the above discussion it may be concluded that at standstill there can be no cross-field, only the pulsating stator field, and therefore, the single-phase induction motor cannot start itself i.e. the inherent starting torque of a single-phase induction motor is zero. If, however, the rotor is made to run by some external means, then it will continue to develop torque as a single-phase induction motor in the direction of rotation.

20.5 EQUIVALENT CIRCUIT (OR CIRCUIT MODEL) OF A SINGLE-PHASE INDUCTION MOTOR

The equivalent circuit of a single-phase induction motor can be developed on the basis of double revolving field theory. To develop the equivalent circuit it is first necessary to consider standstill or blocked rotor conditions. The motor with a blocked

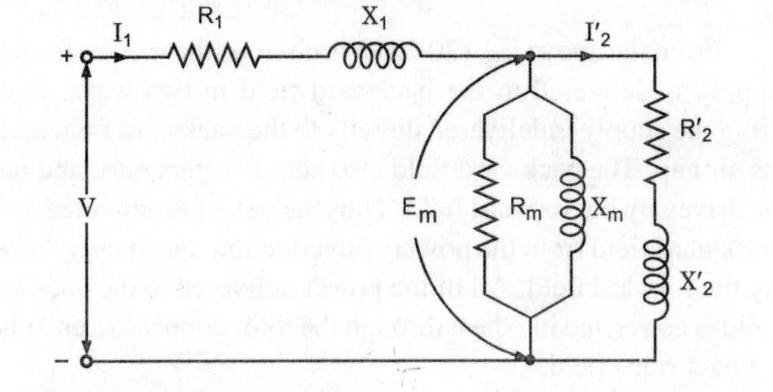

(a) Equivalent Circuit of a Single-Phase Induction Motor at Standstill

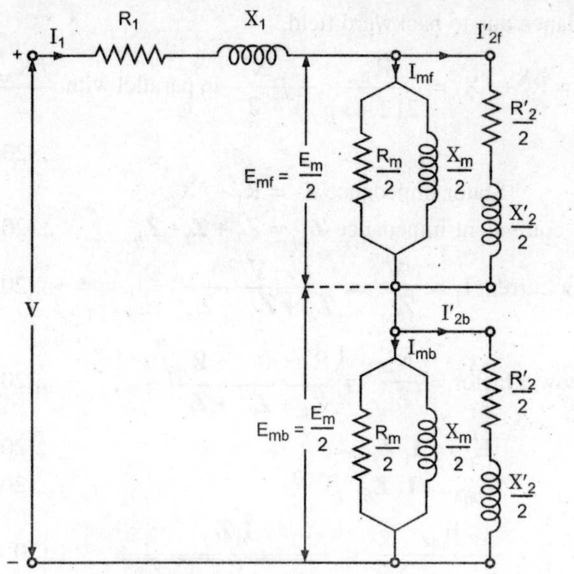

(b) Equivalent Circuit of a Single-Phase Induction Motor at Standstill on The Basis of Double Revolving Field Theory

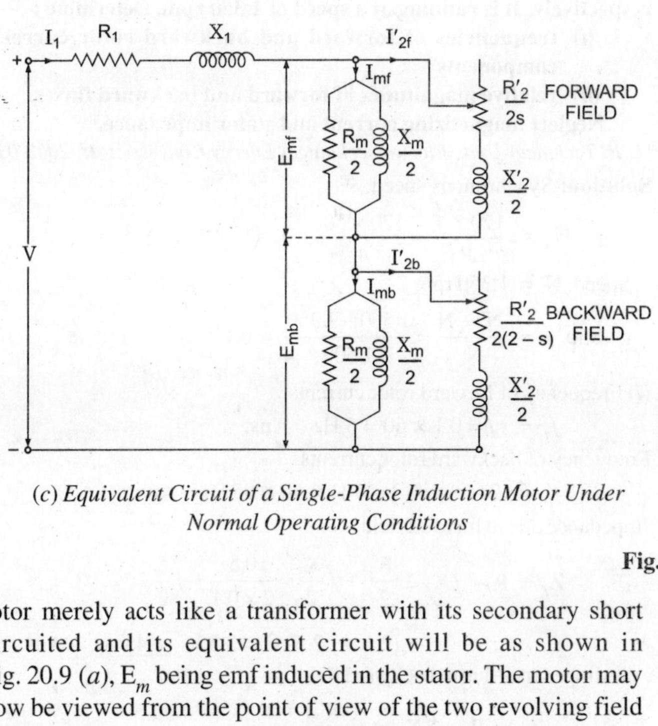

(c) Equivalent Circuit of a Single-Phase Induction Motor Under Normal Operating Conditions

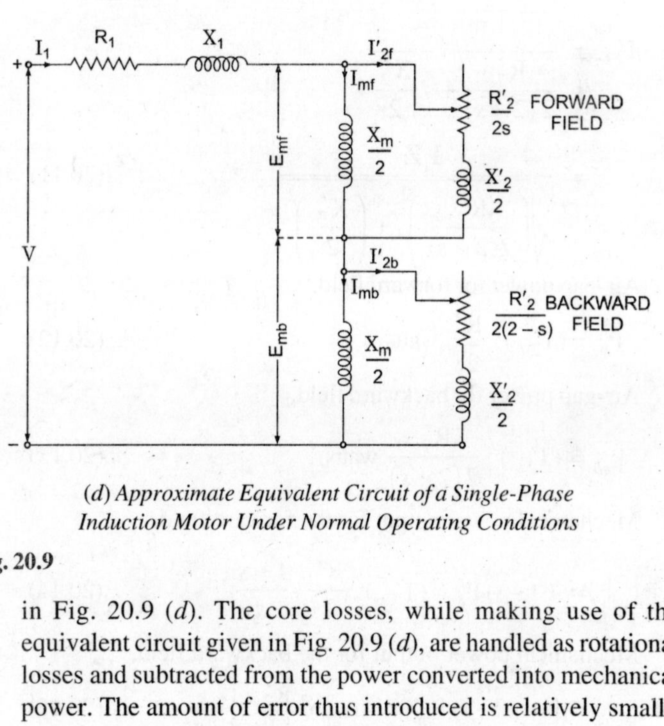

(d) Approximate Equivalent Circuit of a Single-Phase Induction Motor Under Normal Operating Conditions

Fig. 20.9

rotor merely acts like a transformer with its secondary short circuited and its equivalent circuit will be as shown in Fig. 20.9 (*a*), E_m being emf induced in the stator. The motor may now be viewed from the point of view of the two revolving field theory. The two flux components induce in the stator winding the respective emf E_{mf} and E_{mb}. The phasor sum of E_{mf} and E_{mb} equals the applied voltage V (less the voltage drops in stator resistance R_1 and leakage reactance X_1). Since at standstill the two oppositely rotating fields are of same strength, the magnetizing and rotor impedances are divided into two equal halves connected in series, as illustrated in Fig. 20.9 (*b*). Thus, the equivalent circuit of a single-phase induction motor at standstill, on the basis of the double revolving field theory will be that given in Fig. 20.9 (*b*).

When the rotor is running at speed N with respect to forward field, the slip is *s* with respect to forward field and (2 – *s*) with respect to backward rotating field and the equivalent circuit becomes as given in Fig. 20.9 (*c*). For simplification, if the core losses are neglected, the equivalent circuit is modified as given in Fig. 20.9 (*d*). The core losses, while making use of the equivalent circuit given in Fig. 20.9 (*d*), are handled as rotational losses and subtracted from the power converted into mechanical power. The amount of error thus introduced is relatively small.

20.6 PERFORMANCE CHARACTERISTICS

The performance characteristics of a single-phase induction motor, when running with only its main winding energized, can be determined from the equivalent circuit diagram for different values of slip. The calculations will be based on the same principles as were applied to the equivalent circuit for the polyphase induction motor. In calculation for single-phase induction motor the core losses will be taken as rotational losses.

Impedance due to forward field

$$\mathbf{Z}_f = R_f + jX_f = \frac{R'_2}{2s} + j\frac{X'_2}{2} \text{ in parallel with } \frac{jX_m}{2} \quad ...(20.3)$$

Impedance due to backward field,

$$\mathbf{Z}_b = R_b + jX_b = \frac{R'_2}{2(2-s)} + j\frac{X'_2}{2} \text{ in parallel with } \frac{jX_m}{2} \quad ...(20.4)$$

Stator impedance $\mathbf{Z}_1 = R_1 + jX_1$

Total equivalent impedance $\mathbf{Z}_{eq} = \mathbf{Z}_1 + \mathbf{Z}_f + \mathbf{Z}_b$ ...(20.5)

$$\text{Motor current } I_1 = \frac{\mathbf{V}}{\mathbf{Z}_{eq}} = \frac{\mathbf{V}}{\mathbf{Z}_1 + \mathbf{Z}_f + \mathbf{Z}_b} \quad ...(20.6)$$

$$\text{Power factor} = \frac{R_{eq}}{\mathbf{Z}_{eq}} = \frac{R_1 + R_f + R_b}{\mathbf{Z}_1 + \mathbf{Z}_f + \mathbf{Z}_b} \quad ...(20.7)$$

$$\mathbf{E}_{mf} = \mathbf{I}_1 \mathbf{Z}_f \quad ...(20.8)$$

$$\mathbf{E}_{mb} = \mathbf{I}_1 \mathbf{Z}_b \quad ...(20.9)$$

$$I'_{2f} = \frac{E_{mf}}{\frac{R'_2}{2s} + j\frac{X'_2}{2}} = \frac{\mathbf{I}_1\mathbf{Z}_f}{\sqrt{\left(\frac{R'_2}{2s}\right)^2 + \left(\frac{X'_2}{2}\right)^2}} \quad ...(20.10)$$

$$I'_{2b} = \frac{E_{mb}}{\frac{R'_2}{2(2-s)} + j\frac{X'_2}{2}}$$

$$= \frac{\mathbf{I}_1\mathbf{Z}_b}{\sqrt{\left(\frac{R'_2}{2(2-s)}\right)^2 + \left(\frac{X'_2}{2}\right)^2}} \quad ...(20.11)$$

Air-gap power for forward field,

$$P_{gf} = (I'_{2f})^2 \frac{R'_2}{2s} \text{ watts} \quad ...(20.12)$$

Air-gap power for backward field,

$$P_{gb} = (I'_{2b})^2 \frac{R'_2}{2(2-s)} \text{ watts} \quad ...(20.13)$$

Mechanical power output for the forward field,

$$P_{mech\,f} = (1-s)P_{gf} = (I'_{2f})^2 \frac{R'_2}{2} \cdot \frac{1-s}{s} \quad ...(20.14)$$

Mechanical power output for the backward field,

$$P_{mech\,b} = [1-(2-s)P_{gb}] = -(I'_{2b})^2 \frac{R'_2}{2} \cdot \frac{1-s}{2-s} \quad ...(20.15)$$

Mechanical power output,

$$P_{mech\,net} = P_{mech\,f} + P_{mech\,b}$$

$$= (1-s)P_{gf} + [1-(2-s)P_{gb}]$$

$$= (1-s)[P_{gf} - P_{gb}] \quad ...(20.16)$$

Torque developed for the forward field

$$= \frac{P_{gf}}{2\pi N_s/60} = \frac{9.55 P_{gf}}{N_s} \quad ...(20.17)$$

Torque developed for the backward field

$$= \frac{9.55 P_{gb}}{N_s} \quad ...(20.18)$$

$$\text{Net torque developed} = \frac{9.55}{N_s}\left(P_{gf} - P_{gb}\right) \quad ...(20.19)$$

Net power output = $P_{mech\,net}$ – friction and windage losses – core losses ...(20.20)

$$\text{Motor efficiency, } \eta = \frac{\text{Output power}}{\text{Input power}} \times 100 \quad ...(20.21)$$

Power delivered to the backward field

$$= (2-s)P_{gb}$$

$$= P_{gb} + (1-s)P_{gb} = P_{gb} - P_{mech\,b} \quad ...(20.22)$$

From the above Eq. (20.22) it is obvious that power from the supply is delivered to the backward field in two ways. Power from the supply is delivered directly to the backward field across its air gap. The backward field also acts as a generator and must be driven by the forward field. Thus the power is delivered to the backward field from the power converted into mechanical power by the forward field. All of the power delivered to the backward field is converted into heat through the rotor copper losses caused by backward field.

Example 20.1. A 230 V, 4-pole, 50 Hz, 1-phase induction motor has an effective rotor resistance and leakage reactance of 0.5 Ω and 5 Ω respectively. It is running at a speed of 1,350 rpm. Determine :

(i) frequencies of forward and backward rotor current components

(ii) relative magnitudes of forward and backward fluxes.

Neglect magnetizing current and stator impedance.

[*U.P. Technical Univ. Electromechanical Energy Conversion-II, 2002-03*]

Solution: Synchronous speed,

$$N_s = \frac{120 \times f}{P} = \frac{120 \times 50}{4} = 1{,}500 \text{ rpm}$$

Speed, N = 1,350 rpm

$$\text{Slip, } s = \frac{N_s - N}{N_s} = \frac{1{,}500 - 1{,}350}{1{,}500} = 0.1$$

(*i*) Frequency of forward rotor currents,

$f'_f = sf = 0.1 \times 50 = 5$ Hz **Ans.**

Frequency of backward rotor currents,

$f'_b = (2-s)f = (2-0.1) \times 50 = 95$ Hz **Ans.**

Impedance due to forward field,

$$\mathbf{Z}_f = R_f + jX_f = \frac{R'_2}{2s} + j\frac{X'}{2} = \frac{0.5}{2 \times 0.1} + j\frac{5}{2}$$

$= (2.5 + j\,2.5)\ \Omega = 3.536\ \angle 45°\ \Omega$ **Ans.**

Impedance due to backward field,

$$Z_b = R_b + jX_b = \frac{R'_2}{2(2-s)} + j\frac{X'}{2} = \frac{0.5}{2(2-0.1)} + j\frac{5}{2}$$

$= (0.1316 + j\,2.5)\ \Omega = 2.5035\ \angle 86.99°\ \Omega$

(*ii*) Now $\dfrac{\phi_f}{\phi_b} = \dfrac{E_f}{E_b} = \dfrac{I_1 Z_f}{I_1 Z_b} = \dfrac{Z_f}{Z_b} = \dfrac{3.536}{2.5035} = 1.4$ **Ans.**

Example 20.2. A 125-W, 4-pole, 110-V, 50-Hz single-phase induction motor delivers rated output at a slip of 6%. The total copper loss at full load is 25 watts. Calculate the full-load efficiency and the rotor copper loss caused by the backward field. Rotational losses may be assumed to be 25 watts. Neglect stator copper loss.

[*U.P. Technical Univ. Electromechanical Energy Conversion-II, 2010-11*]

Solution: Output of motor = 125 watts

Rotational losses = 25 watts

Mechanical power developed by rotor,

P_{mech} = Output + rotational losses = 125+25 = 150 watts

Full-load slip, s_f = 6% or 0.06

$$\text{Power input to rotor} = P_{gf} - P_{gb} = \frac{P_{mech}}{1-s_f} = \frac{150}{1-0.06} = 160 \text{ watts} \quad ...(i)$$

$$\text{and total copper losses} = s\,P_{gf} + (2-s)\,P_{gb}$$

$$\text{or } 25 = 0.06\,P_{gf} + 1.94\,P_{gb}$$

$$\text{or } 0.06\,P_{gf} + 1.94\,P_{gb} = 25 \text{ watts} \quad ...(ii)$$

Solving Eqs. (*i*) and (*ii*) we have rotor copper loss caused by backward field

$$1.94\,P_{gb} = \frac{25-(0.06\times 160)}{2}\times 1.94 = 14.94 \text{ watts } \textbf{Ans.}$$

Input power to motor

= Output power of motor + rotational losses

+ rotor copper losses + stator copper losses

+ core losses

Neglecting stator copper losses and core losses

Input power to motor = 125 + 25 + 25 = 175 watts

Full-load efficiency,

$$\eta = \frac{\text{Output}}{\text{Input}}\times 100 = \frac{125}{175}\times 100 = 71.43\% \textbf{ Ans.}$$

Example 20.3. A 230 V, 50 Hz, 4-pole single-phase induction motor has the following equivalent circuit impedances.

$$\mathbf{R_{1m} = 2.2\ \Omega, \quad R'_2 = 4.5\ \Omega}$$

$$\mathbf{X_{1m} = 3.1\ \Omega, \quad X'_2 = 2.6\ \Omega, \quad X_m = 80\ \Omega}$$

Friction, windage and core loss = 40 W

For a slip of 0.03 pu, calculate (*i*) input current (*ii*) power factor (*iii*) developed power (*iv*) output power (*v*) efficiency.

[*U.P. Technical Univ. Electromechanical Energy Conversion-II, 2008-09; W.B. Univ. of Technology Electrical Machines-II, 2012-13*]

Solution: Resistance of stator main winding, $R_1 = 2.2\ \Omega$

Leakage reactance of main winding, $X_1 = 3.1\ \Omega$

Magnetising reactance of stator main winding,

$$X_m = 80\ \Omega$$

Rotor resistance, $R'_2 = 4.5\ \Omega$

Rotor standstill reactance, $X'_2 = 2.6\ \Omega$

slip, $s = 0.03$

$(2-s) = (2-0.03) = 1.97$

Stator impedance, $\mathbf{Z}_1 = (2.2 + j\,3.1)\ \Omega$

Forward field impedance,

$$Z_f = \frac{j\dfrac{X_m}{2}\left[\dfrac{R'_2}{2s}+j\dfrac{X'_2}{2}\right]}{\dfrac{R'_2}{2s}+j\dfrac{X_m+X'_2}{2}}$$

$$= \frac{j\dfrac{80}{2}\left[\dfrac{4.5}{2\times 0.03}+j\dfrac{2.6}{2}\right]}{\dfrac{4.5}{2\times 0.03}+j\dfrac{80+2.6}{2}}$$

$$= \frac{j40(75+j1.3)}{75+j41.3} = \frac{40\angle 90^\circ\times 75.011\angle 0.993^\circ}{85.619\angle 28.84^\circ}$$

$$= 35.044\angle 62.153^\circ\ \Omega = (16.37 + j30.98)\ \Omega$$

Backward field impedance,

$$Z_b = \frac{j\dfrac{X_m}{2}\left[\dfrac{R'_2}{2(2-s)}+j\dfrac{X'_2}{2}\right]}{\dfrac{R'_2}{2(2-s)}+j\dfrac{X_m+X'_2}{2}}$$

$$= \frac{j40\left[\dfrac{4.5}{2\times 1.97}+j\dfrac{2.6}{2}\right]}{\dfrac{4.5}{2\times 1.97}+j\dfrac{80+2.6}{2}}$$

$$= \frac{j40\,(1.142+j\,1.3)}{1.142+j41.3}$$

$$= \frac{40\angle 90^\circ\times 1.73\angle 48.7^\circ}{41.316\angle 88.42^\circ} = 1.675\angle 50.28^\circ\ \Omega$$

$$= (1.07 + j\,1.288)\ \Omega$$

Total equivalent impedance,

$$Z_{eq} = \mathbf{Z}_1 + \mathbf{Z}_f + \mathbf{Z}_b$$

$$= (2.2 + j\,3.1) + (16.37 + j\,30.98) + (1.07 + j\,1.288)$$

$$= 19.64 + j\,35.368 = 40.455\angle 60.956\ \Omega$$

(*i*) Star main winding current,

$$I_m = \frac{V_m}{Z} = \frac{230\angle 0^\circ}{40.455\angle 60.956^\circ}$$

$$= 5.685\angle -60.956^\circ \text{ A} \quad \textbf{Ans.}$$

(*ii*) Power factor, $\cos\phi = \text{Cos}(-60.956^\circ) = 0.4855$ (lagging) **Ans.**

(*iii*) Power developed, $P_d = I_m^2 (R_f - R_b)(1-s)$

$= (5.685)^2 (16.37 - 1.07)(1 - 0.03)$

$= 479.65$ W **Ans.**

(*iv*) Output power, $P_{out} = P_d -$ friction, windage and core loss

$= 479.65 - 40 = 439.65$ W

Input power, $P_{in} = V\,I_m \cos\phi$

$= 230 \times 5.685 \times 0.4855 = 634.8$ W

(*v*) Motor efficiency, $\eta = \dfrac{P_{out}}{P_{in}}\times 100 = \dfrac{439.65}{634.8}\times 100$

$= 69.26\%$ **Ans.**

Example 20.4. Draw and explain the equivalent circuit of 1-phase induction motor. A single-phase 230 V, 50 Hz, 6 pole induction motor has the following parameters :

$$\mathbf{r_1 = 0.12\ \Omega,\ r'_2 = 0.14\ \Omega,\ x_1 = x'_2 = 0.25\ \Omega,\ x_m = 15\ \Omega.}$$

If the core loss is 250 W and friction and windage losses are 500 W, determine the efficiency and torque at *s* = 0.05.

Solution: Core loss current, $I_e = \dfrac{250}{230} = 1.087$ A

Half value of core-loss equivalent resistance,

$$\frac{R_m}{2} = \frac{230}{1.087} = 211.6\ \Omega$$

For Forward Field [Fig. 20.9 (*c*)]

Conductance of core-loss branch

$$= \frac{1}{R_m/2} = \frac{1}{211.6} = 0.004726 \text{ S}$$

Susceptance of magnetising branch

$$= \frac{-j}{\dfrac{X_m}{2}} = \frac{-j}{15/2} = -j0.13333$$

$$\text{Rotor admittance} = \frac{1}{\dfrac{r_2'}{2s}+\dfrac{jx_2'}{2}} = \frac{1}{\dfrac{0.14}{2\times 0.05}+j\dfrac{0.25}{2}}$$

$$= \frac{1}{1.4+j0.125}$$

$$= (0.70864 - j0.0327)\ \text{S}$$

$$\text{Admittance, } \mathbf{Y}_f = (0.004726 - j0.13333 + 0.70864 - j0.0327)\ \text{S}$$

$$= 0.71366 - j0.166033\ \text{S}$$

$$\text{Impedance, } \mathbf{Z}_f = \frac{1}{\mathbf{Y}_f} = \frac{1}{0.71366 - j0.166033}$$

$$= (1.3293 + j0.30926)\ \Omega$$

$$= 1.3648 \angle 13.097°\ \Omega$$

For Backward Field

$$\text{Rotor admittance} = \frac{1}{\dfrac{r_2'}{2(2-s)}+\dfrac{jx_2'}{2}}$$

$$= \frac{1}{\dfrac{0.14}{2(2-0.05)}+\dfrac{j0.25}{2}}$$

$$= \frac{1}{0.0359+j0.125} = (2.12255 - j7.39)\ \text{S}$$

$$\text{Admittance, } \mathbf{Y}_b = (0.004726 - j0.133333 + 2.12255 - j7.39)$$

$$= (2.127276 - j7.52333)\ \text{S}$$

$$\text{Impedance, } \mathbf{Z}_b = \frac{1}{\mathbf{Y}_b} = \frac{1}{2.127276 - j7.52333}$$

$$= (0.0348 + j0.12308)\ \Omega$$

$$= 0.1279 \angle 74.21°\ \Omega$$

Equivalent impedance,

$$\mathbf{Z}_{eq} = \mathbf{Z}_1 + \mathbf{Z}_f + \mathbf{Z}_b$$

$$= (0.12 + j0.25) + (1.3293 + j0.30926) + (0.0348 + j0.12308)\ \Omega$$

$$= (1.4841 + j0.68234)\ \Omega = 1.6334 \angle 24.69°\ \Omega$$

$$\text{Input current, } I_1 = \frac{V}{Z_{eq}} = \frac{230}{1.6334} = 140.8\ \text{A}$$

$$\text{Power input} = V_1 I_1 \cos\phi = 230 \times 140 \cos 24.69° = 29{,}256\ \text{W}$$

$$E_{mf} = I_1 \times Z_f = 140.8 \times 1.3648 = 192.173\ \text{V}$$

$$E_{mb} = I_1 \times Z_b = 140.8 \times 0.1279 = 18\ \text{V}$$

$$I'_{2f} = \frac{E_{mf}}{\sqrt{\left(\dfrac{r_2'}{2s}\right)^2 + \left(\dfrac{x_2'}{2}\right)^2}} = \frac{192.173}{\sqrt{(1.4)^2 + (0.125)^2}}$$

$$= 136.72\ \text{A}$$

$$I'_{2b} = \frac{E_{mb}}{\sqrt{\left(\dfrac{r_2'}{2(2-s)}\right)^2 + \left(\dfrac{x_2'}{2}\right)^2}}$$

$$= \frac{18}{\sqrt{0.0359^2 + 0.125^2}} = 138.4\ \text{A}$$

Air-gap power for forward field,

$$P_{gf} = (I'_{2f})^2 \times \frac{r_2'}{2s} = (136.72)^2 \times \frac{0.14}{2\times 0.05} = 26{,}170\ \text{W}$$

Air-gap power for backward field,

$$P_{gb} = (I'_{2b})^2 \times \frac{r_2'}{2(2-s)} = (138.4)^2 \times \frac{0.14}{2\times 1.95} = 688\ \text{W}$$

Power developed,

$$P_{mech} = (1-s)(P_{gf} - P_{gb})$$

$$= (1 - 0.05)(26{,}170 - 688) = 2{,}4208\ \text{W}$$

$$\text{Shaft power} = P_{mech} - \text{friction and windage losses}$$

$$= 2{,}4208 - 500 = 23{,}708\ \text{W}$$

$$\text{Efficiency, } \eta = \frac{\text{Shaft power}}{\text{Power input}} \times 100 = \frac{23{,}708}{29{,}256} \times 100 = 81\%\ \textbf{Ans.}$$

$$\text{Torque, T} = \frac{\text{Shaft power}}{2\pi N/60}$$

$$= \frac{23{,}708 \times 60}{2\pi \times \dfrac{120 \times 50}{6}(1-0.05)} = \frac{23{,}708 \times 60}{2\pi \times 950}$$

$$= 238.3\ \text{N-m}\ \textbf{Ans.}$$

Example 20.5. A 230 V, 6-pole, 50 Hz, 375 W single-phase induction motor has the following parameters at standstill:

$r_1 = 11.4\ \Omega$, $x_1 = 14.3\ \Omega$, $X_0 = 275\ \Omega$

$r_2' = 13.8\ \Omega$, $x_2' = 14.3\ \Omega$, Ignore R_0

Find the input current, power and power factor when the motor is operating at $s = 0.06$.

[*U.P.S.C. I.E.S. Electrical Engineering-II, 2010*]

Solution: Slip, $s = 0.06$

$$(2 - s) = (2 - 0.06) = 1.94$$

$$\text{Stator impedance, } \mathbf{Z}_1 = (11.4 + j\,14.3)\ \Omega$$

Forward field impedance,

$$\mathbf{Z}_f = \frac{j\dfrac{x_m}{2}\left[\dfrac{r_2'}{2s}+\dfrac{jx_2'}{2}\right]}{\dfrac{r_2'}{2s}+j\left(\dfrac{x_m + x_2'}{2}\right)} = \frac{j\dfrac{275}{2}\left[\dfrac{13.8}{2\times 0.06}+j\dfrac{14.3}{2}\right]}{\dfrac{13.8}{2\times 0.06}+j\left(\dfrac{275+14.3}{2}\right)}$$

$$\text{or } \mathbf{Z}_f = \frac{-983.125 + j15{,}812.5}{115 + j144.65} = \frac{15{,}843 \angle 93.56°}{184.7935 \angle 51.51°}$$

$$= 85.74 \angle 42.05°\ \Omega$$

Similarly backward field impedance,

$$\mathbf{Z}_b = \frac{\dfrac{j\,x_m}{2}\left[\dfrac{r_2'}{2-s}+\dfrac{j\,x_2'}{2}\right]}{\dfrac{r_2'}{2s}+j\left(\dfrac{x_m + x_2'}{2}\right)} = \frac{j\dfrac{275}{2}\left[\dfrac{13.8}{2\times(2-0.06)}+j\dfrac{14.3}{2}\right]}{\dfrac{13.8}{2\times(2-0.06)}+j\dfrac{275+14.3}{2}}$$

$$\mathbf{Z}_b = \frac{-983.125 + j\,489.046}{3.5567 + j\,144.65} = \frac{1{,}098.044 \angle 153.55°}{144.694 \angle 88.59°}$$

$$= 7.59 \angle 64.96°\ \Omega$$

Total equivalent impedance

$$\mathbf{Z}_{eq} = \mathbf{Z}_1 + \mathbf{Z}_f + \mathbf{Z}_b$$

$$= (11.4 + j\,14.3) + 85.74 \angle 42.05° + 7.59 \angle 64.96°$$

$$= (11.4 + j\,14.3) + (63.667 + j\,57.427) + (3.212 + j\,6.877)$$

$$= (78.279 + j\,78.604) = 110.9 \angle 45.12°$$

Input current,

$$\mathbf{I} = \frac{\mathbf{V}}{\mathbf{Z}_{eq}} = \frac{230 \angle 0°}{110.9 \angle 45.12°} = 2.074 \angle -45.12°\ \text{A}$$

So input current, I = 2.074 A **Ans.**

Input power factor, $\cos\phi = \text{Cos}(-45.12°) = 0.706$ (lagging) **Ans.**

$$\text{Input power} = VI\cos\phi$$

$$= 230 \times 2.074 \times 0.706 = 336.8\ \text{W}\ \textbf{Ans.}$$

Example 20.6. The constants of a $\frac{1}{4}$ hp, 230 V, 4-pole, 60 Hz, single phase induction motor are as follows :

Stator resistance, $R_1 = 10.0\ \Omega$

Stator reactance, $X_1 = 12.8\ \Omega$

Rotor resistance referred to stator, $R'_2 = 11.65\ \Omega$

Rotor reactance referred to stator, $X_2' = 12.8\ \Omega$

Magnetising reactance, $X_m = 258.0\ \Omega$

The total load is such that the machine runs at 3% slip, when the applied voltage is at 210 V. The iron losses are 35.5 watts at 210 V

Calculate (*i*) input current, (*ii*) power developed (*iii*) shaft power (if mechanical losses are 7 watts) and (*iv*) efficiency.

Solution: Core loss current, $I_e = \dfrac{35.5}{210} = 0.169$ A

$$\frac{R_m}{2} = \frac{210}{0.169} = 1{,}242\ \Omega$$

For forward Field [Fig. 20.9 (*c*)]

Conductance of core-loss branch

$$= \frac{1}{\frac{R_m}{2}} = \frac{1}{1{,}242} = 0.0008 \text{ S}$$

Susceptance of magnetising branch

$$= \frac{-j}{\frac{X_m}{2}} = \frac{1}{\frac{258}{2}} = -j\,0.00775 \text{ S}$$

$$\text{Rotor admittance} = \frac{1}{\frac{R_2'}{2s} + \frac{jX_2'}{2}} = \frac{1}{\frac{11.65}{2\times0.03} + \frac{j12.8}{2}}$$

$$= \frac{1}{194.17 + j6.4}$$

$$= (0.005145 - j\,0.00017) \text{ S}$$

Admittance,

$$Y_f = (0.0008 - j\,0.00775 + 0.005145 - j\,0.00017) \text{ S}$$
$$= (0.005945 - j\,0.00792)\ s$$

Impedance,

$$Z_f = \frac{1}{Y_f} = \frac{1}{0.005945 - j\,0.00792}$$

$$= 60.62 + j\,80.759 = 100.979\ \angle 53.1^\circ\ \Omega$$

For backward field

Rotor admittance

$$= \frac{1}{\frac{R_2'}{2(2-s)} + \frac{jX_2'}{2}} = \frac{1}{\frac{11.65}{2(2-0.03)} - \frac{j12.8}{2}}$$

$$= \frac{1}{2.957 + j6.4} = (0.0595 - j0.012876) \text{ S}$$

$$\text{Admittance, } Y_b = (0.0008 - j\,0.00775 + j\,0.0595 - j\,0.12876) \text{ S}$$
$$= (0.0603 - j\,0.1365) \text{ S}$$

$$\text{Impedance, } Z_b = \frac{1}{Y_b} = \frac{1}{0.0603 - j0.1365}$$

$$= 2.708 + j\,6.13 = 6.70\ \angle 66.17^\circ\ \Omega$$

Equivalent impedance,

$$Z_{eq} = Z_1 + Z_f + Z_b$$
$$= (10 + j\,12.8) + (60.62 + j\,80.759) + j\,(2.708 + j\,6.13)$$
$$= 73.323 + j\,99.689 = 123.75\ \angle 53.66^\circ\ \Omega$$

(*i*) Input current, $I_1 = \dfrac{V}{Z_{eq}} = \dfrac{230}{123.75} = 1.8586$ A **Ans.**

$$\text{Power input} = VI_1 \cos\phi = 230 \times 1.8586 \times \cos 53.66^\circ = 253.3 \text{ W}$$
$$E_{mf} = I_1 \times Z_f = 1.8586 \times 100.979 = 187.68 \text{ V}$$
$$E_{mb} = I_1 \times Z_b = 1.8586 \times 6.70 = 12.45 \text{ V}$$

$$I'_{2f} = \frac{E_{mf}}{\sqrt{\left(\frac{R_2'}{2s}\right)^2 + \left(\frac{X_2'}{2}\right)^2}} = \frac{187.68}{\sqrt{(194.17)^2 + (6.4)^2}} = 0.966 \text{ A}$$

$$I'_{2b} = \frac{E_{mb}}{\sqrt{\left(\frac{R_2'}{2(2-s)}\right)^2 + \left(\frac{X_2'}{2}\right)^2}} = \frac{12.45}{\sqrt{(2.2957)^2 + (6.4)^2}} = 1.831 \text{ A}$$

Air-gap power for forward field,

$$P_{gf} = (I'_{2f})^2 \times \frac{R_2'}{2s} = (0.966)^2 \times \frac{11.65}{2\times0.03} = 181.2 \text{ W}$$

Air-gap power for backward field,

$$P_{gb} = (I'_{2b})^2 \times \frac{R_2'}{2(2-s)} = (1.831)^2 \times \frac{11.65}{2\times1.97} = 9.9 \text{ W}$$

(*ii*) Power developed

$$P_{mech\ net} = (1-s)(P_{gf} - P_{gb})$$
$$= (1 - 0.03)(181.2 - 9.9) = 166.2 \text{ W } \textbf{Ans.}$$

(*iii*) Shaft power $= P_{mech\ net}$ – mechanical losses

$$= 166.2 - 7 = 159.2 \text{ W } \textbf{Ans.}$$

$$(iv) \text{ Efficiency, } \eta = \frac{\text{Power output}}{\text{Power input}} \times 100$$

$$= \frac{159.2}{253.3} \times 100 = 62.85\% \textbf{ Ans.}$$

Example 20.7. A 220 V, 6-pole, 50 Hz, single winding, single-phase induction motor has the following equivalent circuit parameters as referred to the stator.

$$r_{1m} = 3\ \Omega \quad x_{1m} = 5\ \Omega \quad r_2 = 1.5\Omega \quad x_2 = 2\ \Omega$$

Neglect the magnetizing current. When the motor runs at 97% of synchronous speed, compute the ratio E_{mf}/E_{mb} and ratio T_f/T_b.

[U.P. Technical Univ. Electromechanical Energy Conversion-II, 2006-07]

Solution: Referring to Fig. 20.9(*d*)

$$R_1 = 3\ \Omega \quad X_1 = 5\ \Omega \quad R_2' = 1.5\ \Omega \quad X_2' = 2\ \Omega$$

$$\text{Slip, } s = \frac{N_s - N}{N_s} = \frac{N_s - 0.97N_s}{N_s} = 0.03$$

Impedance due to forward field,

$$Z_f = \frac{R_2'}{2s} + \frac{jX_2'}{2} \text{ in parallel with } \frac{jX_m}{2}$$

Since magnetizing current is negligible, $X_m \to \infty$ and

$$Z_f = \frac{R_2'}{2s} + \frac{jX_2'}{2} = \frac{1.5}{2\times0.03} + \frac{j2}{2} = 25 + j1 = 25.02\ \angle 2.3^\circ\ \Omega$$

Similarly impedance due to backward field,

$$Z_b = \frac{R_2'}{2(2-s)} + \frac{jX_2'}{2} = \frac{1.5}{2(2-0.03)} + \frac{j2}{2}$$

$$= 0.3807 + j1 = 1.07 \angle 69.15°\Omega$$

$$\text{Ratio } \frac{E_{mf}}{E_{mb}} = \frac{I_1 Z_f}{I_1 Z_b} = \frac{Z_f}{Z_b} = \frac{25.02}{1.07} = 23.383 \quad \textbf{Ans.}$$

$$\text{Ratio } \frac{T_f}{T_b} = \frac{P_{gf}}{P_{gb}} = \frac{(I_{2f}')^2 \frac{R_2'}{2s}}{(I_{2b}')^2 \frac{R_2'}{2(2-s)}} = \frac{2-s}{s} \quad \because \text{ Here } I_{2f}' = I_{2b}' = I_1$$

$$= \frac{2-0.03}{0.03} = 65.67 \quad \textbf{Ans.}$$

20.7 DETERMINATION OF EQUIVALENT CIRCUIT PARAMETERS BY TESTS

The performance (efficiency, losses and power factor) of single-phase induction motors is obtained by applying load to it and measuring output, input, speed and current, as such motors are usually of small ratings. However, to determine the constants of the machine it is necessary to perform various tests. The most common tests performed on single-phase induction motors for determination of various constants are:

(*a*) **Measurement of Stator Winding Resistance.** The resistance of each stator winding is measured separately and in case of capacitor start motor the condenser is omitted.

(*b*) **Blocked Rotor Test.** This test is performed with main winding and starting winding separately and input voltage, amperes, and watts are measured in each case. This test is sometimes performed at reduced voltage (about 40 per cent of rated voltage) to avoid excessive short-circuit current and heating.

(*c*) **No-Load Test.** This test is performed at rated voltage, exciting the main winding only and input current, voltage and power, as well as the speed, are measured.

(*d*) **Determination of Friction and Windage Losses.** In this test the voltage applied to the motor is reduced until the motor will just run and the power input is measured. The power input, just sufficient to keep the motor running on no load, gives approximately the friction and windage losses.

When the rotor is blocked

Slip, $s = 1.0$. In the equivalent circuit $X_m/2$ can be neglected because R_2', $X_2' < X_m$. Further iron loss is also negligible as flux during blocked rotor test is small. The equivalent circuit on blocked rotor test with instruments is reduced to that shown in Fig. 20.10.

If the readings of the ammeter, voltmeter and wattmeter are I, V and P respectively,

$$\text{then } R_2' + R_1 = \frac{P}{I^2} \qquad \text{... (20.23)}$$

R_1 is measured separately with dc by ammeter-voltmeter method. This dc value is usually taken as the ac value because skin effect at normal supply frequency is negligible in small single phase induction motors.

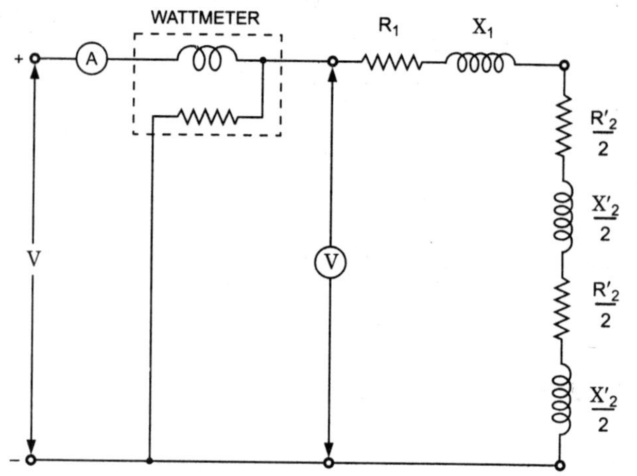

Fig. 20.10

$$\text{So } R_2' = \frac{P}{I^2} - R_1 \qquad \text{... (20.24)}$$

Now equivalent impedance,

$$Z_{eq} = \sqrt{(R_1 + R_2')^2 + (X_1 + X_2')^2} = \frac{V}{I} \qquad \text{... (20.25)}$$

$$\text{or } X_1 + X_2' = \sqrt{Z_{eq}^2 - (R_1 + R_2')^2} \qquad \text{...(20.26)}$$

Assuming $X_1 = X_2'$ we have

$$X_1 = X_2' = \frac{X_1 + X_2'}{2} \qquad \text{... (20.27)}$$

In no-load test

Slip, s is very small, so $R_2'/2s$ in the equivalent circuit [Fig. 20.9 (*d*)] may be considered as infinity and the corresponding rotor circuit may be taken as open circuit. $X_m/2$ in the backward field circuit may also be omitted because $X_m > R_2'/2$, X_2' and $\frac{R_2'}{2(2-s)}$ may be taken as $\frac{R_2'}{4}$ as s is very small. Thus the equivalent circuit on no-load test with the measuring instruments becomes as given in Fig. 20.11. A fictitious resistance R' is added in the circuit shown in Fig. 20.11 to represent stray power loss*, which occur in addition to power losses in the resistances R_1 and $R_2'/4$. However, current through R' negligibly small, because R' is very high.

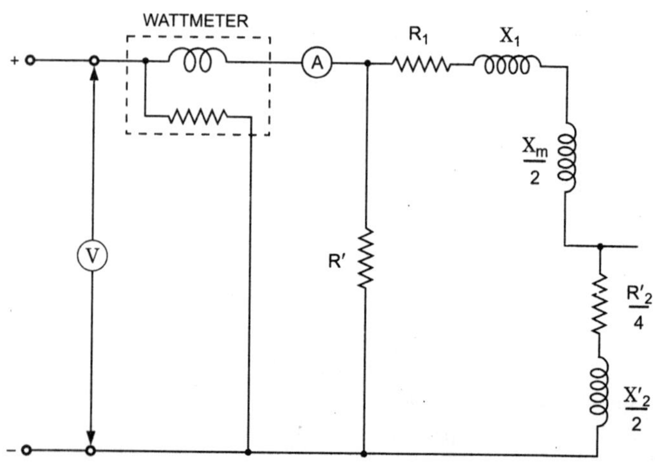

Fig. 20.11

* Stray power losses are disregarded in the equivalent circuit of the motor.

If the readings of voltmeter, ammeter and wattmeter are V_0, I_0 and P_0 respectively, then

$$\text{Stray power loss, } P_s = P_0 - I_0^2\left[R_1 + \frac{R'_2}{4}\right] \quad \text{... (20.28)}$$

The equivalent motor impedance, excluding R′, which carries negligibly small current in comparison to I, can be computed from the relation

$$Z_0 = \frac{V_0}{I_0} \quad \text{... (20.29)}$$

$$\text{and } X_1 + \frac{X'_2}{2} + \frac{X_m}{2} = \sqrt{Z_0^2 - \left(R_1 + \frac{R'_2}{4}\right)^2} \quad \text{... (20.30)}$$

Since X_1 and X'_2 are known from Eq. (20.27), so

$$X_m = 2\sqrt{Z_0^2 - \left(R_1 + \frac{R'_2}{4}\right)^2} - 2\left(X_1 + \frac{X'_2}{2}\right) \quad \text{... (20.31)}$$

Example 20.8. A 220 V, 1-phase induction motor gave the following test results:

Blocked-rotor test: 120 V, 9.6 A, 460 W

No-load test: 220 V, 4.6 A, 125 W

The stator winding resistance is 1.5 Ω and during the blocked rotor test, the starting winding is open. Determine the equivalent circuit parameters and core, friction and windage losses.

[*G.B. Technical Univ. Electromechanical Energy Conversion-II, 2012-13*]

Solution: Stator winding resistance,

$$R_1 = 1.5\ \Omega$$

Equivalent resistance of motor windings,

$$R_{eq} = R'_2 + R_1 = \frac{P}{I^2} = \frac{460}{(9.6)^2} = 5\ \Omega \text{ (From blocked rotor test)}$$

or $R'_2 = 5 - 1.5 = 3.5\ \Omega$ **Ans.**

Equivalent impedance of motor windings,

$$Z_{eq} = \frac{V}{I} = \frac{120}{9.6} = 12.5\ \Omega \quad \text{(From blocked rotor test)}$$

Equivalent reactance of motor windings,

$$X_{eq} = \sqrt{Z_{eq}^2 - R_{eq}^2} = \sqrt{(12.5)^2 - 5^2} = 11.46\ \Omega$$

Assuming $X_1 = X'_2$

Rotor winding reactance referred to stator,

$$X'_2 = X_1 = \frac{X_{eq}}{2} = \frac{11.46}{2} = 5.73\ \Omega \textbf{ Ans.}$$

$$Z_o = \frac{V_o}{I_o} = \frac{220}{4.6} = 47.8\ \Omega \quad \text{(From no-load test)}$$

and Magnetising reactance,

$$X_m = \sqrt{Z_o^2 - \left(R_1 + \frac{R'_2}{4}\right)^2} - 2\left(X_1 + \frac{X'_2}{2}\right)$$

$$= 2\sqrt{(47.8)^2 - \left(1.5 + \frac{3.5}{4}\right)^2} - 2\left(5.73 + \frac{5.73}{2}\right)$$

$$= 84.3\ \Omega \textbf{ Ans.}$$

Core, friction and windage losses

$$= \text{Power input to motor on no load} - I_0^2\left(R_1 + \frac{R'_2}{4}\right)$$

$$= 125 - (4.6)^2\left[1.5 + \frac{3.5}{4}\right] = 74.745 \text{ W } \textbf{Ans.}$$

20.8 STARTING METHODS AND TYPES OF SINGLE-PHASE INDUCTION MOTORS

As already discussed the single-phase induction motor is not self-starting and, therefore, it is necessary to employ some means for making it self-starting. Commercial single-phase induction motors employ the principle of 'phase split', described below, for the purpose of starting.

20.8.1. Principle of Phase Split. It is known that when two windings, spaced 90 electrical degrees apart on the stator of a motor, are excited by two alternating emfs that are 90° displaced in time phase, a rotating magnetic field is produced. If two windings so placed are connected in parallel to a single-phase source, the field produced will alternate but will not revolve since the two windings are equivalent to but one single-phase winding. If, however, an impedance (resistance, inductance or capacitance) is connected in series with one of these windings, the currents may be made to differ in time phase. By proper selection of such an impedance the currents may be made to differ by as much as 90°*, thereby producing a rotating field very much like the field of a two-phase motor. This is the principle of *phase splitting* and the single-phase induction motors employing this principle for starting are known as *split-phase motors*. Split-phase motors are of the following types:

(*a*) Resistance-start motor or simply split-phase motor

(*b*) Capacitor-start motor

(*c*) Permanent capacitor or single-value capacitor motor and

(*d*) Capacitor-start, capacitor-run or two-value capacitor motor

20.8.2. Constructional Features. All of the single-phase induction motors mentioned above, have one construction feature in common, *viz.*, the *auxiliary or starting winding*. The starting winding is provided so that, together with the main or working winding, the motor can stimulate a *two-phase motor* to develop torque at start. In the case of the permanent capacitor motor, the auxiliary winding remains in the circuit after starting, and thus closely approximates a two-phase motor also when it is operating under load. Hence, the term '*split-phase*' may rightly be applied to all of the above mentioned single-phase induction motors. In practice, however, it is the high resistance starting winding type that retains the name, while those that use a capacitor to split the single phase are called by the particular capacitor circuit required.

Except for the single-value capacitor motor, all the motors are provided with a cut-out for the auxiliary winding. Despite the fact that other circuits are now being designed and employed,

* If the phase difference between the two currents is less than 90° and if the two mmfs are not equal, starting torque will be small but in many applications it is sufficient to start the motor.

the centrifugal switch remains an important element in the starting winding circuit. One important advantage of a centrifugal switch is that it opens at a predetermined speed, and subsequently closes at a lower speed when the line switch is opened and the motor is brought to rest. It is thus not subjected to fluctuations in line voltage, or to current variations for sudden change in load.

Centrifugal Switch and its Importance. It is mounted inside the motor. It consist, of a stationary part and a rotating part. The stationary part, consisting of one stationary and the other movable contact, is located in one of the housing or end bell, and does not rotate. The movable contact is held against the stationary one by spring action. It is thus normally closed (NC). Connections to the starting winding and line or capacitor (if employed) are made from this end bell. The rotating part of the switch has no electrical connection. It contains the actuating mechanism, the truely centrifugally operated component. It is mounted on the shaft and held close to it, usually by spring action, when the motor is at rest. When the motor is energized, the rotor speed increases, and, at approximately 75 per cent of rated speed, there is sufficient centrifugal force on the actuating mechanism to overcome its spring tension. It then makes contact with the actual switch, opening it, and holding it open, unless the speed drops to approximately 50 per cent of rated one.

Nevertheless it is the centrifugal switch that is the major cause of trouble and breakdown of motor in the operation of the single-phase induction motor. Should failure of the centrifugal switch not be detected, the motor may get permanently damaged. In the case of a capacitor-start or a high resistance-start motors, the auxiliary winding is not designed to remain energized for more than a few seconds. If the centrifugal switch remains open after the motor has been operating, little harm should occur if the motor is provided with overload protection or if an operator is present. The motor simply won't start on the next attempt, and will be disconnected from service. But in case of remotely located motors, such as in case of heating systems having motors for the oil pump, circulating pump, and one or more blowers, the motor may attempt to start many times during a day, and the excessive cycling may cause damage to the protective relay, if not the motor itself.

On the other hand, should the switch remain closed, the motor will start and run. The starting winding will get quickly overheated, and, if not soon detected, or cut out by a protective relay, this winding may cause damage to the main winding due to their proximity.

With a two-value capacitor motor (or capacitor-start, capacitor-run motor), the starting winding is designed to remain energized after starting, but one capacitor is not. Should the centrifugal switch remain closed, this starting capacitor may quickly get damaged. Also, when running, the higher capacitance in series with the inductive winding may actually reduce the total impedance of the winding or phase, thereby increasing its current above normal. The added capacitor will cause a change in the overall power factor, thereby negating one of the reasons for the motor's use. Should an auto-transformer be used to have the higher capacitance for starting, similar effects occur, although the capacitor may not get damaged.

From the above discussions the importance of centrifugal switch is obvious and it is never repaired. *It should always be replaced.*

In single-phase induction motors, the main winding usually occupies only $\frac{2}{3}$rd of a pole pitch and this makes it possible to achieve a winding factor of 0.827. Filling the other slots with the winding is not rational, since this leads to such a small winding factor (up to $\frac{2}{\pi}$ *i.e.* 0.638) that the copper of the stator winding will be poorly utilized. The remaining empty one-third of the stator slots can be used for the auxiliary winding, whose axis is shifted in space through 90° with respect to the axis of the main winding. A two-winding system of this type can be had from a star-connected three-phase winding by disconnecting one phase winding and using it as the starting winding, the remaining two series connected phase windings serving as the main winding.

In order to provide high resistance and low inductance, the starting winding is wound with fewer turns of fine wire and placed on the top of the stator slots. The main winding, on the other hand, is wound with thicker wire and placed at the bottom of the stator slots. The idea behind it is that the current through the starting winding is merely nearly in phase with the line voltage than is that through the main winding. This gives the effect of splitting the single phase applied to the motor terminals.

With the increase in speed of motor, the torque developed increases and above 85 per cent of synchronous speed, the torque developed by the main winding alone is actually greater than that developed by both windings, and it might be advantageous to open the auxiliary or starting winding circuit at this cross-over point . To allow for individual variations among motors and switches, however, the contacts are usually designed to open at 75 per cent of synchronous speed. This does not affect the operation, because the main winding alone usually develops approximately twice full-load torque at this speed.

The starting winding is not designed for continuous operation and care should be taken that it does not remain connected to the supply after it should have been disconnected by the switch. The series switch is usually centrifugally operated. In case of motors that are hermetically sealed, such as in refrigeration units, electromagnetic type relay is used.

Since split-phase motors are usually started directly from the line hence these motors draw large current (4 to 6 times the rated value) at start. This line current has two components, one of them (the larger of the two) passing through the main winding, while the other passes through the auxiliary winding. Advantage is taken of the high starting current in connection with the use of an electromagnetic type relay that performs the same function as the centrifugal switch. This relay, shown in Fig. 20.12, has a coil connected in series with the main winding, and a pair of contacts, normally open, in the auxiliary winding circuit. At the start, when large current flows through the main winding, the contacts close so that the auxiliary winding circuit is energized. When the motor picks up the speed (75 or 80% of synchronous speed), the main winding current drops to a value that is low enough to cause the contacts to open. Such relays are located outside the motor, where they can be serviced or replaced, a situation that would be impossible if an internally mounted centrifugal switch were used.

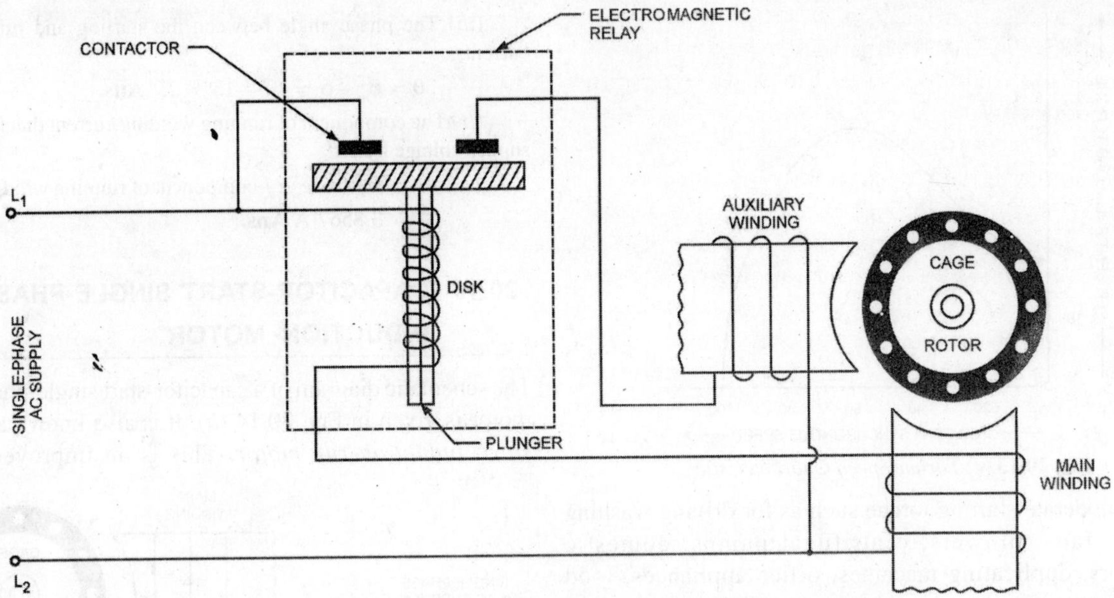

Fig. 20.12

Reversal of Direction of Split-Phase Induction Motor. The direction of rotation of a split-phase induction motor can be reversed simply by reversing the line connections of either the main winding or the auxiliary winding. The noteworthy point is that reversal of a split -phase induction motor takes place only when it is at rest and the starting switch is in its normally closed de-energized position. If reversal is attemped under normal operating conditions, nothing will happen because under normal operating conditions, the starting switch is in open position and the motor is running as a pure single-phase induction motor and developing torque in the direction of rotation.

20.9 RESISTANCE-START SINGLE-PHASE INDUCTION MOTOR

It is a form of split-phase motor having a high resistance connected in series with the starting winding as illustrated in Fig. 20.13 (*a*). Because of low resistance and high inductive reactance of main winding and high resistance and low inductive reactance of starting winding, the starting winding current I_s does not lag so much as the main winding current I_m. At starting the main winding current I_m lags behind the applied voltage V by 65-75 degrees, whereas the starting winding current I_s lags by 35-45 degrees. Thus the phase displacement between main winding current and starting winding current is about 20-30 degrees as shown in phasor diagram [Fig. 20.13 (*b*)]. Since the currents in the two windings are different in magnitude and constants of two windings are different, therefore, the arrangement forms imperfect two phase motor, that produces a rotating magnetic field that is not uniform in either time or space, but is sufficient for starting the motor. The best starting condition will be obtained if I_s and I_m differ in time by about 90°, but this condition is not possible in the motor of this type.

A typical torque-speed characteristic is shown in Fig. 20.13(*c*). The starting torque is 1.5 times to twice the full-load starting torque and starting current is 6 to 8 times full-load current. The speed regulation of a resistance–start single-phase induction motor is very good. The per cent slip in most fractional kilowatt split-phase motors is about 4-6 per cent. Such a motor may operate with a power factor of 0.55-0.65 and efficiency of 60-65 per cent.

Owing to their low cost, resistance-start split-phase motors are most popular single-phase motors. These motors are made in fractional kilowatt $\left(\frac{1}{20} \text{ to } \frac{1}{2} \text{ kW}\right)$ ratings with speed ranging from 2,875 to 700 rpm. Such motors find wide applications for low inertia loads, continuous operating loads and applications

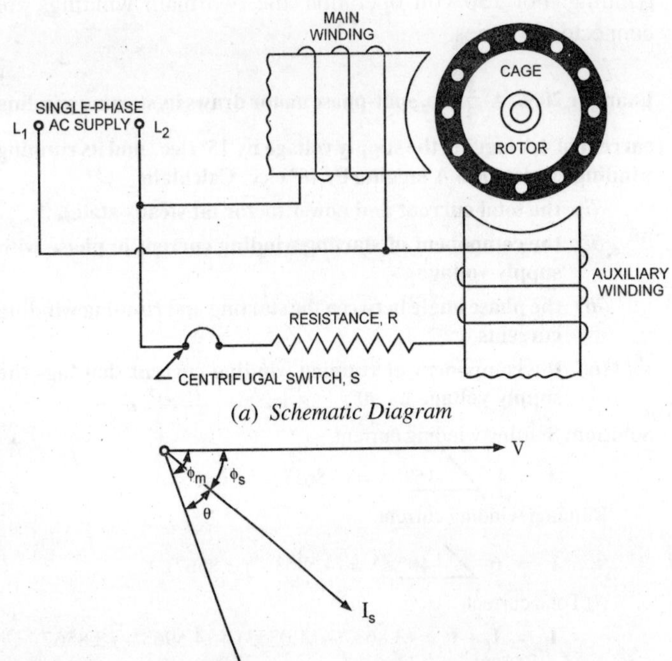

(*a*) *Schematic Diagram*

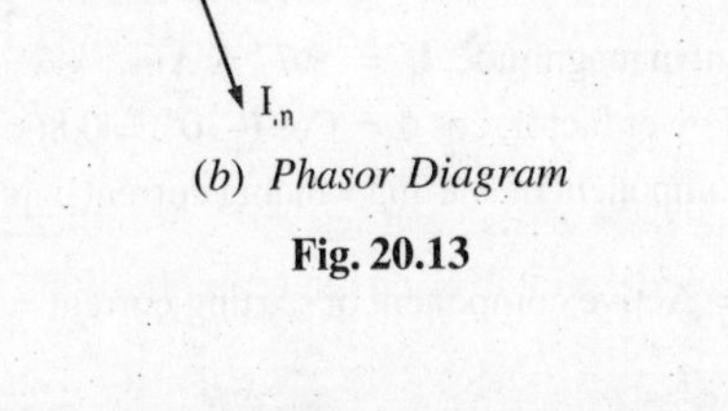

(*b*) *Phasor Diagram*

Fig. 20.13

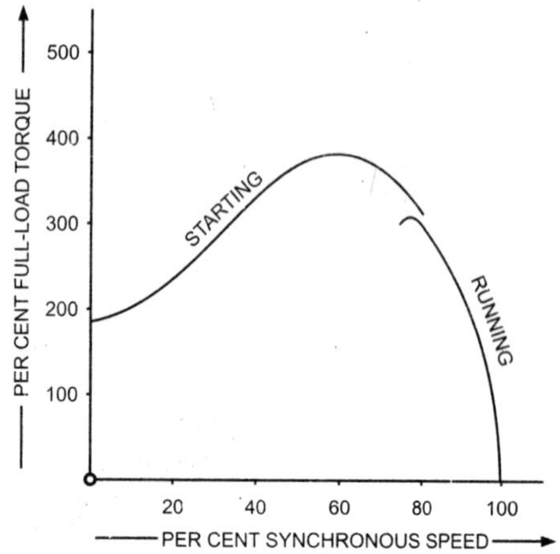

Fig. 20.13 (*c*) *Torque Speed Characteristic*

requiring moderate starting torque such as for driving washing machines, fans, blowers, centrifugal pumps, domestic refrigerators, duplicating machines, office appliances, food processing machines, woodworking tools, grinders, oil burners etc. Because of low starting torque these motors are seldom employed in sizes larger than $\frac{1}{2}$ kW.

20.9.1. Two-Speed, Two-Winding Motors-Dual Voltage Motors. Resistance-start split-phase motors can be constructed as that two speeds are available. This is done by providing two sets of windings, one for each speed.

Split-phase motors are also convertible for use with either a 115 V or 230 V source of supply merely by changing the circuit connections. For motor operation at 115 V, the two main windings are connected in parallel, thereby providing 115 volts to each winding. For 230 volt operation, the two main windings are connected in series.

Example 20.9. A $\frac{1}{4}$ hp, split-phase motor draws its starting winding current of 4 A lagging the supply voltage by 15° elec. and its running winding current is 6 A lagging by 40° elec. Calculate

(*i*) the total current and power factor (at steady state),

(*ii*) the component of starting winding current in phase with supply voltage

(*iii*) the phase angle between the starting and running winding currents

(*iv*) the component of running winding current that lags the supply voltage by 90°.

Solution: Starting winding current,

$$\mathbf{I}_s = 4\ \angle{-15^\circ}\ \text{A} = (3.8637 - j\,1.0353)\ \text{A}$$

Running winding current,

$$\mathbf{I}_m = 6\ \angle{-40^\circ}\ \text{A} = (4.5963 - j\,3.8567)\ \text{A}$$

(*i*) Total current

$$\mathbf{I}_L = \mathbf{I}_s + \mathbf{I}_m = (3.8637 - j\,1.0353) + (4.5963 - j\,3.8567)$$

$$= (8.46 - j\,4.892)\ \text{A or } 9.77\ \angle{-30^\circ}\ \text{A}$$

Total current in magnitude, $\mathbf{I}_L$ = 9.77 A **Ans.**

Power factor, $\cos\phi$ = Cos (–30°) = 0.866 (lag) **Ans.**

(*ii*) The component of starting winding current in phase with supply voltage

= Active component of starting current = 3.8637 A **Ans.**

(*iii*) The phase angle between the starting and running winding currents,

$$\theta = \phi_m - \phi_s = 40^\circ - 15^\circ = 25^\circ \textbf{ Ans.}$$

(*iv*) The component of running winding current that lags behind the supply voltage by 90°

= Reactive or *j*-component of running winding current

= 3.8567 A **Ans.**

20.10 CAPACITOR-START SINGLE-PHASE INDUCTION MOTOR

The schematic diagram of a capacitor-start single-phase induction motor is given in Fig. 20.14 (*a*). It is also known as *capacitor-start, induction-run motor.* This is an improved form of a

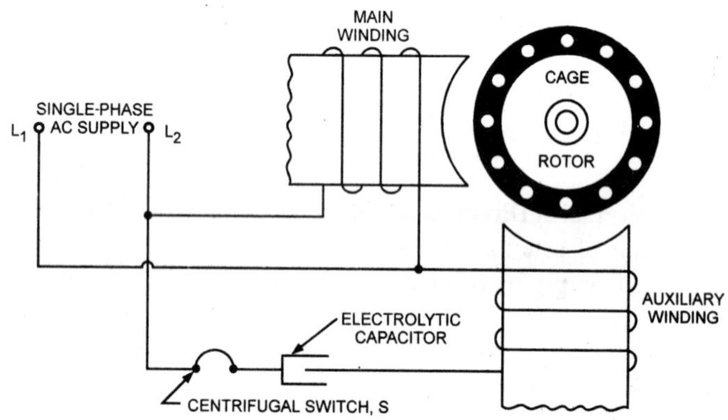

(*a*) *Schematic Diagram*

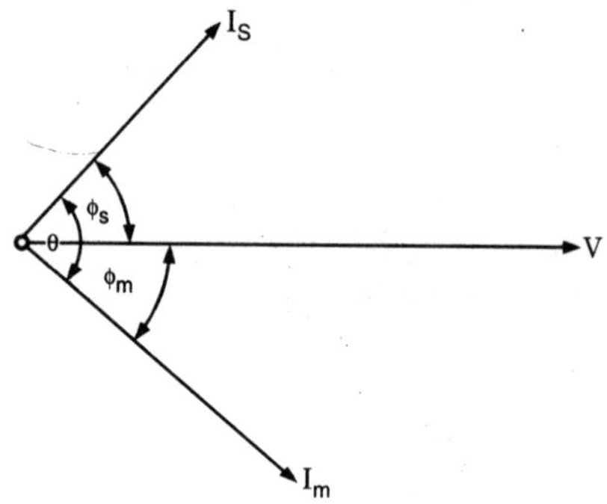

(*b*) *Phasor Diagram*

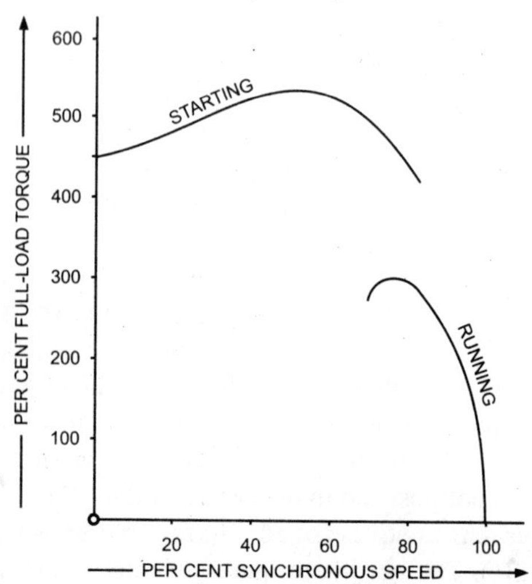

(*c*) *Torque-Speed Characteristic*

Fig. 20.14 *Capacitor-Start Single Phase Induction Motor* resistance-start split-phase motor. Fundamentally this motor is very similar to the resistance-start split -phase motor, except that the starting winding has a few more turns and consists of a heavier wire than the starting winding of a resistance-start motor; there is also a large electrolytic capacitor in place of resistor, connected in series with the starting winding. By using a capacitor of proper value, the starting winding current I_s at standstill, can be made to lead the main winding current I_m by 90º (*i.e.* $\theta = 90º$). The phasor diagram of a capacitor-start motor is given in Fig. 20.14 (*b*).

The capacitor is usually mounted in a metal casing on the top of the motor although it may be mounted in any convenient external position on the motor frame. In some cases, the capacitor is mounted inside the motor housing. The capacitor used in the capacitor-start motor is of the electrolytic type and is designed for extremely short-duty service and is guaranteed for not more than 20 periods of operation per hour, each period not to exceed 3 seconds. The size of the capacitor depends upon the rating of the motor and has a value around 50 μF to a few hundred microfarads. It is primarily due to the development of the inexpensive yet reliable ac electrolytic* capacitor that the capacitor-start motor has largely replaced all other single-phase motors where high starting torque is the essential requirement.

The capacitor-start motor, like resistance–start motor, has the starting winding disconnected by means of a centrifugal switch as the motor picks up speed. Because of short-duty service rating of electrolytic capacitor used in capacitor-start motor, proper operation of the starting (or centrifugal) switch is doubly important. A faulty switch may keep the capacitor in the circuit for too long time during each starting period, and thereby appreciably shorten its life. Of course, the starting winding itself is affected by the switch in the same manner as in the resistance-start motor.

The direction of rotation of a capacitor-start induction motor may be reversed by interchanging the connections to the supply of either the main or starting winding, but it is subject to the same limitations, as given in Art. 20.8.

Since the torque developed by a phase- split motor is proportional to the sine of phase displacement angle θ between the starting winding current I_s and main winding current I_m, it is obvious that the increase in phase displacement angle θ (from 30º to 80º) alone increases the starting torque twice the value developed by a resistance-start motor. The increase in starting winding turns causes decrease in starting current on one hand and produces a large rotating flux, and therefore a further increase in the starting torque on the other hand. Finally it is seen that, for the same magnitudes of two stator winding currents, I_s and I_m the line current I_L is less in the capacitor-start motor, because of the greater angle between them. In addition, the starting winding power factor is also improved. In short it can be said that because of the closer approximation to a true two-phase motor, a greater number of turns on the starting winding, and more allowable current in both of the windings due to lower line current, the capacitor-start motor develops a greater starting torque than does an equally rated resistance-start motor. The line current of a resistance-start motor was found to be two-third higher than line current of a corresponding capacitor-start motor, while the starting torque of the capacitor-start motor was twice that of resistance-start motor.

It is noteworthy that the capacitor-start motor differs from the resistance-start motor only during the starting period, for it develops more starting torque (about 3.5-4.5 times full-load torque) with a lower inrush current. After the machine attains normal speed and the starting winding is cut out, its operating characteristics are identical to those of a corresponding resistance-start motor.

Capacitor-start motors are manufactured in rating ranging from $\frac{1}{10}$ kW to $\frac{3}{4}$ kW, but larger sizes (as large as 6 kW) are also available. They are widely used for heavy duty general purpose applications like refrigerator units, air-conditioners, compressors, jet pumps, sump pumps, centrifugal pumps, compressors, conveyors, fans, blowers, farm and home workshop tools, oil burners, etc.

Cautions:

1. Capacitor-start motors should not be used for applications having starting periods of longer duration. This is due to the fact that electrolytic capacitors are rated for short-duty service only and will get damaged if the starting load does not allow the motor to pick up speed quickly.
2. Electrolytic capacitors have maximum voltage rating and if subjected to voltage exceeding 25 per cent greater than such rating, they will get damaged. This voltage limitation means that a centrifugal switch must positively open at its preset value of speed; if the spring is not properly adjusted or if the load keeps the speed near the point at which the switch flutters, *i.e.*, opens and closes rapidly, it is possible for the capacitor voltage to be twice rated value for an instant, in which even the capacitor will fail. This double voltage may be caused if the capacitor is fully charged at the instant the switch opens and then closes on the ac voltage wave when it is a maximum in the opposite direction.
3. The capacitor-start motor should not be started more than 20 times an hour. If the motor is started too frequently in a short time interval, the dielectric of the capacitor will get damaged by the frequent current surges. As a result, the dielectric will rupture or break down and short the capacitor plates and cause the fuse protection for the branch motor circuit to blow. If the fuse ratings are quite high and do not interrupt the power supply to the motor, the starting winding may burn out.

Example 20.10. A 250 W, 230 V, 50 Hz capacitor start motor has the following impedances at standstill.

Main winding, $Z_m = (7 + j5)\Omega$

Auxiliary winding, $Z_a = (11.5 + j5)\Omega$

Find the value of capacitor to be connected in series with auxiliary winding to give a phase displacement of 90° between the currents in two windings. Draw the circuit and phasor diagram for motor.

[*Rajasthan Univ. Electrical Machines-II, 2004*]

Solution: Main winding impedance, $Z_m = (7 + j5) = 8.60 \angle 35.54°$

Obviously main winding current I_m lags behind the applied voltage V by 35.54°

* AC electrolytic capacitor is not the same as the dc electrolytic capacitor often employed in a filter in radios, amplifiers, and TV sets etc.

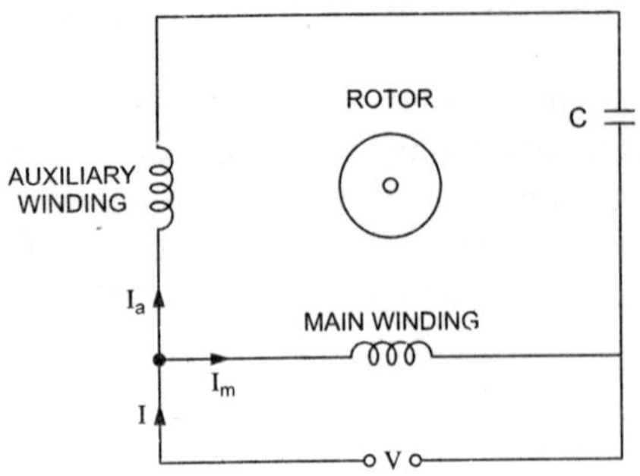

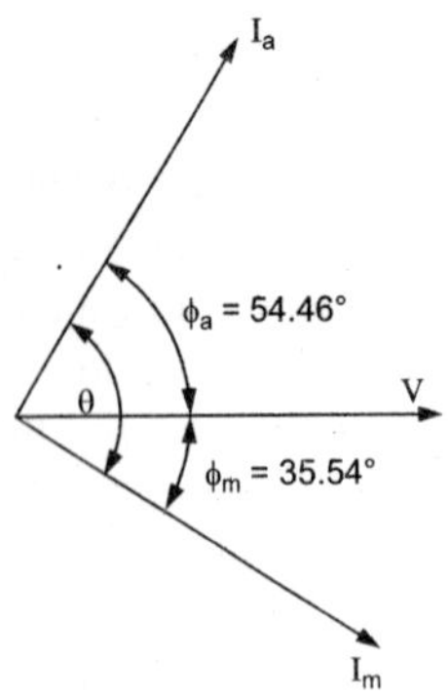

(*a*) *Circuit Diagram* (*b*) *Phasor Diagram*

Fig. 20.15

Auxiliary winding impedance, $\mathbf{Z}_o = (11.5 + j5)\ \Omega$

Since, time phase angle between auxiliary winding current I_a and main winding current I_m is 90°, auxiliary winding current I_a must lead the applied voltage by (90° – 35.54°) or 54.46°.

If X_C is the capacitive reactance of the capacitor C connected in series with the auxiliary winding, then impedance of the auxiliary winding will be given as

$$\mathbf{Z}_a = (11.5 + j5 - jX_C) \text{ or } [11.5 + j(5 - X_C)]$$

For auxiliary winding

$$\text{Tan } \phi_a = \frac{5 - X_C}{11.5}$$

$$\text{or} \quad X_C = 5 - 11.5 \tan \phi_a$$

$$= 5 - 11.5 \tan(-54.46°) = 5 + 16.1 = 21.1\ \Omega$$

Capacitance of the capacitor,

$$C = \frac{1}{2\pi f \times 21.1} = \frac{1}{2\pi \times 50 \times 21.1} = 150.87\ \mu F \quad \textbf{Ans.}$$

Example 20.11. The following data shown result on a 230 V, 50 Hz capacitor-start single-phase induction motor at standstill :

Main winding : 100 V, 2 A, 40 W

Auxiliary winding : 80 V, 1.0 A, 50 W

Determine the value of capacitance for obtaining the maximum starting torque.

[*U.P. Technical Univ. Electromechanical Energy Conversion II, 2005-06*]

Solution: Equivalent impedance with main winding,

$$Z_{01m} = \frac{V_m}{I_m} = \frac{100}{2} = 50\ \Omega$$

Equivalent resistance with main winding,

$$R_{01m} = \frac{P_m}{I_m^2} = \frac{40}{2^2} = 10\ \Omega$$

Assuming rotor resistance and reactance referred to stator, R_2' and X_2' equal to stator resistance R_1 and reactance X_1 respectively

$$R_{1m} = R_2' = \frac{R_{01m}}{2} = \frac{10}{2} = 5\ \Omega$$

$$\text{and } X_{1m} = X_2' = \frac{X_{01m}}{2} = \frac{\sqrt{50^2 - 10^2}}{2} = 24.5\ \Omega$$

Angle of lag of main winding current,

$$\phi_m = \text{Tan}^{-1} \frac{X_{1m}}{R_{1m}} = \text{Tan}^{-1} \frac{24.5}{5} = 78.46°$$

Equivalent impedance with auxiliary winding,

$$Z_{01a} = \frac{V_a}{I_a} = \frac{80}{1} = 80\ \Omega$$

Equivalent resistance with auxiliary winding,

$$R_{01a} = \frac{P_a}{I_a^2} = \frac{50}{1^2} = 50\ \Omega$$

Equivalent reactance with auxiliary winding,

$$X_{01a} = \sqrt{80^2 - 50^2} = 62.45\ \Omega$$

Auxiliary winding resistance

$$R_a = R_{01a} - R_2' = 50 - 5 = 45\ \Omega$$

Auxiliary winding reactance,

$$X_a = X_{01a} - X_2' = 62.45 - 24.5 = 37.95\ \Omega$$

Torque developed at start will be maximum, if the phase angle between main winding current I_m and auxiliary winding current I_a is 90° or the auxiliary winding current leads the applied voltage by (90° – 78.46°) or 11.54°.

If X_c is the capacitive reactance of the capacitor connected in series with the auxiliary winding, then impedance of the auxiliary winding will be given as

$$\mathbf{Z}_a = R_a + X_a - X_c = 45 + j\,37.95 - j\,X_c$$

$$= [45 + j\,(37.95 - X_c)] \text{ ohms}$$

For auxiliary winding

$$\text{Tan } \phi_a = \frac{37.95 - X_c}{45}$$

$$\text{or } X_c = 37.95 - 45 \tan \phi_a = 37.95 - 45 \tan(-11.54°)$$

$$= 37.95 + 9.19 = 47.138\ \Omega$$

$$\text{Capacitance of capacitor, } C = \frac{1}{2\pi f X_c} = \frac{1}{2\pi \times 50 \times 47.138}$$

$$= 67.5\ \mu F \ \textbf{Ans.}$$

20.11 CAPACITOR-START, CAPACITOR-RUN SINGLE-PHASE INDUCTION MOTOR

In order to have the advantages that result from the use of a capacitor during the normal running operation as well as starting period, the capacitor-start, capacitor-run motors have been developed. The capacitor-start, capacitor-run motor is similar to capacitor-start, induction-run motor, except that the auxiliary winding and a capacitor remain connected in the circuit at all times. Since the value of the capacitance which will result in optimum running performance is not the same as the value which will given the best starting torque conditions and also the capacitor employed for normal running should be of continuous-duty rating and that employed for starting should be of short-duty rating, therefore, in such a motor two different capacitors—one small value oil impregnated paper continuous rating capacitor C_R and another much larger value electrolytic short-duty capacitor C_S are employed. It is also necessary to provide a centrifugal switch to take out the electrolytic capacitor out of auxiliary winding circuit when the motor has attained a predetermined speed. The schematic diagram of a capacitor-start, capacitor-run motor is given in Fig. 20.16 (*a*). When the motor attains 75 per cent of synchronous speed, the starting capacitor C_S is taken out of the circuit by the operation of centrifugal or transfer switch, which is normally closed.

It is to be noted that when the centrifugal switch is closed during the starting period, the two capacitors C_S and C_R are in parallel, so that total capacitance in the auxiliary winding circuit is sum of their values; after the switch opens, only the continuous oil type capacitor C_R remains in the auxiliary winding circuit. In practice starting capacitor C_S is about 10-15 times as large as the running capacitor C_R.

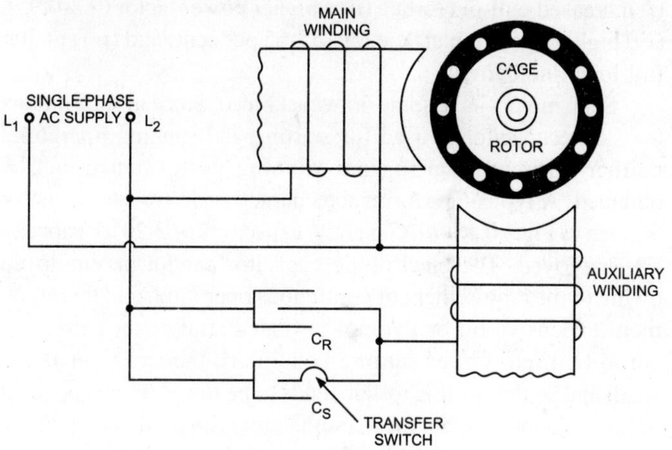

(a) Schematic Diagram

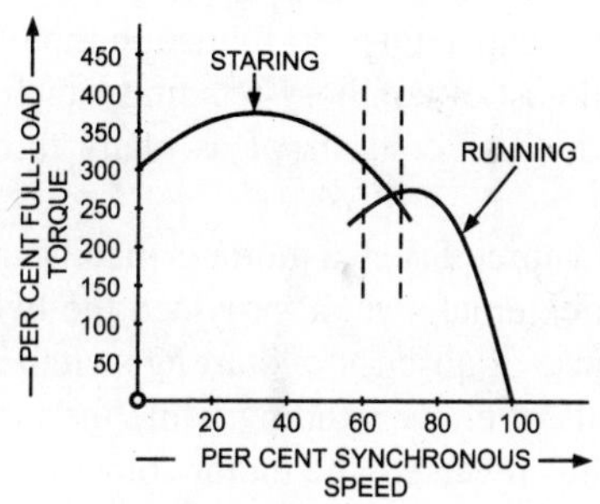

(b) Torque-Speed Characteristic

Fig. 20.16 *Capacitor-Start, Capacitor-Run Single Phase Induction Motor*

Such a motor operates as a two-phase motor from single-phase supply, thereby, producing a constant torque and not a pulsating torque, as in other single-phase motors. Besides their ability to start heavy loads they are extremely quiet in operation, have better efficiency (55-65%) and power factor (0.8-0.95) when loaded and develop up to 25% greater overload capacities. They are indeed splendid machines where the load requirements are severe as in case of compressors, refrigerators and fire strokers, pumps, conveyors etc. The disadvantage of such machines is only high cost. These motors are often employed requiring a quiet operating motor such as in hospitals, studios etc.

In recent years, home air-conditioners have become extremely popular. In many cases, the electrical system in the home may not have the capacity to handle the electrical load of the air-conditioner(s). For this reason, capacitor-start capacitor-run motor have found more wide spread use in the compressors. Both the starting and running currents are low—full-load current may be $\frac{3}{4}$th of that for an equivalent capacitor-start motor.

Such motors have become increasingly popular and are now manufactured in the larger sized single-phase motors where previously only repulsion-start motors were available. A typical torque-speed characteristic is shown in Fig. 20.16 (*b*). The direction of rotation of motor may be reversed by interchanging the connections to the supply of either the main or auxiliary winding.

20.11.1. Capacitor-Start, Capacitor-Run Motor With an Auto-Transformer. Capacitor-start, capacitor-run motors are also manufactured in which an auto-transformer and a single capacitor unit are used. The auto-transformer and a capacitor are sealed in a rectangular iron box and mounted on the top of the motor. The idea is based on the principle that a capacitor of value C, connected across the secondary of a step-up transformer, appears to the primary as though it had a value of K^2C where K is a voltage transformation ratio. For example, if the actual capacitor connected across the high voltage secondary is of 6 μF and the voltage transformation ratio is 4, the low-voltage primary acts as though it had a $(4^2 \times 6)$ or 96 μF capacitor connected to its terminals. Thus the effectiveness of a low value capacitor is increased largely by the use of a step-up transformer. Figure 20.17 illustrates how this is done in one design in which the stepping-up is accomplished by means of an auto-transformer. The transfer switch is now a single-pole double-throw switch, rather than a simple on-off switch. As the motor picks up the speed, the transfer switch shifts the capacitor from one voltage tap to another so that the voltage transformation ratio changes from higher value at starting to a lower value for running. It is noteworthy here that the capacitor used must be capable to withstand a continuous ac voltage of 330 V and an intermittent voltage of 800 V. The capacitor used is usually of the paper-tin foil construction immersed in high grade insulation like wax or mineral oil, or in some other special compound such as Pyranol or Inerteen.

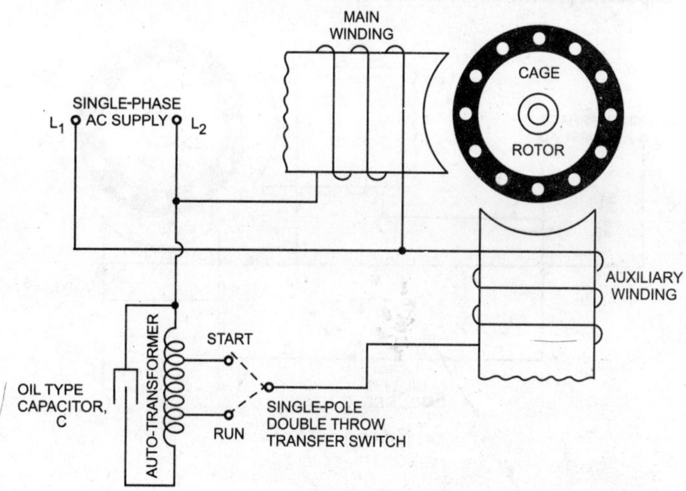

Fig. 20.17 *Capacitor-Start, Capacitor-Run Motor With An Auto-Transformer*

Since an auxiliary winding is permanently connected in the motor circuit, a faulty starting switch will not harm it. If, however, the switch fails to open at proper time, the electrolytic capacitor connected in auxiliary winding circuit may get destroyed. In the case of the transformer capacitor type motor, the voltage across the capacitor may become dangerously high as the motor picks up speed and the dielectric may breakdown. If the switch remains in the open position, the motor using two capacitors will have only low-value oil-filled capacitor in the circuit at starting and thus low starting torque will be developed. In case of a transformer capacitor motor, if the transfer switch remains in the running position, the motor will give low starting torque.

The capacitor-start, capacitor-run motor with an auto-transformer has the advantage of greater design flexibility because any desired capacitance values can be had by a proper choice of transformer turn-ratio. The drawbacks of such a motor are the size and the cost of the auto-transformer and the increased cost of the transfer switch.

Both types of the motors, described above, operate with both the windings while running, and so the direction of rotation can be reversed while running. This, however, causes the transfer switch to return to the starting position, and then again to the running position, when the motor rotates in the opposite direction. This is not desirable.

20.12 PERMANENT-CAPACITOR SINGLE-PHASE INDUCTION MOTOR

Permanent-capacitor single-phase induction motor, also sometimes called the *single-value capacitor–run motor*, has two stator windings placed mutually 90 electrical degrees apart. The main or running winding is connected directly across the single-phase ac supply. A capacitor in series with the auxiliary winding is also connected across the supply lines. There is no centrifugal switch, since the auxiliary winding is energized at all times when the motor is in operation. The auxiliary winding in such motors usually has almost the same size of wire and almost as many turns as the main winding; in fact, in some motors the two windings are identical. The schematic diagram of a permanent capacitor motor is given is Fig. 20.18(*a*).

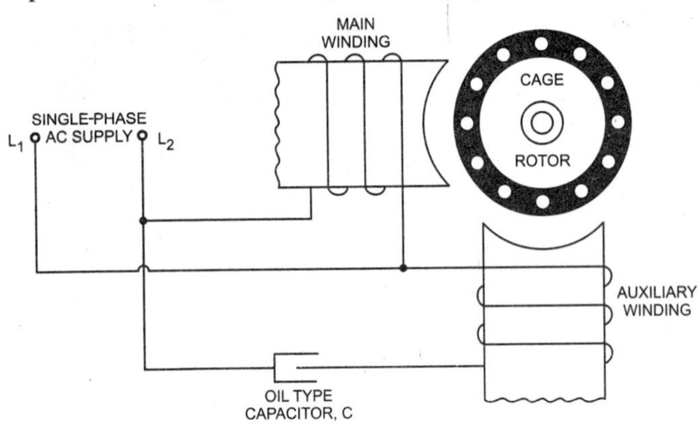

(*a*) *Schematic Diagram*

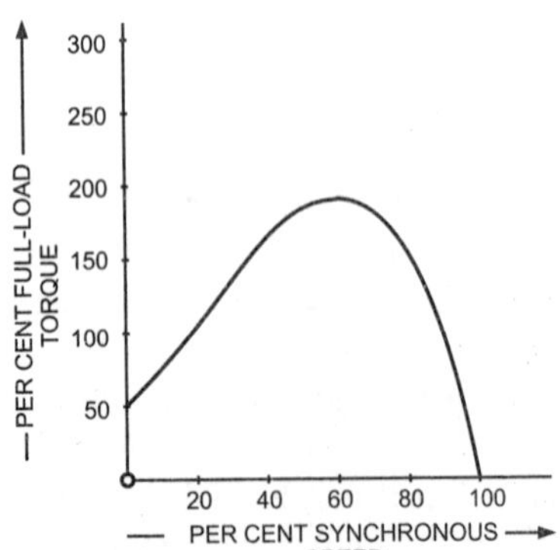

(*b*) *Torque-Speed Characteristic*

Fig. 20.18 *Permanent Capacitor Single Phase Induction Motor*

The auxiliary winding is always in the motor circuit, and therefore, the operation of the motor when loaded resembles more closely to that of a 2-phase motor. One advantage of such operation is that the rotating field developed by the two stator windings is a more nearly uniform, and therefore, the motor is quieter in operation. The other advantages of such a motor are (*i*) increased pull-out torque (*ii*) a higher power factor (0.8-0.95), (*iii*) higher operation efficiency (55-65 per cent) and (*iv*) smaller full-load line current.

Such motors have some drawbacks also. Since in such motors the same capacitor is used for starting and running, therefore, neither optimum starting nor running performance can be obtained. A typical performance characteristic for such a motor is given in Fig. 20.18 (*b*). Generally capacitors of 2-20 μF capacity are employed. The electrolytic capacitor cannot be employed owing to the requirement of continuous operation, and therefore, more expensive oil or Pyranol insulated foil-paper capacitors are to be used. Under starting conditions, the capacitor is too small and so the starting torque tends to be low (50 to 80 per cent of full-load torque); but the manufacturers can partly compensate for this by using a high resistance squirrel cage. The permanent-capacitor induction motor is more expensive than the equivalent resistance-start or capacitor-start induction motor mainly owing to the increased cost of auxiliary winding which is now wound with thicker wire and consists of as many turns as the main winding.

A unique feature of such a motor is that it can be very easily reversed by an external switch, provided the two windings are identical. Because of absence of starting switch it is possible to reverse the leads of either winding while the motor is running. The torque is thus reversed, the motor comes to a standstill, and then starts in the opposite direction. If the two stator windings are identical, it is possible to place the capacitor in series with either of the two windings, making that particular winding the auxiliary or leading one. Circuit for such an operation is given in Fig. 20.19. When the SPDT switch is in position 1 winding A serves as a main winding and winding B as an auxiliary winding. When the SPDT switch is moved in position 2, winding A becomes the auxiliary winding and winding B the main one.

Such motors are widely used in table and ceiling fans, blowers, oil burners, room coolers, portable tools, other domestic and commercial electrical appliances, where low starting torque is required.

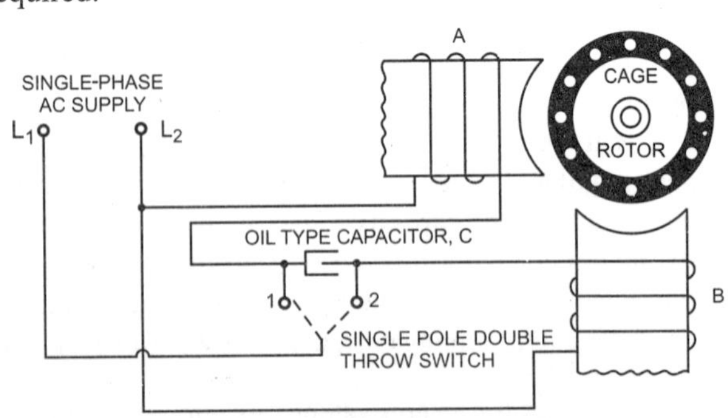

Fig. 20.19 *Reversing Permanent Capacitor Motor*

Because of the simplicity of reversal of motor, such motors are often used to operate devices that must be moved back and forth frequently such as induction regulators, rheostats, furnace controls, valves and arc-welding controls.

Because of the absence of starting switch, it is possible to run a permanent- (or single-value) capacitor run motor over a wide range of speeds, and such a motor has many applications where adjustable speed is required. Such a motor can be arranged for an adjustable speed by the use of a tapped winding or an auto-transformer regulator. It is known that the torque developed in an induction motor is proportional to the square of the applied voltage. For a given load, therefore, if the applied voltage is reduced, the motor will operate with increased slip to develop the required torque. Thus the speed can be controlled by varying the voltage applied to the motor.

20.13 SHADED-POLE MOTOR

Shaded-pole motor is a split-phase type single-phase induction motor. The stator has salient poles, each provided with its own exciting coil. A part $\left(\text{about } \frac{1}{4} \text{ to } \frac{1}{3}\right)$ of each pole is wrapped by a copper strap forming a closed loop, known as a *shading coil,* as shown in Fig. 20.20 (*a*). Stator usually has 2 or 4 poles giving synchronous speed of 3,000 or 1,500 rpm for 50 Hz power supply. The rotor is a simple squirrel cage type with a 60° skew for optimum starting torque and for limiting the torque dip during run-up. A low cage resistance impairs both the starting torque and the dip, and usually the resistance is greater than for any other type of single-phase cage motor, giving a starting torque in the range of 40-80 per cent of full-load torque and pull-out torque of about 110-140 per cent of full-load torque. The gap length is of the order of 0.25-0.5 mm. Too short air gap may result in starting torque variations due to rotor slotting. It has no commutator, brushes, collector rings, contactors, capacitors or moving switch parts, so it is relatively cheaper, simpler and extremely rugged in construction and reliable (absence of centrifugal switch or starting mechanical auxiliary eliminates the possibility of motor failure due to faulty centrifugal switch mechanisms).

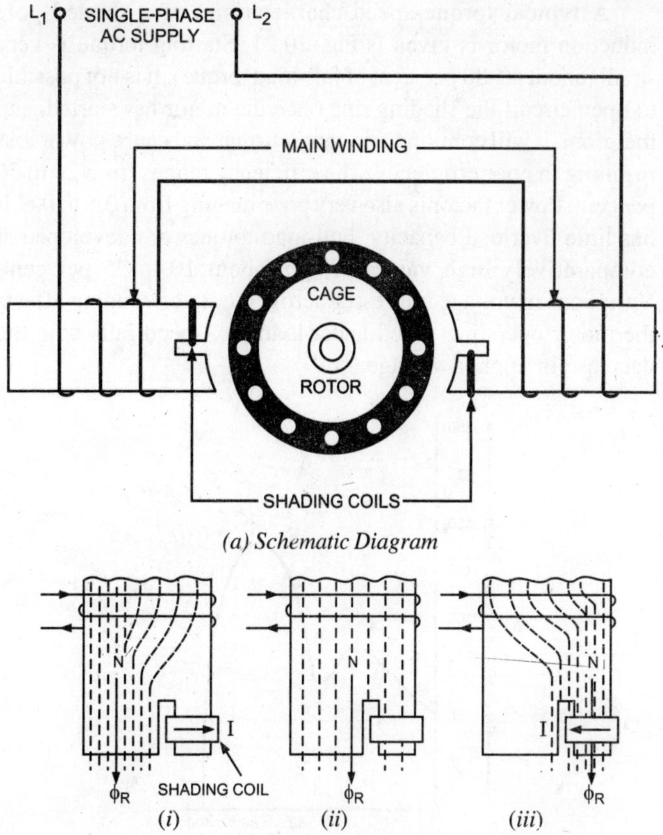

(*a*) *Schematic Diagram*

(*b*) *Shifting of Air-Gap Flux*

Fig. 20.20 *Shaded-Pole Single-Phase Induction Motor*

Action. When the exciting winding is connected to a single-phase ac supply source an alternating current flows through it and sinusoidally varying flux is set up in the core. This change of flux induces short-circuit currents in the shading coil in such a direction as to oppose the core flux. The flux in the shaded part of the pole thus lags behind the flux in the main part of the pole. At the same time, the main flux and shaded-pole flux are obviously displaced in space, although by less than 90°. Since two fields have both space and time displacement, the conditions for setting up a rotating field are present, no matter how imperfectly. Torque is thus developed in the squirrel cage rotor.

The movement of the flux around the stator may be more clearly illustrated in the following discussion. In the following discussion the exciting current will be assumed to be varying sinusoidally. The discussion will be made for three parts of current (*i*) from zero to a nearly positive maximum (*ii*) in the region of maximum current and (*iii*) from a nearly positive maximum to zero.

As the current in the exciting winding increases, a voltage is induced in the shading coil by transformer action. The current in the low resistance shading coil is comparatively high and is in such a direction as to oppose the rapidly changing flux, created by it, in the shaded portion. Thus, when the exciting current increases, the greater portion of flux passes through the unshaded part of each pole and the resultant lies near the centre of the unshaded part of the pole, as illustrated in Fig. 20.20 (*b*)(*i*).

When the exciting (or stator) current and flux attain their maximum values, the rate of change in either current or flux is zero, thereby the emf and so the current induced in the shading coil becomes zero. Consequently the flux distribution is uniform over the entire pole face and the resultant flux lies at the centre of the pole, as illustrated in Fig. 20.20 (*b*)(*ii*).

When the exciting current and flux decrease, induced emf and the current in the shading coil sets up a mmf which aids the stator field. Consequently, the flux concentrates in the shaded portion of the pole and weakens in the unshaded portion, as shown in Fig. 20.20 (*b*)(*iii*). The resultant flux is thus shifted to the centre of the shaded portion of the pole.

The effect of shading coil is to cause the field flux to shift across the pole face from the unshaded portion to shaded portion. This shifting flux is like a rotating weak field moving in the direction from unshaded portion to the shaded portion of the pole. Consequently a small starting torque is developed. As soon as the rotor starts rotating under the influence of starting torque, an additional torque is developed by single-phase induction motor action. The motor ultimately attains a speed slightly less than the synchronous speed and runs as a single phase induction motor.

Operating Characteristics and Applications. The motor performance is a function of the number of main winding turns, pole-arc / pole-pitch ratio, the proportion of the pole-arc in the shaded region, the shading coil resistance, and the rotor cage resistance and the slot skew.

A typical torque-speed characteristic of a shaded-pole induction motor is given is Fig. 20.21. Starting torque is very small (about 40-80 per cent of full-load torque). It is not possible to open circuit the shading ring once the motor has started, and therefore, it will continuously carry current and cause power loss resulting in poor efficiency. The efficiency ranges from 20 to 50 per cent. Power factor is also very poor ranging from 0.5 to 0.6. It has little overload capacity. Full-load torques are developed at comparatively high values of slip, about 10 to 25 per cent. Variations in voltage impressed across the stator winding affects the motor operating speed under load *i.e.*, speed falls with the decrease in applied voltage.

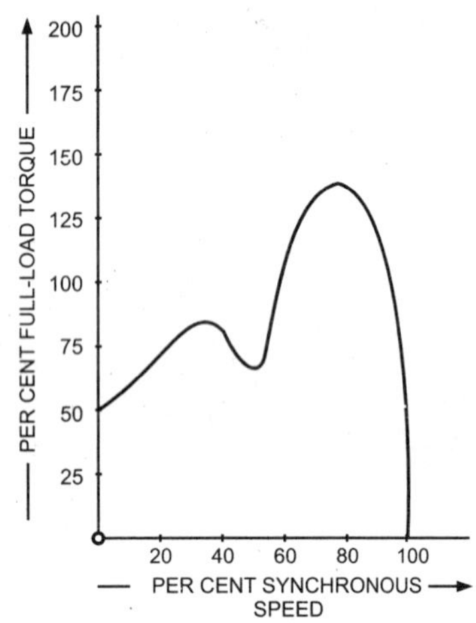

Fig. 20.21 *Torque-Speed Characteristic of Shaded Pole Motor*

The direction of rotation of motor depends upon the position of the shading coil on the pole i.e which portion of the pole is encircled by the shading coil. Unless the machine is constructed so that the shading coil can be shifted to the other side of the pole, the direction of rotation of such a machine cannot be reversed.

The speed of a shaded pole motor can be varied by varying the voltage applied across its stator winding. The voltage applied across the stator winding can be varied in three ways viz (*i*) by using an auto-transformer tapped for several voltages (*ii*) by using a tapped reactance coil (sometimes a resistance) that incurs a voltage drop in the line and (*iii*) by using tapped exciting winding, so that the constant voltage source may be impressed across the entire winding or some part of it.

Because of their poor starting torque, low power factor and poor efficiency, such motors are only suitable for low power applications such as for toys, small fans, electric clocks, hairdryers, time phonographs, slide projectors, humidifiers, small business machines (photocopying machines, vending machines, advertising displays etc.) and other similar applications. They are manufactured in small sizes ranging from below 5 W to about 100-150 W. They are available with built-in gear reducers to give almost any speed (as low as below 1 rpm.)

Example 20.12. The name plate of a 1-phase, 4-pole induction motor gives the following data:

Output 373 W; 230 V; Frequency 50 Hz; Input current 2.9 A; Power factor 0.71; Speed 1,410 rpm. Calculate (*a*) the efficiency and (*b*) the slip of the motor when delivering the rated output.

Solution: Power output, $P_0 = 373$ watts

Power input, $P_{in} = VI \cos \phi = 230 \times 2.9 \times 0.71 = 473.57$ W

(*a*) Motor efficiency, $\eta = \dfrac{P_0}{P_{in}} \times 100 = \dfrac{373}{473.57} \times 100 = 78.76\%$ **Ans.**

Synchronous speed, $N_s = \dfrac{120\,f}{P} = \dfrac{120 \times 50}{4} = 1{,}500$ rpm

(*b*) Slip, $s = \dfrac{N_s - N}{N_s} \times 100$

$= \dfrac{1{,}500 - 1{,}410}{1{,}500} \times 100 = 6\%$ **Ans**

20.14 SPEED CONTROL OF SINGLE-PHASE INDUCTION MOTORS

The stator winding of a single-phase induction motor can easily be arranged to give two synchronous speeds, one double the other. In Fig. 20.22 the connections for two pole motors are shown in which two coils A and B have been connected in series with their mmfs helping each other. If the connections to coil B are reversed so that both the coils are again in series but their mmfs opposing each other, as shown in Fig. 20.23, 4 poles are produced and the synchronous speed is reduced to half. Similarly a four-pole motor can be re-connected to give 8 poles. The change from one speed to the other is easily and quickly accomplished by means of a two-pole double-throw switch.

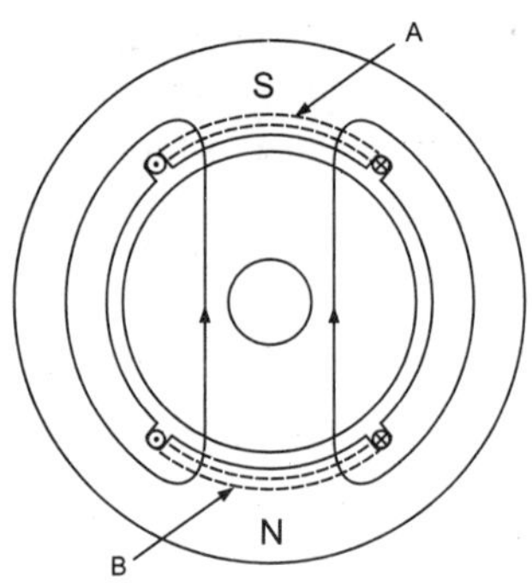

Fig. 20.22 *2-Pole Connection*

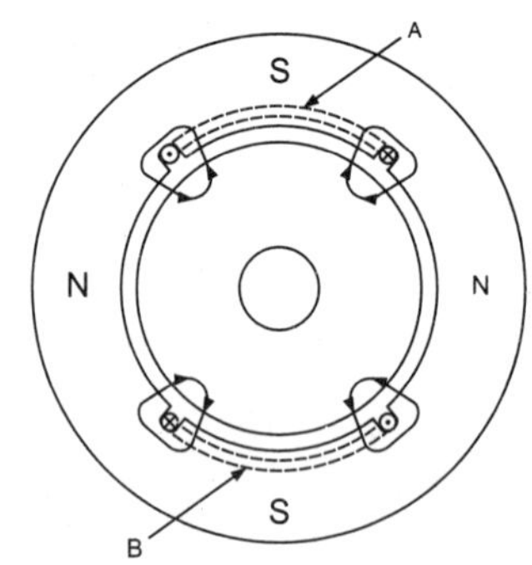

Fig. 20.23 *4-Pole Connection*

Furthermore, because of the comparatively small power handled by single-phase induction motors, excessive slip to give a slight reduction in speed is not serious, from the standpoint of waste of power, as is case of a 3-phase induction motors. For that reason in various cases, such as for fan and blower applications where speeds of 1,100 or 1,300 rpm might be required, it is not unusual to employ 4-pole motor with high slip. This can be accomplished (*i*) by using high rotor resistance, in which case no variation in speed is introduced, but the motor displays a drooping speed curve; or (*ii*) by reducing the voltage applied to the stator winding. For this purpose variable resistance or tapped reactance coil in series with the motor can be employed for speed regulation. This method is inefficient, gives large variations in speed with the variations in load, but it is much used inspite of these drawbacks. Alternatively, the voltage applied to the motor stator winding can be varied either by means of a transformer with taps on its secondary or by means of a variac, which provides the maximum possible number of taps and thus gives finest speed adjustment. This is one of the simplest methods of speed control of fractional kilowatt motors. The speed of the single-phase induction motor can also be controlled by employing tapped exciting winding, so that the constant voltage source may be impressed across the entire winding or some part of it. By this means, a normal speed and a single reduced speed are provided.

Speed control because of its more general use, has brought about the standardization of fan speeds, allowing the use of high slip motors. A squirrel cage motor with 8-10 per cent slip and with low pull-out torque operates satisfactorily not only for constant speed drives, but also for adjustable speed drives through the use of voltage control on the stator winding. With this method of speed control, the torque-speed characteristics resembles with those of a wound rotor motor with different number of external resistances in the motor circuit.

20.15 AC COMMUTATOR MOTORS

It will be appreciated that the 3-ϕ induction motor, although possessing many excellent features, is not a variable speed motor. Moreover, the single-phase induction motor is not self-starting and has a poor power factor and efficiency. AC commutator motors have been developed in an attempt to provide improved starting torque, good power factor and better speed control.

AC commutator motors may be divided broadly into two classes, viz., single-phase and 3-phase machines. The more common single-phase commutator motors are the series motors, universal motors, repulsion motors and the repulsion-induction motors, with various modifications and combination of these types being in use.

20.16 AC SERIES MOTORS

Direct current shunt or series motors rotate in the same direction regardless polarity of the supply. If the line terminals be reversed, both the field current and the armature current are reversed, the torque developed remains in the same direction and therefore, the motor continues to rotate in the direction in which it was rotating before the line terminals were reversed. From this it seems that any dc motor would operate satisfactorily when connected to single-phase ac supply. However, it is not true.

It has been found that the shunt motor develops but little torque when it is connected to an ac supply. The field winding of shunt motor is wound with many turns of fine wire so as to limit the field current but to give high flux. These many turns of wires cause a significant inductive effect, resulting in a high reactance and so a high impedance. The result is a very low field current, much lower than that required to develop sufficient starting torque. There is another adverse effect of high inductance of the shunt field. The high inductance of the shunt field causes the field current and, therefore, the flux to lag nearly 90° in time with respect to the line voltage. The rotor (or armature) current also lags behinds the applied voltage, but because of its fewer turns, the angle of lag cannot be large. Time phase difference between the flux and the rotor current, therefore, is considerable, and since torque developed in an ac motor is proportional to flux , rotor current, and the cosine of the angle between the two, the torque developed is further reduced. For these reasons, viz the reduction of field current and time-phase difference between the flux and the rotor current, a shunt motor will not develop sufficient starting or running torque when connected to single-phase ac supply.

However, in case of a series motor, the field is wound with relatively few turns of heavy wire and, therefore, the inductive effect is less and the field current and the flux are almost in phase with each other. The rotor current is, of course, the same in both field and rotor, and there is no time lag between the flux produced by the field coils and the current in the rotor conductors. Thus the two objections to the operation of a shunt motor on ac supply are overcome in the case of a series motor, and it may operate on both ac and dc.

It is found, however, that an ordinary dc series motor, when connected to single-phase ac supply, operates, but not very satisfactorily. The reasons are as follows :

1. The torque developed is not of constant magnitude, as in dc series motors, but pulsates between zero and maximum values each half cycle because the field and rotor (or armature) currents both reverse every half cycle.
2. The alternating field flux would induce eddy currents in the solid parts of the field structure, such as the yoke and cores, causing excessive heating and lowering of efficiency.
3. From the schematic diagram of an ordinary series motor, shown in Fig. 20.24 it is clear that there is no inductive coupling between the field (stator) and armature (rotor) winding, as they are placed in space quadrature. As a consequence, both the windings act as open-circuited primaries of the transformers and, therefore, each winding has large reactance. This results in small current and abnormal voltage drop and so reduces the output and power factor to such low values as to make the motor impractical.

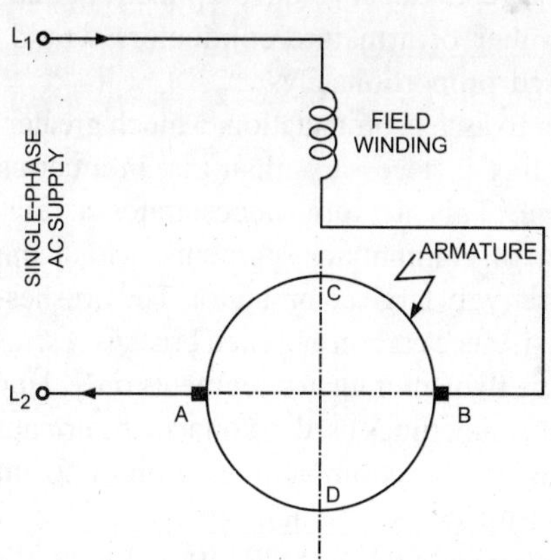

Fig 20.24 *Schematic Diagram of an Ordinary Series Motor*

4. When a series motor operates on dc supply, it develops rotational emf in conductors placed along the brush or q-axis (conductors AB in Fig. 20.24) due to the armature reaction flux. Self-induced emf, often called the reactance voltage, is developed in these conductors due to the reversal of the current in them on crossing the brush. In case of ac operation, in addition to rotational emf and reactance voltage, transformer emf is also developed which will be maximum for conductors A and B (conductors lying midway between the pole centres) because the armature coil with the greatest emf is the one whose magnetic axis is parallel to the main field. Consequently, at the commutator

the voltages developed by transformer action are greatest between segments just where the brushes are situated, at the so-called neutral points. The result would be destructive sparking if measures were not taken to limit the current passing between segments short-circuited by the brushes.

5. Because of large voltage drop in the machine due to its high inductance, the speed for ac operation is much less than that for dc operation for a given torque (refer to Fig. 20.31).
6. Starting torque is also poor due to low power factor at starting.

Hence some modifications are necessary in a dc series motor that is to operate satisfactorily on ac.

20.16.1. Modifications and Improvements in Design of AC Series Motor

1. In order to reduce the eddy current losses, the whole magnetic circuit is laminated, the field magnet system as well as the armature core. For this reason salient or projecting pole construction is unusual; the field magnet system is smooth cored like the stator of an induction motor. Slots are cut axially into the inner face of the stator to accommodate the main, commutating and compensating windings.
2. In order to reduce the reactance of the series field, ac series motors are built with as few turns as possible. Fewer field turns per pole means lower flux per pole which improves the commutation by reducing the transformer emf in the commutation coils.
3. Reduction in number of turns on the field winding would also reduce the flux Φ for a given current and would tend to increase the speed and reduce the torque available for a given current. Hence to develop the required load torque the number of armature conductors would have to be increased proportionately.
4. In order to assist commutation, a much greater subdivision of winding is necessary than that in a corresponding dc machine. This, in turn, necessitates a relatively large number of commutator segments, which implies also a comparatively large commutator. The brushes are made of high resistance carbon and each brush is restricted in width to bridge two commutator segments only. High resistance leads are sometimes used to connect the armature (or rotor) winding to the commutator, in order to minimize the circulating current which flows when a coil is short circuited by a brush (Fig. 20.29).
5. The increase in armature conductors causes increase in the inductive reactance of the armature and thus there is no reduction in the net reactance of the machine. In order to overcome this difficulty the machine is provided with a compensating winding, which neutralizes, as effectively as possible, the armature mmf.
6. However, there is a limit to the number of armature conductors that can be accommodated on a given core, this limit being fixed by the consideration of heating. The only way to develop the desired torque in motors of considerable ratings is to use numerous poles (with lesser flux per pole). The use of numerous poles, each with relatively few turns, has the further advantage of reducing the field winding inductance in addition to the main advantage of improving commutation.
7. To obtain sufficient flux with few ampere-turns per pole, the reluctance of the magnetic circuit must be reduced to minimum. This is accomplished by operating the iron at low flux density and, therefore, at high permeability and by using a very short air gap. Because of the small number of field ampere-turns and the very low flux density a short pole is required.
8. The operating voltage is kept low in order to reduce the inductance. Because high voltage motor with proportionately low current would require a large number of turns to produce the given flux.
9. Since inductive reactance is directly proportional to frequency, ac series motor operating characteristics are better at low frequency. Some series motors are built in large sizes for traction services and are designed to operate at low frequency, 25 Hz or less. However, fractional kilowatt motors may be designed to operate satisfactorily at commercial frequency of 50 Hz.
10. Interpoles are provided in ac series motors to eliminate the armature-resistance leads, which occupy valuable space and lower the rating of the motor by their heating.

20.16.2. Commutation. In single-phase ac series motors, as already explained, two emfs viz rotational (or speed) emf and transformer emf, are induced in the armature conductors. Whatever the brush position may be, there is always an emf induced in the short-circuited element of the armature winding. Thus when the brushes are in normal position *i.e.* perpendicular to the main field axis, transformer induced emf is in the short-circuited element and when the brushes are along field axis, dynamically induced emf is in it. Hence motor gives more sparking than an ordinary dc series motor. Excessive sparking and poor power factor are prevented by providing compensating winding, which carries the armature current and neutralize the armature flux. In Fig. 20.25 the field flux is produced along the axis CD and armature flux is produced along axis AB. Thus the compensating winding must produce a flux along AB and in a direction opposite to that of the armature flux at every instant. The compensating winding is placed in the stator slots, and its axis is kept at 90 electrical degrees to the main field axis. It may be connected in series with both of the armature and series field, as illustrated in Fig. 20.25. The number of turns per pole in the

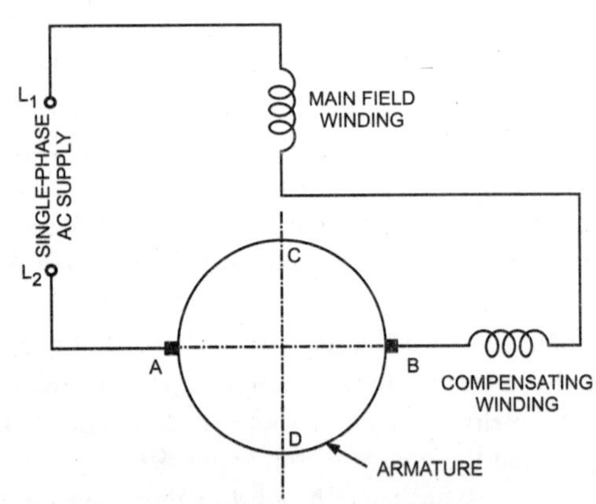

Fig. 20.25 *Conductively Compensated Series Motor*

compensating winding is so chosen that the mmf produced by it is equal and opposite to that of the armature. In this case the compensating winding carries current equal to that flowing in the armature but in opposite direction due to conductive effect, so it is known as *conductively compensated series motor*.

It is possible to combine the main field and the compensating field by employing a single distributed winding which produces its flux in the direction of the resultant flux of the main field and compensating windings, as illustrated in Fig. 20.26.

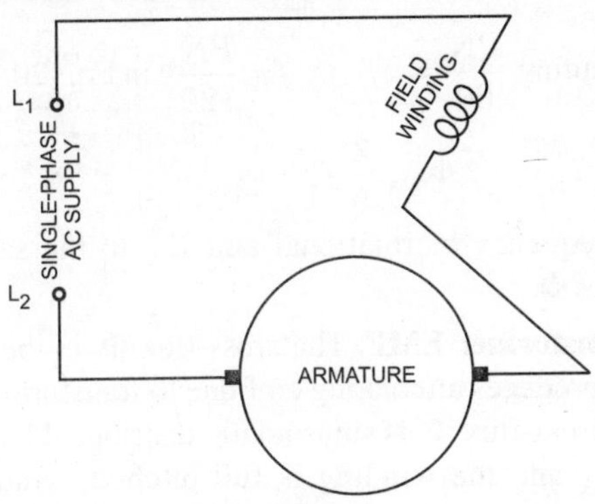

Fig. 20.26

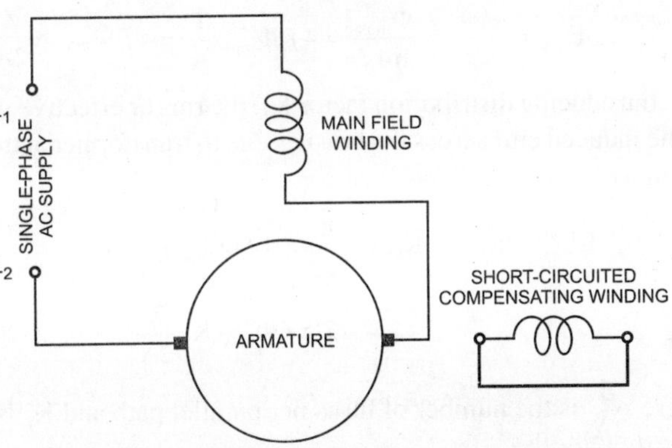

Fig. 20.27 *Inductively Compensated Series Motor*

Another method is to short circuit the compensating winding, as shown in Fig. 20.27, so that it behaves like the short-circuited secondary of a transformer, of which the primary is the armature winding. Since in this case the compensating winding is excited inductively so the motor with short-circuited compensating winding is known as *inductively compensated series motor*. The current in the compensating winding will be almost directly proportional to the armature current and practically 180° out of phase with it. Therefore, the action of the compensating winding will be almost identical with that of a conductively compensating winding. The resistance and inductive reactance characteristics of the armature circuit of the inductively compensated motor will be increased over their values for the armature alone by a certain amount through the coupling of the compensating winding to the armature in accordance with the transformer action. Thus, the total resistance and reactance characteristics of the inductively compensated motor are practically the same as those for an equivalent conductively compensated motor. In fact, the performance characteristics are practically indentical for the two types of ac series motors.

20.16.3. Interpoles. Generally, all ac series motors are provided with commutating poles for improving commutation (as in case of dc motors) but commutating poles alone do not produce satisfactory commutation, unless something is done to neutralize the transformer emf.

The transformer emf in the coils undergoing commutation lags the main flux by 90° and hence lags the current by nearly 90°. The flux in the usual series connected interpole is practically proportional to the current and in phase with it. Hence the rotational or speed emf due to the commutating pole flux is practically in phase with the current and thus must lead the transformer emf by nearly 90°. The speed emf, therefore, cannot neutralize the transformer emf but adds vectorially. However, as in the dc machines, there is a speed emf due to the commutated coils cutting the armature leakage cross-flux, as well as an emf of self-induction. These emfs are in phase with the current but are proportional to the speed. Hence, the interpoles can neutralize them at one speed only.

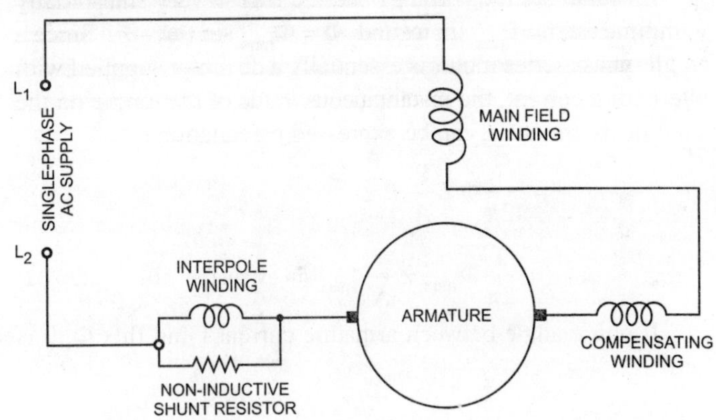

Fig. 20.28 *Series Motor With Interpoles*

The speed emf and the transformer emf in the commutated coils add in quadrature to give a resultant emf. By shunting the interpoles with a non-inductive shunt resistor, as shown in Fig. 20.28, the phase of the interpole flux can be made such as to neutralize the resultant. The neutralization can be made at one speed only, and accordingly the interpoles are only partly effective at starting. However, their use enables the circulating current to be reduced to lower values over range of speeds and load, so that good commutation is achieved for normal running conditions. Commutating poles increase the efficiency of the machine. Output of the machine with a given size of armature core is also increased by employing commutating poles.

The earliest method of reducing the sparking due to transformer emf was by use of high resistance leads from the winding tapping points to the commutator segments, as shown in Fig. 20.29. Such high resistance leads form paths for both, main current as well as circulating current, so copper loss due to

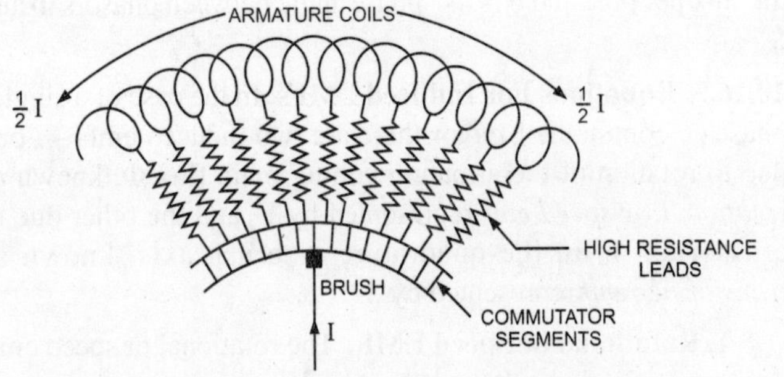

Fig. 20.29

main current is increased, but that due to circulating current is reduced. Therefore, the value of this resistance used to be so decided that it may result in minimum copper loss.

Now this method has been replaced by using interpoles shunted by non-inductive resistors, as shown in Fig. 20.28.

20.16.4. Torque Equation. A schematic diagram of a single-phase series motor with interpole and compensating windings is given in Fig. 20.28.

Since the field winding is connected in series with the armature, the armature and field currents are in phase. But the flux Φ, which is developed by the field current and passes through the armature, lags behind the armature current I by a small angle θ owing to the influence of the steel losses and particularly, of the currents in the short-circuited armature winding coil sections electromagnetically linked with the field winding of the motor.

Assume that the current I and the flux Φ vary sinusoidally with time i.e, $i = I_{max} \sin \omega t$ and $\Phi = \Phi_{max} \sin (\omega t - \theta)$. Since a single-phase series motor is essentially a dc motor supplied with alternating current, the instantaneous value of the torque on the shaft of its armature, can be expressed by equation

$$T = \frac{1}{2\pi} \Phi Z \frac{P}{A} i$$

$$= \frac{1}{2\pi} \Phi_{max} Z \frac{P}{A} I_{max} \sin \omega t \sin (\omega t - \theta) \quad ...(20.32)$$

If phase angle between armature current I and flux Φ, θ is zero

$$T = \frac{1}{2\pi} Z \frac{P}{A} \Phi_{max} I_{max} \sin^2 \omega t \quad ... (20.33)$$

Hence the torque of a single-phase series motor remains positive during the whole period varying sinusoidally during the half cycle from zero to maximum value given by the equation

$$T_{max} = \frac{1}{2\pi} Z \frac{P}{A} \Phi_{max} I_{max}$$

When the phase angle θ is not equal to zero, the torque remains positive during the time interval corresponding to the angle $(\pi - \theta)$ and has negative values during the interval corresponding to the angle θ. The average value of the torque on the motor shaft is given as

$$T_{av} = \frac{1}{\pi} \int_0^{\pi} \frac{1}{2\pi} Z \frac{P}{A} \Phi_{max} I_{max} \sin \omega t \sin (\omega t - \theta) \, d\omega t.$$

$$= \frac{1}{2\sqrt{2}\pi} Z \frac{P}{A} \Phi_{max} I \cos \theta \quad ...(20.34)$$

where I is the effective value of current, Φ_{max} is the peak value of the flux per pole and θ is the phase angle between phasors Φ and I.

20.16.5. Equations For Induced EMFs. In the case of a single-phase ac commutator motor there are two induced emfs — one due to rotation of the armature cutting main flux Φ, known as *rotational or speed emf* represented by E_r and the other due to cross–flux Φ′ in the quadrature or brush axis, known as *transformer emf* represented by E_r

1. Rotational or Speed EMF. The rotational or speed emf, as in case of a dc machine, is given as

$$E = \Phi Z \frac{N}{60} \times \frac{P}{A} \text{ volts}$$

Now when the flux per pole is alternating and the brushes are in the neutral plane, then effective or rms value of rotational emf is given as

$$E_r = \frac{\Phi_{max}}{\sqrt{2}} \frac{ZN}{60} \times \frac{P}{A} \text{ volts} \quad ...(20.35)$$

where Φ_{max} is the peak value of flux pole

Substituting $\frac{PN}{60} = 2f_r \; (\because f = \frac{PN}{120})$ in Eq. (20.35), we get

$$E_r = \sqrt{2} \, \Phi_{max} \frac{Z}{A} f_r \quad ...(20.36)$$

The frequency of rotational emf E_r in the same as the frequency of Φ.

2. Transformer EMF. The cross-flux Φ′ in the neutral or brush axis produces alternating emf due to transformer action.

If the cross-flux Φ′ is sinusoidally distributed having, peak value Φ'_{max} and the winding is full-pitched, winding being commutator winding, the magnitude of induced emf is given as

$$E_{tav} = \frac{d\Phi}{dt} = \frac{\Phi'_{max}}{1/4f} = 4f \Phi'_{max} \frac{T}{A} = 4f \Phi'_{max} \times \frac{Z}{2A}$$

Introducing distribution factor K_b, the rms or effective value of the induced emf across the brushes due to transformer action is given as

$$E_{trms} = K_f E_{av.} K_b = \frac{\pi}{2\sqrt{2}} \times 4f \Phi'_{max} \frac{Z}{2A} K_b$$

$$= \sqrt{2} \pi f \Phi'_{max} K_b \frac{Z}{2A} \quad ...(20.37)$$

where $\frac{Z}{2A}$ is the number of turns per parallel path and K_b is the distribution factor, given as

$$K_b = \frac{\sin n \frac{\alpha}{2}}{n \sin \frac{\alpha}{2}} = \frac{2}{\pi}, \text{ as the winding is uniformly distributed}$$

$$\text{So, } E_t = \sqrt{2} \pi f \Phi'_{max} \times \frac{2}{\pi} \times \frac{Z}{2A} = \sqrt{2} \Phi'_{max} \frac{Z}{A} f \text{ volts} \quad ...(20.38)$$

The frequency of transformer emf E_t is the same as that of cross-flux Φ′.

20.16.6. Series Motor Phasor Diagram. The phasor diagram of a single-phase ac series motor is given in Fig. 20.30. Flux Φ is produced by current I. Flux Φ lags behind the current I by an amount too small to be important, so it has been shown in phase with current I. The resistive drops IR_{se}, IR_p, IR_c and IR_a due to resistances of series field, interpole circuit, compensating winding and of armature respectively are in phase with current I. The reactive drops IX_{se}, IX_p, IX_c and IX_a due to reactances of series field, interpole circuit, compensating winding and of armature respectively are in quadrature with current I and leading. Here it is noteworthy that reactive drop due to series field is much greater than that of either armature or the compensating field even all the measures to reduce series field inductance, as mentioned in Art 20.16.1, have been taken. Unlike the armature cross-flux Φ′, the flux that causes this reactive drop cannot be neutralized

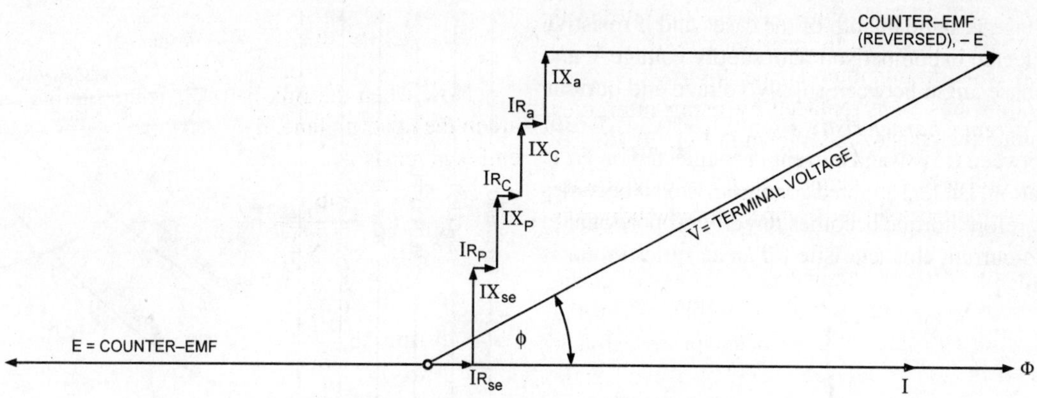

Fig 20.30 *Series Motor Phasor Diagram*

because it is the necessity for the development of torque. The interpole circuit consists of the interpole winding in parallel with the non-inductive shunt, as illustrated in Fig. 20.28. R_p and X_p are the respective equivalent resistance and reactance of the interpole parallel circuit.

The phase angle of E, in the phasor diagram is most easily determined by considering instantaneous values. When the alternating field flux Φ is at its peak value, the armature conductors cut the maximum flux, and the rotational or speed emf is, therefore, a maximum. When the field flux is zero, the rotational or speed emf, E is zero. Since the graphs of flux Φ and generated emf E pass through zero at the same instant, and also through maximum at the same instant, the phasor E must be either exactly in phase with the phasor of flux Φ or exactly in phase opposition. The first possibility is readily eliminated by noting that in case of a motor the generated voltage is in opposition to the direction of flow of current, and, therefore, E is in phase opposition with current and is called the counter–emf.

The terminal voltage V is the phasor sum of the counter-emf reversed and all the resistive and reactive volt drops.

20.16.7. Series Motor Characteristics.

1. *Power output Characteristic.* Mechanical power developed is given by the product of counter-emf E and the armature current I. Though counter-emf E decreases slightly with the increase in current, but if it is neglected in comparison with the magnitude of counter-emf E, mechanical power developed increases almost in proportion with current I.

Power available at shaft or power output is equal to the mechanical power developed less rotational losses. Power output characteristic is shown in Fig. 20.34.

2. *Power Factor Characteristic.* The cosine of the phase angle ϕ is the power factor of the motor.

From phasor diagram shown in Fig. 20.30

$$\sin\phi = \frac{I\left(X_{se} + X_p + X_c + X_a\right)}{V} = \frac{IX}{V}$$

where $X = X_{se} + X_p + X_c + X_a$ and = Effective reactance of the motor

$$\text{or power factor, } \cos\phi = \sqrt{1 - \left(\frac{IX}{V}\right)^2} \qquad ...(20.39)$$

In order to have high power factor the reactance drops must be low and counter-emf high. The reactance drops are lowest and the counter-emf is highest at light loads, and therefore, the power factor of the single-phase series motor is highest (nearly unity) at light loads and falls with the increase in current. This is the reason that power factor of a series motor is low at overloads and at starting. This is the reverse of the power factor relation that exists in case of the induction motor and transformer.

3. *Speed-Current Characteristic.* Speed of a commutator motor is proportional to counter–emf E (E_r or E_b) or proportional to supply voltage less voltage drops because

$$E \;\alpha\; \Phi N \text{ and } E = V - \text{voltage drops}$$

The current causes resistive drops in series field and armature in case of dc operation but in case of ac operation it also causes large reactance drops in series field, armature, compensating windings and interpole circuit in addition to resistive drops in these windings. Hence with ac operation, counter-emf E developed is much less with dc operation. It means speed-current characteristic for ac series motors are more drooping as compared to those for dc series motors, as illustrated in Fig. 20.31.

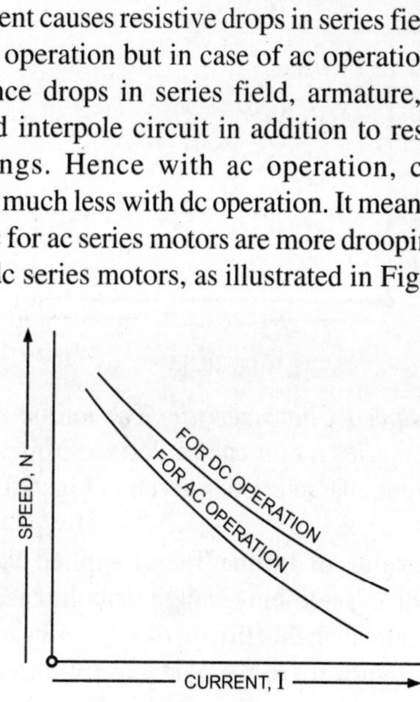

Fig. 20.31

Let the same series commutator motor be operated on dc as well as on ac and with same current I.

The back emf developed in case of dc operation is given as

$$E_b = \frac{\Phi Z N_{dc}}{60} \times \frac{P}{A} \text{ volts and from Eq. (20.35).}$$

$$E_r = \frac{\Phi_{max}}{\sqrt{2}} \frac{Z N_{ac}}{60} \times \frac{P}{A} \text{ volts}$$

Neglecting saturation $\Phi_{max} = \sqrt{2}\,\Phi$ as the peak value of ac flux Φ_{max} is set up by the peak value of current $\sqrt{2}\,I$ while dc flux Φ is set up by current I . So from above two equations

$$\frac{N_{ac}}{N_{dc}} = \frac{E_r}{E_b} = \frac{V\cos\phi - IR}{V - IR} = \cos\phi \qquad ...(20.40)$$

if the supply voltage is same in both of the cases and if resistive drops IR are neglected in comparison with supply voltage V and where ϕ is the phase angle between supply voltage and current

4. *Torque-Current Characteristic.* From Eq. (20.34) $T \propto \Phi I$ if phase angle between flux Φ and current I is neglected or $T \propto I^2$ neglecting saturation. Under saturated condition, flux Φ becomes constant and, therefore, torque becomes directly proportional to current I. Torque-current characteristic for an ac series motor is shown in Fig. 20.32.

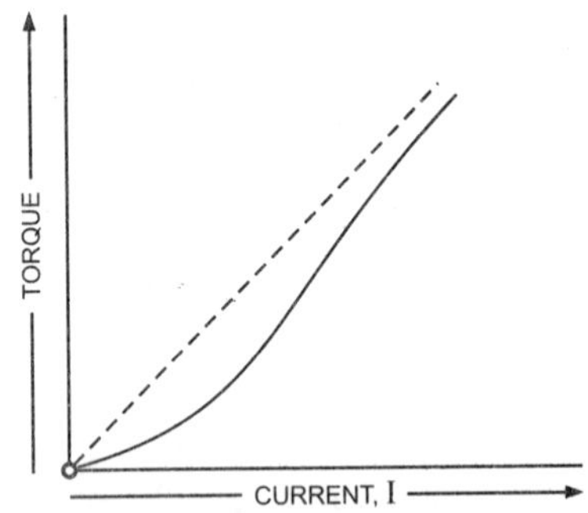

Fig. 20.32

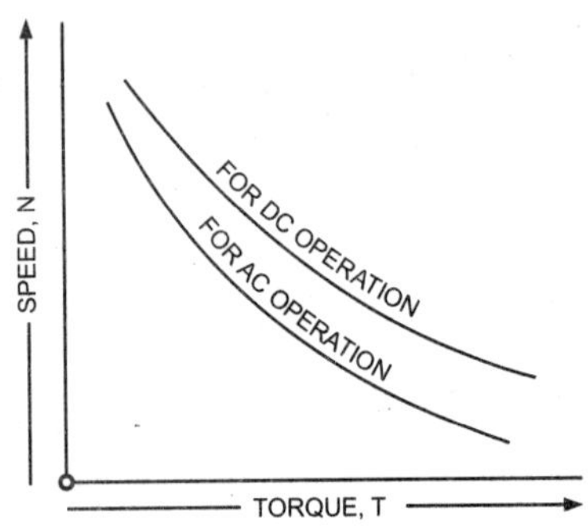

Fig. 20.33

5. *Torque-Speed Characteristic.* The torque-speed characteristic for an ac series motor can be derived from speed-current and torque-current characteristics given in Figs 20.31 and 20.32 respectively.

For given value of torque T and applied voltage V, the armature current is same but voltage drop in case of ac series motor is much more than that in case of a dc series motor so speed of an ac series motor for a given developed torque is less than that of a dc series motor, as illustrated in Fig. 20.33.

The single-phase ac series motor has practically the same operating characteristics as the dc series motor. The torque, or tractive effort, varies nearly as the square of the current and the speed varies inversely as the current. This is illustrated in Fig. 20.34.

However, in case of an ac series motor (*i*) power factor is very low at starting and on overloads on account of high inductive nature of the series field and armature circuits (*ii*) efficiency is not as good as in a corresponding dc machine due to eddy current losses and effects of power factor and (*iii*) starting torque is low due to poor power factor at starting.

For a given kilowatt rating ac series motor is 1.5-2 times in size and weight of the corresponding dc series motor. The construction cost of an ac series motor is much more than that of a dc series motor.

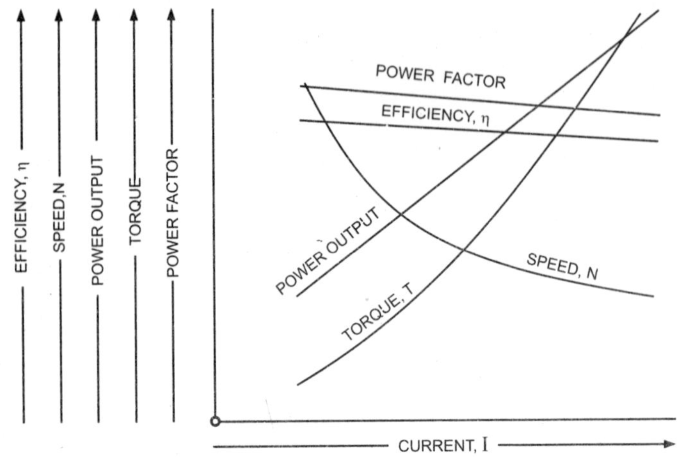

Fig. 20.34

The speed of an ac series motor may be controlled efficiently by taps on a transformer, which is not possible in case of a dc series motor.

20.16.8. AC Series Motor Circle Diagram. From phasor diagram shown in Fig. 20.30,

$$\text{Sin}\,\phi = \frac{IX}{V}$$

$$\text{and } \cos\phi = \frac{E+IR}{V}$$

$$\text{So} \quad \left(\frac{IX}{V}\right)^2 + \left(\frac{E+IR}{V}\right)^2 = \sin^2\phi + \cos^2\phi = 1$$

$$\text{or } (IX)^2 + (E+IR)^2 = V^2 \qquad ...(20.41)$$

The above equation is an equation of a circle of diameter V on appropriate scale.

If a ΔOAB is drawn so as to represent applied voltage V by side OA, speed emf plus resistance drop (i.e. E + IR) by the side AB and reactance drop IX by the side OB, then the locus of point B is a semicircle with OA as diameter because the $\angle ABO$ is always right angle. The current phasor is along AB because counter-emf E and resistance drop IR are in phase with current I. Thus $\angle OAB$ is the power factor angle cos ϕ. The current I, in magnitude, is proportional to OB, reactance X being fixed.

Let the voltage scale be 1mm = K volts. Then current scale will be $\frac{K}{X}$ amperes.

If resistance drop IR represented by BB′ is deducted from AB, then AB′ will represent the counter or rotational emf E. With the increase in load, speed will decrease and therefore, AB′ representing counter-emf E will become smaller.

$$\text{Now } \angle OB'A = \angle OBA + \angle BOB' = 90° + \beta$$

$$\text{where } \beta = \tan^{-1}\frac{BB'}{OB} = \tan^{-1}\frac{IR}{IX} = \tan^{-1}\frac{R}{X}$$

Thus angle $\angle OB'A$ is constant so the point B' lies on another circle, shown dotted in the figure.

Now from ΔOBD in Fig. 20.35.

$$BD = OB\cos\phi = I\cos\phi$$

$\because$ OB represents current on current scale

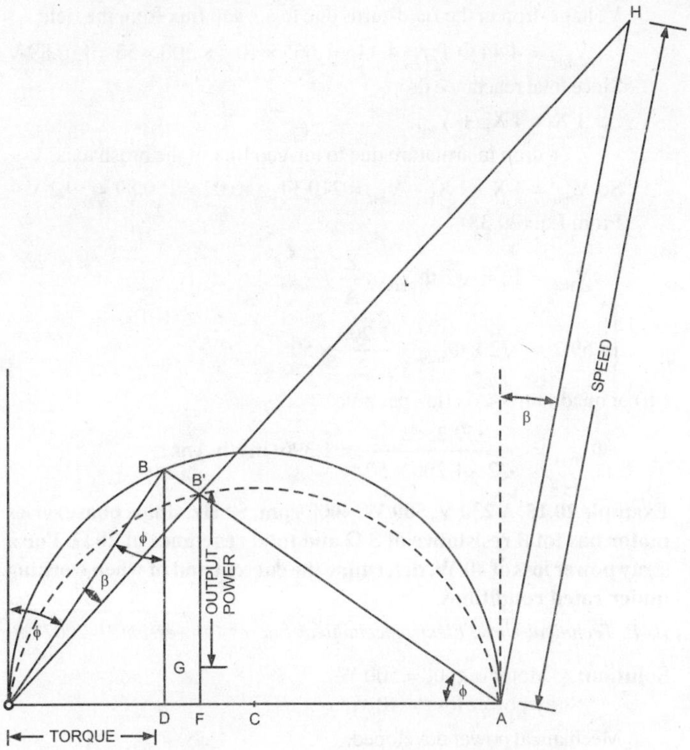

Fig. 20.35 *AC Series Motor Circle Diagram*

Power input being equal to VI cos ϕ is proportional to I cos ϕ, so BD can be taken as the power input to the scale 1 mm $= \frac{KV}{X}$ watts.

Further mechanical power developed,

$$P_{mech} = \text{Power input} - \text{copper losses} = VI \cos\phi - I^2 R$$

Power input to the motor

$$= VI \cos \phi = OA \times BD = 2 \times \text{area of } \Delta\, OAB,$$

$$\text{Copper losses} = I^2 R = I \times IR$$

$$= OB \times BB' = 2 \times \text{area of } \Delta\, OBB'$$

So, *mechanical power developed,*

$$P_{mech} = 2\,[\text{area of } \Delta\, OAB - \text{area of } \Delta\, OBB']$$

$$= 2 \text{ area of } \Delta\, OAB' = B'F \times OA$$

Since OA, being equal to supply voltage V, is constant so B'F represents the mechanical power developed by the motor armature with the same power scale, already determined above.

Since *output power*

$$= P_{mech} - \text{stray power losses} = B'F - FG = B'G$$

where FG represents the stray power losses.

Since torque, T α I^2 neglecting saturation

$$\text{So } T \propto OB^2 = OD \times OA \qquad \because \angle OBA = 90^\circ$$

Since OA is fixed, being equal to supply voltage V, OD represents torque to the proper scale. The torque scale may be determined by computing torque for a particular current from Eq. (20.34) with θ = 0

Now a line is drawn from point A, making an angle β with vertical and intersecting the line OB' extended at H.

Since Δs OHA and OB'A are similar,

$$\text{So } \frac{AB'}{OB'} = \frac{AH}{OA}$$

$$\text{or } \frac{AB'}{OB} = \frac{AH}{OA} \times \frac{OB'}{OB} = \frac{AH}{OA} \sec\beta$$

Since rotational or counter-emf E α flux Φ × speed N

$$\text{or speed, } N \propto \frac{E}{\Phi} \propto \frac{AB'}{OB} \propto \frac{AH}{OA} \sec\beta$$

∵ E is represented by AB' and flux Φ being proportional to I (neglecting saturation) is represented by OB

Since $\frac{\sec\beta}{OA}$ is constant so the speed is represented by AH whose scale may be determined by computing speed for any particular value of current from Eq. (20.35).

Thus the performance characteristics of an ac series motor can be predicted from a circle diagram.

20.16.9. Applications. The torque-speed characteristic of the single-phase series motor is similar to that of the dc series motor i.e. high starting torque and decrease in speed with increase in load making it to have self-relieving property from heavy excessive load, so such a machine is particularly useful for traction services.

Example 20.13. A 75 watt, 2-pole, 50 Hz, 230 V single phase series motor having salient poles has the following data :

Total resistance R= 25 Ω, total reactance 261 Ω, mutual reactance = 121 Ω on the direct axis, stray power loss = 15 W.

Calculate current, pf and output when running at 6,000 rpm.

Solution: If X_{md} and X_{mq} are the mutual reactances in the field and brush-axes respectively and X_L is the total leakage reactance of all the windings, the induced voltage in the field due to air-gap flux in the field or direct axis is $X_{md}I$ and that in the armature due to air-gap flux in the brush or quadrature axis is $X_{mq}I$. In case of fully compensated motor X_{mq} is neglected and total reactance is, therefore, $X = X_L + X_{md}$ and total mutual reactance $X_m = X_{md}$. However, in case of uncompensated series motor, $X = X_L + X_{md} + X_{mq}$.

Rotational or counter-emf E can be expressed in terms of direct-axis mutual reactance X_{md}. If the brushes were shifted to the pole axis, its induced emf due to transformer action of the pole flux would have been $X_{md}I$. The same pole flux induces rotational emf E_r in the armature with the brushes in the normal position i.e., in the interpolar axis and its magnitude is $\frac{f_r}{f}$ times X_{md} as evident from Eqs (20.36) and (20.38).

So counter-or rotational emf,

$$E = X_{md}\, I \cdot \frac{f_r}{f}$$

Rotational frequency ,

$$f_r = \frac{PN}{120} = \frac{2 \times 6{,}000}{120} = 100$$

So counter-emf,

$$E = X_{md} \times I \times \frac{f_r}{f} = 121 \times I \times \frac{100}{50} = 242 \text{ I volts}$$

So supply volts,

$$V = IR + jIX + E$$

$$= IR + IX_{md} \times \frac{f_r}{f} + jIX$$

$$= I\left(R + X_{md} \times \frac{f_r}{f} + jX\right)$$

$$= I\,(25 + 242 + j\,261) = 373.4\, I \angle 44.35^\circ \text{ V}$$

$$\text{or Current, I} = \frac{V}{373.4} = \frac{230}{373.4} = 0.616 \text{ A } \textbf{Ans.}$$

$$\text{Power factor} = \text{Cos } \phi = \text{Cos } 44.35^0 = 0.715 \text{ lagging } \textbf{Ans.}$$

$$\text{Power output} = \text{E I} - \text{stray power losses}$$
$$= 242 \times 0.616 \times 0.616 - 15 = 76.83 \text{ W } \textbf{Ans.}$$

Example 20.14. The following data refer to a 240-V, 2-pole, uncompensated series motor when run on a 50 Hz single-phase supply.

Applied voltage = 240 V (rms) : Input current = 1.5 A; Input watts = 198.75 W.

Field winding resistance = 8.8 Ω; Armature resistance = 4.2 Ω; Speed = 4,800 rpm. The number of turns of the field winding is 300 and the number of armature conductors is 1,200. Leakage reactance drop is 20 per cent of total reactive drop in the motor. Draw the phasor diagram and compute (i) induced voltage in the armature (*ii*) field flux per pole (*iii*) armature (quadrature axis) flux per pole.

Solution: From given datas and phasor diagram shown in Fig. 20.36

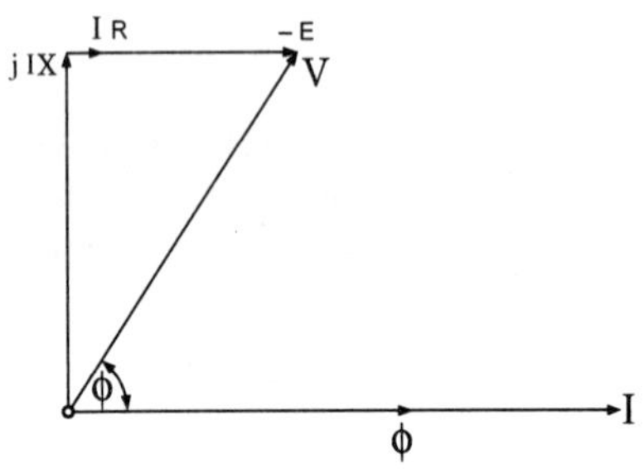

Fig. 20.36

$$\text{Current, I} = 1.5 \text{ A}$$
$$\text{IR drop} = 1.5\,(8.8 + 4.2) = 19.5 \text{ V}$$
$$\text{Power input} = 198.75 \text{ W}$$
$$\text{Supply voltage, V} = 240 \text{ V}$$

$$\text{Power factor, } \cos\phi = \frac{\text{Power input}}{\text{VI}} = \frac{198.75}{240 \times 1.5} = 0.552$$

$$\text{Phase angle, } \phi = \text{Cos}^{-1}\, 0.552 = 56.5°$$

$$\text{and } \sin\phi = \frac{\text{IX}}{\text{V}}$$

so total reactive drop, $\text{IX} = \text{V} \sin\phi = 240 \sin 56.5° = 200.11$ V

(*i*) Induced voltage in armature,

$$\text{E} = \sqrt{\text{V}^2 - (\text{IX})^2} - \text{IR} = \sqrt{240^2 - 200.11^2} - 19.5 = 113 \text{ V}$$

Since from Eq. (20.35) rotational emf E_r is given as

$$E_r = \frac{\Phi_{max}}{\sqrt{2}} \times \frac{ZN}{60} \times \frac{P}{A}$$

$$\text{so } E_r = E = 113 = \frac{\Phi_{max}}{\sqrt{2}} \times \frac{1{,}200 \times 4{,}800}{60} \times \frac{2}{2}$$

(*ii*) or flux per pole,

$$\Phi_{max} = \frac{113 \times \sqrt{2} \times 60}{1{,}200 \times 4{,}800} = 1.665 \text{ m Wb } \textbf{Ans.}$$

Drop due to leakage reactance,

$$\text{IX}_L = \frac{20}{100} \times \text{IX} = \frac{20}{100} \times 200.11 = 40.02 \text{ V}$$

Voltage drop in the field-turns due to air-gap flux from the field

$$V_{mf} = 4.44\, \Phi\, T_f f = 4.44 \times 1.665 \times 10^{-3} \times 300 \times 50 = 110.89 \text{ V}$$

Since total reactance drop,

$$\text{I X} = \text{I X}_L + V_{mf}$$

+ drop in armature due to air-gap flux in the brush axis, V_{ma}

So $V_{ma} = \text{I X} - \text{I X}_L - V_{mf} = 210.11 - 40.02 - 110.89 = 59.2$ V

From Eq. (20.38)

$$V_{ma} = E_t = \sqrt{2}\, \Phi'_{max} \times \frac{Z}{A} f$$

$$\text{or } 59.2 = \sqrt{2} \times \Phi'_{max} \times \frac{1{,}200}{2} \times 50$$

(*iii*) or quadrature–axis flux per pole,

$$\Phi'_{max} = \frac{59.2 \times 2}{\sqrt{2} \times 1{,}200 \times 50} = 1.396 \text{ mWb } \textbf{Ans.}$$

Example 20.15. A 230 V, 500 W, 4000 rpm, 50 Hz, single phase series motor has total resistance of 3 Ω and total reactance of 20 Ω. For a stray power loss of 40 W, determine the current and pf when working under rated conditions.

[*U.P. Technical Univ. Electromechanical Energy Conversion-II, 2007-08*]

Solution: Motor output = 500 W

Stray power loss = 40 W

Mechanical power developed,

$$P_d = 500 + 40 = 540 \text{ W}$$

Rotational or speed emf,

$$E_r = \frac{\text{Mechanical power developed, } P_d}{\text{Supply current, I}} \qquad \because P_d = E_r I$$

From phasor diagram shown in Fig. 20.37

$$V^2 = (E_r + IR)^2 + (IX)^2$$

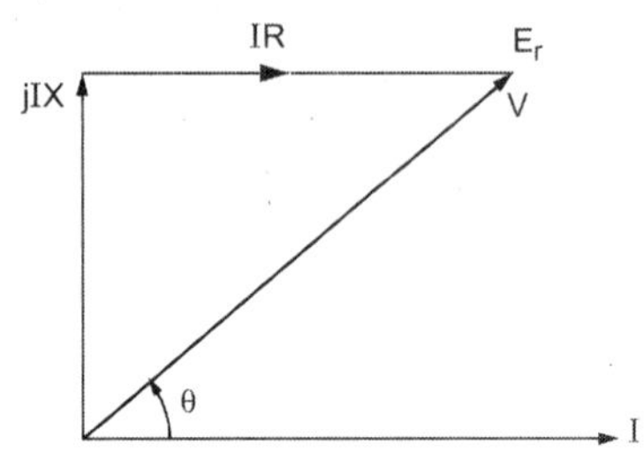

Fig. 20.37

Substituting $E_r = \frac{P_d}{I}$ in above equation we have

$$V^2 = \left(\frac{P_d}{I} + IR\right)^2 + (IX)^2$$

or $(R^2 + X^2)\, I^4 - (V^2 - 2\, P_d R)\, I^2 + P_d^{\,2} = 0$

Substituting the values of V, R, X and P_d is above equation we have

$$(3^2 + 20^2)\, I^4 - (230^2 - 2 \times 540 \times 3)\, I^2 + 540^2 = 0$$
$$\text{or } I^4 - 121.42\, I^2 + 712.96 = 0$$

Solving for I^2 we have

$$I^2 = \frac{121.42 \pm \sqrt{(121.42)^2 - 4 \times 712.96}}{2} = 6.19 \text{ A}^2$$

or $I = \sqrt{6.19} = 2.488$ A **Ans.**

discarding the larger value of current which is very large

$$\text{Sin } \theta = \frac{IX}{V} = \frac{2.488 \times 20}{230} = 0.21635$$

Phase angle $\theta = \text{Sin}^{-1}\, 0.21635 = 12.49°$

Power factor = Cos θ = Cos 12.49° = 0.976 (lagging) **Ans.**

20.17 UNIVERSAL MOTORS

For some applications, it is desirable to employ a motor that operates on either dc or ac. It is possible to build, by a compromise design, small series motors up to about $\frac{1}{2}$ kW rating to operate satisfactorily on dc or ac at 115 or 230 V. Such motors are called the *universal motors.*

A universal motor is a specifically designed series wound motor, that operates at approximately the same speed and output on either dc or ac of approximately same voltage.

Owing to reactance voltage drop, which is present on ac but not on dc, the motor speed is somewhat lower for the same load in ac operation, especially at high loads. On ac, however, the increased saturation of the magnetic circuit at the peak of the sine wave of current may materially reduce the flux below the dc value, and this tends to increase the speed in ac operation. It is possible, therefore, to design small series motors, called the *universal motors*, to have approximately the same torque-speed performance over the operating range, for all frequencies from 0 to 50 Hz.

Because of the difficulty in obtaining like performance on dc and ac at low speeds, most universal motors are designed to operate at speeds exceeding 3,500 rpm. Motors operating at speeds 8,000 to 10,000 rpm are common. High speed results in small size and weight of the motor.

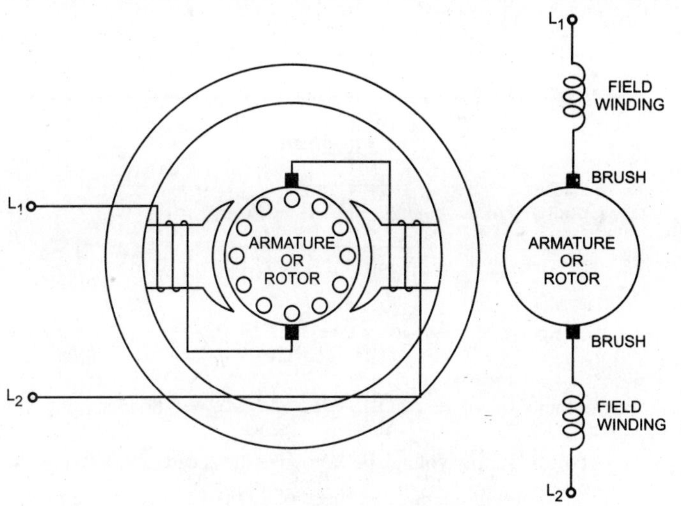

Fig. 20.38 *Universal Motor*

20.17.1. Construction. Basically, a universal motor is similar to a dc series motor in construction. However, it is constructed with few series field turns, laminated armature and field circuits, low-reluctance magnetic path, increased armature conductors and commutator segments and using low flux densities. This is done to minimize the adverse effects caused by high field reactance, eddy current and hysteresis losses.

The commutation on ac is much poorer than on dc, owing to current induced in the short-circuited armature coils, and this provides a definite limitation on their size and utility. If the wide brushes are used, the short-circuit currents are excessive, and the motor starting torque is reduced; while if narrow brushes are used, there may be excessive brush chatter at high speeds, causing short brush life. Good design, therefore, needs careful proportioning of commutator and brush rigging to meet conflicting electrical, mechanical and thermal requirements. Brushes employed are high resistance carbon ones so as to aid commutation.

Compensating winding is sometimes used in large rating motors to improve commutation but in a small rating and high speed universal motor this winding needs not to be provided because in such a motor armature current is so small to cause serious commutation troubles.

Universal motors may be either compensated (distributed field) or uncompensated (concentrated field) type, the latter type being used for the higher speeds and smaller output ratings (usually not exceeding 200 W) only.

A motor of the concentrated field type is usually a salient pole machine with two poles and a winding of relatively few turns to give opposite magnetic polarity. It consists essentially of a laminated core having either straight or skewed slots and a commutator to which the leads of the windings are connected.

Distributed field universal motors are of two types viz; single-field compensated motor and two-field compensated motor. The field windings of a 2-pole, single-field compensated motor resemble the stator winding of a 2-pole, split-phase ac motor. A two-field compensated motor has a stator winding consisting of a main winding and a compensating winding spaced 90 electrical degrees apart.

The frames of universal motors are usually made of aluminium, cast iron, and rolled steel.

The uncompensated motor is simpler and less expensive than the compensated motor but gives poorer speed regulation than a motor with compensation. The distributed-field compensated motor is employed for larger output ratings at high speeds, such as $\frac{3}{4}$ kW at 7,500 rpm, or for lower output ratings at low speeds, such as 200 watts at 3,000 rpm.

20.17.2. Operation. In a series wound motor, the same current flows through field windings and armature, being connected in series with each other, when motor is connected to either dc or ac supply. The magnetic fields developed by the series field winding and armature currents react with each other and hence develop unidirectional torque.

20.17.3. Characteristics. The torque-speed characteristic of a universal motor is quite similar to that of a series wound dc motor i.e., high starting torque and high no-load speed. In a very small series motor, the losses are usually large enough at no load to restrict the speed to a definite value. Where there is a danger of high no-load speed, a governor may be employed to keep speed within specified limits. The governor consists of a centrifugal switch mounted on the motor shaft. The tension of the springs of the switch is adjusted so that the switch opens at a predetermined speed and thus place an external resistor R in series with the armature, thereby reducing the speed. When the speed drops because of increase in load, the switch contacts close, thereby shorting out the series resistor R, thus again raising the speed. The connections are shown in Fig. 20.39. The capacitor C is placed across the governor to reduce sparking.

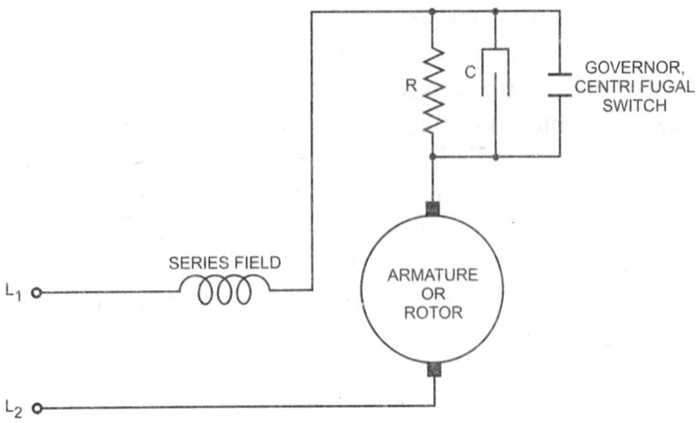

Fig. 20.39

Universal motors being high speed motors, are smaller in size and lighter in weight as compared to other motors of the same output. Full-load power factor is high (approximately 0.9) but it is poor at start and on overloads.

20.17.4. Reversal of Direction of Rotation. The direction of rotation of any series wound motor can be reversed by reversing the direction of the current flow in either the field or armature circuit (but not through both). Universal motors are sensitive to brush position and severe arcing at the brushes occur if the direction of rotation is reversed without shifting the brushes to the neutral plane. This is the reason that the universal motors are usually wound for operation in only one direction.

20.17.5. Speed Control. The factors that determine the speed for any dc motor are the same as those for the ac series or universal motor, viz flux and the generated voltage. Thus the speed of the universal motor, for any given load, can be changed by changing either or both of these quantities. Until recent times, change of generated voltage alone has rarely been employed for speed control. Instead, line voltage has been varied. This has been accomplished by means of tapped resistor, rheostat in series with the line, with an accompanying power loss, or instead by using a tapped auto-transformer. The desired change in speed can be had by using any of the above methods but torque is also reduced with the decrease in speed.

Another method is by using a tapped field, thereby reducing the flux and hence raising the speed. This may be accomplished by using any one of the following two arrangements:

(*i*) By using field poles wound in various sections with wires of different sizes and bringing out the taps from each section.

(*ii*) By using tapped nichrome wire coils wound over a single field pole.

In the field control method, the lesser field that remains in the circuit causes reduction in voltage drop across the field, and therefore, further increase in speed. In this method also change in speed is obtained at the expense of torque i.e. torque decreases with the increase in speed.

20.17.6. Applications. The very small power output rating universal motors, which usually does not exceed 5 or 10 watts, are employed in such equipment as sewing machines, fans, portable hand tools, hairdryers, motion-picture projectors, and electric shavers. The higher rating (5-500 W) universal motors are used in vacuum cleaners, electric typewriters, food mixers and blenders, motion-picture projectors, cameras, adding machines, and calculating machines.

The small series motors are frequently furnished as motor parts, i.e., consisting of bare stators and rotors (with shaft) but without bearings or feet. They can then be compactly "built in" to the power using devices.

Example 20.16. The resistance and total inductance of a single-phase fractional horse power series motor are 30 Ω and 0.5 H respectively. It draws 0.8 A current and runs at 2,000 rpm when connected to a 250 V dc supply. Calculate the speed and power factor when connected to a 250 V, 50 Hz supply and takes the same load current. How much voltage is required for getting 2,000 rpm with ac supply. Assume resistance and inductance remain constant.

Solution: Back emf developed when connected to dc supply,

$$E_b = V - IR = 250 - 0.8 \times 30 = 226 \text{ V}$$

$$\text{Speed, } N_{dc} = 2{,}000 \text{ rpm}$$

When connected to ac supply

$$\text{Resistance drop} = IR = 0.8 \times 30 = 24 \text{ V}$$

$$\text{Reactance drop} = IX = 0.8 \times 2\,\pi \times 50 \times 0.5 = 125.66 \text{ V}$$

From phasor diagram shown in Fig. 20.40.

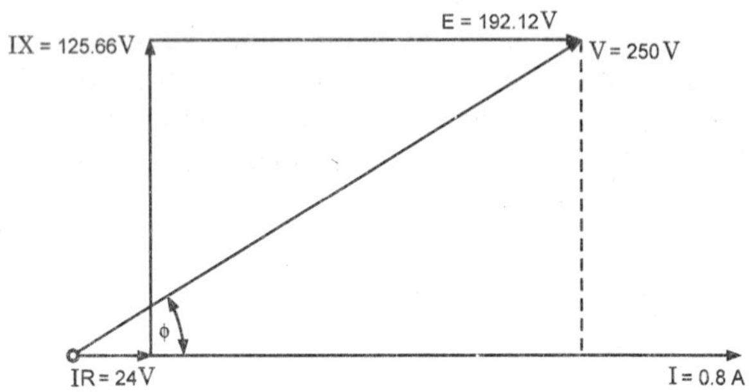

Fig. 20.40

$$\text{Counter-emf, } E = \sqrt{V^2 - (IX)^2} - IR$$

$$= \sqrt{(250)^2 - (125.66)^2} - 24 = 192.12 \text{ V}$$

$$\text{Speed, } N_{ac} = \frac{N_{dc}}{E_b} \times E = \frac{2{,}000}{226} \times 192.12 = 1{,}700 \text{ rpm } \textbf{Ans.}$$

$$\text{Power factor, } \cos\phi = \frac{E + IR}{V} = \frac{192.12 + 24}{250} = 0.8645 \text{ lagging } \textbf{Ans.}$$

Let the ac supply voltage be V′ to give the speed of 2,000 rpm.

The motor wiil operate at a speed of 2,000 when counter emf, E′ = 226 V

$$\text{and } E' = \sqrt{V'^2 - (IX)^2} - IR$$

$$\text{or } (E' + IR)^2 = V'^2 - (IX)^2$$

$$\text{or supply voltage, } V = \sqrt{(E' + IR)^2 + (IX)^2}$$

$$= \sqrt{(226 + 0.8 \times 30)^2 + (0.8 \times 2\pi \times 50 \times 0.5)^2}$$

$$= 279.8 \text{ V } \textbf{Ans.}$$

Example 20.17. A universal motor has a resistance of 30 Ω and an inductance of 0.5 H. When connected to a 250 V dc supply and loaded to 0.8 A, it runs at 2,000 rpm. Determine the speed, torque and power factor when connected to 250 V, 50 Hz ac supply and loaded to take same amount of current.

[*U.P. Technical Univ. Electromechanical Energy Conversion-II, 2005-06; Electrical Machines, 2010-11*]

Solution: Speed, N_{ac} = 1,700 rpm **Ans.**, as worked out in Example 20.16

Power factor, $\cos \phi$ = 0.8645 lagging **Ans.**, as worked out in Example 20.16

Counter emf, E_s = 192.12 V as worked out in Example 20.16

$$\text{Torque developed, } T = \frac{E \times I}{\frac{2\pi N}{60}} = \frac{192.12 \times 0.8 \times 60}{2\pi \times 1{,}700}$$

$$= 0.8633 \text{ N-m } \textbf{Ans.}$$

Example 20.18. A universal motor (ac operated) has a 2-pole armature with 960 conductors. At a certain load, the motor speed is 5,000 rpm and the armature current is 4.6 A, the armature terminal voltage and input are respectively 100 V and 300 W. Compute effective armature reactance and maximum value of useful flux/pole, assuming an armature resistance of 3.5 Ω.

[U.P. Technical Univ. Electromechanical Energy Conversion II, 2006-07; G.B. Technical Univ. Special Electrical Machines, 2012-13]

Solution: Supply voltage, V = 100 V

Power input, P_{in} = 300 W

Current drawn, I = 4.6 A

Armature resistance, R = 3.5 Ω

$$\text{Phase angle, } \phi = \text{Cos}^{-1} \frac{P_{in}}{VI}$$

$$= \text{Cos}^{-1} \frac{300}{100 \times 4.6} = 49.3° \text{ (lagging)}$$

because motor circuit is inductive

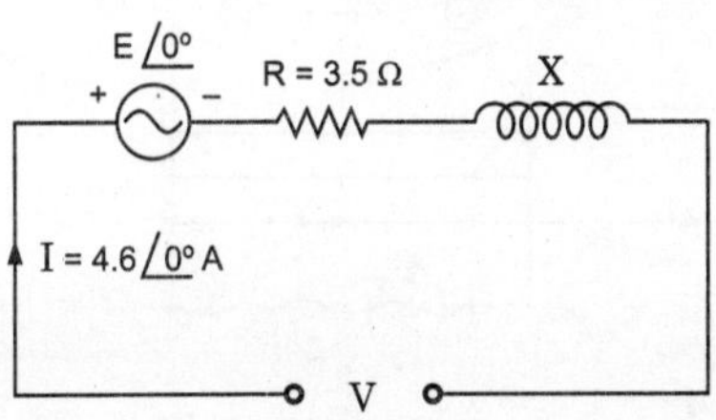

Fig. 20.41

From the circuit given in Fig. 20.41.

Counter emf, E = **V** – **I**(R + *j*X)

$$\text{or } E\angle 0° = 100 \angle 49.3° - 4.6 \angle 0° (3.5 + jX)$$

∵ Counter emf is in phase with I

$$\text{or } E + j0 = 65.21 + j75.81 - 16.1 - j4.6X$$

Equating real and imaginary components we have

$$E = 65.21 - 16.1 = 49.11 \text{ V}$$

$$\text{and } X = \frac{75.81}{4.6} = 16.48\ \Omega \quad \textbf{Ans.}$$

$$\text{Counter emf, } E = \frac{\Phi_{max}}{\sqrt{2}} \times \frac{ZN}{60} \times \frac{P}{A}$$

$$\text{or } 49.11 = \frac{\Phi_{max}}{\sqrt{2}} \times \frac{960 \times 5{,}000}{60} \times \frac{2}{2}$$

$$\text{or } \Phi_{max} = \frac{49.11\sqrt{2} \times 60}{960 \times 5{,}000} = 0.8682 \text{ mWb} \quad \textbf{Ans.}$$

Example 20.19. A 240 V, 50 Hz, 2-pole universal motor operates at a speed of 10,000 rpm on full load and draws a current of 7.12 A at 0.92 lagging power factor. The motor parameters are r_a = 6.34 Ω, x_a = 9.62 Ω, r_s = 4.89 Ω, x_s = 3.6 Ω. Calculate efficiency of the motor if rotational losses are 60 W.

[U.P.S.C. I.E.S. Electrical Engineering-II, 2013]

Solution: Impedance of motor, **Z** = (6.34 + 4.89) + *j* (9.62 + 3.6)

= (11.23 + *j* 13.22) Ω

Taking motor current as reference phasor we have

$$\mathbf{I} = 7.12 \angle 0° \text{ A}$$

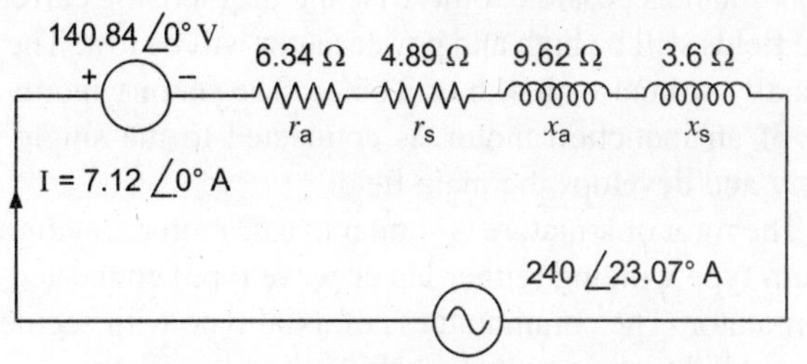

Fig. 20.42

Power factor, $\cos \phi$ = 0.92

Phase angle, $\phi = \text{Cos}^{-1}$ 0.92 = 23.07° (lagging)

So supply voltage, V = 240 ∠23.07° V

= 240(0.92 + *j*0.392) = (220.8 + *j* 94.06)

Power input, P_{in} = VI cos φ = 240 × 7.12 × 0.92

= 1,572.1 W

Counter-emf, E = **V** – $\mathbf{I}_a$ **Z**

= (220.8 + *j* 94.06) – 7.12(11.23 + *j* 13.22)

= (220.8 + *j* 94.06) – (79.96 + *j* 94.12)

= 140.84 V

Power developed by motor = EI = 140.84 × 7.12 = 1,002.8 W

Net power output, P_{out} = Power developed by rotor – rotational losses

= 1,002.8 – 60 = 942.8 W

$$\text{Hence, motor efficiency, } \eta = \frac{P_{out}}{P_{in}} \times 100 = \frac{942.8}{1572.1} \times 100 = 59.94\% \ \textbf{Ans.}$$

20.18 REPULSION TYPE MOTORS

As discussed in Art 20.16, in ac series motors, the rotor and the stator windings are conductively coupled i.e., rotor current is obtained by conduction from the stator. *Repulsion motors* are similar to series motors except that the rotor and the stator windings are inductively coupled i.e., the rotor current is obtained by transformer action from the stator.

There are various types of repulsion motors, each employing the repulsion principle for starting. Repulsion motors include *repulsion motor* which starts and runs on the principle of magnetic repulsion; *repulsion-start induction-run motor*, which starts on the repulsion principle but at a predetermined speed operates as a single-phase induction motor; and the *repulsion-induction motor*, which employs the repulsion-start principle but has an additional armature winding that provides additional running torque.

Though the three types of motors, mentioned above, have similarity in names but they are different in construction, in operating characteristics, and in their applications to the industrial uses.

20.19 REPULSION MOTORS

20.19.1. Construction. Practically all repulsion motors are made with non-salient poles. The stator or field windings used are usually of the distributed type, similar to the main or running winding of a split-phase single-phase induction motor. The fact that the reluctance to the main flux and to the transformer field flux must be kept the minimum possible makes it desirable to use non-salient poles and semi-closed slots and to make the air gap as small as possible, otherwise the magnetising currents for these fields will be high and power factor will be low. The stator is usually wound with 4, 6 or 8 poles. The stator winding, as in case of an induction motor, is connected to the single phase supply and develops the main field.

The rotor or armature is similar to a dc motor armature, with a drum type winding (either lap or wave type) connected to the commutator. The commutator is of axial type with segments or bars parallel to the armature shaft.

There are two cast steel end shields that house the motor bearings which are secured to the motor frame.

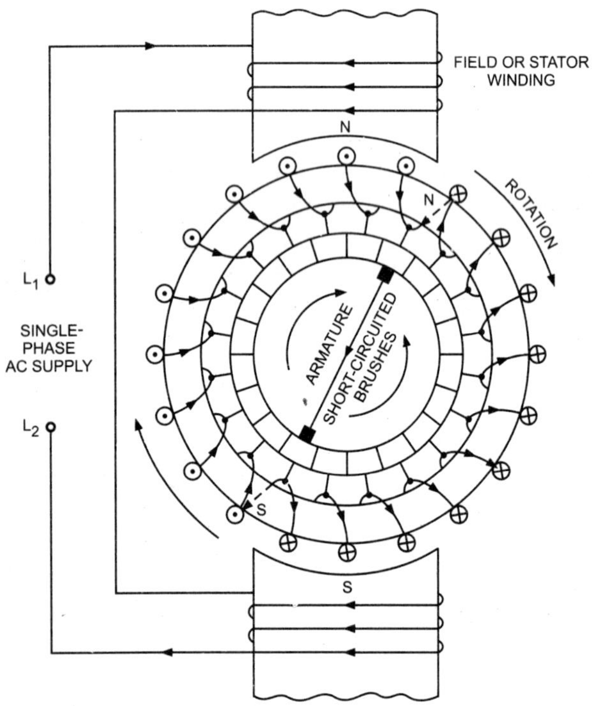

Fig. 20.43 *Simplified Schematic Diagram of a Repulsion Motor*

Carbon brushes, in contact with commutator surface, are held in place by a brush holder assembly mounted on one of the end shields. The brushes directly opposite to each other, as illustrated in Fig. 20.43, are not connected to the supply but are connected to each other by a connector of negligible resistance or short circuit. The brush holder assembly may be moved so that the brushes can make contact with the commutator surface at different points so as to provide the correct direction of rotation and maximum torque output.

Stator core is pressed into a cast steel frame and the armature shaft is supported by the two bearings (either sleeve bearing or ball bearing units).

Simplified schematic diagram of a repulsion motor is given in Fig. 20.43.

20.19.2. Operation. When the stator winding is connected to a single-phase ac supply, a field is developed by the current in the stator windings and this field induces a voltage in the armature and thus causes a flow of current in it. If the brushes are fixed at a small angle (from 15 to 25 electrical space degrees) from the field axis, induced currents in the armature windings develop an armature field with the poles in the neighbourhood of the brushes. The instantaneous polarity of the rotor poles is the same as that of the adjacent stator poles, a repulsion torque is set up that causes the rotation of the motor armature. Thus an application of this principle, that like poles repel each other, gives the motor its name of "*repulsion motor*".

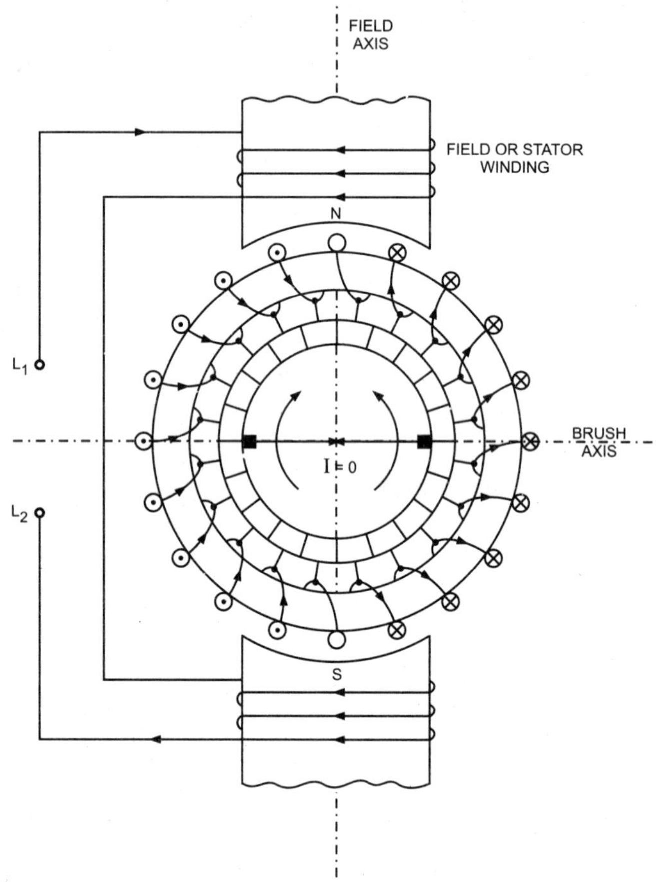

Fig. 20.44 *Brushes in Geometrical Neutral Plane*

Let us consider, for sake of simplicity and easy understanding, the gramme-ring armature and its commutator operating in a bipolar salient pole field, as shown in Figs. 20.44, 20.45, 20.46, 20.43. In such an arrangement the flux paths are simple and the windings and their currents are easy to follow. The basic function of the machine remains the same with either type of construction (salient or non-salient pole). The main field is developed by the stator windings when connected to single-phase ac supply mains.

In Fig. 20.44, the brushes are shown as being placed at right angles to the stator poles (i.e. in the geometrical neutral plane) and short circuited. Each brush is at the midpoint of its transformer winding. As the total emfs in the two windings are the same and the windings are connected in parallel, each point must be at the same potential. The brushes short circuit two points at the same potential, therefore, no current flows between the brushes. Thus there is no current in the rotor or armature windings, and hence, no flux is produced by the armature windings. As a consequence, no torque is developed.

The condition is similar to that of an open-circuited transformer and the motor current will be quite low as the motor impedance is high. This brush position is called as the *high impedance* or *open-circuited* or *neutral position.*

With voltage applied to the stator, in this condition, the motor hums softly and develops no torque. This brush position is, therefore, also called the *soft neutral position.* This brush position is less dangerous than that shown in Fig. 20.45, because motor failure to rotate is not accompanied by high rotor current.

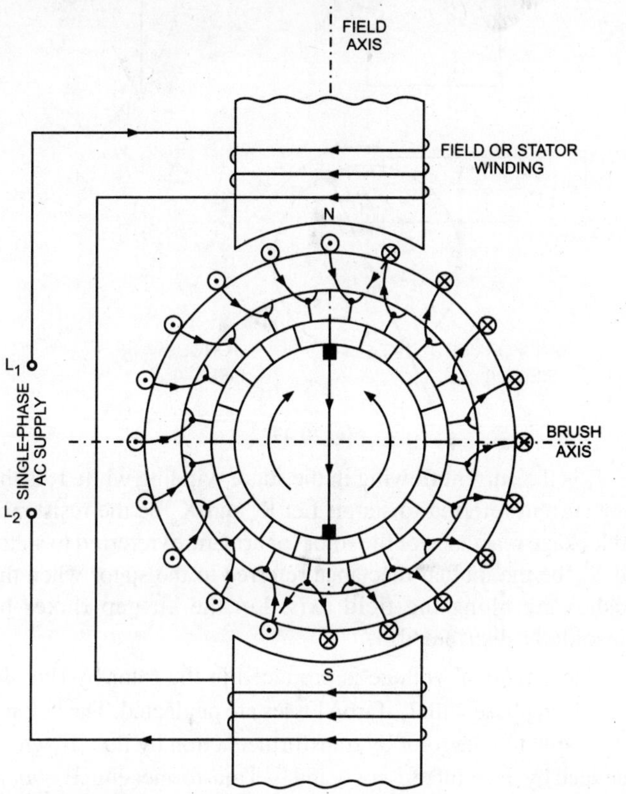

Fig. 20.45 *Brushes Along Field Axis*

In Fig. 20.45, the brushes have been shifted along the field axis, the conditions existing in the field and armature remaining the same as illustrated in Fig. 20.44. The brushes now short circuit the points of the rotor windings across which the maximum pd exists. Current, therefore, flows between the brushes from both sides of the rotor, and in this brush position the current in the rotor is maximum. The motor again develops no torque with the brushes in this position for the following reasons :

Two conditions are necessary for the development of torque. *The angle between the space position of the field axis and the brush axis must be greater than zero. For maximum torque, this* angle should be 90º. *There must be a component of the current in time phase with the flux.* If there is 90º time lag between the current and flux, the current is maximum at the instant the flux is zero, etc. and the average torque is zero. With flux, rotor current, and brush position all fixed, the highest value of torque occurs when the flux and rotor current are in time phase with each other.

Under the conditions shown in Fig. 20.45, the brush axis is parallel to the field axis i.e. angle between the flux and brush axis is zero. Although in this condition rotor current is nearly in time phase with the flux, no torque is developed because of zero space angle between the brush and field axis.

The condition shown in Fig. 20.45 is similar to that of a transformer on short circuit with a low impedance and heavy current. This brush position is, therefore, known as *low impedance* or *short-circuit position.*

In this brush position, the motor growls harshly and, therefore, this brush position is called the *harsh neutral position.*

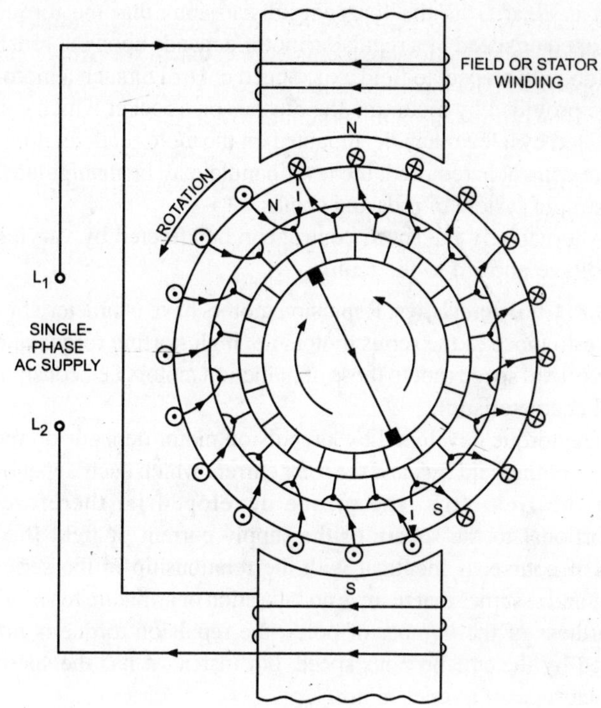

Fig. 20.46

In Fig. 20.46, the brushes have been shifted by an angle α in a counter-clockwise direction from the field axis. As a result the brushes short circuit the points of winding between which a pd exists; therefore, currents flow in the rotor winding and also the rotor current has a component in time phase with the flux. Under these conditions, therefore, the motor develops torque, and if allowed to do so the rotor will rotate. The torque developed in this position of brushes, will be in counter-clockwise direction.

If the brushes are shifted on the opposite side of the field axis i.e., in clockwise direction from the field, as shown in Fig. 20.43, the torque developed will be in clockwise direction.

The fact is that a repulsion motor develops a torque in the direction in which the brushes are shifted from the field axis.

The torque developed by a repulsion motor should be a maximum, theoretically, when the space angle between the pole axis and brush axis, α is 45º, but in practice the angle of inclination α is about 15-25 electrical degrees.

One obvious advantage of the repulsion motor over the series motor is that the rotor of the former can be designed for any convenient low voltage, to ease the problem of commutation, and no brush-yoke insulation is required. It is due to the fact that the armature or rotor is not connected to the supply mains.

The repulsion motor has better commutation than the series motor at speeds below synchronism and poor commutation at very high speeds, owing to greater values of currents induced in the short-circuited coils under the latter conditions. These induced currents provide a braking effect which serves to limit the high no-load speed.

The direction of rotation of a repulsion motor, as already explained, depends upon the position of brushes with respect to the field axis, and so can be reversed by shifting the brushes, round the commutator, on the other side of the field axis. The direction of rotation of the motor can be reversed even while the motor is running.

It is clear from the foregoing discussions that the torque developed or speed of a repulsion motor depends upon the brush position with respect to field axis. Speed control of such a motor can be provided by mounting the brushes on a rocker which can be rotated by a lever handle mounted on the motor end-shield. If remote control is required, the lever handle may be manipulated by a simple system of rods and cranks.

Alternatively the speed control can be affected by varying the voltage applied to the motor.

20.19.3. Characteristics. Repulsion motors have characteristics similar to those of the series motors i.e. high starting torque and high no-load speed (not to those of induction motors i.e., constant speed characteristic).

The torque developed by a repulsion motor depends on the product of the field flux and the rotor current which itself depends upon the field flux, the torque developed is, therefore, proportional to the square of the supply current or field flux. This, of course, is identical with the relationship in the series motor and assumes that there is no saturation or armature reaction. Regardless of the number of poles, the repulsion torque is not limited by the synchronous speed, but instead it has the speed characteristic of a series motor.

The repulsion motor has high starting torque (about 3-5 times full-load torque) and moderate starting current (about 3-4 times full-load current) but poor speed regulation. Shifting the brushes during operation gives a wide range of speed control, as high as 6:1 ratio, and yet provides a continuous variation. The upper speed is not limited by frequency. The motor is a reversing type, and the direction may be changed during rotation.

20.19.4. Disadvantages. The drawbacks of repulsion motors are (*i*) speed variations with the variations in loads, dangerously high at no load (*ii*) low power factor, except at high speeds, (*iii*) tendency to spark at the brushes, sparking at the brushes is negligible at the rated speed which usually occurs near synchronous speed (*iv*) higher cost and (*v*) more attention and maintenance is required.

20.19.5. Applications. The repulsion motor has never enjoyed popularity. The motor is used where sturdy motor with large starting torque and adjustable speed is required. Most common use of this type of motor is in the coil winders in which the operator adjusts the speed by shifting the brushes; the motor is equipped with a special lever mechanism, that shifts the brushes when a foot treadle is pressed.

The ratings in which the repulsion motors are built is limited due to commutation problems. The usual rating of the repulsion motor does not exceed 5 kW.

20.19.6. Performance Characteristics. Let the brush axis be shifted making an angle α (electrical degrees) in clockwise direction from the field axis, as illustrated in Fig. 20.47. The stator field Φ_s may be resolved into two components, a transformer or current-producing field, Φ_q along the brush axis, and a main, or torque-producing field, Φ_d perpendicular to the brush axis, as illustrated in Fig. 20.47.

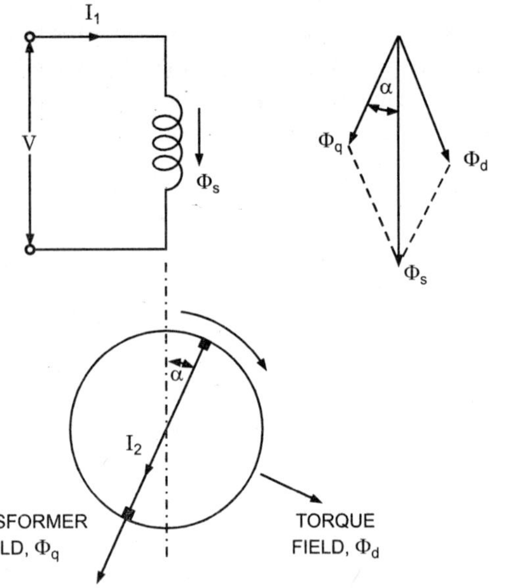

Fig. 20.47

I_1 is the current flowing in the stator winding while I_2 is the rotor current referred to stator. Let R_a and X_a be the resistance and leakage reactance of the rotor (or armature) referred to stator and X_m be the mutual reactance referred to the stator when the brushes are along the field axis. Let the air-gap fluxes be sinusoidally distributed.

The rotational voltage is produced in the rotor by flux Φ_d which is in phase with I_1 if iron losses are neglected. The voltage is developed in the rotor by transformer action by flux Φ_q which is created by the mmf of $I_1 \cos\alpha$ and I_2. Transformer emf E_t would obviously lag the phasor $(I_1 \cos\alpha + I_2)$ by 90°, as in a transformer having negligible iron loss.

Thus the rotor voltage equation may be given as

$$\mathbf{E}_r + \mathbf{E}_t = \mathbf{I}_2 (R_a + j X_a) \qquad ...(20.42)$$

Now flux linking with the stator winding due to flux Φ_d is $\Phi_d \sin\alpha$ and Φ_d is proportional to $I_1 \sin\alpha$. Thus the voltage drop in stator winding due to flux Φ_d, neglecting saturation is $jX_m \sin^2\alpha$. Voltage drop due to flux Φ_q would have been $-E_t$ if the brushes were along the field axis. But because of the displacement of the brushes from the field axis, $\Phi_q \cos\alpha$ links with the stator winding and, therefore, voltage drop due to flux Φ_q is $-E_t \cos\alpha$. Voltage applied to the stator winding equals the sum of voltage drops in stator resistance R_s, stator leakage reactance X_s and voltage drops caused (or emfs induced) by flux Φ_d and Φ_q i.e.

$$\mathbf{V} = \mathbf{I}_1 R_s + j \mathbf{I}_1 X_s + j \mathbf{I}_1 X_m \sin^2\alpha - \mathbf{E}_t \cos\alpha$$

$$= (R_s + j X_s + j X_m \sin^2\alpha)\, \mathbf{I}_1 - \mathbf{E}_t \cos\alpha \qquad ...(20.43)$$

The phasor diagrams for the rotor and stator can be drawn on the basis of the Eqs. (20.42) and (20.43) as illustrated in Figs. 20.48 (*a*) and 20.48 (*b*) respectively.

If the brush axis were perpendicular to field axis, flux Φ_d would have been created by mmf I_1 and the rotational emf would have been $\frac{N}{N_s} X_m I_1$ where N is the actual rotational speed of the

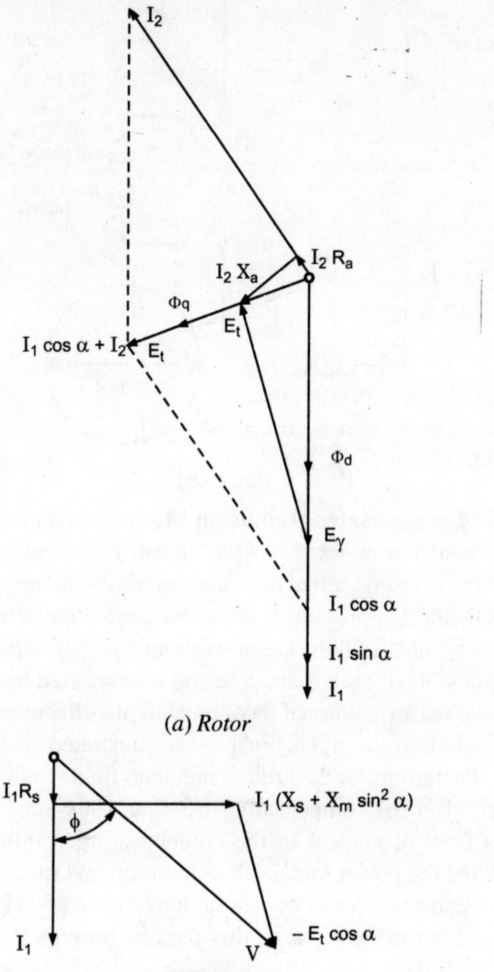

(*a*) *Rotor*

(*b*) *Stator*

Fig. 20.48 *Phasor Diagram For a Repulsion Motor*

rotor and N_s is the synchronous speed. But in the present condition Φ_d is created by $I_1 \sin\alpha$ so rotational emf is given as

$$\mathbf{E}_r = \frac{N}{N_s} X_m I_1 \sin\alpha \qquad \text{...(20.44)}$$

The voltage developed in armature by transformer action is given as

$$\mathbf{E}_t = -jX_m(\mathbf{I}_1 \cos\alpha + \mathbf{I}_2) \qquad \text{...(20.45)}$$

Substituting values of E_r and E_t from Eqs. (20.44) and (20.45) in Eqs. (20.42) and (20.43), we have

$$\frac{N}{N_s} X_m \mathbf{I}_1 \sin\alpha - jX_m (\mathbf{I}_1 \cos\alpha + \mathbf{I}_2) = \mathbf{I}_2 (R_a + jX_a) \qquad \text{...(20.46)}$$

$$\text{and } V = (R_s + jX_s + jX_m \sin^2\alpha)\mathbf{I}_1 + jX_m(\mathbf{I}_1 \cos\alpha + \mathbf{I}_2)\cos\alpha$$

$$= (R_s + jX_s + jX_m) I_1 + jX_m I_2 \cos\alpha \qquad \text{...(20.47)}$$

From Eq. (20.46) rotor equivalent current

$$\mathbf{I}_2 = \frac{\frac{N}{N_s} X_m \sin\alpha - jX_m \cos\alpha}{R_a + jX_a + jX_m} \mathbf{I}_1 \qquad \text{... (20.48)}$$

Substituting the value of I_2 from above Eq. (20.48) in Eq. (20.47) we have

Supply voltage,

$$V = \left[R_s + jX_s + jX_m + jX_m^2 \frac{\frac{N}{N_s}\sin\alpha\cos\alpha - j\cos^2\alpha}{R_a + jX_a + jX_m}\right] I_1 \qquad \text{... (20.49)}$$

For any rotation speed N, the brush angle α, the current I_1 and power factor angle can be determined from Eq. (20.47). With I_1 so computed in complex notation I_2 can be determined from Eq. (20.48).

Mechanical power developed,

$$P_{mech} = V I \cos\phi - I_1^2 R_s - I_2^2 R_a \qquad \text{... (20.50)}$$

Mechanical power output, $P_{out} = P_{mech}$ – stray power losses.

If rotor resistance and reactance are neglected then the Eq. (20.49) becomes

$$\mathbf{V} = \left[R_s + jX_s + jX_m \sin^2\alpha + \frac{N}{N_s} X_m \sin\alpha\cos\alpha\right] \mathbf{I}_1 \qquad \text{... (20.51)}$$

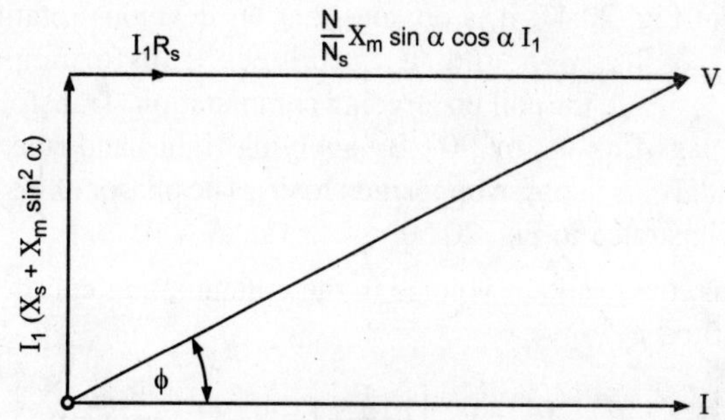

Fig. 20.49

Now the phasor diagram drawn on the basis of the above Eq. (20.51) is given in Fig. 20.49.

From phasor diagram shown in Fig. 20.49,

Mechanical power developed,

$$P_{mech} = VI\cos\phi - I_1^2 R_s = (V\cos\phi - I_1 R_s) I_1$$

$$= \left(I_1 R_s + \frac{N}{N_s} X_m \sin\alpha\cos\alpha - I_1 R_s\right) I_1^2$$

$$= \frac{N}{N_s} X_m I_1^2 \sin\alpha\cos\alpha$$

Torque developed,

$$T = \frac{P_{mech}}{\frac{N}{N_s}} = I_1^2 X_m \sin\alpha\cos\alpha \text{ synchronous watts} \qquad \text{...(20.52)}$$

The expression for starting current can be determined by substituting N = 0 in Eq. (20.51) so

Starting current,

$$I_{st} = \frac{V}{\sqrt{R_s^2 + \left(X_s + X_m \sin^2\alpha\right)^2}} \qquad \text{... (20.53)}$$

and starting torque, T_{st} from Eq. (20.52) is given as

$$T_{st} = I_1^2 X_m \sin\alpha\cos\alpha$$

$$= \frac{1}{2} I_1^2 X_m \sin 2\alpha$$

$$= \frac{1}{2} \frac{V^2 X_m \sin 2\alpha}{R_s^2 + \left(X_s + X_m \sin^2\alpha\right)^2} \text{ synchronous watts} \qquad \text{...(20.54)}$$

From phasor diagram shown in Fig. 20.49.

Power factor,

$$\text{Cos}\,\phi = \frac{\frac{N}{N_s} I_1 X_m \sin\alpha\cos\alpha + I_1 R_s}{V} \qquad \text{...(20.55)}$$

20.19.7. Commutation. From Eq. (20.42), neglecting rotor resistance and leakage reactance, $\mathbf{E}_r + \mathbf{E}_t = 0$ *i.e.* . E_r and E_t are equal in magnitude but opposite in phase, as shown in Fig. 20.50.

From Eqs. (20.36) and (20.38)

$$E_r = \sqrt{2}\,\Phi_{d\,max}\frac{Z}{A}f_r$$

$$\text{and } E_t = \sqrt{2}\,\Phi_{q\,max}\frac{Z}{A}f$$

$$\text{So}\quad \frac{\Phi_{q\,max}}{\Phi_{d\,max}} = \frac{f_r}{f} = \frac{N}{N_s} \qquad ...(20.56)$$

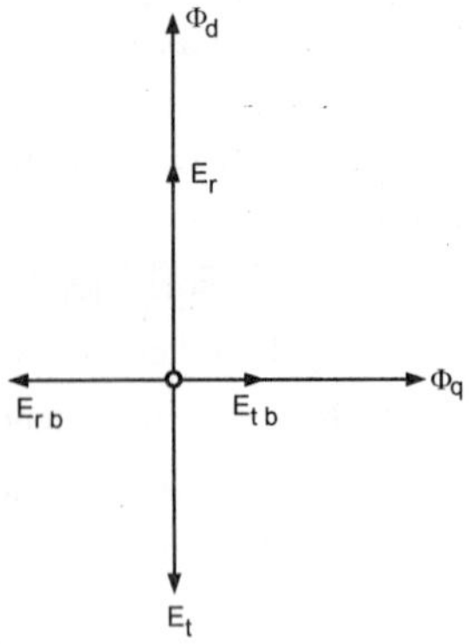

Fig. 20.50

From Fig. 20.47, it is obvious that Φ_q develops rotational emf $E_{rb}\ \alpha\ \Phi_{q\,max} f_r$ and Φ_d produces transformer emf $E_{tb}\ \alpha\ \Phi_{d\,max} f$ in the coil undergoing commutation. Transformer emf E_{tb} lags flux Φ_d by 90°. By applying right hand rule it is found that E_{rb} is in phase opposition to Φ_q. The phasor of E_{rb} and E_{tb} are illustrated in Fig. 20.50.

Thus, the resultant voltage in the commutating coil due to air-gap flux,

$$E'_b = E_{tb} - E_{rb} = E_{tb}\left[1 - \frac{E_{rb}}{E_{tb}}\right] = E_{tb}\left[1 - \frac{\Phi_{q\,max}}{\Phi_{d\,max}}\frac{N}{N_s}\right]$$

Substituting from Eq. (20.56) $\dfrac{\Phi_{q\,max}}{\Phi_{d\,max}} = \dfrac{N}{N_s}$ we have

$$E'_b = E_{tb}\left[1 - \frac{N^2}{N_s^{\,2}}\right] \qquad ... (20.57)$$

Thus it is obvious from Eq. (20.57) that when the machine runs at synchronous speed $E_b' = 0$ and the voltage existing in the coil undergoing commutation is the only small reactance voltage and, therefore, commutation is trouble free. At speeds other than synchronous speed E'_b is high and there may be sparking at the brushes.

The commutation problem can be analyzed otherwise also. At synchronous speed, $N = N_s$ and so from Eq. (20.56) $\Phi_{d\,max} = \Phi_{q\,max}$. These two fluxes Φ_d and Φ_q being equal in magnitude and spaced in quadrature, constitutes pure rotating field in the air gap which rotates at the synchronous speed, as in a balance two-phase motor. Since the magnitude and direction of the speed of rotation of the armature and the rotating field is the same, so no air-gap flux is cut by the commutating coil. Thus the voltage induced due to air-gap flux is zero and commutation obtained is smooth at synchronous speed.

If rotor resistance and reactance are taken into account, which have been neglected in above discussions, the commutation may be smooth and trouble free at a speed of approximately 80 per cent of synchronous speed.

20.19.8. Two Stator Winding Repulsion Motor. Some of the repulsion motors are provided with two windings in stator at right angles to each other and connected in series, as illustrated in Fig. 20.51. One advantage available with this arrangement is that the direction of rotation can be changed by reversing the connections to one of these two stator windings. Instead of moving the brush rocker, it is necessary only to throw the reversing switch, as illustrated in Fig. 20.51.

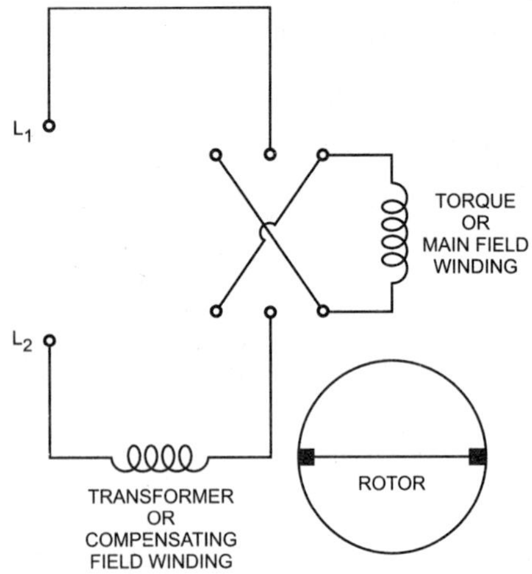

Fig. 20.51

20.19.9. Compensated Repulsion Motor. It is a modified form of the repulsion motor discussed above. It is provided with an additional winding, called the *compensating winding*, to improve power factor and provide better speed regulation. This winding is much smaller than the stator winding and is usually wound in the inner slots of each main pole and is connected in series with rotor through an additional set of brushes placed midway between the usual short-circuited brushes, as illustrated in Fig. 20.52. When the current flows through the main field winding, voltage is induced in the compensating winding. This induced voltage causes flow of current in the compensating winding thereby improving the power factor (close to unity) and speed regulation (almost constant speed) by neutralizing the cross-magnetization produced in the rotor due to flux passing through it.

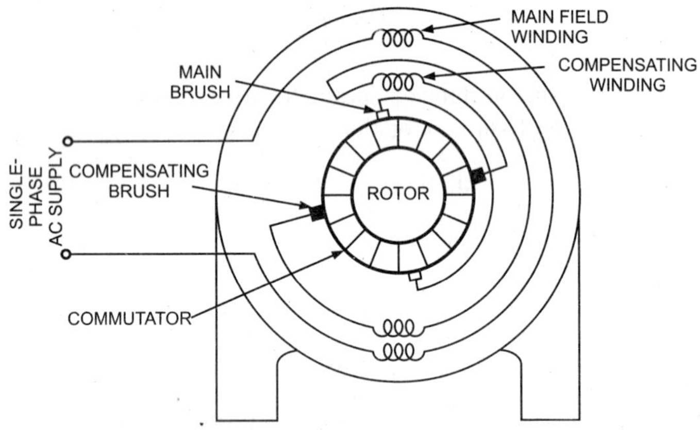

Fig. 20.52 *Compensated Repulsion Motor*

Example 20.20. A 2 kW, 4-pole, 230 V, single winding repulsion motor has the following stator referred parameters:

Mutual reactance = 50 Ω; Stator resistance = 2 Ω, Stator leakage reactance = 5 Ω; Armature resistance 2 Ω, Armature leakage reactance= 6 Ω; Brush shift from low-impedance position = 20°; Stray power loss= 25 watts.

Calculate the load torque when the motor runs at 1,350 rpm.

Solution: Brush shift from field axis (or low impedance position),

$$\alpha = 20°$$

$$\text{Synchronous speed, } N_s = \frac{120\,f}{P} = \frac{120 \times 50}{4} = 1{,}500 \text{ rpm}$$

$$\text{Actual rotational speed, } N = 1{,}350 \text{ rpm}$$

$$\text{Sin } \alpha = \text{Sin } 20° = 0.342$$

$$\text{Cos } \alpha = \text{Cos } 20° = 0.9397$$

From Eq. (20.49), Supply voltage,

$$V = \left[R_s + jX_s + jX_m + jX_m^2 \frac{\frac{N}{N_s}\sin\alpha\cos\alpha - j\cos^2\alpha}{R_a + jX_a + jX_m} \right] I_1$$

$$= \left[2 + j5 + j50 + j50^2 \frac{\frac{1,350}{1,500}\times 0.342\times 0.9397 - j(0.9397)^2}{2 + j6 + j50} \right] I_1$$

$$= \left[2 + j\,55 + j\,2,500 \frac{0.2892 - j\,0.883}{2 + j\,56} \right] I_1$$

$$= \left[2 + j\,55 + j\,2,500 \;\angle 90^\circ \times \frac{0.929 \;\angle -71.86^\circ}{56.0357 \;\angle 87.95^\circ} \right] I_1$$

$$= \left[2 + j\,55 + 41.447 \;\angle -69.81^\circ \right] I_1$$

$$= [2 + j\,55 + 14.305 - j\,38.9]\, I_1 = (16.305 - j\,16.1)\, I_1$$

or supply current,

$$I_1 = \frac{V}{(16.305 - j16.1)} = \frac{230}{22.914 \;\angle -44.64^\circ} = 10.04 \;\angle 44.64^\circ \text{ A}$$

From Eq. (20.48)
Rotor equivalent current,

$$I_2 = \left[\frac{\frac{N}{N_s} X_m \sin\alpha - j\,X_m \cos\alpha}{R_a + jX_a + jX_m} \right] I_1$$

$$= \frac{\frac{1,350}{1,500}\times 50\times 0.342 - j50\times 0.9397}{2 + j6 + j50} \times 10.04 \;\angle 44.64^\circ$$

$$= \frac{15.39 - j46.985}{56.0357 \;\angle 87.95^\circ} \times 10.04 \;\angle 44.64^\circ$$

$$= 8.858 \;\angle -115.17^\circ = (-3.77 - j\,8.017) \text{ A}$$

Mechanical power developed,

$$P_{mech} = V I_1 \cos\phi - I_1^2 R_s - I_2^2 R_a$$

$$= 230 \times 10.04 \cos 44.64^\circ - (10.04)^2 \times 2 - 8.858^2 \times 2$$

$$= 1,643 - 201.6 - 156.9 = 1,284.5 \text{ watts}$$

Power output

$$= P_{mech} - \text{stray power loss} = 1,284.5 - 25 = 1,259.5 \text{ watts}$$

Load torque

$$= \frac{\text{Power output}}{2\pi \frac{N}{60}} = \frac{1,259.5}{2\pi \times 1,350/60} = 8.91 \text{ N-m} \textbf{ Ans.}$$

20.20 REPULSION- START INDUCTION MOTOR

As the name implies, the repulsion-start induction motor starts as a repulsion motor and runs as an induction motor.

20.20.1. Construction. The general construction of a repulsion-start induction motor is quite similar to that of a repulsion motor (i.e. a laminated stator core complete with one winding similar to the main winding of a split-phase motor and a rotor, also called the *armature,* consisting of slots in which a winding is placed and connected to a commutator). The only difference is that in addition to the basic repulsion-motor construction it is equipped with a centrifugal device which operates at about 75-80 per cent of synchronous speed and short circuits all of the commutator segments. In some motors the centrifugal device also lifts the brushes from the commutator in order to reduce wear of the brushes and commutator bars and to make the running operation more quite. It makes no difference electrically whether the brushes are in contact or not after the motor attains the normal speed.

20.20.2. Operation. A repulsion-start induction motor starts in exactly the same manner as a repulsion motor. However, when the motor has accelerated to approximately 75 per cent of rated speed, the torque developed due to induction motor action is more than due to repulsion motor action, commutator segments get short circuited and brushes lifted from the commutator to enable to run as an induction motor.

It may be noted that greater the speed at which change-over action takes place, the lower will be the induction motor current at that instant and, consequently the less disturbance to the line.

After the commutator has been short circuited, the brushes do not carry the current and therefore, may be lifted from the commutator, but lifting is not absolutely necessary.

If the short-circuiting mechanism operates before the repulsion curve crosses the induction motor curve (Fig. 20.53), and if the torque of the induction motor is less than that required to accelerate the load, the motor may slow down until the short circuit is removed from the commutator, in which case the motor will again operate as a repulsion motor. The rotor will speed up until the commutator is again short circuited, after which the rotor will slow down until it again becomes repulsion. This cycle will be repeated over and over again until some change takes place.

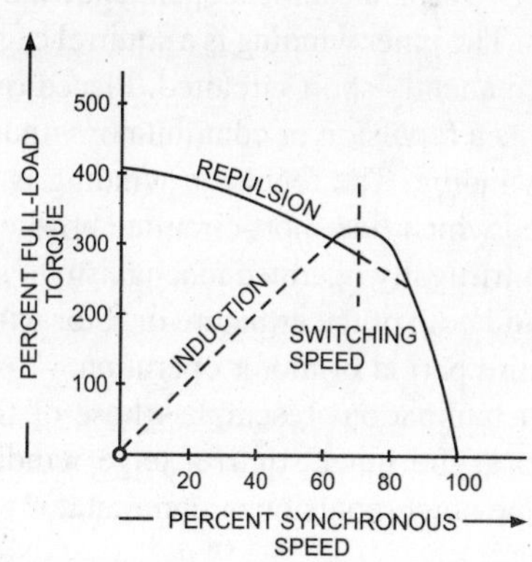

Fig. 20.53 *Torque-Speed Characteristic For Repulsion Start Induction Motor*

20.20.3. Characteristics. The repulsion-start induction motor combines the desirable starting characteristics of the repulsion motor (i.e. high starting torque and moderate starting current) with operating characteristic (constant speed running) of the induction motor. A typical performance characteristic of a repulsion-start induction motor is given in Fig. 20.53. The starting torque is 3 to 6 times of full-load torque, and the starting current is approximately 3-4 times of full-load current.

The direction of rotation of such a motor may be reversed by shifting the brushes, but the brush rigging is not readily accessible, and the method is poorer than with any of the capacitor type motors.

The efficiency and the maximum operating torque of the repulsion-start induction motor are usually less than those of a squirrel cage induction motor of comparative size i.e. a repulsion-start induction motor must be larger than a squirrel cage motor of the same rating to give the same performance.

As compared to capacitor motors, these motors (i) need more maintenance because of commutator, brushes and centrifugal switch, (*ii*) are relatively more expensive, (*iii*) cause radio interference during starting, and (*iv*) are noisier, especially when the brushes continue to ride on commutator.

However, despite all the drawbacks, these motors are still used in integral kilowatt ratings because they (*i*) develop higher starting torque with low starting current and (ii) can withstand longer starting periods than capacitor type motors.

Such motors are suitable for commercial refrigerators, compressors, pumps, hoists and other constant speed drives, particularly those which have a high inertia and prolonged starting period. The usual rating is from $\frac{1}{3}$ kW to 12 kW but for special applications, ratings as high as 30 kW are available.

20.21 REPULSION INDUCTION MOTOR

Such a motor is a combination of repulsion and induction types and operates on the combined principle of both repulsion and induction. In such a motor, the desirable starting characteristics of the repulsion motor and the constant-speed characteristic of the induction motor are obtained.

20.21.1. Construction. The repulsion induction motor has a single-phase stator winding, as a repulsion start induction motor has, but it has two separate and independent windings on rotor in common slots. The inner winding is a squirrel cage winding with rotor bars permanently short circuited. Placed over the squirrel cage winding is a repulsion or commutator winding similar to a dc armature winding. The repulsion winding is connected to a commutator on which ride short-circuited brushes. In this motor there is no centrifugally operated mechanism to short circuit the commutator and both of the armature or rotor windings function during the entire period of motor operation.

The rotor laminations resemble those of the double cage rotor, therefore, the inner squirrel cage winding has a high inductance. The outer repulsion or commutator winding has low inductance.

The schematic diagram of a repulsion-induction motor is given in Fig. 20.54.

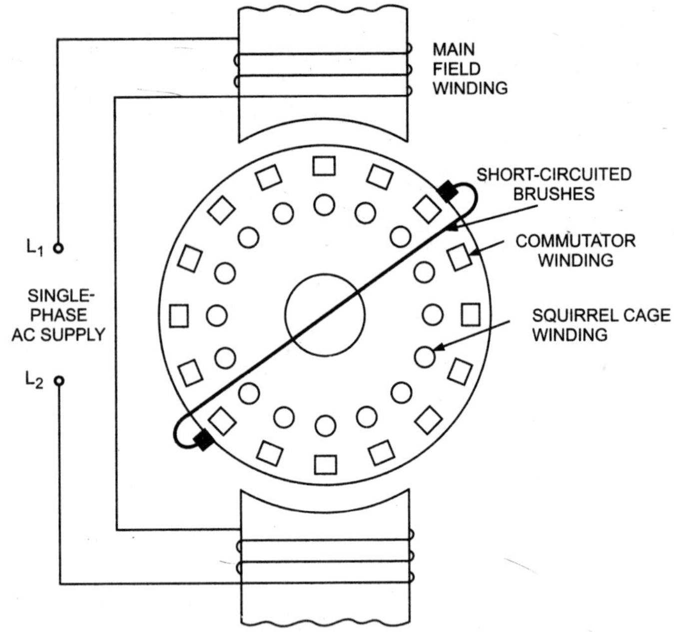

Fig. 20.54 *Repulsion Induction Motor*

20.21.2. Operation. When the motor starts, the squirrel cage winding develops no torque, because due to its high leakage reactance it carries little current, and the motor starts as a repulsion motor giving high starting torque, because due to its low leakage reactance it carries the greater part of the rotor current. As the motor speeds up, the current shifts from the outer winding to the inner winding owing to decreasing reactance of the inner winding with the increasing speed, so that at normal running speed, the squirrel cage winding carries a greater part of the rotor current and thereby contributes considerable torque to the rotor. Both of the windings develop torque, and the output of the motor is the combined output of both of the rotor windings.

The repulsion torque is not limited by the synchronous speed and therefore, this motor operates above synchronous speed, especially at no load. Above synchronous speed, however, the cage winding acts as a generator because of cutting flux in a direction opposite to the below synchronous speed, and therefore, produces a torque opposing the rotation. Thus instead of being unstable at no load, there is a definite upper speed limit. When the motor is loaded, both windings contribute torque and therefore, speed does not fall too much. The speed regulation is usually less than 10 per cent.

20.21.3. Characteristics. The high reactance of the squirrel cage winding causes it to draw only a small additional current at starting, so the starting characteristics are only slightly inferior to the pure repulsion motor. The starting torque is about 2.5 to 3.0 times of the full-load torque and the starting current is about 3 to 4 times full-load current. The commutation of the motor is good at all speeds. No-load speed is above synchronous speed because of effect of the repulsion winding but normal running speed is somewhat below synchronous speed since it runs like an induction motor. No-load current is rather high, sometimes greater than full-load current. The torque-speed characteristic of a repulsion induction motor is given in Fig. 20.55.

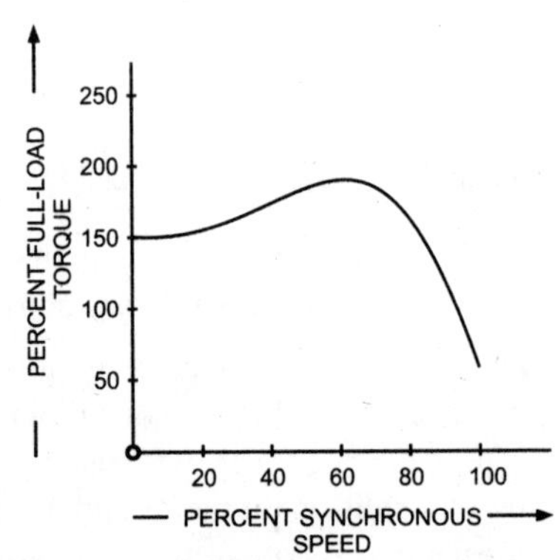

Fig. 20.55 *Torque-Speed Characteristic of a Repulsion Induction Motor*

The motor besides being costlier, has brushes always on the commutator, involving sparking and maintenance problems.

Since its starting current is low, the repulsion induction motor may be operated from lighting circuits when driving frequently started devices.

This motor is particularly suitable where the load can be removed entirely by the declutching or by loose pulley.

Such a motor is used for applications requiring a high starting torque with essentially a constant running speed. The common ratings are $\frac{1}{6}$ to 4 kW although in large ratings are made. Common applications are household refrigerators, garage air pumps, gasoline pumps, compressors etc.

20.22 COMPARISON OF METHODS OF STARTING OF SINGLE-PHASE INDUCTION MOTORS

The various types of single phase induction motors, classified on the basis of methods of their starting, are arranged in ascending order of capital cost as follows.

1. Shaded pole 2. Resistance-start 3. Permanent capacitor 4. Capacitor-start 5. Capacitor-start, capacitor-run 6. Repulsion-start.

If the motors placed at no. 2 and 3 are interchanged in the above list, the motors are then in the ascending order of starting torque, except that motors placed at no. 4 and 5 should be equal.

The repulsion-start induction motor was at one time the most widely used single-phase induction motor, but it has now been almost completely replaced in the fractional kilowatt ratings by the capacitor motors. However, with the increase in output rating, the increase in cost of repulsion start motors is not so rapid as in case of capacitor motors and above the ratings of 1 kW the repulsion-start motors are cheaper than capacitor-start and capacitor-start, capacitor-run motors. The repulsion-start motors are, therefore, still used in ratings above 1 kW, although they have the drawback of being somewhat noisier than the capacitor motors.

20.23 SYNCHRONOUS MOTORS

There are many timing devices where small motors having exact constant speed characteristic will be very advantageous. Very small motors with constant speed characteristic have been developed. They operate from a single-phase ac supply. Because of their exact constant speed characteristic, they are known as *single-phase synchronous motors.* They do not require a dc source of power supply for excitation. The principal applications of such single-phase synchronous motors are for driving of electric clocks, phonographs, record player turntables, tape and other timing devices.

The most commonly used types of single-phase synchronous motors are the *reluctance motors* and the *hysteresis motors*. The efficiency and torque developing ability of these motors is low. The output of most of the commercial motors available is only a few watts. It is practical to design hysteresis motors up to approximately $\frac{1}{8}$ kW.

20.23.1. Reluctance Motors. Reluctance motors operate on the following principle :

"Whenever a piece of ferromagnetic material is located in a magnetic field, a force is exerted upon the material, tending to bring it into the position of the densest portion of the field. The force tends to align the specimen of material so that the reluctance of the magnetic path passing through the material will be at a minimum."

Construction. Reluctance motor is a split-phase induction motor with properly designed salient (shaped) poles. It consists of a stator carrying both the main and auxiliary (or starting) windings for developing a synchronously rotating magnetic field. Except for the permanent-split capacitor motor, all the motors are provided with a starting switch in series with the auxiliary winding. This switch is set to operate at a speed below that at which pull-in torque is exerted.

The usual method of constructing a rotor for a reluctance motor is to assemble it from standard squirrel cage parts, except that some of the teeth are removed. By removing some of the teeth of a normal squirrel cage rotor, salient poles are produced, which offer low reluctance to the stator flux and thereby strongly magnetized. The number of salient poles created on the stator must be equal to the number of poles on stator. Rotor punching for a 4-pole reluctance type synchronous motor is shown in Fig. 20.56.

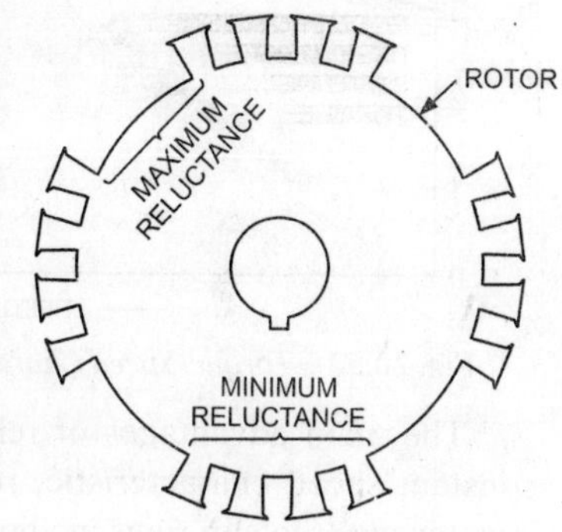

Fig. 20.56 *Reluctance Motor Rotor*

As in a dc excited synchronous motor, pulling into synchronism is to be facilitated when the motor, operating as an induction motor, attains the speed very close to synchronous speed, the rotor should be lighter in weight and have low resistance. For these reasons large-section aluminium bars connected at the ends through a pair of end rings as in a squirrel cage induction motor, are used in rotors.

Operation. When the motor is switched on to a single-phase ac supply, the motor starts as an induction motor, accelerates and attains speed very close to synchronous speed. At about 75% of synchronous speed, a centrifugal switch disconnects the auxiliary winding and the motor continues to run with the main winding only. Since the mechanical load is comparatively small, the slip is negligible, in which case the revolving field permanently magnetizes the projecting rotor poles. The rotor poles then 'lock in step' with the revolving field poles of opposite polarity and continue to rotate at synchronous speed, the speed of the revolving field. The motor adjusts its torque angle for a change in load.

The motor operates at a constant speed up to a little over 200 per cent of its full-load torque. If the load becomes excessive (beyond the value of its pull-out torque), the motor loses synchronism but continues to run as a single-phase induction motor over 500 per cent of its rated torque.

The reluctance of the magnetic path is dependent upon the air gap. Hence more the air gap, more is the reluctance and less the air gap, less is the reluctance. The reluctance is maximum when the stator magnetic field aligns with the cut out portion of the rotor while it is minimum when it aligns with the centre portion of the rotor teeth. Thus, the reluctance of the magnetic circuit depends upon the rotor position with respect to the axis of the stator magnetic field.

Typical torque-speed characteristic of reluctance motor is shown in Fig. 20.57. The starting torque of this type of motor is very high is solely dependent on the rotor position. The starting torque is 3 to 4 times its full-load torque.

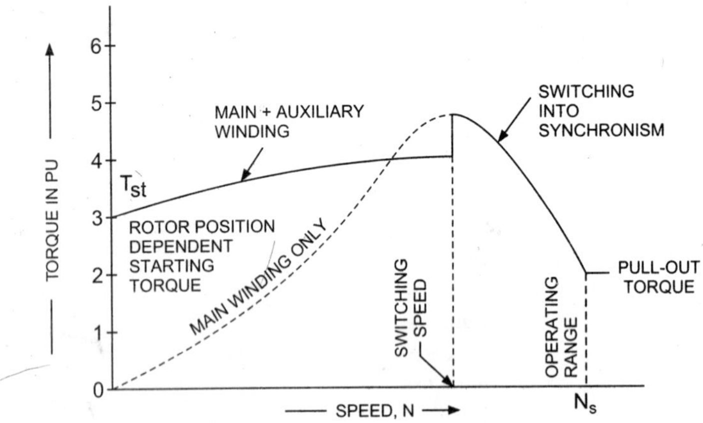

Fig. 20.57 *Torque-Speed Characteristic of Reluctance Motor*

The main advantages of reluctance motor are low cost, constant speed characteristic, robust construction and little maintenance (no slip rings, no brushes, no dc field winding)

Such motors have poor power factor, low efficiency and poor torque. They cannot accelerate high inertia loads to synchronous speed. The pull-in and pull-out torques of such motors are small. The maximum output is largely reduced due to absence of dc field excitation. So, the size of a reluctance motor is larger than that of an equivalent synchronous motor. Such motors are subject to "cogging" because the locked rotor torque varies with the rotor position, but the effect may be minimized by skewing the rotor bars and by not having the number of rotor slots equal to an exact multiple of the number of poles.

Reluctance motors are generally made in the fractional kilowatt ratings and employ the conventional split-phase stator and the centrifugal switch to open the auxiliary winding. When they are built for heavier loads, sometimes in integral kilowatt ratings, the permanent split capacitor construction with no centrifugal switch is preferred.

When the number of salient poles on the rotor is greater, by some multiple, than the number of electrical poles on the stator, the motor will operate at a constant average speed that is submultiple of the apparent synchronous speed; it is then called a *sub-synchronous reluctance motor.*

Reversal of direction of rotation is obtained as in any single phase induction motor.

Applications. Such motors are widely used for absolute constant speed applications, such as in timing devices, signalling devices, recording instruments, teleprinters, phonograph turntables, control apparatus etc.

20.23.2. Hysteresis Motors. Hysteresis motor is a synchronous motor with uniform air gap but with no dc excitation. Its operation depends upon the effect of hysteresis-magnetization developed in a ferromagnetic material, lags behind the magnetizing force.

Construction. The basic construction required for the hysteresis type of synchronous motor consists of a non-salient-pole construction similar to a split-phase induction motor but no winding on the rotor. Such a motor consists of a stator designed to develop a synchronously revolving field from a single-phase ac supply. This is accomplished by using permanent capacitor type construction. The value of capacitance is so adjusted as to give approximately balanced two-phase synchronously revolving field. Alternatively shaded-pole construction is employed to give shaded-pole hysteresis motor. Obviously, in either construction, no centrifugal device is required.

The rotor of such a motor should be built up of such a magnetic material that has high retentivity and permeability i.e. large hysteresis loop so that there is hysteresis effect. The rotor of a hysteresis motor, is therefore, built up of special magnetic material such as cobalt or chrome steel which is mounted on a non-magnetic arbor such as aluminium to make the rotor light. The rotor may be a smooth solid* cylindrical in shape and does not carry any winding or bar.

A 2-pole shaded-pole type hysteresis motor employed for driving ordinary electric clocks is shown in Fig. 20.58.

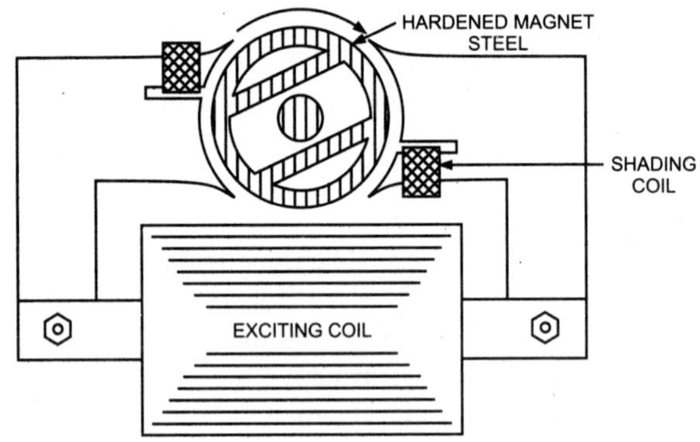

Fig. 20.58

Operation. When the motor is connected to single-phase ac supply, a synchronously revolving field (say in counter-clockwise direction) is developed. The revolving stator flux magnetises the rotor but due to effect of hysteresis, the axis of magnetization of rotor lags behind the axis of stator field by hysteresis lag angle δ, as illustrated in Fig. 20.59 (*a*).

Now the rotor poles get attracted towards the rotating stator field poles and thus the rotor is subjected to torque known as *hysteresis torque*, which is constant at all speeds. The strength of rotor poles remains constant when the stator field axis moves forward.

Instantaneous magnetic conditions in the air gap and rotor are shown in Fig. 20.59 (*a*). The angle δ being hysteresis dependent remains constant, at all rotor speeds. The interaction torque (hysteresis torque) between stator and rotor field, being proportional to the product of stator flux Φ_s, rotor flux Φ_r and sine of hysteresis angle δ, is constant at all speeds of rotor, as illustrated in Fig. 20.59 (*b*). This feature is one of the advantages. Another torque due to eddy current loss is simultaneously created in the motor at starting. The motor starts under the influence of

* To cause a saving in expensive magnetic material instead a thick annular ring of hard steel can be placed on an ordinary cylindrical steel core.

hysteresis torque and eddy current torque and finally attains the synchronous speed. At synchronous speed there is no relative motion between stator field and the rotor and as such there is no eddy current and no torque due to it. So at synchronous speed the motor operates exclusively due to hysteresis torque. At this speed the rotor develops magnetic poles similar to dc excited synchronous motor or the reluctance motor and adjusts its torque angle so as to develop the torque required by the load.

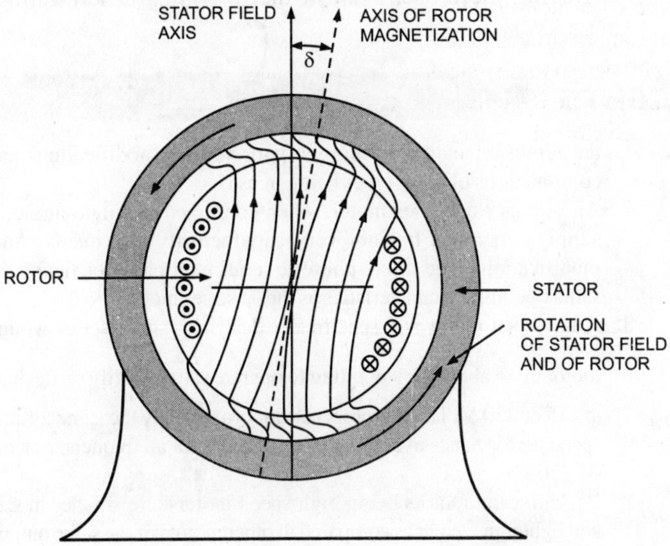

(a) Magnetic Field in a Hysteresis Motor

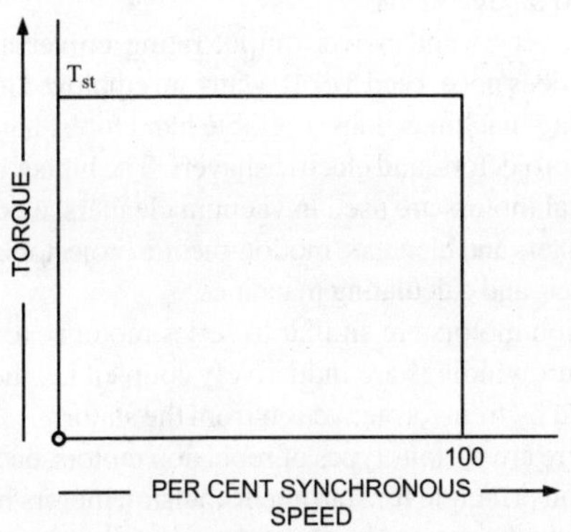

(b) Idealized Torque-Speed Characteristic of a Hysteresis Motor

Fig. 20.59

Magnetic locking between rotor and stator is ensured by high retentivity. The motor either rotates at synchronous speed or not at all due to the principle of magnetic locking.

The hysteresis and eddy current torques developed in a hysteresis motor are derived as below :

The hysteresis loss is given as

$$P_h = K_h (B_{max})^{1.6} f_2 = K_h s f (B_{max})^{1.6} \quad ...(20.58)$$

where B = maximum flux density and f_2, the rotor frequency equals slip s multiplied by the supply frequency f.

Power across air gap,

$$P_g = \frac{P_h}{s} = K_h f (B_{max})^{1.6} \quad ...(20.59)$$

Torque developed,

$$T_h = \frac{P_g}{\omega_s} = \frac{K_h f (B_{max})^{1.6}}{\omega_s} = \text{Constant} \quad ...(20.60)$$

The eddy current loss is given as

$$P_e = K_e f_2^2 (B_{max})^2 = K_e s^2 f^2 (B_{max})^2 \quad ...(20.61)$$

and torque developed,

$$T_e = \frac{P_e}{s w_s} = \left[\frac{K_e f^2 (B_{max})^2}{\omega_s}\right] s \quad ...(20.62)$$

From above Eq. (20.62) it is obvious that eddy current torque is of maximum value at start and reduces linearly with slip vanishing at synchronous speed. This torque aids the hysteresis torque at starting, endowing excellent.

The torque-speed characteristic of hysteresis motor is shown in Fig. 20.59(b). The torque-speed characteristic of a hysteresis motor is quite different from that of an induction motor. In an ideal hysteresis motor, the torque developed by it remains constant at all speeds including synchronous speed while in case of an induction motor the torque developed becomes zero at synchronous speed. Thus, the characteristic of a hysteresis motor reveals that locked rotor, starting and pull-out torques are all equal.

Characteristics and Applications. A hysteresis motor can synchronise any load which it can accelerate, no matter how great the inertia. This is due to development of uniform torque from standstill to synchronous speed, as obvious from Eq. (20.60). The fact that the rotor has no teeth or winding of any sort makes the motor extremely quite in operation; it is not subject to mechanical and magnetic vibrations that must always be present when there are teeth and cutaway sections in the rotor. Resilient mountings further reduce other possible noises so that, virtually hysteresis motors are noiseless.

Because of their noiseless operation and ability to drive high-inertia loads, hysteresis motors are particularly well suited to drive timing devices, electric clocks, tape-decks, turntables and other precision audio equipment.

Commercial motors, being 2-pole motors, run at 3,000 rpm and so for driving an electric clock and other indicating devices, gear train is connected to the motor shaft for reducing the speed.

Rotor being unwound reacts to create the same number of poles as on the stator. Thus by changing the number of stator poles through pole-changing connections, a set of synchronous speeds can be obtained for the motor.

Usual rating is 2-4 watts, but in certain cases 16 watts.

20.23.3. Permanent Magnet Synchronous Motor. A permanent magnet synchronous motor is constructed by embedding

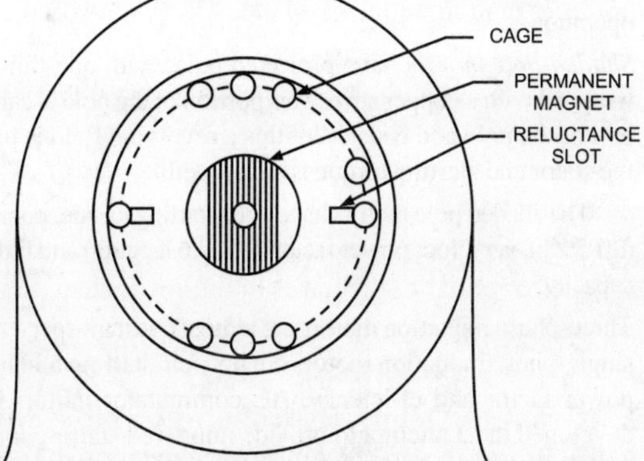

Fig. 20.60 *Permanent Magnet Synchronous Motor*

permanent magnets in the rotor as illustrated in Fig. 20.60. The rotor itself has a squirrel cage construction to provide starting torque. The magnetic path through the annular core carrying the cage winding serves to protect the magnet, but it must be slotted to prevent flux from permanent magnet being short circuited in normal operation.

The rotating stator field may be obtained by means of any one of winding configurations previously discussed viz. shaded pole or permanent capacitor construction. The latter construction (i.e. permanent capacitor one) provides a true two-phase stator near full load and, therefore, a virtually uniform rotating stator field and is commonly employed.

When the motor is connected to a single-phase ac supply, it starts as induction motor, attains nearly synchronous speed and locks into synchronism with the rotating stator field near synchronous speed. It thus operates as a synchronous motor.

Such a motor is quieter in operation and has high power factor and efficiency approaching that of a polyphase motor. It is thus finding more use, even in the low-integral kW ratings (0.5-1.5 kW)

HIGHLIGHTS

1. A single-phase split-induction motor has a squirrel cage rotor and a stator which has two windings. One is running winding or main winding and the other is the starting or auxiliary winding which is disconnected from the supply by a centrifugal switch when rotor attains 70 or 80 per cent of its rated speed.
2. *Operating Principle.* When a single-phase induction motor is connected to a single-phase supply, two fields of magnitude $\Phi_{max}/2$ are developed which rotate in opposite directions. An equal and opposite torque is set up by the two fields and the resultant torque is zero at start. Thus single-phase induction motor is inherently not a self-starting motor. However, if the rotor is rotated in either direction by some external means, torque will be developed and the rotor will pick up the speed in that direction.

   To obtain starting torque in a single-phase induction motor, the single winding is splitted into two windings displaced in space by 90° and having different resistance and inductance. They carry currents of different phase angle which develop a resultant field revolving in space at synchronous speed.
3. Single-phase split-phase induction motors are of two types viz resistance-start induction motor and capacitor-start induction motor.

   The *resistance-start induction motor* has a relatively poor starting torque because the phase angle between the two winding currents is only about 30–50 electrical degrees.

   The capacitor-start induction motor has a comparatively good starting torque because phase angle between the currents flowing in the two windings (running and starting winding) can be as large as 90°.
4. *Capacitor motors* may be (*i*) capacitor-start (*ii*) capacitor-start, capacitor-run and (*iii*) permanent capacitor motors.

   The capacitor-start capacitor-run motor is similar to capacitor-start motor except that the auxiliary winding and a capacitor remain connected in the circuit at all times.

   The permanent capacitor motor has no centrifugal switch and the auxiliary winding remains energized at all times when the motor is in operation.
5. *Shaded-pole motors* have projected poles with one-third pole part wrapped with a copper strip. This portion of the pole is called shaded part of the pole and because of this a revolving field is produced in the stator and starting torque is developed.

   The shaded pole motors have poor starting torque, poor efficiency (20-50%), very poor power factor (0.5–0.6 lagging) and little overload capacity.
6. Three-phase induction motors are almost constant speed motors and single-phase induction motors are not self-starting and have a poor power factor and efficiency. AC commutator motors have been developed in an attempt to provide improved starting torque, good power factor and better speed control. Single-phase commutator motors are the series motors, universal motors, repulsion motors and the repulsion induction motors, with various modifications and combinations of these types being in use.
7. An ordinary dc series motor, when connected to single-phase ac supply, operates, but not very satisfactorily and needs some modifications. The single-phase ac series motor has practically the same operating characteristics as the dc series motor.
8. A universal motor is a specifically designed, small series wound motor up to about $\frac{1}{2}$ kW rating to operate satisfactorily on dc or ac at 115 or 230 V. These motors have approximately the same torque-speed performance over the operating range, for all frequencies from 0 to 50 Hz.

   Universal motors being high speed motors, are smaller in size and lighter in weight as compared to other motors of the same output. Full-load power factor is high (approximately 0.9) but it is poor at start and on overloads.

   The very small power output rating universal motors, which usually does not exceed 5 or 10 watts are employed in such equipment as sewing machines, fans, portable hand tools, hairdryers, motion-picture projectors, and electric shavers. The higher rating (5–500 W) universal motors are used in vacuum cleaners, electric typewriters, food mixers and blenders, motion-picture projectors, cameras, adding machines, and calculating machines.
9. Repulsion motors are similar to series motors except that the rotor and stator windings are inductively coupled i.e. the rotor current is obtained by transformer action from the stator.

   There are various types of repulsion motors, each employing the repulsion principle for starting. Repulsion motors include *repulsion motor* which starts and runs on the principle of magnetic repulsion; *repulsion-start induction-run motor*, which starts on the repulsion principle but at a predetermined speed operates as a single-phase induction motor; and the *repulsion-induction motor*, which employs the repulsion-start principle but has an additional armature winding that provides additional running torque.
10. There are many timing devices where small motors having exact constant speed characteristic will be very advantageous. Very small motors with constant speed characteristic have been developed. They operate from a single-phase ac supply. Because of their exact constant speed characteristic, they are known as *single-phase synchronous motors.* They do not require a dc source of power supply for excitation. The principal applications of such single-phase synchronous motors are for driving of electric clocks, phonographs, record player turntables, tape and other timing devices.

    The most commonly used types of single-phase synchronous motors are the *reluctance motors* and the *hysteresis motors*. The efficiency and torque developing ability of these motors is low. The output of most of the commercial motors available is only a few watts. It is practical to design hysteresis motors up to approximately $\frac{1}{8}$ kW.

EXERCISES

1. Prove that a single-phase motor winding when excited by a single-phase supply produces two equal and opposite revolving fields. [R.G. Technical Univ. Electrical Machine-I, June-2014]
2. Explain the double revolving field theory for operation of single-phase induction motor. [Anna Univ. Electrical Machines-II, Nov./Dec.-2012]
3. Explain double revolving field theory of a single-phase induction motor. [M.D. Univ. Electromechanical Energy Conversion, December-2012; Gujarat Technological Univ. Electrical Machines-II, Summer-2014]
4. Using double revolving field theory explain the working of a single-phase induction motor. [U.P.S.C. I.E.S. Electrical Engineering-II, 2002]
5. Why is it advantageous to use double revolving field theory for determining the running performance of a single-phase induction motor?

   Draw torque-speed characteristics of a single-phase induction motor based on double-revolving field theory and discuss about the magnitude of torque at zero speed and synchronous speed.

   Sketch waveform of rotor current under normal running conditions of the single-phase induction motor. [U.P.S.C. I.E.S. Electrical Engineering-II, 2001]
6. Give the reasons why a single-phase induction motor has poor performance as compared to a 3-phase induction motor. [U.P.S.C. I.E.S. Electrical Engineering-II, 2009]
7. State the reasons for the inferior performance of single-phase motors over three-phase motors. [J.N. Technological Univ. Electrical Machines-III, May-2013]
8. Explain the cross-field theory as applied to a single-phase induction motor.
9. With the help of the torque-slip characteristic explain why single-phase induction motor is not inherently self-starting? [G.G.S.I.P. Univ. Electromechanical Energy Conversion-II, May-2011]
10. Draw and explain the equivalent circuit of a single-phase induction motor on the basis of double revolving field theory. Also draw torque-slip characteristics of it. [U.P. Technical Univ. Electrical Machines, 2010-11]
11. Starting from the expression of sinusoidal supply current, mathematically derive the two revolving field theory model and the equivalent circuit of a single-phase induction motor. [W.B. University of Technology Electrical Machines-II, 2011-12]
12. Describe no-load test and blocked rotor test on a 1-ϕ induction motor and explain how you obtain data for parameters of equivalent circuit of 1-ϕ induction motor. Draw the equivalent circuit of 1-ϕ induction motor both at no-load condition and full-load condition. [U.P. Technical Univ. Electromechanical Energy Conversion II, 2003-04]
13. Explain the following tests that are performed on 1-ϕ IMs:
    (*i*) Stator resistance dc test
    (*ii*) No-load test
    (*iii*) Blocked-rotor test.
    Also mention their utility. [U.P. Technical Univ. Electromechanical Energy Conversion-II, 2013-14]
14. What do you mean by "FORWARD" and "BACKWARD" fields in 1-ϕ IMs? Explain why in case of 1-ϕ IMs, the flux is pulsating in nature. Explain why, 1-ϕ IMs are not self-starting? Enlist the methods of starting of 1-ϕ IMs. Also mention its domestic applications. [U.P. Technical Univ. Electromechanical Energy Conversion-II, 2013-14]
15. Prove that single-phase induction motor is not self-starting. How it is made self starting? Explain the methods. [M.D. Univ. Electrical Machines-II, December-2011]
16. Draw the equivalent circuit of a single phase induction motor. Explain the constructional features and principle of operation of the motor. [Rajasthan Technical Univ. Electrical Machines-II, 2012]
17. Explain different methods of starting 1-phase induction motor. [Gujarat Technological Univ. Electrical Machines-II, Summer-2014]
18. Explain various types of single-phase induction motors alongwith the techniques used for their starting. [G.G.S.I.P. Univ. Electromechanical Energy Conversion-II, May-2011]
19. Describe the construction and principle of operation of split-phase motor. [J.N. Technological Univ. Electrical Machines-III, May-2013]
20. Why single-phase induction motor is not self-starting? Discuss construction and working of capacitor-start, induction-run motor. [Pb. Technical Univ. Asynchronous Machines, May-2013]
21. Explain the operation of single-phase capacitor-start motor. [U.P.S.C. I.E.S. Electrical Engineering-II, 1999]
22. Draw the connection diagram of a capacitor-start induction motor showing starting and main windings. [R.G. Technical Univ. Electrical Machine-I, June-2014]
23. What are the different methods of starting a single-phase induction motor? With the help of appropriate circuit diagram and phasor diagram. Explain the principle of capacitor-start, capacitor-run 1-phase induction motor. Briefly indicate how to design the value of capacitor be used. [W.B. Univ. of Technology Electrical Machines-II, 2011-12]
24. What are the methods adopted to start a single-phase induction motor? Explain with the help of connection diagram the principle of operation of a capacitor-run motor. [W.B. Univ. of Technology Electrical Machines-II, 2012-13]
25. Draw neat phasor diagrams and torque-speed characteristics of single-phase capacitor-run, capacitor-start capacitor-run, and capacitor-start motors. Compare their performance on the basis of these characteristics. [U.P.S.C. I.E.S. Electrical Engineering-II, 2013]
26. Describe the construction, principle of operation of shaded-pole induction motor. Also give some of the applications. [J.N. Technological Univ. Electrical Machines-III, May-2013]
27. Explain the construction, working principle and torque-speed characteristics of shaded pole induction motor. [G.B. Technical Univ. Special Electrical Machines, 2012-13]
28. Explain, with the help of sketch and phasor diagram, how rotating magnetic field is produced in a single-phase shaded-pole motor. Discuss the direction of rotor rotation. [U.P.S.C. I.E.S. Electrical Engineering-II, 2012]
29. Indicate the slip-torque characteristics of different types of single-phase induction motors in one diagram and compare. State reasons for their deviations. [R.G. Technical Univ. Electrical Machine-I, June-2014]
30. Classify 1-phase induction motors by the methods of starting and compare their performance and mention typical application areas. [U.P.S.C. I.E.S. Electrical Engineering-II, 2010]
31. Explain the principle of operation of a single-phase commutator motor. [U.P. Technical Univ. Electrical Machines, 2010-11]
32. Discuss construction and working of an ac commutator motor. [M.D. Univ. Electromechanical Energy Conversion, December-2012]
33. Describe, with the help of neat sketch, ac series motor. Discuss the principle of operation of ac series motor. Explain, why ac shunt motor is not feasible. [U.P. Technical Univ. Electromechanical Energy Conversion II, 2003-04]
34. Explain the principle of operation of ac series motor with a neat diagram. Discuss its applications. [J.N. Technological Univ. Electrical Machines-III, May-2013]
35. Discuss importance of single-phase series motor. Draw its operating characteristics. [Pb. Technical Univ. Asynchronous Machines, May-2013]
36. Draw the phasor diagram of ac series motor, explaining the significance of each phasor element. [G.G.S.I.P. Univ. Electrical Machines, December-2011]

37. Mention the problems usually encountered when a dc series motor is operated on ac. What design modifications are to be incorporated for its satisfactory operation on ac? Compare the speed for dc and ac operations.

38. What are uncompensated and compensated single-phase series commutator motors ? Develop their performance equations and related phasor diagrams.

39. What are the two kinds of emfs induced in the armature of ac commutator machine ? Derive expression for these emfs.

[U.P. Technical Univ. Electromechanical Energy Conversion II, 2002-03]

40. Draw the phasor diagram of a single-phase series motor and show that its performance characteristics can be predicted from a circle diagram. Sketch the typical performance characteristics (torque, speed, power factor and kW output) against current. What are the special design features when compared to a dc series motor ?

41. Explain the construction and working principle of universal motor.

[Gujarat Technological Univ. Electrical Machines-II, Summer-2014]

42. Describe the construction and working of a universal motor and give its applications.

[G.G.S.I.P. Univ. Electromechanical Energy Conversion-II, May-2011]

43. Describe a series (universal) motor. Give its x-sectional view when connected to ac supply. Describe its performance characteristics.

[U.P.S.C. I.E.S. Elec. Engineering II, 1995]

44. Why are the compensating winding and the interpole winding used in a universal motor for ac operation ? Draw the schematic connection diagram for all the stator windings and the armature for ac-operated series (universal) motor. **[U.P.S.C. I.E.S. Elec. Engineering II, 1996]**

45. Discuss the various methods of speed control of universal motors.

[Bharathiyar Univ. Elec. Machines-II, April-1997]

46. Discuss the salient features of repulsion motor.

[Pb. Technical Univ. Asynchronous Machines, May-2013]

47. Give construction and explain principle of operation of a repulsion motor and compare it with a single-phase induction motor.

[U.P. Technical Univ. Electromechanical Energy Conversion II, 2004-05]

48. What are repulsion motors ? Why are they so named ? Develop the phasor diagram for the performance calculations of a single-phase repulsion motor for a brush shift of α electrical degrees.

49. With neat sketch, explain briefly about the construction and working principle of repulsion motor.

[G.B. Technical Univ. Special Electrical Machines, 2012-13]

50. Discuss the constructional features, principle of operation and torque-speed characteristics of repulsion-start induction motor.

[M.G. Univ. Electrical Machines-III, November-2005]

51. Describe the principle of operation of a reluctance motor giving suitable construction diagrams. State its various applications.

[G.G.S.I.P. Univ. Electromechanical Energy Conversion-II, May-2011]

52. Describe the construction of a synchronous hysteresis motor and show that it develops a running torque both at synchronous and asynchronous speed of the rotor.

53. With neat sketch, explain the construction, operating principle and characteristics of hysteresis motor.

[G.B. Technical Univ. Special Electrical Machines, 2012-13]

54. Explain the principle of operation of a permanent magnet fractional horse power synchronous motor. How can the motor be made self-starting ?

55. Write short notes on the following:

(*i*) Double revolving field theory. **[J.N. Technological Univ. May-2013]**

(*ii*) Cross-field theory of 1-phase induction motor. **[M.D. Univ., December-2010]**

(*iii*) Starting of single-phase induction motor. **[U.P. Technical Univ., 2010-11]**

(*iv*) Capacitor motors. **[U.P. Technical Univ., 2010-11]**

(*v*) Shaded-pole induction motor. **[Anna Univ. Nov./Dec.-2012]**

(*vi*) Series motor. **[J.N. Technological Univ., May-2013]**

(*vii*) Universal motor. **[U.P. Technical Univ., 2010-11; W.B. Univ. of Technological, 2011-12; J.N. Technology Univ., May-2013]**

(*viii*) Repulsion motor. **[R.G. Technical Univ., December-2013]**

(*ix*) Reluctance motor. **[J.N. Technological Univ., May-2013]**

(*x*) Hysteresis motor.

[Anna Univ., Nov./Dec.-2012, W.B. Univ. of Technology, 2012-13]

SHORT ANSWER TYPE QUESTIONS WITH ANSWERS

Q. 1. Name the different types of fractional kilowatt motor.

[G.B. Technical Univ., 2012-13]

Ans. Fractional kilowatt motors may be classified as follows:

1. **Single-Phase Induction Motors:**

   (*a*) Resistance split type (*b*) Capacitor split of type (*c*) Shaded-pole type (*d*) Repulsion start.

2. **Single-Phase Commutator Motors:**

   (*a*) Series motors (*b*) Universal motors (*c*) Repulsion motors and repulsion induction motors.

3. **Single-Phase Synchronous Motors:**

   (*a*) Reluctance motors (*b*) Hysteresis motors.

4. **Special Purpose Single-Phase Motors:**

   (*a*) Stepper motors (*b*) Servomotors

Q. 2. How will you resolve a single pulsating field ?

[Bharathidhasan Univ., Nov.-1995; Bharathiyar Univ., April-1997]

Ans. A single pulsating field can be resolved into two components of half its amplitude and rotating in opposite directions with equal speed.

Q. 3. Name the two theories regarding single-phase induction motor.

[Bharathidhasan Univ., Nov.-1996]

Ans. The two theories regarding single-phase induction motor are (*i*) double revolving field theory and (*ii*) cross-field theory.

Q. 4. State the principle that the double revolving field theory makes use of. **[Bharathiyar Univ., Nov.-1995]**

Ans. The double revolving field theory makes use of the principle that pulsating field produced in a single-phase motor can be resolved into two components of half its amplitude and rotating in opposite directions with synchronous speed.

Q. 5. Is the starting torque of a single-phase induction motor zero ?

Ans. Yes, the starting torque of a single-phase induction motor is zero.

Q. 6. Is the torque developed by a single-phase induction motor at synchronous speed zero ?

Ans. No, it is negative.

Q. 7. What is the effect of increasing rotor resistance in a single-phase induction motor ?

Ans. Increase in rotor resistance of a single-phase induction motor reduces its breakdown torque, lowers the efficiency and increases the slip corresponding to the maximum torque.

Q. 8. Why single-phase induction motors are usually set on rubber spring mounts ?

Ans. In a single-phase induction motor, pulsating torque is developed as a direct consequence of the pulsating power in a single-phase circuit.

Q. 9. Draw equivalent circuit of single-phase induction motor. **[Pb. Technical Univ., May-2013]**

Ans. Refer to Figs. 20.9 (*a*), (*b*), (*c*), (*d*).

Q. 10. Why single-phase induction motor is not self starting? **R.G. Technical Univ., June-2014]**

Ans. When a single-phase induction motor is connected to a single-phase supply, two fields of magnitude $\Phi_{max}/2$ are developed which rotate in opposite directions. An equal and opposite torque is set up by the two fields and the resultant torque is zero at start. Thus single-phase induction motor is inherently not a self-starting motor.

Q. 11. Name some of the starting methods of single-phase induction motors. **[J.N. Technological Univ., May-2013; Anna Univ., Nov./Dec.-2012]**

Ans. Commercial single-phase induction motors make use of the principle of phase split for the purpose of starting. Split-phase motors are of the following types:

(*a*) Resistance-start motor or simply split-phase motor

(*b*) Capacitor–start motor

(*c*) Permanent capacitor or single-value capacitor motor and

(*d*) Capacitor-start capacitor-run or two value capacitor motor

Q. 12. What is a split-phase motor ?

Ans. This is a single-phase induction motor equipped with an auxiliary winding, also called the starting winding, connected in parallel with the main winding on stator and differing from it in both phase and space. The starting winding is usually opened by a centrifugally operated device when the motor attains a predetermined speed.

Q. 13. Why an auxiliary winding is required for the operation of a single-phase induction motor? **[G.G.S.I.P. Univ., May-2011]**

Ans. Auxiliary winding, also known as *starting winding*, is provided so that, together with the main winding, the motor can stimulate a two-phase motor to develop torque at start.

Q. 14. A single-phase induction motor is provided with a main winding and an auxiliary winding. Which of the two windings should be more resistive ?

Ans. Auxiliary winding should be more resistive.

Q. 15. In the design of resistance-start single-phase induction motor the starting winding is disconnected once the slip reaches 0.25 or less. Explain.

Ans. The starting winding of a resistance-start motor is not designed for continuous operation and so it is necessary to disconnect it from the motor circuit when the motor attains a definite speed. The starting winding is usually taken out of circuit at a speed of about 75% of synchronous speed because at this speed the torque developed by the main winding alone is greater than that developed by both windings.

Q. 16. How is the direction of rotation of a single-phase induction motor reversed ? **[M.K. Univ., Nov.-1996; Madras Univ., April-1997; Bharathiyar Univ., April-1997]**

Ans. The direction of rotation of a single-phase induction motor (split-phase type) is reversed by reversing the leads to the main or starting winding, but not both.

Q. 17. How does a capacitor-start motor differ from a resistance-start motor ?

Ans. Fundamentally the capacitor-start motor is quite similar to a resistance-start motor, except that its starting winding has a few more turns and is wound with comparatively thick wire than the starting winding of a resistance-start motor and also the external resistance, connected in series with the starting winding, is replaced by a large electrolytic capacitor.

Q. 18. How the direction of rotation of a capacitor-start induction-run motor can be reversed? **[R.G. Technical Univ., December-2011]**

Ans. The direction of rotation of a capacitor-start, induction run motor is reversed by reversing either the starting or running winding leads, not both, to the line.

Q. 19. Explain why a capacitor is connected in one of the windings of single-phase induction motor. **[B.P. Univ. of Technology, 2005]**

Ans. Single-phase induction motors are not self starting. Single-phase induction motors can be made self-starting by using the principle of phase splitting. The capacitor is used in single-phase capacitor motors (capacitor-start motor, capacitor start-capacitor-run motor and permanent capacitor motor) to cause phase splitting. By using a capacitor of suitable value in starting or auxiliary winding, the starting winding current at standstill can be made to lead the main winding current by 90°.

Q. 20. What happens to a capacitor-start induction motor if the auxiliary winding is left in the circuit after it has been used to start the motor ?

Ans. The auxiliary winding of a capacitor-start induction motor is not designed for continuous operation and if left in the circuit after it has been used to start the motor, it will burn out. In addition to it, the short-duty rated electrolytic capacitor connected in series with the starting winding will also get damaged.

Q. 21. Are capacitor-start induction-run motors usually dual voltage or single voltage motors ?

Ans. Capacitor-start, induction-run motors are usually dual voltage motors.

Q. 22. If a capacitor-start motor is rated as 115/230-V motor (the capacitor is usually rated at 115 V), how is it connected, when the motor is used on 230 V ?

Ans. The main or running winding of a dual voltage motor is in two sections so that they may be connected in parallel for 115-V operation and in series for 230-V operation. The capacitor is connected in series with the starting winding; therefore, when operating at 115 V, the winding leads are connected across the line at start, but when operating on 230 V, one lead is connected to the one side of the line and the other lead to the mid point of the running windings.

Q. 23. What is the advantage of using a capacitor-start motor over a resistance-start split-phase motor ? **[Bharathiyar Univ., Nov.-1996]**

Ans. The capacitor-start motor develops more starting torque than does an equally rated resistance-start split-phase motor with a lower inrush current.

Q. 24. Compare the performance of capacitor-start single-phase induction motor and resistance split-phase motor. **[G.G.S.I.P. Univ., December-2011]**

Ans. Since the torque developed by a phase-split motor is proportional to the sine of phase displacement angle θ between the starting winding current I_s and main winding current I_m, it is obvious that the increase in phase displacement angle θ (from 30° to 80°) alone increases the starting torque twice the value developed by a resistance-start motor. The increase in starting winding turns causes decrease in starting current on one hand and produces a large rotating flux, and therefore, a further increase in the starting torque on the other hand. Finally it is seen that, for the same magnitudes of two stator winding currents, I_s and I_m the line current I_L is less in the capacitor-start motor, because of the greater angle between them. In addition, the starting winding power factor is also improved. In short it can be said that because of the closer approximation to a true two-phase motor, a greater number of turns on the starting winding, and more allowable current in both of the windings due to lower line current, the capacitor-start motor develops a greater starting torque than does an equally rated resistance-start motor. The line current of a resistance-start motor was found to be two-third higher than line current of a corresponding capacitor-start motor, while the starting torque of the capacitor-start motor was twice that of resistance-start motor.

It is to be noteworthy that the capacitor-start motor differs from the resistance-start motor only during the starting period, for it develops more starting torque (about 3.5-4.5 times full-load torque) with a lower inrush current. After the machine attains normal speed and the starting winding is cut out, its operating characteristics are identical to those of a corresponding resistance-start motor.

Q. 25. What is a capacitor-start, capacitor-run motor ?

Ans. This is the same motor as the capacitor-start motor, except that it has an extra oil type capacitor connected in series with the starting winding, which always remains in the circuit.

Q. 26. Sketch the circuit diagram of capacitor-start, capacitor-run induction motor and draw speed-torque characteristic. **[G.B. Technical Uni.., 2012-13]**

Ans. Refer to Figs. 20.16 (*a*) and (*b*).

Q. 27. In a 2-phase capacitor motor which capacitor is of fairly high value and to what order ?

Ans. In a 2-phase capacitor motor the starting capacitor is about 10 – 15 times as large as the running capacitor.

Q. 28. What happens when the auxiliary winding of a capacitor motor is disconnected during running condition ?

Ans. The motor will continue running but will develop small torque. Once it stops, it will not start again.

Q. 29. Name the motor being used in a ceiling fan ?

Ans. Split-phase capacitor type single-phase induction motor.

Q. 30. What is a shaded-pole motor ?

Ans. Shaded-pole motor is a salient-pole, split-phase type, squirrel cage rotor single-phase induction motor provided with one or more auxiliary short-circuited stator windings displaced in magnetic position from the main winding.

Q. 31. Does a shaded-pole motor have much starting torque ?

Ans. No, it has very little starting torque.

Q. 32. In what direction do shaded-pole motors turn ?

Ans. Shaded-pole motor rotates in a direction from the unshaded part of the pole to the shaded part.

Q. 33. What role do 'magnetic bridges' play in the operation of a shaded-pole motor ?

Ans. The magnetic bridges cause the flux in the shaded portion of the pole to lag behind the flux in the unshaded portion. This gives, in effect a motion of flux across the pole face and under the influence of this moving flux a starting torque is developed.

Q. 34. State the applications of shaded-pole single-phase induction motors.

Ans. Shaded-pole motor develops very little starting torque and, therefore, it is used only for small fans, electric clocks, hairdryers, time phonographs, advertising displays, toys, instruments and such other applications.

Q. 35. How can a shaded-pole motor be reversed in direction of rotation ?

Ans. Unless the machine is constructed so that the shading coil can be shifted to the other side of the pole, the direction of rotation of such a machine cannot be reversed.

Q. 36. Which is preferable, a small or a large number of rotor slots, in case of single-phase motors ?

Ans. Large number of slots are preferable in the stator of a single-phase motor.

Q. 37. Why is the normal full-load slip of a single-phase induction motor higher than that of a three-phase motor ?

Ans. The normal full-load slip of a single-phase induction motor is higher than that of a 3-ϕ motor, owing to development of backward rotating field. The power is to be delivered to the backward field from the power converted into mechanical power by the forward field.

Q. 38. What is the necessity of having laminated yoke in an ac series motor ?

Ans. Yoke and the pole and armature cores of ac series motors are laminated so that eddy current losses are reduced and so efficiency is improved and heating is reduced.

Q. 39. What happens to the speed and power factor of a single-phase series motor when the load increases ?

Ans. Speed and power factor, both, decrease with the increase in load.

Q. 40. Does the power factor of a single-phase series motor increase or decrease with increasing load torque.

Ans. The power factor of a single-phase series motor decreases with the increase in load torque.

Q. 41. Why should a series motor never be operated on no load ?

Ans. The series motors are never operated on no load because on no load they may attain disastrous high speed and fly into pieces.

Q. 42. What is a universal motor ? **[M.S. Univ., April-1996]**

Ans. A universal motor is a series wound motor, compensated or uncompensated, specifically designed to operate on direct or single-phase alternating current of the same voltage at approximately the same speed and output.

Q. 43. Why should a motor be named as universal motor ? **[M.K. Univ., Nov.-1996]**

Ans. Because the motor runs on dc as well as on ac, it is named as universal motor.

Q. 44. What is the principle of operation of universal motor ? **[M.K. Univ., Nov.-1995]**

Ans. DC motors rotate in the same direction regardless polarity of the supply i.e. if the polarity of the line terminals of any dc motor are reversed, both the direction of the flux and direction of the current in the armature conductors are reversed and, therefore, torque developed remains in the same direction, and the motor continues to rotate in the direction in which it was rotating before the line terminals were reversed.

Q. 45. Why compensating winding is provided in a universal motor for ac operation ?

Ans. The compensating winding is provided in a universal motor for ac operation to neutralize reactance voltage set up by the alternating flux due to transformer action.

Q. 46. How can the direction of rotation of a universal motor be reversed?

Ans. The direction of rotation of universal motor can be reversed by reversing either the field or armature leads, but not both.

Q. 47. Why are high speeds often desirable in operation of universal motors ? **[Bharathiyar Univ., Nov.-1996]**

Ans. Universal motors are designed for operation at high speeds (exceeding 3,500 rpm) owing to the difficulty in obtaining like performance on dc and ac at lower speeds.

Q. 48. Which single-phase motor has the poorest speed regulation?

Ans. Single-phase series motor (uncompensated type) has the poorest speed regulation.

Q. 49. Name three ways in which the speed of a universal motor can be varied.

Ans. Resistance method, tapping field method and centrifugal mechanism method.

Q. 50. What are the characteristics of universal motor. **[G.B. Technical Univ., 2012-13]**

Ans. The torque-speed characteristic of a universal motor is quite similar to that of a series wound dc motor i.e., high starting torque and high no-load speed. In a very small series motor, the losses are usually large enough at no load to restrict the speed to a definite value. Where there is a danger of high no-load speed, a governor may be employed to keep speed within specified limits. The governor consists of a centrifugal switch mounted on the motor shaft. The tension of the springs of the switch is adjusted so that the switch opens at a predetermined speed and thus place an external resistor R in series with the armature, thereby reducing the speed. When the speed drops because of increase in load, the switch contacts close, thereby shorting out the series resistor R, thus again raising the speed.

Universal motors being high speed motors, are smaller in size and lighter in weight as compared to other motors of the same output. Full-load power factor is high (approximately 0.9) but it is poor at start and on overloads.

Q. 51. Suggest suitable single-phase induction motors for the following applications and give reasons for your choice.
(*i*) Ceiling fan (*ii*) Portable drilling machine (*iii*) Motor of domestic refrigerators (*iv*) Sewing machine.
[M.D. Univ. Electromechanical Energy Conversion, December-2006]

Ans. (*i*) **Ceiling Fan:** Permanent capacitor single-phase induction motor. Because of the absence of starting switch, it is possible to run a permanent capacitor single-phase induction motor over a wide range of speeds. Its speed can be controlled by varying the voltage applied to the fan motor.

(*ii*) **Portable Drilling Machine:** Resistance-start split-phase motors because of their low cost and moderate starting torque.

(*iii*) **Motor of Domestic Refrigerators:** Capacitor-start single-phase induction motors because of their higher starting torque.

(*iv*) **Sewing Machine:** Universal motors due to their high no-load speed. They are always built-in the sewing machines with little danger of motor runaway.

Q. 52. In a repulsion motor, which position of the brushes is known as the high impedance position?

Ans. The position at right angles to the stator poles or neutral axis is known as the high impedance position.

Q. 53. What happens when the shaft torque of a repulsion motor with two stator windings increases ?

Ans. With the increase in torque on a repulsion motor stator current increases, and speed and power factor of the motor decreases.

Q. 54. What happens when the brush axis of a repulsion motor is aligned with its stator pole axis?

Ans. The brushes join the point of maximum potential difference, enormous current would flow between the brushes but no torque would be developed as the torque angle would be zero. The motor will growl harshly.

Q. 55. At around what speed (in terms of synchronous speed) commutation of a repulsion motor is best ?

Ans. At an approximately 80 per cent of synchronous speed, the commutation of a repulsion motor is the best.

Q. 56. State whether repulsion motor has series speed-torque characteristics or shunt speed-torque characteristics.

Ans. Series speed-torque characteristics.

Q. 57. What are the two positions of brushes at starting in which the repulsion motor does not develop the starting torque ?

Ans. Field axis and quadrature to field axis positions of the brushes are such in which repulsion motor does not develop any starting torque.

Q. 58. How speed variation is affected in a repulsion motor ?

Ans. Speed variation is affected either by varying the impressed voltage or by changing the position of brushes.

Q. 59. How will you change the direction of rotation of a repulsion motor ?

Ans. The direction of rotation of a repulsion motor is changed by shifting the brushes around the commutator.

Q. 60. Why centrifugal switch is provided in repulsion-start induction motors ?

Ans. Centrifugal switch is provided in a repulsion-start induction motor to short circuit all the commutator segments at about 75-80 per cent of synchronous speed and in some motors also to lift the brushes from the commutator.

Q. 61. Can a repulsion-induction motor ever run at super-synchronous speed ?

Ans. Yes; at no load.

Q. 62. Give applications of repulsion motor. **[Pb. Technical Univ., May-2013]**

Ans. The repulsion motor has never enjoyed popularity. The motor is used where sturdy motor with large starting torque and adjustable speed is required. Most common use of this type of motor is in the coil winders in which the operator adjusts the speed by shifting the brushes; the motor is equipped with a special lever mechanism, that shifts the brushes when a foot treadle is pressed.

The ratings in which the repulsion motors are built is limited due to commutation problems. The usual rating of the repulsion motor does not exceed 5 kW.

Q. 63. Name a small size synchronous motor that abviates the need for dc excitation.

Ans. Reluctance motor.

Q. 64. Why the reluctance motor has low efficiency ?

Ans. Because of varying reluctance.

Q. 65. In a motor running at synchronous speed with no dc excitation, the load torque is independent of the rotor speed. Identify the motor.

Ans. Hysteresis motor.

Q. 66. Does the starting torque of a hysteresis motor have the hysteresis torque as the only component ?

Ans. No; eddy current torque in addition to hysteresis torque.

PROBLEMS

1. A 150 W, 4-pole, 110 V, 50 Hz, 1-ϕ induction motor delivers rated output at a speed of 1,425 rpm. The total copper loss at full load is 30 W. Determine the full-load slip and full-load efficiency. The rotational losses equal 25 watts and neglect stator copper loss. Also determine the copper losses caused by the two fields.
[Ans. 5 %, 73.2 %, 9.73 W, 20.27 W**]**

2. Discuss the double revolving field theory of single phase induction motors. Find the mechanical power output at a slip of 0.05 of the 185 W, 4-pole, 110 V, 60 Hz single-phase induction motor, whose constants are given below:
Resistance of stator main winding, $R_1 = 1.86\ \Omega$
Reactance of stator main winding, $X_1 = 2.56\ \Omega$

Magnetizing reactance of the stator main winding, $X_m = 53.5\ \Omega$

Rotor resistance at standstill, $R_2 = 3.56\ \Omega$

Rotor reactance at standstill, $X_2 = 2.56\ \Omega$

[Nagpur Univ. Elec. Machines, 1991]

[**Ans.** 201.5 watts]

3. Find the mechanical power output of 185 W, 4-pole, 110 V, 60 Hz single-phase induction motor, whose constants are given below at a slip of 0.05

$R_1 = 1.86\ \Omega$, $X_1 = 2.56\ \Omega$, $X_\phi = 53.5\ \Omega$, $R_2 = 3.56\ \Omega$, $X_2 = 2.56\ \Omega$, core loss = 35.0 W, Friction and windage loss = 13.5 W

[**Ans.** 183 watts]

4. Estimate the parameters of the equivalent circuit of a single-phase induction motor from the following data:

Light running test : 220 V, 1.3 A, 60 W

Blocked rotor test : 220 V, 5.5 A, 840 W

Stator resistance : 11 Ω (dc)

Rotor resistance (at standstill/at running) : 1.2.

[Pb. Univ. Elec. Machines-II, Dec.-1993]

[**Ans** $R_1 = 11\ \Omega$; $X_1 = 24.347\ \Omega$; $X_m = 263.5\ \Omega$; $R_2 = 13.97\ \Omega$; $X_2 = 24.347\ \Omega$]

5. A 230 V, 380 W, 50-Hz, 4-pole, 1-phase induction motor gave the following test results:

No-load test:	230 V,	84 W,	2.8 A
Blocked rotor test :	110 V,	460 W,	6.2 A

The stator winding resistance is 4.6 Ω and during the blocked rotor test, the auxiliary winding is open. Determine the equivalent circuit parameters. **[Pb. Univ. Elec. Machines II, May-1995]**

[**Ans.** $R_2' = 7.067\ \Omega$; $X_2' = X_1 = 6.68\ \Omega$, $X_m = 143.75\ \Omega$]

6. A 220 V, 1-ϕ induction motor gave the following test results:

Blocked rotor test :	120 V	9.6 A	460 W
No-load test :	220 V	4.6 A	125 W

The stator winding resistance is 1.5 Ω and during blocked rotor test, the starting winding is open. Determine equivalent circuit parameters.

[U.P. Technical Univ. Electromechanical Energy Conversion II, 2003-04]

[**Ans.** 3.5 Ω, 5.73 Ω, 84.3 Ω]

7. A 1-ϕ induction motor has stator windings in space quadrature and is supplied with a single phase voltage of 200 V at 50 Hz. The standstill impedance of the main winding is $(5.2 + j\,10.1)\ \Omega$ and that of the auxiliary winding is $(19.7 + j\,14.2)\ \Omega$. Find the value of capacitance to be inserted in the auxiliary winding for maximum starting torque.

[Madurai Kamarajar Univ. Elec. Machines II, Nov.-1996]

[**Ans.** 131 µF]

8. A 230-V, 4-pole , 50-Hz single phase induction motor has the following data at standstill.

Main winding $(1.5 + j\,4.0)\ \Omega$; starting winding $(2.2 + j\,5.5)\ \Omega$. For making the motor develop maximum starting torque, find the value of (*i*) resistor (*ii*) capacitor in series with the starting winding.

[Pb. Univ. Elec. Machines-II, May-1994]

[**Ans.** (*i*) 5.74 Ω (*ii*) 503 µF]

[**Hint :** starting torque in case of a resistance-start motor will be maximum when $\phi_s = \dfrac{\phi_m}{2}$]

9. A 250 watt, 230-V, 50-Hz single-phase capacitor-start induction motor has the following constants for the main and auxiliary windings. Main winding $Z_m = (4.5 + j\ 3.7)\ \Omega$, auxiliary winding $Z_a = (9.5 + j\,3.5)$ ohm. Determine the value of the starting capacitor that will place the main and auxiliary winding currents in quadrature at starting. [**Ans.** 211.5 µF]

10. The main and auxiliary winding impedances of a 50 Hz, capacitor-start single-phase induction motor are:

Main winding = $3 + j\,2.7$

Auxiliary winding = $7 + j\,3$

Determine the value of the capacitor to be connected in series to achieve a phase difference of $\alpha = 90°$ between the currents of two windings at start. **[G.B. Technical Univ. Special Electrical Machines, 2012-13]**

[**Ans.** 295.34 µF]

11. A 100 W, 2-pole, 50 Hz, 230 V, single-phase series motor with salient poles has a total resistance of 15 Ω, leakage reactance of 40 Ω, mutual reactance of 80 Ω (in *d*-axis) and 50 Ω (in *q*-axis). If the stray power losses are 20 watts, calculate the current, speed and power factor of the motor at full-load.

[**Ans.** 11,980 rpm, 0.613 A, 0.89 lagging]

12. The following data refer to a 220 V, 2-pole uncompensated series motor when run on a 50 Hz single phase supply. Applied volts = 220 V (rms); Input current = 1.5 A; Input watts = 198.75 W, Field winding resistance = 8.8 Ω; Armature resistance 4.2 Ω; Speed 4,800 rpm. The number of turns on the field winding is 300, and the number of armature conductors 1,200, leakage reactance drop is 20 per cent of the total reactive drop in the motor.

Compute (*i*) induced volts in armature (*ii*) Field flux per pole and (*iii*) armature (quadrature axis) flux per pole.

[**Ans.** (*i*) 113 V (*ii*) 1.665 m Wb (peak value) (*iii*) 0.684 mWb (peak value)]

13. A 300 V, 50 Hz, 2-pole single-phase series motor has 800 turns and 350 turns in armature and field respectively. The total effective resistance is 6 Ω and total leakage reactance is 14 Ω. For a power input of 350 W, the motor draws 2 A while the speed is 3,600 rpm. (*i*) Determine the peak value of field flux per pole (*ii*) If the motor is conductively compensated and draws the same current, find the power factor and speed assuming the motor has full compensation winding of impedance $(1 + j2)\ \Omega$.

[**Ans.** (*i*) 2.4 m Wb (*ii*) 0.7688 (lag) 4.785 rpm]

14. A universal motor has a resistance of 30 Ω and inductance of 0.5 H. When connected to a 250 V dc supply and loaded to take 0.8 A it runs at 2,000 rpm. Estimate its speed and power factor when connected to a 250-V, 50-Hz supply and loaded to take the same current.

[Bharathiyar Univ. Elec. Machines II, Nov.-1996]

[**Ans.** 1,700 rpm; 0.8645 lagging]

15. A 240 V universal motor runs at 1,500 rpm when loaded and connected to 240 V dc supply and it draws a current of 0.8 A. It has resistance of 30 Ω and inductance of 0.5 H. Find its speed and power factor when connected to 240 V, 50 Hz ac supply.

[U.P. Technical Univ. Electromechanical Energy Convevsion-II, 2004-05]

[**Ans.** 1,253 rpm; 0.852 (lagging)]

❑❑❑

CHAPTER 21

Special Electrical Machines

INSIDE THIS CHAPTER

21.1 INDUCTION GENERATOR

If a polyphase induction motor while connected to a constant voltage and frequency mains is mechanically coupled to a prime mover and is driven by it at a speed higher than synchronous speed, the rotor overtakes the rotating magnetic field and the rotor conductors cut the magnetic flux in a reverse direction to motor rotation. As a result, the emfs and currents in the rotor will reverse their direction. Consequently the reaction force between rotating field and rotor current also changes its direction and opposes the rotation. This all has been illustrated in Fig. 21.2. To maintain this rotation mechanical energy is required to be delivered to the rotor; the magnetising current remaining the same, since the rotating field conditions are the same in an induction generator and motor. The power of the machine under such conditions is negative i.e. the machine delivers electrical energy to the mains instead of taking it from the mains. The slip of an induction machine operating as a generator is negative* and the direction of its emf now coincides with the active component of current instead of being opposed to it, as is the case in a motor. It is to be noted that this latter circumstance may serve as a characteristic feature of a machine passing from motor to generator action.

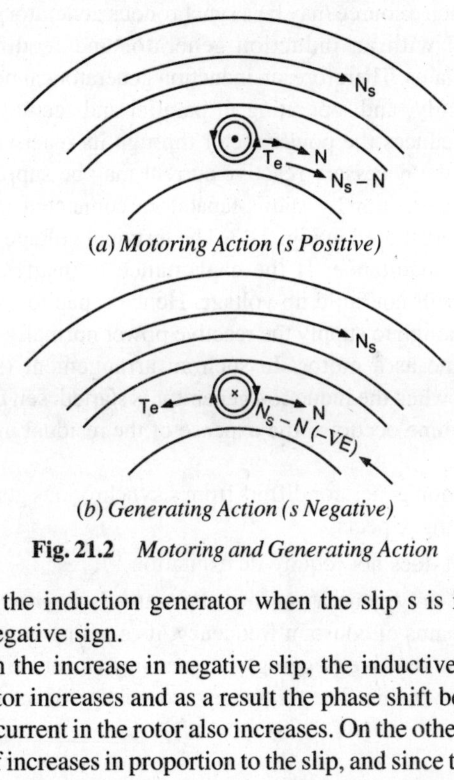

Fig. 21.2 *Motoring and Generating Action*

At super-synchronous speeds, the conditions of an induction generator are electrically similar to those at subsynchronous speeds for induction motor. In this connection it is to be noted that the equivalent circuit and all the characteristic equations of the induction motor also apply to the induction generator when the slip s is introduced with a negative sign.

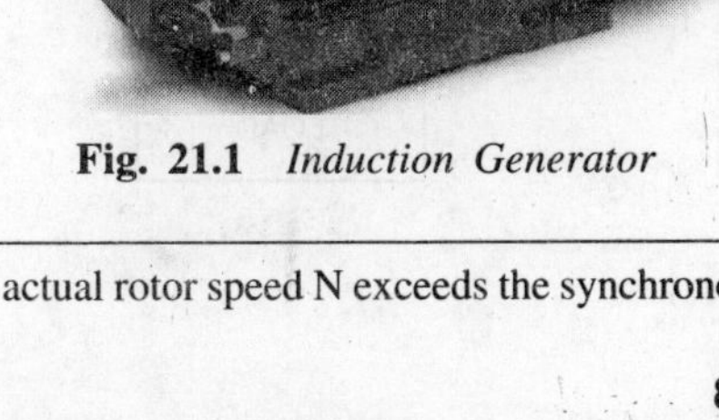

Fig. 21.1 *Induction Generator*

With the increase in negative slip, the inductive reactance of the rotor increases and as a result the phase shift between the emf and current in the rotor also increases. On the other hand, the rotor emf increases in proportion to the slip, and since the braking torque of an induction generator is expressed in the same way as the torque of a motor

$$T \propto \Phi_2 I_2 \cos \phi_2$$

it follows that in the same way as for a motor with slip corresponding to the condition $s = R_2/X_2$ there will be a maximum torque above which the operation of the machine will be unstable. The maximum value of torque which the induction generator can develop, thus, limits its range of operation. The torque-slip characteristics of the machine as a generator resemble to those of the motor [Fig. 21.2 (*c*)].

* In generator operation, the actual rotor speed N exceeds the synchronous speed N_s so the slip s, by equation $s = N_s - N$, becomes negative.

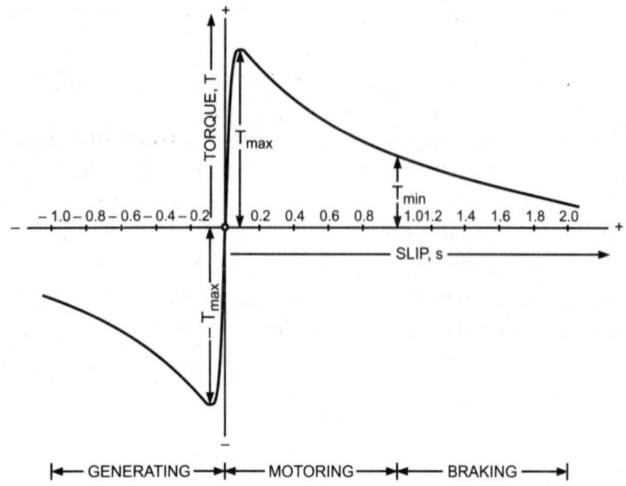

Fig. 21.2 *(c) Torque-Slip Characteristics of Induction Machine*

An induction generator supplies an active current to the mains, but at the same time, it receives from the mains a lagging reactive current (magnetising current), in the same way as a motor. That is why, it requires a source of reactive energy to obtain this current. Such a source may be a synchronous generator operating in parallel with an induction generator and feeding into a common mains. Therefore, an induction generator cannot operate independently, and, operating in parallel and feeding into the mains, it reduces the power factor through its reactive current. However, the necessary reactive current may be supplied to an induction generator by static capacitors connected in parallel with it, as illustrated in Fig. 21.3. The terminal voltage increases with the capacitance. If the capacitance is insufficient, the generator will not build up voltage. Hence capacitor bank must be large enough to supply the reactive power normally drawn by the machine as a motor. In such an arrangement (shown in Fig. 21.3), when the induction generator is started, self excitation of the machine occurs at the expense of the residual magnetism in it.

Induction generator differs from a synchronous generator in the following aspects:

1. It does not require dc excitation.
2. It only generates when the stator is connected to the mains of constant frequency, its exciting current being the reactive (lagging), magnetising current drawn from the mains. The frequency and voltage of the induction generator are those of the line to which it is connected irrespective of load.
3. The frequency of the magnetising current fixes that of the alternating current supplied by the induction generator. Thus the frequency is independent of the speed of the induction generator.
4. The induction generator does not have definite speed for a given frequency, as the synchronous machine has, but the speed with constant frequency varies with the load, the load (active power) being practically proportional to the slip, s. The prime mover must be run at speed slightly more than synchronous speed. The slip speed is usually the same as that when the machine operates as a motor i.e. 30 rpm. So it is essential that the prime mover driving the induction generator should have a rising speed characteristic. The prime mover must be provided with automatic control to increase the speed of the generator when it is required to supply the increased load.
5. No synchronising is required, since the machine cannot generate any emf until it is connected to the supply mains.
6. A unique feature is that the power factor of the output is fixed in value by the generator characteristics and is always leading and independent of the external circuit.

Equivalent Circuit. In the equivalent circuit shown in Fig. 18.30, the induction motor is depicted as a transformer delivering power to an equivalent load resistance $R_2'\left(\frac{1}{s}-1\right)$ which represents the power converted into mechanical work. For negative value of s this load resistance is also negative being equal to $R_2'\left(\frac{1}{-s}-1\right)$ i.e. $-R'_2\left(\frac{1+s}{s}\right)$, that is, it no longer absorbs power but is really a source or in other words there is a transfer of energy from mechanical form to electrical form. The equivalent circuit of an induction machine operating as a generator is given in Fig. 21.4.

Phasor Diagram. The phasor diagram of an induction machine operating as a generator is given in Fig. 21.5. Comparing phasor diagrams (Figs. 18.29 and 21.5) for an induction machine operating as a motor and as a generator we note that in the

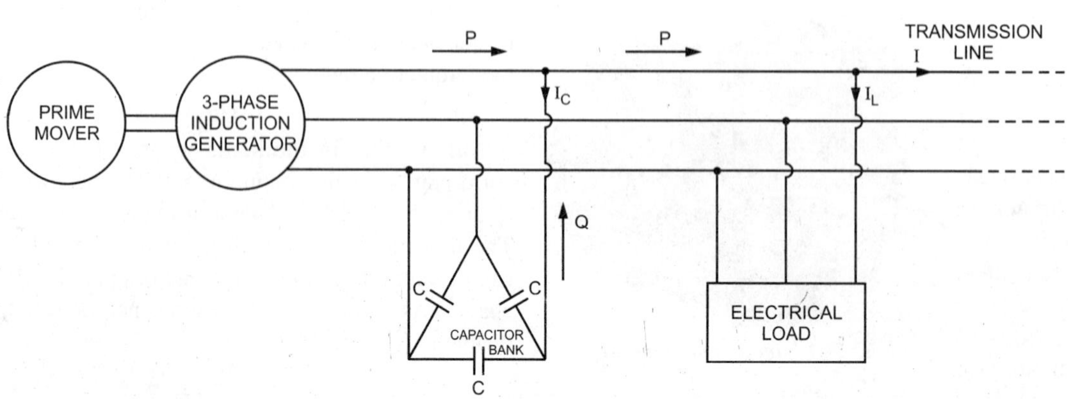

Fig. 21.3

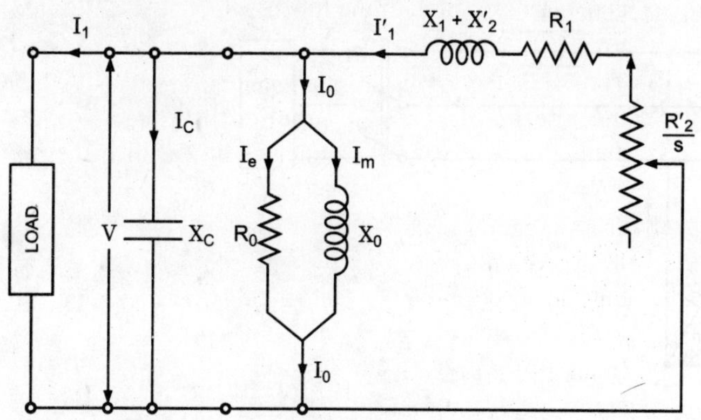

Fig. 21.4 *Approximate Equivalent Circuit of An Induction Generator*

generating case induced emf $\mathbf{E}_1$ is the phasor sum of reversed supply voltage (i.e. $-\mathbf{V}$) and impedance drop in stator winding whereas in case of motoring the applied voltage $\mathbf{V}$ is the phasor sum of reversed induced emf in the stator winding (i.e. $-\mathbf{E}_1$) and impedance drop in the stator winding.

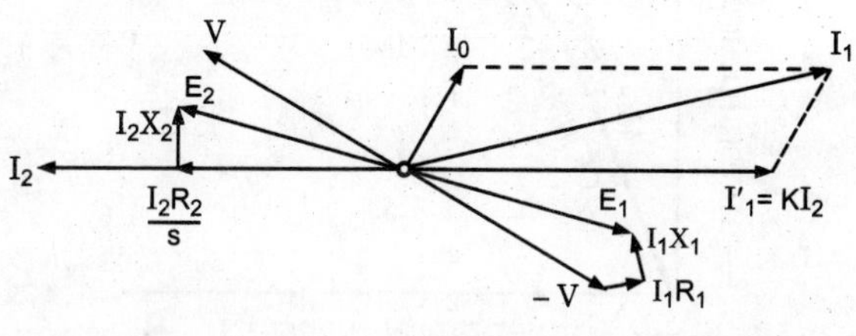

Fig. 21.5 *Phasor Diagram For an Induction Generator*

Circle Diagram. The locus of stator current has been obtained by inverting the impedance locus of a constant reactance in series with a variable resistance. This must now be extended for negative values of R_2/s. The diagram is given in Fig. 21.6.

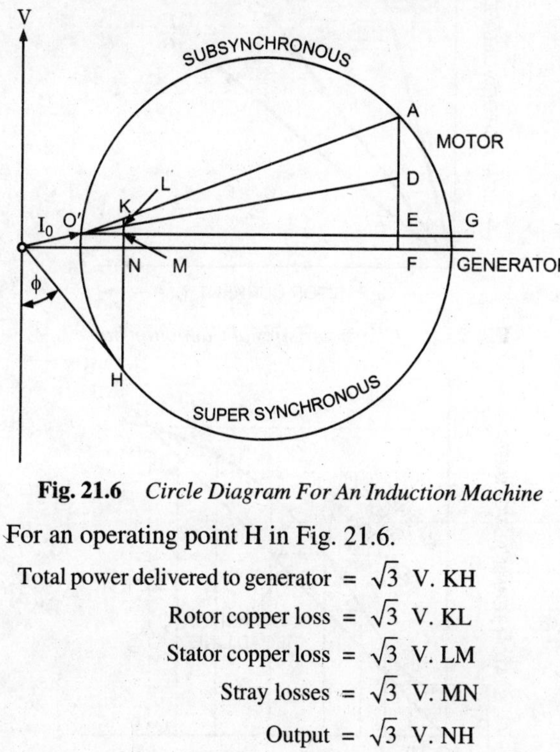

Fig. 21.6 *Circle Diagram For An Induction Machine*

For an operating point H in Fig. 21.6.

Total power delivered to generator = $\sqrt{3}$ V. KH

Rotor copper loss = $\sqrt{3}$ V. KL

Stator copper loss = $\sqrt{3}$ V. LM

Stray losses = $\sqrt{3}$ V. MN

Output = $\sqrt{3}$ V. NH

Slips, s = KL/LH

$$\text{Power factor } \cos\phi = \frac{NH}{OH}$$

Advantages, Disadvantages and Applications. The distinct advantage of the induction generator lies in the fact that it does not hunt or drop out of synchronism, it is simple and rugged in construction, cheaper in cost, easy in maintenance, and when short circuited it delivers little or no sustained power,* because its excitation quickly becomes zero. However, this type of generator is little used to supply commercial power because of the distinct superiority of the synchronous generators. Its main drawbacks are :

1. As already mentioned, it cannot be operated independently and the method of providing self-excitation by connecting suitable capacitors across its terminals is of limited practical importance in view of the magnitude of capacitors required.
2. The other drawback is that it can deliver only leading current whereas most commercial loads require lagging current. Considerable kVA capacity synchronous apparatus is required to supply the total quadrature (lagging) current.
3. Induction generators are not helpful in system stability than the addition of parallel reactors with a rating equal to the generator magnetising reactance.
4. Dangerously high voltages may occur with induction generators over long transmission lines if the synchronous machines at the far end become disconnected and the line capacitance excites the induction machines. Motors with power factor correction capacitors may generate if the capacitors remain connected to the running machine when for some reason the supply is cut off.

It is highly important to prevent shock loads from exceeding the maximum generating torque of the machine or causing the overcurrent trips to function during periods of regeneration.

The induction generator is useful when the prime mover does not run at constant speed such as in hydroelectric power stations having a variable low-head water supply. Where the generator and turbine run submerged in a river to utilise tidal power, induction generators might be considered. The utilisation of wind power is another suitable application for this type of generator since the stator frequency depends on that of paralleled synchronous machines, not on the rotor speed.

The induction generator is very useful for installation in small power stations where it can be operated in parallel and feeding into a common mains without attendent. It does not require synchronising and requires no separate exciter but floats on the line delivering power when the water supply is sufficient to drive it above synchronous speed, otherwise running idle as a motor.

The induction generator is very useful for braking purposes in railway work. If the induction motor be left connected across the line on a downgrade; any tendency of the train to drive it above synchronous speed will cause generator action. In addition to providing effective braking the energy is fed back to the ac mains and thus main generating station is relieved from some of

* An induction generator delivers an instantaneous 3-phase short-circuit current equal to terminal voltage divided by its standstill reactance, but its rate of decay is much faster than that of a synchronous generator of the same rating, and its sustained short-circuit current is zero.

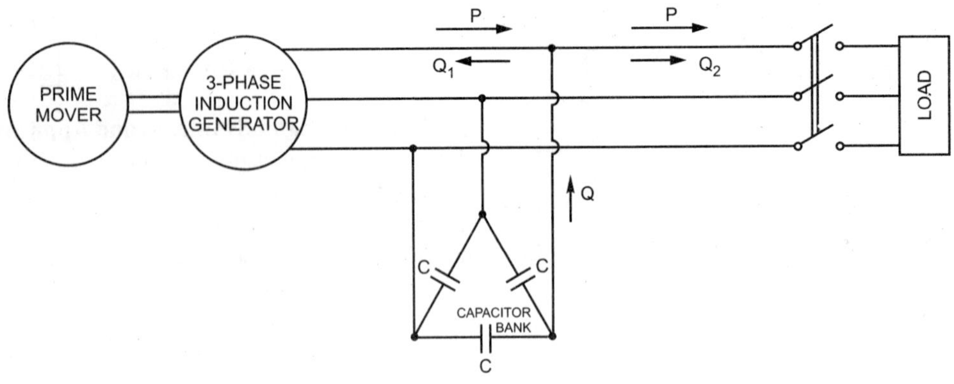

Fig. 21.7

its load. The machine requires no complicated control apparatus when used for regenerative braking, such as is required in case of dc motors operating under similar conditions.

When used on such applications as hoists, the polyphase induction machine supplies very effective overspeed brake since a comparatively small increase in speed causes full generator action to take place exerting a strong braking action, besides furnishing reverse power to the supply line. Since the retarding torque ceases at synchronous speed, the braking from synchronous speed to standstill must be accomplished by some other means.

21.1.1. Self-Excited or Isolated Induction Generator. An isolated induction generator supplying a load is shown in Fig. 21.7. The capacitor bank connected across the stator terminals of induction generator provides the lagging reactive power of both the induction generator as well as load as shown in Fig. 21.7. The terminal voltage increases with the increase in capacitance C. If the capacitance is insufficient, the generator will not build up voltage. Hence the capacitor bank must be large enough to supply the reactive power normally drawn by the machine as motor. In such an arrangement (shown in Fig. 21.7), when the induction generator is started, self-excitation of the machine occurs at the expense of residual magnetism in it.

Capacitance reactive power,

Q = Reactive power required by 3-phase induction generator, Q_1 + Reactive power required by the load, Q_2

In a self-excited induction generator, stator terminal voltage depends upon its magnetization curve as shown in Fig. 21.8. The reactive current of a capacitor is dependent on the voltage across it terminals *i.e.* capacitor voltage $V_C = \omega CI$. In Fig. 21.9, load lines of capacitor banks having different capacitances $C_1 > C_2 > C_3$ are drawn.

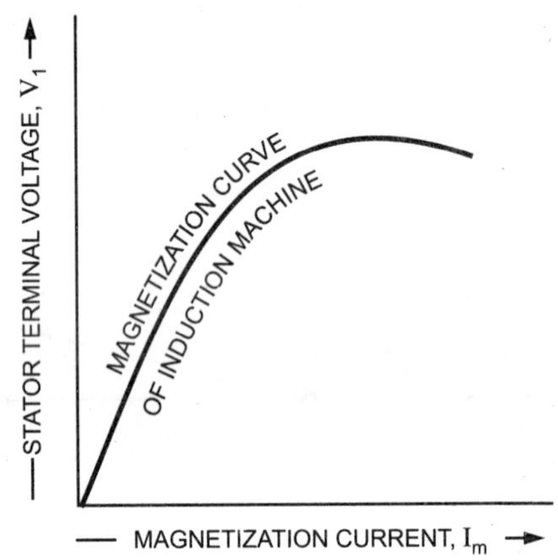

Fig. 21.8 *Characteristic of Induction Generator*

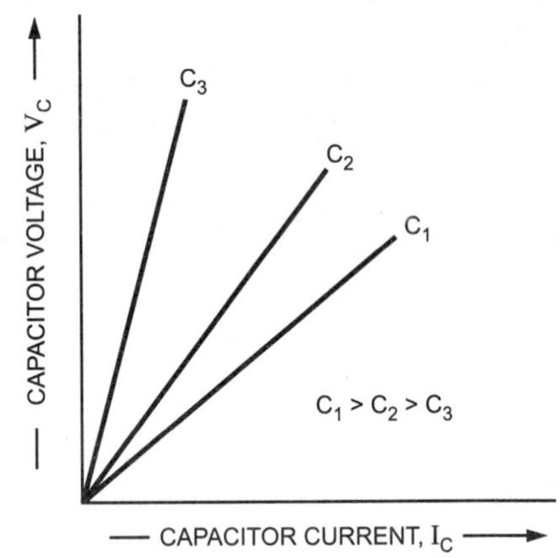

Fig. 21.9 *Characteristics of Capacitor Bank*

When the induction generator is started and run at the required speed, the residual magnetism present in the rotor iron induces a small terminal voltage *oa* across its stator terminals, as shown in Fig. 21.10. This voltage results in a capacitor current *ob*. This current ob produces a flux which aids the residual flux, thus creating more flux and therefore more generated voltage *bc* across the stator terminals. Due to this voltage *bc* capacitor current increases and becomes *od* which eventually induces voltage *de*. This cumulative process of voltage build up continues till the capacitor load line intersects the saturation curve of the induction generator at point *f*, thus giving a no-load

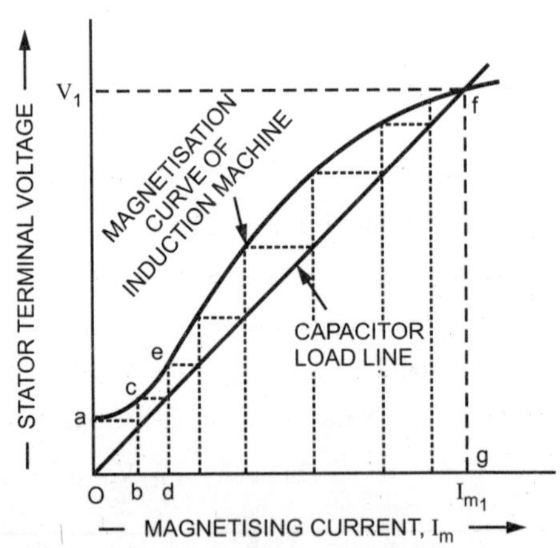

Fig. 21.10 *Building-up Voltage in Induction Generator at No Load*

induced emf *gf* for magnetizing current I_{m1}. The noteworthy point is that the induction generator builds up voltage across its stator terminals like a dc shunt generator.

In case of absence of residual flux in the rotor iron of induction generator there would be no voltage build up. This problem can, however, be solved by operating the induction machine as a 3-phase induction motor for sometime to induce residual magnetism.

The voltage build up depends upon the value of capacitor as shown in Fig. 21.11. Higher the value of capacitor, more is the voltage build up. In case the capacitor load line does not intersect the magnetization curve of induction generator, there would be no voltage build up, as in case of capacitor C_4.

Induction generator is quite suitable for generation of electrical energy in coastal areas, where wind energy is available in abundance.

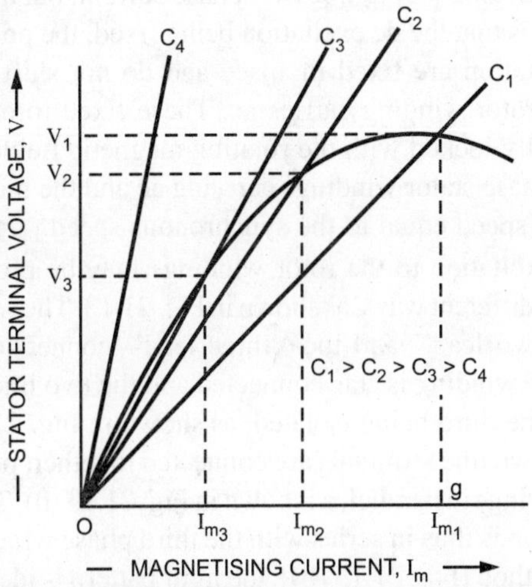

Fig. 21.11 *No-load Voltages For Different Capacitances*

Example 21.1. A 440-V, 6-pole, 970 rpm, 20 kW, 3-phase induction machine is to be employed as an induction generator. The rated current of the machine is 30 A and full-load pf is 80%. Determine (*i*) the speed of the prime mover for generating at a frequency of 50 Hz (*ii*) the capacitance required per phase if the capacitors are connected in delta.

Solution: Synchronous speed of the induction machine,

$$N_s = \frac{120 f}{P} = \frac{120 \times 50}{6} = 1{,}000 \text{ rpm}$$

The prime mover must run at a speed slightly higher than synchronous speed. The slip speed is kept usually the same as that when the machine operates as a motor i.e. 30 rpm.

(*i*) Hence speed of the prime mover,

$$N = N_s + \text{slip speed} = 1{,}000 + 30 = 1{,}030 \text{ rpm } \textbf{Ans.}$$

$$\text{Machine kVA, } S = \frac{\sqrt{3}\,VI}{1{,}000} = \frac{\sqrt{3} \times 440 \times 30}{1{,}000} = 22.863$$

$$\text{Power, } P = S \times \cos\phi = 22.863 \times 0.8 = 18.29 \text{ kW}$$

$$\text{Required power, } Q = \sqrt{S^2 - P^2} = \sqrt{22.863^2 - 18.29^2}$$

$$= 13.718 \text{ kVAR}$$

Hence the kVAR to be supplied by the Δ-connected capacitor bank

$$= \frac{13.718}{3} = 4.5726/\text{phase}$$

$$\text{Capacitor current per phase, } I_C = \frac{4.5726 \times 1{,}000}{440} = 10.39$$

(*ii*) Hence capacitance required,

$$C = \frac{I_C}{2\pi f V} = \frac{10.39}{2\pi \times 50 \times 440}$$

$$= 75 \ \mu\text{F } \textbf{Ans.}$$

Example 21.2. An induction generator having a synchronous speed of 750 rpm is driven at 800 rpm and with the rotor short-circuited, the stator output is 9 kW. Determine (a) the rotor copper loss (b) the rotor resistance per phase, given that the voltage across slip rings at standstill equals 260 V. Neglect stator resistance and rotor reactance. (c) By the addition of external resistance in the rotor circuit, the stator output is reduced to 4 kW. Determine the value of the external resistance inserted per phase.

Solution: Alternator speed, N = 800 rpm

Synchronous speed, N_s = 750 rpm

$$\text{Slip, } s = \frac{N_s - N}{N_s} = \frac{750 - 800}{750} = -\frac{1}{15}$$

Stator output = 9 kW = 9,000 watts

Rotor output = Stator output + stator copper losses + stray losses

Since stator resistance is negligible, therefore, stator copper losses are negligible and also neglecting stray losses

Rotor output = Stator output = 9,000 watts

$$\therefore \text{Slip, } s = \frac{\text{Rotor copper losses}}{\text{Rotor output}}$$

∴ (*a*) Rotor copper losses

$$= \text{Rotor output} \times \text{slip} = 9{,}000 \times \frac{1}{15} = 600 \text{ watts } \textbf{Ans.}$$

(*b*) Induced emf per phase in the rotor at standstill,

$$E_2 = \frac{260}{\sqrt{3}} \text{ V assuming rotor star-connected}$$

$$\text{Rotor current, } I_2 = \frac{E_2}{\sqrt{\left(\frac{R_2}{s}\right)^2 + (X_2)^2}} = \frac{E_2}{R_2 / s}$$

∵ rotor reactance is negligible

$$\text{Rotor copper losses} = 3\, I_2^2\, R_2 = 3 \times \frac{E_2^2}{R_2^2 / s^2} \times R_2 = \frac{3\, E_2^2\, s^2}{R_2}$$

$$\text{or } 600 = \frac{3 \times \left(\frac{260}{\sqrt{3}}\right)^2 \times \frac{1}{225}}{R_2}$$

$$\text{or } R_2 = 3 \times \left(\frac{260}{\sqrt{3}}\right)^2 \times \frac{1}{225} \times \frac{1}{600} = 0.5\ \Omega$$

∴ Rotor resistance per phase,

$$R_2 = 0.5\ \Omega \textbf{ Ans.}$$

(*c*) Let the external resistance inserted in each phase of rotor be R ohms

$$\text{Rotor copper losses} = \frac{3\, E_2^2\, s^2}{R_2 + R} = \frac{3 \times \left(\frac{260}{\sqrt{3}}\right)^2 \times \left(\frac{1}{15}\right)^2}{R_2 + R}$$

$$= \frac{300}{R_2 + R} \text{ watts} \quad \ldots (i)$$

Also rotor copper losses = Slip × rotor output

$$= \frac{1}{15} \times 4{,}000 = 266.66 \text{ watts} \quad \ldots (ii)$$

∵ rotor output = stator output, stator losses being negligible

Equating expressions (*i*) and (*ii*) we have

$$\frac{300}{R_2 + R} = 266.66$$

$$\text{or } R_2 + R = \frac{300}{266.66} = 1.125\ \Omega$$

or external resistance to be inserted in each phase,

$$R = 1.125 - R_2 = 1.125 - 0.5 = 0.625\ \Omega$$

Example 21.3. A 440 V, 50 Hz, 6-pole, 960 rpm star-connected induction machine has the following per phase parameters referred to the stator:

$$\mathbf{R_s = 0.6\ \Omega;\quad R'_r = 0.3\ \Omega;\quad X_s = X'_r = 1\ \Omega}$$

The magnetizing reactance is very high and is neglected. The machine is connected to the 440 V, 50 Hz supply and a certain mechanical load is coupled to it. It is found that the magnitude of the stator current is equal to the rated current of the machine but the machine is running at a speed higher than its rated speed. Find the speed at which the machine is running. Also find the torque developed by the machine. *[GATE, 2002]*

Solution: Equivalent circuit of an induction machine referred to stator, neglecting magnetizing reactance, can be drawn as shown in Fig. 21.12.

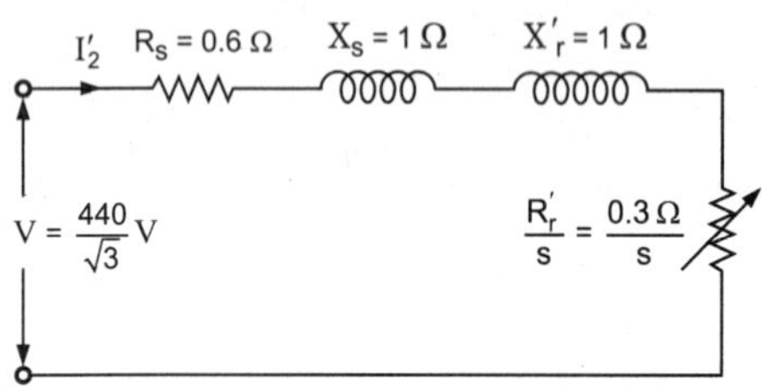

Fig. 21.12

Synchronous speed,

$$N_s = \frac{120f}{P} = \frac{120 \times 50}{6} = 1{,}000 \text{ rpm}$$

$$\text{Slip, } s = \frac{N_s - N}{N_s} = \frac{1{,}000 - 960}{1{,}000} = 0.04$$

$$\text{Current, } I'_2 = \frac{V}{\sqrt{\left(R_s + \dfrac{R'_r}{s}\right)^2 + (X_s + X'_r)^2}}$$

$$= \frac{440/\sqrt{3}}{\sqrt{\left(0.6 + \dfrac{0.3}{0.04}\right)^2 + (1+1)^2}} = \frac{440}{\sqrt{3(65.61 + 4)}}$$

$$= 30.4478 \text{ A}$$

Torque developed by machine

$$= \frac{60}{2\pi N_s} \times 3(I'_2)^2 \times \frac{R'_r}{s} = \frac{60}{2\pi \times 1{,}000} \times 3 \times (30.4478)^2 \times \frac{0.3}{0.04}$$

$$= 199.2 \text{ N-m } \textbf{Ans.}$$

For the same rated current with speed higher than its rated speed, the machine will operate as an induction generator and slip will be same in magnitude but negative.

$$\text{i.e.} -0.04 = \frac{1{,}000 - N}{1{,}000}$$

$$\text{or } N = 1{,}040 \text{ rpm} \quad \textbf{Ans.}$$

21.2 SYNCHRONOUS INDUCTION MOTORS

The synchronous induction motor, as its name implies, is a machine which is capable of running both as an induction motor and as a synchronous motor, the former being its mode of operation during the starting period, and the latter its mode of operation during normal running.

The secondary winding of a synchronous induction motor consists of a polyphase winding, almost invariably of the three-phase type. Since the rotor is uniformly slotted around the whole periphery, there is no saliency, and thus it is a true round rotor machine.

For the synchronous mode of operation it is necessary to supply the rotor winding with direct current. When the rotor windings of a wound rotor induction motor are excited by a direct current, the mmf distribution develops alternate N and S poles in the same way as does a 3-phase current, but the essential difference is that the dc excitation being fixed, the pole axes due to dc excitation are fixed in space and do not shift as in case when the rotor winding carries ac. These fixed rotor poles get magnetically locked with the rotating magnetic field developed by the 3-phase stator windings carrying ac and the motor runs at a constant speed equal to the synchronous speed.

DC excitation to the rotor windings may be provided in a number of different ways, as shown in Fig. 21.13. The two simplest are the "two lead" and the "three lead" connections. In the former, the winding is star-connected and the two line terminals are used, the third being isolated, as shown in Fig. 21.13 (a). In the latter, two line terminals are connected together, putting their phase windings in parallel, as shown in Fig. 21.13 (b). The parallel combination is thus in series with the third phase winding. In the second method [Fig. 21.13 (b)], the mmf pattern is identical with that of a normal 3-phase excitation at the instant that the current in one phase is at its peak value.

As can be inferred from the current distribution in Fig. 21.13, unbalance secondary heating is unavoidable with a 3-phase secondary, but a 2-phase arrangement, as in 21.13 (e) or 21.13 (f), equalizes the heating. However, there is, in Fig. 21.13 (e) , no closed secondary loop apart from that through the exciter, so that the damping is not as good as in 21.13 (f). Though two phase arrangements shown in Fig. 21.13 (e) and 21.13 (f) give uniform heating but develop rather large harmonics and noise. A reduction in exciting current can be achieved by use of a multi-turn secondary, but consequently the induced secondary emf at standstill is high, causing insulation problems.

For an ordinary induction motor, the value of magnetising current is kept as small as possible to ensure a good power factor. A very short air gap, such as would be expected in plain induction motor, would result in a very heavy armature reaction when operating as a synchronous motor because the total field ampere-turns would be reduced for a given number of armature ampere-turns. The actual fact is that a synchronous motor needs a heavy excitation to enable it to operate on a leading power factor. The relatively larger air gap though results in a low power factor when operating as an induction motor, but at normal speed this is immaterial since the power factor can be controlled by variation in dc excitation. The larger air gap and the increase

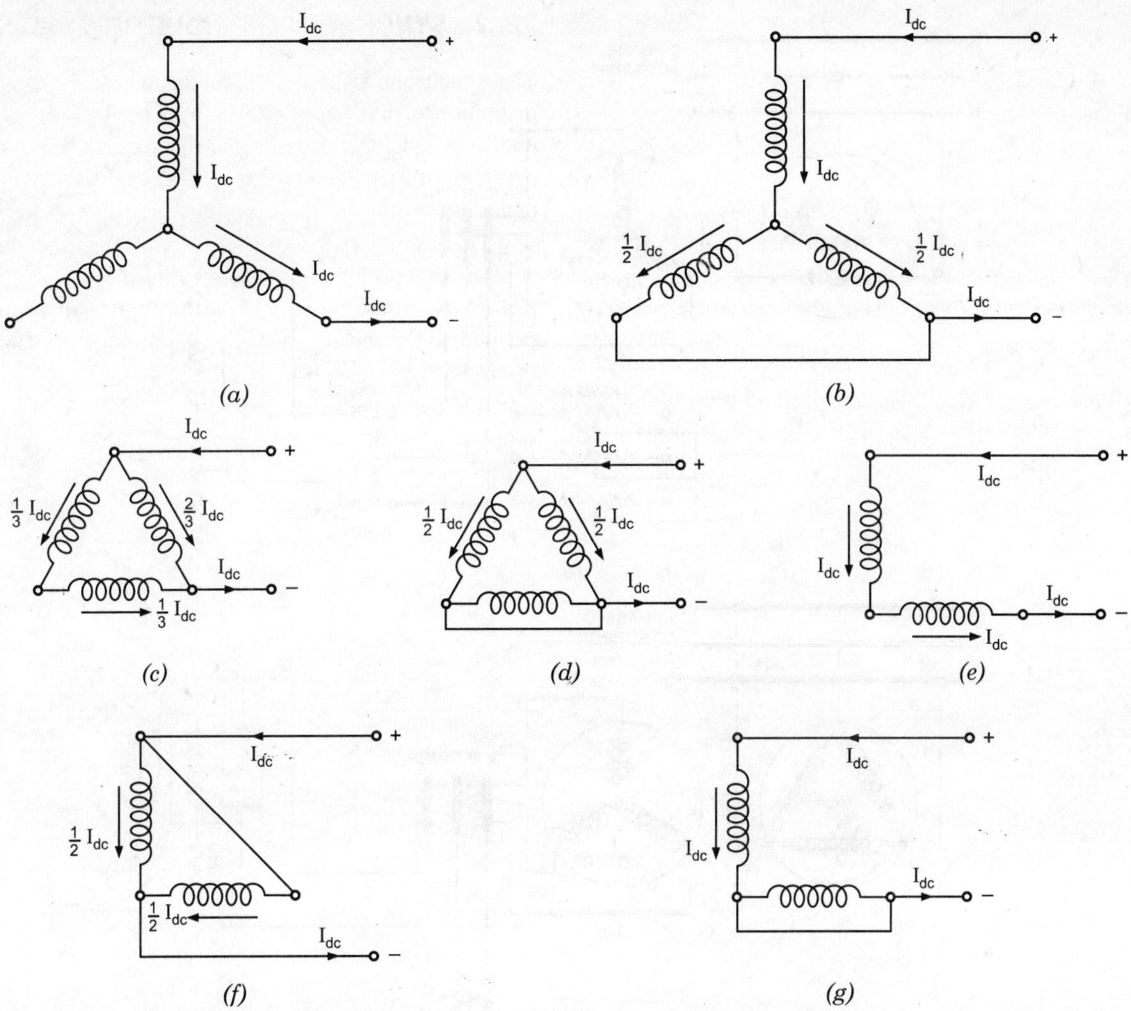

Fig. 21.13 *Rotor Connections For DC Excitation*

in $\frac{\text{excitation emf}}{\text{synchronous impedance}}$ results in a higher peak synchronous torque. So the synchronous induction motor is provided with a large air gap as provided in a plain synchronous motor. A long air gap gives a stiffer machine with a larger overload capacity. The rotor slots are made fewer in number and larger in size. These machines are provided with a heavy rotor winding in order to have a low slip, which facilitates in pulling it into synchronism. Also in order that the induced emf in the field at the starting may not be too high, the field turns provided are few in number and the excitation voltage is kept low. As the exciter winding serves the purpose of damper winding, so there is no need of providing separate damper winding with synchronous induction motors.

The schematic arrangement of starting a synchronous induction motor is shown in Fig. 21.14 (a). The synchronous induction motor is started as a slip-ring induction motor by inserting resistances in the rotor circuit. When the additional resistances are completely cut out and the motor attains the normal induction motor speed, the rotor is disconnected from the starting resistances and connected to the exciter which is usually mounted on the same shaft. Motor will now be operating as a synchronous motor.

The exciter can be left permanently in series with one of the leads, as shown in Fig. 21.14 (*b*), provided the induced voltage at standstill is not too high. At the starting instant the exciter is unexcited, and when the starting resistances are cut out, the machine runs as a normal induction motor. After a short delay, the exciter voltage builds up, and field current is caused to flow through one phase of the rotor winding to the star point where it divides between the other two phases in parallel, each of which carries half the exciter current. The rotor mmf is almost sinusoidally distributed and as this field is established the machine pulls into step and the machine operates as a synchronous motor.

With secondary rheostats for starting, such a motor gives low starting current and high starting torque of the wound rotor induction motor and an improved power factor under load. Hence the synchronous induction motor is essentially a motor, having the induction motor features like high starting torque with low starting current combined with the synchronous motor features like constant speed and power factor control.

The interesting point about this motor is that its peak torque as an induction motor exceeds that as a synchronous motor. If it is momentarily overloaded as a synchronous motor, it may continue to operate as an induction motor-albeit with considerable pulsations of current and torque due to dc until the load falls and it is able to re-synchronise.

However, the extra cost, low efficiency as compared to that of standard types, the small copper space available with a

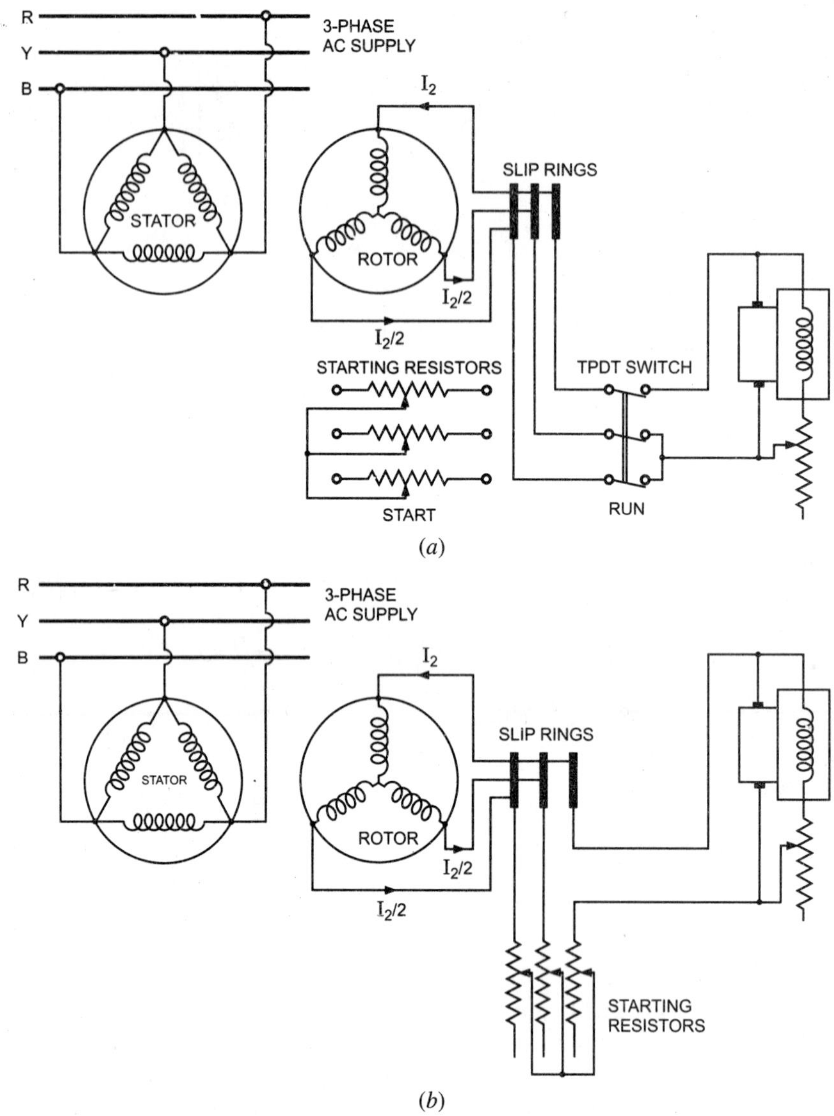

Fig. 21.14 *Typical Circuit For Synchronous Induction Motor*

distributed rotor winding, and the necessary compromise between excessive ring voltage at start, or excessive rotor current in operation are severe handicaps,

A synchronous induction motor can also be made of salient pole type combining the high efficiency and low excitation loss of the salient pole synchronous motor and the excellent starting characteristics of the induction motor. In such motors, the dc field winding is of normal construction; and an independent three-phase starting winding connected to the slip rings is placed in slots in the pole shoes. Each winding can be designed almost independently to fulfil its own function, with consequent improvement in synchronous running and exciter-circuit conditions.

There are three important characteristic torques viz starting torque in the induction mode, pull-in i.e. synchronising from the induction to the synchronous mode, and pull-out torque, corresponding to maximum torque in the synchronous mode. There is naturally a close relation between pull-out torque and the full-load power factor. A motor excited to operate at a pf of 0.9 leading when developing full-load torque might have a pull-out torque 1.5 times of full-load torque, but if it were designed for unity power factor operation while developing full-load torque, the pull-out torque and the necessary exciter rating would both be considerably reduced. The typical curves of stator current, reactive power and power factor against active power output for a machine excited to unity pf on rated active power output are given in Fig. 21.15 (*a*) while Fig. 21.15 (*b*) shows the same quantities for a motor with a strong excitation. The strongly excited motor has a large overload capacity but over the range of normal load its pf is reduced and electrical loss is increased.

The synchronous induction motor is rarely built for ratings below 25 kW because of the relatively higher cost of the exciter. Synchronous induction motors are used where a high starting torque and constant speed operation are required. Typical applications of synchronous induction motors include fans, pumps, blowers, generators, air-compressors, ammonia compressors, machinery and line shafting in industrial works such as cement mills, rolling mills, flour mills, paper mills, rubber works and textile mills.

Synchronous induction motors are very often installed alongwith other induction motors, so that they may improve the overall power factor of the system. Their leading kVAR capacity

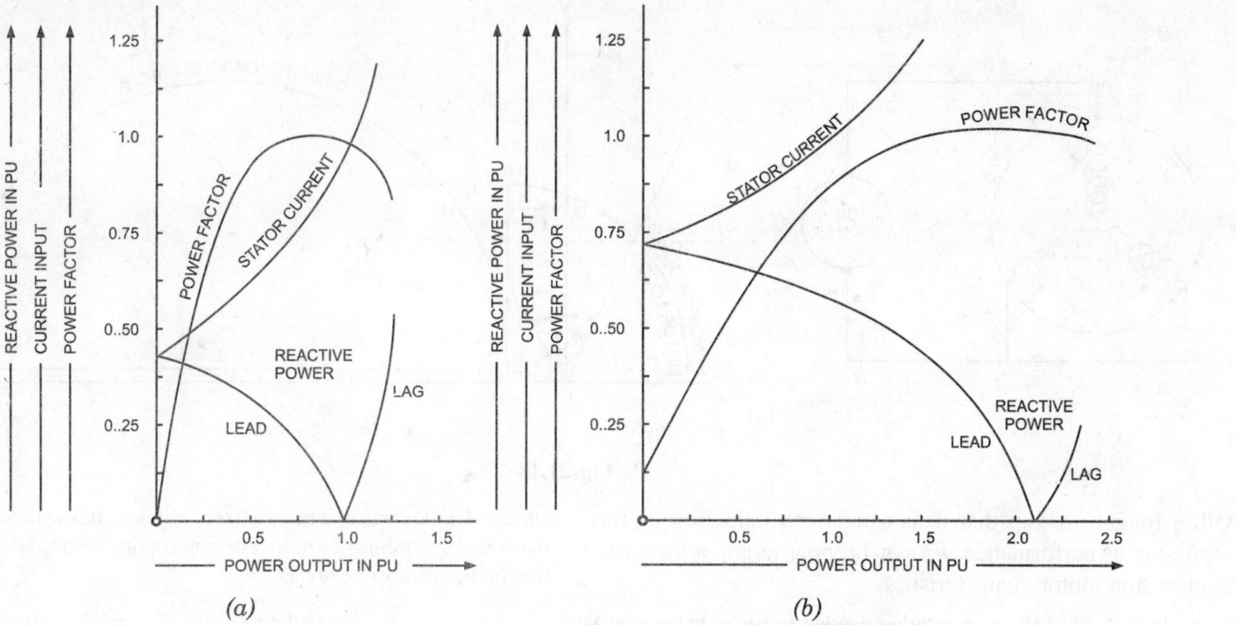

Fig. 21.15 *Performance Characteristics of Synchronous Induction Motor*

is designed to offset the lagging kVAR demand from the induction motors. They have been made for ratings as high as 30 MW.

21.2.1. Circle Diagram For Synchronous Induction Motor. At the instant of synchronising of a synchronous motor the induced emf across the stator circuit of the motor, E is equal to the applied voltage V but opposite in direction as shown in Fig. 16.7 (*a*) vectorially. After synchronising when the external driving means is removed the motor tends to slow down and the induced emf falls back by an angle δ called the load angle or torque angle (refer to Art. 16.4). The resultant voltage E_R causes a current I to flow in the stator circuit according to the phasor diagram shown in Fig. 21.16, where Z_s is the synchronous impedance. Stator current I lags behind E_R by an angle θ, where $\theta = \text{Tan}^{-1} \frac{X_s}{R_e}$ (nearly equal to $\frac{\pi}{2}$ radians or 90°). X_s and R_e being the stator synchronous reactance and effective resistance respectively.

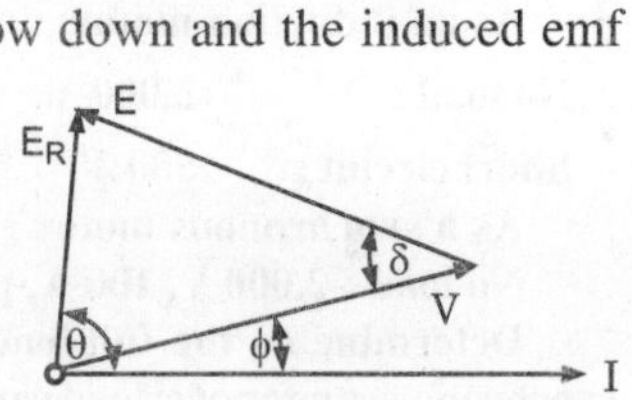

Fig. 21.16

$$i.e. \text{ Stator current, } \mathbf{I} = \frac{E_R}{Z_s} = \frac{V}{Z_s} + \frac{E}{Z_s}$$

Thus the stator current I can be regarded as superposition of two short-circuit currents $\frac{V}{Z_s}$ and $\frac{E}{Z_s}$, taking $\frac{V}{Z_s}$ and $\frac{E}{Z_s}$ on current scale and drawing V/Z_s inclined by the impedance angle θ to V and E/Z_s inclined by an angle δ to V/Z_s. The resultant of these two phasors represents the stator current in phase and magnitude Fig. (21.17). For constant supply voltage V and constant excitation the load angle δ varies with the load and the locus of extremity of the stator current I is thus a circle of radius $\frac{E}{Z_s}$ with its centre at the point O′.

Now in this case $\frac{V}{Z_s}$ is the current drawn by the machine when running at synchronous speed with applied voltage V but without excitation. This current is practically I_0, the no-load current of the machine as an induction motor. The current $\frac{E}{Z_s}$ is the short-circuit current in the stator circuit I_x when driven at synchronous speed and thus depends upon the excitation.

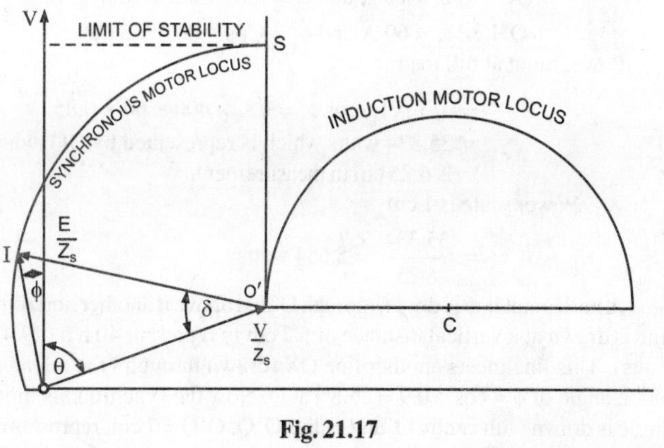

Fig. 21.17

In Fig. 21.17, this locus is superimposed upon the circle diagram for induction motor operation derived by the methods explained in chapter 18. The locus indicates that at light load the synchronous motor will operate at a leading power factor and if load is increased, excitation remaining unchanged, the power factor will become unity at approximately full load and will be lagging under overload conditions. If the load becomes excessive, the limit of stability is reached and the machine pulls out of step but continues to run as an induction motor until the excess load is removed when it should synchronise automatically.

The overload capacity of the machine as a synchronous motor (i.e. the position of the limit of stability) is seen to depend on the value of I_x, which is almost proportional to the excitation. The value of I_0 must also be large unless considerable variation of power factor and load angle can be tolerated over the working range. For this reason a synchronous induction motor is provided

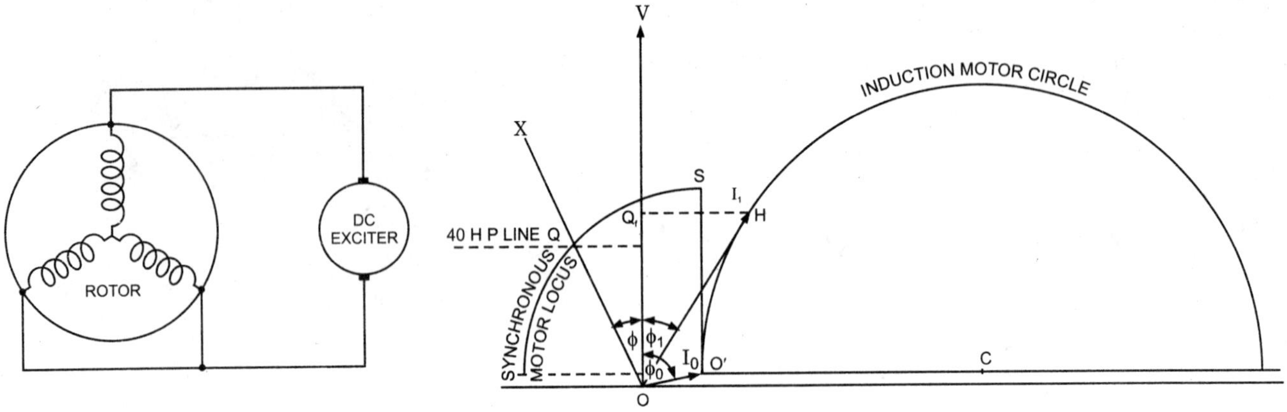

(a) (b)

Fig. 21.18

with a longer air gap than it is usual with induction motors, improving its performance as a synchronous motor at the cost of its induction motor characteristics.

Example 21.4. Find the dc excitation needed to run a 40 hp, 400 V, 3-phase induction motor on full load at synchronous speed at a power factor of 0.9 leading, if its no-load current as an induction motor is 16 A at a power factor of 0.2 and its full-load current is 60 A at a pf of 0.85. The ratio of stator to rotor turns per phase is 2. Find also the pull-out torque in synchronous watts.

Solution: The circle diagram of this machine as an induction motor is drawn, as discussed in Art 18.35, on current scale of 1 cm = 8 A.

In this diagram [Fig. 21.18 (b)]

$$OO' = I_0 = 16 \text{ A and } \phi_0 = \text{Cos}^{-1}\, 0.2 = 78.5°$$

$$OH = I_1 = 60 \text{ A and } \phi_1 = \text{Cos}^{-1}\, 0.85 = 31.8°$$

Power input at full load

$$= \sqrt{3}\ VI_1, \cos\phi_1 = \sqrt{3} \times 400 \times 60 \times 0.85$$

= 35,334 watts which is represented by OQ_f which is 6.25 cm in measurement.

∴ Power scale is 1 cm

$$= \frac{35,334}{6.25} = 5,654 \text{ watts}$$

A horizontal line is drawn through O′ and above it another horizontal line is drawn at a vertical distance of 5.2 cm to represent 40 h p (29,420 watts). This line meets another line OX (drawn through O at a leading phase angle of $\phi = \cos^{-1} 0.9 = 25.8°$) at Q. Now the synchronous motor circle is drawn with centre O′ and radius O′Q. O′Q = 7 cm, representing 56 amperes. This is the ac rotor current equivalent to the dc exciting current. The dc excitation bears a fixed relation to this ac depending on the method of connecting the rotor. For connections assumed [Fig. 21.18 (a)] the equivalent dc excitation is

$$I_{dc} = \sqrt{2}\ I_{ac} \times \frac{\text{stator turns}}{\text{rotor turns}} = \sqrt{2} \times 56 \times 2 = 158 \text{ A}$$

i.e. dc excitation required = 158 A **Ans.**

To obtain the pull-out torque a vertical line is drawn through O′ meeting the synchronous motor circle at S. O′S is the pull-out torque to scale in synchronous watts from the diagram

O′S = 7.0 cm = 7.0 × 5,654 = 39,578 watts

Therefore, pull-out torque = 39,578 synchronous watts **Ans.**

Example 21.5. The following test figures relate to a 400 kW, 2,000 volt, 50 Hz, 6-pole, 3-φ synchronous-induction motor.

As an induction motor:

No load :	**2,000 V,**	**100 A,**	**power factor 0.05**
Short circuit :	**500 V,**	**200 A,**	**power factor 0.15**

As a synchronous motor :

No load : 2,000 V, 100 A, power factor 0.1

Determine (i) the full-load current and power factor (ii) the synchronous motor overload capacity ; and (iii) the maximum torque in N-m against which the motor is capable of starting from rest. Assume rotor copper loss equal to stator copper loss.

Solution : From no-load and short-circuit figures for the machine as an induction motor the circle diagram (Fig. 21.19) is drawn as discussed in Art 18.35. D divides AE since rotor copper loss is equal to stator copper loss. Now O′A is the output line and O′D is the torque line for the machine operating as an induction motor.

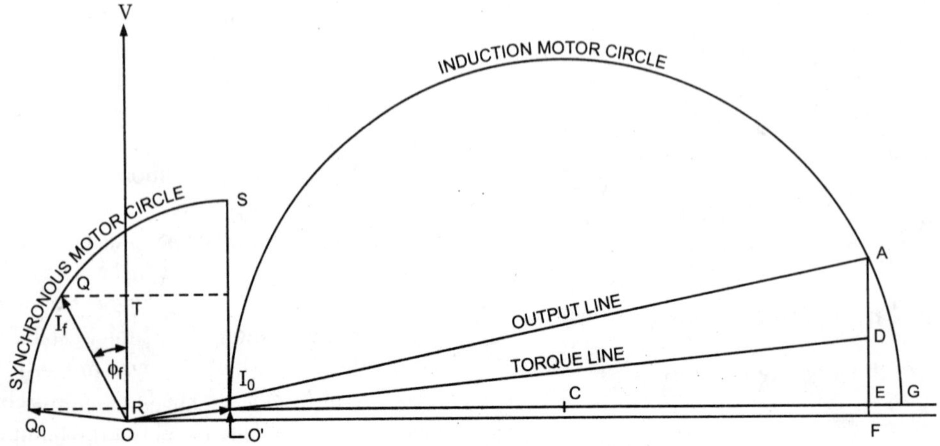

Fig. 21.19

The scales assumed for this diagram are :

Current 1 mm = 5 amperes

Power 1 mm = $\sqrt{3} \times 2{,}000 \times 5 = 17{,}320$ watts

Torque 1 mm = 17,320 synchronous watts.

As a synchronous motor OQ_0 represents the no-load condition where $OQ_0 = 100$ A and $\angle VOQ_0 = \cos^{-1} 0.1 = 84.3°$. The synchronous motor current locus is drawn with centre O′ and radius $O'Q_0$. The horizontal line Q_0R is drawn and above it and parallel to it another line through T is drawn so that RT represents full-load output of 400 kW to the power scale

$$\text{i.e. RT} = \frac{400{,}000}{17{,}320} = 23.1 \text{ mm.}$$

(*i*) The horizontal line through T cuts the synchronous motor current-locus (circle) at Q. Then OQ represents full-load current in magnitude and phase.

∴ Full-load current = OQ × 5 = 28 × 5 = 140 A **Ans.**

Full-load power factor

= Cos ∠ VOQ = Cos 21.5° = 0.93 leading **Ans.**

(*ii*) The pull-out load i.e. the load beyond which the motor fails to run in synchronism, is represented by O′S where S is the point where the synchronous motor circle cuts the vertical line drawn from O′.

$$\therefore \quad \text{Pull-out load} = \text{O}'\text{S} \times 17{,}320 \text{ watts} = \frac{40 \times 17{,}320}{1{,}000} = 693 \text{ kW}$$

Synchronous motor overload capacity

$$= \frac{693 - 400}{400} \times 100 = 73.25\% \textbf{ Ans.}$$

(*iii*) The maximum torque against which the motor is capable of starting from rest is determined from circle diagram of the machine as an induction motor. It is represented by AD.

∴ Maximum torque at start

= AD × 17,320 synchronous watts

= 12 × 17,320 = 2,07,840, synchronous watts

$$= \frac{2{,}07{,}840}{2\pi N_s} = \frac{2{,}07{,}840}{2\pi \times \frac{120 f}{P}} = \frac{2{,}07{,}840}{2\pi \times \frac{120 \times 50}{6}} = 33 \text{ N-m } \textbf{Ans.}$$

i.e. Motor will start from rest against a maximum torque of 33 N-m **Ans.**

21.3 THE VARIABLE SPEED COMMUTATOR TYPE 3-PHASE INDUCTION MOTOR (SCHRAGE MOTOR)

This motor having shunt characteristics and an adjustable speed was invented in 1911 by H. K. Schrage of the Swedish Co ASEA and, therefore, called the Schrage motor. In this motor speed is controlled by injecting a voltage of slip frequency in the secondary circuit. The connection diagram of this machine is shown in Fig. 21.20 while the general arrangement is given in Fig. 21.21.

It is provided with three windings known as primary, secondary and tertiary (adjusting or regulating) windings. The primary winding is placed in the lower part of the slots of rotor and is connected to the 3-phase ac supply by means of slip rings and brushes. It produces the working flux in the machine. The secondary winding is phase wound and located on the stator. End of each secondary phase winding is connected to one of a pair of brushes arranged on the commutator as shown in Fig. 21.20. The reversal of location of primary and secondary windings makes no difference, of course, in the basic induction

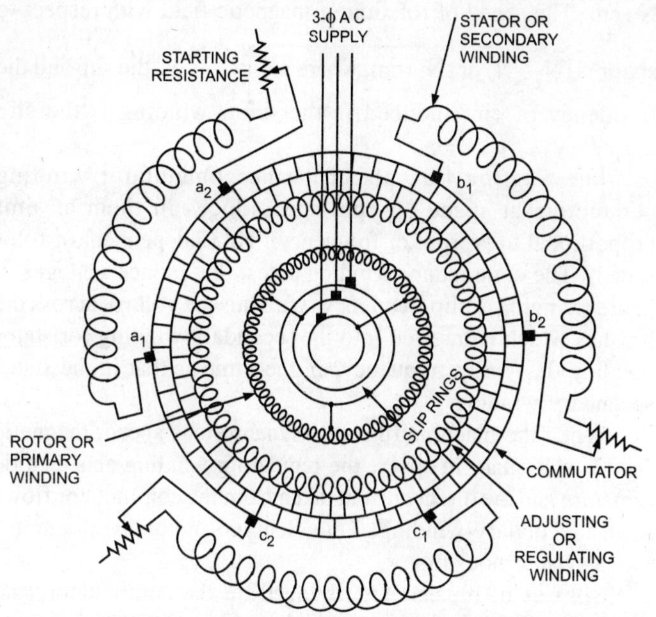

Fig. 21.20 *Connection Diagram of The Schrage Motor*

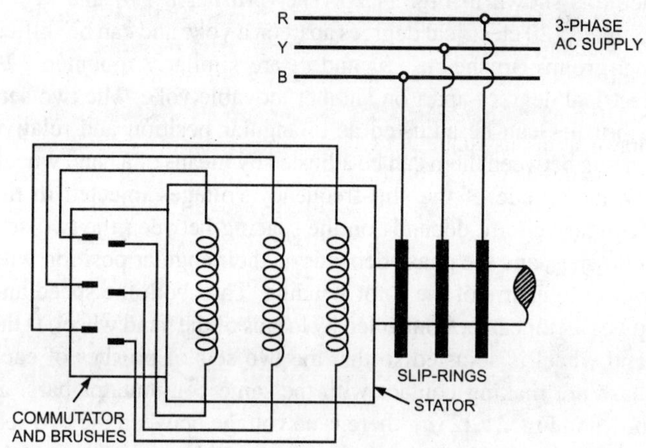

Fig. 21.21 *General Arrangement of The Schrage Motor*

motor action. Another winding called the *tertiary, compensating, adjusting* or *regulating* winding is placed in the same slots as the primary winding on the rotor and is connected to the commutator like the armature winding of a dc motor. The conductors of the third winding are placed in the upper part of the rotor slots to reduce its reactance and so the reactance voltage in the commutating coil. The third winding is magnetically coupled to the primary winding and the voltages induced in each coil of the regulating winding are essentially constant for a given primary voltage and of line frequency. The commutator, acting as a frequency changer, converts this line frequency voltage of the regulating winding to the slip frequency for supplying it into the stator secondary winding as explained below:

Let the supply frequency be f and the corresponding synchronous speed $N_s = \frac{120 f}{P}$ where P is the number of poles. Now when the primary is connected to a balanced 3-phase ac supply, a magnetic field rotating at synchronous speed N_s with respect to the rotor is produced. According to the induction motor theory, the rotor runs in such a direction so as to oppose flux cutting by the stator conductors. Thus the rotor rotates in the direction opposite to that of the magnetic field, say at a speed of N rpm. The speed of rotation of magnetic field with respect to stator is $(N_s - N)$ or sN_s rpm, where $s = \frac{N_s - N}{N_s}$ is the slip and the frequency of emf induced in the stator winding is the slip frequency (sf).

The rotating flux also cuts the commutator winding conductors at speed N_s rpm and induces in them an emf proportional to N_s rpm or frequency f Hz (independent of rotor speed). The commutator winding poles, that induce voltages in it, are changing at slip frequency (sf). Thus the voltage across the brushes which is injected into the secondary winding (or stator winding) has a slip frequency, sf, the same as that in the stator secondary winding.

When the machine runs at a synchronous speed, magnetic field is stationary in space, the regulating winding acts as a dc armature and the direct current taken from the commutator flows in the secondary winding. Thus Schrage motor then operates like a synchronous motor.

Six sets of brushes are arranged on the commutator, and each phase of the secondary winding is connected to a pair of them, as shown in Fig. 21.20. The brushes a_1, b_1 and c_1 are mounted 120 electrical degrees apart on a yoke and can be shifted as a group. Brushes a_2, b_2 and c_2 are similarly mounted 120 electrical degrees apart on another movable yoke. The two sets of brushes can be adjusted as to angular position and relative spacing between them can be adjusted by means of a hand wheel. The magnitude of the slip-frequency voltages injected in the secondary circuit depends on the spacing between the two sets of brushes, and the phase depends on their angular position with respect to centre of the rotor winding. Thus both the speed and power factor can be controlled by means of the hand wheel. If the hand wheel is adjusted so that the two sets of brushes of each phase are making contact with the same commutator bars, as shown in Fig. 21.22 (a), there is no voltage between the two sets of the brushes of any one phase. The secondary winding is short circuited through the commutator segments and the motor operates as an induction motor running at a speed slightly less than synchronous speed and inducing a small voltage in each of the secondary phase windings. If the sets of the brushes are separated from the neutral position, as shown in Fig. 21.22 (b), so that the voltage injected into the secondary winding by the regulating winding is in phase opposition to the voltage induced in the secondary from the primary winding, the motor runs at subsynchronous speeds (speeds below synchronous speed). If the sets of the brushes are separated from neutral position in the opposite direction, the voltage injected into the secondary winding will be in phase with the secondary induced voltage and the speed of motor will increase. As the sets of brushes are separated in this second direction, a position is reached when the motor operates at synchronous speed. Further separation of brush sets as shown in Fig. 21.22 (c) will give speeds above synchronous. The commutator provides maximum voltage when the brushes are separated by one pole pitch. Thus any speed, within reasonable limits, can be obtained by rotating the brush gear in one direction or the other. The normal speed range of these motors is 3 to 1, but much greater speed range up to 15 to 1 or higher may be obtained.

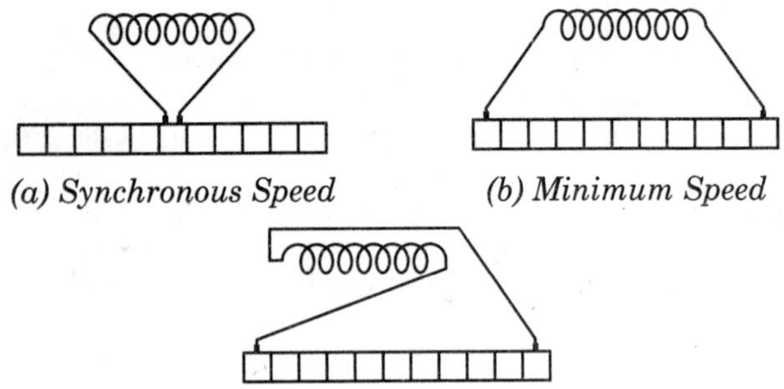

(a) *Synchronous Speed* (b) *Minimum Speed*

(c) *Maximum Speed*

Fig. 21.22

No-load speed of the motor is given as $N = N_s\,(1 - k \sin \frac{\beta}{2})$ where k is a constant depending upon the turn ratio of the secondary and regulating windings and β is brush separation in electrical degrees.

The no-load speeds of such machines range from 7 to 10% of their full-load speeds with the brushes in the maximum speed position and from 20 to 40% above the full-load speeds with the brushes in the minimum speed position. The no-load speed brush separation curve for a 4 pole, 50 Hz Schrage motor is shown in Fig. 21.23.

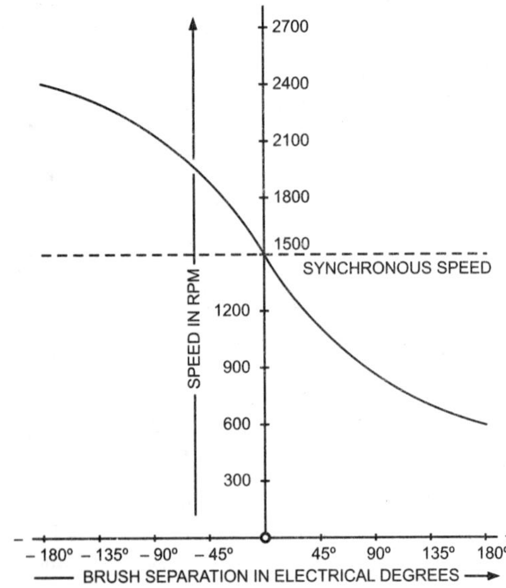

Fig. 21.23

The motor is started with the brushes set in the minimum speed position, when it is necessary only to close the main switch. If the speed range is 2 to 1 or greater the current drawn from the line is limited to about 1.5 times the full-load current. If the speed range is less than 2 to 1, it may be necessary to insert resistances in the secondary circuit for starting purposes as in an ordinary induction motor otherwise, the starting torque may be inadequate and the starting current excessive.

To ensure that the brushes are in the minimum speed position when starting, it is usual to provide a small switch operated by the brush gear and interlocked with the starting switch. The inclusion of an overload tripping coil in the primary of such a machine does not provide complete protection, which, however, may be obtained by incorporating the tripping device in the secondary circuit.

One aspect in which the Schrage motor differs from the induction motor is the power factor; this is very high at hyper-synchronous speeds, but relatively low at subsynchronous speeds. The low power factor at reduced speeds can be improved by an unsymmetrical brush displacement; this can be accomplished by allowing one brush rocker to move faster than the other, so that the centre line of the brushes is rotated around the commutator. If the injected voltage has a quadrature component leading the induced rotor voltage, the power factor of the motor will be improved. Motors which are designed with unsymmetrical brush displacement can be run in one direction only, but by a simple reconnection and readjustment of the brushes the direction of rotation can be changed.

Reversible machines, such as crane and hoisting motors, are designed with symmetrical brush displacement to make the brush mechanism as simple as possible. To reverse such a machine it is necessary simply to interchange any two primary supply leads.

Torque-speed characteristic curves are shown in Fig. 21.24. The output power available from the machine is directly proportional to the operating speed.

Their temperature is 40 °C when operating from maximum to 50 per cent of maximum speed and 50 °C when operating at lower speeds. Maximum running torque ranges from 140 to 250% of full-load depending upon the rating and brush position.

To minimise the difficulties of commutation, such motors are provided with large number of commutator bars and brushes, which together with the special windings result in relatively high cost.

The main advantages of the Schrage motor are (*i*) continuous speed regulation within the required range (*ii*) high power factor for high speed settings (*iii*) high efficiency at all speeds except synchronous speed and very much high at lower speeds.

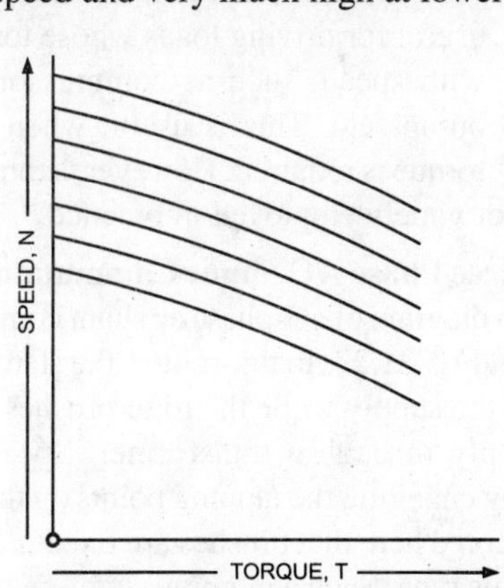

Fig. 21.24

One major drawback of the Schrage motor is that its operating voltage is limited about to 700 V, because the power is to be supplied to the motor through the slip rings. Other disadvantages are its low power factor at low speed settings and its poor commutation and high cost.

Schrage motor is used only where adjustable speed is required as in bakery, machinery, stokers, printing machines, calenders etc.

One important application of the Schrage motor is to drive the hosiery-knitting and ring-spinning machines, where an automatically controlled speed variation of 10 to 30% is often required. They are also used for a wide range of industrial services, driving fans, pumps, conveyors, packaging machinery, paper mills etc. Sizes up to several hundred kW have been built. They are normally rated on a constant torque basis, the kW output varying in direct proportion to the speed.

Example 21.6. A 4-pole, 50 Hz Schrage motor has minimum and maximum no-load speeds of 500 and 2,000 rpm, for brush separation of one pole pitch. What is the brush frequency at these two speeds?

[Pb. Univ. Elec. Machines II, May 1995]

Solution:

$$\text{Synchronous speed, } N_s = \frac{120\,f}{P} = \frac{120 \times 50}{4} = 1{,}500 \text{ rpm}$$

$$\text{Minimum speed, } N_1 = 500 \text{ rpm}$$

$$\text{Slip at minimum speed, } s_1 = \frac{N_s - N}{N_s} = \frac{1{,}500 - 500}{1{,}500} = 0.667$$

$$\text{Brush frequency} = s_1 f = 0.667 \times 50 = 33.33 \text{ Hz } \textbf{Ans.}$$

$$\text{Maximum speed, } N_2 = 2{,}000 \text{ rpm}$$

$$\text{Slip at maximum speed, } s_2 = \frac{N_s - N}{N_s} = \frac{1{,}500 - 2000}{1{,}500} = -0.333$$

$$\text{Brush frequency} = s_2 f = 0.333 \times 50 = 16.67 \text{ Hz } \textbf{Ans.}$$

21.4 THE OSNOS (NO-LAG) MOTOR

General arrangement of the Osnos (no-lag) motor is shown in Fig. 21.25. No-lag motor was developed earlier and Schrage motor is actually modified form of the no-lag motor. This motor is quite similar to the Schrage motor in construction. In this motor phase of the emf can be varied but not the magnitude. EMF from the commutator winding is injected into the secondary winding at an angle leading the secondary emf at 90° and, therefore, improving the power factor.

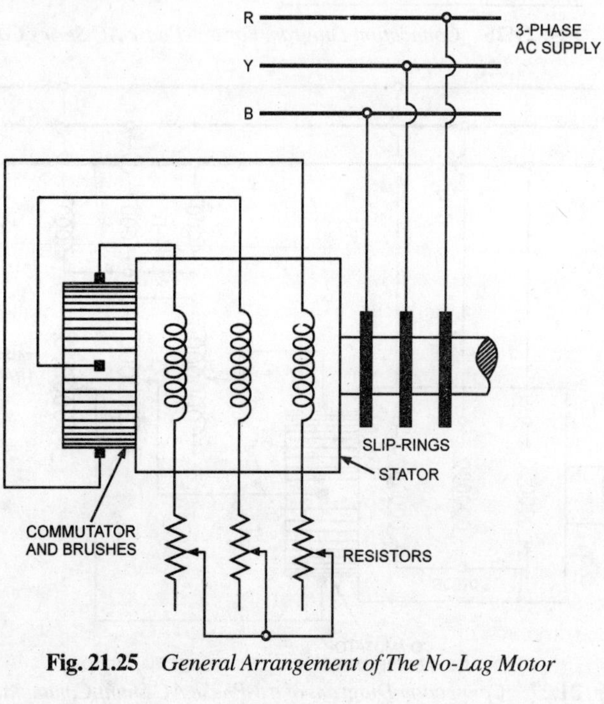

Fig. 21.25 *General Arrangement of The No-Lag Motor*

21.5 THREE-PHASE AC COMMUTATOR MOTORS

Three-phase ac commutator motor may be either series or shunt type.

21.5.1. Three-Phase AC Series Commutator Motor. The connection diagram of a 3-phase ac series commutator motor is shown in Fig. 21.26. The machine has a 3-phase stator winding similar to that of an ordinary induction motor and may be wound for high voltage if required. The rotor carries a winding similar to that of a dc armature and its commutator is provided with three brush sets per pair of poles spaced 120 electrical degrees apart. The stator winding is connected to the ac supply through the primary of a transformer whose secondary is connected to the brushes. Thus, in effect, the stator and rotor windings are in series so that the working flux is dependent on the current giving series speed characteristic. A transformer has been inserted between the stator and rotor in order to reduce the voltage applied to the commutator winding and thus avoid the commutation difficulties.

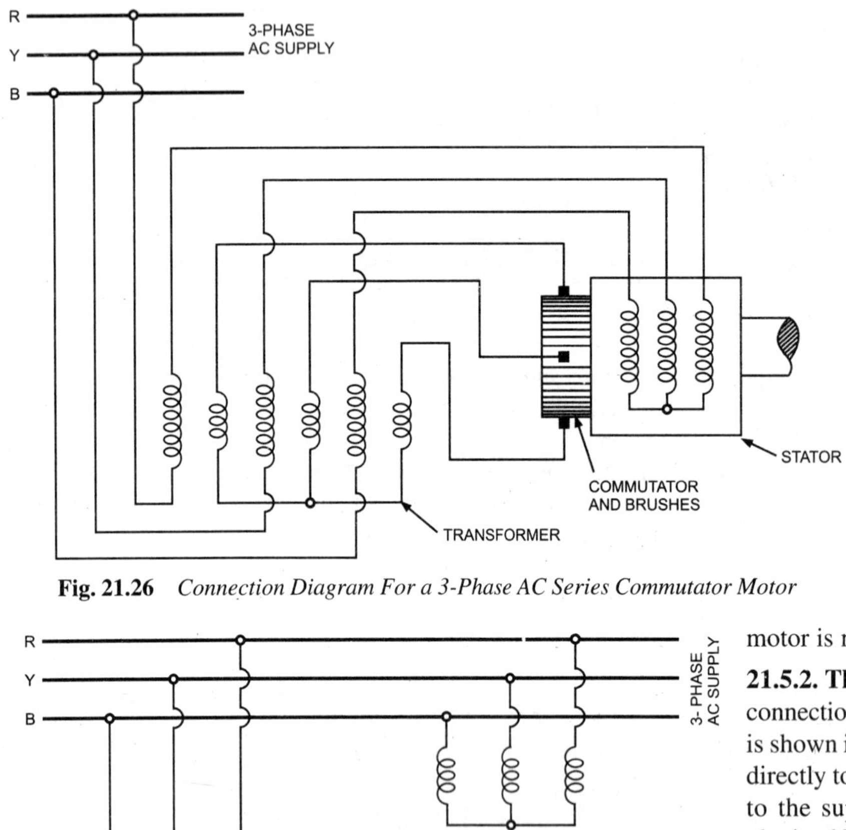

Fig. 21.26 *Connection Diagram For a 3-Phase AC Series Commutator Motor*

The phase relationship of the emfs induced in the stator and rotor depends upon the position of brushes, any movement of which varies the phase of the rotor emf without any change in its magnitude. The brush gear is moved backwards (i.e. in a direction opposite to that of rotation of the rotor) for increasing the speed. For any given position of brushes, the speed falls with the increase in load, as in case of a dc series motor.

With the brushes in the limiting high speed position, the speed does not increase rapidly with the decrease in load, and the characteristic under such a condition approaches that of a compound wound dc motor. So there is no need of providing a centrifugal cut-out device for protection of motor against excessive speed.

The regulation of speed solely by brush displacement makes external resistance unnecessary and avoids the power losses consequent upon their use, and therefore, the efficiency of the motor remains high over the whole operating range. The power factor is also high, approaching unity for speeds near and above synchronous speeds.

If the transformer inserted between stator and rotor of the machine is phase shifting type, it can be used for speed control as a phase shifting device giving the same effect as the movement of the brushes.

The motor is usually started by switching on with the brushes in the low-speed position, when the starting torque is about 50 per cent of full load. As the brushes are moved against the direction of rotation of motor the torque increases and the motor can be brought up to the desired speed.

The speed range is limited between 40 and 150 per cent of synchronous speed owing to commutation difficulties. Such a motor is very useful for driving loads whose torque increases with speed, such as compressors, centrifugal pumps etc. This is useful when large starting torque is required. However, a three phase series motor is not widely employed in practice.

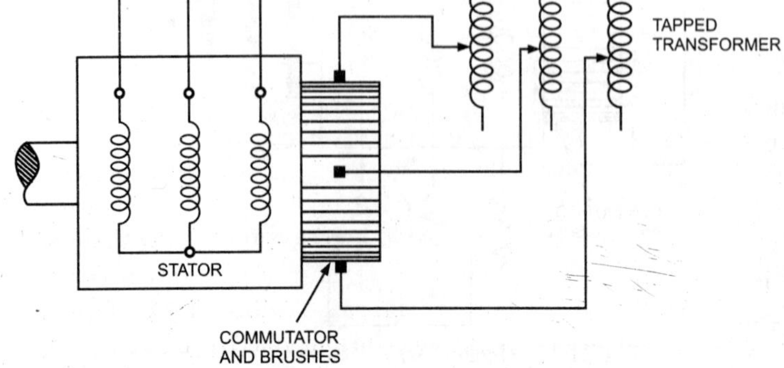

Fig. 21.27 *Connection Diagram of a 3-Phase AC Shunt Commutator Motor*

21.5.2. Three Phase AC Shunt Commutator Motor. The connection diagram of a 3-phase ac shunt commutator motor is shown in Fig. 21.27. In this motor, the stator is connected directly to the supply while the rotor brushes are connected to the supply through a transformer. Speed variation is obtained by changing the tapping points on the transformer. It is to be noted here that brushes are fixed in position in the motor of this type. Such a motor gives more or less constant speed at all loads, for a given tapping on the transformer.

21.6 PHASE ADVANCER

Phase advancers are used to improve the power factor of induction motors. The low power factor of an induction motor is due to the fact that its stator winding draws exciting current which lags behind the supply voltage by 90 electrical degrees. If the exciting ampere turns can be provided from some other ac source, then the stator winding will be relieved of exciting current and the power factor of the motor can be improved. This job is accomplished by the phase advancer which is simply an *ac exciter*. The phase advancer is mounted on the shaft as the main motor and is connected in the rotor circuit of the motor. It provides exciting ampere turns to the rotor circuit at slip frequency. By providing more ampere turns than required, the induction motor can be made to operate on leading power factor like an overexcited synchronous motor. The kVA capacity of the phase advancer need be only 5% or less of kVA correction affected in the main supply circuit. The full corrective kVA could have to be provided if a static or rotary capacitor were used in the supply circuit. The phase advancer is thus very advantageous for use with large motors.

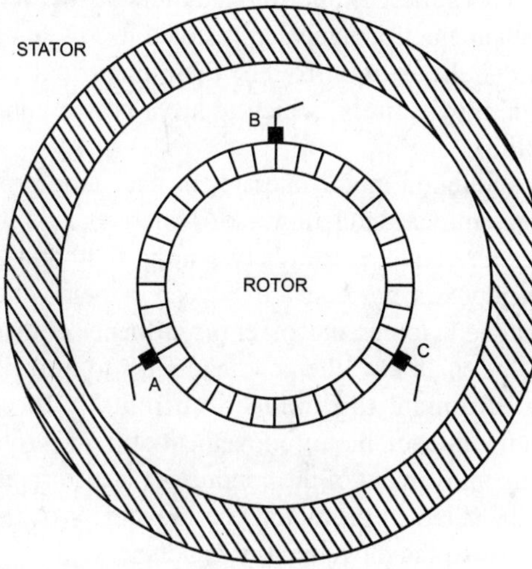

Fig. 21.28 *Phase Advancer*

In its simplest form the phase advancer consists of a rotor built as a dc machine armature and provided with a commutator, and of a stator assembled of steel laminations and acting only as the magnetic circuit for the flux of the phase advancer. Three brushes are arranged on the commutator, 120° apart. A schematic diagram of a 2-pole, 3-phase advancer is shown in Fig. 21.28.

The armature (or rotor) is driven by an auxiliary motor or it may be belt driven from or direct coupled to the main induction motor if its speed is high. At the starting instant the phase advancer is taken out of the circuit and the induction motor slip rings are connected to an ordinary rotor starter by means of a 3-pole change-over switch. When the motor attains rated speed the change-over switch is thrown over to the brushes of the phase advancer, 3-phase low-frequency slip current from the main rotor then flows to the armature of the phase advancer and creates a rotating field. By driving the phase advancer at speed higher than the synchronous speed corresponding to this rotating field, the advancer can be made to supply all the magnetising current for the main induction motor which then operates at unity power factor. If the speed of the phase advancer be increased, the main motor supplies leading current to the network in which it is connected.

Induction motor of about 40 kW have been built with a phase advancer of this type on an extension of the main motor shaft within one of the end casing of the combined set. The cost of a separate phase advancer is not likely to be justified unless the main motor is of 100 kW or more and normally operating at 0.85 or lower power factor. By varying the speed of the phase advancer unity power factor can be maintained down to one-third rated load. The power required to drive it is not considerable since the phase advancer delivers purely reactive power to the rotor of main motor. The active power required is only that absorbed by friction and other losses and is often less than the reduction in the losses of the main motor. The maximum torque of the latter is greatly increased by using phase advancer.

If the advancer emf acts in conjunction with the rotor emf of the main motor the rotor slip decreases and the phase advancer delivers energy to the main motor. On the other hand, if the advancer emf opposes the rotor *emf* the rotor slip is increased and the advancer runs as a motor driven by energy drawn from the rotor circuit. Thus it offers a convenient and economical means of speed control of the main induction motor. Compensating windings on the stator of the advancer ensure good commutation.

21.7 ACTION OF COMMUTATOR AS FREQUENCY CHANGER

Consider a commutator winding fitted with both slip rings and a commutator (Fig. 21.29). If polyphase currents at a frequency *f* are fed to the slip rings with the winding stationary, a rotating field is created rotating at speed of 120 *f*/P rpm. This field cuts the conductors and set up in them emfs of frequency *f*. Also the field moves at a synchronous speed N_s relative to the fixed brushes and the emfs at the brushes will be at a corresponding frequency, i.e. *f*. Slip-ring currents, conductor currents and brush currents and the corresponding emfs are, therefore, all at frequency *f*.

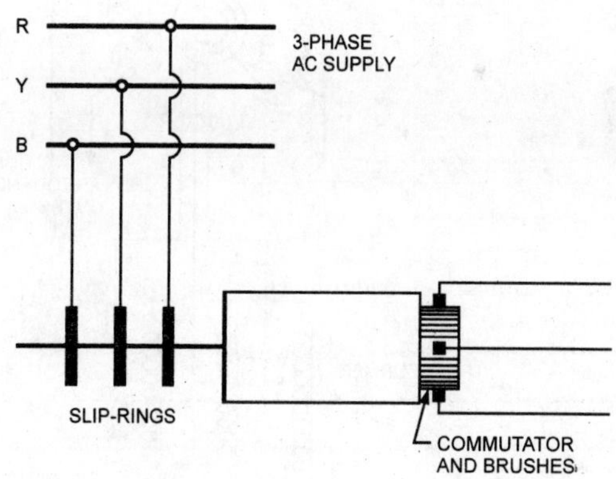

Fig. 21.29 (*a*) *Commutator As Frequency Changer*

Now let the winding be rotated at a speed of N rpm. The speed of the field relative to the conductors will still be N_s since currents at frequency *f* are being fed to fixed points in the winding. The frequency of conductor emfs and currents is thus still *f*. The field is, however, moving at $N_s \pm N$ relative to the fixed brushes, and the frequency of the emfs appearing at the brushes, and therefore, of the currents flowing from them will be $f \pm f_1$

The commutator thus changes the frequency *f* of the currents in the conductors at slip rings to a frequency $f \pm f_1$ at the brushes i.e. from a value in the conductors corresponding to the speed of the field past the conductors to a value at the brushes corresponding to the speed of the field past the brushes. This feature of the commutator is of fundamental importance in the behaviour of commutator machines.

The device can be used as an independent source of variable frequency, usually a low frequency, in which case it is driven by a variable speed motor, just large enough to provide the mechanical losses. The armature speed affects the brush frequency directly, since this is equal to $f - \frac{PN}{120}$ Hz. The relative field/conductor speed, as already explained is $\frac{120\,f}{P}$ so that the generated emf is independent of the rotational speed N. This permits the generation of low frequency with adequate output voltage.

For control of speed and power factor of induction motor slip-frequency currents may be obtained by directly coupling a 3-phase commutator frequency convertor to the shaft of an induction motor, its slip rings being fed from the same a c source and the winding being arranged for the same number of poles as the induction motor. The speed is automatically $\frac{120\,f}{P}(1-s)$ and so the output frequency at the brushes is sf, irrespective of the speed of the motor. Voltage control can be achieved by supplying the frequency convertor slip rings through a 3-phase tapped transformer as illustrated in Fig. 21.29 (*b*). For phase control the commutator brushes are "rocked" round to a new position, such that the space wave of flux travelling around the periphery, traverses the winding section between the brushses at the desired instant in the supply voltage cycle.

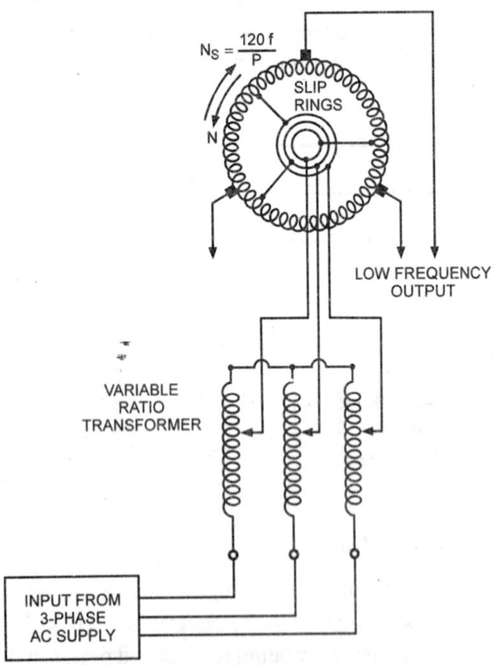

Fig. 21.29 (*b*)

21.8 LINEAR INDUCTION MOTOR

Linear induction motor, abbreviated as LIM, is basically a special purpose motor that is in use to achieve rectilinear motion rather than rotational motion as in case of conventional motors. This is quite an engineering marvel, to convert a general motor for a special purpose with more or less similar working principle thus enhancing its versatility of operation. In a conventional electric motor, the rotor (rotating part) spins inside the stator (stationary part); in a linear motor, the stator is unwrapped and laid out flat and the rotor moves past in a straight line. A linear induction motor is effectively an ac induction motor that has been cut open and unwrapped. The stator is laid out in the form of a track of flat coils made from aluminium or copper and is known as the *primary* and the rotor takes the form of moving platform known as the *secondary*.

The main problem with linear induction motors has been the cost and difficulty of developing suitable electromagnets. Enormously powerful electromagnets are required to levitate (lift) and move something as big as a train, and these typically consume substantial amount of electric power. Linear induction motors often now use superconducting magnets to overcome this problem. When electromagnets are cooled to low temperatures using liquid helium or nitrogen, their electrical resistance disappears almost entirely, which reduces power consumption considerably.

Linear induction motor operates on the same principle on which a conventional induction motor operates i.e. "whenever there occurs a relative motion between the field and the short-circuited conductors, currents are induced in them which results in electromagnetic forces and under the influence of these forces, according to Lenz's law, the conductors try to move in such a way as to eliminate the induced currents". In case of a conventional induction motor, movement of field is rotary about an axis so the movement of the conductors is also rotary. But in case of a linear induction motor, the movement of the field is rectilinear and so the movement of conductors.

In its simplest form, a linear induction motor consists of field system having a 3-phase distributed winding placed in slots as shown in Fig. 21.30. The field system may be single primary system or double primary system (Fig. 21.30). The secondary of the linear induction motor is normally a conducting plate made of either copper or aluminium in which interaction currents are induced. Either member can be the *stator*, the other being the *runner* in accordance with the particular requirements imposed by the duty for which motor is intended.

In a single primary system a ferromagnetic plate is usually placed on the other side of the conducting plate to provide a path of low reluctance to the main flux. However, the ferromagnetic plate gets attracted towards the primary on energisation of the field and this causes unequal gap length on two sides of the conducting plate. This problem can be overcome by employing double primary system. [Fig. 21.30(*b*)].

Which of the two primary and secondary will be shorter in length compared to the other depends upon the use of the motor. When the operating distance is large, the primary is made shorter than the secondary because it is uneconomical to wind a very long 3-phase primary. The short secondary form [Fig. 21.30 (*c*)] is useful with limited operating distance.

When the three phase primary winding of a linear induction motor is energized from a balanced three-phase source, a magnetic field moving in a straight line from one end to the other at a linear synchronous speed v_s is developed. The linear synchronous speed is given as

$$v_s = 2\,\tau f \text{ metres/second} \quad ...(21.1)$$

where τ is the pole pitch in metres and f is the supply frequency in Hz. *It is to be noted here that the synchronous speed does not*

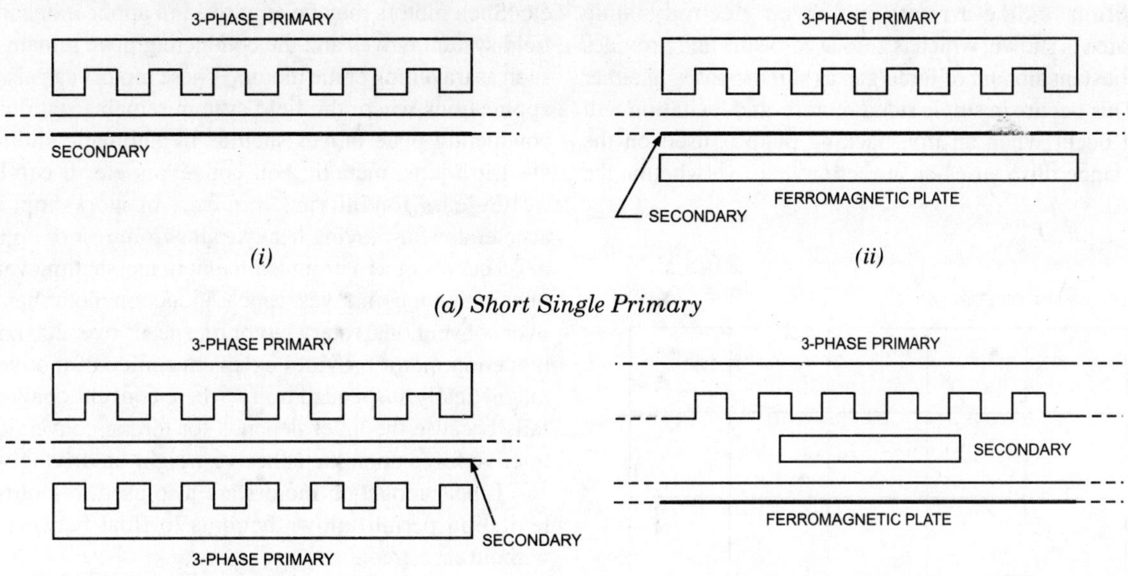

(i) (ii)

(a) Short Single Primary

(b) Short Double Primary

(c) Short Secondary

Fig. 21.30

depend on the number of poles, but only on the pole pitch and the stator supply frequency.

As the flux moves linearly, it drags rotor plate alongwith it in the same direction. This reduces the relative speed of travel of the flux with respect to rotor plate. If the speed of the rotor plate is equal to that of the magnetic field, latter would be stationary when viewed from rotor plate. This is corresponding to the synchronous speed of induction motor. If the rotor plate is moved faster than this speed, the direction of the force would be reversed and a form of regenerative braking based on the principle of induction generator will come into being.

Slip of a linear induction motor is given as

$$s = \frac{v_s - v}{v_s} \quad \text{... (21.2)}$$

where v is the actual speed of rotor plate.

Thrust or force or tractive effort is given as

$$F = \frac{P_2}{v_s} \quad \text{... (21.3)}$$

where P_2 is the actual power supplied to the rotor

Active power flow is similar to that in a conventional rotary induction motor i.e.

$$\text{Copper losses in rotor} = sP_2 \quad \text{... (21.4)}$$

$$\text{and mechanical power developed, } P_{mech} = (1 - s)\, P_2 \quad \text{... (21.5)}$$

Tractive effort will be a function of slip s i.e. $(v_s - v)$. The thrust (or tractive effort) — speed characteristics of a linear induction motor, as shown in Fig. 21.31, are similar to the torque-speed characteristics of a conventional rotary induction motor.

Tractive effort F can be controlled by varying both frequency and voltage simultaneously so that induction density remains constant.

There are two peculiar effects, which are encountered in a linear induction motor but not in a conventional rotary induction motor. These effects are *transverse edge effect* and *end effect.*

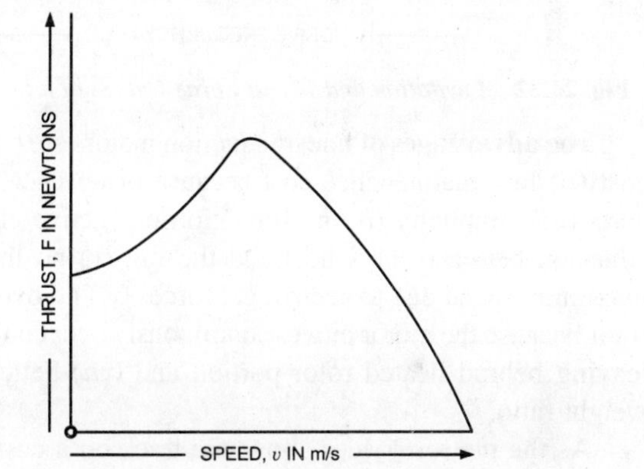

Fig. 21.31

The paths of the induced currents in the secondary are not well defined because the secondary of a linear induction motor is a solid conducting plate. The portion of the current paths parallel to the direction of motion of secondary does not make any contribution towards the production of useful thrust but only contributes towards losses. This effect reduces the effective thrust and increases the losses and is known as the *transverse edge effect* because the current paths parallel to the direction of motion are more towards the edges of the conducting plate.

Unlike a conventional induction motor, a linear induction motor shows 'end effects'. These end effects include losses in performance and efficiency that are believed to be caused by magnetic energy being carried away and lost at the end of the primary by the relative movement of the primary and secondary. With a short secondary, the behaviour is almost identical to a rotary machine, provided it is at least two long poles but with a short primary reduction in thrust occurs at small slip (below about 0.3) until it is eight poles or longer.

However, because of end effect, linear motors cannot run light-normal induction motors are able to run at a near synchronous speed under low load conditions. Due to end effect this creates much more significant losses with linear motors.

In addition, unlike a rotary motor, an electrodynamic levitation force is shown, which is zero at zero slip, and provides a roughly constant amount of force/gap as slip increases in either direction. This occurs in single sided motors, and levitation will usually not occur when an iron backing plate is used on the secondary, since this causes an attraction that overwhelms the lifting force.

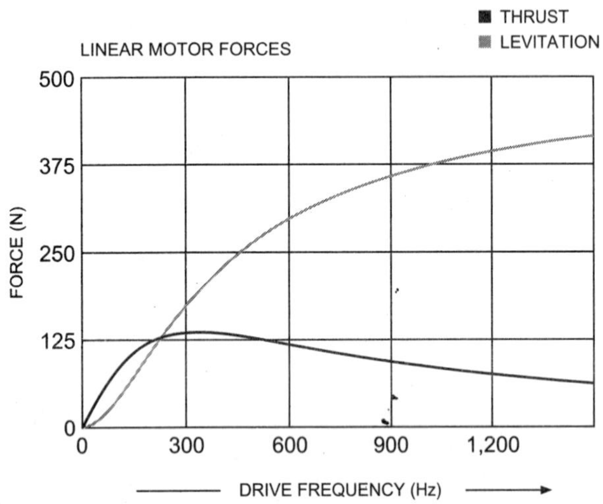

Fig. 21.32 *Levitation and Thrust Force Curves of a Linear Motor*

The **advantages** of linear induction motors are (*i*) low initial cost (*ii*) low maintenance cost because of absence of rotating parts (*iii*) simplicity (*iv*) no limitation of tractive effort due to adhesion between the wheel and the rail (*v*) no limitation of maximum speed due to centrifugal forces (*vi*) no overheating of rotor because the motor moves continuously over cool rotor plate leaving behind heated rotor portion and (*vii*) better power to weight ratio.

As the platform rides above the track on a cushion of air, there is no loss of energy due to friction or vibration but because of larger air gap in a linear induction motor, more power is required and the efficiency is reduced. The lack of an intermediate gearbox to convert rotational motion into linear motion saves energy. Finally, as both acceleration and braking are achieved through electromagnetism, linear motors are much quiter than conventional motors.

The **disadvantages** of linear induction motors are (*i*) poor utilisation of motor due to transverse edge effect and end effect (*ii*) larger air gap and nonmagnetic reaction rail (rotor plate) need more magnetizing current resulting in poor efficiency and low pf (*iii*) very high capital cost of reaction rail fixed along the centre line of the track (*iv*) complications and high cost involved in providing three-phase collector system along the track (*v*) difficulties encountered in maintaining adequate clearances at points and crossings.

APPLICATIONS. Linear induction motors are often used where contactless force is required, where low maintenance is desirable, or where the duty cycle is low. Their practical uses include magnetic levitation, linear propulsion, and linear actuators.

Linear induction motors can be used in conveyors, travelling cranes, haulers, electromagnetic pumps, high-speed rail traction etc. Such motors may be employed in applications in which the field system moves and the conducting plate remains stationary such as travelling crane motors. These motors can also be used in applications where the field system remains stationary and the conducting plate moves such as in automatic sliding doors in electric trains, metallic belt conveyors etc. It can be used on trolley cars for internal transport in workshop, as booster accelerator for moving heavy trains from rest or up the inclines or on curves or as a propulsion unit in marshalling yards in place of shunting locomotives. Linear induction motor has superiority over conventional rotary motor for speeds over 200 kmph. Linear induction motor provides excellent source of motive power for magnetically suspended trains where conventional rotary motor fails because the latter depends for torque conversion to linear tractive force upon the adhesive weight on driving wheels.

Linear induction motors are also used in looms, magnetic levitation permit allows bobbins to float between the fibres without direct contact.

Linear induction motor technology is also used in some roller coasters.

Example 21.7. A single-sided linear induction motor has 98 poles, and its pole pitch is 50 cm. The motor is used for propelling an electric vehicle. Determine the linear synchronous velocity and the vehicle speed in km per hour if the frequency is 50 Hz and the slip is 0.25.

Solution: Supply frequency, $f = 50$ Hz

Pole pitch, $\tau = 50 \text{ cm} = 0.5 \text{ m}$

Slip, $s = 0.25$

Linear synchronous velocity

$$v_s = 2\tau f = 2 \times 0.5 \times 50$$

$$= 50 \text{ m/s} = \frac{50 \times 60 \times 60}{1{,}000} = 180 \text{ kmph } \textbf{Ans.}$$

Vehicle speed, $v = v_s(1 - s) = 180(1 - 0.25) = 135$ kmph **Ans.**

21.9 SERVOMECHANISM

Before going into details of construction and working principle of servomotor it is necessary to understand the basics of servomechanism. So, first of all, let us explain what is servo?

In modern usage the term *servo* or *servomechanism* is restricted to a *feedback control system* in which the controlled variable is mechanical position or time derivatives of position such as velocity and acceleration.

A *feedback control system* is a control system which tends to maintain a prescribed relationship between a controlled quantity and a reference quantity by comparing their functions and using the difference as a means of control.

Electrical feedback control systems rely mainly on electrical energy for operation. Important characteristics usually required for such a control are (*i*) fast response (*ii*) high accuracy, (*iii*) unattended control and (*iv*) remote operation.

The essentials of such a control are (*i*) an error detecting device, (*ii*) an amplifier, and (*iii*) error correcting device, as illustrated in Fig. 21.33. Each element serves a functional purpose in matching the regulated or controlled quantity to the reference quantity. The error-detecting device determines when the regulated quantity is different from the reference quantity. It

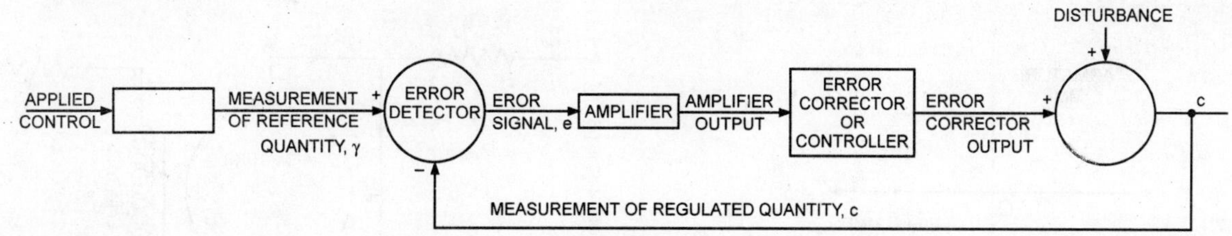

Fig. 21.33

then sends out an error signal to the amplifier which in turn supplies power to the error-correcting device. With this power the error correcting device changes the regulated quantity so that it matches the reference input. The closed loop at feed-back control system is shown in Fig. 21.33.

21.10 SERVOMOTORS

The motors which respond to the error signal abruptly and accelerate the load quickly are called the *servomotors.* The torque-inertia (T/J) ratio is an important figure of a servomotor because the load is to be accelerated by it abruptly.

Servomotors, unlike industrial motors, are not employed for continuous energy conversion but only for precise speed and precise position control at high torques. Such motors are used in feedback control systems as output actuators. No doubt, they operate on the same basic principle on which other electromagnetic motors operate. However, they differ in design, construction and mode of operation. They are of power ratings varying from a fraction of watt to a few hundred watts. They have high speed response owing to their low rotor inertia. The rotors of servomotors are designed with relatively long length and smaller diameter so as keep rotor inertia small. Again modern technology has developed small weight rotors and ironless rotors, electronic rotors. Their operating speeds are very small, and sometimes zero speed. They are of larger size than that of conventional motors of similar power ratings. Servomotors have the following special features.

1. They are capable of developing high torque at all speeds (including zero speed) and holding a static (no motion) position.
2. Because of low inertia, they are able to accelerate, decelerate and reverse in direction of rotation quickly.
3. They do not overheat at standstill or lower speeds.
4. They are able to return to a given position without any drift.

They find wide use in radar, process controllers, tracking and guide systems, computers, robotics, CNC machinery or automated manufacturing.

The fundamental characteristics to be sought in any servomotors are :

(*i*) The motor output torque should be proportional to the voltage applied to it (i.e. control voltage developed by the amplifier in response to an error signal).

(*ii*) The direction of the torque developed by the servomotor should depend upon the instantaneous polarity of the control voltage.

Servomotors are of two types namely dc servomotors and ac servomotors. AC servomotors are generally preferred to dc servomotors except for use in very high power systems. For very high power systems dc motors are preferred because they operate more efficiently than comparable ac servomotors. This enables them to stay cooler. An efficient motor also prevents excessive power waste, although power waste is generally not a prime concern in servomechanism.

The drawbacks of ac motors are: (*i*) nonlinear characteristics and (*ii*) difficult position control through continuous speed or torque changes.

No doubt, the dc motors do not have the above drawbacks, but have the following drawbacks.

(*i*) Maximum torque is restricted by commutation and not by heating as with other motors.

(*ii*) Higher cost owing to the requirements of brush and commutator.

(*iii*) Occurrence of arc during collection of armature current by brush through commutator. In course of time, pitting is caused by this sparking.

Modern science has advanced a lot to reduce wear and tear and maintenance costs.

21.10.1. DC Servomotors. For dc servomotors, armature is deliberately designed to have large resistance so as to provide linear torque-speed characteristics with a large negative slope (Fig. 21.34). The negative slope provides viscous damping for the servo drive system. Figure 21.35 shows that the armature mmf and the excitation field mmf are in quadrature. Because of this fact, we have a fast torque response as torque and flux become decoupled. Accordingly, a step change in the armature voltage/current produces a quick change in the position or speed of motor.

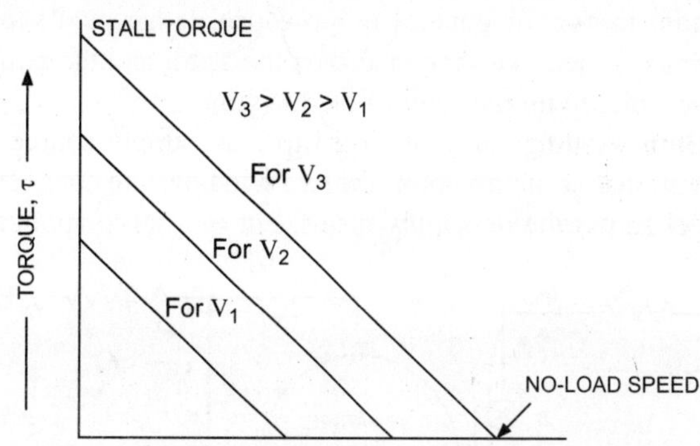

Fig. 21.34 *Torque-Speed Characteristics For Different Control Voltages*

Among the various dc servomotors, are the series motors, split series motors, shunt control motor, and the permanent magnet (fixed excitation) shunt motor. These units develop high

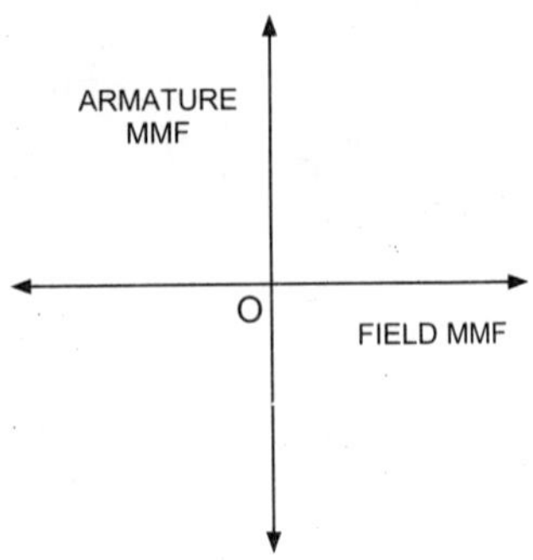

Fig. 21.35

output power in a given size and in the case of the field controlled shunt motor, little control power is required.

The series motor has a high starting torque and draws large current and has poor speed regulation. Reversal can be obtained by reversing field voltage polarity with split-series field winding i.e. one winding for each direction of rotation. The latter reduces motor efficiency.

Small fracional kilowatt, split-series motor (dc series motor with split-field) may be operated as a separate excited field-controlled motor, as illustrated in Fig. 21.36. The armature may be supplied from a constant current source.

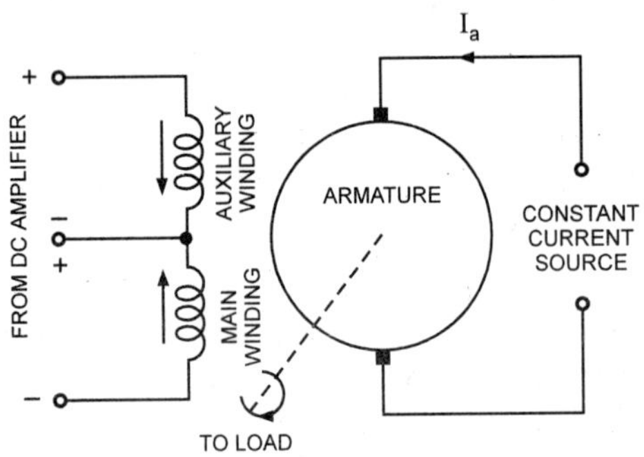

Fig. 21.36

A typical torque-speed curve shows a high stall torque and rapid reduction in torque with the increase in speed. This results in good damping and large velocity error.

A dc shunt type servomotor is not different from any other dc shunt motor for general purpose use. It has two separate windings — field winding placed on the stator and the armature winding placed on the rotor of the machine.

Both windings are connected to a dc supply source. In a conventional dc shunt motor, the two windings are connected in parallel across the dc supply mains, but in a servo application, the windings are driven by separate dc supplies, as illustrated in Figs. 21.37 and 21.38.

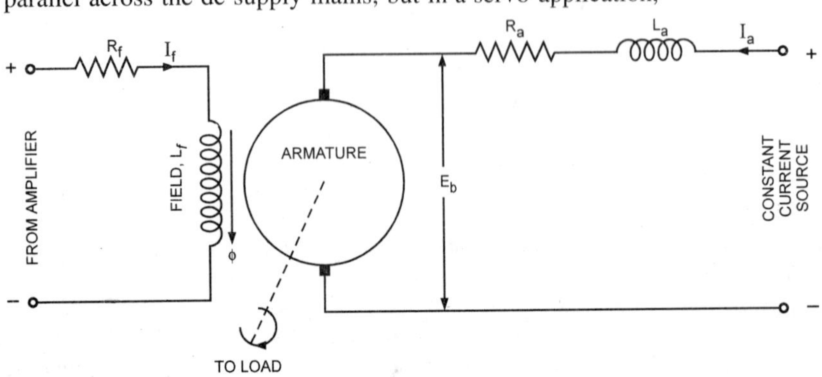

Fig. 21.37 *Schematic Diagram For Field Controlled DC Motor*

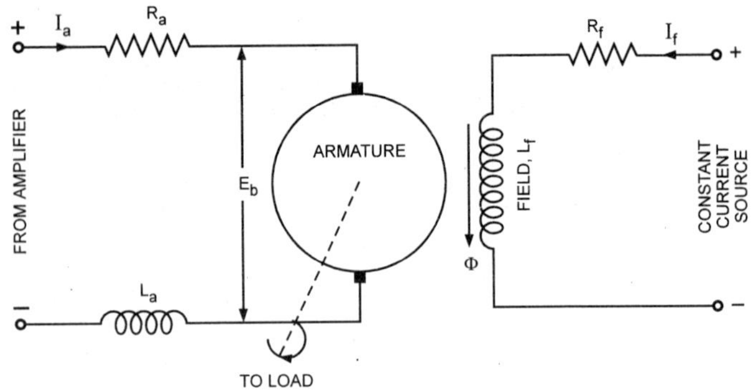

Fig. 21.38 *Schematic Diagram For Armature Controlled DC Motor*

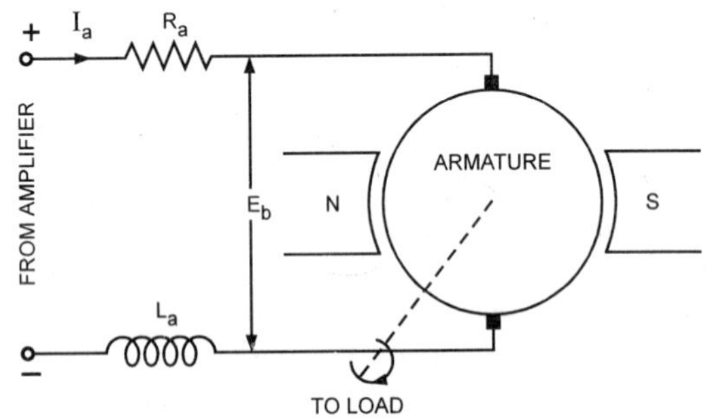

Fig. 21.39 *Schematic Diagram of a Permanent Magnet Armature Controlled DC Motor*

Fig. 21.37 gives the schematic diagram for a field controlled dc motor. In this arrangement the field is excited by the amplified error signal and the armature winding is energized from a constant current source. Torque developed is proportional to the field current up to saturation.

This arrangement is employed in small servomotors because its dynamic response is slower than the armature controlled dc motor, illustrated in Fig. 21.38, owing to the longer time constant of the highly inductive field circuit.

The direction of rotation of motor is reversed if polarity of the field is reversed. Fig 21.38 gives the schematic diagram for the armature controlled dc motor. The motor armature is energized by the amplified error signal and the field is supplied from a constant current source. The field of this motor is normally operated well beyond the knee of the saturation characteristic to keep the torque less sensitive to slight change in field current. High field flux density also increases the torque sensitivity of the motor, because $T \alpha \Phi I_a$, for the small change in armature current. Dynamic response is faster than that of field controlled motor because the armature circuit is essentially a resistive circuit and has shorter time constant. If the polarity of the error signal is reversed, the motor rotates in reversed direction.

The permanent magnet motor is a fixed excitation shunt motor where the field is actually supplied by a permanent magnet as shown in Fig. 21.39. Performance is similar to that of armature controlled fixed field motor.

Comparison between armature and field controlled dc motors is given below in tabular form.

S.No.	Armature Control	Field Control
1.	Large current requirement.	Small current requirement.
2.	Damping by back emf.	Damping by motor and load.
3.	Smaller time constant because of smaller inductance of armature winding.	Longer time constant because of highly inductive field.
4.	Fast speed response to changing current.	Slow speed response to changing current.
5.	Easy to provide constant field current.	Difficult to provide constant current to armature owing to back emf.
6.	High efficiency.	Low efficiency.

21.10.2. AC Servomotors. AC servomotors, as mentioned above, are best suited for low power applications for the same reasons that ac motors are always preferred to dc motors:

1 Squirrel cage rotors are simple and rugged in construction compared to the complex armature windings found on dc machines.
2. AC servomotors have no brush-to-commutator contacts which need frequent inspection and maintenance.
3. There is no insulation around the armature conductors as there is on a dc motor, so the armature can dissipate heat much better.
4. Because the armature has no complicated insulated windings on it, its diameter can be made quite small to reduce rotor inertia; this helps prevent overshoot in a servomechanism.
5. Use of a drift-free ac amplifier in control circuitry.

An ac servomotor is basically a two-phase induction motor except for certain special design features.

There is one important difference between a standard split-phase motor and an ac servomotor. It is that the servomotor has thinner conducting bars in the squirrel cage rotor, so that the motor resistance is higher. It is necessary to have linear torque-speed characteristic, as shown by the curve II is Fig. 21.40. If a conventional induction motor with high $\frac{X}{R}$ ratio is used for servo applications, then because of the positive slope for part of the characteristic, the system employing such a motor becomes unstable. The rotor construction is usually of squirrel-cage or drag-cup* type, as illustrated in Fig. 21.41. The diameter of the rotor is kept small but long length in order to reduce inertia and thus to obtain good accelerating characteristics. Drag-cup construction is used for very low inertia applications.

Such motors normally operate on a frequency of 50 Hz or 400 Hz (for airborne systems). The stator has two distributed windings displaced from each other by 90° (electrical). The two windings are referred to as the main winding and control winding. Sometimes the main winding is called the *fixed* or *reference winding*. The main winding is supplied from a constant voltage source, $V_m = A\angle 0°$. The control winding is supplied with a variable voltage of the same frequency as the reference phase but is displaced in phase by 90° (electrical). The control phase voltage

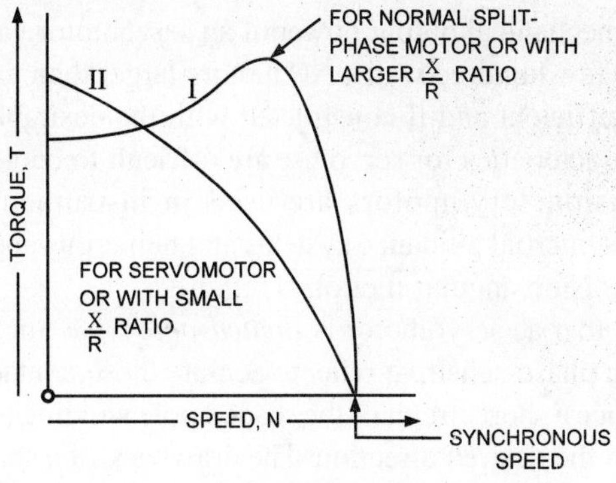

Fig. 21.40

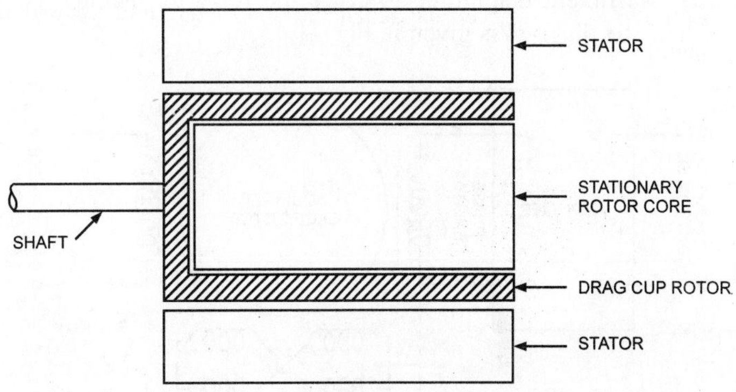

Fig. 21.41 *Drag-Cup Rotor Servomotor*

is shifted in phase by 90° from the reference phase voltage by means of phase-shifting networks included in the voltage amplifications stages of the amplifier. The motor torque gets reversed by phase reversal of the control phase voltage.

With no error signal, the rotor is at standstill. Motor rotation is produced in such a direction so as to reduce the error signal, and the motor ceases to rotate when a null or zero error signal is produced by the control winding. The output torque of the motor is roughly proportional to the applied control voltage, and the direction of torque is determined by the polarity of the control voltage. Whenever the 'error' of a simple proportional servo reverses its direction the actuating signal reverses the polarity and the motor reverses its direction. The fixed or main winding is excited from a source, which is shifted in phase by 90° using a capacitor, and the variable voltage connected to the servo amplifier through a transformer. Schematic diagram is shown in Fig. 21.42.

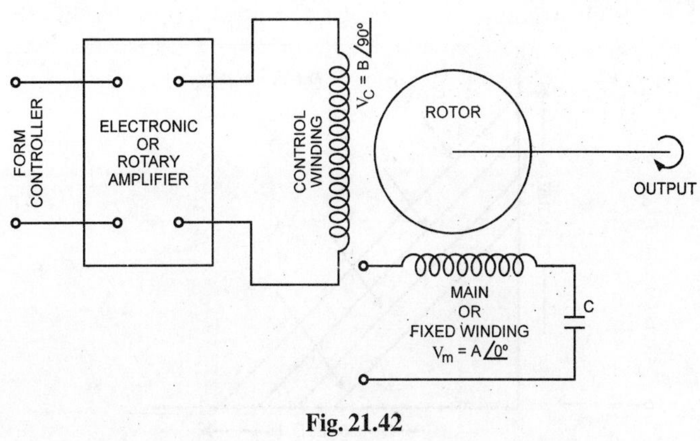

Fig. 21.42

* In a drag-cup rotor, only a light cup is rotating and the rotor core is stationary, therefore, inertia is quite small.

The mechanical output power of ac servomotor varies from 2 watts to few hundred watts. AC motors larger than this ratings are too inefficient and if constructed with the desirable torque-speed characteristics for servo use are difficult to cool.

Precision servomotors are used in instrument servos, computers, inertial guidance systems and numerous applications where precision angular motion is required.

The other ac servomotor is *shaded-pole type.* Such a motor employs a phase-sensitive relay to actuate those contacts which will produce a short circuit of the shaded pole winding to produce rotation in the desired direction. The drawback of a shaded-pole servomotor is that it responds only when the amplifier error signal is of sufficient magnitude to cause the relay to operate. The schematic diagram is given in Fig. 21.43.

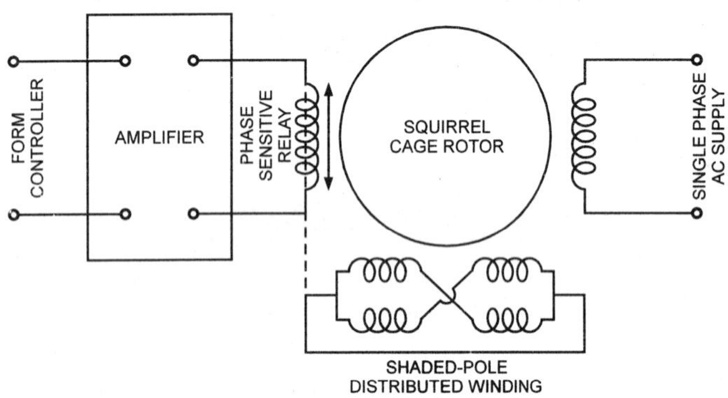

Fig. 21.43

A single-phase ac field winding is connected to the ac supply as shown. If the error signal is of sufficient magnitude as to activate the relay, one pair of shaded-pole windings is shorted and the rotor starts rotating and continues to rotate until the null is produced or the error signal becomes weak and the relay drops out. An error signal of opposite polarity actuates the phase-sensitive relay to short circuit the other pair of shaded-pole distributed windings, causing rotation in the reverse direction.

The torque-speed characteristic of a two-phase induction motor, as already mentioned, depends upon the ratio of resistance to reactance. For high resistance and low reactance, the characteristic is linear and for large ratio of X to R becomes nonlinear as shown in Fig. 21.40. The torque-speed characteristics for various control voltages are almost linear as shown in Fig. 21.44.

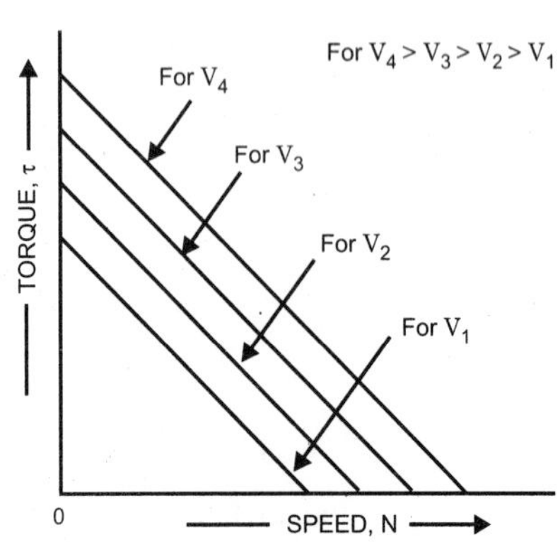

Fig. 21.44 *Torque-Speed Characteristics For Different Voltages*

21.10.3. Three-Phase AC Servomotor. Normally a 3-phase squirrel cage induction motor is a highly nonlinear coupled circuit device. Recently, it has been possible to operate a three-phase squirrel cage induction motor as a linear decoupled machine by making use of a controlled method, known as *vector* or *field oriented control.* In this control method, the currents supplied to the motor are controlled in a way to decouple its torque and flux as in a dc machine. This results in high-speed response and high-torque response. Three-phase squirrel cage induction motors with vector control are being increasingly employed as servomotors for applications in high power servo systems like dc servomotors.

21.11 AC TACHOMETERS

An ac tachometer is a 2-phase induction motor having two windings displaced from each other by 90° (electrical). The two windings are referred to as the main or input winding and the auxiliary or output winding. The main winding is supplied from the carrier frequency, while the auxiliary or output winding is left open circuited as shown in Fig. 21.45. The rotor construction is of drag-cup type so as to keep inertia low. Voltage across output winding will be proportional to rotor speed while its phase shift will be close to 90° (electrical). Output voltage-speed relationship is linear with very small or very large rotor $\frac{X_2}{R_2}$ ratio. Low $\frac{X_2}{R_2}$ provides low speed sensitivity but a wide linear range and vice versa. An intermediate value of $\frac{X_2}{R_2}$ usually meets most of the needs. The phase shift is slightly less than 90° but is quite insensitive to speed. High precision construction is required to maintain concentricity and avoid any direct coupling between the two-phase windings. Pick-up from stray fields is avoided by providing soft-iron shields.

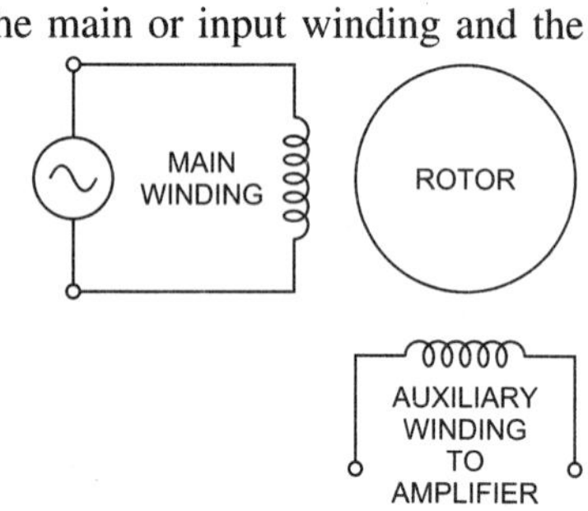

Fig. 21.45 *Schematic Diagram of a 2-Phase AC Tachometer*

AC tachometers are commonly used in 400 Hz control systems which need feedback signal of carrier frequency and of magnitude proportional to speed.

21.12 STEPPER MOTORS

There are many industrial situations which call for precise positioning of an object or precise control of speed without having to resort to closed-loop feedback. Stepper motors are ideally suited to such tasks. A stepper motor rotates a precise angular distance, one step, for each pulse that is delivered to its drive circuit. For example, if a particular stepper motor has a step angle of 7.5° per pulse and the shaft is to be turned through 225° in order to place an object in a certain target position, then it is only necessary to supply 225/7.5 i.e. 30 pulses to the motor's drive circuit. Because the step pulses can be counted digitally and stopped when the desired number have been delivered, stepping motors lend themselves to digital control by a programmable controller or on-line microcomputer. These motors can also provide precise open-loop speed control, since their rotational speed is determined solely by the step pulse frequency, independent of load.

A *stepper motor*, sometimes also called the *stepping motor* or *step motor* is a brushless synchronous motor which divides a full rotation into a number of steps. Unlike a brushless dc motor which rotates continuously when a fixed dc voltage is applied to it, a step motor rotates in discrete steps angles. The stepper motors are manufactured with steps per revolution of 12, 24, 72, 144, 180 and 200 resulting in stepping angles 30, 15, 5, 2.5, 2 and 1.8 degrees per step. The step angles can be as small as 0.72° or as large as 90°.

A stepper motor has the extraordinary ability to operate at very high stepping rate (up to 20,000 steps per second in some motors). When the pulse rate is high, the shaft rotation seems continuous. Operation at high speeds is known as 'slewing'. When in the slewing range, the motor usually emits an audible whine having a fundamental frequency equal to the stepping rate.

In case the stepping rate is increased too fast, the motor loses synchronism and stops. Same things occurs at high speed operation, if command pulses are suddenly stopped instead of being slowed progressively.

Stepper motors are designed to operate for long periods with the rotor held in a fixed position and with rated current flowing in the stator windings. Hence there is no stalling problem in such motors.

Explosive growth of computer industry is mainly responsible for great demand of stepper motors nowadays. They are very popular due to the fact that they can be controlled directly by microprocessors, computers and programmable controller.

There are two main advantages of using stepper motors. One is their compatibility with digital systems and secondly no sensors are required for sensing of position and speed as these are directly obtained by counting input pulses and periodic counting if information about speed is required.

Computer controlled stepper motors are a type of motion-control positioning system. They are typically digitally controlled as part of an open-loop system for use in holding or positioning applications.

In the field of lasers and optics they are frequently employed in precision positioning equipment such as linear actuators, linear stages, rotation stages, gonimeters, and mirror mounts. Other applications are in packaging machinery, and positioning of valve pilot stages for fluid control systems.

Commercially, stepper motors are used in floppy disk drives, flatbed scanners, computer printers, plotters, slot machines, image scanners, compact disc drives, intelligent lighting, camera lenses, CNC machines and more recently they are used in 3D printers.

The range of applications of such motors is increasing as these are becoming available in large power ratings and with reducing cost.

Step Angle. The angle through which the motor turns or the shaft moves for each pulse is known as the *step angle* which is expressed in degrees. The resolution of positioning of a stepper motor is decided by the step angle. Smaller the step angle, greater the number of steps per revolution and higher the resolution of positioning of the motor. Higher the resolution, greater is the accuracy of positioning of objects by the motor.

Step angle can be expressed in terms of the number of stator phases and the number of rotor teeth

$$\text{i.e. step angle, } \alpha = \frac{360^\circ}{\text{Number of phases} \times \text{number of rotor teeth}}$$

$$= \frac{360}{m N_r} \qquad \text{...(21.6)}$$

Alternatively the step angle can be expressed in terms of the rotor and stator poles (teeth) N_r and N_s respectively.

$$\text{i.e. Step angle } \alpha = \frac{N_s - N_r}{N_s N_r} \times 360^\circ \qquad \text{...(21.7)}$$

21.12.1. Stepper Motor System. A stepper motor system consists of three basic elements, often combined with some type of user interface (host computer, PLC or dumb terminal):

1. Indexers. The indexer (or controller) is a microprocessor capable of generating step pulses and direction signals for the driver. In addition, the indexer is typically required to perform many other sophisticated command functions.

2. Drivers. The driver (or amplifier) converts the indexer command signals into the power necessary to energize the motor windings. There are numerous types of drivers, with different voltage and current ratings and construction technology. Not all drivers are suitable to run all motors, so when designing a motion control system the driver selection process is critical.

3. Stepper Motors.

21.12.2. Operating Principle of Stepper Motor. Stepper motors work on the principle of *electromagnetism.* There is a soft iron or magnetic rotor shaft surrounded by the electromagnetic stators. The rotor and stator have poles which may be teethed or not depending upon the types of stepper motor. When the stators are energized the rotor moves to align itself along with the stator (in case of a variable reluctance stepper motor). This way the stators are energized in a sequence to rotate the stepper motor.

The principle of operation of a stepper motor can be easily explained by considering a series of solenoids or electromagnets arranged in a circle and an iron bar (rotor), as illustrated in Fig. 21.46. When these solenoids are energized in sequence, the

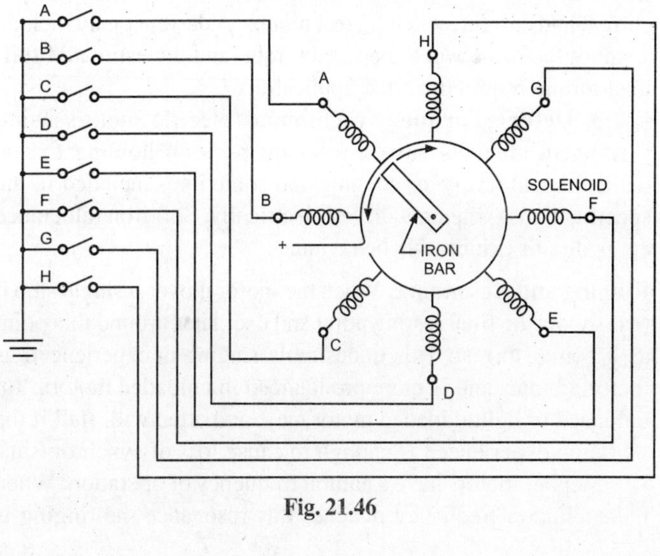

Fig. 21.46

mmf developed in them interact with the iron bar (or rotor) and causes it to turn either in clockwise or counter-clockwise direction, depending upon the switching sequence.

21.12.3. Constructional Features and Working. The construction of the stepper motor is quite simple and consists of a slotted stator equipped with a multi-pole, multi-phase winding that is not unlike the windings of the conventional machines and a rotor structure that carries no winding. They typically use three- and four-phase windings, the number of poles being determined by the required angular change per input pulse. The rotors are either of the *permanent magnet type* or the *variable reluctance* type. Stepper motors operate with an external drive logic circuit; as a train of pulse is applied to the input of the drive circuit, the circuit delivers appropriate currents to the stator windings of the motor to make the axis of the air-gap field around in coincidence with the input pulses. Depending upon the pulse rate and load torque, including inertia effects, the rotor follows the axis of air-gap magnetic field by virtue of the permanent magnet torque and/or the reluctance torque.

By construction the stepper motors come into three broad classes:

1. Permanent magnet stepper motor.
2. Variable reluctance stepper motor.
3. Hybrid stepper motor.

21.12.4. Stepper Motor Torques

1. Pull-in Torque. This is a measure of torque developed by a stepper motor when it is operated without an acceleration state. At low speeds the stepper motor can synchronise itself with an applied step frequency, and this pull-in torque must overcome friction and inertia. It is important to make sure that the load on the motor is frictional rather than inertial as the friction reduces any unwanted oscillations. The pull-in curve defines an area called the start/stop region. Into this region, the motor can be started/stopped instantaneously with a load applied and without loss of synchronism.

2. Pull-out Torque. The pull-out torque of the stepper motor is measured by accelerating the motor to the desired speed and then increasing the torque loading until the motor stalls or misses steps. This measurement is taken across a wide range of speeds and the results are used to generate the stepper motors dynamic performance curve. This curve is affected by drive voltage, drive current and current switching techniques. A designer may include a safety factor between the rated torque and the estimated full-load torque required for the application.

3. Detent Torque. Synchronous electric motors using permanent magnets have a resonant position holding torque (called detent torque or cogging and sometimes included in the specifications) when not driven electrically. Soft iron reluctance cores do not exhibit this behaviour.

Ringing and Resonance. When the motor moves a single step it overshoots the final resting point and oscillates around this point as it comes to rest. This undesirable ringing is experienced as motor vibrates and is more pronounced in unloaded motors. An unloaded or lightly loaded motor may, and often will, stall if the vibration experienced is enough to cause loss of synchronism.

Stepper motors have a natural frequency of operation. When the excitation frequency matches this resonance the ringing is more pronounced, steps may be missed, and stalling is more likely.

21.12.5. Permanent Magnet Stepper Motor. The representative structure of a permanent magnet type stepper motor having 4 stator poles and 6 rotor poles producing a full step angle of 30° is shown in Fig. 21.47. The end terminals of all the windings are brought out to the terminal box for dc excitation. The rotor of such a motor has even number of poles made of high retentivity steel alloy (Alnico). Both rotor and stator may employ salient or non-salient pole construction. Usually the stepper motors having small stepping angles are of non-salient pole construction. In the motor illustrated in Fig. 21.47, the stator pole windings are wrapped in such a way that closure of any winding's control switch causes that winding's pole to become magnetically north. If the six pole rotor is permanently magnetized as shown, and if only a single stator winding is energized at one time, then the rotor will always orient so that the nearest rotor south pole is aligned with the energized stator pole.

For instance if switch A is closed, energizing stator pole winding A, the rotor will align as shown in the figure. If switch A is opened and the switch B is then closed, so that only winding B is energized, the rotor will step 30° clockwise. This step will incur due to the tendency of the south pole nearest the winding B to align with the north stator pole.

If the switch B is then opened and switch C subsequently closed, the rotor will make a second 30° clockwise step, due to attractive tendency of the north stator pole (C) for the rotor pole nearest it. It is not hard to see that when the switching sequence ABCD is repeated continually, the rotor shaft will continue to turn clockwise, one 30° step per switch closure. If the switching sequence is reversed, to A D C B, the rotor shaft would turn in opposite i.e. counter-clockwise direction.

Many stepping motor designs allow the selection of full steps or half steps. This is accomplished by altering the switching sequence to simultaneously energize two stator coils during every second timing interval. Simultaneous energization of two stator windings creates the effects of moving the stator's magnetic north to a position between the two poles.

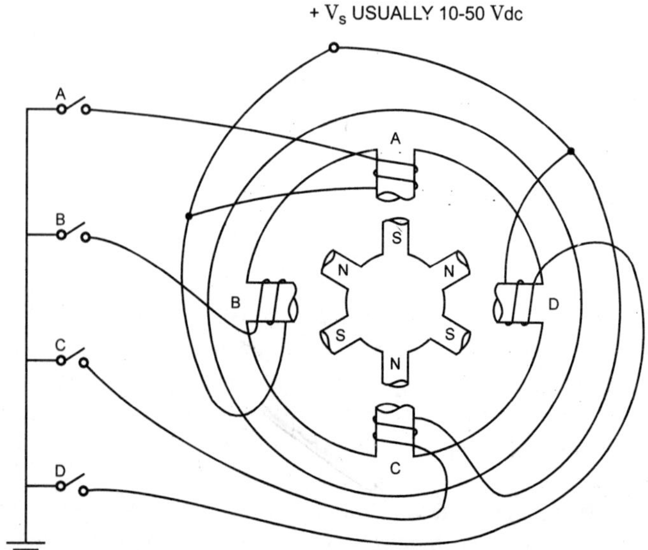

Fig. 21.47 *4-Stator Pole, 6 Rotor Pole Permanent Magnet Stepper Motor*

In reality, stator winding switching is accomplished by transistors rather than mechanical switching. A typical stepper motor needs 1 to 10 A of winding current in order to develop its rated torque. Therefore, the switching transistors must be high power devices.

Permanent magnet stepper motor do not require external exciting current and, therefore, power requirement is low but detent torque is higher as compared to a variable reluctance motor. These motors have large inertia and, therefore, smaller acceleration. However, such motor develops more torque per ampere of stator current than a variable reluctance motor. The major drawback of such motors is that as small permanent magnet rotor cannot be constructed with large number of poles and, therefore, the step size in such motors is relatively large ranging from 30° to 90°. However, disc type permanent magnet stepper motors are available which are magnetized axially to provide a small step size and low inertia. Permanent magnet stepper motor can have maximum step rate of 300 pulses per second while variable reluctance stepper motor may have as high as 1,200 pulses per second.

21.12.6. Variable Reluctance (VR) Stepper Motor. The stator of a variable reluctance stepper motor is similar to that of a permanent magnet stepper motor. It has no permanent magnet on the rotor and the rotor employed is a ferromagnetic multi-toothed one.

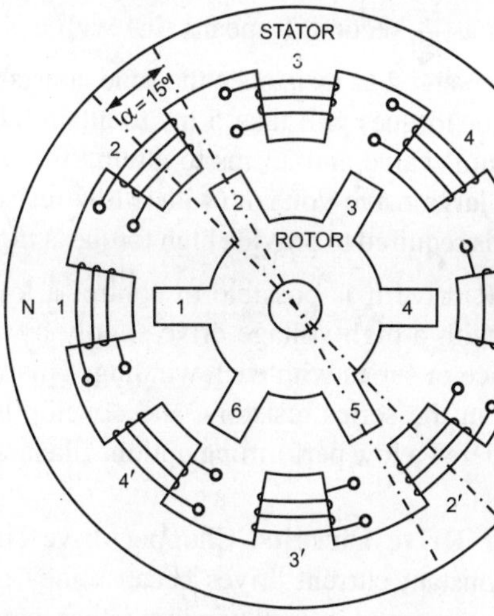

Fig. 21.48 *4-Phase, 8-Pole Stator, 6-Tooth Rotor Variable Reluctance Stepper Motor*

The torque is developed due to large differences in magnetic reluctances that exist between the direct and quadrature axis. The stationary field developed by the direct current in some stator coil tends to develop a torque which causes the rotor to move to the position where the reluctance of the flux path is minimum.

A 4-phase, 8-stator pole, 6-tooth rotor variable reluctance stepper motor is shown schematically in Fig. 21.48. The stator is provided with 4-phase winding and the rotor gives the effects of 6-poles.

Stepping angle is given as, irrespective the type of stepper motor (i.e. permanent magnet type or reluctance type)

$$\alpha = \frac{360°}{\text{Number of phases} \times \text{number of poles or rotor}} = \frac{360}{mp} \quad ...(21.8)$$

In the above case

$$\alpha = \frac{360°}{4 \times 6} = 15°.$$

When the phase 1-1' is energized, keeping other coils unexcited, the rotor aligns as shown in the figure. If now the phase 1-1' is switched off and phase 2-2' is switched on to develop north pole at coil 2 and south pole at coil 2', a reluctance torque is produced to align the axes of rotor poles 2-5 with the axis of stator poles of phase 2-2'. The rotor advances 15° counter-clockwise. Successive excitation of phases 3, 4, 1' yields a total advance of 60° i.e. on rotor pole pitch. Six-times repetition of this switching sequence completes one revolution of the shaft.

Rotor can be made to rotate in clockwise direction by reversing the switching sequence i.e. in sequence 1, 4', 3', 2', 1'.

Multistack VR Stepper Motor. The VR stepper motor discussed above is single-stack VR stepper motor. Multistack VR stepper motors, that provide smaller step angles, are also available. A multistack VR motor may be considered to be made of m identical single stack VR stepper motors with their rotors mounted on a single shaft. The number of poles (or teeth) is same on the stators and rotors and therefore tooth pitch is same. In case of m-stack VR stepper motor, the stator poles in all m stacks are aligned, but the rotor poles are displaced by $\frac{1}{m}$ of the pole pitch angle from one another. All the stator pole windings in a given stack are energized simultaneously and, therefore, the stator winding of each stack forms one phase. Thus the number of phases equal to number of stacks.

Cross-sectional view of a multistack VR stepper motor is shown in Fig. 21.49. The motor is divided into a number of magnetic isolated sections, known as *stacks*, along its axial length. In case of m-stack VR motor having number of rotor poles (or teeth) equal to N_r, the step angle is given as

$$\alpha = \frac{360°}{m N_r} \quad ...(21.9)$$

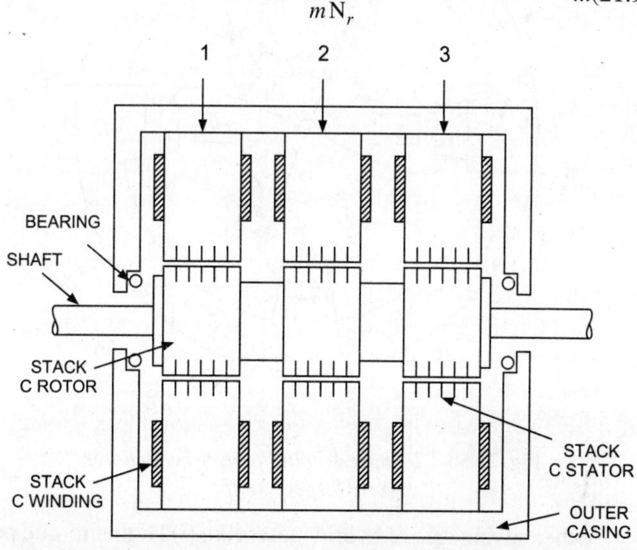

Fig. 21.49 *Cross-Sectional View of Multistack VR Stepper Motor*

Three-stack motors are most common although motors with up to seven stacks (phases) are available. They have the step angles in the range of 2° to 15°. For instance, in a 3 stack VR stepper motor having 8 rotor teeth, the step angle is $\frac{360°}{3 \times 8} = 15°$.

VR stepper motors (single-stack and multistack, both) have the advantages of (*i*) high torque to inertia ratio, (*ii*) fast dynamic response (*iii*) high acceleration rate (*iv*) efficient cooling arrangement (*v*) simple and robust construction due to absence of brushes and (*vi*) lower cost.

21.12.7. Hybrid Stepper Motor. This is in fact a permanent magnet stepper motor with constructional features of toothed rotor used in variable reluctance stepper motor. A 2-phase, 4-pole stator, 6-tooth rotor hybrid stepper motor is shown in Fig. 21.50. Stepping angle is $\frac{360}{2 \times 6} = 30°$.

When phase 1-1' is switched on, keeping phase 2-2' unexcited, the rotor aligns as shown in Fig. 21.50. If now the phase 1-1′ is switched off and phase 2-2' is closed such that lower pole becomes south pole and upper pole becomes north, this would cause rotor to turn in counter-clockwise direction by 30° and the rotor will get locked into the new position. The rotor will turn in opposite i.e. clockwise direction if the phase 2-2' were oppositely excited.

The main advantage of the hybrid motor is that if the stator excitation is removed, the rotor continues to remain locked into the same position, as before removal of excitation. This is due to the reason that the rotor is prevented to move in either direction by the torque because of the permanent magnet excitation.

Hybrid type stepper motor is used in applications where stepping angle is small (e.g.1.5, 2.5 degrees).

The variable reluctance stepper motor's step angles are small, making possible finer resolution than that is obtainable with the permanent magnet type. Maximum stepping rates usually are higher than in permanent magnet stepper motor.

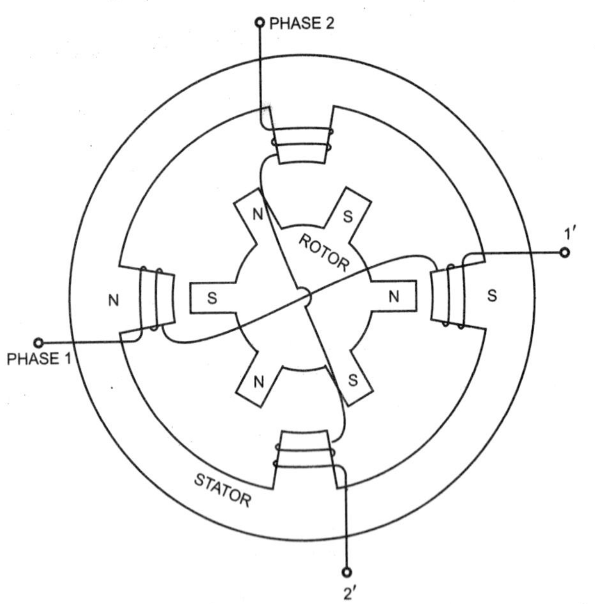

Fig. 21.50 *2-Phase 4-Pole Stator, 6-Tooth Rotor Hybrid Stepper Motor*

Other advantages of hybrid motors are (*i*) Higher torque per unit volume (higher than that of reluctance motor) (*ii*) Availability of detent torque due to permanent magnet (*iii*) Less tendency to resonate (*iv*) High efficiency at lower speeds and lower stepping rates.

Hybrid stepper motors have the drawbacks of (*i*) higher inertia and weight because of rotor magnet (*ii*) higher cost than that of VR stepper motor and (*iii*) affecting performance due to change in magnetic strength.

21.12.8. Stepper Motor Drive Circuits. The performance of a stepper motor strongly depends on the *drive circuit*. Torque curves may be extended to greater speeds if the stator poles can be reversed more quickly, the limiting factor being winding resistance R and inductance L. In order to overcome the inductance L and switch the winding quickly, the drive voltage is to be increased. This leads further to the necessity of limiting the current that these high voltages may otherwise induce.

1. L/R Drive Circuits. Such circuits are also referred to as *constant voltage drives* because a constant voltage (positive or negative) is applied to each winding to set the step positions. However, it is the winding current, not voltage that applies to the stepper motor shaft. The current I in each winding depends on the applied voltage V, winding resistance R and winding inductance. The winding resistance R determines the maximum current from Ohm's law i.e. $I = \frac{V}{R}$. The inductance L determines the maximum rate of change of the current in the winding according to the relation $\frac{di}{dt} = \frac{V}{L}$. Thus in L/R drive, the maximum speed of a stepper motor is restricted by its winding inductance since at some speed the voltage will be changing faster than the current I can keep up. The time constant of the winding is given as $\frac{L}{R}$ second. It means that with a winding of 10 mH inductance and 2 Ω resistance the time constant is 5 ms i.e. the current (or torque) will take 5 ms to attain 0.632 of its final (or maximum) value and 23 ms to attain 99% of its final value. Hence, a large drive voltage (with a low resistance and low inductance) is required to provide high torque at high speeds.

With an L/R drive it is possible to control a low voltage resistive motor with a high voltage drive simply by adding an external resistance in series with each winding. This will cause waste of power in the series resistors, and develop heat. It is, therefore, considered a low performing option, albeit simple and cheap.

2. Chopper Drive Circuits. Chopper drive circuits are referred to as constant current drives because they generate a somewhat constant current in each winding rather than applying a constant voltage. On each new step, a very high voltage is applied to the winding initially. This causes the current in the winding to rise quickly since $dI/dt = V/L$ where V is very large. The current in each winding is monitored by the controller, usually by measuring the voltage across a small sense resistor in series with each winding. When the current exceeds a specified current limit, the voltage is *turned off* or "*chopped*", typically using power transistors. When the winding current drops below the specified limit, the voltage is *turned on* again. In this way, the current is held relatively constant for a particular step position. This requires additional electronics to sense winding currents, and control the switching, but it allows stepper motors to be driven with higher torque at higher speeds than L/R drives. Integrated electronics for this purpose are widely available.

21.12.9. Phase Current Waveforms. A stepper motor is a polyphase synchronous motor, and it is ideally driven by sinusoidal current. A full step waveform is a gross approximation of a sinusoid, and is the reason why the motor exhibits so much vibration. Various drive techniques have been developed to approximate a sinusoidal drive waveform: these are half stepping and microstepping.

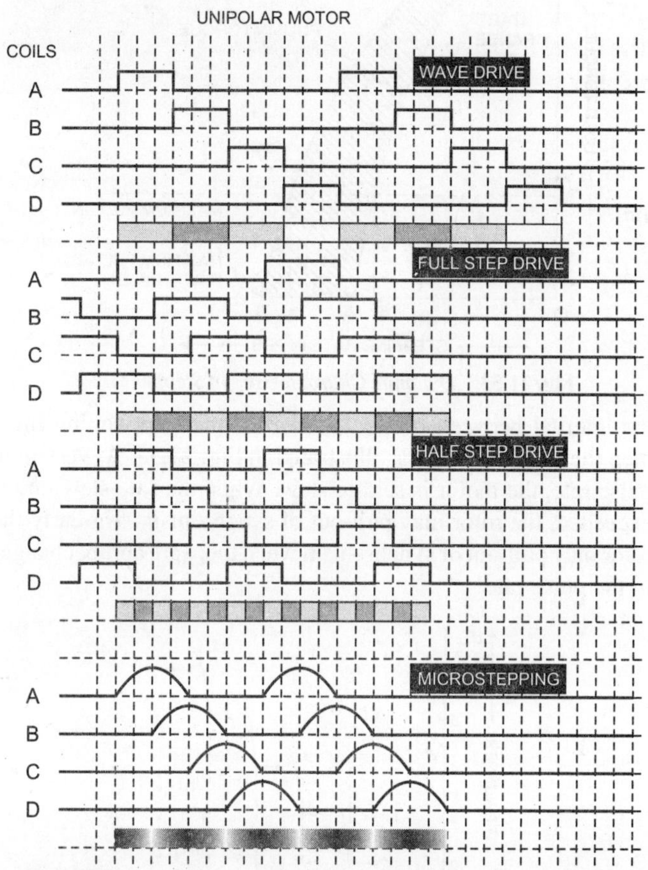

Fig. 21.51 *Different Drive Modes Showing Coil Current on a 4-Phase Unipolar Stepper Motor*

1. Wave Drive or Full-Step Drive (One Phase On). In this drive method only a single phase is activated at a time. It has the same number of steps as the full step drive, but the motor will have significantly less than rated torque. It is rarely used. The animated figure shown above is a wave drive motor. In the animation, rotor has 25 teeth and it takes 4 steps to rotate by one teeth position. So there will be 25 × 4 = 100 steps per full rotation and each step will be 360/100 = 3.6 degrees.

Full-Step Drive (Two Phase On). This is the usual method for full step driving the motor. Two phases are always *on* so the motor will provide its maximum rated torque. As soon as one phase is turned *off*, another one is turned *on*. Wave drive and single-phase full step are both one and the same, with same number of steps but difference in torque.

Half Stepping. In *half stepping*, the drive alternates between two phases *on* and a single phase *on*. This increases the angular resolution. The motor also has less torque (approx 70%) at the full-step position (where only a single phase is *on*). This may be mitigated by increasing the current in the active winding to compensate. The advantage of half stepping is that the drive electronics need not change to support it. In animated figure shown above, if we change it to half stepping, then it will take 8 steps to rotate by 1 teeth position. So there will be 25 × 8 = 200 steps per full rotation and each step will be 360/200 = 1.8°. Its angle per step is half of the full step.

Microstepping. What is commonly referred to as *microstepping* is often "sine cosine microstepping" in which the winding current approximates a sinusoidal ac waveform. Sine cosine microstepping is the most common form, but other waveforms can be used. Regardless of the waveform used, as the microsteps become smaller, motor operation becomes more smooth, thereby greatly reducing resonance in any parts the motor may be connected to, as well as the motor itself. Resolution will be limited by the mechanical stiction, backlash, and other sources of error between the motor and the end device. Gear reducers may be used to increase resolution of positioning.

Step size repeatability is an important step motor feature and a fundamental reason for their use in positioning.

Example. Many modern hybrid step motors are rated such that the travel of every full step (say 1.8 degrees per full step or 200 full steps per revolution) will be within 3% or 5% of the travel of every other full step, as long as the motor is operated within its specified operating ranges. Several manufacturers show that their motors can easily maintain the 3% or 5% equality of step travel size as step size is reduced from full stepping down to 1/10 stepping. Then, as the microstepping divisor number grows, step size repeatability degrades. At large step size reductions it is possible to issue many microstep commands before any motion occurs at all and then the motion can be a "jump" to a new position.

21.12.10. Ratings and Specifications of Stepper Motors. Stepper motors nameplates typically give only the winding current and occasionally the voltage and winding resistance. The rated voltage will produce the rated winding current at dc but this is mostly a meaningless rating, as all modern drives are current limiting and the drive voltages greatly exceed the motor rated voltage.

A stepper's low speed torque will vary directly with current. How quickly the torque falls off at faster speeds depends on the winding inductance and the drive circuitry it is attached to, especially the driving voltage.

Steppers motors should be sized according to published torque curve, which is specified by the manufacturer at particular drive voltages or using their own drive circuitry.

Stepper motors adapted to harsh environments are often referred to as IP65 rated.

21.12.11. Characteristics of Stepper Motors

Static Characteristics. Torque-displacement and torque-current characteristics, known as static characteristics, are shown in Fig. 21.52.

Dynamic Characteristic. The torque-pulse rate characteristic, known as *dynamic characteristic*, is shown in Fig. 21.53. This characteristic provides the relation between electromagnetic torque τ and stepping rate in pulses per second (pps). A stepper motor is usually described by two characteristic curves, known as pull-in torque and pull-out torque curves. Pull-in torque characteristic indicates the maximum stepping rate at which the motor can start, synchronise, stop or reverse for different values of load torque. Pull-out torque characteristic indicates

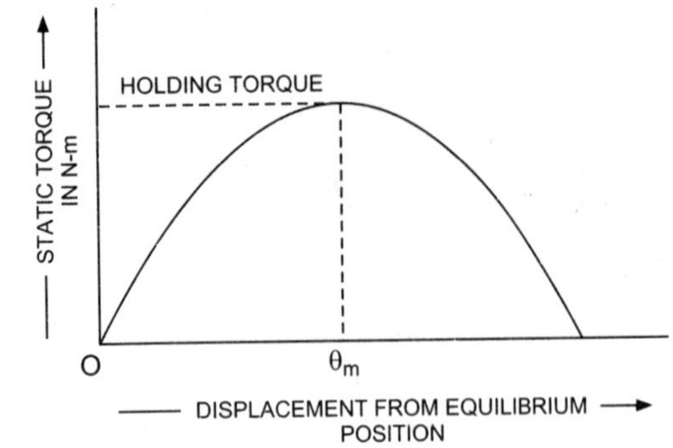

(*a*) *Torque-Displacement Characteristic*

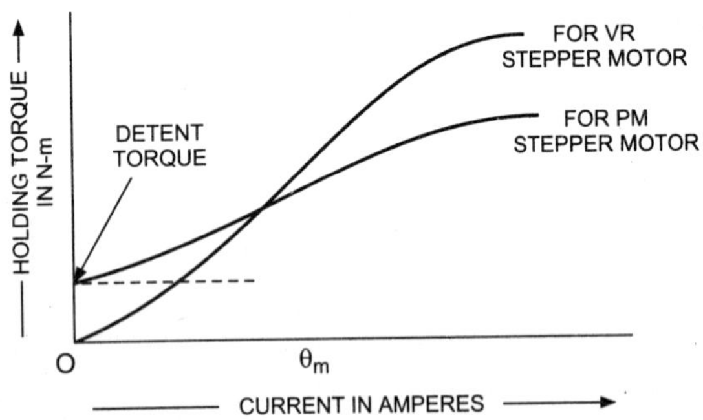

(*b*) *Torque-Current Characteristic*

Fig. 21.52 *Static Characteristics of Stepper Motors*

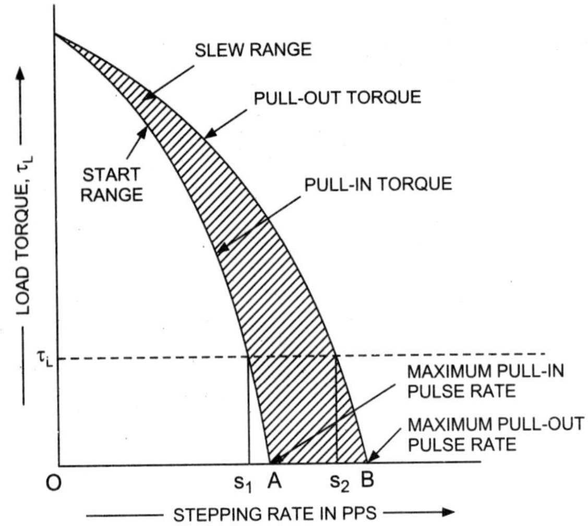

Fig. 21.53 *Dynamic Characteristic of Stepper Motor*

the maximum stepping rate at which the motor can run for different values of load torque if already synchronised, but it cannot start, stop or reverse on command at this rate. For instance, for a load torque of τ_L, the motor can start, synchronise, stop or reverse without missing a pulse, if the pulse rate does not exceed s_1. For load torque τ_1, after starting and synchronisation, the stepping rate can be increased up to s_2 without missing a step or without losing synchronism. In case the stepping rate exceed s_2, the motor will lose synchronism. Thus, the shaded area, area between pull-in torque and pull-out torque curves represents, for different values of load torque, the range of stepping rate which the motor follows without losing synchronism provided that it has already been started and synchronised. This area is known as the slew range. Slew range is useful for speed control and it is not suitable for position control. The area between the zero-pulse rate vertical line and pull-in torque curve is known as the start range. In start range the load position follows the pulses without losing steps. The slew range is one in which the load velocity follows the pulse rate without losing a step, but cannot start, stop or reverse on command. Maximum slew rate increases with the decrease in load torque. Beyond a certain pulse rate, the rotor cannot follow the command and would begin to miss pulses. Step skipping may also occur if motor oscillation amplitude about the locking position is too large.

21.12.12. Operational Features of Stepper Motors. The stepper motor's performance can be visualised from Fig. 21.54. Every time a pulse is supplied to the motor, the rotor steps forward or backward by a definite angle, known as the step angle. It is seen that the rotor overshoots its final position and comes to rest after series of oscillations. This performance is true for low stepping rate. With the increase in pulse rate, the average velocity of the rotor increases and when the pulse rate becomes very high, i.e. the interval between pulses falls below the rotor setting time, then the rotor behaves, as illustrated in Fig. 21.55. At very high pulse rate, the motor is said to slew. When the safe slew rate is exceeded, the rotor may pull-out of synchronism. Similarly the rotor may pull-out of synchronism when there are abrupt changes in the pulse rate.

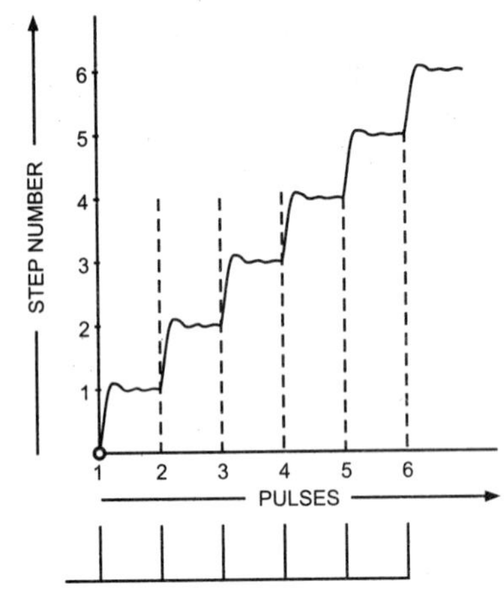

(*a*)

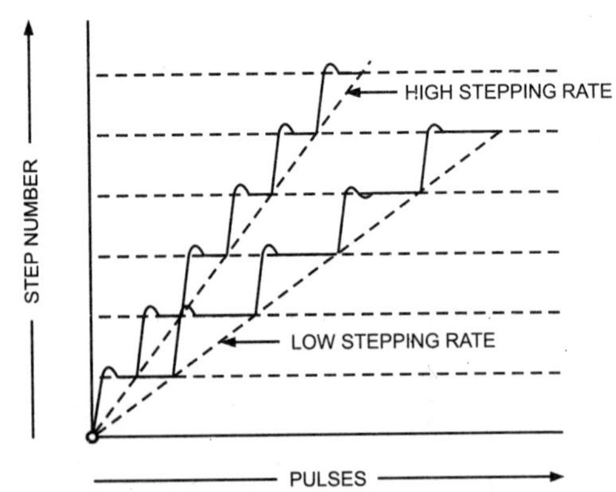

(*b*)

Fig. 21.54

For applications such as machine-tool control, it is necessary to achieve a high slewing rate to start successive machining

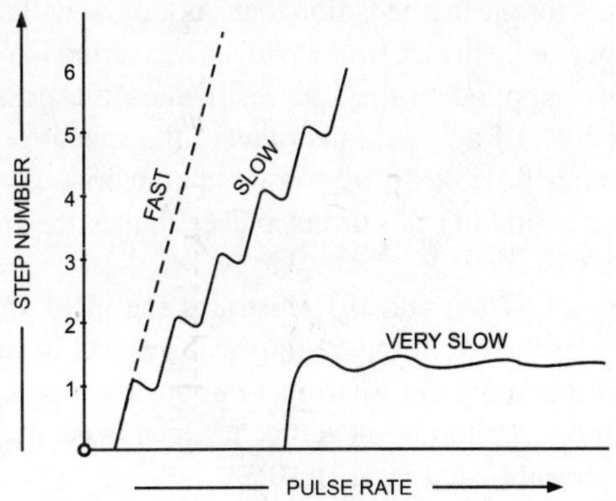

Fig. 21.55

cycles, and the tendency to error or stalling can be eliminated by incorporating the stepper motor into a closed-loop control system.

Example 21.8. Calculate the stepping angle for a 3-stack, 16–tooth variable reluctance motor.

Solution: Number of stacks or phases, $n = 3$

Number of teeth or poles, $p = 16$

$$\text{Stepping angle, } \alpha = \frac{360}{np} = \frac{360}{3 \times 16} = 7.5^\circ \textbf{ Ans.}$$

21.13 SERVOMOTORS vs STEPPER MOTORS

A servomotor consumes power as it rotates to the command position and then rests but stepper motors run warm to the touch because they continue to consume power to lock in and hold the commanded position.

Servomotors are generally used as a high performance alternative to the stepper motor. Stepper motors have some inherent ability to control position, as they have built-in output steps. This often allows them to be used as an open-loop position control, without any feedback encoder, as their drive signal specifies the number of steps of movement to rotate, but for this the controller needs to know the position of stepper motor on power up. So, on first power up, the controller will have to activate the stepper motor and turn it to known position e.g. until it activates an end limit switch. This can be observed when switching on an inkjet printer; the controller will move the inkjet carrier to the extreme left and right to establish the end positions. A servomotor will immediately turn to whatever angle the controller instructs it to, regardless of the initial position at power up.

The lack of feedback of a stepper motor limits its performance, as the stepper motor can only drive a load that is well within its capacity, otherwise missed steps under load may lead to position errors and the system may have to be restarted or calibrated. The encoder and controller of a servomotor are an addition cost, but they optimise the performance of the overall system (for all of speed, power and accuracy) relative to the capacity of the basic motor. With larger systems, where a powerful motor represents an increasing proportion of the system cost, servomotors have the advantage.

Many applications, such as laser cutting machines, may be offered in two ranges, the low-priced range stepper motors and the high-performance range using servomotors.

21.14 SWITCHED RELUCTANCE MOTOR

The increasing availability and progressive improvement of power semiconductor devices as well as intensive computer aided study of the geometry of conventional reluctance motors for optimum torque production have led to the introduction of yet another type of drive called the *switched reluctance motor drive.* It is becoming a serious competitor to converter supplied dc and ac variable speed drive systems. The switched reluctance motor is being considered, nowadays, for applications ranging from low power servomotors to high power traction drives. Motors of power ratings varying from 4 to 22 kW are commercially available at present for many applications.

Construction. The switched reluctance motor (SRM) has both salient pole stator and rotor like a variable reluctance stepper motor. Figure 21.56 shows a one phase winding of a 4-phase switched reluctance motor having 8 poles on stator and 6 poles on rotor. While the rotor has no windings, each stator pole has a concentrated winding around it and each pair of diametrically opposite coils comprise one phase of the motor. These motors are designed for applications different from those for which stepper motors are designed. A stepper motor is designed suitable for open-loop position and speed control in small power applications, where efficiency is not significant. On the other hand a switched reluctance motor (SRM) is employed in variable speed drives and designed to operate efficiently for wide range of speed and torque and needs rotor position sensing. It is, also, quite different from a standard synchronous reluctance motor, described in Art. 20.23, in two aspects. A synchronous reluctance motor has the same number of poles on stator and rotor but the rotor of a SRM has lesser poles than the stator, which is an essential feature to provide self-starting capability and

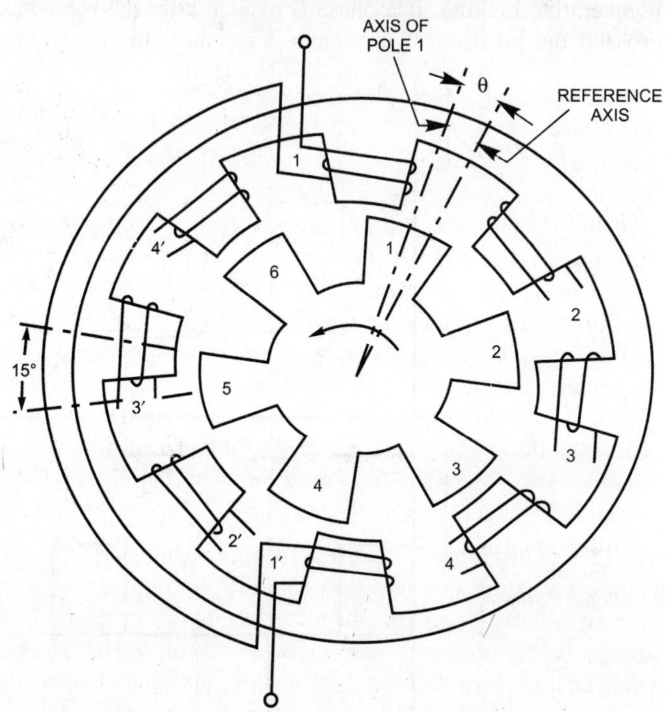

Fig. 21.56 *One Phase Winding of a 4-Phase 8/6 Pole Switched Reluctance Motor*

bidirectional control. They are also different in stator construction—the synchronous reluctance motor has a cylindrical stator with distributed winding, while the SRM has a salient pole stator with concentrated coils like a dc motor.

Though various combinations of stator and rotor pole numbes are possible in SRMs, the commonly used are 8/6 and 6/4. As already mentioned, the stator has concentrated coils and diametrically opposite coils are connected in series or parallel to provide one phase. Thus, motors with pole numbers 6/4 and 8/6 will have three and four phases respectively.

Operation. Reluctance torque is produced when a stator phase is excited by means of unidirectinal currents. This results in the magnetic attraction of an adjacent rotor pole as it tends to align into a position of minimum reluctance. When the number of stator and rotor poles differ, the sequential switching of the excitation from one set of stator poles to the next, in synchronism with the rotor position, produces an almost constant torque resulting in an uniform rotation of rotor. The synchronisation of the switching on the excitation with rotor position can be accomplished with simple rotor position feedback.

Neglecting nonlinearity of the magnetic circuit, the instantaneous torque produced in such machines may be given as

$$T = \frac{1}{2} i^2 \frac{dL}{d\theta} \quad ...(21.10)$$

where i is the instantaneous current in the exciting winding and L is the self-inductance of that winding varying as a function of the angular position of the rotor. It may be noted that the torque developed is independent of the direction of current flow in the windings, so that unidirectional currents can be used to control the motor. The direction of rotation can be reversed by exciting the phases in the reverse sequence. The motor can also provide regenerative braking. If a phase is excited after the rotor has crossed the position of minimum reluctance, the rotor will experience a torque in opposition to its motion, it will decelerate, mechanical energy drawn from it will be converted into electrical energy and supplied to the source. In fact, the possibility of operating in all of the four quadrants of the speed-torque plane and obtaining flexible speed-torque characteristics simply by appropriate switching of current pulses makes the motor very versatile.

Figure 21.57 (*a*) and (*b*) illustrates the ideal variation of inductance of the exciting winding with respect to the angular position of the rotor over a periphery of one rotor pole pitch and the corresponding torque developed, for an assumed value of constant current [using Eq. (21.10)].

Modes of Operation of Motor. There are two distinct modes of operation corresponding to low or high speed. Monitoring of exciting current during low speed operation is essential because of long duration of each phase period and needs chopping of energization to restrict each phase current within the semiconductor ratings. Moreover, the developed torque is controlled by varying the average phase current. Hence, accurate monitoring of the exciting current is required for obtaining high degree of controllability possible. During high speed operation, current control is not essential because the inductance of the winding and the motional back emf induced restrict the excitation to single pulses of current. Torque is controlled by optimal positioning of these pulses rather than the current level. Current monitoring, however, is retained for the sake of protection.

Typical phase current waveforms of the two modes (low speed and high speed) of operation are shown in Figs. 21.58 (*a*) and (*b*).

The complete drive system, comprising a SRM coupled to a load, a power converter and a control system involving rotor position transducer and current sensor is shown in Fig. 21.59.

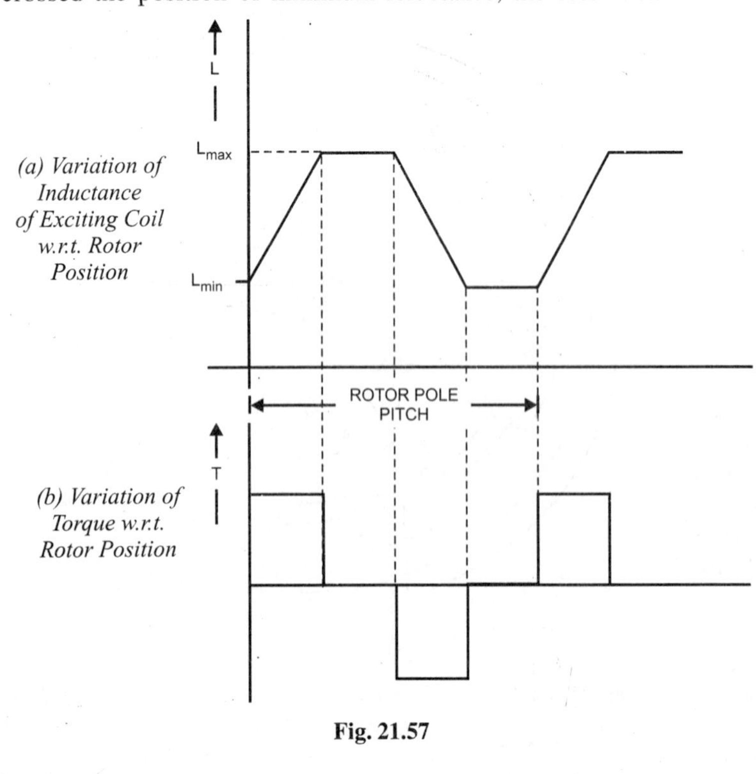

Fig. 21.57

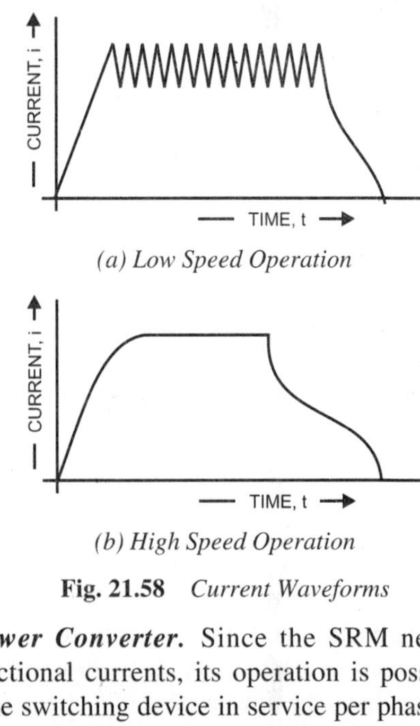

Fig. 21.58 *Current Waveforms*

Power Converter. Since the SRM needs only unidirectional currents, its operation is possible with only one switching device in service per phase, instead of two in series in each phase leg of an inverter for an ac drive. So, the power converter circuits employed for energization of SRMs have few semiconducting devices

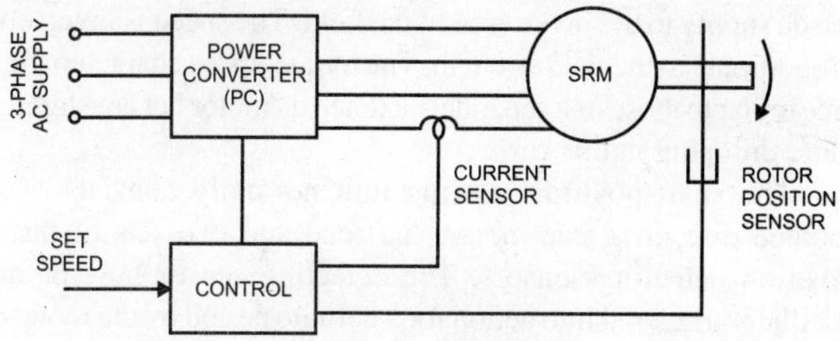

Fig. 21.59 *Block Diagram of a SRM Drive System*

than the inverters supplying ac motors and those devices have only one forward voltage drop in series per phase so that the power losses may, in general, be lower than in conventional inverters. Because of these facts, other factors being the common, results in reduction in the physical size of converter and increase in its reliability.

Different circuit configurations are available for the power converters for SRM drive. All of these have two essential elements—a controlled switch or switches to connect the dc voltage source to the exciting winding to build up current and an alternative path for the current to take when the switch is turned off. The alternative path is provided by a diode or diodes such that the winding experiences a reverse voltage to collapse the current. Three alternatives for one phase of a SRM power converter are depicted in Fig. 21.60. A flexible circuit using two switches per phase is shown in Fig. 21.60 (*a*), whilst the use of bifilar wound motor (as in stepper motors) or a centre tapped supply permits the use of only a single switch per phase as depicted in Figs. 21.60 (*b*) and (*c*). The correct choice of power circuit configuration will depend on the power level of the drive, the supply voltage and the application.

Advantages and Disadvantages of Switched Reluctance (SR) Motor. In SR motor, power is delivered to the stator windings rather than rotor. This greatly simplifies mechanical design as

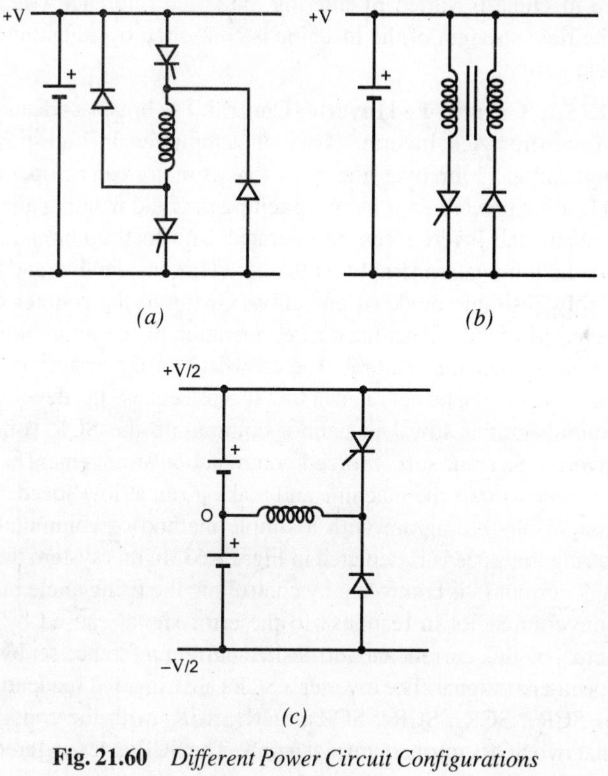

Fig. 21.60 *Different Power Circuit Configurations*

power does not have to be delivered to a moving part, but it complicates the electrical design as some sort of switching system needs to be used to deliver power to the different windings. With modern electronic devices this is not a problem. The laminated rotor has no windings/slip rings and is, therefore cheap to manufacture and extremely robust. The main drawback of SR motor is torque ripple.

The SR motor possesses the advantages of (*i*) higher efficiency (*ii*) more power per unit weight/volume (*iii*) very robust construction (*iv*) possibility of operation at very high speeds (as high as 30,000 rpm) in hazardous atmospheres (*v*) versatile and flexible drive features and (*vi*) possibility of four-quadrant operation with appropriate drive circuitry.

However, the SR motor have the drawbacks of being (*i*) unproven (*ii*) noisy (*iii*) not well-suited for development of smooth torque.

SR motor finds wide applications such as (*i*) traction (*ii*) general purpose industrial drives (*iii*) office and business equipment and (*iv*) domestic appliances like vacuum cleaners, food processors, washing machines etc.

21.15 BRUSHLESS DC (BLDC) MOTOR

A dc motor can be thought of as an ac synchronous machine in which the field is stationary and the armature with its multiphase ac winding is rotating. The armature receives ac supply from a dc source through brushes and commutators. The brushes and commutator constitutes an inverter sensitive to the shaft position. In a similar way, a synchronous motor may be considered to operate as a dc motor. In a synchronous machine the field is rotating whereas the armature is stationary but it should be supplied by an inverter controlled by shaft-position sensing signals. The static inverter with the shaft-position-sensitive controller can very well be regarded as an *electronic commutator* serving the same function as does the mechanical commutator. This facilitates the operation of a synchronous motor as a versatile variable-speed drive like a dc motor but having no mechanical commutator and brushes. This is, no doubt, a great advantage.

No doubt, the brushless dc motors are generally more expensive for the same kW rating than conventional (commutator and brush) motors but possess certain advantages too such as little or no maintenance, much longer operating life, no risk of explosion or possibility of RF radiation due to arcing, higher efficiency, rapid response etc. These motors produce no brush or commutator particles or gases as by- products of operation. These motors are capable of operation submerged in fluids, combustible gases and may even be hermetically sealed. They provide a fairly linear output torque vs input current characteristic, which lends itself to servo applications.

The stator winding of a brushless dc motor may be conventional three-phase winding of a synchronous motor or the conventional armature winding of a dc motor. However, in both cases, the stator winding has to be supplied from a static inverter triggered by shaft-position sensitive signals so that the supply frequency is proportional to the shaft speed. The dc field winding is placed on the rotor and supplied from a static dc source through slip rings mounted on the motor shaft. The equivalent block diagram of a brushless dc motor is shown in Fig. 21.61.

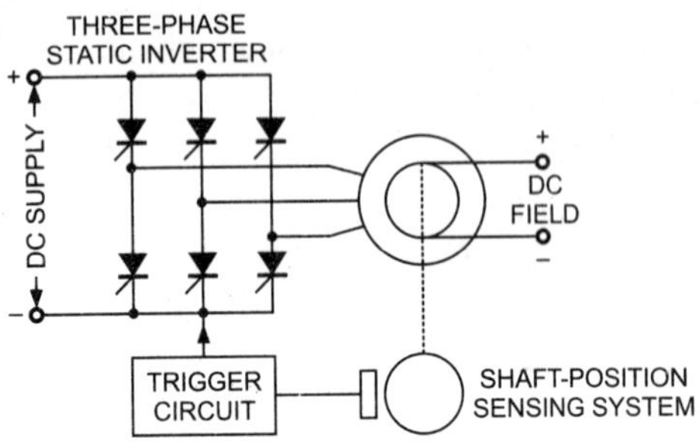

Fig. 21.61 *Equivalent Block Diagram of Brushless DC Motor*

It has already been explained that static inverter together with the shaft-position sensitive trigger circuit is equivalent to the mechanical commutator of a dc machine. A 3-phase inverter is equivalent to a three-segment commutator. If the stator winding is similar to the armature winding of an ordinary dc machine, six symmetrical tappings from the winding can be taken out and the stator winding may be supplied from a six-phase SCR bridge inverter. This is equivalent to six segment commutator. There may be many other variations of the electronic commutation arrangement. For instance, as individual stator coils may be controlled by series SCR switches. However, three segment commutation with a three-phase static inverter is widely used for drives.

It is, therefore, obvious that a 3-phase synchronous motor when supplied by a 3-phase inverter behaves like a simple dc motor but the SCRs of the inverter should be triggered in proper sequence and instant proportional to the position of the rotor shaft. The SCRs may be turned off naturally owing to the nature of the load (synchronous motor). The synchronous motor may be operated at any power factor (leading, lagging behind or unity) by variation of the dc field excitation. When the motor is operated at leading power factor, the inverter can be operated with load commutation i.e. the SCRs are turned off naturally. Otherwise, a force commutated inverter should be employed.

Since the system behaves like a conventional separately-excited dc motor, the speed can be controlled by the variation of the dc supply to the inverter or to the field. The speed is inversely proportional to the field strength. The torque-speed characteristics are similar to those of a separately excited dc motor but are slightly more drooping in this case.

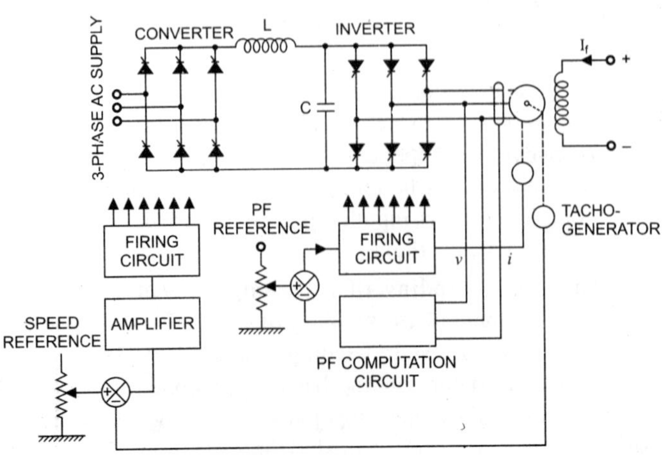

Fig. 21.62 *Block Diagram of Closed-Loop Control of a Brushless DC Motor With Voltage-Fed Inverter*

The shaft-position sensing unit normally consists of a toothed disc on a free motor shaft end and two sets of three position detecting sensors. The detecting sensor may be an oscillator triggered into action for a definite period by the toothed disc or may be an optical unit using phototransistor, light sources producing pulses at definite intervals or may be a Hall device for sensing a magnetic pulse for the required period of time.

The brushless dc motor may be controlled either by a voltage fed inverter, a current fed inverter or a cycloconverter. Open-loop control is simple. To achieve higher accuracy closed-loop control should be employed.

21.15.1. Voltage-Fed Inverter Control. The block schematic diagram of the closed-loop control of a brushless dc motor with a voltage-fed inverter is depicted in Fig. 21.62. This circuit has two feedback loops—the speed loop and the power factor loop. The shaft-position sensing unit provides the triggering pulse for the inverter. SCRs of the inverter are commutated off by a usual force commutation circuit arrangement, not shown in the figure. Signals proportional to the machine input voltage and current are fed to a circuit marked for machine-power factor computation. The signal from the circuit is compared with a fixed reference. The output of the comparator is applied to the phase-shifting firing circuit of the inverter. Thus the power factor of the motor is maintained constant at a desired value. Another signal proportional to the machine speed is compared with speed reference voltage and the output of the comparator is amplified and applied in the phase-control triggering circuit of the converter. Thus the machine speed is maintained at a fixed value by controlling the input dc voltage of the inverter. The speed of the machine is varied by altering the speed reference voltage. The field strength of the machine is controlled by adjusting the field current.

21.15.2. Current-Fed Inverter Control. The brushless dc motor with current-fed inverter is widely employed in high-power applications. Moreover, the synchronous motor can be operated at leading power factor by overexciting the field winding and, as a result, the inverter can be operated with load commutation. Load commutation simplifies the inverter considerably and is a highly desirable mode of operation. Owing to the counter-emf developed by the machine during operation the commutation of the SCRs becomes natural. The drawback of the system is that the drive cannot be operated at low speeds because the developed counter-emf is low and cannot commutate the SCRs of the inverter. So some sort of forced-commutation arrangement has to be made to start the machine and make it run at low speed. The system block diagram with a simple method of commutation during low speeds is depicted in Fig. 21.63. In this system the dc link current I_C is controlled by controlling the firing angle of the converter SCRs in response to the error signal caused by the actual dc link current and the desired current reference set by the speed-error signal. The inverter's SCRs are triggered sequentially as SCR_1, SCR_6, SCR_3, SCR_2, SCR_5, SCR_4 with the condition that two SCRs must conduct at a time. The SCRs are triggered by the angle-control unit which generates a set of six pulses for each electrical revolution of the machine. The pulses are generated at a definite location relative to the shaft position by the shaft-position sensor which measures the shaft angle. These pulses are shifted in phase by a controllable amount determined by the angle command which is a function of the speed error. Thus the SCRs of the inverter are triggered at a particular rotor position and the angle between the stator and field mmfs is controlled. The rotor field current is controlled in the same way, as depicted in the figure.

With the above system the field current, stator current and the rotor angles can be controlled and hence the power factor can be maintained at the desired value so that the load commutation can be achieved.

At low speeds or at the starting instant when the rotor is at standstill, the back emf of the motor is negligibly small and the SCRs of the inverter are not commutated off naturally by the back emf of the motor. Hence, an auxiliary forced-commutation arrangement is necessary for the starting period and can be put off when the motor attains sufficient speed and load commutation is effective. This is accomplished with the help of the capacitor C and two auxiliary SCRs such as SCR_7 and SCR_8.

Current-fed inverters are more suitable for the speed control of a synchronous motor. Apart from the load-commutated inverter, other types of forced commutation are also popular. The speed of the motor can be precisely controlled by the method described above. Reversal of direction of rotation can be achieved by reversing the phase sequence of the SCR switching as the frequency goes through zero value. Reversal of power flow can be achieved by reversing the polarity of the voltage in the dc link. Thus, a four-quadrant operation of the drive is possible.

21.15.3. Brushless Vs Brushed Motors. Brushed dc motors have been in commercial use since 1886. Brushless motors, on the other hand, did not become commercially viable until 1962.

Brushed dc motors develop a maximum torque when stationary, linearly decreasing as speed increases. Some limitations of brushed motors can be overcome by brushless motors; they include higher efficiency and a lower susceptibility to mechanical wear. These benefits come at the cost of potentially less rugged, more complex, and more expensive control electronics.

A typical brushless motor has permanent magnets which rotate around a fixed armature, eliminating problems associated with connecting current to the moving armature. An electronic controller replaces the brush/commutator assembly of the brushed dc motor, which continually switches the phase to the windings to keep the motor turning. The controller performs similar timed power distribution by using a solid-state circuit rather than the brush/commutator system.

Brushless motors offer several advantages over brushed dc motors, including more torque per unit weight, more torque per watt (increased efficiency), increased reliability, reduced noise, longer lifetime (no brush and commutator erosion), elimination

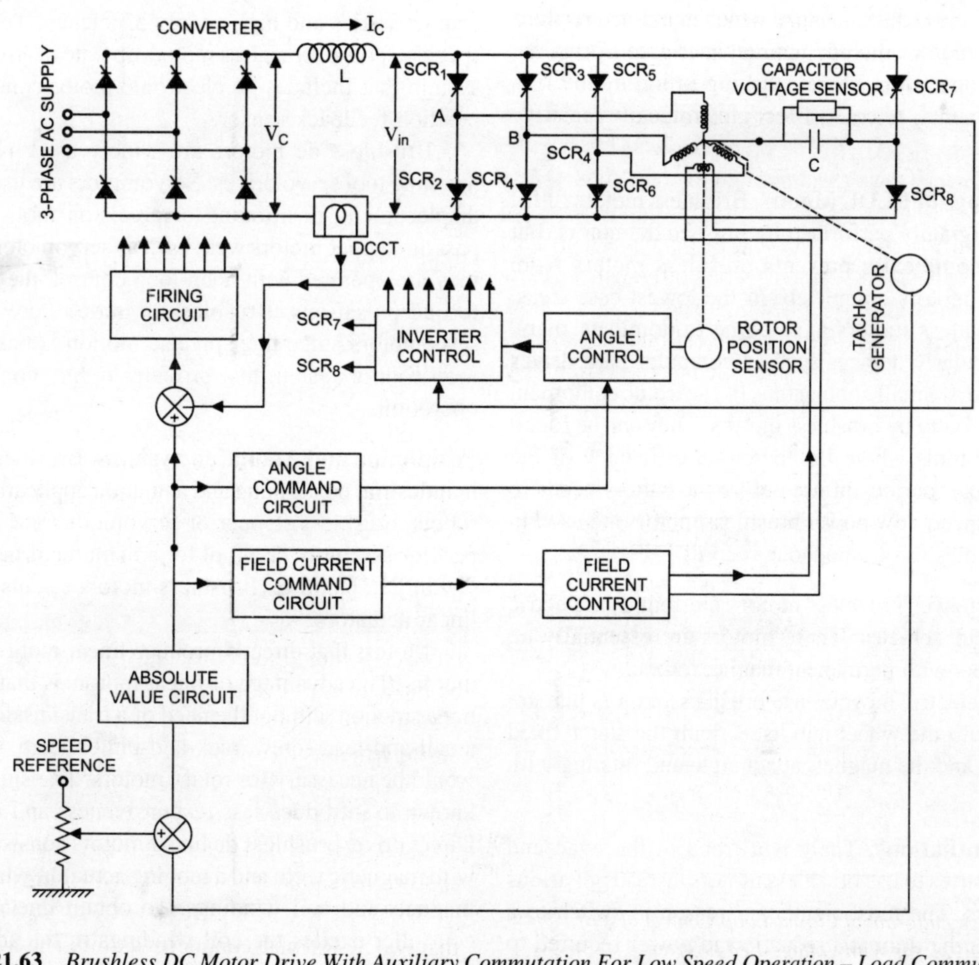

Fig. 21.63 *Brushless DC Motor Drive With Auxiliary Commutation For Low Speed Operation – Load Commutation*

of ionizing sparks from the commutator, and overall reduction of electromagnetic interference (EMI). With no windings on the rotor, they are not subjected to centrifugal forces, and because the windings are supported by the housing, they can be cooled by conduction, requiring no airflow inside the motor for cooling. This in turn means that the motor's internals can be entirely enclosed and protected from dirt or other foreign matter.

Brushless motor commutation can be implemented in software using a microcontroller or microprocessor computer, or may alternatively be implemented in analogue hardware, or in digital firmware using an FPGA. Commutation with electronics instead of brushes allows for greater flexibility and capabilities not available with brushed dc motors, including speed limiting, "micro stepped" operation for slow and/or fine motion control, and a holding torque when stationary.

The maximum power that can be applied to a brushless motor is limited almost exclusively by heat; too much heat weakens the magnets and may damage the winding's insulation.

When converting electricity into mechanical power, brushless motors are more efficient than brushed motors. This improvement is largely due to the brushless motor's speed being determined by the frequency at which the electricity is switched, not the voltage. Additional gains are due to the absence of brushes, which reduces mechanical energy loss due to friction. The enhanced efficiency is greatest in the no-load and low-load region of the motor's performance curve. Under high mechanical loads, brushless motors and high-quality brushed motors are comparable in efficiency.

Environments and requirements in which manufacturers use brushless-type dc motors include maintenance-free operation, high speeds, and operation where sparking is hazardous (i.e. explosive environments) or could affect electronically sensitive equipment.

21.15.4. Applications of BLDC Motors. Brushless motors fulfill many functions originally performed by brushed dc motors, but cost and control complexity prevents brushless motors from replacing brushed motors completely in the lowest-cost areas. Nevertheless, brushless motors have come to dominate many applications particularly devices such as computer hard drives and CD/DVD players. Small cooling fans in electronic equipment are powered exclusively by brushless motors. They can be found in cordless power tools where the increased efficiency of the motor leads to longer periods of use before the battery needs to be charged. Low speed, low power brushless motors are used in direct-drive turntables for gramophone records.

Transport. High power brushless motors are found in electric vehicles and hybrid vehicles. These motors are essentially *ac* synchronous motors with permanent magnet rotors.

A number of electric bicycles use brushless motors that are sometimes built into the wheel hub itself, with the stator fixed solidly to the axle and the magnets attached to and rotating with the wheel.

Heating and Ventilations. There is a trend in the hvac and refrigeration industries to use brushless motors instead of various types of ac motors. The most significant reason to switch to a brushless motor is the dramatic reduction in power required to operate them versus a typical ac motor. While shaded-pole and permanent split-capacitor motors once dominated as the fan motor of choice, many fans are now run using a brushless motor. Some fans use brushless motors also in order to increase overall system efficiency.

In addition to the brushless motor's higher efficiency, certain hv ac systems (especially those featuring variable-speed and/or load modulation) use brushless motors because the built-in microprocessor allows for programmability, better control over airflow, and serial communication.

Industrial Engineering. The application of brushless dc motors within industrial engineering primarily focuses on manufacturing engineering or industrial automation design. In manufacturing, brushless motors are primarily used for motion control, positioning or actuation systems.

Brushless motors are ideally suited for manufacturing applications because of their high power density, good speed-torque characteristics, high efficiency and wide speed ranges and low maintenance. The most common uses of brushless dc motors in industrial engineering are linear motors, servomotors, actuators for industrial robots, extruder drive motors and feed drives for CNC machine tools.

Motion Control Systems. Brushless motors are commonly used as pump, fan and spindle drives in adjustable or variable speed applications. They can develop high torque with good speed response. In addition, they can be easily automated for remote control. Due to their construction, they have good thermal characteristics and high energy efficiency. To obtain a variable speed response, brushless motors operate in an electromechanical system that includes an electronic motor controller and a rotor position feedback sensor.

Brushless dc motors are widely used as servomotors for machine tool servo drives. Servomotors are used for mechanical displacement, positioning or precision motion control. In the past dc stepper motors were used as servomotors; however, since they are operated with open-loop control, they typically exhibit torque pulsations. Brushless dc motors are more suitable as servomotors since their precise motion is based upon a closed loop control system that provides tightly controlled and stable operation.

Positioning and Actuation Systems. Brushless motors are used in industrial positioning and actuation applications. For assembly robots, brushless stepper or servomotors are used to position a part for assembly or a tool for a manufacturing process, such as welding or painting. Brushless motors can also be used to drive linear actuators.

Motors that directly produce linear motion are called linear motors. The advantage of linear motors is that they can produce linear motion without the need of a transmission system, such as a ball-and-lead screw, rack-and-pinion, cam, gears or belts, that would be necessary for rotary motors. Transmission systems are known to introduce less responsiveness and reduced accuracy. Direct drive, brushless dc linear motors consist of a slotted stator with magnetic teeth and a moving actuator, which has permanent magnets and coil windings. To obtain linear motion, a motor controller excites the coil windings in the actuator causing an interaction of the magnetic fields resulting in linear motion. Tubular linear motors are another form of linear motor design operated in a similar way.

Fig. 21.64 *A Microprocessor-Controlled BLDC Motor Powering a Micro Radio-Controlled Airplane. This External Rotor Motor Weighs 5 grams, Consumes Approximately 11 Watts and Produces Thrust of More Than Twice the Weight of the Plane*

Model Engineering. Brushless motors are a popular motor choice for model aircraft including helicopters. Their favorable power-to-weight ratios and large range of available sizes, from under 5 gram to large motors rated at well into the kilowatt output range, have revolutionized the market for electric-powered model flight, displacing virtually all brushed electric motors. They have also encouraged a growth of simple, lightweight electric model aircraft, rather than the previous internal combustion engines powering larger and heavier models. The large power-to-weight ratio of modern batteries and brushless motors allows models to ascend vertically, rather than climb gradually. The low noise and lack of mess compared to small glow fuel internal combustion engines is another reason for their popularity.

Legal restrictions for the use of combustion engine driven model aircraft in some countries have also supported the shift to high-power electric systems.

Radio Controlled Cars. Their popularity has also risen in the radio controlled car area. Brushless motors have been legal in North American RC car racing in accordance to ROAR since 2006. These motors provide a great amount of power to RC racers and, if paired with appropriate gearing and high-discharge Li-Po (lithium polymer) or considerably safer $LiFePO_4$ batteries, these cars can achieve speeds over 160 kilometres per hour.

Brushless motors are capable of producing more torque and have a larger peak rpm compared to nitro or gasoline powered engines. Nitro engines climax to peak at around 26,000 rpm and 1.25 hp, while a smaller brushless motor having generally maximum torque start then tapering off can reach 50,000 rpm and 5 hp.

21.16 LOW-INERTIA DC MOTORS

Low-inertia motors are so designed as to make their armature very light in weight so that they are capable to start, stop and change direction very quickly. This will make them suitable for instrumentation applications. The common types of low inertia motors are (*i*) cup or shell-type permanent magnet dc motors and (*ii*) printed circuit or pancake motors.

21.16.1. Cup or Shell Type PM DC Motors. These motors offer low inertia and low inductance as well as high acceleration characteristics, making them useful in many servo applications. They have hollow cylindrical armatures made of aluminium or copper coils bonded by polymer resin and fiberglass to form a rigid *ironless cup*, which is fastened to an axial shaft. Fig. 21.65 depicts a cutaway view of this class of servomotor. Because the armature has no iron core, it, like the disk motor, has extremely low inertia and a very high torque-to-inertia ratio. This permits the motor to accelerate rapidly for the quick response required in many motion-control applications. The armature rotates in an air gap within very high magnetic flux density. The magnetic field from the stationary magnets is completed through the cup-type armature and a stationary ferrous cylindrical core connected to the motor frame. The shaft rotates within the core, which extends into the rotating cup. Spring-loaded brushes commutate these motors.

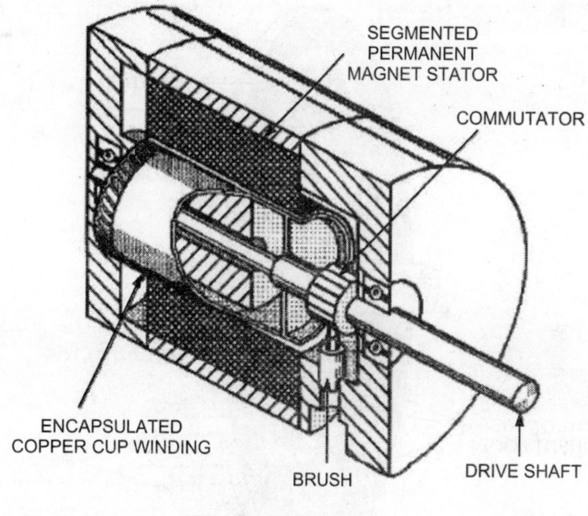

Fig. 21.65 *Cutaway View of a PMDC Servomotor With a Cup-Type Armature*

21.16.2. Printed Circuit or Pancake Motors. The printed circuit motor is an example of an ironless or coreless motor with several unique features. The pancake construction uses an axial magnetic field to achieve the short flat construction. Radial field PCB motors are also possible.

Construction. The rotor windings are printed, stamped or welded onto a thin, disc shaped glass fibre circuit board which rotates in the air gap between pairs of permanent magnets arranged around the periphery of the disk. The windings fan out in a series of radial loops around the surface of the disk. The magnets are arranged alternatively north and south so that the magnetic fields in the air gaps of adjacent magnet pairs are in opposite directions. The magnets are held in place by two iron end caps in a compact "pancake" shaped block to complete the magnetic circuit. Current is fed to the rotor windings via brushes through precious metal commutator segments printed on the disc.

Operating Principle. Traditional electric motors have a radial magnetic field or flux with the rotor current flowing axially along the length of the rotor. In typical *printed circuit motors* the construction is reversed. The magnetic field is axial (oriented along the axis of the machine) and the current flows radially from the axis to the edge of the disc and back again. A tangential force on the disk is created by the current passing through the magnetic fields in the air gaps between the pole pairs of the permanent magnets. So that the return current does not cancel out the effect of the outgoing current, the return wire is physically separated or displaced to one side from the outgoing wire by the width of the magnet. In this way it interacts with the magnetic field of the adjacent magnet which is in the opposite direction and thus reinforces the tangential force on the disk.

In many ways it is similar to Faraday's 1831 disk or homopolar motor which used a single magnet and was driven by a unidirectional current fed by brushes at the centre and on the periphery of the disk.

Main Features: (*i*) Very small rotational inertia (*ii*) High torque/inertia ratio (*iii*) High overload current capability (*iv*) Linear speed-torque characteristic (*v*) Smooth torque down to near-zero speed (*vi*) Very suitable for direct drive applications.

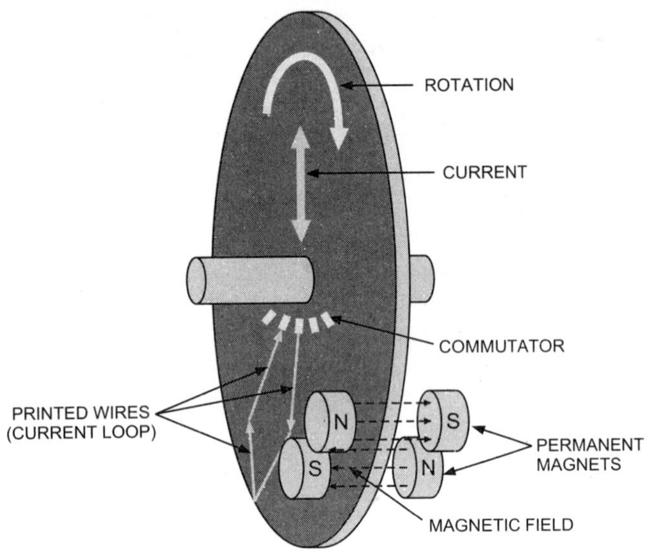

Fig. 21.66 *Printed Circuit Motor*

Advantages: 1. Efficient-no hysteresis or iron loss.

2. Light weight.
3. Simplified armature construction.
4. Minimum radio and TV interference owing to low-voltage design.
5. Flat, so fits into confined space.

Disadvantages: 1. Expensive to make.

2. Armature, being of little mass, can overheat quickly.
3. Restricted to low voltages only.
4. Suited to intermittent duty cycle only because armature overheats quickly.

Applications. The printed circuit motor is a very compact and light weight design making it useful in confined spaces. Since the rotor does not have drag a lump of iron around, it has very low inertia and can run up to speed very quickly. Because of the many commutator segments and the low current capability of the windings, the PCB motor is only suitable for low power applications and is not suitable for continuous operation. It is however ideal for servo systems and industrial controls and automotive applications such as electric window winders.

21.17 SYNCHROS AND RESOLVERS

Synchros and resolvers are basically ac electromechanical, variable-coupling transformer devices. These devices are primarily used for angular data transmission and computing systems. The angular shaft position data is transmitted from one location to another through converted electrical signals.

These electric motor like devices are usually rugged, reliable performers with sufficient resolution for many machine applications. Because synchros and resolvers give an analog output, therefore, converters are required for providing a digital format in order to interface with digital numerical control systems — which are very important today as the strong trend toward digital systems continues. Earlier these converters were comparatively larger in size and expensive in cost. Now these problems have been solved because of technological advancement in electronics and these devices equipped with synchro-to-digital converters are practical approach to overall digital positioning systems and are preferred over encoders for many applications.

21.17.1. Synchros. The basic electromagnetic structure of a synchro consists of a wound rotor and a wound stator, concentrically arranged to give adjustable mutual inductance (coupling) between the windings of rotor and stator. The stator is made up of laminated steel in order to reduce the core losses. It is slotted to accommodate a balanced three phase winding, with the axis of each stator winding displaced by 120° or $\frac{2\pi}{3}$ radians in space. The three stator windings are identical and connected in a Y-configuration. The rotor has a salient pole construction (dumbbell shaped) and is wound with a single phase concentrated winding. In short it can be conveniently said that synchro is similar in construction to that of wound rotor induction motor.

Synchro systems are of two types namely *control* or *error detecting type* and *torque transmission type.*

Torque transmission type synchro systems are employed only for driving light loads, such as pointers, because torque transmission systems give very little output torque. For applications requiring large torques and high accuracies *control* or *error detecting type synchros are* employed.

1. *Control Type Synchro Systems.* These systems are employed as error detectors in positional control systems.

Two different types of synchros, the *synchro control transmitter* and the *synchro control transformer,* are employed in such systems. The synchro control transformer is similar in construction to the synchro control transmitter except that the transmitter has a salient pole rotor while the control transformer has a cylindrical rotor. It is done because in cylindrical construction of rotor the air gap length remains constant irrespective of the position of the rotor and so the input impedance of the control transformer remains unchanged whatever may be the position of rotor.

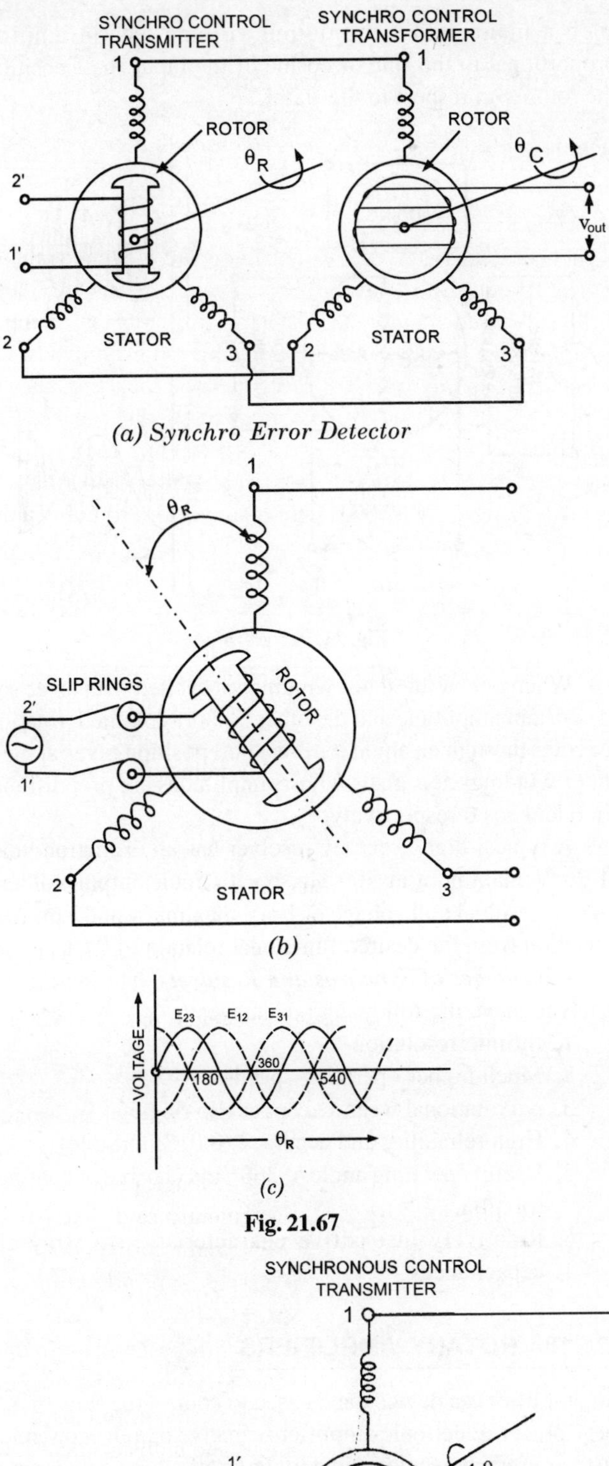

(a) *Synchro Error Detector*

(b)

(c)

Fig. 21.67

The arrangement for control type synchro system is illustrated in Fig. 21.67(*a*).

The coil arrangement of the transmitter alone is illustrated in Fig. 21.67(*b*). When an ac voltage is applied to the rotor through two slip rings, voltages are induced in the stator coils due to transformer action between the rotor and stator windings. For a given stator coil, the open-circuit voltage is sinusoidal in time and varies in amplitude with rotor position, also sinusoidally as shown in Fig. 21.67 (*c*). Thus the three voltage signals from the stator coils uniquely define the angular position of the rotor. When these three voltages are applied to the stator coils of a control transformer, they produce mmf aligned in the same direction as that of the transmitter rotor. The control transformer rotor acts as a "search coil" in detecting the direction of its stator field. If the axis of this coil is aligned with the field, the maximum voltage is induced into the rotor coil of the control transformer. If the axis is perpendicular to the field, zero voltage is induced, giving the null position mentioned above. The output voltage amplitude actually varies sinusoidally with the misalignment angle, but for small angles the sine of the angle is equal to the angle, giving a linear output.

Thus the synchro transmitter — control transformer pair acts as a detector and gives a voltage signal, at the rotor terminals of the control transformer, proportional to the difference between the transmitter and control transformer shaft positions.

Synchros for angle measurement are most widely employed as components of servomechanisms (automatic motion-control feedback systems) where they are employed for measuring and comparing the actual rotational position of load with its commanded position, as in Fig. 21.68. The error voltage signal v_e is an ac voltage of the same frequency as the excitation and of amplitude proportional to the error angle ($\theta_R - \theta_C$) [for small error angles only]. Its phase changes by 180° at the null point; thus the direction of the error is detected. When $\theta_R = \theta_C$, the error voltage and thus the amplifier output or input to the ac motor is zero and the system is at rest. If a command rotation θ_R is now put

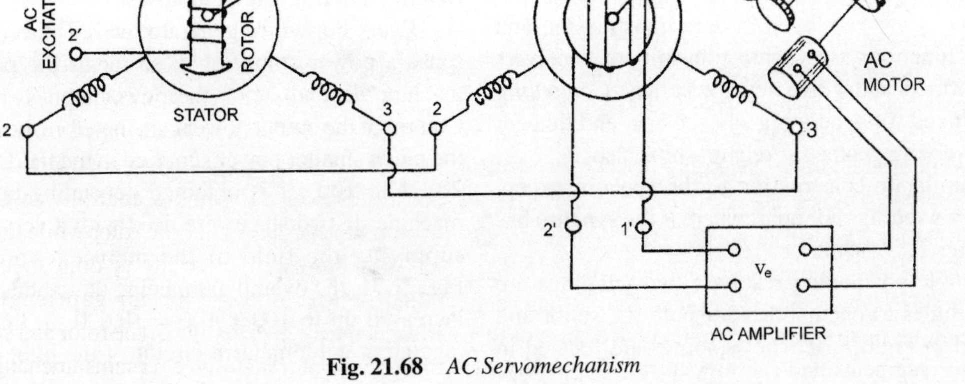

Fig. 21.68 *AC Servomechanism*

in, there will be an error voltage v_e which will cause the motor to rotate so as to return θ_C to correspondence with θ_R.

2. *Torque Transmission Type Synchro Systems.* In instrumentation systems, the synchros are usually employed in the torque transmission mode. In a typical practical setup, a pair of synchros are employed: one acts as a *transmitter* and the other as a *receiver*. The rotors and stators of synchro transmitter and synchro receiver are connected in torque transmission configuration, as illustrated in Fig. 21.69.

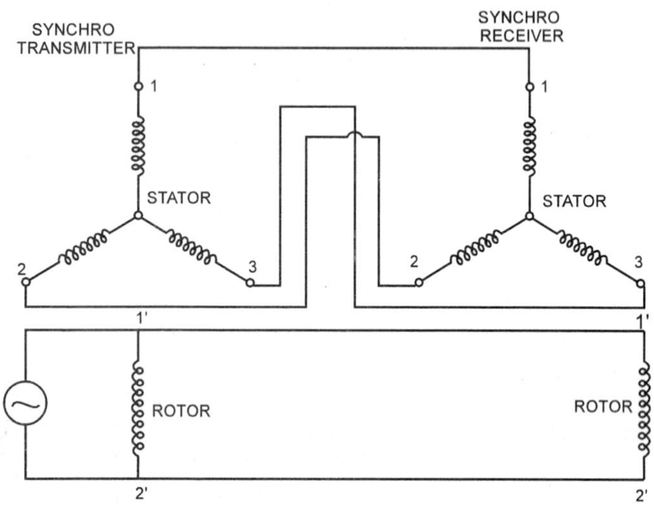

Fig. 21.69 *Torque Transmission Type Synchro System*

In this synchro system, the input is the angular position of the synchro transmitter.

When the rotors of the torque transmitter and torque receiver are in the same angular positions, the emfs induced in the stator windings of both of these are equal, and there is no flow of current between the stator windings. Now if the rotor of transmitter synchro is moved to a new position, there will be an imbalance of emfs induced in the stator windings of synchro transmitter and the synchro-receiver. This voltage imbalance between the stator windings causes currents to flow between the windings. This current flow in the stator winding of receiver produces a torque on the receiver rotor till it assumes the same angular position as that of the transmitter rotor. Thus a synchro can be employed for determination of magnitude and direction of an angular displacement.

The indicator used in such a system is a simple pointer attached to the rotor shaft of the synchro receiver.

21.17.2. Resolvers. Resolvers are of two types namely *synchro resolvers* and *computing resolvers.*

Synchro resolvers are employed in data transmission and perform the same functions as synchro transmitters, receivers and control transformers and with a better accuracy. *Computing resolvers* are employed for generating sine, cosine, and tangent functions and for solving geometric relationships also.

Resolver is similar in construction to the synchro except that it has two stator windings 90° apart whereas the synchro has three stator windings 120° apart.

In a typical angle-positioning resolver, two windings are provided at right angles to one another for both the stator and rotor, as illustrated in Fig. 21.70. The windings are located in such a manner that the output voltage has an amplitude proportional to the sine or cosine of the input shaft position of the rotor with respect to the stator.

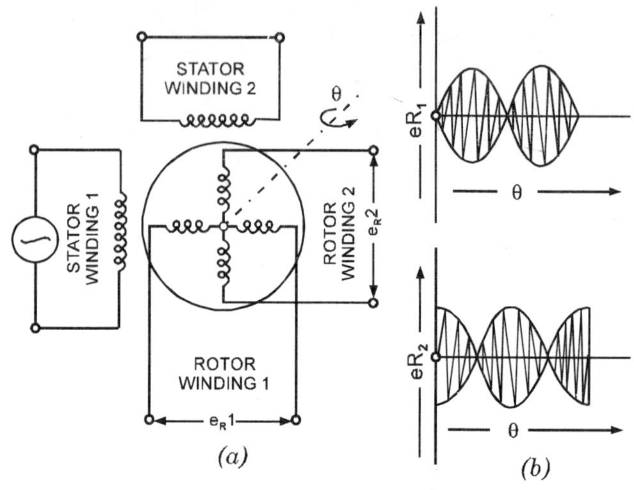

Fig. 21.70 *Resolver*

When one of the stator windings is excited with an ac signal of constant amplitude and the other is short-circuited, rotation of the rotor through an angle θ from a null position gives at the two stator windings ac signals whose amplitudes are proportional to sin θ and cos θ respectively.

A typical high-accuracy resolver has an excitation voltage of 26 V maximum at 400 Hz, open-circuit output voltage of 0-26 V, residual null voltage of 1 mV maximum, and a maximum deviation from the desired functional relation of 0.01 per cent.

Advantages of Synchros and Resolvers. The synchros and resolvers have the following advantages:

1. Infinite resolution.
2. Much higher operating speeds.
3. No rotational wear, except at slip rings.
4. High reliability and accuracy (0.01% feasible).
5. Useful operating angle of 360° and capable of continuous rotation.
6. Relatively insensitive characteristics to stray cable capacitances.

21.18 ROTARY AMPLIFIERS

An amplifier is a device that is used to control the flow of a very large power. Electronic amplifiers can give output economically up to a certain power level and for large amplified power outputs, rotating amplifiers are employed.

Every normal dc generator driven at a constant speed is in a sense a power amplifier in so far as the power output of the machine depends upon the power input to the field windings. Control of the output power can, therefore, be obtained by varying the much smaller power supplied to the field. A power gain from 20 : 1 to 100 : 1 is obtained depending upon the size of the machine. If two stages are used with a very small pilot exciter supplying the field of the main exciter, as illustrated in Fig. 21.71, the overall gain, being the product of the two, can be increased up to 100 × 100 = 10,000. In this arrangement, the regulating equipment in circuit of the pilot exciter has only to deal a very small power, possibly less than 1 watt with normal machines. The transient response of the system is very sluggish, because of large field winding inductance, although such a system is employed in certain applications but it is not suitable with automatic control circuits of the closed loop type. The possibility of combining the two machines in one and so retaining the high power gains, at the same time considerably reducing the time lags, has led to the development of special types of machines, now classified as *rotating amplifiers*.

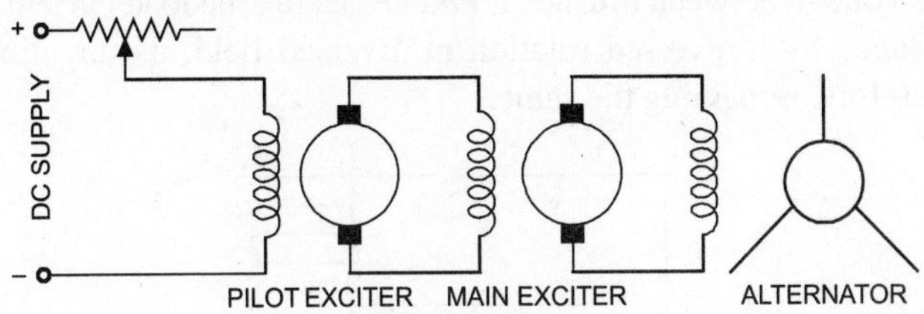

Fig. 21.71

There are two types of such machines (*i*) cross-field generators and (*ii*) rototrol (abbreviated form of rotating control).

21.19 CROSS-FIELD GENERATORS

In an ordinary dc generator the excitation is produced by the mmfs of the field windings. The armature mmf does not contribute to the excitation but opposes the main field and distorts the flux distribution. In the ordinary generator the armature winding is not designed for the purpose of creating excitation for the machine. Its armature mmf may be thought of as an unnecessary evil. It is possible to modify the design so that special generator can be constructed in which the function of a component of armature current is to act as a source of excitation. Such generators are called the *armature-excited* or *cross-field generators*, among which the Rosenberg generator, amplidyne and metadyne are the most important. All these three generators operate on the same principle, called the cross-field principle, explained below, though they differ in construction slightly.

21.19.1. Cross-Field Principle. A separately-excited 2-pole dc generator, consisting of many conductors rotating in a magnetic field produced by field coils, with no load connected to it is shown in Fig. 21.72 (*a*). The brushes are located at right angles to the main field flux or on the magnetic neutral and the voltage induced in the rotating armature conductors is available at the brushes. Since armature conductors do not carry any current (load switch being open), the only magnetic field in the machine is produced by the field coils, and the direction of the field developed is illustrated in Fig. 21.72 (*a*). On closing of switch current will flow through the armature conductors. The armature current sets up a magnetic field at right angles to the main field. Both the flux are illustrated in Fig. 21.72 (*b*).

When the brushes are short circuited, as illustrated in Fig. 21.72 (*c*) the armature current will be very large and the armature winding may get damaged. In order to retain the short circuit and yet prevent the armature winding being damaged, the input to the field is reduced until the armature carries rated current. Since the impedance of the short-circuited armature is very small, very small field current will be required to induce voltage between brushes sufficient to cause flow of rated armature current through armature conductors. So under these conditions the flux due to the main field is reduced to roughly $\frac{1}{100}$ of the former value, while the rated armature current produces a value of flux approximately the original value of the main flux. Thus a small input to the main field controls the large flux created by the armature current.

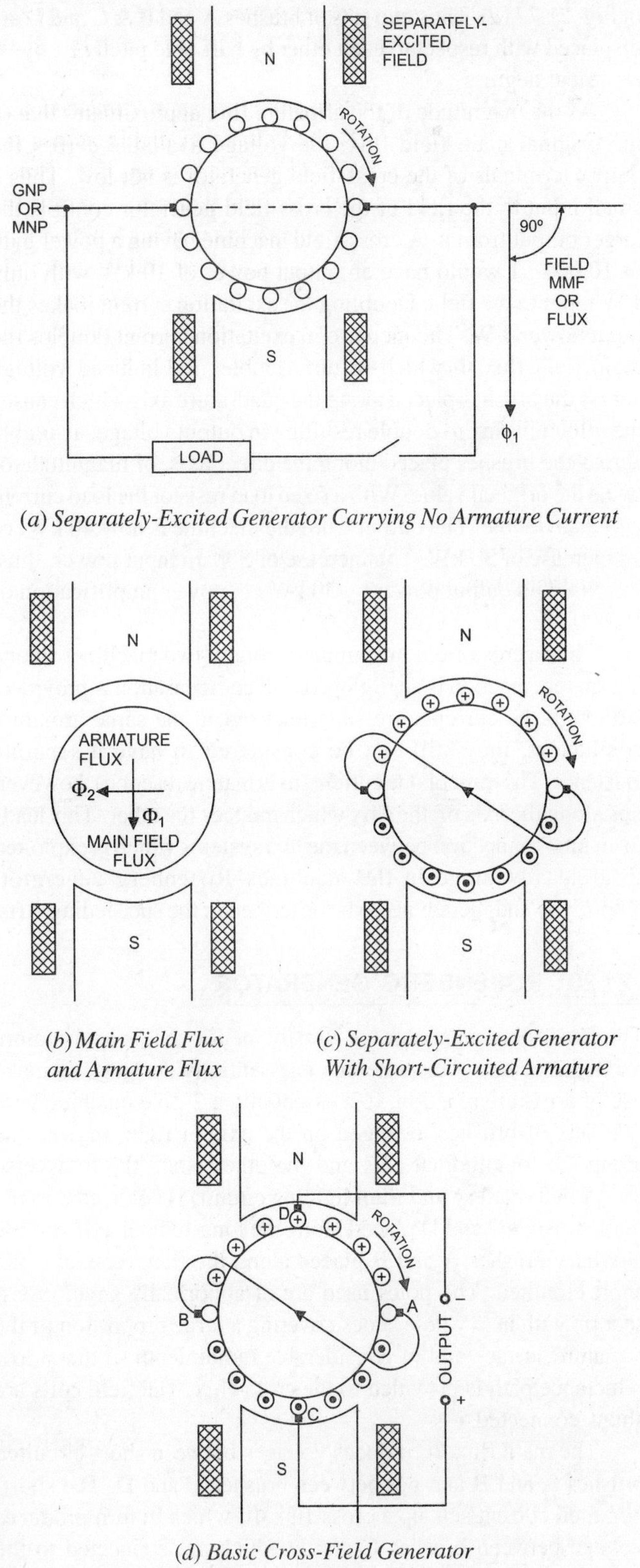

(a) Separately-Excited Generator Carrying No Armature Current

(b) Main Field Flux and Armature Flux

(c) Separately-Excited Generator With Short-Circuited Armature

(d) Basic Cross-Field Generator

Fig. 21.72

Two voltages are induced in the armature as it rotates. One voltage is owing to cutting of main flux by rotating armature conductors and is available between the brushes A and B placed along the *interpolar* or *quadrature axis*. The second voltage induced in the armature conductors is due to cutting of the armature flux and is available between second set of brushes C and D placed along *pole axis* , also known as *direct axis*, and connected to the output terminals of the generator, as illustrated in Fig. 21.72 (*d*). Thus two sets of brushes A and B & C and D are displaced with respect to each other by half pole pitch *i.e.*, by 90 electrical degrees.

As the magnitude of the armature flux approximates that of the original main field flux, the voltage available across the output terminals of the cross-field generator is not low. Thus a small input to the field of the cross-field generator controls the larger output from it. A cross-field machine having a power gain of 10,000 : 1 would have an output power of 10 kW with only 1 W input to the field. Doubling the excitation current makes the input power 4 W. The increase in excitation current doubles the main field flux Φ which in turn doubles the induced voltage across the brushes placed along the quadrature axis which causes the armature flux to double resulting in output voltage, available across the brushes placed along the direct axis, of magnitude of twice the original value. With a fixed load resistor the load current will also double. The output from the machine is now 40 kW, or an increase of 30 kW. The increase of 3 W in input power, thus, increased an output power by 30 kW *i.e.* power amplification of 10,000 : 1.

Thus a cross-field machine is virtually two machines in one if armature reaction is negligible or full compensation is provided. Although the currents are superimposed in the same armature conductors, they still can be considered to have a separate existence. The mmf of one current (its armature reaction), however, lies along the axis of the flux which induces the other. This leads to an interconnection between the two systems which is exploited to great advantage in the machines–Rosenberg generator, amplidyne and metadyne, to be described in the succeeding Arts.

21.20 ROSENBERG GENERATOR

The Rosenberg generator, as already mentioned, is an armature excited machine. This machine is essentially designed to make use of armature reaction. It is essentially a 2-pole machine with two sets of brushes arranged on the axis at right angles, one along the longitudinal axis and the other along the transverse axis. Load (battery and train lighting circuit) is connected to the main brushes C and D placed along the longitudinal axis and the auxiliary brushes A and B placed along the transverse axis are short circuited. The poles used are of abnormally small cross-section with heavy pole shoes covering a large proportion of the armature surface and of considerable radial depth so that a low reluctance path is provided to the cross-flux. The field coils are shunt connected.

The main flux Φ_1 induces voltages between short-circuited brushes A and B and not between brushes C and D. The short-circuited current sets up a cross-flux Φ_2 which in turn produces voltage between brushes C and D which are connected to the load. The resulting load current sets up a flux Φ_3 which opposes the main flux Φ_1. Thus the current in the external circuit bears a controlling influence over the flux cut by the armature inductors enabling the load current to be maintained at or near its rated value.

It should be noted that the direction of the load current is independent of direction of rotation. If the generator is revolved in opposite direction, the polarity of the short-circuited brushes will be reversed, and this will reverse the armature excitation current. This will reverse the direction of the cross-flux Φ_2. Thus the voltage between brushes C and D causing the load current is induced by a reversed rotation in reversed field, its direction, therefore, remaining the same.

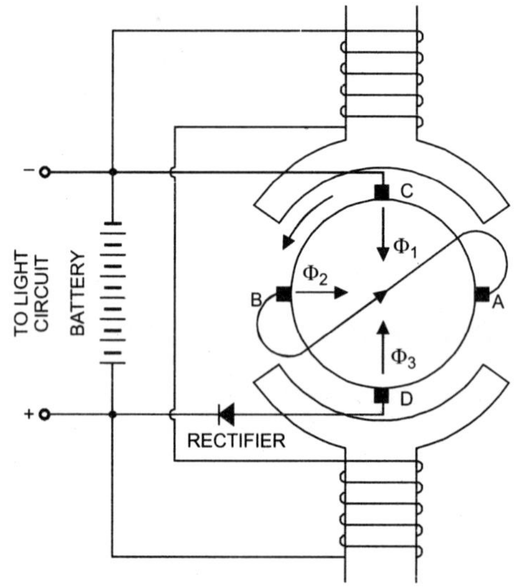

Fig. 21.73 *Rosenberg Generator*

The output current—speed characteristic for a Rosenberg generator is shown in Fig. 21.74, from which it is obvious that no-load component of armature current can flow until a certain speed is obtained and that after this speed is attained, all changes in operation depend upon the increasing speed. The increase in speed causes flux to be cut at a greater rapidity, but the consequent tendency for the output terminal voltage to rise and produce more load component of armature current reacts to reduce the flux cut by the armature inductors and, thereby, the output terminal voltage. Through proper design these two effects can be made to counterbalance each other after a certain speed has been attained. Thus, the Rosenberg generator can be designed to give the desired output current-speed characteristic as illustrated in Fig. 21.74.

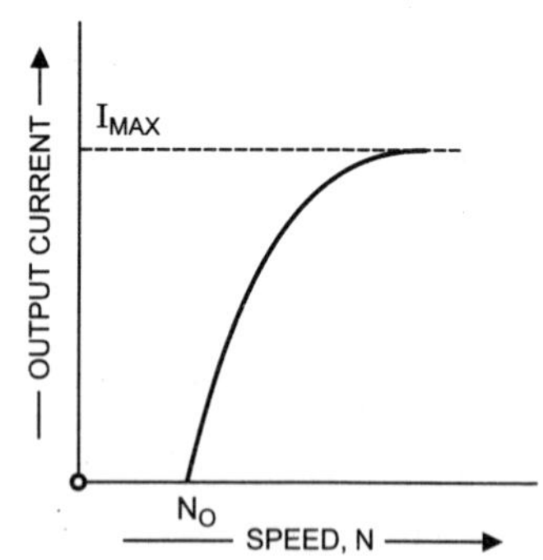

Fig. 21.74 *Output Current-Speed Characteristic For Rosenberg Generator*

Such generators are used in automobile and train services. The generator of a railway coach must be driven from the axle of the car, since there is no prime mover on the coach. A battery is also required to be provided to supply the lighting when the coach is stationary and this battery must be charged by the generator while the train is running. Since the coach travels both forward and backward, the polarity of the generator must be independent of the direction of rotation. Again, since the speed of the coach varies widely, it is necessary for the generator to charge the battery at low speeds but not to overcharge it at high speeds i.e. there must be an inherent current limit in the generator which will not be exceeded whatever the speed or the state of the battery. Both of these requirements are met by the Rosenberg generator, as explained above.

To prevent feedback supply from the batteries to the generator, at low speeds giving generated voltage less than the battery voltage, either some form of automatic reverse current cut-out or a rectifying element is connected in series with the power output terminals of the generator. When lights are in use, they are supplied from the generator when this is in operation, with the cutout closed, or from the battery when the cutout is open and the generator is disconnected.

21.21 AMPLIDYNE AND METADYNE

These machines use the cross-field principle and are development of the Rosenberg generator, described in preceding Art 21.20. In the Rosenberg generator no compensation is provided for the armature reaction mmf set up by the load current as it was essential for giving constant current characteristic when operating under appropriate conditions. In these machines it is essential to neutralize the armature reaction mmf set up by the load current and this is achieved by connecting a compensating winding in series with the power output brush terminals. Commutating poles are also required in these machines for satisfactory commutation. Fully compensated machine is called the *Amplidyne* and under compensated machine is called the *Metadyne*.

21.21.1. Amplidyne (Rotary Amplifier). The amplidyne is a rotary amplifier in which a small amount of power supplied to excite the field is amplified in the generator output. Though the ordinary separately-excited generator operated at low flux densities can be used as a power amplifier since the power required for exciting the field is only 1 to 3 per cent of the rated output of the generator and a small change in the power to the field can be made to produce a large proportionate change in the power output. However, a much larger power amplification can be had by using the amplidyne, since in this machine power is amplified in two stages. Rotary amplifiers are generally built for outputs up to 100 kW.

This machine uses the cross-field principle and a single two-pole armature winding.

The basic constructional features for a 2-pole armature amplidyne are shown in Fig. 21.75 (*a*). The poles, however, are divided and give the appearance of a four-pole field though actually the poles are in pairs with the same polarity. The poles may be salient ones as used in ordinary dc machines or may be slotted so as to take a distributed winding. The machine is sometimes termed as a split-field machine because it has divided poles. Commutating poles may be arranged in all the interpolar gaps. Control windings and compensating windings are arranged, as illustrated in Fig. 21.75 (*a*). Compensating winding in the form of a concentrated coil wound around the field cores, as shown in Fig. 21.75 (*a*) is used in many cases but such a winding does not completely neutralize the armature load current mmf and some distortion of the flux, therefore, exist. For better compensation a distributed type of compensating winding, properly located in slots in the pole faces must be used. The magnetic circuit may then consist of slotted punchings. This form does not provide as much space for a variety of field windings as the form with polar projections, and the various control inputs are more usually combined in an electronic or magnetic amplifier providing the control excitations instead of in the machine itself. Magnetic circuit should be of laminated construction throughout.

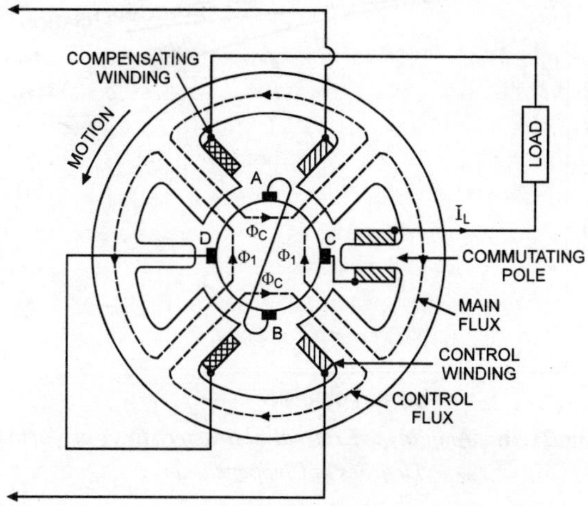

(*a*) *Constructional Feature of Amplidyne*

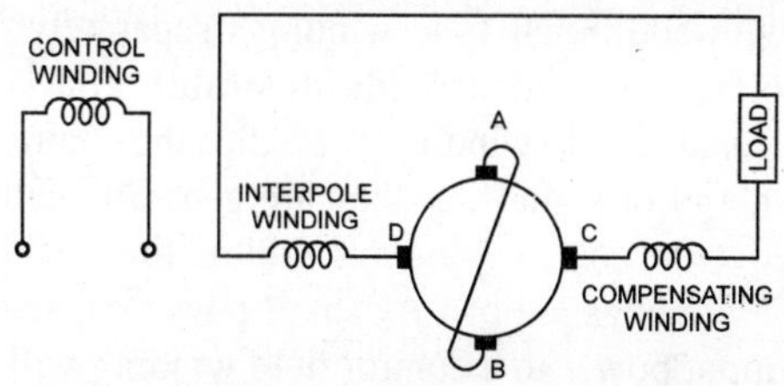

(*b*) *Connection of Amplidyne*

Fig. 21.75

The action of the machine is now explained. The control windings produce a small control flux Φ_c in the horizontal axis. When the machine is driven, a voltage is developed between the short-circuited brushes A and B and, therefore, current flows between them (the control winding is provided with few ampere-turns in order to limit the current through the short-circuited armature winding). Now the current I_1 through the short-circuited armature winding produces armature reaction flux at right angles to the control flux Φ_c which in turn develops emf between the output brushes C and D. In the figure separate flux paths have been shown. Although the fluxes are superimposed in the poles and parts of the armature and yoke, in the absence of saturation they are mutually independent.

When the amplidyne is loaded, load current I_L flows which in turn sets up the flux opposing the control flux Φ_c. It is essential, therefore, for the armature reaction mmf set up by the output current I_L to be neutralised if the machine is to operate as a high gain amplifier. A compensating winding is, therefore, connected in series with the power output brush terminals. The machine operated in this manner is called the amplidyne. This device is virtually two machines in one. Current through the short-circuited brushes which form the load for the first stage, acts also as the field excitation for the second stage. Mechanical input at the shaft is converted into an electrical output, modulated by the control flux. With full compensation, gains of the order of 20,000 : 1 are possible and the power supplied to the control winding may be fraction of a watt even in a 10-20 kW rated machine. Hence the amplidyne can be fed directly from an electronic amplifier in an automatic control scheme.

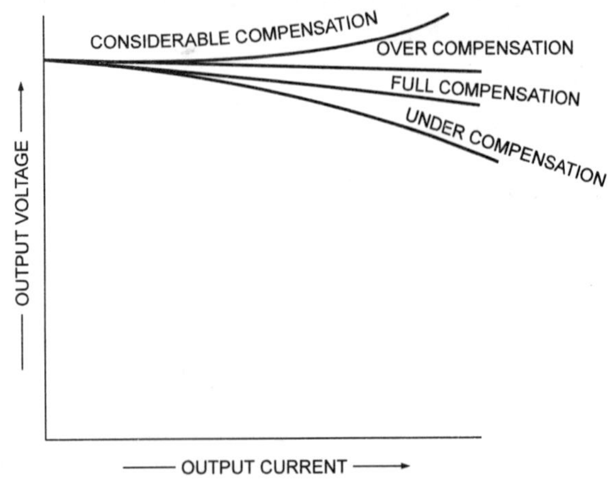

Fig. 21.76 *Amplidyne External Characteristics For Various Degrees of Compensation*

The basic machine, described above, had only one control winding, which was separately excited. Actual amplidynes are provided with additional field windings separately excited as well as self-excited to provide desirable characteristics. Separately-excited field windings are called the control windings since the energy or signals, as they are generally called, which are supplied to the control windings, control the machine output. Since these windings are of very small power capacity, a small change of input power to a control field winding will develop a large change in the output voltage across brushes C and D and, therefore, a large change in power output. It is feasible to construct amplidynes with four separate control field windings so that the output of the machine may be controlled from several independent signals. Commutating poles are fitted at the brush axis to assist commutation, but it is not usual to fit them at the cross axis, except in large machines owing to the small value of I_1, the exciting current.

The external characteristic of an amplidyne is similar to that of an ordinary separately-excited generator, but its slope can be regulated by acting on the compensating winding of the amplidyne. This winding is usually designed to provide magnetizing force somewhat greater than the demagnetising force of armature reaction of the amplidyne *i.e.* to overcompensate for the armature reaction, which is necessary under certain amplidyne operating conditions. But usually the compensating winding is shunted by a rheostat to reduce the magnetising force. This allows the slope of the external characteristic to be regulated. In the case of undercompensation, the voltage drop, with increasing load, will be due not only to the voltage drop in the armature, but also to armature reaction. In the case of full compensation of armature reaction a fall in the amplidyne voltage with increasing load will be due only to voltage drop in the armature. With slight overcompensation, an increase in flux with the load can be made to compensate exactly for the increase in voltage drop in the amplidyne armature, in which case the voltage of the amplidyne will not vary with the load. However, in the case of extreme overcompensation, undesirable self excitation of the amplidyne *i.e.* spontaneous rise in voltage and load current at constant control current (or with the control circuit open) can occur. Such self excitation (similar to that of generator) can continue until saturation of the amplidyne magnetic system becomes effective. Self excitation can cause considerable overloading of the amplidyne and totally disrupt operation of the system containing the amplidyne.

Applications of Amplidynes. An amplidyne is widely used for various purposes, particularly in feedback control systems where the regulated power requirement is in the range of 1 to 50 kW. It may sometimes be found economical to employ it as a voltage regulator in the excitation system of a large alternator to buck or boost voltage in series with the field winding of the main exciter. It can also be employed as a generator in a Ward-Leonard system of speed or position control. In modern control systems these functions of an amplidyne have been taken over by SCR circuitry.

Amplidynes are employed in installations for controlling reversing rolling mills, cold rolling mills, mine hoists, metal cutting and paper machines. However, amplidynes possess a number of shortcomings: poor commutation, which sometimes disrupts the operating conditions of the installation, and a low efficiency of operation. There is a trend in industry to replace amplidynes by static devices, such as magnetic amplifiers operating in conjunction with transistor amplifiers.

21.21.2. Metadyne. In partial compensated machine the opposing flux produced by the output current I_L would neutralize most of the control flux. In control engineering terms, it is a negative current feedback tending to suppress any increase of output current. The action is sufficient to maintain the output current almost constant regardless of change in load resistance. The machine operated in this way is called the Metadyne. In such cases electrical input and output powers are the same, neglecting copper losses and the only mechanical power that is required to overcome the mechanical losses. The output current is controlled by the current through control windings. The output characteristics of Metadyne with different control currents are shown in Fig. 21.77.

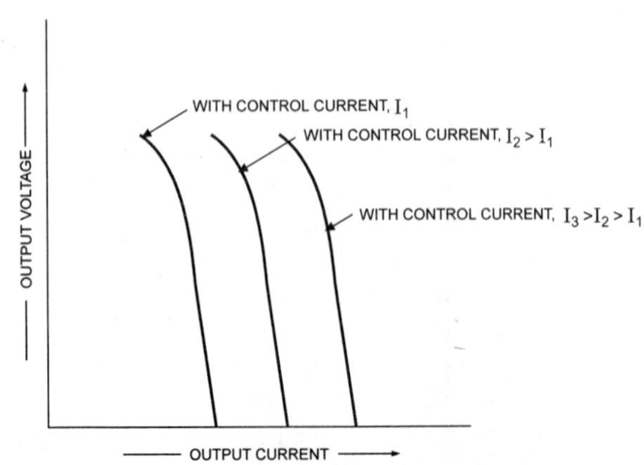

Fig. 21.77 *Metadyne External Characteristics For Various Control Currents*

Note: From the characteristics shown in Figs. 21.76 and 21.77 for amplidyne and metadyne respectively it is obvious that an amplidyne should never be operated on short circuit and a metadyne on open circuit.

Metadyne Generator Rating. The rated output power of a metadyne is appreciably less than that obtainable from the same armature in a normal generator. First the output current has to be reduced by about 10 per cent, owing to the heating caused by the quadrature current I_1 flowing in the armature. Secondly the provision of the commutating zones for two sets of brushes reduces the effective pole-face area, and hence the total flux and terminal voltage by about one-third. The rated output power of the armature is thus about $0.9 \times \frac{2}{3} = 0.6$ of that of the same armature in an ordinary dc generator. For this reason it is not economical to use metadynes for large output power, and, above 15 to 20 kW separately-excited dc generators are employed with metadynes as exciters to provide similar results as from a metadyne generator.

Metadyne Control. The essential part of the metadyne control is metadyne converter. The metadyne converter is a cross-field machine which behaves like a transformer on direct current. The transformation ratio of a metadyne can be varied continuously. It takes power at constant voltage and variable current and delivers the same at constant current and varying voltage.

The metadyne converter essentially consists, in its simplest form, of a 2 pole dc armature with two pairs of brushes and a four pole field magnet, as shown in Fig. 21.78. One pair of the brushes (say A and C) are connected across constant voltage dc supply while the other pair (B and D) are connected to the load (normally a dc series motor).

For understanding the working of a metadyne converter consider first an ordinary dc machine with two poles and two brushes supplied with a current flowing in the direction shown in Fig. 21.79 (*a*). It will cause armature current distribution, as illustrated in the figure with corresponding cross flux, mainly confined to the poles.

Now consider that metadyne converter (a dc machine with two pairs of brushes and two pairs of poles) is running at constant speed and drawing a current I_1 from the dc supply main, which flows through the armature conductors via the brushes A and C, as shown in Fig. 21.79 (*b*). An armature reaction flux Φ_1 set up in usual way is provided with a fairly low reluctance path through

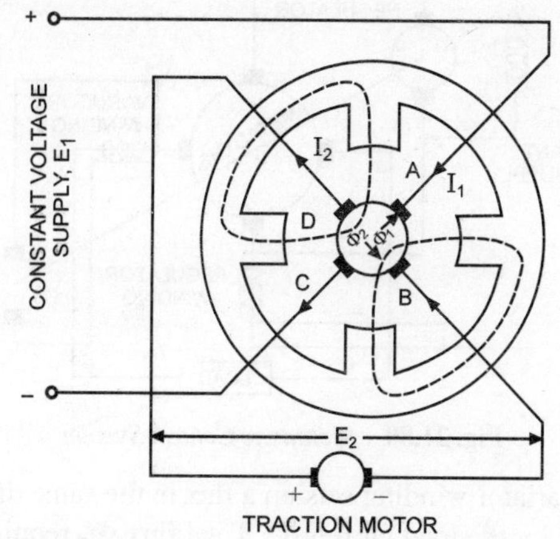

Fig. 21.78 *Simple Metadyne Converter*

the yoke by the four poles, as shown in the figure. Due to rotation of armature conductors in this primary flux, an emf $E_2 = K\,I_1$ is set up between the brushes B and D. When these brushes B and D are connected to a load, a current I_2 flows through the load. The load current I_2 sets up another flux Φ_2 known as *secondary flux*, at the right angles to the first, the distribution is shown in Fig. 21.79 (*c*). This secondary flux Φ_2 causes an emf, $E_1 = K\,I_2$ between brushes A and C opposing the applied voltage. As the applied voltage is constant, the resistance drop is negligible so the back emf E_1 opposing applied voltage and the current I_2 producing E_1 are also constant.

$$\text{Since input} = E_1 I_1 = K I_2 I_1 = K I_2 \times \frac{E_2}{K} = E_2 I_2 = \text{output},$$

therefore, power required to drive the metadyne is very small being equal to the running losses of the machine.

This simple metadyne converter transforms the constant voltage supply into a constant current variable voltage supply to feed the load. The arrangement, therefore, is quite suitable for starting dc traction motors. With this arrangement the load current I_2 and supply voltage V remain constant and as the load increases on account of building of back emf in dc traction motor E_2 and I_1 increases to meet with the increased load.

The metadyne described above has no winding on the poles and is capable of delivering only a single value of constant current but for supplying dc traction motors, after the motor has gained speed, the load current I_2 has to be reduced to the running value. For this purpose the field magnet poles are provided with variator and regulator windings, as shown in Fig. 21.80.

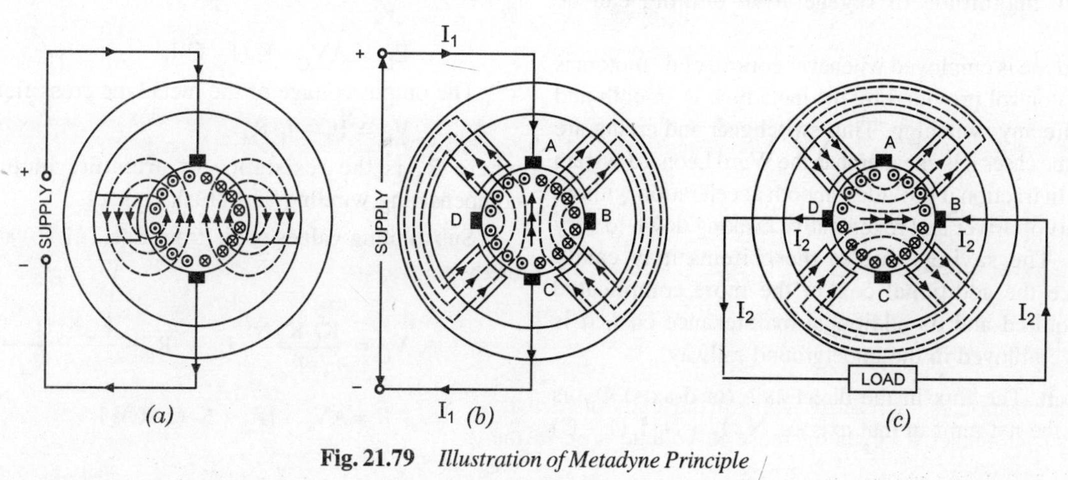

Fig. 21.79 *Illustration of Metadyne Principle*

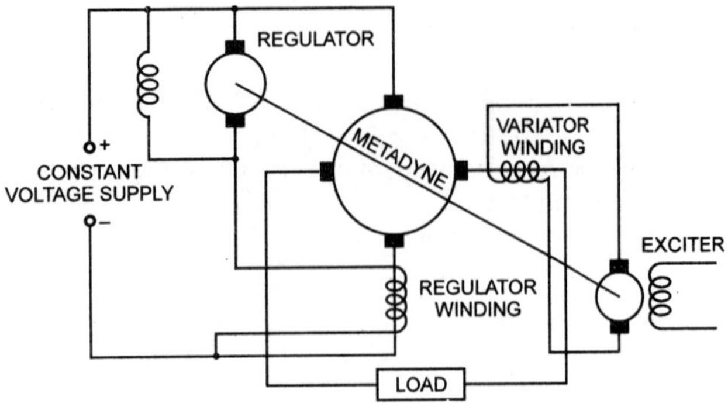

Fig. 21.80 *Metadyne Control System*

The variator winding sets up a flux in the same direction as that set up by the load current I_2. Total flux Φ_2 required in this axis being constant in order to produce a back emf E_1 equal to the constant supply voltage, therefore, if some of this flux is set up by a separate winding, known as *variator winding*, the load current I_2 will decrease and can be of a smaller constant value. Similarly the load current can be increased by causing the current to flow in the variator winding in the opposite direction.

If the output current I_2 is, say, reduced in this way, the voltage remaining the constant, the total output will be reduced but input will remain the same and, therefore, the set will speed up. In order to keep the speed of the metadyne converter constant, an additional winding known as *regulator winding* is provided. By adjusting the current in the regulator winding, the input current can be varied and, therefore, input power can be adjusted equal to output power, the speed of the converter remaining the same.

The regulator winding is supplied from a small dc shunt generator mounted on the shaft of the metadyne. Any tendency towards a change in speed of the metadyne will cause corresponding change in the emf set up by the shunt machine and as it acts in the opposite direction of the supply voltage, so corresponding change in the regulator winding current will result in.

The variator winding is supplied excitation from an exciter mounted on the same shaft, as shown in Fig. 21.80.

With metadyne converter, regenerative braking can be accomplished very easily by reversing the field of the traction motor. This causes the reversal of direction of induced emf E_2 which in turn will change the direction of current I_1. Thus current I_1 can be supplied back to the supply source. By controlling the magnitude of reversed excitation of traction motors supplied by metadyne, the magnitude of regenerative braking can be regulated.

The metadyne is employed whenever control of dc motors is required. The control provided by the metadyne is smooth and does not require any switching. Thus switchgear and arcing are avoided. In some cases it is cheaper than the Ward Leonard system in initial cost. In traction it provides smooth acceleration without skill on the part of driver and regenerative braking down to very slow speeds. The savings due to these items may easily counterbalance the additional cost of the more complicated equipment required and its additional maintenance cost. It is already being employed in the underground railway.

EMF Equation. The flux in the direct axis (or d-axis) Φ_d is developed by the net mmf in that axis *i.e.* $N_C I_C - N_a I_L(1 - C)$ where N_C and N_a are the effective turns of the control field coils and armature respectively, I_C is the current flowing through the control field coils, I_L is the current supplied by the brushes CD and C is the ratio of effective compensation turns to effective armature turns. Thus

$$\text{Flux in d-axis, } \Phi_d = K_1 [N_C I_C - N_a I_L (1 - C)] \quad ...(21.11)$$

if saturation is negligible and K_1 is a constant depending upon the magnetic reluctance of the d-axis.

The voltage induced between brushes A and B,

$$E_q = K_a \Phi_d N = [K_C NI_C - (1 - C) K_q NI_L] \quad ...(21.12)$$

where $K_C N$ is the quadrature axis voltage coefficient at a speed of N rpm in terms of volts per control field ampere and $K_q N$ is the quadrature axis voltage coefficient at a speed of N rpm in terms of volts per unit *d*-axis armature-ampere without any compensation. The above two voltage coefficients are sometimes written as $K_q c$ and $K_q d$ respectively.

The quadrature axis (*q*-axis) armature current

$$I_q = \frac{E_q}{R_q} = \frac{1}{R_q}[K_C NI_C - (1 - C)K_q NI_L] \quad ...(21.13)$$

where R_q is the total resistance in the quadrature axis armature circuit including external resistance connected across brushes A and B, if any.

The flux in quadrature axis,

$$\Phi_q = K_2 N_a I_q \quad ...(21.14)$$

The voltage induced between brushes C and D,

$$E_L = K_a \Phi_q N = K_d NI_q \quad ...(21.15)$$

where K_d is the direct axis voltage coefficient at a speed of N rpm expressed as volts per unit quadrature axis armature-ampere. This is sometimes expressed as K_{dq}.

Substituting the value of quadrature axis armature current I_q from Eq. (21.13) in Eq. (21.15) we get

$$E_L = \frac{K_C K_d N^2}{R_q} I_C - \frac{K_d K_q (1 - C)N^2}{R_q} I_L \quad ...(12.16)$$

At no load, $I_L = 0$, therefore, output voltage at no load,

$$E_{L0} = \frac{K_C K_d N^2}{R_q} I_C = \frac{K_C K_d}{R_q} N^2 \times \frac{V_C}{R_C} = AV_C \quad ...(21.17)$$

where A is the voltage amplification at no-load, V_C is the control field voltage and R_C is the control field circuit resistance.

Let $\frac{K_d K_q N^2}{R_q}$ be equal to K then

$$E_L = AV_C - K(1 - C)I_L \quad ...(21.18)$$

The output voltage of the metadyne cross-field generator,

$$V_L = E_L - I_L R_a$$

where R_a is the resistance of armature including that of compensating winding and brush contact.

Substituting values of E_L from Eqs. (21.16) and (21.18) we have

$$V_L = \frac{K_C K_d N^2}{R_q} I_C - \left[R_a + \frac{K_d K_q (1 - C)N^2}{R_q}\right] I_L$$

$$= AV_C - [R_a + K (1 - C)] I_L \quad ...(21.19)$$

Also $V_L = R_L I_L$ ...(21.20)

where R_L is the load resistance.

Substituting

$V_L = R_L I_L$ in Eq. (21.19) we have

$$R_L I_L = \frac{K_C K_d N^2}{R_q} I_C - \left[R_a + \frac{K_d K_q (1-C) N^2}{R_q}\right] I_L$$

$$I_L = \frac{K_C K_d I_C}{(R_L + R_a)\frac{R_q}{N^2} + K_d K_q (1-C)} \quad ...(21.21)$$

$$\text{or } I_L = \frac{AV_C}{(R_a + R_L) + K(1-C)} \quad ...(21.22)$$

If C = 0 as in case of a Rosenberg generator, load current I_L tends to attain a constant value of $\frac{K_C}{K_q} I_C$ because at higher speeds $\frac{(R_L + R_a) R_q}{N^2}$ tends to become zero in comparison to $K_d K_q$.

The external characteristics (output voltage-output current) for a constant control field current I_C and constant speed N can be plotted from Eq. (21.19) for various degrees of compensation C. The nature of characteristics is illustrated in Fig. 21.81 where no-load voltage E_{LO} is given by Eq. (21.17). From Fig. 21.81 it is obvious that the amplidyne type cross-field generator, which has full compensation (*i.e.* C = 1) has the highest output voltage and so the maximum amplification factor.

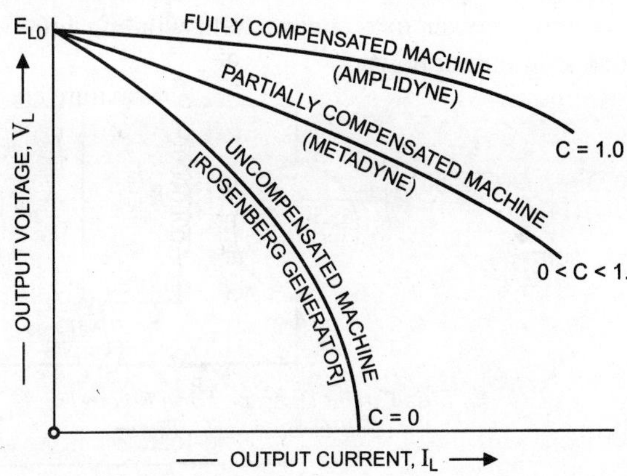

Fig. 21.81 *External Characteristics of Cross-Field Generator For Various Degrees of Compensation at Constant Control Field Current and Constant Speed*

The amplification factor of a cross-field generator can be increased (*i*) either by increasing speed of rotation of the machine but it is limited due to high sparking at the brushes and mechanical stresses in the armature conductors or (*ii*) by reducing the control field ampere-turns which is possible by reducing the air gap in order to create desired flux with smaller control-field ampere-turns. In order to avoid saturation in this direction, lower flux densities are used in the magnetic circuit of the direct-axis field flux.

Example 21.9. A cross - field generator has an input field winding of 100 Ω and 500 turns per pole. It has two pole armature with 400 conductors. The effective resistance of the quadrature circuit is 4 Ω and the internal resistance of the output circuit is 6 Ω including compensating winding. The speed is 3,000 rpm and excitation of 1 AT per pole in the armature winding on any axis or in the control field winding produces a flux of 2 × 10⁻⁵ Wb. Assume 90 per cent compensation and resistance of load is 19 Ω. Neglect nonlinearity. Calculate (*i*) power amplification and (*ii*) number of turns on the compensating winding.

Also assume $I_f = 1$ A and $I_d = 5$ A

Solution: Control field resistance, R_C = 100 Ω

Number of turns on control field coils, N_C = 500

Control current, $I_C = I_f = 1.0$ A

Total resistance in the quadrature axis armature circuit,

R_q = 4 Ω

Resistance of armature including that of compensating winding

R_a = 6 Ω

Load resistance, R_L = 19 Ω

Speed, N = 3,000 rpm.

Armature conductors, Z = 400

Ratio of effective compensation turns to effective armature-turns

C = 0.9

For 1 A of control current

MMF = 500 × 1 = 500 AT/pole

and control flux in direct-axis.

$$\Phi_C = 2 \times 10^{-5} \times 500 = 0.01 \text{ Wb}$$

Quadrature-axis voltage coefficient in terms of volts per control field ampere,

$$K_C N = \Phi_C Z \frac{N}{60} = 0.01 \times 400 \times \frac{3,000}{60} = 200$$

Armature mmf for load current of 1 ampere

$$= \frac{ZI_L}{2AP} = \frac{400 \times 1}{2 \times 2 \times 2} = 50 \text{ AT/pole}$$

So flux, $\Phi_d = 50 \times 2 \times 10^{-5} = 0.001$ Wb

Quadrature-axis voltage coefficient in terms of volts per unit *d*-axis armature ampere without any compensation,

$$K_q N = \Phi_d Z \frac{N}{60} = 0.001 \times 400 \times \frac{3,000}{60} = 20$$

Further $K_d N = K_q N = 20$

Voltage amplification,

$$A = \frac{K_C K_d}{R_q} \times \frac{N^2}{R_C} = \frac{(K_C N)(K_d N)}{R_q R_C} = \frac{200 \times 20}{4 \times 100} = 10$$

$$\text{Constant, } K = \frac{K_d K_q N^2}{R_q} = \frac{(K_d N)(K_q N)}{R_q} = \frac{20 \times 20}{4} = 100$$

$$\text{Now load current, } I_L = \frac{AV_C}{(R_a + R_L) + K(1-C)}$$

$$= \frac{10 \times V_C}{6 + 19 + 100(1 - 0.9)} = 0.286 \, V_C$$

Load voltage, $V_L = R_L I_L = 19 \times 0.286 \, V_C = 5.4286 \, V_C$

Again, $I_C = \frac{V_C}{R_C} = \frac{V_C}{100} = 0.01 \, V_C$

(*i*) Power amplification $= \frac{V_L I_L}{V_C I_C} = \frac{5.4286 \times 0.286}{0.01} = 155.1$ **Ans.**

Effective armature AT/pole per ampere $= \frac{Z}{2AP} = \frac{400}{2 \times 2 \times 2} = 50$

Compensating turns per pole = 0.9 × 50 = 45 turns **Ans.**

Example 21.10. A 2.5 kW, 200 V, metadyne generator driven by a synchronous motor at 1500 rpm, has the following constants.

Control field resistance R_f = 20 Ω

Voltage constant, K_{qf} = 200 V/field A

Voltage constant, $K_{qd} = K_{dq}$ = 100 V/A

Armature resistance, R_a = 5 Ω

The metadyne supplies a 20-ohm resistance at a voltage of 200 V. Find the power input to the control field and power amplification for 50% compensation.

Solution: Resistance of Control field, R_C = 20 Ω

Armature resistance, R_a = 5 Ω

Load resistance, R_L = 20 Ω

Load voltage, V_L = 200 V

Load current,

$$I_L = \frac{V_L}{R_L} = \frac{200}{20} = 10 \text{ A}$$

$$K_C N = \text{Voltage constant } K_{qf} \text{ (or } K_{qC}) = 200 \text{ V/field A}$$

$$K_d N = K_q N = \text{Voltage constant } K_q d = K_{dq} = 100 \text{ V/A}$$

Voltage amplification,

$$A = \frac{(K_C N)(K_d N)}{R_q R_C} = \frac{200 \times 100}{5 \times 20} = 200$$

$$\text{Constant, } K = \frac{(K_d N)(K_q N)}{R_q} = \frac{100 \times 100}{5} = 2{,}000$$

From Eq. (21.22)

Control field voltage,

$$V_C = \frac{I_L}{A}\left[(R_a + R_L) + K(1 - C)\right]$$

$$= \frac{10}{200}\left[(5 + 20) + 2{,}000(1 - 0.5)\right] = 51.25 \text{ V}$$

Control field current,

$$I_C = \frac{V_C}{R_C} = \frac{51.25}{20} = 2.5625 \text{ A}$$

$$\text{Power input} = V_C I_C = 51.25 \times 2.5625 = 131.33 \text{ watts}$$

Power amplification

$$= \frac{V_L I_L}{V_C I_C} = \frac{200 \times 10}{51.25 \times 2.5625} = 15.23 \textbf{ Ans.}$$

21.22 ROTOTROL (REGULEX)

Rototrol is a trade name of rotary amplifier manufactured by the Westingbouse Electric Corporation. This can be used for regulating any electrical quantity such as voltage, current, power or any quantity which can be measured electrically such as speed, torque or tension.

The Rototrol is essentially a small dc generator, similar in mechanical and electrical construction to the standard dc generator of the same size, provided with a number of field windings which can be connected in several ways depending on the requirements. Its armature is driven at constant speed. Input is given to one or more of the control field windings and output is obtained from the armature circuit. All the field windings except the self-excited winding are control windings. The self-excited field winding is commonly designed to be employed as a series field winding connected in series with the output circuit, but it is possible to have the self-excited field winding designed for use as a shunt field connected across the output terminals. The function of self-excited field winding is to provide necessary field strength to furnish the required armature output of the machine and that of control winding is to hold the operation of the machine at the desired points on its magnetisation characteristic curve. The control field windings are of two types—the *pattern field winding* and the *pilot field winding*. The pattern field winding is separately-excited control winding and is used as a calibration or standard of comparison. It is connected in series with a rheostat to a constant voltage source. The pilot field windings are the control field windings which are used for measuring directly or indirectly the quantity to be regulated.

The Rototrol operates on the principle of the unstable action of a self-excited dc generator when its field line coincides with the practically straight portion of its open-circuit characteristic curve. The field line gives the relation between the generated voltage and the Rototrol armature current. The resistor R, known as tuning resistor and connected in series with the self-excited field winding, is adjusted until the resistance of the field circuit is at its critical value. In so far as the self exciting field is concerned the operation of the Rototrol will now be unstable and its output voltage may be any value from zero to maximum value at which the field line and OCC coincide. The voltage at which the Rototrol will operate is stabilized, however, through the control fields. When the control fields are energized, the net mmf produced by them either assists or bucks mmf produced by the self energizing field. Thus only small value of net control-field mmf will stabilize the operation. Thus in Rototrol a small change of input power to a control field winding produces a large change in output power.

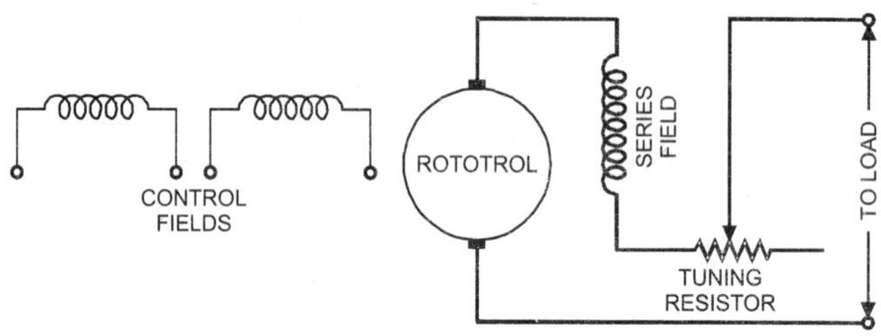

(a) Base Circuit Diagram For a Rototrol With Self-Excited Series Field

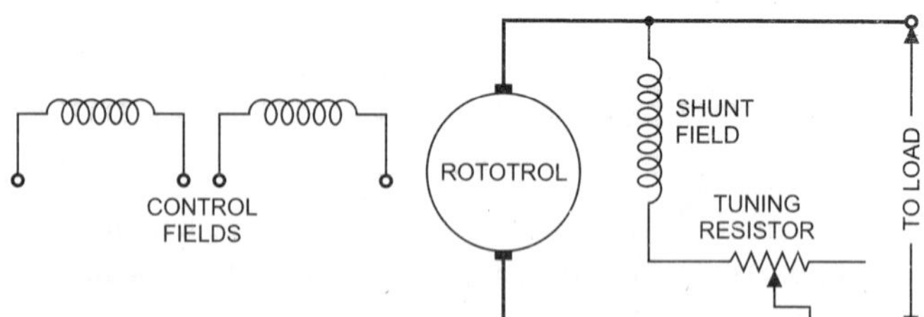

(b)

Fig. 21.82

Circuit diagram of Rototrol for regulating the terminal voltage of a dc generator is shown in Fig. 21.83 In the circuit diagram shown in Fig. 21.83 pattern field winding A is connected in series with a rheostat to the constant voltage source, pilot field winding B is connected in series with a rheostat across the voltage to be regulated and the self-excited field winding C is connected in series with the Rototrol armature, generator field F and a tuning resistor R.

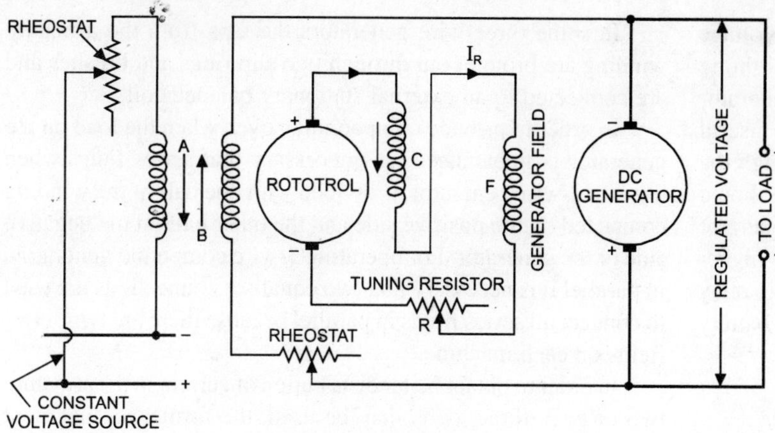

Fig. 21.83

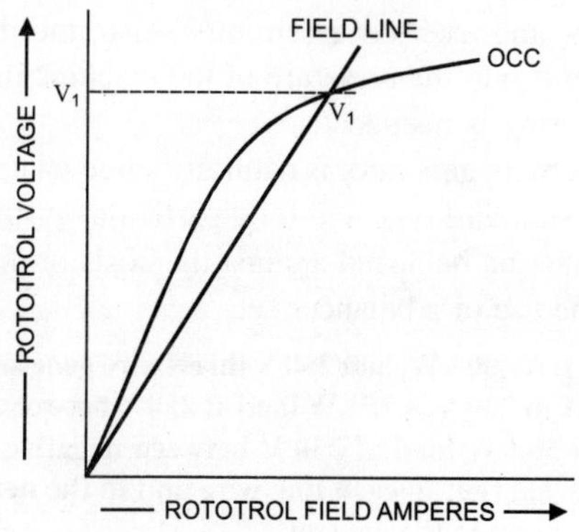

Fig. 21.84

Suppose the tuning resistor R is adjusted until the field line intersects the OCC to give terminal voltage V_1. All the mmf is now supplied by the self exciting field C and the resultant mmf produced by the control windings A and B is zero.

Any decrease in terminal voltage of the generator, that may be due to increase in load connected to it, will reduce the current in field B proportionately and the resultant mmf produced by fields A and B will no longer be zero, but that produced by field A will predominate. This will increase the mmf of the Rototrol field, resulting increase in Rototrol terminal voltage and accordingly increase in current through field C. This action will continue until the generator terminal voltage reaches its regulated value and the resultant mmf produced by windings A and B again becomes zero.

On the other hand increase in generator terminal voltage will cause increase in current through field winding B reducing the Rototrol voltage and accordingly the current in field C until the resultant mmf produced by windings A and B again becomes zero.

As a matter of fact the Rototrol operates satisfactorily if the resultant mmf produced by the control field windings A and B is not exactly zero but does not exceed 5 per cent of the total mmf, when equilibrium is reached.

For controlling current or torque the pilot field B can be connected across the commutating pole field of the regulated machine so that the current in the pilot field B is then proportional to armature current.

For controlling speed the pilot field B may be supplied by a small pilot exciter driven by the motor whose speed is to be regulated.

21.23 THIRD BRUSH GENERATOR

Such generators are used to charge the batteries and lighting system in automobiles and trains. As already said the main requirement for such a service is that the generator must maintain its load current practically constant regardless of speed variations.

Third brush generator is illustrated in Fig. 21.85 In construction it is a simple shunt generator, with the exception that the shunt field is connected to the positive brush A and an auxiliary brush C about 120 electrical degrees away in the direction of rotation and the negative brush B is usually grounded to the frame of the car or engine. The shunt field current obviously depends upon the voltage between positive and auxiliary brushes, the latter being adjustable in position relative to the former and, therefore, charging current can be adjusted–to increase the charging current the auxiliary brush C is required to be shifted towards the grounded brush B in the direction of rotation.

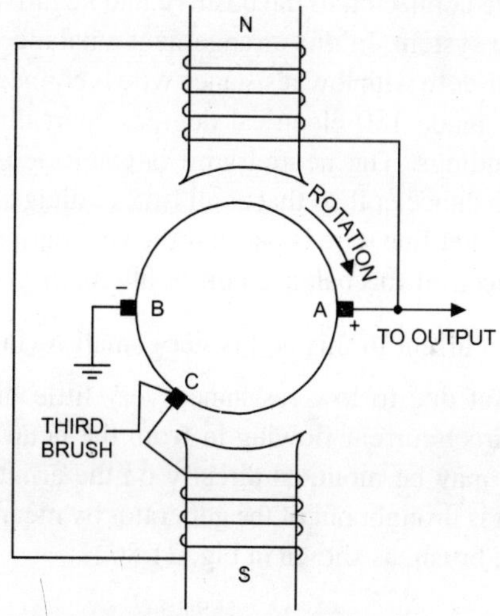

Fig. 21.85 *Third Brush Generator*

In this generator advantage is taken of armature reaction to regulate the charging current. The flux due to armature current produces a distorting effect resulting in weakening of field near the leading pole tips and, therefore, the voltage generated by the inductors between positive brush and auxiliary brush will be reduced by an increase in armature current. This, in turn, reduces the field flux and with it the voltage between the positive and grounded brushes. In other words since induced emf is proportional to the product of flux and speed *i.e.* $E \alpha \Phi N$, at low speeds flux Φ is comparatively high while at high speed flux Φ is correspondingly lower; for a wide range in speed, therefore, the terminal voltage remains nearly constant. Third brush generators are generally designed for weak shunt field and a strong armature reaction field so that the latter may have a substantial effect upon the former one.

With proper setting of the auxiliary brush C, the current supplied to the battery and lamps can be maintained at or near a predetermined constant value.

At low speeds the generated voltage is less than the battery voltage and the battery would also tend to feed current to the generator in a direction opposite to that of the generated voltage of the machine in addition to supply of power to the train lighting system. Such a current would cause motor action in the generator and would put a heavy strain on the batteries with no useful purpose. To prevent this discharging of the battery through the low resistance of the armature circuit of the generator there should be a cut out relay which connects the battery to the generator only after the generator has built upto a voltage slightly higher than that of the battery. If the generator slows down, this relay opens when the current reverses and begins to flow from battery to generator.

21.24 THREE-WIRE GENERATOR

Such generators are used to give 3-wire dc distribution system in which two different voltages are available – high voltage between the two outers for power load and low voltage between either of the outers and middle wire for domestic power and light loads.

It is an ordinary dc generator with special provision being for connecting the neutral wire to the armature windings. The two outers are connected to the positive and negative brushes as in a two wire system. In this arrangement a balance coil wound around an iron core with low resistance wire is connected between the two taps made 180 electrical degrees apart directly to the armature windings. The neutral wire is connected to the mid-point of the balance coil so that at all times voltage between the neutral and either line wire is one half the generator voltage. The voltage induced in the balance coil is alternating at frequency $\frac{PN}{120}$ and the current in this coil is very small owing to its high reactance. But due to low resistance very little hinderence is offered to direct current flowing in from the neutral wire. The balance coil may be mounted directly on the armature and the neutral point is brought out of the generator by means of a single slip-ring and brush, as shown in Fig. 21.86.

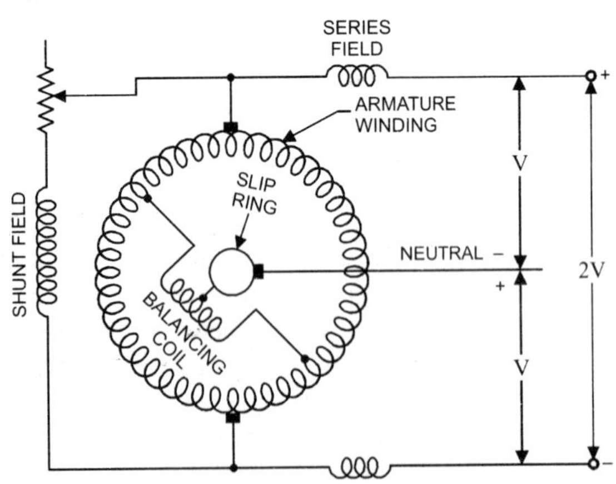

Fig. 21.86 *Three-Wire Generator*

When both sides of the system are equally loaded *i.e.* $I_1 = I_2$; no current flows through the neutral wire. In case of unbalanced loading *i.e.*, $I_1 \neq I_2$ the difference of currents I_1 and I_2 *i.e.* $I_1 - I_2$ flows through the neutral wire. This current divides into two parts in the balancing coil, which enters the armature and then passes through the positive and negative brushes.

In some three-wire generators the taps from the armature winding are brought out through two slip rings and brushes and are connected to an external stationary balance coil.

In order to provide compounding even when the load on the generator is unbalanced it is necessary that series field, when used on 3-wire generator, is divided with one half of the winding connected on the positive side and the other half on the negative side of the generator. For operating 3-wire compound generators in parallel it is necessary that two equaliser connections are used to connect all series fields in parallel because there are two series fields on each machine.

In order to obtain better distribution of current in the armature two or even three, coils may be used, the former arrangement requiring four slip rings and the latter three slip rings, corresponding to two-phase and star-connected three-phase arrangements, and used on ac circuits. Also, the inductance is sometimes built into the armature of the machine in which case only one slip-ring is needed.

The three-wire generator is naturally somewhat more expensive than the standard type and for a particular system this additional cost must be balanced against the costs or disadvantages incident to the use of a balancer set.

Example 21.11. A 400 kW, 500/250 V three-wire generator delivers a load of 250 kW at 500 V, a 75 kW load at 250 V between positive and neutral and a 50 kW load at 250 V between negative and neutral. Determine the current in each line wire and in the neutral and the direction of current in the neutral.

Solution: Current in line wire due to load of 250 kW at 500

$$I = \frac{250 \times 1{,}000}{500} = 500 \text{ A}$$

Current in positive line wire due to load of 75 kW at 250 V,

$$I'_1 = \frac{75 \times 1{,}000}{250} = 300 \text{ A}$$

Current in positive line wire,

$$I_1 = I + I'_1 = 500 + 300 = 800 \text{ A } \textbf{Ans.}$$

Current in negative line wire due to load of 50 kW at 250 V,

$$I'_2 = \frac{50 \times 1{,}000}{250} = 200 \text{ A}$$

Current in negative line wire,

$$I_2 = I + I'_2 = 500 + 200 = 700 \text{ A } \textbf{Ans.}$$

Current in neutral wire,

$$I_N = I_1 - I_2$$

$$= 800 - 700 = 100 \text{ A toward the balancing coil } \textbf{Ans.}$$

21.25 ELECTRIC WELDING GENERATOR

Electric welding is now widely used in the fabrication of iron and steel, repair of fractured castings and the fillings by the deposition of new metal of worn parts. Either dc or ac sources and equipment may be applied to arc welding operations. Because of the steady stream of ions between anode and cathode a dc arc is more easily maintained than an ac arc and is preferred for many types of welding. The direct current is also preferred from the point of view of safety against shuck to the operator.

The main requirements, which must be met by the welding generators, are–they must withstand the short-circuit conditions which occur when the welder short circuits the electrodes (as, for example, when striking the arc), they must provide a more or less constant value of current for a varying arc resistance (varying arc length). These, conditions are met by the external characteristic of the welding generator having a sharply drooping nature. This requires the use of separate excitation, and a small exciter generator is usually included in the arc welding set. A typical circuit diagram is shown in Fig. 21.87 where a dual excitation system, composed of a separately excited shunt field in differential combination with a series field, is used. There are also other types of welding generators designed to give required characteristics.

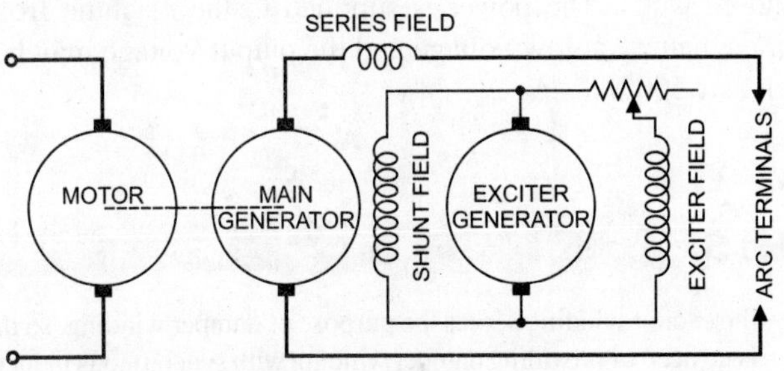

Fig. 21.87 *Circuit Diagram For Arc Welding Generator*

In case of multiple-operator equipment an adjustable series resistor is usually included for each arc circuit, the power being supplied from a constant voltage source. This method is not economical for heavy duty or continuous work as the energy is lost in series resistors. For heavy duty service arc welding generator, described above, is used.

21.26 SERIES BOOSTER

When a series generator is operated at low flux densities so that its terminal voltage rises in direct proportion to the load current ($E \alpha \Phi N$, where speed N is constant) and is used to compensate the line drop in a transmission system it is called a *series booster.*

The wiring diagram of a series booster is shown in Fig. 21.88. The series booster is driven by a shunt motor connected to the bus-bars. Since booster generator is to compensate the line drop only, its output therefore, need not to be as much as that of the load connected to the feeder inspite of the fact that it has to carry the feeder total current.

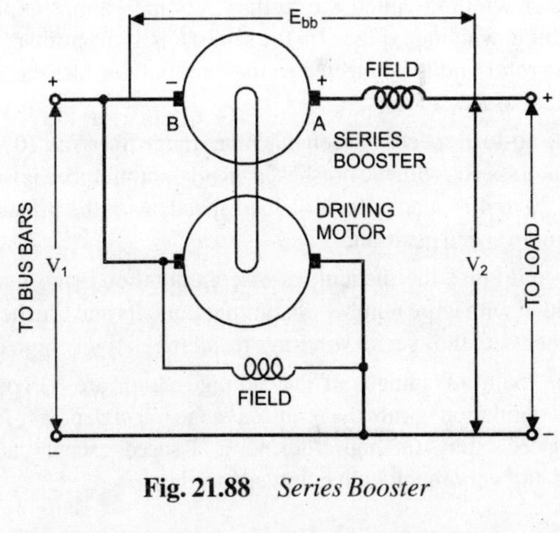

Fig. 21.88 *Series Booster*

In operation, such a motor-generator set must be provided protection against the possibility of getting the driving motor accidently disconnected from the main busbars. Should this happen, the series generator would immediately start working as a series motor without load, reversing its direction of rotation and attempt to race. How it happens is explained below.

When the machine is working as a generator, booster voltage E_{bb} adds to V_1 to give V_2; brush A is then positive and brush B is negative. However, when the booster acts as a motor because of losing its motive power, brush B becomes positive and brush A negative. Finally, since the field current has not changed direction, the booster rotates in reversed direction and tends to run away as unloaded series motor.

21.27 DYNAMOTORS

For obtaining voltage much different from that available from the supply source *dynamotors* are used. Though the same purpose can be served by use of a motor-generator set in which the motor is supplied with power from the available source and the driven generator develops the required voltage, but such an arrangement is expensive and inefficient since two machines are required.

Dynamotor is a single machine that functions both as a motor and a generator. It has a single field structure and an armature with two armature windings and two commutators. Two independent armature windings are usually placed in the same slots, one above the other (in some cases alternate sets of slots are used for each winding) and are connected to separate commutators one at each end of the shaft. Special dynamotors have been constructed with as many as four armature windings and four commutators, two at each end of the shaft, in such an arrangement one winding is connected to the available source and the remaining three windings develop three different voltages required. Dynamotors are less expensive and more efficient than motor-generator sets; the higher efficiency results because the rotational (stray) losses consisting of the friction, windage and iron losses, and the field copper losses are due to a single machine. The dynamotors have the drawback, however, that the generated voltage cannot be controlled and is fixed by the voltage supplied to the motor end.

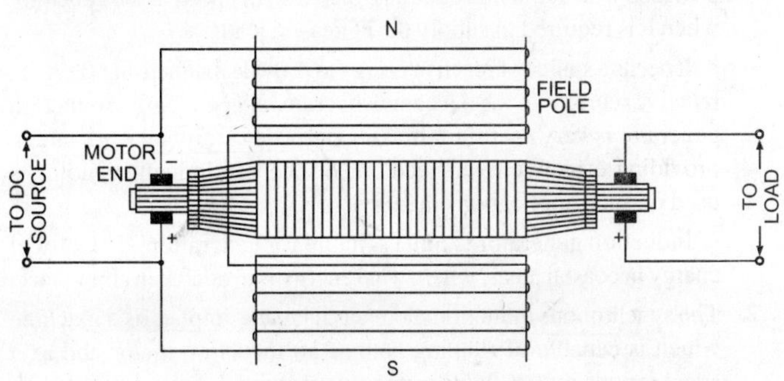

Fig. 21.89 *Dynamotor*

Commutation in dynamotors is usually excellent because the armature reaction effect results from two sets of ampere-turns in opposite directions. This is true because the current in the motor winding is always opposite to that in the generator winding; the net armature ampere-turns are exactly those required to provide sufficient torque for the rotational losses. In addition, the shaft has to transmit but little torque.

When interpoles are fitted to such a machine, the windings then would be relatively feeble under normal conditions, as armature reaction is not to be neutralized. For preventing sparking at sudden momentary overloads on the *lt* side, it is well to have powerful windings on the interpoles, and the reluctance of the interpole magnetic circuit may with advantage be increased to permit of this, by inserting non-magnetic distance pieces between the interpole and the yoke.

On analysing the dynamotor it is found that the induced emf per conductor is the same for either windings because the same flux is cut at the same speed. If the two windings are similar (both lap or wave), the counter-emf in the motor and the generated emf in the generator winding will be directly proportional to their respective number of turns. This implies that any change in speed or any change in flux affects the two windings equally and there will always be a definite ratio between the output voltage to the input voltage if armature drops are neglected. It can be explained in another way also, if the field is weakened the motor speeds up but since the generator armature winding cuts less flux, the product of speed and flux remains constant so that the generated voltage remains constant. On the other hand, if the field is strengthened, the motor slows down but since the armature winding cuts more flux, the product of flux and speed remains again constant and, therefore, generated voltage remains constant. The emf induced in the generator winding is given as the voltage applied to the motor multiplied by the ratio of number of conductors i.e. $V = \frac{Z_G}{Z_M}$ where V is the voltage applied to the motor end Z_G is the number of conductors in generating winding and Z_M is the number of conductors in motoring winding.

Dynamotors are employed mostly for radio and other communication services when rating is of the order of a few hundred watts. The power is supplied to the machine from a storage battery at low voltage and the output voltage may be as high as 1,500 V.

HIGHLIGHTS

1. A polyphase induction machine connected to a constant voltage and frequency mains and driven at a speed higher than synchronous one, operates as an induction generator. Induction generator is often called asynchronous generator, because its speed is not in synchronism with the line frequency.

   At supersynchronous speeds, the conditions of an induction generator are electrically similar to those at subsynchronous speeds for induction motor. In this connection it is to be noted that the equivalent circuit and all the characteristic equations of the induction motor also apply to the induction generator when the slip s is introduced with a negative sign.

   The induction generator does not have definite speed for a given frequency, as the synchronous machine has, but the speed with constant frequency varies with the load, the load (active power) being practically proportional to the slip, s. The prime mover must be run at speed slightly more than synchronous speed. The slip speed is usually the same as that when the machine operates as a motor i.e. 30 rpm. So it is essential that the prime mover driving the induction generator should have a rising speed characteristic. The prime mover must be provided with automatic control to increase the speed of the generator when it is required to supply the increased load.

   It needs a source of reactive energy to provide magnetising (lagging reactive) current to it, so it cannot operate independently. Induction generator is very useful for braking purpose in railway work and in providing effective overspeed brake when an induction motor is used on such applications in hoists.

   Induction generator is quite suitable for generation of electrical energy in coastal areas, where wind energy is available in abundance.

2. The synchronous induction motor, as its name implies, is a machine which is capable of running both as an induction motor and as a synchronous motor, the former being its mode of operation during the starting period, and the latter its mode of operation during normal running.

   Synchronous induction motor is provided with a long air gap so as to be stiff machine with a larger overload capacity. These machines are provided with a heavy rotor winding in order to have a small slip, which facilitates in pulling it into synchronism. Also in order that the induced emf in the field at the start may not be too high, the field turns provided are few in number and the excitation voltage is kept low. As the exciter winding serves the purpose of damper winding, so there is no need of providing damper winding with synchronous induction motor.

   Synchronous induction motor is essentially a motor, having the induction motor feature like high starting torque with low starting current combined with the synchronous motor features like constant speed and power factor control.

   The synchronous induction motor is rarely built for ratings below 25 kW because of the relatively higher cost of the exciter. Synchronous induction motors are used where a high starting torque and constant speed operation are required. Typical applications of synchronous induction motors include fans, pumps, blowers, generators, air-compressors, ammonia compressors, machinery and line shafting in industrial works such as cement mills, rolling mills, flour mills, paper mills, rubber works and textile mills.

3. The variable speed commutator type 3-phase induction motor, also known as *Schrage motor*, has shunt characteristics and adjustable speed. In this motor speed is controlled by injecting a voltage of slip frequency in the secondary circuit.

   This motor is provided with three windings—primary, secondary and tertiary windings. The primary winding is placed in the lower part of the slots of rotor and is connected to the 3-phase ac supply mains by means of slip rings and brushes and produces the working flux. The secondary winding is phase wound and located on stator. Another winding called the tertiary, compensating, adjusting or regulating winding is placed in the same slots as the primary winding on the rotor and is connected to the commutator like the armature winding of a dc motors.

   The no-load speeds of such machines range from 7 to 10% of their full-load speeds with the brushes in the maximum speed position and from 20 to 40% above the full-load speeds with the brushes in the minimum speed position.

   To minimise the difficulties of commutation, such motors are provided with large number of commutator bars and brushes, which together with the special windings result in relatively high cost.

   The main advantages of the Schrage motor are (*i*) continuous speed regulation within the required range (*ii*) high power factor for high speed settings (*iii*) high efficiency at all speeds except synchronous speed and very much high at lower speeds.

One major drawback of the Schrage motor is that its operating voltage is limited about to 700 V, because the power is to be supplied to the motor through the slip rings. Other disadvantages are its low power factor at low speed settings and its poor commutation and high cost.

Schrage motor is used only where adjustable speed is required as in bakery, machinery, stokers, printing machines, calenders etc.

One important application of the Schrage motor is to drive the hosiery-knitting and ring-spinning machines, where an automatically controlled speed variation of 10 to 30% is often required. They are also used for a wide range of industrial services, driving fans, pumps, conveyors, packaging machinery, paper mills etc. Sizes up to several hundred kW have been built. They are normally rated on a constant torque basis, the kW output varying in direct proportion to the speed.

4. Three-phase ac commutator motor may be either series or shunt type. The three-phase ac series commutator motor has a 3-phase stator winding similar to that of an ordinary induction motor and may be wound for high voltage if required. The rotor carries a winding similar to that of a dc armature and its commutator is provided with three brush sets per pair of poles spaced 120 electrical degrees apart. The stator winding is connected to the ac supply through the primary of a transformer whose secondary is connected to the brushes. Thus, in effect, the stator and rotor windings are in series so that the working flux is dependent on the current giving series speed characteristic. A transformer has been inserted between the stator and rotor in order to reduce the voltage applied to the commutator winding and thus avoid the commutation difficulties.

With the brushes in the limiting high speed position, the speed does not increase rapidly with the decrease in load, and the characteristic under such a condition approaches that of a compound wound dc motor. So there is no need of providing a centrifugal cut-out device for protection of motor against excessive speed.

The regulation of speed solely by brush displacement makes external resistance unnecessary and avoids the power losses consequent upon their use, and therefore, the efficiency of the motor remains high over the whole operating range. The power factor is also high, approaching unity for speeds near and above synchronous speeds.

The speed range is limited between 40 and 150 per cent of synchronous speed owing to commutation difficulties. Such a motor is very useful for driving loads whose torque increases with speed, such as compressors, centrifugal pumps etc. This is useful when large starting torque is required. However, a three-phase series motor is not widely employed in practice.

In a 3-phase ac shunt commutator motor, the stator is connected directly to the supply while the rotor brushes are connected to the supply through a transformer. Speed variation is obtained by changing the tapping points on the transformer. It is to be noted here that brushes are fixed in position in the motor of this type. Such a motor gives more or less constant speed at all loads, for a given tapping on the transformer.

5. Phase advancer is a particular type of ac exciter which may be connected in the rotor circuit of an induction motor to improve the power factor. The principle used is that of injection through the slip rings of the motor a current which is leading with regard to the rotor voltage. This current relieves the stator circuit of the duty of magnetising the machine and as leading current is supplied to the rotor at the low voltage corresponding to the slip of the rotor, the kVA capacity of the phase advancer need be only 5% or less of kVA correction affected in the main supply circuit. The full corrective kVA could have to be provided if a static or rotary capacitor were used in the supply circuit. The phase advancer is thus very advantageous for use with large motors.

6. Linear induction motors are induction motors that provides motion in a straight line rather than rotational motion. In a conventional electric motor, the rotor spins inside the stator; in a linear motor the stator is unwrapped and laid out flat and the rotor moves past in the a straight line. Linear motors often use superconducting magnets, which are cooled to low temperatures to reduce power consumption.

A linear induction motor is not that widespread compared to a conventional motor, taking its economic aspects and versatility of usage into consideration. But there are quite a few instances where the linear induction motor is indeed necessary for some specialised applications such as (*i*) automatic sliding doors in electric trains (*ii*) mechanical handling equipment, like propulsion of a train of tubes along a certain route (*iii*) metallic conveyor belts.

7. The motors which respond to the error signal abruptly and accelerate the load quickly are called the *servomotors*. The torque-inertia (T/J) ratio is an important figure of a servomotor because the load is to be accelerated by it abruptly.

The fundamental characteristics to be sought in any servomotor are :

(*i*) The motor output torque should be proportional to the voltage applied to it (i.e. control voltage developed by the amplifier in response to an error signal).

(*ii*) The direction of the torque developed by the servomotor should depend upon the instantaneous polarity of the control voltage.

Servomotors are of two types namely dc servomotors and ac servomotors. AC servomotors are generally preferred to dc servomotors except for use in very high-power systems. For very high power systems dc motors are preferred because they operate more efficiently than comparable ac servo-motors. This enables them to stay cooler. An efficient motor also prevents excessive power waste, although power waste is generally not a prime concern in servomechanism.

8. A *stepper motor*, sometimes also called the *stepping motor* or *step motor*, is a brushless synchronous motor which divides a full rotation into a number of steps. Unlike a brushless dc motor which rotates continuously when a fixed ac voltage is applied to it, a step motor rotates in discrete steps angles. The stepper motors are manufactured with steps per revolution of 12, 24, 72, 144, 180 and 200 resulting in stepping angles 30, 15, 5, 2.5, 2 and 1.8 degrees per step. The step angles can be as small as 0.72° or as large as 90°.

The angle through which the motor turns or the shaft moves for each pulse is known as the *step angle* which is expressed in degrees. The resolution of positioning of a stepper motor is decided by the step angle. Smaller the step angle, greater the number of steps per revolution and higher the resolution of positioning of the motor. Higher the resolution, greater is the accuracy of positioning of objects by the motor.

Stepper motors work on the principle of *electromagnetism.* There is a soft iron or magnetic rotor shaft surrounded by the electromagnetic stators. The rotor and stator have poles which may be teethed or not depending upon the types of stepper motor. When the stators are energized the rotor moves to align itself along with the stator (in case of a variable reluctance stepper motor). This way the stators are energized in a sequence to rotate the stepper motor.

By construction the stepper motors come into three broad classes:

1. Permanent magnet stepper motor.
2. Variable reluctance stepper motor.
3. Hybrid stepper motor.

Stepper motors have rotors of two types—either of the permanent magnet type or the variable reluctance type. Both types of stepper motors, i.e., permanent magnet and variable reluctance type operate on the same basic magnetic principle.

Hybrid stepper motor is in fact a permanent magnet stepper motor with constructional features of toothed rotor used in variable reluctance stepper motor.

9. The switched reluctance (SR) motor is a type of stepper motor, a motor that operates on the same basic principle as a VR motor. Unlike a conventional synchronous motor, both the stator and rotor of a SR motor have salient poles. This doubly, salient arrangement is very effective for electromagnetic energy conversion.

The SR motor has wound field coils as in a dc motor for the stator windings. The rotor, however, has no magnets or coil attached. It is a solid salient-pole rotor made of soft magnetic material (laminated steel). When power is applied to the stator windings, the rotor's magnetic reluctance develops a force that attempts to align the rotor pole with the nearest stator pole. In order to maintain rotation, an electronic control system switches on the windings of successive stator poles in sequence so that the magnetic field of the stator "leads" the rotor pole, pulling it forward.

Rather than using a troublesome high-maintenance mechanical commutator to switch the winding current as in conventional motors, the SR motor makes uses of an electronic position sensor to determine the angle of the rotor shaft and solid state electronics to switch the stator windings, which also offers the opportunity for dynamic control of pulse timing and shaping.

10. Brushless dc motors (BLDC motors or BL motors), also known as *electronically commutated motors* (ECMs, EC motors), are synchronous motors that are powered by a dc electric source via an integrated inverter/switching power supply, which produces an ac signal to drive the motor.

The rotor part of a brushless motor is often a permanent magnet synchronous motor, but can also be a switched reluctance motor, or induction motor.

Brushless motors offer several advantages over brushed dc motors, including more torque per unit weight, more torque per watt (increased efficiency), increased reliability, reduced noise, longer lifetime (no brush and commutator erosion), elimination of ionizing sparks from the commutator, and overall reduction of electromagnetic interference (EMI). With no windings on the rotor, they are not subjected to centrifugal forces, and because the windings are supported by the housing, they can be cooled by conduction, requiring no airflow inside the motor for cooling. This in turn means that the motor's internals can be entirely enclosed and protected from dirt or other foreign matter.

11. Cup (or shell)-type PM dc motors and printed circuit (or pancake) motors are low inertia dc motors, are capable to start, stop and change direction very quickly and are, therefore, suitable for instrumentation applications.

Cup (or shell)-type PM dc motors have hollow cylindrical armatures made of aluminium or copper coils bonded by polymer resin and fiber glass to form a rigid ironless cup, which is fastened to axial shaft.

The printed circuit motor is an example of an ironless or coreless motor with several unique features. The pancake construction uses an axial magnetic field to achieve the short flat construction. Radial field PCB motors are also possible.

These motors have special features of (*i*) very small rotational inertia (*ii*) high torque/inertia ratio (*iii*) high overload capability (*iv*) linear speed-torque characteristic (*v*) smooth torque down to near zero speed and (*vi*) suitability for direct drive applications.

12. Electronic amplifiers can provide output economically up to a certain power level and for large amplified power outputs, rotating amplifiers are employed. Every normal dc generator driven at a constant speed is in a sense a power amplifier in so far as the power output of the machine depends upon the power input to the field windings. Control of the output power can, therefore, be obtained by varying the much smaller power supplied to the field. A power gain from 20 : 1 to 100 : 1 is obtained depending upon the size of the machine.

13. A cross-field machine is virtually two machines in one if armature reaction is negligible or full compensation is provided.

14. Rosenberg generator is an armature-excited machine. Such generators are employed in automobiles and train services.

15. The amplidyne is a rotary amplifier in which small amount of power supplied to excite the field is amplified in the generator output. This machine uses the cross-field principle and is virtually two machines in one.

An amplidyne is widely used for various purposes, particularly in feedback control systems where the regulated power requirement is in the range of 1 to 50 kW.

Amplidynes are employed in installations for controlling reversing rolling mills, cold rolling mills, mine hoists, metal cutting and paper machines. However, amplidynes possess a number of shortcomings: poor commutation, which sometimes disrupts the operating conditions of the installation, and a low efficiency of operation. There is a trend in industry to replace amplidynes by static devices, such as magnetic amplifiers operating in conjunction with transistor amplifiers.

16. In partial compensated machine the opposing flux produced by the output current would neutralize most of the control flux. In control engineering terms, it is a negative current feedback tending to suppress any increase of output current. The action is sufficient to maintain the output current almost constant regardless of change in load resistance. The machine operated in this way is called the Metadyne. In such cases electrical input and output powers are the same, neglecting copper losses and the only mechanical power that is required to overcome the mechanical losses. The output current is controlled by the current through control windings.

17. Third-brush generators are used to charge the batteries and lighting system in automobiles and trains. The main requirement for such a service is that the generator must maintain its load current practically constant regardless of speed variations.

18. Three-wire generators are used to give 3-wire dc distribution system in which two different voltages are available – high voltage between the two outers for power load and low voltage between either of the outers and middle wire for domestic power and light loads.

It is an ordinary dc generator with special provision being for connecting the neutral wire to the armature windings.

19. The main requirements, which must be met by the welding generators, are–they must withstand the short-circuit conditions which occur when the welder short-circuits the electrodes (as, for example, when striking the arc), they must provide a more or less constant value of current for a varying arc resistance (varying arc length). These, conditions are met by the external characteristic of the welding generator having a sharply drooping nature. This requires the use of separate excitation, and a small exciter generator is usually included in the arc welding set.

20. For obtaining voltage much different from that available from the supply source *dynamotors* are used. Though the same purpose can be served by use of a motor-generator set in which the motor is supplied with power from the available source and the driven generator develops the required voltage, but such an arrangement is expensive and inefficient since two machines are required.

Dynamotors are employed mostly for radio and other communication services when rating is of the order of a few hundred watts. The power is supplied to the machine from a storage battery at low voltage and the output voltage may be as high as 1,500 V.

EXERCISES

1. Explain the principle of operation of an induction generator. Can it be classified as a synchronous generator or as an asynchronous generator? **[W.B. Univ. of Technology Electrical Machines-II, 2011-12]**
2. Explain the principle of operation of an induction generator. What are its limitations? **[M.D. Univ. Electrical Machines-II, May-2009; December-2010]**
3. Discuss construction and working principle of an induction generator. **[W.B. Univ. of Technology Electrical Machines-II, 2012-13]**
4. Describe with a neat diagram the principle of operation of induction generator. **[M.D. Univ. Electromechanical Energy Conversion, December-2013]**
5. Discuss the working of induction generator. Mention its advantages and disadvantages. **[Pb. Univ. Elec. Machines-II, December-1995]**
6. When and how a 3-phase induction machine can operate as a motor, as generator and a brake.
7. 'The performance of the induction generator at constant voltage and constant frequency requires a prime mover with rising speed characteristic.' Justify the statement.
8. Explain the equivalent circuit representation of self-excited induction generator. **[Bharathidhasan Univ. Elec. Machines-II, Nov;-1995]**
9. Describe the induction generator regarding circuit and circle diagram. How it is different from dc shunt generator? **[R.G. Technical Univ. Electrical Machines-I, June-2013]**
10. Explain the working of an induction generator. How does the capacitor bank effect its operation? Also discuss its limitations. **[Rajasthan Univ. Electrical Machines-II, 2007]**
11. Explain the operation of induction generator under (*i*) self-excited and (*ii*) line-excited condition. **[M.G. Univ. Electrical Machines-III, November-2005]**
12. Explain the principle of operation of a self-excited induction generator. Give the condition under which this generator may fail to build up. **[J.N. Technological Univ. Electrical Machines-II, January-2014]**
13. What is a synchronous induction motor? **[M.G. Univ. Electrical Machines-III, November-2005]**
14. Describe the construction and working of a synchronous induction motor.
15. Describe the advantages of synchronous induction motor over an ordinary synchronous motor. Indicate various methods of providing excitation for this type of motor.
16. Explain synchronous induction motor. **[Rajasthan Technical Univ. Electrical Machines-II, 2013]**
17. Explain the construction and working principle of Schrage motor. **[Gujarat Technological Univ. Electrical Machines-II, Summer-2014]**
18. Show that in addition to speed control, power factor control is also possible in Schrage motor. **[M.G. Univ. Electrical Machines-III, November-2005]**
19. Discuss how the speed and the power factor of a Schrage motor can be varied by an emf injected into its secondary winding. Compare the frequency of the emfs induced in and injected into the secondary winding.
20. Describe the salient features of the construction of Schrage motor. Explain how are speeds below and above synchronous speed obtained. **[Pb. Univ. Elec. Machines-II, December-1994]**
21. Describe the constructional features of a 3-phase ac series commutator motor. How is the speed control affected in such a motor ?

    How does a 3-phase ac shunt commutator motor differ from a 3-phase ac series commutator motor in construction and operation ?
22. Give the principle of operation, construction and applications of phase advancer.
23. Explain the behaviour of commutator as a frequency convertor.
24. Briefly describe working and applications of linear inductors motor. **[Pb. Technical Univ. Asynchronous Machines, May-2013]**
25. Explain the principle of operation of a linear induction motor. **[M.G. Univ. Electrical Machines-III, November-2005]**
26. Explain the operating principle of a linear induction motor. What is equivalent to synchronous speed of a rotating induction motor ? Mention some of its applications.
27. Explain the principle of operation of a linear induction motor. Draw its characteristics. State its important applications. **[Madurai Kamarajar Univ. Elec. Machines-II, Nov.-1996]**
28. For small and sensitive servomechanism give four reasons why ac servomotors are generally preferred to dc servomotors.
29. What is two phase servomotor ? What are its applications ? Draw its torque-speed characteristics for various control voltages. **[U.P. Technical Univ. Electrical Machines, 2009-10]**
30. Explain the construction, working principle and torque-speed characteristics of two-phase ac servomotor. **[G.B. Technical Univ. Special Electrical Machines, 2012-13]**
31. Explain the working of two-phase ac servomotor and its various applications. **[U.P. Technical Univ. Electrical Machines and Automatic Control, 2009-10]**
32. What is a two-phase servomotor? Draw its torque-speed characteristics for various control voltages. **[U.P. Technical Univ. Electrical Machines and Automatic Control, 2010-11; 2013-14]**
33. Explain the operation of a stepper motor. What are the merits and demerits of stepper motors ? **[U.P. Technical Univ. Electrical Machines, 2008-09]**
34. Explain the principle of operation of stepper motor. What are their applications ? **[U.P. Technical Univ. Electrical Machines, 2009-10]**
35. What do you know about stepper motors? What are the areas of applications? **[U.P. Technical Univ. Electromechanical Energy Conversion-II, 2009-10]**
36. Give classification of stepper motors and briefly discuss their constructional features, working and applications. **[Pb. Technical Univ. Asynchronous Machines, May-2013]**
37. Describe the construction and working principle of any two types of stepper motors. **[G.G.S.I.P. Electrical Machines, December-2011]**
38. Describe the constructional features of 'variable-reluctance' type and 'permanent magnet' type stepper motors. Compare them. Explain the working principle of any one. Discuss the effects of the stepping rate of pulses on the stability of the stepper motor.
39. Explain the working principle of permanent magnet stepper motor with constructional diagram. **[U.P.Technical Univ. Electromechanical Energy Conversion-II, 2006-07]**
40. Explain the operation of variable reluctance type stepper motor. Why does it fall out of step if the pulsating rate of logic controller is increased? **[W.B. Univ. of Technology Electrical Machines-II, 2011-12]**
41. Explain the construction and principle of operation of Hybrid stepper motor. **[G.B. Technical Univ. Special Electrical Machines, 2012-13]**
42. With neat sketch, explain any of drive circuit of stepper motor. **[G.B. Technical Univ. Special Electrical Machines, 2012-13]**
43. Explain the working of single and multistack configured stepping motors. **[Anna Univ. Special Electrical Machines, May/June-2013]**
44. Explain switched reluctance motor. Also write its advantages and disadvantages. **[R.G. Technical Univ. Electrical Machine-II, December-2012]**
45. What is switched reluctance motor? Describe its basic operating principle with regard to a 4-phase, 8-pole stator and 6-pole radial flux (conventional) SR motor. **[W.B. Univ. of Technology Electrical Machines-II, 2011-12]**

46. Briefly explain the construction and principle of operation of a switched reluctance motor. [G.B. Technical Univ. Electric Drives, 2010-11]

47. Describe working principle of switched reluctance motor. Discuss any one converter circuit used. What are its advantages over other ac motor drives? [U.P. Technical Univ. Electric Drives, 2005-06]

48. With neat sketch, explain the construction, operating principle and characteristics of brushless dc motor. [G.B. Technical Univ. Special Electrical Machines, 2012-13]

49. With appropriate circuit diagram explain the operation of a brushless dc motor. Can it run without motor position feedback? [W.B. Univ. of Technology Electrical Machines-II, 2011-12]

50. What are the similarities between a brushless dc motor and a conventional dc motor? Give the advantages of brushless dc motor over conventional dc motor.

51. What is printed circuit motor? Give its construction, operating principle, main features, advantages, disadvantages and applications.

52. What is synchro? Give its construction. Explain with neat diagrams the control type synchro systems and torque transmission type synchro systems?

53. What are resolvers? How are they classified? Give their applications. Give the winding configuration and principle of action of an electromagnetic resolver. [U.P.S.C. I.E.S. Electrical Engineering-I, 1996]

54. What is an amplifier ? What has led to the development of rotating amplifiers ? Name the two machines which fall in the category of rotating amplifiers. Does an ordinary dc generator fall in the category of rotary amplifiers ?

55. Explain the cross-field principle. Name the three machines which operate on this principle.

56. Explain with a neat diagram the essential features of a Rosenberg generator. Describe its applications.

57. What is a cross-field machine? Explain the working principle of amplidyne. [Pb. Univ. Electrical Machines-II, Dec.-1995]

58. In cross-field machines, one pole consists of two polar projection. Explain the reason. [Pb. Univ. Elec. Machines-II, May-1995]

59. Discuss cross-field machine operation as amplidyne. [Pb. Univ. Elec. Machines-II, Dec.-1994]

60. Explain the operation and working of an amplidyne machine as a power amplifier.

Indicate how this machine may be changed into metadyne.

61. Explain the principle of cross-field generator of the metadyne type.

62. Explain with a diagram the principle of operation of a rototrol. How can it be used to regulate the voltage of a dc generator ?

63. What is third brush generator ? Explain its principle of operation.

64. What is a 3-wire generator ? What function does the balancing coil have in operation of 3-wire generator? Why must the series field of 3-wire generators be divided ?

65. Give the main requirements, which must be met by welding generators.

66. Give the principle of operation and application of a series booster.

67. What is a dynamotor ? Describe the constructional features of a dynamotor. Why is it more efficient than the motor-generator set ?

68. Write short notes on the following:

(*i*) Synchronous induction motor.
(*ii*) Schrage motor.
(*iii*) Three-phase ac commutator motors.
(*iv*) Phase advancer.
(*v*) Linear induction motor. [R.G. Technical Univ., June-2013; W.B. Univ. of Technology, 2011-12]
(*vi*) Servomotor. [R.G. Technical Univ, June-2013]
(*vii*) AC servomotor. [W.B. Univ. of Technology, 2011-12; 2012-13]
(*viii*) Stepper motor. [M.D. Univ., December-2004, May-2005; W.B. Univ. of Technology, 2012-13]
(*ix*) Switched reluctance motor. [U.P. Technical Univ., 2013-14]
(*x*) Brushless dc motor. [G.B. Technical Univ., 2012-13]
(*xi*) Printed circuit motor.
(*xii*) Synchro. [W.B. Univ. of Technology, 2012-13]
(*xiii*) Rotary amplifiers.
(*xiv*) Amplidyne.
(*xv*) Booster.

SHORT ANSWER TYPE QUESTIONS WITH ANSWERS

Q. 1. Can an induction generator be classified as a synchronous generator or as an asynchronous generator?

Ans. An induction generator is classified as an asynchronous generator because its speed is not in synchronism with the line frequency.

Q. 2. Why the induction generator is often called an asynchronous generator ?

Ans. Because the speed of an induction generator is not in synchronism with line frequency, so it is often called an asynchronous generator.

Q. 3. What are the limitations of induction generators?

Ans. The main limitations of induction generators are:

1. It cannot be operated independently and the method of providing self-excitation by connecting suitable capacitors across its terminals is of limited practical importance in view of the magnitude of capacitors required.
2. The other drawback is that it can deliver only leading current whereas most commercial loads require lagging current. Considerable KVA capacity synchronous apparatus is required to supply the total quadrature (lagging) current.
3. Induction generators are not helpful in system stability than the addition of parallel reactors with a rating equal to the generator magnetising reactance.
4. Dangerously high voltages may occur with induction generators over long transmission lines if the synchronous machines at the far end become disconnected and the line capacitance excites the induction machines. Motors with power factor correction capacitors may generate if the capacitors remain connected to the running machine when for some reason the supply is cut off.

It is highly important to prevent shock loads from exceeding the maximum generating torque of the machine or causing the overcurrent trips to function during periods of regeneration.

Q. 4. Why an induction generator cannot be operated independently ?

Ans. An induction generator needs a source of reactive energy to provide magnetising (lagging reactive) current to it, so it cannot operate independently.

Q. 5. Is circle diagram valid for induction generator ?

Ans. Yes.

Q. 6. Give the condition under which induction generator may fail to build up.

Ans. An induction generator needs magnetising current (a lagging reactive current), in the same way as an induction motor. So it requires a source of reactive energy which may be a synchronous generator operating in parallel with an induction generator and feeding into a common mains or static capacitors connected in parallel with it. If the capacitor bank is not sufficient to meet the needs of induction generator, the generator may fail to build up.

Q. 7. Can the operating principle of a self-excited induction generator be used for braking the induction motor ?

Ans. Yes, In this case the induction machine receives its excitation from the bank of capacitors and operates as an induction generator.

Q. 8. What is the application for induction generator ?

Ans. Induction generator is very useful for braking purpose in railway work and in providing effective overspeed brake when an induction motor is used on such applications as hoists.

Q. 9. What is a synchronous induction motor. **[M.G. Univ., November-2005]**

Ans. The synchronous induction motor, as its name implies, is a machine which is capable of running both as an induction motor and as a synchronous motor, the former being its mode of operation during the starting period, and the latter its mode of operation during normal running.

The synchronous induction motor is essentially a motor, having the induction motor feature, like high starting torque with low starting current combined with the synchronous motor features like constant speed and power factor control.

Q. 10. An ordinary induction motor is provided with a very short air gap, while in a synchronous induction motor it is necessary to provide relatively larger air gap. Why?

Ans. For an ordinary induction motor, the value of magnetising current is kept as small as possible to ensure a good power factor. A very short air gap, such as would be expected in plain induction motor, would result in a very heavy armature reaction when operating as a synchronous motor because the total field ampere-turns would be reduced for a given number of armature ampere-turns. The actual fact is that a synchronous motor needs a heavy excitation to enable it to operate on a leading power factor. The relatively larger air gap though results in a low power factor when operating as an induction motor, but at normal speed this is immaterial since the power factor can be controlled by variation in dc excitation. The larger air gap and the increase in $\frac{\text{excitation emf}}{\text{synchronous impedance}}$ results in a higher peak synchronous torque. So the synchronous induction motor is provided with a large air gap as provided in a plain synchronous motor. A long air gap gives a stiffer machine with a larger overload capacity.

Q. 11. What are the limitations of synchronous induction motor?

Ans. The extra cost, low efficiency as compared to that of standard types, the small copper space available with a distributed rotor winding, and the necessary compromise between excessive ring voltage at start, or excessive rotor current in operation are severe limitations.

Q. 12. In what rating synchronous induction motors are available ?

Ans. Synchronous induction motors are available in large sizes (ratings ranging from 25 kW to 30,000 kW).

Q. 13. The synchronous induction motor is rarely built for ratings below 25 kW. Why?

Ans. The synchronous induction motor is rarely built for ratings below 25 kW because of the relatively higher cost of exciter.

Q. 14. Give the applications of synchronous induction motors.

Ans. Synchronous induction motors are used where a high starting torque and constant speed operation are required. Typical applications of synchronous induction motors include fans, pumps, blowers, generators, air-compressors, ammonia compressors, machinery and line shafting in industrial works such as cement mills, rolling mills, flour mills, paper mills, rubber works and textile mills.

Synchronous induction motors are very often installed alongwith other induction motors, so that they may improve the overall power factor of the system. Their leading kVAR capacity is designed to offset the lagging kVAR demand from the induction motors. They have been made for ratings as high as 30 MW.

Q. 15. A 3- phase, 6-pole, 50 Hz Schrage motor has the ratio injected emf to induced emf in the secondary winding as 0.33. What is the no-load speed of the motor ?

Ans. 670 rpm because the ratio of no-load speed to the synchronous speed is (I – K) where K is the ratio of the injected voltage to the secondary standstill voltage.

Q. 16. Why the Schrage motor is provided with a large number of commutator bars and brushes ?

Ans. The Schrage motor is provided with a large number of commutator segments and brushes in order to overcome commutation difficulties.

Q. 17. Why is primary winding in Schrage motor provided on its rotor and secondary on stator ?

Ans. Primary and tertiary windings are provided on the rotor so that both of these windings are acted upon by the same magnetic field, the induced voltage in the secondary and the voltage injected into the secondary are of same frequency (equal to slip frequency) and phase and magnitude of injected voltage is controllable.

Q. 18. What happens to the power factor of a Schrage motor when operating at subsynchronous speed if the brush shift is given against the rotation of the motor ?

Ans. The pf will improve.

Q. 19. Give the advantages of Schrage motor.

Ans. The main advantages of the Schrage motor are (*i*) continuous speed regulation within the required range (*ii*) high power factor for high speed settings (*iii*) high efficiency at all speeds except synchronous speed and very much high at lower speeds.

Q. 20. What are the drawbacks of Schrage motor?

Ans. One major drawback of the Schrage motor is that its operating voltage is limited about to 700 V, because the power is to be supplied to the motor through the slip rings. Other disadvantages are its low power factor at low speed settings and its poor commutation and high cost.

Q. 21. Give the applications of Schrage motor.

Ans. Schrage motor is used only where adjustable speed is required as in bakery, machinery, stokers, printing machines, calenders etc.

One important application of the Schrage motor is to drive the hosiery-knitting and ring-spinning machines, where an automatically controlled speed variation of 10 to 30% is often required. They are also used for a wide range of industrial services, driving fans, pumps, conveyors, packaging machinery, paper mills etc. Sizes up to several hundred kW have been built. They are normally rated on a constant torque basis, the kW output varying in direct proportion to the speed.

Q. 22. How is speed control affected in a 3-phase ac series commutator motor ?

Ans. In a 3-phase ac series commutator motor, the speed control is affected by moving all brushes simultaneously round the commutator by means of a hand wheel.

Q. 23. 3-phase ac series commutator motor does not need any protection against excessive speed. Why ?

Ans. With the brushes in the limiting high speed position, the speed does not increase rapidly with the decrease in load, so in a

3-phase ac series commutator motor excessive speed protection needs not to be provided.

Q. 24. How is speed control affected in a 3-phase ac shunt commutator motor ?

Ans. In a 3-phase ac shunt commutator motor the speed is adjusted by changing the tapping points on a transformer inserted between the supply mains and the brushes, which are in fixed position.

Q. 25. What is phase advancer.

Ans. Phase advancers are used to improve the power factor of induction motors. The low power factor of an induction motor is due to the fact that its stator winding draws exciting current which lags behind the supply voltage by 90 electrical degrees. If the exciting ampere turns can be provided from some other ac source, then the stator winding will be relieved of exciting current and the power factor of the motor can be improved. This job is accomplished by the phase advancer which is simply an *ac exciter*. The phase advancer is mounted on the shaft as the main motor and is connected in the rotor circuit of the motor. It provides exciting ampere turns to the rotor circuit at slip frequency. By providing more ampere turns than required, the induction motor can be made to operate on leading power factor like an overexcited synchronous motor.

Q. 26. What is a linear induction motor?

Ans. Linear induction motors are induction motors that provides motion in a straight line rather than rotational motors. In a conventional electric motor, the rotor spins inside the stator; in a linear motor, the stator is unwrapped and laid out flat and the rotor moves past in the a straight line. Linear motors often use superconducting magnets, which are cooled to low temperatures to reduce power consumption.

Q. 27. How can the speed of a linear induction motor be controlled ? **[Bharathiyar Univ., Nov.-1995]**

Ans. The speed of a linear induction motor can be controlled by varying both frequency and voltage simultaneously.

Q. 28. What is the principle of operation of a linear induction motor ?

Ans. The principle of operation of a linear induction motor is the same as that of a conventional rotary induction motor. It is given below:

Whenever there occurs a relative motion between the field and the short-circuited conductors, currents are induced in them which results in electromagnetic forces. Under the influence of the electromagnetic forces developed, the conductors, according to Lenz's law, try to move in such a way as to eliminate the induced currents.

Q. 29. Give the advantages of linear induction motor.

Ans. The **advantages** of linear induction motors are (*i*) low initial cost (*ii*) low maintenance cost because of absence of rotating parts (*iii*) simplicity (*iv*) no limitation of tractive effort due to adhesion between the wheel and the rail (*v*) no limitation of maximum speed due to centrifugal forces (*vi*) no overheating of rotor because the motor moves continuously over cool rotor plate leaving behind heated rotor portion and (*vii*) better power to weight ratio.

Q. 30. Give the disadvantages of LIMs.

Ans. The **disadvantages** of linear induction motors are (*i*) poor utilisation of motor due to transverse edge effect and end effect (*ii*) larger air gap and nonmagnetic reaction rail (rotor plate) need more magnetizing current resulting in poor efficiency and low pf (*iii*) very high capital cost of reaction rail fixed along the centre line of the track (*iv*) complications and high cost involved in providing three phase collector system along the track (*v*) difficulties encountered in maintaining adequate clearances at points and crossings.

Q. 31. What is a servomotor?

Ans. A servomotor is a rotary actuator that allows for precise control of angular position, velocity and acceleration. It consists of a suitable motor (dc or ac) coupled to a sensor for position feedback. It also needs a relatively sophisticated controller, of a dedicated module designed specifically for use with servomotors.

Q. 32. The rotors of servomotors are designed with relatively long length and smaller diameter?

Ans. The rotors of servomotors are designed with relatively long length and smaller diameter so as to keep the rotor inertia small. The rotor with smaller inertia results in high speed response which is an essential requirement for servomotor.

Q. 33. Give the special features of servomotors.

Ans. Servomotors have the following special features:

1. They are capable of developing high torque at all speeds (including zero speed) and holding a static (no motion) position.
2. Because of low inertia, they are able to accelerate, decelerate and reverse in direction of rotation quickly.
3. They do not overheat at standstill or lower speeds.
4. They are able to return to a given position without any drift.

Q. 34. Which types of servomotors generally lend themselves to small horse power requirements ?

Ans. Two-phase and the shaded pole type induction motors.

Q. 35. Give one advantage of the two-phase servomotor over the shaded-pole type.

Ans. Better response to very small control signals.

Q. 36. Why are 'drag-cup-constructions' used in rotors of certain fhp machines for control purposes ?

Ans. In drag-cup-construction all heavy iron laminations remain stationary and only a light cup rotates, so the inertia is quite small.

Q. 37. Give the applications of servomotors.

Ans. Servomotors find wide use in radar, process controllers, tracking and guide systems, computers, robotics, CNC machinery or automated manufacturing.

Q. 38. What is a stepper motor?

Ans. A *stepper motor*, sometimes also called the *stepping motor* or *step motor* is a brushless synchronous motor which divides a full rotation into a number of steps. Unlike a brushless dc motor which rotates continuously when a fixed ac voltage is applied to it, a step motor rotates in discrete steps angles. The stepper motors are manufactured with steps per revolution of 12, 24, 72, 144, 180 and 200 resulting in stepping angles 30, 15, 5, 2.5, 2 and 1.8 degrees per step. The step angles can be as small as 0.72° or as large as 90°.

Q. 39. Define stepping angle in a stepping motor. **[Anna Univ., May/June-2013]**

Ans. The angle through which the motor turns or the shaft moves for each pulse is known as the *step angle* which is expressed in degrees. The resolution of positioning of a stepper motor is decided by the step angle. Smaller the step angle, greater the number of steps per revolution and higher the resolution of positioning of the motor. Higher the resolution, greater is the accuracy of positioning of objects by the motor.

Q. 40. What is the function of drive circuit in a stepping motor? **[Anna Univ., May/June-2013]**

Ans. The function of drive circuit is to convert the controller command signals into the power necessary to energize the motor windings.

Q. 41. Give the limitations of stepper motor. **[Pb. Technical Univ., May-2013]**

Ans. The lack of feedback of a stepper motor limits its performance, as the stepper motor can only drive a load that is well within its capacity, otherwise missed steps under load may lead to positioning errors and the system may have to be restarted or recalibrated.

Q. 42. What are the two types of stepper motors ? **[M.S. Univ; April 1996; Bharathiyar Univ., April-1997]**

Ans. Permanent-magnet type and variable-reluctance type motors.

Q. 43. For the same number of poles which stepper motor produces (*i*) largest stepping angle, (*ii*) smallest stepping angle ?

Ans. (*i*) Permanent magnet type—largest stepping angle
(*ii*) Variable reluctance type—smallest stepping angle.

Q. 44. Where is stepper motor used ? **[Bharathidhasan Univ., Nov.-1996]**

Ans. Stepper motors are used in industrial situations which call for precise positioning of an object or precise control of speed without having to resort to closed-loop feedback.

Q. 45. Why do hybrid stepper motors have sometimes more than six phases?

Ans. Hybrid stepper motors have sometimes more than six phases so as to have smaller step angles.

Q. 46. Is there any disadvantage(s) of having more phases?

Ans. Yes, there are minor disadvantages of having more phases such as increase in the number of leads to be brought out from the motor, requirement of interconnections to the drive circuit and requirement of switching devices.

Q. 47. Do stepper motors need fans (internal or external) for cooling?

Ans. No. Because the heat produced in the stator winding of the stepper motor is conducted through the stator iron into the case which is cooled by conduction, convection and radiation.

Q. 48. What is the main advantage of multistack variable reluctance motor?

Ans. The main advantage of multistack variable reluctance motor is that it is well-suited to high stepping rates.

Q. 49. Will there be any harm if the rotor of hybrid stepper motor is pulled out of its stator?

Ans. Yes. The rotor may become partially demagnetized and, on reassembling, may provide less holding torque.

Q. 50. What is switched reluctance motor?

Ans. SR motor is a type of stepper motor that operates on the same basic principle as a VR motor. A stepper motor is designed suitable for open-loop position and speed control in small power applications, where efficiency is not significant. On the other hand a SR motor is employed in variable speed drives and designed to operate efficiently for wide range of speed and needs rotor position sensing.

Q. 51. Give the difference between synchronous reluctance motor and switched reluctance motor. **[Anna Univ., May/June-2013]**

Ans. SRM is quite different from a standard reluctance motor in two aspects. A synchronous reluctance motor has the same number of poles on stator and rotor but the rotor of a SRM has lesser poles than the stator, which is an essential feature to provide self-starting capability and bidirectional control. They are also different in stator construction—the synchronous reluctance motor has a cylindrical stator with distributed winding, while the SRM has a salient pole stator with concentrated coils like a dc motor.

Q. 52. Write the torque equation of switched reluctance motor. **[Anna Univ., May/June-2013]**

Ans. Neglecting nonlinearity of the magnetic circuit, the instantaneous torque produced in such machines may be given as

$$T = \frac{1}{2} i^2 \frac{dL}{d\theta}$$

where *i* is the instantaneous current in the exciting winding and L is the self-inductance of that winding varying as a function of the angular position of the rotor. It may be noted that the torque developed is independent of the direction of current flow in the windings, so that unidirectional currents can be used to control the motor. The direction of rotation can be reversed by exciting the phases in the reverse sequence.

Q. 53. Give the applications of SR motor.

Ans. SR motor finds wide applications such as (*i*) traction (*ii*) general purpose industrial drives (*iii*) office and business equipment and (*iv*) domestic appliances like vacuum cleaners, food processors, washing machines etc.

Q. 54. What are brushless dc motors?

Ans. Brushless dc motors are synchronous motors that are powered by a dc electric source via an integrated inverter/switching power supply which produces an ac signal to drive the motor. In this context, alternating current does not imply a sinusoidal waveform, but rather a bi-directional current with no restriction on waveform. Additional sensors and electronics control the inverter output amplitude and waveform and frequency.

The rotor part of a brushless motor is often a permanent magnet synchronous motor, but can also be switched reluctance motor or induction motor.

Q. 55. What are the advantages of brushless motors?

Ans. Brushless motors offer several advantages over brushed dc motors, including more torque per unit weight, more torque per watt (increased efficiency), increased reliability, reduced noise, longer lifetime (no brush and commutator erosion), elimination of ionizing sparks from the commutator, and overall reduction of electromagnetic interference (EMI). With no windings on the rotor, they are not subjected to centrifugal forces, and because the windings are supported by the housing, they can be cooled by conduction, requiring no airflow inside the motor for cooling. This in turn means that the motor's internals can be entirely enclosed and protected from dirt or other foreign matter.

Q. 56. What is printed circuit motor?

Ans. The printed circuit motor is an example of an ironless or coreless motor with several unique features. The pancake construction uses an axial magnetic field to achieve the short flat construction. Radial field PCB motors are also possible.

These motors have special features of (*i*) very small rotational inertia (*ii*) high torque/inertia ratio (*iii*) high overload capability (*iv*) linear speed-torque characteristic (*v*) smooth torque down to near zero speed and (*vi*) suitability for direct drive applications.

Q. 57. Give the applications of printed circuit motors.

Ans. Printed circuit motors have been developed specifically to have high performance characteristics when used in direct-drive applications like (*i*) oscillograph (*ii*) high speed paper tape readers (*iii*) X-Y recorders (*iv*) positioning servomotors (*v*) in-built optical position encoder.

It is also manufactured for heavy-duty drives such as battery-driven vehicles, lawn mowers etc.

Q. 58. What are resolvers? **[U.P.S.C. I.E.S., 1996]**

Ans. Resolver is a synchro whose rotor is mechanically driven to translate rotor angle into electrical information corresponding to the sine and cosine of rotor angle. It can interchange rectangular and polar coordinates. It is also called a sine-cosine generator and synchro resolver.

Q. 59. On what principle do Rosenberg generator, amplidyne and metadyne operate ?

Ans. Rosenberg generator, amplidyne and metadyne operate on the principle of cross-field.

Q. 60. How is feedback supply from the batteries to the Rosenberg generator prevented ?

Ans. Some form of automatic reverse current cut-out or a rectifying element is connected in series with the output terminals of the generator to prevent feedback supply from the batteries to the Rosenberg generator.

Q. 61. Why is some form of automatic reverse current cut-out or a rectifying element connected in series with the power output terminals of the Rosenberg generator ?

Ans. To prevent the discharging of the battery through the low resistance of the armature circuit of the generator.

Q. 62. What is difference between Rosenberg generator and the amplidyne?

Ans. Both of the Rosenberg generator and amplidyne are cross-field or armature-excited machines. The main difference between the two machines is that amplidyne is the fully-compensated machine whereas the Rosenberg generator is uncompensated machine.

Q. 63. What is difference between amplidyne and metadyne ?

Ans. Amplidyne is fully compensated armature-excited machine whereas metadyne is a partially compensated armature-excited machine.

Q. 64. Why is series compensating winding employed in the amplidyne?

Ans. Compensating winding is connected in series with the power output brush terminals of the amplidyne to neutralize the armature reaction mmf set up by the load current and increase amplification factor.

Q. 65. Why are commutating poles required in amplidynes ?

Ans. Commutating poles are required in amplidynes to give satisfactory commutation.

Q. 66. Why are amplidynes not extremely overcompensated ?

Ans. In case of extreme overcompensation, undesirable self excitation of the amplidyne i.e. spontaneous rise in voltage and load current at constant control current (or with the control circuit open) can occur. Self excitation can cause considerable overloading of the amplidyne and totally disrupt operation of the system containing the amplidyne.

Q. 67. Why amplidynes are provided with split-poles ?

Ans. Amplidynes are provided with split-poles to provide space for interpoles.

Q. 68. Does a cross-field generator produce only voltage gain, only power gain or both voltage and power gains ?

Ans. Both voltage and power gains.

Q. 69. What is the purpose of compensating winding in cross-field generators ?

Ans. To neutralize the armature reaction mmf set up by the load current and increase amplification factor.

Q. 70. If the number of poles in a cross-field generator is 2, what should be the number of interpoles ?

Ans. Four.

Q. 71. What are the main requirements to be met by the generators used for charging of batteries and lighting systems in automobiles and trains ?

Ans. The main requirements to be met by the generators used for charging of batteries and lighting system in automobiles and trains are that the generator must maintain its load current constant regardless of speed variations and the direction of output current must be independent of direction of rotation.

Q. 72. What is rototrol ?

Ans. Rototrol is a trade name of rotary amplifier manufactured by Westinghouse Electric Corporation.

Q. 73. Why third brush generators are usually designed for weak shunt field and a strong armature reaction field ?

Ans. Third brush generators are usually designed for weak shunt field and a strong armature reaction field so that the latter may have a substantial effect upon the former one.

Q. 74. Where are three-wire generators used ?

Ans. Three-wire generators are used to give 3-wire dc distribution system in which two different voltages are available.

Q. 75. What function does the balance coil have in the operation of a 3-wire generator ?

Ans. The function of the balance coil in the operation of a 3-wire generator is to maintain voltage between the neutral and either line wire one half of the generator voltage.

Q. 76. Is the inductor in a three-wire generator air-cored or iron-cored ?

Ans. Iron-cored.

Q. 77. Why the inductor of a 3-wire generator should be iron-cored ?

Ans. To restrict the flow of alternating current through it.

Q. 78. What are the main requirements, which must be fulfilled by a welding generator ?

Ans. The main requirements to be fulfilled by the electric welding generators are—they must withstand the short-circuit conditions (when the welder short-circuits the electrodes), they must provide approximately constant current for a varying arc resistance.

Q. 79. For what purpose is used series booster ?

Ans. Series booster is used to compensate the line drop.

Q. 80. Why is the output of a series booster not as much as that of the load connected to the feeder inspite of the fact that it carries the total feeder current ?

Ans. It is due to the fact that series booster is to compensate the line drop only rather than supplying total load on rated voltage.

Q. 81. What is the main drawback of a dynamotor ?

Ans. The output voltage of a dynamotor is fixed by the voltage applied to it and cannot be controlled.

Q. 82. Why commutation in dynamotors is usually excellent ?

Ans. Because the armature reaction effect is caused from two sets of ampere-turns in opposite directions.

❑❑❑

SECTION VI

ELECTRONIC CONTROL OF ELECTRIC MOTORS AND OBJECTIVE QUESTIONS

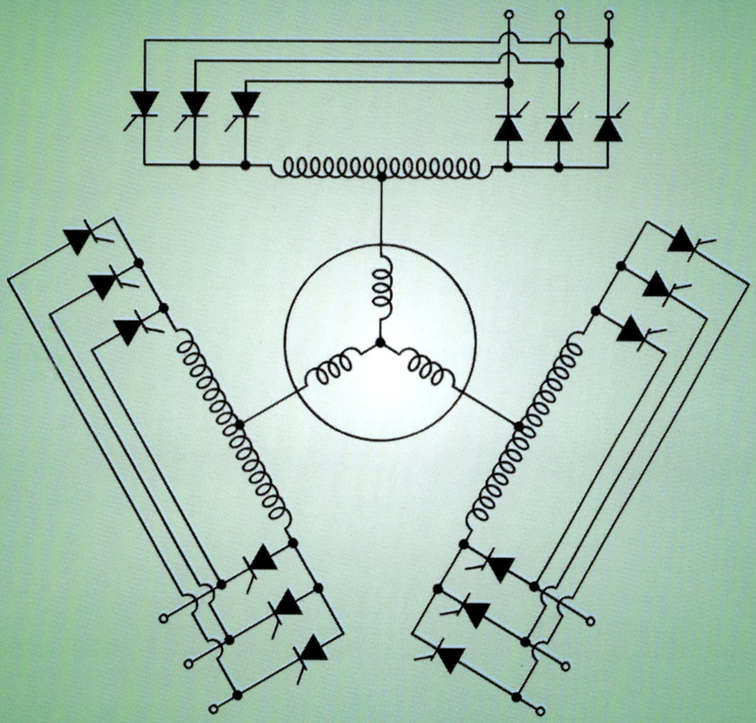

CHAPTERS COVERED IN THIS SECTION

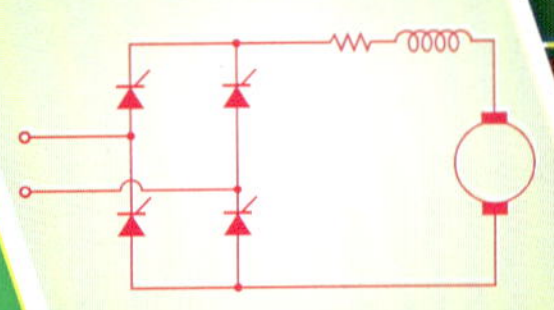

ELECTRONIC CONTROL OF ELECTRIC MOTORS

OBJECTIVE TYPE QUESTIONS

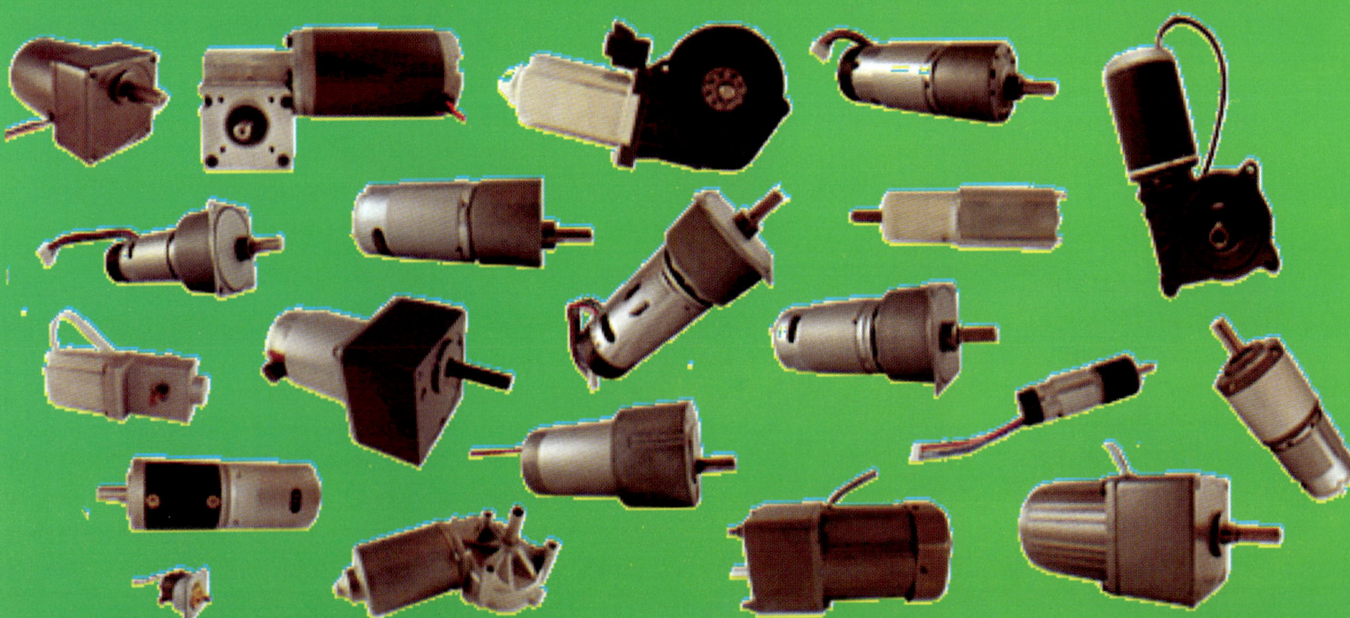

CHAPTER 22

Electronic Control of Electric Motors

INSIDE THIS CHAPTER

22.1 INTRODUCTION

Electric motors are employed in a very large power range, from a few watts to thousands of kilowatts. Many applications require very precise position adjustments (as in robotics). In many applications optimum performance and efficiency are the main concern. Variable speed drive (VSD) systems help in optimisation of process so as to reduce investment costs, operational costs and maintenance costs. Energy saving is another big advantage of variable speed drives.

The advent of thyristors capable of handling large currents has revolutionized the field of electric power control. Thyratrons, ignitrons, mercury-arc rectifiers, magnetic amplifiers motor-generator sets have all been replaced by solid state circuits employing semiconductor diodes and thyristors. Thyristor controlled drives employing both dc and ac motors find wide application in industry as variable speed drives. In the 1960's, ac power used to be converted to dc power for direct control of drive motors with solid state devices (high power silicon diodes and silicon controlled rectifiers). Initially saturable reactors were employed in conjunction with high power silicon rectifiers for dc drives. Nowadays, thyristors are extensively used for ac-to-dc conversion.

Thyristorized control has the following advantages and disadvantages:

Advantages. 1. The response of the control device is faster as it eliminates the time lag introduced by the inductances of the generator field and the armature.

2. Due to low voltage drop across the thyristor, the efficiency of the control system is high.
3. The control device is smaller in size, lighter in weight, cheaper in cost, requiring less space and minimal maintenance.
4. Simple and reliable operation.

Disadvantages. 1. Because of higher ripple content in the converter output, motor heating and commutation problems are serious.

2. Due to switching action of thyristors and nonsinusoidal nature of current, there is more possibility of interference with the communication networks.

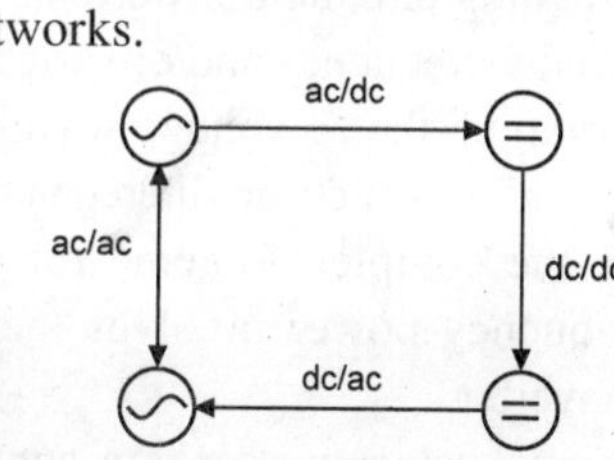

Fig. 22.1 *Four Basic Forms of Power Conversion*

In all variable speed drive systems power electronic converter acts as an interface which accepts electric power from the existing source and converts it in a controlled manner into a suitable form compatible with the particular load or the process for which it is employed. The main sources for electric power are: Single- or three-phase 50 Hz ac from utility systems and dc from storage batteries or solar cells. The four basic forms of power conversion required are: ac to dc, dc to dc, dc to ac and ac to ac. The modern converters are compact, cheap, reliable, durable, flexible, completely controllable. They need reduced maintenance too. They are suitable for all the four basic forms of power conversion mentioned above through rectifiers (ac/dc), choppers (dc/dc), inverters (dc/ac) and cycloconverters or ac regulators (ac/ac).

For dc motor control, controlled dc power from a constant voltage ac supply is obtained by means of *controlled rectifiers* (usually termed as *converters*) using thyristors and diodes. The control of dc voltage is achieved by varying the phase angle at which the thyristors are fired relative to the applied alternating voltage waveform. This scheme of control is known as the *phase control*. In another control system, known as the *integral cycle control*, the current is gated to flow from the ac supply for a number of complete cycles and is then quenched for a further few cycles, the process being repeated continuously. Control is applied by adjusting the ratio of *on* and *off* durations. This method is suitable for the control of fractional kW output dc motors.

Phase controlled converters are simple to operate and are less expensive as they do not require additional circuitry for commutation process. In such converters natural commutation is achieved, *i.e.*, when an incoming thyristor is turned on, it immediately reverse biases the outgoing thyristor and turns it off.

Phase control and integral cycle control methods are also applicable for ac motors wherein the converter circuitry is not required.

Control of dc motors supplied from dc supply is achieved by means of a thyristor switching circuit called the *chopper*. In chopper circuits, the control of the average voltage is achieved by changing the on-to-off time duration ratio for which the dc supply voltage is applied to the motor. This provides an efficient and stepless control of dc motors. The motor can also be made to operate in the regenerative braking mode. Instead of converter circuits, it is possible to use an uncontrolled rectifier, which provides a constant direct voltage, followed by a chopper to provide a variable mean direct output voltage. The chopper controller needs forced commutation of the thyristor.

For the control of ac motors supplied from dc supply, thyristor, transistor or MOSFET based inverters are used. Such switching circuits transfer energy from dc supply to ac load of variable frequency and/or variable voltage. Because of switching operation, the ac voltage waveforms are stepped, the harmonic contents of which are filtered out. As the power supply is normally ac, the complete scheme for obtaining variable-voltage and frequency power involves the use of both an inverter and a converter.

A *cycloconverter* is a control unit for providing variable-voltage and frequency power directly from a fixed frequency supply without the necessity of an intermediate dc stage. The mechanism of voltage and frequency control is a combination of those employed in the phase-controlled converter and the pulse-modulated inverters. Cycloconverters inspite of their attraction of direct ac to ac conversion, suffer from certain drawbacks because of which they did not become popular. Some of these drawbacks are : cycloconverters can generate only a sub-frequency output; they produce output with large harmonic content and have a low input power factor. Cycloconverters are employed for low-speed drives and for controlling linear motors in high-speed transportation systems.

22.2 GENERAL CONFIGURATION OF A MOTOR DRIVE

Figure 22.2 shows a block diagram illustrating the control of a motor drive. The main constituents are power electronic converter, motor, processes, process control computer and controller.

The required characteristics of the motor drive are determined by the process. The process may need a variable-speed drive or a servo drive etc. The motor is selected on the basis of these

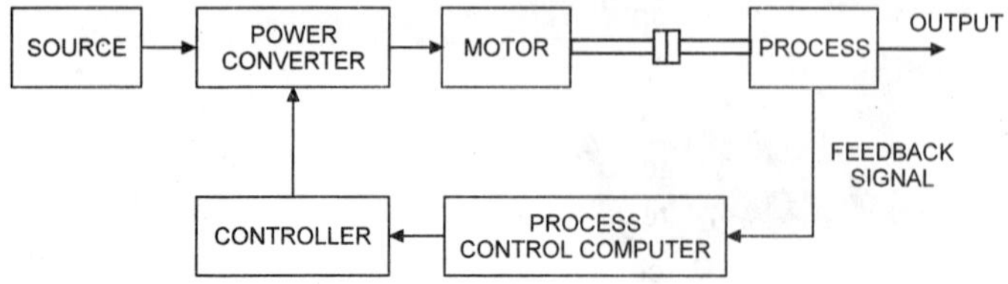

Fig. 22.2 *Block Diagram of a Motor Drive*

requirements. The power electronic converter converts ac single-phase or 3-phase input into a supply suitable to provide desired characteristics of the motor. The process control computer receives a feedback from the process as regards the extent to which the requirements are met. This feedback may be of speed, position etc. The computer feeds an error signal to the controller which, in turn, does the corrective action. The corrective action may be, say, the change in the firing angle of thyristors of the power electronic converter. In some situations the accuracy and response time of the motor to adjust the new speed is of utmost importance. In some other situations the requirements may not be so critical.

22.3 CLASSIFICATION OF POWER CONVERTERS

Solid state power converters are employed for obtaining the appropriate form of electrical energy such as direct current or adjustable-frequency alternating current (required to operate most electronic circuits and motor drives) from fixed-frequency alternating current.

According to functions, the static power converters may be classified in the four basic categories namely ac/dc converters, ac/ac converters, dc/dc converters and dc/ac converters, already discussed in Art. 22.1 [refer to Fig. 22.1].

The inverters (dc/ac converters) may be further categorised as per their source nature as voltage source and current source inverters. A voltage source inverter (VSI) is fed by a constant voltage source system while a current source inverter (CSI) is supplied from a constant current source. A voltage source inverter is one in which dc source has small or negligible impedance. Because of a low internal impedance, the terminal voltage of a VSI remains almost constant with variations in load. It is, therefore, equally suitable to single motor and multimotor drives. On the other hand, CSI is supplied with a controlled current from a dc source of high impedance. Because of a large impedance, the terminal voltage of a CSI varies substantially with a change in load. CSIs are not suitable for multimotor drives because a change in load on any motor affects other motors.

The static power converters may also be categorised into two categories viz. transistor converters and thyristor converters. The transistor converters, which do not require additional commutation circuitry as needed in thyristor converters, have almost completely replaced thyristor converters below 75 kW. The availability of low-cost high-power BJTs has also made possible the use of pulse width modulation (PWM) for improving the motor current waveform.

According to the method of commutation the thyristor converters can mainly be categorised in two types viz. naturally commutated and force commutated converters. In case of ac circuits, ac line voltage is available across the device. When the current in the SCR passes through a natural zero, the device is turned off. This process is known as natural commutation process and the converters based on this principle are known as *line commutated* or *naturally commutated converters.* AC/DC converters, cycloconverters, and ac/ac voltage regulators using SCRs are of the naturally commutated type.

In case of dc circuits, since the supply voltage does not pass through the zero point, some external source is required to commutate the device. This process is known as forced commutation process and the inverters based on this principle are known as force commutated inverters. As the device is to be commutated forcefully, such inverters need complicated commutation circuitries.

Converters may also be grouped into one-quadrant, two-quadrant and four-quadrant converters. A semiconverter is a one-quadrant converter and it has one polarity of output voltage and current. A full converter is a two-quadrant converter and polarity of its output voltage can be either positive or negative. However, its output current has one polarity only. A dual converter is a four-quadrant converter and both its output voltage and current can be either positive or negative.

22.3.1. AC/DC Converters. Rectification is a process of converting alternating current or voltage into a direct current or voltage. This conversion can be achieved by a variety of circuits based on and using switching devices. The widely used switching devices are diodes, thyristors, power transistors, power MOS etc. The rectifier circuits using only diodes as the rectifying elements are known as *uncontrolled rectifiers* because for a fixed value of ac input voltage, the output dc voltage is fixed and cannot be controlled. For some low-power devices where control of output voltage is not required, rectifier circuits using only diodes may be employed. Such circuits have the advantages of low cost and simple circuitry.

Rectifier circuits using thyristors are known as *controlled rectifiers* or *ac-dc converters.* The basic principle of a controlled rectifier is to control the point in time at which the thyristors are allowed to conduct during each cycle. Thus, it is possible to select the time segments of the ac voltage waves which appear at the dc terminals, and the mean output voltage is controlled continuously. This process of control is called the phase control and the controlled rectifiers are, therefore, also called the phase-controlled rectifiers. Since thyristors are available in high voltage and large current ratings, controlled rectifiers using thyristors are employed in many high-power devices. A controlled rectifier may be a single-phase or 3-phase depending on whether the input is single-phase or 3-phase ac.

Controlled rectifiers are of two types namely fully-controlled and half-controlled. The fully-controlled rectifier uses thyristors as the rectifying elements and the dc output voltage is a function of the amplitude of the ac supply voltage and the point-on-wave at which the thyristors are triggered (called firing angle α). The half-controlled rectifiers consist of a mixture of diodes and thyristors, allowing a more limited control over the dc output voltage level than the fully-controlled rectifier. The half-controlled rectifier is cheaper than a fully-controlled rectifier of the same ratings but has operational limitations.

Uncontrolled and half-controlled rectifiers permit power to flow only from the ac system to the dc load and are, therefore, referred to as unidirectional, one-quadrant or semiconverters. However, with a fully-controlled rectifier it is possible, by control of the point-on-wave at which switching occurs, to allow power to be transferred from the dc side of the converter back into the ac system. When this occurs, operation is said to be in the inverting mode. The fully-controlled converters may, therefore, be referred to as bidirectional or two-quadrant converters. Thus it is possible for the phase-controlled converters to provide either a one-quadrant, two-quadrant or four-quadrant operation at its dc terminals. In a four-quadrant operation, two fully-controlled converters, connected back-to-back with one another, are used. Four-quadrant converters or dual converters provide the facility for bidirectional current flow through the load.

The converter type depends on the power to be handled and how much voltage ripple is tolerable. For low powers (below 20 kW), single-phase circuits are adequate but they themselves can take different forms. For high powers (above 20 kW), three-phase circuits are invariably employed.

22.3.2. AC/AC Converters. Thyristor-based ac voltage controllers using the phase control principle convert fixed alternating voltage directly to variable alternating voltage without a change in frequency. Applications of ac voltage controllers or regulators include domestic and industrial heating, transformer tap changing, lighting control, speed control of single-phase and three-phase induction motors etc. Earlier, the devices used for such applications were auto-transformers, tap-changing transformers, magnetic amplifiers, saturable reactors etc. But these devices are now being replaced by ac regulators using thyristors and triacs because of their high efficiency, flexibility in control, compact size and less maintenance. However, ac regulators using thyristors and triacs introduce objectionable harmonics in the circuit. AC regulators are classified as single phase or three phase. Each of these may be half-wave (unidirectional) or full-wave (bidirectional). Because of ac input, ac regulators are always line or naturally commutated and as such no complex commutation circuitry is required in these controllers.

Cycloconverter is defined as a power controller which converts input power at one frequency to output power at a different frequency with one-state conversion. Thus it is an ac/ac converter without any intermediate dc link. Cycloconverters can be classified as single phase-to-single phase, three phase-to-single phase, 3-phase-to-three phase devices. They may also be classified as step-down and step-up. In step-down cycloconverters, the output frequency is lower than the supply or input frequency. In step-up cycloconverters, output frequency is higher than supply frequency. Step-down cycloconverters use line or natural commutation while step-up cycloconverters need forced commutation. Generally step-down cycloconverters are in much use. The desired amplitude of output voltage can be had by changing firing angle of thyristors used in the circuit. With the advent of high-power thyristors cycloconverters are becoming more and more popular. At present, the applications of cycloconverters include speed control of high-power ac drives, induction heating, static VAR compensation, conversion of variable-speed alternator voltage to constant frequency output voltage for use as power supply in aircraft or shipboards.

22.3.3. DC/DC Converters. A dc chopper, also known as dc-dc converter, is a static device (switch) used to obtain variable dc voltage from a source of constant dc voltage. Thus, chopper may be thought of as dc equivalent of an ac transformer as it behaves in an identical manner. Besides the saving in power, the dc chopper offers greater efficiency, fast response, lower maintenance, small size, smooth control, and, for many applications, lower cost than motor-generator sets or gas tubes approaches.

Solid state choppers due to their inherent advantages are widely employed in trolley cars, battery-operated vehicles, traction motor control, control of a large number of dc motors from a common dc bus, marine hoists, forklift trucks, mine haulers etc. They can also be used for regenerative braking in electric traction.

22.3.4. DC/AC Converters. The dc to ac power converters are known as inverters. In other words, an inverter is a circuit which converts a dc power into an ac power at desired output voltage and frequency. The ac output voltage could be fixed at a fixed or variable frequency. The dc input supply may come from a battery, solar cells, fuel cells etc.

The conversion can be achieved either by transistors or thyristors. For low and medium power outputs transistorised inverters are used but for high power outputs, thyristors should be used. For low power outputs, self-oscillating transistorised inverters are suitable but for high power outputs, driven inverters are more common than self-oscillating ones. Moreover for multiphase ac output, there is no alternative other than driven inverters. The driven inverters have better frequency stability because a separate master oscillator is used for this purpose. For the applications in inverters, transistors have some edge over thyristors regarding the switching speed, simplicity in control circuits, higher efficiency and greater reliability. This is mainly due to the fact that thyristor inverters need extra circuits to turn thyristors off, moreover, additional complex logic circuits may be required to avoid false triggering and provide proper commutation timings. Thyristors can handle much higher load currents than transistors and consequently, for high power outputs, thyristors become more desirable than transistors.

The output voltage waveforms of an ideal inverter should be sinusoidal, but the output voltage waveforms of practical inverters are nonsinusoidal and contain certain harmonics. Square wave or quasi-square wave voltages may be acceptable for low and medium power applications, but for high power applications low-distorted sinusoidal waveforms are required. The output frequency of an inverter is determined by the rate at which the semiconductor devices are switched on and off by the inverter control circuitry and consequently, an adjustable frequency ac output is readily provided. The harmonic contents of output voltage can be reduced significantly by switching techniques of available high speed power semiconductor devices. The filtering of harmonics is not feasible when the output frequency varies over a wide range, and the generation of ac waveform with low harmonic content is important. When the ac output voltage of an inverter is supplied to a transformer or ac motor, this output voltage must be varied in conjunction with frequency to maintain the proper magnetic conditions. Some of the important applications of inverters are: variable speed ac motor drives, induction heating, aircraft power supplies uninterruptable power supplies (UPS), high-voltage dc transmission lines etc.

22.4 MATCHING POWER ELECTRONIC CONVERTER AND MOTOR

The motor has to be selected to match the load requirements. The power electronic converter has to match the motor. Some important considerations in matching power electronic converter and motor are as under.

1. ***Current Rating.*** The torque developed by an electric motor depends on the current drawn by it. During most of the operating duration a motor is required to deliver its rated torque, and therefore, the current drawn is also equal to its rated current. But a motor is capable of developing much higher torque (as compared to its rated torque) provided the duration of the torque is small as compared to thermal constant of the motor. Since torque depends on the current, a motor for developing a large torque, requires a large current which has to be supplied by the converter. However, the overcurrent capability of electronic devices is very small. A large current leads to high junction temperature which may destroy the electronic converter. The thermal time constants of the power electronic devices are very small in comparison to those of motors. The current of the converter must be selected properly on the basis of average, rms and peak currents of the motor.

2. ***Voltage Rating.*** Every motor develops a back emf e_b that opposes the applied voltage v. The rate of change of torque depends on the rate of change of current which, in turn, can be expressed as

$$\frac{di}{dt} = \frac{v - e_b}{L} \qquad ...(22.1)$$

where $\frac{di}{dt}$ is the rate of change of current and L is the inductance of the motor. For a reasonably large value of $\frac{di}{dt}$, the output voltage v of converter should be much larger than e_b. Since e_b depends on the motor speed, it is necessary that the voltage rating of the converter is selected on the basis of maximum speed of the motor.

3. ***Switching Frequency and Motor Inductance.*** The steady-state ripple content in the motor current depends on motor inductance L. Large ripple content leads to large motor losses. So the steady-state ripple content in the motor current should be small. This is possible if inductance L is large. But large inductance means that the motor cannot respond quickly to the change in load demand. To meet these opposing requirements, the switching frequency of converter should be high (so that the steady-state ripple content is small). However, a high switching frequency leads to large power losses in the converter. Therefore, a proper selection of switching frequency and inductance is essential.

22.5 DC MOTOR CONTROL THROUGH CONVERTERS

DC motors are widely used in adjustable-speed drives and position control applications. Their speeds below base speed can be controlled by armature voltage control while speeds above base speed are obtained by field flux control. DC motors are preferred over ac motors because these are inherently most suitable for smooth, efficient, wide-range speed control and quick reversals.

Phase-controlled converters provide an adjustable dc output voltage from a fixed ac input voltage. DC choppers also provide adjustable dc output voltage from a fixed dc input voltage. When available supply is ac, an ac-dc converter (phase-controlled converter) is ideally suited to convert the constant ac voltage into a variable dc voltage. Phase-controlled converters are simple to operate and are less expensive as they do not require additional circuitry for commutation process. In such converters natural commutation is achieved, *i.e.*, when an incoming thyristor is turned on, it immediately reverse biases the outgoing thyristor and turns it off. The power conversion efficiency of these converters is more than 95% because of low losses in thyristors. Thus, dc motors can be controlled conveniently and efficiently through phase-controlled converters.

The dc motors used in conjunction with power converters are dc separately-excited motors and dc series motors. The power converters are used in speed control of fractional kW dc motors as well as in large sized motors used in variable-speed reversing drives for rolling mills with motor ratings as large as several MWs.

22.6 THYRISTOR CONTROL OF SEPARATELY-EXCITED DC MOTOR DRIVE

22.6.1. Single-Phase Drives. The single-phase ac is converted into dc by a controlled rectifier or converter and the dc is supplied to the dc motor. By varying the firing angle the voltage applied to the motor can be varied and thus the speed of the motor can be controlled. Single-phase drives can be further classified as

1. Single-phase half-wave converter drives.
2. Single-phase semiconverter drives.
3. Single-phase full converter drives.
4. Single-phase dual converter drives.

1. *Single-Phase Half-Wave Converter Drives.* Figure 22.3 (*a*) shows a single-phase half-wave converter for controlling a separately-excited dc motor. It needs a single thyristor and a freewheeling diode (D_{FW}). Freewheeling diode, also sometimes known as *bypass diode* or *commutating diode*, is used to improve the waveshape of load current and power factor. Freewheeling diode is connected across the motor terminals to allow for dissipation of energy stored in the motor inductance and to provide for continuity of motor current when the thyristors are blocked. It also provides protection against transient overvoltages. Separate converters are employed for the armature and field circuits. It is desirable that the supply to the field winding is provided through a semiconverter or full converter. If half-wave converter is used for supply to field circuit, the high ripple content in the field circuit would cause increase in the iron losses of the machine. It is one quadrant drive, as illustrated in Fig. 22.3 (*b*).

Because of inductance of the field and armature, the thyristor would not turn off at $\omega t = \pi$. Therefore it is desirable to have freewheeling diode (D_{FW}), as shown.

Average value of voltage applied across armature is given by

$$V_a = \frac{1}{2\pi}\left[\int_{\alpha_a}^{\pi} V_{max} \sin \omega t \, d(\omega t) + 0\right]$$

$$= \frac{V_{max}}{2\pi}(1 + \cos \alpha_a) \quad \text{for } 0 \le \alpha_a \le \pi \qquad ...(22.2)$$

where α_a is the firing angle of converter in the armature circuit.

In this circuit the motor current is always discontinuous, resulting in poor performance. This type of drive is used only for small dc motors of rating up to 500 W or so.

2. *Single-Phase Semiconverter Drives.* Figure 22.4 (*a*) shows the circuit of a semiconverter feeding a separately-excited dc motor. Both armature and field circuits are supplied from single-phase ac supply through semiconverters. Because of inductance in the armature and field circuits, freewheeling diodes are required in both the circuits. The voltages applied to the armature and field are

$$V_a = \frac{V_{max}}{\pi}(1 + \cos \alpha_a) \text{ for } 0 \le \alpha_a \le \pi \qquad ...(22.3)$$

$$V_f = \frac{V_{max}}{\pi}(1 + \cos \alpha_f) \text{ for } 0 \le \alpha_f \le \pi \qquad ...(22.4)$$

where α_a and α_f are the firing angles of converters in the armature and field circuits respectively.

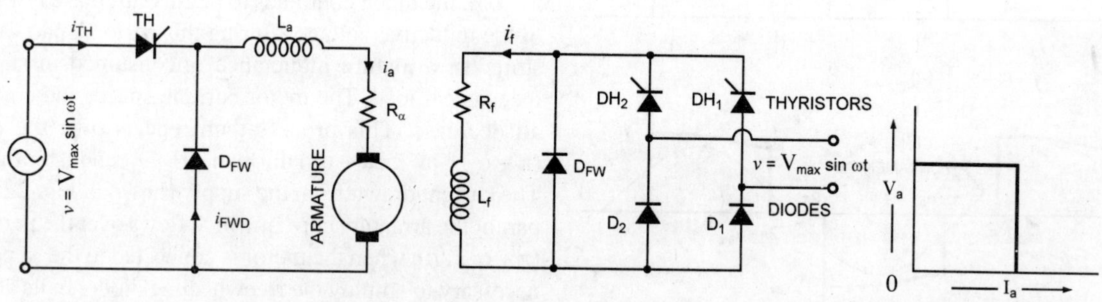

(*a*) Circuit Diagram (*b*) Quadrant of Operation

Fig. 22.3 *Separately-Excited DC Motor Fed From Single-Phase AC Supply Through Half-Wave Converter*

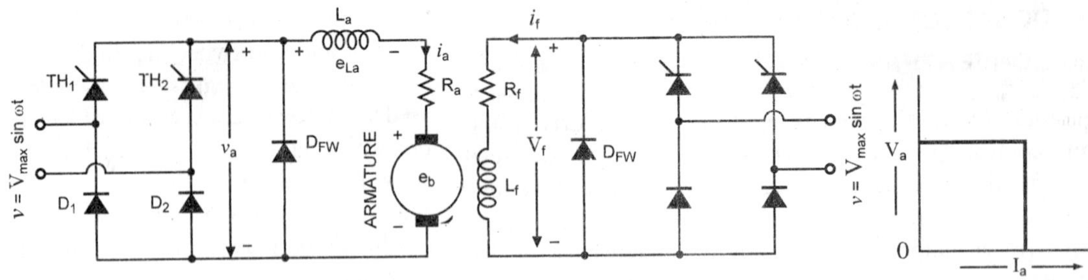

(a) Circuit Diagram (b) Quadrant of Operation

Fig. 22.4 *Separately Excited DC Motor Fed From Single-Phase AC Supply Through Semiconverters*

This is also a one quadrant drive [Fig. 22.4 (*b*)] which provides voltage and current of one polarity at dc terminals. It, therefore, does not provide for regenerative braking, *i.e.*, power flow from dc motor to the ac supply. Where regeneration is not required, this converter is used for reasons of economy. This drive is used for motors up to 15 kW rating.

Typical steady-state voltage and current waveforms are illustrated in Fig. 22.5. The thyristor TH_1 is fired at angle α_a while TH_2 at angle $\pi + \alpha_a$ with respect to supply voltage v and process is repeated continuously.

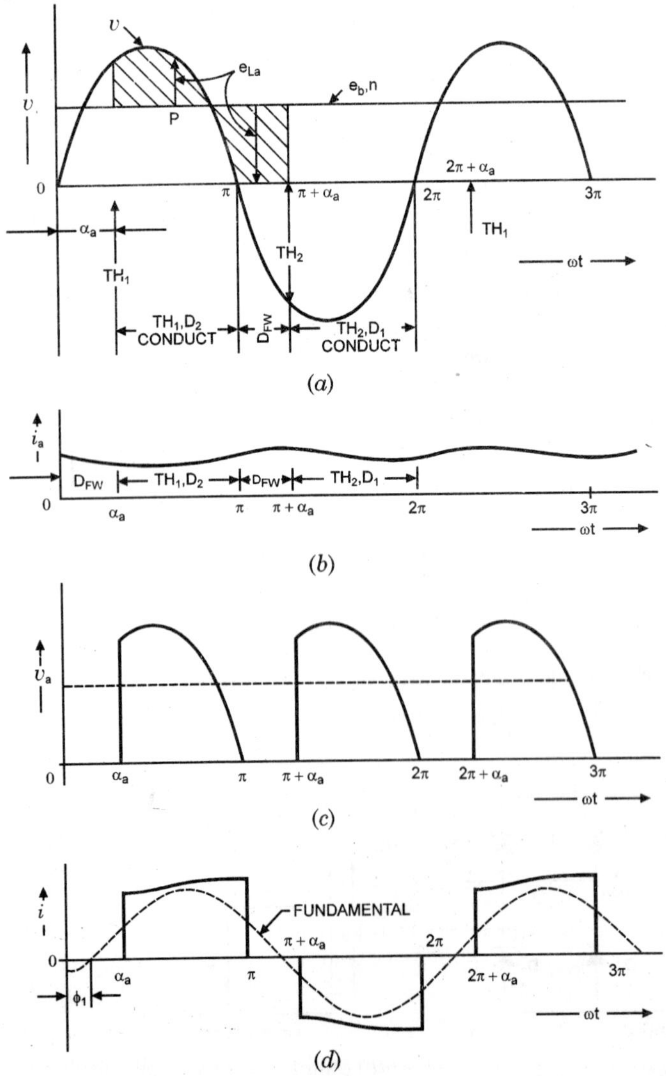

Fig. 22.5 *Voltage and Current Waveforms of Semiconverter Feeding Separately-Excited DC Motor; Continuous Armature Current*

Under steady conditions, as thyristor TH_1 is fired ($\omega t = \alpha_a$), TH_1 and diode D_2 conduct and the motor gets supply, *i.e.*, v_a. At $\omega t = \pi$, v_a tends to become negative as the input voltage reverses in polarity. This makes freewheeling diode to become forward biased and armature current flowing through TH_1 D_2 is transferred to D_{FW} for the freewheeling period $\pi < \omega t < \pi + \alpha_a$ providing for continuity of the armature current during this period when the motor remains disconnected from the supply. At $\omega t = \pi + \alpha_a$, the thyristor TH_2 is fired and TH_2D_1 conduct, causing D_{FW} to become reverse biased, and therefore, open circuited. The motor is once again connected positively to the supply for the next period of $\pi + \alpha_a < \omega t < 2\pi$. This process repeats continuously.

Different voltage and current waveforms of a separately-excited dc motor supplied through a semiconverter are illustrated in Fig. 22.5. Though the voltage across motor terminals [Fig. 22.5 (*c*)] contains harmonics over and above a steady dc value, it is rightly assumed here that the motor does not respond to these harmonics and thus runs at constant speed of *n rps* and has constant induced back emf e_b. As TH_1 fires at $\omega t = \alpha_a$, the motor current is given as

$$\frac{1}{L_a} \int_{\alpha_a}^{\omega t} (v - e_b)\, d(\omega t)\,;$$

assuming R_a negligibly small

up to the point P illustrated in Fig. 22.5 (*a*) ; $v > e_b$ so that the motor current increases. So does the motor back emf e_b. During this period, apart from energy being delivered to the load, energy is also being stored in the motor armature inductance L_a. Beyond the point P, voltage v becomes lesser than induced back emf e_b and the motor current starts to drop. This also implies the reversal of voltage across motor armature inductance which now feeds energy into the system. During the freewheeling period ($\pi < \omega t < \pi + \alpha_a$), the diode continues to be forward biased by the reversal of the inductive voltage. During this period a part of the energy stored in armature inductance is consumed in supplying the mechanical load. The motor current, speed, and emf, therefore, all decrease. This process then repeats over the next period ($\pi + \alpha_a < \omega t < 2\pi + \alpha_a$) through TH_2 D_1, and later through D_{FW}. The current drawn from the supply shown in Fig. 22.5 (*d*) is that part of the armature current which flows over the periods (α_a, π), ($\pi + \alpha_a$, 2π), when the motor is connected to the supply. It is not necessary to employ the freewheeling diode. In its absence at $\omega t = \pi$, diode D_1 becomes forward biased so that freewheeling occurs through TH_1 D_1 till TH_2 is fired. At $\omega t = 2\pi$ freewheeling occurs through TH_2 D_2 and so on.

Discontinuous Armature Current. The armature current becomes discontinuous for large values of the firing angle, high speed and low values of torque. Discontinuous armature current results in deterioration of motor performance. The ratio of peak to average and rms to average value of armature current increases. Thus it is desirable to operate the motor in the continuous current mode. This can be achieved by using an external armature circuit choke, which reduces the rate of decay of current during the freewheeling operation.

The voltage and current waveforms for semiconverter with discontinuous current are illustrated in Fig. 22.6. The motor is connected to the supply through TH_1 and D_2 for the period $\alpha_a < \omega t < \pi$. Beyond π, the motor is shorted through the freewheeling diode D_{FW}. The armature current falls to zero at angle β (extinction angle) $< \pi + \alpha_a$ *i.e.,* before the thyristor TH_2 is fired, thereby making the armature current discontinuous, During α_a to π, the conduction period through TH_1 and D_2, the motor terminal voltage is the same as the input voltage. During π to β, the motor terminal voltage is zero, motor terminals being shorted by the freewheeling diode D_{FW}. From β to $\pi + \alpha_a$, the motor coasts and, therefore, its terminal voltage is the same as its induced emf. It should be noted here that the fundamental of the current drawn from the mains lags behind the voltage by an angle ϕ, $(< \alpha_a)$.

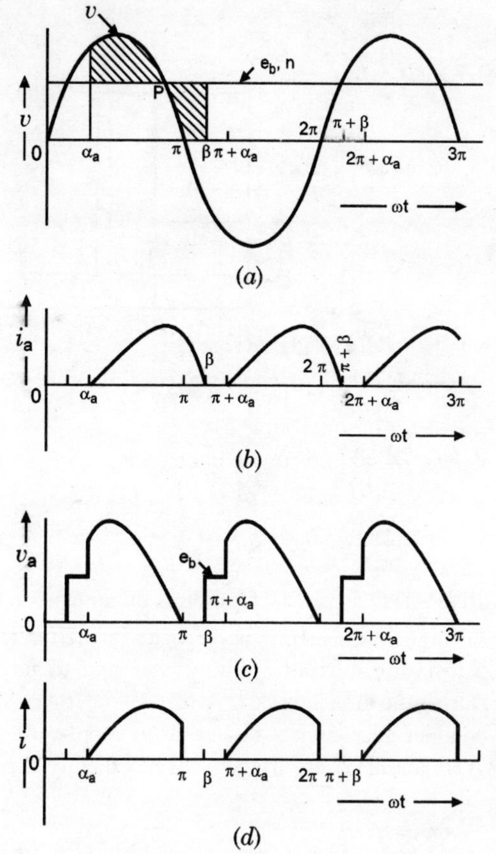

Fig. 22.6 *Voltage and Current Waveforms of Semiconverter Feeding Separately-Excited DC Motor: Discontinuous Armature Current*

Torque-Speed Characteristics. The emf induced in armature conductors of a motor, known as back emf, is given by

$$E_b = k_e \Phi N \qquad ...(22.5)$$

where K_e is emf constant and is equal to $\frac{PZ}{60\,A}$.

In a separately-excited dc motor, the field winding is excited from a separate source. So voltage applied to the armature is given by

$$V_a = E_b + I_a R_a \qquad ...(22.6)$$

where I_a is armature current in amperes and R_a is the armature resistance in ohms.

Thus, from Eqs. (22.5) and (22.6), we have

$$\text{Back emf, } E_b = V - I_a R_a = k_e N \Phi$$

$$\text{or speed of motor, } N = \frac{V - I_a R_a}{k_e \Phi} \qquad ...(22.7)$$

Since field is excited separately, field current I_f and flux Φ are constant. Moreover voltage drop in armature, $I_a R_a$ is negligible. So motor speed N is constant if voltage applied to the armature is constant, and speed varies directly in proportion to the applied voltage *i.e.,* $N \propto V_a$.

Electromagnetic torque developed in a dc motor is given by

$$T_e = k_T \Phi I_a \qquad ...(22.8)$$

where k_T is torque constant and is equal to $\frac{1}{2\pi}\frac{PZ}{A}$

$$\text{and torque constant, } k_T = \frac{1}{2\pi}\frac{PZ}{A} = \frac{60}{2\pi} \times \frac{PZ}{60\,A} = \frac{60}{2\pi} k_e \qquad ...(22.9)$$

From Eq. (22.8), we have

$$\text{Armature current, } I_a = \frac{T_e}{k_T \Phi} \qquad ...(22.10)$$

From Eqs. (22.7) and (22.10), we have

$$N = \frac{V_a}{k_e \Phi} - \frac{T_e R_a}{k_T k_e \Phi^2} \qquad ...(22.11)$$

With a semiconverter,

$$V_a = \frac{V_{max}}{\pi}(1 + \cos \alpha_a) \qquad \text{... Refer to Eq. (22.3)}$$

$$\text{So } N = \frac{V_{max}(1 + \cos \alpha_a)}{\pi k_e \Phi} - \frac{T_e R_a}{k_T k_e \Phi^2} \qquad ...(22.12)$$

The first term of RHS of Eq. (22.12) represents the theoretical speed and the second term represents the speed drop due to voltage drop in armature circuit.

The theoretical no-load speed can be controlled by varying the firing angle α_a.

3. *Single-Phase Full-Converter Drives.* The circuit diagram is illustrated in Fig. 22.7 (*a*). A fullconverter needs four thyristors but no freewheeling diode. Full converters are employed for both the armature and field supply. As illustrated in Fig. 22.7 (*b*), two quadrant operation is possible. The converter in the armature circuit provides $+V_a$ or $-V_a$ thus allowing operation in first and fourth quadrants. Current remains unidirectional because of the unidirectional thyristors. When operating in the fourth quadrant regenerative braking is possible and motor feeds back energy to the source. If α_a and α_f are the firing angles for the converters used in the armature and field circuits, the armature voltage V_a and field voltage V_f are given as

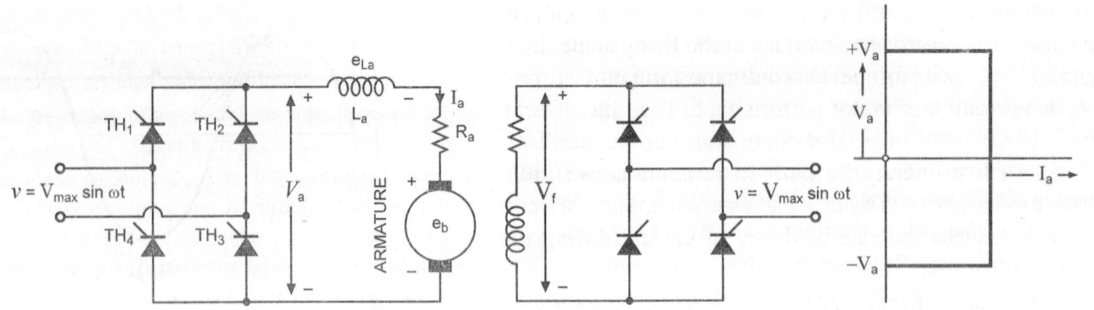

(a) Circuit Diagram *(b) Quadrant of Operation*

Fig. 22.7 *Separately-Excited DC Motor Supplied From Single-Phase AC Supply Through Full Converters*

$$V_a = \frac{2V_{max}}{\pi} \cos \alpha_a \text{ for } 0 \le \alpha_a \le \pi \qquad ...(22.13)$$

$$V_f = \frac{2V_{max}}{\pi} \cos \alpha_f \text{ for } 0 \le \alpha_f \le \pi \qquad ...(22.14)$$

This drive is also employed for motors of rating up to 15 kW.

The voltage and current waveforms are illustrated in Fig. 22.8; armature current i_a being assumed to be almost constant. During the time interval $\alpha_a < \omega t < \pi + \alpha_a$, thyristors TH_1 and TH_3 conduct and connect the motor to supply. At $\pi + \alpha_a$, thyristors TH_2 and TH_4 are triggered. Immediately the supply voltage appears in reverse bias across thyristors TH_1 and TH_3 and turns them off. This is *natural* or *line commutation.* The motor current is transferred from thyristors TH_1 and TH_3 to thyristors TH_2 and TH_4. During the time interval α_a to π energy is supplied to the motor (both v and i are positive and, therefore, are v_a and i_a). But during time interval π to $\pi + \alpha_a$, some of the motor energy is fed back to the supply (v and i and so the v_a and i_a being of opposite polarity). The noteworthy point is that the fundamental of the current drawn from the supply mains lags behind the applied voltage by angle $\phi_1 = \alpha_a$.

The voltage and current waveforms for $\alpha_a > 90°$ are illustrated in Fig. 22.9. The average motor terminal voltage is now negative. On reversing of the motor terminals, it will operate as a generator supplying power back to the ac supply. This is the *inversion*

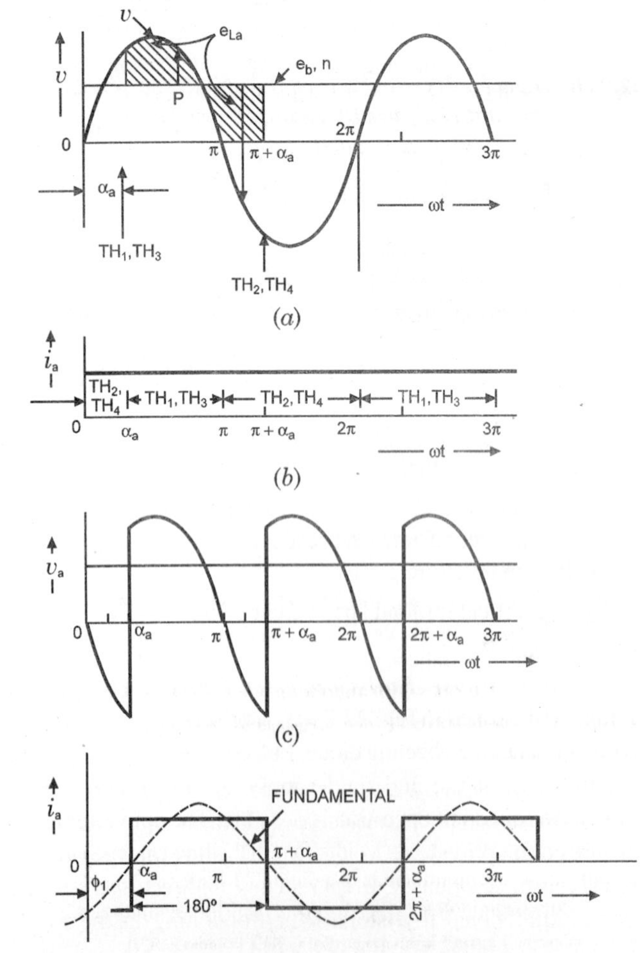

Fig. 22.8 *Voltage and Current Waveforms of Full converter Supplying Separately-Excited DC Motor; Continuous Armature Current (Almost Constant):* $\alpha < 90°$

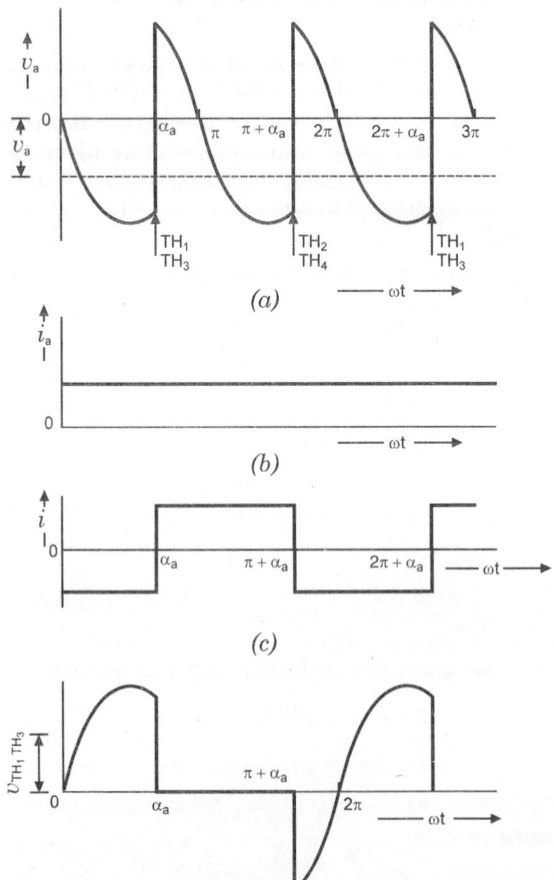

Fig. 22.9 *Voltage and Current Waveforms For Full Converter:* $\alpha_a > 90°$

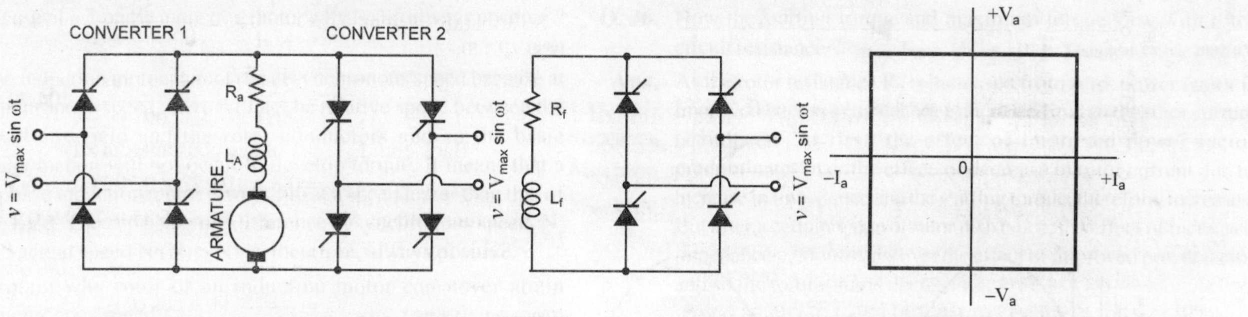

(a) Circuit Diagram *(b) Quadrant of Operation*

Fig. 22.10 *Separately-Excited DC Motor Supplied From Single-Phase AC Supply Through Dual Converters*

operation of the converter and is employed in regenerative braking of the motor. The noteworthy point here is that during the conduction period of either TH_1 TH_3 or TH_2 TH_4 as the supply voltage becomes negative, the armature current begins to drop, causing the inductance polarity to reverse and thus conducting thyristors continue to be forward biased.

From Eqs. (22.11) and (22.13) we have

$$\text{Speed N} = \frac{2\,V_{max} \cos \alpha_a}{\pi k_e \phi} - \frac{T_e R_a}{k_T k_e \phi^2} \quad \text{...(22.15)}$$

In the case of discontinuous current, the average voltage at motor terminals depends upon the extinction angle β which itself depends on the average motor speed N, average armature current I_a and firing angle α_a. We need not to go in its analytical treatment.

4. *Single-Phase Dual Converter Drives*. The circuit diagram is illustrated in Fig. 22.10 (*a*). In this drive, the armature is supplied from two full converters while the field is supplied from another full converter. Converter 1 provides operation in the first and fourth quadrants while converter 2 provides operation in second and third quadrants. Thus it is a *four quadrant drive* and allows all four modes of operation, *i.e.*, forward motoring, forward regenerative braking, reverse motoring and reverse regenerative braking. When converter 1 is operating, a positive voltage V_a is applied to the armature. When converter 2 is operating a negative voltage $-V_a$ is applied to the armature. It is employed for motors of rating up to 15 kW. If α_{a_1} is the firing angle of converter 1, the armature voltage V_a is given as

$$V_a = \frac{2\,V_{max}}{\pi} \cos \alpha_{a_1}, \text{ for } 0 \le \alpha_{a_1} \le \pi \quad \text{...(22.16)}$$

The firing angle α_{a_2} of converter 2 will be equal to $(\pi - \alpha_{a_1})$ and when converter 2 is operating, armature voltage is given as

$$V_a = \frac{2\,V_{max}}{\pi} \cos \alpha_{a_2} \text{ for } 0 \le \alpha_{a_2} \le \pi \quad \text{...(22.17)}$$

The voltage applied to the field is given as

$$V_f = \frac{2\,V_{max}}{\pi} \cos \alpha_f \text{ for } 0 \le \alpha_f \le \pi \quad \text{...(22.18)}$$

where α_f is the firing angle of converter supplying the field. Figure 22.10 (*b*) illustrates the four quadrants of operation.

Example 22.1. A single-phase half-wave converter is operated from a 120 V, 60 Hz supply. If the load is resistive, R = 10 Ω and the delay angle $\alpha = \frac{\pi}{3}$, determine:

(*i*) the efficiency (*ii*) the form factor (*iii*) the ripple factor (*iv*) the transformer utilisation factor ; and (*v*) peak inverse voltage of the thyristor.

Solution: In a half-wave converter, negative half waves are eliminated. The average voltage across load is

$$V_{dc} = \frac{1}{2\pi}\int_{\alpha}^{\pi} V_{max} \sin \omega t \; d(\omega t) = \frac{V_{max}}{2\pi}(1 + \cos \alpha)$$

$$= \frac{120\sqrt{2}}{2\pi}\left(1 + \cos\frac{\pi}{3}\right)$$

$$= \frac{120\sqrt{2}}{2\pi} \times 1.5 = 40.5 \text{ V}$$

$$I_{dc} = \frac{V_{dc}}{R} = \frac{40.5}{10} = 4.05 \text{ A}$$

RMS voltage across the load,

$$V_{rms} = \left[\frac{1}{2\pi}\int_{\alpha}^{\pi} (V_{max} \sin \omega t)^2 \, d(\omega t)\right]^{1/2}$$

$$= V_{max}\left[\frac{1}{4\pi}\int_{\alpha}^{\pi} (1 - \cos 2\omega t)\, d(\omega t)\right]^{1/2}$$

$$= V_{max}\left[\frac{1}{4\pi}\left|\left(\omega t - \frac{\sin 2\omega t}{2}\right)\right|_{\alpha}^{\pi}\right]^{1/2}$$

$$= V_{max}\left[\frac{\pi - \alpha}{4\pi} + \frac{\sin 2\alpha}{8\pi}\right]^{1/2}$$

$$= 120\sqrt{2}\left[\frac{\pi - \pi/3}{4\pi} + \frac{\sin 2\pi/3}{8\pi}\right]^{1/2} = 76.1 \text{ V}$$

$$I_{rms} = \frac{V_{rms}}{R} = \frac{76.1}{10} = 7.61 \text{ A}$$

DC output, $P_{dc} = V_{dc}\, I_{dc} = 40.5 \times 4.05 = 164$ W

AC input, $P_{ac} = V_{rms}\, I_{rms} = 76.1 \times 7.61 = 580$ W

(*i*) Rectification efficiency, $\eta = \frac{P_{dc}}{P_{ac}} = \frac{164}{580} = 0.283$ **Ans.**

(*ii*) Form factor, $K_f = \frac{\text{RMS voltage}}{\text{Average voltage}} = \frac{76.1}{40.5}$

$= 1.879$ **Ans.**

(*iii*) Ripple factor, $\gamma = \sqrt{(K_f^2 - 1)} = \sqrt{1.879^2 - 1}$

$= 1.59$ **Ans.**

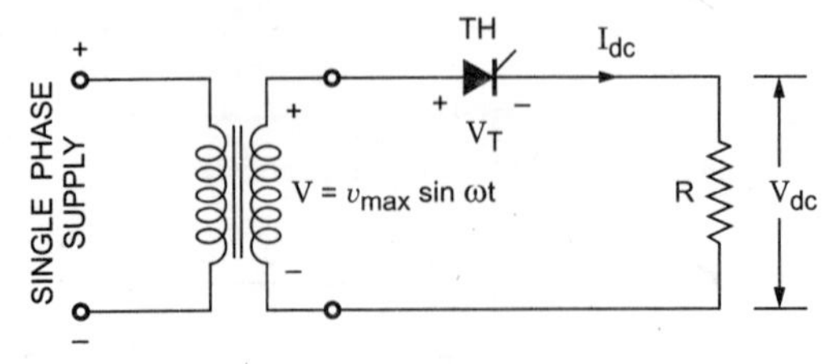

(*a*) *Circuit Diagram*

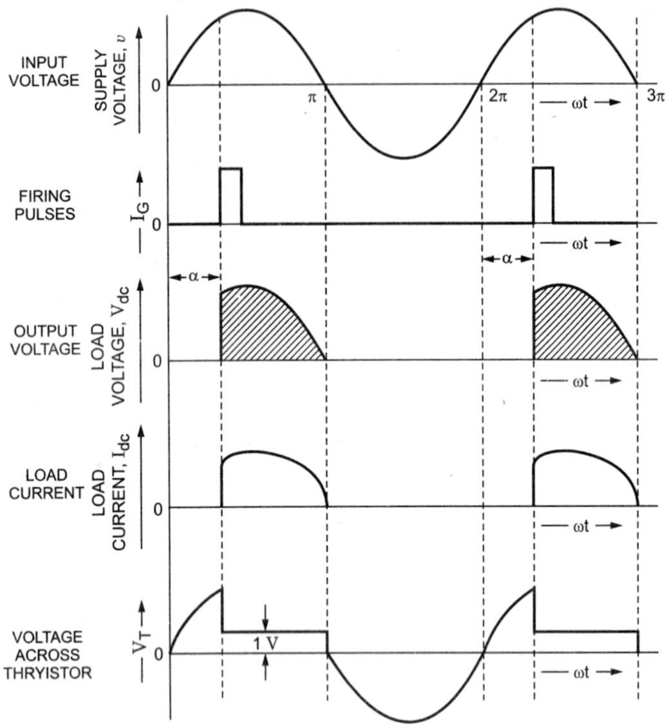

(*b*) *Waveforms of Input Voltage, Firing Pulses, Output Voltage, Current and Voltage Drop Across Thyristor*

Fig. 22.11 *Single-Phase Half-Wave Converter*

Volt-ampere rating of transformer $= 120 \times 7.61 = 913.2$

(*iv*) Transformer utilisation factor $= \dfrac{P_{dc}}{\text{VA rating of transformer}}$

$$= \frac{164}{913.2} = 0.18 \quad \textbf{Ans.}$$

(*v*) Peak inverse voltage, PIV $= V_{max} = 120\sqrt{2} = 169.7$ V **Ans.**

Example 22.2. A single-phase, 220 V, 50 Hz supply feeds a separately-excited dc motor through two single-phase semiconverters, one for the field and another for armature. The firing angle for the field semiconverter is zero. The field resistance is 250 Ω and armature resistance is 0.2 Ω. The load torque is 50 N-m at 1,000 rpm. The voltage constant is 0.8 V/A-rad/s and the torque constant is 0.8 N-m/A². Assuming armature and field currents to be continuous and neglecting losses determine (*i*) field current (*ii*) firing angle of the converter in the armature circuit and (*iii*) power factor of the converter of the armature circuit.

Solution: Field circuit voltage,

$$V_f = \frac{V_{max}}{\pi}(1 + \cos\alpha_f) = \frac{220\sqrt{2}}{\pi} \times (1 + \cos 0°) = 198 \text{ V}$$

(*i*) Field current, $I_f = \dfrac{V_f}{R_f} = \dfrac{198}{250} = 0.792$ A **Ans.**

(*ii*) Since torque developed, $T_d = K_f I_f I_a$ where K_f is torque constant

So armature current, $I_a = \dfrac{T_d}{K_f I_f} = \dfrac{50}{0.8 \times 0.792} = 78.9$ A

$$\omega = \frac{2\pi N}{60} = \frac{2\pi \times 1{,}000}{60} = 104.72 \text{ rad/s}$$

Back emf, E_b = Voltage constant × ω × I_f

$= 0.8 \times 104.72 \times 0.792 = 66.35$ V

Armature voltage, $V_a = E_b + I_a R_a = 66.35 + 78.9 \times 0.2 = 82.13$ V

Since from Eq. (22.3),

$$V_a = \frac{V_{max}}{\pi}(1 + \cos\alpha_a)$$

So $82.13 = \dfrac{220\sqrt{2}}{\pi}(1 + \cos\alpha_a)$

or $\alpha_a = 99.83°$ **Ans.**

(*iii*) Power output of converter in the armature circuit,

$= V_a I_a = 82.13 \times 78.9 = 6{,}480$ W

$$\text{Input current} = \left[\frac{2}{2\pi}\int_{\alpha_a}^{\pi} I_a^2 \, d(\omega t)\right]^{1/2}$$

$$= I_a\left[\frac{\pi - \alpha_a}{\pi}\right]^{1/2} = 78.9\left[\frac{180 - 99.83}{180}\right]^{1/2}$$

$= 52.66$ A

Input volt-amperes $= 220 \times 52.66 = 11{,}585.2$

Input power factor $= \dfrac{6{,}480}{11{,}585.2} = 0.56$ lagging **Ans.**

Example 22.3. A single-phase full converter is used to control the speed of 5 kW, 220 V, 1,000 rpm separately-excited dc motor. The armature resistance is 0.2 Ω and the rated armature current is 30 A. The ac supply voltage is 230 V. The motor voltage constant is $K_a\phi$ = 1.668 volt/radian/second. Determine (*i*) the speed of the motor and (*ii*) motor torque for a firing angle α = 45° and rated motor current. Assume continuous and ripple free armature current.

Solution: Voltage across the armature,

$$V_a = \frac{2V_{max}}{\pi}\cos\alpha_a \qquad \text{Refer to Eq. (22.13)}$$

$$= \frac{2 \times 230\sqrt{2}}{\pi}\cos 45° = 146.42 \text{ V}$$

Back emf, $E_b = V_a - I_a R_a = 146.42 - 30 \times 0.2 = 140.42$ V

(*i*) So speed, $\omega = \dfrac{E_b}{K_a\Phi} = \dfrac{140.42}{1.668} = 84.186$ radians/second

or $N = \dfrac{\omega}{2\pi} \times 60 = \dfrac{84.186}{2\pi} \times 60 = 804$ rpm **Ans.**

(*ii*) Motor torque, $T = K_a \Phi I_a = 1.668 \times 30 = 50.04$ N-m **Ans.**

Example 22.4. A 200 V, 875 rpm, 150 A separately-excited dc motor has an armature resistance of 0.06 Ω. It is fed from a single-phase fully controlled rectifier with an ac source voltage of 220 V, 50 Hz. Assuming continuous conduction calculate the firing angle for rated motor torque at 750 rpm.

Solution: Back emf of motor when operated from 200 V dc supply and runs at 875 rpm,

$$E_{b_1} = V_{dc} - I_a R_a = 200 - 150 \times 0.06 = 191 \text{ V}$$

Back emf of motor when fed from 220 V 50 Hz single-phase supply through fully controlled rectifier and runs at 750 rpm (assuming flux Φ to remain constant),

$$E_{b_2} = E_{b_1} \times \frac{N_2}{N_1} = 191 \times \frac{750}{875} = 163.71 \text{ V}$$

Armature current $I'_a = I_a = 150$ A

∵ torque remains the same and flux is assumed to remain constant

Armature voltage, $V_a = E_b + I_a R_a = 163.71 + 150 \times 0.06 = 172.71$ V

Since for a fully controlled single phase rectifier voltage across armature is given as

$$V_a = \frac{2 V_{max}}{\pi} \cos \alpha_a$$

So $$172.71 = \frac{2 \times 220\sqrt{2}}{\pi} \cos \alpha_a$$

or $\alpha_a = 29.31°$ **Ans.**

Example 22.5. A single-phase fully controlled thyristor converter is connected to a load comprising 2 Ω resistance and 0.3 H inductance. The supply voltage is 230 V at 50 Hz. Estimate the average load voltage, average load current and input power factor for triggering angle of 30°. Assume continuous and ripple free load current.

[*U.P. Technical Univ. Energy Conversion 2005-06*]

Solution: Supply voltage $V_{(rms\ value)} = 230$ V

$$V_{max} = 230\sqrt{2} \text{ V}$$

Average load voltage, $$V_{L\,av} = \frac{2V_{max}}{\pi} \cos \alpha = \frac{2 \times \sqrt{2} \times 230 \cos 30°}{\pi} = 179.33 \text{ V } \textbf{Ans.}$$

Average load current, $$I_{L\,av} = \frac{V_{L\,av}}{R} = \frac{179.33}{2} = 89.665 \text{ A } \textbf{Ans.}$$

As load current is ripple free,

RMS value of load current, $I_{rms} = I_{L\,av} = 89.665$ A

Active power of load, $P = V_{L\,av} \times I_{L\,av} = 179.33 \times 89.665 = 16{,}080$ W

Reactive power of load, $$Q = \frac{2 V_{max}}{\pi} I_{L\,av} \sin \alpha = \frac{2 \times \sqrt{2} \times 230}{\pi} \times 89.665 \sin 30° = 9{,}283.6 \text{ VAR}$$

Input power factor

$$= \text{Cos} \tan^{-1} \frac{Q}{P} = \text{Cos} \tan^{-1} \frac{9283.6}{16{,}080} = \text{Cos } 30° = 0.866 \text{ (lagging) } \textbf{Ans.}$$

Example 22.6. A separately-excited dc motor is controlled by varying its armature voltage using a single-phase full-converter bridge as shown in Fig. 22.12. The field current is kept constant at the rated value. The motor has an armature resistance of 0.2 Ω, and the motor voltage constant is 2.5 V/rad/sec. The motor is driving a mechanical load having a constant torque of 140 N-m. The triggering angle of the converter is 60°. The armature current can be assumed to be continuous and ripple free.

(*a*) Calculate the motor armature current.

(*b*) Evaluate the motor speed in rad/sec. [*GATE 2001*]

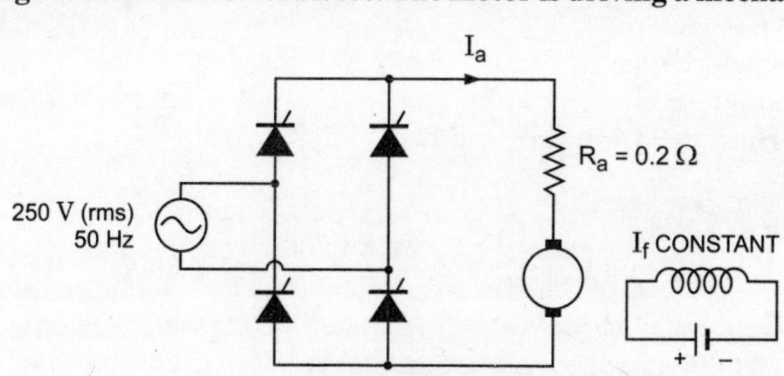

Fig. 22.12

Solution: (*a*) Voltage across the motor armature

$$V_a = \frac{2V_{max}}{\pi} \cos \alpha = \frac{2 \times 250\sqrt{2}}{\pi} \cos 60° = 112.5 \text{ V}$$

∵ $V_{max} = 250\sqrt{2}$ V and $\alpha = 60°$

Now $T\omega = E_b I_a = K_a \omega I_a$ ∵ $E_b = K_a \omega$ ...(*i*)

or Armature current, $$I_a = \frac{T}{K_a} = \frac{140}{2.5} = 56 \text{ A } \textbf{Ans.}$$

(*b*) Back emf, $E_b = V - I_a R_a = 112.5 - 56 \times 0.2 = 101.34$ V

From Eq. (*i*) Speed, $$\omega = \frac{E_b I_a}{T} = \frac{101.34 \times 56}{140} = 40.536 \text{ radians/second } \textbf{Ans.}$$

Example 22.7. A 220 V, 1500 rpm, 10 A separately-excited dc motor is fed from a single-phase fully controlled rectifier with an ac source voltage of 230 V, 50 Hz, R_a = 2 Ω. Assuming continuous conduction, find the firing angles for :

(*i*) Half the rated torque 500 rpm

(*ii*) Rated motor torque and (–1000) rpm.

[*U.P. Technical Univ. Electric Drives 2009-10*]

Solution: Back emf of motor when operated from 220 V dc supply and runs at 1,500 rpm

$$E_{b1} = V_{dc} - I_{a1} R_a = 220 - 10 \times 2 = 200 \text{ V}$$

(*i*) Back emf of motor when fed from 230 V, 50 Hz single-phase supply through fully controlled rectifier and runs at 500 rpm (assuming flux Φ to remain constant)

$$E_{b2} = E_{b_1} \times \frac{N_2}{N_1} = 200 \times \frac{500}{1{,}500} = 66.67 \text{ V}$$

Armature current, $$I_{a2} = \frac{1}{2} I_{a1} = \frac{1}{2} \times 10 = 5 \text{ A}$$

∵ Torque is half and flux is assumed to remain constant

Armature voltage, $V_{a1} = E_{b2} + I_{a2} R_a = 66.67 + 5 \times 2 = 76.67$ V

Since for a fully controlled single phase rectifier voltage across armature is given as

$$V_a = \frac{2 V_{max}}{\pi} \cos \alpha_a$$

or $$76.67 = \frac{2 \times \sqrt{2} \times 230}{\pi} \cos \alpha_{a_1}$$

or $$\text{Cos } \alpha_{a_1} = \frac{76.67 \times \pi}{2\sqrt{2} \times 230} = 0.37$$

or $\alpha_{a_1} = \text{Cos}^{-1} 0.37 = 68.27°$ **Ans.**

(*ii*) Back emf of motor when fed from 230 V, 50 Hz single phase supply through fully controlled rectifier and runs at –1,000 rpm (assuming flux *f* to remain constant)

$$E_{b3} = E_{b_1} \times \frac{N_3}{N_1} = 200 \times \frac{-1{,}000}{1{,}500} = -133.33 \text{ V}$$

Armature current, $I_{a3} = I_{a_1} = 10$ A ∵ Torque remains the same and flux is assumed to remain constant

Armature voltage, $V_{a2} = E_{b_3} + I_{a_3} R_a$

$$= -133.33 + 10 \times 2 = -113.33 \text{ V}$$

and $V_{a2} = \dfrac{2 V_{max}}{\pi} \cos \alpha_{a2}$

So $-113.33 = \dfrac{2 \times \sqrt{2} \times 230}{\pi} \cos \alpha_{a2}$

or $\cos \alpha_{a2} = \cos^{-1}(-0.5473) = 123.18°$ **Ans.**

Example 22.8. A 220 V single-phase 50 Hz supply feeds the armature and field circuits of a separately-excited dc motor through two full converters. The firing angle of field circuit converter is zero. The armature and field resistances are 0.2 Ω and 200 Ω respectively. The torque and voltage constants are 1.12. The firing angle of the armature circuit converter is 30° and armature current is 40 A. Find torque developed and motor speed. Assume that field and armature currents are continuous and constant and neglect brush contact drop.

Solution: Field circuit voltage,

$$V_f = \frac{2 V_{max}}{\pi} \cos \alpha_f = \frac{2 \times 220\sqrt{2}}{\pi} \cos 0° = 198 \text{ V}$$

Field circuit current,

$$I_f = \frac{V_f}{R_f} = \frac{198}{200} = 0.99 \text{ A}$$

Armature circuit voltage,

$$V_a = \frac{2 V_{max} \cos \alpha_a}{\pi} = \frac{2 \times 220\sqrt{2}}{\pi} \cos 30° = 171.53 \text{ V}$$

Back emf, $E_b = V_a - I_a R_a = 171.53 - 40 \times 0.2 = 163.53$ V

Torque developed,

$T_d = K_t I_a I_f = 1.12 \times 40 \times 0.99 = 44.352$ N-m **Ans.**

$$\text{Speed, } N = \frac{60\omega}{2\pi} = \frac{60}{2\pi} \times \frac{\text{back emf}}{K_t I_f} = \frac{60}{2\pi} \times \frac{163.53}{1.12 \times 0.99}$$

$= 1{,}408$ rpm **Ans.**

22.6.2. Three-Phase Drives. For large sized (rating exceeding 15 kW) motors 3-phase drive is employed. The output voltage of a 3-phase converter has less ripple contents than that of single-phase converter. So the filtering requirements for smoothing out the armature current are less and the current is mostly continuous. Three-phase converters could be half-wave, semiconverter, full-converter and dual converter. The 3-phase half-wave converter drives are only of theoretical importance and are generally not employed in industrial applications. This is because of presence of dc component in supply currents. Semi- and full-converters are widely used. A dual converter is employed in case of reversible drives with a power rating up to several MW.

1. *Three-Phase Semiconverter Drives.* It is a one quadrant drive and is employed for motors of rating up to about 100 kW. The field converter may be 3-phase or single-phase. The armature voltage V_a is given as

$$V_a = \frac{3\sqrt{3} V_{max}}{2\pi} (1 + \cos \alpha_a) \text{ for } 0 \le \alpha_a \le \pi \qquad ...(22.19)$$

where V_{max} is the peak value of line-to-neutral voltage and α_a is the firing angle of the converter in the armature circuit.

The circuit diagram for a separately-excited dc motor supplied from 3-phase ac supply through a three-phase semiconverter is shown in Fig. 22.13.

2. *Three-Phase Full Converter Drives.* Three-phase full converter drive is a two quadrant drive and is employed up to

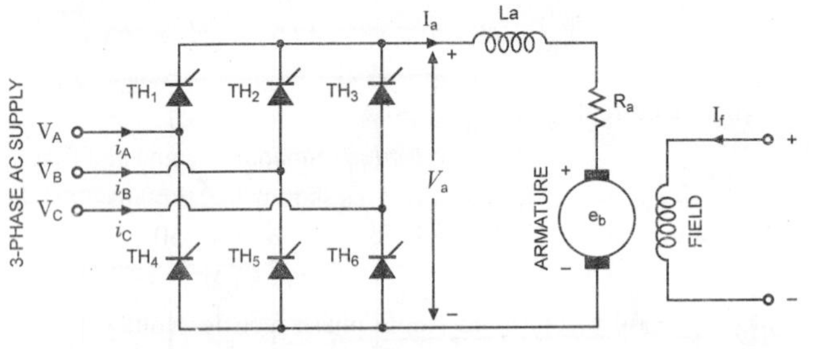

Fig. 22.13 *Three-Phase Semiconverter Supplying Separately-Excited DC Motor*

1,500 kW motors. The converter in the field circuit may be single-phase or 3-phase. When 3-phase full converters are used for both armature and field, the armature and field voltages are given as

$$V_a = \frac{3\sqrt{3} V_{max}}{\pi} \cos \alpha_a \text{ for } 0 \le \alpha_a \le \pi \qquad ...(22.20)$$

$$V_f = \frac{3\sqrt{3} V_{max}}{\pi} \cos \alpha_f \text{ for } 0 \le \alpha_f \le \pi \qquad ...(22.21)$$

3. *Three-Phase Dual Converter Drives.* 3-phase dual converter consists of two converters and can be operated in all the four quadrants. Either converter 1 is operated to supply armature voltage V_a or converter 2 is operated to supply armature voltage $-V_a$. It is used for motors of rating up to 1,500 kW. The armature voltage supplied by converter 1 is given as

$$V_a = \frac{3\sqrt{3} V_{max}}{\pi} \cos \alpha_{a_1}, \text{ for } 0 \le \alpha_{a_1} \le \pi \qquad ...(22.22)$$

where α_{a_1} is the firing angle for converter 1.

For converter 2, the voltage V_a is given as

$$V_a = \frac{3\sqrt{3} V_{max}}{\pi} \cos \alpha_{a_2} \text{ for } 0 \le \alpha_{a_2} \le \pi \qquad ...(22.23)$$

where $\alpha_{a_2} = \pi - \alpha_{a_1}$.

If the field is also supplied by a 3-phase full converter,

$$V_f = \frac{3\sqrt{3} V_{max}}{\pi} \cos \alpha_f \text{ for } 0 \le \alpha_f < \pi \qquad ...(22.24)$$

Example 22.9. A 400 V, 3-phase ac supply feeds a separately-excited dc motor through two 3-phase full converters, one for armature and the other for the field. The firing angle of the field circuit converter is zero. The armature and field resistances are 0.2 Ω and 250 Ω respectively. The armature current is 60 A. The motor voltage constant is 1.3 V/A-rad/second. The speed is 1,200 rpm. Determine the firing angle of the armature circuit converter.

Solution: Field circuit voltage,

$$V_f = \frac{3\sqrt{3}}{\pi} V_{max} \cos \alpha_f = \frac{3\sqrt{3}}{\pi} \times \frac{400}{\sqrt{3}} \times \sqrt{2} \cos 0° = 540.2 \text{ V}$$

...Refer to Eq. (22.21)

Field circuit current,

$$I_f = \frac{V_f}{R_f} = \frac{540.2}{250} = 2.16 \text{ A}$$

Back emf developed,

$$E_b = K I_f \omega = 1.3 \times 2.16 \times \frac{2\pi \times 1{,}200}{60} = 352.86 \text{ V}$$

Armature circuit voltage,

$$V_a = E_b + I_a R_a = 352.86 + 60 \times 0.2 = 364.86 \text{ V}$$

Also armature circuit voltage

$$= \frac{3\sqrt{3}}{\pi} V_{max} \cos \alpha_a \quad \text{...Refer to Eq. (22.20)}$$

$$\text{So} \quad 364.86 = \frac{3\sqrt{3}}{\pi} \times \frac{400}{\sqrt{3}} \times \sqrt{2} \cos \alpha_a$$

or $\alpha_a = 47.52°$ **Ans.**

Example 22.10. A 3-phase half-controlled thyristor bridge with 400 V, 3-phase, 50 Hz supply is feeding a separately-excited dc motor. Armature resistance is 0.2 Ω, armature rated current is 100 A and back emf constant is 0.25 V/rpm. Determine no-load speed if no-load armature current is 5 A and firing angle is 45°. Also determine firing angle to obtain a speed of 1,500 rpm at rated current.

Solution: Armature circuit voltage at no load,

$$V_{a_0} = \frac{3\sqrt{3}}{2\pi} V_{max} (1 + \cos \alpha_a) \quad \text{...Refer to Eq. (22.19)}$$

$$= \frac{3\sqrt{3}}{2\pi} \times \frac{400\sqrt{2}}{\sqrt{3}} (1 + \cos 45°) = 461 \text{ V}$$

Back emf at no load, $E_{b_0} = V_{a_0} - I_{a_0} R_a = 461 - 5 \times 0.2 = 460$ V

Speed at no load, $N_0 = \frac{460}{0.25} = 1{,}840$ rpm **Ans.**

Back emf developed at a speed of 1,500 rpm,

$$E_b = 1{,}500 \times 0.25 = 375 \text{ V}$$

Input voltage to motor, $V_a = E_b + I_a R_a = 375 + 100 \times 0.2 = 395$ V

$$\text{Also} \quad V_a = \frac{3\sqrt{3}}{2\pi} V_{max} (1 + \cos \alpha_a)$$

$$\therefore \quad 395 = \frac{3\sqrt{3}}{2\pi} \times \frac{400\sqrt{2}}{\sqrt{3}} (1 + \cos \alpha_a)$$

or $\alpha_a = 62.45°$ **Ans.**

Example 22.11. A separately-excited dc motor is supplied from 3-phase, 400 V, 50 Hz supply through half controlled 3-phase bridge rectifier. The motor parameters are inductance 0.012 H, resistance 0.8 Ω, armature constant 2 V/rad/s (N-m/A). Neglecting rectifier losses and supply impedance, determine speed-torque characteristics up to 80 N-m at fixed firing angle of 90°.

Solution: Because of armature inductance, the armature current can be assumed to remain constant.

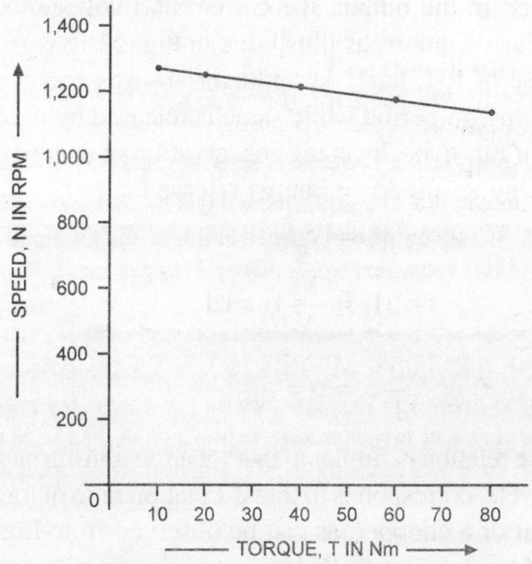

Fig. 22.14

Armature circuit voltage,

$$V_a = \frac{3\sqrt{3}}{2\pi} V_{max} (1 + \cos \alpha_a)$$

$$= \frac{3\sqrt{3}}{2\pi} \times \frac{400}{\sqrt{3}} \times \sqrt{2} (1 + \cos 90°) = 270 \text{ V}$$

Let us determine torque and speed for armature current of 5, 10, 20, 30, 40 A, as tabulated below :

Armature Current in Amperes, I_a	*Torque in N-m* $= K_a I_a = 2 \times I_a$	*Back emf,* $E_b = V_a - I_a R_a = 270 - I_a \times 0.8$ *in volts*	*Speed, N in RPM* $= \frac{E_b}{K_t} \times \frac{60}{2\pi} = \frac{E_b}{2} \times \frac{30}{\pi}$
5	$2 \times 5 = 10$	$270 - 5 \times 0.8 = 266$	$\frac{266}{2} \times \frac{30}{\pi} = 1{,}270$
10	$2 \times 10 = 20$	$270 - 10 \times 0.8 = 262$	$\frac{262}{2} \times \frac{30}{\pi} = 1{,}250$
20	$2 \times 20 = 40$	$270 - 20 \times 0.8 = 254$	$\frac{254}{2} \times \frac{30}{\pi} = 1{,}213$
30	$2 \times 30 = 60$	$270 - 30 \times 0.8 = 246$	$\frac{246}{2} \times \frac{30}{\pi} = 1{,}175$
40	$2 \times 40 = 80$	$270 - 40 \times 0.8 = 238$	$\frac{238}{2} \times \frac{30}{\pi} = 1{,}136$

Speed-torque characteristic is shown in Fig. 22.14.

22.7 CHOPPER CONTROL OF DC MOTORS

In many cases, conversion of dc supply source voltage to different levels is required. For example, subway cars, trolley buses, or battery operated vehicles are fed from a fixed voltage dc supply source. But for their speed control we need conversion of fixed voltage dc source to a variable voltage dc source. Conventional methods for obtaining the variable dc voltage from a fixed dc voltage source are given below:

1. Resistance Control. In this method, a variable resistance is introduced between the load and the source. The drawbacks of this method are wastage of lot of energy and for a given output voltage, requirement of resistances of different values for different values of load current. This method is still employed for older traction installations.

2. Motor-Generator Set (M-G Set). Separately-excited dc generator provides a voltage which can be varied from zero to rated value with either polarity. It has disadvantages of higher cost, large bulk, low efficiency and sluggish response because of the generator field time constant.

3. AC Link Chopper (Inverter-Rectifier). In this method, the dc supply is first converted to ac supply by an inverter (dc to ac converter). The obtained ac supply is then stepped up or down by a transformer and then rectified back to dc by a rectifier. Because of two stages of conversion, the set-up is costly, bulky and less efficient.

4. DC Chopper (DC to DC Power Converter). As already mentioned in Art. 22.3.3, a dc chopper is a static switch to provide variable dc voltage from a source of constant dc voltage. Choppers

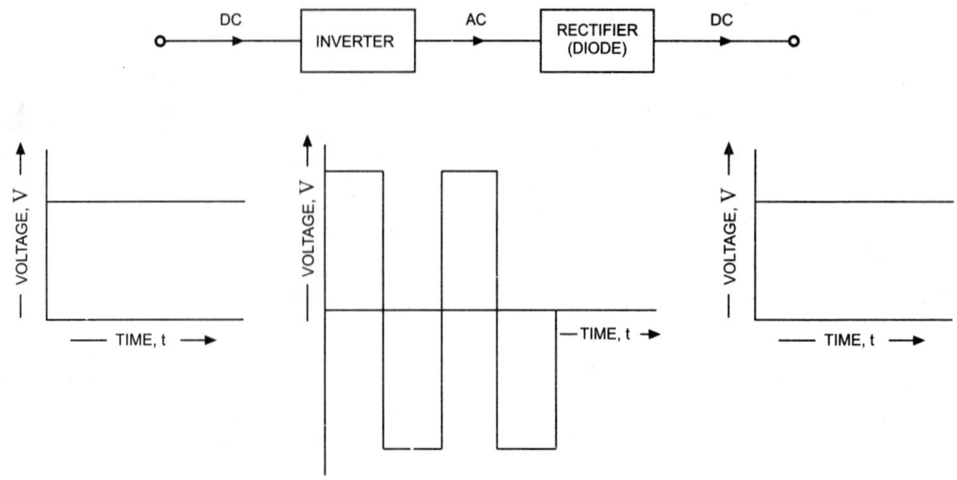

Fig. 22.15 *AC-Link Chopper*

are now being used all over the world for rapid transit systems. These are also employed in trolley cars, marine hoists, forklift trucks and mine haulers. The future electric automobiles are likely to use choppers for their speed control and braking. Chopper systems offer smooth control, high efficiency, fast response and regeneration.

22.7.1. Principle of Operation of Chopper. A *dc chopper* is a high speed on/off semiconductor switch that connects the load to and disconnects it from the supply and produces a chopped load voltage from a constant input supply voltage. As it is an on/off switch between source and load, it can be symbolically represented by a switch S inside a dotted rectangle, as shown in Fig. 22.16. Actually, a chopper consists of main power semiconductor device together with their turn-on and turn-off mechanisms. In low-power chopper circuits, power transistors, GTOs etc. are used widely. In high-power chopper circuits, however, thyristors are in common use.

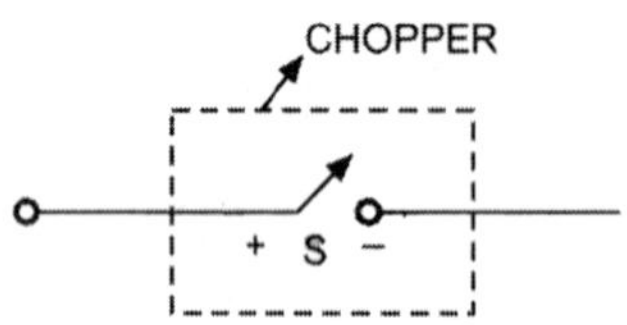

Fig. 22.16 *Symbol of Chopper*

Figure 22.17 illustrates the principle of operation of a chopper. The switch may be closed or opened by supplying on/off pulses to it. Here the mechanism of sending pulses to

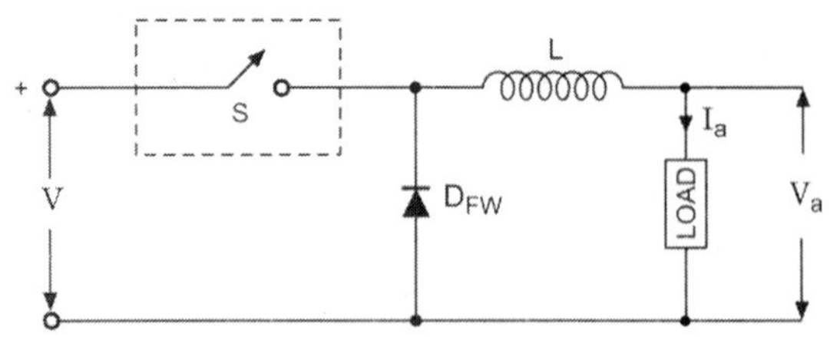

Fig. 22.17 *Elementary Chopper Circuit*

a switch has not been shown. A shunting diode D_{FW} is provided across the load for freewheeling the load current when the switch S is off. When the switch S is closed, the source voltage V, assuming that voltage drop across switch S in on-state is negligible, appears across the load for the duration the switch is closed. This duration is designated T_{on}. After the pulse to switch is removed the load is disconnected from the source. The energy stored in the inductor circulates a current through the freewheeling diode D_{FW}. As a result the load is short circuited and voltage across it falls to zero but the current through the load falls exponentially, as illustrated in Fig. 22.18. The duration for which switch is off is called off time T_{off}. The total time of a pulse is, therefore, given by $T = T_{on} + T_{off}$.

Thus, a dc chopped voltage is produced at the load terminals [Fig. 22.18 (*a*)] and the load current is continuous as shown in Figs. 22.18 (*b*) and (*c*).

From Fig. 22.18 (*a*), the average load voltage V_a is given by

$$V_a = V \frac{T_{on}}{T_{on} + T_{off}} = V \frac{T_{on}}{T} = \alpha V \qquad ...(22.25)$$

where $\alpha = \frac{T_{on}}{T}$ is called *duty cycle* of the circuit. It is analogous to transformation ratio of ac transformer. The value of α ranges from 0 to 1. Hence the output voltage varies between zero and supply voltage V. In the present case the output voltage is less than the supply voltage. It is, therefore, called **step-down** chopper. In choppers, called step-up choppers, the output voltage is more than supply voltage.

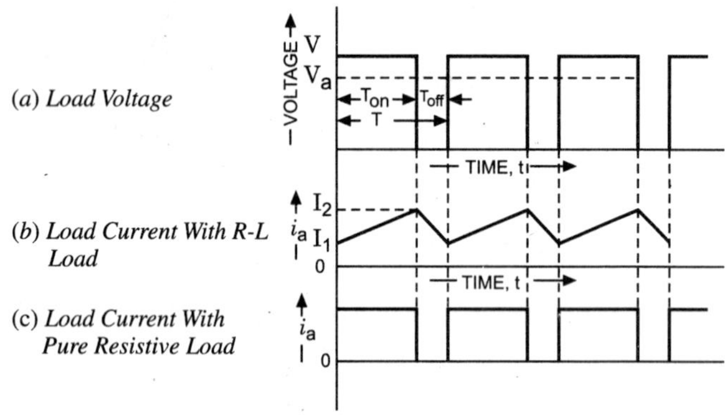

Fig. 22.18 *Voltage and Current Waveforms*

The load current and voltage waveforms, as shown in Figs. 22.18 (*b*) and (*a*) are for R-L load. However, in case, resistive load is used in the output, the current and voltage waveforms have similar variations as illustrated in Fig. 22.18 (*c*) and (*a*).

In case of R-L load, the inductor absorbs energy from the source during on-period while same is released by it during off-period. We can write, by using conservation of energy principle;

$$\text{Energy absorbed} = \text{Energy released}$$

$$\text{or} \quad VI = V_a I_a$$

$$\text{or} \quad I = \frac{V_a}{V} I_a = \alpha I_a$$

$$\text{or} \quad \frac{I}{I_a} = \alpha = \frac{V_a}{V} \qquad ...(22.26)$$

Above relation is similar to that for an ac transformer. Here α, the duty cycle, corresponds to transformation ratio of transformer. The output of a chopper, as can be observed from Eqs. (22.25) and (22.26), can be controlled by a variation in α, the duty cycle.

Control Strategies. It is seen from Eq. (22.25), that, average value of output voltage, V_a can be controlled by periodic opening and closing of the semiconductor switch. There are two types of control strategies for operating the switches that can be used in dc choppers. These control strategies are time-ratio control (TRC) and current limit control.

1. Time-Ratio Control. In this control strategy, the value of $\frac{T_{on}}{T}$ is varied. This is affected in two ways namely variable frequency control and constant frequency control.

(*a*) **Variable Frequency Control (FM).** Under such scheme the chopping frequency *f* is kept variable and for this purpose either (*a*) *on* time, T_{on} or (*b*) *off* time, T_{off} is kept constant. This scheme is, therefore, called frequency modulation scheme and is illustrated in Fig. 22.19.

In Fig. 22.19 (*a*) and (*b*) *on* time T_{on} is kept constant while *off* time T_{off} of the switch is variable. The duty cycles in two figures are $\alpha_1 = \frac{T_{on}}{T_{on} + T_{off}} = \frac{3}{4}$ and $\alpha_2 = \frac{1}{2}$ respectively.

Figure 22.19 (*c*) and (*d*) are for variable on time T_{on} and constant off time T_{off}. The duty cycle in the two cases are respectively $\frac{1}{4}$ and $\frac{2}{5}$. This technique is suitable for switches which require forced commutation to turn off.

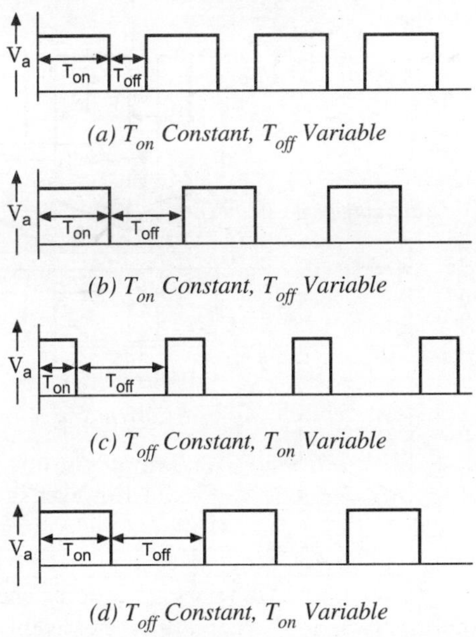

(*a*) T_{on} *Constant*, T_{off} *Variable*

(*b*) T_{on} *Constant*, T_{off} *Variable*

(*c*) T_{off} *Constant*, T_{on} *Variable*

(*d*) T_{off} *Constant*, T_{on} *Variable*

Fig. 22.19 *Output Voltage Waveforms For Variable Frequency Control*

(*b*) **Constant Frequency Control.** Under this control scheme the chopping period T remains constant but the pulse width T_{on} is varied to control the output. This is, therefore, called pulse width modulation (PWM) control. The scheme is illustrated in Fig. 22.20. In the figure, time $T = T_{on} + T_{off}$ is constant but T_{on} and T_{off} both are variable. The duty cycles of $\alpha = 0.75$ and 0.50 are shown, respectively, in Figs. 22.20 (*a*) and (*b*).

Variable frequency control has some disadvantage as compared to constant frequency control method which are:

(*i*) Chopping frequency is required to be varied over a large range to control the output. Thus the output has wide range of harmonics. The design of a filter to remove such a wide range of harmonics is quite difficult.

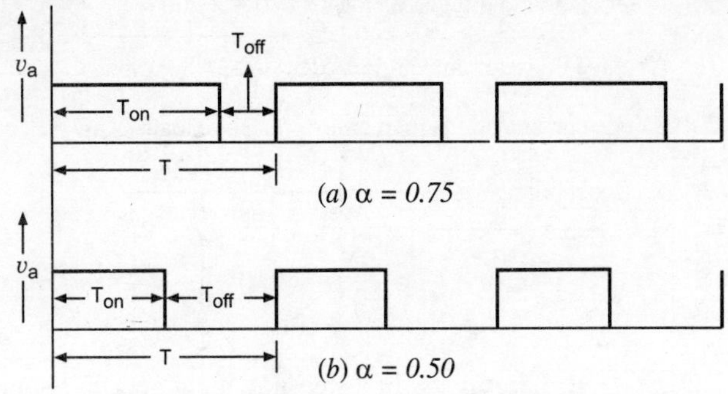

(*a*) $\alpha = 0.75$

(*b*) $\alpha = 0.50$

Fig. 22.20 *Constant Frequency Control*

(*ii*) For the control of duty cycle the off period may be so large that load current falls to zero between two consecutive pulses. This leads to discontinuity in the load current which is not desirable.

(*iii*) For the control of duty cycle α, frequency variations would be wide. As such, there is a possibility of interference with signalling and telephone lines in frequency modulation control.

For above reasons PWM control is preferred over FM control in chopper circuits.

2. Current Limit Control. Under this scheme the chopper is switched *on* and *off* between upper and lower limits of load current. As the load current exceeds upper limit the chopper is turned *off*. The load current now freewheels until it reaches the lower value prescribed. As the load current attains lower limit, the chopper is switched *on*. As the freewheeling operation is involved in this scheme, a storage element is a necessity in the load. As the load current varies between two prescribed limits, this scheme is applicable only either at constant frequency or at constant turn *on* time T_{on}. The waveforms for current limit control are shown in Fig. 22.21. In this case the load current is continuous and difference between upper and lower values of current decides the switching frequency. At smaller values of difference in current, the switching frequency is higher. As a result the ripples in the output are reduced. As an example the switching frequencies of the chopper for $I'_{a\,max}$ and $I_{a\,max}$ as upper limits of the current are, respectively $\frac{1}{T'}$ and $\frac{1}{T}$ where $I'_{a\,max} < I_{a\,max}$ and $T' < T$. By further reducing the difference between upper and lower limits of the current the switching frequency can be increased correspondingly which leads to reduction in the ripple content in the output.

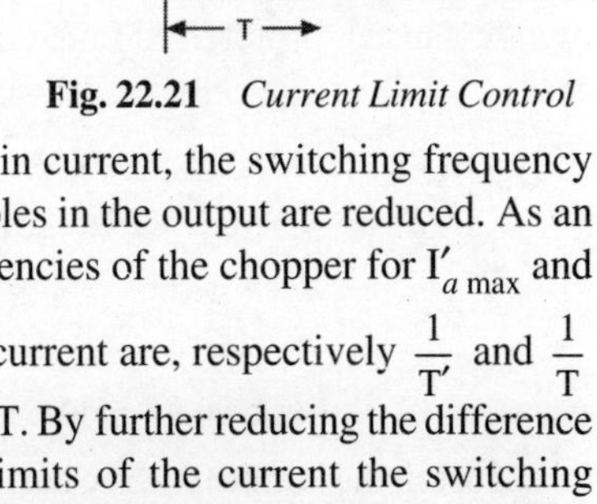

Fig. 22.21 *Current Limit Control*

Example 22.12. For a dc to dc converter shown in Fig. 22.17, the source voltage is 200 volts, load resistance 10 Ω and duty cycle α = 0.4. If load is resistive, calculate the following (*a*) Average load voltage (*b*) Thyristor average current (*c*) Average and rms freewheeling diode currents (*d*) Effective input resistance of chopper.

Solution: For resistive load, the load voltage and load current are in phase:

(*a*) Average load voltage, $V_a = \alpha V_{dc} = 0.4 \times 200 = 80$ volts **Ans.**

(*b*) Average thyristor current $= V_a/R = \frac{80}{10} = 8$ Amps

(*c*) Diode currents are zero in case of resistive load. **Ans.**

(*d*) Effective input resistance $= \frac{\text{DC source voltage}}{\text{Average source current}}$

$$= \frac{V}{\alpha \frac{V}{R}} = \frac{R}{\alpha} = \frac{10}{0.4} = 25\ \Omega \quad \textbf{Ans.}$$

22.7.2. Step-up Choppers. In a step-down chopper the output voltage lies between zero and supply voltage. However, there are techniques by which the output of a chopper obtained is higher than the supply voltage. Such types of choppers are called step-up choppers. The circuit diagram of a step-up chopper is shown in Fig. 22.22 (*a*). In figure, the chopper is represented by a switch S.

In this chopper, a large inductor in series with the supply voltage source is essential, as shown in Fig. 22.22 (*a*). When the chopper is on, the current flows through inductance L, chopper S and power supply source, as shown in Fig. 22.22 (*b*). During this on-period T_{on}, the inductor L stores energy.

When the chopper is off, the inductor current is forced to flow through the diode and load for a period T_{off} because the inductor current cannot die instantaneously. As the current tends to decrease, polarity of the emf induced in the inductor is reversed, as illustrated in Fig. 22.22 (*c*).

Assuming negligible voltage drop across diode D, the voltage equation for outer loop may be written as

$$-V_a + e_L + V = 0$$

$$\text{or } V_a = V + e_L \qquad \text{...(22.27)}$$

If the rate of fall of current through inductor is $\frac{di}{dt}$, Eq. (22.27) may be written as

$$V_a = V + L\frac{di}{dt} \qquad \text{...(22.28)}$$

It means that the inductor voltage adds to the source voltage to force the inductor current into the load. In this manner, the energy stored in inductor is released to load. Here, higher value of inductance L is preferred for getting lesser ripple in the output.

During the time T_{on}, when the chopper is on, the energy stored in the inductor is given by

$$W_1 = \text{Voltage} \times \text{average current through L} \times T_{on}$$

$$= V \times I_a \times T_{on} \qquad \text{...(22.29)}$$

Above Eq. (22.29) is based on the assumption that the variation of output current is linear.

During the period T_{on}, the current through the inductor rises to peak value I_2 while during off period T_{off} it falls to value I_1 and rises again to I_2 in the next on state of the chopper. It is shown in Fig. 22.22 (*d*). Thus, in a cycle the average value of current is given by $I_a = \frac{I_1 + I_2}{2}$. Hence the energy stored in the inductor L is given by

$$W_1 = V \times \frac{I_1 + I_2}{2} \times T_{on} \qquad \text{...(22.30)}$$

and this energy is released by the inductor to the load during the off-period, T_{off}. Thus energy released by inductor L during off period is given by

$$W_0 = \text{Voltage across inductor} \times \text{average current through inductor} \times T_{off}$$

$$= (V_a - V)\left(\frac{I_1 + I_2}{2}\right) T_{off} \qquad \text{...(22.31)}$$

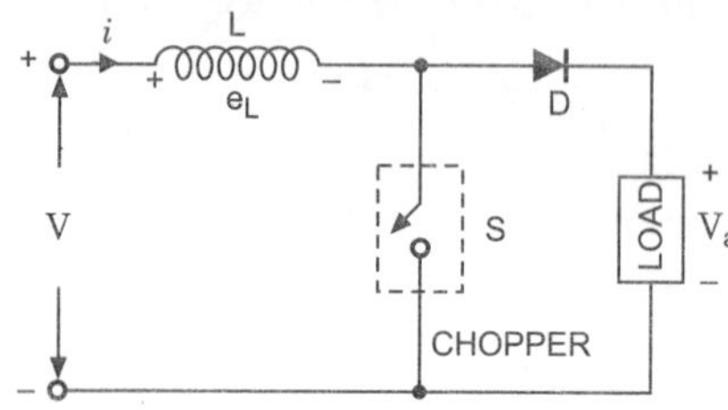

(a) Step-up Chopper

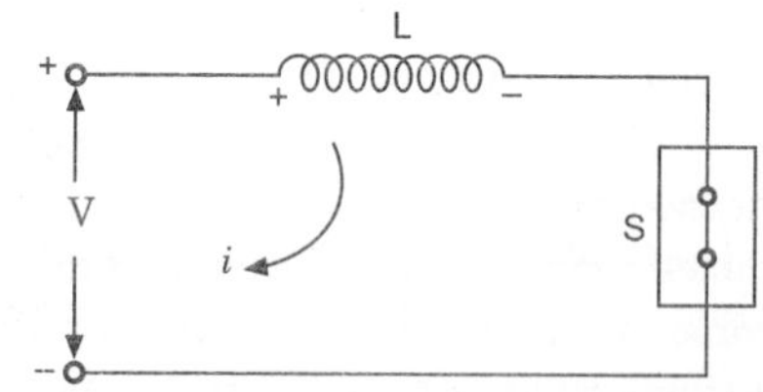

(b) Equivalent Circuit During On-Period

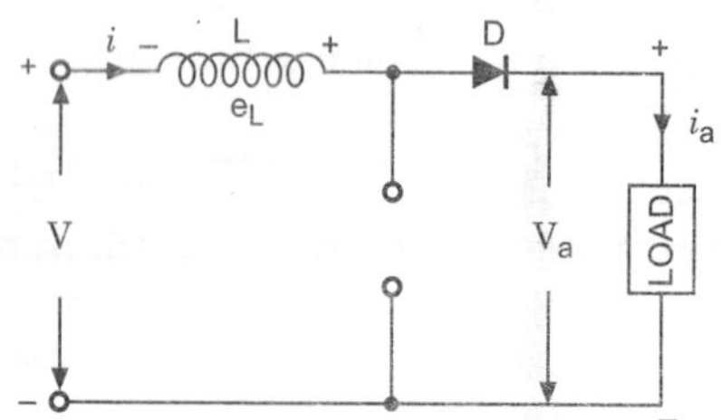

(c) Equivalent Circuit During off-Period

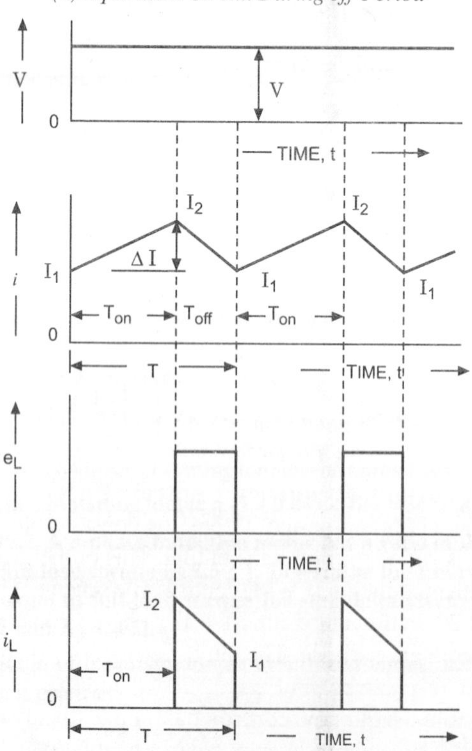

(d) Voltage and Current Waveforms

Fig. 22.22 *Step-Up Chopper*

Assuming the system to be lossless, in the steady-state the two energies will be equal *i.e.*,

$$W_1 = W_0$$

$$\text{or } V \times \frac{I_1 + I_2}{2} \times T_{on} = (V_a - V)\left(\frac{I_1 + I_2}{2}\right) T_{off}$$

$$\text{or } V(T_{on} + T_{off}) = V_a T_{off}$$

$$\text{or } V_a = \frac{V(T_{on} + T_{off})}{T_{off}}$$

$$= \frac{V \times T}{T - T_{on}} = \frac{V}{1 - \frac{T_{on}}{T}} = \frac{V}{1-\alpha} \quad \text{...(22.32)}$$

The value of α, the duty cycle lies between zero and unity. Corresponding average values of load voltage V_a lies between V and infinity. Thus the output voltage of step-up chopper is never less than the supply voltage V *i.e.*, for $0 \le \alpha \le 1$ the output voltage of step-up chopper lies in the range $V \le V_a \le \alpha$.

The step-up chopper can be employed for regenerative braking of dc motors even at low operating speeds. Let V represent the motor armature voltage and V_a the dc source voltage in Fig. 22.22 *(a)*. Regenerative braking takes place when $\left(V + L\frac{di}{dt}\right)$ exceeds V_a. Even at decreasing motor speeds, duty cycle α can be so adjusted that $\left(V + L\frac{di}{dt}\right)$ exceeds fixed supply voltage V_a.

Example 22.13. A step-up chopper shown in Fig. 22.22 *(a)* is fed by a dc source of 220 volts. The inductance has value of 30 mH. If load voltage is 250 volts across a resistance of 10 ohms, calculate average current and duty cycle.

Solution: Average value of load voltage is

$$V_a = \frac{V}{1-\alpha} = 250 \text{ volts}$$

$$\text{or } 250 = \frac{220}{1-\alpha}$$

$$\text{or } \alpha = 1 - \frac{220}{250} = 0.12 \textbf{ Ans.}$$

$$\text{Average load current} = \frac{V_a}{R} = \frac{250}{10} = 25 \text{ A } \textbf{Ans.}$$

22.7.3. Step-up/down Chopper. A chopper can also be used for stepping up or down the supply voltage by varying its duty cycle. The principle of operation is illustrated in Fig. 22.23. As shown, the output voltage polarity is opposite to that of input voltage V.

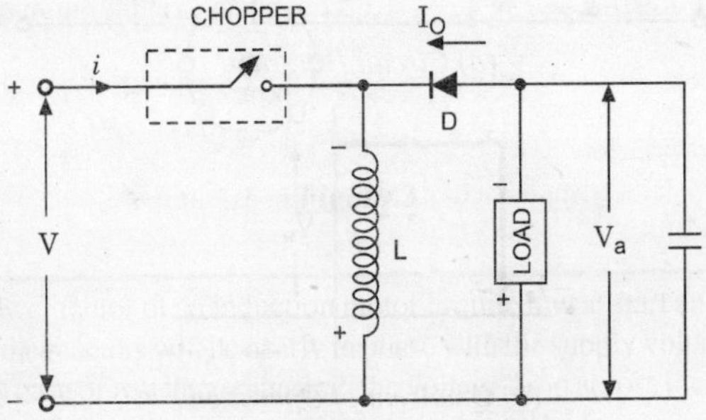

Fig. 22.23 *Step-Up/Down Chopper*

When the chopper is on, the supply current flows through the path V_+ – chopper – inductor L – V_-. As a result energy is stored in the inductor. When the chopper is made off, the current through the inductor L decreases. As a result an emf is induced in the inductor with lower end positive and the energy stored in the inductor is released through the load and diode and the voltage across the load is negative as illustrated. It is thus used to reverse the polarity of load voltage w.r.t. source voltage.

The energy stored in inductor L during on-period T_{on} is given by

$$W_1 = V \times \text{average current} \times T_{on}$$

But the average current through the inductor, as in case of step-up chopper discussed in Art. 22.7.2, is

$$I_a = \frac{I_1 + I_2}{2} \text{ giving}$$

$$W_1 = V \times \frac{I_1 + I_2}{2} T_{on} \quad \text{...(22.33)}$$

During *off* period T_{off}, the energy released by inductor to the load is

$$W_0 = V_a \times \frac{I_1 + I_2}{2} T_{off} \quad \text{...(22.34)}$$

Under steady-state condition, assuming no loss of power.

Energy stored in inductor = Energy released to load

$$\text{or } V \times \frac{I_1 + I_2}{2} T_{on} = V_a \times \frac{I_1 + I_2}{2} T_{off}$$

$$\text{or } V_a = V \times \frac{T_{on}}{T_{off}} = \frac{V \times T_{on}}{T - T_{on}} = \frac{V}{\frac{T}{T_{on}} - \frac{T_{on}}{T_{on}}}$$

Substituting $\frac{T_{on}}{T} = \alpha$ in above equation, we have

$$V_a = \frac{V}{\frac{1}{\alpha} - 1} = \frac{\alpha}{1-\alpha} V \quad \text{...(22.35)}$$

Equation (22.35) can be used for step-up and step-down operation of chopper. In the range $0 < \alpha < 0.5$ the circuit operates in step-down mode while for duty cycle in the range $0.5 < \alpha < 1$ it acts as a step-up chopper.

Example 22.14. A step-up/step-down chopper has an input voltage of 125 V and output voltage of 200 V. If the period of conduction is 1 ms, determine the frequency of switching pulse.

Solution: Average output voltage, V_a = 200 V

Supply voltage, V = 125 V

Period of conduction, T_{on} = 1 ms = 0.001 s

Since output voltage of step-up/step-down chopper is given by

$$V_a = \frac{\alpha V}{1-\alpha},$$

$$200 = \frac{\alpha \times 125}{1-\alpha}$$

$$\text{or } \alpha = \frac{8}{13}$$

Frequency of switching pulse,

$$f = \frac{1}{T} = \frac{1}{T_{on}/\alpha} = \frac{\alpha}{T_{on}} = \frac{8/13}{1 \times 10^{-3}} = \frac{8{,}000}{13} \text{ Hz} = 615.4 \text{ Hz} \quad \textbf{Ans.}$$

22.7.4. Chopper Control of Separately-Excited DC Motor Drive. Chopper drives can be operated in the following modes :

1. Motoring mode.
2. Regenerative braking mode.
3. Rheostatic braking mode.

1. *Motoring Mode.* Figure 22.24 (*a*) illustrates this mode of operation with a one quadrant chopper feeding the motor from dc supply of voltage V, the field being excited from a separate source. Figure 22.24 (*b*) shows the quadrant of operation. When the chopper conducts, voltage across the motor armature is supply voltage V, and when it blocks, voltage across the motor armature is zero. Therefore, both average voltage and current are positive and power can flow from supply source to the load, and thus, represents *motoring operation*. The voltage across the motor armature is given as

$$V_a = \alpha V \qquad ...(22.36)$$

where α is the duty cycle of the chopper.

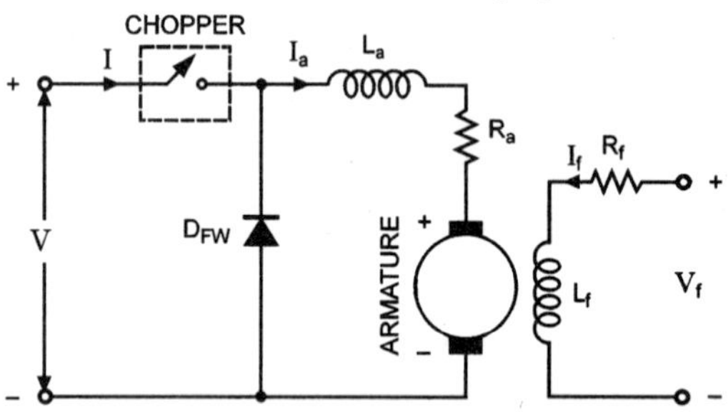

(*a*) Circuit Diagram

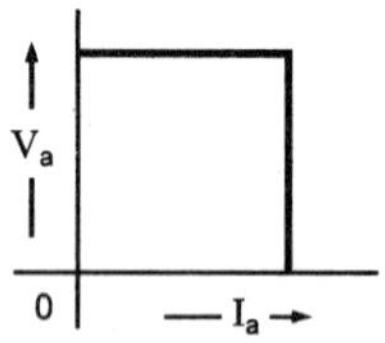

(*b*) Quadrant of Operation

Fig. 22.24 *Separately-Excited DC Motor Supplied by a Chopper*

Freewheeling diode D_{FW} is necessary so that the armature current may flow when the chopper is off. The power supplied to the motor is given as

$$P = V_a I_a = \alpha V I_a \qquad ...(22.37)$$

The current drawn from the supply is

$$I = \alpha I_a \qquad ...(22.38)$$

By varying the duty cycle α, the speed of the motor can be controlled.

2. *Regenerative Braking Mode.* When the motor is to be stopped, brakes are to be applied. In regenerative braking the motor is made to operate as a generator and the kinetic energy of the motor is converted into electrical energy and returned to the supply. Fig. 22.25 (*a*) shows the circuit diagram for this mode of operation. During regenerative braking mode, the chopper is gated so that the motor armature is short circuited through the chopper. Kinetic energy of the motor is partly dissipated in armature resistance R_a and partly stored in inductance L_a. When chopper is turned off, freewheeling diode D_{FW} is turned on and the energy stored in inductance L_a is returned to the supply source (provided the source is receptive).

The operation is in second quadrant, as illustrated in Fig. 22.25 (*b*).

The average voltage V_{ch} is given as

$$V_{ch} = (1-\alpha) V \qquad ...(22.39)$$

Regenerated power P_g is given as

$$P_g = V_{ch} \times I_a = (1-\alpha) V I_a \qquad ...(22.40)$$

where I_a is the armature current

Generated voltage E_g is given as

$$E_g = V_{ch} + I_a R_a = (1-\alpha) V + I_a R_a \qquad ...(22.41)$$

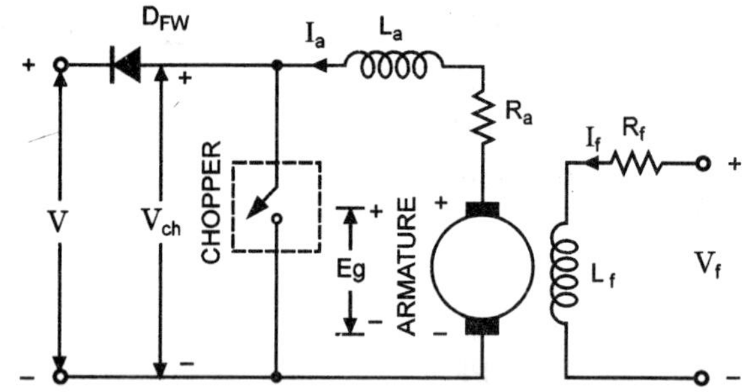

(*a*) Circuit Diagram

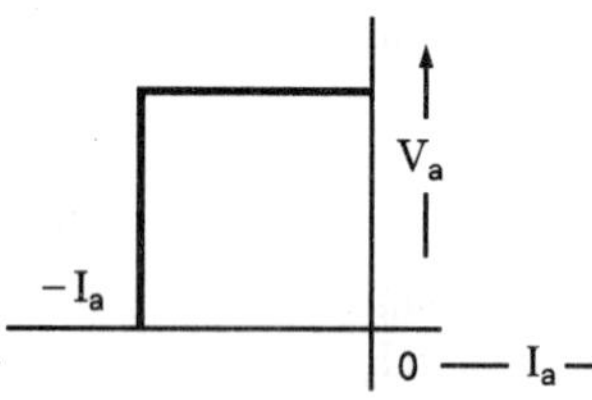

(*b*) Quadrant of Operation

Fig. 22.25 *Regenerative Braking of Separately-Excited DC Motor*

3. *Rheostatic Braking Mode.* Figure 22.26 (*a*) illustrates the circuit for this mode of operation. In this mode also the motor is made to operate as a generator and the kinetic energy of the motor is converted into electrical energy. This electrical energy is dissipated in braking resistance R_b connected to the terminals, as illustrated in the figure. When this mode is to be used, chopper is turned on by a gate pulse.

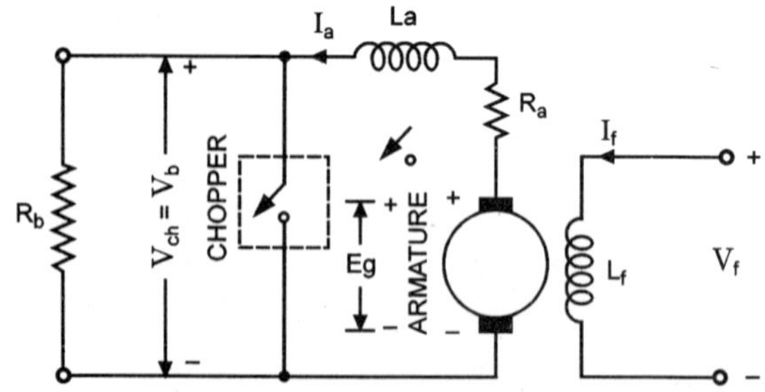

(*a*) Circuit Diagram

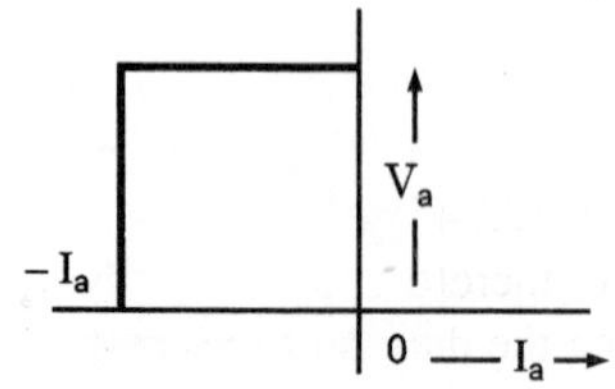

(*b*) Quadrant of Operation

Fig. 22.26 *Rheostatic Braking of Separately-Excited DC Motor*

The kinetic energy is partly dissipated in armature resistance R_a and partly stored in armature inductance L_a. When chopper is turned off, the energy stored in inductance L_a is transferred to braking resistance R_b and dissipated as heat.

The average current in braking resistance R_b is given as

$I_b = (1 - \alpha) I_a$ ...(22.42) where I_a is the armature current.

Voltage across braking resistance R_b,

$$V_b = I_b R_b = R_b I_a (1 - \alpha) \quad ...(22.43)$$

Power dissipated in braking resistance R_b,

$$P_b = I_a^2 R_b (1 - \alpha)^2 \quad ...(22.44)$$

In this mode also the operation is in second quadrant, as illustrated in Fig. 22.26 (*b*).

It is also possible to combine the regenerative and rheostatic braking modes. Regenerative braking is used when the supply system is receptive. During remaining period the rheostatic braking is used.

22.7.5. Step-Down Chopper With R-L-E_b Load. The choppers are widely employed for speed control of dc motors in industrial and traction drives. Figure 22.27 (*a*) shows the basic chopper circuit for the control of a dc series motor. The chopper is shown to consist of a force-commutated thyristor, it could well be a transistor switch. It offers one-quadrant drive [Fig. 22.27 (*b*)]. Armature current is assumed continuous and ripple free. The waveforms for the source voltage V, armature terminal voltage v_a, armature current i_a, supply current i and freewheeling-diode current i_{DFW} are shown in Fig. 22.27 (*c*). From these waveforms we have

Average value of voltage across motor,

$$V_a = \frac{T_{on}}{T} V = \alpha V = f T_{on} V \quad ...\text{Refer to Eq. (22.25)}$$

where α = Duty cycle = $\frac{T_{on}}{T}$ and f = Chopping frequency = $\frac{1}{T}$

Power delivered to load (motor)

= Average value of voltage across motor × average motor current

$= V_a I_a = \alpha V I_a$

Average supply current

$= \frac{T_{on}}{T} I_a = \alpha I_a$

Input power to chopper

= Average supply voltage × average source current = $V \alpha I_a$

For the motor armature circuit,

$$V_a = \alpha V = E_b + I_a (R_a + R_{se}) = k_e \omega_m + I_a (R_a + R_{se})$$

or Motor speed,

$$\omega_m = \frac{\alpha V - I_a (R_a + R_{se})}{k_e} \quad ...(22.45)$$

From above Eq. (22.45) it is obvious that the armature terminal voltage and therefore, speed of the dc motor can be controlled by varying the duty cycle α of the chopper.

So far, armature current i_a has been assumed ripple free and accordingly waveforms in Fig. 22.27 (*c*) are drawn. But in practice, the motor armature current increases during chopper on

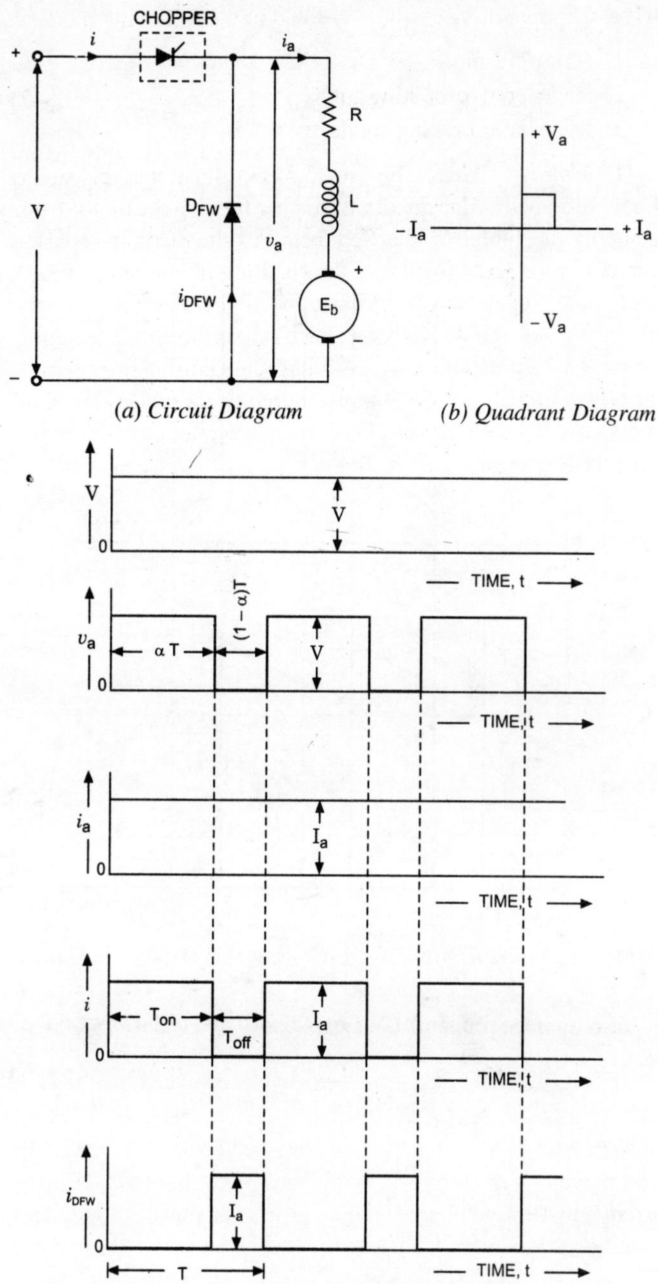

(*a*) Circuit Diagram (*b*) Quadrant Diagram

(*c*) Waveforms

Fig. 22.27 *Step-Down Chopper With R-L-E_b Load*

period and decreases during chopper off period, as illustrated in Fig. 22.28.

For R-L-E_b type of load, E_b is load voltage (back emf of motor). When the chopper is switched on for a period T_{on} and power is supplied to R-L-E_b load. By using Kirchhoff's voltage law (KVL) in the loop consisting of V, chopper and R-L-E_b load, we have

$$V - E_b - iR - L\frac{di}{dt} = 0$$

$$\text{or } V = L\frac{di}{dt} + iR + E_b \text{ for } 0 \le t \le T_{on} \quad ...(22.46)$$

When chopper is off, the load current continues flowing through the freewheeling diode due to energy stored in the inductor. The KVL for the loop containing R-L-E_b and D_{FW} yields

$$-iR - L\frac{di}{dt} - E_b = 0$$

$$\text{or } L\frac{di}{dt} + iR + E_b = 0 \quad \text{for } T_{on} \le t \le T \qquad ...(22.47)$$

The above Eqs. (22.46) and (22.47) can be solved by using Laplace transform. It is seen from Fig. 22.28 that initial value of load current is I_{min} for Eq. (22.46) and I_{max} for Eq. (22.47).

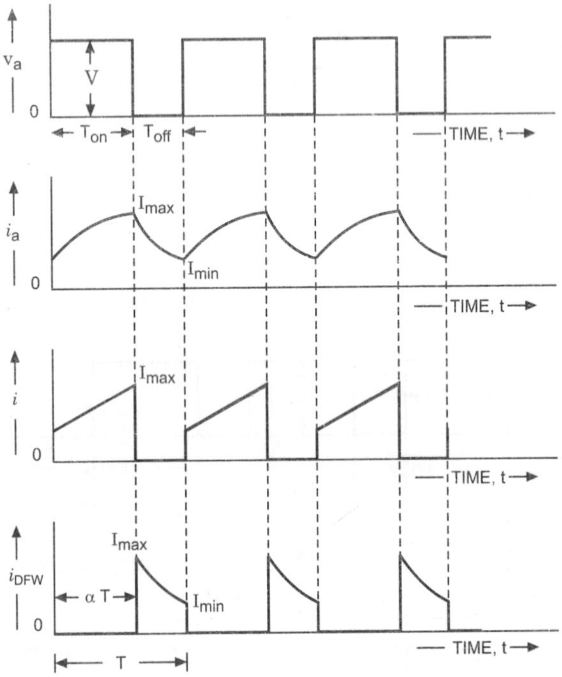

Fig. 22.28 *Waveforms For DC Chopper Drive of Fig. 22.27 (a)*

So Laplace transform of Eqs. (22.46) and (22.47) we have

$$RI(s) + L\,[s\,I(s) - I_{min}] = \frac{V - E_b}{s} \qquad ...(22.48)$$

$$\text{and } RI(s) + L[s\,I(s) - I_{max}] = \frac{-E_b}{s} \qquad ...(22.49)$$

From Eq. (22.48), we have

$$I(s) = \frac{V - E_b}{s(R + Ls)} + \frac{L\,I_{min}}{R + sL} = \frac{V - E_b}{sL\left(s + \frac{R}{L}\right)} + \frac{I_{min}}{s + \frac{R}{L}}$$

Laplace inverse of above equation is

$$i(t) = \frac{V - E_b}{R}\left(1 - e^{\frac{-R}{L}t}\right) + I_{min}\,e^{\frac{-R}{L}t} \qquad ...(22.50)$$

Similarly the time-domain expression for current from Eq. (22.49) is

$$i(t') = -\frac{E_b}{R}\left(1 - e^{\frac{-R}{L}t'}\right) + I_{max}\,e^{\frac{-R}{L}t'} \qquad ...(22.51)$$

Here R = Armature resistance R_a + series field resistance R_{se}

and L = Armature inductance + series field inductance L_{se}.

The values of maximum current I_{max} and minimum current I_{min} can be determined by using the fact that $i(t) = I_{min}$ when $t = 0$ and $i(t') = I_{max}$ when $t' = 0$. The values are found to be

$$I_{max} = \frac{V}{R}\left[\frac{1 - e^{\frac{-R\alpha T}{L}}}{1 - e^{\frac{-RT}{L}}}\right] - \frac{E_b}{R} \qquad ...(22.52)$$

$$\text{and } I_{min} = \frac{V}{R}\left[\frac{e^{\alpha TR/L} - 1}{e^{RT/L} - 1}\right] - \frac{E_b}{R} \qquad ...(22.53)$$

Example 22.15. An R-L-E_b load is supplied from 400 V through a chopper circuit. If R = 0, L = 0.5 H and α = 0.25, calculate the chopping frequency to limit the ripple current to 10 A.

Solution: Average load voltage, $V_a = \alpha V = 0.25 \times 400 = 100$ V

In R-L-E_b load, the difference between input voltage and average voltage is responsible for creating magnetic field in the coil. So

$$L\frac{di}{dt} = V - V_a = 400 - 100 = 300 \text{ V}$$

$$L = 0.5 \text{ H}$$

$$\text{and } \frac{di}{dt} = \frac{10 \text{ A}}{T_{on}}$$

$$\text{So } \frac{0.5 \times 10}{T_{on}} = 300$$

$$\text{or } T_{on} = \frac{0.5 \times 10}{300} = \frac{1}{60} \text{ second}$$

$$\text{and } T = \frac{T_{on}}{\alpha} = \frac{1/60}{0.25} = \frac{1}{15} \text{ second}$$

$$\text{Chopping frequency } f = \frac{1}{T} = 15 \text{ Hz.} \quad \textbf{Ans.}$$

Example 22.16. The speed of a separately-excited dc motor is controlled by a chopper. The supply voltage is 120 V, armature circuit resistance = 0.5 Ω, armature circuit inductance = 20 mH and motor constant 0.05 V/rpm. The motor drives a constant load torque requiring an average current of 20 A. Assume motor current is continuous. Calculate *(a)* the range of speed control *(b)* the range of duty cycle. *[Pb. Univ., Nov.-1990]*

Solution: The motor will operate at maximum speed (or base speed) when the ratio of time-on (T_{on}) to cycle (T), α = 1, In this case

Voltage across the motor armature, $V_a = V = 120$ V

Back emf, $E_b = V_a - I_a R_a = 120 - 20 \times 0.5 = 110$ V

$$\text{Motor speed, } N = \frac{E_b}{\text{Motor constant}} = \frac{110}{0.05} = 2{,}200 \text{ rpm}$$

The minimum speed is zero and at zero speed

Back emf, $E_{b0} = 0$

Voltage across the motor armature,

$$V_a = E_b + I_a R_a = 0 + 20 \times 0.5 = 10 \text{ V}$$

$$\text{Since, } V_a = V \times \frac{T_{on}}{T}$$

$$\therefore \text{ Duty cycle, } \alpha = \frac{T_{on}}{T} = \frac{V_a}{V} = \frac{10}{120} = \frac{1}{12}$$

Hence *(a)* Range of speed control is from 0 to 2,200 rpm **Ans.**

(b) Range of duty cycle is from $\frac{1}{12}$ to 1 **Ans.**

Example 22.17. A 200 V, 940 rpm, 100 A separately-excited dc motor has an armature resistance of 0.02 Ω. The motor is fed from a single quadrant step-down chopper for motoring operation. The input to

the chopper is 200 V dc. Assuming continuous load current, calculate the duty cycle of the chopper for motoring at rated torque and 500 rpm. [*U.P. Technical Univ. Electric Drives, 2004-05*]

Solution: Back emf induced in the motor armature when it is drawing a current of 100 A from 200 V supply and running at 940 rpm

$$E_{b1} = V - I_a R_a = 200 - 100 \times 0.02 = 198 \text{ V}$$

Back emf induced at a speed of 500 rpm

$$E_{b2} = E_{b_1} \times \frac{N_2}{N_1} = 198 \times \frac{500}{940} = 105.32 \text{ V}$$

Voltage across armature when running at 500 rpm

$$V_a = E_{b2} + I_a R_a = 105.32 + 100 \times 0.02$$
$$= 107.32 \text{ V} \quad \because \text{Torque is rated one}$$

$$\text{Duty ratio, } \alpha = \frac{T_{on}}{T} = \frac{V_a}{V} = \frac{107.32}{200} = 0.5366 \quad \textbf{Ans.}$$

Example 22.18. A dc chopper fed from 400 V supply runs a separately-excited motor. The armature resistance R_a = 0.1 Ω. The motor voltage constant is 4 Vs/radian, the average armature current I_a = 150 A. The armature current is continuous and has negligible ripple. Determine (*i*) the input power, (*ii*) the motor speed and (*iii*) the developed torque for a duty cycle of 60%. Neglect the losses in chopper. If the duty cycle of the chopper varies between 10% and 90%, find the minimum and maximum speeds.

Solution: (*i*) Power input

$$= \frac{\alpha \times V \times I_a}{1{,}000} = \frac{0.6 \times 400 \times 150}{1{,}000}$$
$$= 36 \text{ kW} \quad \textbf{Ans.}$$

Voltage across armature circuit, $V_a = \alpha V = 0.6 \times 400 = 240$ V

Back emf developed, $E_b = V_a - I_a R_a$

$$= 240 - 150 \times 0.1 = 225 \text{ V}$$

$$(ii)\ \text{Speed, } N = \frac{60}{2\pi}\omega = \frac{60}{2\pi} \times \frac{E_b}{\text{voltage constant}} = \frac{60}{2\pi} \times \frac{225}{4}$$
$$= 537 \text{ rpm} \quad \textbf{Ans.}$$

$$(iii)\ \text{Torque, } T = \frac{E_b I_a}{2\pi N/60} = \frac{225 \times 150 \times 60}{2\pi \times 537} = 600 \text{ N-m} \quad \textbf{Ans.}$$

For duty cycle of 10%,

$$E_b = \alpha V - I_a R_a = 0.1 \times 400 - 150 \times 0.1 = 25 \text{ V}$$

$$\text{Speed, } N = \frac{60}{2\pi} \times \frac{25}{4} = 60 \text{ rpm}$$

For duty cycle of 90%,

$$E_b = \alpha V - I_a R_a = 0.9 \times 400 - 150 \times 0.1 = 345 \text{ V}$$

$$\text{Speed, } N = \frac{60}{2\pi} \times \frac{345}{4} = 824 \text{ rpm}$$

Minimum speed will be at 10% duty and will be equal to 60 rpm while the maximum speed will be at 90% duty and will be equal to 824 rpm. **Ans.**

Example 22.19. A one quadrant chopper is used for rheostatic braking of a separately-excited dc motor. Armature resistance is 0.1 Ω, braking resistance is 7.5 Ω, voltage constant 1.6 V/A–rad/s, armature current 150 A and field current is 1.5 A. The duty cycle of the chopper is 0.4. Determine (*i*) voltage across the chopper (*ii*) power dissipated in the braking resistance and (*iii*) motor speed.

Solution: (*i*) Average voltage across chopper,

$$V_{ch} = V_b = (1 - \alpha) R_b I_a \quad \text{... Refer to Eq. (22.43)}$$
$$= (1 - 0.4) \times 7.5 \times 150 = 675 \text{ V} \quad \textbf{Ans.}$$

(*ii*) Power dissipated in braking resistance,

$$P_b = I_a^2 R_b (1 - \alpha)^2 \quad \text{... Refer to Eq. (22.44)}$$
$$= 150^2 \times 7.5 (1 - 0.4)^2 = 60{,}750 \text{ W or } 60.75 \text{ kW} \quad \textbf{Ans.}$$

Generated emf, $E_g = V_a + I_a R_a = 675 + 150 \times 0.1 = 690$ V

$$(iii)\ \text{Speed, } N = \frac{60}{2\pi} \times \omega$$
$$= \frac{60}{2\pi} \times \frac{E_g}{\text{voltage constant} \times \text{field current}}$$
$$= \frac{60}{2\pi} \times \frac{690}{1.6 \times 1.5} = 2{,}745 \text{ rpm} \quad \textbf{Ans.}$$

Example 22.20. The motor given in Example 22.8 is to be employed for regenerative braking by reversing the direction of field current. Find the firing angle of the armature circuit converter for the given value of armature current. Also determine the power supplied back to the supply source.

Solution: When the field current is reversed, the direction of back emf is reversed. The motor is now acting as a generator.

Generated emf, $E_g = -163.53$ V

Voltage across armature circuit,

$$V_a = E_g + I_a R_a = -163.53 + 40 \times 0.2 = -155.53 \text{ V}$$

$$\text{Since } V_a = \frac{2 V_{max}}{\pi} \cos \alpha_a$$

$$\therefore \quad -155.53 = \frac{2 \times 220 \times \sqrt{2}}{\pi} \cos \alpha_a$$

$$\text{or} \quad \alpha_a = 141.74° \quad \textbf{Ans.}$$

Power supplied back to the source

$$= V_a I_a = 155.53 \times 40$$
$$= 6{,}221.2 \text{ W or } 6.2212 \text{ kW} \quad \textbf{Ans.}$$

Example 22.21. A constant frequency system is used to control the speed of a dc traction motor from 220 V dc supply. The combined armature and field resistance is 0.2 Ω. The average current in the circuit is 25 A and chopper frequency is 200 Hz. Determine the pulse width, if the average value of back emf is 100 V.

Solution: Average voltage across chopper,

$$V_{ch} = V_a = E_b + I_a (R_a + R_{se}) = 100 + 25 \times 0.2 = 105 \text{ V}$$

$$\text{Also } V_a = V \times \frac{T_{on}}{T} = V \times T_{on} \times f$$

$$\text{So } 105 = 220 \times T_{on} \times 200$$

$$\text{or } T_{on} = \frac{105}{220 \times 200} = 0.00239 \text{ second or } 2.39 \text{ ms} \quad \textbf{Ans.}$$

22.8 COMPARISON OF CONVERTER FED AND CHOPPER FED DC DRIVES

For speed control of a dc motor it is necessary that it is supplied with an adjustable voltage which can be obtained either from a controlled converter or from a chopper. The relative advantages and disadvantages of these two are given below :

1. When available supply is ac, an ac-dc converter is ideally suited to convert the constant ac voltage into a variable dc voltage. If we use chopper, which converts a constant dc voltage into a variable dc voltage, diode rectifier will be required to convert constant ac voltage into a constant dc voltage for feeding it to chopper.

2. The ripple in the output of a chopper is much less than that in the output of a converter. Less ripple means lesser losses.
3. The chances of discontinuous operation in a chopper drive are lesser than that in a converter drive. So speed regulation and transient response of chopper drive are better than that in converter drive.
4. For reducing ripple in the output a converter with higher pulse number is required and a converter with higher pulse number has lower utilisation factor and more cost.

22.9 SPECIAL FEATURES OF THYRISTOR DRIVE MOTORS

The most commonly used dc motors for thyristor drive are separately-excited and series-excited dc motors. The thyristor drive motors usually differ from conventional dc motors in construction. Before discussing the special features incorporated in thyristor drive motors, discussion on the effects of thyristor power supply on performance of a dc motor is necessary.

22.9.1. Effects of Thyristor Power Supply on The Performance of a DC Motor.

1. The output voltage from thyristor converter consists of a dc component and ac harmonic components.
2. The output voltage can change very rapidly in comparison to that of a motor-generator set owing to the absence of field time constants associated with the generator.
3. An abnormally high value of armature current may rise in the event of thyristor fault when operating in the inverter mode.
4. The magnitude of harmonics decreases with the increase in frequency of harmonic for all converters. The magnitude of the harmonics increase as the firing angle is increased and dc voltage is reduced for all bidirectional converters. The higher the armature inductance, the lesser will be the harmonic currents.
5. Torque is developed by the dc component of the current whereas heating is developed by the effective (or rms) value of current. The form factor (ratio of effective value to average value) for half-wave three-phase thyristors may be taken as 1.2 while for full-wave three-phase thyristors it is 1.1. This increases electrical losses and, therefore, heating is 5 to 7 per cent more for three-phase full-bridge converters while for three-phase half-bridge converters it is from 15 to 20 per cent.
6. The commutating ability is seriously affected by the presence of harmonic currents. The peak value of the current is increased, interpole flux will be reduced in magnitude and a time lag will be introduced between the interpole current and the flux due to eddy currents generated in the iron path of the inter polar flux.

The other effects of thyristor power supply on motor performance are heating of interpole winding, saturation of interpole magnetic circuit, transformer voltage at the brushes and increase in voltage commutator segment.

22.9.2. Special Features of Thyristor Drive Motors. The thyristor drive dc motors are designed with the following special features in order to improve their performance.

1. The thyristor drive dc motors are made with larger diameter armature and larger size poles of reduced height.
2. The commutators are made larger in order to provide extra insulation to withstand larger and rapid voltage fluctuations.
3. The yoke as well as the main and commutating poles are laminated to reduce the eddy current effects.
4. Low inertia armatures are employed for improving the response.
5. Compensating windings are used in large motors to reduce armature reaction effect.
6. Split brushes of good commutating quality are used for reducing the effect of transformer voltage in the coils undergoing commutation.
7. The use of a laminated yoke instead of a solid yoke improves the commutation to a greater extent.
8. The use of a large number of commutator bars reduces the voltage between commutator segments and improves the commutation.
9. The use of an octagonal, rather than circular, shape for the frame accommodates more material and gives a larger rating for the same frame size.
10. The current densities used for the armature and interpole winding are reduced compared to conventional dc motors of the same frame size and rating in order to reduce the effect of heating armature and interpoles.
11. A better class of insulation (class F materials as standard insulation) is used to allow higher temperature rise and dissipation of more losses from a given frame.
12. The ratio pole arc/pole pitch is reduced in order to reduce the ratio of commutating zone to neutral zone.
13. Armature inductance is increased to reduce the ripple current. It may also increase reactance voltage. The number of turns per armature coil is kept at the minimum since the reactance voltage is proportional to the square of the turns per coil. Use of dummy coils is avoided.
14. Forced cooling by an auxiliary ac motor is widely used for improving cooling of the motor at reduced speeds.
15. Great care is required to be taken for accurate construction : spacing of brushes, poles and manufacture of commutators.

22.10 SPEED CONTROL OF DC MOTOR USING FEEDBACK LOOPS

The systems of speed control discussed in Art. 22.6 are open-loop systems. An open-loop system can be converted into a closed-loop system by employing a feedback loop, as illustrated in Fig. 22.29. The tacho-generator is a permanent magnet dc generator whose output emf is directly proportional to the speed. The difference of input speed setting and speed feedback provides the error signal. The error signal is amplified by the amplifier and supplied to the firing circuit of the converter. Thus the firing of converter thyristors is advanced or delayed depending on whether the actual speed is less than or more than the desired speed.

Let the desired speed of motor be 1,200 rpm, error signal be 0.1 V, converter output be 240 V and output of tacho-generator be 10 V per 1,000 rpm. If the input setting is 12.1 V, the converter

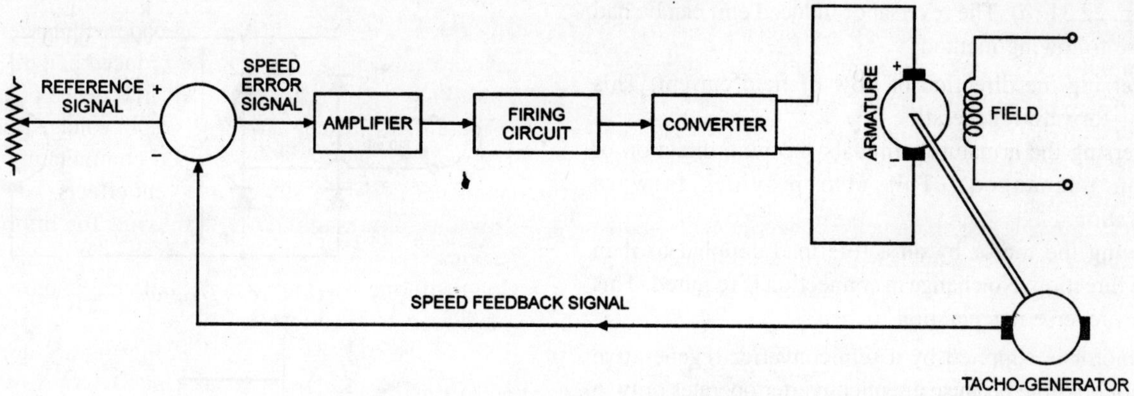

Fig. 22.29 *Speed Control of DC Motor With Negative Feedback*

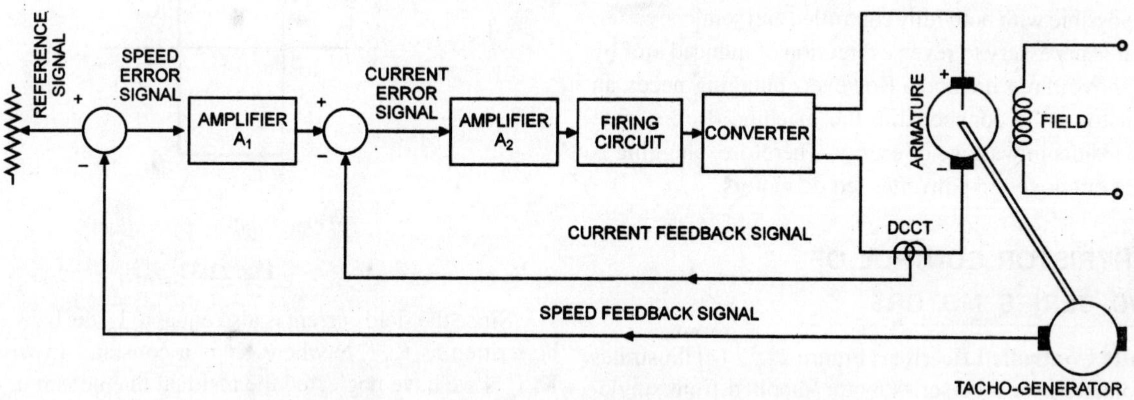

Fig. 22.30 *Speed Control of DC Motor With Current Limiting*

output will be 240 V and motor speed will be 1,200 rpm. If the load torque causes the decrease in speed and let the speed become 1,190 rpm, the output of tacho-generator will be 11.9 V increasing the error signal to 0.2 V. The converter output voltage will increase causing the speed to increase to 1,200 rpm. If the speed tends to increase beyond 1,200 rpm, the error signal will become less than 0.1 V and the output voltage of the converter will decrease, thus reducing the motor speed. Thus negative feedback is employed for control of motor speed.

A change in motor speed causes a temporary increase in motor and thyristor currents. Excessive current, however, can be avoided by an automatic current limiting circuit using another feedback loop, as shown in Fig. 22.30. The dc current transformer (dc CT), measures the converter output current and develops a voltage proportional to current. This signal is fed back, as illustrated at a position between the two amplifiers. If the converter current tends to increase, the output of dc CT increases thus reducing the current error. Thus the output current of the converter is reduced to the safe value.

22.11 ELECTRIC BRAKING OF RECTIFIER CONTROLLED SEPARATELY-EXCITED DC MOTOR

Figure 22.31 (*a*) illustrates a simplified representation of a single-phase fully controlled rectifier supplying a separately-excited dc motor. The polarities of voltages and directions of currents are also marked. The firing angle of the rectifier is in the range of $0 \le \alpha \le 90°$.

For regenerative braking, the direction of power flow is to be reversed. The motor has to operate as a generator and supply power to converter which now operates as an inverter and feeds back power to the supply mains. The circuit diagram for this operation is illustrated in Fig. 22.31 (*b*). Since the inverter output is connected to the supply mains, natural commutation is possible. It is seen that direction of flow of current in both of the circuits is the same. It has to be so since the current through a thyristor can flow only from anode to cathode. For this mode of operation $90° \le \alpha \le 180°$.

It is seen from Fig. 22.31 that direction of induced emf (back emf) in circuit shown in Fig. 22.31 (*b*) is reverse of that in circuit

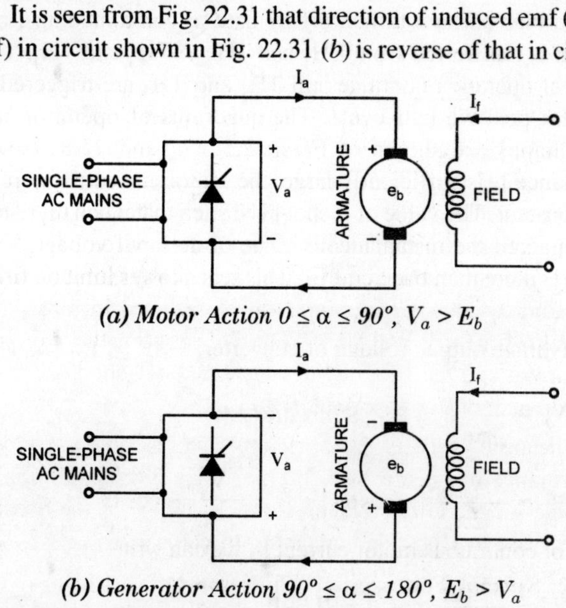

(a) Motor Action $0 \le \alpha \le 90°$, $V_a > E_b$

(b) Generator Action $90° \le \alpha \le 180°$, $E_b > V_a$

Fig. 22.31 *Electric Braking of Separately-Excited DC Motor* shown in Fig. 22.31 (*a*). The reversal of induced emf can be had by any of the following methods :

1. By reversing the direction of flow of field current. This provides forward regeneration.
2. By reversing the armature terminals (keeping direction of rotation unchanged). This also provides forward regeneration.
3. By driving the motor by an active load coupled to it in reverse direction. No change in connection is required. This provides reverse regeneration.

If the motor is supplied by a semiconverter regenerative braking is not possible because a semiconverter operates only in a first quadrant (*i.e.*, cannot operate as an inverter). However, plugging is possible with both fully controlled and semiconverter. In this case it is necessary to reverse direction of induced emf by any of the above three methods. However, plugging needs an external resistor to be connected to the machine (to dissipate energy) and results in wastage of energy. Therefore, plugging is generally not employed in converter fed dc motors.

22.12 THYRISTOR CONTROL OF DC SERIES MOTORS

22.12.1. Fully Controlled Rectifier. Figure 22.32 (*a*) illustrates the circuit diagram for a dc series motor supplied from single-phase ac mains through a fully controlled rectifier. The motor armature and field windings have resistance as well as inductance. Since the output current of a rectifier is not a perfect dc, the role of inductance also comes in. R is resistance of armature including that of field and L is inductance of armature including that of field. During positive half cycle, thyristors TH_1 and TH_2 are forward biased and start conducting at $\omega t = \alpha$. The load current flows through TH_1, motor and TH_2. At $\omega t = \pi$, the supply voltages reverses. Because of inductance L thyristors TH_1 and TH_2 continue to conduct beyond $\omega t = \pi$. From $\omega t = \pi + \alpha$ to $\omega t = 2\pi$, thyristors TH_3 and TH_4 are forward biased. When TH_3 and TH_4 are triggered at $\omega t = \pi + \alpha$, thyristors TH_1 and TH_2 are subjected to reverse bias and are turned off by natural commutation. Load current is transferred from TH_1 and TH_2 to TH_3 and TH_4. This mode of operation continues till TH_1 and TH_2 are triggered in the next positive half cycle. The quadrants of operation and waveshapes are shown in Figs. 22.32 (*b*) and 22.32 (*c*). If inductance L is sufficiently large, the motor current is more or less constant. The value of α should be such that when thyristors are triggered the instantaneous value of ac input voltage $V_{max} \sin \omega t$ is more than back emf E_b. This sets a lower limit on firing angle α.

Average output voltage of converter,

$$V_a = \frac{2}{2\pi}\int_{\alpha}^{\pi+\alpha} V_{max} \sin\omega t \, d(\omega t)$$

$$= \frac{2V_{max}}{2\pi}\left| -\cos\omega t \right|_{\alpha}^{\pi+\alpha} = \frac{2V_{max}}{\pi}\cos\alpha$$

For continuous motor current I_a, we can write

$$V_a = \frac{2V_{max}}{\pi}\cos\alpha = RI_a + E_b$$

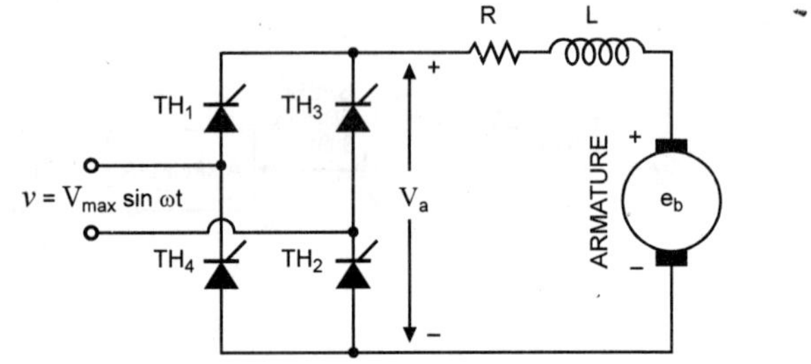

(*a*) Circuit Diagram

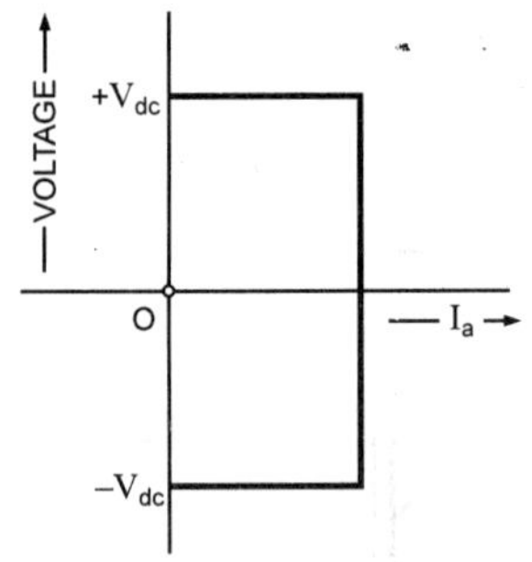

(*b*) Quadrants of Operation

Fig. 22.32

Since the field current is also equal to I_a, the back emf E_b can be written as $K_1 I_a N$ where K_1 is a constant. In writing E_b as $K_1 I_a N$ we have neglected the residual magnetism in the motor.

So $$\frac{2V_{max}}{\pi}\cos\alpha = RI_a + K_1 I_a N$$

$$\text{or } N = \frac{\frac{2V_{max}}{\pi}\cos\alpha - RI_a}{K_1 I_a} \quad \text{...(22.54)}$$

$$\text{and torque, } T \propto \Phi I_a = K_2 I_a^2 \quad \text{...(22.55)}$$

Substituting value of I_a from Eq. (22.54) in Eq. (22.55), we have

$$T = K_2\left[\frac{\frac{2V_{max}\cos\alpha}{\pi}}{R + K_1 N}\right]^2$$

$$\text{or } T = \frac{4K_2 V_{max}^2 \cos^2\alpha}{\pi^2 (R + K_1 N)^2} \quad \text{...(22.56)}$$

Using above Eq. (22.56) speed-torque characteristics for different values of α can be plotted, as illustrated in Fig. 22.32 (*d*).

22.12.2. Semicontrolled Rectifier. A dc series motor supplied from single-phase ac supply mains through a semicontrolled rectifier is shown in Fig. 22.33 (*a*). The rectifier has two thyristors TH_1 and TH_2 and two diodes D_1 and D_2. The freewheeling diode D_{FW} helps in conduction of current when thyristor is not conducting. R is the resistance of armature and field and L is the inductance of armature and field windings.

During positive half cycle thyristor TH_1 is triggered at $\omega t = \alpha$ and starts conducting. The value of α should be such that $V_{max} \sin \alpha > E_b$. The current flows through TH_1, motor and diode D_1 from $\omega t = \alpha$ to $\omega t = \pi$. At $\omega t = \pi$ the input voltage becomes

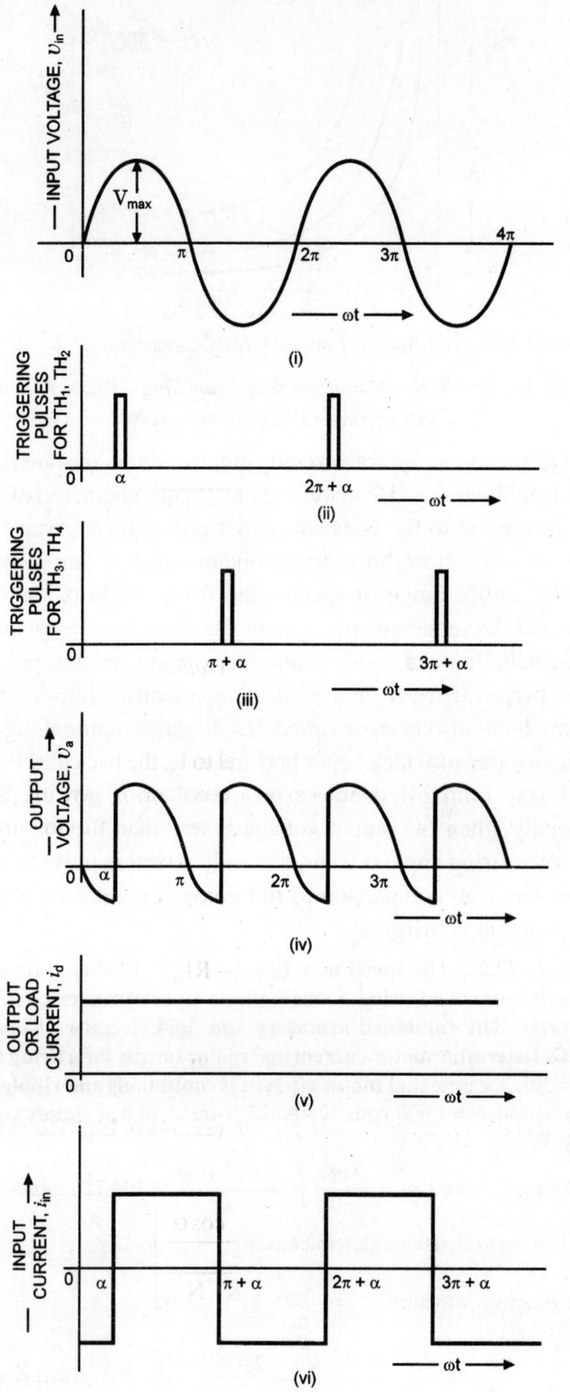

(c) Waveshape

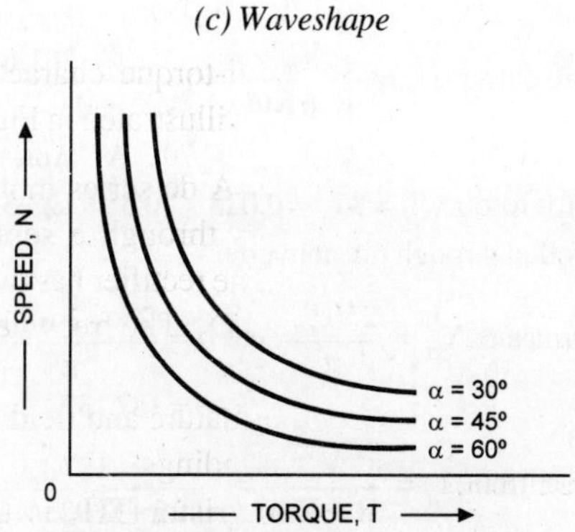

(d) Speed-Torque Characteristics

Fig. 22.32 *DC Series Motor Supplied From Single-Phase AC Supply Through Fully Controlled Rectifier*

negative and freewheeling diode D_{FW} is forward biased. Thus at $\omega t = \pi$, thyristor TH_1 and diode D_1 stop conducting and current is transferred to D_{FW}. During negative half cycle thyristor TH_2 is forward biased and when it is triggered at $\omega t = \pi + \alpha$, the freewheeling diode D_{FW} stops conducting and current is transferred to $TH_2 - D_2$ combination. TH_2 and D_2 conduct from $\omega t = \pi + \alpha$ to $\omega t = 2\pi$. At $\omega t = 2\pi$, thyristor TH_2 and diode D_2 get turned off and current freewheels through D_{FW} from $\omega t = 2\pi$ to $\omega t = 2\pi + \alpha$. At $\omega t = 2\pi + \alpha$, thyristor TH_1 is triggered again and next cycle of operation starts. Thus the circuit operation is: D_{FW} conducts for $0 \le \omega t \le \alpha$; TH_1 and D_1 conduct for $\alpha \le \omega t \le \pi$; D_{FW} conducts for $\pi \le \omega t \le \pi + \alpha$; and TH_2 and D_2 conduct for $\pi + \alpha \le \omega t \le 2\pi$.

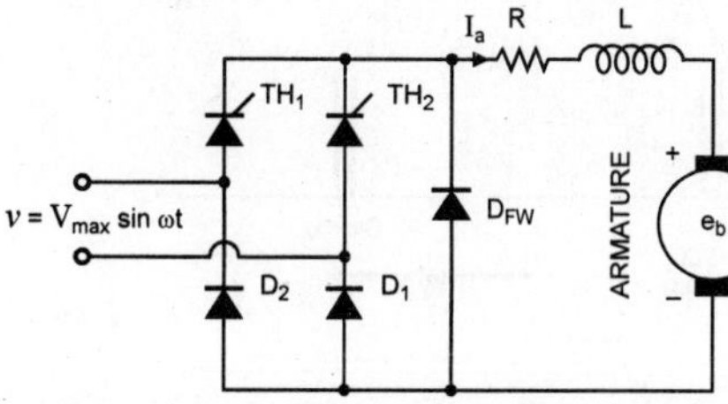

(a) Circuit Diagram

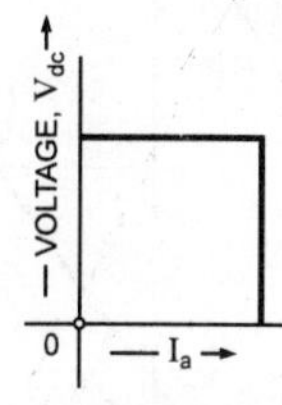

(b) Quadrant of Operation

Fig. 22.33

The quadrant of operation and waveshapes are shown in Figs. 22.33 (*b*) and 22.33 (*c*) respectively.

The average output voltage of semiconverter is given as

$$V_a = \frac{2}{2\pi}\int_{\alpha}^{\pi} V_{max} \sin \omega t \, d(\omega t)$$

$$= \frac{2\,V_{max}}{2\pi}\left| -\cos \omega \right|_{\alpha}^{\pi} = \frac{V_{max}}{\pi}\,(1 + \cos \alpha)$$

The back emf E_b can be written as $K_1 I_a N$, neglecting residual magnetism

$$\text{So } V_a = \frac{V_{max}}{\pi}\,(1 + \cos \alpha) = RI_a + K_1 I_a N$$

$$\text{or } N = \frac{\frac{V_{max}}{\pi}(1 + \cos \alpha) - RI_a}{K_1 I_a} \qquad \text{...(22.57)}$$

and torque $T \propto \phi I_a = K_2 I_a^2$

Substituting the value of I_a from Eq. (22.57) in above equation, we have

$$T = \frac{K_2 V_{max}^2 (1 + \cos \alpha)^2}{\pi^2 (R + K_1 N)^2} \qquad \text{...(22.58)}$$

Using above Eq. (22.58) speed-torque characteristics for different values of α are plotted, as illustrated in Fig. 22.33 (*d*).

During operation of dc separately-excited and dc series motors, it is observed that the region of discontinuous current is smaller in case of dc series motors. The use of semiconverter provides continuous current over almost the entire operating range except when the load torque is very small. If inductance is employed in the armature circuit, continuous armature current can be obtained over the entire range of operation.

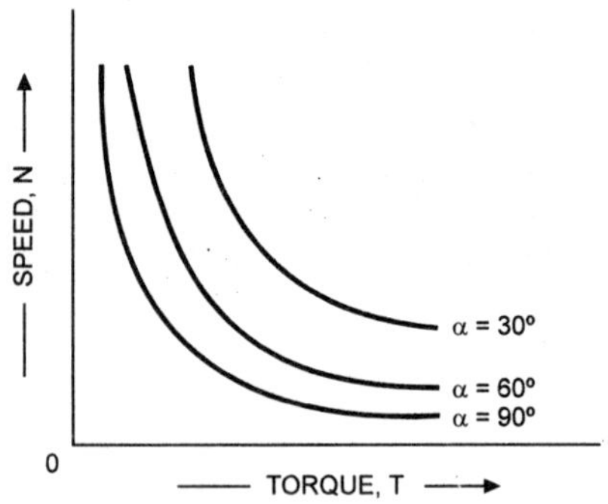

(d) Speed-Torque Characteristics

Fig. 22.33 *DC Series Motor Supplied From Single Phase AC Supply Through Single Phase Semiconverter*

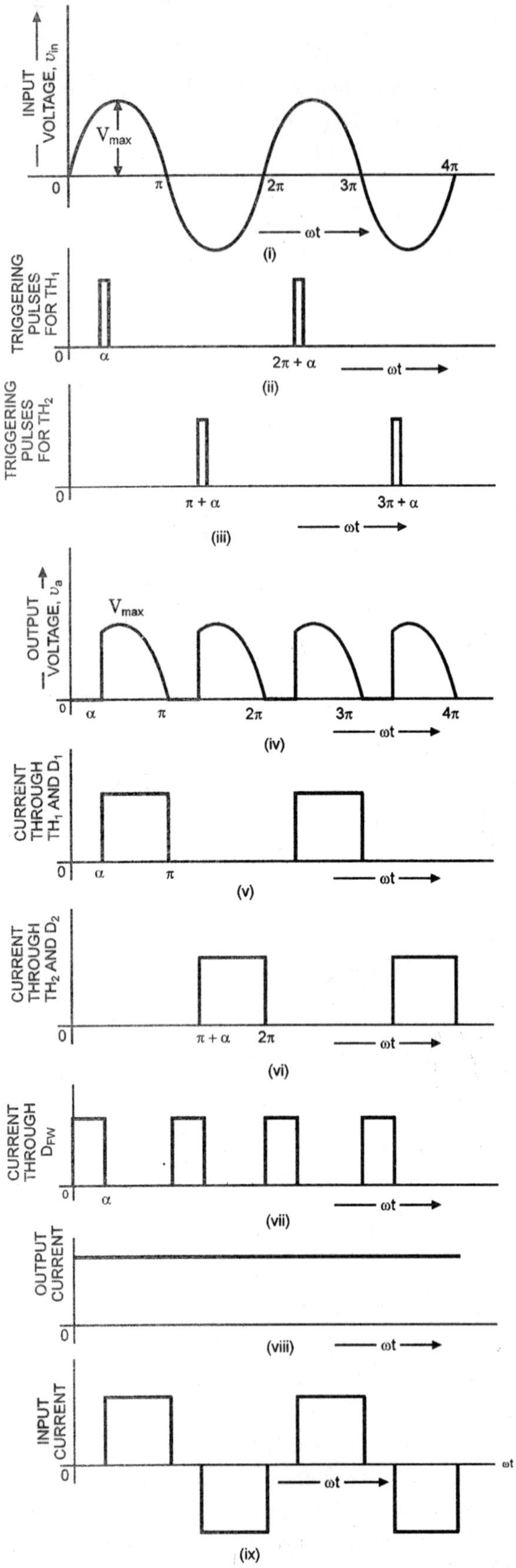

(c) Waveshapes

Fig. 22.33

DC series motors are normally employed for constant output power applications. However, speed-torque characteristics do not correspond to the constant output power for a given firing angle and, therefore, in order to obtain constant output power over the entire range of speeds, the firing angle α has to be adjusted. The semiconverter system, because of its freewheeling action, helps in maintaining continuous current and thus provides better performance to motors in comparison to full-converter system. It has also been observed that dc series motor along with semiconverter provides better performance.

Phase controlled converters have poor power factor especially when the output voltage is less than the maximum, *i.e.,* when firing angle α is large. Semiconverters provide better power factor in comparison to full converters even though the improvement is marginal.

Example 22.22. The speed of a 10 kW, 220 V, 1000 rpm dc series motor is controlled using a single-phase (*i*) semiconverter (*ii*) full converter. The combined armature and field circuit resistance is 0.25 Ω. Determine motor current and motor torque for a firing angle of α = 30°. Assume that motor current is continuous and ripple free, motor speed, N = 1,000 rpm, K = 0.025 N-m/A², and ac supply voltage = 240 V.

Solution: Speed, $\omega = \dfrac{2\pi N}{60} = \dfrac{2\pi \times 1{,}000}{60} = 104.72$ radians/s

(*i*) When controlled through semiconverter

$$\text{Voltage across armature, } V_a = \frac{V_{max}}{\pi}(1 + \cos\alpha)$$

$$= \frac{\sqrt{2} \times 240}{\pi}(1 + \cos 30°) = 201.6 \text{ V}$$

$$\text{Armature current, } I_a = \frac{V_a}{R + K\omega} = \frac{201.6}{0.25 + 0.025 \times 104.72}$$

$$= 70.3 \text{ A} \quad \textbf{Ans.}$$

$$\text{Motor torque, } T = KI_a^2 = 0.025 \times 70.3^2 = 123.53 \text{ N-m} \quad \textbf{Ans.}$$

(*ii*) When controlled through full converter

$$\text{Voltage across armature, } V_a = \frac{2V_{max}}{\pi}\cos\alpha = \frac{2 \times \sqrt{2} \times 240}{\pi}\cos 30°$$

$$= 187.13 \text{ V}$$

$$\text{Armature current, } I_a = \frac{V_a}{R + K\omega} = \frac{187.13}{0.25 + 0.025 \times 104.72}$$

$$= 65.25 \text{ A} \quad \textbf{Ans.}$$

$$\text{Motor torque, } T = KI_a^2 = 0.025 \times 65.25^2 = 106.43 \text{ N-m} \quad \textbf{Ans.}$$

Example 22.23. A dc series motor is supplied from a single-phase, 230 V, 50 Hz supply through a full-wave diode rectifier. The armature and field resistance is 1.0 Ω and torque constant K_t is 0.3 Nm/A². If load torque is 30 N-m, find average motor current and speed. Assume that motor current is constant and emf constant is 0.3 V/A-rad/second.

Solution: DC output voltage,

$$V_{dc} = \frac{1}{\pi}\int_0^{\pi} V_{max} \sin \omega t \, d(\omega t) = \frac{\sqrt{2} \times 230}{\pi}\int_0^{\pi} \sin \omega t \, d(\omega t)$$

$$= \frac{\sqrt{2} \times 230}{\pi} \times 2 = 207 \text{ V}$$

Applying Kirchhoff's second or voltage law to the motor circuit, we have

$$RI_a + (L_a + L_f)\frac{dI_a}{dt} + K_e \omega I_a = V_{max} \sin \omega t$$

where I_a is the armature current, L_a is the armature winding inductance, L_f is the field winding inductance, R is the combined resistance of field and armature windings, ω is the motor angular speed and K_e is the emf constant.

Since armature current I_a is constant, $\frac{dI_a}{dt} = 0$

So $1.0 \times I_a + 0.3\, \omega I_a = 230\sqrt{2} \sin \omega t$

Integrating above equation over one half cycle, we have

$$I_a + 0.3 \omega I_a = 207 \qquad ...(i)$$

Moreover, load torque, $T_L = K_t I_a^2$

So $30 = 0.3 I_a^2$

or $I_a = \sqrt{\frac{30}{0.3}} = 10$ A **Ans.**

Substituting $I_a = 10$ A in Eq. (*i*), we have

$$\omega = \frac{207 - I_a}{0.3 I_a} = \frac{207 - 10}{0.3 \times 10} = 65.667 \text{ rad/second}$$

and speed, $N = \frac{\omega \times 60}{2\pi} = \frac{65.667 \times 60}{2\pi} = 627$ rpm **Ans.**

Example 22.24. A dc series motor is coupled to a load whose torque varies as square of the speed and is supplied from a transformer through a semiconverter. The secondary voltage of the transformer is 675 V. When the firing angle is 100°, the motor speed and armature current are 400 rpm and 36 A. It is desired to raise the speed to 600 rpm. Determine the armature current and firing angle of semiconverter. Armature and field resistance is 0.4 Ω. Assume that magnetic circuit is unsaturated.

Solution: Torque developed by a dc motor is given as

$T = K_t \Phi I_a$

or $T \propto I_a^2$ ∵ magnetic circuit is unsaturated and, therefore, $\Phi \propto I_a$

Also load torque, $T \propto N^2$

Under steady-state conditions

$I_a^2 \propto N^2$

or $I_a \propto N$

So $I_{a2} = I_{a1} \times \frac{N_2}{N_1} = 36 \times \frac{600}{400} = 54$ A **Ans.**

Voltage across armature when firing angle $\alpha_1 = 100°$

$$V_{a_1} = \frac{V_{max}}{\pi}(1 + \cos \alpha_1) = \frac{675\sqrt{2}}{\pi}(1 + \cos 100°) = 251.09 \text{ V}$$

Back emf, $E_{b_1} = V_{a_1} - I_{a_1}(R_a + R_f) = 251.09 - 36 \times 0.4 = 236.69$ V

Let V_{a_2} be the output voltage of semi-converter when speed is 600 rpm

Back emf, $E_{b_2} = V_{a_2} - I_{a_2}(R_a + R_f) = V_{a_2} - 54 \times 0.4 = (V_{a_2} - 21.6)$ volts

Since $E_b \propto \Phi N \propto I_a N$

So $E_{b_2} = E_{b_1} \times \frac{I_{a_2} N_2}{I_{a_1} N_1} = \frac{236.69 \times 54 \times 600}{36 \times 400} = 532.56$ V

Also $E_{b_2} = V_{a_2} - 21.6$

or $V_{a_2} = E_{b_2} + 21.6 = 532.56 + 21.6 = 554.16$ V

Also $V_{a_2} = \frac{V_{max}}{\pi}(1 + \cos \alpha)$

or $554.16 = \frac{675\sqrt{2}}{\pi}(1 + \cos \alpha)$

or $\alpha = 34.54°$ **Ans.**

22.13 THYRISTOR CONTROL OF THREE-PHASE INDUCTION MOTORS

As already stated in Art. 19.3, the speed of an induction motor is given as

$$N = \frac{120 f}{P}(1 - s)$$

Thus the speed of an induction motor with fixed number of poles depends upon the supply frequency *f* and the slip *s* which in turn depends upon the voltage or current supplied to the motor. Following methods are available for speed control of 3-phase induction motors using thyristors.

1. Stator voltage control or variable voltage constant frequency control.
2. Variable voltage and variable frequency control.
3. Rotor resistance control.
4. Secondary foreign voltage control.

Stator voltage variation is accomplished by means of *ac regulators*, which control the rms value of the ac voltage applied to the motor by introducing thyristors connected back-to-back in each supply line. Variable frequency power is obtained by means of a *cycloconverter*, which directly converts fixed frequency ac to variable frequency ac or by an inverter which converts dc to ac. The effective value of an external resistance introduced into the rotor circuit can be controlled by connecting a high frequency chopper across the resistance and varying the time for which the chopper is *on* during the cycle. Static frequency converters are being employed for replacing the auxiliary machines in the Scherbius system. The Kramer scheme has also been changed by making use of a diode bridge rectifier in place of rotor converter, but a dc motor is still required for converting the rectified slip power to mechanical power.

22.13.1. AC Regulators. An ac regulator converts a constant ac voltage into a variable ac voltage of the same frequency. No doubt, ac voltage level can be changed by auto-transformer, tap-changing transformer, saturable reactors etc. These devices have been in use for long time and still are in use. But ac regulators using thyristors and triacs are becoming more and more popular because of their high efficiency, fast control and compact size. However, ac regulators using thyristors and triacs introduce objectionable harmonics in the circuits. AC regulators are classified as single-phase or three-phase. Each of these may be half-wave (*i.e.*, unidirectional) or full-wave (*i.e.*, bidirectional).

Since the input to an ac regulator is ac, it is always line commutated. Therefore, forced commutation is not required. As such the circuits of ac regulators are quite simple. Two types of controls are used in ac regulators. These are known as *integral cycle control* and *phase control*.

In an *integral cycle control*, also known as *on-off control*, the thyristors are used as switches to connect the motor to the supply source for a certain number of cycles of the source voltage and then to disconnect it for another certain number of cycles. Each of the on- and off-times consists of an integral number of cycles. Thyristors are switched on by gate pulses at zero voltage crossing of input voltage.

In *phase control*, the thyristors are employed as switches for connecting the motor to the supply for a certain portion of each cycle of the supply voltage. Most ac regulators use phase control. The power circuit configurations for integral cycle control and phase control do not differ in any manner.

Smooth variation of three-phase ac voltage can be realized by means of different configurations of the power circuit.

The three-phase regulators, as already said, may be half-wave or full-wave. Circuit of *three-phase half-wave regulator* for delta-connected motors or star-connected motors in which neutral point is not accessible is shown in Fig. 22.34. This circuit uses three thyristors and three diodes. Though half-wave ac regulator illustrated in Fig. 22.34 affects a saving in the cost of semiconductor devices and does not give rise to dc components in any part of the system but it introduces more harmonics into the line current than does the full-wave regulator. Half-wave circuit is not used in actual practice.

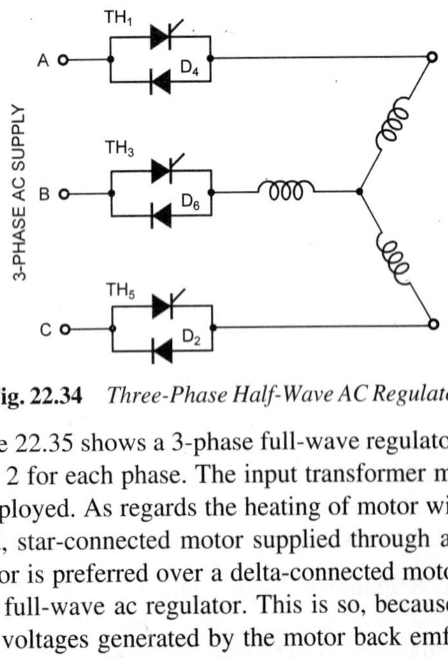

Fig. 22.34 *Three-Phase Half-Wave AC Regulator*

Figure 22.35 shows a 3-phase full-wave regulator. It uses 6 thyristors, 2 for each phase. The input transformer may or may not be employed. As regards the heating of motor windings are concerned, star-connected motor supplied through a full-wave ac regulator is preferred over a delta-connected motor supplied through a full-wave ac regulator. This is so, because any third harmonic voltages generated by the motor back emf can cause circulating currents in case of delta-connected motor.

For delta-connected load circuits in which each end of each phase is accessible, the arrangement illustrated in Fig. 22.36 is employed. This arrangement has the advantage of reducing the current of the device as it has to carry now $1/\sqrt{3}$ of the current if

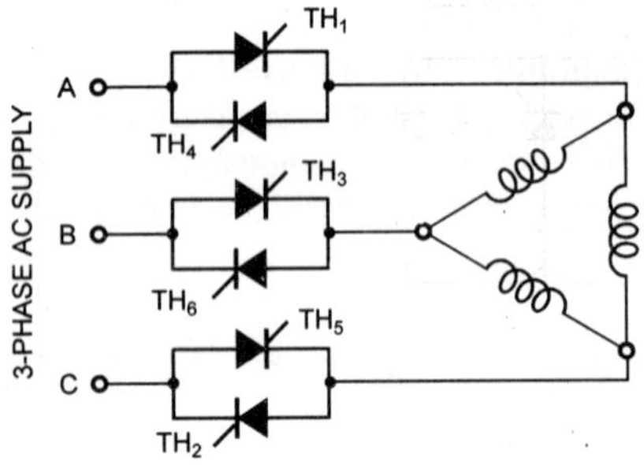

Fig. 22.35 *Three-Phase Full-Wave AC Regulator*

they were connected in the line of the delta winding. Once the phase current wave is known, the line current wave can be constructed by superposition.

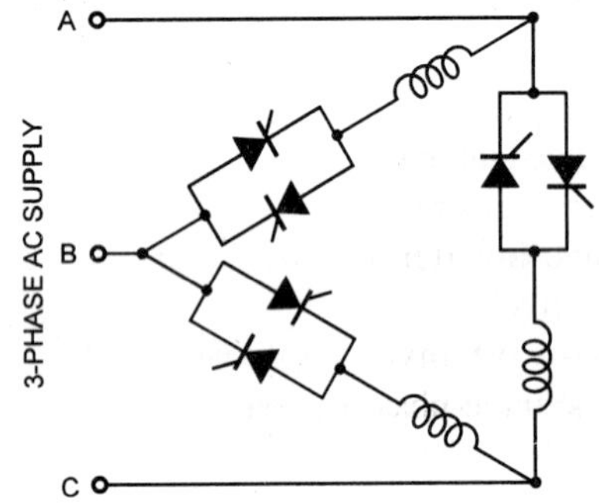

Fig. 22.36 *Delta-Connected Speed Controller*

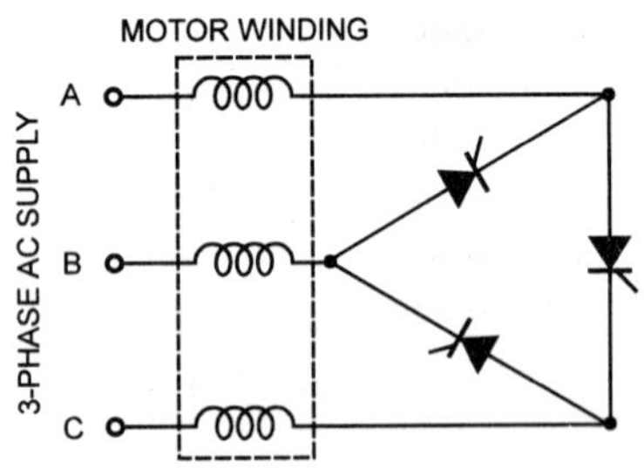

Fig. 22.37 *Neutral Connected Regulator*

For star-connected load circuits in which neutral point is accessible and can be opened, the arrangement illustrated in Fig. 22.37 may be used. In this arrangement, the number of thyristors required reduces to three and control circuitry becomes considerably simplified. The power consumption of the motor may be as high 100% greater than that with sine wave voltage control especially at reduced speeds.

22.13.2. Variable Voltage and Variable Frequency Control. If only frequency is changed and stator voltage is kept constant, the stator flux will not be at its rated value. The operation with flux below or above the rated value is not desirable. For constant flux operation, it is necessary that the induced emf increases or decreases linearly with applied frequency. At higher voltages and at high frequency operation stator drops are very small and thus constant flux operation is obtained by keeping V/f ratio constant. Speed-torque characteristics of a 3-phase induction motor with variable voltage variable frequency (constant V/f control) are illustrated in Fig. 19.12. It may be seen from the Fig. 19.12 that the maximum torque does not alter because the flux per pole remains constant. This mode of operation ensures the ability of developing the highest possible torque per unit

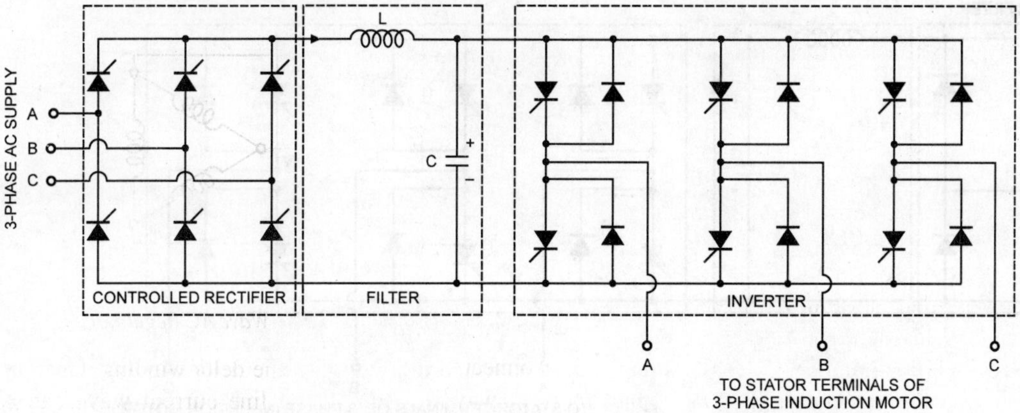

Fig. 22.38 *Variable Voltage Variable Frequency Control of 3-Phase Induction Motor (Square-Wave Inverter Power Circuit)*

the speed ratio is usually restricted to 10 : 1.

The noteworthy point is that electrical machine is designed corresponding to near saturation point on the magnetization curve (or B-H curve). This is done from the point of view of full utilisation of the core. If the stator frequency is reduced, keeping the stator voltage constant, motor will operate in the saturation region and therefore, the motor will draw large stator current and, therefore, results in the best possible utilisation of the available current capability of the drive.

The variable stator voltage and frequency can be obtained from the systems illustrated in Fig. 22.38 or in Fig. 22.41, known as square-wave inverter and pulse width modulated (PWM) inverter respectively.

A square-wave inverter power circuit is illustrated in Fig. 22.38. The three-phase ac supply is converted into dc by a controlled rectifier. The output of the rectifier is supplied to the filter circuit to remove the harmonics. The dc output from filter is fed to a controlled inverter which provides variable voltage variable frequency output. This supply is fed to the stator of the 3-phase induction motor whose speed is to be controlled.

Figure 22.39 shows the phase voltage waveforms V_{AN}, V_{BN}, V_{CN} and line- to-line voltage waveforms V_{AB}, V_{BC} and V_{CA}. Each line voltage waveform is displaced in time phase by 120 electrical degrees from one another and is of quasi-square wave with 120° width. The noteworthy point is that the inverter thyristors are force commutated because induction motor is a lagging power factor load. The feedback diodes help in circulation of load reactive power with filter capacitor and maintain the output voltage waves fixed to the level of dc link voltage.

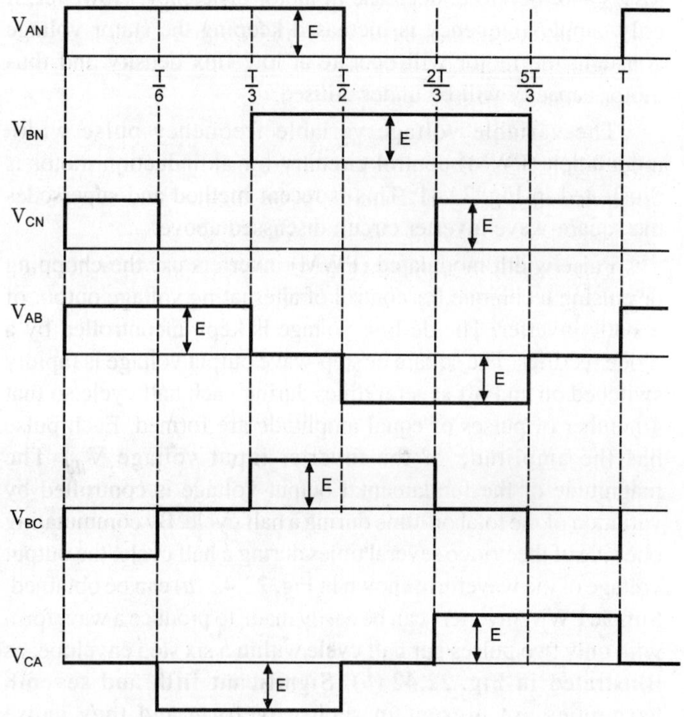

Fig. 22.39 *Phase and Line Voltage Waveforms at the Motor Terminals*

Desired voltage-frequency relationship of induction motor is illustrated in Fig. 22.40. When the frequency is less than the normal frequency, the voltage is reduced by the same proportion so as to keep V/*f* constant. At very low frequencies, as the reactance drop becomes smaller in comparison to stator resistance drop (ω L < R), additional voltage is required to be applied to compensate this effect. It means a higher V/*f* ratio. When the frequency exceeds the normal frequency, the torque decreases with the decrease in air-gap flux and the motor now operates in the constant power region, as illustrated in Fig. 22.40. This is equivalent to field weakening mode of speed control of a dc motor.

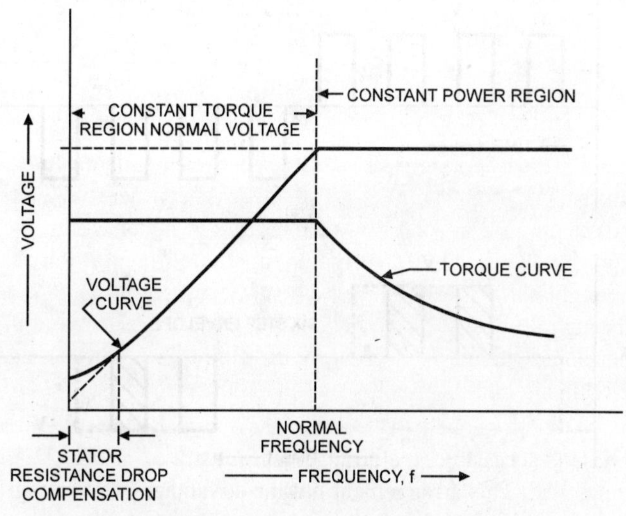

Fig. 22.40 *Voltage-Frequency Relationship*

When operating at reduced voltage the converter voltage decreases and, therefore, the commutation capability of the capacitor also decreases. So the inverter is usually provided with an auxiliary constant voltage dc supply for commutation purposes.

The inverter mentioned above cannot return power back to the ac supply lines unless another phase controlled rectifier is added to form a reversing system. This method of speed control is used in low and medium size 3-phase induction motors where

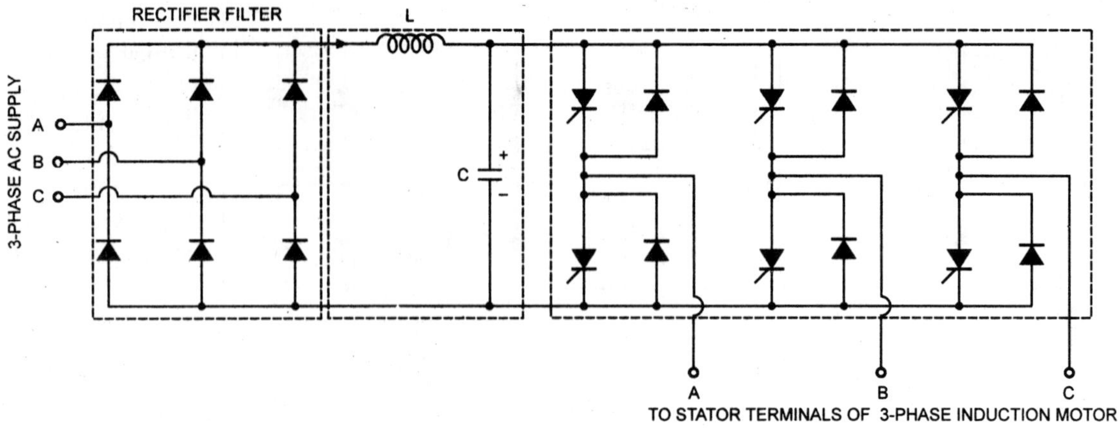

Fig. 22.41 *Variable Voltage Variable Frequency Control of 3-Phase Induction Motor (PWM Control)*

magnetising current causing increase in core and stator copper losses and therefore, decrease in motor efficiency. However, if only supply frequency is increased keeping the stator voltage constant, the motor will operate at low flux density and thus motor capacity will be under utilised.

The variable voltage variable frequency pulse width modulation (PWM) control circuitry for an induction motor is illustrated in Fig. 22.41. This is recent method and supersedes the square-wave inverter circuit discussed above.

Pulse width modulated (PWM) inverters use the chopping or pulsing technique for control of alternating voltage output of a static inverter. The dc link voltage is kept uncontrolled by a diode rectifier. The square or step-wave output voltage is rapidly switched on and off several times during each half cycle so that a number of pulses of equal amplitude are formed. Each pulse has the amplitude of the inverter input voltage V_{dc}. The magnitude of the fundamental output voltage is controlled by variation of the total on-time during a half cycle. By commutating one side of the bridge several times during a half cycle, the output voltage of the waveform shown in Fig. 22.42 (*a*) can be obtained. Simple PWM inverters can be easily made to produce a waveform with only two pulses per half cycle within a six step envelope, as illustrated in Fig. 22.42 (*b*). Significant fifth and seventh harmonics are present in such waveform and they cause appreciable deterioration in the low-speed performance of the ac motor. For elimination of low order harmonics, more refined techniques of PWM are employed, in which high frequency pulsing occurs throughout the half cycle.

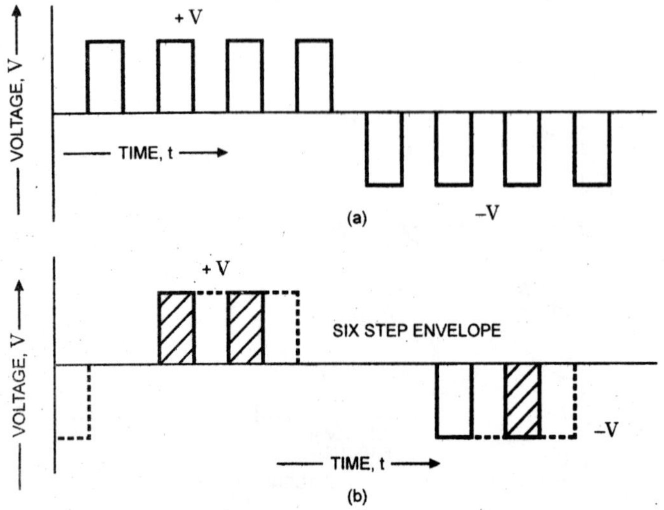

Fig. 22.42 *Basic Waveform of PWM*

In sophisticated PWM systems, the pulse width is varied throughout the half cycle in a sinusoidal manner, as illustrated in Fig. 22.43. Actually, the pulses should be regularly spaced and the pulse width at a particular position should be proportional to the area under the sine wave at that position. In the PWM waveform, the lowest harmonic frequency is at the pulse repetition frequency and, if this is much higher than the fundamental frequency, adequate filtering is provided by the inductance of the machine. Such waveforms are usually produced by means of a control circuit in which a high frequency triangular waveform is mixed with a sinusoidal waveform of the desired frequency. Voltage control is obtained by varying the widths of all pulses without affecting the sinusoidal relationship.

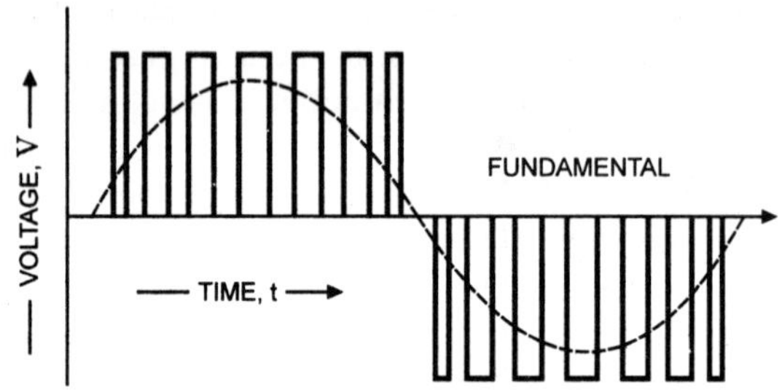

Fig. 22.43 *Output Voltage Waveform of a PWM Inverter With Sinusoidal Modulation of The Pulse Width*

Figure 22.44 illustrates sinusoidal PWM technique where an isosceles triangular wave is compared with the sinusoidal wave signal and the points of commutation are determined by the crossover points. If modulation index is found to be less than unity, only carrier frequency harmonics with fundamental frequency related side bands appears at the output. Such waveform generates less harmonic heating and torque pulsation in comparison to that of the square wave. When the modulation index exceeds unity, maximum voltage is obtained in the square-wave mode. Thus, PWM mode is applicable in the constant torque region while in the constant power region the operation is similar to that of the square-wave mode.

The transistorized PWM control, shown in Fig. 22.45, is used for control of low to medium size motors. No doubt power transistors cost much more than that of the thyristors of the same

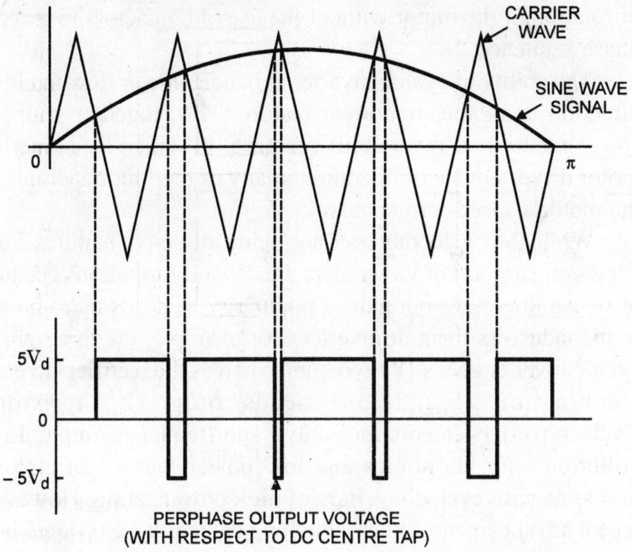

Fig. 22.44 *Principle of Pulse Width Modulation (PWM)*

capability but savings due to elimination of commutation circuit and corresponding commutation losses, the circuit proves to be more economical and efficient. Also, the transistors operate faster, PWM is possible at higher frequency. This cuts down further the machine losses.

22.13.3. Variable Current Variable Frequency Control. A variable current variable frequency control circuit for an induction motor is illustrated in Fig. 22.46. Variable dc voltage provided by a phase controlled rectifier is converted to a current source by connecting a large inductor in series. The large inductance maintains the current constant. The voltage available at the stator terminals of the 3-phase induction motor is almost sinusoidal with superimposed voltage spikes due to commutation. The converter employed is a line commutated while the inverter is forced commutated because induction motor operates at lagging power factor. The phase controlled converter can be replaced by a diode rectifier followed by a dc chopper.

The circuit has the following ***advantages*** :

(*i*) Since the input current is constant, misfiring of devices and short circuits do not pose any problem.

(*ii*) Less number of components in inverter circuit and less commutation loss.

(*iii*) Rugged and reliable power circuit.

(*iv*) Simpler and more reliable control circuit. This is because only 6 thyristors are to be controlled.

(*v*) Peak current of devices is limited.

(*vi*) It can handle reactive or regenerative loads without freewheeling diodes.

Disadvantages : (*i*) Somewhat sluggish response of the drive.

(*ii*) Somewhat bulky and expensive inverter. This is due to large size of the inductance and the commutation capacitors.

(*iii*) Low frequency range of the inverter.

(*iv*) It cannot operate under no-load condition. This is because some minimum load current is necessary for the satisfactory commutation of the inverter.

22.13.4. Cycloconverters. A cycloconverter converts ac at one frequency to an ac of another frequency. Cycloconverters can be

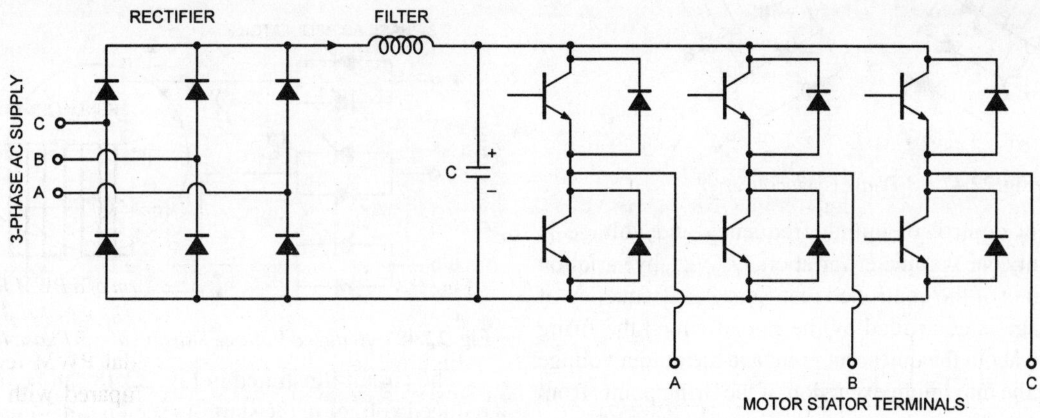

Fig. 22.45 *Transistorized PWM Control*

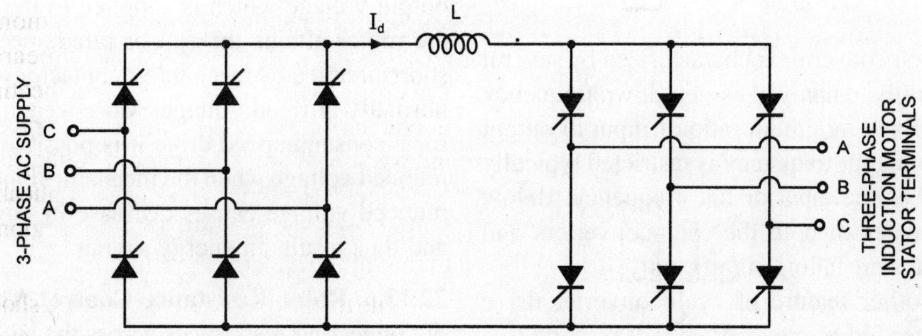

Fig. 22.46 *Variable Current Variable Frequency Control*

classified as single-phase to single-phase, three-phase to single-phase and three-phase to 3-phase devices. They can also be classed as step-up and step-down cycloconverters. A step-up cycloconverter provides an output whose frequency is higher than that of input while a step-down cycloconverter provides an output of frequency lower than that of input. Step-down cycloconverter uses line or natural commutation. Cycloconverters were developed initially for electric traction systems operating at 25 Hz and 16 $\frac{2}{3}$ Hz. In their early stages of use, the cycloconverters used mercury-arc rectifiers. With the development of thyristors the applications of cycloconverters have increased.

The basic power circuit scheme of a 3-phase cycloconverter is shown in Fig. 22.47.

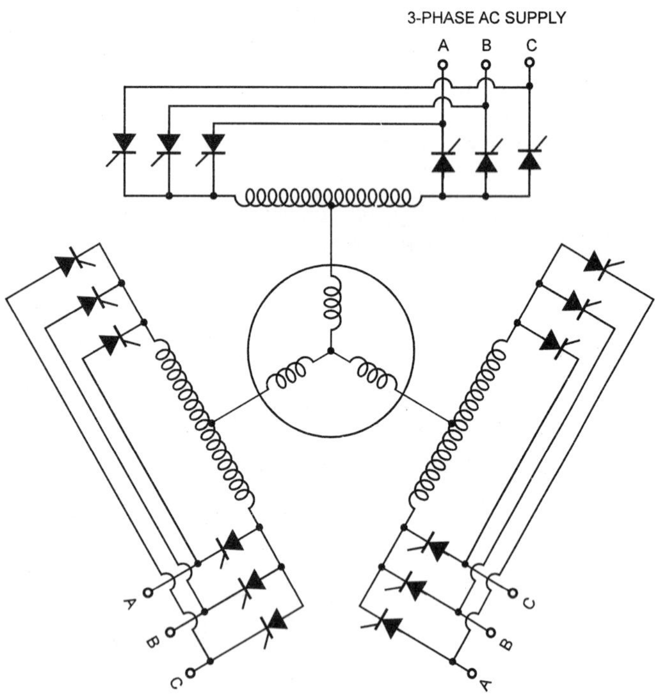

Fig. 22.47 *3-Phase Cycloconverter*

Independent control of output frequency and voltage is obtained with only one parameter variation, *i.e.*, viz., by variation of the firing points of the controlled rectifiers. The frequency of the output voltage is controlled by the rate at which the firing points are varied about the quiescent point and the output voltage is controlled by the maximum excursion of the firing points from the quiescent point. The cycloconverter, with its associated firing circuit, produces an output voltage that is replica of the reference voltage.

The operation of cycloconverter is characterized by several features. They are, generally, employed as step-down frequency converters. There is no fixed minimum ratio of input to output frequency; however, the output frequency is restricted typically to one-third or one-half of the input or line frequency. Below these ratios, the efficiencies of both the cycloconverters and motors supplied by them start falling significantly.

Reversibility is another feature of cycloconverter drive systems. A cycloconverter fed ac motor drive will respond to a change in polarity of the input signals by changing the direction of rotation of the motor without the use of contactors to reverse phase sequence.

The ability of cycloconverter to handle power flow in either direction is another important feature. This, together with the above mentioned reversibility feature, provides an induction motor drive capable of operating in any of the four quadrants of the motor's speed-torque curve.

While the cycloconverter has many attractive features from a theoretical point of view, there are several limitations because of which they have not gained popularity. It needs more power semiconductors than an inverter. For example, the three-phase cycloconverter needs 18 thyristors, whereas the rectifier-inverter combination (Fig. 22.38) needs only 12 thyristors. Cycloconverters can produce only a sub-frequency output. Line pollution with harmonics and low power factor can also be problems with cycloconverters of high power rating. However, recent advancements in fast switching devices have resulted into devices known as *forced commutated direct frequency changers* (FCDFC) which operate at high efficiency and have low harmonic content.

Cycloconverters drives are normally used for large size motors as the cost and complexity of the power and control circuits prohibit their use for general applications. Cycloconverters have been employed in diesel electric locomotives where a high frequency alternator coupled to the engine shaft provides power at the input. These have also been employed in gearless cement mill or ball mill drives.

22.13.5. Reduced Voltage Starting (Soft Start). The starting line current at full voltage of an induction motor may be about 6 times the rated full-load current. Such a high current may cause severe voltage dip in the network supplying the induction motor.

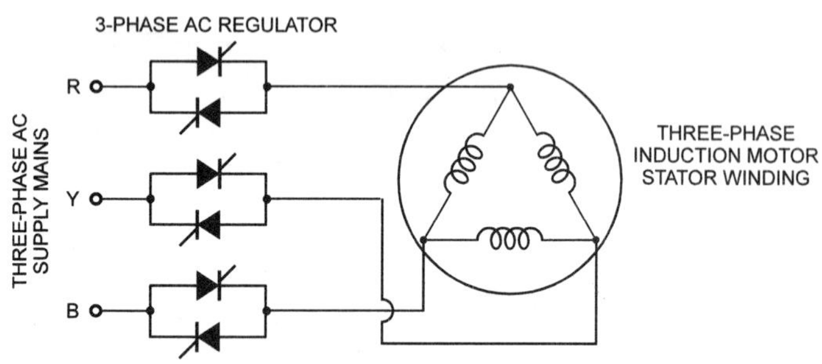

Fig. 22.48 *Reduced Voltage Starting of a 3-Phase Induction Motor*

The circuit illustrated in Fig. 22.48 can be used for feeding a reduced voltage at the start. As seen it is a 3-phase ac regulator. By proper control of firing angle the regulator provides a low output voltage which is supplied to the induction motor. When the motor attains the full or rated speed, the regulator can be short circuited by mechanical contactor so that the motor operates normally at rated voltage. Moreover, if the motor is employed for a constant speed drive it is possible to operate the motor at reduced voltage when the mechanical load is small. Operation at reduced voltage causes decrease in power losses in the motor and thus results in energy saving.

22.13.6. Rotor Resistance Control. The conventional method of connecting resistances across the slip rings of a wound rotor induction motor is a form of rotor voltage control [Art. 19.3.4].

The main drawback of this method of speed control is its poor efficiency because of power wasted in the external resistors.

Figure 22.49 (*a*) shows a 3-phase diode rectifier and gate-turn-off thyristor (GTO)* connected in the rotor circuit of a wound rotor induction motor. The GTO operating as a chopper varies the resistance R as per the duty cycle α. The effective resistance R_e is given by

$$R_e = R\,(1 - \alpha) \qquad ...(22.59)$$

Thus the speed of a wound rotor induction motor is controlled by varying the ratio of on-time to off-time.

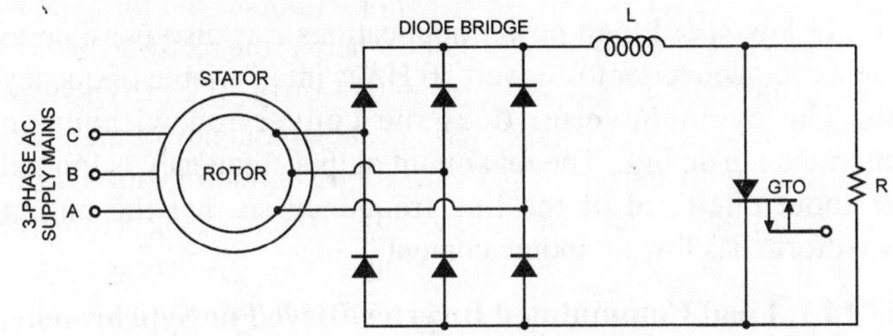

Fig. 22.49 (*a*) *Rotor Voltage Control of a 3-Phase Wound Rotor Induction Motor (Chopper Control)*

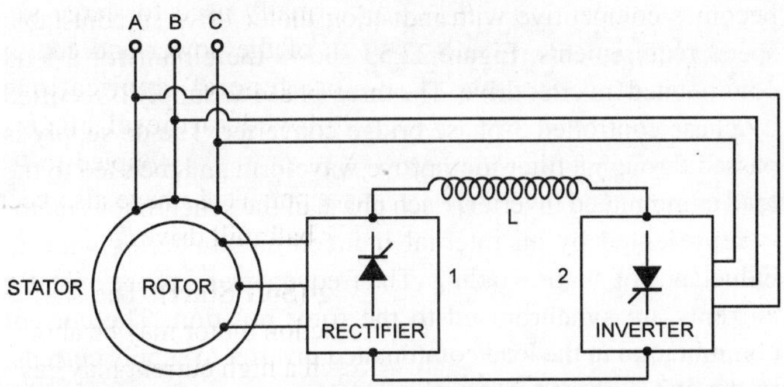

Fig. 22.49 (*b*) *Rotor Voltage Control of a 3-Phase Wound Rotor Induction Motor (Slip-Power Recovery Scheme)*

22.13.7. Slip-Power Recovery Scheme. Figure 22.49 (*b*) shows a scheme known as *slip-power* recovery scheme. The rotor terminals are connected to the 3-phase ac supply mains through two fully controlled thyristor bridges. Bridge 1 acts as a rectifier (or converter) and bridge 2 acts as an inverter. Power output from rotor can be supplied back to the supply source. Since the frequency of rotor currents is slip frequency, this method is known as *slip-power recovery scheme.* By controlling the firing angles of the two bridges, the rotor power output can be varied. Thus the motor slip and speed (for the same torque) will also change. However, the drawback of this scheme is that both the bridges draw reactive power from the supply mains. Therefore, the overall power factor of the motor is poor. If speeds only below synchronous are desired, bridge 1 may be uncontrolled and, thus, may consists of diodes. If both bridges are controlled, the operation of the two bridges can be reversed also to get speeds above synchronous one.

Infact, the slip-power is either returned to the supply network as in Scherbius scheme or used to drive an auxiliary motor which is mechanically coupled to the induction motor shaft, as in Kramer scheme.

1. *Static Scherbius Drive.* The static Scherbius drive also uses the principle of slip-power recovery. The schematic diagram is shown in Fig. 22.50. For achieving both subsychronous and supersynchronous speed control, converters 1 and 2 must be fully-controlled thyristor bridges, one functioning at slip frequency as a rectifier or inverter while the other operating at supply system frequency as an inverter or rectifier. The cost of converters is quite appreciable and a slip-frequency gating circuit is also required. Moreover, at speeds near synchronous, when the slip-frequency emfs are quite small for natural commutation, special connections for forced commutation methods are required. If converter 1 is taken uncontrolled one (diode bridge) the converter cascade and the control unit would become economical and simple, but only subsynchronous speed control would then be available.

The 3-phase transformer between the supply source and inverter 2 is meant to bring the rotor circuit voltage up to the value corresponding to the voltage of the supply source. The main drawback of subsynchronous cascade drive is its poor power factor particularly at reduced speeds.

This drive has applications in large power fan and pump drives which need speed control in a narrow range only. Power ratings of diode bridge, inverter and transformer is just maximum slip times the power rating of the motor resulting in a low cost drive. This drive provides a constant torque control. Constant power control can be obtained from Kramer drive discussed below.

2. *Static Kramer Drive.* Figure 22.51 shows the schematic diagram of the Kramer cascade with static converter. The rotor circuit of a slip-ring induction motor feeds the slip power, rectified by a diode bridge,

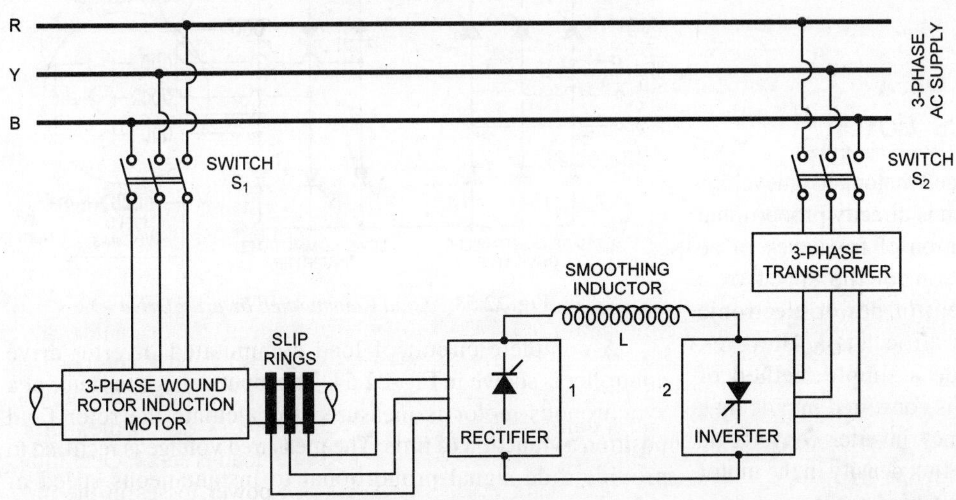

Fig. 22.50 *Static Scherbius Drive*

* A gate turn-off thyristor (GTO) is a special thyristor which can be turned on by a positive gate signal and can be turned off by a negative gate signal. Evidently the use of a GTO in a power electronics circuit eliminates the need of forced commutation circuit because turn-off is achieved by applying a negative gate signal.

to the armature of a separately excited dc motor mechanically coupled to the induction motor. Speed control is achieved by varying the field current of the dc motor. An emf proportional to the back emf of the dc motor may be considered to be injected into the rotor circuit of the induction motor to cause variation in the speed of the system.

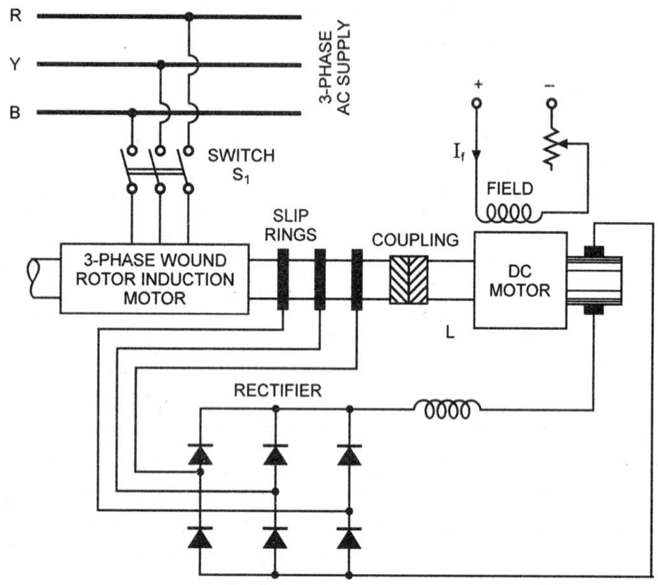

Fig. 22.51. ***Static Kramer Drive***

For achieving larger speed range, diode bridge will need replacement by a thyristor bridge. With thyristor bridge speed can be controlled up to standstill.

Static Kramer drive possesses no line-commutated inverter, it takes less reactive power and introduces smaller harmonic contents of currents than a static Scherbius drive. However, it has maintenance problems which are posed by commutator and brushes of the auxiliary dc motor. It has also drawback of large moment of inertia.

Static Kramer drive systems are used in large power pumps and compressor type loads where speed control is within narrow range and below synchronous speed.

22.14 THYRISTOR CONTROL OF A 3-PHASE SYNCHRONOUS MOTOR

The synchronous motor is a constant speed motor and it develops torque only at synchronous speed which is directly proportional to supply frequency. Therefore, variation of frequency of ac supply is a convenient method to control the speed of a synchronous motor. Before the advent of power electronic devices change of frequency was a very difficult task. However, the power electronic circuits provide a simple method of changing the frequency. The 50 Hz ac is converted into dc by a controlled rectifier. A variable frequency inverter converts dc into variable frequency ac. To keep the flux density in the motor constant voltage and frequency must be varied in the same ratio so that voltage/frequency ratio is constant.

Figure 22.52 shows a system of speed control of a synchronous motor. The 3-phase 50 Hz ac is rectified by a controlled 3-phase rectifier. The filter removes harmonics from the output of the rectifier. The variable frequency inverter feeds a variable frequency ac to the synchronous motor. Thus the speed of a synchronous motor can be controlled. The dc output from filter is also used to excite the rotor winding.

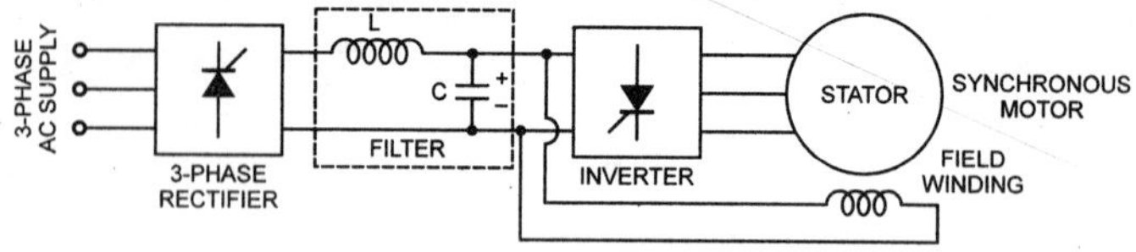

Fig. 22.52 *Speed Control of a Synchronous Motor*

In low-speed high power applications it is also possible to use a cycloconverter to convert 50 Hz ac into variable frequency ac. The cycloconverter does the conversion without an intermediate dc link. The maximum output frequency is limited to about one-third of the line frequency so that the output waveform has low harmonic content.

22.14.1. Load Commutated Inverter Drive For Synchronous Motor. For large sized motors (motors having output rating of more than 750 kW), load commutated synchronous motor drive becomes competitive with induction motor drive in adjustable speed requirements. Figure 22.53 shows the circuit for a load commutated inverter drive. The three-phase ac supply is rectified by phase controlled 3-phase bridge converter. The dc supply is passed through a filter to improve waveform and then fed to the load commutated inverter. Each phase of the synchronous motor is represented by an internal induced emf in series with an inductance of stator winding. The frequency and phase of stator currents are synchronised to the rotor position. The current commutation in the load commutated inverter to supply currents to the stator phases in an appropriate sequence is provided by the stator induced emfs. The presence of 3-phase induced emf in the motor stator winding causes the commutation of the thyristor. A control of firing angle of converter controls its dc output voltage and, therefore, the current.

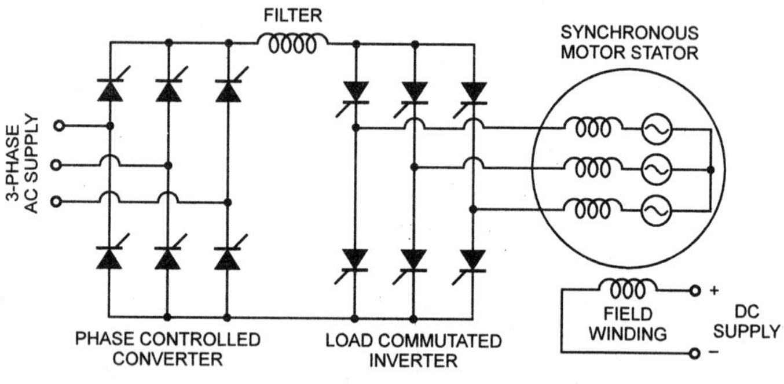

Fig. 22.53 *Load Commutated Inverter Drive*

A complete circuit of load commutated inverter drive controller is shown in Fig. 22.54. The input voltage to stator of a synchronous motor is measured to calculate the rotor field position as function of time. The measured voltage is rectified to provide a dc signal proportional to instantaneous speed of synchronous motor. The turn-off time T_{off} available to thyristor in the inverter is kept constant. Keeping field current and T_{off} constant, the actual speed is compared with the reference speed. The error signal is amplified to provide reference I_d. If actual current I_d is less than reference, the rectifier increases the voltage supplied to inverter, thus increasing I_d and the motor torque. Based on current I_d and measured voltage, the fire pulses to the gates of thyristors of inverter are provided to keep T_{off} constant.

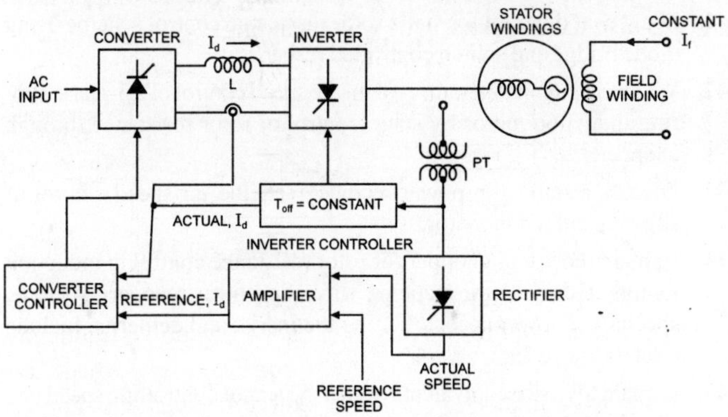

Fig. 22.54 *Load Commutated Inverter Drive For a Synchronous Motor*

22.15 SPEED CONTROL OF SINGLE-PHASE INDUCTION MOTOR USING TRIAC (FAN REGULATOR)

The conventional speed regulator for a fan uses a resistance regulator. The regulator resistance is in series with the fan motor. The speed of the fan is reduced, whenever desired, by increasing the regulator resistance. Thus the voltage drop across the regulator resistance increases and, therefore, voltage applied to the motor is reduced and so the speed. The resistance regulator causes loss of energy. This loss of energy becomes significant at low speeds.

Figure 22.55 shows a speed control circuit using a diac-triac pair. The diac is meant to trigger the triac into conduction. It is preferable to use a matched diac-triac pair. The R-C circuit forms the triggering network. By adjusting the resistor R, the voltage across capacitor C can be adjusted. When voltage across capacitor C exceeds breakdown voltage of diac, it is triggered into conduction and sends a triggering signal to the triac gate. When triac is turned on the motor starts. By varying resistance R, The firing angle of triac can be altered and thus the voltage applied to the motor is changed and consequently the motor speed is changed. Since resistance R, carries very small current, the loss of energy is very small. Sometimes an R-C snubber circuit is added in parallel with the triac to protect it from high voltage transients and high $\frac{dv}{dt}$.

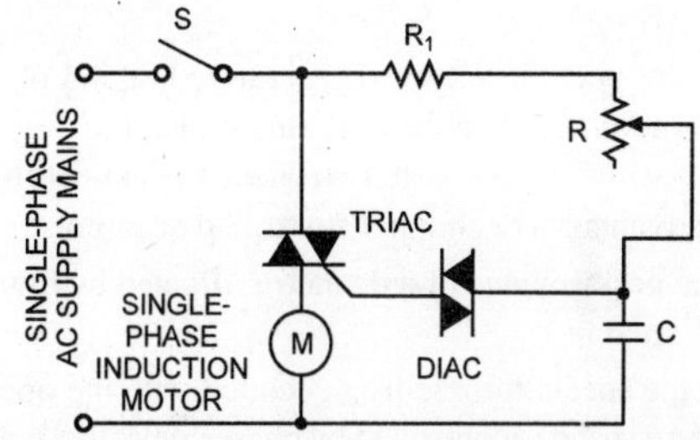

Fig. 22.55 *Speed Control of Single Phase Induction Motor Using Triac*

EXERCISES

1. Why thyristor controlled drives employing both dc and ac motors find wide application in industry ? Explain.
2. Give the classification of static power converters.
3. What are the factors that govern the choice of current rating, voltage rating and switching frequency of the power electronic converter for electric motor drive ?
4. Explain working principle of single-phase half controlled bridge converter fed separately-excited dc motor and draw waveforms of motor terminal voltage, motor current and supply current.

   Compare its merits and demerits with respect to when fed by fully controlled bridge converter. **[U.P. Technical Univ. Electric Drives, 2005-06]**
5. Explain how the speed control of a separately-excited dc motor can be achieved by a fully controlled single-phase thyristor bridge converter. **[Madurai Kamaraj Univ., November-1996]**
6. Draw the complete drive circuit of a single-phase fully controlled rectifier controlling a separately-excited dc motor. The circuit should include the single-phase supply, controlled rectifier (including the thyristors) and the motor. Draw typical waveforms of supply voltage, armature voltage and current at a fixed speed under continuous conduction. Derive an expression for motor speed in terms of supply voltage, firing angle and motor constants. **[U.P. Technical Univ. Electric Drives, 2007-08]**
7. Compare and contrast a half-controlled converter with a fully controlled converter for speed control of a separately-excited dc motor. In which case supply power factor is better and why ?
8. Compare the performance of a separately-excited dc motor when fed from a 1-phase fully controlled bridge converter with that of the same motor fed from a 1-phase half controlled bridge. Draw the relevant waveforms in each case and sketch the variation of average output voltage *vs* firing angle.
9. Describe the operation of a single-phase half controlled bridge converter fed separately-excited dc motor and draw waveforms of motor terminal voltage, armature and supply currents assuming continuous current conduction. List its merits and demerits in comparison to that of with fully controlled bridge rectifier. **[U.P. Technical Univ. Electric Drives, 2008-09]**
10. How do you change from motoring to braking operation in forward direction of a dc separately-excited motor when two fully controlled thyristor converters are used in back-to-back connection? Explain in detail.
11. Draw the circuit arrangements for a dual converter and explain its operation. **[Madras Univ., April-1997]**
12. Describe, with power circuit diagram, the dual converter. What is the relationship between firing angles of the two full converters of the dual converter under continuous current conduction mode ? Explain Why ?
13. Explain two modes in which a dual converter is used to control the dc motor speed. Which of the two methods is better and why ? Explain. **[U.P. Technical Univ. Electric Drives, 2009-10]**
14. A three-phase fully controlled bridge converter is feeding a separately-excited dc motor. Assuming continuous load current, derive the equation relating speed and torque of a motor as a function of triggering delay 'α'. **[U.P. Technical Unv. Electric Drives, 2004-05]**
15. Enumerate the different methods of obtaining variable dc voltage from a fixed dc voltage source. Give their advantages and disadvantages.

16. What is a dc chopper ? Explain the principle of operation of a chopper giving its circuit diagram and output voltage waveforms.
17. Explain the time ratio control (TRC) and current limit control used for choppers.
18. What is the analogy between a chopper and transformer ? Describe the working and obtain an expression for average load voltage of step-up chopper.
19. Discuss in full, the chopper control of a dc separately-excited motor. **[Bharathiyar Univ. Electrical Drives and Control, Oct.-1996]**
20. Draw the block diagram for chopper fed dc motor drive. Explain the working.
21. Draw the power circuit of a single quadrant dc chopper and explain its working.
22. Draw the circuit for the chopper control of a separately-excited dc motor and derive an expression for speed in terms of motor constants, supply voltage, duty ratio and load torque. **[U.P. Technical Univ. Electric Drives, 2007-08]**
23. Regenerative braking is energy efficient braking, explain in view of chopper drives.
24. Discuss a chopper circuit for regenerative braking of a separately excited dc motor. Explain the operating principle. **[U.P. Technical Univ. Electric Drives, 2004-05]**
25. State the advantages of chopper controlled dc motors.
26. How is the performance of a dc motor affected by thyristor power supply ? Explain.
27. What are the special problems associated with the operation of dc drives where speed is controlled by phase controlled converters and choppers ?
28. Explain the speed control of a dc series motor using thyristors.
29. Explain in brief, on the importance of supply harmonics, power factor and ripples in the motor current, in the performance of a converter fed dc motor drive. **[G.B. Technical Univ. Electric Drives, 2010-11]**
30. Explain the chopper control techniques for each of a separately-excited dc motor and dc series motor. **[G.B. Technical Univ. Electric Drives, 2010-11]**
31. How ac motor speed control is achieved by using voltage source inverter ?
32. Explain, giving block diagram, the operation of a dc link voltage source inverter for speed control of 3-phase induction motors.
33. What is 'pulse width modulation (PWM)' ? Explain the purpose.
34. State the advantages of ac voltage control by PWM technique. Explain the sinusoidal PWM technique together with relevant waveforms.
35. What is meant by VVVF control of induction motors ? Explain with a block diagram and with reference to both constant torque and constant power modes of operation.
36. Draw the power circuit of a 3-phase current source inverter and explain its working. State also its inherent advantages over voltage source inverter for speed control of ac motor.
37. Distinguish between voltage source and current source inverters. Compare the two inverters for speed control of 3-phase induction motor.
38. Discuss the advantages and disadvantages of cycloconverter for speed control of 3-phase induction motors.
39. What is a cycloconverter ? Enumerate some of its industrial applications.
40. What is meant by "soft start" ? State and explain the soft start schemes (including electronic means, if any) employed for induction motors.
41. Draw neat diagram for static rotor resistance control scheme using diode bridge and switch controlled resistor.
42. With neat circuit diagram, explain the speed control of a 3-phase slip-ring induction motor by static control of rotor resistance through chopper.
43. Discuss a static slip-power recovery scheme for speed control of slip-ring induction motor.
44. Explain the use of chopper for rotor resistance control in induction motors. Compare this scheme with slip-power recovery scheme, specially at lower speeds, stating the merits and demerits. Include circuits as needed.
45. Explain thyristor equivalent Kramer system of controlling speed of a slip-ring induction motor and regenerative operation.
46. Explain any one slip power recovery scheme for the control of three-phase induction motor drive. Can supersynchronous speeds be achieved through it ? Explain. **[G.B. Technical Univ. Electric Drives, 2010-11]**
47. Explain the slip-power recovery scheme of a three-phase induction motor control. What are its advantages and disadvantages ? **[U.P. Technical Univ. Electric Drives, 2009-10]**
48. Explain working of static slip-power recovery control scheme returning power to supply mains for speed control of three-phase slip-ring induction motor. Why is supply power factor poor and how can it be improved ? **[U.P. Technical Univ. Electric Drives, 2008-09]**
49. Discuss static Scherbius scheme for speed control of a slip-ring induction motor. Draw a neat circuit diagram of the complete scheme. Mention one main advantage of this scheme compared to rotor resistance control. **[U.P. Technical Univ. Electric Drives, 2004-05]**
50. Discuss with the help of a suitable schematic diagram, the operation of a load commutated-fed synchronous motor drive. Why is this drive suitable for high speed and high power applications?
51. Explain the self controlled scheme for the synchronous motor control. **[G.B. Technical Univ. Electric Drives, 2010-11]**
52. Describe self-controlled and load commutated inverter controlled synchronous motor drives in detail and compare them. **[U.P. Technical Univ. Electric Drives, 2009-10]**
53. Explain static speed control of a single-phase induction motor. **[U.P. Technical Univ. Electric Drives, 2005-06]**
54. Write short notes on the following :
    (*i*) Half controlled converters. **[Madurai Kamaraj Univ., Nov.-1996]**
    (*ii*) Full controlled converters.
    (*iii*) Dual converter drives.
    (*iv*) Chopper drives.
    (*v*) Comparison of converter fed and chopper fed dc drives.
    (*vi*) Thyristor control of dc series motors.
    (*vii*) Control circuits for ac drives. **[Bharathidasan Univ., Nov.-1996]**
    (*viii*) Cycloconverters.
    (*ix*) Slip-power recovery scheme for speed control of slip-ring induction motor.
    (*x*) Load commutated inverter drive for synchronous motor.

SHORT ANSWER TYPE QUESTIONS WITH ANSWERS

Q. 1. Why the thyristor control is preferred over Ward Leonard system of speed control ?

Ans. The thyristor control has some distinct advantages over Ward Leonard system of speed control such as low installation cost, ease of maintenance, higher accuracy, greater reliability, quick response, higher efficiency of operation, lesser floor space requirements, operation at a wide range of temperature etc.

Q. 2. What are freewheeling diodes ? **[Bharathiyar Univ., April-1997]**

Ans. Freewheeling diode is a bypass or commutating diode. It is connected across motor terminals to allow for dissipation of energy stored in motor inductance and to provide for continuity of motor current when the thyristors are blocked. It also provides protection against transient overvoltages.

Q. 3. Why half-wave converter is not used for supply to the field circuit of a dc motor ?

Ans. If half-wave converter is used for supply to the field circuit of a dc motor, the high ripple content in the field circuit would cause increase in iron losses of the machine.

Q. 4. What is the relationship of V_{dc} and V_{ac} in terms of firing angle α for half-controlled and full-controlled converters? **[Madurai Kamaraj Univ., Nov.-1996]**

Ans. For half-controlled converter,

$$V_{dc} = \frac{\sqrt{2}\,V_{ac}}{\pi}\,(1 + \cos\alpha) \text{ for } 0 \le \alpha \le \pi.$$

For full-controlled converter,

$$V_{dc} = \frac{2\sqrt{2}\,V_{ac}}{\pi}\cos\alpha \text{ for } 0 \le \alpha \le \pi.$$

Q. 5. What are the drawbacks of half-wave rectifier circuits ? **[Bharathiyar Univ., April-1997]**

Ans. The output voltage wave of a half-wave rectifier is not a perfect dc. It is a discontinuous current and contains various harmonics. So half-wave converter drives have poor performance and their use is limited to small drives (rating up to 500 W).

Q. 6. What is the necessity of dc choke coil and freewheeling diode in a converter circuit ? **[Madurai Kamaraj Univ., Nov.-1996]**

Ans. Freewheeling diode allows for dissipation of energy stored in the motor inductance and dc choke coil which reduces the rate of decay of current during the freewheeling operation when thyristors are blocked and thus motor current is made continuous. Discontinuous armature current results in deterioration of motor performance.

Q. 7. Why three-phase half-wave drives are generally not employed in industrial applications ?

Ans. Because of presence of dc component in supply currents.

Q. 8. What is a chopper ?

Ans. A chopper is a static switch to provide variable dc voltage from a source of constant dc voltage source.

Q. 9. Give the applications of dc choppers.

Ans. DC choppers are employed in trolley cars, marine hoists, forklift trucks, mine haulers and electric traction.

Q. 10. What is meant by duty ratio of a dc chopper ? **[Madras Univ., April-1997]**

Ans. The ratio of on-period to the time period is known as duty ratio, *i.e.,*

$$\text{Duty ratio, } \alpha = \frac{T_{on}}{T} \text{ or } T_{on} \times \text{frequency or } \frac{T_{on}}{T_{on} + T_{off}}.$$

Q. 11. What are the advantages of chopper fed drive over converter fed drive ?

Ans. The advantages of chopper fed drive over converter fed drive are:

1. The ripple in the output of chopper is much less than that in the output of a converter. Lesser ripple means lesser losses.
2. The chances of discontinuous operation in chopper drive are lesser than that in converter drive. So speed regulation and transient response of chopper drive are better than that in converter drive.

Q. 12. Enumerate the control strategies used in dc choppers.

Ans. There are two types of control strategies for operating the switches that can be used in dc choppers. These control strategies are time ratio control (TRC) and current limit control.

Q. 13. What is TRC ? In how many ways can it be affected ?

Ans. TRC (time-ratio control) is a control strategy in which the value of $\frac{T_{on}}{T}$ is varied. This is affected in two ways namely variable frequency control (FM control) and constant frequency control (PWM).

Q. 14. What quantity is varied in PWM and frequency modulation ?

Ans. In pulse width modulation (PWM), the chopping period T is kept constant but the pulse width T_{on} is varied to control the output. In frequency modulation the chopping frequency is varied and for this purpose either T_{on} or T_{off} is varied.

Q. 15. What is a step-up chopper ?

Ans. A step-up chopper, like a step-up transformer, raises the level of input dc voltage.

Q. 16. On what principle does a step-up chopper operate ?

Ans. A step-up chopper operates on the principle of storing energy in an inductance to provide output voltage higher than source voltage.

Q. 17. What is steady-state ripple ?

Ans. When chopper feeds R-L or R-L-E_b load, the current varies between I_{max} and I_{min}. The difference of I_{max} and I_{min} is known as steady-state ripple.

Q. 18. When is steady-state ripple maximum.

Ans. When α = 0.5, steady-state ripple is maximum.

Q. 19. What factors can make the current discontinuous in a chopper feeding an R-L-E_b load ?

Ans. Higher L/R ratio, low duty cycle and higher back emf can make the current discontinuous.

Q. 20. The commutators of the thyristor controlled dc motors are made larger than those for ordinary dc motors. Why ?

Ans. The commutators of the thyristor controlled dc motors are made larger in order to provide extra insulation to withstand larger and rapid voltage fluctuations.

Q. 21. Semiconverter system is preferred for dc series motor drive. Why ?

Ans. The semiconverter system, because of its freewheeling action helps in maintaining continuous current and thus provides better performance to dc series motors in comparison to full converter system.

Q. 22. Why are ac regulators preferred over conventional methods of providing ac variable voltage of constant frequency for ac motor drives ?

Ans. AC regulators using thyristors are becoming popular because of high efficiency, fast control and compact size.

Q. 23. What is difference between on-off control and phase control in the case of ac voltage controller ? **[Madras Univ., October-1996]**

Ans. In on-off control, also known as integral cycle control, the thyristors are used to connect the motor to the supply source for a certain number of cycles and then to disconnect it for another certain number of cycles; each of the on and off times consist of an integral number of cycles. On the other hand in phase control, the thyristors are employed as switches for connecting the motor to the supply for a certain portion of each cycle of the supply voltage.

Q. 24. What is the significance of maintaining the ratio of supply voltage to frequency constant in variable frequency drives of induction motors. **[U.P.T.U. Electric Drives, 2007-08]**

Ans. If only frequency is changed and stator voltage is kept constant, the stator flux will not be at its rated value. The operation with flux below or above the rated value is not desirable.

The noteworthy point is that electrical machine is designed corresponding to near saturation point on the magnetization curve (or B-H curve). This is done from the point of view of full utilisation of the core. If the stator frequency is reduced, keeping the stator voltage constant, motor will operate in the saturation region and therefore, the motor will draw large magnetising current causing increase in stator core and copper losses and therefore, decrease in motor efficiency. However, if only supply frequency is increased keeping the stator voltage constant, the motor will operate at low flux density and thus motor capacity will be under utilised.

For constant flux operation, it is necessary that the induced emf varies linearly with applied frequency. At higher voltages and at high frequency operation stator voltage drops are very small and constant flux operation may be obtained by keeping V/f ratio constant.

Q. 25. State two advantages of current source inverter.

[Madurai Kamaraj Univ., Nov.-1996]

Ans. Current source inverter has the following two main advantages:

1. It can handle reactive or regenerative loads without freewheeling diodes.
2. Misfiring of devices and short circuits, because of constant input current, do not pose any problem.

Q. 26. Mention any two drawbacks of cycloconverters.

[Bharathiyar Univ., April-1997]

Ans. The two drawbacks of cycloconverters are :

1. It needs more power semiconductors than an inverter.
2. They can produce only a sub-frequency output.

Q. 27. Give some applications of cycloconverter.

[Madurai Kamaraj Univ., Nov.-1996]

Ans. Cycloconverter drives are normally used for large sized motors. They have also been employed in gearless cement mill or ball mill drives.

Q. 28. What is meant by GTO ? [Madras Univ., April-1997]

Ans. Gate-turn-off switch is a special thyristor which can be turned on by a positive gate signal and can be turned-off by a negative gate signal.

PROBLEMS

1. A single-phase 230 V, 50 Hz supply feeds a separately-excited dc motor through single-phase semiconverters. The firing angle of the field semiconverter is zero. The armature and field resistances are 0.1 Ω and 150 Ω respectively. The load torque is 150 N-m at 750 rpm. The voltage constant is 0.9 V/A radian/s and torque constant is 0.9 N-m/A^2. Determine (*i*) field current (*ii*) firing angle of converter in armature circuit and (*iii*) power factor of armature circuit converter. Assume continuous current operation and neglect losses.
[**Ans.** (*i*) 1.38 A (*ii*) 96.6° (*iii*) 0.585 lagging]
2. A 230 V single-phase ac supply is used to supply a separately-excited dc motor through two full converters. The firing angle of converter in the field circuit is kept at zero. The field and armature resistances are 150 Ω and 0.15 Ω respectively. The firing angle of converter in the armature circuit is 30° and armature current is 80 A. The torque and voltage constants are 1.2 N-m/A^2 and 1.2 V/A-radian/second respectively. Determine torque developed and motor speed.
[**Ans.** 132.48 N-m; 964.7 rpm]
3. A 220 V, 1500 rpm, 10 A, separately-excited dc motor with an armature resistance of 3 Ω is fed from a single-phase fully controlled rectifier connected to an ac source voltage of 230 V, 50 Hz. Assuming continuous conduction, calculate the firing angle for half the motor torque at 600 rpm. [U.P. Technical Univ. Electric Drives 2007-08]
[**Ans.** 63.93°]
4. A 400 V 3-phase supply feeds a separately-excited dc motor through two three-phase full converters, one for armature and the other for field. The firing angle of the field converter is zero. The armature and field resistances are 0.2 Ω and 200 Ω respectively. The armature current is equal to rated value of 60 A. The motor voltage constant is 1.05 V/A-rad/second. The speed is 1,000 rpm. Find firing angle of armature circuit converter. [**Ans.** 56.65°]
5. A 3-phase fully controlled bridge converter is used to control the speed of a 40 kW, 400 V, 1,000 rpm separately-excited dc motor. The converter is fed from a 3-phase, 400 V, 50 Hz ac power supply. The induced emf constant is 0.3 V/rpm and the armature resistance is 0.5 Ω. Determine (*i*) rated torque (*ii*) firing angle when the motor operates at full load and rated speed. (*iii*) firing angle when the motor should operate at half load and at 500 rpm.
[U.P. Technical Univ. Electric Drives, 2008-09]
[**Ans.** (*i*) 382 N-m (*ii*) 42.23° (*iii*) 68.27°]
6. A 230 V ac supply feeds the armature and field circuits of a separately-excited dc motor through two full converters. The firing angle of the field converter is zero. The armature and field resistances are 0.5 Ω and 300 Ω respectively. The torque and voltage constants are 1.35 each. Determine motor speed and torque for armature current of 60 A and firing angle of armature circuit converter of 45°.
[**Ans.** 1,173 rpm; 55.89 N-m]
7. A dc chopper is supplying a resistive load. If the input and output voltages are 200 V and 128 V and chopping frequency is 800 Hz, find the periods of conduction and blocking in each cycle.
[**Ans.** 0.8×10^{-3} s and 0.45×10^{-3} s]
8. A chopper circuit is operating on TRC principle at a frequency of 4 kHz on 200 V dc supply. If the load voltage is 160 V, compute the conducting and blocking period of thyristor in each cycle.
[**Ans.** 0.2 ms and 0.05 ms]
9. A step-up chopper has an ac input voltage of 100 V and an output voltage of 250 V. The period of blocking in each cycle is 0.6 ms. Find period of conduction and chopping frequency.
[**Ans.** 0.36 ms; 1.6667 kHz]
10. A step-up chopper has input voltage of 200 V and output voltage of 600 V. If the conducting period of chopper is 50 ms, compute the pulse width of output voltage. In case output voltage pulse width is halved for constant frequency operation, determine the average value of new output voltage. [**Ans.** 25 ms; 1,200 V]
11. A separately-excited dc motor is supplied from a 500 V source through a chopper. Armature current is 200 A and armature resistance is 0.1 Ω. The voltage and torque constants are 1.4 V/A rad/s and 1.4 Nm/A^2. The field current is 2.5 A. The duty cycle of the chopper is 0.5. Determine (*i*) input power (*ii*) speed and (*iii*) torque.
[**Ans.** (*i*) 50 kW (*ii*) 627.5 rpm (*iii*) 700 N-m]
12. The chopper used for control of a dc separately-excited motor has supply voltage of 220 V, on-period is 10 ms and off-period is 15 ms. Determine the average load current when the speed of the motor is 1,200 rpm and has a voltage constant of 0.5 V/rad per second. The armature resistance is 2.5 Ω. [**Ans.** 10.07 A]
13. A one quadrant chopper is used for rheostatic braking of a separately-excited dc motor. Armature and braking resistances are 0.2 Ω and 6 Ω respectively. Armature and field currents are 160 A and 1.9 A respectively. Duty cycle of chopper is 0.4 and voltage constant is 1.1 V/A-rad/second. Determine (*i*) average voltage across chopper (*ii*) speed of motor and (*iii*) power dissipated in braking resistance.
[**Ans.** (*i*) 576 V (*ii*) 2,778 rpm (*iii*) 92.16 kW]
14. A dc series motor is coupled to a load whose torque varies as the square of speed. The motor is supplied through a semiconverter with input voltage of 700 V. For firing angle of 90° the armature current is 25 A and motor speed is 400 rpm. Determine the firing angle and armature current if the motor speed is to be raised to 500 rpm. Assume unsaturated magnetic circuit and the combined resistance of armature and field to be 0.4 Ω. [**Ans.** 56.46°; 31.25 A]

❑❑❑

Objective Type Questions

MULTIPLE CHOICE TYPE QUESTIONS

Note: Each of the following questions comprises multiple responses (two or more than two answers). Select the correct answers. In case you feel that there is more than one correct answers, select the answer which you consider the best. In any case, choose ONLY ONE answer or response for each question.

1. A straight conductor of length l moving with a velocity v in the presence of a magnetic field of flux density B directed at an angle θ with the direction of v experiences a force. Which of the following statement(s) is/are true for the magnitude of the force ?

1. It is independent of θ. 2. It is proportional to l^2
3. It is proportional to B. 4. It is independent of v.

Select the correct answer from the codes given below.

Codes:

(a) 1, 2 and 3. (b) 4 alone.
(c) 3 alone. (d) 2 and 4.

[U.P.S.C. I.E.S. E.E.-I, 1994]

2. Consider the following:
Lorentz force $\vec{F} = e(\vec{v} \times \vec{B})$, where e, $\vec{v}$ and $\vec{B}$ are respectively the charge of the particle, velocity of the particle and flux density of uniform magnetic field.
Which one of the following statements is not correct?

(a) Acceleration is normal to the plane containing the particle path and $\vec{B}$.
(b) If the direction of the particle path is normal to B , the acceleration is maximum.
(c) If the particle is at rest, the field will deflect the particle.
(d) If the particle path is in the same direction of B , there will be no acceleration. **[U.P.S.C. I.E.S. E.E. 2007]**

3. An air gap is usually inserted in magnetic circuits so as to

(a) prevent saturation. (b) increase in mmf.
(c) increase in flux. (d) increase in inductance.

4. A straight conductor of circular x-section carries a current. Which one of the following statements is true in this regard ?

(a) No force acts on the conductor at any point.
(b) An axial force acts on the conductor tending to increase its length.
(c) A radial force acts towards the axis tending to reduce its cross-section.
(d) A radial force acts away from the axis tending to increase its cross-section. **[U.P.S.C. I.E.S. E.E.-I, 1994]**

5. Two parallel wires separated by a distance d are carrying a current I in the same direction. The magnetic field along a line running parallel to these wires and midway between them

(a) depends upon I. (b) is zero.
(c) depends upon d.
(d) depends upon the permeability of medium between the wires. **[GATE E.E. 1999]**

6. The laws of electromagnetic induction (Faraday's and Lenz's laws) are summarized in the following equation :

(a) $e = \text{LR}$ (b) $e = \text{L}\dfrac{di}{dt}$
(c) $e = -\dfrac{d\psi}{dt}$ (d) none of the above

[GATE E.E. 1998]

7. A coil of 300 turns is wound on a non-magnetic core having a mean circumference of 300 mm and a cross-sectional area of 300 mm^2. The inductance of the coils corresponding to a magnetizing current of 3 A will be
(Given that $\mu_0 = 4\pi \times 10^{-7}$ H/m)

(a) 37.68 μH (b) 113.04 μH
(c) 3.768 μH (d) 1.1304 μH **[GATE E.E. 2008]**

8. The equivalent inductance of two coils A and B connected as in the given figure is given by

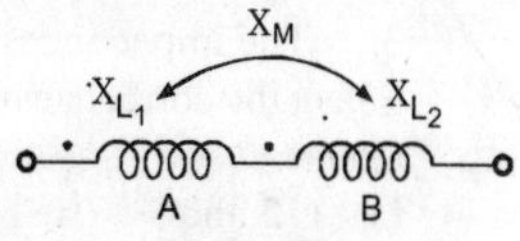

(a) $X_{L1} + X_{L2} - 2X_M$
(b) $X_{L1} + X_{L2} + X_M$
(c) $X_{L1} + X_{L2} - X_M$
(d) $X_{L1} + X_{L2} + 2X_M$ **[U.P.S.C. I.E.S. E.E.-I, 2000]**

9. Two magnetically uncoupled inductive coils have Q factors q_1 and q_2 at chosen frequency. Their respective resistances are R_1 and R_2. When connected in series their effective Q factor at the same frequency is

(a) $q_1R_1 + q_2R_2$ (b) $\dfrac{q_1}{R_1} + \dfrac{q_2}{R_2}$
(c) $\dfrac{q_1R_1 + q_2R_2}{R_1 + R_2}$ (d) $q_1R_1 + q_2R_1$ **[GATE 2013]**

10. Two conducting thin coils X and Y (identical except for a thin cut in coil Y) are placed in a uniform magnetic field which is decreasing at a constant rate. If the plane of the coils is perpendicular to the field lines, which of the following statement is correct? As a result, emf is inducted in

(a) both the coils. (b) coil Y only.
(c) coil X only. (d) none of the two coils.

[U.P.S.C. I.E.S. E.T.E.-I, 2007]

Answers: **1.** (c), **2.** (c), **3.** (a), **4.** (c), **5.** (b), **6.** (c), **7.** (b), **8.** (d), **9.** (c), **10.** (a),

11. Magnetostriction is a phenomenon of
   (a) generation of electricity in ferromagnetic materials.
   (b) generation of magnetism in conductors.
   (c) change in permeability of ferromagnetic materials during magnetisation.
   (d) change in physical dimensions of ferromagnetic materials during magnetisation. [U.P.S.C. I.E.S. E.E.-I, 2005]

12. A coil of 1,000 turns is wound on a core. A current of 1 A flowing through the coil creates a core flux 1 m Wb. The energy stored in the magnetic field is
   (a) 0.25 J. (b) 0.5 J. (c) 1 J. (d) 2 J.
   [U.P.S.C. I.E.S. E.E.-II, 2003; 2005]

13. The voltage applied to a circuit is $100\sqrt{2}\cos(100\pi t)$ volts and the circuit draws a current of $10\sqrt{2}\sin(100\pi t + \pi/4)$ amperes. Taking the voltage as the reference phasor, the phasor representation of the current in amperes is
   (a) $10\sqrt{2}\angle-\pi/4$ (b) $10\angle-\pi/4$
   (c) $10\angle+\pi/4$ (d) $10\sqrt{2}\angle+\pi/4$
   [GATE E.E. 2011]

14. For an ac circuit, if $v(t) = 160\sin(\omega t + 10°)$ and $i(t) = 5\sin(\omega t - 20°)$, then reactive power absorbed by the circuit is
   (a) 100 VARs (b) 200 VARs
   (c) 300 VARs (d) 400 VARs
   [U.P.S.C. I.E.S. E.E.-I, 2010]

15. A single-phase load is supplied by a single-phase voltage source. If the current flowing from the load to source is $10\angle-150°$ and the voltage at the load terminals is $10\angle 60°$, then the
   (a) load absorbs real power and delivers reactive power.
   (b) load absorbs real power and absorbs reactive power.
   (c) load delivers real power and delivers reactive power.
   (d) load delivers real power and absorbs reactive power.
   [GATE 2013]

16. Which of the following are the necessary conditions for an entire 3-ϕ system to be balanced ?
   1. The line voltages are equal in magnitude.
   2. The phase differences between successive line voltages are equal.
   3. The impedances in each of the phases are identical.
   Select the correct answer using the codes given below:
   Codes:
   (a) 1, 2 and 3 (b) 1 and 3 (c) 1 and 2 (d) 2 and 3
   [U.P.S.C. I.E.S. E.E.-I, 1996]

17. The rated voltage of a 3-phase power system is given as
   (a) rms phase voltage. (b) peak phase voltage.
   (c) rms line-to-line voltage. (d) peak line-to-line voltage.
   [GATE E.E. 2004]

18. E_{oa}, E_{ob} and E_{oc} are three phase voltages while E_{ab}, E_{bc} and E_{ca} are three line voltages of a balanced three-phase system having *a-b-c* phase sequence. In relation to E_{oc}, E_{bc} would
   (a) lag by 30°. (b) lead by 30°.
   (c) have the same phase.
   (d) have no definite phase relationship.
   [U.P.S.C. I.E.S. E.E.-I, 1997]

19. For a 3-phase load balanced condition, each phase has the same value of ________.
   (a) impedance (b) resistance
   (c) power factor (d) all of these
   [G.B.T.U. E.E. Even Semester 2010-11]

20. For the three-phase circuit shown in the figure the ratio of the currents $I_R : I_Y : I_B$ is given by

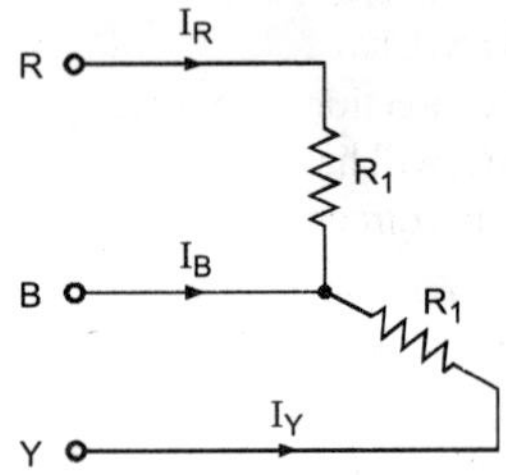

   (a) $1:1:\sqrt{3}$ (b) 1 : 1 : 2 (c) 1 : 1 : 0 (d) $1:1:\sqrt{3/2}$
   [GATE E.E. 2005]

21. A 400 V, 50 Hz, three-phase balanced source supplies power to a star-connected load whose rating is $12\sqrt{3}$ kVA, 0·8 pf (lag). The rating (in kVAR) of the delta-connected (capacitive) reactive power bank necessary to bring the pf to unity is
   (a) 28·78 (b) 21·60 (c) 16·60 (d) 12·47
   [GATE E.E. 2006]

22. In an unbalanced three phase system, phase current $I_a = 1\angle(-90°)$ pu, negative sequence current $I_{b2} = 4\angle(-150°)$ pu, zero sequence current $I_{c0} = 3\angle 90°$ pu. The magnitude of phase current I_b in pu is
   (a) 1.00 (b) 7.81 (c) 11.53 (d) 13.00
   [GATE 2014]

23. The developed electromagnetic force and/or torque in electro-mechanical energy conversion system act in a direction that tends
   (a) to increase the stored energy at constant flux.
   (b) to decrease the stored energy at constant flux.
   (c) to decrease the coenergy at constant mmf.
   (d) to decrease the stored energy at constant mmf. [GATE E.E. 1992]

24. For a linear electromagnetic circuit, the following statement is true
   (a) Field energy is equal to the coenergy.
   (b) Field energy is greater than the coenergy.
   (c) Field energy is lesser than the coenergy.
   (d) Coenergy is zero. [GATE E.E. 2004]

25. In an electromechanical energy conversion device, the developed torque depends upon
   (a) stator field strength and torque angle.
   (b) stator field and rotor field strengths.
   (c) stator field and rotor field strengths and the torque angle.
   (d) stator field strength only. [U.P.S.C. I.E.S. E.E.-II 2001]

26. An electromechanical energy conversion device is shown in the figure. The instantaneous values of armature induced emf, busbar voltage and armature current are respectively *e*, *v* and *i*. T_e, T_a and ω_r represent external torque, armature developed torque and angular velocity of shaft respectively. Assume that the given directions of parameters in the figure are positive. If the device is working in generating mode then

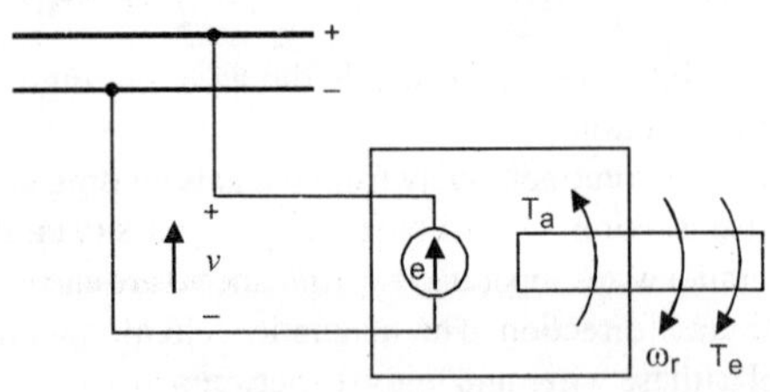

Answers: **11.** (d), **12.** (b), **13.** (b), **14.** (b), **15.** (b), **16.** (a), **17.** (c), **18.** (b), **19.** (d), **20.** (a), **21.** (d), **22.** (c), **23.** (b), **24.** (a), **25.** (c), **26.** (a),

(a) all parameters will have the positive direction.
(b) e, v, T_e and ω_r will have negative direction and others will remain in positive direction.
(c) T_a and i will have negative direction and others will remain in positive direction.
(d) T_e and ω_r' will have negative direction and others will remain in positive direction. **[U.P.S.C. I.E.S. E.E.-II 1999]**

27. In electromechanical energy conversion devices (e.g. generators and motors), a small air gap is left between the stator and the rotor in order to
(a) reduce the reluctance of the magnetic path.
(b) increase flux density in the air gap.
(c) permit mechanical clearance.
(d) avoid saturation of field.

28. When the stator and rotor windings of a 2-pole rotating electrical machine are excited, each would produce a sinusoidal mmf distribution in the air gap with peak values F_s and F_r respectively. The rotor mmf lags stator mmf by a space angle δ at any instant as shown in the figure. Thus, half of stator and rotor surfaces will form one pole with the other half forming the second pole. Further, the direction of torque acting on the rotor can be clockwise or counter-clockwise.

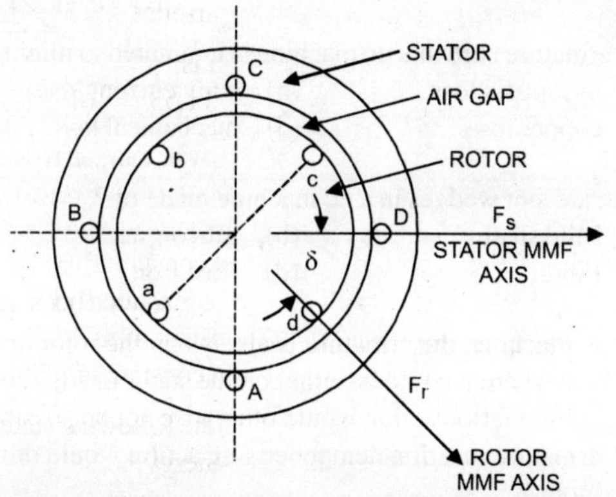

The following table gives four sets of statements as regards poles and torque. Select the correct set corresponding to the mmf axis as shown in the figure.

	Stator Surface ABC forms	*Stator Surface CDA forms*	*Rotor Surface abc forms*	*Rotor Surface cda forms*	*Torque is*
(a)	North Pole	South Pole	North Pole	South Pole	Clockwise
(b)	South Pole	North Pole	North Pole	South Pole	Counter-clockwise
(c)	North Pole	South Pole	South Pole	North Pole	Counter-clockwise
(d)	South Pole	North Pole	South Pole	North Pole	Clockwise

[GATE E.E. 2003]

29. Following are some of the properties of rotating electrical machines.
P Stator winding current is dc, rotor winding current is ac.
Q Stator winding current is ac, rotor winding current is dc.
R Stator winding current is ac, rotor winding current is ac.
S Stator has salient poles and rotor has commutator.
T Rotor has salient poles and slip-rings and stator is cylindrical.
U Both stator and rotor have polyphase windings.

DC machines, synchronous machines and induction machines exhibit some of the above properties as given in the following table. Indicate the correct combination from this table:

	DC Machines	*Synchronous Machines*	*Induction Machines*
(a)	PS	QT	RU
(b)	QU	PT	RS
(c)	PS	RU	QT
(d)	RS	QU	PT

[GATE E.E. 2003]

30. If a current carrying coil is placed in a uniform magnetic field direction of magnetic induction, then
(a) the net force and torque on the coil are both zero.
(b) the net force is zero but torque is finite.
(c) the net force is finite but torque is zero.
(d) the net force and torque are both finite.

31. The general expression for the instantaneous value of the torque of dynamo-electric machines is given by

$$T = \frac{1}{2} i_1^2 \frac{dL_1}{d\theta} + \frac{1}{2} i_2^2 \frac{dL_2}{d\theta} + i_1 i_2 \frac{dM}{d\theta}$$

where i_1 and i_2 are rotor and stator currents respectively, L_1 and L_2 are rotor and stator windings self inductances.
M is the mutual inductance between stator and rotor windings and θ is the angular displacement of rotor.
For a synchronous machine (cylindrical rotor), the torque is given by only the third term of the equation because
(a) the average value of i_1^2 and i_2^2 each is zero.
(b) the first and second terms cancel each other.
(c) $\frac{dL_1}{d\theta}$ and $\frac{dL_2}{d\theta}$ have opposite signs and $\frac{dL_1/d\theta}{dL_2/d\theta} = \frac{i_2^2}{i_1^2}$.
(d) L_1 and L_2 do not vary with respect to θ. **[U.P.S.C. I.E.S. E.E.-II 1997]**

32. Three singly-excited electromagnetic structures are shown in the figures labelled I, II and III (R-Rotor, S-stator). The reluctance torque can be developed in

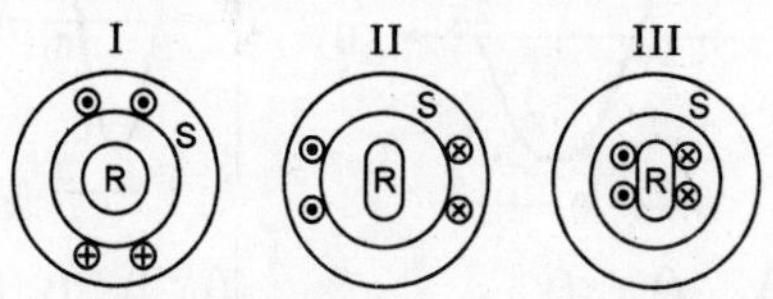

(a) I and II. (b) I alone. (c) II and III. (d) II alone. **[U.P.S.C. I.E.S. E.E.-II 1999]**

33. A rotating electrical machine having its self inductances of both the stator and the rotor windings, independent of rotor position will definitely not develop
(a) starting torque. (b) synchronising torque.
(c) hysteresis torque. (d) reluctance torque. **[GATE F.E. 2004]**

34. EMF induced in a coil rotating in a uniform magnetic field will be maximum when the
(a) flux linking with the coil is maximum.
(b) rate of change of flux linkage is minimum.
(c) rate of change of flux linkage is maximum.
(d) rate of cutting flux by the coil sides is minimum.

35. For a P-pole machine, the relation between electrical and mechanical angles is given by

Answers: 27. (c), **28.** (c), **29.** (a), **30.** (a), **31.** (d), **32.** (d), **33.** (b), **34.** (c), **35.** (d),

(a) $\theta_c = \frac{2}{P}\theta_m$ (b) $\theta_c = P\theta_m$

(c) $\theta_c = \frac{4}{P}\theta_m$ (d) $\theta_c = \frac{P}{2}\theta_m$

[W.B. Univ. of Technology, 2013]

36. What is the form of an air gap flux density waveform of an uncompensated 2-pole dc shunt motor at full load?

(a) 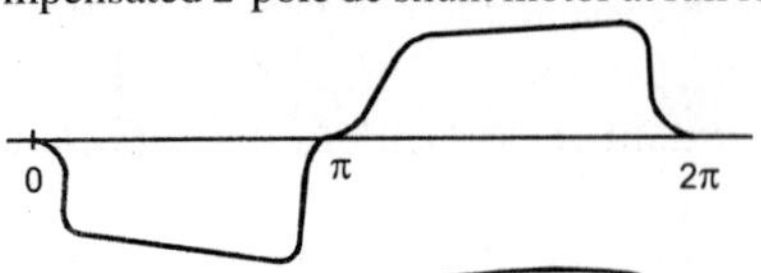

(b) 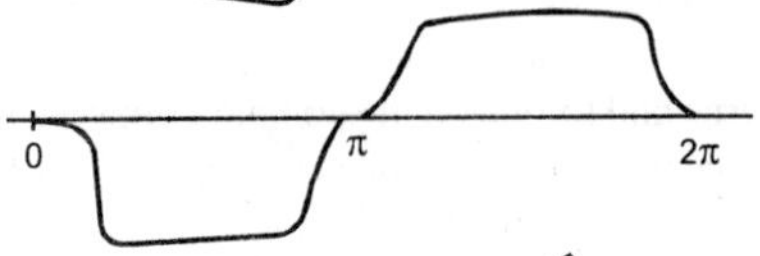

(c) 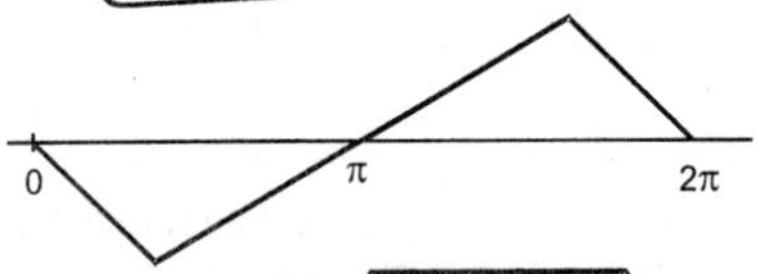

(d) 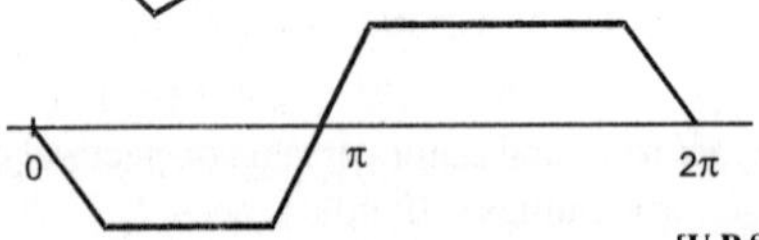

[U.P.S.C. I.E.S. E.E.-II, 2006]

37. An elementary cylindrical machine has one full-pitch coil in the stator, but the rotor may have (*i*) two poles or (*ii*) four poles of permanent magnets. The time varying voltage that would be induced in the stator coil for one rotation of the rotor, while the rotor is revolving at a constant speed, are shown in figures A, B, C and D above.

Select the correct combination from the code given below:

	2-pole	4-pole
(a)	A	D
(b)	A	B
(c)	C	D
(d)	B	C

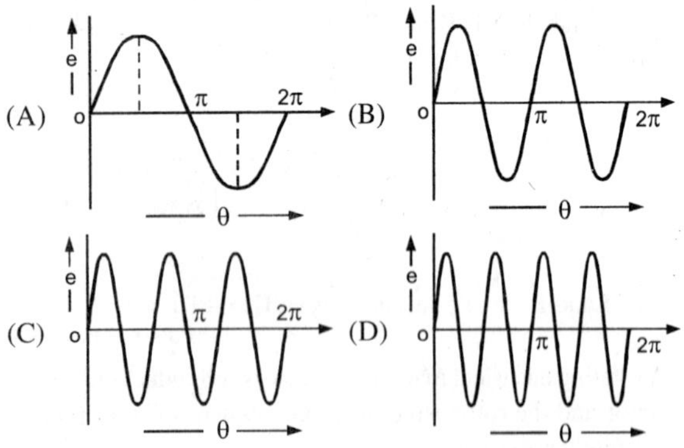

[U.P.S.C. I.E.S. E.E.-II, 2006]

38. Two magnetic poles revolve around a stationary armature carrying two coils $(c_1 - c_1', c_2 - c_2')$ as shown in the figure. Consider the instant when the poles are in a position as shown. Identify the correct statement regarding the polarity of the induced emf at this instant in coil sides c_1 and c_2.

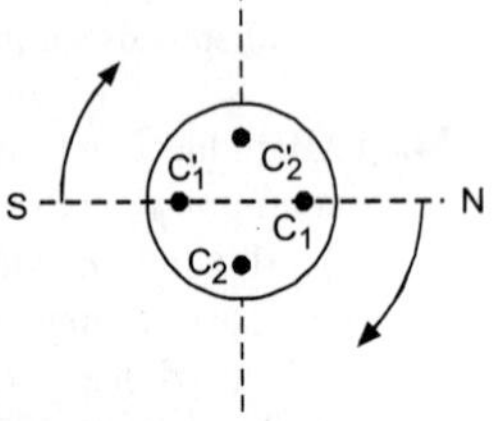

(a) ⊙ in χ_1, no emf in χ_2. (b) ⊗ in χ_1, no emf in χ_2.

(c) ⊙ in χ_2, no emf in χ_1. (d) ⊗ in χ_2, no emf in χ_1.

[GATE E.E. 2005]

39. Stray losses are sum of:

(a) Iron and mechanical losses.

(b) Copper and iron losses.

(c) Copper and mechanical losses.

(d) None of these. [G.B.T.U. Even Semester 2010-11]

40. Generally the no-load losses of an electrical machine is represented in its equivalent circuit by a

(a) parallel resistance with a low value.

(b) series resistance with a low value.

(c) parallel resistance with a high value.

(d) series resistance with a high value. [U.P.S.C. I.E.S. E.E.-II 2002]

41. Neglecting all losses, how is the developed torque (T) of a dc separately excited motor, operating under constant terminal voltage, related to its output power (P)?

(a) $T \propto \sqrt{P}$ (b) $T \propto P$ (c) $T^2 \propto P^3$

(d) T is independent of P. [GATE E.E. 1992; U.P.S.C. I.E.S. E.E.-II, 2007]

42. Consider the following parts of a dc machine:

1. Yoke 2. Armature core

3. Brushes 4. Pole core

Which of the above parts are subjected to iron loss?

(a) 1 and 2 only. (b) 2 only.

(c) 1 only. (d) 1, 2, 3 and 4.

[U.P.S.C. I.E.S. E.E.-II, 2010]

43. The armature core of a dc machine is laminated to minimise

(a) hysteresis loss. (b) eddy current loss.

(c) copper loss. (d) mechanical loss.

[W.B. Univ. of Technology, 2011]

44. What are slot wedges in a dc machine made of ?

(a) Mild steel. (b) Silicon steel.

(c) Fibre. (d) Cast iron.

[U.P.S.C. I.E.S. E.E.-II, 2007]

45. In a dc machine, the armature is always on the rotor unlike an ac machine where it could be either on the stator or rotor, because

(a) commutation action would otherwise not be possible.

(b) armature reaction demagnetising action would otherwise be more.

(c) otherwise it would not be possible to place compensating winding in pole shoes.

(d) self-starting of dc motor would not be possible otherwise.

[U.P.S.C. I.E.S. E.E.-II 2000]

46. The material used for brush is

(a) graphite. (b) aluminium.

(c) mica. (d) wood.

[W.B. Univ. of Technology, 2011]

47. Wave winding is employed in a dc machine of

(a) high current and low voltage rating.

(b) low current and high voltage rating.

(c) high current and high voltage rating.

(d) low current and low voltage rating. [U.P.S.C. I.E.S. E.E.-II, 2006]

48. The number of parallel paths for simple lap winding in a dc machine is

(a) 2 (b) 4 (c) 6

(d) number of poles [W.B. Univ. of Technology, 2013]

49. If two 8-pole dc machines of identical armatures are wound, one with lap winding and the other with wave winding, then

(a) wave wound machine will have more rated current and more voltage.

(b) lap would machine will have more rated voltage and more current.

(c) lap wound machine will have more rated voltage and less current.

Answers: 36. (a), **37.** (B), **38.** (a), **39.** (a), **40.** (c), **41.** (b), **42.** (b), **43.** (b), **44.** (c), **45.** (a), **46.** (a), **47.** (b), **48.** (d), **49.** (d),

(d) wave wound machine will have more rated voltage and less current. **[U.P.S.C. I.E.S. E.E.-II, 1999]**

50. The armature resistance in case of lap wound armature having Z conductors each conductor of length l metres, x-sectional area a m², resistivity and P poles will be

(a) $\rho\frac{l}{a}\frac{Z}{P^2}$. (b) $\rho\frac{l}{a}Z$. (c) $\rho\frac{l}{a}ZP$. (d) $\rho\frac{l}{a}ZP^2$.

51. A 4 pole generator with 16 coils has a two layer lap winding. The pole pitch is

(a) 32 (b) 16 (c) 8 (d) 4

[GATE E.E. 1996]

52. In a dc machine, for the same value of ϕ, Z and N; which one of the following statements is correct?

(a) Armature emf is more with wave winding than with lap winding.
(b) Armature emf is less with wave winding than with lap winding.
(c) Armature emf depends on whether the machine is running as a motor or a generator.
(d) Armature emf is same as long as the flux density in the air gap remains the same. **[U.P.S.C. I.E.S. E.E.-II 2004]**

53. The dummy coils in dc machines are useful to

(a) increase the efficiency.
(b) improve the commutation.
(c) reduce the cost of the machine.
(d) maintain mechanical balance of armature.

[U.P.S.C. I.E.S. E.E.-II 2003]

54. How many segments will be there for the commutator of a 6-pole dc machine having a simple wave-wound armature with 72 slots?

(a) 73 (b) 72
(c) 71 (d) 70 **[U.P.S.C. I.E.S. E.E.-II, 2011]**

55. A 4-pole lap wound dc generator has a developed power of *P* watts and voltage of *E* volts. The adjacent brushes of the machine are removed as they are worn out. If the machine operates with the remaining brushes, the developed voltage and power that can be obtained from the machine are

(a) E, P (b) E/2, P/2 (c) E, P/4 (d) E, P/2

[GATE E.E. 1999]

56. The emf induced in a conductor of machine driven at 600 rpm, the peak value of flux density is 1.0 Wb/m², diameter of machine 2.0 meter and length of machine 0.30 m is:

(a) 41.83 V (b) 29.58 V
(c) 9.42 V (d) 18.84 V

[U.P.S.C. I.E.S. E.E.-II, 2011]

57. The waveform of armature mmf in a dc machine is

(a) square (b) rectangular
(c) triangular (d) sinusoidal

[W.B. Univ. of Technology, 2010, 2013]

58. The armature mmf of a dc machine has

(a) triangular space distribution and rotates at the speed of armature.
(b) trapezoidal space distribution and is stationary in space.
(c) stepped distribution and rotates at the speed of armature.
(d) triangular space distribution and is stationary in space.

[U.P.S.C. I.E.S. E.E.-II 2000]

59. A 6-pole lap wound dc machine has 720 conductors and it draws 50 A from supply mains. What is the flux distribution produced by the armature reaction per pole?

(a) Rectangular in wave shape with a peak of 500 AT.
(b) Rectangular in wave shape with a peak of 100 AT.
(c) Triangular in wave shape with a peak of 500 AT.
(d) Triangular in wave shape with a peak of 1000 AT.

[U.P.S.C. I.E.S. E.E.-II, 2005]

60. Figure shows the extended view of a 2 pole dc machine with 10 armature conductors. Normal brush positions are shown by A and B, placed at the interpolar axis. If the brushes are now shifted, in the direction of rotation, to A′ and B′ as shown, the voltage waveform $V_{A'B'}$ will resemble

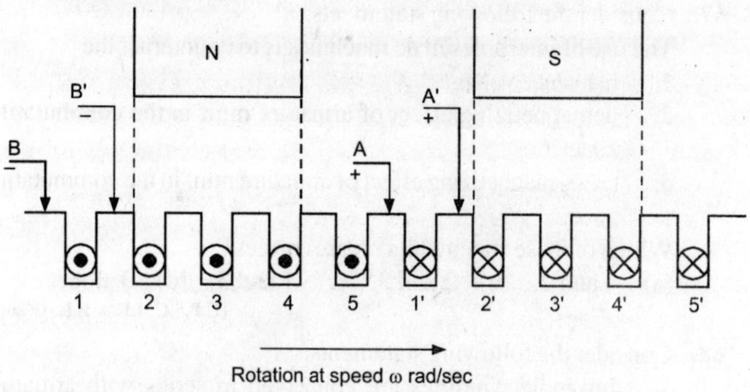

(a)

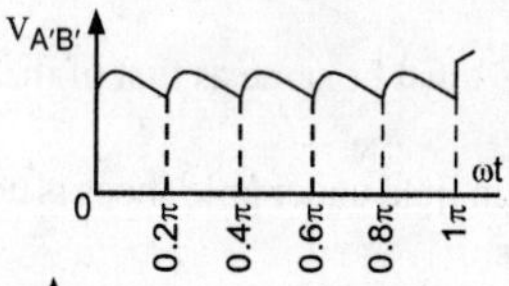

(b)

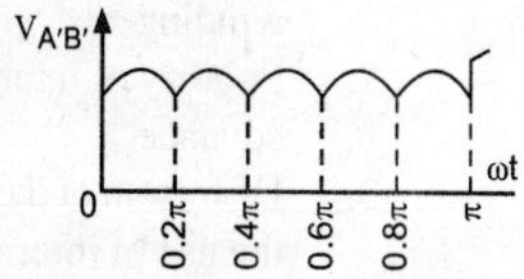

(c)

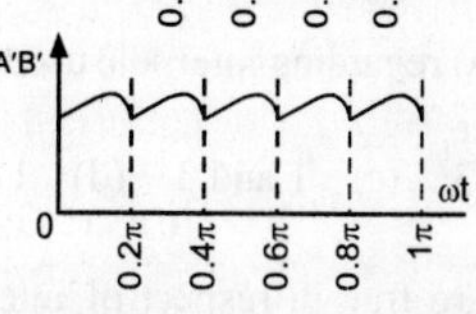

(d)

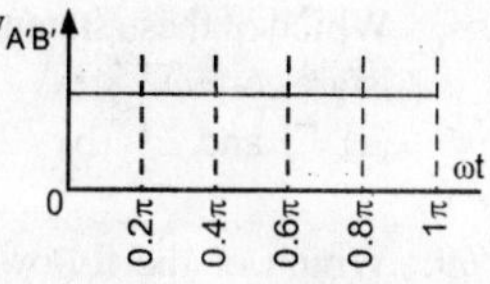

[GATE E.E. 2009]

61. Armature reaction in a dc shunt generator, running at full load with the brushes not shifted from the geometrical neutral plane and saturation neglected, is

(a) absent. (b) cross-magnetising.
(c) demagnetising. (d) magnetising.

[U.P.S.C. I.E.S. E.E.-II, 2008]

62. Air gap at the pole tips of a dc machine is kept more than that at the centre of the pole mainly to reduce

(a) reactance voltage. (b) effect of armature reaction.
(c) losses of armature core. (d) noise of the machine.

63. In a dc shunt generator working on load, the brushes are moved forward in the direction of rotation, as a result of this, commutation will

(a) improve but terminal voltage will fall.
(b) worsen and terminal voltage will fall.
(c) improve and terminal voltage will rise.
(d) worsen and terminal voltage will rise.

[U.P.S.C. I.E.S. Elec. Engineering-II, 1997]

64. If the armature current is increased to double its previous value, and the time of commutation is halved, how will the reactance voltage vary ?

(a) It will be halved. (b) It will remain the same.
(c) It will be doubled. (d) It will become four times.

[U.P.S.C. I.E.S. E.E.-II, 1993, 2007]

65. To have sparkless commutation, the armature reaction effect in a dc machine is neutralized by

(a) compensating winding and commutating poles.
(b) shifting the brush axis from geometrical neutral axis to magnetic neutral axis.
(c) fixing the brush axis in line with the pole axis.
(d) increasing the field excitation. **[U.P.S.C. I.E.S. E.E.-II, 1997]**

66. The interpoles in dc machines have a tapering shape in order to

(a) reduce the overall weight.

Answers: 50. (a), **51.** (c), **52.** (a), **53.** (d), **54.** (b), **55.** (c), **56.** (d), **57.** (c), **58.** (d), **59.** (c), **60.** (a), **61.** (b), **62.** (b), **63.** (a), **64.** (d), **65.** (a), **66.** (b),

(b) reduce the saturation in the interpole.
(c) economic on the material required for interpoles and their windings.
(d) increase the acceleration of commutation.

[U.P.S.C. I.E.S. E.E.-II 1998]

67. Consider the following statements:
The use of interpoles in dc machines is to counteract the
1. reactance voltage.
2. demagnetizing effect of armature mmf in the commutating zone.
3. cross-magnetizing effect of armature mmf in the commutating zone.

Which of these statement(s) is/are correct?
(a) 1 and 2. (b) 2 and 3. (c) 1 and 3. (d) 3 alone.

[U.P.S.C. I.E.S. E.E.-II 2001]

68. Consider the following statements:
1. Interpole windings are connected in series with armature winding.
2. Polarity of interpole must be same as that of the main pole in advance.
3. Distortion of the main field under pole shoes is not affected by the use of interpoles.

Which of these statement(s) regarding interpole used in dc motors is/are correct?
(a) 1 and 2 (b) 2 and 3 (c) 1 and 3 (d) 1 alone.

[U.P.S.C. I.E.S. E.E.-II 2000]

69. Which of the following are true in respect of interpoles in dc machines ?
1. Reduce armature reaction effects in interpolar region.
2. Windings are connected in series with armature.
3. Have the same dimensions as main poles.

Select the correct answer using the code given below :
(a) 1 and 2 only. (b) 2 and 3 only.
(c) 1 and 3 only. (d) 1, 2 and 3.

[U.P.S.C. I.E.S. E.E.-II, 2007]

70. Consider the following statements about commutating poles which are fitted on most large dc shunt motors.
1. The commutating poles are placed in geometric neutral planes and their number is usually equal to the number of main poles.
2. The winding on the commutating poles is connected in series with the shunt field winding on the main poles.
3. The polarity of the commutating pole must be that of the next main pole further ahead.
4. The commutating poles neutralise the reactance voltage produced in the coil undergoing commutation.

(a) 1, 2 and 3 are correct. (b) 1 and 4 are correct.
(c) 2, 3 and 4 are correct. (d) 1, 2 and 4 are correct.

[U.P.S.C. I.E.S. E.E.-II, 1996]

71. In a dc machine, which of the following statements is true?
(a) Compensating winding is used for neutralizing armature reaction while interpole winding is used for producing residual flux.
(b) Compensating winding is used for neutralizing armature reaction while interpole winding is used for improving commutation.
(c) Compensating winding is used for improving commutation while interpole winding is used for neutralizing armature reaction.
(d) Compensating winding is used for improving commutation while interpole winding is used for producing residual flux.

[GATE E.E., 2006]

72. Consider the following statements in respect of a dc generator:
1. Compensating windings are located in teeth cut out in main poles.
2. Compensating windings are series connected so that the armature reaction is aided at any load.
3. Compensating windings are series connected for cancellation of the armature reaction at any load.
4. Compensating windings are shunt connected to aid the main poles at any load.

Which of the statements given above are correct?
(a) 1 and 2. (b) 1, 2 and 3. (c) 1 and 3. (d) 2, 3 and 4.

[U.P.S.C. I.E.S. E.E.-II, 2006]

73. The compensating winding in a dc machine
(a) is located in armature slots for compensation of the armature reaction.
(b) is located on commutating poles for improving commutation.
(c) is located on pole shoes for avoiding the flashover at the commutator surface.
(d) is located on pole shoes to avoid the sparking at the brushes.

[GATE E.E. 2000]

74. In a dc series generator, the terminal voltage with the increase in load will
(a) decrease. (b) increase.
(c) remain constant. (d) none of these.

[W.B. Univ. of Technology, 2010]

75. Consider the following statements related to a shunt generator :
1. The full-load voltage is lower than no-load voltage.
2. The above lowering is partly due to voltage drop in armature resistance.
3. The field current does not change with load and has no effect on generated voltage.

Which of the statements given above are correct ?
(a) 1, 2 and 3. (b) Only 1 and 2.
(c) Only 2 and 3. (d) Only 1 and 3.

[U.P.S.C. I.E.S. E.T.E-I, 2006]

76. Consider the following statements regarding the magnetisation characteristics of a dc generator:
1. It represents the variation of field flux with field current.
2. Variation of open-circuit terminal voltage with field current is independent of speed.
3. It is nonlinear from zero to rated field current.
4. It helps to determine critical field resistance at a given speed.
5. It always starts from the origin.

Which of these statements are correct?
(a) 1 and 4. (b) 1, 2 and 5.
(c) 2, 3 and 4. (d) 3and 5.

[U.P.S.C. I.E.S. E.E.-II, 1999]

77. Out of the following factors for a dc machine
(i) Interpole (ii) Armature resistance
(iii) Armature (iv) Reduction in field current
(v) Effect of armature reaction

The factors that are responsible for decrease in terminal voltage of a shunt generator are
(a) (i), (ii) and (iv). (b) (ii), (iii) and (v).
(c) (i), (iv) and (v). (d) (ii), (iv) and (v).

[GATE E.E. 1996]

78. The graph represent which characteristic of a dc shunt generator?
(a) Internal characteristic.
(b) External characteristic.
(c) Open-circuit characteristic.
(d) Magnetic characteristic.

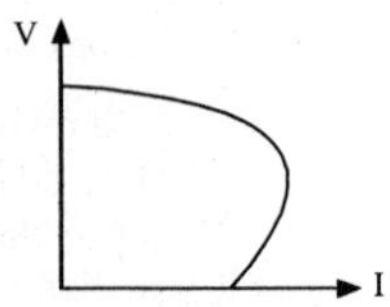

[U.P.S.C. I.E.S E.E.-II, 2008]

Answers: **67.** (c), **68.** (d), **69.** (a), **70.** (b), **71.** (b), **72.** (c), **73.** (c), **74.** (b), **75.** (b), **76.** (a), **77.** (d), **78.** (b),

79. Consider the following statements
1. In a dc series generator, full-load voltage is more than the no-load voltage.
2. In a separately excited dc generator full-load voltage is more than the no-load voltage.
3. In a dc shunt generator, full-load voltage is less than the no-load voltage.

Of these statements.
(a) 1,2 and 3 are correct. (b) 1 and 2 are correct.
(c) 2 and 3 are correct. (d) 1 and 3 are correct.
[U.P.S.C. I.E.S. E.E. II, 1994]

80. Consider the following statements :
For a level compounded dc generator to run at constant speed, the series field mmf must effectively compensate
1. armature reaction mmf.
2. armature resistance voltage drop.
3. brush contact voltage drop.

Which of these statements is/are correct ?
(a) 2 and alone. (b) 1 and 2.
(c) 1 and 3. (d) 1, 2 and 3.
[U.P.S.C. I.E.S E.E.-II, 2000]

81. In a dc compound generator, "flat compound" characteristic, required for certain applications, may be obtained by connecting a variable resistance:
(a) across the series field. (b) in series with the series field.
(c) in parallel with the shunt field.
(d) in series with the shunt field. **[U.P.S.C. I.E.S. E.E.-II, 2005]**

82. A self excited dc shunt generator driven by its prime mover at the rated speed fails to build up voltage across its terminals at no load. What reason can be assigned for this?
(a) The field circuit resistance is higher than the critical resistance.
(b) The initial shunt field mmf does not assist the residual magnetism.
(c) One of the interpole connection is reversed.
(d) The brush-axis shifts slightly from the geometrical neutral axis of the machine. **[U.P.S.C. I.E.S. E.E.-II, 2006]**

83. A cumulatively compounded dc generator is supplying 20 A at 200 V. Now if the series field winding is short circuited, the terminal voltage
(a) will remain unaltered at 200 V.
(b) will rise to 220 V.
(c) will shoot upto a very high value.
(d) will become less than 200 V. **[U.P.S.C. I.E.S. E.E. I, 1992]**

84. A dc shunt generator when driven without connecting field winding shows an open-circuit terminal voltage of 12 V. When field winding is connected and excited the terminal voltage drops to zero because
(a) field resistance is higher than critical resistance.
(b) there is no residual magnetism in the field circuit.
(c) field winding has got wrongly connected.
(d) there is a fault in armature circuit. **[U.P.S.C. I.E.S. E.E.-II 2001]**

85. A dc shunt generator, when driven at its rated speed, is found to be not generating any voltage. Which of the following would account for this?
1. There is no residual magnetism.
2. The connection of the field winding is not proper with respect to armature terminals.
3. The resistance of the field circuit is greater than the critical field resistance.
4. The load resistance is less than the critical armature resistance.
Select the correct answer using the codes given below:

Codes:
(a) 3 and 4. (b) 1, 3 and 4.
(c) 1, 2 and 3. (d) 1, 2, 3 and 4.
[U.P.S.C. I.E.S. E.E.-II, 1997]

86. The type of dc generator used for arc welding purposes is a
(a) series generator. (b) shunt generator.
(c) cumulatively compounded generator.
(d) differentially compounded generator.
[U.P.S.C. I.E.S. E.E. II, 1995]

87. A pair of similar dc shunt generators operate in parallel and supply a common load. It is required to switch off machine no. 1 and allow machine no. 2 to supply the circuit load. The following operations are to be used to achieve this
1. Switch off the main switch of machine no. 1.
2. Reduce the field current of machine no. 1.
3. Increase the field current of machine no. 2.
4. Ensure that machine no. 1 just floats.

The correct sequence of these operations is
(a) 4, 3, 2, 1. (b) 2, 3, 4, 1. (c) 2, 4, 3, 1. (d) 3, 2, 4, 1.
[U.P.S.C. I.E.S. E.E. II, 1996]

88. In the block diagram of a separately excited dc motor, how does the armature induced emf appear as ?
(a) Positive feedback. (b) Negative feedback.
(c) Disturbance input. (d) Output.
[U.S.P.C. I.E.S. E.E.-II, 2005]

89. The armature current drawn by a 120 V dc motor with back emf 110 V and armature resistance of 0.4 Ω is
(a) 4A (b) 25A (c) 200 A (d) 10A
[W.B. Univ. of Technology, 2013]

90. A series motor (dc) is running drawing a load current of 1 A. If the load is reduced such that the current drawn is halved; the speed of the motor would be approximately
(a) unchanged. (b) reduced by 50%.
(c) increased by 50%. (d) increased by 100%.
[W.B. Univ. of Technology, 2013]

91. If the load on a dc shunt motor is increased, its speed decreases primarily due to
(a) increase in its flux. (b) decrease in back emf.
(c) increase in armature current.
(d) increase in brush drop. **[W.B. Univ. of Technology, 2011]**

92. The direction of rotation of a dc series motor can be reversed
(a) by interchanging supply terminals.
(b) by interchanging field terminals.
(c) either by interchanging supply terminals or by interchanging field terminals.
(d) by interchanging supply terminals as well as field terminals.
[U.P.S.C. I.E.S. E.E.-II 2003]

93. When is the mechanical power developed by a dc motor maximum?
(a) Back emf is equal to applied voltage.
(b) Back emf is equal to zero.
(c) Back emf is equal to half the applied voltage.
(d) None of the above. **[U.P.S.C. I.E.S. E.E.-II, 2008]**

94. The dc motor, which can provide zero speed regulation at full load without any controller is
(a) series. (b) shunt.
(c) cumulative compound. (d) differential compound.
[GATE E.E. 2007]

95. What happens to the speed when the flux is reduced by 10% in a 200 V dc shunt motor having an armature resistance of 0.2 ohm carrying a current of 50 A and running at 960 rpm prior to weakening of field. The total torque may be assumed constant. Neglect losses.
(a) 1250 rpm. (b) 1066 rpm.
(c) 920 rpm. (d) 576 rpm.
[U.P.S.C. I.E.S. E.E.-II, 2007]

Answers: 79. (d), **80.** (b), **81.** (a), **82.** (b), **83.** (d), **84.** (c), **85.** (d), **86.** (d), **87.** (b), **88.** (b), **89.** (b), **90.** (d), **91.** (b), **92.** (b), **93.** (c), **94.** (d), **95.** (b),

96. Consider the following statements:
The armature torque in a dc motor is a function of
1. Field flux 2. Armature current
3. Speed 4. Damping
Which of the above statements is/are correct?
(a) 1, 2, 3 and 4. (b) 3 and 4 only.
(c) 1 and 2 only. (d) 4 only.
[U.P.S.C. I.E.S. E.E.-II, 2010]

97. Torque developed by the armature of a dc motor is proportional to which one of the following ?
(a) (Back emf) × (Armature current).
(b) (Magnetic flux per pole) × (Armature current).
(c) (Back emf) × (Magnetic flux per pole).
(d) (Back emf) / (Magnetic flux per pole).
[U.P.S.C. I.E.S E.E.-II, 2007]

98. Armature torque of a dc motor is a function of which of the following factors?
1. Speed. 2. Field flux.
3. Armature current. 4. Residual magnetism.
Select the correct answer using the codes given below:
Codes:
(a) 2 and 3. (b) 1 and 4. (c) 3 and 4. (d) 1 and 2.
[U.P.S.C. I.E.S. E.E.-II 2003]

99. A 200 V, 2,000 rpm, 10 A, separately excited dc motor has an armature resistance of 2 Ω. Rated dc voltage is applied to both the armature and field windings of the motor. If the armature draws 5 A from the source, the torque developed by the motor is
(a) 4.30 Nm (b) 4.77 Nm (c) 0.45 Nm (d) 0.50 Nm
[GATE E.E. 2002]

100. A PMDC commutator motor has a no-load speed of 6,000 rpm when connected to a 120 V dc supply. The armature resistance is 2.5 Ω and other losses may be neglected. The speed of the motor, when supply voltage of 60 V developing a torque of 0.5 Nm, is
(a) 3,000 rpm. (b) 2,673 rpm.
(c) 2,836 rpm. (d) 5,346 rpm. **[GATE E.E. 2000]**

101. A 240 V dc shunt motor with an armature resistance of 0.5 Ω has a full-load current of 40 A. Find the ratio of the stalling torque to the full-load torque when a resistance of 1 Ω is connected in series with the armature
(a) 4 (b) 12
(c) 6 (d) none of the above.
[GATE E.E. 1998]

102. Which of the following curves represents the speed-torque characteristic of a dc series motor ?

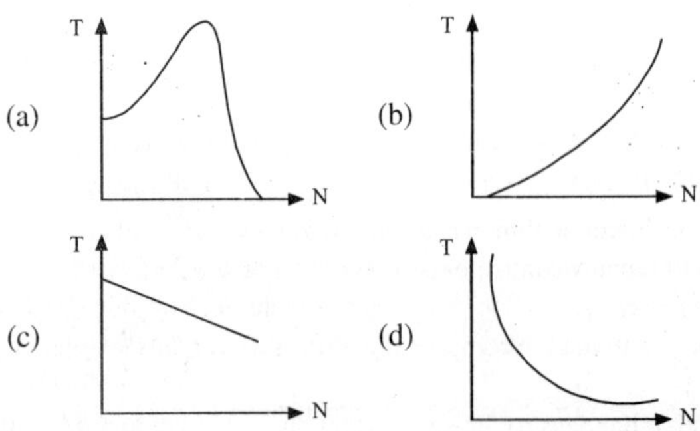

[U.P.S.C. I.E.S. E.E.-II, 2008]

103. A 240 V dc series motor takes 40 A when giving its rated output at 1,500 rpm. Its resistance is 0.3 Ω. The value of resistance which must be added to obtain rated torque at 1,000 rpm is
(a) 6 Ω. (b) 5.7 Ω. (c) 2.2 Ω. (d) 1.9 Ω.
[GATE E.E. 2000]

104. If the armature of a dc motor is supplied from a constant current source, and its field from a constant voltage source, then the torque-speed characteristic of the motor will be

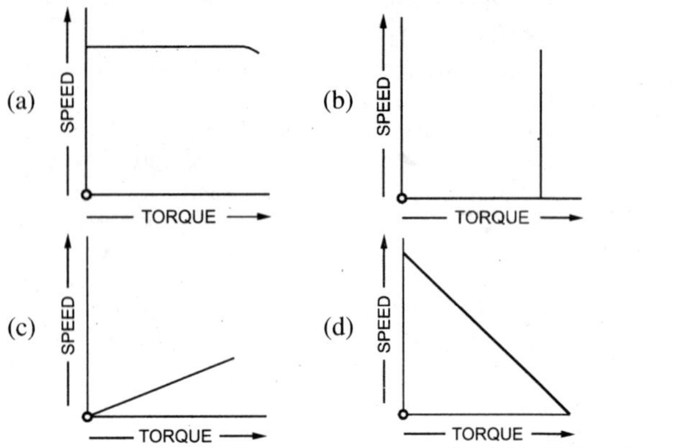

[U.P.S.C. I.E.S. E.E. II 1994]

105. An electric motor with "constant output power" will have a torque-speed characteristic in the form of a
(a) straight line through the origin.
(b) straight-line parallel to speed axis.
(c) circle about the origin.
(d) rectangular hyperbola. **[GATE E.E. 2001]**

106. A dc series motor fed from rated supply voltage is overloaded and its magnetic circuit is saturated. The torque-speed characteristic of this motor will be approximately represented by which curve of figure given:
(a) Curve *A*. (b) Curve *B*.
(c) Curve *C*. (d) Curve *D*.

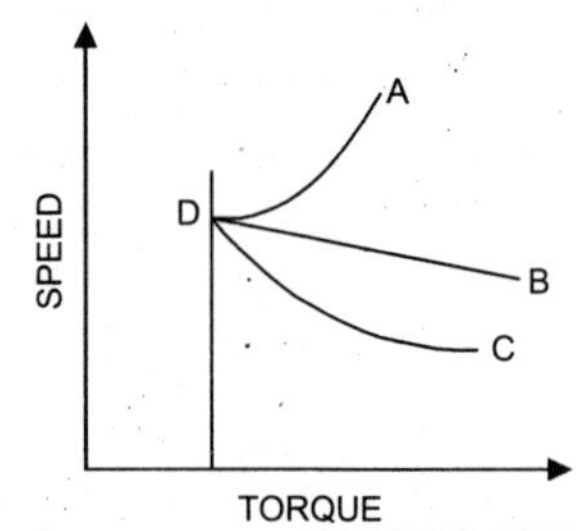

[GATE E.E. 2002]

107. An electric train employing a dc series motor is running at a fixed speed, when a sudden slight drop in the mains voltage occurs. This would result in
(a) drop in speed and rise in current.
(b) rise in speed and drop in current.
(c) rise in speed and rise in current.
(d) drop in speed with current unaltered.
[U.P.S.C. I.E.S. E.E. II, 1995]

108. Consider the following statements :
If the terminal voltage of a dc shunt motor is halved with the load torque varying as the square of speed, then
1. speed is halved.
2. speed remains unaltered.
3. armature current is doubled.
4. armature current remains unaltered.
Of these statements
(a) 1 and 3 are correct. (b) 1 and 4 are correct.
(c) 2 and 3 are correct. (d) 2 and 4 are correct.
[U.P.S.C. I.E.S. E.E.-II 1993]

109. DC shunt motors are used for driving
(a) machine tools. (b) lathes.
(c) centrifugal pumps. (d) line shafts.
(e) blowers and fans. (f) all of these.

110. A cumulative compound long shunt motor is driving a load at rated torque and rated speed. If the series field is shunted by a resistance equal to the resistance of the series field, keeping the torque constant.
(a) the armature current increases.

Answers: **96.** (c), **97.** (b), **98.** (a), **99.** (a), **100.** (b), **101.** (a), **102.** (d), **103.** (d), **104.** (b), **105.** (d), **106.** (b), **107.** (a), **108.** (c), **109.** (f), **110.** (e),

(b) the motor speed increases.
(c) the armature current decreases.
(d) the motor speed decreases.
(e) both (a) and (b).

111. A belt-driven cumulative compounded dc generator is delivering power to the dc mains. If the belt snaps then the machine will run as a
(a) cumulative compounded motor in the same direction.
(b) differentially compounded motor in the same direction.
(c) cumulative compounded motor in the opposite direction.
(d) differentially compounded motor in the opposite direction.
[U.P.S.C. I.E.S. E.E. II, 1994]

112. Consider the following statements.
A cumulatively compounded dc motor runs at a speed N_1 rpm drawing an armature current of I amperes. If, keeping the load torque constant, the series field winding is short-circuited, then the
1. speed of the motor will increase.
2. armature current will increase.
3. armature current will decrease.
4. speed of the motor will decrease.

Of these statements
(a) 3 and 4 are correct. (b) 1 and 4 are correct.
(c) 2 and 4 are correct. (d) 1 and 2 are correct.
[U.P.S.C. I.E.S. E.E.-II, 1996]

113. A differentially compounded dc motor with interpoles and with brushes on the neutral axis is to be driven as a generator in the same direction with the same polarity of the terminal voltage. It will then
(a) be a cumulatively compounded generator but the interpole coil connections are to be reversed.
(b) be a cumulatively compounded generator without reversing the interpole coil connections.
(c) be a differentially compounded generator without reversing the interpole coil connections.
(d) be a differentially compounded generator but the interpole coil connections are to be reversed. **[GATE E.E. 1995]**

114. Which motor should not be used for centrifugal pumps ?
(a) Shunt. (b) Series.
(c) Cumulatively compound.
(d) Differentially compound. **[U.P.S.C. I.E.S. E.E. II, 1992]**

115. The motor used in a lift of a highrise building is
(a) dc series motor. (b) dc shunt motor.
(c) dc compound motor. (d) ac synchronous motor.
[W.B. Univ. of Technology, 2011]

116. Consider the following statements:
PMDC motors used in cassette tape recorders have
1. magnets on stator and armature on the rotor.
2. magnets on rotor and armature on the stator.
3. electronic commutation and no brushes.
4. mechanical commutation.
5. automatic speed governors.

Of these statements
(a) 1, 3 and 5 are correct. (b) 1, 4 and 5 are correct.
(c) 2, 3 and 5 are correct. (d) 1 and 4 are correct.
[U.P.S.C. I.E.S. E.E.-II 1998]

117. The speed of a dc shunt motor may be varied by varying
1. Field current. 2. Supply voltage.
3. Armature circuit.

Select the correct answer using the codes given below :
Codes :
(a) 1, 2, and 3. (b) 1 and 2. (c) 1 and 3. (d) 2 and 3.
[U.P.S.C. I.E.S. E.E.-II, 2005]

118. Consider the following statements:
The speed of a dc motor can be controlled by the variation of
1. armature voltage. 2. field current.
3. armature circuit resistance.
4. angle of brush shift.

Which of the statements given above are correct?
(a) 1, 2, 3. (b) 2, 3, 4.
(c) 1, 3, 4. (d) 1, 2, 4.
[U.P.S.C. I.E.S. Elec. Engineering-II, 2004]

119. The field flux method of speed control of dc shunt motor is generally used for which one of the following ?
(a) Constant hp drive below the base speed.
(b) Constant torque drive above the base speed.
(c) Constant torque drive below the base speed.
(d) Constant hp drive above the base speed.
[U.P.S.C. I.E.S. E.E.-II, 2007]

120. Following controls are considered for dc motors
I. Control flux.
II. Armature resistance control.
III. Supply voltage control.

Which of the above controls play significant role in the speed control of dc motors.
(a) I only. (b) II and III only.
(c) I and III only. (d) I and II.
[U.P.S.C. I.E.S. E.E. II, 1992]

Common Data for Q. (121-122)

A separately excited DC motor runs at 1,500 rpm under no load with 200 V applied to the armature. The field voltage is maintained at its rated value. The speed of the motor, when it delivers a torque of 5 Nm, is 1,400 rpm as shown in the figure. The rotational losses and armature reaction are neglected.

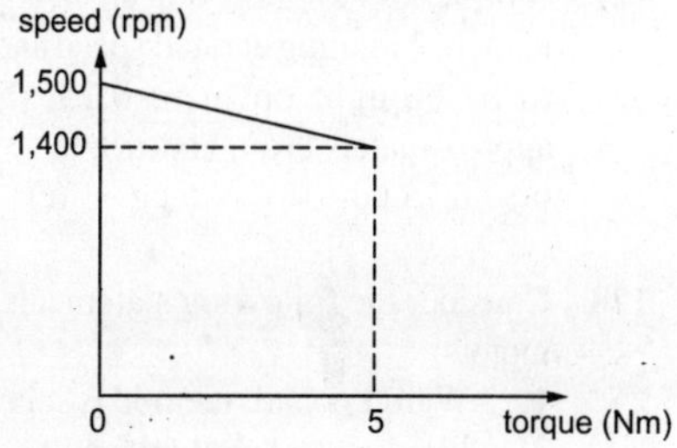

121. The armature resistance of the motor is
(a) 2 Ω (b) 3.4 Ω
(c) 4.4 Ω (d) 7.7 Ω **[GATE E.E., 2010]**

122. For the motor to deliver a torque of 2.5 Nm at 1,400 rpm, the armature voltage to be applied is
(a) 125.5 V (b) 193.3 V
(c) 200 V (d) 241.7 V **[GATE E.E., 2011]**

123. For a constant supply voltage, what are the effects of inserting a series resistance in the field circuit of a dc shunt motor, on its speed and torque?
(a) Speed will decrease and the torque will decrease.
(b) Speed will increase and the torque will increase.
(c) Speed will increase and the torque will decrease.
(d) Speed will decrease and the torque will increase.
[U.P.S.C. I.E.S. E.E.-II, 2006]

124. A 50 kW dc shunt motor is loaded to draw rated armature current at any given speed. When driven
(*i*) at half the rated speed by armature voltage control and
(*ii*) at 1.5 times the rated speed by field control, the respective output powers delivered by the motor are approximately.
(a) 25 kW in (*i*) and 75 kW in (*ii*).
(b) 25 kW in (*i*) and 50 kW in (*ii*).
(c) 50 kW in (*i*) and 75 kW in (*ii*).
(d) 50 kW in (*i*) and 50 kW in (*ii*).
[U.P.S.C. I.E.S., 2006; GATE E.E. 2005]

Answers: 111. (b), **112.** (d), **113.** (b), **114.** (b), **115.** (a), **116.** (c), **117.** (c), **118.** (a), **119.** (d), **120.** (d), **121.** (b), **122.** (b), **123.** (c), **124.** (b),

125. In a coal mine, Ward Leonard method of speed control, as shown in the given figure, is used to lift coal from the pit through a crane and bucket arrangement. After lifting the coal to desired level and emptying at the dumping ground, the bucket has to be lowered into the pit. The action to be taken at this stage is to reverse the

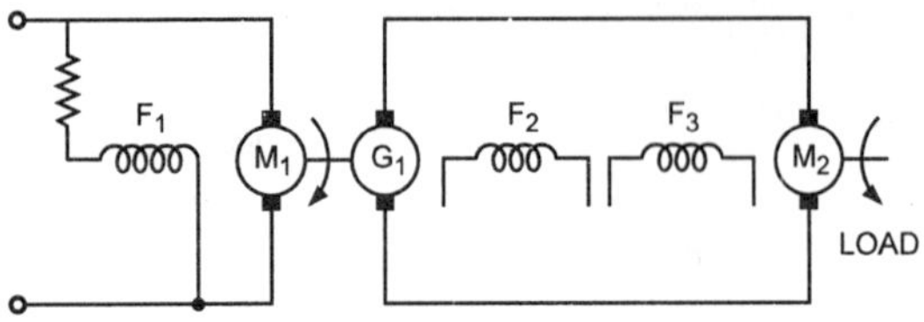

(a) direction of current in F_2 and slowly increase it.
(b) current in F_3 and slowly increase it.
(c) supply terminal of M_1.
(d) direction of current in F_1 and slowly increase it.
[U.P.S.C. I.E.S. E.E. II 1995]

126. For a given amount of kinetic energy to be released by the flywheel of Ward-Leonard-Ilgner control system at a given percentage reduction in speed, the mass of the flywheel would depend upon its radius of gyration r_g and its initial peripheral speed U_{p1}.
The mass of the flywheel is
(a) directly proportional to both r_g^2, and U_{p1}^2
(b) directly proportional to r_g^2, but inversely to U_{p1}^2
(c) inversely proportional to r_g^2, but directly to U_{p1}^2
(d) inversely proportional to both r_g^2 and U_{p1}^2
[U.P.S.C. I.E.S. E.E.-II, 2001]

127. A dc series motor driving an electric train faces a constant power load. It is running at rated speed and rated voltage. If the speed has to be brought down to 0.25, the supply voltage has to be approximately brought down to
(a) 0.75 pu (b) 0.5 pu (c) 0.25 pu (d) 0.125 pu
[GATE E.E. 2003]

128. Consider the following statements regarding speed control of dc motors:
1. Ward Leonard method is suitable for constant torque drive.
2. Field control method facilitates speed control below the base speed.
3. Armature resistance control method is more efficient when compared to Ward Leonard method.
4. Field control method is suitable for constant power drives.

Of these statements
(a) 1, 2 and 3 are correct. (b) 1, 3 and 4 are correct.
(c) 2, 3 and 4 are correct. (d) 1 and 4 are correct.
[U.P.S.C. I.E.S. E.E. II 1993]

129. Consider the following statements regarding the speed control of dc motors:
1. Ward-Leonard method is suitable for constant torque drives.
2. Ward-Leonard method is suitable for constant power drives.
3. Field-control method facilitates speed control below base speed.
4. Armature resistance control method is more efficient than Ward-Leonard method.
5. Field control method is suitable for constant torque drives.
6. Armature resistance control method is suitable for constant torque drives.

From these the correct answer is
(a) 1, 2, 6 (b) 1, 2, 4 (c) 1, 6 (d) 1, 3, 4
[U.P.S.C. I.E.S. E.E. II 2001]

130. For a given torque, reducing diverter resistance of a dc series motor
(a) increases its speed but armature current remains the same.
(b) increases its speed demanding more armature current.
(c) decreases its speed demanding less armature current.
(d) decreases its speed but armature current remains the same.
[U.P.S.C. I.E.S. E.E.-II 2002]

131. A dc series motor is operating at rated speed with rated excitation. It has two resistances R_1 and R_2 connected across the field and the armature respectively. Speeds above the rated speed can be obtained by
(a) increasing R_1 only. (b) reducing R_2 only.
(c) reducing R_1 and increasing R_2.
(d) increasing R_1 and reducing R_2.

132. A cumulatively compounded long shunt motor is driving a load at rated torque and rated speed. If the series field is shunted by a resistance equal to the resistance of the series field, keeping the torque constant,
(a) the armature current increases.
(b) the motor speed increases.
(c) the armature current decreases.
(d) the motor speed decreases. **[GATE E.E. 1993]**

133. Consider the following statements regarding the starting of a dc shunt motor:
1. External field resistance should be minimum.
2. External field resistance should be maximum.
3. Reduced voltage should be applied to the armature.
4. Increased voltage should be applied to the armature.

Of these statements
(a) 1 and 3 are correct. (b) 1 and 4 are correct.
(c) 2 and 3 are correct. (d) 2 and 4 are correct.
[U.P.S.C.I.E.S. E.E. II, 1994]

134. Which one of the following statements is correct for the ideal field excitation during starting in case of shunt and compound motors?
(a) Maximum, because motor torque required to overcome friction and load torque.
(b) Maximum, because voltage should be built slowly.
(c) Maximum, because small excitation will lead to low operating speed during starting.
(d) It can be anything because it does not matter in case of these motors. **[U.P.S.C. I.E.S. E.E.-II, 2011]**

135. Consider the following statements:
At starting, the field excitation of a dc shunt motor is kept at its maximum value to reduce
1. sparking at brushes. 2. acceleration time.
3. starting current. 4. voltage dip in supply.

Of these statements
(a) 1 and 2 are correct. (b) 2, 3 and 4 are correct.
(c) 1, 3, and 4 are correct. (d) 1, 2, 3, and 4 are correct.
[U.P.S.C. I.E.S. E.E.-II, 1998]

136. A dc shunt motor has external resistances of R_a and R_f in the armature and field circuits respectively. Armature current at starting can be reduced by keeping
(a) R_f maximum and R_a minimum.
(b) R_a maximum and R_f maximum.
(c) R_a minimum and R_f maximum.
(d) R_a maximum and R_f minimum. **[U.P.S.C. I.E.S. E.E. II, 1996]**

137. Which one of the following figures best represents the variation in armature current and speed with time during the starting of a dc shunt motor ?

(a)
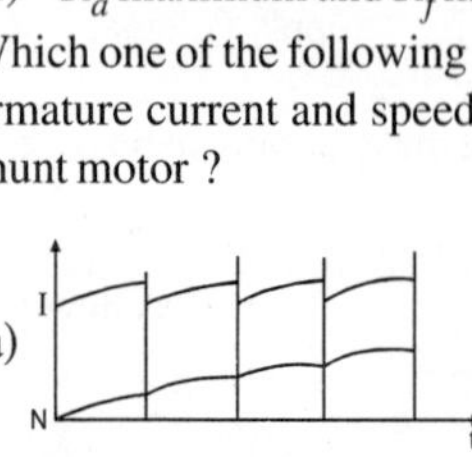

(b)
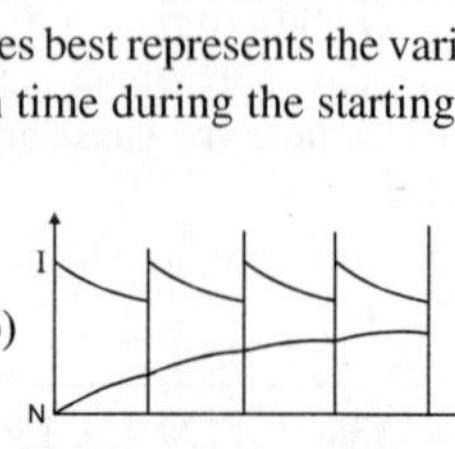

Answers: **125.** (a), **126.** (d), **127.** (b), **128.** (d), **129.** (a), **130.** (b), **131.** (c), **132.** (a), **133.** (a), **134.** (a), **135.** (b), **136.** (d), **137.** (b),

(c) 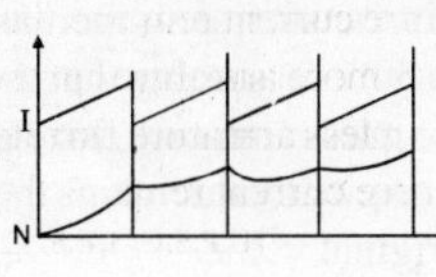

(d) 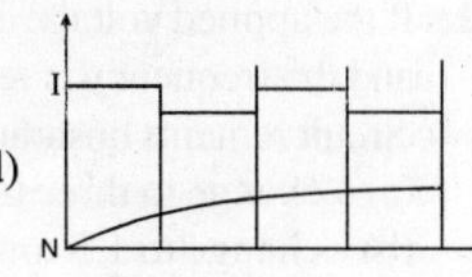

[U.P.S.C. I.E.S. E.E.-II, 1997]

138. Plugging of dc machine is normally executed by
(a) reversing the field polarity.
(b) reversing the armature polarity.
(c) reversing both field and armature polarity.
(d) connecting the resistance across armature terminals.

139. The electric braking system commonly employed in rolling mills, elevators and printing presses is
(a) plugging. (b) rheostatic.
(c) dynamic. (d) regenerative.

140. A 240 V, dc shunt motor draws 15 A while supplying the rated load at a speed of 80 rad/s. The armature resistance is 0.5 Ω and the field winding resistance is 80 Ω.
The net voltage across the armature resistance at the time of plugging will be
(a) 6 V (b) 234 V (c) 240 V (d) 474 V
[GATE E.E. 2008]

Statement for Linked Answer Questions (141-142).

A 240 V, dc shunt motor draws 15 A while supplying the rated load at a speed of 80 rad/s. The armature resistance is 0.5 Ω and the field winding resistance is 80 Ω.

141. The net voltage across the armature resistance at the time of plugging will be
(a) 6 V (b) 234 V (c) 240 V (d) 474 V
[GATE E.E., 2008]

142. The external resistance to be added in the armature circuit to limit the armature current to 125% of its rated value is
(a) 31.1 Ω (b) 31.9 Ω
(c) 15.1 Ω (d) 15.9 Ω **[GATE E.E., 2008]**

143. Regenerative braking
(a) can be easily applied to dc shunt motors.
(b) can be easily applied to dc series motors.
(c) can be used for stopping the motor.
(d) cannot be used when the load on motor has overhauling characteristics.

144. Consider the following statements:
In a dc machine, iron loss occurs in
1. armature core. 2. yoke.
3. pole cores. 4. pole shoes.
Which of the statements given above are correct?
(a) 1 and 4. (b) 1 and 2. (c) 1 and 3. (d) 2 and 3.
[U.P.S.C. I.E.S. E.E.-II 2004]

145. Two series motors are mechanically coupled. One machine is running as a motor and the other as a generator. The iron and friction losses of the machines will be identical when
(a) their speeds are identical.
(b) their speeds and excitations are identical.
(c) their speeds are equal but back emfs are half the supply voltages.
(d) their ratings and armature sizes are identical.
[U.P.S.C. I.E.S. E.E.-II 1992]

146. The maximum efficiency occurs in a separately excited dc generator when the terminal voltage is 220 V and the induced emf is 240 V, the stray losses, if the armature resistance is 0.2 Ω, will be:
(a) 1,000 W (b) 2,000 W (c) 3,000 W (d) 4,000 W
[U.P.S.C. I.E.S. E.E.-II, 2011]

147. The curve giving relation between efficiency and useful power output in kW for a dc motor is

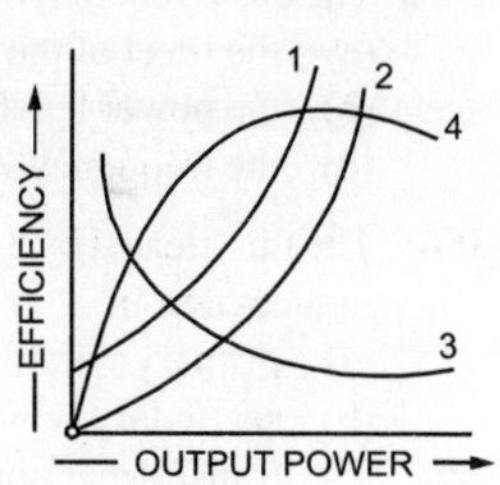

(a) 1
(b) 2
(c) 3
(d) 4

148. The efficiency of a dc machine will be maximum when
(a) $I = \sqrt{\dfrac{P_C}{R_a}}$ (b) $I = \dfrac{\sqrt{P_C}}{R_a}$
(c) $I = \dfrac{P_C}{R_a^2}$ (d) $I = \dfrac{P_C}{R_a}$

149. In Swinburne's test of a dc machine
(a) no-load losses are calculated and copper losses are measured.
(b) no-load losses are measured and the copper losses are calculated.
(c) both the no-load losses and the copper losses are calculated.
(d) both the no-load losses and the copper losses are measured.

150. In Hopkinson's efficiency test for dc machines
(a) both the machines are mechanically coupled.
(b) both the machines are operated at full load.
(c) one machine acts as a generator and another as a motor.
(d) the armature current of the machine operating as a motor is more than that of the machine operating as a generator.
(e) the field current of the machine operating as a generator is more than that of the machine operating as a motor.
(f) all of the above.

151. Which of the following tests can be conducted on other than shunt machines ?
(a) Swinburne's test. (b) Retardation test.
(c) Field's test. (d) Back-to-back test.
[U.P.S.C. I.E.S. Elec. Engg-II, 1992]

152. To conduct load test on a dc shunt motor, it is coupled to a generator which is identical to the motor. The field of the generator is also connected to the same supply source as the motor. The armature of generator is connected to a load resistance. The armature resistance is 0.2 pu. Armature reaction and mechanical losses can be neglected. With rated voltage across the motor, the load resistance across the generator is adjusted to obtain rated armature current in both motor and generator. The pu value of this load resistance is
(a) 1.0 (b) 0.98 (c) 0.96 (d) 0.94
[GATE E.E. 2003]

153. Growler is used to detect
(a) condition of insulation used.
(b) working condition of bearing.
(c) working condition of bearing.
(d) none of the above.

Answers: 138. (b), **139.** (a), **140.** (d), **141.** (d), **142.** (a), **143.** (a), **144.** (a), **145.** (b), **146.** (b), **147.** (d), **148.** (a), **149.** (b), **150.** (f), **151.** (c), **152.** (c), **153.** (a),

154. The sparking at brushes may occur due to
(a) trouble in brushes. (b) trouble in commutator.
(c) trouble in armature. (d) excessive load.
(e) any of the above causes.

155. The basic function of a transformer is to change
(a) the level of the voltage.
(b) the power level. (c) the power factor.
(d) the frequency.

156. The flux created by the current flowing through the primary winding induces emf in
(a) primary winding only.
(b) secondary winding only.
(c) transformer core only.
(d) both primary and secondary windings.

157. The iron core in a transformer provides a _______ path to the main flux.
(a) low reluctance (b) high reluctance
(c) low resistance (d) high conductivity

158. For an ideal transformer the windings should have
(a) maximum resistance on primary side and least resistance on secondary side.
(b) least resistance on primary side and maximum resistance on secondary side.
(c) equal resistance on primary and secondary sides.
(d) no ohmic resistance on either side.
[U.P.S.C. I.E.S. E.E.-II, 1992]

159. For an ideal transformer shown in the given figure

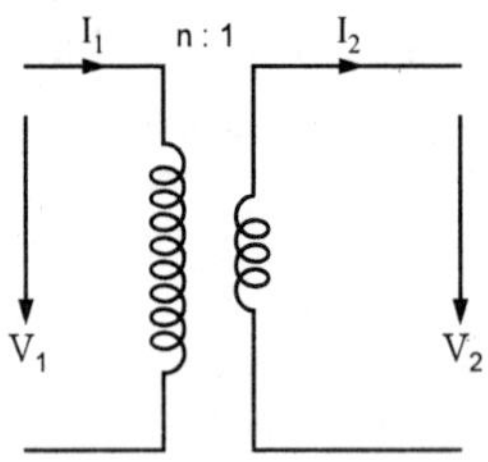

(a) $V_1 = nV_2, I_2 = -n I_1$ (b) $V_2 = nV_1, I_2 = -n I_1$
(c) $V_1 = nV_2, I_2 = \frac{1}{n} I_2$ (d) $V_1 = nV_2, I_2 = -\frac{1}{n} I_1$
[U.P.S.C. I.E.S. E.E.-I, 1995]

160. A single-phase air core transformer, fed from a rated sinusoidal supply, is operating at no load. The steady state magnetizing current drawn by the transformer from the supply will have the waveform

(a)
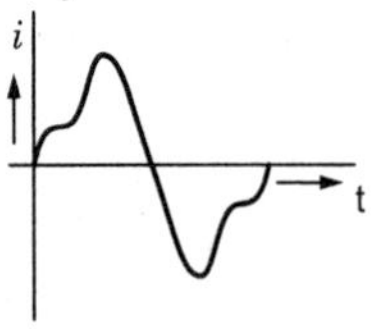

(b)
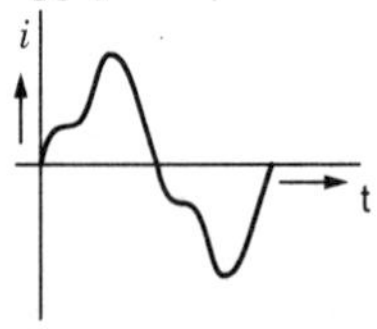

(c)
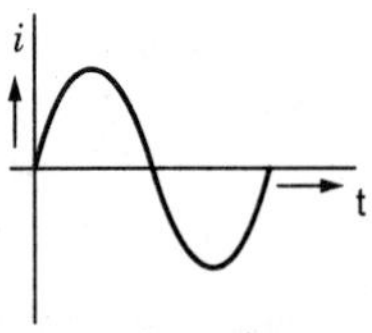

(d)
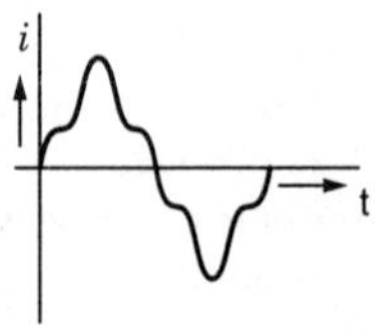

[GATE E.E., 2011]

161. If the applied voltage to a transformer primary is increased by keeping the V/*f* ratio fixed, then the magnetizing current and the core loss will, respectively:
(a) decrease and remain the same.
(b) increase and decrease. (c) both remain the same.
(d) remain the same and increase. **[U.P.S.C. I.E.S. E.E.-II, 2006]**

162. If the applied voltage of a certain transformer is increased by 50% and the frequency is reduced to 50% (assuming that the magnetic circuit remains unsaturated), the maximum core flux density will
(a) change to three times the original value.
(b) change to 1.5 times the original value.
(c) change to 0.5 times the original value.
(d) remain the same as the original value.
[U.P.S.C. I.E.S. E.E.-II, 1997]

163. The applied voltage of a certain transformer is increased by 75%; while the frequency of the applied voltage is reduced by 25%. The maximum core flux density will:
(a) increase by seven times. (b) increase by three times.
(c) reduce to one quarter. (d) remain the same.
[U.P.S.C. I.E.S. E.E.-II, 2011]

164. If the supply frequency of a transformer is increased, keeping the supply voltage constant, then
(a) both magnetising component and core loss component of current will decrease.
(b) both magnetising component and core loss component of current will increase.
(c) magnetising component of current will increase while core loss component of current will decrease.
(d) magnetising component of current will decrease while core loss component of current will increase.
[U.P.S.C. I.E.S. E.E.-II, 1996]

165. A transformer, designed for a supply frequency of 50 Hz is supplied with 60 Hz system of the same voltage,
(a) the eddy current and hysteresis loss will increase.
(b) eddy current loss will increase and hysteresis loss will increase.
(c) eddy current loss will not change but hysteresis loss will decrease.
(d) eddy current loss will be same but hysteresis loss will increase.
[W.B. Univ. of Technology, 2013]

166. For a single phase, two winding transformer, the supply frequency and voltage are both increased by 10%. The percentage changes in the hysteresis loss and eddy current loss, respectively, are
(a) 10 and 21 (b) −10 and 21
(c) 21 and 10 (d) −21 and 10 **[GATE 2014]**

167. The core loss of a single phase, 230/115 V, 50 Hz power transformer is measured from 230 V side by feeding the primary (230 V side) from a variable voltage variable frequency source while keeping the secondary open circuited. The core loss is measured to be 1050 W for 230 V, 50 Hz input. The core loss is again measured to be 500 W for 138 V, 30 Hz input. The hysteresis and eddy current losses of the transformer for 230 V, 50 Hz input are respectively,
(a) 508 W and 542 W. (b) 468 W and 582 W.
(c) 498 W and 552 W. (d) 488 W and 562 W.
[GATE 2014]

168. A 50 Hz transformer having equal hysteresis and eddy current losses at rated excitation is operated at 45 Hz at 90% of its rated voltage. Compared to rated operating point, the core losses under this condition.
(a) reduce by 10%. (b) reduce by 19%.
(c) reduce by 14.5%. (d) remain unchanged.
[U.P.S.C. I.E.S. E.E.-II, 1998; GATE E.E. 1998]

169. Can a 50 Hz transformer be used for 25 Hz, if the input voltage is maintained constant at the rated value corresponding to 50 Hz ?
(a) Yes, since the voltage is constant current level will not change.
(b) No, flux will be doubled which will drive the core to excessive saturation.

Answers: 154. (b), **155.** (a), **156.** (d), **157.** (a), **158.** (d), **159.** (a), **160.** (c), **161.** (d), **162.** (a), **163.** (b), **164.** (a), **165.** (c), **166.** (a), **167.** (a), **168.** (c), **169.** (b),

(c) No, owing to decreased resistance of transformer, input current will be doubled at the load.
(d) Yes, at constant voltage, insulation will not be overstressed. **[U.P.S.C. I.E.S. E.E.-II, 1993]**

170. In a transformer core, third and fifth harmonic components of fluxes are respectively 10% and 4% of the fundamental flux. The third and fifth harmonic induced emfs in the winding, in terms of the fundamental induced emfs are respectively
(a) 30%, 20% (b) 10%, 12%
(c) 50%, 20% (d) 50%, 12%
[U.P.S.C. I.E.S. E.E.-II, 2001]

171. If a sinusoidal voltage source is connected to a power transformer, its no-load current would be
(a) sinusoidal and lagging the voltage by 90º.
(b) sinusoidal and lagging the voltage by less than 90º.
(c) rich in third harmonic and its fundamental would lag the voltage by 90º.
(d) rich in third harmonic and its fundamental would lag the voltage by less than 90º. **[U.P.S.C. I.E.S. E.E.-II, 1999]**

172. A single-phase transformer has a turns ratio of 1 : 2 and is connected to a purely resistive load as shown in the figure. The magnetizing current drawn is 1 A, and the secondary currents is 1 A. If core losses and leakage reactances are neglected, the primary current is

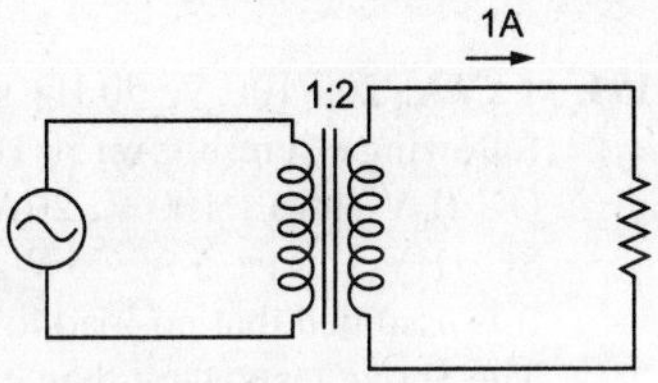

(a) 1.41 A (b) 2 A (c) 2.24 A (d) 3 A
[GATE E.E., 2010]

173. At which condition of the transformer the equivalent circuit will be as shown in the figure?

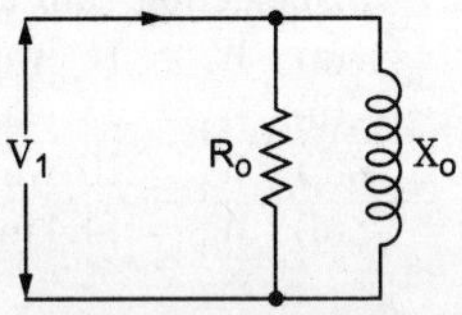

(a) Under short circuit.
(b) Under rated load.
(c) Under open circuit.
(d) Under load and no load.
[U.P.S.C. I.E.S. E.E.-II, 2009]

174. At which condition of the transformer the equivalent circuit will be as shown in the figure?

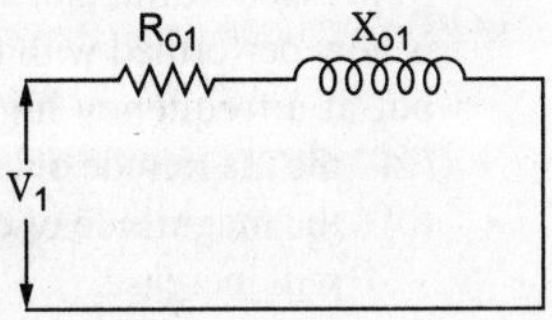

(a) Under short circuit. (b) Under open circuit.
(c) Under no load.
(d) Under rated load. **[U.P.S.C. I.E.S. E.E.-II, 2009]**

175. The core flux of a practical transformer with a resistance load
(a) is strictly constant with load changes.
(b) increases linearly with load.
(c) increases as the square root of the load.
(d) decreases with increase of load. **[GATE E.E. 2001]**

176. Figure shows an ideal single-phase transformer. The primary and secondary coils are wound on the core as shown. Turn-ratio $N_1/N_2 = 2$. The correct phasors of voltages E_1, E_2 currents I_1, I_2 and core flux ϕ are as shown in

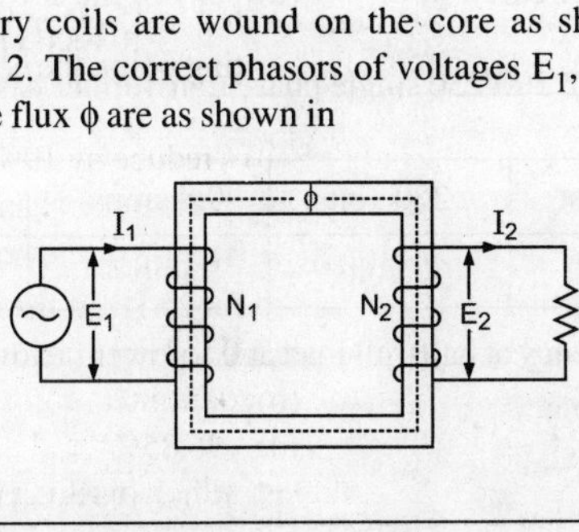

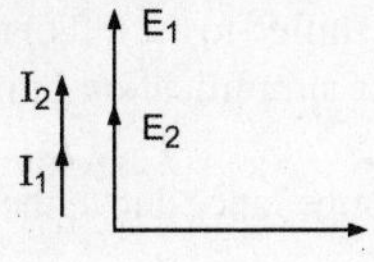

Fig. A

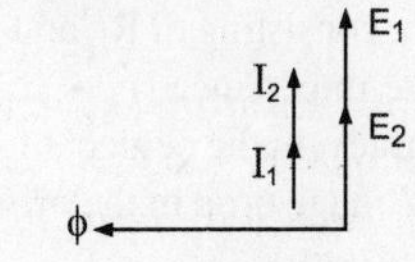

Fig. B

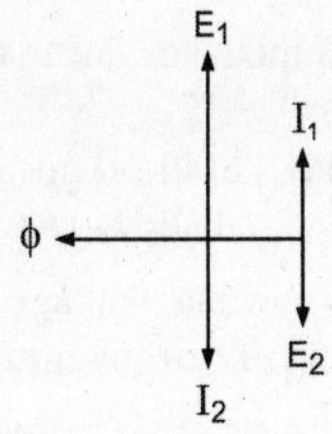

Fig. C

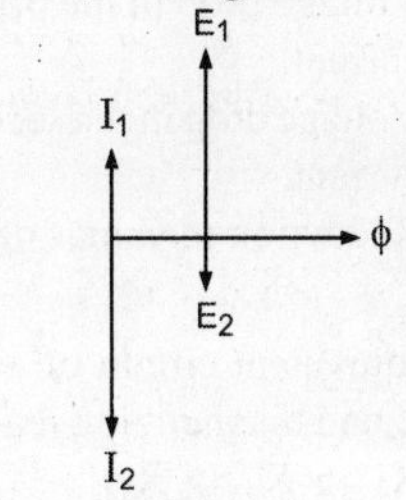

Fig. D

(a) Fig. *A*. (b) Fig. *B*.
(c) Fig. *C*. (d) Fig. *D*. **[GATE E.E. 2003]**

177. The inductive reactance of a transformer depends on
(a) electromotive force. (b) magnetomotive force.
(c) magnetic flux. (d) leakage flux.
[U.P.S.C. I.E.S. E.E.-II 1992]

178. Consider the circuit shown in the given figure. For maximum power transfer to the load, the primary to secondary turn-ratio must be

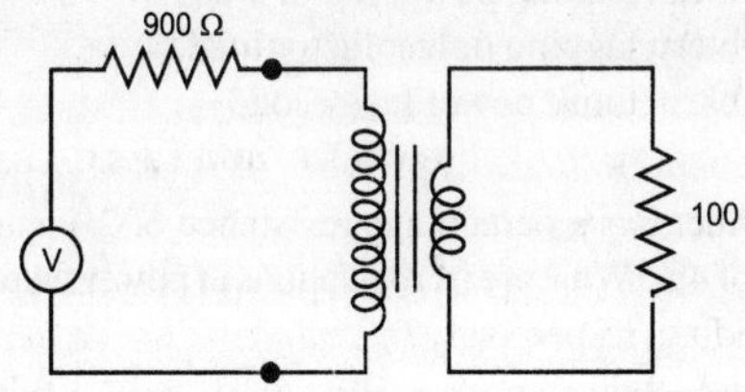

(a) 9 : 1. (b) 3 : 1 (c) 1 : 3. (d) 1 : 9.
[U.P.S.C. I.E.S. E.E-I 1995]

179. For the purpose of analysis exact equivalent circuit of a transformer is usually replaced by an approximate equivalent circuit. In doing so errors introduced due to winding copper loss and core loss are of different nature. Due to this, the analysis by approximate equivalent circuit gives fairly satisfactory results. Under the circumstances, which one of the following statements is correct in respect of losses referred to approx. equivalent circuit as compared to exact equivalent circuit ?
(a) This accounts for somewhat lesser primary winding copper loss and less core loss.
(b) This accounts for somewhat lesser primary winding copper loss and more core loss.
(c) This accounts for somewhat greater secondary winding copper loss and less core loss.
(d) This accounts for somewhat lesser secondary winding copper loss and core loss. **[U.P.S.C. I.E.S. E. E.-II 1995]**

180. The exact equivalent circuit of a two-winding transformer is given in the figure. For affecting simplification, the parallel magnetising

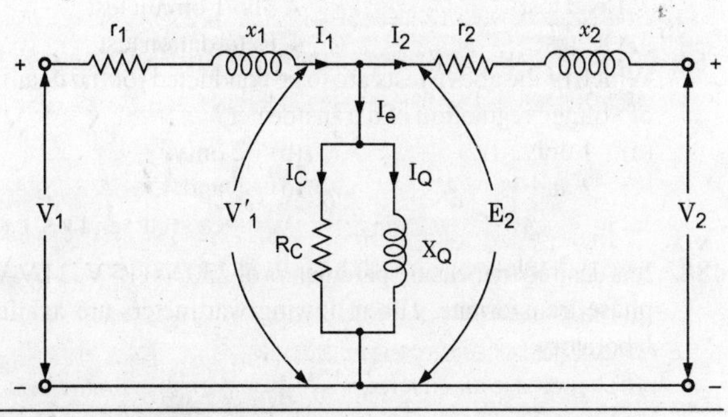

Answers: 170. (a), **171.** (d), **172.** (c), **173.** (c), **174.** (a), **175.** (a), **176.** (d), **177.** (d), **178.** (b), **179.** (b), **180.** (b),

branch, consisting of R_C and X_Q is shifted to the left of the primary leakage impedance $(r_1 + jx_1)$. This simplification introduces the inaccuracy, in the neglect of
(a) Voltage drop in the primary impedance due to the secondary current.
(b) Voltage drop in the primary impedance due to the exciting current.
(c) Voltage drop in the secondary impedance due to the exciting current.
(d) Reduction in values of R_C and X_Q of the exciting current.
[U.P.S.C. I.E.S. E.E. II, 2005]

181. The equivalent circuit of a transformer has leakage reactances X_1, X_2', and magnetizing reactance X_M. Their magnitudes satisfy
(a) $X_1 >> X_2' >> X_M$ (b) $X_1 << X_2' << X_M$
(c) $X_1 \approx X_2' >> X_M$ (d) $X_1 \approx X_2' << X_M$
[GATE E.E. 2005; U.P.S.C. I.E.S. E.E.-II, 2006]

182. Full-load voltage regulation of a power transformer is zero when power factor of the load is near
(a) unity and leading. (b) zero and leading.
(c) zero and lagging. (d) unity and lagging.
[U.P.S.C. I.E.S. E.E.-II, 1999]

183. In a transformer, zero voltage regulation at full load is:
(a) not possible.
(b) possible at leading power factor load.
(c) possible at lagging power factor load.
(d) possible at unity power factor load.
[GATE E.E. 2007; U.P.S.C. I.E.S. E.E.-II, 2011]

184. A transformer has a percentage resistance of 2% and percentage reactance of 4%. What are its regulations at power factor 0.8 lagging and 0.8 leading, respectively?
(a) 4% and – 0.8% (b) 3.2% and – 1.6%
(c) 1.6% and – 3.2% (d) 4.8% and – 0.6%
[U.P.S.C. I.E.S. E.E. II, 2008]

185. The losses in a transformer are :
I Copper loss
II Eddy current loss
III Hysteresis loss.
The constant power loss of a transformer is given by
(a) I only. (b) I and II only.
(c) II and III only. (d) I, II and III.
[U.P.S.C. I.E.S. E.E.-II, 1992]

186. In transformers, which of the following statements is valid?
(a) In an open-circuit test, copper losses are obtained while in short-circuit test, core losses are obtained.
(b) In an open-circuit test, current is drawn at high power factor.
(c) In a short-circuit, current is drawn at zero power factor.
(d) In an open-circuit test, current is drawn at low power factor.
[GATE, E.E. 2006]

187. Consider the following tests:
1. Load test 2. Short-circuit test
3. OC test 4. Retardation test
Which of the above tests are to be conducted for the determination of voltage regulation of a transformer?
(a) 1 only. (b) 2 only.
(c) 2 and 3. (d) 3 and 4.
[U.P.S.C. I.E.S. E.E.-II, 2009]

188. It is desired to measure parameters of 230 V/115 V, 2 kVA, single-phase transformer. The following wattmeters are available in a laboratory:
W_1 : 250 V, 10 A, Low Power Factor.
W_2 : 250 V, 5 A, Low Power Factor.
W_3 : 150 V, 10 A, High Power Factor.
W_4 : 150 V, 5 A, High Power Factor.
The wattmeters used in open-circuit test and short-circuit test of the transformer will respectively be
(a) W_1 and W_2 (b) W_2 and W_4
(c) W_1 and W_4 (d) W_2 and W_3 [GATE E.E. 2008]

189. In a 100 kVA, 1100/220 V, 50 Hz single phase transformer with 2,000 turns on high-voltage side, the open-circuit test result gives 220 V, 91 A, 5 kW on low-voltage side. The core-loss component of current is, approximately
(a) 9.1 A (b) 22.7 A (c) 45.0 A (d) 91 A
[U.P.S.C. I.E.S. E.E.-II 2002]

190. While performing the open-circuit and short-circuit tests on a transformer to determine parameters the status of the low voltage (LV) and high voltage (HV) windings will be such that
(a) in OC, LV is open and in SC HV is shorted.
(b) in OC, HV is open and in SC LV is shorted.
(c) in OC, LV is open and in SC LV is shorted.
(d) In OC, HV is open and in SC HV is shorted.
[U.P.S.C. I.E.S. E.E.-II, 1996]

191. 1 kVA, 200/100 V, 50 Hz single phase transformer gave the following test results on 50 Hz:
OC (LV side) : 100 V, 20 Watts.
SC (HV side) : 5 A, 25 Watts.
It is assumed that no-load loss components are equally divided. The above tests were then conducted on the same transformer at 40 Hz. Test results were:
OC (HV) : 160 V, W_1 Watts.
SC (LV) : 10 A, W_2 Watts.
Neglecting skin effect, W_1 and W_2 will be
(a) W_1 = 16 watts, W_2 = 25 watts.
(b) W_1 = 25 watts, W_2 = 31.25 watts.
(c) W_1 = 20 watts, W_2 = 20 watts.
(d) W_1 = 14.4 watts, W_2 = 25 watts.
[U.P.S.C. I.E.S. E.E.-II, 2003]

192. Short-circuit test is performed on a transformer with a certain impressed voltage at rated frequency. If the short-circuit test is now performed with the same magnitude of impressed voltage, but at a frequency higher than the rated frequency, then
(a) the magnitude of current and power factor will both increase.
(b) the magnitude of current will decrease but the power factor will increase.
(c) the magnitude of current will increase but power factor will decrease.
(d) the magnitude of current as well as the power factor will decrease.
[U.P.S.C. I.E.S. E.E.-II, 1993]

193. The efficiency of two identical transformers under load conditions can be determined by
(a) back to back test. (b) open-circuit test.
(c) short-circuit test. (d) any of the above.
[U.P.S.C I.E.S. E.E.-II, 1992]

194. A 10 kVA, 2,500/250 single phase transformer has the following results:

O.C. Test	250 volts	0.8 amps	50 W
S.C. Test	60 volts	3 amps	45 W

Then efficiency at half-full load at 0.8 power factor will be:
(a) 98.49% (b) 97.68%
(c) 98.28% (d) 96.85%
[U.P.S.C. I.E.S. E.E.-II, 2011]

Answers: **181.** (d), **182.** (a), **183.** (b), **184.** (a), **185.** (c), **186.** (d), **187.** (c), **188.** (d), **189.** (b), **190.** (b), **191.** (d), **192.** (d), **193.** (a), **194.** (c),

195. The efficiency of a 100 kVA transformer is 0.98 at full as well as half load. For this transformer at full-load the copper loss
(a) is less than core less. (b) is equal to core loss.
(c) is more than core loss. (d) none of the above.
[GATE E.E. 1998]

196. If P_c and P_{sc} represent core and full-load ohmic losses respectively, the maximum kVA delivered to load corresponding to maximum efficiency is equal to rated kVA multiplied by
(a) $\frac{P_c}{P_{sc}}$ (b) $\sqrt{\frac{P_c}{P_{sc}}}$ (c) $\left(\frac{P_c}{P_{sc}}\right)^2$ (d) $\left(\frac{P_{sc}}{P_c}\right)^2$
[U.P.S.C. I.E.S. E.E.-II, 2001]

197. A single-phase transformer has no-load loss of 64 W, as obtained from an open-circuit test. When a short-circuit test is performed on it with 90% of the rated currents flowing in its both LV and HV windings, the measured loss is 81 W. The transformer has maximum efficiency when operated at
(a) 50% of the rated current (b) 64% of the rated current
(c) 80% of the rated current (d) 88% of the rated current
[GATE 2013]

198. When a short-circuit test on a transformer is performed at 25 V, 50 Hz, the drawn current 'I_1' is at a lagging pf angle of ϕ_1. If the test is performed by 25 V, 25 Hz and the drawn current 'I_2' is at a lagging power factor angle of ϕ_2 then
(a) $I_2 > I_1$ and $\phi_2 < \phi_1$. (b) $I_2 < I_1$ and $\phi_2 < \phi_1$.
(c) $I_2 > I_1$ and $\phi_2 > \phi_1$. (d) $I_2 < I_1$ and $\phi_2 > \phi_1$.
[U.P.S.C. I.E.S. E.E.-II, 1999]

199. A single phase transformer is to be switched to the supply to have minimum inrush current. The switch should be closed at
(a) maximum supply voltage.
(b) zero supply voltage.
(c) $\frac{1}{\sqrt{2}}$ maximum supply voltage.
(d) $\frac{1}{2}$ maximum supply voltage. **[GATE E.E. 2001]**

200. The desirable properties of transformer core material are
(a) low permeability and low hysteresis loss.
(b) high permeability and high hysteresis loss.
(c) high permeability and low hysteresis loss.
(d) low permeability and high hysteresis loss.
[U.P.S.C. I.E.S. E.E.-II, 1992]

201. The use of higher flux density in the transformer design
(a) reduces the weight per kVA.
(b) increases the weight per kVA.
(c) has no relation with the weight of transformer.
(d) increases the weight per kW. **[U.P.S.C. I.E.S. E.E.-II, 2003]**

202. What does the use of higher flux density value in a transformer design lead to ?
(a) Increase in weight per kVA.
(b) Decrease in weight per kVA.
(c) Reduced iron losses.
(d) Reduced copper losses. **[U.P.S.C. I.E.S. E.E.-II, 2007]**

203. What is the core loss in a high frequency ferrite core transformer used in SMPS power supply?
(a) 10% of rated power. (b) 5% of rated power.
(c) 2% of rated power. (d) 1% of rated power.
[U.P.S.C. I.E.S. E.E.-II, 2009]

204. For a specified input voltage and frequency, if the equivalent radius of the core of a transformer is reduced by half, the factor by which the number of turns in the primary should change to maintain the same no-load current is
(a) 1/4 (b) 1/2 (c) 2 (d) 4 **[GATE 2014]**

205. If the iron core of a transformer is replaced by an air core, then the hysteresis losses in the transformer will
(a) increase. (b) decrease.
(c) remain unchanged. (d) become zero.
[U.P.S.C. I.E.S. E.E.-II, 2006]

206. Which of the following will improve the mutual coupling between primary and secondary circuits ?
(a) Transformer oil of high breakdown voltage.
(b) High reluctance magnetic core.
(c) Winding material of high resistivity.
(d) Low reluctance magnetic core. **[U.P.S.C. I.E.S. E.E.-II, 1992]**

207. When are eddy-current losses in a transformer reduced ?
(a) If laminations are thick.
(b) If the number of turns in primary winding is reduced.
(c) If the number of turns in secondary winding is reduced.
(d) If laminations are thin. **[U.P.S.C. I.E.S. E.E.-II, 2008]**

208. The joints in the transformer core laminations are staggered so as to
(a) avoid continuous gap causing increase in magnetizing current.
(b) increase the mechanical strength of the assembled core.
(c) avoid undue humming noise.
(d) facilitate assembly of core after putting preformed coils on the core.
(e) all of the above.

209. For CRGOS mitred overlap is preferred for the core-yoke joints as it
(a) reduces the magnetizing current and also the core losses.
(b) improves the mechanical strength.
(c) reduces magnetostriction.
(d) makes better use of core space.

210. In a large capacity transformer, 5% of the turns at the end of hv winding are provided with extra insulation so as to provide protection
(a) against corona. (b) against lightening.
(c) due to surges occurring during switching operations.
(d) all of the above.

211. When a transformer winding suffers a short circuit, the adjoining turns of the same winding experience
(a) an attractive force. (b) a repulsive force.
(c) no force. **[GATE E.E. 1994]**

212. The function of oil in a transformer is
(a) to provide insulation and cooling.
(b) to provide protection against lightning.
(c) to provide protection against short-circuit.
(d) to provide lubrication.
[GATE E.E. 1996; U.P.S.C. I.E.S. E.E.-II, 2003]

213. Consider the following statements relating to the constructional features of a large power transformer:
1. The conservator is used to maintain the level of oil in the transformer tank.
2. The bushing is used to protect transformer insulation against lightning voltages.
3. The Buchholz relay is an overcurrent relay.
4. Silica gel is used to absorb moisture.

Which of these statements are correct?
(a) 1, 2, 3 and 4. (b) 2 and 3.
(c) 1 and 4. (d) 1, 2 and 4.
[U.P.S.C. I.E.S. E.E.-II, 2003]

214. In a large power transformer, a "conservator" drum is provided above the completely oil filled transformer tank and connected to it by a short pipe. The conservator drum is linked to external air

Answers: **195.** (c), **196.** (b), **197.** (c), **198.** (a), **199.** (a), **200.** (c), **201.** (a), **202.** (b), **203.** (d), **204.** (d), **205.** (d), **206.** (d), **207.** (d), **208.** (e), **209.** (a), **210.** (c), **211.** (a), **212.** (a), **213.** (c), **214.** (c), through a breather. What is the purpose of providing the conservator?

(a) To store reserve oil to make up oil losses due leakage.
(b) To prevent explosion due to rise in oil pressure inside the tank during a fault.
(c) To accommodate change in oil-level during the "load - cycle" of the transformer load.
(d) To exert additional pressure by the conservator-oil on the oil inside the main tank to prevent disintegration of oil at high temperature. [U.P.S.C. I.E.S. E.E.-II, 2007]

215. Consider the following statements regarding transformers:

1. The function of the magnetizing component of no-load current is to sustain the alternating flux in the core.
2. Short-circuit test is performed to find core losses only.
3. The function of the breather in transformer is to arrest flow of moisture when outside air enters the transformer.

Which of the statements given above are correct?

(a) 1 and 2 only. (b) 1 and 3 only.
(c) 2 and 3 only. (d) 1, 2 and 3.
[U.P.S.C. I.E.S. E.E.-II, 2007]

216. Sludge formation in transformer oil is due to which one of the following ?

(a) Ingress of dust particles and moisture in the oil.
(b) Appearance of small fragments of paper, varnish, cotton and other organic materials in the oil.
(c) Chemical reaction of transformer oil with the insulating materials.
(d) Oxidation of transformer oil. [U.P.S.C. I.E.S. E.E.-II, 2008]

217. A Buchholz relay is used for

(a) protection of a transformer against all internal faults.
(b) protection of a transformer against external faults.
(c) protection of a transformer against both internal and external faults.
(d) protection of induction motors. [U.P.S.C. I.E.S. E.E.-II, 2005]

218. Distribution transformers have core losses

(a) > copper loss. (b) < copper losses.
(c) = copper loss. (d) $= \frac{1}{2}$ (copper losses).

219. The low-voltage winding of a core type transformer is subdivided into two equal halves, each of half the original width of the single winding with the high-voltage winding in between (instead of having the usual construction of low voltage winding adjacent to the core and surrounded by the high-voltage winding). Such an interlacing of coils would make the combined primary and secondary leakage reactance (in terms of the primary) nearly

(a) twice. (b) equal. (c) half. (d) one-fourth.
[U.P.S.C. I.E.S. E.E.-II, 1997]

220. Transformers are rated in kVA instead of kW because

(a) load pf is often not known.
(b) kVA is fixed where kW depends on load pf.
(c) total transformer loss depends on volt-amperes.
(d) it has become customary. [R.G.T.U. B.E.E. June 2007]

221. The single phase, 50 Hz iron core transformer in the circuit has both the vertical arms of cross-sectional area 20 cm^2 and both the horizontal arms of cross-sectional area 10 cm^2. If the two windings shown were wound instead on opposite horizontal arms, the mutual inductance will

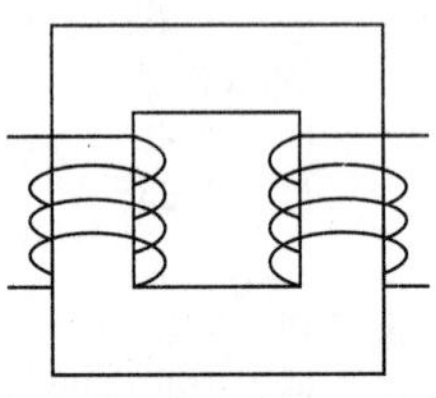

(a) double. (b) remain same.
(c) be halved. (d) become one quarter.
[GATE E.E. 2009]

222. The three limbed non-ideal core shown in the figure has three windings with nominal inductances L each when measured individually with a single phase ac source. The inductance of the windings as connected will be

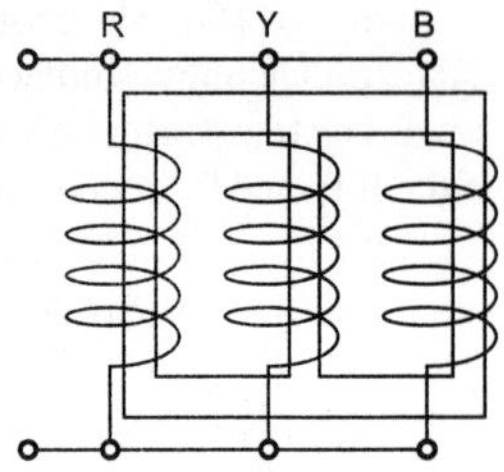

(a) very low. (b) L/3.
(c) 3L. (d) very high. [GATE E.E. 2006]

223. In the figure, transformer T_1 has two secondaries, all three windings having the same number of turns and with polarities as indicated. One secondary is shorted by a 10 Ω resistor R, and the other by a 15 mF capacitor. The switch SW is opened ($t = 0$) when the capacitor is charged to 5V with the left plate as positive. At ($t = 0^+$) the voltage V_p and current I_R are

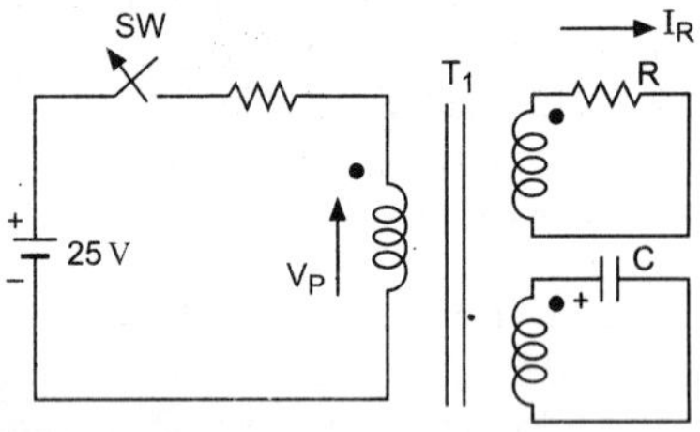

(a) –25 V, 0.0 A
(b) very large voltage, very large current.
(c) 5.0 V, 0.5 A (d) –5.0 V, –0.5 A [GATE E.E. 2007]

224. The core of a two-winding transformer is subjected to a magnetic flux variation as indicated in the figure.

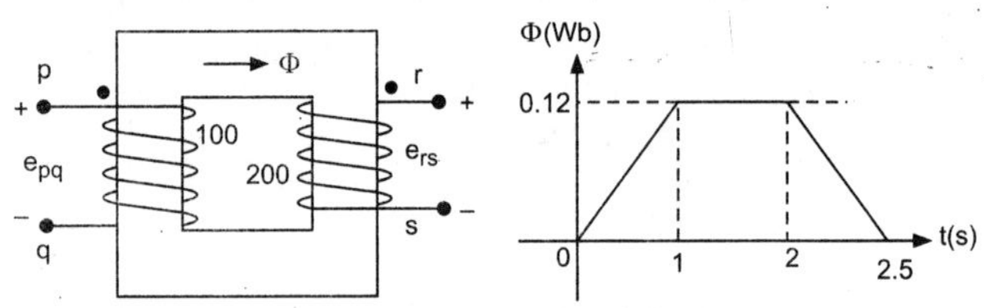

The induced emf (e_{rs}) in the secondary winding as a function of time will be of the form

(a)

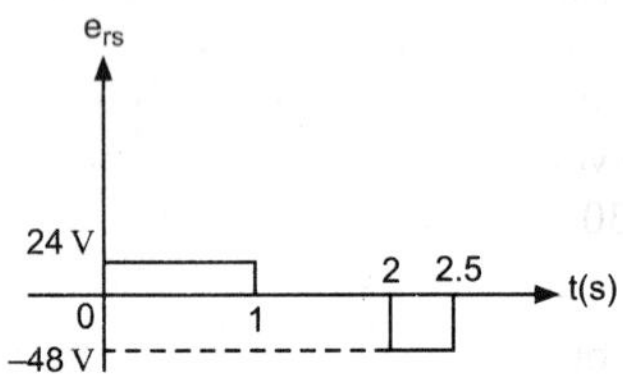

(b)

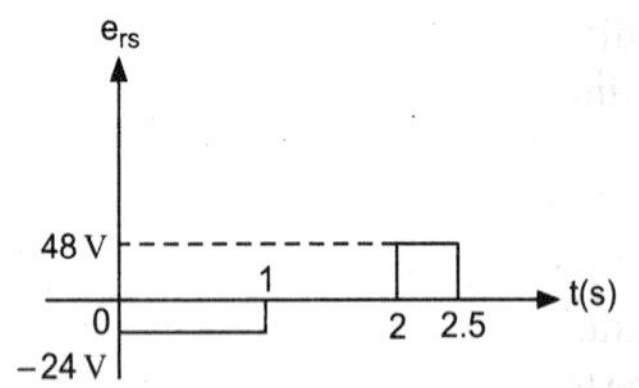

(c)

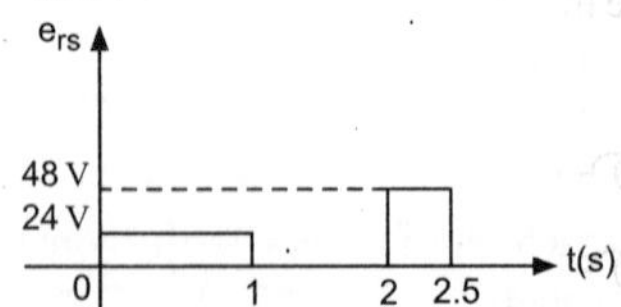

(d)

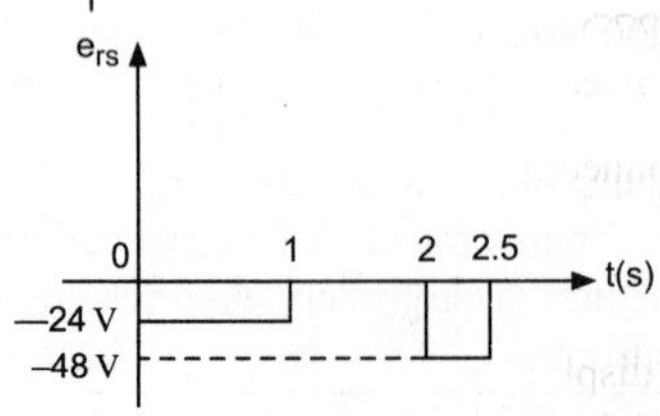

[GATE E.E. 2008]

Answers: 215. (b), **216.** (d), **217.** (a), **218.** (b), **219.** (c), **220.** (c), **221.** (c), **222.** (a), **223.** (d), **224.** (b),

Common Data for Questions 225 and 226

The circuit diagram shows a two-winding, lossless transformer with no leakage flux, excited from a current source, $i(t)$, whose waveform is also shown. The transformer has a magnetizing inductance of $400/\pi$ mH.

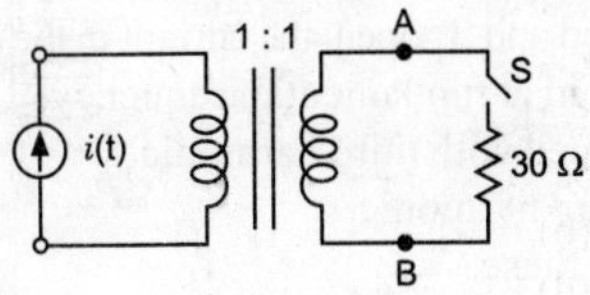

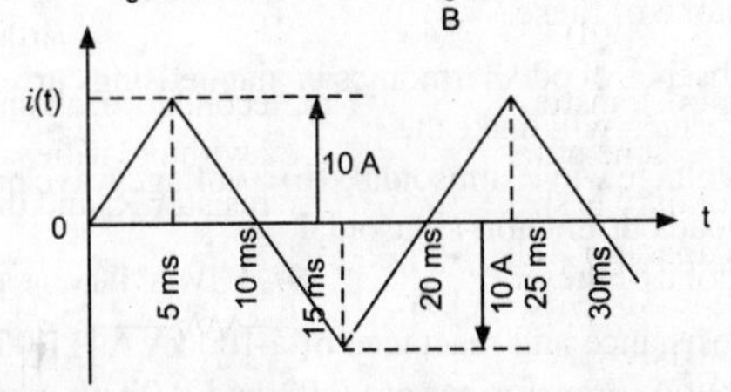

225. The peak voltage across A and B, with S open is

(a) $\frac{400}{\pi V}$ (b) 800 V (c) $\frac{4,000}{\pi V}$ (d) $\frac{800}{\pi V}$

[GATE E.E. 2009]

226. If the waveform of $i(t)$ is changed to $i(t) = 10 \sin(100\,\pi t)$ A, the peak voltage across A and B with S closed is

(a) 400 V (b) 240 V (c) 320 V (d) 160 B

[GATE E.E. 2009]

227. If the waveform of the voltage impressed on the primary of a Y-Δ bank contains 5th harmonics, what are the waveforms of the resultant voltages of the primary and secondary?

	Primary	Secondary
(a)	Peaked	Peaked
(b)	Peaked	Flat-topped
(c)	Flat-topped	Peaked
(d)	Flat-topped	Flat-topped

[U.P.S.C. I.E.S. E.E.-II, 2005]

228. Two 3-limb, 3-phase delta-star connected transformers are supplied from the same source. One of the transformers is Dy1 and the other is Dy11 connection. The phase difference between the corresponding phase voltage of the secondaries would be

(a) 0° (b) 30° (c) 60° (d) 120°

[U.P.S.C. I.E.S. E.E.-II, 1997]

229. Which three-phase connection can be used in a transformer to introduce a phase difference of 30° between its output and corresponding input line voltages ?

(a) Star- Star. (b) Star-Delta.
(c) Delta-Delta. (d) Delta- Zigzag.

[GATE E.E. 2005; U.P.S.C. I.E.S. E.E.-II, 2006; 2011]

230. Three single-phase transformers are connected to form 3-phase transformer bank. The transformers are connected in the following manner:

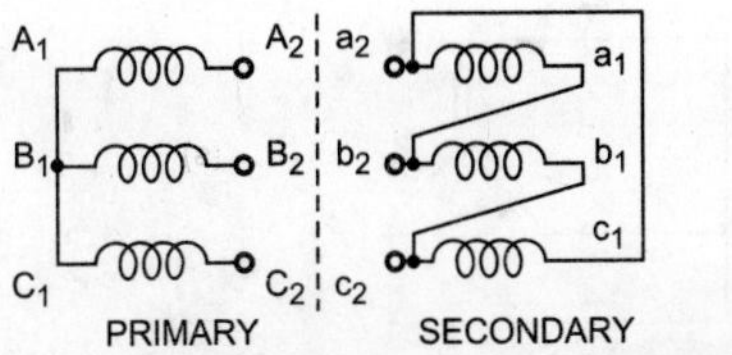

The transformer connection will be represented by

(a) Yd0 (b) Yd1
(c) Yd6 (d) Yd11 **[GATE E.E. 2008]**

231. What is the phase displacement between primary and secondary voltages for a star-delta, 3-phase transformer connection shown

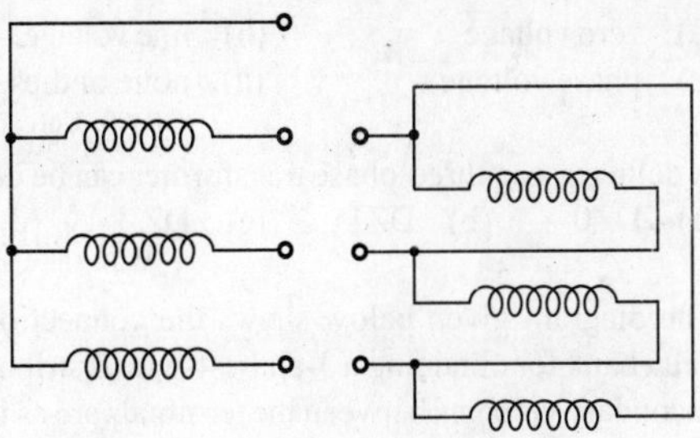

(a) 30° lagging. (b) 30° leading.
(c) 0° (d) 180° **[U.P.S.C. I.E.S. E.E.-II, 2005]**

232. In a 3-phase Δ/Y transformer shown in the figure, the phase displacement of secondary line voltage with corresponding primary line voltage will be

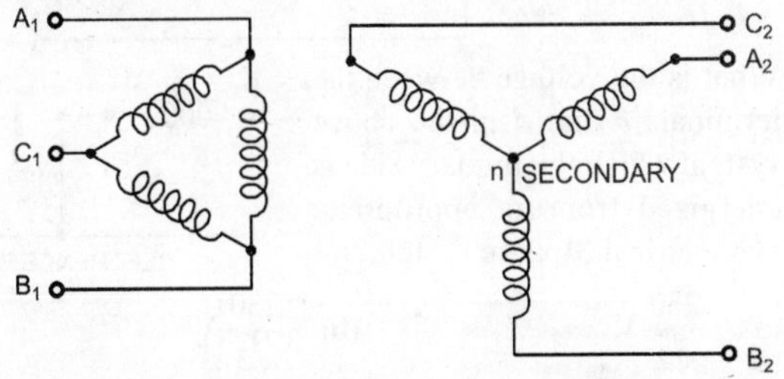

(a) zero. (b) 30° lag. (c) 30° lead. (d) 180°

[U.P.S.C. I.E.S. E.E.-II, 1998]

233. Which of the following connection of transformer will give the highest secondary voltage ?

(a) Delta primary, delta secondary.
(b) Delta primary, star secondary.
(c) Star primary, star secondary.
(d) Star primary, delta secondary. **[U.P.S.C. I.E.S. E.E.-II, 1992]**

234. Three single phase 11,000/220 V transformers are connected to form 3-phase transformer bank. High voltage side is connected in star, and low voltage side is connected in delta. What are the voltage ratings and turn ratio of 3-phase transformer?

(a) 19,052/220 V, 50 (b) 19052/220 V, $50\sqrt{3}$
(c) 11,000/381 V, $50\sqrt{3}$ (d) 11,000/220 V, 50

[U.P.S.C. I.E.S. E.E.-I-II, 2005]

235. A delta/star transformer has a phase-to-phase voltage transformation ratio of K $\left[K = \frac{\text{delta phase voltage}}{\text{star phase voltage}}\right]$. The line-to-line voltage ratio of star/delta connections is given by

(a) $K/\sqrt{3}$ (b) K (c) $K\sqrt{3}$ (d) $\sqrt{3}/K$

[U.P.S.C. I.E.S. E.E.-II, 2002]

236. In the statement, "In a star/delta transformer, there exists a phase difference of "X" between the corresponding phases of primary and secondary "Y". "X" and "Y" stand respectively for

(a) 60° and voltages. (b) 60° and currents.
(c) 30° and voltages. (d) 30° and currents.

[U.P.S.C. I.E.S. E.E.-II, 2000]

237. Two single phase transformers with turn-ratios 1 and 2 respectively are connected to a 3-phase supply on the primary side as shown in the figure. The voltmeter V_2 will read

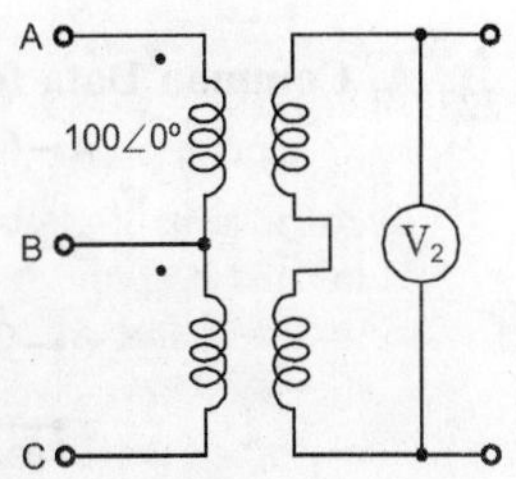

(a) 100 V (b) 173 V
(c) 200 V (d) 265 V

[UP.S.C. I.E.S. E.E. II 1998]

238. In a properly connected delta winding of a 3-phase transformer, a voltmeter connected by opening a node of delta will show

Answers: 225. (d), **226.** (a), **227.** (b), **228.** (c), **229.** (b), **230.** (b), **231.** (a), **232.** (c), **233.** (b), **234.** (a), **235.** (d), **236.** (c), **237.** (b), **238.** (b),

(a) zero voltage (b) line voltage
(c) phase voltage (d) none of these

[W.B. Univ. of Technology, 2011]

239. A delta zig-zag three-phase transformer can be designated as
(a) DZ0 (b) DZ1 (c) DZII (d) none of these.

[W.B. Univ. of Technology, 2011]

240. The diagram given below shows the connection of a four-wire delta bank for obtaining a 3-phase 4-wire distribution system. The secondary voltages between the terminals are as indicated.

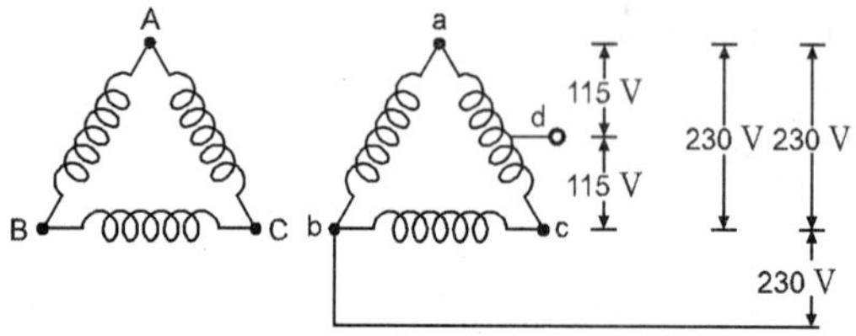

What is the voltage between the terminals *b* and *d* in the above system which the primary side is energized from an appropriate symmetrical 3-phase system?

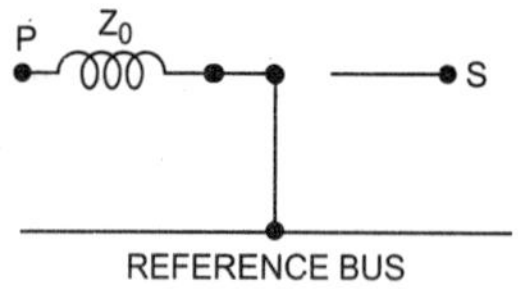

(a) $\frac{230}{\sqrt{2}}$ V (b) $\frac{230}{\sqrt{3}}$ V
(c) $115\times\sqrt{3}$ V (d) $115\times\sqrt{2}$ V

[U.P.S.C. I.E.S. E.E.-II, 2004]

241. A 3-phase transformer having zero-sequence impedance Z_0 has zero-sequence network as represented in the figure. The connections of its windings are

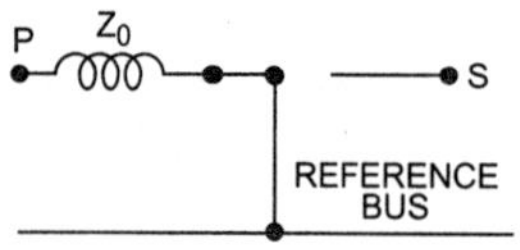

(a) star with isolated neutral-delta.
(b) star with grounded neutral-delta.
(c) delta-star with grounded neutral.
(d) delta-delta.

[U.P.S.C. I.E.S. E.E.-II, 2002]

242.

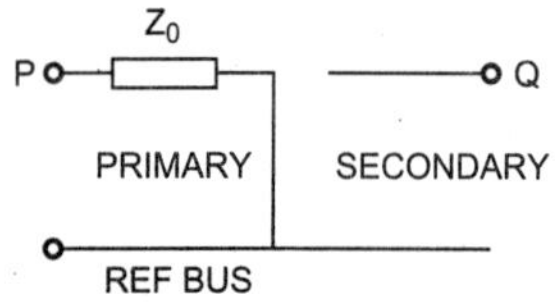

A zero-sequence network of a 3-phase transformer is given above. Z_0 is the zero-sequence impedance of the transformer. What is the type of transformer connections ?

(a)

(b)
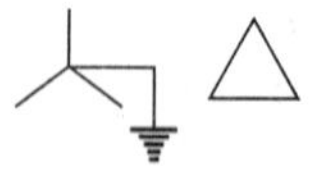

(c)
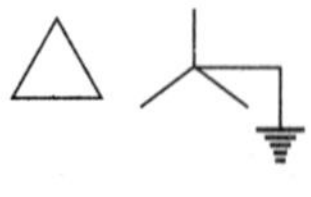
(d)
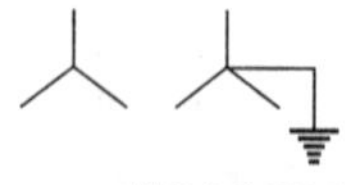

[U.P.S.C. I.E.S. E.E.-II, 2005]

Common Data for Questions 243 and 244:

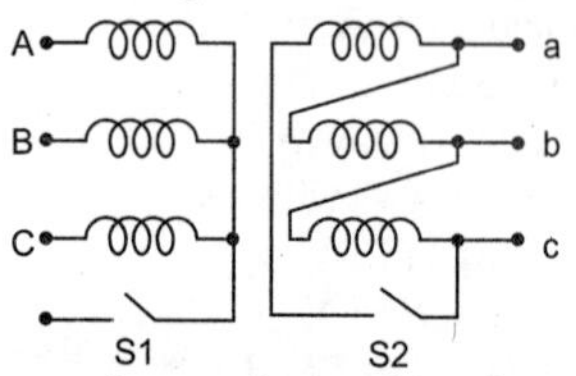

The star-delta transformer shown above is excited on the star side with balanced, 4-wire, 3-phase, sinusoidal voltage supply of rated magnitude. The transformer is under no load condition.

243. With both S_1 and S_2 open, the core flux waveform will be
(a) a sinusoid at fundamental frequency.
(b) flat-topped with third harmonic.
(c) peaky with third harmonic.
(d) none of these. **[GATE E.E. 2009]**

244. With S_2 closed and S_1 open, the current in the delta winding will be
(a) a sinusoid at fundamental frequency.
(b) flat-topped with third harmonic.
(c) only third harmonic.
(d) none of these. **[GATE E.E. 2009]**

245. The absence of odd harmonics in magnetising current of a 3-phase transformer, will make the
(a) voltage wave sinusoidal. (b) voltage wave non-sinusoidal.
(c) load current non-sinusoidal.
(d) none of these. **[W.B. Univ. of Technology, 2010]**

246. The resistance and reactance of a 100 kVA, 11,000/400 V, Δ-Y distribution transformer are 0.02 and 0.07 pu respectively. The phase impedance of the transformer referred to the primary is
(a) $(0.02 + j\,0.07)\ \Omega$ (b) $(0.55 + j\,1.925)\ \Omega$
(c) $(15.125 + j\,52.94)\ \Omega$ (d) $(72.6 + j\,254.1)\ \Omega$

[GATE E.E. 2004]

247. For ΔY transformer, connections and terminal markings are shown in the below figure. If 1 and 2 represents positive and negative sequence, then the correct solution in per unit values for VA_1 and VA_2 is

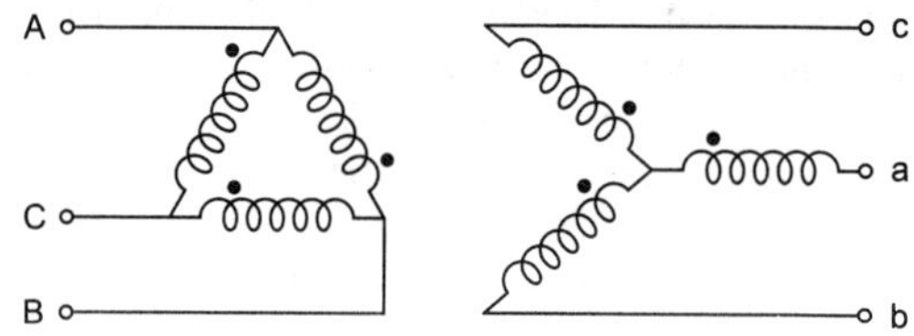

(a) jVa_1 and $-jVa_2$ (b) Va_1 and $-jVa_2$
(c) jVa_1 and $-Va_2$ (d) Va_1 and $-Va_2$

[U.P.S.C. I.E.S. E.E.-II, 2010]

248. A balanced star-connected and purely resistive load is connected at the secondary of a star-delta transformer as shown in the figure. The line-to-line voltage rating of the transformer is 110 V/220 V. Neglecting the non-idealities of the transformer, the impedance 'Z' of the equivalent star-connected load, referred to the primary side of the transformer, is

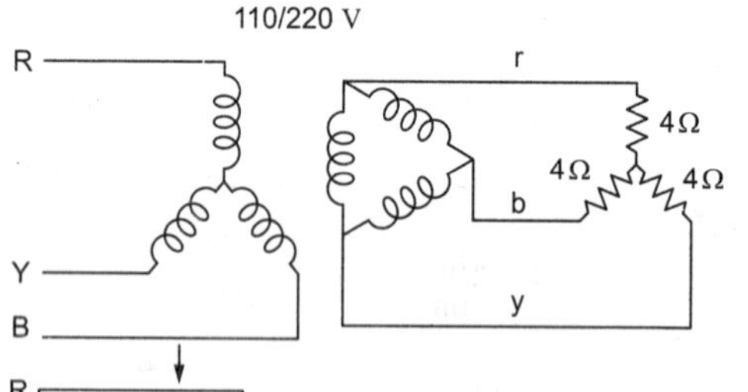

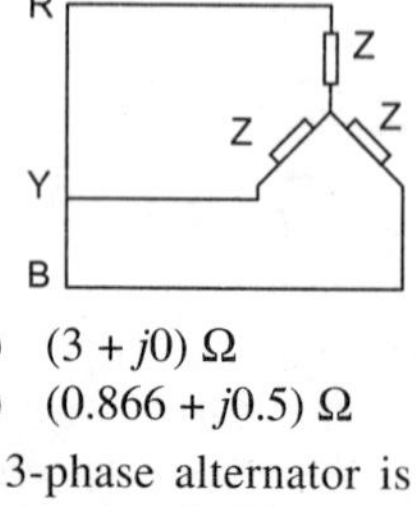

(a) $(3 + j0)\ \Omega$ (b) $(0.866 - j0.5)\ \Omega$
(c) $(0.866 + j0.5)\ \Omega$ (d) $(1 + j0)\ \Omega$ **[GATE E.E., 2010]**

249. A 3-phase alternator is connected to a Dd transformer. The hysteresis and eddy current losses of the transformer are respectively 300 W and 400 W. If the speed of the alternator is reduced by 10%,

Answers: 239. (a), **240.** (c), **241.** (b), **242.** (b), **243.** (b), **244.** (a), **245.** (a), **246.** (d), **247.** (d), **248.** (d), **249.** (b),

then the hysteresis and eddy current losses of the transformer will be respectively
(a) 228 W and 262.44 W. (b) 270 W and 324 W.
(c) 243 W and 360 W. (d) 270 W and 400 W.
[U.P.S.C. I.E.S. E.E.-II, 2000]

250. The windings of a Q kVA, $\frac{V_1}{V_2}$ volts, three phase Dd core type transformer are reconnected to work as a single phase transformer. The power rating and the maximum voltage ratio of the new configuration will be respectively
(a) $\frac{Q}{3}$ kVA and $\frac{V_1}{V_2}$ volts. (b) 3 Q kVA and $\frac{3V_1}{3V_2}$ volts.
(c) 2 Q kVA and $\frac{\sqrt{3}V_1}{\sqrt{3}V_2}$ volts. (d) $\frac{2}{3}$ Q kVA, $\frac{2V_1}{2V_2}$ volts.
[U.P.S.C. I.E.S. E.E.-II 2003; GATE E.E. 1999]

251. When one transformer is removed from a Δ-Δ bank of 30 kVA transformer, the capacity of the resulting 3-phase transformer in V-V connection will be:
(a) 11.5 kVA (b) 17.3 kVA
(c) 20 kVA (d) 25.9 kVA
[U.P.S.C. I.E.S. E.E.-II, 2011]

252. An 11 kV/400V, 1000 kVA, Y/Y transformer is reconnected as Δ/Y with the high voltage side connected in delta. The rating for new connection will be
(a) $\frac{11}{\sqrt{3}}$ kV/400 V, 1000 kVA
(b) 11 kV/400 V, 1732 kVA
(c) $11\sqrt{3}$ kV/400 V, 1000 kVA
(d) 11 kV/400, $\sqrt{3}$ V, $\frac{1000}{\sqrt{3}}$ kVA **[U.P.S.C. I.E.S. E.E.-II, 1993]**

253. Scott connections are used for
(a) single phase to three-phase transformation.
(b) three phase to single-phase transformation.
(c) three phase to two-phase transformation.
(d) three phase to six-phase transformation.
[W.B. Univ. of Technology, 2013]

254. In Scott connection, if the turns ratio of main transformer is K, then the teaser transformer has turns ratio of
(a) $\frac{2K}{\sqrt{3}}$ (b) $\frac{\sqrt{3}K}{2}$ (c) $\frac{K}{\sqrt{3}}$ (d) $\frac{K}{2}$
[U.P.S.C. I.E.S. E.E.-II, 2010]

255. Supply to one terminal of a delta-wye connected three phase core type transformer which is on no load fails. Assuming magnetic circuit symmetry, voltages on the secondary side will be
(a) 230, 230, 115 (b) 230, 115, 115
(c) 345, 115, 115 (d) 345, 0, 345 **[GATE E.E. 1995]**

256. In a transformer fed from a fundamental frequency voltage source, the source of harmonics is the
(a) overload. (b) poor insulation.
(c) iron loss. (d) saturation of core.
[U.P.S.C. I.E.S. E.E.-II, 1997]

257. Zero sequence currents can flow from a line into a transformer bank if the windings are
(a) grounded star/delta (b) delta/star.
(c) star/grounded star. (d) delta/delta.
[U.P.S.C.I.E.S. E.E.-II, 1997]

258. The magnetizing current in a transformer is rich in
(a) 3rd harmonic. (b) 5th harmonic.
(c) 7th harmonic. (d) 13th harmonic.
[GATE E.E. 1998]

259. Tertiary winding is used in transformer connected in
(a) delta/delta. (b) delta/star.
(c) star/star. (d) star/zig-zag.
[W.B. Univ. of Technology, 2011]

260. A 400 V/200 V/200 V, 50 Hz three-winding transformer is connected as shown in the figure. The reading of the voltmeter V will be

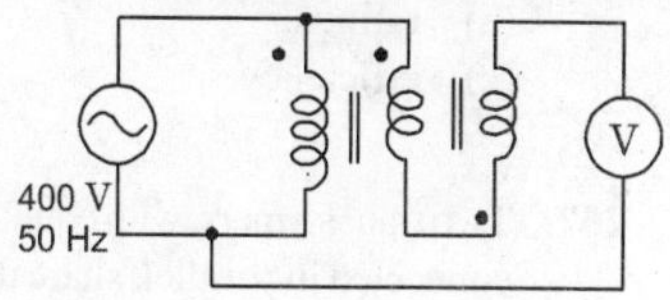

(a) zero V (b) 400 V (c) 600 V (d) 800 V
[GATE E.E. 2000; 2002]

261. Figure shows an ideal three-winding transformer. The three windings 1, 2, 3 of the transformer are wound on the same core as shown. The turn-ratio $N_1 : N_2 : N_3$ is 4 : 2 : 1. A resistor of 10 Ω is connected across winding 2. A capacitor of reactance 2.5 Ω is connected across winding 3. Winding 1 is connected across a 400 V, ac supply. If the supply voltage phasor $V_1 = 400\angle 0°$, the supply current phasor I is given by

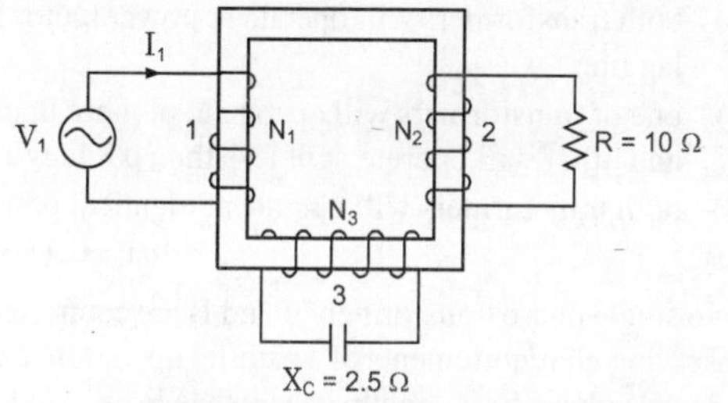

(a) (– 10 + *j* 10) A (b) (– 10 – *j* 10) A
(c) (10 + *j* 10) A (d) (10 – *j* 10) A
[GATE E.E. 2003]

262. Conditions for parallel operation of transformer are:
1. The same voltage ratio.
2. The same per unit impedance.
3. The same polarity.
4. The same phase sequence.
5. The relative phase displacement.
Which are the absolutely essential conditions?
(a) 1, 2, 3, 4 and 5. (b) 2, 3 and 4.
(c) 3, 4 and 5. (d) 1, 2 and 3.
[U.P.S.C. I.E.S. E.E.-II, 2011]

263. Possible three-to-three phase transformer connection for parallel operation is
(a) Δ-Y to Δ-Y (b) Δ-Δ to Δ-Y
(c) Y-Y to Δ-Y (d) Δ-Y to Y-Δ
[U.P.S.C. I.E.S. E.E.-II, 2002]

264. Two 3-phase transformers are to be connected for parallel operation. Which one of the following arrangements is impossible?
(a) Transformer A; primary Y; secondary Y
Transformer B; primary Δ; secondary Δ
(b) Transformer A; primary Δ; secondary Y
Transformer B; primary Δ; secondary Δ
(c) Transformer A; primary Y; secondary Δ
Transformer B: primary Δ; secondary Y
(d) Transformer A; primary Δ; secondary Δ
Transformer B; primary Δ; secondary Δ
[U.P.S.C. I.E.S. E.E.-II, 1995]

Answers: **250.** (d), **251.** (b), **252.** (a), **253.** (c), **254.** (b), **255.** (b), **256.** (d), **257.** (a), **258.** (a), **259.** (c), **260.** (a), **261.** (d), **262.** (c), **263.** (a), **264.** (b),

265. Two transformers when operating in parallel will share the load depending upon which of the following?
(a) Magnetizing current. (b) Leakage reactance.
(c) Per unit impedance. (d) Efficiency.
[U.P.S.C. I.E.S. E.E.-II, 2007]

266. Two transformers operating in parallel will share the load depending upon their
(a) ratings. (b) leakage reactance.
(c) efficiency. (d) per unit impedance.
[U.P.S.C. I.E.S. E.E.-II, 2003]

267. Two transformers, with equal rating and negligible excitation current, connected in parallel, share the load in the ratio of their kVA ratings only, if their pu impedances (based on their own kVA) are
(a) equal.
(b) in the inverse ratio of their ratings.
(c) in the direct ratio of their ratings.
(d) pure reactive. [U.P.S.C. I.E.S. E.E.-II, 2002]

268. Two single phase 100 kVA transformers, each having different leakage impedances are connected in parallel. When a load of 150 kVA at 0.8 pf lagging is applied
(a) both transformers will operate at power factor more than 0.8 lagging.
(b) both transformers will operate at power factor less than 0.8 lagging.
(c) one of transformers will operate at pf more than 0.8 lagging and other will operate at pf less than 0.8 lagging.
(d) both transformers will operate at identical power factors.
[U.P.S.C. I.E.S. E.E.-II, 1992]

269. Two single phase transformers A and B are connected in parallel, observing all requirements of a parallel operation, except that the induced voltage E_a is slightly greater than E_b; Z_{ea} and Z_{eb} being the equivalent impedances of A and B, both referred to the secondary side.

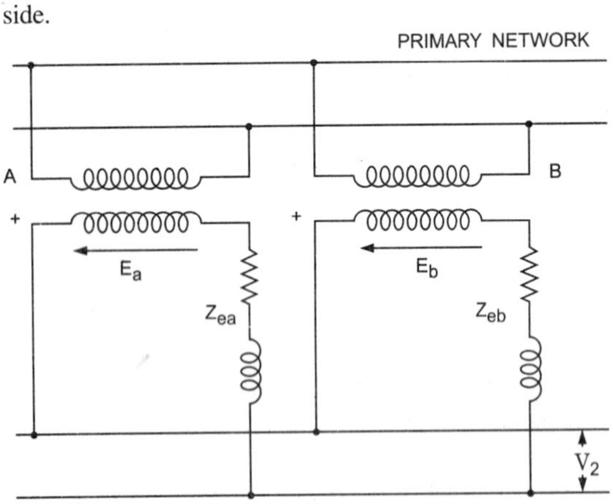

Under this operating condition with the primary bus-bars being energized, a circulating current will flow
(a) only in the secondary windings of A and B.
(b) in both the primary and secondary windings of A and B.
(c) in both the primary and secondary windings of A and B, as well as in the primary side network.
(d) in the primary and secondary windings of A and B, and boost the voltages in the secondary side of both A and B.
[U.P.S.C. I.E.S. E.E.-II, 2005]

270. If per unit impedances of two transformers connected in parallel are not equal which one of the following statements is correct?
(a) The power factor of the two transformers will be different from that of the connected load.
(b) Transformers will get overloaded.
(c) Dead short circuit occurs.
(d) The transformer with higher per unit impedance will share more load. [U.P.S.C. I.E.S. E.E.-II, 2004]

271. In an auto-transformer, power is transferred, through
(a) Conduction process only.
(b) Induction process only.
(c) Both conduction and induction processes.
(d) Mutual coupling. [U.P.S.C. I.E.S. E.E.-II, 2010]

272. A single-phase, 10 kVA, 2000/200 V, 50 Hz transformer is connected to form an auto-transformer as shown in the figure. What are the values of V_1 and I_2, respectively?

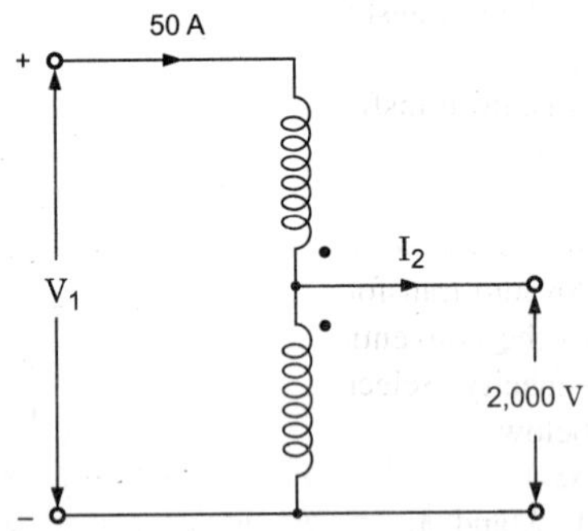

(a) 2,200 V, 55 A. (b) 2200 V, 45 A.
(c) 2000 V, 45 A. (d) 1800 V, 45 A.
[U.P.S.C. I.E.S. E.E.-II, 2005]

273. A two-winding transformer is converted into an auto-transformer. If we apply additive polarity and subtractive polarity for the connections, then the secondary voltage is 2640 V and 2160 V, respectively. What is the ratio of primary to secondary voltage of the original transformer?
(a) 66 : 54. (b) 54 : 66. (c) 10 : 1. (d) 1 : 10.
[U.P.S.C. I.E.S. E.E.-II, 2006]

274. A 50 kVA, 3300/230 V single phase transformer is connected as an auto-transformer shown in figure. The nominal rating of the auto-transformer will be

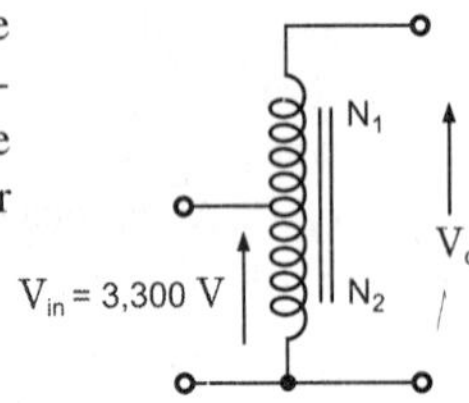

(a) 50.0 kVA (b) 53.5 kVA
(c) 717.4 kVA (d) 767.4 kVA
[GATE E.E. 2004]

275. Neglecting losses, if the power transformed inductively is equal to power transformed conductively in an auto-transformer, then the secondary to primary ratio of transformer is
(a) 0.5 (b) 2 (c) 1.5 (d) 1.25
[U.P.S.C. I.E.S. E.E.-II, 2010]

276. It is advisable to use auto-transformer if the transformation ratio is
(a) greater than 1 (b) near to 1
(c) 0.25 (d) 0.5 [U.P.S.C. I.E.S. E.E.-II, 2010]

Statement for Linked Answer Questions 277 and 278

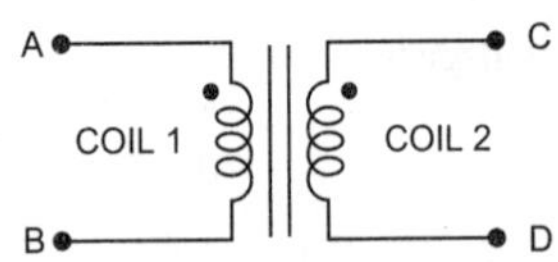

The figure above shows coils 1 and 2, with dot markings as shown, having 4,000 and 6,000 turns respectively. Both the coils have a rated current of 25 A. Coil 1 is excited with single phase, 400 V, 50 Hz supply.

Answers: 265. (c), **266.** (d), **267.** (a), **268.** (c), **269.** (b), **270.** (a), **271.** (c), **272.** (d), **273.** (c), **274.** (d), **275.** (a), **276.** (b),

277. The coils are to be connected to obtain a single-phase, $\frac{400}{1,000}$ V, auto-transformer to drive a load of 10 kVA. Which of the options given should be exercised to realize the required auto-transformer?
(a) Connect A and D; Common B.
(b) Connect B and D; Common C.
(c) Connect A and C; Common B.
(d) Connect A and C; Common D. **[GATE E.E. 2009]**

278. In the auto-transformer obtained in Question 270, the current in each coil is
(a) Coil-1 is 25 A and Coil-2 is 10 A.
(b) Coil-1 is 10 A and Coil-2 is 25 A.
(c) Coil-1 is 10 A and Coil-2 is 15 A.
(d) Coil-1 is 15 A and Coil-2 is 10 A. **[GATE E.E. 2009]**

279. In case of auto-transformers, which of the following statements are correct?
1. An auto-transformer requires less copper as compared to a conventional, 2-winding transformer of the same capacity.
2. An auto-transformer provides isolation between the primary and secondary windings.
3. An auto-transformer has less leakage reactance as compared to the conventional, 2-winding transformer of the same capacity. Select the correct answer using the codes given below:

Codes:
(a) 1, 2 and 3. (b) 1 and 2.
(c) 1 and 3. (d) 2 and 3.
[U.P.S.C. I.E.S. E.E.-II, 2001]

280. What is the efficiency of an auto-transformer in comparison to that of a two-winding transformer of the same rating?
(a) Slightly less than that of a two-winding transformer.
(b) Same as that of a two-winding transformer.
(c) More than that of a two-winding transformer.
(d) As low as 1/5th of the efficiency of a two-winding transformer.
[U.P.S.C. I.E.S. E.E.-II, 2006]

281. The following arrangement consists of an ideal transformer and an attenuator which attenuates by a factor of 0.8. An ac voltage V_{WX_1} = 100 V is applied across WX to get an open-circuit voltage V_{YZ_1} across YZ. Next an ac voltage V_{YZ_2} = 100 V is applied across YZ to get an open-circuit voltage WX_2 across WX. Then V_{YZ_1}/V_{WX_1}, V_{WX_2}/V_{YZ_2} are respectively

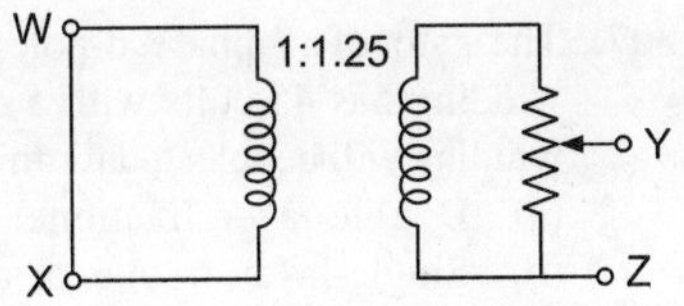

(a) 125/100 and 100/80 (b) 100/100 and 80/100
(c) 100/100 and 100/100 (d) 80/100 and 80/100
[GATE 2013]

282. Tappings of a transformer are provided
(a) at the phase end of lv side.
(b) at the phase end of hv side.
(c) at the neutral end of hv side.
(d) at the middle of hv side.

283. In a constant voltage transformer (CVT), the output voltage remains constant due to
(a) capacitor. (b) input inductor.
(c) saturation. (d) tapped windings.
[GATE E.E. 2000]

284. In a single phase induction regulator, compensating winding is provided in order to
(a) increase efficiency.
(b) reduce the voltage drop due to reactance of the secondary winding to negligible value.
(c) give wider variation of output voltage.
(d) all of the above.

285. In a 3-phase induction regulator, the output line voltages are in phase with the supply line voltage in
(a) maximum buck position only.
(b) maximum boost position only.
(c) both maximum boost and maximum buck positions.
(d) all positions.

286. A single phase induction regulator is a constant input transformer to obtain smooth variation of the output voltage by varying the
(a) ratio of turns between primary and secondary windings.
(b) frequency. (c) flux density in the core.
(d) angle between the magnetic axes of the primary and secondary windings. **[U.P.S.C. I.E.S. E.E.-II, 1998]**

287. The three-phase transformers each rated at 75 MVA, 132 kV/11 kV, have the following different methods of cooling:
1. Self cooled. 2. Forced oil cooled.
3. Forced air cooled.
The correct sequence in ascending order in terms of the weights of these transformers is
(a) 1, 2, 3 (b) 2, 3, 1 (c) 3, 1, 2 (d) 3, 2, 1
[U.P.S.C. I.E.S. E.E.-II, 1998]

288. In operating a 400 Hz transformer at 50 Hz
(a) only voltage is reduced in the same proportion as the frequency.
(b) only kVA rating is reduced in the same proportion as the frequency.
(c) both voltage and kVA rating are reduced in the same proportions as the frequency.
(d) none of the above.

289. Repeated switching-in of transformers located near the generating stations may cause
(a) excessive temperature rise.
(b) mechanical distortion of the windings and subsequent turn-to-turn insulation failure.
(c) intense local eddy currents in the core.
(d) surface flash-over of the bushings.

290. Dust should never be allowed to accumulate on the core and windings in a dry type transformer as it is likely to
(a) reduce dissipation of heat.
(b) absorb oil and grease.
(c) short circuit the windings.
(d) corrode the metal surface.

291. Which one of the following is the primary reason for placing field on rotor in an alternator?
(a) Small power in the field circuit.
(b) Insulation of high voltage is made easy on stator than on rotor.
(c) Large power in stator.
(d) Large current in stator. **[U.P.S.C. I.E.S. E.E.-II, 2004]**

292. Cylindrical rotor synchronous machines are mainly used in
(a) hydel power plants. (b) steam power plants.
(c) both hydel and steam power plants.
(d) none of these. **[W.B. Univ. of Technology, 2012-13]**

293. Four important parameters of alternators have comparatively larger or smaller values, depending upon the type of the alternator. In comparison to a steam turbine driven alternator, a hydraulically driven machine will have which one of the following combinations?

Answers: **277.** (a), **278.** (d), **279.** (c), **280.** (c), **281.** (b) **282.** (d), **283.** (c), **284.** (b), **285.** (c), **286.** (d), **287.** (d), **288.** (c), **289.** (b), **290.** (a), **291.** (b), **292.** (b), **293.** (b),

	Number of armature conductors	Axial length of armature conductors	Number of poles	Operating Speed
(a)	Smaller	Larger	Smaller	Higher
(b)	Larger	Smaller	Larger	Lower
(c)	Larger	Larger	Smaller	Lower
(d)	Smaller	Smaller	Larger	Higher

[U.P.S.C. I.E.S. E.E.-II, 2004]

294. Consider the following statements. The synchronous generators used in thermal power stations have
1. Cylindrical rotors.
2. Slip rings and brushes.
3. Laminated rotor.
4. Stator slots in the multiples of 6.
5. 3-phase stator-connected windings.

Which of these statements are correct?
(a) 1, 2, 3 and 4 (b) 1, 2, 3 and 5
(c) 3, 4 and 5 (d) 1, 2, 4 and 5
[U.P.S.C. I.E.S. E.E.-II, 1999]

295. The field system of an alternator is usually excited at
(a) 250/250 V dc. (b) 110/220 V ac.
(c) 3-phase, 50 Hz 400 V. (d) 230 V ac.

296. Number of slip rings required in a conventional synchronous machine is
(a) 0 (b) 1 (c) 2 (d) 3
[W.B. Univ. of Technology, 2011-12]

297. In a salient pole synchronous machine, the air gap is
(a) uniform under the whole pole shoe.
(b) least under the middle of the pole shoe and increases outwards.
(c) largest under the middle of the pole shoe and decreases outwards.
(d) least at one end of the pole shoe and increases to the maximum value at the other end. [U.P.S.C. IE.S. E.E.-II 2004]

298. A 400 V, 60 Hz synchronous machine has 6 poles. The synchronous speed in rpm is
(a) 1,200 rpm. (b) 3,000 rpm.
(c) 1,000 rpm. (d) 1,500 rpm.
[W.B. Univ. of Technology, 2012-13]

299. In turbo-alternators, smooth cylindrical rotors have very long axial length because it
(a) reduces friction and windage loss.
(b) reduces centrifugal force.
(c) number of conductors being less, they have to be necessarily long for generating the required voltage.
(d) none of the above.

300. Two mechanically coupled alternators deliver power at 50 Hz and 60 Hz respectively. The highest speed of the alternators is
(a) 3,600 rpm. (b) 3,000 rpm.
(c) 600 rpm. (d) 500 rpm.
[U.P.S.C. I.E.S. E.E.-II, 1994]

301. Consider the following statements regarding the design of the distributed armature winding in a 3-phase alternator:
1. It reduces the phase belt harmonics.
2. It increases utilization of armature iron and copper.
3. It increases rigidity and mechanical strength of the winding.
4. It reduces copper in the overhang of the winding.

Which of these statements are correct?
(a) 1, 3 and 4. (b) 2, 3 and 4.
(c) 1, 2 and 3. (d) 1, 2, 3 and 4.
[U.P.S.C. I.E.S. E.E.-II, 2000]

302. If the space flux distribution is non-sinusoidal, emf induced in the distributed winding will be
(a) more sinusoidal in comparison to the flux distribution.
(b) less sinusoidal in comparison to the flux distribution.
(c) of the same shape as that of flux distribution.
(d) none of the above.

303. The armature winding of a 2-pole, 3-phase alternator for each phase is distributed in a number of slots per phase. The rms value of the voltage per phase is less than the rms value of the voltage per coil multiplied by the number of coils in series because the
(a) rms value of the voltage in different coils of the phase is different.
(b) equal rms voltages in different coils of the phase has mutual phase difference.
(c) maximum values of the induced voltages in different coils of the phase are different.
(d) different coils of the phase pass through different saturated regions of the magnetic circuit. [U.P.S.C. I.E.S. E.E.-II, 1993]

304. Consider the following statements:
The effects of slot harmonics are effectively reduced by
1. skewing the slots. 2. fractional slot winding.
3. short-chorded winding. 4. distributed winding.

Of these statements
(a) 1 and 2 are correct. (b) 2, 3 and 4 are correct.
(c) 1, 3 and 4 are correct. (d) 1 and 3 are correct.
[U.P.S.C. I.E.S. E.E.-II, 1998]

305. In a polyphase ac machine 'short pitching' of coils leads to
(a) high terminal voltage. (b) higher efficiency.
(c) high power factor. (d) better induced emf.
[W.B. Univ. of Technology, 2011-12]

306. Skew of rotor bar eliminates
(a) the effect of space harmonics.
(b) the entire effect of crawling.
(c) magnetic noise.
(d) virbration due to unequal force developed on rotor.

307. The stator of a 3-phase, 6-pole a c machine has 45 slots. The stator winding has 45 coils with a coil span of 6 slots. What type of winding will be selected for this machine?
(a) Double-layer, fractional slot, short-pitched winding.
(b) Single-layer, fractional slot, short-pitched winding.
(c) Single-layer, integral slot, full-pitch winding.
(d) Double-layer, fractional slot, full-pitch winding.
[U.P.S.C. I.E.S. E.E.-II, 2008]

308. It is desirable to eliminate 5th harmonic voltage from the phase voltage of an alternator, the coils should be short pitched by an electrical angle of
(a) 30° (b) 36°
(c) 18° (d) none of these.
[W.B. Univ. of Technology, 2011-12]

309. Distributed winding and short chording employed in ac machines will result in
(a) increase in emf and reduction in harmonics.
(b) reduction in emf and increase in harmonics.
(c) increase in both emf and harmonics.
(d) reduction in both emf and harmonics. [GATE E.E. 2008]

310. In a synchronous generator, a divided winding rotor is preferable to a conventional winding rotor because of
(a) higher efficiency.
(b) increased steady-state stability limit.
(c) higher short-circuit ratio.
(d) better damping. [U.P.S.C. I.E.S. E.E.-II, 1997]

Answers: **294.** (d), **295.** (a), **296.** (c), **297.** (b), **298.** (a), **299.** (c), **300.** (c), **301.** (c), **302.** (a), **303.** (b), **304.** (b), **305.** (d), **306.** (a), **307.** (a), **308.** (b), **309.** (d), **310.** (b),

311. A 3-phase, 4-pole alternator has 48 stator slots carrying the 3-phase distributed winding. Each coil of the winding is short chorded by one slot pitch. The winding factor is given by

(a) $\frac{1}{16}$ cot (7.5°) (b) $\frac{1}{8}$cot(7.5°)

(c) $\frac{1}{16 \sin (7.5°)}$ (d) cos (7.5°)

[U.P.S.C. I.E.S. E. E. II, 1995]

312. In a cylindrical rotor synchronous machine, the phasor addition of stator and rotor mmfs is possible because

(a) two mmfs are rotating in opposite directions.
(b) two mmfs are rotating in same direction at different speeds.
(c) two mmfs are stationary with respect to each other.
(d) one mmf is stationary and the other mmf is rotating.

[U.P.S.C. I.E.S. E.E.-II, 2001]

313. If the dimensions of all the parts of a synchronous generator, and the number of field and armature turns are doubled, then the generated voltage will change by a factor of

(a) 1 (b) 2 (c) 4 (d) 8

[U.P.S.C. I.E.S. E.E.-II, 1998]

314. In a 3-phase star-connected generator, if the fundamental and the third harmonic rms voltages in each phase are respectively V_1 and V_3, the line voltage of the generator will be

(a) $\sqrt{3(V_1^2 + V_3^2)}$ (b) $\sqrt{3}\, V_1$

(c) $\sqrt{3}\left(V_1 + \frac{V_3}{3}\right)$ (d) $\sqrt{V_1^2 + \frac{V_3^2}{3}}$

[U.P.S.C. I.E.S. E.E.-II, 1999]

315. A 3-phase alternator is wound with a 60 degree phase-spread armature windings and develops 300 kVA. If the armature is reconnected utilizing all the coils for single phase operation with a phase spread of 180 degrees, the new rating of the machine is

(a) 100 kVA (b) 200 kVA
(c) 250 kVA (d) 300 kVA **[GATE E.E. 1999]**

316. A 20-pole alternator is having 180 identical stator slots with 6 conductors in each slot. All the coils of a phase are in series. If the coils are connected to realize single-phase winding, the generated voltage is V_1. If the coils are reconnected to realize three-phase star-connected winding, the generated phase voltage is V_2. Assuming full pitch, single-layer winding, the ratio V_1/V_2 is

(a) $\frac{1}{\sqrt{3}}$ (b) $\frac{1}{2}$ (c) $\sqrt{3}$ (d) 2 **[GATE 2014]**

317. The phase sequence of a three-phase alternator will reverse if

(a) the field current is reversed keeping the direction of rotation same.
(b) the field current remains the same but the direction of rotation is reversed.
(c) the field current is reversed and the number of poles is doubled.
(d) the number of poles is doubled without reversing the field current. **[GATE E.E. 2000]**

318. How can the reactive power delivered by a synchronous generator be controlled?

(a) By changing the prime mover input.
(b) By changing the excitation.
(c) By changing the direction of rotation.
(d) By changing the prime mover speed.

[U.P.S.C. I.E.S E.E.-II, 2007]

319. Figure shows the magnetization curves of an alternator at rated armature current, unity power factor and also at no load. The magnetization curve for rated armature current, 0.8 power factor leading is given by

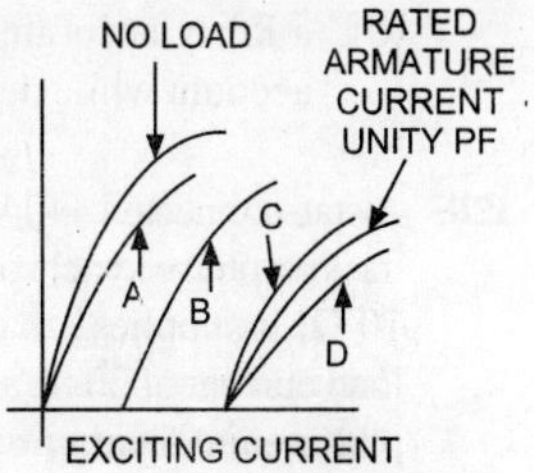

(a) curve A (b) curve B
(c) curve C (d) curve D

[GATE E.E. 2001]

320. Armature reaction effect is/are ______ on the main field flux.

(a) distorting (or cross-magnetizing) as well as demagnetizing.
(b) only assisting (c) only demagnetizing
(d) only distorting

321. Armature reaction AT of a synchronous generator at rated voltage zero power actor lagging is

(a) magnetizing (b) demagnetizing
(c) cross magnetizing
(d) both magnetizing and cross magnetizing.

[W.B. Univ. of Technology, 2012-13]

322. The synchronous reactance is the

(a) reaction due to armature reaction of the machine.
(b) reactance due to leakage flux.
(c) combined reactance due to leakage flux and armature reaction.
(d) reactance either due to armature reaction or leakage flux.

[U.P.S.C. I.E.S. E.E. II 2003]

323. What does the SCR (short-circuit ratio) of a synchronous machine yield ?

(a) $\frac{1}{X_s \text{(unsaturated) pu}}$ (b) $\frac{1}{X_s \text{(unsaturated) in Ohm}}$

(c) $\frac{1}{X_s \text{(adjusted) pu}}$ (d) $\frac{1}{X_s \text{(adjusted) in Ohm}}$

[U.P.S.C. I.E.S E.E.-II, 2007]

324. A 100 kVA, 415 V (line), star-connected synchronous machine generates rated open circuit voltage of 415 V at a field current of 15 A. The short-circuit armature current at a field current of 10 A is equal to the rated armature current. The per unit saturated synchronous reactance is

(a) 1.731 (b) 1.5 (c) 0.666 (d) 0.577

[GATE E.E. 2007]

325. A 1000 kVA, 6.6 kV, 3-phase star-connected cylindrical pole synchronous generator has a synchronous reactance of 20 Ω. Neglect armature resistance and consider operation at full load and unity power factor.

(*i*) The induced emf (line-to-line) is close to
(a) 5.5 kV (b) 7.2 kV (c) 9.6 kV (d) 12.5 kV

(*ii*) The power (or torque) angle is close to
(a) 13.9° (b) 18.3° (c) 24.6° (d) 33.0°

[GATE E.E 2005]

326. The leakage reactance of a three-phase alternator is determined by performing

(a) open-circuit and zero power factor tests.
(b) zero power factor and slip tests.
(c) open- and short-circuit tests.
(d) short-circuit and slip tests. **[U.P.S.C. I.E.S. E.E.-II, 2003]**

327. The 'synchronous impedance method' of finding the voltage regulation of a cylindrical rotor alternator is generally considered as

(a) a pessimistic method because saturation is not considered.
(b) an optimistic method because saturation is not considered.
(c) a fairly accurate method even if power factor is not taken into account while determining synchronous impedance.

Answers: 311. (b), **312.** (c), **313.** (c), **314.** (a), **315.** (b), **316.** (d), **317.** (b), **318.** (b), **319.** (c), **320.** (a), **321.** (b), **322.** (c), **323.** (c), **324.** (a), **325.** (i) (b), (ii) (c), **326.** (a), **327.** (a),

(d) a fairly accurate method when power factor is taken into account while determining synchronous impedance. **[U.P.S.C. I.E.S. E.E.-II, 2002]**

328. A star-connected 440 V, 50 Hz alternator has per phase synchronous reactance of 10 Ω. It supplies a balanced capacitive load current of 20 A, as shown in the per phase equivalent circuit of figure. It is desirable to have zero voltage regulation. The load power factor should be

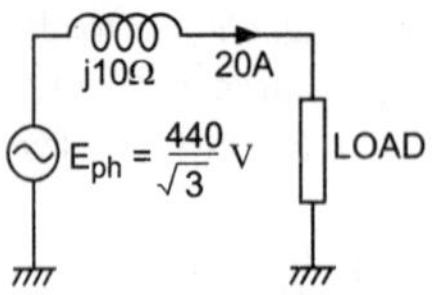

(a) 0.82 (b) 0.47 (c) 0.39 (d) 0.92 **[GATE E.E. 2001]**

329. Curves *X* and *Y* in the figure denote open-circuit and full-load zero power factor (Z pf) characteristics of a synchronous generator. *Q* is a point on the Z pf characteristic at 1.0 pu voltage. The vertical distance *PQ* in the figure gives the voltage drop across

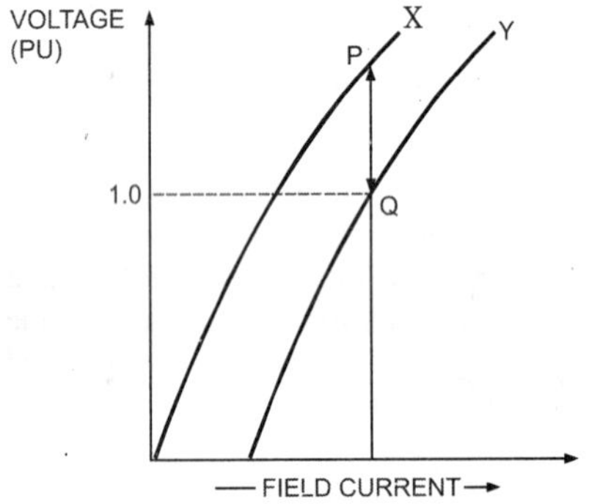

(a) synchronous reactance. (b) magnetising reactance.
(c) Potier reactance. (d) leakage reactance. **[GATE E.E. 2003]**

330. Consider the following:
1. LPF method. 2. EMF method.
3. ZPF method. 4. MMF method.

Which of the above methods are correct for determination of voltage regulation of an alternator?
(a) 1, 2 and 3. (b) 2, 3 and 4.
(c) 2 and 3 only. (d) 3 and 4 only. **[U.P.S.C. I.E.S. E.E.-II, 2009]**

331. The four methods of calculation of voltage regulation of a 3-phase alternator are
1. EMF method.
2. Saturated synchronous reactance method.
3. New ASA method. 4. MMF method.

The correct sequence of the ascending order of the values of regulation obtained by these methods is
(a) 3, 4, 2, 1 (b) 4, 3, 1, 2
(c) 3, 4, 1, 2 (d) 4, 3, 2, 1 **[U.P.S.C. I.E.S. E. E. II 1993]**

332. A synchronous motor is operated from a bus voltage of 1.0 pu and is drawing 1.0 pu zero power factor leading current. Its synchronous reactance is 0.5 pu. What is the excitation emf of the motor?
(a) 2.0. (b) 1.5. (c) 1.0. (d) 0.5. **[U.P.S.C. I.E.S. E.E.-II, 2005]**

333. Which one of the following statements is not correct in respect of synchronous machines?
(a) In salient pole machines, the direct-axis synchronous reactance is greater than the quadrature axis synchronous reactance.
(b) The damper bars help the motor to self start.
(c) Short-circuit ratio is the ratio of field current required to produce the rated voltage on open circuit to the rated armature current.
(d) The V-curve of a synchronous motor represents the variation in armature current with field excitation at a given output power. **[U.P.S.C. I.E.S. E.E.-II, 2006; GATE E.E. 2005]**

334. In a salient pole synchronous machine where X_d = *d*-axis synchronous reactance, X_q = quadrature-axis synchronous reactance,
(a) $X_q = X_d$ (b) $X_q > X_d$
(c) $X_q < X_d$ (d) $X_q = 0$ **[W.B. Univ. of Technology, 2011-12, 2012-13]**

335. For maximum current during slip test on a synchronous machine, the armature mmf aligns along
(a) *d*-axis. (b) *q*-axis.
(c) 45° to *d*-axis. (d) 45° to *q*-axis. **[U.P.S.C. I.E.S. E.E.-II, 2002]**

336. While conducting a "slip" test for determination of direct- and quadrature-axis synchronous reactance X_d and X_q of salient pole synchronous machine, the rotor of the machine is run with a slip '*s*' and stator supply frequency '*f*'. The frequency of
1. voltage induced across open-field terminals
2. envelope of armature terminal voltage
3. envelope of armature current and
4. armature current

will be respectively
(a) *sf*, *sf*, *sf* and *f*. (b) *sf*, *f*, *sf* and *f*.
(c) *f*, *sf*, *f* and *sf*. (d) *f*, (1 – *s*) *f*, (2 – *s*) *f* and *sf*. **[U.P.S.C. I.E.S. E.E.-II, 1997]**

337. In which one of the following is reluctance power developed?
(a) Salient pole alternator.
(b) Non-salient pole alternator.
(c) Squirrel cage induction motor.
(d) Transformer. **[U.P.S.C. I.E.S. E.E.-II, 2006]**

338. Which is the value of load angle when the power output of a salient pole synchronous generator is maximum?
(a) 0° (b) 45° (c) 90° (d) none of the above. **[U.P.S.C. I.E.S. E.E.-II, 2006]**

339. A salient pole synchronous machine has maximum power output when power angle δ is
(a) 90° (b) 45°
(c) less than 90° (d) more than 90°. **[W.B. Univ. of Technology, 2012-13]**

340. The two-reaction phasor diagram of a salient pole synchronous generator is considered, where

I_a = armature current per phase
X_L = armature leakage reactance per phase
X_d = direct-axis synchronous reactance
X_q = quadrature-axis synchronous reactance and
V = generator terminal voltage, taken as reference.

The voltage in-phase with the direct-axis will be the phasor sum of
(a) $V + I_aR_a + I_aX_L$ (b) $V + I_aR_a + I_aX_d$
(c) $V + I_aR_a + I_aX_q$ (d) $V + I_aR_a + I_a(X_d + X_q)$ **[U.P.S.C. I.E.S. E.E.-II, 1999]**

341. Higher synchronous reactance is preferred in the present day alternators, because one can have
(a) reduced subtransient current.
(b) reduced harmonic currents.
(c) reduced transient currents.
(d) higher voltage regulation with load. **[GATE E.E. 1999]**

342. When a 3-phase alternator is suddenly short-circuited at its terminals, the initial value of short-circuit current is limited by which one of the following:

Answers: 328. (a), **329.** (a), **330.** (b), **331.** (d), **332.** (d), **333.** (c), **334.** (c), **335.** (b), **336.** (a), **337.** (a), **338.** (d), **339.** (c), **340.** (b), **341.** (c), **342.** (a),

(a) Subtransient reactance X''_d.
(b) Transient reactance X'_d.
(c) Synchronous reactance X_S.
(d) Sum of X''_d, X'_d and X_S. **[U.P.S.C. I.E.S. E.E.-II, 2004]**

343. X_d, X'_d and X''_d are steady-state *d*-axis synchronous reactance, transient *d*-axis reactance and subtransient *d*-axis reactance of a synchronous machine respectively. Which of the following statements is true?
(a) $X_d > X'_d > X''_d$ (b) $X''_d > X'_d > X_d$
(c) $X'_d > X''_d > X_d$ (d) $X_d > X''_d > X'_d$
[GATE E.E. 2001]

344. The phasor diagram of a synchronous machine connected to an infinite bus is shown in the figure: The machine is acting as a

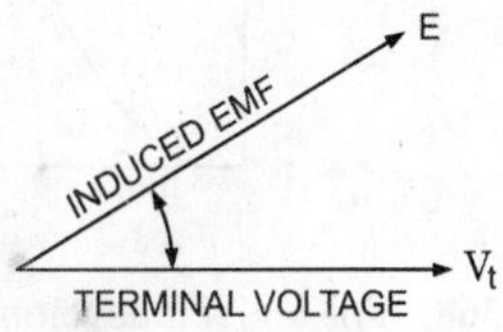

(a) generator and operating at a lagging power factor.
(b) generator and operating at a leading power factor.
(c) motor operating at a leading power factor.
(d) motor and operating at a lagging power factor.
[U.P.S.C. I.E.S. E.E.-II, 1999]

345. A round rotor synchronous generator has a leakage reactance of 10%, armature reaction reactance of 90% and negligible armature resistance. With the machine initially running at rated speed and terminal voltage of 1.0 pu, a 3-phase short-circuit is applied. The sustained armature current will be
(a) 1.25 pu (b) 1.11 pu (c) 1.0 pu (d) 0.9 pu
[U.P.S.C. I.E.S. E.E.-II, 1993]

346. A 3-phase synchronous generator with constant steam input supplies power to an infinite bus at a lagging power factor. If the excitation is increased
(a) both power angle and power factor decrease.
(b) both power angle and power factor increase.
(c) the power angle decreases while power factor increases.
(d) the power angle increases while power factor decreases.
[U.P.S.C. I.E.S. E.E.-II, 2005]

347. If the prime mover of the alternator supplying load to an infinite bus is suddenly shunt down, then it will
(a) stop. (b) continue to run as an alternator.
(c) continue to run as a synchronous motor in the reverse direction.
(d) continue to run as a synchronous motor in the same direction.
[W.B. Univ. of Technology, 2011-12]

348. A stand alone engine driven synchronous generator is feeding a partly inductive load. A capacitor is now connected across the load to completely nulify the inductive current. For this operating condition
(a) the field current and fuel input have to be reduced.
(b) the field current and fuel input have to be increased.
(c) the field current has to be increased and fuel input left unaltered.
(d) the field current has to be reduced and fuel input left unaltered.
[GATE E.E. 2003]

349. When a balanced 3-phase distributed type armature winding is carrying 3-phase, balanced currents, the strength of the resultant rotating magnetic field is
(a) three times the amplitude of each constituent pulsating magnetic field.
(b) equal to the amplitude of each constituent pulsating magnetic field.
(c) half the amplitude of each constituent pulsating magnetic field.
(d) one and half times the amplitude of each constituent of pulsating magnetic field. **[U.P.S.C. I.E.S. E.E.-II, 1994]**

350. An ideal synchronous motor has no starting torque because the
(a) rotor is made up of salient poles.
(b) relative velocity between the stator and the rotor mmfs is zero.
(c) relative velocity between the stator and rotor mmfs is not zero.
(d) rotor winding is highly reactive. **[U.P.S.C. I.E.S. E.E.-II, 1998]**

351. A synchronous machine is revolving armature and stationary field type. Under steady running conditions, the air gap field
(a) rotates at synchronous speed with respect to stator.
(b) rotates at synchronous speed in the direction of rotation of rotor.
(c) remains stationary with respect to stator.
(d) remains stationary with respect to rotor.

352. What is the angle between the induced voltage and supply voltage of a synchronous motor under running condition?
(a) Zero. (b) Greater than zero but ≤ 90°.
(c) Between 90° and 180°. (d) > 180. **[U.P.S.C. I.E.S. E.E.-II, 2007]**

353. The flux per pole in a synchronous motor with the field circuit ON and the stator disconnected from the supply is found to be 25 m Wb. When the stator is connected to the rated supply with the field excitation unchanged, the flux per pole in the machine is found to be 20 m Wb while the motor is running on no load. Assuming no-load losses to be zero, the no-load current by the motor from the supply
(a) lags behind the supply voltage.
(b) leads the supply voltage.
(c) is in phase with the supply voltage.
[GATE E.E. 2002]

354. Synchronizing power of a synchronous motor varies
(a) directly as synchronous reactance, X_s.
(b) directly as the square of the synchronous reactance, X_s^2.
(c) inversely as the synchronous reactance, $1/X_s$.
(d) none of the above.

355. A 3-phase synchronous motor connected to an infinite bus is operating at half-full load with normal excitation. When the load on the synchronous motor is suddenly increased
(a) its speed will first decrease and then become synchronous.
(b) its speed will first increase and then become synchronous.
(c) its speed will fluctuate around synchronous speed and then become synchronous.
(d) its speed will remain unchanged. **[U.P.S.C. I.E.S. E.E-II, 1994]**

356. A synchronous generator connected to an infinite bus is over-excited. Considering only the reactive power, from the point of view of the system, the machine acts as
(a) capacitor. (b) inductor.
(c) resistor. (d) none of the above.
[GATE E.E. 1998]

357. A synchronous machine connected to a power system grid bus-bar is operating as a generator. To make the machine operate as a motor, the
(a) direction of rotation is to be reversed.
(b) phase-sequence is to be changed.
(c) field excitation is to be decreased.
(d) mechanical input is to be less than the losses at the shaft.
[U.P.S.C. I.E.S. E.E.-II, 1993]

358. What is the effect of the field failure of salient pole synchronous motor connected with infinite bus?

Answers: 343. (a), **344.** (b), **345.** (c), **346.** (a), **347.** (d), **348.** (d), **349.** (d), **350.** (b), **351.** (c), **352.** (c), **353.** (b), **354.** (c), **355.** (c), **356.** (a), **357.** (d), **358.** (d),

(a) Reduce motor torque and speed.
(b) Not charge motor torque and speed.
(c) Stop the motor.
(d) Reduce motor torque but motor will continue to run at synchronous speed. **[U.P.S.C. I.E.S. E.E.-II, 2009]**

359. When the excitation of normally operating unloaded salient-pole synchronous motor suddenly gets disconnected, it continues to run as a
(a) Scharge motor. (b) Spherical motor.
(c) Switched-reluctance motor.
(d) Variable-reluctance motor. **[U.P.S.C. I.E.S. E.E.-II, 2002]**

360. A synchrnous motor, connected to an infinite bus, is working at a leading power factor. Its excitation emf E_f and terminal voltage V_f are related as under
(a) $E_f > V_t$ and E_f lags V_t (b) $E_f < V_t$ and E_f lags V_t
(c) $E_f > V_t$ and E_f leads V_t (d) $E_f < V_t$ and E_f leads V_t

361. A synchronous motor is floating on infinite mains at no load. If its excitation is increased, it will draw
(a) unity power factor current.
(b) zero power factor lagging current.
(c) zero power factor leading current.
(d) no current. **[W.B. Univ. of Technology, 2012-13]**

362. A synchronous motor is operating on no load at unity power factor. If the field current is increased, the pf will become
(a) leading and current will decrease.
(b) lagging and current will increase.
(c) lagging and current will decrease.
(d) leading and current will increase.
[W.B. Univ. of Technology, 2011-12]

363. A three-phase, salient pole synchronous motor is connected to an infinite bus. It is operated at no load at normal excitation. The field excitation of the motor is first reduced to zero and then increased in the reverse direction gradually. Then the armature current
(a) increases continuously.
(b) first increases and then decreases steeply.
(c) first decreases and then increases steeply.
(d) remains constant. **[GATE E.E., 2011]**

364. A synchronous motor operates at 0.8 pf lagging. If the field current of the motor is continuously increased
(a) the power factor decreases upto a certain value of the field current and there-after it increases.
(b) the armature current increases upto a certain value of the field current and thereafter it decreases.
(c) the power factor increases upto a certain value of field current and thereafter it decreases.
(d) the armature current decreases upto a certain value of field current and there after it increases.
(e) both (c) and (d). **[GATE E.E. 1993]**

365. The torque angle of a synchronous machine operating from a constant voltage bus, is usually defined as the space angle between
(a) rotor mmf wave and stator mmf wave.
(b) rotor mmf wave and resultant flux density wave.
(c) stator mmf wave and resultant flux density wave.
(d) stator mmf wave and resultant mmf wave. **[GATE E.E. 1992]**

366. A synchronous motor on load draws a current at a leading power factor angle ϕ. If the internal power angle ______ which is the phase angle between the excitation emf and the current in the phasor diagram is ψ, then the air-gap excitation mmf lags behind the armature mmf.
(a) ψ (b) $\frac{\pi}{2}+\psi$ (c) $\frac{\pi}{2}-\psi$ (d) $\psi+\phi$
[GATE E.E. 1995]

367. Which of the following graphs represents the speed-torque characteristic of a synchronous motor ?

(a)
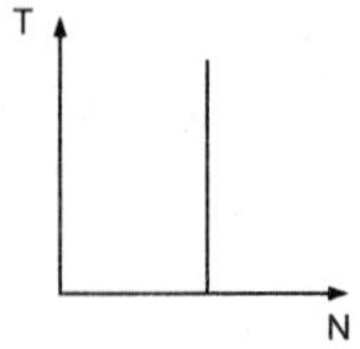

(b)
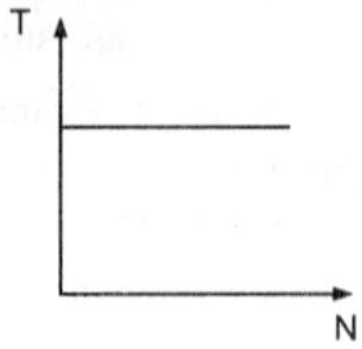

(c)
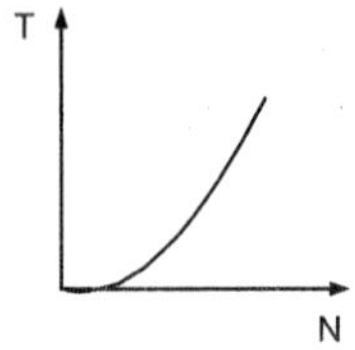

(d)
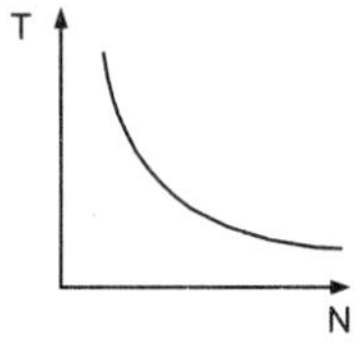

[U.P.S.C. I.E.S. E.E.-II, 2003]

368. For a given developed power, a synchronous motor operating from a constant voltage and constant frequency supply, will draw the minimum and maximum armature currents, I_{min} and I_{max} respectively, corresponding to
(a) I_{min} at unity pf, but i_{max} at zero pf.
(b) I_{max} at unity pf, but I_{min} at zero pf.
(c) both I_{min} and I_{max} at unity pf.
(d) both I_{min} and I_{max} at zero pf. **[U.P.S.C. I.E.S. E.E.-II, 1997]**

369. A synchronous motor operating at rated voltage draws 1.0 pu current at 1.0 power factor. The machine parameters are synchronous reactance 1.0 pu; armature resistance negligible. Apart from supplying this rated power, if the motor has to supply an additional reactive power of 0.8 pu, then the field current has to be increased by
(a) 42% (b) 46% (c) 52% (d) 69%
[U.P.S.C. I.E.S. E.E.-II, 1996]

Statement for Linked Answer Questions 370 and 371

A synchronous motor is connected to an infinite bus at 1.0 pu voltage and draws 0.6 pu current at unity power factor. Its synchronous reactance is 1.0 pu and resistance is negligible.

370. The excitation voltage (E) and load angle (δ) will respectively be
(a) 0.8 pu and 36.86° lag. (b) 0.8 pu and 36.86° lead.
(c) 1.17 pu and 30.96° lead. (d) 1.17 pu and 30.96° lag.
[GATE E.E. 2008]

371. Keeping the excitation voltage same, the load on the motor is increased such that the motor current increases by 20%. The operating power factor will become
(a) 0.995 lagging. (b) 0.995 leading.
(c) 0.791 lagging. (d) 0.848 leading. **[GATE E.E. 2008]**

372. An induction motor and synchronous motor are connected to a common feeder line. To operate the feeder line at unity pf, the synchronous motor should be
(a) underexcited. (b) overexcited.
(c) normally excited.
(d) disconnected from the common terminals.
[U.P.S.C. I.E.S. E.E.-II 2000]

373. The power factor of a synchronous motor
(a) improves with increase in excitation and may even become leading at higher excitation.
(b) decreases with increase in excitation.
(c) is independent of excitation.
(d) increases with loading for a given excitation.
[U.P.S.C. I.E.S. E.E.-II, 2002]

Answers: 359. (d), **360.** (c), **361.** (c), **362.** (d), **363.** (b), **364.** (e), **365.** (a), **366.** (d), **367.** (a), **368.** (a), **369.** (b), **370.** (c), **371.** (d), **372.** (b), **373.** (a),

374. Synchronous condenser means
(a) a synchronous motor with capacitor connected across terminals to improve pf.
(b) a synchronous motor operating at full load with leading pf.
(c) an overexcited synchronous motor partially supplying mechanical load, and also improving pf of the system to which it is connected.
(d) an overexcited synchronous motor operating at no load with leading pf used in large power stations for improvement of pf. **[U.P.S.C. I.E.S. E.E.-II 2001]**

375. When a 3-phase synchronous motor is running at synchronous speed, the damper winding produces
(a) damping torque. (b) eddy current torque.
(c) torque aiding the developed torque.
(d) no torque. **[W.B. Univ. of Technology, 2011-12]**

376. During hunting of synchronous motor
(a) negative phase sequence currents are generated.
(b) harmonics are developed in the armature circuit.
(c) damper bar develops torque.
(d) field excitation increases. **[GATE E.E. 1996]**

377. A cylindrical rotor synchronous motor is switched on to the dc supply with its field windings shorted on themselves. It will
(a) not start.
(b) start but not run at synchronous speed.
(c) start as an induction motor and then run as a synchronous motor.
(d) start and run as a synchronous motor. **[GATE E.E. 1996]**

378. When a salient pole 3-phase synchronous motor is started by induction motor action and its field is connected across a field discharge resistance, starting torque is developed by
(a) reluctance torque due to saliency of the rotor.
(b) eddy current and hysteresis torque in pole faces.
(c) induction motor torque in field and damper windings.
(d) all of the above.

379. In figure given below, the characteristic that corresponds to the variation of a synchronous reactance of a synchronous motor with field current is

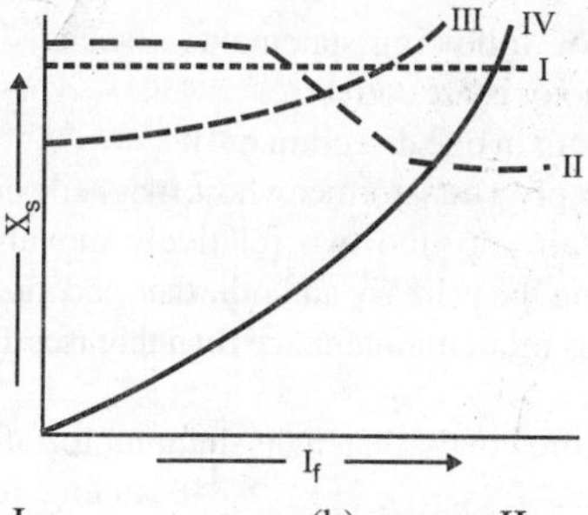

(a) curve I. (b) curve II.
(c) curve III. (d) curve IV. **[GATE E.E. 1996]**

380. A 440 V, 3-phase, 10 pole and 50 Hz synchronous motor delivering a torque of $\frac{50}{\pi}$ Nm, delivers a power of:
(a) 50 W (b) 500 W
(c) 1000 W (d) 2000 W
[U.P.S.C. I.E.S. E.E.-II, 2011]

381. The negative phase sequence in a 3-phase synchronous motor exists when the motor is
(a) underloaded. (b) overloaded.
(c) supplied with unbalanced voltage.
(d) hot. **[U.P.S.C. I.E.S. E.E.-II, 1992]**

382. Which of the following synchronous motors has cost comparable to that of an induction motor ?
(a) High kW output high speed.
(b) High kW output low speed.
(c) Low kW output low speed.
(d) Low kW output high speed.

383. Which of the following conditions are to be satisfied for proper synchronization of alternators?
1. Equal terminal voltage. 2. Same frequency.
3. Same phase sequence. 4. Same kVA rating.
5. Same phase displacement

Select the correct answer using the code given below:
(a) 1, 3 and 4 only. (b) 1, 2, 4 and 5 only.
(c) 2, 3, 4 and 5 only. (d) 1, 2, 3 and 5 only.
[U.P.S.C I.E.S. E.E.-II, 2008]

384. Which one of the following is not a necessary condition to be satisfied for synchronising an incoming alternator to an already operating alternator?
(a) Same voltage magnitude.
(b) Same frequency.
(c) Same prime mover speed.
(d) Same phase sequence. **[U.P.S.C. I.E.S. E.E.-II 2004]**

385. A three phase synchronous generator is to be connected to the infinite bus. The lamps are connected as shown in the figure for the synchronization. The phase sequence of bus voltage is R-Y-B and that of incoming generator voltage is R′-Y′-B′.

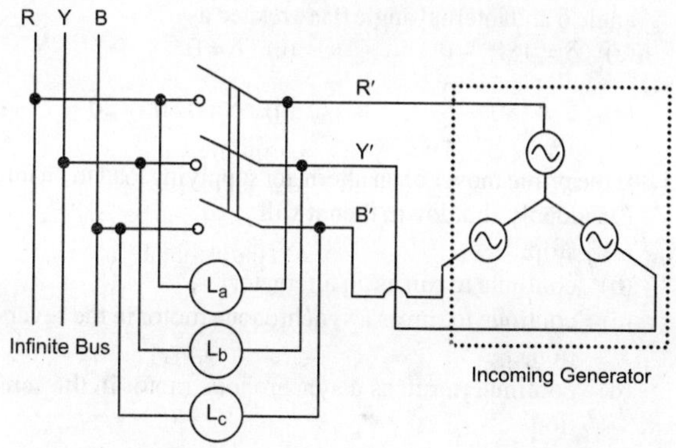

It was found that the lamps are becoming dark in the sequence L_a-L_b-L_c. It means that the phase sequence of incoming generator is
(a) opposite to infinite bus and its frequency is more than infinite bus.
(b) opposite to infinite bus but its frequency is less than infinite bus.
(c) same as infinite bus and its frequency is more than infinite bus.
(d) same as infinite bus and its frequency is less than infinite bus. **[GATE 2014]**

386. The electrical stiffness of a synchronous generator connected to a very large grid can be increased by
(a) increasing the excitation or power angle of the machine.
(b) reducing the excitation or the synchronous reactance of the machine.
(c) increasing the synchronous reactance of the machine.
(d) operating the generator at a much lower MW level compared to the steady state limit. **[U.P.S.C. I.E.S. E.E.-II 2002]**

387. Two alternators A and B are sharing an inductive load equally. If the excitation of alternator A is increased
(a) alternator B will deliver more current and alternator A will deliver less current.

Answers: 374. (d), **375.** (d), **376.** (c), **377.** (b), **378.** (d), **379.** (b), **380.** (c), **381.** (c), **382.** (b), **383.** (d), **384.** (c), **385.** (a), **386.** (c), **387.** (b),

(b) alternator B will deliver less current and alternator A will deliver more current.
(c) both will continue to share load equally.
(d) both will deliver more current. [U.P.S.C. I.E.S. E.E.II-1992]

388. Two similar synchronous generators are working in parallel to supply a common load demand with identical excitations and steam supplies to their prime movers. Now, if the steam supply to the prime mover of one of the generators is increased compared to the other, with field excitation kept unchanged, then
(a) its active power component will remain the same but the reactive power contribution will increase.
(b) its active power will decrease while the reactive power will increase.
(c) both active and reactive components of power will increase.
(d) its active power contribution will increase but reactive power contributions of both will remain unchanged.
[U.P.S.C. I.E.S. E.E.-II, 1993]

389. A 3-phase, 11 kV, 5 MVA alternator has synchronous reactance of 10 Ω per phase. Its excitation is such that the generated emf is 14 kV. If the alternator is connected to infinite bus bar, the maximum output at the given excitation is
(a) 15,400 kW (b) 8,000 kW
(c) 6,200 kW (d) 5,135 kW
[U.P.S.C. I.E.S. E.E.-II, 2010]

390. An alternator connected to an infinite bus, for a given excitation, will develop maximum electromagnetic power when the power angle δ and internal angle θ are related as
(a) $\delta = 180° - \theta$ (b) $\delta = \theta$
(c) $\delta = 180° - \frac{\theta}{2}$ (d) $\delta = 180° - 2\theta$

391. If the prime mover of an alternator supplying load to an infinite bus is suddenly shutdown, then it will
(a) stop.
(b) continue to run as an alternator.
(c) continue to run as a synchronous motor in the reverse direction.
(d) continue to run as a synchronous motor in the same direction.

392. A salient pole synchronous generator delivering power to an infinite bus through a reactive tie line reaches its steady state stability limit. What is the power angle of the generator relative to the infinite bus voltage reference?
(a) Greater than 90 degrees.
(b) Equal to 90 degrees.
(c) Less than 90 degrees.
(d) Zero. [U.P.S.C. I.E.S. E.E.-II, 2009]

393. A large ac generator supplying power to an infinite bus, has a sudden short circuit occurring at its terminals. Assuming the prime mover input and the voltage behind the transient reactance to remain constant immediately after the fault, the acceleration of the generator rotor is
(a) inversely proportional to the moment of inertia of the machine.
(b) inversely proportional to the square of the voltage.
(c) directly proportional to the square of the short-circuit current.
(d) directly proportional to the short-circuit power.
[U.P.S.C. I.E.S. E.E.-II, 1994]

394. Four identical alternators each rated for 20 MVA, 11 kV having a subtransient reactance of 16% are working in parallel. The short-circuit level at the bus-bars is
(a) 500 MVA (b) 400 MVA
(c) 125 MVA (d) 80 MVA
[U.P.S.C. I.E.S. E.E.-II, 1996]

395. A power station consists of two synchronous generators A and B of ratings 250 MVA and 500 MVA with inertia 1.6 pu and 1 pu respectively on their own base MVA ratings. The equivalent per unit inertia constant for the system on 100 MVA common base is
(a) 2.6 (b) 0.615 (c) 1.625 (d) 9.0
[GATE E.E. 1998]

396. A loss-less cylindrical rotor synchronous generator is floating on the system bus-bar after being properly synchronised and is running at no-load. Now, if the prime mover is acting as a mechanical load on the machine, then
(a) the torque developed by the machine will act in the same direction as the prime mover torque.
(b) the torque developed by the machine will act in the opposite direction to that of the prime mover torque.
(c) the machine rotor would fall-back and the induced emf would lag behind the terminal voltage.
(d) the machine rotor would overspeed and the induced emf would lead the terminal voltage. [U.P.S.C. I.E.S. E.E.-II, 2006]

397. Two identical synchronous machines A and B, running at the same speed, are linked through an inductor, as shown

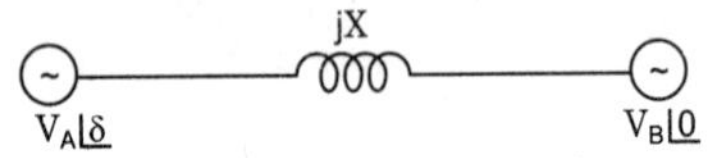

Machine A will supply active and reactive power to machine B when δ is
(a) positive and $V_A < V_B$ (b) negative and $V_A > V_B$
(c) negative and $V_A < V_B$ (d) positive and $V_A > V_B$
[U.P.S.C I.E.S. E.E.-II, 2008]

398. The angle δ in the swing equation of a synchronous generator is the
(a) angle between stator voltage and current.
(c) angular displacement of the rotor with respect to the stator.
(c) angular displacement of the stator mmf with respect to synchronously rotating axis.
(d) an angular displacement of an axis fixed to the rotor with respect to a synchronously rotating axis. [GATE 2013]

399. Which of the following statements associated with polyphase induction motor is/are correct ?
(a) It can run in one direction only.
(b) It is simply a transformer whose magnetic circuit is separated by an air gap into two relatively movable portions, one carrying the primary and other the secondary winding.
(c) It needs more maintenance than that required by a dc shunt motor.
(d) It is more costly than a dc shunt motor of the same output rating.

400. In a 3-phase slip-ring induction motor, the rotor winding terminals are brought out through slip rings to
(a) connect extra resistance across them during starting.
(b) connect them either in star or in delta as per need.
(c) connect to 3-phase ac supply.
(d) close the rotor circuit externally.

401. A smaller air gap in a polyphase induction motor helps to
(a) reduce the chance of crawling.
(b) increase the starting torque.
(c) reduce the chance of cogging.
(d) reduce the magnetizing current. [U.P.S.C. I.E.S. E.E.-II, 2004]

402. In an induction motor, if the air gap is increased
(a) its speed will reduce. (b) its efficiency will improve.

Answers: 388. (d), **389.** (a), **390.** (a), **391.** (d), **392.** (c), **393.** (a), **394.** (a), **395.** (d), **396.** (c), **397.** (d), **398.** (d), **399.** (b), **400.** (a), **401.** (d), **402.** (c)

(c) its power factor will reduce.
(d) its breakdown torque will reduce.

[U.P.S.C. I.E.S. E.E.-II, 2004]

403. Which of the following parameters in an induction motor influences the magnetizing reactance to the maximum extent?
(a) Axial length of the rotor stack.
(b) Axial length of the stator stack.
(c) Radial length of air gap. (d) Number of slots on the stator.

[U.P.S.C I.E.S. E.E.-II, 2006]

404. The rotor slots are slightly skewed in squirrel-cage induction motor to
(a) increase the strength of rotor bars.
(b) reduce the magnetic hum and locking tendency of rotor.
(c) economise on the copper to be used.
(d) provide ease of fabrication. **[U.P.S.C. I.E.S. E.E.-II, 2003]**

405. Skewing of the rotor in a three-phase squirrel-cage induction motor reduces
(a) noise, parasitic torque, starting torque and pull-out torque.
(b) noise and parasitic torque, but increases starting torque and pull-out torque.
(c) noise and pullout torque, but increases parasitic torque and starting torque.
(d) noise, parasitic torque and starting torque, but increases pull-out torque. **[U.P.S.C. I.E.S. E.E.-II, 2001]**

406. Consider the following statements:
Skewing of rotor slots in a 3-phase induction motor (cage rotor) may
1. introduce additional leakage reactance.
2. eliminate slot harmonics.
Which of the statements given above is/are correct?
(a) 1 only. (b) 2 only.
(c) Both 1 and 2. (d) Neither 1 nor 2.

[U.P.S.C I.E.S. E.E.-II, 2008]

407. The principle of operation of a 3-phase induction motor is almost similar to that of
(a) synchronous motor.
(b) repulsion start induction motor.
(c) transformer with a shorted secondary.
(d) capacitor-start induction motor. **[R.G.T.U. B.E.E. June 2007]**

408. If any two leads from slip rings are interchanged in a 3-phase induction motor, the motor will
(a) continue running in the same direction as before.
(b) run in a direction opposite to previous one.
(c) not run.
(d) get damaged.

409. The voltage actually used for setting up of the useful flux in the air gap of 3-phase induction motor is
(a) = applied voltage. (b) > applied voltage.
(c) < applied voltage. (d) = rotor induced emf.

[U.P.S.C. I.E.S. E.E.-II, 2006]

410. Leakage flux in an induction motor is
(a) flux that leaks through the machine.
(b) flux that links both stator and rotor windings.
(c) flux that links none of the windings.
(d) flux that links the stator winding or rotor winding but not both. **[GATE 2013]**

411. At 50 Hz, the speed of rotating magnetic field for a 4-pole, 3-ϕ induction motor is
(a) 1,500 rpm (b) 3,000 prm
(c) 750 rpm (d) none of these.

[W.B. Univ. of Technology, 2011]

412. If N_s is the synchronous speed, N is the rotor speed and s is the slip then relation is
(a) $N_s = (1 - s)$ N (b) N = sN_s
(c) N = $(s - 1)$ N_s (d) None of these.

[G.B.T.U. Even Semester 2010-11]

413. A balanced 3-phase induction motor runs at slip s. If ω_s is its synchronous speed, what is the relative speed between the stator mmf and rotor mmf ?
(a) $s\omega_s$ (b) $(1 - s)\omega_s$ (c) ω_s (d) zero

[U.P.S.C I.E.S. E.E.-II, 2008]

414. The rotor of a three phase, 5 kW, 400 V, 50 Hz, slip ring induction motor is wound for 6 poles while its stator is wound for 4 poles. The approximate average no load steady state speed when this motor is connected to 400 V, 50 Hz supply is
(a) 1,500 rpm. (b) 500 rpm.
(c) 0 rpm. (d) 1,000 rpm. **[GATE E.E. 2002]**

415. The direction of rotation of a 3-phase induction motor is clockwise when it is supplied with 3-phase sinusoidal voltage having phase sequence *A-B-C*. For counterclockwise rotation of the motor, the phase sequence of the power supply should be
(a) B-C-A (b) C-A-B
(c) A-C-B (d) B-C-A and C-A-B

[GATE E.E. 2004]

416. The mmf produced by the rotor currents of a 3-phase induction motor
(a) rotates at the speed of rotor in the air gap.
(b) is at standstill with respect to stator mmf.
(c) rotates at slip speed with respect to stator mmf.
(d) rotates at synchronous speed with respect to rotor.

[U.P.S.C. I.E.S. E.E.-II, 2010]

417. A 6-pole, 50 Hz wound rotor induction motor when supplied at the rated voltage and frequency with slip-rings open circuited, developed a voltage of 100 V between any two slip rings. If the rotor is driven by an external means at 1,000 rpm opposite to the direction of stator field, the frequency of voltage across slip rings will be
(a) zero. (b) 50 Hz (c) 100 Hz (d) 200 Hz

[U.P.S.C. I.E.S. E.E.-II, 2000]

418. A 3-phase induction motor is applied with a 400 V, 3-ϕ, 50 Hz supply and the motor runs at 1,470 rpm. The rotor current will have a frequency of
(a) 2.0 Hz (b) 1.0 Hz (c) 4.0 Hz (d) 4.8 Hz

[W.B. Univ. of Technology, 2013]

419. A 3-phase induction motor is operating at slip s. If its two supply leads are interchanged, then its slip at that instant will be
(a) $2 - s$ (b) $2 + s$ (c) $1 + s$ (d) $1 - s$

420. A voltmeter gives 120 oscillations per minute when connected to the rotor of an induction motor. The frequency is 50 Hz. The slip of the motor is
(a) 2% (b) 4% (c) 5% (d) 25%

[U.P.S.C. I.E.S. E.E.-II, 1995, 2000]

421. If E_2 is the standstill rotor phase emf, I_2 is the standstill rotor phase current and cos ϕ_2 is the rotor power factor then torque developed by a 3-phase induction motor varies as
(a) $E_2 I_2$ (b) $E_2I_2 \cos \phi_2$
(c) $E_2I_2/\cos \phi_2$ (d) $E_2/I_2 \cos \phi_2$

422. For a more accurate analysis of the torque-slip relationship, the equivalent circuit of a 3-phase induction motor

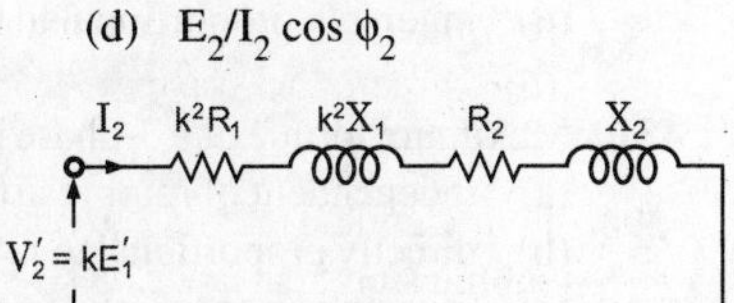

Answers: **403.** (c), **404.** (b), **405.** (a), **406.** (c), **407.** (c), **408.** (a), **409.** (a), **410.** (d), **411.** (a), **412.** (d), **413.** (d), **414.** (c), **415.** (c), **416.** (b), **417.** (c), **418.** (b), **419.** (a), **420.** (b), **421.** (b), **422.** (a), for running condition is to be derived from the given standstill condition:

K = ratio of transformation

$E_1 = V_1$ (applied voltage - $I_\mu X_L$)

At running condition, if the two reactances in the equivalent circuit are denoted by *A* and *B*, then the values of *A* and *B* will be

(a) K^2R_1 and R_2 (b) sK^2R_1 and R_2
(c) K^2R_1 and sR_2 (d) sK^2R_1 and sR_2

[U.P.S.C. I.E.S. E.E.-II, 1998]

423. The torque developed in an induction motor is nearly proportional to

(a) $\frac{1}{V}$ (b) V
(c) V^2 (d) none of these

[U.P.T.U. Even Semester 2009-10]

424. If a 400 V, 50 Hz star-connected, 3-phase squirrel cage induction motor is operated from a 400 V, 75 Hz supply, the torque that the motor can now provide while drawing rated current from the supply

(a) reduces. (b) increases.
(c) remains the same.
(d) increases or reduces depending upon the rotor resistance.

[GATE E.E. 2002]

425. When the applied voltage per phase is reduced to one-half, the starting torque of a three-phase squirrel cage induction motor becomes

(a) 1/2 of the initial value. (b) 1/4 of the initial value.
(c) twice of the initial value. (d) 4 times of the initial value.

[U.P.S.C. I.E.S. E.E.-II, 2006]

426. If the supply voltage decreases by 4%, the torque in a 3-phase induction motor would decrease by

(a) 4% (b) 8% (c) 16% (d) 6%

[W.B. Univ. of Technology, 2013]

427. The condition for maximum starting torque in an induction motor is

(a) $R_2 = \frac{X_2}{2}$ (b) $R_2 = \sqrt{R_1^2 + (X_1 + X_2)^2}$
(c) $R_2 = \frac{X_1 + X_2}{2}$ (d) $R_2 = R_1 + X_1 + X_2$

[W.B. Univ. of Technology, 2013]

428. For maximum starting torque in an induction motor

(a) $R_2 = 0.5\,X_2$ (b) $R_2 = X_2$
(c) $R_2 = 2X_2$ (d) $X_2 = 0$

[W.B. Univ. of Technology, 2011]

429. An induction motor has a rotor resistance of 0.002 ohm/phase. If the resistance is increased to 0.004 ohm/phase then the maximum torque will

(a) reduce to half. (b) increase by 100%.
(c) increase by 200%. (d) remain unaltered.

[U.P.S.C. I.E.S. E.E.-II, 1993]

430. Breakdown torque of a 3-phase induction motor of negligible stator impedance is

(a) directly proportional to the rotor resistance.
(b) inversely proportional to the rotor resistance.
(c) directly proportional to the reactance.
(d) inversely proportional to the rotor leakage reactance.

[U.P.S.C. I.E.S. E.E.-II, 1993, 2008]

431. Maximum torque of a 3-phase induction motor is

(a) independent of rotor resistance r_2'
(b) directly proportional to r_2'
(c) inversely proportional to r_2'
(d) proportional to r_2'

[W.B. Univ. of Technology, 2010]

432. In a 3-phase induction motor, the torque developed is maximum when the rotor circuit resistance per phase is equal to

(a) rotor leakage reactance per phase at standstill.
(b) slip times the rotor leakage reactance per phase at standstill.
(c) stator resistance per phase.
(d) stator leakage reactance per phase. [U.P.S.C. I.E.S. E.E.-II 2004]

433. A 3 phase, 50 Hz, six pole induction motor has a rotor resistance of 0.1 Ω and reactance of 0.92 Ω. Neglect the voltage drop in stator and assume that the rotor resistance is constant. Given that the full load slip is 3%, the ratio of maximum torque to full load torque is

(a) 1.567 (b) 1.712 (c) 1.948 (c) 2.134

[GATE 2014]

434. A 3-phase induction motor is driving a constant torque load of 1 pu at 5% slip. It has a maximum torque of 2 pu at 10% slip. Torque-speed variation in stable zone is assumed to be linear. For a stable operation of the motor, the minimum per unit supply voltage is approximately

(a) 0.25 (b) 0.50 (c) 0.707 (d) 0.80

[U.P.S.C. I.E.S. E.E.-II, 1999]

435. The torque-slip characteristic of a polyphase induction motor becomes almost linear at small values of slip, because in this range of slips

(a) the effective rotor circuit resistance is very large compared to the rotor reactance.
(b) the rotor resistance is equal to the stator resistance.
(c) the rotor resistance is equal to the rotor reactance.
(d) the rotor resistance is equal to the stator reactance.

[U.P.S.C. I.E.S. E.E.-II, 1993]

436. A 3-phase squirrel cage induction motor supplied from a balanced 3-phase source drives a mechanical load. The torque-speed characteristics of the motor (solid curve) and of the load (dotted curve) are shown. Of the two equilibrium points A and B, which of the following options correctly describes the stability of A and B?

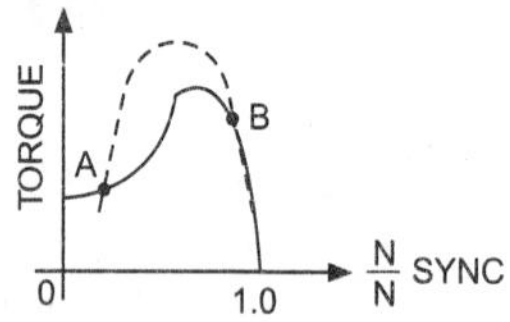

(a) A is stable B is unstable. (b) A is unstable B is stable.
(c) Both are stable. (d) Both are unstable.

[GATE E.E. 2009]

437. What is the ratio of starting torque and maximum torque of a 3- phase, 50 Hz, 4 pole induction motor for a maximum torque at 1,200 rpm?

(a) 0.421 (b) 0.384
(c) 0.6 (d) 0.5 [U.P.S.C. I.E.S. E.E.-II, 2009]

438. In an induction motor, what is the ratio of rotor copper loss and rotor input ?

(a) $1/s$ (b) s (c) $(1 - s)$ (d) $s/(1 - s)$

[U.P.S.C I.E.S. E.E.-II, 2007]

439. For an induction motor, operating at slip *s*, the ratio of gross power output to air gap power is equal to

(a) $(1 - s)^2$ (b) $(1 - s)$
(c) $\sqrt{(1-s)}$ (d) $(1-\sqrt{s})$ [GATE E.E. 2005]

440. In the equivalent circuit of a 3-phase induction motor, the mechanical load on the motor can be represented by a resistance of value (R = rotor resistance, *s* = slip)

(a) R_2/s (b) $\frac{R_2}{1-s}$ (c) $R_2\frac{1-s}{s}$ (d) $\frac{R_2}{s(1-s)}$

[W.B. Univ. of Technology, 2013]

Answers: **423.** (c), **424.** (a), **425.** (b), **426.** (b), **427.** (b), **428.** (b), **429.** (d), **430.** (d), **431.** (a), **432.** (b), **433.** (c), **434.** (c), **435.** (a), **436.** (b), **437.** (b), **438.** (b), **439.** (b), **440.** (c),

441. An induction motor has rotor resistance R_r, standstill rotor induced voltage E_{ss}, and stator to rotor effective turn-ratio of m. In an equivalent circuit of this machine, the rotor circuit resistance is shown as m^2R_r/s, where s is the slip. This implies that the value of the equivalent rotor circuit voltage will be
(a) E_{ss}. (b) sE_{ss}. (c) mE_{ss}. (d) $m\,s\,E_{ss}$.
[U.P.S.C. I.E.S. E. E.-II, 1993]

442. The power factor of an induction motor operating at no load will have a value around
(a) 0.9 lag. (b) 0.2 lead. (c) 0.2 lag. (d) 0.9 lead.
[U.P.S.C. I.E.S. E.E.-II, 2010]

443. The power factor of a lightly loaded induction motor is quite low because
(a) the current drawn is largely a magnetising current due to air gap.
(b) of the current due to air gap.
(c) the current drawn is largely a magnetising component due to laminated core.
(d) the current drawn is largely an energy component due to laminated core.

444. No-load test on a 3-phase induction motor was conducted at different supply voltages and a plot of input power versus voltage was drawn. This curve was extrapolated to intersect the y-axis. This intersection point yields
(a) core loss. (b) stator copper loss.
(c) stray load loss. (d) friction and windage loss.
[GATE E.E. 2003]

445. What is the shunt resistance component in equivalent circuit obtained by no-load test of an induction motor representative of?
(a) Windage and frictional losses only.
(b) Core losses only.
(c) Core, windage and frictional losses.
(d) Copper losses. **[U.P.S.C. I.E.S. E.E.-II, 2005; GATE E.E. 2003]**

446. Under no-load condition, if the applied voltage to an induction motor is reduced from the rated voltage to half the rated value.
(a) the speed decreases and the stator current increases.
(b) both the speed and the stator current decrease.
(c) the speed and the stator current remain practically constant.
(d) there is negligible change in the speed but the stator current decreases. **[GATE E.E. 2005]**

447. Blocked rotor test on a 3-phase induction motor helps to find out
(a) short circuit power factor.
(b) fixed losses.
(c) motor resistance referred to stator.
[W.B. Univ. of Technology, 2010]

448. Which one of the following is correct statement?
The output line in an induction motor circle diagram is the line joining the tip of the
(a) no load current phasor to the point corresponding to slip = 0.
(b) no load current phasor to the point corresponding to slip = 1.
(c) short circuit current phasor to the point corresponding to slip = 1.
(d) short circuit current phasor to the point corresponding to slip = 0. **[U.P.S.C. I.E.S. E.E.-II, 2007]**

449. Consider the following curve:

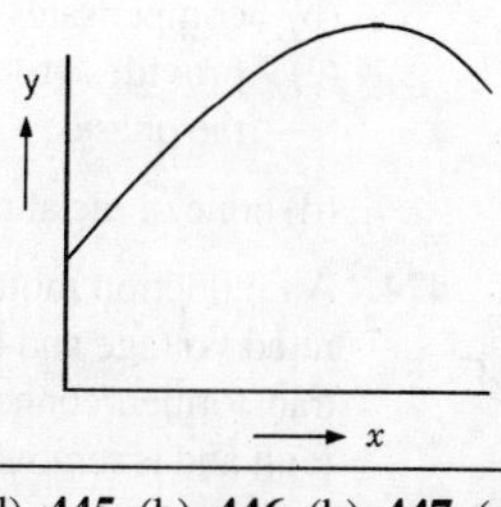

Which of the following characteristics of the induction motor is represented by the above curve?

	x-axis	y-axis
(a)	Output	pf
(b)	Speed	pf
(c)	Load	efficiency
(d)	Speed	torque

[U.P.S.C. I.E.S. E.E.-II, 2004]

450. If the rotor power factor of a 3-phase induction motor is 0.866, the spatial displacement between the stator magnetic field and the rotor magnetic field will be
(a) 30° (b) 90° (c) 120° (d) 150°
[U.P.S.C. I.E.S. E.E.-II, 2000]

451. A 3-phase induction motor when started picks up speed but runs stably at about half the normal speed. This is because of
(a) unbalance in the supply voltages.
(b) non-sinusoidal nature of the supply voltage.
(c) stator circuit asymmetry.
(d) rotor circuit asymmetry. **[U.P.S.C. I.E.S. E.E.-II, 1998]**

452. An induction motor is said to be crawling when
(a) it runs at one-seventh of rated speed.
(b) it accelerates too fast.
(c) it is subjected to fluctuating loads.
(d) it is started on full load.

453. Zero sequence impedance of a 3-phase cage induction motor is due to
(a) stator circuit 3rd time harmonic.
(b) stator circuit 3rd space harmonic.
(c) rotor circuit 3rd time harmonic.
(d) rotor circuit 3rd space harmonic. **[U.P.S.C. I.E.S. E.E.-II, 1998]**

454. The stator of a 6-pole, 3-phase induction motor is fed from a 3-phase 50 Hz supply which contains a pronounced fifth time harmonic. The speed of the fifth space harmonic field produced by the fifth time harmonic in the stator supply will be
(a) 200 rpm. (b) 1,500 rpm.
(c) 1,000 rpm. (d) 5,000 rpm.
[U.P.S.C. I.E.S. E.E.-II, 2000]

455. The synchronous speed for the seventh space harmonic mmf wave of a 3-phase, 8-pole, 50 Hz induction machine is
(a) 107.14 rpm in forward direction.
(b) 107.14 rpm in reverse direction.
(c) 5,250 rpm in forward direction.
(d) 5,250 rpm in reverse direction. **[GATE E.E. 2004]**

456. An induction motor when started on load does not accelerate upto full speed but runs at 1/7th of the rated speed. The motor is said to be
(a) locking. (b) plugging. (c) crawling. (d) cogging.
[U.P.S.C. I.E.S. E.E.-II, 2001]

457. Crawling in an induction motor is due to
(a) time harmonics in supply.
(b) slip ring rotor.
(c) space harmonics produced by winding currents.
(d) insufficient starting torque. **[U.P.S.C. I.E.S. E.E.-II, 1999]**

458. The phenomenon of crawling in a 3-phase induction motor may be due to
(a) unbalanced supply voltage.
(b) 7th space harmonics of air-gap field.
(c) 7th time harmonics of voltage wave.
(d) 5th space harmonics. **[U.P.S.C. I.E.S. E.E.-II, 1998]**

459. In an induction motor, when the number of stator slots is equal to an integral multiple of rotor slots,
(a) there may be a discontinuity in torque-slip characteristics.

Answers: 441. (c), **442.** (c), **443.** (a), **444.** (d), **445.** (b), **446.** (b), **447.** (a), **448.** (b), **449.** (a), **450.** (c), **451.** (b), **452.** (a), **453.** (b), **454.** (d), **455.** (a), **456.** (c), **457.** (c), **458.** (b), **459.** (d),

(b) a high starting torque will be available.
(c) the maximum torque will be high.
(d) the machine will fail to start. [U.P.S.C. I.E.S. E.E.-II, 2003]

460. Cogging and crawling are phenomenon associated with
(a) cage induction machines and they are essentially the same.
(b) squirrel cage induction machines, the former during starting and the latter at a fraction of its rated speed.
(c) squirrel cage induction machines, the former at a fraction of its rated speed and the latter during starting.
(d) wound rotor induction machines and they are reduced by skewing, chording and distribution of windings.
[U.P.S.C. I.E.S. E.E.-II, 1995]

461. In an induction motor, if the air gap is increased
(a) speed will reduce. (b) efficiency will improve.
(c) power factor will be lowered.
(d) breakdown torque will reduce. [GATE E.E. 1996]

462. If two induction motors A and B are identical except that the air-gap of motor 'A' is 50% greater than that of motor 'B' then
(a) the no-load power factor of A will be better than that of B.
(b) the no-load power factor of A will be poorer than that of B.
(c) the core losses of A will be more than those of B.
(d) the operating flux of A will be smaller than that of B.
[U.P.S.C. I.E.S. E.E.-II, 1997]

463. Semiclosed or totally closed slots are used in induction motors essentially to
(a) improve pull-out torque. (b) increase pull-out torque.
(c) increase efficiency.
(d) reduce magnetizing current and improve power factor.
[U.P.S.C. I.E.S. E.E.-II, 1994]

464. ''Cogging'' in induction motor occurs when
(a) number of stator teeth – number of rotor teeth = odd number.
(b) number of stator teeth – number of rotor teeth = even number.
(c) number of stator teeth – number of rotor teeth = 0.
(d) number of stator teeth – number of rotor teeth = negative number. [U.P.S.C. I.E.S. E.E.-II, 1992]

465. The required load torque line intersects the resultant torque-speed characteristic of a 3-phase squirrel cage induction motor at points P, Q and R as shown in the figure. Which is/are the stable operating point(s)?

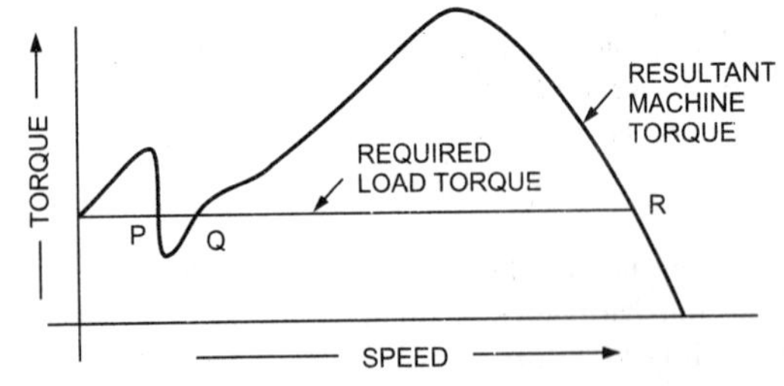

(a) P and Q. (b) Q and R.
(c) P and R. (d) Only R.
[U.P.S.C. I.E.S. E.E.-II, 2006]

466. Which of the following statements regarding skewing of rotor bars in a squirrel cage induction motor are correct ?
1 It prevents cogging.
2 It produces more uniform torque.
3 It increases starting torque.
4 It reduces motor 'hum' during its operation.
Select the answer using the codes given below
(a) 2, 3, 4. (b) 1, 2, 3.
(c) 1, 3, 4. (d) 1, 2, and 4.
[U.P.S.C. I.E.S. E.E.-II, 1997]

467. In the equivalent circuit of a double-cage induction motor, the two rotor cages can be considered
(a) to be in parallel. (b) to be in series-parallel.
(c) to be in series. (d) to be in parallel with stator.
[U.P.S.C I.E.S. E.E.-II, 2008]

468. The deep bar rotor of an induction motor is considered equivalent to a double cage rotor for the purpose of starting and performance. At the time of starting, the current distribution in the deep bar rotor slot will be as shown

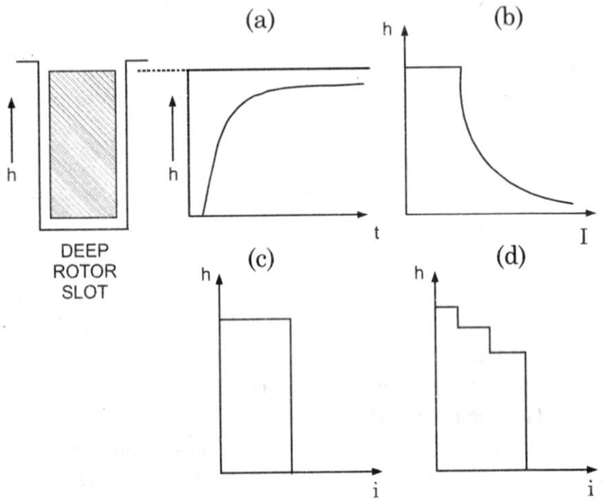

[U.P.S.C. I.E.S. E.E.-II, 1998]

469. The advantage of the double squirrel-cage induction motor over single cage rotor is that its
(a) efficiency is higher. (b) power factor is higher.
(c) slip is larger. (d) starting current is lower.
[U.P.S.C. I.E.S. E.E.-II 2001]

470. Motor A has shallow and wider slots. Motor B has deeper and narrow slots. If both are 3-phase 400 V, 50 Hz, 1440 rpm induction motors. It can be concluded that
(a) Motor A has more starting torque as compared to motor B.
(b) Motor B has more starting torque as compare to motor A.
(c) Motor A has more pull-out torque as compared to motor B.
(d) Motor B has more pull-out torque as compared to motor A.
[U.P.S.C. I.E.S. E.E.-II, 1992]

471. Unbalanced supply voltage given to a 3-phase, delta-connected induction motor will cause
(a) zero sequence currents. (b) less heating of motor.
(c) negative sequence component current.
(d) all the above. [GATE E.E. 1996]

472. A 4-pole induction motor, supplied by a slightly unbalanced three-phase, 50 Hz source, is rotating at 1,440 rpm. The electrical frequency in Hz of the induced negative sequence current in rotor is
(a) 100 (b) 98 (d) 52 (d) 48
[GATE 2013]

473. Single phase preventer
(a) suppresses negative sequence currents.
(b) compensates for voltage drops.
(c) provides protection in the event of non-availability of one of the phases.
(d) none of the above.

474. An induction motor is fed from a balanced three phase supply at rated voltage and frequency through a bank of three single phase transformers connected in delta-delta. One unit of the bank develops fault and is removed. Then

Answers: 460. (b), **461.** (c), **462.** (b), **463.** (d), **464.** (c), **465.** (c), **466.** (c), **467.** (a), **468.** (a), **469.** (d), **470.** (b), **471.** (c), **472.** (b), **473.** (c), **474.** (a),

(a) single phasing will occur and the machine fails to start.
(b) single phasing will not occur but the motor terminal voltages will become unbalanced and the machine can be loaded to the extent of 57.7% of its rating.
(c) the machine can be loaded to the extent of 57.7% of its rating with balanced supply at its terminals.
(d) the machine can be loaded to the extent of $66\frac{2}{3}$ % of its rating with balanced supply at its terminals. **[GATE E.E. 1995]**

475. Consider the following statements:
In a 3-phase induction motor connected to a 3-phase supply; if one of the lines suddenly gets disconnected, then the
1. motor will come to a standstill.
2. motor will continue to run at the same speed with line current unchanged.
3. motor will continue to run at a slightly reduced speed with increase in line current.
4. rotor current will have both of sf and $(2–s)f$ component frequencies where s is the slip and f is the supply frequency.

Of these statements
(a) 1 and 4 are correct. (b) 1 and 2 are correct.
(c) 3 and 4 are correct. (d) 2 and 3 are correct.
[U.P.S.C. I.E.S. E.E.-II, 1995]

476. The following starting method for an induction motor is inferior from the point of view of poor starting torque per ampere of the line current drawn
(a) Direct-on-line starting.
(b) Auto transformer method of starting.
(c) Series inductor method of starting.
(d) Star-delta method of stating. **[GATE E.E. 1999]**

477. The starting current of an induction motor is 5 times the full-load current while the full-load slip is 4%. What is the ratio of starting torque to full-load torque?
(a) 0.6 (b) 0.8
(c) 1.0 (d) 1.2 **[U.P.S.C. I.E.S. E.E.-II, 2009]**

478. A 3-phase, 4 pole, 400 V, 50 Hz, star-connected induction motor has following circuit parameters

$$r_1 = 1.0\ \Omega,\ r_2' = 0.5\ \Omega,\ x_1 = x_2' = 1.2\ \Omega,\ x_m = 35\ \Omega$$

The starting torque when the motor is started direct-on-line is (use approximate equivalent circuit model)
(a) 63.6 Nm (b) 74.3 Nm (c) 190.8 Nm (c) 222.9 Nm
[GATE E.E. 2006]

479. Star-delta starting of a polyphase induction motor is equivalent to auto-transformer starting with
(a) 85% tapping. (b) 58% tapping.
(c) 52% tapping. (d) 33% tapping.
[W.B. Univ. of Technology, 2010]

480. Consider the following statements:
Star-delta starter is used in 3-phase induction motor because it
1. prevents heating of the motor windings.
2. ensures permissible minimum starting current.
3. is regulated by electricity authorities.
4. ensures smooth run-up to full load. Of these statements.

(a) 1, 2 and 3 are correct. (b) 2, 3 and 4 are correct.
(c) 1, 3 and 4 are correct. (d) 1 and 2 are correct.
[U.P.S.C. I.E.S. E.E.-II, 1998]

481. A 3-phase slip-ring induction motor having negligible stator impedance drives a constant torque load. If an additional resistance is included in the rotor circuit, what does the motor experience?
(a) Increase in both the stator current and the slip.
(b) No change in the stator current and increase in the slip.
(c) Increase in the ststor current and no change in the slip.
(d) Decrease in the stator current and increase in the slip.
[U.P.S.C I.E.S. E.E.-II, 2008]

482. Which of the following methods are suitable for the speed control of squirrel cage induction motors?
1. Voltage control. 2. Rotor resistance control.
3. Frequency control. 4. Pole changing method.

Select the correct answer using the codes given below:
(a) 2, 3 and 4. (b) 1, 3 and 4.
(c) 1, 2 and 3. (d) 2 and 4.
[U.P.S.C. I.E.S. E.E.-II, 2005]

483. In a constant V/f control of induction motor, the ratio V/f is maintained constant from 0 to base frequency, where V is the voltage applied to the motor at fundamental frequency f. Which of the following statements relating to low frequency operation of the motor is TRUE?
(a) At low frequency, the stator flux increases from its rated value.
(b) At low frequency, the stator flux decreases from its rated value.
(c) At low frequency, the motor saturates.
(d) At low frequency, the stator flux remains unchanged at its rated value. **[GATE 2014]**

484. A 3-phase induction motor fed from a 3-phase voltage regulator is suitable for driving loads whose torque is
(a) constant irrespective of speed.
(b) inversely proportional to the square of the speed.
(c) directly proportional to the square of the speed.
(d) inversely proportional to speed. **[U.P.S.C. I.E.S. E.E.-II, 1998]**

485. While operating on variable frequency supplies, the ac motor requires variable voltage as well in order to
(a) protect the insulation. (b) avoid the effect of saturation.
(c) improve the capabilities of the inverter.
(d) protect the thyristor from $\frac{dv}{dt}$. **[U.P.S.C. I.E.S. E.E.-II, 1993]**

486. An ac induction motor is used for a speed control application. It is driven from an inverter with a constant V/f control. The motor name plate details are as follows:
V : 415 V Ph : 3
f : 50 Hz N : 2,850 rpm
The motor is run with the inverter output frequency set at 40 Hz, and with half the rated slip. The running speed of the motor is
(a) 2,400 rpm. (b) 2,280 rpm.
(c) 2,340 rpm. (d) 2,790 rpm. **[GATE E.E. 2003]**

487. The speed of a 4-pole induction motor is controlled by varying the supply frequency while maintaining the ratio of supply voltage to supply frequency (V/f) constant. At rated frequency of 50 Hz and rated voltage of 400 V its speed is 1,440 rpm. Find the speed at 30 Hz, if the load torque is constant.
(a) 882 rpm. (b) 864 rpm. (c) 840 rpm. (d) 828 rpm.
[GATE E.E. 2006]

488. A 3-phase induction motor is driving a constant torque load at rated voltage and frequency. If both voltage and frequency are halved, following statements relate to new condition if stator resistance, leakage reactance and core loss are ignored.
P. The difference between synchronous speed and actual speed remains the same.
Q. The air gap flux remains same.
R. The stator current remains same.
S. The pu slip remains same.

Answers: 475. (c), **476.** (c), **477.** (c), **478.** (a), **479.** (b), **480.** (a), **481.** (d), **482.** (b), **483.** (b), **484.** (c), **485.** (b), **486.** (c), **487.** (c), **488.** (b).

Among the above, correct statements are
(a) All. (b) P, Q and R.
(c) Q, R and S. (d) P and S. [GATE E.E. 2003]

489. The absolute speed of magnetic field in space of a 3-phase rotor fed induction motor is
(a) synchronous speed N_s. (b) rotor speed N_r.
(c) $(N_s - N_r)$. (d) $(N_s + N_r)$.
[U.P.S.C. I.E.S. E.E.-II, 1996]

490. Consider the following statements regarding speed control of induction motors by means of external rotor resistors :
1. Regulation in speed is accompanied by reduced efficiency.
2. With a large resistance in the rotor circuit, the speed would vary considerably with variation in torque.
3. The method is very complicated.

The disadvantages of such a method of speed control would include
(a) 1 and 2. (b) 2 and 3. (c) 1 and 3. (d) 1,2 and 3.
[U.P.S.C. I.E.S. E.E.-II, 1997]

491. The injected emf in the rotor of induction motor must have
(a) the same frequency as the stator slip frequency.
(b) the same phase as the rotor emf.
(c) a high value for satisfactory speed control.
(d) the same phase as the rotor emf and a high value for satisfactory speed control. [U.P.S.C. I.E.S. E.E.-II, 2003]

492. When will a slip-ring induction motor run at super synchronous speed?
(a) If a voltage is injected in the rotor circuit in phase opposition to the rotor induced emf.
(b) If an emf is injected in the rotor circuit in phase with the rotor induced emf.
(c) If motor is coupled with active load.
(d) If motor is coupled with passive load.
[U.P.S.C. I.E.S. E.E.-II, 2009]

493. A voltage source inverter is used to control the speed of a three-phase, 50 Hz, squirrel cage induction motor. Its slip for rated torque is 4%. The flux is maintained at rated value. If the stator resistance and rotational losses are neglected, then the frequency of the impressed voltage to obtain twice the rated torque at starting should be
(a) 10 Hz (b) 5 Hz
(c) 4 Hz (d) 2 Hz [GATE E.E., 2007]

494. A three-phase slip-ring induction motor, provided with a commutator winding, is shown in the figure. The motor rotates in clockwise direction when the rotor windings are closed.

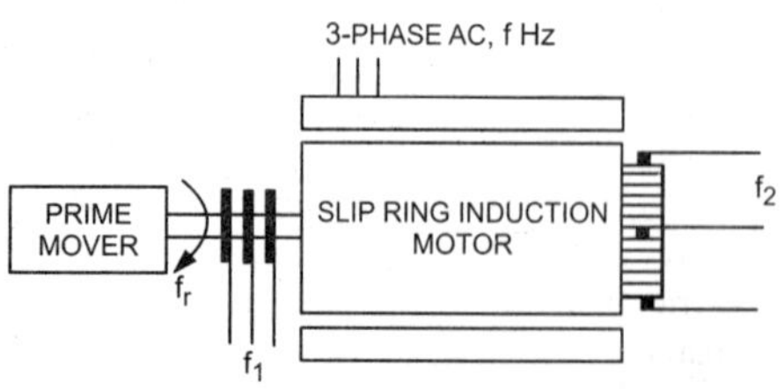

If the rotor winding is open circuited and the system is made to run at rotational speed f_r with the help of prime-mover in anti-clockwise direction, then the frequency of voltage across slip rings is f_1 and frequency of voltage across commutator brushes is f_2. The values of f_1 and f_2 respectively are
(a) $f+f_r$ and f (b) $f-f_r$ and f
(c) $f-f_r$ and $f+f_r$ (d) $f+f_r$ and $f-f_r$ [GATE 2014]

495. A separately excited dc machine is coupled to a 50 Hz, three-phase, 4-pole induction machine as shown in the figure. The dc machine is energized first and the machines rotate at 160 rpm. Subsequently the induction machine is also connected to a 50 Hz, three-phase source, the phase sequence being consistent with the direction of rotation. In steady state,

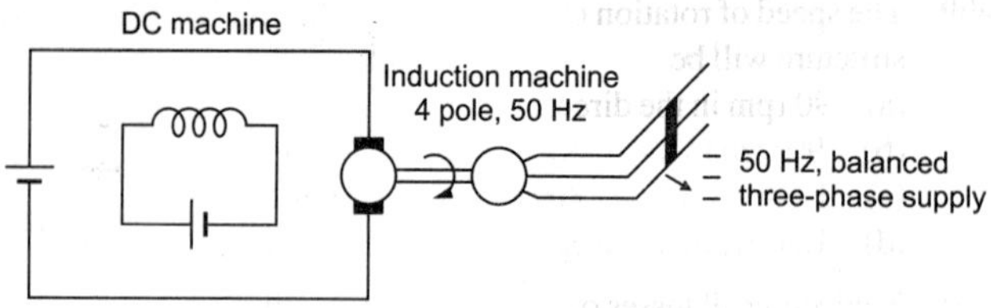

(a) both machines act as generators.
(b) the dc machine acts as a generator, and the induction machine acts as a motor.
(c) the dc machine acts as a motor, and the induction machine acts as a generator.
(d) both machines act as motors. [GATE E.E., 2010]

496. Cascade method of speed control involves the use of two coupled induction motors. The necessary condition for speed control is that
(a) both the motors are of the wound rotor type having the same number of poles.
(b) both the motors are of the squirrel cage rotor type having different number of poles.
(c) one motor is of the slip-ring type but both the motors have the same number of poles.
(d) one motor is of the slip-ring type and the two motors have different number of poles. [U.P.S.C. I.E.S. E.E.-II, 1993]

497. In a slip power recovery scheme for a 3-phase induction motor, if slip power is
(a) returned to supply, constant power drive is obtained.
(b) added to the main shaft, constant power drive is obtained.
(c) subtracted from the main shaft, constant torque drive is obtained.
(d) obtained from the supply, constant torque drive is obtained.

498. A three phase, wound rotor induction motor is to be operated with slip energy recovery in the constant torque mode, when it delivers an output power P_0 at slip s. Then theoretically, the maximum power that is available for recovery at the rotor terminals, is equal to

(a) P_0 (b) $P_0 s$ (c) $\dfrac{P_0}{(1-s)}$ (d) $\dfrac{P_0 s}{(1-s)}$
[GATE E.E. 2000]

Statement for Linked Answer Questions 499 and 500

A 3-phase, 440 V, 50 Hz, 4 pole, slip ring induction motor is fed from the rotor side through an auto transformer and the stator is connected to a variable resistance as shown in the figure.

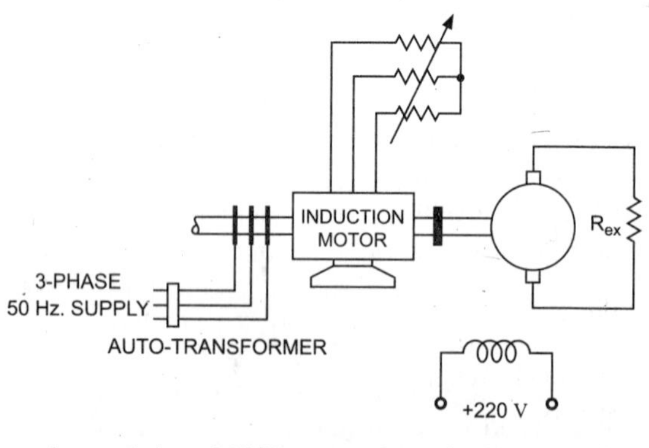

The motor is coupled to a 220 V, separately excited dc generator feeding power to fixed resistance of 10 Ω. Two wattmeter method is used to

Answers: **489.** (c), **490.** (a), **491.** (a), **492.** (a), **493.** (c), **494.** (a), **495.** (c), **496.** (d), **497.** (b), **498.** (b), measure the input power to induction motor. The variable resistance is adjusted such that motor runs at 1,410 rpm and the following readings were recorded W_1 = 1,800 W, W_2 = –200 W.

499. The speed of rotation of stator magnetic field with respect to rotor structure will be
(a) 90 rpm in the direction of rotation.
(b) 90 rpm in the opposite direction of rotation.
(c) 1,500 rpm in the direction of rotation.
(d) 1,500 rpm in the opposite direction of rotation. **[GATE E.E. 2008]**

500. Neglecting all losses of both the machines, the dc generator power output and the current through resistance (R_{ex}) will respectively be
(a) 96 W, 3.10 A (b) 120 W, 3.46 A
(c) 1,504 W, 12.26 A (d) 1,880 W, 13.71 A
[GATE E.E. 2008]

501. In dynamic braking of 3-phase induction motors
(a) the supply terminals of any two stator phases are interchanged.
(b) any two stator terminals are earthed.
(c) the stator terminals are switched over to a dc source from the ac supply.
(d) a dc voltage is injected in the rotor circuit.

502. In a 3-phase induction motor, the regenerative braking occur when
(a) the load is lowered by a hoisting machine.
(b) the load is raised by a hoisting machine.
(c) the number of poles is reduced in a pole-changing motor.
(d) the motor speed falls due to overload.

503. The stator of a $\frac{2}{4}$ pole changing cage motor is initially wound for 2-poles. The reconnection of the stator winding to 4-poles through a change over switch, while the motor is running would result in
(a) constant torque drive. (b) constant hp drive.
(c) plugging to standstill.
(d) regenerative braking to half the original speed.
[U.P.S.C. I.E.S. E.E.-II, 1992]

504. A 3-phase wound rotor induction motor, when started with load connected to its shaft, was found to start but settle down at about half synchronous speed. If the rotor winding as well as stator winding were star-connected, the cause of the malfunctioning could be attributed to
(a) one of the stator phase winding being short-circuited.
(b) one of the supply fuses being blown.
(c) one of the rotor phase being open-circuited.
(d) two of the rotor phases being open-circuited.
[U.P.S.C. I.E.S. E.E.-II, 1997]

505. The rotor slots in an induction motor are, usually not kept parallel to the shaft as it
(a) helps the rotor teeth to remain under the stator teeth.
(b) helps in reducing the tendency of the rotor teeth to remain under the stator teeth.
(c) improves pf. (d) improves efficiency.

506. When a single phase induction motor is excited with single phase ac voltage, the magnetic field set up is equivalent to
(a) two fields, rotating in opposite directions with different speeds.
(b) two fields, rotating at synchronous speed in opposite directions.
(c) two fields, rotating at synchronous speed.
(d) two fields, rotating in the same direction but at different speeds. **[U.P.S.C. I.E.S. E.E.-II 2004]**

507. A single phase induction motor with only the main winding excited would exhibit the following response at synchronous speed
(a) Rotor current is zero.
(b) Rotor current is non-zero and is at slip frequency.
(c) Forward and backward totaling fields are equal.
(d) Forward rotating field is more than backward rotating field.
[GATE E.E. 2003]

508. The backward rotor slip in a single-phase induction motor is equal to
(a) $1 - s$ (b) s (c) $2 - s$ (d) $s/2$
[W.B. Univ. of Technology, 2011-12]

509. An 8-pole single phase induction motor is running at 690 rpm. What is its slip with respect to forward and backward fields, respectively?
(a) 0.08, 2.0 (b) 0.08, 1.92
(c) 1.92, 0.08 (d) 2.0, 0.08
[U.P.S.C. I.E.S. E.E.-II, 2007]

510. A 230 V, 50 Hz, 4 pole, single-phase induction motor is rotating in the clockwise (forward) direction at a speed of 1,425 rpm. If the rotor resistance at standstill is 7.8 Ω, then the effective rotor resistance in the backward branch of the equivalent circuit will be
(a) 2 Ω (b) 4 Ω (c) 78 Ω (d) 156 Ω
[GATE E.E. 2008]

511. When the main winding of a single-phase induction motor is connected to 220 V, 50-Hz source, it takes a current of 50 $\angle -60°$ A at standstill. Neglecting stator leakage impedance and magnetizing current, the rotor circuit parameters r_2 and x_2 in ohms are respectively
(a) 2, 3.464 (b) 3.464, 2 (c) 4, 6.928 (d) 6.928, 4
[W.B. Univ. of Technology, 2011-12]

512. Phase splitting can be accomplished in a single phase induction motor
(a) only by adding a capacitor in series with the auxiliary winding.
(b) only by causing the auxiliary winding to have high reactance.
(c) only by causing the auxiliary winding to have low resistance.
(d) by any one of the above three methods.

513. Why is a centrifugal switch used in a single-phase induction motor?
(a) To protect the motor from overloading.
(b) To improve the starting performance of the motor.
(c) To cut off the starting winding at an appropriate instant.
(d) To cut in the capacitor during running conditions.
[U.P.S.C. I.E.S. E.E.-II, 2008]

514. The torque-speed characteristic of two phase induction motor is largely affected by
(a) voltage. (b) speed.
(c) $\frac{X}{R}$ ratio. (d) supply frequency.
[U.P.S.C. I.E.S. E.E.-II, 1992]

515. In a capacitor-start single-phase induction motor, a capacitor is connected
(a) in series with the main winding.
(b) in parallel with the main winding.
(c) in series with the auxiliary winding.
(d) in parallel with the auxiliary winding.
[W.B. Univ. of Technology, 2012-13]

516. In a single phase induction motor driving a fan load, the reason for having a high resistance rotor is to achieve
(a) low starting torque. (b) quick acceleration.
(c) high efficiency. (d) reduced size. **[GATE E.E. 2005]**

517. Capacitor in a single-phase induction motor is used for
(a) improving the power factor.
(b) improving the starting torque.
(c) starting the motor.
(d) reducing the harmonics. **[U.P.S.C. I.E.S. E.E.-II, 2002]**

Answers: 499. (a), **500.** (c), **501.** (c), **502.** (a), **503.** (d), **504.** (c), **505.** (b), **506.** (b), **507.** (d), **508.** (c), **509.** (b), **510.** (b), **511.** (a), **512.** (d), **513.** (c), **514.** (c), **515.** (c), **516.** (b), **517.** (c),

518. For a single phase capacitor start induction motor which of the following statements is valid?
(a) The capacitor is used for power factor improvement.
(b) The direction of rotation can be changed by reversing the main winding terminals.
(c) The direction of rotation cannot be changed.
(d) The direction of rotation can be changed by interchanging the supply terminals. **[GATE E.E. 2006]**

519. Which one of the following capacitor-start split-phase motor will have the largest value of capacitance ?
(a) $\frac{1}{2}$ hp; 3,450 rpm. (b) $\frac{1}{4}$ hp; 1,725 rpm.
(c) $\frac{1}{2}$ hp; 1,140 rpm. (d) $\frac{3}{4}$ hp; 1,140 rpm.
[U.P.S.C. I.E.S. E.E.-II, 1993]

520. A single phase, 230 V, 50 Hz, 4-pole, capacitor-start induction motor has the following standstill impedances.
Main winding $Z_m = 6.0 + j\,4.0\ \Omega$
Auxiliary winding, $Z_a = 8.0 + j\,6.0\ \Omega$
The value of the starting capacitor required to produce 90° phase difference between the currents in the main and auxiliary windings will be
(a) 176.84 μF. (b) 187.24 μF.
(c) 265.26 μF. (d) 280.86 μF. **[GATE E.E 2004]**

521. In a capacitor-start and run 1-phase induction motor, starting capacitor C_1 and permanent capacitor C_2 are as under:
(a) C_1 : electrolytic type; C_2 : paper type.
(b) C_1 : electrolytic type; C_2 : mica type.
(c) C_1 : mica type; C_2 : electrolytic type.
(d) C_1 : paper type; C_2 : electrolytic type.

522. Which one of the following statements regarding a capacitor-start, capacitor-run single phase induction motor is correct?
(a) One of the capacitors is in the circuit during starting and both are in the circuit during running.
(b) Both the capacitors are in the circuit during starting and one of them in the circuit during running.
(c) Both the capacitors are in the circuit during starting and running.
(d) One of the capacitor is in the circuit during starting and other is in the circuit during running.
[U.P.S.C. I.E.S. E.E.-II, 1999]

523. A ceiling fan uses
(a) split phase motor.
(b) capacitor-start, capacitor-run motor.
(c) universal motor. (d) capacitor-start motor.
[W.B. Univ. of Technology, 2012-13]

524. In a shaded-pole motor, shading coils are used to
(a) reduce winding losses. (b) reduce friction losses.
(c) produce rotating magnetic field.
(d) protect against sparking. **[U.P.S.C. I.E.S. E.E.-II, 1992]**

525. In a shaded-pole induction motor, the rotor runs from the
(a) shaded portion to the unshaded portion of the pole while the flux in the former leads that of the latter.
(b) shaded portion to the unshaded portion of the pole while the flux in the former lags that in the latter.
(c) unshaded portion to the shaded portion while the flux in the former leads that in the latter.
(d) unshaded portion to the shaded portion while the flux in the former lags that in the latter. **[U.P.S.C. I.E.S. E.E.-II 2001]**

526. Consider the following statements regarding fractional horse power shaded pole motor;
1. Its direction of rotation is from unshaded to shaded portion of the poles.
2. Its direction of rotation is from shaded to unshaded portion of the poles.
3. It can remain stalled for short periods without any harm.
4. It has a very poor power factor.
Of these statements
(a) 1,3 and 4 are correct. (b) 2,3 and 4 are correct.
(c) 2 and 4 are correct. (d) 1 and 3 are correct.
[U.P.S.C. I.E.S. E.E.-II, 1997]

527. Which of the following are the features of a shaded pole motor?
1. Salient pole stator. 2. Uniform air gap.
3. Two stator windings one of which is a short-circuited ring.
4. Squirrel cage rotor.
Select the correct answer using the codes given below:
Codes:
(a) 1 and 4. (b) 2 and 4.
(c) 1, 3 and 4. (d) 1, 2 and 3.
[U.P.S.C. I.E.S. E.E.-II, 1998]

528. The type of single-phase induction motor having the highest power factor at full load is
(a) shaded pole type. (b) split-phase type.
(c) capacitor- start type. (d) capacitor-run type.
[GATE E.E. 2004]

529. Consider the following types of single phase motors:
1. Capacitor start induction motor.
2. Capacitor start and run induction motor.
3. Permanent split capacitor motor.
4. Shaded pole motor.
The correct sequence of these in the ascending order of magnitude of starting torque is
(a) 1-2-3-4. (b) 2-3-4-1. (c) 3-4-2-1. (d) 4-3-2-1.
[U.P.S.C. I.E.S. E.E.-II, 2003]

530. Consider the following single-phase motors
I Capacitor start motor II Capacitor start and run motor
III Permanent split capacitor motor
IV Shaded pole motor
The correct sequence of the increasing order of their cost is
(a) IV, III, II, I. (b) IV, III, I, II.
(c) III, IV, II, I. (d) III, IV, I, II.
[U.P.S.C. I.E.S. E.E.-II, 1993]

531. In hand tool applications, which one of the following single-phase motors is used?
(a) Shaded pole motor. (b) Capacitor start motor.
(c) Capacitor run motor. (d) AC series motor.
[U.P.S.C. I.E.S. E.E.-II, 2005]

532. A 200 V, 50 Hz, single-phase induction motor has the following connection diagram and winding orientations shown. MM′ is the axis of the main stator winding (M_1M_2) and AA′ is that of the auxiliary winding (A_1A_2). Directions of the winding axis indicate direction of flux when currents in the windings are in the directions shown. Parameters of each winding are indicated. When switch S is closed, the motor

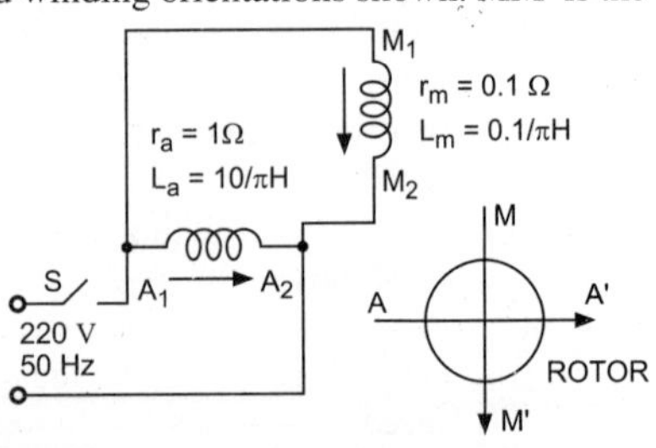

Answers: 518. (b), **519.** (d), **520.** (a), **521.** (a), **522.** (b), **523.** (a), **524.** (c), **525.** (c), **526.** (a), **527.** (a), **528.** (d), **529.** (d), **530.** (b), **531.** (d), **532.** (b),

(a) rotates clockwise. (b) rotates anticlockwise.
(c) does not rotate.
(d) rotates momentarily and comes to a halt. **[GATE E.E. 2009]**

533. A 6-pole, 50 Hz, single-phase induction motor runs at a speed of 900 rpm. The frequency of current in the cage rotor will be
(a) 5 Hz (b) 55 Hz (c) 95 Hz (d) 100 Hz
[W.B. Univ. of Technology, 2012-13]

534. A dc series motor is accidently connected to single phase ac supply. The torque produced will be
(a) of zero average value. (b) oscillating.
(c) steady and unidirectional.
(d) pulsating and unidirectional. **[U.P.S.C. I.E.S. E.E.-II, 2002]**

535. A 230 V dc series motor is connected is a 230 V ac supply
(a) the motor will vibrate violently.
(b) the motor will run with less efficiency and more sparking.
(c) the motor will not run.
(d) the fuse will be blown. **[W.B. Univ. of Technology, 2011-12]**

536. In an ac series motor, compensation winding is provided to
(a) neutralize the armature mmf.
(b) improve commutation and reduce sparking at the brushes.
(c) reduce reactance drop.
(d) reduce armature heating and vibrations.
(e) all of the above except (d).

537. For improvement of commutation in ac series motors shunted commutating poles are commonly used and commutating pole windings are shunted by a
(a) non-inductive resistor to make the commutating pole current to lead the main current.
(b) non-inductive resistor to make the commutating pole current to lag behind the main current.
(c) reactor to make the commutating pole current to lag behind the main current.
(d) given statements are false.

538. A universal motor runs at
(a) higher speed with dc supply and with less sparking.
(b) higher speed with ac supply and with less sparking.
(c) same speed with both ac and dc supplies.
(d) higher speed with ac supply but with increased sparking at the brushes. **[U.P.S.C.I.E.S. E. E.-II, 1994]**

539. The speed of a universal motor can be controlled by
(a) introducing a variable resistance in series with the motor.
(b) tapping the field at various points.
(c) centrifugal mechanism.
(d) all of the above steps. **[U.P.S.C. I.E.S. E.E.-II, 1993]**

540. In universal motors, normally the ratio of width of brushes to the width of commutator segments is
(a) 2 : 1 (b) 1 : 1 (c) 1 : 2 (d) 4 : 1

541. Repulsion motor
(a) receives its armature current by induction.
(b) is similar to a series motor except that rotor and stator windings are inductively coupled.
(c) can be directly switched on to supply line.
(d) has high starting torque and high speed on no load.
(e) all of the above.

542. The commutation of repulsion motor is at its best
(a) during starting.
(b) when runs near synchronous speed.
(c) when runs at 50% of synchronous speed.
(d) when runs at 150% of synchronous speed.

543. The torque-speed characteristic of a repulsion motor resembles which of the following dc motor characteristic ?
(a) Separately excited. (b) Shunt.
(c) Series. (d) Compound. **[GATE E.E. 1996]**

544. In a repulsion motor by shifting the brush position
(a) speed and direction of rotation can be controlled.
(b) only speed can be controlled.
(c) speed, torque and direction of rotation can be controlled.
(d) only torque can be controlled by varying the speed at constant output power.

545. Which of the following statements regarding repulsion start induction motor is/are correct ?
(a) It runs below synchronous speeds on all loads.
(b) It needs more maintenance of commutator and other mechanical devices.
(c) It makes quite a bit noise on starting.
(d) It starts as a repulsion motor with its brushes set to give maximum torque.
(e) all of the above.

546. The repulsion induction motor does not have the drawback of
(a) higher cost. (b) careful maintenance.
(c) poor starting torque. (d) tendency of sparking.

547. A single phase reluctance motor
(a) has salient pole rotor structure and runs at subsynchronous speed.
(b) has salient pole rotor structure and runs at super-synchronous speed.
(c) has salient pole rotor structure and runs at synchronous speed.
(d) has non-salient pole rotor structure and runs at synchronous speed.

548. Reluctance torque is the
(a) torque developed in a given synchronous motor as a function of field excitation and the sine of the torque angle.
(b) torque developed by salient pole synchronous or synchronous induction motors due to variations in air gap produced by armature reaction.
(c) torque developed by non-salient pole asynchronous motor.
(d) torque developed by none of the above.

549. For a reluctance type motor, if the stator magnetic field angular velocity is ω and the actual rotor angular velocity is ω_r, then which of the following equations is satisfied if the average electromagnetic torque is not zero?
(a) $\omega = \omega_r/2$ (b) $\omega = \omega_r$
(c) $\omega = 2\omega_r$ (d) $\omega = 4\omega_r$
[U.P.S.C. I.E.S. E.E.-II, 2004]

550. The torque of a reluctance motor can be effectively increased by
(a) increasing reluctance of the magnetic circuit along the direct axis.
(b) decreasing reluctance of the magnetic circuit along the quadrature axis.
(c) increasing the ratio of quadrature axis reluctance to direct-axis reluctance.
(d) decreasing the ratio of quadrature axis reluctance to direct-axis reluctance. **[U.P.S.C. I.E.S. E.E.-II, 2001]**

551. In a single phase hysteresis motor
(a) starting torque is caused by both eddy current and hysteresis while the running torque is caused by hysteresis.
(b) starting as well as running torques are caused by both eddy current and hysteresis.

Answers: **533.** (a), **534.** (d), **535.** (b), **536.** (e), **537.** (b), **538.** (a), **539.** (d), **540.** (a), **541.** (e), **542.** (b), **543.** (c), **544.** (c), **545.** (e), **546.** (c), **547.** (c), **548.** (b), **549.** (b), **550.** (c), **551.** (a),

(c) starting torque is caused by only eddy current while running torque is caused by only hysteresis.
(d) starting as well as running torques both are caused by only hysteresis.

552. Hysteresis motors are used for Hi-Q record players because of their
(a) constant (synchronous) speed.
(b) extremely steady torque.
(c) insensitivity to supply voltage fluctuations.
(d) non-dependence on centrifugal switch requirement.
[U.P.S.C. I.E.S. E.E.-II, 1998]

553. The main reason for using a hysteresis motor for high quality tape recorders and record players is that
(a) its speed is constant (synchronous).
(b) it develops extremely steady torque.
(c) it requires no centrifugal switch.
(d) its operation is not affected by mechanical vibrations.
[U.P.S.C. I.E.S. E.E.-II, 1995]

554. A fluctuating voltage supply is detrimental to a refrigerator motor but not to a ceiling fan, although both are single phase induction motors because, the refrigerator motor
(a) is made more robust than the fan motor.
(b) is subjected to short duty cycle but the fan motor is subjected to continuous duty.
(c) is enclosed in a sealed unit while the fan motor is open to the environment.
(d) load is constant, but the fan motor load is voltage dependent.
[U.P.S.C. I.E.S. E.E.-II, 1993]

555. An electric motor gets overheated. This may be due to
(a) overloading. (b) shorted stator winding.
(c) worn-out or dry bearings.
(d) low or high voltage. (e) any of the above.

556. If the centrifugal switch fails to open when the motor attains 70 to 80 per cent of synchronous speed, it would result in
(a) overloading of main winding.
(b) damage to centrifugal switch.
(c) overheating of auxiliary winding due to continuous flow of current and may get damaged.
(d) none of the above.

557. The stator and rotor pole number may be different in a
(a) pole changing induction motor.
(b) reluctance motor. (c) repulsion motor.
(d) synchronous motor. **[U.P.S.C. I.E.S. E.E.-II, 1993]**

558. In a self excited induction generator, to keep the frequency of generated voltage constant with the increase in load, the speed of the induction machine should be
(a) increased. (b) decreased.
(c) maintained less than the rated synchronous speed.
(d) maintained more than the rated synchronous speed.
[U.P.S.C. I.E.S. E.E.-II, 1995]

559. Induction generator runs at
(a) supersynchronous speed.
(b) subsynchronous speed.
(c) synchronous speed. (d) none of these.
[W.B. Univ. of Technology, 2011-12, 2012-13]

560. Consider the following statements; If a 3-phase squirrel cage induction machine operates at a slip of - 0.05 (*i.e.* minus 0.05), then the machine will
1. draw electrical power from the mains.
2. draw mechanical power through the shaft.
3. deliver electrical power to the mains.

Of these statements
(a) 1,2 and 3 are correct. (b) 1 and 2 are correct.
(c) 2 and 3 are correct. (d) 1 and 3 are correct.
[U.P.S.C. I.E.S. E.E.-II, 1996]

561. A three-phase, 4-pole, self-excited induction generator is feeding power to a load at a frequency f_1. If the load is partially removed, the frequency becomes f_2. If the speed of the generator is maintained at 1500 rpm in both the cases, then
(a) $f_1, f_2 > 50$ Hz and $f_1 > f_2$ (b) $f_1 < 50$ Hz and $f_2 > 50$ Hz
(c) $f_1, f_2 < 50$ Hz and $f_2 > f_1$ (d) $f_1 > 50$ Hz and $f_2 < 50$ Hz
[GATE 2014]

562. A 3-phase induction machine draws active power '*P*' and reactive power '*Q*' from the grid. If it is operated as a generator, *P* and *Q* will respectively be
(a) positive and negative. (b) negative and negative.
(c) positive and positive. (d) negative and positive.
[U.P.S.C. I.E.S. E.E.-II 1998]

563. In a 3-phase induction machine, motoring, generating and braking operations take place in the range of slip ''*s*'' given by
(a) motoring: $1 > s > 0$; generating : $0 > s > -1$; braking : $s > 1$.
(b) motoring : $s > 1$; generating $1 > s > 0$; braking ; $0 > s > -1$.
(c) motoring: $s > 1$; generating $0 > s > -1$ braking : $1 > s > 0$.
(d) motoring: $0 > s > -1$; generating : $s > 1$; braking : $1 > s > 0$. **[U.P.S.C. I.E.S. E.E.-II, 1994]**

564. In a 3-phase induction machine, motoring, generating and braking operations take place in the range of slip "S" is
(a) $1 > s > 0$, $0 > s > -2$ and $s > 1$
(b) $s > 1$, $1 > s > -1$ and $0 > s > -1$
(c) $s > 1$, $0 > s > -1$ and $1 > s > 0$
(d) $0 > s > -1$, $s > 1$ and $1 > s > 0$ **[U.P.S.C. I.E.S. E.E.-II, 2010]**

565. Which of the following statements about synchronous-induction motors is incorrect?
(a) It is basically a wound rotor induction motor with fewer and larger slots on rotor.
(b) It has a very small air gap as expected in a plain induction motor.
(c) It is connected to 3-phase ac supply on stator side and to dc on rotor side.
(d) It starts as an induction motor and runs as a synchronous motor.
(e) It is provided with a heavy rotor winding.

566. Synchronous-induction motors
(a) are used where high starting torque is required.
(b) can be operated on full load on any desired pf.
(c) are very often installed alongwith other induction motors in order to improve the overall power factor of the system.
(d) have been made for rating up to 30,000 kW.
(e) all of the above statements regarding synchronous induction motors are correct.

567. The secondary of a linear induction motor normally consists of a
(a) concentrated single phase winding.
(b) distributed single phase winding.
(c) solid conducting plate.
(d) distributed three phase winding.

568. A Schrage motor is a variable speed commutator type three phase induction motor and has three windings,
(a) all the three windings located in stator.
(b) all the three windings located in rotor.
(c) two windings in rotor and one winding in stator.
(d) two windings in stator and one winding on rotor.

569. Which of the following statements about Schrage motor is incorrect?

Answers: **552.** (b), **553.** (b), **554.** (d), **555.** (e), **556.** (c), **557.** (b), **558.** (a), **559.** (a), **560.** (c), **561,** (c), **562.** (d), **563.** (a), **564.** (a), **565.** (b), **566.** (e), **567.** (c), **568.** (c), **569.** (b),

(a) it gives normally speed range of 3 to 1 but much speed ranges upto 15 to 1 and higher may also be obtained.
(b) It is cheaper than the slip-ring induction motor.
(c) Its efficiency is higher than that of a slip-ring induction motor at all speeds except synchronous speed and is very much high at lower speeds.
(d) While operating exactly at synchronous speed it behaves, in a sense, like a synchronous motor with compensating winding functioning as a dc exciter.

570. A 3-phase slip-ring induction motor is fed from the rotor side with stator winding short circuited. The frequency of the currents in the short-circuited stator is
(a) slip frequency. (b) supply frequency.
(c) frequency corresponding to rotor speed.
(d) zero. **[GATE E.E. 1993]**

571. In particular motor there is a stator phase winding and a short-circuited rotor winding. The motor runs as synchronous speed but does not necessitate any dc excitation. The motor is a
(a) 3-phase induction motor.
(b) 3-phase synchronous motor.
(c) 3-phase reluctance motor.
(d) hysteresis motor.

572. Phase advancers are employed with large induction motors to
(a) regulate their speeds. (b) reduce copper losses.
(c) improve power factor. (d) reduce noise and vibrations.

573. Which of the following statements about a phase advancer is incorrect ?
(a) It is a particular type of ac exciter which may be connected in the rotor circuit of an induction motor to improve power factor.
(b) It uses the principle of injection through slip-rings of the motor a current leading with regard to the rotor voltage.
(c) It resembles with an induction motor in construction.
(d) Its kVA capacity needs be only 5% or less of kVA correction affected in the main supply circuit.
(e) it has compensating windings on the stator to ensure good commutation.

574. Consider the following statements regarding ac servomotor:
1. The torque-speed curve has negative slope.
2. It is sensitive to noise.
3. The rotor has high resistance and low inertia.
4. It has slow acceleration.
Which of the following are the characteristics of ac servomotor as control component?
(a) 1 and 2. (b) 2 and 3. (c) 1 and 3. (d) 2 and 4. **[U.P.S.C I.E.S. E.E.-I, 2002]**

575. Given that D=rotor diameter and L=axial length, a high performance a c servomotor is characterized by
(a) large D, large L. (b) large D, small L.
(c) small D, small L. (d) small D, large L. **[U.P.S.C.I.E.S. E.E.-II, 1993]**

576. While a 2-phase ac servomotor is in operation, if the voltage across the control field winding becomes zero, then the motor has a tendency to run as a single phase induction motor. To prevent this
(a) rotor having high mass moment of inertia is used for such a motor.
(b) drag cup type of light rotor and high resistance is preferred.
(c) a low resistance is used.
(d) the number of turns in the control field winding used is less than the main reference winding. **[U.P.S.C. I.E.S. E.E.-II, 1993]**

577. A linear servomotor must have
(a) high rotor resistance. (b) high rotor reactance.
(c) a large air gap.
(d) both high rotor resistance and reactance. **[U.P.S.C. I.E.S. E.E.-II, 2005]**

578. A two-phase servomotor develops maximum torque at
(a) forward speed of one-half of the synchronous speed.
(b) backward speed of one-half of the synchronous speed.
(c) synchronous speed.
(d) a speed of twice the synchronous speed. **[U.P.S.C. I.E.S. E.E.-II, 1999]**

579. Ratio of the rotor reactance X to the rotor resistance R for a 2-phase servomotor
(a) is equal to that of a normal induction motor.
(b) is less than that of a normal induction motor.
(c) is greater than that of a normal induction motor.
(d) may be less or greater than that of a normal induction motor. **[GATE E.E 2000]**

580. A stepper motor is a
(a) dc motor. (b) single-phase ac motor.
(c) a two-phase motor. (d) multiphase motor. **[W.B. Univ. of Technology, 2012-13]**

581. In a stepper motor, the detent torque means
(a) minimum of the static torque with the phase winding excited.
(b) maximum of the static torque with the phase winding excited.
(c) minimum of the static torque with the phase winding unexcited.
(d) maximum of the static torque with the phase winding unexcited. **[GATE E.E. 2008]**

582. For a given stepper motor, the following torque has the highest numerical value
(a) Detent torque. (b) Pull-in torque.
(c) Pull-out torque. (d) Holding torque. **[GATE E.E. 2004]**

583. Which one among the following has the highest numerical value in a stepper motor?
(a) detent torque. (b) Holding torque.
(c) Dynamic torque. (d) Ripple torque. **[U.P.S.C. I.E.S. E.E.-II, 2006]**

584. Switched reluctance motor means
(a) salient pole synchronous motor without excitation winding.
(b) a stepper motor with salient poles.
(c) synchronous motor with salient poles on stator and rotor.
(d) a stepper motor with closed loop control and with rotor position sensor. **[U.P.S.C. I.E.S. E.E.-II, 2001]**

585. The following motor definitely has a permanent magnet rotor:
(a) DC commutator motor.
(b) Brushless dc motor.
(c) Stepper motor. (d) Reluctance motor. **[GATE E.E. 2004]**

586. For a 1.8°, 2-phase bipolar stepper motor, the stepping rate is 100 steps/second. The rotational speed of the motor in rpm is
(a) 15 (b) 30 (c) 60 (d) 90 **[GATE E.E. 2004]**

587. A 1.8° step, 4-phase stepper motor has a total of 40 teeth on 8 poles of stator. The number of rotor teeth for this motor will be
(a) 40 (b) 50 (c) 100 (d) 80 **[GATE E.E. 2000]**

588. A stepping motor has four concentric coils on its 4-pole stator. It has an unwound variable reluctance motor with 5 teeth. The step angle for this motor is
(a) 36° (b) 18° (c) 9° (d) 72°

Answers: 570. (a), **571.** (c), **572.** (c), **573.** (c), **574.** (c), **575.** (d), **576.** (b), **577.** (a), **578.** (a), **579.** (b), **580.** (d), **581.** (c), **582.** (c), **583.** (b), **584.** (d), **585.** (c), **586.** (b), **587.** (b), **588.** (b),

589. Stepper motors are mostly used for
(a) high power requirements.
(b) control system applications.
(c) very high speed of operation.
(d) very low speed of operation. [U.P.S.C. I.E.S. E.E.-II, 2002]

590. Stepper motors are widely used because of
(a) wide speed range. (b) large rating.
(c) no need for field control.
(d) compatibility with digital systems. [U.P.S.C. I.E.S. E.E.-II, 2002]

591. Which one of the following types of motors is most suitable for a computer printer drive?
(a) Reluctance motor. (b) Hysteresis motor.
(c) Shaded pole motor. (d) Stepper motor.
[U.P.S.C. I.E.S. E.E.-II 2004; GATE E.E. 1996]

592. A 3 ϕ, 3 stack, variable reluctance step motor has 20 poles on each rotor and stator stack. The step angle of this step motor is
(a) 3° (b) 6° (c) 9° (d) 18°
[GATE E.E. 2007]

593. Group I lists different applications and group II lists the motors for these applications. Match the application with most suitable motor and choose the right combination among the choices given thereafter.

Group I		Group II
P	Food mixer	1. PMDC motor
Q	Cassette tape recorder induction motor	2. Single phase induction motor
		3. Universal motor
R	Domestic water pump induction motor	4. Three-phase induction motor
		5. DC series motor
S	Escalator	6. Stepper motor

(a) P-3, Q-6, R-4, S-5. (b) P-1, Q-3, R-2, S-4.
(c) P-3, Q-1, R-2, S-4. (d) P-3, Q-2, R-1, S-4.
[GATE E.E. 2003]

594. In brushless dc motor we have
(a) no mechanical commutator.
(b) no brushes. (c) no arcing.
(d) all of these. [W.B. Univ. of Technology, 2011-12]

595. A synchro has
(a) 3-phase winding on rotor and 1-phase winding on stator.
(b) 3-phase winding on stator and commutator winding on rotor.
(c) 3-phase winding on stator and 1-phase winding on rotor.
(d) 1-phase winding on stator and commutator winding on rotor.
[W.B. Univ. of Technology, 2012-13]

596. Position synchros normally have ______ phase stator winding and ________ phase rotor winding.
(a) three, single (b) single, three
(c) three, three (d) single, single

597. When the stators of two slip-ring induction motors are fed from a common 3-phase ac supply in parallel and the rotor windings are connected in opposition, they are called
(a) power selsyns. (b) position selsyns.
(c) power synchros. (d) position synchros.

598. In case of a cross-field machine, the amplification ratio may be as high as
(a) 10,000:1 (b) 2,000:1 (c) 500:1 (d) 100:1

599. In cross-field generators
(a) the reversal of speed does not reverse the polarity of the output.
(b) compensating winding is provided in order to increase the amplification factor.
(c) the number of brush studs is double of that of poles.
(d) all of the above.

600. In Rosenberg generators,
(a) use of armature reaction is made.
(b) the field coils are shunt connected.
(c) the direction of load current is independent of direction of rotation and remains almost constant between wider limits of speed.
(d) all of the above.

601. Which of the following statement(s) associated with a Rosenberg generator is/are incorrect ?
(a) It is an armature excited high speed machine.
(b) It provides constant current over wider limits of speed.
(c) It is a constant voltage generator at low speeds.
(d) It is essentially a 2-pole machine with two sets of brushes arranged on the axis at right angles to each other.
(e) Its voltage polarity is independent of direction of rotation.

602. An amplidyne is provided with split poles so as to
(a) increase amplification factor.
(b) provide space for interpoles.
(c) increase efficiency.
(d) dampout mechanical oscillations.

603. Which of the following statements associated with Amplidyne and Metadyne are correct ?
(a) These are cross-field machines.
(b) Amplidyne is fully compensated machine and Metadyne is under compensated machine.
(c) Commutating poles are also provided in these machines.
(d) Compensating winding is connected in series with the load output brush terminals.
(e) all of the above.

604. Which of the following statements associated with a 3-wire generator are correct?
(a) It is an ordinary dc generator with a special provision for connecting the neutral wire to the armature winding.
(b) It is somewhat more expensive than tne standard one.
(c) For operating 3-wire compound generators in parallel it is necessary that two equalizer connections are used to connect all series fields in parallel.
(d) There are two series fields on each 3-wire compound generator.
(e) all of the above.

605. If the number of poles in the first stage of a Rototrol (Regulex) is 2, the number of poles in the second stage will be
(a) 2 (b) 4 (c) 6 (d) 8

606. The proper alignment of a machine is essential otherwise
(a) excessive wear and tear will take place on the bearings.
(b) the load current will become excessively high.
(c) the speed of the motor may become abnormal.
(d) the motor may not operate at all.

607. Before connecting the motor to the supply mains it is necessary to ensure that
(a) there is a proper insulation between field winding, armature winding and earth.
(b) protective devices have been adjusted at their proper settings.
(c) all the necessary connections show continuity.
(d) supply voltage is appropriate.
(e) all of the above.

608. For preventive maintenance of a dc machine it is to be ensured that
(a) insulation resistances of the field and armature windings are proper.
(b) commutator and brushes are in good condition.
(c) bearings are properly greased.
(d) motor and its control gear are free from oil and grease.
(e) all of the above.

Answers: **589.** (b), **590.** (d), **591.** (d), **592.** (b), **593.** (c), **594.** (b), **595.** (c), **596.** (a), **597.** (a), **598.** (a), **599.** (d), **600.** (d), **601.** (c), **602.** (b), **603.** (e), **604.** (e), **605.** (b), **606.** (a), **607.** (e), **608.** (e),

609. Which of the following statement(s) associated with installation and preventive maintenance of dc machines is incorrect ?
(a) Accumulation of carbon dust on the commutator surface may result in flash-over.
(b) The commutator should be lubricated in order to reduce brush friction.
(c) Accumulation of dust on the machine may result in overheating of the machine.
(d) Sufficient space is necessary all around the motor for movement of operational and maintenance personnels.
(e) Resistance of earth continuity conductor is to be checked periodically in order to ensure the effectiveness of the earthing system.

610. An amplidyne has split poles to
(a) provide space for interpoles.
(b) increase amplification factor.
(c) increase efficiency. (d) damp out oscillations.

611. In a single-phase full-wave controlled bridge rectifier, minimum output voltage and maximum output voltage are obtained at which conduction angles ?
(a) 0°, 180° respectively. (b) 180°, 0° respectively.
(c) 0°, 0° respectively. (d) 180°, 180° respectively.
[U.P.S.C. I.E.S. E.E.-II, 2007]

612. In a single-phase semiconverter with resistive load and for a firing angle α, each SCR conduction and freewheeling action take place respectively, for
(a) α, 0° (b) $\pi - \alpha, \alpha$ (c) $\pi + \alpha, \alpha$ (d) $\pi - \alpha$, 0°
[U.P.S.C. I.E.S. E.E.-II, 1998]

613. The advantage of using a freewheeling diode with bridge type ac/dc converter is
(a) regenerative braking. (b) reliable speed control.
(c) improved power factor. (d) reduced cost of the system.

614. A single-phase full convertor is connected across 250 V ac. Its output voltage when firing angle is 30° is given by
(a) 210 V. (b) 195 V.
(c) 250 V. (d) none of the above.

615. In a dual converter, the circulating current
(a) allows smooth reversal of load current, but increases the response time.
(b) allows smooth reversal of load current with improved speed of response.
(c) does not allow smooth reversal of load current, but reduces the response time.
(d) flows if there is no interconnecting inductor.
[U.P.S.C. I.E.S. E.E.-II, 2006]

616. A separately-excited dc motor is started using a 3-phase ac/dc controlled rectifier using 'soft starting'. For limiting the starting current, it is required that firing angle should be
(a) gradually increased from 0° to 180°
(b) fixed at 30°
(c) gradually reduced from 180° to 30°
(d) gradually increased from 30° to 180°
[U.P.S.C. I.E.S. E.E.-II, 2010]

617. A step-down chopper is operated in the continuous conduction mode in steady state with a constant duty ratio D. If V_0 is the magnitude of the dc output voltage and if V_s is the magnitude of the input voltage, the ratio V_o/V_s is given by
(a) D. (b) 1 − D. (c) $\dfrac{1}{1-D}$. (d) $\dfrac{D}{1-D}$.
[GATE E.E. 2002]

618. A 220 V, 20 A, 1,000 rpm, separately-excited dc motor has an armature resistance of 2.5 Ω. The motor is controlled by a step-down chopper with a frequency of 1 kHz. The input dc voltage to the chopper is 250 V. The duty cycle of the chopper for the motor to operate at a speed of 600 rpm delivering the rated torque will be
(a) 0.518 (b) 0.608 (c) 0.852 (d) 0.902
[GATE E.E. 2008]

619. The separately-excited dc motor in the figure below has a rated armature current of 20 A and a rated armature voltage of 150 V. An ideal chopper switching at 5 kHz is used to control the armature voltage. If L_a = 0.1 mH R_a = 1 Ω, neglecting armature reaction, the duty ratio of the chopper to obtain 50% of the rated torque at the rated speed and the rated field current is

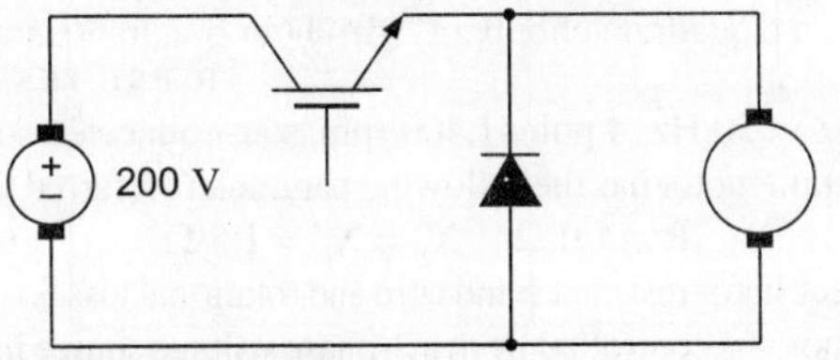

(a) 0.4 (b) 0.5 (c) 0.6 (d) 0.7
[GATE 2013]

620. A dc chopper is used in regenerative braking mode of a dc series motor. The dc supply is 600 V, the duty cycle is 70%. The average value of armature current is 100 A. It is continuous and ripple free. What is the value of power fed back to the supply?
(a) 3 kW (b) 9 kW (c) 18 kW (d) 35 kW
[U.P.S.C. I.E.S. E.E.-II, 2009]

621. A load resistance of 10 Ω is fed through a single-phase voltage controller from a voltage source of 200 sin 314*t*. If the power delivered to the load is 1 kW, the firing angle delay is
(a) 30° (b) 60° (c) 45° (d) 90°

622. A single-phase SCR based ac regulator is feeding power to a load consisting of 5 Ω resistance and 16 mH inductance. The input supply is 230 V, 50 Hz ac. The maximum firing angle at which the voltage across the device becomes zero all throughout and the rms value of current through SCR, under this operating condition, are
(a) 30° and 46 A (b) 30° and 23 A
(c) 45° and 23 A (d) 45° and 32 A **[GATE 2014]**

623. A single-phase ac voltage regulator is fed from at 50 Hz supply system. If it supplies a load comprising a resistance of 2 Ω connected in series with an inductance of 6.36 mH, then the range of firing angle 'α' providing controlled voltage would be
(a) $0° < \alpha < 180°$. (b) $45° < \alpha < 180°$.
(c) $90° < \alpha < 180°$. (d) $0° < \alpha < 45°$.
[U.P.S.C. I.E.S. E.E.-II, 2000]

624. What is the power factor of a single-phase ac regulator feeding a resistive load ?
(a) (Per unit power)^2 (b) (Per unit power)$^{1/2}$
(c) $\dfrac{(\text{Per unit power})^2}{\sqrt{2}}$ (d) $\dfrac{(\text{Per unit power})^{1/2}}{2}$
[U.P.S.C. I.E.S. E.E.-II, 2006]

625. An advantage of a cycloconverter is:
(a) very good power factor.
(b) requires few number of thyristors.
(c) commutation failure does not short circuit the source.
(d) load commutation is possible. **[U.P.S.C. I.E.S. E.E.-II, 2011]**

626. Compared to a single-phase half-bridge inverter, the output power of a single phase full-bridge inverter is higher by a factor of

Answers: **609.** (b), **610.** (a), **611.** (b), **612.** (b), **613.** (c), **614.** (a), **615.** (b), **616.** (c), **617.** (a), **618.** (b), **619.** (d), **620.** (c), **621.** (d), **622.** (c), **623.** (a), **624.** (b), **625.** (d), **626.** (d),

(a) 12. (b) 8. (c) 4. (d) 2.
[U.P.S.C. I.E.S. E.E.-II, 2002]

627. Chopper circuits are used to control the speed of dc motors which are fed from fixed dc voltage sources. Variable voltage output can be obtained by which of the following methods.
1. varying the 'on' to 'off' time ratio.
2. adjusting the wave shape.
3. adjusting the frequency of switching.
4. using a saturable core reactor in series with the motor.

The correct answer is
(a) 1 and 2 (b) 1 and 3 (c) 2 and 3 (d) 2 and 4

628. Which of the following configurations is used for both motoring and regenerative braking ?
(a) First quadrant chopper. (b) Second quadrant chopper.
(c) Two quadrant chopper. (d) Four quadrant chopper.
[U.P.S.C. I.E.S. E.E.-II, 2006]

629. A 400 V, 50 Hz, 4 pole, 1,400 rpm, star-connected squirrel cage induction motor has the following parameters referred to the stator:
$$R = 1.0\ \Omega,\quad X_S = X_r' = 1.5\ \Omega$$
Neglect stator resistance and core and rotational losses of the motor. The motor is controlled from a 3-phase voltage source inverter with constant V/f control. The stator line-to-line voltage (rms) and frequency to obtain the maximum torque at starting will be :
(a) 20.6 V, 2.7 Hz. (b) 133.3 V, 16.7 Hz.
(c) 266.6 V, 33.3 Hz. (d) 323.3 V, 40.3 Hz.
[GATE E.E. 2008]

630. The operation of an inverter fed induction motor can be shifted from motoring to regenerative braking by
(a) reversing phase sequence.
(b) reducing inverter voltage.
(c) decreasing inverter frequency.
(d) increasing inverter frequency. [U.P.S.C. I.E.S. E.E.-II, 2002]

631. A 3-phase induction motor is used as an adjustable speed drive from zero to 2.0 per unit speed using a variable frequency inverter. In the speed range of 1.0 to 2.0 per unit, in order to ensure satisfactory operation, which one of the following sets of quantities is to be maintained approximately constant ?
(a) Voltage and power. (b) Flux and torque.
(c) Voltage and torque. (d) Flux and power.
[U.P.S.C. I.E.S. E.E.-II, 1998]

632. In a self-controlled synchronous motor fed from a variable frequency inverter
(a) The rotor poles invariably have damper windings.
(b) There are stability problems.
(c) The speed of the rotor decides stator frequency.
(d) The frequency of the stator decides the rotor speed.
[U.P.S.C. I.E.S. E.E.-II, 2003]

633. A 3-phase wound rotor induction motor is controlled by a chopper-controlled resistance in its rotor circuit. A resistance of 2 Ω is connected in the rotor circuit and a resistance of 4 Ω is additionally connected during OFF periods of the chopper. The OFF period of the chopper is 4 ms. The average resistance in the rotor circuit for the chopper frequency of 200 Hz is
(a) $\frac{26}{5}\Omega$ (b) $\frac{24}{5}\Omega$ (c) $\frac{18}{5}\Omega$ (d) $\frac{16}{5}\Omega$
[U.P.S.C. I.E.S. E.E.-II, 2001]

634. An electric motor, developing a starting torque of 15 Nm, starts with a load torque of 7 Nm on its shaft. If the acceleration at start is 2 rad/sec^2, the moment of inertia of the systems must be (neglecting viscous and Coulomb friction)
(a) 0.25 kg m^2. (b) 0.25 Nm^2.
(c) 4 kg m^2. (d) 4 Nm^2. [GATE E.E. 2005]

Answers: 627. (b), **628.** (c), **629.** (b), **630.** (c), **631.** (a), **632.** (d), **633.** (d), **634.** (c).

MATCHING TYPE QUESTIONS

Note: Each of the following questions consists of two lists. Match the items given in list 1 and those in list 2. Select your answers to these questions using the codes given in each of the questions.

1.

	List 1 (DC Motors)		**List 2 (Characteristics)**
A	Cumulatively compounded motor.	1.	Fairly constant speed irrespective of the load.
B	Differentially compounded motor.	2.	It may start in reverse directtion.
C	Series motor.	3.	Definite no-load speed.
D	Shunt motor.	4.	Never started without load.

Codes:	A	B	C	D
(a)	2	3	1	4
(b)	3	2	4	1
(c)	1	3	2	4
(d)	4	1	3	2

[U.P.S.C. I.E.S. E.E. II, 1994; 2005]

2.

	List 1 (DC Machine Quantity)		**List 2 (Relation)**
A.	Developed power.	1.	$\propto N\phi$
B.	Torque.	2.	$\propto E_b I_a$
C.	Generated emf.	3.	$\propto E_b/\phi$
D.	Speed.	4.	$\propto I_a\phi$

Codes:	A	B	C	D
(a)	2	4	1	3
(b)	3	1	4	2
(c)	2	1	4	3
(d)	3	4	1	2

[U.P.S.C. I.E.S. E.E.-II., 2008]

3.

	List 1 (DC Motor)	**List 2 (Characteristic Labelled 1, 2, 3 and 4)**
A.	Shunt motor.	1.
B.	Series motor.	2.
C.	Cumulative compound motor.	3.
D.	Differential compound motor.	4.

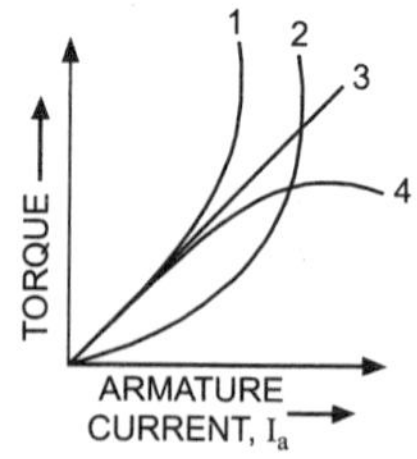

Codes:	A	B	C	D
(a)	3	2	1	4
(b)	4	1	2	3
(c)	3	1	2	4
(d)	4	2	1	3

[U.P.S.C. I.E.S. E.E.-II, 2001]

4.

	List 1 (Type of Machine)		**List 2 (Characteristic/Application)**
A.	DC shunt generator.	1.	Electric traction.
B.	DC series motor.	2.	Has good voltage regulation.

Answers: 1. (b), **2.** (a), **3.** (c), **4.** (d),

C. Level compounded dc. generator. — 3. Must have residual flux.
D. DC series generator. — 4. Used as boosters.

Codes:	A	B	C	D
(a)	4	1	2	3
(b)	3	2	1	4
(c)	4	2	1	3
(d)	3	1	2	4

[U.P.S.C. I.E.S. E.E.-II., 2001]

5. **List 2** — **List 2**

A. Field Control — 1. Below base speed
B. Armature Control — 2. Above base speed
3. Above base torque
4. Below base torque

Codes:	A	B
(a)	1	3
(b)	2	1
(c)	2	3
(d)	1	4

[GATE E.E. 2003]

6. **List 1** — **List 2**

A. Voltage drop test — 1. Efficiency.
B. Hopkinson's test — 2. Separation of iron and friction losses.
C. Swinburne's test — 3. Open and short-circuited armature coils.
D. Retardation test — 4. Temperature rise.

Codes :	A	B	C	D
(a)	3	1	4	2
(b)	2	4	1	3
(c)	3	4	1	2
(d)	2	1	4	3

[U.P.S.C. I.E.S. E.E.-II. 2000]

7. **List 1 (Test)** — **List 2 (Quantities)**

A. OC Test — 1. Copper loss and iron loss.
B. SC Test — 2. Total losses.
C. Sumpner's Test — 3. Iron loss.
D. Load Test — 4. Copper loss.

Codes:	A	B	C	D
(a)	3	4	1	2
(b)	2	1	4	3
(c)	3	1	4	2
(d)	2	4	1	3

[U.P.S.C. I.E.S. E.E.-II, 2005]

8. **List 1 (Connections of 3-phase transformers)** — **List 2 (Utilization factor)**

A. Delta-zig zag. — 1. 0.5
B. Star-zig zag. — 2. 0.667
C. Open delta. — 3. 0.866
4. 1.0

Codes:	A	B	C
(a)	3	3	3
(b)	1	2	3
(c)	2	3	1
(d)	2	2	2

9. **List 1 (Type of AC Windings)** — **List 2 (Features of AC Windings)**

A. Single layer — 1. Coil span and pole pitch are equal.
B. Double layer — 2. Coil span is shorter than pole pitch.
C. Full Pitch — 3. Number of slots is equal to number of coils.
D. Chorded — 4. Number of slots is twice the number of coils.

Codes:	A	B	C	D
(a)	4	3	1	2
(b)	3	4	2	1
(c)	4	3	2	1
(d)	3	4	1	2

[U.P.S.C. I.E.S. E.E.-II, 2003]

10. **List 1 (Name of Test)** — **List 2 (Result)**

A. Open circuit and short circuit tests. — 1. Leakge reactance.
B. Open circuit and zero power factor tests. — 2. Direct axis synchronous reactance.
C. Slip test. — 3. Quadrature axis synchronous reactance.
D. Maximum lagging current test. — 4. Ratio of direct axis synchronous reactance to quadrature axis sychronous reactance.

Codes:	A	B	C	D
(a)	1	2	4	3
(b)	1	2	3	4
(c)	2	1	4	3
(d)	2	1	3	3

[U.P.S.C. I.E.S. E.E.-II, 2002]

11. **List 1 (Machine Components)** — **List 2 (Type of Machine)**

A. Amortisseur winding — 1. Squirrel cage induction motor
B. Breather — 2. DC motor
C. End-Rings — 3. Transformer
D. Commutator — 4. Synchronous motor

Code:	A	B	C	D
(a)	2	3	1	4
(b)	4	3	1	2
(c)	2	1	3	4
(d)	4	1	3	2

[U.P.S.C. I.E.S. E.E.-II, 2009]

12. **List 1 (Parts of Electrical Machine)** — **List 2 (Approximate Nature of Airgap mmf Produced in them)**

A. DC machine stator — 1. Sinusoidal
B. DC machine rotor — 2. Trapezoidal
C. Salient pole synchronous machine stator — 3. Triangular
D. Squirrel cage rotor of induction motor

Codes:	A	B	C	D
(a)	1	2	1	3
(b)	3	1	3	1
(c)	2	3	1	3
(d)	2	3	1	1

[U.P.S.C. I.E.S. E.E.-II, 1997]

13. **List 1 (Machines)** — **List 2 (Tests)**

A. DC motors — 1. Slip test
B. Single phase transformers — 2. Blocked rotor test
C. 3-phase induction motors — 3. Swinburne's test
D. Salient pole alternators — 4. Sumpner's test

Codes:	A	B	C	D
(a)	1	2	4	3
(b)	3	4	2	1
(c)	1	4	2	3
(d)	3	2	4	1

[U.P.S.C. I.E.S. E.E.-II, 2005]

Answers: 5. (b), **6.** (c), **7.** (a), **8.** (a), **9.** (a), **10.** (c), **11.** (b), **12.** (d), **13.** (b),

14.

List 1 (Methods of control)	**List 2 (Applications)**
A. Injection of voltage into the rotor circuit of a 3 phase induction motor.	1. Steel rolling mills.
B. Variation of applied voltage to a 3 phase induction motor at constant frequency.	2. Textile mills.
C. Series-parallel control of dc series motors.	3. Fans and blowers.
D. Ward-Leonard control of dc motors.	4. Traction.
	5. Position control in a servosystem.

Codes :	A	B	C	D
(a)	1	2	3	4
(b)	4	5	2	1
(c)	1	5	2	4
(d)	2	3	4	1

[U.P.S.C. I.E.S. E.E.-II, 1994]

15.

List 1 (Method of speed control of 3-phase wound-type induction motor)	**List 2 (Performance achieved)**
A. Stator voltage control.	1. Both speed and pf can be controlled.
B. Rotor resistance control.	2. Maximum torque remains constant.
C. Constant volts/Hz control.	3. Starting torque increases.
D. Injection of voltage in rotor circuit.	4. Starting torque decreases.

Codes:	A	B	C	D
(a)	2	1	4	3
(b)	4	3	2	1
(c)	2	3	4	1
(d)	4	1	2	3

[U.P.S.C. I.E.S. E.E.-II, 2008]

16.

List 1	**List 2**
A. Single-phase induction motor.	1. Rotor resistance starting.
B. 3-phase slip ring induction motor.	2. Not self-starting.
C. 3-phase squirrel cage induction. motor.	3. Auto-transformer starting.

Codes:	A	B	C
(a)	2	1	3
(b)	1	2	3
(c)	2	3	1
(d)	3	1	2

[U.P.S.C. I.E.S. E.E.-II, 1999]

17.

List 1	**List 2**
A. General purpose split phase FHP motor.	1. Refrigerators
B. General purpose capacitor start FHP motor.	2. Hair dryers.
C. Permanent split capacitor start FHP motor.	3. Unit heaters.
D. Shaded pole FHP motor.	4. Fans, blowers.

Codes:	A	B	C	D
(a)	1	2	4	3
(b)	1	2	3	4
(c)	4	1	2	3
(d)	4	1	3	2

[U.P.S.C. I.E.S. E.E.-II, 2007]

18.

List 1 (Motors)	**List 2 (Applications)**
A. Permanent magnet dc motor.	1. Cassette tape recorder.
B. Stepper motor.	2. Ceiling fan.
C. Single phase induction motor.	3. Hand drill.
D. Universal motor.	4. Digital control.

Codes :	A	B	C	D
(a)	1	4	2	3
(b)	1	3	2	4
(c)	4	1	3	2
(d)	1	2	4	3

[U.P.S.C. I.E.S. E. E.-II, 1993]

19.

List 1 (Type of motor) — **List 2 (Torque-speed characteristics)**

A. Hysteresis motor.
B. Capacitor start split phase motor.

1.

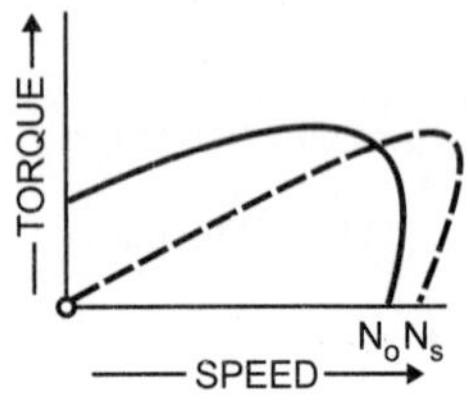

C. Reluctance motor.

2.
3.

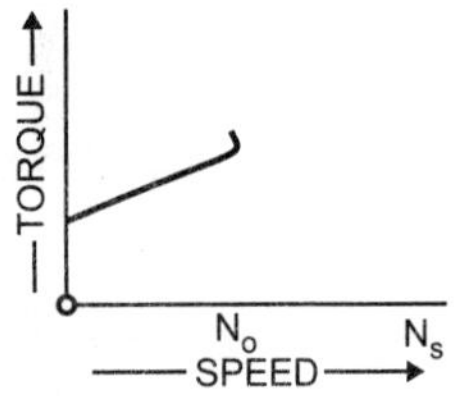

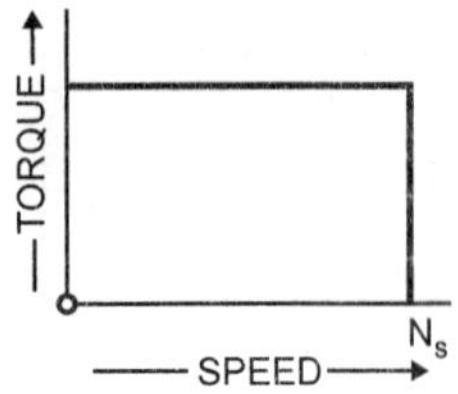

Codes:	A	B	C
(a)	3	1	2
(b)	2	3	1
(c)	2	1	3
(d)	1	3	2

[U.P.S.C. I.E.S. E.E.-II, 1993]

20.

List 1 (Types of Motors)	**List 2 (Applications)**
A. Shaded pole motor	1. Robotics
B. Capacitor motor	2. Rockets
C. Permanent magnet motor	3. Ceiling fan
D. Steppe motor	4. Table fan

Codes:	A	B	C	D
(a)	4	3	1	2
(b)	3	4	2	1
(c)	3	4	1	2
(d)	4	3	2	1

[U.P.S.C. I.E.S. E.E.-II, 1998]

Answers: 14. (d), **15.** (b), **16.** (a), **17.** (a), **18.** (a), **19.** (a), **20.** (a),

TRUE/FALSE TYPES QUESTIONS

Note: Identify whether the following statements are TRUE (T) or FALSE (F)

1. An electromagnetic field is said to be conservative when divergence of the field is zero.
2. The reluctance of a material is defined as its ability to conduct magnetic flux. **[R.G.T.U. M.P., December-2010]**
3. A minus sign is required to be placed before the right hand side of the expression for emf induced just to indicate the phenomenon explained by Lenz.
4. Hysteresis loss is the amount of energy dissipated as heat in the magnetic material.
5. Eddy current losses are reduced by using thin plates of high resistivity material.
6. The relation between apparent power (kVA), true power (kW) and reactive power (kVAR) is $\text{kVA} = \sqrt{(\text{kW})^2 + (\text{kVAR})^2}$.
7. The quality factor Q is defined as

$$Q = \frac{1}{2\pi} \times \frac{\text{Total energy stored}}{\text{Energy dissipated per cycle}}$$

8. Shunt capacitors are employed for power factor improvement. **[Mahamaya Technical Univ., Even Semester 2010-11]**
9. After very long time, a capacitor behaves as short circuit. **[Mahamaya Technical Univ. Even Semester 2010-11]**
10. A capacitor with zero initial conditions at $t = 0^+$ acts as current source.
11. For a linear electromagnetic circuit, field energy is equal to the coenergy.
12. In electromechanical energy conversion devices (*e.g.* generators and motors) a small air gap is left between the stator and rotor in order to reduce the reluctance of magnetic path.
13. Hysteresis loss can be reduced by laminating the iron core.
14. Iron losses are function of both flux and speed.
15. Windage losses largely depend upon the speed of the motor.
16. A body cools down from 60°C to 50°C in 10 minutes. The body will take the same *i.e.* 10 minutes in further cooling down from 50°C to 40°C, the external conditions remaining unchanged.
17. 15 minute rated motors are suitable for light duty cranes.
18. The magnetic field of a dc generator is connected to a 50 Hz sinusoidal supply, the voltage induced in the armature will be of 50 Hz sinusoidal.
19. The yoke of a dc generator is made of cast iron as it gives mechanical protection to the machine.
20. The brushes of a dc machine are made of carbon. **[R.G.T.U. M.P., December, 2010]**
21. The pole shoe acts as a support to field coils and spreads out the flux in the air gap.
22. Two slip rings put together face to face form a split ring.
23. The mica insulation between commutator segments is harder than the commutator material itself.
24. Armature core is laminated to reduce hysteresis loss.
25. The armature shaft is invariably made of cast steel.
26. Ventilating ducts are provided in a machine for reducing iron and friction losses.
27. Brush rocker assembly cannot be moved back and forth.
28. Wave winding is preferred for welding generator.
29. Equalizer rings are provided only in case of wave wound dc machines.
30. Frog-leg winding is combined lap and wave winding on a single rotor.
31. In a dc motor, the reluctance of the air gap is negligible in comparison to that of the iron.
32. In a dc machine the armature mmf is stationary with respect to field poles but rotating with respect to the armature.
33. Brushes are to be shifted to the magnetic neutral axis in order to avoid induction of reactance voltage in the coils undergoing commutation.
34. Larger air gaps can be provided on non-interpole machines.
35. Interpoles are never made of laminated construction.
36. Commutating poles or interpoles create a flux in the interpolar region in a direction opposite that produced due to armature reaction.
37. Internal characteristic of a dc generator is also referred to as the performance characterisitc.
38. In self-excited dc generators, the residual magnetism is of the order of 2.5% of normal magnetism.
39. The series fields must have resistance in direct proportion to the ratings of the generators.
40. In comparison to a self excited dc generator, separately excited dc generator has better voltage regulation.
41. Critical resistance in a dc shunt generator is a function of speed.
42. In case of compound wound dc generators the degree of compounding may be adjusted by means of divertor which shunts the series field.
43. DC series generators are most suitable for parallel operation.
44. DC motors are still favoured in applications where wide range of speed variation is required.
45. In general, dc motors are preferred where frequent starting, stopping, speed variation and reversal is required.
46. A dc motor develops unidirectional torque with the help of commutator.
47. There exists two magnetic fields in an electric rotating machine.
48. Two magnetic fields in an electric machine always try to align with each other.
49. The angle between the axis of two magnetic fields in an electric machine is called the torque angle.
50. The equation $T\omega = ei$ holds good for both generating and motoring action.
51. With the increase in load on the generator, the electromagnetic torque developed decreases.
52. In a dc motor polarity of the interpole should be same as main pole behind.
53. A cumulatively compounded long shunt generator, when operated as a motor, it would operate as a differentially compounded long shunt motor.

Answers: **1.** (F), **2.** (T), **3.** (T), **4.** (T), **5.** (T), **6.** (T), **7.** (F), **8.** (T), **9.** (F), **10.** (F), **11.** (T), **12.** (F), **13.** (F), **14.** (T), **15.** (T), **16.** (F), **17.** (T), **18.** (T), **19.** (F), **20.** (T), **21.** (T), **22.** (F), **23.** (T), **24.** (F), **25.** (F), **26.** (F), **27.** (F), **28.** (F), **29.** (F), **30.** (T), **31.** (F), **32.** (T), **33.** (T), **34.** (T), **35.** (F), **36.** (T), **37.** (F), **38.** (T), **39.** (F), **40.** (T), **41.** (T), **42.** (T), **43.** (F), **44.** (T), **45.** (T), **46.** (T), **47.** (T), **48.** (T), **49.** (T), **50.** (T), **51.** (F), **52.** (T), **53.** (T),

54. Negative speed regulation means drop of speed with the increase in load.
55. Torque-armature current characteristic is also called the mechanical characteristic.
56. The torque-armature current characteristic for a dc series motor is of the form of a parabola.
57. DC series motor is used for traction. **[R.G.T.U. M.P., December, 2010]**
58. DC shunt motors are preferred for belt driven loads.
59. DC shunt motors are used for driving constant speed air blowers.
60. In a cumulative compound dc motor, the maximum speed is fixed by the shunt field.
61. Differential compound dc motor, when started on heavy load, may start rotating in reverse direction.
62. In Ward Leonard system, the lower limit is imposed by residual magnetism of the generator.
63. In Ward Leonard system, the motor of the M-G set is of variable speed type.
64. The speed of a dc series motor can be varied by controlling the terminal voltage.
65. In paralleling field coil method of speed control, several speeds can be obtained by regrouping of field coils.
66. At the starting instant when a dc motor is connected to the dc supply mains, back emf is zero and so the full supply voltage acts across the armature.
67. Three point starters are not suitable for applications where speed variation by flux control is required because the motor may stop at very high speed.
68. Drum type controller is employed with a dc motor of 10kW or more requiring frequent starting, stopping, reversing and speed control.
69. DC series motors are unsuitable for dynamic braking with large load inertia.
70. Electromechanical or friction brakes are operated by electro-magnets or electric operated thyristors.
71. Hopkinson test is a regenerative test.
72. Transformer is a static electric device. **[R.G.T.U. M.P., December, 2010]**
73. A transformer can be connected to dc. **[U.P. Technical Univ. Even Semester 2009-10]**
74. For step up transformer, the transformer ratio is greater than unity. **[R.G.T.U. M.P., December, 2010]**
75. Mutual flux is the flux in the core which links with both the primary and secondary windings.
76. In a transformer of given voltage rating, the magnetizing current increases with the increase in frequency.
77. The secondary terminal voltage and the secondary induced emf are in phase opposition.
78. A good transformer must have regulation as high as possible. **[R.G.T.U. M.P., December, 2010]**
79. The core losses in a transformer are independent of power factor.
80. The commercial efficiency of a transformer is higher at lower power factor loads.
81. The all day efficiency of a transformer can be computed if its iron losses, full-load copper losses, rated output and load cycle is known.
82. At low frequencies, thin laminations are used.
83. In a transformer, the magnetic coupling between primary and secondary can be increased by employing low reluctance magnetic core.
84. Spacers are provided between adjacent coils in a transformer to insulate the coils from each other.
85. Leakage fluxes in a transformer may be minimized by sectionalizing and interleaving the primary and secondary windings.
86. In sandwich type winding parts of hv and lv windings alternate along the height of the limb.
87. For system voltages exceeding 66 kV condenser bushings are used for economic reasons.
88. Conservator consists of an air tight metal drum fixed at the top of the transformer tank.
89. Sludging of transformer oil means formation of semisolid hydrocarbon due to heat and oxidation.
90. In a shell type transformer, the primary and secondary windings are wound on separate formers.
91. In shell type transformers, the iron core surrounds the windings and there are two parallel magnetic paths in which the magnetic flux divides.
92. Scott-connections are employed for 2-phase to 3-phase conversion.
93. A V-V connected transformer can be operated in parallel with a Δ–Δ connected transformer but not with a Δ-Y connected transformer.
94. A Dy1 transformer can be operated in parallel with YD11 transformer having the same voltage ratings.
95. The tertiary winding in a transformer supplies the substation auxiliaries at a voltage different from those of primary and secondary windings.
96. In star-star transformers distortion on voltage wave shape due to third harmonic flux and voltage unbalance on single phase loads are minimum if they are bank of single phase transformers.
97. In an auto-transformer the electrical isolation between primary and secondary sides is lost.
98. Auto-transformers are preferred over conventional two winding transformers in power applications only for economic reasons.
99. In case of a current transformer, the overall power factor is high when both of the windings are close together and falls as the distance between them increases.
100. High leakage transformers can be designed to give open-circuit voltage between 20 and 30 kV and short-circuit current ranging from 1 to 2 m A.
101. In a tap-changing transformer, the taps are usually provided on the hv side.
102. Variac is a continuously variable auto-transformer employing carbon brush on a rotary shaft making contact with transformer winding exposed turns.
103. Three-phase induction regulators are unable to balance the line voltages when supplying unbalanced loads.
104. The potential transformer is a step-down two winding transformer operating on no-load.
105. In a 3-phase synchronous machine, the rotor rotates in the direction opposite to that of the stator field.
106. The pole face in a salient pole type rotor is so shaped that the radial air-gap length increases from the pole centre to pole tips.

Answers: **54.** (F), **55.** (F), **56.** (T), **57.** (T), **58.** (T), **59.** (T), **60.** (T), **61.** (T), **62.** (T), **63.** (F), **64.** (T), **65.** (T), **66.** (T), **67.** (T), **68.** (T), **69.** (F), **70.** (T), **71.** (T), **72.** (T), **73.** (F), **74.** (T), **75.** (T), **76.** (F), **77.** (F), **78.** (F), **79.** (T), **80.** (F), **81.** (T), **82.** (F), **83.** (T), **84.** (F), **85.** (T), **86.** (T), **87.** (T), **88.** (T), **89.** (T), **90.** (F), **91.** (T), **92.** (F), **93.** (T), **94.** (T), **95.** (T), **96.** (F), **97.** (T), **98.** (T), **99.** (T), **100.** (T), **101.** (T), **102.** (T), **103.** (T), **104.** (T), **105.** (F), **106.** (T),

107. Cylindrical rotor synchronous machines have always horizontal configuration.

108. Slip rings are made of aluminium.

[U.P. Technical Univ. Even Semester 2009-10]

109. In an alternator, a fractional rather than integral number of slots per pole is often used in order to eliminate harmonics in the waveform.

110. For a double layer winding the number of slots in the stator of a synchronous machine is a multiple of the number of poles and the number of phases.

111. The sum of the armature mmfs in the three-phases of the stator winding of a synchronous generator is zero.

112. Synchronous reactance can be determined from open-circuit and short-circuit tests.

113. In an alternator with a load which is partially inductive or capacitive, there is a cross-magnetization in addition to direct demagnetization or magnetization.

114. Under short-circuit conditions, the armature reaction in the alternator is having magnetising effect.

115. A small short-circuit ratio indicates small armature reaction in an alternator.

116. Direct-axis synchronous reactance of a salient pole synchronous machine can be estimated fairly from short-circuit and open-circuit tests.

117. In a pulsating field, the polarity reverses at the rate of supply frequency but the field is stationary in nature.

118. The relation between synchronous speed and number of poles is

$$N_s = \frac{120\,f}{P}$$

[R.G.T.U. M.P., December, 2010]

119. The rotor poles of a synchronous machine are magnetically locked with its stator poles when the machine is running.

120. The working of synchronous motor is similar to transmission of mechanical power by shaft.

121. An increase in field current of a synchronous motor causes a decrease in armature current. Armature current lags the terminal voltage.

122. In two identical cylindrical rotor and salient pole synchronous machines, the power developed for the same load angle is greater in salient pole machine.

123. A salient pole synchronous motor at no load is connected to an infinite bus. The field circuit is accidently opened. Machine continues to run drawing lagging current from the bus.

124. Pull-out torque and maximum torque are same in a synchronous motor.

125. It is difficult to synchronise salient pole rotor synchronous machine in comparison to smooth cylindrical rotor machine.

126. A synchronous motor when operated overexcited and without mechanical load can be employed for improving voltage regulation of a transmission line.

127. A 3-phase synchronous motor provided with damper grids is self-starting.

128. Before synchronising a 3-phase synchronous motor to the supply it has to be run up to or near its synchronous speed by some external means.

129. The field winding of a 3-phase synchronous motor is energized from a dc supply source just before its rotor attains synchronous speed in order to avoid slipping of poles.

130. Large 3-phase synchronous motors, specially those designed for starting under full load, use three phase wound rotor type winding.

131. The control equipment of a synchronous motor is more complex in comparison to that of an induction motor.

132. Synchronous motors are widely used for line shaft drives in small workshops.

133. Synchronisation process by lamp method is more accurate than that by synchroscope.

134. Load shared by a synchronous generator is governed by its excitation.

135. The number of terminals in the terminal box of an induction motor is six.

136. Single layer concentric windings are employed in case of large induction motors operating at high voltages.

137. Skewing of rotor slots increases rotor resistance.

138. The rotor of an induction motor may be considered as an analog of a mechanical clutch.

139. At infinite slip induction motor develops no torque.

140. The no-load current in an induction motor is dependent on its kW rating.

141. The torque developed in an induction motor is nearly proportional to supply voltage.

142. In a 3-phase induction motor, the current varies directly as the voltage whereas the torque varies as the square of applied voltage.

143. Power factor of an induction motor in loaded condition is more than that in no-load condition.

144. When the load on a 3-phase induction motor is increased from no-load to full load, its stator pf increases while rotor pf decreases.

145. The maximum torque developed in an induction motor does not depend upon the rotor resistance.

146. The magnetizing component of the no-load current in a 3-phase induction motor is much less than power loss component at no load.

147. The power output of an induction motor will be maximum when slip is zero.

148. The resistance R_0 of the exciting branch of the equivalent circuit of a 3-phase induction motor represents the friction and windage losses only.

149. Blocked rotor test and no-load run test of an induction motor fall under the category of type test.

150. No-load test at unity slip is performed to determine equivalent circuit of induction motor.

151. In a 3-phase induction motor, the mechanical power developed in terms of air-gap power P_s is $(1 - s)\,P_s$.

152. The synchronous speed of the *n*th order harmonic is 1/*n*th of the synchronous speed of fundamental.

153. In case of 3-phase induction motor the slip of rotor with respect to fifth harmonic is given by 6 + 5 s.

154. The number of stator slots should be equal to the number of rotor slots to avoid cogging.

155. Cogging and crawling are phenomena associated with squirrel cage motor, the former at a fraction of its rated speed and the latter during starting.

156. The skewing of slots has no effect on the production of harmonics in any rotor.

Answers: **107.** (T), **108.** (F), **109.** (T), **110.** (T), **111.** (T), **112.** (T), **113.** (T), **114.** (F), **115.** (F), **116.** (F), **117.** (T), **118.** (T), **119.** (T), **120.** (T), **121.** (T), **122.** (T), **123.** (T), **124.** (T), **125.** (F), **126.** (T), **127.** (T), **128.** (T), **129.** (T), **130.** (T), **131.** (T), **132.** (F), **133.** (F), **134.** (F), **135.** (T), **136.** (T), **137.** (T), **138.** (T), **139.** (F), **140.** (F), **141.** (F), **142.** (T), **143.** (T), **144.** (T), **145.** (T), **146.** (F), **147.** (F), **148.** (F), **149.** (F), **150.** (F), **151.** (T), **152.** (T), **153.** (F), **154.** (F), **155.** (F), **156.** (F),

157. Deep bar or double cage construction is used for improving the starting performance of the motor.
158. At the start of the motor the lower cage offers very low impedance to the flow of current in comparison to upper cage.
159. During starting of a double cage induction motor, the torque developed by a lower cage is much more than that developed by the upper cage.
160. Three phase induction motor has a low efficiency.
[U.P. Technical Univ. Even Semester 2009-10]
161. Induction motor cannot be used in electric traction.
162. Unbalance rotor resistance may cause a slip-ring induction motor run at about half the synchronous speed.
163. Single phasing of induction motor means its operation when one of its supply lines gets disconnected.
164. A 3-phase induction motor coupled to a pump is operating at normal speed. If one line gets disconnected, the motor stops.
[GATE E.E. 1994]
165. An auto-transformer with x % tap is used to start an induction motor. Supply current is reduced to $x/100$ of the value under direct-on-line.
166. The speed of a 3-phase induction motor can be controlled very smoothly over a wide range.
167. Stator voltage control of speed control of induction motor is suitable for constant load drive.
168. Stator voltage control is employed for machine tools.
169. Stator voltage control is most suitable for fan type loads.
170. The applications where speed ratio other than 1 : 2 is required, speed control of induction motor can be done by variation of number of stator poles using multiple stator windings.
171. The main disadvantage of pole changing method of speed control is limited speed range.
172. In variable frequency induction motor drives the ratio V/f is increased at low speed.
173. For constant V/f operation, maximum torque remains constant at any frequency.
174. V/f control provides high torque to current ratio during starting.
175. Rotor resistance control is inefficient at low values of speed.
176. Slip-power recovery scheme is very economical for very low power induction motors.
177. A 3-phase induction motor can be dried by blocking the rotor and circulating current at a voltage around 10% of rated voltage.
178. During plugging of induction motors, slip is positive.
179. The motors built in fractional-horse power sizes are technically termed small motors.
180. The stator of a single phase induction motor carries one winding only.
181. In a split-phase induction motor the auxiliary winding has a lower resistance to reactance ratio.
182. In case of 1-phase induction motor, inductance may be used in series with main winding for getting starting torque.
183. In a resistance split-phase motor the auxiliary winding has higher R/X ratio as compared to the main winding.
184. In a capacitor-start single phase induction motor, capacitor is used in series with the motor to improve power factor.
185. Induction motors are used in air-conditioners.
186. Two value capacitor motor is preferred for tape recorder.
187. In a shaded-pole motor the direction of rotation is from shaded portion to unshaded portion.
188. In a shaded-pole motor, the effect of the shading ring is to reduce the flux in the shaded portion and, at the same time, to shift it in time phase behind the flux in the unshaded portion.
189. Shaded-pole motors develop higher starting torque than 1-phase capacitor start motors.
190. Shaded-pole induction motors are used in refrigerator compressors.
191. AC series motors need not have laminated yokes.
192. When a dc series motor of normal design is connected across single phase ac supply unidirectional torque is developed.
193. Large ac series motors are usually designed for operation on high frequencies.
194. A single-phase compensated series motor exhibits shunt characteristics.
195. When the load torque on a single phase ac series motor is increased, its speed and power factor will decrease.
196. AC series motors have high torque to weight ratio than dc series motor of equal ratings.
197. The direction of rotation of a universal motor can be reversed by reversing the supply terminals.
198. Universal motors are generally used with speed reducing gears.
199. Universal motors are used in food mixer grinder.
200. A repulsion motor runs in the direction of brush shift.
201. The speed of repulsion-start induction motor can be varied by brush-shifting.
202. Repulsion motors are unsuitable for high speed running.
203. The commutation of a repulsion motor is at its best when it runs at about 50% of synchronous speed.
204. Repulsion motor can be used for high inertia load requiring high starting torque.
205. The air-gap reluctance of a reluctance motor is uniform.
206. In a reluctance motor, when the load increases to an extent that it cannot maintain synchronous speed it will run as an induction motor.
207. Efficiency and power factor of reluctance motor are high as compared to single phase induction motor of same size.
208. Uniform air gap is used to produce a rotating magnetic field in reluctance motor.
209. Reluctance motors have starting torques.
210. An hysteresis motor is a smooth noiseless motor.
211. Hysteresis motors can pull-into step just about any load they can carry, regardless of the connected inertia of that load.
212. Hysteresis loss is minimum in hysteresis motor at standstill.
213. The hysteresis motor develops a constant hysteresis torque at all speeds from standstill up to but not including synchronous speed.

Answers: **157.** (T), **158.** (F), **159.** (F), **160.** (F), **161.** (F), **162.** (F), **163.** (T), **164.** (T), **165.** (F), **166.** (F), **167.** (F), **168.** (F), **169.** (T), **170.** (F), **171.** (T), **172.** (F), **173.** (T), **174.** (T), **175.** (T), **176.** (F), **177.** (T), **178.** (T), **179.** (T), **180.** (F), **181.** (F), **182.** (F), **183.** (T), **184.** (F), **185.** (T), **186.** (F), **187.** (F), **188.** (T), **189.** (F), **190.** (F), **191.** (F), **192.** (T), **193.** (F), **194.** (F), **195.** (T), **196.** (F), **197.** (F), **198.** (T), **199.** (T), **200.** (T), **201.** (T), **202.** (F), **203.** (F), **204.** (T), **205.** (F), **206.** (T), **207.** (F), **208.** (F), **209.** (T), **210.** (T), **211.** (T), **212.** (T), **213.** (F),

214. Hysteresis motor is normally free from mechanical and magnetic vibrations.

215. Eddy current motors operate on dc supply only.

216. A permanent magnet motor is a variable reluctance motor.

217. Permanent magnet motors are also called brushless motors.

218. If any two phases of a 3-phase induction motor are interchanged it will operate as an induction generator.

219. The torque-speed characteristics of the machine as a generator resemble to those of the motor.

220. When used in parallel with synchronous machine, an induction generator can deliver only lagging loads.

221. Induction motor can work as induction generator with slip-power recovery scheme.

222. The equivalent rotor circuit resistance R_2 is negative for induction generator.

223. Induction generators are designed with high rotor resistance so that they can be started against loads.

224. Induction generator is often used with wind mills.

225. At synchronous speed secondary windings of Schrage motor carry dc.

226. A long air gap in a synchronous induction motor makes it a stiffer machine with a larger overload capability.

227. The operating principle of the linear induction motor is the same as that of a conventional rotary motor.

228. The speed of a linear induction motor can be increased increasing the number of poles, the supply frequency and pole-pitch being kept fixed.

229. 'Emf effect' in a flat linear induction motor increases with the increase in number of poles.

230. Commutator motors are recommended for use in mines.

231. The brush shift against the direction of rotation of a Schrage motor gives improved power factor for both subsynchronous and supersynchronous operation.

232. Amplidyne is considered to be a dc amplifier.

233. Third brush generators are usually designed for weak shunt field and a strong armature reaction field.

234. Can the metadyne be used as a constant current source?

235. Average output voltage from a full wave thyristor controlled rectifier is $\frac{V_{max}}{\pi}\cos\frac{\alpha}{2}$ where α is firing delay angle.

236. For four-quadrant operation dual converter is required.

237. Thyristor drive provides higher accuracy, greater reliability, high operating efficiency, and fast response.

238. The thyristor drive motors have laminated yoke as well as main and commutating poles.

239. The thyristor drive motors have armatures of smaller diameter in comparison to ordinary dc motors.

240. A single-phase cycloconverter with centre input transformer needs 4 thyristors.

Answers: **214.** (T), **215.** (F), **216.** (F), **217.** (F), **218.** (F), **219.** (T), **220.** (F), **221.** (T), **222.** (T), **223.** (F), **224.** (T), **225.** (T), **226.** (T), **227.** (T), **228.** (F), **229.** (F), **230.** (F), **231.** (F), **232.** (T), **233.** (T), **234.** (yes). **235.** (F), **236.** (T), **237.** (T), **238.** (T), **239.** (F), **240.** (T).

FILL-IN-THE BLANK TYPE QUESTIONS

Note: Fill in the blanks with appropriate word or words.

1. ________ force drives or tends to drive flux through a magnetic circuit.

2. The ratio of total flux (flux in the iron path) to useful flux (flux in the air gap) is known as ________ factor.

3. The product of the number of turns and flux linking with the coil is known as ________.

4. The inductance of a coil is proportional to the ________ of the number of turns, all other parameters remaining the same. **[GATE E.E. 1994]**

5. The inductance has dimensions of ________.

6. Two identical coils of negligible resistance when connected in series across a 200 V, 50 Hz source draws, a current of 10 A. When the terminals of one of the coils are reversed, the current drawn is 8 A. The coefficient of coupling between the two coils is ________. **[GATE E.E. 1997]**

7. Two identical coupled inductors are connected in series. The measured inductances for the two possible series connections are 380 μH and 240 μH. Their mutual inductance in μH is ________. **[GATE 2014]**

8. The equivalent inductance seen at terminals A-B in figure given below is ________ H. **[GATE E.E. 1992]**

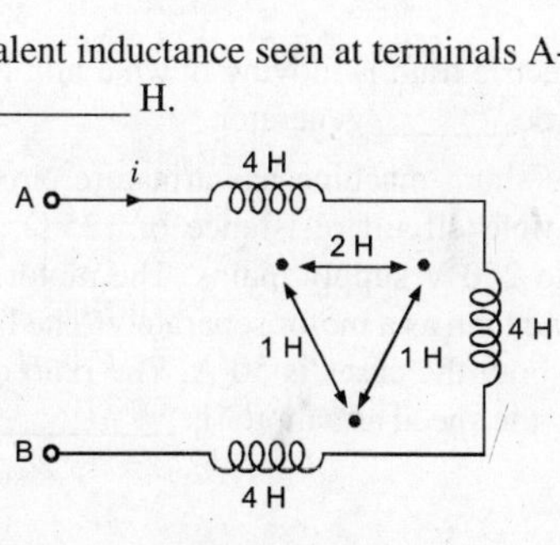

9. The lagging of magnetic flux density B behind the magnetic field intensity is known as magnetic ________.

10. The ________ currents produce a magnetic field of their own which opposes the main magnetic field.

11. The voltage and current waveforms for an element are shown in figure. The circuit element is ________ and its value is ________.

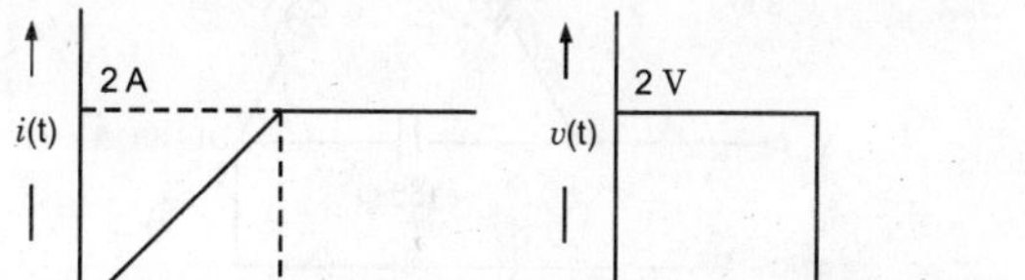

12. An electromagnetic field is said to be conservative when curl of the field is ________.

13. If the static magnetic field density is $\vec{B}$, then $\nabla\vec{B}$ = ________.

14. The force on a charge q moving with velocity V under the influence of electric and magnetic fields is given by ________.

15. The emf induced in a coil rotating in a uniform magnetic field is maximum when the flux linkage by the coil is ________.

16. A sinusoidal current having rms value of 8 ∠0° A is added to another sinusoidal current of rms value 6 ∠90° A. The rms value of the resultant current is ________. **[U.P. Technical Univ. Odd Semester 2009-10]**

17. In the circuit below the voltage **Vs** has a phase angle of ________ with respect to V_L. [GATE E.E. 1994]

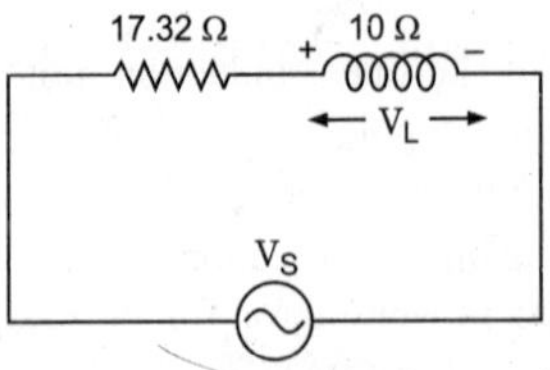

18. The increase in resistance due to non-uniform distribution of current in a conductor is known as________ effect. [GATE E.E.. 1994]

19. The power factor in a RLC series circuit will be lagging if ______. [Mahamaya Technical Univ. Even Semester 2010-11]

20. A series R-L-C circuit has the following parameter values R = 10 Ω, L = 0.01 H, C = 100 μF. The Q-factor of the circuit at resonance is ________. [GATE E.E. 1995]

21. In case of a resonant circuit the formula relating the resonance frequency ω_0, the quality factor Q_0 and the bandwidth $\Delta\omega$ is
($\omega_0 = Q_0 \times \Delta\omega$, $Q_0 = \omega_0 \times \Delta\omega$, $\Delta\omega = Q_0\omega_0$) [G.G.S.I.P. Univ. Delhi May 2010]

22. In delta-connected generator, the sum of instantaneous voltage around the delta is equal to _______. [Mahamaya Technical Univ. Even Semester 2010-11]

23. Three resistors of 10 Ω each forming a delta-connected load are connected across a 3-Φ balanced supply line to line voltage of 400 V. The total power consumed by this load is________ kW.

24. If W_1, W_2 and W_3 are the readings of three wattmeters used to measure the power in 3-phase, 4-wire circuit, the total power of load circuit will be _______. [U.P. Technical Univ. Odd Semester 2008-09; M.T.U. Even Semester 2010-11]

25. In two wattmeter method, if the readings are equal with opposite sign then the power factor of the load is ______. [G.B. Technical Univ. Odd Semester 2010-11

26. A set of 3 equal resistors, each of value R_X, connected in star across RYB of fig shown consumes the same power as the unbalanced delta-connected load shown. The value of R_X is _______Ω [GATE E.E. 1994]

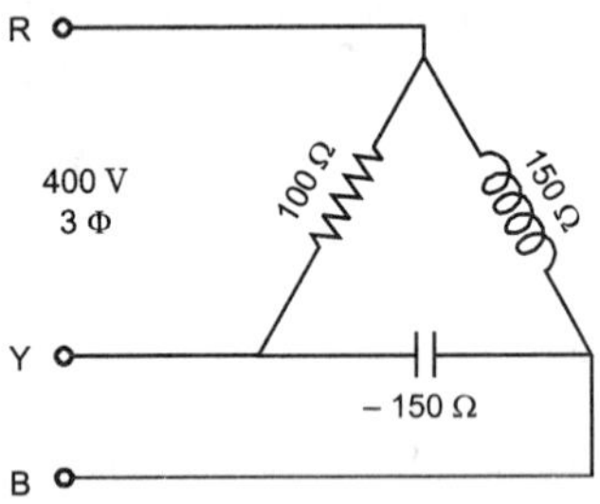

27. A three phase star-connected load is drawing power at a voltage of 0.9 pu and 0.8 power factor lagging. The three phase base power and base current are 100 MVA and 437.38 A respectively. The line-to-line load voltage in kV is _________. [GATE 2014]

28. Lamps of 40 W, 100 W and 200 W are connected in different phases of a 3-phase, 4-wire supply. If the neutral wire breaks, then the lamp likely to fuse first will be_________. (40/100/200 W)

29. For a particular motor, the cooling time constant is _______ than heating time constant.

30. The output indicated on the name plate of the motor is always the output power _______.

31. The emf induced in a coil rotating in a uniform magnetic field is maximum when the flux linkage by the coil is _______.

32. The yoke of a large machine is invariably made of fabricated steel because of its _______.

33. A dc machine armature is laminated to reduce _______ loss.

34. If the flux in the air gap is Φ, then the flux in the armature core section of a dc machine is _______.

35. The component of a dc generator which plays vital role in providing dc is _______.

36. Mica is _______ than copper.

37. The axial slots on the armature surface are provided to accommodate the _______.

38. The armature shaft is invariably made of _______.

39. Lap winding is suitable for _______ current but _______ voltage generators.

40. Wave winding is suitable for _______ current but _______ voltage machines.

41. A 4-pole lap wound dc generator generates 200 V at 1,000 rpm. If this generator is now wave wound and runs at 500 rpm, the generated voltage will be ______. [U.P. Technical Univ. Odd Semester 2009-10]

42. In dc machines, the field flux axis and armature mmf axis are respectively along _______ axis and _______ axis.

43. In a dc machine, the armature reaction and the inductance of commutating coils causes over/under/linear/sinusoidal commutation.

44. Commutation in a dc machine can be improved by increasing the _______ resistance.

45. A 200 V, 10 kW lap-wound dc generator has 10 poles and 500 conductors on its armature. If the pole face covers 80% of the pole pitch, the pole face conductors required to fully compensate for armature reaction will be _______ conductors per pole. [GATE 1997]

46. The series field of a short shunt generator is excited by _______ current.

47. An external resistance added in the field circuit of a dc shunt generator will cause _______ in terminal voltage.

48. A differentially compounded dc generator is delivering full-load current at a terminal voltage of 220 V. If the series field winding is short circuited its terminal voltage will _______.

49. If the shunt field is connected across the armature only in a dc compound generator, the generator is said to be _______ shunt compound dc generator.

50. The back emf of dc motor is given by E_b _______. [G.B. Technical Univ. Odd Semester 2010-11]

51. The current drawn by a 120 V dc motor of armature resistance 0.4 Ω and back emf 112 V is _______. [Mahamaya Technical Univ. Even Semester 2010-11]

52. The direction of rotation of a dc compound motor can be reversed by interchanging _______ connections.

53. The efficiency of a dc motor when developing maximum power will be _______ than 50%.

54. When the field of a dc motor is weakened, the speed of the motor _______.

55. When an electric train is moving down a hill, the dc motor will operate as a dc _______ generator.

56. A 250 V dc shunt machine has armature circuit resistance of 0.6 Ω and field circuit resistance of 125 Ω. The machine is connected to 250 V supply mains. The motor is operated as a generator and then as a motor separately. The line current of the machine in both the cases is 50 A. The ratio of the speed as a generator to the speed as a motor is ___________. [GATE 2014]

57. A 15 kW, 230 V dc shunt motor has armature circuit resistance of 0.4 Ω and field circuit resistance of 230 Ω. At no load and rated voltage, the motor runs at 1400 rpm and the line current drawn by the motor is 5 A. At full load, the motor draws a line current of 70 A. Neglect armature reaction. The full load speed of the motor in rpm is ______. **[GATE 2014]**

58. A separately excited dc motor has to be run at 125% rated speed when delivering 50% torque. The excitation will be ______ per cent. **[GATE E.E. 1991]**

59. A separately excited dc motor has an armature resistance of 0.5 Ω. It runs off at 250 V dc supply drawing an armature current of 20 A at 1,500 rpm. The torque developed for an armature current of 10 A will be ______ for the same field current. **[GATE E.E. 1992]**

60. ______ motor has self load relieving properties.
[U.P. Technical Univ. Even Semester 2009-10; M.T.U. Even Semester 2010-11]

61. ______ motor will be preferred for elevators.
[U.P. Technical Univ. Even Semester 2009-10; M.T.U. Even Semester 2010-11]

62. Hoist and cranes require ______ starting torque.

63. ______ wound dc motor has self relieving property.

64. In dc motors, field control method is used to have speeds ______ rated one.

65. A dc shunt motor is fitted with a field regulator for speed control. For a constant torque load the speed will be minimum for a regulator resistance of ______ ohms.

66. A variable resistor in series with a shunt field winding in a dc machine is called ______.

67. The armature voltage control is suitable if the dc machine is driven at constant ______.

68. Series-parallel control of dc series motors widely used in electric traction gives a speed range of approximately ______.

69. The field coil of a two-pole dc series motor is made up of two identical sections. In one case (*i*) two sections of the field coil are connected in series, and in another case (*ii*) the two sections are connected in parallel. If the motor takes the rated current in both the cases, then torque (*i*) : torque (*ii*) is ______ and speed (*i*) : speed (*ii*) is ______ (Neglect the effect of magnetic saturation). **[GATE E.E. 1997]**

70. A DOL starter is employed for starting electric motors of rating up to ______.

71. Overload release coil in a dc motor starter carries ______ load current.

72. In plugging of a dc shunt motor, the resistance is inserted in the armature circuit to limit the ______.

73. The electric braking system commonly used in rolling mills, elevators and printing presses is ______.

74. In a properly designed dc generator of 5 kW capacity, the total losses may be of the order of ______ watts.

75. The armature copper losses are proportional to the square of the armature current and are about ______ to ______ of full load losses.

76. Field copper losses are ______ to ______ per cent of the full-load losses.

77. DC machines are designed to have maximum efficiency around ______.

78. A 5 kW, 200 V dc shunt motor has an armature resistance of 1 Ω and shunt field resistance of 100 Ω. At no load the motor draws 6 A from a 200 V supply and runs at 1,000 rpm. The rotational loss of the machine is ______ W and the no-load torque is ______ Nm. **[GATE E.E. 1997]**

79. Brake test is performed on small dc machines for determination of ______.

80. ______ voltage is applied while performing Swinburne's test on dc machines.

81. Transformer action results when the magnetic flux in its core is ______.

82. The mmf produced by the secondary winding will ______ the primary mmf.

83. One transformer has leakage impedance of $(1 + j\,4)$ Ω and $(3 + j\,11)$ Ω for its primary and secondary windings respectively. This transformer has ______ (lv/hv) primary and ______ (lv/hv) secondary.

84. Maximum regulation occurs when the power factor of the load is ______.

85. Two windings of a single phase transformer are indicated by terminals *AB* and *CD* as shown in figure. When a voltage of 100 V is applied across *AB* with *BD* short-circuited, a voltage of 200 V appears across *AC*. The turns ratio from *CD* to *AB* is ______

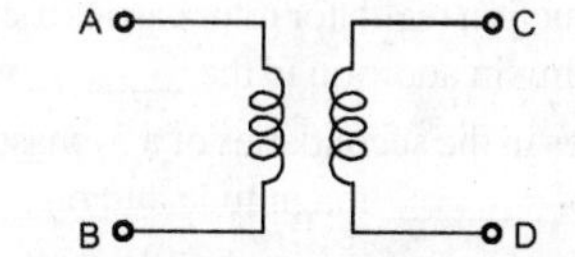

86. A single phase transformer is working at maximum efficiency. The copper losses are 100 W, then iron losses would be ______.
[U.P. Technical Univ. Odd Semester 2008-09]

87. A 100 kVA single phase transformer operating at 0.9 power factor has 90% maximum efficiency. The iron loss will be ______.
[U.P. Technical Univ. Odd Semester 2009-10]

88. The cross sections of small transformer core are ______.

89. The core of a transformer is assembled with laminated sheets so as to reduce ______ loss.

90. In sandwich type winding the leakage flux is ______ than that in cylindrical type winding.

91. The humming noise in a transformer depends on ______.

92. Conservator is not found in ______ cooled transformers.

93. Elaborate cooling arrangement is very essential for ______ rating power transformers.

94. Three identical single-phase transformers are connected in star-star configuration without neutral wire. Assuming that the supply is 3-phase balanced at 50 Hz, the frequency of the voltage appearing between the supply neutral and the primary star point is nearly ______ Hz.

95. The percentage impedance of a 100 kVA, 11 kV/400 V, Δ/*Y*, 50 Hz transformer is 4.5%. For the circulation of half the full-load current during short-circuit test, with low voltage terminal shorted, the applied voltage on the high voltage side will be ______. **[GATE E.E. 1995]**

96. A 3-phase transformer bank consists of three identical 2,300/230, 15 kVA single phase transformers connected in Δ/Δ. The bank supplies a 20 kVA, unity pf 3-phase load. If one of the single phase transformer develops a fault, and is removed, the load carried by each of the transformers now operating in delta will be ______ kVA. **[GATE E.E 1997]**

97. Two transformers A and B of equal ratings are connected in parallel. If the impedance of transformer A is more than that of B, transformer A will supply ______ load than that being supplied by B.

98. For parallel operation of two three phase transformers, it is essential that they have ______ phase sequence.

99. An auto-transformer is a transformer having a part of its winding ______ to the primary and secondary circuits.

100. A single phase, 50 kVA, 1000V/100 V two winding transformer is connected as an autotransformer as shown in the figure.

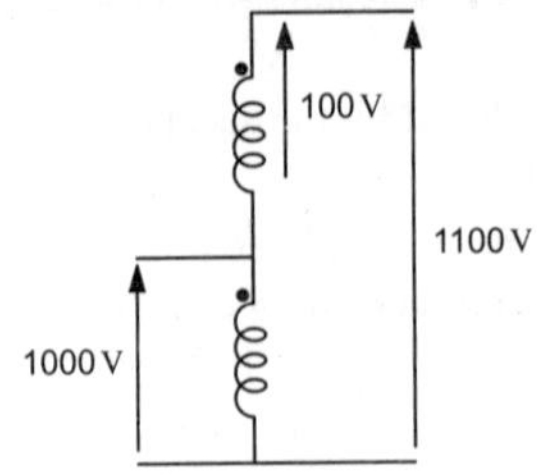

The kVA rating of the autotransformer is ____________. **[GATE 2014]**

101. Constant current transformer is used for _______ lighting.

102. Tappings are provided on transformers for varying the _______ voltage.

103. In a 1-phase induction regulator rotor carries a short-circuited compensating winding in addition to the _______ winding.

104. Induced voltages in the secondaries of a 3-phase induction voltage regulators are of _______ magnitude and of _______ phase.

105. If the voltage of a transformer is reduced, may be due to any reason, its output volt-amperes will be reduced in the _______ proportion.

106. The highest speed of 50 Hz ac generator can be _______. **[Mahamaya Technical Univ. Even Semester 2010-11]**

107. The field coils of synchronous machine are _______ insulated and _______ impregnated.

108. The frequency of generated emf by a synchronous generator can be adjusted by regulating the _______ of the prime mover.

109. The coils are so connected that emfs induced in them _______ each other.

110. In a fractional pitch winding the coil span is less than _______ electrical degrees.

111. It is desirable that the 5th harmonic voltage should be eliminated from the output of an alternator. The armature windings of the alternator should be short chorded by an angle of _______.

112. The distribution factor for a 36 slot stator with three phase, 8-pole winding, having 120° phase spread is _______. **[GATE E.E 1995]**

113. The synchronous reactance of a 200 MVA, 10 kV, 3-ϕ, 50 Hz generator is 1.0 pu at its own base. Its pu reactance at 100 MVA, 20 kV base will be _______ pu. **[GATE E.E 1997]**

114. A single phase alternator has a synchronous reactance of 2 Ω and negligible resistance. If it supplies 10 A to a purely capacitive load at 200 V, the generated emf will be _______ V and the regulation will be _______ %. **[GATE E.E 1997]**

115. A star connected 400 V, 50 Hz, 4 pole synchronous machine gave the following open circuit and short circuit test results:

Open circuit test: V_{oc} = 400 V (rms, line-to-line) at field current, I_f = 2.3 A

Short circuit test: I_{sc} = 10 A (rms, phase) at field current, I_f = 1.5 A

The value of per phase synchronous impedance in Ω at rated voltage is _________. **[GATE 2014]**

116. The quadrature axis synchronous reactance of a salient pole alternator is _______ than the direct axis synchronous reactance.

117. $X_d - X_d' - X_d''$.

118. A synchronous generator is connected to an infinite bus with excitation voltage E_f = 1.3 pu. The generator has a synchronous reactance of 1.1 pu and is delivering real power (P) of 0.6 pu to the bus. Assume the infinite bus voltage to be 1.0 pu. Neglect stator resistance. The reactive power (Q) in pu supplied by the generator to the bus under this condition is _________. **[GATE 2014]**

119. Thermal conductivity of hydrogen is nearly _______ times of that of air while its density is about _______ per cent of that of air.

120. In a 3-phase winding three windings are _______ electrical degrees apart from each other.

121. The resultant flux set up by a three phase winding makes _______ revolution(s) per cycle.

122. The percentage full-load slip in case of a synchronous motor is _______.

123. The maximum torque that a synchronous motor can develop without loss of synchronism is known as _______ torque.

124. For 100% excitation in a synchronous motor the counter emf is _______ applied voltage.

125. The synchronous motor operating underexcited will have _______ power factor.

126. A synchronous motor operated at power factors ranging from lagging through unity to leading without mechanical load is also called the _______ reactor.

127. The process of connecting an alternator in parallel with another alternator or with common bus-bars is called the _______.

128. When two alternators are operating in proper synchronism, the synchronising power will be _______.

129. In a system there are two generators operating in parallel. One generator, of rating 250 MVA, has an inertia constant of 6 MJ/MVA while the other generator of 150 MVA has an inertia constant of 4 MJ/MVA. The inertia constant for the combined system on 100 MVA common base is _______ MJ/MVA. **[GATE E.E 1994]**

130. There are two generators in a power system. No-load frequencies of the generators are 51.5 Hz and 51 Hz, respectively, and both are having droop constant of 1 Hz/MW. Total load in the system is 2.5 MW. Assuming that the generators are operating under their respective droop characteristics, the frequency of the power system in Hz in the steady state is __________. **[GATE 2014]**

131. An alternator synchronised to the infinite bus delivers lagging power to the bus when the field is _______ (overexcited, under-excited, normally excited).

132. The stator slots used in a 3-phase induction motor are usually of _______ type.

133. Under operating conditions the rotor winding of a wound type induction motor is _______ at the slip rings.

134. For an induction motor and synchronous induction motor of same rating, air gap of synchronous motor _______ air-gap of induction motor.

135. The slip of induction motor at synchronous speed will be _______.

136. An induction motor runs stably under constant torque load at 1,250 rpm at a 50 Hz supply. Its number of poles is _______. **[GATE E.E 1995]**

137. The rotor speed of a 6-pole 50 Hz induction motor is 960 rpm, the percentage slip is _____. **[Mahamaya Technical Univ. Even Semester 2010-11]**

138. Percentage no-load current of a transformer _______ percentage of no load current of an induction motor of comparable rating.

139. A 3-phase induction motor can achieve the maximum value of power factor to be _______.

140. An induction motor at starting behaves like a _______ circuited transformer.

141. The speed characteristic of an induction motor is somewhat similar to that of a _______.

142. An 8-pole, 3-phase, 50 Hz induction motor is operating at a speed of 700 rpm. The frequency of the rotor current of the motor in Hz is ___________. **[GATE 2014]**

143. In a 3-phase induction motor iron loss in rotor is _______ in comparison to that of a stator.

144. Skewing of rotor slots _______ rotor resistance.

145. Cogging is avoided in induction motor by ________

146. Crawling in induction motor is caused due to ________

147. Double cage induction motors are designed to achieve high _______ torque.

148. Very large three phase induction motors are started by direct on line/by star-delta starters/by auto-transformer starter/by none of the above.

149. Pole changing method of speed control is applicable only to _______ induction motor.

150. The stator of a 3-phase, 6-pole wound rotor induction motor is connected to 50 Hz source but its rotor is energised from 20 Hz. The rotor of this motor would run at speed of _______ rpm and _______ rpm.

151. When started by means of an auto-transformer with 50% tapping, supply current at start of an induction motor is reduced to _______ of that when started by means of a star-delta starter. **[GATE E.E. 1995]**

152. In a variable frequency induction motor drive, the voltage must be varied _______ to the frequency. **[GATE E.E. 1994]**

153. A 3-phase, 20 KW, 400 V, 1470 rpm, 50 Hz squirrel cage induction motor develops a torque of 100 Nm at a speed of 1,400 rpm. If the motor is connected to a 30 Hz supply, for keeping the same air gap flux, the supply voltage should be _______ and for the same load torque, the new speed will be _______ rpm. **[GATE E.E. 1997]**

154. A six pole, 50 Hz induction motor rotating at 1,440 rpm is in _______ mode. **[GATE E.E. 1994]**

155. If the air gap of the induction motor is increased, the magnetizing current of the motor will _______.

156. If a polyphase induction motor while connected to a constant voltage and frequency mains is mechanically coupled to a prime mover and is driven at a speed higher than synchronous speed. It will operate as an induction _______.

157. The linear synchronous speed of a linear motor is given as v_s = _______ where τ is the pole pitch in metres and f is the stator supply frequency in Hz.

158. The Schrage motor is started with the brushes set in the _______ speed position.

159. For a given frame size and temperature rise the output of single phase motor is about _______ per cent of that of a 3-phase motor.

160. For the same rating, the size of a single phase induction motor is about three times/the same as/1.5 times/one-third that of the corresponding three phase induction motor.

161. The main winding and the auxiliary winding placed on the stator of a single phase induction motor are spaced _______ electrical degrees apart.

162. In a shaded pole motor the necessary phase splitting is produced by _______.

163. A single phase induction motor draws 12 MW power at 0.6 lagging power. A capacitor is connected in parallel to the motor to improve the power factor of the combination of motor and capacitor to 0.8 lagging. Assuming that the real and reactive power drawn by the motor remains same as before, the reactive power delivered by the capacitor in MVAR is _________. **[GATE 2014]**

164. Repulsion motor has _______ efficiency in comparison to other single phase motors.

165. The direction of rotation of a single phase repulsion motor can be reversed by _______

166. The hysteresis motor rotor is made of magnetic material having _______ hysteresis loop area.

167. Fully compensated cross-field generator is called the amplidyne and undercompensated cross-field generator is called the _______.

168. The power input to the amplidyne is of the order of _______.

169. Metadynes should _______ be operated on open circuit.

170. Amplidynes should never be operated on _______.

171. The output rating of the armature of a metadyne is about _______ times of that of the same armature in an ordinary dc generator.

172. The _______ type cross-field generator which has full compensation has the _______ output voltage.

173. Third brush generators are usually designed for _______ shunt field and a _______ armature reaction field.

174. A 3-phase ac-to -dc diode bridge rectifier is supplied from a three-phase, 440 V source. The rectifier supplies a purely resistive load. The average dc voltage across the load will be _______ V. **[GATE 1995]**

175. A single-phase fully controlled dual converter uses _______ thyristors.

176. In a single-phase converter using a centre tapped transformer and with a highly inductive load, the firing angle is α. Each thyristor conducts for a period _______.

177. The figure shows the circuit diagram of a rectifier. The load consists of a resistance 10 Ω and an inductance 0.05 H connected in series. Assuming ideal thyristor and ideal diode, the thyristor firing angle (in degree) needed to obtain an average load voltage of 70 V is _______.

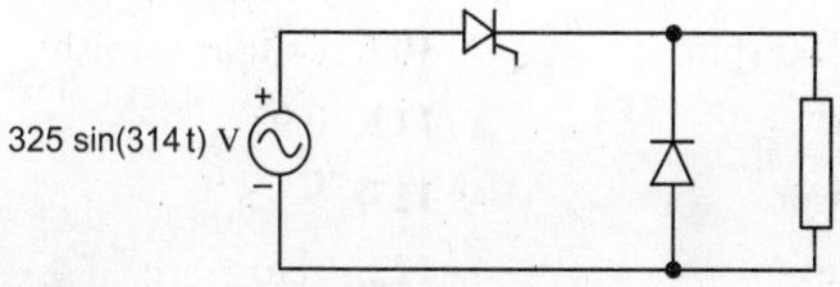

[GATE 2014]

178. A fully controlled converter bridge feeds a highly inductive load with ripple free load current. The input supply (v_s) to the bridge is a sinusoidal source. Triggering angle of the bridge converter is α = 300. The input power factor of the bridge is _________.

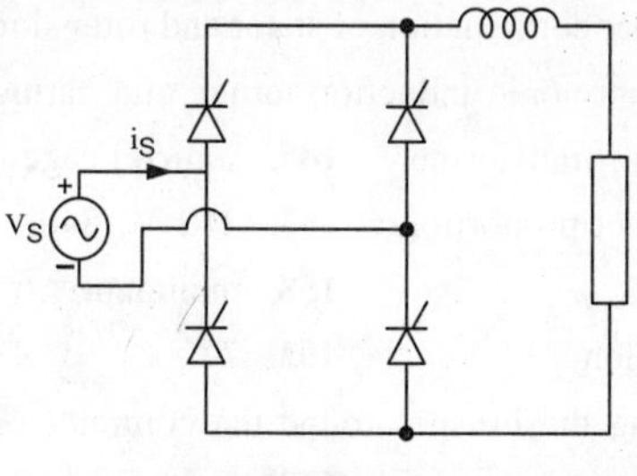

[GATE 2014]

179. A dc chopper has a T_{on} of 1ms and its frequency is 500 Hz. Its duty cycle will be _______ percent. (25/50/75/100)

180. A step-up chopper is used to feed a load at 400 V dc from a 250 V dc source. The inductor current is continuous. If the 'off' time of the switch is 20 μs, the switching frequency of the chopper in kHz is ___________. **[GATE 2014]**

181. A single-phase ac regulator uses _______ thyristors connected in _______.

182. The speed of a dc motor can be controlled by using _______. (transistors/thyristors/diodes)

FILL-IN-THE BLANK TYPE QUESTIONS

1. Magnetomotive	**2.** leakage	**3.** flux linkage	**4.** square	**5.** flux per current
6. 1/9	**7.** 35	**8.** 8	**9.** hysteresis	**10.** eddy
11. inductor, 2 H	**12.** zero	**13.** 0	**14.** $q(E + V \times B)$	**15.** zero
16. $10 \angle 38.87°$ A	**17.** 60°	**18.** skin	**19.** $\omega L > \dfrac{1}{\omega C}$	**20.** 1.0
21. $\omega_0 = Q_0 \times \Delta\omega$	**22.** zero	**23.** 48.0	**24.** $W_1 + W_2 + W_3$	**25.** zero
26. 100	**27.** 118.8	**28.** 40 W	**29.** greater	**30.** at the shaft
31. minimum	**32.** high permeability	**33.** eddy current	**34.** $\Phi/2$	**35.** commutator
36. harder	**37.** winding	**38.** mild steel	**39.** high; low	**40.** low; high
41. 200 V	**42.** pole, magnetic neutral	**43.** under	**44.** brush	**45.** 4
46. load	**47.** decrease	**48.** increase	**49.** short	**50.** $= \dfrac{\Phi ZN}{60} \times \dfrac{P}{A}$
51. 20 A	**52.** armature	**53.** less	**54.** increases	**55.** series
56. 1.27	**57.** 1,241	**58.** 80	**59.** 15.28 N-m	**60.** DC series
61. DC series	**62.** high	**63.** Series	**64.** above	**65.** zero
66. shunt field regulator	**67.** torque	**68.** 4 : 1	**69.** 2, 1/2	**70.** 3.75 kW
71. full	**72.** armature current	**73.** plugging	**74.** 300	**75.** 30, 40
76. 20; 30	**77.** full load	**78.** 784, 7.5	**79.** net output	**80.** Rated
81. alternating	**82.** oppose	**83.** *hv, lv*	**84.** lagging	**85.** 1 or 3
86. 100 W	**87.** 5 kW	**88.** rectangular or square	**89.** eddy current	**90.** less
91. flux density	**92.** air	**93.** large	**94.** 150	**95.** 247.5 V
96. 10	**97.** less	**98.** same	**99.** common	**100.** 550
101. series	**102.** output	**103.** primary	**104.** fixed; varying	**105.** same
106. 3,000 rpm	**107.** cotton; varnish	**108.** speed	**109.** help	**110.** 180
111. 36°	**112.** 0.97	**113.** 0.125	**114.** 180; –10	**115.** 15.06
116. lesser	**117.** $>;>$	**118.** 1.092	**119.** 7.5; 7	**120.** 120
121. one	**122.** zero	**123.** pull-out	**124.** equal to	**125.** lagging
126. synchronous	**127.** synchronising	**128.** zero	**129.** 21	**130.** 50
131. overexcited	**132.** semi-closed	**133.** short circuited	**134.** $>$	**135.** zero
136. 4	**137.** 4%	**138.** $<$	**139.** 0.9	**140.** short-
141. dc shunt motor	**142.** 3.333	**143.** negligible	**144.** increases	
145. a proper combination of stator and rotor slots.				
146. both harmonic induction torque and harmonic synchronous torque.			**147.** starting	
148. by auto-transformer	**149.** squirrel cage	**150.** 1,400; 600	**151.** 3/4	
152. in direct proportion	**153.** 240 V; 840	**154.** generating	**155.** increase	**156.** generator
157. $2\tau f$	**158.** minimum	**159.** 50	**160.** 1.5 times	**161.** 90
162. induction	**163.** 7	**164.** high		
165. shifting the brushes round the commutator on the other side of the field axis			**166.** Large	
167. metadyne	**168.** few watts	**169.** never	**170.** short circuit	**171.** 0.6
172. amplidyne; highest	**173.** weak; strong	**174.** 594	**175.** 8	**176.** π
177. 69.310	**178.** 0.7797 (lagging)	**179.** 50	**180.** 31.25	**181.** 2; antiparallel
182. thyristors				

Appendices

APPENDIX A

STANDARD PREFIXES

Multiple	*SI Prefix*	*Symbol*
10^{12}	tera	T
10^{9}	giga	G
10^{6}	mega	M
10^{3}	kilo	k
10^{2}	hecto	h
10	deka	da
10^{-1}	deci	d
10^{-2}	centi	c
10^{-3}	milli	m
10^{-6}	micro	μ
10^{-9}	nano	n
10^{-12}	pico	p
10^{-15}	femto	f
10^{-18}	atto	a

APPENDIX B

GREEK ALPHABETS

Letter	*Capital*	*Small*
Alpha	A	α
Beta	–	β
Gamma	–	γ
Delta	Δ	δ
Epsilon	–	ε, ∈
Zeta	–	ς
Eta	–	η
Theta	–	θ
Iota	–	ι
Lambda	Λ	λ
Mu	–	μ
Nu	–	ν
Xi	–	ξ
Pi	–	π
Rho	–	ρ
Sigma	Σ	σ
Phi	Φ	ϕ
Psi	Ψ	ψ
Omega	Ω(ohm)	ω

APPENDIX C

CONSTANTS

Gravitational acceleration, $g = 9.807$ m/s^2

Permittivity of free space, $\varepsilon_0 = 8.854 \times 10^{-12}$ F/m

Permeability of free space, $\mu = 4\pi \times 10^{-7}$ H/m

$\pi = 3.141592654$

APPENDIX D

CONVERSION FACTORS

Quantity	*SI Unit*	*English Unit*	*C.G.S. Unit*
Length	1 m	3.28084 ft, 39.3701 in	100 cm
Mass	1 kg	2.20462 lb (mass)	1,000 gm
Force	1 N	0.2248 lb	10^5 dynes
Torque	1 N-m	0.7376 lb-ft	10^7 dyne-cm
Power	1 W, 1 J/s	1.341×10^{-3} hp	–
Energy	1 J, W-s	0.7376 ft-lb	10^7 erg
Moment of Inertia	1 kg-m^2	23.73 lb-ft^2	10^7 g cm^2
Magnetic flux	1 Wb	10^8 lines	10^8 maxwells
Magnetic flux density	1 Wb/m^2, 1 T	64.5×10^3 lines/in^2	10^4 gauss
Magnetic field intensity	1 A/m, 1 A-T/m	0.0254 A/in	0.004 π oersted

Additional Conversion Factors

1 US gallon = 0.8327 UK gallon

1 Ton = 1,016.05 kg = 2,240 lb = 20 cwt

1 N-m^2 = 2.420 lb.ft^2

1 torr = 1.333 m bar

1 pascal = 1 N/m^2

1 kW = 1,000 W = 1.34102 hp

1 hp = 745.7 W = 0.7457 kW

1 hp (metric) = 735.5 W

1 kWh = 3.600×10^6 joules = 859 845 calories = 3,412.1 BThu

1 cal = 4.1868 J = 1.163×10^{-6} kWh

Index

A

B

H

I

K

L

M

N

O

P

R

S

T

U

V

W

Z

❑❑❑

Win Prizes!

ATTENTION READERS

We request you for your frank assessment regarding some of the aspects of the book given as under:

THEORY AND PERFORMANCE

of

ELECTRICAL MACHINES

By

J.B. Gupta

Please fill up the given space in neat capital letters.

(i) What topics of your syllabus that are important from you examination point of view are not covered in the book ?

....

....

....

....

(ii) What are the chapters wherein the treatment of the subject matter is not systematic or organised or updated ?

....

....

....

....

(iii) Have you come across misprints/mistakes/factual inaccuracies in the book? Please specify the chapters and the page numbers.

....

....

....

....

(iv) What other books on the same subject have you found/heard better that the present book? Please specify in terms of price and quality.

....

....

....

....

(v) Further suggestions and comments foı the improvement of the book:

....

....

....

....

(PS: If need be please attach a separate sheet to write your views.)

Other Details :

(i) Who recommended you the book ? (Please tick in the box near the option relevant to you.)

☐ Teachers ☐ Friends ☐ Bookseller

(ii) Name of the recommending teacher, his designation and address.

....

....

....

....

(iii) Name and address of the bookseller you purchased the book form :

....

....

....

....

(iv) Name and address of your institution (Please mention the University or Board, as the case may be)

....

....

....

....

(v) Your name and complete postal address :

....

....

....

....

(vi) Course you are enrolled in :

....

....

....

....

The best assessment will be awarded each month. The award will be in the form of our publications as decided by the Editorial Board.

Please mail the filled up coupon at your earliest to:

S.K. Kataria & Sons®

4885/109, Prakash Mahal, Dr. Subhash Bhargav Lane,
Opposite Delhi Medical Association, Daryaganj, New Delhi-110002 (INDIA)
Phone: +91-11-43551243 Mobile: +91-9871775858, 7678425209
e-mail: katariabook@gmail.com; katariabooks@yahoo.com
Website: www.skkatariaandsons.com